Solar System Data

Body	Mass (kg)	Mean Radius (m)	Period (s)	Distance from Sun (m)
Mercury	3.18×10^{23}	2.43×10^{6}	7.60×10^{6}	5.79×10^{10}
Venus	4.88×10^{24}	6.06×10^{6}	1.94×10^{7}	1.08×10^{11}
Earth	5.98×10^{24}	6.37×10^{6}	3.156×10^{7}	1.496×10^{11}
Mars	6.42×10^{23}	3.37×10^{6}	5.94×10^{7}	2.28×10^{11}
Jupiter	1.90×10^{27}	6.99×10^{7}	3.74×10^{8}	7.78×10^{11}
Saturn	5.68×10^{26}	5.85×10^{7}	9.35×10^{8}	1.43×10^{12}
Uranus	8.68×10^{25}	2.33×10^{7}	2.64×10^{9}	2.87×10^{12}
Neptune	1.03×10^{26}	2.21×10^{7}	5.22×10^{9}	4.50×10^{12}
Pluto	$\approx 1.4 \times 10^{22}$	$\approx 1.5 \times 10^{6}$	7.82×10^{9}	5.91×10^{12}
Moon	7.36×10^{22}	1.74×10^{6}	—	—
Sun	1.991×10^{30}	6.96×10^{8}	—	—

Values of Physical Data Often Used[a]

Acceleration due to gravity	$9.80\ m/s^2$
Average earth-moon distance	3.84×10^{8} m
Average earth-sun distance	1.496×10^{11} m
Average radius of the earth	6.37×10^{6} m
Density of air (20°C and 1 atm)	$1.20\ kg/m^3$
Density of water (20°C and 1 atm)	$1.00 \times 10^{3}\ kg/m^3$
Mass of the earth	5.98×10^{24} kg
Mass of the moon	7.36×10^{22} kg
Mass of the sun	1.99×10^{30} kg
Standard atmospheric pressure	1.013×10^{5} Pa

[a] These are the values of the constants as used in the text.

Some Prefixes for Powers of Ten

Power	Prefix	Abbreviation	Power	Prefix	Abbreviation
10^{-18}	atto	a	10^{1}	deka	da
10^{-15}	femto	f	10^{2}	hecto	h
10^{-12}	pico	p	10^{3}	kilo	k
10^{-9}	nano	n	10^{6}	mega	M
10^{-6}	micro	μ	10^{9}	giga	G
10^{-3}	milli	m	10^{12}	tera	T
10^{-2}	centi	c	10^{15}	peta	P
10^{-1}	deci	d	10^{18}	exa	E

PHYSICS

For Scientists & Engineers

Third edition

Isaac Newton's manuscripts on the aberration of light and the prism he used for his experiments. (Erich Lessing/MAGNUM)

PHYSICS

For Scientists & Engineers

Third edition

Raymond A. Serway
James Madison University

SAUNDERS GOLDEN SUNBURST SERIES
Saunders College Publishing
Philadelphia Fort Worth Chicago
San Francisco Montreal Toronto
London Sydney Tokyo

Text Typeface: Caledonia
Compositor: Progressive Typographers, Inc.
Acquisitions Editor: John Vondeling
Developmental Editor: Ellen Newman
Managing Editor: Carol Field
Project Editor: Sally Kusch
Copy Editor: Will Eaton
Manager of Art and Design: Carol Bleistine
Art and Design Coordinator: Doris Bruey
Text Designer: Edward A. Butler
Cover Designer: Lawrence R. Didona
Text Artwork: Tom Mallon
Layout Artist: Tracy Baldwin
Director of EDP and Production Manager: Tim Frelick

Cover Credit: © THE EXPLORATORIUM by S. Schwartzenberg

Printed in the United States of America

Physics for Scientists & Engineers, 3rd edition

0-03-030258-7

Library of Congress Catalog Card Number: 89-043325

12 032 9876543

Preface

This textbook is intended for a course in introductory physics for students majoring in science or engineering. The entire contents of the text could be covered in a three-semester course, but it is possible to use the material in shorter sequences with the omission of selected chapters and sections. The mathematical background of the student taking this course should ideally include one semester of calculus. If that is not possible, the student should be enrolled in a concurrent course in introduction to calculus.

Star trails around the south celestial pole. (© Anglo-Australian Telescope Board 1980)

OBJECTIVES

The main objectives of this introductory physics textbook are twofold: to provide the student with a clear and logical presentation of the basic concepts and principles of physics, and to strengthen an understanding of the concepts and principles through a broad range of interesting applications to the real world. In order to meet these objectives, emphasis is placed on sound physical arguments. At the same time, I have attempted to motivate the student through practical examples that demonstrate the role of physics in other disciplines.

COVERAGE

The material covered in this book is concerned with fundamental topics in classical physics and an introduction to modern physics. The book is divided into six parts. In the first volume, Part I (Chapters 1–15) deals with the fundamentals of Newtonian mechanics and the physics of fluids; Part II (Chapters 16–18) covers wave motion and sound; Part III (Chapters 19–22) is concerned with heat and thermodynamics. In the second volume, Part IV (Chapters 23–34) treats electricity and magnetism, Part V (Chapters 35–38) covers light and optics, and Part VI (Chapters 39–40) deals with relativity and an introduction to quantum physics.

Features Most instructors would agree that the textbook selected for a course should be the student's major "guide" for understanding and learning the subject matter. Furthermore, a textbook should be easily accessible and should be styled and written for ease in instruction. With these points in mind, I have included many pedagogic features in the textbook which are intended to enhance its usefulness to both the student and instructor. These are as follows:

Organization The book is divided into six parts: mechanics, wave motion and sound, heat and thermodynamics, electricity and magnetism, light and optics, and modern physics. Each part includes an overview of the subject matter to be covered in that part and some historical perspectives.

Style As an aid for rapid comprehension, I have attempted to write the book in a style that is clear, logical, and succinct. The writing style is somewhat informal and relaxed, which I hope students will find appealing and enjoyable to read. New terms are carefully defined, and I have tried to avoid jargon.

Previews Most chapters begin with a chapter preview, which includes a brief discussion of chapter objectives and content.

Important Statements and Equations Most important statements and definitions are set in blue print for added emphasis and ease of review. Important equations are highlighted with a tan screen for review or reference.

Marginal Notes Comments and marginal notes set in blue print are used to locate important statements, equations, and concepts in the text.

Illustrations The readability and effectiveness of the text material and worked examples are enhanced by the large number of figures, diagrams, photographs, and tables. Full color is used to add clarity to the artwork and to make it as realistic as possible. For example, vectors are color-coded, and curves in *xy*-plots are drawn in color. Three-dimensional effects are produced with the use of color airbrushed areas, where appropriate. The color photographs have been carefully selected, and their accompanying captions have been written to serve as an added instructional tool. A complete description of the pedagogical use of color appears on p. xvii.

Experiencing circular motion on a roller coaster. (© Zerschling, Photo Researchers)

Mathematical Level Calculus is introduced gradually, keeping in mind that a course in calculus is often taken concurrently. Most steps are shown when basic equations are developed, and reference is often made to mathematical appendices at the end of the text. Vector products are introduced later in the text where they are needed in physical applications. The dot product is introduced in Chapter 7, "Work and Energy." The cross product is introduced in Chapter 11, which deals with rotational dynamics.

Worked Examples A total of 355 worked examples of varying difficulty are presented as an aid in understanding concepts. This represents an increase of 10% over the 2nd edition. In many cases, these examples will serve as models for solving the end-of-the-chapter problems. The examples are set off in a blue box, and the solution answers are highlighted with a tan screen. Most examples are given titles to describe their content.

Worked Example Exercises Many of the worked examples are followed immediately by exercises with answers. These exercises are intended to make the textbook more interactive with the student, and to test immediately the student's understanding of problem-solving techniques. The exercises represent extensions of the worked examples and are numbered in case the instructor wishes to assign them.

Units The international system of units (SI) is used throughout the text. The British engineering system of units (conventional system) is used only to a limited extent in the chapters on mechanics, heat, and thermodynamics.

Biographical Sketches Throughout the text I have added short biographies of important scientists to give the third edition a more historical emphasis.

Problem-Solving Strategies and Hints As an added feature of the third edition, I have included general strategies for solving the types of problems featured in both the examples and in the end-of-chapter problems. It is my hope that this feature will help the students identify the steps in solving a problem and eliminate any uncertainty they might have. This feature is highlighted by a light blue screen for emphasis and ease of location.

Summaries Each chapter contains a summary which reviews the important concepts and equations discussed in that chapter. The summaries are highlighted with a gold screen.

Thought Questions A list of 922 questions requiring verbal answers (an increase of 41% over the 2nd edition) is given at the end of each chapter. Some questions provide the student with a means of self-testing the concepts presented in the chapter. Others could serve as a basis for initiating classroom discussions. Answers to most questions are included in the Student Study Guide With Computer Exercises that accompanies the text.

Problems An extensive set of problems is included at the end of each chapter, with a total of 2910 problems. This represents an increase of 42.5% of new or revised problems over the 2nd edition. Answers to odd-numbered problems are given at the end of the book; these pages have colored edges for ease of location. For the convenience of both the student and the instructor, about two thirds of the problems are keyed to specific sections of the chapter. The remaining problems, labeled "Additional Problems," are not keyed to specific sections. In my opinion, assignments should consist mainly of the keyed problems to help build self-confidence in students.

In general, the problems within a given section are presented so that the straightforward problems are first, followed by problems of increasing difficulty. For ease in identifying the intermediate-level problems, the problem number is printed in blue. I have also included a small number of challenging problems, which are indicated by a problem number printed in magenta.

Spreadsheet Problems and Examples Selected problems and examples throughout the text have been keyed with a color box to indicate that there are spreadsheets on a separate disk that accompany these problems. This is an option; that is, these problems can also be solved analytically without using the software. The spreadsheets will help the student work some of the difficult problems; instructions for use of the spreadsheets are found in Appendix F. The data disk is available free to instructors who adopt the text and can be used with a variety of programs (e.g., Lotus 1-2-3) that the instructor may already have.

Calculator/Computer Problems Numerical problems that can best be solved with the use of programmable calculators or a computer are given in a selected number of chapters. These will be useful in those courses where the instructor wishes to put programming skills to practice.

Guest Essays I have included 11 essays, written by guest authors, on topics of current interest to scientists and engineers. Eight of the essays are entirely new; the remaining three have been revised and updated. The essays are intended as supplemental readings for the student and are marked with a blue bar at the edge of the page so they can be located easily.

Special Topics Many chapters include special topic sections which are intended to expose the student to various practical and interesting applications of physical principles. Most of these are considered optional, and as such are labeled with an asterisk (*).

Appendices and Endpapers Several appendices are provided at the end of the text, including the new appendix with instructions for problem-solving with spreadsheets. Most of the appendix material represents a review of mathematical techniques used in the text, including scientific notation, algebra, geometry, trigonometry, differential calculus, and integral calculus. Reference to these appendices is made throughout the text. Most mathematical review sections include worked examples and exercises with answers. In addition to the mathematical reviews, the appendices contain tables of physical data, conversion factors, atomic masses, and the SI units of physical quantities, as well as a periodic chart. Other useful information, including fundamental constants and physical data, planetary data, a list of standard prefixes, mathematical symbols, the Greek alphabet, and standard abbreviation of units appears on the endpapers.

Hurricane Elena photographed from space. (NASA)

CHANGES IN THE THIRD EDITION

A number of changes and improvements have been made in preparing the third edition of this text. Many of these changes are in response to comments and suggestions offered by users of the text and reviewers of the manuscript. The following represent the major changes in the third edition:

1. Substantial revision of the problems to include more at the intermediate level (about 40% of the problems are new or revised). In addition, the end-of-chapter questions have been revised and increased by 40%. *All problems in the text have been reviewed for accuracy.*
2. Chapter 44 (Superconductivity) and Chapter 47 (Particle Physics and Cosmology) are *new* chapters which have been added to the extended version.
3. Increase in the number of worked examples and student exercises (with answers) that follow the examples. This feature makes the text more interactive with the student.
4. Addition of Problem-Solving Strategies and Hints where appropriate provides additional information and guidelines to help students solve problems.
5. Many entirely new and updated guest essays. As a new feature, these essays now include both questions and problems in case the instructor wishes to assign them.
6. Some sections have been rewritten to improve clarity, especially in the chapters on fluids, relativity, and thermodynamics.
7. Revised and updated coverage in the modern physics chapters.
8. New full-color design, which is used to highlight important concepts throughout the text and in the figures. In addition, about 200 color photographs have been carefully selected, and their accompanying captions have been written to serve as an added instructional tool.
9. Approximately 75 problems and examples in the text have been keyed to indicate that there are spreadsheets on a separate disk that accompany these problems. This is an optional feature since these problems can also be solved analytically without using the software. The spreadsheets will help the students work some of the more difficult problems; instructions for use of the spreadsheets appear in Appendix F. The data disk will be available free to instructors who adopt the text by returning a response card in the Instructor's Manual.
10. Biographical Sketches of many prominent physicists have been added to give the third edition a more historical emphasis.

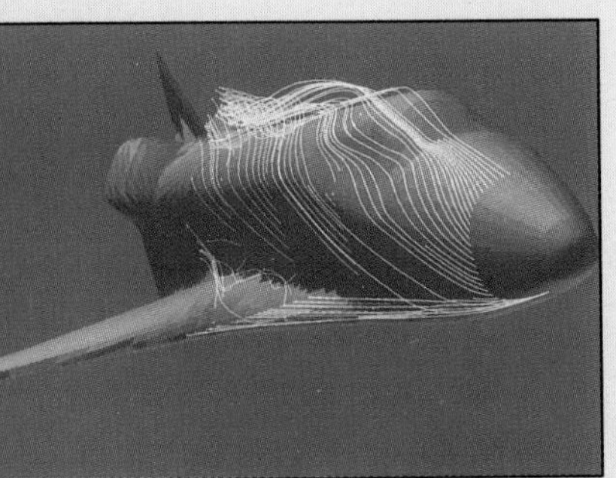

Computer simulation of the pattern of air flow around a space shuttle. (NASA)

ANCILLARIES

The following ancillaries are available with this text:

Instructor's Manual with Solutions This manual consists of complete, worked-out solutions to all the problems in the text and answers to even-numbered problems.

The instructor will be able to obtain the Spreadsheet Data Disk (IBM) that contains problems and examples keyed to the text by returning the response card in the back of the Instructor's Manual. The Spreadsheet Disk is a data disk that can be used with a variety of spreadsheet programs (e.g., Lotus 1-2-3) that the instructor may already have. Use of the Spreadsheet Disk is optional.

Student Study Guide with Computer Exercises (IBM or Macintosh) This unique student aid combines the value of a problem-solving-oriented study guide with a select group of integrated and interactive computer exercises. The Study Guide contains chapter objectives, a skills section that reviews mathematical techniques, and suggested approaches to problem-solving methodology; notes from selected chapter sections, which include a glossary of important terms, theorems, and concepts; answers to selected end-of-chapter questions; solutions to selected end-of-chapter problems; and programmed exercises. The Study Guide also includes the option of using a select group of computer programs (presented in special computer modules) that are interactive in nature. The student's input will have direct and immediate effect on the output. This feature will enable students to work through many challenging numerical problems and experience the power of the computer in scientific computations. The courseware disk is available for the IBM or Macintosh computers and is packaged with the Study Guide.

Computerized Test Bank Available for the IBM PC, Apple II, and Macintosh computers, this test bank contains over 2300 multiple choice questions, representing every chapter of the text. The test bank enables the instructor to create many unique tests and permits the editing of questions as well as addition of new questions. The software program solves all problems and prints each answer on a separate grading key.

Printed Test Bank This printed test bank contains the multiple choice questions from the software disk. It is provided as another source of test questions and is helpful for the instructor who does not have access to a computer.

Overhead Transparency Acetates This collection of transparencies consists of 300 full-color figures from the text and features large print for easy viewing in the classroom.

Physics Laboratory Manual To supplement the learning of basic physical principles while introducing laboratory procedures and equipment, each chapter of the laboratory manual includes a pre-laboratory assignment, objectives, an equipment list, the theory behind the experiment, experimental procedure, calculations, graphs, and questions. In addition, a laboratory report is provided for each experiment so the student can record data, calculations, and experimental results.

Instructor's Manual for Physics Laboratory Manual Each chapter contains a discussion of the experiment, teaching hints, answers to selected questions, and a post-laboratory quiz with short answer and essay questions. We have also included a list of the suppliers of scientific equipment and a summary of the equipment needed for all the laboratory experiments in the manual.

TEACHING OPTIONS

This book is structured in the following sequence of topics: classical mechanics, matter waves, heat and thermodynamics, electricity and magnetism, light waves, optics, relativity, and an introduction to quantum physics. This presentation is a more traditional sequence, with the subject of matter waves presented before electricity and magnetism. Some instructors may prefer to cover this material after completing electricity and magnetism (after Chapter 34). The chapter on relativity was placed near the end of the text because this topic is often treated as an introduction to the era of "modern physics." If time permits, instructors may choose to cover Chapter 39 after completing Chapter 14, which concludes the material on Newtonian mechanics.

For those instructors teaching a two-semester sequence, some sections and chapters could be deleted without any loss in continuity. I have labeled these with asterisks (*) in the Table of Contents and in the appropriate sections of the text. For student enrichment, some of these sections or chapters could be given as extra reading assignments. The guest essays could also serve the same purpose.

ACKNOWLEDGMENTS

The third editon of this textbook was prepared with the guidance and assistance of many professors who reviewed part or all of the manuscript. I wish to acknowledge the following scholars and express my sincere appreciation for their suggestions, criticisms, and encouragement:

George Alexandrakis, University of Miami; Bo Casserberg, University of Minnesota; Soumya Chakravarti, California State Polytechnic University; Edward Chang, University of Massachusetts, Amherst; Hans Courant, University of Minnesota; F. Paul Esposito, University of Cincinnati; Clark D. Hamilton, National Bureau of Standards; Mark Heald, Swarthmore College; Paul Holoday, Henry Ford Community College; Larry Kirkpatrick, Montana State University; Barry Kunz, Michigan Technological University; Douglas A. Kurtze, Clarkson University; Robert Long, Worcester Polytechnic Institute; Nolen G. Massey, University of Texas at Arlington; Charles E. McFarland, University of Missouri at Rolla; James Monroe, The Pennsylvania State University, Beaver Campus; Fred A. Otter, University of Connecticut; Eric Peterson, Highland Community College; Jill Rugare, DeVry Institute of Technology; Charles Scherr, University of Texas at Austin; John Shelton, College of Lake County; Kervork Spartalian, University of Vermont; Robert W. Stewart, University of Victoria; James Stith, United States Military Academy; Carl T. Tomizuka, University of Arizona; Som Tyagi, Drexel University; James Walker, Washington State University; George Williams, University of Utah; and Edward Zimmerman, University of Nebraska, Lincoln.

Special thanks go to the many people who provided me with useful comments and suggestions for improvement during the development of this third edition. These include Albert A. Bartlett, David C. Currot, Chelcie Liu, Howard C. McAllister, A. J. Slavin, J. C. Sprott, and William W. Wood.

I would also like to thank the following professors for their suggestions during the development of the prior editions of this textbook:

Elmer E. Anderson, University of Alabama; Wallace Arthur, Fairleigh Dickinson University; Duane Aston, California State University at Sacramento; Richard Barnes, Iowa State University; Marvin Blecher, Virginia Polytechnic Institute and State University; William A. Butler, Eastern Illinois University; Don Chodrow, James Madison University; Clifton Bob Clark, University of North Carolina at Greensboro; Lance E. De Long, University of Kentucky; Jerry S. Faughn, Eastern Kentucky University; James B. Gerhart, University of Washington; John R. Gordon, James Madison University; Herb Helbig, Clarkson University; Howard Herzog, Broome Community College; Larry Hmurcik, University of Bridgeport; William Ingham, James Madison University; Mario Iona, University of Denver; Karen L. Johnston, North Carolina State University; Brij M. Khorana, Rose-Hulman Institute of Technology; Carl Kocher, Oregon State University; Robert E. Kribel, Jacksonville State University; Fred Lipschultz, University of Connecticut; Francis A. Liuima, Boston College; Charles E. McFarland, University of Missouri, Rolla; Clem Moses, Utica College; Curt Moyer, Clarkson University; Bruce Morgan, U.S. Naval Academy; A. Wilson Nolle, The University of Texas at Austin; Thomas L. O'Kuma, San Jacinto College North; George Parker, North Carolina State University; William F. Parks, University of Missouri, Rolla; Philip B. Peters, Virginia Military Institute; Joseph W. Rudmin, James Madison University; James H. Smith, University of Illinois at Urbana-Champaign; Edward W. Thomas, Georgia Institute of Technology; Gary Williams, University of California, Los Angeles; George A. Williams, University of Utah; and Earl Zwicker, Illinois Institute of Technology.

I would like to thank the following people for contributing many interesting new problems and questions to the text: Ron Canterna, University of Wyoming; Paul Feldker, Florissant Valley Community College; Roger Ludin, California Polytechnic State University; Richard Reimann, Boise State University; Jill Rugare, DeVry Institute of Technology; Stan Shepard, The Pennsylvania State University; Som Tyagi, Drexel University; Steve Van Wyk, Chapman College; and James Walker, Washington State University.

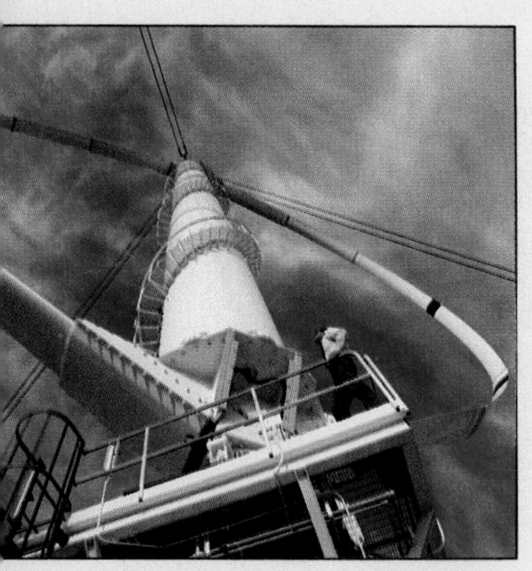

Vertical-axis wind generator. (U.S. Department of Energy)

Special thanks go to the following people for writing guest essays: Isaac D. Abella, University of Chicago; Albert A. Bartlett, University of Colorado at Boulder; Gordon Batson, Clarkson University; Leon Blitzer, University of Arizona; Roger A. Freedman and Paul K. Hansma, University of California, Santa Barbara; Robert G. Fuller, University of Nebraska; Clark D. Hamilton, National Bureau of Standards; Laurent D. Hodges, Iowa State University; A. Jayaraman, AT&T/Bell Laboratories; Edward Lacy; Samson A. Marshall, Michigan Technological Institute; John D. Meakin, University of Delaware; Philip Morrison, Massachusetts Institute of Technology; Brian B. Schwartz, Brooklyn College, C.U.N.Y., and the American Physical Society; Clifford Will, Washington University; Sidney C. Wolff, National Optical Astronomy Observatory; Dean A. Zollman, Kansas State University; Alma C. Zook, Pomona College. I appreciate the assistance of Carl T. Tomizuka in coordinating the essays.

A surfer "riding the pipe" on a wave. (© Doug Peebles/Index Stock International)

I am especially grateful to the following people for their careful accuracy reviews of all the problems and examples in the text: Stanley Bashkin, University of Arizona; Jeffrey J. Braun, University of Evansville; Louis H. Cadwell, Providence College, Ralph V. McGrew, Broome Community College; Charles D. Teague, Eastern Kentucky University; and Steve Van Wyk, Chapman College.

I appreciate the assistance of Jeffrey J. Braun, Charles Teague and Steve Van Wyk in reorganizing the problem sets. My grateful thanks also go to Steve Van Wyk and Louis H. Cadwell for the preparation of the Instructor's Manual that accompanies the text. I am indebted to my colleague and friend John R. Gordon for his many contributions during the development of this text, his continued encouragement and support, and for his expertise in revising the Student Study Guide. I am grateful to David Oliver for developing the computer software that accompanies the Student Study Guide and to David Stetser for developing the Spreadsheet Data Disks that accompany the Instructor's Manual and the Spreadsheet Appendix in the back of this text. Support for David Stetser's work has been provided by Miami University—Middletown and the Department of Physics and Astronomy, Center for Advanced Studies, University of New Mexico, Albuquerque, New Mexico. I thank David Loyd for preparing the Physics Laboratory Manual and accompanying Instructor's Manual that can be used with this text. I appreciate the assistance of Louis H. Cadwell in preparing the answers that appear at the end of the text, and in preparing some of the test questions for the Computerized Test Bank and Printed Test Bank. I also thank the staff of the Physics Department at Georgia Tech for providing many of the questions for this test bank. I am grateful to Mario Iona for making many excellent suggestions for improving the figures in the text. I thank Sarah Evans, Ellen Newman, Henry Leap, and Jim Lehman for locating and/or providing many excellent photographs. I thank my son Mark for writing many of the biographical sketches included in this edition. I thank Agatha Brabon, Linda Delosh, Mary Thomas, Georgina Valverde, and Linda Miller for an excellent job in typing various stages of the original manuscript. During the development of this textbook, I have benefited from valuable discussions with many people including Subash Antani, Gabe Anton, Randall Caton, Don Chodrow, Jerry Faughn, John R. Gordon, Herb Helbig, Lawrence Hmurcik, William Ingham, David Kaup, Len Ketelsen, Henry Leap, H. Kent Moore, Charles McFarland, Frank Moore, Clem Moses, Curt Moyer, William Parks, Dorn Peterson, Joe Rudmin, Joe Scaturro, Alex Serway, John Serway, Georgio Vianson, and Harold Zimmerman. Special recognition is due to my mentor and friend, Sam Marshall, a gifted teacher and scientist who helped me sharpen my writing skills while I was a graduate student.

Special thanks and recognition go to the professional staff at Saunders College Publishing for their fine work during the development and production of this text, especially Ellen Newman, Senior Developmental Editor; Sally Kusch, Senior Project Manager; and Carol Bleistine, Manager of Art and Design. I thank John Vondeling, Associate Publisher, for his great enthusiasm for the project, his friendship, and his confidence in me as an author. I am most appreciative of the intelligent copyediting by Will Eaton, the excellent artwork by Tom Mallon and the excellent design work by Edward A. Butler.

A special note of appreciation goes to the hundreds of students at Clarkson University who used the first edition of this text in manuscript form during its development. I am most grateful for the supportive environment provided by James Madison University. I also wish to thank the many users of the second edition who submitted suggestions and pointed out errors. With the help of such cooperative efforts, I hope to have achieved my main objective; that is, to provide an effective textbook for the student.

And last, I thank my wonderful family for their continued patience and understanding. The completion of this enormous task would not have been possible without their endless love and faith in me.

Raymond A. Serway
James Madison University
Harrisonburg, Virginia

To the Student

I feel it is appropriate to offer some words of advice which should be of benefit to you, the student. Before doing so, I will assume that you have read the preface, which describes the various features of the text that will help you through the course.

Computer simulation of gaseous flow around a post in a rocket engine. (U.S. Department of Energy)

HOW TO STUDY

Very often instructors are asked "How should I study physics and prepare for examinations?" There is no simple answer to this question, but I would like to offer some suggestions based on my own experiences in learning and teaching over the years.

First and foremost, maintain a positive attitude towards the subject matter, keeping in mind that physics is the most fundamental of all natural sciences. Other science courses that follow will use the same physical principles, so it is important that you understand and be able to apply the various concepts and theories discussed in the text.

CONCEPTS AND PRINCIPLES

It is essential that you understand the basic concepts and principles *before* attempting to solve assigned problems. This is best accomplished through a careful reading of the textbook before attending your lecture on that material. In the process, it is useful to jot down certain points which are not clear to you. Take careful notes in class, and then ask questions pertaining to those ideas that require clarification. Keep in mind that few people are able to absorb the full meaning of scientific material after one reading. Several readings of the text and notes may be necessary. Your lectures and laboratory work should supplement the text and clarify some of the more difficult material. You should reduce memorization of material to a minimum. Memorizing passages from a text, equations, and derivations does not necessarily mean you understand the material. Your understanding of the material will be enhanced through a combination of efficient study habits, discussions with other students and instructors, and your ability to solve the problems in the text. Ask questions whenever you feel it is necessary.

STUDY SCHEDULE

It is important to set up a regular study schedule, preferably on a daily basis. Make sure to read the syllabus for the course and adhere to the schedule set by your instructor. The lectures will be much more meaningful if you read the corresponding textual material before attending the lecture. As a general rule, you should devote about two hours of study time for every hour in class. If you are having trouble with the course, seek the advice of the instructor or students who have taken the course. You may find it necessary to seek further instruction from experienced students. Very often, instructors will offer review sessions in addition to regular class periods. It is important that you avoid the practice of delaying study until a day or two before an exam. More often than not, this will lead to disastrous results. Rather than an all-night study session, it is better to briefly review the basic concepts and equations, followed by a good night's rest. If you feel in need of additional help in understanding the concepts, preparing for exams, or in problem-solving, we suggest that you acquire a copy of the Student Study Guide which accompanies the text, which should be available at your college bookstore.

USE THE FEATURES

You should make full use of the various features of the text discussed in the preface. For example, marginal notes are useful for locating and describing important equations and concepts, while important statements and definitions are highlighted in color. Many useful tables are contained in appendices, but most are incorporated in the text where they are used most often. Appendix B is a convenient review of mathematical techniques. Answers to odd-numbered problems are given at the end of the text, and answers to most end-of-chapter questions are provided in the study guide. Exercises (with answers), which follow some worked examples, represent extensions of those examples, and in most cases you are expected to perform a simple calculation. Their purpose is to test your problem-solving skills as you read through the text. Problem-Solving Strategies and Hints are included in selected chapters throughout the text to give you additional information to help you solve problems. An overview of the entire text is given in the table of contents, while the index will enable you to locate specific material quickly. Footnotes are sometimes used to supplement the discussion or to cite other references on the subject. Many chapters include problems that require the use of programmable calculators or computers. Problems and examples with boxes can be solved either analytically or with the use of spreadsheets available from your instructor. These are intended for those courses that place some emphasis on numerical methods. You may want to develop appropriate programs for some of these problems even if they are not assigned by your instructor.

After reading a chapter, you should be able to define any new quantities introduced in that chapter, and discuss the principles and assumptions that were used to arrive at certain key relations. The chapter summaries and the review sections of the study guide should help you in this regard. In some cases, it will be necessary to refer to the index of the text to locate certain topics. You should be able to correctly associate with each physical quantity a symbol used to represent that quantity and the unit in which the quantity is specified. Furthermore, you should be able to express each important relation in a concise and accurate prose statement.

A model steam engine. (Courtesy of CENCO)

THE IMPORTANCE OF PROBLEM SOLVING

R.P. Feynman, Nobel laureate in physics, once said, "You do not know anything until you have practiced." In keeping with this statement, I strongly advise that you develop the skills necessary to solve a wide range of problems. Your ability to solve problems will be one of the main tests of your knowledge of physics, and therefore you should try to solve as many problems as possible. It is essential that you understand basic concepts and principles before attempting to solve problems. It is good practice to try to find alternate solutions to the same problem. For example, problems in mechanics can be solved using Newton's laws, but very often an alternative method using energy considerations is more direct. You should not deceive yourself into thinking you understand the problem after seeing its solution in class. You must be able to solve the problem and similar problems on your own.

The method of solving problems should be carefully planned. A systematic plan is especially important when a problem involves several concepts. First, read the problem several times until you are confident you understand what is being asked. Look for any key words that will help you interpret the problem, and perhaps allow you to make certain assumptions. Your ability to interpret the question properly is an integral part of problem solving. You should acquire the habit of writing down the information given in a problem, and decide what quantities need to be found. You might want to construct a table listing quantities given, and quantities to be found. This procedure is sometimes used in the worked examples of the text. After you have decided on the method you feel is appropriate for the situation, proceed with your solution. General problem-solving strategies of this type are included in the text and are highlighted by a light blue screen.

I often find that students fail to recognize the limitations of certain formulas or physical laws in a particular situation. It is very important that you understand and remember the assumptions which underlie a particular theory or formalism. For example, certain equations in kinematics apply only to a particle moving with constant acceleration. These equations are not valid for situations in which the acceleration is not constant, such as the motion of an object connected to a spring, or the motion of an object through a fluid.

GENERAL PROBLEM-SOLVING STRATEGY

Most courses in general physics require the student to learn the skills of problem solving, and examinations are largely composed of problems that test such skills. This brief section describes some useful ideas which will enable you to increase your accuracy in solving problems, enhance your understanding of physical concepts, eliminate initial panic or lack of direction in approaching a problem, and organize your work. One way to help accomplish these goals is to adopt a problem-solving strategy. Many chapters in this text will include a section labeled "Problem-Solving Strategies and Hints" which should help you through the "rough spots."

In developing problem-solving strategies, five basic steps are commonly used.

1. Draw a suitable diagram with appropriate labels and coordinate axes if needed.
2. As you examine what is being asked in the problem, identify the basic physical principle (or principles) that are involved, listing the knowns and unknowns.
3. Select a basic relationship or derive an equation that can be used to find the unknown, and solve the equation for the unknown symbolically.
4. Substitute the given values along with the appropriate units into the equation.
5. Obtain a numerical value for the unknown. The problem is verified and receives a check mark if the following questions can be properly answered: Do the units match? Is the answer reasonable? Is the plus or minus sign proper or meaningful?

Diagram

Given Data

Basic Equation

Working Equation

Evaluation and Check

A menu for problem solving.

One of the purposes of this strategy is to promote accuracy. Properly drawn diagrams can eliminate many sign errors. Diagrams also help to isolate the physical principles of the problem. Symbolic solutions and carefully labeled knowns and unknowns will help eliminate other careless errors. The use of symbolic solutions should help you think in terms of the physics of the problem. A check of units at the end of the problem can indicate a possible algebraic error. The physical layout and organization of your problem will make the final product more understandable and easier to follow. Once you have developed an organized system for examining problems and extracting relevant information, you will become a more confident problem solver.

EXAMPLE A person driving in a car at a speed of 20 m/s applies the brakes and stops in a distance of 100 m. What was the acceleration of the car?

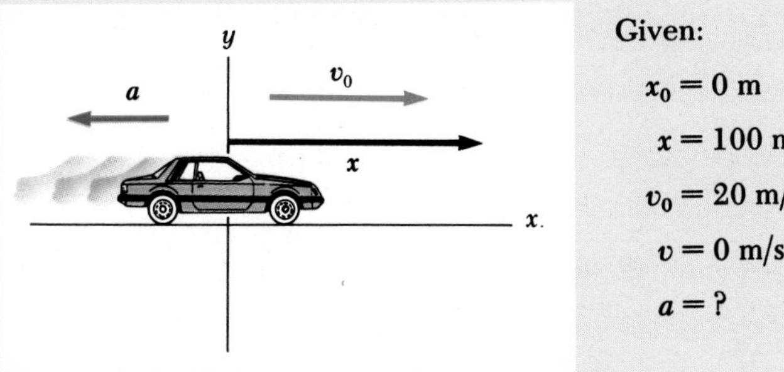

Given:

$x_0 = 0$ m

$x = 100$ m

$v_0 = 20$ m/s

$v = 0$ m/s

$a = ?$

$$v^2 = v_0^2 + 2a(x - x_0) \qquad a = \frac{(0\ \text{m/s})^2 - (20\ \text{m/s})^2}{2(100\ \text{m})} = -2\ \text{m/s}^2$$

$$v^2 - v_0^2 = 2a(x - x_0)$$

$$a = \frac{v^2 - v_0^2}{2(x - x_0)} \qquad \frac{\text{m}^2/\text{s}^2}{\text{m}} = \frac{\text{m}}{\text{s}^2}$$

EXPERIMENTS

Physics is a science based upon experimental observations. In view of this fact, I recommend that you try to supplement the text through various type of "hands-on" experiments, either at home or in the laboratory. These can be used to test ideas and models discussed in class or in the text. For example, the common "Slinky" toy is excellent for studying traveling waves; a ball swinging on the end of a long string can be used to investigate pendulum motion; various masses attached to the end of a vertical spring or rubber band can be used to determine their elastic nature; an old pair of Polaroid sunglasses and some discarded lenses and magnifying glass are the components of various experiments in optics; you can get an approximate measure of the acceleration of gravity by dropping a ball from a known height by simply measuring the time of its fall with a stopwatch. The list is endless. When physical models are not available, be imaginative and try to develop models of your own.

AN INVITATION TO PHYSICS

It is my sincere hope that you too will find physics an exciting and enjoyable experience, and that you will profit from this experience, regardless of your chosen profession. Welcome to the exciting world of physics.

The scientist does not study nature because it is useful; he studies it because he delights in it, and he delights in it because it is beautiful. If nature were not beautiful, it would not be worth knowing, and if nature were not worth knowing, life would not be worth living.

Henri Poincaré

Pedagogical Use of Color

The various colors that you will see in the illustrations of this text are used to improve clarity and understanding. Many figures with three-dimensional representations are air-brushed in various colors to make them as realistic as possible.

Color coding has been used in various parts of the book to identify specific physical quantities. The following schemes should be noted.

Part I (Chapters 1–15) covers topics in classical mechanics:

- Displacement and position vectors
- Velocity vectors
- Force vectors
- Acceleration vectors
- Torque and angular momentum vectors
- Linear or rotational motion directions or

Part IV (Chapters 23–34) deals with topics in electricity and magnetism:

- Electric fields
- Magnetic fields
- Positive charges
- Negative charges
- Resistors, batteries, and switches
- Capacitors and inductors (coils)
- Voltmeters
- Ammeters
- Ground symbol

Part V (Chapters 35–38) discusses geometrical and physical optics:

- Light rays
- Lenses
- Mirrors
- Objects
- Images

Most graphs are presented with curves plotted in either red or blue, and coordinate axes are in black. Several colors are used in those graphs where many physical quantities are plotted simultaneously, or in those cases where different processes may be occurring and need to be distinguished.

Finally, motional shading effects have been incorporated in many figures to remind the readers that they are dealing with a dynamic system rather than a static system. These figures will appear to be somewhat like a "multiflash" photograph of a moving system, with faint images of the "past history" of the system's path. In some figures, a broad, colored arrow is used to indicate the direction of motion of the system.

In addition to the use of color in the figures, the pedagogy in the text has been enhanced with color as well. We have used the following color-coded system:

Important equations	tan screen
Important statements	blue type
Marginal notes	blue type
Problem-Solving Strategy and Hints	light blue screen
Examples	blue box
Summaries	gold screen
Intermediate-level problems	problem number in blue type
Challenging problems	problem number in magenta type
Spreadsheet problems	black or color box around problem number

The Guest Essays are marked with a blue bar on the edge of the page so they can be located easily.

Introducing the Book

5
The Laws of Motion

A sailboat with its spinnaker billowing runs before the wind. What force or forces are causing the forward motion? Does the water exert a force on the boat? (© Doug Peebles/Index Stock International)

In the previous two chapters on kinematics, we described the motion of particles based on the definition of displacement, velocity, and acceleration. However, we would like to be able to answer specific questions related to the causes of motion, such as "What mechanism causes motion?" and "Why do some objects accelerate at a higher rate than others?" In this chapter, we shall describe the change in motion of particles using the concepts of force, mass, and momentum. We shall then discuss the three basic laws of motion, which are based on experimental observations and were formulated nearly three centuries ago by Sir Isaac Newton.

5.1 INTRODUCTION TO CLASSICAL MECHANICS

The purpose of classical mechanics is to provide a connection between the motion of a body and the forces acting on it. Keep in mind that classical mechanics deals with objects that are large compared with the dimensions of atoms

95

PHOTOGRAPHS

Chapter opening photographs have been carefully selected to relate to the principles discussed in the chapter. The captions serve as an added instructional tool.

CHAPTER PREVIEWS

Include a brief discussion of chapter objectives and content.

COLOR-CODED LINE ART

Color is used to add clarity and three-dimensional effects. A complete description of the color system used appears on page xvii.

$$\sum F_x = T = ma_x \qquad \text{or} \qquad a_x = \frac{T}{m}$$

In this situation, there is no acceleration in the y direction. Applying $\Sigma F_y = ma_y$ with $a_y = 0$ gives

$$N - W = 0 \qquad \text{or} \qquad N = W$$

That is, the normal force is equal to and opposite the weight.

If $\boldsymbol{T}$ is a *constant* force, then the acceleration, $a_x = T/m$, is also a constant. Hence, the equations of kinematics from Chapter 3 can be used to obtain the displacement, Δx, and velocity, v, as functions of time. Since $a_x = T/m =$ constant, these expressions can be written

$$\Delta x = v_0 t + \tfrac{1}{2}\left(\frac{T}{m}\right)t^2$$

$$v = v_0 + \left(\frac{T}{m}\right)t$$

where v_0 is the velocity at $t = 0$.

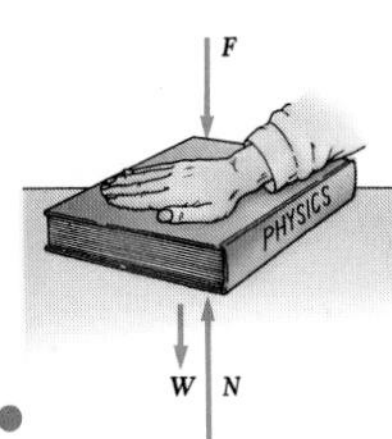

Figure 5.8 When one pushes downward on an object with a force $\boldsymbol{F}$, the normal force $\boldsymbol{N}$ is greater than the weight. That is, $\boldsymbol{N} = \boldsymbol{W} + \boldsymbol{F}$.

In the example just presented, we found that the normal force $\boldsymbol{N}$ was equal in magnitude and opposite the weight $\boldsymbol{W}$. *This is not always the case.* For example, suppose you were to push down on a book with a force $\boldsymbol{F}$ as in Figure 5.8. In this case, $\Sigma F_y = 0$ gives $N - W - F = 0$, or $N = W + F$. Other examples in which $N = W$ will be presented later.

Consider a lamp of weight W suspended from a chain of negligible weight fastened to the ceiling, as in Figure 5.9a. The free-body diagram for the lamp is shown in Figure 5.9b, where the forces on it are the weight, $\boldsymbol{W}$, acting downward, and the force of the chain on the lamp, $\boldsymbol{T}$, acting upward. The force $\boldsymbol{T}$ is the constraint force in this case. (If we cut the chain, $\boldsymbol{T} = 0$ and the body executes free fall.)

If we apply the first law to the lamp, noting that $\boldsymbol{a} = 0$, we see that since there are no forces in the x direction, the equation $\Sigma F_x = 0$ provides no helpful information. The condition $\Sigma F_y = 0$ gives

$$\sum F_y = T - W = 0 \qquad \text{or} \qquad T = W$$

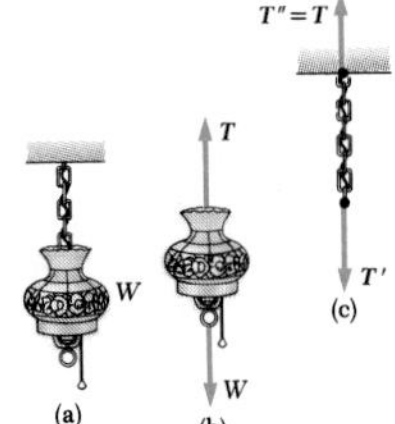

Figure 5.9 (a) A lamp of weight $\boldsymbol{W}$ suspended by a light chain from a ceiling. (b) The forces acting on the lamp are the force of gravity, $\boldsymbol{W}$, and the tension in the chain, $\boldsymbol{T}$. (c) The forces acting on the chain are $\boldsymbol{T}'$, that exerted by the lamp, and $\boldsymbol{T}''$, that exerted by the ceiling.

Note that $\boldsymbol{T}$ and $\boldsymbol{W}$ are *not* action-reaction pairs. The reaction to $\boldsymbol{T}$ is $\boldsymbol{T}'$, the force exerted on the chain by the lamp, as in Figure 5.9c. The force $\boldsymbol{T}'$ acts downward and is transmitted to the ceiling. That is, the force of the chain on the ceiling, $\boldsymbol{T}'$, is *downward* and equal to W in magnitude. The ceiling exerts an equal and opposite force, $\boldsymbol{T}'' = \boldsymbol{T}$, on the chain, as in Figure 5.9c.

PROBLEM-SOLVING STRATEGY

The following procedure is recommended when dealing with problems involving the application of Newton's laws:

1. Draw a simple, neat diagram of the system.
2. Isolate the object of interest whose motion is being analyzed. Draw a free-body diagram for this object, that is, a diagram showing *all external forces acting on the object.* For systems containing more than one object, draw *separate* diagrams for each object. *Do not* include forces that the object exerts on its surroundings.

PROBLEM-SOLVING STRATEGIES AND HINTS

Outlines general strategies for solving the types of problems featured in both the examples and the end-of-chapter problems. These steps are highlighted with a light blue screen so they can be easily located.

3. Establish convenient coordinate axes for each body and find the components of the forces along these axes. Now, apply Newton's second law, $\Sigma \boldsymbol{F} = m\boldsymbol{a}$, in *component* form. Check your dimensions to make sure that all terms have units of force.
4. Solve the component equations for the unknowns. Remember that you must have as many independent equations as you have unknowns in order to obtain a complete solution.
5. It is a good idea to check the predictions of your solutions for extreme values of the variables. You can often detect errors in your results by doing so.

EXAMPLE 5.2 A Traffic Light at Rest

A traffic light weighing 100 N hangs from a cable tied to two other cables fastened to a support, as in Figure 5.10a. The upper cables make angles of 37° and 53° with the horizontal. Find the tension in the three cables.

Solution First we construct a free-body diagram for the traffic light, as in Figure 5.10b. The tension in the vertical cable, T_3, supports the light, and so we see that $T_3 = W = 100$ N. Now we construct a free-body diagram for the knot that holds the three cables together, as in Figure 5.10c. This is a convenient point to choose because all forces in question act at this point. We choose the coordinate axes as shown in Figure 5.10c and resolve the forces into their x and y components:

Force	x component	y component
T_1	$-T_1 \cos 37°$	$T_1 \sin 37°$
T_2	$T_2 \cos 53°$	$T_2 \sin 53°$
T_3	0	-100 N

The first condition for equilibrium gives us the equations

$$(1) \quad \sum F_x = T_2 \cos 53° - T_1 \cos 37° = 0$$

$$(2) \quad \sum F_y = T_1 \sin 37° + T_2 \sin 53° - 100 \text{ N} = 0$$

From (1) we see that the horizontal components of $\boldsymbol{T}_1$ and $\boldsymbol{T}_2$ must be equal in magnitude, and from (2) we see that the sum of the vertical components of $\boldsymbol{T}_1$ and $\boldsymbol{T}_2$ must balance the weight of the light. We can solve (1) for T_2 in terms of T_1 to give

$$T_2 = T_1 \left(\frac{\cos 37°}{\cos 53°} \right) = 1.33T_1$$

This value for T_2 can be substituted into (2) to give

$$T_1 \sin 37° + (1.33T_1)(\sin 53°) - 100 \text{ N} = 0$$

$$T_1 = 60.0 \text{ N}$$

$$T_2 = 1.33T_1 = 79.8 \text{ N}$$

Exercise 2 In what situation will $T_1 = T_2$?
Answer When the supporting cables make equal angles with the horizontal support.

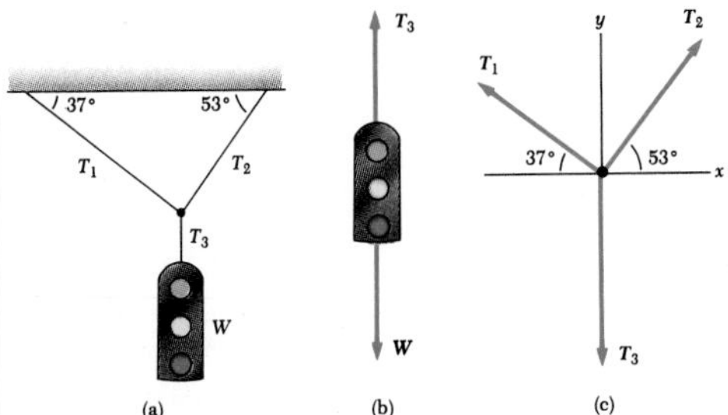

Figure 5.10 (Example 5.2) (a) A traffic light suspended by cables. (b) Free-body diagram for the traffic light. (c) Free-body diagram for the knot.

WORKED EXAMPLE EXERCISES

Many of the worked examples are followed immediately by exercises with answers to make the textbook more interactive and to test the student's understanding of problem-solving techniques.

WORKED EXAMPLES

Presented as an aid in understanding concepts, these examples serve as models for solving the end-of-chapter problems. The examples are set off in a blue box, and the solution answer is highlighted with a tan screen.

SUMMARY

Newton's first law

Newton's first law states that a body at rest will remain at rest or a body in uniform motion in a straight line will maintain that motion unless an external resultant force acts on the body.

Newton's second law

Newton's second law states that the time rate of change of momentum of a body is equal to the resultant force acting on the body. If the mass of the body is constant, the net force equals the product of the mass and its acceleration, or $\Sigma \mathbf{F} = m\mathbf{a}$.

Newton's first and second laws are valid in an inertial frame of reference. An **inertial frame** is one in which an object, subject to no net external force, moves with constant velocity including the special case of $v = 0$.

Inertial frame

Mass is a scalar quantity. The mass that appears in Newton's second law is called **inertial mass**.

Weight

The **weight** of a body is equal to the product of its mass and the acceleration of gravity, or $\mathbf{W} = m\mathbf{g}$.

Newton's third law

Newton's third law states that if two bodies interact, the force exerted on body 1 by body 2 is equal to and opposite the force exerted on body 2 by body 1. Thus, an isolated force cannot exist in nature.

Forces of friction

The n
surface is
mum forc
$f_s \leq \mu_s N$, w
When a b
opposite t
nitude of t
friction. U

More on F

As we hav
ing Newto
recognize
construct
the free-b
of mechan
free-body
to constru
lems. Whe
you constr
As usu
notes a no

SUMMARY

Reviews important concepts and equations discussed in the chapter.

MARGINAL NOTES

Comments and notes in blue print are used to locate important statements, equations, and concepts in the chapter.

PROBLEMS KEYED TO SECTIONS

Allows students to refer back to the sections to look at a worked example for help in solving problems.

its speed increases to 7.0×10^5 m/s in a distance of 5.0 cm. Assuming its acceleration is constant, (a) determine the force on the electron and (b) compare this force with the weight of the electron, which we neglected.

25. A 15-lb block rests on the floor. (a) What force does the floor exert on the block?(b) If a rope is tied to the block and run vertically over a pulley and the other end attached to a free-hanging 10-lb weight, what is the force of the floor on the 15-lb block? (c) If we replace the 10-lb weight in (b) by a 20-lb weight, what is the force of the floor on the 15-lb block?

Section 5.8 Some Applications of Newton's Laws

26. Find the tension in each cord for the systems described in Figure 5.22. (Neglect the mass of the cords.)

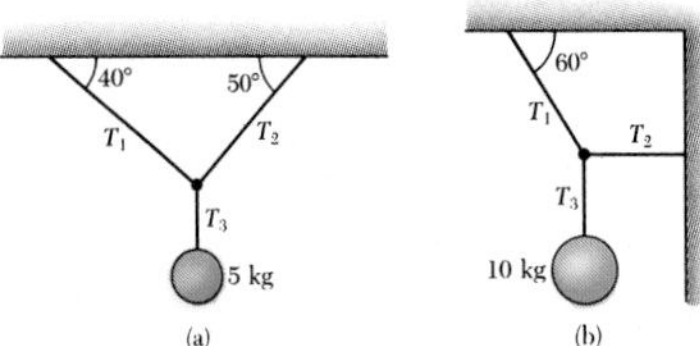

Figure 5.22 (Problem 26).

27. Three 10-kg objects are suspended as shown (Fig. 5.23). Determine each of the labeled tension forces.

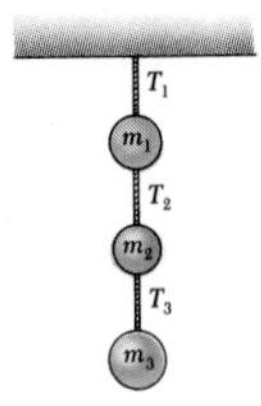

Figure 5.23 (Problem 27).

28. A 200-N weight is tied to the middle of a strong rope, and two people pull at opposite ends of the rope in an attempt to lift the weight. (a) What force $\mathbf{F}$ must each person apply to suspend the weight as shown in Figure 5.24? (b) Can they pull in such a way as to make the rope horizontal? Explain.

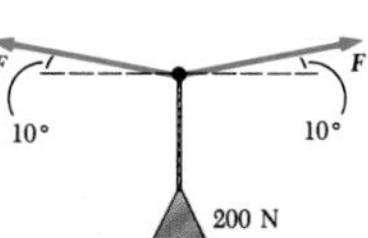

Figure 5.24 (Problem 28).

29. A 50-kg mass hangs from a rope 5 m in length, which is fastened to the ceiling. What horizontal force applied to the mass will deflect it 1 m sideways from the vertical and maintain it in that position?

30. The systems shown in Figure 5.25 are in equilibrium. If the spring scales are calibrated in N, what do they read in each case? (Neglect the mass of the pulleys and strings, and assume the incline is smooth.)

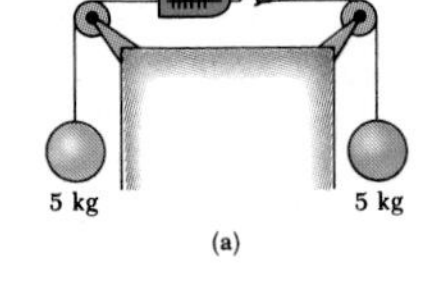

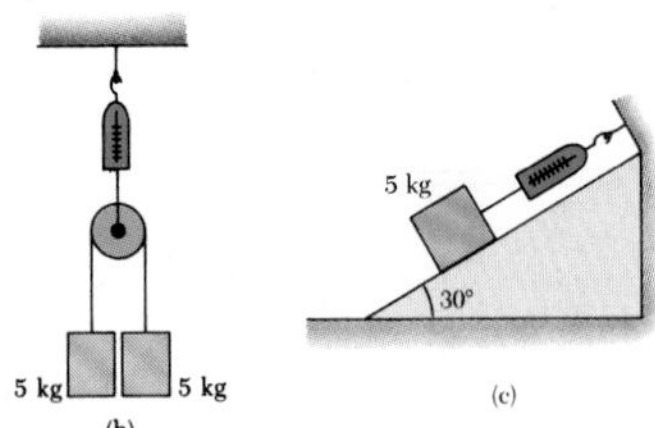

Figure 5.25 (Problem 30).

31. A bag of cement hangs from three wires as shown (Fig. 5.26). Two of the wires make angles θ_1 and θ_2 with the horizontal. If the system is in equilibrium, (a) show that

$$T_1 = \frac{W\cos(\theta_2)}{\sin(\theta_1 + \theta_2)}$$

(b) Given that $W = 200$ N, $\theta_1 = 10°$ and $\theta_2 = 25°$, find the tensions T_1, T_2, and T_3 in the wires.

32. A woman at an airport is pulling her 20-kg suitcase at constant speed by pulling on a strap at an angle θ

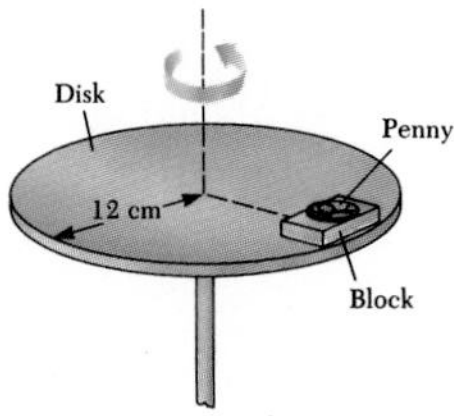

Figure 6.24 (Problem 50).

through its center, as shown in Figure 6.25. The period of revolution of the hoop is T. The bead has mass m and is at rest relative to the hoop at the angle θ. Answer the following in terms of the given parameters and the acceleration due to gravity g. (a) Find the normal force $\mathbf{N}$ exerted on the bead by the wire. (b) Find the angle between the normal force and the axis of rotation. (c) Find the centripetal force exerted on the bead.

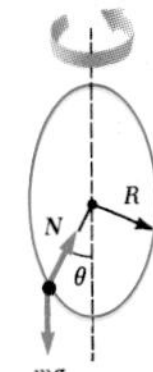

Figure 6.25 (Problem 51).

52. An amusement park ride consists of a rotating circular platform 8 m in diameter from which "bucket seats" are suspended at the end of 2.5-m chains. (For simplicity only one chain is shown [Figure 6.26], though for stability at least two would be used.) When the system rotates the chains holding the seats make an angle $\theta = 28°$ with the vertical. (a) What is the speed of the seat? (b) If a child of mass 40 kg sits in the 10-kg seat, what is the tension in the chain?

53. The following experiment is performed in a spacecraft that is at rest where the net gravitational field is zero. A small sphere is injected into a viscous medium with initial velocity $\mathbf{v}_0$. The sphere experiences a resistive force $\mathbf{R} = -b\mathbf{v}$. Find the velocity of the sphere as a function of time. (*Hint:* Apply Newton's second law, write a as dv/dt, separate the variables, and integrate the equation.)

CALCULATOR/COMPUTER PROBLEMS

54. A hailstone of mass 4.8×10^{-4} kg and radius 0.5 cm falls through the atmosphere and experiences a net force given by Equation 6.9. This expression can be written in the form

$$m\frac{dv}{dt} = mg - Kv^2$$

where $K = \frac{1}{2}C\rho A$, $\rho = 1.29$ kg/m^3 and $C = 0.5$. (a) What is the terminal velocity of the hailstone? (b) Use a method of numerical integration to find the velocity and position of the hailstone at 1-s intervals, taking $v_0 = 0$. Continue your calculation until terminal velocity is reached.

55. A 0.5-kg block slides down a 30° incline of length 1 m. The coefficient of kinetic friction between the block and the incline varies with the block's velocity according to the expression

$$\mu = 0.3 + 1.2\sqrt{v}$$

where v is in m/s. (a) Use a numerical method to find the velocity of the block at intervals of 10 cm during its motion. (b) If the length of the plane is extended to several km, will the block reach terminal velocity? If so, what is its terminal velocity, and at what point does it occur on the incline?

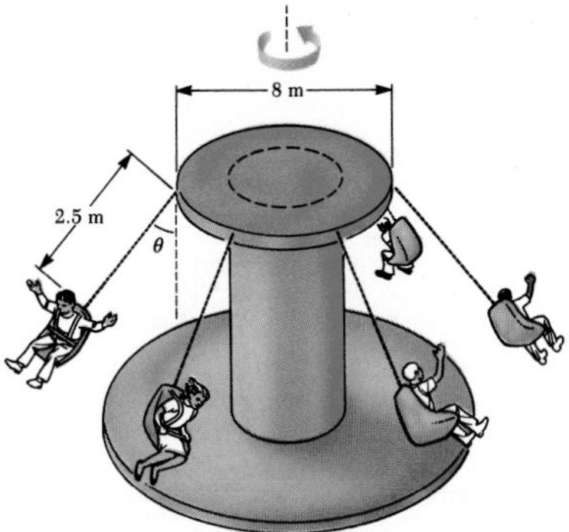

Figure 6.26 (Problem 52).

SPREADSHEET PROBLEMS

Spreadsheets on a separate disk available from your instructor accompany these problems. These problems can also be solved analytically without using the software.

PROBLEMS GRADED BY LEVEL OF DIFFICULTY

Intermediate-level problems are indicated by printing the problem number in blue print; challenging problems have a magenta problem number.

The fictional traveler Lemuel Gulliver spent a busy time in a kingdom called Lilliput, where all living things—men, cattle, trees, grass—were exactly similar to our world, except that they were all built on the scale of one inch to the foot. Lilliputians were a little under 6 inches high, on the average, and built proportionately just as we are. Gulliver also visited Brobdingnag, the country of the giants, who were exactly like men but 12 times as tall. As Swift described it, daily life in both kingdoms was about like ours (in the 18th century). His commentary on human behavior is still worth reading, but we shall see that people of such sizes just could not have been as he described them.

Long before Swift lived, Galileo understood why very small or very large models of man could not be like us, but apparently Dean Swift had never read what Galileo wrote. One character in Galileo's "Two New Sciences" says, "Now since . . . in geometry, . . . mere size cuts no figure, I do not see that the properties of circles, triangles, cylinders, cones, and other solid figures will change with their size. . . ." But his physicist friend replies, "The common opinion is here absolutely wrong." Let us see why.

We start with the strength of a rope. It is easy to see that if one man who pulls with a certain strength can almost break a certain rope, two such ropes will just withstand the pull of two men. A single large rope with the same total area of cross-section as the two smaller ropes combined will contain just double the number of fibers of one of the small ropes, and it will also do the job. In
or rope is proportional to its area of cro
Experience and theory agree in this c
holds, not only for ropes or cables supp
supporting a thrust. The thrust which a c
a given material, is also proportional to t

Now the body of a man or an animal i
skeleton—supported by various braces
But the weight of the body which must b
flesh and bone present, that is, to the vo

Let us now compare Gulliver with th
Since the giant is exactly like Gulliver in
sions is 12 times the corresponding one
columns and braces is proportional to the
of their linear dimension (strength ∝ L^2),
as Gulliver's. Because his weight is propo
12^3 or 1728 times as great as Gulliver's.
ratio a dozen times smaller than ours. Just
much trouble as we should have in carry

In reality, of course, Lilliput and Br
effects of a difference in scale if we com
The smaller ones are not scale models o
sponding leg bones of two closely rela
gazelle, the other a bison. Notice that the
geometrically to that of the smaller. It is
ing the scale change, which would make

Galileo wrote very clearly on this ve
dingnag, or of any normal-looking giants
giant the same proportion of limb as that
a harder and stronger material for making
strength in comparison with men of me

*Adapted from PSSC PHYSICS, 2nd edition,
Development Center, Inc., Newton, MA.

ESSAY

Scaling—the Physics of Lilliput*

Philip Morrison
Massachusetts Institute of Technology

GUEST ESSAYS

Intended as supplemental readings for the student. The blue bar at the edges of the pages help students locate them easily. The essays have been placed at the ends of chapters to avoid interrupting the main textual material.

very biggest things we can make which have some roundness, which are fully three-dimensional, are buildings and great ships. These lack a good deal of being a thousand times larger than men in their linear dimensions.

Within our present technology our scaling arguments are important. If we design a new large object on the basis of a small one, we are warned that new effects too small to detect on our scale may enter and even become the most important things to consider. We cannot just scale up and down blindly, geometrically, but by scaling in the light of physical reasoning, we can sometimes foresee what changes will occur. In this way we can employ scaling in intelligent airplane design, for example, and not arrive at a jet transport that looks like a bee—and won't fly.

Suggested Readings

Galileo, Galilei, "Dialogues Concerning Two New Sciences," trans. by Henry Crew and Alphonso De Salvio, Evanston, Northwestern University Press, 1946, pp. 1–6, 125–128.

Haldane, J. B. S., "On Being the Right Size." *World of Mathematics,* Vol. II, edited by James R. Newman. New York, Simon & Schuster, 1956.

Holcomb, Donald F. and Philip Morrison, *My Father's Watch—Aspects of the Physical World,* Englewood Cliffs, NJ, Prentice Hall, 1974, pp. 68–83.

Smith, Cyril S., "The Shape of Things," *Scientific American,* January, 1954, p. 58.

Thompson, D'Arcy W., "On Magnitude," in *On Growth and Form,* Cambridge University Press, 1952 and 1961.

Essay Questions

1. The leg bones of one animal are twice as strong as those of another closely related animal of similar shape. (a) What would you expect to be the ratio of these animals' heights? (b) What would you expect to be the ratio of their weights?
2. A hummingbird must eat very frequently and even then must have a highly concentrated form of food such as sugar. What does the concept of scaling tell you about the size of a hummingbird?
3. About how many Lilliputians would it take to equal the mass of one citizen of Brobdingnag?

Essay Problems

1. The total surface area of a rectangular solid is the sum of the areas of the six faces. If each dimension of a given rectangular solid is doubled, what effect does this have on the total surface area?
2. A hollow metal sphere has a wall thickness of 2 cm. If you increase both the diameter and thickness of this sphere so that the overall volume is three times the original overall volume, how thick will the shell of the new sphere be?
3. If your height and all your other dimensions were doubled, by what factor would this change (a) your weight? (b) the ability of your leg bones to support your weight?
4. According to the zoo, an elephant of mass 4.0×10^3 kg consumes 3.4×10^2 times as much food as a guinea pig of mass 0.70 kg. They are both warm-blooded, plant-eating, similarly shaped animals. Find the ratio of their surface areas, which is approximately the ratio of their heat losses, and compare it with the known ratio of food consumed.
5. A rectangular water tank is supported above the ground by four pillars 5 m long whose diameters are 20 cm. If the tank were made 10 times longer, wider, and deeper, what diameter pillars would be needed? How much more water would the tank hold?
6. How many state maps of scale 1 : 1 000 000 would you need to cover the state with those maps?

SUGGESTED READINGS, ESSAY QUESTIONS, AND PROBLEMS

Most of the essays contain suggested readings, questions, and problems for added flexibility in covering these topics.

Contents Overview

Thermogram produced with an infrared scanner. (VANSCAN® Thermogram by Daedalus Enterprises, Inc.)

Contents

° These sections are optional.

Coronet formed by a drop of milk.
(© Harold E. Edgerton. Courtesy of Palm Press, Inc.)

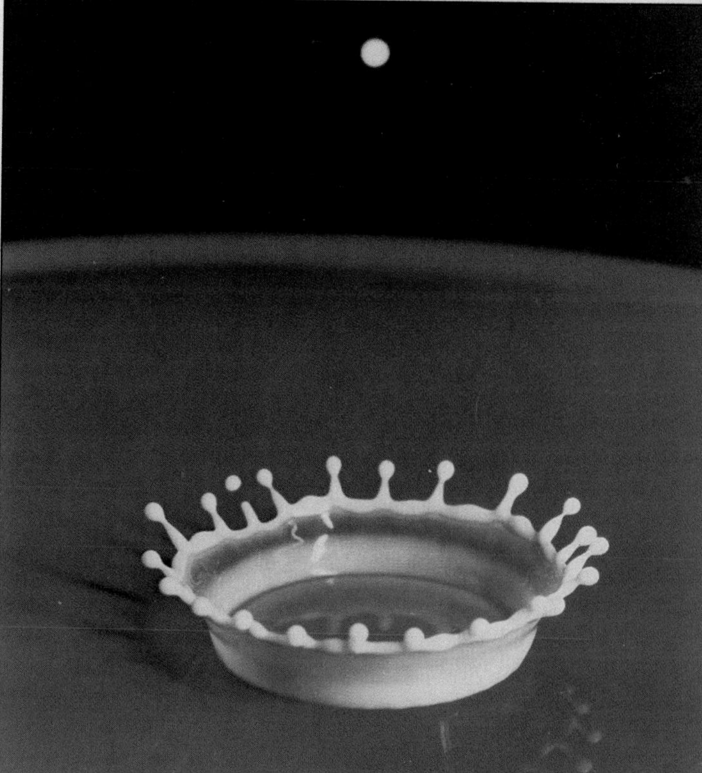

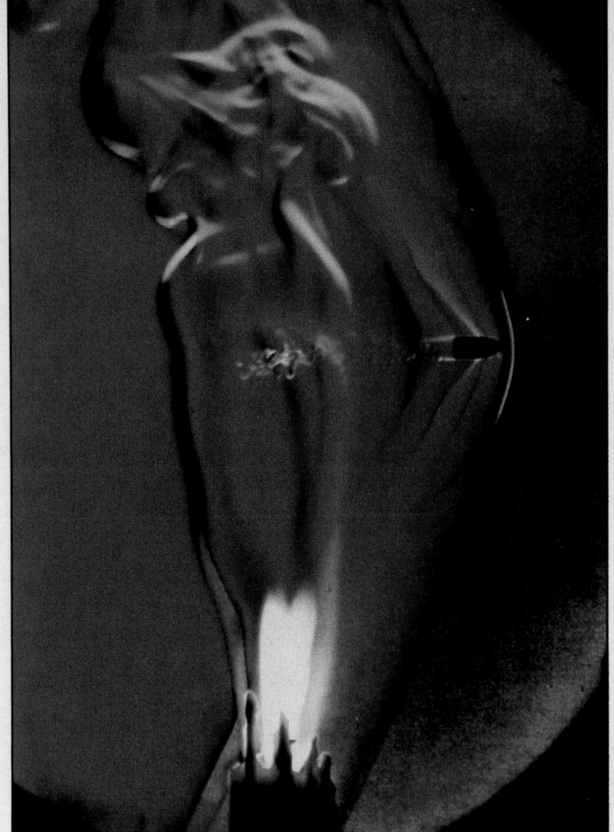

Supersonic candlelight. (Bullet through Flame, 1973. © Harold E. Edgerton. Courtesy of Palm Press, Inc.)

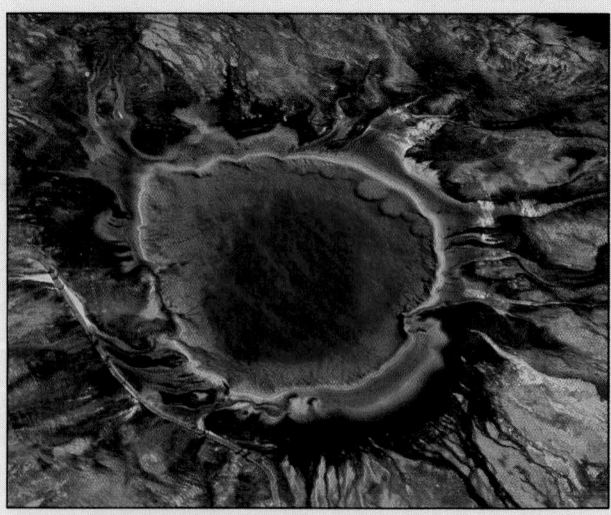

Grand Prismatic Spring, Yellowstone National Park. (© Paul Chesley/Photographers Aspen)

The first integrated circuit. (Courtesy of Texas Instruments)

Aurora Borealis seen from inside a dome on the western shore of Hudson Bay. (© David Hiser/Photographers Aspen)

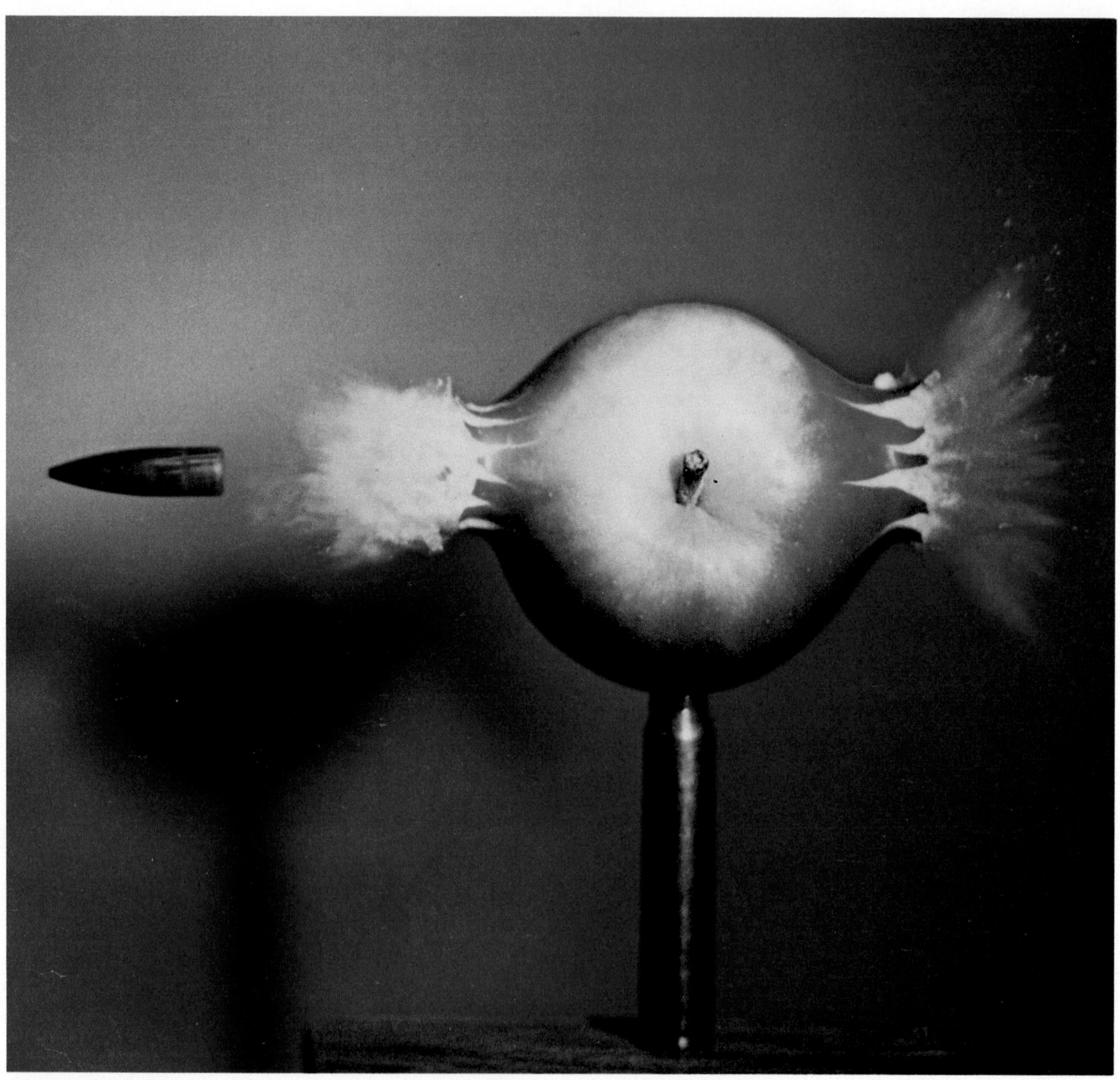

What would William Tell think? An apple being pierced by a 30-caliber bullet traveling at a supersonic speed of 900 m/s. This collision was photographed with a microflash stroboscope using an exposure time of 0.33 μs. Shortly after the photograph was taken, the apple disintegrated completely. Note that the points of both entry and exit of the bullet are visually explosive. (Shooting the Apple, 1964. © Harold Edgerton. Courtesy of Palm Press, Inc.)

PART I
Mechanics

Facts which at first seem improbable will, even in scant explanation, drop the cloak which has hidden them and stand forth in naked and simple beauty.
GALILEO GALILEI

Nature and Nature's laws lay hid in night: God said, Let Newton be! and all was light.
ALEXANDER POPE

Physics, the most fundamental physical science, is concerned with the basic principles of the universe. It is the foundation upon which the other physical sciences—astronomy, chemistry, and geology—are based. The beauty of physics lies in the simplicity of the fundamental physical theories and in the manner in which just a small number of fundamental concepts, equations, and assumptions can alter and expand our view of the world around us.

The myriad physical phenomena in our world are a part of one or more of the following five areas of physics:

1. Mechanics, which is concerned with the motion of material objects
2. Thermodynamics, which deals with heat, temperature, and the behavior of a large number of particles
3. Electromagnetism, which involves the theory of electricity, magnetism, and electromagnetic fields
4. Relativity, which is a theory describing particles moving at any speed less than the speed of light
5. Quantum mechanics, a theory dealing with the behavior of particles at the submicroscopic level as well as the macroscopic world

The first part of this textbook deals with mechanics, sometimes referred to as classical mechanics or newtonian mechanics. This is an appropriate place to begin an introductory text since many of the basic principles used to understand mechanical systems can later be used to describe such natural phenomena as waves and heat transfer. Furthermore, the laws of conservation of energy and momentum introduced in mechanics retain their importance in the fundamental theories that follow, including the theories of modern physics.

The first serious attempts to develop a theory of motion were provided by the Greek astronomers and philosophers. Although they devised a complex model to describe the motions of heavenly bodies, their model lacked correlation between such motions and the motions of objects on earth. The study of mechanics was enhanced by a number of careful astronomical investigations by Copernicus, Brahe, and Kepler in the 16th century. In the 16th and 17th centuries, Galileo attempted to relate the motion of falling bodies and projectiles to the motion of planetary bodies, and Sevin and Hooke were studying forces and their relation to motion. A major development in the theory of mechanics was provided by Newton in 1687 when he published his *Principia.* Newton's elegant theory, which remained unchallenged for more than 200 years, was based on Newton's hypothesis of universal gravitation and contributions made by Galileo and others.

Today, mechanics is of vital importance to students from all disciplines. It is highly successful in describing the motions of material bodies, such as the planets, rockets, and baseballs. In the first part of the text, we shall describe the laws of mechanics and examine a wide range of phenomena that can be understood with these fundamental ideas.

1
Introduction: Physics and Measurement

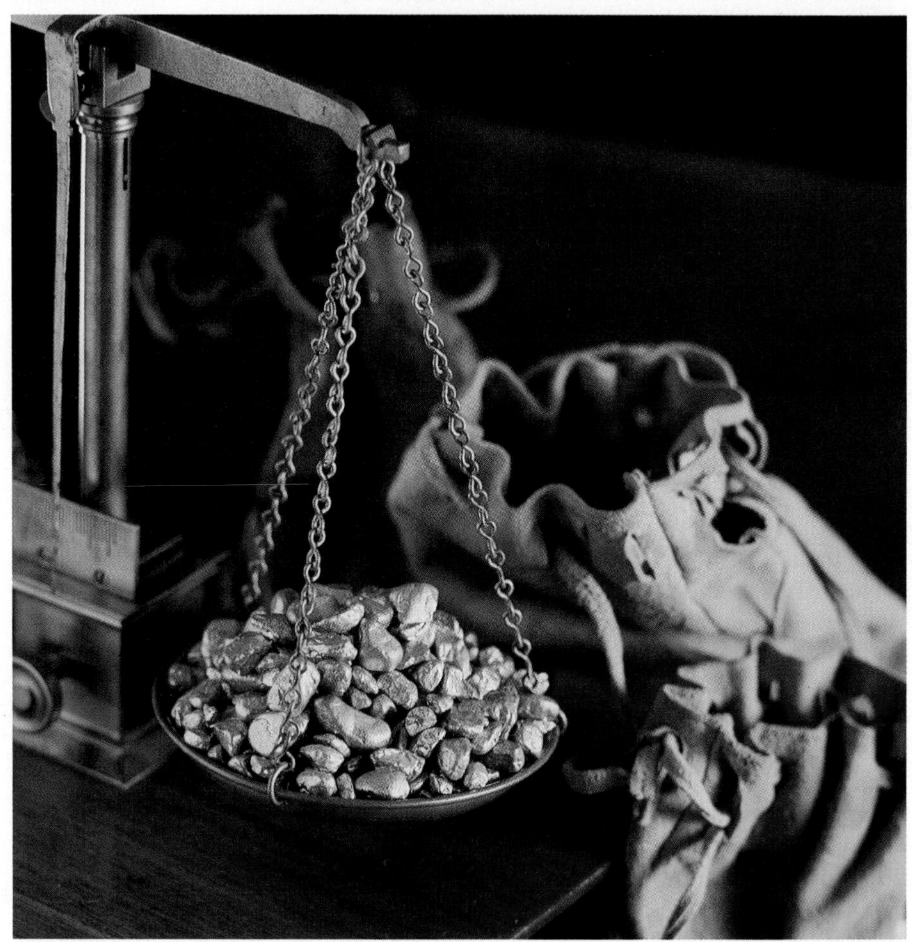

Quantitative measurement is one of the tools of physicists. Here gold is shown being measured on a scale. (© R. Pleasant, F.P.G. International.)

Physics is a fundamental science concerned with understanding the natural phenomena that occur in our universe. It is a science based on experimental observations and quantitative measurements. The main objective of the scientific approach is to develop physical theories based on fundamental laws that will predict the results of some experiments. Fortunately, it is possible to explain the behavior of many physical systems with a limited number of fundamental laws. These fundamental laws are expressed in the language of mathematics, the tool that provides a bridge between theory and experiment.

Whenever a discrepancy arises between theory and experiment, new theories and experiments must be formulated to remove the discrepancy.

Many times a theory is satisfactory under limited conditions; a more general theory might be satisfactory without such limitations. A classic example is Newton's laws of motion, which accurately describe the motion of bodies at normal speeds but do not apply to objects moving at speeds comparable to the speed of light. The special theory of relativity developed by Albert Einstein (1879–1955) successfully predicts the motion of objects at speeds approaching the speed of light and hence is a more general theory of motion.

Classical physics, developed prior to 1900, includes the theories, concepts, laws, and experiments in three major disciplines: (1) classical mechanics, (2) thermodynamics (heat transfer, temperature, and the behavior of a large number of identical particles), and (3) electromagnetism (the study of electric and magnetic phenomena, optics and radiation).

Galileo Galilei (1564–1642) made significant contributions to classical mechanics through his work on the laws of motion with constant acceleration. In the same era, Johannes Kepler (1571–1630) used astronomical observations to develop empirical laws for the motion of planetary bodies.

The most important contributions to classical mechanics were provided by Isaac Newton (1642–1727), who developed classical mechanics as a systematic theory and was one of the originators of the calculus as a mathematical tool. Major developments in classical physics continued in the 18th century. However, thermodynamics and electricity and magnetism were not developed until the latter part of the 19th century, principally because the apparatus for controlled experiments was either too crude or unavailable until then. Although many electric and magnetic phenomena had been studied earlier, it was the work of James Clerk Maxwell (1831–1879) that provided a unified theory of electromagnetism. In this text we shall treat the various disciplines of classical physics in separate sections; however, we will see that the disciplines of mechanics and electromagnetism are basic to all the branches of classical and modern physics.

A new era in physics, usually referred to as *modern physics*, began near the end of the 19th century. Modern physics developed mainly because of the discovery that many physical phenomena could not be explained by classical physics. The two most important developments in this modern era were the theories of relativity and quantum mechanics. Einstein's theory of relativity completely revolutionized the traditional concepts of space, time, and energy. Among other things, Einstein's theory corrected Newton's laws of motion for describing the motion of objects moving at speeds comparable to the speed of light. The theory of relativity also assumes that the speed of light is the upper limit of the speed of an object or signal and shows the equivalence of mass and energy. The formulation of quantum mechanics by a number of distinguished scientists provided a description of physical phenomena at the atomic level.

Scientists are constantly working at improving our understanding of fundamental laws, and new discoveries are being made every day. In many research areas, there is a great deal of overlap between physics, chemistry, and biology. The many technological advances in recent times are the result of the efforts of many scientists, engineers, and technicians. Some of the most notable recent developments are (1) unmanned space missions and manned moon landings, (2) microcircuitry and high-speed computers, and (3) sophisticated imaging techniques used in scientific research and medicine. The impact of such developments and discoveries on our society has indeed been great, and it is very likely that future discoveries and developments will be exciting, challenging, and of great benefit to humanity.

1.1 STANDARDS OF LENGTH, MASS, AND TIME

The laws of physics are expressed in terms of basic quantities that require a clear definition. For example, such physical quantities as force, velocity, volume, and acceleration can be described in terms of more fundamental quantities that in themselves are defined in terms of measurements or comparison with established standards. In mechanics, the three fundamental quantities are length (L), time (T), and mass (M). All other physical quantities in mechanics can be expressed in terms of these.

Obviously, if we are to report the results of a measurement to someone who wishes to reproduce this measurement, a standard must be defined. It would be meaningless if a visitor from another planet were to talk to us about a length of 8 "glitches" if we do not know the meaning of the unit "glitch." On the other hand, if someone familiar with our system of measurement reports that a wall is 2 meters high and our unit of length is defined as 1 meter, we then know that the height of the wall is twice our fundamental unit of length. Likewise, if we are told that a person has a mass of 75 kilograms and our unit of mass is defined as 1.0 kilogram, then that person is 75 times as massive as our fundamental unit of mass.[1]

In 1960, an international committee established rules to decide on a set of standards for these fundamental quantities. The system that was established is an adaptation of the metric system, and it is called the **International System** (SI) of units. The abbreviation SI comes from its French name "Système Internationale." In this system, the units of mass, length, and time are the kilogram, meter, and second, respectively. Other standard SI units established by the committee are those for temperature (the *kelvin*), electric current (the *ampere*), luminous intensity (the *candela*) and for the amount of substance (the *mole*, see Sec. 1.2). These seven fundamental units are the basic SI units. In the study of mechanics, however, we will be concerned only with the units of mass, length, and time. Definitions of units are under constant review and are changed from time to time.

Mass

> The SI unit of mass, the **kilogram**, is defined as the mass of a specific platinum-iridium alloy cylinder kept at the International Bureau of Weights and Measures at Sèvres, France.

This mass standard was established in 1887, and there has been no change since that time because platinum-iridium is an unusually stable alloy. The Sèvres cylinder is 3.9 centimeters in diameter and 3.9 centimeters in height. A duplicate is kept at the National Bureau of Standards in Gaithersburg, Md.

[1] The need for assigning numerical values to various physical quantities through experimentation was expressed by Lord Kelvin (William Thomson) as follows: "I often say that when you can measure what you are speaking about, and express it in numbers, you should know something about it, but when you cannot express it in numbers, your knowledge is of a meager and unsatisfactory kind."

(*Left*) The National Standard Kilogram No. 20, an accurate copy of the International Standard Kilogram kept at Sèvres, France, is housed under a double bell jar in a vault at the National Bureau of Standards. (*Right*) The primary frequency standard (an atomic clock) at the National Bureau of Standards. This device keeps time with an accuracy of about 3 millionths of a second per year. (Photos courtesy of National Bureau of Standards, U.S. Department of Commerce)

Time

Before 1960, the standard of time was defined in terms of the *mean solar day* for the year 1900.[2] Thus, the *mean solar second*, representing the basic unit of time, was originally defined as $(\frac{1}{60})(\frac{1}{60})(\frac{1}{24})$ of a mean solar day. Time that is referenced to the rotation of the earth about its axis is called *universal time*. The rotation of the earth is now known to vary substantially with time.

In 1967, the second was redefined to take advantage of the high precision that could be obtained using a device known as an *atomic clock*. In this device, the frequencies associated with certain atomic transitions (which are extremely stable and insensitive to the clock's environment) can be measured to a precision of one part in 10^{12}. This is equivalent to an uncertainty of less than one second every 30 000 years. Such frequencies are highly insensitive to changes in the clock's environment. Thus, in 1967 the SI unit of time, the *second*, was redefined using the characteristic frequency of a particular kind of cesium atom as the "reference clock":

> **One second**—the time required for a cesium-133 atom to undergo 9 192 631 770 vibrations.

This new standard has the distinct advantage of being "indestructible" and reproducible.

[2] A solar day is the time interval between successive appearances of the sun at the highest point it reaches in the sky each day.

TABLE 1.1 Mass of Various Bodies (Approximate Values)

	Mass (kg)
Milky Way Galaxy	7×10^{41}
Sun	2×10^{30}
Earth	6×10^{24}
Moon	7×10^{22}
Shark	1×10^{3}
Human	7×10^{1}
Frog	1×10^{-1}
Mosquito	1×10^{-5}
Bacterium	1×10^{-15}
Hydrogen atom	1.67×10^{-27}
Electron	9.11×10^{-31}

Length

When the metric system was first established in France in 1792, the meter was defined to be 10^{-7} times the distance from the equator to the north pole through Paris. This standard was eventually abandoned for practical reasons. Until 1960, the standard for length, the *meter,* was defined as the distance between two lines on a specific platinum-iridium bar stored under controlled conditions. This standard was abandoned for several reasons, a principal one being that the limited accuracy with which the separation between the lines on the bar can be determined does not meet the present requirements of science and technology. Until recently, the meter was defined as 1 650 763.73 wavelengths of orange-red light emitted from a krypton-86 lamp. However, in October 1983, the meter was redefined as follows:

> **One meter** — the distance traveled by light in vacuum during a time of 1/299 792 458 second.

In effect, this latest definition establishes that the speed of light in vacuum is 299 792 458 meters per second.

The orders of magnitude (approximate values) of various masses, lengths, and time intervals are presented in Tables 1.1 to 1.3. Note the wide range of these quantities.[3] You should study these tables and get a feel for what is meant by a kilogram of mass, for example, or by a time interval of 10^{10} seconds. Systems of units commonly used are the SI or *mks* system, in which the units of mass, length, and time are the kilogram (kg), meter (m), and second (s), respectively; the *cgs* or Gaussian system, in which the units of length, mass, and time are the centimeter (cm), gram (g), and second, respectively; and the British engineering system (sometimes called the conventional system), in which the

TABLE 1.2 Approximate Values of Some Measured Lengths

	Length (m)
Distance from earth to most remote quasar known	1.4×10^{26}
Distance from earth to most remote known normal galaxies	4×10^{25}
Distance from earth to nearest large galaxy (M 31 in Andromeda)	2×10^{22}
Distance from sun to nearest star (Proxima Centauri)	4×10^{16}
One lightyear	9.46×10^{15}
Mean orbit radius of the earth	1.5×10^{11}
Mean distance from earth to moon	3.8×10^{8}
Distance from the equator to the north pole	1×10^{7}
Mean radius of the earth	6.4×10^{6}
Typical altitude of orbiting earth satellite	2×10^{5}
Length of a football field	9.1×10^{1}
Length of a housefly	5×10^{-3}
Size of smallest dust particles	1×10^{-4}
Size of cells of most living organisms	1×10^{-5}
Diameter of a hydrogen atom	1×10^{-10}
Diameter of an atomic nucleus	1×10^{-14}

[3] If you are unfamiliar with the use of powers of ten (scientific notation), you should review Section B.2 of the mathematical appendix at the back of this book.

TABLE 1.3 Approximate Values of Some Time Intervals

	Interval (s)
Age of the universe	5×10^{17}
Age of the earth	1.3×10^{17}
Average age of a college student	6.3×10^{8}
One year	3.2×10^{7}
One day (time for one revolution of earth about its axis)	8.6×10^{4}
Time between normal heartbeats	8×10^{-1}
Period[a] of audible sound waves	1×10^{-3}
Period of typical radio waves	1×10^{-6}
Period of vibration of an atom in a solid	1×10^{-13}
Period of visible light waves	2×10^{-15}
Duration of a nuclear collision	1×10^{-22}
Time for light to cross a proton	3.3×10^{-24}

[a] Period is defined as the time interval of one complete vibration.

units of length, mass, and time are the foot (ft), slug, and second, respectively. Throughout most of this text we shall use SI units since they are almost universally accepted in science and industry. We will make some limited use of conventional units in the study of classical mechanics.

Some of the most frequently used prefixes for the various powers of ten and their abbreviations are listed in Table 1.4. For example, 10^{-3} m is equivalent to 1 millimeter (mm), and 10^3 m is 1 kilometer (km). Likewise, 1 kg is 10^3 g, and 1 megavolt (MV) is 10^6 volts.

TABLE 1.4 Some Prefixes for Powers of Ten

Power	Prefix	Abbreviation
10^{-18}	atto	a
10^{-15}	femto	f
10^{-12}	pico	p
10^{-9}	nano	n
10^{-6}	micro	μ
10^{-3}	milli	m
10^{-2}	centi	c
10^{-1}	deci	d
10^{3}	kilo	k
10^{6}	mega	M
10^{9}	giga	G
10^{12}	tera	T
10^{15}	peta	P
10^{18}	exa	E

1.2 DENSITY AND ATOMIC MASS

A fundamental property of any substance is its **density** ρ (Greek letter rho), defined as *mass per unit volume* (a table of the letters in the Greek alphabet is provided at the back of the book):

$$\rho \equiv \frac{m}{V} \tag{1.1}$$

For example, aluminum has a density of 2.70 g/cm^3, and lead has a density of 11.3 g/cm^3. Therefore, a piece of aluminum of volume 10 cm^3 has a mass of 27.0 g, while an equivalent volume of lead would have a mass of 113 g. A list of densities for various substances is given in Table 1.5.

The difference in density between aluminum and lead is due, in part, to their different *atomic weights;* the atomic weight of lead is 207 and that of aluminum is 27. However, the ratio of atomic weights, $207/27 = 7.67$, does not correspond to the ratio of densities, $11.3/2.70 = 4.19$. The discrepancy is due to the difference in atomic spacings and atomic arrangements in their crystal structures.

All ordinary matter consists of atoms, and each atom is made up of electrons and a nucleus. Practically all of the mass of an atom is contained in the nucleus, which consists of protons and neutrons. Thus we can understand why the atomic weights of the various elements differ. The mass of a nucleus is

TABLE 1.5 Densities of Various Substances

Substance	Density ρ (kg/m^3)
Gold	19.3×10^3
Uranium	18.7×10^3
Lead	11.3×10^3
Copper	8.93×10^3
Iron	7.86×10^3
Aluminum	2.70×10^3
Magnesium	1.75×10^3
Water	1.00×10^3
Air	0.0013×10^3

measured relative to the mass of an atom of the carbon-12 isotope (this isotope of carbon has six protons and six neutrons).

The mass of ^{12}C is defined to be exactly 12 atomic mass units (u), where $1\ \text{u} = 1.6605402 \times 10^{-27}$ kg. In these units, the proton and neutron have masses of about 1 u. More precisely,

$$\text{mass of proton} = 1.0073\ \text{u}$$

$$\text{mass of neutron} = 1.0087\ \text{u}$$

The mass of the nucleus of ^{27}Al is *approximately* 27 u. In fact, a more precise measurement shows that the nuclear mass is always slightly *less* than the combined mass of the protons and neutrons making up the nucleus. The processes of nuclear fission and nuclear fusion are based on this mass difference.

One **mole** of any element (or compound) consists of Avogadro's number, N_A, of molecules of the substance. Avogadro's number is defined so that one mole of carbon-12 atoms has a mass of exactly 12 g. Its value has been found to be $N_A = 6.02 \times 10^{23}$ molecules/mole. For example, one mole of aluminum has a mass of 27 g, and one mole of lead has a mass of 207 g. Although the two have different masses, one mole of aluminum contains the same number of atoms as one mole of lead. Since there are 6.02×10^{23} atoms in one mole of any element, the mass per atom is given by

Atomic mass

$$m = \frac{\text{atomic weight}}{N_A} \tag{1.2}$$

For example, the mass of an aluminum atom is

$$m = \frac{27\ \text{g/mole}}{6.02 \times 10^{23}\ \text{atoms/mole}} = 4.5 \times 10^{-23}\ \text{g/atom}$$

Note that 1 u is equal to $N_A{}^{-1}$ g.

EXAMPLE 1.1 How Many Atoms in the Cube?
A solid cube of aluminum (density 2.7 g/cm³) has a volume of 0.2 cm³. How many aluminum atoms are contained in the cube?

Solution Since the density equals mass per unit volume, the mass of the cube is

$$\rho V = (2.7\ \text{g/cm}^3)(0.2\ \text{cm}^3) = 0.54\ \text{g}.$$

To find the number of atoms, N, we can set up a proportion using the fact that one mole of aluminum (27 g) contains 6.02×10^{23} atoms:

$$\frac{6.02 \times 10^{23}\ \text{atoms}}{27\ \text{g}} = \frac{N}{0.54\ \text{g}}$$

$$N = \frac{(0.54\ \text{g})(6.02 \times 10^{23}\ \text{atoms})}{27\ \text{g}} = 1.2 \times 10^{22}\ \text{atoms}$$

1.3 DIMENSIONAL ANALYSIS

The word *dimension* has a special meaning in physics. It usually denotes the physical nature of a quantity. Whether a distance is measured in units of feet or meters or furlongs, it is a distance. We say its dimension is *length*.

The symbols that will be used to specify length, mass, and time are L, M, and T, respectively. We will often use brackets [ ] to denote the dimensions of a physical quantity. For example, in this notation the dimensions of velocity,

TABLE 1.6 Dimensions of Area, Volume, Velocity, and Acceleration

System	Area (L^2)	Volume (L^3)	Velocity (L/T)	Acceleration (L/T^2)
SI	m^2	m^3	m/s	m/s^2
cgs	cm^2	cm^3	cm/s	cm/s^2
British engineering (conventional)	ft^2	ft^3	ft/s	ft/s^2

v, are written $[v] = L/T$, and the dimensions of area, A, are $[A] = L^2$. The dimensions of area, volume, velocity, and acceleration are listed in Table 1.6, along with their units in the three common systems. The dimensions of other quantities, such as force and energy, will be described as they are introduced in the text.

In many situations, you may be faced with having to derive or check a specific formula. Although you may have forgotten the details of the derivation, there is a useful and powerful procedure called *dimensional analysis* that can be used to assist in the derivation or to check your final expression. This procedure should always be used and should help minimize the rote memorization of equations. Dimensional analysis makes use of the fact that *dimensions can be treated as algebraic quantities.* That is, quantities can be added or subtracted only if they have the same dimensions. Furthermore, the terms on each side of an equation must have the same dimensions. By following these simple rules, you can use dimensional analysis to help determine whether or not an expression has the correct form, because the relationship can be correct only if the dimensions on each side of the equation are the same.

To illustrate this procedure, suppose you wish to derive a formula for the distance x traveled by a car in a time t if the car starts from rest and moves with constant acceleration a. In Chapter 3, we shall find that the correct expression for this special case is $x = \frac{1}{2}at^2$. Let us check the validity of this expression from a dimensional analysis approach.

The quantity x on the left side has the dimension of length. In order for the equation to be dimensionally correct, the quantity on the right side must also have the dimension of length. We can perform a dimensional check by substituting the basic units for acceleration, L/T^2, and time, T, into the equation. That is, the dimensional form of the equation $x = \frac{1}{2}at^2$ can be written as

$$L = \frac{L}{\cancel{T^2}} \cdot \cancel{T^2} = L$$

The units of time cancel as shown, leaving the unit of length.

A more general procedure of dimensional analysis is to set up an expression of the form

$$x \propto a^n t^m$$

when n and m are exponents that must be determined and the symbol $\propto$ indicates a proportionality. This relationship is only correct if the dimensions of both sides are the same. Since the dimension of the left side is length, the dimension of the right side must also be length. That is,

$$[a^n t^m] = L = LT^0$$

Since the dimensions of acceleration are L/T^2 and the dimension of time is T,

$$(L/T^2)^n T^m = L$$

or

$$L^n T^{m-2n} = L$$

Since the exponents of L and T must be the same on both sides, we see that $n = 1$ and $m = 2$. Therefore, we conclude that

$$x \propto at^2$$

This result differs by a factor of 2 from the correct expression, which is given by $x = \frac{1}{2}at^2$.

EXAMPLE 1.2 Analysis of an Equation

Show that the expression $v = v_0 + at$ is dimensionally correct, where v and v_0 represent velocities, a is acceleration, and t is a time interval.

Solution Since

$$[v] = [v_0] = L/T$$

and the dimensions of acceleration are L/T^2, the dimensions of at are

$$[at] = (L/T^2)(T) = L/T$$

and the expression is dimensionally correct. On the other hand, if the expression were given as $v = v_0 + at^2$, it would be dimensionally *incorrect.* Try it and see!

EXAMPLE 1.3 Analysis of a Power Law

Suppose we are told that the acceleration of a particle moving in a circle of radius r with uniform velocity v is proportional to some power of r, say r^n, and some power of v, say v^m. How can we determine the powers of r and v?

Solution Let us take a to be

$$a = kr^n v^m$$

where k is a dimensionless constant. With the known dimensions of a, r, and v, we see that the dimensional equation must be

$$L/T^2 = L^n(L/T)^m = L^{n+m}/T^m$$

This dimensional equation is balanced under the conditions

$$n + m = 1 \qquad \text{and} \qquad m = 2$$

Therefore, $n = -1$ and we can write the acceleration

$$a = kr^{-1}v^2 = k\frac{v^2}{r}$$

When we discuss uniform circular motion later, we shall see that $k = 1$ if compatible units are used. Note that k would not equal 1 if, for example, v were in km/h and you wanted a in m/s^2.

1.4 CONVERSION OF UNITS

Sometimes it is necessary to convert units from one system to another. Conversion factors between the SI and conventional units of length are as follows:

$$1 \text{ mile} = 1609 \text{ m} = 1.609 \text{ km} \qquad 1 \text{ ft} = 0.3048 \text{ m} = 30.48 \text{ cm}$$

$$1 \text{ m} = 39.37 \text{ in.} = 3.281 \text{ ft} \qquad 1 \text{ in.} = 0.0254 \text{ m} = 2.54 \text{ cm}$$

A more complete list of conversion factors can be found in Appendix A. Units can be treated as algebraic quantities that can cancel each other. For

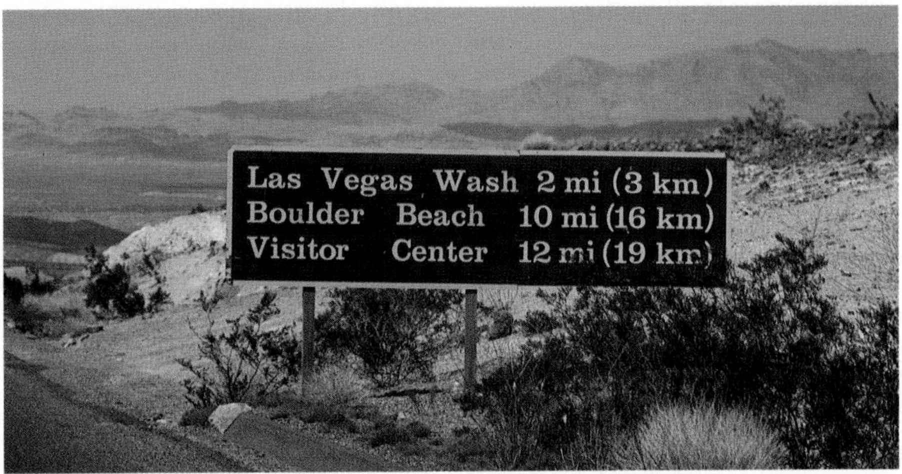

Conversion of miles to kilometers. (Courtesy of Ohio Department of Transportation)

example, suppose we wish to convert 15.0 in. to centimeters. Since 1 in. = 2.54 cm (exactly), we find that

$$15.0 \text{ in.} = (15.0 \cancel{\text{in.}}) \left(2.54 \frac{\text{cm}}{\cancel{\text{in.}}}\right) = 38.1 \text{ cm}$$

EXAMPLE 1.4 The Density of a Cube
The mass of a solid cube is 856 g and each edge has a length of 5.35 cm. Determine the density ρ of the cube in SI units.

Solution Since 1 g = 10^{-3} kg and 1 cm = 10^{-2} m, the mass, m, and volume, V, in SI units are given by

$$m = 856 \text{ g} \times 10^{-3} \text{ kg/g} = 0.856 \text{ kg}$$

$$V = L^3 = (5.35 \text{ cm} \times 10^{-2} \text{ m/cm})^3$$
$$= (5.35)^3 \times 10^{-6} \text{ m}^3 = 1.53 \times 10^{-4} \text{ m}^3$$

Therefore

$$\rho = \frac{m}{V} = \frac{0.856 \text{ kg}}{1.53 \times 10^{-4} \text{ m}^3} = 5.60 \times 10^3 \text{ kg/m}^3$$

1.5 ORDER-OF-MAGNITUDE CALCULATIONS

It is often useful to compute an approximate answer to a given physical problem even where little information is available. Such results can then be used to determine whether or not a more precise calculation is necessary. These approximations are usually based on certain assumptions, which must be modified if more precision is needed. Thus, we will sometimes refer to an *order of magnitude* of a certain quantity as the power of ten of the number that describes that quantity. Usually, when an order-of-magnitude calculation is made, the results are reliable to within a factor of 10. If a quantity increases in value by three orders of magnitude, this means that its value is increased by a factor of $10^3 = 1000$.

The spirit of attempting order-of-magnitude calculations, sometimes referred to as "guesstimates" or "ball-park figures," is given in the following quotation: "Make an estimate before every calculation, try a simple physical argument . . . before every derivation, guess the answer to every puzzle. Courage: no one else needs to know what the guess is."[4]

EXAMPLE 1.5 The Number of Atoms in a Solid
Estimate the number of atoms in 1 cm^3 of a solid.

Solution From Table 1.2 we note that the diameter of an atom is about 10^{-10} m. Thus, if in our model we assume that the atoms in the solid are solid spheres of this diameter, then the volume of each sphere is about 10^{-30} m^3 (more precisely, volume $= 4\pi r^3/3 = \pi d^3/6$, where $r = d/2$). Therefore, since 1 $cm^3 = 10^{-6}$ m^3, the number of atoms in the solid is of the order of $10^{-6}/10^{-30} = 10^{24}$ atoms.

A more precise calculation would require knowledge of the density of the solid and the mass of each atom. However, our estimate agrees with the more precise calculation to within a factor of 10. (This approach should be used for Problem 38.)

EXAMPLE 1.6 How Much Gas Do We Use?
Estimate the number of gallons of gasoline used by all U.S. cars each year.

Solution Since there are about 200 million people in the United States, an estimate of the number of cars in the country is 40 million (assuming one car and five people per family). We must also estimate that the average distance traveled per year is 10 000 miles. If we assume a gasoline consumption of 20 mi/gal, each car uses about 500 gal/year. Multiplying this by the total number of cars in the United States gives an estimated total consumption of 2×10^{10} gal. This corresponds to a yearly consumer expenditure of over 20 billion dollars! This is probably a low estimate since we haven't accounted for commercial consumption and for such factors as two-car families.

1.6 SIGNIFICANT FIGURES

When one performs measurements on certain quantities, the measured values are only known to within the limits of the experimental uncertainty. The value of the uncertainty can depend on various factors such as the quality of the apparatus, the skill of the experimenter, and the number of measurements performed. Suppose that in a laboratory experiment we are asked to measure the area of a rectangular plate using a meter stick as a measuring instrument. Let us assume that the accuracy to which we can measure a particular dimension of the plate is ± 0.1 cm. If the length of the plate is measured to be 16.3 cm, we can only claim that its length lies somewhere between 16.2 cm and 16.4 cm. In this case, we say that the measured value has three significant figures. Likewise, if its width is measured to be 4.5 cm, the actual value lies between 4.4 cm and 4.6 cm. (This measured value has only two significant figures.) Note that the significant figures include the first estimated digit. Thus, we could write the measured values as 16.3 ± 0.1 cm and 4.5 ± 0.1 cm.

Suppose now that we would like to find the area of the plate by multiplying the two measured values together. If we were to claim that the area is (16.3 cm)(4.5 cm) $= 73.35$ cm^2, our answer would be unjustifiable since it contains four significant figures, which is greater than the number of significant figures in either of the measured lengths. A good "rule of thumb" to use as a guide in determining the number of significant figures that can be claimed is as follows: **When multiplying several quantities, the number of significant figures in the final answer is the same as the number of significant figures in**

[4] E. Taylor and J.A. Wheeler, *Spacetime Physics*, San Francisco, W.H. Freeman, 1966, p. 60.

the *least* accurate of the quantities being multiplied, where "least accurate" means "having the lowest number of significant figures." The same rule applies to division.

Applying this rule to the multiplication example above, we see that the answer for the area can have only two significant figures since the dimension of 4.5 cm has only two significant figures. Thus, we can only claim the area to be 73 cm^2, realizing that the value can range between (16.2 cm)(4.4 cm) = 71 cm^2 and (16.4 cm)(4.6 cm) = 75 cm^2.

The presence of zeros in an answer may be misinterpreted. For example, suppose the mass of an object is measured to be 1500 g. This value is ambiguous because it is not known whether the last two zeros are being used to locate the decimal point or whether they represent significant figures in the measurement. In order to remove this ambiguity, it is common to use scientific notation to indicate the number of significant figures. In this case, we would express the mass as 1.5×10^3 g if there are two significant figures in the measured value and 1.50×10^3 g if there are three significant figures. Likewise, a number such as 0.00015 should be expressed in scientific notation as 1.5×10^{-4} if it has two significant figures or as 1.50×10^{-4} if it has three significant figures. The three zeros between the decimal point and the digit 1 in the number 0.00015 are not counted as significant figures because they are present only to locate the decimal point. In general, a **significant figure** is a reliably known digit (other than a zero used to locate the decimal point).

For addition and subtraction, the number of decimal places must be considered. **When numbers are added (or subtracted), the number of decimal**

Could this be the result of poor data analysis? (Photo Roger Viollet, Mill Valley, CA, University Science Books, 1982)

places in the result should equal the smallest number of decimal places of any term in the sum. For example, if we wish to compute $123 + 5.35$, the answer would be 128 and not 128.35. As another example, if we compute the sum $1.0001 + 0.0003 = 1.0004$, the result has five significant figures, even though one of the terms in the sum, 0.0003, only has one significant figure. Likewise, if we perform the subtraction $1.002 - 0.998 = 0.004$, the result has three decimal places in accordance with the rule, but only one significant figure.

Throughout this book, *we shall generally assume that the given data are precise enough to yield an answer having three significant figures.* Thus, if we state that a jogger runs a distance of 5 m, it is to be understood that the distance covered is 5.00 m. Likewise, if the speed of a car is given as 23 m/s, its value is understood to be 23.0 m/s.

EXAMPLE 1.7 The Area of a Rectangle
A rectangular plate has a length of (21.3 ± 0.2) cm and a width of (9.80 ± 0.10) cm. Find the area of the plate and the uncertainty in the calculated area.

Solution

$$\text{Area} = \ell w = (21.3 \pm 0.2)\text{ cm} \times (9.80 \pm 0.10)\text{ cm}$$

$$\approx (21.3 \times 9.80 \pm 21.3 \times 0.10 \pm 9.80 \times 0.2)\text{ cm}^2$$

$$\approx (209 \pm 4)\text{ cm}^2$$

Note that the input data were given only to three significant figures, so we cannot claim any more in our result. Furthermore, you should realize that the uncertainty in the product (2%) is *approximately* equal to the sum of the uncertainties in the length and width (each uncertainty is about 1%).

EXAMPLE 1.8 Installing a Carpet
A carpet is to be installed in a room whose length is measured to be 12.71 m (four significant figures) and whose width is measured to be 3.46 m (three significant figures). Find the area of the room.

Solution If you multiply 12.71 m by 3.46 m on your calculator, you will get an answer of 43.9766 m^2. How many of these numbers should you claim? Our rule of thumb for multiplication tells us that you can only claim the number of significant figures in the least accurate of the quantities being measured. In this example, we have only three significant figures in our least accurate measurement, so we should express our final answer as 44.0 m^2. Note that in the answer given, we used a general rule for rounding off numbers which states that the last digit retained is to be increased by 1 if the first digit dropped was equal to 5 or greater.

1.7 MATHEMATICAL NOTATION

Many mathematical symbols will be used throughout this book, some of which you are surely aware of, such as the symbol $=$ to denote the equality of two quantities.

The symbol $\propto$ is used to denote a proportionality. For example, $y \propto x^2$ means that y is proportional to the square of x.

The symbol $<$ means *less than,* and $>$ means *greater than.* For example, $x > y$ means x is greater than y.

The symbol $\ll$ means *much less than,* and $\gg$ means *much greater than.*

The symbol $\approx$ is used to indicate that two quantities are *approximately equal* to each other.

The symbol $\equiv$ means *is defined as.* This is a stronger statement than a simple $=$.

It is convenient to use a symbol to indicate the change in a quantity. For example, Δx (read delta x) means the *change in the quantity* x. (It does not mean the product of Δ and x). For example, if x_i is the initial position of a particle and x_f is its final position, then the *change in position* is written

$$\Delta x = x_f - x_i$$

We will often have occasion to sum several quantities. A useful abbreviation for representing such a sum is the Greek letter Σ (capital sigma). Suppose we wish to sum a set of five numbers represented by x_1, x_2, x_3, x_4, and x_5. In the abbreviated notation, we would write the sum

$$x_1 + x_2 + x_3 + x_4 + x_5 \equiv \sum_{i=1}^{5} x_i$$

where the subscript i on a particular x represents any one of the numbers in the set. For example, if there are five masses in a system, m_1, m_2, m_3, m_4, and m_5, the *total* mass of the system $M = m_1 + m_2 + m_3 + m_4 + m_5$ could be expressed

$$M = \sum_{i=1}^{5} m_i$$

Finally, the *magnitude* of a quantity x, written $|x|$, is simply the absolute value of that quantity. The magnitude of x is *always positive*, regardless of the sign of x. For example, if $x = -5$, $|x| = 5$; if $x = 8$, $|x| = 8$.

A list of these symbols and their meanings is given on the back endsheet.

SUMMARY

Mechanical quantities can be expressed in terms of three fundamental quantities, *mass, length,* and *time,* which have the units *kilograms* (kg), *meters* (m), and *seconds* (s), respectively, in the SI system. It is often useful to use the *method of dimensional analysis* to check equations and to assist in deriving expressions.

The **density** of a substance is defined as its *mass per unit volume.* Different substances have different densities mainly because of differences in their atomic masses and atomic arrangements.

The number of atoms in one mole of any element or compound is called **Avogadro's number, N_A**, which has the value 6.02×10^{23} atoms/mole.

QUESTIONS

1. What types of natural phenomena could serve as alternative time standards?
2. The height of a horse is sometimes given in units of "hands." Why is this a poor standard of length?
3. Express the following quantities using the prefixes given in Table 1.4: (a) 3×10^{-4} m, (b) 5×10^{-5} s, (c) 72×10^{2} g.
4. Does a dimensional analysis give any information on constants of proportionality that may appear in an algebraic expression? Explain.
5. Suppose that two quantities A and B have different dimensions. Determine which of the following arithmetic operations *could* be physically meaningful: (a) $A + B$, (b) A/B, (c) $B - A$, (d) AB.
6. What accuracy is implied in an order-of-magnitude calculation?
7. Apply an order-of-magnitude calculation to an everyday situation you might encounter. For example, how far do you walk or drive each day?
8. Estimate your age in seconds.

9. Estimate the masses of various objects around you in grams or in kilograms. If a scale is available, check your estimates.

10. Is it possible to use length, density, and time as three fundamental units rather than length, mass, and time? If so, what could be used as a standard of density?

PROBLEMS

Section 1.2 Density and Atomic Mass

1. Calculate the density of a solid cube that measures 5 cm on each side and has a mass of 350 g.

2. A solid sphere is to be made out of copper, which has a density of 8.93 g/cm^3. If the mass of the sphere is to be 475 g, what radius must it have?

3. How many grams of copper are required to make a hollow spherical shell with an inner radius of 5.70 cm and an outer radius of 5.75 cm? The density of copper is 8.93 g/cm^3.

4. A hollow cylindrical container has a length of 800 cm and an inner radius of 30 cm. If the cylinder is completely filled with water, what is the mass of the water? Assume 1.0 g/cm^3 as the density of water.

5. Calculate the mass of an atom of (a) helium, (b) iron, and (c) lead. Give your answers in atomic mass units and in grams. The atomic weights are 4, 56, and 207, respectively, for the atoms given.

6. A small particle of iron in the shape of a cube is observed under a microscope. The edge of the cube is 5×10^{-6} cm long. Find (a) the mass of the cube and (b) the number of iron atoms in the particle. The atomic weight of iron is 56, and its density is 7.86 g/cm^3.

7. Compute the ratio of the atomic mass of lead to that of mercury and compare this ratio to the ratio of their densities. Use Table 15.1 in Chapter 15 and Appendix C.

8. A flat circular plate of copper has a radius of 0.243 m and a mass of 62 kg. What is the thickness of the plate?

Section 1.3 Dimensional Analysis

9. Show that the expression $x = vt + \frac{1}{2}at^2$ is dimensionally correct, where x is a coordinate and has units of length, v is velocity, a is acceleration, and t is time.

10. The displacement of a particle when moving under uniform acceleration is some function of the time and the acceleration. Suppose we write this displacement $s = ka^mt^n$, where k is a dimensionless constant. Show by dimensional analysis that this expression is satisfied if $m = 1$ and $n = 2$. Can this analysis give the value of k?

11. The square of the speed of an object undergoing a uniform acceleration a is some function of a and the displacement s, according to the expression given by $v^2 = ka^ms^n$, where k is a dimensionless constant. Show by dimensional analysis that this expression is satisfied only if $m = n = 1$.

12. Show that the equation $v^2 = v_0{}^2 + 2ax$ is dimensionally correct, where v and v_0 represent velocities, a is acceleration, and x is a distance.

13. Which of the equations below are dimensionally correct?
(a) $v = v_0 + ax$
(b) $y = (2 \text{ m})\cos(kx)$, where $k = 2 \text{ m}^{-1}$.

14. The period T of a simple pendulum is measured in time units and is given by

$$T = 2\pi\sqrt{\frac{\ell}{g}}$$

where ℓ is the length of the pendulum and g is the acceleration due to gravity in units of length divided by the square of time. Show that this equation is dimensionally consistent.

15. Suppose that the displacement of a particle is related to time according to the expression $s = ct^3$. What are the dimensions of the constant c?

16. Show that if $x = At^3 + Bt$ is dimensionally correct when x has units of length and t has units of time, then the derivative $dx/dt = 3At^2 + B$ has units of length divided by time.

17. (a) One of the fundamental laws of motion states that the acceleration of an object is directly proportional to the resultant force on it and inversely proportional to its mass. From this statement, determine the dimensions of force. (b) The newton (N) is the SI unit of force. According to your results for (a), how can you express a newton using the SI fundamental units of mass, length, and time?

Section 1.4 Conversion of Units

18. Convert the volume 8.50 in.^{3} to m^3, recalling that 1 in. = 2.54 cm and 1 cm = 10^{-2} m.

19. A rectangular building lot is 100.0 ft by 150.0 ft. Determine the area of this lot in m^2.

20. An object in the shape of a rectangular parallelepiped measures 2.0 in. $\times$ 3.5 in. $\times$ 6.5 in. Determine the volume of the object in m^3.

21. A creature moves at a speed of 5 furlongs per fortnight (not a very common unit of speed). Given that 1 furlong = 220 yards and 1 fortnight = 14 days, determine the speed of the creature in m/s. (The creature is probably a snail.)

22. A section of land has an area of 1 square mile and contains 640 acres. Determine the number of square meters there are in 1 acre.

23. A solid piece of lead has a mass of 23.94 g and a volume of 2.10 cm^3. From these data, calculate the density of lead in SI units (kg/m^3).

24. A quart container of ice cream is to be made in the form of a cube. What should be the length of a side in cm? (Use the conversion 1 gallon = 3.786 liter.)

25. Estimate the age of the earth in years using the data in Table 1.3 and the appropriate conversion factors.

26. The proton, which is the nucleus of the hydrogen atom, can be pictured as a sphere whose diameter is 3×10^{-13} cm having a mass of 1.67×10^{-24} g. Determine the density of the proton in SI units and compare this number with the density of lead, which has the value 11.3×10^3 kg/m^3.

27. Using the fact that the speed of light in free space is about 3.00×10^8 m/s, determine how many miles a pulse from a laser beam will travel in one hour.

28. A painter is to cover the walls in a room 8 ft high and 12 ft along each side. What surface area in square meters must he cover?

29. (a) Find a conversion factor to convert from mi/h to km/h. (b) Until recently, federal law mandated that highway speeds would be 55 mi/h. Use the conversion factor of part (a) to find the speed in km/h. (c) The maximum highway speed has been raised to 65 mi/h in some places. In km/h, how much increase is this over the 55 mi/h limit?

30. (a) How many seconds are there in a year? (b) If one micrometeorite (a sphere with a diameter of 10^{-6} m) strikes each square meter of the moon each second, how many years would it take to cover the moon to a depth of 1 m? (*Hint:* Consider a cubic box on the moon 1 m on a side, and find how long it will take to fill the box.)

31. A house is 50 ft long, 26 ft wide, and has 8-ft high ceilings. What is the volume of the house in cubic meters and in cubic centimeters?

32. The base of a pyramid covers an area of 13 acres (1 acre = 43 560 ft^2) and has a height of 481 ft. If the volume of a pyramid is given by the expression $V = (1/3)Bh$, where B is the area of the base and h is the height, find the volume of this pyramid in cubic meters.

33. The pyramid described in Problem 32 contains approximately two million stone blocks that average 2.50 tons each. Find the weight of this pyramid in pounds.

34. A part for an aircraft engine is manufactured from steel (same density as iron) and has a mass of 4.86 kg. If, instead, a magnesium-aluminum alloy is used ($\rho = 2.55$ g/cm^3), what will be the mass of the part?

35. Radio waves are electromagnetic and travel at a speed of about 3.0×10^8 m/s in free space. Use this fact and the data in Table 1.2 to determine the time it would take an electromagnetic pulse to make a round trip from the earth to Proxima Centauri, the star nearest the sun.

36. The mean radius of the earth is 6.37×10^6 m, and that of the moon is 1.74×10^8 cm. From these data calculate (a) the ratio of the earth's surface area to that of the moon and (b) the ratio of the earth's volume to that of the moon. Recall that the surface area of a sphere is $4\pi r^2$ and the volume of a sphere is $\frac{4}{3}\pi r^3$.

37. From the fact that the average density of the earth is 5.5 g/cm^3 and its mean radius is 6.37×10^6 m, compute the mass of the earth.

38. The mass of a copper atom is 1.06×10^{-22} g, and the density of copper is 8.9 g/cm^3. Determine the number of atoms in 1 cm^3 of copper and compare the result with the estimate in Example 1.5.

39. Aluminum is a very lightweight metal, with a density of 2.7 g/cm^3. What is the weight in pounds of a solid sphere of aluminum of radius 50 cm? The result might surprise you. (*Note:* A 1-kg mass corresponds to a weight of 2.2 pounds.)

Section 1.5 Order-of-Magnitude Calculations

40. Estimate the total number of times the heart of a human beats in an average lifetime of 70 years. (See Table 1.3 for data.)

41. Estimate the number of Ping–Pong balls that would fit into an average-size room (without crushing them).

42. Estimate the amount of motor oil used by all cars in the United States each year and its cost to the consumers.

43. Soft drinks are commonly sold in aluminum containers. Estimate the number of such containers thrown away each year by U.S. consumers. Approximately how many tons of aluminum does this represent?

44. Approximately how many raindrops fall on a 1-acre lot during a 1-in. rainfall?

45. Determine the approximate number of bricks needed to face all four sides of an average-size home.

46. A particular hamburger chain advertises that it has sold more than 50 billion hamburgers. Estimate how many pounds of hamburger meat must have been used by the restaurant chain and how many head of cattle were required to furnish the meat.

47. Imagine yourself to be the equipment manager of a professional baseball team. One of your jobs is to keep baseballs on hand for games. One of the ways by which balls are lost is by players hitting them into the stands either as home runs or foul balls. Estimate how many baseballs you have to buy per season in order to account for such losses. Assume your team plays an 81-game home schedule in a season.

48. Assume that you watch every pitch of every game of a 162 game major league baseball season. Approximately how many pitches would you see thrown?

49. Estimate the number of piano tuners living in New York City. This problem was posed by the physicist Enrico Fermi, who was well known for making order-of-magnitude calculations.

Section 1.6 Significant Figures

50. Determine the number of significant figures in the following numbers: (a) 23 cm (b) 3.589 s (c) 4.67×10^3 m/s (d) 0.0032 m.
51. Calculate (a) the circumference of a circle of radius 3.5 cm and (b) the area of a circle of radius 4.65 cm.
52. Carry out the following arithmetic operations: (a) the sum of the numbers 756, 37.2, 0.83, and 2.5; (b) the product 3.2×3.563; (c) the product $5.6 \times \pi$.
53. If the length and width of a rectangular plate are measured to be (15.30 ± 0.05) cm and (12.80 ± 0.05) cm, respectively, find the area of the plate and the approximate uncertainty in the calculated area.
54. The *radius* of a solid sphere is measured to be (6.50 ± 0.20) cm, and its mass is measured to be (1.85 ± 0.02) kg. Determine the density of the sphere in kg/m^3 and the uncertainty in the density.
55. How many significant figures are there in (a) 78.9 ± 0.2, (b) 3.788×10^9, (c) 2.46×10^{-6}, (d) 0.0053?
56. A farmer measures the distance around a rectangular field. The length of the long sides of the rectangle is found to be 38.44 m, and the length of the short sides is found to be 19.5 m. What is the total distance around the field?
57. A sidewalk is to be constructed around a swimming pool that measures (10.0 ± 0.1) m $\times$ (17.0 ± 0.1) m. If the sidewalk is to measure (1.00 ± 0.01) m wide by (9.0 ± 0.1) cm thick, what volume of concrete is needed, and what is the approximate uncertainty of this volume?

Section 1.7 Mathematical Notation

58. Compute the value of $\sum_{i=1}^{4} x_i$ if $x_i = (2i + 1)$.

59. Determine whether the expression $\sum_{i=1}^{4} i^2$ is equal to $\left(\sum_{i=1}^{4} i\right)^2$.

"All right—now convert the whole thing to metric."

ESSAY

Scaling—the Physics of Lilliput*

Philip Morrison
Massachusetts Institute of Technology

The fictional traveler Lemuel Gulliver spent a busy time in a kingdom called Lilliput, where all living things—men, cattle, trees, grass—were exactly similar to our world, except that they were all built on the scale of one inch to the foot. Lilliputians were a little under 6 inches high, on the average, and built proportionately just as we are. Gulliver also visited Brobdingnag, the country of the giants, who were exactly like men but 12 times as tall. As Swift described it, daily life in both kingdoms was about like ours (in the 18th century). His commentary on human behavior is still worth reading, but we shall see that people of such sizes just could not have been as he described them.

Long before Swift lived, Galileo understood why very small or very large models of man could not be like us, but apparently Dean Swift had never read what Galileo wrote. One character in Galileo's "Two New Sciences" says, "Now since . . . in geometry, . . . mere size cuts no figure, I do not see that the properties of circles, triangles, cylinders, cones, and other solid figures will change with their size. . . ." But his physicist friend replies, "The common opinion is here absolutely wrong." Let us see why.

We start with the strength of a rope. It is easy to see that if one man who pulls with a certain strength can almost break a certain rope, two such ropes will just withstand the pull of two men. A single large rope with the same total area of cross-section as the two smaller ropes combined will contain just double the number of fibers of one of the small ropes, and it will also do the job. In other words, the breaking strength of a wire or rope is proportional to its area of cross-section, or to the square of its diameter. Experience and theory agree in this conclusion. Furthermore, the same relation holds, not only for ropes or cables supporting a pull, but also for columns or struts supporting a thrust. The thrust which a column will support, comparing only those of a given material, is also proportional to the cross-sectional area of the column.

Now the body of a man or an animal is held up by a set of columns or struts—the skeleton—supported by various braces and cables, which are muscles and tendons. But the weight of the body which must be supported is proportional to the amount of flesh and bone present, that is, to the volume.

Let us now compare Gulliver with the Brobdingnagian giant, 12 times his height. Since the giant is exactly like Gulliver in construction, every one of his linear dimensions is 12 times the corresponding one of Gulliver's. Because the strength of his columns and braces is proportional to their cross-sectional area and thus to the square of their linear dimension (strength $\propto L^2$), his bones will be 12^2 or 144 times as strong as Gulliver's. Because his weight is proportional to his volume and thus to L^3, it will be 12^3 or 1728 times as great as Gulliver's. So the giant will have a strength-to-weight ratio a dozen times smaller than ours. Just to support his own weight, he would have as much trouble as we should have in carrying 11 men on our back.

In reality, of course, Lilliput and Brobdingnag do not exist. But we can see real effects of a difference in scale if we compare similar animals of very different sizes. The smaller ones are not scale models of the larger ones. Figure 1 shows the corresponding leg bones of two closely related animals of the deer family: one a tiny gazelle, the other a bison. Notice that the bone of the large animal is not at all similar geometrically to that of the smaller. It is much thicker for its length, thus counteracting the scale change, which would make a strictly similar bone too weak.

Galileo wrote very clearly on this very point, disproving the possibility of Brobdingnag, or of any normal-looking giants: ". . . if one wishes to maintain in a great giant the same proportion of limb as that found in an ordinary man he must either use a harder and stronger material for making the bones, or he must admit a diminution of strength in comparison with men of medium stature; for if his height be increased

(Continued on next page)

*Adapted from PSSC PHYSICS, 2nd edition, 1965; D.C. Heath and Company with Education Development Center, Inc., Newton, MA.

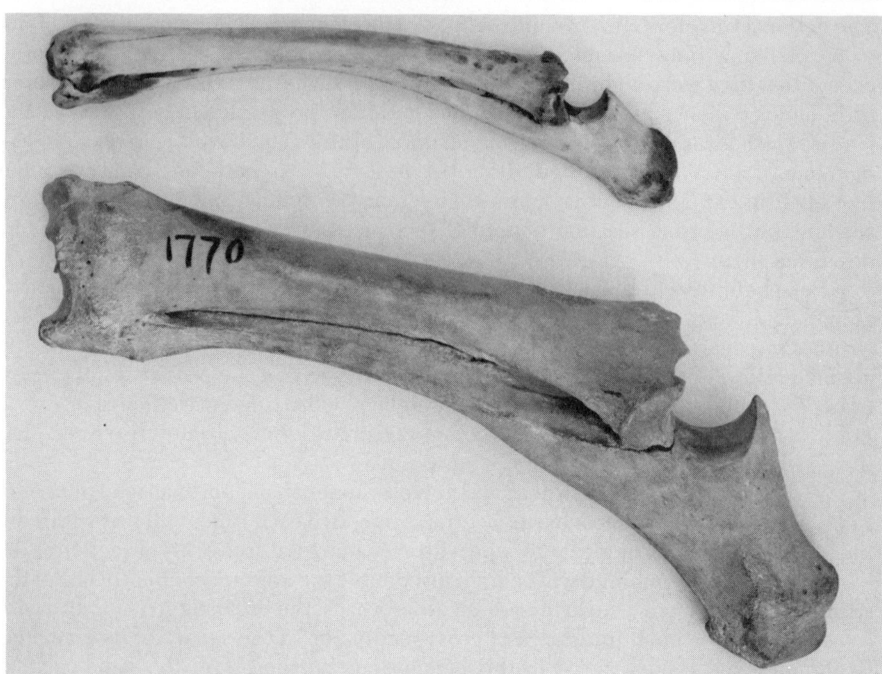

Figure 1a The front leg bones of a bison and a gazelle. The animals are related, but the gazelle is much smaller. The photos show the approximate relative sizes of the bones.

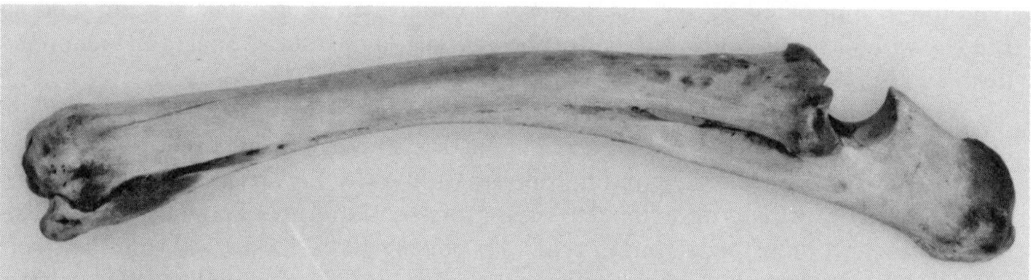

Figure 1b The leg bone of a gazelle enlarged to the same length as the bison bone. Note that the bone of the larger animal is much thicker in comparison to its length than that of the gazelle. The small deer is generally more lightly and gracefully built. Can you visualize how much different Lilliputians must have been from men of normal size?

inordinately he will fall and be crushed under his own weight. Whereas, if the size of a body be diminished, the strength of that body is not diminished in the same proportion; indeed, the smaller the body the greater its relative strength. Thus a small dog could probably carry on his back two or three dogs of his own size; but I believe that a horse could not carry even one of his own size." The sketch of Figure 2 is taken from Galileo, who drew it to illustrate the paragraph just quoted.

An elephant is already so large that his limbs are clumsily thickened. However, a whale, the largest of all animals, may weigh 40 times as much as an elephant; yet the whale's bones are not proportionately thickened. They are strong enough because the whale is supported by water. What is the fate of a stranded whale? His ribs break. Some of the dinosaurs of old were animals of whalelike size; how did *they* get along?

Following Galileo, we have investigated the problems of scaling up to giants. Now let's take a look at some of the problems that arise when we scale down.

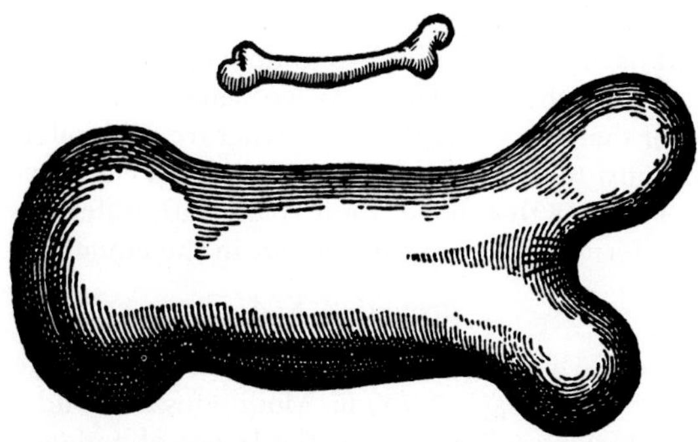

Figure 2 Galileo's drawing illustrating scaling. Over 300 years ago, Galileo wrote concerning the fact that a bone of greater length must be increased in thickness in greater proportion in order to be comparably strong.

When you climb dripping wet out of a pool there is a thin film of water on your skin. Your fingers are no less wet than your forearm; the thickness of the water film is much the same over most of your body. Roughly, at least, the amount of water you bring out is proportional to the surface area of your body. You can express this by the relation

$$\text{amount of water} \propto L^2,$$

where L is your height. The original load on your frame is as before, proportional to your volume. So, the ratio *extra load/original load* is proportional to L^2/L^3, or to $1/L$. Perhaps you carry out of the pool a glassful or so, which amounts to about a 1% increase in what you have to move about. But a Lilliputian will bring out about 12% of his weight, which would be equivalent to a heavy winter suit of clothing with an overcoat. Getting out of the pool would be no fun! If a fly gets wet, his body load doubles, and he is all but imprisoned by the drop of water.

There is a still more important effect of the scale of a living body. Your body loses heat mainly through the skin (and some through breathing out warm air). It is easy to believe— and it can be checked by experiment—that the heat loss is proportional to the surface area, so that

$$\text{heat loss} \propto L^2,$$

keeping other factors, like the temperature, nature of skin, and so on, constant. The food taken in must supply this heat, as well as the surplus energy we use in moving about. So minimum food needs go as L^2. If a man like Gulliver can live off a leg of lamb and a loaf of bread for a day or two, a Lilliputian with the same body temperature will require a volume of food only $(\frac{1}{12})^2$ as large. But his leg of lamb, scaled down to his world, will be smaller in volume by a factor of $(\frac{1}{12})^3$. Therefore, he would need a dozen of his roasts and loaves to feel as well fed as Gulliver did after one. Lilliputians must be a hungry lot, restless, active, graceful, but easily waterlogged. You can recognize these qualities in many small mammals, like a mouse.

We can see why there are no warm-blooded animals much smaller than the mouse. Fish and frogs and insects can be very much smaller because their temperature is not higher than their surroundings. In accord with the scaling laws of area and volume, small, warm-blooded animals need relatively a great deal of food; really small ones could not gather or even digest such an enormous amount. Certainly the agriculture of the Lilliputians could not have supported a kingdom like the one Gulliver described.

(Continued on next page)

Now we see that neither Brobdingnag nor Lilliput can really be a scale model of our world. But what have these conclusions to do with physics?

Let's start again with the very large. As we scale up any system, the load will eventually be greater than the strength of the structure. This effect applies to every physical system, not just to animals, of course. Buildings can be very large because their materials are stronger than bone, their shapes are different, and they do not move. These facts determine the constants like K in the equation

$$\text{strength} = KL^2$$

but the same laws hold. No building can be made which will look like the Empire State but be as high as a mountain, say 10 000 m. Mountains are solid structures, for the most part, without interior cavities. Just as the bones of a giant must be thick, an object of mountainous size on the earth must be all but solid, or else built of new materials yet unknown.

Our arguments are not restricted to the surface of the earth. We can imagine building a tremendous structure far out in space away from the gravitational pull of the earth. The load then is not given by the earth's gravitational pull, but as the structure is built larger and larger each part pulls gravitationally on every other and soon the outside of the structure is pulled in with great force. The inside, built of ordinary materials, is crushed, and large protuberances on the surface break off or sink in. As a result any large structure like a planet has a simple shape, and if it is large enough, the shape is close to a sphere. Any other shape will be unable to support itself. Here is the essential reason why the planets and the sun tend to be spherical. The pull of gravity is important for us on earth, but as we extend the range of dimensions which we study, it becomes absolutely dominant in the very large. Only motion can change this result. The great masses of gas which are nebulae, for example, are changing in time, and hence the law that large objects must be simple in shape is modified.

When we go from our size to the very small, gravitational effects cease to be important. But as we saw in investigating Lilliput, surface effects become significant. If we go far enough toward the very small, surfaces no longer appear smooth, but are so rough that we have difficulty in defining a surface. Other descriptions must be used. In any case, it will not come as a complete surprise that in the domain of the atom, the very small, scale factors demonstrate that the dominant pull is one which is not easily observed in everyday experience.

Such arguments as these run through all of physics. Like order-of-magnitude measurements, they are extremely valuable when we begin the study of any physical system. How the behavior of a system will change with changes in the scale of its dimensions, its motion, and so on, is often the best guide to a detailed analysis.

Even more, it is by the study of systems built on many unusual scales that physicists have been able to uncover unsuspected physical relations. When changing scale, one aspect of the physical world may be much emphasized and another one may be minimized. In this way we may discover, or at least get a clearer view of, things which are less obvious on our normal scale of experience. It is largely for this reason that physicists examine, in and out of their laboratories, the very large and the very small, the slow and the rapid, the hot and the cold, and all the other unusual circumstances they can contrive. In examining what happens in these circumstances we use instruments both to produce the unusual circumstances and to extend our senses in making measurements.

It is hard to resist pointing out how much the scale of man's own size affects the way he sees the world. It has been largely the task of physics to try to form a picture of the world which does not depend upon the way we happen to be built. But it is hard to get rid of these effects of our own scale. We can build big roads and bridges which are long and thin, but are essentially not three-dimensional, complex structures. The very biggest things we can make which have some roundness, which are fully three-dimensional, are buildings and great ships. These lack a good deal of being a thousand times larger than men in their linear dimensions.

Within our present technology our scaling arguments are important. If we design a new large object on the basis of a small one, we are warned that new effects too small to detect on our scale may enter and even become the most important things to consider. We cannot just scale up and down blindly, geometrically, but by scaling in the light of physical reasoning, we can sometimes foresee what changes will occur. In this way we can employ scaling in intelligent airplane design, for example, and not arrive at a jet transport that looks like a bee—and won't fly.

Suggested Readings

Galileo, Galilei, "Dialogues Concerning Two New Sciences," trans. by Henry Crew and Alphonso De Salvio, Evanston, Northwestern University Press, 1946, pp. 1–6, 125–128.

Haldane, J. B. S., "On Being the Right Size." *World of Mathematics,* Vol. II, edited by James R. Newman. New York, Simon & Schuster, 1956.

Holcomb, Donald F. and Philip Morrison, *My Father's Watch—Aspects of the Physical World,* Englewood Cliffs, NJ, Prentice Hall, 1974, pp. 68–83.

Smith, Cyril S., "The Shape of Things," *Scientific American,* January, 1954, p. 58.

Thompson, D'Arcy W., "On Magnitude," in *On Growth and Form,* Cambridge University Press, 1952 and 1961.

Essay Questions

1. The leg bones of one animal are twice as strong as those of another closely related animal of similar shape. (a) What would you expect to be the ratio of these animals' heights? (b) What would you expect to be the ratio of their weights?
2. A hummingbird must eat very frequently and even then must have a highly concentrated form of food such as sugar. What does the concept of scaling tell you about the size of a hummingbird?
3. About how many Lilliputians would it take to equal the mass of one citizen of Brobdingnag?

Essay Problems

1. The total surface area of a rectangular solid is the sum of the areas of the six faces. If each dimension of a given rectangular solid is doubled, what effect does this have on the total surface area?
2. A hollow metal sphere has a wall thickness of 2 cm. If you increase both the diameter and thickness of this sphere so that the overall volume is three times the original overall volume, how thick will the shell of the new sphere be?
3. If your height and all your other dimensions were doubled, by what factor would this change (a) your weight? (b) the ability of your leg bones to support your weight?
4. According to the zoo, an elephant of mass 4.0×10^3 kg consumes 3.4×10^2 times as much food as a guinea pig of mass 0.70 kg. They are both warm-blooded, plant-eating, similarly shaped animals. Find the ratio of their surface areas, which is approximately the ratio of their heat losses, and compare it with the known ratio of food consumed.
5. A rectangular water tank is supported above the ground by four pillars 5 m long whose diameters are 20 cm. If the tank were made 10 times longer, wider, and deeper, what diameter pillars would be needed? How much more water would the tank hold?
6. How many state maps of scale 1 : 1 000 000 would you need to cover the state with those maps?

2

Vectors

Weathervanes can be used to determine the direction of the wind velocity vector at any instant. (© Pat Le Croix, The IMAGE Bank)

Mathematics is the basic tool used by scientists and engineers to describe the behavior of physical systems. Physical quantities that have both numerical and directional properties are represented by vectors. Some examples of vector quantities are force, velocity, and acceleration. This chapter is primarily concerned with vector algebra and with some general properties of vectors. The addition and subtraction of vectors will be discussed, together with some common applications to physical situations. Discussion of the products of vectors will be delayed until these operations are needed.[1]

Vectors will be used throughout this text, and it is imperative that you master both their graphical and algebraic properties.

[1] The dot, or scalar, product is discussed in Section 7.3, and the cross, or vector, product is introduced in Section 11.2.

2.1 COORDINATE SYSTEMS AND FRAMES OF REFERENCE

Many aspects of physics deal in some form or other with locations in space. For example, the mathematical description of the motion of an object requires a method for describing the position of the object. Thus, it is perhaps fitting that we first discuss how one describes the position of a point in space. This is accomplished by means of coordinates. A point on a line can be described with one coordinate. A point in a plane is located with two coordinates, and three coordinates are required to locate a point in space.

A coordinate system used to specify locations in space consists of

1. A fixed reference point O, called the origin
2. A set of specified axes or directions with an appropriate scale and labels on the axes
3. Instructions that tell us how to label a point in space relative to the origin and axes.

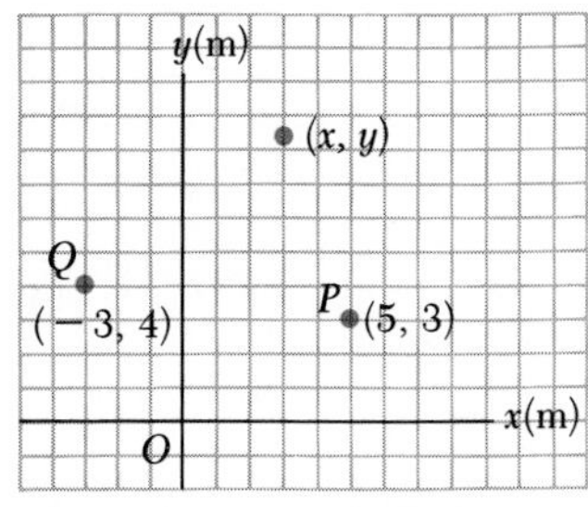

Figure 2.1 Designation of points in a cartesian coordinate system. Every point is labeled with coordinates (x, y).

One convenient coordinate system that we will use frequently is the *cartesian coordinate system*, sometimes called the *rectangular coordinate system*. Such a system in two dimensions is illustrated in Figure 2.1. An arbitrary point in this system is labeled with the coordinates (x, y). Positive x is taken to the right of the origin, and positive y is upward from the origin. Negative x is to the left of the origin, and negative y is downward from the origin. For example, the point P, which has coordinates $(5, 3)$, may be reached by first going 5 meters to the right of the origin and 3 meters above the origin. Similarly, the point Q has coordinates $(-3, 4)$, corresponding to going 3 meters to the left of the origin and 4 meters above the origin.

Sometimes it is more convenient to represent a point in a plane by its *plane polar coordinates*, (r, θ), as in Figure 2.2a. In this coordinate system, r is the distance from the origin to the point having cartesian coordinates (x, y) and θ is the angle between r and a fixed axis, usually measured counterclockwise from the positive x axis. From the right triangle in Figure 2.2b, we find $\sin\theta = y/r$ and $\cos\theta = x/r$. (A review of trigonometric functions is given in Appendix B.4.) Therefore, starting with plane polar coordinates, one can obtain the cartesian coordinates through the equations

$$x = r\cos\theta \tag{2.1}$$

$$y = r\sin\theta \tag{2.2}$$

Furthermore, it follows that

$$\tan\theta = y/x \tag{2.3}$$

and

$$r = \sqrt{x^2 + y^2} \tag{2.4}$$

You should note that these expressions relating the coordinates (x, y) to the coordinates (r, θ) apply only when θ is defined as in Figure 2.2a, where positive θ is an angle measured *counterclockwise* from the positive x axis. If the reference axis for the polar angle θ is chosen to be other than the positive x axis, or the sense of increasing θ is chosen differently, then the corresponding expressions relating the two sets of coordinates will change.

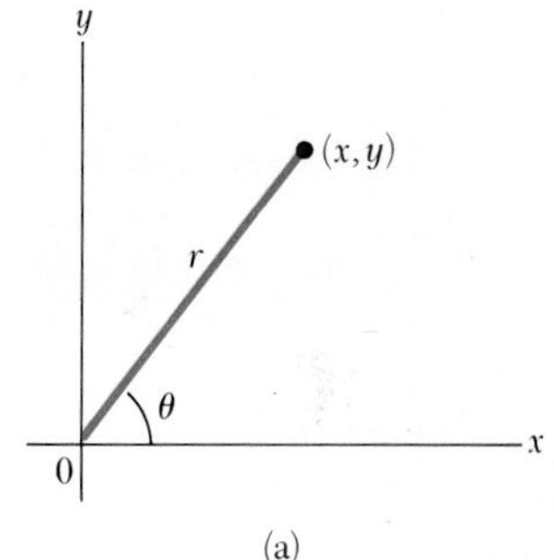

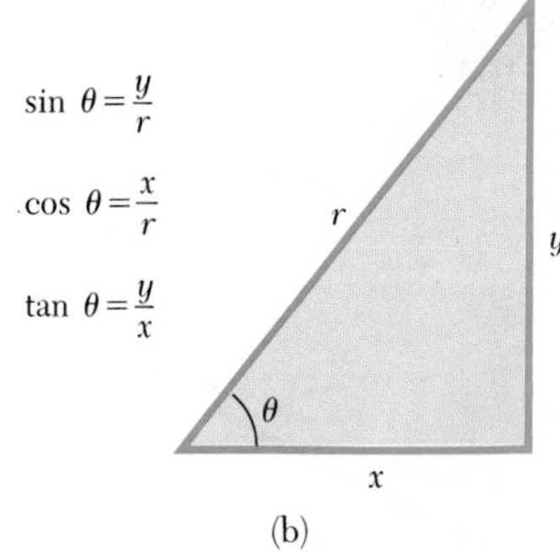

Figure 2.2 (a) The plane polar coordinates of a point are represented by the distance r and the angle θ. (b) The right triangle used to relate (x, y) to (r, θ).

EXAMPLE 2.1 Polar Coordinates
The cartesian coordinates of a point are given by $(x, y) = (-3.5, -2.5)$ meters as in Figure 2.3. Find the polar coordinates of this point.

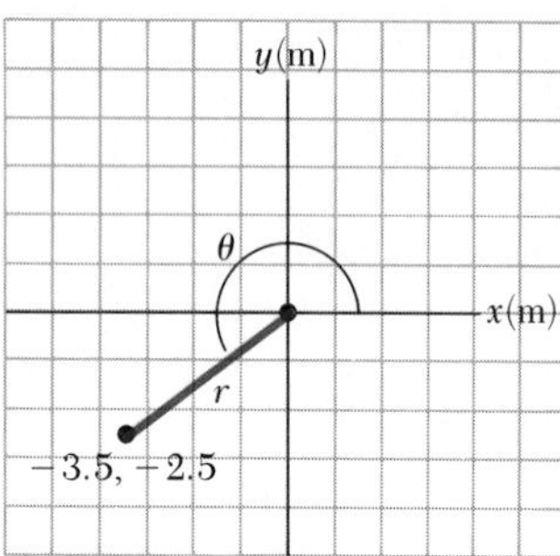

Figure 2.3 (Example 2.1).

Solution

$$r = \sqrt{x^2 + y^2} = \sqrt{(-3.5)^2 + (-2.5)^2} = \boxed{4.3 \text{ meters}}$$

$$\tan\theta = \frac{y}{x} = \frac{-2.5}{-3.5} = 0.714$$

$$\theta = \boxed{216°}$$

Note that you must use the signs of x and y to find that θ is in the third quadrant of the coordinate system. That is, $\theta = 216°$ and not 36°.

Most scientific calculators provide conversions between cartesian and polar coordinates.

2.2 VECTORS AND SCALARS

The physical quantities that we shall encounter in this text can be placed in one or the other of two categories: they are either scalars or vectors. A scalar is a quantity that is completely specified by a number with appropriate units. That is,

Jennifer pointing in the right direction. (Photo by Raymond A. Serway)

> A **scalar** has only magnitude and no direction. On the other hand, a **vector** is a physical quantity that must be specified by both magnitude and direction.

The number of apples in a basket is an example of a scalar quantity. If you are told there are 38 apples in the basket, this completely specifies the required information; no direction is required. Other examples of scalars are temperature, volume, mass, and time intervals. The rules of ordinary arithmetic are used to manipulate scalar quantities.

Force is an example of a vector quantity. To describe completely the force on an object, we must specify both the direction of the applied force and a number to indicate the force's magnitude. When the motion (velocity) of an object is described, we must specify both how fast it is moving and the direction of its motion.

Another simple example of a vector quantity is the **displacement** of a particle, defined as the *change in the position* of the particle. Suppose the particle moves from some point O to the point P along a straight path, as in Figure 2.4. We represent this displacement by drawing an arrow from O to P, where the tip of the arrow represents the direction of the displacement and the length of the arrow represents the magnitude of the displacement. If the particle travels along some other path from O to P, such as the broken line in Figure 2.4, its displacement is still $\mathbf{OP}$. The vector displacement along any indirect path from O and P is defined as being equivalent to the displacement for the direct path from O to P. Thus, *the displacement of a particle is completely known if its initial and final coordinates are known.* The path need not

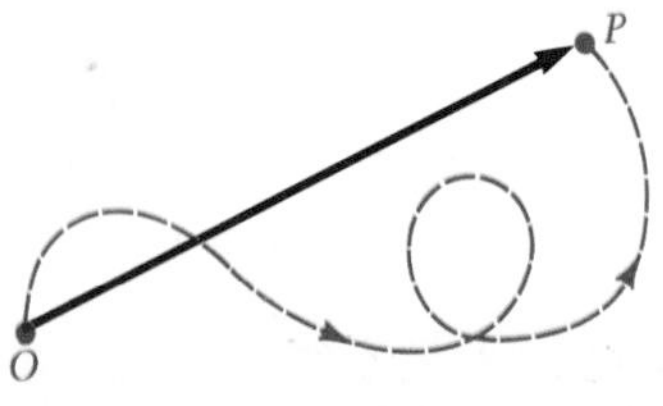

Figure 2.4 As a particle moves from O to P along the broken line, its displacement vector is the arrow drawn from O to P.

be specified. In other words, *the displacement is independent of the path,* if the end points of the path are fixed.

It is important to note that the distance traveled by a particle is distinctly different from its displacement. This distance traveled (a scalar quantity) is the length of the path, which in general can be much greater than the magnitude of the displacement (see Fig. 2.4). Also, the magnitude of the displacement is the shortest distance between the end points.

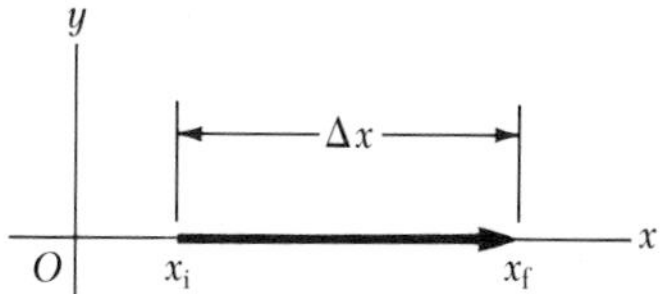

Figure 2.5 A particle moving along the x axis from x_i to x_f undergoes a displacement $\Delta x = x_f - x_i$.

If the particle moves along the x axis from position x_i to position x_f, as in Figure 2.5, its displacement is given by $x_f - x_i$. As mentioned in Chapter 1, we use the Greek letter delta (Δ) to denote the *change* in a quantity. Therefore, we write the change in the position of the particle (the displacement)

Definition of displacement along a line

$$\Delta x = x_f - x_i \tag{2.5}$$

From this definition, we see that Δx is positive if x_f is greater than x_i and negative if x_f is less than x_i. For example, if a particle changes its position from $x_i = -3$ units to $x_f = 5$ units, its displacement is 8 units.

There are many physical quantities in addition to displacement that are vectors. These include velocity, acceleration, force, and momentum, all of which will be defined in later chapters. In this text, we will use boldface letters, such as **A**, to represent an arbitrary vector. Another common method for vector notation that you should be aware of is to use an arrow over the letter: $\vec{A}$. The magnitude of the vector **A** is written A or, alternatively, $|\mathbf{A}|$. The magnitude of a vector has physical units, such as m for displacement or m/s for velocity, as discussed in Chapter 1. Vectors combine according to special rules, which will be discussed in Sections 2.3 and 2.4.

2.3 SOME PROPERTIES OF VECTORS

Equality of Two Vectors Two vectors **A** and **B** are defined to be equal if they have the same magnitude and point in the same direction. That is, **A** = **B** only if $A = B$ *and* they act along parallel direction. For example, all the vectors in Figure 2.6 are equal even though they have different starting points. This property allows us to translate a vector parallel to itself in a diagram without affecting the vector. In fact, any true vector can be moved parallel to itself without affecting the vector.

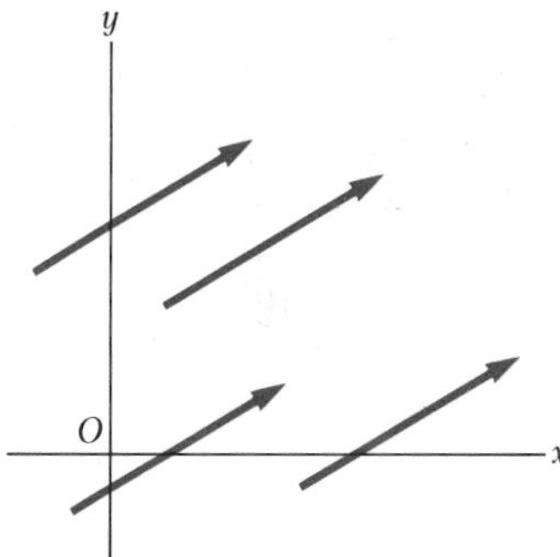

Figure 2.6 Four representations of the same vector.

Addition When two or more vectors are added together, *all* vectors involved *must* have the same units. For example, it would be meaningless to add a velocity vector to a displacement vector since they are different physical quantities. Scalars also obey the same rule. For example, it would be meaningless to add time intervals and temperatures.

The rules for vector sums are conveniently described by geometric methods. To add vector **B** to vector **A**, first draw vector **A**, with its magnitude represented by a convenient scale, on graph paper and then draw vector **B** to the same scale with its tail starting from the tip of **A**, as in Figure 2.7. The *resultant vector* **R** = **A** + **B** is the blue vector drawn from the tail of **A** to the tip of **B**. This is known as the *triangle method of addition.* An alternative graphical procedure for adding two vectors, known as the *parallelogram rule of addition,* is shown in Figure 2.8a. In this construction, the tails of the two vectors **A** and **B** are together and the resultant vector **R** is the diagonal of a parallelogram formed with **A** and **B** as its sides.

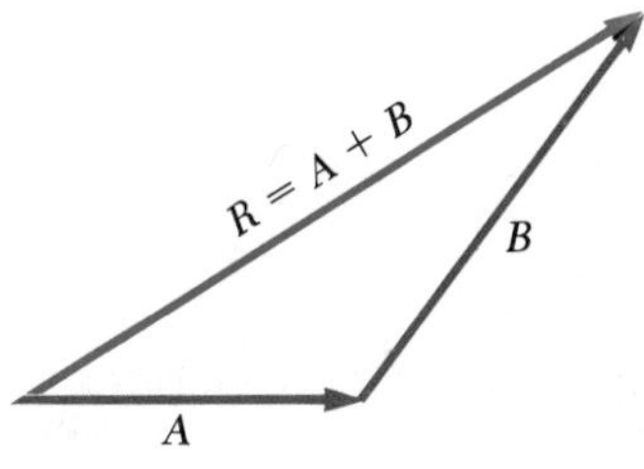

Figure 2.7 When vector **A** is added to vector **B**, the resultant **R** is the blue vector that runs from the tail of **A** to the tip of **B**.

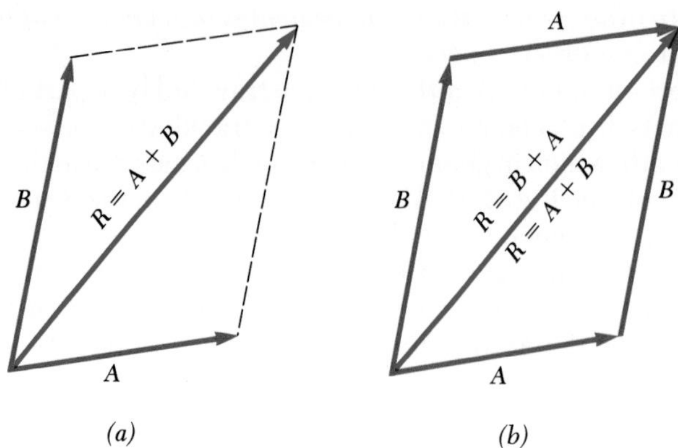

Figure 2.8 (a) In this construction, the resultant ***R*** is the diagonal of a parallelogram with sides ***A*** and ***B***. (b) This construction shows that ***A*** + ***B*** = ***B*** + ***A***.

When two vectors are added, the sum is independent of the order of the addition. This can be seen from the geometric construction in Figure 2.8b and is known as the **commutative law of addition:**

Commutative law

$$\mathbf{A} + \mathbf{B} = \mathbf{B} + \mathbf{A} \tag{2.6}$$

If three or more vectors are added, their sum is independent of the way in which the individual vectors are grouped together. A geometric proof of this for three vectors is given in Figure 2.9. This is called the **associative law of addition:**

Associative law

$$\mathbf{A} + (\mathbf{B} + \mathbf{C}) = (\mathbf{A} + \mathbf{B}) + \mathbf{C} \tag{2.7}$$

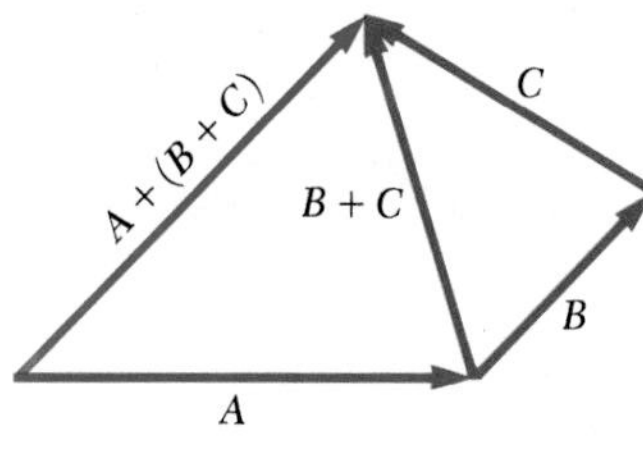

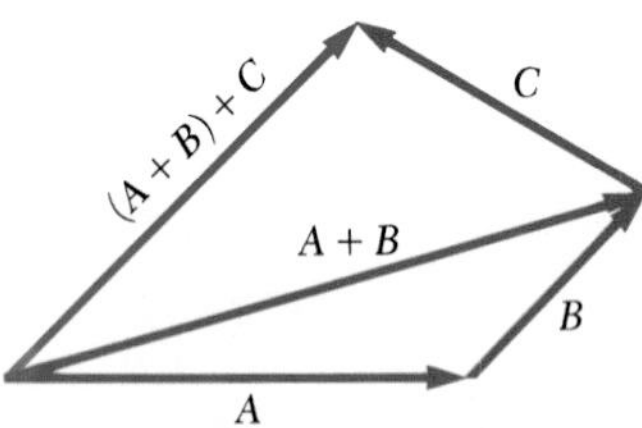

Figure 2.9 Geometric constructions for verifying the associative law of addition.

Geometric constructions can also be used to add more than three vectors. This is shown in Figure 2.10 for the case of four vectors. The resultant vector sum ***R*** = ***A*** + ***B*** + ***C*** + ***D*** is *the vector that completes the polygon.* In other words, ***R*** is *the vector drawn from the tail of the first vector to the tip of the last vector.* Again, the order of the summation is unimportant.

Thus we conclude that *a vector is a quantity that has both magnitude and direction and also obeys the laws of vector addition* as described in Figures 2.7 to 2.10.

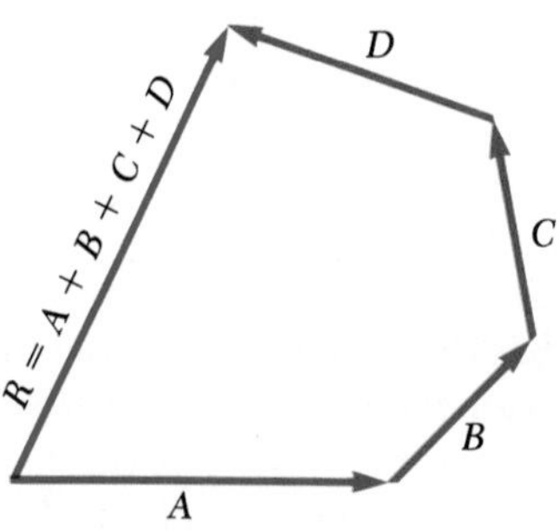

Figure 2.10 Geometric construction for summing four vectors. The resultant vector ***R*** in blue completes the polygon.

Negative of a Vector The negative of the vector $\mathbf{A}$ is defined as the vector that when added to $\mathbf{A}$ gives zero for the vector sum. That is, $\mathbf{A} + (-\mathbf{A}) = 0$. The vectors $\mathbf{A}$ and $-\mathbf{A}$ have the same magnitude but point in opposite directions.

Subtraction of Vectors The operation of vector subtraction makes use of the definition of the negative of a vector. We define the operation $\mathbf{A} - \mathbf{B}$ as vector $-\mathbf{B}$ added to vector $\mathbf{A}$:

$$\mathbf{A} - \mathbf{B} = \mathbf{A} + (-\mathbf{B}) \tag{2.8}$$

The geometric construction for subtracting two vectors is shown in Figure 2.11.

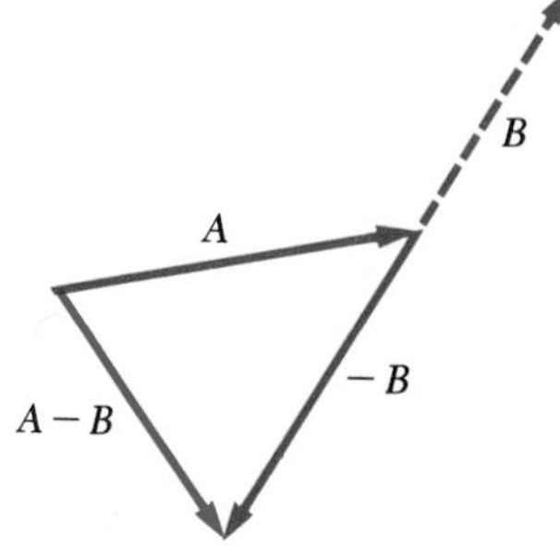

Figure 2.11 This construction shows how to subtract vector $\mathbf{B}$ from vector $\mathbf{A}$. The vector $-\mathbf{B}$ is equal in magnitude and opposite the vector $\mathbf{B}$.

Multiplication of a Vector by a Scalar If a vector $\mathbf{A}$ is multiplied by a positive scalar quantity m, the product $m\mathbf{A}$ is a vector that has the same direction as $\mathbf{A}$ and magnitude mA. If m is a negative scalar quantity, the vector $m\mathbf{A}$ is directed opposite $\mathbf{A}$. For example, the vector $5\mathbf{A}$ is five times as long as $\mathbf{A}$ and points in the same direction as $\mathbf{A}$. On the other hand, the vector $-\frac{1}{3}\mathbf{A}$ is one third the length of $\mathbf{A}$ and points in the direction opposite $\mathbf{A}$ (because of the negative sign).

EXAMPLE 2.2 A Vacation Trip

A car travels 20.0 km due north and then 35.0 km in a direction 60° west of north, as in Figure 2.12. Find the magnitude and direction of the car's resultant displacement.

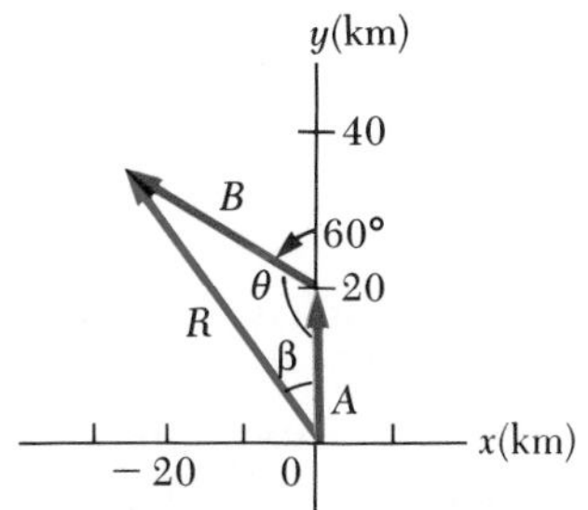

Figure 2.12 (Example 2.2) Graphical method for finding the resultant displacement $\mathbf{R} = \mathbf{A} + \mathbf{B}$.

Solution The problem can be solved geometrically using graph paper and a protractor, as shown in Figure 2.12. The resultant displacement $\mathbf{R}$ is the sum of the two individual displacements $\mathbf{A}$ and $\mathbf{B}$.

An algebraic solution for the magnitude of $\mathbf{R}$ can be obtained using the law of cosines from trigonometry as applied to the obtuse triangle (Appendix B.4). Since $\theta = 180° - 60° = 120°$ and $R^2 = A^2 + B^2 - 2AB\cos\theta$, we find that

$$\begin{aligned} R &= \sqrt{A^2 + B^2 - 2AB\cos\theta} \\ &= \sqrt{(20\text{ km})^2 + (35\text{ km})^2 - 2(20\text{ km})(35\text{ km})\cos 120°} \\ &= \boxed{48.2\text{ km}} \end{aligned}$$

The direction of $\mathbf{R}$ measured from the northerly direction can be obtained from the law of sines from trigonometry:

$$\frac{\sin\beta}{B} = \frac{\sin\theta}{R}$$

$$\sin\beta = \frac{B}{R}\sin\theta = \frac{35.0\text{ km}}{48.2\text{ km}}\sin 120° = 0.629$$

or

$$\beta = \boxed{38.9°}$$

Therefore, the resultant displacement of the car is 48.2 km in a direction 38.9° west of north.

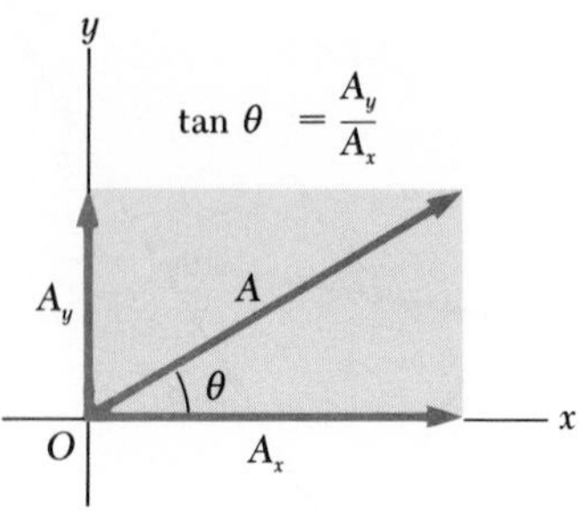

Figure 2.13 Any vector **A** lying in the *xy* plane can be represented by its rectangular vector components $\mathbf{A}_x$ and $\mathbf{A}_y$, where $\mathbf{A} = \mathbf{A}_x + \mathbf{A}_y$.

2.4 COMPONENTS OF A VECTOR AND UNIT VECTORS

The geometric method of adding vectors is not the recommended procedure in situations where high precision is required or in three-dimensional problems. In this section, we describe a method of adding vectors that makes use of the *projections* of a vector along the axes of a rectangular coordinate system. These projections are called the **components** of the vector. Any vector can be completely described by its components.

Consider a vector **A** lying in the *xy* plane and making an arbitrary angle θ with the positive *x* axis, as in Figure 2.13. The vector **A** can be expressed as the sum of two other vectors $\mathbf{A}_x$ and $\mathbf{A}_y$, called the **vector components** of **A**. The vector component $\mathbf{A}_x$ represents the projection of **A** along the *x* axis, while $\mathbf{A}_y$ represents the projection of **A** along the *y* axis. From Figure 2.13, we see that $\mathbf{A} = \mathbf{A}_x + \mathbf{A}_y$. We will often refer to the magnitudes of $\mathbf{A}_x$ and $\mathbf{A}_y$, namely A_x and A_y, as the **components** of **A**. The components of a vector are scalars with units that can be positive or negative. The component A_x is positive if $\mathbf{A}_x$ points along the positive *x* axis and is negative if $\mathbf{A}_x$ points along the negative *x* axis. The same is true for the component A_y.

From Figure 2.13 and the definition of the sine and cosine of an angle, we see that $\cos\theta = A_x/A$ and $\sin\theta = A_y/A$. Hence, the rectangular components of **A** are given by

Components of the vector *A*

$$A_x = A\cos\theta$$

and (2.9)

$$A_y = A\sin\theta$$

These components form two sides of a right triangle the hypotenuse of which has a magnitude *A*. Thus, it follows that the magnitude of **A** and its direction are related to its rectangular components through the expressions

Magnitude of *A*

$$A = \sqrt{A_x^{\,2} + A_y^{\,2}} \tag{2.10}$$

and

Direction of *A*

$$\tan\theta = \frac{A_y}{A_x} \tag{2.11}$$

II	*y* I
A_x negative	A_x positive
A_y positive	A_y positive
	x
A_x negative	A_x positive
A_y negative	A_y negative
III	IV

Figure 2.14 The signs of the rectangular components of a vector **A** depend on the quadrant in which the vector is located.

To solve for θ, we can write $\theta = \tan^{-1}(A_y/A_x)$, which is read "$\theta$ equals the angle the tangent of which is the ratio A_y/A_x." *Note that the signs of the rectangular components A_x and A_y depend on the angle θ.* For example, if $\theta = 120°$, A_x is negative and A_y is positive. On the other hand, if $\theta = 225°$, both A_x and A_y are negative. Figure 2.14 summarizes the signs of the components when **A** lies in the various quadrants.

If you choose reference axes or an angle other than what is shown in Figure 2.13, the components of a vector must be modified accordingly. In many applications it is more convenient to express the components of a vector in a coordinate system having axes that are not horizontal and vertical, but still perpendicular to each other. Suppose a vector $\boldsymbol{B}$ makes an angle θ with the x' axis defined in Figure 2.15. The rectangular components of $\boldsymbol{B}$ along these axes are given by $B_{x'} = B \cos \theta$ and $B_{y'} = B \sin \theta$, as in Equation 2.9. The magnitude and direction of $\boldsymbol{B}$ are obtained from expressions equivalent to Equations 2.10 and 2.11. Thus, we can express the components of a vector in *any* coordinate system that is convenient for a particular situation.

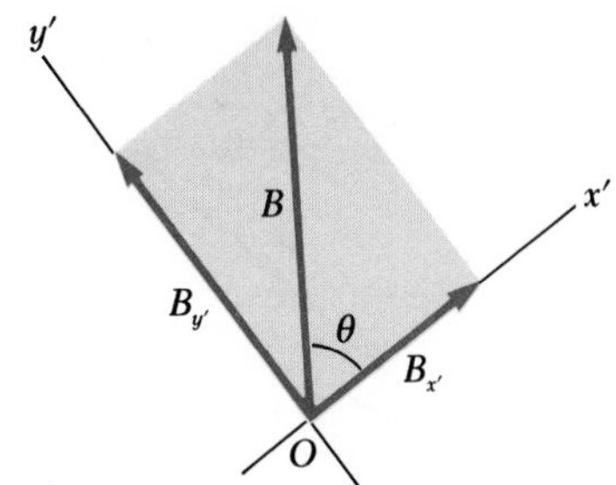

Figure 2.15 The vector components of a vector $\boldsymbol{B}$ in a coordinate system that is tilted.

The components of a vector, such as a displacement, are different when viewed from different coordinate systems. Furthermore, the components of a vector can change with respect to a fixed coordinate system if the vector changes in magnitude, orientation, or both.

Vector quantities are often expressed in terms of unit vectors. A **unit vector** is *a dimensionless vector one unit in length* used to specify a given *direction.* Unit vectors have no other physical significance. They are used simply as a convenience in describing a direction in space. We will use the symbols $\boldsymbol{i}$, $\boldsymbol{j}$, and $\boldsymbol{k}$ to represent unit vectors pointing in the x, y, and z directions, respectively. Thus, the unit vectors $\boldsymbol{i}$, $\boldsymbol{j}$, and $\boldsymbol{k}$ form a set of mutually perpendicular vectors as shown in Figure 2.16a, where the magnitude of the unit vectors equals unity; that is, $|\boldsymbol{i}| = |\boldsymbol{j}| = |\boldsymbol{k}| = 1$.

Consider a vector $\boldsymbol{A}$ lying in the xy plane, as in Figure 2.16b. The product of the component A_x and the unit vector $\boldsymbol{i}$ is the vector $A_x\boldsymbol{i}$ parallel to the x axis with magnitude A_x. Likewise, $A_y\boldsymbol{j}$ is a vector of magnitude A_y parallel to the y axis. Thus, the unit-vector notation for the vector $\boldsymbol{A}$ is written

$$\boldsymbol{A} = A_x\boldsymbol{i} + A_y\boldsymbol{j} \tag{2.12}$$

The vectors $A_x\boldsymbol{i}$ and $A_y\boldsymbol{j}$ are the component vectors of $\boldsymbol{A}$. These should not be confused with A_x and A_y, which we shall always refer to as the components of $\boldsymbol{A}$.

Now suppose we wish to add vector $\boldsymbol{B}$ to vector $\boldsymbol{A}$, where $\boldsymbol{B}$ has components B_x and B_y. The procedure for performing this sum is to simply add the x and y components separately. The resultant vector $\boldsymbol{R} = \boldsymbol{A} + \boldsymbol{B}$ is therefore given by

$$\boldsymbol{R} = (A_x + B_x)\boldsymbol{i} + (A_y + B_y)\boldsymbol{j} \tag{2.13}$$

Thus, the rectangular components of the resultant vector are given by

$$\begin{aligned} R_x &= A_x + B_x \\ R_y &= A_y + B_y \end{aligned} \tag{2.14}$$

The magnitude of $\boldsymbol{R}$ and the angle it makes with the x axis can then be obtained from its components using the relationships

$$R = \sqrt{R_x^2 + R_y^2} = \sqrt{(A_x + B_x)^2 + (A_y + B_y)^2} \tag{2.15}$$

and

$$\tan \theta = \frac{R_y}{R_x} = \frac{A_y + B_y}{A_x + B_x} \tag{2.16}$$

The procedure just described for adding two vectors $\boldsymbol{A}$ and $\boldsymbol{B}$ using the component method can be checked using a geometric construction, as in

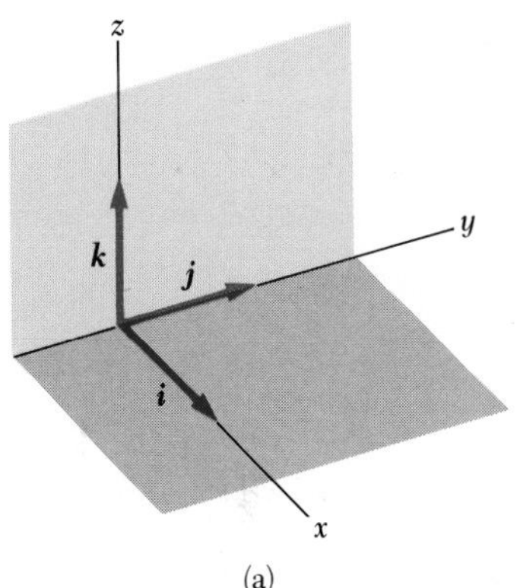

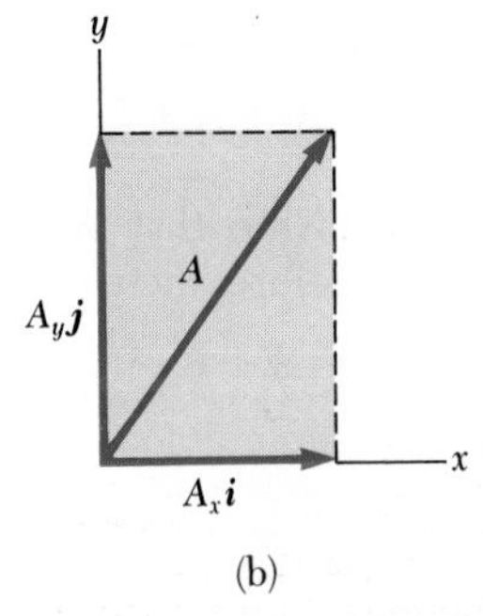

Figure 2.16 (a) The unit vectors $\boldsymbol{i}$, $\boldsymbol{j}$, and $\boldsymbol{k}$ are directed along the x, y, and z axes, respectively. (b) A vector $\boldsymbol{A}$ lying in the xy plane has component vectors $A_x\boldsymbol{i}$ and $A_y\boldsymbol{j}$ where A_x and A_y are the rectangular components of $\boldsymbol{A}$.

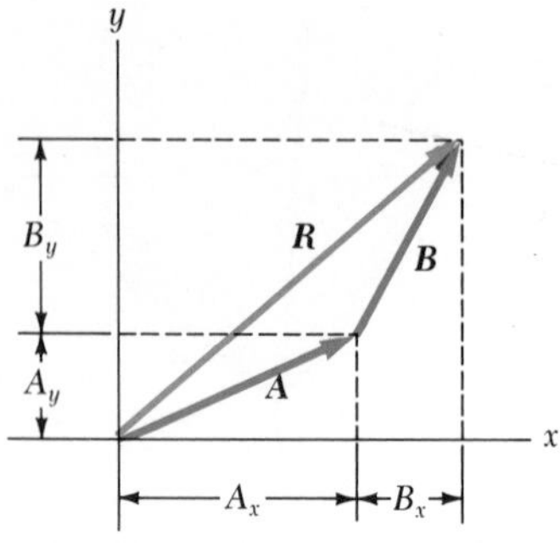

Figure 2.17 Geometric construction showing the relation between the components of the resultant $\boldsymbol{R}$ of two vectors and the individual vector components.

Figure 2.17. Again, you must take note of the *signs* of the components when using either the algebraic or the geometric method.

The extension of these methods to three-dimensional vectors is straightforward. If $\boldsymbol{A}$ and $\boldsymbol{B}$ both have x, y, and z components, we express them in the form

$$\boldsymbol{A} = A_x\boldsymbol{i} + A_y\boldsymbol{j} + A_z\boldsymbol{k} \tag{2.17}$$

$$\boldsymbol{B} = B_x\boldsymbol{i} + B_y\boldsymbol{j} + B_z\boldsymbol{k} \tag{2.18}$$

The sum of $\boldsymbol{A}$ and $\boldsymbol{B}$ is given by

$$\boldsymbol{R} = \boldsymbol{A} + \boldsymbol{B} = (A_x + B_x)\boldsymbol{i} + (A_y + B_y)\boldsymbol{j} + (A_z + B_z)\boldsymbol{k} \tag{2.19}$$

Thus, the resultant vector also has a z component given by $R_z = A_z + B_z$. The same procedure can be used to sum up three or more vectors.

PROBLEM-SOLVING STRATEGY

When two or more vectors are to be added, the following step-by-step procedure is recommended:

1. Select a coordinate system.
2. Draw a sketch of the vectors to be added (or subtracted), with a label on each vector.
3. Find the x and y components of all vectors.
4. Find the resultant components (the algebraic sum of the components) in both the x and y directions.
5. Use the Pythagorean theorem to find the magnitude of the resultant vector.
6. Use a suitable trigonometric function to find the angle the resultant vector makes with the x axis.

EXAMPLE 2.3 The Sum of Two Vectors □

Find the sum of two vectors $\boldsymbol{A}$ and $\boldsymbol{B}$ lying in the xy plane and given by

$$\boldsymbol{A} = 2\boldsymbol{i} + 2\boldsymbol{j} \qquad \text{and} \qquad \boldsymbol{B} = 2\boldsymbol{i} - 4\boldsymbol{j}$$

Solution Note that $A_x = 2$, $A_y = 2$, $B_x = 2$, and $B_y = -4$. Therefore, the resultant vector $\boldsymbol{R}$ is given by

$$\boldsymbol{R} = \boldsymbol{A} + \boldsymbol{B} = (2 + 2)\boldsymbol{i} + (2 - 4)\boldsymbol{j} = 4\boldsymbol{i} - 2\boldsymbol{j}$$

or

$$R_x = 4, R_y = -2$$

The magnitude of $\boldsymbol{R}$ is given by

$$R = \sqrt{R_x^2 + R_y^2} = \sqrt{(4)^2 + (-2)^2} = \sqrt{20} = 4.47$$

Many examples in this text will be followed by an exercise. The purpose of these exercises is to test your understanding of the example by asking you to do a calculation or answer some other question related to the example. Answers to these exercises will be provided at the end of the exercise, when appropriate. Here is your first exercise, related to Example 2.3.

Exercise 1 Find the angle θ that the resultant vector $\boldsymbol{R}$ makes with the positive x axis.
Answer 333°.

EXAMPLE 2.4 The Resultant Displacement

A particle undergoes three consecutive displacements given by $\boldsymbol{d}_1 = (\boldsymbol{i} + 3\boldsymbol{j} - \boldsymbol{k})$ cm, $\boldsymbol{d}_2 = (2\boldsymbol{i} - \boldsymbol{j} - 3\boldsymbol{k})$ cm, and $\boldsymbol{d}_3 = (-\boldsymbol{i} + \boldsymbol{j})$ cm. Find the resultant displacement of the particle.

Solution

$$\begin{aligned}\boldsymbol{R} &= \boldsymbol{d}_1 + \boldsymbol{d}_2 + \boldsymbol{d}_3\\ &= (1 + 2 - 1)\boldsymbol{i} + (3 - 1 + 1)\boldsymbol{j} + (-1 - 3 + 0)\boldsymbol{k}\\ &= (2\boldsymbol{i} + 3\boldsymbol{j} - 4\boldsymbol{k})\text{ cm}\end{aligned}$$

That is, the resultant displacement has components $R_x =$ 2 cm, $R_y = 3$ cm, and $R_z = -4$ cm. Its magnitude is

$$R = \sqrt{R_x^2 + R_y^2 + R_z^2} = \sqrt{(2\text{ cm})^2 + (3\text{ cm})^2 + (-4\text{ cm})^2}$$
$$= 5.39\text{ cm}$$

EXAMPLE 2.5 Taking a Hike

A hiker begins a trip by first walking 25 km due southeast from her base camp. On the second day, she walks 40 km in a direction 60° north of east, at which point she discovers a forest ranger's tower.
(a) Determine the rectangular components of the hiker's displacements for the first and second days.

Solution If we denote the displacement vectors on the first and second days by **A** and **B**, respectively, and use the camp as the origin of coordinates, we get the vectors shown in Figure 2.18. Displacement **A** has a magnitude of 25.0 km and is 45° southeast. Its rectangular components are

$$A_x = A\cos(-45°) = (25\text{ km})(0.707) = \boxed{17.7\text{ km}}$$

$$A_y = A\sin(-45°) = -(25\text{ km})(0.707) = \boxed{-17.7\text{ km}}$$

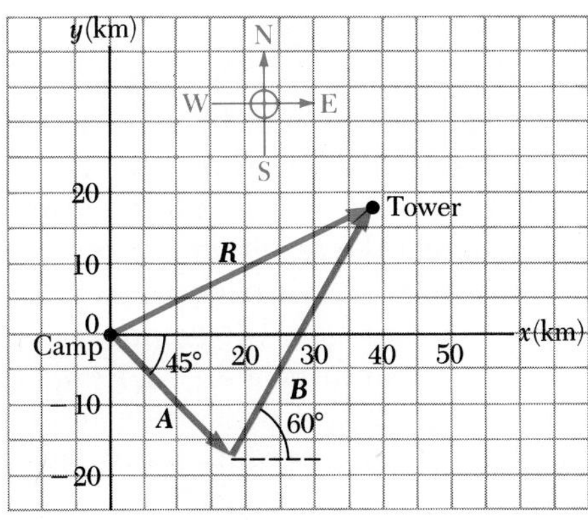

Figure 2.18 (Example 2.5) The total displacement of the hiker is the vector **R** = **A** + **B**.

The negative value of A_y indicates that the y coordinate had decreased for this displacement. The signs of A_x and A_y are also evident from Figure 2.18. The second displacement, **B**, has a magnitude of 40.0 km and is 60° north of east. Its rectangular components are

$$B_x = B\cos 60° = (40\text{ km})(0.50) = \boxed{20.0\text{ km}}$$

$$B_y = B\sin 60° = (40\text{ km})(0.866) = \boxed{34.6\text{ km}}$$

(b) Determine the rectangular components of the hiker's total displacement for the trip.

Solution The resultant displacement for the trip, **R** = **A** + **B**, has components given by

$$R_x = A_x + B_x = 17.7\text{ km} + 20.0\text{ km} = \boxed{37.7\text{ km}}$$

$$R_y = A_y + B_y = -17.7\text{ km} + 34.6\text{ km} = \boxed{16.9\text{ km}}$$

In unit-vector form, we can write the total displacement **R** = (37.7***i*** + 16.9***j***) km.

Exercise 2 Determine the magnitude and direction of the total displacement.
Answer 41.3 km, 24.1° north of east from the base camp.

EXAMPLE 2.6 Let's Fly Away

A commuter airplane starts from an airport and takes the route shown in Figure 2.19. First, it flies to city A located 175 km in a direction 30° north of east. Next, it flies 150 km 20° west of north to city B. Finally, it flies 190 km due west to city C. Find the location of city C relative to the location of the starting point.

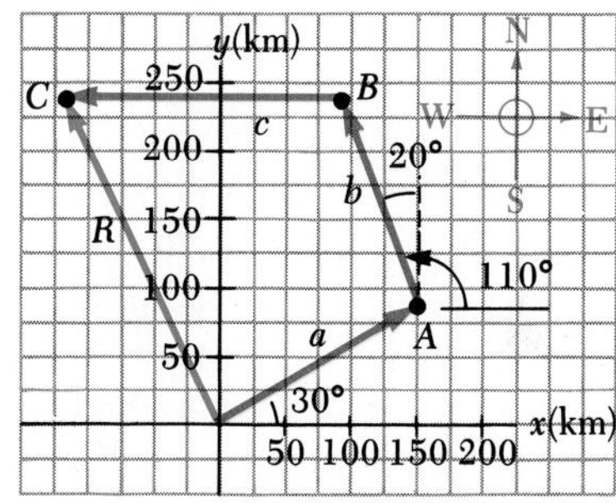

Figure 2.19 (Example 2.6) The airplane starts at the origin, flies to *A* and *B*, and ends its trip at *C*.

Solution As in the previous example, it is convenient to choose the coordinate system shown in Figure 2.19, where the x axis points to the east and the y axis points to the north. Let us denote the three consecutive displacements by the vectors **a**, **b**, and **c**. The first displacement ***a*** has a magnitude of 175 km and has rectangular components

$$a_x = a\cos(30°) = (175\text{ km})(0.866) = 152\text{ km}$$

$$a_y = a\sin(30°) = (175\text{ km})(0.500) = 87.5\text{ km}$$

The second displacement b whose magnitude is 150 km has rectangular components

$$b_x = b\cos(110°) = (150\text{ km})(-0.342) = -51.3\text{ km}$$

$$b_y = b\sin(110°) = (150 \text{ km})(0.940) = 141 \text{ km}$$

Finally, the third displacement **c** whose magnitude is 190 km has rectangular components

$$c_x = c\cos(180°) = (190 \text{ km})(-1) = -190 \text{ km}$$

$$c_y = c\sin(180°) = 0$$

Therefore, the components of the position vector ***R*** from the starting point to city C are

$$R_x = a_x + b_x + c_x = 152 \text{ km} - 51.3 \text{ km} - 190 \text{ km}$$

$$= -89.7 \text{ km}$$

$$R_y = a_y + b_y + c_y = 87.5 \text{ km} + 141 \text{ km} + 0$$

$$= 228 \text{ km}$$

In unit vector notation, $\boldsymbol{R} = (-89.7\boldsymbol{i} + 228\boldsymbol{j})$ km. That is, city C can be reached from the starting point by first traveling 89.7 km due west followed by a trip 228 km due north.

Exercise 3 Find the magnitude and direction of the final position vector ***R***.
Answer 245 km, 21.4° west of north

SUMMARY

Vectors are quantities that have both magnitude and direction and obey the vector law of addition. **Scalars** are quantities that have only magnitude.

Two vectors ***A*** and ***B*** can be added using either the triangle method or the parallelogram rule. In the triangle method (Figure 2.20a), the vector $\boldsymbol{C} = \boldsymbol{A} + \boldsymbol{B}$ runs from the tail of ***A*** to the tip of ***B***. In the parallelogram method (Figure 2.20b), ***C*** is the diagonal of a parallelogram having ***A*** and ***B*** as its sides.

The x component, A_x, of the vector ***A*** is equal to its projection along the x axis of a coordinate system as in Figure 2.21, where $A_x = A\cos\theta$. Likewise, the y component, A_y, of ***A*** is its projection along the y axis, where $A_y = A\sin\theta$. The resultant of two or more vectors can be found by resolving all vectors into their x and y components, adding their resultant x and y components, and then using the Pythagorean theorem to find the magnitude of the resultant vector. The angle that the resultant vector makes with respect to the x axis can be found by use of a suitable trigonometric function.

If a vector ***A*** has an x component equal to A_x and a y component equal to A_y, the vector can be expressed in unit-vector form as $\boldsymbol{A} = A_x\boldsymbol{i} + A_y\boldsymbol{j}$. In this notation, ***i*** is a unit vector pointing in the positive x direction and ***j*** is a unit vector in the positive y direction. Since ***i*** and ***j*** are unit vectors, $|\boldsymbol{i}| = |\boldsymbol{j}| = 1$.

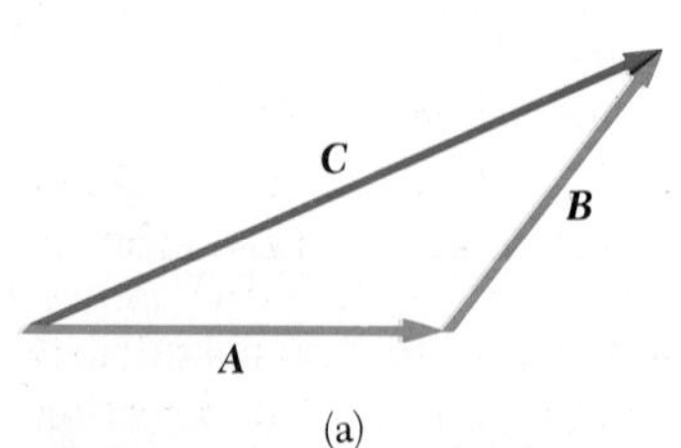

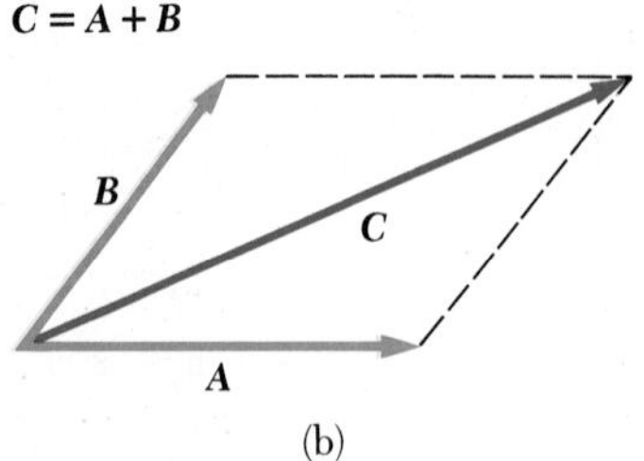

Figure 2.20 (a) Vector addition using the triangle method. (b) Vector addition using the parallelogram rule.

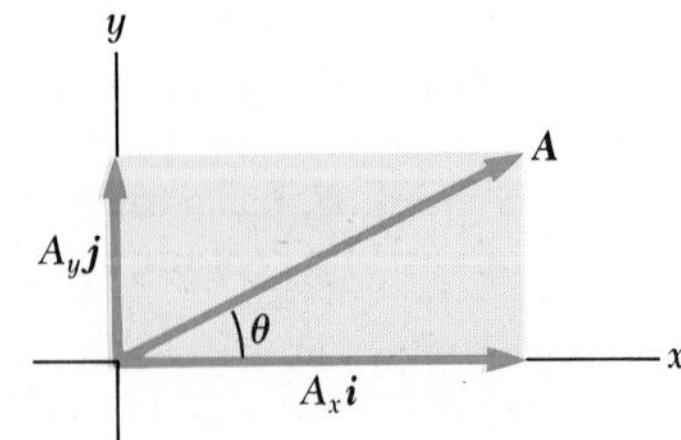

Figure 2.21 The x and y components of a vector ***A*** are A_x and A_y.

QUESTIONS

1. A book is moved once around the perimeter of a table of dimensions 1 m × 2 m. If the book ends up at its initial position, what is its displacement? What is the distance traveled?
2. If **B** is added to **A**, under what condition does the resultant vector have a magnitude equal to $A + B$? Under what conditions is the resultant vector equal to zero?
3. Can the magnitude of a particle's displacement be greater than the distance traveled? Explain.
4. The magnitudes of two vectors **A** and **B** are $A = 5$ units and $B = 2$ units. Find the largest and smallest values possible for the resultant vector $\boldsymbol{R} = \boldsymbol{A} + \boldsymbol{B}$.
5. A vector **A** lies in the xy plane. For what orientations of **A** will both of its rectangular components be negative? For what orientations will its components have opposite signs?
6. Can a vector have a component equal to zero and still have a nonzero magnitude? Explain.
7. If one of the components of a vector is not zero, can its magnitude be zero? Explain.
8. If the component of vector **A** along the direction of vector **B** is zero, what can you conclude about the two vectors?
9. If $\boldsymbol{A} = \boldsymbol{B}$, what can you conclude about the components of **A** and **B**?
10. Can the magnitude of a vector have a negative value? Explain.
11. If $\boldsymbol{A} + \boldsymbol{B} = 0$, what can you say about the components of the two vectors?
12. Which of the following are vectors and which are not: force, temperature, the volume of water in a can, the ratings of a TV show, the height of a building, the velocity of a sports car, the age of the universe?
13. Under what circumstances would a nonzero vector lying in the xy plane have components that are equal in magnitude?
14. Is it possible to add a vector quantity to a scalar quantity? Explain.
15. Two vectors have unequal magnitudes. Can their sum be zero? Explain.
16. (a) What is the resultant displacement of a walk of 80 m followed by a walk of 125 m when both displacements are in the eastward direction? (b) What is the resultant displacement in a situation in which the 125-m walk is in the direction opposite the 80-m walk?
17. While traveling along a straight interstate highway you notice that the mile marker reads 260. You travel until you reach the 150-mile marker and then retrace your path to the 175-mile marker. What is the magnitude of your resultant displacement from the 260-mile marker?
18. A submarine dives at an angle of 30° with the horizontal and follows a straight-line path for a total distance of 50 m. How far is the submarine below the surface of the water?
19. A roller coaster travels 135 ft at an angle of 40° above the horizontal. How far does it move horizontally and vertically?

PROBLEMS

Section 2.1 Coordinate Systems and Frames of Reference

1. Two points in the xy plane have cartesian coordinates (2.0, −4.0) and (−3.0, 3.0), where the units are in m. Determine (a) the distance between these points and (b) their polar coordinates.
2. A point in the xy plane has cartesian coordinates (−3.0, 5.0) m. What are the polar coordinates of this point?
3. The polar coordinates of a point are $r = 5.50$ m and $\theta = 240°$. What are the cartesian coordinates of this point?
4. Two points in a plane have polar coordinates (2.50 m, 30°) and (3.80 m, 120°). Determine (a) the cartesian coordinates of these points and (b) the distance between them.
5. A certain corner of a room is selected as the origin of a rectangular coordinate system. If a fly is crawling on an adjacent wall at a point having coordinates (2.0, 1.0), where the units are in meters, what is the distance of the fly from the corner of the room?
6. Express the location of the fly in Problem 5 in polar coordinates.
7. A point is located in a polar coordinate system by the coordinates $r = 2.5$ m and $\theta = 35°$. Find the x and y coordinates of this point, assuming the two coordinate systems have the same origin.

Section 2.2 Vectors and Scalars and Section 2.3 Some Properties of Vectors

8. A shopper pushing a cart through a store moves 40 m down one aisle then makes a 90° turn and moves 15 m. He then makes another 90° turn and moves 20 m. How far is the shopper away from his original position in magnitude and direction? The direction moved in any of the 90° turns is not given. As a result, could there be more than one answer?

9. A man pushing a mop across a floor causes it to undergo two displacements. The first has a magnitude of 150 cm and makes an angle of 120° with the positive x axis. The resultant displacement has a magnitude of 140 cm and is directed at an angle of 35° to the positive x axis. Find the magnitude and direction of the second displacement.
10. A pedestrian moves 6 km east and 13 km north. Find the magnitude and direction of the resultant displacement vector using the graphical method.
11. Vector $\boldsymbol{A}$ is 3 units in length and points along the positive x axis. Vector $\boldsymbol{B}$ is 4 units in length and points along the negative y axis. Use graphical methods to find the magnitude and direction of the vectors (a) $\boldsymbol{A}+\boldsymbol{B}$, (b) $\boldsymbol{A}-\boldsymbol{B}$.
12. Vector $\boldsymbol{A}$ is 6 units in length and at an angle of 45° to the x axis. Vector $\boldsymbol{B}$ is 3 units in length and is directed along the positive x axis ($\theta = 0$). Find the resultant vector $\boldsymbol{A}+\boldsymbol{B}$ using (a) graphical methods and (b) the law of cosines.
13. A person walks along a circular path of radius 5 m, around one half of the circle. (a) Find the magnitude of the displacement vector. (b) How far did the person walk? (c) What is the magnitude of the displacement if the circle is completed?
14. A particle undergoes three consecutive displacements such that its *total* displacement is zero. The first displacement is 8 m westward. The second is 13 m northward. Find the magnitude and direction of the third displacement using the graphical method.
15. Each of the displacement vectors $\boldsymbol{A}$ and $\boldsymbol{B}$ shown in Figure 2.22 has a magnitude of 3 m. Find graphically (a) $\boldsymbol{A}+\boldsymbol{B}$, (b) $\boldsymbol{A}-\boldsymbol{B}$, (c) $\boldsymbol{B}-\boldsymbol{A}$, (d) $\boldsymbol{A}-2\boldsymbol{B}$.

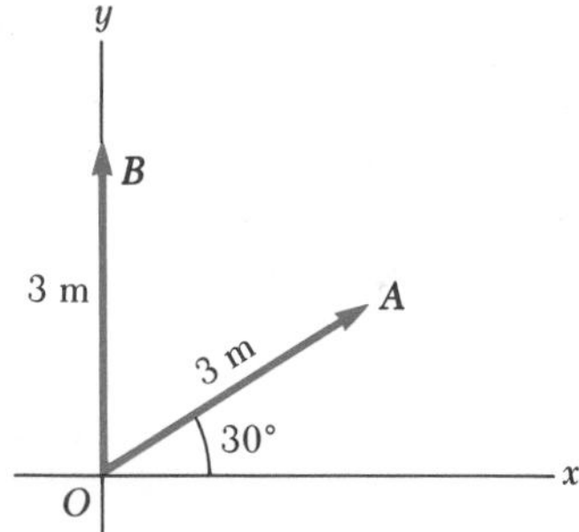

Figure 2.22 (Problems 15 and 37).

16. A dog searching for a bone walks 3.5 m south, then 8.2 m at an angle 30° north of east, and finally 15 m west. Find the dog's resultant displacement vector using graphical techniques.
17. A roller coaster moves 200 ft horizontally, then rises 135 ft at an angle of 30° above the horizontal. It then travels 135 ft at an angle of 40° downward. What is its displacement from its starting point at the end of this movement? Use graphical techniques.
18. A taxi driver travels due south for 10 km and then moves 6 km in a direction 30° north of east. Find the magnitude and direction of the car's resultant displacement.
19. Find the horizontal and vertical components of the 100-m displacement of a superhero who flies from the top of a tall building following the path shown in Figure 2.23.

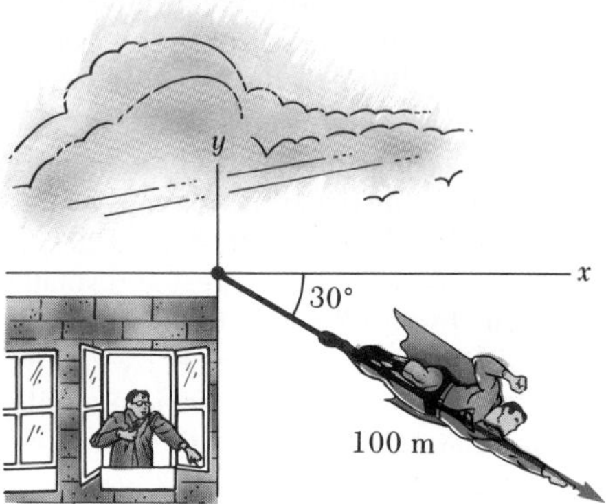

Figure 2.23 (Problem 19).

20. A man lost in a maze makes three consecutive displacements such that at the end of the walks he is right back where he started. The first displacement is 8 m westward, and the second is 13 m northward. Find the magnitude and direction of the third displacement using the graphical method.
21. A jogger runs 100 m due west, then changes direction for the second leg of the run. At the end of the run, she is 175 m away from the starting point at an angle of 15° north of west. What was the direction and length of her second displacement? Use graphical techniques.
22. While exploring a cave, a spelunker starts at the entrance and moves the following distances. She goes 75 m north, 250 m east, 125 m at an angle 30° north of east, and 150 m south. Find the resultant displacement from the cave entrance.

Section 2.4 Components of a Vector and Unit Vectors

23. A vector has an x component of -25 units and a y component of 40 units. Find the magnitude and direction of this vector.
24. A displacement vector $\boldsymbol{A}$ makes an angle θ with the positive x axis, as in Figure 2.13. Find the rectangular components of $\boldsymbol{A}$ for the following values of A and θ: (a) $A = 8$ m, $\theta = 60°$; (b) $A = 6$ ft, $\theta = 120°$; (c) $A = 12$ cm, $\theta = 225°$.
25. Vector $\boldsymbol{A}$ lies in the xy plane. Construct a table of the signs of the x and y components of $\boldsymbol{A}$ when the vector lies in the first, second, third, and fourth quadrants.
26. A displacement vector lying in the xy plane has a magnitude of 50 m and is directed at an angle of 120° to the positive x axis. What are the rectangular components of this vector?

27. Find the magnitude and direction of the resultant of three displacements having components (3, 2) m, (−5, 3) m, and (6, 1) m.

28. Vector **A** has x and y components of −8.7 cm and 15 cm, respectively; vector **B** has x and y components of 13.2 cm and −6.6 cm, respectively. If $\mathbf{A} - \mathbf{B} + 3\mathbf{C} = 0$, what are the components of **C**?

29. Two vectors are given by $\mathbf{A} = 3\mathbf{i} - 2\mathbf{j}$ and $\mathbf{B} = -\mathbf{i} - 4\mathbf{j}$. Calculate (a) $\mathbf{A} + \mathbf{B}$, (b) $\mathbf{A} - \mathbf{B}$, (c) $|\mathbf{A} + \mathbf{B}|$, (d) $|\mathbf{A} - \mathbf{B}|$, (e) the direction of $\mathbf{A} + \mathbf{B}$ and $\mathbf{A} - \mathbf{B}$.

30. Three vectors are given by $\mathbf{A} = \mathbf{i} + 3\mathbf{j}$, $\mathbf{B} = 2\mathbf{i} - \mathbf{j}$, and $\mathbf{C} = 3\mathbf{i} + 5\mathbf{j}$. Find (a) the sum of the three vectors and (b) the magnitude and direction of the resultant vector.

31. Obtain expressions for the position vectors with polar coordinates (a) 12.8 m, 150°; (b) 3.3 cm, 60°; (c) 22 in., 215°.

32. Vectors **A** and **B** have components $A_x = -5.0$ cm, $A_y = 1.1$ cm, $A_z = -3.5$ cm and $B_x = 8.8$ cm, $B_y = -6.3$ cm, $B_z = 9.2$ cm. Determine the components of the vectors (a) $\mathbf{A} + \mathbf{B}$, (b) $\mathbf{B} - \mathbf{A}$, (c) $3\mathbf{B} + 2\mathbf{A}$. (d) Express the vector $\mathbf{B} - \mathbf{A}$ in unit-vector notation.

33. A particle undergoes the following consecutive displacements: 3.5 m south, 8.2 m northeast, and 15.0 m west. What is the resultant displacement?

34. A quarterback takes the ball from the line of scrimmage, runs backward for 10 yards, then sideways parallel to the line of scrimmage for 15 yards. At this point, he throws a 50-yard forward pass straight downfield perpendicular to the line of scrimmage. What is the magnitude of the football's resultant displacement?

35. An airplane flies from city A to city B in a direction due east for 800 miles. In the next part of the trip the airplane flies from city B to city C in a direction 40° north of east for 600 miles. What is the resultant displacement of the airplane between city A and city C?

36. A particle undergoes three consecutive displacements. The first is to the east and has a magnitude of 25 m. The second is to the north and has a magnitude of 42 m. If the resultant displacement has a magnitude of 38 m and is directed at an angle of 30° north of east, what are the magnitude and direction of the third displacement?

37. Find the x and y components of the vectors **A** and **B** shown in Figure 2.22. Derive an expression for the resultant vector $\mathbf{A} + \mathbf{B}$ in unit-vector notation.

38. A particle undergoes two displacements. The first has a magnitude of 150 cm and makes an angle of 120° with the positive x axis. The *resultant* displacement has a magnitude of 140 cm and is directed at an angle of 35° to the positive x axis. Find the magnitude and direction of the second displacement.

39. The vector **A** has x, y, and z components of 8, 12, and −4 units, respectively. (a) Write a vector expression for **A** in unit-vector notation. (b) Obtain a unit-vector expression for a vector **B** one fourth the length of **A** pointing in the same direction as **A**. (c) Obtain a unit-vector expression for a vector **C** three times the length of **A** pointing in the direction opposite **A**.

40. Two vectors are given by $\mathbf{A} = -2\mathbf{i} + \mathbf{j} - 3\mathbf{k}$ and $\mathbf{B} = 5\mathbf{i} + 3\mathbf{j} - 2\mathbf{k}$. (a) Find a third vector **C** such that $3\mathbf{A} + 2\mathbf{B} - \mathbf{C} = 0$. (b) What are the magnitudes of **A**, **B**, and **C**?

41. A vector **A** has a magnitude of 35 units and makes an angle of 37° with the positive x axis. Describe (a) a vector **B** that is in the direction opposite **A** and is one fifth the size of **A**, and (b) a vector **C** that when added to **A** will produce a vector twice as long as **A** pointing in the negative y direction.

42. A vector **A** has a positive x component of 4 units and a negative y component of 2 units. What second vector **B** when added to **A** will produce a *resultant* vector three times the magnitude of **A** directed in the positive y direction?

43. A vector **A** has a negative x component 3 units in length and a positive y component 2 units in length. (a) Determine an expression for **A** in unit-vector notation. (b) Determine the magnitude and direction of **A**. (c) What vector **B** when added to **A** gives a resultant vector with no x component and a negative y component 4 units in length?

44. A particle moves in the xy plane from the point (3, 0) m to the point (2, 2) m. (a) Determine a vector expression for the resultant displacement. (b) What are the magnitude and direction of this displacement vector?

45. Find the magnitude and direction of a displacement vector having x and y components of −5 m and 3 m, respectively.

46. Three vectors are given by $\mathbf{A} = 6\mathbf{i}$, $\mathbf{B} = 9\mathbf{j}$, and $\mathbf{C} = (-3\mathbf{i} + 4\mathbf{j})$. (a) Find the magnitude and direction of the resultant vector. (b) What vector must be added to these three to make the resultant vector zero?

47. Three vectors are oriented as shown in Figure 2.24 where $|\mathbf{A}| = 20$, $|\mathbf{B}| = 40$, and $|\mathbf{C}| = 30$ units. Find (a) the x and y components of the resultant vector and (b) the magnitude and direction of the resultant vector.

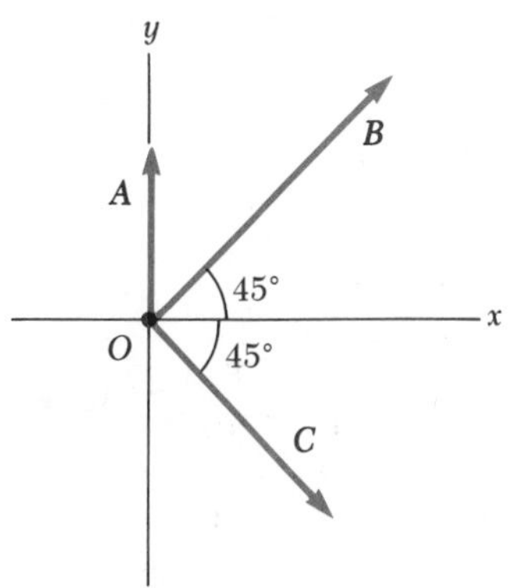

Figure 2.24 (Problem 47).

ADDITIONAL PROBLEMS

48. A vector is given by $\boldsymbol{R} = 2\boldsymbol{i} + \boldsymbol{j} + 3\boldsymbol{k}$. Find (a) the magnitudes of the x, y, and z components, (b) the magnitude of $\boldsymbol{R}$, and (c) the angles between $\boldsymbol{R}$ and the x, y, and z axes.

49. A person going for a walk follows the path shown in Figure 2.25. The total trip consists of four straight-line paths. At the end of the walk, what is the person's resultant displacement measured from the starting point?

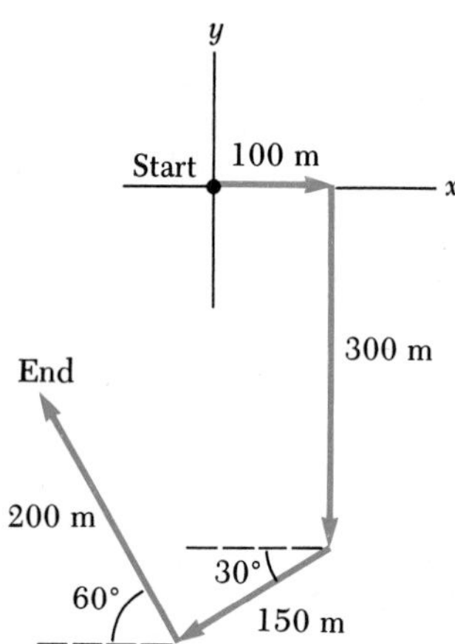

Figure 2.25 (Problem 49).

50. Two people pull on a stubborn mule as seen from the helicopter view shown in Figure 2.26. Find (a) the single force which is equivalent to the two forces shown, and (b) the force that a third person would have to exert on the mule to make the net force equal to zero.

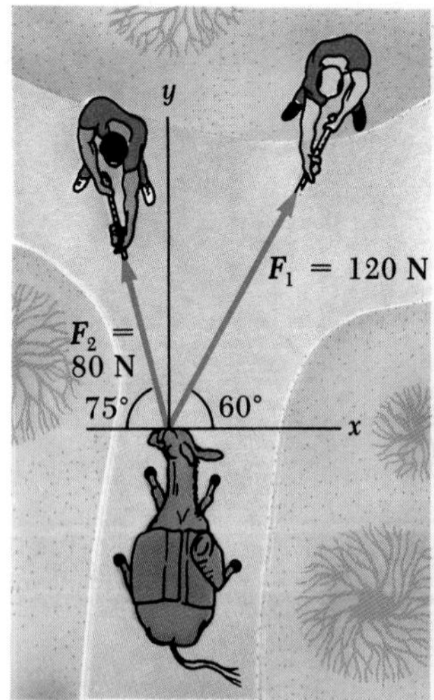

Figure 2.26 (Problem 50).

51. A particle moves from a point in the xy plane having cartesian coordinates $(-3, -5)$ m to a point with coordinates $(-1, 8)$ m. (a) Write vector expressions for the position vectors in unit-vector form for these two points. (b) What is the displacement vector? (See Problem 54 for definition.)

52. A rectangular parallelepiped has dimensions a, b, and c, as in Figure 2.27. (a) Obtain a vector expression for the face diagonal vector $\boldsymbol{R}_1$. What is the magnitude of this vector? (b) Obtain a vector expression for the body diagonal vector $\boldsymbol{R}_2$. What is the magnitude of this vector?

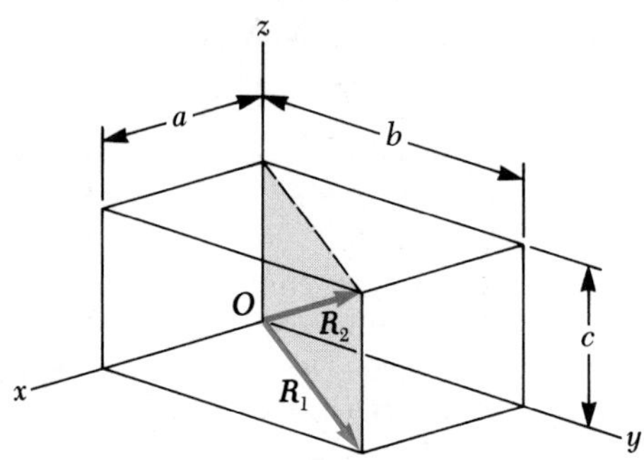

Figure 2.27 (Problem 52).

53. A point P is described by the coordinates (x, y) with respect to the normal cartesian coordinate system shown in Figure 2.28. Show that (x', y'), the coordinates of this point in the rotated $x'y'$ coordinate system, are related to (x, y) and the rotation angle α by the expressions

$$x' = x \cos \alpha + y \sin \alpha$$

and

$$y' = -x \sin \alpha + y \cos \alpha$$

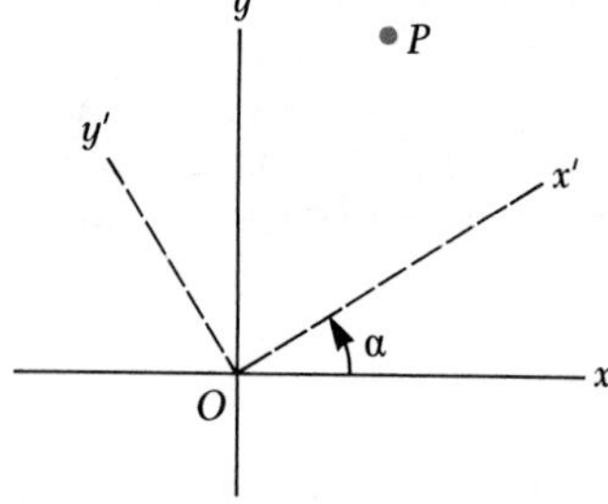

Figure 2.28 (Problem 53).

54. (a) Show that a point lying in the xy plane and having coordinates (x, y) can be described by the position vector $\boldsymbol{r} = x\boldsymbol{i} + y\boldsymbol{j}$. (b) Show that the magnitude of this vector is $r = \sqrt{x^2 + y^2}$. (c) Show that the displacement vector for a particle moving from (x_1, y_1) to (x_2, y_2) is given by $\boldsymbol{d} = (x_2 - x_1)\boldsymbol{i} + (y_2 - y_1)\boldsymbol{j}$. (d) Plot the position vectors $\boldsymbol{r}_1$ and $\boldsymbol{r}_2$ and the displacement vector $\boldsymbol{d}$, and verify by the graphical method that $\boldsymbol{d} = \boldsymbol{r}_2 - \boldsymbol{r}_1$.

3

Motion in One Dimension

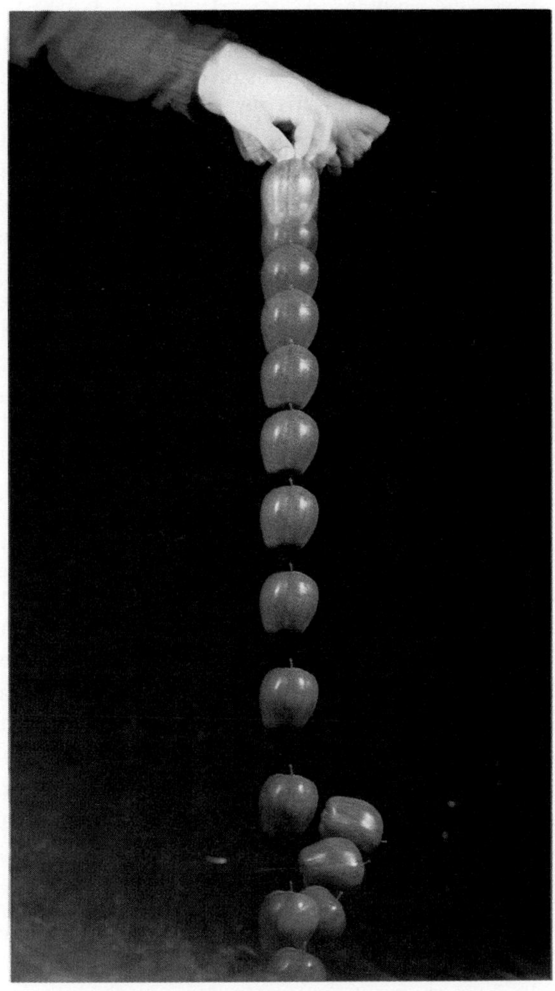

Newton's apple, a multiflash study of motion in one dimension. (© Harold E. Edgerton. Courtesy of Palm Press Inc.)

Dynamics is concerned with the study of the motion of an object and the relation of this motion to such physical concepts as force and mass. It is convenient to describe motion using the concepts of space and time, without regard to the causes of the motion. This portion of mechanics is called *kinematics.* In this chapter we shall consider motion along a straight line, that is, one-dimensional motion. In the next chapter we shall extend our discussion to two-dimensional motion. Starting with the concept of displacement discussed in the previous chapter, we shall define velocity and acceleration. Using these concepts, we shall proceed to study the motion of objects undergoing constant acceleration. The subject of *dynamics,* which is concerned with the causes of motion and relationships between motion, forces, and the properties of moving objects, will be discussed in Chapters 5 and 6.

From everyday experience we recognize that motion represents the continuous change in the position of an object. The movement of an object through space may be accompanied by the rotation or vibration of the object. Such motions can be quite complex. However, it is sometimes possible to simplify matters by temporarily neglecting the internal motions of the moving object. In many situations, an object can be treated as a *particle* if the only motion being considered is one of translation through space. An idealized particle is a mathematical point with no size. For example, if we wish to describe the motion of the earth around the sun, we can treat the earth as a particle and obtain reasonable accuracy in a prediction of the earth's orbit. This approximation is justified because the radius of the earth's orbit is large compared with the dimensions of the earth and sun. On the other hand, we could not use the particle description to explain the internal structure of the earth and such phenomena as tides, earthquakes, and volcanic activity. On a much smaller scale, it is possible to explain the pressure exerted by a gas on the walls of a container by treating the gas molecules as particles. However, the particle description of the gas molecules is generally inadequate for understanding those properties of the gas that depend on the internal motions of the gas molecules, namely, rotations and vibrations.

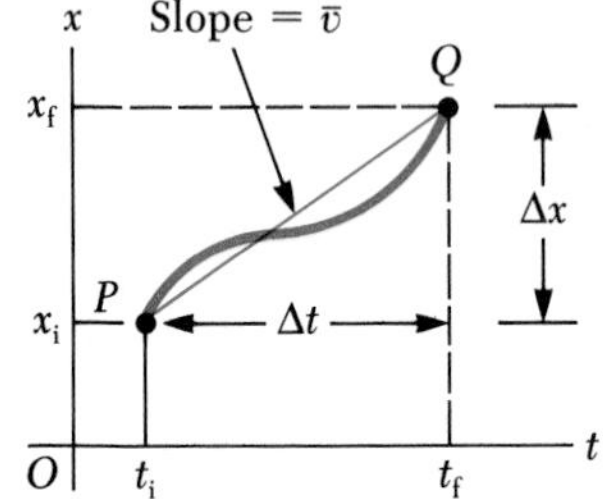

Figure 3.1 Position-time graph for a particle moving along the x axis. The average velocity $\bar{v}$ in the interval $\Delta t = t_f - t_i$ is the slope of the straight line connecting the points P and Q.

3.1 AVERAGE VELOCITY

The motion of a particle is completely known if its position in space is known at all times. Consider a particle moving along the x axis from point P to point Q. Let its position at point P be x_i at some time t_i, and let its position at point Q be x_f at time t_f. (The indices i and f refer to the initial and final values.) At times other than t_i and t_f, the position of the particle between these two points may vary as in Figure 3.1. Such a plot is often called a *position-time graph.* In the time interval $\Delta t = t_f - t_i$, the displacement of the particle is $\Delta x = x_f - x_i$. (Recall that the displacement is defined as the change in the position of the particle, which equals its final minus its initial position value.)

The x-component of the **average velocity** of the particle, $\bar{v}$, is defined as the ratio of its displacement, Δx, and the time interval, Δt:

Average velocity

$$\bar{v} \equiv \frac{\Delta x}{\Delta t} = \frac{x_f - x_i}{t_f - t_i} \tag{3.1}$$

From this definition, we see that the average velocity has the dimensions of length divided by time, or m/s in SI units and ft/s in conventional units. The average velocity is *independent* of the path taken between the points P and Q. This is true because the average velocity is proportional to the displacement, Δx, which in turn depends only on the initial and final coordinates of the particle. It therefore follows that if a particle starts at some point and returns to the same point via any path, its average velocity for this trip is zero, since its displacement along such a path is zero. The displacement should not be confused with the distance traveled, since the distance traveled for any motion is clearly nonzero. Thus, average velocity gives us no details of the motion between points P and Q. (How we evaluate the velocity at some instant in time is discussed in the next section.) Finally, note that the average velocity in one dimension can be positive or negative, depending on the sign of the displacement. (The time interval, Δt, is always positive.) If the coordinate of the particle increases in time (that is, if $x_f > x_i$), then Δx is positive and $\bar{v}$ is positive. This corresponds to a velocity in the positive x direction. On the other hand, if the coordinate decreases in time ($x_f < x_i$), Δx is negative; hence $\bar{v}$ is negative. This corresponds to a velocity in the negative x direction.

The average velocity can also be interpreted geometrically by drawing a straight line between the points P and Q in Figure 3.1. This line forms the hypotenuse of a triangle of height Δx and base Δt. The slope of this line is the ratio $\Delta x/\Delta t$. Therefore, we see that the *average* velocity of the particle during the time interval t_i to t_f is equal to the "slope" of the straight line joining the initial and final points on the space-time graph. (The word *slope* will often be used when referring to the graphs of physical data. Regardless of what data are plotted, the word *slope* will represent the ratio of the change in the quantity represented on the vertical axis to the change in the quantity represented on the horizontal axis.)

EXAMPLE 3.1 Calculate the Average Velocity
A particle moving along the x axis is located at $x_i = 12$ m at $t_i = 1$ s and at $x_f = 4$ m at $t_f = 3$ s. Find its displacement and average velocity during this time interval.

Solution The displacement is given by

$$\Delta x = x_f - x_i = 4\text{ m} - 12\text{ m} = -8\text{ m}$$

The average velocity is

$$\bar{v} = \frac{\Delta x}{\Delta t} = \frac{x_f - x_i}{t_f - t_i} = \frac{4\text{ m} - 12\text{ m}}{3\text{ s} - 1\text{ s}} = -\frac{8\text{ m}}{2\text{ s}} = -4\text{ m/s}$$

Since the displacement and average velocity are negative for this time interval, we conclude that the particle has moved to the left, toward decreasing values of x.

3.2 INSTANTANEOUS VELOCITY

We would like to be able to define the velocity of a particle at a particular instant of time, rather than just over a finite interval of time. The velocity of a particle at any instant of time, or at some point on a space-time graph, is called the **instantaneous velocity.** This concept is especially important when the average velocity in different time intervals is *not constant.*

Consider the motion of a particle between the two points P and Q on the space-time graph shown in Figure 3.2. As the point Q is brought closer and closer to the point P, the time intervals ($\Delta t_1, \Delta t_2, \Delta t_3, \ldots$) get progressively smaller. The average velocity for each time interval is the slope of the appropriate dotted line in Figure 3.2. As the point Q approaches P, the time interval approaches zero, but at the same time the slope of the dotted line approaches that of the blue line tangent to the curve at the point P. The slope of the line tangent to the curve at P is defined to be the *instantaneous velocity* at the time t_i. In other words,

> the instantaneous velocity, v, equals the limiting value of the ratio $\Delta x/\Delta t$ as Δt approaches zero[1]:

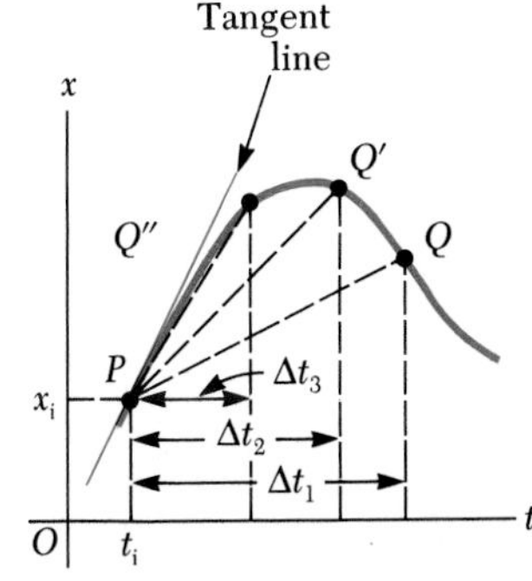

Figure 3.2 Position-time graph for a particle moving along the x axis. As the time intervals starting at t_i get smaller and smaller, the average velocity for that interval approaches the slope of the line tangent at P. The instantaneous velocity at P is defined as the slope of the blue tangent line at the time t_i.

[1] Note that the displacement, Δx, also approaches zero as Δt approaches zero. However, as Δx and Δt become smaller and smaller, the ratio $\Delta x/\Delta t$ approaches a value equal to the *true* slope of the line tangent to the x versus t curve.

$$v \equiv \lim_{\Delta t \to 0} \frac{\Delta x}{\Delta t} \tag{3.2}$$

In the calculus notation, this limit is called the *derivative* of x with respect to t, written dx/dt:

Definition of the derivative

$$v \equiv \lim_{\Delta t \to 0} \frac{\Delta x}{\Delta t} = \frac{dx}{dt} \tag{3.3}$$

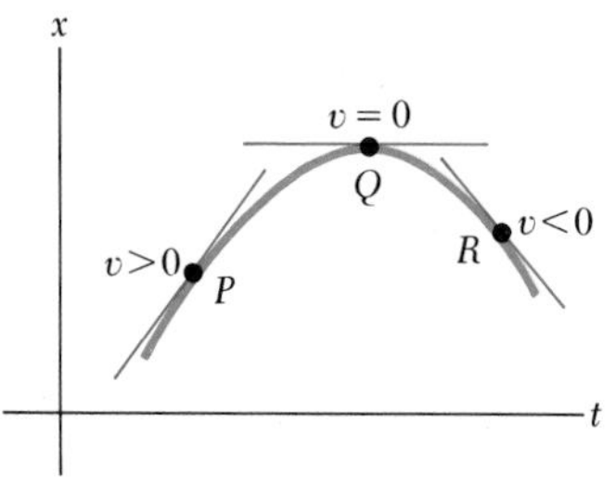

Figure 3.3 In the position-time graph shown here, the velocity is positive at P, where the slope of the tangent line is positive; the velocity is zero at Q, where the slope of the tangent line is zero; and the velocity is negative at R, where the slope of the tangent line is negative.

The instantaneous velocity can be positive, negative, or zero.

When the slope of the space-time graph is positive, such as at the point P in Figure 3.3, v is positive. At point R, v is negative since the slope is negative. Finally, the instantaneous velocity is zero at the peak Q (the turning point), where the slope is zero. *From here on, we shall usually use the word* velocity *to designate instantaneous velocity.*

The **instantaneous speed** of a particle is defined as the magnitude of the instantaneous velocity vector. Hence, by definition, *speed* can never be negative.

It is also possible to find the displacement of a particle if its velocity is known as a function of time using a mathematical technique called integration. Because this procedure may not be familiar to many students, the topic is treated in Section 3.6, which is optional, for general interest and for those courses that cover this material.

EXAMPLE 3.2 Average and Instantaneous Velocity □

A particle moves along the x axis. Its x coordinate varies with time according to the expression $x = -4t + 2t^2$, where x is in m and t is in s. The position-time graph for this motion is shown in Figure 3.4. Note that the particle first moves in the negative x direction for the first second of motion, stops instantaneously at $t = 1$ s, and then heads back in the positive x direction for $t > 1$ s. (a) Determine the displacement of the particle in the time intervals $t = 0$ to $t = 1$ s and $t = 1$ s to $t = 3$ s.

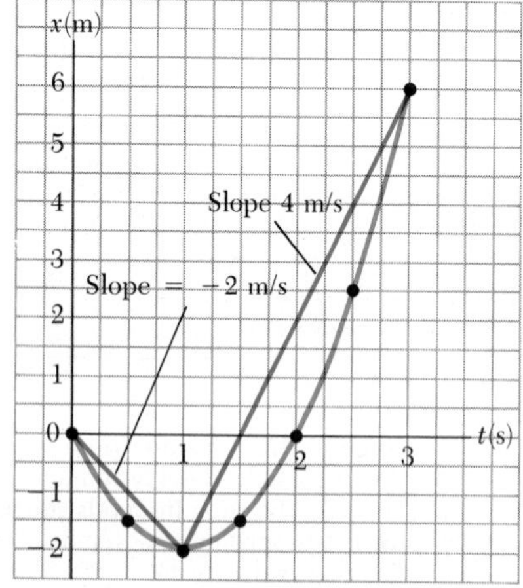

t(s)	x(m)
0	0
0.5	−1.5
1	−2
1.5	−1.5
2	0
2.5	2.5
3	6

Figure 3.4 (Example 3.2) Position-time graph for a particle having an x coordinate that varies in time according to $x = -4t + 2t^2$. Note that $\bar{v}$ is *not* the same as $v = -4 + 4t$.

In the first time interval, we set $t_i = 0$ and $t_f = 1$ s. Since $x = -4t + 2t^2$, we get for the first displacement

$$\begin{aligned}\Delta x_{01} &= x_f - x_i \\ &= [-4(1) + 2(1)^2] - [-4(0) + 2(0)^2] \\ &= -2 \text{ m}\end{aligned}$$

Likewise, in the second time interval we can set $t_i = 1$ s and $t_f = 3$ s. Therefore, the displacement in this interval is

$$\begin{aligned}\Delta x_{13} &= x_f - x_i \\ &= [-4(3) + 2(3)^2] - [-4(1) + 2(1)^2] \\ &= 8 \text{ m}\end{aligned}$$

Note that these displacements can also be read directly from the position-time graph (Fig. 3.4).

(b) Calculate the average velocity in the time intervals $t = 0$ to $t = 1$ s and $t = 1$ s to $t = 3$ s.

In the first time interval, $\Delta t = t_f - t_i = 1$ s. Therefore, using Equation 3.1 and the results from (a) gives

$$\bar{v}_{01} = \frac{\Delta x_{01}}{\Delta t} = \frac{-2\text{ m}}{1\text{ s}} = \boxed{-2\text{ m/s}}$$

Likewise, in the second time interval, $\Delta t = 2$ s; therefore

$$\bar{v}_{13} = \frac{\Delta x_{13}}{\Delta t} = \frac{8\text{ m}}{2\text{ s}} = \boxed{4\text{ m/s}}$$

These values agree with the slopes of the lines joining these points in Figure 3.4.

(c) Find the instantaneous velocity of the particle at $t = 2.5$ s.

By measuring the slope of the position-time graph at $t = 2.5$ s, we find that $v = 6$ m/s. (You should show that the velocity is -4 m/s at $t = 0$ and zero at $t = 1$ s.) Do you see any symmetry in the motion? For example, does the speed ever repeat itself?[2]

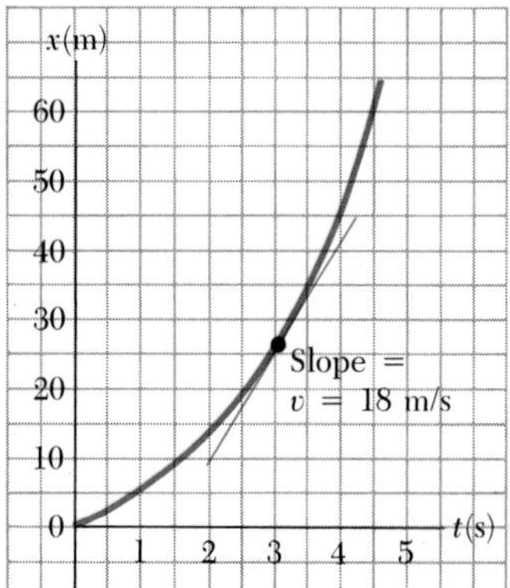

Figure 3.5 (Example 3.3) Position-time graph for a particle having an x coordinate that varies in time according to $x = 3t^2$. Note that the instantaneous velocity at $t = 3$ s equals the slope of the blue line tangent to the curve at this point.

EXAMPLE 3.3 The Limiting Process

The position of a particle moving along the x axis varies in time according to the expression $x = 3t^2$, where x is in m, 3 is in m/s^2, and t is in s. Find the velocity at any time.

Solution The position-time graph for this motion is shown in Figure 3.5. We can compute the velocity at any time t by using the definition of the instantaneous velocity. If the initial coordinate of the particle at time t is $x_i = 3t^2$, then the coordinate at a later time $t + \Delta t$ is

$$x_f = 3(t + \Delta t)^2 = 3[t^2 + 2t\,\Delta t + (\Delta t)^2]$$
$$= 3t^2 + 6t\,\Delta t + 3(\Delta t)^2$$

Therefore, the displacement in the time interval Δt is

$$\Delta x = x_f - x_i = 3t^2 + 6t\,\Delta t + 3(\Delta t)^2 - 3t^2$$
$$= 6t\,\Delta t + 3(\Delta t)^2$$

The average velocity in this time interval is

$$\bar{v} = \frac{\Delta x}{\Delta t} = 6t + 3\,\Delta t$$

To find the instantaneous velocity, we take the limit of this expression as Δt approaches zero. In doing so, we see that the term $3\,\Delta t$ goes to zero, therefore

$$v = \lim_{\Delta t \to 0} \frac{\Delta x}{\Delta t} = \boxed{6t\text{ m/s}}$$

Notice that this expression gives us the velocity at *any* general time t. It tells us that v is increasing linearly in time. It is then a straightforward matter to find the velocity at some specific time from the expression $v = 6t$. For example, at $t = 3$ s, the velocity is $v = 6(3) = 18$ m/s. Again, this can be checked from the slope of the graph (the blue line) at $t = 3$ s.

The limiting process can also be examined numerically. For example, we can compute the displacement and average velocity for various time intervals beginning at $t = 3$ s, using the expressions for Δx and $\bar{v}$. The results of such calculations are given in Table 3.1. Notice that as the time intervals get smaller and smaller, the average velocity more nearly approaches the value of the instantaneous velocity at $t = 3$ s, namely, 18 m/s.

TABLE 3.1 Displacement and Average Velocity for Various Time Intervals for the Function $x = 3t^2$ (the intervals begin at $t = 3$ s)

Δt (s)	Δx (m)	$\Delta x/\Delta t$ (m/s)
1.00	21	21
0.50	9.75	19.5
0.25	4.69	18.8
0.10	1.83	18.3
0.05	0.9075	18.15
0.01	0.1803	18.03
0.001	0.018003	18.003

[2] We could also use the rules of differential calculus to find the velocity from the displacement. That is, $v = \frac{dx}{dt} = \frac{d}{dt}(-4t + 2t^2) = 4(-1 + t)$ m/s. Therefore, at $t = 2.5$ s, $v = 4(-1 + 2.5) = 6$ m/s. A review of basic operations in the calculus is provided in Appendix B.6.

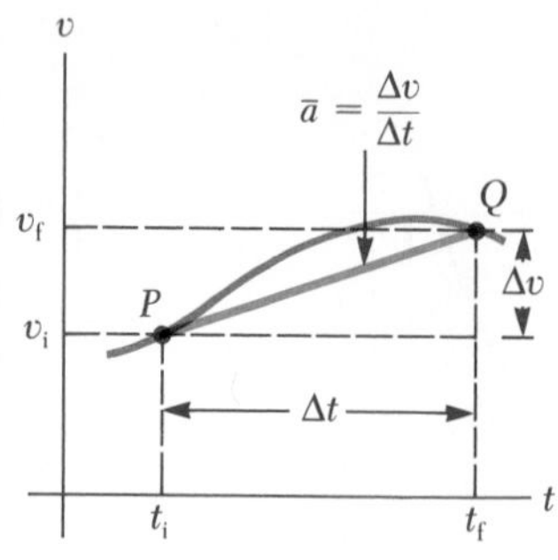

Figure 3.6 Velocity-time graph for a particle moving in a straight line. The slope of the line connecting the points P and Q is defined as the average acceleration in the time interval $\Delta t = t_f - t_i$.

3.3 ACCELERATION

When the velocity of a particle changes with time, the particle is said to be *accelerating*. For example, the velocity of a car will increase when you "step on the gas." The car will slow down when you apply the brakes. However, we need a more precise definition of acceleration than this.

Suppose a particle moving along the x axis has a velocity v_i at time t_i and a velocity v_f at time t_f, as in Figure 3.6.

The **average acceleration** of the particle in the time interval $\Delta t = t_f - t_i$ is defined as the ratio $\Delta v/\Delta t$, where $\Delta v = v_f - v_i$ is the *change* in velocity in this time interval:

$$\bar{a} \equiv \frac{v_f - v_i}{t_f - t_i} = \frac{\Delta v}{\Delta t} \tag{3.4}$$

Acceleration is a vector quantity having dimensions of length divided by $(\text{time})^2$, or L/T^2. Some of the common units of acceleration are meters per second per second (m/s^2) and feet per second per second (ft/s^2).

In some situations, the value of the average acceleration may be different over different time intervals. It is therefore useful to define the **instantaneous acceleration** as the limit of the average acceleration as Δt approaches zero. This concept is analogous to the definition of instantaneous velocity discussed in the previous section. If we imagine that the point Q is brought closer and closer to the point P in Figure 3.6 and take the limit of the ratio $\Delta v/\Delta t$ as Δt approaches zero, we get the *instantaneous acceleration:*

$$a \equiv \lim_{\Delta t \to 0} \frac{\Delta v}{\Delta t} = \frac{dv}{dt} \tag{3.5}$$

That is, the instantaneous acceleration equals the derivative of the velocity with respect to time, which by definition is the slope of the velocity-time graph. One can interpret the derivative of the velocity with respect to time as the *time rate of change of velocity*. Again you should note that if a is positive, the acceleration is in the positive x direction, whereas negative a implies acceleration in the negative x direction. *From now on we shall use the term* acceleration *to mean instantaneous acceleration.* Average acceleration is seldom used in physics.

Since $v = dx/dt$, the acceleration can also be written

$$a = \frac{dv}{dt} = \frac{d}{dt}\left(\frac{dx}{dt}\right) = \frac{d^2x}{dt^2} \tag{3.6}$$

That is, the acceleration equals the *second derivative* of the coordinate with respect to time.

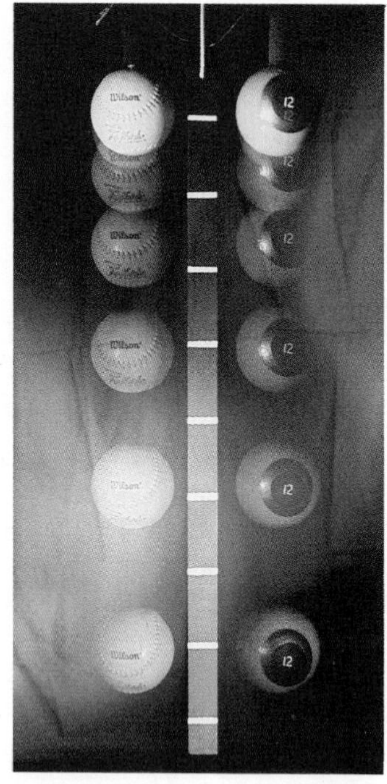

A multiflash photograph of a freely falling baseball (mass 0.23 kg) and shotput (mass 5.4 kg) taken at a flash rate of 1/15 s. The spacing between markers is 10 cm. Note that the two objects fall at the same rate. Why is this so, in view of the fact that they have different masses? (Courtesy of Henry Leap)

Figure 3.7 shows how the acceleration-time curve can be derived from the velocity-time curve. In these sketches, the acceleration at any time is simply the slope of the velocity-time graph at that time. Positive values of the acceleration correspond to those points where the velocity is increasing in the positive x direction. The acceleration reaches a maximum at time t_1, when the slope of the velocity-time graph is a maximum. The acceleration then goes to zero at time t_2, when the velocity is a maximum (that is, when the velocity is

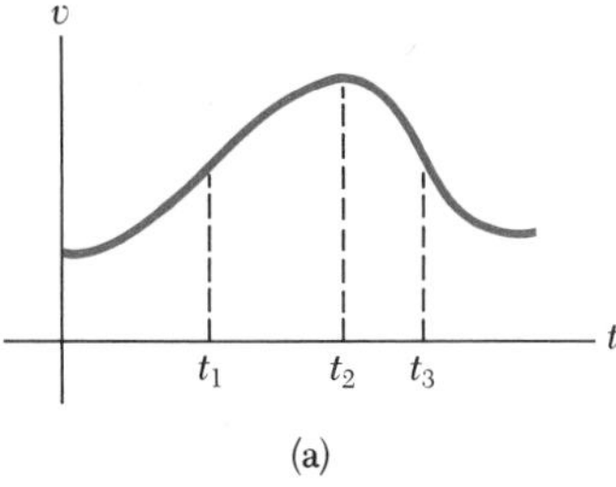

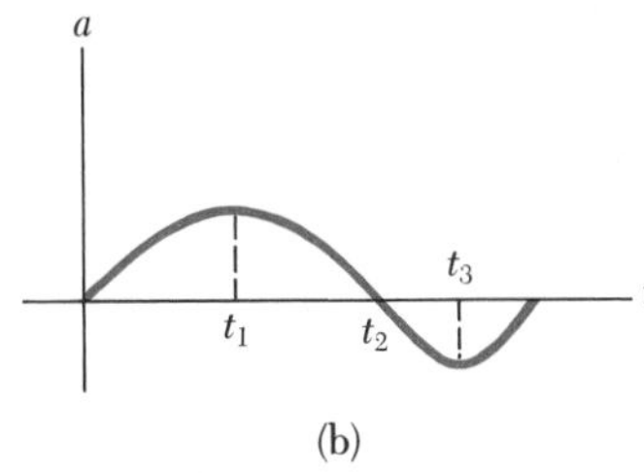

Figure 3.7 The instantaneous acceleration can be obtained from the velocity-time graph (a). At each instant, the acceleration in the a versus t graph (b) equals the slope of the line tangent to the v versus t curve.

momentarily not changing and the slope of the v versus t graph is zero). Finally, the acceleration is negative when the velocity in the positive x direction is decreasing in time.

EXAMPLE 3.4 Average and Instantaneous Acceleration

The velocity of a particle moving along the x axis varies in time according to the expression $v = (40 - 5t^2)$ m/s, where t is in s. (a) Find the average acceleration in the time interval $t = 0$ to $t = 2$ s.

The velocity-time graph for this function is given in Figure 3.8. The velocities at $t_i = 0$ and $t_f = 2$ s are found by substituting these values of t into the expression given for the velocity:

$$v_i = (40 - 5t_i^2)\ \text{m/s} = [40 - 5(0)^2]\ \text{m/s} = 40\ \text{m/s}$$

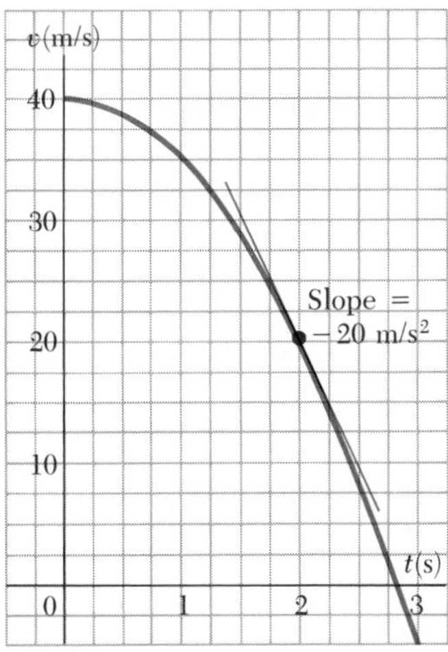

Figure 3.8 (Example 3.4) The velocity-time graph for a particle moving along the x axis according to the relation $v = (40 - 5t^2)$ m/s. Note that the acceleration at $t = 2$ s is equal to the slope of the blue tangent line at that time.

$$v_f = (40 - 5t_f^2)\ \text{m/s} = [40 - 5(2)^2]\ \text{m/s} = 20\ \text{m/s}$$

Therefore, the average acceleration in the specified time interval $\Delta t = t_f - t_i = 2$ s is given by

$$\bar{a} = \frac{v_f - v_i}{t_f - t_i} = \frac{(20 - 40)\ \text{m/s}}{(2 - 0)\ \text{s}} = \boxed{-10\ \text{m/s}^2}$$

The negative sign is consistent with the fact that the slope of the line joining the initial and final points on the velocity-time graph is negative.

(b) Determine the acceleration at $t = 2$ s.

The velocity at time t is given by $v_i = (40 - 5t^2)$ m/s, and the velocity at time $t + \Delta t$ is given by

$$v_f = 40 - 5(t + \Delta t)^2 = 40 - 5t^2 - 10t\,\Delta t - 5(\Delta t)^2$$

Therefore, the change in velocity over the time interval Δt is

$$\Delta v = v_f - v_i = [-10t\,\Delta t - 5(\Delta t)^2]\ \text{m/s}$$

Dividing this expression by Δt and taking the limit of the result as Δt approaches zero, we get the acceleration at *any* time t:

$$a = \lim_{\Delta t \to 0} \frac{\Delta v}{\Delta t} = \lim_{\Delta t \to 0} (-10t - 5\,\Delta t) = -10t\ \text{m/s}^2$$

Therefore, at $t = 2$ s, we find that

$$a = (-10)(2)\ \text{m/s}^2 = \boxed{-20\ \text{m/s}^2}$$

This result can also be obtained by measuring the slope of the velocity-time graph at $t = 2$ s. Note that the acceleration is not constant in this example. Situations involving constant acceleration will be treated in the next section.

So far we have evaluated the derivatives of a function by starting with the definition of the function and then taking the limit of a specific ratio. Those of you familiar with the calculus should recognize that there are specific rules for taking the derivatives of various functions. These rules, which are listed in Appendix B.6, enable us to evaluate derivatives quickly.

Suppose x is proportional to some power of t, such as

$$x = At^n$$

where A and n are constants. (This is a very common functional form.) The derivative of x with respect to t is given by

$$\frac{dx}{dt} = nAt^{n-1}$$

Applying this rule to Example 3.3, where $x = 3t^2$, we see that $v = dx/dt = 6t$, in agreement with our result of taking the limit explicitly. Likewise, in Example 3.4, where $v = 40 - 5t^2$, we find that $a = dv/dt = -10t$. (Note that the rate of change of any constant quantity is zero.)

3.4 ONE-DIMENSIONAL MOTION WITH CONSTANT ACCELERATION

If the acceleration of a particle varies in time, the motion can be complex and difficult to analyze. A very common and simple type of one-dimensional motion occurs when the acceleration is constant, or uniform. Because the acceleration is constant, the average acceleration equals the instantaneous acceleration. Consequently, the velocity increases or decreases at the same rate throughout the motion.

If we replace $\bar{a}$ by a in Equation 3.4, we find that

$$a = \frac{v_f - v_i}{t_f - t_i}$$

For convenience, let $t_i = 0$ and t_f be any arbitrary time t. Also, let $v_i = v_0$ (the initial velocity at $t = 0$) and $v_f = v$ (the velocity at any arbitrary time t). With this notation, we can express the acceleration as

$$a = \frac{v - v_0}{t}$$

or

Velocity as a function of time

$$v = v_0 + at \qquad \text{(for constant } a) \qquad (3.7)$$

This expression enables us to predict the velocity at *any* time t if the initial velocity, acceleration, and elapsed time are known. A graph of velocity versus time for this motion is shown in Figure 3.9a. The graph is a straight line the slope of which is the acceleration, a, consistent with the fact that $a = dv/dt$ is a constant. From this graph and from Equation 3.7, we see that the velocity at any time t is the sum of the initial velocity, v_0, and the change in velocity, at. The graph of acceleration versus time (Fig. 3.9b) is a straight line with a slope of zero, since the acceleration is constant. Note that if the acceleration were negative (the particle is slowing down), the slope of Figure 3.9a would be negative.

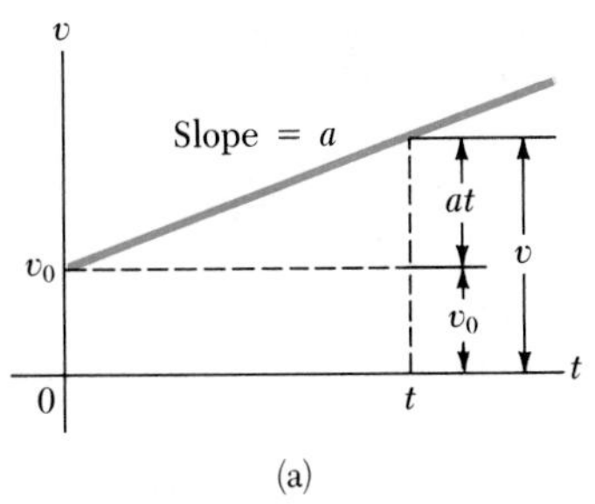

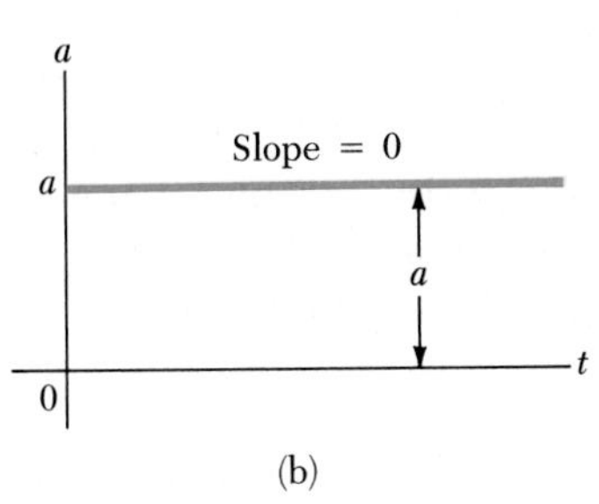

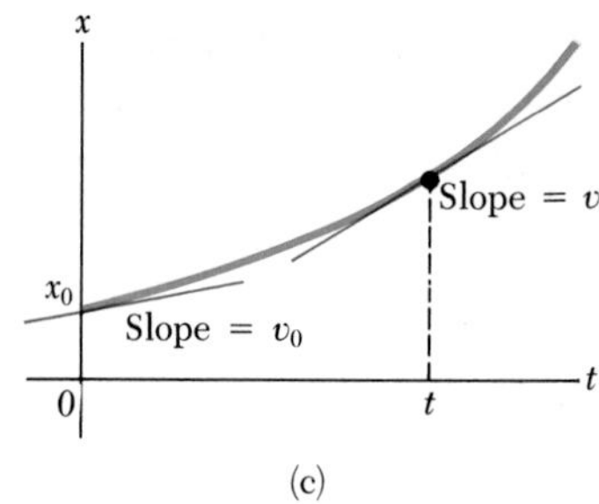

Figure 3.9 A particle moving along the x axis with constant acceleration a; (a) the velocity-time graph, (b) the acceleration-time graph, and (c) the space-time graph.

Because the velocity varies linearly in time according to Equation 3.7, we can express the average velocity in any time interval as the arithmetic mean of the initial velocity, v_0, and the final velocity, v:

$$\bar{v} = \frac{v_0 + v}{2} \qquad \text{(for constant } a\text{)} \qquad (3.8)$$

Note that this expression is only useful when the acceleration is constant, that is, when the velocity varies linearly with time.

We can now use Equations 3.1 and 3.8 to obtain the displacement as a function of time. Again, we choose $t_i = 0$, at which time the initial position is $x_i = x_0$. This gives

$$\Delta x = \bar{v}\,\Delta t = \left(\frac{v_0 + v}{2}\right) t$$

or

$$x - x_0 = \tfrac{1}{2}(v + v_0)t \qquad \text{(for constant } a\text{)} \qquad (3.9)$$

We can obtain another useful expression for the displacement by substituting Equation 3.7 into Equation 3.9:

$$x - x_0 = \tfrac{1}{2}(v_0 + v_0 + at)t$$

$$x - x_0 = v_0 t + \tfrac{1}{2}at^2 \qquad \text{(for constant } a\text{)} \qquad (3.10)$$

The validity of this expression can be checked by differentiating it with respect to time, to give

$$v = \frac{dx}{dt} = \frac{d}{dt}\left(x_0 + v_0 t + \frac{1}{2}at^2\right) = v_0 + at$$

Finally, we can obtain an expression that does not contain the time by substituting the value of t from Equation 3.7 into Equation 3.9. This gives

$$x - x_0 = \tfrac{1}{2}(v_0 + v)\left(\frac{v - v_0}{a}\right) = \frac{v^2 - {v_0}^2}{2a}$$

Velocity as a function of displacement

$$v^2 = v_0{}^2 + 2a(x - x_0) \qquad \text{(for constant } a\text{)} \qquad (3.11)$$

A position-time graph for motion under constant acceleration assuming positive a is shown in Figure 3.9c. Note that the curve representing Equation 3.10 is a parabola. The slope of the tangent to this curve at $t = 0$ equals the initial velocity, v_0, and the slope of the tangent line at any time t equals the velocity at that time.

If motion occurs in which the acceleration is *zero*, then we see that

$$\left.\begin{aligned} v &= v_0 \\ x - x_0 &= vt \end{aligned}\right\} \text{ when } a = 0$$

That is, when the acceleration is zero, the velocity is a constant and the displacement changes linearly with time.

Equations 3.7 through 3.11 are five *kinematic expressions that may be used to solve any problem in one-dimensional motion with constant acceleration.* Keep in mind that these relationships were derived from the definition of velocity and acceleration, together with some simple algebraic manipulations and the requirement that the acceleration be constant. It is often convenient to choose the initial position of the particle as the origin of the motion, so that $x_0 = 0$ at $t = 0$. In such a case, the displacement is simply x.

The four kinematic equations that are used most often are listed in Table 3.2 for convenience.

The choice of which kinematic equation or equations you should use in a given situation depends on what is known beforehand. Sometimes it is necessary to use two of these equations to solve for two unknowns, such as the displacement and velocity at some instant. For example, suppose the initial velocity, v_0, and acceleration, a, are given. You can then find (1) the velocity after a time t has elapsed, using $v = v_0 + at$, and (2) the displacement after a time t has elapsed, using $x - x_0 = v_0t + \frac{1}{2}at^2$. You should recognize that the quantities that vary during the motion are velocity, displacement, and time.

You will get a great deal of practice in the use of these equations by solving a number of exercises and problems. Many times you will discover that there is more than one method for obtaining a solution.

TABLE 3.2 Kinematic Equations for Motion in a Straight Line Under Constant Acceleration

Equation	Information Given by Equation
$v = v_0 + at$	Velocity as a function of time
$x - x_0 = \frac{1}{2}(v + v_0)t$	Displacement as a function of velocity and time
$x - x_0 = v_0t + \frac{1}{2}at^2$	Displacement as a function of time
$v^2 = v_0{}^2 + 2a(x - x_0)$	Velocity as a function of displacement

Note: Motion is along the x axis. At $t = 0$, the position of the particle is x_0 and its velocity is v_0.

EXAMPLE 3.5 The Supercharged Sportscar

A certain automobile manufacturer claims that its super-deluxe sportscar will accelerate uniformly from rest to a speed of 87 mi/h in 8 s. (a) Determine the acceleration of the car.

First note that $v_0 = 0$ and the velocity after 8 s is 87 mi/h = 128 ft/s. (It is useful to note that 60 mi/h = 88 ft/s exactly.) Because we are given v_0, v, and t, we can use $v = v_0 + at$ to find the acceleration:

$$a = \frac{v - v_0}{t} = \frac{128 \text{ ft/s}}{8 \text{ s}} = 16 \text{ ft/s}^2$$

(b) Find the distance the car travels in the first 8 s.

Let the origin be at the original position of the car, so that $x_0 = 0$. Using Equation 3.9 we find that

$$x = \tfrac{1}{2}(v_0 + v)t = \tfrac{1}{2}(128 \text{ ft/s})(8 \text{ s}) = 512 \text{ ft}$$

(c) What is the velocity of the car 10 s after it begins its motion, assuming it continues to accelerate at the rate of 16 ft/s²?

Again, we can use $v = v_0 + at$, with $v_0 = 0$, $t = 10$ s, and $a = 16$ ft/s². This gives

$$v = v_0 + at = 0 + (16 \text{ ft/s}^2)(10 \text{ s}) = 160 \text{ ft/s}$$

which corresponds to 109 mi/h.

EXAMPLE 3.6 Accelerating an Electron

An electron in a cathode ray tube of a TV set enters a region where it accelerates uniformly from a speed of 3×10^4 m/s to a speed of 5×10^6 m/s in a distance of 2 cm. (a) How long is the electron in this region where it accelerates?

Taking the direction of motion to be along the x axis, we can use Equation 3.9 to find t, since the displacement and velocities are known:

$$x - x_0 = \tfrac{1}{2}(v_0 + v)t$$

$$t = \frac{2(x - x_0)}{v_0 + v} = \frac{2(2 \times 10^{-2} \text{ m})}{(3 \times 10^4 + 5 \times 10^6) \text{ m/s}}$$

$$= 7.95 \times 10^{-9} \text{ s}$$

(b) What is the acceleration of the electron in this region?

To find the acceleration, we can use $v = v_0 + at$ and the results from (a):

$$a = \frac{v - v_0}{t} = \frac{(5 \times 10^6 - 3 \times 10^4) \text{ m/s}}{7.95 \times 10^{-9} \text{ s}}$$

$$= 6.25 \times 10^{14} \text{ m/s}^2$$

We also could have used Equation 3.11 to obtain the acceleration, since the velocities and displacement are known. Try it! Although a is very large in this example, the acceleration occurs over a very short time interval and is a typical value for such charged particles in acceleration.

EXAMPLE 3.7 Watch out for the Speed Limit

A car traveling at a constant speed of 30 m/s (≈67 mi/h) passes a trooper hidden behind a billboard. One second after the speeding car passes the billboard, the trooper sets in chase after the car with a constant acceleration of 3.0 m/s². How long does it take the trooper to overtake the speeding car?

Solution To solve this problem algebraically, let us write expressions for the position of each vehicle as functions of time. It is convenient to choose the origin at the position of the billboard and take $t = 0$ as the time the trooper begins his motion. At that instant, the speeding car has already traveled a distance of 30 m since it travels at a constant speed of 30 m/s. Thus, the initial position of the speeding car is given by $x_0 = 30$ m. Since the car moves with constant speed, its acceleration is zero, and applying Equation 3.10 gives

$$x_C = 30 \text{ m} + (30 \text{ m/s})t$$

Note that at $t = 0$, this expression does give the car's correct initial position $x_C = x_0 = 30$ m. Likewise, for the trooper who starts from the origin at $t = 0$, we have $x_0 = 0$, $v_0 = 0$, and $a = 3.0$ m/s². Hence, the position of the trooper versus time is given by

$$x_T = \tfrac{1}{2}at^2 = \tfrac{1}{2}(3.0 \text{ m/s}^2)t^2$$

The trooper overtakes the car at the instant that $x_T = x_C$, or

$$\tfrac{1}{2}(3.0 \text{ m/s}^2)t^2 = 30 \text{ m} + (30 \text{ m/s})t$$

This gives the quadratic equation

$$1.5t^2 - 30t - 30 = 0$$

whose positive solution is $t = 21$ s. Note that in this time interval, the trooper travels a distance of about 660 m.

Exercise 1 This problem can also be easily solved graphically. On the *same* graph, plot the position versus time for *each vehicle*, and from the intersection of the two curves determine the time at which the trooper overtakes the speeding car.

3.5 FREELY FALLING BODIES

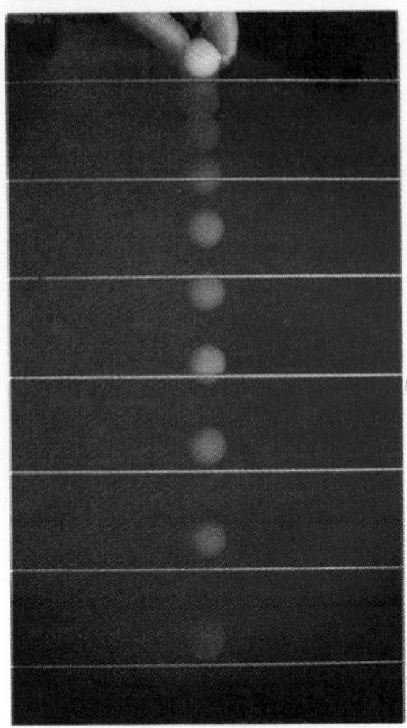

A multiflash photograph of a freely falling ball. The time interval between flashes is 1/15 s, and the separation between the horizontal markers is 10 cm. Can you estimate g from these data? (Courtesy of Henry Leap)

It is well known that all objects, when dropped, will fall toward the earth with nearly constant acceleration. There is a legendary story that Galileo Galilei first discovered this fact by observing that two different weights dropped simultaneously from the Leaning Tower of Pisa hit the ground at approximately the same time. Although there is some doubt that this particular experiment was carried out, it is well established that Galileo did perform many systematic experiments on objects moving on inclined planes. Through careful measurements of distances and time intervals, he was able to show that the displacement of an object starting from rest is proportional to the square of the time the object is in motion. This observation is consistent with one of the kinematic equations we derived for motion under constant acceleration (Eq. 3.10). Galileo's achievements in the science of mechanics paved the way for Newton in his development of the laws of motion.

You might want to try the following experiment. Drop a coin and a crumpled-up piece of paper simultaneously from the same height. In the absence of air resistance, both will experience the same motion and hit the floor at the same time. (In a real experiment, air resistance cannot be neglected.) In the idealized case, where air resistance *is* neglected, such motion is referred to as *free fall.* If this same experiment could be conducted in a good vacuum, where air friction is truly negligible, the paper and coin would fall with the same acceleration, regardless of the shape of the paper. On August 2, 1971, such an experiment was conducted on the moon by astronaut David Scott. He simultaneously released a geologist's hammer and a falcon's feather, and in unison they fell to the lunar surface. This demonstration would have surely pleased Galileo!

Acceleration due to gravity $g = 9.80\ \text{m/s}^2$

We shall denote the *acceleration due to gravity* by the symbol **g**. The magnitude of g decreases with increasing altitude. Furthermore, there are slight variations in g with latitude. The vector **g** is directed downward toward the center of the earth. At the earth's surface, the magnitude of g is approxi-

Galileo Galilei (1564–1642), an Italian physicist and astronomer, investigated the motion of objects in free fall (including projectiles) and the motion of an object on an inclined plane, established the concept of relative motion, and noted that a swinging pendulum could be used to measure time intervals. Following the invention of the telescope, he said, "I now have visual proof of what I already knew through my intellect." Galileo made several major discoveries in astronomy; he discovered four moons of Jupiter and many new stars, investigated the nature of the moon's surface, discovered sun spots and the phases of Venus, and proved that the Milky Way consists of an enormous number of stars. (Courtesy of AIP Niels Bohr Library)

mately 9.80 m/s², or 980 cm/s², or 32 ft/s². Unless stated otherwise, we shall use this value for g when doing calculations.

Definition of free fall

When we use the expression *freely falling object,* we do not necessarily refer to an object dropped from rest. A freely falling object is any object moving freely under the influence of gravity, *regardless* of its initial motion. Objects thrown upward or downward and those released from rest are all falling freely once they are released. Furthermore, it is important to recognize that any freely falling object experiences an acceleration directed *downward.* This is true regardless of the initial motion of the object.

An object thrown upward (or downward) will experience the same acceleration as an object released from rest. Once they are in free fall, all objects will have an acceleration downward, equal to the acceleration due to gravity.

If we neglect air resistance and assume that the gravitational acceleration does not vary with altitude, then the motion of a freely falling body is equivalent to motion in one dimension under constant acceleration. Therefore our kinematic equations for constant acceleration can be applied. We shall take the vertical direction to be the y axis and call y positive upward. With this choice of coordinates, we can replace x by y in Equations 3.7, 3.9, 3.10, and 3.11. Furthermore, since positive y is upward, the acceleration is negative (downward) and given by $a = -g$. The negative sign simply indicates that the acceleration is downward. With these substitutions, we get the following expressions:[3]

Kinematic equations for a freely falling body

$$v = v_0 - gt \tag{3.12}$$

$$y - y_0 = \tfrac{1}{2}(v + v_0)t \tag{3.13}$$

$$y - y_0 = v_0 t - \tfrac{1}{2}gt^2 \tag{3.14}$$

(for constant a; $a = -g$)

$$v^2 = v_0^2 - 2g(y - y_0) \tag{3.15}$$

You should note that *the negative sign for the acceleration is already included in these expressions.* Therefore, when using these equations in any free-fall problem, you should simply substitute $g = 9.80\ \text{m/s}^2$.

Consider the case of a particle thrown vertically upward from the origin with a velocity v_0. In this case, v_0 is *positive* and $y_0 = 0$. Graphs of the displacement and velocity as functions of time are shown in Figure 3.10. Note that the

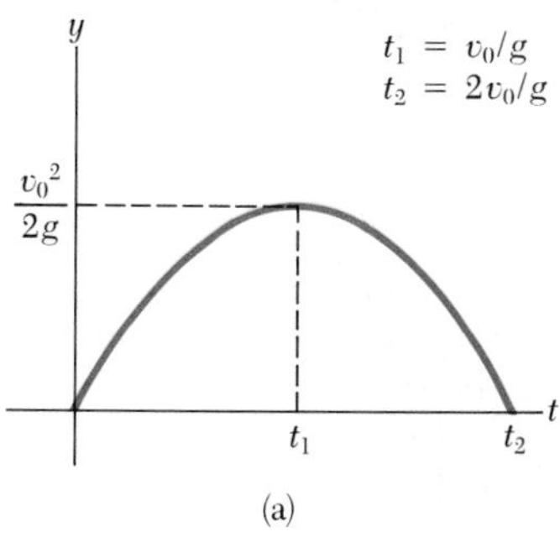

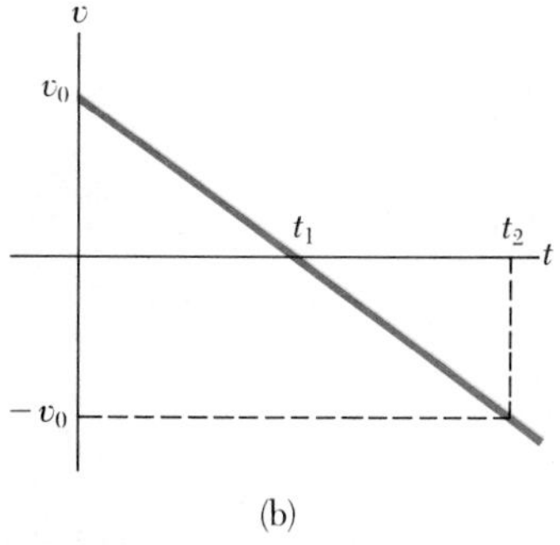

Figure 3.10 Graphs of (a) the displacement versus time and (b) the velocity versus time for a freely falling particle, where y and v are taken to be positive upward. Note the symmetry in (a) about $t = t_1$.

[3] One can also take y positive downward, in which case $a = +g$. The results will be the same, regardless of the convention chosen.

velocity is initially positive, but decreases in time and goes to zero at the peak of the path. From Equation 3.12, we see that this occurs at the time $t_1 = v_0/g$. At this time, the displacement has its largest positive value, which can be calculated from Equation 3.14 with $t = t_1 = v_0/g$. This gives $y_{max} = v_0^2/2g$.

At the time $t_2 = 2t_1 = 2v_0/g$, we see from Equation 3.14 that the displacement is again zero, that is, the particle has returned to its starting point. Furthermore, at time t_2 the velocity is given by $v = -v_0$. (This follows directly from Equation 3.12.) Hence, there is symmetry in the motion. In other words, both the displacement and the magnitude of the velocity repeat themselves in the time interval $t = 0$ to $t = 2v_0/g$.

In the examples that follow, we shall, for convenience, assume that $y_0 = 0$ at $t = 0$. Notice that this does not affect the solution to the problem. If y_0 is nonzero, then the graph of y versus t (Fig. 3.10a) is simply shifted upward or downward by an amount y_0, while the graph of v versus t (Fig. 3.10b) remains unchanged.

EXAMPLE 3.8 Look Out Below!
A golf ball is released from rest from the top of a very tall building. Neglecting air resistance, calculate the position and velocity of the ball after 1, 2, and 3 s.

Solution We choose our coordinates such that the starting point of the ball is at the origin ($y_0 = 0$ at $t = 0$) and remember that we have defined y to be positive upward. Since $v_0 = 0$, Equations 3.12 and 3.14 become

$$v = -gt = -(9.80 \text{ m/s}^2)t$$

$$y = -\tfrac{1}{2}gt^2 = -\tfrac{1}{2}(9.80 \text{ m/s}^2)t^2$$

where t is in s, v is in m/s, and y is in m. These expressions give the velocity and displacement at any time t after the ball is released. Therefore, at $t = 1$ s,

$$v = -(9.80 \text{ m/s}^2)(1 \text{ s}) = \boxed{-9.80 \text{ m/s}}$$

$$y = -\tfrac{1}{2}(9.80 \text{ m/s}^2)(1 \text{ s})^2 = \boxed{-4.90 \text{ m}}$$

Likewise, at $t = 2$ s, we find that $v = -19.6$ m/s and $y = -19.6$ m. Finally, at $t = 3$ s, $v = -29.4$ m/s and $y = -44.1$ m. The minus signs for v indicate that the velocity vector is directed downward, and the minus signs for y indicate displacement in the negative y direction.

Exercise 2 Calculate the position and velocity of the ball after 4 s.
Answer -78.4 m, -39.2 m/s.

EXAMPLE 3.9 Not a Bad Throw for a Rookie
A stone is thrown from the top of a building with an initial velocity of 20 m/s straight upward. The building is 50 m high, and the stone just misses the edge of the roof on its way down, as in Figure 3.11. Determine (a) the time needed for the stone to reach its maximum height, (b) the maximum height, (c) the time needed for the stone to return to the level of the thrower, (d) the velocity of the stone at this instant, and (e) the velocity and position of the stone at $t = 5$ s.

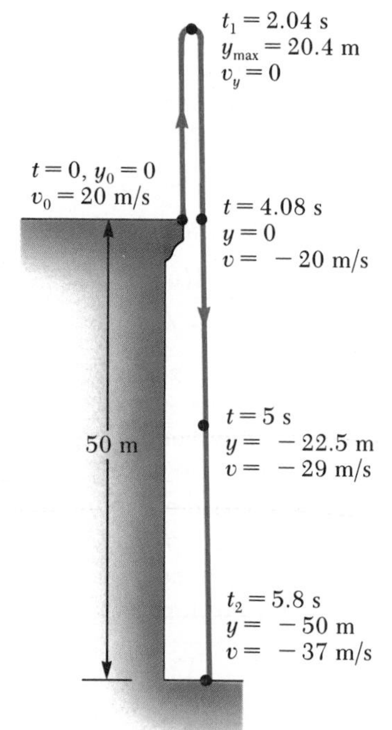

Figure 3.11 (Example 3.9) Position and velocity versus time for a freely falling particle thrown initially upward with a velocity $v_0 = 20$ m/s.

Solution (a) To find the time necessary to reach the maximum height, use Equation 3.12, $v = v_0 - gt$, noting that $v = 0$ at maximum height:

$$20\text{ m/s} - (9.80\text{ m/s}^2)t_1 = 0$$

$$t_1 = \frac{20\text{ m/s}}{9.80\text{ m/s}^2} = 2.04\text{ s}$$

(b) This value of time can be substituted into Equation 3.14, $y = v_0t - \frac{1}{2}gt^2$, to give the maximum height as measured from the position of the thrower:

$$y_{max} = (20\text{ m/s})(2.04\text{ s}) - \tfrac{1}{2}(9.80\text{ m/s}^2)(2.04\text{ s})^2$$

$$= 20.4\text{ m}$$

(c) When the stone is back at the height of the thrower, the y coordinate is zero. From the expression $y = v_0t - \frac{1}{2}gt^2$ (Eq. 3.14), with $y = 0$, we obtain the expression

$$20t - 4.9t^2 = 0$$

This is a quadratic equation and has two solutions for t. (For some assistance in solving quadratic equations, see Appendix B.2.) The equation can be factored to give

$$t(20 - 4.9t) = 0$$

One solution is $t = 0$, corresponding to the time the stone starts its motion. The other solution is $t = 4.08$ s, which is the solution we are after.

(d) The value for t found in (c) can be inserted into $v = v_0 - gt$ (Eq. 3.12) to give

$$v = 20\text{ m/s} - (9.80\text{ m/s}^2)(4.08\text{ s}) = -20.0\text{ m/s}$$

Note that the velocity of the stone when it arrives back at its original height is equal in magnitude to its initial velocity but opposite in direction. This indicates that the motion is symmetric.

(e) From $v = v_0 - gt$ (Eq. 3.12), the velocity after 5 s is

$$v = 20\text{ m/s} - (9.80\text{ m/s}^2)(5\text{ s}) = -29.0\text{ m/s}$$

We can use $y = v_0t - \frac{1}{2}gt^2$ (Eq. 3.14) to find the position of the particle at $t = 5$ s:

$$y = (20\text{ m/s})(5\text{ s}) - \tfrac{1}{2}(9.80\text{ m/s}^2)(5\text{ s})^2 = -22.5\text{ m}$$

Exercise 3 Find (a) the velocity of the stone just before it hits the ground and (b) the total time the stone is in the air.
Answer (a) −37.1 m/s (b) 5.83 s.

*3.6 KINEMATIC EQUATIONS DERIVED FROM CALCULUS

This is an optional section that assumes that the reader is familiar with the techniques of integral calculus. If you have not studied integration in your calculus course as yet, this section should be skipped or covered at some later time after you become familiar with integration.

The velocity of a particle moving in a straight line can be obtained from a knowledge of its position as a function of time. Mathematically, the velocity equals the derivative of the coordinate with respect to time. It is also possible to find the displacement of a particle if its velocity is known as a function of time. In the calculus, this procedure is referred to as integration, or the antiderivative. Graphically, it is equivalent to finding the area under a curve.

Suppose the velocity versus time plot for a particle moving along the x axis is as shown in Figure 3.12. Let us divide the time interval $t_f - t_i$ into many small intervals of duration Δt_n. From the definition of average velocity, we see that the displacement during any small interval such as the shaded one in Figure 3.12 is given by $\Delta x_n = \bar{v}_n \Delta t_n$, where $\bar{v}_n$ is the average velocity in that interval. Therefore, the displacement during this small interval is simply the area of the shaded rectangle. The total displacement for the interval $t_f - t_i$ is the sum of the areas of all the rectangles:

$$\Delta x = \sum_n \bar{v}_n \Delta t_n$$

where the sum is taken over all the rectangles from t_i to t_f. Now, as each interval is made smaller and smaller, the number of terms in the sum increases and the sum approaches a value equal to the area under the velocity-time

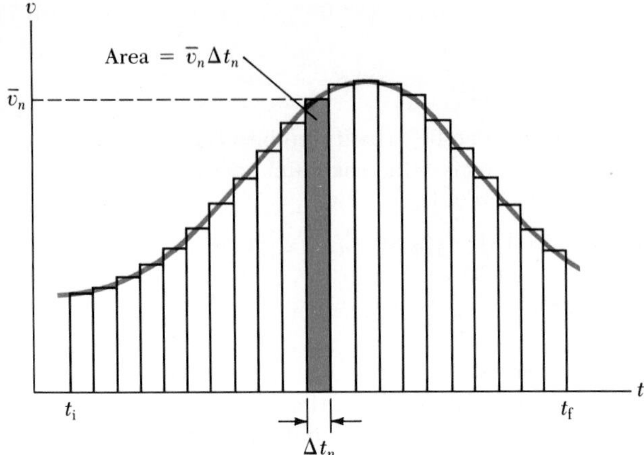

Figure 3.12 Velocity versus time curve for a particle moving along the x axis. The area of the shaded rectangle is equal to the displacement Δx in the time interval Δt_n, while the total area under the curve is the total displacement of the particle.

graph. Therefore, in the limit $n \rightarrow \infty$, or $\Delta t_n \rightarrow 0$, we see that the displacement is given by

$$\Delta x = \lim_{\Delta t_n \to 0} \sum_n v_n \, \Delta t_n \tag{3.16}$$

or

Displacement = area under the velocity-time graph

Note that we have replaced the average velocity $\bar{v}_n$ by the instantaneous velocity v_n in the sum. As you can see from Figure 3.12, this approximation is clearly valid in the limit of very small intervals. We conclude that if the velocity-time graph for motion along a straight line is known, the displacement during any time interval can be obtained by measuring the area under the curve.

The limit of the sum in Equation 3.16 is called a **definite integral** and is written

Definite integral

$$\lim_{\Delta t_n \to 0} \sum_n v_n \, \Delta t_n = \int_{t_i}^{t_f} v(t)\, dt \tag{3.17}$$

where $v(t)$ denotes the velocity at any time t. If the explicit functional form of $v(t)$ is known, the specific integral can be evaluated.

If a particle moves with a constant velocity v_0 as in Figure 3.13, its displacement during the time interval Δt is simply the area of the shaded rectangle, that is,

$$\Delta x = v_0 \, \Delta t \qquad \text{(when } v = v_0 = \text{constant)}$$

As another example, consider a particle moving with a velocity that is proportional to t, as in Figure 3.14. Taking $v = at$, where a is the constant of proportionality (the acceleration), we find that the displacement of the particle during the time interval $t = 0$ to $t = t_1$ is the area of the shaded triangle in Figure 3.14:

$$\Delta x = \tfrac{1}{2}(t_1)(at_1) = \tfrac{1}{2}at_1^{\,2}$$

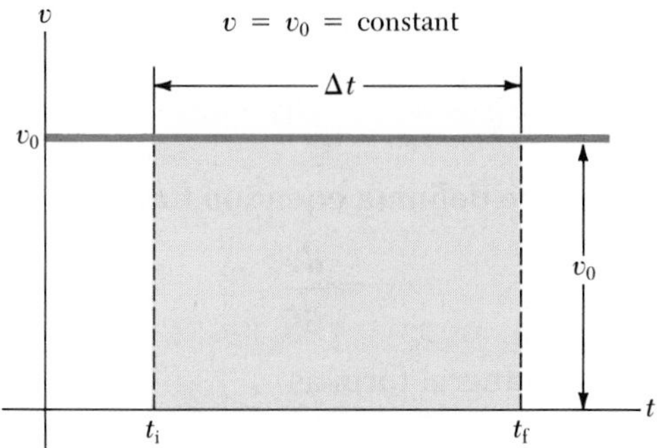

Figure 3.13 The velocity versus time curve for a particle moving with constant velocity v_0.

Kinematic Equations

We will now make use of the defining equations for acceleration and velocity to derive the kinematic equations.

The defining equation for acceleration,

$$a = \frac{dv}{dt}$$

may also be written in terms of an integral (or antiderivative) as

$$v = \int a\, dt + C_1$$

where C_1 is a constant of integration. For the special case where the acceleration a is a constant, this reduces to

$$v = at + C_1$$

The value of C_1 depends on the initial conditions of the motion. If we take $v = v_0$ when $t = 0$, and substitute these into the last equation, we have

$$v_0 = a(0) + C_1$$

or

$$C_1 = v_0$$

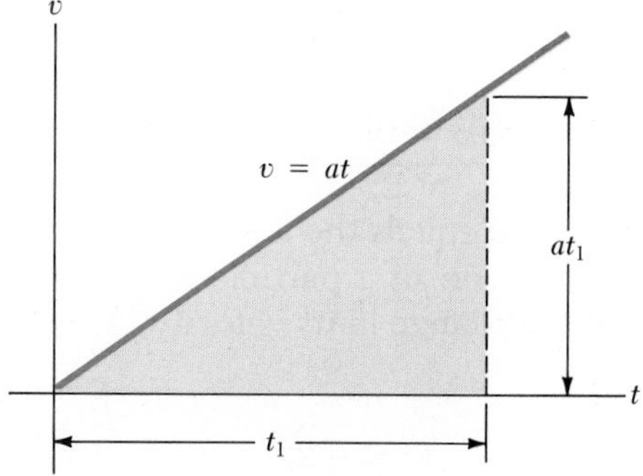

Figure 3.14 The velocity versus time curve for a particle moving with a velocity that is proportional to the time.

Hence, we obtain the first kinematic equation:

$$v = v_0 + at \qquad \text{(for constant } a\text{)}$$

Now let us consider the defining equation for velocity:

$$v = \frac{dx}{dt}$$

We can also write this in integral form as

$$x = \int v \, dt + C_2$$

where C_2 is another constant of integration. Since $v = v_0 + at$, this becomes

$$x = \int (v_0 + at) \, dt + C_2$$

$$x = \int v_0 \, dt + \int at \, dt + C_2$$

$$x = v_0 t + \tfrac{1}{2}at^2 + C_2$$

To find C_2, we make use of the initial condition that $x = x_0$ when $t = 0$. This gives $C_2 = x_0$. Therefore, we have

$$x = x_0 + v_0 t + \tfrac{1}{2}at^2 \qquad \text{(for constant } a\text{)}$$

This is the second equation of kinematics. Recall that $x - x_0$ is equal to the displacement of the object, where x_0 is its initial position.

SUMMARY

The **average velocity** of a particle during some time interval is equal to the ratio of the displacement, Δx, and the time interval, Δt:

Average velocity

$$\bar{v} \equiv \frac{\Delta x}{\Delta t} \tag{3.1}$$

The **instantaneous velocity** of a particle is defined as the limit of the ratio $\Delta x/\Delta t$ as Δt approaches zero. By definition, this equals the derivative of x with respect to t, or the time rate of change of the position:

Instantaneous velocity

$$v \equiv \lim_{\Delta t \to 0} \frac{\Delta x}{\Delta t} = \frac{dx}{dt} \tag{3.3}$$

The **speed** of a particle equals the absolute value of the velocity.

The **average acceleration** of a particle during some time interval is defined as the ratio of the change in its velocity, Δv, and the time interval, Δt:

Average acceleration

$$\bar{a} \equiv \frac{\Delta v}{\Delta t} \tag{3.4}$$

The **instantaneous acceleration** is equal to the limit of the ratio $\Delta v/\Delta t$ as $\Delta t \rightarrow 0$. By definition, this equals the derivative of v with respect to t, or the time rate of change of the velocity:

$$a \equiv \lim_{\Delta t \to 0} \frac{\Delta v}{\Delta t} = \frac{dv}{dt} \tag{3.5}$$

Instantaneous acceleration

The slope of the tangent to the x versus t curve at any instant equals the instantaneous velocity of the particle.

The slope of the tangent to the v versus t curve equals the instantaneous acceleration of the particle.

The area under the v versus t curve in any time interval equals the displacement of the particle in that interval.

The **equations of kinematics** for a particle moving along the x axis with uniform acceleration a (constant in magnitude and direction) are

Equations of kinematics

$$v = v_0 + at \tag{3.7}$$

$$x - x_0 = \tfrac{1}{2}(v_0 + v)t \tag{3.9}$$

(constant a only)

$$x - x_0 = v_0 t + \tfrac{1}{2}at^2 \tag{3.10}$$

$$v^2 = v_0^2 + 2a(x - x_0) \tag{3.11}$$

Freely falling body

A body falling freely in the presence of the earth's gravity experiences a gravitational acceleration directed toward the center of the earth. If air friction is neglected, and if the altitude of the motion is small compared with the earth's radius, then one can assume that the acceleration of gravity, g, is constant over the range of motion, where g is equal to 9.80 m/s^2, or 32 ft/s^2. Assuming y positive upward, the acceleration is given by $-g$, and the equations of kinematics for a body in free fall are the same as those given above, with the substitutions $x \rightarrow y$ and $a \rightarrow -g$.

QUESTIONS

1. Average velocity and instantaneous velocity are generally different quantities. Can they ever be equal for a specific type of motion? Explain.
2. If the average velocity is nonzero for some time interval, does this mean that the instantaneous velocity is never zero during this interval? Explain.
3. If the average velocity equals zero for some time interval Δt and if $v(t)$ is a continuous function, show that the instantaneous velocity must go to zero some time in this interval. (A sketch of x versus t might be useful in your proof.)
4. Is it possible to have a situation in which the velocity and acceleration have opposite signs? If so, sketch a velocity-time graph to prove your point.
5. If the velocity of a particle is nonzero, can its acceleration ever be zero? Explain.
6. If the velocity of a particle is zero, can its acceleration ever be nonzero? Explain.
7. Can the equations of kinematics (Eqs. 3.7 through 3.11) be used in a situation where the acceleration varies in time? Can they be used when the acceleration is zero?
8. A ball is thrown vertically upward. What are its velocity and acceleration when it reaches its maximum altitude? What is its acceleration just before it strikes the ground?
9. A stone is thrown upward from the top of a building. Does the stone's displacement depend on the location of the origin of the coordinate system? Does the stone's velocity depend on the origin? (Assume that the coordinate system is stationary with respect to the building.) Explain.
10. A child throws a marble in the air with an initial velocity v_0. Another child drops a ball at the same instant. Compare the accelerations of the two objects while they are in flight.

11. A student at the top of a building of height h throws one ball upward with an initial speed v_0 and then throws a second ball downward with the same initial speed. How do the final velocities of the balls compare when they reach the ground?
12. Can the instantaneous velocity of an object ever be greater in magnitude than the average velocity? Can it ever be less?
13. If a car is traveling eastward, can its acceleration be westward? Explain.
14. If the average velocity of an object is zero in some time interval, what can you say about the displacement of the object for that interval?
15. A rapidly growing plant doubles in height each week. At the end of the 25th day, the plant reaches the height of a building. At what time was the plant one-fourth the height of the building?
16. Two cars are moving in the same direction in parallel lanes along a highway. At some instant, the velocity of car A exceeds the velocity of car B. Does this mean that the acceleration of A is greater than that of B? Explain.
17. A ball is thrown upward. While the ball is in the air, (a) does its acceleration increase, decrease, or remain constant? (b) Describe what happens to its velocity.
18. Car A traveling south from New York to Miami has a speed of 25 m/s. Car B traveling west from New York to Chicago also has a speed of 25 m/s. Are their velocities equal? Explain.
19. An apple is dropped from some height above the earth's surface as in the photograph to the right. Neglecting air resistance, how much does its speed increase each second during its fall?

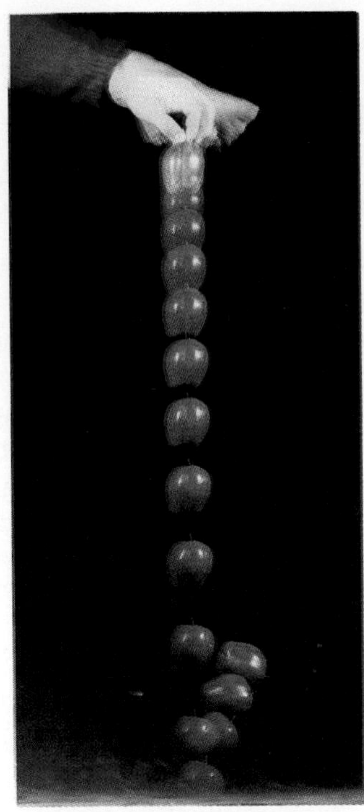

(Question 19)

20. An object released from rest falls freely with a constant downward acceleration of 9.8 m/s^2. After it has dropped for 3 seconds, (a) how far has it traveled? (b) what is its velocity? (c) what is its acceleration?
21. A pebble is dropped in a water well, and the "splash" is heard 2 s later. What is the *approximate* depth of the well?

PROBLEMS

Section 3.1 Average Velocity

1. The position of a pine wood derby car was observed at various times and the results are summarized in the table below. Find the average velocity of the car for (a) the first second, (b) the last 3 seconds, and (c) the entire period of observation.

x (m)	0	2.3	9.2	20.7	36.8	57.5
t (s)	0	1.0	2.0	3.0	4.0	5.0

2. A motorist drives north for 35 minutes at 85 km/h and then stops for 15 minutes. He then continues north, traveling 130 km in 2 h. (a) What is his total displacement? (b) What is his average velocity?
3. The displacement versus time for a certain particle moving along the x axis is shown in Figure 3.15. Find the average velocity in the time intervals (a) 0 to 2 s, (b) 0 to 4 s, (c) 2 s to 4 s, (d) 4 s to 7 s, (e) 0 to 8 s.

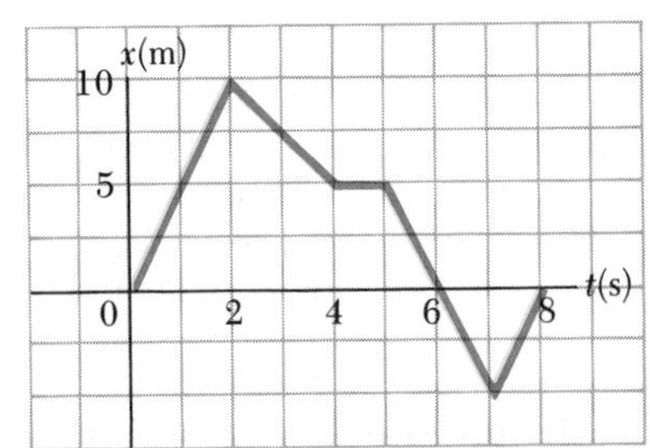

Figure 3.15 (Problem 3).

4. A jogger runs in a straight line with an average velocity of 5 m/s for 4 min, and then with an average velocity of 4 m/s for 3 min. (a) What is her total displacement? (b) What is her average velocity during this time?
5. An athlete swims the length of a 50-m pool in 20 s and makes the return trip to the starting position in 22 s.

Determine his average velocity in (a) the first half of the swim, (b) the second half of the swim, and (c) the round trip.

6. The position of a particle along the x axis is given by $x = 3t^3 - 7t$ where x is in meters and t is in seconds. What is the average velocity of the particle during the interval from $t = 2.0$ s to $t = 5.0$ s?

7. A car makes a 200-km trip at an average speed of 40 km/h. A second car starting 1 h later arrives at their mutual destination at the same time. What was the average speed of the second car?

Section 3.2 Instantaneous Velocity

8. Using Figure 3.16, determine (a) the average velocity between $t = 2.0$ s and $t = 5.0$ s, and (b) the instantaneous velocity at $t = 3.0$ s.

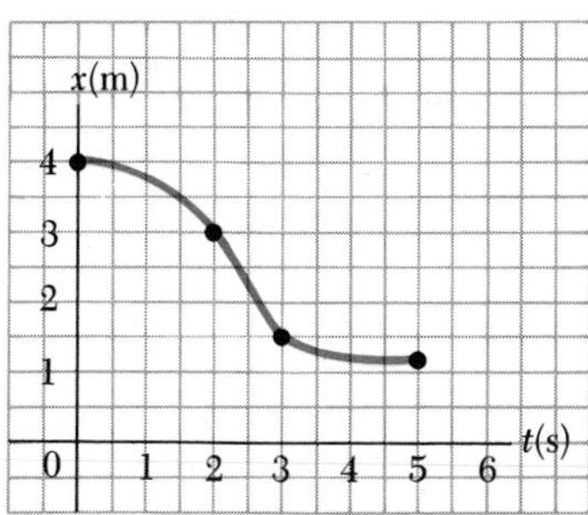

Figure 3.16 (Problem 8).

9. The position-time graph for a particle moving along the x axis is as shown in Figure 3.17. (a) Find the average velocity in the time interval $t = 1.5$ s to $t = 4$ s. (b) Determine the instantaneous velocity at $t = 2$ s by measuring the slope of the tangent line shown in the graph. (c) At what value of t is the velocity zero?

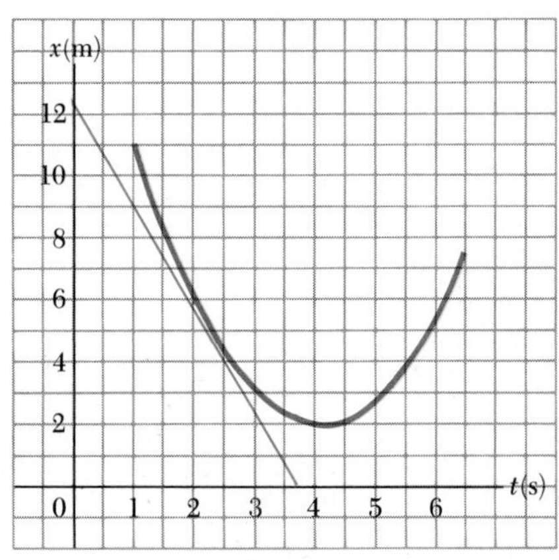

Figure 3.17 (Problem 9).

10. From the data in Problem 1, find the instantaneous velocity at $t = 2.0$ s.

11. At $t = 1$ s, a particle moving with constant velocity is located at $x = -3$ m, and at $t = 6$ s, the particle is located at $x = 5$ m. (a) From this information, plot the position as a function of time. (b) Determine the velocity of the particle from the slope of this graph.

12. Use the data in Problem 1 to (a) construct a graph of position versus time. (b) By constructing tangents to the $x(t)$ curve, find the instantaneous velocity of the car at several instants. (c) Plot the instantaneous velocity versus time and, from this, determine the average acceleration of the car. (d) What was the initial velocity of the car?

13. Find the instantaneous velocity of the particle described in Figure 3.15 at the following times: (a) $t = 1$ s, (b) $t = 3$ s, (c) $t = 4.5$ s, and (d) $t = 7.5$ s.

14. The position-time graph for a particle moving along the z axis is as shown in Figure 3.18. Determine whether the velocity is positive, negative, or zero at times (a) t_1, (b) t_2, (c) t_3, (d) t_4.

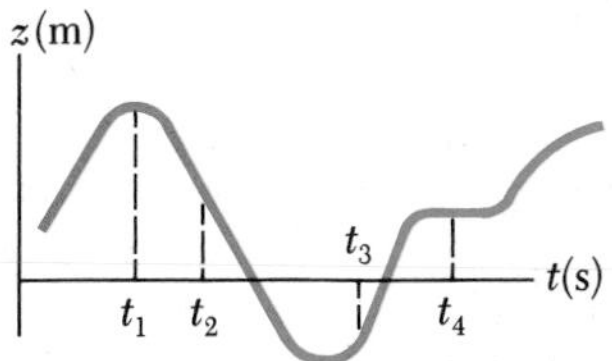

Figure 3.18 (Problem 14).

Section 3.3 Acceleration

15. A particle is moving with a velocity $v_0 = 60$ m/s at $t = 0$. Between $t = 0$ and $t = 15$ s the velocity decreases uniformly to zero. What was the average acceleration during this 15-s interval? What is the significance of the sign of your answer?

16. A car traveling in a straight line has a velocity of 30 m/s at some instant. Two seconds later its velocity is 25 m/s. What is its average acceleration in this time interval?

17. A particle moves along the x axis according to the equation $x = 2t + 3t^2$, where x is in m and t is in s. Calculate the instantaneous velocity and instantaneous acceleration at $t = 3$ s.

18. A particle moving in a straight line has a velocity of 8 m/s at $t = 0$. Its velocity at $t = 20$ s is 20 m/s. (a) What is its average acceleration in this time interval? (b) Can the average velocity be obtained from the information presented? Explain.

19. The velocity-time graph for an object moving along the x axis is as shown in Figure 3.19. (a) Plot a graph of the acceleration versus time. (b) Determine the average acceleration of the object in the time intervals $t = 5$ s to $t = 15$ s and $t = 0$ to $t = 20$ s.

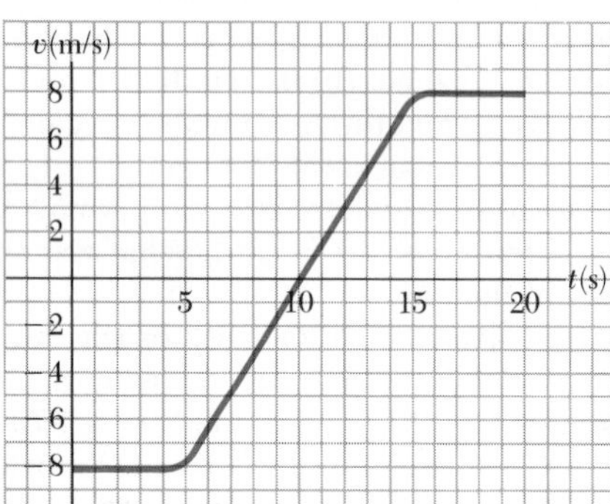

Figure 3.19 (Problem 19).

20. The velocity of a particle as a function of time is shown in Figure 3.20. At $t = 0$, the particle is at $x = 0$. (a) Sketch the acceleration as a function of time. (b) Determine the average acceleration of the particle in the time interval $t = 2$ s to $t = 8$ s. (c) Determine the instantaneous acceleration of the particle at $t = 4$ s.

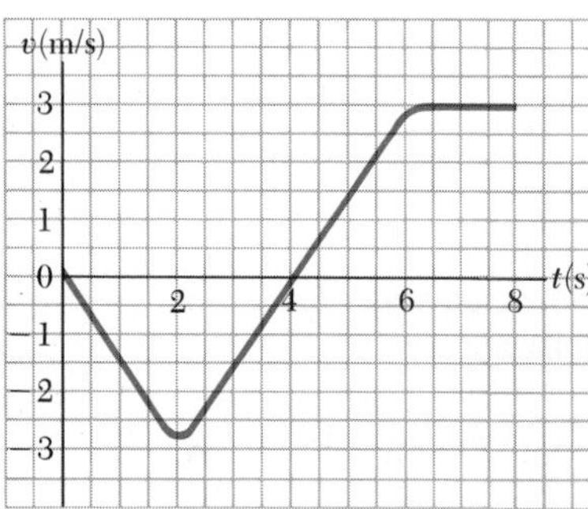

Figure 3.20 (Problem 20).

21. A particle moves along the x axis according to the equation $x = 2 + 3t - t^2$, where x is in m and t is in s. At $t = 3$ s, find (a) the position of the particle, (b) its velocity, and (c) its acceleration.
22. The velocity of a particle moving along the x axis varies in time according to $v = (15 - 8t)$ m/s. Find (a) the acceleration of the particle, (b) its velocity at $t = 3$ s, and (c) its average velocity in the time interval $t = 0$ to $t = 2$ s.
23. The velocity of a certain particle as a function of time is plotted in Figure 3.21. (a) Sketch the acceleration as a function of time. (b) Does the particle ever travel with constant acceleration? Explain. (c) Estimate the acceleration of the particle at $t = 6$ s.

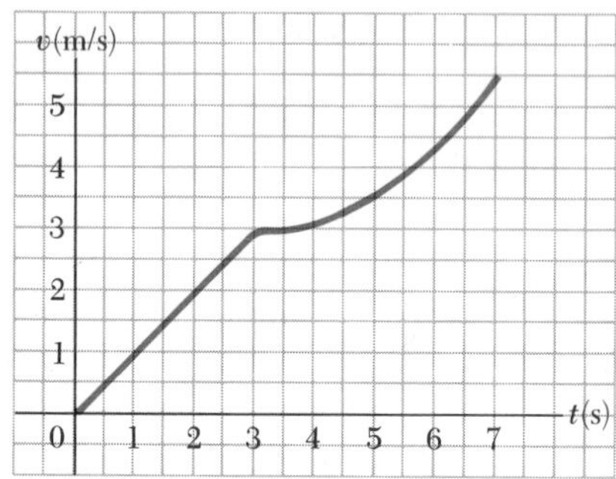

Figure 3.21 (Problem 23).

Section 3.4 One-Dimensional Motion with Constant Acceleration

24. A particle travels in the positive x direction for 10 s at a constant speed of 50 m/s. It then accelerates uniformly to a speed of 80 m/s in the next 5 s. Find (a) the average acceleration of the particle in the first 10 s, (b) its average acceleration in the interval $t = 10$ s to $t = 15$ s, (c) the total displacement of the particle between $t = 0$ and $t = 15$ s, and (d) its average speed in the interval $t = 10$ s to $t = 15$ s.
25. A body moving with uniform acceleration has a velocity of 12 cm/s when its x coordinate is 3 cm. If its x coordinate 2 s later is -5 cm, what is the magnitude of its acceleration?
26. An automobile traveling initially at a speed of 60 m/s is accelerated uniformly to a speed of 85 m/s in 12 s. How far does the automobile travel during the 12-s interval?
27. The initial velocity of an electron is -2×10^5 m/s. If its acceleration is 4.2×10^{14} m/s^2, (a) how far does it travel before momentarily coming to rest? (b) how long will it take to undergo a displacement of 5.0 cm?
28. Figure 3.22 represents part of the performance data of a car owned by a proud physics student. (a) Calculate from the graph the total distance traveled. (b) What distance does it travel between the times $t = 10$ s and $t = 40$ s? (c) Draw a graph of its acceleration versus time between $t = 0$ and $t = 50$ s. (d) Write an equation for x as a function of time for each phase of the motion represented by (i) Oa, (ii) ab, (iii) bc. (e) What is the average velocity of the car between $t = 0$ and $t = 50$ s?

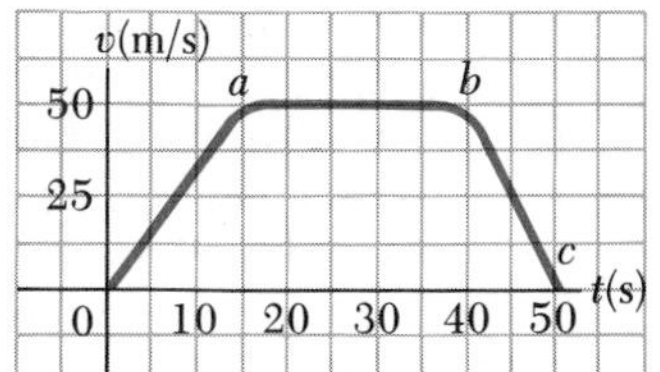

Figure 3.22 (Problem 28).

29. The initial speed of a body is 5.2 m/s. What is its speed after 2.5 s if it (a) accelerates uniformly at 3.0 m/s^2 and (b) accelerates uniformly at -3.0 m/s^2 (that is, it accelerates in the negative x direction)?

30. A hockey puck sliding on a frozen lake comes to rest after traveling 200 m. If its initial velocity is 3.0 m/s, (a) What is its acceleration if it is assumed constant? (b) How long is it in motion? (c) What is its speed after traveling 150 m?

31. A jet plane lands with a velocity of 100 m/s and can accelerate at a maximum rate of -5.0 m/s^2 as it comes to rest. (a) From the instant it touches the runway, what is the minimum time needed before it comes to rest? (b) Can this plane land on a small tropical island airport where the runway is 0.80 km long?

32. A sports car is traveling along a straight road at 140 km/h. When the brakes are applied, it undergoes a uniform negative acceleration, slowing to 70 km/h in a distance of 200 m. (a) What is the acceleration of the sports car? (b) If it continues to accelerate at this rate, how far will it travel while its speed decreases from 70 to 35 km/h? (c) How far does it travel as its speed changes from 140 km/h to zero?

33. A racing car reaches a speed of 50 m/s. At this instant, it undergoes a uniform negative acceleration using a parachute and braking system and comes to rest 5 s later. (a) Determine the acceleration of the car. (b) How far does the car travel after "turning on the brakes"?

34. A motorist driving at 60 mi/h along a straight road sees a 35 mi/h zone ahead. He slows down at a constant rate and in 3 s reaches the 35 mi/h zone traveling at 45 mi/h. (a) What is his acceleration? (b) How far is he in the 35 mi/h zone before his velocity is at the legal limit?

35. A particle starts from rest from the top of an inclined plane and slides down with constant acceleration. The inclined plane is 2.0 m long, and it takes 3.0 s for the particle to reach the bottom. Find (a) the acceleration of the particle, (b) its speed at the bottom of the incline, (c) the time it takes the particle to reach the middle of the incline, and (d) its speed at the midpoint.

36. Two express trains started 5 minutes apart. Starting from rest, each is capable of a maximum speed of 160 km/h after uniformly accelerating over a distance of 2.0 km. (a) What is the acceleration of each train? (b) How far ahead is the first train when the second one starts? (c) How far apart are they when they are both traveling at maximum speed?

37. A body is traveling at 16 m/s and comes to rest after undergoing a uniform negative acceleration for 40 m. (a) What is the acceleration of this body? (b) How long does it take to come to rest?

38. A go-cart travels the first half of a 100-m track with a constant speed of 5 m/s. In the second half of the track, it experiences a mechanical problem and slows down at 0.2 m/s^2. How long does it take the go-cart to travel the 100-m distance?

39. A car moving at a constant speed of 30 m/s suddenly stalls at the bottom of a hill. The car undergoes a constant acceleration of -2 m/s^2 (opposite its motion) while ascending the hill. (a) Write equations for the position and the velocity as functions of time, taking $x = 0$ at the bottom of the hill where $v_0 = 30$ m/s. (b) Determine the maximum distance traveled by the car up the hill after stalling.

40. An electron has an initial velocity of 3.0×10^5 m/s. If it undergoes an acceleration of 8.0×10^{14} m/s^2, (a) how long will it take to reach a velocity of 5.4×10^5 m/s and (b) how far has it traveled in this time?

41. A railroad car is released from a locomotive on an incline. When the car reaches the bottom of the incline, it has a speed of 30 mi/h, at which point it passes through a retarder track that slows it down. If the retarder track is 30 ft long, what negative acceleration must it produce to bring the car to rest?

42. An indestructible bullet, 2 cm long, is fired straight through a board which is 10.0 cm thick. The bullet strikes the board with a speed of 420 m/s and emerges with a speed of 280 m/s. (a) What is the average acceleration of the bullet through the board? (b) What is the total time that the bullet is in contact with the board? (c) How many thicknesses of board (calculated to the nearest tenth cm) would it take to completely stop the bullet?

43. Until recently, the world's land speed record was held by Colonel John P. Stapp, USAF. On March 19, 1954, he rode a rocket-propelled sled that moved down the track at 632 mi/h. He and the sled were safely brought to rest in 1.4 s. Determine (a) the negative acceleration he experienced and (b) the distance he traveled during this negative acceleration.

44. A hockey player is standing on his skates on a frozen pond when an opposing player skates by with the puck, moving with a uniform speed of 12 m/s. After 3 s, the first player makes up his mind to chase his opponent. If he accelerates uniformly at 4 m/s^2, (a) how long does it take him to catch the opponent?

(b) How far has he traveled in this time? (Assume the player with the puck remains in motion at constant speed.)

Section 3.5 Freely Falling Bodies

45. A woman is reported to have fallen 144 ft from the 17th floor of a building, landing on a metal ventilator box, which she crushed to a depth of 18 in. She suffered only minor injuries. Neglecting air resistance, calculate (a) the speed of the woman just before she collided with the ventilator, (b) her acceleration while in contact with the box, and (c) the time it took to crush the box.
46. A ball is thrown directly downward with an initial velocity of 8 m/s from a height of 30 m. When does the ball strike the ground?
47. A student throws a set of keys vertically upward to her sorority sister in a window 4.0 m above. The keys are caught 1.5 s later by the sister's outstretched hand. (a) With what initial velocity were the keys thrown? (b) What was the velocity of the keys just before they were caught?
48. A hot air balloon is traveling vertically upward at a constant speed of 5.0 m/s. When it is 21.0 m above the ground, a package is released from the balloon. (a) How long after being released is the package in the air? (b) What is the velocity of the package just before impact with the ground? (c) Repeat (a) and (b) for the case of the balloon descending at 5.0 m/s.
49. A ball is thrown vertically upward from the ground with an initial speed of 15 m/s. (a) How long does it take the ball to reach its maximum altitude? (b) What is its maximum altitude? (c) Determine the velocity and acceleration of the ball at $t = 2$ s.
50. A ball thrown vertically upward is caught by the thrower after 20 s. Find (a) the initial velocity of the ball and (b) the maximum height it reaches.
51. A ball is thrown straight up from the ground with a speed of 4.0 m/s. (a) How much time elapses between the two moments when its velocity has a magnitude of 2.5 m/s? (b) How far is it from the ground at those times?
52. An object is thrown vertically upward in such a way that it has a speed of 19.6 m/s when it reaches one half its maximum altitude. What are (a) its maximum altitude, (b) its velocity 1 s after it is thrown, and (c) its acceleration when it reaches its maximum altitude?
53. A stone is thrown upwards from the edge of a cliff 18 m high. It just misses the cliff on the way down and hits the ground below with a speed of 18.8 m/s. (a) With what velocity was it released? (b) What is its maximum distance from the ground during its flight?
54. A stone falls from rest from the top of a high cliff. A second stone is thrown downward from the same height 2.0 s later with an initial speed of 30 m/s. If both stones hit the ground below simultaneously, how high is the cliff?
55. A ball is thrown vertically upward with an initial speed of 10 m/s. One second later, a stone is thrown vertically upward with an initial speed of 25 m/s. Determine (a) the time it takes the stone to reach the same height as the ball, (b) the velocity of the ball and stone when they are at the same height, and (c) the total time each is in motion before returning to the original height.

*3.6 Kinematic Equations Derived from Calculus

56. The displacement of a body (in m) as a function of time (in s) is $x = 5 + 4t - 3t^2$. (a) What is the velocity as a function of time? What are the velocity and acceleration at (b) $t = 0.1$ s? (c) $t = 0$?
57. The acceleration of a motorboat as a function of time is given by the equation $a = Bt - Ct^2$, where the units of a are m/s^2. (a) What are the units of B and C? (b)What is the velocity as a function of time if the motorboat starts from rest at $t = 0$? (c) At what time $T > 0$ is the acceleration zero? (d) What is the velocity at the time T found in (c)?
58. A particle moves along the x axis with an acceleration that is proportional to the time, according to the expression $a = 30t$, where a is in m/s^2. Initially the particle is at rest at the origin. Find (a) the instantaneous velocity and (b) the instantaneous position as functions of time.
59. A particle is moving along the x axis. Its velocity as a function of time is given by $v = 5 + 10t$, where v is in m/s. The position of the particle at $t = 0$ is 20 m. Find (a) the acceleration as a function of time, (b) the position as a function of time, and (c) the velocity of the particle at $t = 0$.
60. The acceleration of a marble in a certain fluid is proportional to its velocity squared, and is given (in m/s^2) by $a = -3v^2$ for $v > 0$. If the marble enters this fluid with a speed of 1.50 m/s, how long will it take before the marble's speed is reduced to half of its initial value?

ADDITIONAL PROBLEMS

61. Given the position function

$$x(t) = \tfrac{1}{4}at^2 + 2bt$$

where $a = 1$ m/s^2 and $b = 1$ m/s, evaluate the average velocity $\bar{v}$ during elapsed time intervals starting at $t_a = 2$ s for which $\Delta t = 1$ s, 0.5 s, 0.1 s, 0.01 s, and 0.001 s.
62. A motorist is traveling at 18.0 m/s when he sees a deer in the road 38.0 m ahead. (a) If the maximum negative acceleration of the vehicle is -4.5 m/s^2, what is the maximum reaction time Δt of the motorist that will

allow him to avoid hitting the deer? (b) If his reaction time is 0.30 s, how fast will he be traveling when he hits the deer?

63. An inquisitive physics student and mountain climber climbs a 50-m cliff that overhangs a calm pool of water. He throws two stones vertically downward 1 s apart and observes that they cause a single splash. The first stone has an initial velocity of 2 m/s. (a) At what time after release of the first stone will the two stones hit the water? (b) What initial velocity must the second stone have if they are to hit simultaneously? (c) What will the velocity of each stone be at the instant they hit the water?

64. A particle moves along the positive x axis in such a way that its coordinate varies in time according to the expression $x = 4 + 2t - 3t^2$, where x is in m and t is in s. (a) Make a graph of x versus t for the interval $t = 0$ to $t = 2$ s. (b) Determine the initial position and initial velocity of the particle. (c) Determine at what time the particle reaches a *maximum* position coordinate. (*Note:* At this time, $v = 0$.) (d) Calculate the coordinate, velocity, and acceleration at $t = 2$ s.

65. A "superball" is dropped from a height of 2 m above the ground. On the first bounce the ball reaches a height of 1.85 m, where it is caught. Find the velocity of the ball (a) just as it makes contact with the ground and (b) just as it leaves the ground on the bounce. (c) Neglecting the time the ball spends in contact with the ground, find the total time required for the ball to go from the dropping point to the point where it is caught.

66. A Cessna 150 aircraft has a lift-off speed of about 125 km/h. (a) What minimum constant acceleration does this require if the aircraft is to be airborne after a take-off run of 250 m? (b) What is the corresponding take-off time? (c) If the aircraft continues to accelerate at this rate, what speed will it reach 25 s after it begins to roll?

67. A ball is dropped from rest from a window 8 m above a sidewalk at the same instant that a falling water drop passes the window with a speed of 15 m/s. How much time passes between the two impacts on the sidewalk? Neglect air resistance.

68. A falling object requires 1.50 s to travel the last 30 m before hitting the ground. From what height above the ground did it fall?

69. A young woman named Kathy Kool buys a superdeluxe sports car that can accelerate at the rate of 16 ft/s^2. She decides to test the car by dragging with another speedster, Stan Speedy. Both start from rest, but experienced Stan leaves 1 s before Kathy. If Stan moves with a constant acceleration of 12 ft/s^2 and Kathy maintains an acceleration of 16 ft/s^2, find (a) the time it takes Kathy to overtake Stan, (b) the distance she travels before she catches him, and (c) the velocities of both cars at the instant she overtakes him.

70. A hockey player takes a slap shot at a puck at rest on the ice. The puck glides over the ice for 10 ft without friction, at which point it runs over rough ice. The puck then accelerates opposite its motion at a uniform rate of -20 ft/s^2. If the velocity of the puck is 40 ft/s after traveling 100 ft from the point of impact, (a) what is the average acceleration imparted to the puck as it is struck by the hockey stick? (Assume that the time of contact is 0.01 s.) (b) How far in all does the puck travel before coming to rest? (c) What is the total time the puck is in motion, neglecting contact time?

71. A particle is observed at time $t = 0$ to be at the coordinate position $x_0 = 5$ m and moving with a velocity $v_0 = 20$ m/s. The particle undergoes a constant negative acceleration (i.e., an acceleration in the direction opposite to the velocity). If, 10 s after the initial observation, the particle has a velocity $v = 2$ m/s, what is the acceleration? What is the position function? How long a time will elapse before the particle returns to $x = 5$ m?

72. A truck traveling at a constant speed of 80 km/h passes a more slowly moving car. The instant the truck passes the car, the car begins to accelerate at a constant rate of 1.2 m/s^2 and passes the truck 0.5 km farther down the road. What is the speed of the car when it passes the truck?

73. A student stands on the edge of a building 100 ft above the ground and throws a baseball upward with some initial velocity. The ball is blown slightly sideways by a crosswind and then falls to the ground, just missing the building. If the total time for the ball to travel from the student's hand to the ground is 6 s, find (a) the initial velocity of the ball, (b) its final velocity as it hits the ground, and (c) its velocity after 3 s.

74. A rock, starting from rest and falling freely, falls the last 1/2 of the total distance of fall in 3.0 s. Find (a) the height from which the rock was released, and (b) the total time of fall.

75. A person sees a lightning bolt passing close to an airplane flying in the distance. The person hears thunder 5 s after seeing the bolt and sees the airplane overhead 10 s after hearing the thunder. If the speed of sound in air is 1100 ft/s, (a) find the distance the airplane is from the person at the instant of the bolt. (Neglect the time it takes the light to travel from the bolt to the eye.) (b) Assuming the plane travels with a constant speed toward the person, find the velocity of the airplane. (c) Look up the speed of light in air and defend the approximation used in (a).

76. A rocket is fired vertically upward with an initial velocity of 80 m/s. It accelerates upward at 4 m/s^2 until it reaches an altitude of 1000 m. At that point, its en-

gines fail and the rocket goes into free flight, with acceleration -9.80 m/s^2. (a) How long is the rocket in motion? (b) What is its maximum altitude? (c) What is its velocity just before it collides with the earth? (*Hint:* Consider the motion while the engine is operating separate from the free-flight motion.)

77. In a 100-m race, Maggie and Judy cross the finish line in a dead heat, both taking 10.2 s. Accelerating uniformly, Maggie takes 2.0 s and Judy 3.0 s to attain maximum speed, which they maintain for the rest of the race. (a) What is the acceleration of each sprinter? (b) What are their respective maximum speeds? (c) Which sprinter is ahead at the 6-s mark, and by how much?

78. A train travels in time in the following manner. In the first hour, it travels with a speed v, in the next half hour it has a speed $3v$, in the next 90 min it travels with a speed $v/2$, and in the final 2 h it travels with a speed $v/3$. (a) Plot the speed-time graph for this trip. (b) How far does the train travel in this trip? (c) What is the average speed of the train over the entire trip?

79. A commuter train can minimize the time t between two stations by accelerating ($a_1 = 0.1 \text{ m/s}^2$) for a time t_1 then undergoing a negative acceleration ($a_2 = -0.5 \text{ m/s}^2$) by using his brakes for a time t_2. Since the stations are only 1 km apart, the train never reaches its maximum velocity. Find the minimum time of travel t, and the time t_1.

80. In order to protect his food from hungry bears, a boy scout raises his food pack, mass m, with a rope that is thrown over a tree limb of height h above his hands. He walks away from the vertical rope with constant velocity v_0 holding the free end of the rope in his hands (see Fig. 3.23). (a) Show that the velocity v of the food pack is $x(x^2 + h^2)^{-1/2}v_0$ where x is the distance he has walked away from the vertical rope. (b) Show that the acceleration a of the food pack is $h^2(x^2 + h^2)^{-3/2}v_0{}^2$. (c) What values do the acceleration and velocity v have shortly after the boy scout leaves the vertical rope? (d) What values do the velocity and acceleration approach as the distance x continues to increase?

Figure 3.23 (Problem 80).

81. Two objects A and B are connected by a rigid rod that has a length L. The objects slide along perpendicular guide rails, as shown in Figure 3.24. If A slides to the left with a constant speed v, find the velocity of B when $\alpha = 60°$.

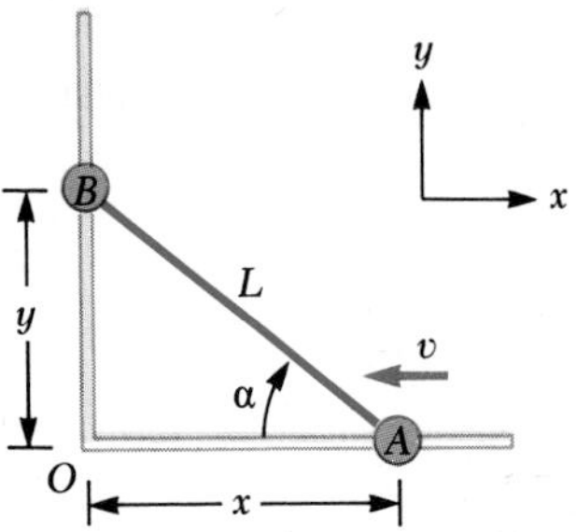

Figure 3.24 (Problem 81).

CALCULATOR/COMPUTER PROBLEMS

82. In Problem 80 let the height h equal 6 m and the velocity v_0 equal 2 m/s. Assume that the food pack starts from rest on a ledge over a cliff 6 m below the boy scout's hands. (a) Tabulate and graph the velocity-time graph. (b) Tabulate and graph the acceleration-time graph. (Let the range of time be from 0 s to 6 s and the time intervals be 0.5 s.)

83. **A particle undergoes a varying acceleration. The velocity is measured at 0.5 s intervals and is tabulated below. (a) Determine the average acceleration in each interval. (b) Use a numerical integration procedure to determine the position of the particle at the end of each time interval. Assume the initial position of the particle is zero.**

t (s)	0	0.5	1.0	1.5	2.0	2.5	3.0	3.5	4.0	4.5	5.0
v (m/s)	0	1	3	4.5	7.0	9.5	10.5	12	14	15	17.5

84. The acceleration of a particle moving along the x axis varies with position according to the expression

$$a = a_0 e^{-bx}$$

where $a_0 = 3 \text{ m/s}^2$ and $b = 1 \text{ m}^{-1}$. If the particle starts at rest from the origin, use a numerical integration method to find the position of the particle at $t = 2.37$ s. The accuracy of your calculation should be at least 1%.

85. A particle moving along the x axis undergoes an acceleration given by $a = \sqrt{3 + t^3} \text{ m/s}^2$. Use a numerical integration method to find the position and velocity of the particle at $t = 5.7$ s to within 1% accuracy.

4
Motion in Two Dimensions

Multiflash exposure of a tennis player executing a backhand swing. Such photographs can be used to study the quality of sports equipment and the performance of an athlete. (© Zimmerman, FPG International)

In this chapter we deal with the kinematics of a particle moving in a plane, or two-dimensional motion. Some common examples of motion in a plane are the motion of projectiles and satellites and the motion of charged particles in uniform electric fields. We begin by showing that velocity and acceleration are vector quantities. As in the case of one-dimensional motion, we shall derive the kinematic equations for two-dimensional motion from the fundamental definitions of displacement, velocity, and acceleration. As special cases of motion in two dimensions, we shall treat motion in a plane with constant acceleration and uniform circular motion.

4.1 THE DISPLACEMENT, VELOCITY, AND ACCELERATION VECTORS

In the previous chapter we found that the motion of a particle moving along a straight line is completely known if its coordinate is known as a function of time. Now let us extend this idea to the motion of a particle in the xy plane. We

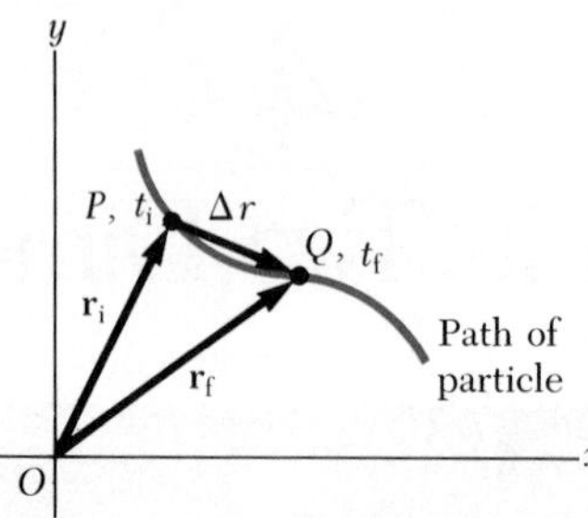

Figure 4.1 A particle moving in the xy plane is located with the position vector $\boldsymbol{r}$ drawn from the origin to the particle. The displacement of the particle as it moves from P to Q in the time interval $\Delta t = t_f - t_i$ is equal to the vector $\Delta \boldsymbol{r} = \boldsymbol{r}_f - \boldsymbol{r}_i$.

begin by describing the position of a particle with a *position vector* $\boldsymbol{r}$, drawn from the origin of some reference frame to the particle located in the xy plane, as in Figure 4.1. At time t_i, the particle is at the point P, and at some later time t_f, the particle is at Q. As the particle moves from P to Q in the time interval $\Delta t = t_f - t_i$, the position vector changes from $\boldsymbol{r}_i$ to $\boldsymbol{r}_f$, where the indices i and f refer to initial and final values. Because $\boldsymbol{r}_f = \boldsymbol{r}_i + \Delta \boldsymbol{r}$, the **displacement vector** for the particle is given by

Definition of the displacement vector

$$\Delta \boldsymbol{r} \equiv \boldsymbol{r}_f - \boldsymbol{r}_i \tag{4.1}$$

The direction of $\Delta \boldsymbol{r}$ is indicated in Figure 4.1. Note that the displacement vector equals the difference between the final position vector and the initial position vector. As we see from Figure 4.1, the magnitude of the displacement vector is *less* than the distance traveled along the curved path.

We now define the **average velocity** of the particle during the time interval Δt as the ratio of the displacement to the time interval for this displacement:

Average velocity

$$\bar{\boldsymbol{v}} \equiv \frac{\Delta \boldsymbol{r}}{\Delta t} \tag{4.2}$$

Since the displacement is a vector and the time interval is a scalar, we conclude that the average velocity is a *vector* quantity directed along $\Delta \boldsymbol{r}$. Note that the average velocity between points P and Q is *independent of the path* between the two points. This is because the average velocity is proportional to the displacement, which in turn depends only on the initial and final position vectors. As we did in the case of one-dimensional motion, we conclude that if a particle starts its motion at some point and returns to this point via any path, its average velocity is zero for this trip since its displacement is zero.

Consider again the motion of a particle between two points in the xy plane, as in Figure 4.2. As the time intervals become smaller and smaller, the displacements, $\Delta \boldsymbol{r}_1, \Delta \boldsymbol{r}_2, \Delta \boldsymbol{r}_3, \ldots$, get progressively smaller and the direction of the displacement approaches that of the line tangent to the path at the point P.

The **instantaneous velocity**, $\boldsymbol{v}$, is defined as the limit of the average velocity, $\Delta \boldsymbol{r}/\Delta t$, as Δt approaches zero:

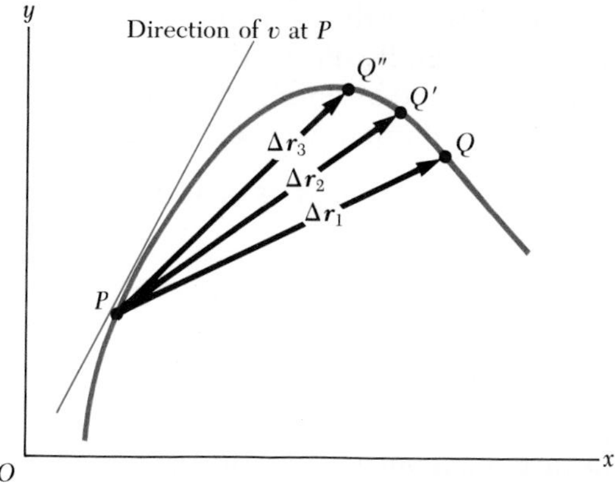

Figure 4.2 As a particle moves between two points, its average velocity is in the direction of the displacement vector $\Delta \boldsymbol{r}$. As the point Q moves closer to P, the direction of $\Delta \boldsymbol{r}$ approaches that of the line tangent to the curve at P. By definition, the instantaneous velocity at P is in the direction of this tangent line.

$$\boldsymbol{v} \equiv \lim_{\Delta t \to 0} \frac{\Delta \boldsymbol{r}}{\Delta t} = \frac{d\boldsymbol{r}}{dt} \tag{4.3}$$

Instantaneous velocity

That is, the instantaneous velocity equals the derivative of the position vector with respect to time. The direction of the velocity vector is along a line that is tangent to the path of the particle and in the direction of motion. This is illustrated in Figure 4.3 for two points along the path. The magnitude of the instantaneous velocity vector is called the *speed.* Note that Equation 4.3 is a logical generalization of differentiation as developed in the study of calculus.

As the particle moves from P to Q along some path, its instantaneous velocity vector changes from $\boldsymbol{v}_i$ at time t_i to $\boldsymbol{v}_f$ at time t_f (Figure 4.3).

The **average acceleration** of the particle as it moves from P to Q is defined as the ratio of the change in the instantaneous velocity vector, $\Delta \boldsymbol{v}$, to the elapsed time, Δt:

$$\bar{\boldsymbol{a}} \equiv \frac{\boldsymbol{v}_f - \boldsymbol{v}_i}{t_f - t_i} = \frac{\Delta \boldsymbol{v}}{\Delta t} \tag{4.4}$$

Average acceleration

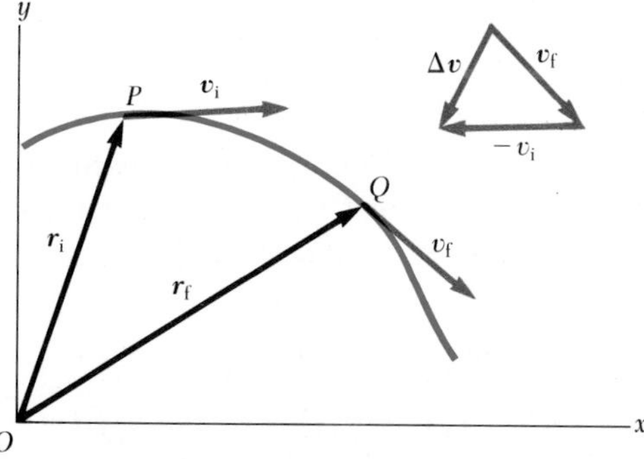

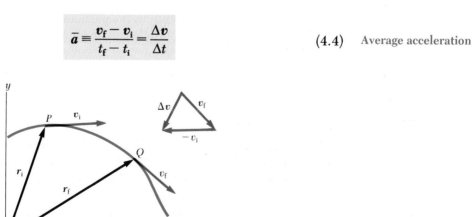

Figure 4.3 The average acceleration vector, $\bar{\boldsymbol{a}}$, for a particle moving from P to Q is in the direction of the change in the velocity, $\Delta \boldsymbol{v} = \boldsymbol{v}_f - \boldsymbol{v}_i$.

Since the average acceleration is the ratio of a vector, $\Delta \boldsymbol{v}$, and a scalar, Δt, we conclude that $\bar{\boldsymbol{a}}$ is a vector quantity directed along $\Delta \boldsymbol{v}$. As is indicated in Figure 4.3, the direction of $\Delta \boldsymbol{v}$ is found by adding the vector $-\boldsymbol{v}_i$ (the negative of $\boldsymbol{v}_i$) to the vector $\boldsymbol{v}_f$, since by definition $\Delta \boldsymbol{v} = \boldsymbol{v}_f - \boldsymbol{v}_i$.

The **instantaneous acceleration,** $\boldsymbol{a}$, is defined as the limiting value of the ratio $\Delta \boldsymbol{v}/\Delta t$ as Δt approaches zero:

Instantaneous acceleration

$$\boldsymbol{a} \equiv \lim_{\Delta t \to 0} \frac{\Delta \boldsymbol{v}}{\Delta t} = \frac{d\boldsymbol{v}}{dt} \qquad (4.5)$$

In other words, the instantaneous acceleration equals the first derivative of the velocity vector with respect to time.

It is important to recognize that a particle can accelerate for several reasons. First, the magnitude of the velocity vector (the speed) may change with time as in one-dimensional motion. Second, a particle accelerates when the direction of the velocity vector changes with time (a curved path) even though its speed is constant. Finally, the acceleration may be due to a change in both the magnitude and the direction of the velocity vector.

4.2 MOTION IN TWO DIMENSIONS WITH CONSTANT ACCELERATION

Let us consider the motion of a particle in two dimensions with constant acceleration. That is, we assume that the magnitude and direction of the acceleration remain unchanged during the motion.

A particle in motion can be described by its position vector $\boldsymbol{r}$. The position vector for a particle moving in the xy plane can be written

$$\boldsymbol{r} = x\boldsymbol{i} + y\boldsymbol{j} \qquad (4.6)$$

where x, y, and $\boldsymbol{r}$ change with time as the particle moves. If the position vector is known, the velocity of the particle can be obtained from Equations 4.3 and 4.6, which give

$$\boldsymbol{v} = \frac{d\boldsymbol{r}}{dt} = \frac{dx}{dt}\boldsymbol{i} + \frac{dy}{dt}\boldsymbol{j}$$

$$\boldsymbol{v} = v_x\boldsymbol{i} + v_y\boldsymbol{j} \qquad (4.7)$$

Because $\boldsymbol{a}$ is a constant, its components a_x and a_y are also constants. Therefore, we can apply the equations of kinematics to both the x and y components of the velocity vector. Substituting $v_x = v_{x0} + a_x t$ and $v_y = v_{y0} + a_y t$ into Equation 4.7 gives

$$\boldsymbol{v} = (v_{x0} + a_x t)\boldsymbol{i} + (v_{y0} + a_y t)\boldsymbol{j}$$

$$= (v_{x0}\boldsymbol{i} + v_{y0}\boldsymbol{j}) + (a_x\boldsymbol{i} + a_y\boldsymbol{j})t$$

Velocity vector as a function of time

$$\boldsymbol{v} = \boldsymbol{v}_0 + \boldsymbol{a}t \qquad (4.8)$$

This result states that the velocity of a particle at some time t equals the vector sum of its initial velocity, $\boldsymbol{v}_0$, and the additional velocity $\boldsymbol{a}t$ acquired in the time t as a result of its constant acceleration.

Similarly, from kinematics we know that the x and y coordinates of a particle moving with constant acceleration are given by

$$x = x_0 + v_{x0}t + \tfrac{1}{2}a_x t^2 \qquad \text{and} \qquad y = y_0 + v_{y0}t + \tfrac{1}{2}a_y t^2$$

Substituting these expressions into Equation 4.6 gives

$$\begin{aligned}\boldsymbol{r} &= (x_0 + v_{x0}t + \tfrac{1}{2}a_x t^2)\boldsymbol{i} + (y_0 + v_{y0}t + \tfrac{1}{2}a_y t^2)\boldsymbol{j} \\ &= (x_0\boldsymbol{i} + y_0\boldsymbol{j}) + (v_{x0}\boldsymbol{i} + v_{y0}\boldsymbol{j})t + \tfrac{1}{2}(a_x\boldsymbol{i} + a_y\boldsymbol{j})t^2\end{aligned}$$

or

$$\boldsymbol{r} = \boldsymbol{r}_0 + \boldsymbol{v}_0 t + \tfrac{1}{2}\boldsymbol{a}t^2 \tag{4.9}$$

This equation says that the displacement vector $\boldsymbol{r} - \boldsymbol{r}_0$ is the vector sum of a displacement $\boldsymbol{v}_0 t$, arising from the initial velocity of the particle, and a displacement $\frac{1}{2}\boldsymbol{a}t^2$, resulting from the uniform acceleration of the particle. Graphical representations of Equations 4.6 and 4.7 are shown in Figures 4.4a and 4.4b. For simplicity in drawing the figure, we have taken $\boldsymbol{r}_0 = 0$ in Figure 4.4b. That is, we assume that the particle is at the origin at $t = 0$. Note from Figure 4.4b that $\boldsymbol{r}$ is generally not along the direction of $\boldsymbol{v}_0$ or $\boldsymbol{a}$, since the relation between these quantities is a vector expression. For the same reason, from Figure 4.4a we see that $\boldsymbol{v}$ is generally not along the direction of $\boldsymbol{v}_0$ or $\boldsymbol{a}$. Finally, if we compare Figures 4.4a and 4.4b we see that $\boldsymbol{v}$ and $\boldsymbol{r}$ are not in the same direction. This is because $\boldsymbol{v}$ is linear in t, while $\boldsymbol{r}$ is quadratic in t. It is also important to recognize that since Equations 4.8 and 4.9 are *vector* expressions having one or more components (in general, three components), we may write the component forms of these expressions along the x and y axes with $\boldsymbol{r}_0 = 0$

$$\boldsymbol{v} = \boldsymbol{v}_0 + \boldsymbol{a}t \qquad \begin{cases} v_x = v_{x0} + a_x t \\ v_y = v_{y0} + a_y t \end{cases}$$

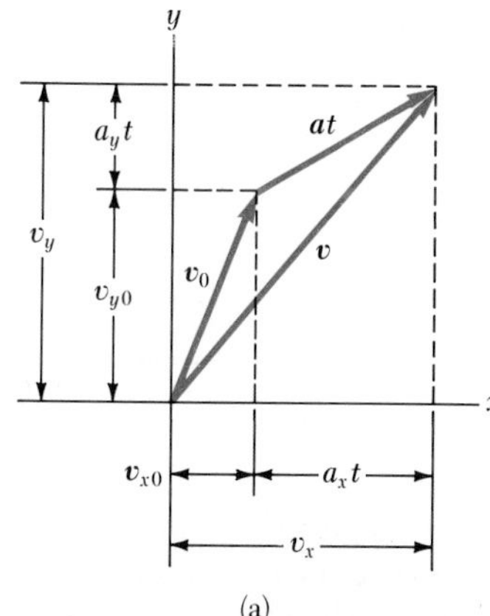

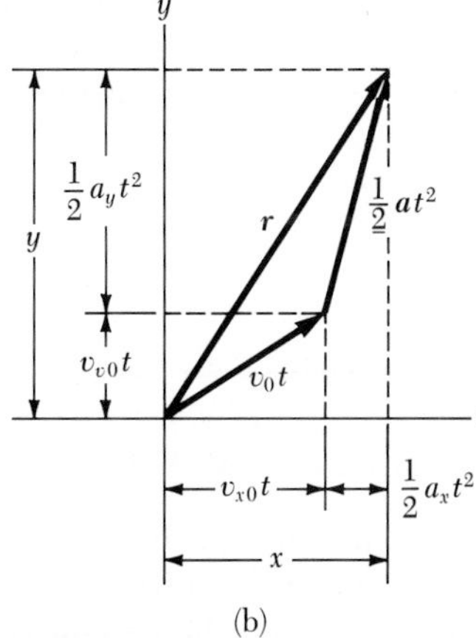

Figure 4.4 Vector representations and rectangular components of (a) the velocity and (b) the displacement of a particle moving with a uniform acceleration $\boldsymbol{a}$.

$$\boldsymbol{r} = \boldsymbol{v}_0 t + \tfrac{1}{2}\boldsymbol{a}t^2 \qquad \begin{cases} x = v_{x0}t + \tfrac{1}{2}a_x t^2 \\ y = v_{y0}t + \tfrac{1}{2}a_y t^2 \end{cases}$$

These components are illustrated in Figure 4.4. In other words, two-dimensional motion with constant acceleration is equivalent to two independent motions in the x and y directions with constant accelerations a_x and a_y.

EXAMPLE 4.1 Motion in a Plane □

A particle moves in the xy plane with an x component of acceleration only, given by $a_x = 4$ m/s². The particle starts from the origin at $t = 0$ with an initial velocity having an x component of 20 m/s and a y component of -15 m/s. (a) Determine the components of velocity as a function of time and the total velocity vector at any time.

Since $v_{x0} = 20$ m/s and $a_x = 4$ m/s², the equations of kinematics give

$$v_x = v_{x0} + a_x t = (20 + 4t) \text{ m/s}$$

Also, since $v_{y0} = -15$ m/s and $a_y = 0$,

$$v_y = v_{y0} = -15 \text{ m/s}$$

Therefore, using the above results and noting that the velocity vector $\boldsymbol{v}$ has two components, we get

$$\boldsymbol{v} = v_x\boldsymbol{i} + v_y\boldsymbol{j} = [(20 + 4t)\boldsymbol{i} - 15\boldsymbol{j}] \text{ m/s}$$

We could also obtain this result using Equation 4.8 directly, noting that $\boldsymbol{a} = 4\boldsymbol{i}$ m/s² and $\boldsymbol{v}_0 = (20\boldsymbol{i} - 15\boldsymbol{j})$ m/s. Try it!

(b) Calculate the velocity and speed of the particle at $t = 5$ s.

With $t = 5$ s, the result from (a) gives

$$\boldsymbol{v} = \{[20 + 4(5)]\boldsymbol{i} - 15\boldsymbol{j}\} \text{ m/s} = (40\boldsymbol{i} - 15\boldsymbol{j}) \text{ m/s}$$

That is, at $t = 5$ s, $v_x = 40$ m/s and $v_y = -15$ m/s. The speed is defined as the magnitude of $\boldsymbol{v}$, or

$$v = |\boldsymbol{v}| = \sqrt{v_x^2 + v_y^2} = \sqrt{(40)^2 + (-15)^2} \text{ m/s}$$

$$= 42.7 \text{ m/s}$$

(*Note:* v is larger than v_0. Why?)

The angle θ that $\boldsymbol{v}$ makes with the x axis can be calculated using the fact that $\tan\theta = v_y/v_x$, or

$$\theta = \tan^{-1}\left(\frac{v_y}{v_x}\right) = \tan^{-1}\left(\frac{-15}{40}\right) = -20.6°$$

(c) Determine the x and y coordinates at any time t and the displacement vector at this time.

Since at $t = 0$, $x_0 = y_0 = 0$, the expressions for the x and y coordinates, the equations of kinematics give

$$x = v_{x0}t + \tfrac{1}{2}a_x t^2 = (20t + 2t^2) \text{ m}$$

$$y = v_{y0}t = (-15t) \text{ m}$$

Therefore, the displacement vector at any time t is given by

$$\boldsymbol{r} = x\boldsymbol{i} + y\boldsymbol{j} = [(20t + 2t^2)\boldsymbol{i} - 15t\boldsymbol{j}] \text{ m}$$

Alternatively, we could obtain $\boldsymbol{r}$ by applying Equation 4.9 directly, with $\boldsymbol{v}_0 = (20\boldsymbol{i} - 15\boldsymbol{j})$ m/s and $\boldsymbol{a} = 4\boldsymbol{i}$ m/s². Try it!

Thus, for example, at $t = 5$ s, $x = 150$ m and $y = -75$ m, or $\boldsymbol{r} = (150\boldsymbol{i} - 75\boldsymbol{j})$ m. It follows that the distance of the particle from the origin to this point is the magnitude of the displacement, or

$$|\boldsymbol{r}| = r = \sqrt{(150)^2 + (-75)^2} \text{ m} = 168 \text{ m}$$

Note that this is *not* the distance that the particle travels in this time! Can you determine this distance from the available data?

4.3 PROJECTILE MOTION

Anyone who has observed a baseball in motion (or, for that matter, any object thrown into the air) has observed projectile motion. For an arbitrary direction of the initial velocity, the ball moves in a curved path. This very common form of motion is surprisingly simple to analyze if the following two assumptions are made: (1) the acceleration due to gravity, **g**, is constant over the range of

Assumptions of projectile motion

motion and is directed downward,[1] and (2) the effect of air resistance is negligible.[2] With these assumptions, we shall find that the path of a projectile, which we call its *trajectory*, is *always* a parabola. *We shall use these assumptions throughout this chapter.*

If we choose our reference frame such that the y direction is vertical and positive upward, then $a_y = -g$ (as in one-dimensional free fall) and $a_x = 0$ (since air friction is neglected). Furthermore, let us assume that at $t = 0$, the projectile leaves the origin ($x_0 = y_0 = 0$) with a velocity v_0, as in Figure 4.5. If the vector $\boldsymbol{v}_0$ makes an angle θ_0 with the horizontal, as in Figure 4.5, then from the definitions of the cosine and sine functions we have

$$\cos \theta_0 = v_{x0}/v_0 \qquad \text{and} \qquad \sin \theta_0 = v_{y0}/v_0$$

Therefore, the initial x and y components of velocity are given by

$$v_{x0} = v_0 \cos \theta_0 \qquad \text{and} \qquad v_{y0} = v_0 \sin \theta_0$$

Substituting these expressions into Equations 4.8 and 4.9 with $a_x = 0$ and $a_y = -g$ gives the velocity components and coordinates for the projectile at any time t:

$$v_x = v_{x0} = v_0 \cos \theta_0 = \text{constant} \tag{4.10}$$

Horizontal velocity component

$$v_y = v_{y0} - gt = v_0 \sin \theta_0 - gt \tag{4.11}$$

Vertical velocity component

$$x = v_{x0}t = (v_0 \cos \theta_0)t \tag{4.12}$$

Horizontal position component

$$y = v_{y0}t - \tfrac{1}{2}gt^2 = (v_0 \sin \theta_0)t - \tfrac{1}{2}gt^2 \tag{4.13}$$

Vertical position component

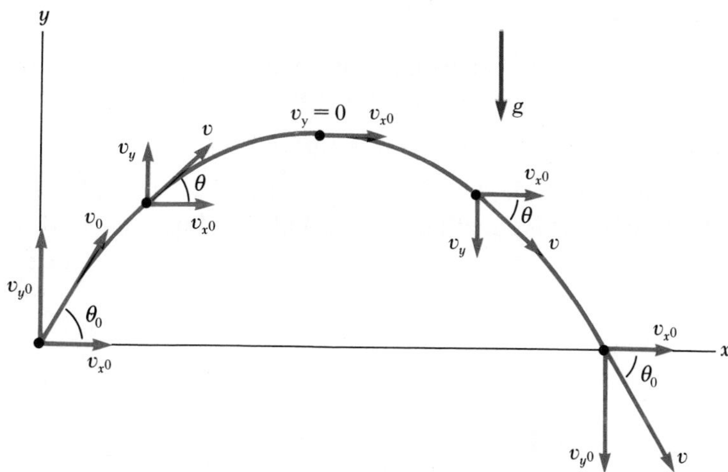

Figure 4.5 The parabolic trajectory of a projectile that leaves the origin with a velocity v_0. Note that the velocity vector, $\boldsymbol{v}$, changes with time. However, the x component of velocity, v_{x0}, remains constant in time. Also, $v_y = 0$ at the peak.

[1] This approximation is reasonable as long as the range of motion is small compared with the radius of the earth (6.4×10^6 m). In effect, this approximation is equivalent to assuming that the earth is flat over the range of motion considered.

[2] This approximation is generally *not* justified, especially at high velocities. In addition, the spin of a projectile, such as a baseball, can give rise to some very interesting effects associated with aerodynamic forces (for example, a curve thrown by a pitcher).

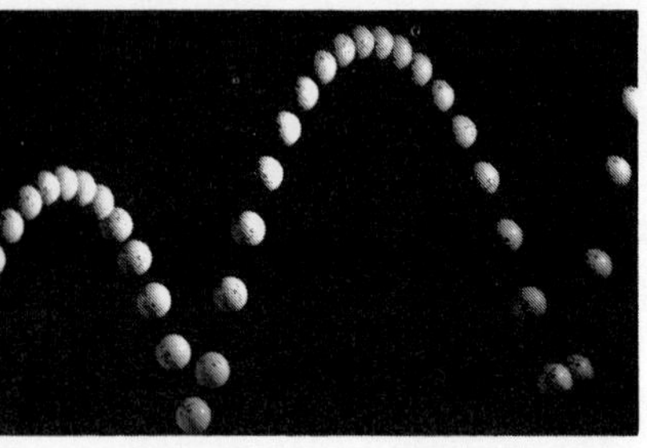

A golfball bouncing off a hard surface. Note the parabolic path of the ball following each bounce. (© Harold E. Edgerton. Courtesy of Palm Press Inc.)

From Equation 4.10 we see that v_x remains constant in time and is equal to the initial x component of velocity, since there is no horizontal component of acceleration. Also, for the y motion we note that v_y and y are identical to the expressions for the freely falling body discussed in Chapter 3. In fact, *all* of the equations of kinematics developed in Chapter 3 are applicable to projectile motion.

If we solve for t in Equation 4.12 and substitute this expression for t into Equation 4.13, we find that

$$y = (\tan \theta_0)x - \left(\frac{g}{2v_0^2 \cos^2 \theta_0}\right) x^2 \qquad (4.14)$$

which is valid for the angles in the range $0 < \theta_0 < \pi/2$. This is of the form $y = ax - bx^2$, which is the equation of a parabola that passes through the origin. Thus, we have proved that the trajectory of a projectile is a parabola. Note that the trajectory is *completely* specified if v_0 and θ_0 are known.

One can obtain the speed, v, as a function of time for the projectile by noting that Equations 4.10 and 4.11 give the x and y components of velocity at any instant. Therefore, by definition, since v is equal to the magnitude of $\boldsymbol{v}$,

$$v = \sqrt{v_x^2 + v_y^2} \qquad (4.15)$$

Also, since the velocity vector is tangent to the path at any instant, as shown in Figure 4.5, the angle θ that $\boldsymbol{v}$ makes with the horizontal can be obtained from v_x and v_y through the expression

Angle of trajectory

$$\tan \theta = \frac{v_y}{v_x} \qquad (4.16)$$

The vector expression for the position vector as a function of time for the projectile follows directly from Equation 4.9, with $\boldsymbol{a} = \boldsymbol{g}$:

$$\boldsymbol{r} = \boldsymbol{v}_0 t + \tfrac{1}{2}\boldsymbol{g}t^2$$

This expression is equivalent to Equations 4.12 and 4.13 and is plotted in Figure 4.6. Note that this equation is consistent with Equation 4.13, since the

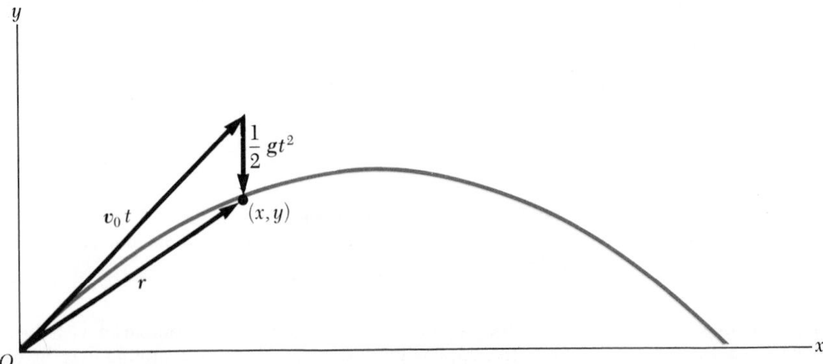

Figure 4.6 The displacement vector, $\boldsymbol{r}$, of a projectile having an initial velocity at the origin of $\boldsymbol{v}_0$. The vector $\boldsymbol{v}_0 t$ would be the displacement of the projectile if gravity were absent, and the vector $\frac{1}{2}\boldsymbol{g}t^2$ is its vertical displacement due to gravity in the time t.

expression for $\boldsymbol{r}$ is a vector equation and $\boldsymbol{a} = \mathbf{g} = -g\boldsymbol{j}$ when the upward direction is taken to be positive. It is interesting to note that the motion can be considered the superposition of the term $\boldsymbol{v}_0 t$, which is the displacement if no acceleration were present, and the term $\frac{1}{2}\mathbf{g}t^2$, which arises from the acceleration due to gravity. In other words, if there were no gravitational acceleration, the particle would continue to move along a straight path in the direction of $\boldsymbol{v}_0$. Therefore, the vertical distance, $\frac{1}{2}gt^2$, through which the particle "falls" measured from the straight line is that of a freely falling body. *We conclude that projectile motion is the superposition of two motions: (1) the motion of a freely falling body in the vertical direction with constant acceleration and (2) uniform motion in the horizontal direction with constant velocity.*

Horizontal Range and Maximum Height of a Projectile Let us assume that a projectile is fired from the origin at $t = 0$ with a positive v_y component, as in Figure 4.7. There are two special points that are interesting to analyze: the peak with cartesian coordinates labeled $(R/2, h)$ and the point with coordinates $(R, 0)$. The distance R is called the *horizontal range* of the projectile, and h is its *maximum height.* Let us find h and R in terms of v_0, θ_0, and g.

We can determine the maximum height, h, reached by the projectile by noting that at the peak, $v_y = 0$. Therefore, Equation 4.11 can be used to determine the time t_1 it takes to reach the peak:

$$t_1 = \frac{v_0 \sin \theta_0}{g}$$

Substituting this expression for t_1 into Equation 4.13 gives h in terms of v_0 and θ_0:

Maximum height of projectile

$$h = (v_0 \sin \theta_0) \frac{v_0 \sin \theta_0}{g} - \tfrac{1}{2}g\left(\frac{v_0 \sin \theta_0}{g}\right)^2$$

$$h = \frac{{v_0}^2 \sin^2 \theta_0}{2g} \qquad (4.17)$$

The range, R, is the horizontal distance traveled in twice the time it takes to reach the peak, that is, in a time $2t_1$. (This can be seen by setting $y = 0$ in Equation 4.13 and solving the quadratic for t. One solution of this quadratic is

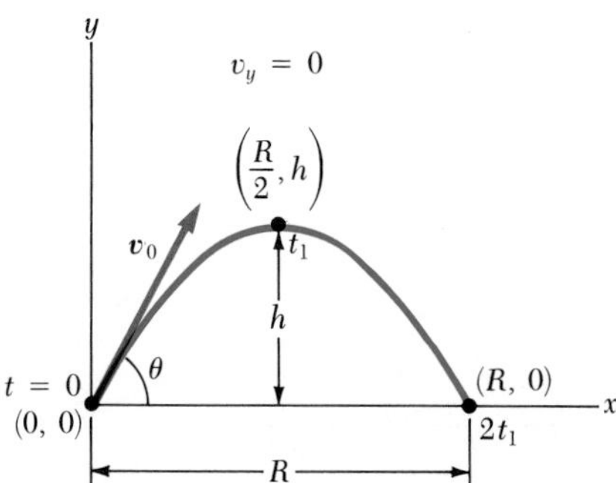

Figure 4.7 A projectile fired from the origin at $t = 0$ with an initial velocity $\boldsymbol{v}_0$. The maximum height of the projectile is h, and its horizontal range is R.

$t = 0$, and the second is $t = 2t_1$.) Using Equation 4.12 and noting that $x = R$ at $t = 2t_1$, we find that

$$R = (v_0 \cos \theta_0)2t_1 = (v_0 \cos \theta_0) \frac{2v_0 \sin \theta_0}{g}$$

$$R = \frac{2v_0^2 \sin \theta_0 \cos \theta_0}{g}$$

Since $\sin 2\theta = 2 \sin \theta \cos \theta$, R can be written in the more compact form

Range of projectile

$$R = \frac{v_0^2 \sin 2\theta_0}{g} \tag{4.18}$$

Keep in mind that Equations 4.17 and 4.18 are useful only for calculating h and R if v_0 and θ_0 are known and only for a symmetric path, as shown in Figure 4.7 (which means that only $\mathbf{v}_0$ has to be specified). The general expressions given by Equations 4.10 through 4.13 are the *most important* results, since they give the coordinates and velocity components of the projectile at *any* time t.

You should note that the maximum value of R from Equation 4.18 is $R_{max} = v_0^2/g$. This result follows from the fact that the maximum value of $\sin 2\theta_0$ is unity, which occurs when $2\theta_0 = 90°$. Therefore, we see that R is a maximum when $\theta_0 = 45°$, as you would expect if air friction is neglected.

Figure 4.8 illustrates various trajectories for a projectile of a given initial speed. As you can see, the range is a maximum for $\theta_0 = 45°$. In addition, for any θ_0 other than 45°, a point with coordinates $(R, 0)$ can be reached with *two* complementary values of θ_0, such as 75° and 15°. Of course, the maximum height and time of flight will be different for these two values of θ_0.

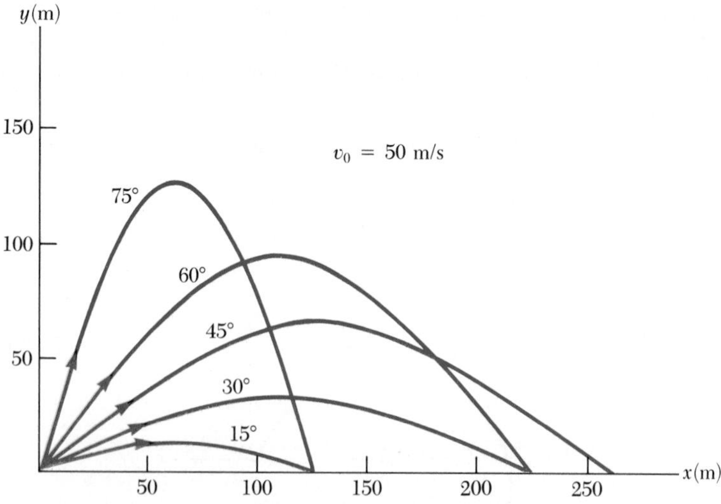

Figure 4.8 A projectile fired from the origin at an initial speed of 50 m/s at various angles of projection. Note that a point along the x axis can be reached at any two complementary values of θ_0.

EXAMPLE 4.2 The Long-jump
A long-jumper leaves the ground at an angle of 20° to the horizontal and at a speed of 11 m/s. (a) How far does he jump? (Assume that the motion of the long-jumper is equivalent to that of a particle.)

Solution His horizontal motion is described by using Equation 4.12:

$$x = (v_0 \cos \theta_0)t = (11 \text{ m/s})(\cos 20°)t$$

The value of x can be found if t, the total time of the jump, is known. We are able to find t using the expression $v_y = v_0 \sin \theta_0 - gt$ by noting that at the top of the jump the vertical component of velocity goes to zero:

$$v_y = v_0 \sin \theta_0 - gt$$

$$0 = (11 \text{ m/s}) \sin 20° - (9.80 \text{ m/s}^2)t_1$$

$$t_1 = 0.384 \text{ s}$$

Note that t_1 is the time interval to reach the *top* of the jump. Because of the symmetry of the vertical motion, an identical time interval passes before the jumper returns to the ground. Therefore, the *total time* in the air is $t = 2t_1 = 0.768$ s. Substituting this into the expression for x gives

$$x = (11 \text{ m/s})(\cos 20°)(0.768 \text{ s}) = \boxed{7.94 \text{ m}}$$

(b) What is the maximum height reached?

Solution The maximum height reached is found using Equation 4.13, with $t = t_1 = 0.384$ s:

$$y_{max} = (v_0 \sin \theta_0)t_1 - \tfrac{1}{2}gt_1^2$$

$$y_{max} = (11 \text{ m/s})(\sin 20°)(0.384 \text{ s}) - \tfrac{1}{2}(9.80 \text{ m/s}^2)(0.384 \text{ s})^2$$

$$= \boxed{0.722 \text{ m}}$$

The assumption that the motion of the long-jumper is that of a projectile is an oversimplification of the situation. Nevertheless, the values obtained are reasonable. Note that we also could have used Equations 4.17 and 4.18 to find the maximum height and horizontal range. However, the method used in our solution is more instructive.

EXAMPLE 4.3 It's A Bull's Eye Every Time
In a very popular lecture demonstration, a projectile is fired at a falling target in such a way that the projectile leaves the gun at the same time the target is dropped from rest, as in Figure 4.9. Let us show that if the gun is initially aimed at the target, the projectile will hit the target.[3]

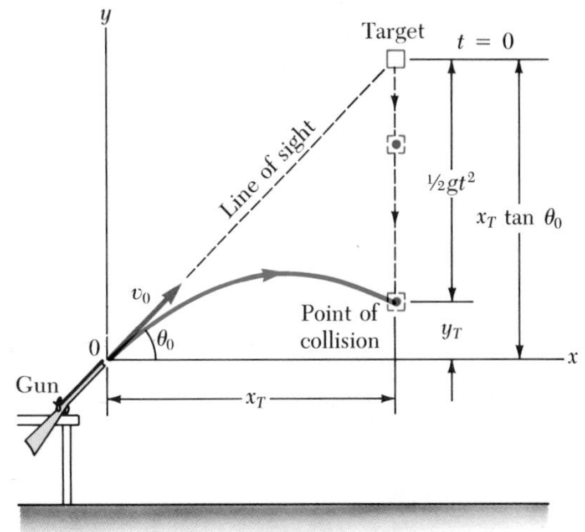

Figure 4.9 (Example 4.3) Schematic diagram of the projectile-and-target demonstration. If the gun is aimed directly at the target and is fired at the same instant the target begins to fall, the projectile will hit the target. Both fall through the same vertical distance in a time t, since both experience the same acceleration, $a_y = -g$.

Solution We can argue that a collision will result under the conditions stated by noting that both the projectile and the target experience the *same* acceleration, $a_y = -g$, as soon as they are released. First, note from Figure 4.9 that the initial y coordinate of the target is $x_T \tan \theta_0$ and that it falls through a distance $\frac{1}{2}gt^2$ in a time t. Therefore, the y coordinate of the target as a function of time is

$$y_T = x_T \tan \theta_0 - \tfrac{1}{2}gt^2$$

Now if we write equations for x and y for the projectile path over time, using Equations 4.12 and 4.13 simultaneously, we get

$$y_P = x_P \tan \theta_0 - \tfrac{1}{2}gt^2$$

Thus, when $x_P = x_T$, we see by comparing the two equations above that $y_P = y_T$ and a collision results.

The result could also be arrived at with vector methods, using expressions for the position vectors for the projectile and target.

You should also note that a collision will *not* always take place. There is the further restriction that a collision

[3] In one variation of the demonstration, the target is a tin can held by an electromagnet energized with a small battery. At the instant the projectile leaves the gun, a small switch at the top of the gun is opened by the moving projectile. This opens the circuit containing the electromagnet, allowing the target to fall.

will result only when $v_0 \sin \theta_0 \geq \sqrt{gd/2}$, where d is the initial elevation of the target above the *floor*, as in Figure 4.9. If $v_0 \sin \theta_0$ is less than this value, the projectile will strike the floor before reaching the target.

EXAMPLE 4.4 That's Quite an Arm

A stone is thrown from the top of a building upward at an angle of 30° to the horizontal and with an initial speed of 20 m/s, as in Figure 4.10. If the height of the building is 45 m, (a) how long is the stone "in flight"?

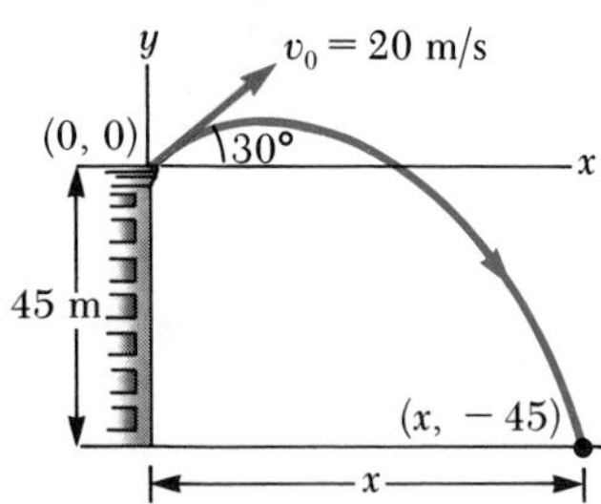

Figure 4.10 (Example 4.4).

Solution The initial x and y components of the velocity are

$$v_{x0} = v_0 \cos \theta_0 = (20 \text{ m/s})(\cos 30°) = 17.3 \text{ m/s}$$

$$v_{y0} = v_0 \sin \theta_0 = (20 \text{ m/s})(\sin 30°) = 10 \text{ m/s}$$

To find t, we can use $y = v_{y0}t - \frac{1}{2}gt^2$ (Equation 4.13) with $y = -45$ m and $v_{y0} = 10$ m/s (we have chosen the top of the building as the origin, as shown as in Figure 4.10):

$$-45 \text{ m} = (10 \text{ m/s})t - \tfrac{1}{2}(9.80 \text{ m/s}^2)t^2$$

Solving the quadratic equation for t gives, for the positive root, $t = 4.22$ s. Does the negative root have any physical meaning? (Can you think of another way of finding t from the information given?)

(b) What is the speed of the stone just before it strikes the ground?

Solution The y component of the velocity just before the stone strikes the ground can be obtained using the equation $v_y = v_{y0} - gt$ (Equation 4.11) with $t = 4.22$ s:

$$v_y = 10 \text{ m/s} - (9.80 \text{ m/s}^2)(4.22 \text{ s}) = 31.4 \text{ m/s}$$

Since $v_x = v_{x0} = 17.3$ m/s, the required speed is given by

$$v = \sqrt{v_x^2 + v_y^2} = \sqrt{(17.3)^2 + (-31.4)^2} \text{ m/s} = 35.9 \text{ m/s}$$

Exercise 1 Where does the stone strike the ground?
Answer 73 m from the base of the building.

EXAMPLE 4.5 The Stranded Explorers

An Alaskan rescue plane drops a package of emergency rations to a stranded party of explorers, as shown in Figure 4.11. If the plane is traveling horizontally at 40 m/s at a height of 100 m above the ground, where does the package strike the ground relative to the point at which it was released?

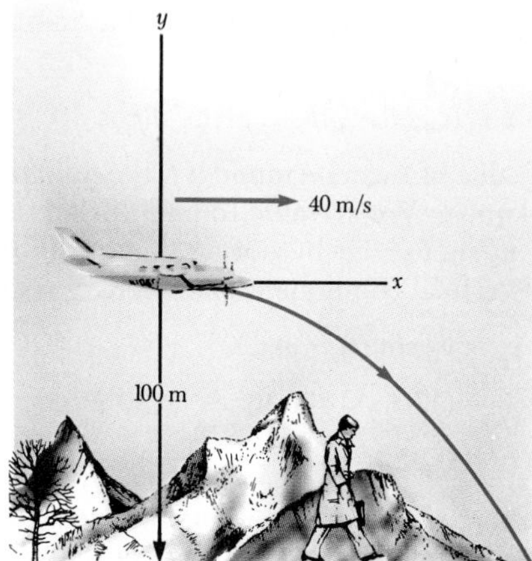

Figure 4.11 (Example 4.5) To an observer on the ground, a package released from the rescue plane travels along the path shown.

Solution The coordinate system for this problem is selected as shown in Figure 4.11, with the positive x direction to the right and the positive y direction upward.

Consider first the horizontal motion of the package. The only equation available to us is $x = v_{x0}t$

The initial x component of the package velocity is the same as the velocity of the plane when the package was released, 40 m/s. Thus, we have

$$x = (40 \text{ m/s})t$$

If we know t, the length of time the package is in the air, we can determine x, the distance traveled by the package along the horizontal. To find t, we move to the equations for the vertical motion of the package. We know that at the instant the package hits the ground its y coordinate is -100 m. We also know that the initial velocity of the package in the vertical direction, v_{y0}, is zero because the package was released with only a horizontal component of velocity. From Equation 4.13, we have

$$y = -\tfrac{1}{2}gt^2$$

$$-100 \text{ m} = -\tfrac{1}{2}(9.80 \text{ m/s}^2)t^2$$

$$t^2 = 20.4 \text{ s}^2$$

$$t = 4.51 \text{ s}$$

The value for the time of flight substituted into the equation for the x coordinate gives

$$x = (40 \text{ m/s})(4.51 \text{ s}) = 180 \text{ m}$$

Exercise 2 What are the horizontal and vertical components of the velocity of the package just before it hits the ground?
Answer $v_x = 40$ m/s; $v_y = -44.1$ m/s.

EXAMPLE 4.6 The End of the Ski Jump
A ski jumper travels down a slope and leaves the ski track moving in the horizontal direction with a speed of 25 m/s as in Figure 4.12. The landing incline below him falls off with a slope of 35°. (a) Where does he land on the incline?

Solution It is convenient to select the origin ($x = y = 0$) at the beginning of the jump. Since $v_{x0} = 25$ m/s, and $v_{y0} = 0$ in this case, Equations 4.12 and 4.13 give

$$(1) \qquad x = v_{x0}t = (25 \text{ m/s})t$$

$$(2) \qquad y = v_{y0}t - \tfrac{1}{2}gt^2 = -\tfrac{1}{2}(9.80 \text{ m/s}^2)t^2$$

Taking d to be the distance he travels along the incline before landing, then from the right triangle in Figure 4.12, we see that his x and y coordinates at the point of landing are given by $x = d \cos 35°$ and $y = -d \sin 35°$. Substituting these relations into (1) and (2) gives

$$(3) \qquad d \cos 35° = (25 \text{ m/s})t$$

$$(4) \qquad -d \sin 35° = -\tfrac{1}{2}(9.80 \text{ m/s}^2)t^2$$

Eliminating t from these equations gives $d = 109$ m. Hence, the x and y coordinates of the point at which he lands are

$$x = d \cos 35° = (109 \text{ m}) \cos 35° = 89.3 \text{ m}$$

$$y = -d \sin 35° = -(109 \text{ m}) \sin 35° = -62.5 \text{ m}$$

Exercise 3 Determine how long the ski jumper is airborne, and his vertical component of velocity just before he lands on the slope.
Answer 3.57 s; $v_y = -35.0$ m/s

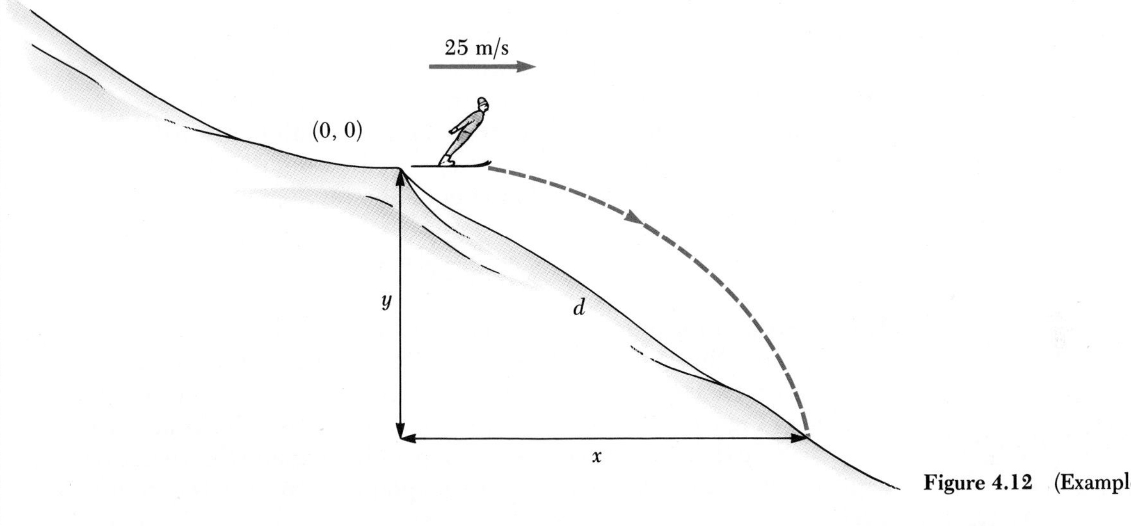

Figure 4.12 (Example 4.6).

4.4 UNIFORM CIRCULAR MOTION

Figure 4.13a shows an object moving in a circular path with *constant linear speed v*. It is often surprising to students to find that *even though the object moves at a constant speed, it still has an acceleration.* To see why this occurs, consider the defining equation for average acceleration, $\bar{\mathbf{a}} = \Delta\mathbf{v}/\Delta t$.

Note that the acceleration depends on *the change in the velocity vector.* Because velocity is a vector, there are two ways in which an acceleration can be produced: by a change in the *magnitude* of the velocity and by a change in the *direction* of the velocity. It is the latter situation that is occurring for an object moving in a circular path with constant speed. The velocity vector is always tangent to the path of the particle and in this case is perpendicular to $\mathbf{r}$.

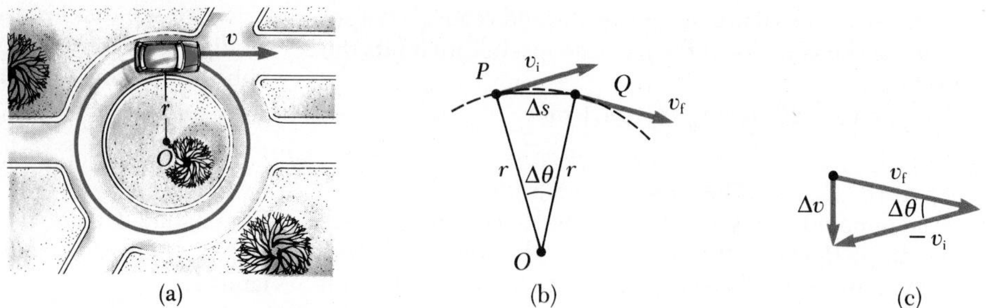

Figure 4.13 (a) Circular motion of an object moving with a constant speed. (b) As the particle moves from P to Q, the direction of its velocity vector changes from $\boldsymbol{v}_i$ to $\boldsymbol{v}_f$. (c) The construction for determining the direction of the change in velocity, $\Delta\boldsymbol{v}$, which is toward the center of the circle.

We shall show that the acceleration vector in this case is perpendicular to the path and always points toward the center of the circle. An acceleration of this nature is called a **centripetal acceleration** (center-seeking) and its magnitude is given by

Centripetal acceleration

$$a_r = \frac{v^2}{r} \tag{4.19}$$

To derive Equation 4.19, consider Figure 4.13b. Here an object is seen first at point P with velocity $\boldsymbol{v}_i$ at time t_i and then at point Q with velocity $\boldsymbol{v}_f$ at a later time t_f. Let us also assume here that $\boldsymbol{v}_i$ and $\boldsymbol{v}_f$ differ only in direction; their magnitudes are the same (that is, $v_i = v_f = v$). In order to calculate the acceleration, let us begin with the defining equation for average acceleration:

$$\overline{\boldsymbol{a}} = \frac{\boldsymbol{v}_f - \boldsymbol{v}_i}{t_f - t_i} = \frac{\Delta \boldsymbol{v}}{\Delta t}$$

This equation indicates that we must vectorially subtract $\boldsymbol{v}_i$ from $\boldsymbol{v}_f$, where $\Delta\boldsymbol{v} = \boldsymbol{v}_f - \boldsymbol{v}_i$ is the change in the velocity. That is, $\Delta\boldsymbol{v}$ is obtained by adding to $\boldsymbol{v}_f$ the vector $-\boldsymbol{v}_i$. This can be accomplished graphically as shown by the vector triangle in Figure 4.13c. Note that when Δt is very small, Δs and $\Delta\theta$ are also very small. In this case, $\boldsymbol{v}_f$ will be almost parallel to $\boldsymbol{v}_i$ and the vector $\Delta\boldsymbol{v}$ will be approximately perpendicular to them, pointing toward the center of the circle.

Now consider the triangle in Figure 4.13b, which has sides Δs and r. This triangle and the one with sides Δv and v in Figure 4.13c are similar. (Two triangles are similar if the angle between any two sides is the same for both triangles and if the ratio of lengths of these sides is the same.) This enables us to write a relationship between the lengths of the sides:

$$\frac{\Delta v}{v} = \frac{\Delta s}{r}$$

This equation can be solved for Δv and the expression so obtained can be substituted into $\overline{a} = \Delta v/\Delta t$ to give $\overline{a}\,\Delta t = v\,\Delta s/r$, or

$$\overline{a} = \frac{v}{r}\frac{\Delta s}{\Delta t}$$

Now imagine that points P and Q in Figure 4.13b become extremely close together. In this case $\Delta \boldsymbol{v}$ would point toward the center of the circular path, and because the acceleration is in the direction of $\Delta \boldsymbol{v}$, it too is toward the center. Furthermore, as the two points P and Q approach each other, Δt approaches zero, and the ratio $\Delta s/\Delta t$ approaches the velocity v. Hence, in the limit $\Delta t \rightarrow 0$, the acceleration is

$$a_r = \frac{v^2}{r}$$

Thus we conclude that in uniform circular motion, the acceleration is directed inward toward the center of the circle and has a magnitude given by v^2/r. You should show that the dimensions of a_r are [L]/[T²], as required because this is a true acceleration. We shall return to the discussion of circular motion in Section 6.1.

4.5 TANGENTIAL AND RADIAL ACCELERATION IN CURVILINEAR MOTION

Let us consider the motion of a particle along a curved path where the velocity changes both in direction and in magnitude, as described in Figure 4.14. In this situation, the velocity of the particle is always tangent to the path; however, the acceleration vector $\boldsymbol{a}$ is now at some angle to the path. As the particle moves along the curved path in Figure 4.14, we see that the direction of the total acceleration vector, $\boldsymbol{a}$, changes from point to point. This vector can be resolved into two component vectors: a radial component vector, $\boldsymbol{a}_r$, and a tangential component vector, $\boldsymbol{a}_t$. That is, the *total* acceleration vector, $\boldsymbol{a}$, can be written as the vector sum of these component vectors:

$$\boldsymbol{a} = \boldsymbol{a}_r + \boldsymbol{a}_t \qquad (4.20)$$

Total acceleration

The tangential acceleration arises from the change in the speed of the particle, and its absolute magnitude is given by

$$a_t = \frac{d|\boldsymbol{v}|}{dt} \qquad (4.21)$$

Tangential acceleration

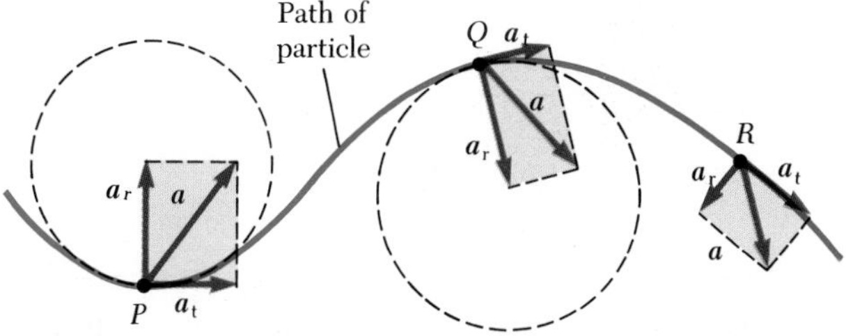

Figure 4.14 The motion of a particle along an arbitrary curved path lying in the xy plane. If the velocity vector $\boldsymbol{v}$ (always tangent to the path) changes in direction and magnitude, the component vectors of the acceleration of the particle are a tangential vector, $\boldsymbol{a}_t$, and a radial vector, $\boldsymbol{a}_r$.

The radial acceleration is due to the time rate of change in direction of the velocity vector and has an absolute magnitude given by

Centripetal acceleration

$$a_r = \frac{v^2}{r} \tag{4.22}$$

where r is the radius of curvature of the path at the point in question. Since $\boldsymbol{a}_r$ and $\boldsymbol{a}_t$ are perpendicular component vectors of $\boldsymbol{a}$, it follows that $a = \sqrt{a_r^2 + a_t^2}$. As in the case of uniform circular motion, $\boldsymbol{a}_r$ always points toward the center of curvature, as shown in Figure 4.14. Also, at a given speed, a_r is large when the radius of curvature is small (as at points P and Q in Fig. 4.14) and small when r is large (such as at point R). The direction of $\boldsymbol{a}_t$ is either in the same direction as $\boldsymbol{v}$ (if v is increasing) or opposite $\boldsymbol{v}$ (if v is decreasing).

Note that in the case of uniform circular motion, where v is constant, $a_t = 0$ and the acceleration is always radial, as we described in Section 4.4. Furthermore, if the direction of $\boldsymbol{v}$ doesn't change, then there is no radial acceleration and the motion is one-dimensional ($a_r = 0$, $a_t \neq 0$).

It is convenient to write the acceleration of a particle moving in a circular path in terms of unit vectors. We can do this by defining the unit vectors $\hat{\boldsymbol{r}}$ and $\hat{\boldsymbol{\theta}}$, where $\hat{\boldsymbol{r}}$ is *a unit vector directed radially outward along the radius vector,* from the center of curvature, and $\hat{\boldsymbol{\theta}}$ is *a unit vector tangent to the circular path,* as in Figure 4.15a. The direction of $\hat{\boldsymbol{\theta}}$ is in the direction of increasing θ, where θ is measured counterclockwise from the positive x axis. Note that both $\hat{\boldsymbol{r}}$ and $\hat{\boldsymbol{\theta}}$ "move along with the particle" and so vary in time relative to a stationary observer. Using this notation, we can express the total acceleration as

$$\boldsymbol{a} = \boldsymbol{a}_t + \boldsymbol{a}_r = \frac{d|\boldsymbol{v}|}{dt}\hat{\boldsymbol{\theta}} - \frac{v^2}{r}\hat{\boldsymbol{r}} \tag{4.23}$$

These vectors are described in Figure 4.15b. The negative sign for $\boldsymbol{a}_r$ indicates that it is always directed radially inward, *opposite* the unit vector $\hat{\boldsymbol{r}}$.

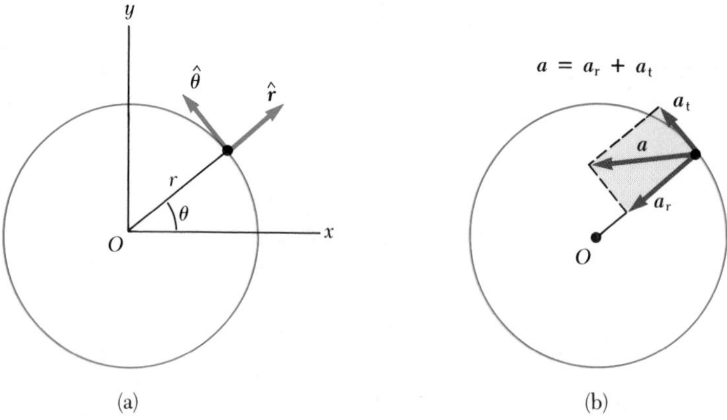

Figure 4.15 (a) Description of the unit vectors $\hat{\boldsymbol{r}}$ and $\hat{\boldsymbol{\theta}}$. (b) The total acceleration $\boldsymbol{a}$ of a particle rotating in a circle consists of a radial component vector, $\boldsymbol{a}_r$, directed toward the center of rotation, and a tangential component vector, $\boldsymbol{a}_t$. The component vector $\boldsymbol{a}_t$ is zero if the speed is constant.

EXAMPLE 4.7 The Rotating Ball

A ball tied to the end of a string 0.5 m in length swings in a vertical circle under the influence of gravity, as in Figure 4.16. When the string makes an angle of $\theta = 20°$ with the vertical, the ball has a speed of 1.5 m/s. (a) Find the radial component of acceleration at this instant.

Since $v = 1.5$ m/s and $r = 0.5$ m, we find that

$$a_r = \frac{v^2}{r} = \frac{(1.5 \text{ m/s})^2}{0.5 \text{ m}} = 4.5 \text{ m/s}^2$$

(b) When the ball is at an angle θ to the vertical, it has a tangential acceleration of magnitude $g \sin\theta$ (the component of **g** tangent to the circle). Therefore, at $\theta = 20°$, we find that $a_t = g \sin 20° = 3.36$ m/s². Find the magnitude and direction of the *total* acceleration at $\theta = 20°$.

Since $\mathbf{a} = \mathbf{a}_r + \mathbf{a}_t$, the magnitude of $\mathbf{a}$ at $\theta = 20°$ is given by

$$a = \sqrt{a_r^2 + a_t^2} = \sqrt{(4.5)^2 + (3.36)^2} \text{ m/s}^2 = 5.62 \text{ m/s}^2$$

If ϕ is the angle between $\mathbf{a}$ and the string, then

$$\phi = \tan^{-1}\frac{a_t}{a_r} = \tan^{-1}\left(\frac{3.36 \text{ m/s}^2}{4.5 \text{ m/s}^2}\right) = 36.7°$$

Note that all of the vectors—$\mathbf{a}$, $\mathbf{a}_t$, and $\mathbf{a}_r$—change in direction *and* magnitude as the ball swings through the circle. When the ball is at its lowest elevation ($\theta = 0$), $a_t = 0$, since there is no tangential component of **g** at this angle, and a_r is a *maximum*, since v is a maximum. When the ball is at its highest position ($\theta = 180°$), a_t is again zero but a_r is a minimum, since v is a minimum. Finally, in the two horizontal positions, ($\theta = 90°$ and $270°$), $|\mathbf{a}_t| = g$ and a_r is somewhere between its minimum and maximum values.

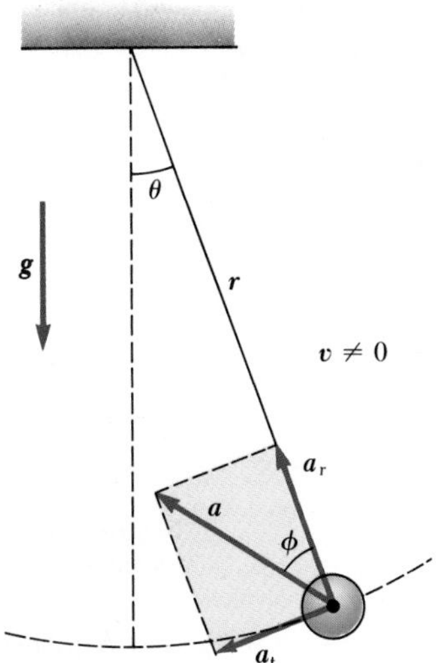

Figure 4.16 (Example 4.7) Circular motion of a ball tied on a string of length r. The ball swings in a vertical plane, and its acceleration, $\mathbf{a}$, has a radial component vector, $\mathbf{a}_r$, and a tangential component vector, $\mathbf{a}_t$.

4.6 RELATIVE VELOCITY AND RELATIVE ACCELERATION

In this section, we describe how observations made by different observers in different frames of reference are related to each other. We shall find that observers in different frames of reference may measure different displacements, velocities, and accelerations for a particle in motion. That is, two observers moving with respect to each other will generally not agree on the outcome of a measurement.

For example, if two cars are moving in the same direction with speeds of 50 mi/h and 60 mi/h, a passenger in the slower car will claim that the speed of the faster car relative to that of the slower car is 10 mi/h. Of course, a stationary observer will measure the speed of the faster car to be 60 mi/h. This simple example demonstrates that velocity measurements differ in different frames of reference.

Next, suppose a person riding on a moving vehicle (observer A) throws a ball straight up in the air according to his frame of reference, as in Figure 4.17a. According to observer A, the ball will move in a vertical path. On the other hand, a stationary observer (B) will see the path of the ball as a parabola, as illustrated in Figure 4.17b.

Figure 4.17 (a) Observer A in a moving vehicle throws a ball upward and sees a straight-line path for the ball. (b) A stationary observer B sees a parabolic path for the same ball.

Another simple example is to imagine a package being dropped from an airplane flying parallel to the earth with a constant velocity. An observer on the airplane would describe the motion of the package as a straight line toward the earth. On the other hand, an observer on the ground would view the trajectory of the package as a parabola. Relative to the ground, the package has a vertical component of velocity (resulting from the acceleration of gravity and equal to the velocity measured by the observer in the airplane) *and* a horizontal component of velocity (given to it by the airplane's motion). If the airplane continues to move horizontally with the same velocity, the package will hit the ground directly beneath the airplane (assuming that friction is neglected)!

In a more general situation, consider a particle located at the point P in Figure 4.18. Imagine that the motion of this particle is being described by two observers, one in reference frame S, fixed with respect to the earth, and another in reference frame S′, moving to the right relative to S with a constant velocity $\boldsymbol{u}$. (Relative to an observer in S′, S moves to the left with a velocity $-\boldsymbol{u}$.) The location of an observer in his own frame of reference is irrelevant in this discussion, but to be definite the observer can be placed at the origin.

We label the position of the particle with respect to the S frame with the position vector $\boldsymbol{r}$ and label its position relative to the S′ frame with the vector $\boldsymbol{r}'$, at some time t. If the origins of the two reference frames coincide at $t = 0$,

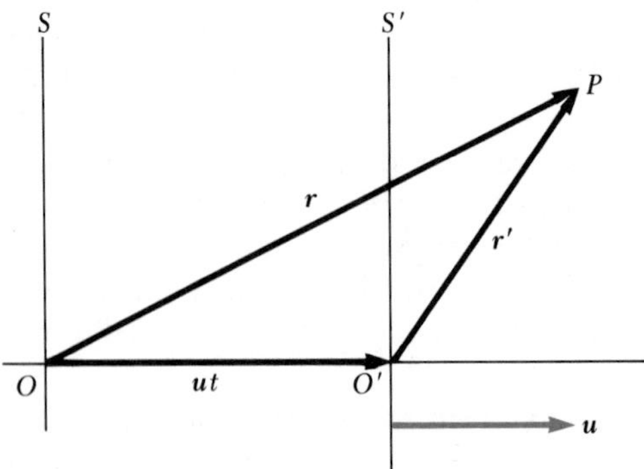

Figure 4.18 A particle located at the point P is described by two observers, one in the fixed frame of reference, S, the other in the frame S′, which moves with a constant velocity $\boldsymbol{u}$ to the right. The vector $\boldsymbol{r}$ is the particle's position vector relative to S, and $\boldsymbol{r}'$ is the position vector relative to S′.

then the vectors $\boldsymbol{r}$ and $\boldsymbol{r}'$ are related to each other through the expression $\boldsymbol{r} = \boldsymbol{r}' + \boldsymbol{u}t$, or

$$\boldsymbol{r}' = \boldsymbol{r} - \boldsymbol{u}t \tag{4.24}$$

Galilean coordinate transformation

That is, in a time t the S′ frame is displaced to the right by an amount $\boldsymbol{u}t$.

If we differentiate Equation 4.24 with respect to time and note that $\boldsymbol{u}$ is constant, we get

$$\frac{d\boldsymbol{r}'}{dt} = \frac{d\boldsymbol{r}}{dt} - \boldsymbol{u}$$

$$\boldsymbol{v}' = \boldsymbol{v} - \boldsymbol{u} \tag{4.25}$$

Galilean velocity transformation

where $\boldsymbol{v}'$ is the velocity of the particle observed in the S′ frame and $\boldsymbol{v}$ is the velocity observed in the S frame. Equations 4.24 and 4.25 are known as *Galilean transformation equations.*

Although observers in the two different reference frames will measure different velocities for the particles, they will measure the *same acceleration* when $\boldsymbol{u}$ is constant. This can be seen by taking the time derivative of Equation 4.25, which gives

$$\frac{d\boldsymbol{v}'}{dt} = \frac{d\boldsymbol{v}}{dt} - \frac{d\boldsymbol{u}}{dt}$$

But $du/dt = 0$, since $\boldsymbol{u}$ is constant. Therefore, we conclude that $\boldsymbol{a}' = \boldsymbol{a}$ since $\boldsymbol{a}' = d\boldsymbol{v}'/dt$ and $\boldsymbol{a} = d\boldsymbol{v}/dt$. That is, *the acceleration of the particle measured by an observer in the earth's frame of reference will be the same as that measured by any other observer moving with* constant velocity *with respect to the first observer.*

EXAMPLE 4.8 A Boat Crossing a River

A boat heading due north crosses a wide river with a speed of 10 km/h relative to the water. The river has a uniform speed of 5 km/h due east. Determine the velocity of the boat with respect to a stationary ground observer.

Solution The moving reference frame, S′, is attached to a cork floating on the river, and the observer is in the stationary reference frame, S (the earth). The vectors $\boldsymbol{u}$, $\boldsymbol{v}$, and $\boldsymbol{v}'$ are defined as follows:

$\boldsymbol{u}$ = velocity of the water with respect to the earth

$\boldsymbol{v}$ = velocity of the boat with respect to the earth

$\boldsymbol{v}'$ = velocity of the boat with respect to the water

In this example, $\boldsymbol{u}$ is to the right, $\boldsymbol{v}'$ is straight up, and $\boldsymbol{v}$ is at the angle θ_1, as defined in Figure 4.19a. Since these three vectors form a right triangle, the speed of the boat with respect to the earth is

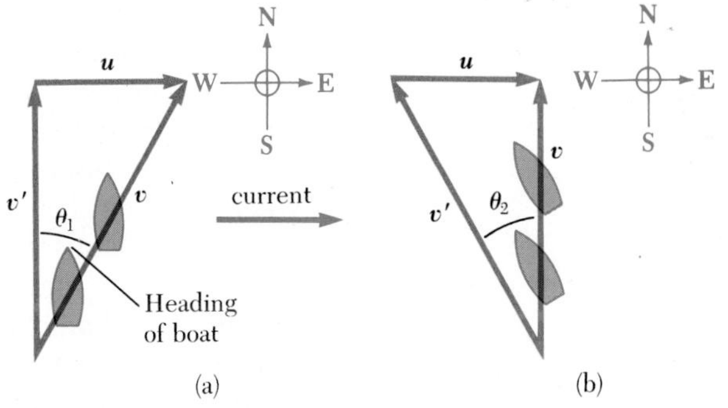

Figure 4.19 (Examples 4.8 and 4.9) (a) If the boat heads north, the motion of the boat relative to the earth is northeast along $\boldsymbol{v}$ when the river flows eastward. (b) If the boat is to travel north, it must have a northwest heading as shown. In both cases, $\boldsymbol{v} = \boldsymbol{v}' + \boldsymbol{u}$ and the heading of the boat is parallel to $\boldsymbol{v}'$.

$$v = \sqrt{(v')^2 + u^2} = \sqrt{(10)^2 + (5)^2} \text{ km/h} = \boxed{11.2 \text{ km/h}}$$

and the direction of $\boldsymbol{v}$ is

$$\theta_1 = \tan^{-1}\left(\frac{u}{v'}\right) = \tan^{-1}\left(\frac{5}{10}\right) = 26.6°$$

Therefore, the boat will be traveling 63.4° north of east with respect to the earth.

EXAMPLE 4.9 Which Way Should We Head?
If the boat in Example 4.8 travels with the same speed of 10 km/h relative to the water and is to travel due north, as in Figure 4.19b, what should be its heading?

Solution Intuitively, we know that the boat must head upstream. For this example, the vectors $\boldsymbol{u}$, $\boldsymbol{v}$, and $\boldsymbol{v}'$ are oriented as shown in Figure 4.19b, where $\boldsymbol{v}'$ is now the hypotenuse of the right triangle. Therefore, the boat's speed relative to the earth is

$$v = \sqrt{(v')^2 - u^2} = \sqrt{(10)^2 - (5)^2}\ \text{km/h} = 8.66\ \text{km/h}$$

$$\theta_2 = \tan^{-1}\left(\frac{u}{v}\right) = \tan^{-1}\left(\frac{5}{8.66}\right) = 30°$$

where θ_2 is west of north.

EXAMPLE 4.10 Airplane Flying in a Crosswind
An airplane is headed due east at an airspeed of 400 km/h. At the same time, a wind blows northward with a speed of 75 km/h with respect to the earth. Find the magnitude and direction of the airplane's velocity relative to the earth.

Solution The airplane's velocity, $\boldsymbol{v}_{pe}$, relative to the earth is the vector sum of the two velocities given in the problem, and as indicated in the vector diagram in Figure 4.20. Since the airplane's velocity relative to the air is $\boldsymbol{v}_{pa} = (400\ \text{km/h})\boldsymbol{i}$, while the wind's velocity relative to the earth is $\boldsymbol{v}_{ae} = (75\ \text{km/h})\boldsymbol{j}$, we find

$$|\boldsymbol{v}_{pe}| = \sqrt{(400\ \text{km/h})^2 + (75\ \text{km/h})^2} = 407\ \text{km/h}$$

The angle θ that $\boldsymbol{v}_{pe}$ makes with the horizontal is

$$\tan\theta = \frac{75\ \text{km/h}}{400\ \text{km/h}} = 0.1875 \qquad \text{or} \qquad \theta = 10.6°$$

Hence, the airplane's velocity relative to the earth is 407 km/h in the direction 10.6° north of east.

Exercise 4 If the airplane moves at the same airspeed of 400 km/h, in what direction must it head in order to move due east relative to the earth?
Answer 10.8° south of east.

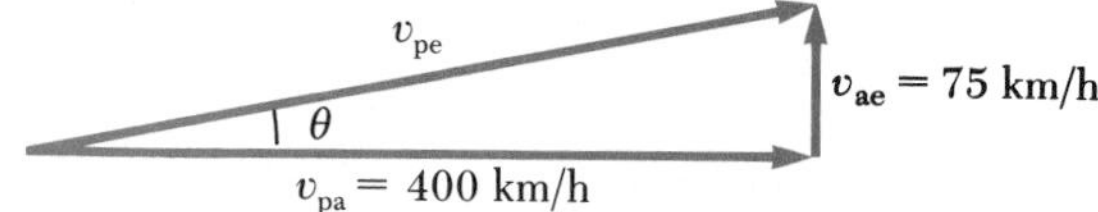

Figure 4.20 (Example 4.10).

*4.7 RELATIVE MOTION AT HIGH SPEEDS

As we mentioned in the previous section, the Galilean transformation equations given by Equations 4.24 and 4.25 relate the coordinates and velocity of a particle as measured in the earth's reference frame to those measured in a frame moving with uniform motion with respect to the earth. However, it is important to note that these transformations are valid *only* at particle speeds (relative to both observers) that are small compared with the speed of light, c (where $c \approx 3 \times 10^8$ m/s). When the particle speed according to either observer approaches the speed of light, these transformation equations must be replaced by the equations used by Einstein in his special theory of relativity. Although we shall discuss the theory of relativity in Chapter 40, a forecast of some of its predictions is in order. As it turns out, the relativistic transformation equations reduce to the Galilean transformation equations when the particle speed is small compared with the speed of light. This is in keeping with the **correspondence principle** first proposed by Niels Bohr, which in effect states that if an old theory accurately describes a number of physical phenomena, then any new theory *must* account for the same phenomena over the range of validity of the old theory.

You might wonder how one can test the validity of the transformation equations. The Galilean transformation equations are used in Newtonian mechanics, while Einstein's theory uses relativistic transformations. From experiments on high-speed particles such as electrons and protons in particle accelerators, one finds that Newtonian mechanics *fails* at particle speeds approaching the speed of light. On the other hand, Einstein's theory of special relativity is in agreement with experiment at *all* speeds. Finally, Newtonian mechanics places no upper limit on speed of a particle. In contrast, the relativistic velocity transformation equations predict particle speeds that *can never exceed the speed of light.* Electrons and protons accelerated through very high voltages can acquire speeds close to the speed of light, but never reach this value. Hence, experimental results are in complete agreement with the theory of relativity.

SUMMARY

If a particle moves with *constant* acceleration $\boldsymbol{a}$ and has a velocity $\boldsymbol{v}_0$ and position $\boldsymbol{r}_0$ at $t = 0$, its velocity and position at some later time t are given by

Velocity vector as a function of time

$$\boldsymbol{v} = \boldsymbol{v}_0 + \boldsymbol{a}t \tag{4.8}$$

Position vector as a function of time

$$\boldsymbol{r} = \boldsymbol{r}_0 + \boldsymbol{v}_0 t + \tfrac{1}{2}\boldsymbol{a}t^2 \tag{4.9}$$

For two-dimensional motion in the xy plane under constant acceleration, these vector expressions are equivalent to two component expressions, one for the motion along x with an acceleration $\boldsymbol{a}_x$ and one for the motion along y with an acceleration $\boldsymbol{a}_y$.

Projectile motion is two-dimensional motion under constant acceleration, where $a_x = 0$ and $a_y = -g$. In this case, if $x_0 = y_0 = 0$, the components of Equations 4.8 and 4.9 reduce to

Projectile motion equations

$$v_x = v_{x0} = \text{constant} \tag{4.10}$$

$$v_y = v_{y0} - gt \tag{4.11}$$

$$x = v_{x0}t \tag{4.12}$$

$$y = v_{y0}t - \tfrac{1}{2}gt^2 \tag{4.13}$$

where $v_{x0} = v_0 \cos \theta_0$, $v_{y0} = v_0 \sin \theta_0$ is the initial speed of the projectile, and θ_0 is the angle v_0 makes with the positive x axis. Note that these expressions give the velocity components (and hence the velocity vector) and the coordinates (and hence the position vector) at *any* time t that the projectile is in motion.

As you can see from Equations 4.10 through 4.13, it is useful to think of projectile motion as the superposition of two motions: (1) uniform motion in the x direction, where v_x remains constant, and (2) motion in the vertical direction, subject to a constant downward acceleration of magnitude $g = 9.80\ \text{m/s}^2$. Hence, one can analyze the motion in terms of separate horizontal and vertical components of velocity, as in Figure 4.21.

A particle moving in a circle of radius r with constant speed v undergoes a centripetal (or radial) acceleration, $\boldsymbol{a}_r$, because the direction of $\boldsymbol{v}$ changes in time. The magnitude of $\boldsymbol{a}_r$ is given by

Centripetal acceleration

$$a_r = \frac{v^2}{r} \tag{4.19}$$

and its direction is always toward the center of the circle.

If a particle moves along a curved path in such a way that the magnitude and direction of $\boldsymbol{v}$ change in time, the particle has an acceleration vector that can be described by two component vectors: (1) a radial component vector, $\boldsymbol{a}_r$, arising from the change in direction of $\boldsymbol{v}$, and (2) a tangential component vector, $\boldsymbol{a}_t$, arising from the change in magnitude of $\boldsymbol{v}$. The magnitude of $\boldsymbol{a}_r$ is v^2/r, and the magnitude of $\boldsymbol{a}_t$ is $d|\boldsymbol{v}|/dt$.

The velocity of a particle, $\boldsymbol{v}$, measured in a fixed frame of reference, S, is related to the velocity of the same particle, $\boldsymbol{v}'$, measured in a moving frame of reference, S′, by

Galilean velocity transformation

$$\boldsymbol{v}' = \boldsymbol{v} - \boldsymbol{u} \tag{4.25}$$

where $\boldsymbol{u}$ is the velocity of S′ relative to S.

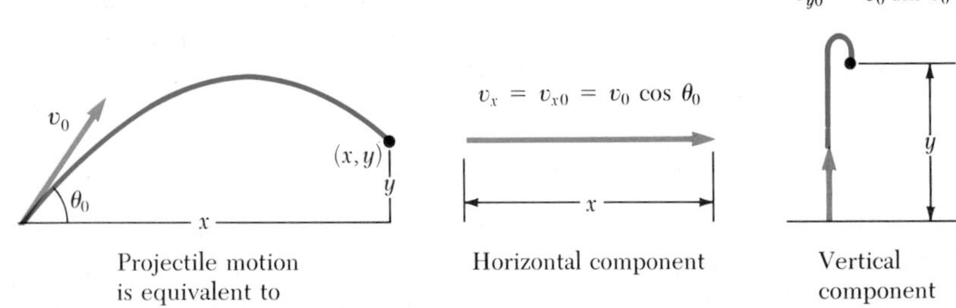

Figure 4.21 Analyzing motion in terms of the horizontal and vertical components of velocity.

QUESTIONS

1. If the average velocity of a particle is zero in some time interval, what can you say about the displacement of the particle for that interval?
2. If you know the position vectors of a particle at two points along its path and also know the time it took to get from one point to the other, can you determine the particle's instantaneous velocity? its average velocity? Explain.
3. Describe a situation in which the velocity of a particle is perpendicular to the position vector.
4. Can a particle accelerate if its speed is constant? Can it accelerate if its velocity is constant? Explain.
5. Explain whether or not the following particles have an acceleration: (a) a particle moving in a straight line with constant speed and (b) a particle moving around a curve with constant speed.
6. Correct the following statement: "The racing car rounds the turn at a constant velocity of 90 miles per hour."
7. Determine which of the following moving objects would exhibit an approximate parabolic trajectory: (a) a ball thrown in an arbitrary direction, (b) a jet airplane, (c) a rocket leaving the launching pad, (d) a rocket a few minutes after launch with failed engines, (e) a tossed stone moving to the bottom of a pond.
8. A student argues that as a satellite orbits the earth in a circular path, it moves with a constant velocity and therefore has no acceleration. The professor claims that the student is wrong since the satellite must have a centripetal acceleration as it moves in its circular orbit. What is wrong with the student's argument?
9. What is the fundamental difference between the unit vectors $\hat{\boldsymbol{r}}$ and $\hat{\boldsymbol{\theta}}$ defined in Figure 4.15 and the unit vectors $\boldsymbol{i}$ and $\boldsymbol{j}$?
10. At the end of its arc, the velocity of a pendulum is zero. Is its acceleration also zero at this point?
11. If a rock is dropped from the top of a sailboat's mast, will it hit the deck at the same point whether the boat is at rest or in motion at constant velocity?
12. A stone is thrown upward from the top of the building. Does the stone's displacement depend on the location of the origin of the coordinate system? Does the stone's velocity depend on the location of the origin?
13. Inspect the multiple image photograph (Fig. 4.22) of two golf balls released simultaneously under the conditions indicated. Explain why both balls hit the floor simultaneously.
14. Is it possible for a vehicle to travel around a curve without accelerating? Explain.

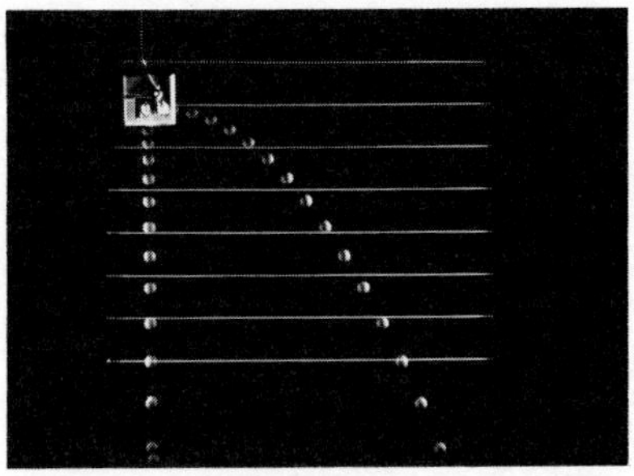

Figure 4.22 This multiple image photograph of two golf balls released simultaneously illustrates both free fall and projectile motion. The right ball was projected horizontally with an initial velocity of 2.0 m/s. The light flashes were 1/30 s apart, and the white parallel lines (actually strings) were placed $15\frac{1}{4}$ cm apart.

15. A baseball is thrown with an initial velocity of $(10\boldsymbol{i} + 15\boldsymbol{j})$ m/s. When it reaches the top of its trajectory, what is (a) its velocity and (b) its acceleration. Neglect air resistance.
16. An object moves in a circular path with constant speed v. (a) Is the velocity of the object constant? (b) Is its acceleration constant? Explain.
17. As a projectile moves in its parabolic path, is there any point along its path where the velocity and acceleration are (a) perpendicular to each other? (b) parallel to each other?
18. A projectile is fired at some angle to the horizontal with some initial speed v_0, and air resistance is neglected. Is the projectile a freely falling body? What is its acceleration in the vertical direction? What is its acceleration in the horizontal direction?
19. A projectile is fired at an angle of 30° with the horizontal with some initial speed. If a second projectile is fired with the same initial speed, what other projectile angle would give the same range? Neglect air resistance.
20. A projectile is fired on the earth with some initial velocity. Another projectile is fired on the moon with the *same* initial velocity. Neglecting air resistance, which projectile has the greater range? Which reaches the greater altitude? (Note that the acceleration due to gravity on the moon is about 1.6 m/s^2.)
21. As a projectile moves through its parabolic trajectory, which of the quantities, if any, remain constant? (a) speed, (b) acceleration, (c) horizontal component of velocity, (d) vertical component of velocity.
22. The maximum range of a projectile occurs when it is launched at an angle of 45° with the horizontal if air resistance is neglected. If air resistance is not neglected, will this optimum angle be greater or less than 45°? Explain.
23. A ball is tossed upwards in the air by a passenger on a train moving with a constant velocity. Describe the path of the ball as seen by the passenger. Describe its path as seen by a stationary observer outside the train. How would these observations change if the train were accelerating along the track?
24. A person drops a spoon on a train moving with constant velocity. What is the acceleration of the spoon relative to (a) the train, and (b) the earth?

PROBLEMS

Section 4.1 The Displacement, Velocity, Acceleration Vectors

1. Suppose that the trajectory of a particle is given by $\boldsymbol{r}(t) = x(t)\boldsymbol{i} + y(t)\boldsymbol{j}$ with $x(t) = at^2 + bt$ and $y(t) = ct + d$, where a, b, c, and d are constants that have appropriate dimensions. What displacement does the particle undergo between $t = 1$ s and $t = 3$ s?
2. Suppose that the position vector function for a particle is given as $\boldsymbol{r}(t) = x(t)\boldsymbol{i} + y(t)\boldsymbol{j}$, with $x(t) = at + b$ and $y(t) = ct^2 + d$, where $a = 1$ m/s, $b = 1$ m, $c = 1/8$ m/s^2, and $d = 1$ m. (a) Calculate the average velocity during the time interval from $t = 2$ s to $t = 4$ s. (b) Determine the velocity and the speed at $t = 2$ s.
3. Find the magnitude and direction of the average velocity vector of the tip of a 5.0-cm-long minute hand as the time changes from 4:15 to 4:30.
4. A cockroach on an apartment table is crawling with constant acceleration given by $\boldsymbol{a} = (0.3\boldsymbol{i} - 0.2\boldsymbol{j})$ cm/s^2. It starts from a point $(-4, 2)$ cm at $t = 0$ with velocity $\boldsymbol{v}_0 = 1.0\boldsymbol{j}$ cm/s. (a) What are the components of its velocity and position vectors at any time t? (b) What are the magnitude and direction of the velocity and position vectors at $t = 10.0$ s?

Section 4.2 Motion in Two Dimensions with Constant Acceleration

5. At $t = 0$, a particle moving in the xy plane with constant acceleration has a velocity of $\boldsymbol{v}_0 = (3\boldsymbol{i} - 2\boldsymbol{j})$ m/s at the origin. At $t = 3$ s, its velocity is $\boldsymbol{v} = (9\boldsymbol{i} + 7\boldsymbol{j})$ m/s. Find (a) the acceleration of the particle and (b) its coordinates at any time t.
6. A particle starts from rest at $t = 0$ at the origin and moves in the xy plane with a constant acceleration of $\boldsymbol{a} = (2\boldsymbol{i} + 4\boldsymbol{j})$ m/s^2. After a time t has elapsed, determine (a) the x and y components of velocity, (b) the coordinates of the particle, and (c) the speed of the particle.
7. A fish swimming in a horizontal plane has velocity $\boldsymbol{v}_0 = (4\boldsymbol{i} + \boldsymbol{j})$ m/s at a point in the ocean whose distance from a certain rock is $\boldsymbol{r}_0 = (10\boldsymbol{i} - 4\boldsymbol{j})$ m. After swimming with constant acceleration for 20.0 s, its velocity is $\boldsymbol{v} = (20\boldsymbol{i} - 5\boldsymbol{j})$ m/s. (a) What are the com-

ponents of the acceleration? (b) What is the direction of the acceleration with respect to unit vector $\boldsymbol{i}$? (c) Where is the fish at $t = 25$ s and in what direction is it moving?

8. The vector position of a particle varies in time according to the expression $\boldsymbol{r} = (3\boldsymbol{i} - 6t^2\boldsymbol{j})$ m. (a) Find expressions for the velocity and acceleration as functions of time. (b) Determine the particle's position and velocity at $t = 1$ s.

9. A particle initially located at the origin has an acceleration of $\boldsymbol{a} = 3\boldsymbol{j}$ m/s^2 and an initial velocity of $\boldsymbol{v}_0 = 5\boldsymbol{i}$ m/s. Find (a) the vector position and velocity at any time t and (b) the coordinates and speed of the particle at $t = 2$ s.

Section 4.3 Projectile Motion

(Neglect air resistance in all problems.)

10. A student stands at the edge of a cliff and throws a stone horizontally over the edge with a speed of 18 m/s. The cliff is 50 m above a flat horizontal beach, as shown in Figure 4.23. How long after being released does the stone strike the beach below the cliff? With what speed and angle of impact does it land?

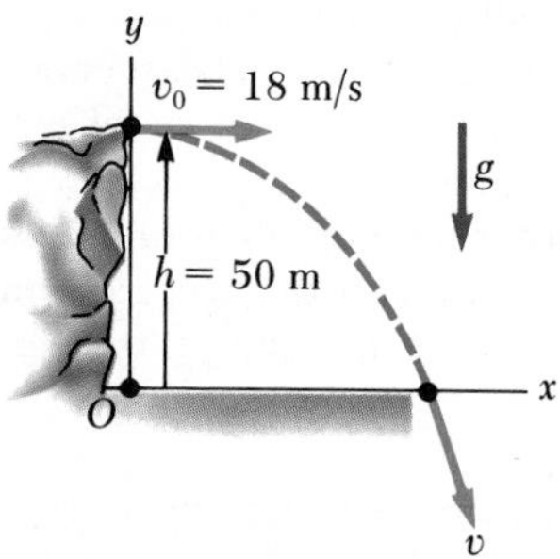

Figure 4.23 (Problem 10).

11. In a local bar, a customer slides an empty beer mug on the counter for a refill. The bartender is momentarily distracted and does not see the mug, which slides off the counter and strikes the floor 1.4 m from the base of the counter. If the height of the counter is 0.86 m, (a) with what velocity did the mug leave the counter, and (b) what was the direction of the mug's velocity just before it hit the floor?

12. A student decides to measure the muzzle velocity of the pellets from his BB gun. He points the gun horizontally. On a vertical wall a distance x away from the gun, a target is placed. The shots hit the target a vertical distance y below the gun. (a) Show that the position of the pellet when traveling through the air is given by $y = Ax^2$, where A is a constant. (b) Express the constant A in terms of the initial velocity and the acceleration due to gravity. (c) If $x = 3.0$ m and $y = 0.21$ m, what is the speed of the BB?

13. A projectile is fired from level ground with velocity $\boldsymbol{v} = (12.0\boldsymbol{i} + 24.0\boldsymbol{j})$ m/s. (a) What are the vertical and horizontal velocity components after 4 s? (b) What are the coordinates of the point at which the height is maximum?

14. A football, kicked at an angle of 50° to the horizontal, travels a horizontal distance of 20 m before hitting the ground. Find (a) the initial speed of the football, (b) the time it is in the air, and (c) the maximum height it reaches.

15. A ball is thrown horizontally from the top of a building 35 m high. The ball strikes the ground at a point 80 m from the base of the building. Find (a) the time the ball is in flight, (b) its initial velocity, and (c) the x and y components of velocity just before the ball strikes the ground.

16. An athlete releases a shot from his outstretched arm 2.3 m above the ground at an angle of 60° with the horizontal. The shot hit the ground 20.5 m away, 0.60 m short of the state record. (a) What are the velocity components when it hits the ground? (b) What would be the range if he threw it at 45° from a height of 2.2 m? (Assume the initial speed is unchanged.)

17. A place kicker must kick a football from a point 36 m (about 40 yards) from the goal and the ball must clear the crossbar, which is 3.05 m high. When kicked, the ball leaves the ground with a speed of 20.0 m/s at an angle of 53° to the horizontal. (a) By how much does the ball clear or fall short of clearing the crossbar? (b) Does the ball approach the crossbar while still rising or while falling?

18. Show that the horizontal range of a projectile with a fixed initial speed will be the same for any two complementary angles, such as 30° and 60°.

19. It has been said that in his youth George Washington threw a silver dollar across a river. Assuming that the river was 75 m wide, (a) what *minimum initial* speed was necessary to get the coin across the river and (b) how long was the coin in flight?

20. A rifle is aimed horizontally through its bore at the center of a large target 200 m away. The initial velocity of the bullet is 500 m/s. (a) Where does the bullet strike the target? (b) To hit the center of the target, the barrel must be at an angle above the line of sight. Find the angle of elevation of the barrel.

21. During World War I, the Germans had a gun called Big Bertha that was used to shell Paris. The shell had an initial speed of 1700 m/s (approximately 5 times the speed of sound) at an initial inclination of 55° to the horizontal. In order to hit the target, adjustments were made for air resistance and other effects. If we ignore those effects, (a) how far away did the shell hit? (b) How long was it in the air?

22. The maximum horizontal distance a certain baseball player is able to hit the ball is 150 m. On one pitch,

this player hits the ball in such a way that it has the same initial speed as his maximum-distance hit, but makes an angle of 20° with the horizontal. Where will this ball strike the ground with respect to home plate?

23. A projectile is fired in such a way that its horizontal range is equal to three times its maximum height. What is the angle of projection?

24. A flea can jump a vertical height h. (a) What is the maximum horizontal distance it can jump? (b) What is the time in the air in both cases?

25. An astronaut on a strange planet finds that she can jump a *maximum* horizontal distance of 30 m if her initial speed is 9 m/s. What is the acceleration of gravity on the planet?

26. The projectile fired from a pump air gun has an initial speed of 30.0 m/s at an angle of 30° above the horizontal. If the air pressure is reduced to produce a projectile speed of 27.0 m/s, at what angle must the gun be fired to achieve the same range? Assume that it returns to its initial height in both cases.

Section 4.4 Uniform Circular Motion

27. Find the acceleration of a particle moving with a constant speed of 8 m/s in a circle 2 m in radius.

28. Young David who slew Goliath experimented with slings before tackling the giant. He found that with a sling of length 0.6 m, he could revolve the sling at the rate of 8 rev/s. If he increased the length to 0.9 m, he could revolve the sling only 6 times per second. (a) Which rate of rotation gives the larger linear speed? (b) What is the centripetal acceleration at 8 rev/s? (c) What is the centripetal acceleration at 6 rev/s?

29. A hunter uses a stone attached to the end of a rope as a crude weapon. The stone is swung overhead in a horizontal circle 1.6 m in diameter at the rate of 3 revolutions per second. What is the centripetal acceleration of the stone?

30. From information in the front cover of this book, compute the radial acceleration of a point on the surface of the earth at the equator.

31. The orbit of the moon about the earth is approximately circular, with a mean radius of 3.84×10^8 m. It takes 27.3 days for the moon to complete one revolution about the earth. Find (a) the mean orbital speed of the moon and (b) its centripetal acceleration.

32. In the spin cycle of a washing machine, the tub of radius 0.30 m develops a speed of 630 rpm. What is the maximum linear speed with which water leaves the machine?

33. A particle moves in a circular path 0.4 m in radius with constant speed. If the particle makes five revolutions in each second of its motion, find (a) the speed of the particle and (b) its acceleration.

34. A tire 0.5 m in radius rotates at a constant rate of 200 revolutions per minute. Find the speed and acceleration of a small stone lodged in the tread of the tire (on its outer edge).

Section 4.5 Tangential and Radial Acceleration in Curvilinear Motion

35. Figure 4.24 represents the total acceleration of a particle moving clockwise in a circle of radius 2.5 m at a given instant of time. At this instant of time, find (a) the centripetal acceleration, (b) the speed of the particle, and (c) its tangential acceleration.

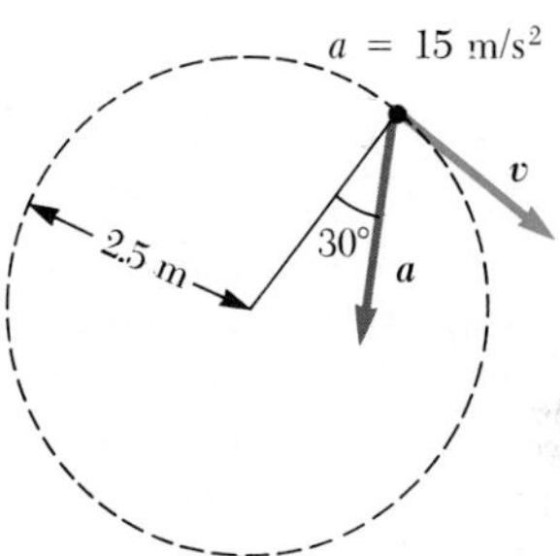

Figure 4.24 (Problem 35).

36. The speed of a particle moving in a circle 2 m in radius increases at the constant rate of 3 m/s². At some instant, the magnitude of the total acceleration is 5 m/s². At this instant, find (a) the centripetal acceleration of the particle and (b) its speed.

37. A train slows down as it rounds a sharp horizontal turn, slowing from 90 km/h to 50 km/h in the 15 s that it takes to round the bend. The radius of the curve is 150 m. Compute the acceleration at the moment the train speed reaches 50 km/h.

38. A pendulum of length 1 m swings in a vertical plane (Figure 4.16). When the pendulum is in the two horizontal positions ($\theta = 90°$ and $\theta = 270°$), its speed is 5 m/s. (a) Find the magnitude of the centripetal acceleration and tangential acceleration for these positions. (b) Draw vector diagrams to determine the direction of the total acceleration for these two positions. (c) Calculate the magnitude and direction of the total acceleration.

39. A student swings a ball attached to the end of a string 0.6 m in length in a vertical circle. The speed of the ball is 4.3 m/s at its highest point and 6.5 m/s at its lowest point. Find the acceleration of the ball at (a) its highest point and (b) its lowest point.

40. At some instant of time, a particle moving counterclockwise in a circle of radius 2 m has a speed of 8 m/s and its total acceleration is directed as shown in Figure 4.25. At this instant, determine (a) the centripetal

acceleration of the particle, (b) the tangential acceleration, and (c) the magnitude of the total acceleration.

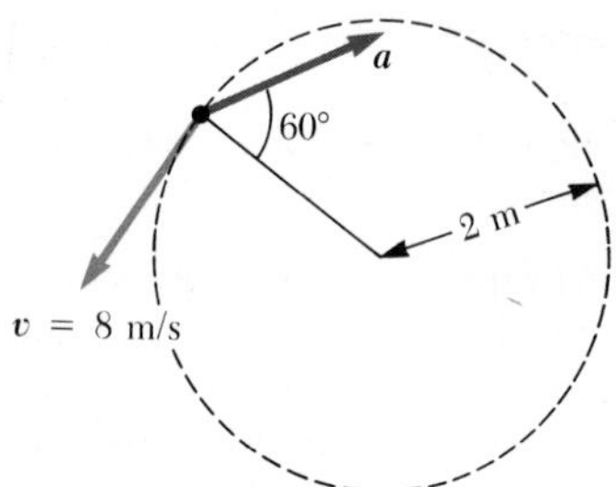

Figure 4.25 (Problem 40).

Section 4.6 Relative Velocity and Relative Acceleration

41. A car travels north with a speed of 60 km/h on a straight highway. A truck travels in the opposite direction with a speed of 50 km/h. (a) What is the velocity of the car relative to the truck? (b) What is the velocity of the truck relative to the car?

42. A motorist traveling west on Interstate Route 80 at 80 km/h is being chased by a police car traveling at 95 km/h. (a) What is the velocity of the motorist relative to the police car? (b) What is the velocity of the police car relative to the motorist?

43. A river has a steady speed of 0.5 m/s. A student swims upstream a distance of 1 km and returns to the starting point. If the student can swim at a speed of 1.2 m/s in still water, how long does the trip take? Compare this with the time the trip would take if the water were still.

44. Two canoeists in identical canoes exert the same effort paddling in a river. One paddles directly upstream (and moves upstream), whereas the other paddles directly downstream. An observer on the riverbank reckons their speeds to be 1.2 m/s and 2.9 m/s. How fast is the river flowing?

45. A shopper in a department store can walk up a stationary (stalled) escalator in 30 s. If the normally functioning escalator can carry the standing shopper to the next floor in 20 s, how long would it take the shopper to walk up the moving escalator? Assume the same walking effort for the shopper whether the escalator is stalled or moving.

46. A boat crosses a river with a width $w = 160$ m in which the current flows with a uniform speed of 1.5 m/s. The steersman maintains a bearing (i.e., the direction in which his boat points) perpendicular to the river and a throttle setting to give a constant speed of 2 m/s with respect to the water. (a) What is the velocity of the boat relative to a stationary shore observer? (b) How far downstream from the initial position is the boat when it reaches the opposite shore?

47. The pilot of an airplane notes that the compass indicates a heading due west. The airplane's speed relative to the air is 150 km/h. If there is a wind of 30 km/h toward the north, find the velocity of the airplane relative to the ground.

48. The pilot of an aircraft wishes to fly due west in a wind blowing at 50 km/h toward the south. If the speed of the aircraft in the absence of a wind is 200 km/h, (a) in what direction should the aircraft head and (b) what should its speed be relative to the ground?

49. A car travels due east with a speed of 50 km/h. Rain is falling vertically with respect to the earth. The traces of the rain on the side windows of the car make an angle of 60° with the vertical. Find the velocity of the rain with respect to (a) the car and (b) the earth.

50. A child in danger of drowning in a river is being carried downstream by a current that flows uniformly with a speed of 2.5 km/h. The child is 0.6 km from shore and 0.8 km upstream of a boat landing when a rescue boat sets out. (a) If the boat proceeds at its maximum speed of 20 km/h with respect to the water, what heading relative to the shore should the boatman take? (b) What angle does the boat velocity $\boldsymbol{v}$ make with the shore? (c) How long will it take the boat to reach the child?

51. A bolt drops from the ceiling of a train car which is accelerating northward at a rate of 2.5 m/s^2. What is the acceleration of the bolt with respect to (a) the train car? (b) the stationary train station?

52. A science student is riding on a flatcar of a train traveling along a straight horizontal track at a constant speed of 10 m/s. The student throws a ball into the air along a path that he judges to make an initial angle of 60° with the horizontal and to be in line with the track. The student's professor, who is standing on the ground nearby, observes the ball to rise vertically. How high does the ball rise?

ADDITIONAL PROBLEMS

53. At $t = 0$ a particle leaves the origin with a velocity of 6 m/s in the positive y direction. Its acceleration is given by $\boldsymbol{a} = (2\boldsymbol{i} - 3\boldsymbol{j})$ m/s^2. When the particle reaches its *maximum* y coordinate, its y component of velocity is zero. At this instant, find (a) the velocity of the particle and (b) its x and y coordinates.

54. A boy throws a ball into the air as hard as he can and then runs as fast as he can under the ball in order to catch it. If his maximum speed in throwing the ball is 20 m/s and his best time for a 20-m dash is 3 s, how high does the ball rise?

55. A car is parked on a steep incline overlooking the ocean, where the incline makes an angle of 37° with the horizontal. The negligent driver leaves the car in neutral, and the parking brakes are defective. The car rolls from rest down the incline with a constant accel-

eration of 4 m/s^2 and travels 50 m to the edge of the cliff. The cliff is 30 m above the ocean. Find (a) the speed of the car when it reaches the cliff and the time it takes to get there, (b) the velocity of the car when it lands in the ocean, (c) the total time the car is in motion, and (d) the position of the car relative to the base of the cliff when the car lands in the ocean.

56. A cannon is fired at an angle of 30° above the horizontal from a cliff that is 20 m above a flat river bottom. What is the initial speed of the projectile if it is found to land 40 m from the base of the cliff?

57. A batter hits a pitched baseball 1 m above the ground, imparting to the ball a speed of 40 m/s. The resulting line drive is caught on the fly by the left fielder 60 m from home plate with his glove 1 m above the ground. If the shortstop, 45 m from home plate and in line with the drive, were to jump straight up to make the catch, instead of allowing the left fielder to make the play, how high above the ground would his glove have to be?

58. The initial speed of a cannonball is 200 m/s. If it is fired at a target that is at a horizontal distance of 2 km from the cannon, find (a) the two projected angles that will result in a hit and (b) the total time of flight for each of the two trajectories found in (a).

59. A cannon fires a shell with initial speed v_0 at angle of elevation θ_1. When the angle is raised to θ_2, the range of the shell is increased by a distance b. Show that

$$b = \frac{v_0^2}{g}(\sin 2\theta_2 - \sin 2\theta_1)$$

60. What would be the length of the second hand of a clock if the centripetal acceleration of its tip is $0.01g$?

61. A stone at the end of a sling is whirled in a vertical circle of radius 1.2 m at a constant speed $v_0 = 1.5$ m/s as in Figure 4.26. The center of the string is 1.5 m from the ground. What is the range of the stone if it is released when the sling is inclined at 30° with the horizontal (a) at A? (b) at B? What is the acceleration of the stone (c) just before it is released at A? (d) just after it is released at A?

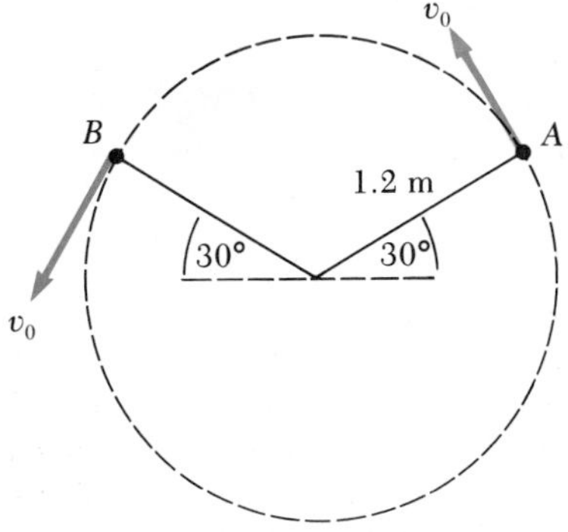

Figure 4.26 (Problem 61).

62. A truck is moving due north with a constant velocity of 10 m/s on a horizontal stretch of road. A boy riding on the back of the truck wishes to throw a ball while the truck is moving and to catch the ball after the truck has gone 20 m. (a) Neglecting air resistance, at what angle to the vertical should the ball be thrown? (b) What should be the initial speed of the ball? (c) What is the shape of the path of the ball as seen by the boy? (d) An observer on the ground watches the boy throw the ball up and catch it. In this observer's fixed frame of reference, determine the general shape of the ball's path and the initial velocity of the ball.

63. A dart gun is fired while being held horizontally at a height of 1 m above ground level. With the gun at rest relative to the ground, the dart from the gun travels a horizontal distance of 5 m. A child holds the same gun in a horizontal position while sliding down a 45° incline at a constant speed of 2 m/s. How far will the dart travel if the gun is fired when it is 1 m above the ground?

64. A rocket is launched at an angle of 53° to the horizontal with an initial speed of 100 m/s. It moves along its initial line of motion with an acceleration of 30 m/s^2 for 3 s. At this time its engines fail and the rocket proceeds to move as a free body. Find (a) the maximum altitude reached by the rocket, (b) its total time of flight, and (c) its horizontal range.

65. A boat requires 2 min to cross a river that is 150 m wide. The boat's speed relative to the water is 3 m/s and the river current flows at a speed of 2 m/s. At what possible upstream or downstream points does the boat reach the opposite shore?

66. A home run in a baseball game is hit in such a way that the ball just clears a wall 21 m high, located 130 m from home plate. The ball is hit at an angle of 35° to the horizontal, and air resistance is negligible. Find (a) the initial speed of the ball, (b) the time it takes the ball to reach the wall, and (c) the velocity components and the speed of the ball when it reaches the wall. (Assume the ball is hit at a height of 1 m above the ground.)

67. A daredevil is shot out of a cannon at 45° to the horizontal with an initial speed of 25 m/s. A net is located at a horizontal distance of 50 m from the cannon. At what height above the cannon should the net be placed in order to catch the daredevil?

68. The position of a particle moving in the xy plane varies in time according to the equation $\mathbf{r} = 3\cos 2t\mathbf{i} + 3\sin 2t\mathbf{j}$, where $\mathbf{r}$ is in m and t is in s. (a) Show that the path of the particle is a circle 3 m in radius centered at the origin. (*Hint:* Let $\theta = 2t$.) (b) Calculate the velocity and acceleration vectors. (c) Show that the acceleration vector always points toward the origin (opposite $\mathbf{r}$) and has a magnitude of v^2/r.

69. A bomber is flown horizontally with a ground speed of 275 m/s at an altitude of 3000 m over level terrain. Neglect the effects of air resistance. (a) How far from the point vertically under the point of release will a bomb hit the ground? (b) If the plane maintains its original course and speed, where will it be when the bomb hits the ground? (c) For the above conditions, at what angle from the vertical at the point of release must the telescopic bomb sight be set so that the bomb will hit the target seen in the sight at the time of release?

70. A football is thrown toward a receiver with an initial speed of 20 m/s at an angle of 30° above the horizontal. At that instant, the receiver is 20 m from the quarterback. In what direction and with what constant speed should the receiver run in order to catch the football at the level at which it was thrown?

71. A flea is at point A on a turntable 10 cm from the center. The turntable is rotating at $33\frac{1}{3}$ rev/min in the clockwise direction. The flea jumps vertically upward to a height of 5 cm and lands on the turntable at point B. Place the coordinate origin at the center of the turntable with the positive x axis fixed in space through the position from which the flea jumped. (a) Find the linear displacement of the flea. (b) Find the position of point A when the flea lands. (c) Find the position of point B when the flea lands.

72. A student who is able to swim at a speed of 1.5 m/s in still water wishes to cross a river that has a current of velocity 1.2 m/s toward the south. The width of the river is 50 m. (a) If the student starts from the west bank of the river, in what direction should she head in order to swim directly across the river? How long will this trip take? (b) If she heads due east, how long will it take to cross the river? (*Note:* The student travels farther than 50 m in this case.)

73. A rifle has a maximum range of 500 m. (a) For what angles of elevation would the range be 350 m? What is the range when the bullet leaves the rifle (b) at 14°? (c) at 76°?

74. A river flows with a uniform velocity v. A person in a motorboat travels 1 km upstream, at which time a log is seen floating by. The person continues to travel upstream for one more hour at the same speed and then returns downstream to the starting point, where the same log is seen again. Find the velocity of the river. (*Hint:* The time of travel of the boat after it meets the log equals the time of travel of the log.)

75. A fisherman wishes to cross a river 1 km in width in which the current is 5 km/h toward the north. The fisherman is on the west bank. His boat is propelled with a speed of 4 km/h relative to the water. (a) In what direction should he head in order to make the crossing in minimum time? (b) How long will the crossing take? (c) Determine the velocity of the boat with respect to a stationary ground observer. (c) Find the final downstream displacement.

76. The fisherman in Problem 75 desires to cross the same river in the same boat starting from the west bank. This time he wants to cross the river such that his downstream displacement is minimized. (a) In what direction should he head? (b) Find the final downstream displacement of the boat.

77. Two soccer players, Mary and Jane, begin running from approximately the same point at the same time. Mary runs in an easterly direction at 4.0 m/s, while Jane takes off in a direction 60° N of E at 5.4 m/s. (a) How long is it before they are 25 m apart? (b) What is the velocity of Jane relative to Mary? (c) How far apart are they after 4.0 s?

78. After delivering his toys in the usual manner, Santa decides to have some fun and slide down an icy roof, as in Figure 4.27. He starts from rest at the top of the roof, which is 8 m in length, and accelerates at the rate of 5 m/s². The edge of the roof is 6 m above a soft snowbank, which Santa lands on. Find (a) Santa's velocity components when he reaches the snowbank, (b) the total time he is in motion, and (c) the distance d between the house and the point where he lands in the snow.

Figure 4.27 (Problem 78).

79. A skier leaves the ramp of a ski jump with a velocity of 10 m/s, 15° above the horizontal, as in Figure 4.28. The slope is inclined at 50°, and air resistance is negli-

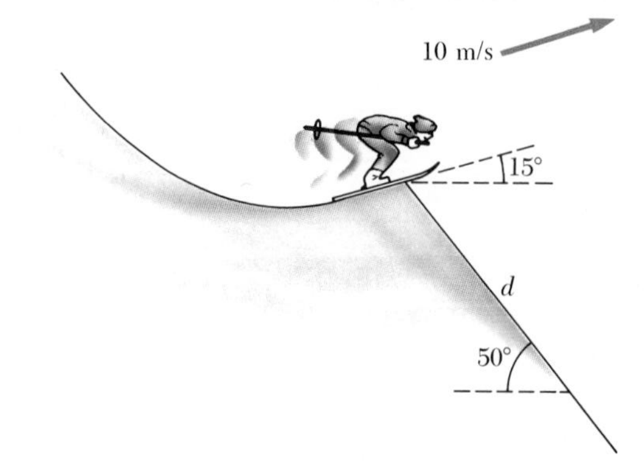

Figure 4.28 (Problem 79).

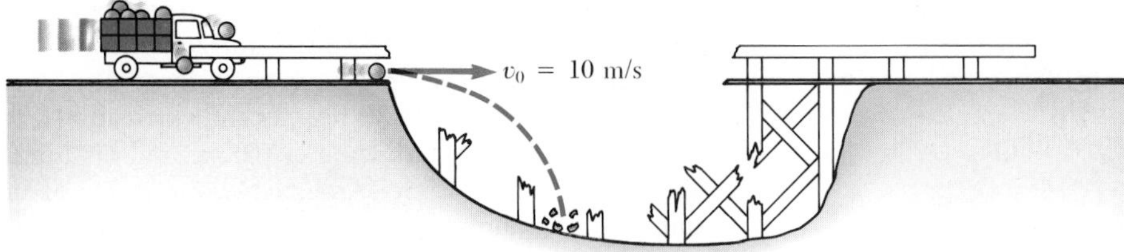

Figure 4.29 (Problem 81).

gible. Find (a) the distance that the jumper lands down the slope and (b) the velocity components just before landing. (How do you think the results might be affected if air resistance were included? Note that jumpers lean forward in the shape of an airfoil with their hands at their sides to increase their distance. Why does this work?)

80. A golf ball leaves the ground at an angle θ and hits a tree while moving horizontally at height h above the ground. If the tree is a horizontal distance of b from the point of projection, show that (a) $\tan \theta = 2\ h/b$. (b) What is the initial velocity of the ball in terms of b and h?

81. A truck loaded with cannonball watermelons stops suddenly to avoid running over the edge of a washed-out bridge (see Figure 4.29). The quick stop causes a number of melons to fly off the truck. One melon rolls over the edge with an initial velocity $v_0 = 10$ m/s in the horizontal direction. What are the x and y coordinates of the melon when it splatters on the bank, if a cross-section of the bank has the shape of a parabola ($y^2 = 16x$ where x and y are measured in meters) with its vertex at the edge of the road?

82. An enemy ship is on the east side of a mountain island as shown in Figure 4.30. The enemy ship can maneuver to within 2500 m of the 1800-m-high mountain peak and can shoot projectiles with an initial speed of 250 m/s. If the western shoreline is horizontally 300 m from the peak, what are the distances from the western shore at which a ship can be safe from the bombardment of the enemy ship?

83. A hawk is flying horizontally at 10.0 m/s in a straight line 200 m above the ground. A mouse it was carrying is released from its grasp. The hawk continues on its path at the same speed for two seconds before attempting to retrieve its prey. To accomplish the retrieval, it dives in a straight line at constant speed and recaptures the mouse 3.0 m above the ground. Assuming no air resistance (a) find the diving speed of the hawk. (b) What angle did the hawk make with the horizontal during its descent? (c) For how long did the mouse "enjoy" free flight?

84. The determined coyote is out once more to try to capture the elusive road runner. The coyote wears a pair of Acme jet-powered roller skates, which provide a constant horizontal acceleration of 15 m/s^2 (Fig. 4.31). The coyote starts off at rest 70 m from the edge of a cliff at the instant the road runner zips by in the direction of the cliff. (a) If the road runner moves with constant speed, determine the minimum speed he must have in order to reach the cliff before the coyote. (b) If the cliff is 100 m above the base of a canyon, determine where the coyote lands in the canyon (assume his skates are still in operation when he is in "flight"). (c) Determine the coyote's velocity components just before he lands in the canyon. (As usual, the road runner is saved by making a sudden turn at the cliff.)

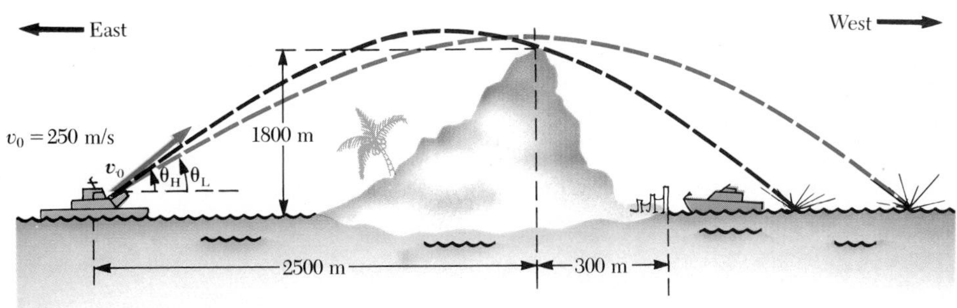

Figure 4.30 (Problem 82).

Figure 4.31 (Problem 84).

85. A U.S. Olympic decathlon star, who happens to be a bright physics student, is trapped on the roof of a burning building with a pencil, paper, pocket calculator, and his favorite physics textbook. He has about 15 min to decide whether to jump to the next building by either running at top speed horizontally off the edge or by using the long-jump technique. The next building is horizontally 30 ft away and vertically 10 ft below. His 100-m dash time is 10.3 s, and his long-jump distance is 25.5 ft. (Assume he long-jumps at an angle of 45° above the horizontal.) Perform calculations to decide which method (if any) he can use to reach the other building safely.

CALCULATOR/COMPUTER PROBLEM

86. A projectile is fired from the origin with an initial speed of v_0 at an angle θ_0 to the horizontal. Write programs that will enable you to tabulate the projectile's x and y coordinates, displacement, x and y components of velocity, and its speed as functions of time. Tabulate the above values for the following inputs of $v_0 = 50$ m/s, $\theta_0 = 60°$ at time intervals of 0.2 s until a total time of 4.4 s is reached.

5

The Laws of Motion

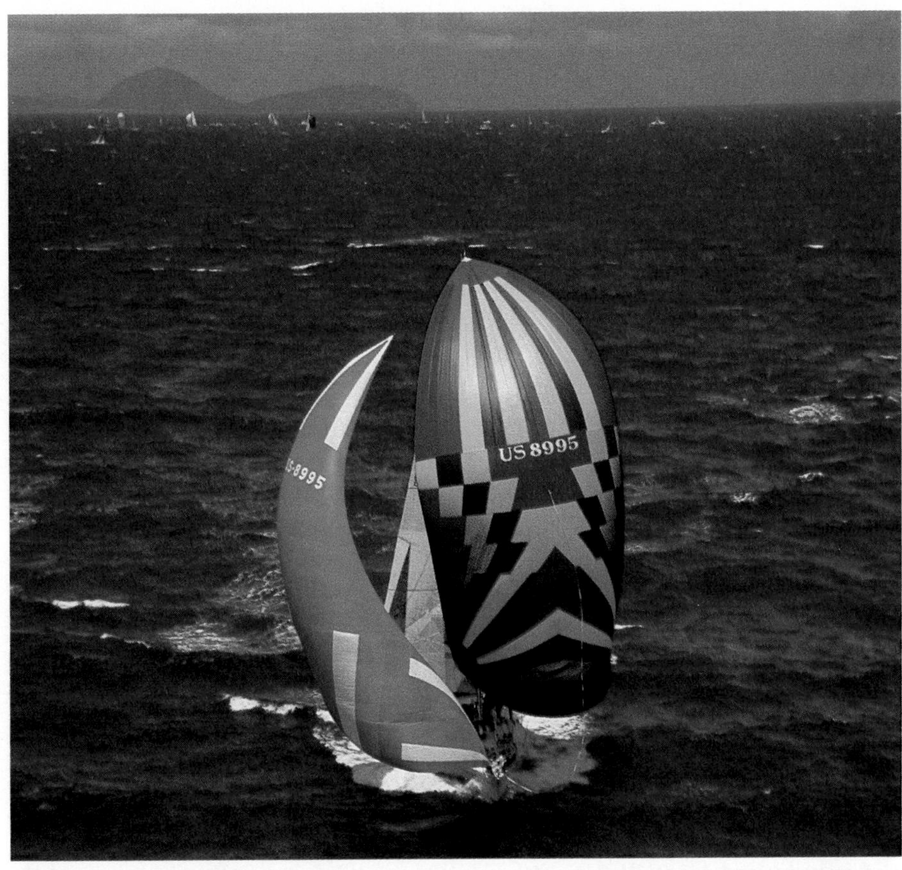

A sailboat with its spinnaker billowing runs before the wind. What force or forces are causing the forward motion? Does the water exert a force on the boat? (© Doug Peebles/ Index Stock International)

In the previous two chapters on kinematics, we described the motion of particles based on the definition of displacement, velocity, and acceleration. However, we would like to be able to answer specific questions related to the causes of motion, such as "What mechanism causes motion?" and "Why do some objects accelerate at a higher rate than others?" In this chapter, we shall describe the change in motion of particles using the concepts of force, mass, and momentum. We shall then discuss the three basic laws of motion, which are based on experimental observations and were formulated nearly three centuries ago by Sir Isaac Newton.

5.1 INTRODUCTION TO CLASSICAL MECHANICS

The purpose of classical mechanics is to provide a connection between the motion of a body and the forces acting on it. Keep in mind that classical mechanics deals with objects that are large compared with the dimensions of atoms

($\approx 10^{-10}$ m) and move at speeds that are much less than the speed of light (3×10^8 m/s).

We shall see that it is possible to describe the acceleration of an object in terms of the resultant force acting on it and the mass of the object. This force represents the interaction of the object with its environment. The mass of an object is a measure of the object's inertia, that is, the tendency of the object to resist an acceleration when a force acts on it.

We shall also discuss *force laws,* which describe the quantitative method of calculating the force on an object if its environment is known. We shall see that although the force laws are rather simple in form, they successfully explain a wide variety of phenomena and experimental observations. These force laws, together with the laws of motion, are the foundations of classical mechanics.

5.2 THE CONCEPT OF FORCE

Everyone has a basic understanding of the concept of force from everyday experiences. When you push or pull an object, you exert a force on it. You exert a force when you throw or kick a ball. In these examples, the word *force* is associated with the result of muscular activity and some change in the state of motion of an object. Forces do not always cause an object to move. For example, as you sit reading this book, the force of gravity acts on your body, and yet you remain stationary. You can push on a block of stone and not move it.

What force (if any) causes a distant star to drift freely through space? Newton answered such questions by stating that the change in velocity of an object is caused by forces. Therefore, if an object moves with uniform motion (constant velocity), no force is required to maintain the motion. Since only a force can cause a change in velocity, we can think of force as that which causes a body to accelerate.

A body accelerates due to an external force

Now consider a situation in which several forces act simultaneously on an object. In this case, the object will accelerate only if the *net force* acting on it is not equal to zero. We shall often refer to the net force as the *resultant force,* or the *unbalanced force. If the net force is zero, the acceleration is zero and the velocity of the object remains constant.* That is, if the net force acting on the object is zero, either the object will be at rest or it will move with constant velocity. *When the velocity of a body is constant or if the body is at rest, it is said to be in equilibrium.*

Definition of equilibrium

Whenever a force is exerted on an object, its shape can change. For example, when you squeeze a soft rubber ball or strike a punching bag with your fist, the objects will be deformed to some extent. Even more rigid objects such as an automobile are deformed under the action of external forces. The deformations can be permanent if the forces are large enough, as in the case of a collision between vehicles.

In this chapter, we shall be concerned with the relation between the force on an object and the acceleration of that object. If you pull on a coiled spring, as in Figure 5.1a, the spring stretches. If the spring is calibrated, the distance that it stretches can be used to measure the strength of the force. If you pull hard enough on a cart to overcome friction, as in Figure 5.1b, it will move. Finally, when a football is kicked, as in Figure 5.1c, the football is both deformed and set in motion. These are all examples of a class of forces called *contact forces.* That is, they represent the result of physical contact between

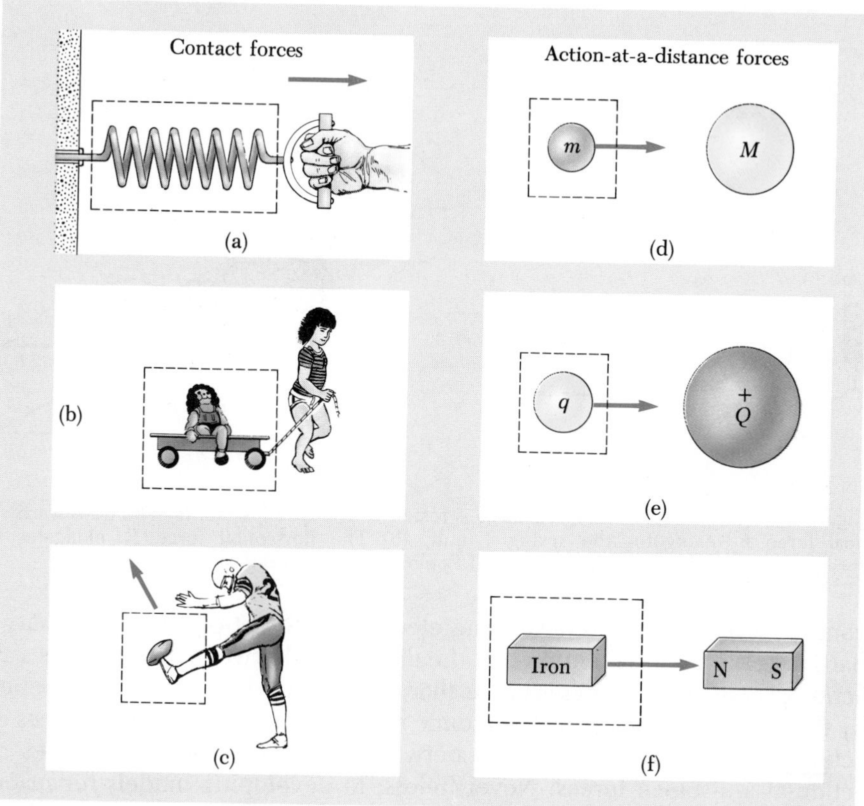

Figure 5.1 Some examples of forces applied to various objects. In each case a force is exerted on the particle or object within the boxed area. The environment external to the boxed area provides the force on the object.

two objects. Other examples of contact forces include the force of a gas on the walls of a container (the result of the collisions of molecules with the walls) and the force of our feet on the floor.

Another class of forces, which do not involve physical contact between two objects but act through empty space, are known as *action-at-a-distance forces.* The force of gravitational attraction between two objects is an example of this class of force, illustrated in Figure 5.1d. This force keeps objects bound to the earth and gives rise to what we commonly call the *weight* of an object. The planets of our solar system are bound under the action of gravitational forces. Another common example of an action-at-a-distance force is the electric force that one electric charge exerts on another electric charge, as in Figure 5.1e. These might be an electron and proton forming the hydrogen atom. A third example of an action-at-a-distance force is the force that a bar magnet exerts on a piece of iron, as shown in Figure 5.1f. The forces between atomic nuclei are also action-at-a-distance forces but are usually very short-range. They are the dominating interaction for particle separations of the order of 10^{-15} m.

Compression of football as player's kick sets it in motion. (© Harold E. Edgerton. Courtesy of Palm Press Inc.)

Early scientists, including Newton himself, were uneasy with the concept of a force acting at a distance. To overcome this conceptual problem, Michael Faraday (1791–1867) introduced the concept of a *field.* According to this approach, when a mass m_1 is placed at some point P near a mass m_2, one can say that m_1 interacts with m_2, by virtue of the gravitational field that exists at P. The field at P is produced by mass m_2. In Chapter 23, we shall see that the field

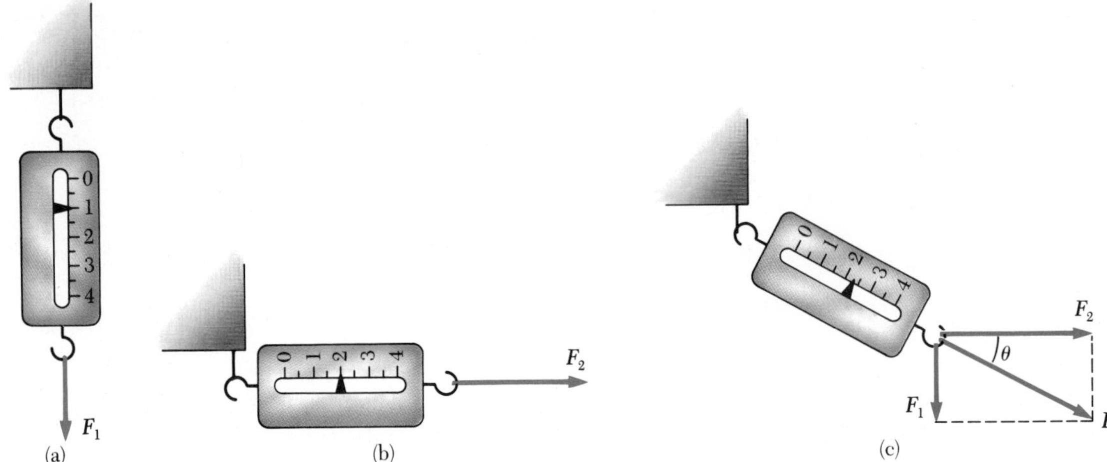

Figure 5.2 The vector nature of a force is tested with a spring scale. (a) The downward vertical force $\boldsymbol{F}_1$ elongates the spring 1 unit. (b) The horizontal force $\boldsymbol{F}_2$ elongates the spring 2 units. (c) The combination $\boldsymbol{F}_1$ and $\boldsymbol{F}_2$ elongates the spring $\sqrt{1^2 + 2^2} = \sqrt{5}$ units.

concept is also useful in describing electrical interactions between charged particles. We should mention that the distinction between contact forces and action-at-a-distance forces is not as sharp as you may have been led to believe by the above discussion. At the atomic level, the so-called contact forces are actually due to repulsive forces between charges, which themselves are action-at-a-distance forces. Nevertheless, in developing models for macroscopic phenomena, it is convenient to use both classifications of forces. However, the only known *fundamental* forces in nature are (1) gravitational attraction between objects because of their masses, (2) electromagnetic forces between charges at rest or in motion, (3) strong nuclear forces between subatomic particles, and (4) weak nuclear forces (the so-called weak interaction), which arise in certain radioactive decay processes. In classical physics, we shall be concerned only with gravitational and electromagnetic forces.

Fundamental forces in nature

It is convenient to use the deformation of a spring to measure force. Suppose a force is applied vertically to a spring with a fixed upper end, as in Figure 5.2a. We can calibrate the spring by defining the unit force, $\boldsymbol{F}_1$, as the force that produces an elongation of 1 cm. If a force $\boldsymbol{F}_2$, applied horizontally as in Figure 5.2b, produces an elongation of 2 cm, the magnitude of $\boldsymbol{F}_2$ is 2 units. If the two forces $\boldsymbol{F}_1$ and $\boldsymbol{F}_2$ are applied simultaneously, as in Figure 5.2c, the elongation of the spring is found to be $\sqrt{5} = 2.24$ cm. The single force, $\boldsymbol{F}$, that would produce this same elongation is the vector sum of $\boldsymbol{F}_1$ and $\boldsymbol{F}_2$, as described in Figure 5.2c. That is, $|\boldsymbol{F}| = \sqrt{F_1{}^2 + F_2{}^2} = \sqrt{5}$ units, and its direction is $\theta = \arctan(-0.5) = -26.6°$. *Because forces are vectors, you must use the rules of vector addition to get the resultant force on a body.* Springs that elongate in proportion to an applied force are said to obey *Hooke's law.* Such springs can be constructed and calibrated to measure unknown forces.

5.3 NEWTON'S FIRST LAW AND INERTIAL FRAMES

Before we state Newton's first law, consider the following simple experiment. Suppose a book is lying on a table. Obviously, the book will remain at rest in the absence of any influences. Now imagine that you push the book with a horizontal force large enough to overcome the force of friction, which is usually present, between the book and table. The book can then be set in motion with constant velocity if the applied force is equal in magnitude to the force of friction and in the direction opposite the friction force. If the applied force exceeds the force of friction, the book accelerates. If the book is released, it stops sliding after moving a short distance since the force of friction retards its motion (or causes a negative acceleration). Now imagine that the book is pushed across a smooth, highly waxed floor. The book will again come to rest, but not as quickly as before. If you could imagine the possibility of a floor so highly polished that friction is completely absent, the book, once set in motion, will slide until it hits the wall.

Isaac Newton (1642–1727), an English physicist and mathematician, was one of the most brilliant scientists in history. Before the age of 30, he formulated the basic concepts and laws of mechanics, discovered the law of universal gravitation, and invented the mathematical methods of calculus. As a consequence of his theories, Newton was able to explain the motion of the planets, the ebb and flow of the tides, and many special features of the motion of the moon and earth. He also interpreted many fundamental observations concerning the nature of light. His contributions to physical theories dominated scientific thought for two centuries and remain important today. (Courtesy AIP Niels Bohr Library)

Before about 1600, scientists felt that the natural state of matter was the state of rest. Galileo was the first to take quite a different approach to motion and the natural state of matter. He devised thought experiments, such as an object moving on a frictionless surface, and concluded that it is not the nature of an object to stop once set in motion: rather, it is its nature to resist deceleration and acceleration.

This new approach to motion was later formalized by Newton in a form that has come to be known as **Newton's first law of motion:**

> An object at rest will remain at rest and an object in motion will continue in motion with a constant velocity (that is, constant speed in a straight line) unless it experiences a net external force (or resultant force).

In simpler terms, we can say that *when the net force on a body is zero, its acceleration is zero.* That is, when $\Sigma \boldsymbol{F} = 0$, then $\boldsymbol{a} = 0$. From the first law, we conclude that an isolated body (a body that does not interact with its environment) is either at rest or moving with constant velocity. Actually, Newton was not the first to state this law. Several decades earlier Galileo wrote, "Any velocity once imparted to a moving body will be rigidly maintained as long as the external causes of retardation are removed."

Another example of uniform motion on a nearly frictionless plane is the motion of a light disk on a column of air (the lubricant), as in Figure 5.3. If the disk is given an initial velocity, it will coast a great distance before coming to rest. This idea is used in the game of air hockey, where the disk makes many collisions with the walls before coming to rest.

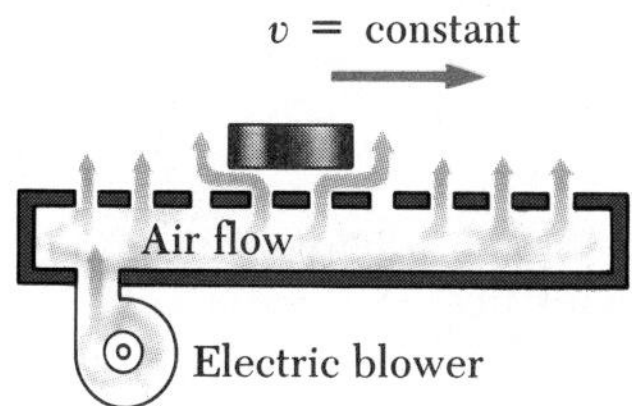

Figure 5.3 A disk moving on a column of air is an example of uniform motion, that is, motion in which the acceleration is zero.

Finally, consider a spaceship traveling in space and far removed from any planets or other matter. The spaceship requires some propulsion system to *change* its velocity. However, if the propulsion system is turned off when the spaceship reaches a velocity $\boldsymbol{v}$, the spaceship will "coast" in space with the same velocity, and the astronauts get a "free ride" (that is, no propulsion system is required to keep them moving at the velocity $\boldsymbol{v}$).

Inertial Frames

Newton's first law, sometimes called the *law of inertia,* defines a special set of reference frames called *inertial frames.*

Inertial frame

> An **inertial frame of reference** is one in which an object, subject to no force, moves with constant velocity. That is, a reference frame in which Newton's first law is valid is called an inertial frame.

A reference frame that moves with constant velocity relative to the distant stars is the best approximation of an inertial frame. The earth is not an inertial frame because of its orbital motion about the sun and rotational motion about its own axis. As the earth travels in its nearly circular orbit about the sun, it experiences a centripetal acceleration of about 4.4×10^{-3} m/s^2 toward the sun. In addition, since the earth rotates about its own axis once every 24 h, a point on the equator experiences an additional centripetal acceleration of 3.37×10^{-2} m/s^2 toward the center of the earth. However, these are small compared with g and can often be neglected. In most situations *we shall assume that the earth is an inertial frame.*

Thus, if an object is in uniform motion ($\boldsymbol{v}$ = constant) an observer in one inertial frame (say, one at rest with respect to the object) will claim that the acceleration and the resultant force on the object are zero. An observer in *any other* inertial frame will also find that $\boldsymbol{a} = 0$ and $\boldsymbol{F} = 0$ for the object. According to the first law, a body at rest and one moving with constant velocity are equivalent. Unless stated otherwise, we shall usually write the laws of motion with respect to an observer "at rest" in an inertial frame.

5.4 INERTIAL MASS

Inertia

If you attempt to change the state of motion of any body, the body will resist this change. **Inertia** is the property of matter that relates to the tendency of an object to remain at rest or in uniform motion. For instance, consider two large, solid cylinders of equal size, one being balsa wood and the other steel. If you were to push the cylinders along a horizontal, rough surface, it would certainly take more effort to get the steel cylinder rolling. Likewise, once they are in motion, it would require more effort to bring the steel cylinder to rest. Therefore, we say that the steel cylinder has more inertia than the balsa wood cylinder.

Mass is a term used to measure inertia, and the **SI** unit of mass is the kilogram. The greater the mass of a body, the less it will accelerate (change its state of motion) under the action of an applied force. For example, if a given force acting on a 3-kg mass produces an acceleration of 4 m/s^2, the same force when applied to a 6-kg mass will produce an acceleration of 2 m/s^2. This idea will be used to obtain a quantitative description of the concept of mass.

Mass and weight are different quantities

It is important to point out that mass should not be confused with weight. *Mass and weight are two different quantities.* The weight of a body is equal to the force of gravity acting on the body and varies with location. For example, a person who weighs 180 lb on earth weighs only about 30 lb on the moon. On the other hand, the mass of a body is the same everywhere, regardless of location. An object having a mass of 2 kg on earth will also have a mass of 2 kg on the moon.

A quantitative measurement of mass can be made by comparing the accelerations that a given force will produce on different bodies. Suppose a force acting on a body of mass m_1 produces an acceleration $\boldsymbol{a}_1$, and the *same force* acting on a body of mass m_2 produces an acceleration $\boldsymbol{a}_2$. The ratio of the two masses is defined as the *inverse* ratio of the magnitudes of the accelerations produced by the same force:

$$\frac{m_1}{m_2} \equiv \frac{a_2}{a_1} \tag{5.1}$$

If one of these is a standard known mass of, say, 1 kg, the mass of an unknown can be obtained from acceleration measurements. For example, if the standard 1-kg mass undergoes an acceleration of 3 m/s^2 under the influence of some force, a 2-kg mass will undergo an acceleration of 1.5 m/s^2 under the action of the same force.

Mass is an inherent property of a body and is independent of the body's surroundings and of the method used to measure the mass. It is an experimental fact that *mass is a scalar quantity*. Finally, *mass is a quantity that obeys the rules of ordinary arithmetic*. That is, several masses can be combined in a simple numerical fashion. For example, if you combine a 3-kg mass with a 5-kg mass, their total mass would be 8 kg. This can be verified experimentally by comparing the acceleration of each object produced by a known force with the acceleration of the combined system using the same force.

5.5 NEWTON'S SECOND LAW

Newton's first law explains what happens to an object when the resultant of all external forces on it is zero. In such instances, the object either remains at rest or moves in a straight line with constant speed. Newton's second law answers the question of what happens to an object that has a nonzero resultant force acting on it.

Imagine a situation in which you are pushing a block of ice across a smooth horizontal surface, such that frictional forces can be neglected. When you exert some horizontal force $\boldsymbol{F}$, the block moves with some acceleration $\boldsymbol{a}$. If you apply a force twice as large, the acceleration doubles. Likewise, if the applied force is increased to $3\boldsymbol{F}$, the acceleration is tripled, and so on. From such observations, we can conclude that *the acceleration of an object is directly proportional to the resultant force acting on it.* The acceleration of an object also depends on its mass. This can be understood by considering the following set of experiments. If you apply a force $\boldsymbol{F}$ to a block of ice on a frictionless surface, it will undergo some acceleration $\boldsymbol{a}$. If the mass of the block is doubled, the same applied force will produce an acceleration $\boldsymbol{a}/2$. If the mass is tripled, the same applied force will produce an acceleration $\boldsymbol{a}/3$, and so on. According to this observation, we conclude that *the acceleration of an object is inversely proportional to its mass.*

These observations are summarized in **Newton's second law,** which states that

> the acceleration of an object is directly proportional to the resultant force acting on it and inversely proportional to its mass.

Note that if the resultant force is zero, then $\boldsymbol{a} = 0$, which corresponds to the equilibrium situation where $\boldsymbol{v}$ is equal to a constant. Thus we can relate mass and force through the following mathematical statement of Newton's second law:[1]

$$\sum \boldsymbol{F} = m\boldsymbol{a} \qquad (5.2)$$

Newton's second law

You should note that Equation 5.2 is a *vector* expression and hence is equivalent to the following three component equations:

[1] Equation 5.2 is valid only when the speed of the particle is much less than the speed of light. We will treat the relativistic situation in Chapter 39.

TABLE 5.1 Units of Force, Mass, and Acceleration[a]

System of Units	Mass	Acceleration	Force
SI	kg	m/s^2	$N = kg \cdot m/s^2$
cgs	g	cm/s^2	$dyne = g \cdot cm/s^2$
British engineering (conventional)	slug	ft/s^2	$lb = slug \cdot ft/s^2$

[a] $1\ N = 10^5$ dyne $= 0.225$ lb

Newton's second law—component form

$$\sum F_x = ma_x \qquad \sum F_y = ma_y \qquad \sum F_z = ma_z \tag{5.3}$$

Units of Force and Mass

The SI unit of force is the **newton,** which is defined as the force that, when acting on a 1-kg mass, produces an acceleration of 1 m/s^2.

From this definition and Newton's second law, we see that the newton can be expressed in terms of the following fundamental units of mass, length, and time:

Definition of newton

$$1\ N \equiv 1\ kg \cdot m/s^2 \tag{5.4}$$

The unit of force in the cgs system is called the **dyne** and is defined as that force that, when acting on a 1-g mass, produces an acceleration of 1 cm/s^2:

Definition of dyne

$$1\ dyne \equiv 1\ g \cdot cm/s^2 \tag{5.5}$$

In the British engineering system, the unit of force is the **pound,** defined as the force that, when acting on a 1-slug mass,[2] produces an acceleration of 1 ft/s^2:

Definition of pound

$$1\ lb = 1\ slug \cdot ft/s^2 \tag{5.6}$$

Since 1 kg $= 10^3$ g and 1 m $= 10^2$ cm, it follows that 1 N $= 10^5$ dynes. It is left as a problem to show that 1 N $= 0.225$ lb. The units of force, mass, and acceleration are summarized in Table 5.1.

[2] The *slug* is the *unit of mass* in the British engineering system and is that system's counterpart of the SI *kilogram.* When we speak of going on a diet to lose a few pounds, we really mean that we want to lose a few slugs, that is, we want to reduce our mass. When we lose those few slugs, the force of gravity (pounds) on our reduced mass decreases (since $\mathbf{W} = m\mathbf{g}$) and that is how we "lose a few pounds." Since most of the calculations we shall carry out in our study of classical mechanics will be in SI units, the slug will seldom be used in this text.

EXAMPLE 5.1 An Accelerating Hockey Puck

A hockey puck with a mass of 0.3 kg slides on the horizontal frictionless surface of an ice rink. Two forces act on the puck as shown in Figure 5.4. The force $\mathbf{F}_1$ has a magnitude of 5 N, and $\mathbf{F}_2$ has a magnitude of 8 N. Determine the acceleration of the puck.

Solution The resultant force in the x direction is

$$\sum F_x = F_{1x} + F_{2x} = F_1 \cos 20° + F_2 \cos 60°$$
$$= (5\ N)(0.940) + (8\ N)(0.500) = 8.70\ N$$

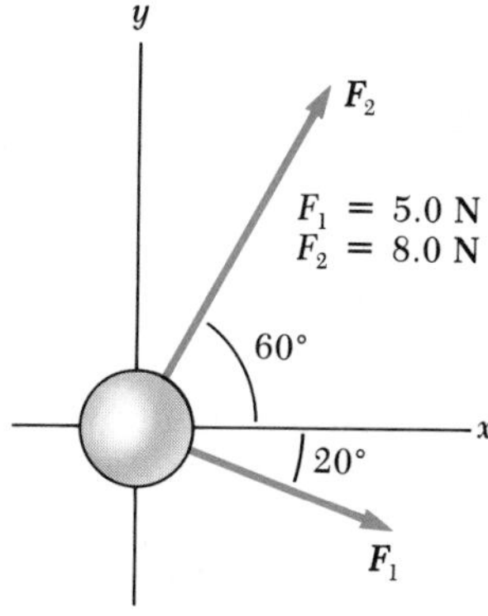

Figure 5.4 (Example 5.1) An object moving on a frictionless surface will accelerate in the direction of the *resultant* force, $\mathbf{F}_1 + \mathbf{F}_2$.

The resultant force in the y direction is

$$\sum F_y = F_{1y} + F_{2y} = -F_1 \sin 20° + F_2 \sin 60°$$
$$= -(5 \text{ N})(0.342) + (8 \text{ N})(0.866) = 5.22 \text{ N}$$

Now we can use Newton's second law in component form to find the x and y components of acceleration:

$$a_x = \frac{\Sigma F_x}{m} = \frac{8.70 \text{ N}}{0.3 \text{ kg}} = 29.0 \text{ m/s}^2$$

$$a_y = \frac{\Sigma F_y}{m} = \frac{5.22 \text{ N}}{0.3 \text{ kg}} = 17.4 \text{ m/s}^2$$

The acceleration has a magnitude of

$$a = \sqrt{(29.0)^2 + (17.4)^2} \text{ m/s}^2 = \boxed{33.8 \text{ m/s}^2}$$

and its direction is

$$\theta = \tan^{-1}(a_y/a_x) = \tan^{-1}(17.4/29.0) = \boxed{31.0°}$$

relative to the positive x axis.

Exercise 1 Determine the components of a third force that when applied to the puck will cause it to be in equilibrium.
Answer $F_x = -8.70$ N, $F_y = -5.22$ N.

5.6 WEIGHT

We are well aware of the fact that bodies are attracted to the earth. The force exerted by the earth on a body is called the **weight** of the body W. This force is directed toward the center of the earth.[3] More precisely, the weight of an object is defined as the resultant gravitational force on the object due to all other bodies in the universe!

We have seen that a freely falling body experiences an acceleration g acting toward the center of the earth. Applying Newton's second law to the freely falling body, with $\mathbf{a} = \mathbf{g}$ and $\mathbf{F} = \mathbf{W}$, gives

$$\mathbf{W} = m\mathbf{g} \tag{5.7}$$

Since the weight depends on g, it varies with geographic location. Bodies weigh less at higher altitudes than at sea level. This is because g decreases with increasing distance from the center of the earth. Hence, weight, unlike mass, is not an inherent property of a body. Therefore, you should not confuse mass with weight. For example, if a body has a mass of 70 kg, then the magnitude of its weight in a location where $g = 9.80$ m/s^2 is $mg = 686$ N (about 154 lb). At the top of a mountain where $g = 9.76$ m/s^2, this weight would be 683 N. This corresponds to a decrease in weight of about 0.4 lb. Therefore, if you want to lose weight without going on a diet, climb a mountain or weigh yourself at 30 000 ft during a flight on a jet airplane.

Since $\mathbf{W} = m\mathbf{g}$, we can compare the masses of two bodies by measuring their weights using a spring scale or a chemical balance. That is, the ratio of the weights of two bodies equals the ratio of their masses at a given location.

Astronaut Edwin E. Aldrin, Jr., walking on the moon after the Apollo 11 lunar landing. The weight of this astronaut on the moon is less than it is on earth, but his mass remains the same. (Courtesy of NASA)

[3] This ignores the fact that the mass distribution of the earth is not perfectly spherical.

5.7 NEWTON'S THIRD LAW

Newton's third law states that if two bodies interact, the force exerted on body 1 by body 2 is equal to and opposite the force exerted on body 2 by body 1. That is,

$$\boldsymbol{F}_{12} = -\boldsymbol{F}_{21} \tag{5.8}$$

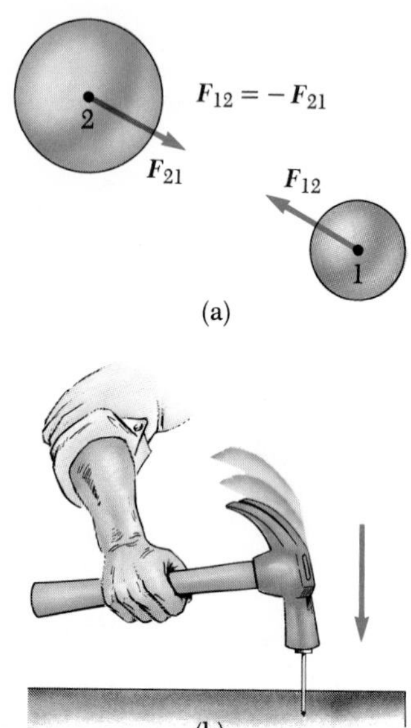

Figure 5.5 Newton's third law. (a) The force of body 1 on body 2 is equal to and opposite the force of body 2 on body 1. (b) The force of the hammer on the nail is equal to and opposite the force of the nail on the hammer.

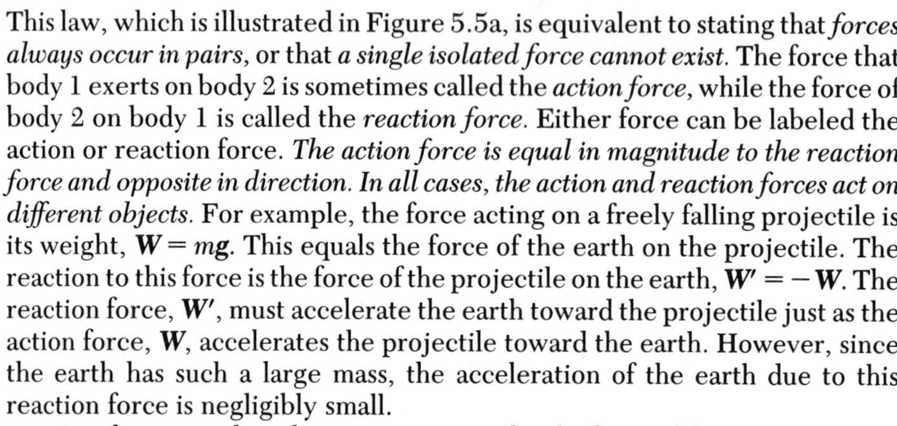

This law, which is illustrated in Figure 5.5a, is equivalent to stating that *forces always occur in pairs*, or that *a single isolated force cannot exist*. The force that body 1 exerts on body 2 is sometimes called the *action force*, while the force of body 2 on body 1 is called the *reaction force*. Either force can be labeled the action or reaction force. *The action force is equal in magnitude to the reaction force and opposite in direction. In all cases, the action and reaction forces act on different objects.* For example, the force acting on a freely falling projectile is its weight, $\mathbf{W} = m\mathbf{g}$. This equals the force of the earth on the projectile. The reaction to this force is the force of the projectile on the earth, $\mathbf{W}' = -\mathbf{W}$. The reaction force, $\mathbf{W}'$, must accelerate the earth toward the projectile just as the action force, $\mathbf{W}$, accelerates the projectile toward the earth. However, since the earth has such a large mass, the acceleration of the earth due to this reaction force is negligibly small.

Another example is shown in Figure 5.5b. The force of the hammer on the nail (the action) is equal to and opposite the force of the nail on the hammer (the reaction). You directly experience Newton's third law if you slam your fist against a wall or if you kick a football with your bare foot. You should be able to identify the action and reaction forces in these cases.

The weight of a body, $\mathbf{W}$, has been defined as the force the earth exerts on the body. If the body is a block at rest on a table, as in Figure 5.6a, the reaction force to $\mathbf{W}$ is the force the block exerts on the earth, $\mathbf{W}'$. The block does not accelerate since it is held up by the table. The table, therefore, exerts an upward action force, $\mathbf{N}$, on the block, called the **normal force.**[4] The normal

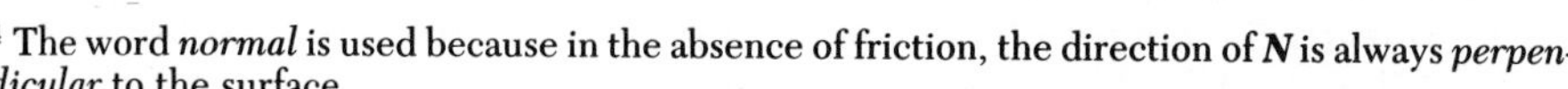

[4] The word *normal* is used because in the absence of friction, the direction of $\boldsymbol{N}$ is always *perpendicular* to the surface.

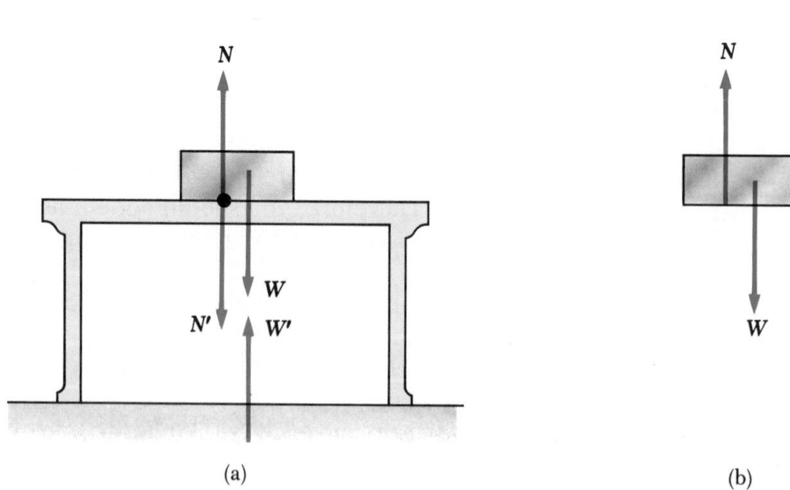

Figure 5.6 When a block is lying on a table, the forces acting on the block are the normal force, $\boldsymbol{N}$, and the force of gravity, $\boldsymbol{W}$, as illustrated in (b). The reaction to $\boldsymbol{N}$ is the force of the block on the table $\boldsymbol{N}'$. The reaction to $\boldsymbol{W}$ is the force of the block on the earth, $\boldsymbol{W}'$.

force is the force that prevents the block from falling through the table, and can have any value needed, up to the point of breaking the table. The normal force balances the weight and provides equilibrium. The reaction to $\mathbf{N}$ is the force of the block on the table, $\mathbf{N}'$. Therefore, we conclude that

$$\mathbf{W} = -\mathbf{W}' \quad \text{and} \quad \mathbf{N} = -\mathbf{N}'$$

Note that the forces acting on the block are $\mathbf{W}$ and $\mathbf{N}$, as in Figure 5.6b. We shall be interested only in such external forces when treating the motion of a body. From the first law, we see that since the block is in equilibrium ($\mathbf{a} = 0$), it follows that $W = N = mg$.

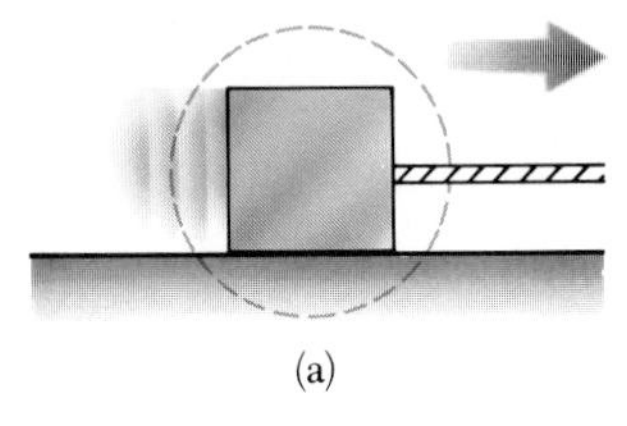

(a)

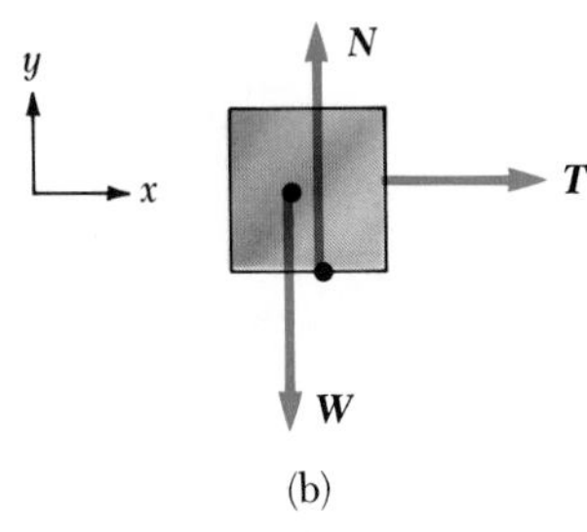

(b)

Figure 5.7 (a) A block being pulled to the right on a smooth surface. (b) The free-body diagram that represents the external forces on the block.

5.8 SOME APPLICATIONS OF NEWTON'S LAWS

In this section we present some simple applications of Newton's laws to bodies that are either in equilibrium ($\mathbf{a} = 0$) or moving linearly under the action of constant external forces. As our model, we shall assume that the bodies behave as particles so that we need not worry about rotational motions. In this section, we shall also neglect the effects of friction for those problems involving motion. This is equivalent to stating that the surfaces are *smooth*. Finally, we shall usually neglect the mass of any ropes involved in a particular problem. In this approximation, the magnitude of the force exerted at any point along the rope is the same at all points along the rope.

When we apply Newton's laws to a body, we shall be interested only in those external forces that act *on the body*. For example, in Figure 5.6 the only external forces acting on the block are $\mathbf{N}$ and $\mathbf{W}$. The reactions to these forces, $\mathbf{N}'$ and $\mathbf{W}'$, act on the table and on the earth, respectively, and do not appear in Newton's second law as applied to the block.

Tension

When an object such as a block is being pulled by a rope attached to it, the rope exerts a force on the object. The **tension** in the rope is defined as the force that the rope exerts on the object. In general, tension in a rope is the force that the rope exerts on what is attached to it.

Consider a block being pulled to the right on the smooth, horizontal surface of a table, as in Figure 5.7a. Suppose you are asked to find the acceleration of the block and the force of the table on the block. First, note that the horizontal force being applied to the block acts through the string. The force that the string exerts on the block is denoted by the symbol $\mathbf{T}$. The magnitude of $\mathbf{T}$ is equal to the tension in the string. A dotted circle is drawn around the block in Figure 5.7a to remind you to isolate the block from its surroundings. Since we are interested only in the motion of the block, we must be able to *identify all external forces acting on it.* These are illustrated in Figure 5.7b. In addition to the force $\mathbf{T}$, the force diagram for the block includes the weight, $\mathbf{W}$, and the normal force, $\mathbf{N}$. As before, $\mathbf{W}$ corresponds to the force of gravity pulling down on the block and $\mathbf{N}$ is the upward force of the table on the block. Such a force diagram is referred to as a **free-body diagram.** The construction of such a diagram is an important step in applying Newton's laws. The *reactions* to the forces we have listed, namely, the force of the string on the hand, the force of the block on the earth, and the force of the block on the table, are not included in the free-body diagram since they act on *other* bodies and not on the block.

Free-body diagrams are important when applying Newton's laws

We are now in a position to apply Newton's second law in component form to the system. The only force acting in the x direction is $\mathbf{T}$. Applying $\Sigma F_x = ma_x$ to the horizontal motion gives

$$\sum F_x = T = ma_x \qquad \text{or} \qquad a_x = \frac{T}{m}$$

In this situation, there is no acceleration in the y direction. Applying $\Sigma F_y = ma_y$ with $a_y = 0$ gives

$$N - W = 0 \qquad \text{or} \qquad N = W$$

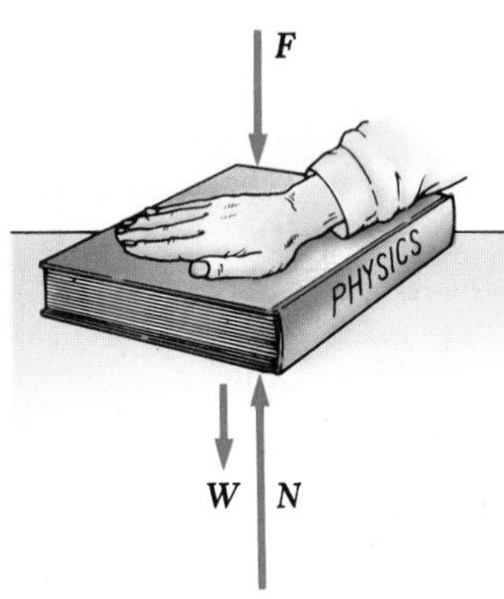

Figure 5.8 When one pushes downward on an object with a force $\boldsymbol{F}$, the normal force $\boldsymbol{N}$ is greater than the weight. That is, $\boldsymbol{N} = \boldsymbol{W} + \boldsymbol{F}$.

That is, the normal force is equal to and opposite the weight.

If $\boldsymbol{T}$ is a *constant* force, then the acceleration, $a_x = T/m$, is also a constant. Hence, the equations of kinematics from Chapter 3 can be used to obtain the displacement, Δx, and velocity, v, as functions of time. Since $a_x = T/m =$ constant, these expressions can be written

$$\Delta x = v_0 t + \tfrac{1}{2}\left(\frac{T}{m}\right)t^2$$

$$v = v_0 + \left(\frac{T}{m}\right)t$$

where v_0 is the velocity at $t = 0$.

In the example just presented, we found that the normal force $\boldsymbol{N}$ was equal in magnitude and opposite the weight $\boldsymbol{W}$. *This is not always the case.* For example, suppose you were to push down on a book with a force $\boldsymbol{F}$ as in Figure 5.8. In this case, $\Sigma F_y = 0$ gives $N - W - F = 0$, or $N = W + F$. Other examples in which $N = W$ will be presented later.

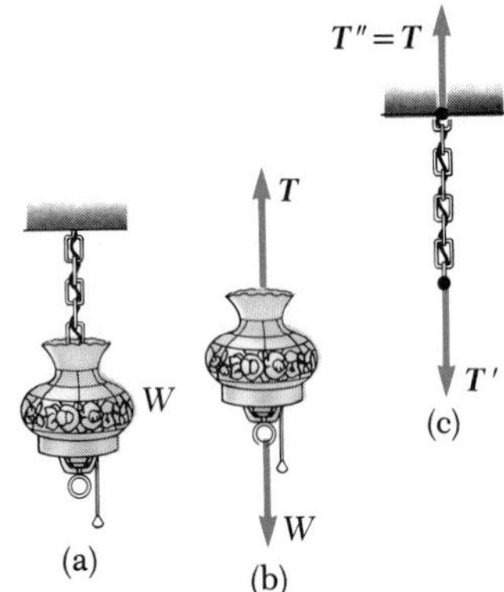

Figure 5.9 (a) A lamp of weight $\boldsymbol{W}$ suspended by a light chain from a ceiling. (b) The forces acting on the lamp are the force of gravity, $\boldsymbol{W}$, and the tension in the chain, $\boldsymbol{T}$. (c) The forces acting on the chain are $\boldsymbol{T}'$, that exerted by the lamp, and $\boldsymbol{T}''$, that exerted by the ceiling.

Consider a lamp of weight W suspended from a chain of negligible weight fastened to the ceiling, as in Figure 5.9a. The free-body diagram for the lamp is shown in Figure 5.9b, where the forces on it are the weight, $\boldsymbol{W}$, acting downward, and the force of the chain on the lamp, $\boldsymbol{T}$, acting upward. The force $\boldsymbol{T}$ is the constraint force in this case. (If we cut the chain, $\boldsymbol{T} = 0$ and the body executes free fall.)

If we apply the first law to the lamp, noting that $\boldsymbol{a} = 0$, we see that since there are no forces in the x direction, the equation $\Sigma F_x = 0$ provides no helpful information. The condition $\Sigma F_y = 0$ gives

$$\sum F_y = T - W = 0 \qquad \text{or} \qquad T = W$$

Note that $\boldsymbol{T}$ and $\boldsymbol{W}$ are *not* action-reaction pairs. The reaction to $\boldsymbol{T}$ is $\boldsymbol{T}'$, the force exerted on the chain by the lamp, as in Figure 5.9c. The force $\boldsymbol{T}'$ acts downward and is transmitted to the ceiling. That is, the force of the chain on the ceiling, $\boldsymbol{T}'$, is *downward* and equal to W in magnitude. The ceiling exerts an equal and opposite force, $\boldsymbol{T}'' = \boldsymbol{T}$, on the chain, as in Figure 5.9c.

PROBLEM-SOLVING STRATEGY

The following procedure is recommended when dealing with problems involving the application of Newton's laws:

1. Draw a simple, neat diagram of the system.
2. Isolate the object of interest whose motion is being analyzed. Draw a free-body diagram for this object, that is, a diagram showing *all external forces acting on the object.* For systems containing more than one object, draw *separate* diagrams for each object. *Do not* include forces that the object exerts on its surroundings.

3. Establish convenient coordinate axes for each body and find the components of the forces along these axes. Now, apply Newton's second law, $\Sigma \boldsymbol{F} = m\boldsymbol{a}$, in *component* form. Check your dimensions to make sure that all terms have units of force.
4. Solve the component equations for the unknowns. Remember that you must have as many independent equations as you have unknowns in order to obtain a complete solution.
5. It is a good idea to check the predictions of your solutions for extreme values of the variables. You can often detect errors in your results by doing so.

EXAMPLE 5.2 A Traffic Light at Rest

A traffic light weighing 100 N hangs from a cable tied to two other cables fastened to a support, as in Figure 5.10a. The upper cables make angles of 37° and 53° with the horizontal. Find the tension in the three cables.

Solution First we construct a free-body diagram for the traffic light, as in Figure 5.10b. The tension in the vertical cable, T_3, supports the light, and so we see that $T_3 = W = 100$ N. Now we construct a free-body diagram for the knot that holds the three cables together, as in Figure 5.10c. This is a convenient point to choose because all forces in question act at this point. We choose the coordinate axes as shown in Figure 5.10c and resolve the forces into their x and y components:

Force	x component	y component
T_1	$-T_1 \cos 37°$	$T_1 \sin 37°$
T_2	$T_2 \cos 53°$	$T_2 \sin 53°$
T_3	0	-100 N

The first condition for equilibrium gives us the equations

$$(1) \qquad \sum F_x = T_2 \cos 53° - T_1 \cos 37° = 0$$

$$(2) \qquad \sum F_y = T_1 \sin 37° + T_2 \sin 53° - 100 \text{ N} = 0$$

From (1) we see that the horizontal components of $\boldsymbol{T}_1$ and $\boldsymbol{T}_2$ must be equal in magnitude, and from (2) we see that the sum of the vertical components of $\boldsymbol{T}_1$ and $\boldsymbol{T}_2$ must balance the weight of the light. We can solve (1) for $\boldsymbol{T}_2$ in terms of $\boldsymbol{T}_1$ to give

$$T_2 = T_1 \left(\frac{\cos 37°}{\cos 53°} \right) = 1.33 T_1$$

This value for T_2 can be substituted into (2) to give

$$T_1 \sin 37° + (1.33 T_1)(\sin 53°) - 100 \text{ N} = 0$$

$$T_1 = \boxed{60.0 \text{ N}}$$

$$T_2 = 1.33 T_1 = \boxed{79.8 \text{ N}}$$

Exercise 2 In what situation will $T_1 = T_2$?
Answer When the supporting cables make equal angles with the horizontal support.

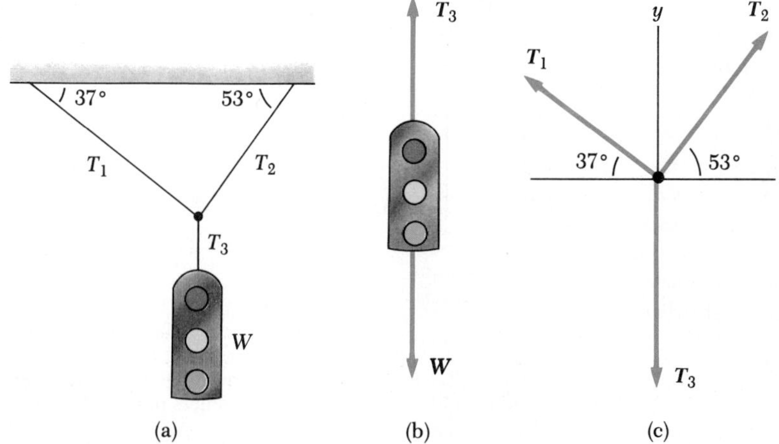

Figure 5.10 (Example 5.2) (a) A traffic light suspended by cables. (b) Free-body diagram for the traffic light. (c) Free-body diagram for the knot.

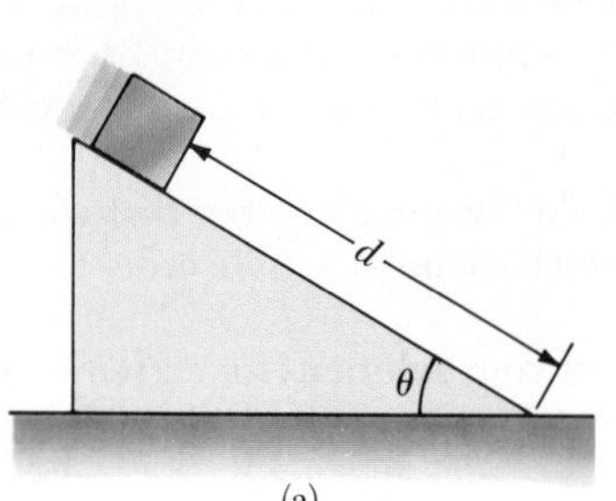

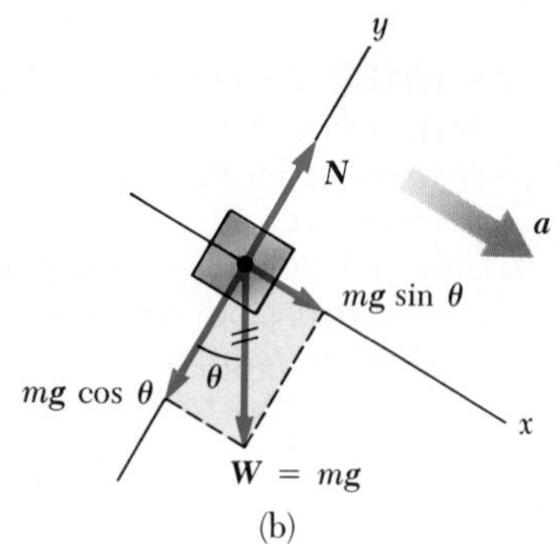

Figure 5.11 (Example 5.3) (a) A block sliding down a smooth incline. (b) The free-body diagram for the block. Note that its acceleration along the incline is $g \sin \theta$.

EXAMPLE 5.3 Block on a Smooth Incline

A block of mass m is placed on a smooth, inclined plane of angle θ, as in Figure 5.11a. (a) Determine the acceleration of the block after it is released.

The free-body diagram for the block is shown in Figure 5.11b. The only forces on the block are the normal force, **N**, acting perpendicular to the plane and the weight, **W**, acting vertically downward. *It is convenient to choose the coordinate axes with x along the incline and y perpendicular to it.* Then, we replace the weight vector by a component of magnitude $mg \sin \theta$ along the *positive* x axis and another of magnitude $mg \cos \theta$ in the *negative* y direction. Applying Newton's second law in component form while noting that $a_y = 0$ gives

$$(1) \qquad \sum F_x = mg \sin \theta = ma_x$$

$$(2) \qquad \sum F_y = N - mg \cos \theta = 0$$

From (1) we see that the acceleration along the incline is provided by the component of weight down the incline:

$$(3) \qquad a_x = g \sin \theta$$

From (2) we conclude that the component of weight perpendicular to the incline is *balanced* by the normal force, or $N = mg \cos \theta$. The acceleration given by (3) is found to be *independent* of the mass of the block! It depends only on the angle of inclination and on g!

Special Cases We see that when $\theta = 90°$, $a = g$ and $N = 0$. This corresponds to the block in free fall. Also, when $\theta = 0$, $a_x = 0$ and $N = mg$ (its maximum value).

(b) Suppose the block is released from rest at the top, and the distance from the block to the bottom is d. How long does it take the block to reach the bottom, and what is its speed just as it gets there?

Since $a_x =$ constant, we can apply the kinematic equation $x - x_0 = v_{x0}t + \frac{1}{2}a_x t^2$ to the block. Since the displacement $x - x_0 = d$ and $v_{x0} = 0$, we get

$$d = \tfrac{1}{2}a_x t^2$$

or

$$(4) \qquad t = \sqrt{\frac{2d}{a_x}} = \sqrt{\frac{2d}{g \sin \theta}}$$

Also, since $v_x{}^2 = v_{x0}{}^2 + 2a_x(x - x_0)$ and $v_{x0} = 0$, we find that

$$v_x{}^2 = 2a_x d$$

$$(5) \qquad v_x = \sqrt{2a_x d} = \sqrt{2gd \sin \theta}$$

Again, t and v_x are *independent* of the mass of the block. This suggests a simple method of measuring g using an inclined air track or some other smooth incline. Simply measure the angle of inclination, the distance traveled by the block, and the time it takes to reach the bottom. The value of g can then be calculated from (4) and (5).

EXAMPLE 5.4 Atwood's Machine

When two unequal masses are hung vertically over a light, frictionless pulley as in Figure 5.12a, the arrange-

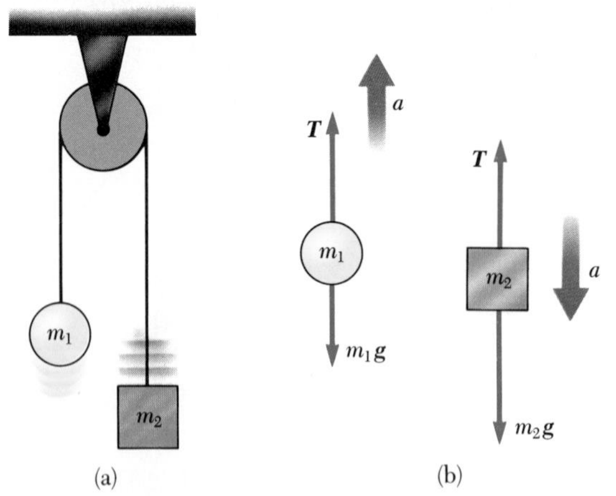

Figure 5.12 (Example 5.4) Atwood's machine. (a) Two masses connected by a light string over a frictionless pulley. (b) Free-body diagrams for m_1 and m_2.

ment is called *Atwood's machine.* The device is sometimes used in the laboratory to measure the acceleration of gravity. Determine the acceleration of the two masses and the tension in the string.

Solution The free-body diagrams for the two masses are shown in Figure 5.12b, where we assume that $m_2 > m_1$. When Newton's second law is applied to m_1, with a upwards for this mass, we find

$$(1) \qquad \sum F_y = T - m_1 g = m_1 a$$

Similarly, for m_2 we find

$$(2) \qquad \sum F_y = T - m_2 g = -m_2 a$$

The negative sign on the right-hand side of (2) indicates that m_2 accelerates downwards, in the negative y direction.

When (2) is subtracted from (1), T drops out and we get

$$-m_1 g + m_2 g = m_1 a + m_2 a$$

or

$$(3) \qquad a = \left(\frac{m_2 - m_1}{m_1 + m_2} \right) g$$

If (3) is substituted into (1), we get

$$(4) \qquad T = \left(\frac{2m_1 m_2}{m_1 + m_2} \right) g$$

Special Cases Note that when $m_1 = m_2$, $a = 0$ and $T = m_1 g = m_2 g$, as we would expect for the balanced case. Also, if $m_2 \gg m_1$, $a \approx g$ (a freely falling body) and $T \approx 2m_1 g$.

Exercise 3 Find the acceleration and tension of an Atwood's machine in which $m_1 = 2$ kg and $m_2 = 4$ kg.
Answer $a = 3.27\ \text{m/s}^2$, $T = 26.1$ N.

EXAMPLE 5.5 Two Connected Objects

Two unequal masses are attached by a light string that passes over a light, frictionless pulley as in Figure 5.13a. The block of mass m_2 lies on a smooth incline of angle θ. Find the acceleration of the two masses and the tension in the string.

Solution Since the two masses are connected by a string (which we assume doesn't stretch), they both have accelerations of the same magnitude. The free-body diagrams for the two masses are shown in Figures 5.13b and 5.13c. Applying Newton's second law in component form to m_1 while *assuming* that $\boldsymbol{a}$ is upward for this mass gives

Equations of motion for m_1:

$$(1) \qquad \sum F_x = 0$$

$$(2) \qquad \sum F_y = T - m_1 g = m_1 a$$

Note that in order for a to be positive, it is necessary that $T > m_1 g$.

Now, for m_2 it is convenient to choose the positive x' axis along the incline as in Figure 5.13c. Applying Newton's second law in component form to m_2 gives

Equations of motion for m_2:

$$(3) \qquad \sum F_{x'} = m_2 g \sin\theta - T = m_2 a$$

$$(4) \qquad \sum F_{y'} = N - m_2 g \cos\theta = 0$$

Expressions (1) and (4) provide no information regarding the acceleration. However, if we solve (2) and (3) simultaneously for the unknowns a and T, we get

$$(5) \qquad a = \frac{m_2 g \sin\theta - m_1 g}{m_1 + m_2}$$

When this is substituted into (2), we find

$$(6) \qquad T = \frac{m_1 m_2 g(1 + \sin\theta)}{m_1 + m_2}$$

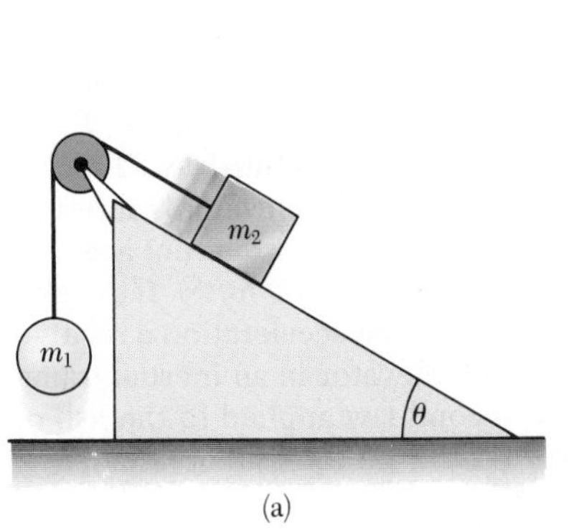

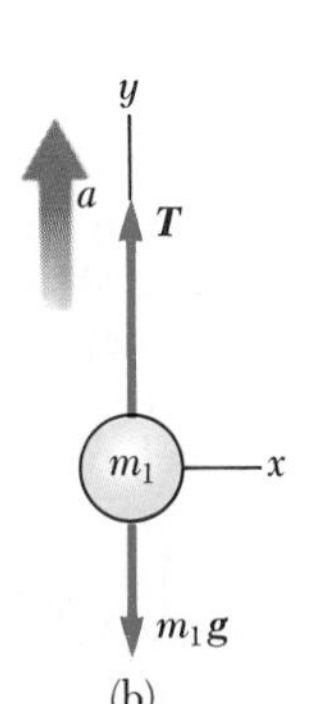

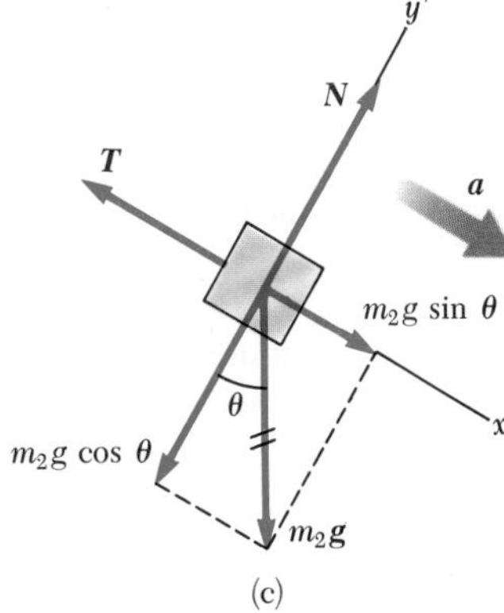

Figure 5.13 (Example 5.5) (a) Two masses connected by a light string over a frictionless pulley. (b) The free-body diagram for m_1. (c) The free-body diagram for m_2 (the incline is smooth).

Note that m_2 accelerates down the incline if $m_2 \sin \theta$ exceeds m_1 (that is, if a is positive as we assumed). If m_1 exceeds $m_2 \sin \theta$, the acceleration of m_2 is up the incline and downward for m_1. You should also note that the result for the acceleration, (5), can be interpreted as the resultant unbalanced force on the system divided by the total mass of the system.

Exercise 4 If $m_1 = 10$ kg, $m_2 = 5$ kg, and $\theta = 45°$, find the acceleration.
Answer $a = -4.22$ m/s^2, where the negative sign indicates that m_2 accelerates up the incline.

EXAMPLE 5.6 One Block Pushes Another
Two blocks of masses m_1 and m_2 are placed in contact with each other on a smooth, horizontal surface as in Figure 5.14a. A constant horizontal force $\boldsymbol{F}$ is applied to m_1 as shown. (a) Find the acceleration of the system.

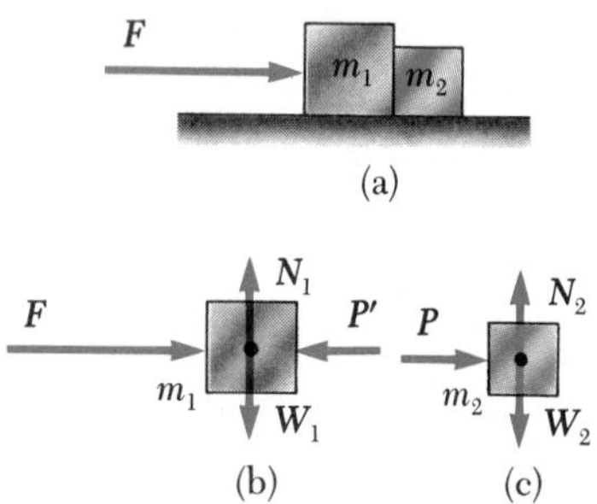

Figure 5.14 Example 5.6.

Solution Both blocks must experience the *same* acceleration since they are in contact with each other. Since the force F is the *only* horizontal force on the system (the two blocks), we have

$$\sum F_x(\text{system}) = F = (m_1 + m_2)a$$

or

$$(1) \qquad a = \frac{F}{m_1 + m_2}$$

(b) Determine the magnitude of the contact force between the two blocks.

Solution To solve this part of the problem, it is necessary to first construct free-body diagrams for each block shown in Figures 5.14b and 5.14c, where the contact force is denoted by $\boldsymbol{P}$. From Figure 5.14c, we see that the *only horizontal* force acting on m_2 is the contact force $\boldsymbol{P}$ (the force of m_1 acting on m_2) which is to the right. Applying Newton's second law to m_2 gives

$$(2) \qquad \sum F_x = P = m_2 a$$

Substituting the value of the acceleration a from (1) into (2) gives

$$(3) \qquad P = m_2 a = \left(\frac{m_2}{m_1 + m_2}\right) F$$

From this result, note that the contact force $\boldsymbol{P}$ is *less* than the applied force $\boldsymbol{F}$. This is consistent with the fact that the force required to accelerate m_2 alone must be *less* than the force required to produce the same acceleration for the system of two blocks.

It is instructive to check this expression for P by considering the forces acting on m_1, shown in Figure 15.14b. In this case, the horizontal forces acting on m_1 are the applied force $\boldsymbol{F}$ to the right, and the contact force $\boldsymbol{P'}$ to the left (the force of m_2 on m_1). From Newton's third law, $\boldsymbol{P'}$ is the reaction to $\boldsymbol{P}$, so that $|\boldsymbol{P'}| = |\boldsymbol{P}|$. Applying Newton's second law to m_2 gives

$$(4) \qquad \sum F_x = F - P' = F - P = m_2 a$$

Substituting the value of a from (2) into (4) gives

$$P = F - m_2 a = F - \frac{m_1 F}{m_1 + m_2} = \left(\frac{m_2}{m_1 + m_2}\right) F$$

This agrees with (3), as it must.

Exercise 5 If $m_1 = 4$ kg, $m_2 = 3$ kg, and $F = 9$ N, find the acceleration of the system and the magnitude of the contact force.
Answer $a = 1.29$ m/s^2; $P = 3.86$ N

EXAMPLE 5.7 Weighing a Fish in an Elevator
A person weighs a fish on a spring scale attached to the ceiling of an elevator, as shown in Figure 5.15. Show that if the elevator accelerates or decelerates, the spring scale reads a weight different from the true weight of the fish.

Solution The external forces acting on the fish are its true weight, $\boldsymbol{W}$, and the upward constraint force, $\boldsymbol{T}$, exerted on it by the scale. By Newton's third law, T is also the reading of the spring scale. If the elevator is at rest or moving at constant velocity, then the fish is not accelerating and $T = W = mg$ (where $g = 9.80$ m/s^2). If the elevator accelerates upward with an acceleration $\boldsymbol{a}$ relative to an observer outside the elevator in an inertial frame (Fig. 5.15a), then the second law applied to the fish of mass m gives the total force $\boldsymbol{F}$ on the fish:

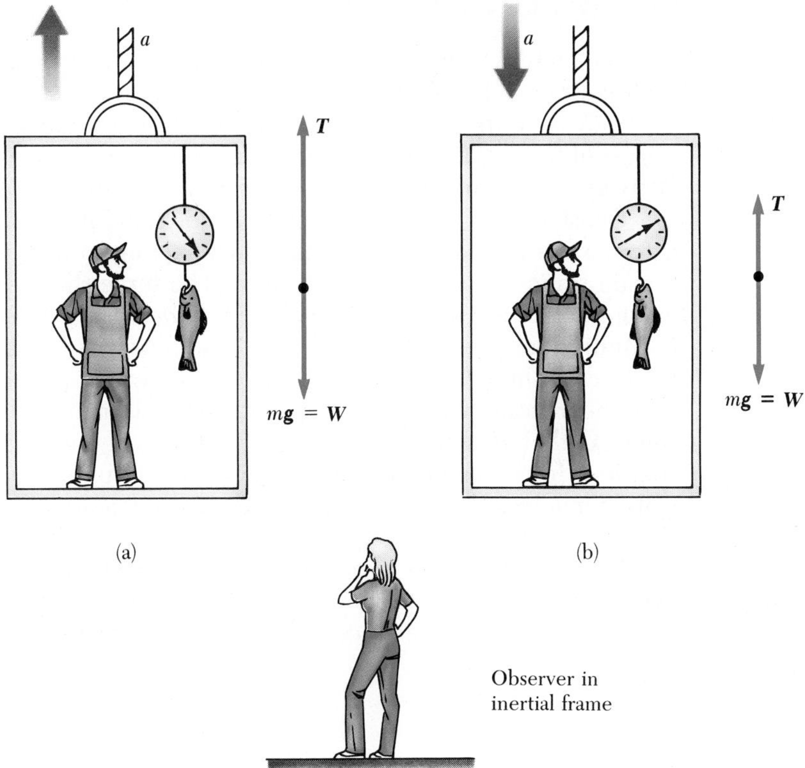

Figure 5.15 (Example 5.7) Apparent weight versus true weight. (a) When the elevator accelerates *upward* the spring scale reads a value *greater* than the true weight. (b) When the elevator accelerates *downward* the spring scale reads a value *less* than the true weight. The spring scale reads the *apparent weight.*

$$(1) \qquad \sum F = T - W = ma \qquad (\text{if } \boldsymbol{a} \text{ is upward})$$

Likewise, if the elevator accelerates downward as in Figure 5.15b, Newton's second law applied to the fish becomes

$$(2) \qquad \sum F = T - W = -ma \qquad (\text{if } \boldsymbol{a} \text{ is downward})$$

Thus, we conclude from (1) that the scale reading, T, is greater than the true weight, W, if $\boldsymbol{a}$ is upward. From (2) we see that T is less than W if $\boldsymbol{a}$ is downward.

For example, if the true weight of the fish is 40 N, and $\boldsymbol{a}$ is 2 m/s² upward, then the scale reading is

$$T = ma + mg = mg\left(\frac{a}{g} + 1\right)$$

$$= W\left(\frac{a}{g} + 1\right) = (40\text{ N})\left(\frac{2\text{ m/s}^2}{9.80\text{ m/s}^2} + 1\right)$$

$$= 48.2\text{ N}$$

If a is 2 m/s² downward, then

$$T = -ma + mg = mg\left(1 - \frac{a}{g}\right)$$

$$= W\left(1 - \frac{a}{g}\right) = (40\text{ N})\left(1 - \frac{2\text{ m/s}^2}{9.80\text{ m/s}^2}\right)$$

$$= 31.8\text{ N}$$

Hence, if you buy a fish in an elevator, make sure the fish is weighed while the elevator is at rest or accelerating downward! Furthermore, note that from the information given here, one cannot determine the *direction* of motion of the elevator.

Special Cases If the elevator cable breaks, then the elevator falls freely and $a = -g$. Since $W = mg$, we see from (1) that the apparent weight, T, is zero, that is, the fish appears to be weightless. If the elevator accelerates *downward* with an acceleration *greater* than g, the fish (along with the person in the elevator) will eventually hit the ceiling since its acceleration will still be that of a freely falling body relative to an outside observer.

5.9 FORCES OF FRICTION

When a body is in motion on a rough surface, or through a viscous medium such as air or water, there is resistance to motion because of the interaction of the body with its surroundings. We call such resistance a **force of friction.** Forces of friction are very important in our everyday lives. Forces of friction allow us to walk or run and are necessary for the motion of wheeled vehicles.

Consider a block on a horizontal table, as in Figure 5.16a. If we apply an external horizontal force $\boldsymbol{F}$ to the block, acting to the right, the block will remain stationary if $\boldsymbol{F}$ is not too large. The force that keeps the block from moving acts to the left and is called the *frictional force*, $\boldsymbol{f}$. As long as the block is in equilibrium, $f = F$. Since the block is stationary, we call this frictional force the *force of static friction*, $\boldsymbol{f}_s$. Experiments show that this force arises from the roughness of the two surfaces, so that contact is made only at a few points, as shown in the "magnified" view of the surfaces in Figure 5.16a. Actually, the frictional force is much more complicated than presented here since it ultimately involves the electrostatic forces between atoms or molecules where the surfaces are in contact.

If we increase the magnitude of $\boldsymbol{F}$, as in Figure 5.16b, the block will eventually slip. When the block is on the verge of slipping, f_s is a maximum. When F *exceeds* $f_{s,max}$ the block moves and accelerates to the right. When the

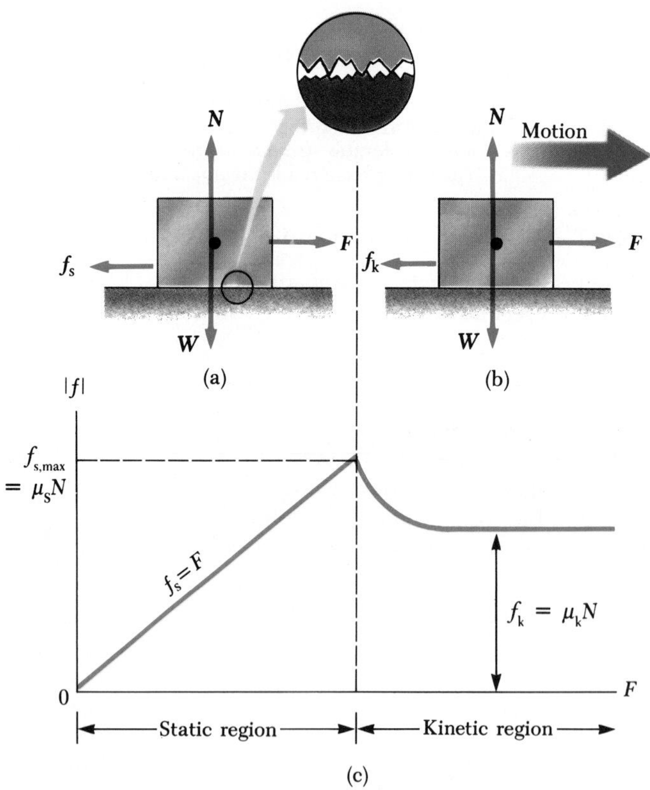

Figure 5.16 The force of friction, $\boldsymbol{f}$, between a block and a rough surface is opposite the applied force, $\boldsymbol{F}$. (a) The force of static friction equals the applied force. (b) When the applied force exceeds the force of kinetic friction, the block accelerates to the right. (c) A graph of the magnitude of the frictional force versus the applied force. Note that $f_{s,max} > f_k$.

block is in motion, the retarding frictional force becomes *less* than $f_{s,max}$ (Fig. 5.16c). When the block is in motion, we call the retarding force the *force of kinetic friction*, f_k. The unbalanced force in the x direction, $F - f_k$, produces an acceleration to the right. If $F = f_k$ the block moves to the right with constant speed. If the applied force is removed, then the frictional force acting to the left decelerates the block and eventually brings it to rest.

In a simplified model, we can imagine that the force of kinetic friction is less than $f_{s,max}$ because of the reduction in roughness of the two surfaces when the object is in motion.

Experimentally, one finds that both f_s and f_k are *proportional to the normal force acting on the block.* The experimental observations can be summarized by the following laws of friction:

1. The force of static friction between any two surfaces in contact is opposite the applied force and can have values given by

$$f_s \leq \mu_s N \qquad (5.9)$$

Force of static friction

where the dimensionless constant μ_s is called the **coefficient of static friction** and N is the normal force. The equality in Equation 5.9 holds when the block is on the *verge* of slipping, that is, when $f_s = f_{s,max} = \mu_s N$. The inequality holds when the applied force is *less* than this value.

2. The force of kinetic friction acting on an object is opposite to the direction of motion of the object and is given by

$$f_k = \mu_k N \qquad (5.10)$$

Force of kinetic friction

where μ_k is the **coefficient of kinetic friction.**

3. The values of μ_k and μ_s depend on the nature of the surfaces, but μ_k is generally less than μ_s. Typical values of μ range from around 0.05 for smooth surfaces to 1.5 for rough surfaces. Table 5.2 lists some reported values.

TABLE 5.2 Coefficients of Friction[a]

	μ_s	μ_k
Steel on steel	0.74	0.57
Aluminum on steel	0.61	0.47
Copper on steel	0.53	0.36
Rubber on concrete	1.0	0.8
Wood on wood	0.25–0.5	0.2
Glass on glass	0.94	0.4
Waxed wood on wet snow	0.14	0.1
Waxed wood on dry snow	—	0.04
Metal on metal (lubricated)	0.15	0.06
Ice on ice	0.1	0.03
Teflon on Teflon	0.04	0.04
Synovial joints in humans	0.01	0.003

[a] All values are approximate.

Finally, the coefficients of friction are nearly independent of the area of contact between the surfaces. Although the coefficient of kinetic friction varies with speed, we shall neglect any such variations.

EXAMPLE 5.8 Experimental Determination of μ_s and μ_k

In this example we describe a simple method of measuring the coefficients of friction between an object and a rough surface. Suppose the object is a small block placed on a surface inclined with respect to the horizontal, as in Figure 5.17. The angle of the inclined plane is increased until the block slips. By measuring the angle θ_c at which this slipping just occurs, we obtain μ_s directly. We note that the only forces acting on the block are its weight, mg, the normal force, **N**, and the force of static friction, f_s. Taking x parallel to the plane and y perpendicular to the plane, Newton's second law applied to the block gives

Static case: (1) $\sum F_x = mg \sin\theta - f_s = 0$

(2) $\sum F_y = N - mg\cos\theta = 0$

We can eliminate mg by substituting $mg = N/\cos\theta$ from (2) into (1) to get

(3) $$f_s = mg\sin\theta = \left(\frac{N}{\cos\theta}\right)\sin\theta = N\tan\theta$$

When the inclined plane is at the critical angle, θ_c (called the angle of repose), $f_s = f_{s,\max} = \mu_s N$, and so at this angle, (3) becomes

$$\mu_s N = N\tan\theta_c$$

Static case: $\mu_s = \tan\theta_c$

For example, if we find that the block just slips at $\theta_c = 20°$, then $\mu_s = \tan 20° = 0.364$. Once the block starts to move at $\theta \geq \theta_c$, it will accelerate down the incline and the force of friction is $f_k = \mu_k N$. However, if θ is reduced below θ_c, an angle θ_c' can be found such that the block moves down the incline with constant speed ($a_x = 0$). In this case, using (1) and (2) with f_s replaced by f_k gives

Kinetic case: $\mu_k = \tan\theta_c'$

where $\theta_c' < \theta_c$.

You should try this simple experiment using a coin as the block and a notebook as the inclined plane. Also, you can try taping two coins together to prove that you still get the same critical angles as with one coin.

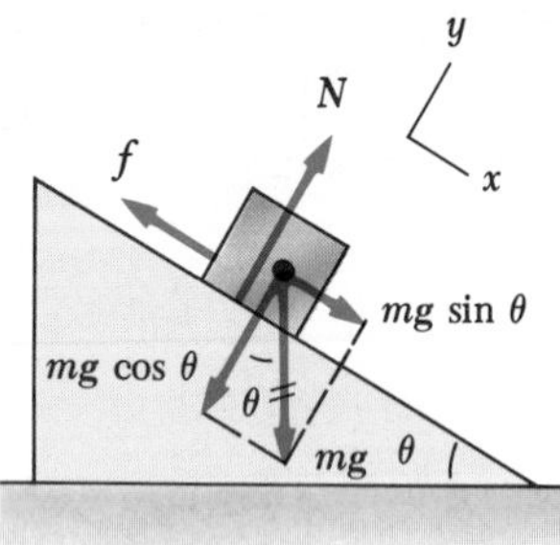

Figure 5.17 (Example 5.8) The external forces acting on a block lying on a rough incline are the weight, mg, the normal force, **N**, and the force of friction, f. Note that the weight vector is resolved into a component along the incline, $mg\sin\theta$, and a component perpendicular to the incline, $mg\cos\theta$.

EXAMPLE 5.9 The Sliding Hockey Puck

A hockey puck on a frozen pond is hit and given an initial speed of 20 m/s. If the puck always remains on the ice and slides a distance of 120 m before coming to rest, determine the coefficient of kinetic friction between the puck and the ice.

Solution The forces acting on the puck after it is in motion are shown in Figure 5.18. If we assume that the force of friction, f_k, remains constant, then this force produces a uniform negative acceleration of the puck. Applying Newton's second law to the puck in component form gives

(1) $\sum F_x = -f_k = ma$

(2) $\sum F_y = N - mg = 0 \quad (a_y = 0)$

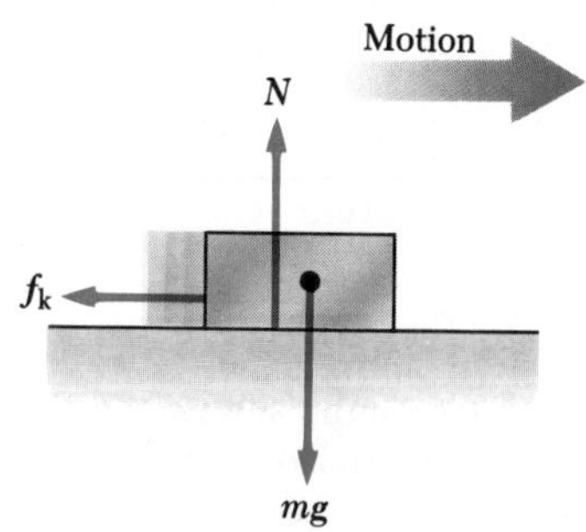

Figure 5.18 (Example 5.9) *After* the puck is given an initial velocity, the external forces acting on it are the weight, mg, the normal force, **N**, and the force of kinetic friction, f_k.

But $f_k = \mu_k N$, and from (2) we see that $N = mg$. Therefore, (1) becomes

$$-\mu_k N = -\mu_k mg = ma$$

$$a = -\mu_k g$$

The negative sign means that the acceleration is to the left, corresponding to a negative acceleration of the puck. Also, the acceleration is independent of the mass of the puck and is *constant* since we are assuming that μ_k remains constant.

Since the acceleration is constant, we can use the kinematic equation $v^2 = v_0^2 + 2ax$, with the final speed $v = 0$. This gives

$$v_0^2 + 2ax = v_0^2 - 2\mu_k gx = 0$$

$$\mu_k = \frac{v_0^2}{2gx}$$

In our example, $v_0 = 20$ m/s and $x = 120$ m:

$$\mu_k = \frac{(20 \text{ m/s})^2}{2(9.80 \text{ m/s}^2)(120 \text{ m})} = 0.170$$

Note that μ_k has no dimensions.

EXAMPLE 5.10 Connected Objects with Friction □ A block of mass m_1 on a rough, horizontal surface is connected to a second mass m_2 by a light cord over a light, frictionless pulley as in Figure 5.19a. A force of magnitude F is applied to mass m_1 as shown. The coefficient of kinetic friction between m_1 and the surface is μ. Determine the acceleration of the masses and the tension in the cord.

Solution First we draw the free-body diagrams of m_1 and m_2 as in Figures 5.19b and 5.19c. Note that the force $\mathbf{F}$ has components $F_x = F\cos\theta$ and $F_y = F\sin\theta$. Therefore, in this case N is *not* equal to $m_1 g$. Applying Newton's second law to both masses and *assuming* the motion of m_1 is to the right, we get

Motion of m_1: $\sum F_x = F\cos\theta - f_k - T = m_1 a$

(1) $\sum F_y = N + F\sin\theta - m_1 g = 0$

Motion of m_2: $\sum F_x = 0$

(2) $\sum F_y = T - m_2 g = m_2 a$

But $f_k = \mu N$, and from (1), $N = m_1 g - F\sin\theta$; therefore

(3) $f_k = \mu(m_1 g - F\sin\theta)$

That is, the frictional force is *reduced* because of the positive y component of $\mathbf{F}$. Substituting (3) and the value of T from (2) into (1) gives

$$F\cos\theta - \mu(m_1 g - F\sin\theta) - m_2(a + g) = m_1 a$$

Solving for a, we get

$$(4) \qquad a = \frac{F(\cos\theta + \mu\sin\theta) - g(m_2 + \mu m_1)}{m_1 + m_2}$$

We can find T by substituting this value of a into (2). Note that the acceleration for m_1 can be either to the right or left,[5] depending on the sign of the numerator in (4). If the motion of m_1 is to the *left*, we must reverse the sign of f_k since the frictional force *opposes* the motion. In this case, the value of a is the same as in (4) with μ replaced by $-\mu$.

[5] A close examination of (4) shows that when $\mu m_1 > m_2$, there is a range of values of F for which no motion occurs at a given angle θ.

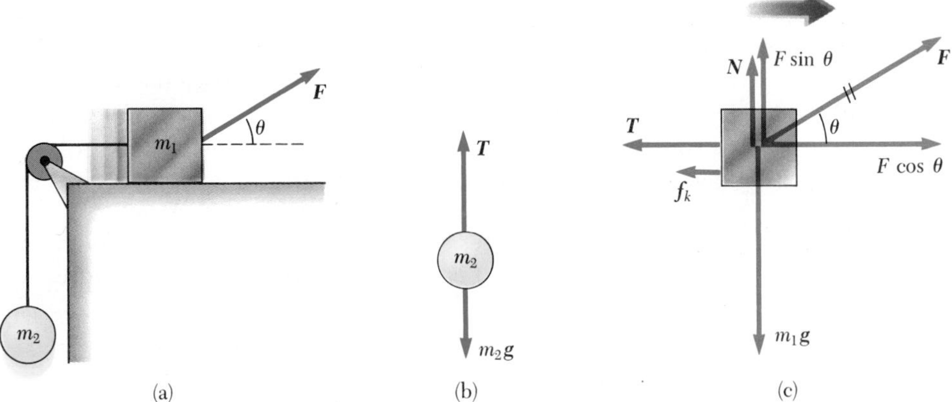

Figure 5.19 (Example 5.10) (a) The external force, $\mathbf{F}$, applied as shown can cause m_1 to accelerate to the right. (b) and (c) The free-body diagrams assuming that m_1 accelerates to the right while m_2 accelerates upward. Note that the force of kinetic friction in this case is given by $f_k = \mu_k N = \mu_k(m_1 g - F\sin\theta)$.

SUMMARY

Newton's first law

Newton's first law states that a body at rest will remain at rest or a body in uniform motion in a straight line will maintain that motion unless an external resultant force acts on the body.

Newton's second law

Newton's second law states that the time rate of change of momentum of a body is equal to the resultant force acting on the body. If the mass of the body is constant, the net force equals the product of the mass and its acceleration, or $\Sigma \mathbf{F} = m\mathbf{a}$.

Inertial frame

Newton's first and second laws are valid in an inertial frame of reference. An **inertial frame** is one in which an object, subject to no net external force, moves with constant velocity including the special case of $v = 0$.

Mass is a scalar quantity. The mass that appears in Newton's second law is called **inertial mass.**

Weight

The **weight** of a body is equal to the product of its mass and the acceleration of gravity, or $\mathbf{W} = m\mathbf{g}$.

Newton's third law

Newton's third law states that if two bodies interact, the force exerted on body 1 by body 2 is equal to and opposite the force exerted on body 2 by body 1. Thus, an isolated force cannot exist in nature.

Forces of friction

The **maximum force of static friction, f_s,** between a body and a rough surface is proportional to the normal force acting on the body. This maximum force occurs when the body is on the verge of slipping. In general, $f_s \le \mu_s N$, where μ_s is the *coefficient of static friction* and $\mathbf{N}$ is the normal force. When a body slides over a rough surface, the *force* of *kinetic friction*, $\mathbf{f}_k$, is opposite the motion and is also proportional to the normal force. The magnitude of this force is given by $f_k = \mu_k N$, where μ_k is the *coefficient of kinetic friction.* Usually, $\mu_k < \mu_s$.

More on Free-Body Diagrams

As we have seen throughout this chapter, in order to be successful in applying Newton's second law to a mechanical system you must first be able to recognize all the forces acting on the system. That is, you must be able to construct the correct free-body diagram. The importance of constructing the free-body diagram cannot be overemphasized. In Figure 5.20 a number of mechanical systems are presented together with their corresponding free-body diagrams. You should examine these carefully and then proceed to construct free-body diagrams for other systems described in the problems. When a system contains more than one element, it is important that you construct a free-body diagram for *each* element.

As usual, $\mathbf{F}$ denotes some applied force, $\mathbf{W} = m\mathbf{g}$ is the weight, $\mathbf{N}$ denotes a normal force, $\mathbf{f}$ is frictional force, and $\mathbf{T}$ is the force of tension.

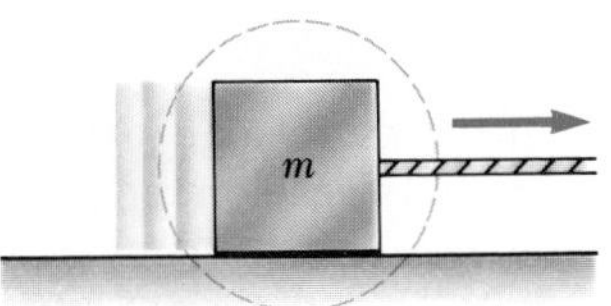

A block pulled to the right on a *rough*, horizontal surface

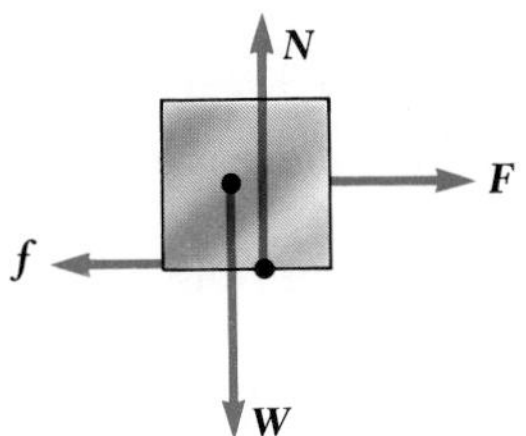

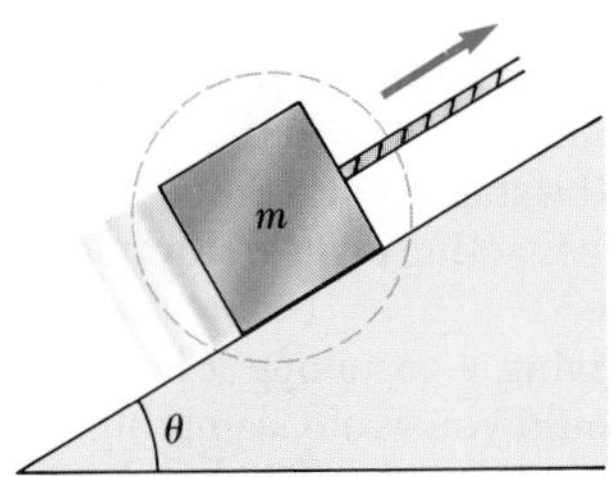

A block being pulled up a *rough* incline

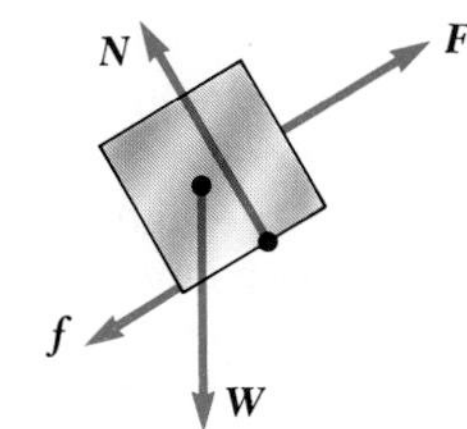

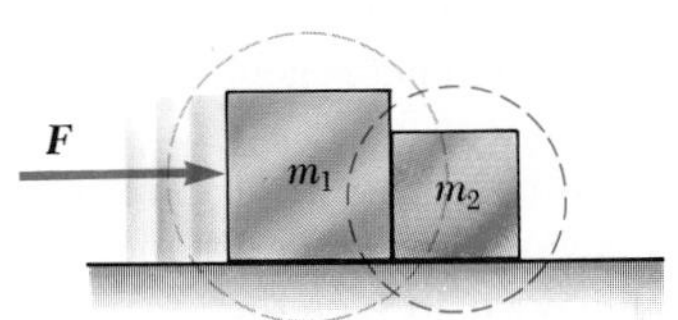

Two blocks in contact, pushed to the right on a smooth surface

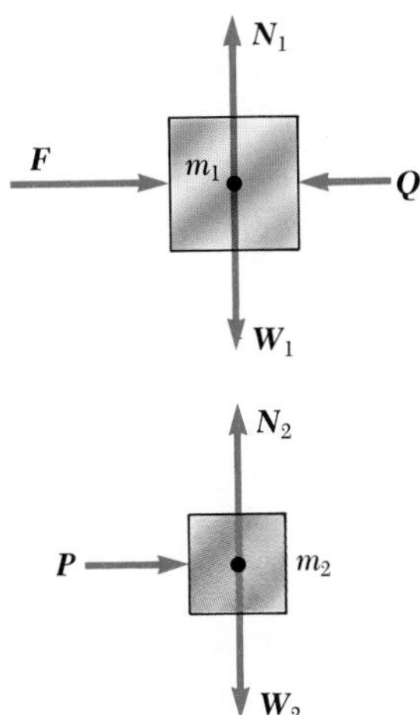

Note: $\boldsymbol{P} = -\boldsymbol{Q}$ since they are an action-reaction pair

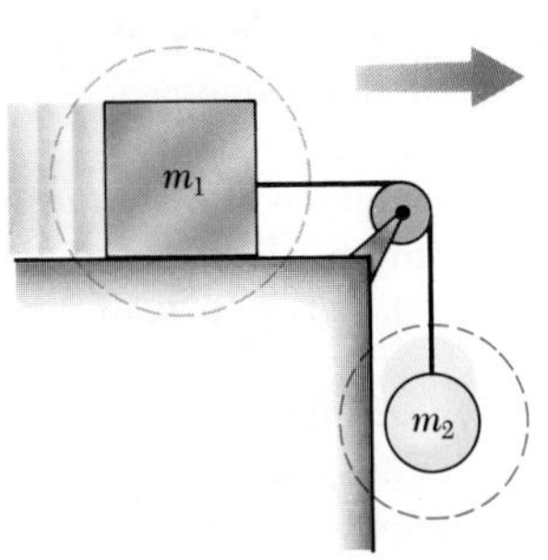

Two blocks connected by a light cord. The surface is rough and the pulley is smooth

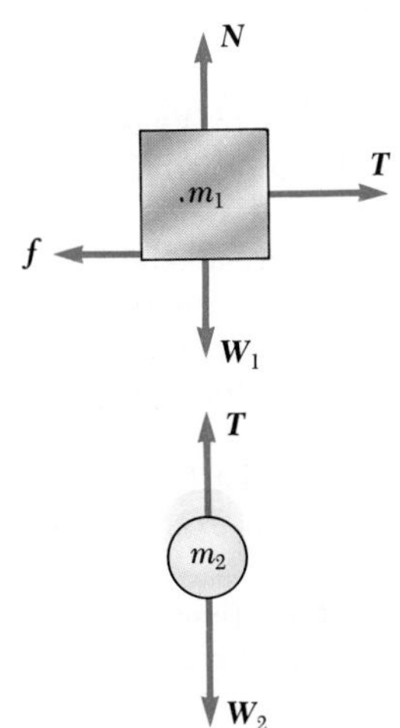

Figure 5.20 Various mechanical configurations (left) and the corresponding free-body diagrams (right).

QUESTIONS

1. If an object is at rest, can we conclude that there are no external forces acting on it?
2. If gold were sold by weight, would you rather buy it in Denver or in Death Valley? If sold by mass, at which of the two locations would you prefer to buy it? Why?
3. A passenger sitting in the rear of a bus claims that he was injured when the driver slammed on the brakes, causing a suitcase to come flying toward the passenger from the front of the bus. If you were the judge in this case, what disposition would you make? Why?
4. A space explorer is in a spaceship moving through space far from any planet or star. She notices a large rock, taken as a specimen from an alien planet, floating around the cabin of the spaceship. Should she push it gently toward a storage compartment or kick it toward the compartment? Why?
5. How much does an astronaut weigh out in space, far from any planet?
6. Although the frictional force between two surfaces may decrease as the surfaces are smoothed, the force will again increase if the surfaces are made extremely smooth and flat. How do you explain this?
7. Why is it that the frictional force involved in the rolling of one body over another is less than for a sliding motion?
8. A massive metal object on a rough metal surface may undergo contact welding to that surface. Discuss how this affects the frictional forces between the object and the surface.
9. The observer in the elevator of Example 5.7 would claim that the "weight" of the fish is T, the scale reading. This is obviously wrong. Why does this observation differ from that of a person outside the elevator at rest with respect to the elevator?
10. Identify the action-reaction pairs in the following situations: a man takes a step; a snowball hits a girl in the back; a baseball player catches a ball; a gust of wind strikes a window.
11. While a football is in flight, what forces act on it? What are the action-reaction pairs while the football is being kicked and while it is in flight?
12. A ball is held in a person's hand. (a) Identify all the external forces acting on the ball and the reaction to each. (b) If the ball is dropped, what force is exerted on it while it is falling? Identify the reaction force in this case. (Neglect air resistance.)
13. Identify all the action-reaction pairs that exist for a horse pulling a cart. Include the earth in your examination.
14. If a car is traveling westward with a constant speed of 20 m/s, what is the resultant force acting on it?
15. A large crate is placed on the bed of a truck without being tied to the truck. (a) As the truck accelerates forward, the crate remains at rest relative to the truck. What force causes the crate to accelerate? (b) If the truck driver slams on the brakes, what could happen to the crate?
16. A child pulls a wagon with some force, causing it to accelerate. Newton's third law says that the wagon exerts an equal and opposite reaction force on the child. How can the wagon accelerate?
17. A rubber ball is dropped onto the floor. What force causes the ball to bounce back into the air?
18. What is wrong with the statement, "Since the car is at rest, there are no forces acting on it"? How would you correct this sentence?
19. Suppose you are driving a car along a highway at a high speed. Why should you avoid slamming on your brakes if you want to stop in the shortest distance?
20. If you have ever taken a ride in an elevator of a high-rise building, you may have experienced the nauseating sensation of "heaviness" and "lightness" depending on the direction of $\boldsymbol{a}$. Explain these sensations. Are we truly weightless in free fall?
21. A 0.15-kg baseball is thrown upward with an initial speed of 20 m/s. If air resistance is neglected, what is the net force on the ball when it reaches half its maximum height?
22. The driver of a speeding empty truck slams on the brakes and skids to a stop through a distance d. (a) If the truck carried a heavy load such that its mass were doubled, what would be its "skidding distance"? (b) If the initial speed of the truck is halved, what would be its "skidding distance"?
23. Does it make sense to say that an object possesses force? Explain.
24. In an attempt to define Newton's third law, a student states that the action and reaction forces are equal and opposite each other. If this is the case, how can there ever be a net force on an object?
25. In a tug-of-war between two athletes, each athlete pulls on the rope with a force of 200 N. What is the tension in the rope? What force does each athlete exert against the ground?
26. If you push on a heavy box which is at rest, it requires some force $\boldsymbol{F}$ to start its motion. However, once it is sliding, it requires a *smaller* force to maintain that motion. Why is this so?
27. What causes a rotary lawn sprinkler to turn?
28. The force of gravity is twice as great on a 20-N rock as it is on a 10-N rock. Why doesn't the 20-N rock have a greater free-fall acceleration?
29. Is it possible to have motion in the absence of a force? Explain.
30. Is there any relation between the net force acting on an object and the direction in which it moves? Explain.

PROBLEMS

Section 5.1 through Section 5.7

1. A force, $\boldsymbol{F}$, applied to an object of mass m_1 produces an acceleration of 3 m/s^2. The same force applied to a second object of mass m_2 produces an acceleration of 1 m/s^2. (a) What is the value of the ratio m_1/m_2? (b) If m_1 and m_2 are combined, find their acceleration under the action of the force $\boldsymbol{F}$.
2. A 6-kg object undergoes an acceleration of 2 m/s^2. (a) What is the magnitude of the resultant force acting on it? (b) If this same force is applied to a 4-kg object, what acceleration will it produce?
3. A force of 10 N acts on a body of mass 2 kg. What is (a) the acceleration of the body, (b) its weight in N, and (c) its acceleration if the force is doubled?
4. A 3-kg particle starts from rest and moves a distance of 4 m in 2 s under the action of a single, constant force. Find the magnitude of the force.
5. A 5.0-g bullet leaves the muzzle of a rifle with a speed of 320 m/s. What average force is exerted on the bullet while it is traveling down the 0.82-m-long barrel of the rifle?
6. A pitcher releases a baseball of weight 1.4 N with a speed of 32 m/s by uniformly accelerating his arm for 0.09 s. If the ball starts from rest, (a) through what distance does the ball accelerate before release? (b) What average force is exerted on the ball to produce this acceleration?
7. A 3-kg mass undergoes an acceleration given by $\boldsymbol{a} = (2\boldsymbol{i} + 5\boldsymbol{j})$ m/s^2. Find the resultant force, $\boldsymbol{F}$, and its magnitude.
8. Verify the following conversions: (a) 1 N = 10^5 dynes, (b) 1 N = 0.225 lb.
9. A person weighs 120 lb. Determine (a) her weight in N and (b) her mass in kg.
10. On a hypothetical planet, the acceleration due to gravity at its surface is 0.072 g. If an object weighs 25 N on the earth, what will it weigh on the surface of the hypothetical planet?
11. What is the mass of an astronaut whose weight on the moon is 115 N? The acceleration due to gravity on the moon is 1.63 m/s^2.
12. If a man weighs 900 N on earth, what would he weigh on Jupiter, where the acceleration due to gravity is 25.9 m/s^2?
13. On planet X, an object weighs 10 N. On planet B where the acceleration due to gravity is 1.6g, the object weighs 27 N. What is the mass of the object and what is the acceleration due to gravity (in m/s^2) on planet X?
14. One or more external forces are exerted on each object shown in Figure 5.1. Clearly identify the reaction to all of these forces. (*Note:* The reaction forces act on other objects.)
15. A brick of weight $\boldsymbol{W}$ rests on top of a vertical spring of weight $\boldsymbol{W}_s$. The spring rests on a table. (a) Draw a free-body diagram of the brick and label all forces acting on it. (b) Repeat (a) for the spring. (c) Identify all action-reaction pairs of forces in this brick-spring-table-earth system.
16. Two forces $\boldsymbol{F}_1$ and $\boldsymbol{F}_2$ act on a body of mass 28 kg. $F_1 = 8$ N in a direction due east while $F_2 = 10$ N in a direction 50° west of north. What are the magnitude and direction of the acceleration of this body?
17. Two forces $\boldsymbol{F}_1 = (3\boldsymbol{i} - 5\boldsymbol{j})$ N and $\boldsymbol{F}_2 = (2\boldsymbol{i} + \boldsymbol{j})$ N act on a 1.5-kg body. (a) What are the magnitude and direction of the acceleration? (b) If the body is initially resting at the origin, what are the coordinates of its position at $t = 4.0$ s?
18. Two forces $\boldsymbol{F}_1$ and $\boldsymbol{F}_2$ act on a 5-kg mass. If $F_1 = 20$ N and $F_2 = 15$ N, find the acceleration in (a) and (b) of Figure 5.21.

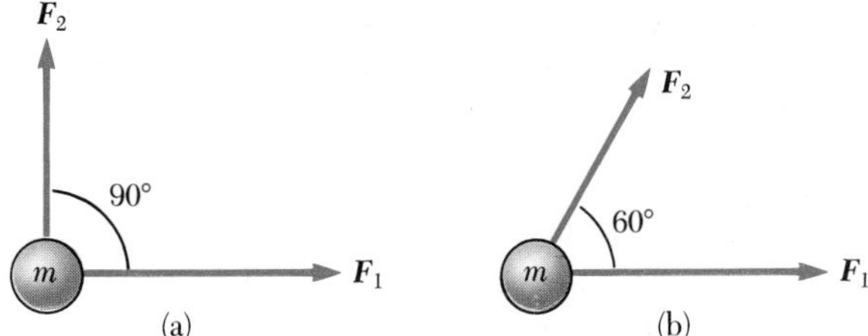

Figure 5.21 (Problem 18).

19. A 2-kg particle moves along the x axis under the action of a single, constant force. If the particle starts from rest at the origin at $t = 0$ and is observed to have a velocity of $-6.0\boldsymbol{i}$ m/s at $t = 4$ s, what are the magnitude and direction of the force?
20. A 9-kg object undergoes an acceleration of 2 m/s^2 to the right under the action of two forces, $\boldsymbol{F}_1$ and $\boldsymbol{F}_2$. $\boldsymbol{F}_1$ acts to the right and has a magnitude of 25 N. What are the magnitude and direction of $\boldsymbol{F}_2$?
21. A 4-kg object has a velocity of $3\boldsymbol{i}$ m/s at one instant. Eight seconds later, its velocity is $(8\boldsymbol{i} + 10\boldsymbol{j})$ m/s. Assuming the object was subject to a constant net force, find (a) the components of the force and (b) its magnitude.
22. A horse of mass 240 kg attempts to pull a 100-kg cart. What horizontal force must the ground exert on the horse to produce an acceleration of 2.0 m/s^2? (The cart has frictionless wheels.)
23. A 2-ton truck provides an acceleration of 3 ft/s^2 to a 5-ton trailer. If the truck exerts the same pull on a 15-ton trailer, what acceleration will result?
24. An electron of mass 9.1×10^{-31} kg has an initial speed of 3.0×10^5 m/s. It travels in a straight line, and

its speed increases to 7.0×10^5 m/s in a distance of 5.0 cm. Assuming its acceleration is constant, (a) determine the force on the electron and (b) compare this force with the weight of the electron, which we neglected.

25. A 15-lb block rests on the floor. (a) What force does the floor exert on the block? (b) If a rope is tied to the block and run vertically over a pulley and the other end attached to a free-hanging 10-lb weight, what is the force of the floor on the 15-lb block? (c) If we replace the 10-lb weight in (b) by a 20-lb weight, what is the force of the floor on the 15-lb block?

Section 5.8 Some Applications of Newton's Laws

26. Find the tension in each cord for the systems described in Figure 5.22. (Neglect the mass of the cords.)

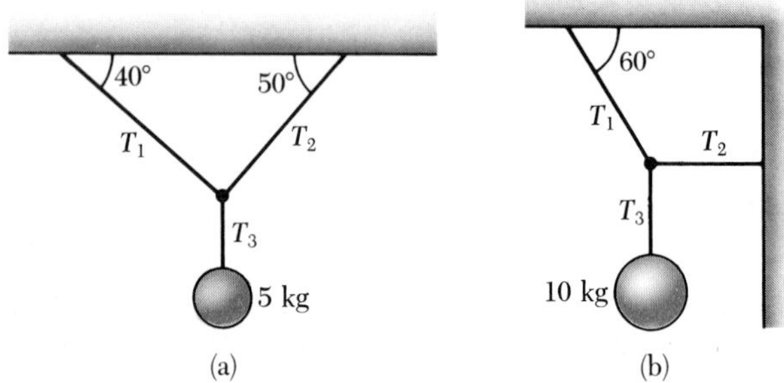

Figure 5.22 (Problem 26).

27. Three 10-kg objects are suspended as shown (Fig. 5.23). Determine each of the labeled tension forces.

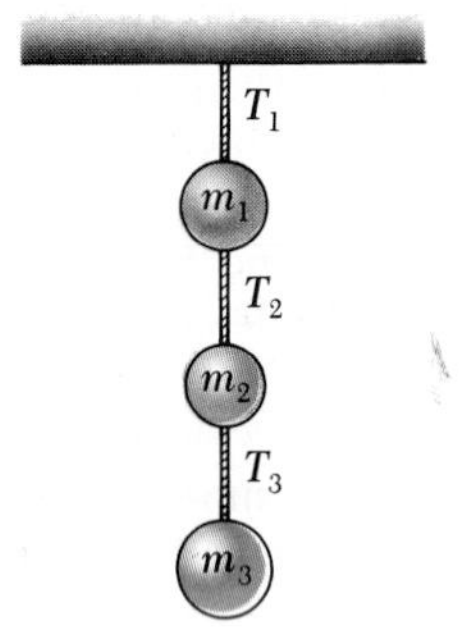

Figure 5.23 (Problem 27).

28. A 200-N weight is tied to the middle of a strong rope, and two people pull at opposite ends of the rope in an attempt to lift the weight. (a) What force $\boldsymbol{F}$ must each person apply to suspend the weight as shown in Figure 5.24? (b) Can they pull in such a way as to make the rope horizontal? Explain.

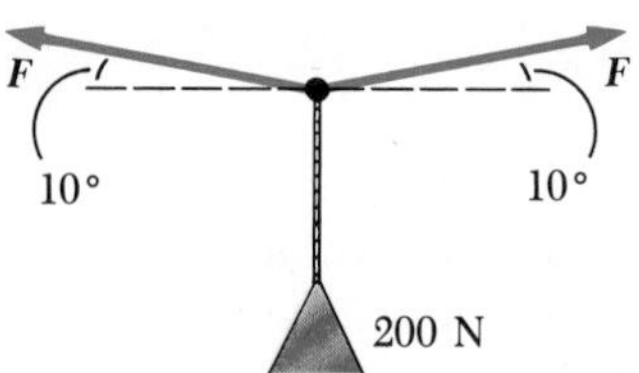

Figure 5.24 (Problem 28).

29. A 50-kg mass hangs from a rope 5 m in length, which is fastened to the ceiling. What horizontal force applied to the mass will deflect it 1 m sideways from the vertical and maintain it in that position?

30. The systems shown in Figure 5.25 are in equilibrium. If the spring scales are calibrated in N, what do they read in each case? (Neglect the mass of the pulleys and strings, and assume the incline is smooth.)

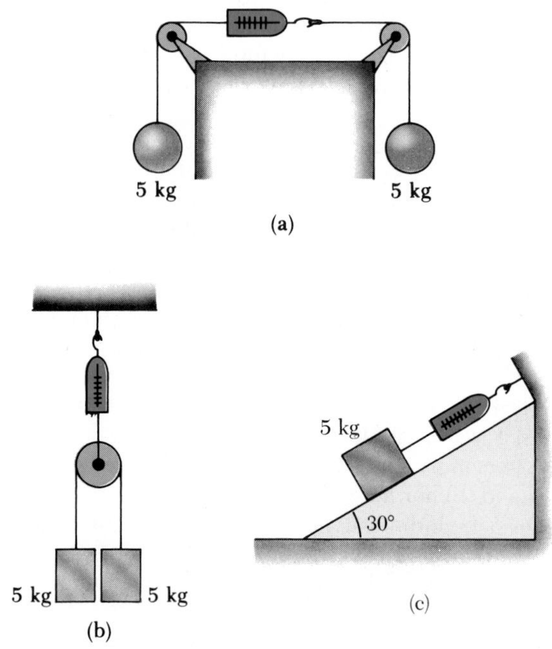

Figure 5.25 (Problem 30).

31. A bag of cement hangs from three wires as shown (Fig. 5.26). Two of the wires make angles θ_1 and θ_2 with the horizontal. If the system is in equilibrium, (a) show that

$$T_1 = \frac{W\cos(\theta_2)}{\sin(\theta_1 + \theta_2)}$$

(b) Given that $W = 200$ N, $\theta_1 = 10°$ and $\theta_2 = 25°$, find the tensions T_1, T_2, and T_3 in the wires.

32. A woman at an airport is pulling her 20-kg suitcase at constant speed by pulling on a strap at an angle θ

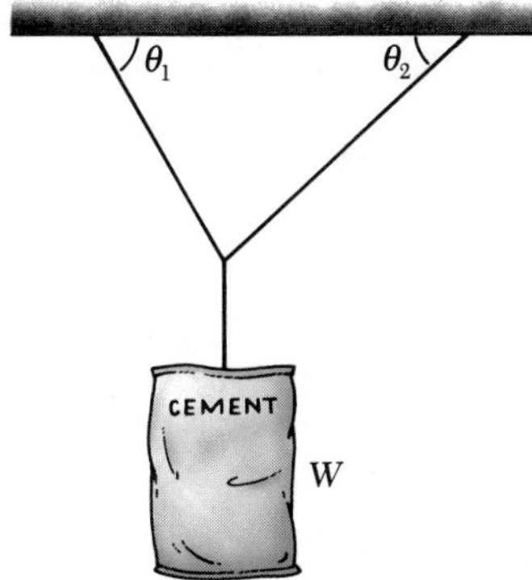

Figure 5.26 (Problem 31).

above the horizontal (Fig. 5.27). She pulls on the strap with a 35-N force and the friction force on the suitcase is 20 N. (a) What angle does the strap make with the horizontal? (b) What normal force does the ground exert on the suitcase?

Figure 5.27 (Problem 32).

33. The parachute on a race car of weight 8820 N opens at the end of a quarter-mile run when the car is traveling at 55 m/s What is the total retarding force required to stop the car in a distance of 1000 m in the event of a brake failure?
34. A rifle bullet with a mass of 12 g, traveling with a speed of 400 m/s, strikes a large wooden block, which it penetrates to a depth of 15 cm. Determine the magnitude of the frictional force (assumed constant) that acts on the bullet.
35. An automobile ($m = 2200$ kg) moves with a velocity $v = 32$ m/s on a level road. If a constant braking force of 6000 N is applied to the automobile wheels, what distance will the vehicle travel after the brakes are applied?
36. A rope with a mass $m = 0.8$ kg is attached to a block with a mass $M = 4$ kg. The rope-block combination is pulled across a horizontal frictionless surface by a 12-N force that is applied to the free end of the rope. What is the acceleration of the system? What force does the rope exert on the block?
37. In the system shown (Fig. 5.28), a horizontal force $\boldsymbol{F}_x$ acts on the 8-kg mass. (a) For what values of $\boldsymbol{F}_x$ will the 2-kg mass accelerate upward? (b) For what values of $\boldsymbol{F}_x$ will the tension in the cord be zero? (c) Plot the acceleration of the 8-kg mass versus $\boldsymbol{F}_x$. Include values of $\boldsymbol{F}_x$ from -100 N to $+100$ N.

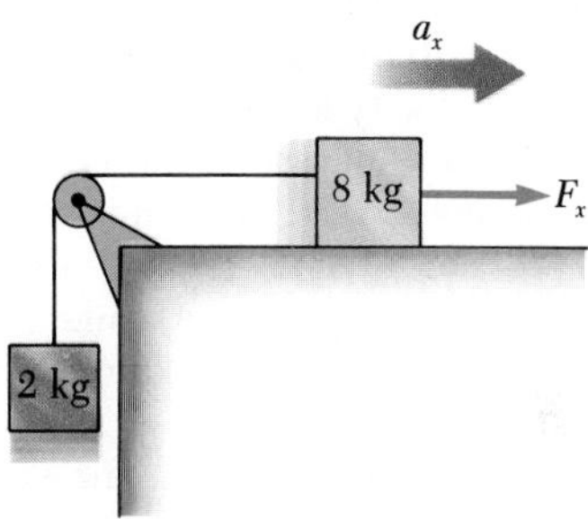

Figure 5.28 (Problem 37).

38. Two masses, m_1 and m_2, situated on a frictionless, horizontal surface are connected by a light string. A force, $\boldsymbol{F}$, is exerted on one of the masses to the right (Fig. 5.29). Determine the acceleration of the system and the tension, T, in the string.

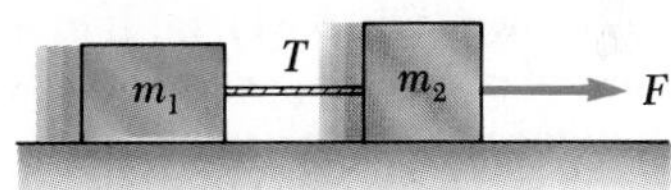

Figure 5.29 (Problems 38 and 55).

39. Two masses of 3 kg and 5 kg are connected by a light string that passes over a smooth pulley as in Figure 5.12. Determine (a) the tension in the string, (b) the acceleration of each mass, and (c) the distance each mass moves in the first second of motion if they start from rest.
40. A block slides down a smooth plane having an inclination of $\theta = 15°$ (Fig. 5.30). If the block starts from rest

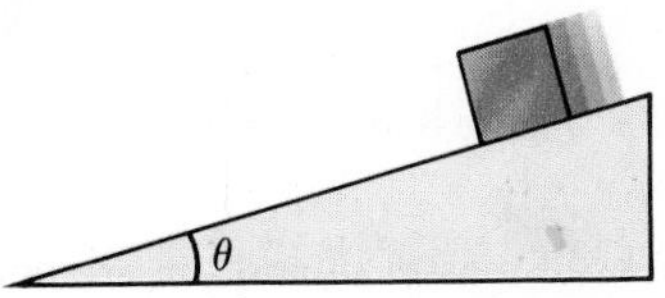

Figure 5.30 (Problems 40 and 41).

at the top and the length of the incline is 2 m, find (a) the acceleration of the block and (b) its speed when it reaches the bottom of the incline.

41. A block is given an initial velocity of 5 m/s up a smooth 20° incline (Fig. 5.30). How far up the incline does the block slide before coming to rest?

42. Two masses are connected by a light string that passes over a smooth pulley as in Figure 5.13. If the incline is frictionless and if $m_1 = 2$ kg, $m_2 = 6$ kg, and $\theta = 55°$, find (a) the acceleration of the masses, (b) the tension in the string, and (c) the speed of each mass 2 s after they are released from rest.

43. A 72-kg man stands on a spring scale in an elevator. Starting from rest, the elevator ascends, attaining its maximum velocityof 1.2 m/s in 0.8 s. It travels with this constant velocity for the next 5.0 s. The elevator then undergoes a uniform *negative* acceleration for 1.5 s and comes to rest. What does the spring scale register (a) before the elevator starts to move? (b) during the first 0.8 s? (c) while the elevator is traveling at constant velocity? (d) during the negative acceleration period?

44. A 3.0-kg mass is moving in a plane with its x and y coordinates given by $x = 5t^2 - 1$ and $y = 3t^3 + 2$, where x and y are given in meters while t is in seconds. Find the magnitude of the net force acting on this mass at $t = 2.0$ s.

45. A net horizontal force $F = A + Bt^3$ acts on a 2-kg object, where $A = 5.0$ N and $B = 2.0$ N/s^3. What is the horizontal velocity of this object 4 seconds after it starts from rest?

46. Mass m_1 on a smooth horizontal table is connected to mass m_2 through a very light pulley P_1 and a light fixed pulley P_2 as shown (Fig. 5.31). (a) If a_1 and a_2 are the accelerations of m_1 and m_2, respectively, what is the relation between these accelerations? Express the (b) tensions in the strings, and (c) the accelerations a_1 and a_2 in terms of the masses m_1, m_2, and g.

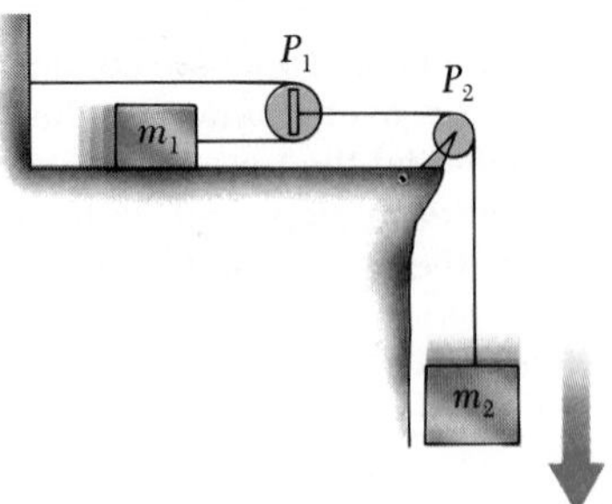

Figure 5.31 (Problem 46).

47. Two blocks of masses m and $4m$ are connected by a light string passing over a fixed pulley as shown (Fig. 5.32). The $4m$ mass is on a smooth horizontal surface and is pushed toward the left by a constant horizontal force F. (a) What value of F will give the system an acceleration of $0.4g$? (b) Find the tension in the string when F has the value determined in (a). (c) What happens when $F > 4mg$?

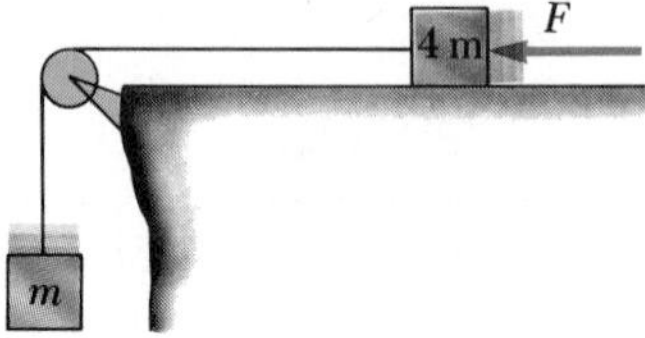

Figure 5.32 (Problem 47).

Section 5.9 Forces of Friction

48. A 25-kg block is initially at rest on a rough, horizontal surface. A horizontal force of 75 N is required to set the block in motion. After it is in motion, a horizontal force of 60 N is required to keep the block moving with constant speed. Find the coefficients of static and kinetic friction from this information.

49. The coefficient of static friction between a 5-kg block and a horizontal surface is 0.4. What is the *maximum* horizontal force that can be applied to the block before it slips?

50. A racing car accelerates uniformly from 0 to 80 mi/h in 8 s. The external force that accelerates the car is the frictional force between the tires and the road. If the tires do not spin, determine the *minimum* coefficient of friction between the tires and the road.

51. In a game of shuffleboard, a disk is given an initial speed of 5 m/s. It slides a distance of 8 m before coming to rest. What is the coefficient of kinetic friction between the disk and the surface?

52. A car is traveling at 50 mi/h on a horizontal highway. (a) If the coefficient of friction between the road and tires on a rainy day is 0.1, what is the *minimum* distance in which the car will stop? (b) What is the stopping distance when the surface is dry and $\mu = 0.6$? (c) Why should you avoid "slamming on" your brakes if you want to stop in the shortest distance?

53. A boy drags his 60-N sled at constant speed up a 15° hill. He does so by pulling with a 25-N force on a rope attached to the sled. If the rope is inclined at 35° to the horizontal, (a) what is the coefficient of kinetic friction between the sled and the snow? (b) At the top of the hill, he jumps on the sled and slides down the hill. What is his acceleration down the slope?

54. A block moves up a 45° incline with constant speed under the action of a force of 15 N applied *parallel* to the incline. If the coefficient of kinetic friction is 0.3, determine (a) the weight of the block and (b) the minimum force required to allow the block to move *down* the incline at constant speed.

55. Two blocks connected by a light rope are being dragged by a horizontal force F (Fig. 5.29). Suppose that $F = 50$ N, $m_1 = 10$ kg, $m_2 = 20$ kg, and the coefficient of kinetic friction between each block and the surface is 0.1. (a) Draw a free-body diagram for each block. (b) Determine the tension, T, and the acceleration of the system.

56. A hockey puck is hit on a frozen lake and starts moving with a velocity of 12.0 m/s. At $t = 5.0$ s, its velocity is 6.0 m/s. (a) What is the average acceleration of the puck? (b) What is the average value of the coefficient of friction between the puck and the ice? (c) How far does the puck travel during this 5-s interval?

57. A 3-kg block starts from rest at the top of a 30° incline and slides a distance of 2 m down the incline in 1.5 s. Find (a) the acceleration of the block, (b) the coefficient of kinetic friction between the block and the plane, (c) the frictional force acting on the block, and (d) the speed of the block after it has slid 2 m.

58. A block slides on a *rough* incline. The coefficient of kinetic friction between the block and the plane is μ_k. (a) If the block accelerates *down* the incline, show that the acceleration of the block is given by $a = g(\sin\theta - \mu_k \cos\theta)$. (b) If the block is projected *up* the incline, show that its acceleration is $a = -g(\sin\theta + \mu_k\cos\theta)$.

59. In order to determine the coefficients of friction between rubber and various surfaces, a student uses a rubber eraser and an incline. In one experiment the eraser slips down the incline when the angle of inclination is 36° and then moves down the incline with constant speed when the angle is reduced to 30°. From these data, determine the coefficients of static and kinetic friction for this experiment.

60. A crate of weight $\mathbf{W}$ is pushed by a force $\mathbf{F}$ on a rough horizontal floor. If the coefficient of static friction is μ_s and F is directed at angle ϕ *below* the horizontal (a) show that the minimum value of F that will move the crate is given by

$$F = \frac{\mu_s W \sec\phi}{1 - \mu_s \tan\phi}$$

(b) Find the minimum value of F which can produce motion when $\mu_s = 0.4$, $W = 100$ N, and $\phi = 0°$, 15°, 30°, 45°, and 60°.

61. Two masses are connected by a light string, which passes over a frictionless pulley as in Figure 5.13. The incline is rough. When $m_1 = 3$ kg, $m_2 = 10$ kg, and $\theta = 60°$, the 10-kg mass moves down the incline and accelerates *down* the incline at the rate of 2 m/s². Find (a) the tension in the string and (b) the coefficient of kinetic friction between the 10-kg mass and the plane.

62. A box rests on the back of a truck. The coefficient of static friction between the box and the surface is 0.3. (a) When the truck accelerates, what force accelerates the box? (b) Find the *maximum* acceleration the truck can have before the box slides.

63. A block slides down a 30° incline with *constant* acceleration. The block starts from rest at the top and travels 18 m to the bottom, where its speed is 3 m/s. Find (a) the coefficient of kinetic friction between the block and the incline and (b) the acceleration of the block.

ADDITIONAL PROBLEMS

64. Two masses on a rough horizontal surface are connected by a light, rigid bar. m_1 is to the left of m_2. A horizontal force $\mathbf{F}$ is applied to the mass m_1 toward m_2 causing the system to accelerate to the right. The coefficient of kinetic friction between the blocks and the surface is μ. (a) Draw a free-body diagram for *each* mass. Identify all the forces in your diagrams. (b) Write a statement of Newton's second law in the horizontal and vertical direction for each mass in *symbolic* form. (c) Find the contact force between the bar and each block in terms of m_1, m_2, and $\mathbf{F}$. (d) Find the acceleration of the system in terms of the given parameters and g.

65. A mass M is held in place by an applied force $\mathbf{F_A}$ and a pulley system as shown in Figure 5.33. The pulleys are massless and frictionless. Find (a) the tension in each section of rope, T_1, T_2, T_3, T_4, and T_5 and (b) the applied force $\mathbf{F_A}$.

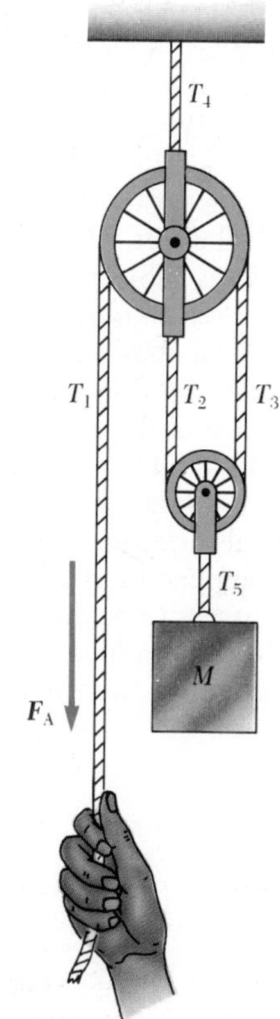

Figure 5.33 (Problem 65).

66. "Big" Al remembered from high school physics that pulleys can be used to aid in lifting heavy objects. Al designed the pulley system, shown in Figure 5.34, to lift a safe to a second-floor office. The safe weighs 400 lb, and Al can pull with a force of 240 lb. (a) Will Big Al be able to raise the safe? (b) What is the maximum weight Big Al can lift using his pulley system? (*Note:* The large pulley is fastened by a yoke to the rope that Big Al is pulling.)

Figure 5.34 (Problem 66).

67. The largest caliber antiaircraft gun operated by the Luftwaffe during World War II was the 12.8-cm Flak 40. This weapon fired a 25.8-kg shell with a muzzle velocity of 880 m/s. What propulsive force was necessary to attain the muzzle velocity within the 6.0-m barrel? (Assume constant acceleration and neglect the earth's gravitational effect.)

68. A 0.50-kg particle is subject to the earth's gravitational force and a *constant horizontal* force of 15 N. It is held in place 5 cm above a table top, and 20 cm from the edge as shown (Fig. 5.35). When released, will it miss the edge of the table? If not, how far in from the edge does it hit?

69. Three blocks are in contact with each other on a frictionless, horizontal surface as in Figure 5.36. A horizontal force $\boldsymbol{F}$ is applied to m_1. If $m_1 = 2$ kg, $m_2 = 3$ kg, $m_3 = 4$ kg, and $F = 18$ N, find (a) the acceleration of the blocks, (b) the *resultant* force on each block, and (c) the magnitude of the contact forces between the blocks.

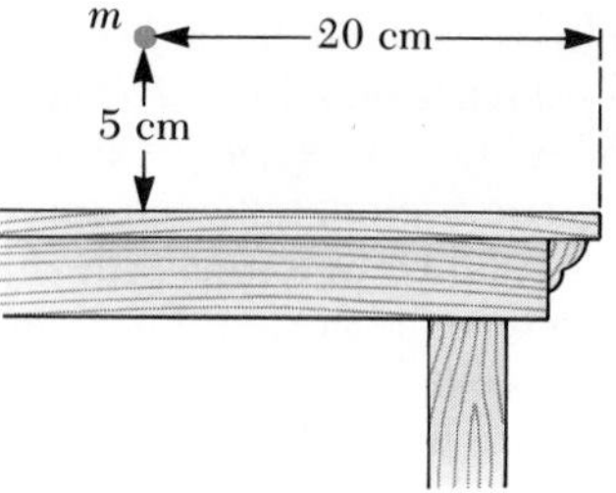

Figure 5.35 (Problem 68).

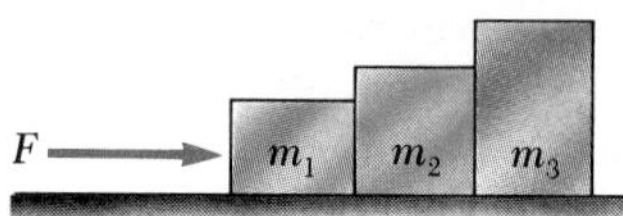

Figure 5.36 (Problems 69 and 70).

70. Repeat Problem 69 given that the coefficient of kinetic friction between the blocks and the surface is 0.1. Use the data given in Problem 69.

71. Three baggage carts of masses m_1, m_2, and m_3 are towed by a tractor of mass M along an airport apron. The wheels of the tractor exert a total frictional force F on the ground as shown (Fig. 5.37). In the following, express your answers in terms of F, M, m_1, m_2, m_3, and g. (a) What are the magnitude and direction of the horizontal force exerted on the tractor by the ground? (b) What is the smallest value of the coefficient of static friction, μ_s, that will prevent the wheels from slipping? Assume that each of the two drive wheels on the tractor bears 1/3 of the tractor's weight. (c) What is the acceleration, a, of the system (tractor plus baggage carts)? (d) What are the tensions T_1, T_2, and T_3 in the connecting cables? (e) What is the net force on the cart of mass m_2?

72. A horizontal force $\boldsymbol{F}$ is applied to a frictionless pulley of mass m_2 as in Figure 5.38. The horizontal surface is smooth. (a) Show that the acceleration of the block of mass m_1 is *twice* the acceleration of the pulley. Find (b) the acceleration of the pulley and the block and (c) the tension in the string. A constant supporting force is applied to the axle of the pulley equal to its weight.

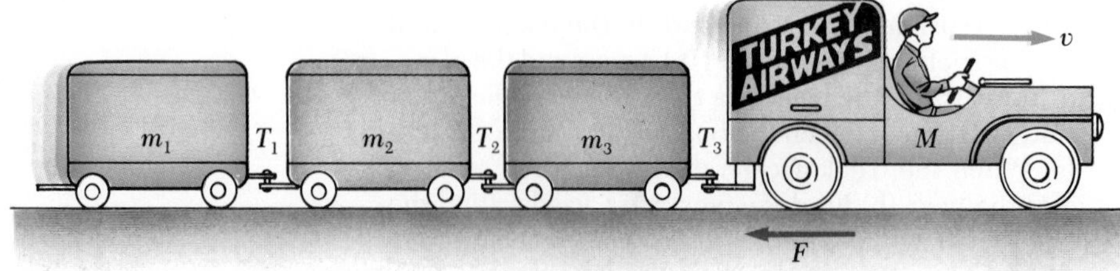

Figure 5.37 (Problem 71).

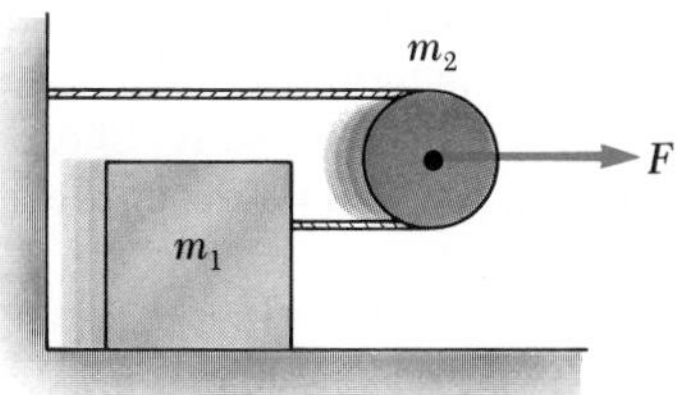

Figure 5.38 (Problem 72).

73. A 2-kg block is placed on top of a 5-kg block as in Figure 5.39. The coefficient of kinetic friction between the 5-kg block and the surface is 0.2. A horizontal force $\boldsymbol{F}$ is applied to the 5-kg block. (a) Draw a free-body diagram for each block. What force accelerates the 2-kg block? (b) Calculate the force necessary to pull both blocks to the right with an acceleration of 3 m/s^2. (c) Find the minimum coefficient of static friction between the blocks such that the 2-kg block does not slip under an acceleration of 3 m/s^2.

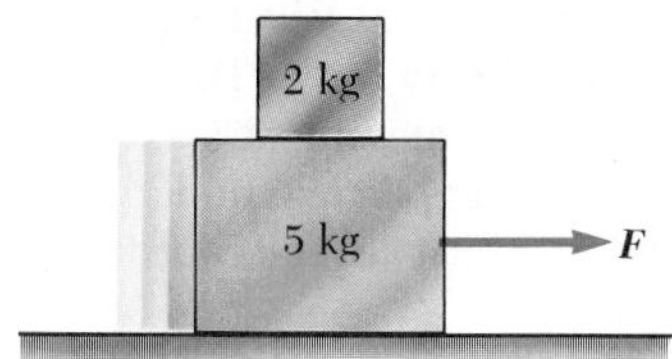

Figure 5.39 (Problem 73).

74. A 5-kg block is placed on top of a 10-kg block (Fig. 5.40). A horizontal force of 45 N is applied to the 10-kg block, while the 5-kg block is tied to the wall. The coefficient of kinetic friction between the moving surfaces is 0.2. (a) Draw a free-body diagram for each block and identify the action-reaction forces between the blocks. (b) Determine the tension in the string and the acceleration of the 10-kg block.

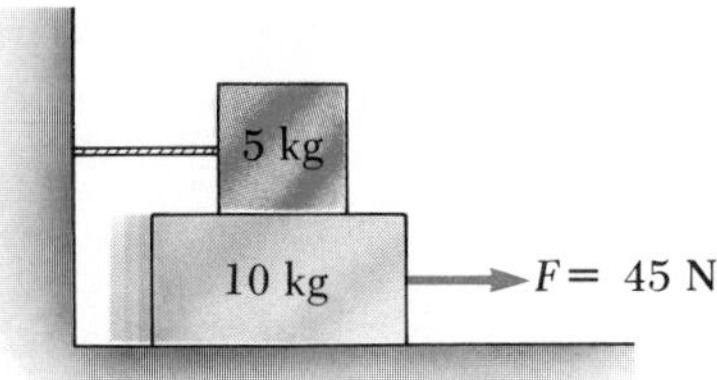

Figure 5.40 (Problem 74).

75. An inventive child named Pat wants to reach an apple in a tree without climbing the tree. Sitting in a chair connected to a rope that passes over a frictionless pulley (Fig. 5.41), Pat pulls on the loose end of the rope with such a force that the spring scale reads 250 N. Pat's true weight is 320 N and the chair weighs 160 N. (a) Draw free-body diagrams for Pat and the chair considered as separate systems, and another diagram for Pat and the chair considered as one system. (b) Show that the acceleration of the system is *upward* and find its magnitude. (c) Find the force that Pat exerts on the chair.

Figure 5.41 (Problem 75).

76. Consider a system consisting of a horse pulling a sled. According to Newton's third law, the force exerted by the horse on the sled is equal to and opposite the force exerted by the sled on the horse. Therefore, one might argue that the system can never move. Explain, using complete force diagrams on the horse and sled, that motion in this system is possible despite Newton's third law. Be sure to identify all of your forces.

77. Two blocks are fastened to the top of an elevator as in Figure 5.42. The elevator accelerates upward at 4

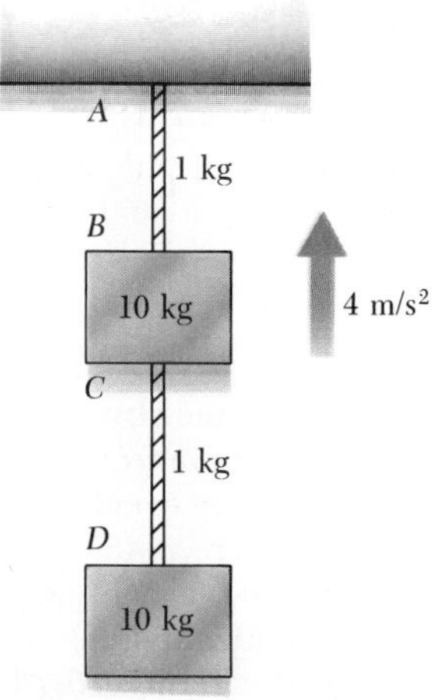

Figure 5.42 (Problem 77).

m/s^2. Each rope has a mass of 1 kg. Find the tensions in the ropes at points A, B, C, and D.

78. Two masses m and M are attached with strings as shown in Figure 5.43. If the system is in equilibrium, show that $\tan \theta = 1 + \dfrac{2M}{m}$.

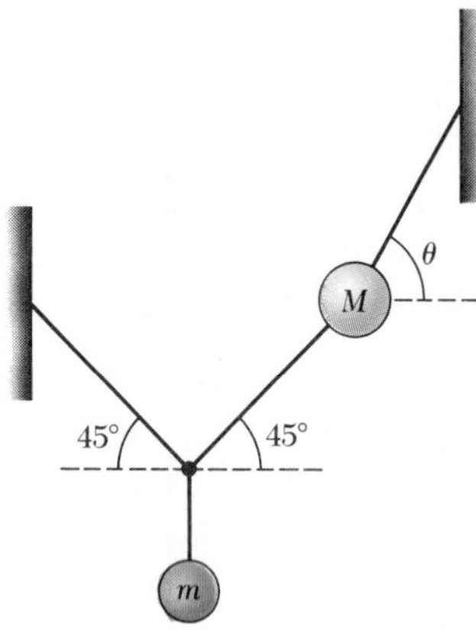

Figure 5.43 (Problem 78).

79. A wire ABC supports a body of weight W as shown (Fig. 5.44). The wire passes over a fixed pulley at B and is firmly attached to a vertical wall at A. AB makes an angle ϕ with the vertical while the pulley at B exerts a force of magnitude F inclined at angle θ with the horizontal on the wire. (a) Show that if the system is in equilibrium, $\theta = (1/2)\phi$. (b) Show that $F = 2W \sin(\phi/2)$. (c) Sketch a graph of F as ϕ increases from 0° to 180°.

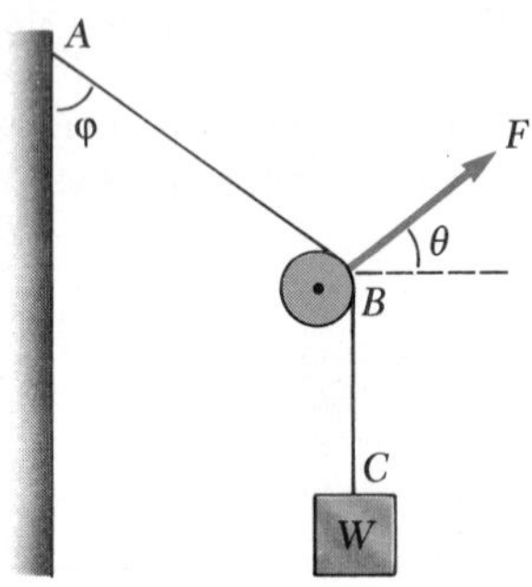

Figure 5.44 (Problem 79).

80. A butterfly mobile is formed by supporting four metal butterflies of equal mass m from a string of length L. The points of support are evenly spaced a distance ℓ apart as shown in Figure 5.45. The string forms an angle θ_1 with the ceiling at each end point. The center section of string is horizontal. (a) Find the tension in each section of string in terms of θ_1, m, and g. (b) Find the angle θ_2, in terms of θ_1, that the sections of string between the outside butterflies and the inside butterflies form with the horizontal. (c) Show that the distance D between the end points of the string is

$$D = \frac{L}{5}\left(2 \cos \theta_1 + 2 \cos \left[\tan^{-1}\left(\tfrac{1}{2} \tan \theta_1\right)\right] + 1\right)$$

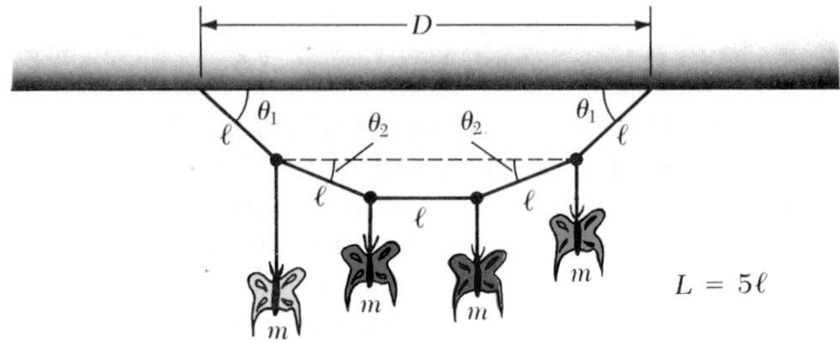

Figure 5.45 (Problem 80).

81. Two forces $\boldsymbol{F}_1 = (-6\boldsymbol{i} - 4\boldsymbol{j})$ N and $\boldsymbol{F}_2 = (-3\boldsymbol{i} + 7\boldsymbol{j})$ N act on a particle of mass 2 kg initially at rest at coordinates (−2 m, +4 m). (a) What are the components of the particle's velocity at $t = 10$ s? (b) In what direction is the particle moving at $t = 10$ s? (c) What displacement does the particle undergo during the first 10 seconds? (d) What are the coordinates of the particle at $t = 10$ s?

82. A bowling ball attached to a spring scale is suspended from the ceiling of an elevator as in Figure 5.15. (The ball replaces the fish!) The scale reads 16 lb when the elevator is at rest. (a) What will the scale read if the elevator accelerates *upward* at the rate of 8 ft/s²? (b) What will the scale read if the elevator accelerates *downward* at the rate of 8 ft/s²? (c) If the supporting rope can withstand a maximum tension of 25 lb and the weight of the scale is neglected, what is the maximum acceleration the elevator can have before the rope breaks? (d) If the spring scale weighs 5 lb, which rope breaks first? Why?

83. What horizontal force must be applied to the cart shown in Figure 5.46 in order that the blocks remain *stationary* relative to the cart? Assume all surfaces, wheels, and pulley are frictionless. (*Hint:* Note that the tension in the string accelerates m_1.)

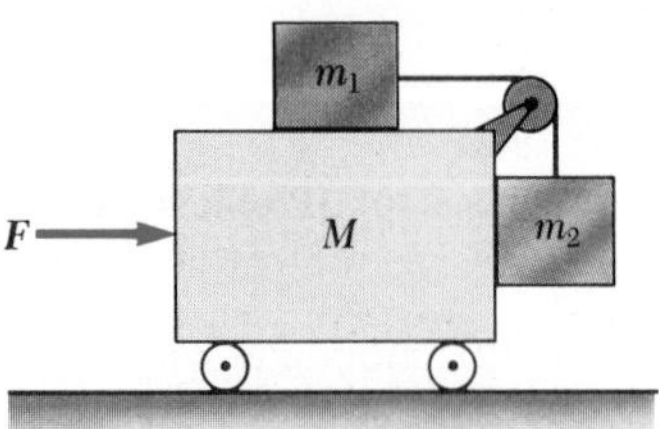

Figure 5.46 (Problems 83 and 84).

84. Initially the system of masses shown in Figure 5.46 is held motionless. All surfaces, pulley, and wheels are frictionless. In this case let the force $\boldsymbol{F}$ be zero and assume that m_2 can only move vertically. At the *instant after* the system of masses is released find: (a) the tension T in the string, (b) the acceleration of m_2, (c) the acceleration of M, and (d) the acceleration of m_1. (*Note:* The pulley accelerates along with the cart.)

85. The three blocks in Figure 5.47 are connected by light strings that pass over frictionless pulleys. The acceleration of the system is 2 m/s^2 to the left and the surfaces are rough. Find (a) the tensions in the strings and (b) the coefficient of kinetic friction between blocks and surfaces. (Assume the same μ for both blocks.)

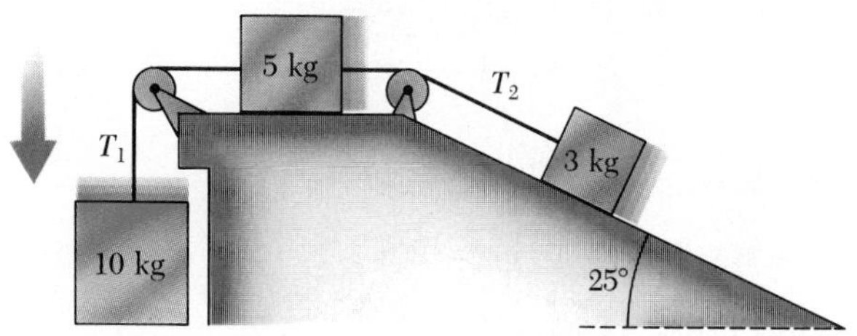

Figure 5.47 (Problem 85).

86. In Figure 5.48, the coefficient of kinetic friction between the 2-kg and 3-kg blocks is 0.3. The horizontal surface and the pulleys are frictionless and the masses are released from the rest. (a) Draw free-body diagrams for each block. (b) Determine the acceleration of each block. (c) Find the tension in the strings.

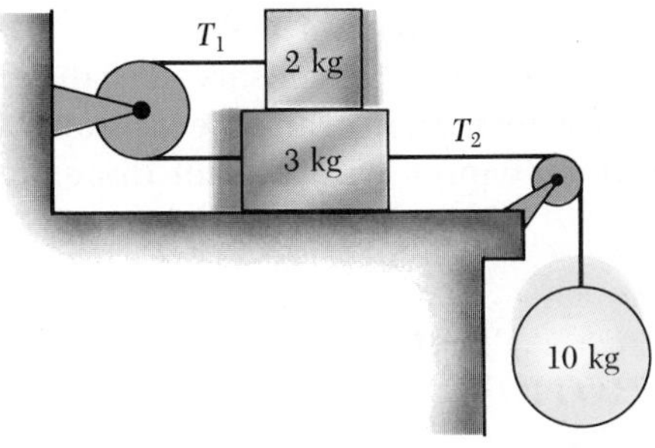

Figure 5.48 (Problem 86).

87. Two blocks of mass 2 kg and 7 kg are connected by a light string that passes over a frictionless pulley (Fig. 5.49). The inclines are smooth. Find (a) the acceleration of each block and (b) the tension in the string.

88. The system described in Figure 5.49 is observed to have an acceleration of 1.5 m/s^2 when the inclines are rough. Assume the coefficients of kinetic friction between each block and the inclines are the same. Find (a) the coefficient of kinetic friction and (b) the tension in the string.

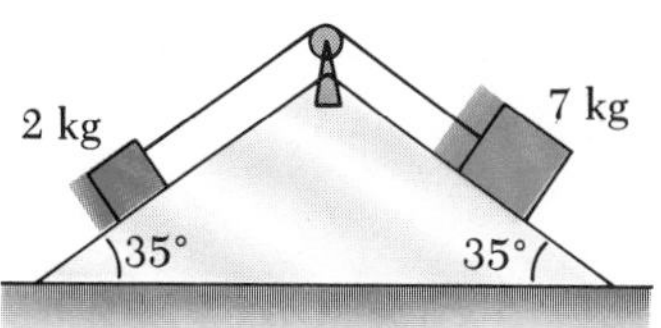

Figure 5.49 (Problems 87 and 88).

89. The wedge shown (Fig. 5.50) is moving on a smooth horizontal surface with an acceleration of 2 m/s^2. A 5-kg block rests on the wedge and is attached by a light string at A. There is no friction between the wedge and the block. (a) What is the tension in the string? (b) What normal force does the wedge exert on the block? (c) Compare your answers for (a) and (b) to the values obtained when the wedge is at rest.

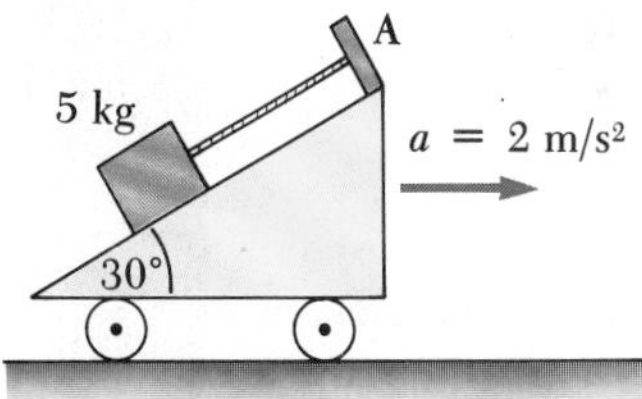

Figure 5.50 (Problem 89).

90. Before 1960 it was commonly believed that the maximum attainable coefficient of static friction of an automobile tire was less than 1. Then about 1962, three companies independently developed racing tires with coefficients of 1.6. Since then, tires have improved, as illustrated in the following problem. According to the 1982 *Guinness Book of Records,* the fastest $\frac{1}{4}$ mile covered by a wheel-powered car from a standing start is 5.64 s. This record elapsed time was set by Don Garlits in October 1975. (a) Assuming that the torque applied by Garlits' rear wheels nearly lifted his front wheels off the pavement, what is the lowest value of μ_s his tires could have had? (b) Suppose Garlits were able to double his engine power, keeping other things equal. How would this affect his elapsed time?

91. A magician attempts to pull a tablecloth from under a 200-g mug. The mug is located 30 cm from the edge of the cloth. If there is a small frictional force of 0.1 N exerted on the mug by the cloth, and the cloth is pulled with a constant acceleration of 3.0 m/s^2, how far will the mug move on the tabletop before the cloth is completely out from under it? (*Hint:* The cloth moves more than 30 cm before it is out from under the mug!)

6

Circular Motion and Other Applications of Newton's Laws

Passengers on a roller coaster at the Texas State Fair experience the thrill of circular motion. (© Zerschling, Photo Researchers)

In the previous chapter we introduced Newton's laws of motion and applied them to situations involving linear motion. In this chapter we shall apply Newton's laws of motion to the dynamics of circular motion. We shall also discuss the motion of an object when observed in an accelerated or noninertial frame of reference and the motion of an object through a viscous medium. Finally, we conclude this chapter with a brief discussion of the fundamental forces in nature.

6.1 NEWTON'S SECOND LAW APPLIED TO UNIFORM CIRCULAR MOTION

In Section 4.4 we found that a particle moving in a circular path of radius r with uniform speed v experiences an acceleration that has a magnitude

$$a_r = \frac{v^2}{r}$$

Because the velocity vector, $\boldsymbol{v}$, changes its direction continuously during the motion, the acceleration vector, $\boldsymbol{a}_r$, is directed toward the center of the circle and, hence, is called centripetal acceleration. Furthermore, $\boldsymbol{a}_r$ is always perpendicular to the velocity vector, $\boldsymbol{v}$.

Consider a ball of mass m tied to a string of length r and being whirled in a horizontal circular path on a table top as in Figure 6.1. Let us assume that the ball moves with constant speed. The inertia of the ball tends to maintain the motion in a straight-line path; however, the string prevents this motion along a straight line by exerting a force on the ball to make it follow its circular path. This force is directed along the length of the string toward the center of the circle, as shown in Figure 6.1, and is an example of a class of forces called **centripetal forces.** If we apply Newton's second law along the radial direction, we find that the required centripetal force is

$$F_r = ma_r = m\frac{v^2}{r} \tag{6.1}$$

Uniform circular motion

Like the centripetal acceleration, the centripetal force acts toward the center of the circular path followed by the particle. Because they act toward the center of rotation, centripetal forces cause a change in the direction of the velocity. Centripetal forces are no different from any other forces we have encountered. The term *centripetal* is used simply to indicate that *the force is directed toward the center of a circle.* In the case of a ball rotating at the end of a string, the tension force is the centripetal force. For a satellite in a circular orbit around the earth, the force of gravity is the centripetal force. The centripetal force acting on a car rounding a curve on a flat road is the force of friction between the tires and the pavement, and so forth.

Regardless of the example used, if the centripetal force acting on an object should vanish, the object would no longer move in its circular path; instead it would move along a straight-line path tangent to the circle. This idea is illustrated in Figure 6.2 for the case of the ball whirling in a circle at the end of a string. If the string breaks at some instant, the ball will move along the straight-line path tangent to the circle at the point where the string broke.

In general, a body can move in a circular path under the influence of such forces as friction, the gravitational force, or a combination of forces. Let us consider some examples of uniform circular motion. In each case, be sure to recognize the external force (or forces) that causes the body to move in its circular path.

A ten-hour exposure showing circular star trails around the south celestial pole. (© Anglo-Australian Telescope Board 1980)

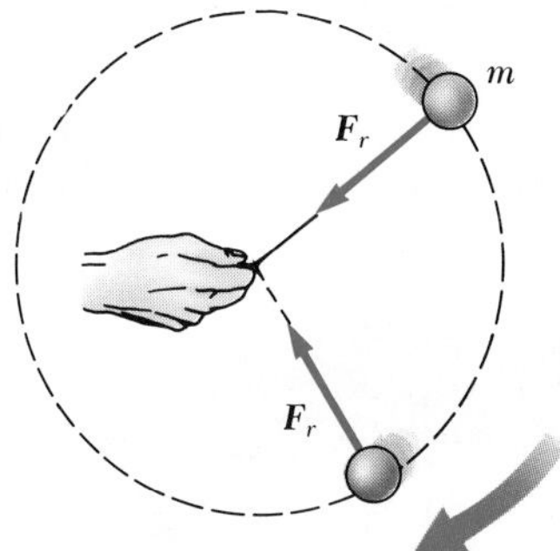

Figure 6.1 A ball moving in a circular path. A force $\mathbf{F}_r$ directed toward the center of the circle keeps the ball moving in the circle with a constant speed.

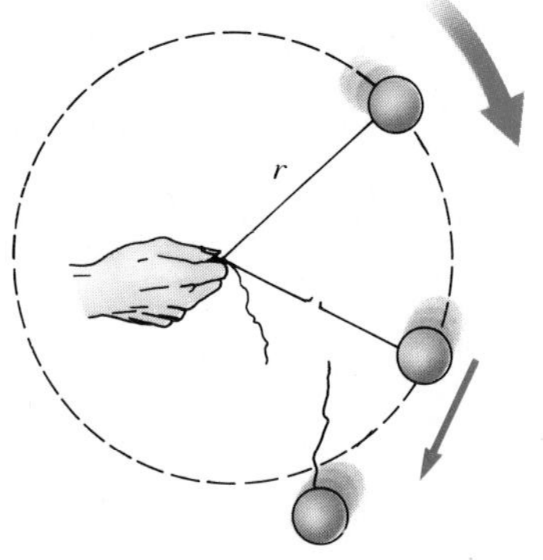

Figure 6.2 When the string breaks, the ball moves in the direction tangent to the circular path.

EXAMPLE 6.1 How Fast Can It Spin?
A ball of mass 0.5 kg is attached to the end of a cord whose length is 1.5 m. The ball is whirled in a horizontal circle as in Figure 6.2. If the cord can withstand a maximum tension of 50 N, what is the maximum speed the ball can have before the cord breaks?

Solution Because the centripetal force in this case is the tension $\boldsymbol{T}$ in the cord, Equation 6.1 gives

$$T = m\frac{v^2}{r}$$

Solving for v, we have

$$v = \sqrt{\frac{Tr}{m}}$$

The maximum speed that the ball can have will correspond to the maximum value of the tension. Hence, we find

$$v_{max} = \sqrt{\frac{T_{max}r}{m}} = \sqrt{\frac{(50\text{ N})(1.5\text{ m})}{0.5\text{ kg}}} = 12.2\text{ m/s}$$

Exercise 1 Calculate the tension in the cord if the speed of the ball is 5 m/s.
Answer 8.33 N.

EXAMPLE 6.2 The Conical Pendulum
A small body of mass m is suspended from a string of length L. The body revolves in a horizontal circle of radius r with constant speed v, as in Figure 6.3. Since the string sweeps out the surface of a cone, the system is known as a *conical pendulum*. Find the speed of the body and the period of revolution, T_P.

Solution The free-body diagram for the mass m is shown in Figure 6.3, where the tension, $\boldsymbol{T}$, has been resolved into a vertical component, $T\cos\theta$, and a component $T\sin\theta$ acting toward the center of rotation. Since the body does not accelerate in the vertical direction, the vertical component of the tension must balance the weight. Therefore,

$$(1) \qquad T\cos\theta = mg$$

Since the centripetal force in this example is provided by the component $T\sin\theta$, from Newton's second law we get

$$(2) \qquad T\sin\theta = ma_r = \frac{mv^2}{r}$$

By dividing (2) by (1), we eliminate T and find that

$$\tan\theta = \frac{v^2}{rg}$$

But from the geometry, we note that $r = L\sin\theta$, therefore

$$v = \sqrt{rg\tan\theta} = \sqrt{Lg\sin\theta\tan\theta}$$

The period of revolution, T_P (not to be confused with the tension T), is given by

$$(3) \qquad T_P = \frac{2\pi r}{v} = \frac{2\pi r}{\sqrt{rg\tan\theta}} = 2\pi\sqrt{\frac{L\cos\theta}{g}}$$

The intermediate algebraic steps used in obtaining (3) are left to the reader. Note that T_P is independent of m! If we take $L = 1.00$ m and $\theta = 20°$, we find using (3) that

$$T_P = 2\pi\sqrt{\frac{(1.00\text{ m})(\cos 20°)}{9.80\text{ m/s}^2}} = 1.95\text{ s}$$

Is it physically possible to have a conical pendulum with $\theta = 90°$?

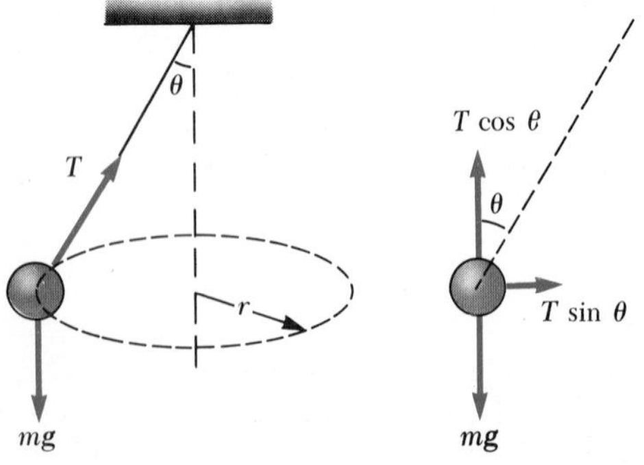

Figure 6.3 (Example 6.2) The conical pendulum and its free-body diagram.

EXAMPLE 6.3 What is the Maximum Speed of the Car?
A 1500-kg car moving on a flat road negotiates a curve whose radius is 35 m as in Figure 6.4. If the coefficient of

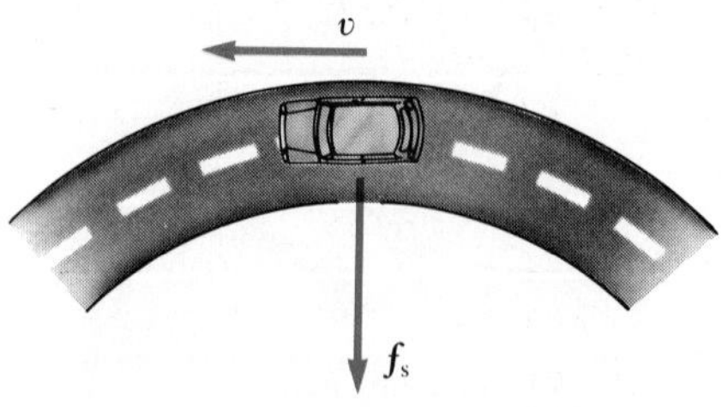

Figure 6.4 (Example 6.3) The force of static friction directed towards the center of the circular arc keeps the car moving in a circle.

static friction between the tires and the dry pavement is 0.50, find the maximum speed the car can have in order to make the turn successfully.

Solution In this case, the centripetal force which enables the car to remain in its circular path is the force of static friction. Hence, from Equation 6.1 we have

$$(1) \qquad f_s = m\frac{v^2}{r}$$

The maximum speed that the car can have around the curve corresponds to the speed at which it is on the verge of skidding outwards. At this point, the friction force has its maximum value given by

$$f_{s\,max} = \mu N$$

Because the normal force equals the weight in this case, we find

$$f_{s\,max} = \mu mg = (0.5)(1500 \text{ kg})(9.80 \text{ m/s}^2) = 7350 \text{ N}$$

Substituting this value into (1), we find that the maximum speed is

$$v_{max} = \sqrt{\frac{f_{s\,max}r}{m}} = \sqrt{\frac{(7350 \text{ N})(35 \text{ m})}{1500 \text{ kg}}} = 13.1 \text{ m/s}$$

Exercise 2 On a wet day, the car described in this example begins to skid on the curve when its speed reaches 8 m/s. What is the coefficient of static friction in this case?
Answer 0.187.

EXAMPLE 6.4 The Banked Exit Ramp
An engineer wishes to design a curved exit ramp for a tollroad in such a way that a car will not have to rely on friction to round the curve without skidding. Suppose that a typical car rounds the curve with a speed of 30 mi/h (13.4 m/s) and that the radius of the curve is 50 m. At what angle should the curve be banked?

Solution On a level road, the centripetal force must be provided by a force of friction between car and road. However, if the road is banked at an angle θ, as in Figure 6.5, the normal force, **N**, has a horizontal component $N \sin \theta$ pointing toward the center of the circular path followed by the car. We assume that only the component $N \sin \theta$ furnishes the centripetal force. Therefore, the banking angle we calculate will be one for which *no frictional force is required*. In other words, a car moving at the correct speed (13.4 m/s) can negotiate the curve even on an icy surface. Newton's second law written for the radial direction gives

$$(1) \qquad N \sin \theta = \frac{mv^2}{r}$$

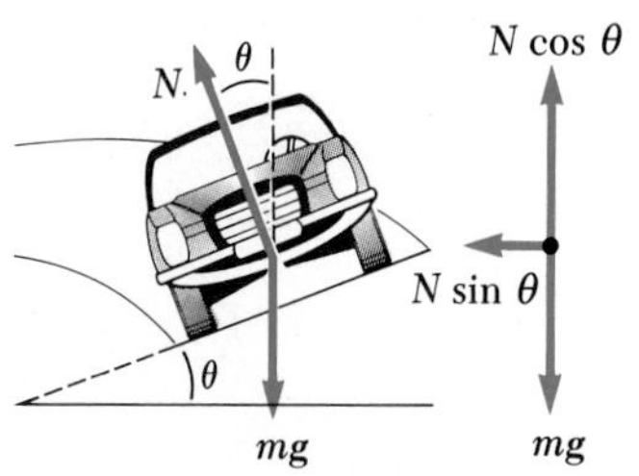

Figure 6.5 (Example 6.4) End view of a car rounding a curve on a road banked at an angle θ to the horizontal. The centripetal force is provided by the horizontal component of the normal force when friction is neglected. Note that **N** is the *sum* of the forces which the road exerts on the wheels of the car.

The car is in equilibrium in the vertical direction. Thus, from $\Sigma F_y = 0$, we have

$$(2) \qquad N \cos \theta = mg$$

Dividing (1) by (2) gives

$$\tan \theta = \frac{v^2}{rg}$$

$$\theta = \tan^{-1}\left[\frac{(13.4 \text{ m/s})^2}{(50 \text{ m})(9.80 \text{ m/s}^2)}\right] = 20.1°$$

If a car rounds the curve at a speed lower than 13.4 m/s, the driver will have to rely on friction to keep from sliding down the incline. A driver who attempts to negotiate the curve at a speed higher than 13.4 m/s will have to depend on friction to keep from sliding up the ramp.

Exercise 3 Write Newton's second law applied to the radial direction for the car in a situation in which a frictional force f is directed *down* the slope of the banked road.
Answer $N \sin \theta + f \cos \theta = mv^2/r$.

EXAMPLE 6.5 Satellite Motion □
This example treats the problem of a satellite moving in a circular orbit about the earth. In order to understand this problem, we must first note that the gravitational force between two particles having masses m_1 and m_2, separated by a distance r, is attractive and has a magnitude given by

$$F = G\frac{m_1 m_2}{r^2} \qquad (6.2)$$

where $G = 6.672 \times 10^{-11}$ N · m²/kg². This is Newton's universal law of gravity, which we shall discuss in detail in Chapter 14.

Now consider a satellite of mass m moving in a circular orbit about the earth at a constant speed v and at an

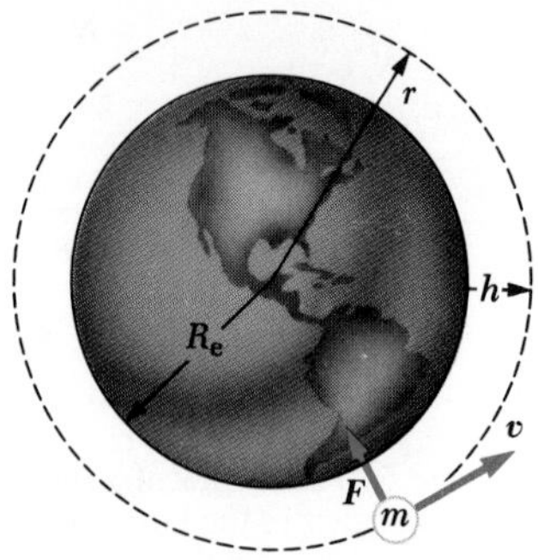

Figure 6.6 (Example 6.5) A satellite of mass m moving in a circular orbit of radius r and with constant speed v around the earth. The centripetal force is provided by the gravitational force between the satellite and the earth.

altitude h above the earth's surface as in Figure 6.6. (a) Determine the speed of the satellite in terms of G, h, R_e (the radius of the earth), and M_e (the mass of the earth).

Solution Because the only external force on the satellite is the force of gravity, which acts toward the center of the earth, we have

$$F_r = G\frac{M_e m}{r^2}$$

From Newton's second law, and the fact that $r = R_e + h$, we get

$$G\frac{M_e m}{r^2} = m\frac{v^2}{r}$$

or

$$v = \sqrt{\frac{GM_e}{r}} = \sqrt{\frac{GM_e}{R_e + h}} \qquad (6.3)$$

(b) Determine the satellite's period of revolution, T_p (the time for one revolution about the earth).

Solution Since the satellite travels a distance of $2\pi r$ (the circumference of the circle) in a time T_p, we find using Equation 6.3 that

$$T_p = \frac{2\pi r}{v} = \frac{2\pi r}{\sqrt{GM_e/r}} = \left(\frac{2\pi}{\sqrt{GM_e}}\right) r^{3/2} \qquad (6.4)$$

The planets move around the sun in approximately circular orbits. The radii of these orbits can be calculated from Equation 6.4 with M_e replaced by the mass of the sun. The fact that the square of the period is proportional to the cube of the radius of the orbit was first recognized as an empirical relation based on planetary data. We shall return to this topic in Chapter 14.

Exercise 4 A satellite is in a circular orbit at an altitude of 1000 km. The radius of the earth is 6.37×10^6 m. Find the speed of the satellite and the period of its orbit. **Answer** 7.35×10^3 m/s; 6.31×10^3 s $= 105$ min.

EXAMPLE 6.6 Let's Go Loop-the-loop

A pilot of mass m in a jet aircraft executes a "loop-the-loop" maneuver as illustrated in Figure 6.7a. In this flying pattern, the aircraft moves in a vertical circle of radius 2.70 km at a *constant speed* of 225 m/s. Determine the force of the seat on the pilot at (a) the bottom of the loop and (b) the top of the loop. Express the answers in terms of the weight of the pilot, mg.

Solution (a) The free-body diagram for the pilot at the bottom of the loop is shown in Figure 6.7b. The only forces acting on the pilot are the downward force of gravity, mg, and the upward force $\mathbf{N}_{bot}$ exerted by the seat. Since the net force upwards which provides the centripetal acceleration is $N_{bot} - mg$, Newton's second law for the radial direction gives

$$N_{bot} - mg = m\frac{v^2}{r}$$

or

$$N_{bot} = mg + m\frac{v^2}{r} = mg\left[1 + \frac{v^2}{rg}\right]$$

Substituting the values given for the speed and radius, $v = 225$ m/s and $r = 2.70 \times 10^3$ m/s, gives

$$N_{bot} = mg\left[1 + \frac{(225\ \text{m/s})^2}{(2.70 \times 10^3\ \text{m/s})(9.80\ \text{m/s}^2)}\right]$$

$$= 2.91mg$$

Hence, the force of the seat on the pilot is *greater* than his true weight, mg, by a factor of 2.91. In such situations, the pilot experiences an "apparent weight" which is greater than his true weight by the factor 2.91. This is discussed further in Section 6.4. (b) The free-body diagram for the pilot at the top of the loop is shown in Figure 6.7c. At this point, both the weight and the force of the seat on the pilot, $\mathbf{N}_{top}$, act *downwards*, so the net force downwards which provides the centripetal acceleration has a magnitude $N_{top} + mg$. Applying Newton's second law gives

$$N_{top} + mg = m\frac{v^2}{r}$$

or

$$N_{top} = m\frac{v^2}{r} - mg = mg\left[\frac{v^2}{rg} - 1\right]$$

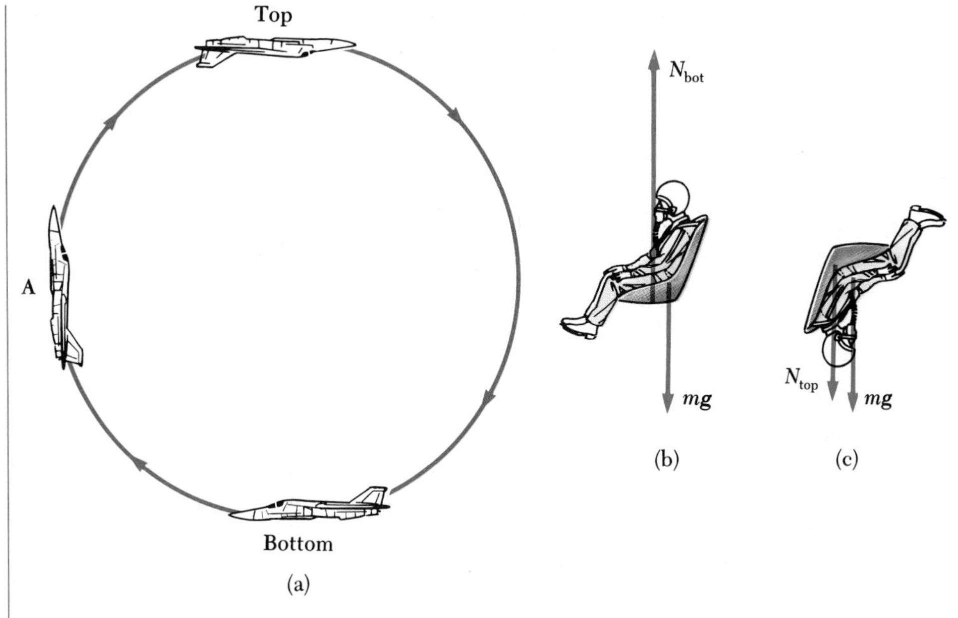

Figure 6.7 Example 6.6.

$$N_{top} = mg\left[\frac{(225 \text{ m/s})^2}{(2.70 \times 10^3 \text{ m/s})(9.80 \text{ m/s}^2)} - 1\right]$$

$$= 0.91mg$$

In this case, the force of the seat on the pilot is *less* than the true weight by a factor of 0.91. Hence, the pilot will feel lighter at the top of the loop.

Exercise 5 Calculate the force of the seat on the pilot if the aircraft is at point A in Figure 6.7a, midway up the loop.
Answer $N_A = 1.91mg$ directed to the right.

6.2 NONUNIFORM CIRCULAR MOTION

In Chapter 4 we found that if a particle moves with varying speed in a circular path, there is, in addition to the centripetal component of acceleration, a tangential component of magnitude dv/dt. Therefore, the force acting on the particle must also have a tangential and a radial component. That is, since the total acceleration is given by $\boldsymbol{a} = \boldsymbol{a}_r + \boldsymbol{a}_t$, the total force is given by $\boldsymbol{F} = \boldsymbol{F}_r + \boldsymbol{F}_t$, as shown in Figure 6.8. The vector component, $\boldsymbol{F}_r$, is directed toward the center of the circle and is responsible for the centripetal acceleration. The

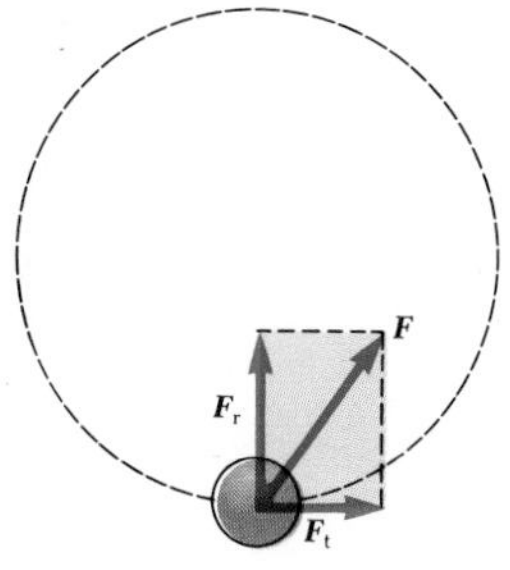

Figure 6.8 When the force acting on a particle has a tangential component F_t, its speed changes. The total force on the particle in this case is the vector sum of the tangential force and the centripetal force. That is, $\boldsymbol{F} = \boldsymbol{F}_t + \boldsymbol{F}_r$.

vector component, $\mathbf{F}_t$, tangent to the circle is responsible for the tangential acceleration, which causes the speed of the particle to change with time. The following example demonstrates this type of motion.

EXAMPLE 6.7 Follow the Rotating Ball

A small sphere of mass m is attached to the end of a cord of length R, which rotates in a *vertical* circle about a fixed point O, as in Figure 6.9a. Let us determine the tension in the cord at any instant that the speed of the sphere is v when the cord makes an angle θ with the vertical.

Solution First we note that the speed is *not* uniform since there is a tangential component of acceleration arising from the weight of the sphere. From the free-body diagram in Figure 6.9a, we see that the only forces acting on the sphere are the weight, $m\mathbf{g}$, and the constraint force (or tension), $\mathbf{T}$. Now we resolve $m\mathbf{g}$ into a tangential component, $mg \sin \theta$, and a radial component, $mg \cos \theta$. Applying Newton's second law to the forces in the tangential direction gives

$$\sum F_t = mg \sin \theta = ma_t$$

$$(1) \qquad a_t = g \sin \theta$$

This component causes v to change in time, since $a_t = dv/dt$. Applying Newton's second law to the forces in the radial direction and noting that both $\mathbf{T}$ and $\mathbf{a}_r$ are directed toward O, we get

$$\sum F_r = T - mg \cos \theta = \frac{mv^2}{R}$$

$$(2) \qquad T = m\left(\frac{v^2}{R} + g \cos \theta\right)$$

Limiting Cases At the *top* of the path, where $\theta = 180°$, we see from (2) that since $\cos 180° = -1$,

$$T_{\text{top}} = m\left(\frac{v_{\text{top}}^2}{R} - g\right)$$

This is the *minimum* value of T. Note that at this point $a_t = 0$, and therefore the acceleration is radial and directed downward, as in Figure 6.9b.

At the *bottom* of the path, where $\theta = 0$, again from (2) we see that since $\cos 0 = 1$,

$$T_{\text{bot}} = m\left(\frac{v_{\text{bot}}^2}{R} + g\right)$$

This is the *maximum* value of T. Again, at this point $a_t = 0$, and the acceleration is radial and directed upward.

Exercise 6 At what orientation of the system would the cord most likely break if the average speed increased?
Answer At the bottom of the path, where T has its maximum value.

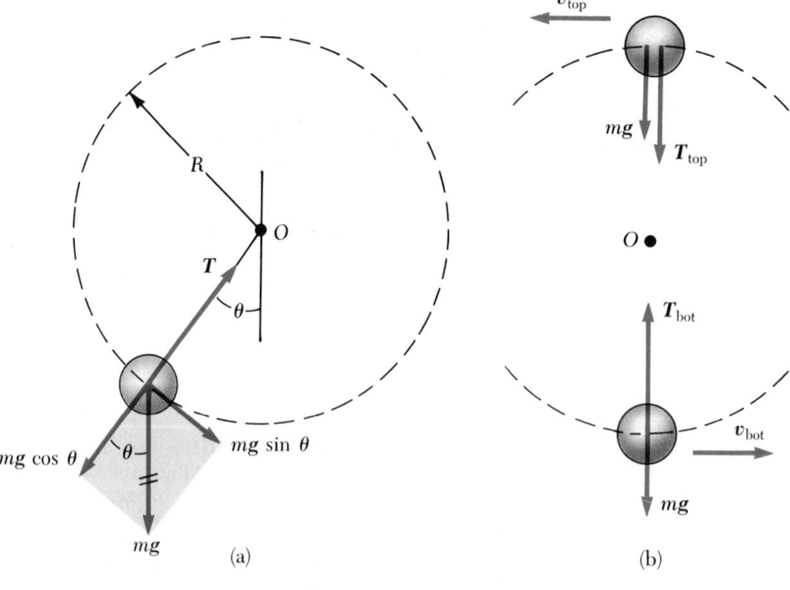

Figure 6.9 (Example 6.7) (a) Forces acting on a mass m connected to a string of length R and rotating in a vertical circle centered at O. (b) Forces acting on m when it is at the top and bottom of the circle. Note that the tension at the bottom is a maximum and the tension at the top is a minimum.

*6.3 MOTION IN ACCELERATED FRAMES

When Newton's laws of motion were introduced in Chapter 5, we emphasized that the laws are valid when observations are made in an *inertial* frame of reference. In this section, we shall analyze how an observer in a noninertial frame of reference (one that is accelerating) would attempt to apply Newton's second law.

If a particle moves with an acceleration $\boldsymbol{a}$ relative to an observer in an inertial frame, then the inertial observer may use Newton's second law and correctly claim that $\Sigma \boldsymbol{F} = m\boldsymbol{a}$. If an observer in an accelerated frame (the noninertial observer) tries to apply Newton's second law to the motion of the particle, the noninertial observer must introduce *fictitious* forces to make Newton's second law work in that frame. Sometimes, these fictitious forces are referred to as **inertial forces.** These forces "invented" by the noninertial observer *appear* to be real forces in the accelerating frame. However, we emphasize that these fictitious forces *do not* exist when the motion is observed in an inertial frame. The fictitious forces are used only in an accelerating frame but *do not* represent "real" forces on the body. (By "real" forces, we mean the interaction of the body with its environment.) If the fictitious forces are properly defined in the accelerating frame, then the description of motion in this frame will be equivalent to the description by an inertial observer who considers only real forces. Usually, motions are analyzed using inertial reference frames, but there are cases in which an accelerating frame is more convenient.

Fictitious or inertial forces

In order to understand better the motion of a rotating system, consider a car traveling along a highway at a high speed and approaching a curved exit ramp, as in Figure 6.10. As the car takes the sharp left turn onto the ramp, a person sitting in the passenger seat slides to the right across the seat and hits the door. At that point, the force of the door keeps him from being ejected from the car. What causes the passenger to move toward the door? A popular, but *improper*, explanation is that some mysterious force pushes him outward. (This is often called the "centrifugal" force, but we shall not use this term since it often creates confusion.) The passenger invents this fictitious force in order to explain what is going on in his accelerated frame of reference.

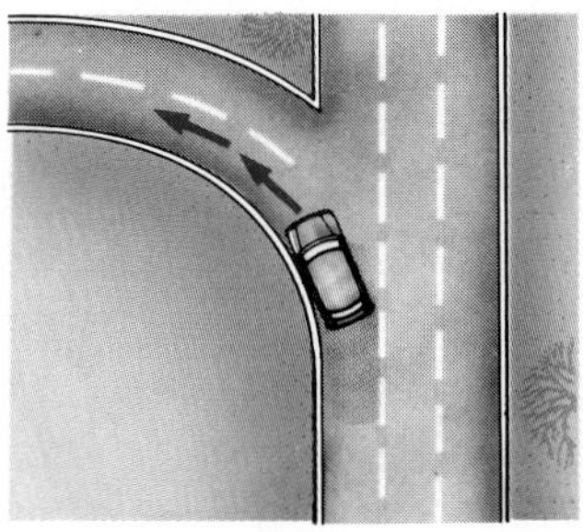

Figure 6.10 A car approaching a curved exit ramp.

The phenomenon is correctly explained as follows. Before the car enters the ramp, the passenger is moving in a straight-line path. As the car enters the ramp and travels a curved path, the passenger, because of inertia, tends to move along the original straight-line path. This is in accordance with Newton's first law: the natural tendency of a body is to continue moving in a straight line. However, if a sufficiently large centripetal force (toward the center of curvature) acts on the passenger, he will move in a curved path along with the car. The origin of this centripetal force is the force of friction between the passenger and the car seat. If this frictional force is not large enough, the passenger will slide across the seat as the car turns under him. Eventually, the passenger encounters the door, which provides a large enough centripetal force to enable the passenger to follow the same curved path as the car. The passenger slides toward the door not because of some mysterious outward force but because *there is no centripetal force large enough to allow him to travel along the circular path followed by the car.*

In summary, one must be very careful to distinguish real forces from fictitious ones in describing motion in an accelerating frame. An observer in a car rounding a curve is in an accelerating frame and invents a fictitious outward force to explain why he or she is thrown outward. A stationary observer outside the car, however, considers only real forces on the passenger. To this observer, the mysterious outward force *does not exist!* The only real external force on the passenger is the centripetal (inward) force due to friction or the normal force of the door.

EXAMPLE 6.8 Linear Accelerometer

A small sphere of mass m is hung from the ceiling of an accelerating boxcar, as in Figure 6.11. According to the inertial observer at rest (Figure 6.11a), the forces on the sphere are the tension $\mathbf{T}$ and the weight $m\mathbf{g}$. The inertial observer concludes that the acceleration of the sphere of mass m is the same as that of the boxcar and that this acceleration is provided by the horizontal component of $\mathbf{T}$. Also, the vertical component of $\mathbf{T}$ balances the weight. Therefore, the inertial observer writes Newton's second law as $\mathbf{T} + m\mathbf{g} = m\mathbf{a}$, which in component form becomes

$$\text{Inertial observer} \begin{cases} (1) & \sum F_x = T \sin\theta = ma \\ (2) & \sum F_y = T\cos\theta - mg = 0 \end{cases}$$

Thus, by solving (1) and (2) simultaneously, the inertial observer can determine the acceleration of the car through the relation

$$a = g \tan\theta$$

Therefore, since the deflection of the string from the vertical serves as a measure of the acceleration of the car, *a simple pendulum can be used as an accelerometer.*

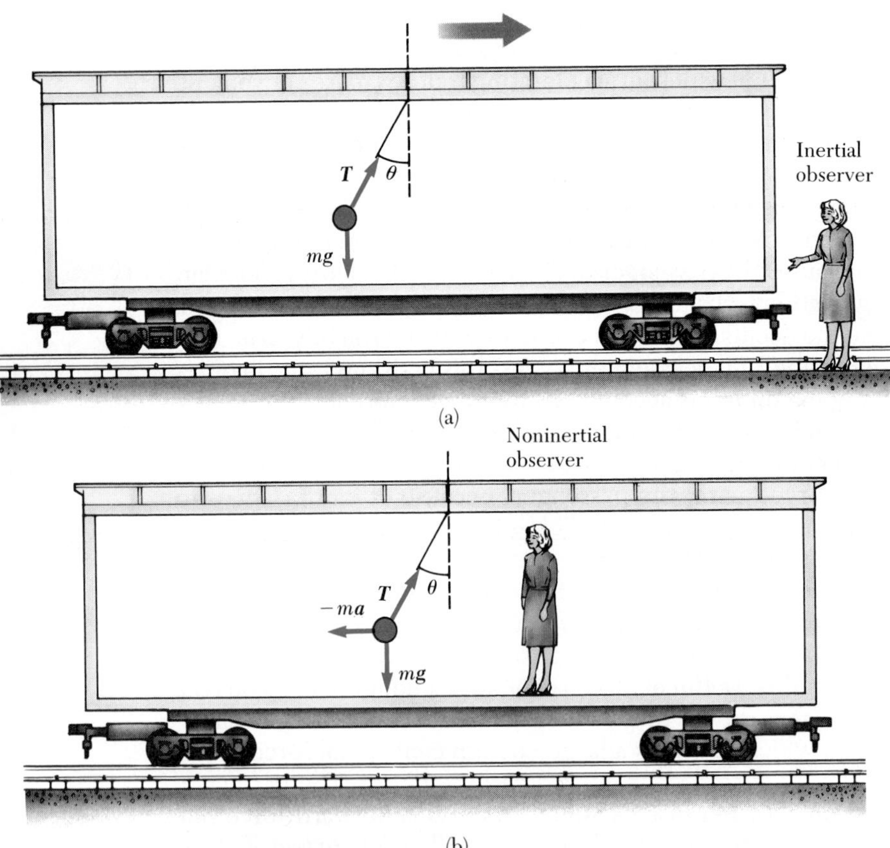

Figure 6.11 (Example 6.8) (a) A ball suspended from the ceiling of a boxcar accelerating to the right is deflected as shown. The inertial observer at rest outside the car claims that the acceleration of the ball is provided by the horizontal component of $\mathbf{T}$. (b) A noninertial observer riding in the car says that the net force on the ball is zero and that the deflection of the string from the vertical is due to a fictitious force, $-m\mathbf{a}$, which balances the horizontal component of $\mathbf{T}$.

According to the noninertial observer riding in the car, described in Figure 6.11b, the sphere is at rest and the acceleration is zero. Therefore, the noninertial observer introduces a *fictitious force*, $-m\boldsymbol{a}$, to balance the horizontal component of $\boldsymbol{T}$ and claims that the net force on the sphere is *zero!* In this noninertial frame of reference, Newton's second law in component form gives

$$\text{Noninertial observer} \begin{cases} \sum F_x' = T \sin \theta - ma = 0 \\ \sum F_y' = T \cos \theta - mg = 0 \end{cases}$$

These expressions are equivalent to (1) and (2); therefore the noninertial observer gets the same mathematical results as the inertial observer. However, the physical interpretation of the deflection of the string *differs* in the two frames of reference. Note that even though a pendulum is used, it does not oscillate in this application.

EXAMPLE 6.9 Fictitious Force in a Rotating System
An observer in a rotating system is another example of a noninertial observer. Suppose a block of mass m lying on a horizontal, frictionless turntable is connected to a string as in Figure 6.12. According to an inertial observer, if the block rotates uniformly, it undergoes a centripetal acceleration v^2/r, where v is its tangential speed. The inertial observer concludes that this centripetal acceleration is provided by the force of tension in the string, $\boldsymbol{T}$, and writes Newton's second law $T = mv^2/r$.

According to a noninertial observer attached to the turntable, the block is at rest. Therefore, in applying Newton's second law, this observer introduces a fictitious *outward* force called the *centrifugal force*, of magnitude mv^2/r. According to the noninertial observer, this "centrifugal" force balances the force of tension and therefore $T - mv^2/r = 0$.

You should be careful when using fictitious forces to describe physical phenomena. Remember that fictitious forces, such as centrifugal force, are used *only* in noninertial, or accelerated, frames of reference. When solving problems, it is generally best to use an inertial frame.

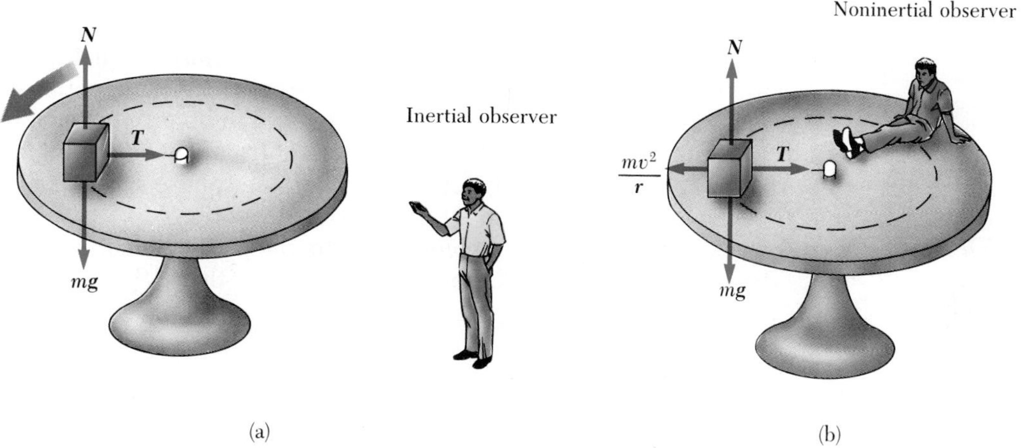

Figure 6.12 (Example 6.9) A block of mass m connected to a string tied to the center of a rotating turntable. (a) The inertial observer claims that the centripetal force is provided by the force of tension, $\boldsymbol{T}$. (b) The noninertial observer claims that the block is not accelerating and therefore he introduces a fictitious centrifugal force mv^2/r, which acts outward and balances the tension.

*6.4 MOTION IN THE PRESENCE OF RESISTIVE FORCES

In the previous chapter we discussed the force of sliding friction, that is, the resistive force on an object moving along a rough, solid surface. Such forces are nearly independent of velocity, and matters are simplified by assuming them to be constant in magnitude. Now let us consider what happens when an object moves through a liquid or gas. In such situations, the medium exerts a resistive force $\boldsymbol{R}$ on the object. The magnitude of this force depends on the velocity of the object, and its direction is always opposite the direction of motion of the object relative to the medium. The magnitude of the resistive force is generally found to increase with increasing velocity. Some examples of such resistive forces are the air resistance to flying airplanes and moving cars and the viscous forces on objects moving through a liquid.

In general, the resistive force can have a complicated velocity dependence. In the following discussions, we will consider two situations. First, we will assume that the resistive force is proportional to the velocity. Objects falling through a fluid and very small objects, such as particles of dust moving through air, experience such a force. Second, we will treat situations for which the resistive force is assumed to be proportional to the square of the speed of the object. Large objects, such as a skydiver moving through air in free fall in the presence of gravity, experience such a force.

Resistive Force Proportional to Velocity

When an object moves at low speeds through a viscous medium, it experiences a resistive drag force that is proportional to the velocity of the object. Let us assume that the resistive force, $\boldsymbol{R}$, has the form

$$\boldsymbol{R} = -b\boldsymbol{v} \tag{6.5}$$

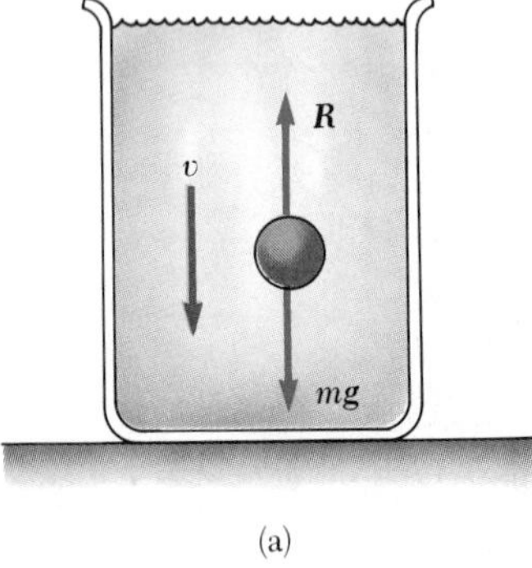

(a)

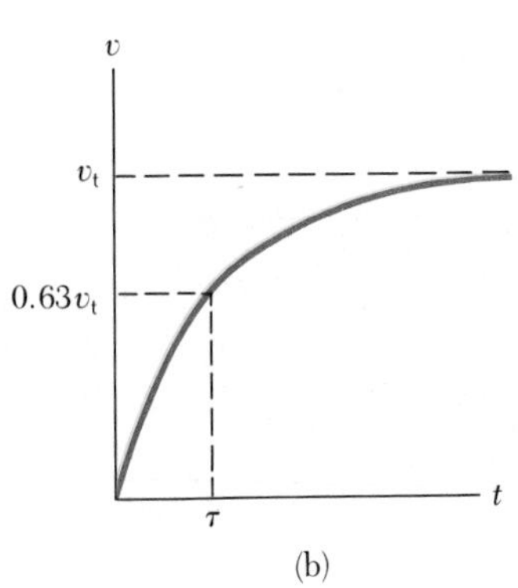

(b)

Figure 6.13 (a) A small sphere falling through a viscous fluid. (b) The velocity-time graph for an object falling through a viscous medium. The object reaches a maximum, or terminal, velocity, v_t, and τ is the time it takes to reach $0.63v_t$.

where $\boldsymbol{v}$ is the velocity of the object and b is a constant that depends on the properties of the medium and on the shape and dimensions of the object. If the object is a sphere of radius r, then b is found to be proportional to r.

Consider a sphere of mass m released from rest in a fluid, as in Figure 6.13a. Assuming the only forces acting on the sphere are the resistive force, $-b\boldsymbol{v}$, and the weight, $m\boldsymbol{g}$, let us describe its motion.[1]

Applying Newton's second law to the vertical motion, choosing the downward direction to be positive, and noting that $\Sigma F_y = mg - bv$, we get

$$mg - bv = m\frac{dv}{dt}$$

where the acceleration is downward. Simplifying the above expression gives

$$\frac{dv}{dt} = g - \frac{b}{m}v \tag{6.6}$$

Equation 6.6 is called a *differential equation*, and the methods of solving such an equation may not be familiar to you as yet. However, note that initially, when $v = 0$, the resistive force is zero and the acceleration, dv/dt, is simply g. As t increases, the resistive force increases and the acceleration *decreases*. Eventually, the acceleration becomes zero when the resistive force *equals* the weight. At this point, the body continues to move with zero acceleration, and it reaches its *terminal velocity*, v_t. The terminal velocity can be obtained from Equation 6.6 by setting $a = dv/dt = 0$. This gives

$$mg - bv_t = 0 \qquad \text{or} \qquad v_t = mg/b$$

[1] There is also a *buoyant* force, which is constant and equal to the weight of the displaced fluid. This will only change the weight of the sphere by a constant factor. We shall discuss such buoyant forces in Chapter 15.

The expression for v that satisfies Equation 6.6 with $v = 0$ at $t = 0$ is

$$v = \frac{mg}{b}(1 - e^{-bt/m}) = v_t(1 - e^{-t/\tau}) \qquad (6.7)$$

This function is plotted in Figure 6.13b. The time $\tau = m/b$ is the time it takes the object to reach 63% of its terminal velocity. We can check that Equation 6.7 is a solution to Equation 6.6 by direct differentiation:

$$\frac{dv}{dt} = \frac{d}{dt}\left(\frac{mg}{b} - \frac{mg}{b}e^{-bt/m}\right) = -\frac{mg}{b}\frac{d}{dt}e^{-bt/m} = ge^{-bt/m}$$

Substituting this expression and Equation 6.7 into Equation 6.6 shows that our solution satisfies the differential equation.

The high cost of fuel has prompted many truck owners to install wind deflectors on their cabs to reduce drag. (Courtesy of Henry Leap)

EXAMPLE 6.10 Sphere Falling in Oil
A small sphere of mass 2 g is released from rest in a large vessel filled with oil. The sphere reaches a terminal velocity of 5 cm/s. Determine the time constant τ and the time it takes the sphere to reach 90 percent of its terminal velocity.

Solution Since the terminal velocity has a magnitude given by $v_t = mg/b$, the coefficient b is

$$b = \frac{mg}{v_t} = \frac{(2\text{ g})(980\text{ cm/s}^2)}{5\text{ cm/s}} = 392\text{ g/s}$$

Therefore, the time τ is given by

$$\tau = \frac{m}{b} = \frac{2\text{ g}}{392\text{ g/s}} = 5 \times 10^{-3}\text{ s}$$

The velocity of the sphere as a function of time is given by Equation 6.7, $v = v_t(1 - e^{-t/\tau})$. To find the time t it takes the sphere to reach a velocity of $0.90\ v_t$, we set $v = 0.90\ v_t$ into this expression and solve for t:

$$0.90\ v_t = v_t(1 - e^{-t/\tau})$$

$$1 - e^{-t/\tau} = 0.90$$

$$e^{-t/\tau} = 0.10$$

$$-\frac{t}{\tau} = -2.30$$

$$t = 2.30\ \tau = 2.30(5.10 \times 10^{-3}\text{ s})$$

$$= 11.7 \times 10^{-3}\text{ s} = 11.7\text{ ms}$$

Exercise 7 Calculate the speed that the sphere would have at $t = 11.7$ ms if air resistance were not present and compare it to the true speed at that instant.
Answer 11.5 cm/s compared to the true speed of 4.50 m/s.

Air Drag

We have seen that an object moving through a fluid experiences a resistive drag force. If the object is small and moves at low speeds, the drag force is proportional to the velocity, as we have already discussed. However, for larger objects moving at high speeds through air, such as airplanes, skydivers, and baseballs, the drag force is approximately proportional to the *square* of the speed. In these situations, the magnitude of the drag force can be expressed as

$$R = \tfrac{1}{2}C\rho Av^2 \qquad (6.8)$$

where ρ is the density of air, A is the cross-sectional area of the falling object measured in a plane perpendicular to its motion, and C is a dimensionless empirical quantity called the *drag coefficient.* The drag coefficient has a value of about 0.5 for spherical objects but can be as high as about 2 for irregularly shaped objects.

Figure 6.14 By spreading their arms and legs out from the sides of their bodies and by keeping the plane of their bodies parallel to the ground, skydivers experience maximum air drag resulting in a specific terminal speed. (U.S. Air Force Academy photo by SSgt. West C. Jacobs, 94ATS)

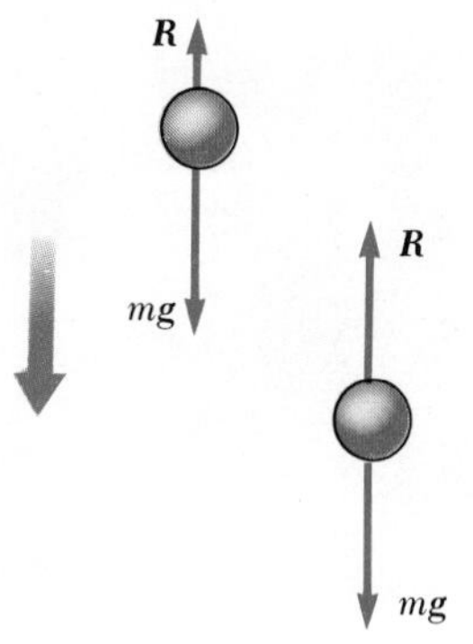

Figure 6.15 An object falling through air experiences a drag force, **R**, and the force of gravity, **mg**. The object reaches terminal velocity (on the right) when the net force is zero, that is, when $R = mg$. Before this occurs, the acceleration varies with speed according to Equation 6.10.

Consider an airplane in flight experiencing such a drag force. Equation 6.8 shows that the drag force is proportional to the density of air and hence decreases with decreasing air density. Since air density decreases with increasing altitude, the drag force on a jet airplane flying at a given speed must also decrease with increasing altitude. Furthermore, if the plane's speed is doubled, the drag force increases by a factor of 4. In order to maintain this increased speed, the propulsive force also increases by a factor of 4 and the power required (force times speed) must increase by a factor of 8.

Now let us analyze the motion of a mass in free fall subject to an upward air drag force given by $R = \frac{1}{2}C\rho Av^2$. Suppose a mass m is released from rest from the position $y = 0$ as in Figure 6.15. The mass experiences two external forces: the weight, $m\mathbf{g}$, downward and the drag force, $\mathbf{R}$, upward. There is also an upward buoyant force which we will neglect. Hence, the magnitude of the net force is given by

$$F_{\text{net}} = mg - \tfrac{1}{2}C\rho Av^2 \tag{6.9}$$

Substituting $F_{\text{net}} = ma$ into Equation 6.9, we find that the mass has a downward acceleration of magnitude

$$a = g - \left(\frac{C\rho A}{2m}\right)v^2 \tag{6.10}$$

Again, we can calculate the terminal velocity, v_t, using the fact that when the weight is balanced by the drag force, the net force is zero and therefore the acceleration is zero. Setting $a = 0$ in Equation 6.10 gives

$$g - \left(\frac{C\rho A}{2m}\right)v_t^2 = 0$$

Terminal velocity

$$v_t = \sqrt{\frac{2mg}{C\rho A}} \tag{6.11}$$

Using this expression, we can determine how the terminal speed depends on the dimensions of the object. Suppose the object is a sphere of radius r. In this case, $A \propto r^2$ and $m \propto r^3$ (since the mass is proportional to the volume). Therefore, $v_t \propto \sqrt{r}$. That is, as r increases, the terminal speed increases with the square root of the radius.

Table 6.1 lists the terminal speeds for several objects falling through air.

TABLE 6.1 Terminal Speed for Various Objects Falling Through Air

Object	Mass (kg)	Area (m²)	v_t(m/s)[a]
Skydiver	75	0.7	60
Baseball (radius 3.66 cm)	0.145	4.2×10^{-3}	33
Golf ball (radius 2.1 cm)	0.046	1.4×10^{-3}	32
Hailstone (radius 0.5 cm)	4.8×10^{-4}	7.9×10^{-5}	14
Raindrop (radius 0.2 cm)	3.4×10^{-5}	1.3×10^{-5}	9

[a] The drag coefficient, C, is assumed to be 0.5 in each case.

6.5 THE FUNDAMENTAL FORCES OF NATURE

In the previous chapter, we described a variety of forces that are experienced in our everyday activities, such as the force of gravity which acts on all bodies at or near the earth's surface, and the force of friction as one surface slides over another. Other forces we have encountered are the force of tension in a rope, the normal force acting on an object that is in contact with the floor or some other object, and the drag force as an object moves through air or some other medium. Other forces that we shall encounter include the restoring force in a deformed spring, the electrostatic force between two charged objects, and the magnetic force between a magnet and a piece of iron.

Forces also act in the atomic and subatomic world. For example, atomic forces within the atom are responsible for holding its constituents together, and nuclear forces act on different parts of the nucleus to keep its parts from separating.

Until recently, physicists believed that there were four fundamental forces in nature: the gravitational force, the electromagnetic force, the strong nuclear force, and the weak nuclear force.

The **gravitational force** is the mutual force of attraction between all masses. We have already encountered the gravitational force when describing the weight of an object. Although gravitational forces can be very significant between macroscopic objects, the gravitational force is actually the weakest of the four fundamental forces. This statement is based on the relative strengths of the four forces when considering the interaction between elementary particles. For example, the gravitational force between the electron and proton in the hydrogen atom is only about 10^{-47} N, whereas the electrostatic force between the two particles is about 10^{-7} N. Thus, we see that the strength of the gravitational force is insignificant in comparison to that of the electrostatic force. We shall learn more about the nature of the gravitational force in Chapter 14.

The **electromagnetic force** is an attraction or repulsion between two charged particles that are in relative motion. Later in this text we shall see that electric and magnetic forces are closely related. In fact, the magnetic force can be viewed as an additional electric force that acts whenever the interacting charges are in motion. Although the electric force between two charged elementary particles is much stronger than the gravitational force between them, the electric force is of medium strength. The force that causes a rubbed comb to attract bits of paper, and the force that a magnet exerts on an iron nail are examples of electromagnetic forces. It is interesting to note that essentially all forces in our macroscopic world (apart from the gravitational force) are manifestations of the electromagnetic force when examined very closely. For example, friction forces, contact forces, tension forces, and forces in stretched springs or other deformed bodies are essentially the consequence of electromagnetic forces between charged particles in close proximity.

The **strong nuclear force** is responsible for the stability of nuclei. This force represents the "glue" that holds the nuclear constituents (called nucleons) together. It is the strongest of all the fundamental forces. For separations of about 10^{-15} m (a typical nuclear dimension), the strong nuclear force is one to two orders of magnitude stronger than the electromagnetic force. However, the strong nuclear force decreases rapidly with increasing separation and is negligible for separations greater than about 10^{-14} m.

Finally, the **weak nuclear force** is a short range nuclear force that tends to produce instability in certain nuclei. Most radioactive decay reactions are caused by the weak nuclear force. The weak nuclear force is about 12 orders of magnitude weaker than the electric force.

In 1979, physicists predicted that the electromagnetic force and the weak force are manifestations of one and the same force called the **electroweak** force. This prediction was confirmed experimentally in 1984. Thus, the current view is that there are only three fundamental forces in nature.

Physicists and cosmologists believe that the fundamental forces of nature are closely related to the origin of the universe. The big bang theory of the creation of the universe states that the universe erupted from a pointlike singularity on the order of 15 to 20 billion years ago. According to this theory, the first moments ($\approx 10^{-10}$ s) after the big bang saw such extremes of energy that all four fundamental forces were unified. Scientists continue in their search for a possible connection among the three fundamental forces, as Einstein himself dreamed.

SUMMARY

Newton's second law applied to a particle moving in **uniform circular motion** states that the net force in the radial direction must equal the product of the mass and the centripetal acceleration:

Uniform circular motion

$$F_r = ma_r = \frac{mv^2}{r} \tag{6.1}$$

The force that provides the centripetal acceleration could be, for example, the force of gravity (as in satellite motion), the force of friction, or the force of tension (as in a string). A particle moving in nonuniform circular motion has both a centripetal (or radial) acceleration and a nonzero tangential component of acceleration. In the case of a particle rotating in a vertical circle, the force of gravity provides the tangential acceleration and part or all of the centripetal acceleration.

Fictitious forces

An observer in a noninertial (accelerated) frame of reference must introduce **fictitious forces** when applying Newton's second law in that frame. If these fictitious forces are properly defined, the description of motion in the noninertial frame will be equivalent to that made by an observer in an inertial frame. However, the observers in the two different frames will not agree on the causes of the motion.

A body moving through a liquid or gas experiences a **resistive force** that is velocity dependent. This resistive force, which opposes the motion, generally increases with velocity. The force depends on the shape of the body and the properties of the medium through which the body is moving. In the limiting case for a falling body, when the resistive force equals the weight ($a = 0$), the body reaches its **terminal velocity.** There are only four fundamental forces in nature: the gravitational force, the electromagnetic force, the strong nuclear force, and the weak nuclear force.

QUESTIONS

1. Because the earth rotates about its axis and about the sun, it is a noninertial frame of reference. Assuming the earth is a uniform sphere, why would the *apparent weight* of an object be greater at the poles than at the equator?
2. Explain why the earth is not spherical in shape and bulges at the equator.
3. How would you explain the force that pushes a rider toward the side of a car as the car rounds a corner?
4. When an airplane does an inside "loop-the-loop" in a vertical plane, at what point would the pilot appear to be heaviest? What is the constraint force acting on the pilot?
5. A skydiver in free fall reaches terminal velocity. After the parachute is opened, what parameters change to decrease this terminal velocity?
6. Why is it that an astronaut in a space capsule orbiting the earth experiences a feeling of weightlessness?
7. Why does mud fly off a rapidly turning wheel?
8. A pail of water can be whirled in a vertical path such that none is spilled. Why does the water stay in, even when the pail is above your head?
9. Imagine that you attach a heavy object to one end of a spring and then whirl the spring and object in a horizontal circle (by holding the free end of the spring). Does the spring stretch? If so, why? Discuss this in terms of centripetal force.
10. It has been suggested that rotating cylinders about 10 mi in length and 5 mi in diameter be placed in space and used as colonies. The purpose of the rotation is to simulate gravity for the inhabitants. Explain this concept for producing an effective gravity.
11. Why does a pilot tend to black out when pulling out of a steep dive?
12. Cite an example of a situation in which an automobile driver can have a centripetal acceleration but no tangential acceleration.
13. Is it possible for a car to move in a circular path in such a way that it has a tangential acceleration but no centripetal acceleration?
14. Analyze the motion of a rock dropped into water in terms of its speed and acceleration as it falls. Assume that there is a resistive force acting on the rock that increases as the velocity increases.
15. Consider a skydiver falling through air *before* reaching terminal velocity. As the velocity of the skydiver increases, what happens to her acceleration.
16. Centrifuges are often used in dairies to separate the cream from the milk. Which remains on the bottom?
17. We often think of the brakes and the gas pedal on a car as the devices which accelerate the car. Could a steering wheel also fall into this category? Explain.
18. Suppose that a baseball and a softball are dropped from an airplane. Which has the higher terminal velocity? Which experiences the greater acceleration before reaching terminal velocity, say one second after they are released?
19. Consider a small raindrop and a large raindrop falling through the atmosphere. Compare their terminal velocities. What are their accelerations when they reach terminal velocity?

PROBLEMS

Section 6.1 Newton's Second Law Applied to Uniform Circular Motion

1. A toy car completes one lap around a circular track (a distance of 200 m) in 25 s. (a) What is the average speed? (b) If the mass of the car is 1.5 kg, what is the magnitude of the centripetal force that keeps it in a circle?
2. In a cyclotron (one type of particle accelerator), a deuteron (of atomic mass 2 u) reaches a final velocity of 10% of the speed of light while moving in a circular path of radius 0.48 m. The deuteron is maintained in the circular path by a magnetic force. What magnitude of force is required?
3. What centripetal force is required to keep a 1.5 kg mass moving in a circle of radius 0.4 m at a speed of 4 m/s?
4. In a hydrogen atom, the electron in orbit about the proton feels an attraction of about 8.20×10^{-8} N. If the radius of the orbit is 5.3×10^{-11} m, what is the frequency in revolutions per second? See the front cover for additional data.
5. A 3-kg mass attached to a light string rotates in circular motion on a horizontal, frictionless table. The radius of the circle is 0.8 m, and the string can support a mass of 25 kg before breaking. What range of speeds can the mass have before the string breaks?
6. A satellite of mass 300 kg is in a circular orbit about the earth at an altitude equal to the earth's mean radius (see Example 6.5). Find (a) the satellite's orbital speed, (b) the period of its revolution, and (c) the gravitational force acting on it.
7. (a) What is the radius of the planet Venus if a satellite 2.0×10^6 m from its surface moves in circular orbits around Venus with speed 6.35×10^3 m/s and centripetal acceleration 5.01 m/s^2? (b) What is the period of the satellite around Venus? (The period T is the time for one complete orbit.)

8. (a) What is the speed of a stone at the end of a string 1.2 m long if it moves in a circle with centripetal acceleration 2.5 m/s^2? (b) The string is likely to break if its tension exceeds 3.0 N. What is the maximum safe speed of a 0.2-kg mass attached at the end of it?
9. A crate of eggs is located in the middle of the flatbed of a pickup truck as the truck negotiates a curve in the road. The curve may be regarded as an arc of a circle of radius 35 m. If the coefficient of static friction between the crate and the flatbed of the truck is 0.6, what must be the maximum speed of the truck if the crate is not to slide during the maneuver?
10. A highway curve has a radius of 150 m and is designed for a traffic speed of 40 mi/h (17.9 m/s). (a) If the curve is not banked, determine the minimum coefficient of friction between the car and the road. (b) At what angle should the curve be banked if friction is neglected (Figure 6.5)?
11. An air puck of mass 0.25 kg is tied to a string and allowed to revolve in a circle of radius 1.0 m on a horizontal frictionless table. The other end of the string passes through a hole in the center of the table and a mass of 1.0 kg is tied to it. The suspended mass remains in equilibrium while the puck on the table top revolves. (a) What is the tension in the string? (b) What is the centripetal force acting on the puck? (c) What is the speed of the puck?
12. The speed of the tip of the minute hand of a town clock is 1.75×10^{-3} m/s. (a) What is the speed of the tip of the second hand of the same length? (b) What is the centripetal acceleration of the tip of the second hand?
13. A coin is placed 30 cm from the center of a rotating, horizontal turntable. The coin is observed to slip when its speed is 50 cm/s. (a) What provides the centripetal force when the coin is stationary relative to the turntable? (b) What is the coefficient of static friction between the coin and the turntable?

Section 6.2 Nonuniform Circular Motion

14. A car traveling on a straight road at 9.0 m/s goes over a hump in the road. The hump may be regarded as an arc of a circle of radius 11.0 m. (a) What is the apparent weight of a 60-N woman in the car as she rides over the hump? (b) What must be the speed of the car over the hump if she is to experience weightlessness? (The apparent weight must be zero.)
15. A pail of water is rotated in a vertical circle of radius 1 m. What is the minimum speed of the pail at the top of the circle if no water is to spill out?
16. A 0.5-kg mass attached to the end of a string swings in a vertical circle of radius $R = 2$ m (Figure 6.9). When $\theta = 20°$, the speed of the mass is 8 m/s. At this instant, find (a) the tension in the string, (b) the tangential and radial components of acceleration, and (c) the magnitude of the total acceleration.
17. A 40-kg child sits in a conventional swing of length 3 m, supported by two chains. If the tension in each chain at the lowest point is 350 N, find (a) the child's speed at the lowest point, and (b) the force of the seat on the child at the lowest point. (Neglect the mass of the seat.)
18. A ball attached to the end of a string 0.8 m in length is rotated in a vertical circle (Fig. 6.9). Determine the minimum speed of the ball at the top of its path if it maintains a circular path. (*Note:* Below this speed, the tension in the string is zero at the top.)
19. A Ferris wheel with radius 20 m makes 1 revolution every 9.0 seconds. What force does a 55-kg passenger exert on the seat when she is at the top of the Ferris wheel?
20. A roller-coaster vehicle has a mass of 500 kg when fully loaded with passengers (Fig. 6.16). (a) If the vehicle has a speed of 20 m/s at point *A*, what is the force of the track on the vehicle at this point? (b) What is the maximum speed the vehicle can have at *B* in order that it remain on the track?

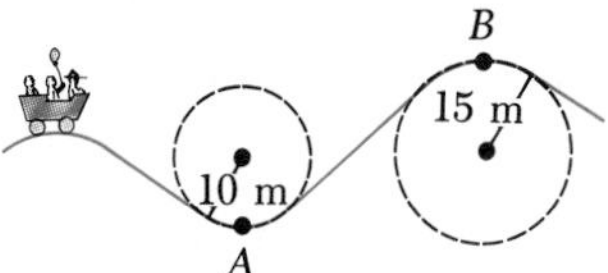

Figure 6.16 (Problem 20).

21. Tarzan ($m = 85$ kg) tries to cross a river by swinging from a vine. The vine is 10 m long, and his speed at the bottom of the swing (as he just clears the water) is 8 m/s. Tarzan doesn't know that the vine has a breaking strength of 1000 N. Does he make it safely across the river?
22. At the Six Flags Great America amusement park in Gurnee, Illinois, there is a roller coaster that incorporates some of the latest design technology and some basic physics. The vertical loop, instead of being circular, is shaped like a teardrop (Fig. 6.17). This produces higher speeds at the top of the loop. The cars ride on the inside of the loop at the top, and the higher speeds ensure that the cars remain on the track. The biggest loop is 40 m high (about 130 ft), with a maximum speed of 31 m/s (nearly 70 mph) at the bottom. (*New York Times*, Aug. 2, 1988.) Suppose the speed at the top is 13.0 m/s and the corresponding centripetal acceleration is $2g$. (a) What is the radius of the teardrop of the arc at the top? (b) If the total mass of the roller coaster at the top of the loop is M, what force does the rail exert on it at the top? (c) Suppose instead, the roller coaster had a circular loop of radius 20 m. If the cars have the same speed, 13 m/s at the top, what is the centripetal acceleration at the top? (d) Comment on the normal force at the top in this situation.

Figure 6.17 (Problem 22).

*Section 6.3 Motion in Accelerated Frames

23. A block is attached to a string, which in turn is connected to a peg at the center of a rotating turntable, as in Figure 6.12. If the turntable is *rough*, describe the forces on the block as observed by (a) someone on the turntable and (b) an observer at rest relative to the turntable. (c) For a given velocity, does the tension in the string increase, decrease, or remain the same as the turntable is made smoother?

24. A ball is suspended from the ceiling of a moving car by a string 25 cm in length. An observer in the car notes that the ball deflects 6 cm from the vertical toward the rear of the car. What is the acceleration of the car?

25. A 0.5-kg object is suspended from the ceiling of an accelerating boxcar as in Figure 6.11. If $a = 3\ \text{m/s}^2$, find (a) the angle that the string makes with the vertical and (b) the tension in the string.

26. A 5-kg mass attached to a spring scale rests on a smooth, horizontal surface as in Figure 6.18. The spring scale, attached to the front end of a boxcar, reads 18 N when the car is in motion. (a) If the spring scale reads zero when the car is at rest, determine the acceleration of the car. (b) What will the spring scale read if the car moves with constant velocity? (c) Describe the forces on the mass as observed by someone in the car and by someone at rest outside the car.

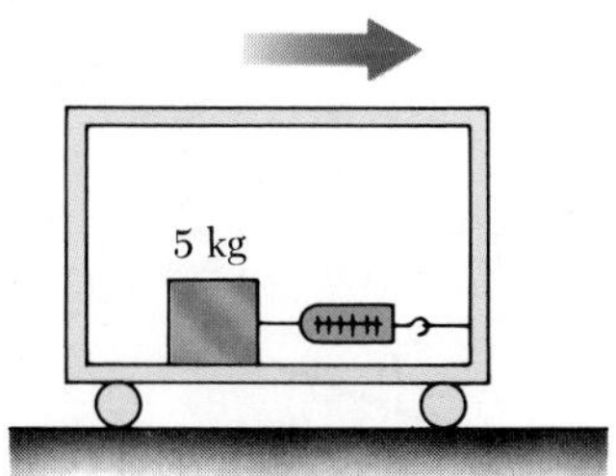

Figure 6.18 (Problem 26).

27. A mass m rests on spring scales situated on the floor of an elevator. The elevator moves down with a constant downward acceleration a. (a) In the noninertial frame of reference of the elevator, draw the free-body diagram and apply Newton's second law to find the apparent weight indicated by the scale. (b) Repeat part (a) in a fixed (inertial) frame of reference.

28. A plumb bob does not hang exactly along a line directed to the center of the earth's rotation. How much does the plumb bob deviate from a radial line at 35° north latitude?

*Section 6.4 Motion in the Presence of Resistive Forces

29. A skydiver of mass 80 kg jumps from a slow-moving aircraft and reaches a terminal speed of 50 m/s. (a) What is the acceleration of the skydiver when her speed is 30 m/s? What is the drag force on the diver when her speed is (b) 50 m/s and (c) 30 m/s?

30. A small piece of Styrofoam packing material is dropped from a height of 2.0 m above the ground. Until the terminal velocity is reached, the acceleration of the piece of Styrofoam is given by $a = g - bv$, where g is the acceleration due to gravity, v the velocity, and b a constant. When it has fallen 0.5 m the terminal velocity is reached, and it takes an extra 5 s to reach the ground. (a) What is the numerical value of the constant b? (b) What is the acceleration at $t = 0$? (c) What is the acceleration when the velocity is 0.15 m/s?

31. A motor boat cuts its engine when its speed is 10 m/s and coasts to rest. The equation governing the motion of the motorboat during this period is $v = v_0 e^{-ct}$, where v is the velocity at time t, v_0 is the initial velocity, and c is a constant. At $t = 20$ s the velocity is 5 m/s. (a) Find the constant c. (b) What is the velocity at $t = 40$ s? (c) Differentiate the expression above for $v(t)$ and thus show that the acceleration of the boat is proportional to the velocity at any time.

32. (a) Estimate the terminal velocity of a wooden sphere (density 0.83 g/cm^3) moving in air if its radius is 8.0 cm. (b) From what height would a freely-falling object reach this speed in the absence of air resistance?

33. A small, spherical bead of mass 3 g is released from rest at $t = 0$ in a bottle of liquid shampoo. The terminal velocity, v_t, is observed to be 2 cm/s. Find (a) the value of the constant b in Equation 6.7, (b) the time, τ, it takes to reach $0.63v_t$, and (c) the value of the retarding force when the bead reaches terminal velocity.

ADDITIONAL PROBLEMS

34. A spinning ball of radius 5.0 cm slows uniformly from 30 rev/min to rest in 0.3 s. Compute the radial, tangential, and net acceleration of a point on the equator of the ball at the beginning of this time period.

35. In the Bohr model of the hydrogen atom the velocity of the electron is approximately 2.2×10^6 m/s. Find (a) the centripetal force acting on the electron as it revolves in a circular orbit of radius 0.53×10^{-10} m, (b) the centripetal acceleration of the electron, and (c) the number of revolutions per second made by the electron.

36. A 0.40-kg pendulum bob passes through the lowest part of its path with a speed of 8.2 m/s. What is the tension in the pendulum cable if the pendulum is 80 cm long?

37. A 100-gram bead is free to slide along a piece of string ABC 0.8 m long. The ends of the string are attached to a vertical rod at points A and B, which are 0.4 m apart as in Figure 6.19. When the rod is rotated, BC is horizontal and equal to 0.3 m. (a) What is the tension in the string? (b) What is the speed of the bead at C?

38. A small turtle, appropriately named "Dizzy," is placed on a horizontal, rotating turntable at a distance of 20 cm from its center. Dizzy's mass is 50 g, and the coefficient of static friction between his feet and the turntable is 0.3. Find (a) the *maximum* number of revolutions per second the turntable can have if Dizzy is to remain stationary relative to the turntable and (b) Dizzy's speed and radial acceleration when he is on the verge of slipping.

39. Because of the earth's rotation about its axis, a point on the equator experiences a centripetal acceleration of 0.034 m/s^2, while a point at the poles experiences no centripetal acceleration. (a) Show that at the equator the gravitational force on an object (the true weight) must *exceed* the object's apparent weight. (b) What is the apparent weight at the equator and at the poles of a person having a mass of 75 kg? (Assume the earth is a uniform sphere and take $g =$ 9.800 m/s^2.)

40. A piece of putty is initially located at point A on the rim of a grinding wheel rotating about a horizontal axis. The putty is dislodged from point A when the diameter through A is horizontal. The putty rises vertically and returns to A the instant the wheel completes one revolution. (a) Find the velocity of a point on the rim of the wheel in terms of the acceleration due to gravity and the radius R of the wheel. (b) If the mass of the putty is m, what is the magnitude of the force that held it to the wheel?

41. A railroad track has a curve of 400 m radius. The tracks are banked toward the inside at an angle of 6°. For trains of what speed was this track designed? (Assume that the correct speed requires only the normal force to keep the train on the track.)

42. A car rounds a banked curve as in Figure 6.5. The radius of curvature of the road is R, the banking angle is θ, and the coefficient of static friction is μ. (a) Determine the *range* of speeds the car can have without slipping up or down the road. (b) Find the minimum value for μ such that the minimum speed is zero. (c) What is the range of speeds possible if $R = 100$ m, $\theta = 10°$, and $\mu = 0.1$ (slippery conditions)?

43. A model airplane of a mass 0.75 kg flies in a horizontal circle at the end of a 60-m control wire, with a speed of 35 m/s. Compute the tension in the wire if it makes a constant angle of 20° with the horizontal. The airplane is acted upon by the tension in the control line, its weight, and the aerodynamic lift, which acts at 20° inward from the vertical as shown in Fig. 6.20.

44. A child's toy consists of a small wedge that has an acute angle θ (Figure 6.21). The sloping side of the wedge is smooth and a mass m on it remains at the same height if the wedge is spun at a constant speed. The wedge is spun by rotating a rod which is firmly attached to the wedge at one end. Show that when the wedge rises up the plane a distance L the speed of the mass m is given by

$$v = \sqrt{gL \sin \theta}$$

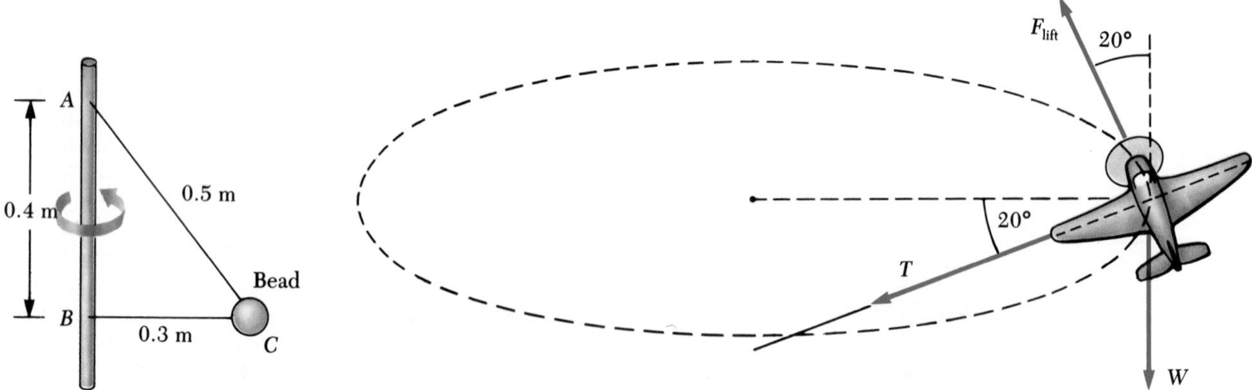

Figure 6.19 (Problem 37).

Figure 6.20 (Problem 43).

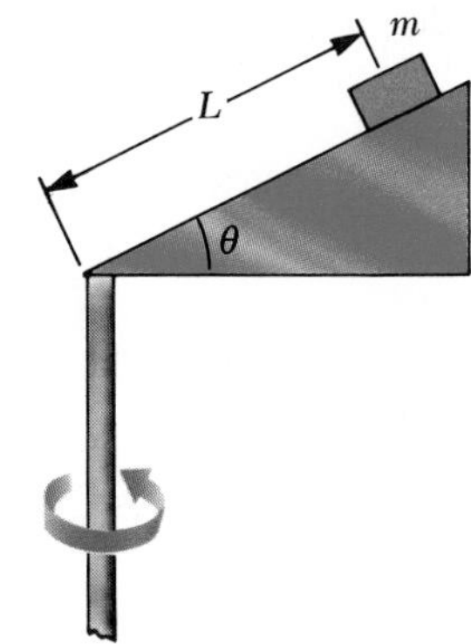

Figure 6.21 (Problem 44).

45. The pilot of an airplane executes a constant-speed loop-the-loop maneuver in a vertical plane. The speed of the airplane is 300 mi/h, and the radius of the circle is 1200 ft. (a) What is the pilot's apparent weight at the lowest point if his true weight is 160 lb? (b) What is his apparent weight at the highest point? (c) Describe how the pilot could experience weightlessness if both the radius and velocity can be varied. (*Note:* His apparent weight is equal to the force that the seat exerts on his body.)
46. A 4-kg mass is attached to a *horizontal* rod by two strings, as in Figure 6.22. The strings are under tension when the rod rotates about its axis. If the speed of the mass is 4 m/s when observed at the following positions, find the tension in the string when the mass is (a) at its lowest point, (b) in the horizontal position, and (c) at its highest point.

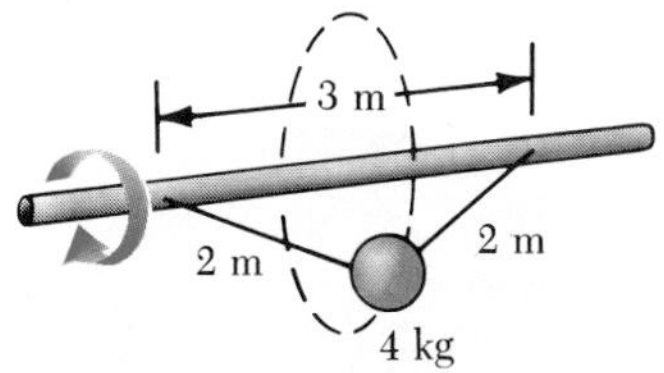

Figure 6.22 (Problems 46 and 47).

47. Suppose the rod in the system shown in Figure 6.22 is made *vertical* and rotates about this axis. If the mass rotates at a constant speed of 6 m/s in a horizontal plane, determine the tensions in the upper and lower strings.
48. A student builds and calibrates an accelerometer, which she uses to determine the speed of her car around a certain highway curve. The accelerometer is a simple pendulum with a protractor which she attaches to the roof of her car. Her friend observes that the pendulum hangs at an angle of 15° from the vertical when the car has a speed of 23.0 m/s. (a) What is the centripetal acceleration of the car rounding the curve? (b) What is the radius of the curve? (c) What is speed of the car if the pendulum deflection is 9.0° while rounding the same curve?
49. An amusement park ride consists of a large vertical cylinder that spins about its axis fast enough that any person inside is held up against the wall when the floor drops away (Figure 6.23). The coefficient of static friction between the person and the wall is μ_s, and the radius of the cylinder is R. (a) Show that the *maximum* period of revolution necessary to keep the person from falling is $T = (4\pi^2 R\mu_s/g)^{1/2}$. (b) Obtain a numerical value for T if $R = 4$ m and $\mu_s = 0.4$. How many revolutions per minute does the cylinder make?

Figure 6.23 (Problem 49).

50. A penny of mass 3.1 g rests on a small 20-g block supported by a spinning disk (Figure 6.24). If the coefficient of friction between the block and the disk is 0.75 (static) and 0.64 (kinetic) while for the penny and block it is 0.45 (kinetic) and 0.52 (static), what is the maximum speed of the disk in rpm without the block or penny sliding on the disk?
51. A bead is threaded on a frictionless vertical wire hoop of radius R. The hoop rotates about a vertical axis

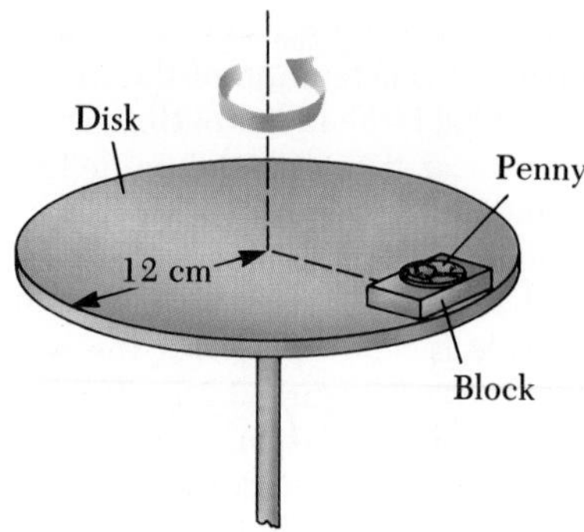

Figure 6.24 (Problem 50).

through its center, as shown in Figure 6.25. The period of revolution of the hoop is T. The bead has mass m and is at rest relative to the hoop at the angle θ. Answer the following in terms of the given parameters and the acceleration due to gravity g. (a) Find the normal force $\boldsymbol{N}$ exerted on the bead by the wire. (b) Find the angle between the normal force and the axis of rotation. (c) Find the centripetal force exerted on the bead.

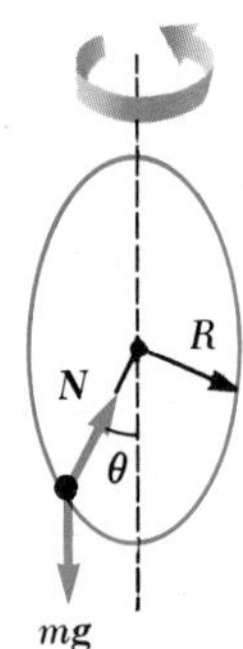

Figure 6.25 (Problem 51).

52. An amusement park ride consists of a rotating circular platform 8 m in diameter from which "bucket seats" are suspended at the end of 2.5-m chains. (For simplicity only one chain is shown [Figure 6.26], though for stability at least two would be used.) When the system rotates the chains holding the seats make an angle $\theta = 28°$ with the vertical. (a) What is the speed of the seat? (b) If a child of mass 40 kg sits in the 10-kg seat, what is the tension in the chain?

53. The following experiment is performed in a spacecraft that is at rest where the net gravitational field is zero. A small sphere is injected into a viscous medium with initial velocity $\boldsymbol{v}_0$. The sphere experiences a resistive force $\boldsymbol{R} = -b\boldsymbol{v}$. Find the velocity of the sphere as a function of time. (*Hint:* Apply Newton's second law, write a as dv/dt, separate the variables, and integrate the equation.)

CALCULATOR/COMPUTER PROBLEMS

54. A hailstone of mass 4.8×10^{-4} kg and radius 0.5 cm falls through the atmosphere and experiences a net force given by Equation 6.9. This expression can be written in the form

$$m \frac{dv}{dt} = mg - Kv^2$$

where $K = \frac{1}{2}C\rho A$, $\rho = 1.29$ kg/m³ and $C = 0.5$. (a) What is the terminal velocity of the hailstone? (b) Use a method of numerical integration to find the velocity and position of the hailstone at 1-s intervals, taking $v_0 = 0$. Continue your calculation until terminal velocity is reached.

55. A 0.5-kg block slides down a 30° incline of length 1 m. The coefficient of kinetic friction between the block and the incline varies with the block's velocity according to the expression

$$\mu = 0.3 + 1.2\sqrt{v}$$

where v is in m/s. (a) Use a numerical method to find the velocity of the block at intervals of 10 cm during its motion. (b) If the length of the plane is extended to several km, will the block reach terminal velocity? If so, what is its terminal velocity, and at what point does it occur on the incline?

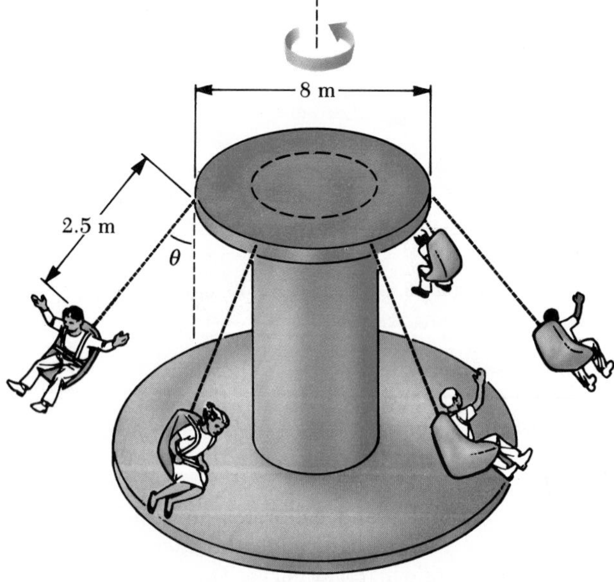

Figure 6.26 (Problem 52).

ESSAY

Dynamics of Satellite Orbits

Leon Blitzer
University of Arizona

Equations 6.3 and 6.4 for the velocity and period of a satellite in a circular orbit about an assumed spherical earth have long been known:

$$v = \sqrt{\frac{GM_e}{r}} \qquad (6.3)$$

$$T_p = \left(\frac{2\pi}{\sqrt{GM_e}}\right) r^{3/2} \qquad (6.4)$$

However, the advent of the Space Age had to await the development of rockets with sufficient thrust to propel the missile into orbit. With the launch of Sputnik I on October 4, 1957, artificial earth satellites became a reality, and since then numerous satellites and space probes have been sent into orbit. These manned, as well as unmanned, space explorations have captured the interest and imagination of the entire world. The unique character of the satellite is that it provides a sustained observing platform outside the atmosphere for studying the earth and its environs, as well as outer space. Today, hundreds of satellites and space probes are in orbit, and their applications encompass just about the entire range of science and engineering research: astronomy, global communication, navigation, weather reconnaissance, planetary studies, geodesy, studies of biological organisms in a weightless environment, cosmic ray and solar physics, ocean and land surveillance for marine and mineral resources, agricultural surveys, location of sites for archaeological explorations, etc. There is no doubt that the future will see an increasing number of satellites and space probes being used for ever-expanding applications.

You should note that Equations 6.3 and 6.4 have application beyond artificial earth satellites, for they are valid for any satellite (moon) moving in a circular orbit about its parent planet, or any planet moving in a circular orbit about the sun. More generally, satellite and planetary orbits are elliptical, the circle being a special case of the ellipse.°

The Satellite Orbit Paradox

Clearly, forces are required to overcome the earth's gravitational attraction and to propel a satellite into orbit, with higher orbits requiring greater forces. Consider a satellite moving in a given circular orbit. If the satellite is subjected to some force in the *same direction* as its motion, it will be propelled into a higher orbit and will travel at a slower speed according to Equation 6.3. Conversely, if the satellite is subjected to some force in a direction opposite to its motion, it will be driven into a lower orbit and will move at a greater speed according to Equation 6.3.

It is known that even at altitudes of 500 km and more above the earth there are sufficient atmospheric particles to create a significant frictional (drag) force on the rapidly moving satellite. Since the frictional force is in a direction opposite to its motion, it will cause the satellite to shift into a lower orbit and move with greater velocity. Hence, we have the so-called "drag paradox"; namely, atmospheric *friction* causes the orbital velocity to *increase.* Indeed, all satellites moving within the earth's atmosphere slowly spiral inward at ever-increasing speed until they burn up or impact the earth (see Fig. 1). Moreover, this is also the case for satellites moving in elliptical orbits.†

Note that the paradox is not limited to drag, for *any* force acting in the same direction as the motion of the satellite causes the missile to shift into a higher orbit and move with slower speed, while any retarding force actually results in an increase in speed.†

° This is apart from perturbations due to the attraction of other planets, tidal forces, radiation pressure, particle drag, electromagnetic forces, nonspherical shapes, etc.

† See L. Blitzer, "Satellite Orbit Paradox: A General View," *Amer. J. Physics* **39**:882, 1971.

(continued on next page)

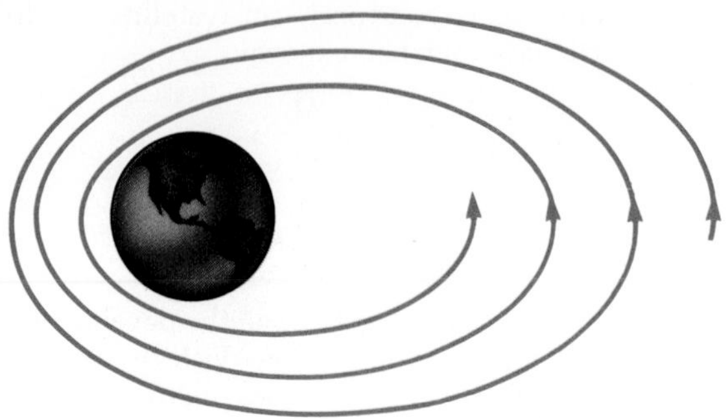

Figure 1 Shrinking of the orbit under atmospheric drag.

Geostationary SYNCOM Satellites

Consider a satellite moving in a circular orbit in the plane of the equator at such a distance that its orbital period is synchronous with the rotational period of the earth, namely one sidereal day. Such a satellite will then be at a fixed geographic longitude, and is referred to as **geostationary.** Figure 2 shows three uniformly spaced satellites in synchronous equatorial orbits. This configuration of three geostationary satellites, when equipped with radio transponders, can provide line-of-sight global communication between any two points on earth. Practically all satellites currently used for communication are in such 24-hour synchronous orbits and hence referred to as SYNCOMS.

Libration of Synchronous Satellites

Because of its nonuniform shape and mass distribution — oceans, mountains, variations in density — the earth is far from spherical. In fact, it is flattened at the poles and bulges at the equator, with the equatorial cross-section being very nearly elliptical.

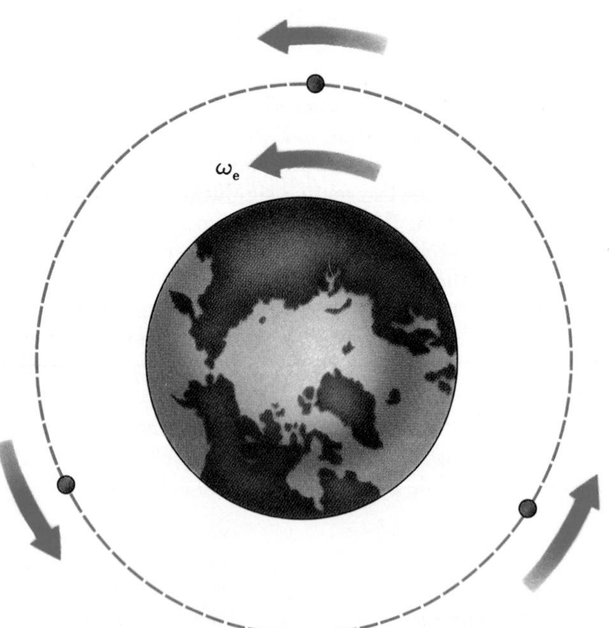

Figure 2 Three satellites uniformly spaced in synchronous equatorial orbits. (Not drawn to scale.)

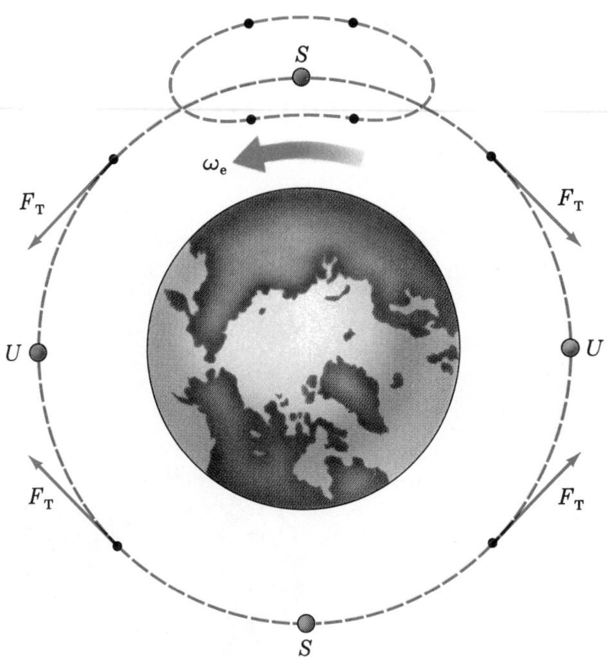

Figure 3 Equatorial section of Earth (looking south along polar axis). $\boldsymbol{F}_T$ is the net tangential force on the satellite at the positions shown. S is the stable equilibrium position, U, unstable. The blue dashed curve shows the librational path of a 24-hour satellite.

(On the basis of satellite tracking data the difference between the major and minor axes of the equatorial ellipse is about 130 m.)

Let us examine the motion of a 24-hour satellite in a frame of reference rotating with the earth (see Fig. 3). It is clear from symmetry that when the satellite is on the extension of either principal axis of the equatorial ellipse, at positions S or U, the gravitational force is purely radial. These must then be equilibrium positions, or stationary points, in the rotating frame. On the other hand, at off-axis positions, the greater gravitational attraction will be toward the nearest major axis. Hence, there is a net tangential force $\boldsymbol{F}_T$ toward the nearest major axis, as indicated for various positions of the satellite in Figure 3.

At first thought, one might expect the satellite to accelerate in the direction of $\boldsymbol{F}_T$. However, in accordance with the orbit paradox, the satellite will actually drift in the opposite direction toward the nearest equilibrium position S on the minor axis. Since it acquires momentum, the satellite will drift past S, the direction of $\boldsymbol{F}_T$ will then be reversed, and the drift will gradually be reversed. Hence the satellite will oscillate, or librate, about the equilibrium position S on the minor axis. Contrary to the drag paradox, there is no friction involved in this process.

The path of one such 24-hour satellite in the rotating earth frame is shown as a dashed curve in Figure 3. The period of libration depends on the amplitude, which, in turn, is determined by the initial position. For small-amplitude librations the period is approximately 2.1 years. Unless provided with propulsion for *repositioning,* all SYNCOM satellites experience such librations.

Suggested Readings

Blitzer, L., "Satellite Orbit Paradox: A General View," *Amer. J. Phys.* **39**:882, 1971.

"Basic Facts About The Satellite Orbits," *Sky and Telescope* **15**:408, 1956.

Dubridge, L.A., "Fun In Space," *Amer. J. Phys.* **28**:719, 1960.

Blitzer, L., "Equilibrium and Stability of a Pendulum in an Orbiting Spaceship," *Amer. J. Phys.* **47**:241, 1979.

7
Work and Energy

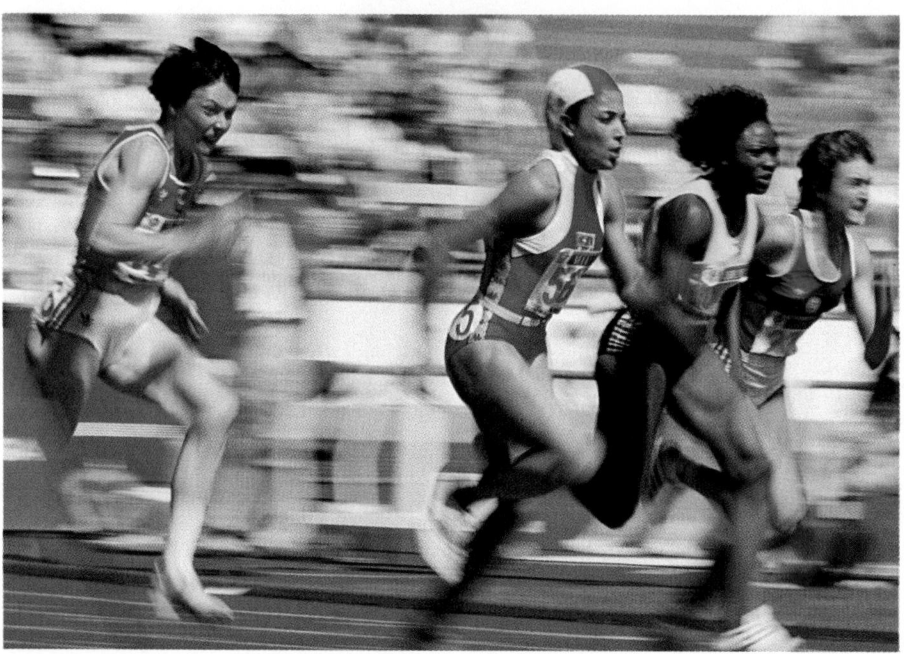

Working hard, sprinters at the 1988 Summer Olympics compete in the 100-meter dash. U.S. athlete Florence Griffith-Joyner, wearing cap, ran the race in 10.62 seconds, setting a new Olympic record. (Wide World Photos)

The concept of energy is one of the most important physical concepts in both contemporary science and engineering practice. In everyday usage, we think of energy in terms of the cost of fuel for transportation and heating, electricity for lights and appliances, and the foods we consume. However, these ideas do not really define energy. They only tell us that fuels are needed to do a job and that those fuels provide us with something we call energy.

Energy is present in various forms, including mechanical energy, electromagnetic energy, chemical energy, thermal energy, and nuclear energy. The various forms of energy are related to each other through the fact that when energy is transformed from one form to another, the total amount of energy remains the same. This is the point that makes the energy concept so useful. That is, if an isolated system loses energy in some form, then the law of conservation of energy says that the system will gain an equal amount of energy in other forms. For example, when an electric motor is connected to a battery, chemical energy is converted to electrical energy, which in turn is converted to mechanical energy. The transformation of energy from one form to another is an essential part of the study of physics, engineering, chemistry, biology, geology, and astronomy.

7.1 INTRODUCTION

In this chapter, we shall be concerned only with the mechanical form of energy. We shall see that the concepts of work and energy can be applied to the dynamics of a mechanical system without resorting to Newton's laws. However, it is important to note that the work-energy concepts are based upon Newton's laws and therefore do not involve any new physical principles.

Although the approach we shall use provides the same results as Newton's laws in describing the motion of a mechanical system, the general ideas of the work-energy concept can be applied to a wide range of phenomena in the fields of electromagnetism and atomic and nuclear physics. In addition, in a complex situation the "energy approach" can often provide a much simpler analysis than the direct application of Newton's second law.

This alternative method of describing motion is especially useful when the force acting on a particle is not constant. In this case, the acceleration is not constant, and we cannot apply the simple kinematic equations we developed in Chapter 3. Often, a particle in nature is subject to a force that varies with the position of the particle. Such forces include gravitational forces and the force exerted on a body attached to a spring. We shall describe techniques for treating such systems with the help of an extremely important development called the *work-energy theorem,* which is the central topic of this chapter.

We begin by defining work, a concept that provides a link between the concepts of force and energy. In Chapter 8, we shall discuss the law of conservation of energy and apply it to various problems.

7.2 WORK DONE BY A CONSTANT FORCE

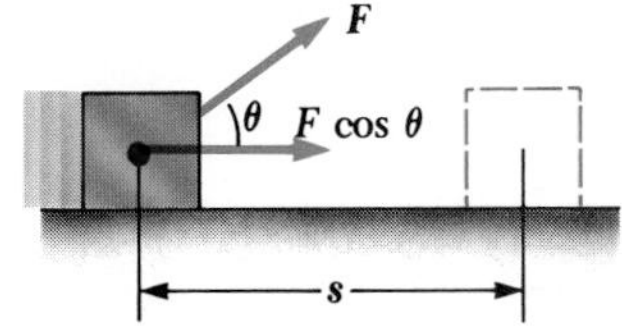

Figure 7.1 If an object undergoes a displacement **s**, the work done by the force **F** is $(F \cos \theta)s$.

Consider an object that undergoes a displacement **s** along a straight line under the action of a constant force **F**, which makes an angle θ with **s**, as in Figure 7.1.

> The work done by the constant force is defined as the product of the component of the force in the direction of the displacement and the magnitude of the displacement.

Since the component of **F** in the direction of **s** is $F \cos \theta$, the work W done by **F** is given by

Work done by a constant force

$$W \equiv (F \cos \theta)s \tag{7.1}$$

According to this definition, work is done by **F** on an object under the following conditions: (1) the object must undergo a displacement and (2) **F** must have a nonzero component in the direction of **s**. From the first condition, we see that a force does no work on an object if the object does not move ($s = 0$). For example, if a person pushes against a brick wall, a force is exerted on the wall but the person does no work since the wall is fixed. However, the person's muscles are contracting (undergoing displacement) in the process so that *internal* energy is being used up. Thus, we see that the meaning of work in physics is distinctly different from its meaning in day-to-day affairs. Likewise, if you hold a weight at arm's length for some period of time, no work is done on the weight (assuming no wiggling or oscillations of the arms). Even though you must exert an upward force to support the weight, the work done by the force

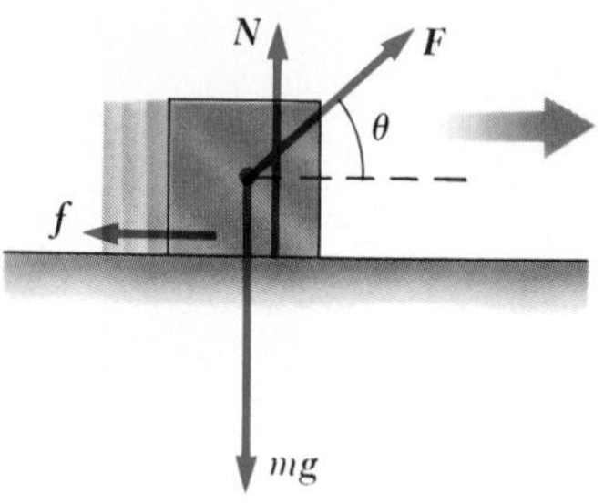

Figure 7.2 When an object is displaced horizontally on a rough surface, the normal force, **N**, and the weight, *m***g**, do *no* work. The work done by **F** is $(F \cos \theta)s$, and the work done by the frictional force is $-fs$.

is zero since the displacement is zero. After holding the weight for a long period of time, your arms would tire and you would claim that the effort required a considerable amount of "work."

From the second condition, note that the work done by a force is also zero when the force is perpendicular to the displacement, since $\theta = 90°$ and $\cos 90° = 0$. For example, in Figure 7.2, both the work done by the normal force and the work done by the force of gravity are zero since both forces are perpendicular to the displacement and have zero components in the direction of **s**.

The sign of the work also depends on the direction of **F** relative to **s**. The work done by the applied force is positive when the vector associated with the component $F \cos \theta$ is in the *same direction* as the displacement. For example, when an object is lifted, the work done by the applied force is positive since the lifting force is upward, that is, in the same direction as the displacement. In this situation, the work done by the gravitational force is negative. When the vector associated with the component $F \cos \theta$ is in the direction *opposite* the displacement, *W is negative.* The factor $\cos \theta$ that appears in the definition of *W automatically* takes care of the sign.

A common example in which W is negative is the work done by a frictional force when a body slides over a rough surface. If the force of sliding friction is **f**, and the body undergoes a linear displacement **s**, the work done by the frictional force is

Work done by a sliding frictional force

$$W_f = -fs \tag{7.2}$$

where the negative sign comes from the fact that $\theta = 180°$ and $\cos 180° = -1$.

Finally, if an applied force **F** acts along the direction of the displacement, then $\theta = 0$, and $\cos 0 = 1$. In this case, Equation 7.1 gives

$$W = Fs \tag{7.3}$$

Work is a scalar quantity, and its units are force multiplied by length. Therefore, the SI unit of work is the **newton · meter** (N · m). Another name for the newton-meter is the **joule** (J). The units of work in the cgs and British engineering systems are **dyne · cm**, which is also called the **erg**, and ft · lb, respectively. These are summarized in Table 7.1. Note that $1 \text{ J} = 10^7$ ergs.

Since work is a scalar quantity we can combine the work done by each of the separate forces to get the total work done. For instance, if there are three forces contributing to the work done, there would be three terms in the sum, each corresponding to the work done by a given force. The following example illustrates this point.

Does the weight lifter do any work as he holds the weight over his head? Does he do any work as he raises the weight? (© M. Brittan/Index Stock International)

TABLE 7.1 Units of Work in the Three Common Systems of Measurement

System	Unit of Work	Name of Combined Unit
SI	newton · meter (N · m)	joule (J)
cgs	dyne · centimeter (dyne · cm)	erg
British engineering (conventional)	foot · pound (ft · lb)	foot · pound (ft · lb)

EXAMPLE 7.1 Dragging a Box
A box is dragged across a rough floor by a constant force of magnitude 50 N. The force makes an angle of 37° above the horizontal. A frictional force of 10 N retards the motion, and the box is displaced a distance of 3 m to the right. (a) Calculate the work done by the 50-N force.

Using the definition of work (Equation 7.1) and given that $F = 50$ N, $\theta = 37°$, and $s = 3$ m,

$$W_F = (F\cos\theta)s = (50\text{ N})(\cos 37°)(3\text{ m})$$

$$= 120\text{ N}\cdot\text{m} = \boxed{120\text{ J}}$$

The vertical component of $\boldsymbol{F}$ does no work.

(b) Calculate the work done by the frictional force.

$$W_f = -fs = (-10\text{ N})(3\text{ m}) = -30\text{ N}\cdot\text{m} = \boxed{-30\text{ J}}$$

(c) Determine the net work done on the box by all forces acting on it.

Since the normal force, $\boldsymbol{N}$, and the force of gravity, $m\boldsymbol{g}$, are both perpendicular to the displacement, they do no work. Therefore, the net work done on the box is the sum of (a) and (b):

$$W_{net} = W_F + W_f = 120\text{ J} - 30\text{ J} = \boxed{90\text{ J}}$$

Later we shall show that the net work done on the body equals the change in its kinetic energy, which establishes the physical significance of W_{net}.

Exercise 1 Find the net work done on the box if it is pulled a distance of 3 m with a horizontal force of 50 N, assuming the frictional force is 15 N.
Answer 105 J.

7.3 THE SCALAR PRODUCT OF TWO VECTORS

We have defined work as a *scalar* quantity given by the product of the magnitude of the displacement and the component of the force in the direction of the displacement. It is convenient to express Equation 7.1 in terms of a **scalar product** of the two vectors $\boldsymbol{F}$ and $\boldsymbol{s}$. We write this scalar product $\boldsymbol{F}\cdot\boldsymbol{s}$. Because of the dot symbol used, the scalar product is often called the *dot product*. Thus, we can express Equation 7.1 as a scalar product:

$$W = \boldsymbol{F}\cdot\boldsymbol{s} = Fs\cos\theta \qquad (7.4)$$

Work expressed as a dot product

In other words, $\boldsymbol{F}\cdot\boldsymbol{s}$ (read "F dot s") is a shorthand notation for $Fs\cos\theta$.

In general, the scalar product of any two vectors $\boldsymbol{A}$ and $\boldsymbol{B}$ is defined as a scalar quantity equal to the product of the magnitudes of the two vectors and the cosine of the angle θ that is included between the directions of $\boldsymbol{A}$ and $\boldsymbol{B}$.

That is, the scalar product (or dot product) of $\boldsymbol{A}$ and $\boldsymbol{B}$ is defined by the relation

$$\boldsymbol{A}\cdot\boldsymbol{B} \equiv AB\cos\theta \qquad (7.5)$$

Scalar product of any two vectors $\boldsymbol{A}$ and $\boldsymbol{B}$

where θ is the angle between $\boldsymbol{A}$ and $\boldsymbol{B}$, as in Figure 7.3, A is the magnitude of $\boldsymbol{A}$, and B is the magnitude of $\boldsymbol{B}$. Note that $\boldsymbol{A}$ and $\boldsymbol{B}$ need not have the same units.

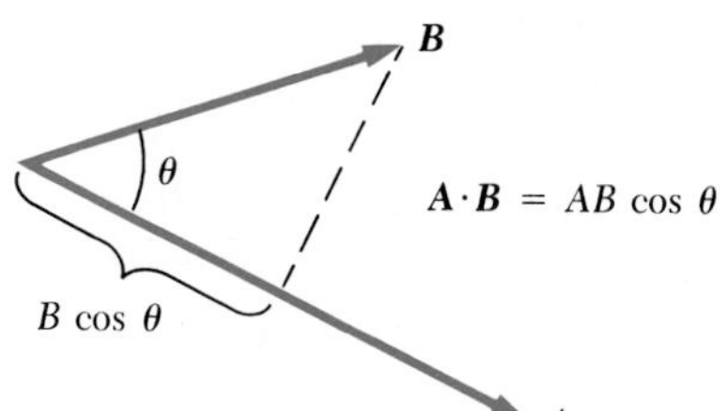

Figure 7.3 The scalar product $\boldsymbol{A}\cdot\boldsymbol{B}$ equals the magnitude of $\boldsymbol{A}$ multiplied by the projection of $\boldsymbol{B}$ onto $\boldsymbol{A}$.

In Figure 7.3 $B \cos \theta$ is the projection of $\mathbf{B}$ onto $\mathbf{A}$. Therefore, the definition of $\mathbf{A} \cdot \mathbf{B}$ as given by Equation 7.5 can be considered as the product of the magnitude of $\mathbf{A}$ and the projection of $\mathbf{B}$ onto $\mathbf{A}$.[1] From Equation 7.5 we also see that the scalar product is *commutative.* That is,

The order of the dot product can be reversed

$$\mathbf{A} \cdot \mathbf{B} = \mathbf{B} \cdot \mathbf{A} \tag{7.6}$$

Finally, the scalar product obeys the *distributive law of multiplication,* so that

$$\mathbf{A} \cdot (\mathbf{B} + \mathbf{C}) = \mathbf{A} \cdot \mathbf{B} + \mathbf{A} \cdot \mathbf{C} \tag{7.7}$$

The dot product is simple to evaluate from Equation 7.5 when $\mathbf{A}$ is either perpendicular or parallel to $\mathbf{B}$. If $\mathbf{A}$ is perpendicular to $\mathbf{B}$ ($\theta = 90°$), then $\mathbf{A} \cdot \mathbf{B} = 0$. Also, $\mathbf{A} \cdot \mathbf{B} = 0$ in the more trivial case when either $\mathbf{A}$ or $\mathbf{B}$ is zero. If $\mathbf{A}$ and $\mathbf{B}$ point in the same direction ($\theta = 0°$), then $\mathbf{A} \cdot \mathbf{B} = AB$. If $\mathbf{A}$ and $\mathbf{B}$ point in opposite directions ($\theta = 180°$), then $\mathbf{A} \cdot \mathbf{B} = -AB$. The scalar product is negative when $90° < \theta < 180°$.

The unit vectors $\mathbf{i}$, $\mathbf{j}$, and $\mathbf{k}$, which were defined in Chapter 2, lie in the positive x, y, and z directions, respectively, of a right-handed coordinate system. Therefore, it follows from the definition of $\mathbf{A} \cdot \mathbf{B}$ that the scalar products of these unit vectors are given by

Dot products of unit vectors

$$\mathbf{i} \cdot \mathbf{i} = \mathbf{j} \cdot \mathbf{j} = \mathbf{k} \cdot \mathbf{k} = 1 \tag{7.8a}$$

$$\mathbf{i} \cdot \mathbf{j} = \mathbf{i} \cdot \mathbf{k} = \mathbf{j} \cdot \mathbf{k} = 0 \tag{7.8b}$$

Two vectors $\mathbf{A}$ and $\mathbf{B}$ can be expressed in component form as

$$\mathbf{A} = A_x\mathbf{i} + A_y\mathbf{j} + A_z\mathbf{k}$$

$$\mathbf{B} = B_x\mathbf{i} + B_y\mathbf{j} + B_z\mathbf{k}$$

Therefore Equations 7.8a and 7.8b reduce the scalar product of $\mathbf{A}$ and $\mathbf{B}$ to

$$\mathbf{A} \cdot \mathbf{B} = A_xB_x + A_yB_y + A_zB_z \tag{7.9}$$

In the special case where $\mathbf{A} = \mathbf{B}$, we see that

$$\mathbf{A} \cdot \mathbf{A} = A_x^2 + A_y^2 + A_z^2 = A^2$$

[1] This is equivalent to stating that $\mathbf{A} \cdot \mathbf{B}$ equals the product of the magnitude of $\mathbf{B}$ and the projection of $\mathbf{A}$ onto $\mathbf{B}$.

EXAMPLE 7.2 The Scalar Product

The vectors $\mathbf{A}$ and $\mathbf{B}$ are given by $\mathbf{A} = 2\mathbf{i} + 3\mathbf{j}$ and $\mathbf{B} = -\mathbf{i} + 2\mathbf{j}$. (a) Determine the scalar product $\mathbf{A} \cdot \mathbf{B}$.

$$\mathbf{A} \cdot \mathbf{B} = (2\mathbf{i} + 3\mathbf{j}) \cdot (-\mathbf{i} + 2\mathbf{j})$$

$$= -2\mathbf{i} \cdot \mathbf{i} + 2\mathbf{i} \cdot 2\mathbf{j} - 3\mathbf{j} \cdot \mathbf{i} + 3\mathbf{j} \cdot 2\mathbf{j}$$

$$= -2 + 6 = 4$$

where we have used the fact that $\mathbf{i} \cdot \mathbf{j} = \mathbf{j} \cdot \mathbf{i} = 0$. The same result is obtained using Equation 7.9 directly, where $A_x = 2$, $A_y = 3$, $B_x = -1$, and $B_y = 2$.

(b) Find the angle θ between $\mathbf{A}$ and $\mathbf{B}$.

The magnitudes of $\mathbf{A}$ and $\mathbf{B}$ are given by

$$A = \sqrt{A_x^2 + A_y^2} = \sqrt{(2)^2 + (3)^2} = \sqrt{13}$$

$$B = \sqrt{B_x^2 + B_y^2} = \sqrt{(-1)^2 + (2)^2} = \sqrt{5}$$

Using Equation 7.5 and the result from (a) gives

$$\cos \theta = \frac{\mathbf{A} \cdot \mathbf{B}}{AB} = \frac{4}{\sqrt{13}\sqrt{5}} = \frac{4}{\sqrt{65}}$$

$$\theta = \cos^{-1} \frac{4}{8.06} = 60.3°$$

EXAMPLE 7.3 Work Done by a Constant Force

A particle moving in the xy plane undergoes a displacement $\mathbf{s} = (2\mathbf{i} + 3\mathbf{j})$ m while a constant force given by $\mathbf{F} = (5\mathbf{i} + 2\mathbf{j})$ N acts on the particle. (a) Calculate the magnitude of the displacement and the force.

$$s = \sqrt{x^2 + y^2} = \sqrt{(2)^2 + (3)^2} = \sqrt{13} \text{ m}$$

$$F = \sqrt{F_x^2 + F_y^2} = \sqrt{(5)^2 + (2)^2} = \sqrt{29} \text{ N}$$

(b) Calculate the work done by the force $\mathbf{F}$.

Substituting the expressions for $\mathbf{F}$ and $\mathbf{s}$ into Equation 7.4 and using Equation 7.8, we get

$$W = \mathbf{F} \cdot \mathbf{s} = (5\mathbf{i} + 2\mathbf{j}) \cdot (2\mathbf{i} + 3\mathbf{j}) \text{ N} \cdot \text{m}$$

$$= 5\mathbf{i} \cdot 2\mathbf{i} + 2\mathbf{j} \cdot 3\mathbf{j} = 16 \text{ N} \cdot \text{m} = 16 \text{ J}$$

Exercise 2 Calculate the angle between $\mathbf{F}$ and $\mathbf{s}$.
Answer 34.5°.

7.4 WORK DONE BY A VARYING FORCE: THE ONE-DIMENSIONAL CASE

Consider an object being displaced along the x axis under the action of a varying force, as in Figure 7.4. The object is displaced along the x axis from $x = x_i$ to $x = x_f$. In such a situation, we cannot use $W = (F \cos \theta)s$ to calculate the work done by the force, since this relationship applies only when $\mathbf{F}$ is constant in magnitude and direction. However, if we imagine that the object undergoes a very small displacement Δx, described in Figure 7.4a, then the x component of the force, F_x, is approximately constant over this interval and we can express the work done by the force for this small displacement as

$$\Delta W = F_x \, \Delta x \tag{7.10}$$

This is just the area of the shaded rectangle in Figure 7.4a. If we imagine that the F_x versus x curve is divided into a large number of such intervals, as in Figure 7.4a, then the total work done for the displacement from x_i to x_f is approximately equal to the sum of a large number of such terms:

$$W \cong \sum_{x_i}^{x_f} F_x \, \Delta x$$

If the displacements are allowed to approach zero, then the number of terms in the sum increases without limit, but the value of the sum approaches a definite value equal to the *true area* under the curve bounded by F_x and the x axis. As you probably have learned in the calculus, this limit of the sum is called an **integral** and is represented by

$$\lim_{\Delta x \to 0} \sum_{x_i}^{x_f} F_x \, \Delta x = \int_{x_i}^{x_f} F_x \, dx$$

The limits on the integral, $x = x_i$ to $x = x_f$, define what is called a **definite integral.** (An *indefinite integral* represents the limit of a sum over an unspecified interval. Appendix B.7 gives a brief description of integration.) This definite integral is numerically equal to the area under the F_x versus x curve between x_i and x_f. Therefore, we can express the work done by F_x for the displacement of the object from x_i to x_f as

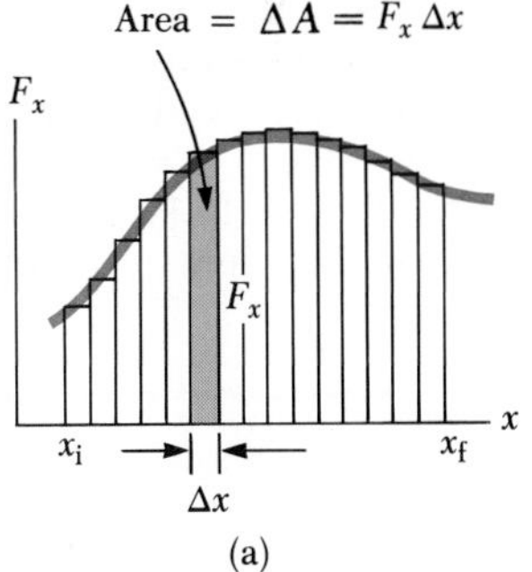

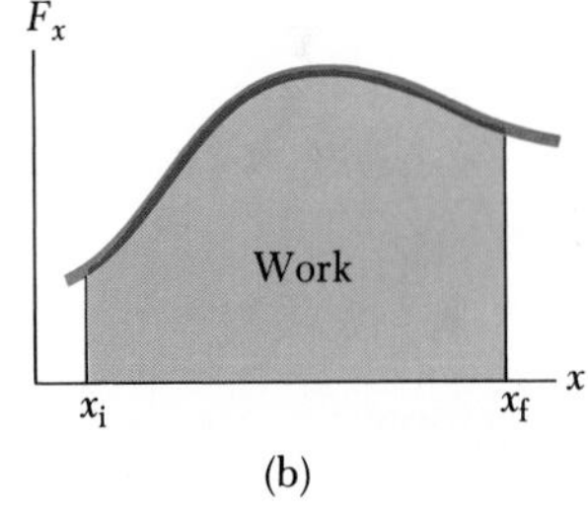

Figure 7.4 (a) The work done by the force F_x for the small displacement Δx is $F_x \, \Delta x$, which equals the area of the shaded rectangle. The total work done for the displacement from x_i to x_f is approximately equal to the sum of the areas of all the rectangles. (b) The work done by the variable force F_x as the particle moves from x_i to x_f is *exactly* equal to the area under this curve.

$$W = \int_{x_i}^{x_f} F_x \, dx \tag{7.11}$$

This equation reduces to Equation 7.1 when $F_x = F \cos \theta$ is constant.

If more than one force acts on the object, the total work done is just the work done by the resultant force. If we express the resultant force in the x direction as ΣF_x in the x direction, then the *net work* done as the object moves from x_i to x_f is

$$W_{\text{net}} = \int_{x_i}^{x_f} \left(\sum F_x \right) dx \tag{7.12}$$

EXAMPLE 7.4

A force acting on an object varies with x as shown in Figure 7.5. Calculate the work done by the force as the object moves from $x = 0$ to $x = 6$ m.

Solution The work done by the force is equal to the total area under the curve from $x = 0$ to $x = 6$ m. This area is equal to the area of the rectangular section from $x = 0$ to $x = 4$ m plus the area of the triangular section from $x = 4$ m to $x = 6$ m. The area of the rectangle is $(4)(5)$ N · m = 20 J, and the area of the triangle is equal to $\frac{1}{2}(2)(5)$ N · m = 5 J. Therefore, the total work done is 25 J.

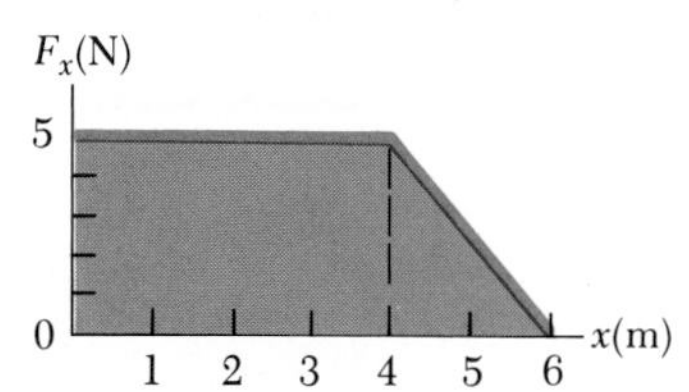

Figure 7.5 (Example 7.4) The force acting on a particle is constant for the first 4 m of motion and then decreases linearly with x from $x = 4$ m to $x = 6$ m. The net work done by this force is the area under this curve.

Work Done by a Spring

A common physical system for which the force varies with position is shown in Figure 7.6. A body on a horizontal, smooth surface is connected to a helical spring. If the spring is stretched or compressed a small distance from its unstretched, or equilibrium, configuration, the spring will exert a force on the body given by

Spring force

$$F_s = -kx \tag{7.13}$$

where x is the displacement of the body from its unstretched ($x = 0$) position and k is a positive constant called the *force constant* of the spring. As we learned in Chapter 5, this force law for springs is known as **Hooke's law.** Note that Hooke's law is only valid in the limiting case of small displacements. The value of k is a measure of the stiffness of the spring. Stiff springs have large k values, and soft springs have small k values.

The negative sign in Equation 7.13 signifies that the force exerted by the spring is always directed *opposite* the displacement. For example, when $x > 0$ as in Figure 7.6a, the spring force is to the left, or negative. When $x < 0$ as in Figure 7.6c, the spring force is to the right, or positive. Of course, when $x = 0$ as in Figure 7.6b, the spring is unstretched and $F_s = 0$. Since the spring force always acts toward the equilibrium position, it is sometimes called a *restoring force.* Once the mass is displaced some distance x_m from equilibrium and then

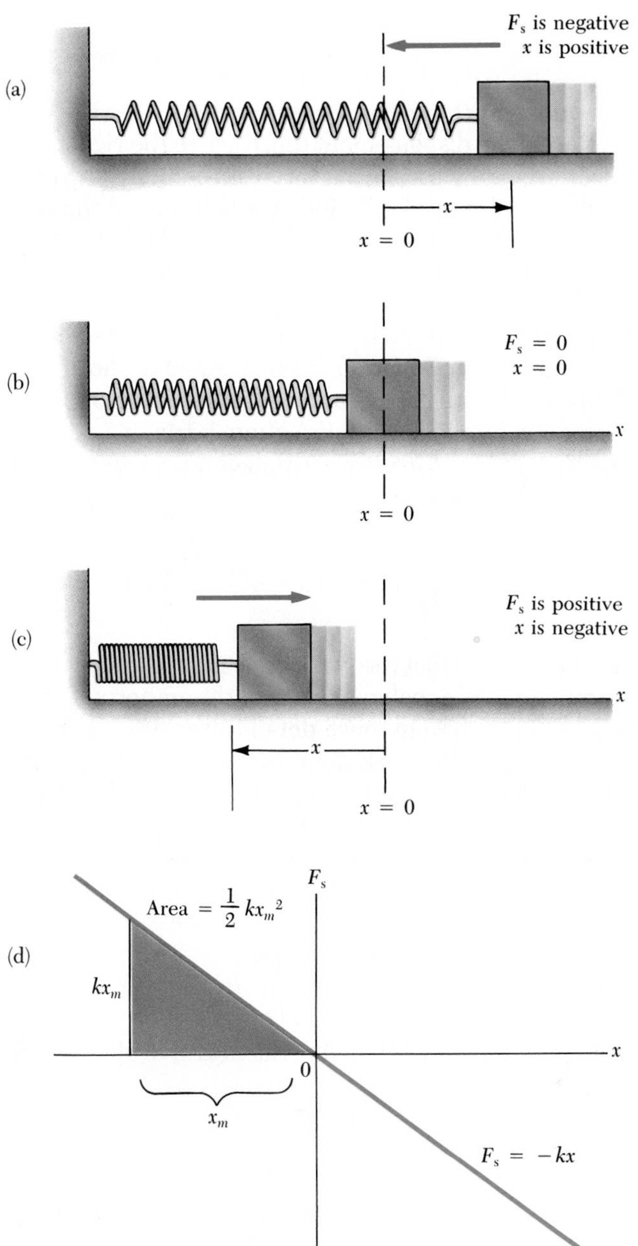

Figure 7.6 The force of a spring on a block varies with the block's displacement from the equilibrium position $x = 0$. (a) When x is positive (stretched spring), the spring force is to the left. (b) When x is zero, the spring force is zero (natural length of the spring). (c) When x is negative (compressed spring), the spring force is to the right. (d) Graph of F_s versus x for systems described above. The work done by the spring force as the block moves from $-x_m$ to 0 is the area of the shaded triangle, $\frac{1}{2}kx_m^2$.

released, it will move from $-x_m$ through zero to $+x_m$. The details of the ensuing oscillating motion will be given in Chapter 13.

Suppose that the block is pushed to the left a distance x_m from equilibrium, as in Figure 7.6c, and then released. Let us calculate the *work done by the spring force* as the body moves from $x_i = -x_m$ to $x_f = 0$. Applying Equation 7.11, we get

Work done by a spring

$$W_s = \int_{x_i}^{x_f} F_s \, dx = \int_{-x_m}^{0} (-kx) \, dx = \tfrac{1}{2}kx_m^2 \tag{7.14a}$$

That is, the work done by the spring force is positive since the spring force is in the same direction as the displacement (both are to the right). However, if we consider the work done by the spring force as the body moves from $x_i = 0$ to $x_f = x_m$, we find that $W_s = -\frac{1}{2}kx_m^2$, since for this part of the motion, the displacement is to the right and the spring force is to the left. Therefore, the *net* work done by the spring force as the body moves from $x_i = -x_m$ to $x_f = x_m$ is *zero*.

If we plot F_s versus x as in Figure 7.6d, we arrive at the same results. Note that the work calculated in Equation 7.14a is equal to the area of the shaded triangle in Figure 7.6d, with base x_m and height kx_m. The area of this triangle is $\frac{1}{2}kx_m^2$, the work done by the spring, Equation 7.14a.

If the mass undergoes an *arbitrary* displacement from $x = x_i$ to $x = x_f$, the work done by the spring force is given by

$$W_s = \int_{x_i}^{x_f} (-kx) \, dx = \tfrac{1}{2}kx_i^2 - \tfrac{1}{2}kx_f^2 \tag{7.14b}$$

From this equation, we see that the work done is zero for any motion that ends where it began ($x_i = x_f$). We shall make use of this important result in describing the motion of this system in more detail in the next chapter.

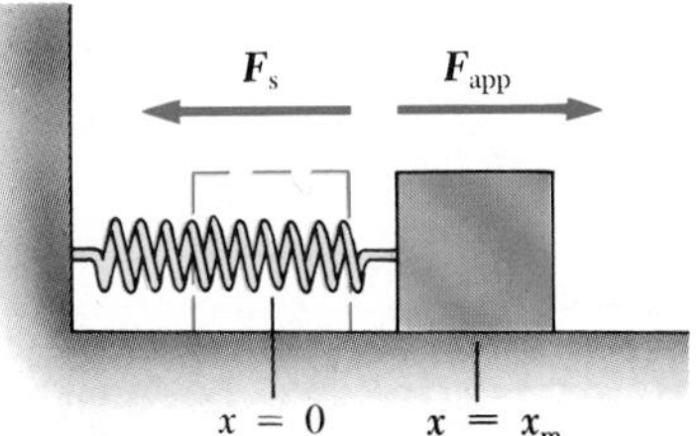

Figure 7.7 A block being pulled to the right on a frictionless surface by a force F_{app} from $x = 0$ to $x = x_m$. If the process is carried out very slowly, the applied force is equal and opposite to the spring force at all times.

Now let us consider the work done by an *external agent* in stretching a spring *very slowly* from $x_i = 0$ to $x_f = x_m$, as in Figure 7.7. This work can be easily calculated by noting that the *applied force*, $\mathbf{F}_{app}$, is equal to and opposite the spring force, $\mathbf{F}_s$, at any value of the displacement, so that $F_{app} = -(-kx) = kx$. Therefore, the work done by this applied force (the external agent) is given by

$$W_{F_{app}} = \int_0^{x_m} F_{app} \, dx = \int_0^{x_m} kx \, dx = \tfrac{1}{2}kx_m^2$$

You should note that this work is equal to the negative of the work done by the spring force for this displacement.

EXAMPLE 7.5 The Spring Force Does Work

A block lying on a smooth, horizontal surface is connected to a spring with a force constant of 80 N/m. The spring is compressed a distance of 3.0 cm from equilibrium as in Figure 7.6c. Calculate the work done by the spring force as the block moves from $x_i = -3.0$ cm to its unstretched position, $x_f = 0$.

Solution Using Equation 7.14a with $x_m = -3.0 \text{ cm} = -3 \times 10^{-2}$ m, we get

$$W_s = \tfrac{1}{2}kx_m^2 = \tfrac{1}{2}\left(80 \, \frac{\text{N}}{\text{m}}\right)(-3 \times 10^{-2} \text{ m})^2$$

$$= 3.6 \times 10^{-2} \text{ J}$$

EXAMPLE 7.6 Measuring *k* for a Spring

A common technique used to measure the force constant of a spring is described in Figure 7.8. The spring is hung vertically as shown in Figure 7.8a. A body of mass m is then attached to the lower end of the spring as in Figure 7.8b. The spring stretches a distance d from its equilibrium position under the action of the "load" mg. Since the spring force is upward, it must balance the weight mg downward when the system is at rest. In this case, we can apply Hooke's law to give $|\mathbf{F}_s| = kd = mg$, or

$$k = mg/d$$

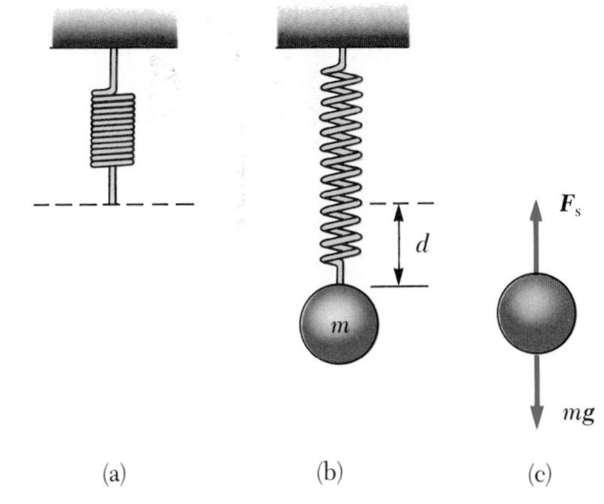

Figure 7.8 (Example 7.6) Determination of the force constant of a helical spring. The elongation d of the spring is due to the weight mg. Since the spring force upward balances the weight, it follows that $k = mg/d$.

For example, if a spring is stretched a distance of 2.0 cm by a mass of 0.55 kg, the force constant of the spring is

$$k = \frac{mg}{d} = \frac{(0.55 \text{ kg})(9.80 \text{ m/s}^2)}{2.0 \times 10^{-2} \text{ m}} = 2.7 \times 10^2 \text{ N/m}$$

EXAMPLE 7.7 Work Done in Moving a Car

A sports car on a horizontal surface is pushed by a horizontal force that varies with position according to the graph shown in Figure 7.9. Determine an approximate value for the total work done in moving the car from $x = 0$ to $x = 20$ m.

Solution We can obtain the result from the graph by dividing the total displacement into many small displacements. For simplicity, we choose to divide the total displacement into ten consecutive displacements, each 2 m in length, as shown in Figure 7.9. The work done during each small displacement is *approximately* equal to the area of the dotted rectangle. For example, the work done for the first displacement, from $x = 0$ to $x = 2$ m, is the area of the smallest rectangle, (2 m)(5 N) = 10 J; the work done for the second displacement, from $x = 2$ m to $x = 4$ m, is the area of the second rectangle, equal to (2 m)(12 N) = 24 J. Continuing in this fashion, we get the areas indicated in Figure 7.9, the *sum* of which gives the total work done from $x = 0$ to $x = 20$ m. This result is

$$W_{\text{total}} \approx 442 \text{ J}$$

The accuracy of the result will of course improve as the widths of the intervals are made smaller.

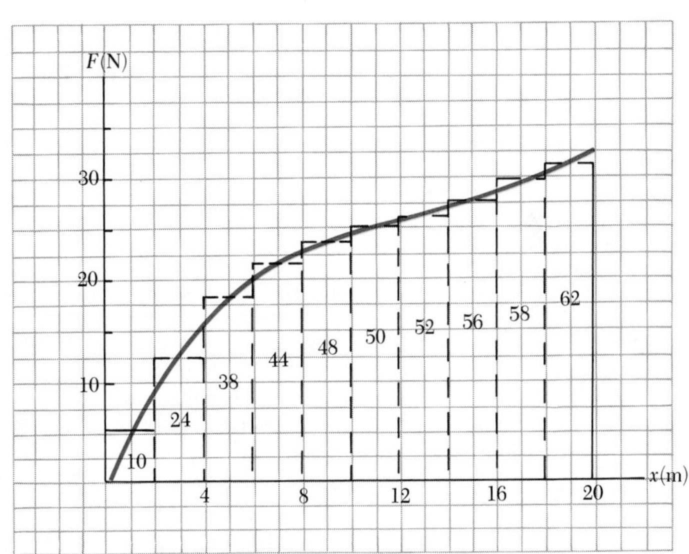

Figure 7.9 (Example 7.7) A graph of force versus position for a car moving along the x axis. The numbers within the rectangles represent the work done (area of rectangle) during that interval.

7.5 WORK AND KINETIC ENERGY

In Chapter 5 we found that a particle accelerates when the resultant force on it is not zero. Consider a situation in which a constant force $\boldsymbol{F}_x$ acts on a particle of mass m moving in the x direction. Newton's second law states that $\boldsymbol{F}_x = m\boldsymbol{a}$, where $\boldsymbol{a}_x$ is constant since $\boldsymbol{F}_x$ is constant. If the particle is displaced from $x_i = 0$ to $x_f = s$ the work done by the force $\boldsymbol{F}_x$ is

$$W = F_x s = (ma_x)s \tag{7.15}$$

However, in Chapter 3 we found that the following relationships are valid when a particle undergoes constant acceleration:

$$s = \tfrac{1}{2}(v_i + v_f)t \qquad a_x = \frac{v_f - v_i}{t}$$

where v_i is the velocity at $t = 0$ and v_f is the velocity at time t. Substituting these expressions into Equation 7.15 gives

$$W = m\left(\frac{v_f - v_i}{t}\right)\tfrac{1}{2}(v_i + v_f)t$$

$$W = \tfrac{1}{2}mv_f^2 - \tfrac{1}{2}mv_i^2 \tag{7.16}$$

The product of one half the mass and the square of the speed is defined as the **kinetic energy** of the particle.

That is, the kinetic energy, K, of a particle of mass m and speed v is defined as

Kinetic energy is energy associated with the motion of a body

$$K \equiv \tfrac{1}{2}mv^2 \tag{7.17}$$

Kinetic energy is a scalar quantity and has the same units as work. For example, a 1-kg mass moving with a speed of 4.0 m/s has a kinetic energy of 8.0 J. Table 7.2 lists the kinetic energies for various objects. We can think of kinetic energy as energy associated with the motion of a body. It is often convenient to write Equation 7.16

Work-energy theorem

$$W = K_f - K_i = \Delta K \tag{7.18}$$

TABLE 7.2 Kinetic Energies for Various Objects

Object	Mass (kg)	Speed (m/s)	Kinetic energy (J)
Earth orbiting the sun	5.98×10^{24}	2.98×10^{4}	2.65×10^{33}
Moon orbiting the earth	7.35×10^{22}	1.02×10^{3}	3.82×10^{28}
Rocket moving at escape velocity[a]	500	6.18×10^{5}	9.5×10^{13}
Automobile at 55 mi/h	2000	25	6.3×10^{5}
Running athlete	70	10	3.5×10^{3}
Stone dropped from 10 m	1	14	9.8×10^{1}
Golf ball at terminal speed	0.046	32	2.4×10^{1}
Raindrop at terminal speed	3.5×10^{-5}	9	1.4×10^{-3}
An oxygen molecule in air	5.3×10^{-26}	500	6.6×10^{-21}

[a] Escape velocity is the minimum velocity an object must reach near the earth's surface in order to escape the earth's gravitational force.

That is,

the work done by the resultant constant force F in displacing a particle equals the change in kinetic energy of the particle.

Work done on a particle equals the change in its kinetic energy

The change here means the final minus the initial value of the kinetic energy.

Equation 7.18 is an important result known as the **work-energy theorem.** This theorem was derived for the case where the force is constant, but we can show that it is valid even when the force is varying: If the resultant force acting on a body in the x direction is $\Sigma \boldsymbol{F}_x$, then Newton's second law states that $\Sigma \boldsymbol{F}_x = m\boldsymbol{a}$. Thus, we can use Equation 7.12 and express the net work done as

$$W_{\text{net}} = \int_{x_i}^{x_f} \left(\sum F_x\right) dx = \int_{x_i}^{x_f} ma\, dx$$

Because the resultant force varies with x, the acceleration and velocity also depend on x. We can now use the following chain rule to evaluate W_{net}:

$$a = \frac{dv}{dt} = \frac{dv}{dx}\frac{dx}{dt} = v\frac{dv}{dx}$$

Substituting this into the expression for W gives

$$W_{\text{net}} = \int_{x_i}^{x_f} mv\frac{dv}{dx}\, dx = \int_{v_i}^{v_f} mv\, dv = \tfrac{1}{2}mv_f^2 - \tfrac{1}{2}mv_i^2 \qquad (7.19)$$

The limits of the integration were changed because the variable was changed from x to v.

The work-energy theorem given by Equation 7.18 is also valid in the more general case when the force varies in direction and magnitude while the particle moves along an arbitrary curved path in three dimensions. In this situation, we express the work as

$$W = \int_{i}^{f} \boldsymbol{F} \cdot d\boldsymbol{s} \qquad (7.20)$$

General expression for work done by a force $\boldsymbol{F}$

where the limits i and f represent the initial and final coordinates of the particle. The integral given by Equation 7.20 is called a *line integral.* Because the infinitesimal displacement vector can by expressed as $d\boldsymbol{s} = dx\boldsymbol{i} + dy\boldsymbol{j} + dz\boldsymbol{k}$ and because $\boldsymbol{F} = F_x\boldsymbol{i} + F_y\boldsymbol{j} + F_z\boldsymbol{k}$, Equation 7.20 reduces to

$$W = \int_{x_i}^{x_f} F_x\, dx + \int_{y_i}^{y_f} F_y\, dy + \int_{z_i}^{z_f} F_z\, dz \qquad (7.21)$$

This is the general expression[2] that is used to calculate the work done by a force when a particle undergoes a displacement from the point with coordinates (x_i, y_i, z_i) to the point with coordinates (x_f, y_f, z_f).

Thus, we conclude that

the work done on a particle by the resultant force acting on it is equal to the change in the kinetic energy of the particle.

Work can be positive, negative, or zero

[2] In the general expression, Equation 7.21, the component F_x can depend on y and z as well as on x. However, y and z are treated as constants in the integration over x; similarly for F_y and F_z.

The work-energy theorem also says that the speed of the particle will increase ($K_f > K_i$) if the net work done on it is positive, whereas its speed will decrease ($K_f < K_i$) if the net work done on it is negative. That is, the speed and kinetic energy of a particle will change only if work is done on the particle by the net external force. Because of this connection between work and change in kinetic energy, we can also think of the kinetic energy of a body as the work the body can do in coming to rest.

EXAMPLE 7.8 A Block Pulled on a Smooth Surface
A 6-kg block initially at rest is pulled to the right along a horizontal smooth surface by a constant, horizontal force of 12 N, as in Figure 7.10a. Find the speed of the block after it moves a distance of 3 m.

Solution The weight is balanced by the normal force, and neither of these forces does work since the displacement is horizontal. Since there is no friction, the resultant external force is the 12-N force. The work done by this force is

$$W_F = Fs = (12\text{ N})(3\text{ m}) = 36\text{ N}\cdot\text{m} = 36\text{ J}$$

Using the work-energy theorem and noting that the initial kinetic energy is zero, we get

$$W_F = K_f - K_i = \tfrac{1}{2}mv_f^2 - 0$$

$$v_f^2 = \frac{2W_F}{m} = \frac{2(36\text{ J})}{6\text{ kg}} = 12\text{ m}^2/\text{s}^2$$

$$v_f = 3.46\text{ m/s}$$

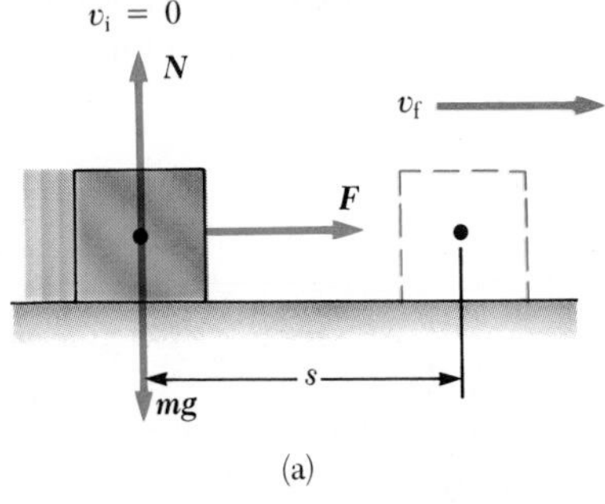

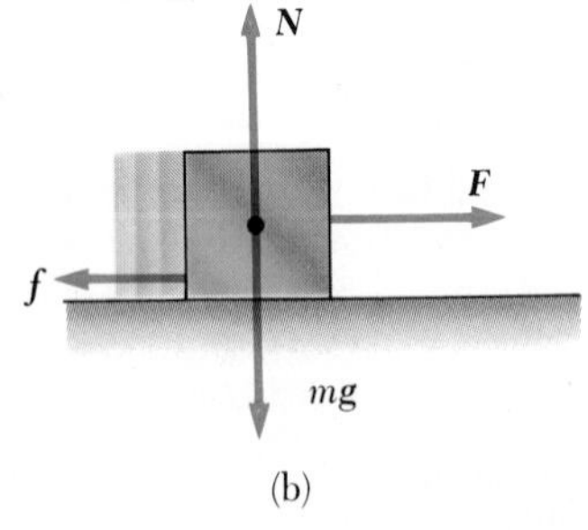

Figure 7.10 (a) Example 7.8. (b) Example 7.9.

Exercise 3 Find the acceleration of the block, and determine the final speed of the block using the kinematic equation $v_f^2 = v_i^2 + 2as$.
Answer $a = 2\text{ m/s}^2$; $v_f = 3.46\text{ m/s}$.

EXAMPLE 7.9 A Block Pulled on a Rough Surface
Find the final speed of the block described in Example 7.8 if the surface is rough and the coefficient of kinetic friction is 0.15.

Solution In this case, we must calculate the net work done on the block, which equals the sum of the work done by the applied 12-N force and the frictional force f, as in Figure 7.10b. Since the frictional force opposes the displacement, the work this force does is *negative.* The magnitude of the frictional force is given by $f = \mu N = \mu mg$; therefore, the work done by this force is this force multiplied by the displacement (see Eq. 7.2) or

$$W_f = -fs = -\mu mgs = (-0.15)(6\text{ kg})(9.80\text{ m/s}^2)(3\text{ m})$$
$$= -26.5\text{ J}$$

Therefore, the net work done on the block is

$$W_{net} = W_F + W_f = 36.0\text{ J} - 26.5\text{ J} = 9.50\text{ J}$$

Applying the work-energy theorem with $v_i = 0$ gives

$$W_{net} = \tfrac{1}{2}mv_f^2$$

$$v_f^2 = \frac{2W_{net}}{m} = \frac{19}{6}\text{ m}^2/\text{s}^2$$

$$v_f = 1.78\text{ m/s}$$

Exercise 4 Find the acceleration of the block from Newton's second law, and determine the final speed of the block using kinematics.
Answer $a = 0.530\text{ m/s}^2$; $v_f = 1.78\text{ m/s}$.

EXAMPLE 7.10 A Mass-Spring System
A block of mass 1.6 kg is attached to a spring with a force constant of 10^3 N/m, as in Figure 7.6. The spring is compressed a distance of 2.0 cm and the block is released from rest. (a) Calculate the speed of the block as it passes

through the equilibrium position $x = 0$, if the surface is frictionless.

Using Equation 17.4a, the work done by the spring with $x_m = -2.0 \text{ cm} = -2 \times 10^{-2}$ m is

$$W_s = \tfrac{1}{2}kx_m^2 = \tfrac{1}{2}\left(10^3 \frac{\text{N}}{\text{m}}\right)(-2 \times 10^{-2} \text{ m})^2 = 0.20 \text{ J}$$

Using the work-energy theorem with $v_i = 0$ gives

$$W_s = \tfrac{1}{2}mv_f^2 - \tfrac{1}{2}mv_i^2$$

$$0.20 \text{ J} = \tfrac{1}{2}(1.6 \text{ kg})v_f^2 - 0$$

$$v_f^2 = \frac{0.4 \text{ J}}{1.6 \text{ kg}} = 0.25 \text{ m}^2/\text{s}^2$$

$$v_f = 0.50 \text{ m/s}$$

(b) Calculate the speed of the block as it passes through the equilibrium position if a constant frictional force of 4.0 N retards its motion.

The work done by the frictional force for a displacement of 2×10^{-2} m is given by

$$W_f = -fs = -(4 \text{ N})(2 \times 10^{-2} \text{ m}) = -0.08 \text{ J}$$

The net work done on the block is the work done by the spring plus the work done by friction. In part (a), we found $W_s = 0.20$ J, therefore

$$W_{net} = W_s + W_f = 0.20 \text{ J} - 0.08 \text{ J} = 0.12 \text{ J}$$

Applying the work-energy theorem gives

$$\tfrac{1}{2}mv_f^2 = W_{net}$$

$$\tfrac{1}{2}(1.6 \text{ kg})v_f^2 = 0.12 \text{ J}$$

$$v_f^2 = \frac{0.24 \text{ J}}{1.6 \text{ kg}} = 0.15 \text{ m}^2/\text{s}^2$$

$$v_f = 0.39 \text{ m/s}$$

Note that this value for v_f is *less* than that obtained in the frictionless case. Is this result sensible?

EXAMPLE 7.11 Block Pushed Along an Incline

A block of mass m is pushed up a rough incline by a constant force $\mathbf{F}$ acting parallel to the incline, as in Figure 7.11a. The block is displaced a distance d up the incline. (a) Calculate the work done by the force of gravity for this displacement.

The force of gravity is downward but has a component *down* the plane. This is given by $-mg \sin\theta$ if the positive x direction is chosen to be up the plane (Figure 7.11b). Therefore, the work done by gravity for the displacement d is

$$W_g = (-mg \sin\theta)d = -mgh$$

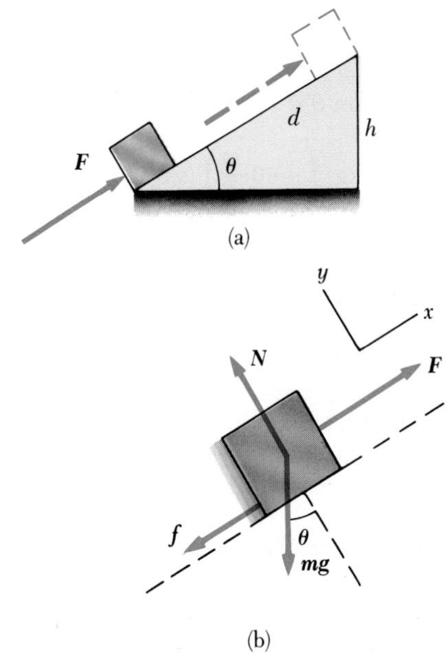

Figure 7.11 (Example 7.11) A block is pushed up a rough incline by a constant force $\mathbf{F}$.

where $h = d \sin\theta$ is the *vertical* displacement. That is, the work done by gravity has a magnitude equal to the force of gravity multiplied by the *upward vertical displacement.* In the next chapter, we shall show that this result is valid in general for any particle displaced between two points. Furthermore, the result is independent of the path taken between these points.

(b) Calculate the work done by the applied force $\mathbf{F}$.

Since $\mathbf{F}$ is in the same direction as the displacement, we get

$$W_F = \mathbf{F} \cdot \mathbf{s} = Fd$$

(c) Find the work done by the force of kinetic friction if the coefficient of friction is μ.

The magnitude of the force of friction is $f = \mu N = \mu mg \cos\theta$. Since the direction of this force is *opposite* the direction of the displacement, we find that

$$W_f = -fd = -\mu mgd \cos\theta$$

(d) Find the net work done on the block for this displacement.

Using the results to (a), (b), and (c), we get

$$W_{net} = W_g + W_F + W_f$$
$$= -mgd \sin\theta + Fd - \mu mgd \cos\theta$$

or

$$W_{net} = Fd - mgd(\sin\theta + \mu \cos\theta)$$

For example, if we take $F = 15$ N, $d = 1.0$ m, $\theta = 25°$, $m = 1.5$ kg, and $\mu = 0.30$, we find that

$$W_g = -(mg\sin\theta)d$$

$$= -(1.5\text{ kg})\left(9.80\ \frac{\text{m}}{\text{s}^2}\right)(\sin 25°)(1.0\text{ m})$$

$$= -6.2\text{ J}$$

$$W_F = Fd = (15\text{ N})(1\text{ m}) = 15\text{ J}$$

$$W_f = -\mu mgd\cos\theta$$

$$= -(0.30)(1.5\text{ kg})\left(9.80\ \frac{\text{m}}{\text{s}^2}\right)(1.0\text{ m})(\cos 25°)$$

$$= -4.0\text{ J}$$

$$W_{\text{net}} = W_g + W_F + W_f = 4.8\text{ J}$$

EXAMPLE 7.12 Minimum Stopping Distance

An automobile traveling at 48 km/h can be stopped in a minimum distance of 40 m by applying the brakes. If the same automobile is traveling at 96 km/h, what is the minimum stopping distance?

Solution We shall assume that when the brakes are applied, the car does not skid. To get the minimum stopping distance, d, we take the frictional force f between the tires and road to be a *maximum*. The work done by this frictional force, $-fd$, must equal the change in kinetic energy of the automobile. Since the kinetic energy has a final value of zero and an initial value of $\frac{1}{2}mv^2$, we get

$$W_f = K_f - K_i$$

$$-fd = 0 - \tfrac{1}{2}mv^2$$

$$d = \frac{mv^2}{2f}$$

If we assume f is the same for the two initial speeds, we can take m and f as constants. Therefore, the ratio of stopping distances is given by

$$\frac{d_2}{d_1} = \left(\frac{v_2}{v_1}\right)^2$$

Taking $v_1 = 48$ km/h, $v_2 = 96$ km/h, and $d_1 = 40$ m gives

$$\frac{d_2}{d_1} = \left(\frac{96}{48}\right)^2 = 4$$

$$d_2 = 4d_1 = 4(40\text{ m}) = 160\text{ m}$$

This shows that the *minimum stopping distance varies as the square of the ratio of speeds.* If the speed is doubled, as it is in this example, the distance increases by a factor of 4.

7.6 POWER

From a practical viewpoint, it is interesting to know not only the work done on an object, but also the rate at which the work is being done. **Power** is defined as *the time rate of energy transfer.*

If an external force is applied to an object, and if the work done by this force is ΔW in the time interval Δt, then the **average power** during this interval is defined as the ratio of the work done to the time interval:

Average power

$$\bar{P} \equiv \frac{\Delta W}{\Delta t} \tag{7.22}$$

According to the work-energy theorem, this work done on the object contributes to increasing the energy of the object, so we can also say that power is the time rate of energy transfer. The **instantaneous power,** P, is the limiting value of the average power as Δt approaches zero:

$$P \equiv \lim_{\Delta t \to 0} \frac{\Delta W}{\Delta t} = \frac{dW}{dt} \tag{7.23}$$

From Equation 7.4, we can express the work done by a force $\boldsymbol{F}$ for a displacement $d\boldsymbol{s}$, since $dW = \boldsymbol{F} \cdot d\boldsymbol{s}$. Therefore, the instantaneous power can be written

$$P = \frac{dW}{dt} = \boldsymbol{F} \cdot \frac{d\boldsymbol{s}}{dt} = \boldsymbol{F} \cdot \boldsymbol{v} \quad (7.24)$$

Instantaneous power

where we have used the fact that $\boldsymbol{v} = d\boldsymbol{s}/dt$.

The unit of power in the SI system is J/s, which is also called a *watt*, W (after James Watt):

$$1 \text{ W} = 1 \text{ J/s} = 1 \text{ kg} \cdot \text{m}^2/\text{s}^3$$

The watt

The symbol W for watt should not be confused with the symbol for work.

The unit of power in the British engineering system is the horsepower (hp), where

$$1 \text{ hp} = 550 \text{ ft} \cdot \text{lb/s} = 746 \text{ W}$$

A new unit of energy (or work) can now be defined in terms of the unit of power. One kilowatt-hour (kWh) is the energy converted or consumed in 1 h at the constant rate of 1 kW. The numerical value of 1 kWh is

$$1 \text{ kWh} = (10^3 \text{ W})(3600 \text{ s}) = 3.6 \times 10^6 \text{ J}$$

It is important to realize that a kWh is a unit of energy, not power. When you pay your electric bill, you are buying energy, and the amount of electricity used is usually in multiples of kWh. For example, an electric bulb rated at 100 W would "consume" 3.6×10^5 J of energy in 1 h.

Although the W and the kWh are commonly used only in electrical applications, they can be used in other scientific areas. For example, an automobile engine can be rated in kW as well as in hp. Likewise, the power consumption of an electrical appliance can be expressed in hp.

EXAMPLE 7.13 Power Delivered by an Elevator Motor

An elevator has a mass of 1000 kg and carries a maximum load of 800 kg. A constant frictional force of 4000 N retards its motion upward, as in Figure 7.12. (a) What must be the minimum power delivered by the motor to lift the elevator at a constant speed of 3 m/s?

The motor must supply the force $\boldsymbol{T}$ that pulls the elevator upward. From Newton's second law and from the fact that $a = 0$ since v is constant, we get

$$T - f - Mg = 0$$

where M is the *total* mass (elevator plus load), equal to 1800 kg. Therefore,

$$\begin{aligned} T &= f + Mg \\ &= 4 \times 10^3 \text{ N} + (1.8 \times 10^3 \text{ kg})(9.80 \text{ m/s}^2) \\ &= 2.16 \times 10^4 \text{ N} \end{aligned}$$

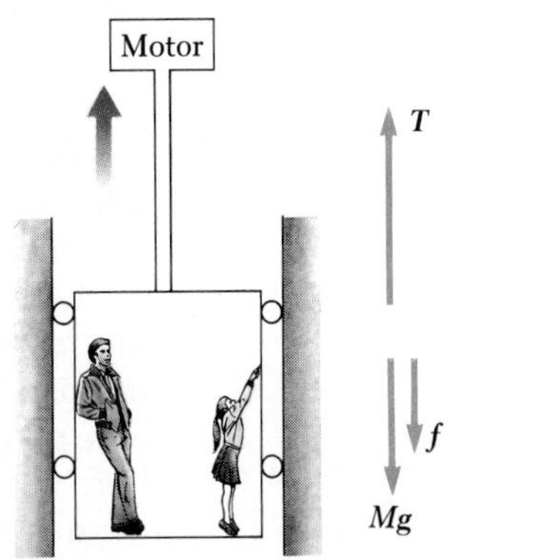

Figure 7.12 (Example 7.13) A motor provides a force $\boldsymbol{T}$ upward on the elevator. A frictional force $\boldsymbol{f}$ and the total weight $\boldsymbol{Mg}$ act downward.

Using Equation 7.24 and the fact that $\mathbf{T}$ is in the same direction as $\mathbf{v}$ gives

$$P = \mathbf{T} \cdot \mathbf{v} = Tv$$

$$= (2.16 \times 10^4 \text{ N})(3 \text{ m/s}) = 6.49 \times 10^4 \text{ W}$$

$$= 64.9 \text{ kW} = 87.0 \text{ hp}$$

(b) What power must the motor deliver at any instant if it is designed to provide an upward acceleration of 1.0 m/s²?

Applying Newton's second law to the elevator gives

$$T - f - Mg = Ma$$

$$T = M(a + g) + f$$

$$= (1.8 \times 10^3 \text{ kg})(1.0 + 9.80) \text{ m/s}^2 + 4 \times 10^3 \text{ N}$$

$$= 2.34 \times 10^4 \text{ N}$$

Therefore, using Equation 7.24 we get for the required power

$$P = Tv = (2.34 \times 10^4\, v) \text{ W}$$

where v is the instantaneous speed of the elevator in m/s. Hence, the power required increases with increasing speed.

*7.7 ENERGY AND THE AUTOMOBILE

Automobiles powered by gasoline engines are known to be very inefficient machines. Even under ideal conditions, less than 15% of the available energy in the fuel is used to power the vehicle. The situation is much worse under stop-and-go driving in the city. The purpose of this section is to use the concepts of energy, power, and forces of friction to analyze some factors that affect automobile fuel consumption.

There are many mechanisms that contribute to the energy losses in a typical automobile.[3] About two thirds of the energy available from the fuel is lost in the engine. Part of this energy ends up in the atmosphere via the exhaust system, and part is used in the engine's cooling system. As we shall see in Chapter 22, the large power loss in the exhaust and cooling system is not easy to overcome because of some fundamental laws of thermodynamics. About 10% of the available energy is lost in the automobile's drive-train mechanism; this loss includes friction in the transmission, drive shaft, wheel and axle bearings, and differential. Friction in other moving parts such as in the motor dissipates about 6% of the energy, and 4% of the available energy is used to operate fuel and oil pumps and such accessories as power steering, air conditioning, power brakes, and electrical components. Finally, about 14% of the available energy is used to propel the automobile. This energy is used mainly to overcome road friction and air resistance.

Table 7.3 lists the power losses for an automobile with an available fuel power of 136 kW. These data apply to a typical 1450-kg "gas-guzzler" with a gas consumption rate of 15 liters per 100 km (about 15 mi/gal).

Let us examine the power requirements to overcome road friction and air drag in more detail. The coefficient of rolling friction μ between the tires and the road is about 0.016. For a 1450-kg car, the weight is 14 200 N and the force of rolling friction is $\mu N = \mu W = 227$ N. As the speed of the car increases, there is a small reduction in the normal force as a result of a reduction in air pressure as air flows over the top of the car. This causes a slight reduction in the force of rolling friction f_r with increasing speed, as shown in Table 7.4.

Now let us consider the effect of air friction, that is, the drag force that results from air moving past the various surfaces of the car. The drag force

[3] An excellent article on this subject is the one by G. Waring in *The Physics Teacher*, Vol. 18 (1980), p. 494. The data in Tables 7.3 and 7.4 were taken from this article.

TABLE 7.3 Power Losses in a Typical Automobile Assuming a Total Available Power of 136 kW

Mechanism	Power Loss (kW)	Power Loss (%)
Exhaust (heat)	46	33
Cooling system	45	33
Drive train	13	10
Internal friction	8	6
Accessories	5	4
Propulsion of vehicle	19	14

associated with air friction for large objects is proportional to the square of the speed (in m/s) (Section 6.4) and may be written

$$f_a = \tfrac{1}{2}CA\rho v^2 \qquad (7.25)$$

where C is the drag coefficient, A is the cross-sectional area of the moving object, and ρ is the density of air. This expression can be used to calculate the values in Table 7.4 using $C = 0.5$, $\rho = 1.293\ \text{kg/m}^3$, and $A \approx 2\ \text{m}^2$.

The magnitude of the total frictional force, f_t, is given by the sum of the rolling friction force and the air drag force:

$$f_t = f_r + f_a \approx \text{constant} + \tfrac{1}{2}CA\rho v^2 \qquad (7.26)$$

At low speeds, road resistance and air drag are comparable, but at high speeds air drag is the predominant resistive force, as shown in Table 7.4. Road friction can be reduced by reducing tire flexing (increase the air pressure slightly above recommended values) and using radial tires. Air drag can be reduced by using a smaller cross-sectional area and streamlining the car. Though driving a car with the windows open does create more air drag, resulting in a 3% decrease in mpg, driving with the windows closed and the air conditioner running results in a 12% decrease in mileage.

The total power needed to maintain a constant speed v equals the product $f_t v$. This must equal the power delivered to the wheels. For example, from Table 7.4 we see that at $v = 26.8$ m/s, the required power is

$$P = f_t v = (683\ \text{N})\left(26.8\ \frac{\text{m}}{\text{s}}\right) = 18.3\ \text{kW}$$

TABLE 7.4 Frictional Forces and Power Requirements for a Typical Car

v (m/s)	N (N)	f_r (N)	f_a (N)	f_t (N)	$P = f_t v$ (kW)
0	14 200	227	0	227	0
8.9	14 100	226	51	277	2.5
17.8	13 900	222	204	426	7.6
26.8	13 600	218	465	683	18.3
35.9	13 200	211	830	1041	37.3
44.8	12 600	202	1293	1495	66.8

In this table, N is the normal force, f_r is road friction, f_a is air friction, f_t is total friction, and P is the power delivered to the wheels.

This can be broken into two parts: (1) the power needed to overcome road friction, $f_r v$, and (2) the power needed to overcome air drag, $f_a v$. At $v = 26.8$ m/s, these have the values

$$P_r = f_r v = (218\text{ N})\left(26.8\ \frac{\text{m}}{\text{s}}\right) = 5.8\text{ kW}$$

$$P_a = f_a v = (465\text{ N})\left(26.8\ \frac{\text{m}}{\text{s}}\right) = 12.5\text{ kW}$$

Note that $P = P_r + P_a$.

On the other hand, at $v = 44.7$ m/s, we find that $P_r = 9.0$ kW, $P_a = 57.8$ kW, and $P = 66.8$ kW. This shows the importance of air drag at high speeds.

EXAMPLE 7.14 Gas Consumed by Compact Car

A compact car has a mass of 800 kg, and its efficiency is rated at 14%. (That is, 14% of the available fuel energy is delivered to the wheels.) Find the amount of gasoline used to accelerate the car from rest to 60 mi/h (27 m/s). Use the fact that the energy equivalent of one gallon of gasoline is 1.3×10^8 J.

Solution The energy required to accelerate the car from rest to a speed v is its kinetic energy, $\frac{1}{2}mv^2$. For this case,

$$E = \tfrac{1}{2}mv^2 = \tfrac{1}{2}(800\text{ kg})\left(27\ \frac{\text{m}}{\text{s}}\right)^2 = 2.9 \times 10^5\text{ J}$$

If the engine were 100% efficient, each gallon of gasoline would supply an energy 1.3×10^8 J. Since the engine is only 14% efficient, each gallon delivers only $(0.14)(1.3 \times 10^8\text{ J}) = 1.8 \times 10^7$ J. Hence, the number of gallons used to accelerate the car is

$$\text{Number of gal} = \frac{2.9 \times 10^5\text{ J}}{1.8 \times 10^7\text{ J/gal}} = \boxed{0.016\text{ gal}}$$

At this rate, a gallon of gas would be used after 62 such accelerations. This demonstrates the severe energy requirements for extreme stop-and-start driving.

EXAMPLE 7.15 Power Delivered to Wheels

Suppose the car described in Example 7.14 has a mileage rating of 35 mi/gal when traveling at 60 mi/h. How much power is delivered to the wheels?

Solution From the given data, we see that the car consumes 60/35 = 1.7 gal/h. Using the fact that each gallon is equivalent to 1.3×10^8 J, we find that the total power used is

$$P = \frac{(1.7\text{ gal/h})(1.3 \times 10^8\text{ J/gal})}{3.6 \times 10^3\text{ s/h}}$$

$$= \frac{2.2 \times 10^8\text{ J}}{3.6 \times 10^3\text{ s}} = \boxed{62\text{ kW}}$$

Since 14% of the available power is used to propel the car, we see that the power delivered to the wheels is $(0.14)(62\text{ kW}) = 8.7$ kW. This is about one half the value obtained for the large 1450-kg car discussed in the text. Size is clearly an important factor in power-loss mechanisms.

EXAMPLE 7.16 Car Accelerating Up a Hill

Consider a car of mass m accelerating up a hill, as in Figure 7.13. Assume that the magnitude of the drag force is given by

$$|\mathbf{f}| = (218 + 0.70v^2)\text{ N}$$

where v is the speed in m/s. Calculate the power that the engine must deliver to the wheels.

Solution The forces on the car are shown in Figure 7.13, where $\mathbf{F}$ is the force of static friction that propels the car and the remaining forces have their usual meaning. Newton's second law applied to the motion along the road surface gives

$$\sum F_x = F - |\mathbf{f}| - mg\sin\theta = ma$$

$$F = ma + mg\sin\theta + |\mathbf{f}|$$

$$= ma + mg\sin\theta + (218 + 0.70v^2)$$

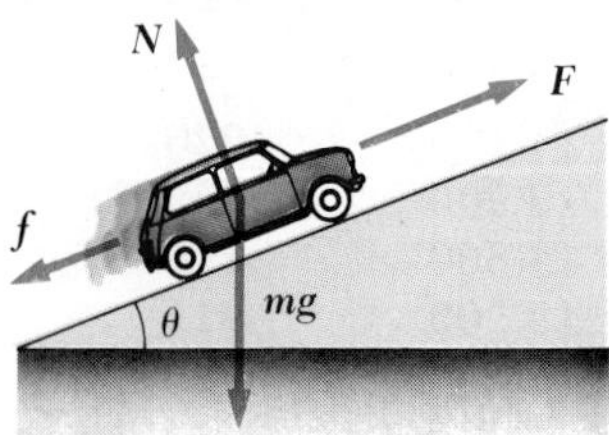

Figure 7.13 (Example 7.16).

Therefore, the power required for propulsion is

$$P = Fv = mva + mvg \sin\theta + 218v + 0.70v^3$$

In this expression, the term mva represents the power the engine must deliver to accelerate the car. If the car moves at constant speed, this term is zero and the power requirement is reduced. The term $mvg \sin\theta$ is the power required to overcome the force of gravity as the car moves up the incline. This term would be zero for motion on a horizontal surface. The term $218v$ is the power required to counterbalance rolling friction. Finally, the term $0.70v^3$ is the power needed to overcome air drag.

If we take $m = 1450$ kg, $v = 27$ m/s (= 60 mi/h), $a = 1$ m/s^2, and $\theta = 10°$, the various terms in P are calculated to be

$$mva = (1450 \text{ kg})(27 \text{ m/s})(1 \text{ m/s}^2)$$
$$= 39 \text{ kW} = 52 \text{ hp}$$
$$mvg \sin\theta = (1450 \text{ kg})(27 \text{ m/s})(9.80 \text{ m/s}^2)(\sin 10°)$$
$$= 67 \text{ kW} = 89 \text{ hp}$$
$$218v = 218(27) = 5.9 \text{ kW} = 7.9 \text{ hp}$$
$$0.70v^3 = 0.70(27)^3 = 14 \text{ kW} = 18 \text{ hp}$$

Hence, the total power required is 126 kW, or 167 hp. Note that the power requirements for traveling at *constant* speed on a horizontal surface are only 20 kW, or 26 hp (the sum of the last two terms). Furthermore, if the mass is halved (as in compact cars), the power required is also reduced by almost the same factor.

*7.8 RELATIVISTIC KINETIC ENERGY

The laws of Newtonian mechanics are only valid for describing the motion of particles moving at speeds that are small compared with the speed of light, $c(\approx 3 \times 10^8$ m/s). When the particle speeds are comparable to c, the equations of Newtonian mechanics must be replaced by the more general equations predicted by the theory of relativity. One consequence of the theory of relativity is that the kinetic energy of a particle of mass m moving with a speed v is no longer given by $K = mv^2/2$. Instead, one must use the relativistic form of the kinetic energy given by

$$K = mc^2\left(\frac{1}{\sqrt{1-(v/c)^2}} - 1\right) \quad (7.27)$$

Relativistic kinetic energy

According to this expression, speeds greater than c are not allowed since K would be imaginary for $v > c$. Furthermore, as v approaches c, K approaches ∞. This is consistent with experimental observations on subatomic particles such as electrons and protons, which have shown that no particles travel at speeds greater than c. (That is, c is the ultimate speed.) From the point of view of the work-energy theorem, v can only approach c, since it would take an infinite amount of work to attain the speed $v = c$.

All formulas in the theory of relativity *must* reduce to those in Newtonian mechanics at low particle speeds. It is instructive to show that this is the case for the kinetic energy relation by analyzing Equation 7.27 when v is small compared to c. In this case, we expect that K should reduce to the Newtonian expression, $K = mv^2/2$. We can check this by using the binomial expansion applied to the quantity $[1-(v/c)^2]^{-1/2}$, with $v/c \ll 1$. If we let $x = (v/c)^2$, the expansion gives

$$\frac{1}{(1-x)^{1/2}} = 1 + \frac{x}{2} + \frac{3}{8}x^2 + \cdots$$

Making use of this expansion in Equation 7.27 gives

$$K = mc^2\left(1 + \frac{v^2}{2c^2} + \frac{3}{8}\frac{v^4}{c^4} + \cdots - 1\right)$$

$$= \frac{1}{2}mv^2 + \frac{3}{8}m\frac{v^4}{c^2} + \cdots$$

$$\approx \frac{1}{2}mv^2 \qquad \text{for} \qquad \frac{v}{c} \ll 1$$

Thus, we see that the relativistic kinetic energy expression does indeed reduce to the Newtonian expression for speeds that are small compared with c. We shall return to the subject of relativity in more depth in Chapter 39.

SUMMARY

The **work** done by a *constant* force $\boldsymbol{F}$ acting on a particle is defined as the product of the component of the force in the direction of the particle's displacement and the magnitude of the displacement. If the force makes an angle θ with the displacement $\boldsymbol{s}$, the work done by $\boldsymbol{F}$ is

Work done by a constant force

$$W \equiv Fs \cos\theta \qquad (7.1)$$

The scalar, or dot, **product** of any two vectors $\mathbf{A}$ and $\mathbf{B}$ is defined by the relationship

Scalar product

$$\mathbf{A} \cdot \mathbf{B} \equiv AB \cos\theta \qquad (7.5)$$

where the result is a scalar quantity and θ is the included angle between the directions of the two vectors. The scalar product obeys the commutative and distributive laws.

The *work* done by a *varying* force acting on a particle moving along the x axis from x_i to x_f is given by

Work done by a varying force

$$W \equiv \int_{x_i}^{x_f} F_x\, dx \qquad (7.11)$$

where F_x is the component of force in the x direction. If there are several forces acting on the particle, the net work done by all forces is the sum of the individual work done by each force.

The **kinetic energy** of a particle of mass m moving with a speed v (where v is small compared with the speed of light) is defined as

Kinetic energy

$$K \equiv \tfrac{1}{2}mv^2 \qquad (7.17)$$

The **work-energy theorem** states that the net work done on a particle by external forces equals the change in kinetic energy of the particle:

Work-energy theorem

$$W_{\text{net}} = K_f - K_i = \tfrac{1}{2}mv_f^2 - \tfrac{1}{2}mv_i^2 \qquad (7.18)$$

The **instantaneous power** is defined as the time rate of doing work. If an agent applies a force $\boldsymbol{F}$ to an object moving with a velocity $\boldsymbol{v}$, the power delivered by that agent is given by

$$P \equiv \frac{dW}{dt} = \boldsymbol{F} \cdot \boldsymbol{v} \qquad (7.24)$$

Instantaneous power

When objects move at speeds comparable to the speed of light, c, their kinetic energy must be calculated using the relativistic expression given by

$$K = mc^2\left(\frac{1}{\sqrt{1-(v/c)^2}} - 1\right) \qquad (7.27)$$

Relativistic kinetic energy

QUESTIONS

1. When a particle rotates in a circle, a *centripetal force* acts on it directed toward the center of rotation. Why is it that this force does no work on the particle?
2. Explain why the work done by the force of sliding friction is negative when an object undergoes a displacement on a rough surface.
3. Is there any direction associated with the dot product of two vectors?
4. If the dot product of two vectors is positive, does this imply that the vectors must have positive rectangular components?
5. As the load on a spring hung vertically is increased, one would not expect the F_s versus x curve to always remain linear as in Figure 7.6d. Explain qualitatively what you would expect for this curve as m is increased.
6. Can the kinetic energy of an object have a negative value?
7. If the speed of a particle is doubled, what happens to its kinetic energy?
8. What can be said about the speed of an object if the net work done on that object is zero?
9. Using the work-energy theorem, explain why the force of kinetic friction always has the effect of *reducing* the kinetic energy of a particle.
10. Can the average power ever equal the instantaneous power? Explain.
11. In Example 7.13, does the required power increase or decrease as the force of friction is reduced?
12. An automobile sales representative claims that a "souped-up" 300-hp engine is a necessary option in a compact car (instead of a conventional 130-hp engine). Suppose you intend to drive the car within speed limits ($\leq$55 mi/h) and on flat terrain. How would you counteract this sales pitch?
13. One bullet has twice the mass of a second bullet. If both are fired such that they have the same velocity, which has more kinetic energy? What is the ratio of kinetic energies of the two bullets?
14. When a punter kicks a football, is he doing any work on the ball while his toe is in contact with it? Is he doing any work on the ball after it loses contact with his toe? Are there any forces doing work on the ball while it is in flight?
15. Discuss the work done by a pitcher throwing a baseball. What is the approximate distance through which the force acts as the ball is thrown?
16. Estimate the time it takes you to climb a flight of stairs. Then approximate the power required to perform this task. Express your value in horsepower.
17. Do frictional forces always reduce the kinetic energy of a body? If your answer is no, give examples which illustrate the effect.
18. Cite two examples in which a force is exerted on an object without doing any work on the object.
19. Two sharpshooters fire .30-caliber rifles using identical shells. The barrel of rifle A is 2 cm longer than that of rifle B. Which rifle will have the higher muzzle velocity? (*Hint:* The force of the expanding gases in the barrel accelerates the bullets.)
20. A team of furniture movers wishes to load a truck using a ramp from the ground to the rear of the truck. One of the movers claims that less work would be done in loading the truck if the length of the ramp were increased, in order to decrease the angle of the ramp with respect to the horizontal. Is his claim valid? Explain.
21. As a simple pendulum swings back and forth, the forces acting on the suspended mass are the force of gravity, the tension in the supporting cord, and air resistance. (a) Which of these forces, if any, do no work on the pendulum? (b) Which of these forces does negative work at all times during its motion? (c) Describe the work done by the force of gravity while the pendulum is swinging.
22. The kinetic energy of an object depends on the frame of reference in which its motion is measured. Give an example to illustrate this point.

PROBLEMS

Section 7.2 Work Done by a Constant Force

1. If a man lifts a 20-kg bucket from a well and does 6 kJ of work, how deep is the well? Assume the speed of the bucket remains constant as it is lifted.
2. A 65-kg woman climbs a flight of 20 stairs, each 23 cm high. How much work was done against the force of gravity in the process?
3. A tugboat exerts a constant force of 5000 N on a ship moving at constant speed through a harbor. How much work does the tugboat do on the ship in a distance of 3 km?
4. Verify the following energy unit conversions: (a) 1 J $= 10^7$ ergs, (b) 1 J $= 0.737$ ft · lb.
5. A cheerleader lifts his 50-kg partner straight up off the ground a distance of 0.6 m before releasing her. If he does this 20 times, how much work has he done on her?
6. A 100-kg sled is dragged by a team of dogs a distance of 2 km over a horizontal surface at a constant velocity. If the coefficient of friction between the sled and the snow is 0.15, find the work done by (a) the team of dogs and (b) the force of friction.
7. A horizontal force of 150 N is used to push a 40-kg box on a rough, horizontal surface through a distance of 6 m. If the box moves at constant speed, find (a) the work done by the 150-N force, (b) the work done by friction, and (c) the coefficient of kinetic friction.
8. A 15-kg block is dragged over a rough, horizontal surface by a constant force of 70 N acting at an angle of 20° above the horizontal. The block is displaced 5 m, and the coefficient of kinetic friction is 0.3. Find the work done by (a) the 70-N force, (b) the force of friction, (c) the normal force, and (d) the force of gravity. (e) What is the net work done on the block?
9. A man moved a broom horizontally by pushing on it with a force of 80 N directed at 50° below the horizontal. When he stopped moving, he had done 450 J of work. How far did he push the broom?
10. Batman, whose mass is 80 kg, is holding onto the free end of a 12-m rope, the other end of which is fixed to a tree limb above. He is able to get the rope in motion as only Batman knows how, eventually getting it to swing enough that he can reach a ledge when the rope makes a 60° angle with the downward vertical. How much work was done against the force of gravity in this maneuver?
11. A cart loaded with bricks has a total mass of 18 kg and is pulled at constant speed by a rope. The rope is inclined at 20° above the horizontal and the cart moves on a horizontal plane. The coefficient of kinetic friction between the ground and the cart is 0.5. (a) What is the tension in the rope? When the cart is moved 20 m, (b) how much work is done on the cart by the rope? (c) How much work is done by the friction force?

Section 7.3 The Scalar Product of Two Vectors

12. Two vectors are given by $\mathbf{A} = 4\mathbf{i} + 3\mathbf{j}$ and $\mathbf{B} = -\mathbf{i} + 3\mathbf{j}$. Find (a) $\mathbf{A} \cdot \mathbf{B}$ and (b) the angle between $\mathbf{A}$ and $\mathbf{B}$.
13. A vector is given by $\mathbf{A} = -2\mathbf{i} + 3\mathbf{j}$. Find (a) the magnitude of $\mathbf{A}$ and (b) the angle that $\mathbf{A}$ makes with the positive y axis. [In (b), use the definition of the scalar product.]
14. Vector $\mathbf{A}$ has a magnitude of 5 units, and $\mathbf{B}$ has a magnitude of 9 units. The two vectors make an angle of 50° with each other. Find $\mathbf{A} \cdot \mathbf{B}$.
15. Given two arbitrary vectors $\mathbf{A}$ and $\mathbf{B}$, show that $\mathbf{A} \cdot \mathbf{B} = A_xB_x + A_yB_y + A_zB_z$. (*Hint:* Write $\mathbf{A}$ and $\mathbf{B}$ in unit vector form and use Eq. 7.8.)
16. For the three vectors $\mathbf{A} = 3\mathbf{i} + \mathbf{j} - \mathbf{k}$, $\mathbf{B} = -\mathbf{i} + 2\mathbf{j} + 5\mathbf{k}$, and $\mathbf{C} = 2\mathbf{j} - 3\mathbf{k}$, find $\mathbf{C} \cdot (\mathbf{A} - \mathbf{B})$.
17. A force $\mathbf{F} = (6\mathbf{i} - 2\mathbf{j})$ N acts on a particle that undergoes a displacement $\mathbf{s} = (3\mathbf{i} + \mathbf{j})$ m. Find (a) the work done by the force on the particle and (b) the angle between $\mathbf{F}$ and $\mathbf{s}$.
18. Vector $\mathbf{A}$ is 2 units long and points in the positive y direction. Vector $\mathbf{B}$ has a negative x component 5 units long, a positive y component 3 units long, and no z component. Find $\mathbf{A} \cdot \mathbf{B}$ and the angle between the vectors.
19. The scalar product of vectors $\mathbf{A}$ and $\mathbf{B}$ is 6 units. The magnitude of each vector is 4. Find the angle between the vectors.
20. Find the angle between the vectors $\mathbf{A} = -5\mathbf{i} - 3\mathbf{j} + 2\mathbf{k}$ and $\mathbf{B} = -2\mathbf{j} - 2\mathbf{k}$.
21. Using the definition of the scalar product, find the angles between the following pairs of vectors: (a) $\mathbf{A} = 3\mathbf{i} - 2\mathbf{j}$ and $\mathbf{B} = 4\mathbf{i} - 4\mathbf{j}$, (b) $\mathbf{A} = -2\mathbf{i} + 4\mathbf{j}$ and $\mathbf{B} = 3\mathbf{i} - 4\mathbf{j} + 2\mathbf{k}$, (c) $\mathbf{A} = \mathbf{i} - 2\mathbf{j} + 2\mathbf{k}$ and $\mathbf{B} = 3\mathbf{j} + 4\mathbf{k}$.

Section 7.4 Work Done by a Varying Force: The One-Dimensional Case

22. The object moves from $x = 0$ to $x = 3$ m. If the resultant force acting on that object is in the x direction and varies as shown in Figure 7.14, determine the total work done on that object.
23. A body is subject to a force F_x that varies with position as in Figure 7.15. Find the work done by the force on the body as it moves (a) from $x = 0$ to $x = 5$ m, (b) from $x = 5$ m to $x = 10$ m, and (c) from $x = 10$ m to $x = 15$ m. (d) What is the total work done by the force over the distance $x = 0$ to $x = 15$ m?

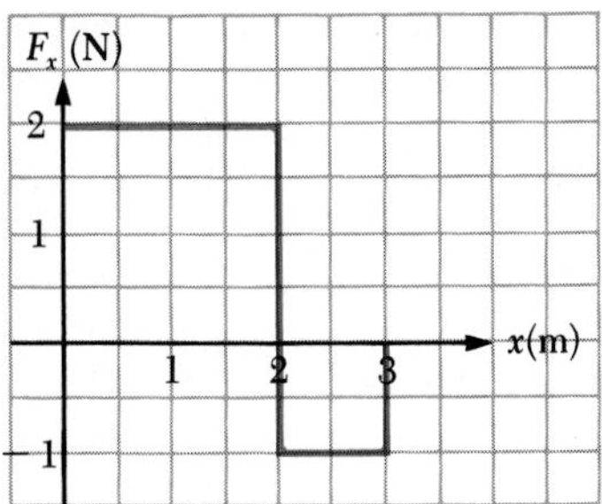

Figure 7.14 (Problem 22).

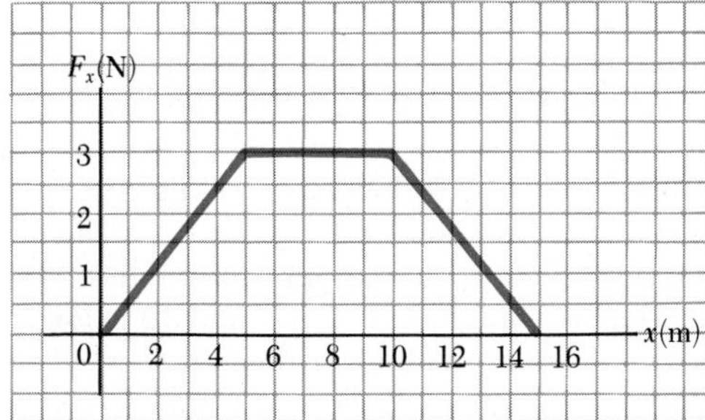

Figure 7.15 (Problems 23 and 38).

24. The force acting on a particle varies as in Figure 7.16. Find the work done by the force as the particle moves (a) from $x = 0$ to $x = 8$ m, (b) from $x = 8$ m to $x = 10$ m, and (c) from $x = 0$ to $x = 10$ m.

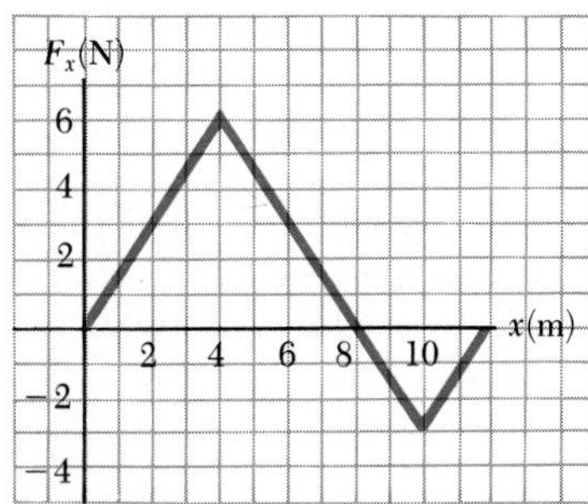

Figure 7.16 (Problems 24 and 39).

25. An archer pulls his bow string back 0.4 m by exerting a force that increases uniformly from zero to 230 N. (a) What is the equivalent spring constant of the bow? (b) How much work is done in pulling the bow?

26. The force acting on a particle is given by $F_x = (8x - 16)$ N, where x is in m. (a) Make a plot of this force versus x from $x = 0$ to $x = 3$ m. (b) From your graph, find the net work done by this force as the particle moves from $x = 0$ to $x = 3$ m.

27. When a 4-kg mass is hung vertically on a certain light spring that obeys Hooke's law, the spring stretches 2.5 cm. If the 4-kg mass is removed, (a) how far will the spring stretch if a 1.5-kg mass is hung on it, and (b) how much work must an external agent do to stretch the same spring 4.0 cm from its unstretched position?

28. A marine in the jungle finds himself in the middle of a swamp. It is estimated that the force F_x he must exert in the x direction as he struggles to get out is $F_x = (1000 - 50x)$ N, where x is in m. (a) Sketch the graph of F_x versus x. (b) What is the average force in traveling from zero to x? (c) If he travels a distance $x = 20$ m to get completely free of the swamp, how much work does he do?

29. If an applied force varies with position according to $F_x = 3x^3 - 5$, where x is in m, how much work is done by this force on an object that moves from $x = 4$ m to $x = 7$ m?

30. Show that the net work done by a force of the form $F = A \sin(kx)$ is zero for any integral number of cycles. (*Note:* Each "cycle" corresponds to an increase in x by the amount $x = 2\pi/k$.)

31. How much work is done on an object by a force that varies according to $F = 3 \sin(4x)$ in one half cycle? (*Note:* One "cycle" corresponds to an increase of 2π radians in the value of $4x$.)

Section 7.5 Work and Kinetic Energy

32. A 0.6-kg particle has a speed of 2 m/s at point **A** and kinetic energy of 7.5 J at point **B**. What is (a) its kinetic energy at point **A**? (b) its velocity at point **B**? (c) the total work done on the particle as it moves from **A** to **B**?

33. A 0.3-kg ball has a speed of 15 m/s. (a) What is its kinetic energy? (b) If its speed is doubled, what is its kinetic energy?

34. Calculate the kinetic energy of a 1000-kg satellite orbiting the earth at a speed of 7×10^3 m/s.

35. A mechanic pushes a 2500-kg car from rest to a speed v doing 5000 J of work in the process. During this time, the car moves 25 m. Neglecting friction between the car and the road, (a) what is the final speed, v, of the car? (b) What is the horizontal force exerted on the car?

36. A 3-kg mass has an initial velocity $\mathbf{v}_0 = (6\mathbf{i} - 2\mathbf{j})$ m/s. (a) What is its kinetic energy at this time? (b) Find the *change* in its kinetic energy if its *velocity changes* to $(8\mathbf{i} + 4\mathbf{j})$ m/s. (*Hint:* Remember that $v^2 = \mathbf{v} \cdot \mathbf{v}$.)

37. A 40-kg box initially at rest is pushed a distance of 5 m along a rough, horizontal floor with a constant applied force of 130 N. If the coefficient of friction between the box and floor is 0.3, find (a) the work done by the applied force, (b) the work done by friction, (c) the change in kinetic energy of the box, and (d) the final speed of the box.

38. A 4-kg particle is subject to a force that varies with position as shown in Figure 7.15. The particle starts from rest at $x = 0$. What is the speed of the particle at (a) $x = 5$ m, (b) $x = 10$ m, (c) $x = 15$ m?

39. The force acting on a 6-kg particle varies with position as shown in Figure 7.16. If its velocity is 2 m/s at $x = 0$, find its speed and kinetic energy at (a) $x = 4$ m, (b) $x = 8$ m, (c) $x = 10$ m.

40. In a ballistics demonstration, an FBI scientist fires a 55-g bullet horizontally at close range into a trough of sand. The bullet is released with speed 350.0 m/s and comes to rest in the sand after traveling 18 cm. Assuming that only 25% of the energy expended by the bullet appears as mechanical work (the remaining 75% goes into heat), (a) how much mechanical energy is given to the sand? (b) What average force does the sand exert on the bullet?

41. A 6-kg mass is lifted vertically through a distance of 5 m by a light string with a tension of 80 N. Find (a) the work done by the force of tension, (b) the work done by gravity, and (c) the final speed of the mass if it starts from rest.

42. In the giant roller coaster in Gurnee, Illinois (see Problem 22, Chapter 6), the speed of the cars is 13 m/s at the top of the loop 40 m in height. What is its speed at the bottom if we neglect any friction effects?

43. An Atwood's machine consists of a light fixed pulley with a light inextensible string over it (Fig. 5.12). The ends of the string support masses of 0.2 kg and 0.3 kg. The masses are held at rest beside each other and then released. Neglecting any friction, what is the speed of each mass the instant they have both moved 0.4 m?

44. A 2-kg block is attached to a light spring of force constant 500 N/m as in Figure 7.6. The block is pulled 5 cm to the right of equilibrium and released from rest. Find the speed of the block as it passes through equilibrium if (a) the horizontal surface is frictionless and (b) the coefficient of friction between the block and surface is 0.35.

45. A sled of mass m is given a kick on a frozen pond, imparting to it an initial speed $v_0 = 2$ m/s. The coefficient of kinetic friction between the sled and ice is $\mu_k = 0.1$. Use the work-energy theorem to find the distance the sled moves before coming to rest.

46. A block of mass 12 kg slides from rest down a frictionless 35° incline and is stopped by a strong spring with $k = 3.0 \times 10^4$ N/m. The block slides a total distance $d = 3.0$ m from the point of release to the point where it comes to rest against the spring. When the block comes to rest, how far has the spring been compressed?

47. A crate of mass 10 kg is pulled up a rough incline with an initial speed of 1.5 m/s. The pulling force is 100 N parallel to the incline, which makes an angle of 20° with the horizontal. If the coefficient of kinetic friction is 0.4, and the crate is pulled a distance of 5 m, (a) how much work is done against gravity? (b) How much work is done against friction? (c) How much work is done by the 100-N force? (d) What is the change in kinetic energy of the crate? (d) What is the speed of the crate after being pulled 5 m?

48. A block of mass 0.6 kg slides 6.0 m down a frictionless ramp inclined at 20° to the horizontal. It then travels on a rough horizontal surface where $\mu_k = 0.5$. (a) What is the speed of the block at the end of the incline? (b) What is its speed after traveling 1.0 m on the rough plane? (c) What distance does it travel on this horizontal plane before coming to rest?

49. A 4-kg block is given an initial speed of 8 m/s at the bottom of a 20° incline. The frictional force that retards its motion is 15 N. (a) If the block is directed *up* the incline, how far will it move before it stops? (b) Will it slide back down the incline?

50. "Winkin, Blinkin and Nod one day . . ." abandoned their boat and took to the slopes, complete with sled. They found a hill of slope 2.87° and 30 m long. Starting from rest at the top of the hill they slid to the bottom where their speed was 5.0 m/s. If the combined mass of the three plus sled was 70 kg, what was the average frictional force on the slope?

51. A 3-kg block is moved up a 37° incline under the action of a constant *horizontal* force of 40 N. The coefficient of kinetic friction is 0.1, and the block is displaced 2 m up the incline. Calculate (a) the work done by the 40-N force, (b) the work done by gravity, (c) the work done by friction, and (d) the *change* in kinetic energy of the block. (*Note:* The applied force is *not* parallel to the incline.)

52. A 4-kg block attached to a string 2 m in length rotates in a circle on a horizontal surface. (a) If the surface is frictionless, identify all the forces on the block and show that the work done by each force is zero for any displacement of the block. (b) If the coefficient of friction between the block and surface is 0.25, find the work done by the force of friction in each revolution of the block.

Section 7.6 Power

53. A 700-N marine in basic training climbs a 10-m vertical rope at uniform speed in 8 s. What is his power output?

54. Water flows over a section of the Niagara Falls at a rate of 1200 kg/s and falls 50 m. How many 60-W bulbs can be lit with this power?

55. A weightlifter lifts 250 kg through 2 m in 1.5 s. What is his power output?

56. A 200-kg crate is pulled along a level surface by an engine. The coefficient of friction between the crate and surface is 0.4. (a) How much power must the en-

gine deliver to move the crate at a constant speed of 5 m/s? (b) How much work is done by the engine in 3 min?

57. A 1500-kg car accelerates uniformly from rest to a speed of 10 m/s in 3 s. Find (a) the work done on the car in this time, (b) the average power delivered by the engine in the first 3 s, and (c) the instantaneous power delivered by the engine at $t = 2$ s.

58. A certain automobile engine delivers a power of 30 hp (2.24×10^4 W) to its wheels when moving at a constant speed of 27 m/s (≈ 60 mi/h). What is the resistive force acting on the automobile at that speed?

59. A speedboat requires 130 hp to move at a constant speed of 15 m/s (≈ 33 mi/h). Calculate the resistive force due to the water at that speed.

60. A car of weight 2500 N operating at a rate of 130 kW develops a maximum speed of 31 m/s on a level, horizontal road. Assuming that the resistive force (due to friction and air resistance) remains constant, (a) what is its maximum speed on an incline of 1 in 20? (i.e., if θ is the angle of the incline with the horizontal, $\sin\theta = 1/20$). (b) What is its power output on a 1 in 10 incline if it is traveling at 10 m/s?

61. A 65-kg athlete runs a distance of 600 m up a mountain inclined at 20° to the horizontal. He performs this feat in 80 s. Assuming that air resistance is negligible, (a) how much work does he perform and (b) what is his power output during the run?

62. A single, constant force $\mathbf{F}$ acts on a particle of mass m. The particle starts at rest at $t = 0$. (a) Show that the instantaneous power delivered by the force at any time t is equal to $(F^2/m)t$. (b) If $F = 20$ N and $m = 5$ kg, what is the power delivered at $t = 3$ s?

*Section 7.7 Energy and the Automobile

63. The car described in Table 7.4 travels at a constant speed of 35.9 m/s. At this speed, determine (a) the power needed to overcome air drag, and (b) the total power delivered to the wheels.

64. A passenger car carrying two people has a fuel economy of 25 mi/gal. It travels a distance of 3000 miles. A jet airplane making the same trip with 150 passengers has a fuel economy of 1 mi/gal. Compare the fuel consumed per passenger for the two modes of transportation.

65. A small car is designed with a drag coefficient of 0.40 and an equivalent frontal area of 1.7 m^2. The car with one occupant has a mass of 760 kg and an efficiency of 15%. Estimate the amount of gas required to move the car 1.4 km up a 10° incline at a constant speed of 64 km/h.

66. Suppose the empty car described in Table 7.4 has a fuel economy of 6.4 km/liter (15 mi/gal) when traveling at a speed of 26.8 m/s (60 mi/h). Assuming constant efficiency, determine the fuel economy of the car if the total mass of the passengers and the driver is 350 kg.

67. When an air conditioner is added to the car described in Problem 66, the additional output power required to operate the air conditioner is 1.54 kW. If the fuel economy is 6.4 km/liter without the air conditioner, what is the fuel economy when the air conditioner is operating?

*Section 7.8 Relativistic Kinetic Energy

68. An electron moves with a speed of $0.995c$, where c is the speed of light. (a) What is the kinetic energy of the electron? (b) If you use the classical expression to calculate its kinetic energy, what percentage error would result?

69. A proton in a high-energy accelerator moves with a speed of $c/2$, where c is the speed of light. Use the work-energy theorem to find the work required to increase its speed to (a) $0.75c$, (b) $0.995c$.

ADDITIONAL PROBLEMS

70. A 15-kg box is dragged at uniform speed up an incline that is 8.0 m long and makes a 15° angle with the horizontal. If the coefficient of friction between the box and the incline is 0.40, how much work is done by (a) the applied force, (b) the frictional force, (c) the normal force, and (d) the gravitational force?

71. A bartender in a western saloon slides a bottle of rye on the horizontal counter to a cowboy at the other end of the bar 7 m away. With what speed did he release the bottle if the coefficient of sliding friction is 0.1 and the bottle comes to rest in front of the cowboy?

72. A woman raises a 10-kg flag from the ground to the top of a 10-m flagpole at constant velocity, 0.25 m/s. (a) Find the work done by the woman while raising the flag. (b) Find the work done by gravity. (c) What is the power output of the woman while raising the flag?

73. Three vectors that form a closed triangle satisfy the condition $\mathbf{C} = \mathbf{A} - \mathbf{B}$ (Fig. 7.17). Use this fact and the definition of the scalar product to derive the law of cosines in trigonometry,

$$C^2 = A^2 + B^2 - 2AB\cos\theta$$

(*Hint:* Find the scalar product $\mathbf{C} \cdot \mathbf{C}$ in terms of A and B.)

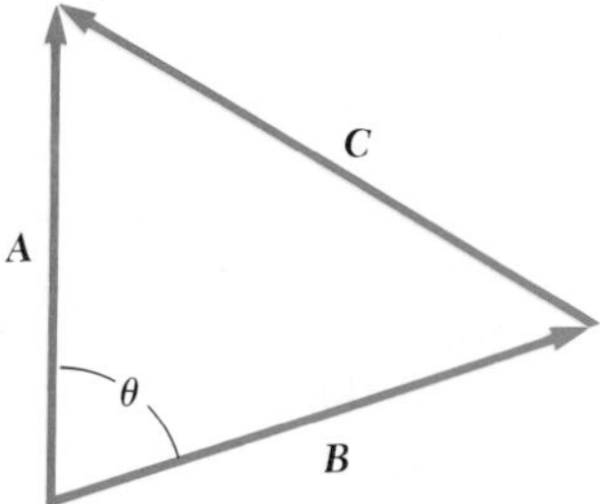

Figure 7.17 (Problem 73).

74. The direction of an arbitrary vector **A** can be completely specified with the angles α, β, and γ that the vector makes with the x, y, and z axes, respectively. If $\mathbf{A} = A_x\mathbf{i} + A_y\mathbf{j} + A_z\mathbf{k}$, (a) find expressions for $\cos\alpha$, $\cos\beta$, and $\cos\gamma$ (these are known as *direction cosines*) and (b) show that these angles satisfy the relation $\cos^2\alpha + \cos^2\beta + \cos^2\gamma = 1$. (*Hint:* Take the scalar product of **A** with **i**, **j**, and **k** separately.)

75. A 4-kg particle moves along the x-axis. Its position varies with time according to $x = t + 2t^3$, where x is in m and t is in s. Find (a) the kinetic energy at any time t, (b) the acceleration of the particle and the force acting on it at time t, (c) the power being delivered to the particle at time t, and (d) the work done on the particle in the interval $t = 0$ to $t = 2$ s. (*Note:* $P = dW/dt$.)

76. An ideal Atwood's machine has a 3-kg and 2-kg mass at the ends of the string (Fig. 5.12). The 2-kg mass is released from rest on the floor, 4 m below the 3-kg mass. (a) If the pulley is frictionless, what will be the speed of the masses when they pass each other? (b) Suppose now that the pulley does not rotate and the string must slide over it. If the total frictional force between the pulley and string is 5 N, what are their speeds when the masses pass each other?

77. The resultant force acting on a 2-kg particle moving along the x axis varies as $F_x = 3x^2 - 4x + 5$, where x is in m and F_x is in N. (a) Find the net work done on the particle as it moves from $x = 1$ m to $x = 3$ m. (b) If the speed of the particle is 5 m/s at $x = 1$ m, what is its speed at $x = 3$ m?

78. Prove the work-energy theorem, $W = \Delta K$, for a general three-dimensional displacement. (*Note:* $\mathbf{F} = m\, d\mathbf{v}/dt$ and $d\mathbf{s} = \mathbf{v}\, dt$.)

79. A 200-g block is pressed against a spring of force constant 1.4 kN/m until the block compresses the spring 10 cm. The spring rests at the bottom of a ramp inclined at 60° to the horizontal. Use the work-energy theorem to determine how far up the incline the block moves before momentarily coming to rest (a) if there is no friction between the block and the ramp, and (b) if the coefficient of kinetic friction is 0.4.

80. A block of mass m is attached to a light spring of force constant k as in Figure 7.6. The spring is compressed a distance d from its equilibrium position and released from rest. (a) If the block comes to rest when it first reaches the equilibrium position, what is the coefficient of friction between the block and surface? (b) If the block first comes to rest when the spring is *stretched* a distance of $d/2$ from equilibrium, what is μ?

81. A 0.4-kg particle slides on a horizontal, circular track 1.5 m in radius. It is given an initial speed of 8 m/s. After one revolution, its speed drops to 6 m/s because of friction. (a) Find the work done by the force of friction in one revolution. (b) Calculate the coefficient of kinetic friction. (c) What is the total number of revolutions the particle will make before coming to rest?

82. A light, inextensible string is attached to a mass of 0.25 kg lying on a rough, horizontal table. The string passes over a light frictionless pulley and is attached to a 0.4-kg mass hanging vertically from it. The coefficient of sliding friction between block and plane is 0.2. Use the work-energy theorem to determine (a) the velocity of the blocks after each has moved 20 m from rest and (b) the mass that must be added to the 0.25-kg mass, so that, given an initial velocity, the blocks continue to move at a constant speed. (c) What mass must be removed from the 0.4 kg mass to accomplish the same thing as in (b)?

83. A projectile, mass m, is shot horizontally with initial velocity $\mathbf{v}_0$ from a height h above a flat desert floor. The instant before the projectile hits the desert floor find (a) the work done on the projectile by gravity, (b) the change in kinetic energy since the projectile was fired, and (c) the final kinetic energy of the projectile.

84. Referring to Problem 83, find (a) the instantaneous rate at which work is being done on the projectile and (b) if the mass of the projectile is 10 kg and the initial height is 40 m, the instantaneous rate that work is being done after 1 s, 2 s, and 3 s. (*Note:* Be careful of the elapsed time.)

85. A 60-kg load is raised by a two-pulley arrangement as shown (Fig. 7.18). How much work is done by the force **F** to raise the load 3 m if there is a frictional force of 20 N in each pulley? (The pulleys do not rotate, but the rope slides across each surface.)

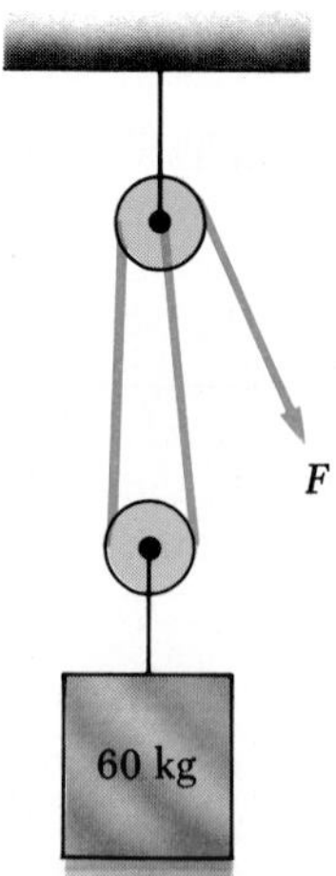

Figure 7.18 (Problem 85).

86. A small sphere of mass m hangs from a string of length L as in Figure 7.19. A variable horizontal force **F** is applied to the mass in such a way that it moves slowly

from the vertical position until the string makes an angle θ with the vertical. Assuming the sphere is always in equilibrium, (a) show that $F = mg \tan \theta$. (b) Make use of Equation 7.20 to show that the work done by the force $\boldsymbol{F}$ is equal to $mgL(1 - \cos \theta)$. (*Hint:* Note that $s = L\theta$, and so $ds = L\, d\theta$.

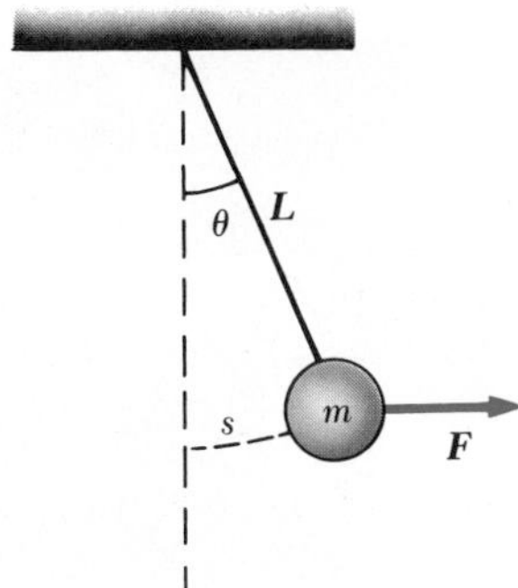

Figure 7.19 (Problem 86).

87. One end of a light spring of force constant 400 N/m is fixed in a metal tube that performs like a spring gun. When the spring is its natural length, the free end is flush with the open end of the tube. The spring is compressed 0.25 m in the metal tube, which points in the air at 60° to the horizontal. On top of the compressed spring is a 200-g steel ball which leaves the tube when the spring is released. Assume negligible air resistance and no friction in the tube. (a) What is the speed of the ball as it leaves the tube? (b) What is its speed at the highest point? (c) What is the maximum height (above where it left the tube) that the ball ever attains?

88. A 10-kg block is attached to one end of a spring having a force constant k. The other end of the spring is fixed to a rigid support with the block and spring sitting on a horizontal table. The coefficient of kinetic friction between the block and table is 0.5. With the spring initially unstretched, the block is pushed by a horizontal force that *compresses* the spring a distance L. (a) Obtain an expression for the total work done in terms of L and k. (b) Find the value of the spring constant if, after pushing the block 7 cm, the work done against friction is equal to the work done against the spring force. (c) If 60% of the energy expended has been used to compress the spring, how much total work has been done?

89. A car of mass m travels with constant speed v on a level road for a distance d. According to actual tests, the drag force is *approximately* given by $\boldsymbol{f} = -Km\boldsymbol{v}$, where $K = 0.018\ \text{s}^{-1}$. (a) Show that the work done by the engine to overcome the drag force is given by $Kmvd$. (b) Show that the power that the engine must deliver to the wheels to maintain this speed is Kmv^2. (c) Obtain numerical values for the work done and power delivered taking $m = 1500$ kg, $v = 27$ m/s, and $d = 100$ km. (d) If the car has a fuel economy of 15 mi/gal, what is the efficiency of the engine? (In this case, we define efficiency as the work done divided by the energy consumed.)

90. A passenger car of mass 1500 kg accelerates from rest to 97 km/h in 10 s. (a) Find the acceleration of the car. (b) Show that the coefficient of friction between the rear tires and road must be at least 0.55. (c) Determine the limiting frictional force on the car. (d) Find the average power delivered by the engine. (Assume that the normal force on each tire is $\frac{1}{4}mg$.)

91. Suppose a car is modeled as a cylinder moving with a speed v, as in Figure 7.20. In a time Δt, a column of air of mass Δm must be moved a distance $v\,\Delta t$ and hence must be given a kinetic energy $\frac{1}{2}(\Delta m)v^2$. Using this model, show that the power loss due to air resistance is $\frac{1}{2}\rho A v^3$ and the drag force is $\frac{1}{2}\rho A v^2$, where ρ is the density of air.

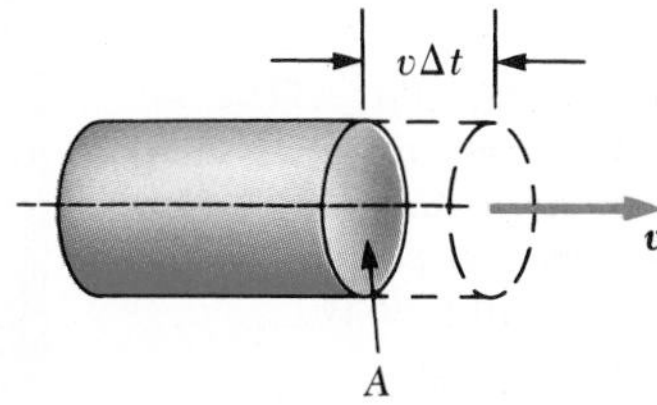

Figure 7.20 (Problem 91).

CALCULATOR/COMPUTER PROBLEMS

92. A 5-kg particle starts at the origin and moves along the x axis. The net force acting on the particle is measured at intervals of 1 m to be: 27.0, 28.3, 36.9, 34.0, 34.5, 34.5, 46.9, 48.2, 50.0, 63.5, 13.6, 12.2, 32.7, 46.6, 27.9 (in newtons). Determine the total work done on the particle over this interval.

93. A 0.178-kg particle moves along the x axis from $x = 12.8$ m to $x = 23.7$ m under the influence of a force given by

$$F = \frac{375}{x^3 + 3.75x}$$

where F is in newtons and x is in meters. Use a method of numerical integration to estimate the total work done by this force during this displacement. Your calculations should have an accuracy of at least 2%.

8

Potential Energy and Conservation of Energy

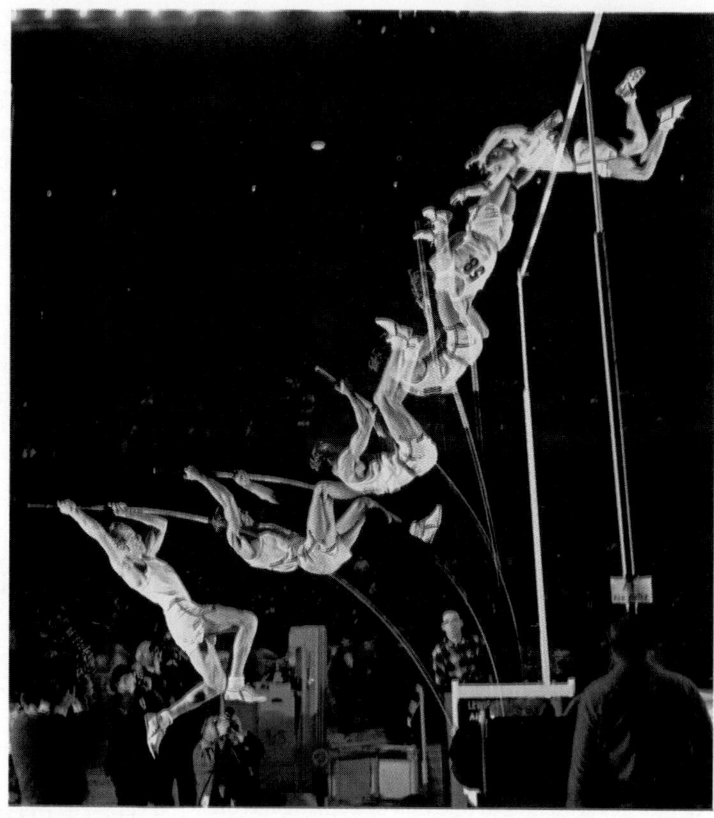

This multiflash photograph illustrates various forms of energy which you will study in this chapter. The energy stored temporarily in the bent pole is a form of potential energy. How many other forms of energy can you identify in this picture? (© Harold E. Edgerton. Courtesy of Palm Press Inc.)

In Chapter 7 we introduced the concept of kinetic energy, which is associated with the motion of an object. We found that the kinetic energy of an object can change only if work is done on the object. In this chapter we introduce another form of mechanical energy, called *potential energy*, associated with the position or configuration of an object. We shall find that the potential energy of a system can be thought of as stored energy that can be converted to kinetic energy or do work. The potential energy concept can be used only when dealing with a special class of forces called *conservative forces*. When only internal conservative forces, such as gravitational or spring forces, act on a system, the kinetic energy gained (or lost) by the system as its members change their relative positions is compensated by an equal energy loss (or gain) in the form of potential energy. This is known as the *law of conservation of mechanical energy*. A more general energy conservation law applies to an isolated system when all forms of energy and energy transformations are taken into account.

8.1 CONSERVATIVE AND NONCONSERVATIVE FORCES

Conservative Forces

In Example 7.11 we found that the work done by the gravitational force acting on a particle equals the weight of the particle multiplied by its vertical displacement, assuming that g is constant over the range of the displacement. As we shall see in Section 8.4, this result is valid for an arbitrary displacement of the particle. That is, the work done by gravity depends only on the initial and final coordinates and is independent of the path taken between these points. When a force exhibits these properties, it is called a **conservative force.** In addition to the gravitational force, other examples of conservative forces are the electrostatic force and the restoring force in a spring.

In general, a force is conservative if the work done by that force acting on a particle moving between two points is independent of the path the particle takes between the points.

That is, the work done on a particle by a conservative force depends only on the initial and final coordinates of the particle. With reference to the *arbitrary* paths shown in Figure 8.1a, we can write this condition

Property of a conservative force

$$W_{PQ}\text{ (along 1)} = W_{PQ}\text{ (along 2)}$$

A conservative force has another property, which can be derived from the above condition. Suppose the particle moves from P to Q along path 1, and then from Q to P along path 2, as in Figure 8.1b. The work done by a conservative force in the reverse path 2 from Q to P is equal to the *negative* of the work done from P to Q along path 2. Therefore, we can write the condition of a conservative force

$$W_{PQ}\text{ (along 1)} = -W_{QP}\text{ (along 2)}$$

$$W_{PQ}\text{ (along 1)} + W_{QP}\text{ (along 2)} = 0$$

Hence, a conservative force also has the property that

the total work done by a conservative force on a particle is zero when the particle moves around any closed path and returns to its initial position.

We can interpret this property of a conservative force in the following manner. The work-energy theorem says that the net work done on a particle displaced between two points equals the change in its kinetic energy. Therefore, if all the forces acting on the particle are conservative, then $W = 0$ for a round trip. This means that the particle will return to its starting point with the same kinetic energy it had when it started its motion.

To illustrate that the force of gravity is conservative, recall that the work done by the gravitational force as a particle of mass m moves between two points of elevation y_i and y_f is given by

$$W_g = -mg(y_f - y_i)$$

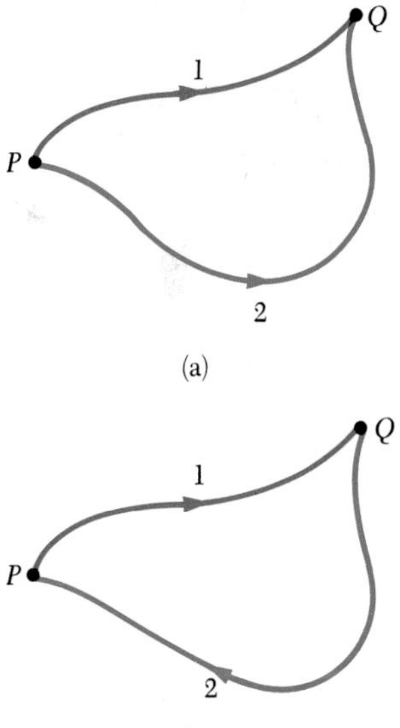

Figure 8.1 (a) A particle moves from P to Q along two different paths. The work done by a conservative force acting on the particle is the same along each path. If the force is nonconservative, the work done by this force differs along the two paths. (b) A particle moves from P to Q and then from Q back to P along a different path. That is, it moves in a closed path.

That is, the work done by the gravitational force mg (in the negative y direction) equals the force multiplied by the displacement in the y direction. From this expression, we first note that W_g depends only on the initial and final y coordinates and is *independent* of the path taken. Furthermore, if y_i and y_f are at the same elevation or if the particle makes a round trip, then $y_i = y_f$ and $W_g = 0$. For example, if a ball is thrown vertically upward with an initial speed v_i, and if air resistance is neglected, the ball must return to the thrower's hand with the same speed (and same kinetic energy) it had at the start of its motion.

Another example of a conservative force is the force of a spring on a block attached to the spring, where the restoring force is given by $F_s = -kx$. In the previous chapter, we found that the work done by the spring on the block is

Work done by the spring force

$$W_s = \tfrac{1}{2}kx_i^2 - \tfrac{1}{2}kx_f^2$$

where the initial and final coordinates of the block are measured from the equilibrium position of the block, $x = 0$. We see that W_s again depends only on the initial and final x coordinates. In addition, $W_s = 0$ for a round trip, where $x_i = x_f$.

Nonconservative Forces

A force is **nonconservative** if the work done by that force on a particle moving between two points depends on the path taken.

That is, the work done by a nonconservative force in taking a particle from P to Q in Figure 8.1a will differ for paths 1 and 2. We can write this

Property of a nonconservative force

$$W_{PQ} \text{ (along 1)} \neq W_{PQ} \text{ (along 2)}$$

Furthermore, from this condition we can show that if a force is nonconservative, the work done by that force on a particle that moves through any closed path is *not necessarily zero*. Since the work done in going from P to Q along path 2 is equal to the negative of the work done in going from Q to P along path 2, it follows from the first condition of a nonconservative force that

$$W_{PQ} \text{ (along 1)} \neq -W_{QP} \text{ (along 2)}$$

$$W_{PQ} \text{ (along 1)} + W_{QP} \text{ (along 2)} \neq 0$$

The force of sliding friction is a good example of a nonconservative force. If an object is moved over a rough, horizontal surface between two points along various paths, the work done by the frictional force certainly depends on the path. The negative work done by the frictional force along any particular path between two points will equal the force of friction multiplied by the length of the path. Paths of different lengths involve different amounts of work. The absolute magnitude of the least work done by the frictional force will correspond to a straight-line path between the two points. Furthermore, for a closed path you should note that the total work done by friction is nonzero since the force of friction opposes the motion along the entire path.

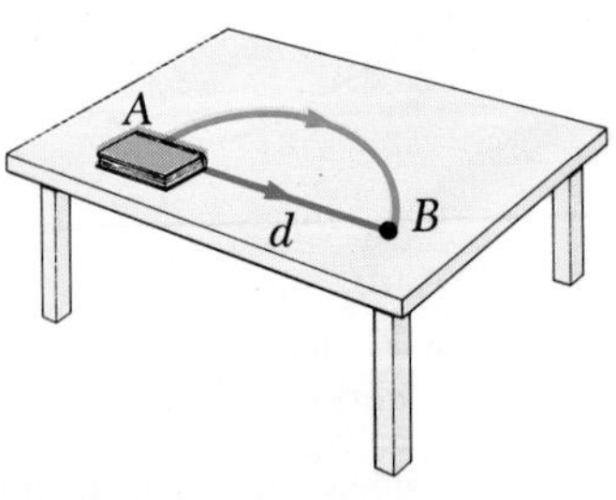

Figure 8.2 The work done by the force of friction depends on the path taken as the book is moved from A to B.

As an instructive example, suppose you were to displace a book between two points on a rough, horizontal surface such as a table. If the book is displaced in a straight line between two points, A and B in Figure 8.2, the work done by friction is simply $-fd$, where d is the distance between the points.

However, if the book is moved along *any other* path between the two points, the work done by friction would be *greater* (in absolute magnitude) than $-fd$. For example, the work done by friction along the semicircular path in Figure 8.2 is equal to $-f(\pi d/2)$, where d is the diameter of the circle. Finally, if the book is moved through any closed path (such as a circle), the work done by friction would clearly be nonzero since the frictional force opposes the motion.

In the example of a ball thrown vertically in the air with an initial speed v_i, careful measurements would show that because of air resistance, the ball would return to the thrower's hand with a speed less than v_i. Consequently, the final kinetic energy is less than the initial kinetic energy. The presence of a nonconservative force has reduced the ability of the system to do work by virtue of its motion. We shall sometimes refer to a nonconservative force as a *dissipative force*. For this reason, frictional forces are often referred to as being dissipative.

8.2 POTENTIAL ENERGY

In the previous section we found that the work done by a conservative force does not depend on the path taken by the particle and is independent of the particle's velocity. The work done is a function only of the particle's initial and final coordinates. For these reasons, we can define a potential energy function U such that the work done equals the decrease in the potential energy. That is, the work done by a conservative force $\boldsymbol{F}$ as the particle moves along the x axis is[1]

$$W_c = \int_{x_i}^{x_f} F_x\,dx = -\Delta U = U_i - U_f \tag{8.1}$$

That is, *the work done by a conservative force equals the negative of the change in the potential energy associated with that force*, where the change in the potential energy is defined as $\Delta U = U_f - U_i$. We can also express Equation 8.1 as

$$\Delta U = U_f - U_i = -\int_{x_i}^{x_f} F_x\,dx \tag{8.2}$$

Change in potential energy

where F_x is the component of $\boldsymbol{F}$ in the direction of the displacement.

It is often convenient to establish some particular location, x_i, to be a reference point and to then measure all potential energy differences with respect to this point. With this understanding, we can define the potential energy function as

$$U_f(x) = -\int_{x_i}^{x_f} F_x\,dx + U_i \tag{8.3}$$

[1] For a general displacement, the work done in two or three dimensions also equals $U_i - U_f$, where $U = U(x, y, z)$. We write this formally $W = \int_i^f \boldsymbol{F}\cdot d\boldsymbol{s} = U_i - U_f$.

Furthermore, the value of U_i is often taken to be zero at some arbitrary reference point. It really doesn't matter what value we assign to U_i, since it only shifts $U_f(x)$ by a constant, and it is only the *change* in potential energy that is physically meaningful. (In the next section, we shall see that the change in the particle's potential energy is related to a change in its kinetic energy.) If the conservative force is known as a function of position, we can use Equation 8.3 to calculate the change in potential energy of a body as it moves from x_i to x_f. It is interesting to note that in the one-dimensional case, a force is *always* conservative if it is a function of position only. This is generally not true for motion involving two- or three-dimensional displacements.

The work done by a nonconservative force does depend on the path as a particle moves from one position to another and could also depend on the particle's velocity or on other quantities. Therefore, the work done is not simply a function of the initial and final coordinates of the particle. We conclude that there is no potential energy *function* associated with a nonconservative force.

8.3 CONSERVATION OF MECHANICAL ENERGY

Suppose a particle moves along the x axis under the influence of only *one* conservative force, $\mathbf{F}_x$. If this is the only force acting on the particle, then the work-energy theorem tells us that the work done by that force equals the change in kinetic energy of the particle:

$$W_c = \Delta K$$

Since the force is conservative, according to Equation 8.1 we can write $W_c = -\Delta U$. Hence,

$$\Delta K = -\Delta U$$

$$\Delta K + \Delta U = \Delta(K + U) = 0 \tag{8.4}$$

This is the **law of conservation of mechanical energy,** which can be written in the alternative form

Conservation of mechanical energy

$$K_i + U_i = K_f + U_f \tag{8.5}$$

If we now define the total mechanical energy of the system, E, as the sum of the kinetic energy and potential energy, we can express the conservation of mechanical energy as

$$E_i = E_f \tag{8.6a}$$

where

Total mechanical energy

$$E \equiv K + U \tag{8.6b}$$

The law of conservation of mechanical energy states that the total mechanical energy of a system remains constant if the only force that does work is a conservative force. This is equivalent to the statement that if the kinetic energy of a conservative system increases (or decreases) by some amount, the potential energy must decrease (or increase) by the same amount.

If more than one conservative force acts on the system, then there is a potential energy function associated with *each* force. In such a case, we can write the law of conservation of mechanical energy

$$K_i + \sum U_i = K_f + \sum U_f \tag{8.7}$$

Conservation of mechanical energy

where the number of terms in the sums equals the number of conservative forces present. For example, if a mass connected to a spring oscillates vertically, two conservative forces act on it: the spring force and the force of gravity. We will discuss this situation later in a worked example.

8.4 GRAVITATIONAL POTENTIAL ENERGY NEAR THE EARTH'S SURFACE

When an object moves in the presence of the earth's gravity, the gravitational force can do work on that object. In the case of a freely falling object, the work done by gravity is a function of the *vertical* displacement of the object. This result is also valid in the more general case where the object undergoes both a horizontal and vertical displacement, such as in the case of a projectile.

Consider a particle being displaced from P to Q along various paths in the presence of a constant gravitational force[2] (Fig. 8.3). The work done along the path PAQ can be broken into two segments. The work done along PA is $-mgh$ (since $\mathbf{mg}$ is opposite to this displacement), and the work done along AQ is zero (since $\mathbf{mg}$ is perpendicular to this path). Hence, $W_{PAQ} = -mgh$. Likewise, the work done along PBQ is also $-mgh$, since $W_{PB} = 0$ and $W_{BQ} = -mgh$. Now consider the general path described by the solid line from P to Q. The curve is broken down into a series of horizontal and vertical steps. There is no work done by the force of gravity along the horizontal steps, since $\mathbf{mg}$ is perpendicular to these elements of displacement. Work is done by the force of gravity only along the vertical displacements, where the work done in the nth vertical step is $-mg\ \Delta y_n$. Thus, the total work done by the force of gravity as the

[2] The assumption that the force of gravity is constant is a good one as long as the vertical displacement is small compared with the earth's radius.

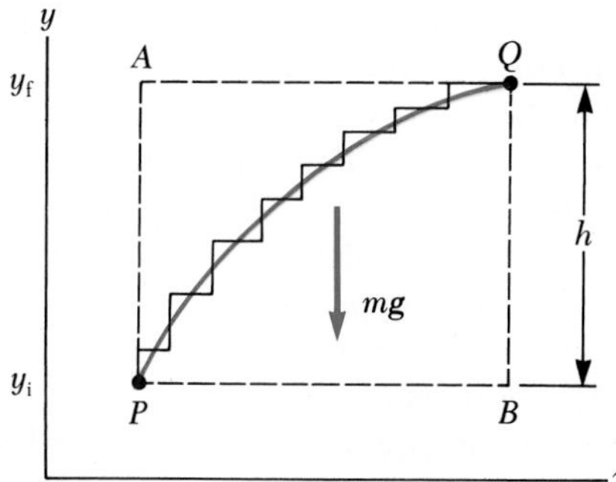

Figure 8.3 A particle that moves between the points P and Q under the influence of the force of gravity can be envisioned as moving along a series of horizontal and vertical steps. The work done by the force of gravity along each horizontal element is zero, and the net work done by the force of gravity is equal to the sum of the work done along each of the vertical displacements.

particle is displaced upward a distance h is the sum of the work done along each vertical displacement. Summation of all such terms gives

$$W_g = -mg \sum_n \Delta y_n = -mgh$$

Since $h = y_f - y_i$, we can express W_g as

$$W_g = mgy_i - mgy_f \tag{8.8}$$

We conclude that since the work done by the force of gravity is independent of the path, the gravitational force is a conservative force.

Since the force of gravity is conservative, we can define a **gravitational potential energy function U_g** as

Gravitational potential energy

$$U_g \equiv mgy \tag{8.9}$$

where we have chosen to take $U_g = 0$ at $y = 0$. Note that this function depends on the choice of origin of coordinates and is valid only when the displacement of the object in the vertical direction is small compared with the earth's radius. A general expression for the gravitational potential energy will be developed in Chapter 14.

Substituting the definition of U_g (Eq. 8.9) into the expression for the work done by the force of gravity (Eq. 8.8) gives

$$W_g = U_i - U_f = -\Delta U_g \tag{8.10}$$

That is, *the work done by the force of gravity is equal to the initial value of the potential energy minus the final value of the potential energy.* We conclude from Equation 8.10 that when the displacement is upward, $y_f > y_i$, and therefore $U_i < U_f$ and the work done by gravity is negative. This corresponds to the case where the force of gravity is *opposite* the displacement. When the object is displaced downward, $y_f < y_i$, and so $U_i > U_f$ and the work done by gravity is positive. In this case, $m\mathbf{g}$ is in the *same* direction as the displacement.

The term *potential energy* implies that the object has the potential, or capability, of gaining kinetic energy or doing work when released from some point under the influence of gravity. The choice of the origin of coordinates for measuring U_g is completely arbitrary, since only differences in potential energy are important. However, it is often convenient to choose the surface of the earth as the reference position $y_i = 0$.

If the force of gravity is the *only* force acting on a body, then the total mechanical energy of the body is conserved (Eq. 8.5). Therefore, the law of conservation of mechanical energy for a freely falling body can be written

Conservation of mechanical energy for a freely falling body

$$\tfrac{1}{2}mv_i^2 + mgy_i = \tfrac{1}{2}mv_f^2 + mgy_f \tag{8.11}$$

EXAMPLE 8.1 Ball in Free Fall □

A ball of mass m is dropped from a height h above the ground as in Figure 8.4. (a) Determine the speed of the ball when it is at a height y above the ground, neglecting air resistance.

Since the ball is in free fall, the only force acting on it is the gravitational force. Therefore, we can use the law of conservation of mechanical energy. When the ball is released from rest at a height h above the ground, its kinetic energy is $K_i = 0$ and its potential energy is $U_i = mgh$, where the y coordinate is measured from ground level. When the ball is at a distance y above the ground, its kinetic energy is $K_f = \frac{1}{2}mv_f^2$ and its potential energy relative to the ground is $U_f = mgy$. Applying Equation 8.11, we get

$$K_i + U_i = K_f + U_f$$

$$0 + mgh = \tfrac{1}{2}mv_f^2 + mgy$$

$$v_f^2 = 2g(h - y)$$

$$v_f = \sqrt{2g(h-y)}$$

(b) Determine the speed of the ball at y if it is given an initial speed v_i at the initial altitude h.

In this case, the initial energy includes kinetic energy equal to $\frac{1}{2}mv_i^2$ and Equation 8.11 gives

$$\tfrac{1}{2}mv_i^2 + mgh = \tfrac{1}{2}mv_f^2 + mgy$$

$$v_f^2 = v_i^2 + 2g(h - y)$$

$$v_f = \sqrt{v_i^2 + 2g(h-y)}$$

Note that this result is consistent with an expression from kinematics, $v_y^2 = v_{y0}^2 - 2g(y - y_0)$, where $y_0 = h$. Furthermore, this result is valid even if the initial velocity is at an angle to the horizontal (the projectile situation).

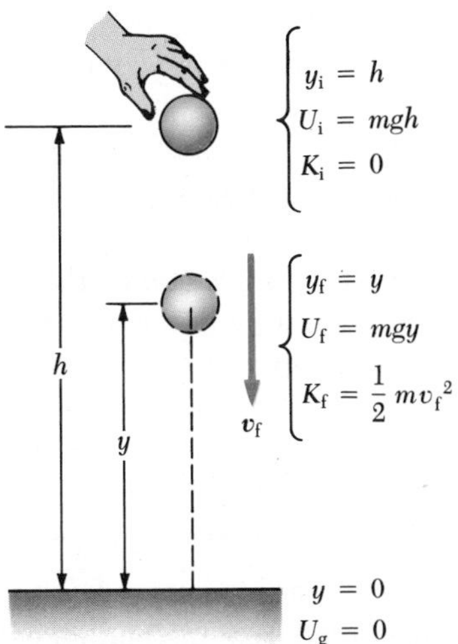

Figure 8.4 (Example 8.1) A ball is dropped from a height h above the floor. Initially, its total energy is potential energy, equal to mgh relative to the floor. At the elevation y, its total energy is the sum of kinetic and potential energies.

EXAMPLE 8.2 The Pendulum

A pendulum consists of a sphere of mass m attached to a light cord of length L as in Figure 8.5. The sphere is released from rest when the cord makes an angle θ_0 with the vertical, and the pivot at 0 is frictionless. (a) Find the speed of the sphere when it is at the lowest point, b.

The only force that does work on m is the force of gravity, since the force of tension is always perpendicular to each element of the displacement and hence does no work. Since the force of gravity is a conservative force, the total mechanical energy is conserved. Therefore, as the pendulum swings, there is a continuous transfer between potential and kinetic energy. At the instant the pendulum is released, the energy is entirely potential energy. At point b, the pendulum has kinetic energy but has lost some potential energy. At point c, the pendulum has regained its initial potential energy and its kinetic energy is again zero. If we measure the y coordinates from the center of rotation, then $y_a = -L \cos \theta_0$ and $y_b = -L$. Therefore, $U_a = -mgL \cos \theta_0$ and $U_b = -mgL$. Applying the principle of conservation of mechanical energy gives

$$K_a + U_a = K_b + U_b$$

$$0 - mgL \cos \theta_0 = \tfrac{1}{2}mv_b^2 - mgL$$

$$(1) \qquad v_b = \sqrt{2gL(1 - \cos \theta_0)}$$

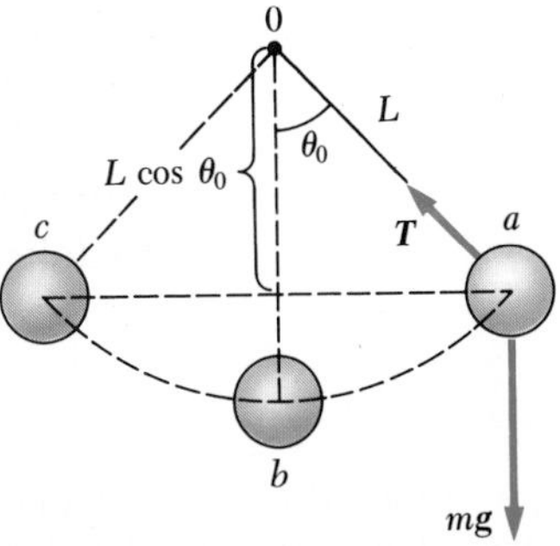

Figure 8.5 (Example 8.2) If the pendulum is released from rest at the angle θ_0, it will never swing above this position during its motion. At the start of the motion, position a, its energy is entirely potential energy. This is transformed into kinetic energy at the lowest elevation, position b.

(b) What is the tension $\mathbf{T}$ in the cord at b?

Since the force of tension does no work, it cannot be determined using the energy method. To find $\mathbf{T}_b$, we can apply Newton's second law to the radial direction. First, recall that the centripetal acceleration of a particle moving in a circle is equal to v^2/r directed toward the center of rotation. Since $r = L$ in this example, we get

(2) $\sum F_r = T_b - mg = ma_r = mv_b^2/L$

Substituting (1) into (2) gives for the tension at point b

(3) $T_b = mg + 2mg(1 - \cos\theta_0) = mg(3 - 2\cos\theta_0)$

Exercise 1 A pendulum of length 2.0 m and mass 0.5 kg is released from rest when the supporting cord makes an angle of 30° with the vertical. Find the speed of the sphere and the tension in the cord when the sphere is at its lowest point.
Answer 2.29 m/s; 6.21 N.

8.5 NONCONSERVATIVE FORCES AND THE WORK-ENERGY THEOREM

In real physical systems, nonconservative forces, such as friction, are usually present. Therefore, the total mechanical energy is not a constant. However, we can use the work-energy theorem to account for the presence of nonconservative forces. If W_{nc} represents the work done on a particle by all nonconservative forces and W_c is the work done by all conservative forces, we can write the work-energy theorem

$$W_{nc} + W_c = \Delta K$$

Since $W_c = -\Delta U$ (Eq. 8.1), this equation reduces to

$$W_{nc} = \Delta K + \Delta U = (K_f - K_i) + (U_f - U_i) \tag{8.12}$$

That is,

> the work done by all nonconservative forces equals the change in kinetic energy plus the change in potential energy.

Since the total mechanical energy is given by $E = K + U$, we can also express Equation 8.12 as

Work done by nonconservative forces

$$W_{nc} = (K_f + U_f) - (K_i + U_i) = E_f - E_i \tag{8.13}$$

That is, *the work done by all nonconservative forces equals the change in the total mechanical energy of the system.* Of course, when there are no nonconservative forces present, it follows that $W_{nc} = 0$ and $E_i = E_f$; that is, the total mechanical energy is conserved.

EXAMPLE 8.3 Block Moving on Incline

A 3-kg block slides down a rough incline 1 m in length as in Figure 8.6a. The block starts from rest at the top and experiences a constant force of friction of magnitude 5 N; the angle of inclination is 30°. (a) Use energy methods to determine the speed of the block when it reaches the bottom of the incline.

Since $v_i = 0$, the initial kinetic energy is zero. If the y coordinate is measured from the bottom of the incline, then $y_i = 0.50$ m. Therefore, the total mechanical energy of the block at the top is potential energy given by

$$E_i = U_i = mgy_i = (3\text{ kg})\left(9.80\ \frac{\text{m}}{\text{s}^2}\right)(0.50\text{ m}) = 14.7\text{ J}$$

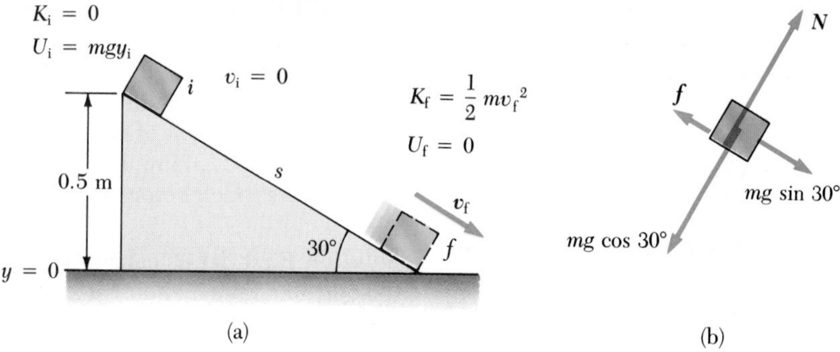

Figure 8.6 (Example 8.3) (a) A block slides down a rough incline under the influence of gravity. Its potential energy decreases while its kinetic energy increases. (b) Free-body diagram for the block.

When the block reaches the bottom, its kinetic energy is $\frac{1}{2}mv_f^2$, but its potential energy is *zero* since its elevation is $y_f = 0$. Therefore, the total mechanical energy at the bottom is $E_f = \frac{1}{2}mv_f^2$. However, we cannot say that $E_i = E_f$ in this case, because there is a nonconservative force that does work on the block, namely, the force of friction, $W_{nc} = -fs$, where s is the displacement along the plane. (Recall that the forces normal to the plane do no work on the block since they are perpendicular to the displacement.) In this case, $f = 5$ N and $s = 1$ m, therefore,

$$W_{nc} = -fs = (-5 \text{ N})(1 \text{ m}) = -5 \text{ J}$$

That is, some mechanical energy is lost because of the presence of the retarding force. Applying the work-energy theorem in the form of Equation 8.13 gives

$$W_{nc} = E_f - E_i$$

$$-fs = \tfrac{1}{2}mv_f^2 - mgy_i$$

$$\tfrac{1}{2}mv_f^2 = 14.7 \text{ J} - 5 \text{ J} = 9.7 \text{ J}$$

$$v_f^2 = \frac{19.4 \text{ J}}{3 \text{ kg}} = 6.47 \text{ m}^2/\text{s}^2$$

$$v_f = 2.54 \text{ m/s}$$

(b) Check the answer to (a) using Newton's second law to first find the acceleration.

Summing the forces along the plane gives

$$mg \sin 30° - f = ma$$

$$a = g \sin 30° - \frac{f}{m} = (9.80 \text{ m/s}^2)(0.500) - \frac{5 \text{ N}}{3 \text{ kg}}$$

$$= 3.23 \text{ m/s}^2$$

Since the acceleration is constant, we can apply the expression $v_f^2 = v_i^2 + 2as$, where $v_i = 0$:

$$v_f^2 = 2as = 2(3.23 \text{ m/s}^2)(1 \text{ m}) = 6.46 \text{ m}^2/\text{s}^2$$

$$v_f = 2.54 \text{ m/s}$$

Exercise 2 If the inclined plane is assumed to be frictionless, find the final speed of the block and its acceleration along the incline.
Answer 3.13 m/s; 4.90 m/s².

EXAMPLE 8.4 Motion on a Curved Track

A child of mass m takes a ride on an irregularly curved slide of height h, as in Figure 8.7. The child starts from rest at the top. (a) Determine the speed of the child at the bottom, assuming there is no friction present.

First, note that the normal force, **N**, does no work on the child since this force is always perpendicular to each element of the displacement. Furthermore, since there is no friction, $W_{nc} = 0$ and we can apply the law of conservation of mechanical energy. If we measure the y co-

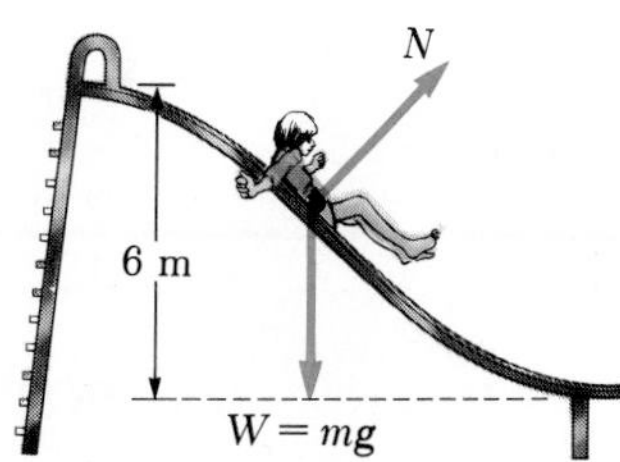

Figure 8.7 (Example 8.4) If the slide is frictionless, the speed of the child at the bottom depends only on the height of the slide and is independent of the shape of the slide.

ordinate from the bottom of the slide, then $y_i = h$, $y_f = 0$, and we get

$$K_i + U_i = K_f + U_f$$

$$0 + mgh = \tfrac{1}{2}mv_f^2 + 0$$

$$v_f = \sqrt{2gh}$$

Note that this result is the same as if the child had fallen vertically through a distance h! For example, if $h = 6$ m, then

$$v_f = \sqrt{2gh} = \sqrt{2\left(9.80\ \frac{\text{m}}{\text{s}^2}\right)(6\ \text{m})} = \boxed{10.8\ \text{m/s}}$$

(b) If there were a frictional force acting on the child, what would be the work done by this force?

In this case, $W_{nc} \neq 0$ and mechanical energy is *not* conserved. We can use Equation 8.13 to find the work done by friction, assuming the final velocity at the bottom is known:

$$W_{nc} = E_f - E_i = \tfrac{1}{2}mv_f^2 - mgh$$

For example, if $v_f = 8.0$ m/s, $m = 20$ kg, and $h = 6$ m, we find that

$$W_{nc} = \tfrac{1}{2}(20\ \text{kg})(8.0\ \text{m/s})^2 - (20\ \text{kg})\left(9.80\ \frac{\text{m}}{\text{s}^2}\right)(6\ \text{m})$$

$$= \boxed{-536\ \text{J}}$$

Again, W_{nc} is negative since the *work done by sliding friction is always negative.* Note, however, that because the slide is curved, the normal force changes in magnitude and direction during the motion. Therefore, the frictional force, which is proportional to N, also changes during the motion. Do you think it would be possible to determine μ from these data?

EXAMPLE 8.5 Let's Go Skiing

A skier starts from rest at the top of a smooth incline of height 20 m as in Figure 8.8. At the bottom of the incline, the skier encounters a horizontal rough surface where the coefficient of kinetic friction between the skis and snow is 0.21. How far does the skier travel on the horizontal surface before coming to rest?

Solution First, let us calculate the speed of the skier at the bottom of the incline. Since the incline is smooth, we can apply conservation of mechanical energy (as in Example 8.4a) to find

$$v = \sqrt{2gh} = \sqrt{2(9.80\ \text{m/s}^2)(20\ \text{m})} = 19.8\ \text{m/s}$$

Now we apply the work-energy theorem as the skier moves along the rough horizontal surface. The work done by the friction force along the horizontal is $W_{nc} = -fs$, where s is the horizontal displacement. Therefore,

$$W_{nc} = -fs = K_f - K_i$$

To find the distance the skier travels before coming to rest, we take $K_f = 0$. Since $v_i = 19.8$ m/s, and the friction force is given by $f = \mu N = \mu mg$, we get

$$-\mu mgs = -\tfrac{1}{2}mv_i^2$$

or

$$s = \frac{v_i^2}{2\mu g} = \frac{(19.8\ \text{m/s})^2}{2(0.21)(9.80\ \text{m/s}^2)} = \boxed{95.2\ \text{m}}$$

Exercise 3 Find the horizontal distance the skier travels before coming to rest if the incline is also rough with a coefficient of kinetic friction equal to 0.21. **Answer** 40.3 m

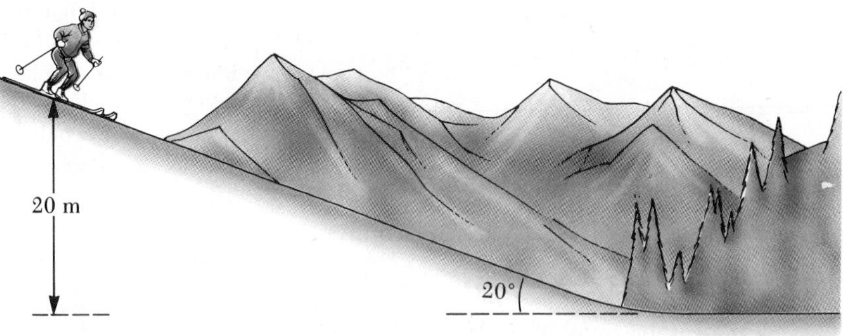

Figure 8.8 Example 8.5.

8.6 POTENTIAL ENERGY STORED IN A SPRING

Now let us consider another mechanical system that is conveniently described using the concept of potential, or stored, energy. A block of mass m slides on a frictionless, horizontal surface with constant velocity v_i and collides with a

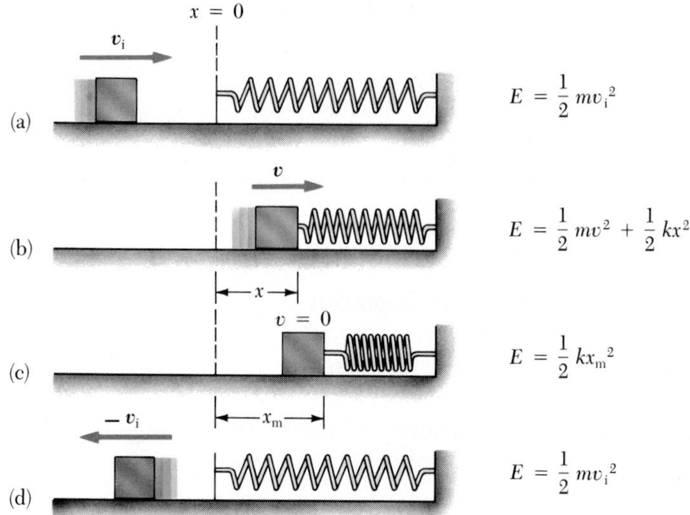

Figure 8.9 A block sliding on a smooth, horizontal surface collides with a light spring. (a) Initially the mechanical energy is all kinetic energy. (b) The mechanical energy is the sum of the kinetic energy of the block and the elastic potential energy in the spring. (c) The energy is entirely potential energy. (d) The energy is transformed back to the kinetic energy of the block. The total energy remains constant.

light coiled spring as in Figure 8.9. The description that follows is greatly simplified by assuming that the spring is very light and therefore its kinetic energy is negligible. The spring exerts a force on the block to the left as the spring is compressed, and eventually the block comes to rest (Fig. 8.9c). The initial energy in the system (block + spring) is the initial kinetic energy of the block. When the block comes to rest after colliding with the spring, its kinetic energy is zero. Because the spring force is conservative and because there are no external forces that can do work on the system (including gravity), the total mechanical energy of the system must remain constant. Thus, there is a transfer of energy from kinetic energy of the block to potential energy stored in the spring. Eventually, the block moves in the opposite direction and regains all of its initial kinetic energy, as described in Figure 8.8d.

To describe the potential energy stored in the spring, recall from the previous chapter that the work done by the spring on the block as the block moves from $x = x_i$ to $x = x_f$ is

$$W_s = \tfrac{1}{2}kx_i^2 - \tfrac{1}{2}kx_f^2$$

The quantity $\frac{1}{2}kx^2$ is defined as the **elastic potential energy** stored in the spring, denoted by the symbol U_s:

Potential energy stored in a spring

$$U_s = \tfrac{1}{2}kx^2 \qquad (8.14)$$

The elastic potential energy stored in the spring is zero when the spring is unstretched, or undeformed, ($x = 0$). Furthermore, U_s is a *maximum* when the spring has reached its maximum compression (Fig. 8.9c). Finally, U_s is *always* positive since it is proportional to x^2.

The total mechanical energy of the block-spring system can be expressed as

$$E = \tfrac{1}{2}mv_i^2 + \tfrac{1}{2}kx_i^2 = \tfrac{1}{2}mv_f^2 + \tfrac{1}{2}kx_f^2 \quad (8.15)$$

Applying this expression to the system described in Figure 8.9 and noting that $x_i = 0$, we get

$$E = \tfrac{1}{2}mv_i^2 = \tfrac{1}{2}mv_f^2 + \tfrac{1}{2}kx_f^2 \quad (8.16)$$

This expression says that for any displacement x_f, when the speed of the block is v_f, the sum of the kinetic and potential energies is equal to a *constant E*, which equals the total energy. In this case, the total energy is the initial kinetic energy of the block.

Now suppose there are nonconservative forces acting on the block-spring system. In this case, we can apply the work-energy theorem in the form of Equation 8.13, which gives

$$W_{nc} = (\tfrac{1}{2}mv_f^2 + \tfrac{1}{2}kx_f^2) - (\tfrac{1}{2}mv_i^2 + \tfrac{1}{2}kx_i^2) \quad (8.17)$$

That is, the total mechanical energy is not a constant of the motion when nonconservative forces act on the system. Again, if W_{nc} is due to a force of friction, then W_{nc} is *negative* and the final energy is less than the initial energy.

PROBLEM-SOLVING STRATEGIES

As we have seen, many problems in physics can be solved using the principle of conservation of mechanical energy. Other examples involving springs as part of the system will follow. The following steps should be followed in applying this principle:

1. Define your system, which may consist of more than one object.
2. Select a reference position for the zero point of potential energy (both gravitational and spring), and use this throughout your analysis. If there is more than one conservative force, then remember to write expressions for the potential energy associated with each force. (Examples 8.6 and 8.8 below, for instance, involve systems with springs and with objects whose gravitational potential energy changes.)
3. Determine whether or not friction forces are present. Remember that if friction or air resistance is present, mechanical energy *is not conserved.*
4. If mechanical energy is conserved, then you can proceed to write the total initial energy E_i at some point as the sum of the kinetic and potential energy at that point. Then, write an expression for the total final energy $E_f = K_f + U_f$ at the final point that is of interest. Since mechanical energy is conserved, you can equate the two total energies and solve for the quantity that is unknown.
5. If friction forces are present, then you should first write expressions for the total initial and total final energies. In this case, however, the total final energy differs from the total initial energy, the difference being the work done by the nonconservative forces. That is, you should apply $W_{nc} = E_f - E_i$ as in Examples 8.4b and 8.5.

EXAMPLE 8.6 The Spring-Loaded Popgun

The launching mechanism of a toy gun consists of a spring of unknown spring constant as shown in Figure 8.10a. By compressing the spring a distance of 0.12 m, the gun is able to launch a 20-g projectile to a maximum height of 20 m when fired vertically from rest. Neglecting all resistive forces, (a) determine the value of the spring constant.

Solution Since the projectile starts at rest, the initial kinetic energy in the system is zero. If the reference of gravitational potential energy is taken at the lowest position of the projectile, then its initial gravitational potential energy is also zero. Hence, the total initial energy of the system is the elastic potential energy stored in the spring, which is $kx^2/2$, where $x = 0.12$ m. Since the projectile rises to a maximum height $h = 20$ m, its final gravitational potential energy is equal to mgh, its final kinetic energy is zero, and the final elastic potential energy is zero. Since there are no nonconservative forces, conservation of mechanical energy can be applied to give

$$\tfrac{1}{2}kx^2 = mgh$$

$$\tfrac{1}{2}k(0.12\text{ m})^2 = (0.02\text{ kg})(9.80\text{ m/s}^2)(20\text{ m})$$

or

$$k = 544\text{ N/m}$$

(b) Find the speed of the projectile as it moves through the equilibrium position of the spring (where $x = 0$) as shown in Figure 8.10b.

Solution Using the same reference level for the gravitational potential energy as in part (a), we see that the initial energy of the system is still the elastic potential energy $kx^2/2$. The final energy of the system when the projectile moves through the unstretched position of the spring consists of the kinetic energy of the projectile, $mv^2/2$, and the gravitational potential energy of the projectile, mgx. Hence, conservation of energy in this case gives

$$\tfrac{1}{2}kx^2 = \tfrac{1}{2}mv^2 + mgx$$

Solving for v gives

$$v = \sqrt{\frac{kx^2}{m} - 2gx}$$

$$v = \sqrt{\frac{(544\text{ N/m})(0.12\text{ m})^2}{(0.02\text{ kg})} - 2(9.80\text{ m/s}^2)(0.12\text{ m})}$$

$$= 19.7\text{ m/s}$$

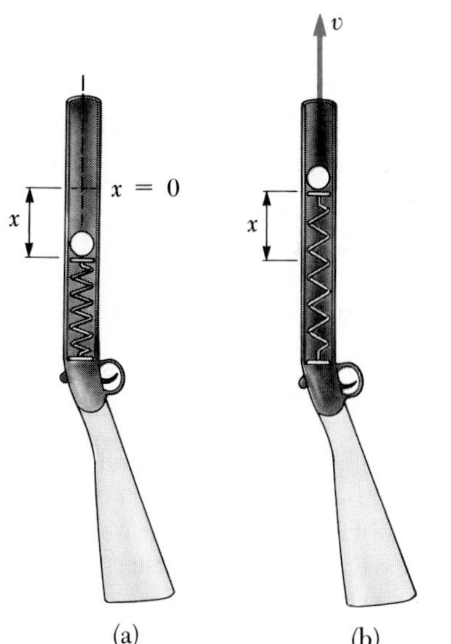

Figure 8.10 Example 8.6.

Exercise 4 What is the speed of the projectile when it is at a height of 10 m?
Answer 14.0 m/s

EXAMPLE 8.7 Mass-Spring Collision

A mass of 0.80 kg is given an initial velocity $v_i = 1.2$ m/s to the right and collides with a light spring of force constant $k = 50$ N/m, as in Figure 8.9. (a) If the surface is frictionless, calculate the initial maximum compression of the spring after the collision.

Solution The total mechanical energy is conserved since $W_{nc} = 0$. Applying Equation 8.15 to this system with $v_f = 0$ gives

$$\tfrac{1}{2}mv_i^2 + 0 = 0 + \tfrac{1}{2}kx_f^2$$

$$x_f = \sqrt{\frac{m}{k}}\,v_i = \sqrt{\frac{0.8\text{ kg}}{50\text{ N/m}}}(1.2\text{ m/s}) = 0.152\text{ m}$$

(b) If a constant force of friction acts between the block and the surface with $\mu = 0.5$ and if the speed of the block just as it collides with the spring is $v_i = 1.2$ m/s, what is the maximum compression in the spring?

Solution In this case, the mechanical energy of the system is *not* conserved because of the presence of friction, which does negative work on the system. The magnitude of the frictional force is

$$f = \mu N = \mu mg = 0.5(0.80\text{ kg})\left(9.80\,\frac{\text{m}}{\text{s}^2}\right) = 3.92\text{ N}$$

Therefore, the work done by the force of friction as the block is displaced from $x_i = 0$ to $x_f = x$ is

$$W_{nc} = -fx = (-3.92x)\ \text{J}$$

Substituting this into Equation 8.17 gives

$$W_{nc} = (0 + \tfrac{1}{2}kx^2) - (\tfrac{1}{2}mv_i^2 + 0)$$

$$-3.92x = \frac{50}{2}x^2 - \frac{1}{2}(0.80)(1.2)^2$$

$$25x^2 + 3.92x - 0.576 = 0$$

Solving the quadratic equation for x gives $x = 0.0924$ m and $x = -0.249$ m. The physically acceptable root is $x = 0.0924\ \text{m} = 9.24$ cm. The negative root is unacceptable since the block must be displaced to the right of the origin after coming to rest. Note that 9.24 cm is *less* than the distance obtained in the frictionless case (a). This result is what we should expect, since friction retards the motion of the system.

EXAMPLE 8.8 Connected Blocks in Motion

Two blocks are connected by a light string that passes over a frictionless pulley as in Figure 8.11. The block of mass m_1 lies on a rough surface and is connected to a spring of force constant k. The system is released from rest when the spring is unstretched. If m_2 falls a distance h before coming to rest, calculate the coefficient of kinetic friction between m_1 and the surface.

Solution In this situation there are two forms of potential energy to consider: the gravitational potential energy and the elastic potential energy stored in the spring. We can write the work-energy theorem

$$(1) \qquad W_{nc} = \Delta K + \Delta U_g + \Delta U_s$$

where ΔU_g is the *change* in the gravitational potential energy and ΔU_s is the *change* in the elastic potential energy of the system. In this situation, $\Delta K = 0$ since the initial and final speeds of the system are zero. Also, W_{nc} is the work done by friction, given by

$$(2) \qquad W_{nc} = -fh = -\mu m_1 gh$$

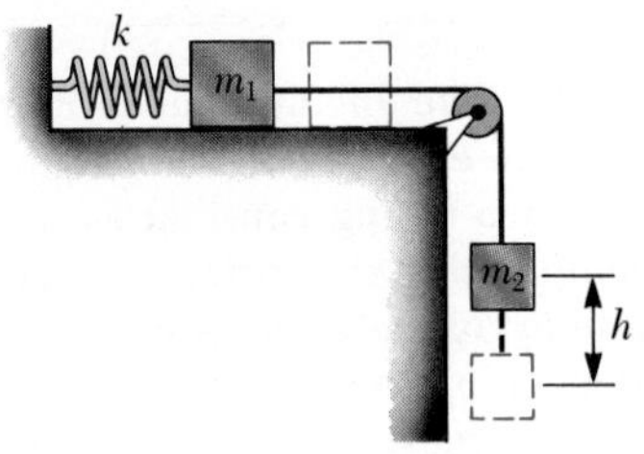

Figure 8.11 (Example 8.8) As the system moves from the highest to the lowest elevation of m_2, the system loses gravitational potential energy but gains elastic potential energy in the spring. Some mechanical energy is lost because of the presence of the nonconservative force of friction between m_1 and the surface.

The change in the gravitational potential energy is associated only with m_2 since the vertical coordinate of m_1 does not change. Therefore, we get

$$(3) \qquad \Delta U_g = U_f - U_i = -m_2 gh$$

where the coordinates have been measured from the lowest position of m_2. The change in the elastic potential energy in the spring is given by

$$(4) \qquad \Delta U_s = U_f - U_i = \tfrac{1}{2}kh^2 - 0$$

Substituting (2), (3), and (4) into (1) gives

$$-\mu m_1 gh = -m_2 gh + \tfrac{1}{2}kh^2$$

$$\mu = \frac{m_2 g - \frac{1}{2}kh}{m_1 g}$$

This represents a possible experimental technique for measuring the coefficient of kinetic friction. For example, if $m_1 = 0.50$ kg, $m_2 = 0.30$ kg, $k = 50$ N/m, and $h = 5.0 \times 10^{-2}$ m, we find that

$$\mu = \frac{(0.30\ \text{kg})\left(9.80\ \frac{\text{m}}{\text{s}^2}\right) - \frac{1}{2}\left(50\ \frac{\text{N}}{\text{m}}\right)(5.0 \times 10^{-2}\ \text{m})}{(0.50\ \text{kg})\left(9.80\ \frac{\text{m}}{\text{s}^2}\right)}$$

$$= 0.345$$

8.7 RELATIONSHIP BETWEEN CONSERVATIVE FORCES AND POTENTIAL ENERGY

In the previous sections we saw that the concept of potential energy is related to the configuration, or coordinates, of a system. In a few examples, we showed how to obtain the potential energy from a knowledge of the conservative force. (Remember that one can associate a potential energy function only with a conservative force.)

According to Equation 8.1, the change in the potential energy of a particle under the action of a conservative force equals the negative of the work done by the force. If the system undergoes an infinitesimal displacement, dx, we can express the infinitesimal change in potential energy, dU, as

$$dU = -F_x\, dx$$

Therefore, the conservative force is related to the potential energy function through the relationship

$$F_x = -\frac{dU}{dx} \qquad (8.18)$$

Relation between force and potential energy

That is, *the conservative force equals the negative derivative of the potential energy with respect to x.*[3]

We can easily check this relationship for the two examples already discussed. In the case of the deformed spring, $U_s = \frac{1}{2}kx^2$, and therefore

$$F_s = -\frac{dU_s}{dx} = -\frac{d}{dx}\left(\tfrac{1}{2}kx^2\right) = -kx$$

which corresponds to the restoring force in the spring. Since the gravitational potential energy function is given by $U_g = mgy$, it follows from Equation 8.18 that $F_g = -mg$.

We now see that U is an important function, since the conservative force can be derived from it. Furthermore, Equation 8.18 should clarify the fact that adding a constant to the potential energy is unimportant.

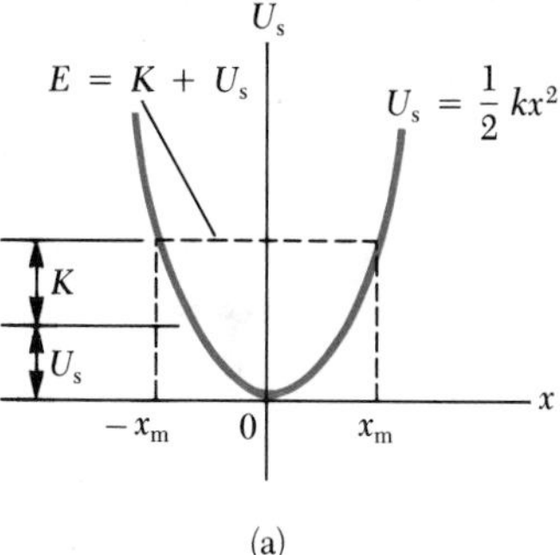

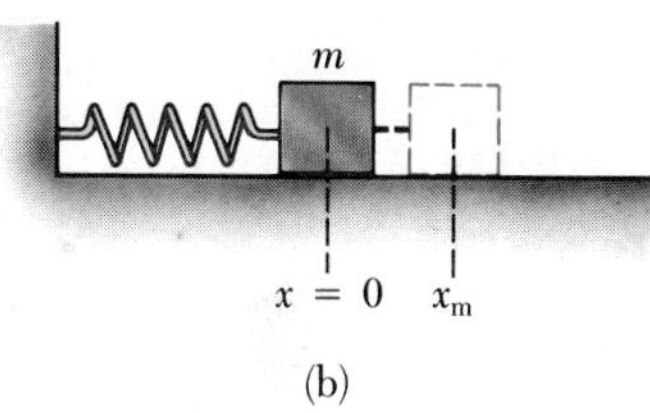

Figure 8.12 (a) The potential energy as a function of x for the block-spring system described in (b). The block oscillates between the turning points, which have the coordinates $x = \pm x_m$. Note that the restoring force of the spring always acts toward $x = 0$, the position of stable equilibrium.

*8.8 ENERGY DIAGRAMS AND STABILITY OF EQUILIBRIUM

The qualitative behavior of the motion of a system can often be understood through an analysis of its potential energy curve. Consider the potential energy function for the mass-spring system, given by $U_s = \frac{1}{2}kx^2$. This function is plotted versus x in Figure 8.12a. The force is related to U through the expression

$$F_s = -\frac{dU_s}{dx} = -kx$$

That is, the force is equal to the negative of the *slope* of the U versus x curve. When the mass is placed at rest at the equilibrium position ($x = 0$), where $F = 0$, it will remain there unless some external force acts on it. If the spring is stretched from equilibrium, x is positive and the slope dU/dx is positive; therefore F_s is negative and the mass accelerates back toward $x = 0$. If the spring is compressed, x is negative and the slope is negative; therefore F_s is positive and again the mass accelerates toward $x = 0$.

From this analysis, we conclude that the $x = 0$ position is one of **stable equilibrium.** That is, any movement away from this position results in a force

Stable equilibrium

[3] In a three-dimensional problem, where U depends on x, y, z, the force is related to U through the expression $\boldsymbol{F} = -\boldsymbol{i}\,\partial U/\partial x - \boldsymbol{j}\,\partial U/\partial y - \boldsymbol{k}\,\partial U/\partial z$, where $\partial/\partial x$, etc., are partial derivatives. In the language of vector calculus, $\boldsymbol{F}$ is said to equal the negative of the gradient of the scalar quantity $U(x, y, z)$.

that is directed back toward $x = 0$. In general, *positions of stable equilibrium correspond to those points for which U(x) has a minimum value.*

From Figure 8.12 we see that if the mass is given an initial displacement x_m and released from rest, its total energy initially is the potential energy stored in the spring, given by $\frac{1}{2}kx_m^2$. As motion commences, the system acquires kinetic energy at the expense of losing an equal amount of potential energy. Since the total energy must remain constant, the mass oscillates between the two points $x = \pm x_m$, called the *turning points.* In fact, because there is no energy loss (no friction), the mass will oscillate between $-x_m$ and $+x_m$ forever. (We shall discuss these oscillations further in Chapter 13.) From an energy viewpoint, the energy of the system cannot exceed $\frac{1}{2}kx_m^2$; therefore the mass must stop at these points and, because of the spring force, accelerate toward $x = 0$.

Another simple mechanical system that has a position of stable equilibrium is that of a ball rolling about in the bottom of a spherical bowl. If the ball is displaced from its lowest position, it will always tend to return to that position when released.

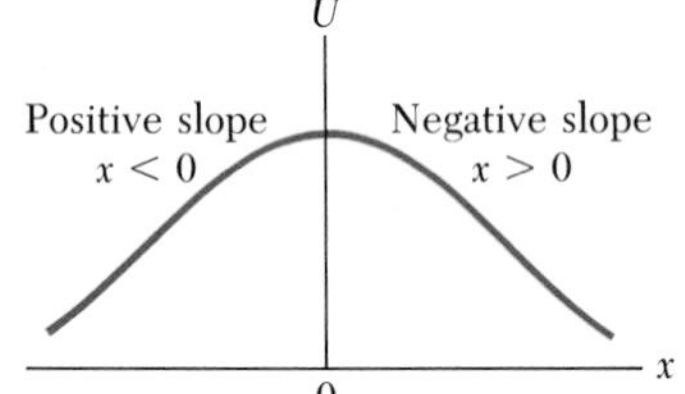

Figure 8.13 A plot of U versus x for a system that has a position of unstable equilibrium, located at $x = 0$. In this case, the force on the system for finite displacements is directed away from $x = 0$.

Now consider an example where the U versus x curve is as shown in Figure 8.13. In this case, $F_x = 0$ at $x = 0$, and so the particle is in equilibrium at this point. However, this is a position of **unstable equilibrium** for the following reason. Suppose the particle is displaced to the *right* ($x > 0$). Since the slope is negative for $x > 0$, $F_x = -dU/dx$ is positive and the particle will accelerate away from $x = 0$. Now suppose that the particle is displaced to the left ($x < 0$). In this case, the force is *negative* since the slope is positive for $x < 0$. Therefore, the particle will again accelerate away from the equilibrium position. Therefore, the $x = 0$ position in this situation is called a position of *unstable equilibrium,* since for any displacement from this point, the force pushes the particle farther away from equilibrium. In fact, the force pushes the particle toward a position of lower potential energy. A ball placed on the top of an inverted spherical bowl is obviously in a position of unstable equilibrium. That is, if the ball is displaced slightly from the top and released, it will surely roll off the bowl. In general, *positions of unstable equilibrium correspond to those points for which U(x) has a maximum value.*[4]

Neutral equilibrium

Finally, a situation may arise where U is constant over some region, and hence $F = 0$. This is called a position of **neutral equilibrium.** Small displacements from this position produce neither restoring nor disrupting forces. A ball lying on a flat horizontal surface is an example of an object in neutral equilibrium.

8.9 CONSERVATION OF ENERGY IN GENERAL

We have seen that the total mechanical energy of a system is conserved when only conservative forces act on the system. Furthermore, we were able to associate a potential energy function with each conservative force. In other words, mechanical energy is lost when nonconservative forces, such as friction, are present.

We can generalize the energy conservation principle to include all forces acting on the system, both conservative and nonconservative. In the study of thermodynamics we shall find that mechanical energy can be transformed into

[4] Mathematically, you can test whether an extreme of U is stable or unstable by examining the sign of d^2U/dx^2.

thermal energy. For example, when a block slides over a rough surface, the mechanical energy lost is transformed into internal energy temporarily stored in the block, as evidenced by a measurable increase in its temperature. On a submicroscopic scale, we shall see that this internal energy is associated with the vibration of atoms about their equilibrium positions. Since this internal atomic motion has kinetic and potential energy, one can say that frictional forces arise fundamentally from conservative atomic forces.[5] Therefore, if we include this increase in the internal energy of the system in our work-energy theorem, the total energy is conserved.

This is just one example of how you can analyze a system and always find that the total energy of an isolated system does not change, as long as you account for all forms of energy. That is, *energy can never be created or destroyed. Energy may be transformed from one form to another, but the total energy of an isolated system is always constant.* From a universal point of view, we can say that the *total energy of the universe is constant.* Therefore, if one part of the universe gains energy in some form, another part must lose an equal amount of energy. No violation of this principle has been found.

Total energy is always conserved

Other examples of energy transformations include the energy carried by sound waves resulting from the collision of two objects, the energy radiated by an accelerating charge in the form of electromagnetic waves (a radio antenna), and the elaborate sequence of energy conversions in a thermonuclear reaction.

In subsequent chapters, we shall see that the energy concept, and especially transformations of energy between various forms, join together the various branches of physics. In other words, one cannot really separate the subjects of mechanics, thermodynamics, and electromagnetism. Finally, from a practical viewpoint, all mechanical and electronic devices rely on some forms of energy transformation.

*8.10 MASS-ENERGY EQUIVALENCE

This chapter has been concerned with the important principle of energy conservation and its application to various physical phenomena. Another important principle, called the **law of conservation of mass,** says that in any kind of ordinary physical or chemical process, *matter is neither created nor destroyed.* That is, the mass of the system before the process equals the mass of the system after the process.

At this point, it would appear that energy and mass are two quantities that are separately conserved. However, in 1905 Einstein made the incredible discovery that energy can be transformed into mass and that mass can be transformed into energy. Mass and energy are not conserved separately, but are conserved as a single entity called *mass-energy.* Hence, *energy and mass are considered to be equivalent concepts.* The relation between energy and mass is given by Einstein's most famous formula

$$E_0 = mc^2 \tag{8.19}$$

[5] By introducing the nonconservative force, friction, we are able to limit the system we are studying. We have, in effect, avoided the complex problem of describing the dynamics of 10^{23} molecules and their interactions.

where c is the speed of light ($c \approx 3 \times 10^8$ m/s), and E_0 is the energy equivalent of a mass m. In relativity, the mass m associated with an object at rest is called the **rest mass**, while the corresponding energy E_0 is called the **rest energy.** As we shall see in Chapter 39, mass increases with speed; however, this dependence is insignificant for $v \ll c$. Hence, the masses that we use for describing situations in our everyday experiences are taken to be rest masses.

The rest energy associated with a small amount of matter is enormous. For example, the rest energy of 1 kg of *any* substance is calculated to be

$$E_0 = mc^2 = (1 \text{ kg})(3 \times 10^8 \text{ m/s})^2 = 9 \times 10^{16} \text{ J}$$

This is equivalent to the energy content of about 15 million barrels of crude oil (about one day's consumption in the entire United States)! If this energy could easily be converted to useful work, our energy resources would be unlimited.

In reality, matter is not freely converted to energy. However, in some instances, a significant fraction of the rest energy can be converted to work. The exchange between mass and energy is most notable in nuclear reactions, where fractional changes in mass of about 10^{-3} are routinely observed. A good example is the enormous energy released when the uranium-235 nucleus undergoes fission (splits) into smaller fragments. This happens because the ^{235}U nucleus is more massive than the sum of the masses of the fission products. The awesome nature of the energy released in such reactions is vividly demonstrated in the explosion of a nuclear weapon.

Since the mass and the energy of a system change together, both can be regarded as aspects of one quantity, mass-energy. The conservation law that describes any transformation or process states that *the mass-energy of an isolated system remains constant.*

Another aspect of the energy-mass relationship given by Equation 8.19 is that *energy has mass.* Whenever the energy of an object changes in any way, its mass changes. If ΔE is the change in energy of an object, its change in mass is given by

$$\Delta m = \frac{\Delta E}{c^2} \tag{8.20}$$

In ordinary experiments, if energy ΔE in any form (such as kinetic, potential, or thermal energy) is supplied to an object, the change in the mass of the object would be $\Delta m = \Delta E/c^2$. However, since c^2 is so large, the changes in mass in any ordinary mechanical experiment (or chemical reaction) are too small to be detected.

*8.11 QUANTIZATION OF ENERGY

As you may have learned in your chemistry course, all ordinary matter consists of atoms, and each atom consists of a collection of electrons and a nucleus. Thus, on the fine scale of the atomic world, we see that mass comes in discrete quantities corresponding to the atomic masses. In the language of modern physics, we say that mass is *quantized.* As we shall learn in the extended version of this text, many more physical quantities including energy are quantized. The quantized nature of energy is especially revealing in the atomic and subatomic world.

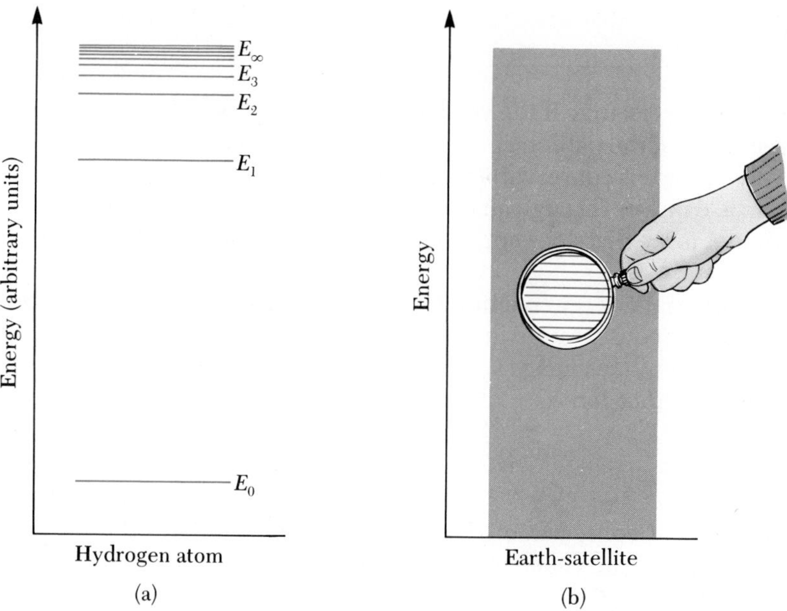

Figure 8.14 (a) Quantum states of the hydrogen atom. The lowest state, E_0, is the ground state. (b) The energy levels of an earth satellite are also quantized, but are so close together that they cannot be distinguished from each other.

First, consider the energy levels of the hydrogen atom (an electron orbiting around a proton). The electron in the atom can only occupy certain energy levels, called *quantum states*, as shown in Figure 8.14a. The atom cannot have any energy values lying between these quantum states. The lowest energy level, labeled E_0, is called the *ground state* of the atom. The ground state corresponds to the state that the atom would usually occupy if it were isolated. The atom could move to higher energy states by absorbing energy from some external source or by colliding with other atoms. The highest energy on the scale shown in Figure 8.14a, E_∞, corresponds to the energy of the atom when the electron is completely removed from the proton, and is called the *ionization energy*. Note that the energy levels get closer together at the high end of the scale.

Next, consider a satellite in orbit about the earth. If you were asked to describe the possible energies that the satellite could have, it would be reasonable (but incorrect) to say that the satellite could have any arbitrary energy you choose. However, just like the hydrogen atom, *the energy of the satellite is also quantized.* If you were to construct an energy level diagram for the satellite showing its allowed energies, the levels would be very close together and essentially form a continuum of states as in Figure 8.14b. In fact, the levels are so close to one another that it is impossible to tell they are not continuous. In other words, we have no way of experiencing quantization of energy in the macroscopic world; hence we can ignore it in describing everyday experiences.

SUMMARY

A force is **conservative** if the work done by that force acting on a particle is independent of the path the particle takes between two given points. Alternatively, a force is conservative if the work done by that force is zero when the particle moves through an arbitrary closed path and returns to its initial position. A force that does not meet these criteria is said to be **nonconservative.**

A **potential energy** function U can be associated only with a conservative force. If a conservative force $\boldsymbol{F}$ acts on a particle that moves along the x axis from x_i to x_f, *the change in the potential energy equals the negative of the work done by that force:*

Change in potential energy

$$U_f - U_i = -\int_{x_i}^{x_f} F_x\,dx \tag{8.2}$$

The **law of conservation of mechanical energy** states that if the only force acting on a mechanical system is conservative, the total mechanical energy is conserved:

Conservation of mechanical energy

$$K_i + U_i = K_f + U_f \tag{8.5}$$

The **total mechanical energy of a system** is defined as the sum of the kinetic energy and potential energy:

Total mechanical energy

$$E \equiv K + U \tag{8.6b}$$

The **gravitational potential energy** of a particle of mass m that is elevated a distance y near the earth's surface is given by

Gravitational potential energy

$$U_g \equiv mgy \tag{8.9}$$

The **work-energy theorem** states that the work done by all nonconservative forces acting on a system equals the change in the total mechanical energy of the system:

Work done by nonconservative forces

$$W_{nc} = E_f - E_i \tag{8.13}$$

The **elastic potential energy** stored in a spring of force constant k is

Potential energy stored in a spring

$$U_s \equiv \tfrac{1}{2}kx^2 \tag{8.14}$$

QUESTIONS

1. A bowling ball is suspended from the ceiling of a lecture hall by a strong cord. The bowling ball is drawn away from its equilibrium position and released from rest at the tip of the demonstrator's nose. If the demonstrator remains stationary, explain why she will not be struck by the ball on its return swing. Would the demonstrator be safe if the ball were given a push from this position?
2. Can the gravitational potential energy of an object ever have a negative value? Explain.
3. A ball is dropped by a person from the top of a building, while another person at the bottom observes its motion. Will these two people agree on the value of the ball's potential energy? on the *change* in potential energy of the ball? on the kinetic energy of the ball?
4. When a person runs in a track event at constant velocity, is any work done? (*Note:* Although the runner may move with constant velocity, the legs and arms undergo acceleration.) How does air resistance enter into the picture?

5. Our body muscles exert forces when we lift, push, run, jump, etc. Are these forces conservative?
6. When nonconservative forces act on a system, does the total mechanical energy remain constant?
7. If three different conservative forces and one nonconservative force act on a system, how many potential energy terms will appear in the work-energy theorem?
8. A block is connected to a spring that is suspended from the ceiling. If the block is set in motion and air resistance is neglected, will the total energy of the system be conserved? How many forms of potential energy are there for this situation?
9. Consider a ball fixed to one end of a rigid rod with the other end pivoted on a horizontal axis so that the rod can rotate in a vertical plane. What are the positions of stable and unstable equilibrium?
10. A ball rolls on a horizontal surface. Is the ball in stable, unstable, or neutral equilibrium?
11. Is it physically possible to have a situation where $E - U < 0$?
12. What would the curve of U versus x look like if a particle were in a region of neutral equilibrium?
13. Explain the energy transformations that occur during the following athletic events: (a) the pole vault, (b) the shotput, (c) the high jump. What is the source of energy in each case?
14. Discuss all the energy transformations that occur during the operation of an automobile.
15. A ball is thrown straight up into the air. At what position is its kinetic energy a maximum? At what position is its gravitational potential energy a maximum?
16. Three identical balls are thrown from the top of a building, all with the same initial speed. One ball is thrown horizontally, the second at some angle above the horizontal, and the third at some angle below the horizontal. Neglecting air resistance, compare the speeds of the balls as they reach the ground.
17. Consider the earth to be a perfect sphere. By how much does your potential energy change if you (a) walk from the north pole to the equator? (b) jump through a "tunnel" from the north pole to the south pole passing through the center of the earth?
18. Does a single external force acting on a particle necessarily change (a) its kinetic energy? (b) its velocity?
19. Is the work done by a nonconservative force always negative?
20. In the high jump, is any portion of the running athlete's kinetic energy converted into potential energy during the jump?
21. In the pole vault or high jump, why does the athlete attempt to keep his or her center of gravity as low as possible near the top of the jump?

PROBLEMS

Section 8.1 Conservative and Nonconservative Forces

1. A 4-kg particle moves from the origin to the position having coordinates $x = 5$ m and $y = 5$ m under the influence of gravity acting in the negative y direction (Fig. 8.15). Using Equation 7.21, calculate the work done by gravity in going from O to C along the following paths: (a) OAC, (b) OBC, (c) OC. Your results should all be identical. Why?

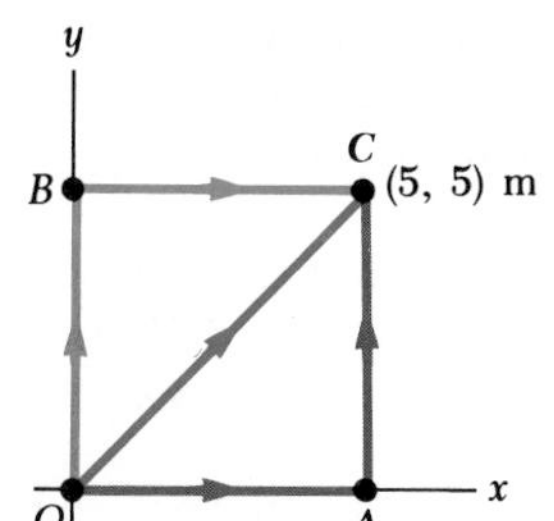

Figure 8.15 (Problems 1, 2, 3, and 5).

2. (a) Starting with Equation 7.20 for the definition of work, show that *any constant force* is conservative. (b) As a special case, suppose a particle of mass m is under the influence of force $\mathbf{F} = (3\mathbf{i} + 4\mathbf{j})$ N and moves from O to C in Figure 8.15. Calculate the work done by $\mathbf{F}$ along the three paths OAC, OBC, and OC, and show that they are identical.
3. A particle moves in the xy plane in Figure 8.15 under the influence of a frictional force that opposes its displacement. If the frictional force has a magnitude of 3 N, calculate the total work done by friction along the following *closed* paths: (a) the path OA followed by the return path AO, (b) the path OA followed by AC and the return path CO, and (c) the path OC followed by the return path CO. (d) Your results for the three closed paths should all be different and nonzero. What is the significance of this?
4. A single conservative force acting on a particle varies as $\mathbf{F} = (-Ax + Bx^2)\mathbf{i}$ N, where A and B are constants and x is in m. (a) Calculate the potential energy associated with this force, taking $U = 0$ at $x = 0$. (b) Find the change in potential energy and change in kinetic energy as the particle moves from $x = 2$ m to $x = 3$ m.
5. A force acting on a particle moving in the xy plane is given by $\mathbf{F} = (2y\mathbf{i} + x^2\mathbf{j})$ N, where x and y are in m. The particle moves from the origin to a final position having coordinates $x = 5$ m and $y = 5$ m, as in Figure 8.15. Calculate the work done by $\mathbf{F}$ along (a) OAC, (b) OBC, (c) OC. (d) Is $\mathbf{F}$ conservative or nonconservative? Explain.

Section 8.3 Conservation of Mechanical Energy

6. A 4-kg particle moves along the x axis under the influence of a single conservative force. If the work done on the particle is 80 J as the particle moves from $x = 2$ m to $x = 5$ m, find (a) the change in the particle's kinetic energy, (b) the change in its potential energy, and (c) its speed at $x = 5$ m if it starts at rest at $x = 2$ m.
7. A single conservative force $F_x = (2x + 4)$ N acts on a 5-kg particle, where x is in m. As the particle moves along the x axis from $x = 1$ m to $x = 5$ m, calculate (a) the work done by this force, (b) the change in the potential energy of the particle, and (c) its kinetic energy at $x = 5$ m if its speed at $x = 1$ m is 3 m/s.
8. At time t_i, the kinetic energy of a particle is 30 J and its potential energy is 10 J. At some later time t_f, its kinetic energy is 18 J. (a) If only conservative forces act on the particle, what is its potential energy at time t_f? What is its total energy? (b) If the potential energy at time t_f is 5 J, are there any nonconservative forces acting on the particle? Explain.
9. A single constant force $\boldsymbol{F} = (3\boldsymbol{i} + 5\boldsymbol{j})$ N acts on a 4-kg particle. (a) Calculate the work done by this force if the particle moves from the origin to the point with vector position $\boldsymbol{r} = (2\boldsymbol{i} - 3\boldsymbol{j})$ m. Does this result depend on the path? Explain. (b) What is the speed of the particle at $\boldsymbol{r}$ if its speed at the origin is 4 m/s? (c) What is the change in the potential energy of the particle?
10. Use conservation of energy to determine the final speed of a mass of 5.0 kg attached to a light cord over a massless, frictionless pulley and attached to another mass of 3.5 kg when the 5.0 kg mass has fallen (starting from rest) a distance of 2.5 m. (See Figure 8.16.)

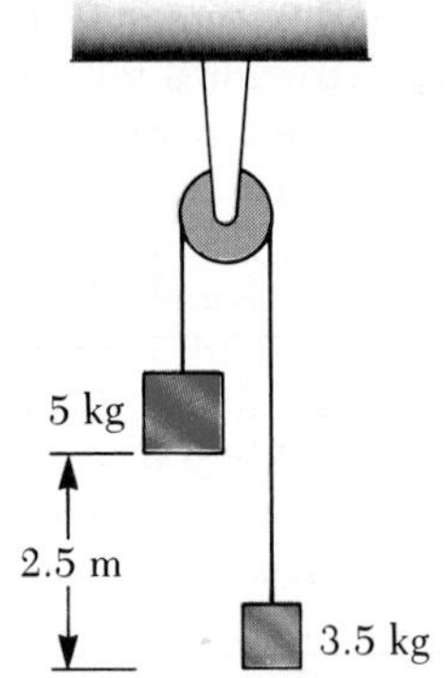

Figure 8.16 (Problem 10).

11. A bead slides without friction around a loop-the-loop (Fig. 8.17). If the bead is released from a height $h = 3.5R$, what is its speed at point A? How large is the normal force on it if its mass is 5.0 g?
12. A particle of mass 0.5 kg is shot from P as shown in Figure 8.18 with an initial velocity $\boldsymbol{v}_0$, having a horizontal component of 30 m/s. The particle rises to a maximum height of 20 m above P. Using the conservation of energy, determine: (a) the vertical component of $\boldsymbol{v}_0$, (b) the work done by the gravitational force on the particle during its motion from P to B, and (c) the horizontal and the vertical components of the velocity vector when the particle reaches B.

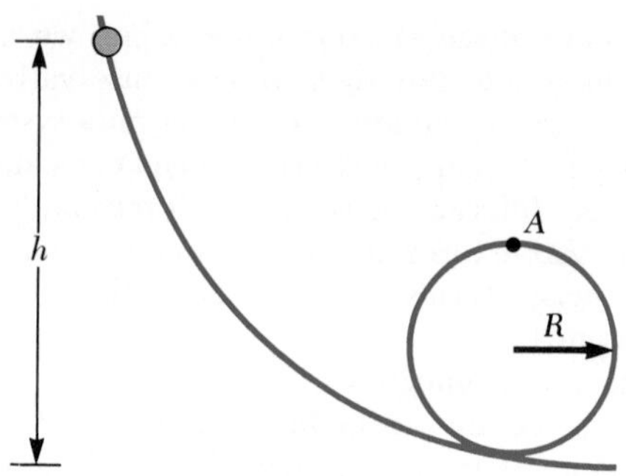

Figure 8.17 (Problem 11).

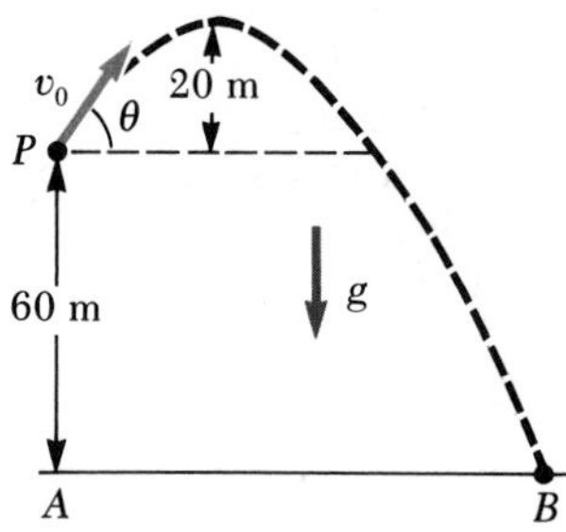

Figure 8.18 (Problem 12).

Section 8.4 Gravitational Potential Energy Near the Earth's Surface

13. A rocket is launched at an angle of 53° to the horizontal from an altitude h with a speed v_0. (a) Use energy methods to find the speed of the rocket when its altitude is $h/2$. (b) Find the x and y components of velocity when the rocket's altitude is $h/2$, using the fact that $v_x = v_{x0} =$ constant (since $a_x = 0$) and the results to part (a).
14. A 5-kg mass is attached to a light string of length 2 m to form a pendulum (Fig. 8.5). The mass is given an initial speed of 4 m/s at its lowest position. When the string makes an angle of 37° with the vertical, find (a) the *change* in the potential energy of the mass, (b) the speed of the mass, and (c) the tension in the string. (d) What is the maximum height reached by the mass above its lowest position?
15. A 0.5-kg ball is thrown vertically upward with an initial speed of 16 m/s. Assuming its initial potential energy is zero, find its kinetic energy, potential energy, and total mechanical energy (a) at its initial position,

(b) when its height is 5 m, and (c) when it reaches the top of its flight. (d) Find its maximum height using the law of conservation of energy.

16. A 0.4-kg ball is thrown into the air and reaches a maximum altitude of 20 m. Taking its initial position as the point of zero potential energy and using energy methods, find (a) its initial speed, (b) its total mechanical energy, and (c) the ratio of its kinetic energy to its potential energy when its altitude is 10 m.

17. Two masses are connected by a light string passing over a light frictionless pulley as shown in Figure 8.19. The 5-kg mass is released from rest. Using the law of conservation of energy, (a) determine the velocity of the 3-kg mass just as the 5-kg mass hits the ground. (b) Find the maximum height to which the 3-kg mass will rise.

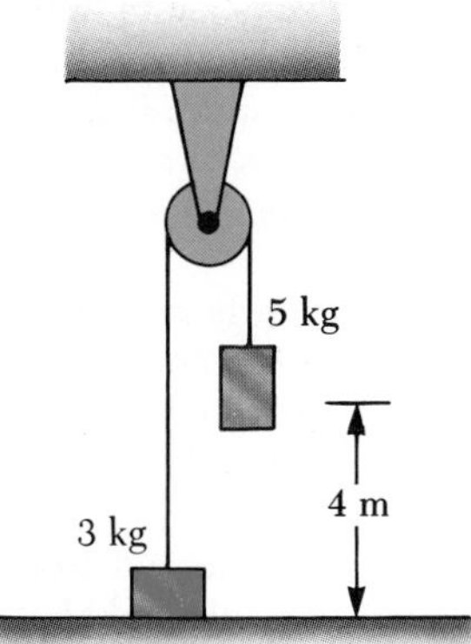

Figure 8.19 (Problem 17).

18. A child slides down the frictionless slide shown in Figure 8.20. In terms of R and H, at what height h will he lose contact with the section of radius R?

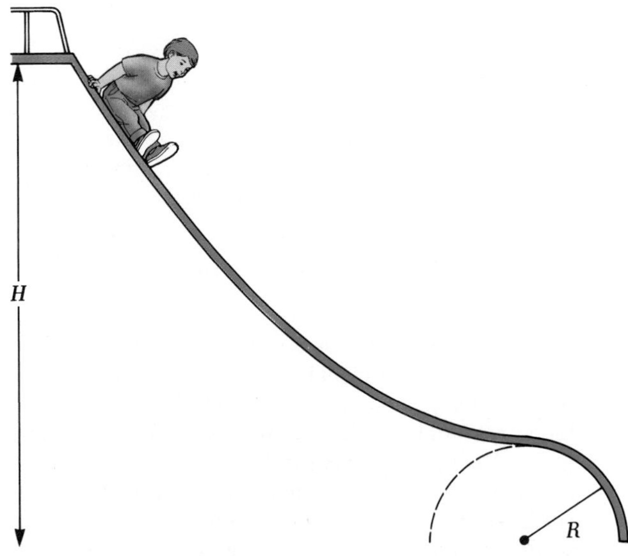

Figure 8.20 (Problem 18).

Section 8.5 Nonconservative Forces and the Work-Energy Theorem

19. A 5-kg block is set into motion up an inclined plane as in Figure 8.21 with an initial speed of 8 m/s. The block comes to rest after traveling 3 m along the plane, as shown in the diagram. The plane is inclined at an angle of 30° to the horizontal. (a) Determine the change in kinetic energy. (b) Determine the change in potential energy. (c) Determine the frictional force on the block (assumed to be constant). (d) What is the coefficient of kinetic friction?

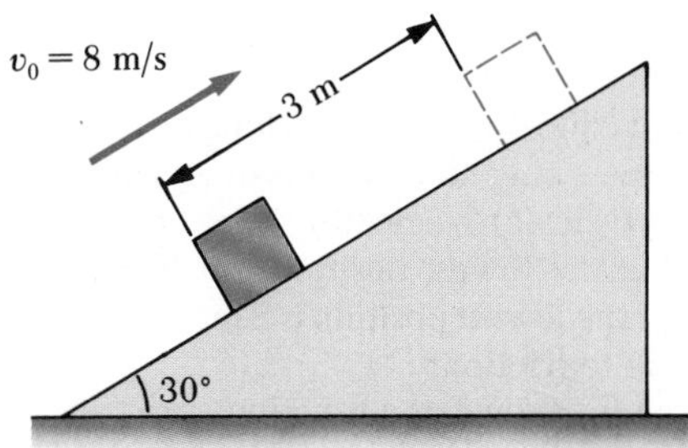

Figure 8.21 (Problem 19).

20. A block with a mass of 3 kg starts at a height $h =$ 60 cm on a plane with an inclination angle of 30°, as shown in Figure 8.22. Upon reaching the bottom of the ramp, the block slides along a horizontal surface. If the coefficient of friction on both surfaces is $\mu_k = 0.20$, how far will the block slide on the horizontal surface before coming to rest? [*Hint:* Divide the path into two straight-line parts.]

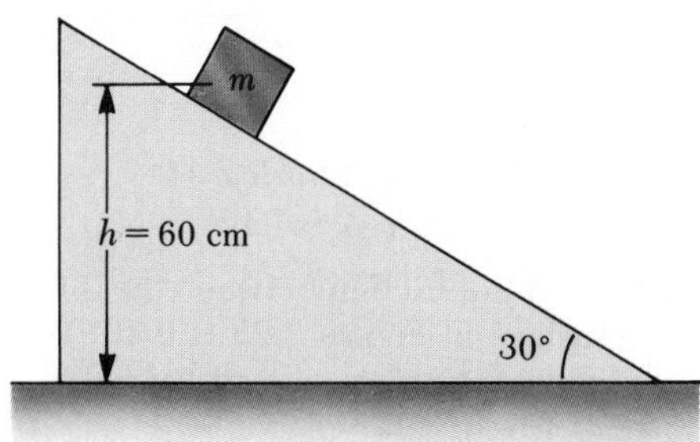

Figure 8.22 (Problem 20).

21. A child starts from rest at the top of a slide of height $h = 4$ m (Fig. 8.7). (a) What is her speed at the bottom if the incline is frictionless? (b) If she reaches the bottom with a speed of 6 m/s, what percentage of her total energy is lost as a result of friction?

22. A 0.5-kg bead slides on a curved wire, starting from rest at point A in Figure 8.23. The segment from A to B is frictionless, and the segment from B to C is rough. (a) Find the speed of the bead at B. (b) If the bead comes to rest at C, find the total work done by friction in going from B to C. (c) What is the net work done by nonconservative forces as the bead moves from A to C?

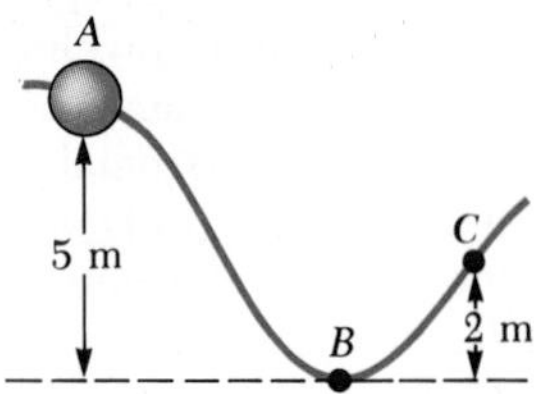

Figure 8.23 (Problem 22).

23. A 25-kg child on a swing 2 m long is released from rest when the swing supports make an angle of 30° with the vertical. (a) Neglecting friction, find the child's speed at the lowest position. (b) If the speed of the child at the lowest position is 2 m/s, what is the energy loss due to friction?

24. A force F_x, shown as a function of distance in Figure 8.24, acts on a 5-kg mass. If the particle starts from rest at $x = 0$ m, determine the speed of the particle at $x = 2$, 4, and 6 m.

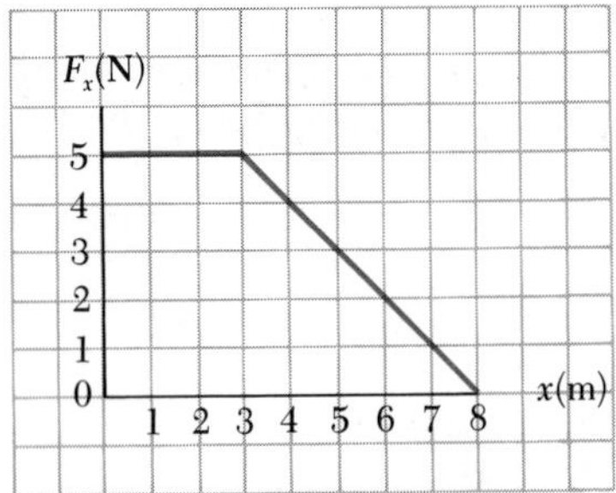

Figure 8.24 (Problem 24).

25. The coefficient of friction between the 3.0-kg object and the surface in Figure 8.25 is 0.40. What is the speed of the 5.0-kg mass when it has fallen a vertical distance of 1.5 m?

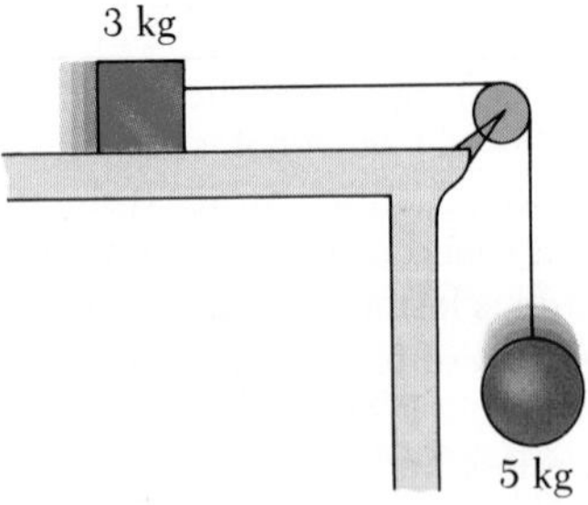

Figure 8.25 (Problem 25).

26. A toy gun uses a spring to project a 5.3-g soft rubber sphere. The spring constant is 8.0 N/m, the barrel of the gun is 15 cm long, and there is a constant frictional force of 0.032 N between the barrel and the projectile. With what speed is the projectile launched from the barrel of the gun if the spring was compressed 5.0 cm?

27. A mass of 2.5 kg is attached to a light spring with $k = 65$ N/m. The spring is stretched and allowed to oscillate freely on a frictionless horizontal surface. When the spring is stretched 10 cm, the kinetic energy of the attached mass and the elastic potential energy are equal. What is the maximum speed of the mass?

Section 8.6 Potential Energy Stored in a Spring

28. A spring has a force constant of 300 N/m. How much work must be done on the spring to stretch it (a) 4 cm from its equilibrium position and (b) from $x = 2$ cm to $x = 4$ cm, where $x = 0$ is its equilibrium position? (In the unstretched position, the potential energy is defined to be zero.)

29. A spring has a force constant of 500 N/m. What is the elastic potential energy stored in the spring when (a) it is stretched 4 cm from equilibrium, (b) it is compressed 3 cm from equilibrium, and (c) it is unstretched?

30. A block of mass m is released from rest and slides down a frictionless track of height h above a table (Fig. 8.26). At the bottom of the track, where the surface is horizontal, the block strikes and sticks to a light spring. (a) Find the maximum distance the spring is compressed. (b) Obtain a numerical value for this distance if $m = 1.0$ kg, $h = 2$ m, and $k = 400$ N/m.

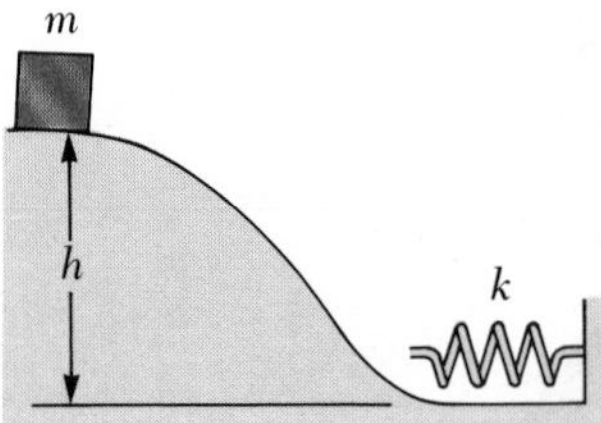

Figure 8.26 (Problem 30).

31. An 8-kg block travels on a rough, horizontal surface and collides with a spring as in Figure 8.9. The speed of the block *just before* the collision is 4 m/s. As the block rebounds to the left with the spring uncompressed, its speed as it leaves the spring is 3 m/s. If the coefficient of kinetic friction between the block and surface is 0.4, determine (a) the work done by friction while the block is in contact with the spring and (b) the maximum distance the spring is compressed.

32. A 3-kg mass is fastened to a light spring that passes over a pulley (Fig. 8.27). The pulley is frictionless, and the mass is released from rest when the spring is unstretched. If the mass drops a distance of 10 cm before coming to rest, find (a) the force constant of the spring, and (b) the speed of the mass when it is 5 cm below its starting point.

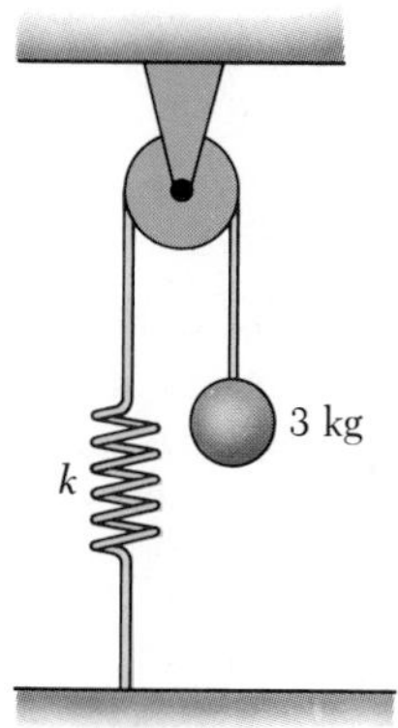

Figure 8.27 (Problem 32).

33. A block of mass 0.25 kg is placed on a vertical spring of constant $k = 5000$ N/m and is pushed downward, compressing the spring a distance of 0.1 m. As the block is released it leaves the spring and continues to travel upward. To what maximum height above the point of release does the block rise?

34. A block of mass 2 kg is kept at rest by compressing a horizontal massless spring having a spring constant k = 100 N/m by 10 cm. As the block is released it travels on a rough horizontal surface a distance of 0.25 m before it stops. Calculate the coefficient of kinetic friction between the horizontal surface and the block.

35. A 10-kg block is released from point *A* on a track *ABCD* as shown in Fig. 8.28. The track is frictionless except for the portion *BC*, of length 6 m. The block travels down the track and hits a spring of force constant $k = 2250$ N/m and compresses it a distance of 0.3 m from its equilibrium position before coming to rest momentarily. Determine the coefficient of kinetic friction between the track portion *BC* and the block.

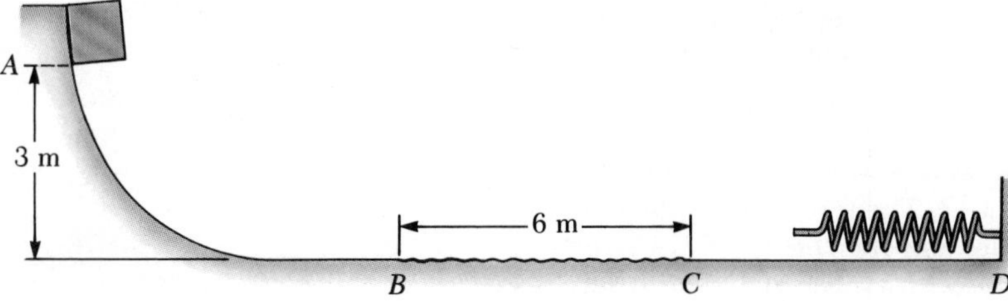

Figure 8.28 (Problem 35).

36. A 6-kg mass, when attached to a light vertical spring of length 10 cm, stretches the spring by 0.3 cm. The mass is now pulled downward, stretching the spring to a length of 10.7 cm, and released. Find the speed of the oscillating mass when the spring's length is 10.4 cm.

Section 8.7 Relationship Between Conservative Forces and Potential Energy

37. The potential energy of a two-particle system separated by a distance r is given by $U(r) = A/r$, where A is a constant. Find the radial force $\mathbf{F}_r$.

38. The potential energy function for a system is given by $U = ax^2 - bx$, where a and b are constants. (a) Find the force $\mathbf{F}_x$ associated with this potential energy function. (b) At what value of x is the force zero?

39. A potential energy function for a two dimensional force is of the form

$$U = 3x^3y - 7x$$

Find the force that acts at the point (x, y).

*Section 8.8 Energy Diagrams and Stability of Equilibrium

40. Consider the potential energy curve $U(x)$ versus x shown in Figure 8.29. (a) Determine whether the force F_x is positive, negative, or zero at the various points indicated. (b) Indicate points of stable, unsta-

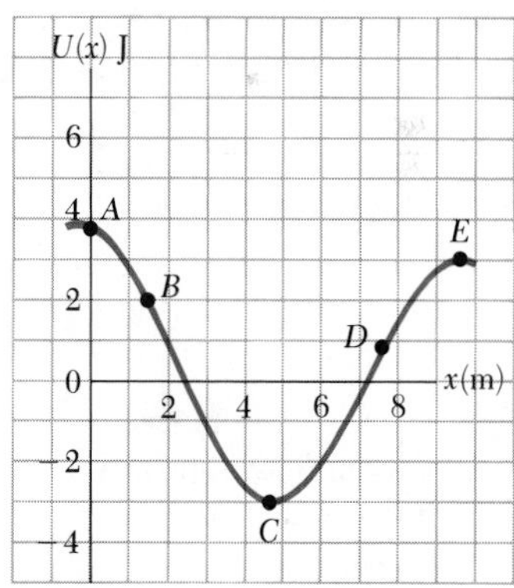

Figure 8.29 (Problem 40).

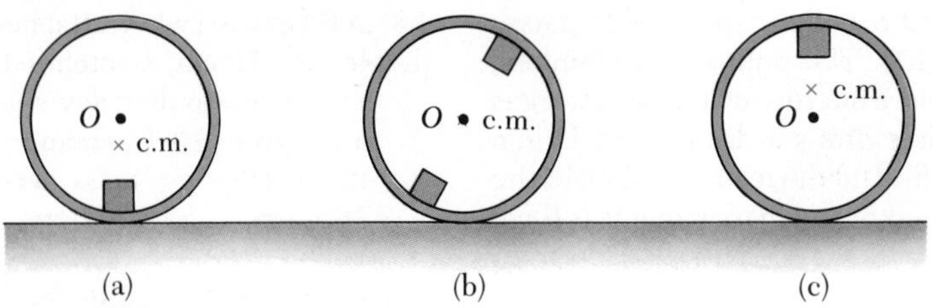

Figure 8.30 (Problem 42).

ble, or neutral equilibrium. (c) With reference to the potential energy curve in Figure 8.29, make a rough sketch of the curve F_x versus x from $x = 0$ to $x = 8$ m.

41. A right circular cone can be balanced on a horizontal surface in three different ways. Sketch these three equilibrium configurations, and identify them as being positions of stable, unstable, or neutral equilibrium.

42. A hollow pipe has one or two extra weights attached to its inner surface as shown in Figure 8.30(a),(b),(c). Explain why one is in unstable equilibrium, one in neutral equilibrium, and one in stable equilibrium. (In each diagram, O is the center of curvature and c.m. is the center of mass.)

43. A particle of mass $m = 5$ kg is released from point A on a frictionless track shown in Figure 8.31. Determine: (a) the speed of mass m at points B and C, and (b) the net work done by the force of gravity in moving the particle from A to C.

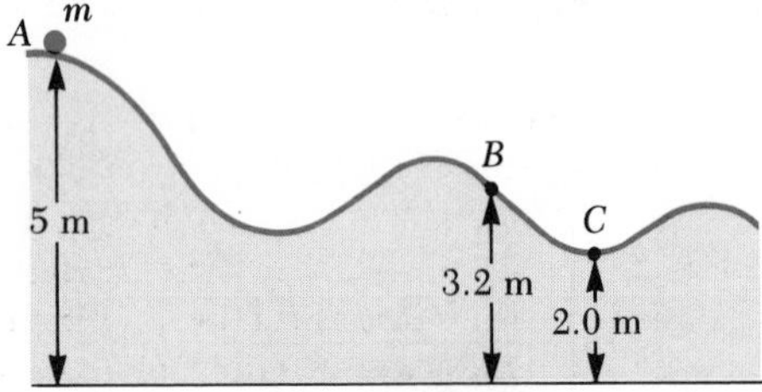

Figure 8.31 (Problem 43).

*Section 8.10 Mass–Energy Equivalence

44. Find the energy equivalence of the following objects. (a) An electron of mass 9.11×10^{-31} kg, (b) a uranium atom of mass 4.0×10^{-25} kg, (c) a paper clip of mass 2 g, (d) the earth of mass 5.99×10^{24} kg.

45. The correct expression for the kinetic energy of a particle moving with a speed v is given by Equation 7.27, which can be written as $K = \gamma mc^2 - mc^2$, where the factor $\gamma = [1 - (v/c)^2]^{-1/2}$. The term γmc^2 is the *total energy* of the particle, and the term mc^2 is its *rest energy*. A proton moves with a speed of $0.990c$, where c is the speed of light. Find (a) its rest energy, (b) its total energy, and (c) its kinetic energy.

ADDITIONAL PROBLEMS

46. A 200-g particle is released from rest at point A along the diameter on the inside of a smooth hemispherical bowl of radius $R = 30$ cm (Fig. 8.32). Calculate (a) its gravitational potential energy at point A relative to point B, (b) its kinetic energy at point B, (c) its speed at point B, and (d) its kinetic energy and potential energy at point C.

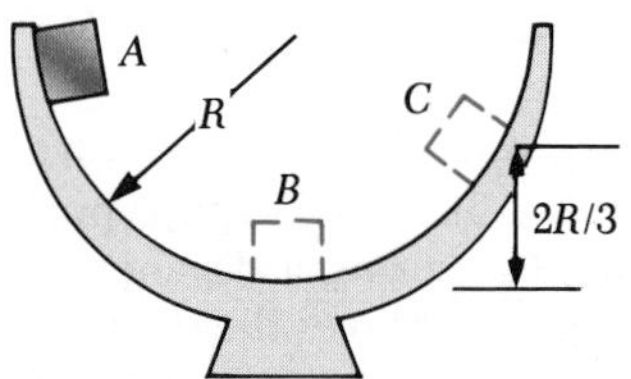

Figure 8.32 (Problems 46 and 47).

47. The particle described in Problem 46 (Fig. 8.32) is released from point A at rest, and the surface of the bowl is rough. The speed of the particle at point B is 1.5 m/s. (a) What is its kinetic energy at B? (b) How much energy is lost as a result of friction as the particle goes from A to B? (c) Is it possible to determine μ from these results in any simple manner? Explain.

48. A child's toy consists of a piece of plastic attached to a spring (Fig. 8.33). The spring is compressed against the floor a distance of 2 cm, and the toy is released. If the mass of the toy is 100 g and it rises to a maximum height of 60 cm, estimate the force constant of the spring.

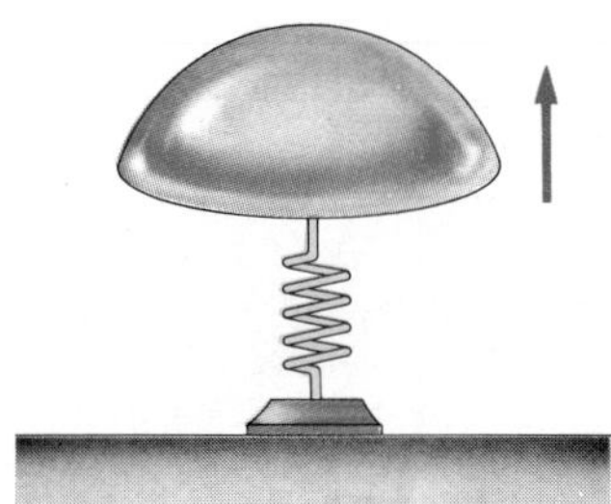

Figure 8.33 (Problem 48).

49. A frictionless roller coaster is given an initial speed v_0 at a height h, as in Figure 8.34. The radius of curvature of the track at point A is R. (a) Find the *maximum* value of v_0 necessary in order that the roller coaster *not* leave the track at A. (b) Using the value of v_0 calculated in (a), determine the value of h' necessary if the roller coaster is to make it just to point B.

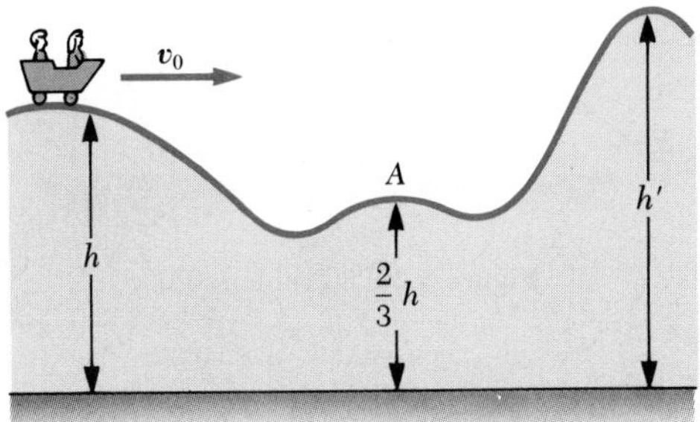

Figure 8.34 (Problem 49).

50. A 2-kg mass is suspended by means of a light string that passes over a frictionless pulley as shown in Figure 8.35. The other end of the string is connected to a 1-kg mass that rests upon a horizontal frictionless surface. The system starts in motion with the string connected to the 1-kg mass making an angle of 30° with the horizontal. When the string makes an angle of 45° with the horizontal, how much work has been done on the 1-kg mass? (The pulley is 2 m above the surface and the surface is frictionless.)

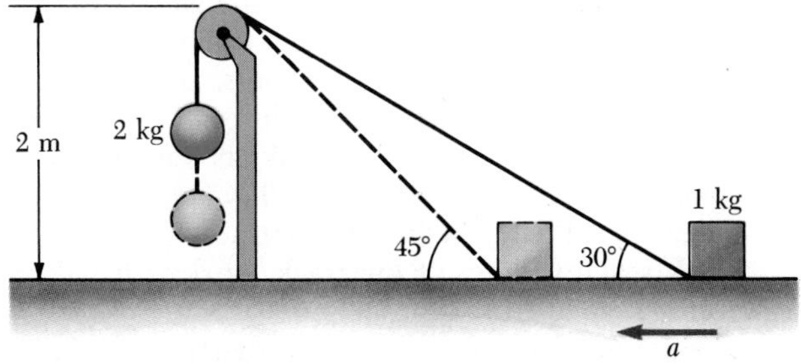

Figure 8.35 (Problem 50).

51. The masses of the javelin, the discus, and the shot are 0.8 kg, 2.0 kg, and 7.2 kg, respectively, and record throws in the track events using these objects are about 89 m, 69 m, and 21 m, respectively. Neglecting air resistance, (a) calculate the minimum initial kinetic energies that would produce these throws, and (b) estimate the average force exerted on each object during the throw assuming the force acts over a distance of 2 m. (c) Do your results suggest that air resistance is an important factor?

52. An olympic high jumper whose height is 2 m makes a record leap of 2.3 m over a horizontal bar. Estimate the speed with which he must leave the ground to perform this feat. (*Hint:* Estimate the position of his center of gravity before jumping, and assume he is in a horizontal position when he reaches the peak of his jump.)

53. Prove that the following forces are conservative and find the change in potential energy corresponding to these forces taking $x_i = 0$ and $x_f = x$: (a) $F_x = ax + bx^2$, (b) $F_x = Ae^{\alpha x}$. (a, b, A, and α are all constants.)

54. Find the forces corresponding to the following potential energy functions: (a) K/y, (b) bx^3, (c) e^{-ar}/r. (K, b, and a are all constants.)

55. A 2-kg block situated on a rough incline is connected to a spring of neglible mass having a spring constant of 100 N/m (Fig. 8.36). The block is released from rest when the spring is unstretched and the pulley is frictionless. The block moves 20 cm down the incline before coming to rest. Find the coefficient of kinetic friction between the block and the incline.

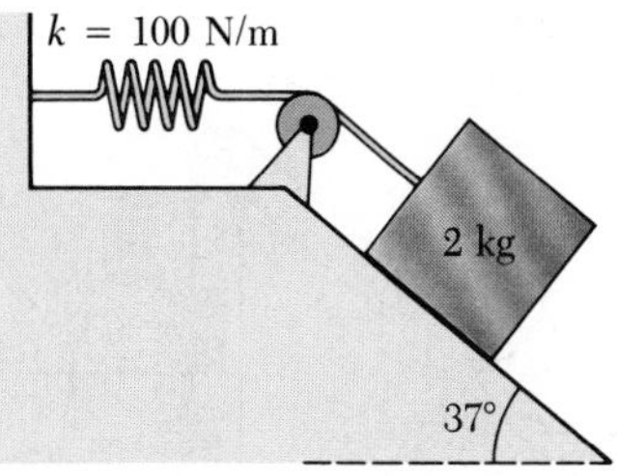

Figure 8.36 (Problems 55 and 56).

56. Suppose the incline is *smooth* for the system described in Problem 55 (Fig. 8.36). The block is released from rest with the spring initially unstretched. (a) How far does it move down the incline before coming to rest? (b) What is the acceleration of the block when it reaches its lowest point? Is the acceleration constant? (c) Describe the energy transformations that occur during the descent of the block.

57. A ball whirls around in a vertical circle at the end of a string. If the ball's total energy remains constant, show that the tension in the string at the bottom is greater than the tension at the top by six times the weight of the ball.

58. A pendulum of length L swings in the vertical plane. The string hits a peg located a distance d below the point of suspension (Fig. 8.37). (a) Show that if the pendulum is released at a height *below* that of the peg, it will return to this height after striking the peg. (b) Show that if the pendulum is released from the horizontal position ($\theta = 90°$) and the pendulum is to swing in a complete circle centered on the peg, then the minimum value of d must be $3L/5$.

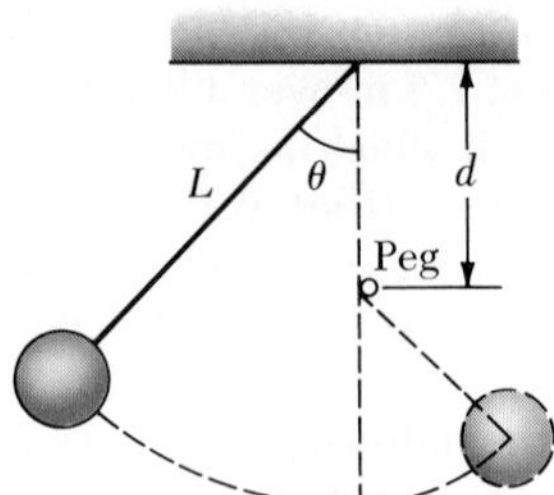

Figure 8.37 (Problem 58).

59. A 20-kg block is connected to a 30-kg block by a string that passes over a frictionless pulley. The 30-kg block is connected to a spring of neglible mass with force constant 250 N/m, as in Figure 8.38. The spring is unstretched when the system is as shown in the figure, and the incline is smooth. The 20-kg block is pulled a distance of 20 cm down the incline (so that the 30-kg block is 40 cm above the floor) and is released from rest. Find the speed of each block when the 30-kg block is 20 cm above the floor (that is, when the spring is unstretched).

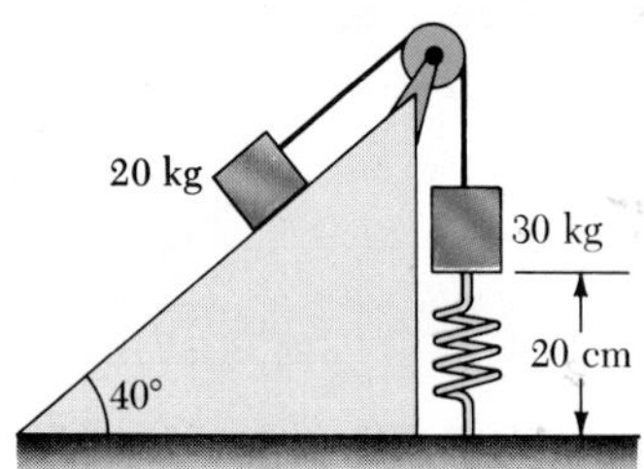

Figure 8.38 (Problem 59).

60. A potential energy function for a system is given by $U(x) = -x^3 + 2x^2 + 3x$. (a) Determine the force F_x as a function of x. (b) For what values of x is the force equal to zero? (c) Plot $U(x)$ versus x and F_x versus x and indicate points of stable and unstable equilibrium.

61. A crane is to lift 2000 kg of material to a height of 150 m in 1 min at a uniform rate. What electric power is required to drive the crane motor if 35 percent of the electric power is converted to mechanical power? (Is it reasonable to neglect the kinetic energy in this process?)

62. A block of mass m is dropped from rest at a height h directly above the top of a vertical spring having a force constant k. Find the *maximum* distance the spring will be compressed.

63. A force $\boldsymbol{F}$ acts on a block of mass 50 kg as shown in Figure 8.39. The block moves with a constant speed of 10 m/s up the incline a distance of 20 m. The coefficient of kinetic friction between the block and the incline is $\mu_k = 0.2$. Calculate the work done on the block by (a) the force $\boldsymbol{F}$, (b) the force of friction, and (c) the force of gravity.

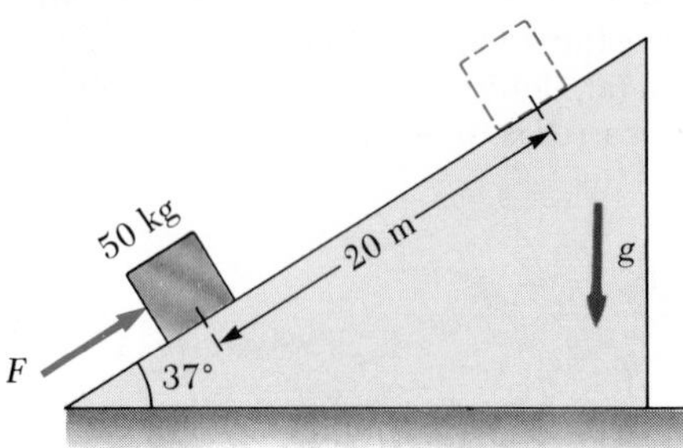

Figure 8.39 (Problem 63).

64. A 5-kg block free to move on a horizontal frictionless surface is attached to a spring. The spring is compressed 0.1 m from equilibrium and released. The speed of the block is 1.2 m/s when it passes the equilibrium position of the spring. The same experiment is now repeated with the frictionless surface replaced by a surface with a coefficient of kinetic friction $\mu_k = 0.3$. Determine the speed of the block at the equilibrium position of the spring.

65. A block of mass 0.5 kg is pushed against a horizontal spring of negligible mass, compressing the spring a distance of Δx (see Fig. 8.40). The spring constant is 450 N/m. When released the block travels along a frictionless horizontal surface to point B, the bottom of a vertical circular track of radius $R = 1$ m and continues to move up the track. The circular track is not smooth. The speed of the block at the bottom of the vertical circular track is $v_B = 12$ m/s, and the block experiences an average frictional force of 7.0 N while sliding up the curved track. (a) What was the initial compression of the spring? (b) What is the speed of the block at the top of the circular track? (c) Does the block reach the top of the track, or does it fall off before reaching the top?

66. The potential energy function of a particle as a function of distance x is given by $U = Ax^2 - Bx^3$, where A and B are positive constants. (a) Find the expression for the force the particle experiences as a function of x. (b) Locate the point(s) where the force vanishes. (c) Will the equilibrium of the particle be stable at these points? (d) What sort of motion would result if the particle were released from rest a small distance away from each of these points?

67. Two blocks A and B (with mass 50 kg and 100 kg respectively) are connected by a string as shown in Figure 8.41. The pulley is frictionless and of negligible mass. The coefficient of kinetic friction between block A and the incline is $\mu_k = 0.25$. Determine the

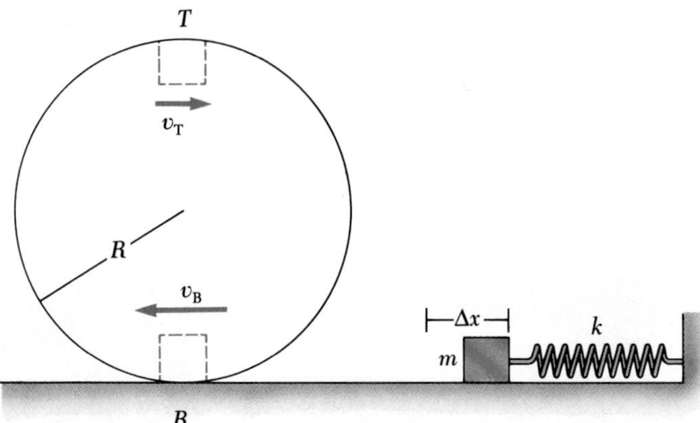

Figure 8.40 (Problem 65).

change in the kinetic energy of block A as it moves from C to D, a distance of 20 m up the incline.

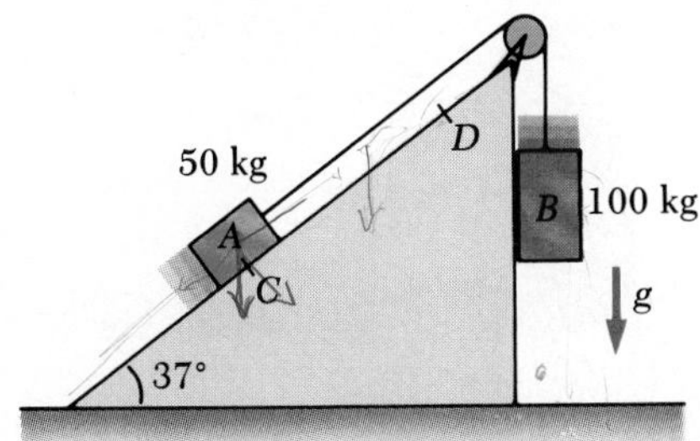

Figure 8.41 (Problem 67).

68. A uniform rope of length L lies on a horizontal smooth table. Part of the rope, of length d, hangs over the table, and the rope is released from rest (Fig. 8.42). Using energy methods, find (a) the velocity of the rope at the instant all of the rope leaves the table and (b) the time it takes this to occur. (*Hint:* Note the motion of the center of gravity of the rope.)

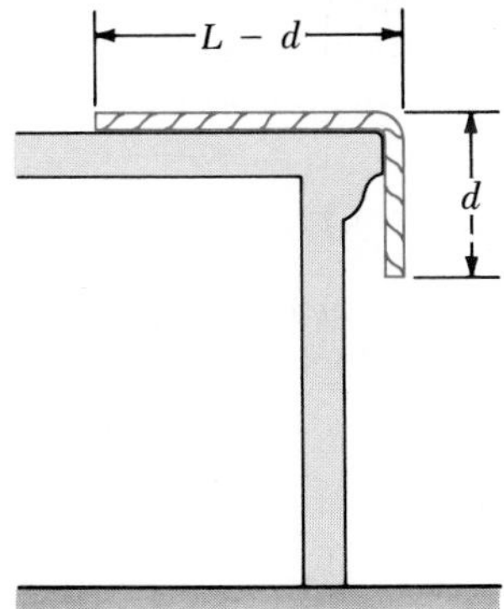

Figure 8.42 (Problem 68).

69. A 1.0-kg mass slides to the right on a surface with coefficient of friction $\mu = 0.25$ (Fig. 8.43). It has a speed of $v_i = 3$ m/s when contact is made with a spring with spring constant $k = 50$ N/m. The mass comes to rest after the spring has been compressed a distance d. The mass is then forced toward the left by the spring and it continues to move in that direction beyond the unstretched position. Finally the mass comes to rest a distance D to the left of the unstretched position. Find the following: (a) the compressed distance d, (b) the velocity v at the unstretched position, and (c) the distance D where the mass will come to rest to the left of the unstretched position.

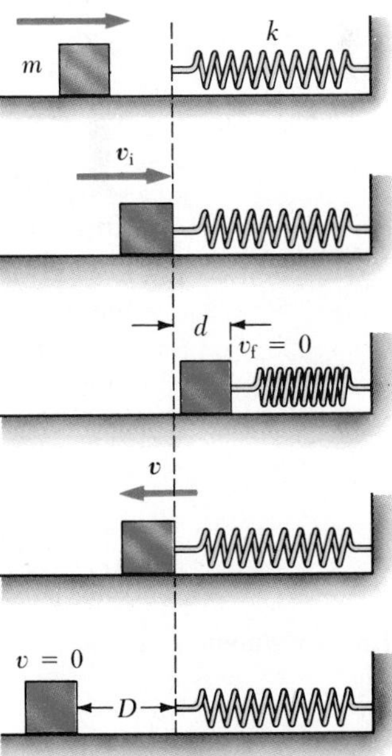

Figure 8.43 (Problem 69).

9

Linear Momentum and Collisions

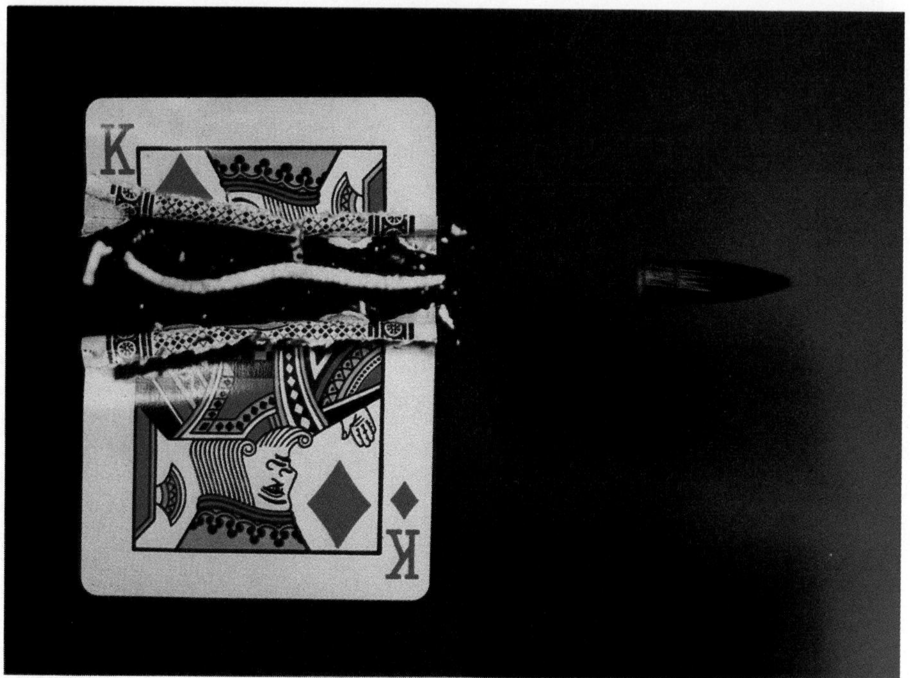

Cutting the card quickly: A bullet traveling at about 900 m/s collides with the king of diamonds. The photograph was taken using a microflash stroboscope at an exposure time of less than 1 μs. (© Harold E. Edgerton. Courtesy of Palm Press, Inc.)

In this chapter we shall analyze the motion of a system containing many particles. We shall introduce the concept of the linear momentum of the system of particles and show that this momentum is conserved when the system is isolated from its surroundings. The law of momentum conservation is especially useful for treating such problems as the collisions between particles and for analyzing rocket propulsion. The concept of the center of mass of a system of particles will also be introduced. We shall show that the overall motion of a system of interacting particles can be represented by the motion of an equivalent particle located at the center of mass.

9.1 LINEAR MOMENTUM AND IMPULSE

The **linear momentum** of a particle of mass m moving with a velocity $\boldsymbol{v}$ is defined to be the product of the mass and velocity:[1]

Definition of linear momentum of a particle

$$\boldsymbol{p} \equiv m\boldsymbol{v} \tag{9.1}$$

[1] This expression is nonrelativistic, and is valid only when $v \ll c$, where c is the speed of light. In relativity, momentum is defined by the relation $\boldsymbol{p} = m\boldsymbol{v}/(1 - v^2/c^2)^{1/2}$.

Momentum is a vector quantity since it equals the product of a scalar, m, and a vector, $\boldsymbol{v}$. Its direction is along $\boldsymbol{v}$, and it has dimensions of ML/T. In the SI system, momentum has the units kg·m/s.

If a particle is moving in an arbitrary direction, $\boldsymbol{p}$ will have three components and Equation 9.1 is equivalent to the component equations given by

$$p_x = mv_x \qquad p_y = mv_y \qquad p_z = mv_z \tag{9.2}$$

We can relate the linear momentum to the force acting on the particle using Newton's second law of motion: *The time rate of change of the momentum of a particle is equal to the resultant force on the particle.*[2] That is,

$$\boldsymbol{F} = \frac{d\boldsymbol{p}}{dt} \tag{9.3}$$

Newton's second law for a particle

From Equation 9.3 we see that if the resultant force is zero, the momentum of the particle must be constant. In other words, the linear momentum of a particle is conserved when $\boldsymbol{F} = 0$. Of course, if the particle is *isolated* (that is, if it does not interact with its environment), then by necessity, $\boldsymbol{F} = 0$ and $\boldsymbol{p}$ remains unchanged. This result can also be obtained directly through the application of Newton's second law in the form $\boldsymbol{F} = m\, d\boldsymbol{v}/dt$. That is, when the force is zero, the acceleration of the particle is zero and the velocity remains constant.

Equation 9.3 can be written

$$d\boldsymbol{p} = \boldsymbol{F}\, dt \tag{9.4}$$

We can integrate this expression to find the change in the momentum of a particle. If the momentum of the particle changes from $\boldsymbol{p}_i$ at time t_i to $\boldsymbol{p}_f$ at time t_f, then integrating Equation 9.4 gives

$$\Delta\boldsymbol{p} = \boldsymbol{p}_f - \boldsymbol{p}_i = \int_{t_i}^{t_f} \boldsymbol{F}\, dt \tag{9.5}$$

The quantity on the right side of Equation 9.5 is called the *impulse* of the force $\boldsymbol{F}$ for the time interval $\Delta t = t_f - t_i$. Impulse is a vector defined by

$$\boldsymbol{I} \equiv \int_{t_i}^{t_f} \boldsymbol{F}\, dt = \Delta\boldsymbol{p} \tag{9.6}$$

Impulse of a force

That is,

the **impulse** of the force $\boldsymbol{F}$ equals the change in the momentum of the particle.

This statement, known as the **impulse-momentum theorem,** is equivalent to Newton's second law. From this definition, we see that impulse is a vector quantity having a magnitude equal to the area under the force-time curve, as

Impulse-momentum theorem

[2] The formula $\boldsymbol{F} = d\boldsymbol{p}/dt$ is also valid in relativity provided that we use the relation $\boldsymbol{p} = m\boldsymbol{v}/(1 - v^2/c^2)^{1/2}$ for the momentum. We shall return to the relativistic treatment of motion in Chapter 39.

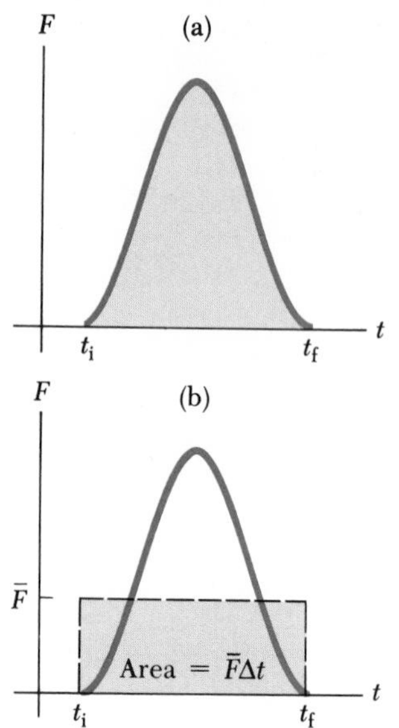

Figure 9.1 (a) A force acting on a particle may vary in time. The impulse is the area under the force versus time curve. (b) The average force (horizontal line) would give the same impulse to the particle in the time Δt as the real time-varying force described in (a).

described in Figure 9.1a. In this figure, it is assumed that the force varies in time in the general manner shown and is nonzero in the time interval $\Delta t = t_f - t_i$. The direction of the impulse vector is the same as the direction of the change in momentum. Impulse has the dimensions of momentum, that is, ML/T. Note that impulse is *not* a property of the particle itself, but is a quantity that measures the degree to which an external force changes the momentum of the particle. Therefore, when we say that an impulse is given to a particle, it is implied that momentum is transferred from an external agent to that particle.

Since the force can generally vary in time as in Figure 9.1a, it is convenient to define a time-averaged force $\overline{\boldsymbol{F}}$, given by

$$\overline{\boldsymbol{F}} \equiv \frac{1}{\Delta t}\int_{t_i}^{t_f} \boldsymbol{F}\, dt \tag{9.7}$$

where $\Delta t = t_f - t_i$. Therefore, we can express Equation 9.6 as

$$\boldsymbol{I} = \Delta \boldsymbol{p} = \overline{\boldsymbol{F}}\, \Delta t \tag{9.8}$$

This average force, described in Figure 9.1b, can be thought of as the constant force that would give the same impulse to the particle in the time interval Δt as the actual time-varying force gives over this same interval.

In principle, if $\boldsymbol{F}$ is known as a function of time, the impulse can be calculated from Equation 9.6. The calculation becomes especially simple if the force acting on the particle is constant. In this case, $\overline{\boldsymbol{F}} = \boldsymbol{F}$ and Equation 9.8 becomes

$$\boldsymbol{I} = \Delta \boldsymbol{p} = \boldsymbol{F}\, \Delta t \tag{9.9}$$

In many physical situations, we shall use the so-called **impulse approximation.** In this approximation, *we assume that one of the forces exerted on a particle acts for a short time but is much larger than any other force present.* This approximation is especially useful in treating collisions, where the duration of the collision is very short. When this approximation is made, we refer to the force as an *impulsive force.* For example, when a baseball is struck with a bat, the time of the collision is about 0.01 s, and the average force the bat exerts on the ball in this time is typically several thousand newtons. This is much greater than the force of gravity, so the impulse approximation is justified. When we use this approximation, it is important to remember that $\boldsymbol{p}_i$ and $\boldsymbol{p}_f$ represent the momenta *immediately* before and after the collision, respectively. Therefore, in the impulse approximation there is very little motion of the particle during the collision.

EXAMPLE 9.1 Teeing Off

A golf ball of mass 50 g is struck with a club (Fig. 9.2). The force on the ball varies from zero when contact is made up to some maximum value (where the ball is deformed) back to zero when the ball leaves the club. Thus, the force-time curve is qualitatively described by Figure 9.1. Assuming that the ball travels a distance of 200 m, (a) estimate the impulse due to the collision.

Solution Neglecting air resistance, we can use the expression for the range of a projectile (Chapter 4) given by

$$R = \frac{{v_0}^2}{g}\sin 2\theta_0$$

Figure 9.2 A golf ball being struck by a club. (© Harold E. Edgerton. Courtesy of Palm Press, Inc.)

Let us assume that the launch angle is 45°, which provides the maximum range for any given launch speed. The initial velocity of the ball is then estimated to be

$$v_0 = \sqrt{Rg} = \sqrt{(200\text{ m})(9.80\text{ m/s}^2)} = 44\text{ m/s}$$

Since $v_i = 0$ and $v_f = v_0$ for the ball, the magnitude of the impulse imparted to the ball is

$$I = \Delta p = mv_0 = (50 \times 10^{-3}\text{ kg})\left(44\ \frac{\text{m}}{\text{s}}\right) = \boxed{2.2\text{ kg}\cdot\text{m/s}}$$

(b) Estimate the time of the collision.

Solution From Figure 9.2, it appears that a reasonable estimate of the distance the ball travels while in contact with the club is the radius of the ball, about 2 cm. The time it takes the club to move this distance (the contact time) is then

$$\Delta t = \frac{\Delta x}{v_0} = \frac{2 \times 10^{-2}\text{ m}}{44\text{ m/s}} = \boxed{4.5 \times 10^{-4}\text{ s}}$$

Exercise 1 Estimate the magnitude of the average force exerted on the ball during the collision with the club.
Answer 4.9×10^3 N. This force is extremely large compared with the weight (gravity force) of the ball, which is only 0.49 N.

EXAMPLE 9.2 How Good are the Bumpers?
In a particular crash test, an automobile of mass 1500 kg collides with a wall as in Figure 9.3. The initial and final velocities of the automobile are $v_i = -15.0$ m/s and $v_f = 2.6$ m/s, respectively. If the collision lasts for 0.150 s, find the impulse due to the collision and the average force exerted on the automobile.

Solution The initial and final momenta of the automobile are given by

$$p_i = mv_i$$
$$= (1500\text{ kg})(-15.0\text{ m/s}) = -2.25 \times 10^4\text{ kg}\cdot\text{m/s}$$
$$p_f = mv_f = (1500\text{ kg})(2.6\text{ m/s}) = 0.39 \times 10^4\text{ kg}\cdot\text{m/s}$$

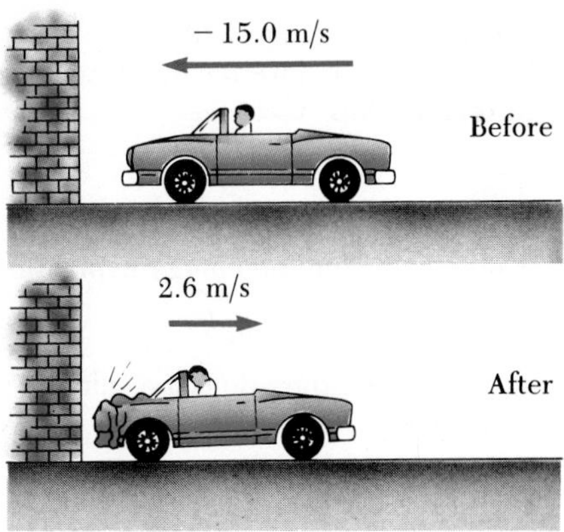

Figure 9.3 (Example 9.2).

Hence, the impulse, which equals the change in momentum, is

$$I = \Delta p = p_f - p_i$$
$$= 0.39 \times 10^4\text{ kg}\cdot\text{m/s} - (-2.25 \times 10^4\text{ kg}\cdot\text{m/s})$$
$$I = \boxed{2.64 \times 10^4\text{ kg}\cdot\text{m/s}}$$

The average force exerted on the automobile is

$$\overline{F} = \frac{\Delta p}{\Delta t} = \frac{2.64 \times 10^4\text{ kg}\cdot\text{m/s}}{0.150\text{ s}} = \boxed{1.76 \times 10^5\text{ N}}$$

EXAMPLE 9.3 Follow the Bouncing Ball
A ball of mass 100 g is dropped from a height $h = 2$ m above the floor (Fig. 9.4). It rebounds vertically to a height $h' = 1.5$ m after colliding with the floor. (a) Find the momentum of the ball immediately before and after the ball collides with the floor.

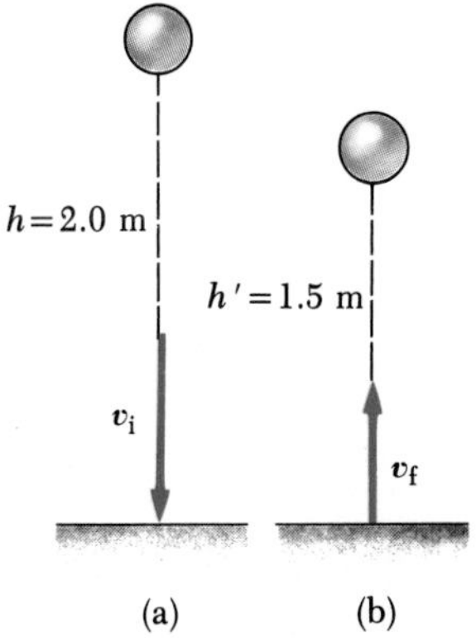

Figure 9.4 (Example 9.3) (a) The ball is dropped from a height h and reaches the floor with a velocity $\mathbf{v}_i$. (b) The ball rebounds from the floor with a velocity $\mathbf{v}_f$ and reaches a height h'.

Solution Using the energy methods, we can find $\boldsymbol{v}_i$, the velocity of the ball just before it collides with the floor, through the relationship

$$\tfrac{1}{2}mv_i^2 = mgh$$

Likewise, v_f, the ball's velocity right after colliding with the floor, is obtained from the energy expression

$$\tfrac{1}{2}mv_f^2 = mgh'$$

Substituting into these expressions the values $h = 2.0$ m and $h' = 1.5$ m gives

$$v_i = \sqrt{2gh} = \sqrt{(2)(9.80)(2)} \text{ m/s} = 6.26 \text{ m/s}$$

$$v_f = \sqrt{2gh'} = \sqrt{(2)(9.80)(1.5)} \text{ m/s} = 5.42 \text{ m/s}$$

Since $m = 0.1$ kg, the vector expressions for the initial and final linear momenta are given by

$$\boldsymbol{p}_i = m\boldsymbol{v}_i = -0.626\boldsymbol{j} \text{ kg}\cdot\text{m/s}$$

$$\boldsymbol{p}_f = m\boldsymbol{v}_f = 0.542\boldsymbol{j} \text{ kg}\cdot\text{m/s}$$

(b) Determine the average force exerted by the floor on the ball. Assume the time of the collision is 10^{-2} s (a typical value).

Using Equation 9.5 and the definition of $\overline{\boldsymbol{F}}$, we get

$$\Delta\boldsymbol{p} = \boldsymbol{p}_f - \boldsymbol{p}_i = \overline{\boldsymbol{F}}\,\Delta t$$

$$\overline{\boldsymbol{F}} = \frac{[0.542\boldsymbol{j} - (-0.626\boldsymbol{j})] \text{ kg}\cdot\text{m/s}}{10^{-2}\text{ s}} = 1.17 \times 10^2\boldsymbol{j} \text{ N}$$

Note that this average force is much greater than the force of gravity ($mg \approx 1.0$ N). That is, the impulsive force due to the collision with the floor overwhelms the gravitational force. In this inelastic collision, the energy lost by the ball is transformed into thermal energy.

9.2 CONSERVATION OF LINEAR MOMENTUM FOR A TWO-PARTICLE SYSTEM

Consider two particles that can interact with each other but are isolated from their surroundings (Fig. 9.5). That is, the particles exert forces on each other, but no external forces are present.[3] Suppose that at some time t, the momentum of particle 1 is $\boldsymbol{p}_1$ and the momentum of particle 2 is $\boldsymbol{p}_2$. We can apply Newton's second law to each particle and write

$$\boldsymbol{F}_{12} = \frac{d\boldsymbol{p}_1}{dt} \quad \text{and} \quad \boldsymbol{F}_{21} = \frac{d\boldsymbol{p}_2}{dt}$$

where $\boldsymbol{F}_{12}$ is the force on particle 1 due to particle 2 and $\boldsymbol{F}_{21}$ is the force on particle 2 due to particle 1. These forces could be gravitational forces, electrostatic forces, or have some other origin. This really isn't important for the present discussion. However, Newton's third law tells us that $\boldsymbol{F}_{12}$ and $\boldsymbol{F}_{21}$ are equal in magnitude and opposite in direction. That is, they form an action-reaction pair and $\boldsymbol{F}_{12} = -\boldsymbol{F}_{21}$. We can also express this condition as

$$\boldsymbol{F}_{12} + \boldsymbol{F}_{21} = 0$$

or

$$\frac{d\boldsymbol{p}_1}{dt} + \frac{d\boldsymbol{p}_2}{dt} = \frac{d}{dt}(\boldsymbol{p}_1 + \boldsymbol{p}_2) = 0$$

Since the time derivative of the total momentum, $\boldsymbol{P} = \boldsymbol{p}_1 + \boldsymbol{p}_2$, is *zero*, we conclude that the *total* momentum, $\boldsymbol{P}$, must remain constant, that is,

$$\boldsymbol{P} = \boldsymbol{p}_1 + \boldsymbol{p}_2 = \text{constant} \tag{9.10}$$

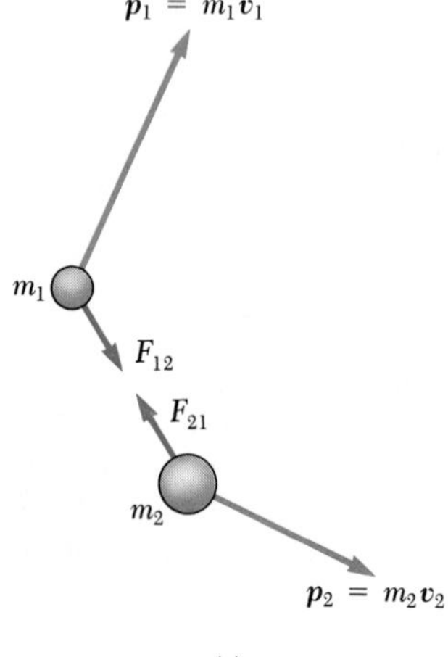

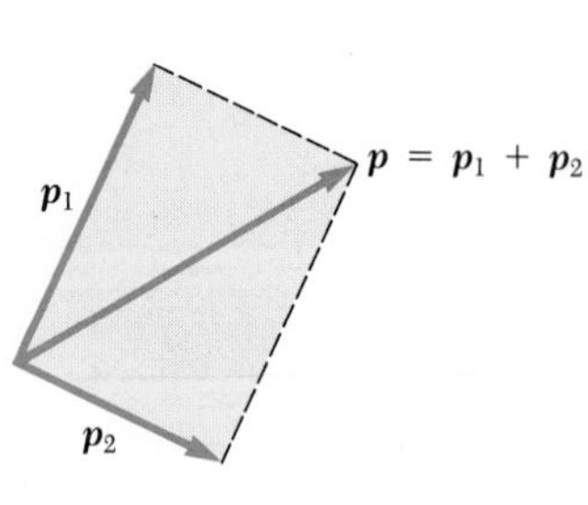

Figure 9.5 (a) At some instant, the momentum of m_1 is $\boldsymbol{p}_1 = m_1\boldsymbol{v}_1$ and the momentum of m_2 is $\boldsymbol{p}_2 = m_2\boldsymbol{v}_2$. Note that $\boldsymbol{F}_{12} = -\boldsymbol{F}_{21}$. (b) The total momentum of the system, $\boldsymbol{P}$, is equal to the vector sum $\boldsymbol{p}_1 + \boldsymbol{p}_2$.

[3] A truly isolated system cannot be achieved in the earth-bound laboratory, since gravitational forces and friction will always be present.

This vector equation is equivalent to three component equations. In other words, Equation 9.10 in component form says that the total momenta in the x, y, and z directions are all independently conserved, or

$$P_{ix} = P_{fx} \qquad P_{iy} = P_{fy} \qquad P_{iz} = P_{fz}$$

We can state this law, known as **the conservation of linear momentum,** as follows:

> If two particles of masses m_1 and m_2 form an isolated system, then the total momentum of the system is conserved, regardless of the nature of the force between them. More simply, whenever two particles collide their total momentum remains constant, provided they are isolated.

Suppose $\mathbf{v}_{1i}$ and $\mathbf{v}_{2i}$ are the initial velocities of particles 1 and 2, and $\mathbf{v}_{1f}$ and $\mathbf{v}_{2f}$ are their velocities at some later time. Applying Equation 9.10, we can express the conservation of linear momentum of this isolated system in the form

$$m_1\mathbf{v}_{1i} + m_2\mathbf{v}_{2i} = m_1\mathbf{v}_{1f} + m_2\mathbf{v}_{2f} \tag{9.11}$$

or

$$\mathbf{p}_{1i} + \mathbf{p}_{2i} = \mathbf{p}_{1f} + \mathbf{p}_{2f} \tag{9.12}$$

Conservation of momentum

That is, *the total momentum of the isolated system at all times equals its initial total momentum.* We can also describe the law of conservation of momentum in another way. Since we require that the system be isolated, the only forces acting must be internal to the system (the action-reaction pair). In other words, if there are no external forces present, the total momentum of the system remains constant. Therefore, momentum conservation for an isolated system is an alternative and equivalent statement of Newton's third law.

The law of conservation of momentum is considered to be one of the most important laws of mechanics. Mechanical energy is only conserved for an isolated system when conservative forces alone act on a system. On the other hand, momentum is conserved for an isolated two-particle system *regardless* of the nature of the internal forces. In fact, in Section 9.7 we shall show that the law of conservation of linear momentum also applies to an isolated system of n particles.

EXAMPLE 9.4 The Recoiling Cannon

A 3000-kg cannon rests on a frozen pond as in Figure 9.6. The cannon is loaded with a 30-kg cannonball and is fired horizontally. If the cannon recoils to the right with a velocity of 1.8 m/s, what is the velocity of the cannonball just after it leaves the cannon?

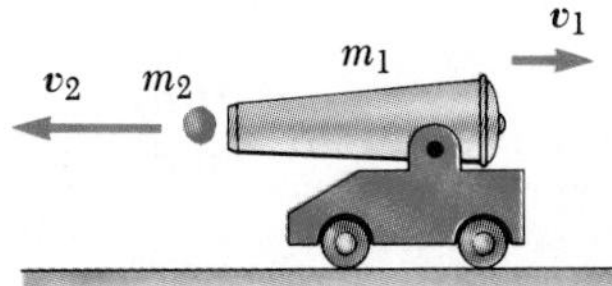

Figure 9.6 (Example 9.4) When the cannonball is fired to the right, it recoils to the right.

Solution We take the system to consist of the cannonball and the cannon. The system is not really isolated because of the force of gravity and the normal force. However, both of these forces are directed perpendicular to the motion of the system. Therefore, momentum is conserved in the x direction since there are no external forces in this direction (assuming the surface is frictionless).

The total momentum of the system before firing is zero. Therefore, the total momentum after firing must be zero, or

$$m_1v_1 + m_2v_2 = 0$$

With $m_1 = 3000$ kg, $v_1 = 1.8$ m/s, and $m_2 = 30$ kg, solving for v_2, the velocity of the cannonball, gives

$$v_2 = -\frac{m_1}{m_2}v_1 = -\left(\frac{3000\text{ kg}}{30\text{ kg}}\right)(1.8\text{ m/s}) = -180\text{ m/s}$$

The negative sign for v_2 indicates that the ball is moving to the left after firing, in the direction opposite to the movement of the cannon.

9.3 COLLISIONS

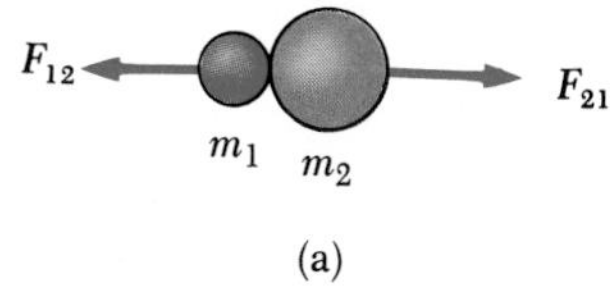

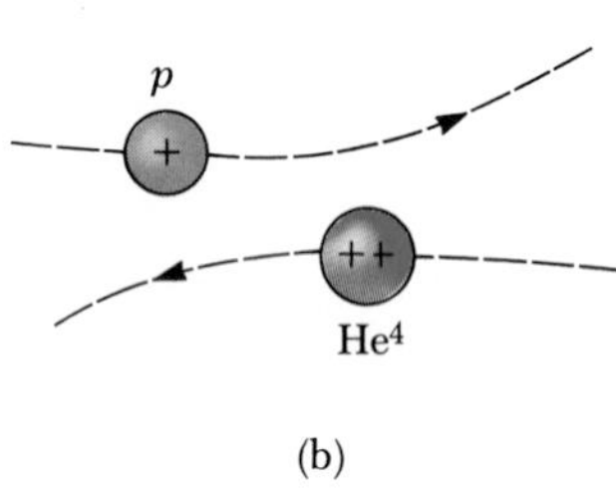

Figure 9.7 (a) The collision between two objects as the result of direct contact. (b) The collision between two charged particles.

In this section we shall use the law of conservation of momentum to describe what happens when two particles collide with each other. We shall use the term **collision** to represent the event of two particles coming together for a short time, producing impulsive forces on each other. *The impulsive force due to the collision is assumed to be much larger than any external forces present.*

The collision process may be the result of physical contact between two objects, as described in Figure 9.7a. This is a common observation when two macroscopic objects, such as two billiard balls or a baseball and a bat, collide. The notion of what we mean by a collision must be generalized since "contact" on a submicroscopic scale is ill-defined and meaningless. More accurately, impulsive forces arise from the electrostatic interaction of the electrons in the surface atoms of the two bodies.

To understand this on a more fundamental basis, consider a collision on an atomic scale (Fig. 9.7b), such as the collision of a proton with an alpha particle (the nucleus of the helium atom). Since the two particles are positively charged, they repel each other because of the strong electrostatic force between them at close separations. Such a process is commonly called a *scattering process.*

When the two particles of masses m_1 and m_2 collide as in Figure 9.7, the impulse forces may vary in time in a complicated way such as described in Figure 9.8. If $\boldsymbol{F}_{12}$ is the force on m_1 due to m_2, then the change in momentum of m_1 due to the collision is given by

$$\Delta \boldsymbol{p}_1 = \int_{t_i}^{t_f} \boldsymbol{F}_{12}\, dt$$

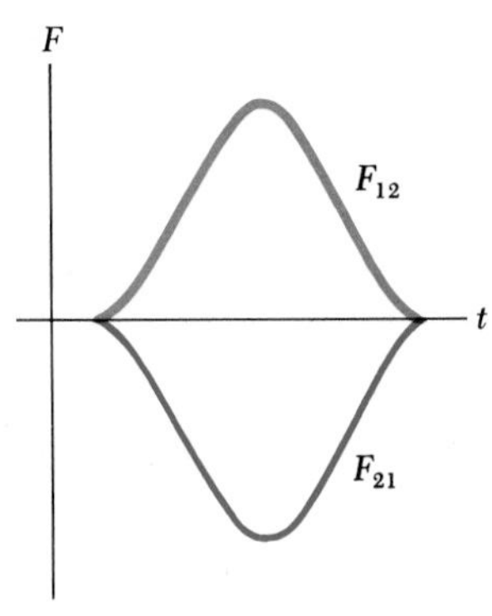

Figure 9.8 The force as a function of time for the two colliding particles described in Figure 9.6a. Note that $\boldsymbol{F}_{12} = -\boldsymbol{F}_{21}$.

Likewise, if $\boldsymbol{F}_{21}$ is the force on m_2 due to m_1, the change in momentum of m_2 is given by

$$\Delta \boldsymbol{p}_2 = \int_{t_i}^{t_f} \boldsymbol{F}_{21}\, dt$$

However, Newton's third law states that the force on m_1 due to m_2 is equal to and opposite the force on m_2 due to m_1, or $\boldsymbol{F}_{12} = -\boldsymbol{F}_{21}$. (This is described graphically in Fig. 9.8.) Hence, we conclude that

$$\Delta \boldsymbol{p}_1 = -\Delta \boldsymbol{p}_2$$

$$\Delta \boldsymbol{p}_1 + \Delta \boldsymbol{p}_2 = 0$$

Since the total momentum of the system is $\boldsymbol{P} = \boldsymbol{p}_1 + \boldsymbol{p}_2$, we conclude that the *change* in the momentum of the system due to the collision is zero, that is,

$$\boldsymbol{P} = \boldsymbol{p}_1 + \boldsymbol{p}_2 = \text{constant}$$

This is precisely what we expect if there are no external forces acting on the system (Section 9.2). However, the result is also valid if we consider the motion just before and just after the collision. Since the impulsive forces due to the collision are internal, they do not affect the total momentum of the system. Therefore, we conclude that

> for any type of collision, the total momentum of the system just before the collision equals the total momentum of the system just after the collision.

Momentum is conserved for any collision

Whenever a collision occurs between two bodies, we have seen that *the total momentum is always conserved.* However, the total kinetic energy is generally *not* conserved when a collision occurs because some of the kinetic energy is converted into thermal energy and internal elastic potential energy when the bodies are deformed during the collision.

Inelastic collision

We define an **inelastic collision** as a *collision in which momentum is conserved but kinetic energy is not.* For a general inelastic collision, we can apply the law of conservation of momentum in the form given by Equation 9.11. The collision of a rubber ball with a hard surface is inelastic since some of the kinetic energy of the ball is lost when it is deformed while it is in contact with the surface. When two objects collide and stick together after the collision, the collision is called **perfectly inelastic.** This is an extreme case of an inelastic collision. For example, if two pieces of putty collide, they stick together and move with some common velocity after the collision. If a meteorite collides with the earth, it becomes buried in the earth and the collision is perfectly inelastic. However, not all of the initial kinetic energy is necessarily lost even in a perfectly inelastic collision because the combined bodies may not be stationary after the collision.

Elastic collision

An **elastic collision** is defined as a *collision in which both momentum and kinetic energy are conserved.* Billiard ball collisions and the collisions of air molecules with the walls of a container at ordinary temperatures are highly elastic. In reality, collisions in the macroscopic world, such as those between billiard balls, can be only approximately elastic because in such collisions there is always some permanent deformation of the objects; hence there is always some loss of kinetic energy. However, truly elastic collisions do occur between atomic and subatomic particles. Elastic and perfectly inelastic collisions are *limiting* cases, and most collisions are cases in between.

Properties of inelastic and elastic collision

1. An **inelastic collision** is one in which momentum is conserved, but kinetic energy is not.
2. A **perfectly inelastic collision** between two objects is an inelastic collision in which the two objects stick together after the collision, so their final velocities are the same.
3. An **elastic collision** is one in which both momentum and kinetic energy are conserved.

9.4 COLLISIONS IN ONE DIMENSION

In this section, we treat collisions in one dimension and consider two extreme types of collisions: (1) perfectly inelastic and (2) elastic. The important distinction between these two types of collisions is that *momentum is conserved in both cases, but kinetic energy is conserved only in the case of an elastic collision.*

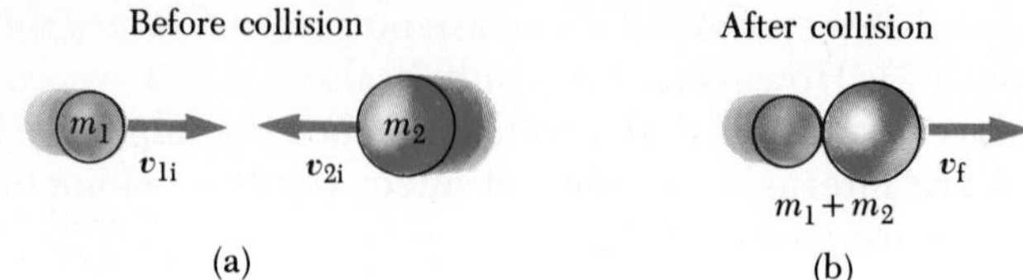

Figure 9.9 Schematic representation of a perfectly inelastic head-on collision between two particles: (a) before the collision and (b) after the collision.

Perfectly Inelastic Collisions

Consider two particles of masses m_1 and m_2 moving with initial velocities $\mathbf{v}_{1i}$ and $\mathbf{v}_{2i}$ along a straight line, as in Figure 9.9. If the two particles stick together and move with some common velocity $\mathbf{v}_f$ after the collision, then only the linear momentum of the system is conserved. Therefore, we can say that the total momentum before the collision equals the total momentum of the composite system after the collision, that is,

$$m_1\mathbf{v}_{1i} + m_2\mathbf{v}_{2i} = (m_1 + m_2)\mathbf{v}_f \tag{9.13}$$

$$\mathbf{v}_f = \frac{m_1\mathbf{v}_{1i} + m_2\mathbf{v}_{2i}}{m_1 + m_2} \tag{9.14}$$

EXAMPLE 9.5 The Cadillac Versus the "Beetle"
A large luxury car with a mass of 1800 kg stopped at a traffic light is struck from the rear by a compact car with a mass of 900 kg. The two cars become entangled as a result of the collision. (a) If the compact car was moving at 20 m/s before the collision, what will be the velocity of the entangled mass after the collision?

Solution The momentum before the collision is that of the compact car alone because the large car was initially at rest. Thus, we have for the momentum before the collision

$$p_i = m_1v_i = (900\text{ kg})(20\text{ m/s}) = 1.80 \times 10^4\text{ kg}\cdot\text{m/s}$$

After the collision, the mass that moves is the sum of the masses of the large car and the compact car. The momentum of the combination is

$$p_f = (m_1 + m_2)v_f = (2700\text{ kg})(v_f)$$

Equating the momentum before to the momentum after and solving for v_f, the velocity of the wreckage, we have

$$v_f = \frac{p_i}{m_1 + m_2} = \frac{1.80 \times 10^4\text{ kg}\cdot\text{m/s}}{2700\text{ kg}} = 6.67\text{ m/s}$$

(b) How much kinetic energy is lost in the collision?

Solution Since the luxury car is at rest before the collision, $v_{2i} = 0$, the initial kinetic energy (before the collision) is

$$K_i = \tfrac{1}{2}m_1v_{1i}^2 + \tfrac{1}{2}m_2v_{2i}^2$$
$$= \tfrac{1}{2}(900\text{ kg})(20\text{ m/s})^2 + 0 = 1.80 \times 10^5\text{ J}$$

Because the vehicles move with a common velocity v_f after the collision, the final kinetic energy (after the collision) is

$$K_f = \tfrac{1}{2}(m_1 + m_2)v_f^2 = \tfrac{1}{2}(900\text{ kg} + 1800\text{ kg})(6.67\text{ m/s})^2$$
$$= 0.60 \times 10^5\text{ J}$$

Hence, the *loss* in kinetic energy is

$$K_i - K_f = 1.20 \times 10^5\text{ J}$$

EXAMPLE 9.6 The Ballistic Pendulum
The ballistic pendulum (Fig. 9.10) is a system used to measure the velocity of a fast-moving projectile, such as a bullet. The bullet is fired into a large block of wood suspended from some light wires. The bullet is stopped by the block, and the entire system swings through a height h. Since the collision is perfectly inelastic and momentum is conserved, Equation 9.14 gives the veloc-

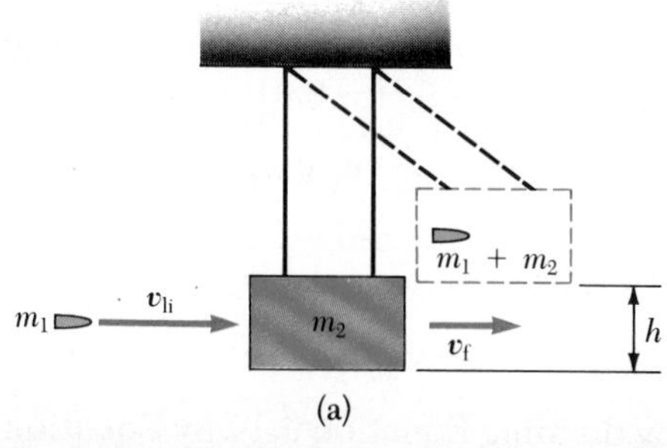

(a)

(b)

Figure 9.10 (Example 9.6) (a) Diagram of a ballistic pendulum. Note that $\mathbf{v}_f$ is the velocity of the system right after the perfectly inelastic collision. (b) Multiflash photo of a ballistic pendulum used in the laboratory. (Courtesy of CENCO)

ity of the system *right after* the collision in the impulse approximation. The kinetic energy *right after* the collision is given by

$$(1) \qquad K = \tfrac{1}{2}(m_1 + m_2)v_f^2$$

With $v_{2i} = 0$, Equation 9.14 becomes

$$(2) \qquad v_f = \frac{m_1 v_{1i}}{m_1 + m_2}$$

Substituting this value of v_f into (1) gives

$$K = \frac{m_1^2 v_{1i}^2}{2(m_1 + m_2)}$$

where v_{1i} is the initial velocity of the bullet. Note that this kinetic energy is *less* than the initial kinetic energy of the bullet. However, *after* the collision, energy is conserved and the kinetic energy at the bottom is transformed into potential energy in the bullet and in the block at the height h; that is,

$$\frac{m_1^2 v_{1i}^2}{2(m_1 + m_2)} = (m_1 + m_2)gh$$

$$v_{1i} = \left(\frac{m_1 + m_2}{m_1}\right)\sqrt{2gh}$$

Hence, it is possible to obtain the initial velocity of the bullet by measuring h and the two masses. Why would it be incorrect to equate the initial kinetic energy of the incoming bullet to the final gravitational energy of the bullet-block combination?

Exercise 2 In a ballistic pendulum experiment, suppose that $h = 5$ cm, $m_1 = 5$ g, and $m_2 = 1$ kg. Find (a) the initial speed of the projectile, and (b) the loss in energy due to the collision.
Answer 199 m/s; 98.5 J.

Elastic Collisions

Now consider two particles that undergo an elastic head-on collision (Fig. 9.11). In this case, both momentum and kinetic energy are conserved; therefore we can write these conditions

$$m_1 v_{1i} + m_2 v_{2i} = m_1 v_{1f} + m_2 v_{2f} \qquad (9.15)$$

$$\tfrac{1}{2}m_1 v_{1i}^2 + \tfrac{1}{2}m_2 v_{2i}^2 = \tfrac{1}{2}m_1 v_{1f}^2 + \tfrac{1}{2}m_2 v_{2f}^2 \qquad (9.16)$$

where v is positive if a particle moves to the right and negative if it moves to the left.

In a typical problem involving elastic collisions, there will be two unknown quantities and Equations 9.15 and 9.16 can be solved simultaneously to find these. However, an alternative approach, one that involves a little mathematical manipulation of Equation 9.16, often simplifies this process. To see this, let's cancel the factor of $\frac{1}{2}$ in Equation 9.16 and rewrite it as

$$m_1(v_{1i}^2 - v_{1f}^2) = m_2(v_{2f}^2 - v_{2i}^2)$$

The force from a nitrogen-propelled, hand-controlled device allows an astronaut to move about freely in space without restrictive tethers. (Courtesy of NASA).

Here we have moved the terms containing m_1 to one side of the equation and those containing m_2 to the other. Next, let us factor both sides of the equation:

$$m_1(v_{1i} - v_{1f})(v_{1i} + v_{1f}) = m_2(v_{2f} - v_{2i})(v_{2f} + v_{2i}) \tag{9.17}$$

We now separate the terms containing m_1 and m_2 in the equation for the conservation of momentum (Eq. 9.15) to get

$$m_1(v_{1i} - v_{1f}) = m_2(v_{2f} - v_{2i}) \tag{9.18}$$

Our final result is obtained by dividing Equation 9.17 by Equation 9.18 to get

$$v_{1i} + v_{1f} = v_{2f} + v_{2i}$$

or

$$v_{1i} - v_{2i} = -(v_{1f} - v_{2f}) \tag{9.19}$$

This equation, in combination with the equation for conservation of momentum, can be used to solve problems dealing with perfectly elastic collisions. According to Equation 9.19, the relative velocity of the two objects before the collision, $v_{1i} - v_{2i}$, equals the negative of the relative velocity of the two objects after the collision, $-(v_{1f} - v_{2f})$.

Suppose that the masses and the initial velocities of both particles are known. Equations 9.15 and 9.16 can be solved for the final velocities in terms of the initial velocities, since there are two equations and two unknowns. Solving for v_{1f} and v_{2f} gives

Elastic collision: relations between final and initial velocities

$$v_{1f} = \left(\frac{m_1 - m_2}{m_1 + m_2}\right)v_{1i} + \left(\frac{2m_2}{m_1 + m_2}\right)v_{2i} \tag{9.20}$$

$$v_{2f} = \left(\frac{2m_1}{m_1 + m_2}\right)v_{1i} + \left(\frac{m_2 - m_1}{m_1 + m_2}\right)v_{2i} \tag{9.21}$$

It is important to remember that the appropriate signs for v_{1i} and v_{2i} must be included in Equations 9.20 and 9.21 since velocities are vectors. For example, if m_2 is moving to the left initially, as in Figure 9.11, then v_{2i} is negative.

Let us consider some special cases: If $m_1 = m_2$, then we see that $v_{1f} = v_{2i}$ and $v_{2f} = v_{1i}$. That is, the particles exchange velocities if they have equal masses. This is what one observes in billiard ball collisions.

If m_2 is initially at rest, $v_{2i} = 0$, and Equations 9.20 and 9.21 become

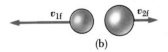

$$v_{1f} = \left(\frac{m_1 - m_2}{m_1 + m_2}\right)v_{1i} \tag{9.22}$$

$$v_{2f} = \left(\frac{2m_1}{m_1 + m_2}\right)v_{1i} \tag{9.23}$$

Before collision

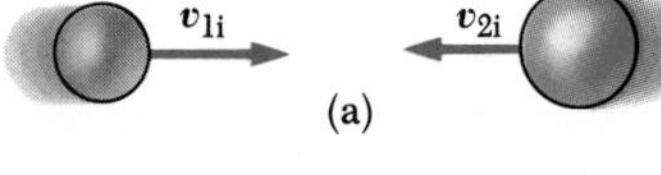

(a)

After collision

v_{1f} v_{2f}

(b)

Figure 9.11 Schematic representation of an elastic head-on collision between two particles; (a) before the collision and (b) after the collision.

If m_1 is very large compared with m_2, we see from Equations 9.22 and 9.23 that $v_{1f} \approx v_{1i}$ and $v_{2f} \approx 2v_{1i}$. That is, when a very heavy particle collides head-on with a very light one initially at rest, the heavy particle continues its motion unaltered after the collision, while the light particle rebounds with a velocity equal to about twice the initial velocity of the heavy particle. An example of such a collision would be the collision of a moving heavy atom, such as uranium, with a light atom, such as hydrogen.

If m_2 is much larger than m_1, and m_2 is initially at rest, then we find from Equations 9.22 and 9.23 that $v_{1f} \approx -v_{1i}$ and $v_{2f} \approx 0$. That is, when a very light particle collides head-on with a very heavy particle initially at rest, the light particle will have its velocity reversed, while the heavy particle will remain approximately at rest. For example, imagine what happens when a marble hits a stationary bowling ball.

EXAMPLE 9.7 A Two-body Collision with Spring

A block of mass $m_1 = 1.60$ kg moving to the right with a speed of 4.00 m/s on a frictionless horizontal track collides with a spring attached to a second block of mass $m_2 = 2.10$ kg moving to the left with a speed of 2.50 m/s, as in Figure 9.12a. The spring has a spring constant of 600 N/m. At the instant when m_1 is moving to the right with a speed of 3.00 m/s, determine (a) the velocity of m_2 and (b) the distance x that the spring is compressed.

Solution (a) First, note that the initial velocity of m_2 is -2.50 m/s because its direction is to the left. Since the total momentum of the system is conserved, we have

$$m_1v_{1i} + m_2v_{2i} = m_1v_{1f} + m_2v_{2f}$$

$$(1.60 \text{ kg})(4.00 \text{ m/s}) + (2.10 \text{ kg})(-2.50 \text{ m/s}) = (1.60 \text{ kg})(3.00 \text{ m/s}) + (2.10 \text{ kg})v_{2f}$$

$$v_{2f} = -1.74 \text{ m/s}$$

The negative value for v_{2f} means that m_2 is still moving towards the left at that instant.

(b) To determine the compression in the spring, x, shown in Figure 9.12b, we can make use of conservation of

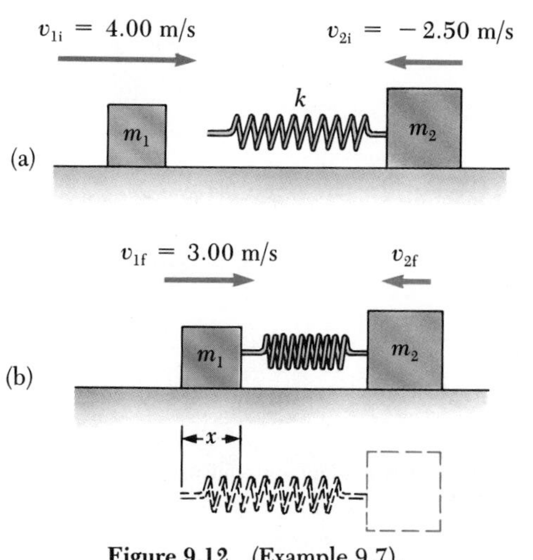

Figure 9.12 (Example 9.7).

energy since there are no friction forces acting on the system. Thus, we have

$$\tfrac{1}{2}m_1v_{1i}^2 + \tfrac{1}{2}m_2v_{2i}^2 = \tfrac{1}{2}m_1v_{1f}^2 + \tfrac{1}{2}m_2v_{2f}^2 + \tfrac{1}{2}kx^2$$

Substituting the given values and the result to part (a) into this expression gives

$$x = 0.173 \text{ m}$$

Exercise 3 Find the velocity of m_1 and the compression in the spring at the instant that m_2 is at rest.

Answer 0.719 m/s; 0.251 m.

EXAMPLE 9.8 Slowing Down Neutrons by Collisions

In a nuclear reactor, neutrons are produced when the isotope ${}^{235}_{92}U$ undergoes fission. These neutrons are moving at high speeds (typically 10^7 m/s) and must be slowed down to about 10^3 m/s. Once the neutrons have slowed down, they have a high probability of producing another fission event and hence a sustained chain reaction. The high-speed neutrons can be slowed down by passing them through a solid or liquid material called a *moderator*. The slowing-down process involves elastic collisions. Let us show that a neutron can lose most of its kinetic energy if it collides elastically with a moderator containing light nuclei, such as deuterium and carbon. Hence, the moderator material is usually heavy water (D_2O) or graphite (which contains carbon nuclei).

Solution Let us assume that the moderator nucleus of mass m_2 is at rest initially and that the neutron of mass m_1 and initial velocity v_{1i} collides head-on with it. Since momentum and energy are conserved, Equations 9.22 and 9.23 apply to the head-on collision of a neutron with the moderator nucleus. The initial kinetic energy of the neutron is

$$K_i = \tfrac{1}{2}m_1v_{1i}^2$$

After the collision, the neutron has a kinetic energy given by $\tfrac{1}{2}m_1v_{1f}^2$, where v_{1f} is given by Equation 9.22. We can express this energy as

$$K_1 = \tfrac{1}{2}m_1v_{1f}^2 = \frac{m_1}{2}\left(\frac{m_1 - m_2}{m_1 + m_2}\right)^2 v_{1i}^2$$

Therefore, the *fraction* of the total kinetic energy possessed by the neutron *after* the collision is given by

$$(1) \qquad f_1 = \frac{K_1}{K_i} = \left(\frac{m_1 - m_2}{m_1 + m_2}\right)^2$$

From this result, we see that the final kinetic energy of the neutron is small when m_2 is close to m_1 and is zero when $m_1 = m_2$.

We can calculate the kinetic energy of the moderator nucleus after the collision using Equation 9.23:

$$K_2 = \tfrac{1}{2}m_2 v_{2f}^2 = \frac{2m_1^2 m_2}{(m_1 + m_2)^2} v_{1i}^2$$

Hence, the fraction of the total kinetic energy transferred to the moderator nucleus is given by

$$(2) \qquad f_2 = \frac{K_2}{K_i} = \frac{4m_1 m_2}{(m_1 + m_2)^2}$$

Since the total energy is conserved, (2) can also be obtained from (1) with the condition that $f_1 + f_2 = 1$, so that $f_2 = 1 - f_1$.

Suppose that heavy water is used for the moderator. Collisions of the neutrons with deuterium nuclei in D_2O ($m_2 = 2m_1$) predict that $f_1 = 1/9$ and $f_2 = 8/9$. That is, 89% of the neutron's kinetic energy is transferred to the deuterium nucleus. In practice, the moderator efficiency is reduced because head-on collisions are very unlikely to occur. How would the result differ if graphite were used as the moderator?

9.5 TWO-DIMENSIONAL COLLISIONS

In the previous section and in Section 9.2, it was shown that the total momentum of a system of two particles is conserved when the system is isolated. For a general collision of two particles, this implies that the total momentum in *each* of the directions x, y, and z is conserved (Eq. 9.12). Thus, for a three-dimensional problem we would get three component equations for the conservation of momentum.

Let us consider a two-dimensional problem in which a particle of mass m_1 collides with a particle of mass m_2, where m_2 is initially at rest (Fig. 9.13). The collision is not head-on, but glancing. After the collision, m_1 moves at an angle θ with respect to the horizontal and m_2 moves at an angle ϕ with respect to the horizontal. Applying the law of conservation of momentum in component form, $P_{xi} = P_{xf}$ and $P_{yi} = P_{yf}$, and noting that $P_{yi} = 0$, we get

x component of momentum

$$m_1 v_{1i} = m_1 v_{1f} \cos\theta + m_2 v_{2f} \cos\phi \qquad (9.24a)$$

y component of momentum

$$0 = m_1 v_{1f} \sin\theta - m_2 v_{2f} \sin\phi \qquad (9.24b)$$

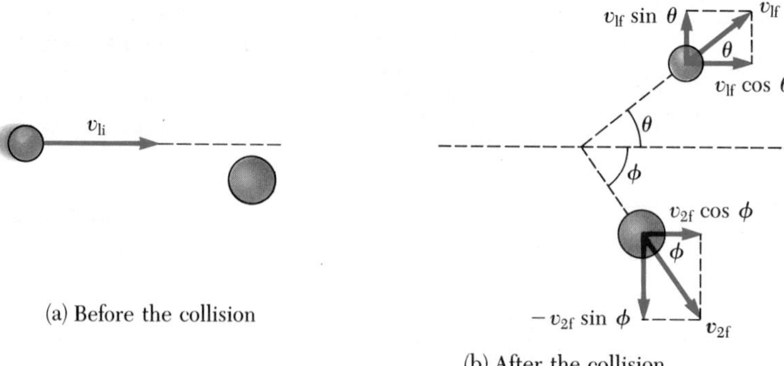

Figure 9.13 Schematic representation of an elastic glancing collision between two particles: (a) before the collision and (b) after the collision.

Now let us assume that the collision is elastic, in which case we can also write a third equation for the conservation of kinetic energy, in the form

$$\tfrac{1}{2}m_1v_{1i}^2 = \tfrac{1}{2}m_1v_{1f}^2 + \tfrac{1}{2}m_2v_{2f}^2 \qquad (9.25)$$

Conservation of energy

If we know the initial velocity, v_{1i}, and the masses, we are left with four unknowns. Since we only have three equations, one of the four remaining quantities (v_{1f}, v_{2f}, θ, or ϕ) must be given to determine the motion after the collision from conservation principles alone.

Again, it is important to note that if the collision is inelastic, kinetic energy is *not* conserved, and Equation 9.25 does *not* apply. Examples of both inelastic and elastic collisions will now be presented.

PROBLEM SOLVING STRATEGY: COLLISIONS

The following procedure is recommended when dealing with problems involving collisions between two objects.

1. Set up a coordinate system and define your velocities with respect to that system. It is convenient to have the x axis coincide with one of the initial velocities.
2. In your sketch of the coordinate system, draw all velocity vectors with labels and include all the given information.
3. Write expressions for the x and y components of the momentum of each object before and after the collision. Remember to include the appropriate signs for the components of the velocity vectors. For example, if an object is moving in the negative x direction, its x component of velocity must be taken to be negative. It is essential that you pay careful attention to signs.
4. Now write expressions for the *total* momentum in the x direction *before* and *after* the collision and equate the two. Repeat this procedure for the total momentum in the y direction. These steps follow from the fact that because the momentum of the *system* is conserved in any collision, the total momentum along any direction must be conserved. It is important to emphasize that it is the momentum of the *system* (the two colliding objects) that is conserved, not the momentum of the individual objects.
5. If the collision is inelastic, kinetic energy is *not* conserved, and you should then proceed to solve the momentum equations for the unknown quantities.
6. If the collision is elastic, kinetic energy is also conserved, so you can equate the total kinetic energy before the collision to the total kinetic energy after the collision. This gives an additional relationship between the various velocities.

EXAMPLE 9.9 Collision at an Intersection

A 1500-kg car traveling east with a speed of 25 m/s collides at an intersection with a 2500-kg van traveling north at a speed of 20 m/s, as shown in Figure 9.14. Find the direction and magnitude of the velocity of the wreckage after the collision, assuming that the vehicles undergo a perfectly inelastic collision (that is, they stick together).

Solution Let us choose east to be along the positive x direction and north to be along the positive y direction, as in Figure 9.14. Before the collision, the only object having momentum in the x direction is the car. Thus, the total initial momentum of the system (car plus van) in the x direction is

$$\sum p_{xi} = (1500\ \text{kg})(25\ \text{m/s}) = 37\ 500\ \text{kg}\cdot\text{m/s}$$

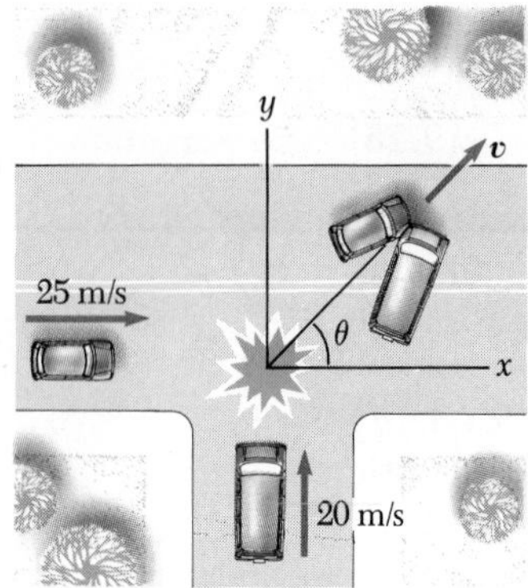

Figure 9.14 (Example 9.9) Top view of a car colliding with a van.

Now let us assume that the wreckage moves at an angle θ and speed v after the collision, as in Figure 9.14. The total momentum in the x direction after the collision is

$$\sum p_{xf} = (4000 \text{ kg})(v \cos \theta)$$

Because momentum is conserved in the x direction, we can equate these two equations to get

$$(1) \qquad 37\,500 \text{ kg}\cdot\text{m/s} = (4000 \text{ kg})(v \cos \theta)$$

Similarly, the total initial momentum of the system in the y direction is that of the van, which has the value (2500 kg)(20 m/s). Applying conservation of momentum to the y direction, we have

$$\sum p_{yi} = \sum p_{yf}$$

$$(2500 \text{ kg})(20 \text{ m/s}) = (4000 \text{ kg})(v \sin \theta)$$

$$(2) \qquad 50\,000 \text{ kg}\cdot\text{m/s} = (4000 \text{ kg})(v \sin \theta)$$

If we divide (2) by (1), we get

$$\tan \theta = \frac{50\,000}{37\,500} = 1.33$$

$$\theta = \boxed{53.1^\circ}$$

When this angle is substituted into (2)—or alternatively into (1)—the value of v is

$$v = \frac{50\,000 \text{ kg}\cdot\text{m/s}}{(4000 \text{ kg})(\sin 53^\circ)} = \boxed{15.6 \text{ m/s}}$$

EXAMPLE 9.10 Proton–Proton Collision

A proton collides in a perfectly elastic fashion with another proton initially at rest. The incoming proton has an initial speed of 3.5×10^5 m/s and makes a glancing collision with the second proton, as in Figure 9.13. (At close separations, the protons exert a repulsive electrostatic force on each other.) After the collision, one proton is observed to move at an angle of 37° to the original direction of motion, and the second deflects at an angle ϕ to the same axis. Find the final speeds of the two protons and the angle ϕ.

Solution Since $m_1 = m_2$, $\theta = 37^\circ$, and we are given $v_{1i} = 3.5 \times 10^5$ m/s, Equations 9.24 and 9.25 become

$$v_{1f} \cos 37^\circ + v_{2f} \cos \phi = 3.5 \times 10^5$$

$$v_{1f} \sin 37^\circ - v_{2f} \sin \phi = 0$$

$$v_{1f}^2 + v_{2f}^2 = (3.5 \times 10^5)^2$$

Solving these three equations with three unknowns simultaneously gives

$$v_{1f} = \boxed{2.80 \times 10^5 \text{ m/s}} \qquad v_{2f} = \boxed{2.11 \times 10^5 \text{ m/s}}$$

$$\phi = \boxed{53.0^\circ}$$

It is interesting to note that $\theta + \phi = 90^\circ$. This result is *not* accidental. *Whenever two equal masses collide elastically in a glancing collision and one of them is initially at rest, their final velocities are always at right angles to each other.* The next example illustrates this point in more detail.

EXAMPLE 9.11 Billiard Ball Collision

In a game of billiards, the player wishes to "sink" the target ball in the corner pocket, as shown in Figure 9.15. If the angle to the corner pocket is 35°, at what angle θ is the cue ball deflected? Assume that friction and rotational motion ("English") are unimportant, and assume the collision is elastic.

Solution Since the target is initially at rest, $v_{2i} = 0$, and conservation of kinetic energy gives

$$\tfrac{1}{2}m_1 v_{1i}^2 = \tfrac{1}{2}m_1 v_{1f}^2 + \tfrac{1}{2}m_2 v_{2f}^2$$

But $m_1 = m_2$, so that

$$(1) \qquad v_{1i}^2 = v_{1f}^2 + v_{2f}^2$$

Applying conservation of momentum to the two-dimensional collision gives

$$(2) \qquad \mathbf{v}_{1i} = \mathbf{v}_{1f} + \mathbf{v}_{2f}$$

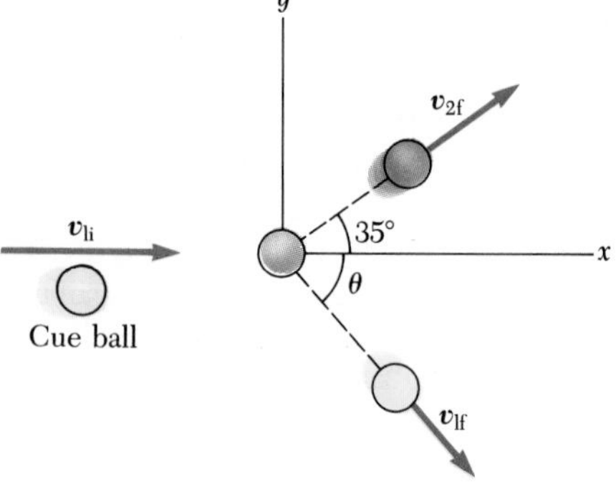

Figure 9.15 (Example 9.11).

If we square both sides of (2), we get

$$v_{1i}^2 = (\boldsymbol{v}_{1f} + \boldsymbol{v}_{2f}) \cdot (\boldsymbol{v}_{1f} + \boldsymbol{v}_{2f})$$
$$= v_{1f}^2 + v_{2f}^2 + 2\boldsymbol{v}_{1f} \cdot \boldsymbol{v}_{2f}$$

But $\boldsymbol{v}_{1f} \cdot \boldsymbol{v}_{2f} = v_{1f}v_{2f} \cos(\theta + 35°)$, and so

$$(3) \qquad v_{1i}^2 = v_{1f}^2 + v_{2f}^2 + 2v_{1f}v_{2f} \cos(\theta + 35°)$$

Subtracting (1) from (3) gives

$$2v_{1f}v_{2f} \cos(\theta + 35°) = 0$$
$$\cos(\theta + 35°) = 0$$

$$\theta + 35° = 90° \quad \text{or} \quad \theta = 55.0°$$

Again, this shows that whenever two equal masses undergo a glancing elastic collision and one of them is initially at rest, they will move at right angles to each other after the collision.

9.6 THE CENTER OF MASS

In this section we describe the overall motion of a mechanical system in terms of a very special point called the *center of mass* of the system. The mechanical system can be either a system of particles or an extended object. We shall see that the mechanical system moves as if all its mass were concentrated at the center of mass. Furthermore, if the resultant external force on the system is $\boldsymbol{F}$ and the total mass of the system is M, the center of mass moves with an acceleration given by $\boldsymbol{a} = \boldsymbol{F}/M$. That is, the system moves as if the resultant external force were applied to a single particle of mass M located at the center of mass. This result was implicitly assumed in earlier chapters since nearly all examples referred to the motion of extended objects.

Consider a mechanical system consisting of a pair of particles connected by a light, rigid rod (Fig. 9.16). The center of mass is located somewhere on the line joining the particles and is closer to the larger mass. If a single force is applied at some point on the rod closer to the smaller mass, the system will rotate clockwise (Fig. 9.16a). If the force is applied at a point on the rod closer

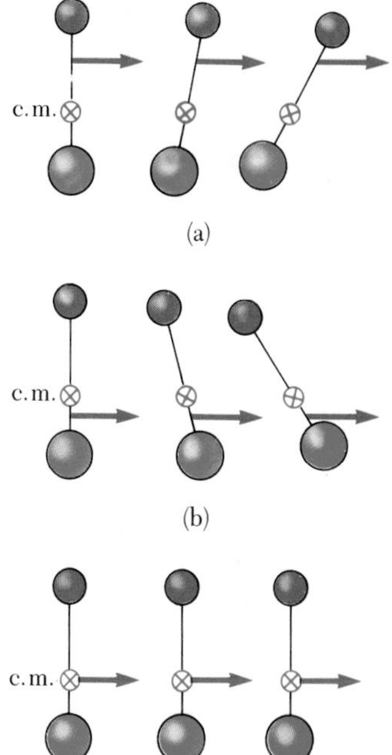

Figure 9.16 Two unequal masses are connected by a light, rigid rod. (a) The system rotates clockwise when a force is applied above the center of mass. (b) The system rotates counterclockwise when a force is applied below the center of mass. (c) The system moves in the direction of $\boldsymbol{F}$ without rotating when a force is applied at the center of mass.

Ballet dancers appear to float horizontally through the air as they execute a *grand jeté* (great leap). Their centers of mass are nonetheless following a parabolic path. (© John Terence Turner, FPG International)

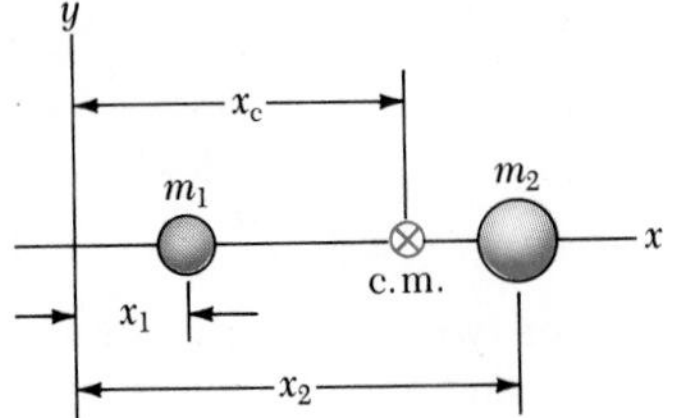

Figure 9.17 The center of mass of two particles on the x axis is located at x_c, a point between the particles, closer to the larger mass.

to the larger mass, the system will rotate in the counterclockwise direction (Fig. 9.16b). If the force is applied at the center of mass, the system will move in the direction of $\boldsymbol{F}$ without rotating (Fig. 9.16c). Thus, the center of mass can be easily located.

One can describe the position of the center of mass of a system as being the *average position* of the system's mass. For example, the center of mass of the pair of particles described in Figure 9.17 is located on the x axis and lies somewhere between the particles. The x coordinate of the center of mass in this case is defined to be

$$x_c \equiv \frac{m_1x_1 + m_2x_2}{m_1 + m_2} \tag{9.26}$$

For example, if $x_1 = 0$, $x_2 = d$, and $m_2 = 2m_1$, we find that $x_c = \frac{2}{3}d$. That is, the center of mass lies closer to the more massive particle. If the two masses are equal, the center of mass lies midway between the particles.

We can extend the center of mass concept to a system of many particles in three dimensions. The x coordinate of the center of mass of n particles is defined to be

$$x_c \equiv \frac{m_1x_1 + m_2x_2 + m_3x_3 + \cdots + m_nx_n}{m_1 + m_2 + m_3 + \cdots + m_n} = \frac{\Sigma m_ix_i}{\Sigma m_i} \tag{9.27}$$

where x_i is the x coordinate of the ith particle and Σm_i is the *total mass* of the system. For convenience, we shall express the total mass as $M = \Sigma m_i$, where the sum runs over all n particles. The y and z coordinates of the center of mass are similarly defined by the equations

$$y_c \equiv \frac{\Sigma m_iy_i}{M} \quad \text{and} \quad z_c \equiv \frac{\Sigma m_iz_i}{M} \tag{9.28}$$

The center of mass can also be located by its position vector, $\boldsymbol{r}_c$. The rectangular coordinates of this vector are x_c, y_c, and z_c, defined in Equations 9.27 and 9.28. Therefore,

$$\boldsymbol{r}_c = x_c\boldsymbol{i} + y_c\boldsymbol{j} + z_c\boldsymbol{k}$$

$$= \frac{\Sigma m_ix_i\boldsymbol{i} + \Sigma m_iy_i\boldsymbol{j} + \Sigma m_iz_i\boldsymbol{k}}{M} \tag{9.29}$$

Vector position of the center of mass for a system of particles

$$\boldsymbol{r}_c \equiv \frac{\Sigma m_i\boldsymbol{r}_i}{M} \tag{9.30}$$

where $\boldsymbol{r}_i$ is the position vector of the ith particle, defined by

$$\boldsymbol{r}_i \equiv x_i\boldsymbol{i} + y_i\boldsymbol{j} + z_i\boldsymbol{k}$$

Although the location of the center of mass for a rigid body is somewhat more cumbersome, the basic ideas we have discussed still apply. We can think of a general rigid body as a system of a large number of particles (Fig. 9.18). The particle separation is very small, and so the body can be considered to have a continuous mass distribution. By dividing the body into elements of mass Δm_i, with coordinates x_i, y_i, z_i, we see that the x coordinate of the center of mass is approximately

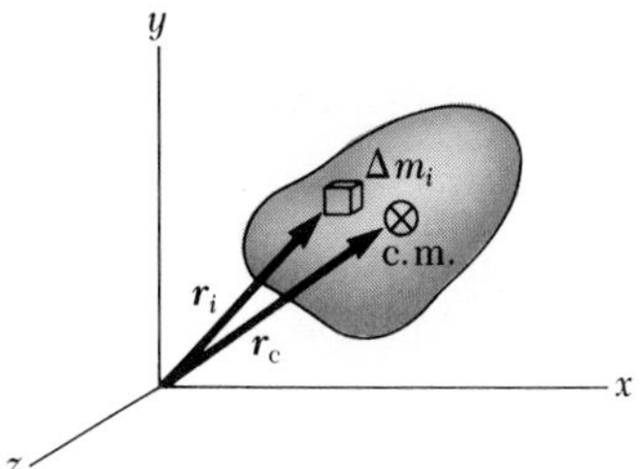

Figure 9.18 A rigid body can be considered a distribution of small elements of mass Δm_i. The center of mass is located at the vector position $\boldsymbol{r}_c$, which has coordinates x_c, y_c, and z_c.

$$x_c \approx \frac{\Sigma x_i \, \Delta m_i}{M}$$

with similar expressions for y_c and z_c. If we let the number of elements, n, approach infinity, then x_c will be given precisely. In this limit, we replace the sum by an integral and replace Δm_i by the differential element dm, so that

$$x_c = \lim_{\Delta m_i \to 0} \frac{\Sigma x_i \, \Delta m_i}{M} = \frac{1}{M} \int x \, dm \tag{9.31}$$

Likewise, for y_c and z_c we get

$$y_c = \frac{1}{M} \int y \, dm \quad \text{and} \quad z_c = \frac{1}{M} \int z \, dm \tag{9.32}$$

We can express the vector position of the center of mass of a rigid body in the form

$$\boldsymbol{r}_c = \frac{1}{M} \int \boldsymbol{r} \, dm \tag{9.33}$$

where this is equivalent to the three scalar expressions given by Equations 9.31 and 9.32.

The center of mass of various homogeneous, symmetric bodies must lie on an axis of symmetry. For example, the center of mass of a homogeneous rod must lie on the rod, midway between its ends. The center of mass of a homogeneous sphere or a homogeneous cube must lie at its geometric center. One can determine the center of mass of an irregularly shaped planar body experimentally by suspending the body from two different points (Fig. 9.19). The body is first hung from point A, and a vertical line AB is drawn (which can be established with a plumb bob) when the body is in equilibrium. The body is then hung from point C, and a second vertical line, CD, is drawn. The center of mass coincides with the intersection of these two lines. In fact, if the body is hung freely from any point, the vertical line through this point must pass through the center of mass.

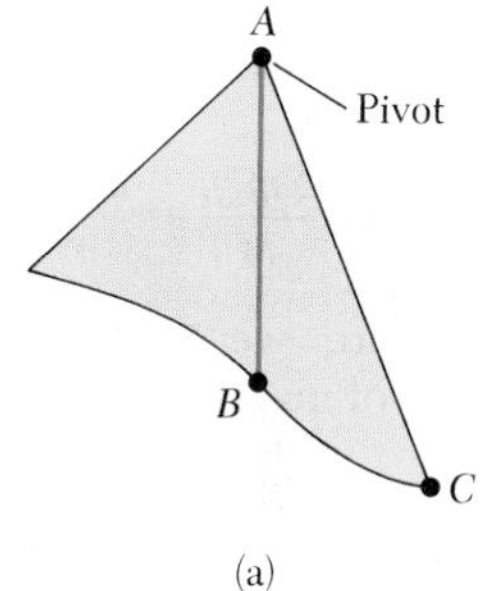

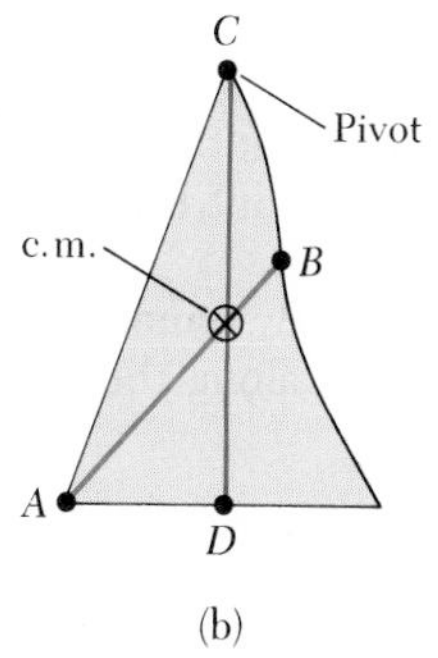

Figure 9.19 An experimental technique for determining the center of mass of an irregular planar object. The object is hung freely from two different pivots, A and C. The intersection of the two vertical lines AB and CD locates the center of mass.

Since a rigid body is a continuous distribution of mass, each portion is acted upon by the force of gravity. The net effect of all of these forces is equivalent to the effect of a single force, $M\boldsymbol{g}$, acting through a special point, called the **center of gravity.** If $\boldsymbol{g}$ is constant over the mass distribution, then the center of gravity coincides with the center of mass. If a rigid body is pivoted at its center of gravity, it will be balanced in any orientation.

EXAMPLE 9.12 The Center of Mass of Three Particles

A system consists of three particles located at the corners of a right triangle as in Figure 9.20. Find the center of mass of the system.

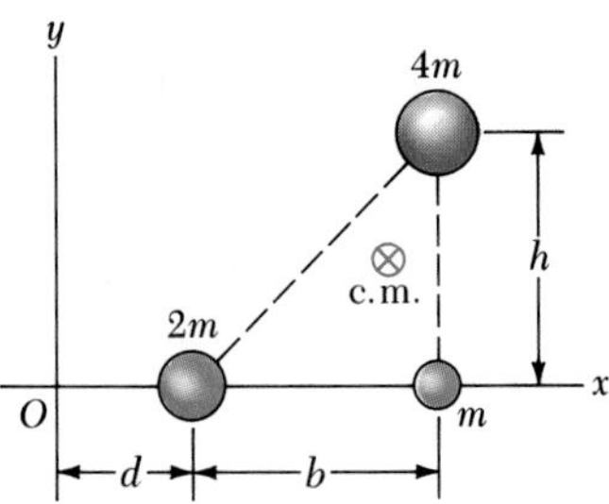

Figure 9.20 (Example 9.12) The center of mass of the three particles is located inside the triangle.

Solution Using the basic defining equations for the coordinates of the center of mass, and noting that $z_c = 0$, we get

$$x_c = \frac{\Sigma m_i x_i}{M} = \frac{2md + m(d+b) + 4m(d+b)}{7m}$$

$$= d + \frac{5}{7}b$$

$$y_c = \frac{\Sigma m_i y_i}{M} = \frac{2m(0) + m(0) + 4mh}{7m} = \frac{4}{7}h$$

Therefore, we can express the position vector to the center of mass measured from the origin as

$$\mathbf{r}_c = x_c\mathbf{i} + y_c\mathbf{j} = \left(d + \frac{5}{7}b\right)\mathbf{i} + \frac{4}{7}h\mathbf{j}$$

EXAMPLE 9.13 The Center of Mass of a Rod

(a) Show that the center of mass of a uniform rod of mass M and length L lies midway between its ends (Fig. 9.21).

Solution By symmetry, we see that $y_c = z_c = 0$ if the rod is placed along the x axis. Furthermore, if we call the mass per unit length λ (the linear mass density), then $\lambda = M/L$ for a uniform rod. If we divide the rod into elements of length dx, then the mass of each element is $dm = \lambda\, dx$. Since an arbitrary element is at a distance x from the origin, Equation 9.31 gives

$$x_c = \frac{1}{M}\int_0^L x\, dm = \frac{1}{M}\int_0^L x\lambda\, dx = \frac{\lambda}{M}\frac{x^2}{2}\bigg]_0^L = \frac{\lambda L^2}{2M}$$

Because $\lambda = M/L$, this reduces to

$$x_c = \frac{L^2}{2M}\left(\frac{M}{L}\right) = \frac{L}{2}$$

One can also argue that by symmetry, $x_c = L/2$.

(b) Suppose the rod is *nonuniform* and the mass per unit length varies linearly with x according to the expression $\lambda = \alpha x$, where α is a constant. Find the x coordinate of the center of mass as a fraction of L.

Solution In this case, we replace dm by $\lambda\, dx$, where λ is *not* constant. Therefore, x_c is given by

$$x_c = \frac{1}{M}\int_0^L x\, dm = \frac{1}{M}\int_0^L x\lambda\, dx = \frac{\alpha}{M}\int_0^L x^2\, dx = \frac{\alpha L^3}{3M}$$

We can also eliminate α by noting that the total mass of the rod is related to α through the relation

$$M = \int dm = \int_0^L \lambda\, dx = \int_0^L \alpha x\, dx = \frac{\alpha L^2}{2}$$

Substituting this into the expression for x_c gives

$$x_c = \frac{\alpha L^3}{3\alpha L^2/2} = \frac{2}{3}L$$

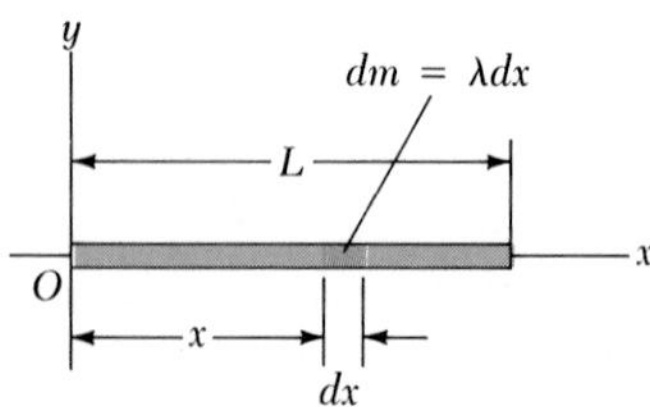

Figure 9.21 (Example 9.13) The center of mass of a uniform rod of length L is located at $x_c = L/2$.

EXAMPLE 9.14 □

An object of mass M is in the shape of a right triangle whose dimensions are shown in Figure 9.22. Locate the coordinates of the center of mass, assuming the object has a uniform mass per unit area.

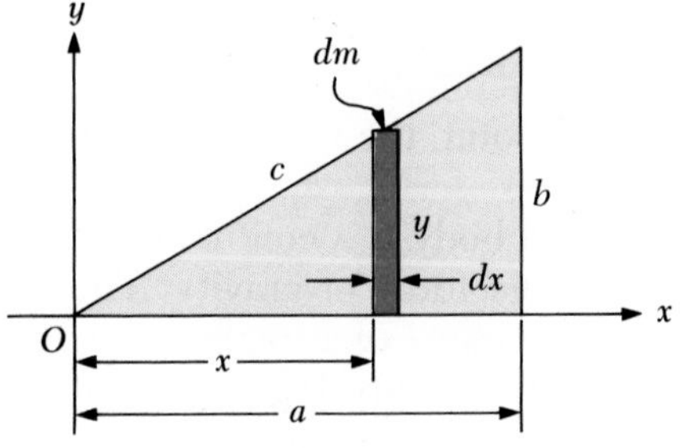

Figure 9.22 (Example 9.14).

Solution To evaluate the x coordinate of the center of mass, we divide the triangle into narrow strips of width dx and height y as in Figure 9.22. The mass of this strip dm can be expressed as

$$dm = \frac{\text{Total mass}}{\text{Total area}} \times \text{Area of strip}$$

or

$$dm = \frac{M}{\frac{1}{2}ab}\,(y\,dx) = \left(\frac{2M}{ab}\right) y\,dx$$

Therefore, the x coordinate of the center of mass is

$$x_c = \frac{1}{M}\int x\,dm = \frac{1}{M}\int_0^a x\left(\frac{2M}{ab}\right) y\,dx = \frac{2}{ab}\int_0^a xy\,dx$$

In order to evaluate this integral, we must express the variable y in terms of the variable x. From similar triangles in Figure 9.22, we see that

$$\frac{y}{x} = \frac{b}{a} \quad \text{or} \quad y = \frac{b}{a}x$$

With this substitution, x_c becomes

$$x_c = \frac{2}{ab}\int_0^a x\left(\frac{b}{a}x\right) dx = \frac{2}{a^2}\int_0^a x^2\,dx = \frac{2}{a^2}\left[\frac{x^3}{3}\right]_0^a = \frac{2}{3}a$$

By a similar calculation, you can easily show that the y coordinate of the center of mass is given by $y_c = \frac{1}{3}b$. Thus, the results are

$$x_c = \frac{2}{3}a \quad \text{and} \quad y_c = \frac{1}{3}b$$

Exercise 4 Show that the y coordinate of the center of mass is given by $y_c = b/3$.

9.7 MOTION OF A SYSTEM OF PARTICLES

We can begin to understand the physical significance and utility of the center of mass concept by taking the time derivative of the position vector of the center of mass, $\boldsymbol{r}_c$, given by Equation 9.30. Assuming that M remains constant, that is, no particles enter or leave the system, we get the following expression for the **velocity of the center of mass:**

$$\boldsymbol{v}_c = \frac{d\boldsymbol{r}_c}{dt} = \frac{1}{M}\sum m_i \frac{d\boldsymbol{r}_i}{dt} = \frac{\Sigma m_i \boldsymbol{v}_i}{M} \qquad (9.34)$$

Velocity of the center of mass

where $\boldsymbol{v}_i$ is the velocity of the ith particle. Rearranging Equation 9.34 gives

$$M\boldsymbol{v}_c = \sum m_i\boldsymbol{v}_i = \sum \boldsymbol{p}_i = \boldsymbol{P} \qquad (9.35)$$

Total momentum of a system of particles

The right side of Equation 9.35 equals the total momentum of the system. Therefore, we conclude that *the* **total momentum of the system** *equals the total mass multiplied by the velocity of the center of mass.* In other words, the total momentum of the system is equal to that of a single particle of mass M moving with a velocity $\boldsymbol{v}_c$.

If we now differentiate Equation 9.34 with respect to time, we get the **acceleration of the center of mass:**

$$\boldsymbol{a}_c = \frac{d\boldsymbol{v}_c}{dt} = \frac{1}{M}\sum m_i \frac{d\boldsymbol{v}_i}{dt} = \frac{1}{M}\sum m_i\boldsymbol{a}_i \qquad (9.36)$$

Acceleration of the center of mass

Rearranging this expression and using Newton's second law, we get

$$M\boldsymbol{a}_c = \sum m_i\boldsymbol{a}_i = \sum \boldsymbol{F}_i \qquad (9.37)$$

where $\boldsymbol{F}_i$ is the force on particle i.

The forces on any particle in the system may include both external forces (from outside the system) and internal forces (from within the system). However, by Newton's third law, the force of particle 1 on particle 2, for example, is equal to and opposite the force of particle 2 on particle 1. Thus, when we sum over all internal forces in Equation 9.37, they cancel in pairs and the net force on the system is due *only* to external forces. Thus, we can write Equation 9.37 in the form

Newton's second law for a system of particles

$$\sum \boldsymbol{F}_{\text{ext}} = M\boldsymbol{a}_c = \frac{d\boldsymbol{P}}{dt} \tag{9.38}$$

That is, the resultant external force on the system of particles equals the total mass of the system multiplied by the acceleration of the center of mass. If we compare this to Newton's second law for a single particle, we see that

> the center of mass moves like an imaginary particle of mass M under the influence of the resultant external force on the system.

In the absence of external forces, the center of mass moves with uniform velocity as in the case of the rotating wrench shown in Figure 9.23.

Finally, we see that if the resultant external force is zero, then from Equation 9.38 it follows that

$$\frac{d\boldsymbol{P}}{dt} = M\boldsymbol{a}_c = 0$$

so that

$$\boldsymbol{P} = M\boldsymbol{v}_c = \text{constant} \qquad (\text{when } \sum \boldsymbol{F}_{\text{ext}} = 0) \tag{9.39}$$

That is, the total linear momentum of a system of particles is conserved if there are no external forces acting on the system. Therefore, it follows that for an *isolated* system of particles, both the total momentum and velocity of the center of mass are constant in time. This is a generalization to a many-particle system of the law of conservation of momentum that was derived in Section 9.2 for a two-particle system.

Suppose an isolated system consisting of two or more members is at rest. The center of mass of such a system will remain at rest unless acted upon by an

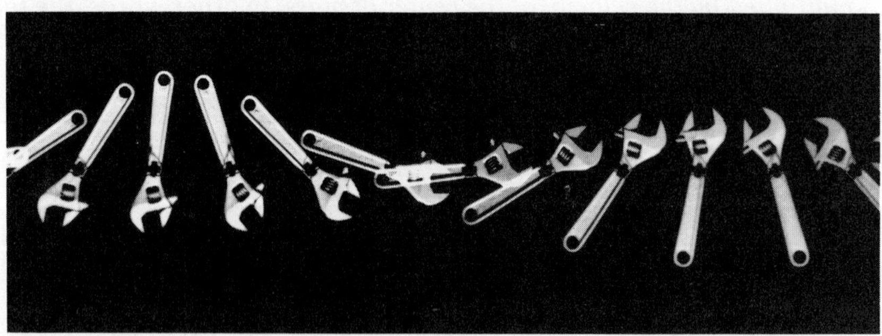

Figure 9.23 Multiflash photograph of a wrench moving on a horizontal surface. The center of mass of the wrench moves in a straight line as the wrench rotates about this point shown by the black marker. (Education Development Center, Newton, Mass.)

external force. For example, consider a system made up of a swimmer and a raft, with the system initially at rest. When the swimmer dives off the raft, the center of mass of the system will remain at rest (if we neglect the friction between raft and water). Furthermore, the momentum of the diver will be equal in magnitude to the momentum of the raft, but opposite in direction.

As another example, suppose an unstable atom initially at rest suddenly decays into two fragments of masses M_1 and M_2, with velocities $\mathbf{v}_1$ and $\mathbf{v}_2$, respectively. (An example of such a radioactive decay is that of the uranium-238 nucleus, which decays into an alpha particle—the helium nucleus—and the thorium-234 nucleus.) Since the total momentum of the system before the decay is zero, the total momentum of the system after the decay must also be zero. Therefore, we see that $M_1\mathbf{v}_1 + M_2\mathbf{v}_2 = 0$. If the velocity of one of the fragments after the decay is known, the recoil velocity of the other fragment can be calculated. Can you explain the origin of the kinetic energy of the fragments?

EXAMPLE 9.15 Exploding Projectile

A projectile is fired into the air and suddenly explodes into several fragments (Fig. 9.24). What can be said about the motion of the center of mass of the fragments after the collision?

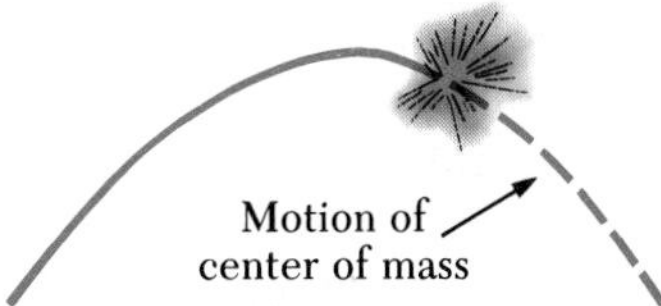

Figure 9.24 (Example 9.15) When a projectile explodes into several fragments, the center of mass of the fragments follows the same parabolic path the projectile would have taken had there been no explosion.

Solution The only external force on the projectile is the force of gravity. Thus, the projectile follows a parabolic path. If the projectile did not explode, it would continue to move along the parabolic path indicated by the broken line in Figure 9.24. Since the forces due to the explosion are internal, they do not affect the motion of the center of mass. Thus, after the explosion the center of mass of the fragments follows the *same* parabolic path the projectile would have followed if there had been no explosion.

EXAMPLE 9.16 The Exploding Rocket

A rocket is fired vertically upward. It reaches an altitude of 1000 m and a velocity of 300 m/s. At this instant, the rocket explodes into three equal fragments. One fragment continues to move upward with a speed of 450 m/s right after the explosion. The second fragment has a speed of 240 m/s moving in the easterly direction right after the explosion. (a) What is the velocity of the third fragment right after the explosion?

Solution Let us call the total mass of the rocket M; hence the mass of each fragment is $M/3$. The total momentum just before the explosion must equal the total momentum of the fragments right after the explosion since the forces of the explosion are internal to the system and cannot affect the total momentum of the system.

Before the explosion:

$$\mathbf{P}_i = M\mathbf{v}_0 = 300M\mathbf{j}$$

After the explosion:

$$\mathbf{P}_f = 240\left(\frac{M}{3}\right)\mathbf{i} + 450\left(\frac{M}{3}\right)\mathbf{j} + \frac{M}{3}\mathbf{v}$$

where $\mathbf{v}$ is the unknown velocity of the third fragment. Equating these two expressions gives

$$M\frac{\mathbf{v}}{3} + 80M\mathbf{i} + 150M\mathbf{j} = 300M\mathbf{j}$$

$$\mathbf{v} = (-240\mathbf{i} + 450\mathbf{j})\ \text{m/s}$$

(b) What is the position of the center of mass relative to the ground 3 s after the explosion? (Assume the rocket engine is nonoperative after the explosion.)

Solution The center of mass of the fragments moves as a freely falling body since the explosion doesn't affect the motion of the center of mass (Example 9.15). If $t = 0$ is the time of the explosion, then $y_0 = 1000$ m and $v_0 = 300$ m/s for the center of mass. Using an expression from kinematics, we get for the y coordinate of the center of mass

$$y_c = y_0 + v_0 t - \tfrac{1}{2}gt^2 = 1000 + 300t - 4.9t^2$$

Thus, at $t = 3$ s,

$$y_c = [1000 + 300\,(3) - 4.9(3)^2]\ \text{m} = \boxed{1.86\ \text{km}}$$

Note that the x coordinate of the center of mass doesn't change. That is, in a given time interval the second fragment moves to the right by the same distance that the third fragment moves to the left.

*9.8 ROCKET PROPULSION

When ordinary vehicles, such as automobiles, boats, and locomotives, are propelled, the driving force for the motion is one of friction. In the case of the automobile, the driving force is the force of the road on the car. A locomotive "pushes" against the tracks; hence the driving force is the force of the tracks on the locomotive. However, a rocket moving in space has no air, tracks, or water to "push" against. Therefore, the source of the propulsion of a rocket must be different. Figure 9.25 is a dramatic photograph of a spacecraft at lift-off. *The operation of a rocket depends upon the law of conservation of momentum as applied to a system of particles, where the system is the rocket plus its ejected fuel.*

The propulsion of a rocket can be understood by first considering the mechanical system consisting of a machine gun mounted on a cart on wheels. As the machine gun is fired, each bullet receives a momentum $m\boldsymbol{v}$ in some direction where $\boldsymbol{v}$ is measured with respect to a stationary earth frame. For each bullet that is fired, the gun and cart must receive a compensating momentum in the opposite direction (as in Example 9.4). That is, the reaction force of the bullet on the gun accelerates the cart and gun. If there are n bullets fired each second, then the average force on the gun is equal to $\boldsymbol{F}_{av} = nm\boldsymbol{v}$.

Figure 9.25 Lift-off of the space shuttle Columbia. Massive amounts of thrust are generated by the shuttle's liquid-fueled engines, aided by the two solid fuel boosters. Many physical principles are involved in the areas of mechanics, thermodynamics, and electricity and magnetism. (Courtesy of NASA)

In a similar manner, as a rocket moves in free space (a vacuum), *its momentum changes when some of its mass is released in the form of ejected gases* (Fig. 9.25). *Since the ejected gases acquire some momentum, the rocket receives a compensating momentum in the opposite direction.* Therefore, *the rocket is accelerated as a result of the "push," or thrust, from the exhaust gases.* In free space, the center of mass of the entire system moves uniformly, independent of the propulsion process. It is interesting to note that the rocket and machine gun represent cases of the *inverse* of an inelastic collision; that is, momentum is conserved, but the kinetic energy of the system is *increased* (at the expense of internal energy).

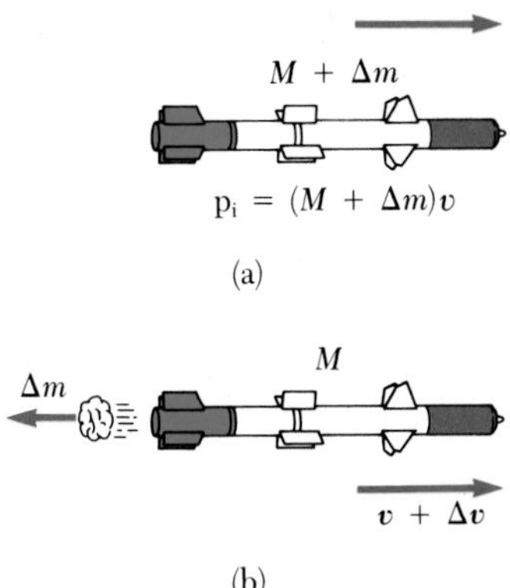

Figure 9.26 Rocket propulsion. (a) The initial mass of the rocket is $M + \Delta m$ at a time t, and its speed is v. (b) At a time $t + \Delta t$, the rocket's mass has reduced to M, and an amount of fuel Δm has been ejected. The rocket's speed increases by an amount Δv.

Suppose that at some time t, the momentum of the rocket plus the fuel is $(M + \Delta m)v$ (Fig. 9.26a). At some short time later, Δt, the rocket ejects some fuel of mass Δm and the rocket's speed therefore increases to $v + \Delta v$ (Fig. 9.26b). If the fuel is ejected with a velocity v_e *relative to the rocket*, then the velocity of the fuel relative to a stationary frame of reference is $v - v_e$. Thus, if we equate the total initial momentum of the system to the total final momentum, we get

$$(M + \Delta m)v = M(v + \Delta v) + \Delta m(v - v_e)$$

Simplifying this expression gives

$$M\,\Delta v = v_e\,\Delta m$$

We also could have arrived at this result by considering the system in the center of mass frame of reference; that is, a frame whose velocity equals the center of mass velocity. In this frame, the total momentum is zero; therefore if the rocket gains a momentum $M\,\Delta v$ by ejecting some fuel, the exhaust gases obtain a momentum $v_e\,\Delta m$ in the *opposite* direction, and so $M\,\Delta v - v_e\,\Delta m = 0$. If we now take the limit as Δt goes to zero, then $\Delta v \rightarrow dv$ and $\Delta m \rightarrow dm$. Furthermore, the increase in the exhaust mass, dm, corresponds to an equal decrease in the rocket mass, so that $dm = -dM$. Note that dM is given a negative sign because it represents a decrease in mass. Using this fact, we get

$$M\,dv = -v_e\,dM \qquad (9.40)$$

Integrating this equation, and taking the initial mass of the rocket plus fuel to be M_i and the final mass of the rocket plus its remaining fuel to be M_f, we get

$$\int_{v_i}^{v_f} dv = -v_e \int_{M_i}^{M_f} \frac{dM}{M}$$

$$v_f - v_i = v_e \ln\left(\frac{M_i}{M_f}\right) \qquad (9.41)$$

Expression for rocket propulsion

This is the basic expression of rocket propulsion. First, it tells us that the increase in velocity is proportional to the exhaust velocity, v_e. Therefore, the exhaust velocity should be very high. Second, the increase in velocity is proportional to the logarithm of the ratio M_i/M_f. Therefore, this ratio should be as large as possible, which means that the rocket should carry as much fuel as possible.

The *thrust* on the rocket is the force exerted on the rocket by the ejected exhaust gases. We can obtain an expression for the thrust from Equation 9.40:

$$\text{Thrust} = M\frac{dv}{dt} = \left|v_e \frac{dM}{dt}\right| \tag{9.42}$$

Here we see that the thrust increases as the exhaust velocity increases and as the rate of change of mass (burn rate) increases.

EXAMPLE 9.17 A Rocket in Space

A rocket moving in free space has a speed of 3×10^3 m/s. Its engines are turned on, and fuel is ejected in a direction opposite the rocket's motion at a speed of 5×10^3 m/s relative to the rocket. (a) What is the speed of the rocket once its mass is reduced to one half its mass before ignition?

Applying Equation 9.41, we get

$$v_f = v_i + v_e \ln\left(\frac{M_i}{M_f}\right)$$

$$= 3 \times 10^3 + 5 \times 10^3 \ln\left(\frac{M_i}{0.5M_i}\right)$$

$$= 6.47 \times 10^3 \text{ m/s}$$

(b) What is the thrust on the rocket if it burns fuel at the rate of 50 kg/s?

$$\text{Thrust} = \left|v_e \frac{dM}{dt}\right| = \left(5 \times 10^3 \frac{\text{m}}{\text{s}}\right)\left(50 \frac{\text{kg}}{\text{s}}\right)$$

$$= 2.50 \times 10^5 \text{ N}$$

SUMMARY

The **linear momentum** of a particle of mass m moving with a velocity $\boldsymbol{v}$ is defined to be

$$\boldsymbol{p} \equiv m\boldsymbol{v} \tag{9.1}$$

Impulse

The **impulse** of a force $\boldsymbol{F}$ on a particle is equal to the change in the momentum of the particle and is given by

$$\boldsymbol{I} = \Delta\boldsymbol{p} = \int_{t_i}^{t_f} \boldsymbol{F}\, dt \tag{9.6}$$

Impulsive forces are forces that are very strong compared with other forces on the system, and usually act for a very short time, as in the case of collisions.

Conservation of momentum

The **law of conservation of momentum** for two interacting particles states that if two particles form an isolated system, their total momentum is conserved regardless of the nature of the force between them. Therefore, the total momentum of the system at all times equals its initial total momentum, or

$$\boldsymbol{p}_{1i} + \boldsymbol{p}_{2i} = \boldsymbol{p}_{1f} + \boldsymbol{p}_{2f} \tag{9.12}$$

Elastic and inelastic collision

When two particles collide, the total momentum of the system before the collision always equals the total momentum after the collision, regardless of the nature of the collision. An **inelastic collision** is a collision for which the mechanical energy is not conserved, but momentum is conserved. A perfectly inelastic collision corresponds to the situation where the colliding bodies stick together after the collision. An **elastic collision** is one in which both momentum and kinetic energy are conserved.

In a two- or three-dimensional collision, the components of momentum in each of the three directions (x, y, and z) are conserved independently.

The **vector position of the center of mass of a system of particles** is defined as

Center of mass for a system of particles

$$\boldsymbol{r}_c \equiv \frac{\Sigma m_i \boldsymbol{r}_i}{M} \quad (9.30)$$

where $M = \Sigma m_i$ is the total mass of the system and $\boldsymbol{r}_i$ is the vector position of the ith particle.

The *vector position of the center of mass of a rigid body* can be obtained from the integral expression

Center of mass for a rigid body

$$\boldsymbol{r}_c = \frac{1}{M}\int \boldsymbol{r}\, dm \quad (9.33)$$

The **velocity of the center of mass for a system of particles** is given by

Velocity of the center of mass

$$\boldsymbol{v}_c = \frac{\Sigma m_i \boldsymbol{v}_i}{M} \quad (9.34)$$

The total momentum of a system of particles equals the total mass multiplied by the velocity of the center of mass, that is, $\boldsymbol{P} = M\boldsymbol{v}_c$.

Newton's second law applied to a system of particles is given by

Newton's second law for a system of particles

$$\sum \boldsymbol{F}_{ext} = M\boldsymbol{a}_c = \frac{d\boldsymbol{P}}{dt} \quad (9.38)$$

where $\boldsymbol{a}_c$ is the acceleration of the center of mass and the sum is over all external forces. Therefore, the center of mass moves like an imaginary particle of mass M under the influence of the resultant external force on the system. It follows from Equation 9.38 that the total momentum of the system is conserved if there are no external forces acting on it.

QUESTIONS

1. If the kinetic energy of a particle is zero, what is its linear momentum? If the total energy of a particle is zero, is its linear momentum necessarily zero? Explain.
2. If the velocity of a particle is doubled, by what factor is its momentum changed? What happens to its kinetic energy?
3. If two particles have equal kinetic energies, are their momenta necessarily equal? Explain.
4. Does a large force always produce a larger impulse on a body than a smaller force? Explain.
5. An isolated system is initially at rest. Is it possible for parts of the system to be in motion at some later time? If so, explain how this might occur.
6. If two objects collide and one is initially at rest, is it possible for both to be at rest after the collision? Is it possible for one to be at rest after the collision? Explain.
7. Explain why momentum is conserved when a ball bounces from a floor.
8. Is it possible to have a collision in which all of the kinetic energy is lost? If so, cite an example.
9. In a perfectly elastic collision between two particles, does the kinetic energy of each particle change as a result of the collision?
10. When a ball rolls down an incline, its momentum increases. Does this imply that momentum is not conserved? Explain.
11. Consider a perfectly inelastic collision between a car and a large truck. Which vehicle loses more kinetic energy as a result of the collision?
12. Can the center of mass of a body lie outside the body? If so, give examples.
13. A boy stands at one end of a floating raft that is stationary relative to the shore. He then walks to the opposite end of the raft, away from the shore. What hap-

pens to the center of mass of the system (boy + raft)? Does the raft move? Explain.

14. Three balls are thrown into the air simultaneously. What is the acceleration of their center of mass while they are in motion?
15. A meter stick is balanced in a horizontal position with the index fingers of the right and left hands. If the two fingers are brought together, the stick remains balanced and the two fingers always meet at the 50-cm mark regardless of their original positions (try it!). Carefully explain this observation.
16. A researcher tranquilizes a polar bear on a glacier. How might the researcher, knowing her own weight, be able to *estimate* the weight of the polar bear using a measuring tape and a rope?
17. A sharpshooter fires a rifle while standing with the butt of the gun against his shoulder. If the forward momentum of a bullet is the same as the backward momentum of the gun, why isn't it as dangerous to be hit by the gun as by the bullet?
18. A piece of mud is thrown against a brick wall and sticks to the wall. What happens to the momentum of the mud? Is momentum conserved? Explain.
19. Early in this century, Robert Goddard proposed sending a rocket to the moon. Critics took the position that in a vacuum, such as exists between the earth and the moon, the gases emitted by the rocket would have nothing to push against to propel the rocket. According to *Scientific American* (January 1975), Goddard placed a gun in a vacuum and fired a blank cartridge from it. (A blank cartridge fires only the wadding and hot gases of the burning gunpowder.) What happened when the gun was fired?
20. A pole vaulter falls from a height of 15 ft onto a foam rubber pad. Could you calculate his velocity just before he reaches the pad? Would you be able to calculate the force exerted on him due to the collision? Explain.
21. As a ball falls toward the earth, its momentum increases. How would you reconcile this fact with the law of conservation of momentum?
22. A man is at rest sitting at one end of a boat in the middle of a lake. If he walks to the opposite end of the boat toward the east, why does the boat move west? What can you say about the center of mass of the boat-man system?
23. Explain how you would use a balloon to demonstrate the mechanism responsible for rocket propulsion.
24. Explain the maneuver of decelerating a spacecraft. What other maneuvers are possible?
25. Does the center of mass of a rocket in free space accelerate? Explain. Can the speed of a rocket exceed the exhaust velocity of the fuel? Explain.
26. A ball is dropped from a tall building in New York. Identify the system whose linear momentum is conserved.
27. A bomb, initially at rest, explodes into several pieces. In this process (a) is the linear momentum conserved? (b) Is the kinetic energy conserved? Explain.
28. Is kinetic energy always lost in an inelastic collision? Explain.
29. A skater is standing still on a frictionless ice rink. Her friend throws a Frisbee straight at her. In which of the following cases is the largest momentum transferred to the skater? (a) The skater catches the Frisbee and holds it, (b) catches it momentarily but drops it vertically down, (c) catches the Frisbee momentarily and then throws it back to her friend.
30. The moon revolves around the earth as viewed by us on the earth. Is the moon's linear momentum conserved? Is its kinetic energy conserved? Assume, for simplicity, that the moon's orbit is perfectly circular.
31. A large sheet is held at its edges by two students in such a way that it forms a nearly vertical "net" to catch an object. A third student, who happens to be the star pitcher on the baseball team, is asked to throw a raw egg into the sheet. Explain why the egg does not break in the process, regardless of the initial speed of the egg. (If you try this one, make sure the pitcher hits the target near its center, and do not allow the the egg to fall on the floor after being caught.)
32. If a raw egg is dropped, it falls apart upon impact with the floor. However, if you drop a raw egg onto a thick foam rubber cushion from a height of about 1 m, the egg will rebound without breaking. Why is this possible? (In this demonstration, be sure to catch the egg after the first bounce.)

PROBLEMS

Section 9.1 Linear Momentum and Impulse

1. A 3-kg particle has a velocity of $(3\mathbf{i} - 4\mathbf{j})$ m/s. Find its x and y components of momentum and the magnitude of its total momentum.
2. The momentum of a 1250-kg car is equal to the momentum of a 5000-kg truck traveling at a speed of 10 m/s. What is the speed of the car?
3. A 1500-kg automobile travels eastward at a speed of 8 m/s. It makes a 90° turn to the north in a time of 3 s and continues with the same speed. Find (a) the impulse delivered to the car as a result of the turn and (b) the average force exerted on the car during the turn.
4. A 0.3-kg ball moving along a straight line has a velocity of $5\mathbf{i}$ m/s. It collides with a wall and rebounds with

a velocity of $-4\mathbf{i}$ m/s. Find (a) the change in its momentum and (b) the average force exerted on the wall if the ball is in contact with the wall for 5×10^{-3} s.

5. The force $\boldsymbol{F}_x$ acting on a 2-kg particle varies in time as shown in Figure 9.27. Find (a) the impulse of the force, (b) the final velocity of the particle if it is initially at rest, and (c) the final velocity of the particle if it is initially moving along the x axis with a velocity of -2 m/s.

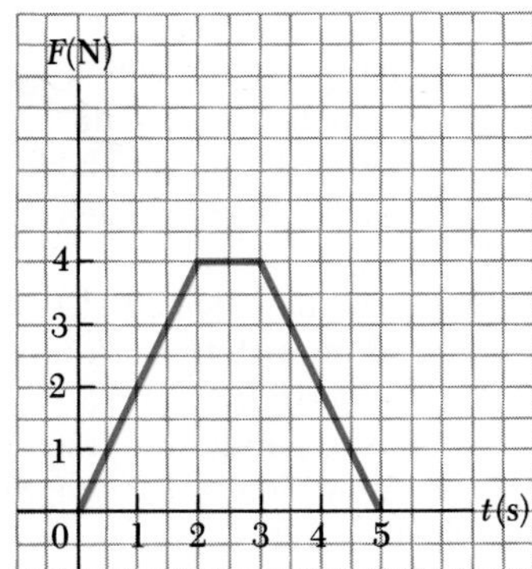

Figure 9.27 (Problems 5 and 6).

6. Find the average force exerted on the particle described in Figure 9.27 for the time interval $t_i = 0$ to $t_f = 5$ s.
7. An estimated force-time curve for a baseball struck by a bat is shown in Figure 9.28. From this curve, determine (a) the impulse delivered to the ball, (b) the average force exerted on the ball, and (c) the peak force exerted on the ball.
8. A garden hose is held in the manner shown in Figure 9.29. What force is necessary to hold the nozzle stationary if the discharge rate is 0.60 kg/s with a velocity of 25 m/s?
9. Calculate the magnitude of the linear momentum for the following cases: (a) a proton of mass 1.67×10^{-27} kg moving with a speed of 5×10^6 m/s; (b) a 15-g bullet moving with a speed of 500 m/s; (c) a 75-kg sprinter running at a speed of 12 m/s, and (d) the earth (mass 5.98×10^{24} kg) moving with an orbital speed of 2.98×10^4 m/s.
10. If the momentum of an object is doubled in magnitude, what happens to its kinetic energy? (b) If the kinetic energy of an object is tripled, what happens to its momentum?
11. A 0.5-kg football is thrown with a speed of 15 m/s. A stationary receiver catches the ball and brings it to rest in 0.02 s. (a) What is the impulse delivered to the ball? (b) What is the average force exerted on the receiver?
12. A single constant force of 60 N accelerates a 5-kg object from a speed of 2 m/s to a speed of 8 m/s. Find (a) the impulse acting on the object in this interval and (b) the time interval over which this impulse is delivered.
13. A 0.15 kg baseball is thrown with a speed of 40 m/s. It is hit straight back at the pitcher with a speed of 50 m/s. (a) What is the impulse delivered to the baseball? (b) Find the average force exerted by the bat on the ball if the ball is in contact with the bat for 2×10^{-3} s. Compare this with the weight of the ball and determine whether or not the impulse approximation is valid in the situation.
14. A tennis player receives a shot with the ball (0.06 kg) traveling horizontally at 50 m/s, and returns the shot with the ball traveling horizontally as 40 m/s in the opposite direction. What is the impulse delivered to the ball by the racket?
15. A 3.0-kg steel ball strikes a massive wall with a speed of 10 m/s at an angle of 60° with the surface. It bounces off with the same speed and angle (see Fig. 9.30). If the ball is in contact with the wall for 0.20 s, what is the average force exerted on the ball by the wall?
16. Water falls at a rate of 0.25 liter/s from a height of 60 m without splashing into a 750-g bucket on a scale. If the bucket is originally empty, what does the scale read after 3 s?

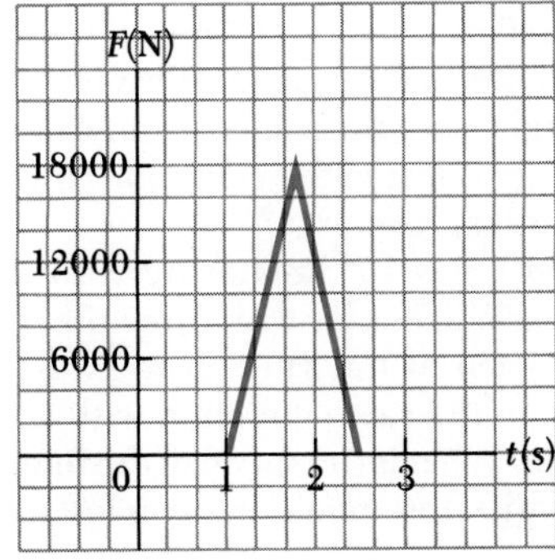

Figure 9.28 (Problem 7).

Figure 9.29 (Problem 8).

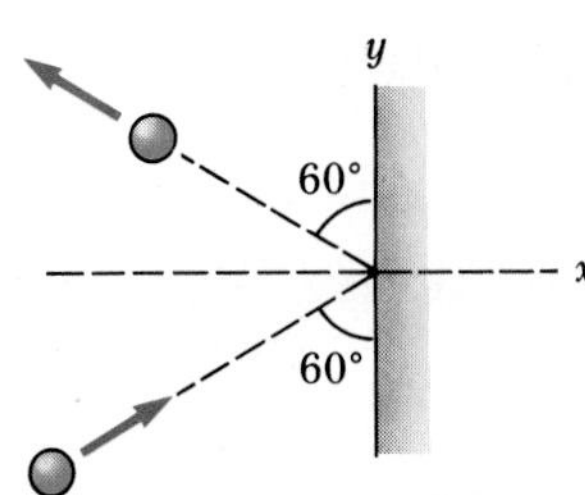

Figure 9.30 (Problem 15).

Section 9.2 Conservation of Linear Momentum for a Two-Particle System

17. A 40-kg child standing on a frozen pond throws a 2-kg stone to the east with a speed of 8 m/s. Neglecting friction between the child and ice, find the recoil velocity of the child.

18. Two blocks of masses M and $3M$ are placed on a horizontal, frictionless surface. A light spring is attached to one of them, and the blocks are pushed together with the spring between them (Fig. 9.31). A string holding them together is burned, after which the block of mass $3M$ moves to the right with a speed of 2 m/s. What is the speed of the block of mass M? (Assume they are initially at rest.)

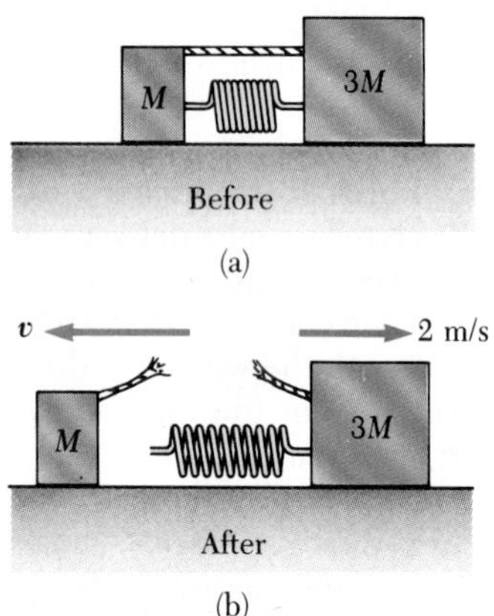

Figure 9.31 (Problem 18).

19. A 60-kg boy and a 40-kg girl, both wearing skates, face each other at rest. The boy pushes the girl, sending her eastward with a speed of 4 m/s. Describe the subsequent motion of the boy. (Neglect friction.)

20. Two particles with masses $m_1 = 2.0$ kg and $m_2 =$ 5.0 kg are free to slide on a straight horizontal frictionless guide wire (Fig. 9.32). The particle with the smaller mass is moving with a speed of 17.0 m/s and overtakes the larger particle, which is moving in the same direction with a speed of 3.0 m/s. The larger particle has an ideal inertia-less spring ($k =$ 4480 N/m) attached to the side being approached by the smaller particle, as shown in the diagram. (a) What is the maximum compression of the spring as the two particles collide? (b) What are the final velocities of the particles?

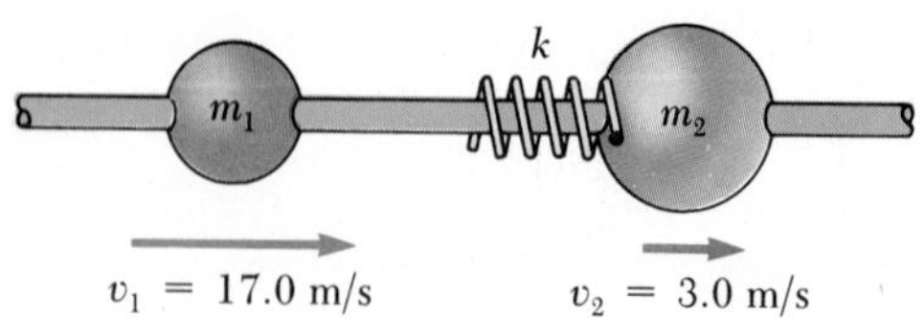

Figure 9.32 (Problem 20).

21. A 45-kg girl is standing on a plank that has a mass of 150 kg. The plank, originally at rest, is free to slide on a frozen lake, which is a flat, frictionless supporting surface. The girl begins to walk along the plank at a constant velocity of 1.5 m/s *relative to the plank*. (a) What is her velocity relative to the ice surface? (b) What is the velocity of the plank relative to the ice surface?

Section 9.3 Collisions and Section 9.4 Examples of Collisions in One Dimension

22. A 2.5-kg mass moving initially with a speed of 10 m/s makes a perfectly inelastic head-on collision with a 5-kg mass initially at rest. (a) Find the final velocity of the composite particle. (b) How much energy is lost in the collision?

23. A 2000-kg meteorite has a speed of 120 m/s just before colliding head-on with the earth. Determine the recoil speed of the earth (mass 5.98×10^{24} kg).

24. Consider the ballistic pendulum described in Example 9.6 and shown in Figure 9.10. (a) Show that the ratio of the kinetic energy after the collision to the kinetic energy before the collision is given by the ratio $m_1/(m_1 + m_2)$, where m_1 is the mass of the bullet and m_2 is the mass of the block. (b) If $m_1 = 8$ g and $m_2 = 2$ kg, what percentage of the original energy is left after the inelastic collision? What accounts for the missing energy?

25. A 10-g bullet is fired into a 2.5-kg ballistic pendulum and becomes embedded in it. If the pendulum rises a vertical distance of 8 cm, calculate the initial speed of the bullet.

26. A 90-kg halfback running north with a speed of 10 m/s is tackled by a 120-kg opponent running south with a speed of 4 m/s. If the collision is assumed to be perfectly inelastic and head-on, (a) calculate the velocity of the players just after the tackle and (b) determine the energy lost as a result of the collision. Can you account for the missing energy?

27. A 1200-kg car traveling initially with a speed of 25 m/s in an easterly direction crashes into the rear end of a 9000-kg truck moving in the same direction at 20 m/s (Fig. 9.33). The velocity of the car right after the collision is 18 m/s to the east. (a) What is the velocity of the truck right after the collision? (b) How much mechanical energy is lost in the collision? How do you account for this loss in energy?

28. A railroad car of mass 2.5×10^4 kg moving with a speed of 4 m/s collides and couples with three other coupled railroad cars each of the same mass as the single car and moving in the same direction with an initial speed of 2 m/s. (a) What is the speed of the four cars after the collision? (b) How much energy is lost in the collision?

Figure 9.33 (Problem 27).

29. A neutron in a reactor makes an elastic head-on collision with the nucleus of a carbon atom initially at rest. (a) What fraction of the neutron's kinetic energy is transferred to the carbon nucleus? (b) If the initial kinetic energy of the neutron is 1 MeV $= 1.6 \times 10^{-13}$ J, find its final kinetic energy and the kinetic energy of the carbon nucleus after the collision. (The mass of the carbon nucleus is about 12 times the mass of the neutron.)
30. A neutron moving with a velocity of $3 \times 10^6 \mathbf{i}$ m/s makes a head-on elastic collision with a stationary helium nucleus (the mass of He is 4 u). Find (a) the final velocity of each particle and (b) the fraction of the initial kinetic energy transferred to the helium nucleus.
31. A 10-g particle moving to the right with a speed of 30 cm/s makes an elastic head-on collision with a 20-g particle initially at rest. Find (a) the final velocity of each particle and (b) the fraction of the total energy transferred to the 20-g particle.
32. Two billiard balls have velocities of 2.0 m/s and -0.5 m/s before they meet in an elastic head-on collision. What are their final velocities?
33. A 2-kg mass moving with an initial speed of 7 m/s collides with and sticks to a 6-kg mass initially at rest. The combined mass then proceeds to collide with and stick to a 2-kg mass also at rest initially. If the collisions are all head-on, find (a) the final speed of the system and (b) the amount of kinetic energy lost.
34. A 0.2-kg ball fastened to the end of a string 1.5 m in length to form a pendulum is released in the horizontal position. At the bottom of its swing, the ball collides with a 0.3-kg block initially resting on a frictionless surface (Fig. 9.34). (a) If the collision is elastic, calculate the speed of the ball and of the block just after the collision. (b) If the collision is completely inelastic (they stick), determine the height that the center of mass rises after the collision.
35. A 12-g bullet is fired into a 100-g wooden block initially at rest on a horizontal surface. After impact, the block slides 7.5 m before coming to rest. If the coefficient of friction between the block and the surface is 0.65, what was the speed of the bullet immediately before impact?

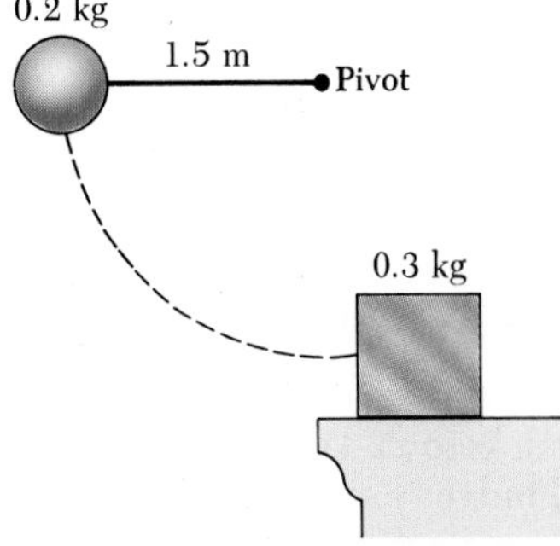

Figure 9.34 (Problem 34).

36. How may head-on collisions will the neutron in Prob. 9.29 need to make to reduce its energy to 1 eV? How will the number of head-on collisions change if carbon is replaced by an element whose nucleus is one third as massive as carbon?
37. Consider a frictionless track *ABC* as shown in Figure 9.35. A block of mass $m_1 = 5$ kg is released from *A*. It makes a head-on elastic collision with a block of mass $m_2 = 10$ kg at *B*, initially at rest. Calculate the maximum height to which m_1 will rise after the collision.

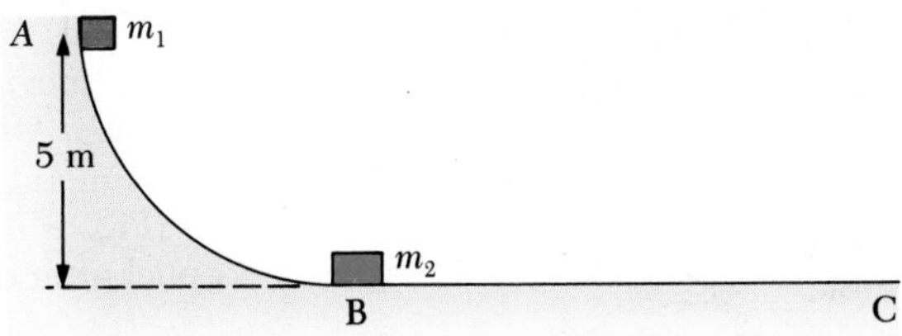

Figure 9.35 (Problem 37).

38. A block of mass $m_1 = 2$ kg starts from rest on an incline at 53° to the horizontal. The coefficient of kinetic friction between the incline and the block is $\mu_k = 0.25$. (a) If the speed of the block at the foot of the incline is 8 m/s to the right, determine the height from which the block was released. (b) Another block of mass $m_2 = 6$ kg is at rest on the smooth horizontal surface. Block m_1 collides with block m_2. After collision, the blocks stick together and move to the right. Determine the speed of the blocks after collision.

39. A bullet of mass $m = 10$ g is fired horizontally with speed v into the block of a ballastic pendulum of mass $M = 1$ kg. The block is attached by a very light string of length 2 m to a rigid support. The block is free to move in a vertical circle. The bullet is stopped and gets embedded in the block. Determine the minimum value of v such that the pendulum will swing through a complete circle.

40. Two blocks of mass $m_1 = 2$ kg and $m_2 = 4$ kg are released from a height of 5 m on a frictionless track as shown in Figure 9.36. The blocks undergo an elastic head-on collision. (a) Determine the velocities of the two blocks just before collision. (b) Using Equations 9.20 and 9.21 determine the velocities of the two blocks immediately after collision. (c) Determine the maximum heights to which m_1 and m_2 will rise after the first collision.

Section 9.5 Two-Dimensional Collisions

41. A 3-kg mass with an initial velocity of $5\mathbf{i}$ m/s collides with and sticks to a 2-kg mass with an initial velocity of $-3\mathbf{j}$ m/s. Find the final velocity of the composite mass.

42. A proton moving with a velocity $v_0\mathbf{i}$ collides elastically with another proton initially at rest. If both protons have the same speed after the collision, find (a) the speed of each proton after the collision in terms of v_0 and (b) the direction of the velocity vectors after the collision.

43. An unstable nucleus of mass 17×10^{-27} kg initially at rest disintegrates into three particles. One of the particles, of mass 5.0×10^{-27} kg, moves along the y axis with a velocity of 6×10^6 m/s. Another particle, of mass 8.4×10^{-27} kg, moves along the x axis with a velocity of 4×10^6 m/s. Find (a) the velocity of the third particle and (b) the total energy given off in the process.

44. A 0.3-kg puck, initially at rest on a horizontal, frictionless surface, is struck by a 0.2-kg puck moving initially along the x axis with a velocity of 2 m/s. After the collision, the 0.2-kg puck has a speed of 1 m/s at an angle of $\theta = 53°$ to the positive x axis (Fig. 9.13). (a) Determine the velocity of the 0.3-kg puck after the collision. (b) Find the fraction of kinetic energy lost in the collision.

45. Two shuffleboard disks of equal mass, one orange and the other yellow, are involved in a perfectly elastic glancing collision. The yellow disk is initially at rest and is struck by the orange disk moving with a speed of 5 m/s. After the collision, the orange disk moves along a direction that makes an angle of 37° with its initial direction of motion and the velocity of the yellow disk is perpendicular to that of the orange disk (after the collision). Determine the final speed of each disk.

46. A bomb initially at rest explodes into two parts, one with three times the mass of the other. If the lighter mass moves with a velocity of $(2\mathbf{i} + 6\mathbf{j})$ m/s, what is the velocity of the second fraction immediately after collision?

47. In the previous problem there are obviously forces present that make the two parts of the bomb fly apart. How do you justify using conservation of linear momentum to solve Problem 46?

48. A 200-g cart moves on a horizontal, frictionless surface with a constant speed of 25 cm/s. A 50-g piece of modeling clay is dropped vertically onto the cart. (a) If the clay sticks to the cart, find the final speed of the system. (b) After the collision, the clay has no momentum in the vertical direction. Does this mean that the law of conservation of momentum is violated?

49. A particle of mass m, moving with speed v, collides obliquely with an identical particle initially at rest. Show that as a consequence of the conservation of kinetic energy and linear momentum, the two particles move at 90° from each other after collision. (*Hint:* $(\mathbf{A} + \mathbf{B})^2 = A^2 + B^2 + 2AB \cos \theta$.)

50. A particle of mass m_1 and initial velocity $\mathbf{v}_1$ collides with a stationary mass m_2. After the collision, m_1 and m_2 are deflected as shown in Figure 9.37. The velocity of m_1 after collision is $\mathbf{v}'_1$. Show that

$$\tan \theta_2 = \frac{v'_1 \sin \theta_1}{v_1 - v_1 \cos \theta_2}$$

From the information given and the result you derived, can you make the assumption that the collision is perfectly elastic?

Figure 9.36 (Problem 40.)

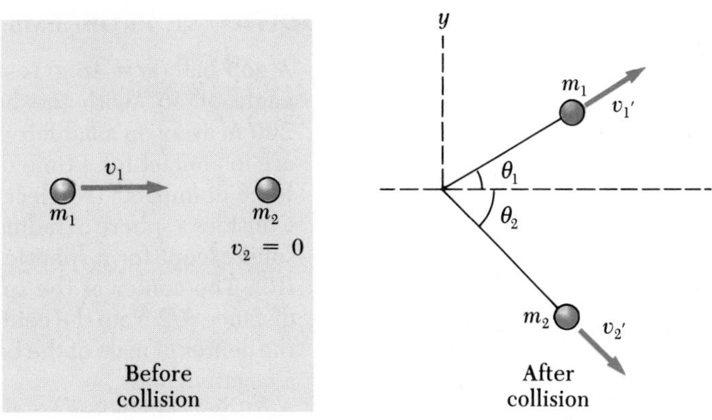

Figure 9.37 (Problem 50).

Section 9.6 The Center of Mass

51. A 3-kg particle is located on the x axis at $x = -5$ m, and a 4-kg particle is on the x axis at $x = 3$ m. Find the center of mass.

52. Three masses located in the xy plane have the following coordinates: a 3-kg mass has coordinates given by $(3, -2)$ m; a 4-kg mass has coordinates $(-2, 4)$ m; a 1-kg mass has coordinates $(2, 2)$ m. Find the coordinates of the center of mass.

53. The mass of the Sun is 329 390 Earth masses and the mean distance from the center of the Sun to the center of the Earth is 1.496×10^8 km. Treating the Earth and Sun as particles, with each mass concentrated at the respective geometric center, how far from the center of the Sun is the C.M. of the Earth-Sun system? Compare this distance with the mean radius of the Sun (6.960×10^5 km).

54. The separation between the hydrogen and chlorine atoms of the HCl molecule is about 1.30×10^{-10} m. Determine the location of the center of mass of the molecule as measured from the hydrogen atom. (Chlorine is 35 times more massive than hydrogen.)

55. A shell is fired from a cannon at level ground to hit an enemy bunker situated R away from the cannon. At the highest point of the trajectory the shell explodes into two equal parts. One part of the shell is found to hit the ground at $R/2$. At what distance from the cannon would the second fragment fall?

56. Compute the location of the center of mass of a piece of plywood cut into the shape of an isosceles *right* triangle if the two equal sides are 1.3 m long. (*Hint:* Use the integral form of the center of mass equation.)

57. A uniform piece of sheet steel is shaped as shown in Figure 9.38. Compute the x and y components of the center of mass of the piece.

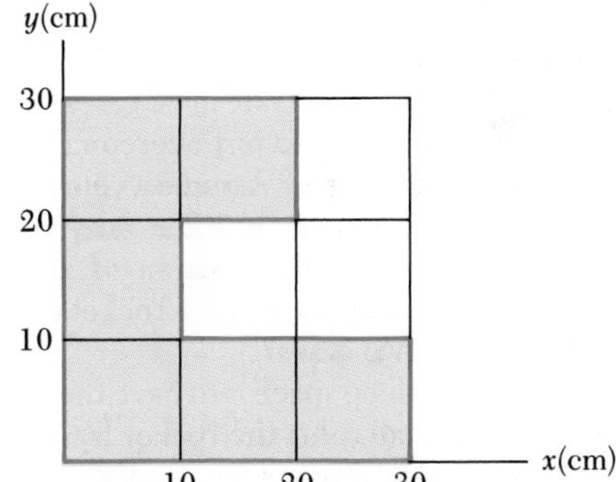

Figure 9.38 (Problem 57.)

Section 9.7 Motion of a System of Particles

58. A 5-kg particle moves along the x axis with a velocity of 3 m/s. A 2-kg particle moves along the x axis with a velocity of -2.5 m/s. Find (a) the velocity of the center of mass and (b) the total momentum of the system.

59. A 2-kg particle has a velocity of $(2\mathbf{i} - 3\mathbf{j})$ m/s, and a 3-kg particle has a velocity of $(\mathbf{i} + 6\mathbf{j})$ m/s. Find (a) the velocity of the center of mass and (b) the total momentum of the system.

60. A 2-kg particle has a velocity of $\mathbf{v}_1 = (2\mathbf{i} - 10t\mathbf{j})$ m/s, where t is in s. A 3-kg particle moves with a constant velocity of $\mathbf{v}_2 = 4\mathbf{i}$ m/s. At $t = 0.5$ s, find (a) the velocity of the center of mass, (b) the acceleration of the center of mass, and (c) the total momentum of the system.

61. (a) A 3.0-g particle is moving toward a 7.0-g particle with a speed of 3.0 m/s. With what speed do each approach the center of mass? (b) What is the momentum of each particle, relative to the center of mass?

62. Romeo entertains Juliet by playing his guitar from the rear of their boat in still water. After the serenade, Juliet carefully moves to the rear of the boat (away from shore) to plant a kiss on Romeo's cheek. If the 80-kg boat is facing shore and the 55-kg Juliet moves 2.7 m toward the 77-kg Romeo, how far does the boat move toward shore?

Section 9.8 Rocket Propulsion

63. A rocket engine consumes 80 kg of fuel per second. If the exhaust velocity is 2.5×10^3 m/s, calculate the thrust on the rocket.
64. The first stage of a Saturn V space vehicle consumes fuel at the rate of 1.5×10^4 kg/s, with an exhaust velocity of 2.6×10^3 m/s. (These are approximate figures.) (a) Calculate the thrust produced by these engines. (b) Find the initial acceleration of the vehicle on the launch pad if its initial mass is 3×10^6 kg. [You must include the force of gravity to solve (b).]
65. A rocket with a total mass of 160 000 kg sits vertically on a launch pad. (a) At what rate must gases be ejected on ignition if the thrust is to just overcome the gravitational force on the rocket? Assume a velocity for the ejected gases of 3000 m/s. (b) What rate is required on ignition if an initial acceleration of $2g$ is to be achieved? (c) What thrust does the rocket engine deliver in the above two cases?
66. A rocket for use in deep space is to have the capability of boosting a payload (plus the rocket frame and engine) of 3.0 metric tons to an achieved speed of 10 000 m/s with an engine and fuel designed to produce an exhaust velocity of 2000 m/s. (a) How much fuel and oxidizer is required? (b) If a different fuel and engine design could give an exhaust velocity of 5000 m/s, what amount of fuel and oxidizer would be required for the same task? Comment.
67. Fuel aboard a rocket has a density of 1.4×10^3 kg/m^3 and is ejected with a speed of 3.0×10^3 m/s. If the engine is to provide a thrust of 2.5×10^6 N, what volume of fuel must be burned per second?
68. A rocket with an initial total mass M_i is launched vertically from the Earth's surface. When the launch fuel has been completely burned, the rocket has reached an altitude small compared to the Earth's radius (so that the acceleration due to gravity may be considered constant during the burn). Show that the final velocity is

$$v = -v_e \ln (M_i/M_f) - gt$$

where the time of burn t_1 is

$$t_1 = (M_i - M_f)(dm/dt)^{-1}$$

(M_f is the final total mass of the rocket, v_e is the exhaust gas velocity, and dm/dt is the constant rate of fuel consumption.)

ADDITIONAL PROBLEMS

69. A golf ball ($m = 46$ g) is struck a blow that makes an angle of 45° with the horizontal. The drive lands 200 m away on a flat fairway. If the golf club and ball are in contact for a time of 7 ms, what is the average force of impact? (Neglect air resistance effects.)
70. Consider a sphere of radius R and mass density ρ that is solid except for a spherical hollow volume of radius $R/2$. The center of the spherical void is located at a distance $R/2$ from the center of the large sphere. Find the center of mass of the body. (*Hint:* Treat the void as a negative mass.)
71. A 30-06 caliber hunting rifle fires a bullet of mass 0.012 kg with a muzzle velocity of 600 m/s to the right. The rifle has a mass of 4.0 kg. (a) What is the recoil velocity of the rifle as the bullet leaves the rifle? (b) If the rifle is stopped by the hunter's shoulder in a distance of 2.5 cm, what is the average force exerted on the shoulder by the rifle? (c) If the hunter's shoulder is partially restricted from recoiling, would the force exerted on the shoulder be the same as in part (b)? Explain.
72. An 8-g bullet is fired into a 2.5-kg block initially at rest at the edge of a frictionless table of height 1 m (Fig. 9.39). The bullet remains in the block, and after impact the block lands 2 m from the bottom of the table. Determine the initial speed of the bullet.

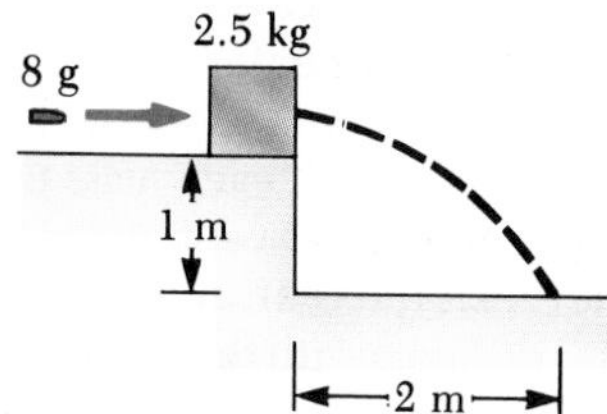

Figure 9.39 (Problem 72).

73. An attack helicopter is equipped with a 20-mm cannon that fires 130-g shells in the forward direction with a muzzle velocity of 800 m/s. The fully loaded helicopter has a mass of 4000 kg. A burst of 160 shells is fired in a 4-s interval. What is the resulting average force on the helicopter and by what amount is its forward speed reduced?
74. A loaded howitzer is rigidly bolted to a railroad flatcar. The mass of the combination is 20 000 kg. The flatcar is observed to be coasting along a level, straight (frictionless) track with a constant speed V. The howitzer is set to an elevation angle of 60° with the horizontal and then fired. After the howitzer is fired, the flatcar and attached howitzer are seen to recoil, moving in the opposite direction along the track with the same speed V. (a) If the muzzle velocity of the howitzer shell is 300 m/s, and if the shell has a mass of

50 kg, find V. (b) How high would this shell rise in the air?

75. A jet aircraft is traveling at 500 mi/h (223 m/s) in horizontal flight. The engine takes in air at a rate of 80 kg/s and burns fuel at a rate of 3 kg/s. If the exhaust gases are ejected at 600 m/s relative to the aircraft, find the thrust of the jet engine and the delivered horsepower.

76. A spacecraft is stationary in deep space when its rocket engine is ignited for a 100-s "burn." Hot gases are ejected at a constant rate of 150 kg/s with a velocity relative to the spacecraft of 3000 m/s. The initial mass of the spacecraft (plus fuel and oxidizer) is 25 000 kg. Determine the thrust of the rocket engine and the initial acceleration in units of g. What is the final velocity of the spacecraft?

77. A 75-kg firefighter slides down a pole while a constant frictional force of 300 N retards his motion. A horizontal 20-kg platform is supported by a spring at the bottom of the pole to cushion the fall. The firefighter starts from rest 4 m above the platform, and the spring constant is 4000 N/m. Find (a) the firefighter's speed just before he collides with the platform and (b) the maximum distance the spring will be compressed. (Assume the frictional force acts during the entire motion.)

78. A 70-kg baseball player jumps vertically up to catch a baseball of mass 0.160 kg traveling horizontally with a speed of 35 m/s. If the vertical speed of the player at the instant of catching the ball is 0.2 m/s, determine the magnitude and direction of the player's velocity just after the catch.

79. An 80-kg astronaut is working on the engines of his ship, which is drifting through space with a constant velocity. The astronaut, wishing to get a better view of the universe, pushes against the ship and later finds himself 30 m behind the ship. Without a thruster, the only way to return to the ship is to throw his 0.5-kg wrench directly away from the ship. If he throws the wrench with a speed of 20 m/s, how long does it take the astronaut to reach the ship?

80. A cannon is rigidly attached to a carriage, which can move along horizontal rails but is connected to a post by a large spring of force constant $k = 2 \times 10^4$ N/m, as shown in Figure 9.40. The cannon fires a 200-kg projectile at a velocity of 125 m/s, directed 45° above the horizontal. (a) If the mass of the cannon and its carriage is 5000 kg, find the recoil velocity of the cannon. (b) Determine the maximum extension of the spring. (c) Consider the "system" to consist of the cannon, carriage, and shell. Is the momentum of this system conserved during the firing? Why or why not?

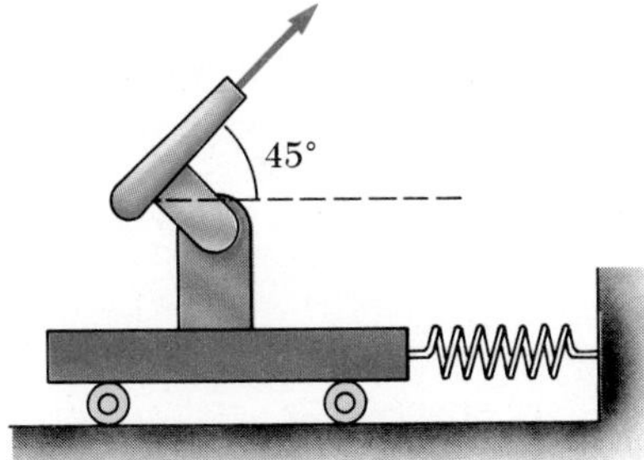

Figure 9.40 (Problem 80).

81. A chain of length L and total mass M is released from rest with its lower end just touching the top of a table, as in Figure 9.41a. Find the force of the table on the chain after the chain has fallen through a distance x, as in Figure 9.41b. (Assume each link comes to rest the instant it reaches the table.)

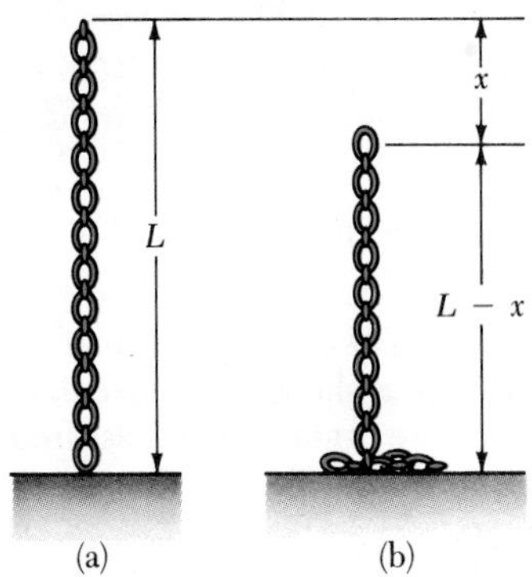

Figure 9.41 (Problem 81).

82. Two gliders are set in motion on an air track. A spring of force constant k is attached to the near side of one glider. The first glider of mass m_1 has velocity $\mathbf{v}_1$ and the second glider of mass m_2 has velocity $\mathbf{v}_2$ as shown in Figure 9.42 ($\mathbf{v}_1 > \mathbf{v}_2$). When m_1 collides with the spring attached to m_2 and compresses the spring to its maximum compression x_m, the velocity of the gliders is $\mathbf{v}$. In terms of $\mathbf{v}_1$, $\mathbf{v}_2$, m_1, m_2, and k, find (a) the velocity $\mathbf{v}$ at maximum compression, (b) the maximum compression x_m, and (c) the velocities of each glider after the first glider has again lost contact with the spring.

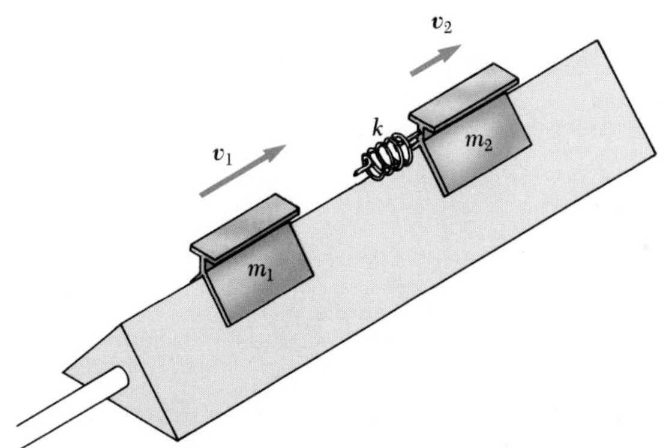

Figure 9.42 (Problem 82).

83. A 40-kg child stands at one end of a 70-kg boat that is 4 m in length (Fig. 9.43). The boat is initially 3 m from the pier. The child notices a turtle on a rock at the far end of the boat and proceeds to walk to that end to catch the turtle. Neglecting friction between the boat and water, (a) describe the subsequent motion of the system (child + boat). (b) Where will the child be *relative to the pier* when he reaches the far end of the boat? (c) Will he catch the turtle? (Assume he can reach out 1 m from the end of the boat.)

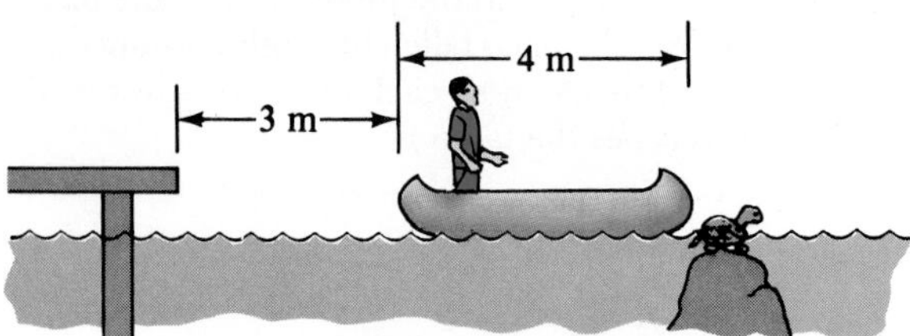

Figure 9.43 (Problem 83).

84. An object of mass M is in the shape of a right triangle with dimensions as shown in Figure 9.44. Locate the coordinates of the center of mass, assuming the object has a uniform mass per unit area.

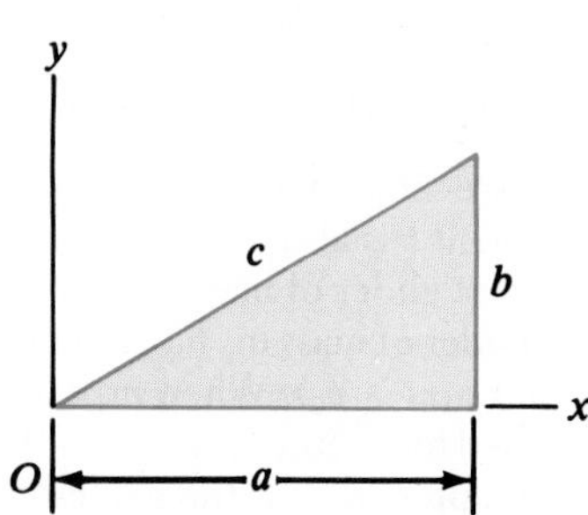

Figure 9.44 (Problem 84).

85. A 5-g bullet moving with an initial speed of 400 m/s is fired into and passes through a 1-kg block, as in Figure 9.45. The block, initially at rest on a frictionless, horizontal surface, is connected to a spring of force constant 900 N/m. If the block moves a distance of 5 cm to the right after impact, find (a) the speed at which the bullet emerges from the block and (b) the energy lost in the collision.

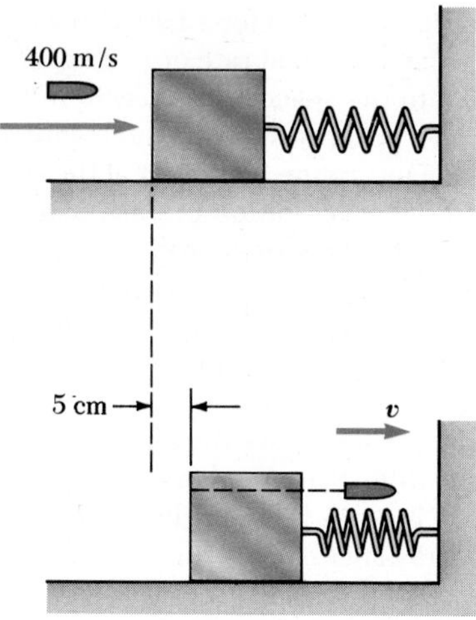

Figure 9.45 (Problem 85).

CALCULATOR/COMPUTER PROBLEM

86. Consider a head-on elastic collision between a moving particle of mass m_1 and an initially stationary particle of mass m_2 (see Example 9.8). (a) Plot f_2, the fraction of energy transferred to m_2, as a function of the ratio m_2/m_1 and show that f_2 reaches a maximum when $m_2/m_1 = 1$. (b) Perform an analytical calculation that verifies that f_2 is a maximum when $m_1 = m_2$.

10

Rotation of a Rigid Body About a Fixed Axis

Rotational motion of a ballet dancer. (© 1986, John Terence Turner, FPG International)

When an extended body, such as a wheel, rotates about its axis, the motion cannot be analyzed by treating the body as a particle, since at any given time different parts of the body have different velocities and accelerations. For this reason, it is convenient to consider an extended object as a large number of particles, each with its own velocity and acceleration.

In dealing with the rotation of a body, analysis is greatly simplified by assuming the body to be rigid. **A rigid body** is defined as a body that is nondeformable, or one in which the separations between all pairs of particles in the body remain constant. All real bodies in nature are deformable to some extent; however, our rigid-body model is useful in many situations where deformation is negligible. In this chapter, we shall treat the rotation of a rigid body about a fixed axis, commonly referred to as *pure rotational motion.*

Rigid body

The vector nature of angular velocity and angular acceleration and the concept of angular momentum will be presented in detail in Chapter 11.

10.1 ANGULAR VELOCITY AND ANGULAR ACCELERATION

Figure 10.1 illustrates a planar rigid body of arbitrary shape confined to the xy plane and rotating about a fixed axis through O perpendicular to the plane of the figure. A particle on the body at P is at a fixed distance r from the origin and

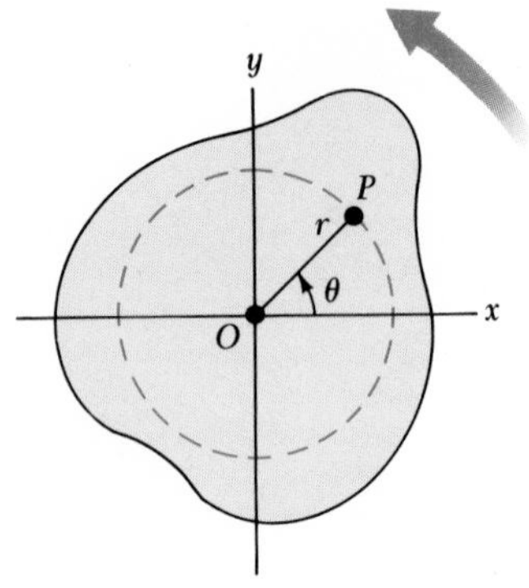

Figure 10.1 Rotation of a rigid body about a fixed axis through O perpendicular to the plane of the figure (the z axis). Note that a particle at P rotates in a circle of radius r centered at O.

rotates in a circle of radius r about O. In fact, *every* particle on the body undergoes circular motion about O. It is convenient to represent the position of the point P with its polar coordinates (r, θ). In this representation, the only coordinate that changes in time is the angle θ; r remains constant. (In rectangular coordinates, both x and y vary in time.) As the particle moves along the circle from the positive x axis ($\theta = 0$) to the point P, it moves through an arc length s, which is related to the angular position θ through the relation

$$s = r\theta \tag{10.1a}$$

$$\theta = s/r \tag{10.1b}$$

It is important to note the units of θ as expressed by Equation 10.1b. The angle θ is the ratio of an arc length and the radius of the circle, and hence is a pure number. However, we commonly refer to the unit of θ as a **radian** (rad), where

one rad is the angle subtended by an arc length equal to the radius of the arc.

Since the circumference of a circle is $2\pi r$, it follows that 360° corresponds to an angle of $2\pi r/r$ rad or 2π rad (one revolution). Hence, 1 rad $= 360°/2\pi \approx 57.3°$. To convert an angle in degrees to an angle in radians, we can use the fact that 2π radians $= 360°$; hence

$$\theta\ (\text{rad}) = \frac{\pi}{180°}\ \theta\ (\text{deg})$$

For example, 60° equals $\pi/3$ rad, and 45° equals $\pi/4$ rad.

As the particle travels from P to Q in Figure 10.2 in a time Δt, the radius vector sweeps out an angle $\Delta\theta = \theta_2 - \theta_1$, which equals the **angular displacement.** We define the **average angular velocity** $\overline{\omega}$ (omega) as the ratio of this angular displacement to the time interval Δt:

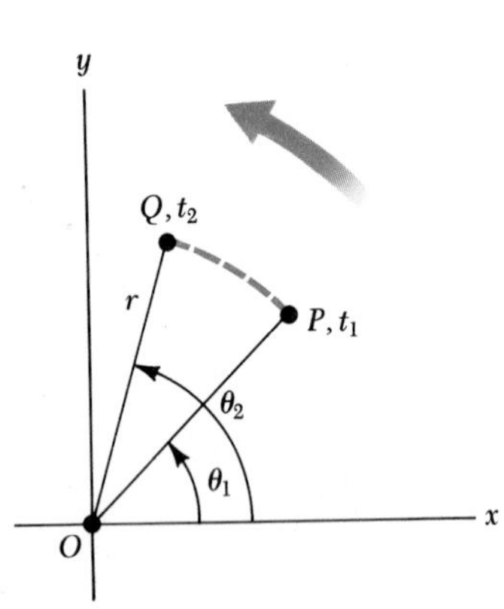

Figure 10.2 A particle on a rotating rigid body moves from P to Q along the arc of a circle. In the time interval $\Delta t = t_2 - t_1$, the radius vector sweeps out an angle $\Delta\theta = \theta_2 - \theta_1$.

$$\overline{\omega} \equiv \frac{\theta_2 - \theta_1}{t_2 - t_1} = \frac{\Delta\theta}{\Delta t} \tag{10.2}$$

In analogy to linear velocity, the **instantaneous angular velocity,** ω, is defined as the limit of the ratio in Equation 10.2 as Δt approaches zero:

$$\omega \equiv \lim_{\Delta t \to 0} \frac{\Delta\theta}{\Delta t} = \frac{d\theta}{dt} \tag{10.3}$$

Angular velocity has units of rad/s, or s^{-1}, since radians are not dimensional. Let us adopt the convention that the fixed axis of rotation for the rigid body is the z axis, as in Figure 10.1. We shall take ω to be positive when θ is increasing (counterclockwise motion) and negative when θ is decreasing (clockwise motion).

If the instantaneous angular velocity of a body changes from ω_1 to ω_2 in the time interval Δt, the body has an angular acceleration. The **average angular acceleration** $\overline{\alpha}$ (alpha) of a rotating body is defined as the ratio of the change in the angular velocity to the time interval Δt:

Average angular acceleration

$$\overline{\alpha} \equiv \frac{\omega_2 - \omega_1}{t_2 - t_1} = \frac{\Delta\omega}{\Delta t} \tag{10.4}$$

In analogy to linear acceleration, the **instantaneous angular acceleration** is defined as the limit of the ratio $\Delta\omega/\Delta t$ as Δt approaches zero:

$$\alpha \equiv \lim_{\Delta t \to 0} \frac{\Delta\omega}{\Delta t} = \frac{d\omega}{dt} \tag{10.5}$$

Instantaneous angular acceleration

Angular acceleration has units of rad/s^2 or s^{-2}. Note that α is positive when ω is increasing in time and negative when ω is decreasing in time.

For rotation about a fixed axis, we see that every *particle on the rigid body has the* same *angular velocity and the* same *angular acceleration.* That is, the quantities ω and α characterize the rotational motion of the *entire* rigid body. Using these quantities, we can greatly simplify the analysis of rigid-body rotation. The angular displacement (θ), angular velocity (ω), and angular acceleration (α) are analogous to linear displacement (x), linear velocity (v), and linear acceleration (a), respectively, for the corresponding motion discussed in Chapter 3. The variables θ, ω, and α differ dimensionally from the variables x, v, and a, only by a length factor.

We have already indicated how the signs for ω and α are determined; however, we have not specified any direction in space associated with these vector quantities.[1] For rotation about a fixed axis, the only direction in space that uniquely specifies the rotational motion is the direction along the axis of rotation. However, we must also decide on the sense of these quantities, that is, whether they point into or out of the plane of Figure 10.1.

As we have already mentioned, the direction of $\boldsymbol{\omega}$ is along the axis of rotation, which is the z axis in Figure 10.1. By convention, we take the direction of $\boldsymbol{\omega}$ to be *out* of the plane of the diagram when the rotation is counterclockwise and *into* the plane of the diagram when the rotation is clockwise. To further illustrate this convention, it is convenient to use the *right-hand rule* shown in Figure 10.3a. The four fingers of the right hand are wrapped in the direction of the rotation. The extended right thumb points in the direction of $\boldsymbol{\omega}$. Figure 10.3b illustrates that $\boldsymbol{\omega}$ is also in the direction of advance of a similarly rotating right-handed screw. Finally, the sense of $\boldsymbol{\alpha}$ follows from its definition as $d\boldsymbol{\omega}/dt$. It is the same as $\boldsymbol{\omega}$ if the angular speed (the magnitude of $\boldsymbol{\omega}$) is increasing in time and antiparallel to $\boldsymbol{\omega}$ if the angular speed is decreasing in time.

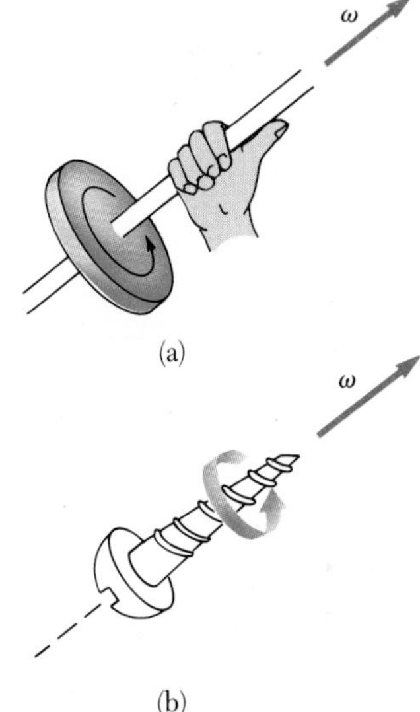

Figure 10.3 (a) The right-hand rule for determining the direction of the angular velocity. (b) The direction of $\boldsymbol{\omega}$ is in the direction of advance of a right-handed screw.

10.2 ROTATIONAL KINEMATICS: ROTATIONAL MOTION WITH CONSTANT ANGULAR ACCELERATION

In the study of linear motion, we found that the simplest form of accelerated motion to analyze is motion under constant linear acceleration (Chapter 3). Likewise, for rotational motion about a fixed axis the simplest accelerated motion to analyze is motion under constant angular acceleration. Therefore, we shall next develop kinematic relations for rotational motion under constant

[1] Although we do not verify it here, the instantaneous angular velocity and instantaneous angular acceleration are vector quantities, but the corresponding average values are not. This is because angular displacement is not a vector quantity for finite rotations.

TABLE 10.1 A Comparison of Kinematic Equations for Rotational and Linear Motion Under Constant Acceleration

Rotational Motion About Fixed Axis with α = Constant. Variables: θ and ω	Linear Motion with a = Constant. Variables: x and v
$\omega = \omega_0 + \alpha t$	$v = v_0 + at$
$\theta = \theta_0 + \omega_0 t + \frac{1}{2}\alpha t^2$	$x = x_0 + v_0 t + \frac{1}{2}at^2$
$\omega^2 = \omega_0{}^2 + 2\alpha(\theta - \theta_0)$	$v^2 = v_0{}^2 + 2a(x - x_0)$

angular acceleration. If we write Equation 10.5 in the form $d\omega = \alpha\, dt$ and let $\omega = \omega_0$ at $t_0 = 0$, we can integrate this expression directly:

Rotational kinematic equations

$$\omega = \omega_0 + \alpha t \qquad (\alpha = \text{constant}) \tag{10.6}$$

Likewise, substituting Equation 10.6 into Equation 10.3 and integrating once more (with $\theta = \theta_0$ at $t_0 = 0$), we get

$$\theta = \theta_0 + \omega_0 t + \tfrac{1}{2}\alpha t^2 \tag{10.7}$$

If we eliminate t from Equations 10.6 and 10.7, we get

$$\omega^2 = \omega_0{}^2 + 2\alpha(\theta - \theta_0) \tag{10.8}$$

Notice that these kinematic expressions for rotational motion under constant angular acceleration are of the *same form* as those for linear motion under constant linear acceleration with the substitutions $x \to \theta$, $v \to \omega$, and $a \to \alpha$. Table 10.1 gives a comparison of the kinematic equations for rotational and linear motion. Furthermore, the expressions are valid for both rigid-body rotation and particle motion about a *fixed* axis.

EXAMPLE 10.1 Rotating Wheel

A wheel rotates with a constant angular acceleration of 3.5 rad/s^2. If the angular velocity of the wheel is 2.0 rad/s at $t_0 = 0$, (a) what angle does the wheel rotate through in 2 s?

$$\theta - \theta_0 = \omega_0 t + \tfrac{1}{2}\alpha t^2$$

$$= \left(2.0\ \frac{\text{rad}}{\text{s}}\right)(2\ \text{s}) + \tfrac{1}{2}\left(3.5\ \frac{\text{rad}}{\text{s}^2}\right)(2\ \text{s})^2$$

$$= 11\ \text{rad} = 630° = 1.75\ \text{rev}$$

(b) What is the angular velocity at $t = 2$ s?

$$\omega = \omega_0 + \alpha t = 2.0\ \text{rad/s} + \left(3.5\ \frac{\text{rad}}{\text{s}^2}\right)(2\ \text{s})$$

$$= 9.0\ \text{rad/s}$$

We could also obtain this result using Equation 10.8 and the results of (a). Try it!

Exercise 1 Find the angle that the wheel rotates through between $t = 2$ s and $t = 3$ s.
Answer 10.8 rad.

10.3 RELATIONSHIPS BETWEEN ANGULAR AND LINEAR QUANTITIES

In this section we shall derive some useful relationships between the angular velocity and acceleration of a rotating rigid body and the linear velocity and acceleration of an arbitrary point in the body. In order to do so, we should keep in mind that when a rigid body rotates about a fixed axis, *every* particle of the body moves in a circle the center of which is the axis of rotation (Fig. 10.4).

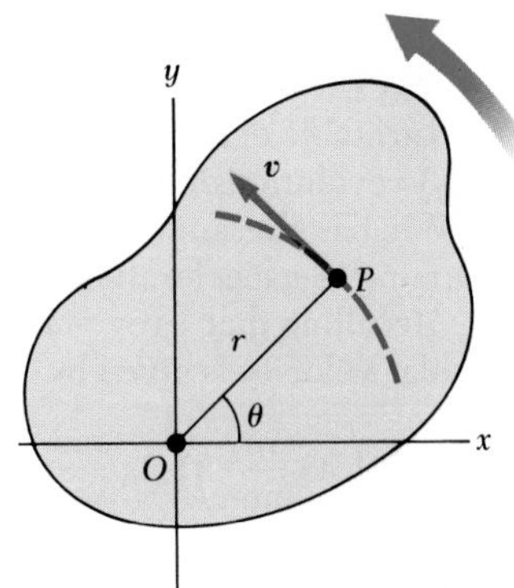

Figure 10.4 As a rigid body rotates about the fixed axis through O, the point P has a linear velocity $\boldsymbol{v}$, which is always tangent to the circular path of radius r.

We can first relate the angular velocity of the rotating body to the tangential velocity, v, of a point P on the body. Since P moves in a circle, the linear velocity vector is always tangent to the circular path, and hence the phrase *tangential velocity*. The magnitude of the tangential velocity of the point P is, by definition, ds/dt, where s is the distance traveled by this point measured along the circular path. Recalling that $s = r\theta$ and noting that r is constant, we get

$$v = \frac{ds}{dt} = r\frac{d\theta}{dt}$$

Relationship between linear and angular speed

$$v = r\omega \tag{10.9}$$

That is, the tangential velocity of a point on a rotating rigid body equals the distance of that point from the axis of rotation multiplied by the angular velocity. Therefore, although every point on the rigid body has the same *angular* velocity, not every point has the same *linear* velocity. In fact, Equation 10.9 shows that the linear velocity of a point on the rotating body increases as one moves outward from the center of rotation toward the rim, as you would intuitively expect.

We can relate the angular acceleration of the rotating rigid body to the tangential acceleration of the point P by taking the time derivative of v:

$$a_t = \frac{dv}{dt} = r\frac{d\omega}{dt}$$

$$a_t = r\alpha \tag{10.10}$$

That is, the tangential component of the linear acceleration of a point on a rotating rigid body equals the distance of that point from the axis of rotation multiplied by the angular acceleration.

In Chapter 4 we found that a point rotating in a circular path undergoes a centripetal, or radial, acceleration of magnitude v^2/r and directed toward the center of rotation (Fig. 10.5). Since $v = r\omega$ for the point P on the rotating body, we can express the centripetal acceleration as

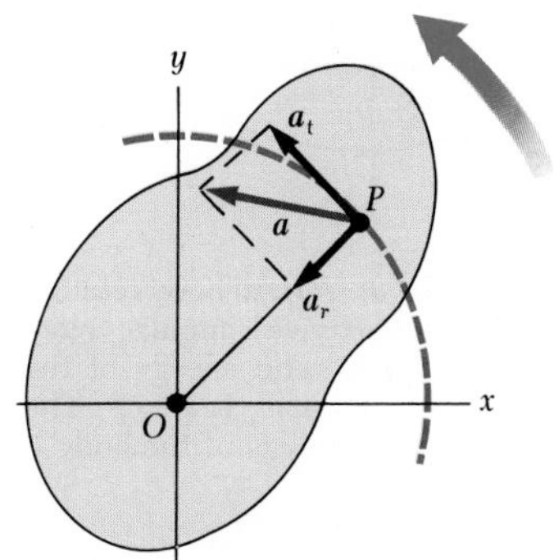

Figure 10.5 As a rigid body rotates about a fixed axis through O, the point P experiences a tangential component of acceleration, $\boldsymbol{a}_t$, and a centripetal component of acceleration, $\boldsymbol{a}_r$. The total acceleration of this point is $\boldsymbol{a} = \boldsymbol{a}_t + \boldsymbol{a}_r$.

$$a_r = \frac{v^2}{r} = r\omega^2 \tag{10.11}$$

The *total linear acceleration* of the particle is $\boldsymbol{a} = \boldsymbol{a}_t + \boldsymbol{a}_r$. Therefore, the magnitude of the total linear acceleration of the point P on the rotating rigid body is given by

$$a = \sqrt{a_t^2 + a_r^2} = \sqrt{r^2\alpha^2 + r^2\omega^4} = r\sqrt{\alpha^2 + \omega^4} \tag{10.12}$$

EXAMPLE 10.2 A Rotating Turntable

The turntable of a record player rotates initially at a rate of 33 revolutions/min and takes 20 s to come to rest. (a) What is the angular acceleration of the turntable, assuming the acceleration is uniform?

Recalling that 1 rev = 2π rad, we see that the initial angular velocity is given by

$$\omega_0 = \left(33\ \frac{\text{rev}}{\text{min}}\right)\left(2\pi\ \frac{\text{rad}}{\text{rev}}\right)\left(\frac{1}{60}\ \frac{\text{min}}{\text{s}}\right) = 3.46\ \text{rad/s}$$

Using $\omega = \omega_0 + \alpha t$ and the fact that $\omega = 0$ at $t = 20$ s, we get

$$\alpha = -\frac{\omega_0}{t} = -\frac{3.46\ \text{rad/s}}{20\ \text{s}} = \boxed{-0.173\ \text{rad/s}^2}$$

where the negative sign indicates a negative angular acceleration (ω is decreasing).

(b) How many rotations does the turntable make before coming to rest?

Using Equation 10.7, we find that the angular displacement in 20 s is

$$\Delta\theta = \theta - \theta_0 = \omega_0 t + \tfrac{1}{2}\alpha t^2$$

$$= [3.46(20) + \tfrac{1}{2}(-0.173)(20)^2]\ \text{rad} = \boxed{34.6\ \text{rad}}$$

This corresponds to $34.6/2\pi$ rev, or 5.50 rev.

(c) What are the magnitudes of the radial and tangential components of the linear acceleration of a point on the rim at $t = 0$?

We can use $a_t = r\alpha$ and $a_r = r\omega^2$, which gives

$$a_t = r\alpha = (14\ \text{cm})\left(0.173\ \frac{\text{rad}}{\text{s}^2}\right) = \boxed{2.42\ \text{cm/s}^2}$$

$$a_r = r\omega_0^2 = (14\ \text{cm})\left(3.46\ \frac{\text{rad}}{\text{s}}\right)^2 = \boxed{168\ \text{cm/s}^2}$$

Exercise 2 If the radius of the turntable is 14 cm, what is the initial linear speed of a point on the rim of the turntable?

Answer 48.4 cm/s.

10.4 ROTATIONAL KINETIC ENERGY

Let us consider a rigid body as a collection of small particles and let us assume that the body rotates about the fixed z axis with an angular velocity ω (Fig. 10.6). Each particle of the body has some kinetic energy, determined by its mass and velocity. If the mass of the ith particle is m_i and its speed is v_i, the kinetic energy of this particle is

$$K_i = \tfrac{1}{2}m_i v_i^2$$

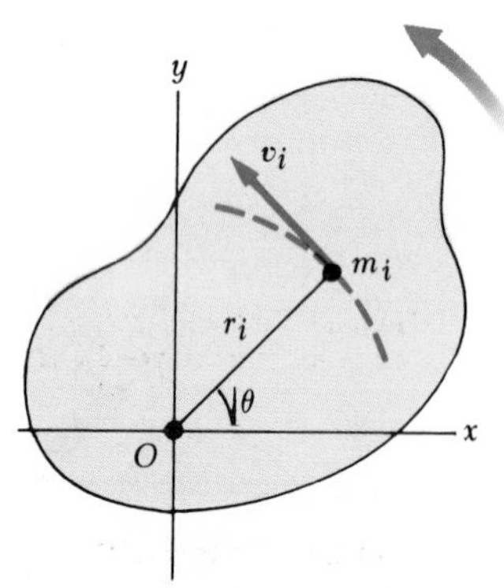

Figure 10.6 A rigid body rotating about the z axis with angular velocity ω. The kinetic energy of the particle of mass m_i is $\frac{1}{2}m_i v_i^2$. The total kinetic energy of the body is $\frac{1}{2}I\omega^2$.

To proceed further, we must recall that although every particle in the rigid body has the same angular velocity, ω, the individual linear velocities depend on the distance r_i from the axis of rotation according to the expression $v_i = r_i\omega$ (Eq. 10.9). The *total* kinetic energy of the rotating rigid body is the sum of the kinetic energies of the individual particles:

$$K = \sum K_i = \sum \tfrac{1}{2}m_i v_i^2 = \tfrac{1}{2}\sum m_i r_i^2 \omega^2$$

$$K = \tfrac{1}{2}\left(\sum m_i r_i^2\right)\omega^2 \tag{10.13}$$

where we have factored ω^2 from the sum since it is common to every particle. The quantity in parentheses is called the **moment of inertia, I**:

$$I = \sum m_i r_i^2 \tag{10.14}$$

Using this notation, we can express the kinetic energy of the rotating rigid body (Eq. 10.13) as

$$K = \tfrac{1}{2}I\omega^2 \tag{10.15}$$

From the definition of moment of inertia, we see that it has dimensions of ML^2 ($kg \cdot m^2$ in SI units and $g \cdot cm^2$ in cgs units). It plays the role of mass in *all* rotational equations. Although we shall commonly refer to the quantity $\frac{1}{2}I\omega^2$ as the **rotational kinetic energy,** it is not a new form of energy. It is ordinary kinetic energy, since it was derived from a sum over individual kinetic energies of the particles contained in the rigid body. However, the form of the kinetic energy given by Equation 10.15 is a convenient one in dealing with rotational motion, providing we know how to calculate I. It is important that you recognize the analogy between kinetic energy associated with linear motion, $\frac{1}{2}mv^2$, and rotational kinetic energy, $\frac{1}{2}I\omega^2$. The quantities I and ω in rotational motion are analogous to m and v in linear motion, respectively. In the next section we shall describe how to calculate moments of inertia for rigid bodies. The following examples illustrate how to calculate moments of inertia and rotational kinetic energy for a distribution of particles.

EXAMPLE 10.3 The Oxygen Molecule

Consider the diatomic molecule oxygen, O_2, which is rotating in the xy plane about the z axis passing through its center, perpendicular to its length. At room temperature, the "average" separation between the two oxygen atoms is 1.21×10^{-10} m (the atoms are treated as point masses). (a) Calculate the moment of inertia of the molecule about the z axis.

Since the mass of an oxygen atom is 2.66×10^{-26} kg and the distance of each atom from the z axis is $d/2$, the moment of inertia about the z axis is

$$I = \sum m_i r_i^2 = m\left(\frac{d}{2}\right)^2 + m\left(\frac{d}{2}\right)^2 = \frac{md^2}{2}$$

$$= \left(\frac{2.66 \times 10^{-26}}{2}\ \text{kg}\right)(1.21 \times 10^{-10}\ \text{m})^2$$

$$= 1.95 \times 10^{-46}\ \text{kg} \cdot \text{m}^2$$

(b) If the angular velocity of the molecule about the z axis is 2.0×10^{12} rad/s, what is its rotational kinetic energy?

$$K = \tfrac{1}{2}I\omega^2$$

$$= \tfrac{1}{2}(1.95 \times 10^{-46}\ \text{kg} \cdot \text{m}^2)\left(2.0 \times 10^{12}\ \frac{\text{rad}}{\text{s}}\right)^2$$

$$= 3.89 \times 10^{-22}\ \text{J}$$

This is about one order of magnitude smaller than the average kinetic energy associated with the translational motion of the molecule at room temperature, which is about 6.2×10^{-21} J.

EXAMPLE 10.4 Four Rotating Particles

Four point masses are fastened to the corners of a frame of negligible mass lying in the xy plane (Fig. 10.7). (a) If the rotation of the system occurs about the y axis with an angular velocity ω, find the moment of inertia about the y axis and the rotational kinetic energy about this axis.

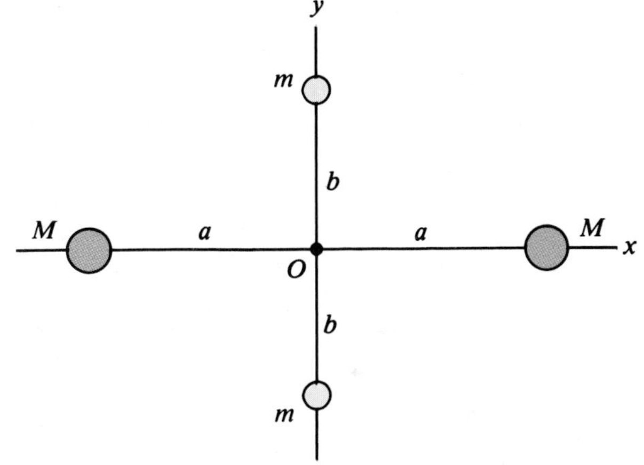

Figure 10.7 (Example 10.4) All particles are at a fixed separation as shown. The moment of inertia depends on the axis about which it is evaluated.

First, note that the particles of mass m that lie on the y axis do not contribute to I_y (that is, $r_i = 0$ for these particles about this axis). Applying Equation 10.14, we get

$$I_y = \sum m_i r_i^2 = Ma^2 + Ma^2 = 2Ma^2$$

Therefore, the rotational kinetic energy about the y axis is

$$K = \tfrac{1}{2}I_y\omega^2 = \tfrac{1}{2}(2Ma^2)\omega^2 = Ma^2\omega^2$$

The fact that the masses m do not enter into this result makes sense, since these particles have no motion about the chosen axis of rotation; hence they have no kinetic energy.

(b) Now suppose the system rotates in the xy plane about an axis through O (the z axis). Calculate the moment of inertia about the z axis and the rotational kinetic energy about this axis.

Since r_i in Equation 10.14 is the *perpendicular* distance to the axis of rotation, we get

$$I_z = \sum m_i r_i^2 = Ma^2 + Ma^2 + mb^2 + mb^2$$

$$= 2Ma^2 + 2mb^2$$

$$K = \tfrac{1}{2} I_z \omega^2 = \tfrac{1}{2}(2Ma^2 + 2mb^2)\omega^2 = (Ma^2 + mb^2)\omega^2$$

Comparing the results for (a) and (b), we conclude that the moment of inertia, and therefore the rotational kinetic energy associated with a given angular speed, depends on the axis of rotation. In (b), we would expect the result to include all masses and distances, since all particles are in motion for rotation in the xy plane. Furthermore, the fact that the kinetic energy in (a) is smaller than in (b) indicates that it would take less effort (work) to set the system into rotation about the y axis than about the z axis.

10.5 CALCULATION OF MOMENTS OF INERTIA

We can evaluate the moment of inertia of an extended body by imagining that the body is divided into volume elements, each of mass Δm. Now we can use the definition $I = \Sigma r^2 \Delta m$ and take the limit of this sum as $\Delta m \rightarrow 0$. In this limit, the sum becomes an integral over the whole body, where r is the perpendicular distance from the axis of rotation to the element Δm. Hence,

$$I = \lim_{\Delta m \rightarrow 0} \sum r^2 \Delta m = \int r^2 \, dm \qquad (10.16)$$

To evaluate the moment of inertia using Equation 10.16, it is necessary to express the element of mass dm in terms of its coordinates. It is common to define a mass density in various forms. For a three-dimensional body, it is appropriate to use the *local volume density,* that is, *mass per unit volume.* In this case, we can write

$$\rho = \lim_{\Delta V \rightarrow 0} \frac{\Delta m}{\Delta V} = \frac{dm}{dV}$$

$$dm = \rho \, dV$$

Therefore, the moment of inertia can be expressed in the form

$$I = \int \rho r^2 \, dV$$

If the body is homogeneous, then ρ is constant and the integral can be evaluated for a known geometry. If ρ is not constant, then its variation with position must be specified. When dealing with an object in the form of a sheet of uniform thickness t, it is convenient to define a surface density $\sigma = \rho t$, which signifies *mass per unit area.* Finally, when mass is distributed along a uniform rod of cross-sectional area A, we sometimes use linear density, $\lambda = \rho A$, where λ is defined as *mass per unit length.*

EXAMPLE 10.5 Uniform Hoop

Find the moment of inertia of a uniform hoop of mass M and radius R about an axis perpendicular to the plane of the hoop, through its center (Fig. 10.8).

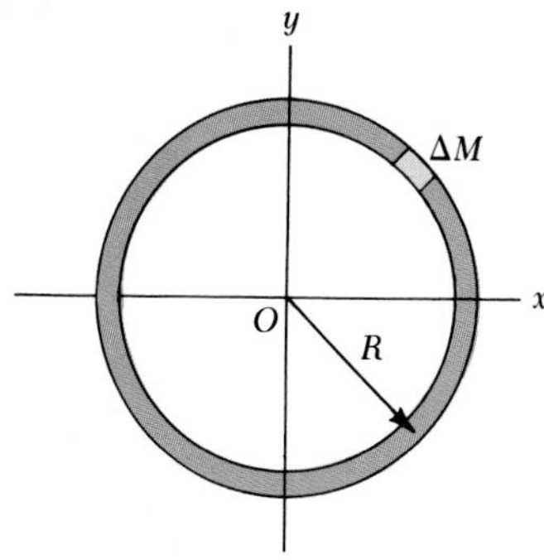

Figure 10.8 (Example 10.5) The mass elements of a uniform hoop are all the same distance from O.

Solution All elements of mass are at the same distance $r = R$ from the axis, so, applying Equation 10.16, we get for the moment of inertia about the z axis through O:

$$I_z = \int r^2\, dm = R^2 \int dm = MR^2$$

EXAMPLE 10.6 Uniform Rigid Rod □

Calculate the moment of inertia of a uniform rigid rod of length L and mass M (Fig. 10.9) about an axis perpendicular to the rod (the y axis) passing through its center of mass.

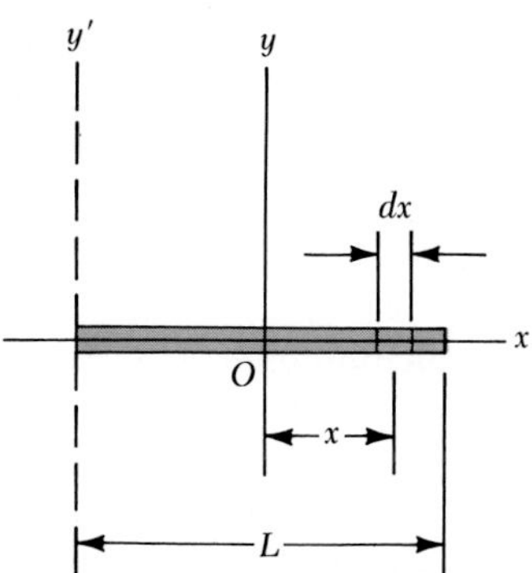

Figure 10.9 (Example 10.6) A uniform rigid rod of length L. The moment of inertia about the y axis is less than that about the y' axis.

Solution The shaded element of width dx has a mass dm equal to the mass per unit length multiplied by the element of length, dx. That is, $dm = \frac{M}{L}\, dx$. Substituting this into Equation 10.16, with $r = x$, we get

$$I_y = \int r^2\, dm = \int_{-L/2}^{L/2} x^2 \frac{M}{L}\, dx = \frac{M}{L} \int_{-L/2}^{L/2} x^2\, dx$$

$$= \frac{M}{L}\left[\frac{x^3}{3}\right]_{-L/2}^{L/2} = \frac{1}{12} ML^2$$

Exercise 3 Calculate the moment of inertia of a uniform rigid rod about an axis perpendicular to the rod through one end (the y' axis). Note that the calculation requires that the limits of integration be from $x = 0$ to $x = L$. *Answer* $\frac{1}{3}ML^2$.

EXAMPLE 10.7 Uniform Solid Cylinder

A uniform solid cylinder has a radius R, mass M, and length L. Calculate the moment of inertia of the cylinder about its axis (the z axis in Fig. 10.10).

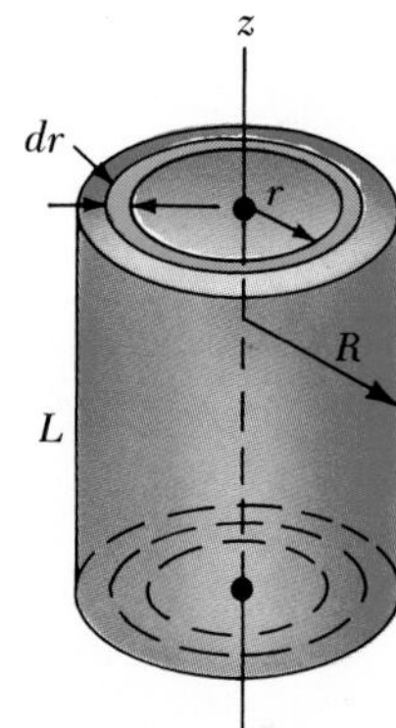

Figure 10.10 (Example 10.7) Calculating I about the z axis for a uniform solid cylinder.

Solution In this example, it is convenient to divide the cylinder into cylindrical shells of radius r, thickness dr, and length L, as in Figure 10.10. In this case, cylindrical shells are chosen because one wants all mass elements dm to have a single value for r, which makes the calculation of I more straightforward. The volume of each shell is its cross-sectional area multiplied by the length, or $dV = dA \cdot L = (2\pi r\, dr)L$. If the *mass per unit volume* is ρ, then the mass of this differential volume element is $dm = \rho dV = \rho\, 2\pi r L\, dr$. Substituting this into Equation 10.16, we get

$$I_z = \int r^2\, dm = 2\pi\rho L \int_0^R r^3\, dr = \frac{\pi\rho L R^4}{2}$$

However, since the total volume of the cylinder is $\pi R^2 L$, $\rho = M/V = M/\pi R^2 L$. Substituting this into the above result gives

$$I_z = \tfrac{1}{2}MR^2$$

TABLE 10.2 Moments of Inertia of Homogeneous Rigid Bodies with Different Geometries

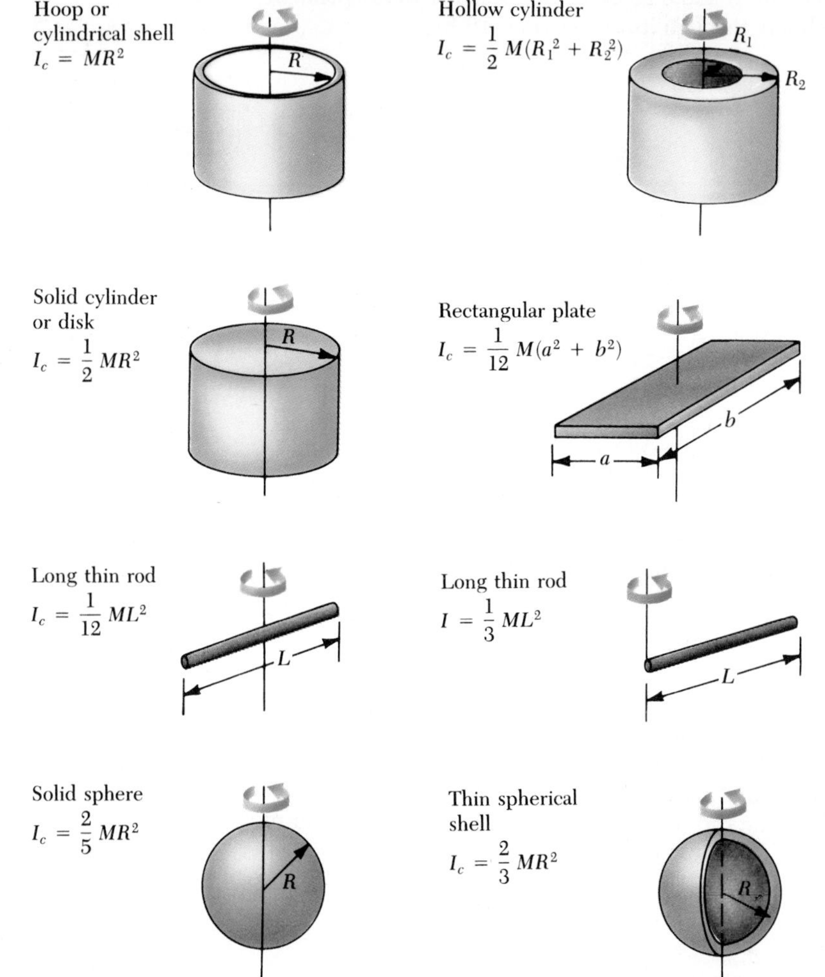

As we saw in the previous examples, the moments of inertia of rigid bodies with simple geometry (high symmetry) are relatively easy to calculate provided the reference axis coincides with an axis of symmetry. Table 10.2 gives the moments of inertia for a number of bodies about specific axes.[2]

The calculation of moments of inertia about an arbitrary axis can be somewhat cumbersome, even for a highly symmetric body, such as a sphere. In this regard, there is an important theorem, called the *parallel-axis theorem*, that often simplifies the calculation of moments of inertia. Suppose the moment of inertia about any axis through the center of mass is I_c. The parallel-axis theorem states that the moment of inertia about any axis that is *parallel* to and a

[2] Civil engineers use the moment of inertia concept to characterize the elastic properties (rigidity) of such structures as loaded beams. Hence, it is often useful even in a nonrotational context.

distance D away from the axis that passes through the center of mass is given by

$$I = I_c + MD^2 \tag{10.17}$$

Parallel-axis theorem

***Proof of the Parallel-axis Theorem** Suppose a body rotates in the xy plane about an axis through O as in Figure 10.11 and the coordinates of the center of mass are x_c, y_c. Let the element Δm have coordinates x, y relative to the origin. Since this element is at a distance $r = \sqrt{x^2 + y^2}$ from the z axis, the moment of inertia about the z axis through O is

$$I = \int r^2 \, dm = \int (x^2 + y^2) dm$$

However, we can relate the coordinates x, y to the coordinates of the center of mass, x_c, y_c, and the coordinates relative to the center of mass, x', y' through the relations $x = x' + x_c$ and $y = y' + y_c$. Therefore,

$$\begin{aligned} I &= \int [(x' + x_c)^2 + (y' + y_c)^2] dm \\ &= \int [(x')^2 + (y')^2] dm + 2x_c \int x' \, dm + 2y_c \int y' \, dm + (x_c^2 + y_c^2) \int dm \end{aligned}$$

The first term on the right is, by definition, the moment of inertia about an axis parallel to the z axis, through the center of mass. The second two terms on the right are zero, since by definition of the center of mass $\int x' \, dm = \int y' \, dm = 0$ (x', y' are the coordinates of the mass element relative to the center of mass). Finally, the last term on the right is simply MD^2, since $\int dm = M$ and $D^2 = x_c^2 + y_c^2$. Therefore, we conclude that

$$I = I_c + MD^2$$

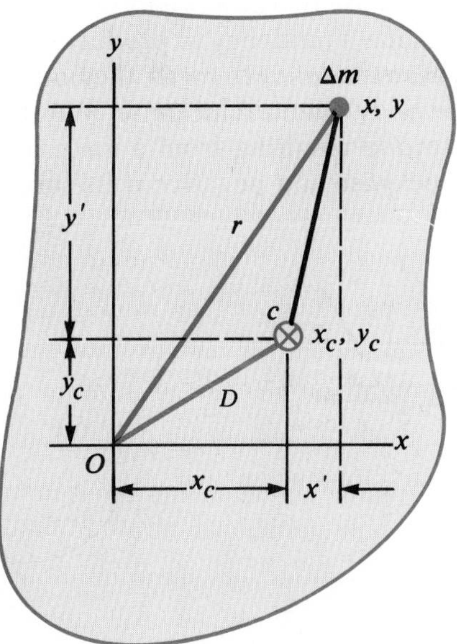

Figure 10.11 The parallel-axis theorem. If the moment of inertia about an axis perpendicular to the figure through the center of mass at c is I_c, then the moment of inertia about the z axis is $I_z = I_c + MD^2$.

EXAMPLE 10.8 Applying the Parallel-axis Theorem Consider once again a uniform rigid rod of mass M and length L as in Figure 10.9. Find the moment of inertia of the rod about an axis perpendicular to the rod through one end (the y' axis in Fig. 10.9).

Solution Since the moment of inertia of the rod about an axis perpendicular to the rod through its center of mass is $ML^2/12$, and the distance between this axis and the parallel axis through its end is $D = L/2$, the parallel-axis theorem gives

$$I = I_c + MD^2 = \tfrac{1}{12}ML^2 + M\left(\frac{L}{2}\right)^2 = \tfrac{1}{3}ML^2$$

Exercise 4 Calculate the moment of inertia of the rod about an axis perpendicular to the rod and through the point $x = L/4$.
Answer $I = \frac{7}{48}ML^2$

10.6 TORQUE

When a force is exerted on a rigid body pivoted about some axis, the body will tend to rotate about that axis. The tendency of a force to rotate a body about some axis is measured by a quantity called the **torque** (τ). Consider the wrench pivoted about the axis through O in Figure 10.12. The applied force $\boldsymbol{F}$ generally can act at an angle ϕ to the horizontal. We define the magnitude of the torque, τ, (Greek letter tau), resulting from the force $\boldsymbol{F}$ by the expression

Definition of torque

$$\tau \equiv rF\sin\phi = Fd \tag{10.18}$$

Moment arm

It is very important that you recognize that *torque is defined only when a reference axis is specified.* The quantity $d = r\sin\phi$, called the **moment arm** (or *lever arm*) of the force $\boldsymbol{F}$, represents the perpendicular distance from the rotation axis to the line of action of $\boldsymbol{F}$. Note that the only component of $\boldsymbol{F}$ that tends to cause a rotation is $F\sin\phi$, the component perpendicular to r. The horizontal component, $F\cos\phi$, passes through O and has no tendency to produce a rotation. If there are two or more forces acting on a rigid body, as in Figure 10.13, then each has a tendency to produce a rotation about the pivot at O. For example, $\boldsymbol{F}_2$ has a tendency to rotate the body clockwise, and $\boldsymbol{F}_1$ has a tendency to rotate the body counterclockwise. We shall use the convention that the sign of the torque resulting from a force is positive if its turning tendency is counterclockwise and negative if its turning tendency is clock-

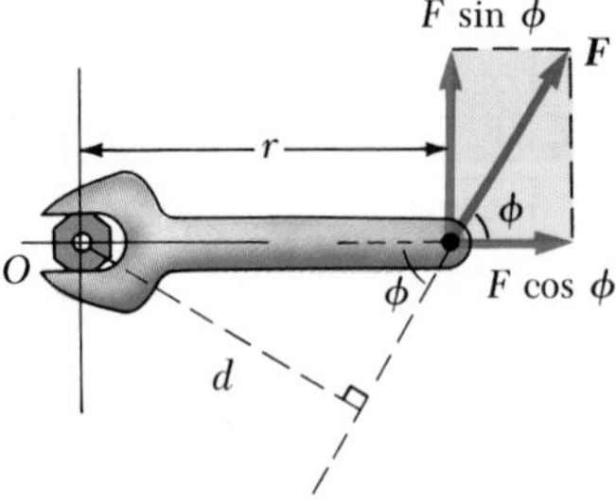

Figure 10.12 The force $\boldsymbol{F}$ has a greater rotating tendency about O as F increases and as the moment arm, d, increases. It is the component $F\sin\phi$ that tends to rotate the system about O.

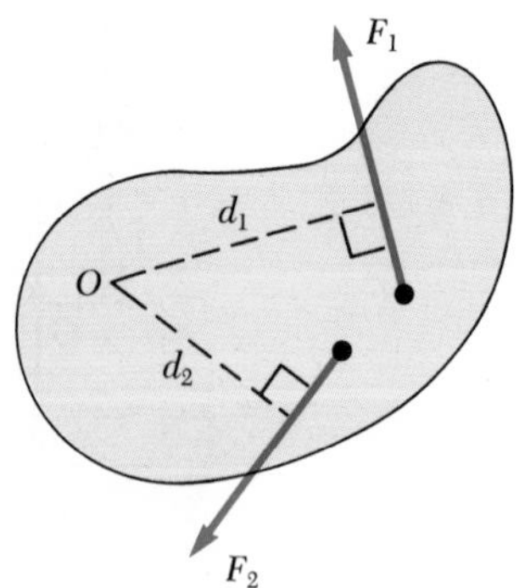

Figure 10.13 The force $\boldsymbol{F}_1$ tends to rotate the body counterclockwise about O, and $\boldsymbol{F}_2$ tends to rotate the body clockwise.

wise. For example, in Figure 10.13, the torque resulting from $\boldsymbol{F}_1$, which has a moment arm of d_1, is *positive* and equal to $+F_1d_1$; the torque from $\boldsymbol{F}_2$ is *negative* and equal to $-F_2d_2$. Hence, the *net* torque acting on the rigid body about O is

$$\tau_{\text{net}} = \tau_1 + \tau_2 = F_1d_1 - F_2d_2$$

From the definition of torque, we see that the rotating tendency increases as $\boldsymbol{F}$ increases and as d increases. For example, it is easier to close a door if we push at the doorknob rather than at a point close to the hinge. *Torque should not be confused with force.* Torque has units of force times length, or N·m in SI units, the same combination that gives work. In Section 10.7 we shall see that the concept of torque is convenient for analyzing the rotational dynamics of a rigid body. The vector nature of torque will be described in detail in the next chapter.

EXAMPLE 10.9 The Net Torque on a Cylinder

A solid cylinder is pivoted about a frictionless axle as in Figure 10.14. A rope wrapped around the outer radius, R_1, exerts a force $\boldsymbol{F}_1$ to the right on the cylinder. A second rope wrapped around another section of radius R_2 exerts a force $\boldsymbol{F}_2$ downward on the cylinder. (a) What is the net torque acting on the cylinder about the z axis through O?

The torque due to $\boldsymbol{F}_1$ is $-R_1F_1$ and is negative because it tends to produce a clockwise rotation. The torque due to $\boldsymbol{F}_2$ is $+R_2F_2$ and is positive because it tends to produce a counterclockwise rotation. Therefore, the net torque is

$$\tau_{\text{net}} = \tau_1 + \tau_2 = \boxed{R_2F_2 - R_1F_1}$$

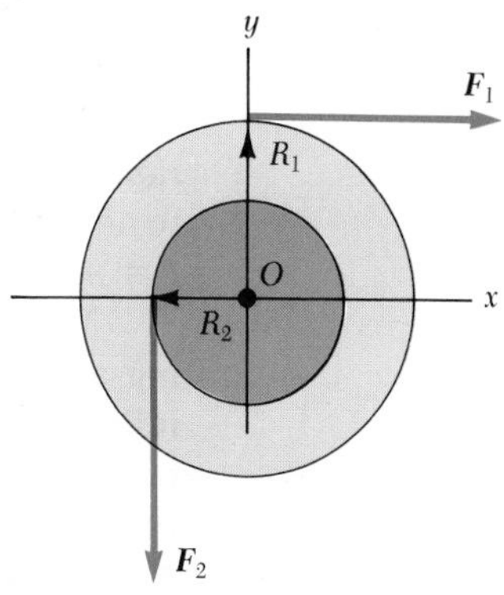

Figure 10.14 (Example 10.9) A solid cylinder pivoted about the z axis through O. The moment arm of $\boldsymbol{F}_1$ is R_1, and the moment arm of $\boldsymbol{F}_2$ is R_2.

(b) Suppose $F_1 = 5$ N, $R_1 = 1.0$ m, $F_2 = 6$ N, and $R_2 = 0.5$ m. What is the net torque and which way will the cylinder rotate?

$$\tau_{\text{net}} = (6\text{ N})(0.5\text{ m}) - (5\text{ N})(1.0\text{ m}) = \boxed{-2\text{ N}\cdot\text{m}}$$

Since the net torque is negative, the cylinder will rotate in the clockwise direction.

10.7 RELATIONSHIP BETWEEN TORQUE AND ANGULAR ACCELERATION

In this section we shall show that the angular acceleration of a rigid body rotating about a fixed axis is proportional to the net torque acting about that axis. Before discussing the more complex case of rigid-body rotation, it is instructive first to discuss briefly the case of a particle rotating about some fixed point under the influence of an external force. The ideas embodied in this situation will then be extended to the case of a rigid body rotating about a fixed axis.

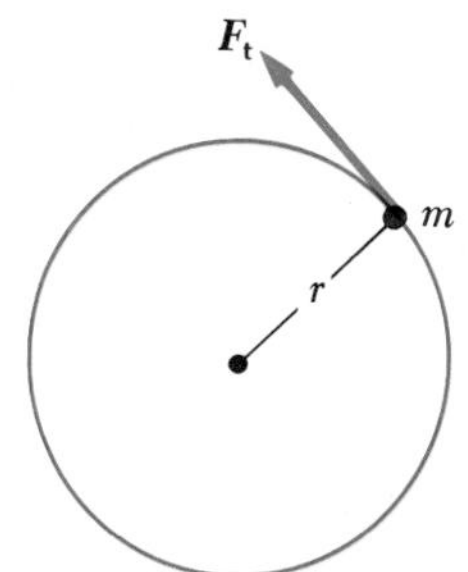

Figure 10.15 A particle rotating in a circle under the influence of a tangential force $\boldsymbol{F}_t$. A centripetal force $\boldsymbol{F}_r$ (not shown) must also be present to maintain the circular motion.

Consider a particle of mass m rotating in a circle of radius r under the influence of a tangential force $\boldsymbol{F}_t$ as in Figure 10.15 and a centripetal force $\boldsymbol{F}_r$ not shown in the figure. (The centripetal force *must* be present to keep the particle moving in its circular path.) The tangential force provides a tangential acceleration $\boldsymbol{a}_t$, and

$$F_t = ma_t$$

The torque about the origin due to the force $\boldsymbol{F}_t$ is the product of the magnitude of the force, F_t, and the moment arm of the force:

$$\tau = F_t r = (ma_t)r$$

Since the tangential acceleration is related to the angular acceleration through the relation $a_t = r\alpha$, the torque can be expressed

$$\tau = (mr\alpha)r = (mr^2)\alpha$$

Recall that the quantity mr^2 is the moment of inertia of the rotating mass about the z axis passing through the origin, so that

Relationship between torque and angular acceleration

$$\tau = I\alpha \tag{10.19}$$

That is, *the torque acting on the particle is proportional to its angular acceleration*, and the proportionality constant is the moment of inertia. It is important to note that $\tau = I\alpha$ is the rotational analogue of Newton's second law of motion, $F = ma$.

Now let us extend this discussion to a rigid body of arbitrary shape rotating about a fixed axis as in Figure 10.16. The body can be regarded as an infinite number of mass elements dm of infinitesimal size. Each mass element rotates in a circle about the origin, and each has a tangential acceleration $\boldsymbol{a}_t$ produced by a tangential force $d\boldsymbol{F}_t$. For any given element, we know from Newton's second law that

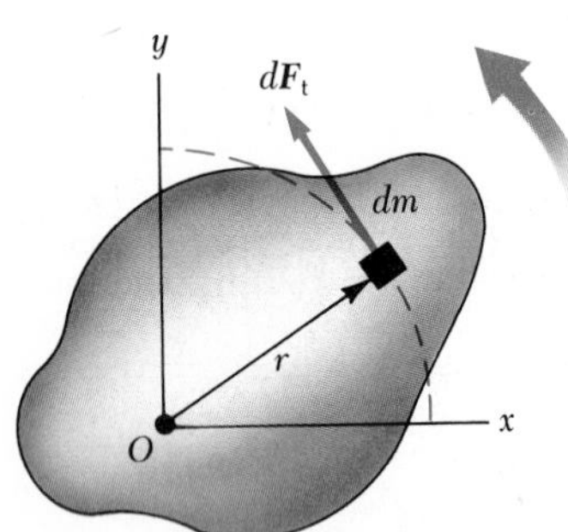

Figure 10.16 A rigid body pivoted about an axis through O. Each mass element dm rotates about O with the same angular acceleration $\boldsymbol{\alpha}$, and the net torque on the body is proportional to $\boldsymbol{\alpha}$.

$$dF_t = (dm)a_t$$

The torque $d\tau$ associated with the force $d\boldsymbol{F}_t$ acting about the origin is given by

$$d\tau = r\, dF_t = (r\, dm)a_t$$

Since $a_t = r\alpha$, the expression for $d\tau$ becomes

$$d\tau = (r\, dm)r\alpha = (r^2\, dm)\alpha$$

It is important to recognize that although each point of the rigid body may have a different $\boldsymbol{a}_t$, all mass elements have the *same* angular acceleration, $\boldsymbol{\alpha}$. With this in mind, the above expression can be integrated to obtain the net torque about O:

$$\tau_{\text{net}} = \int (r^2\, dm)\alpha = \alpha \int r^2\, dm$$

where α can be taken outside the integral since it is common to all mass elements. Since the moment of inertia of the body about the rotation axis through O is defined by $I = \int r^2\, dm$, the expression for τ_{net} becomes

Torque is proportional to angular acceleration

$$\tau_{\text{net}} = I\alpha \tag{10.20}$$

Again we see that the net torque about the rotation axis is proportional to the angular acceleration of the body with the proportionality factor being I, which depends upon the axis of rotation and upon the size and shape of the body.

In view of the complex nature of the system, the important result that $\tau_{net} = I\alpha$ is strikingly simple and in complete agreement with experimental observations. The simplicity of the result lies in the manner in which the motion is described.

Although each point on a rigid body rotating about a fixed axis may not experience the same force, linear acceleration, or linear velocity, every point on the body has the same angular acceleration and angular velocity at any instant. Therefore, at any instant the rotating rigid body as a whole is characterized by specific values for angular acceleration, net torque, and angular velocity.

Every point has the same ω and α

Finally, you should note that the result $\tau_{net} = I\alpha$ would also apply if the forces acting on the mass elements had radial components as well as tangential components. This is because the line of action of all radial components must pass through the axis of rotation, and hence would produce *zero* torque about that axis.

EXAMPLE 10.10 Rotating Rod

A uniform rod of length L and mass M is free to rotate about a frictionless pivot at one end, as in Figure 10.17. The rod is released from rest in the horizontal position. What is the *initial* angular acceleration of the rod and the *initial* linear acceleration of the right end of the rod?

Solution The weight Mg, located at the geometric center of the rod, acts at its center of mass as shown in Figure 10.17. The magnitude of the torque due to this force about an axis through the pivot is

$$\tau = \frac{MgL}{2}$$

The support force at the hinge has zero torque about an axis through the pivot, because this force passes through the axis (hence $r = 0$). Since $\tau = I\alpha$, where $I = \frac{1}{3}ML^2$ for this axis of rotation (see Table 10.2), we get

$$I\alpha = Mg\,\frac{L}{2}$$

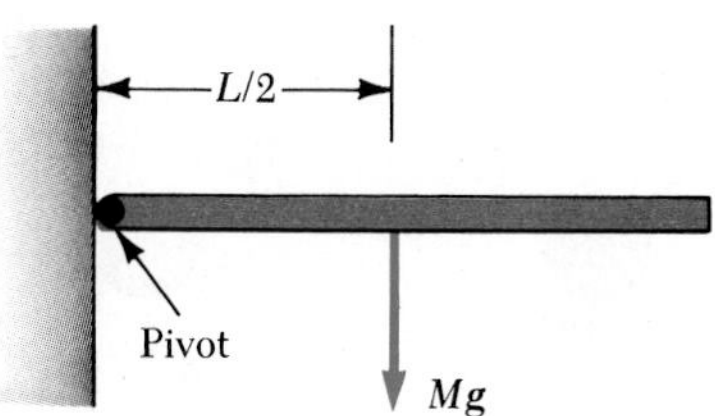

Figure 10.17 (Example 10.10) The uniform rod is pivoted at the left end.

$$\alpha = \frac{Mg(L/2)}{\frac{1}{3}ML^2} = \frac{3g}{2L}$$

This angular acceleration is common to *all* points on the rod.

To find the linear acceleration of the right end of the rod, we use the relation $a_t = R\alpha$, with $R = L$. This gives

$$a_t = L\alpha = \tfrac{3}{2}g$$

This result is rather interesting, since $a_t > g$. That is, the end of the rod has an acceleration *greater* than the acceleration due to gravity. Therefore, if a coin were placed at the end of the rod, the end of the rod would fall faster than the coin when released.

Other points on the rod have a linear acceleration less than $\frac{3}{2}g$. For example, the middle of the rod has an acceleration $\frac{3}{4}g$.

EXAMPLE 10.11 Angular Acceleration of a Wheel

A wheel of radius R, mass M, and moment of inertia I is mounted on a frictionless, horizontal axle as in Figure 10.18. A light cord wrapped around the wheel supports a body of mass m. Calculate the linear acceleration of the suspended body, the angular acceleration of the wheel, and the tension in the cord.

Solution The torque acting on the wheel about its axis of rotation is $\tau = TR$. The weight of the wheel and the normal force of the axle on the wheel pass through the axis of

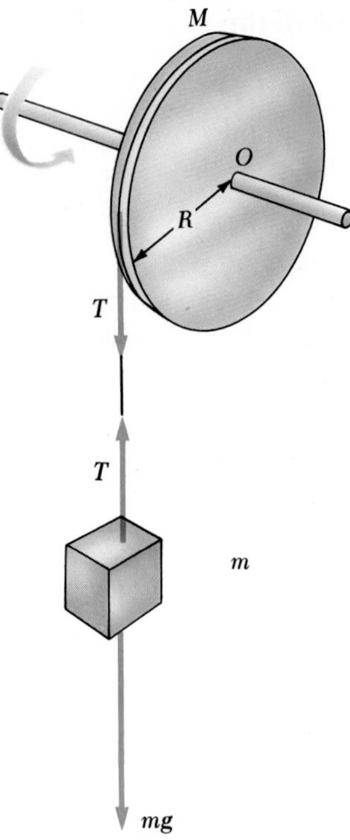

Figure 10.18 (Example 10.11) The cord attached to m is wrapped around the pulley, which produces a torque about the axle through O.

rotation and produce no torque. Since $\tau = I\alpha$, we get

$$\tau = I\alpha = TR$$

$$\alpha = TR/I \quad (1)$$

Now let us apply Newton's second law to the motion of the suspended mass m, making use of the free-body diagram (Fig. 10.18):

$$\sum F_y = T - mg = -ma$$

$$a = \frac{mg - T}{m} \quad (2)$$

The linear acceleration of the suspended mass is equal to the tangential acceleration of a point on the rim of the wheel. Therefore, the angular acceleration of the wheel and this linear acceleration are related by $a = R\alpha$. Using this fact together with (1) and (2) gives

$$a = R\alpha = \frac{TR^2}{I} = \frac{mg - T}{m}$$

$$T = \frac{mg}{1 + \frac{mR^2}{I}}$$

Likewise, solving for a and α gives

$$a = \frac{g}{1 + I/mR^2}$$

$$\alpha = \frac{a}{R} = \frac{g}{R + I/mR}$$

Exercise 5 The wheel in Figure 10.18 is a solid disk of $M = 2.0$ kg, $R = 30$ cm, and $I = 0.09$ kg·m². The suspended object has a mass of $m = 0.5$ kg. Find the tension in the cord and the angular acceleration of the wheel.
Answer 3.27 N; 10.9 rad/s².

EXAMPLE 10.12

Two masses m_1 and m_2 are connected to each other by a light cord which passes over two identical pulleys, each having a moment of inertia I. (See Figure 10.19a.) Find the acceleration of each mass and the tensions $\boldsymbol{T}_1$, $\boldsymbol{T}_2$, and $\boldsymbol{T}_3$ in the cord. (Assume that no slipping occurs between the cord and pulleys.)

Solution First, let us write Newton's second law of motion as applied to each block, taking $m_2 > m_1$. The free-body diagrams are shown in Figure 10.19b.

$$T_1 - m_1 g = m_1 a \quad (1)$$

$$m_2 g - T_3 = m_2 a \quad (2)$$

Next, we must include the effect of the pulleys on the motion. Free-body diagrams for the pulleys are shown in Figure 10.19c.

The net torque about the axle for the pulley on the left is $(T_2 - T_1)R$, while the net torque for the pulley on the right is $(T_3 - T_2)R$. Using the relation $\tau_{\text{net}} = I\alpha$ for

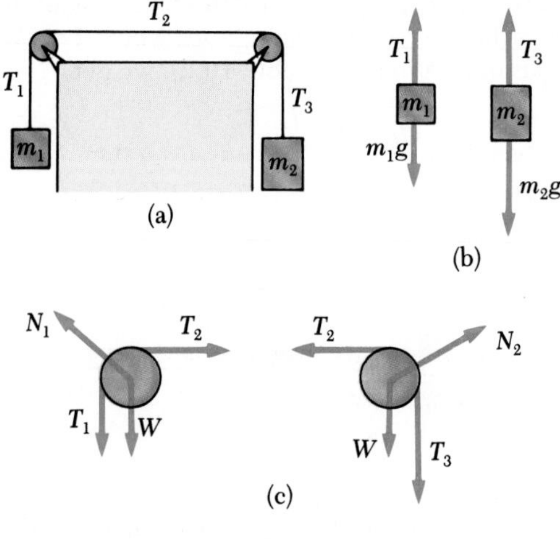

Figure 10.19 (Example 10.12).

each wheel, and noting that each wheel has the same α gives

$$(T_2 - T_1)R = I\alpha \tag{3}$$

$$(T_3 - T_2)R = I\alpha \tag{4}$$

We now have four equations with four unknowns; namely, a, T_1, T_2, and T_3. These can be solved simultaneously. Adding Equations (3) and (4) gives

$$(T_3 - T_1)R = 2I\alpha \tag{5}$$

Adding Equations (1) and (2) gives

$$T_1 - T_3 + m_2 g - m_1 g = (m_1 + m_2)a$$

or

$$T_3 - T_1 = (m_2 - m_1)g - (m_1 + m_2)a \tag{6}$$

Substituting Equation (6) into Equation (5), we have

$$[(m_2 - m_1)g - (m_1 + m_2)a]R = 2I\alpha$$

Since $\alpha = \dfrac{a}{R}$, this can be simplified as follows:

$$(m_2 - m_1)g - (m_1 + m_2)a = 2I\frac{a}{R^2}$$

or

$$a = \frac{(m_2 - m_1)g}{m_1 + m_2 + 2\dfrac{I}{R^2}} \tag{7}$$

This value of a can then be substituted into Equations (1) and (2) to give T_1 and T_3. Finally, T_2 can be found from Equation (3) or Equation (4).

10.8 WORK AND ENERGY IN ROTATIONAL MOTION

The description of a rotating rigid body would not be complete without a discussion of the rotational kinetic energy and how its change is related to the work done by external forces.

Again, we shall restrict our discussion to rotation about a fixed axis located in an inertial frame. Furthermore, we shall see that the important relationship $\tau_{net} = I\alpha$ derived in the previous section can also be obtained by considering the rate at which energy is changing with time.

Consider a rigid body pivoted at the point O in Figure 10.20. Suppose a single external force $\mathbf{F}$ is applied at the point P. The work done by $\mathbf{F}$ as the body rotates through an infinitesimal distance $ds = r\,d\theta$ in a time dt is

$$dW = \mathbf{F}\cdot d\mathbf{s} = (F\sin\phi)r\,d\theta$$

where $F\sin\phi$ is the tangential component of $\mathbf{F}$, or the component of the force along the displacement. Note from Figure 10.20 that *the radial component of* $\mathbf{F}$ *does no work since it is perpendicular to the displacement.*

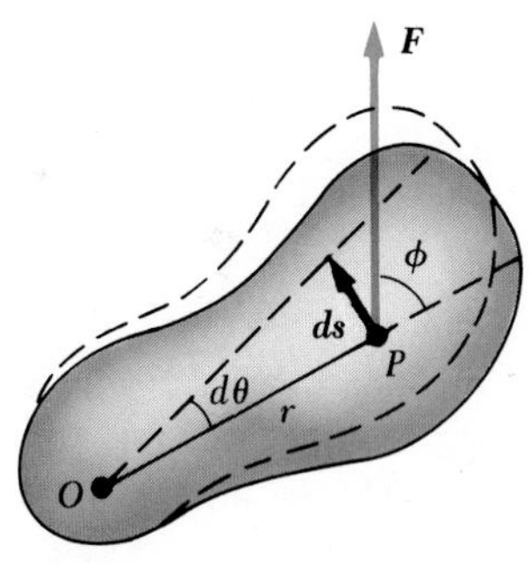

Figure 10.20 A rigid body rotates about an axis through O under the action of an external force $\mathbf{F}$ applied at P.

Since the magnitude of the torque due to $\mathbf{F}$ about the origin was defined as $rF\sin\phi$, we can write the work done for the infinitesimal rotation

$$dW = \tau\,d\theta \tag{10.21}$$

The rate at which work is being done by $\mathbf{F}$ for rotation about the fixed axis is obtained by formally dividing the left and right sides of Equation 10.21 by dt:

$$\frac{dW}{dt} = \tau\frac{d\theta}{dt} \tag{10.22}$$

But the quantity dW/dt is, by definition, the instantaneous power, P, delivered by the force. Furthermore, since $d\theta/dt = \omega$, Equation 10.22 reduces to

Power delivered to a rigid body

$$P = \frac{dW}{dt} = \tau\omega \tag{10.23}$$

This expression is analogous to $P = Fv$ in the case of linear motion, and the expression $dW = \tau\, d\theta$ is analogous to $dW = F_x\, dx$.

The Work-Energy Theorem in Rotational Motion

In linear motion, we found the energy concept, and in particular the work-energy theorem, to be extremely useful in describing the motion of a system. The energy concept can be equally useful in simplifying the analysis of rotational motion. From what we learned of linear motion, we expect that for rotation of a symmetric object (such as a symmetric wheel) about a fixed axis, the work done by external forces will equal the change in the rotational kinetic energy. To show that this is in fact the case, let us begin with $\tau = I\alpha$. Using the chain rule from the calculus, we can express the torque as

$$\tau = I\alpha = I\frac{d\omega}{dt} = I\frac{d\omega}{d\theta}\frac{d\theta}{dt} = I\frac{d\omega}{d\theta}\omega$$

Rearranging the above expression and noting that $\tau\, d\theta = dW$, we get

$$\tau\, d\theta = dW = I\omega\, d\omega$$

Integrating this expression, we get for the total work done

Work-energy theorem for rotational motion

$$W = \int_{\theta_0}^{\theta} \tau\, d\theta = \int_{\omega_0}^{\omega} I\omega\, d\omega = \tfrac{1}{2}I\omega^2 - \tfrac{1}{2}I\omega_0^2 \tag{10.24}$$

where the angular velocity changes from ω_0 to ω as the angular displacement changes from θ_0 to θ. This expression is analogous to the expression for the work-energy theorem in linear motion with m replaced by I and v replaced by ω. That is,

> the net work done by external forces in rotating a symmetric rigid body about a fixed axis equals the change in the body's rotational kinetic energy.

TABLE 10.3 A Comparison of Useful Equations in Rotational and Translational Motion

Rotational Motion About a Fixed Axis	Linear Motion
Angular velocity $\omega = d\theta/dt$	Linear velocity $v = dx/dt$
Angular acceleration $\alpha = d\omega/dt$	Linear acceleration $a = dv/dt$
Resultant torque $\Sigma\tau = I\alpha$	Resultant force $\Sigma F = Ma$
If $\alpha =$ constant $\begin{cases} \omega = \omega_0 + \alpha t \\ \theta - \theta_0 = \omega_0 t + \frac{1}{2}\alpha t^2 \\ \omega^2 = \omega_0^2 + 2\alpha(\theta - \theta_0) \end{cases}$	If $a =$ constant $\begin{cases} v = v_0 + at \\ x - x_0 = v_0 t + \frac{1}{2}at^2 \\ v^2 = v_0^2 + 2a(x - x_0) \end{cases}$
Work $W = \int_{\theta_0}^{\theta} \tau\, d\theta$	Work $W = \int_{x_0}^{x} F_x\, dx$
Kinetic energy $K = \frac{1}{2}I\omega^2$	Kinetic energy $K = \frac{1}{2}mv^2$
Power $P = \tau\omega$	Power $P = Fv$
Angular momentum $L = I\omega$	Linear momentum $p = mv$
Resultant torque $\tau = dL/dt$	Resultant force $F = dp/dt$

Table 10.3 lists the various equations we have discussed pertaining to rotational motion, together with the analogous expressions for linear motion. The last two equations in Table 10.3, involving the concept of angular momentum L, will be discussed in Chapter 11 and are included for completeness. In all cases, note the similarity between the equations of rotational motion and those of linear motion.

EXAMPLE 10.13 Rotating Rod—Revisited

A uniform rod of length L and mass M is free to rotate on a frictionless pin through one end (Fig. 10.21). The rod is released from rest in the horizontal position. (a) What is the angular velocity of the rod when it is at its lowest position?

The question can be easily answered by considering the mechanical energy of the system. When the rod is in the horizontal position, it has no kinetic energy. Its potential energy relative to the lowest position of its center of mass (O') is $MgL/2$. When it reaches its lowest position, the energy is entirely kinetic energy, $\frac{1}{2}I\omega^2$, where I is the moment of inertia about the pivot. Since $I = \frac{1}{3}ML^2$ (Table 10.2) and since mechanical energy is conserved, we have

$$\tfrac{1}{2}MgL = \tfrac{1}{2}I\omega^2 = \tfrac{1}{2}(\tfrac{1}{3}ML^2)\omega^2$$

$$\omega = \sqrt{\frac{3g}{L}}$$

For example, if the rod is a meter stick, we find that $\omega = 5.42$ rad/s.

(b) Determine the linear velocity of the center of mass and the linear velocity of the lowest point on the rod in the vertical position.

$$v_c = r\omega = \frac{L}{2}\omega = \boxed{\tfrac{1}{2}\sqrt{3gL}}$$

The lowest point on the rod has a linear velocity equal to $2v_c = \sqrt{3gL}$.

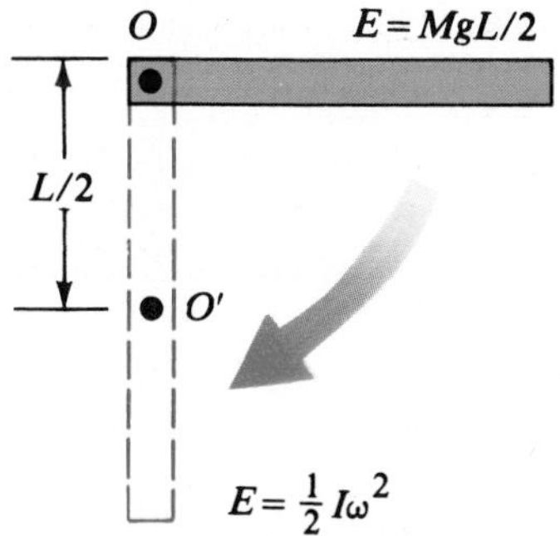

Figure 10.21 (Example 10.13) A uniform rigid rod pivoted at O rotates in a vertical plane under the action of gravity.

EXAMPLE 10.14 Connected Masses

Consider two masses connected by a string passing over a pulley having a moment of inertia I about its axis of rotation, as in Figure 10.22. The string does not slip on the pulley, and the system is released from rest. Find the linear velocities of the masses after m_2 descends through a distance h, and the angular velocity of the pulley at this time.

Solution If we neglect friction in the system, then mechanical energy is conserved and we can state that the increase in kinetic energy of the system equals the decrease in potential energy. Since $K_i = 0$ (the system is initially at rest), we have

$$\Delta K = K_f - K_i = \tfrac{1}{2}m_1v^2 + \tfrac{1}{2}m_2v^2 + \tfrac{1}{2}I\omega^2$$

where m_1 and m_2 have a common speed. But $v = R\omega$, so that

$$\Delta K = \tfrac{1}{2}\left(m_1 + m_2 + \frac{I}{R^2}\right)v^2$$

From Figure 10.22, we see that m_2 loses potential energy while m_1 gains potential energy. That is, $\Delta U_2 = -m_2gh$ and $\Delta U_1 = m_1gh$. Applying the law of conservation of energy in the form $\Delta K + \Delta U_1 + \Delta U_2 = 0$ gives

$$\tfrac{1}{2}\left(m_1 + m_2 + \frac{I}{R^2}\right)v^2 + m_1gh - m_2gh = 0$$

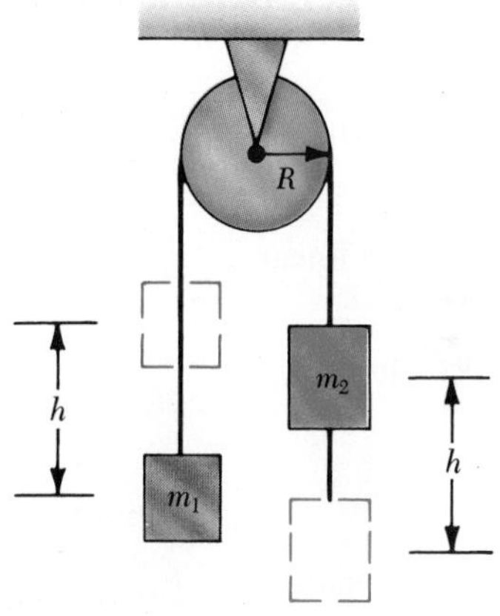

Figure 10.22 (Example 10.14).

$$v = \left[\frac{2(m_2 - m_1)gh}{\left(m_1 + m_2 + \frac{I}{R^2}\right)} \right]^{1/2}$$

Since $v = R\omega$, the angular velocity of the pulley at this instant is given by $\omega = v/R$.

Exercise 6 Repeat the calculation of v in Example 10.14 using $\tau_{net} = I\alpha$ applied to the pulley and Newton's second law applied to m_1 and m_2. Make use of the procedure presented in Examples 10.11 and 10.12.

SUMMARY

The instantaneous angular velocity of a particle rotating in a circle or of a rigid body rotating about a fixed axis is given by

Instantaneous angular velocity

$$\omega = \frac{d\theta}{dt} \tag{10.3}$$

where ω is in rad/s, or s^{-1}.

The instantaneous angular acceleration of a rotating body is given by

Instantaneous angular acceleration

$$\alpha = \frac{d\omega}{dt} \tag{10.5}$$

and has units of rad/s^2, or s^{-2}.

When a rigid body rotates about a fixed axis, every part of the body has the same angular velocity and the same angular acceleration. However, different parts of the body, in general, have different linear velocities and linear accelerations.

If a particle or body undergoes rotational motion about a fixed axis under constant angular acceleration α, one can apply equations of kinematics in analogy with kinematic equations for linear motion under constant linear acceleration:

Rotational kinematic equations

$$\omega = \omega_0 + \alpha t \tag{10.6}$$

$$\theta = \theta_0 + \omega_0 t + \tfrac{1}{2}\alpha t^2 \tag{10.7}$$

$$\omega^2 = \omega_0{}^2 + 2\alpha(\theta - \theta_0) \tag{10.8}$$

When a rigid body rotates about a fixed axis, the angular velocity and angular acceleration are related to the linear velocity and tangential linear acceleration through the relationships

Relationship between linear and angular speed

$$v = r\omega \tag{10.9}$$

Relationship between linear and angular acceleration

$$a_t = r\alpha \tag{10.10}$$

The moment of inertia of a system of particles is given by

Moment of inertia for a system of particles

$$I = \sum m_i r_i{}^2 \tag{10.14}$$

If a rigid body rotates about a fixed axis with angular velocity ω, its kinetic energy can be written

Kinetic energy of a rotating rigid body

$$K = \tfrac{1}{2}I\omega^2 \tag{10.15}$$

where I is the moment of inertia about the axis of rotation.

The **moment of inertia** of a rigid body is given by

$$I = \int r^2\,dm \qquad (10.16)$$

Moment of inertia for a rigid body

where r is the distance from the mass element dm to the axis of rotation.

The **torque** associated with a force $\mathbf{F}$ acting on a body has a magnitude equal to

$$\tau = Fd \qquad (10.18)$$

Torque

where d is the moment arm of the force, which is the perpendicular distance from some origin to the line of action of the force. Torque is a measure of the tendency of the force to rotate the body about some axis.

If a rigid body free to rotate about a fixed axis has a **net external torque** acting on it, the body will undergo an angular acceleration α, where

$$\tau_{\text{net}} = I\alpha \qquad (10.20)$$

Net torque

The rate at which work is being done by external forces in rotating a rigid body about a fixed axis, or the **power delivered**, is given by

$$P = \tau\omega \qquad (10.23)$$

Power delivered to a rigid body

The net work done by external forces in rotating a rigid body about a fixed axis equals the change in the rotational kinetic energy of the body:

$$W = \tfrac{1}{2}I\omega^2 - \tfrac{1}{2}I\omega_0{}^2 \qquad (10.24)$$

Work-energy theorem for rotational motion

This is the **work-energy theorem** applied to rotational motion.

QUESTIONS

1. What is the magnitude of the angular velocity, $\boldsymbol{\omega}$, of the second hand of a clock? What is the direction of $\boldsymbol{\omega}$ as you view a clock hanging vertically? What is the angular acceleration, $\boldsymbol{\alpha}$ of the second hand?
2. A wheel rotates counterclockwise in the xy plane. What is the direction of $\boldsymbol{\omega}$? What is the direction of $\boldsymbol{\alpha}$ if the angular velocity is decreasing in time?
3. Are the kinematic expressions for θ, ω, and α valid when the angular displacement is measured in degrees instead of in radians?
4. A turntable rotates at a constant rate of 45 rev/min. What is the magnitude of its angular velocity in rad/s? What is its angular acceleration?
5. When a wheel of radius R rotates about a fixed axis, do all points on the wheel have the same angular velocity? Do they all have the same linear velocity? If the angular velocity is constant and equal to ω_0, describe the linear velocities and linear accelerations of the points at $r = 0$, $r = R/2$, and $r = R$.
6. Suppose $a = b$ and $M > m$ in the system of particles described in Figure 10.7. About what axis (x, y, or z) does the moment of inertia have the smallest value? the largest value?
7. A wheel is in the shape of a hoop as in Figure 10.8. In two separate experiments, the wheel is rotated from rest to an angular velocity ω. In one experiment, the rotation occurs about the z axis through O; in the other, the rotation occurs about an axis parallel to z through P. Which rotation requires more work?
8. Suppose the rod in Figure 10.9 has a nonuniform mass distribution. In general, would the moment of inertia about the y axis still equal $\frac{1}{12}ML^2$? If not, could the moment of inertia be calculated without knowledge of the manner in which the mass is distributed?
9. Suppose that only two external forces act on a rigid body, and the two forces are equal in magnitude but opposite in direction. Under what condition will the body rotate?
10. Explain how you might use the apparatus described in Example 10.11 to determine the moment of inertia of the wheel. (If the wheel is not a uniform disk the moment of inertia is not necessarily equal to $\frac{1}{2}MR^2$.)

11. Using the results from Example 10.11, how would you calculate the angular velocity of the wheel and the linear velocity of the suspended mass at, say, $t = 2$ s, if the system is released from rest at $t = 0$? Is the relation $v = R\omega$ valid in this situation?
12. If a small sphere of mass M were placed at the end of the rod in Figure 10.21, would the result for ω be greater than, less than, or equal to the value obtained in Example 10.13?
13. Explain why changing the axis of rotation of an object changes its moment of inertia.
14. Is it possible to change the translational kinetic energy of an object without changing its rotational kinetic energy?
15. Two cylinders having the same dimensions are set into rotation about their axes with the same angular velocity. One is hollow, and the other is filled with water. Which cylinder would be easier to stop rotating?
16. Must an object be rotating to have a nonzero moment of inertia?
17. If you see an object rotating, is there necessarily a net torque acting on it?
18. Can a (momentarily) stationary object have a nonzero angular acceleration?
19. A particle is moving in a circle with constant speed. Locate one point about which the particle's angular momentum is constant and another about which it changes with time.
20. The polar diameter of the earth is slightly less than the equatorial diameter. How would the moment of inertia of the earth change if some mass from near the equator were removed and transferred to the polar regions to make the earth a perfect sphere?
21. During a wrecking operation, a tall chimney is toppled by an explosive charge at its base. The chimney is observed to rupture on its lower half while toppling, so that the bottom part reaches the ground before the top part. Explain why this occurs.

PROBLEMS

Section 10.2 Rotational Kinematics: Rotational Motion with Constant Angular Acceleration

1. A wheel starts from rest and rotates with constant angular acceleration to an angular velocity of 12 rad/s in a time of 3 s. Find (a) the angular acceleration of the wheel and (b) the angle in radians through which it rotates in this time.
2. The turntable of a record player rotates at the rate of $33\frac{1}{3}$ rev/min and takes 60 s to come to rest when switched off. Calculate (a) its angular acceleration and (b) the number of revolutions it makes before coming to rest.
3. What is the angular speed in rad/s of (a) the earth in its orbit about the sun and (b) the moon in its orbit about the earth?
4. A wheel rotates in such a way that its angular displacement in a time t is given by $\theta = at^2 + bt^3$, where a and b are constants. Determine equations for (a) the angular speed and (b) the angular acceleration, both as functions of time.
5. An electric motor rotating a workshop grinding wheel at a rate of 100 rev/min is switched off. Assuming constant negative acceleration of magnitude 2 rad/s^2 (a) how long will it take for the grinding wheel to stop? (b) through how many radians has the wheel turned during the time found in (a)?
6. The angular position of a point on a wheel can be described by $\theta = 5 + 10t + 2t^2$ rad. Determine the angular position, speed, and acceleration of the point at $t = 0$ and $t = 3$ s.
7. A grinding wheel, initially at rest, is rotated with constant angular acceleration $\alpha = 5$ rad/s^2 for 8 s. The wheel is then brought to rest with uniform negative acceleration in 10 revolutions. Determine the negative acceleration required and the time needed to bring the wheel to rest.
8. A wheel, starting from rest, rotates with an angular acceleration $\alpha = (10 + 6t)$ rad/s^2, where t is in seconds. Determine the angle in radians through which the wheel has turned in the first four seconds.

Section 10.3 Relationships Between Angular and Linear Quantities

9. A racing car travels on a circular track of radius 250 m. If the car moves with a constant speed of 45 m/s, find (a) the angular speed of the car and (b) the magnitude and direction of the car's acceleration.
10. The racing car described in Problem 9 starts from rest and accelerates uniformly to a speed of 45 m/s in 15 s. Find (a) the average angular speed of the car in this interval, (b) the angular acceleration of the car, (c) the magnitude of the car's linear acceleration at $t = 10$ s, and (d) the total distance traveled in the first 30 s.
11. A wheel 2 m in diameter rotates with a constant angular acceleration of 4 rad/s^2. The wheel starts at rest at $t = 0$, and the radius vector at point P on the rim makes an angle of 57.3° with the horizontal at this time. At $t = 2$ s, find (a) the angular speed of the wheel, (b) the linear velocity and acceleration of the point P, and (c) the position of the point P.
12. A cylinder of radius 0.1 m starts from rest and rotates about its axis with a constant angular acceleration of 5 rad/s^2. At $t = 3$ s, what is (a) its angular velocity, (b) the linear speed of a point on its rim, and (c) the

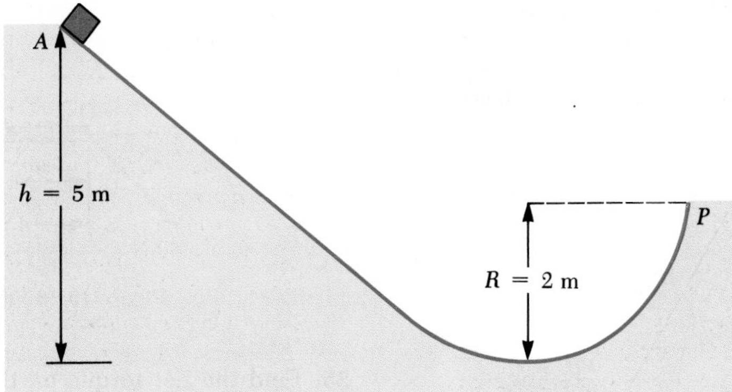

Figure 10.23 (Problem 15).

radial and tangential components of acceleration of a point on its rim?

13. A disk 8 cm in radius rotates at a constant rate of 1200 rev/min about its axis. Determine (a) the angular speed of the disk, (b) the linear speed at a point 3 cm from its center, (c) the radial acceleration of a point on the rim, and (d) the total distance a point on the rim moves in 2 s.
14. A car is travelling at 36 km/h on a straight road. The radius of the tires is 25 cm. Find the angular speed of one of the tires with its axle taken as the axis of rotation.
15. A 6-kg block is released from A on a frictionless track as shown in Figure 10.23. Determine the radial and tangential components of acceleration for the block at P.

Section 10.4 Rotational Kinetic Energy

16. A tire of moment of inertia 80 kg·m^2 rotates about a fixed central axis at the rate of 600 rev/min. What is its kinetic energy?
17. The four particles in Figure 10.24 are connected by rigid rods of negligible mass. If the system rotates in the xy plane about the z axis with an angular velocity of 6 rad/s, calculate (a) the moment of inertia of the system about the z axis and (b) the kinetic energy of the system.

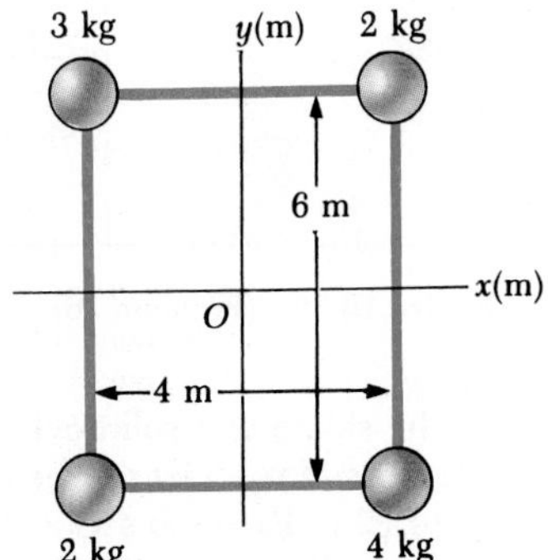

Figure 10.24 (Problems 17 and 18).

18. The system of particles described in Problem 17 (Fig. 10.24) rotates about the y axis. Calculate (a) the moment of inertia about the y axis and (b) the work required to take the system from rest to an angular speed of 6 rad/s.
19. Three particles are connected by rigid rods of negligible mass lying along the y axis (Fig. 10.25). If the system rotates about the x axis with an angular speed of 2 rad/s, find (a) the moment of inertia about the x axis and the total kinetic energy evaluated from $\frac{1}{2}I\omega^2$ and (b) the linear speed of each particle and the total kinetic energy evaluated from $\Sigma\frac{1}{2}m_iv_i^2$.

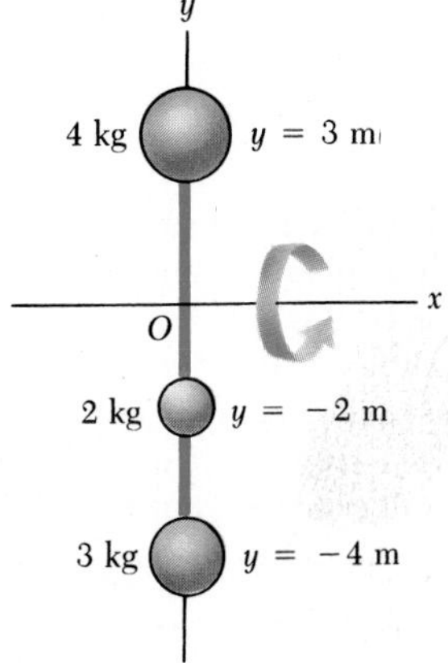

Figure 10.25 (Problem 19).

20. Three particles each of mass M are arranged at the vertices of an equilateral triangle as shown in Figure 10.26. Determine the moment of inertia about the x, y

and z axes. The z axis passes through O and is normal to the xOy plane.

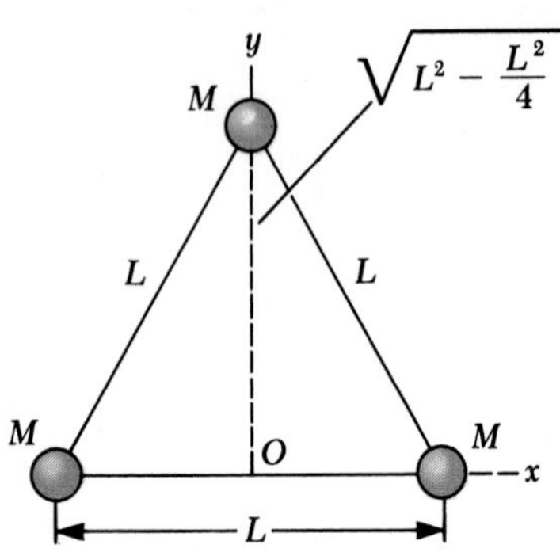

Figure 10.26 (Problem 20).

21. Two masses M and m are connected together by a rigid rod of length L and negligible mass as in Figure 10.27. For an axis perpendicular to the rod, show that the system has the minimum moment of inertia when the axis passes through the center of mass. Show that this moment of inertia is $I = \mu L^2$ where $\mu = mM/(m + M)$.

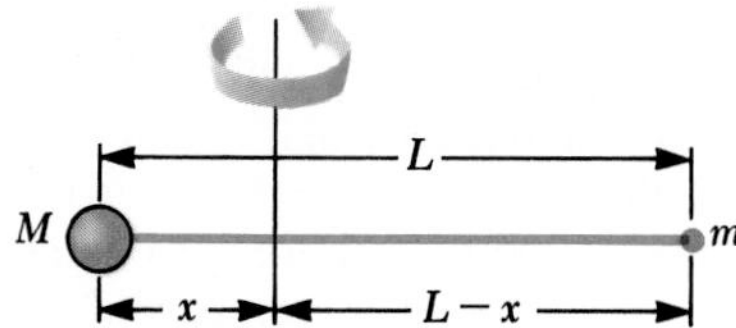

Figure 10.27 (Problem 21).

Section 10.5 Calculation of Moments of Inertia

22. Following the procedure used in Example 10.6, prove that the moment of inertia about the y' axis of the rigid rod in Figure 10.9 is $\frac{1}{3}ML^2$.
23. Use the parallel-axis theorem and Table 10.2 to find the moments of inertia of (a) a solid cylinder about an axis parallel to the center of mass axis and passing through the edge of the cylinder and (b) a solid sphere about an axis tangent to the surface of the sphere.

Section 10.6 Torque

24. Calculate the net torque (magnitude and direction) on the beam shown in Figure 10.28 about (a) an axis through O, perpendicular to the figure and (b) an axis through C, perpendicular to the figure.

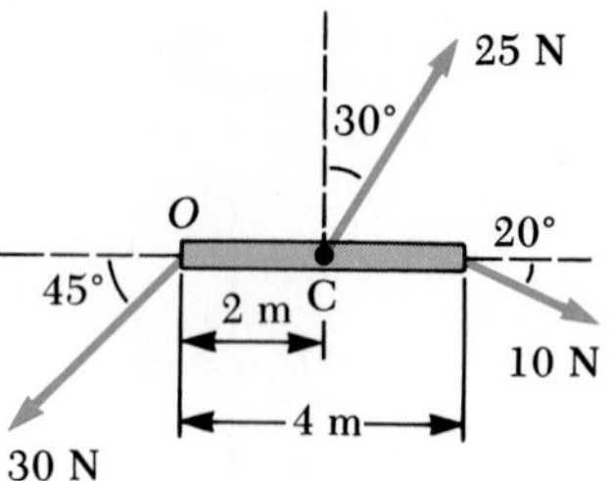

Figure 10.28 (Problem 24).

25. Find the net torque on the wheel in Figure 10.29 about the axle through O if $a = 10$ cm and $b = 25$ cm.

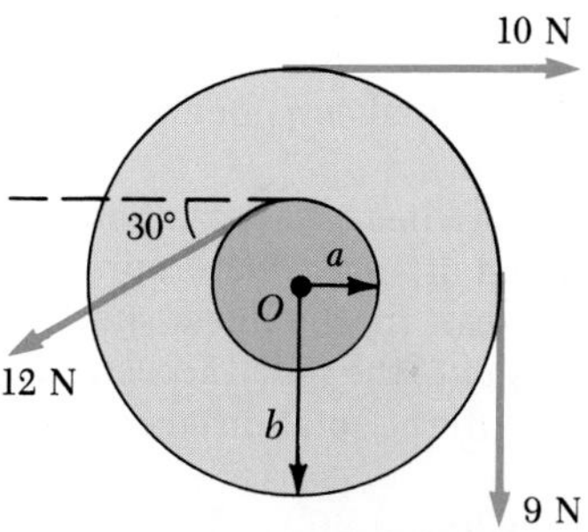

Figure 10.29 (Problem 25).

26. Find the mass m needed to balance the 150-kg truck on the incline shown in Figure 10.30. The angle of inclination θ is 45°. Assume all pulleys are frictionless and massless.

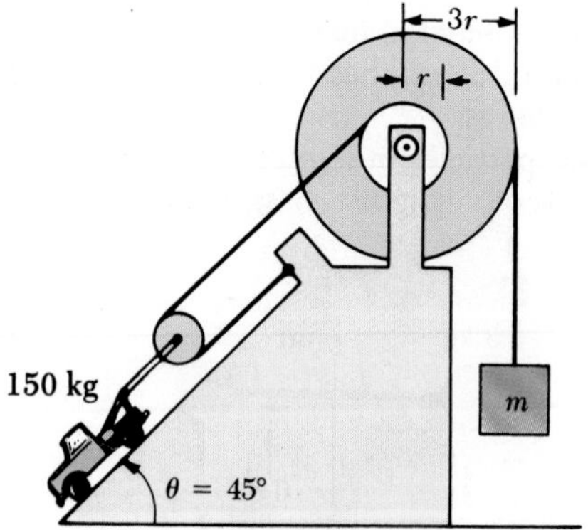

Figure 10.30 (Problem 26).

27. A flywheel in the shape of a solid cylinder of radius $R = 0.6$ m and mass $M = 15$ kg can be brought to an angular speed of 12 rad/s in 0.6 s by a motor exerting a constant torque. After the motor is turned off, the

flywheel makes 20 revolutions before coming to rest because of frictional losses (assumed constant during rotation). What percentage of the power generated by the motor is used to overcome the frictional losses?

Section 10.7 Relationship Between Torque and Angular Acceleration

28. The combination of an applied force and a frictional force produces a constant total torque of 36 N·m on a wheel rotating about a fixed axis. The applied force acts for 6 s, during which time the angular speed of the wheel increases from 0 to 10 rad/s. The applied force is then removed, and the wheel comes to rest in 60 s. Find (a) the moment of inertia of the wheel, (b) the magnitude of the frictional torque, and (c) the total number of revolutions of the wheel.

29. If a motor is to produce a torque of 50 N·m on a wheel rotating at 2400 rev/min, how much power must the motor deliver?

30. The system described in Example 10.11 (Fig. 10.18) is released from rest. After the mass m has fallen through a distance h, find (a) the linear velocity of the mass m and (b) the angular speed of the wheel.

Section 10.8 Work and Energy in Rotational Motion

31. A wheel 1 m in diameter rotates on a fixed, frictionless, horizontal axle. Its moment of inertia about this axis is 5 kg·m^2. A constant tension of 20 N is maintained on a rope wrapped around the rim of the wheel, so as to cause the wheel to accelerate. If the wheel starts from rest at $t = 0$, find (a) the angular acceleration of the wheel, (b) the wheel's angular speed at $t = 3$ s, (c) the kinetic energy of the wheel at $t = 3$ s, and (d) the length of rope unwound in the first 3 s.

32. A 15-kg mass is attached to a cord that is wrapped around a wheel of radius $r = 10$ cm (Fig. 10.31). The acceleration of the mass down the frictionless incline is measured to be 2.5 m/s^2. Assuming the axle of the wheel to be frictionless, determine (a) the tension in the rope, (b) the moment of inertia of the wheel, and (c) the angular speed of the wheel 2 s after it begins rotating, starting from rest.

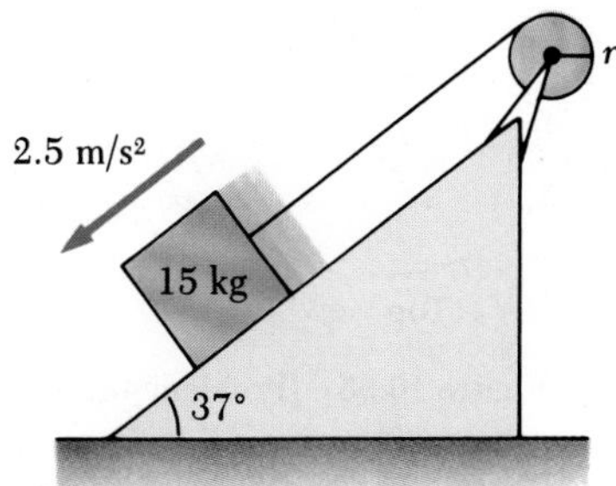

Figure 10.31 (Problem 32).

33. (a) A uniform solid disk of radius R and mass M is free to rotate on a frictionless pivot through a point on its rim (Fig. 10.32). If the disk is released from rest in the position shown by the green circle, what is the velocity of its center of mass when it reaches the position indicated by the dashed circle? (b) What is the speed of the lowest point on the disk in the dashed position? (c) Repeat part (a) if the object is a uniform hoop.

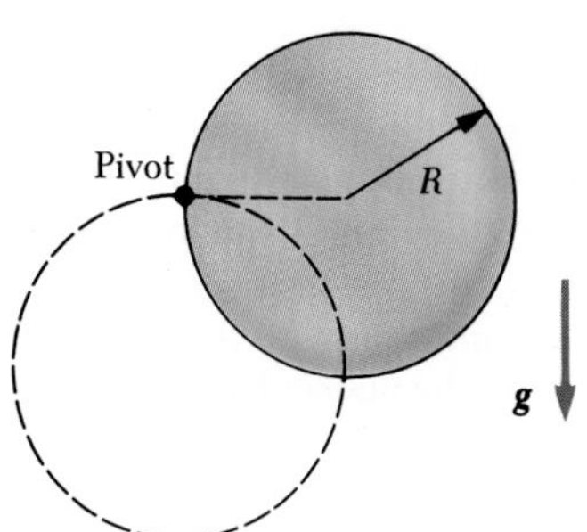

Figure 10.32 (Problem 33).

34. A potter's wheel, a thick stone disk of radius 0.5 m and mass 100 kg, is freely rotating at 50 rev/min. The potter can stop the wheel in 6 s by pressing a wet rag against the rim and exerting a radially inward force of 70 N. Find the effective coefficient of kinetic friction between the wheel and the wet rag.

35. A weight of 50 N is attached to the free end of a light string wrapped around a pulley of radius 0.25 m and mass 3 kg. The pulley is free to rotate in a vertical plane about the horizontal axis passing through its center. The weight is released 6 m above the floor. (a) Determine the tension in the string, the acceleration of the mass, and the velocity with which the weight hits the floor. (b) Find the velocity calculated in part (a) by using the law of conservation of energy.

36. Consider Figure 10.22 (Example 10.14) with $m_1 =$ 12.5 kg, $m_2 = 20$ kg, $R = 0.2$ m and the mass of the pulley $M = 5$ kg. Mass m_1 is resting on the floor and mass m_2 is 4 m above the floor when it is released from rest. Calculate the time taken by m_2 to hit the floor. How would your answer change if the pulley were massless?

ADDITIONAL PROBLEMS

37. A string is wound around a uniform disk of radius 0.25 m and mass 8 kg. The disk starts at rest and is free to turn about its axis. The end of the string is pulled with a constant force of 12 N. At time $t = 2$ s after the constant force is applied, determine (a) the torque exerted on the disk, (b) the angular acceleration of the disk, (c) the acceleration of the end of the string, (d) the angular velocity of the disk, (e) the velocity of the end of the string, (f) the kinetic energy of the disk, (g) the work done on the disk, (h) the angle θ through which the disk has turned, and (i) the length of rope pulled from the disk.

38. Calculate the moment of inertia of a uniform solid sphere of mass M and radius R about a diameter (see Table 10.2). (*Hint:* Treat the sphere as a set of disks of various radii, and first obtain an expression for the moment of inertia of one of these disks about the symmetry axis.)

39. A uniform solid cylinder of mass M and radius R rotates on a horizontal, frictionless axle (Fig. 10.33). Two equal masses hang from light cords wrapped around the cylinder. If the system is released from rest, find (a) the tension in each cord, (b) the acceleration of each mass, and (c) the angular velocity of the cylinder after the masses have descended a distance h.

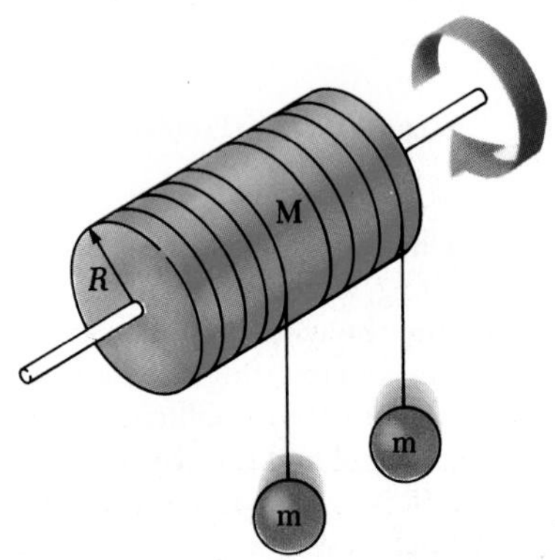

Figure 10.33 (Problem 39).

40. Find by integration the moment of inertia of a hollow cylinder about its symmetry axis. The mass of the cylinder is M, its inner radius is R_1, and its outer radius is R_2. (Check your result against the value given in Table 10.2.)

41. A 4-m length of light nylon cord is wound around a uniform cylindrical spool of radius 0.5 m and 1-kg mass. The spool is mounted on a frictionless axle and is initially at rest. The cord is pulled from the spool with a constant acceleration of 2.5 m/s^2. (a) How much work has been done on the spool, when it reaches an angular speed, $\omega = 8$ rad/s? (b) Assuming there is enough cord on the spool, how long will it take the spool to reach an angular speed of 8 rad/s? (c) Is there enough cord on the spool to enable the spool to reach this angular speed of 8 rad/s?

42. Many machines make use of heavy circular disks, called flywheels, to help maintain uniformity in rotational motion. The rotational inertia of a flywheel smooths fluctuations in rotational velocity incurred during operation, such as between power strokes in a gasoline engine. A particular flywheel of diameter 0.6 m and mass 200 kg is mounted on a frictionless bearing. A motor connected to the flywheel accelerates the flywheel from rest to 1000 rev/min. (a) What is the moment of inertia of the flywheel? (b) How much work is done on the flywheel during this acceleration? (c) After 1000 rpm is achieved the motor is disengaged from the flywheel. A friction brake is used to slow the rotational rate to 500 rpm. How much energy is dissipated as heat within the friction brake?

43. A long uniform rod of length L and mass M is pivoted about a horizontal, frictionless pin through one end. The rod is released from rest in a vertical position as in Figure 10.34. At the instant the rod is horizontal, find (a) the angular velocity of the rod, (b) its angular acceleration, (c) the x and y components of the acceleration of its center of mass, and (d) the components of the reaction force at the pivot.

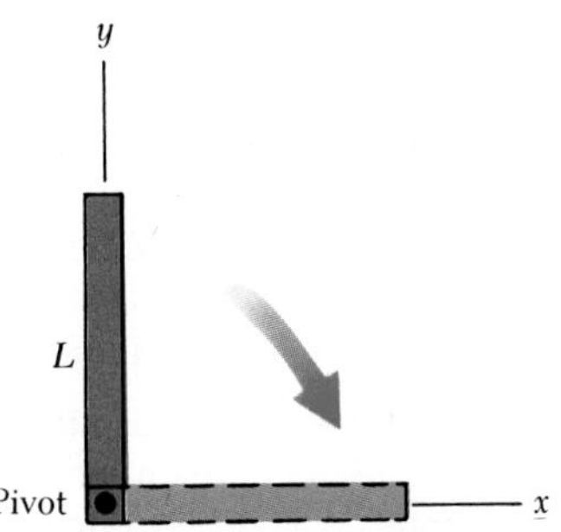

Figure 10.34 (Problem 43).

44. A small flat disk with a mass of 2 kg slides on a frictionless horizontal surface. The disk is held in a circular orbit by a 1.5-m rod of negligible mass, pivoted at one end, as shown in Figure 10.35. Initially, the disk has an orbit speed of 5 m/s. A 1-kg mass of putty is dropped onto the disk from directly above. If the putty sticks to the disk, what is the new period of rotation?

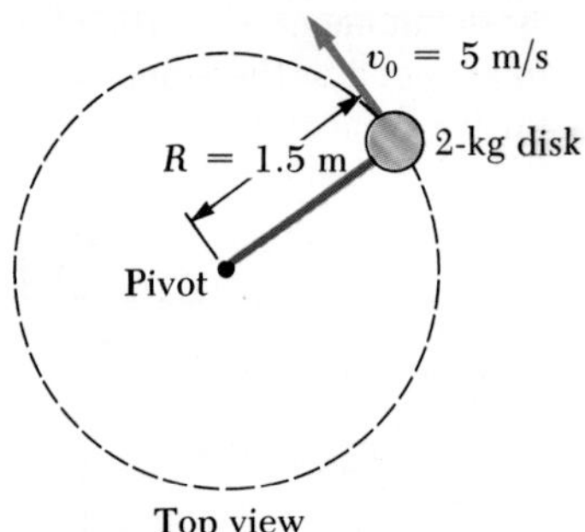

Figure 10.35 (Problem 44).

45. For any given rotational axis, the radius of gyration, K, of a rigid body is defined by the expression $K^2 = I/M$,

where M is the total mass of the body and I is the moment of inertia about the given axis. That is, the radius of gyration is equal to the distance of an imaginary point mass M from the axis of rotation such that I for the point mass about that axis is the same as for the rigid body about the same axis. Find the radius of gyration of (a) a solid disk of radius R, (b) a uniform rod of length L, and (c) a solid sphere of radius R, all three rotating about a central axis.

46. A bright physics student purchases a wind vane for her father's garage. The vane consists of a rooster sitting on top of an arrow. The vane is fixed to a vertical shaft of radius r and mass m that is free to turn in its roof mount as shown in Figure 10.36. The student sets up an experiment to measure the rotational inertia of the rooster and arrow attached to the shaft. String wound about the vertical shaft passes over a pulley and is connected to a mass M hanging over the edge of the garage roof. When the mass M is released, the student determines the time t that the mass takes to fall through a distance h. From these data the student is able to find the rotational inertia I of the rooster and arrow. Find the expression for I in terms of m, M, r, g, h, and t.

Figure 10.36 (Problem 46).

47. A uniform hollow cylindrical spool has an inside radius of $R/2$, outside radius R and mass M (see Fig. 10.37). It is mounted so as to rotate on a rough, fixed horizontal axle. A mass m is connected to the end of a string that is wound around the spool. The mass m is observed to fall from rest through a distance y in time t. Show that the torque due to the frictional forces between the spool and the axle is

$$\tau_F = R\left[m\left(g - \frac{2y}{t^2}\right) - \tfrac{5}{4}M\left(\frac{y}{t^2}\right)\right]$$

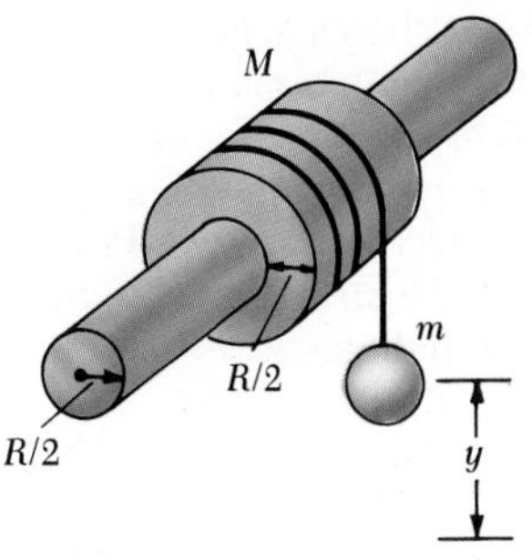

Figure 10.37 (Problem 47).

48. A cord is wrapped around a pulley of mass m and radius r. The free end of the cord is connected to a block of mass M. The block starts from rest and then slides down a rough incline which makes an angle θ with the horizontal. The coefficient of kinetic friction between the block and the incline is μ. (a) Use the work-energy theorem to show that the block's velocity v as a function of displacement d down the incline is

$$v = \left[4gd\left(\frac{M}{m + 2M}\right)(\sin\theta - \mu\cos\theta)\right]^{1/2}$$

(b) Find the acceleration of the block in terms of μ, m, M, g, and θ.

49. A circular hole of radius $R/4$ is cut in a disk of radius R and mass M. The hole is centered at $R/2$ from the center of the disk. Find an expression for the moment of inertia of this object about an axis normal to and passing through the center of the disk. [*Hint:* Consider the hole to be a disk of "negative-mass" material.]

50. A mass m_1 is connected by a light cord to a mass m_2, which slides on a smooth surface (Fig. 10.38). The pulley rotates about a frictionless axle and has a moment of inertia I and radius R. Assuming the cord does not slip on the pulley, find (a) the acceleration of the two masses, (b) the tensions T_1 and T_2, and (c) numerical values for a, T_1, and T_2 if $I = 0.5\ \text{kg}\cdot\text{m}^2$, $R = 0.3$ m, $m_1 = 4$ kg, and $m_2 = 3$ kg. (d) What would your answers be if the inertia of the pulley was neglected?

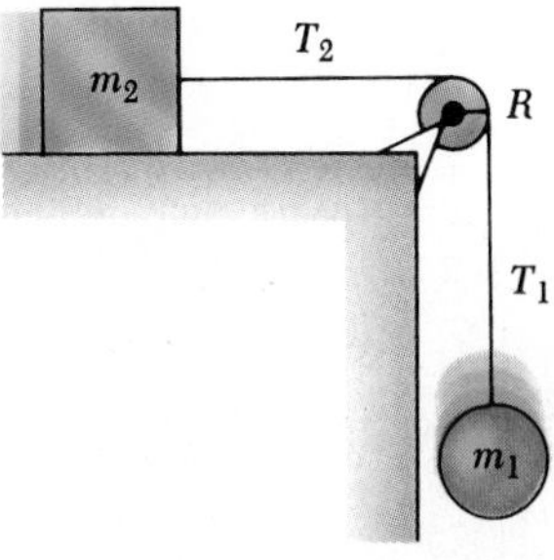

Figure 10.38 (Problem 50).

51. Two blocks, as shown in Fig. 10.39, are connected by a string of negligible mass passing over a pulley of radius 0.25 m and moment of inertia I. The block on the incline is moving up with a constant acceleration of 2 m/s^2. (a) Determine T_1 and T_2, the tensions in the two parts of the string, and (b) find the moment of inertia of the pulley.

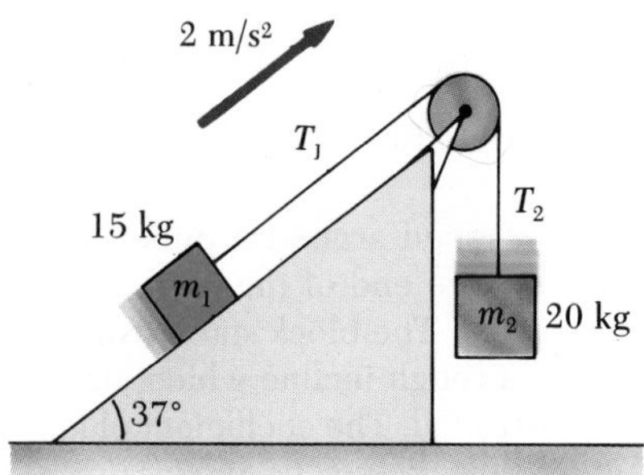

Figure 10.39 (Problem 51).

52. The pulley shown in Figure 10.40 has a radius R and moment of inertia I. One end of the mass m is connected to a spring of force constant k, and the other end is fastened to a cord wrapped around the pulley. The pulley axle and the incline are smooth. If the pulley is wound counterclockwise so as to stretch the spring a distance d from its *unstretched* position and then released from rest, find (a) the angular velocity of the pulley when the spring is again unstretched and (b) a numerical value for the angular velocity at this point if $I = 1\ \text{kg}\cdot\text{m}^2$, $R = 0.3$ m, $k = 50$ N/m, $m = 0.5$ kg, $d = 0.2$ m, and $\theta = 37°$.

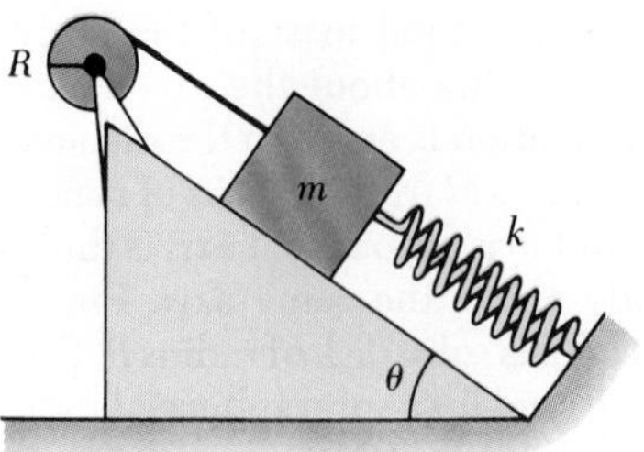

Figure 10.40 (Problem 52).

53. A cylindrical spool of thread of radius 0.1 m and mass 4 kg is mounted so that it can freely rotate about its longitudinal axis of symmetry. The spool starts rotating from rest as the thread is pulled off the spool, maintaining a constant tension of 8 N in the string. Neglect the mass of the unwrapped thread. Determine (a) the torque acting on the spool and the angular acceleration of the spool. At the instant the angular velocity of the spool is 10 rad/s, determine (b) the angle through which the spool has turned from the start, (c) the tangential and the radial acceleration of a point on the rim of the spool, and (d) the rotational kinetic energy.

54. An electrical motor can accelerate a Ferris wheel of moment of inertia $I = 20\ 000\ \text{kg}\cdot\text{m}^2$ from rest to 10 rev/min in 12 s. When the motor is turned off, the Ferris wheel slows down from 10 to 8 rev/min in 10 s due to frictional losses. Determine (a) the torque generated by the motor to bring the wheel to 10 rev/min, and (b) the power needed to maintain the Ferris wheel's rotation speed of 10 rev/min.

11

Rolling Motion, Angular Momentum, and Torque

Hurricane Elena photographed from space. This spiral-shaped, nonrigid mass of air undergoes rotation and has angular momentum. (Courtesy of NASA)

In the previous chapter we learned how to treat the rotation of a rigid body about a fixed axis. This chapter deals in part with the more general case, where the axis of rotation is not fixed in space. We begin by describing the rolling motion of an object such as a cylinder or sphere. Next, we define a vector product. The vector product is a convenient mathematical tool for expressing such quantities as torque and angular momentum. The central point of this chapter is to develop the concept of the angular momentum of a system of particles, a quantity that plays a key role in rotational dynamics. In analogy to the conservation of linear momentum, we shall find that the angular momentum of any isolated system (an isolated rigid body or any other isolated collection of particles) is always conserved. This conservation law is a special case of the result that the time rate of change of the total angular momentum of any system of particles equals the resultant external torque acting on the system.

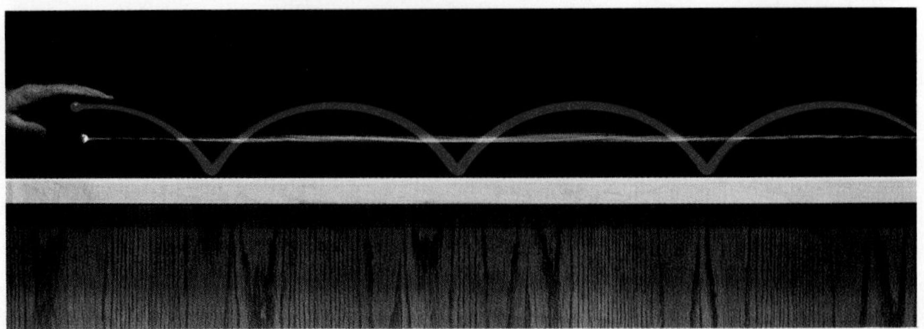

Figure 11.1 Light sources at the center and rim of a rolling cylinder illustrate the different paths these points take. The center moves in a straight line, as indicated by the green line, while a point on the rim moves in the path of a cycloid, as indicated by the red curve. (Courtesy of Henry Leap and Jim Lehman)

11.1 ROLLING MOTION OF A RIGID BODY

In this section we shall treat the motion of a rigid body that is rotating about a moving axis. The general motion of a rigid body in space is very complex. However, we can simplify matters by restricting our discussion to a homogeneous rigid body having a high degree of symmetry, such as a cylinder, sphere, or hoop. Furthermore, we shall assume that the body undergoes rolling motion in a plane.

Suppose a cylinder is rolling on a straight path as in Figure 11.1. The center of mass moves in a straight line, while a point on the rim moves in a more complex path, which corresponds to the path of a cycloid. As we shall see later in this chapter, it is convenient to view this latter motion as a combination of rotation about the center of mass and the translation of the center of mass.

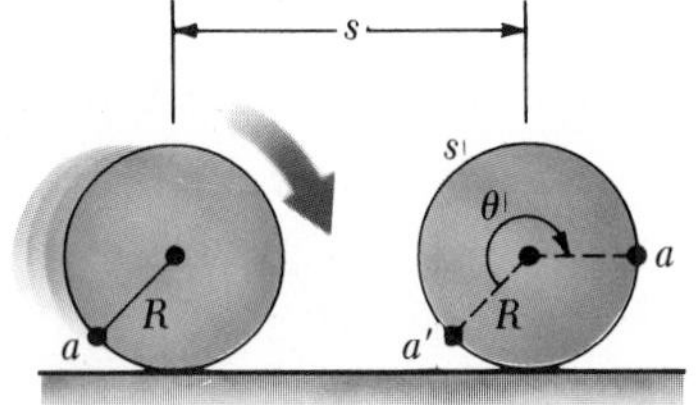

Figure 11.2 For pure rolling motion, as the cylinder rotates through an angle θ, the center of the cylinder moves a distance $s = R\theta$.

Now consider a uniform cylinder of radius R rolling on a rough, horizontal surface (Fig. 11.2). As the cylinder rotates through an angle θ, its center of mass moves a distance $s = R\theta$. Therefore, the velocity and acceleration of the center of mass for *pure rolling motion* are given by

$$v_c = \frac{ds}{dt} = R\frac{d\theta}{dt} = R\omega \tag{11.1}$$

$$a_c = \frac{dv_c}{dt} = R\frac{d\omega}{dt} = R\alpha \tag{11.2}$$

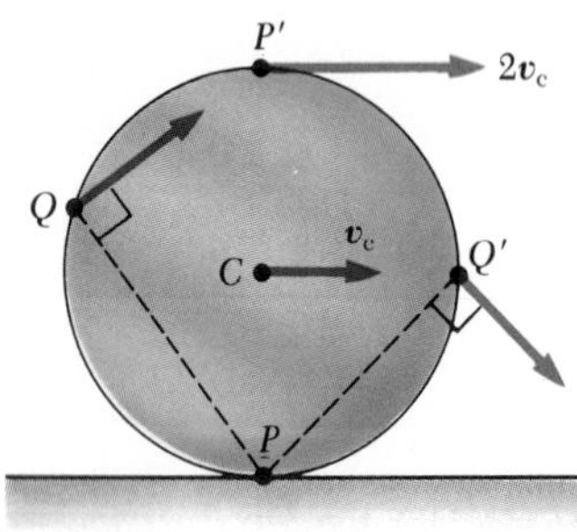

Figure 11.3 All points on a rolling body move in a direction perpendicular to an axis through the contact point P. The center of the body moves with a velocity $\mathbf{v}_c$, while the point P' moves with the velocity $2\mathbf{v}_c$.

The linear velocities of various points on the rolling cylinder are illustrated in Figure 11.3. Note that the linear velocity of any point is in a direction perpendicular to the line from that point to the contact point. At any instant, the point P is at rest relative to the surface since sliding does not occur.

A general point on the cylinder, such as Q, has both horizontal and vertical components of velocity. However, the points P and P' and the point at the center of mass are unique and of special interest. Relative to the surface on which the cylinder is moving, the center of mass moves with a velocity $v_c = R\omega$, whereas the contact point P has zero velocity. The point P' has a velocity equal to $2v_c = 2R\omega$, since all points on the cylinder have the same angular velocity.

We can express the total kinetic energy of the rolling cylinder as

$$K = \tfrac{1}{2}I_P\omega^2 \tag{11.3}$$

where I_P is the moment of inertia about the axis through P. Applying the parallel-axis theorem, we can substitute $I_P = I_c + MR^2$ into Equation 11.3 to get

$$K = \tfrac{1}{2}I_c\omega^2 + \tfrac{1}{2}MR^2\omega^2$$

$$K = \tfrac{1}{2}I_c\omega^2 + \tfrac{1}{2}Mv_c^2 \tag{11.4}$$

Total kinetic energy of a rolling body

where we have used the fact that $v_c = R\omega$.

We can think of Equation 11.4 as follows: The first term on the right, $\frac{1}{2}I_c\omega^2$, represents the rotational kinetic energy about the center of mass, and the term $\frac{1}{2}Mv_c^2$ represents the kinetic energy the cylinder would have if it were just translating through space without rotating. Thus, we can say that

> the total kinetic energy of an object undergoing rolling motion is the sum of a rotational kinetic energy about the center of mass and the translational kinetic energy of the center of mass.

We can use energy methods to treat a class of problems concerning the rolling motion of a rigid body down a rough incline. We shall assume that the rigid body in Figure 11.4 does not slip and is released from rest at the top of the incline. Note that rolling motion is possible only if a frictional force is present between the object and the incline to produce a net torque about the center of mass. Despite the presence of friction, there is no loss of mechanical energy since the contact point is at rest relative to the surface at any instant. On the other hand, if the rigid body were to slide, mechanical energy would be lost as motion progressed.

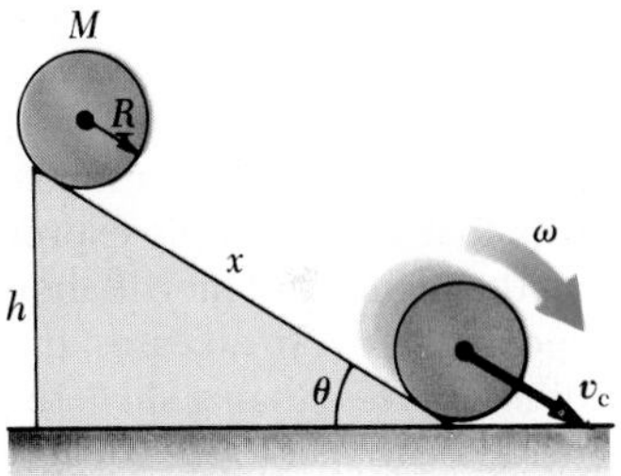

Figure 11.4 A round object rolling down an incline. Mechanical energy is conserved if no slipping occurs and there is no rolling friction.

Using the fact that $v_c = R\omega$ for pure rolling motion, we can express Equation 11.4 as

$$K = \tfrac{1}{2}I_c\left(\frac{v_c}{R}\right)^2 + \tfrac{1}{2}Mv_c^2$$

$$K = \tfrac{1}{2}\left(\frac{I_c}{R^2} + M\right)v_c^2 \tag{11.5}$$

When the rolling cylinder reaches the bottom of the incline, it has lost potential energy Mgh, where h is the height of the incline. If the body starts from rest at the top, its kinetic energy at the bottom, given by Equation 11.5, must equal its potential energy at the top. Therefore, the velocity of the center of mass at the bottom can be obtained by equating the two quantities:

$$\tfrac{1}{2}\left(\frac{I_c}{R^2} + M\right)v_c^2 = Mgh$$

$$v_c = \left(\frac{2gh}{1 + I_c/MR^2}\right)^{1/2} \tag{11.6}$$

EXAMPLE 11.1 Sphere Rolling Down an Incline
If the rigid body shown in Fig. 11.4 is a solid sphere, calculate the velocity of its center of mass at the bottom and determine the linear acceleration of the center of mass of the sphere.

Solution For a uniform solid sphere, $I_c = \frac{2}{5}MR^2$, and therefore Equation 11.6 gives

$$v_c = \left(\frac{2gh}{1 + \frac{\frac{2}{5}MR^2}{MR^2}}\right)^{1/2} = \left(\frac{10}{7}gh\right)^{1/2}$$

The vertical displacement is related to the displacement x along the incline through the relation $h = x \sin\theta$. Hence, after squaring both sides, we can express the equation above as

$$v_c^2 = \frac{10}{7} gx \sin\theta$$

Comparing this with the familiar expression from kinematics, $v_c^2 = 2a_c x$, we see that the acceleration of the center of mass is given by

$$a_c = \tfrac{5}{7} g \sin\theta$$

The results are quite interesting in that both the velocity and acceleration of the center of mass are *independent* of the mass and radius of the sphere! That is, *all homogeneous solid spheres would experience the same velocity and acceleration on a given incline.* If we repeated the calculations for a hollow sphere, a solid cylinder, or a hoop, we would obtain similar results. The constant factors that appear in the expressions for v_c and a_c depend on the moment of inertia about the center of mass for the specific body. In all cases, the acceleration of the center of mass is *less* than g sin θ, the value it would have if the plane were frictionless and no rolling occurred.

EXAMPLE 11.2 Another Look at the Rolling Sphere

In this example, let us consider the solid sphere rolling down an incline and verify the results of Example 11.1 using dynamic methods. The free-body diagram for the sphere is illustrated in Figure 11.5.

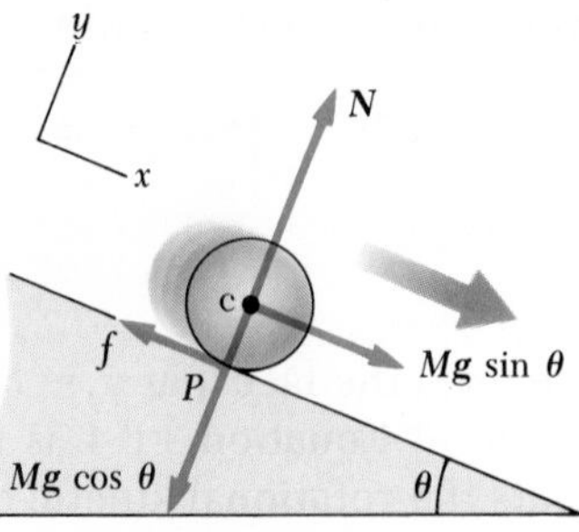

Figure 11.5 (Example 11.2) Free-body diagram for a solid sphere rolling down an incline.

Solution Newton's second law applied to the center of mass motion gives

$$(1) \qquad \sum F_x = Mg\sin\theta - f = Ma_c$$

$$\sum F_y = N - Mg\cos\theta = 0$$

where x is measured downward along the inclined plane. Now let us write an expression for the torque acting on the sphere. A convenient axis to choose is an axis through the center of the sphere, perpendicular to the plane of the figure.[1] Since $\mathbf{N}$ and $M\mathbf{g}$ go through this origin, they have zero moment arms and do not contribute to the torque. However, the force of friction produces a torque about this axis equal to fR in the clockwise direction; therefore

$$\tau_c = fR = I_c\alpha$$

Since $I_c = \frac{2}{5}MR^2$ and $\alpha = a_c/R$, we get

$$(2) \qquad f = \frac{I_c\alpha}{R} = \left(\frac{\frac{2}{5}MR^2}{R}\right)\frac{a_c}{R} = \tfrac{2}{5}Ma_c$$

Substituting (2) into (1) gives

$$a_c = \tfrac{5}{7} g \sin\theta$$

which agrees with the result of Example 11.1. Note that $a_c < g\sin\theta$ because of the retarding frictional force.

[1] You should note that although the point at the center of mass is not an inertial frame, the expression $\tau_c = I\alpha$ still applies in the center of mass frame.

11.2 THE VECTOR PRODUCT AND TORQUE

Consider a force $\mathbf{F}$ acting on a rigid body at the vector position $\mathbf{r}$ (Fig. 11.6). *The origin O is assumed to be in an inertial frame, so that Newton's second law is valid.* The *magnitude* of the torque due to this force relative to the origin is, by definition, equal to $rF\sin\phi$, where ϕ is the angle between $\mathbf{r}$ and $\mathbf{F}$. The axis about which $\mathbf{F}$ would tend to produce rotation is perpendicular to the plane formed by $\mathbf{r}$ and $\mathbf{F}$. If the force lies in the xy plane as in Figure 11.6, then the torque, $\boldsymbol{\tau}$ is represented by a vector parallel to the z axis. The force in Figure 11.6 creates a torque that tends to rotate the body counterclockwise looking down the z axis, so the sense of $\boldsymbol{\tau}$ is toward increasing z, and $\boldsymbol{\tau}$ is in the positive z direction. If we reversed the direction of $\mathbf{F}$ in Figure 11.6, $\boldsymbol{\tau}$ would then be in the negative z direction. The torque involves two vectors, $\mathbf{r}$ and $\mathbf{F}$, and is in fact defined to be equal to the *vector product*, or *cross product*, of $\mathbf{r}$ and $\mathbf{F}$:

$$\boldsymbol{\tau} \equiv \mathbf{r} \times \mathbf{F} \tag{11.7}$$

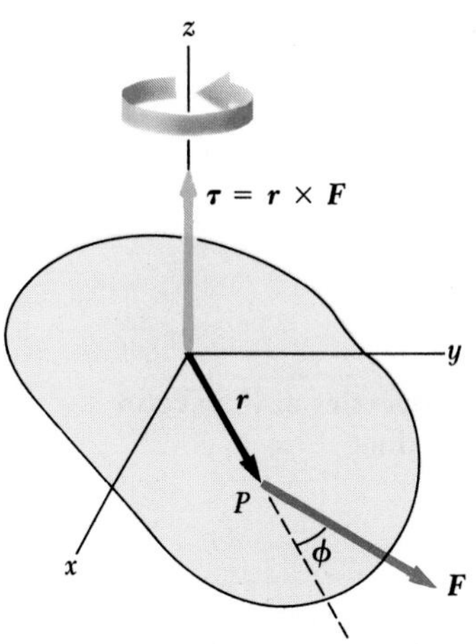

Figure 11.6 The torque vector $\boldsymbol{\tau}$ lies in a direction perpendicular to the plane formed by the position vector $\mathbf{r}$ and the applied force $\mathbf{F}$.

We must now give a formal definition of the vector product. Given any two vectors $\mathbf{A}$ and $\mathbf{B}$, the vector product $\mathbf{A} \times \mathbf{B}$ is defined as a third vector $\mathbf{C}$, the *magnitude* of which is $AB \sin \theta$, where θ is the angle included between $\mathbf{A}$ and $\mathbf{B}$. That is, if $\mathbf{C}$ is given by

$$\mathbf{C} = \mathbf{A} \times \mathbf{B} \tag{11.8}$$

then its magnitude is

$$C \equiv |\mathbf{C}| = |AB \sin \theta| \tag{11.9}$$

Note that the quantity $AB \sin \theta$ is equal to the area of the parallelogram formed by $\mathbf{A}$ and $\mathbf{B}$, as shown in Figure 11.7. The *direction* of $\mathbf{A} \times \mathbf{B}$ is perpendicular to the plane formed by $\mathbf{A}$ and $\mathbf{B}$, as in Figure 11.7, and its sense is determined by the advance of a right-handed screw when turned from $\mathbf{A}$ to $\mathbf{B}$ through the angle θ. A more convenient rule to use for the direction of $\mathbf{A} \times \mathbf{B}$ is the right-hand rule illustrated in Figure 11.7. The four fingers of the right hand are pointed along $\mathbf{A}$ and then "wrapped" into $\mathbf{B}$ through the angle θ. The direction of the erect right thumb is the direction of $\mathbf{A} \times \mathbf{B}$. Because of the notation, $\mathbf{A} \times \mathbf{B}$ is often read "**A** cross **B**"; hence the term *cross product.*

Some properties of the vector product which follow from its definition are as follows:

1. Unlike the scalar product, the order in which the two vectors are multiplied in a cross product is important, that is,

$$\mathbf{A} \times \mathbf{B} = -(\mathbf{B} \times \mathbf{A}) \tag{11.10}$$

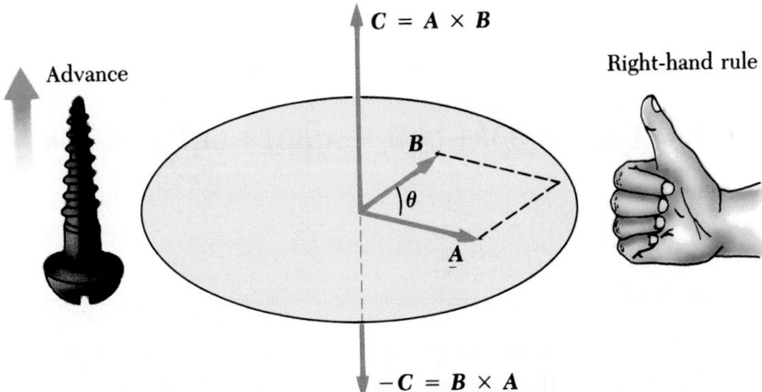

Figure 11.7 The vector product $\mathbf{A} \times \mathbf{B}$ is a third vector $\mathbf{C}$ having a magnitude $AB \sin \theta$ equal to the area of the parallelogram shown. The direction of $\mathbf{C}$ is perpendicular to the plane formed by $\mathbf{A}$ and $\mathbf{B}$, and its sense is determined by the right-hand rule.

Therefore, if you change the order of the cross product, you must change the sign. You could easily verify this relation with the right-hand rule (Fig. 11.7).

2. If $\boldsymbol{A}$ is parallel to $\boldsymbol{B}$ ($\theta = 0°$ or $180°$), then $\boldsymbol{A} \times \boldsymbol{B} = 0$; therefore, it follows that $\boldsymbol{A} \times \boldsymbol{A} = 0$.
3. If $\boldsymbol{A}$ is perpendicular to $\boldsymbol{B}$, then $|\boldsymbol{A} \times \boldsymbol{B}| = AB$.
4. It is also important to note that the vector product obeys the *distributive law*, that is,

Properties of the vector product

$$\boldsymbol{A} \times (\boldsymbol{B} + \boldsymbol{C}) = \boldsymbol{A} \times \boldsymbol{B} + \boldsymbol{A} \times \boldsymbol{C} \tag{11.11}$$

5. Finally, the derivative of the cross product with respect to some variable such as t is given by

$$\frac{d}{dt}(\boldsymbol{A} \times \boldsymbol{B}) = \boldsymbol{A} \times \frac{d\boldsymbol{B}}{dt} + \frac{d\boldsymbol{A}}{dt} \times \boldsymbol{B} \tag{11.12}$$

where it is important to preserve the multiplicative order of $\boldsymbol{A}$ and $\boldsymbol{B}$, in view of Equation 11.10.

It is left as an exercise to show from Equations 11.8 and 11.9 and the definition of unit vectors that the cross products of the rectangular unit vectors $\boldsymbol{i}, \boldsymbol{j}$, and $\boldsymbol{k}$ obey the following expressions:

Cross products of unit vectors

$$\boldsymbol{i} \times \boldsymbol{i} = \boldsymbol{j} \times \boldsymbol{j} = \boldsymbol{k} \times \boldsymbol{k} = 0 \tag{11.13a}$$

$$\boldsymbol{i} \times \boldsymbol{j} = -\boldsymbol{j} \times \boldsymbol{i} = \boldsymbol{k} \tag{11.13b}$$

$$\boldsymbol{j} \times \boldsymbol{k} = -\boldsymbol{k} \times \boldsymbol{j} = \boldsymbol{i} \tag{11.13c}$$

$$\boldsymbol{k} \times \boldsymbol{i} = -\boldsymbol{i} \times \boldsymbol{k} = \boldsymbol{j} \tag{11.13d}$$

Signs are interchangeable. For example, $\boldsymbol{i} \times (-\boldsymbol{j}) = -\boldsymbol{i} \times \boldsymbol{j} = -\boldsymbol{k}$.

The cross product of *any* two vectors $\boldsymbol{A}$ and $\boldsymbol{B}$ can be expressed in the following determinant form:

$$\boldsymbol{A} \times \boldsymbol{B} = \begin{vmatrix} \boldsymbol{i} & \boldsymbol{j} & \boldsymbol{k} \\ A_x & A_y & A_z \\ B_x & B_y & B_z \end{vmatrix}$$

Expanding this determinant gives the result

$$\boldsymbol{A} \times \boldsymbol{B} = (A_yB_z - A_zB_y)\boldsymbol{i} + (A_zB_x - A_xB_z)\boldsymbol{j} + (A_xB_y - A_yB_x)\boldsymbol{k} \tag{11.14}$$

EXAMPLE 11.3 The Cross Product

Two vectors lying in the xy plane are given by the equations $\boldsymbol{A} = 2\boldsymbol{i} + 3\boldsymbol{j}$ and $\boldsymbol{B} = -\boldsymbol{i} + 2\boldsymbol{j}$. Find $\boldsymbol{A} \times \boldsymbol{B}$, and verify explicitly that $\boldsymbol{A} \times \boldsymbol{B} = -\boldsymbol{B} \times \boldsymbol{A}$.

Solution Using Equations 11.13a through 11.13d for the cross product of unit vectors gives

$$\boldsymbol{A} \times \boldsymbol{B} = (2\boldsymbol{i} + 3\boldsymbol{j}) \times (-\boldsymbol{i} + 2\boldsymbol{j})$$

$$= 2\boldsymbol{i} \times 2\boldsymbol{j} + 3\boldsymbol{j} \times (-\boldsymbol{i}) = 4\boldsymbol{k} + 3\boldsymbol{k} = 7\boldsymbol{k}$$

(We have omitted the terms with $\boldsymbol{i}\times\boldsymbol{i}$ and $\boldsymbol{j}\times\boldsymbol{j}$, since they are zero.)

$$\boldsymbol{B}\times\boldsymbol{A} = (-\boldsymbol{i}+2\boldsymbol{j})\times(2\boldsymbol{i}+3\boldsymbol{j})$$
$$= -\boldsymbol{i}\times 3\boldsymbol{j} + 2\boldsymbol{j}\times 2\boldsymbol{i} = -3\boldsymbol{k} - 4\boldsymbol{k} = -7\boldsymbol{k}$$

Therefore, $\boldsymbol{A}\times\boldsymbol{B} = -\boldsymbol{B}\times\boldsymbol{A}$.

As an alternative method for finding $\boldsymbol{A}\times\boldsymbol{B}$, we could use Equation 11.8, with $A_x = 2, A_y = 3, A_z = 0$ and $B_x = -1, B_y = 2, B_z = 0$. This gives

$$\boldsymbol{A}\times\boldsymbol{B} = (0)\boldsymbol{i} + (0)\boldsymbol{j} + [2\times 2 - 3\times(-1)]\boldsymbol{k} = 7\boldsymbol{k}$$

Exercise 1 Use the results to this example and Equation 11.9 to find the angle between $\boldsymbol{A}$ and $\boldsymbol{B}$.
Answer 60.3°.

11.3 ANGULAR MOMENTUM OF A PARTICLE

A particle of mass m, located at the vector position $\boldsymbol{r}$, moves with a velocity $\boldsymbol{v}$ (Fig. 11.8).

The **instantaneous angular momentum** $\boldsymbol{L}$ of the particle relative to the origin O is defined by the cross product of its instantaneous vector position and the instantaneous linear momentum $\boldsymbol{p}$:

$$\boldsymbol{L} \equiv \boldsymbol{r}\times\boldsymbol{p} \tag{11.15}$$

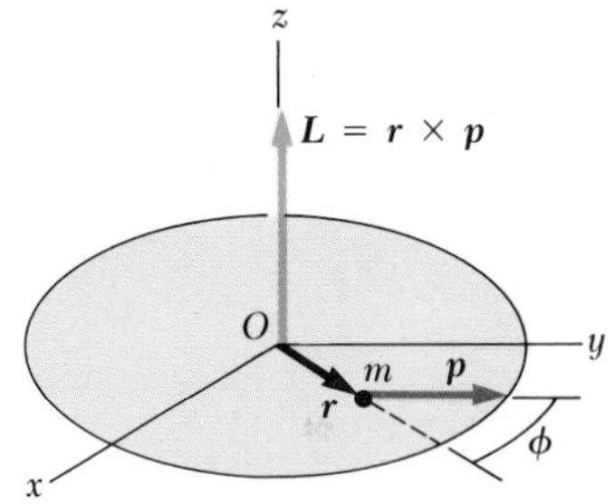

Figure 11.8 The angular momentum $\boldsymbol{L}$ of a particle of mass m and momentum $\boldsymbol{p}$ located at the position $\boldsymbol{r}$ is a vector given by $\boldsymbol{L} = \boldsymbol{r}\times\boldsymbol{p}$. The value of $\boldsymbol{L}$ depends on the origin and is a vector perpendicular to both $\boldsymbol{r}$ and $\boldsymbol{p}$.

The SI units of angular momentum are kg·m²/s. It is important to note that both the magnitude and direction of $\boldsymbol{L}$ depend on the choice of the origin. The direction of $\boldsymbol{L}$ is perpendicular to the plane formed by $\boldsymbol{r}$ and $\boldsymbol{p}$, and its sense is governed by the right-hand rule. For example, in Figure 11.8 $\boldsymbol{r}$ and $\boldsymbol{p}$ are assumed to be in the xy plane, so that $\boldsymbol{L}$ points in the z direction. Since $\boldsymbol{p} = m\boldsymbol{v}$, the magnitude of $\boldsymbol{L}$ is given by

$$L = mvr\sin\phi \tag{11.16}$$

where ϕ is the angle between $\boldsymbol{r}$ and $\boldsymbol{p}$. It follows that L is zero when $\boldsymbol{r}$ is parallel to $\boldsymbol{p}$ ($\phi = 0$ or 180°). In other words, when the particle moves along a line that passes through the origin, it has zero angular momentum with respect to the origin. This is equivalent to stating that it has no tendency to rotate about the origin. On the other hand, if $\boldsymbol{r}$ is perpendicular to $\boldsymbol{p}$ ($\phi = 90°$), then L is a maximum and equal to mrv. In this case, the particle has maximum tendency to rotate about the origin. In fact, at that instant the particle moves exactly as though it were on the rim of a wheel rotating about the origin in a plane defined by $\boldsymbol{r}$ and $\boldsymbol{p}$.

Alternatively, one may note that a particle has nonzero angular momentum about some point if its position vector measured from that point rotates about the point. On the other hand, if the position vector simply increases or decreases in length, the particle moves along a line passing through the origin and therefore has zero angular momentum with respect to that origin.

In the case of the linear motion of a particle, we found that the resultant force on a particle equalled the time rate of change of its linear momentum. We shall now show that Newton's second law implies that the resultant torque acting on a particle equals the time rate of change of its angular momentum. Let us start by writing the torque on the particle in the form

$$\boldsymbol{\tau} = \boldsymbol{r}\times\boldsymbol{F} = \boldsymbol{r}\times\frac{d\boldsymbol{p}}{dt} \tag{11.17}$$

where we have used the fact that $\boldsymbol{F} = d\boldsymbol{p}/dt$. Now let us differentiate Equation 11.15 with respect to time using the rule given by Equation 11.12.

$$\frac{d\boldsymbol{L}}{dt} = \frac{d}{dt}(\boldsymbol{r} \times \boldsymbol{p}) = \boldsymbol{r} \times \frac{d\boldsymbol{p}}{dt} + \frac{d\boldsymbol{r}}{dt} \times \boldsymbol{p}$$

It is important to adhere to the order of terms since $\boldsymbol{A} \times \boldsymbol{B} = -\boldsymbol{B} \times \boldsymbol{A}$.

The last term on the right in the above equation is zero, since $\boldsymbol{v} = d\boldsymbol{r}/dt$ is parallel to $\boldsymbol{p}$. Therefore,

$$\frac{d\boldsymbol{L}}{dt} = \boldsymbol{r} \times \frac{d\boldsymbol{p}}{dt} \qquad (11.18)$$

Comparing Equations 11.17 and 11.18, we see that

Torque equals time rate of change of angular momentum

$$\boldsymbol{\tau} = \frac{d\boldsymbol{L}}{dt} \qquad (11.19)$$

which is the rotational analog of Newton's second law, $\boldsymbol{F} = d\boldsymbol{p}/dt$. This result says that

> the **torque** acting on a particle is equal to the time rate of change of the particle's angular momentum.

It is important to note that Equation 11.19 is valid only if the origins of $\boldsymbol{\tau}$ and $\boldsymbol{L}$ are the *same*. It is left as an exercise to show that Equation 11.19 is also valid when there are several forces acting on the particle, in which case $\boldsymbol{\tau}$ is the *net* torque on the particle. *Furthermore, the expression is valid for any origin fixed in an inertial frame.* Of course, the same origin must be used in calculating all torques as well as the angular momentum.

A System of Particles

The total angular momentum, $\boldsymbol{L}$, of a system of particles about some point is defined as the vector sum of the angular momenta of the individual particles:

$$\boldsymbol{L} = \boldsymbol{L}_1 + \boldsymbol{L}_2 + \cdots + \boldsymbol{L}_n = \sum \boldsymbol{L}_i$$

where the vector sum is over all of the n particles in the system.

Since the individual momenta of the particles may change in time, the total angular momentum may also vary in time. In fact, from Equations 11.17 through 11.18, we find that the time rate of change of the total angular momentum equals the vector sum of *all* torques, including those associated with internal forces between particles and those associated with external forces. However, the net torque associated with internal forces is zero. To understand this, recall that Newton's third law tells us that the internal forces occur in equal and opposite pairs that lie along the line of separation of each pair of particles. Therefore, the torque due to each action-reaction force pair is zero. By summation, we see that *the net internal torque vanishes.* Finally, we conclude that the total angular momentum can vary with time *only* if there is a net *external* torque on the system, so that we have

$$\sum \boldsymbol{\tau}_{\text{ext}} = \sum \frac{d\boldsymbol{L}_i}{dt} = \frac{d}{dt}\sum \boldsymbol{L}_i = \frac{d\boldsymbol{L}}{dt} \qquad (11.20)$$

That is,

> the time rate of change of the total angular momentum of the system about some origin in an inertial frame equals the net external torque acting on the system about that origin.

Note that Equation 11.20 is the rotational analog of $\boldsymbol{F}_{ext} = d\boldsymbol{p}/dt$ for a system of particles (Chapter 9).

EXAMPLE 11.4 Linear Motion

A particle of mass m moves in the xy plane with a velocity $\boldsymbol{v}$ along a straight line (Fig. 11.9). What is the magnitude and direction of its angular momentum with respect to the origin O?

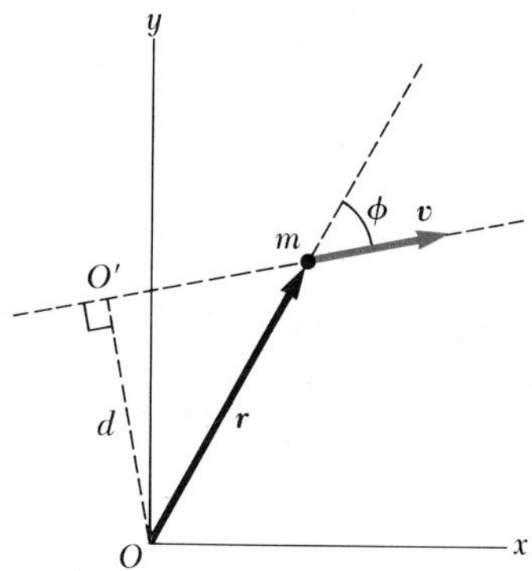

Figure 11.9 (Example 11.4) A particle moving in a straight line with a velocity $\boldsymbol{v}$ has an angular momentum equal in magnitude to mvd relative to O, where $d = r \sin \phi$ is the distance of closest approach to the origin. The vector $\boldsymbol{L} = \boldsymbol{r} \times \boldsymbol{p}$ points *into* the diagram in this case.

Solution From the definition of angular momentum, $\boldsymbol{L} = \boldsymbol{r} \times \boldsymbol{p} = rmv \sin \phi\,(-\boldsymbol{k})$. Therefore the magnitude of $\boldsymbol{L}$ is given by

$$L = mvr \sin \phi = mvd$$

where $d = r \sin \phi$ is the distance of closest approach of the particle from the origin. The direction of $\boldsymbol{L}$ from the right-hand rule is *into* the diagram, and we can write the vector expression $\boldsymbol{L} = -(mvd)\boldsymbol{k}$. The angular momentum relative to the origin O' is zero.

EXAMPLE 11.5 Circular Motion

A particle moves in the xy plane in a circular path of radius r, as in Figure 11.10. (a) Find the magnitude and direction of its angular momentum relative to O when its velocity is $\boldsymbol{v}$.

Since $\boldsymbol{r}$ is perpendicular to $\boldsymbol{v}$, $\phi = 90°$ and the magnitude of $\boldsymbol{L}$ is simply

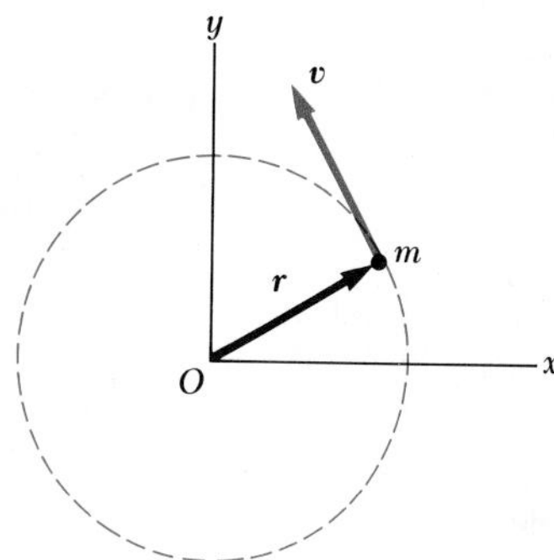

Figure 11.10 (Example 11.5) A particle moving in a circle of radius r has an angular momentum equal in magnitude to mvr relative to the center. The vector $\boldsymbol{L} = \boldsymbol{r} \times \boldsymbol{p}$ points *out* of the diagram.

$$L = mvr \sin 90° = mvr \qquad \text{(for } \boldsymbol{r} \text{ perpendicular to } \boldsymbol{v}\text{)}$$

The direction of $\boldsymbol{L}$ is perpendicular to the plane of the circle, and its sense depends on the direction of $\boldsymbol{v}$. If the sense of the rotation is counterclockwise, as in Figure 11.10, then by the right-hand rule, the direction of $\boldsymbol{L} = \boldsymbol{r} \times \boldsymbol{p}$ is *out* of the paper. Hence, we can write the vector expression $\boldsymbol{L} = (mvr)\boldsymbol{k}$. On the other hand, if the particle were to move clockwise, $\boldsymbol{L}$ would point into the paper.

(b) Find an alternative expression for L in terms of the angular velocity, ω.

Since $v = r\omega$ for a particle rotating in a circle, we can express L as

$$L = mvr = mr^2\omega = I\omega$$

where I is the moment of inertia of the particle about the z axis through O. Furthermore, in this case the angular momentum is in the *same* direction as the angular velocity vector, $\boldsymbol{\omega}$ (see Section 10.1), and so we can write $\boldsymbol{L} = I\boldsymbol{\omega} = I\omega\boldsymbol{k}$.

Exercise 2 A car of mass 1500 kg moves in a circular race track of radius 50 m with a speed of 40 m/s. What is the magnitude of its angular momentum relative to the center of the race track?
Answer 3.00×10^6 kg·m²/s.

11.4 ROTATION OF A RIGID BODY ABOUT A FIXED AXIS

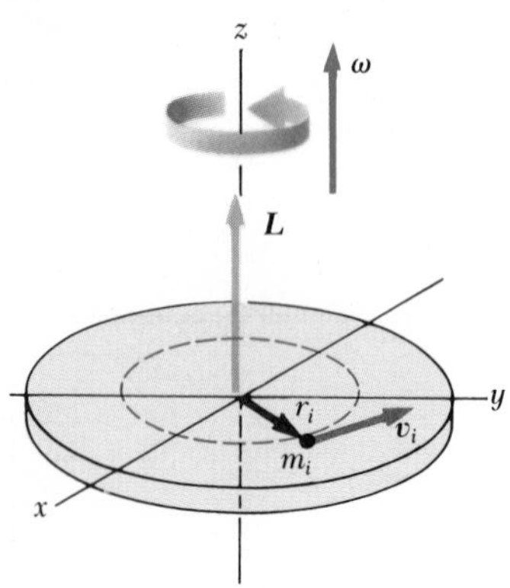

Figure 11.11 When a rigid body rotates about an axis, the angular momentum $\mathbf{L}$ is in the same direction as the angular velocity $\boldsymbol{\omega}$, according to the expression $\mathbf{L} = I\boldsymbol{\omega}$.

Let us consider a rigid body rotating about an axis that is fixed in direction. We shall assume that the z axis coincides with the axis of rotation, as in Figure 11.11. Each particle of the rigid body rotates in the xy plane about the z axis with an angular velocity ω. The magnitude of the angular momentum of the particle of mass m_i is $m_i v_i r_i$ about the origin O. Because $v_i = r_i \omega$, we can express the magnitude of the angular momentum of the ith particle as

$$L_i = m_i r_i^2 \omega$$

The vector $\mathbf{L}_i$ is directed along the z axis, corresponding to the direction of $\boldsymbol{\omega}$.

We can now find the z component of the angular momentum of the rigid body by taking the sum of L_i over all particles of the body:

$$L_z = \sum m_i r_i^2 \omega = \left(\sum m_i r_i^2\right) \omega$$

or

$$L_z = I\omega \tag{11.21}$$

where L_z is the z component of the angular momentum and I is the moment of inertia of the rigid body about the z axis.

Now let us differentiate Equation 11.21 with respect to time, noting that I is constant for a rigid body:

$$\frac{dL_z}{dt} = I\frac{d\omega}{dt} = I\alpha \tag{11.22}$$

where α is the angular acceleration relative to the axis of rotation. Because the product $I\alpha$ is equal to the net torque (see Eq. 11.20), we can express Equation 11.22 as follows:

$$\sum \tau_{\text{ext}} = \frac{dL_z}{dt} = I\alpha \tag{11.23}$$

That is, the net external torque acting on a rigid body rotating about a fixed axis equals the moment of inertia about the axis of rotation multiplied by its angular acceleration relative to that axis.

You should note that if a symmetrical rigid body rotates about a fixed axis passing through its center of mass, one can write Equation 11.21 in vector form, $\mathbf{L} = I\boldsymbol{\omega}$, where $\mathbf{L}$ is its total angular momentum measured with respect to the axis of rotation. Furthermore, the expression is valid for any body, regardless of its symmetry, if $\mathbf{L}$ stands for the component of angular momentum along the axis of rotation.[2]

[2] In general, the expression $\mathbf{L} = I\boldsymbol{\omega}$ is not always valid. If a rigid body rotates about an arbitrary axis, $\mathbf{L}$ and $\boldsymbol{\omega}$ may point in different directions. In fact, in this case, the moment of inertia cannot be treated as a scalar. Strictly speaking, $\mathbf{L} = I\boldsymbol{\omega}$ applies only to rigid bodies of any shape that rotate about one of three mutually perpendicular axes (called *principal axes*) through the center of mass. This is discussed in more advanced texts on mechanics.

EXAMPLE 11.6 Rotating Sphere

A uniform solid sphere of radius $R = 0.50$ m and mass 15 kg rotates about the z axis through its center, as in Figure 11.12. Find its angular momentum when the angular velocity is 3 rad/s.

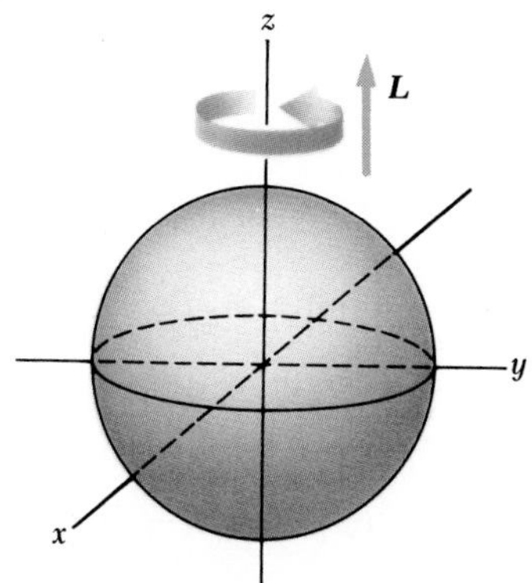

Figure 11.12 (Example 11.6) A sphere that rotates about the z axis in the direction shown has an angular momentum $\mathbf{L}$ in the positive z direction. If the direction of rotation is reversed, $\mathbf{L}$ will point in the negative z direction.

Solution The moment of inertia of the sphere about an axis through its center is

$$I = \tfrac{2}{5}MR^2 = \tfrac{2}{5}(15\text{ kg})(0.5\text{ m})^2 = 1.5\text{ kg}\cdot\text{m}^2$$

Therefore, the magnitude of the angular momentum is

$$L = I\omega = (1.5\text{ kg}\cdot\text{m}^2)(3\text{ rad/s}) = \boxed{4.5\text{ kg}\cdot\text{m}^2/\text{s}}$$

EXAMPLE 11.7 Rotating Rod

A rigid rod of mass M and length ℓ rotates in a vertical plane about a frictionless pivot through its center (Fig. 11.13). Particles of masses m_1 and m_2 are attached at the ends of the rod. (a) Determine the angular momentum when the angular velocity is ω.

The moment of inertia of the system equals the sum of the moments of inertia of the three components: the rod, m_1, and m_2. Using Table 10.2 we find that the total moment of inertia about the z axis through O is

$$I = \tfrac{1}{12}M\ell^2 + m_1\left(\frac{\ell}{2}\right)^2 + m_2\left(\frac{\ell}{2}\right)^2 = \frac{\ell^2}{4}\left(\frac{M}{3} + m_1 + m_2\right)$$

Therefore, when the angular velocity is ω, the angular momentum is given by

$$L = I\omega = \boxed{\frac{\ell^2}{4}\left(\frac{M}{3} + m_1 + m_2\right)\omega}$$

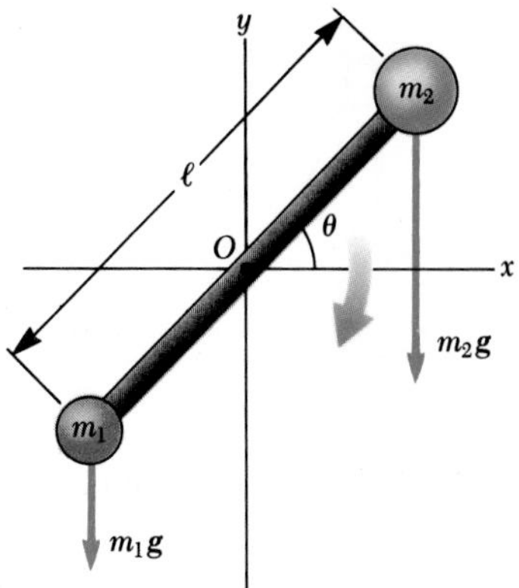

Figure 11.13 (Example 11.7) Since gravitational forces act on the system rotating in a vertical plane, there is in general a net nonzero torque about O when $m_1 \neq m_2$, which in turn produces an angular acceleration according to $\tau_{\text{net}} = I\alpha$.

(b) Determine the angular acceleration of the system when the rod makes an angle θ with the horizontal.

The torque due to the force m_1g about the pivot is

$$\tau_1 = m_1g\frac{\ell}{2}\cos\theta \qquad \text{(out of the plane)}$$

The torque due to the force m_2g about the pivot is

$$\tau_2 = -m_2g\frac{\ell}{2}\cos\theta \qquad \text{(into the plane)}$$

Hence, the net torque about O is

$$\tau_{\text{net}} = \tau_1 + \tau_2 = \tfrac{1}{2}(m_1 - m_2)g\ell\cos\theta$$

You should note that τ_{net} is *out* of the plane if $m_1 > m_2$ and is *into* the plane if $m_1 < m_2$. To find α, we use $\tau_{\text{net}} = I\alpha$, where I was obtained in (a). This gives

$$\alpha = \frac{\tau_{\text{net}}}{I} = \boxed{\frac{2(m_1 - m_2)g\cos\theta}{\ell\left(\frac{M}{3} + m_1 + m_2\right)}}$$

Note that α is zero when θ is $\pi/2$ or $-\pi/2$ (vertical position) and α is a maximum when θ is 0 or π (horizontal position). Furthermore, the angular velocity of the system changes since α varies in time.

Exercise 3 If $m_1 > m_2$, at what value of θ is ω a maximum? Knowing the angular velocity at some instant, how would you calculate the linear velocity of m_1 and m_2?

EXAMPLE 11.8 Two Connected Masses

Two masses, m_1 and m_2, are connected by a light cord that passes over a pulley of radius R and moment of inertia I about its axle as in Figure 11.14. The mass m_2 slides on a frictionless, horizontal surface. Let us determine the

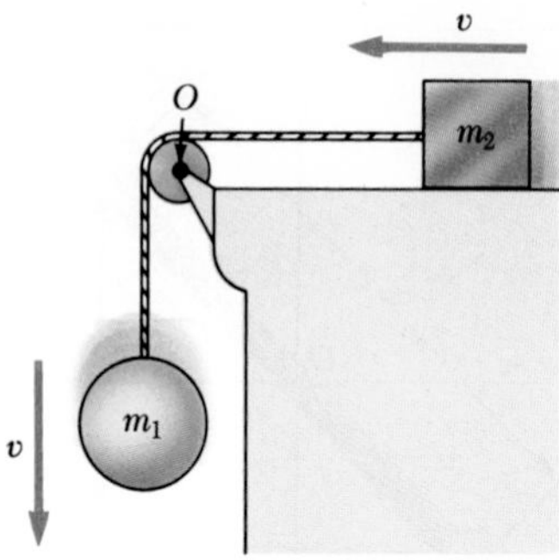

Figure 11.14 (Example 11.8).

acceleration of the two masses using the concepts of angular momentum and torque.

Solution First, let us calculate the angular momentum of the system, which consists of the two masses plus the pulley. We shall calculate the angular momentum about an axis along the axle of the pulley through O. At the instant the masses m_1 and m_2 have a speed v, the angular momentum of m_1 is m_1vR, while the angular momentum of m_2 is m_2vR. At the same instant, the angular momentum of the pulley is $I\omega = Iv/R$. Therefore, the total angular momentum of the system is

$$(1) \qquad L = m_1vR + m_2vR + I\frac{v}{R}$$

Now let us evaluate the total external torque on the system about the axle. Because the force of the axle on the pulley has zero moment arm, it does not contribute to the torque. Furthermore, the normal force acting on m_2 is balanced by its weight $m_2\mathbf{g}$, hence these forces do not contribute to the torque. The external force $m_1\mathbf{g}$ produces a torque about the axle equal in magnitude to m_1gR, where R is the moment arm of the force about the axle. This is the total external torque about O; that is, $\tau_{\text{ext}} = m_1gR$. Using this result, together with (1) and Equation 11.23 gives

$$\tau_{\text{ext}} = \frac{dL}{dt}$$

$$m_1gR = \frac{d}{dt}\left[(m_1 + m_2)Rv + I\frac{v}{R}\right]$$

or

$$(2) \qquad m_1gR = (m_1 + m_2)R\frac{dv}{dt} + \frac{I}{R}\frac{dv}{dt}$$

Because $dv/dt = a$, we can solve this for a to get

$$a = \frac{m_1g}{(m_1 + m_2) + I/R^2}$$

You may wonder why we did not include the forces of tension in evaluating the net torque about the axle. The reason is that the forces of tension are *internal* to the system under consideration. Only the *external* torques contribute to the change in angular momentum.

11.5 CONSERVATION OF ANGULAR MOMENTUM

In Chapter 9 we found that the total linear momentum of a system of particles remains constant when the resultant external force acting on the system is zero. We have an analogous conservation law in rotational motion which states that *the total angular momentum of a system is constant if the resultant external torque acting on the system is zero.* This follows directly from Equation 11.20, where we see that if

$$\sum \tau_{\text{ext}} = \frac{d\mathbf{L}}{dt} = 0 \qquad (11.24)$$

then

$$\mathbf{L} = \text{constant} \qquad (11.25)$$

For a system of particles, we write this conservation law as $\Sigma \mathbf{L}_i = \text{constant}$. If a body undergoes a redistribution of its mass, then its moment of inertia changes and we express this conservation of angular momentum in the form

$$\mathbf{L}_i = \mathbf{L}_f = \text{constant} \qquad (11.26)$$

If the system is a body rotating about a *fixed* axis, such as the z axis, then we can write $L_z = I\omega$, where L_z is the component of $\boldsymbol{L}$ along the axis of rotation and I is the moment of inertia about this axis. In this case, we can express the conservation of angular momentum as

$$I_i\omega_i = I_f\omega_f = \text{constant} \tag{11.27}$$

Conservation of angular momentum

This expression is valid for rotations either about a fixed axis or about an axis through the center of mass of the system as long as the axis remains parallel to itself. We only require the net external torque to be zero.

Although we do not prove it here, there is an important theorem concerning the angular momentum relative to the center of mass. This theorem states that

> the resultant torque acting on a body about the center of mass equals the time rate of change of angular momentum regardless of the motion of the center of mass.

This theorem applies even if the center of mass is accelerating, provided $\boldsymbol{\tau}$ and $\boldsymbol{L}$ are evaluated relative to the center of mass.

In Equation 11.27 we have a third conservation law to add to our list. Furthermore, we can now state that the energy, linear momentum, and angular momentum of an isolated system all remain constant.

There are many examples that demonstrate conservation of angular momentum, some of which should be familiar to you. You may have observed a figure skater undergoing a spin motion in the finale of an act. The angular velocity of the skater increases upon pulling his or her hands and feet close to the body. Neglecting friction between the skater and the ice, we see that there are no external torques on the skater. The change in angular velocity is due to the fact that since angular momentum is conserved, the product $I\omega$ remains constant and a decrease in the moment of inertia of the skater causes an increase in the angular velocity. Similarly, when divers (or acrobats) wish to make several somersaults, they pull their hands and feet close to their bodies in order to rotate at a higher rate. In these cases, the external force due to gravity acts through the center of mass and hence exerts no torque about this point. Therefore, the angular momentum about the center of mass must be conserved, or $I_i\omega_i = I_f\omega_f$. For example, when divers wish to double their angular velocity, they must reduce their moment of inertia to half its initial value.

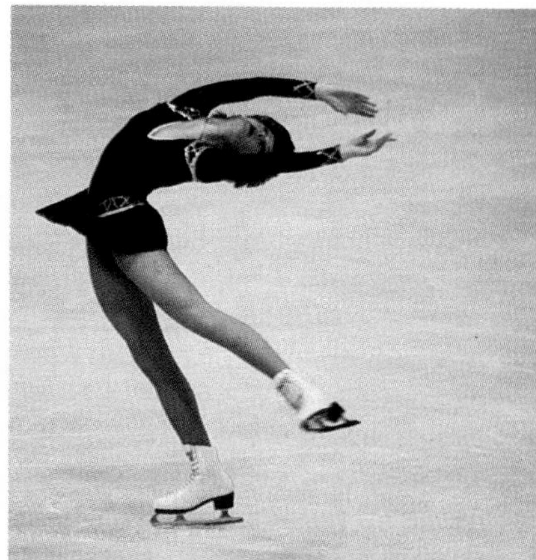
Figure skater in a spin. During a spinning routine, her angular velocity increases as she draws her hands (or feet) closer to her body—one example of conservation of angular momentum. (© G. Aschendorf, Photo Researchers)

EXAMPLE 11.9 A Projectile–Cylinder Collision
A projectile of mass m and velocity $\boldsymbol{v}_0$ is fired at a solid cylinder of mass M and radius R (Fig. 11.15). The cylinder is initially at rest and is mounted on a fixed horizontal axle that runs through the center of mass. The line of motion of the projectile is perpendicular to the axle and at a distance $d < R$ from the center. Find the angular speed of the system after the projectile strikes and adheres to the surface of the cylinder.

Solution Let us evaluate the angular momentum of the system (projectile + cylinder) about the axle of the cylinder. About this point, the net external torque on the system is zero about this axle. Hence, the angular momentum of the system is the same before and after the collision.

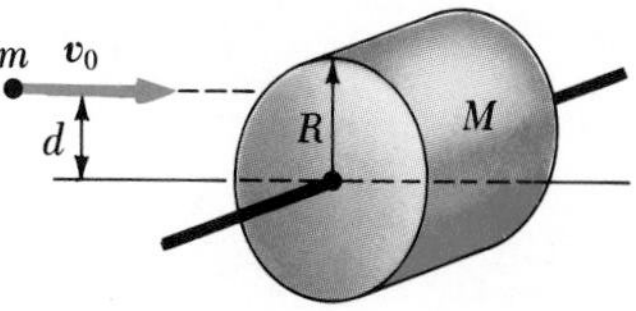

Figure 11.15 (Example 11.9) The angular momentum of the system before the collision equals the angular momentum right after the collision with respect to the center of mass if we neglect the weight of the projectile.

Before the collision, only the projectile has angular momentum with respect to a point on the axle. The magnitude of this angular momentum is mv_0d, and it is directed along the axle into the paper. After the collision, the total angular momentum of the system is $I\omega$, where I is the total moment of inertia about the axle (projectile + cylinder). Since the total angular momentum is conserved, we get

$$mv_0d = I\omega = (\tfrac{1}{2}MR^2 + mR^2)\omega$$

$$\omega = \frac{mv_0d}{\frac{1}{2}MR^2 + mR^2}$$

This suggests another technique for measuring the velocity of a bullet.

Exercise 4 In this example, mechanical energy is *not* conserved, since the collision is inelastic. Show that $\frac{1}{2}I\omega^2 < \frac{1}{2}mv_0^2$. What do you suppose accounts for the energy loss?

EXAMPLE 11.10 The Merry-Go-Round

A horizontal platform in the shape of a circular disk rotates in a horizontal plane about a frictionless vertical axle (Fig. 11.16). The platform has a mass of 100 kg and a radius of 2 m. A student whose mass is 60 kg walks slowly from the rim of the platform toward the center. If the angular velocity of the system is 2 rad/s when the student is at the rim, (a) calculate the angular velocity when the student has reached a point 0.5 m from the center.

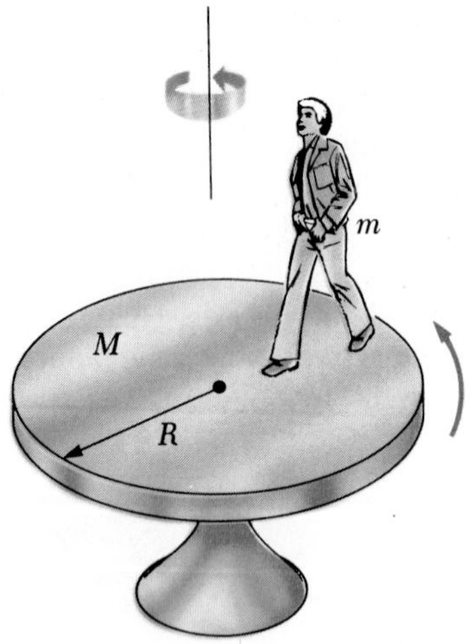

Figure 11.16 (Example 11.10) As the student walks toward the center of the rotating platform, the angular velocity of the system increases since the angular momentum must remain constant.

Let us call the moment of inertia of the platform I_p and the moment of inertia of the student I_s. Treating the student as a point mass m, we can write the *initial* moment of inertia of the system about the axle of rotation

$$I_i = I_p + I_s = \tfrac{1}{2}MR^2 + mR^2$$

where M and R are the mass and radius of the platform, respectively. When the student has walked to the position $r < R$, the moment of inertia of the system *reduces* to

$$I_f = \tfrac{1}{2}MR^2 + mr^2$$

Since there are no external torques on the system (student + platform) about the axis of rotation, we can apply the law of conservation of angular momentum:

$$I_i\omega_i = I_f\omega_f$$

$$(\tfrac{1}{2}MR^2 + mR^2)\omega_i = (\tfrac{1}{2}MR^2 + mr^2)\omega_f$$

$$\omega_f = \left(\frac{\frac{1}{2}MR^2 + mR^2}{\frac{1}{2}MR^2 + mr^2}\right)\omega_i$$

Substituting the values given for M, R, m, and ω_i we get

$$\omega_f = \left(\frac{200 + 240}{200 + 15}\right)(2 \text{ rad/s}) = 4.1 \text{ rad/s}$$

(b) Calculate the initial and final kinetic energies of the system.

$$K_i = \tfrac{1}{2}I_i\omega_i^2 = \tfrac{1}{2}(440 \text{ kg}\cdot\text{m}^2)\left(2\ \frac{\text{rad}}{\text{s}}\right)^2 = 880 \text{ J}$$

$$K_f = \tfrac{1}{2}I_f\omega_f^2 = \tfrac{1}{2}(215 \text{ kg}\cdot\text{m}^2)\left(4.1\ \frac{\text{rad}}{\text{s}}\right)^2 = 1800 \text{ J}$$

Note that the kinetic energy of the system *increases!* Although this result may surprise you, it can be explained as follows: In the process of walking toward the center of the platform, the student has to exert some muscular effort and perform positive work, which in turn is transformed into an increase in kinetic energy of the system. In other words, internal forces within the system do work. The student must exert a centripetal force to keep from flying off the platform. Hence, he has to do work against the reaction to this force.

Exercise 5 Show that the gain in kinetic energy can be accounted for using the work-energy theorem.

EXAMPLE 11.11 Spinning on a Stool

A student sits on a pivoted stool while holding a pair of weights, as in Figure 11.17. The stool is free to rotate about a vertical axis with negligible friction. The student is set in rotating motion with the weights outstretched. Why does the angular velocity of the system increase as the weights are pulled inward?

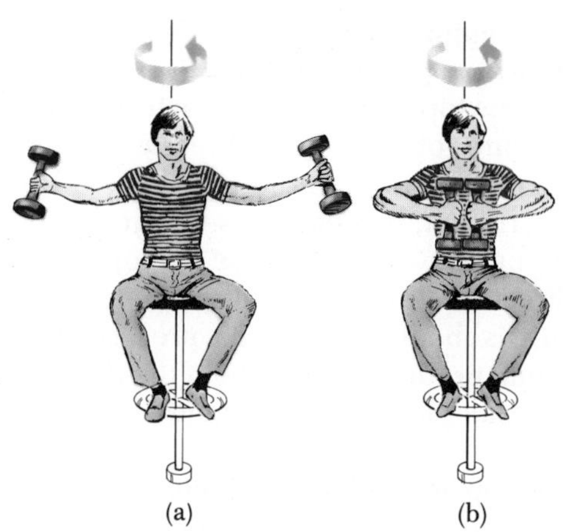

Figure 11.17 (Example 11.11) (a) The student is given an initial angular velocity while holding two masses as shown. (b) When the masses are pulled in close to the body, the angular velocity of the system increases. Why?

Solution The initial angular momentum of the system is $I_i\omega_i$, where I_i refers to the initial moment of inertia of the entire system (student + weights + stool). After the weights are pulled in, the angular momentum of the system is $I_f\omega_f$. Note that $I_f < I_i$ since the weights are now closer to the axis of rotation, reducing the moment of inertia. Since the net external torque on the system is zero, angular momentum is conserved, so $I_i\omega_i = I_f\omega_f$. Therefore, $\omega_f > \omega_i$, or the angular velocity increases. As in the previous example the kinetic energy of the system increases as the weights are pulled inward. The increase in kinetic energy arises from the fact that the student does work in pulling the weights toward the axis of rotation.

Exercise 6 Suppose the student were to drop the weights to his side rather than pull them inward horizontally. What would account for the increase in the kinetic energy of the system in this situation?

EXAMPLE 11.12 The Spinning Bicycle Wheel

In another favorite classroom demonstration, a student holds the axle of a spinning bicycle wheel while seated on a pivoted stool (Fig. 11.18). The student and stool are initially *at rest* while the wheel is spinning in a horizontal plane with an initial angular momentum $\boldsymbol{L}_0$ pointing upward. Explain what happens if the wheel is inverted about its center by 180°.

Figure 11.18 (Example 11.12) The wheel is initially spinning when the student is at rest. What happens when the wheel is inverted?

Solution In this situation, the system consists of the student, wheel, and stool. Initially, the total angular momentum of the system is $\boldsymbol{L}_0$, corresponding to the contribution from the spinning wheel. As the wheel is inverted, a torque is supplied by the student, but this is *internal* to the system. There is *no* external torque acting on the system about the vertical axis. Therefore, *the angular momentum of the system must be conserved.*

Initially, we have

$$\boldsymbol{L}_{\text{system}} = \boldsymbol{L}_0 \qquad \text{(upward)}$$

After the wheel is inverted,

$$\boldsymbol{L}_{\text{system}} = \boldsymbol{L}_0 = \boldsymbol{L}_{\text{student+stool}} + \boldsymbol{L}_{\text{wheel}}$$

In this case, $\boldsymbol{L}_{\text{wheel}} = -\boldsymbol{L}_0$ since it is now rotating in the opposite sense. Therefore

$$\boldsymbol{L}_0 = \boldsymbol{L}_{\text{student+stool}} - \boldsymbol{L}_0$$

$$\boldsymbol{L}_{\text{student+stool}} = 2\boldsymbol{L}_0$$

This shows that *the student and stool will start to turn, acquiring an angular momentum having a magnitude twice that of the spinning wheel and directed upward.*

Exercise 7 How much angular momentum would the student acquire if the wheel is tilted through an angle θ measured from the vertical axis?

Answer $L_0(1 - \cos\theta)$.

*11.6 THE MOTION OF GYROSCOPES AND TOPS

A very unusual and fascinating type of motion that you probably have observed is that of a top spinning about its axis of symmetry as in Fig. 11.19a. If the top spins about its axis very rapidly, the axis will rotate about the vertical

Precessional motion

direction as indicated, thereby sweeping out a cone. The motion of the axis of the top about the vertical, known as **precessional motion,** is usually slow compared with the spin motion of the top. It is quite natural to wonder why the top doesn't fall over. Since the center of mass is not directly above the pivot point O, there is clearly a net torque acting on the top about O due to the weight force $\boldsymbol{Mg}$. From this description, it is easy to see that the top would certainly fall if it were not spinning. However, because the top is spinning, it has an angular momentum $\boldsymbol{L}$ directed along its axis of symmetry. As we shall show, the motion of the rotation axis about the z axis (the precessional motion) arises from the fact that the torque produces a change in the *direction* of the rotation axis. This is an excellent example of the importance of the directional nature of angular momentum.

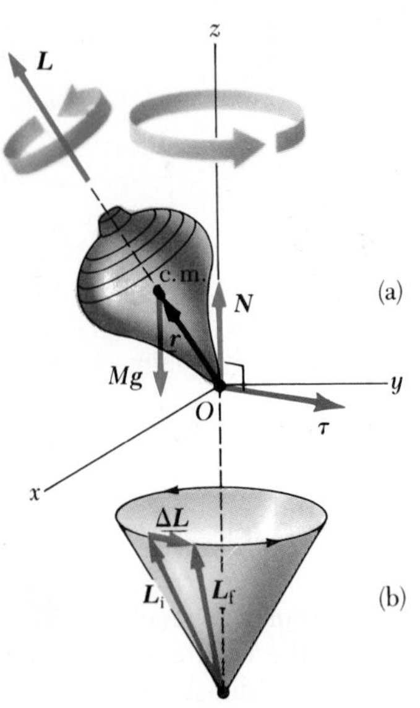

Figure 11.19 Precessional motion of a top spinning about its axis of symmetry. The only external forces acting on the top are the normal force, $\boldsymbol{N}$, and the force of gravity, $\boldsymbol{Mg}$. The direction of the angular momentum, $\boldsymbol{L}$, is along the axis of symmetry.

The two forces acting on the top are the downward force of gravity, $\boldsymbol{Mg}$, and the normal force, $\boldsymbol{N}$, acting upward at the pivot point O. The normal force produces no torque about the pivot since its moment arm is zero. However, the force of gravity produces a torque $\boldsymbol{\tau} = \boldsymbol{r} \times \boldsymbol{Mg}$ about O, where the direction of $\boldsymbol{\tau}$ is perpendicular to the plane formed by $\boldsymbol{r}$ and $\boldsymbol{Mg}$. By necessity, the vector $\boldsymbol{\tau}$ lies in a horizontal plane perpendicular to the angular momentum vector. The net torque and angular momentum of the body are related through the expression

$$\boldsymbol{\tau} = \frac{d\boldsymbol{L}}{dt}$$

From this expression, we see that the nonzero torque produces a *change* in angular momentum $d\boldsymbol{L}$, which is in the same direction as $\boldsymbol{\tau}$. Therefore, like the torque vector, $d\boldsymbol{L}$ must also be at right angles to $\boldsymbol{L}$. Figure 11.19b illustrates the resulting precessional motion of the axis of the top. In a time Δt, the change in angular momentum $\Delta\boldsymbol{L} = \boldsymbol{L}_f - \boldsymbol{L}_i = \boldsymbol{\tau}\,\Delta t$. Note that because $\Delta\boldsymbol{L}$ is perpendicular to $\boldsymbol{L}$ the magnitude of $\boldsymbol{L}$ doesn't change ($|\boldsymbol{L}_i| = |\boldsymbol{L}_f|$). Rather, what is changing is the *direction* of $\boldsymbol{L}$. Since the change in angular momentum is in the direction of $\boldsymbol{\tau}$, which lies in the xy plane, the top undergoes precessional motion. Thus, the effect of the torque is to deflect the angular momentum of the top in a direction perpendicular to its spin axis.

We have presented a rather qualitative description of the motion of a top. In general, the motion of such an object is very complex. However, the essential features of the motion can be illustrated by considering the simple gyroscope shown in Fig. 11.20. This device consists of a wheel free to spin about an axle that is pivoted at a distance h from the center of mass of the wheel. If the wheel is given an angular velocity $\boldsymbol{\omega}$ about its axis, the wheel will have a spin angular momentum $\boldsymbol{L} = I\boldsymbol{\omega}$ directed along the axle as shown. Let us consider the torque acting on the wheel about the pivot O. Again, the force, $\boldsymbol{N}$, of the support on the axle produces no torque about O. On the other hand, the weight $\boldsymbol{Mg}$ produces a torque of magnitude Mgh about O. The direction of this torque is *perpendicular* to the axle (and perpendicular to $\boldsymbol{L}$), as described in Figure 11.20. This torque causes the angular momentum to change in the direction perpendicular to the axle. Hence, the axle moves in the direction of the torque, that is, in the horizontal plane. There is an assumption that we must make in order to simplify the description of the system. The *total* angular momentum of the precessing wheel is actually the sum of the spin angular momentum, $I\boldsymbol{\omega}$, and the angular momentum due to the motion of the center of mass about the pivot. In our treatment, we shall neglect the contribution from

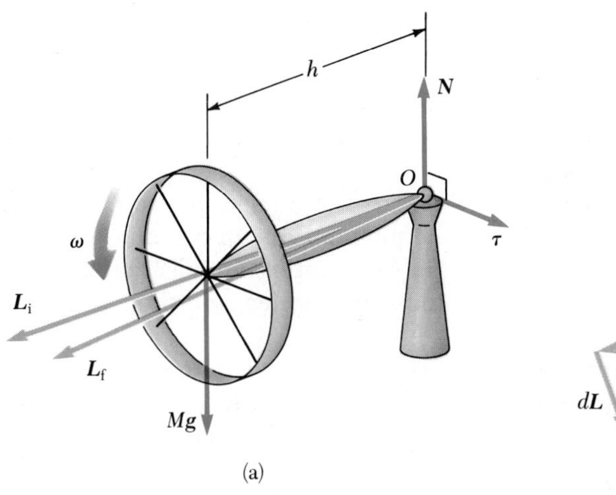

(a)

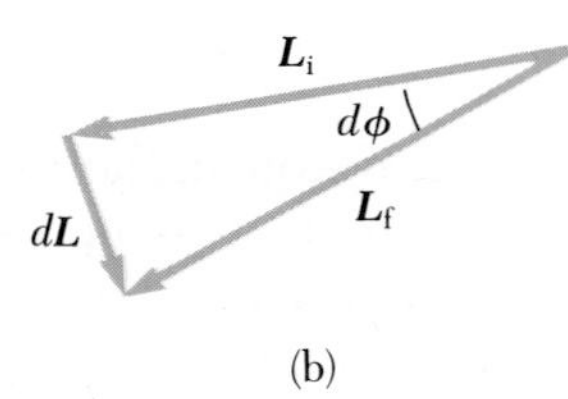

(b)

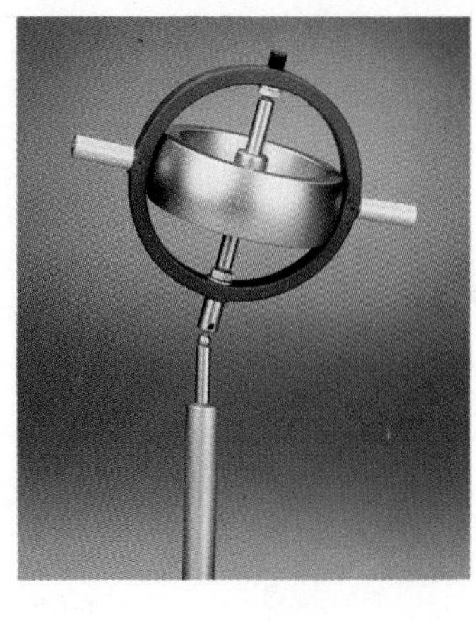

(c)

Figure 11.20 (a) The motion of a simple gyroscope pivoted a distance h from its center of gravity. Note that the weight $M\mathbf{g}$ produces a torque about the pivot that is perpendicular to the axle. (b) This results in a change in angular momentum $d\mathbf{L}$ in the direction perpendicular to the axle. The axle sweeps out an angle $d\phi$ in a time dt. (c) Photograph of a toy-gyroscope. (Courtesy of CENCO)

the center of mass motion and take the total angular momentum to be just $I\boldsymbol{\omega}$. In practice, this is a good approximation if ω is made very large.

In a time dt the torque due to the weight force adds to the system an *additional* angular momentum equal to $d\mathbf{L} = \boldsymbol{\tau}\, dt$ in the direction perpendicular to $\mathbf{L}$. This additional angular momentum, $\boldsymbol{\tau}\, dt$, when added vectorially to the original spin angular momentum, $I\boldsymbol{\omega}$, *causes a shift in the direction of the total angular momentum.* We can express the magnitude of this change in angular momentum as

$$dL = \tau\, dt = (Mgh)\, dt$$

The vector diagram in Fig. 11.20 shows that in the time dt, the angular momentum vector rotates through an angle $d\phi$, which is also the angle through which the axle rotates. From the vector triangle formed by the vectors $\mathbf{L}_i$, $\mathbf{L}_f$, and $d\mathbf{L}$ and from the expression above, we see that

$$d\phi = \frac{dL}{L} = \frac{(Mgh)\, dt}{L}$$

Using $L = I\omega$, we find that the rate at which the axle rotates about the vertical axis is given by

$$\omega_p = \frac{d\phi}{dt} = \frac{Mgh}{I\omega} \qquad (11.28)$$

Precessional frequency

The angular frequency ω_p is called the **precessional frequency.** This result is valid only when $\omega_p \ll \omega$. Otherwise, a much more complicated motion is involved. As you can see from Equation 11.28, the condition that $\omega_p \ll \omega$ is met when $I\omega$ is large compared with Mgh. Furthermore, note that the precessional frequency decreases as ω increases, that is, as the wheel spins faster about its axis of symmetry.

*11.7 ANGULAR MOMENTUM AS A FUNDAMENTAL QUANTITY

We have seen that the concept of angular momentum is very useful for describing the motion of macroscopic systems. However, the concept is also valid on a submicroscopic scale and has been used extensively in the development of modern theories of atomic, molecular, and nuclear physics. In these developments, it was found that the angular momentum of a system is a *fundamental* quantity. The word *fundamental* in this context implies that angular momentum is an inherent property of atoms, molecules, and their constituents.

In order to explain the results of a variety of experiments on atomic and molecular systems, it is necessary to assign discrete values to the angular momentum. These discrete values are some multiple of a fundamental unit of angular momentum, which equals $\hbar = h/2\pi$, where h is called Planck's constant.

$$\text{Fundamental unit of angular momentum} = \hbar = 1.054 \times 10^{-34}\,\frac{\text{kg}\cdot\text{m}^2}{\text{s}^2}$$

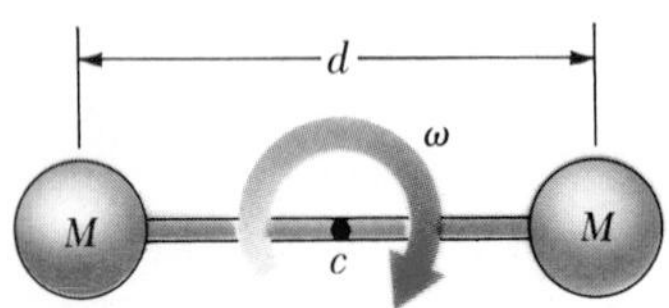

Figure 11.21 The rigid-rotor model of the diatomic molecule. The rotation occurs about the center of mass in the plane of the diagram.

Let us accept this postulate for the time being and show how it can be used to estimate the rotational frequency of a diatomic molecule. Consider the O_2 molecule as a rigid rotor, that is, two atoms separated by a fixed distance d and rotating about the center of mass (Fig. 11.21). Equating the rotational angular momentum to the fundamental unit $\hbar$, we can estimate the lowest rotational frequency:

$$I_c\omega \approx \hbar \qquad \text{or} \qquad \omega \approx \frac{\hbar}{I_c}$$

In Example 10.3, we found that the moment of inertia of the O_2 molecule about this axis of rotation is 2.03×10^{-46} kg·m². Therefore,

$$\omega \approx \frac{\hbar}{I_c} = \frac{1.054 \times 10^{-34}\ \text{kg}\cdot\text{m}^2/\text{s}}{2.03 \times 10^{-46}\ \text{kg}\cdot\text{m}^2} = 5.19 \times 10^{11}\ \text{rad/s}$$

This result is in good agreement with measured rotational frequencies. Furthermore, the rotational frequencies are much lower than the vibrational frequencies of the molecule, which are typically of the order of 10^{13} Hz.

This simple example shows that certain classical concepts and mechanical models might be useful in describing some features of atomic and molecular systems. However, a wide variety of phenomena on the submicroscopic scale can be explained only if one assumes discrete values of the angular momentum associated with a particular type of motion.

Historically, the Danish physicist Niels Bohr (1885–1962) was the first to suggest this radical idea in his theory of the hydrogen atom. Strictly classical models were unsuccessful in describing many properties of the hydrogen atom, such as the fact that the atom emits radiation at discrete frequencies. Bohr postulated that the electron could only occupy circular orbits about the proton for which the orbital angular momentum was equal to $n\hbar$, where n is an integer. That is, Bohr made the bold postulate that the orbital angular momentum is quantized. From this rather simple model, one can estimate the rotational frequencies of the electron in the various orbits (Problem 35).

Although Bohr's theory provided some insight concerning the behavior of matter at the atomic level, it is basically incorrect. Subsequent developments in quantum mechanics from 1924 to 1930 provided models and interpretations that are still accepted. We shall discuss this further in Chapter 40.

Later developments in atomic physics indicated that the electron also possesses another kind of angular momentum, called *spin,* which is also an inherent property of the electron. The spin angular momentum is also restricted to discrete values. We shall return to this important property later in the extended version of this text and discuss its great impact on modern physical science.

SUMMARY

The **total kinetic energy** of a rigid body, such as a cylinder, that is rolling on a rough surface without slipping equals the rotational kinetic energy about its center of mass, $\frac{1}{2}I_c\omega^2$, plus the translational kinetic energy of the center of mass, $\frac{1}{2}Mv_c^2$:

Total kinetic energy of a rolling body

$$K = \tfrac{1}{2}I_c\omega^2 + \tfrac{1}{2}Mv_c^2 \tag{11.4}$$

In this expression, v_c is the velocity of the center of mass and $v_c = R\omega$ for pure rolling motion.

The **torque** $\boldsymbol{\tau}$ due to a force $\boldsymbol{F}$ about an origin in an inertial frame is defined to be

Torque

$$\boldsymbol{\tau} \equiv \boldsymbol{r} \times \boldsymbol{F} \tag{11.7}$$

Given two vectors $\boldsymbol{A}$ and $\boldsymbol{B}$, their **cross product** $\boldsymbol{A} \times \boldsymbol{B}$ is a vector $\boldsymbol{C}$ having a magnitude

Magnitude of the cross product

$$C \equiv |AB \sin \theta| \tag{11.9}$$

where θ is the angle included between $\boldsymbol{A}$ and $\boldsymbol{B}$. The direction of the vector $\boldsymbol{C} = \boldsymbol{A} \times \boldsymbol{B}$ is perpendicular to the plane formed by $\boldsymbol{A}$ and $\boldsymbol{B}$, and its sense is determined by the right-hand rule. Some properties of the cross product include the facts that $\boldsymbol{A} \times \boldsymbol{B} = -\boldsymbol{B} \times \boldsymbol{A}$ and $\boldsymbol{A} \times \boldsymbol{A} = 0$.

The **angular momentum** $\boldsymbol{L}$ of a particle of linear momentum $\boldsymbol{p} = m\boldsymbol{v}$ is given by

Angular momentum of a particle

$$\boldsymbol{L} = \boldsymbol{r} \times \boldsymbol{p} = m\boldsymbol{r} \times \boldsymbol{v} \tag{11.15}$$

where $\boldsymbol{r}$ is the vector position of the particle relative to an origin in an inertial frame. If ϕ is the angle between $\boldsymbol{r}$ and $\boldsymbol{p}$, the magnitude of $\boldsymbol{L}$ is given by

$$L = mvr \sin \phi \tag{11.16}$$

The **net external torque** acting on a particle or rigid body is equal to the time rate of change of its angular momentum:

$$\sum \tau_{\text{ext}} = \frac{d\boldsymbol{L}}{dt} \tag{11.20}$$

Angular momentum of a rigid body about a fixed axis

The *z component of angular momentum* of a rigid body rotating about a fixed axis (the z axis) is given by

$$L_z = I\omega \tag{11.21}$$

where I is the moment of inertia about the axis of rotation, and ω is its angular velocity.

The *net external torque* acting on a rigid body equals the product of its moment of inertia about the axis of rotation and its angular acceleration:

$$\sum \tau_{\text{ext}} = I\alpha \tag{11.23}$$

Conservation of angular momentum

If the net external torque acting on a system is zero, the total angular momentum of the system is constant. Applying this **conservation of angular momentum** law to a body whose moment of inertia changes gives

$$I_i\omega_i = I_f\omega_f = \text{constant} \tag{11.27}$$

QUESTIONS

1. Is it possible to calculate the torque acting on a rigid body without specifying the origin? Is the torque independent of the location of the origin?
2. Is the triple product defined by $\mathbf{A}\cdot(\mathbf{B}\times\mathbf{C})$ equal to a scalar or vector quantity? Explain why the operation $(\mathbf{A}\cdot\mathbf{B})\times\mathbf{C}$ has no meaning.
3. In the expression for torque, $\boldsymbol{\tau} = \mathbf{r}\times\mathbf{F}$, is r equal to the moment arm? Explain.
4. Can a particle moving in a straight line have nonzero angular momentum?
5. If the angular momentum of a particle is constant as measured by one observer, will it be constant according to all observers?
6. If the torque acting on a particle about an *arbitrary* origin is zero, what can you say about its angular momentum about that origin?
7. A particle moves in a straight line, and you are told that the torque acting is zero about some unspecified origin. Does this necessarily imply that the net force on the particle is zero? Can you conclude that its velocity is constant? Explain.
8. Suppose that the velocity vector of a particle is completely specified. What can you conclude about the *direction* of its angular momentum vector with respect to the direction of motion?
9. If the net torque acting on a rigid body is nonzero about some origin, is there any other origin about which the net torque is zero?
10. If a system of particles is in motion, is it possible for the total angular momentum to be zero about some origin? Explain.
11. A ball is thrown in such a way that it does not spin about its own axis. Does this mean that the angular momentum is zero about an arbitrary origin? Explain.
12. Why is it easier to keep your balance on a moving bicycle than on a bicycle at rest?
13. A scientist at a hotel sought assistance from a bellhop to carry a mysterious suitcase. When the unaware bellhop rounded a corner carrying the suitcase, it suddenly moved away from him for some unknown reason. At this point, the alarmed bellhop dropped the suitcase and ran off. What do you suppose might have been in the suitcase?
14. When a cylinder rolls on a horizontal surface as in Figure 11.4, are there any points on the cylinder that have only a vertical component of velocity at some instant? If so, where are they?
15. Three homogeneous rigid bodies—a solid sphere, a solid cylinder, and a hollow cylinder—are placed at the top of an incline (Fig. 11.22). If they all are released from rest at the same elevation and roll without slipping, which reaches the bottom first? Which reaches last? You should try this at home and note that the result is *independent* of the masses and radii.

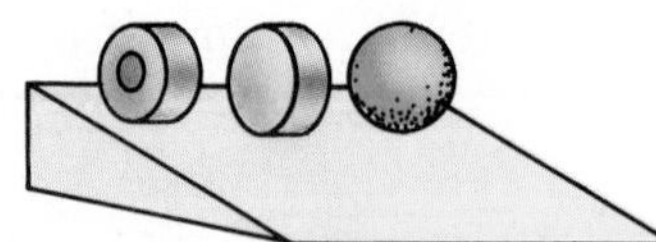

Figure 11.22 (Question 15) Which object wins the race?

16. A mouse is initially at rest on a horizontal turntable mounted on a frictionless vertical axle. If the mouse begins to walk around the perimeter, what happens to the turntable? Explain.

17. Stars originate as large bodies of slowly rotating gas. Because of gravity, these regions of gas slowly decrease in size. What happens to the angular velocity of a star as it shrinks? Explain.
18. Use the principle of conservation of angular momentum to form a hypothesis that explains how a cat can always land on its feet regardless of the position from which it is dropped.
19. Often when a high diver wants to turn a flip in midair, she will draw her legs up against her chest. Why does this make her rotate faster? What should she do when she wants to come out of her flip?
20. As a tether ball winds around a pole, what happens to its angular velocity? Explain.
21. Space colonies have been proposed that consist of large cylinders placed in space. Gravity would be simulated in these cylinders by setting them into rotation about their long axis. Discuss the difficulties that would be encountered in attempting to set the cylinders into rotation.
22. For a particle undergoing uniform circular motion, how are its linear momentum, $\boldsymbol{p}$ and the angular momentum $\boldsymbol{L}$ oriented with respect to each other?
23. If the net force acting on a system is zero, then is it necessarily true that the net torque on it is also zero?
24. Why do tightrope walkers carry a long pole to help balance themselves while walking a tightrope?

PROBLEMS

Section 11.1 Rotation of a Rigid Body

1. A cylinder of mass 10 kg rolls without slipping on a rough surface. At the instant its center of mass has a speed of 10 m/s, determine (a) the translational kinetic energy of its center of mass, (b) the rotational kinetic energy about its center of mass, and (c) its total kinetic energy.
2. A solid sphere has a radius of 0.2 m and a mass of 150 kg. How much work is required to get the sphere rolling with an angular speed of 50 rad/s on a horizontal surface? (Assume the sphere starts from rest and rolls without slipping.)
3. (a) Determine the acceleration of the center of mass of a uniform solid disk rolling down an incline and compare this acceleration with that of a uniform hoop. (b) What is the minimum coefficient of friction required to maintain pure rolling motion for the disk?
4. A uniform solid disk and a uniform hoop are placed side by side at the top of a rough incline of height h. If they are released from rest and roll without slipping, determine their velocities when they reach the bottom. Which object reaches the bottom first?
5. The center of mass of a uniform cylinder of mass M and radius R is moving with speed v on a horizontal surface. The cylinder rolls without slipping. Find the total kinetic energy of the cylinder with respect to a reference frame fixed to the horizontal surface.
6. The centers of mass of a solid sphere and a solid cylinder, each of mass M and radius R, are moving with speed v with respect to the floor. Find the ratios of their total kinetic energies.

Section 11.2 The Vector Product and Torque

7. Two vectors are given by $\boldsymbol{A} = -3\boldsymbol{i} + 4\boldsymbol{j}$ and $\boldsymbol{B} = 2\boldsymbol{i} + 3\boldsymbol{j}$. Find (a) $\boldsymbol{A} \times \boldsymbol{B}$ and (b) the angle between $\boldsymbol{A}$ and $\boldsymbol{B}$.
8. Find $\boldsymbol{A} \times \boldsymbol{B}$ for the vectors (a) $\boldsymbol{A} = 3\boldsymbol{j}$ and $\boldsymbol{B} = 2\boldsymbol{i} + 2\boldsymbol{j}$, (b) $\boldsymbol{A} = 3\boldsymbol{i} - \boldsymbol{j}$ and $\boldsymbol{B} = 4\boldsymbol{k}$, (c) $\boldsymbol{A} = 3\boldsymbol{j} + \boldsymbol{k}$ and $\boldsymbol{B} = -2\boldsymbol{i}$.
9. Vector $\boldsymbol{A}$ is in the negative y direction, and vector $\boldsymbol{B}$ is in the negative x direction. What are the directions of (a) $\boldsymbol{A} \times \boldsymbol{B}$ and (b) $\boldsymbol{B} \times \boldsymbol{A}$?
10. A particle is located at the vector position $\boldsymbol{r} = (\boldsymbol{i} + 3\boldsymbol{j})$ m, and the force acting on it is $\boldsymbol{F} = (3\boldsymbol{i} + 2\boldsymbol{j})$ N. What is the torque about (a) the origin and (b) the point having coordinates (0, 6) m?
11. If $|\boldsymbol{A} \times \boldsymbol{B}| = \boldsymbol{A} \cdot \boldsymbol{B}$, what is the angle between $\boldsymbol{A}$ and $\boldsymbol{B}$?
12. Verify Equation 11.14 for the cross product of any two vectors $\boldsymbol{A}$ and $\boldsymbol{B}$, and show that the cross product may be written in the following determinant form:

$$\boldsymbol{A} \times \boldsymbol{B} = \begin{vmatrix} \boldsymbol{i} & \boldsymbol{j} & \boldsymbol{k} \\ A_x & A_y & A_z \\ B_x & B_y & B_z \end{vmatrix}$$

13. Two forces $\boldsymbol{F}_1$ and $\boldsymbol{F}_2$ act along the two sides of an equilateral triangle as shown in Figure 11.23. Find a

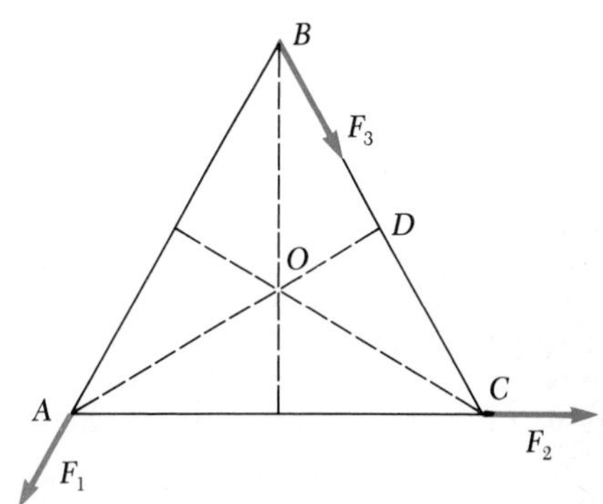

Figure 11.23 (Problem 13).

third force $\boldsymbol{F}_3$ to be applied at B and along BC which will make the net torque about the point of intersection of the altitudes zero. Will the net torque change if $\boldsymbol{F}_3$ were to be applied not at B but at any other point along BC?

14. A particle is located at the vector position $\boldsymbol{r} = (4\boldsymbol{i} + 6\boldsymbol{j})$ m and the force acting on it is given by $\boldsymbol{F} = (3\boldsymbol{i} + 2\boldsymbol{j})$ N. (a) What is the torque acting on the particle about the origin? (b) Determine a point on the y axis about which the torque will be in the opposite direction of (a) and half in magnitude.

Section 11.3 Angular Momentum of a Particle

15. A light rigid rod 1 m in length rotates in the xy plane about a pivot through the rod's center. Two particles of mass 4 kg and 3 kg are connected to its ends (Fig. 11.24). Determine the angular momentum of the system about the origin at the instant the speed of each particle is 5 m/s.

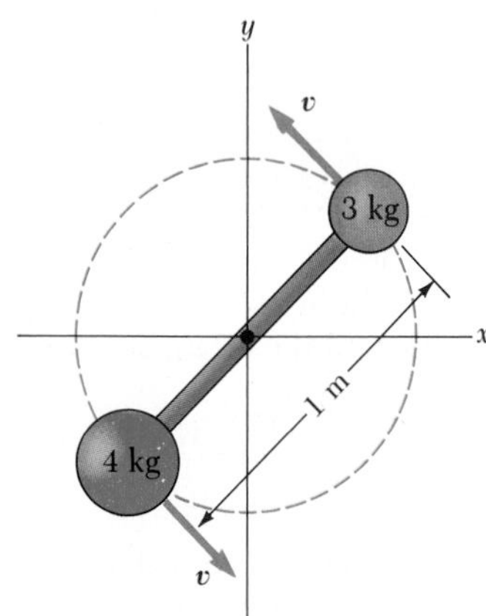

Figure 11.24 (Problem 15).

16. A particle of mass 0.3 kg moves in the xy plane. At the instant its coordinates are (2, 4) m, its velocity is given by $(3\boldsymbol{i} + 2\boldsymbol{j})$ m/s. At this instant, determine the angular momentum of the particle relative to the origin.
17. The position vector of a particle of mass 2 kg is given as a function of time by $\boldsymbol{r} = (6\boldsymbol{i} + 5t\boldsymbol{j})$ m. Determine the angular momentum of the particle as a function of time.
18. Two particles move in *opposite* directions along a straight line (Fig. 11.25). The particle of mass m moves to the right with a speed v while the particle of mass $3m$ moves to the left with a speed v. What is the *total* angular momentum of the system relative to (a) the point A, (b) the point O, and (c) the point B?
19. A particle of mass m moves in a straight line with a constant velocity $\boldsymbol{v} = v\boldsymbol{j}$ along the positive y axis. Determine the angular momentum of the particle (both magnitude and direction) relative to (a) the point having coordinates $(-d, 0)$, (b) the point having coordinates $(2d, 0)$, and (c) the origin.

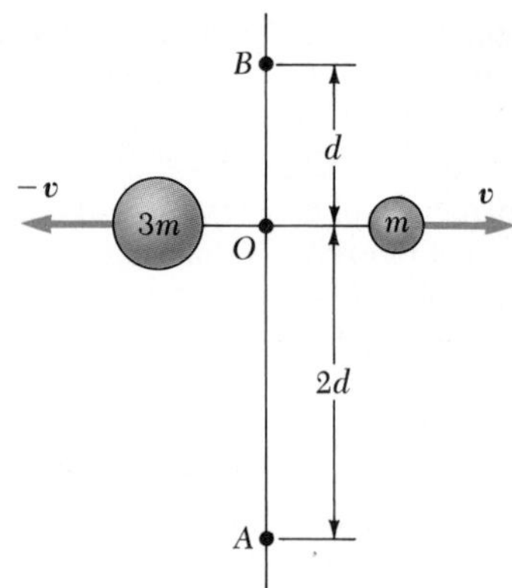

Figure 11.25 (Problem 18).

20. An airplane of mass 12 000 kg flies level to the ground at an altitude of 10 km with a constant speed of 175 m/s relative to the earth. (a) What is the magnitude of the airplane's angular momentum relative to a ground observer who is directly below the airplane? (b) Does this value change as the airplane continues its motion along a straight line?
21. (a) Calculate the angular momentum of the earth due to its spinning motion about its axis. (b) Calculate the angular momentum of the earth due to its orbital motion about the sun and compare this with (a). (Take the earth-sun distance to be 1.49×10^{11} m.)
22. A 4-kg mass is attached to a light cord, which is wound around a pulley (Fig. 10.18). The pulley is a uniform solid cylinder of radius 8 cm and mass 2 kg. (a) What is the net torque on the system about the point O? (b) When the 4-kg mass has a speed v, the pulley has an angular velocity $\omega = v/R$. Determine the total angular momentum of the system about O. (c) Using the fact that $\boldsymbol{\tau} = d\boldsymbol{L}/dt$ and your result from (b), calculate the acceleration of the 4-kg mass.
23. A particle of mass m is shot with an initial velocity $\mathbf{v}_0$ making an angle θ with the horizontal as shown in Figure 11.26. The particle moves in the gravitational field of the earth. Find the angular momentum of the particle about the origin when the particle is (a) at the origin, (b) at the highest point of its trajectory, and (c) just before it hits the ground. (d) What torque causes its angular momentum to change?

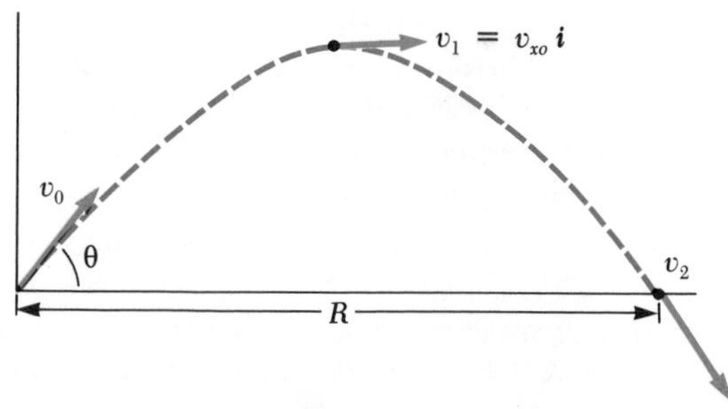

Figure 11.26 (Problem 23).

Section 11.4 Rotation of a Rigid Body About a Fixed Axis

24. A uniform solid disk of mass 3 kg and radius 0.2 m rotates about a fixed axis perpendicular to its face. If the angular frequency of rotation is 6 rad/s, calculate the angular momentum of the disk when the axis of rotation (a) passes through its center of mass and (b) passes through a point midway between the center and the rim.
25. A particle of mass 0.4 kg is attached to the 100-cm mark of a meter stick of mass 0.1 kg. The meter stick rotates on a horizontal, smooth table with an angular velocity of 4 rad/s. Calculate the angular momentum of the system if the stick is pivoted about an axis (a) perpendicular to the table through the 50-cm mark and (b) perpendicular to the table through the 0-cm mark.
26. Two masses M and m connected by a rigid rod of negligible mass are lying on a horizontal frictionless surface. Where should a force normal to the rod be applied so that the masses will not rotate about the point of application of the force?

Section 11.5 Conservation of Angular Momentum

27. A cylinder with moment of inertia I_1 rotates with angular velocity ω_0 about a vertical, frictionless axle. A second cylinder, with moment of inertia I_2 initially not rotating, drops onto the first cylinder (Fig. 11.27). Since the surfaces are rough, the two eventually reach the same angular velocity ω. (a) Calculate ω. (b) Show that energy is lost in this situation and calculate the ratio of the final to the initial kinetic energy.

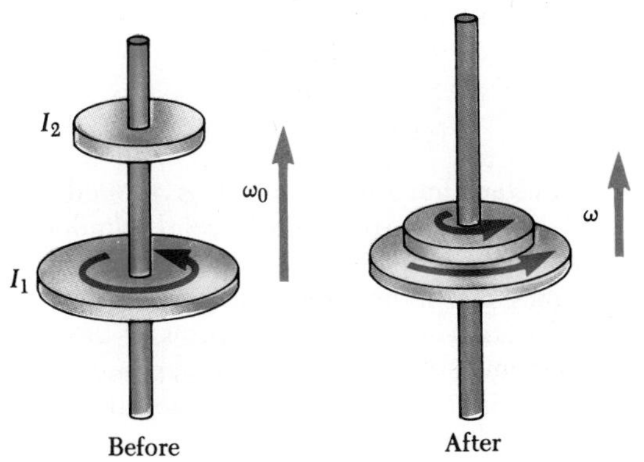

Figure 11.27 (Problem 27).

28. A uniform solid cylinder of mass 2 kg and radius 25 cm rotates about a fixed vertical, frictionless axle with an angular speed of 10 rad/s. A 0.5-kg piece of putty is dropped vertically onto the cylinder at a point 15 cm from the axle. If the putty sticks to the cylinder, calculate the final angular speed of the system. (Assume the putty is a point particle.)
29. A woman whose mass is 60 kg stands at the rim of a horizontal turntable having a moment of inertia of 500 kg · m² and a radius of 2 m. The system is initially at rest, and the turntable is free to rotate about a frictionless, vertical axle through its center. The woman then starts walking around the rim in a clockwise direction (looking downward) at a constant speed of 1.5 m/s relative to the earth. (a) In what direction and with what angular speed does the turntable rotate? (b) How much work does the woman do to set the system into motion?
30. A uniform rod of mass 100 g and length 50 cm rotates in a horizontal plane about a fixed, vertical, frictionless pin through its center. Two small beads, each of mass 30 g, are mounted on the rod such that they are able to slide without friction along its length. Initially the beads are held by catches at positions 10 cm on each side of center, at which time the system rotates at an angular speed of 20 rad/s. Suddenly, the catches are released and the small beads slide outward along the rod. Find (a) the angular speed of the system at the instant the beads reach the ends of the rod, and (b) the angular speed of the rod after the beads fly off the ends.
31. The student in Figure 11.17 holds two weights, each of mass 10 kg. When his arms are extended horizontally, the weights are 1 m from the axis of rotation and he rotates with an angular speed of 2 rad/s. The moment of inertia of the student plus the stool is 8 kg · m² and is assumed to be constant. If the student pulls the weights horizontally to 0.25 m from the rotation axis, calculate (a) the final angular speed of the system and (b) the change in the mechanical energy of the system.
32. A particle of mass $m = 10$ g and speed $v_0 = 5$ m/s collides with and sticks to the edge of a uniform solid sphere of mass $M = 1$ kg and radius $R = 20$ cm (Fig. 11.28). If the sphere is initially at rest and is pivoted about a frictionless axle through O perpendicular to the plane of the paper, (a) find the angular velocity of the system after the collision and (b) determine how much energy is lost in the collision.

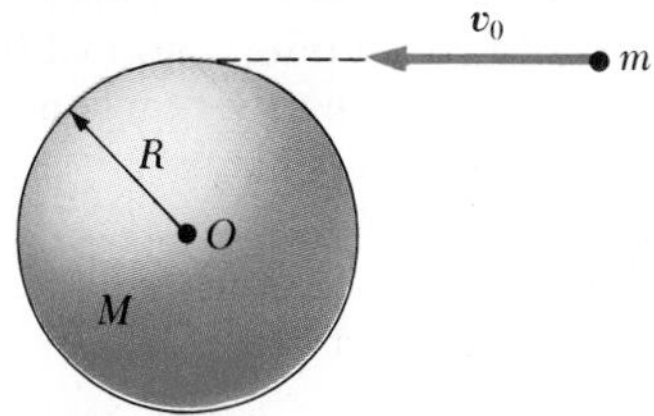

Figure 11.28 (Problem 32).

33. A wooden block of mass M resting on a frictionless horizontal surface is attached to a rigid rod of length ℓ and of negligible mass (Fig. 11.29). The rod is pivoted at the other end. A bullet of mass m traveling parallel to the horizontal surface and normal to the rod with

speed v hits the block and gets embedded in it. (a) What is the angular momentum of the bullet and block system? (b) What fraction of the original kinetic energy is lost in the collision?

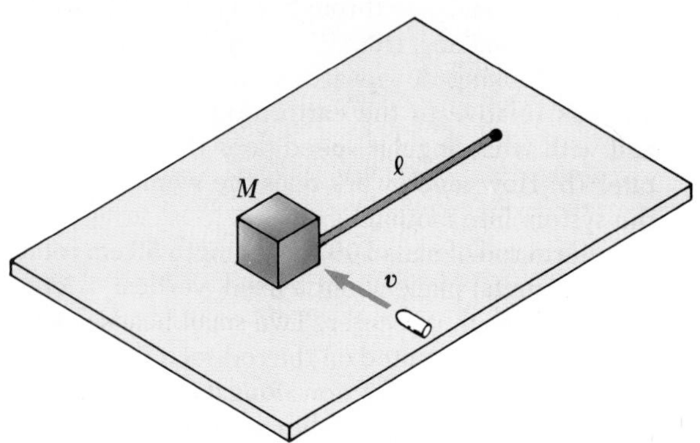

Figure 11.29 (Problems 33 and 34).

34. Consider the previous problem with $\ell = 2$ m, $M = 2$ kg, $m = 10$ g, and $v = 200$ m/s. In this case the bullet leaves the block with speed $v = 25$ m/s horizontal to the surface. (a) Determine the angular momentum of the block just after the bullet exits it. (b) Determine the kinetic energy lost in this collision. Neglect the transit time through the block.

°Section 11.7 Angular Momentum as a Fundamental Quantity

35. In the Bohr model of the hydrogen atom, the electron moves in a circular orbit of radius 0.529×10^{-10} m around the proton. Assuming the orbital angular momentum of the electron is equal to $\hbar$, calculate (a) the orbital speed of the electron, (b) the kinetic energy of the electron, and (c) the angular frequency of the electron's motion.

ADDITIONAL PROBLEMS

36. A uniform solid sphere of radius r is placed on the inside surface of a hemispherical bowl of radius R. The sphere is released from rest at an angle θ to the vertical and rolls without slipping (Fig. 11.30). Determine the angular speed of the sphere when it reaches the bottom of the bowl.

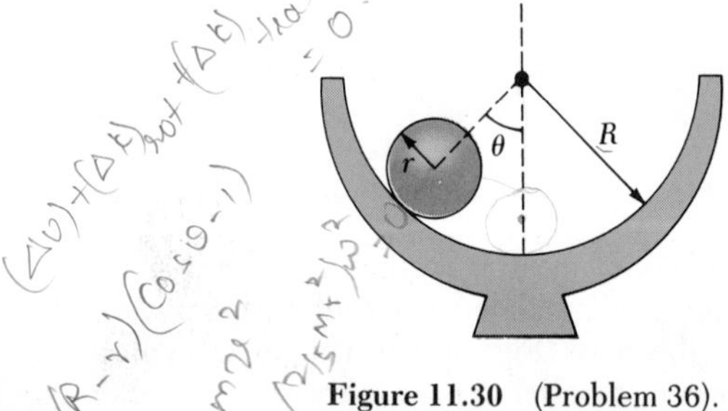

Figure 11.30 (Problem 36).

37. (a) Compute the kinetic energy of the earth due to its annual orbit around the sun. (b) Compute the rotational kinetic energy of the earth due to its daily rotation about its own axis. (c) Compute the ratio $K_{\text{orbit}}/K_{\text{rotation}}$.

38. A thin uniform cylindrical turntable of radius 2 m and mass 30 kg rotates in a horizontal plane with an initial angular velocity, 4π rad/s. The turntable bearing is frictionless. A small clump of clay of mass 0.25 kg is dropped onto the turntable and sticks at a point 1.8 m from the center of rotation. (a) Find the final angular velocity of the clay and turntable. (Treat the clay as a point mass.) (b) Is mechanical energy conserved in this collision? Explain and use numerical results to verify your answer.

39. A string is wound around a uniform disk of radius R and mass M. The disk is released from rest with the string vertical and its top end tied to a fixed support (Fig. 11.31). As the disk descends, show that (a) the tension in the string is one third the weight of the disk, (b) the acceleration of the center of mass is $2g/3$, and (c) the velocity of the center of mass is $(4gh/3)^{1/2}$. Verify your answer to (c) using the energy approach.

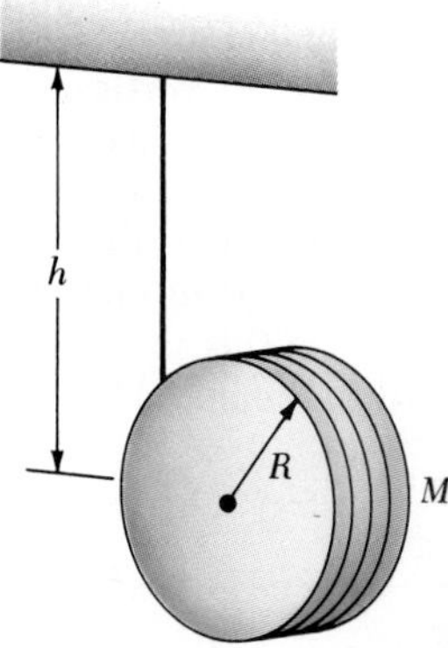

Figure 11.31 (Problem 39).

40. A constant horizontal force $\boldsymbol{F}$ is applied to a lawn roller in the form of a uniform solid cylinder of radius R and mass M (Fig. 11.32). If the roller rolls without slipping on the horizontal surface, show that (a) the acceleration of the center of mass is $2\boldsymbol{F}/3M$ and (b) the minimum coefficient of friction necessary to prevent slipping is $F/3Mg$. (*Hint:* Take the torque with respect to the center of mass.)

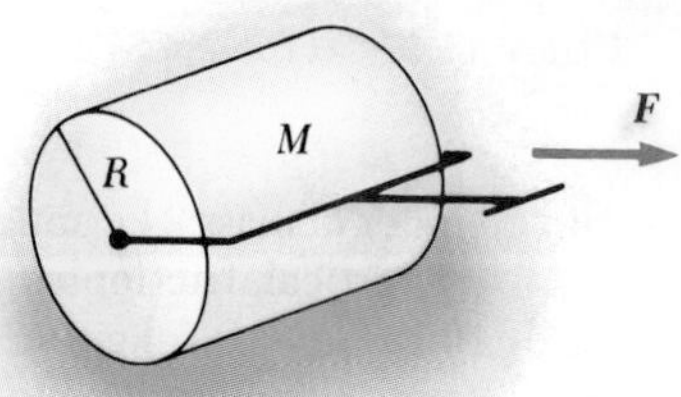

Figure 11.32 (Problem 40).

41. A light rope passes over a light, frictionless pulley. One end is fastened to a bunch of bananas of mass M, and a monkey of mass M clings to the other end of the rope (Fig. 11.33). The monkey climbs the rope in an attempt to reach the bananas. (a) Treating the system as consisting of the monkey, bananas, rope, and pulley, evaluate the net torque about the pulley axis. (b) Using the results to (a), determine the total angular momentum about the pulley axis and describe the motion of the system. Will the monkey reach the bananas?

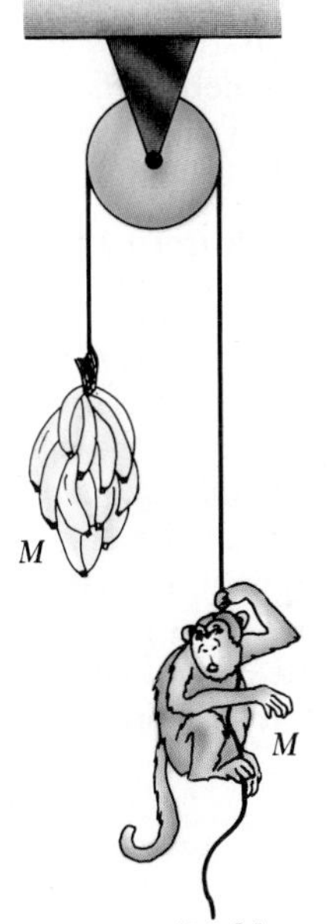

Figure 11.33 (Problem 41).

42. A small, solid sphere of mass m and radius r rolls without slipping along the track shown in Figure 11.34. If it starts from rest at the top of the track at a height h, where h is large compared to r, (a) what is the minimum value of h (in terms of the radius of the loop R) such that the sphere completes the loop? (b) What are the force components on the sphere at the point P if $h = 3R$?

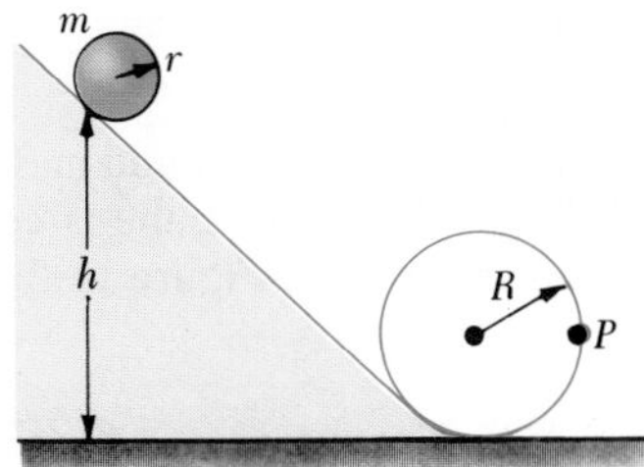

Figure 11.34 (Problem 42).

43. A massless string is wrapped around a uniform solid cylinder (Fig. 11.35) that has a mass $M = 15$ kg and a radius $R = 6$ cm. The cylinder is free to rotate about its axis on frictionless bearings. One end of the string is attached to the cylinder and the free end is pulled tangentially by a force that maintains a constant tension of 2 N in the string. (a) What is the angular acceleration α of the cylinder? (b) What is the angular speed ω of the cylinder 2 s after force is applied if the cylinder was originally at rest?

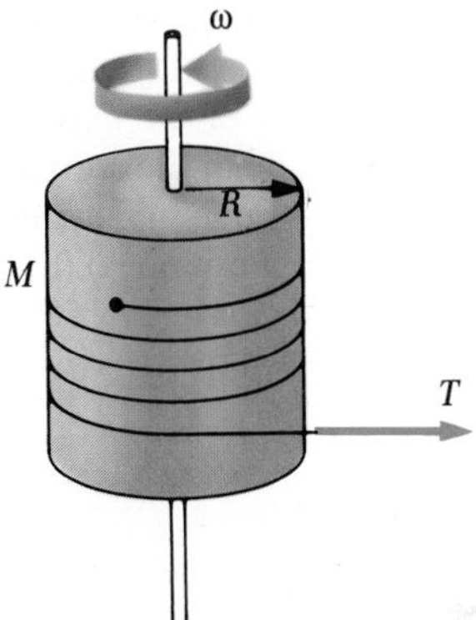

Figure 11.35 (Problem 43).

44. Consider the problem of the solid sphere rolling down an incline as described in Example 11.1. (a) Choose the axis of the origin for the torque equation as the instantaneous axis through the contact point P and show that the acceleration of the center of mass is given by $a_c = \frac{5}{7}g \sin \theta$. (b) Show that the *minimum* coefficient of friction such that the sphere will roll without slipping is given by $\mu_{\text{min}} = \frac{2}{7}\tan \theta$.

45. The position vector of a particle of mass 5 kg is given by $\boldsymbol{r} = (2t^2\boldsymbol{i} + 3\boldsymbol{j})$ m, where t is in seconds. Determine the angular momentum and the torque acting on the particle about the origin.

46. A particle of mass m is located at the vector position $\boldsymbol{r}$ and has a linear momentum $\boldsymbol{p}$. (a) If $\boldsymbol{r}$ and $\boldsymbol{p}$ both have nonzero x, y, and z components, show that the angular momentum of the particle relative to the origin has components given by $L_x = yp_z - zp_y$, $L_y = zp_x - xp_z$, and $L_z = xp_y - yp_x$. (b) If the particle moves only in the xy plane, prove that $L_x = L_y = 0$ and $L_z \neq 0$.

47. A force $\boldsymbol{F}$ acts on the particle described in Problem 46. (a) Find the components of the torque acting on the particle about the origin when the particle is located at the position $\boldsymbol{r}$ and the force has three components. (b) From this result, show that if the particle moves in the xy plane and the force has only x and y components, the torque (and angular momentum) must be in the z direction.

48. It is proposed to power a passenger bus with a massive rotating flywheel that is periodically brought up to its maximum speed (3500 rpm) by means of an electric motor. The flywheel has a mass of 1200 kg, a diameter of 1.8 m, and is in the shape of a solid cylinder. (This is

not an efficient shape for a flywheel that is designed to power a vehicle: can you see why?) (a) What is the maximum amount of kinetic energy that can be stored in the flywheel? (b) If the bus requires an average power of 30 hp, how long will the flywheel rotate?

49. A mass m is attached to a cord passing through a small hole in a frictionless, horizontal surface (Fig. 11.36). The mass is initially orbiting in a circle of radius r_0 with velocity v_0. The cord is then slowly pulled from below, decreasing the radius of the circle to r. (a) What is the velocity of the mass when the radius is r? (b) Find the tension in the cord as a function of r. (c) How much work is done in moving m from r_0 to r? (*Note:* The tension depends on r.) (d) Obtain numerical values for v, T, and W when $r = 0.1$ m, if $m = 50$ g, $r_0 = 0.3$ m, and $v_0 = 1.5$ m/s.

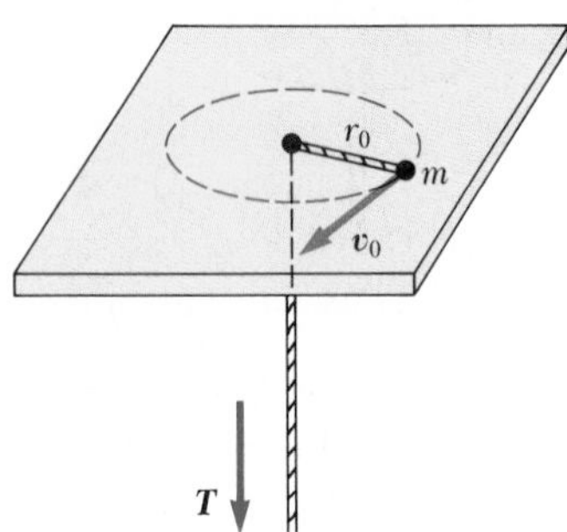

Figure 11.36 (Problem 49).

50. A bowling ball is given an initial speed v_0 on an alley such that it *initially slides without rolling*. The coefficient of friction between the ball and the alley is μ. Show that at the time *pure rolling motion occurs*, (a) the velocity of the ball's center of mass is $5v_0/7$ and (b) the distance it has traveled is $12v_0^2/49\ \mu g$. (*Hint:* When pure rolling motion occurs, $v_c = R\omega$ and $\alpha = a_c/R$. Since the frictional force provides the deceleration, from Newton's second law it follows that $a_c = -\mu g$.)

51. A trailer with loaded weight **W** is being pulled by a vehicle with a force **F**, as in Figure 11.37. The trailer is loaded such that its center of gravity is located as shown. Neglect the force of rolling friction and assume the trailer has an acceleration a. (a) Find the vertical component of **F** in terms of the given parameters. (b) If $a = 2$ m/s^2 and $h = 1.5$ m, what must be the value of d in order that $F_y = 0$ (no vertical load on the vehicle)? (c) Find the values of F_x and F_y given that $W = 1500$ N, $d = 0.8$ m, $L = 3$ m, $h = 1.5$ m, and $a = -2$ m/s^2.

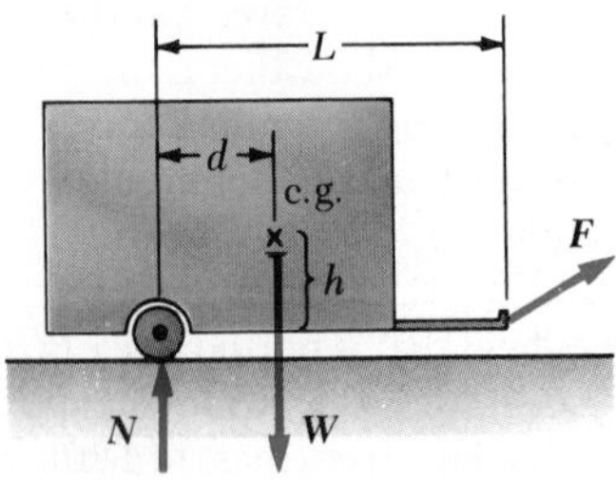

Figure 11.37 (Problem 51).

52. (a) A thin rod of length h and mass M is held vertically up with its lower end resting on a frictionless horizontal surface. The rod is then let go to fall freely. Determine the velocity of its center of mass just before it hits the horizontal surface. (b) Suppose the rod were pivoted at its lower end. Determine the speed of the rod's center of mass just before it hits the surface.

53. A uniform solid disk of mass M rotates about an axis parallel to the symmetry axis through its center, as in Figure 11.38. Show that the angular momentum of the disk is given by

$$\mathbf{L} = I_c\boldsymbol{\omega} + \mathbf{r}_c \times M\mathbf{v}_c$$

where I_c is the moment of inertia about the axis through its center of mass, $\mathbf{r}_c$ is the vector from O to the center of mass, and $\mathbf{v}_c$ is the velocity of the center of mass. The first term on the right side of this expression is called the *spin angular momentum* since it refers to that part of the angular momentum associated with the spin of the system about the center of mass. The second term on the right side is usually referred to as the *orbital angular momentum.* (*Hint:* Use the parallel-axis theorem.)

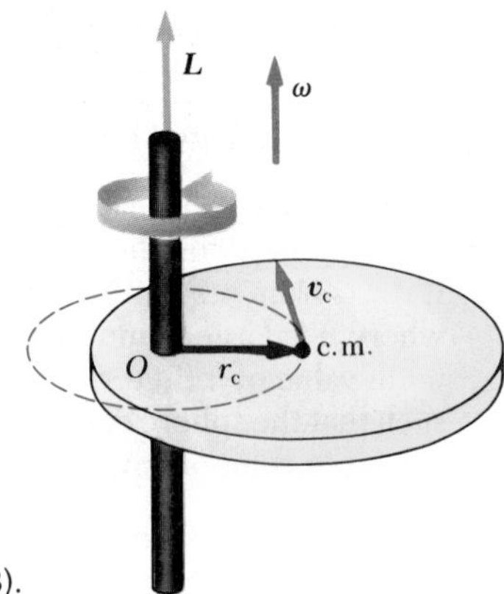

Figure 11.38 (Problem 53).

54. A solid cube of wood of side $2a$ and mass M is resting on a horizontal surface. The cube is constrained to rotate about an axis AB (see Figure 11.39). A bullet of

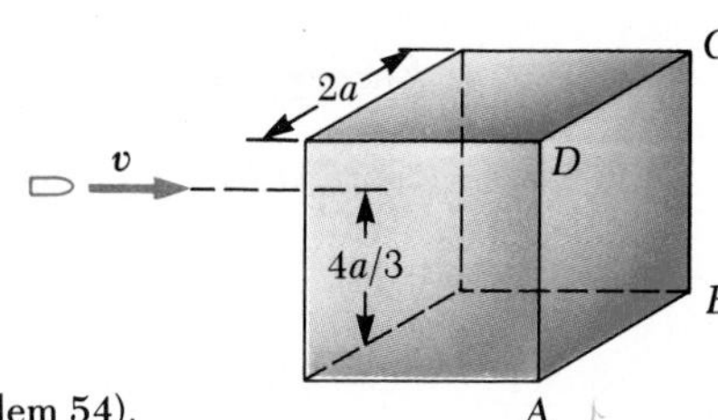

Figure 11.39 (Problem 54).

mass m and speed v is shot at the face opposite to $ABCD$ at a height of $4a/3$. The bullet gets embedded in the block. Find the minimum value of the v required to tip the cube to fall on its face $ABCD$. Assume $m \ll M$.

55. A neutron star is a collection of neutrons bound together by their mutual gravitation. The density of such a star is comparable to that of an atomic nucleus (10^{17} kg/m^3). Assuming the neutron star to be spherical in shape and of uniform mass density, show that the maximum frequency with which it may rotate, without mass flying off at the equator, is $\omega = (4\pi\rho G/3)^{1/2}$, where ρ is the mass density. Calculate ω for $\rho = 10^{17}$ kg/m^3. Determine the kinetic energy of such a star about its axis of rotation. The radius of the star is 10 km.

56. A solid sphere is released from height h from the top of an incline making an angle θ with the horizontal. Calculate the velocity of the sphere when it reaches the bottom of the incline in the case that (a) it rolls without slipping, and (b) it slips frictionlessly without rolling. Compare the times required to reach the bottom in cases (a) and (b).

57. A spool of wire of mass M and radius R is unwound under a constant force $\boldsymbol{F}$ (Fig. 11.40). Assuming the spool is a uniform solid cylinder that *doesn't slip,* show that (a) the acceleration of the center of mass is $4\boldsymbol{F}/3M$ and (b) the force of friction is to the *right* and equal to $\boldsymbol{F}/3$. (c) If the cylinder starts from rest and rolls without slipping, what is the velocity of its center of mass after it has rolled through a distance d? (Assume the force remains constant.)

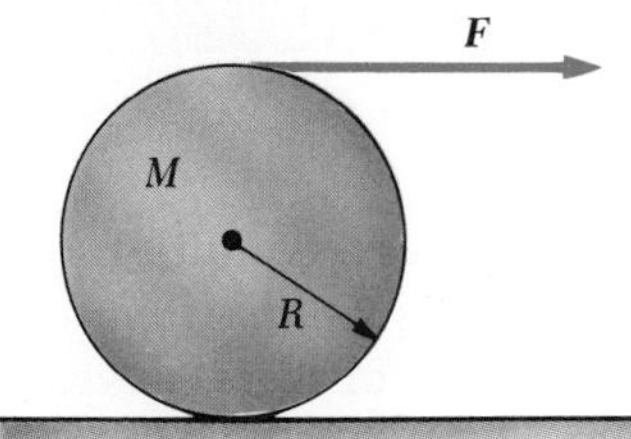

Figure 11.40 (Problem 57).

58. A uniform solid disk is set into rotation about an axis through its center with an angular velocity ω_0. The rotating disk is lowered to a *rough,* horizontal surface with this angular velocity and released as in Figure 11.41. (a) What is the angular velocity of the disk once pure rolling takes place? (b) Find the fractional loss in kinetic energy from the time the disk is released until pure rolling occurs. (*Hint:* Angular momentum is conserved about an axis through the point of contact.)

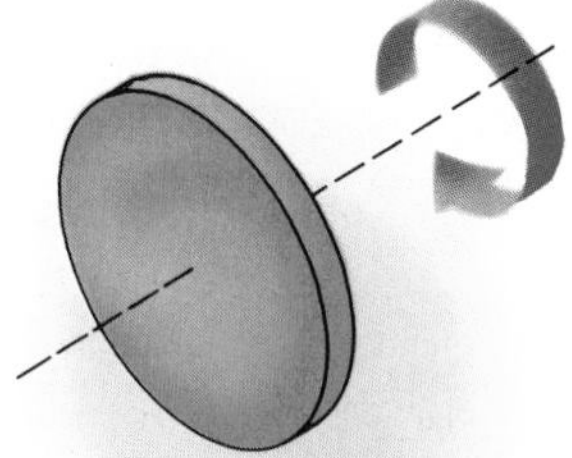

Figure 11.41 (Problems 58 and 59).

59. Suppose a solid disk of radius R is given an angular velocity ω_0 about an axis through its center and is then lowered to a rough, horizontal surface and released, as in Problem 58 (Fig. 11.41). Furthermore, assume that the coefficient of friction between the disk and surface is μ. (a) Show that the *time* it takes pure rolling motion to occur is given by $R\omega_0/3\mu g$. (b) Show that the *distance* the disk travels before pure rolling occurs is given by $R^2\omega_0{}^2/18\,\mu g$. (See hint in Problem 58.)

60. A large, cylindrical roll of tissue paper of initial radius R lies on a long, horizontal surface with the open end of the paper nailed to the surface so that it can unroll easily. The roll is given a *slight* shove ($v_0 \approx 0$) and commences to unroll. (a) Determine the speed of the center of mass of the roll when its radius has diminished to r. (b) Calculate a numerical value for this speed at $r = 1$ mm, assuming $R = 6$ m. (c) What happens to the energy of the system when the paper is completely unrolled? (*Hint:* Assume the roll has a uniform density and apply energy methods.)

61. A solid cube of side $2a$ and mass M is sliding on a frictionless surface with uniform velocity $\boldsymbol{v}_0$ as in Figure 11.42a. It hits a small obstacle at the end of the table, which causes the cube to tilt as in Figure 11.42b. Find the minimum value of $\boldsymbol{v}_0$ such that the cube falls off the table. Note that the moment of inertia of the cube about an axis along one of its edges is $8Ma^2/3$. (*Hint:* The cube undergoes an *inelastic collision* at the edge.)

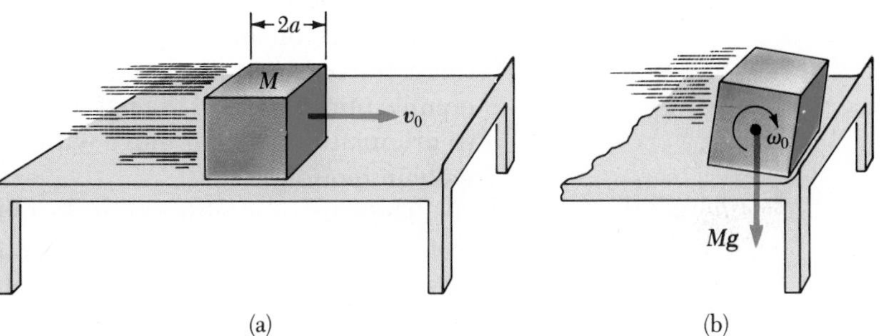

Figure 11.42 (Problem 61).

12

Static Equilibrium and Elasticity

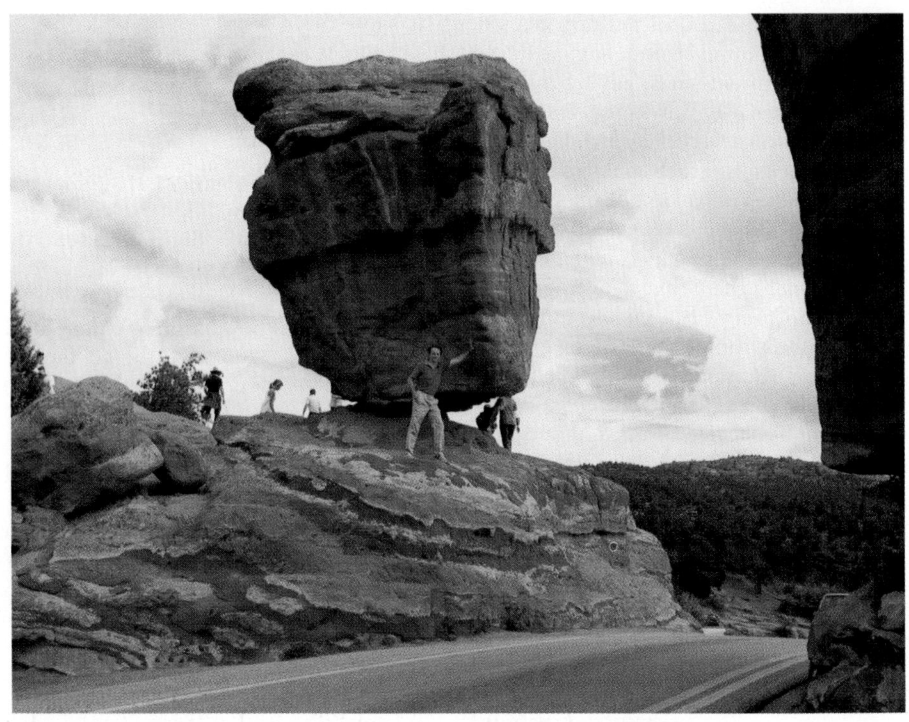

A large balanced rock photographed at the Garden of the Gods in Colorado Springs, Colorado. This is an example of unstable equilibrium. (Photo by David Serway)

Part of this chapter is concerned with the conditions under which a rigid body is in equilibrium. The term *equilibrium* implies that the body is either at rest or that its center of mass moves with constant velocity. We shall deal with bodies at rest, or bodies in *static equilibrium*. This represents a common situation in engineering practice, and the principles involved are of special interest to civil engineers, architects, and mechanical engineers, who deal with various structural designs, such as bridges and buildings. Those of you who are engineering students will undoubtedly take an intensified course in statics in the future.

In Chapter 5 we stated that one necessary condition for equilibrium is that the net force on an object be zero. If the object is treated as a single particle, this is the *only* condition that must be satisfied in order that the particle be in equilibrium. That is, if the net force on the particle is zero, it will remain at rest (if originally at rest) or move with constant velocity in a straight line (if originally in motion).

The situation with real objects is somewhat more complex because objects cannot be treated as particles. An object has a definite size, shape, and mass distribution. In order for an object to be in static equilibrium, the net force on it must be zero *and* the object must have no tendency to rotate. This second condition of equilibrium requires that *the net torque about any origin be zero.* In order to establish whether or not an object is in equilibrium, we must know the size and shape of the object, the forces acting on different parts of the object, and the points of application of the various forces.

In the first part of this chapter, we shall be concerned with objects that are assumed to be rigid. *A rigid object is defined as one that does not deform under the application of external forces.* That is, all parts of a rigid object remain at a fixed separation with respect to each other when subjected to external forces.

The last section of this chapter deals with the realistic situation of objects that deform under load conditions. Deformation is an important consideration in understanding the mechanics of materials and structural designs. Such deformations are usually elastic in nature and will not affect the conditions of equilibrium. By *elastic* we mean that when the deforming forces are removed, the object returns to its original shape. Several elastic constants will be defined, each corresponding to a different type of deformation.

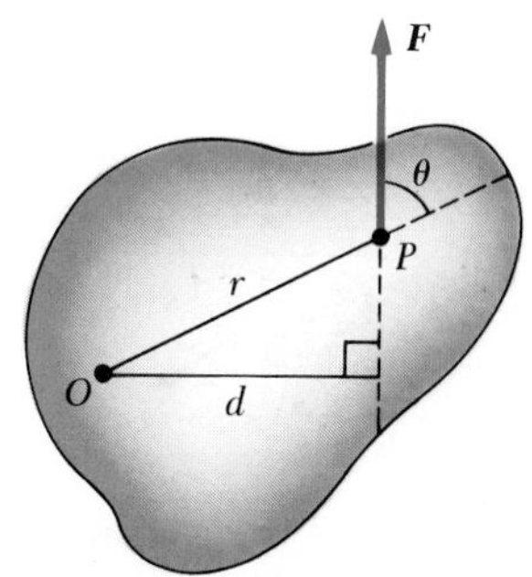

Figure 12.1 A single force $\boldsymbol{F}$ acts on a rigid object at the point P. The moment arm of $\boldsymbol{F}$ relative to O is the perpendicular distance d from O to the line of action of $\boldsymbol{F}$.

12.1 THE CONDITIONS OF EQUILIBRIUM OF A RIGID OBJECT

Consider a single force $\boldsymbol{F}$ acting on a rigid object as in Figure 12.1. The effect of the force on the object depends on its point of application, P. If $\boldsymbol{r}$ is the position vector of this point relative to O, *the torque associated with the force* $\boldsymbol{F}$ *about* O *is given by*

$$\boldsymbol{\tau} = \boldsymbol{r} \times \boldsymbol{F} \tag{12.1}$$

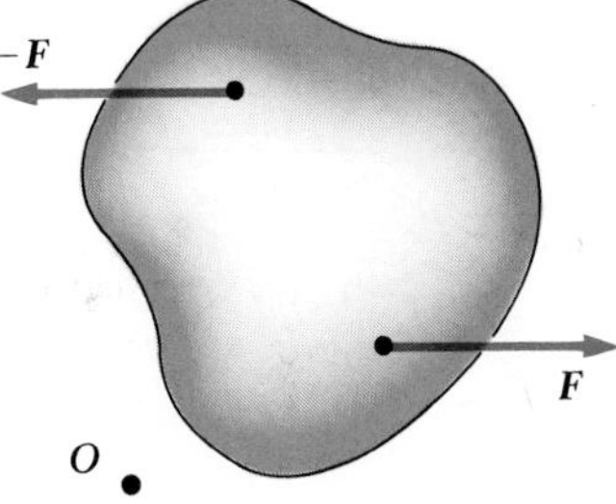

Figure 12.2 The two forces acting on the object are equal in magnitude and opposite in direction, yet the object is not in equilibrium.

Recall that the vector $\boldsymbol{\tau}$ is perpendicular to the plane formed by $\boldsymbol{r}$ and $\boldsymbol{F}$. Furthermore, the sense of $\boldsymbol{\tau}$ is determined by the sense of the rotation that $\boldsymbol{F}$ tends to give to the object. The right-hand rule can be used to determine the direction of $\boldsymbol{\tau}$: Close your right hand such that your four fingers wrap in the direction of rotation that $\boldsymbol{F}$ tends to give the object; your thumb will point in the direction of $\boldsymbol{\tau}$. Hence, in Figure 12.1 $\boldsymbol{\tau}$ is directed *out* of the paper.

As you can see from Figure 12.1, the tendency of $\boldsymbol{F}$ to make the object rotate about an axis through O depends on the moment arm d (the perpendicular distance to the line of action of the force) as well as on the magnitude of $\boldsymbol{F}$. By definition, the magnitude of $\boldsymbol{\tau}$ is given by Fd.

Now suppose two forces, $\boldsymbol{F}_1$ and $\boldsymbol{F}_2$, act on a rigid object. The two forces will have the same effect on the object only if they have the same magnitude, the same direction, and the same line of action. In other words,

Equivalent forces

> two forces $\boldsymbol{F}_1$ and $\boldsymbol{F}_2$ are equivalent if and only if $\boldsymbol{F}_1 = \boldsymbol{F}_2$ and if they have the same torque about any given point.

An example of two equal and opposite forces that are *not* equivalent is shown in Figure 12.2. The force directed toward the right tends to rotate the object clockwise about an axis through O, whereas the force directed toward the left tends to rotate it counterclockwise about that axis.

When an object is pivoted about an axis through its center of mass, the object will undergo an angular acceleration about this axis if there is a nonzero torque acting on the object. As an example, suppose an object is pivoted about an axis through its center of mass as in Figure 12.3. Two equal and opposite

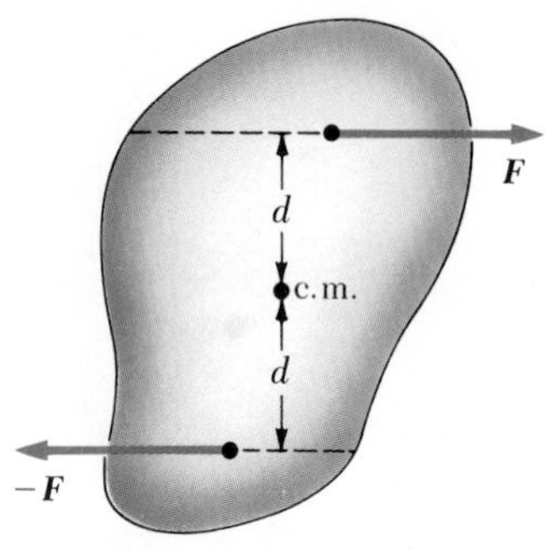

Figure 12.3 Two equal and opposite forces acting on the object form a couple. In this case, the object will rotate clockwise. The net torque on the object about the center of mass is $2Fd$.

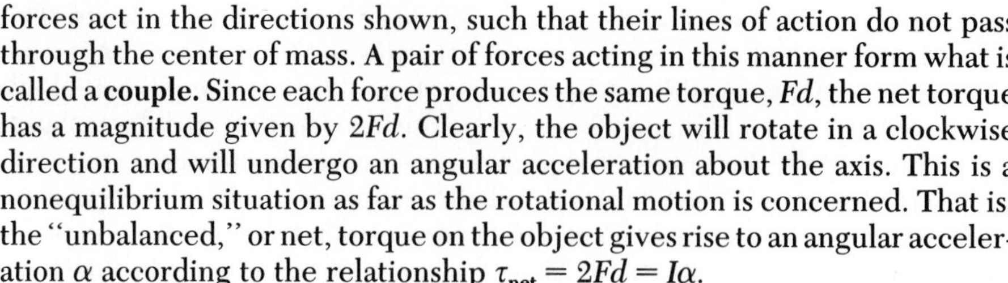

forces act in the directions shown, such that their lines of action do not pass through the center of mass. A pair of forces acting in this manner form what is called a **couple.** Since each force produces the same torque, Fd, the net torque has a magnitude given by $2Fd$. Clearly, the object will rotate in a clockwise direction and will undergo an angular acceleration about the axis. This is a nonequilibrium situation as far as the rotational motion is concerned. That is, the "unbalanced," or net, torque on the object gives rise to an angular acceleration α according to the relationship $\tau_{\text{net}} = 2Fd = I\alpha$.

In general, an object will be in rotational equilibrium only if its angular acceleration $\alpha = 0$. Since $\tau_{\text{net}} = I\alpha$ for rotation about a fixed axis, a necessary condition of equilibrium for an object is that *the net torque about any origin must be zero.* We now have *two necessary conditions for equilibrium of an object, which can be stated as follows:*

1. The resultant external force must equal zero. $$\sum \mathbf{F} = 0 \tag{12.2}$$
2. The resultant external torque must be zero about *any* origin. $$\sum \boldsymbol{\tau} = 0 \tag{12.3}$$

The first condition is a statement of **translational equilibrium,** that is, the linear acceleration of the center of mass of the object must be zero when viewed from an inertial reference frame. The second condition is a statement of **rotational equilibrium,** that is, the angular acceleration about any axis must be zero. In the special case of **static equilibrium,** which is the main subject of this chapter, the object is at rest so that it has no linear or angular velocity (that is, $v_c = 0$ and $\omega = 0$).

The two vector expressions given by Equations 12.2 and 12.3 are equivalent, in general, to six scalar equations. Three of these come from the first condition of equilibrium, and three follow from the second condition (corresponding to x, y, and z components). Hence, in a complex system involving several forces acting in various directions, you would be faced with solving a set of equations with many unknowns. We will restrict our discussion to situations in which all the forces lie in a common plane, which we assume to be the xy plane. Forces whose vector representations are in the same plane are said to be *coplanar.* In this case, we shall have to deal with only *three* scalar equations. Two of these come from balancing the forces in the x and y directions. The third comes from the torque equation, namely, that the net torque about *any* point in the xy plane must be zero. Hence, these two conditions of equilibrium provide the equations

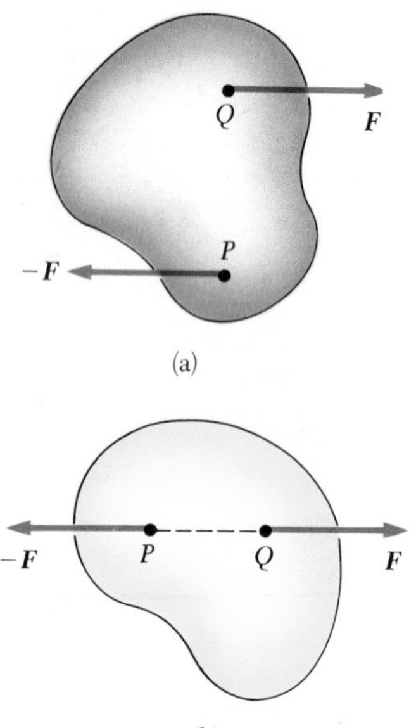

Figure 12.4 (a) The object is not in equilibrium since the two forces do not have the same line of action. (b) The object is in equilibrium since the two forces act along the same line.

$$\sum F_x = 0 \qquad \sum F_y = 0 \qquad \sum \tau_z = 0 \tag{12.4}$$

where the axis of the torque equation is *arbitrary,* as we shall show later.

There are two cases of equilibrium that are often encountered. The first case deals with an object subjected to only two forces, and the second case is concerned with an object subjected to three forces.

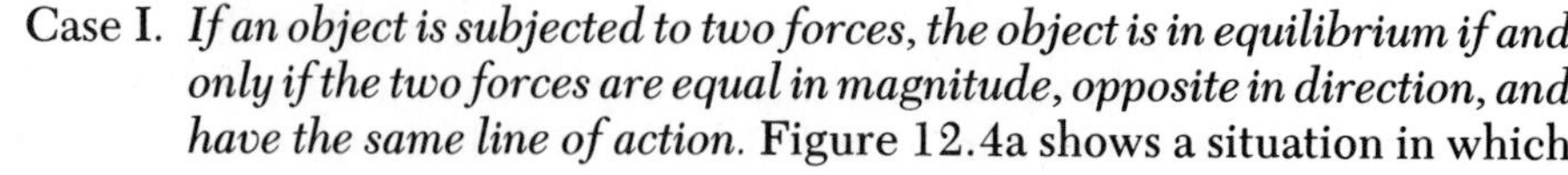

Case I. *If an object is subjected to two forces, the object is in equilibrium if and only if the two forces are equal in magnitude, opposite in direction, and have the same line of action.* Figure 12.4a shows a situation in which the object is not in equilibrium because the two forces are not along the same line. Note that the torque about any axis, such as one through P, is not zero, which violates the second condition of equilibrium. In Figure 12.4b, the object is in equilibrium because the forces have the same line of action. In this situation, it is easy to see that the net torque about any axis is zero.

Case II. *If an object subjected to three forces is in equilibrium, the lines of action of the three forces must intersect at a common point.* That is, the forces must be *concurrent.* (One exception to this rule is the situation in which none of the lines of action intersect. In this situation, the forces must be parallel.) Figure 12.5 illustrates the general rule. The lines of action of the three forces pass through the point S. The conditions of equilibrium require that $\boldsymbol{F}_1 + \boldsymbol{F}_2 + \boldsymbol{F}_3 = 0$ and that the net torque about any axis be zero. Note that as long as the forces are concurrent, the net torque about an axis through S must be zero.

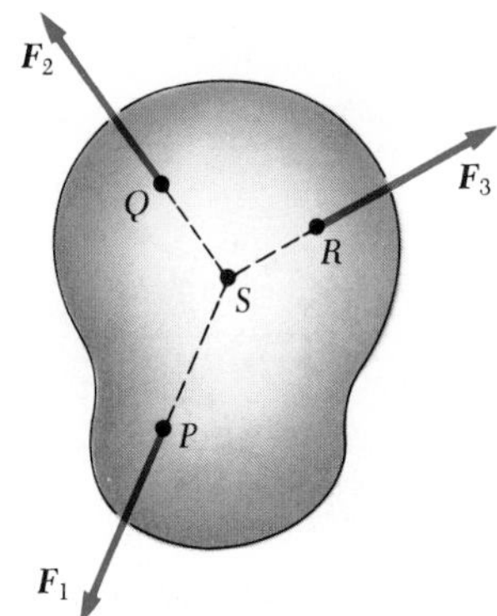

Figure 12.5 If three forces act on an object that is in equilibrium, their lines of action must intersect at a point S (or they must be parallel).

We can easily show that if an object is in translational equilibrium and the net torque is zero with respect to one point, it must be zero about *any* point. The point can be inside or outside the boundaries of the object. Consider an object under the action of several forces such that the resultant force $\Sigma \boldsymbol{F} = \boldsymbol{F}_1 + \boldsymbol{F}_2 + \boldsymbol{F}_3 + \cdots = 0$. Figure 12.6 describes this situation (for clarity, only four forces are shown). The point of application of $\boldsymbol{F}_1$ is specified by the position vector $\boldsymbol{r}_1$. Similarly, the points of application of $\boldsymbol{F}_2, \boldsymbol{F}_3, \ldots$ are specified by $\boldsymbol{r}_2, \boldsymbol{r}_3, \ldots$ (not shown). The net torque about O is

$$\sum \tau_0 = \boldsymbol{r}_1 \times \boldsymbol{F}_1 + \boldsymbol{r}_2 \times \boldsymbol{F}_2 + \boldsymbol{r}_3 \times \boldsymbol{F}_3 + \cdots$$

Now consider another arbitrary point, O', having a position vector $\boldsymbol{r}'$ relative to O. The point of application of $\boldsymbol{F}_1$ relative to this point is identified by the vector $\boldsymbol{r}_1 - \boldsymbol{r}'$. Likewise, the point of application of $\boldsymbol{F}_2$ relative to O' is $\boldsymbol{r}_2 - \boldsymbol{r}'$, and so forth. Therefore, the torque about O' is

$$\sum \tau_{O'} = (\boldsymbol{r}_1 - \boldsymbol{r}') \times \boldsymbol{F}_1 + (\boldsymbol{r}_2 - \boldsymbol{r}') \times \boldsymbol{F}_2 + (\boldsymbol{r}_3 - \boldsymbol{r}') \times \boldsymbol{F}_3 + \cdots$$

$$\sum \tau_{O'} = \boldsymbol{r}_1 \times \boldsymbol{F}_1 + \boldsymbol{r}_2 \times \boldsymbol{F}_2 + \boldsymbol{r}_3 \times \boldsymbol{F}_3 + \cdots - \boldsymbol{r}' \times (\boldsymbol{F}_1 + \boldsymbol{F}_2 + \boldsymbol{F}_3 + \cdots)$$

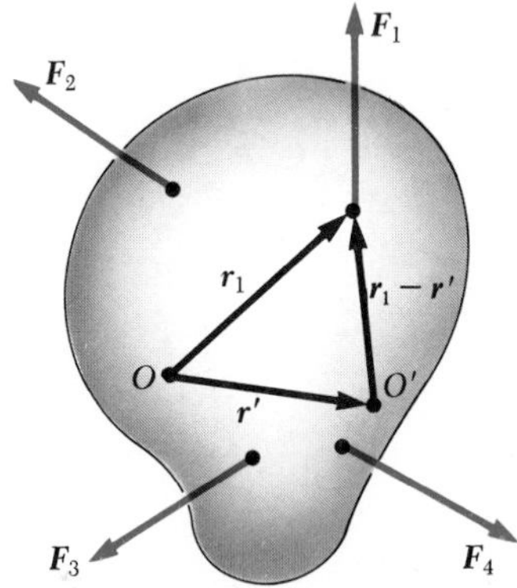

Figure 12.6 Construction for showing that if the net torque about origin O is zero, the net torque about any other origin, such as O', will be zero.

Since the net force is assumed to be zero, the last term in this last expression vanishes and we see that $\Sigma \tau_{O'} = \Sigma \tau_O$. Hence,

> if an object is in translational equilibrium and the net torque is zero about one point, it must be zero about any other point.

12.2 THE CENTER OF GRAVITY

Whenever we deal with rigid objects, one of the forces that must be considered is the weight of the object, that is, the force of gravity acting on the object. In order to compute the torque due to the weight force, all of the weight can be considered as being concentrated at a single point called the *center of gravity.* As we shall see, the center of gravity of an object coincides with its center of mass if the object is in a uniform gravitational field.

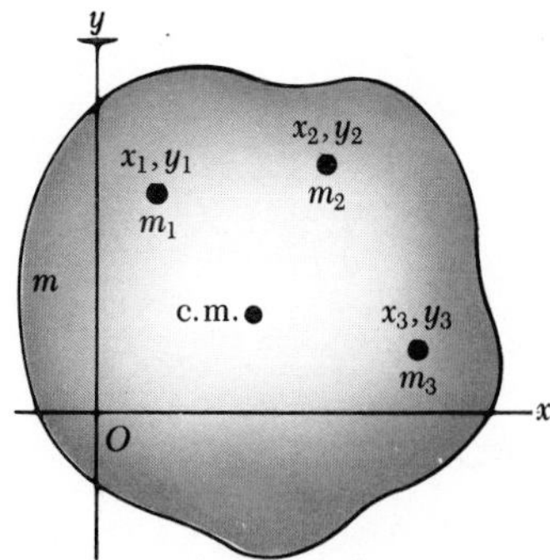

Figure 12.7 An object can be divided into many small particles with specific masses and coordinates. These can be used to locate the center of mass.

Consider an object of arbitrary shape lying in the xy plane, as in Figure 12.7. Suppose the object is divided into a large number of very small particles of masses $m_1, m_2, m_3, \ldots$ having coordinates $(x_1, y_1), (x_2, y_2), (x_3, y_3), \ldots$.

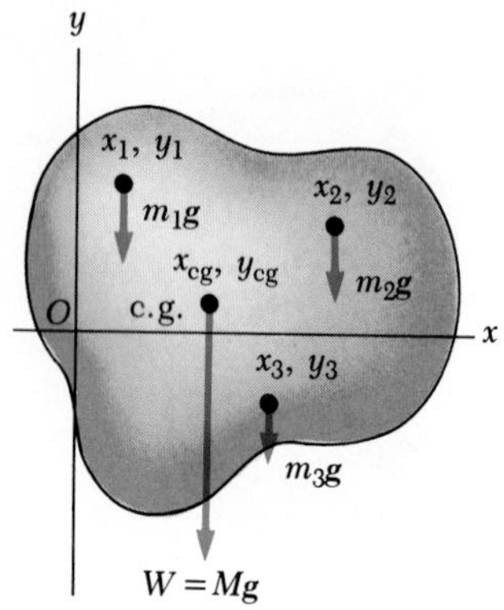

Figure 12.8 The center of gravity of the object is located at the center of mass if the value of **g** is constant over the object.

In Chapter 9 we defined the x coordinate of the center of mass of such an object to be

$$x_c = \frac{m_1x_1 + m_2x_2 + m_3x_3 + \cdots}{m_1 + m_2 + m_3 + \cdots} = \frac{\Sigma m_i x_i}{\Sigma m_i}$$

The y coordinate of the center of mass is similar to this, with x_i replaced by y_i.

Let us now examine the situation from another point of view by considering the weight of each part of the object, as in Figure 12.8. Each particle contributes a torque about the origin equal to its weight multiplied by its moment arm. For example, the torque due to the weight m_1g_1 is $m_1g_1x_1$, and so forth. We wish to locate the one position of the single force **W** (the total weight of the object) whose effect on the rotation of the object is the same as that of the individual particles. This point is called the *center of gravity* of the object. Equating the torque exerted by **W** acting at the center of gravity to the sum of the torques acting on the individual particles gives

$$(m_1g_1 + m_2g_2 + m_3g_3 + \cdots)x_{cg} = m_1g_1x_1 + m_2g_2x_2 + m_3g_3x_3 + \cdots$$

where this expression accounts for the fact that the acceleration of gravity can in general vary over the object. If we assume that g is uniform over the object (as is usually the case), then the g terms in the above equation cancel and we get

$$x_{cg} = \frac{m_1x_1 + m_2x_2 + m_3x_3 + \cdots}{m_1 + m_2 + m_3 + \cdots} \tag{12.5}$$

In other words, *the center of gravity is located at the center of mass as long as the body is in a uniform gravitational field.*

In several examples that will be presented in the next section, we shall be concerned with homogeneous, symmetric objects for which the center of gravity coincides with the geometric center of the object. A rigid object in a uniform gravitational field can be balanced by a single force equal in magnitude to the weight of the object, as long as the force is directed upward through the object's center of gravity.

12.3 EXAMPLES OF RIGID OBJECTS IN STATIC EQUILIBRIUM

In this section we present several examples of rigid objects in static equilibrium. In working such problems, it is important to first recognize *all* external forces acting on the object. Failure to do so will result in an incorrect analysis. The following procedure is recommended when analyzing a body in equilibrium under the action of several external forces:

PROBLEM-SOLVING STRATEGY: OBJECTS IN EQUILIBRIUM

1. Make a sketch of the object under consideration.
2. Draw a free-body diagram and label all external forces acting on the object. Try to guess the correct direction for each force. If you select a direction that leads to a negative sign in your solution for a force, do not be alarmed; this merely means that the direction of the force is the opposite of what you assumed.

3. Resolve all forces into rectangular components, choosing a convenient coordinate system. Then apply the first condition for equilibrium, which balances forces. Remember to keep track of the signs of the various force components.
4. Choose a convenient axis for calculating the net torque on the object. Remember that the choice of the origin for the torque equation is *arbitrary;* therefore choose an origin that will simplify your calculation as much as possible. Becoming adept at this is a matter of practice.
5. The first and second conditions of equilibrium give a set of linear equations with several unknowns. All that is left is to solve the simultaneous equations for the unknowns in terms of the known quantities.

EXAMPLE 12.1 The Seesaw

A uniform board of weight 40 N supports two children weighing 500 N and 350 N, respectively, as shown in Figure 12.9. If the support (called the *fulcrum*) is under the center of gravity of the board and if the 500-N child is 1.5 m from the center, (a) determine the upward force N exerted on the board by the support.

Solution First note that, in addition to N, the external forces acting on the board are the weights of the children and the weight of the board, all of which act downward. We can assume that the center of gravity of the board is at its geometric center because we were told that the board is uniform. Since the system is in equilibrium, the upward force N must balance all the downward forces. From $\Sigma F_y = 0$, we have

$$N - 500\text{ N} - 350\text{ N} - 40\text{ N} = 0 \quad \text{or} \quad N = \boxed{890\text{ N}}$$

It should be pointed out here that the equation $\Sigma F_x = 0$ also applies to this situation, but it is unnecessary to consider this equation because we have no forces acting horizontally on the board.

(b) Determine where the 350-N child should sit to balance the system.

Solution To find this position, we must invoke the second condition for equilibrium. Taking the center of gravity of the board as the axis for our torque equation, we see from $\Sigma\tau = 0$ that

$$(500\text{ N})(1.5\text{ m}) - (350\text{ N})(x) = 0$$

$$x = \boxed{2.14\text{ m}}$$

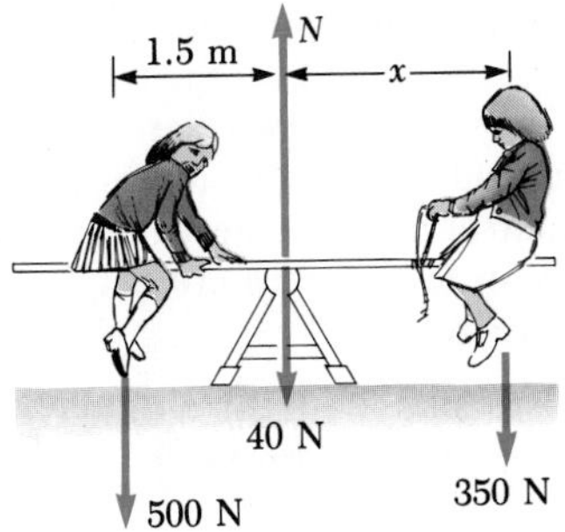

Figure 12.9 (Example 12.1) A balanced system.

Exercise 1 If the fulcrum did not lie under the center of gravity of the board, what other information would you need to solve the problem?

EXAMPLE 12.2 A Weighted Hand

A 50-N weight is held in the hand with the forearm in the horizontal position, as in Figure 12.10a. The biceps muscle is attached 3 cm from the joint, and the weight is 35 cm from the joint. Find the upward force that the biceps exerts on the forearm (made up of the radius and ulna) and the downward force on the upper arm (the humerus) acting at the joint. Neglect the weight of the forearm.

Solution The forces acting on the forearm are equivalent to those acting on a bar of length 35 cm, as shown in Figure 12.10b, where $\boldsymbol{F}$ is the upward force of the biceps and $\boldsymbol{R}$ is the downward force at the joint. From the first condition for equilibrium, we have

$$(1) \qquad \Sigma F_y = F - R - 50\text{ N} = 0$$

From the second condition for equilibrium, we know that the sum of the torques about any point must be zero. With the joint O as the axis, we have

$$Fd - W\ell = 0$$

$$F(3\text{ cm}) - (50\text{ N})(35\text{ cm}) = 0$$

$$F = \boxed{583\text{ N}}$$

This value for F can be substituted into (1) to give $R = 533$ N. These values correspond to $F = 131$ lb and

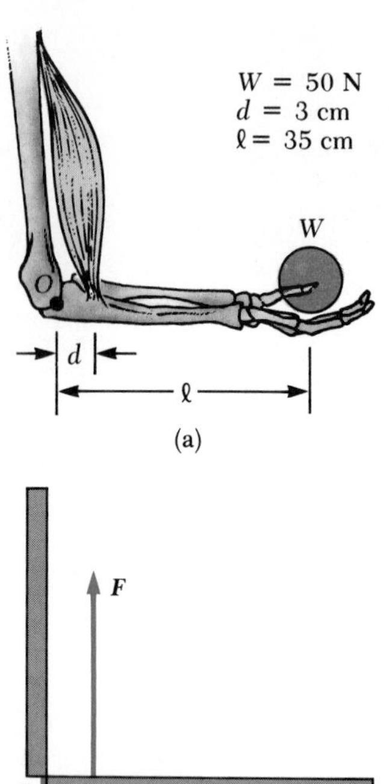

Figure 12.10 (Example 12.2) (a) In this skeletal view, the biceps are pulling (essentially) at right angles to the forearm. (b) The mechanical model for the system described in (a).

$R = 120$ lb. Hence, the forces at joints and in muscles can be extremely large.

Exercise 2 In reality, the biceps makes an angle of 15° with the vertical, so that $\boldsymbol{F}$ has both a vertical and a horizontal component. Find the value of $\boldsymbol{F}$ and the components of $\boldsymbol{R}$ including this fact in your analysis.
Answer $F = 604$ N, $R_x = 156$ N, $R_y = 533$ N.

EXAMPLE 12.3 Standing on a Horizontal Beam

A uniform horizontal beam of length 8 m and weight 200 N is attached to a wall by a pin connection that allows the beam to rotate. Its far end is supported by a cable that makes an angle of 53° with the horizontal (Fig. 12.11a). If a 600-N person stands 2 m from the wall, find the tension in the cable and the force exerted on the beam by the wall.

Solution First we must identify all the external forces acting on the beam. These are its weight, the tension, $\boldsymbol{T}$, in the cable, the force $\boldsymbol{R}$ exerted by the wall at the pivot (the direction of this force is unknown), and the weight of the person on the beam. These are all indicated in the free-body diagram for the beam (Fig. 12.11b). If we resolve $\boldsymbol{T}$ and $\boldsymbol{R}$ into horizontal and vertical components and apply the first condition for equilibrium, we get

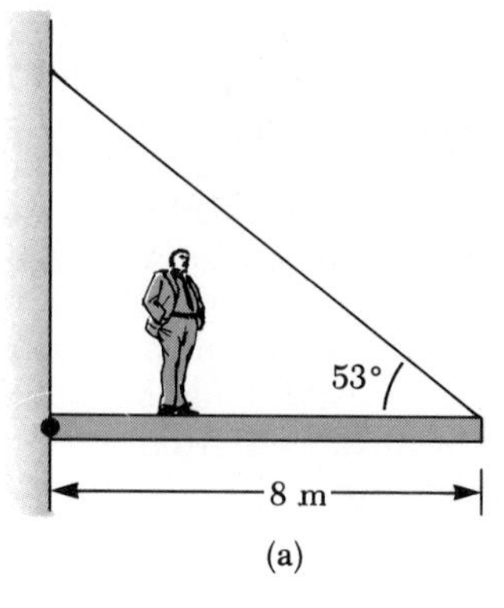

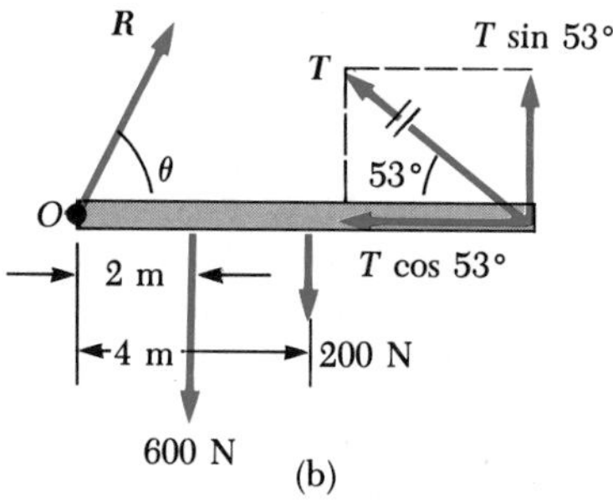

Figure 12.11 (Example 12.3) (a) A uniform beam supported by a cable. (b) The free-body diagram for the beam.

(1) $\sum F_x = R\cos\theta - T\cos 53° = 0$

(2) $\sum F_y = R\sin\theta + T\sin 53° - 600\ \text{N} - 200\ \text{N} = 0$

Because R, T, and θ are all unknown, we cannot obtain a solution from these expressions alone. (The number of simultaneous equations must equal the number of unknowns in order for us to be able to solve for the unknowns.)

Now let us invoke the condition for rotational equilibrium. A convenient axis to choose for our torque equation is the one that passes through the pivot at O. The feature that makes this point so convenient is that the force $\boldsymbol{R}$ and the horizontal component of $\boldsymbol{T}$ both have a lever arm of zero, and hence zero torque, about this pivot. Recalling our convention for the sign of the torque about an axis and noting that the lever arms of the 600-N, 200-N, and $T \sin 53°$ forces are 2 m, 4 m, and 8 m, respectively, we get

(3) $$\sum \tau_O = (T\sin 53°)(8\ \text{m}) - (600\ \text{N})(2\ \text{m}) - (200\ \text{N})(4\ \text{m}) = 0$$

$$T = 313\ \text{N}$$

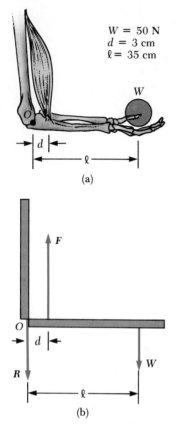

Thus the torque equation with this axis gives us one of the unknowns directly! This value is substituted into (1) and (2) to give

$$R \cos \theta = 188 \text{ N}$$

$$R \sin \theta = 550 \text{ N}$$

We divide these two equations and recall the trigonometric identity $\sin \theta / \cos \theta = \tan \theta$ to get

$$\tan \theta = \frac{550 \text{ N}}{188 \text{ N}} = 2.93$$

$$\theta = \boxed{71.1^\circ}$$

Finally,

$$R = \frac{188 \text{ N}}{\cos \theta} = \frac{188 \text{ N}}{\cos 71.1^\circ} = \boxed{581 \text{ N}}$$

If we had selected some other axis for the torque equation, the solution would have been the same. For example, if the axis were to pass through the center of gravity of the beam, the torque equation would involve both T and R. However, this equation, coupled with (1) and (2), could still be solved for the unknowns. Try it!

When many forces are involved in a problem of this nature, it is convenient to "keep the books straight" by setting up a table of forces, their lever arms, and their torques. For instance, in the example just given, we would construct the following table. Setting the sum of the terms in the last column equal to zero represents the condition of rotational equilibrium.

Force Component	Lever Arm Relative to O (m)	Torque About O (N·m)
$T \sin 53^\circ$	8	$8T \sin 53^\circ$
$T \cos 53^\circ$	0	0
200 N	4	$-4(200)$
600 N	2	$-2(600)$
$R \sin \theta$	0	0
$R \cos \theta$	0	0

EXAMPLE 12.4 The Leaning Ladder

A uniform ladder of length ℓ and weight $W = 50$ N rests against a smooth, vertical wall (Fig. 12.12a). If the coefficient of static friction between the ladder and ground is $\mu = 0.40$, find the *minimum* angle θ_{min} such that the ladder will *not* slip.

Solution The free-body diagram showing all the external forces acting on the ladder is illustrated in Figure

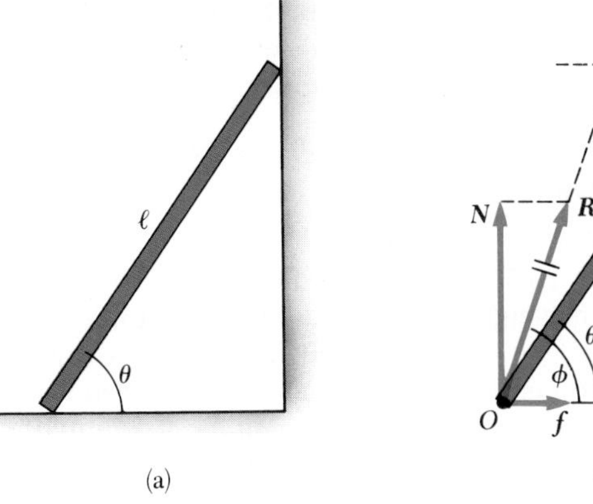

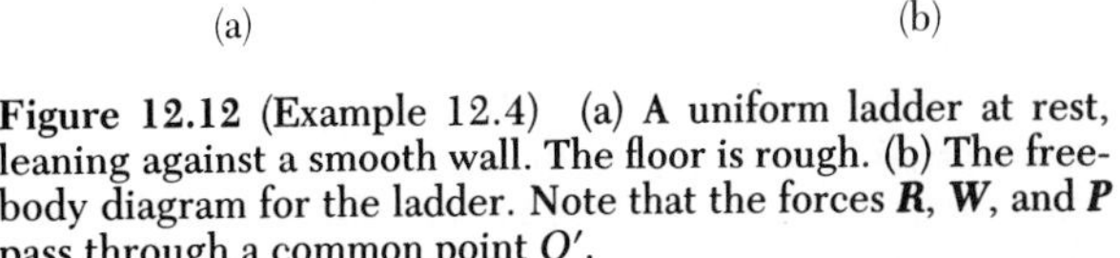

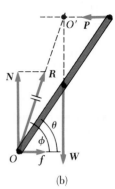

Figure 12.12 (Example 12.4) (a) A uniform ladder at rest, leaning against a smooth wall. The floor is rough. (b) The free-body diagram for the ladder. Note that the forces **R**, **W**, and **P** pass through a common point O'.

12.12b. The reaction force at the ground, $\mathbf{R}$, is the vector sum of a normal force, $\mathbf{N}$, and the force of friction, $\mathbf{f}$. The reaction force at the wall, $\mathbf{P}$, is horizontal, since the wall is smooth. From the first condition for equilibrium applied to the ladder, we have

$$\sum F_x = f - P = 0$$

$$\sum F_y = N - W = 0$$

Since $W = 50$ N, we see from the equation above that $N = W = 50$ N. Furthermore, *when the ladder is on the verge of slipping, the force of friction must be a maximum,* given by $f_{\text{max}} = \mu N = 0.40(50 \text{ N}) = 20$ N. (Recall that $f_s \leq \mu N$.) Thus, at this angle, $P = 20$ N.

To find the value of θ, we must use the second condition of equilibrium. When the torques are taken about the origin O at the bottom of the ladder, we get

$$\sum \tau_O = P\ell \sin \theta - W\frac{\ell}{2} \cos \theta = 0$$

But $P = 20$ N when the ladder is about to slip and $W = 50$ N, so that the expression above gives

$$\tan \theta_{\text{min}} = \frac{W}{2P} = \frac{50 \text{ N}}{40 \text{ N}} = 1.25$$

$$\theta_{\text{min}} = \boxed{51.3^\circ}$$

It is interesting to note that the result does not depend on ℓ!

An alternative approach to analyzing this problem is to consider the intersection O' of the forces **W** and **P**. Since the torque about any origin must be zero, the torque about O' must be zero. This requires that the line of action of $\mathbf{R}$ (the resultant of $\mathbf{N}$ and $\mathbf{f}$) pass through O'! That is, since this is a three-force body, the forces must be concurrent. With this condition, one could then obtain the angle ϕ that $\boldsymbol{R}$ makes with the horizontal (where ϕ is greater than θ), assuming the length of the ladder is known.

Exercise 3 With reference to Figure 12.12, show that $\tan \phi = 2 \tan \theta$.

EXAMPLE 12.5 Raising a Cylinder

A cylinder of weight W and radius R is to be raised onto a step of height h as shown in Figure 12.13. A rope is wrapped around the cylinder and pulled horizontally. Assuming the cylinder doesn't slip on the step, find the *minimum* force $\boldsymbol{F}$ necessary to raise the cylinder and the reaction force at P.

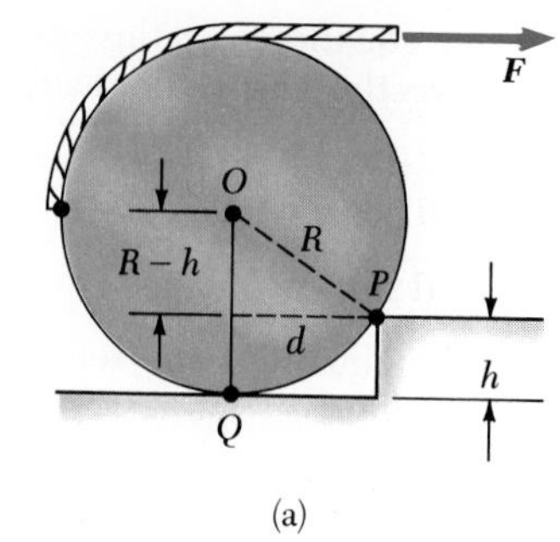

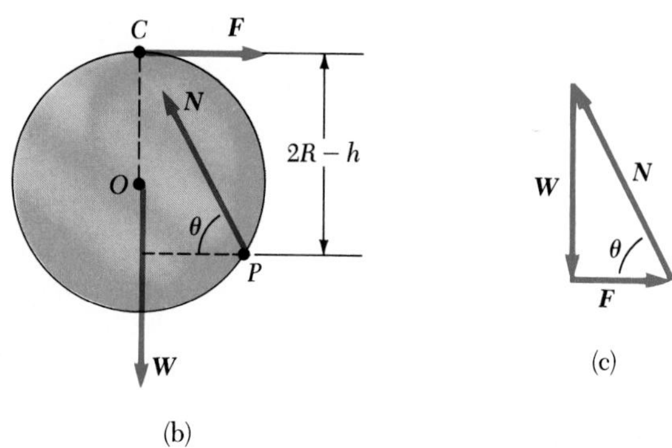

Figure 12.13 (Example 12.5) (a) A cylinder of weight **W** being pulled by a force **F** over a step. (b) The free-body diagram for the cylinder when it is just about to be raised. (c) The *vector* sum of the three external forces is zero.

Solution When the cylinder is just about to be raised, the reaction force at Q goes to zero. Hence, at this time there are only three forces on the cylinder, as shown in Figure 12.13b. From the dotted triangle drawn in Figure 12.13a, we see that the moment arm d of the weight relative to the point P is given by

$$d = \sqrt{R^2 - (R-h)^2} = \sqrt{2Rh - h^2}$$

The moment arm of $\boldsymbol{F}$ relative to P is $2R - h$. Therefore, the net torque acting on the cylinder about P is

$$Wd - F(2R - h) = 0$$

$$W\sqrt{2Rh - h^2} - F(2R - h) = 0$$

$$F = \frac{W\sqrt{2Rh - h^2}}{2R - h}$$

Hence, the second condition of equilibrium was sufficient to obtain the magnitude of $\boldsymbol{F}$. We can determine the components of $\boldsymbol{N}$ by using the first condition of equilibrium:

$$\sum F_x = F - N\cos\theta = 0$$

$$\sum F_y = N\sin\theta - W = 0$$

Dividing gives

$$(1) \qquad \tan\theta = \frac{W}{F}$$

and solving for N gives

$$(2) \qquad N = \sqrt{W^2 + F^2}$$

For example, if we take $W = 500$ N, $h = 0.3$ m, and $R = 0.8$ m, we find that $F = 240$ N, $\theta = 64.3°$, and $N = 551$ N.

Exercise 4 Solve this problem by noting that the three forces acting on the cylinder are concurrent and must pass through the point C. The three forces form the sides of the triangle shown in Figure 12.13c.

12.4 ELASTIC PROPERTIES OF SOLIDS

In our study of mechanics thus far, we assumed that objects remain undeformed when external forces act on them. In reality, all objects are deformable. That is, it is possible to change the shape or size of an object (or both) through the application of external forces. Although these changes are observed as large-scale deformations, the internal forces that resist the deformation are due to short-range forces between atoms.

We shall discuss the elastic properties of solids in terms of the concepts of stress and strain. **Stress** is a quantity that is proportional to the force causing a deformation; more specifically, stress is the external force per unit cross-sectional area acting on an object. **Strain** is a measure of the degree of deformation. It is found that, for sufficiently small stresses, the stress is proportional to the strain; the constant of proportionality depends on the material being deformed and on the nature of the deformation. We call this proportionality constant the **elastic modulus.** The elastic modulus is therefore the ratio of stress to strain:

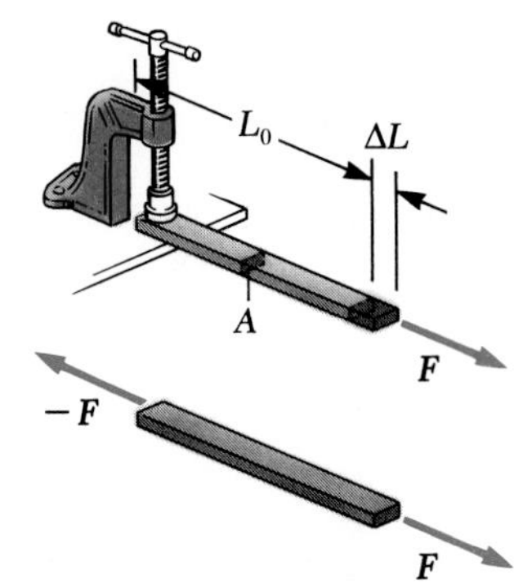

Figure 12.14 A long bar clamped at one end is stretched by an amount ΔL under the action of a force $\boldsymbol{F}$.

$$\text{Elastic modulus} \equiv \frac{\text{stress}}{\text{strain}} \tag{12.6}$$

We shall consider three types of deformation and define an elastic modulus for each:

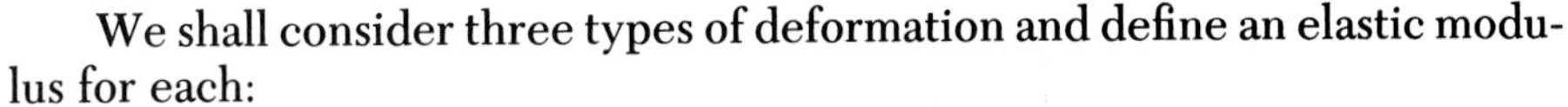

1. **Young's modulus,** which measures the resistance of a solid to a change in its length
2. **Shear modulus,** which measures the resistance to motion of the planes of a solid sliding past each other
3. **Bulk modulus,** which measures the resistance that solids or liquids offer to changes in their volume

Young's Modulus: Elasticity in Length

Consider a long bar of cross-sectional area A and length L_0 that is clamped at one end (Fig. 12.14). When an external force $\boldsymbol{F}$ is applied along the bar and perpendicular to the cross section, internal forces in the bar resist distortion ("stretching"), but the bar attains an equilibrium in which its length is greater and in which the external force is exactly balanced by internal forces. In such a situation, the bar is said to be stressed. We define the **tensile stress** as the ratio of the magnitude of the external force F to the cross-sectional area A. The **tensile strain** in this case is defined as the ratio of the change in length, ΔL, to the original length, L_0, and is therefore a dimensionless quantity. Thus, we can use Equation 12.6 to define **Young's modulus,** Y:

$$Y \equiv \frac{\text{tensile stress}}{\text{tensile strain}} = \frac{F/A}{\Delta L/L_0} \tag{12.7}$$

Young's modulus

This quantity is typically used to characterize a rod or wire stressed under either tension or compression. Note that because the strain is a dimensionless quantity, Y has units of force per unit area. Typical values are given in Table 12.1. Experiments show that (a) the change in length for a fixed applied force is proportional to the original length and (b) the force necessary to produce a given strain is proportional to the cross-sectional area. Both of these observations are in accord with Equation 12.7.

It is possible to exceed the *elastic limit* of a substance by applying a sufficiently large stress (Fig. 12.15). When the stress exceeds the elastic limit, the object is permanently distorted and will not return to its original shape after the stress is removed. Beyond the elastic limit, the stress-strain curve

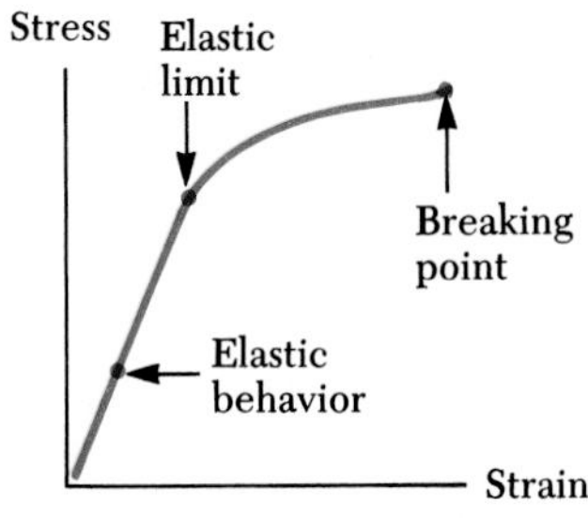

Figure 12.15 Stress versus strain curve for an elastic solid.

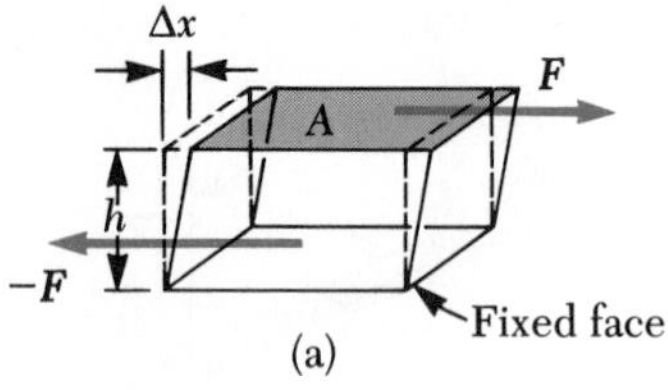

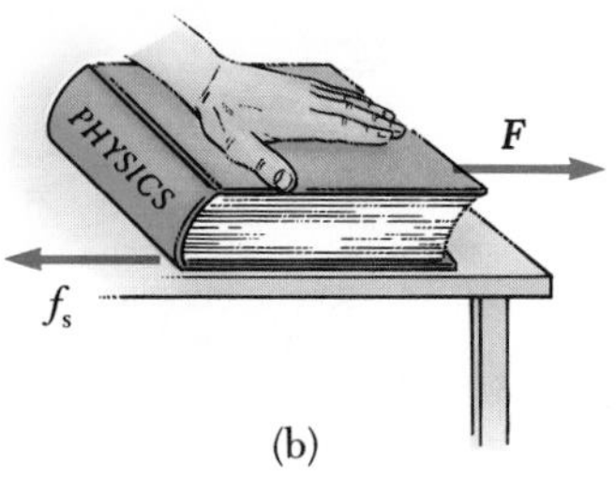

Figure 12.16 (a) A shear deformation in which a rectangular block is distorted by two forces of equal magnitude but opposite directions applied to two parallel faces. (b) A book under shear stress.

TABLE 12.1 Typical Values for Elastic Modulus

Substance	Young's Modulus (N/m²)	Shear Modulus (N/m²)	Bulk Modulus (N/m²)
Aluminum	7.0×10^{10}	2.5×10^{10}	7.0×10^{10}
Brass	9.1×10^{10}	3.5×10^{10}	6.1×10^{10}
Copper	11×10^{10}	4.2×10^{10}	14×10^{10}
Steel	20×10^{10}	8.4×10^{10}	16×10^{10}
Tungsten	35×10^{10}	14×10^{10}	20×10^{10}
Glass	$6.5-7.8 \times 10^{10}$	$2.6-3.2 \times 10^{10}$	$5.0-5.5 \times 10^{10}$
Quartz	5.6×10^{10}	2.6×10^{10}	2.7×10^{10}
Water	—	—	0.21×10^{10}
Mercury	—	—	2.8×10^{10}

departs from a straight line. Hence, its shape is permanently changed. As the stress is increased even further, the material will ultimately break.

Shear Modulus: Elasticity of Shape

Another type of deformation occurs when a body is subjected to a force $\mathbf{F}$ tangential to one of its faces while the opposite face is held in a fixed position by a force of friction, $\mathbf{f}_s$ (Fig. 12.16a). If the object is originally a rectangular block, a shear stress results in a shape whose cross-section is a parallelogram. For this situation, the stress is called a shear stress. A book pushed sideways as in Figure 12.16b is an example of an object under a shear stress. There is no change in volume under this deformation. We define the **shear stress** as F/A, the ratio of the tangential force to the area, A, of the face being sheared. The **shear strain** is defined as the ratio $\Delta x/h$, where Δx is the horizontal distance the sheared face moves and h is the height of the object. In terms of these quantities, the **shear modulus, S,** is

Shear modulus

$$S \equiv \frac{\text{shear stress}}{\text{shear strain}} = \frac{F/A}{\Delta x/h} \tag{12.8}$$

Values of the shear modulus for some representative materials are given in Table 12.1. The units of shear modulus are force per unit area.

Bulk Modulus: Volume Elasticity

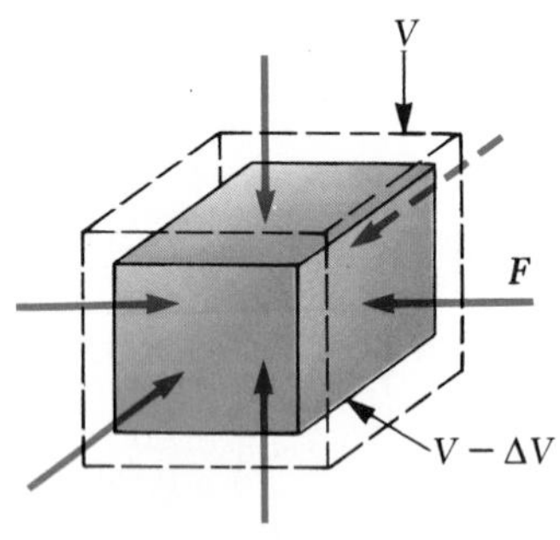

Figure 12.17 When a solid is under uniform pressure, it undergoes a change in volume but no change in shape. This cube is compressed on all sides by forces normal to its surfaces.

Finally, we define the bulk modulus of a substance, which characterizes the response of a substance to uniform squeezing. Suppose that the external forces acting on an object are at right angles to all of its faces (Fig. 12.17) and distributed uniformly over all the faces. As we shall see in Chapter 15, this occurs when an object is immersed in a fluid. A body subject to this type of deformation undergoes a change in volume but no change in shape. The **volume stress, ΔP,** is defined as the ratio of the magnitude of the normal force, F, to the area, A. The quantity $\Delta P = F/A$ is called the **pressure.** The volume strain is equal to the change in volume, ΔV, divided by the original volume, V.

Thus, from Equation 12.6 we can characterize a volume compression in terms of the **bulk modulus,** ***B***, defined as

$$B \equiv \frac{\text{volume stress}}{\text{volume strain}} = -\frac{F/A}{\Delta V/V} = -\frac{\Delta P}{\Delta V/V} \qquad (12.9)$$

Bulk modulus

A negative sign is inserted in this defining equation so that B is a positive number. This is because an increase in pressure (positive ΔP) causes a decrease in volume (negative ΔV) and vice versa.

Table 12.1 lists bulk moduli for some materials. If you look up such values in a different source, you will often find that the reciprocal of the bulk modulus is listed. The reciprocal of the bulk modulus is called the **compressibility** of the material. You should note from Table 12.1 that both solids and liquids have a bulk modulus. However, there is no shear modulus and no Young's modulus for liquids because a liquid will not sustain a shearing stress or a tensile stress (it will flow instead).

Prestressed Concrete

If the stress on a solid object exceeds a certain value, the object will break or fracture. The maximum stress that can be applied before fracture occurs depends on the nature of the material and the type of stress that is applied. For example, concrete has a tensile strength of about 2×10^6 N/m^2, a compressive strength of 20×10^6 N/m^2, and a shear strength of 2×10^6 N/m^2. If the actual stress exceeds these values, the concrete fractures. It is common practice to use large safety factors to prevent failure in concrete structures.

Concrete is normally very brittle when cast in thin sections. Thus, concrete slabs tend to sag and crack at unsupported areas, as in Figure 12.18a. The slab can be strengthened by using steel rods to reinforce the concrete at specific depths, as in Figure 12.18b. Recall that concrete is much stronger under compression than under tension. For this reason, vertical columns of concrete that are under compression can support very heavy loads, whereas horizontal beams of concrete will tend to sag and crack because of their smaller shear strength. A significant increase in shear strength is achieved, however, by prestressing the reinforced concrete, as in Figure 12.18c. As the concrete is being poured, the steel rods are held under tension by external forces. The external forces are released after the concrete cures, which results in a permanent tension in the steel and hence a compressive stress on the concrete. This enables the concrete slab to support a much heavier load.

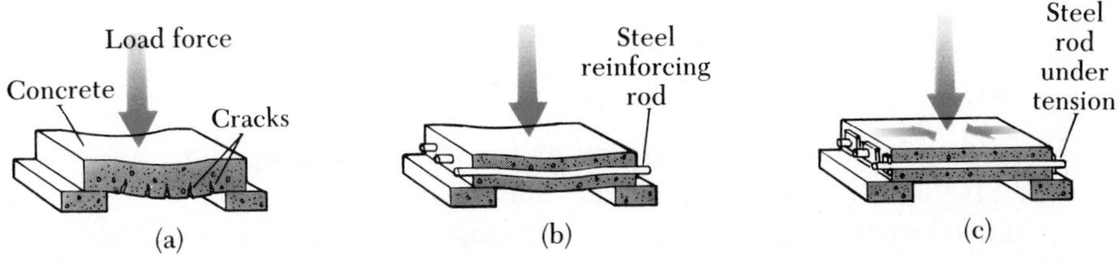

Figure 12.18 (a) A concrete slab with no reinforcement tends to crack under a heavy load. (b) The strength of the concrete slab is increased by using steel reinforcement rods. (c) The slab is further strengthened by prestressing the concrete with steel rods under tension.

Another method that has been successful in strengthening concrete is the use of fibers mixed in the cement and aggregate. Problems such as cracking can be controlled by fibrous materials, such as glass, steel, nylon, polypropylene, and, more recently, glass fibers.

EXAMPLE 12.6 Measuring Young's Modulus
A load of 102 kg is supported by a wire of length 2 m and cross-sectional area 0.1 cm². The wire is stretched by 0.22 cm. Find the tensile stress, tensile strain, and Young's modulus for the wire from this information.

Solution

$$\text{Tensile stress} = \frac{F}{A} = \frac{Mg}{A} = \frac{(102 \text{ kg})(9.80 \text{ m/s}^2)}{0.1 \times 10^{-4} \text{ m}^2}$$

$$= 1.0 \times 10^8 \text{ N/m}^2$$

$$\text{Tensile strain} = \frac{\Delta L}{L_0} = \frac{0.22 \times 10^{-2} \text{ m}}{2 \text{ m}}$$

$$= 0.11 \times 10^{-2}$$

$$Y = \frac{\text{tensile stress}}{\text{tensile strain}} = \frac{1.0 \times 10^8 \text{ N/m}^2}{0.11 \times 10^{-2}}$$

$$= 9.1 \times 10^{10} \text{ N/m}^2$$

Comparing this value for Y with the values in Table 12.1, we conclude that the wire is probably made of brass.

EXAMPLE 12.7 Squeezing a Lead Sphere
A solid lead sphere of volume 0.5 m³ is lowered to a depth in the ocean where the water pressure is equal to 2×10^7 N/m². The bulk modulus of lead is equal to 7.7×10^9 N/m². What is the change in volume of the sphere?

Solution From the definition of bulk modulus, we have

$$B = -\frac{\Delta P}{\Delta V/V}$$

or

$$\Delta V = -\frac{V\Delta P}{B}$$

In this case, the change in pressure, ΔP, has the value 2×10^7 N/m². (This is large relative to atmospheric pressure, 1.01×10^5 N/m².) Taking $V = 0.5$ m³ and $B = 7.7 \times 10^9$ N/m², we get

$$\Delta V = -\frac{(0.5 \text{ m}^3)(2 \times 10^7 \text{ N/m}^2)}{7.7 \times 10^9 \text{ N/m}^2} = -1.3 \times 10^{-3} \text{ m}^3$$

The negative sign indicates a *decrease* in volume.

SUMMARY

A rigid object is in **equilibrium** if and only if the following conditions are satisfied: (1) *the resultant external force must be zero* and (2) *the resultant external torque must be zero about* any *origin*. That is,

Conditions for equilibrium

$$\sum \mathbf{F} = 0 \qquad (12.2)$$

$$\sum \boldsymbol{\tau} = 0 \qquad (12.3)$$

The first condition is the *condition of translational equilibrium*, and the second is the *condition of rotational equilibrium*.

If two forces act on a rigid object, the object is in equilibrium if and only if the forces are equal in magnitude and opposite in direction and have the same line of action.

When three forces act on a rigid object that is in equilibrium, the three forces must be concurrent, that is, their lines of action must intersect at a common point.

The *center of gravity* of an object coincides with the center of mass if the object is in a uniform gravitational field.

The elastic properties of a solid can be described using the concepts of stress and strain. **Stress is a quantity proportional to the force producing a deformation; strain is a measure of the degree of deformation. Stress is proportional to strain, and the constant of proportionality is the elastic modulus:**

$$\text{Elastic modulus} \equiv \frac{\text{stress}}{\text{strain}} \tag{12.6}$$

Three common types of deformation are: (1) the resistance of a solid to elongation under a load, characterized by **Young's modulus,** Y; (2) the resistance of a solid to the motion of planes in the solid sliding past each other, characterized by the **shear modulus,** S; (3) the resistance of a solid (or a liquid) to a volume change, characterized by the **bulk modulus,** B.

QUESTIONS

1. Can a body be in equilibrium if only one external force acts on it? Explain.
2. Can a body be in equilibrium if it is in motion? Explain.
3. Locate the center of gravity for the following uniform objects: (a) a sphere, (b) a cube, (c) a right circular cylinder.
4. The center of gravity of an object may be located outside the object. Give a few examples for which this is the case.
5. You are given an arbitrarily shaped piece of plywood, together with a hammer, nail, and plumb bob. How could you use these items to locate the center of gravity of the plywood? (*Hint:* Use the nail to suspend the plywood.)
6. In order for a chair to be balanced on one leg, where must the center of gravity of the chair be located?
7. Give an example in which the net torque acting on an object is zero and yet the net force is nonzero.
8. Give an example in which the net force acting on an object is zero and yet the net torque is nonzero.
9. Can an object be in equilibrium if the only torques acting on it produce clockwise rotation?
10. A tall crate and a short crate of equal mass are placed side by side on an incline (without touching each other). As the incline angle is increased, which crate will topple first? Explain.
11. A male and a female student are asked to do the following task. Face a wall, step three foot lengths away from the wall, and then lean over and touch the wall with your nose, keeping your hands behind your back. The male usually fails, but the female succeeds. How would you explain this?
12. When lifting a heavy object, why is it recommended to straighten your back as much as possible, lifting from the knees, rather than bending over and lifting from the waist?
13. Would you expect the center of gravity and the center of mass of the Empire State Building to coincide precisely? Explain.
14. Give several examples where several forces are acting on a system in such a way that their sum is zero, but the system is not in equilibrium.
15. If you measure the net torque and the net force on a system to be zero, (a) could the system still be rotating with respect to you? (b) Could it be translating with respect to you?
16. A ladder is resting inclined against a wall. Would you feel safer climbing up the ladder if you were told that the floor is frictionless but the wall is rough or that the wall is smooth but the floor is rough? Justify your answer.
17. What kind of deformation does a cube of Jello exhibit when it "jiggles"?

PROBLEMS

Section 12.1 The Conditions of Equilibrium of a Rigid Object

1. Write the necessary condition of equilibrium for the body shown in Figure 12.19. Take the origin of the torque equation at the point O.

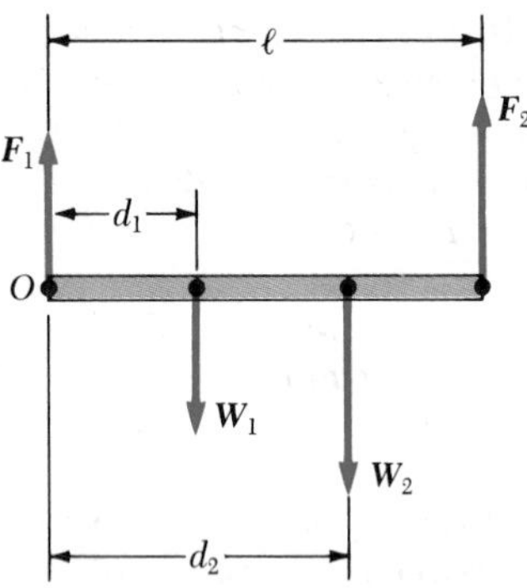

Figure 12.19 (Problem 1).

2. Write the necessary conditions of equilibrium for the body shown in Figure 12.20. Take the origin of the torque equation at the point O.

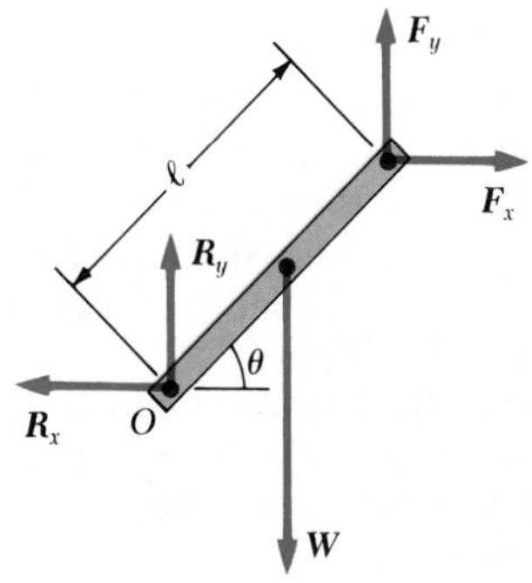

Figure 12.20 (Problem 2).

3. A uniform beam of weight W and length ℓ has weights W_1 and W_2 at two positions, as in Figure 12.21. The beam is resting at two points. For what value of x will the beam be balanced at P such that the normal force at O is zero?

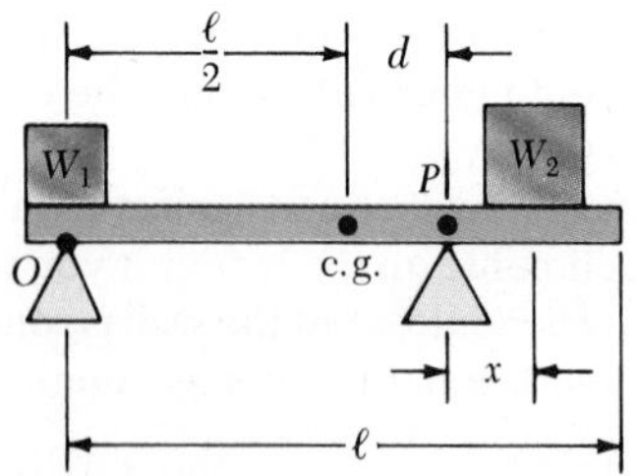

Figure 12.21 (Problems 3 and 4).

4. With reference to Figure 12.21, find x such that the normal force at O will be one half the normal force at P. Neglect the weight of the beam.
5. A ladder of weight 400 N and length 10 m is placed against a smooth vertical wall. A person weighing 800 N stands on the ladder 2 m from the bottom as measured along the ladder. The foot of the ladder is 8 m from the bottom of the wall. Calculate the force exerted by the wall, and the normal force exerted by the floor on the ladder.
6. A student gets his car stuck in a snow drift. Not at a loss, having studied physics, he attaches one end of a stout rope to the vehicle and the other end to the trunk of a nearby tree, allowing for a small amount of slack. The student then exerts a force F on the center of the rope in the direction perpendicular to the car-tree line, as shown in Figure 12.22. If the rope is inextensible and if the magnitude of the applied force is 500 N, what is the force on the car? (Assume equilibrium conditions.)

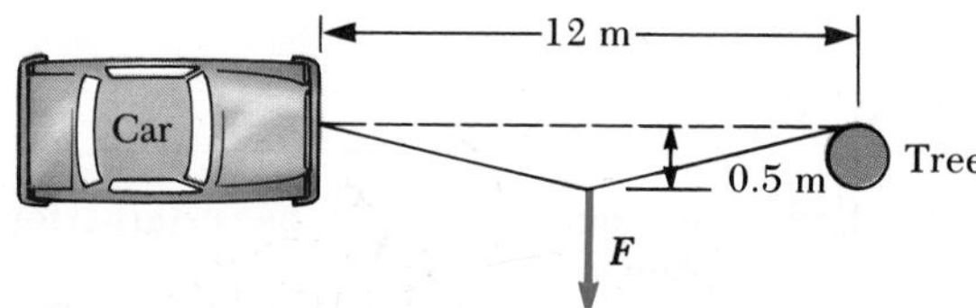

Figure 12.22 (Problem 6).

Section 12.2 The Center of Gravity

7. A flat plate in the shape of a letter T is cut with the dimensions shown in Figure 12.23. Locate the center

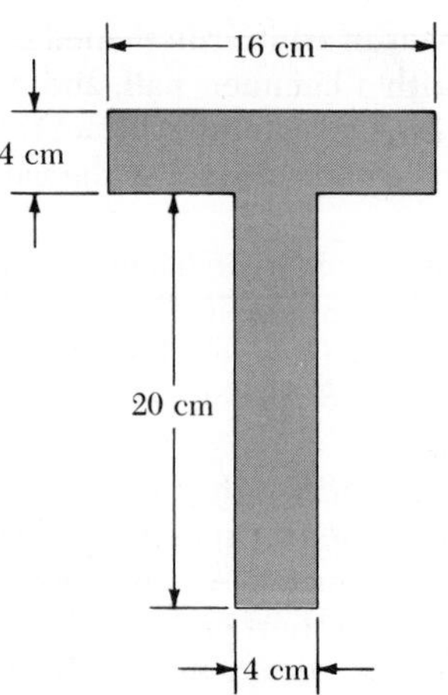

Figure 12.23 (Problem 7).

of gravity. (*Hint:* Note that the weights of the two rectangular parts are proportional to their volumes.)

8. A carpenter's square has the shape of an L, as in Figure 12.24. Locate the center of gravity. (See hint in Problem 7.)

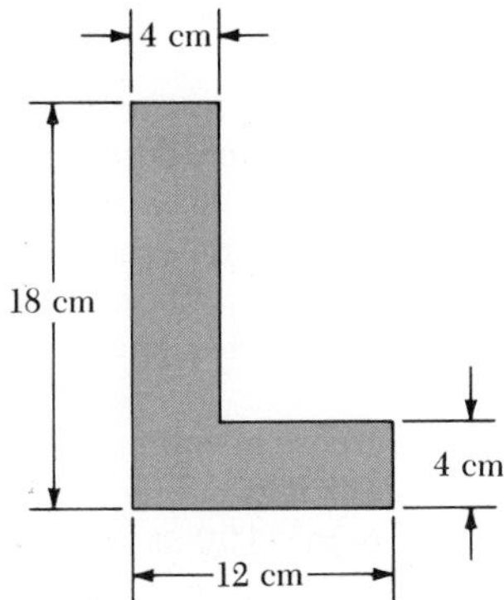

Figure 12.24 (Problem 8).

9. Consider the following mass distribution: 5 kg at (0, 0), 3 kg at (0, 4), and 4 kg at (3, 0). The position coordinates are in meters. Where should a fourth mass of 8 kg be placed so that the center of gravity of the four-mass arrangement will be at (0, 0)?

10. A rod of length L and linear mass density $\lambda(x) = Ax$ kg/m, where A is a constant and x is measured in meters from one end of the rod. Assuming the rod is in a uniform gravitational field, determine the location of the center of gravity of the rod.

Section 12.3 Examples of Rigid Bodies in Static Equilibrium

11. A meter stick supported at the 50-cm mark has masses of 300 g and 200 g hanging from it at the 10-cm and 60-cm marks, respectively. Determine the position at which one would hang a third, 400-g mass to keep the meter stick balanced.

12. Two pans of a balance are 50 cm apart. The fulcrum of the balance has been shifted 1 cm away from the center by a dishonest shopkeeper. By what percentage is the true weight of the goods being marked up by the shopkeeper? The balance itself has negligible mass.

13. A ladder with a uniform density and a mass m rests against a frictionless vertical wall at an angle of 60°. The lower end rests on a flat surface where the coefficient of static friction is $\mu_s = 0.40$. A student with a mass $M = 2\,m$ attempts to climb the ladder. What fraction of the length L of the ladder will the student have reached when the ladder begins to slip?

14. A uniform plank of length 6 m and mass 30 kg rests horizontally on a scaffold, with 1.5 m of the plank hanging over one end of the scaffold. How far can a painter of mass 70 kg walk on the overhanging part of the plank before it tips?

15. An automobile with a mass of 1500 kg has a wheel base (the distance between the axles) of 3.0 m. The center of mass of the automobile is on the center line at a point 1.2 m behind the front axle. Find the force exerted by the ground on each of the four wheels.

16. A 48-kg diver stands at the end of a 3-m-long diving board. What torque does the weight of the diver produce about an axis perpendicular to and in the plane of the diving board through its midpoint?

17. A cue stick strikes a cue ball and delivers a horizontal impulse J in such a way that the ball rolls without slipping as it starts to move. At what height above the ball's center (in terms of the radius of the ball) was the blow struck?

18. A uniform flexible cable has a mass of 100 kg and is suspended between two fixed points, A and B, at the same level, as shown in Figure 12.25. At the support points the cable makes angles of 30°. (a) Find the components of the force exerted on each support by the cable. (b) Find the tension in the cable at the lowest point.

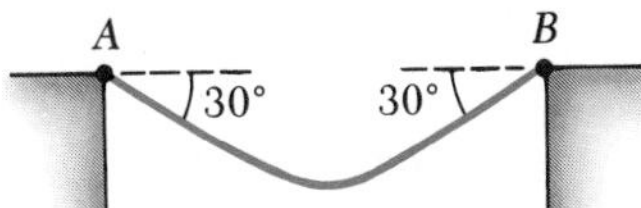

Figure 12.25 (Problem 18).

19. A hemispherical sign 1 m in diameter and of uniform mass density is supported by two strings as shown in Figure 12.26. What fraction of the sign's weight is supported by each string?

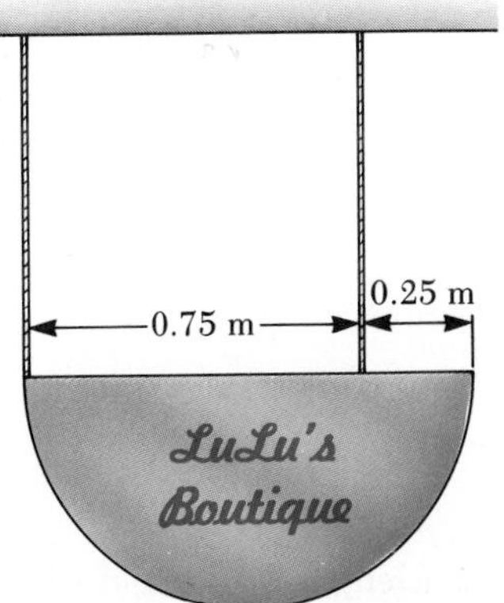

Figure 12.26 (Problem 19).

20. A 10-m-long uniform beam of 100 kg is supported at the two ends by two strings to hang horizontally. If one of the strings can withstand a maximum load of 600 N, at what maximum distance, measured from the stronger string, could a mass of 50 kg be placed so that the weaker string would not break?

21. A uniform rod AB of length 5 m and weight 50 N is pivoted at A and held in equilibrium by a rope as shown in Figure 12.27. A load of 100 N hangs from the rod at a distance x from A. If the breaking strength of the rope is 50 N, find the maximum value of x.

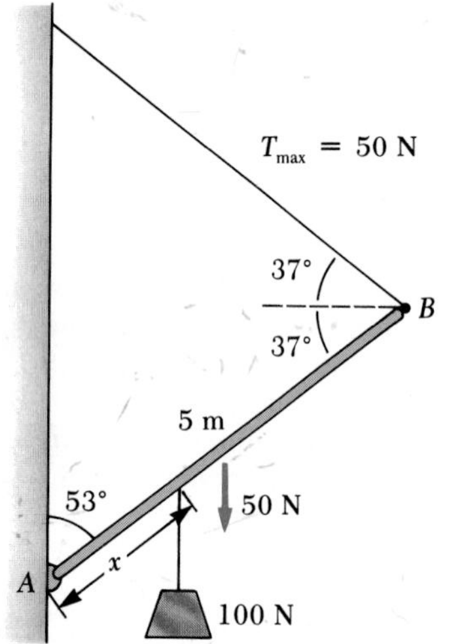

Figure 12.27 (Problem 21).

22. A beam 20 m long and of uniform mass density is resting on two supports as shown in Figure 12.28. The beam weighs 500 N. (a) A person of weight 625 N stands at the geometrical midpoint of the beam. Calculate the force exerted on the beam by the two supports. (b) The person is able to walk a maximum distance x as shown, before the beam tips over. Calculate the distance x.

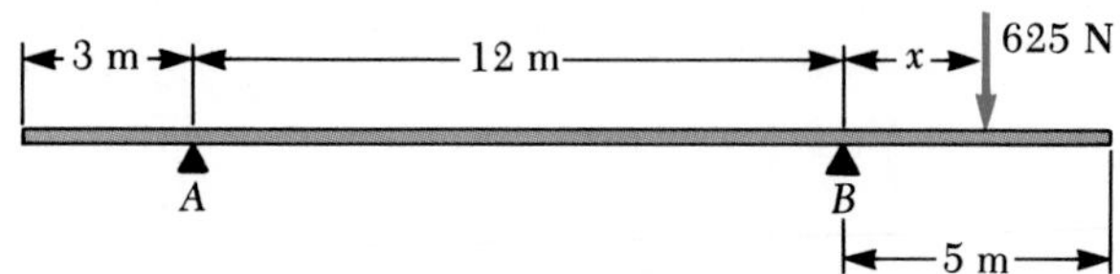

Figure 12.28 (Problem 22).

Section 12.4 Elastic Properties of Solids

23. A steel wire has a length of 10 m and a cross-sectional area of 0.2 cm^2. Under what load will its length increase by 0.1 cm?

24. A mass of 2 kg is supported by a copper wire of length 4 m and diameter 4 mm. Determine (a) the stress in the wire and (b) the elongation of the wire.

25. A cube of steel 4 cm on an edge is subjected to a shearing force of 3000 N while one face is clamped. Find the shearing strain in the cube.

26. The *elastic limit* of a material is defined as the maximum stress that can be applied to the material before it becomes permanently deformed. If the elastic limit of copper is 1.5×10^8 N/m^2, determine the *minimum* diameter a copper wire can have under a load of 10 kg if its elastic limit is not to be exceeded.

27. What increase in pressure is necessary to decrease the volume of a 4-cm-diameter sphere of mercury by 0.1%?

28. Determine the decrease in volume of a cube of copper 10 cm on an edge if it is subjected to a bulk stress (pressure) of 10^7 N/m^2 (100 atm).

29. If the shear stress in steel exceeds about 4.0×10^8 N/m^2, it ruptures. Determine the shearing force necessary to (a) shear a steel bolt 1 cm in diameter and (b) punch a 1-cm-diameter hole in a steel plate that is 0.5 cm thick.

30. Two wires are made of the same metal, but have different dimensions. Wire 1 is four times longer and twice the diameter of wire 2. If they are both under the same load, compare (a) the stresses in the two wires and (b) the elongations of the two wires.

31. By how much will the density of an aluminum block change if it is subjected to a pressure of 2×10^8 N/m^2 on all surfaces? (The density of aluminum at 1 atm is 2700 kg/m^3.)

32. A 5.0-m nylon cable with an effective diameter of 1.0 cm is used to suspend a tire swing. When a 30-kg child sits on the swing, the rope stretches 13 cm. Compute the stress, strain, and Young's modulus for this cable.

ADDITIONAL PROBLEMS

33. A bridge of length 50 m and mass 8×10^4 kg is supported at each end as in Figure 12.29. A truck of mass 3×10^4 kg is located 15 m from one end. What are the forces on the bridge at the points of support?

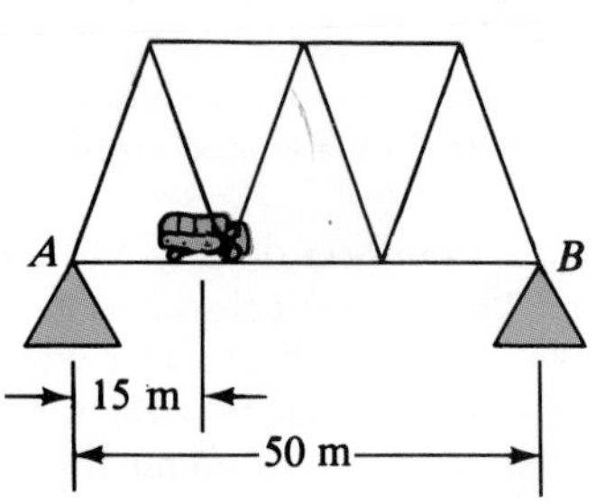

Figure 12.29 (Problem 33).

34. A solid sphere of radius R and mass M is placed in a wedge as shown in Figure 12.30. The inner surfaces of the wedge are frictionless. Determine the forces exerted by the wedge on the sphere at the two contact points.

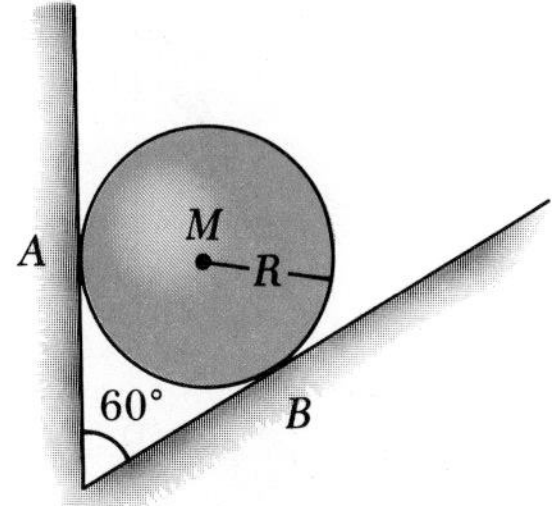

Figure 12.30 (Problem 34).

35. A 10-kg monkey climbs up a 120-N uniform ladder of length ℓ, as in Figure 12.31. The upper and lower ends of the ladder rest on frictionless surfaces. The lower end of the ladder is fastened to the wall by a horizontal rope AB that can support a maximum tension of 110 N. (a) Draw a free-body diagram for the ladder. (b) Find the tension in the rope when the monkey is one third the way up the ladder. (c) Find the maximum distance d the monkey can walk up the ladder before the rope breaks, expressing your answer as a fraction of the length ℓ.

Figure 12.31 (Problem 35).

36. A hungry bear weighing 700 N walks out on a beam in an attempt to retrieve some "goodies" hanging at the end of the beam (Fig. 12.32). The beam is uniform, weighs 200 N, and is 6 m long; the goodies weigh 80 N. (a) Draw a free-body diagram for the beam. (b) When the bear is at $x = 1$ m, find the tension in the wire and the components of the reaction force at the hinge. (c) If the wire can withstand a maximum tension of 900 N, what is the maximum distance the bear can walk before the wire breaks?

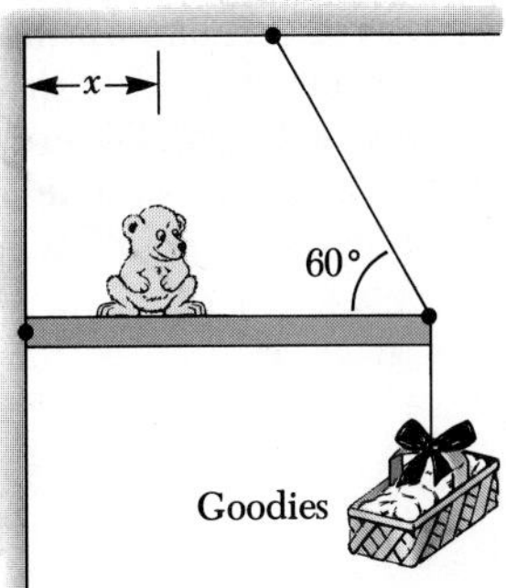

Figure 12.32 (Problem 36).

37. A uniform beam of length 4 m and mass 10 kg supports a 20-kg mass as in Figure 12.33. (a) Draw a free-body diagram for the beam. (b) Determine the tension in the supporting wire and the components of the reaction force at the pivot.

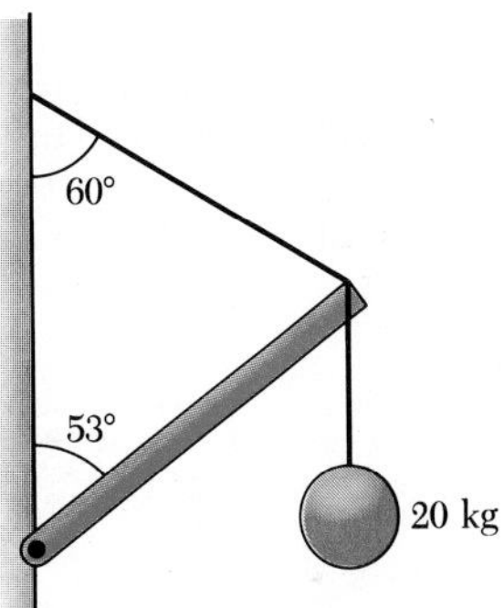

Figure 12.33 (Problem 37).

38. A 1200-N uniform boom is supported by a cable as in Figure 12.34. The boom is pivoted at the bottom, and a 2000-N object hangs from its top. Find the tension in the supporting cable and the components of the reaction force on the boom at the hinge.

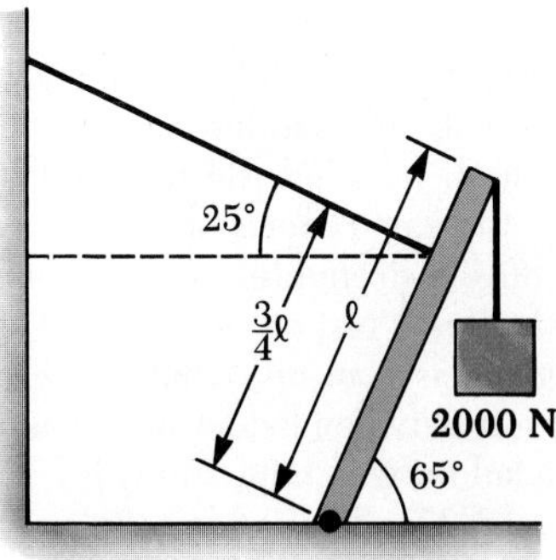

Figure 12.34 (Problem 38).

39. A uniform sign of weight W and width 2ℓ hangs from a light, horizontal beam, hinged at the wall and supported by a cable (Fig. 12.35). Determine (a) the tension in the cable and (b) the components of the reaction force at the hinge in terms of W, d, ℓ, and θ.

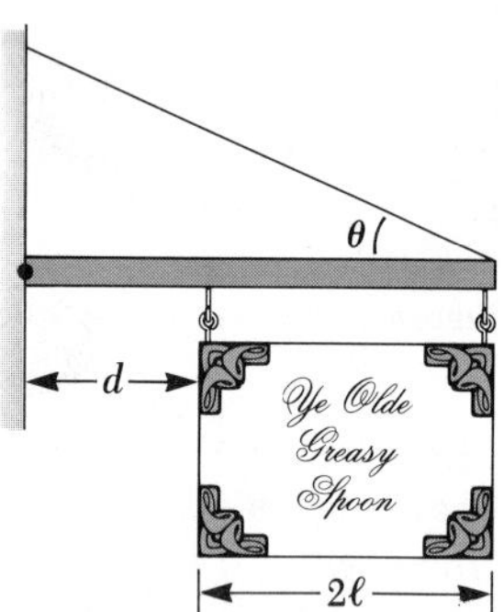

Figure 12.35 (Problem 39).

40. A crane of mass 3000 kg supports a load of 10 000 kg as in Figure 12.36. The crane is pivoted with a smooth pin at A and rests against a smooth support at B. Find the reaction forces at A and B.

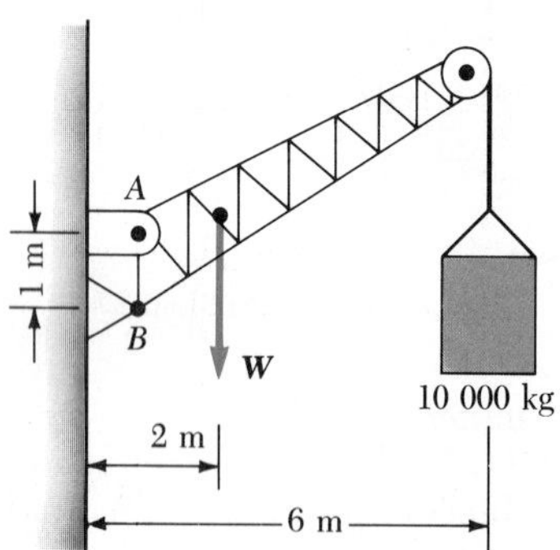

Figure 12.36 (Problem 40).

41. A 15-m uniform ladder weighing 500 N rests against a frictionless wall. The ladder makes a 60° angle with the horizontal. (a) Find the horizontal and vertical forces that the earth exerts on the base of the ladder when an 800-N firefighter is 4 m from the bottom. (b) If the ladder is just on the verge of slipping when the firefighter is 9 m up, what is the coefficient of static friction between ladder and ground?

42. A uniform ladder weighing 200 N is leaning against a wall (Fig. 12.37). The ladder slips when θ is 60°, where θ is the angle between the ladder and the horizontal. Assuming the coefficients of static friction at the wall and the floor *are the same*, obtain a value for μ_s.

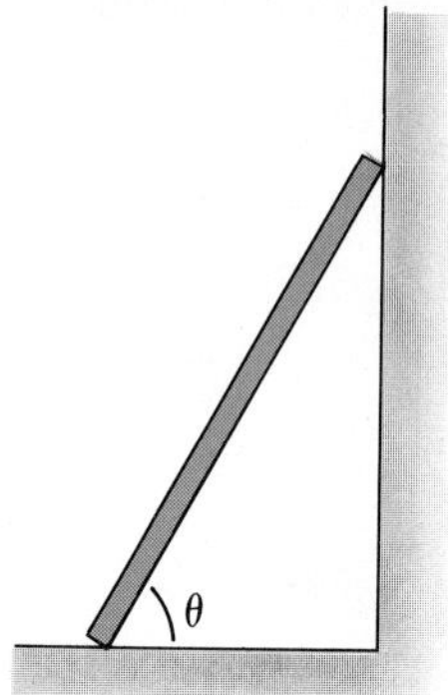

Figure 12.37 (Problem 42).

43. A 10 000-N shark is supported by a cable attached to a 4-m rod that can pivot at the base. Calculate the cable tension needed to hold the system in the position shown in Figure 12.38. Find the horizontal and vertical forces exerted on the base of the rod. (Neglect the weight of the rod.)

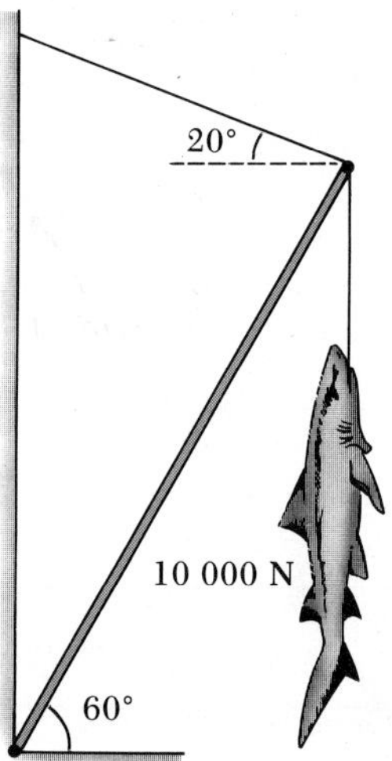

Figure 12.38 (Problem 43).

44. When a person stands on tiptoe (a strenuous position), the position of the foot is as shown in Figure 12.39a. The total weight **W** is supported by the force **N** of the floor on the toe. A mechanical model for the situation is shown in Figure 12.39b, where **T** is the tension in the Achilles tendon and **R** is the force on the foot due to the tibia. Find the values of T, R and θ using the model and dimensions given, with $W = 700$ N.

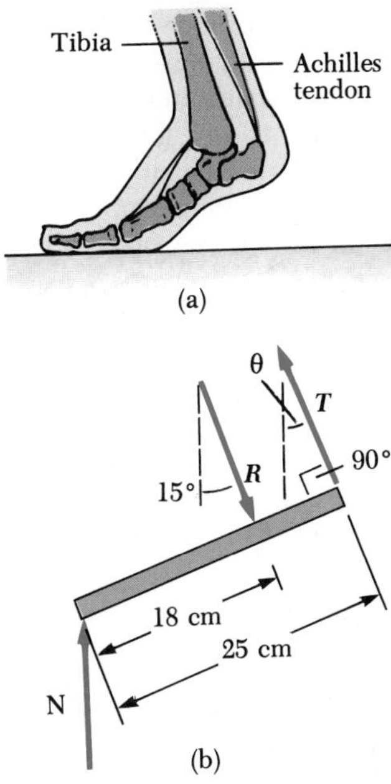

Figure 12.39 (Problem 44).

45. A person bends over and lifts a 200-N object as in Figure 12.40a, with the back in the horizontal position (a terrible way to lift an object). The back muscle attached at a point two thirds up the spine maintains the position of the back, where the angle between the spine and this muscle is 12°. Using the mechanical model shown in Figure 12.40b and taking the weight of the upper body to be 350 N, find the tension in the back muscle and the compressional force in the spine.

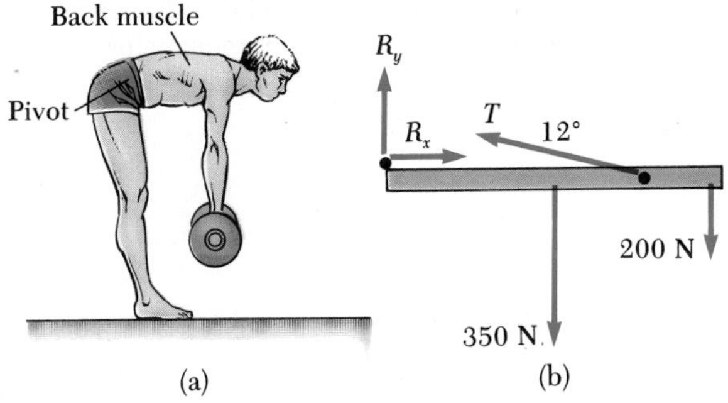

Figure 12.40 (Problem 45).

46. A disk of mass m and of radius r rests on an inclined surface and is supported by a rope that is tangent to the disk and parallel to the inclined surface. The inclined surface makes an angle θ with the horizontal as shown in Figure 12.41. Find (a) the minimum value of the coefficient of static friction, in terms of θ, that will prevent the disk from slipping down the inclined surface and (b) the tension in the rope in terms of m, g, and θ.

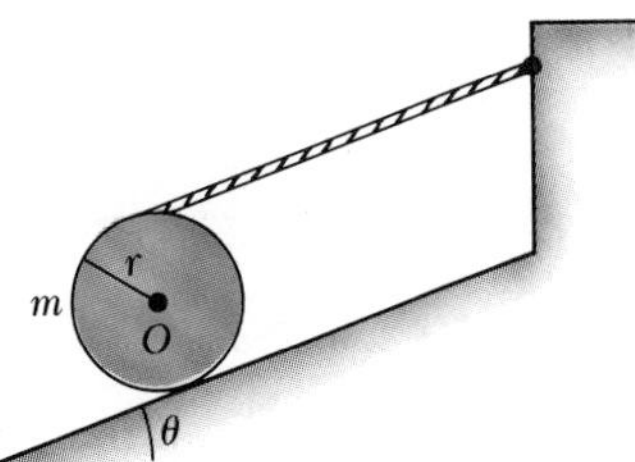

Figure 12.41 (Problem 46).

47. A force $\boldsymbol{F}$ acts on a rectangular block weighing 400 N as in Figure 12.42. (a) If the block slides with constant speed when $F = 200$ N and $h = 0.4$ m, find the coefficient of sliding friction and the position of the resultant normal force. (b) If $F = 300$ N, find the value of h for which the block will just begin to tip from a vertical position.

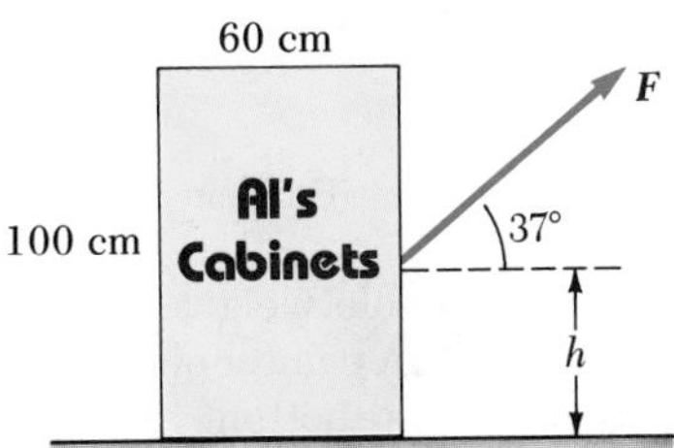

Figure 12.42 (Problem 47).

48. Consider the rectangular block of Problem 47. A force $\boldsymbol{F}$ is applied horizontally at the upper edge. (a) What is the minimum force required to start to tip the block on its edge? (b) What is the minimum coefficient of static friction required for the block to tip with the application of $\boldsymbol{F}$ in (a)? (c) Find the magnitude and the direction of the minimum force required to tip the block if the point of application can be chosen anywhere on the block.

49. A uniform beam of weight w is inclined at an angle θ to the horizontal with its upper end supported by a horizontal rope tied to a wall and its lower end resting on a rough floor (Fig. 12.43). (a) If the coefficient of static friction between the beam and floor is μ_s, determine an expression for the *maximum* weight W that can be suspended from the top before the beam slips. (b) Determine the magnitude of the reaction force at the floor and the magnitude of the force of the beam on the rope at P in terms of w, W and μ_s.

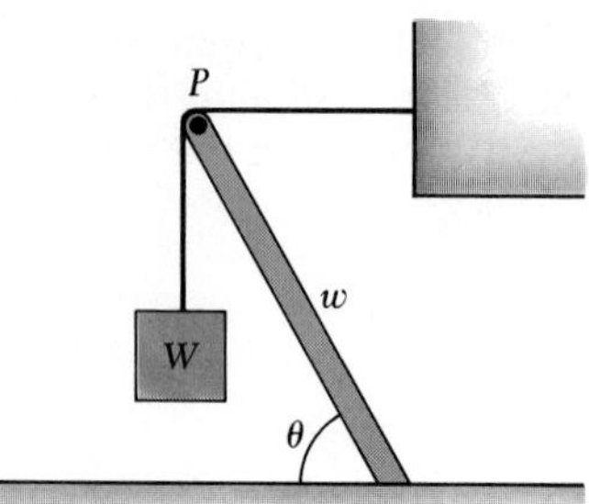

Figure 12.43 (Problem 49).

50. The cylinder shown in Figure 12.44 is held in position by a rope that supplies a force $\boldsymbol{F}$ and by static friction. What is the *minimum* value of μ_s such that the cylinder will remain in equilibrium when $\boldsymbol{F}$ is at the angle θ with the horizontal?

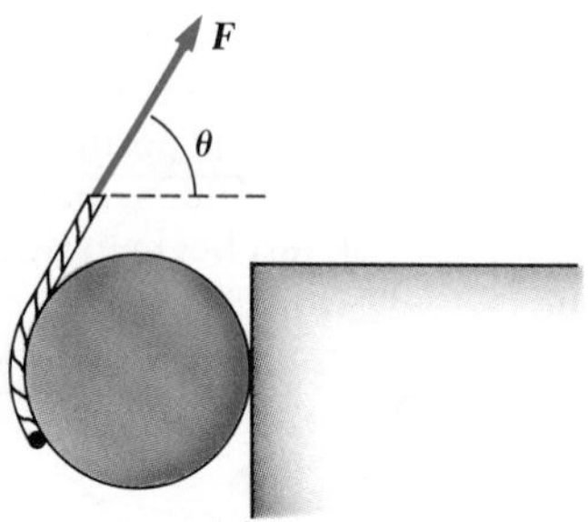

Figure 12.44 (Problem 50).

51. A stepladder of negligible weight is constructed as shown in Figure 12.45. A painter of mass 70 kg stands on the ladder 3 m from the bottom. Assuming the floor is frictionless, find (a) the tension in the horizontal bar connecting the two halves of the ladder, (b) the normal forces at A and B, and (c) the components of the reaction force at the hinge C that the left half of the ladder exerts on the right half. (*Hint:* Treat each half of the ladder separately.)

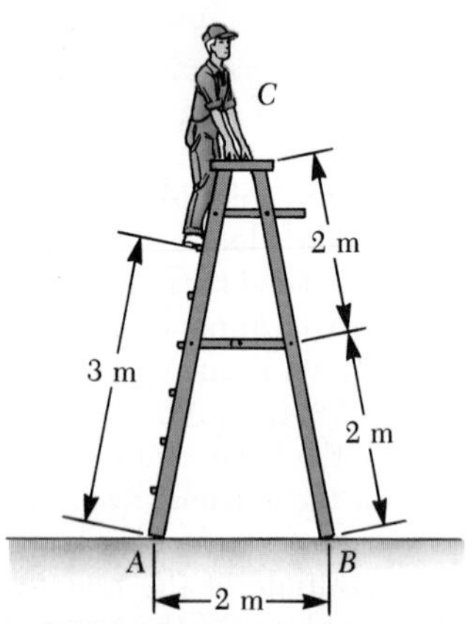

Figure 12.45 (Problem 51).

52. On a hot summer afternoon Kathy Kool is driving in her sports car and enjoying a *thick* vanilla milk shake. She places the milkshake on the carpeted floor to enable her to shift into low gear at a red light. (The height of the cylindrical cup is twice its diameter, and the carpet prevents the cup from sliding.) What is the maximum acceleration the car can have before the cup tips over?

53. One end of a 15-m-long rigid rod of 200 N rests on a frictionless floor. The other end is hinged at a point 9 m above the floor. The tension in the retaining horizontal string (see Fig. 12.46) can be adjusted by changing the counterweight W. Determine the forces exerted on the bar by the floor and the hinge as a function of tension T in the string.

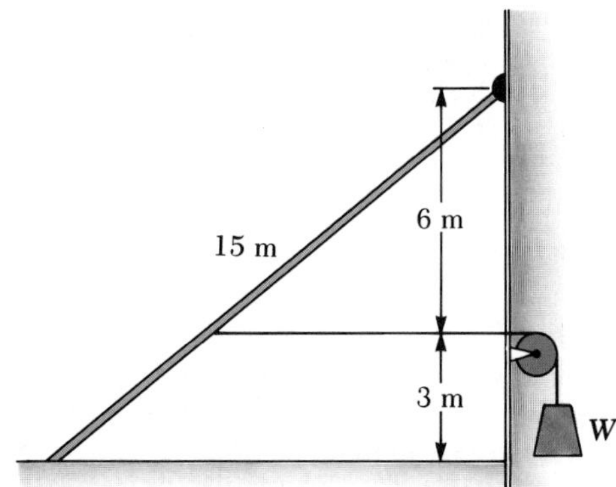

Figure 12.46 (Problem 53).

54. A 10-m-long uniform beam weighing 250 N, hinged at B, is to be kept at 53° to the vertical as shown in Figure 12.47. This is to be done by adjusting the tension in the string, its orientation, or both. Derive a relation between T and θ. For what value of T will θ become 53°? Determine the minimum value of T and its orientation that will keep the beam oriented in the desired direction.

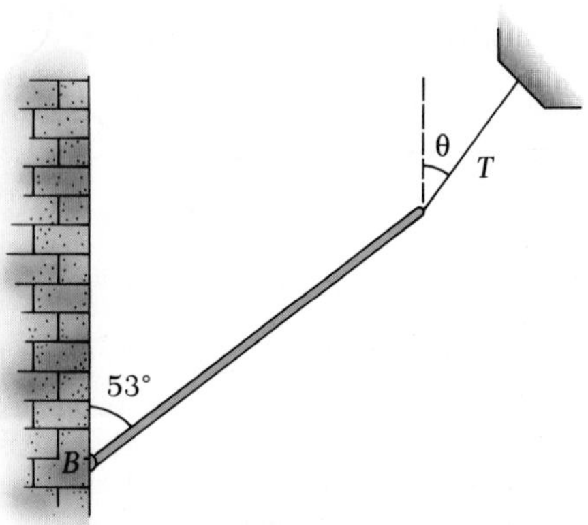

Figure 12.47 (Problem 54).

55. One end of a uniform, 4-m-long rod of weight W is supported by a string at one end. The other end is resting against the wall where it is held by friction (see Fig. 12.48). The coefficient of static friction between the wall and the rod is $\mu_s = 0.5$. Determine the distance of the nearest point from A from which an additional weight W can be hung without the rod slipping from A.

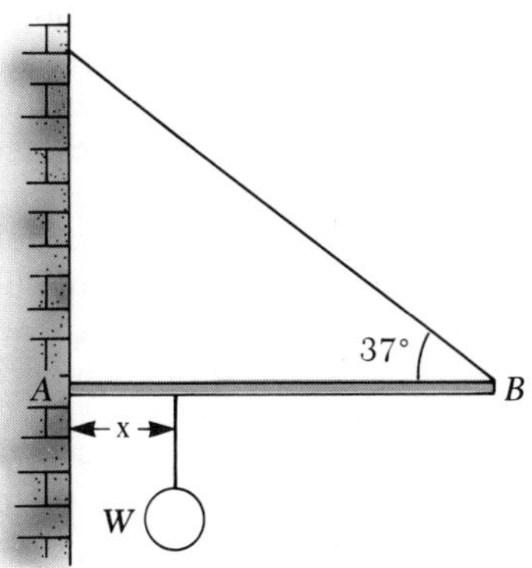

Figure 12.48 (Problem 55).

56. A wire of length L, Young's modulus Y, and cross-sectional area A is stretched elastically by an amount ΔL. By Hooke's law, the restoring force is given by $-k\,\Delta L$. (a) Show that the constant k is given by $k = YA/L$. (b) Show that the work done in stretching the wire by an amount ΔL is given by

$$\text{Work} = \tfrac{1}{2}\frac{YA}{L}(\Delta L)^2$$

57. The distortion of the earth's crustal plates is an example of shear on a large scale. A particular type of crustal rock is determined to have a shear modulus of 1.5×10^{10} N/m^2. What shear stress is involved when a 10-km layer of rock is sheared through a distance of 5 m?

58. A 30-kg hammer strikes a 2.3-cm-diameter steel spike while moving with a speed of 20 m/s. It rebounds with a speed of 10 m/s in 0.11 s. What is the average strain in the spike during the impact?

59. (a) Estimate the force with which a karate master strikes a board if the speed at time of impact is $v = 10$ m/s, decreasing to 1 m/s during a 0.002 s time-of-contact with the board. The mass of coordinated hand and arm is 1 kg. (b) Estimate the shear stress if this force is exerted on a 1-cm-thick pine board which is 10 cm wide. (c) If the maximum shear stress a pine board can receive before breaking is 3.6×10^6 N/m^2, will the board break?

60. A steel cable 3 cm^2 in cross-sectional area has a mass of 2.4 kg per meter length. If 500 m of the cable are hung over a vertical cliff, how much does the cable stretch under its own weight? $Y_{\text{steel}} = 2 \times 10^{11}$ N/m^2.

"I think you should be more explicit here in step two."

ESSAY

Arch Structures

Gordon Batson
Clarkson University, Potsdam N.Y.

Of all structures built for various utilitarian purposes, a bridge and its structural components are the most visible. The load-carrying tasks of the principal structural components can be comprehended easily; the supporting cables of a suspension bridge are under tension induced by the weight and loads on the bridge.

The arch is another type of structure whose shape indicates that the loads are carried by compression. The arch can be visualized as an up-side-down suspension cable.

The stone arch is one of the oldest existing structures found in buildings, walls, and bridges. Other materials, such as timber, may have been used prior to stone, but nothing of these remains today most likely because of fires, warfare, and the decay processes of nature. Although stone arches were constructed prior to the Roman Empire, the Romans constructed some of the largest and most enduring stone arches.

Before the development of the arch, the principal method of spanning a space was the simple post-and-beam construction (Fig. 1a), in which a horizontal beam is supported by two columns. This type of construction was used to build the great Greek temples. The columns of these temples are closely spaced because of the limited length of available stones. Much larger spans can now be achieved using steel beams, but the spans are limited because the beams tend to sag under heavy loads.

The corbeled arch (or false arch) shown in Figure 1b is another primitive structure; it is only a slight improvement over post-and-beam construction. The stability of this false arch depends upon the horizonal projection of one stone over another and the downward weight of stones from above.

The semicircular arch (Fig. 2a) developed by the Romans was a great technological achievement in architectural design. The stability of this true (or voussoir) arch depends on the compression between its wedge-shaped stones. (That is, the stones are forced to squeeze against each other.) This results in horizontal outward forces at the springing of the arch (where it starts curving), which must be supported by the foundation (abutments) on the stone wall shown on the sides of the arch (Fig. 2a). It is common to use very heavy walls (buttresses) on either side of the arch to provide the horizontal stability. If the foundation of the arch should move, the compressive forces between the wedge-shaped stones may decrease to the extent that the arch collapses. The surfaces of the stones used in the semicircular arches constructed by the Romans

Post-and-beam
(a)

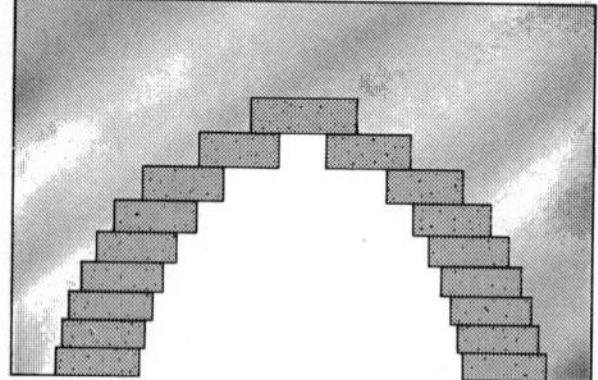

Corbeled (false) arch
(b)

Figure 1 Some methods of spanning space: (a) simple post-and-beam structure and (b) corbeled, or false, arch.

Semicircular arch (Roman)
(a)

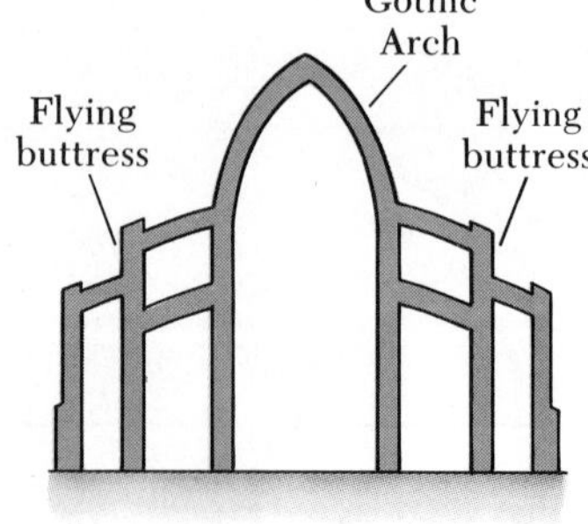

Pointed arch (Gothic)
(b)

Figure 2 (a) The semicircular arch developed by the Romans. (b) Gothic arch with flying buttresses to provide lateral support. (Typical cross-section of a church or cathedral.) The buttresses transfer the speading forces of the arch by vertical loads to the foundation of the structure.

were cut, or "dressed," to make a very tight joint; it is interesting to note that mortar was usually not used in these joints. The resistance to slipping between stones was provided by the compression force and the friction between the stone faces.

Another important architectural innovation was the pointed Gothic arch shown in Figure 2b. This type of stucture was first used in Europe beginning in the 12th century, followed by the construction of several magnificent Gothic cathedrals in France in the 13th century. One of the most striking features of these cathedrals is their extreme height. For example, the cathedral at Chartres rises to 118 ft and the one at Reims has a height of 137 ft. It is interesting to note that such magnificent Gothic structures evolved over a very short period of time, without the benefit of any mathematical theory of structures. However, Gothic arches required flying buttresses to prevent the spreading of the arch supported by the tall, narrow columns. The fact that they have been stable for more than 700 years attests to the technical skill of their builders and architects, which was probably acquired through experience and intuition.

Figure 3 shows how the horizontal force at the base of an arch varies with arch height for an arch hinged at the peak. For a given load $\boldsymbol{P}$, the horizontal force at the base is doubled when the height is reduced by a factor of 2. This explains why the horizontal force required to support a high pointed arch is less than that required for a circular arch. For a given span L, the horizontal force at the base is proportional to the total load P and inversely proportional to the height h. Therefore, in order to minimize the horizontal force at the base, the arch must be made as light and high as possible.

With the advent of more advanced methods of structural analysis, it has become possible to determine the optimum shape of an arch under given load conditions.

One of the most impressive modern arches, the St. Louis Gateway Arch, designed by Eero Saarinen, has a span of 192 m and a height of 192 m (Fig. 4). The largest steel-truss arch bridge, the New River Gorge Bridge in Charleston, West Virginia, has a span of 520 m. Beautiful concrete arch bridges were designed and built in the 1920s and 1930s by Robert Maillart in Switzerland. The Sando Bridge in Sweden, a single arch of reinforced concrete, spans 264 m. Today, the arch is still the most common structure used to span large distances.

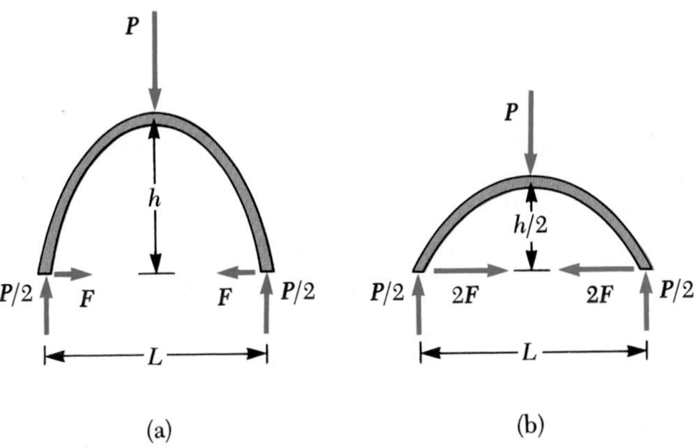

Figure 3 *(Above)* When the height of an arch is reduced by a factor of 2, and the load force $\boldsymbol{P}$ remains the same, the horizontal force at the base is doubled.

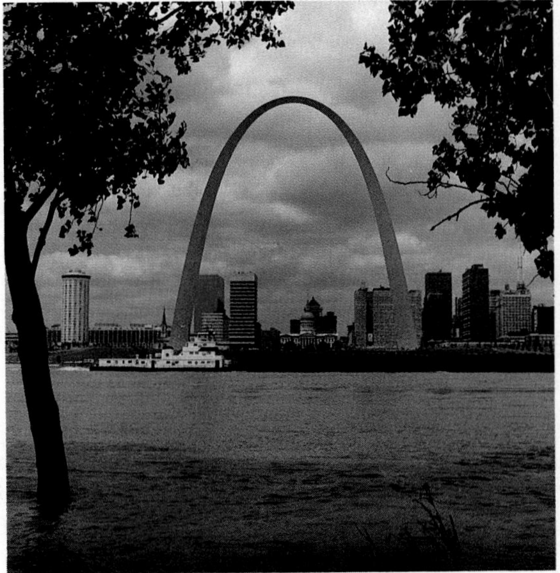

Figure 4 *(Right)* The St. Louis Gateway Arch seen from the Mississippi River. This beautiful structure has the shape of an inverted freely hanging cable, so that all of its members are under compression.

13

Oscillatory Motion

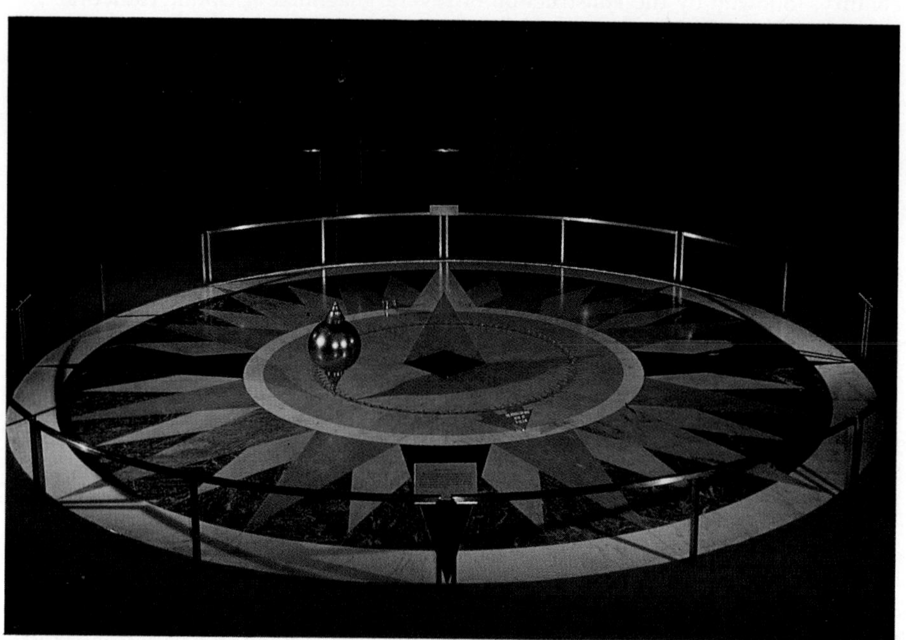

The Foucault pendulum at the Smithsonian Institution in Washington, D.C. This type of pendulum was first used by the French physicist Jean Foucault to verify the earth's rotation experimentally. During its swinging motion, the pendulum's plane of oscillation appears to rotate, as the bob successively knocks over the red indicators arranged in a horizontal circle. In reality, the pendulum's plane of motion is fixed in space, while the earth rotates beneath the swinging pendulum. (Courtesy of the Smithsonian Institution)

The main objectives of the previous chapters was to discover that the motion of a body can be predicted if the initial conditions describing its state of motion and the external forces acting on it are known. If a force varies in time, the velocity and acceleration of the body will also change with time. A very special kind of motion occurs when the force on a body is proportional to the displacement of the body from equilibrium. If this force always acts toward the equilibrium position of the body, a repetitive back-and-forth motion will result about this position. The motion is an example of what is called *periodic* or *oscillatory* motion.

You are most likely familiar with several examples of periodic motion, such as the oscillations of a mass on a spring, the motion of a pendulum, and the vibrations of a stringed musical instrument. The number of systems that exhibit oscillatory motion is extensive. For example, the molecules in a solid oscillate about their equilibrium positions; electromagnetic waves, such as light waves, radar, and radio waves, are characterized by oscillating electric and magnetic field vectors; and in alternating-current circuits, voltage, current, and electrical charge vary periodically with time.

Most of the material in this chapter deals with *simple harmonic motion.* For this type of motion, an object oscillates between two spatial positions for an indefinite period of time, with no loss in mechanical energy. In real mechanical systems, retarding (or frictional) forces are always present. Such forces reduce the mechanical energy of the system as motion progresses, and the oscillations are said to be *damped.* If an external driving force is applied such that the energy loss is balanced by the energy input, we call the motion a *forced oscillation.*

13.1 SIMPLE HARMONIC MOTION

A particle moving along the x axis is said to exhibit **simple harmonic motion** when x, its displacement from equilibrium, varies in time according to the relationship

Displacement versus time for simple harmonic motion

$$x = A\cos(\omega t + \delta) \tag{13.1}$$

where A, ω, and δ are constants of the motion. In order to give physical significance to these constants, it is convenient to plot x as a function of t, as in Figure 13.1. First, we note that A, called the **amplitude** of the motion, is simply the *maximum displacement* of the particle in either the positive or negative x direction. The constant ω is called the *angular frequency* (defined in Eq. 13.4). The constant angle δ is called the **phase constant** (or phase angle) and along with the amplitude A is determined uniquely by the initial displacement and velocity of the particle. The constants δ and A tell us what the displacement was at time $t = 0$. The quantity $(\omega t + \delta)$ is called the **phase** of the motion and is useful in comparing the motions of two systems of particles. Note that the function x is periodic and repeats itself when ωt increases by 2π radians.

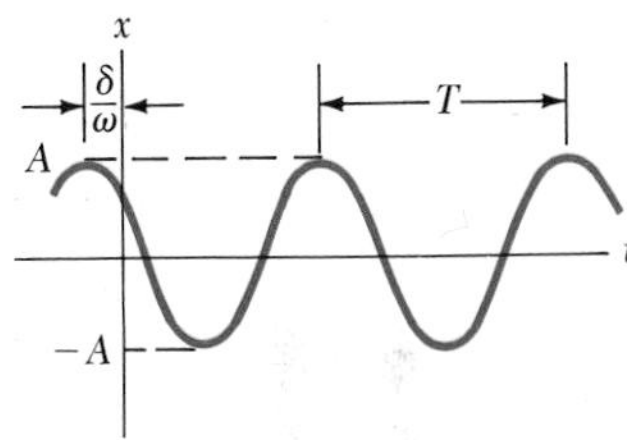

Figure 13.1 Displacement versus time for a particle undergoing simple harmonic motion. The amplitude of the motion is A and the period is T.

The **period,** T, is the time for the particle to go through one full cycle of its motion. That is, the value of x at time t equals the value of x at time $t + T$. We can show that the period of the motion is given by $T = 2\pi/\omega$ by using the fact that the phase increases by 2π radians in a time T:

$$\omega t + \delta + 2\pi = \omega(t + T) + \delta$$

Hence, $\omega T = 2\pi$ or

$$T = \frac{2\pi}{\omega} \tag{13.2}$$

Period

The inverse of the period is called the **frequency** of the motion, f. The frequency represents the *number of oscillations the particle makes per unit time:*

$$f = \frac{1}{T} = \frac{\omega}{2\pi} \tag{13.3}$$

Frequency

The units of f are cycles/s, or hertz (Hz).

Rearranging Equation 13.3 gives

$$\omega = 2\pi f = \frac{2\pi}{T} \tag{13.4}$$

Angular frequency

The constant ω is called the **angular frequency** and has units of rad/s. We shall discuss the geometric significance of ω in Section 13.4.

We can obtain the velocity of a particle undergoing simple harmonic motion by differentiating Equation 13.1 with respect to time:

Velocity in simple harmonic motion

$$v = \frac{dx}{dt} = -\omega A \sin(\omega t + \delta) \tag{13.5}$$

The acceleration of the particle is given by dv/dt:

Acceleration in simple harmonic motion

$$a = \frac{dv}{dt} = -\omega^2 A \cos(\omega t + \delta) \tag{13.6}$$

Since $x = A \cos(\omega t + \delta)$, we can express Equation 13.6 in the form

$$a = -\omega^2 x \tag{13.7}$$

From Equation 13.5 we see that since the sine and cosine functions oscillate between ± 1, the extreme values of v are equal to $\pm \omega A$. Equation 13.6 tells us that the extreme values of the acceleration are $\pm \omega^2 A$. Therefore, the *maximum* values of the velocity and acceleration are given by

Maximum values of velocity and acceleration in simple harmonic motion

$$v_{\max} = \omega A \tag{13.8}$$

$$a_{\max} = \omega^2 A \tag{13.9}$$

Figure 13.2a represents the displacement versus time for an arbitrary value of the phase constant. The projection of a point moving with uniform

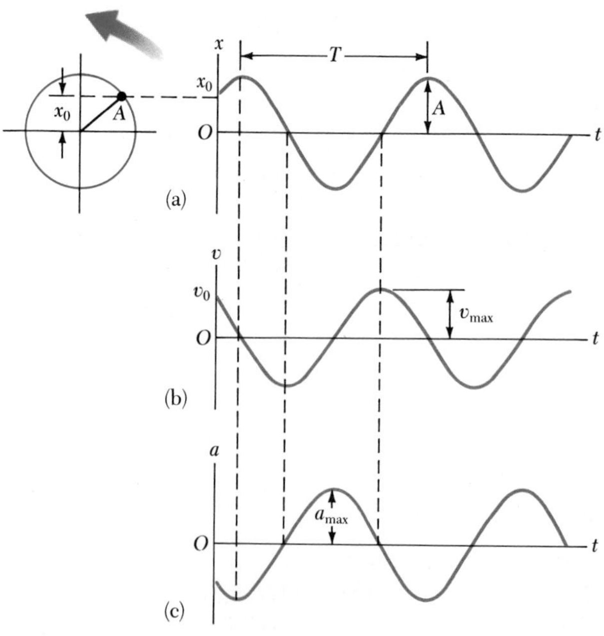

Figure 13.2 Graphical representation of simple harmonic motion: (a) the displacement versus time, (b) the velocity versus time, and (c) the acceleration versus time. Note that the velocity is 90° out of phase with the displacement and the acceleration is 180° out of phase with the displacement.

circular motion on a reference circle of radius A also moves in sinusoidal fashion. This will be discussed in more detail in Section 13.5.

The velocity and acceleration versus time curves are illustrated in Figures 13.2b and 13.2c. These curves show that the phase of the velocity differs from the phase of the displacement by $\pi/2$ rad, or 90°. That is, when x is a maximum or a minimum, the velocity is zero. Likewise, when x is zero, the speed is a maximum. Furthermore, note that the phase of the acceleration differs from the phase of the displacement by π radians, or 180°. That is, when x is a maximum, a is a maximum in the opposite direction.

As we stated earlier, the solution $x = A\cos(\omega t + \delta)$ is a general solution of the equation of motion, where the phase constant δ and the amplitude A must be chosen to meet the initial conditions of the motion. The phase constant is important when comparing the motion of two or more oscillating particles. Suppose that the initial position x_0 and initial velocity v_0 of a single oscillator are given, that is, at $t = 0$, $x = x_0$ and $v = v_0$. Under these conditions, the equations $x = A\cos(\omega t + \delta)$ and $v = -\omega A\sin(\omega t + \delta)$ give

$$x_0 = A\cos\delta \qquad \text{and} \qquad v_0 = -\omega A\sin\delta$$

Dividing these two equations eliminates A, giving

$$\frac{v_0}{x_0} = -\omega\tan\delta$$

$$\tan\delta = -\frac{v_0}{\omega x_0} \tag{13.10a}$$

The phase angle ϕ and amplitude A can be obtained from the initial conditions

Furthermore, if we take the sum $x_0{}^2 + \left(\frac{v_0}{\omega}\right)^2 = A^2\cos^2\delta + A^2\sin^2\delta$ and solve for A, we find that

$$A = \sqrt{x_0{}^2 + \left(\frac{v_0}{\omega}\right)^2} \tag{13.10b}$$

Thus, we see that δ and A are known if x_0, ω, and v_0 are specified. We shall treat a few specific cases in the next section.

We conclude this section by pointing out the following important properties of a particle moving in simple harmonic motion:

Properties of simple harmonic motion

1. The displacement, velocity, and acceleration all vary sinusoidally with time but are not in phase, as shown in Figure 13.2.
2. The acceleration of the particle is proportional to the displacement, but in the opposite direction.
3. The frequency and the period of motion are independent of the amplitude.

EXAMPLE 13.1 An Oscillating Body □

A body oscillates with simple harmonic motion along the x axis. Its displacement varies with time according to the equation

$$x = (4.0\text{ m})\cos\left(\pi t + \frac{\pi}{4}\right)$$

where t is in s, and the angles in the parentheses are in radians. (a) Determine the amplitude, frequency, and period of the motion.

By comparing this equation with the general relation for simple harmonic motion, $x = A\cos(\omega t + \delta)$, we see that $A = 4.0$ m and $\omega = \pi$ rad/s; therefore we find $f = \omega/2\pi = \pi/2\pi = 0.50\ \text{s}^{-1}$ and $T = 1/f = 2.0$ s.

(b) Calculate the velocity and acceleration of the body at any time t.

$$v = \frac{dx}{dt} = -4.0 \sin\left(\pi t + \frac{\pi}{4}\right) \frac{d}{dt}(\pi t)$$

$$= -(4\pi \text{ m/s}) \sin\left(\pi t + \frac{\pi}{4}\right)$$

$$a = \frac{dv}{dt} = -4\pi \cos\left(\pi t + \frac{\pi}{4}\right) \frac{d}{dt}(\pi t)$$

$$= -(4\pi^2 \text{ m/s}^2) \cos\left(\pi t + \frac{\pi}{4}\right)$$

(c) Using the results to (b), determine the position, velocity, and acceleration of the body at $t = 1$ s.

Noting that the angles in the trigonometric functions are in radians, we get at $t = 1$ s

$$x = (4.0 \text{ m}) \cos\left(\pi + \frac{\pi}{4}\right) = (4.0 \text{ m}) \cos\left(\frac{5\pi}{4}\right)$$

$$= (4.0 \text{ m})(-0.707) = -2.83 \text{ m}$$

$$v = -(4\pi \text{ m/s}) \sin\left(\frac{5\pi}{4}\right) = -(4\pi \text{ m/s})(-0.707) = 8.89 \text{ m/s}$$

$$a = -(4\pi^2 \text{ m/s}^2) \cos\left(\frac{5\pi}{4}\right) = -(4\pi^2 \text{ m/s}^2)(-0.707)$$

$$= 27.9 \text{ m/s}^2$$

(d) Determine the maximum speed and maximum acceleration of the body.

From the general relations for v and a found in (b), we see that the maximum values of the sine and cosine functions are unity. Therefore, v varies between $\pm 4\pi$ m/s, and a varies between $\pm 4\pi^2$ m/s^2. Thus, $v_{max} = 4\pi$ m/s and $a_{max} = 4\pi^2$ m/s^2. The same results are obtained using $v_{max} = \omega A$ and $a_{max} = \omega^2 A$, where $A = 4.0$ m and $\omega = \pi$ rad/s.

(e) Find the displacement of the body between $t = 0$ and $t = 1$ s.

The x coordinate at $t = 0$ is given by

$$x_0 = (4.0 \text{ m}) \cos\left(0 + \frac{\pi}{4}\right) = (4.0 \text{ m})(0.707) = 2.83 \text{ m}$$

In (c), we found that the coordinate at $t = 1$ s was -2.8 m; therefore the displacement between $t = 0$ and $t = 1$ s is

$$\Delta x = x - x_0 = -2.83 \text{ m} - 2.83 \text{ m} = -5.66 \text{ m}$$

Because the particle's velocity changes sign during the first second, the magnitude of Δx is *not* the same as the distance traveled in the first second.

(f) What is the phase of the motion at $t = 2$ s?

The phase is defined as $\omega t + \delta$, where in this case $\omega = \pi$ and $\delta = \pi/4$. Therefore, at $t = 2$ s, we get

$$\text{Phase} = (\omega t + \delta)_{t=2} = \pi(2) + \pi/4 = 9\pi/4 \text{ rad}$$

13.2 MASS ATTACHED TO A SPRING

In Chapter 7 we introduced the physical system consisting of a mass attached to the end of a spring, where the mass is free to move on a horizontal, frictionless surface (Fig. 13.3). We know from experience that such a system will oscillate back and forth if disturbed from the equilibrium position $x = 0$, where the spring is unstretched. If the surface is frictionless, the mass will exhibit simple harmonic motion. One possible experimental arrangement that clearly demonstrates that such a system exhibits simple harmonic motion is illustrated in Figure 13.4, in which a mass oscillating vertically on a spring has a marking pen attached to it. While the mass is in motion, a sheet of paper is moved horizontally as shown, and the marking pen traces out a sinusoidal pattern. We can understand this qualitatively by first recalling that when the mass is displaced a small distance x from equilibrium, the spring exerts a force on m given by **Hooke's law,**

$$F = -kx \tag{13.11}$$

where k is the force constant of the spring. We call this a **linear restoring force** since it is linearly proportional to the displacement and is always directed toward the equilibrium position, *opposite* to the displacement. That is, when the mass is displaced to the right in Figure 13.3, x is positive and the restoring force is to the left. When the mass is displaced to the left of $x = 0$, then x is negative and F is to the right. If we now apply Newton's second law to the motion of m in the x direction, we get

$$F = -kx = ma$$

$$a = -\frac{k}{m}x \tag{13.12}$$

that is, *the acceleration is proportional to the displacement of the mass from equilibrium and is in the opposite direction.* If the mass is displaced a maximum distance $x = A$ at some initial time and released from rest, its *initial* acceleration will be $-kA/m$ (that is, it has its extreme negative value). When it passes through the equilibrium position, $x = 0$ and its acceleration is zero. At this instant, its velocity is a maximum. It will then continue to travel to the left of equilibrium and finally reach $x = -A$, at which time its acceleration is kA/m (maximum positive) and its velocity is again zero. Thus, we see that the mass will oscillate between the turning points $x = \pm A$. In one full cycle of its motion, the mass travels a distance $4A$.

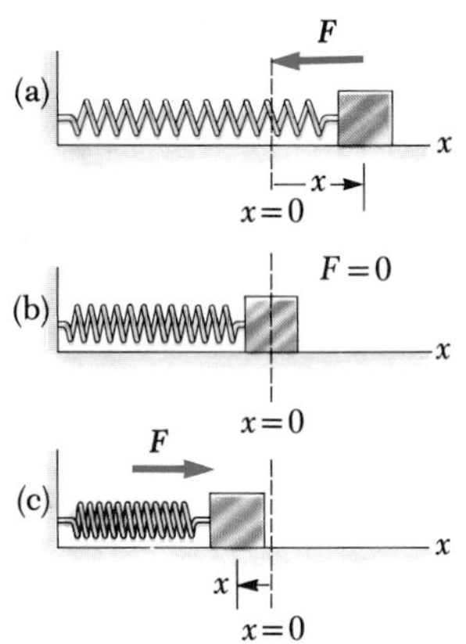

Figure 13.3 A mass attached to a spring on a frictionless surface exhibits simple harmonic motion. (a) When the mass is displaced to the right of equilibrium, the displacement is positive and the acceleration is negative. (b) At the equilibrium position, $x = 0$, the acceleration is zero but the speed is a maximum. (c) When the displacement is negative, the acceleration is positive.

We shall now describe the motion in a quantitative fashion. This can be accomplished by recalling that $a = dv/dt = d^2x/dt^2$. Thus, we can express Equation 13.12 as

$$\frac{d^2x}{dt^2} = -\frac{k}{m}x \tag{13.13}$$

If we denote the ratio k/m by the symbol ω^2,

$$\omega^2 = k/m \tag{13.14}$$

then Equation 13.13 can be written in the form

$$\frac{d^2x}{dt^2} = -\omega^2 x \tag{13.15}$$

What we now require is a solution to Equation 13.15, that is, a function $x(t)$ that satisfies this second-order differential equation. The nature of such a solution $x(t)$ as an algebraic relationship is that it reduces the differential equation to an identity. However, since Equations 13.15 and 13.7 are equivalent, we see that the solution must be that of simple harmonic motion:

$$x(t) = A\cos(\omega t + \delta)$$

To see this explicitly, note that if

$$x = A\cos(\omega t + \delta)$$

then

$$\frac{dx}{dt} = A\frac{d}{dt}\cos(\omega t + \delta) = -\omega A\sin(\omega t + \delta)$$

$$\frac{d^2x}{dt^2} = -\omega A\frac{d}{dt}\sin(\omega t + \delta) = -\omega^2 A\cos(\omega t + \delta)$$

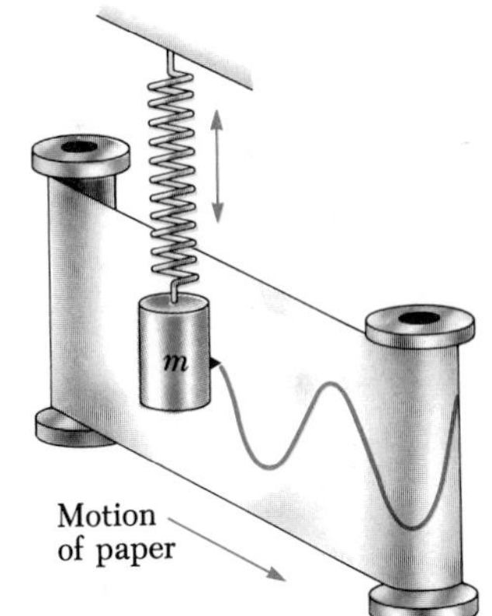

Figure 13.4 An experimental apparatus for demonstrating simple harmonic motion. A pen attached to the oscillating mass traces out a sine wave on the moving chart paper.

Comparing the expressions for x and d^2x/dt^2, we see that $d^2x/dt^2 = -\omega^2 x$ and Equation 13.15 is satisfied.

The following general statement can be made based on the above discussion:

> Whenever the force acting on a particle is linearly proportional to the displacement and in the opposite direction, the particle will exhibit simple harmonic motion.

We shall give additional physical examples in subsequent sections.

Since the period is given by $T = 2\pi/\omega$ and the frequency is the inverse of the period, we can express the period and frequency of the motion for this system as

Period and frequency for mass-spring system

$$T = \frac{2\pi}{\omega} = 2\pi \sqrt{\frac{m}{k}} \tag{13.16}$$

$$f = \frac{1}{T} = \frac{1}{2\pi} \sqrt{\frac{k}{m}} \tag{13.17}$$

That is, the period and frequency depend *only* on the mass and on the force constant of the spring. As we might expect, the frequency is larger for a stiffer spring and decreases with increasing mass.

It is interesting to note that a mass suspended from a vertical spring attached to a fixed support will also exhibit simple harmonic motion. Although there is a gravitational force to consider in this case, the equation of motion still reduces to Equation 13.15, where the displacement is measured from the equilibrium position of the suspended mass. The proof of this is left as a problem (Problem 56).

Special Case I In order to better understand the physical significance of our solution of the equation of motion, let us consider the following special case. Suppose we pull the mass from equilibrium by a distance A and release it from rest from this stretched position, as in Figure 13.5. We must then require that our solution for $x(t)$ obey the *initial conditions* that at $t = 0$, $x_0 = A$ and $v_0 = 0$. These conditions will be met if we choose $\delta = 0$, giving $x = A \cos \omega t$ as our solution. Note that this is consistent with $x = A \cos(\omega t + \delta)$, where $x_0 = A$ and $\delta = 0$. To check this, we see that the solution $x = A \cos \omega t$ satisfies the condition that $x_0 = A$ at $t = 0$, since $\cos 0 = 1$. Thus, we see that A and δ contain the

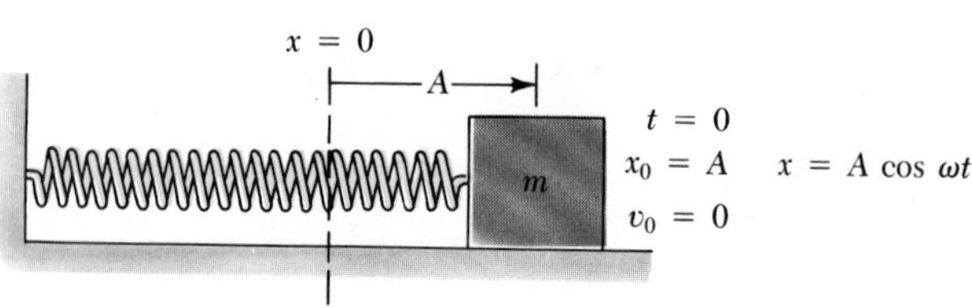

Figure 13.5 A mass-spring system that starts from rest at $x_0 = A$. In this case, $\delta = 0$, and so $x = A \cos \omega t$.

information on initial conditions. Now let us investigate the behavior of the velocity and acceleration for this special case. Since $x = A \cos \omega t$

$$v = \frac{dx}{dt} = -\omega A \sin \omega t$$

and

$$a = \frac{dv}{dt} = -\omega^2 A \cos \omega t$$

From the velocity expression $v = -\omega A \sin \omega t$, we see that at $t = 0$, $v_0 = 0$, as we require. The expression for the acceleration tells us that at $t = 0$, $a = -\omega^2 A$. Physically this makes sense, since the force on the mass is to the left when the displacement is positive. In fact, at this position $F = -kA$ (to the left), and the initial acceleration is $-kA/m$.

We could also use a more formal approach to show that $x = A \cos \omega t$ is the correct solution by using the relation $\tan \delta = -v_0/\omega x_0$ (Eq. 13.10a). Since $v_0 = 0$ at $t = 0$, $\tan \delta = 0$ and so $\delta = 0$.

The displacement, velocity, and acceleration versus time are plotted in Figure 13.6 for this special case. Note that the acceleration reaches extreme values of $\pm\omega^2 A$ when the displacement has extreme values of $\pm A$. Furthermore, the velocity has extreme values of $\pm\omega A$, which both occur at $x = 0$. Hence, the quantitative solution agrees with our qualitative description of this system.

Special Case II Now suppose that the mass is given an initial velocity v_0 to the *right* at the unstretched position of the spring so that at $t = 0$, $x_0 = 0$ and $v = v_0$ (Fig. 13.7). Our particular solution must now satisfy these initial conditions. Since the mass is moving toward positive x values at $t = 0$, and $x_0 = 0$ at $t = 0$, the solution has the form $x = A \sin \omega t$.

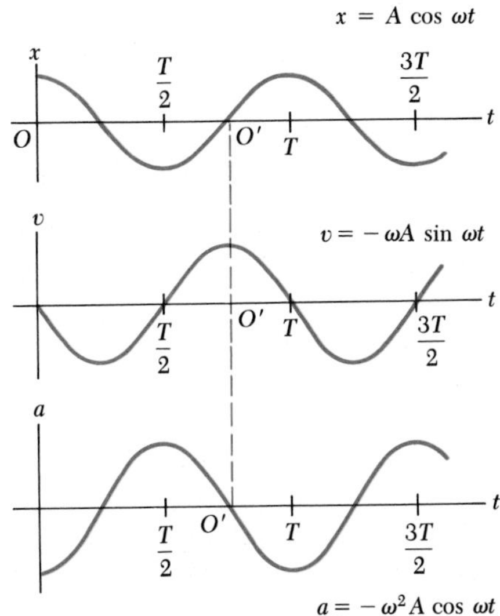

Figure 13.6 Displacement, velocity, and acceleration versus time for a particle undergoing simple harmonic motion under the initial conditions that at $t = 0$, $x_0 = A$ and $v_0 = 0$.

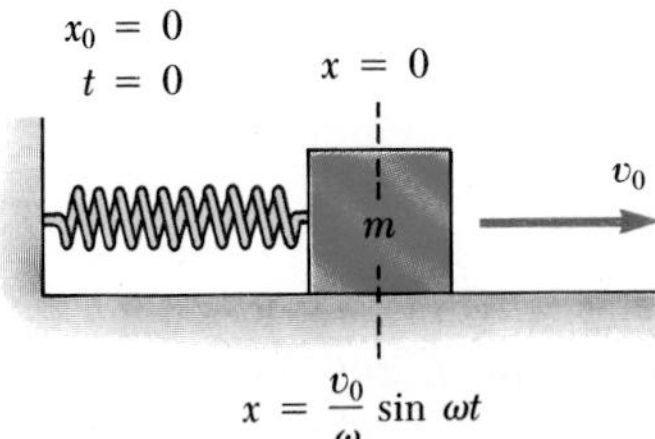

Figure 13.7 The mass-spring system starts its motion at the equilibrium position, $x_0 = 0$ at $t = 0$. If its initial velocity is v_0 to the right, its x coordinate varies as $x = \frac{v_0}{\omega} \sin \omega t$.

Applying $\tan \delta = -v_0/\omega x_0$ and the initial condition that $x_0 = 0$ at $t = 0$ gives $\tan \delta = -\infty$ or $\delta = -\pi/2$. Hence, the solution is $x = A \cos(\omega t - \pi/2)$, which can be written $x = A \sin \omega t$. Furthermore, from Equation 13.10b we see that $A = v_0/\omega$; therefore we can express our solution as

$$x = \frac{v_0}{\omega} \sin \omega t$$

The velocity and acceleration in this case are given by

$$v = \frac{dx}{dt} = v_0 \cos \omega t$$

$$a = \frac{dv}{dt} = \omega v_0 \sin \omega t$$

This is consistent with the fact that the mass always has a maximum speed at $x = 0$, while the force and acceleration are zero at this position. The graphs of these functions versus time in Figure 13.6 correspond to the origin at O'. What would be the solution for x if the mass is initially moving to the left in Figure 13.7?

EXAMPLE 13.2 That Car Needs a New Set of Shocks

A car of mass 1300 kg is constructed using a frame supported by four springs. Each spring has a force constant of 20 000 N/m. If two people riding in the car have a combined mass of 160 kg, find the frequency of vibration of the car when it is driven over a pot hole in the road.

Solution We shall assume that the weight is evenly distributed. Thus, each spring supports one fourth of the load. The total mass supported by the springs is 1460 kg, and therefore each spring supports 365 kg. Hence, the frequency of vibration is

$$f = \frac{1}{2\pi}\sqrt{\frac{k}{m}} = \frac{1}{2\pi}\sqrt{\frac{20\ 000\ \text{N/m}}{365\ \text{kg}}} = 1.18\ \text{Hz}$$

Exercise 1 How long does it take the car to execute two complete vibrations?
Answer 1.70 s.

EXAMPLE 13.3 A Mass-Spring System

A mass of 200 g is connected to a light spring of force constant 5 N/m and is free to oscillate on a horizontal, frictionless surface. If the mass is displaced 5 cm from equilibrium and released from rest, as in Figure 13.5, (a) find the period of its motion.

This situation corresponds to Case I, where $x = A \cos \omega t$ and $A = 5 \times 10^{-2}$ m. Therefore,

$$\omega = \sqrt{\frac{k}{m}} = \sqrt{\frac{5\ \text{N/m}}{200 \times 10^{-3}\ \text{kg}}} = 5\ \text{rad/s}$$

Therefore

$$T = \frac{2\pi}{\omega} = \frac{2\pi}{5} = 1.26\ \text{s}$$

(b) Determine the maximum speed of the mass.

$$v_{\text{max}} = \omega A = (5\ \text{rad/s})(5 \times 10^{-2}\ \text{m}) = 0.250\ \text{m/s}$$

(c) What is the maximum acceleration of the mass?

$$a_{\text{max}} = \omega^2 A = (5\ \text{rad/s})^2(5 \times 10^{-2}\ \text{m}) = 1.25\ \text{m/s}^2$$

(d) Express the displacement, speed, and acceleration as functions of time.

The expression $x = A \cos \omega t$ is our special solution for Case I, and so we can use the results from (a), (b), and (c) to get

$$x = A \cos \omega t = (0.05\ \text{m}) \cos 5t$$

$$v = -\omega A \sin \omega t = -(0.25\ \text{m/s}) \sin 5t$$

$$a = -\omega^2 A \cos \omega t = -(1.25\ \text{m/s}^2) \cos 5t$$

13.3 ENERGY OF THE SIMPLE HARMONIC OSCILLATOR

Let us examine the mechanical energy of the mass-spring system described in Figure 13.6. Since the surface is frictionless, we expect that the total mechanical energy is conserved, as was shown in Chapter 8. We can use Equation 13.5 to express the kinetic energy as

Kinetic energy of a simple harmonic oscillator

$$K = \tfrac{1}{2}mv^2 = \tfrac{1}{2}m\omega^2 A^2 \sin^2(\omega t + \delta) \qquad (13.18)$$

The elastic potential energy stored in the spring for any elongation x is given by $\tfrac{1}{2}kx^2$. Using Equation 13.1, we get

Potential energy of a simple harmonic oscillator

$$U = \tfrac{1}{2}kx^2 = \tfrac{1}{2}kA^2 \cos^2(\omega t + \delta) \qquad (13.19)$$

We see that K and U are *always* positive quantities. Since $\omega^2 = k/m$, we can express the *total energy* of the simple harmonic oscillator as

$$E = K + U = \tfrac{1}{2}kA^2[\sin^2(\omega t + \delta) + \cos^2(\omega t + \delta)]$$

But $\sin^2\theta + \cos^2\theta = 1$, where $\theta = \omega t + \delta$; therefore this equation reduces to

Total energy of a simple harmonic oscillator

$$E = \tfrac{1}{2}kA^2 \qquad (13.20)$$

That is,

the energy of a simple harmonic oscillator is a constant of the motion and proportional to the square of the amplitude.

In fact, the total mechanical energy is just equal to the maximum potential energy stored in the spring when $x = \pm A$. At these points, $v = 0$ and there is no kinetic energy. At the equilibrium position, $x = 0$ and $U = 0$, so that the total energy is all in the form of kinetic energy. That is, at $x = 0$, $E = \tfrac{1}{2}mv_{\text{max}}^2 = \tfrac{1}{2}m\omega^2A^2$.

Plots of the kinetic and potential energies versus time are shown in Figure 13.8a, where we have taken $\delta = 0$. In this situation, both K and U are always

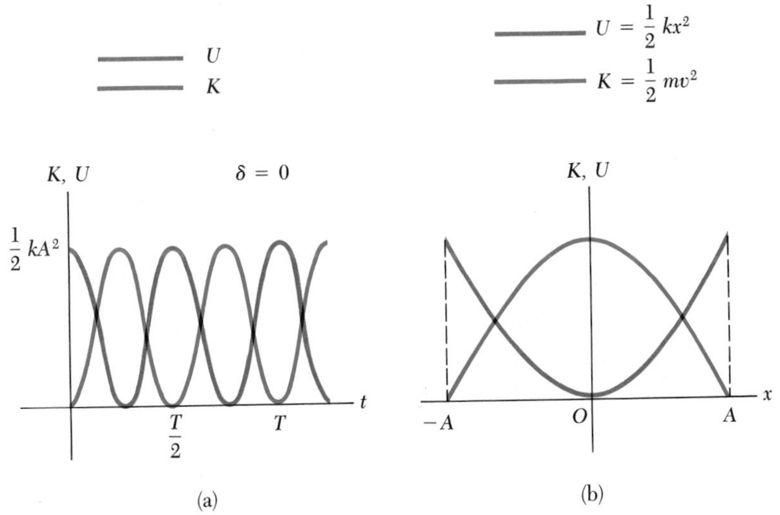

Figure 13.8 (a) Kinetic energy and potential energy versus time for a simple harmonic oscillator with $\delta = 0$. (b) Kinetic energy and potential energy versus displacement for a simple harmonic oscillator. In either plot, note that $K + U =$ constant.

positive and their sum at all times is a constant equal to $\frac{1}{2}kA^2$, the total energy of the system. The variations of K and U with displacement are plotted in Figure 13.8b. Energy is continuously being transferred between potential energy stored in the spring and the kinetic energy of the mass. Figure 13.9 illustrates the position, velocity, acceleration, kinetic energy, and potential energy of the mass-spring system for one full period of the motion. Most of the ideas discussed so far are incorporated in this important figure. We suggest that you study this figure carefully.

Finally, we can use energy conservation to obtain the velocity for an arbitrary displacement x by expressing the total energy at some arbitrary position as

$$E = K + U = \tfrac{1}{2}mv^2 + \tfrac{1}{2}kx^2 = \tfrac{1}{2}kA^2$$

Velocity as a function of position for a simple harmonic oscillator

$$v = \pm\sqrt{\frac{k}{m}(A^2 - x^2)} = \pm\,\omega\sqrt{A^2 - x^2} \qquad (13.21)$$

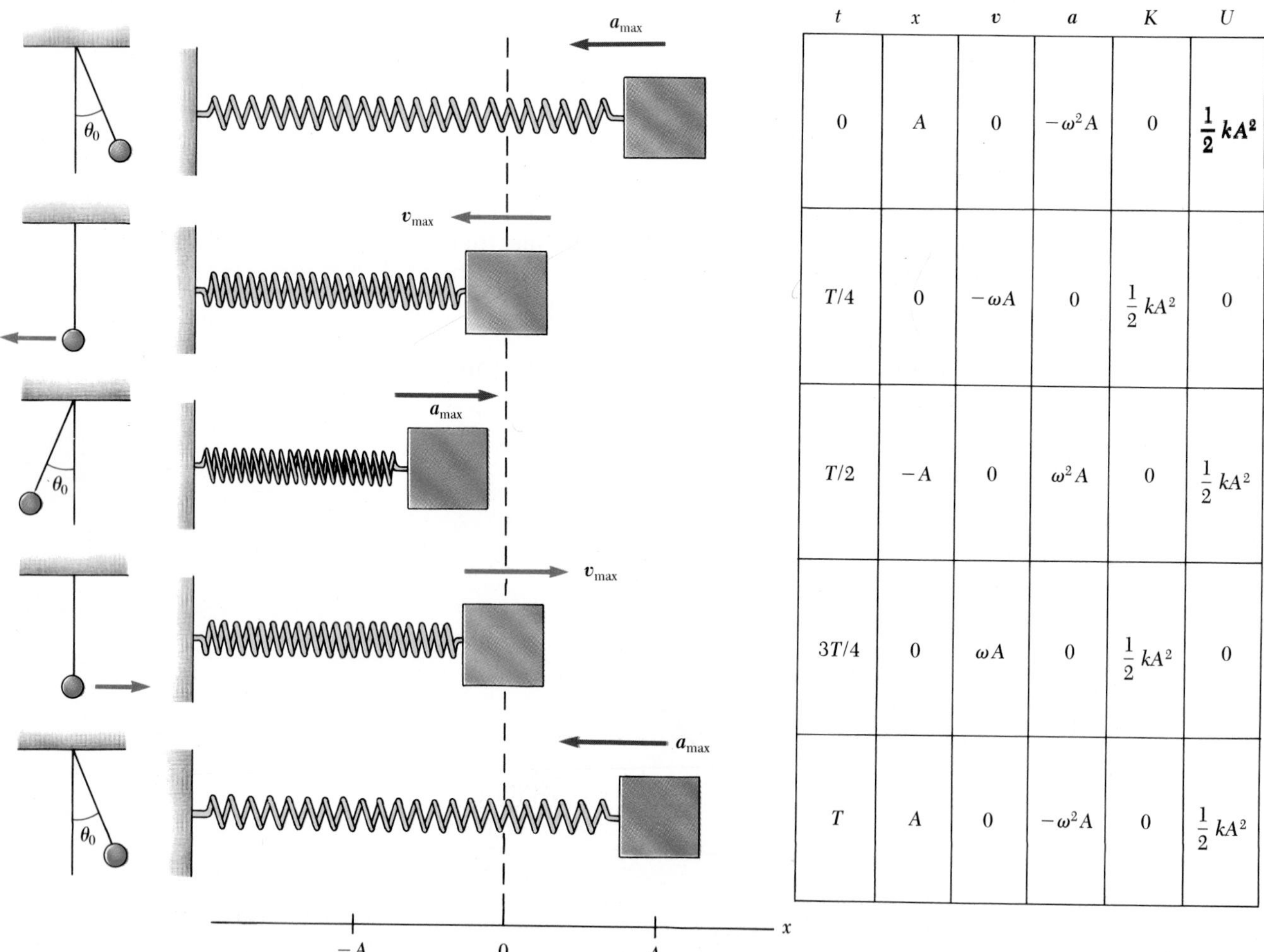

t	x	v	a	K	U
0	A	0	$-\omega^2 A$	0	$\frac{1}{2}kA^2$
$T/4$	0	$-\omega A$	0	$\frac{1}{2}kA^2$	0
$T/2$	$-A$	0	$\omega^2 A$	0	$\frac{1}{2}kA^2$
$3T/4$	0	ωA	0	$\frac{1}{2}kA^2$	0
T	A	0	$-\omega^2 A$	0	$\frac{1}{2}kA^2$

Figure 13.9 Simple harmonic motion for a mass-spring system and its analogy to the motion of a simple pendulum. The parameters in the table at the right refer to the mass-spring system, assuming that at $t = 0$, $x = A$ so that $x = A \cos \omega t$ (Case I).

Again, this expression substantiates the fact that the speed is a maximum at $x = 0$ and is zero at the turning points, $x = \pm A$.

EXAMPLE 13.4 Oscillations on a Horizontal Surface
A mass of 0.5 kg connected to a light spring of force constant 20 N/m oscillates on a horizontal, frictionless surface. (a) Calculate the total energy of the system and the maximum speed of the mass if the amplitude of the motion is 3 cm.

Using Equation 13.20, we get

$$E = \tfrac{1}{2}kA^2 = \tfrac{1}{2}\left(20\ \frac{\text{N}}{\text{m}}\right)(3 \times 10^{-2}\ \text{m})^2$$

$$= 9.00 \times 10^{-3}\ \text{J}$$

When the mass is at $x = 0$, $U = 0$ and $E = \frac{1}{2}mv_{max}^2$; therefore

$$\tfrac{1}{2}mv_{max}^2 = 9.00 \times 10^{-3}\ \text{J}$$

$$v_{max} = \sqrt{\frac{18 \times 10^{-3}\ \text{J}}{0.5\ \text{kg}}} = 0.190\ \text{m/s}$$

(b) What is the velocity of the mass when displacement is equal to 2 cm?

We can apply Equation 13.21 directly:

$$v = \pm\sqrt{\frac{k}{m}(A^2 - x^2)} = \pm\sqrt{\frac{20}{0.5}(3^2 - 2^2) \times 10^{-4}}$$

$$= \pm 0.141\ \text{m/s}$$

The positive and negative signs indicate that the mass could be moving to the right or left at this instant.

(c) Compute the kinetic and potential energies of the system when the displacement equals 2 cm.

Using the result to (b), we get

$$K = \tfrac{1}{2}mv^2 = \tfrac{1}{2}(0.5\ \text{kg})(0.14\ \text{m/s})^2 = 5.00 \times 10^{-3}\ \text{J}$$

$$U = \tfrac{1}{2}kx^2 = \tfrac{1}{2}\left(20\ \frac{\text{N}}{\text{m}}\right)(2 \times 10^{-2}\ \text{m})^2 = 4.00 \times 10^{-3}\ \text{J}$$

Note that the sum $K + U$ equals the total energy, E.

Exercise 2 For what values of x does the speed of the mass equal 0.10 m/s?
Answer ± 2.55 m.

13.4 THE PENDULUM

The Simple Pendulum

The simple pendulum is another mechanical system that exhibits periodic, oscillatory motion. It consists of a point mass m suspended by a light string of length L, where the upper end of the string is fixed as in Figure 13.10. The motion occurs in a vertical plane and is driven by the force of gravity. We shall show that the motion is that of a simple harmonic oscillator, provided the angle θ that the pendulum makes with the vertical is small.

The forces acting on the mass are the tension, $\mathbf{T}$, acting along the string, and the weight $m\mathbf{g}$. The tangential component of the weight, $mg \sin\theta$, always acts toward $\theta = 0$, opposite the displacement. Therefore, the tangential force is a restoring force, and we can write the equation of motion in the tangential direction

$$F_t = -mg \sin\theta = m\frac{d^2s}{dt^2}$$

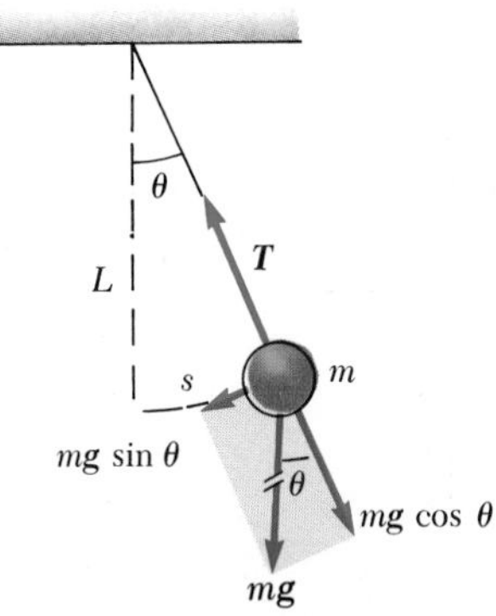

Figure 13.10 When θ is small, the simple pendulum oscillates with simple harmonic motion about the equilibrium position ($\theta = 0$). The restoring force is $mg \sin\theta$, the component of weight tangent to the circle.

where s is the displacement measured along the arc and the minus sign indicates that F_t acts toward the equilibrium position. Since $s = L\theta$ and L is constant, this equation reduces to

$$\frac{d^2\theta}{dt^2} = -\frac{g}{L}\sin\theta$$

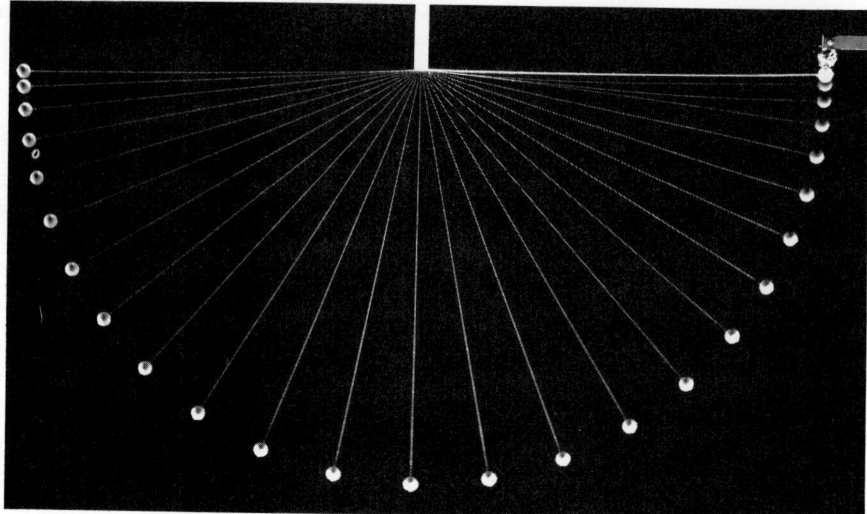

The motion of a simple pendulum captured with multiflash photography. Is the motion simple harmonic in this case? (© Berenice Abbott, Science Source/Photo Researchers)

The right side is proportional to $\sin\theta$, rather than to θ; hence we conclude that the motion is not simple harmonic motion. However, if we assume that θ is *small*, we can use the approximation $\sin\theta \approx \theta$, where θ is measured in *radians*.[1] Therefore, the equation of motion becomes

Equation of motion for the simple pendulum (small θ)

$$\frac{d^2\theta}{dt^2} = -\frac{g}{L}\theta \qquad (13.22)$$

Now we have an expression that is of exactly the same form as Equation 13.15, and so we conclude that the motion is simple harmonic motion. Therefore, θ can be written as $\theta = \theta_0 \cos(\omega t + \delta)$, where θ_0 is the *maximum angular displacement* and the angular frequency ω is given by

Angular frequency of motion for the simple pendulum

$$\omega = \sqrt{\frac{g}{L}} \qquad (13.23)$$

The period of the motion is

Period of motion for the simple pendulum

$$T = \frac{2\pi}{\omega} = 2\pi\sqrt{\frac{L}{g}} \qquad (13.24)$$

In other words, *the period and frequency of a simple pendulum depend only on the length of the string and the acceleration of gravity.* Since the period is *independent* of the mass, we conclude that *all* simple pendulums of equal length at the same location oscillate with equal periods.[2] The analogy between

[1] This approximation can be understood by examining the series expansion for $\sin\theta$, which is $\sin\theta = \theta - \theta^3/3! + \cdots$. For small values of θ, we see that $\sin\theta \approx \theta$. The difference between θ and $\sin\theta$ for $\theta = 15°$ is only about 1%.

[2] The period of oscillation for the simple pendulum with arbitrary amplitude is

$$T = 2\pi\sqrt{\frac{L}{g}}\left(1 + \frac{1}{4}\sin^2\frac{\theta_0}{2} + \frac{9}{64}\sin^4\frac{\theta_0}{2} + \cdots\right)$$

where θ_0 is the maximum angular displacement in radians.

the motion of a simple pendulum and the mass-spring system is illustrated in Figure 13.9.

The simple pendulum can be used as a timekeeper. It is also a convenient device for making precise measurements of the acceleration of gravity. Such measurements are important since variations in local values of **g** can provide information on the location of oil and other valuable underground resources.

EXAMPLE 13.5 What is the Height of That Tower?
A man enters a tall tower. He needs to know the height of the tower. He notes that a long pendulum extends from the ceiling almost to the floor and that its period is 12 s. How tall is the tower?

Solution If we use $T = 2\pi\sqrt{L/g}$ and solve for L, we get

$$L = \frac{gT^2}{4\pi^2} = \frac{(9.80\ \text{m/s}^2)(12\ \text{s})^2}{4\pi^2} = 35.7\ \text{m}$$

Exercise 3 If the pendulum described in this example is taken to the moon, where the acceleration due to gravity is 1.67 m/s², what would its period be there?
Answer 29.1 s.

The Physical Pendulum

A physical, or compound, pendulum consists of any rigid body suspended from a fixed axis that does not pass through the body's center of mass. The system will oscillate when displaced from its equilibrium position. Consider a rigid body pivoted at a point O that is a distance d from the center of mass (Fig. 13.11). The torque about O is provided by the force of gravity, and its magnitude is $mgd \sin\theta$. Using the fact that $\tau = I\alpha$, where I is the moment of inertia about the axis through O, we get

$$-mgd \sin\theta = I\frac{d^2\theta}{dt^2}$$

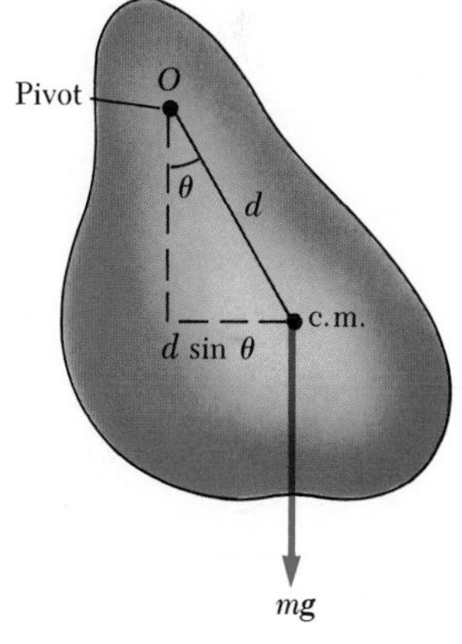

Figure 13.11 The physical pendulum consists of a rigid body pivoted at the point O, and not through the center of mass. At equilibrium, the weight vector passes through O, corresponding to $\theta = 0$. The restoring torque about O when the system is displaced through an angle θ is $mgd \sin\theta$.

The minus sign on the left indicates that the torque about O tends to decrease θ. That is, the force of gravity produces a restoring torque.

If we again assume that θ is small, then the approximation $\sin\theta \approx \theta$ is valid and the equation of motion reduces to

$$\frac{d^2\theta}{dt^2} = -\left(\frac{mgd}{I}\right)\theta = -\omega^2\theta \qquad (13.25)$$

Thus, we note that the equation is of the same form as Equation 13.15, and so the motion is simple harmonic motion. That is, the solution of Equation 13.25 is $\theta = \theta_0 \cos(\omega t + \delta)$, where θ_0 is the maximum angular displacement and

$$\omega = \sqrt{\frac{mgd}{I}}$$

The period is given by

$$T = \frac{2\pi}{\omega} = 2\pi\sqrt{\frac{I}{mgd}} \qquad (13.26)$$

One can use this result to measure the moment of inertia of a planar rigid body. If the location of the center of mass, and hence of d, are known, the moment of inertia can be obtained through a measurement of the period. Finally, note that Equation 13.26 reduces to the period of a simple pendulum (Eq. 13.24) when $I = md^2$, that is, when all the mass is concentrated at the center of mass.

EXAMPLE 13.6 A Swinging Rod

A uniform rod of mass M and length L is pivoted about one end and oscillates in a vertical plane (Fig. 13.12). Find the period of oscillation if the amplitude of the motion is small.

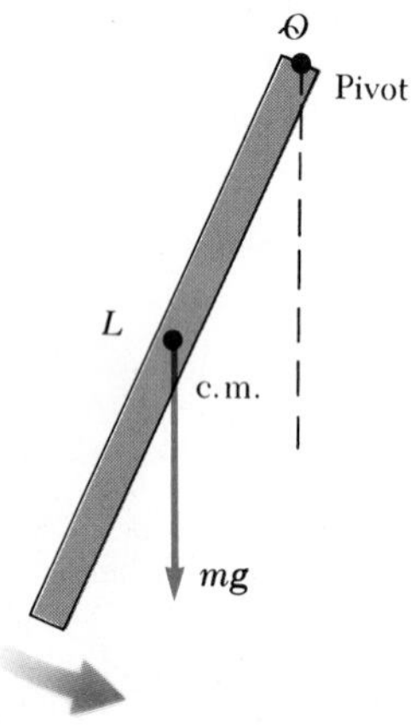

Figure 13.12 (Example 13.6) A rigid rod oscillating about a pivot through one end is a physical pendulum with $d = L/2$ and $I_0 = \frac{1}{3}ML^2$.

Solution In Chapter 10 we found that the moment of inertia of a uniform rod about an axis through one end is $\frac{1}{3}ML^2$. The distance d from the pivot to the center of mass is $L/2$. Substituting these quantities into Equation 13.26 gives

$$T = 2\pi \sqrt{\frac{\frac{1}{3}ML^2}{Mg\frac{L}{2}}} = 2\pi\sqrt{\frac{2L}{3g}}$$

Comment In one of the early moon landings, an astronaut walking on the moon's surface had a belt hanging from his spacesuit, and the belt oscillated as a compound pendulum. A scientist on earth observed this motion on TV and was able to estimate the acceleration of gravity on the moon from this observation. How do you suppose this calculation was done?

Exercise 4 Calculate the period of a meter stick pivoted about one end and oscillating in a vertical plane as in Figure 13.12.

Answer 1.64 s.

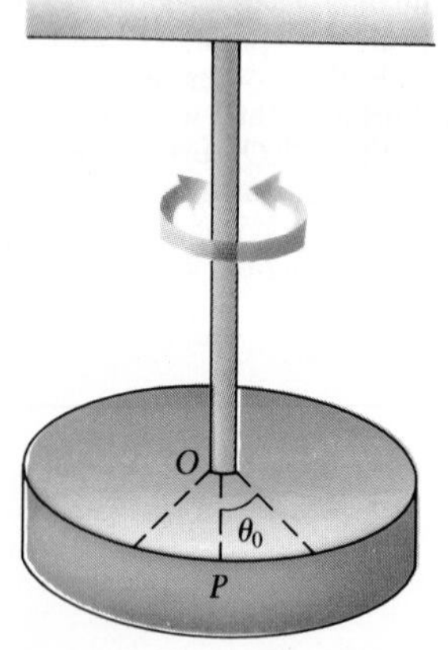

Figure 13.13 A torsional pendulum consists of a rigid body suspended by a wire attached to a rigid support. The body oscillates about the line OP with an amplitude θ_0.

Torsional Pendulum

Figure 13.13 shows a rigid body suspended by a wire attached at the top to a fixed support. When the body is twisted through some small angle θ, the twisted wire exerts a restoring torque on the body proportional to the angular displacement. That is,

$$\tau = -\kappa\theta$$

where κ (the Greek letter kappa) is called the *torsion constant* of the support wire. The value of κ can be obtained by applying a known torque to twist the wire through a measurable angle θ. Applying Newton's second law for rotational motion gives

$$\tau = -\kappa\theta = I\frac{d^2\theta}{dt^2}$$

$$\frac{d^2\theta}{dt^2} = -\frac{\kappa}{I}\theta \qquad (13.27)$$

Again, this is the equation of motion for a simple harmonic oscillator, with $\omega = \sqrt{\kappa/I}$ and a period

$$T = 2\pi\sqrt{\frac{I}{\kappa}} \tag{13.28}$$

This system is called a *torsional pendulum.* There is no small angle restriction in this situation, as long as the elastic limit of the wire is not exceeded. The balance wheel of a watch oscillates as a torsional pendulum, energized by the mainspring. Torsional pendulums are also used in laboratory galvanometers and the Cavendish torsional balance.

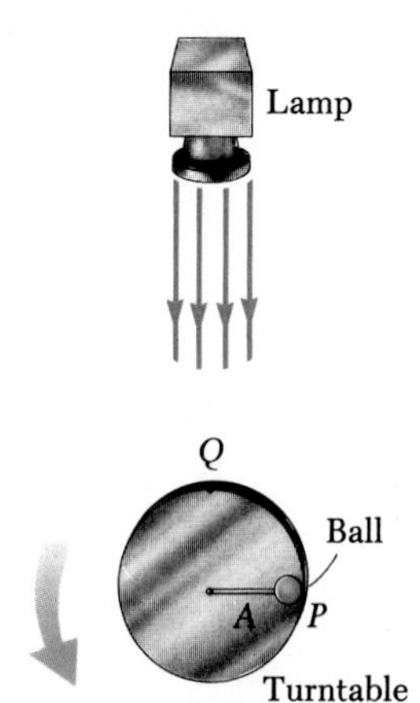

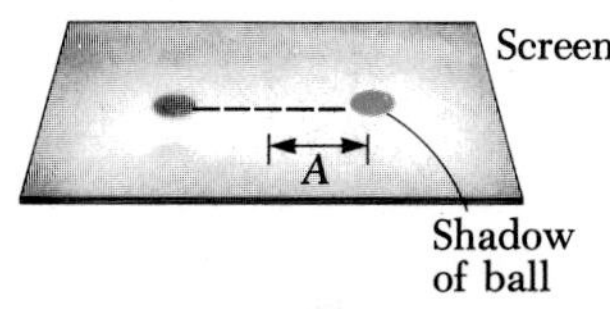

Figure 13.14 Experimental setup for demonstrating the connection between simple harmonic motion and uniform circular motion. As the ball rotates on the turntable with constant angular velocity, its shadow on the screen moves back and fo·th with simple harmonic motion.

*13.5 COMPARING SIMPLE HARMONIC MOTION WITH UNIFORM CIRCULAR MOTION

We can better understand and visualize many aspects of simple harmonic motion along a straight line by looking at its relationship to uniform circular motion. Figure 13.14 shows an experimental arrangement useful for developing this concept. This figure represents a top view of a ball attached to the rim of a phonograph turntable of radius A, illuminated from the side by a lamp. Rather than concentrating on the ball, let us focus our attention on the shadow that the ball casts on the screen. We find that *as the turntable rotates with constant angular velocity, the shadow of the ball moves back and forth with simple harmonic motion.*

Consider a particle at point P moving in a circle of radius A with constant angular velocity ω (Fig. 13.15a). We shall refer to this circle as the *reference circle* for the motion. As the particle rotates, the position vector of the particle rotates about the origin, O. At some instant of time, t, the angle between OP and the x axis is $\omega t + \delta$, where δ is the angle that OP makes with the x axis at $t = 0$. We take this as our reference point for measuring the angular displacement. As the particle rotates on the reference circle, the angle that OP makes with the x axis *changes* with time. Furthermore, the projection of P onto the x

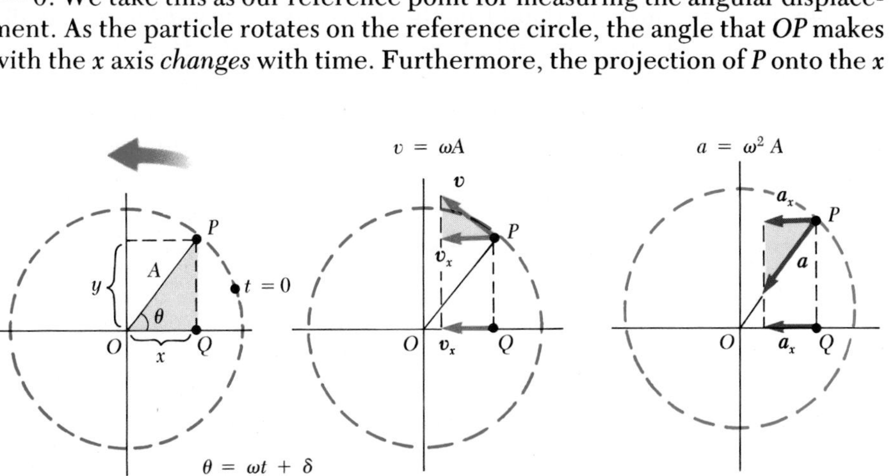

Figure 13.15 Relationship between the uniform circular motion of a point P and the simple harmonic motion of the point Q. A particle at P moves in a circle of radius A with constant angular velocity ω. (a) The x components of the points P and Q are equal and vary in time as $x = A\cos(\omega t + \delta)$. (b) The x component of velocity of P equals the velocity of Q. (c) The x component of the acceleration of P equals the acceleration of Q.

axis, labeled point Q, moves back and forth along a line parallel to the diameter of the reference circle, between the limits $x = \pm A$.

Note that points P and Q have the *same* x coordinate. From the right triangle OPQ, we see that the x coordinate of P and Q is given by

$$x = A \cos(\omega t + \delta) \tag{13.29}$$

This expression shows that the point Q moves with simple harmonic motion along the x axis. Therefore, we conclude that

> simple harmonic motion along a straight line can be represented by the projection of uniform circular motion along a diameter.

By a similar argument, you can see from Figure 13.15a that the projection of P along the y axis also exhibits simple harmonic motion. Therefore, *uniform circular motion can be considered a combination of two simple harmonic motions*, one along x and one along y, where the two differ in phase by 90°.

The geometric interpretation we have presented shows that the time for one complete revolution of the point P on the reference circle is equal to the period of motion, T, for simple harmonic motion between $x = \pm A$. That is, the angular speed of the point P is the same as the angular frequency, ω, of simple harmonic motion along the x axis. The phase constant δ for simple harmonic motion corresponds to the initial angle that OP makes with the x axis. The radius of the reference circle, A, equals the amplitude of the simple harmonic motion.

Since the relationship between linear and angular velocity for circular motion is $v = r\omega$, the particle moving on the reference circle of radius A has a velocity of magnitude ωA. From the geometry in Figure 13.15b, we see that the x component of this velocity is given by $-\omega A \sin(\omega t + \delta)$. By definition, the point Q has a velocity given by dx/dt. Differentiating Equation 13.29 with respect to time, we find that the velocity of Q is the same as the x component of velocity of P.

The acceleration of the point P on the reference circle is directed radially inward toward O and has a magnitude given by $v^2/A = \omega^2 A$. From the geometry in Figure 13.15c, we see that the x component of this acceleration is equal to $-\omega^2 A \cos(\omega t + \delta)$. This also coincides with the acceleration of the projected point Q along the x axis, as you can easily verify from Equation 13.29.

EXAMPLE 13.7 Circular Motion with Constant Speed

A particle rotates counterclockwise in a circle of radius 3.0 m with a constant angular speed of 8 rad/s, as in Figure 13.15. At $t = 0$, the particle has an x coordinate of 2.0 m. (a) Determine the x coordinate as a function of time.

Since the amplitude of the particle's motion equals the radius of the circle and $\omega = 8$ rad/s, we have

$$x = A \cos(\omega t + \delta) = (3.0 \text{ m}) \cos(8t + \delta)$$

We can evaluate δ using the initial condition that $x = 2.0$ m at $t = 0$:

$$2.0 \text{ m} = (3.0 \text{ m}) \cos(0 + \delta)$$

$$\delta = \cos^{-1}(\tfrac{2}{3}) = 48° = 0.841 \text{ rad}$$

Therefore, the x coordinate versus time is of the form

$$x = (3.0 \text{ m}) \cos(8t + 0.841)$$

Note that the angles in the cosine function are in radians.

(b) Find the x components of the particle's velocity and acceleration at any time t.

$$v_x = \frac{dx}{dt} = (-3.0)(8) \sin(8t + 0.84)$$

$$= -(24 \text{ m/s}) \sin(8t + 0.841)$$

$$a_x = \frac{dv_x}{dt} = (-24)(8)\cos(8t + 0.841)$$

$$= -(192 \text{ m/s}^2)\cos(8t + 0.841)$$

From these results, we conclude that $v_{max} = 240$ m/s and $a_{max} = 192$ m/s^2. Note that these values also equal the tangential velocity, ωA, and centripetal acceleration, $\omega^2 A$.

*13.6 DAMPED OSCILLATIONS

The oscillatory motions we have considered so far have dealt with an ideal system, that is, one that oscillates indefinitely under the action of a linear restoring force. In realistic systems, dissipative forces, such as friction, are present and retard the motion of the system. Consequently, the mechanical energy of the system will diminish in time, and the motion is said to be *damped.*

One common type of retarding force, which we discussed in Chapter 6, is proportional to the velocity and acts in the direction opposite the motion. This is often observed for the motion of an object through gases like air. Because the retarding force can be expressed as $\mathbf{R} = -b\mathbf{v}$, where b is a constant, and the restoring force is $-kx$, we can write Newton's second law as

$$\sum F_x = -kx - bv = ma_x$$

$$-kx - b\frac{dx}{dt} = m\frac{d^2x}{dt^2} \qquad (13.30)$$

The solution of this equation requires mathematics that may not be familiar to you as yet, and so it will simply be stated without proof. When the retarding force is small compared with kx, that is, when b is small, the solution to Equation 13.30 is

$$x = A\,e^{-\frac{b}{2m}t}\cos(\omega t + \delta) \qquad (13.31)$$

where the frequency of motion is

$$\omega = \sqrt{\frac{k}{m} - \left(\frac{b}{2m}\right)^2} \qquad (13.32)$$

This can be verified by substitution of the solution into Equation 13.30. Figure 13.16a shows the displacement as a function of time in this case. We see that *when the dissipative force is small compared with the restoring force, the oscillatory character of the motion is preserved but the amplitude of vibration decreases in time,* and the motion will ultimately cease. This is known as an **underdamped oscillator.** The dashed blue line in Figure 13.16a, which is the *envelope* of the oscillatory curve, represents the exponential factor that appears in Equation 13.31. This shows that *the amplitude decays exponentially with time.* For motion with a given spring constant and particle mass, the oscillations dampen more rapidly as the maximum value of the dissipative force approaches the maximum value of the restoring force. One example of a damped harmonic oscillator is a mass immersed in a fluid as in Figure 13.16b.

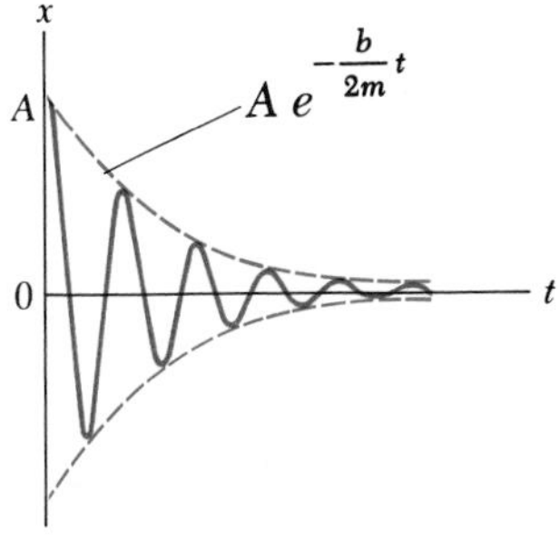

(a)

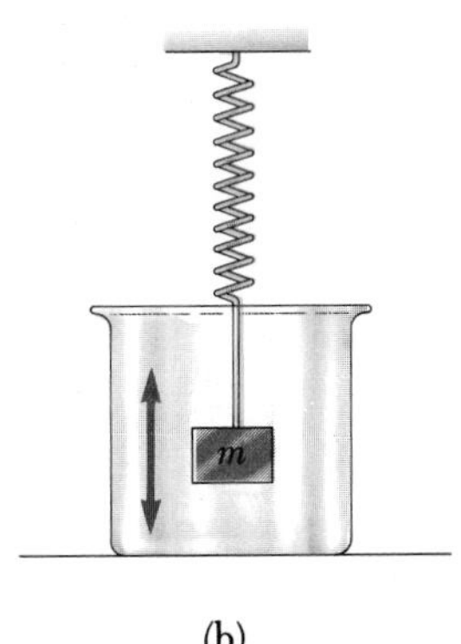

(b)

Figure 13.16 (a) Graph of the displacement versus time for an underdamped oscillator. Note the decrease in amplitude with time. (b) One example of a damped oscillator is a mass on a spring submersed in a liquid.

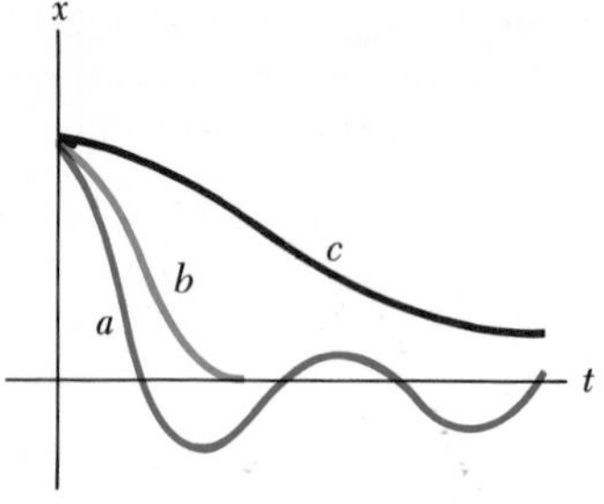

Figure 13.17 Plots of displacement versus time for (*a*) an underdamped oscillator, (*b*) a critically damped oscillator and (*c*) an overdamped oscillator.

It is convenient to express the frequency of vibration in the form

$$\omega = \sqrt{\omega_0{}^2 - \left(\frac{b}{2m}\right)^2}$$

where $\omega_0 = \sqrt{k/m}$ represents the frequency of oscillation in the absence of a resistive force (the undamped oscillator). In other words, when $b = 0$, the resistive force is zero and the system oscillates with its natural frequency, ω_0. As the magnitude of the resistive force approaches the value of the restoring force in the spring, the oscillations dampen more rapidly. When b reaches a critical value b_c such that $b_c/2m = \omega_0$, the system does not oscillate and is said to be **critically damped.** In this case, the system returns to equilibrium in an exponential manner with time, as in Figure 13.17.

If the medium is so viscous that the resistive force is greater than the restoring force, that is, if $b/2m > \omega_0$, the system will be **overdamped.** Again, the displaced system does not oscillate but simply returns to its equilibrium position. As the damping increases, the time it takes the displacement to reach equilibrium also increases, as indicated in Figure 13.17. In any case, when friction is present, the energy of the oscillator will eventually fall to zero. The loss in mechanical energy dissipates into thermal energy in the resistive medium.

*13.7 FORCED OSCILLATIONS

We have seen that the energy of a damped oscillator decreases in time as a result of the dissipative force. It is possible to compensate for this energy loss by applying an external force that does positive work on the system. At any instant, energy can be put into the system by an applied force that acts in the direction of motion of the oscillator. For example, a child on a swing can be kept in motion by appropriately timed "pushes." The amplitude of motion will remain constant if the energy input per cycle of motion exactly equals the energy lost as a result of friction.

A common example of a forced oscillator is a damped oscillator driven by an external force that varies harmonically, such as $F = F_0 \cos \omega t$, where ω is the angular frequency of the force and F_0 is a constant. Adding this driving force to the left side of Equation 13.30 gives

$$F_0 \cos \omega t - b\frac{dx}{dt} - kx = m\frac{d^2x}{dt^2} \tag{13.33}$$

Again, the solution of this equation is rather lengthy and will not be presented. However, after a sufficiently long period of time, when the energy input per cycle equals the energy lost per cycle, a *steady-state* condition is reached in which the oscillations proceed with constant amplitude. At this time, when the system is in steady state, Equation 13.33 has the following solution:

$$x = A\cos(\omega t + \delta) \tag{13.34}$$

where

$$A = \frac{F_0/m}{\sqrt{(\omega^2 - \omega_0{}^2)^2 + \left(\dfrac{b\omega}{m}\right)^2}} \tag{13.35}$$

and where $\omega_0 = \sqrt{k/m}$ is the frequency of the undamped oscillator ($b = 0$). One can argue physically that in a steady state the oscillator must have the same frequency as that of the driving force, so the solution given by Equation 13.34 is expected. In fact, when this solution is substituted into Equation 13.33, one finds that it is indeed a solution provided the amplitude is given by Equation 13.35.

Equation 13.35 shows that the motion of the forced oscillator is not damped since it is being driven by an external force. That is, the external agent provides the necessary energy to overcome the losses due to the resistive force. Note that the mass oscillates at the frequency of the driving force, ω. For small damping, the amplitude becomes large when the frequency of the driving force is near the natural frequency of oscillation, or when $\omega \approx \omega_0$. The dramatic increase in amplitude near the natural frequency is called **resonance,** and the frequency ω_0 is called the **resonance frequency** of the system.

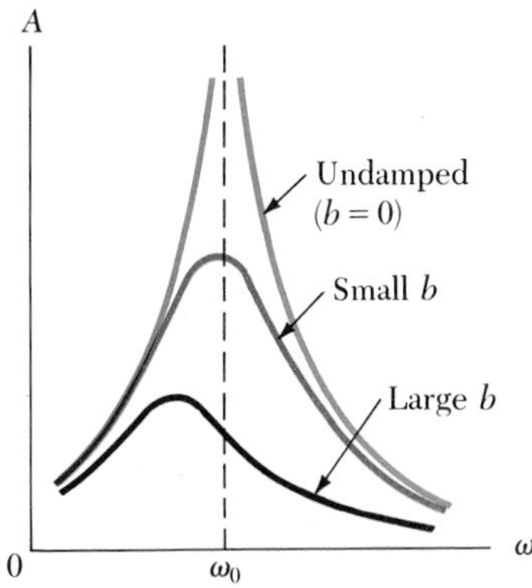

Figure 13.18 Graph of the amplitude versus frequency for a damped oscillator when a periodic driving force is present. When the frequency of the driving force equals the natural frequency, ω_0, resonance occurs. Note that the shape of the resonance curve depends on the size of the damping coefficient, b.

Physically, the reason for large-amplitude oscillations at the resonance frequency is that energy is being transferred to the system under the most favorable conditions. This can be better understood by taking the first time derivative of x, which gives an expression of the velocity of the oscillator. In doing so, one finds that v is proportional to $\sin(\omega t + \delta)$. When the applied force is in phase with v, the rate at which work is done on the oscillator by the force $\mathbf{F}$ (or the power) equals Fv. Since the quantity Fv is always positive when $\mathbf{F}$ and $\mathbf{v}$ are in phase, we conclude that *at resonance the applied force is in phase with the velocity and the power transferred to the oscillator is a maximum.*

A graph of the amplitude as a function of frequency for the forced oscillator with and without a resistive force is shown in Figure 13.18. Note that the amplitude increases with decreasing damping ($b \rightarrow 0$). Furthermore, the resonance curve is broadened as the damping increases. Under steady-state conditions, and at any driving frequency, the energy transferred into the system equals the energy lost because of the damping force; hence the average total energy of the oscillator remains constant. In the absence of a damping force ($b = 0$), we see from Equation 13.35 that the steady-state amplitude approaches infinity as $\omega \rightarrow \omega_0$. In other words, if there are no losses in the system, and we continue to drive an initially motionless oscillator with a sinusoidal force that is in phase with the velocity, the amplitude of motion will build up without limit (Fig. 13.18). This does not occur in practice since some damping will always be present. That is, at resonance the amplitude will be large but finite for small damping.

One experiment that demonstrates a resonance phenomenon is illustrated in Figure 13.19. Several pendulums of different lengths are suspended from a stretched string. If one of them, such as P, is set in sideways motion, the others will begin to oscillate, since they are coupled by the stretched string. Of those that are forced into oscillation by this coupling, pendulum Q, whose length is the same as that of P (and hence the two pendula have the same natural frequency), will oscillate with the greatest amplitude.

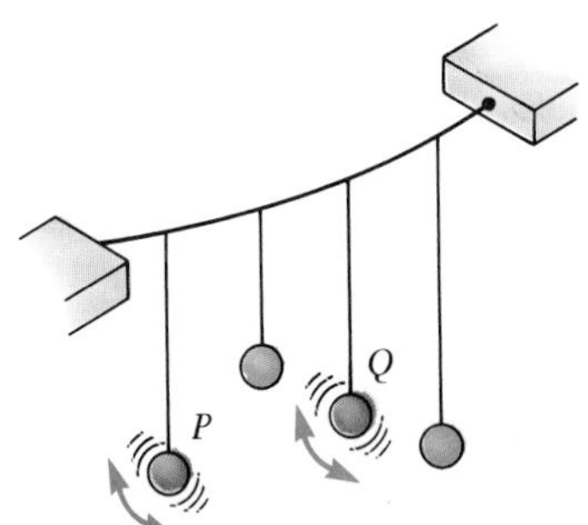

Figure 13.19 If pendulum P is set into oscillation, pendulum Q will eventually oscillate with the greatest amplitude because of the coupling between them and the fact that they have the same natural frequency of vibration.

Later in the text we shall see that the phenomenon of resonance appears in other areas of physics. For example, certain electrical circuits have natural (or resonant) frequencies. A structure such as a bridge has natural frequencies, which can be set into resonance by an appropriate driving force. A striking example of such a structural resonance occurred in 1940, when the Tacoma Narrows bridge in Washington was destroyed by resonant vibrations. It should be noted that although the winds were not particularly strong on that occasion, the bridge ultimately collapsed because vortices (turbulences) generated by the wind blowing through the bridge structure occurred at a frequency which matched the natural frequency of the structure. A more detailed discussion of this dramatic event is given in the essay at the end of this chapter.

Many other examples of resonant vibrations can be cited. A resonant vibration which you may have experienced is the "singing" of telephone wires in the wind. Mechanical machines are often broken apart if one vibrating part is at resonance with some other moving part. Finally, soldiers marching in cadence across bridges have been known to set up resonant vibrations in the structure, causing the bridges to collapse. In one such famous accident, which occurred in France in 1850, a collapsed suspension bridge resulted in the death of 226 soldiers.

SUMMARY

The position of a simple harmonic oscillator varies periodically in time according to the relation

Displacement versus time for simple harmonic motion

$$x = A\cos(\omega t + \delta) \tag{13.1}$$

where A is the amplitude of the motion, ω is the angular frequency, and δ is the phase constant. The value of δ depends on the initial position and velocity of the oscillator.

The time for one complete vibration is called the **period** of the motion, defined by

Period

$$T = \frac{2\pi}{\omega} \tag{13.2}$$

The inverse of the period is the **frequency** of the motion, which equals the number of oscillations per second.

The **velocity** and **acceleration** of a simple harmonic oscillator are given by

Velocity in simple harmonic motion

$$v = \frac{dx}{dt} = -\omega A \sin(\omega t + \delta) \tag{13.5}$$

Acceleration in simple harmonic motion

$$a = \frac{dv}{dt} = -\omega^2 A \cos(\omega t + \delta) \tag{13.6}$$

Thus, the maximum velocity is ωA, and the maximum acceleration is $\omega^2 A$. The velocity is zero when the oscillator is at its turning points, $x = \pm A$, and the speed is a maximum at the equilibrium position, $x = 0$. The magnitude of the acceleration is a maximum at the turning points and is zero at the equilibrium position.

A mass-spring system exhibits simple harmonic motion on a frictionless surface, with a period given by

Period of motion for mass-spring system

$$T = \frac{2\pi}{\omega} = 2\pi\sqrt{\frac{m}{k}} \tag{13.16}$$

where k is the force constant of the spring and m is the mass attached to the spring.

The kinetic energy and potential energy for a simple harmonic oscillator vary with time and are given respectively by

Kinetic and potential energy of a simple harmonic oscillator

$$K = \frac{1}{2}mv^2 = \frac{1}{2}m\omega^2A^2\sin^2(\omega t + \delta) \tag{13.18}$$

$$U = \frac{1}{2}kx^2 = \frac{1}{2}kA^2\cos^2(\omega t + \delta) \tag{13.19}$$

The **total energy** of a simple harmonic oscillator is a constant of the motion and is given by

Total energy of a simple harmonic oscillator

$$E = \frac{1}{2}kA^2 \tag{13.20}$$

The potential energy of a simple harmonic oscillator is a maximum when the particle is at its turning points (maximum displacement from equilibrium) and is zero at the equilibrium position. The kinetic energy is zero at the turning points and is a maximum at the equilibrium position.

A **simple pendulum** of length L exhibits simple harmonic motion for small angular displacements from the vertical, with a *period* given by

Period of motion for a simple pendulum

$$T = 2\pi\sqrt{\frac{L}{g}} \tag{13.24}$$

That is, the period is *independent* of the suspended mass.

A **physical pendulum** exhibits simple harmonic motion about a pivot that does not go through the center of mass. The period of this motion is

Period of motion for a physical pendulum

$$T = 2\pi\sqrt{\frac{I}{mgd}} \tag{13.26}$$

where I is the moment of inertia about an axis through the pivot and d is the distance from the pivot to the center of mass.

Damped oscillations occur in a system in which a dissipative force opposes the linear restoring force. If such a system is set into motion and then left to itself, the mechanical energy decreases in time because of the presence of the nonconservative damping force. It is possible to compensate for this loss in energy by driving the system with an external periodic force that is in phase with the motion of the system. When the frequency of the driving force matches the natural frequency of the *undamped* oscillator that starts its motion from rest, energy is continuously transferred to the oscillator and its amplitude increases without limit.

QUESTIONS

1. What is the total distance traveled by a body executing simple harmonic motion in a time equal to its period if its amplitude is A?
2. If the coordinate of a particle varies as $x = -A\cos\omega t$, what is the phase constant δ in Equation 13.1? At what position does the particle begin its motion?
3. Does the displacement of an oscillating particle between $t = 0$ and a later time t necessarily equal the position of the particle at time t? Explain.
4. Determine whether or not the following quantities can be in the same direction for a simple harmonic oscillator: (a) displacement and velocity, (b) velocity and acceleration, (c) displacement and acceleration.
5. Can the amplitude A and phase constant δ be determined for an oscillator if only the position is specified at $t = 0$? Explain.
6. Describe qualitatively the motion of a mass-spring system if the mass of the spring is not neglected.
7. If a mass-spring system is hung vertically and set into oscillation, why does the motion eventually stop?
8. Explain why the kinetic and potential energies of a mass-spring system can never be negative.

9. A mass-spring system undergoes simple harmonic motion with an amplitude A. Does the total energy change if the mass is doubled but the amplitude is not changed? Do the kinetic and potential energies depend on the mass? Explain.
10. What happens to the period of a simple pendulum if its length is doubled? What happens to the period if the mass that is suspended is doubled?
11. A simple pendulum is suspended from the ceiling of a stationary elevator, and the period is determined. Describe the changes, if any, in the period if the elevator (a) accelerates upward, (b) accelerates downward, and (c) moves with constant velocity.
12. A simple pendulum undergoes simple harmonic motion when θ is small. Will the motion be *periodic* if θ is large? How does the period of motion change as θ increases?
13. Give a few examples of damped oscillations that are commonly observed.
14. Will damped oscillations occur for any values of b and k? Explain.
15. Is it possible to have damped oscillations when a system is at resonance? Explain.
16. At resonance, what does the phase constant δ equal in Equation 13.34? (*Hint:* Compare this with the expression for the driving force, which must be in phase with the velocity at resonance.)
17. A platoon of soldiers marches in step along a road. Why are they ordered to break step when crossing a bridge?
18. Give as many examples as you can in the workings of an automobile where the motion is simple harmonic or damped.
19. If a grandfather clock were running slow, how could we adjust the "length" of the pendulum to correct the time?

PROBLEMS

Section 13.1 Simple Harmonic Motion

1. The displacement of a particle is given by the expression $x = (4\text{ m})\cos(3\pi t + \pi)$, where x is in m and t is in s. Determine (a) the frequency and period of the motion, (b) the amplitude of the motion, (c) the phase constant, and (d) the position of the particle at $t = 0$.
2. For the particle described in Problem 1, determine (a) the velocity at any time t, (b) the acceleration at any time, (c) the maximum velocity and maximum acceleration, and (d) the velocity and acceleration at $t = 0$.
3. A particle oscillates with simple harmonic motion such that its displacement varies according to the expression as $x = (5\text{ cm})\cos(2t + \pi/6)$, where x is in cm and t is in s. At $t = 0$, find (a) the displacement of the particle, (b) its velocity, and (c) its acceleration. (d) Find the period and amplitude of the motion.
4. A particle moving with simple harmonic motion travels a total distance of 20 cm in each cycle of its motion, and its maximum acceleration is 50 m/s^2. Find (a) the angular frequency of the motion and (b) the maximum speed of the particle.
5. The displacement of a body is given by the expression $x = (8.0\text{ cm})\cos(2t + \pi/3)$, where x is in cm and t is in s. Calculate (a) the velocity and acceleration at $t = \pi/2$ s, (b) the maximum speed and the earliest time ($t > 0$) at which the particle has this speed, and (c) the maximum acceleration and the earliest time ($t > 0$) at which the particle has this acceleration.
6. At $t = 0$ a particle moving with simple harmonic motion is at $x_0 = 2$ cm, where its velocity is given by $v_0 = -24$ cm/s. If the period of its motion is 0.5 s and the frequency is 2 Hz, find (a) the phase constant; (b) the amplitude; (c) the displacement, velocity, and acceleration as functions of time; and (d) the maximum speed and maximum acceleration.
7. A particle moving along the x axis with simple harmonic motion starts from the origin at $t = 0$ and moves toward the right. If the amplitude of its motion is 2 cm and the frequency is 1.5 Hz, (a) show that its displacement is given by $x = (2\text{ cm})\sin(3\pi t)$. Determine (b) the maximum speed and the earliest time ($t > 0$) at which the particle has this speed, (c) the maximum acceleration and the earliest time ($t > 0$) at which the particle has this acceleration, and (d) the total *distance* traveled between $t = 0$ and $t = 1$ s.
8. A piston in an automobile engine is in simple harmonic motion. If its amplitude of oscillation from centerline is ± 5 cm, and the mass of the piston is 2 kg, find the maximum velocity and acceleration of the piston when the auto engine is running at the rate of 3600 rev/min.

Section 13.2 Mass Attached to a Spring

Neglect spring masses.

9. A weight of 0.2 N is hung from a spring with a force constant $k = 6$ N/m. How much is the spring displaced?
10. A spring stretches by 3.9 cm when a 10-g mass is hung from it. If a total mass of 25 g attached to this spring oscillates in simple harmonic motion, calculate the period of motion.
11. The frequency of vibration of a mass-spring system is 5 Hz when a 4-g mass is attached to the spring. What is the force constant of the spring?

12. A 1-kg mass attached to a spring of force constant 25 N/m oscillates on a horizontal, frictionless surface. At $t = 0$, the mass is released from rest at $x = -3$ cm. (That is, the spring is compressed by 3 cm.) Find (a) the period of its motion, (b) the maximum values of its speed and acceleration, and (c) the displacement, velocity, and acceleration as functions of time.
13. A simple harmonic oscillator takes 12 s to undergo 5 complete vibrations. Find (a) the period of its motion, (b) the frequency in Hz, and (c) the angular frequency in rad/s.
14. A mass-spring system oscillates such that the displacement is given by $x = (0.25 \text{ m}) \cos (2\pi t)$. (a) Find the speed and acceleration of the mass when $x = 0.10$ m. (b) Determine the maximum speed and maximum acceleration.
15. A 0.5-kg mass attached to a spring of force constant 8 N/m vibrates with simple harmonic motion with an amplitude of 10 cm. Calculate (a) the maximum value of its speed and acceleration, (b) the speed and acceleration when the mass is at $x = 6$ cm from the equilibrium position, and (c) the time it takes the mass to move from $x = 0$ to $x = 8$ cm.
16. A particle that is attached to a vertical spring is pulled down a distance of 4.0 cm below its equilibrium position and is released from rest. The initial upward acceleration of the particle is 0.30 m/s^2. (a) What is the period T of the ensuing oscillations? (b) With what velocity does the particle pass through its equilibrium position? (c) What is the equation of motion for the particle? (Choose the upward direction to be positive.)
17. A particle that hangs from an ideal spring has an angular frequency for oscillations, $\omega_0 = 2.0$ rad/s. The spring is suspended from the ceiling of an elevator car and hangs motionless (relative to the elevator car) as the car descends at a constant velocity of 1.5 m/s. The car then stops suddenly. (a) With what amplitude will the particle oscillate? (b) What is the equation of motion for the particle? (Choose the upward direction to be positive.)

Section 13.3 Energy of the Simple Harmonic Oscillator

Neglect spring masses.

18. A 200-g mass is attached to a spring and executes simple harmonic motion with a period of 0.25 s. If the total energy of the system is 2 J, find (a) the force constant of the spring and (b) the amplitude of the motion.
19. A mass-spring system oscillates with an amplitude of 3.5 cm. If the spring constant is 250 N/m and the mass is 0.5 kg, determine (a) the mechanical energy of the system, (b) the maximum speed of the mass, and (c) the maximum acceleration.
20. A simple harmonic oscillator has a total energy E. (a) Determine the kinetic and potential energies when the displacement equals one half the amplitude. (b) For what value of the displacement does the kinetic energy equal the potential energy?
21. The amplitude of a system moving with simple harmonic motion is doubled. Determine the change in (a) the total energy, (b) the maximum velocity, (c) the maximum acceleration, and (d) the period.
22. A mass-spring system of force constant 50 N/m undergoes simple harmonic motion on a horizontal surface with an amplitude of 12 cm. (a) What is the total energy of the system? (b) What is the kinetic energy of the system when the mass is 9 cm from equilibrium? (c) What is its potential energy when $x = 9$ cm?
23. A particle executes simple harmonic motion with an amplitude of 3.0 cm. At what displacement from the midpoint of its motion will its speed equal one half of its maximum speed?
24. A block with a mass of 0.50 kg is attached to the free end of an ideal spring that provides a restoring force of 40 N per meter of extension. The block is free to slide over a frictionless horizontal surface. The block is set into motion by giving it an initial potential energy of 2 J and an initial kinetic energy of 1.5 J. (a) Plot a graph of the potential energy of the system for x-values in the range $-0.5 \text{ m} \leq x \leq +0.5$ m. (b) Determine the amplitude of the oscillation—first, from the graph, and then by algebraic calculation. (c) What is the speed of the block as it passes through the equilibrium position? (d) At what displacements is the kinetic energy equal to the potential energy? (e) Find the angular frequency ω_0 and the period T of the motion. (f) If the initial displacement was $x_0 > 0$ and if the initial velocity was $v < 0$, determine the initial phase angle ϕ_0. (g) Write down the equation of motion $x(t)$.

Section 13.4 The Pendulum

25. A simple pendulum has a period of 2.50 s. (a) What is its length? (b) What would its period be on the moon where $g_m = 1.67$ m/s^2?
26. Calculate the frequency and period of a simple pendulum of length 10 m.
27. If the length of a simple pendulum is quadrupled, what happens to (a) its frequency and (b) its period?
28. A simple pendulum 2.00 m in length oscillates in a location where $g = 9.80$ m/s^2. How many complete oscillations will it make in 5 min?
29. A simple pendulum has a mass of 0.25 kg and a length of 1 m. It is displaced through an angle of 15° and then released. What is (a) the maximum velocity? (b) the maximum angular acceleration? (c) the maximum restoring force?

30. A uniform rod is pivoted at one end as in Figure 13.12. If the rod swings with simple harmonic motion, what must its length be in order that its period be equal to that of a simple pendulum 1 m long?
31. A simple pendulum has a length of 3.00 m. Determine the *change* in its period if it is taken from a point where $g = 9.80\ \text{m/s}^2$ to a higher elevation, where the acceleration due to gravity decreases to $g = 9.79\ \text{m/s}^2$.
32. A circular hoop of radius R is hung over a knife edge. Show that its period of oscillation is equal to that of a simple pendulum of length $2R$.
33. A physical pendulum in the form of a planar body exhibits simple harmonic motion with a frequency of 1.5 Hz. If the pendulum has a mass of 2.2 kg and the pivot is located 0.35 m from the center of mass, determine the moment of inertia of the pendulum.
34. A thin rod has a mass M and a length $L = 1.6$ m. One end of the rod is attached by a pivot to a fixed support and the hanging rod executes small oscillations about the pivot point. Find the frequency of these oscillations. If a particle with a mass M is added to the lower end of the rod, by what factor will the period change?
35. A clock balance wheel has a period of oscillation of 0.25 s. The wheel is constructed so that 20 g of mass is concentrated around a rim of 0.5 cm radius. What is (a) the wheel's moment of inertia? (b) the torsion constant of the attached spring?

Section 13.6 Damped Oscillations

36. Show that the damping constant, b, has units of kg/s.
37. Show that Equation 13.31 is a solution of Equation 13.30 provided that $b^2 < 4mk$.
38. Show that the time rate of change of mechanical energy for a damped, undriven oscillator is given by $dE/dt = -bv^2$ and hence is *always negative*. (*Hint:* Differentiate the expression for the mechanical energy of an oscillator, $E = \frac{1}{2}mv^2 + \frac{1}{2}kx^2$, and make use of Eq. 13.30.)
39. A pendulum of length 1 m is released from an initial angle of 15°. After 1000 s, its amplitude is reduced by friction to 5.5°. What is the value of $b/2m$?

Section 13.7 Forced Oscillations

40. A 2-kg mass attached to a spring is driven by an external force $F = (3\ \text{N}) \cos(2\pi t)$. If the force constant of the spring is 20 N/m, determine (a) the period and (b) the amplitude of the motion. (*Hint:* Assume that there is *no* damping, that is, $b = 0$, and make use of Eq. 13.35.)
41. Calculate the resonant frequencies of the following systems: (a) a 3-kg mass attached to a spring of force constant 240 N/m, (b) a simple pendulum 1.5 m in length.
42. Consider an *undamped* forced oscillator ($b = 0$), and show that Equation 13.34 is a solution of Equation 13.33, with an amplitude given by Equation 13.35.
43. A weight of 40 N is suspended from a spring with force constant 200 N/m. The system is undamped and is subjected to a harmonic force of frequency 10 Hz, resulting in a forced-motion amplitude of 2 cm. Determine the maximum value of the impressed force.

ADDITIONAL PROBLEMS

44. A car with bad shock absorbers bounces up and down with a period of 1.5 s after hitting a bump. The car has a mass of 1500 kg and is supported by four springs of equal force constant k. Determine a value for k.
45. A large passenger of mass 150 kg sits in the car (Problem 44) with bad shocks. The mass of the car is now 1650 kg. What is the new period of oscillation?
46. A block rests on a flat plate that executes vertical simple harmonic motion with a period of 1.2 s. What is the maximum amplitude of the motion for which the block will not separate from the plate?
47. When the simple pendulum illustrated in Figure 13.20 makes an angle θ with the vertical, its speed is v. (a) Calculate the total mechanical energy of the pendulum as a function of v and θ. (b) Show that when θ is small, the potential energy can be expressed as $\frac{1}{2}mgL\theta^2 = \frac{1}{2}m\omega^2 s^2$. (*Hint:* In part (b), approximate $\cos\theta$ by $\cos\theta \approx 1 - \theta^2/2$.)

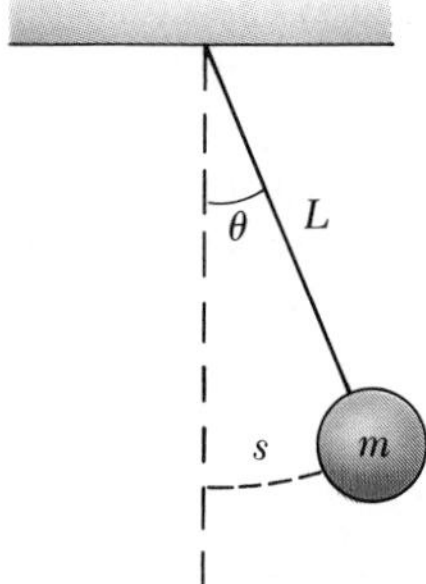

Figure 13.20 (Problem 47).

48. A horizontal platform vibrates with simple harmonic motion in the horizontal direction with a period of 2 s. A body on the platform starts to slide when the amplitude of vibration reaches 0.3 m. Find the coefficient of static friction between the body and the platform.
49. A particle of mass m slides inside a hemispherical bowl of radius R. Show that for small displacements from equilibrium, the particle exhibits simple harmonic motion with an angular frequency equal to that of a simple pendulum of length R. That is, $\omega = \sqrt{g/R}$.
50. A horizontal plank of mass m and length L is pivoted at one end, and the opposite end is attached to a spring of force constant k (Fig. 13.21). The moment of inertia of the plank about the pivot is $\frac{1}{3}mL^2$. If the plank is

displaced a *small* angle θ from the horizontal and released, show that it will move with simple harmonic motion with an angular frequency given by $\omega = \sqrt{3k/m}$.

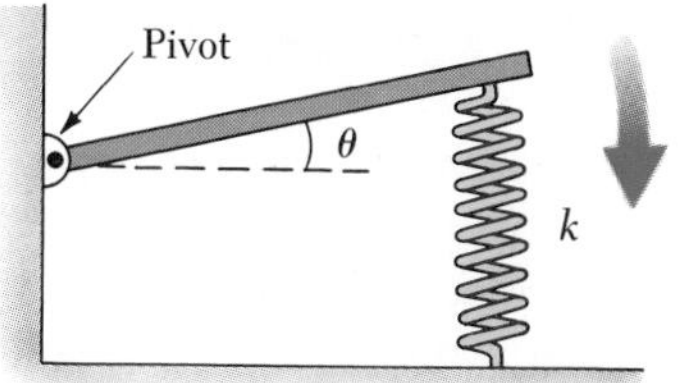

Figure 13.21 (Problem 50).

51. A mass M is attached to the end of a uniform rod of mass M and length L, which is pivoted at the top (Fig. 13.22). (a) Determine the tensions in the rod at the pivot and at the point P when the system is stationary. (b) Calculate the period of oscillation for small displacements from equilibrium, and determine this period for $L = 2$ m. (*Hint:* Assume the mass at the end of the rod is a point mass, and make use of Eq. 13.26.)

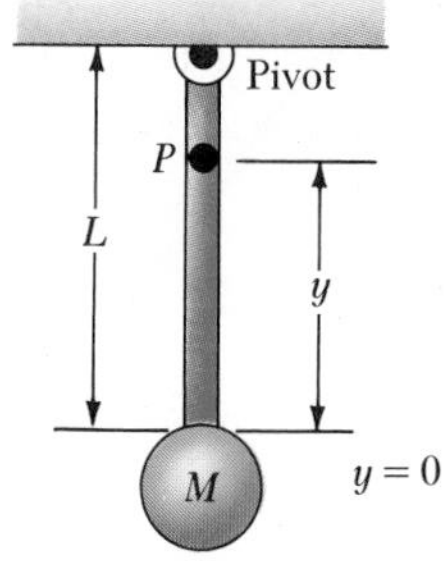

Figure 13.22 (Problem 51).

52. A mass M is connected to a spring of mass m and oscillates in simple harmonic motion on a horizontal, smooth surface (Fig. 13.23). The force constant of the spring is k and the equilibrium length is ℓ. Find (a) the kinetic energy of the system when the mass has a speed v and (b) the period of oscillation. (*Hint:* Assume that all portions of the spring oscillate in phase and that the velocity of a segment dx is proportional to the distance from the fixed end; that is, $v_x = \frac{x}{\ell} v$. Also, note that the mass of a segment of the spring is $dm = \frac{m}{\ell} dx$.)

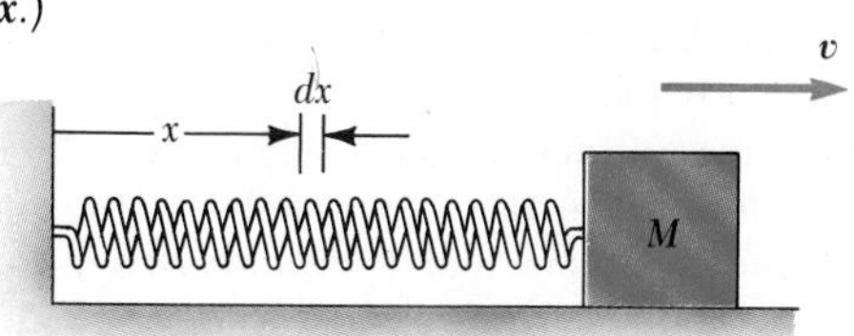

Figure 13.23 (Problem 52).

53. A small thin disk of radius r and mass m is attached rigidly to the face of a second thin disk of radius R and mass M as shown in Figure 13.24. The center of the small disk is located at the edge of the large disk. The large disk is mounted at its center on a frictionless axle. The assembly is rotated through an angle θ from its equilibrium position and released. (a) Show that the speed of the center of the small disk as it passes through the equilibrium position is

$$v = 2\left[\frac{Rg(1 - \cos\theta)}{(M/m) + (r/R)^2 + 2}\right]^{1/2}$$

(b) Show that the period of the motion is

$$T = 2\pi\left[\frac{(M + 2m)R^2 + mr^2}{2mgR}\right]^{1/2}$$

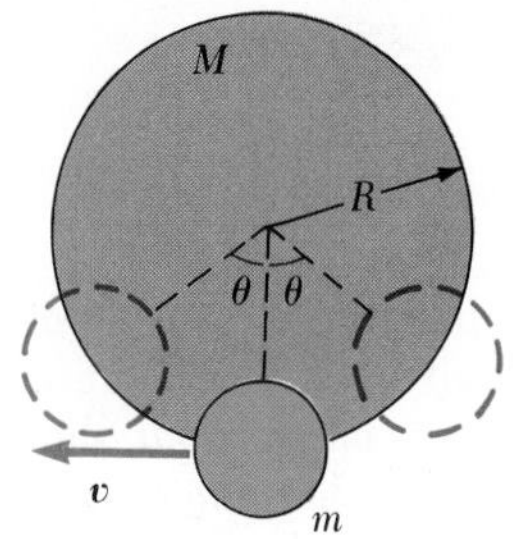

Figure 13.24 (Problem 53).

54. A mass m is connected to two springs of force constants k_1 and k_2 as in Figures 13.25a and 13.25b. In each case, the mass moves on a frictionless table and is displaced from equilibrium and released. Show that in each case the mass exhibits simple harmonic motion with periods (a) $T = 2\pi\sqrt{\frac{m(k_1 + k_2)}{k_1 k_2}}$ and (b) $T = 2\pi\sqrt{\frac{m}{k_1 + k_2}}$

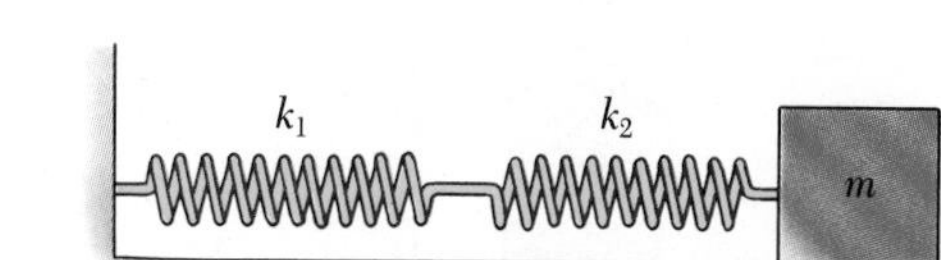

(a)

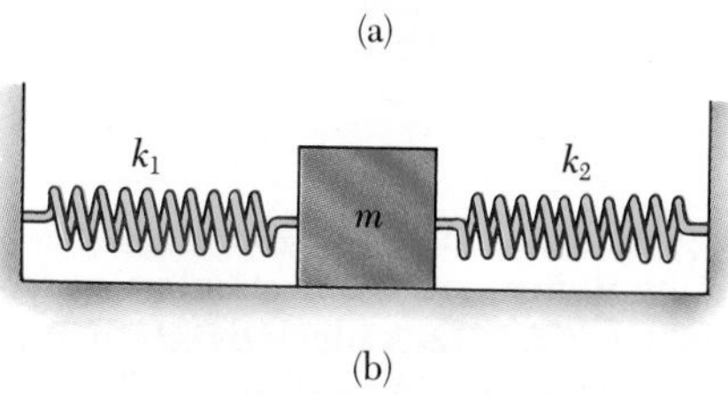

(b)

Figure 13.25 (Problem 54).

55. A pendulum of length L and mass M has a spring of force constant k connected to it at a distance h below its point of suspension (Fig. 13.26). Find the frequency of vibration of the system for small values of the amplitude (small θ). (Assume the vertical suspension of length L is rigid, but neglect its mass.)

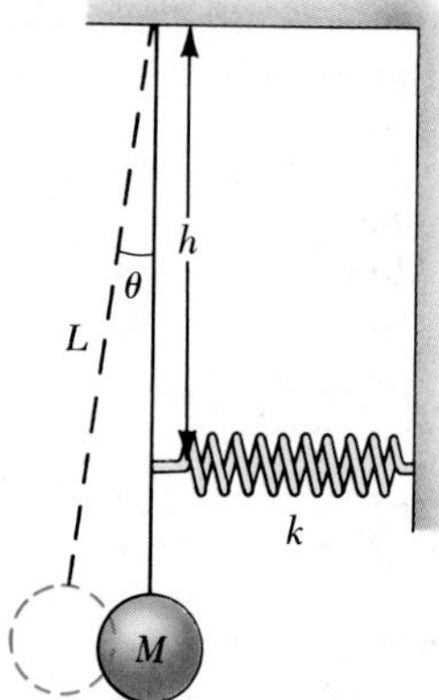

Figure 13.26 (Problem 55).

56. A mass m is attached to a spring of force constant k hanging vertically (Fig. 13.27). If the mass is released when the spring is *unstretched*, show that (a) the system exhibits simple harmonic motion with displacement measured from the unstretched position given by $y = \frac{mg}{k}(1 - \cos \omega t)$, where $\omega = \sqrt{k/m}$, and (b) the maximum tension in the spring is $2mg$.

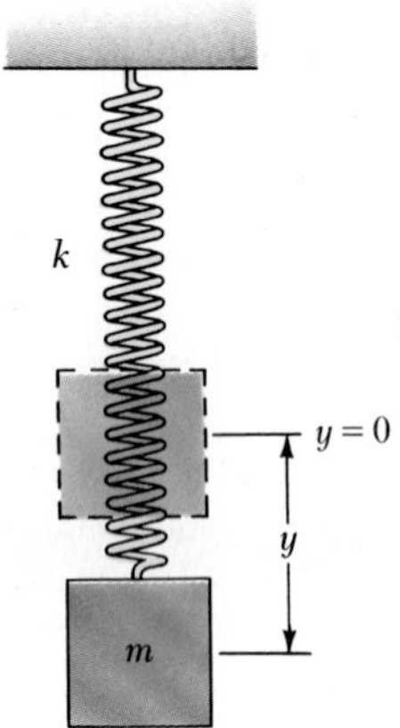

Figure 13.27 (Problem 56).

57. A flat plate P executes horizontal simple harmonic motion by sliding across a frictionless surface with a frequency $f = 1.5$ Hz. A block B rests on the plate, as shown in Figure 13.28, and the coefficient of static friction between the block and the plate is $\mu_s = 0.60$. What maximum amplitude of oscillation can the plate-block system have if the block is not to slip on the plate?

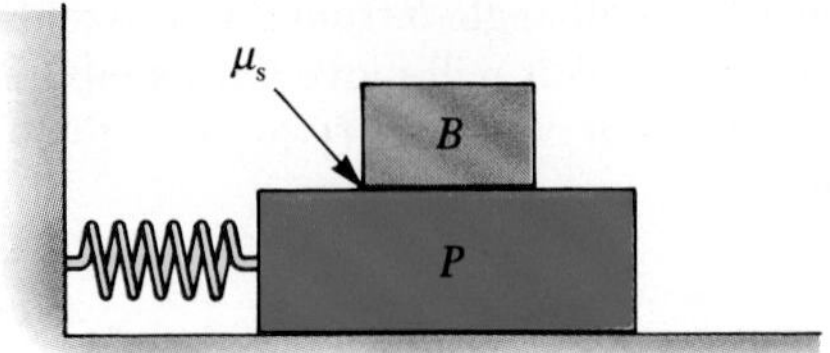

Figure 13.28 (Problem 57).

58. Consider a slender rod with mass $M = 4$ kg and length $\ell = 1.2$ m that is pivoted on a frictionless horizontal bearing at a point $\ell/4$ from one end, as shown in Figure 13.29. (a) Derive (from the definition) the expression for the moment of inertia of the rod about the pivot. (b) Obtain an equation that gives the angular acceleration α of the rod as a function of θ. (c) Determine the period of small amplitude oscillations about the vertical position.

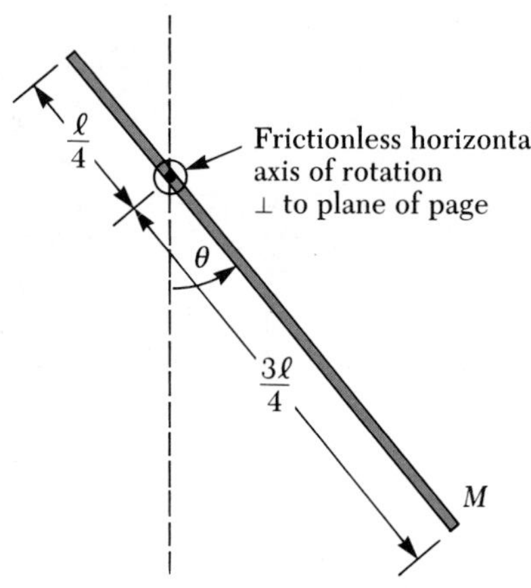

Figure 13.29 (Problem 58).

59. A spring stretches by 3.0 cm from its relaxed length when a force of 7.5 N is applied. A particle with a mass of 0.50 kg is attached to the free end of the spring, which is then compressed horizontally by 5.0 cm from its relaxed length and released from rest at $t = 0$. (a) What is the force constant of the spring, k? (b) Find the period, the frequency, and the angular frequency of the oscillations. (c) Find the amplitude A and the phase constant δ of the motion. (d) Find the maximum velocity v_{max} and the maximum acceleration a_{max}. (e) Find the total energy E of the system. (f) Determine the displacement x, the velocity v, and the acceleration a when $\omega_0 t = \pi/8$.

60. A 50-g mass attached to a spring moves on a horizontal, frictionless surface in simple harmonic motion. Its amplitude is 16 cm, and its period is 4 s. At $t = 0$, the mass is released from rest at $x = 16$ cm, as in Figure 13.5. Find (a) the displacement as a function of time and its value at $t = 0.5$ s, (b) the magnitude and direc-

tion of the force acting on the mass at $t = 0.5$ s, (c) the minimum time required for the mass to reach the position $x = 8$ cm, (d) the velocity at any time t and the speed at $x = 8$ cm, and (e) the total mechanical energy and force constant of the spring.

61. The mass of the deuterium molecule (D_2) is twice that of the hydrogen molecule (H_2). If the vibrational frequency of H_2 is 1.3×10^{14} Hz, what is the vibrational frequency of D_2, assuming that the "spring constant" of attracting forces is the same?

62. Show that if a torsional pendulum is twisted through an angle θ and then held the potential energy is $U = \frac{1}{2}\kappa\theta^2$.

63. A block with a mass of 2 kg hangs without vibrating at the end of a spring ($k = 500$ N/m) that is attached to the ceiling of an elevator car. The car is rising with an upward acceleration of $\frac{1}{3}g$ when the acceleration suddenly ceases (at $t = 0$). (a) What is the angular frequency of oscillation of the block after the acceleration ceases? (b) By what amount is the spring stretched during the time that the elevator car is accelerating? (c) What is the amplitude of the oscillation and the initial phase angle observed by a rider in the car? Take the upward direction to be positive.

64. A spherical mass m of radius R is suspended from a massless rod of length $L - R$ (Fig. 13.30). (a) Determine the moment of inertia for this physical pendulum about the point O using the parallel-axis theorem. (b) Calculate the period for small displacements from equilibrium. (c) If $R \ll L$, show that the period is that of a simple pendulum.

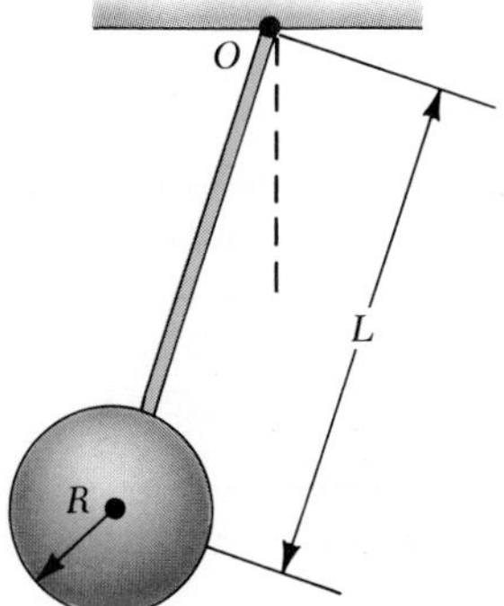

Figure 13.30 (Problem 64).

65. A mass m is connected to two rubber bands of length L, each under tension T, as in Figure 13.31. The mass is displaced by a *small* distance y vertically. Assuming the tension does not change appreciably, show that (a) the restoring force is $-(2T/L)y$ and (b) the system exhibits simple harmonic motion with an angular frequency given by $\omega = \sqrt{2T/mL}$.

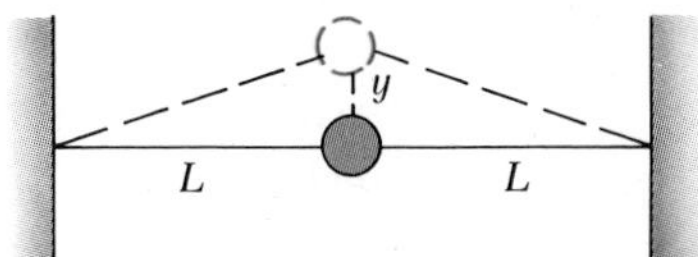

Figure 13.31 (Problem 65).

66. A light cubical container of volume a^3 is initially filled with a fluid of mass density ρ. The cube is initially supported by a light string to form a pendulum of length L_0 measured from the center of mass of the filled container. The fluid is allowed to flow from the bottom of the cube at a constant rate (dM/dt). At any time t the level of the fluid in the container is ℓ and the length of the pendulum is L (measured relative to the instantaneous center of mass). (a) Sketch the apparatus and label the dimensions a, ℓ, L_0, and L. (b) Find the time rate of change of the period as a function of time t. (c) Find the period T as a function of time.

67. What is the period of oscillation of the water in the tube shown in Figure 13.32 if it is initially displaced 6 cm? The total length of fluid in the tube is 60 cm.

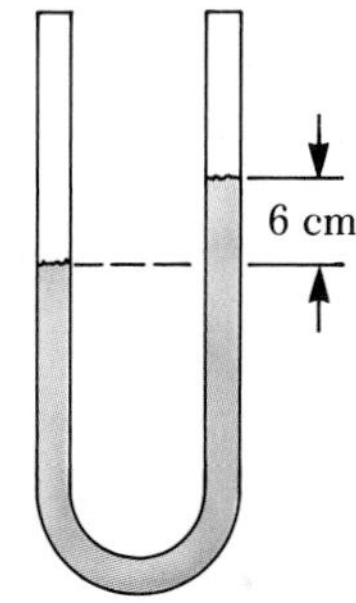

Figure 13.32 (Problem 67).

CALCULATOR/COMPUTER PROBLEMS

68. Using Equations 13.18 and 13.19, plot (a) the kinetic energy versus time and (b) the potential energy versus time for a simple harmonic oscillator. For convenience, take $\delta = 0$. What features do these graphs illustrate?

69. An object attached to the end of a spring vibrates with an amplitude of 20 cm. Find the position of the object at these times: 0, $T/8$, $T/4$, $3T/8$, $T/2$, $5T/8$, $3T/4$, $7T/8$, and T, where T is the period of vibration. Plot your results (position along the vertical axis and time along the horizontal axis).

70. A body oscillates with simple harmonic motion according to the equation $x = (-7 \text{ cm}) \cos (2\pi t)$. (a) Determine the velocity and acceleration as functions of time. (b) Make a table of x, v, and a versus t for the interval $t = 0$ to $t = 1$ s in steps of 0.1 s. (c) Plot x, v, and a versus time for this interval.

ESSAY

Galloping Gertie: The Tacoma Narrows Bridge Collapse

Robert G. Fuller
University of Nebraska–Lincoln

Dean A. Zollman
Kansas State University

The Tacoma Narrows Bridge was not the first suspension bridge to collapse. In fact, a survey of the history of suspension bridges shows that several were destroyed by wind or other oscillating forces (Table 1).

However, the Tacoma Narrows Bridge was by far the longest and most expensive suspension bridge to collapse as a result of interaction with the wind. Perhaps this collapse seemed so striking because nearly 50 years had elapsed since the previous collapse of a bridge.

At the time of the Tacoma Narrows Bridge collapse in 1940, many theories were advanced to explain what had happened. What follows are excerpts from six different explanations. Each presents a slightly different view of the role of design and wind factors in the collapse.

Why Did The Tacoma Narrows Bridge Collapse? (Six Theories)

It is very improbable that resonance with alternating vortices plays an important role in the oscillations of suspension bridges. First, it was found that there is no sharp correlation between wind velocity and oscillation frequency such as is required in case of resonance with vortices whose frequency depends on the wind velocity. Secondly, there is no evidence for the formation of alternating vortices at a cross section similar to that used in the Tacoma Bridge, at least as long as the structure is not oscillating. It seems that it is more correct to say that the vortex formation and frequency is determined by the oscillation of the structure than that the oscillatory motion is induced by the vortex formation. *A Report to the Honorable John M. Carmody, Administrator, Federal Works Agency, Washington, D.C. March 28, 1941.*

The primary cause of the collapse lies in the general proportions of the bridge and the type of stiffening girders and floor. The ratio of the width of the bridge to the length of the main span was so much smaller and the vertical stiffness was so much less than those of previously constructed bridges that forces heretofore not considered became dominant. *Board of Investigation, Tacoma Narrows Bridge, L.J. Sverdrup, Chairman, June 26, 1941.*

Once any small undulation of the bridge is started, the resultant effect of a wind tends to cause a building up of vertical undulations. There is a tendency for the undulations to change to a twisting motion, until the torsional oscillations reach destructive proportions. **Bridges and Their Builders,** *D. Steinman and S. Watson, Putnam's Sons, N.Y., 1941.*

The experimental results described in a (1942) report indicated rather definitely that the motions were a result of vortex shedding. *University of Washington Engineering Experiment Station Bulletin No. 116, 1952.*

TABLE 1 Collapsed Suspension Bridges

Bridge	Designer	Span Length (ft)	Failure Date
Dryburgh (Scotland)	J. and W. Smith	260	1818
Union (England)	Sir Samuel Brown	449	1821
Nassau (Germany)	L. and W.	245	1834
Brighton (England)	Sir Samuel Brown	255	1836
Montrose (Scotland)	Sir Samuel Brown	432	1838
Menai (Wales)	T. Telford	580	1839
Roche (France)	LeBlanc	641	1852
Wheeling (USA)	C. Ellet	1010	1854
Niagara (USA)	E. Serrell	1041	1864
Niagara (USA)	S. Keefer	1260	1889
Tacoma Narrows	L. Moisseff	2800	1940

Summing up the whole bizarre accident, Galloping Gertie tore itself to pieces, because of two characteristics: 1. It was a long, narrow, shallow, and therefore very flexible structure standing in a wind ridden valley; 2. Its stiffening support was a solid girder, which, combined with a solid floor, produced a cross section peculiarly vulnerable to aerodynamic effects. **Bridges and Men,** *J. Gies, Doubleday and Co., 1963.*

Aerodynamic instability was responsible for the failure of the Tacoma Narrows Bridge in 1940. The magnitude of the oscillations depends on the structure shape, natural frequency, a, natural frequency, and damping. The oscillations are caused by the periodic shedding of vortices on the leeward side of the stucture, a vortex being shed first from the upper section and then the lower section. **Wind Forces on Buildings and Structures,** *E. Houghton and N. Carruthers, J. Wiley & Sons, N.Y. 1976.*

Physical Principles

The general principle of the Tacoma Narrows Bridge collapse is straightforward: a resonance effect. However, the details of the physical mechanisms involved are neither trivial nor obvious.

Every system has a natural fundamental vibration frequency. If forces are exerted on that system at the right frequency and phase, sympathetic vibrations can be excited (see Section 13.7). Oscillating forces at the right frequency and phase can cause sympathetic vibrations of catastrophic proportions. The forces applied to the bridge by the wind were applied at a natural frequency of the bridge. Thus, the amplitude of the bridge's oscillations increased until the steel and concrete could no longer stand the stress.

But how does the fluctuating force of just the right frequency arise as the wind blows across the bridge? The first idea that comes to mind is that the gusty wind arrived in pulses, thereby striking the bridge at just the appropriate frequency to cause the large oscillations. Closer examination of this explanation shows it cannot be correct. While all wind speeds fluctuate, these fluctuations tend to be random in phase and variable in frequency. Wind gusts are not the answer. Furthermore, the kinds of forces that must be exerted on the bridge are vertical forces—transverse to the direction of the wind. The wind was blowing across the bridge (from side to side, as shown in Figure 1), while the forces on the bridge were acting vertically. These oscillating vertical forces can be explained by a concept called **vortex shedding.** When a wind that exceeds a minimum speed blows around any object, vortices will be formed on the back side of that object (see Figure 2).

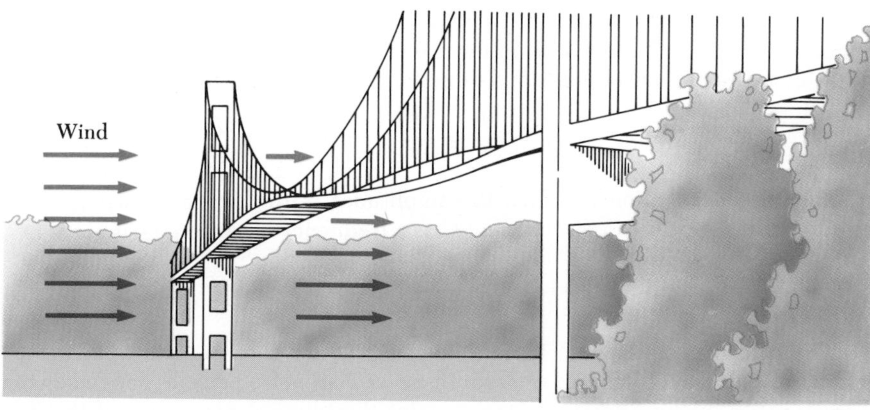

Figure 1 Direction of wind blowing across the bridge.

(Continued on next page)

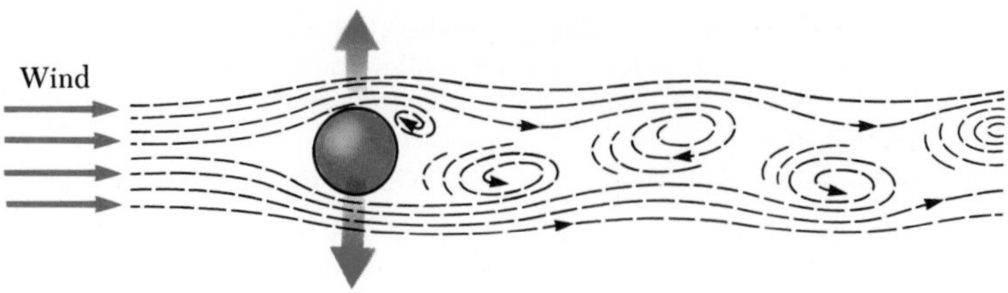

Figure 2 Vortex shedding.

As the wind increases in speed, the vortices form on alternate sides of the downwind side of the object, break loose, and flow downstream. At the time a vortex breaks loose from the back side of the object, a transverse force is exerted on the object. The frequency of these fluctuating eddies is about 20% of the ratio of the velocity of the wind to the width of the object. These lateral forces can be as much as twice as large as the drag forces. Thus, vortex shedding allows us to understand the origin of the fluctuating vertical forces on the Tacoma Narrows Bridge even though the wind was blowing across it in a transverse, horizontal direction.

Model of the Tacoma Narrows Bridge

Let us create a model of the Tacoma Narrows Bridge and treat it as if it were suspended by two springs with equal force constants, as shown in a cross-sectional view in Figure 3.

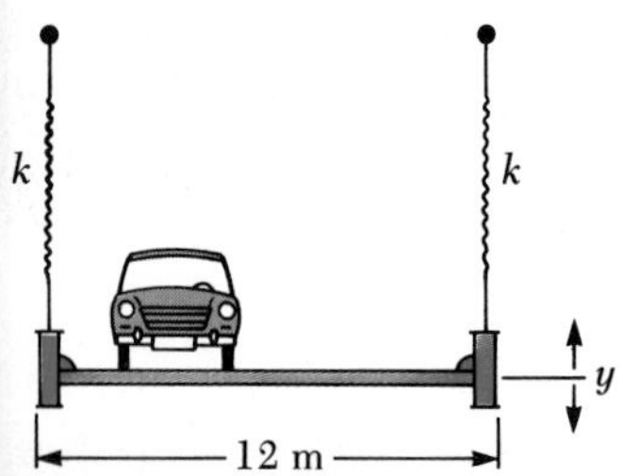

Figure 3 Cross-sectional view of the bridge.

We can then write down the equations of motion using Newton's second law for translation (Eq. 1) and for rotation (Eq. 2).

$$Ma_y = -k\,(y_1 + y_2) \tag{1}$$

$$I\alpha = -\frac{kW}{2}\,(y_2 - y_1) \tag{2}$$

We then make a small angle approximation (Eq. 3)

$$\theta = \frac{y_2 - y_1}{W} \tag{3}$$

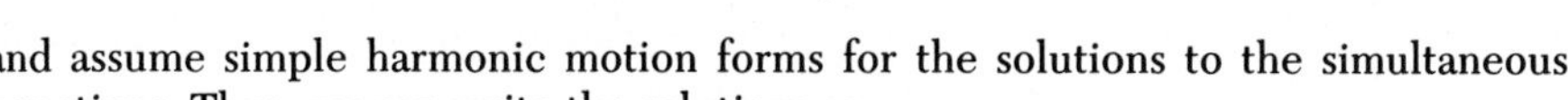

and assume simple harmonic motion forms for the solutions to the simultaneous equations. Thus, we can write the solutions as:

$$y_1 = A_1 \sin(\omega t) \quad \text{and} \quad y_2 = A_2 \sin(\omega t.)$$

We can write down in a standard way the two normal mode solutions for this cross section of the bridge. The *vertical* motion in which the amplitudes of oscillation of the two sides are equal in magnitude and direction has a frequency, ω_v,

$$\omega_1{}^2 = \omega_2{}^2 = \omega_v{}^2 = \frac{2k}{M} \qquad (\text{for } A_1 = A_2) \tag{4}$$

The *torsional* motion in which the amplitudes of the two sides are equal in magnitude but opposite in direction and has a frequency ω_t, where ω_t is given by

$$\omega_t{}^2 = \omega_1{}^2 = \omega_2{}^2 = \frac{kW^2}{2MR^2} \qquad (\text{for } A_1 = -A_2) \tag{5}$$

The latter frequency describes the twisting motion that ultimately caused the bridge to fall down. The exact values for these oscillation frequencies depend on the characteristics of the bridge. On the basis of the physical properties of the first Tacoma Narrows Bridge, we find that the values appropriate for this analysis are:

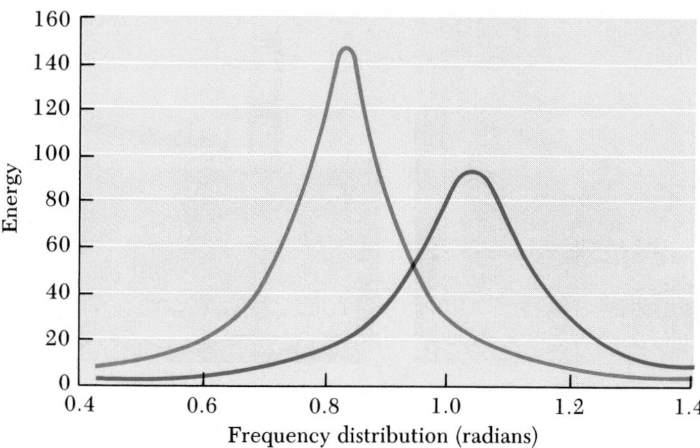

Figure 4 Typical values for $\Delta\omega$.

mass per unit length $= 4.3 \times 10^3$ kg/m, width of the bridge $= 12$ m, radius of gyration of the bridge $= 4.8$ m, effective spring constant $= 1.5 \times 10^3$ N/m. These numerical values result in the vertical normal mode frequency of 8 cycles per minute and the torsional motion of 10 cycles per minute. The approximate equality of these two frequencies played an important role in the fate of the Tacoma Narrows Bridge.

In a real system the normal mode frequency will not be a single frequency, but rather a distribution of frequencies. The energy per unit time that is accepted by a mode of oscillation is given by the following equation:

$$P(\omega) = \frac{1}{(\omega - \omega_0)^2 + \left(\dfrac{\Delta\omega}{2}\right)^2} \qquad (6)$$

where $\Delta\omega$ is the width of the resonance response curve at half maximum. The maximum increases with the tendency of the system to resist oscillations. For the original Tacoma Narrows Bridge, the tendencies of the bridge to resist vertical and rotational, or torsional, motions were different. Hence, the $P(\omega)$ function for vertical and torsional motion has different values for ω_0, the normal mode frequency and for $\Delta\omega$. Using the constants given above, the values for ω_v, and ω_t, for vertical and torsional motion can be computed. Graphs of $P(\omega)_v$ and $P(\omega)_t$ versus ω using typical values for $\Delta\omega$ are shown in Figure 4. The area of overlap of the two curves indicates the tendency of the vertical motion to pump energy into the rotational motion.

As can be seen in Table 2, the ratio of torsional to vertical frequencies for other long bridges is significantly larger than the ratio for the first Tacoma Narrows Bridge.

TABLE 2 Ratio of Torsional to Vertical Frequencies in Suspension Bridges

Bridge	Length (m)	f_v(min^{-1})	f_t(min^{-1})	Ratio f_t/f_v
Verrazano	4260	6.2	11.9	1.9
Golden Gate	4200	5.6	11.0	1.9
Severn	3240	7.7	30.6	3.9
Tacoma Narrows (1st)	2800	8.0	10.0	1.25

(Continued on next page)

(a)

(b)

Figure 5 (a) High winds set up vibrations in the bridge, causing it to oscillate at a frequency near to one of the natural frequencies of the bridge structure. (b) Once established, this resonance condition led to the bridge's collapse. (UPI/Bettmann Newsphotos)

Even before the Tacoma Narrows Bridge was opened, the vortex shedding forces were pumping energy into the vertical motion of the bridge. Vertical oscillations were noticed early, and many people avoided using the bridge. However, the torsional oscillations did not occur until the day of the collapse. On that day a mechanical failure allowed the torsional oscillations to begin. Because this motion was closely coupled to the vertical motion of the bridge, it quickly led to its destruction. The photographs (Fig. 5) show the collapse of the Tacoma Narrows Bridge and provide a vivid demonstration of mechanical resonance.

Public Perceptions

It is interesting to recall some public perceptions involving the bridge collapse. Locals had already noticed for some time that, when driving across this technological marvel, they would experience an undeniable bounce. Some tried to deal with this humorously. For example, the bridge was often referred to as "Galloping Gertie."

An editorial in the *Tacoma Times* dated August 25, 1940 (soon after the bridge was opened) made the following comment regarding the tolls being charged:

> There is no truth to the rumor that part of the Narrows Bridge toll is for the scenic railway effects. The charge is for cross only and the bounce is free.

Before the Tacoma Bridge collapsed, bridges had been considered secure, so much so that a local insurance agent who had arranged a second $800,000 auxiliary policy on the bridge had never bothered to pay the premium. Instead, he pocketed the money and was sent to jail following the disaster.

The collapse of the Tacoma Narrows Bridge was a watershed in the design of suspension bridges. The debates as to who was responsible and whether anything could have been done to prevent the collapse continue.

Suggested Readings

Aerodynamic Stability of Suspension Bridges, University of Washington, Engineering Experiment Station, Bulletin No. 116, Parts I, II, III, IV, and V, University of Washington Press, Seattle, 1949, 1950, 1952, and 1954.

O'Connor, C., *Design of Bridge Superstructures,* New York, John Wiley & Sons, 1971.

The Failure of the Tacoma Narrows Bridge. A Reprint of Original Reports, School of Engineering, Texas Engineering Experiment Station, College Station, Texas, Bulletin No. 78, 1944.

Fuller, Robert G., Dean Zollman, and Thomas C. Campbell, *The Puzzle of the Tacoma Narrows Bridge Collapse,* New York, John Wiley & Sons, 1982.

Houghton, E.L., and N.B. Carruthers, *Wind Forces on Buildings and Structures: An Introduction,* New York, Halsted Press, 1976.

Scigliano, Eric, "Galloping Gertie," *Pacific Northwest,* January 1989.

Simiu, E., and R.H. Scanlan, *Wind Effects on Structures: An Introduction to Wind Engineering,* New York, John Wiley & Sons, 1978.

Steinman, D.B., "Suspension Bridges: The Aerodynamic Problem and its Solution," *American Scientist,* July 1954, pp. 397–438.

Wind Effects on Bridges and Other Flexible Structures, Notes on Applied Science, No. 1, National Physics Laboratory, London, 1955.

Essay Questions

1. The Tacoma Narrows Bridge was a two-lane bridge. How would the bridge have behaved had it been a four-lane bridge?
2. Based on current knowledge, what would have been your advice for a "quick fix" for the bridge which could possibly have averted the disaster?

Essay Problems

1. Derive the frequency relations of Equations (4) and (5) by solving Equations (1), (2), and (3). Assume that the solutions are simple harmonic vibrations.
2. Calculate the effect of stiffening the bridge suspension by increasing the effective spring constant of the Tacoma Narrows Bridge by 50%.
3. How much change in the *total* mass of the Tacoma Narrows Bridge is necessary to bring the vibrational frequency down to that of the Golden Gate Bridge? Assume uniform mass distribution.

14

The Law of Universal Gravitation

This image of Saturn, taken by Voyager I *spacecraft on October 18, 1980, was color-enhanced to increase the visibility of large, bright features in Saturn's North Temperate Belt. The largest violet-colored cloud belt (its true color is brownish) is Saturn's North Equatorial Belt. The distinct color difference between this and other belts and zones may be due to a thicker haze layer covering the northern portion of the belt. Three separate* Voyager I *images taken through ultraviolet, green, and violet filters were used to construct this blue, green, and red color composite. (Courtesy of NASA)*

Prior to 1686, a great mass of data had been collected on the motions of the moon and the planets but a clear understanding of the forces that caused these celestial bodies to move the way they did was not available. In that year, however, Isaac Newton provided the key that unlocked the secrets of the heavens. He knew, from the first law, that a net force had to be acting on the moon. If not, it would move in a straight-line path rather than in its almost circular orbit. Newton reasoned that this force arose as a result of a gravitational attraction that the earth exerted on the moon. He also concluded that there could be nothing special about the earth-moon system or the sun and its planets that would cause gravitational forces to act on them alone. In other words, he saw that the same force of attraction that causes the moon to follow its path also causes an apple to fall to earth from a tree. He wrote, "I deduced that the forces which keep the planets in their orbs must be reciprocally as the squares of their distances from the centers about which they revolve; and thereby compared the force requisite to keep the Moon in her orb with force of gravity at the surface of the Earth; and found them answer pretty nearly."

In this chapter we shall study the law of universal gravitation. Emphasis will be placed on describing the motion of the planets, since astronomical data provide an important test of the validity of the law of universal gravitation. We shall show that the laws of planetary motion developed by Johannes Kepler (1571–1630) follow from the law of universal gravitation and the concept of the conservation of angular momentum. A general expression for the gravitational potential energy will be derived, and the energetics of planetary and satellite motion will be treated. The law of universal gravitation will also be used to determine the force between a particle and an extended body.

14.1 NEWTON'S UNIVERSAL LAW OF GRAVITY

It has been said that Newton was struck on the head by a falling apple while napping under a tree (or some variation of this legend). This supposedly prompted Newton to imagine that perhaps all bodies in the universe are attracted to each other in the same way the apple was attracted to the earth. Newton proceeded to analyze astronomical data on the motion of the moon around the earth. From the analysis of such data, Newton made the bold statement that the law of force governing the motion of planets has the *same* mathematical form as the force law that attracts a falling apple to the earth.

In 1687 Newton published his work on the universal law of gravity in his *Mathematical Principles of Natural Philosophy.* **Newton's law of gravitation** states that

> every particle in the universe attracts every other particle with a force that is directly proportional to the product of their masses and inversely proportional to the square of the distance between them.

If the particles have masses m_1 and m_2 and are separated by a distance r, the magnitude of this gravitational force is

$$F = G\frac{m_1 m_2}{r^2} \qquad (14.1)$$

Universal law of gravity

where G is a universal constant called the *gravitational constant,* which has been measured experimentally. Its value in SI units is

$$G = 6.672 \times 10^{-11}\,\frac{\text{N}\cdot\text{m}^2}{\text{kg}^2} \qquad (14.2)$$

The force law given by Equation 14.1 is often referred to as an **inverse-square law,** since the magnitude of the force varies as the inverse square of the separation of the particles. We can express this force in vector form by defining a unit vector $\hat{\boldsymbol{r}}_{12}$ (Fig. 14.1). Because this unit vector is in the direction of the displacement vector $\boldsymbol{r}_{12}$ directed from m_1 to m_2, the force on m_2 due to m_1 is given by

$$\boldsymbol{F}_{21} = -G\frac{m_1 m_2}{r_{12}^{\;2}}\,\hat{\boldsymbol{r}}_{12} \qquad (14.3)$$

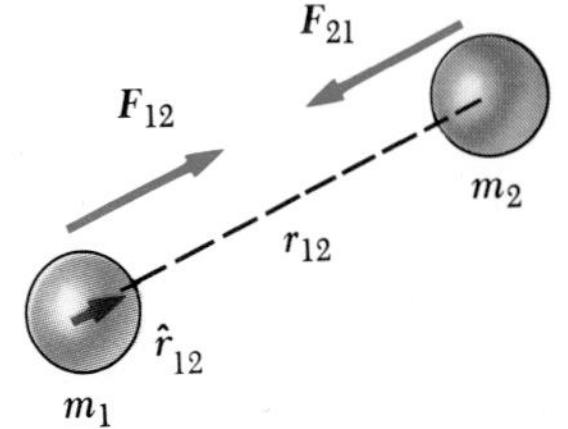

Figure 14.1 The gravitational force between two particles is attractive. The unit vector $\hat{\boldsymbol{r}}_{12}$ is directed from m_1 to m_2. Note that $\boldsymbol{F}_{12} = -\boldsymbol{F}_{21}$.

The minus sign in Equation 14.3 indicates that m_2 is attracted to m_1, and so the force must be directed toward m_1. Likewise, by Newton's third law the force on m_1 due to m_2, designated $\boldsymbol{F}_{12}$, is equal in magnitude to $\boldsymbol{F}_{21}$ and in the opposite direction. That is, these forces form an action-reaction pair, and $\boldsymbol{F}_{12} = -\boldsymbol{F}_{21}$.

Properties of the gravitational force

There are several features of the inverse-square law that deserve some attention. The gravitational force acts as an action-at-a-distance force, which always exists between two particles, regardless of the medium that separates them. The force varies as the inverse square of the distance between the particles and therefore decreases rapidly with increasing separation. Finally, the gravitational force is proportional to the mass of each particle.

Another important fact is that *the gravitational force exerted by a finite-size, spherically symmetric mass distribution on a particle outside the sphere is the same as if the entire mass of the sphere were concentrated at its center.* For example, the force on a particle of mass m at the earth's surface has the magnitude

$$F = G\frac{M_e m}{R_e^{\,2}}$$

where M_e is the earth's mass and R_e is the earth's radius. This force is directed toward the center of the earth.

14.2 MEASUREMENT OF THE GRAVITATIONAL CONSTANT

The gravitational constant, G, was first measured in an important experiment by Sir Henry Cavendish in 1798. The Cavendish apparatus consists of two small spheres each of mass m fixed to the ends of a light horizontal rod suspended by a fine fiber or thin metal wire, as in Figure 14.2. Two large spheres each of mass M are then placed near the smaller spheres. The attractive force between the smaller and larger spheres causes the rod to rotate and twist the wire suspension. If the system is oriented as shown in Figure 14.2, the rod rotates clockwise when viewed from the top. The angle through which the suspended rod rotates is measured by the deflection of a light beam reflected

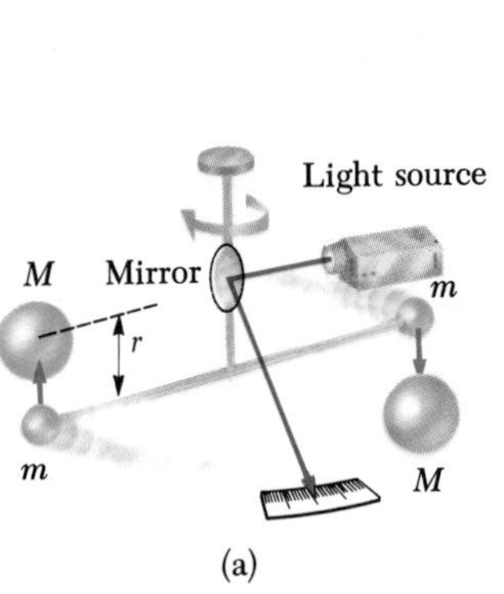

(a)

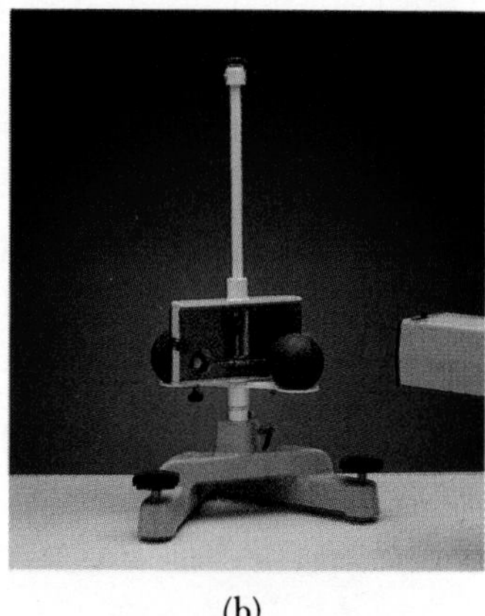

(b)

Figure 14.2 (a) Schematic diagram of the Cavendish apparatus for measuring G. The smaller spheres of mass m are attracted to the large spheres of mass M, and the bar rotates through a small angle. A light beam reflected from a mirror on the rotating apparatus measures the angle of rotation. (b) Photograph of a student Cavendish apparatus. (Courtesy of PASCO Scientific) from a mirror attached to the vertical suspension. The deflected spot of light is an effective technique for amplifying the motion. The experiment is carefully repeated with different masses at various separations. In addition to providing a value for G, the results show that the force is attractive, proportional to the product mM, and inversely proportional to the square of the distance r.

EXAMPLE 14.1 Three Interacting Masses

Three uniform spheres of mass 2 kg, 4 kg, and 6 kg are placed at the corners of a right triangle as in Figure 14.3, where the coordinates are in m. Calculate the resultant gravitational force on the 4-kg mass, assuming the spheres are isolated from the rest of the universe.

Solution First we calculate the individual forces on the 4-kg mass due to the 2-kg and 6-kg masses separately, and then we find the vector sum to get the resultant force on the 4-kg mass.

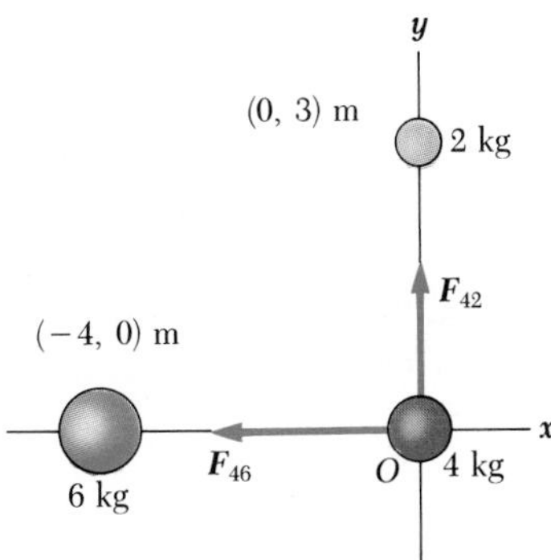

Figure 14.3 (Example 14.1) The *resultant* force on the 4-kg mass is the vector sum $\boldsymbol{F}_{46} + \boldsymbol{F}_{42}$.

The force on the 4-kg mass due to the 2-kg mass is upward and given by

$$\boldsymbol{F}_{42} = G\frac{m_4 m_2}{{r_{42}}^2}\boldsymbol{j}$$

$$= \left(6.67 \times 10^{-11}\,\frac{\text{N}\cdot\text{m}^2}{\text{kg}^2}\right)\frac{(4\text{ kg})(2\text{ kg})}{(3\text{ m})^2}\boldsymbol{j}$$

$$= 5.93 \times 10^{-11}\boldsymbol{j}\text{ N}$$

The force on the 4-kg mass due to the 6-kg mass is to the left and given by

$$\boldsymbol{F}_{46} = G\frac{m_4 m_6}{{r_{46}}^2}(-\boldsymbol{i})$$

$$= \left(-6.67 \times 10^{-11}\,\frac{\text{N}\cdot\text{m}^2}{\text{kg}^2}\right)\frac{(4\text{ kg})(6\text{ kg})}{(4\text{ m})^2}\boldsymbol{i}$$

$$= -10.0 \times 10^{-11}\boldsymbol{i}\text{ N}$$

Therefore, the resultant force on the 4-kg mass is the vector sum of $\boldsymbol{F}_{42}$ and $\boldsymbol{F}_{46}$:

$$\boldsymbol{F}_4 = \boldsymbol{F}_{42} + \boldsymbol{F}_{46} = (-10.0\boldsymbol{i} + 5.93\boldsymbol{j}) \times 10^{-11}\text{ N}$$

The magnitude of this force is 11.6×10^{-11} N, which is only 2.62×10^{-11} lb! The force makes an angle of 149° with the positive x axis.

14.3 WEIGHT AND GRAVITATIONAL FORCE

In Chapter 5 we defined the weight of a body of mass m as simply mg, where g is the magnitude of the acceleration due to gravity. Now, we are in a position to obtain a more fundamental description of g. Since the force on a freely falling body of mass m near the surface of the earth is given by Equation 14.1, we can equate mg to this expression to give

$$mg = G\frac{M_e m}{{R_e}^2}$$

$$g = G\frac{M_e}{{R_e}^2} \qquad (14.4)$$

Acceleration due to gravity

where M_e is the mass of the earth and R_e is the earth's radius. Using the facts that $g = 9.80\text{ m/s}^2$ at the earth's surface and the radius of the earth is approximately 6.38×10^6 m, we find from Equation 14.4 that $M_e = 5.98 \times 10^{24}$ kg.

From this result, the average density of the earth is calculated to be

$$\rho_e = \frac{M_e}{V_e} = \frac{M_e}{\frac{4}{3}\pi R_e^3} = \frac{5.98 \times 10^{24}\ \text{kg}}{\frac{4}{3}\pi(6.38 \times 10^6\ \text{m})^3} = 5.50 \times 10^3\ \text{kg/m}^3$$

Since this value is about twice the density of most rocks at the earth's surface, we conclude that the inner core of the earth has a much higher density.

Now consider a body of mass m a distance h above the earth's surface, or a distance r from the earth's center, where $r = R_e + h$. The magnitude of the gravitational force acting on this mass is given by

$$F = G\frac{M_e m}{r^2} = G\frac{M_e m}{(R_e + h)^2}$$

If the body is in free fall, then $F = mg'$ and we see that g', the acceleration of gravity at the altitude h, is given by

Variation of g with altitude

$$g' = \frac{GM_e}{r^2} = \frac{GM_e}{(R_e + h)^2} \qquad (14.5)$$

Thus, it follows that g' *decreases* with *increasing altitude*. Since the true weight of a body is mg', we see that as $r \to \infty$, the true weight approaches zero.

EXAMPLE 14.2 Variation of g with Altitude h □

Determine the magnitude of the acceleration of gravity at an altitude of 500 km. By what percentage is the weight of a body reduced at this altitude?

Solution Using Equation 14.5 with $h = 500$ km, $R_e = 6.38 \times 10^6$ m, and $M_e = 5.98 \times 10^{24}$ kg gives

$$g' = \frac{GM_e}{(R_e + h)^2}$$

$$= \frac{(6.67 \times 10^{-11}\ \text{N} \cdot \text{m}^2/\text{kg}^2)(5.98 \times 10^{24}\ \text{kg})}{(6.38 \times 10^6 + 0.5 \times 10^6)^2\ \text{m}^2}$$

$$= 8.43\ \text{m/s}^2$$

Since $g'/g = 8.43/9.8 = 0.86$, we conclude that the weight of a body is reduced by about 14% at an altitude of 500 km. Values of g' at other altitudes are listed in Table 14.1.

TABLE 14.1 Acceleration Due to Gravity, g', at Various Altitudes

Altitude h (km)[a]	g' (m/s^2)
1000	7.33
2000	5.68
3000	4.53
4000	3.70
5000	3.08
6000	2.60
7000	2.23
8000	1.93
9000	1.69
10 000	1.49
50 000	0.13
∞	0

[a] All values are distances above the earth's surface.

14.4 KEPLER'S LAWS

The movements of the planets, stars, and other celestial bodies have been observed by people for thousands of years. In early history, scientists regarded the earth as the center of the universe. This so-called geocentric model was elaborated and formalized by the Greek astronomer Claudius Ptolemy in the second century A.D. and was accepted for the next 1400 years. In 1543, the Polish astronomer Nicolaus Copernicus (1473–1543) suggested that the earth and the other planets revolve in circular orbits about the sun (the heliocentric hypothesis).

The Danish astronomer Tycho Brahe (1546–1601) made accurate astronomical measurements over a period of 20 years and provided the basis for the currently accepted model of the solar system. It is interesting to note that these precise observations, made on the planets and 777 stars visible to the naked eye, were carried out with a large sextant and compass because the telescope had not yet been invented.

Johannes Kepler (1571–1630), German astronomer. (The Bettmann Archive)

The German astronomer Johannes Kepler, who was Brahe's student, acquired Brahe's astronomical data and spent about 16 years trying to deduce a mathematical model for the motion of the planets. After many laborious calculations, he found that Brahe's precise data on the revolution of Mars about the sun provided the answer. Such data are difficult to sort out because the earth is also in motion about the sun. Kepler's analysis first showed that the concept of circular orbits about the sun had to be abandoned. He eventually discovered that the orbit of Mars could be accurately described by an ellipse with the sun at one focus. He then generalized this analysis to include the motion of all planets. The complete analysis is summarized in three statements, known as **Kepler's laws.** These empirical laws applied to the solar system are:

1. All planets move in elliptical orbits with the sun at one of the focal points.
2. The radius vector drawn from the sun to any planet sweeps out equal areas in equal time intervals.
3. The square of the orbital period of any planet is proportional to the cube of the semimajor axis of the elliptical orbit.

Kepler's laws

About 100 years later, Newton demonstrated that these laws were the consequence of a simple force that exists between any two masses. Newton's law of universal gravitation, together with his development of the laws of motion, provides the basis for a full mathematical solution to the motion of planets and satellites. More important, Newton's universal law of gravity correctly describes the gravitational attractive force between *any* two masses.

(Left) The earth as viewed from space. *(Right) Voyager I* took this photo of Jupiter and two of its satellites (Io, left, and Europa) on February 13, 1979. Io is about 350,000 km (220,000 miles) above Jupiter's Great Red Spot; Europa is about 600,000 km (375,000 miles) above Jupiter's clouds. (Courtesy of NASA)

14.5 THE LAW OF UNIVERSAL GRAVITATION AND THE MOTION OF PLANETS

In formulating his law of universal gravitation, Newton used the following observation, which suggests that the gravitational force is proportional to the inverse square of the separation. Let us compare the acceleration of the moon in its orbit with the acceleration of an object falling near the earth's surface, such as the legendary apple (Fig. 14.4). Assume that both accelerations have the same cause, namely, the gravitational attraction of the earth. From the inverse-square law, Newton found that the acceleration of the moon toward the earth (centripetal acceleration) should be proportional to $1/r_m^2$, where r_m is the earth-moon separation. Furthermore, the acceleration of the apple toward the earth should vary as $1/R_e^2$, where R_e is the radius of the earth. Using the values $r_m = 3.84 \times 10^8$ m and $R_e = 6.37 \times 10^6$ m, the ratio of the moon's acceleration, a_m, to the apple's acceleration, g, is predicted to be

$$\frac{a_m}{g} = \frac{(1/r_m)^2}{(1/R_e)^2} = \left(\frac{R_e}{r_m}\right)^2 = \left(\frac{6.37 \times 10^6 \text{ m}}{3.84 \times 10^8 \text{ m}}\right)^2 = 2.75 \times 10^{-4}$$

Therefore

The acceleration of the moon

$$a_m = (2.75 \times 10^{-4})(9.80 \text{ m/s}^2) = 2.70 \times 10^{-3} \text{ m/s}^2$$

The centripetal acceleration of the moon can also be calculated kinematically from a knowledge of its orbital period, T, where $T = 27.32$ days $= 2.36 \times 10^6$ s, and its mean distance from the earth, r_m. In a time T, the moon travels a distance $2\pi r_m$, which equals the circumference of its orbit. Therefore, its orbital speed is $2\pi r_m/T$, and its centripetal acceleration is

$$a_m = \frac{v^2}{r_m} = \frac{(2\pi r_m/T)^2}{r_m} = \frac{4\pi^2 r_m}{T^2} = \frac{4\pi^2(3.84 \times 10^8 \text{ m})}{(2.36 \times 10^6 \text{ s})^2} = 2.72 \times 10^{-3} \text{ m/s}^2$$

This agreement provides strong evidence that the inverse-square law of force is correct.

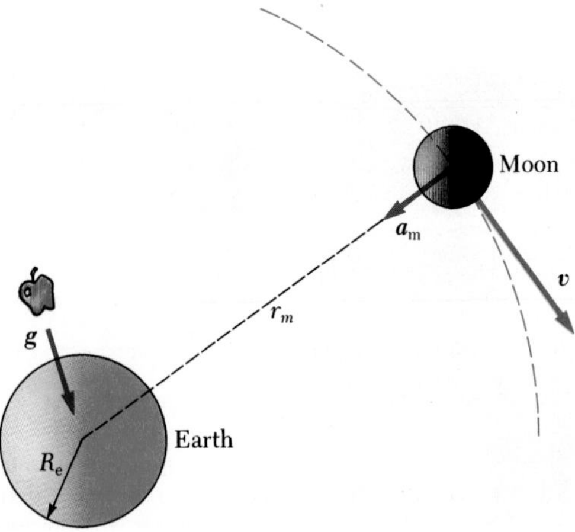

Figure 14.4 As the moon revolves about the earth, the moon experiences a centripetal acceleration a_m directed toward the earth. An object near the earth's surface experiences an acceleration equal to g. (Dimensions are not to scale.)

Although these results must have been very encouraging to Newton, he was deeply troubled by an assumption made in the analysis. In order to evaluate the acceleration of an object at the earth's surface, the earth was treated as if its mass were all concentrated at its center. That is, Newton assumed that the earth acts as a particle as far as its influence on an exterior object is concerned. Several years later, and based on his pioneering work in the development of the calculus, Newton proved this point. (The details of the derivation are given in Section 14.11.) For this reason, and because of Newton's inherent shyness, the publication of the theory of gravitation was delayed for about 20 years.

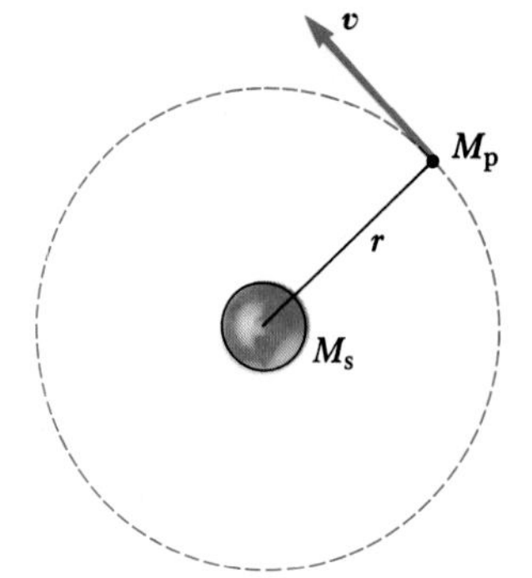

Figure 14.5 A planet of mass M_p moving in a circular orbit about the sun. The orbits of all planets except Mars, Mercury, and Pluto are nearly circular.

Kepler's Third Law

It is informative to show that Kepler's third law can be predicted from the inverse-square law for circular orbits.[1] Consider a planet of mass M_p which is assumed to be moving about the sun of mass M_s in a circular orbit, as in Figure 14.5. Since the gravitational force on the planet is equal to the centripetal force needed to keep it moving in a circle,

$$\frac{GM_sM_p}{r^2} = \frac{M_pv^2}{r}$$

But the orbital velocity of the planet is simply $2\pi r/T$, where T is its period; therefore the above expression becomes

$$\frac{GM_s}{r^2} = \frac{(2\pi r/T)^2}{r}$$

$$T^2 = \left(\frac{4\pi^2}{GM_s}\right)r^3 = K_sr^3 \qquad (14.6)$$

Kepler's third law

where K_s is a constant given by

$$K_s = \frac{4\pi^2}{GM_s} = 2.97 \times 10^{-19}\ \text{s}^2/\text{m}^3$$

Equation 14.6 is Kepler's third law. The law is also valid for elliptical orbits if we replace r by the length of the semimajor axis, a (Fig. 14.6). Note that the constant of proportionality, K_s, is independent of the mass of the planet. Therefore, Equation 14.6 is valid for *any* planet. If we were to consider the orbit of a satellite about the earth, such as the moon, then the constant would have a different value, with the sun's mass replaced by the earth's mass. In this case, the proportionality constant equals $4\pi^2/GM_e$.

A collection of useful planetary data is given in Table 14.2. The last column of this table verifies that T^2/r^3 is a constant whose value is given by $K_s = 4\pi^2/GM_s = 2.97 \times 10^{-19}\ \text{s}^2/\text{m}^3$.

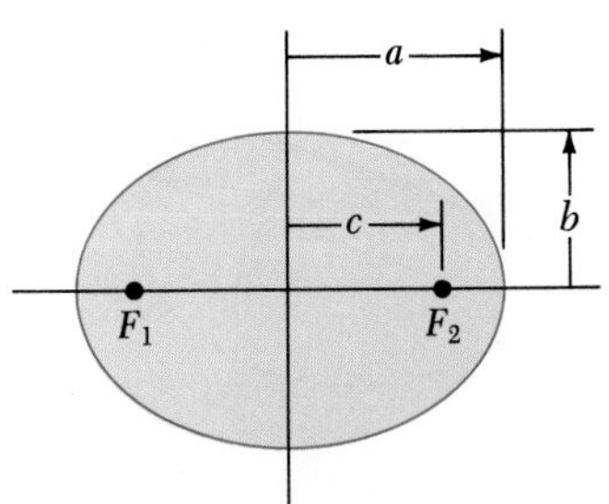

Figure 14.6 Plot of an ellipse. The semimajor axis has a length a, and the semiminor axis has a length b. The focal points are located at a distance c from the center, where $a^2 = b^2 + c^2$, and the eccentricity is defined as $e = c/a$.

[1] The orbits of all planets except Mars, Mercury, and Pluto are very close to being circular. For example, the ratio of the semiminor to the semimajor axis for the earth is $b/a = 0.99986$. Although the orbit of the earth is nearly circular, the sun is *not* at the center of the orbit. Hence, the difference between perihelion (minimum distance from the sun) and aphelion (maximum distance from the sun) is much greater than might be inferred from the comparison of semimajor and semiminor axes.

TABLE 14.2 Useful Planetary Data

Body	Mass (kg)	Mean Radius (m)	Period (s)	Distance from Sun (m)	$\frac{T^2}{r^3}\left(\frac{s^2}{m^3}\right)$
Mercury	3.18×10^{23}	2.43×10^{6}	7.60×10^{6}	5.79×10^{10}	2.97×10^{-19}
Venus	4.88×10^{24}	6.06×10^{6}	1.94×10^{7}	1.08×10^{11}	2.99×10^{-19}
Earth	5.98×10^{24}	6.37×10^{6}	3.156×10^{7}	1.496×10^{11}	2.97×10^{-19}
Mars	6.42×10^{23}	3.37×10^{6}	5.94×10^{7}	2.28×10^{11}	2.98×10^{-19}
Jupiter	1.90×10^{27}	6.99×10^{7}	3.74×10^{8}	7.78×10^{11}	2.97×10^{-19}
Saturn	5.68×10^{26}	5.85×10^{7}	9.35×10^{8}	1.43×10^{12}	2.99×10^{-19}
Uranus	8.68×10^{25}	2.33×10^{7}	2.64×10^{9}	2.87×10^{12}	2.95×10^{-19}
Neptune	1.03×10^{26}	2.21×10^{7}	5.22×10^{9}	4.50×10^{12}	2.99×10^{-19}
Pluto	$\approx 1.4 \times 10^{22}$	$\approx 1.5 \times 10^{6}$	7.82×10^{9}	5.91×10^{12}	2.96×10^{-19}
Moon	7.36×10^{22}	1.74×10^{6}	—	—	—
Sun	1.991×10^{30}	6.96×10^{8}	—	—	—

For a more complete set of data, see, for example, the *Handbook of Chemistry and Physics,* Boca Raton, Florida, The Chemical Rubber Publishing Co.

EXAMPLE 14.3 The Mass of the Sun

Calculate the mass of the sun using the fact that the period of the earth is 3.156×10^7 s and its distance from the sun is 1.496×10^{11} m.

Solution Using Equation 14.6, we get

$$M_s = \frac{4\pi^2 r^3}{GT^2} = \frac{4\pi^2(1.496 \times 10^{11}\ \text{m})^3}{\left(6.67 \times 10^{-11}\ \frac{\text{N}\cdot\text{m}^2}{\text{kg}^2}\right)(3.156 \times 10^7\ \text{s})^2}$$

$$= 1.99 \times 10^{30}\ \text{kg}$$

Note that the sun is 333 000 times as massive as the earth!

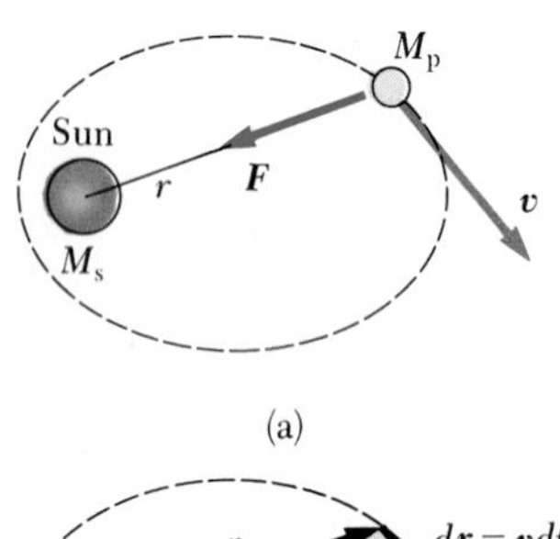

(a)

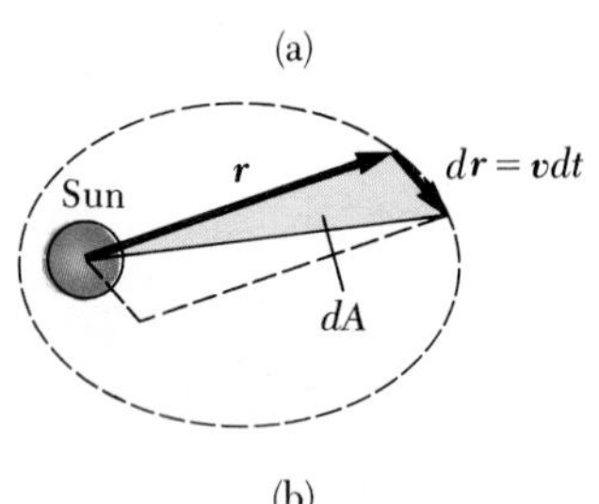

(b)

Figure 14.7 (a) The force acting on a planet acts toward the sun, along the radius vector. (b) As a planet orbits the sun, the area swept out by the radius vector in a time dt is equal to one half the area of the parallelogram formed by the vectors $\boldsymbol{r}$ and $d\boldsymbol{r} = \boldsymbol{v}\,dt$.

Kepler's Second Law and Conservation of Angular Momentum

Consider a planet (or comet) of mass m moving about the sun in an elliptical orbit (Fig. 14.7). The gravitational force acting on the planet is always along the radius vector, directed toward the sun. Such a force directed toward or away from a fixed point (that is, one that is a function of r only) is called a **central force.**[2] The torque acting on the planet due to this central force is clearly zero since $\boldsymbol{F}$ is parallel to $\boldsymbol{r}$. That is,

$$\boldsymbol{\tau} = \boldsymbol{r} \times \boldsymbol{F} = \boldsymbol{r} \times F(r)\hat{\boldsymbol{r}} = 0$$

But recall that the torque equals the time rate of change of angular momentum, or $\boldsymbol{\tau} = d\boldsymbol{L}/dt$. Therefore,

because $\boldsymbol{\tau} = 0$, the angular momentum L of the planet is a constant of the motion:

$$\boldsymbol{L} = \boldsymbol{r} \times \boldsymbol{p} = m\boldsymbol{r} \times \boldsymbol{v} = \text{constant}$$

Since $\boldsymbol{L}$ is a constant of the motion, we see that the planet's motion at any instant is restricted to the plane formed by $\boldsymbol{r}$ and $\boldsymbol{v}$.

[2] Another example of a central force is the electrostatic force between two charged particles.

We can relate this result to the following geometric consideration. The radius vector $\boldsymbol{r}$ in Figure 14.7b sweeps out an area dA in a time dt. This area equals one half the area $|\boldsymbol{r} \times d\boldsymbol{r}|$ of the parallelogram formed by the vectors $\boldsymbol{r}$ and $d\boldsymbol{r}$. Since the displacement of the planet in a time dt is given by $d\boldsymbol{r} = \boldsymbol{v}\,dt$, we get

$$dA = \tfrac{1}{2}|\boldsymbol{r} \times d\boldsymbol{r}| = \tfrac{1}{2}|\boldsymbol{r} \times \boldsymbol{v}\,dt| = \frac{L}{2m}\,dt$$

$$\frac{dA}{dt} = \frac{L}{2m} = \text{constant} \qquad (14.7)$$

Kepler's second law

where L and m are both constants of the motion. Thus, we conclude that

> the radius vector from the sun to any planet sweeps out equal areas in equal times.

It is important to recognize that this result, which is Kepler's second law, is a consequence of the fact that the force of gravity is a central force, which in turn implies conservation of angular momentum. Therefore, the law applies to *any* situation that involves a central force, whether inverse-square or not.

The inverse-square nature of the force of gravity is not revealed by Kepler's second law. Although we do not prove it here, Kepler's first law is a direct consequence of the fact that the gravitational force varies as $1/r^2$. That is, under an inverse-square force law, the orbits of the planets can be shown to be ellipses with the sun at one focus.

EXAMPLE 14.4 Motion in an Elliptical Orbit

A planet of mass m moves in an elliptical orbit about the sun (Fig. 14.8). The minimum and maximum distances of the planet from the sun are called the *perihelion* (indicated by p in Fig. 14.8) and *aphelion* (indicated by a), respectively. If the speed of the planet at p is v_p, what is its speed at a? Assume the distances r_a and r_p are known.

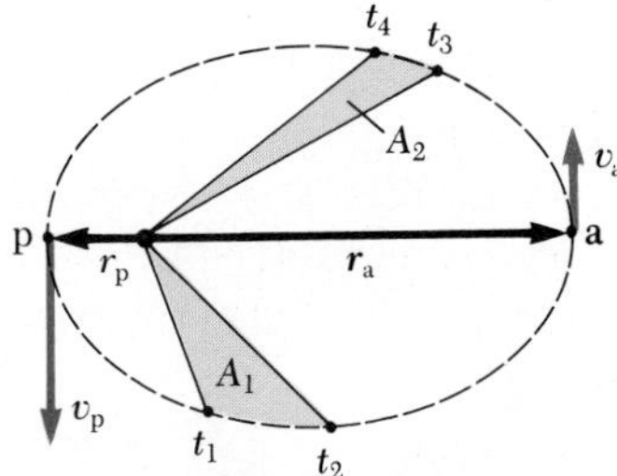

Figure 14.8 (Example 14.4) As a planet moves about the sun in an elliptical orbit, its angular momentum is conserved. Therefore, $mv_ar_a = mv_pr_p$, where the subscripts a and p represent aphelion and perihelion, respectively.

Solution The angular momentum of the planet relative to the sun is $m\boldsymbol{r} \times \boldsymbol{v}$. At the points a and p, $\boldsymbol{v}$ is perpendicular to $\boldsymbol{r}$. Therefore, the magnitude of the angular momentum at these positions is $L_a = mv_ar_a$ and $L_p = mv_pr_p$. The direction of the angular momentum is out of the plane of the paper. Since angular momentum is conserved, we see that

$$mv_ar_a = mv_pr_p$$

$$v_a = \frac{r_p}{r_a}\,v_p$$

14.6 THE GRAVITATIONAL FIELD

The gravitational force between two masses can be considered an action-at-a-distance. That is, the two masses interact even though they are not in contact with each other. An alternative approach in describing the gravitational interaction is to introduce the concept of a **gravitational field, g,** at every point in space. When a particle of mass m is placed at a point where the field is $\mathbf{g}$, the particle experiences a force $\mathbf{F} = m\mathbf{g}$. In other words, the field $\mathbf{g}$ exerts a force on the particle. Hence, the gravitational field is defined by

Gravitational field

$$\mathbf{g} \equiv \frac{\mathbf{F}}{m} \tag{14.8}$$

That is, the gravitational field at any point equals the gravitational force that a test mass experiences divided by that test mass. Consequently, if $\mathbf{g}$ is known at some point in space, a test particle of mass m experiences a gravitational force $m\mathbf{g}$ when placed at that point.

As an example, consider an object of mass m near the earth's surface. The gravitational force on the object is directed toward the center of the earth and has a magnitude mg. Thus we see that the gravitational field that the object experiences at some point has a magnitude equal to the acceleration of gravity at that point. Since the gravitational force on the object has a magnitude GM_em/r^2 (where M_e is the mass of the earth), the field $\mathbf{g}$ at a distance r from the center of the earth is given by

$$\mathbf{g} = \frac{\mathbf{F}}{m} = -\frac{GM_e}{r^2}\,\hat{\mathbf{r}}$$

This expression is valid at all points *outside* the earth's surface, assuming that the earth is spherical and that the earth's rotation can be neglected. At the earth's surface, where $r = R_e$, $\mathbf{g}$ has a magnitude of 9.80 m/s^2.

The field concept is used in many other areas of physics. In fact, the field concept was first introduced by Michael Faraday (1791–1867) in the study of electromagnetism. Later in the text we shall use the field concept to describe

Messier 83 (NGC 5236), a large spiral galaxy in the constellation of Centaurus, is one of the brightest galaxies visible in the southern hemisphere. This whirlpool of stars is about 30,000 lightyears in diameter and 10 million lightyears away from our solar system. It is receding from us at a speed of about 320 km (200 miles) per second. (© Anglo-Australian Telescope Board 1986)

electromagnetic interactions. Gravitational, electrical, and magnetic fields are all examples of *vector fields* since a vector is associated with each point in space. On the other hand, a *scalar field* is one in which a scalar quantity is used to describe each point in space. For example, the variation in temperature over a given region can be described by a scalar temperature field.

14.7 GRAVITATIONAL POTENTIAL ENERGY

In Chapter 8 we introduced the concept of gravitational potential energy, that is, the energy associated with the position of a particle. We emphasized the fact that the gravitational potential energy function, $U = mgy$, is valid only when the particle is near the earth's surface. Since the gravitational force between two particles varies as $1/r^2$, we expect that the correct potential energy function will depend on the amount of separation between the particles.

Before we calculate the specific form for the gravitational potential energy function, we shall first verify that *the gravitational force is conservative.* In order to establish the conservative nature of the gravitational force, we first note that it is a central force. By definition, a central force is one that depends only on the polar coordinate r, and hence can be represented by $F(r)\hat{\mathbf{r}}$, where $\hat{\mathbf{r}}$ is a unit vector directed from the origin to the particle under consideration. Such a force acts from some origin and is directed parallel to the radius vector.

Consider a central force acting on a particle moving along the general path P to Q in Figure 14.9. The central force acts from the point O. This path can be approximated by a series of radial and circular segments. By definition, a central force is always directed along one of the radial segments; therefore the work done along any *radial segment* is given by

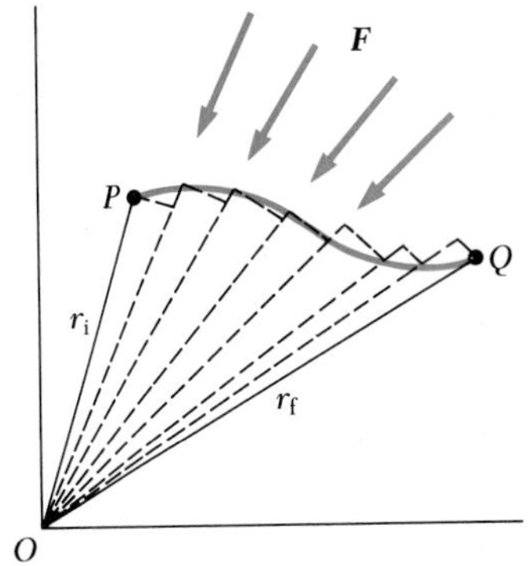

Figure 14.9 A particle moves from P to Q while under the action of a central force $\mathbf{F}$, which is in the radial direction. The path is broken into a series of radial and circular segments. Since the work done along the circular segments is zero, the work done is independent of the path.

$$dW = \mathbf{F} \cdot d\mathbf{r} = F(r)\,dr$$

You should recall that by definition the work done by a force that is perpendicular to the displacement is zero. Hence, the work done along any circular segment is *zero* because $\mathbf{F}$ is perpendicular to the displacement along these segments. Therefore, the total work done by $\mathbf{F}$ is the sum of the contributions along the radial segments:

Work done by a central force

$$W = \int_{r_i}^{r_f} F(r)\,dr$$

where the subscripts i and f refer to the initial and final positions. This result applies to *any* path from P to Q. Therefore, we conclude that *any central force is conservative.* We are now assured that a potential energy function can be obtained once the form of the central force is specified. You should recall from Chapter 8 that the change in the gravitational potential energy associated with a given displacement is defined as the negative of the work done by the gravitational force during that displacement, or

$$\Delta U = U_f - U_i = -\int_{r_i}^{r_f} F(r)\,dr \qquad (14.9)$$

We can use this result to evaluate the gravitational potential energy function. Consider a particle of mass m moving between two points P and Q above

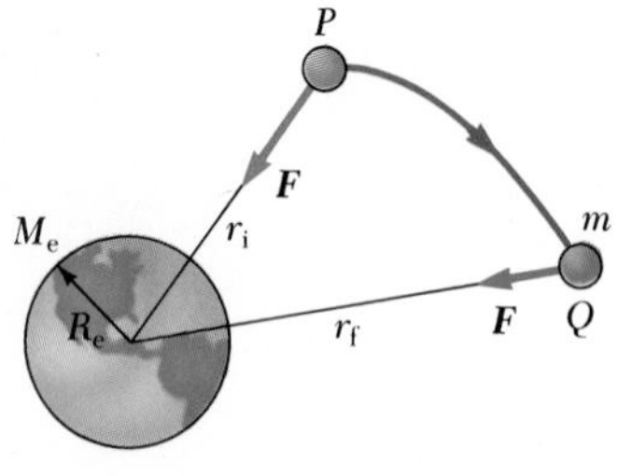

Figure 14.10 As a particle of mass m moves from P to Q above the earth's surface, the potential energy changes according to Equation 14.10.

the earth's surface (Fig. 14.10). The particle is subject to the gravitational force given by Equation 14.1. We can express the force on m in vector form as

$$\boldsymbol{F} = -\frac{GM_em}{r^2}\hat{\boldsymbol{r}}$$

where $\hat{\boldsymbol{r}}$ is a unit vector directed from the earth to the particle and the negative sign indicates that the force is attractive. Substituting this into Equation 14.9, we can compute the change in the gravitational potential energy function:

$$U_f - U_i = GM_em\int_{r_i}^{r_f}\frac{dr}{r^2} = GM_em\left[-\frac{1}{r}\right]_{r_i}^{r_f}$$

$$U_f - U_i = -GM_em\left(\frac{1}{r_f} - \frac{1}{r_i}\right) \qquad (14.10)$$

As always, the choice of a reference point for the potential energy is completely arbitrary. It is customary to choose the reference point where the force is zero. Taking $U_i = 0$ at $r_i = \infty$, we obtain the important result

Gravitational potential energy $r > R_e$

$$U(r) = -\frac{GM_em}{r} \qquad (14.11)$$

This expression applies to the earth-particle system separated by a distance r, provided that $r > R_e$. The result is not valid for particles moving inside the earth, where $r < R_e$. We shall treat this situation in Section 14.10. Because of our choice of U_i, the function $U(r)$ is always negative (Fig. 14.11).

Although Equation 14.11 was derived for the particle-earth system, it can be applied to *any* two particles. That is, the gravitational potential energy associated with *any pair* of particles of masses m_1 and m_2 separated by a distance r is given by

$$U = -\frac{Gm_1m_2}{r} \qquad (14.12)$$

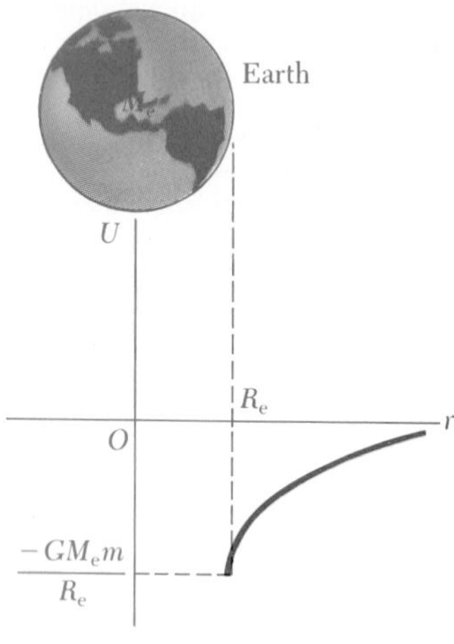

Figure 14.11 Graph of the gravitational potential energy, U, versus r for a particle above the earth's surface. The potential energy goes to zero as r approaches ∞.

This expression shows that the gravitational potential energy for any pair of particles varies as $1/r$, whereas the force between them varies as $1/r^2$. Furthermore, the potential energy is *negative* since the force is attractive and we have taken the potential energy as zero when the particle separation is infinity. Since the force between the particles is attractive, we know that an external agent must do positive work to increase the separation between the two particles. The work done by the external agent produces an increase in the potential energy as the two particles are separated. That is, U becomes less negative as r increases. (Note that part of the work done can also produce a change in kinetic energy of the system. That is, if the work done in separating the particles exceeds the increase in potential energy, the excess energy is accounted for by the increase in kinetic energy of the system.) When the two particles are separated by a distance r, an external agent would have to supply an energy *at least* equal to $+Gm_1m_2/r$ in order to separate the particles by an infinite distance. It is convenient to think of the absolute value of the potential energy as the *binding energy* of the system. If the external agent supplies an energy *greater than* the binding energy, Gm_1m_2/r, the additional energy of the system will be in the form of kinetic energy when the particles are at an infinite separation.

We can extend this concept to three or more particles. In this case, the total potential energy of the system is the sum over all *pairs* of particles.[3] Each pair contributes a term of the form given by Equation 14.12. For example, if the system contains three particles as in Figure 14.12, we find that

$$U_{\text{total}} = U_{12} + U_{13} + U_{23} = -G\left(\frac{m_1m_2}{r_{12}} + \frac{m_1m_3}{r_{13}} + \frac{m_2m_3}{r_{23}}\right) \qquad (14.13)$$

The absolute value of U_{total} represents the work needed to separate the particles by an infinite distance. If the system consists of four particles, there are six terms in the sum, corresponding to the six distinct pairs of interaction forces.

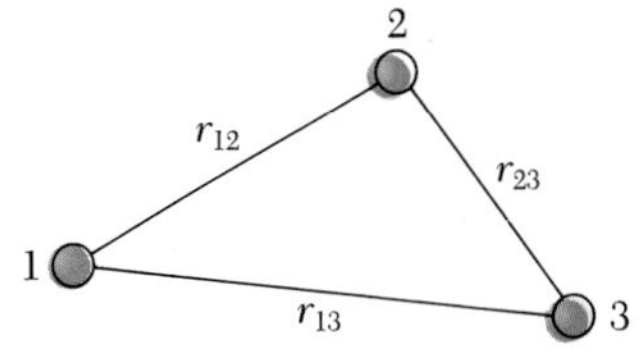

Figure 14.12 Diagram of three interacting particles.

EXAMPLE 14.5 The Change in Potential Energy

A particle of mass m is displaced through a small vertical distance Δy near the earth's surface. Let us show that the general expression for the change in gravitational potential energy given by Equation 14.10 reduces to the familiar relationship $\Delta U = mg\ \Delta y$.

Solution We can express Equation 14.10 in the form

$$\Delta U = -GM_em\left(\frac{1}{r_f} - \frac{1}{r_i}\right) = GM_em\left(\frac{r_f - r_i}{r_ir_f}\right)$$

If both the initial and the final position of the particle are close to the earth's surface, then $r_f - r_i = \Delta y$ and $r_ir_f \approx R_e^2$. (Recall that r is measured from the center of the earth.) Therefore, the *change* in potential energy becomes

$$\Delta U \approx \frac{GM_em}{R_e^2}\,\Delta y = mg\ \Delta y$$

where we have used the fact that $g = GM_e/R_e^2$. Keep in mind that the reference point is arbitrary, since it is the *change* in potential energy that is meaningful.

14.8 ENERGY CONSIDERATIONS IN PLANETARY AND SATELLITE MOTION

Consider a body of mass m moving with a speed v in the vicinity of a massive body of mass M, where $M \gg m$. The system might be a planet moving around the sun or a satellite in orbit around the earth. If we assume that M is at rest in an inertial reference frame, then the total energy E of the two-body system when the bodies are separated by a distance r is the sum of the kinetic energy of the mass m and the potential energy of the system, given by Equation 14.12.[4] That is,

$$E = K + U$$

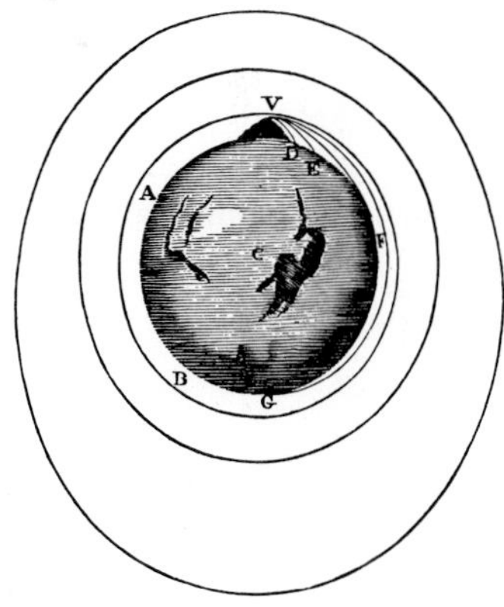

"... the greater the velocity ... with which (a stone) is projected, the farther it goes before it falls to the earth. We may therefore suppose the velocity to be so increased, that it would describe an arc of 1, 2, 5, 10, 100, 1000 miles before it arrived at the earth, till at last, exceeding the limits of the earth, it should pass into space without touching."— Newton, *System of the World.*

[3] The fact that one can add potential energy terms for all pairs of particles stems from the experimental fact that gravitational forces obey the superposition principle. That is, if $\Sigma F = F_{12} + F_{13} + F_{23} + \cdots$ then there exists a potential energy term for each interaction F_{ij}.

[4] You might recognize that we have ignored the acceleration and kinetic energy of the larger mass. To see that this is reasonable, consider an object of mass m falling toward the earth. Since the center of mass of the object-earth system is stationary, it follows that $mv = M_ev_e$. Thus, the earth acquires a kinetic energy equal to

$$\tfrac{1}{2}M_ev_e^2 = \tfrac{1}{2}\frac{m^2}{M_e}v^2 = \frac{m}{M_e}K,$$

where K is the kinetic energy of the object. Since $M_e \gg m$, the kinetic energy of the earth is negligible.

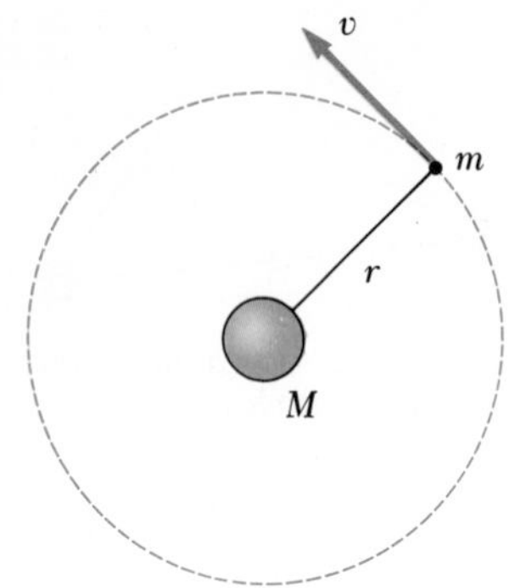

Figure 14.13 A body of mass m moving in a circular orbit about a body of mass M.

$$E = \tfrac{1}{2}mv^2 - \frac{GMm}{r} \tag{14.14}$$

Furthermore, the total energy is conserved if we assume the system is isolated. Therefore as the mass m moves from P to Q in Figure 14.10, the total energy remains constant and Equation 14.14 gives

$$E = \tfrac{1}{2}mv_i^2 - \frac{GMm}{r_i} = \tfrac{1}{2}mv_f^2 - \frac{GMm}{r_f} \tag{14.15}$$

This result shows that E may be positive, negative, or zero, depending on the value of the velocity of the mass m. However, for a bound system, such as the earth and sun, E is necessarily *less than zero.* We can easily establish that $E < 0$ for the system consisting of a mass m moving in a circular orbit about a body of mass M, where $M \gg m$ (Fig. 14.13). Newton's second law applied to the body of mass m gives

$$\frac{GMm}{r^2} = \frac{mv^2}{r}$$

Multiplying both sides by r and dividing by 2 gives

$$\tfrac{1}{2}mv^2 = \frac{GMm}{2r} \tag{14.16}$$

Substituting this into Equation 14.14, we obtain

$$E = \frac{GMm}{2r} - \frac{GMm}{r}$$

Total energy for circular orbits

$$E = -\frac{GMm}{2r} \tag{14.17}$$

This clearly shows that *the total energy must be negative in the case of circular orbits.* Note that *the kinetic energy is positive and equal to one half the magnitude of the potential energy.* The absolute value of E is also equal to the binding energy of the system.

The total mechanical energy is also negative in the case of elliptical orbits.[5] The expression for E for elliptical orbits is the same as Equation 14.17 with r replaced by the semimajor axis length, a.

Both the total energy and the total angular momentum of a planet-sun system are constants of the motion.

[5] This is shown in more advanced mechanics texts. One can also show that if $E = 0$, the mass would move in a parabolic path, whereas if $E > 0$, its path would be hyperbolic. Nothing in Equation 14.14 precludes a particle with $E \geq 0$ from reaching infinitely great distances from the gravitating center (that is, the particle's orbit is unbound). Infinitely great distances are energetically forbidden to a particle with $E < 0$, so its orbit is bound.

EXAMPLE 14.6 Changing the Orbit of a Satellite
Calculate the work required to move an earth satellite of mass m from a circular orbit of radius $2R_e$ to one of radius $3R_e$.

Solution Applying Equation 14.17, we get for the total initial and final energies

$$E_i = -\frac{GM_e m}{4R_e} \qquad E_f = -\frac{GM_e m}{6R_e}$$

Therefore, the work required to increase the energy of the system is

$$W = E_f - E_i = -\frac{GM_e m}{6R_e} - \left(-\frac{GM_e m}{4R_e}\right) = \frac{GM_e m}{12R_e}$$

For example, if we take $m = 10^3$ kg, we find that the work required is $W = 5.2 \times 10^9$ J, which is the energy equivalent of 39 gal of gasoline.

If we wish to determine how the energy is distributed after doing work on the system, we find from Equation 14.16 that the change in kinetic energy is $\Delta K = -GM_e m/12R_e$ (it decreases), while the corresponding change in potential energy is $\Delta U = GM_e m/6R_e$ (it increases). Thus, the work done on the system is given by $W = \Delta K + \Delta U = GM_e m/12R_e$, as we calculated above. In other words, part of the work done goes into increasing the potential energy and part goes into decreasing the kinetic energy.

It is interesting to point out that the process of orbit injection consists of two stages. First, the satellite is placed in an elliptical orbit. Then, its potential energy is maximized, giving it additional kinetic energy.

Escape Velocity

Suppose an object of mass m is projected vertically upward from the earth's surface with an initial speed v_i, as in Figure 14.14. We can use energy considerations to find the minimum value of the initial speed such that the object will escape the earth's gravitational field. Equation 14.15 gives the total energy of the object at any point when its velocity and distance from the center of the earth are known. At the surface of the earth, where $v_i = v$, $r_i = R_e$. When the object reaches its maximum altitude, $v_f = 0$ and $r_f = r_{max}$. Because the total energy of the system is conserved, substitution of these conditions into Equation 14.15 gives

$$\tfrac{1}{2}mv_i^2 - \frac{GM_e m}{R_e} = -\frac{GM_e m}{r_{max}}$$

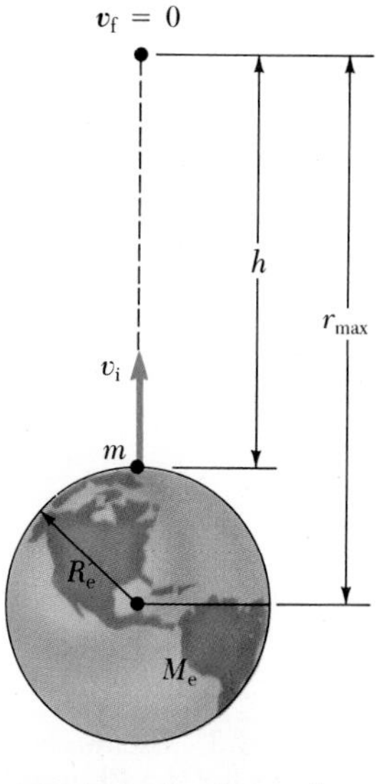

Solving for v_i^2 gives

$$v_i^2 = 2GM_e\left(\frac{1}{R_e} - \frac{1}{r_{max}}\right) \tag{14.18}$$

$v_f = 0$, h, r_{max}, v_i, m, R_e, M_e

Figure 14.14 An object of mass m projected upward from the earth's surface with an initial speed v_i reaches a maximum altitude h.

Therefore, if the initial speed is known, this expression can be used to calculate the maximum altitude h, since we know that $h = r_{max} - R_e$.

We are now in a position to calculate the minimum speed the object must have at the earth's surface in order to escape from the influence of the earth's gravitational field. This corresponds to the situation where the object can *just* reach infinity with a final speed of *zero*. Setting $r_{max} = \infty$ in Equation 14.18 and taking $v_i = v_{esc}$ (the escape velocity), we get

$$v_{esc} = \sqrt{\frac{2GM_e}{R_e}} \tag{14.19}$$

Escape velocity

Note that this expression for v_{esc} is independent of the mass of the object projected from the earth. For example, a spacecraft has the same escape velocity as a molecule. Furthermore, the result is independent of the *direction* of the velocity, provided the trajectory does not intersect the earth. If the object is given an initial speed equal to v_{esc}, its *total* energy is equal to zero. This can be seen by noting that when $r = \infty$, the object's kinetic energy and its potential energy are both zero. If v_i is greater than v_{esc}, the *total* energy will be greater than zero and the object will have some residual kinetic energy at $r = \infty$.

EXAMPLE 14.7 Escape Velocity of a Rocket
Calculate the escape velocity from the earth for a 5000-kg spacecraft, and determine the kinetic energy it must have at the earth's surface in order to escape the earth's field.

Solution Using Equation 14.19 with $M_e = 5.98 \times 10^{24}$ kg and $R_e = 6.37 \times 10^6$ m gives

$$v_{esc} = \sqrt{\frac{2GM_e}{R_e}}$$

$$= \sqrt{\frac{2(6.67 \times 10^{-11}\ \text{N}\cdot\text{m}^2/\text{kg}^2)(5.98 \times 10^{24}\ \text{kg})}{6.37 \times 10^6\ \text{m}}}$$

$$= 1.12 \times 10^4\ \text{m/s}$$

This corresponds to about 25 000 mi/h.

The kinetic energy of the spacecraft is given by

$$K = \tfrac{1}{2}mv_{esc}^2 = \tfrac{1}{2}(5 \times 10^3\ \text{kg})(1.12 \times 10^4\ \text{m/s})^2$$

$$= 3.14 \times 10^{11}\ \text{J}$$

Finally, you should note that Equations 14.18 and 14.19 can be applied to objects projected vertically from *any* planet. That is, in general, the escape velocity from any planet of mass M and radius R is given by

$$v_{esc} = \sqrt{\frac{2GM}{R}}$$

A list of escape velocities for the planets, the moon, and the sun is given in Table 14.3. Note that the values vary from 2.3 km/s for the moon to about 618 km/s for the sun. These results, together with some ideas from the kinetic theory of gases (Chapter 21), explain why some planets have atmospheres and others do not. As we shall see later, a gas molecule has an average kinetic energy which depends on its temperature. Hence, lighter atoms, such as hydrogen and helium, have a higher average velocity than the heavier species. When the velocity of the lighter atoms is not much less than the escape velocity, a significant fraction of the molecules have a chance to escape from the planet. This mechanism also explains why the earth does not retain hydrogen and helium molecules in its atmosphere while much heavier molecules, such as oxygen and nitrogen, do not escape. On the other hand, Jupiter has a very large escape velocity (60 km/s), which enables it to retain hydrogen, the primary constituent of its atmosphere.

TABLE 14.3 Escape Velocities for the Planets, the Moon, and the Sun

Planet	v_{esc} (km/s)
Mercury	4.3
Venus	10.3
Earth	11.2
Moon	2.3
Mars	5.0
Jupiter	60
Saturn	36
Uranus	22
Neptune	24
Pluto	1.1
Sun	618

*14.9 THE GRAVITATIONAL FORCE BETWEEN AN EXTENDED BODY AND A PARTICLE

We have emphasized that the law of universal gravitation given by Equation 14.3 is valid only if the interacting objects are considered as particles. In view of this, how can we calculate the force between a particle and an object having finite dimensions? This is accomplished by treating the *extended* object as a collection of particles and making use of integral calculus. We shall take the approach of first evaluating the potential energy function, from which the force can be calculated.

The potential energy associated with a system consisting of a point mass m and an extended body of mass M is obtained by dividing the body into segments of mass ΔM_i (Fig. 14.15). The potential energy associated with this element and with the particle of mass m is $-Gm\,\Delta M_i/r_i$, where r_i is the distance from the particle to the element ΔM_i. The total potential energy of the system is obtained by taking the sum over all segments as $\Delta M_i \rightarrow 0$. In this limit, we can express U in integral form as

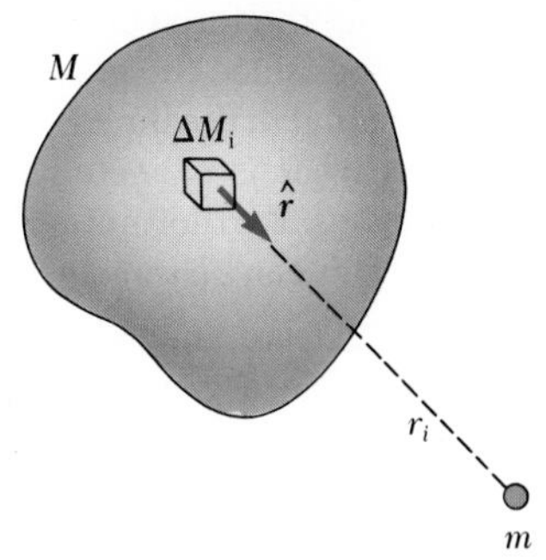

Figure 14.15 A particle of mass m interacting with an extended body of mass M. The potential energy of the system is given by Equation 14.20. The total force on a particle at P due to an extended body can be obtained by taking a vector sum over all forces due to each segment of the body.

$$U = -Gm \int \frac{dM}{r} \tag{14.20}$$

Once U has been evaluated, the force can be obtained by taking the negative derivative of this scalar function (see Section 8.7). If the extended body has spherical symmetry, the function U depends only on r and the force is given by $-dU/dr$. We shall treat this situation in Section 14.10. In principle, one can evaluate U for any specified geometry; however, the integration can be cumbersome.

An alternative approach to evaluating the force between a particle and an extended body is to perform a vector sum over all segments of the body. Using the procedure outlined in evaluating U and the law of universal gravitation (Eq. 14.3), the total force on the particle is given by

Total force between a particle and an extended body

$$\mathbf{F} = -Gm \int \frac{dM}{r^2}\,\hat{\mathbf{r}} \tag{14.21}$$

where $\hat{\mathbf{r}}$ is a unit vector directed from the element dM toward the particle (Fig. 14.15). This procedure is not always recommended, since working with a vector function is more difficult than working with the scalar potential energy function. However, if the geometry is simple, as in the following example, the evaluation of $\mathbf{F}$ can be straightforward.

EXAMPLE 14.8 Force Between a Mass and a Bar □

A homogeneous bar of length L and mass M is at a distance h from a point mass m (Fig. 14.16). Calculate the force on m.

Solution The segment of the bar that has a length dx has a mass dM. Since the mass per unit length is a constant, it then follows that the ratio of masses, dM/M, is equal to the ratio of lengths, dx/L, and so $dM = \frac{M}{L}\,dx$. The variable r in Equation 14.21 is x in our case, and the force on m is to the right; therefore we get

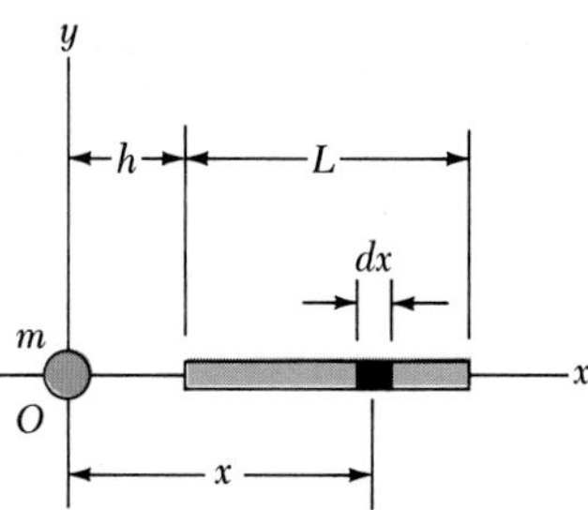

Figure 14.16 (Example 14.8) The force on a particle at the origin due to the bar is to the right. Note that the bar is *not* equivalent to a particle of mass M located at its center of mass.

$$\mathbf{F} = Gm \int_h^{L+h} \frac{M}{L}\frac{dx}{x^2}\,\mathbf{i}$$

$$\mathbf{F} = \frac{GmM}{L}\left[-\frac{1}{x}\right]_h^{L+h}\mathbf{i} = \frac{GmM}{h(L+h)}\,\mathbf{i}$$

We see that the force on m is in the positive x direction, as expected, since the gravitational force is attractive.

Note that in the limit $L \to 0$, the force varies as $1/h^2$, which is what is expected for the force between two point masses. Furthermore, if $h \gg L$, the force also varies as $1/h^2$. This can be seen by noting that the denominator of the expression for F can be expressed in the form $h^2\left(1 + \frac{L}{h}\right)$, which is approximately equal to h^2. Thus, when bodies are separated by distances that are large compared with their characteristic dimensions, they behave like particles.

*14.10 GRAVITATIONAL FORCE BETWEEN A PARTICLE AND A SPHERICAL MASS

In this section we shall describe the gravitational force between a particle and a spherically symmetric mass distribution. We have already stated that a large sphere attracts a particle outside it as if the total mass of the sphere were concentrated at its center. Let us describe the nature of the force on a particle when the extended body is either a spherical shell or a solid sphere, and then apply these facts to some interesting systems.

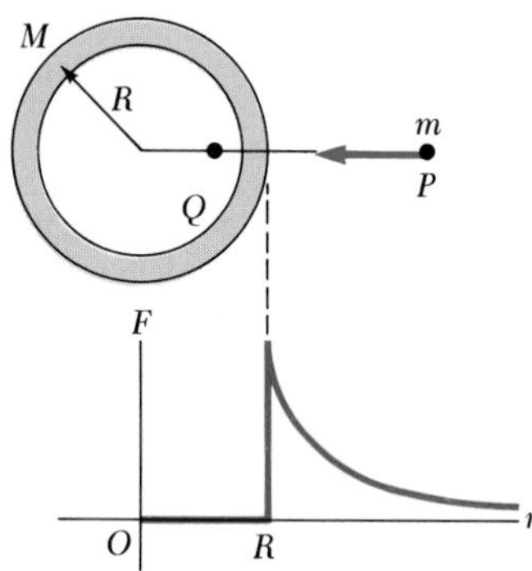

Figure 14.17 The force on a particle when it is outside the spherical shell is given by GMm/r^2 and acts toward the center. The force on the particle is zero everywhere inside the shell.

Spherical Shell

Force on a particle due to a spherical shell

1. If a particle of mass m is located *outside* a spherical shell of mass M (say, point P in Fig. 14.17), the spherical shell attracts the particle as though the mass of the shell were concentrated at its center.

2. If the particle is located *inside* the spherical shell (point Q in Fig. 14.17), the force on it is zero. We can express these two important results in the following way:

$$\mathbf{F} = -\frac{GMm}{r^2}\hat{\mathbf{r}} \qquad \text{for } r > R \tag{14.22a}$$

$$\mathbf{F} = 0 \qquad \text{for } r < R \tag{14.22b}$$

The force as a function of the distance r is plotted in Figure 14.17. Note that the shell of mass does *not* act as a gravitational shield. The particle may experience forces due to other masses outside the shell.

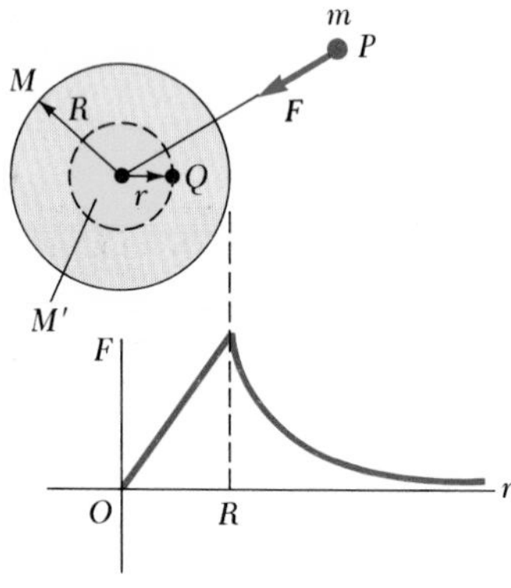

Figure 14.18 The force on a particle when it is outside a uniform solid sphere is given by GMm/r^2 and is directed toward the center. The force on the particle when it is inside such a sphere is proportional to r and goes to zero at the center.

Solid Sphere

1. If a particle of mass m is located *outside* a homogeneous solid sphere of mass M (point P in Fig. 14.18), the sphere attracts the particle as though the mass of the sphere were concentrated at its center. That is, Equation 14.22a applies in this situation. This follows from case 1 above, since a solid sphere can be considered a collection of concentric spherical shells.

2. If a particle of mass m is located *inside* a homogeneous solid sphere of mass M (point Q in Fig. 14.18), the force on m is due *only* to the mass M' contained within the sphere of radius $r < R$, represented by the dotted line in Figure 14.18. In other words,

Force on a particle due to a solid sphere

$$\mathbf{F} = -\frac{GmM}{r^2}\hat{\mathbf{r}} \qquad \text{for } r > R \tag{14.23a}$$

$$\mathbf{F} = -\frac{GmM'}{r^2}\hat{\mathbf{r}} \qquad \text{for } r < R \tag{14.23b}$$

Since the sphere is assumed to have a uniform density, it follows that the ratio of masses M'/M is equal to the ratio of volumes V'/V, where V is the total volume of the sphere and V' is the volume within the dotted surface. That is,

$$\frac{M'}{M} = \frac{V'}{V} = \frac{\frac{4}{3}\pi r^3}{\frac{4}{3}\pi R^3} = \frac{r^3}{R^3}$$

Solving this equation for M' and substituting the value obtained into Equation 14.23b, we get

$$\boldsymbol{F} = -\frac{GmM}{R^3} r\,\hat{\boldsymbol{r}} \qquad \text{for } r < R \tag{14.24}$$

That is, the force goes to zero at the center of the sphere, as we would intuitively expect. The force as a function of r is plotted in Figure 14.18.

3. If a particle is located *inside* a solid sphere having a density ρ that is spherically symmetric but *not* uniform, then M' in Equation 14.23 is given by an integral of the form $M' = \int \rho\, dV$, where the integration is taken over the volume contained *within* the dotted surface. This integral can be evaluated if the radial variation of ρ is given. The integral is easily evaluated if the mass distribution has spherical symmetry, that is, if ρ is a function of r only. In this case, we take the volume element dV as the volume of a spherical shell of radius r and thickness dr, so that $dV = 4\pi r^2\, dr$. For example, if $\rho(r) = Ar$, where A is a constant, it is left as a problem (Problem 63) to show that $M' = \pi A r^4$. Hence we see from Equation 14.23b that F is proportional to r^2 in this case and is zero at the center.

EXAMPLE 14.9 A Free Ride

An object moves in a smooth, straight tunnel dug between two points on the earth's surface (Fig. 14.19). Show that the object moves with simple harmonic motion and find the period of its motion. Assume that the earth's density is uniform throughout its volume.

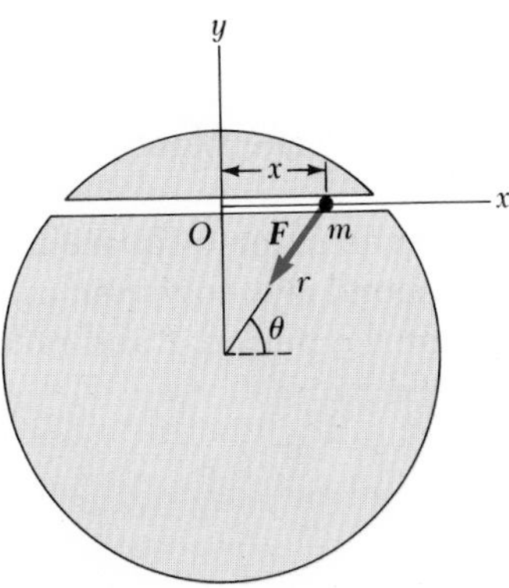

Figure 14.19 A particle moves along a tunnel dug through the earth. The component of the gravitational force $\boldsymbol{F}$ along the x axis is the driving force for the motion. Note that this component always acts toward the origin O.

Solution When the object is in the tunnel, the gravitational force on it acts toward the earth's center and is given by Equation 14.24:

$$F = -\frac{GmM_e}{R_e^3} r$$

The y component of this force is balanced by the normal force exerted by the tunnel wall, and the x component of the force is given by

$$F_x = -\frac{GmM_e}{R_e^3} r \cos\theta$$

Since the x coordinate of the object is given by $x = r\cos\theta$, we can write F_x in the form

$$F_x = -\frac{GmM_e}{R_e^3} x$$

Applying Newton's second law to the motion along x gives

$$F_x = -\frac{GmM_e}{R_e^3} x = ma$$

$$a = -\frac{GM_e}{R_e^3} x = -\omega^2 x$$

But this is the equation of simple harmonic motion with angular velocity ω (Chapter 13), where

$$\omega = \sqrt{\frac{GM_e}{R_e^3}}$$

The period is calculated using the data in Table 14.2 and the above result:

$$T = \frac{2\pi}{\omega} = 2\pi\sqrt{\frac{R_e^3}{GM_e}}$$

$$= 2\pi\sqrt{\frac{(6.37 \times 10^6)^3}{(6.67 \times 10^{-11})(5.98 \times 10^{24})}}$$

$$= 5.06 \times 10^3 \text{ s} = 84.3 \text{ min}$$

This period is the same as that of a satellite in a circular orbit just above the earth's surface. Note that the result is *independent* of the length of the tunnel.

It has been proposed to operate a mass-transit system between any two cities using this principle. A one-way trip would take about 42 min. A more precise calculation of the motion must account for the fact that the earth's density is not uniform as we have assumed. More important, there are many practical problems to consider. For instance, it would be impossible to achieve a frictionless tunnel, and so some auxiliary power source would be required. Can you think of other problems?

SUMMARY

Universal law of gravity

Newton's law of universal gravitation states that the gravitational force of attraction between any two particles of masses m_1 and m_2 separated by a distance r has the magnitude

$$F = G\frac{m_1 m_2}{r^2} \tag{14.1}$$

where G is the universal gravitational constant, which has the value $6.672 \times 10^{-11}\ \text{N}\cdot\text{m}^2/\text{kg}^2$.

Kepler's laws

Kepler's laws of planetary motion state that

1. All planets move in elliptical orbits with the sun at one of the focal points.
2. The radius vector drawn from the sun to any planet sweeps out equal areas in equal time intervals.
3. The square of the orbital period of any planet is proportional to the cube of the semimajor axis for the elliptical orbit.

Kepler's second law is a consequence of the fact that the force of gravity is a *central force,* that is, one that is directed toward a fixed point. This in turn implies that the angular momentum of the planet-sun system is a constant of the motion.

Kepler's third law is consistent with the inverse-square nature of the law of universal gravitation. Newton's second law, together with the force law given by Equation 14.1, verifies that the period T and radius r of the orbit of a planet about the sun are related by

Kepler's third law

$$T^2 = \left(\frac{4\pi^2}{GM_s}\right)r^3 \tag{14.6}$$

where M_s is the mass of the sun. Most planets have nearly circular orbits about the sun. For elliptical orbits, Equation 14.6 is valid if r is replaced by the semimajor axis, a.

The gravitational force is a conservative force, and therefore a potential energy function can be defined. The **gravitational potential energy** associated with two particles separated by a distance r is given by

Gravitational potential energy for a pair of particles

$$U = -\frac{Gm_1m_2}{r} \quad (14.12)$$

where U is taken to be zero at $r = \infty$. The total potential energy for a system of particles is the sum of energies for all pairs of particles, with each pair represented by a term of the form given by Equation 14.12.

If an isolated system consists of a particle of mass m moving with a speed v in the vicinity of a massive body of mass M, the *total energy* of the system is given by

$$E = \tfrac{1}{2}mv^2 - \frac{GMm}{r} \quad (14.14)$$

That is, the energy is the sum of the kinetic and potential energies. The total energy is a constant of the motion.

If m moves in a circular orbit of radius r about M, where $M \gg m$, *the total energy of the system is*

Total energy for circular orbits

$$E = -\frac{GMm}{2r} \quad (14.17)$$

The total energy is negative for any bound system, that is, one in which the orbit is closed, such as an elliptical orbit.

The *potential energy* of gravitational attraction between a particle of mass m and an extended body of mass M is given by

Total potential energy for a particle–extended-body system

$$U = -Gm \int \frac{dM}{r} \quad (14.20)$$

where the integral is over the extended body, dM is an infinitesimal mass element of the body, and r is the distance from the particle to the element.

If a particle is outside a uniform spherical shell or solid sphere with a spherically symmetric internal mass distribution, the sphere attracts the particle as though the mass of the sphere were concentrated at the center of the sphere.

If a particle is inside a uniform spherical shell, the gravitational force on the particle is zero.

If a particle is inside a homogeneous solid sphere, the force on the particle acts toward the center of the sphere and is linearly proportional to the distance from the center to the particle.

QUESTIONS

1. Estimate the gravitational force between you and a person 2 m away from you.
2. Use Kepler's second law to convince yourself that the earth must move faster in its orbit during December, when it is closest to the sun, than it does during June, when it is farthest from the sun.
3. How would you explain the fact that planets such as Saturn and Jupiter have periods much greater than one year?
4. If a system consists of five distinct particles, how many terms appear in the expression for the total potential energy?

5. Is it possible to calculate the potential energy function associated with a particle and an extended body without knowing the geometry or mass distribution of the extended body?
6. Does the escape velocity of a rocket depend on its mass? Explain.
7. Compare the energies required to reach the moon for a 10^5-kg spacecraft and a 10^3-kg satellite.
8. Explain why it takes more fuel for a spacecraft to travel from the earth to the moon than for the return trip. Estimate the difference.
9. Is the magnitude of the potential energy associated with the earth-moon system greater than, less than, or equal to the kinetic energy of the moon relative to the earth?
10. Explain carefully why there is no work done on a planet as it moves in a circular orbit around the sun, even though a gravitational force is acting on the planet. What is the *net* work done on a planet during each revolution as it moves around the sun in an elliptical orbit?
11. A particle is projected through a small hole into the interior of a large spherical shell. Describe the motion of the particle in the interior of the shell.
12. Explain why the force on a particle due to a uniform sphere must be directed toward the center of the sphere. Would this be the case if the mass distribution of the sphere were not spherically symmetric?
13. Neglecting the density variation of the earth, what would be the period of a particle moving in a smooth hole dug through the earth's center?
14. With reference to Figure 14.8, consider the area swept out by the radius vector in the time intervals $t_2 - t_1$ and $t_4 - t_3$. Under what condition is A_1 equal to A_2?
15. If A_1 equals A_2 in Figure 14.8, is the average speed of the planet in the time interval $t_2 - t_1$ less than, equal to, or greater than its average speed in the time interval $t_4 - t_3$?
16. At what position in its elliptical orbit is the speed of a planet a maximum? At what position is the speed a minimum?
17. If you are given the mass and radius of planet X, how would you calculate the acceleration of gravity on the surface of this planet?
18. If a hole could be dug to the center of the earth, do you think that the force on a mass m would still obey Equation 14.1 there? What do you think the force on m would be at the center of the earth?
19. Henry Cavendish, in his 1798 experiment, was said to have "weighed the Earth." Explain this statement.
20. The *Voyager* spacecraft was accelerated toward escape velocity from the sun by the gravitational force from the giant planet Jupiter. How is this possible? Wouldn't the additional speed gained on approach be lost as the spacecraft receded away?
21. How would you find the mass of the moon?
22. The Apollo 13 spaceship developed trouble in the oxygen system about halfway to the moon. Why did the mission continue on around the moon, and then return home, rather than immediately turn back to earth?
23. By how much is the acceleration due to gravity at the earth's equator reduced because of the rotation of the earth? How does this effect vary with latitude?
24. A possible antisatellite weapon has been discussed, where another satellite is launched into a contrary orbit at the same height, but in the opposite direction. If this new satellite, filled with copper wire, were exploded would the copper chaff impacting the original satellite do any damage?

PROBLEMS

Section 14.1 through Section 14.3

1. Two identical, isolated particles, each of mass 2 kg, are separated by a distance of 30 cm. What is the magnitude of the gravitational force of one particle on the other?
2. A 200-kg mass and a 500-kg mass are separated by a distance of 0.40 m. (a) Find the net gravitational force due to these masses acting on a 50-kg mass placed midway between them. (b) At what position (other than infinitely remote ones) would the 50-kg mass experience a net force of zero?
3. Three 5-kg masses are located at the corners of an equilateral triangle having sides 0.25 m in length. Determine the magnitude and direction of the resultant gravitational force on one of the masses due to the other two masses.
4. Two stars of masses M and $4M$ are separated by a distance d. Determine the location of a point measured from M at which the net force on a third mass would be zero.
5. Four particles are located at the corners of a rectangle as in Figure 14.20. Determine the x and y components of the resultant force acting on the particle of mass m.

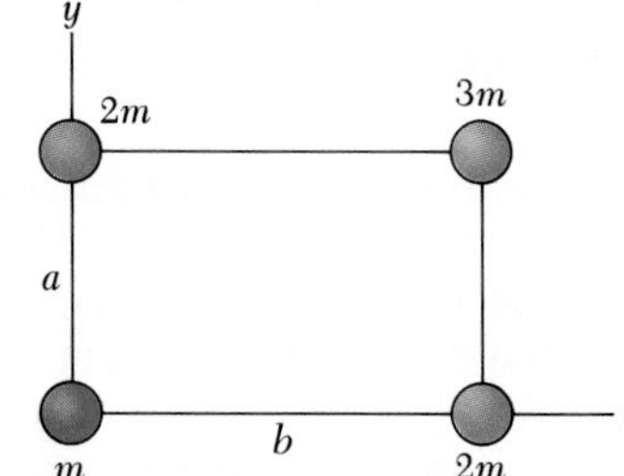

Figure 14.20 (Problem 5).

6. Calculate the acceleration of gravity at a point that is a distance R_e above the surface of the earth, where R_e is the radius of the earth.
7. Using the data given in Figure 14.3, determine a vector expression for the resultant force on the 6-kg mass. What is the magnitude of this force?
8. Two objects attract each other with a gravitational force of 1.0×10^{-8} N when separated by 20 cm. If the total mass of the two objects is 5.0 kg, what is the mass of each?
9. Jupiter has an average density of 1.34×10^3 kg/m^3 and an average diameter of 1.436×10^5 km. What is the acceleration due to gravity at the surface of Jupiter?
10. Consider the earth to be a sphere of uniform density. What would the density need to be to account for the measured acceleration due to gravity at the surface?
11. Plaskett's binary system consists of two stars that revolve in a circular orbit about a center of gravity midway between them. This means that the masses of the two stars are equal (Figure 14.21). If the orbital velocity of each star is 220 km/s and the orbital period of each is 14.4 days, find the mass M of each star. (For comparison, the mass of our sun is 2×10^{30} kg.)

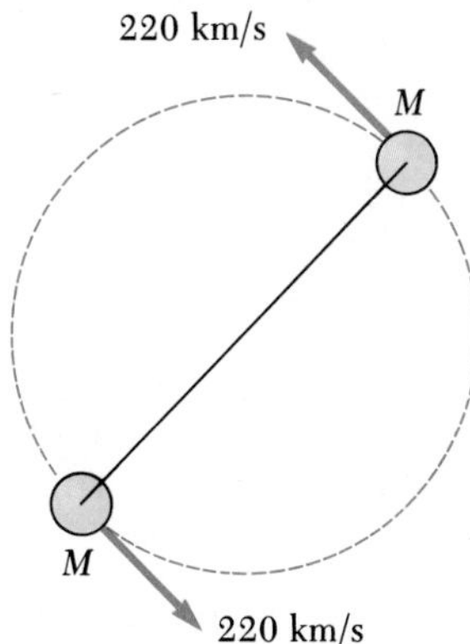

Figure 14.21 (Problem 11).

12. The moon is 384 400 km distant from the earth's center and it completes an orbit in 27.3 days. (a) Determine the moon's orbital velocity. (b) How far does the moon "fall" towards earth in 1 second?

Section 14.4 Kepler's Laws

Section 14.5 The Law of Universal Gravitation and the Motion of Planets

13. Given that the moon's period about the earth is 27.32 days and the earth-moon distance is 3.84×10^8 m, estimate the mass of the earth. Assume the orbit is circular. Why do you suppose your estimate is high?
14. A satellite is in a circular orbit about the earth. (a) Evaluate the constant K that appears in Kepler's third law as applied to this situation. (b) What is the period of the orbit if the satellite is at an altitude of 2×10^6 m?
15. Io, a small moon of the giant planet Jupiter, has an orbital period of 1.77 days and an orbital radius of 4.22×10^5 km. From this data, determine the mass of Jupiter.
16. A satellite of Mars has a period of 459 min. The mass of Mars is 6.42×10^{23} kg. From this information, determine the radius of the satellite's orbit.
17. At its aphelion, the planet Mercury is 6.99×10^{10} m from the sun, and at its perihelion, it is 4.60×10^{10} m from the sun. If its orbital speed is 3.88×10^4 m/s at the aphelion, what is its orbital speed at the perihelion?
18. At what distance from the earth's center will an artificial satellite in an equatorial circular orbit (that is, an orbit that lies entirely in the Earth's equatorial plane) have a period equal to 1 day? Such a *synchronous* (or *geostationary*) satellite moves in synchronism with the Earth, remaining in a fixed position above a point on the equator. Synchronous satellites are used extensively as fixed relay stations in the worldwide communications network.
19. Using the law of universal gravitation, determine the average speed of the earth in its orbit about the sun. (Use data from the inside cover of this book.)
20. Halley's comet approaches the sun to within 0.57 A.U. (1 A.U. = 150×10^6 km), and its orbital period is 75.6 years. How far from the sun will Halley's comet travel before it starts its return journey? (Figure 14.22.)

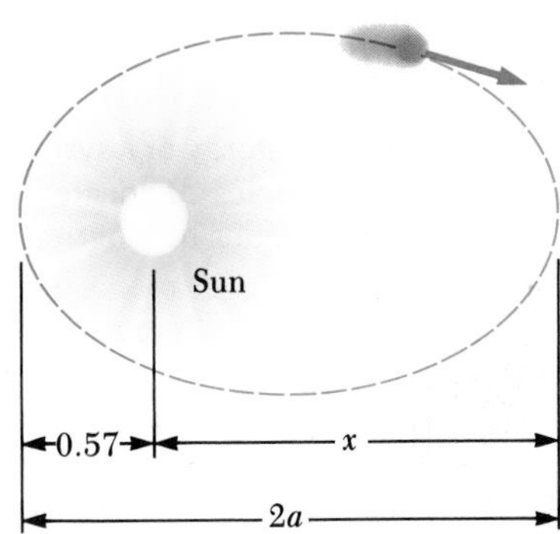

Figure 14.22 (Problem 20).

Section 14.6 The Gravitational Field

21. Compute the magnitude and direction of the gravitational field at a point P on the perpendicular bisector of two equal masses separated by 2a as shown in Figure 14.23.

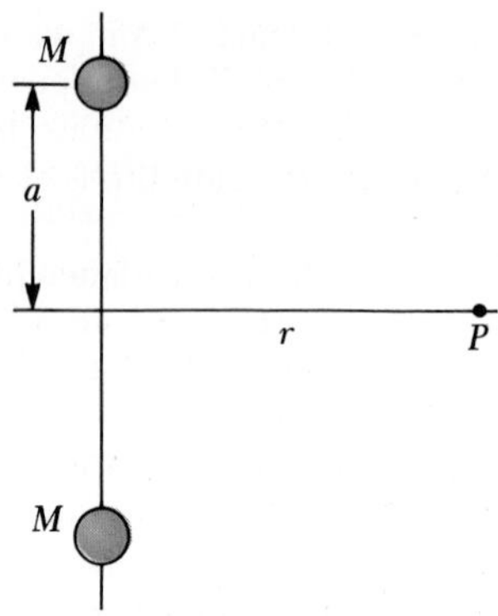

Figure 14.23 (Problem 21).

22. Find the gravitational field at a distance r along the axis of a thin ring of mass M and radius a.
23. At what point along the line connecting the earth and the moon is there zero gravitational force on an object? (Ignore the presence of the sun and the other planets.)

Section 14.7 Gravitational Potential Energy

Assume $U = 0$ at $r = \infty$.

24. A satellite of the earth has a mass of 100 kg and is at an altitude of 2×10^6 m. (a) What is the potential energy of the satellite-earth system? (b) What is the magnitude of the force on the satellite?
25. A system consists of three particles, each of mass 5 g, located at the corners of an equilateral triangle with sides of 30 cm. (a) Calculate the potential energy of the system. (b) If the particles are released simultaneously, where will they collide?
26. How much energy is required to move a 1000-kg mass from the earth's surface to an altitude equal to twice the earth's radius?
27. Show that the potential energy of a system consisting of four equal-mass particles M located at the corners of a square having sides of length d is given by $U = -(GM^2/d)(4 + \sqrt{2})$.

Section 14.8 Energy Considerations in Planetary and Satellite Motion

28. Calculate the escape velocity from the moon using the data in Table 14.2.
29. A spaceship is fired from the earth's surface with an initial speed of 2.0×10^4 m/s. What will its speed be when it is very far from the earth? (Neglect friction.)
30. A 500-kg spaceship is in a circular orbit of radius $2R_e$ about the earth. (a) How much energy is required to transfer the spaceship to a circular orbit of radius $4R_e$? (b) Discuss the change in the potential energy, kinetic energy, and total energy.
31. (a) Calculate the minimum energy required to send a 3000-kg spacecraft from the earth to a distant point in space where earth's gravity is negligible. (b) If the journey is to take three weeks, what *average* power would the engines have to supply?
32. A rocket is fired vertically from the earth's surface and reaches a maximum altitude equal to three earth radii. What was the initial speed of the rocket? (Neglect friction, the earth's rotation, and the earth's orbital motion.)
33. A satellite moves in a circular orbit around a planet just above its surface. Show that the orbital velocity v and escape velocity of the satellite are related by the expression $v_{esc} = \sqrt{2}v$.
34. A satellite is in a circular orbit just above the surface of the moon. (The radius of the moon is 1738 km.) (a) What is the acceleration of the satellite? (b) What is the speed of the satellite? (c) What is the period of the satellite orbit?
35. A satellite with a mass of 500 kg is in a circular orbit at an altitude of 500 km above the earth's surface. Because of air friction, the satellite eventually is brought to the earth's surface, and it hits the earth with a velocity of 2 km/s. How much energy was absorbed by the atmosphere through friction?
36. An artificial earth satellite is "parked" in an equatorial circular orbit at an altitude of 10^3 km. What is the minimum additional velocity that must be imparted to the satellite if it is to escape from earth's gravitational attraction? How does this compare with the minimum escape velocity for leaving from the earth's surface?
37. (a) What is the minimum velocity necessary for a spacecraft to escape the solar system, starting at the earth's orbit? (b) Voyager 1 achieved a maximum velocity of 125 000 km/h on its way to photograph Jupiter. Beyond what distance from the sun is this velocity sufficient to escape the solar system?

*Section 14.9 The Gravitational Force Between an Extended Body and a Particle

38. A uniform rod of mass M is in the shape of a semicircle of radius R (Fig. 14.24). Calculate the force on a point mass m placed at the center of the semicircle.

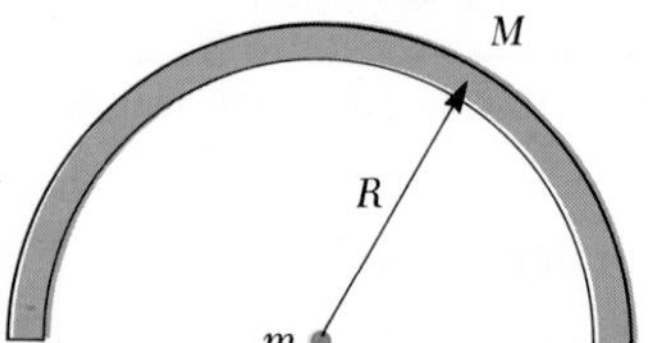

Figure 14.24 (Problem 38).

39. A *nonuniform* rod of length L is placed along the x axis at a distance h from the origin, as in Figure 14.16. The mass per unit length, λ, varies according to the expres-

sion $\lambda = \lambda_0 + Ax^2$, where λ_0 and A are constants. Find the force on a particle of mass m placed at the origin. (*Hint:* An element of the rod has a mass $dM = \lambda\, dx$.)

°Section 14.10 Gravitational Force Between a Particle and a Spherical Mass

40. A spherical shell has a radius of 0.5 m and mass of 80 kg. Find the force on a particle of mass 50 g placed (a) 0.3 m from the center of the shell and (b) outside the shell 1 m from its center.
41. A uniform solid sphere has a radius of 0.4 m and a mass of 500 kg. Find the magnitude of the force on a particle of mass 50 g located (a) 1.5 m from the center of the sphere, (b) at the surface of the sphere, and (c) 0.2 m from the center of the sphere.
42. A uniform solid sphere of mass m_1 and radius R_1 is inside and concentric with a spherical shell of mass m_2 and radius R_2 (Fig. 14.25). Find the force on a particle of mass m located at (a) $r = a$, (b) $r = b$, (c) $r = c$, where r is measured from the center of the spheres.

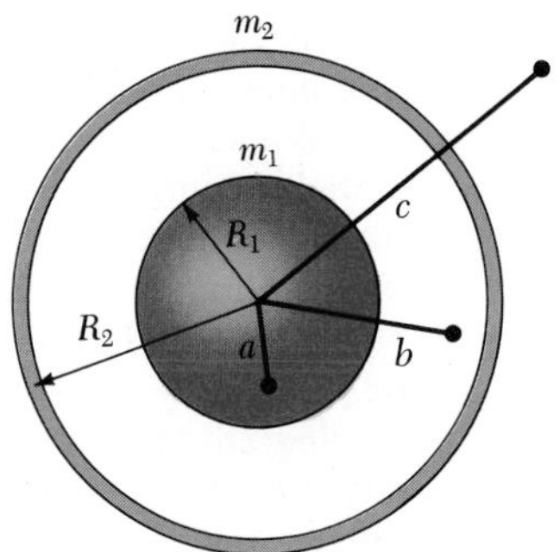

Figure 14.25 (Problem 42).

ADDITIONAL PROBLEMS

43. Calculate the fractional difference $\Delta g/g$ in the acceleration due to gravity at points on the Earth's surface nearest to and farthest from the Moon, taking into account the gravitational effect of the Moon. (This difference is responsible for the occurrence of the *lunar tides* on the Earth.)
44. Voyagers 1 and 2 surveyed the surface of Jupiter's moon Io and photographed active volcanoes spewing liquid sulfur to heights of 70 km above the surface of this moon. Estimate the velocity with which the liquid sulfur left the volcano. Io's mass is 8.9×10^{22} kg, and its radius is 1820 km.
45. A cylindrical habitat in space 6 km in diameter and 30 km long has been proposed (by G.K. O'Neill, 1974). Such a habitat would have cities, land, and lakes on the inside surface and air and clouds in the center. This would all be held in place by rotation of the cylinder about its long axis. How fast would the cylinder have to rotate to produce a 1-g gravity field at the walls of the cylinder?
46. Arranged in a rectangular coordinate system are a 2-kg mass at the origin, a 3-kg mass at the position (0, 2), and a 4-kg mass at (4, 0), where all distances are in m. Find the resultant gravitational force exerted on the mass at the origin by the other two masses.
47. Three masses are aligned along the x axis of a rectangular coordinate system such that a 2-kg mass is at the origin, a 3-kg mass is at (2, 0) m, and a 4-kg mass is at (4, 0) m. (a) Find the gravitational force exerted on the 4-kg mass by the other two masses. (b) Find the magnitude and direction of the gravitational force exerted on the 3 kg mass by the other two.
48. Four objects are located at the corners of a rectangle, as in Figure 14.26. Determine the magnitude and direction of the resultant force acting on the 2-kg mass at the origin.

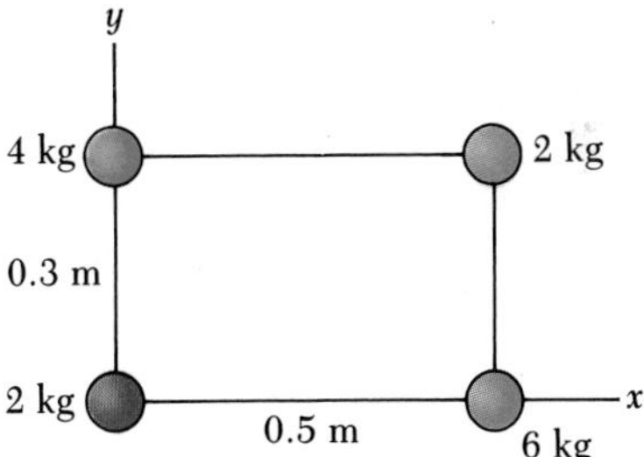

Figure 14.26 (Problem 48).

49. An object of mass m moves in a smooth straight tunnel of length L dug through a chord of the earth as discussed in Example 14.9 (Fig. 14.19). (a) Determine the effective force constant of the harmonic motion and the amplitude of the motion. (b) Using energy considerations, find the maximum speed of the object. Where does this occur? (c) Obtain a numerical value for the maximum speed if $L = 2500$ km.
50. For any planet, comet, or asteroid orbiting the sun, Kepler's third law may be written $T^2 = kr^3$ where T is the orbital period and r is the semimajor axis of the orbit. (a) What is the value of k if T is measured in years and r is measured in A.U.s? One astronomical unit (A.U.) is the mean distance from the earth to the sun. (b) Use this new value of k to quickly find the orbital period of Jupiter if its mean radius from the sun is 5.2 A.U.
51. Two ocean liners, each with a mass of 40 000 metric tons, are moving on parallel courses, 100 m apart. What is the magnitude of the acceleration of one of the liners toward the other due to the mutual gravitational attraction?
52. An airplane in a wide sweeping "outside" loop can create "zero gees" inside the aircraft cabin. What must be the radius of curvature of the flight path for an

aircraft moving at 480 km/h to create a condition of "weightlessness" inside the aircraft?

53. What angular velocity (in rev/min) is needed for a centrifuge to produce an acceleration of 1000g at a radius arm of 10 cm?

54. The maximum distance from the earth to the sun (at the aphelion) is 1.521×10^{11} m, and the distance of closest approach (at the perihelion) is equal to 1.471×10^{11} m. If the earth's orbital speed at the perihelion is 3.027×10^4 m/s, determine (a) the earth's orbital speed at the aphelion, (b) the kinetic and potential energy at the perihelion, and (c) the kinetic and potential energy at the aphelion. Is the total energy conserved? (Neglect the effect of the moon and other planets.)

55. Two hypothetical planets of masses m_1 and m_2 and radii r_1 and r_2, respectively, are at rest when they are an infinite distance apart. Because of their gravitational attraction, they head toward each other on a collision course. (a) When their center-to-center separation is d, find the speed of each planet and their *relative* velocity. (b) Find the kinetic energy of each planet *just* before they collide if $m_1 = 2 \times 10^{24}$ kg, $m_2 = 8 \times 10^{24}$ kg, $r_1 = 3 \times 10^6$ m, and $r_2 = 5 \times 10^6$ m. (*Hint:* Note that both energy and momentum are conserved.)

56. Use the equation $F = mv^2/r$ to calculate the centripetal force needed to make the earth follow its path about the sun. Compare this value for F with the value found by using Newton's universal law of gravity.

57. When the Apollo 11 spacecraft orbited the moon, its mass was 9.979×10^3 kg, its period was 119 min, and its mean distance from the moon's center was 1.849×10^6 m. Assuming its orbit was circular and the moon to be a uniform sphere, find (a) the mass of the moon, (b) the orbital speed of the spacecraft, and (c) the minimum energy required for the craft to leave the orbit and escape the moon's gravity.

58. Studies of the relationship of the sun to the local galaxy—the Milky Way—have revealed that the sun is located near the outer edge of the galactic disc, about 30 000 light years from the center. Furthermore, it has been found that the sun has an orbital velocity of approximately 250 km/s around the galactic center. (a) What is the period of the sun's galactic motion? (b) What is the approximate mass of the Milky Way galaxy? Using the fact that the sun is a typical star, estimate the number of stars in our local galaxy.

59. Using the data in Table 14.2, calculate the total potential energy of the sun-moon-earth system. Assume that the moon and earth are at the same distance from the sun.

60. A hypothetical planet of mass M has three moons of equal mass m, each moving in the same circular orbit of radius R (Fig. 14.27). The moons are equally spaced and thus form an equilateral triangle. Find (a) the total potential energy of the system and (b) the orbital speed of each moon such that they maintain this configuration.

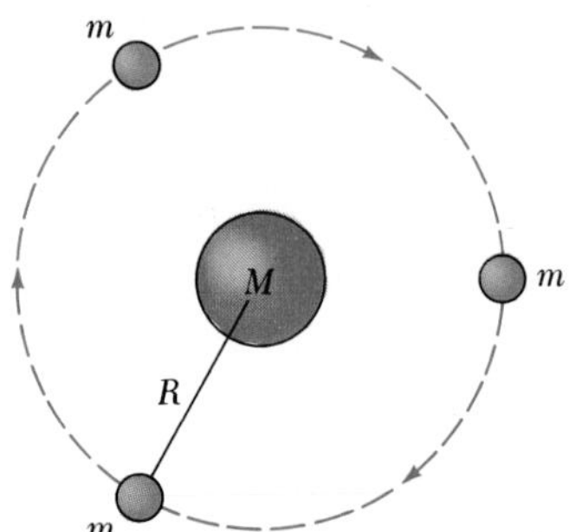

Figure 14.27 (Problem 60).

61. A satellite of mass 200 kg is placed in Earth orbit at a height 200 km above the surface. (a) Assuming a circular orbit, how long does the satellite take to complete one orbit? (b) What is the satellite's speed? (c) What is the minimum energy necessary to place this satellite in orbit (assuming no air friction)?

62. A particle of mass m lies along the symmetry axis of a uniform circular ring of mass M and radius R (Fig. 14.28). (a) Find the force on m if it is at a distance d from the plane of the ring. (b) Show that your result to (a) reduces to what you would intuitively expect (1) when m is at the center of the ring ($d = 0$) and (2) when m is distant from the ring ($d \gg R$).

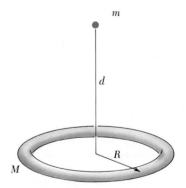

Figure 14.28 (Problem 62).

63. A sphere of mass M and radius R has a *nonuniform* density that varies with r, the distance from its center, according to the expression $\rho = Ar$, for $0 \le r \le R$.

(a) What is the constant A in terms of M and R? (b) Determine the force on a particle of mass m placed *outside* the sphere. (c) Determine the force on the particle if it is *inside* the sphere. (*Hint:* See Section 14.10.)

64. Two stars of masses M and m, separated by a distance d, revolve in circular orbits about their center of mass (Fig. 14.29). Show that each star has a period given by

$$T^2 = \frac{4\pi^2}{G(M+m)}\, d^3$$

(*Hint:* Apply Newton's second law to each star, and note that the center of mass condition requires that $Mr_2 = mr_1$, where $r_1 + r_2 = d$.)

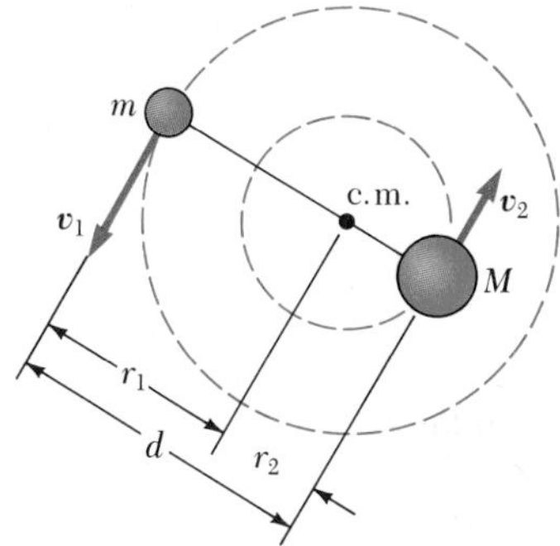

Figure 14.29 (Problem 64).

65. In 1978, astronomers at the U.S. Naval Observatory discovered that the planet Pluto has a moon, called Charon, that eclipses the planet every 6.4 days. If, from observation, the center-to-center separation between Pluto and Charon is 19 700 km, find the total mass $(M+m)$ of Pluto and its moon. (*Hint:* Use the result of Problem 64.)

66. In an effort to explain large meteor collisions with the earth (which seem to occur approximately once every 3×10^7 years), scientists have postulated the existence of a solar companion star which is extremely dim and extremely far from the sun. This star (which some call Nemesis) moves through the axis of the cloud of comets, disturbing their orbits every 3×10^7 years. If this hypothetical star has the above orbital period and a mass of 0.2 times the sun's mass, determine the average distance of this (solar companion) star from the sun. $M_{sun} = 2 \times 10^{30}$ kg.

67. A satellite is in a circular orbit about a planet of radius R. If the altitude of the satellite is h and its period is T, (a) show that the density of the planet is given by

$$\rho = \frac{3\pi}{GT^2}\left(1 + \frac{h}{R}\right)^3$$

(b) Calculate the average density of the planet if the period is 200 min and the satellite's orbit is close to the planet's surface.

68. The acceleration of gravity at an altitude h is given by Equation 14.5. If $h \ll R_e$, show that the acceleration of gravity at h is given *approximately* by

$$g' \approx g\left(1 - 2\frac{h}{R_e}\right)$$

(*Hint:* Start with Equation 14.5, and use the binomial expansion for the denominator.)

69. A particle of mass m is located *inside* a uniform solid sphere of radius R and mass M. If the particle is at a distance r from the center of the sphere, (a) show that the gravitational potential energy of the system is given by $U = (GmM/2R^3)r^2 - 3GmM/2R$. (b) How much work is done by the gravitational force in bringing the particle from the surface of the sphere to its center?

70. (a) A planet has a mass M_r, radius r, and uniform density ρ. Consider a segment of this planet in the form of a thin spherical shell near its surface of mass $dm = \rho 4\pi r^2\, dr$. Show that the energy required to remove the mass in this shell to infinity is given by

$$dW = \frac{GM_r\, dm}{r} = \frac{GM_r}{r}(\rho 4\pi r^2\, dr)$$

(b) If the mass of the planet inside radius r is given by $M_r = \rho \dfrac{4\pi r^3}{3}$, find the total binding energy of the planet if it is built up of spherical shells, starting at radius $r = 0$ and continuing until $r = R_o$, the outer radius, where it has a mass $M_o = \dfrac{4}{3}\rho\pi R_o^{\,3}$.

"I must say it isn't easy adjusting to a 24-second day."

ESSAY

The Big Bang

Sidney C. Wolff
National Optical Astronomy Observatory

Newton's universal law of gravitation — today it is taught in every introductory physics course. It has become so familiar that we can easily forget how profoundly this simple equation changed the nature of science. Newton not only offered a precise mathematical description of the motions of material objects, he also showed that this description applies far beyond the Earth. The same force that causes apples to fall to the ground also keeps the Moon revolving about the Earth and holds the planets in their orbits around the Sun. The heavens were transformed. No longer mysterious and unknowable, the Moon, the planets, and potentially even the stars, all subject to the same physical laws that apply on Earth, became legitimate objects of scientific inquiry.

But is Newton's law of gravitation truly universal? Does it apply to the stars? Newton assumed that it did, but measurements available during his lifetime were not precise enough to offer proof. The influence of gravity is detected by analyzing the *motions* of objects. Apart from their rising and setting, which is a consequence of the Earth's rotation, stars did not appear to the astronomers of Newton's time to have motions of their own. The fact that there are clusters or groupings of stars, like the Pleiades and Hyades, suggested that attractive forces do influence stars. Proof of this hypothesis, however came only a century later.

In January 1782, Sir William Herschel, best known for his discovery of the planet Uranus, published a catalog of 269 pairs of stars that appear to be very close together in the sky. At that time, he did not know whether these double stars form physical systems or simply happen to appear in very nearly the same direction but at very different distances from the Earth. About 20 years later, Herschel repeated the observations of these double stars and discovered that some pairs had changed position relative to each other in a way that could only be explained if they were revolving around each other. Several more decades of measurements by other astronomers proved that these stars do indeed follow orbits that can be described by the inverse square force of Newton's law of gravitation.

The conclusion that the motions of stars are controlled by gravity raises a problem that was recognized by Newton. Up until the beginning of the 20th century it was assumed that the universe is static, that is, that it is neither expanding nor contracting. If the theory of gravitation applies to all of the objects in a static universe, and if the universe is finite in size, then it is difficult to see why the universe has not already collapsed to a single mass as all of the objects are attracted to one another. Newton thought that in an infinite universe it might be impossible for all the matter to fall together into a single large mass. It seemed more likely to him that some matter would aggregate to form one large mass, other matter would form a second large mass, and so on. The result would be an infinite number of large masses scattered through infinite space.

Figure 1 The spiral galaxy M 83, which is about 10 million lightyears distant from the Earth and has a diameter of about 30,000 lightyears. A galaxy like M 83 contains billions of individual stars. (National Optical Astronomy Observatories).

The next major advance in gravitational theory after Newton's work is embodied in Einstein's general theory of relativity. In 1917, Einstein applied his new theory to the universe as a whole and found that not even infinite universes can be static. Since Einstein believed that the universe is neither expanding nor contracting, he introduced a new term into his equations. This term, which is called the cosmological constant, is equivalent to a repulsive force that can balance gravitational attraction over large distances and permit a static universe. There is, however, no evidence for such a repulsion in nature. Einstein introduced the cosmological constant in order to make general relativity conform to the notion, widely held at the time, that the universe is static.

In fact, the universe is *not* static. Observations show that it is *expanding*. This fundamental fact underlies all of modern cosmology. In a classic paper published in 1931, Edwin Hubble and Milton Humason compared distances and velocities of remote galaxies, which are giant systems of billions of individual stars (Figure 1). That comparison established that galaxies are moving away from us with velocities that are

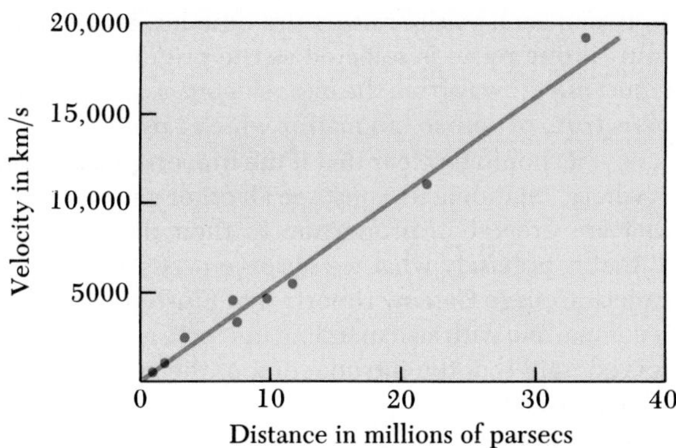

Figure 2 Hubble and Humason's velocity-distance relation, adapted from their 1931 paper in the *Astrophysical Journal.* Distance is measured in parsecs, and one parsec is the distance that light travels in 3.26 years. (From Abell, Morrison, and Wolff, *Exploration of the Universe,* 5th ed., 1987, p. 658)

proportional to their distances from us. This relationship can be expressed as a formula:

$$v = H_0 r,$$

Where v is the velocity with which a given galaxy is moving away from us, r is the distance to the galaxy, and H_0 is the constant of proportionality. This equation is called the *Hubble law* and H_0 is called the *Hubble constant* (Figure 2).

If velocity is proportional to distance, then the universe must be expanding. An analogy can help to show why. Suppose a cook is making raisin bread. Suppose that after yeast has been mixed into the dough and the bread is set aside to rise, it doubles in size during the next hour (Figure 3). All of the raisins move further apart, and in

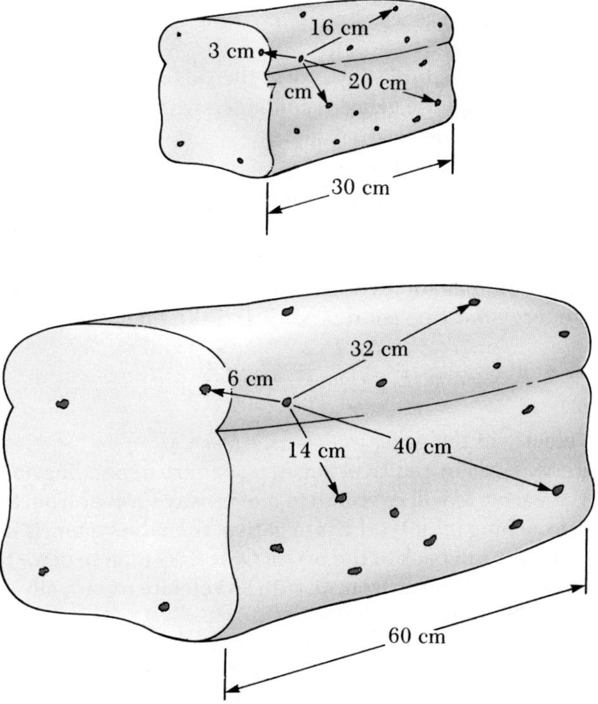

Figure 3 Expanding raisin bread. (Adapted from Abell, George O., David Morrison, and Sidney C. Wolff, *Exploration of the Universe,* 5th ed., Philadelphia, Saunders College Publishing, 1987; p. 659)

(Continued on next page)

fact the distance between each pair of raisins also doubles. Since each distance doubles during the hour, if one raisin is selected as the origin of a coordinate system, every other raisin must move away from the one selected at a speed proportional to its distance. The same is true, of course, no matter which raisin you select.

From this analogy, it should be clear that if the universe is uniformly expanding, all observers everywhere, including us, must see all other galaxies moving away from them at speeds that are greater in proportion to their distances. As Hubble and Humason showed, that is precisely what we observe on Earth.

American physicist George Gamow reports that Einstein, upon learning that his relativity theory is compatible with an expanding universe and that the expansion had been directly observed, said that the introduction of the cosmological constant was "the biggest blunder of my life."

In order to draw any conclusions about the universe as a whole, it is necessary to assume that the parts of the universe that we can actually observe are truly representative. This assumption is known as the cosmological principle, which asserts that, apart from local irregularities, the universe at any given time is the same everywhere. If this principle is valid, then we can reach conclusions about the universe from measurements of a small portion of it that is close enough to study in detail.

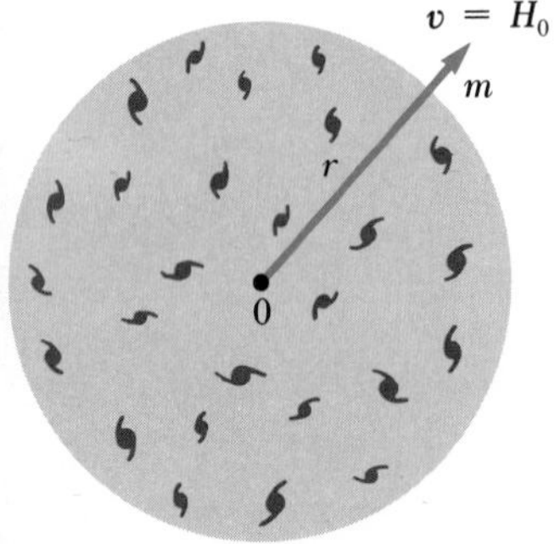

Figure 4 m is the mass of a galaxy located at the surface of a sphere of radius r centered on O. The sphere contains many galaxies with a total mass M.

It is possible to derive some important conclusions about the expanding universe from Newton's law of gravitation, the cosmological principle, and one additional assumption. Imagine a sphere of radius r centered on the observer at point O (Figure 4). The assumption that must be made is that the motion of a galaxy located at the surface of the sphere is the result only of the gravitational forces of the galaxies inside the sphere. The gravitational forces of all the galaxies outside the sphere are assumed to cancel each other and produce no net effect on the galaxy at the surface of the sphere.

This assumption may seem plausible for an infinite universe, but it cannot be proved valid in Newtonian theory. It is only with the general theory of relativity that the calculations that we are about to make about the motion of galaxies can be shown to be correct. Nevertheless, these Newtonian calculations do yield correct results, and those results allow us to reach some interesting conclusions about the past and future of the universe.

Let m be the mass of the typical galaxy at the surface of the sphere in Figure 4. Let M be the total mass of all the galaxies contained within the sphere. This mass can be calculated according to the relationship.

$$M = \text{density} \cdot \text{volume} = \frac{\rho_0 4\pi r^3}{3},$$

where ρ_0 is the density of matter inside the sphere at the present time.

If there are no additional forces other than gravity acting on the galaxy with mass m, then its total energy must be conserved. If E is the total energy, then

$$E = (\tfrac{1}{2})mv^2 - \frac{GMm}{r},$$

where v is the velocity of the galaxy.

The total energy E can be positive, negative, or zero depending on the value of v. If E is positive, the galaxy M will continue to move away forever from the observer at O, eventually approaching infinity. If E is negative, then the system is bound, and the galaxy m will eventually fall back to the origin O. If E is equal to zero, the galaxy will just barely continue to move away from O, with its velocity v gradually slowing so that it approaches zero as r approaches infinity.

The cosmological principle also assures us that any conclusion that we reach about this sphere and the galaxy at the surface of the sphere must apply to all other galaxies throughout the universe. Therefore, positive energy corresponds to a universe that will expand forever. Negative E corresponds to a universe that will eventually stop its expansion and collapse again.

Since $v = H_0 r$ according to the Hubble law, if $E = 0$, then

$$\rho_0 = \frac{3H_0^2}{8\pi G}$$

In other words, if the actual density of the universe now is equal to $3H_0^2/8\pi G$, then the universe is marginally bound and will just barely manage to continue its expansion forever.

The situation is exactly analogous to that of an object launched upward from the surface of the Earth. If the velocity of the object exceeds the escape velocity, the object will never return to the Earth's surface. If the velocity is less than the escape velocity, the object will fall back to the ground. Similarly, if the galaxy has a velocity that is high enough to overcome the attractive force of all the other galaxies within the sphere, it will continue to move away from them, and the expansion of the universe will continue. If the velocity is too low, however, gravity will decelerate the galaxy until its velocity becomes zero, and then the galaxy will reverse direction and begin to move closer to the other galaxies within the sphere.

The obvious question is which of these possibilities describes the future of the universe. The answer depends on how the actual denisty compares with the critical density, that is, on how ρ_0 compares with $3H_0^2/8\pi G$. The value of ρ_0 can be estimated by counting up the mass associated with the galaxies within some volume of space of known size. Most galaxies occur in groups or clusters and can be "weighed" by measuring the gravitational influence that they exert on other nearby galaxies (Figure 5). For example, if two galaxies are in orbit around each other, and if their separations and velocities are known, their combined mass can be estimated from Kepler's third law. There are mathematical procedures that are not much more complex for estimating the total mass of a cluster containing not just two but many galaxies.

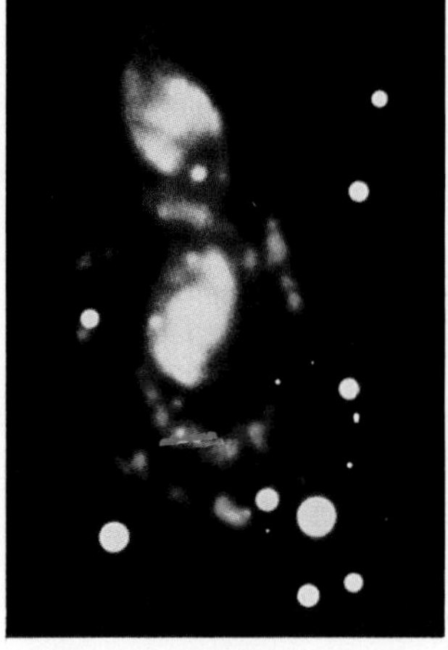

Figure 5 The average density of the universe can be estimated by studying the velocities of galaxies in pairs, small groups, or clusters. This pair of interacting galaxies (NGC 2207 and IC 2103) is in the constellation Canis Major. The blue color indicates regions where there are very young stars. (National Optical Astronomy Observatories).

The best available observational estimates of the average density of the universe suggest that the amount of matter that it contains is no more than 10 to 20 percent of the value required to halt the expansion. There are some theoretical arguments for thinking that the density may be even higher and may just exactly equal the critical density required to slow the expansion to a halt. The negative acceleration is so gradual in this case that an infinite amount of time is required to reduce the expansion velocity to zero. Therefore, both observation and theory suggest that galaxies will never reverse direction and move closer together.

The only way to escape this conclusion would be if astronomers have somehow failed to detect a very large fraction of the total mass of the universe. The missing mass would have to be dark and radiate no light. It would also have to be located in regions that are empty of galaxies so that we have no way to detect the gravitational force of this dark matter by studying its influence on nearby visible objects. Whether or not dark matter exists in large quantities is one of the most hotly disputed issues in all of astronomy.

As the universe expands, galaxies separate from each other. We can also, however, imagine what must have happened in the past. If we extrapolate backward in time, we find all of the galaxies coming together, until some time in the distant past when all matter was crowded to an extreme density—a condition that marks a unique beginning of the universe, or at least of that universe we can know about. At that beginning, the universe suddenly began its expansion with a phenomenon called the **Big Bang**.

With the data now available, we can even estimate when our universe began. The total amount of matter and energy of the universe creates gravitation, whereby all

(Continued on next page)

objects pull on all other objects. This mutual attraction must slow the expansion, which means that in the past the expansion must have been at a greater rate than it is today. How much greater depends on the importance of gravitation in decreasing the rate of expansion. At the extreme, if the total mass-energy density is low enough that gravitation is ineffective (an essentially "empty" universe), the acceleration would be zero, and in that case the universe would always have been expanding at the present rate.

Clearly, that extreme of an empty universe corresponds to the greatest age of the universe (since the Big Bang). If the expansion were faster in the past, galaxies would have reached their present separation in a smaller time. Consequently, we can obtain an estimate of the upper limit to the age of the universe by asking how long it would take for distant galaxies, always moving away from us at their present rates, to have reached those distances. Call that maximum possible age T_0. According to the Hubble law,

$$v = H_0 r.$$

But velocity is just distance divided by time; that is, $v = r/T_0$. Hence,

$$\frac{r}{T_0} = H_0 r,$$

or

$$T_0 = \frac{1}{H_0}.$$

We see, then, that the maximum age of the universe is just the reciprocal of the Hubble constant. The value of the Hubble constant is uncertain, primarily because of the uncertainties in deriving the distances to remote galaxies, but the best estimate is that the Big Bang occurred 10 to 15 billion years ago.

If current ideas are correct, there can be no rejuvenation of the universe. At the time of the Big Bang, the universe was very dense and hot. As it expanded, it cooled. Hydrogen and helium nuclei and then atoms formed. Somehow—theory is not yet very good at explaining this point—hydrogen and helium atoms came together to form concentrations of matter that would become galaxies. Within these protogalaxies stars formed. In many galaxies, star formation has continued to the present. Each star, some in a million years and others only after billions of years, exhausts its sources of nuclear energy. Some stars die gently, others explosively, hurling a portion of their mass into space to be incorporated into a new generation of stars. As they die, most stars also leave behind a portion of their mass in the form of a dense core. The matter in these dense cores is forever lost to the universe. It radiates no energy and cannot, under any conditions that we now believe likely to occur, be broken apart into gas and dust particles that would provide the raw material for new stars. Ultimately, after the gas and the dust are used up, star formation must cease, a final generation of stars will die, and the galaxies will fade into blackness.

Astronomy and physics as we understand them today offer no alternative to this outcome. Today's knowledge, however, is most likely incomplete.

At the end of the 19th century, physics was thought to be totally understood. Theories of gravity, heat, electricity, and magnetism all seemed adequate to explain experimental data. Physicists expected to make no major new breakthroughs but simply to refine and improve the existing mathematical descriptions of physical phenomena. The revolution came in 1900 when Max Planck proposed the quantum principle.

With time, the picture presented here of the origin, evolution, and ultimate fate of the material universe may undergo a comparably radical transformation. Indeed, astronomers are attracted to this area of research precisely because the subject is not fully understood and seems likely to yield its secrets to a systematic scientific attack. To learn something new, something never known before, is the ultimate challenge and satisfaction of scientific research.

Suggested Readings

Ferris, T., *The Red Limit,* 2nd ed., New York, Morrow, Quill, 1983.

Harrison, E.R., *Cosmology,* Cambridge, England, Cambridge University Press, 1981.

Silk, J., *The Big Bang,* New York, W.H. Freeman, 1989.

Tucker, W., and K. Tucker, *The Dark Matter,* New York, William Morrow, 1988.

Wagoner, R., and D. Goldsmith, *Cosmic Horizons: Understanding the Universe,* New York, W.H. Freeman, 1983.

Essay Questions

1. When Hubble and Humason first reported their results for the relation between distance and velocity for galaxies, they thought that the constant of proportionality H_0 was equal to about 200 km/s per million light years. How did their estimate for the age of the universe differ from the modern value based on $H_0 = 20$ km/s per million light years? The oldest stars in the universe are about 15 billion years old. Is this consistent with Hubble and Humason's value for the Hubble constant H_0?
2. In the text, it was explained that it was possible to estimate the masses of two galaxies orbiting about each other by measuring their separations and velocities and applying Kepler's third law. Could this be done by measuring the orbital period directly? Why or why not? (Remember that galaxies contain billions of stars and have diameters of about 100,000 light years.)
3. Explain why the observation that all galaxies are moving away from us does not mean that we are at the center of the universe.

Essay Problems

1. Show that if the total energy of a galaxy at the surface of a sphere in Figure 4 is equal to zero, then $\rho_0 = 3H_0^2/8\pi G$.
2. The maximum age of the universe is equal to $1/H_0$. Show that if the actual density of the universe is equal to $3H_0^2/8\pi G$, then the age of the universe equals $(2/3)/H_0$. (*Hint:* use the fact that $E = 0$ for this density, and derive an expression for $v = dr/dt$. Integrate this equation and assume that at $t = 0$, $r = 0$.)
3. Current estimates for the Hubble constant H_0 range from 15 to 30 km/s per million light years, where one light year is the distance that light travels in one year. What is the corresponding range in the estimated age of the universe?
4. Derive an expression for escape velocity in terms of mean density. Now apply this result to two galaxies, separated by a distance R and moving away from each other at just the escape velocity. Derive an expression for the critical density at which the escape and expansion velocities are equal. Assume that $H_0 = 20$ km/s per million light years. How does your value for the critical density compare with the observed density, which is currently estimated to be less than about 10^{-30} g/cm^3. What does your calculation indicate about the ultimate fact of the universe?

15

Fluid Mechanics

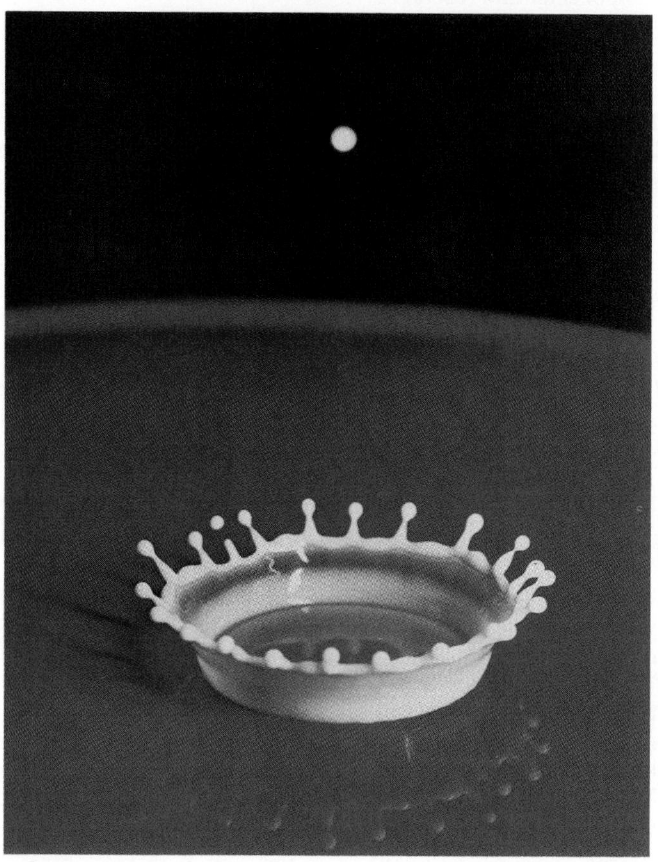

Coronet formed by a drop of milk. (© Harold E. Edgerton. Courtesy of Palm Press, Inc.)

In our treatment of the mechanics of fluids, we shall see that no new physical principles are needed to explain such effects as the buoyant force on a submerged object and the dynamic lift on an airplane wing. First, we shall present a discussion of the various states of matter. Next, we shall consider a fluid at rest and derive an expression for the pressure as a function of its density and depth. We shall then treat fluids in motion, or fluid dynamics. A fluid in motion can be described by a model in which certain simplifying assumptions are made. We shall use this model to analyze some situations of practical importance. An underlying principle known as the *Bernoulli principle* will enable us to determine relations between the pressure, density, and velocity at every point in a fluid. As we shall see, the Bernoulli principle is a result of conservation of energy applied to an ideal fluid. We conclude the chapter with a brief discussion of internal friction in a fluid and turbulent motion.

15.1 STATES OF MATTER

Matter is normally classified as being in one of three states: solid, liquid, or gaseous. Often, this classification is extended to include a fourth state referred to as a plasma.

Everyday experience tells us that a solid has a definite volume and shape. A brick maintains its familiar shape and size day in and day out. We also know that a liquid has a definite volume but no definite shape. For example, when you fill the tank in a car, the gasoline assumes the shape of the tank in the vehicle, but if you have a gallon of gasoline before filling, you will have a gallon after. Finally, a gas has neither definite volume nor definite shape. These definitions help us to picture the states of matter, but they are somewhat artificial. For example, asphalt and plastics are normally considered solids, but over long periods of time they tend to flow like liquids. Likewise, water can be a solid, liquid, or gas (or combinations of these), depending on the temperature and pressure. The response time of the change in shape to an *external* force or pressure determines if we treat the substance as a solid, a very viscous fluid, or another state.

Droplets of mercury lying on a glass surface. Mercury is the only metal that is liquid at room temperature. Note that the small droplets are almost spherical, while the large droplets are flattened. This shows that the effect of surface tension has more influence on the shape of the small (lighter) droplets. (© Charles Steele)

The fourth state of matter can occur when matter is heated to very high temperatures. Under these conditions, one or more electrons surrounding each atom are freed from the nucleus. The resulting substance is a collection of free electrically charged particles: the negatively charged electrons and the positively charged ions. Such an ionized gas with equal amounts of positive and negative charges is called a **plasma.** The plasma state exists inside stars, for example. If we were to take a grand tour of our universe, we would find that there is far more matter in the plasma state than in the more familiar forms of solid, liquid, and gas because there are far more stars around than any other form of celestial matter. However, in this chapter we shall ignore this plasma state and concentrate instead on the more familiar solid, liquid, and gaseous forms that make up the environment on our planet.

All matter consists of some distribution of atoms and molecules. The atoms in a solid are held at specific positions with respect to one another by forces that are mainly electrical in origin. The atoms of a solid vibrate about these equilibrium positions because of thermal agitation. However, at low temperatures, this vibrating motion is slight and the atoms can be considered to be almost fixed. As thermal energy (heat) is added to the material, the amplitude of these vibrations increases. One can view the vibrating motion of the atom as that which would occur if the atom were bound to its equilibrium position by springs attached to neighboring atoms. One such vibrating collection of atoms and imaginary springs is shown in Figure 15.1. If a solid is compressed by external forces, we can picture these external forces as compressing these tiny internal springs. When the external forces are removed, the solid tends to return to its original shape and size. For this reason, a solid is said to have elasticity.

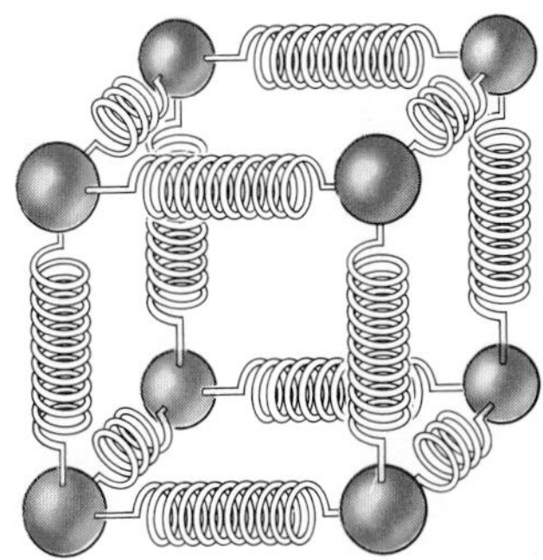

Figure 15.1 A model of a solid. The atoms (spheres) are imagined as being attached to each other by springs, which represent the elastic nature of the interatomic forces.

Solids can be classified as being either crystalline or amorphous. A **crystalline solid** is one in which the atoms have an ordered, periodic structure. For example, in the sodium chloride crystal (common table salt), sodium and chlorine atoms occupy alternate corners of a cube face, as in Figure 15.2a. In an **amorphous solid,** such as glass, the atoms are arranged in a disordered fashion, as in Figure 15.2b.

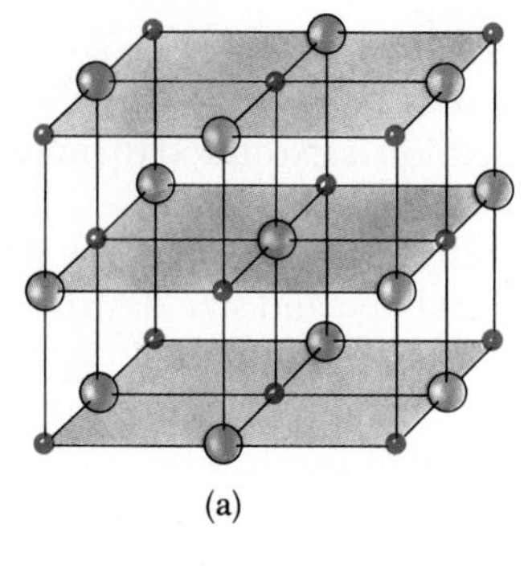

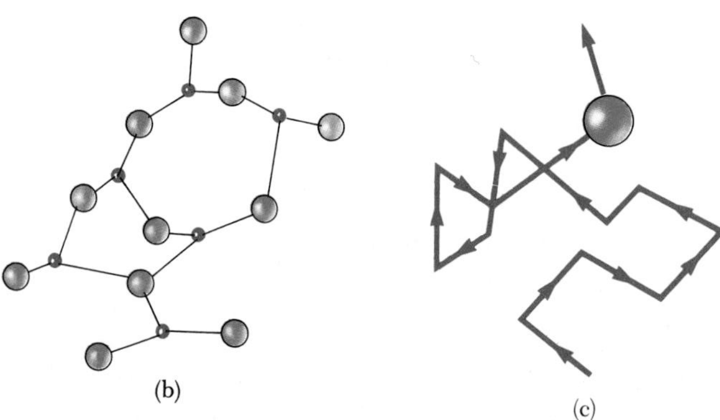

Figure 15.2 (a) The NaCl structure, with Na^+ and Cl^- ions at alternate corners of a cube. The small spheres represent Na^+ ions, and the large spheres represent Cl^- ions. (b) In an amorphous solid, the atoms are arranged in a random fashion. (c) Erratic motion of a molecule in a liquid.

In any given substance, the liquid state exists at a higher temperature than the solid state. Thermal agitation is greater in the liquid state than in the solid state. As a result, the molecular forces in a liquid are not strong enough to keep the molecules in fixed positions, and the molecules wander through the liquid in a random fashion (Fig. 15.2c). Solids and liquids have the following property in common. When one tries to compress a liquid or a solid, strong repulsive atomic forces act internally to resist the deformation.

In the gaseous state, the molecules are in constant random motion and exert only weak forces on each other. The average separation distances between the molecules of a gas are quite large compared with the dimensions of the molecules. Occasionally, the molecules collide with each other; however, most of the time they move as nearly free, noninteracting particles. We shall have more to say about the properties of gases in subsequent chapters.

15.2 DENSITY AND PRESSURE

The **density** of a substance is defined as its mass per unit volume. That is, a substance of mass m and volume V has a density ρ given by

Definition of density

$$\rho \equiv \frac{m}{V} \tag{15.1}$$

TABLE 15.1 Densities of Some Common Substances

Substance	ρ (kg/m³)[a]	Substance	ρ (kg/m³)[a]
Ice	0.917×10^3	Water	1.00×10^3
Aluminum	2.70×10^3	Glycerine	1.26×10^3
Iron	7.86×10^3	Ethyl alcohol	0.806×10^3
Copper	8.92×10^3	Benzene	0.879×10^3
Silver	10.5×10^3	Mercury	13.6×10^3
Lead	11.3×10^3	Air	1.29
Gold	19.3×10^3	Oxygen	1.43
Platinum	21.4×10^3	Hydrogen	8.99×10^{-2}
		Helium	1.79×10^{-1}

[a] All values are at standard atmospheric pressure and temperature (STP), that is, atmospheric pressure and 0°C. To convert to grams per cubic centimeter, multiply by 10^{-3}.

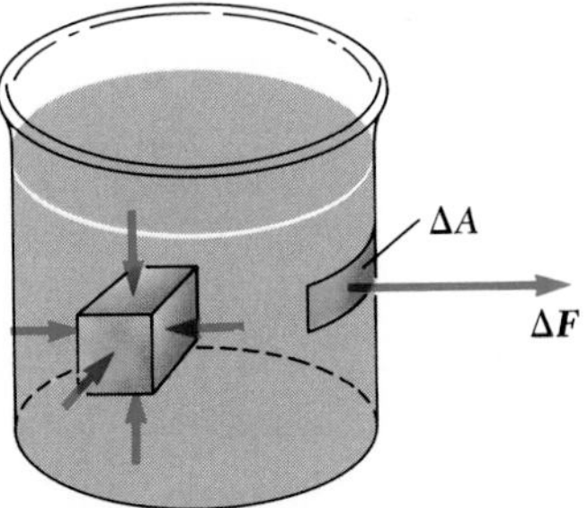

Figure 15.3 The force of the fluid on a submerged object at any point is perpendicular to the surface of the object. The force of the fluid on the walls of the container is perpendicular to the walls at all points.

The units of density are kg/m³ in the SI system and g/cm³ in the cgs system. Table 15.1 lists the densities of various substances. These values vary slightly with temperature, since the volume of a substance is temperature dependent (as we shall see in Chapter 16). Note that under standard conditions (0°C and atmospheric pressure) the densities of gases are about 1/1000 the densities of solids and liquids. This implies that the average molecular spacing in a gas under these conditions is about ten times greater than in a solid or liquid.

The **specific gravity** of a substance is defined as the ratio of its density to the density of water at 4°C, which is 1.0×10^3 kg/m³. By definition, specific gravity is a dimensionless quantity. For example, if the specific gravity of a substance is 3, its density is $3(1.0 \times 10^3 \text{ kg/m}^3) = 3.0 \times 10^3 \text{ kg/m}^3$.

We have seen that fluids do not sustain shearing stresses, and thus the only stress that can exist on an object submerged in a fluid is one that tends to compress the object. The force exerted by the fluid on the object is always perpendicular to the surfaces of the object, as shown in Figure 15.3.

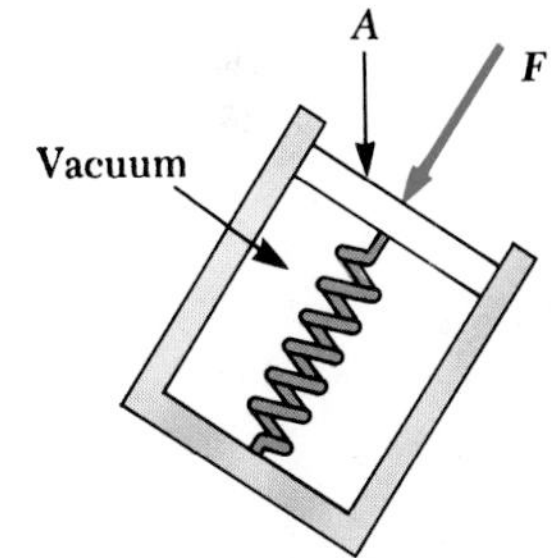

Figure 15.4 A simple device for measuring pressure.

The pressure at a specific point in a fluid can be measured with the device pictured in Figure 15.4. The device consists of an evacuated cylinder enclosing a light piston connected to a spring. As the device is submerged in a fluid, the fluid presses down on the top of the piston and compresses the spring until the inward force of the fluid is balanced by the outward force of the spring. The fluid pressure can be measured directly if the spring is calibrated in advance. This is accomplished by applying a known force to the spring to compress it a given distance.

Definition of pressure

If F is the magnitude of the normal force on the piston and A is the area of the piston, then the **pressure,** P, of the fluid at the level to which the device has been submerged is defined as the ratio of force to area:

$$P \equiv \frac{F}{A} \tag{15.2}$$

The pressure in a fluid is not the same at all points. To define the pressure at a specific point, consider a fluid enclosed as in Figure 15.3. If the normal force exerted by the fluid is ΔF over a surface element of area ΔA, then the **pressure** at that point is

$$P = \lim_{\Delta A \to 0} \frac{\Delta F}{\Delta A} = \frac{dF}{dA} \tag{15.3}$$

As we shall see in the next section, the pressure in a fluid in the presence of the force of gravity varies with depth. Therefore, to get the total force on a plane wall of a container, we have to integrate Equation 15.3 over the surface.

Since pressure is force per unit area, it has units of N/m^2 in the SI system. Another name for the SI unit of pressure is **pascal** (Pa).

$$1 \text{ Pa} \equiv 1 \text{ N/m}^2 \tag{15.4}$$

15.3 VARIATION OF PRESSURE WITH DEPTH

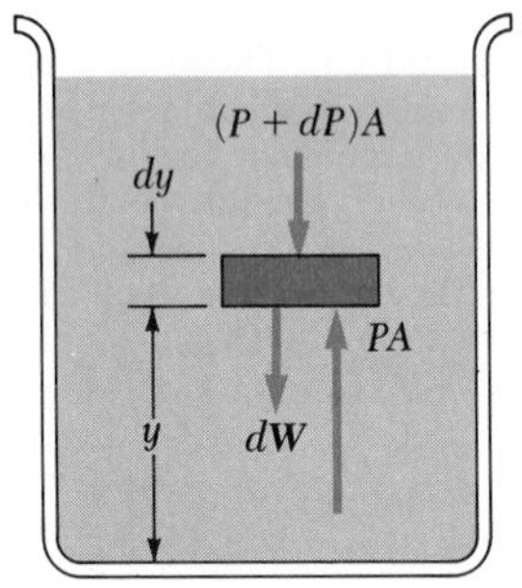

Figure 15.5 The variation of pressure with depth in a fluid. The volume element is at rest, and the forces on it are shown.

Consider a fluid at rest in a container (Fig. 15.5). We first note that *all points at the same depth are at the same pressure.* If this were not the case, a given element of the fluid would not be in equilibrium. To see this, let us select a portion of the fluid contained within an imaginary cylinder of cross-sectional area A and height dy. The upward force on the bottom of the cylinder is PA, and the downward force on the top is $(P + dP)A$. The weight of the cylinder, the volume of which is dV, is given by $dW = \rho g\, dV = \rho g A\, dy$, where ρ is the density of the fluid. Since the cylinder is in equilibrium, the forces must add to zero, and so we get

$$\sum F_y = PA - (P + dP)A - \rho g A\, dy = 0$$

$$\frac{dP}{dy} = -\rho g \tag{15.5}$$

From this result, we see that an increase in elevation (positive dy) corresponds to a decrease in pressure (negative dP). If P_1 and P_2 are the pressures at the elevations y_1 and y_2 above the reference level, and if the density is uniform, then integrating Equation 15.5 gives

$$P_2 - P_1 = -\rho g(y_2 - y_1) \tag{15.6}$$

If the vessel is open at the top (Fig. 15.6), then the pressure at the depth h can be obtained from Equation 15.5. Taking atmospheric pressure to be $P_a = P_2$, and noting that the depth $h = y_2 - y_1$, we find that

$$P = P_a + \rho g h \tag{15.7}$$

where we usually take $P_a \approx 1.01 \times 10^5$ Pa (14.7 lb/in.^{2}). In other words,

the **absolute pressure** P at a depth h below the surface of a liquid open to the atmosphere is *greater* than atmospheric pressure by an amount $\rho g h$.

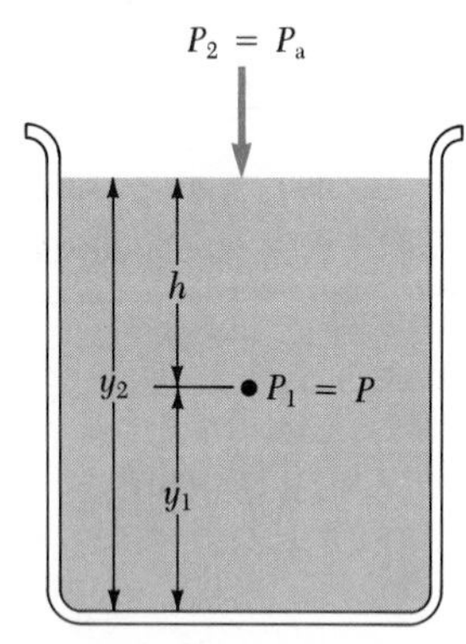

Figure 15.6 The pressure P at a depth h below the surface of a liquid open to the atmosphere is given by $P = P_a + \rho g h$.

This result also verifies that the *pressure is the same at all points having the same elevation.* Furthermore, *the pressure is not affected by the shape of the vessel.*

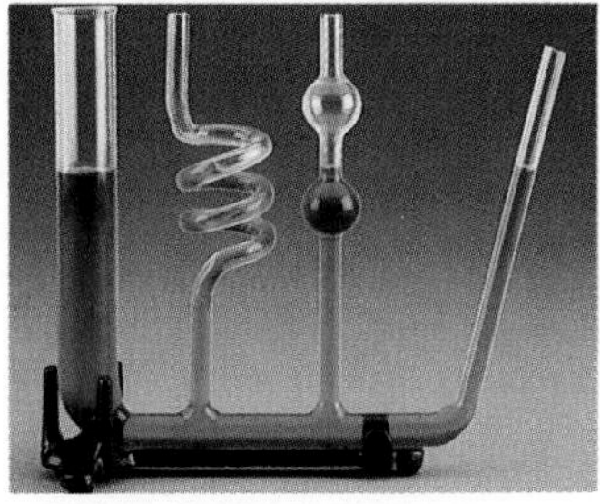

This photograph illustrates that the pressure in a liquid is the same at all points having the same elevation. Note that the shape of the vessel does not affect the pressure. (Courtesy of CENCO)

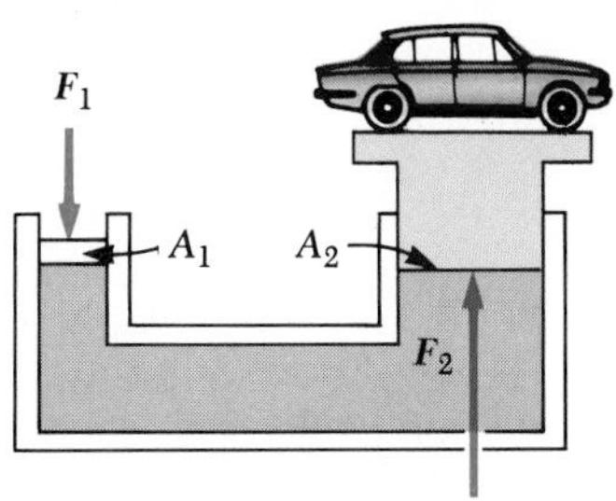

Figure 15.7 Schematic diagram of a hydraulic press. Since the increase in pressure is the same at the left and right sides, a small force $\boldsymbol{F}_1$ at the left produces a much larger force $\boldsymbol{F}_2$ at the right.

In view of the fact that the pressure in a fluid depends only upon depth, any increase in pressure at the surface must be transmitted to every point in the fluid. This was first recognized by the French scientist Blaise Pascal (1623–1662) and is called **Pascal's law:**

> A change in pressure applied to an enclosed fluid is transmitted undiminished to every point of the fluid and the walls of the containing vessel.

An important application of Pascal's law is the hydraulic press illustrated by Figure 15.7. A force $\boldsymbol{F}_1$ is applied to a small piston of area A_1. The pressure is transmitted through a fluid to a larger piston of area A_2. Since the pressure is the same on both sides, we see that $P = F_1/A_1 = F_2/A_2$. Therefore, the force F_2 is larger than F_1 by the multiplying factor A_2/A_1. Hydraulic brakes, car lifts, hydraulic jacks, fork lifts, and so on make use of this principle.

EXAMPLE 15.1 The Car Lift

In a car lift used in a service station, compressed air exerts a force on a small piston having a radius of 5 cm. This pressure is transmitted to a second piston of radius 15 cm. What force must the compressed air exert in order to lift a car weighing 13 300 N? What air pressure will produce this force?

Solution Because the pressure exerted by the compressed air is transmitted undiminished throughout the fluid, we have

$$F_1 = \left(\frac{A_1}{A_2}\right) F_2 = \frac{\pi(5 \times 10^{-2}\ \text{m})^2}{\pi(15 \times 10^{-2}\ \text{m})^2}(13\ 300\ \text{N})$$

$$= \boxed{1.48 \times 10^3\ \text{N}}$$

The air pressure that will produce this force is given by

$$P = \frac{F_1}{A_1} = \frac{1.48 \times 10^3\ \text{N}}{\pi(5 \times 10^{-2}\ \text{m})^2} = \boxed{1.88 \times 10^5\ \text{Pa}}$$

This pressure is approximately twice atmospheric pressure.

Note that the input work (the work done by $\boldsymbol{F}_1$) is equal to the output work (the work done by $\boldsymbol{F}_2$), so that energy is conserved.

EXAMPLE 15.2 The Water Bed

A water bed is 2 m on a side and 30 cm deep. (a) Find its weight.

Solution Since the density of water is 1000 kg/m³, the mass of the bed is

$$M = \rho V = (1000\ \text{kg/m}^3)(1.2\ \text{m}^3) = 1.20 \times 10^3\ \text{kg}$$

and its weight is

$$W = Mg = (1.20 \times 10^3 \text{ kg})(9.80 \text{ m/s}^2)$$

$$= \boxed{1.18 \times 10^4 \text{ N}}$$

This is equivalent to approximately 2640 lb. In order to support such a heavy load, you would be well advised to keep your water bed in the basement or on a sturdy, well-supported floor.

(b) Find the pressure that the water bed exerts on the floor when the bed rests in its normal position. Assume that the entire lower surface of the bed makes contact with the floor.

Solution The weight of the water bed is 1.18×10^4 N. The cross-sectional area is 4 m^2 when the bed is in its normal position. This gives a pressure exerted on the floor of

$$P = \frac{1.18 \times 10^4 \text{ N}}{4 \text{ m}^2} = \boxed{2.95 \times 10^3 \text{ Pa}}$$

Exercise 1 Calculate the pressure that would be exerted on the floor if the bed rests on its side.
Answer Since the area of its side is 0.6 m^2, the pressure is 1.96×10^4 Pa.

EXAMPLE 15.3 Pressure In the Ocean
Calculate the pressure at an ocean depth of 1000 m. Assume the density of water is 1.0×10^3 kg/m^3 and take $P_a = 1.01 \times 10^5$ Pa.

Solution

$$P = P_a + \rho g h$$

$$= 1.01 \times 10^5 \text{ Pa}$$

$$+ (1.0 \times 10^3 \text{ kg/m}^3)(9.80 \text{ m/s}^2)(10^3 \text{ m})$$

$$P = \boxed{9.90 \times 10^6 \text{ Pa}}$$

This is approximately 100 times greater than atmospheric pressure! Obviously, the design and construction of vessels that will withstand such enormous pressures are not trivial matters.

Exercise 2 Calculate the total force exerted on the outside of a circular submarine window of diameter 30 cm at this depth.
Answer 7.00×10^5 N.

EXAMPLE 15.4 The Force on a Dam
Water is filled to a height H behind a dam of width w (Fig. 15.8). Determine the resultant force on the dam.

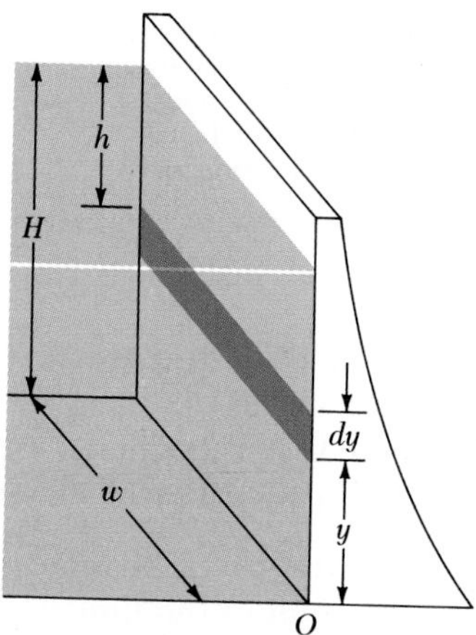

Figure 15.8 (Example 15.4) The total force on a dam must be obtained from the expression $F = \int P\, dA$, where dA is the area of the dark strip.

Solution The pressure at the depth h beneath the surface at the shaded portion is

$$P = \rho g h = \rho g(H - y)$$

(We have left out atmospheric pressure since it acts on both sides of the dam.) Using Equation 15.3, we find the force on the shaded strip to be

$$dF = P\, dA = \rho g(H - y) w\, dy$$

Therefore, the total force on the dam is

$$F = \int P dA = \int_0^H \rho g(H - y) w\, dy = \boxed{\tfrac{1}{2} \rho g w H^2}$$

For example, if $H = 30$ m and $w = 100$ m, we find that $F = 4.4 \times 10^8$ N $= 9.9 \times 10^7$ lb! Note that because the pressure increases with depth, the dam is designed such that its thickness increases with depth, as in Figure 5.8.

15.4 PRESSURE MEASUREMENTS

One simple device for measuring pressure is the open-tube manometer illustrated in Figure 15.9a. One end of a U-shaped tube containing a liquid is open to the atmosphere, and the other end is connected to a system of unknown pressure P. The pressure at point B equals $P_a + \rho gh$, where ρ is the density of the fluid. But the pressure at B equals the pressure at A, which is also the unknown pressure P. Therefore, we conclude that

$$P = P_a + \rho gh$$

The pressure P is called the **absolute pressure,** while $P - P_a$ is called the **gauge pressure.** Thus, if the pressure in the system is greater than atmospheric pressure, h is positive. If the pressure is less than atmospheric pressure (a partial vacuum), h is negative.

Another instrument used to measure pressure is the common barometer, invented by Evangelista Torricelli (1608–1647). A long tube closed at one end is filled with mercury and then inverted into a dish of mercury (Fig. 15.9b). The closed end of the tube is nearly a vacuum, so its pressure can be taken as zero. Therefore, it follows that $P_a = \rho gh$, where ρ is the density of the mercury and h is the height of the mercury column. One atmosphere of pressure is defined to be the pressure equivalent of a column of mercury that is exactly 0.76 m in height at 0°C, with $g = 9.80665\ \text{m/s}^2$. At this temperature, mercury has a density of $13.595 \times 10^3\ \text{kg/m}^3$; therefore

$$P_a = \rho gh = (13.595 \times 10^3\ \text{kg/m}^3)(9.80665\ \text{m/s}^2)(0.7600\ \text{m})$$
$$= 1.013 \times 10^5\ \text{Pa}$$

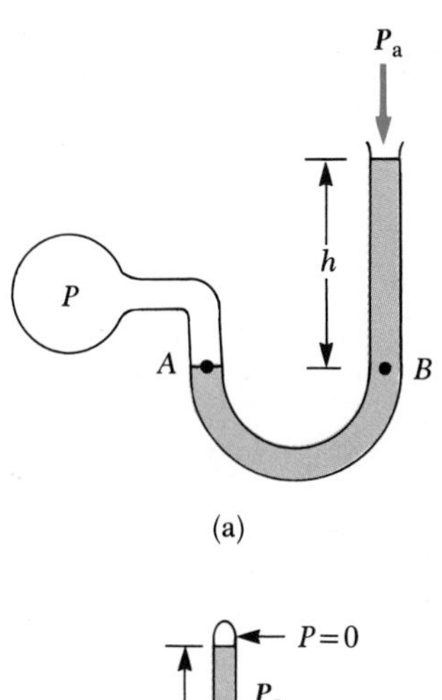

(a)

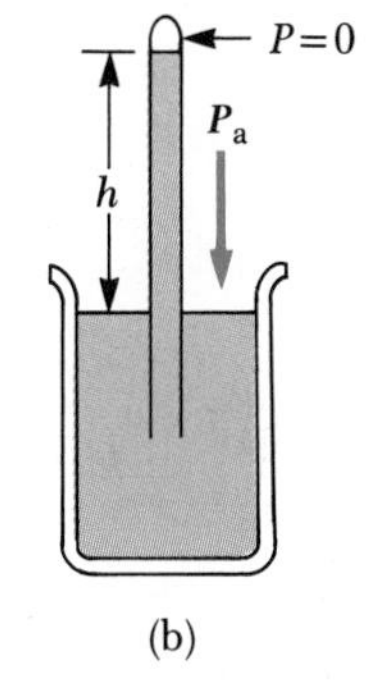

(b)

Figure 15.9 Two devices for measuring pressure: (a) the open-tube manometer; (b) the mercury barometer.

15.5 BUOYANT FORCES AND ARCHIMEDES' PRINCIPLE

Archimedes' principle can be stated as follows:

Archimedes' principle

> Any body completely or partially submerged in a fluid is buoyed up by a force equal to the weight of the fluid displaced by the body.

Everyone has experienced Archimedes' principle. As an example of a common experience, recall that it is relatively easy to lift someone if the person is in a swimming pool whereas lifting that same individual on dry land is much harder. Evidently, water provides partial support to any object placed in it. We say that an object placed in a fluid is buoyed up by the fluid, and we call this upward force the **buoyant force.** According to Archimedes' principle,

> the magnitude of the buoyant force always equals the weight of the fluid displaced by the object.

The buoyant force acts vertically upward through what was the center of gravity of the displaced fluid.

Archimedes' principle can be verified in the following manner. Suppose we focus our attention on the indicated cube of water in the container of Figure 15.10. This cube of water is in equilibrium under the action of the forces on it. One of these forces is the weight of the cube of water. What cancels this downward force? Apparently, the rest of the water inside the container is buoying up the cube and holding it in equilibrium. Thus, the buoyant force, B, on the cube of water is exactly equal in magnitude to the weight of the water inside the cube:

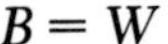

$$B = W$$

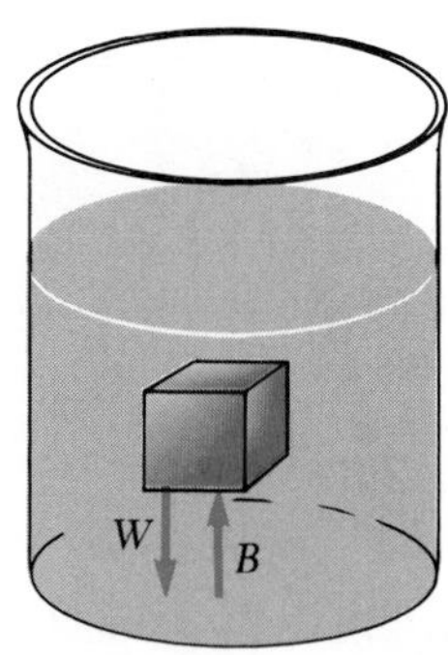

Figure 15.10 The external forces on the cube of water are its weight $\boldsymbol{W}$ and the buoyancy force $\boldsymbol{B}$. Under equilibrium conditions, $B = W$.

Biographical Sketch

Archimedes
(287–212 B.C.)

Archimedes, a Greek mathematician, physicist, and engineer, was perhaps the greatest scientist of antiquity. He was the first to compute accurately the ratio of a circle's circumference to its diameter and also showed how to calculate the volume and surface area of spheres, cylinders, and other geometric shapes. He is well known for discovering the nature of the buoyant force acting on objects and was also a gifted inventor. One of his practical inventions, still in use today, is the Archimedes' screw, an inclined rotating coiled tube used originally to lift water from the holds of ships. He also invented the catapult and devised systems of levers, pulleys, and weights for raising heavy loads. Such inventions were successfully used by the soldiers to defend his native city, Syracuse, during a two-year siege by the Romans.

According to legend, Archimedes was asked by King Hieron to determine whether the king's crown was made of pure gold or had been alloyed with some other metal. The task was to be performed without damaging the crown. Archimedes presumably arrived at a solution while taking a bath, noting a partial loss of weight after submerging his arms and legs in the water. As the story goes, he was so excited about his great discovery that he ran through the streets of Syracuse naked shouting, "Eureka!" which is Greek for "I have found it."

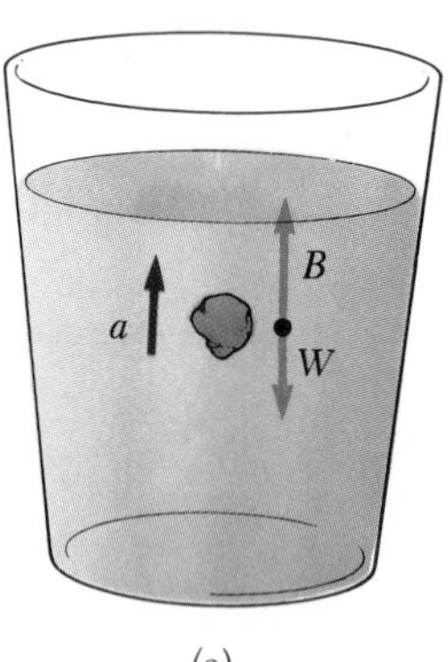

(a)

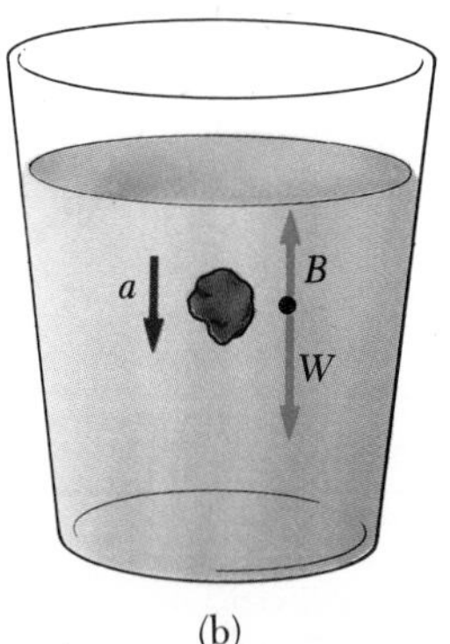

(b)

Figure 15.11 (a) A totally submerged object with a density *less* than the density of water will experience a net upward force. (b) If a totally submerged object has a density greater than the density of the fluid, it will sink.

Now, imagine that the cube of water is replaced by a cube of steel of the same dimensions. What is the buoyant force on the steel? The water surrounding a cube will behave in the same way whether a cube of water or a cube of steel is being buoyed up; therefore, *the buoyant force acting on the steel is the same as the buoyant force acting on a cube of water of the same dimensions.* This result applies for a submerged object of any shape, size, or density.

Let us show explicitly that the buoyant force is equal in magnitude to the weight of the displaced fluid. The pressure at the bottom of the cube in Figure 15.10 is greater than the pressure at the top by an amount $\rho_f gh$, where ρ_f is the density of the fluid and h is the height of the cube. Since the pressure difference, ΔP, is equal to the buoyant force per unit area, that is, $\Delta P = B/A$, we see that $B = (\Delta P)(A) = (\rho_f gh)(A) = \rho_f gV$, where V is the volume of the cube. Since the mass of the water in the cube is $M = \rho_f V$, we see that

$$B = W = \rho_f Vg = Mg \tag{15.8}$$

where W is the weight of the displaced fluid.

Before proceeding with a few examples, it is instructive to compare the forces on a totally submerged object with those acting on a floating object.

Case I A Totally Submerged Object When an object is *totally* submerged in a fluid of density ρ_f, the upward buoyant force is given by $B = \rho_f V_o g$, where V_o is the volume of the object. If the object has a density ρ_o, its weight is equal to $W = Mg = \rho_o V_o g$, and the net force on it is $B - W = (\rho_f - \rho_o)V_o g$. Hence, if the density of the object is *less* than the density of the fluid as in Figure 15.11a, the unsupported object will accelerate upward. If the density of the object is *greater* than the density of the fluid as in Figure 15.11b, the unsupported object will sink.

Case II A Floating Object Now consider an object in static equilibrium *floating* on a fluid; that is, one which is *partially* submerged. In this case, the upward buoyant force is balanced by the downward weight of the object. If

V is the volume of the fluid displaced by the object (which corresponds to that volume of the object which is beneath the fluid level), then the buoyant force has a magnitude given by $B = \rho_f V g$. Since the weight of the object is $W = Mg = \rho_o V_o g$, and $W = B$, we see that $\rho_f V g = \rho_o V_o g$, or

$$\frac{\rho_o}{\rho_f} = \frac{V}{V_o} \tag{15.9}$$

Under normal conditions, the average density of a fish is slightly greater than the density of water. This being the case, a fish would sink if it did not have some mechanism for adjusting its density. The fish accomplishes this by internally regulating the size of its swim bladder. In this manner, fish are able to maintain neutral buoyancy as they swim to various depths.

EXAMPLE 15.5 A Submerged Object
A piece of aluminum is suspended from a string and then completely immersed in a container of water (Fig. 15.12). The mass of the aluminum is 1 kg, and its density is 2.7×10^3 kg/m³. Calculate the tension in the string before and after the aluminum is immersed.

Solution When the piece of aluminum is suspended in air, as in Figure 15.12a, the tension in the string, $\boldsymbol{T}_1$ (the reading on the scale), is equal to the weight, Mg, of the aluminum, assuming that the buoyant force of air can be neglected:

$$T_1 = Mg = (1 \text{ kg})(9.80 \text{ m/s}^2) = \boxed{9.80 \text{ N}}$$

When immersed in water, the aluminum experiences an upward buoyant force $\boldsymbol{B}$, as in Figure 15.2b, which reduces the tension in the string. Since the system is in equilibrium,

$$T_2 + B - Mg = 0$$

$$T_2 = Mg - B = 9.80 \text{ N} - B$$

In order to calculate B, we must first calculate the volume of the aluminum:

$$V_{Al} = \frac{M}{\rho_{Al}} = \frac{1 \text{ kg}}{2.7 \times 10^3 \text{ kg/m}^3} = 3.70 \times 10^{-4} \text{ m}^3$$

Since the buoyant force equals the weight of the water displaced, we have

$$B = M_w g = \rho_w V_{Al} g$$

$$= (1 \times 10^3 \text{ kg/m}^3)(3.7 \times 10^{-4} \text{ m}^3)(9.80 \text{ m/s}^2)$$

$$= 3.63 \text{ N}$$

Therefore,

$$T_2 = 9.80 \text{ N} - B = 9.80 \text{ N} - 3.63 \text{ N} = \boxed{6.17 \text{ N}}$$

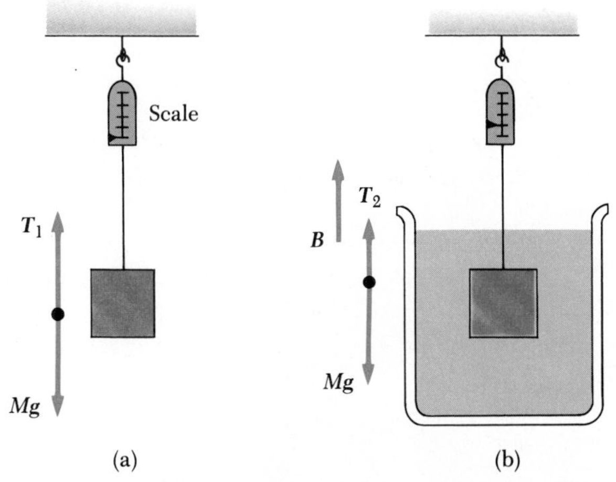

Figure 15.12 (Example 15.5) (a) When the aluminum is suspended in air, the scale reads the true weight, Mg (neglecting the buoyancy of air). (b) When the aluminum is immersed in water, the buoyant force, $\boldsymbol{B}$, reduces the scale reading to $T_2 = Mg - B$.

EXAMPLE 15.6 The Floating Ice Cube □
An ice cube floats in a glass of water as in Figure 15.13. What fraction of the ice cube lies above the water level?

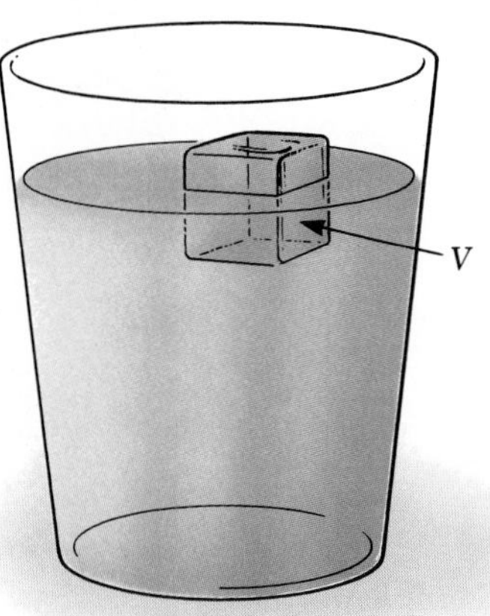

Figure 15.13 (Example 15.6).

Solution This problem corresponds to Case II described in the text. The weight of the ice cube is $W = \rho_i V_i g$, where $\rho_i = 917$ kg/m^3 and V_i is the volume of the ice cube. The upward buoyant force equals the weight of the displaced water; that is, $B = \rho_w V g$, where V is the volume of the ice cube *beneath* the water (the shaded region in Fig. 5.13) and ρ_w is the density of the water, $\rho_w = 1000$ kg/m^3. Since $\rho_i V_i g = \rho_w V g$, we see that the fraction of ice *beneath* water is $V/V_i = \rho_i/\rho_w$. Hence, the fraction of ice that lies *above* the water level is

$$f = 1 - \frac{\rho_i}{\rho_f} = 1 - \frac{917 \text{ kg/m}^3}{1000 \text{ kg/m}^3} = 0.083 \quad \text{or} \quad 8.3\%$$

Exercise 3 What fraction of the volume of an iceberg lies *below* the level of the sea? Note that the density of seawater is 1024 kg/m^3.
Answer 0.896 or 89.6%.

15.6 FLUID DYNAMICS

Thus far, our study of fluids has been restricted to fluids at rest. We now turn our attention to the subject of fluids in motion. Instead of trying to study the motion of each particle of the fluid as a function of time, we shall take the more usual approach of describing the properties of the fluid at each point as a function of time.

Flow Characteristics

When fluid is in motion, its flow can be characterized as being one of two main types. The flow is said to be **steady** if each particle of the fluid follows a smooth path, and the paths of each particle do not cross each other, as in Figure 15.14. Thus, in steady flow, the velocity of the fluid at any point remains constant in time. Above a certain critical speed, fluid flow becomes **nonsteady** or **turbulent.** Turbulent flow is an irregular flow characterized by small whirlpool-like regions as in Figure 15.15. As an example, the flow of water in a stream becomes turbulent in regions where rocks and other obstructions are encountered, often forming "white water" rapids.

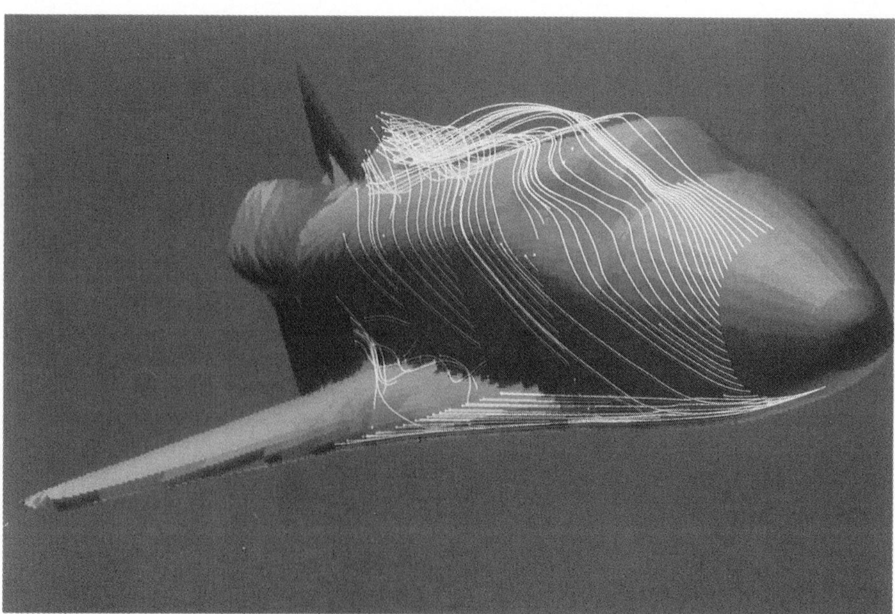

Figure 15.14 Steady flow: a computer simulation of the pattern of air flow around a space shuttle. (Courtesy of NASA)

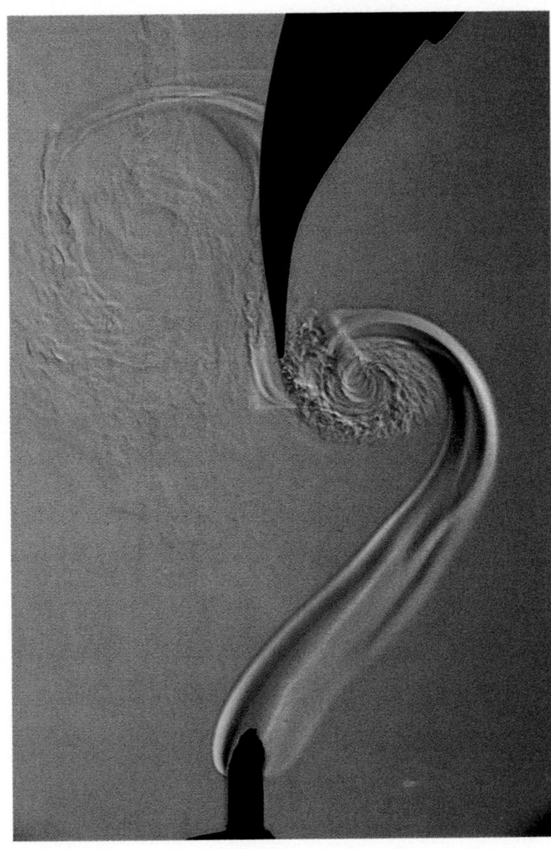

Figure 15.15 Turbulent flow: the tip of a rotating blade (the dark region) forms a vortex in air that is being heated by an alcohol lamp (the wick is at the bottom). Note the air turbulence on both sides of the rotating blade. (© Harold R. Edgerton. Courtesy of Palm Press, Inc.)

The term **viscosity** is commonly used in fluid flow to characterize the degree of internal friction in the fluid. This internal friction is associated with the resistance to two adjacent layers of the fluid to move relative to each other. Because of viscosity, part of the kinetic energy of a fluid is converted to thermal energy. This is similar to the mechanism by which an object sliding on a rough horizontal surface loses kinetic energy.

Because the motion of a real fluid is very complex, and not yet fully understood, we shall make some simplifying assumptions in our approach. As we shall see, many features of real fluids in motion can be understood by considering the behavior of an ideal fluid. In our model of an **ideal fluid,** we make the following four assumptions:

1. **Nonviscous fluid.** In a nonviscous fluid, internal friction is neglected. An object moving through a nonviscous fluid would experience no retarding viscous force.
2. **Steady flow.** In steady flow, we assume that the velocity of the fluid at each point remains constant in time.
3. **Incompressible fluid.** The density of the fluid is assumed to remain constant in time.
4. **Irrotational flow.** Fluid flow is irrotational if there is no angular momentum of the fluid about any point. If a small wheel placed anywhere in the fluid does not rotate about its center of mass, the flow is irrotational. (If the wheel were to rotate, as it would if turbulence were present, the flow would be rotational.)

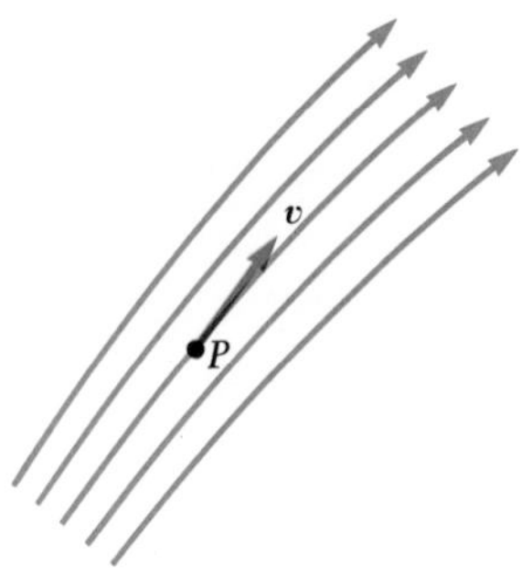

Figure 15.16 This diagram represents a set of streamlines (blue lines). A particle at P follows one of these streamlines, and its velocity is tangent to the streamline at each point along its path.

15.7 STREAMLINES AND THE EQUATION OF CONTINUITY

The path taken by a fluid particle under steady flow is called a **streamline.** The velocity of the fluid particle is always tangent to the streamline at that point, as shown in Figure 15.16. No two streamlines can cross each other, for if they did, a fluid particle could move either way at the crossover point, and the flow would not be steady. A set of streamlines as shown in Figure 15.16 forms what is called a *tube of flow.* Note that fluid particles cannot flow into or out of the sides of this tube, since the streamlines would be crossing each other.

Consider a fluid flowing through a pipe of nonuniform size as in Figure 15.17. The particles in the fluid move along the streamlines in steady flow. At all points the velocity of the particle is tangent to the streamline along which it moves.

In a small time interval Δt, the fluid at the bottom end of the pipe moves a distance $\Delta x_1 = v_1\,\Delta t$. If A_1 is the cross-sectional area in this region, then the mass contained in the shaded region is $\Delta m_1 = \rho_1 A_1\,\Delta x_1 = \rho_1 A_1 v_1\,\Delta t$. Similarly, the fluid that moves through the upper end of the pipe in the time Δt has a mass $\Delta m_2 = \rho_2 A_2 v_2\,\Delta t$. However, since *mass is conserved* and because the flow is steady, the mass that crosses A_1 in a time Δt must equal the mass that crosses A_2 in a time Δt. Therefore $\Delta m_1 = \Delta m_2$, or

$$\rho_1 A_1 v_1 = \rho_2 A_2 v_2 \qquad (15.10)$$

This expression is called the **equation of continuity.**

Since ρ is constant for an *incompressible* fluid, Equation 15.10 reduces to

$$A_1 v_1 = A_2 v_2 = \text{constant} \qquad (15.11)$$

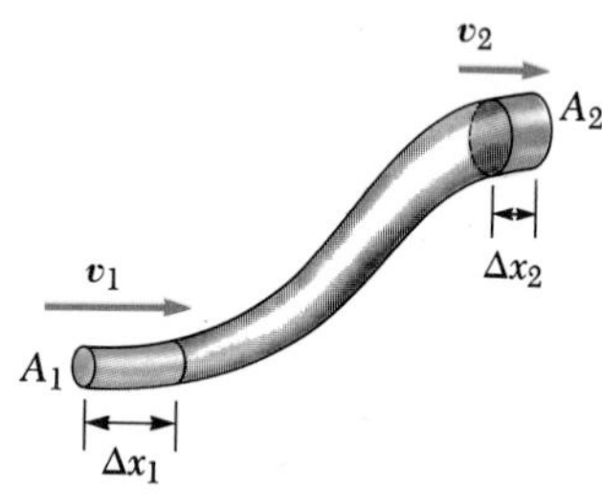

Figure 15.17 An incompressible fluid moving with steady flow through a pipe of varying cross-sectional area. The volume of fluid flowing through A_1 in a time interval Δt must equal the volume flowing through A_2 in the same time interval. Therefore, $A_1 v_1 = A_2 v_2$.

That is,

> the product of the area and the fluid speed at all points along the pipe is a constant.

Therefore, as one would expect, the speed is high where the tube is constricted and low where the tube is wide. The product Av, which has the dimensions of volume/time, is called the *volume flux,* or flow rate. The condition Av = constant is equivalent to the fact that the amount of fluid that enters one end of the tube in a given time interval equals the amount of fluid leaving the tube in the same time interval, assuming no leaks.

EXAMPLE 15.7 Filling a Water Bucket

A water hose 2 cm in diameter is used to fill a 20-liter bucket. If it takes 1 min to fill the bucket, what is the speed v at which the water leaves the hose? (Note that 1 liter $= 10^3$ cm^3.)

Solution The cross-sectional area of the hose is

$$A = \pi\frac{d^2}{4} = \pi\left(\frac{2^2}{4}\right)\text{cm}^2 = \pi\ \text{cm}^2$$

According to the data given, the flow rate is equal to 20 liters/min. Equating this to the product Av gives

$$Av = 20\,\frac{\text{liters}}{\text{min}} = \frac{20\times10^3\ \text{cm}^3}{60\ \text{s}}$$

$$v = \frac{20\times10^3\ \text{cm}^3}{(\pi\ \text{cm}^2)(60\ \text{s})} = 106\ \text{cm/s}$$

Exercise 4 If the diameter of the hose is reduced to 1 cm, what will the speed of the water be as it leaves the hose, assuming the same flow rate?
Answer 424 cm/s.

15.8 BERNOULLI'S EQUATION

As a fluid moves through a pipe of varying cross section and elevation, the pressure will change along the pipe. In 1738 the Swiss physicist Daniel Bernoulli (1700–1782) first derived an expression that relates the pressure to fluid speed and elevation. As we shall see, this result is a consequence of energy conservation as applied to our ideal fluid.

Again, *we shall assume that the fluid is incompressible and nonviscous and flows in an irrotational, steady manner.* Consider the flow through a nonuniform pipe in a time Δt, as illustrated in Figure 15.18. The force on the lower end of the fluid is P_1A_1, where P_1 is the pressure at point 1. The work done by this force is $W_1 = F_1\,\Delta x_1 = P_1A_1\,\Delta x_1 = P_1\,\Delta V$, where ΔV is the volume of the lower shaded region. In a similar manner, the work done on the fluid at the upper end in the time Δt is given by $W_2 = -P_2A_2\,\Delta x_2 = -P_2\,\Delta V$. (The volume that passes through 1 in a time Δt equals the volume that passes through 2 in the same time interval.) This work is negative since the fluid force opposes the displacement. Thus the net work done by these forces in the time Δt is

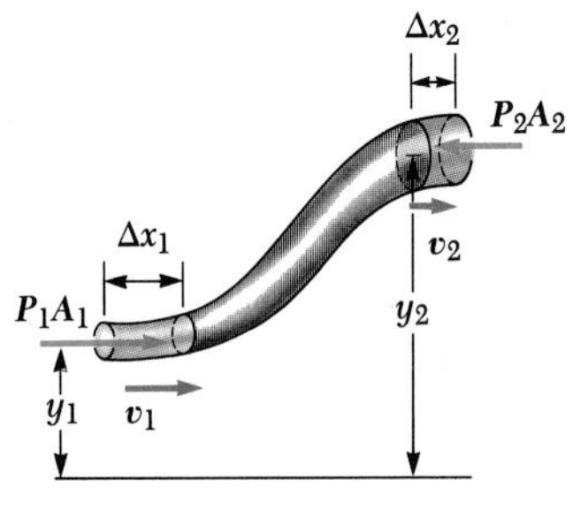

Figure 15.18 An incompressible fluid flowing through a constricted pipe with steady flow. The fluid in the section of length Δx_1 moves to the section of length Δx_2. The volumes of fluid in the two sections are equal.

$$W = (P_1 - P_2)\,\Delta V$$

Part of this work goes into changing the kinetic energy of the fluid, and part goes into changing the gravitational potential energy. If Δm is the mass passing through the pipe in the time Δt, then the change in its kinetic energy is

$$\Delta K = \tfrac{1}{2}(\Delta m)v_2{}^2 - \tfrac{1}{2}(\Delta m)v_1{}^2$$

The change in its potential energy is

$$\Delta U = \Delta mgy_2 - \Delta mgy_1$$

We can apply the work-energy theorem in the form $W = \Delta K + \Delta U$ (Chapter 8) to this volume of fluid to give

$$(P_1 - P_2)\,\Delta V = \tfrac{1}{2}(\Delta m)v_2{}^2 - \tfrac{1}{2}(\Delta m)v_1{}^2 + \Delta mgy_2 - \Delta mgy_1$$

Daniel Bernoulli was a Swiss physicist and mathematician who made important discoveries in hydrodynamics. Born into a family of mathematicians on February 8, 1700, he was the only member of the family to make a mark in physics. He was educated and received his doctorate in Basle, Switzerland.

Bernoulli's most famous work, *Hydrodynamica*, was published in 1738; it is both a theoretical and a practical study of equilibrium, pressure, and velocity of fluids. He showed that as the velocity of fluid flow increases, its pressure decreases. Referred to as "Bernoulli's principle," his work is used to produce a vacuum in chemical laboratories by connecting a vessel to a tube through which water is running rapidly. Bernoulli's principle is an early formulation of the idea of conservation of energy.

Bernoulli's *Hydrodynamica* also attempted the first explanation of the behavior of gases with changing pressure and temperature; this was the beginning of the kinetic theory of gases.

(Photograph courtesy of The Bettmann Archive)

Biographical Sketch

Daniel Bernoulli

(1700–1782)

If we divide each term by ΔV, and recall that $\rho = \Delta m/\Delta V$, the above expression reduces to

$$P_1 - P_2 = \tfrac{1}{2}\rho v_2^2 - \tfrac{1}{2}\rho v_1^2 + \rho g y_2 - \rho g y_1$$

Rearranging terms, we get

$$P_1 + \tfrac{1}{2}\rho v_1^2 + \rho g y_1 = P_2 + \tfrac{1}{2}\rho v_2^2 + \rho g y_2 \qquad (15.12)$$

This is **Bernoulli's equation** as applied to a nonviscous, incompressible fluid in steady flow. It is often expressed as

Bernoulli's equation

$$P + \tfrac{1}{2}\rho v^2 + \rho g y = \text{constant} \qquad (15.13)$$

Bernoulli's equation says that the sum of the pressure, (P), the kinetic energy per unit volume ($\tfrac{1}{2}\rho v^2$), and potential energy per unit volume ($\rho g y$) has the same value at all points along a streamline.

When the fluid is at *rest*, $v_1 = v_2 = 0$ and Equation 15.12 becomes

$$P_1 - P_2 = \rho g(y_2 - y_1) = \rho g h$$

which agrees with Equation 15.6. The following examples represent some interesting applications of Bernoulli's equation.

EXAMPLE 15.8 The Venturi Tube

The horizontal constricted pipe illustrated in Figure 15.19, known as a *Venturi tube*, can be used to measure flow velocities in an incompressible fluid. Let us determine the flow velocity at point 2 if the pressure difference $P_1 - P_2$ is known.

Solution Since the pipe is horizontal, $y_1 = y_2$ and Equation 15.12 applied to points 1 and 2 gives

$$P_1 + \tfrac{1}{2}\rho v_1^2 = P_2 + \tfrac{1}{2}\rho v_2^2$$

From the equation of continuity (Eq. 15.11), we see that $A_1 v_1 = A_2 v_2$ or

$$v_1 = \frac{A_2}{A_1} v_2$$

Substituting this expression into the previous equation gives

$$P_1 + \tfrac{1}{2}\rho \left(\frac{A_2}{A_1}\right)^2 v_2^2 = P_2 + \tfrac{1}{2}\rho v_2^2$$

$$v_2 = A_1 \sqrt{\frac{2(P_1 - P_2)}{\rho(A_1^2 - A_2^2)}} \qquad (15.14)$$

We can also obtain an expression for v_1 using this result and the continuity equation. Note that since $A_2 < A_1$, it follows that P_1 is *greater* than P_2. In other words, the pressure is *reduced* in the constricted part of the pipe. This result is somewhat analogous to the following situation: Consider a very crowded room, where people are squeezed together. As soon as a door is opened and people begin to exit, the squeezing (pressure) is least near the door where the motion (flow) is greatest.

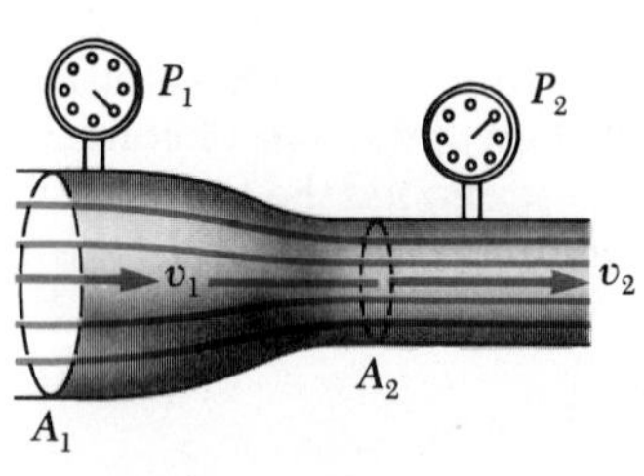

(a)

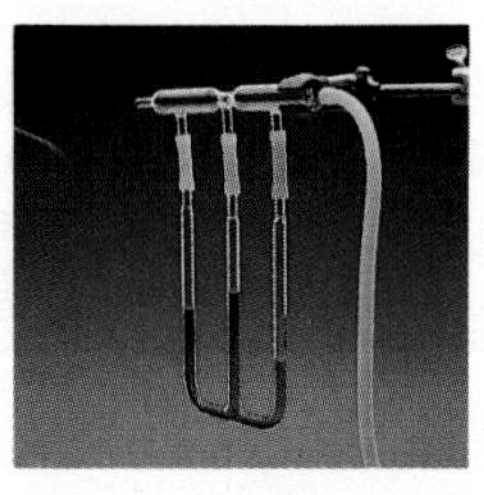

(b)

Figure 15.19 (a) (Example 15.8) Schematic diagram of a Venturi tube. The pressure P_1 is greater than the pressure P_2, since $v_1 < v_2$. This device can be used to measure the speed of fluid flow. (b) Photograph of a Venturi tube. (Courtesy of CENCO)

EXAMPLE 15.9 Torricelli's law (speed of efflux)
A tank containing a liquid of density ρ has a small hole in its side at a distance y_1 from the bottom (Fig. 15.20). The air above the liquid is maintained at a pressure P. Determine the speed at which the fluid leaves the hole when the liquid level is a distance h above the hole.

Solution If we assume the tank is large in cross section compared to the hole ($A_2 \gg A_1$), then the fluid will be approximately at rest at the top, point 2. Applying Bernoulli's equation to points 1 and 2, and noting that at the hole $P_1 = P_a$, we get

$$P_a + \tfrac{1}{2}\rho v_1^2 + \rho g y_1 = P + \rho g y_2$$

But $y_2 - y_1 = h$, and so this reduces to

$$v_1 = \sqrt{\frac{2(P - P_a)}{\rho} + 2gh} \qquad (15.15)$$

If A_1 is the cross-sectional area of the hole, then the flow rate from the hole is given by $A_1 v_1$. When P is large compared with atmospheric pressure (and therefore the term $2gh$ can be neglected), the speed of efflux is mainly a function of P. Finally, if the tank is open to the atmosphere, then $P = P_a$ and $v_1 = \sqrt{2gh}$. In other words, the speed of efflux for an open tank is equal to that acquired by a body falling freely through a vertical distance h. This is known as **Torricelli's law.**

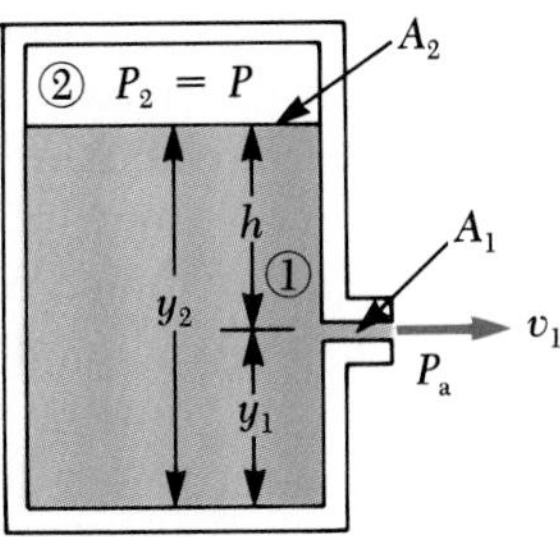

Figure 15.20 (Example 15.9) The speed of efflux, v_1, from the hole in the side of the container is given by $v_1 = \sqrt{2gh}$.

*15.9 OTHER APPLICATIONS OF BERNOULLI'S EQUATION

Many common phenomena can be explained at least in part by Bernoulli's equation. A qualitative description of such phenomena will now be given.

Consider the streamlines that flow around an airplane wing as shown in Figure 15.21. The shapes of airplane wings are designed such that the upper surface has a smaller radius of curvature than the lower surface. Air flowing over the upper surface follows more of a curved path than the air flowing over the lower surface. Since the direction of decreasing pressure is toward the center of curvature, the pressure is *lower* at the upper surface. The approaching air accelerates into this low pressure region; hence the air speed is greater past the upper surface. The net upward force $\boldsymbol{F}$ on the wing, called the **dynamic lift,** depends on several factors, such as the speed of the airplane, the area of the wing, its curvature, and the angle between the wing and the horizontal. As this angle increases, turbulent flow can set in above the wing to reduce the lift predicted by the Bernoulli principle. In practice, wings are usually tilted upward causing the air mass below the wing to be deflected downward. The impact of the air against the lower surface of the wing results in an additional upward force on the wing.

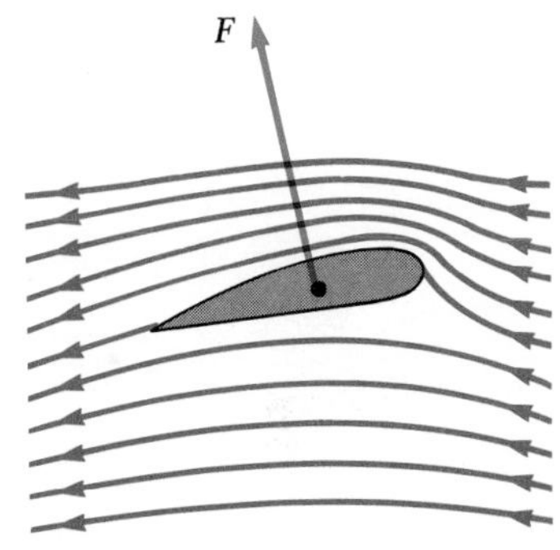

Figure 15.21 Streamline flow around an airplane wing. The pressure above is less than the pressure below, and there is a dynamic lift force upward.

If you place a sheet of paper on a table and blow across its top surface, the paper rises. This is because the faster moving air over the top surface causes a reduction in air pressure above the paper and hence a net upward force.

It is well known that the roofs of buildings are often blown off by strong winds, hurricanes, or tornados. The Bernoulli principle tells us that atmospheric pressure decreases under such conditions. If a building is not well vented, the air pressure inside may be substantially greater than the external air pressure, which may cause the roof to be blown off. A well-vented building is less likely to be damaged in this manner.

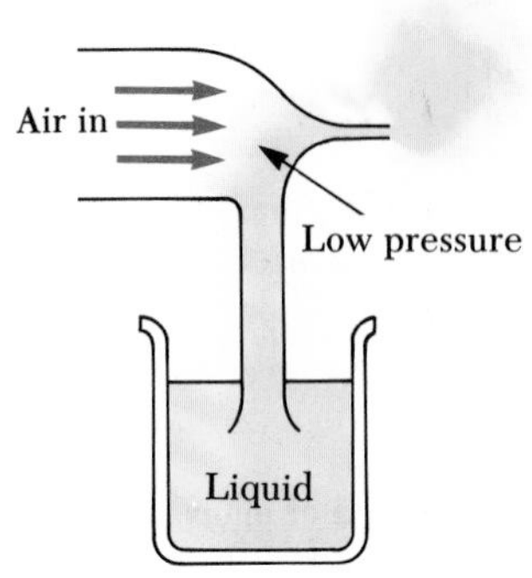

Figure 15.22 A stream of air passing over a tube dipped into a liquid will cause the liquid to rise in the tube as shown.

A number of devices operate in the manner described in Figure 15.22. A stream of air passing over an open tube reduces the pressure above the tube. This reduction in pressure causes the liquid to rise into the air stream. The liquid is then dispersed into a fine spray of droplets. You might recognize that this so-called atomizer is used in perfume bottles and paint sprayers. The same principle is used in the carburetor of a gasoline engine. In this case, the low-pressure region in the carburetor is produced by air drawn in by the piston through the air filter. The gasoline vaporizes, mixes with the air, and enters the cylinder of the engine for combustion.

If a person has advanced arteriosclerosis, the Bernoulli principle produces a sign called vascular flutter. In this situation, the artery is constricted as a result of an accumulation of plaque on its inner walls. In order to maintain a constant flow rate through such a constricted artery, the driving pressure must increase. Such an increase in pressure requires a greater demand on the heart muscle. If the blood velocity is sufficiently high in the constricted region, the artery may collapse under external pressure, causing a momentary interruption in blood flow. At this point, there is no Bernoulli principle and the vessel reopens under arterial pressure. As the blood rushes through the constricted artery, the internal pressure drops and again the artery closes. Such variations in blood flow can be heard with a stethoscope. If the plaque becomes dislodged and ends up in a smaller vessel that delivers blood to the heart, the person can suffer a heart attack.

Finally, consider the circulation of air around a thrown baseball. If the ball is not spinning, as in Figure 15.23a, the motion of the airstream past the ball is nearly streamline. In this figure, the ball is moving from right to left. Hence, from the point of view of the baseball, the airstream is moving from left to right. A symmetric region of turbulence occurs behind the ball as shown. When the ball is spinning counterclockwise, as in Figure 15.23b, layers of air near the surface of the ball are carried in the direction of spin because of viscosity, which will be discussed in Section 15.11. The combined effect of the steady flow of air and the air dragged along due to the spinning motion produces the streamlines shown in 15.23b and an asymmetric turbulence pattern. The velocity of the air below the ball is greater than the velocity above the ball. Thus, from Bernoulli's equation, the air pressure above the ball is greater than the air pressure below the ball, and the ball experiences a sideways deflecting force. This deflecting force is often called the *dynamic lift*. When a pitcher wishes to throw a curve ball that deflects sideways, the spin axis should be vertical (perpendicular to the page in Fig. 15.23b). On the other hand, if he wishes to throw a "sinker," the spin axis should be horizontal. Tennis balls and golf balls with spin also exhibit dynamic lift.

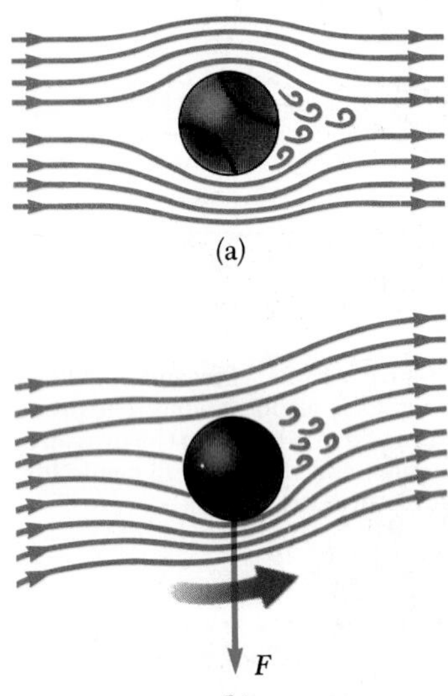

Figure 15.23 (a) The airstream around a nonrotating baseball moving from right to left. The streamlines represent the flow relative to the baseball. Note the symmetric region of turbulence behind the ball. (b) The airstream around a spinning baseball. The ball experiences a deflecting force partly because of the Bernoulli principle.

*15.10 ENERGY FROM THE WIND

The wind as a source of energy is not a new concept — there is some evidence that windmills were used in Babylon and in China as early as 2000 B.C. The kinetic energy carried by the winds originates from solar energy.

Although the wind is a large potential source of energy (about 5 kW per acre in the United States), it has been harnessed only on a small scale. It has been estimated that, on a global scale, the winds account for a total available power of 2×10^{10} kW (three times the world power consumption in 1972). Therefore, if only a small percentage of the available power could be harnessed, wind power would represent a significant fraction of our energy needs. As with all indirect energy resources, wind power systems have some disadvantages, which in this case arise mainly from the variability of wind velocities.

The largest windmill built in the United States was a 1.25-MW generator installed at Grandpa's Knob near Rutland, Vermont. The machine's blades were 175 ft in diameter, and the facility operated intermittently between 1941 and 1945. Unfortunately, one of its two main blades broke off as a result of material fatigue and was never repaired. Despite this failure, the windmill was considered a technological success, since wartime needs limited the quality of available materials. Nevertheless, the project was abandoned because costs were not competitive with hydroelectric power. The U.S. Department of Energy has recently developed wind machines each capable of generating 2.5 MW.

We can use some of the ideas developed in this chapter to estimate wind power. Any wind energy machine involves the conversion of the kinetic energy of moving air to mechanical energy of an object, usually by means of a rotating shaft. The kinetic energy per unit volume of a moving column of air is given by

$$\frac{KE}{\text{volume}} = \tfrac{1}{2}\rho v^2$$

where ρ is the density of air and v is its speed. The rate of flow of air through a column of cross-sectional area A is Av (Fig. 15.24). This can be considered as the volume of air crossing the area each second. In the working machine, A is the cross-sectional area of the wind-collecting system, such as a set of rotating propeller blades. Multiplying the kinetic energy per unit volume by the flow rate gives the rate at which energy is transferred, or, in other words, the power:

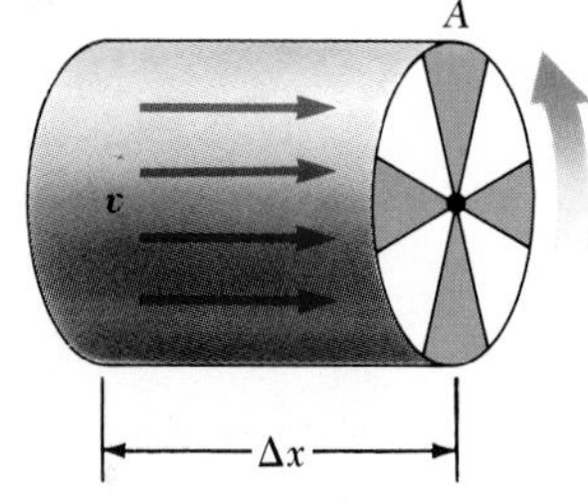

Figure 15.24 Wind moving through a cylindrical column of cross-sectional area A with a speed v.

$$\text{Power} = \frac{KE}{\text{volume}} \times \frac{\text{volume}}{\text{time}} = (\tfrac{1}{2}\rho v^2)(Av) = \tfrac{1}{2}\rho v^3 A \qquad (15.16)$$

Therefore, the available power per unit area is given by

$$\frac{P}{A} = \tfrac{1}{2}\rho v^3 \qquad (15.17)$$

where the symbol P for power is not to be confused with pressure. According to this result, if the moving air column could be brought to rest, a power of $\frac{1}{2}\rho v^3$ would be available for each square meter that was intercepted. For example, if we assume a moderate speed of 12 m/s (27 mi/h) and take $\rho = 1.3\ \text{kg/m}^3$, we find that

$$\frac{P}{A} = \tfrac{1}{2}\left(1.3\ \frac{\text{kg}}{\text{m}^3}\right)\left(12\ \frac{\text{m}}{\text{s}}\right)^3 \approx 1100\ \frac{\text{W}}{\text{m}^2} = 1.1\ \frac{\text{kW}}{\text{m}^2}$$

Since the power per unit area varies as the cube of the velocity, its value doubles if v increases by only 26%. Conversely, the power output would be halved if the velocity decreased by 26%.

This calculation is based on ideal conditions and assumes that all of the kinetic energy is available for power. In reality, the air stream emerges from

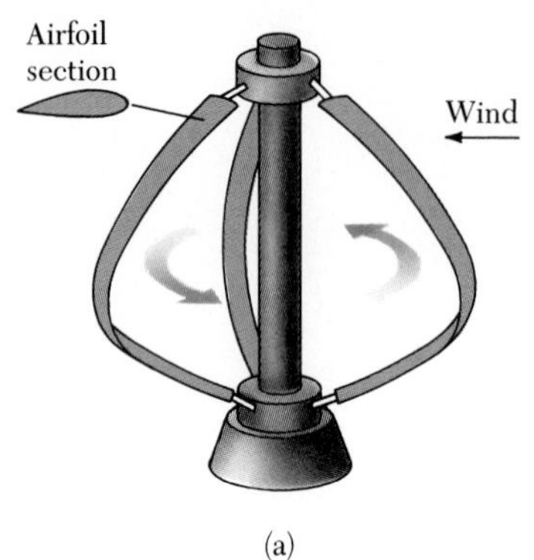

(a)

(b)

(c)

Figure 15.25 (a) A vertical-axis wind generator. (b) A horizontal-axis wind generator. (c) Photograph of a vertical-axis wind generator. (Courtesy of U.S. Department of Energy)

the wind generator with some residual velocity, and more refined calculations show that, at best, one can extract only 59.3% of this quantity.[1] The expression for the maximum available power per unit area for the ideal wind generator is found to be

$$\frac{P_{\text{max}}}{A} = \frac{8}{27}\rho v^3 \tag{15.18}$$

In a real wind machine, further losses resulting from the nonideal nature of the propeller, gearing, and generator reduce the total available power to around 15% of the value predicted by Equation 15.17. Sketches of two types of wind turbines are shown in Figure 15.25.

EXAMPLE 15.10 Power Output of a Windmill
Calculate the power output of a wind generator having a blade diameter of 80 m, assuming a wind speed of 10 m/s and an overall efficiency of 15%.

Solution Since the radius of the blade is 40 m, the cross-sectional area of the propellers is given by

$$A = \pi r^2 = \pi(40 \text{ m})^2 = 5.0 \times 10^3 \text{ m}^2$$

If 100% of the available wind energy could be extracted, the maximum available power would be

$$P_{\text{max}} = \tfrac{1}{2}\rho A v^3 = \tfrac{1}{2}\left(1.2 \frac{\text{kg}}{\text{m}^3}\right)(5.0 \times 10^3 \text{ m}^2)\left(10 \frac{\text{m}}{\text{s}}\right)^3$$

$$= 3.0 \times 10^6 \text{ W} = 3.0 \text{ MW}$$

Since the overall efficiency is 15%, the output power is

$$P = 0.15 P_{\text{max}} = 0.45 \text{ MW}$$

In comparison, a large steam-turbine plant has a power output of about 1 GW. Hence, one would require 2200 such wind generators to equal this output under these conditions. The large number of generators required for reasonable output power is clearly a major disadvantage of wind power. (See Problem 52.)

---

*15.11 VISCOSITY

We have seen that a fluid does not support a shearing stress. However, fluids do offer some degree of resistance to shearing motion. This resistance to shearing motion is a form of internal friction which is called *viscosity*. In the

[1] For more details, see J. H. Krenz, *Energy Conversion and Utilization,* Boston, Allyn and Bacon, 1976, Chapter 8.

case of liquids, the viscosity arises because of a frictional force between adjacent layers of the fluid as they slide past one another. The degree of viscosity of a fluid can be understood with the following example. If two plates of glass are separated by a layer of fluid such as oil, with one plate fixed in position, it is easy to slide one plate over the other (Fig. 15.26). However, if the fluid separating the plates is tar, the task of sliding one plate over the other becomes much more difficult. Thus, we would conclude that tar has a higher viscosity than oil. In Figure 15.26, note that the velocity of successive layers increases linearly from 0 to v as one moves from a layer adjacent to the fixed plate to a layer adjacent to the moving plate.

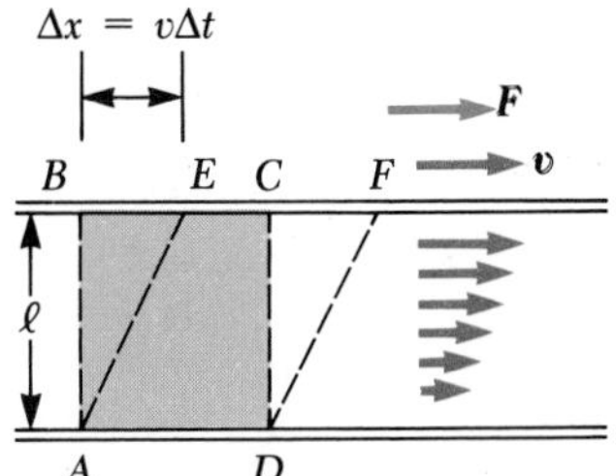

Figure 15.26 A layer of liquid between two solid surfaces in which the lower surface is fixed and the upper surface moves to the right with a velocity $\boldsymbol{v}$.

Recall that in a solid a shearing stress gives rise to a relative displacement of adjacent layers (Section 12.4). In an analogous fashion, adjacent layers of a fluid under shear stress are set into relative motion. Again, consider two parallel layers, one fixed and one moving to the right under the action of an external force $\boldsymbol{F}$ as in Figure 15.26. Because of this motion, a portion of the liquid is distorted from its original shape, *ABCD*, at one instant to the shape *AEFD* after a short time interval. If you refer to Section 12.4, you will recognize that the liquid has undergone a shear strain. By definition, the shear stress on the liquid is equal to the ratio F/A, while the shear strain is defined by the ratio $\Delta x/\ell$. That is,

$$\text{Shear stress} = \frac{F}{A} \qquad \text{Shear strain} = \frac{\Delta x}{\ell}$$

The upper plate moves with a speed v, and the fluid adjacent to this plate has the same speed. Thus, in a time Δt, the fluid at the upper plate moves a distance $\Delta x = v\,\Delta t$, and we can express the shear strain per unit time as

$$\frac{\text{Shear strain}}{\Delta t} = \frac{\Delta x/\ell}{\Delta t} = \frac{v}{\ell}$$

This equation states that the rate of change of shearing strain is v/ℓ.

The **coefficient of viscosity,** η, for the fluid is defined as the ratio of the shearing stress to the rate of change of the shear strain:

$$\eta \equiv \frac{F/A}{v/\ell} = \frac{F\ell}{Av} \tag{15.19}$$

Coefficient of viscosity

The SI unit of viscosity is $N \cdot s/m^2$. The coefficients of viscosity for some substances are given in Table 15.2. The cgs unit of viscosity is $\text{dyne} \cdot s/cm^2$, which is called the **poise.**

TABLE 15.2 The Viscosities of Various Fluids

Fluid	T(°C)	Viscosity η ($N \cdot s/m^2$)
Water	20	1.0×10^{-3}
Water	100	0.3×10^{-3}
Whole blood	37	2.7×10^{-3}
Glycerine	20	830×10^{-3}
10-wt motor oil	30	250×10^{-3}

The expression for η given by Equation 15.19 is only valid if the fluid velocity varies linearly with position. In this case, it is common to say that the velocity gradient, v/ℓ, is uniform. If the velocity gradient is *not* uniform, we must express η in the general form

$$\eta \equiv \frac{F/A}{dv/dy} \tag{15.20}$$

where the velocity gradient dv/dy is the change in velocity with position as measured perpendicular to the direction of velocity.

EXAMPLE 15.11 Measuring the Coefficient of Viscosity

A metal plate of area 0.15 m^2 is connected to an 8-g mass via a string that passes over an ideal pulley (massless and frictionless), as in Figure 15.27. A lubricant with a film thickness of 0.3 mm is placed between the plate and surface. When released, the plate is observed to move to the right with a constant speed of 0.085 m/s. Find the coefficient of viscosity of the lubricant.

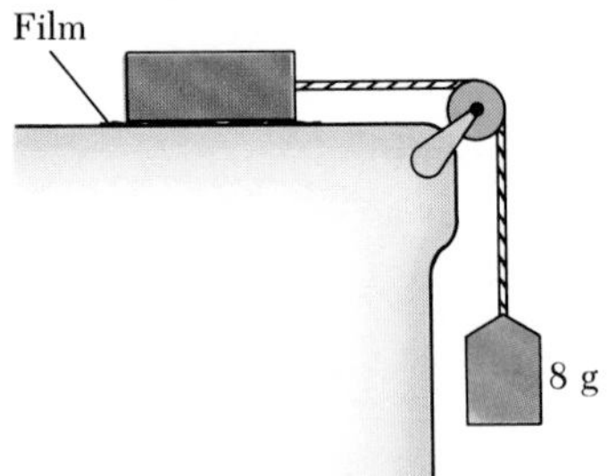

Figure 15.27 (Example 15.11).

Solution Because the plate moves with constant speed, its acceleration is zero. The plate moves to the right under the action of the tension force, T, and the frictional force, f, associated with the viscous fluid. In this case, the tension is equal in magnitude to the suspended weight, therefore,

$$f = T = mg = (8 \times 10^{-3}\text{ kg})(9.80\text{ m/s}^2)$$
$$= 7.84 \times 10^{-2}\text{ N}$$

The lubricant in contact with the horizontal surface is at rest, while the layer in contact with the plate moves at the speed of the plate. Assuming the velocity gradient is uniform, we have

$$\eta = \frac{F\ell}{Av} = \frac{(7.84 \times 10^{-2}\text{ N})(0.3 \times 10^{-3}\text{ m})}{(0.05\text{ m}^2)(0.085\text{ m/s})}$$

$$= 5.53 \times 10^{-3}\text{ N}\cdot\text{s/m}^2$$

*15.12 TURBULENCE

If adjacent layers of a viscous fluid flow smoothly over each other, the stable streamline flow is called *laminar flow*. However, at sufficiently high velocities, the fluid flow changes from laminar flow to a highly irregular and random motion of the fluid called *turbulent flow*. The velocity at which turbulence occurs depends on the geometry of the medium surrounding the fluid and the fluid viscosity.

There are many examples of turbulent flow that can be cited. Water flowing in a rock-filled stream or river, and smoke rising from a chimney on a windy day are turbulent in nature. Likewise, the water in the wake of a speedboat and the air in the wakes left by airplanes and other moving vehicles represent turbulent flow.

Experimentally, it is found that the onset of turbulence is determined by a dimensionless parameter called the **Reynolds number,** RN, given by

Reynolds number

$$RN = \frac{\rho v d}{\eta} \tag{15.21}$$

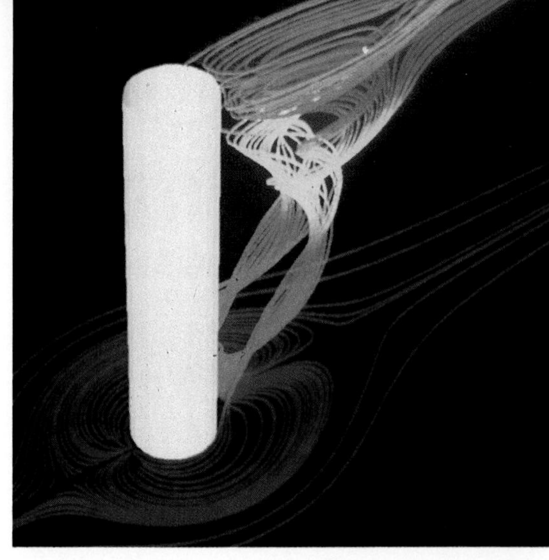

Fluid flow. A Cray-2 (one of the most powerful computers in the world) simulation of gaseous flow around a post or cylinder in a rocket engine. Note the tornado-like vortex wake at the right. Such simulations help explain phenomena observed in research experiments. (Courtesy of U.S. Department of Energy)

where ρ is the fluid density, v is the fluid speed, η is the viscosity, and d is a geometrical length associated with the flow. For flow through a tube, d would be the diameter of the tube. In the case of flow around a sphere, d would be the diameter of the sphere.

Experiments show that if the Reynolds number is below about 2000, the flow of fluid through a tube is laminar; turbulence occurs if the Reynolds number is above about 3000.

SUMMARY

The **density,** ρ, of a substance is defined as its mass per unit volume and has units of kg/m^3 in the SI system:

$$\rho \equiv \frac{m}{V} \qquad (15.1)$$

Density

The **pressure,** P, in a fluid is the force per unit area that the fluid exerts on any surface:

$$P \equiv \frac{F}{A} \qquad (15.2)$$

Average pressure

In the SI system, pressure has units of N/m^2, and $1\ N/m^2 = 1$ pascal (Pa).

The pressure in a fluid varies with depth h according to the expression

$$P = P_a + \rho g h \qquad (15.7)$$

Pressure at any depth h

where P_a is atmospheric pressure ($= 1.01 \times 10^5\ N/m^2$) and ρ is the density of the fluid, assumed uniform.

Pascal's law states that when pressure is applied to an enclosed fluid, the pressure is transmitted undiminished to every point of the fluid and of the walls of the container.

When an object is partially or fully submerged in a fluid, the fluid exerts an upward force on the object called the **buoyant force.** According to **Archimedes' principle,** the buoyant force is equal to the weight of the fluid displaced by the body.

Various aspects of fluid dynamics (fluids in motion) can be understood by assuming that the fluid is nonviscous and incompressible and that the fluid motion is a steady flow with no turbulence.

Using these assumptions, one obtains two important results regarding fluid flow through a pipe of nonuniform size:

1. The flow rate through the pipe is a constant, which is equivalent to stating that the product of the cross-sectional area, A, and the speed, v, at any point is a constant. That is,

Equation of continuity

$$A_1v_1 = A_2v_2 = \text{constant} \tag{15.11}$$

2. The sum of the pressure, kinetic energy per unit volume, and potential energy per unit volume has the same value at all points along a streamline. That is,

Bernoulli's equation

$$P + \tfrac{1}{2}\rho v^2 + \rho gy = \text{constant} \tag{15.13}$$

This is known as **Bernoulli's equation** and is fundamental in the study of fluid dynamics.

The viscosity of a fluid is a measure of its resistance to shearing motion. The **coefficient of viscosity** for a fluid is defined as the ratio of the shearing stress to the rate of change of the shear strain.

The onset of turbulence in fluid flow is determined by a parameter called the **Reynolds number,** whose value depends on the fluid density, the fluid speed, the viscosity, and a geometrical factor.

QUESTIONS

1. Two glass tumblers that weigh the same but have different shapes and different cross-sectional areas are filled to the same level with water. According to the expression $P = P_a + \rho gh$, the pressure is the same at the bottom of both tumblers. In view of this, why does one tumbler weigh more than the other?
2. If the top of your head has an area of 100 cm^2, what is the weight of air above your head?
3. When you drink a liquid through a straw, you reduce the pressure in your mouth and let the atmosphere move the liquid. Explain how this works. Could you use a straw to sip a drink on the moon?
4. Indian fakirs stretch out for a nap on a bed of nails. How is this possible?
5. Pascal used a barometer with water as the working fluid. Why is it impractical to use water for a typical barometer?
6. A person sitting in a boat floating in a small pond throws a heavy anchor overboard. Does the level of the pond rise, fall, or remain the same?
7. Steel is much denser than water. How, then, do boats made of steel float?
8. A helium-filled balloon will rise until its density becomes the same as that of the air. If a sealed submarine begins to sink, will it go all the way to the bottom of the ocean or will it stop when its density becomes the same as that of the surrounding water?
9. A fish rests on the bottom of a bucket of water while the bucket is being weighed. When the fish begins to swim around, does the weight change?
10. Will a ship ride higher in the water of an island lake or in the ocean? Why?
11. If 1 000 000 N of weight were placed on the deck of the World War II battleship North Carolina, the ship would sink only 2.5 cm lower in the water. What is the cross-sectional area of the ship at water level?
12. Lead has a greater density than iron, and both are denser than water. Is the buoyant force on a lead object greater than, less, than, or equal to the buoyant force on an iron object of the same volume?
13. An ice cube is placed in a glass of water. What happens to the level of the water as the ice melts?
14. A woman wearing high-heeled shoes is invited into a home in which the kitchen has a newly installed vinyl floor covering. Why should the homeowner be concerned?

15. A typical silo on a farm has many bands wrapped around its perimeter, as shown in the photograph. Why is the spacing between successive bands smaller at the lower regions of the silo?

(Question 15).

16. The water supply for a city is often provided from reservoirs built on high ground. Water flows from the reservoir, through pipes, and into your home when you turn the tap on your faucet. Why is the water flow more rapid out of a faucet on the first floor of a building than in an apartment on a higher floor?
17. Smoke rises in a chimney faster when a breeze is blowing. Use Bernoulli's principle to explain this phenomenon.
18. Why do many trailer trucks use wind deflectors on the top of their cabs? (See photograph.) How do such devices reduce fuel consumption?

(Question 18) The high cost of fuel has prompted many truck owners to install wind deflectors on their cabs to reduce air drag. (Courtesy of Henry Leap)

19. Consider the cross section of the wing on an airplane. The wing is designed such that the air travels faster over the top than under the bottom. Explain why there is a net upward force (lift) on the wing due to the Bernoulli principle.
20. When a fast-moving train passes a train at rest, the two tend to be drawn together. How does the Bernoulli effect explain this phenomenon?
21. A baseball thrown from left field toward home plate is seen from above to be spinning counterclockwise. In which direction does the ball deflect?
22. A tornado or hurricane will often lift the roof of a house. Use the Bernoulli principle to explain why this occurs. Why should you keep your windows open during these conditions?
23. If you suddenly turn on your shower water at full speed, why is the shower curtain pushed inward?
24. If you hold a sheet of paper and blow across the top surface, the paper rises. Explain.
25. If air from a hair dryer is blown over the top of a Ping-Pong ball, the ball can be suspended in air. Explain how the ball can remain in equilibrium.
26. Two ships passing near each other in a harbor tend to be drawn together and run the risk of a sideways collision. How does the Bernoulli principle explain this?
27. When ski-jumpers are airborne, why do they bend their bodies forward and keep their hands at their sides?

(Question 27). (© Thomas Zimmerman, FPG International)

28. When an object is immersed in a fluid at rest, why is the net force on it in the horizontal direction equal to zero?
29. Explain why a sealed bottle partially filled with a liquid can float.
30. When is the buoyant force on a swimmer the greatest —when the swimmer is exhaling or inhaling?
31. A piece of wood is partially submerged in a container filled with water. If the container is sealed and pressurized above atmospheric pressure, does the wood rise, fall, or remain at the same level? (*Hint:* Wood is porous.)
32. A flat plate is immersed in a fluid at rest. For what orientation of the plate will the pressure on its flat surface be uniform?

33. Because atmospheric pressure is 10^5 N/m^2 and the area of a person's chest is about 0.13 m^2, the force of the atmosphere on one's chest is around 13 000 N. In view of this enormous force, why don't our bodies collapse?
34. How does a bumblebee stay aloft? A helicopter?
35. How would you determine the density of an irregularly shaped rock?
36. Why do airplane pilots prefer to take off into the wind?
37. Baseball pitchers are capable of throwing several interesting kinds of pitches. Can you describe how each of the following pitches is thrown by a right-handed pitcher? (a) a curveball (breaks left), (b) screwball (breaks right), (c) sinker (breaks down), (d) submarine (breaks up), and (e) knuckleball (random deviations).
38. If you release a ball while inside a freely falling elevator, the ball remains in front of you rather than falling to the floor, because the ball, the elevator, and you all experience the same downward acceleration, **g**. What happens if you repeat this experiment with a helium-filled balloon? (This one is tricky.)
39. Two identical ships set out to sea. One is loaded with a cargo of Styrofoam, and the other is empty. Which ship will be more submerged?
40. A small piece of steel is tied to a block of wood. When the wood is placed in a tub of water with the steel on top, half of the block is submerged. If the block is inverted so that the steel is underwater, will the amount of the block submerged increase, decrease, or remain the same? What happens to the water level in the tub when the block is inverted?
41. Prairie dogs ventilate their burrows by building a mound over one of their entrances, which is open to a stream of air. A second entrance at ground level is open to almost stagnant air. How does this construction create an air flow through the burrow?
42. An unopened can of diet cola floats when placed in a tank of water, whereas a can of regular cola of the same brand sinks in the tank. What do you suppose could explain this behavior?
43. The photograph below shows a glass cylinder containing four fluids of different densities. From top to bottom, the fluids are oil (orange), water (yellow), salt water (green), and mercury (silver). The cylinder also contains four objects, which from top to bottom are a Ping-Pong ball, a piece of wood, an egg, and a steel ball. (a) Which of these fluids has the lowest density, and which has the greatest? (b) What can you conclude about the density of each object in the cylinder?

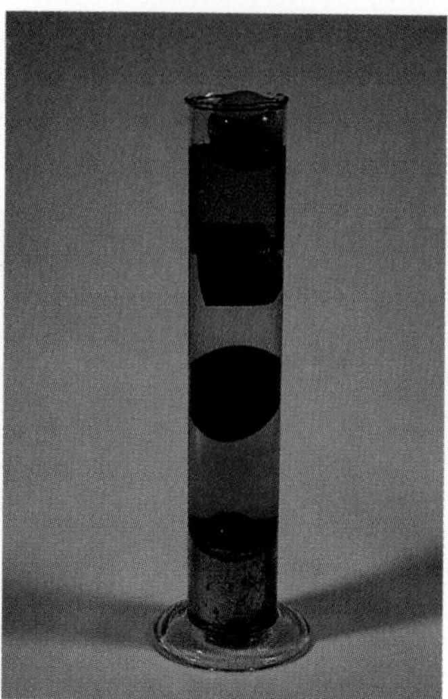

(Question 43). (Courtesy of Henry Leap and Jim Lehman)

44. In the photograph below, an air stream moves from right to left through a tube which is constricted at the middle. Three Ping-Pong balls are levitated in equilibrium above the vertical columns through which the air escapes. (a) Why is the ball at the right at a higher elevation than the one in the middle? (b) Why is the ball at the left at a lower elevation than the ball at the right even though the horizontal tube has the same dimensions at these two points?

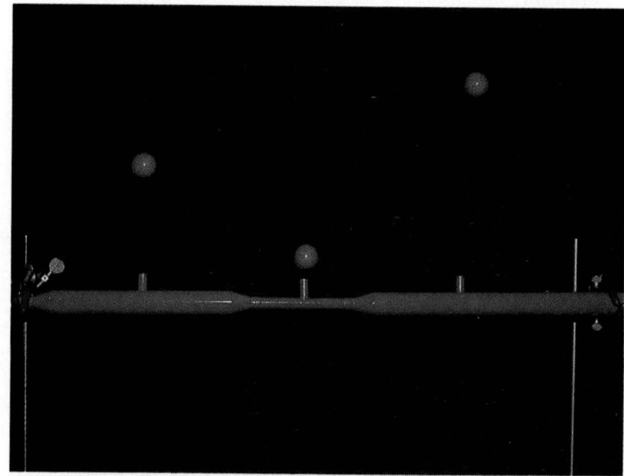

(Question 44). (Courtesy of Henry Leap and Jim Lehman)

PROBLEMS

Section 15.2 Density and Pressure

1. Calculate the mass of a solid iron sphere that has a diameter of 3.0 cm.
2. A small ingot of shiny grey metal has a volume of 25 cm^3 and a mass of 535 g. What is the metal? (See Table 15.1).
3. Estimate the density of the *nucleus* of an atom. What does this result suggest concerning the structure of matter? (Use the fact that the mass of a proton is 1.67×10^{-27} kg and its radius is about 10^{-15} m.)
4. A king orders a gold crown having a mass of 0.5 kg. When it arrives from the metalsmith, the volume of the crown is found to be 185 cm^3. Is the crown made of solid gold?
5. A 50-kg woman balances on one heel of a pair of high-heel shoes. If the heel is circular with radius 0.5 cm, what pressure does she exert on the floor?
6. Use the fact that normal atmospheric pressure is 1.013×10^5 N/m^2 and that the radius of the earth is 6.38×10^6 m to calculate the mass of the earth's atmosphere.
7. Estimate the density of a neutron star. Such an object is thought to have a radius of only 10 km and a mass equal to that of the sun. ($M_{sun} = 2 \times 10^{30}$ kg.)

Section 15.3 Variation of Pressure with Depth

8. Determine the absolute pressure at the bottom of a lake that is 30 m deep.
9. At what depth in a lake is the absolute pressure equal to three times atmospheric pressure?
10. The small piston of a hydraulic lift has a cross-sectional area of 3 cm^2, and the large piston has an area of 200 cm^2 (Fig. 15.7). What force must be applied to the small piston to raise a load of 15 000 N? (In service stations this is usually accomplished with compressed air.)
11. The spring of the pressure gauge shown in Figure 15.4 has a force constant of 1000 N/m, and the piston has a diameter of 2 cm. Find the depth in water for which the spring compresses by 0.5 cm.
12. A swimming pool has dimensions 30 m × 10 m and a flat bottom. When the pool is filled to a depth of 2 m with fresh water, what is the total force due to the water on the bottom? On each end? On each side?
13. What must be the contact area between a suction cup (completely exhausted) and a ceiling in order to support the weight of an 80-kg student?
14. A magician is immersed in 4.0 m of water in a sealed trunk. If the lid of the trunk measures 0.70 m × 2.0 m, what is the force of the water on the trunk lid?
15. What is the fractional change in the density of sea water between the surface (where the pressure is equal to 1 atm) and a depth of 5.2 km (where the pressure is 500 atm)?
16. What is the hydrostatic force on the back of Grand Coulee Dam if the water in the reservoir is 150 m deep and the width of the dam is 1200 m?
17. In some places, the Greenland ice sheet is 1 km thick. Estimate the pressure on the ground underneath the ice. ($\rho_{ice} = 920$ kg/m^3.)

Section 15.4 Pressure Measurements

18. The U-shaped tube in Figure 15.9a contains mercury. What is the absolute pressure, P, on the left side if $h = 20$ cm? What is the gauge pressure?
19. If the fluid in the barometer illustrated in Figure 15.9b is water, what will be the height of the water column in the vertical tube at atmospheric pressure?
20. The open vertical tube in Figure 15.28 contains two fluids of densities ρ_1 and ρ_2, which do not mix. Show that the pressure at the depth $h_1 + h_2$ is given by the expression $P = P_a + \rho_1 g h_1 + \rho_2 g h_2$.

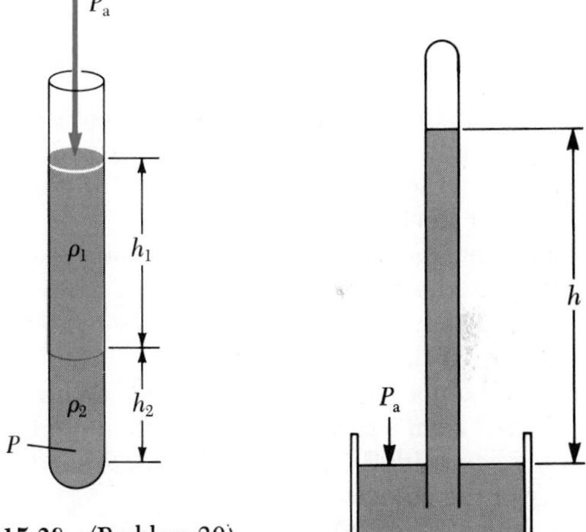

Figure 15.28 (Problem 20).

Figure 15.29 (Problem 21).

21. Blaise Pascal duplicated Torricelli's barometer using (as a Frenchman would) a red Bordeaux wine as the working liquid. The density of the wine he used was 0.984×10^3 kg/m^3. What was the height h of the wine column for normal atmospheric pressure? (Refer to Fig 15.29 and use $g = 9.80$ m/s^2.) Would you expect the vacuum above the column to be as good as for mercury?
22. Normal atmospheric pressure is 1.013×10^5 Pa. The approach of a storm causes the height of a mercury barometer to drop by 20 mm from the normal height. What is the atmospheric pressure? (The density of mercury is 13.59 g/cm^3.)

23. A simple U-tube that is open at both ends is partially filled with water (Figure 15.30). Kerosene ($\rho_k = 0.82 \times 10^3$ kg/m^3) is then poured into one arm of the tube, forming a column 6 cm in height, as shown in the diagram. What is the difference h in the heights of the two liquid surfaces?

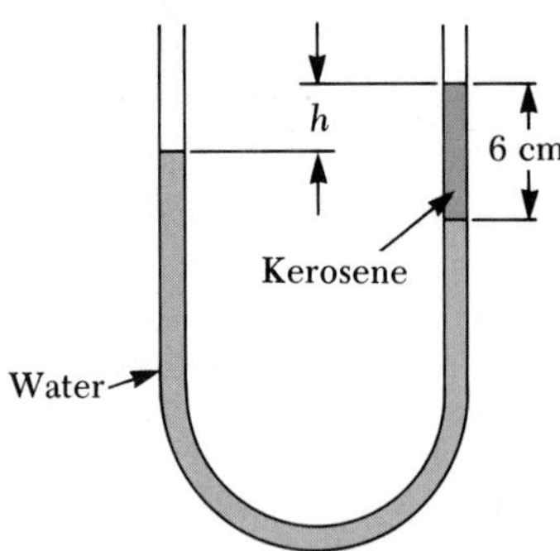

Figure 15.30 (Problem 23).

Section 15.5 Buoyant Forces and Archimedes' Principle

24. Calculate the buoyant force on a solid object made of copper and having a volume of 0.2 m^3 if it is submerged in water. What is the result if the object is made of steel?
25. Show that only 11% of the total volume of an iceberg is above the water level. (Note that sea water has a density of 1.03×10^3 kg/m^3, and ice has a density of 0.92×10^3 kg/m^3.)
26. A solid object has a weight of 5.0 N. When it is suspended from a spring scale and submerged in water, the scale reads 3.5 N (Fig. 15.12b). What is the density of the object?
27. A cube of wood 20 cm on a side and having a density of 0.65×10^3 kg/m^3 floats on water. (a) What is the distance from the top of the cube to the water level? (b) How much lead weight has to be placed on top of the cube so that its top is just level with the water?
28. A balloon filled with helium at atmospheric pressure is designed to support a mass M (payload + empty balloon). (a) Show that the volume of the balloon must be *at least* $V = M/(\rho_a - \rho_{He})$, where ρ_a is the density of air and ρ_{He} is the density of helium. (Ignore the volume of the payload.) (b) If $M = 2000$ kg, what radius should the balloon have?
29. A hollow plastic ball has a radius of 5 cm and a mass of 100 g. The ball has a tiny flap at the top through which lead shot can be inserted. How many grams of lead can be inserted into the ball before it sinks in water? (Assume the ball does not leak when the flap is closed.)
30. A 10-kg block of metal measuring 12 cm × 10 cm × 10 cm is suspended from a scale and immersed in water as in Figure 15.12b. The 12-cm dimension is vertical, and the top of the block is 5 cm from the surface of the water. (a) What are the forces on the top and bottom of the block? (Take $P_a = 1.0130 \times 10^5$ N/m^2.) (b) What is the reading of the spring scale? (c) Show that the buoyant force equals the difference between the forces at the top and bottom of the block.
31. A frog in a hemispherical pod finds that he just floats without sinking in a blue-green sea (density 1.35 g/cm^3). If the pod has a radius of 6 cm and has negligible mass, what is the mass of the frog? (See Fig. 15.31.)

Figure 15.31 (Problem 31).

32. A hollow brass tube (diam. = 4.0 cm) is sealed at one end and filled with lead shot to give a total mass of 0.20 kg. When the tube is floated in pure water, what will be the depth z of the bottom of the tube?
33. A Styrofoam slab has a thickness of 10 cm and a density of 300 kg/m^3. What is the area of the slab if it floats just awash in fresh water when a 75-kg swimmer is aboard?
34. A rectangular air mattress has a length of 2.0 m, a width of 0.50 m, and a thickness of 0.08 m. If the mass of the material is 2.3 kg, what mass can just be supported by the mattress in water?
35. How much helium (in cubic meters) is required to lift a balloon with a 400-kg payload to a height of 8000 m? ($\rho_{He} = 0.18$ kg/m^3.) Assume the balloon maintains a *constant* volume and that the density of air *decereases* with height as $\rho_{air} = \rho_0 e^{-z/8000}$. ($z$ = height in meters, ρ_0 = sea level density = 1.25 kg/m^3.)

Sections 15.6–15.8 Fluid Dynamics and Bernoulli's Equation

36. The rate of flow of water through a horizontal pipe is 2 m^3/min. Determine the velocity of flow at a point where the diameter of the pipe is (a) 10 cm, (b) 5 cm.
37. A large storage tank filled with water develops a small hole in its side at a point 16 m below the water level. If the rate of flow from the leak is 2.5×10^{-3} m^3/min, determine (a) the speed at which the water leaves the hole and (b) the diameter of the hole.

38. Water flows through a constricted pipe at a uniform rate (Fig. 15.18). At one point, where the pressure is 2.5×10^4 Pa, the diameter is 8.0 cm; at another point 0.5 m higher, the pressure is 1.5×10^4 Pa and the diameter is 4.0 cm. (a) Find the speed of flow in the lower and upper sections. (b) Determine the rate of flow through the pipe.
39. Water flows through a horizontal constricted pipe. The pressure is 4.5×10^4 Pa at a point where the speed is 2 m/s and the area is A. Find the speed and pressure at a point where the area is $A/4$.
40. The water supply of a building is fed through a main 6-cm-diameter pipe. A 2-cm-diameter faucet tap located 2 m above the main pipe is observed to fill a 25-liter container in 30 s. (a) What is the speed at which the water leaves the faucet? (b) What is the *gauge pressure* in the 6-cm main pipe? (Assume the faucet is the only "leak" in the building.)
41. Water flows through a fire hose of diameter 6.35 cm at a rate of 0.012 m³/s. The fire hose ends in a nozzle of inner diameter 2.2 cm. What is the velocity with which the water exits the nozzle?
42. The Garfield Thomas water tunnel at Pennsylvania State University has a circular cross-section which constricts from a diameter of 3.6 m to the test section, which is 1.2 m in diameter. If the velocity of flow is 3 m/s in the larger-diameter pipe, determine the velocity of flow in the test section.
43. Old Faithful Geyser in Yellowstone Park erupts at approximately 1-hour intervals, and the height of the fountain reaches 40 m. (a) With what velocity does the water leave the ground? (b) What is the pressure (above atmospheric) in the heated underground chamber?

°Section 15.9 Other Applications of Bernoulli's Equation

44. An airplane has a mass of 16 000 kg, and each wing has an area of 40 m². During level flight, the pressure on the lower wing surface is 7.0×10^4 Pa. Determine the pressure on the upper wing surface.
45. Each wing of an airplane has an area of 25 m². If the speed of the air is 50 m/s over the lower wing surface and 65 m/s over the upper wing surface, determine the weight of the airplane. (Assume the plane travels in level flight at constant speed at an elevation where the density of air is 1 kg/m³. Also assume that all of the lift is provided by the wings.)
46. A Venturi tube may be used as a fluid flow meter (Figure 15.19). If the difference in pressure $P_1 - P_2 =$ 21 000 Pa ($\approx$3 lb/in.²), find the fluid flow rate in m³/s given that the radius of the outlet tube is 1 cm, the radius of the inlet tube is 2 cm, and the fluid is gasoline ($\rho = 700$ kg/m³).
47. A pitot tube can be used to determine the velocity of air flow by measuring the difference between the total pressure and the static pressure (Fig. 15.32). If the fluid in the tube is mercury, density $\rho_{\mathrm{Hg}} =$ 13 600 kg/m³, and $\Delta h = 5$ cm, find the velocity of air flow. (Assume that the air is stagnant at point A and take $\rho_{\mathrm{air}} = 1.25$ kg/m³.)

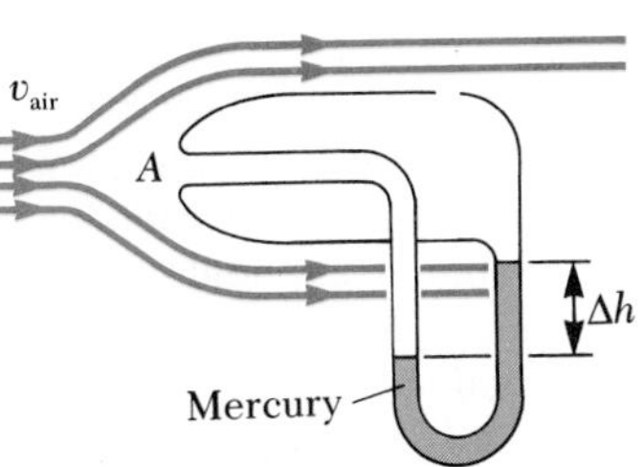

Figure 15.32 (Problem 47).

48. A siphon is used to drain water from a tank, as indicated in Figure 15.33. The siphon has a uniform diameter d. Assume steady flow. (a) Derive an expression for the volume discharge rate at the end of the siphon. (Select the reference level at point 3.) (b) What is the limitation on the height of the top of the siphon above the water surface? (In order to have a continuous flow of liquid, the pressure in the liquid cannot go below atmospheric pressure.)

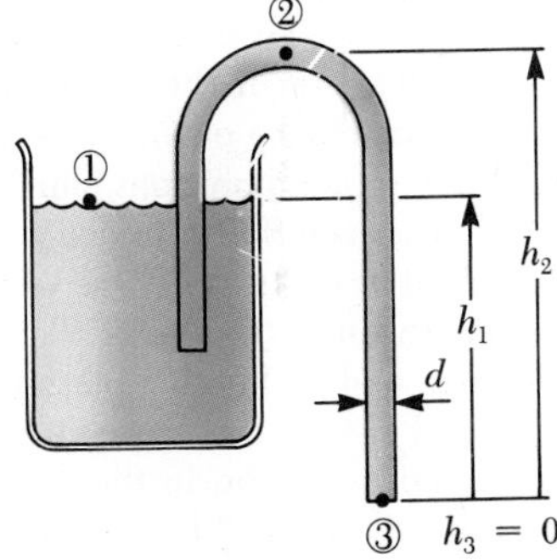

Figure 15.33 (Problem 48).

49. A large storage tank is filled to a height h_0. If the tank is punctured at a height h from the bottom of the tank, (Fig. 15.34) how far from the tank will the stream land?

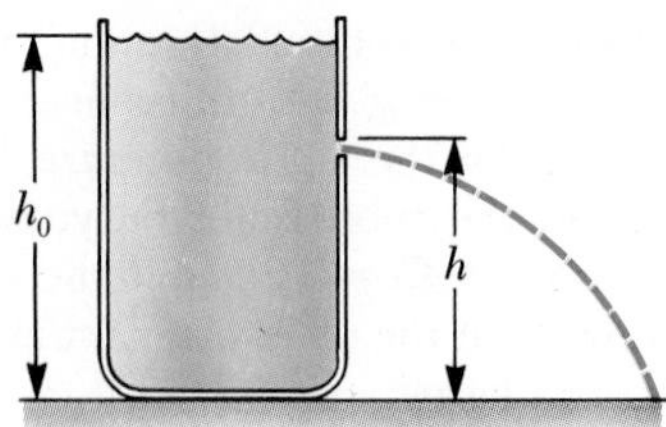

Figure 15.34 (Problem 49).

50. A hole is punched in the side of a 20 cm-tall container, full of water as in the photograph below. If the water is to shoot as far as possible horizontally, (a) how far up the container should the hole be punched? (b) Neglecting friction losses, how far (initially) from the side of the container will the water land?

(Problem 50.) (Photograph courtesy of CENCO)

*Section 15.10 Energy from the Wind

51. Calculate the power output of a windmill having blades 10 m in diameter if the wind speed is 8 m/s. Assume that the efficiency of the system is 20%.
52. According to one rather ambitious plan, it would take 50 000 windmills, each 800 ft in diameter, to obtain an average output of 200 GW. These would be strategically located through the Great Plains, along the Aleutian Islands, and on floating platforms along the Atlantic and Gulf coasts and on the Great Lakes. The annual energy consumption in the United States in 1980 was about 8.3×10^{19} J. What fraction of this could be supplied by the array of windmills?
53. Consider a windmill with blades of cross-sectional area A, as in Figure 15.35, and assume the mill is facing directly into the wind. (a) If the wind speed is v, show that the kinetic energy of the air that passes through the blades in a time Δt is given by the expression $K = \frac{1}{2}\rho A v^3 \Delta t$. (b) What is the maximum available power according to this model? Compare your result with Equation 15.16.

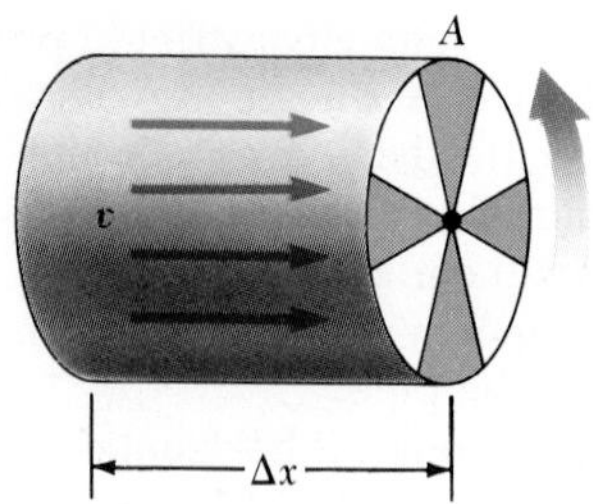

Figure 15.35 (Problem 53).

*Sections 15.11 and 15.12 Viscosity and Turbulence

54. What is the Reynolds number for the flow of liquid in the 1.2-m-diameter Alaska pipeline? The density of crude oil is 850 kg/m^3, its velocity is 3 m/s, and its viscosity is 0.3 Pa·s. Is the flow laminar or turbulent?
55. The viscosity of a certain liquid is determined (at 40° C) by measuring the flow rate through a tube under a known pressure difference between the ends. The tube has a radius of 0.70 mm and a length of 1.50 m. When a pressure difference of $\frac{1}{20}$ atm is applied, a volume of 292 cm^3 is collected in 10 min. What is the viscosity of the liquid? What is the liquid?

ADDITIONAL PROBLEMS

56. A girl of mass 50 kg sits on a 1 m × 1 m × 0.06 m raft made of solid Styrofoam. If the raft *just* supports the girl (that is, the raft is totally submerged), determine the density of the Styrofoam.
57. A sample of copper is to be subjected to a hydrostatic pressure that will increase its density by 0.1 percent. What pressure is required?
58. One side of the U-shaped tube in Figure 15.36 is filled with a liquid of density ρ_1 while the other side contains a liquid of density ρ_2. If the liquids do not mix, show that $\rho_2 = (h_1/h_2)\rho_1$.

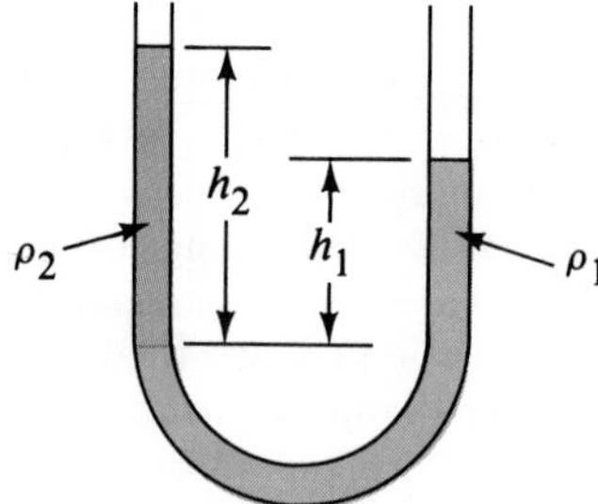

Figure 15.36 (Problem 58).

59. One method of measuring the density of a liquid is illustrated in Figure 15.37. One side of the U-shaped tube is in the liquid being tested; the other side is in water of density ρ_w. When the air is partially removed at the upper part of the tube, show that the density of the liquid on the left is given by $\rho = (h_w/h)\rho_w$.

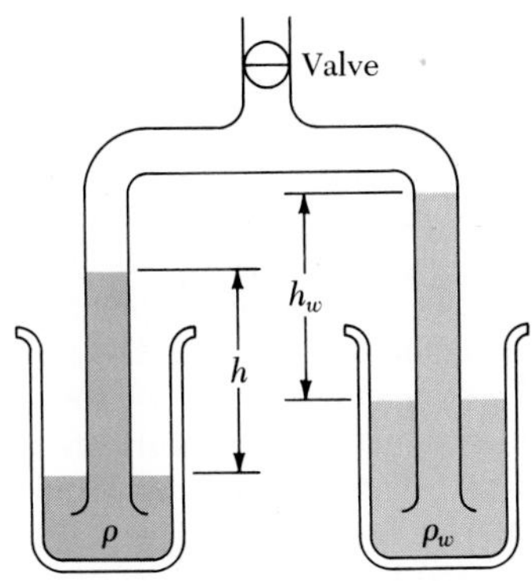

Figure 15.37 (Problem 59).

60. A tank that has a flat bottom of area A and vertical sides is filled to a depth h in the water. There is 1 atm of pressure at the top surface. (a) What is the absolute pressure at the bottom of the tank? (b) Suppose that an object of total mass M (and average density less than the density of water) is placed into the tank and floats there. What is the resulting increase in the absolute pressure at the bottom of the tank? (c) Evaluate your results for a backyard swimming pool ($h =$ 1.50 m; a circular tank with 6-m diameter). If two persons with a combined total mass of 150 kg get into the pool and float quietly there, what is the pressure increase at the bottom of the pool?

61. The true weight of a body is its weight when measured in a vacuum where there are no buoyant forces. A body of volume V is weighed in air on a balance using weights of density ρ. If the density of air is ρ_a and the balance reads W', show that the true weight W is given by

$$W = W' + \left(V - \frac{W'}{\rho g}\right)\rho_a g$$

62. A block of cross-sectional area A, height ℓ, and density ρ is in equilibrium between two fluids of densities ρ_1 and ρ_2 (Fig. 15.38), where $\rho_1 < \rho < \rho_2$. The fluids do not mix. (a) Show that the buoyant force on the block is given by $B = [\rho_1 g y + \rho_2 g(\ell - y)]A$. (b) Show that the density of the block is equal to $\rho = [\rho_1 y + \rho_2(\ell - y)]/\ell$.

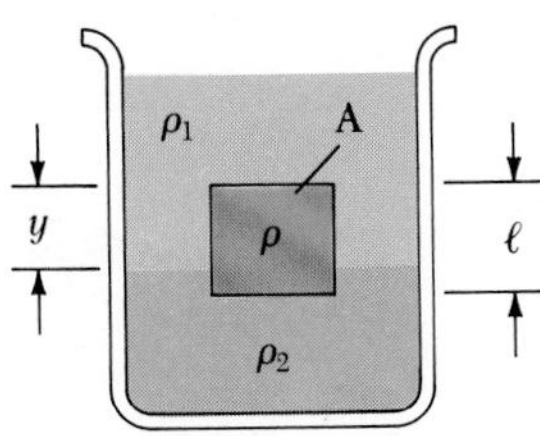

Figure 15.38 (Problem 62).

63. An evacuated steel ball with an outside diameter of 10.0 cm floats in fresh water barely submerged. The density of the steel is 7.86×10^3 kg/m^3. What is the wall thickness of the ball?

64. A horizontal tube of radius 1 cm is joined to a second horizontal tube of radius 0.5 cm. There is a pressure difference of 6660 Pa between the two tubes. What volume of water flows through the tubes per second?

65. If a 1-megaton nuclear weapon is exploded at ground level, the peak overpressure (that is, the pressure increase above normal atmospheric pressure) will be 0.2 atm at a distance of 6 km. What force due to such an explosion will be exerted on the side of a house with dimensions 4.5 m $\times$ 22 m?

66. Water flows at a rate of 2.0×10^{-4} m^3/s in a pipe with a diameter of 3 cm. A small crack (0.1 mm $\times$ 3 mm) develops, and the output flow rate decreases by 1 percent. (a) What is the average velocity of the water that squirts from the crack? (b) What is the average velocity of the water delivered by the pipe?

67. With reference to Figure 15.8, show that the total torque exerted by the water behind the dam about an axis through O is $\frac{1}{6}\rho g w H^3$. Show that the effective line of action of the total force exerted by the water is at a distance $\frac{1}{3}H$ above O.

68. In 1657 Otto von Guericke, inventor of the air pump, evacuated a sphere made of two brass hemispheres. Two teams of eight horses each could pull the hemispheres apart only on some trials, and then "with greatest difficulty." (Fig. 15.39). (a) Show that the force F required to pull the evacuated hemispheres apart is $\pi R^2(P_a - P)$, where R is the radius of the hemispheres and P is the pressure inside the hemispheres, which is much less than P_a. (b) Determine the force if $P = 0.1P_a$ and $R = 0.3$ m.

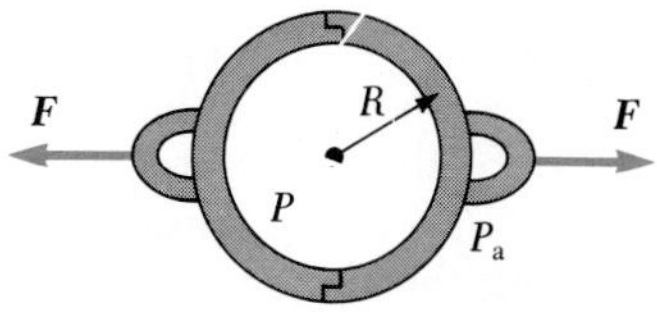

Figure 15.39 (Problem 68).

69. Find the diameter of the largest helium-filled balloon that can be held down using string that snaps when the tension exceeds 40 lb (178 newtons). Use 1.29 kg/m^3 for the density of air and 0.200 kg/m^3 for the density of helium. Assume that the balloon is spherical and ignore the mass of its skin. (*Note:* The density value given here for helium exceeds the value given in Table 15.1. The pressure within the helium balloon has been assumed to be about 10% above atmospheric pressure.)

70. A large water tank has a 1-cm^2 hole in its side at a point 2 m below the surface. What is the mass discharge rate (in kg/s) through the hole?

71. The flow rate of the Columbia River is approximately 3200 m^3/s. What would be the maximum power output of the turbines in a dam if the water were to fall a vertical distance of 160 m?

72. As a first approximation, the earth's continents may be thought of as granite blocks floating in a denser rock (called peridotite) in the same way that ice floats in water. (a) Show that a formula describing this phenomenon is

$$\rho_g t = \rho_p d,$$

where ρ_g is the density of granite (2800 kg/m^3), ρ_p is the density of peridotite (3300 kg/m^3), t is the thickness of a continent, and d is the depth to which a continent floats in the peridotite. (b) If a continent rises 5 km above the surface of the peridotite (this surface may be thought of as the ocean floor), what is the thickness of the continent?

73. Consider a composite "raft" consisting of two square slabs, each of side s, attached face to face. One slab has density ρ_1 and thickness h_1, while the other has density $\rho_2 > \rho_1$ and thickness h_2. (a) Find the average density $\bar{\rho}$ of the raft. (b) Assume that $\bar{\rho} < \rho_w$, so that the raft floats in water. The raft is placed in water with the denser slab on the bottom. Find d, the depth of the bottom surface of the raft. (c) If the raft is placed in water with the denser slab on the *top* find d', the depth of the bottom surface of the raft. Comment on your answer. (d) For which of the orientations described in (b) and (c) is the gravitational potential energy of the entire system (consisting of the raft and the body of water in which it is floating) greater? Find the potential energy difference.

74. A water tank has a conical top surface which slopes upward at an angle α (to the horizontal). The tank is full, but there is a small hole at the apex of the cone, so that the pressure there is 1 atm. If a small hole is opened in the tank wall at a distance s from the apex (measured along the sloping surface), the resulting water stream falls back onto the sloping surface a distance s' down from the leak. (The total distance from the apex to the splash point is thus $s + s'$.) (a) Find s' in terms of s and α. (You may assume that the initial velocity vector of the stream is normal to the tank surface.) (b) Evaluate the ratio s'/s for $\alpha = 45°$. (c) Find the value of α for which $s' = s$. (Give your answer in both radians and in degrees.)

75. A cable of mass density ρ_c and diameter d extends vertically downward a distance h through water, and a block of mass M_b and density ρ_b is hung from the bottom end of the cable. Both ρ_c and ρ_b exceed ρ_w, the density of water. Find (a) the tension T_ℓ at the lower end of the cable, (b) the tension T_u at the upper end of the cable, and (c) the tensions T_ℓ' and T_u' that would exist at the lower and upper ends of the cable if the entire assembly were in air rather than water. (Neglect the buoyant force provided by the air.) (d) Evaluate T_ℓ, T_u, T_ℓ', and T_u' for the case of a 100-meter steel cable supporting a prefabricated concrete object of mass 2.00 metric tons: $\rho_c = 7.86 \times 10^3$ kg/m^3, $d = 2 \times 10^{-2}$ m, $h = 100$ m, $M_b = 2.00 \times 10^3$ kg, and $\rho_b = 2.38 \times 10^3$ kg/m^3.

76. Show that the variation of atmospheric pressure with altitude is given by $P = P_0 e^{-\alpha h}$, where $\alpha = \rho_0 g/P_0$, P_0 is atmospheric pressure at some reference level, and ρ_0 is the atmospheric density at this level. Assume that the decrease in atmospheric pressure with increasing altitude is given by Equation 15.5 and that the density of air is proportional to the pressure.

77. A cube of ice whose edge is 20 mm is floating in a glass of ice-cold water with one of its faces parallel to the water surface. (a) How far below the water surface is the bottom face of the block? (b) Ice-cold ethyl alcohol is gently poured onto the water surface to form a layer 5 mm thick above the water. When the ice cube attains hydrostatic equilibrium again, what will be the distance from the top of the *water* to the bottom face of the block? (c) Additional cold ethyl alcohol is poured onto the water surface until the top surface of the alcohol coincides with the top surface of the ice cube (in hydrostatic equilibrium). How thick is the required layer of ethyl alcohol?

SO MUCH FOR THE QUESTION OF WHETHER OR NOT A CURVEBALL REALLY CURVES.
S. harris

ESSAY

HIGH PRESSURE PHYSICS

A. Jayaraman
AT&T Bell Laboratories
Murray Hill, NJ

Knowledge about the behavior of matter under extreme pressures is of great interest to several scientific disciplines. Scientists working in the field of high pressure research have continually strived to reach higher and higher pressures to expand this knowledge. In recent years a novel pressure-generating device called the *diamond anvil cell* has evolved. With this device, static pressures (sustained application of pressure as opposed to dynamic pressures generated in shock waves, lasting only for microseconds) of over 2 Mbar can be reached under laboratory conditions, and a variety of sophisticated measurements can be performed to probe the behavior of matter under high pressures. [Pressure is force per unit area and is expressed in bars or Pascals; 1 bar = 10^5 Pascals; 1 kbar or kilobar = 10^3 bars; 10 kbar = 1 GPa (gigapascals); 1 Mbar or megabar = 10^6 bars or 100 GPa.]

The late Professor P. W. Bridgman of Harvard University almost single-handedly pioneered high pressure research for over half a century, until his death in 1961. He not only studied an unbelievable number of elements and compounds under pressure, but also invented every technique he used to reach pressures of up to 100 kbar. Bridgman was awarded the Nobel prize in 1946 for the development of pressure-generation techniques and for the discovery of new phenomena under pressure. Generation of very high pressure involves the use of the principle of force multiplication: A force F applied to a larger area A_1 is multiplied by the ratio A_1/A_2 at the delivery end having a smaller area A_2. (See Section 15.4.) This ratio can vary from 10 to 1000 in any practical pressure-generating device. Bridgman used this principle to construct high pressure apparatus.

Diamond has two properties that qualify it as the best material for the containment of high pressure; (1) it is the hardest substance known to science and (2) it is transparent to optical radiation as well as to x-rays. Although these properties of the diamond were well known for a long time, it was only in 1959 that the diamond anvil device was born. A diamond anvil cell capable of generating a pressure of a half a million atmospheres fits easily in the palm of a hand (see Fig. 1 (a) and (b)). One of the greatest advantages of the diamond anvil cell is that pressurized samples can be directly viewed with a microscope. The basic principle of the diamond anvil cell is simple. A sample placed between flat parallel faces of two diamond anvils is subjected to very high pressure by means of a force pushing one anvil against the other. Since

(*Continued on next page*)

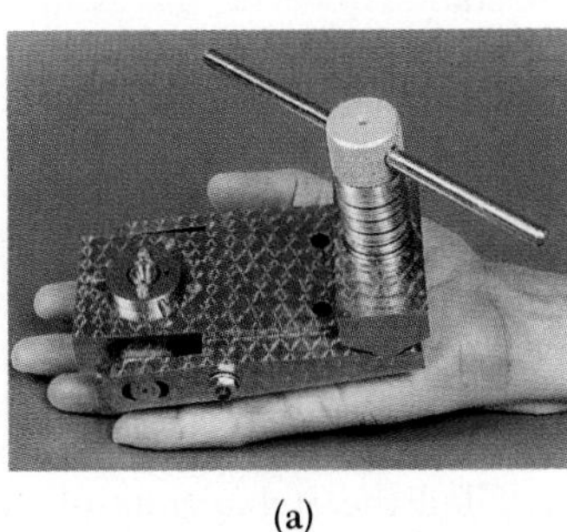

(a)

Figure 1 (a) Photograph of a diamond anvil cell for generating pressure of $\frac{1}{2}$ Mbar.

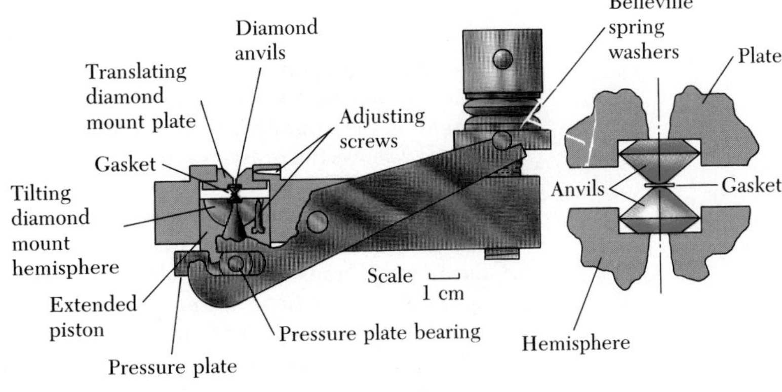

(b)

Figure 1 (b) Diagramatic sketch of a diamond anvil cell according to the National Bureau of Standards design. On the extreme right, the diamond anvil region is shown magnified. To generate hydrostatic pressure, a thin metal gasket with a 200-μm diameter hole is introduced between the anvil faces, and the hole filled with a pressure medium, usually a mixture of methanol and ethanol for hydrostatic pressures up to 10 GPa. Rare gases like argon and xenon can be condensed into the hole for use as a hydrostatic pressure medium to much higher pressures.

the flat faces have a very small area ($\approx 0.1\ mm^2$) there is a large pressure multiplication. A prerequisite to using the diamond anvil cell for pressure generation is that the anvil flats be perfectly aligned axially and set parallel to each other as well. The first objective is easily realized with the help of centering screws, and the second with the help of a tiltable hemispherical mount or a set of two half-cylindrical rocker-mounts. Force on the diamond is applied by compressing the Belleville spring washers with the screws (Fig. 1) or other types of lever-arm arrangement. The diamonds used in the diamond anvil cell must be of high quality and flawless. Usually $\frac{1}{3}$ to $\frac{1}{2}$ carat gem-cut diamonds whose culet has been ground to a small flat surface are used. The pressure is measured by the Ruby fluorescence technique, which is in wide use at the present time. Ruby crystals exhibit a strong red fluorescence when excited by light, such as the blue line of a helium-cadmium laser. This red fluorescent emission consists of two well-defined peaks at wavelengths of 694.2 nm (R_1 line) and 692.7 nm (R_2 line). These shift to higher wavelengths with increasing pressure; the R_1 shift has been calibrated, using the equation of state of standard substances, to construct a pressure-scale. The ruby scale is almost linear up to 300 GPa, the shift being 0.365 nm/GPa. At higher pressures the scale begins to bend somewhat towards the pressure axis.

Diamond cells have been adapted for low-temperature studies down to liquid helium temperatures. For high-temperature investigations heating is accomplished with an yttrium aluminum garnet (YAG) laser. If the sample is absorbing at the wavelength of the laser line, it can be very quickly heated to about 4000°C, without heating the cell. The temperature is usually measured by optical pyrometry or spectroscopy. The diamond anvil cell is an excellent tool for high pressure spectroscopy and is well suited for high pressure x-ray diffraction studies. With modern synchrotron x-ray sources, diffraction data can be collected in a fraction of a second.

One of the most useful basic data for understanding the fundamental properties of a solid is the application of the pressure-volume relationship. This information can be obtained from high pressure x-ray measurements. By fitting the pressure-volume data to a theoretical equation of state, the bulk modulus B (inverse of the compressibility) of a solid ($B = V(\partial P/\partial V)T$ and its pressure derivative B' can be evaluated. One of the equations commonly used in this connection is the Murnaghan equation of state, given by

$$P = \frac{B}{B'}\left[\left(\frac{V_0}{V}\right)^{B'} - 1\right]$$

The above equation of state assumes that dB/dP is a constant. However, this assumption may not be true for highly compressible solids; in that case nonlinear terms should be incorporated into the equation of state. Many such equations are in use. Figure 2 shows the pressure-volume data for silver, fitted to the Murnaghan equation of state.

To a high pressure researcher the most rewarding experience is discovering a pressure-induced phase transition. The fundamental philosophy behind pressure-induced phase transitions is as follows: The total energy of a system in thermodynamic equilibrium must be at its minimum value. (A system in thermodynamic equilibrium is one in which no changes in pressure, volume or temperature can occur without external influences.) This minimum energy state is called the *stable state* of the system. When the volume is decreased by compression, the total energies of the possible states change (see Fig. 3). The system adjusts itself through either structural transformations or changes in its electronic structure, or both, to minimize the total energy. Structural transformations involve geometrical rearrangement of the atoms, while the electronic structure changes are connected with the electronic energy bands in a solid. In many instances these changes profoundly influence the physical properties of solids. Figure 3 shows the structural stability of silicon, a well-known

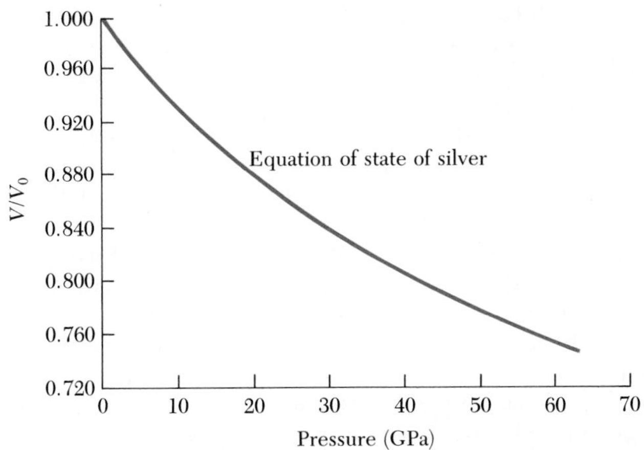

Figure 2 The pressure-volume relationship for silver generated using the Murnaghan equation of state with $B_\beta = 118$ GPa and $B' = 3.8$. The experimental data fit this curve.

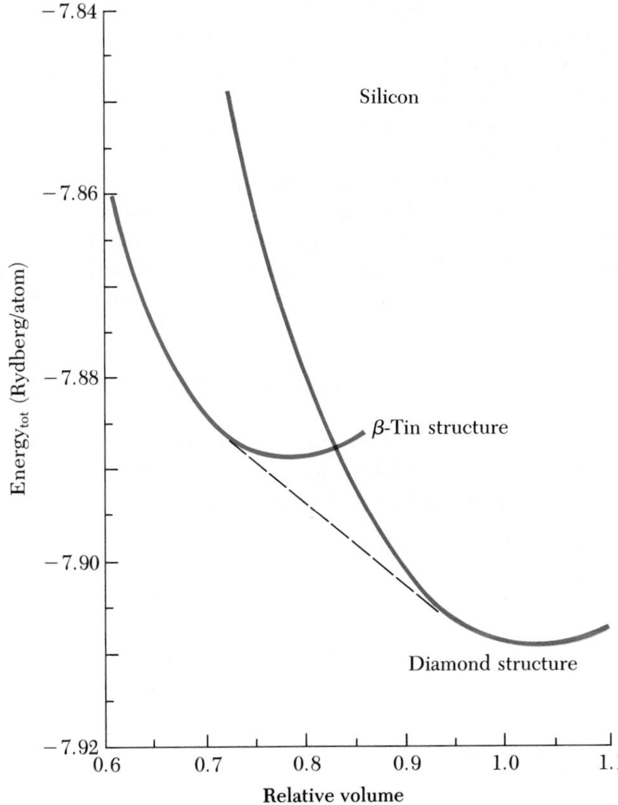

Figure 3 Total energy curves calculated by Yin and Cohen of the University of California, Berkeley, for the diamond and β-tin phases of silicon plotted against relative volume. The dashed line is the common tangent of the energy curves for the two phases. The slope of this line gives the transformation pressure ($dE/dV = P$). The diamond phase of Si transforms to β-tin phase around 12 GPa.

(Continued on next page)

semiconductor, which has the diamond structure at ambient pressure. Upon compression the energy of the diamond phase increases and crosses the β-tin at V/V_0 (0.82). At a pressure of about 10 GPa, Si transforms to the metallic tin structure called β-tin, with a large increase in density. The β-tin phase is metallic and superconducting, a spectacular change indeed. Such behavior under the influence of pressure is shown by germanium and many compound semiconductors as well. One of the exciting predictions of solid state theory is that hydrogen becomes metallic at a pressure of about 3 Mbar. Hydrogen solidifies to a clear transparent solid near a pressure of 57 kbar at 298 K. It is a very good insulator but under the influence of pressure, the localized electron orbitals can overlap, leading to a metallic state. According to solid state theory, metallic hydrogen is likely to be a high temperature superconductor. The high pressure behavior of hydrogen is therefore not only of fundamental interest, but it has implications for the internal structure of the giant planets, which are believed to be largely made up of hydrogen. A rather spectacular pressure-induced semiconductor-to-metal transition occurs at a moderate pressure in a compound known as samarium monosulfide (SmS). This transition is a classic case of electronic structure change. The transition looks like "alchemy," for the dull, black-looking substance glitters like gold when the applied pressure exceeds 7 kbar. In the high pressure phase, one bound electron from the $4f$ level per Sm is squeezed out into the $5d$ conduction band, without causing any change in structure.

The most spectacular application of high pressure research to technology is the synthesis of diamond from graphite. If carbon is subjected to a pressure of 65 kbar at ~1400°C in the presence of nickel or iron, it is spontaneously converted into diamond. Millions of carats of diamonds are synthesized in this manner annually for industrial applications. Perhaps the best scientific application of high pressure research is in the realm of geophysics. Inside the earth both pressure and temperature increase with depth; at the center, pressure is estimated to be 3.5 Mbar and temperature, 4000°C. Furthermore, the interior of the earth is stratified into a crust, mantle and core, with some minor features superimposed. Experiments on silicate minerals (the major components of the earth) have revealed that the major and minor features of the crust and the mantle region are related to phase transitions induced by high pressure in these minerals. A major phase transition in silicate minerals, recently discovered with the diamond anvil cell, is to the so-called "perovskite" phase, which seems to dominate the entire lower mantle region. Thus, the diamond anvil cell is serving as a transparent window right up to the core-mantle boundary, located at a depth of 2900 km.

It is an interesting fact that the recent discovery of the high T_c superconductivity in the compound $YBa_2Cu_3O_7$ (~92 K), was triggered by a high-pressure experiment in lanthanum cuprate. Professor Chu and his colleagues at the University of Houston, while investigating the effect of pressure on the superconducting T_c of barium-substituted La_2CuO_4, found a remarkable increase with pressure in T_c from 30 K to ~50 K, when the compound was pressurized to about 50 kbar. From this they concluded that the substitution of the smaller yttrium ion for lanthanum should exert the same influence as pressure from within the system. Indeed, when such a substitution was tried, the now famous $YBa_2Cu_3O_7$ system with a T_c close to 90°K was discovered.

Since the advent of the diamond anvil cell, research regarding the properties of materials under very high pressures has made a quantum jump. A new level of understanding of the physics and chemistry of solids at high pressure is emerging. The possibility of reaching megabar pressures has stimulated solid state theoretical calculations. These calculations, facilitated by modern computing, are able to predict pressure-induced transitions and the electronic band structure at high pressure to an astonishing degree of precision. Several insulating systems have been driven to the metallic state in recent years. Two spectacular examples of this are solid xenon—a rare gas at normal pressure and temperature—and cesium iodide (CsI), a compound which is a highly insulating substance. Theory had predicted insulator-to-metal transitions in these systems at high pressure, and this has been experimentally realized. Experiments carried out with the diamond cell indicate that solid xenon exhibits optical properties characteristic of a metal at about 1.7 Mbar pressure. Similarly, the optical properties of cesium iodide near 1.2 Mbar are characteristic of the metallic state.

What is the limit of pressure attainable with the diamond anvils? One thing that would limit the maximum pressure would be a phase transition in diamond. A recent theoretical calculation shows that the diamond structure would be stable up to at least 10 Mbar. However, the ultimate pressure that can be reached in a diamond anvil cell would be determined by the yield strength of diamond. From plasticity theory this is estimated to be about 5 Mbar. In recent years, pressures in the neighborhood of 5 Mbar have been attained by two groups in the United States. X-ray diffraction experiments have been actually carried out by the Cornell University group to 2.5 Mbar, with rhenium as the gasket material and as a sample. If such pressures could be attained, high pressure physics would leap to even greater heights.

Suggested Readings

Jayaraman, A., "Diamond Anvil Cell and High Pressure Physical Investigations," *Reviews of Modern Physics,* Volume 58 (1983), pp. 65–108.

Jayaraman, A., "The Diamond-Anvil High Pressure Cell," *Scientific American,* April 1984.

Jayaraman, A., "Ultrahigh Pressure," *Review of Scientific Instruments,* Volume 57 (1986), pp. 1013–1031.

A surfer "riding the pipe" on an ocean wave. (© Doug Peebles/Index Stock International)

PART II
Mechanical Waves

The impetus is much quicker than the water, for it often happens that the wave flees the place of its creation, while the water does not; like the waves made in a field of grain by the wind, where we see the waves running across the field while the grain remains in place.

LEONARDO DA VINCI

As we look around us, we find many examples of objects that vibrate or oscillate: a pendulum, the strings of a guitar, an object suspended on a spring, the piston of an engine, the head of a drum, the reed of a saxophone. Most elastic objects will vibrate when an impulse is applied to them. That is, once they are distorted, their shape tends to be restored to some equilibrium configuration. Even at the atomic level, the atoms in a solid vibrate about some position as if they were connected to their neighbors by some imaginary springs.

Wave motion is closely related to the phenomenon of vibration. Sound waves, earthquake waves, waves on stretched strings, and water waves are all produced by some source of vibration. As a sound wave travels through some medium, such as air, the molecules of the medium vibrate back and forth; as a water wave travels across a pond, the water molecules vibrate up and down. As waves travel through a medium, the particles of the medium move in repetitive cycles. Therefore, the motion of the particles bears a strong resemblance to the periodic motion of a vibrating pendulum or a mass attached to a spring.

There are many other phenomena in nature whose explanation requires us to understand first the concepts of vibrations and waves. Although many large structures, such as skyscrapers and bridges, appear to be rigid, they actually vibrate, a fact that must be taken into account by the architects and engineers who design and build them (see the essay in Chapter 13). To understand how radio and television work, we must understand the origin and nature of electromagnetic waves and how they propagate through space. Finally, much of what scientists have learned about atomic structure has come from information carried by waves. Therefore, we must first study waves and vibrations in order to understand the concepts and theories of atomic physics.

16

Wave Motion

Interference of water waves produced in ripple tank. The sources of the waves are two objects that vibrate perpendicular to the surface of the tank. (Courtesy of CENCO)

Most of us experienced waves as children when we dropped a pebble into a pond. The disturbance created by the pebble excites ripple waves, which move outward, finally reaching the shore of the pond. If you were to examine carefully the motion of a leaf floating near the disturbance, you would see that it moves up and down and sideways about its original position, but does not undergo any net displacement away or toward the source of the disturbance. That is, the water wave (or disturbance) moves from one place to another, *yet the water is not carried with it.*

An excerpt from a book by Einstein and Infeld gives the following remarks concerning wave phenomena.[1]

> A bit of gossip starting in Washington reaches New York very quickly, even though not a single individual who takes part in spreading it travels between these two cities. There are two quite different motions involved, that of the rumor, Washington to New York, and that of the persons who spread the rumor. The wind, passing over a field of grain, sets up a wave which spreads out across

[1] Albert Einstein and Leopold Infeld, *The Evolution of Physics*, New York, Simon and Schuster, 1961. Excerpt from *What is a Wave?*

> the whole field. Here again we must distinguish between the motion of the wave and the motion of the separate plants, which undergo only small oscillations. . . . The particles constituting the medium perform only small vibrations, but the whole motion is that of a progressive wave. The essentially new thing here is that for the first time we consider the motion of something which is not matter, but energy propagated through matter.

Water waves represent only one example of a wide variety of physical phenomena that have wavelike characteristics. The world is full of waves: sound waves; mechanical waves, such as a wave on a string; earthquake waves; shock waves generated by supersonic aircraft; and electromagnetic waves, such as visible light, radio waves, television signals, and x-rays. In the present chapter, we shall confine our attention to mechanical waves, that is, waves that travel only in a material substance.

The wave concept is rather abstract. When we observe what we call a water wave, what we see is a rearrangement of the water's surface. Without the water, there would be no wave. A wave traveling on a string would not exist without the string. Sound waves travel through air as a result of pressure variations from point to point. In such cases, what we interpret as a wave corresponds to the disturbance of a body or medium. Therefore, we can consider a wave to be the *motion of a disturbance.* The motion of the disturbance (that is, the wave itself, or the state of the medium) is not to be confused with the motion of the particles. The mathematics used to describe wave phenomena is common to all waves. In general, we shall find that mechanical wave motion is described by specifying the positions of all points of the disturbed medium as a function of time.

16.1 INTRODUCTION

The mechanical waves discussed in this chapter require (1) some source of disturbance, (2) a medium that can be disturbed, and (3) some physical connection or mechanism through which adjacent portions of the medium can influence each other. We shall find that all waves carry energy. The amount of energy transmitted through a medium and the mechanism responsible for the transport of energy will differ from case to case. For instance, the power of ocean waves during a storm is much greater than the power of sound waves generated by a single human voice.

Three physical characteristics are important in characterizing waves: the wavelength, the frequency, and the wave velocity. One **wavelength** is the *minimum distance between any two points on a wave that behave identically.* For example, in the case of water waves, the wavelength is the distance between adjacent crests or between adjacent troughs.

Most waves are periodic in nature. The **frequency** of such periodic waves is *the rate at which the disturbance repeats itself.*

Waves travel, or *propagate,* with a specific velocity, which depends on the properties of the medium being disturbed. For instance, sound waves travel through air at 20°C with a speed of about 344 m/s (781 mi/h), whereas the speed of sound in most solids is higher than 344 m/s. A special class of waves that do not require a medium in order to propagate is electromagnetic waves, which travel very swiftly through a vacuum with a speed of about 3×10^8 m/s (186 000 mi/s). We shall discuss electromagnetic waves further in Chapter 34.

16.2 TYPES OF WAVES

One way to demonstrate wave motion is to flip one end of a long rope that is under tension and has its opposite end fixed, as in Figure 16.1. In this manner, a single wave bump (or pulse) is formed that travels (to the right in Fig. 16.1) with a definite speed. This type of disturbance is called a **traveling wave.** Figure 16.1 represents four consecutive "snapshots" of the traveling wave. As we shall see later, the speed of the wave depends on the tension in the rope and on the properties of the rope. The rope is the *medium* through which the wave travels. The shape of the wave pulse changes very little as it travels along the rope.[2]

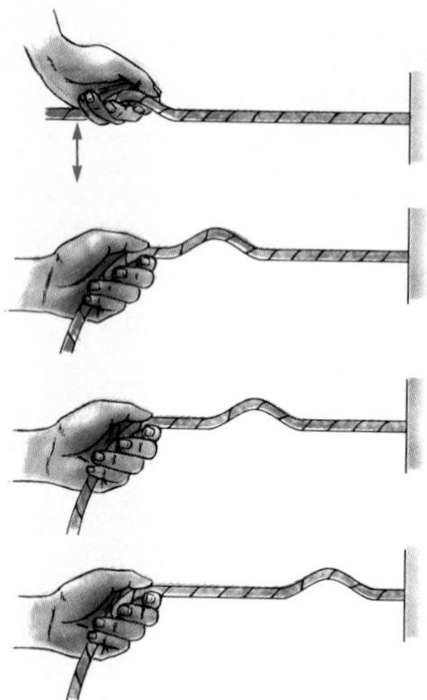

Figure 16.1 A wave pulse traveling down a stretched rope. The shape of the pulse is approximately unchanged as it travels along the rope.

Note that, as the wave pulse travels along the rope, *each segment of the rope that is disturbed moves in a direction perpendicular to the wave motion.* Figure 16.2 illustrates this point for one particular segment, labeled *P*. Note that there is no motion of any part of the rope in the direction of the wave.

A traveling wave such as this, in which the particles of the disturbed medium move perpendicular to the wave velocity, is called a **transverse wave.**[3]

In another class of waves, called **longitudinal waves,** the particles of the medium undergo displacements in a direction *parallel* to the direction of wave motion.

Sound waves, which we shall discuss in Chapter 17, are longitudinal waves that result from the disturbance of the medium. The disturbance corresponds to a series of high- and low-pressure regions that travel through air or through any material medium with a certain velocity. A longitudinal pulse can be easily produced in a stretched spring, as in Figure 16.3. The left end of the spring is given a sudden jerk (consisting of a brief push to the right and equally brief pull to the left) along the length of the spring; this creates a sudden compression of the coils. The compressed region *C* (pulse) travels along the spring, and so we see that the disturbance is parallel to the wave motion. Region *C* is followed by a region *R*, where the coils are extended.[4]

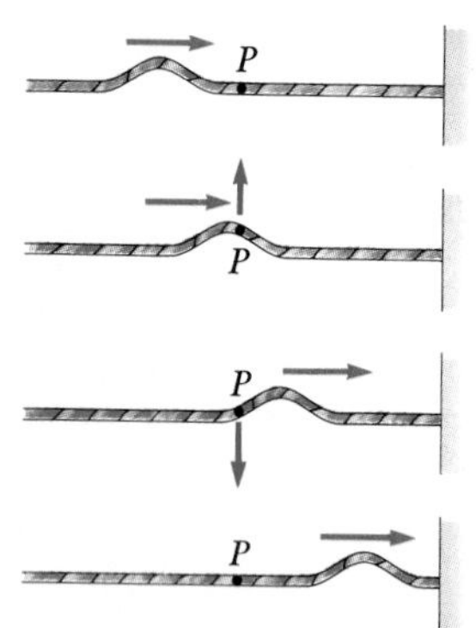

Figure 16.2 A pulse traveling on a stretched rope is a transverse wave. That is, any element *P* on the rope moves in a direction *perpendicular* to the wave motion.

Some waves in nature are neither transverse nor longitudinal, but a combination of the two. Surface water waves are a good example. When a water wave travels on the surface of deep water, water molecules at the surface move in nearly circular paths, as shown in Figure 16.4, where the water surface is drawn as a series of crests and troughs. Note that the disturbance has both transverse and longitudinal components. As the wave passes, water molecules at the crests move in the direction of the wave, and molecules at the troughs move in the opposite direction. Hence, there is no *net* displacement of a water molecule after the passage of any number of complete waves.

[2] Strictly speaking, the pulse will change its shape and gradually spread out during the motion. This effect is called *dispersion* and is common to many mechanical waves.

[3] Other examples of transverse waves are electromagnetic waves, such as light, radio, and television waves. At a given point in space, the electric and magnetic fields of an electromagnetic wave are perpendicular to the direction of the wave and to each other, and vary in time as the wave passes. As we shall see later, electromagnetic waves are produced by accelerating charges.

[4] In the case of longitudinal pressure waves in a gas, each compressed area is a region of higher-than-average pressure and density, and each extended region is a region of lower-than-average pressure and density.

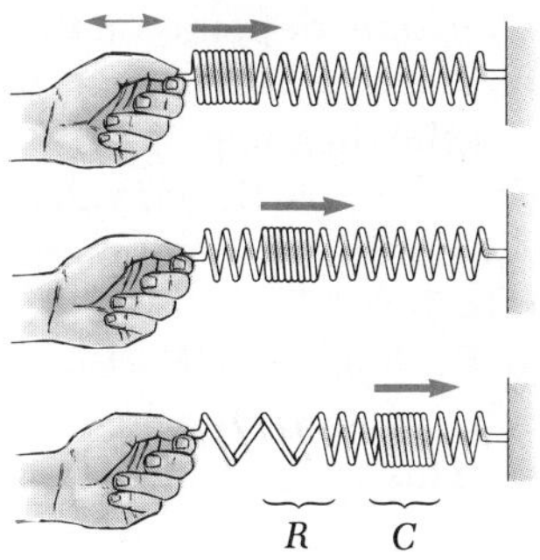

Figure 16.3 A longitudinal pulse along a stretched spring. The disturbance of the medium (the displacement of the coils) is in the direction of the wave motion. For the starting motion described in the text, the compressed region C is followed by an extended region R.

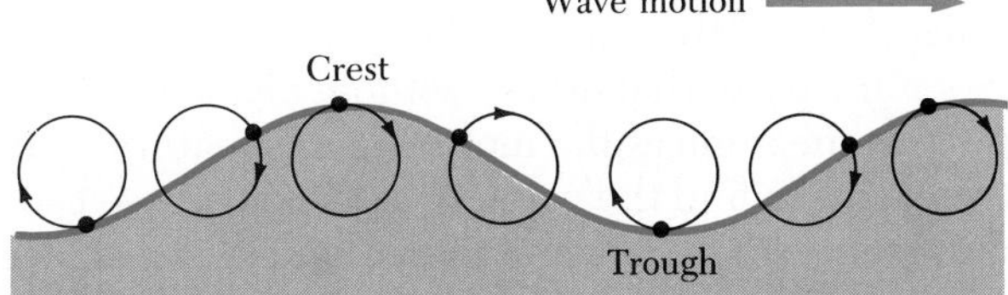

Figure 16.4 Wave motion on the surface of water. The particles at the water's surface move in nearly circular paths. Each particle is displaced horizontally and vertically from its equilibrium position, represented by circles.

16.3 ONE-DIMENSIONAL TRAVELING WAVES

So far we have given only a verbal and graphical description of a traveling wave. Let us now give a mathematical description of a one-dimensional traveling wave. Consider again a wave pulse traveling to the right on a long stretched string with constant speed v, as in Figure 16.5. The pulse moves along the x axis (the axis of the string), and the transverse displacement of the string is measured with the coordinate y.

Figure 16.5a represents the shape and position of the pulse at time $t = 0$. At this time, the shape of the pulse, whatever it may be, can be represented as $y = f(x)$. That is, y is some definite function of x. The *maximum displacement*, y_m, is called the **amplitude** of the wave. Since the speed of the wave pulse is v, it travels to the right a distance vt in a time t (Fig. 16.5b).

If the shape of the wave pulse doesn't change with time, we can represent the displacement y for all later times measured in a stationary frame with the origin at 0 as

$$y = f(x - vt) \tag{16.1}$$

Wave traveling to the right

Similarly, if the wave pulse travels to the *left*, its displacement is given by

$$y = f(x + vt) \tag{16.2}$$

Wave traveling to the left

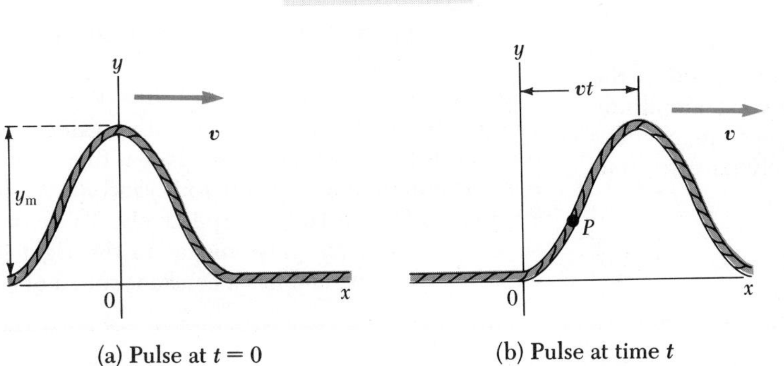

Figure 16.5 A one-dimensional wave pulse traveling to the right with a speed v. (a) At $t = 0$, the shape of the pulse is given by $y = f(x)$. (b) At some later time t, the shape remains unchanged and the vertical displacement is given by $y = f(x - vt)$.

The displacement y, sometimes called the *wave function*, depends on the two variables x and t. For this reason, it is often written $y(x, t)$, which is read "y as a function of x and t." It is important to understand the meaning of y.

Consider a particular point P on the string, identified by a particular value of its coordinates. As the wave passes the point P, the y coordinate of this point will increase, reach a maximum, and then decrease to zero. Therefore, the **wave function** *y represents the y coordinate of any point P at any time t.* Furthermore, if t is fixed, then the wave function y as a function of x *defines a curve representing the actual shape of the pulse at this time.* This is equivalent to a "snapshot" of the wave at this time.

For a pulse that moves without changing its shape, the velocity of a wave pulse is the same as the motion of any feature along the pulse profile, such as the crest. To find the velocity of the pulse, we can calculate how far the crest moves in a short time and then divide this distance by the time interval. The crest of the pulse corresponds to that point for which y has its maximum value. In order to follow the motion of the crest, some particular value, say x_0, must be substituted for $x - vt$. (The value x_0 is called the *argument* of the function y.) Regardless of how x and t change individually, we must require that $x - vt = x_0$ in order to stay with the crest. This, therefore, represents the equation of motion of the crest. At $t = 0$, the crest is at $x = x_0$; at a time dt later, the crest is at $x = x_0 + v\,dt$. Therefore, in a time dt, the crest has moved a distance $dx = (x_0 + v\,dt) - x_0 = v\,dt$. Clearly, the wave speed, often called the **phase velocity,** is given by

Phase velocity

$$v = dx/dt \tag{16.3}$$

The wave velocity, or phase velocity, must not be confused with the transverse velocity (which is in the y direction) of a particle in the medium.

The following example illustrates how a specific wave function is used to describe the motion of a traveling wave pulse.

EXAMPLE 16.1 A Pulse Moving to the Right □

A traveling wave pulse moving to the right along the x axis is represented by the wave function

$$y(x, t) = \frac{2}{(x - 3t)^2 + 1}$$

where x and y are measured in cm and t is in s. Let us plot the waveform at $t = 0$, $t = 1$ s, and $t = 2$ s.

Solution First, note that this function is of the form $y = f(x - vt)$. By inspection, we see that the speed of the wave is $v = 3$ cm/s. Furthermore, the wave amplitude (the maximum value of y) is given by $y_m = 2$ cm. At times $t = 0$, $t = 1$ s, and $t = 2$ s, the wave function expressions are

$$y(x, 0) = \frac{2}{x^2 + 1} \qquad \text{at } t = 0$$

$$y(x, 1) = \frac{2}{(x - 3)^2 + 1} \qquad \text{at } t = 1 \text{ s}$$

$$y(x, 2) = \frac{2}{(x - 6)^2 + 1} \qquad \text{at } t = 2 \text{ s}$$

We can now use these expressions to plot the wave function versus x at these times. For example, let us evaluate $y(x, 0)$ at $x = 0.5$ cm:

$$y(0.5, 0) = \frac{2}{(0.5)^2 + 1} = 1.60 \text{ cm}$$

Likewise, $y(1, 0) = 1.0$ cm, $y(2, 0) = 0.40$ cm, etc. A continuation of this procedure for other values of x yields the waveform shown in Figure 16.6a. In a similar manner, one obtains the graphs of $y(x, 1)$ and $y(x, 2)$, shown in Figures 16.6b and 16.6c, respectively. These snapshots show that the wave pulse moves to the right without changing its shape and has a constant speed of 3 cm/s.

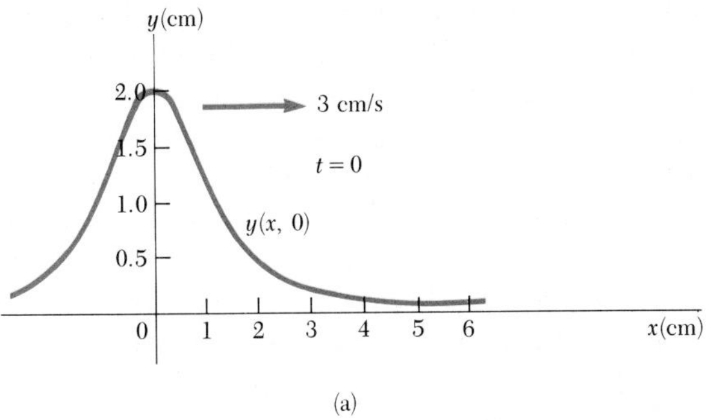

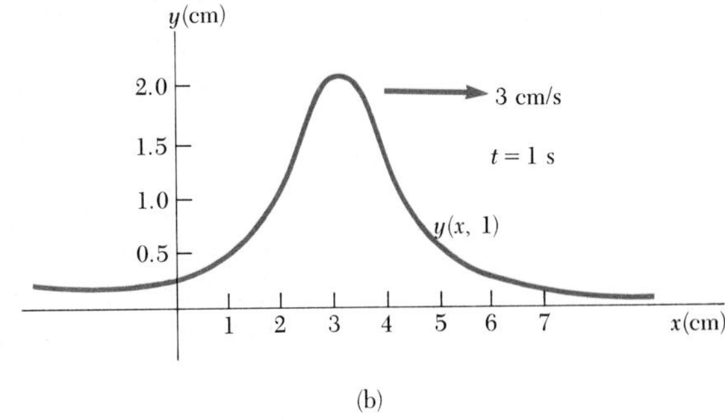

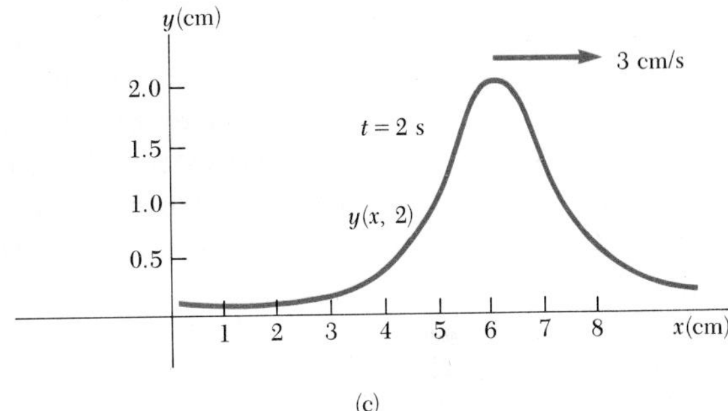

Figure 16.6 (Example 16.1) Graphs of the function $y(x, t) = 2/[(x - 3t)^2 + 1]$. (a) $t = 0$, (b) $t = 1$ s, and (c) $t = 2$ s.

16.4 SUPERPOSITION AND INTERFERENCE OF WAVES

Many interesting wave phenomena in nature cannot be described by a single moving pulse. Instead, one must analyze complex waveforms in terms of a combination of many traveling waves. To analyze such wave combinations, one can make use of the **superposition principle:**

> If two or more traveling waves are moving through a medium, the resultant wave function at any point is the algebraic sum of the wave functions of the individual waves.

Linear waves obey the superposition principle

This rather striking property is exhibited by many waves in nature. Waves that obey this principle are called *linear waves*, and they are generally characterized by small wave amplitudes. Waves that violate the superposition principle are called *nonlinear waves* and are often characterized by large amplitudes. In this book, we shall deal only with linear waves.

One consequence of the superposition principle is the observation that *two traveling waves can pass through each other without being destroyed or even altered.* For instance, when two pebbles are thrown into a pond, the expanding circular surface waves do not destroy each other. In fact, the ripples pass through each other. The complex pattern that is observed can be viewed as two independent sets of expanding circles. Likewise, when sound waves from two sources move through air, they also can pass through each other. The resulting sound one hears at a given point is the resultant of both disturbances.

A simple pictorial representation of the superposition principle is obtained by considering two pulses traveling in opposite directions on a stretched string, as in Figure 16.7. The wave function for the pulse moving to the right is y_1, and the wave function for the pulse moving to the left is y_2. The pulses have the same speed, but different shapes. Each pulse is assumed to be symmetric (although this is not a necessary condition), and both displacements are taken to be positive. When the waves begin to overlap (Fig. 16.7b), the resulting complex waveform is given by $y_1 + y_2$. When the crests of the pulses exactly coincide (Fig. 16.7c), the resulting waveform $y_1 + y_2$ is symmetric. The two pulses finally separate and continue moving in their original directions (Fig. 16.7d). Note that the final waveforms remain unchanged, as if the two pulses never met! The combination of separate waves in the same region of space to produce a resultant wave is called *interference*. For the two pulses shown in Figure 16.7, the displacements of the individual pulses are in the same direction, and the resultant waveform (when the pulses overlap) exhibits a displacement greater than those of the individual pulses.

Now consider two identical pulses traveling in opposite directions on an infinitely long string, where one is inverted relative to the other, as in Figure

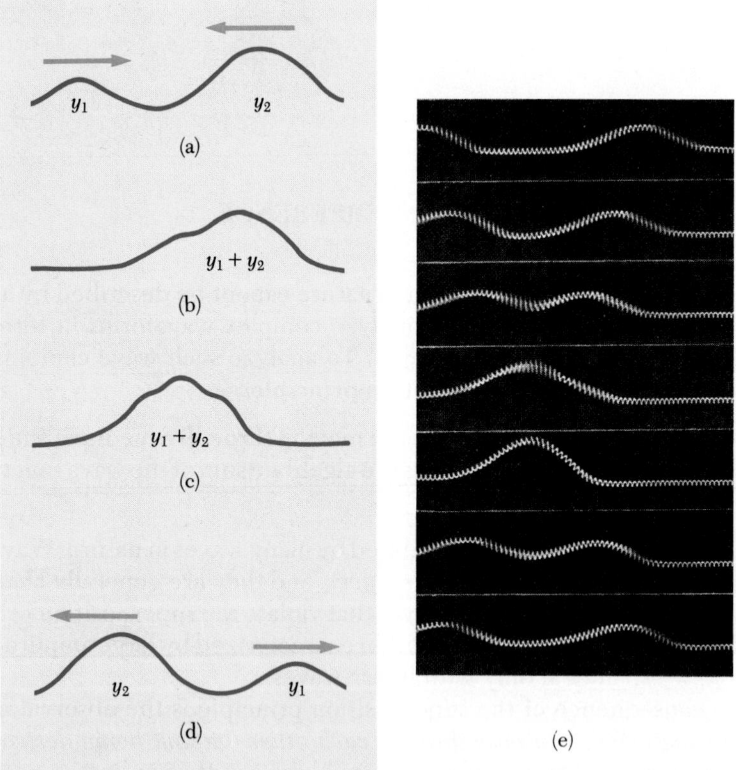

Figure 16.7 (Left) Two wave pulses traveling on a stretched string in opposite directions pass through each other. When the pulses overlap, as in (b) and (c), the net displacement of the string equals the sum of the displacements of each pulse. Since the pulses both have positive displacements, we refer to their superposition as *constructive interference*. (Right) Photograph of superposition of two equal and symmetric pulses traveling in opposite directions on a stretched string. (Photo, Education Development Center, Newton, Mass.)

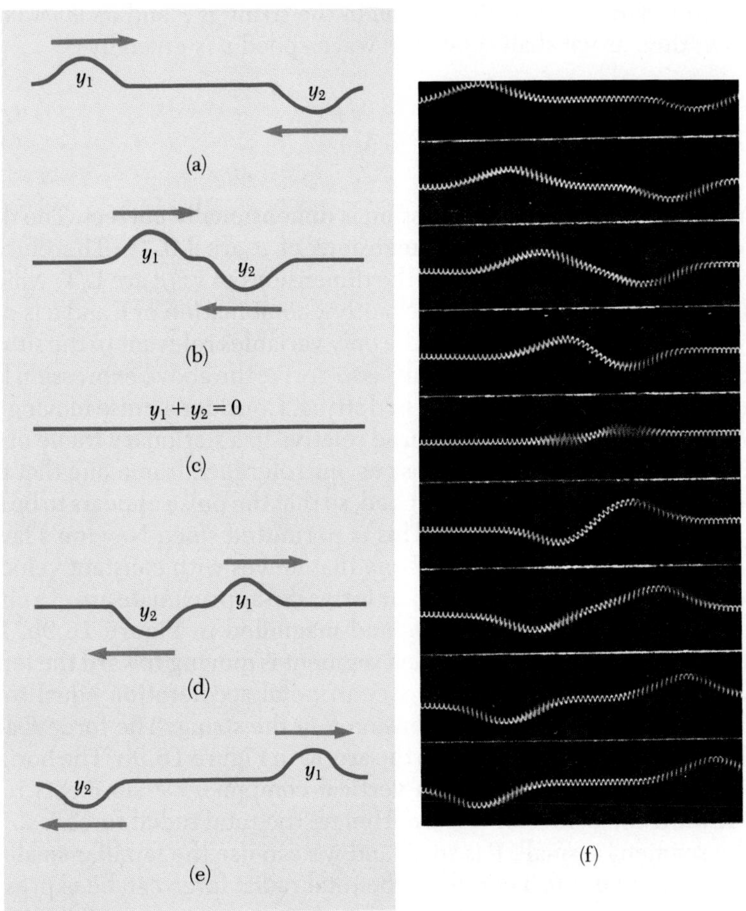

Figure 16.8 (Left) Two wave pulses traveling in opposite directions with equal but opposite displacements. When the two overlap, their displacements subtract from each other. In frame (c), the displacement is zero for all values of x. (Right) Photograph of superposition of two symmetric pulses traveling in opposite directions, where one is inverted relative to the other. (Photo, Education Development Center, Newton, Mass.)

16.8. In this case, when the pulses begin to overlap, the resultant waveform is the *algebraic difference* between the two separate displacements. Again, the two pulses pass through each other as indicated. When the two pulses exactly overlap, they *cancel* each other (assuming the upper positive displacement of the pulse y_1 is equal in magnitude to that of the inverted pulse y_2). At this time, the string is horizontal and the energy associated with the disturbance is contained in the kinetic energy of the string, where the string segments move *vertically*. That is, when the two pulses exactly overlap, the segments of the string on either side of the crossover point are moving vertically, but in opposite directions.

16.5 THE VELOCITY OF WAVES ON STRINGS

For linear waves, *the velocity of mechanical waves depends only on the properties of the medium through which the disturbance travels.* In this section, we shall focus our attention on determining the speed of a transverse pulse travel-

Speed of a wave on a stretched string

ing on a stretched string. If the *tension* in the string is F and its *mass per unit length* is μ, then as we shall show, the wave speed v is given by

$$v = \sqrt{\frac{F}{\mu}} \tag{16.4}$$

First, we verify that this expression is dimensionally correct. The dimensions of F are MLT^{-2}, and the dimensions of μ are ML^{-1}. Therefore, the dimensions of F/μ are L^2/T^2; hence the dimensions of $\sqrt{F/\mu}$ are L/T, which are indeed the dimensions of velocity. No other combination of F and μ is dimensionally correct, assuming they are the only variables relevant to the situation.

Now let us use a mechanical analysis to derive the above expression for the speed of a pulse traveling on a stretched string. Consider a pulse moving to the right with a uniform speed v, measured relative to a stationary frame of reference. It is more convenient to choose as our reference frame one that moves along with the pulse with the same speed, so that the pulse appears to be at rest in this frame, as in Figure 16.9a. This is permitted since Newton's laws are valid in either a stationary frame or one that moves with constant velocity. A *small* segment of the string of length Δs forms the approximate arc of a circle of radius R, as shown in Figure 16.9a and magnified in Figure 16.9b. In the pulse's frame of reference, the shaded segment is moving toward the left with a speed v. This small segment has a centripetal acceleration equal to v^2/R, which is supplied by the force of tension $\mathbf{F}$ in the string. The force $\mathbf{F}$ acts on each side of the segment, tangent to the arc, as in Figure 16.9b. The horizontal components of $\mathbf{F}$ cancel, and each vertical component $F \sin \theta$ acts radially inward toward the center of the arc. Hence, the total radial force is $2F \sin \theta$. Since the segment is small, θ is small and we can use the familiar small-angle approximation $\sin \theta \approx \theta$. Therefore, the total radial force can be expressed as

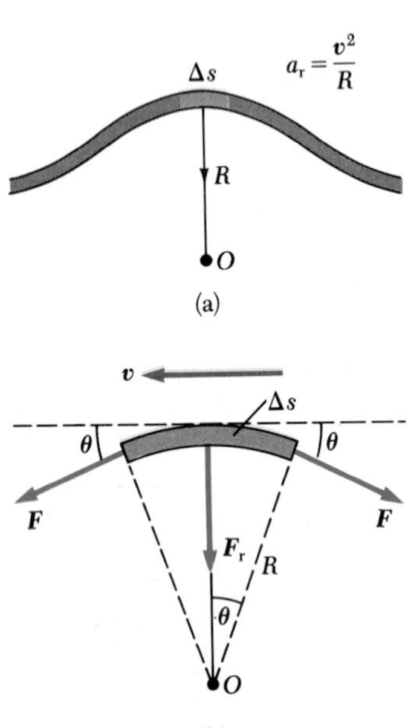

Figure 16.9 (a) To obtain the speed v of a wave on a stretched string, it is convenient to describe the motion of a small segment of the string in a moving frame of reference. (b) The net force on a small segment of length Δs is in the radial direction. The horizontal components of the tension force cancel.

$$F_r = 2F \sin \theta \approx 2F\theta$$

The small segment has a mass given by $m = \mu \Delta s$, where μ is the mass per unit length of the string. Since the segment forms part of a circle and subtends an angle 2θ at the center, $\Delta s = R(2\theta)$, and hence

$$m = \mu \, \Delta s = 2\mu R\theta$$

If we apply Newton's second law to this segment, the radial component of motion gives

$$F_r = mv^2/R \qquad \text{or} \qquad 2F\theta = 2\mu R\theta v^2/R$$

where F_r is the force which supplies the centripetal acceleration of the segment and maintains the curvature at this point.

Solving for v gives

$$v = \sqrt{\frac{F}{\mu}}$$

Notice that this derivation is based on the assumption that the pulse height is small relative to the length of the string. Using this assumption, we were able to use the approximation that $\sin \theta \approx \theta$. Furthermore, the model assumes that the tension F is not affected by the presence of the pulse, so that F is the same at all points on the string. Finally, this proof does *not* assume any particular shape for the pulse. Therefore, we conclude that a pulse of *any shape* will travel on the string with speed $v = \sqrt{F/\mu}$ without changing its shape.

EXAMPLE 16.2 The Speed of a Pulse on a Cord
A uniform cord has a mass of 0.3 kg and a length of 6 m. Tension is maintained in the cord by suspending a 2-kg mass from one end (Fig. 16.10). Find the speed of a pulse on this cord.

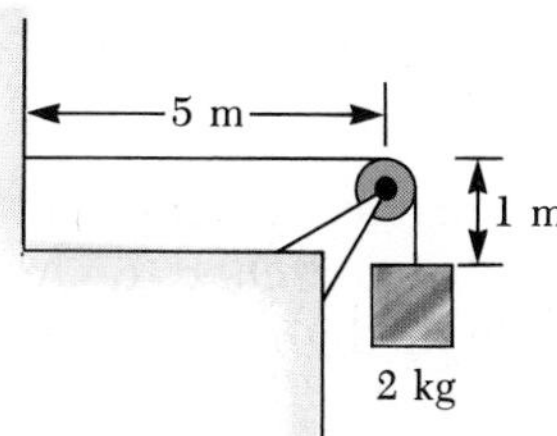

Figure 16.10 (Example 16.2) The tension F in the cord is maintained by the suspended mass. The wave speed is calculated using the expression $v = \sqrt{F/\mu}$.

Solution The tension F in the cord is equal to the weight of the suspended 2-kg mass:

$$F = mg = (2 \text{ kg})(9.80 \text{ m/s}^2) = 19.6 \text{ N}$$

(This calculation of the tension neglects the small mass of the cord. Strictly speaking, the cord can never be exactly horizontal, and therefore the tension is not uniform.)

The mass per unit length μ is

$$\mu = \frac{m}{\ell} = \frac{0.3 \text{ kg}}{6 \text{ m}} = 0.05 \text{ kg/m}$$

Therefore, the wave speed is

$$v = \sqrt{F/\mu} = \sqrt{19.6 \text{ N}/0.05 \text{ kg/m}} = 19.8 \text{ m/s}$$

Exercise 1 Find the time it takes the pulse to travel from the wall to the pulley.
Answer 0.253 s.

16.6 REFLECTION AND TRANSMISSION OF WAVES

Whenever a traveling wave reaches a boundary, part or all of the wave will be reflected. For example, consider a pulse traveling on a string fixed at one end (Fig. 16.11). When the pulse reaches the fixed wall, it will be reflected. Since the support attaching the string to the wall is assumed to be rigid, the pulse does not transmit any part of the disturbance to the wall.

Note that the reflected pulse is inverted. This can be explained as follows. When the pulse meets the end of the string that is fixed at the support, the string produces an upward force on the support. By Newton's third law, the support must then exert an equal and opposite (downward) reaction force on the string. This downward force causes the pulse to invert upon reflection.

Now consider another case where the pulse arrives at the end of a string that is free to move vertically, as in Figure 16.12. The tension at the free end is maintained by tying the string to a ring of negligible mass that is free to slide vertically on a smooth post. Again, the pulse will be reflected, but this time its displacement is not inverted. As the pulse reaches the post, it exerts a force on the free end, causing the ring to accelerate upward. In the process, the ring "overshoots" the height of the incoming pulse by a factor of 2 and is then returned to its original position by the downward component of the tension. This produces a reflected pulse that is not inverted, whose amplitude is the same as that of the incoming pulse.

Finally, we may have a situation in which the boundary is intermediate between these two extreme cases, that is, one in which the boundary is neither

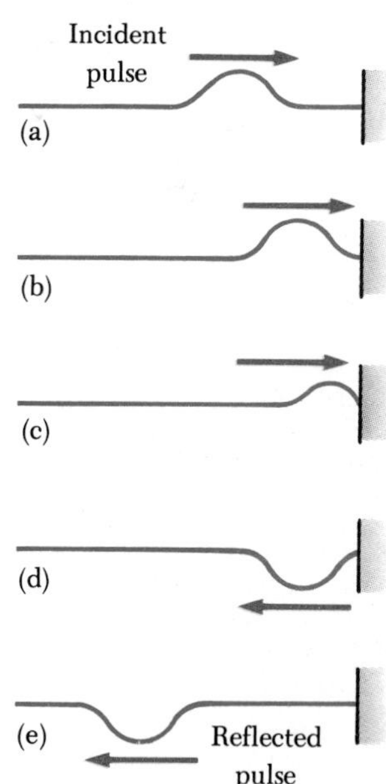

Figure 16.11 The reflection of a traveling wave at the fixed end of a stretched string. Note that the reflected pulse is inverted, but its shape remains the same.

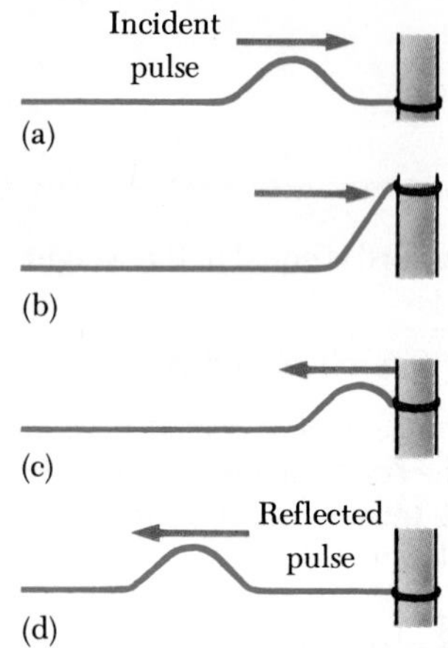

Figure 16.12 The reflection of a traveling wave at the free end of a stretched string. In this case, the reflected pulse is not inverted.

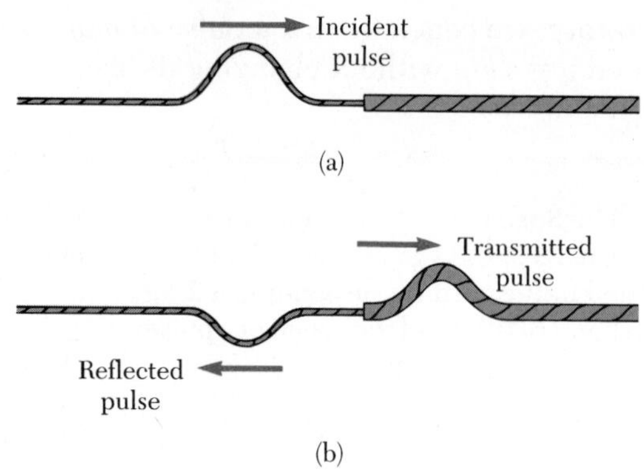

Figure 16.13 (a) A pulse traveling to the right on a light string tied to a heavier string. (b) Part of the incident pulse is reflected (and inverted), and part is transmitted to the heavier string.

rigid nor free. In this case, part of the incident energy is transmitted and part is reflected. For instance, suppose a light string is attached to a heavier string as in Figure 16.13. When a pulse traveling on the light string reaches the knot, part of it is reflected and inverted and part of it is transmitted to the heavier string. As one would expect, the reflected pulse has a smaller amplitude than the incident pulse, since part of the incident energy is transferred to the pulse in the heavier string. The inversion in the reflected wave is similar to the behavior of a pulse meeting a rigid boundary, where it is totally reflected.

When a pulse traveling on a heavy string strikes the boundary of a lighter string, as in Figure 16.14, again part is reflected and part is transmitted. However, in this case, the reflected pulse is not inverted. In either case, the relative heights of the reflected and transmitted pulses depend on the relative densities of the two strings.

If the strings are identical, there is no discontinuity at the boundary, and hence no reflection takes place.

In the previous section, we found that the speed of a wave on a string increases as the density of the string decreases. A pulse travels more slowly on a heavy string than on a light string if both are under the same tension. The following general rules apply to reflected waves: *When a wave pulse travels from medium A to medium B and $v_A > v_B$ (that is, when B is denser than A), the pulse is inverted upon reflection. When a wave pulse travels from medium A to*

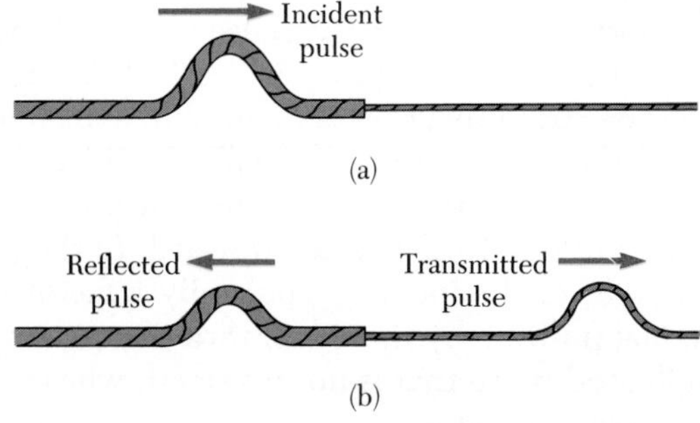

Figure 16.14 (a) A pulse traveling to the right on a heavy string tied to a lighter string. (b) The incident pulse is partially reflected and partially transmitted. In this case, the reflected pulse is not inverted.

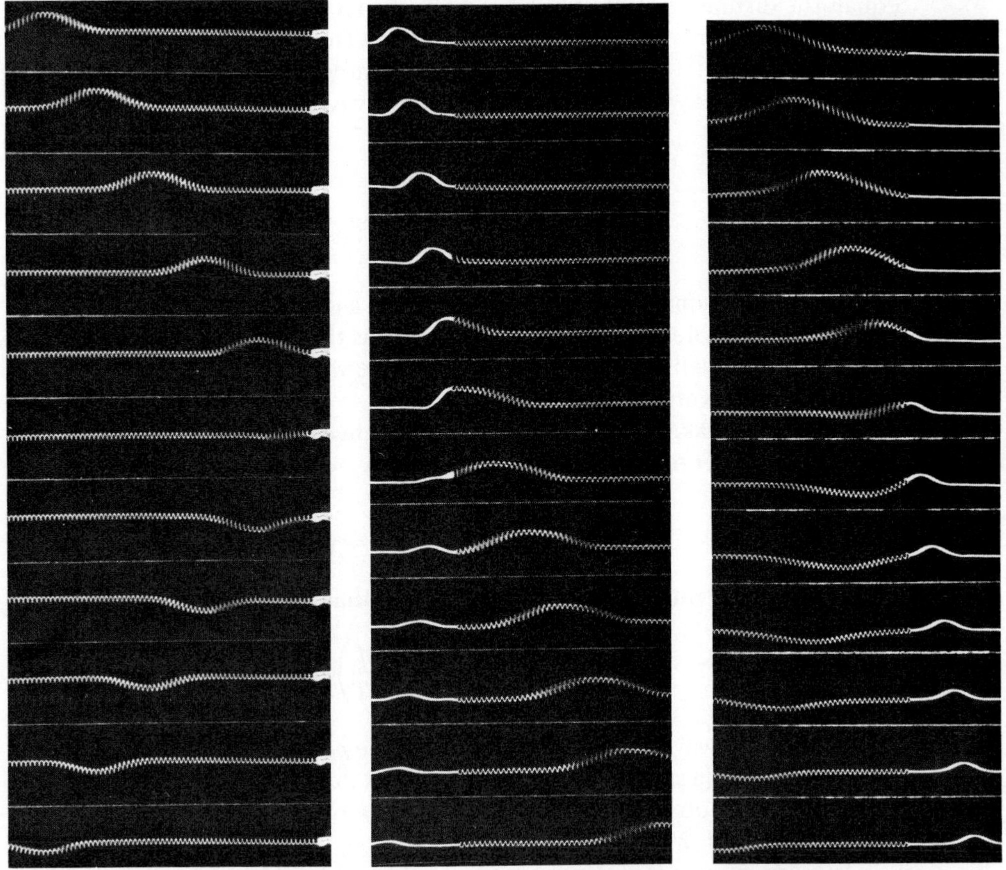

Photographs showing: (Left) Reflection of a pulse from a fixed end. The reflected pulse is inverted. (Center) A pulse passing from a heavy spring to a light spring. At the junction the pulse is partially transmitted and partially reflected. The reflected pulse is not inverted. (Right) A pulse passing from a light spring to a heavy spring. At the junction the pulse is partially transmitted and partially reflected. Note that the reflected pulse is inverted. (Photos, Education Development Center, Newton, Mass.)

medium B and $v_A < v_B$ (A is denser than B), it is not inverted upon reflection. Similar rules apply to other waveforms such as harmonic waves described in the following section.

16.7 HARMONIC WAVES

In this section, we introduce an important waveform known as a **harmonic wave.** *A harmonic wave has a sinusoidal shape,* as shown in Figure 16.15. The red curve represents a snapshot of the traveling harmonic wave at $t = 0$, and the blue curve represents a snapshot of the wave at some later time t. At $t = 0$, the displacement of the curve can be written

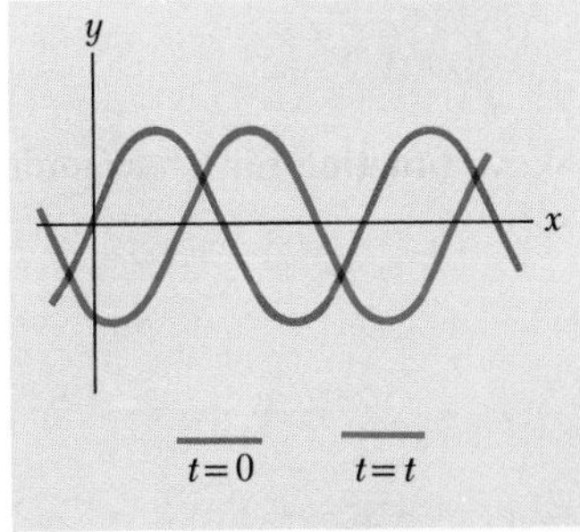

Figure 16.15 A one-dimensional harmonic wave traveling to the right with a speed v. The red curve represents a snapshot of the wave at $t = 0$, and the blue curve at some later time t.

$$y = A \sin\left(\frac{2\pi}{\lambda} x\right) \quad (16.5)$$

The constant A, called the **amplitude** of the wave, represents the *maximum* value of the displacement. The constant λ, called the **wavelength** of the wave, equals the distance between two successive maxima, or *crests*, or between any two adjacent points that have the same phase. Thus, we see that the displacement repeats itself when x is increased by any integral multiple of λ. If the wave moves to the right with a phase velocity v, the wave function at some later time t is given by

$$y = A \sin\left[\frac{2\pi}{\lambda}(x - vt)\right] \tag{16.6}$$

That is, the harmonic wave moves to the right a distance vt in the time t, as in Figure 16.15. Note that the wave function has the form $f(x - vt)$ and represents a wave traveling to the right. If the wave were traveling to the left, the quantity $x - vt$ would be replaced by $x + vt$.

The time it takes the wave to travel a distance of one wavelength is called the **period,** T. Therefore, the phase velocity, wavelength, and period are related by

$$v = \lambda/T \qquad \text{or} \qquad \lambda = vT \tag{16.7}$$

Substituting this into Equation 16.6, we find that

$$y = A \sin\left[2\pi\left(\frac{x}{\lambda} - \frac{t}{T}\right)\right] \tag{16.8}$$

This form of the wave function clearly shows the *periodic* nature of y. That is, at any given time t (a snapshot of the wave), y has the *same* value at the positions x, $x + \lambda$, $x + 2\lambda$, etc. Furthermore, at any given position x, the value of y at times t, $t + T$, $t + 2T$, etc.

We can express the harmonic wave function in a convenient form by defining two other quantities, called the **wave number** k and the **angular frequency** ω:

Wave number

$$k \equiv 2\pi/\lambda \tag{16.9}$$

Angular frequency

$$\omega \equiv 2\pi/T \tag{16.10}$$

Using these definitions, we see that Equation 16.8 can be written in the more compact form

Wave function for a harmonic wave

$$y = A \sin(kx - \omega t) \tag{16.11}$$

We shall use this form most frequently.

The **frequency** of a harmonic wave equals the number of times a crest (or any other point on the wave) passes a *fixed* point each second. The frequency is related to the period by the relationship

Frequency

$$f = \frac{1}{T} \tag{16.12}$$

The most common unit for f is s^{-1}, or hertz (Hz). The corresponding unit for T is s.

Using Equations 16.9, 16.10, and 16.12, we can express the phase velocity v in the alternative forms

$$v = \frac{\omega}{k} \tag{16.13}$$

Velocity of a harmonic wave

$$v = \lambda f \tag{16.14}$$

The wave function given by Equation 16.11 assumes that the displacement y is zero at $x = 0$ and $t = 0$. This need not be the case. If the transverse displacement is not zero at $x = 0$ and $t = 0$, we generally express the wave function in the form

$$y = A \sin(kx - \omega t - \phi) \tag{16.15}$$

General relation for a harmonic wave

where ϕ is called the **phase constant.** This constant can be determined from the initial conditions.

EXAMPLE 16.3 A Traveling Sinusoidal Wave

A sinusoidal wave traveling in the positive x direction has an amplitude of 15 cm, a wavelength of 40 cm, and a frequency of 8 Hz. The displacement of the wave at $t = 0$ and $x = 0$ is also 15 cm, as shown in Figure 16.16. (a) Find the wave number, period, angular frequency, and phase velocity of the wave.

Solution Using Equations 16.9, 16.10, 16.12, and 16.14 and given the information that $\lambda = 40$ cm and $f = 8$ Hz, we find the following:

$$k = 2\pi/\lambda = 2\pi/40 \text{ cm} = 0.157 \text{ cm}^{-1}$$

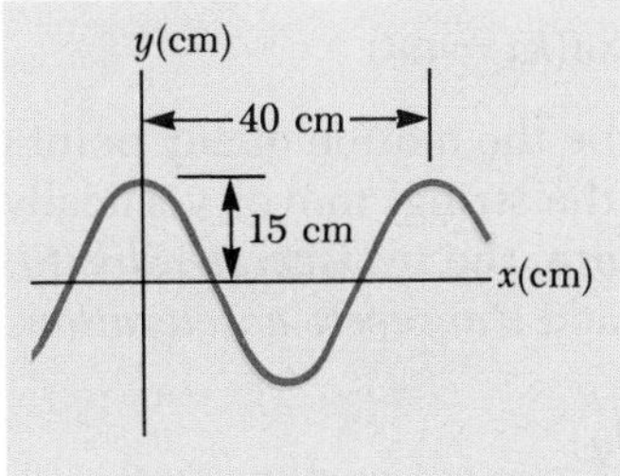

Figure 16.16 (Example 16.3) A harmonic wave of wavelength $\lambda = 40$ cm and amplitude $A = 15$ cm. The wave function can be written in the form $y = A\cos(kx - \omega t)$.

$$T = 1/f = 1/8 \text{ s}^{-1} = 0.125 \text{ s}$$

$$\omega = 2\pi f = 2\pi(8 \text{ s}^{-1}) = 50.3 \text{ rad/s}$$

$$v = f\lambda = (8 \text{ s}^{-1})(40 \text{ cm}) = 320 \text{ cm/s}$$

(b) Determine the phase constant ϕ, and write a general expression for the wave function.

Solution Since the amplitude $A = 15$ cm and since it is given that $y = 15$ cm at $x = 0$ and $t = 0$, substitution into Equation 16.15 gives

$$15 = 15 \sin(-\phi) \qquad \text{or} \qquad \sin(-\phi) = 1$$

Since $\sin(-\phi) = -\sin\phi$, we see that $\phi = -\pi/2$ rad (or $-90°$). Hence, the wave function is of the form

$$y = A \sin\left(kx - \omega t + \frac{\pi}{2}\right) = A\cos(kx - \omega t)$$

This can be seen by inspection, noting that the cosine argument is displaced by 90° from the sine function. Substituting the values for A, k, and ω into this expression gives

$$y = (15 \text{ cm})\cos(0.157x - 50.3t)$$

Harmonic Waves on Strings

One method of producing a wave on a very long string is shown in Figure 16.17. One end of the string is connected to a blade that is set into vibration. As the blade oscillates vertically with simple harmonic motion, a traveling wave moving to the right is set up on the string. Figure 16.17 represents snapshots of the wave at intervals of one quarter of a period. Note that *each particle of the*

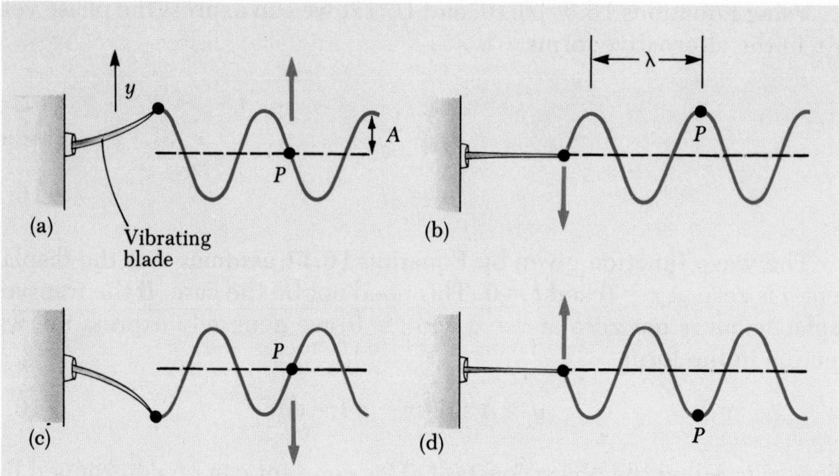

Figure 16.17 One method for producing harmonic waves on a continuous string. The left end of the string is connected to a blade that is set into vibration. Note that every segment, such as P, oscillates with simple harmonic motion in the vertical direction.

string, such as P, oscillates vertically in the y direction with simple harmonic motion. This must be the case because each particle follows the simple harmonic motion of the blade. Therefore, every segment of the string can be treated as a simple harmonic oscillator vibrating with a frequency equal to the frequency of vibration of the blade that drives the string.[5] Note that although each segment oscillates in the y direction, the wave (or disturbance) travels in the x direction with a speed v. Of course, this is the definition of a transverse wave. In this case, the energy carried by the traveling wave is supplied by the vibrating blade. (In reality, the oscillations would gradually decrease in amplitude because of air resistance and the friction internal to the string.)

If the waveform at $t = 0$ is as described in Figure 16.17b, then the wave function can be written

$$y = A \sin(kx - \omega t)$$

We can use this expression to describe the motion of any point on the string. The point P (or any other point on the string) moves vertically, and so *its x coordinate remains constant.* Therefore, the *transverse velocity,* v_y (not to be confused with the wave velocity v), and *transverse acceleration,* a_y, are given by

$$v_y = \left.\frac{dy}{dt}\right]_{x=\text{constant}} = \frac{\partial y}{\partial t} = -\omega A \cos(kx - \omega t) \tag{16.16}$$

$$a_y = \left.\frac{dv_y}{dt}\right]_{x=\text{constant}} = \frac{\partial v_y}{\partial t} = -\omega^2 A \sin(kx - \omega t) \tag{16.17}$$

The *maximum* values of these quantities are simply the absolute values of the coefficients of the cosine and sine functions:

$$(v_y)_{\text{max}} = \omega A \tag{16.18}$$

$$(a_y)_{\text{max}} = \omega^2 A \tag{16.19}$$

[5] In this arrangement, we are assuming that the mass always oscillates in a vertical line. The tension in the string would vary if the mass were allowed to move sideways. Such a motion would make the analysis very complex.

You should recognize that the transverse velocity and transverse acceleration do not reach their maximum values simultaneously. In fact, the transverse velocity reaches its maximum value (ωA) when the displacement $y = 0$, whereas the transverse acceleration reaches its maximum value ($\omega^2 A$) when $y = -A$. Finally, Equations 16.18 and 16.19 are identical to the corresponding equations for simple harmonic motion.

EXAMPLE 16.4 A Harmonically Driven String

The string shown in Figure 16.17 is driven at one end at a frequency of 5 Hz. The amplitude of the motion is 12 cm, and the wave speed is 20 m/s. Determine the angular frequency and wave number for this wave, and write an expression for the wave function.

Solution Using Equations 16.10, 16.12, and 16.13 gives

$$\omega = 2\pi/T = 2\pi f = 2\pi(5 \text{ Hz}) = 31.4 \text{ rad/s}$$

$$k = \omega/v = \frac{31.4 \text{ rad/s}}{20 \text{ m/s}} = 1.57 \text{ m}^{-1}$$

Since $A = 12$ cm $= 0.12$ m, we have

$$y = A \sin(kx - \omega t) = (0.12 \text{ m}) \sin(1.57x - 31.4t)$$

Exercise 2 Calculate the maximum values for the transverse velocity and transverse acceleration of any point on the string.
Answer 3.77 m/s; 118 m/s^2.

16.8 ENERGY TRANSMITTED BY HARMONIC WAVES ON STRINGS

As waves propagate through a medium, they transport energy. This is easily demonstrated by hanging a mass on a stretched string and then sending a pulse down the string, as in Figure 16.18. When the pulse meets the suspended mass, the mass will be momentarily displaced, as in Figure 16.18b. In the process, energy is transferred to the mass since work must be done in moving it upward.

In this section, we describe the rate at which energy is transported along a string. We shall assume a sinusoidal wave when we calculate the power transferred for this one-dimensional wave. Later, we shall extend these ideas to three-dimensional waves.

Consider a harmonic wave traveling on a string (Fig. 16.19). The source of the energy is some external agent at the left end of the string, which does work in producing the oscillations. Let us focus our attention on an element of the string of length Δx and mass Δm. Each such segment moves vertically with simple harmonic motion. Furthermore, each segment has the same frequency, ω, and the same amplitude, A. As we found in Chapter 13, the total energy E associated with a particle moving with simple harmonic motion is given by

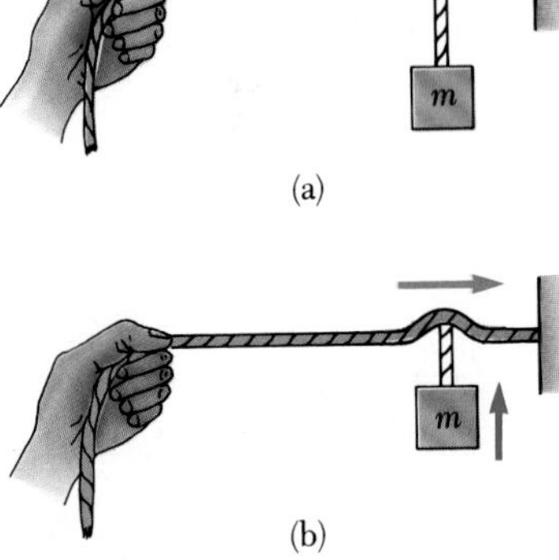

Figure 16.18 (a) A pulse traveling to the right on a stretched string on which a mass has been suspended. (b) Energy and momentum are transmitted to the suspended mass when the pulse arrives.

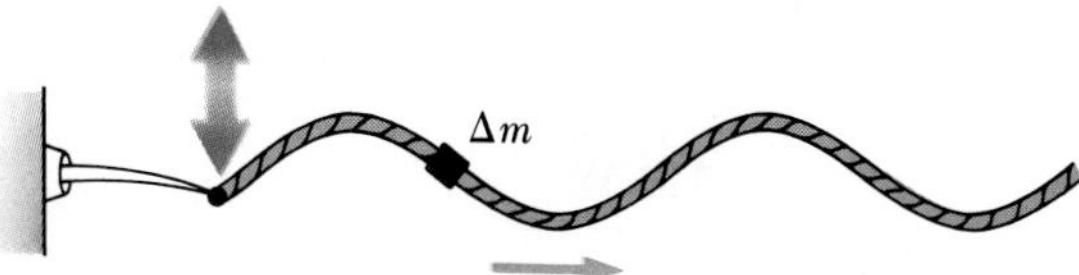

Figure 16.19 A harmonic wave traveling along the x axis on a stretched string. Every segment moves vertically, and each has the same total energy. The power transmitted by the wave equals the energy contained in one wavelength divided by the period of the wave.

$E = \frac{1}{2}kA^2 = \frac{1}{2}m\,\omega^2A^2$, where k is the equivalent force constant of the restoring force. If we apply this to the element of length Δx, we see that the total energy of this element is

$$\Delta E = \tfrac{1}{2}(\Delta m)\omega^2A^2$$

If μ is the mass per unit length of the string, then the element of length Δx has a mass Δm that is equal to $\mu\,\Delta x$. Hence, we can express the energy ΔE as

$$\Delta E = \tfrac{1}{2}(\mu\,\Delta x)\omega^2A^2 \qquad (16.20)$$

If the wave travels from left to right as in Figure 16.19, the energy ΔE arises from the work done on the element Δm by the string element to the left of Δm. Similarly, the element Δm does work on the element to its right, so we see that energy is transmitted to the right. The rate at which energy is transmitted along the string, or the power, is given by dE/dt. If we let Δx approach 0, Equation 16.20 gives

$$\text{Power} = \frac{dE}{dt} = \tfrac{1}{2}\left(\mu\,\frac{dx}{dt}\right)\omega^2A^2$$

Since dx/dt is equal to the wave speed, v, we have

Power

$$\text{Power} = \tfrac{1}{2}\mu\omega^2A^2v \qquad (16.21)$$

This shows that the power transmitted by a harmonic wave on a string is proportional to (a) the wave speed, (b) the square of the frequency, and (c) the square of the amplitude. In fact, *all* harmonic waves have the following general property: *The power transmitted by any harmonic wave is proportional to the square of the frequency and to the square of the amplitude.*

Thus, we see that a wave traveling through a medium corresponds to energy transport through the medium, with no net transfer of matter. An oscillating source provides the energy and produces a harmonic disturbance of the medium. The disturbance is able to propagate through the medium as the result of the interaction between adjacent particles. In order to verify Equation 16.20 by direct experiment, one would have to design some device at the far end of the string to extract the energy of the wave without producing any reflections.

EXAMPLE 16.5 Power Supplied to a Vibrating Rope
A stretched rope having mass per unit length of $\mu = 5 \times 10^{-2}$ kg/m is under a tension of 80 N. How much power must be supplied to the rope to generate harmonic waves at a frequency of 60 Hz and an amplitude of 6 cm?

Solution The wave speed on the stretched rope is given by

$$v = \sqrt{\frac{T}{\mu}} = \left(\frac{80\text{ N}}{5 \times 10^{-2}\text{ kg/m}}\right)^{1/2} = 40\text{ m/s}$$

Since $f = 60$ Hz, the angular frequency ω of the harmonic waves on the string has the value

$$\omega = 2\pi f = 2\pi(60\text{ Hz}) = 377\text{ s}^{-1}$$

Using these values in Equation 16.21 for the power, with $A = 6 \times 10^{-2}$ m, gives

$$P = \tfrac{1}{2}\mu\omega^2A^2v$$
$$= \tfrac{1}{2}(5 \times 10^{-2}\text{ kg/m})(377\text{ s}^{-1})^2(6 \times 10^{-2}\text{ m})^2\ (40\text{ m/s})$$
$$= 512\text{ W}$$

*16.9 THE LINEAR WAVE EQUATION

Earlier in this chapter, we introduced the concept of the wave function to represent waves traveling on a string. All wave functions $y(x, t)$ represent solutions of an equation called the *linear wave equation.* This equation gives a complete description of the wave motion, and from it one can derive an expression for the wave velocity. Furthermore, the wave equation is basic to many forms of wave motion. In this section, we shall derive the wave equation as applied to waves on strings.

Consider a small segment of a string of length Δx and tension F, on which a traveling wave is propagating (Fig. 16.20). Let us assume that the ends of the segment make small angles θ_1 and θ_2 with the x axis.

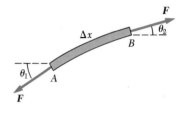

Figure 16.20 A segment of a string under tension **F**. Note that the slopes at points A and B are given by $\tan\theta_1$ and $\tan\theta_2$, respectively.

The net force on the segment in the vertical direction is given by

$$\sum F_y = F\sin\theta_2 - F\sin\theta_1 = F(\sin\theta_2 - \sin\theta_1)$$

Since we have assumed that the angles are small, we can use the small-angle approximation $\sin\theta \approx \tan\theta$ and express the net force as

$$\sum F_y \approx F(\tan\theta_2 - \tan\theta_1)$$

However, the tangents of the angles at A and B are defined as the slope of the curve at these points. Since the slope of a curve is given by $\partial y/\partial x$, we have[6]

$$\sum F_y \approx F[(\partial y/\partial x)_B - (\partial y/\partial x)_A] \tag{16.22}$$

We now apply Newton's second law, $\Sigma F_y = ma_y$, to the segment, where m is the mass of the segment, given by $m = \mu\,\Delta x$. This gives

$$\sum F_y = ma_y = \mu\,\Delta x(\partial^2 y/\partial t^2) \tag{16.23}$$

where we have used the fact that $a_y = \partial^2 y/\partial t^2$. Equating Equation 16.23 to Equation 16.22 gives

$$\mu\,\Delta x(\partial^2 y/\partial t^2) = F[(\partial y/\partial x)_B - (\partial y/\partial x)_A]$$

$$\frac{\mu}{F}\frac{\partial^2 y}{\partial t^2} = \frac{[(\partial y/\partial x)_B - (\partial y/\partial x)_A]}{\Delta x} \tag{16.24}$$

The right side of Equation 16.24 can be expressed in a different form if we note that the derivative of any function is defined as

$$\frac{\partial f}{\partial x} = \lim_{\Delta x \to 0} \frac{f(x + \Delta x) - f(x)}{\Delta x}$$

If we associate $f(x + \Delta x)$ with $(\partial y/\partial x)_B$ and $f(x)$ with $(\partial y/\partial x)_A$, we see that in the limit $\Delta x \to 0$, Equation 16.24 becomes

$$\frac{\mu}{F}\frac{\partial^2 y}{\partial t^2} = \frac{\partial^2 y}{\partial x^2} \tag{16.25}$$

Linear wave equation

This is the linear wave equation as it applies to waves on a string.

[6] It is necessary to use partial derivatives because y depends on both x and t.

We shall now show that the harmonic wave function represents a solution of this wave equation. If we take the harmonic wave function to be of the form $y(x, t) = A \sin(kx - \omega t)$, the appropriate derivatives are

$$\partial^2 y/\partial t^2 = -\omega^2 A \sin(kx - \omega t)$$

$$\partial^2 y/\partial x^2 = -k^2 A \sin(kx - \omega t)$$

Substituting these expressions into Equation 16.25 gives

$$k^2 = (\mu/F)\omega^2$$

Using the relation $v = \omega/k$ in the above expression, we see that

$$v^2 = \omega^2/k^2 = F/\mu$$

$$v = \sqrt{F/\mu}$$

This represents another proof of the expression for the wave velocity on a stretched string.

The linear wave equation given by Equation 16.25 is often written in the form

Linear wave equation in general

$$\frac{\partial^2 y}{\partial x^2} = \frac{1}{v^2}\frac{\partial^2 y}{\partial t^2} \tag{16.26}$$

This expression applies in general to various types of waves moving through nondispersive media. For waves on strings, y represents the vertical displacement. For sound waves, y corresponds to variations in the pressure or density of a gas. In the case of electromagnetic waves, y corresponds to electric or magnetic field components.

We have shown that the harmonic wave function is one solution of the linear wave equation. Although we do not prove it here, the linear wave equation is satisfied by *any* wave function having the form $y = f(x \pm vt)$. Furthermore, we have seen that the wave equation is a direct consequence of Newton's second law applied to any segment of the string. Similarly, the wave equation in electromagnetism can be derived from the fundamental laws of electricity and magnetism. This will be discussed further in Chapter 34.

SUMMARY

Transverse wave

A **transverse wave** is a wave in which the particles of the medium move in a direction *perpendicular* to the direction of the wave velocity. An example is a wave on a stretched string.

Longitudinal wave

Longitudinal waves are waves for which the particles of the medium move in a direction *parallel* to the direction of the wave velocity. Sound waves are longitudinal.

Any one-dimensional wave traveling with a speed v in the positive x direction can be represented by a wave function of the form $y = f(x - vt)$. Likewise, the wave function for a wave traveling in the negative x direction has the form $y = f(x + vt)$. The shape of the wave at any instant (a snapshot of the wave) is obtained by holding t constant.

The **superposition principle** says that when two or more waves move through a medium, the resultant wave function equals the algebraic sum of the individual wave functions. Waves that obey this principle are said to be *linear*. When two waves combine in space, they interfere to produce a resultant wave. The *interference* may be *constructive* (when the individual displacements are in the same direction) or *destructive* (when the displacements are in opposite directions).

Superposition principle

The **speed** of a wave traveling on a stretched string of mass per unit length μ and tension F is

$$v = \sqrt{\frac{F}{\mu}} \qquad (16.4)$$

Speed of a wave on a stretched string

When a pulse traveling on a string meets a fixed end, the pulse is reflected and inverted. If the pulse reaches a free end, it is reflected but not inverted.

The **wave function** for a one-dimensional harmonic wave traveling to the right can be expressed as

$$y = A \sin[(2\pi/\lambda)(x - vt)] = A \sin(kx - \omega t) \qquad (16.6, 16.11)$$

Wave function for a harmonic wave

where A is the amplitude, λ is the wavelength, k is the wave number, and ω is the angular frequency. If T is the period (the time it takes the wave to travel a distance equal to one wavelength) and f is the frequency, then v, k and ω can be written

$$v = \lambda/T = \lambda f \qquad (16.7, 16.14)$$

$$k = 2\pi/\lambda \qquad (16.9)$$

Wave number

$$\omega = 2\pi/T = 2\pi f \qquad (16.10, 16.12)$$

Angular frequency

The **power** transmitted by a harmonic wave on a stretched string is given by

$$P = \tfrac{1}{2}\mu\omega^2 A^2 v \qquad (16.21)$$

Power

The wave function $y(x, t)$ for many kinds of waves satisfies the following **linear wave equation:**

$$\frac{\partial^2 y}{\partial x^2} = \frac{1}{v^2}\frac{\partial^2 y}{\partial t^2} \qquad (16.26)$$

Linear wave equation in general

QUESTIONS

1. Why is a wave pulse traveling on a string considered a transverse wave?
2. How would you set up a longitudinal wave in a stretched spring? Would it be possible to set up a transverse wave in a spring?
3. By what factor would you have to increase the tension in a stretched string in order to double the wave speed?
4. When a wave pulse travels on a stretched string, does it always invert upon reflection? Explain.
5. Can two pulses traveling in opposite directions on the same string reflect from each other? Explain.
6. Does the transverse velocity of a segment on a stretched string depend on the wave velocity?
7. If you were to shake one end of a stretched rope periodically three times each second, what would be the period of the harmonic waves set up in the rope?
8. Harmonic waves are generated on a string under constant tension by a vibrating source. If the power deliv-

ered to the string is doubled, by what factor does the amplitude change? Does the wave velocity change under these circumstances?

9. Consider a wave traveling on a stretched rope. What is the difference, if any, between the speed of the wave and the speed of a small section of the rope?
10. If a long rope is hung from a ceiling and waves are sent up the rope from its lower end, the waves do not ascend with constant speed. Explain.
11. What happens to the wavelength of a wave on a string when the frequency is doubled? Assume the tension in the string remains the same.
12. What happens to the velocity of a wave on a string when the frequency is doubled? Assume the tension in the string remains the same.
13. How do transverse waves differ from longitudinal waves?
14. When all the strings on a guitar are stretched to the same tension, will the velocity of a wave along the more massive bass strings be faster or slower than the velocity of a wave on the lighter strings?
15. If you stretch a rubber hose and pluck it, you can observe a pulse traveling up and down the hose. What happens to the speed if you stretch the hose tighter? If you fill the hose with water?
16. In a longitudinal wave in a spring, the coils move back and forth in the direction the wave travels. Does the speed of the wave depend on the maximum velocity of each of the coils?
17. When two waves interfere, can the resultant wave be larger than either of the two original waves? Under what conditions?
18. A solid may transport a longitudinal wave as well as a transverse wave, but a fluid may only transport a longitudinal wave. Why?
19. In an earthquake both S (shear) and P (pressure) waves are sent out. The S (transverse) waves travel through the earth slower than the P (longitudinal) waves (5 km/s versus 9 km/s). By detecting the time of arrival of the waves, how may one determine how far away the epicenter of the quake was? How many detection centers are necessary to pinpoint the location of the epicenter?

PROBLEMS

Section 16.3 One-Dimensional Traveling Waves

1. At $t = 0$, a transverse wave pulse in a wire is described by the function

$$y = \frac{6}{x^2 + 3}$$

where x and y are in m. Write the function $y(x, t)$ that describes this wave if it is traveling in the positive x direction with a speed of 4.5 m/s.

2. Two wave pulses A and B are moving in *opposite* directions along a stretched string with a speed of 2 cm/s. The amplitude of A is twice the amplitude of B. The pulses are shown in Figure 16.21 at $t = 0$. Sketch the shape of the string at $t = 1$, 1.5, 2, 2.5, and 3 s.

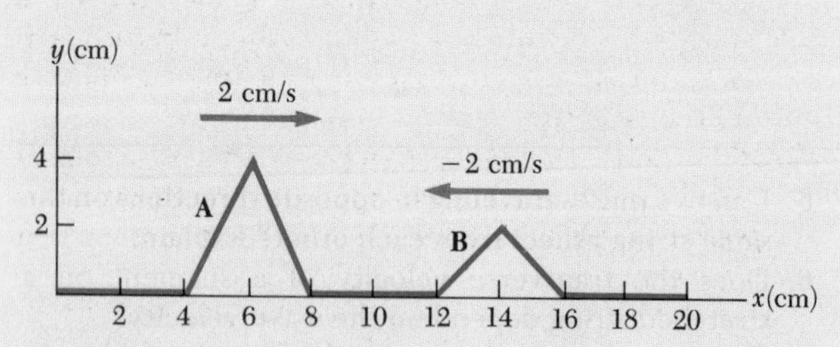

Figure 16.21 (Problem 2).

3. A traveling wave pulse moving to the right along the x axis is represented by the following wave function:

$$y(x, t) = \frac{4}{2 + (x - 4t)^2},$$

where x and y are measured in cm and t is in s. Plot the shape of the waveform at $t = 0$, 1, and 2 s.

4. A traveling wave pulse moving to the left is described by the following function:

$$y = \frac{10}{5 + (x + 2t)^4}$$

where x and y are in cm and t is in s. (a) Compute the speed of this wave. (b) Plot this wave at two different times and show that it is traveling to the left.

5. In a relationship describing a harmonic wave, the expression $(kx \pm \omega t)$, called the *phase,* appears in some form. When the phase is held constant, one travels along with the wave. Show that to hold the phase constant one must move at a speed of v in the positive x direction when the − sign is used and one must move in the negative x direction when the + sign is used.

Section 16.4 Superposition and Interference of Waves

6. Two waves in one string are described by the following relationships:

$$y_1 = 3\cos(4x - 5t) \qquad y_2 = 4\sin(5x - 2t)$$

Find the superposition of the waves $y_1 + y_2$ at the points (a) $x = 1$, $t = 1$, (b) $x = 1$, $t = 0.5$, (c) $x = 0.5$, $t = 0$. (Remember that the arguments of the trigonometric functions are in radians.)

7. Two harmonic waves in a string are defined by the following functions:

$$y_1 = 2 \text{ cm} \sin(20x - 30t) \; y_2 = 2 \text{ cm} \sin(25x - 40t),$$

where the y's and x are in cm and t is in s. (a) What is the phase difference between these two waves at the point $x = 5$ cm at $t = 2$ s? (b) What is the positive x value closest to the origin for which the two phases will differ by $\pm\pi$ at $t = 2$ s? (This is where the two waves will add to zero.)

8. Two waves are described by $y_1 = 1/(ax - bt)$ and $y_2 = 1/(ax + bt)$ where a and b are constants. Find a relationship for the superposition of the two waves $y_1 + y_2$. What points along this wave would be physically unrealistic and why?

9. One wave pulse in a string is described by the equation

$$y_1 = \frac{5}{(3x - 4t)^2 + 2}$$

A second wave pulse in the same string is described by

$$y_2 = \frac{-5}{(3x + 4t - 6)^2 + 2}$$

(a) In which direction does each pulse travel? (b) At what time will the two waves exactly cancel everywhere? (c) At what point do the two waves always cancel?

Section 16.5 The Velocity of Waves on Strings

10. Transverse waves with a speed of 50 m/s are to be produced in a stretched string. A 5-m length of string with a total mass of 0.06 kg is used. What is the required tension in the string?

11. Calculate the wave speed in the string described in Problem 10 if the tension in the string is 8 N.

12. The tension in a cord 15 m in length is 20 N. The measured transverse wave speed in the cord is 60 m/s. Calculate the total mass of the cord.

13. Transverse waves travel with a speed of 20 m/s in a string under a tension of 6 N. What tension is required for a wave speed of 30 m/s in the same string?

14. Tension is maintained in a horizontal string as shown in Figure 16.10. The observed wave speed is 24 m/s when the suspended mass is 3 kg. What is the linear density of the string?

15. The elastic limit of a piece of steel wire is equal to 2.7×10^9 Pa. What is the maximum velocity at which transverse wave pulses can propagate along this wire without exceeding this stress? (The density of the steel is 7.86 g/cm^3.)

16. Transverse pulses travel with a speed of 200 m/s along a taut copper wire that has a diameter of 1.5 mm. What is the tension in the wire? (The density of copper is given in Table 15.1.)

17. A 30-m steel wire and a 20-m copper wire, both with 1-mm diameters, are connected end-to-end and stretched to a tension of 150 N. How long will it take a transverse wave to travel the entire length of the two wires?

18. A light string with a mass per unit length 8 g/m has its ends tied to two walls which are separated by a distance equal to 3/4 the length of the string. (See Fig. 16.22.) A mass m is suspended from the center of the string, putting a tension in the string. (a) Find an expression for the transverse wave velocity in the string as a function of the hanging mass. (b) How much mass should be suspended from the string to have a wave speed of 60 m/s?

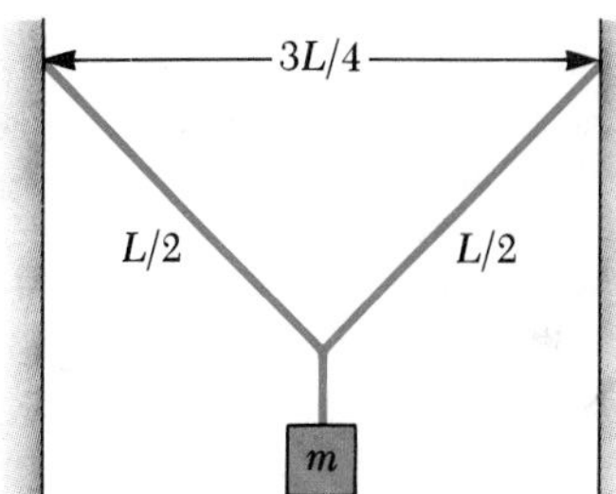

Figure 16.22 (Problem 18).

Section 16.7 Harmonic Waves

19. (a) Plot y versus t at $x = 0$ for a harmonic wave of the form $y = (15 \text{ cm}) \cos(0.157x - 50.3t)$, where x and y are in cm and t is in s. (b) Determine the period of vibration from this plot and compare your result with the value found in Example 16.3.

20. For a certain transverse wave, it is observed that the distance between two successive maxima is 1.2 m. It is also noted that eight crests, or maxima, pass a given point along the direction of travel every 12 s. Calculate the wave speed.

21. A harmonic wave is traveling along a rope. It is observed that the oscillator that generates the wave completes 40 vibrations in 30 s. Also, a given maximum travels 425 cm along the rope in 10 s. What is the wavelength?

22. When a particular wire is vibrating with a frequency of 4 Hz, a transverse wave of wavelength 60 cm is produced. Determine the speed of wave pulses along the wire.

23. One form of the wave function for a harmonic wave is given by Equation 16.11. The displacement y is expressed as a function of x and t in terms of the wave number k and the angular frequency ω. Write equivalent equations in which y is shown as a function of x and t in terms of (a) k and v, (b) λ and v, (c) λ and f, and (d) f and v.

24. A harmonic wave train is described by

$$y = (0.25 \text{ m})\sin(0.3x - 40t),$$

where x and y are in m and t is in s. Determine for this wave the (a) amplitude, (b) angular frequency, (c) wave number, (d) wavelength, (e) wave speed, and (f) direction of motion.

25. Determine the quantities (a) through (f) of Problem 24 when the wave train is described by $y = (0.2 \text{ m})\sin 4\pi(0.4x + t)$. Again x and y are in m and t is in s.

26. In Example 16.3 the harmonic wave was found to be described by $y = (15 \text{ cm})\cos(0.157x - 50.3t)$, where x and y are in cm and t is in s. (a) Plot y versus x at $t = 0$ and $t = 0.050$ s. (b) Determine the wave speed from this plot and compare your result with the value found in Example 16.3.

27. (a) Write the expression for y as a function of x and t for a sinusoidal wave traveling along a rope in the *negative x* direction with the following characteristics: $y_{max} = 8$ cm, $\lambda = 80$ cm, $f = 3$ Hz, and $y(0, t) = 0$ at $t = 0$. (b) Write the expression for y as a function of x for the wave in (a) assuming that $y(x, 0) = 0$ at the point $x = 10$ cm.

28. Consider a wave in a string described by

$$y = (15 \text{ cm})\sin[(\pi/16)(2x - 64t)],$$

where x and y are in cm and t is in s. (a) Calculate the maximum transverse velocity of a point on the string. (b) Calculate the transverse velocity of the point at $x = 6$ cm when $t = 0.25$ s.

29. A transverse harmonic wave has a period $T = 25$ ms and travels in the negative x direction with a speed of 30 m/s. At $t = 0$, a particle on the string at $x = 0$ has a displacement of 2.0 cm and a velocity $v = -2.0$ m/s. (a) What is the amplitude of the wave? (b) What is the initial phase angle? (c) What is the magnitude of the maximum transverse velocity? (d) Write the wave function for the wave.

30. (a) Write the expression for y as a function of x and t for a sinusoidal wave traveling along a rope in the positive x direction with the following characteristics: $y_{max} = 8$ cm, $\lambda = 80$ cm, $f = 3$ Hz, and $y(0, t) = y_{max}$ at $t = 0$. (b) Determine the speed and wave number for the wave described in (a).

31. For the wave described by the relationship

$$y = (0.08 \text{ m})\sin(0.24x - 30t)$$

where x and y are in m and t is in s, determine (a) the wave velocity and (b) the maximum transverse velocity.

32. For the wave described in the previous problem determine the wave velocity and the instantaneous transverse velocity at (a) $x = 20$ m, $t = 0.12$ s and (b) $x = 10$ m, $t = 0$ s.

33. A wave in a string has an amplitude of 2 cm. The waves travel in the $+x$ direction with a speed of 128 m/s and it is noted that 5 complete waves fit in 4 m of the string. (a) Write down an equation describing this wave in the wave number–angular-frequency format. (b) Write an equation describing this wave where the wavelength and frequency are explicitly shown.

34. A wave in a string is described by the wave function $y = (0.10 \text{ m})\sin(0.5\,x - 20\,t)$. (a) Show that a particle in the string at $x = 2$ m executes harmonic motion. (b) Determine the frequency of the oscillation and the initial phase angle of this particular point.

Section 16.8 Energy Transmitted by Harmonic Waves on Strings

35. A stretched rope has a mass of 0.18 kg and a length of 3.6 m. What power must be supplied in order to generate harmonic waves having an amplitude of 0.1 m and a wavelength of 0.5 m and traveling with a speed of 30 m/s?

36. A wire of mass 0.24 kg is 48 m long and under a tension of 60 N. An electric vibrator operating at an angular frequency of 80π rad/s is generating harmonic waves in the wire. The vibrator can supply energy to the wire at a maximum rate of 400 J/s. What is the maximum amplitude of the wave pulses?

37. Transverse waves are being generated on a rope under *constant tension.* By what factor will the required power be increased or decreased if (a) the length of the rope is doubled and the angular frequency remains constant, (b) the amplitude is doubled and the angular frequency is halved, (c) both the wavelength and the amplitude are doubled, and (d) both the length of the rope and the wavelength are halved?

38. Harmonic waves 5 cm in amplitude are to be transmitted along a string that has a linear density of 4×10^{-2} kg/m. If the maximum power delivered by the source is 300 W and the string is under a tension of 100 N, what is the highest vibrational frequency at which the source can operate?

39. A harmonic wave in a string is described by the equation:

$$y = (0.15 \text{ m})\sin(0.8x - 50t)$$

where x and y are in m and t is in s. If the mass per unit length of this string is 12 g/m, determine (a) the speed of the wave, (b) the wavelength, (c) the frequency, and (d) the power transmitted to the wave.

40. It is found that a 6-m segment of a long string contains 4 complete waves and has a mass of 180 g. The string is vibrating sinusoidally with a frequency of 50 Hz and a peak to peak displacement of 15 cm. (Peak to

peak means the vertical distance from the farthest positive displacement to the farthest negative displacement.) (a) Write down the function which describes this wave traveling in the positive x direction, and (b) determine the power being supplied to the string.

Section 16.9 The Linear Wave Equation

41. Show that the wave function $y = \ln[A(x - vt)]$ is a solution to Equation 16.25, where A is a constant.

42. Show that the wave function $y = e^{A(x-vt)}$ is a solution of the wave equation (Equation 16.25), where A is a constant.

43. In Section 16.9 it is verified that $y_1 = A \sin(kx - \omega t)$ is a solution to the wave equation. The wave function $y_2 = B \cos(kx - \omega t)$ describes a wave $\pi/2$ radians out of phase with the first. (a) Determine whether or not $y = A \sin(kx - \omega t) + B \cos(kx - \omega t)$ is a solution to the wave equation. (b) Determine if $y = A(\sin kx) \cdot B(\cos \omega t)$ is a solution to the wave equation.

ADDITIONAL PROBLEMS

44. A traveling wave propagates according to the expression $y = (4.0 \text{ cm})\sin(2.0x - 3.0t)$ where x is in cm. Determine (a) the amplitude, (b) the wavelength, (c) the frequency, (d) the period, and (e) the direction of travel of the wave.

45. The wave function for a linearly polarized wave on a taut string is (in SI units)

$$y(x, t) = (0.35 \text{ m})\sin(10\pi t - 3\pi x + \pi/4)$$

(a) What is the velocity of the wave? (Give the speed and the direction.) (b) What is the displacement at $t = 0$, $x = 0.1$ m? (c) What is the wavelength and what is the frequency of the wave? (d) What is the maximum magnitude of the transverse velocity of the string?

46. What is the average rate at which power is transmitted along the string in Problem 45 if $\mu = 75$ g/m? What is the total energy per wavelength in the wave?

47. (a) Determine the speed of transverse waves on a stretched string that is under a tension of 80 N if the string has a length of 2 m and a mass of 5 g. (b) Calculate the power required to generate these waves if they have a wavelength of 16 cm and are 4 cm in amplitude.

48. A harmonic wave in a rope is described by the wave function $y = (0.2 \text{ m})\sin[\pi(0.75x - 18t)]$ where x and y are in m and t is in s. This wave is traveling in a rope that has a linear mass density of 0.25 kg/m. If the tension in the rope is provided by an arrangement like the one illustrated in Figure 16.10, what is the value of the suspended mass?

49. Consider the sinusoidal wave of Example 16.3, for which it was determined that

$$y = (15 \text{ cm})\cos(0.157x - 50.3t)$$

At a given instant, let point A be at the origin and point B be the first point along x that is 60° out of phase with point A. What is the coordinate of point B?

50. A harmonic traveling wave moving in the positive x direction has an amplitude of 2.0 cm, a wavelength of 4.0 cm, and a frequency of 5 Hz. (a) Determine the speed of the wave and (b) write an expression for the transverse displacement as a function of x and t.

51. A transverse wave propagating along the positive x axis has the following properties: $y_{max} = 6$ cm, $\lambda = 8\pi$ cm, $v = 48$ cm/s, and the displacement of the wave at $t = 0$ and $x = 0$ is -2 cm. Determine the (a) wave number, (b) angular frequency, and (c) phase constant for the wave. (d) What is the first positive value of t for which the displacement at $x = 0$ will be $+2$ cm? (e) For this initial condition, find the coordinate of the particle on the positive x axis closest to the origin for which $y = 0$.

52. **A rope of total mass m and length L is suspended vertically. Show that a transverse wave pulse will travel the length of the rope in a time $t = 2\sqrt{L/g}$. (*Hint:* First find an expression for the velocity at any point a distance x from the lower end of the rope, by considering the tension in the rope as resulting from the weight of the segment below that point.)**

53. A wire of density ρ is tapered so that its cross-sectional area varies with x, according to

$$A = \{10^{-3}x + 0.01\}\text{cm}^2$$

(a) If the wire is subject to a tension F, derive a relationship for the velocity of a wave as a function of position. (b) If the wire is aluminum and is subject to a tension of 24 N, determine the velocity at the origin and at $x = 10$ m.

54. A harmonic wave in a string is described by the equation

$$y = (0.15 \text{ m})\sin(0.8x - 50t)$$

where x and y are in m and t is in s. If the mass per unit length of this string is 12 g/m, (a) find the maximum transverse acceleration of a point in this string. (b) Determine the maximum transverse force on a 1-cm segment of the string and compare this force to the tension in the string.

55. An aluminum wire is clamped at each end under zero tension at room temperature (22°C). The tension in the wire is increased by reducing the temperature, which results in a decrease in the wire's equilibrium length. What strain ($\Delta L/L$) will result in a transverse wave speed of 100 m/s? Take the cross-sectional area of the wire to be 5×10^{-6} m². Furthermore, you

should use the following properties of aluminum: density, $\rho = 2.7 \times 10^3$ kg/m^3; and Young's modulus, $Y = 6.8 \times 10^{10}$ N/m^2.

56. (a) Show that the speed of longitudinal waves along a spring of force constant k is $v = \sqrt{kL/\mu}$, where L is the unstretched length of the spring and μ is the mass per unit length. (b) A spring of mass 0.4 kg has an unstretched length of 2 m and a force constant of 100 N/m. Using the results to (a), determine the speed of longitudinal waves along this spring.

57. It is stated in Problem 52 that a wave pulse will travel from the bottom to the top of a rope of length L in a time $t = 2\sqrt{L/g}$. Use this result to answer the following questions. (It is *not* necessary to set up any new integrations.) (a) How long does it take for a wave pulse to travel halfway up the rope of length L? (Give your answer as a fraction of the quantity $(2\sqrt{L/g})$. (b) A pulse starts traveling up the rope. How far has the pulse traveled after a time $\sqrt{L/g}$?

58. A string of length L consists of two distinct sections. The left half has mass density $\mu = \mu_0/2$, while the right half has mass per unit length $\mu' = 3\mu = 3\mu_0/2$. Tension in the string is F_0. Notice that this string has the same total mass as a uniform string of length L and mass per unit length μ_0. (a) Find the speeds v and v' at which transverse wave pulses travel in the two sections of the string. Express the speeds in terms of F_0 and μ_0, and also as multiples of the speed $v_0 = \sqrt{F_0/\mu_0}$. (b) Find the time required for a wave to travel from one end of the string to the other. Give your result as a multiple of $T_0 = L/v_0$.

59. A wave pulse traveling along a string of linear mass density μ is described by the relationship

$$y = [A_0 e^{-bx}]\sin(kx - \omega t)$$

where the factors in brackets before the sine are said to be the amplitude. (a) What is the power $P(x)$ carried by this wave at a point x? (b) What is the power carried by this wave at the origin? (c) Compute the ratio of $\dfrac{P(x)}{P(0)}$.

CALCULATOR/COMPUTER PROBLEM

60. Two transverse wave pulses traveling in opposite directions along the x axis are represented by the following wave functions:

$$y_1(x, t) = \frac{6}{(x - 3t)^2} \qquad y_2(x, t) = -\frac{3}{(x + 3t)^2}$$

where x and y are measured in cm and t is in s. Write a program which will enable you to obtain the shape of the composite waveform $y_1 + y_2$ as a function of time. Use your program and make plots of the waveform at $t = 0$, 0.5, 1, 1.5, 2, 2.5, and 3 s.

17

Sound Waves

Audible waves are produced by this rock band.
(© A. D'Amario, III, FPG International)

This chapter deals with the properties of longitudinal waves traveling through various media. Sound waves are the most important example of longitudinal waves. They can travel through any material medium (that is, gases, solids, or liquids) with a speed that depends on the properties of the medium. As sound waves travel through a medium, the particles in the medium vibrate to produce density and pressure changes along the direction of motion of the wave. This is in contrast to a transverse wave, where the particle motion is perpendicular to the direction of wave motion. The displacements that occur as a result of sound waves involve the longitudinal displacements of individual molecules from their equilibrium positions. This results in a series of high- and low-pressure regions called *condensations* and *rarefactions*, respectively. If the source of the sound waves, such as the diaphragm of a loudspeaker, vibrates sinusoidally, the pressure variations are also sinusoidal. We shall find that the mathematical description of harmonic sound waves is identical to that of harmonic string waves discussed in the previous chapter.

There are three categories of longitudinal mechanical waves that cover different ranges of frequency: (1) *Audible waves* are sound waves that lie within the range of sensitivity of the human ear, typically, 20 Hz to 20 000 Hz. They can be generated in a variety of ways, such as by musical instruments, human vocal cords, and loudspeakers. (2) *Infrasonic waves* are longitudinal waves with frequencies below the audible range. Earthquake waves are an example. (3) *Ultrasonic waves* are longitudinal waves with frequencies above the audible range. For example, they can be generated by inducing vibrations in a quartz crystal with an applied alternating electric field.

Any device that transforms one form of power into another is called a *transducer.* In addition to the loudspeaker and the quartz crystal, ceramic and magnetic phonograph pickups are common examples of sound transducers. Some transducers can generate ultrasonic waves. Such devices are used in the construction of ultrasonic cleaners and for underwater navigation.

17.1 VELOCITY OF SOUND WAVES

Sound waves are compressional waves traveling through a compressible medium, such as air. The compressed region of air which propagates corresponds to a variation in the normal value of the air pressure. The speed of such compressional waves depends on the compressibility of the medium and on the inertia of the medium. If the compressible medium has a bulk modulus B and an equilibrium density ρ, the speed of sound in that medium is

$$v = \sqrt{\frac{B}{\rho}} \tag{17.1}$$

Recall that the **bulk modulus** (Section 12.4) is defined as the ratio of the change in pressure, ΔP, to the resulting fractional change in volume, $-\Delta V/V$:

$$B = -\frac{\Delta P}{\Delta V/V} \tag{17.2}$$

Note that B is always positive, since an increase in pressure (positive ΔP) results in a decrease in volume. Hence, the ratio $\Delta P/\Delta V$ is always negative.

It is interesting to compare Equation 17.1 with the expression for the speed of transverse waves on a string, $v = \sqrt{F/\mu}$, discussed in the previous chapter. In both cases, the wave speed depends on an elastic property of the medium (B or F) and on an inertial property of the medium (ρ or μ). In fact, the speed of *all mechanical waves* follows an expression of the general form

$$v = \sqrt{\frac{\text{elastic property}}{\text{inertial property}}}$$

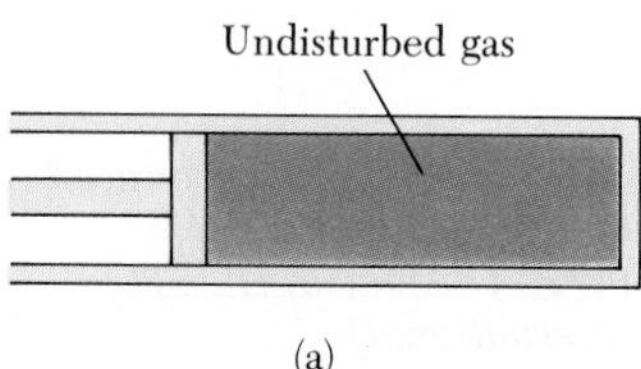

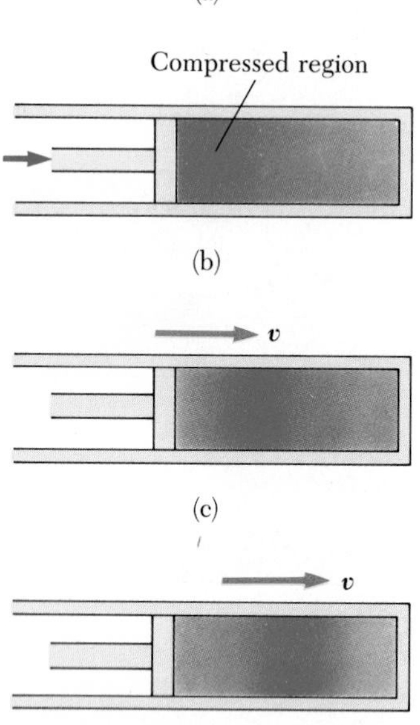

Figure 17.1 Motion of a longitudinal pulse through a compressible medium. The compression (darker region) is produced by the moving piston.

Let us describe pictorially the motion of a longitudinal pulse moving through a long tube containing a compressible gas or liquid (Fig. 17.1). A piston at the left end can be moved to the right to compress the fluid and create the longitudinal pulse. This is a convenient arrangement, since the wave motion is one-dimensional. Before the piston is moved, the medium is undisturbed and of uniform density, as represented by the uniformly shaded region in Figure 17.1a. When the piston is suddenly pushed to the right (Fig. 17.1b), the medium just in front of it is compressed (represented by the heavier shaded region). The pressure and density in this region are higher than normal. When the piston comes to rest (Fig. 17.1c), the compressed region continues to move to the right, corresponding to a longitudinal pulse traveling down the tube with a speed v. Note that the piston speed does *not* equal v.

Furthermore, the compressed region does not "stay with" the piston until the piston stops.

Let us now determine the speed of sound waves in various media with a few examples.

EXAMPLE 17.1 Sound Waves in a Solid Bar
If a solid bar is struck at one end with a hammer, a longitudinal pulse will propagate down the bar with a speed

$$v = \sqrt{\frac{Y}{\rho}} \tag{17.3}$$

where Y is the Young's modulus for the material, defined as the longitudinal stress divided by the longitudinal strain (Chapter 12). Find the speed of sound in an aluminum bar.

Solution Using Equation 17.3 and the available data for aluminum, $Y = 7.0 \times 10^{10}$ N/m^2 and the density $\rho = 2.7 \times 10^3$ kg/m^3, we find that

$$v_{\text{Al}} = \sqrt{\frac{7.0 \times 10^{10}\ \text{N/m}^2}{2.7 \times 10^3\ \text{kg/m}^3}} \approx 5100\ \text{m/s}$$

This is a typical value for the speed of sound in solids. The result is much larger than the speed of sound in gases. This makes sense since the molecules of a solid are close together (in comparison to the molecules of a gas) and hence respond more rapidly to a disturbance.

EXAMPLE 17.2 Speed of Sound in a Liquid
Find the speed of sound in water, which has a bulk modulus of about 2.1×10^9 N/m^2 and a density of about 10^3 kg/m^3.

Solution Using Equation 17.1, we find that

$$v_{\text{water}} = \sqrt{\frac{B}{\rho}} \approx \sqrt{\frac{2.1 \times 10^9\ \text{N/m}^2}{1 \times 10^3\ \text{kg/m}^3}} = 1500\ \text{m/s}$$

This result is much smaller than that for the speed of sound in aluminum, calculated in the previous example. In general, sound waves travel more slowly in liquids than in solids. This is because liquids are more compressible than solids and hence have a smaller bulk modulus.

The speed of sound in various media is given in Table 17.1.

TABLE 17.1 Speed of Sound in Various Media

Medium	v(m/s)
Gases	
Air (0°C)	331
Air (20°C)	343
Hydrogen (0°C)	1286
Oxygen (0°C)	317
Helium (0°C)	972
Liquids at 25°C	
Water	1493
Methyl alcohol	1143
Sea water	1533
Solids	
Aluminum	5100
Copper	3560
Iron	5130
Lead	1322
Vulcanized rubber	54

17.2 HARMONIC SOUND WAVES

If the source of a longitudinal wave, such as a vibrating diaphragm, oscillates with simple harmonic motion, the resulting disturbance will also be harmonic. One can produce a one-dimensional harmonic sound wave in a long, narrow tube containing a gas by means of a vibrating piston at one end, as in Figure 17.2. The darker regions in this figure represent regions where the gas is compressed, and so the density and pressure are *above* their equilibrium values.

A compressed layer is formed at times when the piston is being pushed into the tube. This compressed region, called a **condensation,** moves down the

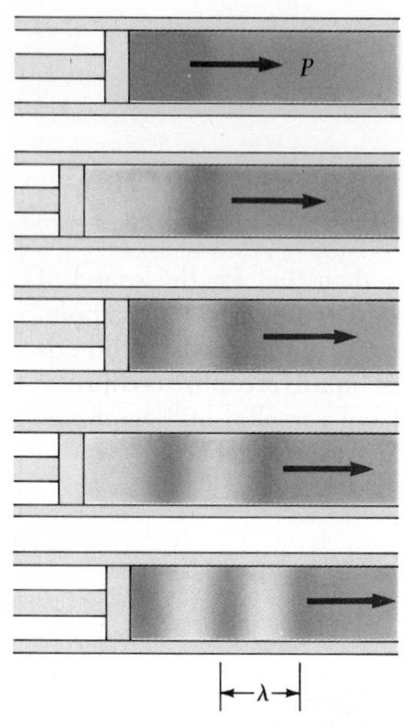

Figure 17.2 A harmonic longitudinal wave propagating down a tube filled with a compressible gas. The source of the wave is a vibrating piston at the left. The high- and low-pressure regions are dark and light, respectively.

tube as a pulse, continuously compressing the layers in front of it. When the piston is withdrawn from the tube, the gas in front of it expands and the pressure and density in this region fall below their equilibrium values (represented by the lighter regions in Figure 17.2). These low-pressure regions, called **rarefactions,** also propagate along the tube, following the condensations. Both regions move with a speed equal to the speed of sound in that medium (about 343 m/s in air at 20°C).

As the piston oscillates back and forth in a sinusoidal fashion, regions of condensation and rarefaction are continuously set up. The distance between two successive condensations (or two successive rarefactions) equals the wavelength, λ. As these regions travel down the tube, any small volume of the medium moves with simple harmonic motion parallel to the direction of the wave. If $s(x, t)$ is the displacement of a small volume element measured from its equilibrium position, we can express this harmonic displacement function as

$$s(x, t) = s_m \cos(kx - \omega t) \tag{17.4}$$

where s_m is the *maximum displacement from equilibrium* (the displacement amplitude), k is the wave number, and ω is the angular frequency of the piston. Note that the displacement is along x, the direction of motion of the sound wave, which of course means we are describing a longitudinal wave. The variation in the pressure of the gas, ΔP, measured from its equilibrium value is also harmonic and given by

$$\Delta P = \Delta P_m \sin(kx - \omega t) \tag{17.5}$$

The derivation of this expression will be given below.

The **pressure amplitude** ΔP_m is the *maximum change in pressure from the equilibrium value.* As we shall show later, the pressure amplitude is proportional to the displacement amplitude, s_m, and is given by

Pressure amplitude

$$\Delta P_m = \rho v \omega s_m \tag{17.6}$$

where ωs_m is the maximum longitudinal velocity of the medium in front of the piston.

Thus, we see that a sound wave may be considered as either a displacement wave or a pressure wave. A comparison of Equations 17.4 and 17.5 shows that *the pressure wave is 90° out of phase with the displacement wave.* Graphs of these functions are shown in Figure 17.3. Note that the pressure variation is a maximum when the displacement is zero, whereas the displacement is a maximum when the pressure variation is zero. Since the pressure is proportional to the density, the variation in density from the equilibrium value follows an expression similar to Equation 17.5.

We shall now give a derivation of Equations 17.5 and 17.6. From Equation 17.2, we see that the pressure variation in a gas is given by

$$\Delta P = -B(\Delta V/V)$$

The volume of a layer of thickness Δx and cross-sectional area A is $V = A\,\Delta x$. The change in the volume ΔV accompanying the pressure change is equal to

$A\,\Delta s$, where Δs is the difference in s between x and $x+\Delta x$. That is, $\Delta s = s(x+\Delta x) - s(x)$. Hence, we can express ΔP as

$$\Delta P = -B\frac{\Delta V}{V} = -B\frac{A\,\Delta s}{A\,\Delta x} = -B\frac{\Delta s}{\Delta x}$$

As Δx approaches zero, the ratio $\Delta s/\Delta x$ becomes $\partial s/\partial x$. (The partial derivative is used here to indicate that we are interested in the variation of s with position at a *fixed* time.) Therefore,

$$\Delta P = -B(\partial s/\partial x) \tag{17.7}$$

If the displacement is the simple harmonic function given by Equation 17.2, we find that

$$\Delta P = -B\frac{\partial}{\partial x}[s_{\text{m}}\cos(kx-\omega t)] = Bs_{\text{m}}k\sin(kx-\omega t)$$

Since the bulk modulus is given by $B = \rho v^2$ (Eq. 17.1), the pressure variation reduces to

$$\Delta P = \rho v^2 s_{\text{m}} k \sin(kx-\omega t)$$

Furthermore, from Equation 16.13, we can write $\omega = kv$; hence ΔP can be expressed as

$$\Delta P = \rho\omega s_{\text{m}} v \sin(kx-\omega t) = \Delta P_{\text{m}}\sin(kx-\omega t) \tag{17.8}$$

where ΔP_{m} is the maximum pressure variation, given by Equation 17.6.

$$\Delta P_{\text{m}} = \rho v\omega s_{\text{m}}$$

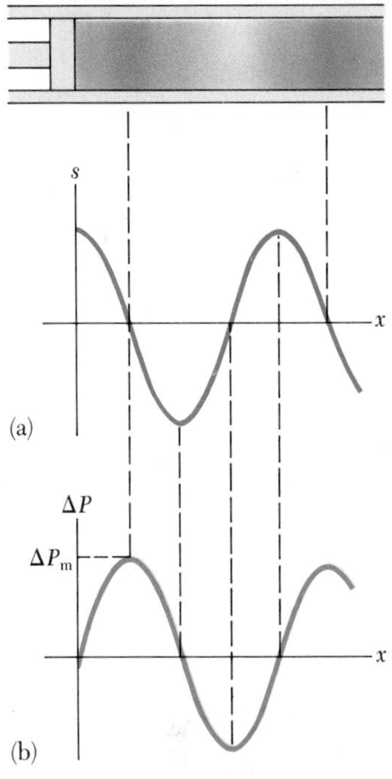

Figure 17.3 (a) Displacement amplitude versus position and (b) pressure amplitude versus position for a harmonic longitudinal wave. The displacement wave is 90° out of phase with the pressure wave.

17.3 ENERGY AND INTENSITY OF HARMONIC SOUND WAVES

In the previous chapter, we showed that waves traveling on stretched strings transport energy. The same concepts are now applied to sound waves. Consider a layer of air of mass Δm and width Δx in front of a piston oscillating with a frequency ω, as in Figure 17.4. The piston transmits energy to the layer of air.[1] Since the average kinetic energy equals the average potential energy in simple harmonic motion (as was shown in Chapter 13), the average total energy of the mass Δm equals its maximum kinetic energy. Therefore, we can express the average energy of the moving layer of gas as

$$\Delta E = \tfrac{1}{2}\Delta m(\omega s_{\text{m}})^2 = \tfrac{1}{2}(\rho A\,\Delta x)(\omega s_{\text{m}})^2$$

where $A\,\Delta x$ is the volume of the layer. The time rate at which energy is transferred (or the power) to each layer is given by

$$\text{Power} = \frac{\Delta E}{\Delta t} = \tfrac{1}{2}\rho A\left(\frac{\Delta x}{\Delta t}\right)(\omega s_{\text{m}})^2 = \tfrac{1}{2}\rho A v(\omega s_{\text{m}})^2$$

where $v = \Delta x/\Delta t$ is the velocity of the disturbance to the right.

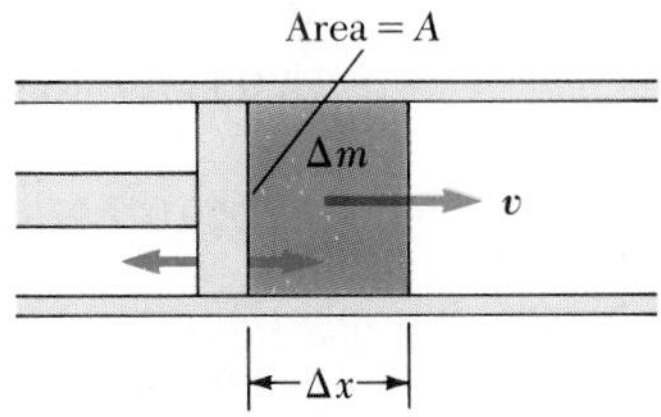

Figure 17.4 An oscillating piston transfers energy to the gas in the tube, causing the layer of width Δx and mass Δm to oscillate with an amplitude s_{m}.

[1] Although it is not proved here, the work done by the piston equals the energy carried away by the wave. For a detailed mathematical treatment of this concept, see Frank S. Crawford, Jr., *Waves*, New York, McGraw-Hill, 1968, Berkeley Physics Course, Volume 3, Chapter 4.

We define the **intensity** I of a wave, or the power per unit area, to be the rate at which sound energy flows through a unit area A perpendicular to the direction of travel of the wave.

In this case, the intensity is given by

Intensity of a sound wave

$$I = \frac{\text{power}}{\text{area}} = \tfrac{1}{2}\rho(\omega s_m)^2 v \tag{17.9}$$

Thus, we see that the intensity of the harmonic sound wave is proportional to the square of the amplitude and to the square of the frequency (as in the case of a harmonic string wave). This can also be written in terms of the pressure amplitude ΔP_m, using Equation 17.6, which gives

$$I = \frac{\Delta P_m{}^2}{2\rho v} \tag{17.10}$$

EXAMPLE 17.3 Hearing Limitations

The faintest sounds the human ear can detect at a frequency of 1000 Hz correspond to an intensity of about 10^{-12} W/m^2 (the so-called *threshold of hearing*). Likewise, the loudest sounds that the ear can tolerate correspond to an intensity of about 1 W/m^2 (the *threshold of pain*). Determine the pressure amplitudes and maximum displacements associated with these two limits.

Solution First, consider the faintest sounds. Using Equation 17.10 and taking $v = 343$ m/s and the density of air to be $\rho = 1.20$ kg/m^3, we get

$$\Delta P_m = (2\rho v I)^{1/2}$$
$$= [2(1.20 \text{ kg/m}^3)(343 \text{ m/s})(10^{-12} \text{ W/m}^2)]^{1/2}$$
$$= 2.87 \times 10^{-5} \text{ N/m}^2$$

Since atmospheric pressure is about 10^5 N/m^2, this means the ear can discern pressure fluctuations as small as 3 parts in 10^{10}! The corresponding maximum displacement can be calculated using Equation 17.6, recalling that $\omega = 2\pi f$:

$$s_m = \frac{\Delta P_m}{\rho \omega v} = \frac{2.87 \times 10^{-5} \text{ N/m}^2}{(1.20 \text{ kg/m}^3)(2\pi \times 10^3 \text{ s}^{-1})(343 \text{ m/s})}$$
$$= 1.11 \times 10^{-11} \text{ m}$$

This is a remarkably small number! If we compare this result for s_m with the diameter of a molecule (about 10^{-10} m), we see that the ear is an extremely sensitive detector of sound waves.

In a similar manner, one finds that the loudest sounds the human ear can tolerate correspond to a pressure amplitude of about 30 N/m^2 and a maximum displacement of 1.1×10^{-5} m. Note that the small pressure amplitudes, called acoustic pressure, correspond to fluctuations taking place above and below atmospheric pressure.

Intensity in Decibels

The previous example illustrates the wide range of intensities that the human ear can detect. For this reason, it is convenient to use a logarithmic intensity scale, where the **intensity level** β is defined by the equation

Intensity in decibels

$$\beta \equiv 10 \log\left(\frac{I}{I_0}\right) \tag{17.11}$$

The constant I_0 is the *reference intensity*, taken to be at the threshold of hearing ($I_0 = 10^{-12}$ W/m^2), and I is the intensity in W/m^2 at the level β, where β is

TABLE 17.2 Decibel Scale Intensity for Some Sources

Source of Sound	β (dB)
Nearby jet airplane	150
Jackhammer; machine gun	130
Siren; rock concert	120
Subway; power mower	100
Busy traffic	80
Vacuum cleaner	70
Normal conversation	50
Mosquito buzzing	40
Whisper	30
Rustling leaves	10
Threshold of hearing	0

measured in decibels (dB).[2] On this scale, the threshold of pain ($I = 1$ W/m^2) corresponds to an intensity level of $\beta = 10 \log(1/10^{-12}) = 10 \log(10^{12}) =$ 120 dB. Likewise, the threshold of hearing corresponds to $\beta = 10 \log(1/1) =$ 0 dB. Nearby jet airplanes can create intensity levels of 150 dB, and subways and riveting machines have levels of 90 to 100 dB. The electronically amplified sounds heard at rock concerts can be at levels of up to 120 dB, the threshold of pain. Prolonged exposure to such high intensity levels may produce serious damage to the ear. Ear plugs are recommended whenever intensity levels exceed 90 dB. Recent evidence also suggests that "noise pollution" may be a contributing factor to high blood pressure, anxiety, and nervousness. Table 17.2 gives some typical values of the sound intensities of various sources.

17.4 SPHERICAL AND PLANE WAVES

If a spherical body pulsates or oscillates periodically such that its radius varies harmonically with time, a sound wave with spherical wave fronts will be produced (Fig. 17.5). The wave moves outward from the source at a constant speed if the medium is uniform.

Since all points on the sphere behave in the same way, we conclude that the energy in a spherical wave will propagate equally in all directions. That is, no one direction is preferred over any other. If P_{av} is the average power emitted by the source, then this power at any distance r from the source must be distributed over a spherical surface of area $4\pi r^2$. Hence, the wave intensity at a distance r from the source is

$$I = P_{av}/A = P_{av}/4\pi r^2 \tag{17.12}$$

Since P_{av} is the same through any spherical surface centered at the source, we see that the intensities at distances r_1 and r_2 are given by

$$I_1 = P_{av}/4\pi r_1^2 \quad \text{and} \quad I_2 = P_{av}/4\pi r_2^2$$

[2] The "bel" is named after the inventor of the telephone, Alexander Graham Bell (1847–1922). The prefix deci- is the metric system scale factor that stands for 10^{-1}.

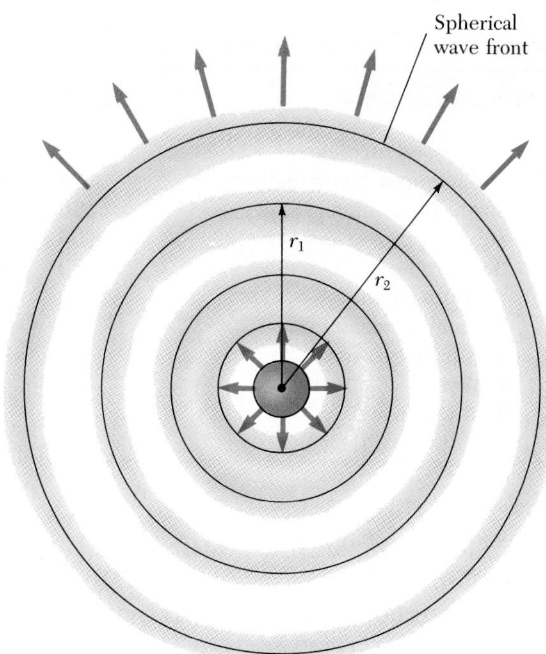

Figure 17.5 A spherical wave propagating radially outward from an oscillating spherical body. The intensity of the spherical wave varies as $1/r^2$.

Therefore, the ratio of intensities on these two spherical surfaces is

$$\frac{I_1}{I_2} = \frac{r_2^{\,2}}{r_1^{\,2}}$$

In Equation 17.9 we found that the intensity was also proportional to $s_m^{\,2}$, the square of the wave amplitude. Comparing this result with Equation 17.12, we conclude that the wave amplitude of a spherical wave must vary as $1/r$. Therefore, we can write the wave function ψ (Greek letter "psi") for an outgoing spherical wave in the form

$$\psi(r, t) = (s_0/r)\sin(kr - \omega t) \tag{17.13}$$

where s_0 is a constant.

It is useful to represent spherical waves by a series of circular arcs concentric with the source, as in Figure 17.6. Each arc represents a surface over which the phase of the wave is constant. We call such a surface of constant phase a **wavefront.** The distance between adjacent wavefronts equals the wavelength, λ. The radial lines pointing outward from the source are called **rays.**

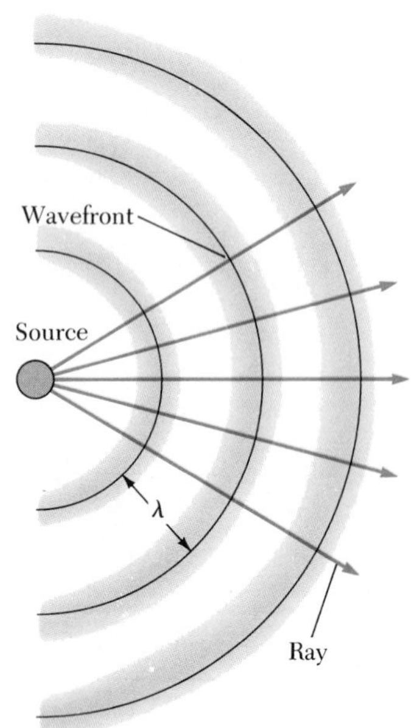

Figure 17.6 Spherical waves emitted by a point source. The circular arcs represent the spherical wavefronts concentric with the source. The rays are radial lines pointing outward from the source perpendicular to the wavefronts.

Now consider a small portion of the wavefronts at *large* distances from the source, as in Figure 17.7. In this case, the rays are nearly parallel and the wavefronts are very close to being planar. Therefore, at distances from the source that are large compared with the wavelength, we can approximate the wavefronts by parallel planes. We call such a wave a **plane wave.** Any small portion of a spherical wave that is far from the source can be considered a plane wave.

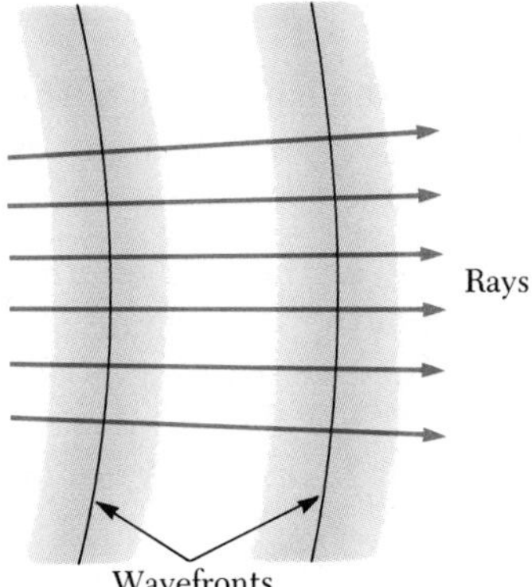

Figure 17.7 At large distances from a point source, the wavefronts are nearly parallel planes and the rays are nearly parallel lines perpendicular to the planes. Hence, a small segment of a spherical wavefront is approximately a plane wave.

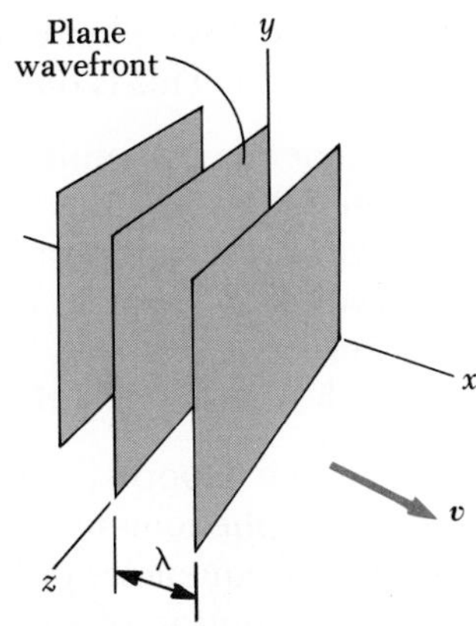

Figure 17.8 Representation of a planar wave moving in the positive x direction. The wavefronts are planes parallel to the yz plane.

Figure 17.8 illustrates a plane wave propagating along the x axis. If x is taken to be the direction of the wave motion (or rays) in Figure 17.8, then the wavefronts are parallel to the yz plane. In this case, the wave function depends only on x and t and has the form

$$\psi(x, t) = s_0 \sin(kx - \omega t) \qquad (17.14)$$

Plane wave representation

That is, the wave function for a plane wave is identical in form to that of a one-dimensional traveling wave. The intensity is the same on successive wavefronts of the plane wave.

EXAMPLE 17.4 Intensity Variations of a Point Source □

A source emits sound waves with a power output of 80 W. Assume the source is a point source. (a) Find the intensity at a distance 3 m from the source.

Solution A point source emits energy in the form of spherical waves (Fig. 17.5). Let P_{av} be the average power output of the source. At a distance r from the source, the power is distributed over the surface area of a sphere, $4\pi r^2$. Therefore, the intensity at a distance r from the source is given by Equation 17.12. Since $P_{av} = 80$ W and $r = 3$ m, we find that

$$I = \frac{P_{av}}{4\pi r^2} = \frac{80 \text{ W}}{4\pi(3 \text{ m})^2} = 0.707 \text{ W/m}^2$$

which is close to the threshold of pain.

(b) Find the distance at which the sound reduces to a level of 40 dB.

Solution We can find the intensity at the 40-dB level by using Equation 17.11 with $I_0 = 10^{-12}$ W/m². This gives

$$\log(I/I_0) = 4$$

$$I = 10^4 I_0 = 10^{-8} \text{ W/m}^2$$

Using this value for I in Equation 17.12 and solving for r, we get

$$r = (P_{av}/4\pi I)^{1/2} = \frac{(80 \text{ W})^{1/2}}{(4\pi \times 10^{-8} \text{ W/m}^2)^{1/2}}$$

$$= 2.52 \times 10^4 \text{ m}$$

which equals about 16 miles!

*17.5 THE DOPPLER EFFECT

When a car or truck is moving while its horn is blowing, the frequency of the sound you hear is higher as the vehicle approaches you and lower as it moves away from you. This is one example of the Doppler effect.[3]

In general, a Doppler effect is experienced whenever there is *relative* motion between the source and the observer. When the source and observer are moving toward each other, the frequency heard by the observer is *higher* than the frequency of the source. When the source and observer move away from each other, the observer hears a frequency which is *lower* than the source frequency.

Although the Doppler effect is most commonly experienced with sound waves, it is a phenomenon common to all harmonic waves. For example, there is a shift in frequencies of light waves (electromagnetic waves) produced by the relative motion of source and observer. The Doppler effect is used in police radar systems to measure the speed of motor vehicles. Likewise, astronomers use the effect to determine the relative motion of stars, galaxies, and other celestial objects.

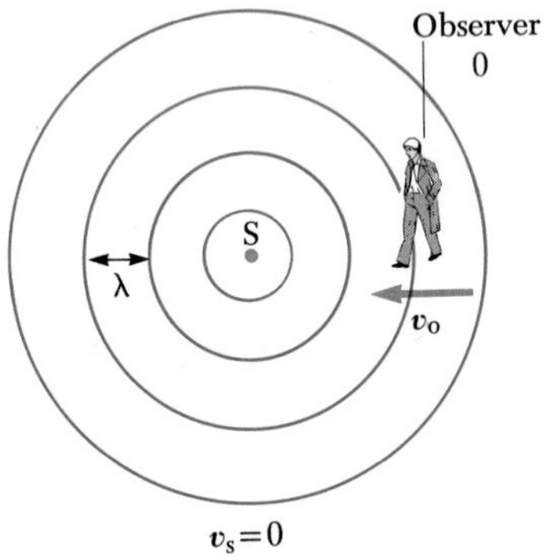

Figure 17.9 An observer O moving with a speed v_o *toward* a stationary point source S hears a frequency f' that is *greater* than the source frequency.

First, let us consider the case where the observer O is moving and the sound source S is stationary. For simplicity, we shall assume that the air is also stationary and that the observer moves directly toward the source. Figure 17.9 describes the situation when the observer moves with a speed v_O toward the source (considered as a point source), which is at rest ($v_S = 0$). In general, "at rest" means at rest with respect to the medium, air.

We shall take the frequency of the source to be f, the wavelength to be λ, and the velocity of sound to be v. If the observer were also stationary, clearly he or she would detect f wavefronts per second. (That is, when $v_O = 0$ and $v_S = 0$, the observed frequency equals the source frequency.) When the observer moves towards the source, the speed of the waves relative to the observer is $v' = v + v_O$, but the wavelength λ is unchanged. Hence, the frequency heard by the observer is *increased* and given by

$$f' = \frac{v'}{\lambda} = \frac{v + v_O}{\lambda}$$

Since $\lambda = v/f$, we can express f' as

$$f' = f\left(1 + \frac{v_O}{v}\right) \quad \text{(Observer moving toward the source)} \quad (17.15)$$

Similarly, if the observer is moving *away* from the source, the speed of the wave relative to the observer is $v' = v - v_O$. The frequency heard by the observer in this case is *lowered* and is given by

$$f' = f\left(1 - \frac{v_O}{v}\right) \quad \text{(Observer moving away from the source)} \quad (17.16)$$

In general, when an observer moves with a speed v_O relative to a stationary source, the frequency heard by the observer is

Frequency heard with an observer in motion

$$f' = f\left(1 \pm \frac{v_O}{v}\right) \quad (17.17)$$

[3] Named after the Austrian physicist Christian Johann Doppler (1803–1853), who discovered the effect for light waves.

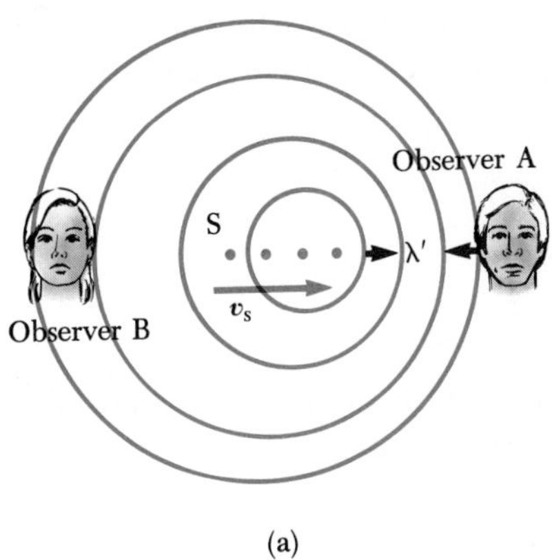

(a)

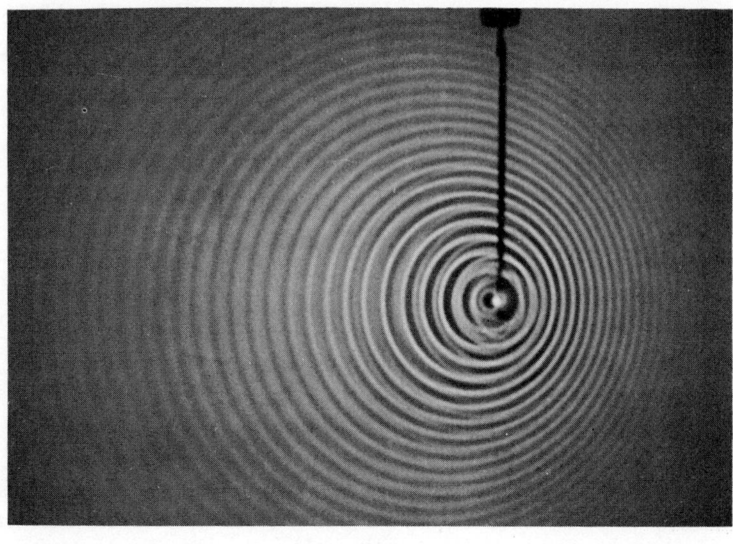

(b)

Figure 17.10 (a) A source S moving with a speed v_S toward a stationary observer A and away from a stationary observer B. Observer A hears an *increased* frequency, and observer B hears a *decreased* frequency. (b) The Doppler effect in water observed in a ripple tank. (Courtesy Educational Development Center, Newton, Mass.)

where the *positive* sign is used when the observer moves *toward* the source and the *negative* sign holds when the observer moves *away* from the source.

Now consider the situation in which the source is in motion and the observer is at rest. If the source moves directly toward observer A in Figure 17.10a, the wavefronts seen by the observer are closer together as a result of the motion of the source in the direction of the outgoing wave. As a result, the wavelength λ' measured by observer A is shorter than the wavelength λ of the source. During each vibration, which lasts for a time T (the period), the source moves a distance $v_ST = v_S/f$ and the wavelength is *shortened* by this amount. Therefore, the observed wavelength λ' is given by

$$\lambda' = \lambda - \Delta\lambda = \lambda - (v_S/f)$$

Since $\lambda = v/f$, the frequency heard by observer A is

$$f' = \frac{v}{\lambda'} = \frac{v}{\lambda - \dfrac{v_S}{f}} = \frac{v}{\dfrac{v}{f} - \dfrac{v_S}{f}}$$

$$f' = f\left(\frac{1}{1 - \dfrac{v_S}{v}}\right) \tag{17.18}$$

"I love hearing that lonesome wail of the train whistle as the magnitude of the frequency of the wave changes due to the Doppler effect."

That is, the observed frequency is *increased* when the source moves toward the observer.

In a similar manner, when the source moves away from an observer B at rest (where observer B is to the left of the source, as in Fig. 17.10a), observer B measures a wavelength λ' that is *greater* than λ and hears a *decreased* frequency given by

$$f' = f\left(\frac{1}{1 + \dfrac{v_S}{v}}\right) \tag{17.19}$$

Combining Equations 17.18 and 17.19, we can express the general relationship for the observed frequency when the source is moving and the observer is at rest as

Frequency heard with source in motion

$$f' = f\left(\frac{1}{1 \mp \frac{v_S}{v}}\right) \tag{17.20}$$

Finally, if both the source and the observer are in motion, one finds the following general relationship for the observed frequency:

Frequency heard with observer and source in motion

$$f' = f\left(\frac{v \pm v_O}{v \mp v_S}\right) \tag{17.21}$$

In this expression, the *upper* signs ($+v_O$ and $-v_S$) refer to motion of one *toward* the other, and the lower signs ($-v_O$ and $+v_S$) refer to motion of one *away* from the other.

A convenient rule to remember concerning signs when working with all Doppler effect problems is the following:

The word *toward* is associated with an *increase* in the observed frequency. The words *away from* are associated with a *decrease* in the observed frequency.

EXAMPLE 17.5 The Moving Train Whistle

A train moving at a speed of 40 m/s sounds its whistle, which has a frequency of 500 Hz. Determine the frequencies heard by a stationary observer as the train approaches and then recedes from the observer.

Solution We can use Equation 17.18 to get the apparent frequency as the train approaches the observer. Taking $v = 343$ m/s for the speed of sound in air gives

$$f' = f\left(\frac{1}{1 - \frac{v_S}{v}}\right) = (500 \text{ Hz})\left(\frac{1}{1 - \frac{40 \text{ m/s}}{343 \text{ m/s}}}\right)$$

$$= 566 \text{ Hz}$$

Likewise, Equation 17.19 can be used to obtain the frequency heard as the train recedes from the observer:

$$f' = f\left(\frac{1}{1 + \frac{v_S}{v}}\right) = (500 \text{ Hz})\left(\frac{1}{1 + \frac{40 \text{ m/s}}{343 \text{ m/s}}}\right)$$

$$= 448 \text{ Hz}$$

EXAMPLE 17.6 The Noisy Siren

An ambulance travels down a highway at a speed of 33.5 m/s (= 75 mi/h). Its siren emits sound at a frequency of 400 Hz. What is the frequency heard by a passenger in a car traveling at 24.6 m/s (= 55 mi/h) in the opposite direction as the car approaches the ambulance and as the car moves away from the ambulance?

Solution Let us take the velocity of sound in air to be $v = 343$ m/s. We can use Equation 17.21 in both cases. As the ambulance and car approach each other, the observed apparent frequency is

$$f' = f\left(\frac{v + v_O}{v - v_S}\right) = (400 \text{ Hz})\left(\frac{343 \text{ m/s} + 24.6 \text{ m/s}}{343 \text{ m/s} - 33.5 \text{ m/s}}\right)$$

$$= 475 \text{ Hz}$$

Likewise, as they recede from each other, a passenger in the car hears a frequency

$$f' = f\left(\frac{v - v_O}{v + v_S}\right) = (400 \text{ Hz})\left(\frac{343 \text{ m/s} - 24.6 \text{ m/s}}{343 \text{ m/s} + 33.5 \text{ m/s}}\right)$$

$$= 338 \text{ Hz}$$

The *change* in frequency as detected by the passenger in the car is 475 − 338 = 137 Hz, which is more than 30% of the actual frequency emitted.

Exercise 1 Suppose that the passenger car is parked on the side of the highway as the ambulance travels down the highway at the speed of 33.5 m/s. What frequency will the passenger in the car hear as the ambulance (a) approaches the parked car and (b) recedes from the parked car?
Answer (a) 443 Hz (b) 364 Hz

Shock Waves

Now let us consider what happens when the source velocity v_S *exceeds* the wave velocity v. This situation is described graphically in Figure 17.11. The circles represent spherical wavefronts emitted by the source at various times during its motion. At $t = 0$, the source is at S_0, and at some later time t, the source is at S_n. In the time t, the wavefront centered at S_0 reaches a radius of vt. In this same interval, the source travels a distance $v_S t$ to S_n. At the instant the source is at S_n, waves are just beginning to be generated and so the wavefront has zero radius at this point. The line drawn from S_n to the wavefront centered on S_0 is tangent to all other wavefronts generated at intermediate times. Thus, we see that the envelope of these waves is a cone whose apex angle θ is given by

$$\sin \theta = v/v_S$$

The ratio v_S/v is referred to as the *Mach number*. The conical wavefront produced when $v_S > v$ (supersonic speeds) is known as a *shock wave*. An interesting analogy to shock waves is the V-shaped wavefronts produced by a boat (the bow wave) when the boat's speed exceeds the speed of the surface water waves.

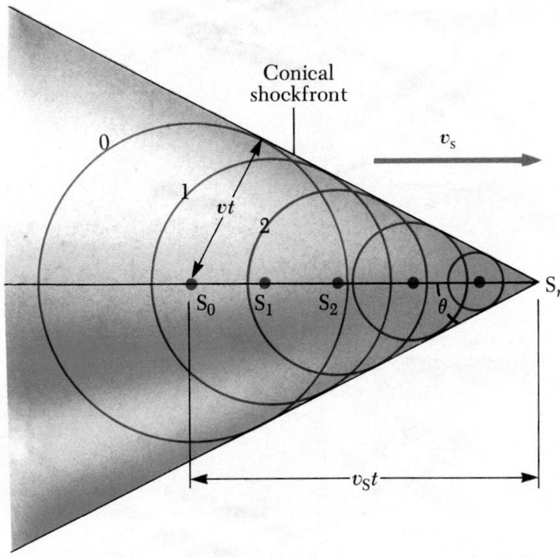

Figure 17.11 Representation of a shock wave produced when a source moves from S_0 to S_n with a speed v_S, which is *greater* than the wave speed v in that medium. The envelope of the wavefronts forms a cone whose apex angle is given by $\sin \theta = v/v_S$.

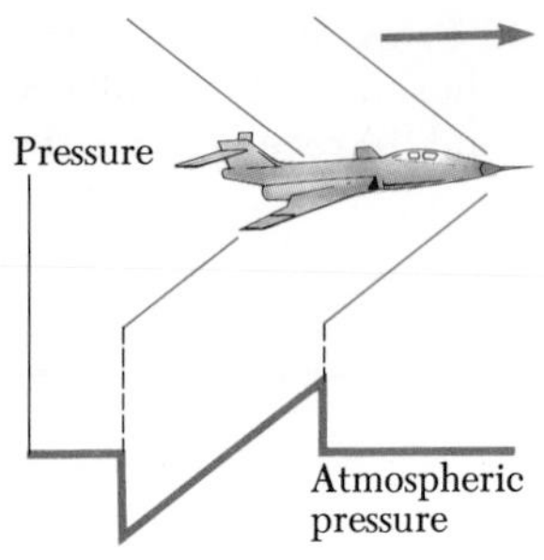

Figure 17.12 Two shock waves produced by the nose and tail of a jet airplane traveling at supersonic speeds.

Jet airplanes traveling at supersonic speeds produce shock waves, which are responsible for the loud explosion, or "sonic boom," one hears. The shock wave carries a great deal of energy concentrated on the surface of the cone, with correspondingly large pressure variations. Such shock waves are unpleasant to hear and can cause damage to buildings when aircraft fly supersonically at low altitudes. In fact, an airplane flying at supersonic speeds produces a double boom because two shock fronts are formed, one from the nose of the plane and one from the tail (Fig. 17.12).

Figure 17.13 is a dramatic photographic of a bullet passing through the flame of a candle. There is a clear outline of the shock wave produced by the bullet moving at supersonic speed.

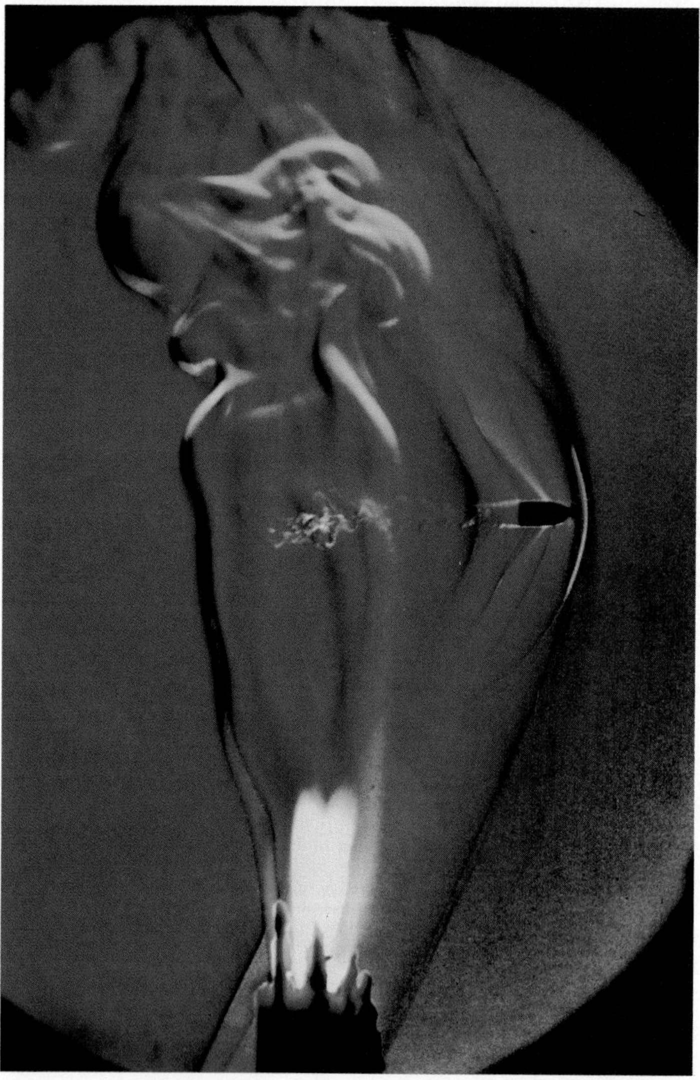

Figure 17.13 Supersonic candlelight. A stroboscopic shadow photograph showing a bullet traveling at supersonic speed passing through the hot air rising above a candle. This type of picture, called a *schlieren,* enables one to observe regions of nonuniform air density. Note the shock wave in the vicinity of the bullet. (Bullet through Flame, 1973. © Harold E. Edgerton. Courtesy of Palm Press, Inc.)

SUMMARY

Sound waves are longitudinal in nature and travel through a compressible medium with a speed that depends on the compressibility and inertia of that medium. The **speed of sound** in a medium of compressibility B and density ρ is

$$v = \sqrt{\frac{B}{\rho}} \tag{17.1}$$

Speed of sound in a compressible medium

In the case of harmonic sound waves, the **variation in pressure** from the equilibrium value is given by

$$\Delta P = \Delta P_m \sin(kx - \omega t) \tag{17.7}$$

Pressure variation

where ΔP_m is the **pressure amplitude.** The pressure wave is 90° out of phase with the displacement wave. If the displacement amplitude is s_m, then ΔP_m has the value

$$\Delta P_m = \rho v \omega s_m \tag{17.6}$$

Pressure amplitude

The **intensity of a harmonic sound wave,** which is the power per unit area, is given by

$$I = \tfrac{1}{2}\rho(\omega s_m)^2 v = \frac{\Delta P_m{}^2}{2\rho v} \tag{17.9, 17.10}$$

Intensity of a sound wave

The **intensity of a spherical wave** produced by a point source is proportional to the average power emitted and inversely proportional to the square of the distance from the source.

The change in frequency heard by an observer whenever there is relative motion between the source and observer is called the **Doppler effect.** If the observer moves with a speed v_O and the source is at rest, the observed frequency f' is

$$f' = f\left(1 \pm \frac{v_O}{v}\right) \tag{17.17}$$

Frequency heard by an observer in motion

where the positive sign is used when the observer moves toward the source and the negative sign refers to motion away from the source.

If the *source moves* with a speed v_S and the observer is at rest, the observed frequency is

$$f' = f\left(\frac{1}{1 \mp \dfrac{v_S}{v}}\right) \tag{17.20}$$

Frequency heard with source in motion

where $-v_S$ refers to motion *toward* the observer and $+v_S$ refers to motion *away* from the observer.

When the *observer and source are both moving,* the observed frequency is

$$f' = f\left(\frac{v \pm v_O}{v \mp v_S}\right) \tag{17.21}$$

Frequency heard with observer and source in motion

QUESTIONS

1. Why are sound waves characterized as longitudinal?
2. As a result of a distant explosion, an observer senses a ground tremor and then hears the explosion. Explain.
3. If an alarm clock is placed in a good vacuum and then activated, no sound will be heard. Explain.
4. Some sound waves are harmonic, whereas others are not. Give an example of each.
5. In Example 17.4, we found that a point source with a power output of 80 W reduces to an intensity level of 40 dB at a distance of about 16 miles. Why do you suppose you cannot normally hear a rock concert going on 16 miles away?
6. If the distance from a point source is tripled, by what factor does the intensity decrease?
7. Explain how the Doppler effect is used with microwaves to determine the speed of an automobile.
8. If you are in a moving vehicle, explain what happens to the frequency of your echo as you move *toward* a canyon wall. What happens to the frequency as you move *away* from the wall?
9. Suppose an observer and a source of sound are both at rest, and a strong wind blows toward the observer. Describe the effect of the wind (if any) on (a) the observed wavelength, (b) the observed frequency, and (c) the wave velocity.
10. Of the following sounds, which is most likely to have an intensity level of 60 dB: a rock concert, the turning of a page in this text, normal conversation, a cheering crowd at a football game, or background noise at a church?
11. Estimate the decibel level of each of the sounds in Question 10.
12. A binary star system consists of two stars revolving about each other. If we observe the light reaching us from one of these stars as it makes one complete revolution about the other, what does the Doppler effect predict will happen to this light?
13. How could an object move with respect to an observer such that the sound from it is not shifted in frequency?
14. Why is it not possible to use sonar (sound waves) to determine the speed of an object traveling faster than the speed of sound in that medium?
15. Why is it so quiet after a snowfall?
16. Why is the intensity of an echo less than that of the original sound?
17. If the wavelength of a sound source is reduced by a factor of 2, what happens to its frequency? Its speed?
18. A sound wave travels in air at a frequency of 500 Hz. If part of the wave travels from the air into water, does its frequency change? Does its wavelength change? Justify your answers.
19. In a recent discovery, a nearby star was found to have a large planet orbiting about it, although the planet could not be seen. In terms of the concept of systems rotating about their center of mass and the Doppler shift for light (which is in many ways similar to that of sound), explain how an astronomer could determine the presence of the invisible planet.
20. Explain how the distance to a lightning bolt may be determined by counting the seconds between the flash and the sound of the thunder. Does the speed of the light signal have to be taken into account?

PROBLEMS

Section 17.1 Velocity of Sound Waves

1. Suppose that you hear a thunder clap 16.2 s after seeing the associated lightning stroke. The speed of sound waves in air is 343 m/s and the speed of light in air is 3.0×10^8 m/s. How far are you from the lightning stroke?
2. A stone is dropped into a deep canyon and is heard to strike the bottom 10.2 s after release. The speed of sound waves in air is 343 m/s. How deep is the canyon? What would be the percentage error in the depth if the time required for the sound to reach the canyon rim were ignored?
3. Find the velocity of sound in mercury, which has a bulk modulus of about 2.8×10^{10} N/m^2 and a density of 13 600 kg/m^3.
4. The ocean floor is underlaid by a layer of basalt that constitutes the crust, or uppermost layer of the earth in this region. Below this crust is found more dense peridotite rock, which forms the earth's mantle. The boundary between these two layers is called the Mohorovičić discontinuity ("Moho" for short). If an explosive charge is set off at the surface of the basalt, it generates a seismic wave that is reflected back at the Moho. If the velocity of this wave in basalt is 6.5 km/s, and the two-way travel time is 1.85 s, what is the thickness of this oceanic crust?
5. (a) What are the SI units of bulk modulus as expressed in Equation 17.2? (b) Show that the SI units of $\sqrt{B/\rho}$ are m/s, as required by Equation 17.1.
6. The density of aluminum is 2.7×10^3 kg/m^3. Use the value for the speed of sound in aluminum given in

Table 17.1 to calculate Young's modulus for this material.

7. You are watching a pier being constructed on the far shore of a salt water inlet when some blasting occurs. You hear the sound in the water 4.5 s before it reaches you through the air. How wide is the inlet? (*Hint:* See Table 17.1. Assume the air temperature is 20°C.)

8. The density of acetone is 792 kg/m^3. Its bulk modulus has a temperature dependence given by the relation $B = -6 \times 10^6 T + 9.02 \times 10^8$ where B is in N/m^2 and T is in °C. What is the speed of sound in acetone at 20°C?

9. The speed of sound through air (in m/s) is given by the following function of temperature:

$$v = 331.5 + 0.607T$$

where T is the Celsius temperature. In dry air the temperature decreases about 1°C for every 150 m rise in altitude. (a) Assuming this change is constant up to an altitude of 9000 m, how long will it take the sound from an airplane flying at 9000 m to reach the ground on a day when the ground temperature is 30°C? (b) Compare this to the time it would take if the air were a constant 30°C. Which time is longer?

Section 17.2 Harmonic Sound Waves

(In this section, use the following values as needed unless otherwise specified: the equilibrium density of air, $\rho = 1.2$ kg/m^3; the velocity of sound in air, $v =$ 343 m/s. Also, pressure variations ΔP are measured relative to atmospheric pressure.)

10. Calculate the pressure amplitude of a 2000-Hz sound wave in air if the displacement amplitude is equal to 2×10^{-8} m.

11. A sound wave in air has a pressure amplitude equal to 4×10^{-3} N/m^2. Calculate the displacement amplitude of the wave at a frequency of 10 kHz.

12. The pressure amplitude corresponding to the threshold of hearing is 2.9×10^{-5} N/m^2. At what frequency will a sound wave in air have this pressure amplitude if the displacement amplitude is 2.8×10^{-10} m?

13. An experimenter wishes to generate in air a sound wave that has a displacement amplitude equal to 5.5×10^{-6} m. The pressure amplitude is to be limited to 8.4×10^{-1} N/m^2. What is the minimum wavelength the sound wave can have?

14. A sound wave in air has a pressure amplitude of 4 N/m^2 and a frequency of 5000 Hz. $\Delta P = 0$ at the point $x = 0$ when $t = 0$. (a) What is ΔP at $x = 0$ when $t = 2 \times 10^{-4}$ s and (b) what is ΔP at $x = 0.02$ m when $t = 0$?

15. A harmonic sound wave can be described in the displacement mode by

$$s(x, t) = (2\ \mu\text{m})\cos[(15.7\ \text{m}^{-1})x - (858\ \text{s}^{-1})t]$$

(a) Find the amplitude, wavelength, and speed of this wave and state what material this sound wave is traveling through. [See Table 17.1.] (b) Determine the instantaneous displacement of the molecules at the position $x = 0.05$ m at $t = 3$ ms. (c) Determine the maximum velocity of the molecules' oscillatory motion.

16. Consider the sound wave whose harmonic displacement is described in Problem 15. What is the pressure variation at $x = 0$ when $t = \pi/2\omega$? Take $\rho =$ 900 kg/m^2.

17. Write an expression that describes the pressure variation as a function of position and time for a harmonic sound wave in air if $\lambda = 0.1$ m and $\Delta P_m = 0.2$ N/m^2.

18. Write the function that describes the displacement wave corresponding to the pressure wave in Problem 17.

Section 17.3 Energy and Intensity of Harmonic Sound Waves

19. Calculate the intensity level in dB of a sound wave that has an intensity of 4 μW/m^2.

20. A vacuum cleaner has a measured sound level of 70 dB. What is the intensity of this sound in W/m^2?

21. (a) Calculate the intensity in W/m^2 of the wave described in Problem 14. (b) Express this intensity in dB.

22. The intensity of a sound wave at a fixed distance from a speaker vibrating at 1000 Hz is 0.6 W/m^2. (a) Determine the intensity if the frequency is increased to 2500 Hz while a *constant* displacement amplitude is maintained. (b) Calculate the intensity if the frequency is reduced to 500 Hz and the displacement amplitude is doubled.

23. Calculate the pressure amplitude corresponding to a sound intensity of 120 dB (a rock concert).

24. An explosive charge is detonated at a height of several kilometers in the atmosphere. At a distance of 400 m from the explosion the acoustic pressure reaches a maximum of 10 N/m^2. Assuming that the atmosphere is homogeneous over the distances considered, what will be the sound intensity level (in dB) at 4 km from the explosion? (Sound waves in air are absorbed at a rate of approximately 7 dB/km.)

25. The sound level 2 m from a source is measured to be 90 dB. (a) What is the sound level 4 m from the source? (b) How far away must one be to measure a sound level of 45 dB?

26. Two sources have sound levels of 75 dB and 80 dB. If they are sounding simultaneously (a) what is the combined sound level? (b) What is their combined intensity in W/m^2?

Section 17.4 Spherical and Plane Waves

27. An experiment requires a sound intensity of 1.2 W/m^2 at a distance of 4 m from a speaker. What power output is required?

28. A source emits sound waves with a uniform power of 100 W. At what distance will the intensity be just below the threshold of pain, which is 1 W/m^2?
29. The sound level at a distance of 3 m from a source is 120 dB. At what distance will the sound level be (a) 100 dB and (b) 10 dB?
30. Spherical waves of wavelength 25 cm are propagating outward from a point source. (a) Compare the wave amplitude at $r = 50$ cm and $r = 200$ cm. (b) Compare the intensity at $r = 50$ cm with the intensity at $r = 100$ cm. (c) Compare the amplitude of the wave function at a specific time at $r = 50$ cm and $r = 75$ cm.
31. A loudspeaker in a stereo system puts out 50 W maximum power. Assuming that the wave radiates from a point source, determine the sound levels and maximum harmonic displacement at a point 5 m from the loudspeaker for waves of frequency (a) 30 Hz, (b) 400 Hz, and (c) 20 000 Hz.
32. A spherical wave is radiating from a point source and is described by the following:

$$\psi(r, t) = \left[\frac{25.0}{r}\right]\sin(1.25r - 1870t)$$

where ψ is in Pa, r in m and t in s. (a) What is the maximum pressure amplitude 4 m from the source? (b) Determine the speed of the wave and hence the material the wave is in. (c) Find the intensity of the wave in dB at a distance 4 m from the source. (d) Find the instantaneous pressure 5 m from the source at 0.08 s.

°Section 17.5 The Doppler Effect

33. A bullet fired from a rifle travels at Mach 1.38 (that is, $v_S/v = 1.38$). What angle does the shock front make with the path of the bullet?
34. At what speed should a supersonic aircraft fly so that the conical wavefront will have an apex angle of 40°?
35. The Concorde flies at Mach 1.5. What is the angle between the direction of propagation of the shock wave and the direction of the plane's velocity?
36. A commuter train passes a passenger platform at a constant speed of 40 m/s. The train horn is sounded at its characteristic frequency of 320 Hz. (a) What change in frequency is observed by a person on the platform as the train passes? (b) What wavelength does a person on the platform observe as the train approaches?
37. Standing at a crosswalk, you hear a frequency of 560 Hz from the siren on an approaching police car. After the police car passes, the observed frequency of the siren is 480 Hz. Determine the car's speed from these observations.
38. A trumpet player traveling in an open convertible plays concert A (440 Hz) on his horn. A stationary observer directly ahead of the automobile hears the note A# (466.16 Hz). What is the speed of the automobile?
39. A train is moving parallel to a highway with a constant speed of 20 m/s. A car is traveling in the same direction as the train with a speed of 40 m/s. As the auto overtakes and passes the train, the car horn sounds at a frequency of 510 Hz and the train horn sounds at a frequency of 320 Hz. (a) What frequency does an occupant of the car observe for the train horn just before passing? (b) What frequency does a train passenger observe for the car horn just after passing?
40. A train passenger hears a frequency of 520 Hz as the train approaches a bell on a trackside safety gate; the bell is actually emitting a signal of 500 Hz. What frequency will the passenger hear just after passing the bell?
41. When high-energy, charged particles move through a transparent medium with a velocity greater than the velocity of light in that medium, a shock wave, or bow wave, of light is produced. This phenomenon is called the *Cerenkov effect* and can be observed in the vicinity of the core of a swimming pool reactor due to high-speed electrons moving through the water. In a particular case, the Cerenkov radiation produces a wavefront with a cone angle of 53°. Calculate the velocity of the electrons in the water. (Use 2.25×10^8 m/s as the velocity of light in water.)
42. A driver traveling northbound on a highway is driving at a speed of 25 m/s. A police car driving southbound at a speed of 40 m/s approaches with its siren sounding at a base frequency of 2500 Hz. (a) What frequency is observed by the driver as the police car approaches? (b) What frequency is detected by the driver after the police car passes him? (c) Repeat (a) and (b) for the case when the police car is traveling northbound.
43. A supersonic jet traveling at Mach 3 at an altitude of 20 000 m is directly overhead at time $t = 0$ as in Figure 17.14. (a) How long will it be before one en-

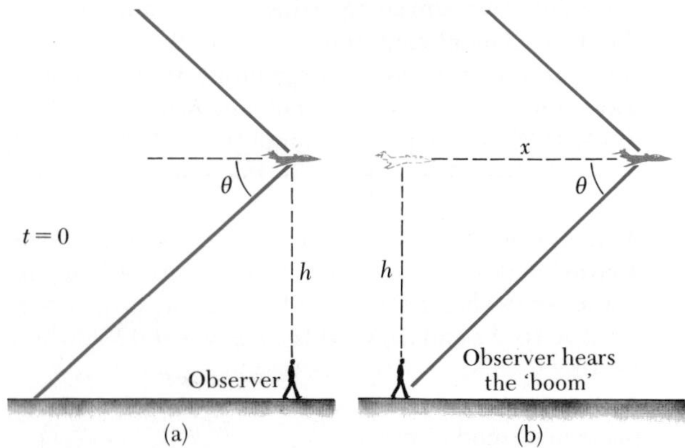

Figure 17.14 (Problem 43).

counters the shock wave? (b) Where will the plane be when it is finally heard? (Assume the speed of sound is uniform at 335 m/s.)

ADDITIONAL PROBLEMS

44. A copper rod is given a sharp compressional blow at one end. The sound of the blow, traveling through air at 0°C, reaches the opposite end of the rod 6.4 ms later than the sound transmitted through the rod. What is the length of the rod? (Refer to Table 17.1.)

45. An earthquake on the ocean floor in the Gulf of Alaska induces a *tsunami* (sometimes called a "tidal wave") that reaches Hilo, Hawaii, 4450 km distant, in a time of 9 h 30 min. Tsunamis have enormous wavelengths (100–200 km), and for such waves the propagation velocity is $v \approx \sqrt{g\bar{d}}$, where $\bar{d}$ is the average depth of the water. From the information given, find the average wave velocity and the average ocean depth between Alaska and Hawaii. (This method was used in 1856 to estimate the average depth of the Pacific Ocean long before soundings were made to give a direct determination.)

46. A sound wave in air at 0°C has an intensity equal to 10^{-6} W/cm^2 (a *loud* sound). (a) What is the amplitude of the acoustic pressure? (b) If the wavelength of the wave is 0.80 m, what is the displacement amplitude?

47. (a) The sound level of a jackhammer is measured as 130 dB and that of a siren as 120 dB. Find the ratio of the intensities of the two sound sources. (b) Two sources have measured intensities of $I_1 = 100\ \mu$W/m^2 and $I_2 = 200\ \mu$W/m^2. By how many dB is source 1 lower than source 2?

48. The measured speed of sound in copper is 3560 m/s, and the density of copper is 8890 kg/m^3. Based on this information, by what percent would you expect a block of copper to decrease in volume when subjected to a uniform external (gauge) pressure of 2 atm?

49. Two ships are moving along a line due east. The trailing vessel has a speed relative to a land-based observation point of 64 km/h, and the leading ship has a speed of 45 km/h relative to that station. The two ships are in a region of the ocean where the current is moving uniformly due west at 10 km/h. The trailing ship transmits a sonar signal at a frequency of 1200 Hz. What frequency is monitored by the leading ship? (Use 1520 m/s as the speed of sound in ocean water.)

50. Consider a longitudinal (compressional) wave of wavelength λ traveling with speed v along the x direction through a medium of density ρ. The *displacement* of the molecules of the medium from their equilibrium position is given by

$$s = s_m \sin(kx - \omega t)$$

Show that the pressure variation in the medium is given by

$$P = -\left(\frac{2\pi\rho v^2}{\lambda} s_m\right)\cos(kx - \omega t)$$

51. A meteoroid the size of a truck enters the earth's atmosphere at a speed of 20 km/s and is not significantly slowed before entering the ocean. (a) What is the Mach angle of the shock wave from the meteoroid in the atmosphere? (Use 331 m/s as the sound speed.) (b) Assuming that the meteoroid survives the impact with the ocean surface, what is the (initial) Mach angle of the shock wave that the meteoroid produces in the water? (Use the wave speed for sea water given in Table 17.1.)

52. (a) Use values from Table 17.2 to determine the resultant intensity in dB when a vacuum cleaner and a power mower are operated against a background of busy traffic. (b) In Table 17.2, a buzzing mosquito is rated at 40 dB and normal conversation at 55 dB. How many buzzing mosquitos are required to equal normal conversation in sound intensity?

53. By proper excitation, it is possible to produce both longitudinal and transverse waves in a long metal rod. A particular metal rod is 150 cm long and has a radius of 0.2 cm and a mass of 50.9 g. Young's modulus for the material is 6.8×10^{10} N/m^2. What must the tension in the rod be if the ratio of the speed of longitudinal waves to the speed of transverse waves is 8?

54. Use the information in Table 12.1 and calculate the velocities of the three different types of wave propagation that are possible in copper.

55. The gas filling the tube shown in Figure 17.2 is air at 20°C and at a pressure of 1.5×10^5 N/m^2. The piston shown is driven at a frequency of 600 Hz. The diameter of the piston is 10 cm, and the amplitude of its motion is 0.1 cm. What power must be supplied to maintain the oscillation of the piston?

56. Consider plane harmonic sound waves propagating in three different media at 0°C: air, water, and iron. Each wave has the same intensity (I_0) and the same angular frequency (ω_0). (a) Compare the values of λ (the wavelength) in the three media. (b) Compare the values of s_m (the displacement amplitude) in the three media. (*Hint:* Refer to Tables 15.1 and 17.1.) (c) Compare the values of ΔP_m (the pressure amplitude) in the three media. (d) For $\omega_0 = 2000\pi$ rad/s and $I_0 = 10^{-6}$ W/m^2 (60 dB), evaluate λ, s_m, and ΔP_m for each of the three media.

57. In order to be able to determine her speed, a skydiver carries a tone generator. A friend on the ground at the landing site has equipment for receiving and analyzing sound waves. While the skydiver is falling at terminal speed, her tone generator emits a steady tone of 1800 Hz. (Assume that the air is calm and that the sound speed is 343 m/s, independent of altitude.)

(a) If her friend on the ground (directly beneath the skydiver) receives waves of frequency 2150 Hz, what is the skydiver's speed of descent? (b) If the skydiver were also carrying sound-receiving equipment sensitive enough to detect waves reflected from the ground, what frequency would she receive?

58. A high-tech model airplane equipped with a sonar range and speed finder is headed straight for a brick wall at constant speed. At $t = 0$, it emits a short burst of waves of frequency f. At $t = T$, it receives the echo; the received frequency is f_r. Let v represent the speed of sound in air and let v_p represent the speed of the airplane. (a) Obtain an equation for v_p in terms of f, f_r, and v. (b) Let d represent the distance between the model and the wall at $t = 0$. Obtain an equation for d in terms of v, v_p, and T, and then use the result of part (a) to write d in terms of f, f_r, v_p, and T. (c) Use the results of parts (a) and (b) to find d_r, the distance between the model and the wall at $t = T$. Express d_r in terms of f, f_r, v, and T. (d) Evaluate v_p, d, and d_r for the following case: $v = 340$ m/s, $f = 5000$ Hz, $f_r = 5280$ Hz, and $T = 0.275$ s.

59. Three metal rods are located relative to each other as shown in Figure 17.15, where $L_1 + L_2 = L_3$. Values of density and Young's modulus for the three materials are $\rho_1 = 2.7 \times 10^3$ kg/m^3, $Y_1 = 7 \times 10^{10}$ N/m^2, $\rho_2 = 11.3 \times 10^3$ kg/m^3, $Y_2 = 1.6 \times 10^{10}$ N/m^2, $\rho_3 = 8.8 \times 10^3$ kg/m^3, and $Y_3 = 11 \times 10^{10}$ N/m^2. (a) If $L_3 = 1.5$ m, what must the ratio L_1/L_2 be if a sound wave is to travel the length of rods 1 *and* 2 in the same time for the wave to travel the length of rod 3? (b) If the frequency of the source is 4000 Hz, determine the phase difference between the wave traveling along rods 1 and 2 and the one traveling along rod 3.

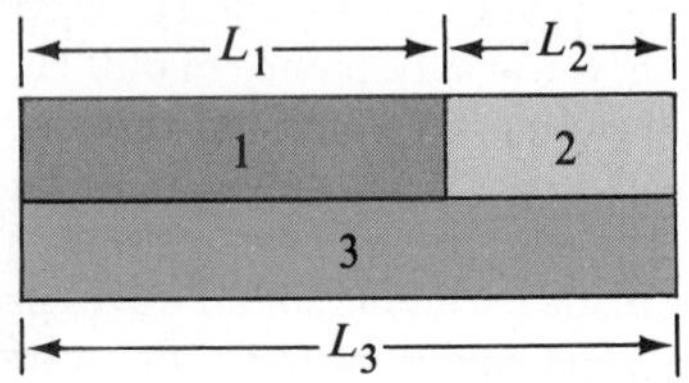

Figure 17.15 (Problem 59).

60. A bat, moving at 5 m/s, is chasing a flying insect. If the bat emits a 40-kHz chirp and receives back an echo at 40.4 kHz, at what velocity is the insect moving toward or away from the bat? (Take the velocity of sound in air to be $v = 340$ m/s.)

(Problem 60). (© Merlin Tuttle, Science Source/Photo Researchers)

61. A supersonic aircraft is flying parallel to the ground. When the aircraft is directly overhead, an observer sees a rocket fired from the aircraft. Ten seconds later the observer hears the sonic boom, followed 2.8 s later by the sound of the rocket engine. What is the Mach number of the aircraft?

62. The volume knob on a radio has what is known as a "logarithmic taper." The electrical device connected to the knob (called a potentiometer) has a resistance R whose logarithm is proportional to the angular position of the knob: that is, $\log R \propto \theta$. If the intensity of the sound I (in W/m^2) produced by the speaker is proportional to the resistance R, show that the intensity level β (in dB) is a linear function of θ.

ESSAY

The Physics of Woodwinds

Alma C. Zook
Pomona College

All wind instruments have the same basic structure: the mouthpiece, which is the initial source of sound waves, is joined to the tube or bore of the instrument, and with it determines the final pitch of the tone being played (Fig. 1). In addition, brass instruments and some woodwinds have a bell at the end of the bore that radiates the sound into the air.

All wind instruments except the flute use some form of reed in the mouthpiece (Fig. 2(a)). For brass instruments, the "reed" is the player's lips; for the woodwinds, the reed is one or more pieces of cane. The discussion that follows describes the behavior of the oboe reed, but it may be applied to any reed. The reed of an oboe is actually a double reed, as indicated in the side view in Fig. 2(b). Before the instrument can be played, the reed must be thoroughly moistened to make the cane flexible. When moist, the cane behaves like a spring with a relatively low spring constant.

Sound production begins when a player places the reed in his or her mouth and pushes air through it. As a result, the speed of the air between the blades of the reed is increased from zero to some value, which in turn causes the air pressure inside the reed to decrease. The air pressure in the player's mouth, however, is equal to or slightly greater than the ambient air pressure. The difference in pressure between the inside and outside of the reed forces the blades together, stopping the flow of air through the reed. Once the air flow is cut off, the air speed inside the reed drops to zero, and the air pressure inside rises to its original value. At this point the pressure difference between the inside and outside of the reed is zero, the reed springs back to its equilibrium (open) configuration, and the cycle repeats itself.

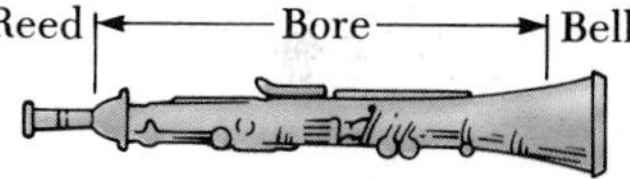

Figure 1 Woodwind structure.

The rapid opening and closing of the reed blades results in alternating compressions and rarefactions inside the bore of the instrument—that is, sound waves. But the vibration of the reed is irregular and not particularly pleasant to hear, because it contains many frequencies with no particular relationship to each other. Converting the sound wave from the vibrating reed into musical sound requires that most of the sound energy be filtered out, leaving only frequencies that are integer multiples of some lowest or *fundamental* frequency. (Such frequencies are referred to as "harmonically related.") That filtering process is carried out by the bore of the instrument.

Although many frequencies are present in the sound generated by the reed, the frequencies that are primarily responsible for the musical sound of the instrument are those that can set up standing waves inside the bore. (Standing waves will be discussed further in the next chapter.) Any other frequencies interfere destructively with themselves as they reflect off the two ends of the instrument. In this respect, the inside of a musical instrument is like a laser cavity, which also uses the occurrence of standing waves to remove all but one desired frequency. With a laser, of course, the wave is light rather than sound.

The particular frequencies that can set up standing waves in a wind instrument

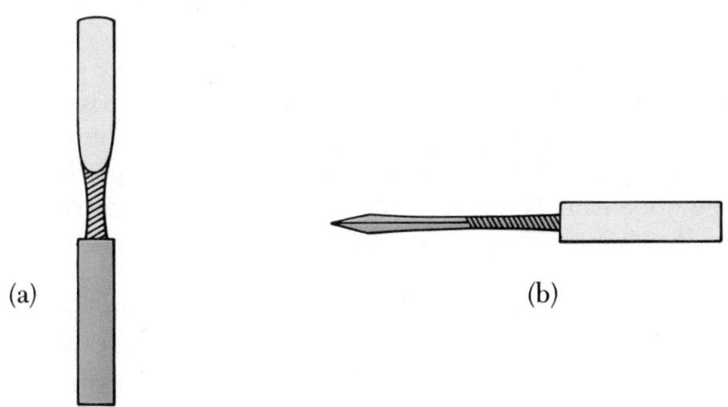

Figure 2 The oboe reed.

(Continued on next page)

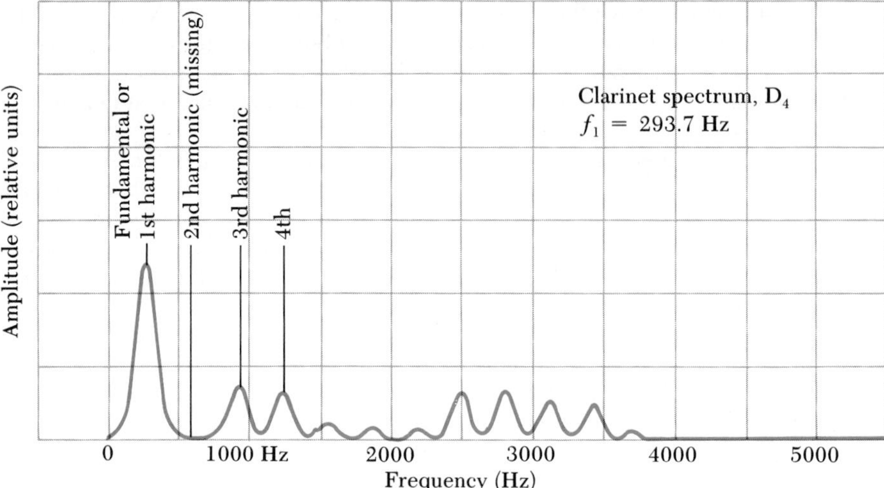

Figure 3 Clarinet spectrum.

depend on its geometry. The clarinet has a cylindrical bore for most of its length and behaves approximately like an air column that is closed on one end (the reed end) and open on the other. Its natural frequencies are therefore given by

$$f_n = (2n - 1)\frac{v}{4L}$$

where n is an integer, v is the speed of sound in air, and L is the distance from the reed to the first open tone hole. The lowest or fundamental frequency corresponding to $n = 1$ is

$$f_1 = \frac{v}{4L},$$

and all other natural frequencies are *odd* multiples of the fundamental frequency. We would therefore expect the sound from a clarinet to have only *odd* harmonics in its spectrum, and this is in fact approximately the case, at least in its lowest register (Fig. 3). The absence of the even harmonics in the spectrum of the clarinet gives its sound a "hollow" quality.

The oboe, on the other hand, has a conical bore, with the reed at the narrower end (which is quite narrow—about 5 mm). The conical bore causes the instrument to act acoustically like an air column that is open at both ends, with natural frequencies

$$f_n = n\frac{v}{2L}$$

where n is an integer. So the oboe differs acoustically from the clarinet in two important respects. Although the oboe and clarinet are about the same length, the oboe's lowest possible note, corresponding to $n = 1$, has a fundamental frequency about twice that of the clarinet. Therefore, the clarinet can play pitches almost an octave *lower* than the oboe can. The oboe makes up for its more restricted pitch range by having a richer harmonic spectrum, containing all integer-multiple frequencies of the fundamental, and not just the odd multiples (Fig. 4).

Clearly, real musical instruments are not just simple cylinders and cones: the bell and tone holes introduce more complicated geometry that in turn leads to the particular balance of the harmonics we associate with a given instrument. In addition, neither the reed nor the air in the bore is a perfectly linear system, which causes the spectrum of an instrument to depend on the loudness at which it is played. Neverthe-

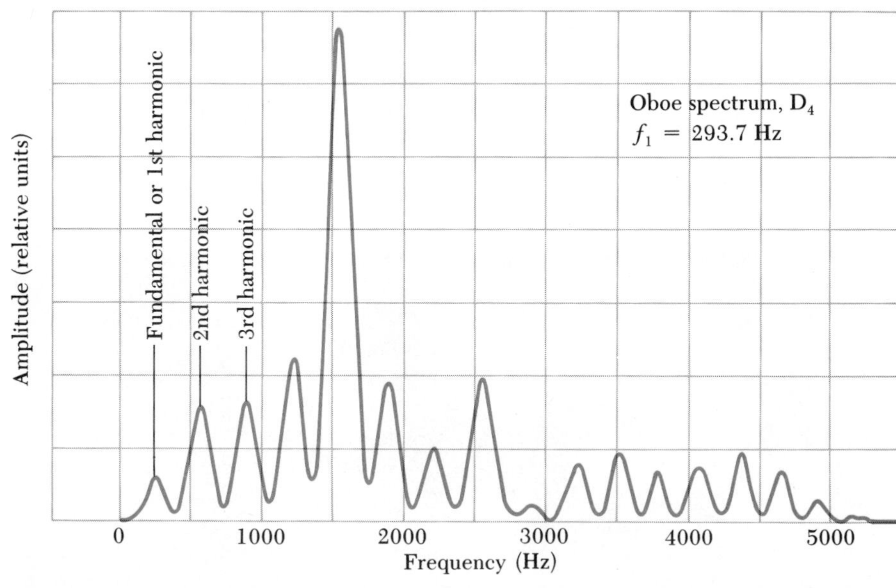

Figure 4 Oboe spectrum.

less, much of the behavior of any wind instrument can be explained using familiar physics.

Suggested Readings

Backus, J. *The Acoustical Foundations of Music,* New York, Norton, 1977.

Benade, A.H., "The Physics of Woodwinds," *Scientific American,* October 1960.

Benade, A.H., "The Physics of Brasses," *Scientific American,* July 1973.

Fletcher, N.H., and S. Thwaites, "The Physics of Organ Pipes," *Scientific American,* January 1983.

Hall, D.E., *Musical Acoustics: An Introduction,* Belmont, CA, Wadsworth, 1980.

Essay Problems

1. The speed of sound changes with temperature. This temperature variation is described approximately by the equation

$$v(T) = 331.5 + 0.607\ T$$

where v is in m/s and T is the Celsius temperature. Therefore, blowing through an instrument raises the temperature of the air inside it, unless the room is very warm. For a certain fingering, a clarinet has an effective length of 24.6 cm. This fingering sounds the pitch F_4 ($f = 349.2$ Hz) at 20°C. During a concert, the air temperature inside the bore rises to 27°C. (a) What new frequency will the clarinet sound for this fingering? (b) What is the percentage of change in the frequency? (Note that a 1% change in frequency would be readily noticeable to a listener.)

2. What is the frequency of the third available harmonic in a closed organ pipe with a length of 2.5 m? (Assume $T = 20$°C.)

3. The English horn is badly named, since it is neither English nor a horn. It is essentially a larger version of the oboe, differing primarily in the shape of the bell, which does not flare the way the bell on the oboe does. For a given finger position, the English horn plays a note that is lower in frequency by a factor of 2/3 compared with the note on the oboe with the same fingering. If the lowest note on an oboe is B_3 ($f = 246.9$ Hz) and the oboe is 66 cm long, how long must the English horn be? The true length of an English horn is 92 cm. Can you account for the discrepancy?

18
Superposition and Standing Waves

Even when silent, this organ at the Mormon Tabernacle conveys a sense of the power of sound waves. (Courtesy of Henry Leap)

An important aspect of waves is the combined effect of two or more waves traveling in the same medium. For instance, what happens to a string when a wave traveling toward its fixed end is reflected back on itself? What is the pressure variation in the air when the instruments of an orchestra sound together?

In a linear medium, that is, one in which the restoring force of the medium is proportional to the displacement of the medium, one can apply the *principle of superposition* to obtain the resultant disturbance. This principle can be applied to many types of waves, including waves on strings, sound waves, surface water waves, and electromagnetic waves. The superposition principle states that the actual displacement of any part of the disturbed medium equals the vector sum of the displacements caused by the individual waves. We discussed this principle in Chapter 16 as it applies to wave pulses. The term *interference* was used to describe the effect produced by combining two waves moving simultaneously through a medium.

This chapter is concerned with the superposition principle as it applies to harmonic waves. If the harmonic waves that combine in a given medium have the same frequency and wavelength, one finds that a stationary pattern, called a *standing wave*, can be produced at certain frequencies under certain circumstances. For example, a stretched string fixed at both ends has a discrete set of oscillation patterns, called *modes of vibration*, which depend upon the tension and mass per unit length of the string. These modes of vibration are found in stringed musical instruments. Other musical instruments, such as the organ and flute, make use of the natural frequencies of sound waves in hollow pipes. Such frequencies depend upon the length of the pipe, its shape, and upon whether one end is open or closed.

We also consider the superposition and interference of waves with different frequencies and wavelengths. When two sound waves with nearly the same frequency interfere, one hears variations in the loudness called *beats*. The beat frequency corresponds to the rate of alternation between constructive and destructive interference. Finally, we describe how any complex periodic waveform can, in general, be described by a sum of sine and cosine functions.

18.1 SUPERPOSITION AND INTERFERENCE OF HARMONIC WAVES

The superposition principle tells us that when two or more waves move in the same linear medium, the net displacement of the medium (the resultant wave) at any point equals the algebraic sum of the displacements of all the waves. Let us apply this superposition principle to two harmonic waves traveling in the same direction in a medium. If the two waves are traveling to the right and have the same frequency, wavelength, and amplitude but differ in phase, we can express their individual wave functions as

$$y_1 = A_0 \sin(kx - \omega t) \qquad \text{and} \qquad y_2 = A_0 \sin(kx - \omega t - \phi)$$

Hence, the resultant wave function y is given by

$$y = y_1 + y_2 = A_0\,[\sin(kx - \omega t) + \sin(kx - \omega t - \phi)]$$

In order to simplify this expression, it is convenient to make use of the following trigonometric identity:

$$\sin a + \sin b = 2\cos\left(\frac{a-b}{2}\right)\sin\left(\frac{a+b}{2}\right)$$

If we let $a = kx - \omega t$ and $b = kx - \omega t - \phi$, we find that the resultant wave y reduces to

$$y = \left(2A_0 \cos\frac{\phi}{2}\right)\sin\left(kx - \omega t - \frac{\phi}{2}\right) \tag{18.1}$$

Resultant of two traveling harmonic waves

There are several important features of this result. The resultant wave function y is also harmonic and has the *same* frequency and wavelength as the individual waves. The amplitude of the resultant wave is $2A_0 \cos(\phi/2)$, and its phase is equal to $\phi/2$. If the phase constant ϕ equals 0, then $\cos(\phi/2) = \cos 0 = 1$ and the amplitude of the resultant wave is $2A_0$. In other words, the amplitude of the resultant wave is twice as large as the amplitude of either individual wave. In this case, the waves are said to be everywhere *in phase* and thus *interfere constructively*. That is, the crests and troughs of the individual

Constructive interference

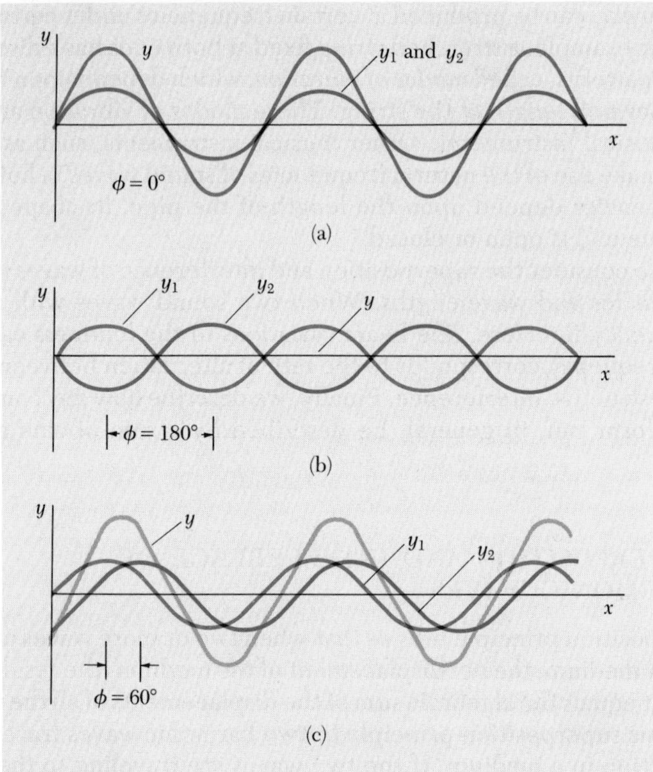

Figure 18.1 The superposition of two waves with amplitudes y_1 and y_2. (a) When the two waves are in phase, the result is constructive interference. (b) When the two waves are 180° out of phase, the result is destructive interference. (c) When the phase angle lies in the range $0 < \phi < 180°$, the resultant y falls somewhere between that shown in (a) and that shown in (b).

waves occur at the same positions, as is shown by the broken lines in Figure 18.1a. In general, constructive interference occurs when $\cos(\phi/2) = \pm 1$, or when $\phi = 0, 2\pi, 4\pi, \ldots$. On the other hand, if ϕ is equal to π radians (or any *odd* multiple of π) then $\cos(\phi/2) = \cos(\pi/2) = 0$ and the resultant wave has *zero* amplitude everywhere. In this case, the two waves *interfere destructively*. That is, the crest of one wave coincides with the trough of the second (Fig. 18.1b) and their displacements cancel at every point. Finally, when the phase constant has an arbitrary value between 0 and π, as in Figure 18.1c, the resultant wave has an amplitude whose value is somewhere between 0 and $2A_0$.

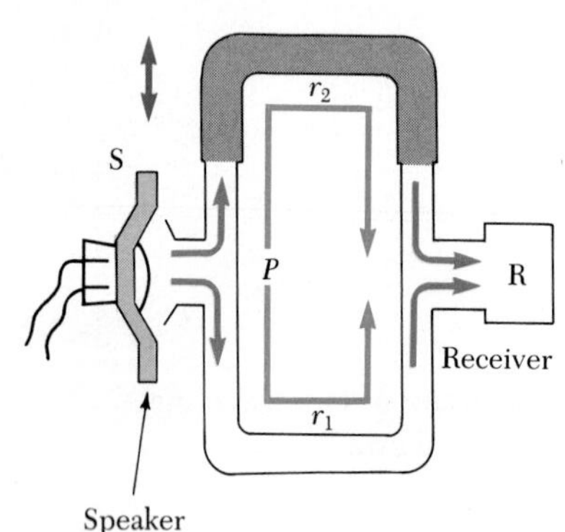

Figure 18.2 An acoustical system for demonstrating interference of sound waves. Sound from the speaker propagates into a tube and splits into two parts at *P*. The two waves, which superimpose at the opposite side, are detected at R. Note that the upper path length, r_2, can be varied by the sliding section.

Interference of Sound Waves

One simple device for demonstrating interference of sound waves is illustrated in Figure 18.2. Sound from a loudspeaker S is sent into a tube at *P*, where there is a T-shaped junction. Half the sound power travels in one direction and half in the opposite direction. Thus, the sound waves that reach the receiver R at the other side can travel along two different paths. The receiver may be a microphone whose output is amplified and fed into earphones or an oscilloscope. The total distance from the speaker to the receiver is called the *path length, r*. The path length for the lower path is fixed at r_1. Along the upper path, the path length r_2 can be varied by sliding the U-shaped tube, similar to that on a slide trombone. When the difference in the path lengths $\Delta r = |r_2 - r_1|$ is either zero or some integral multiple of the wavelength λ, the two waves reaching the receiver are in phase and interfere constructively, as in Figure 18.1a. For this case, a maximum in the sound intensity will be detected at the receiver. If the path length r_2 is adjusted such that the path difference Δr is $\lambda/2$, $3\lambda/2$, . . . , $n\lambda/2$ (for n odd), the two waves are exactly 180° out of phase at the receiver and hence cancel each other. In this case of completely destructive interference, no sound is detected at the receiver. This simple experiment is a striking illustration of the phenomenon of interference. In addition, it demonstrates the fact that a phase difference may arise between two waves generated by the same source when they travel along paths of unequal lengths.

It is often useful to express the path difference in terms of the phase difference ϕ between the two waves. Since a path difference of one wavelength corresponds to a phase difference of 2π radians, we obtain the ratio $\lambda/2\pi = \Delta r/\phi$, or

$$\Delta r = \frac{\lambda}{2\pi}\phi \tag{18.2}$$

Relationship between path difference and phase angle

There are many other examples of interference phenomena in nature. Later, in Chapter 37, we shall describe several interesting interference effects involving light waves.

EXAMPLE 18.1 Two Speakers Driven by the Same Source

A pair of speakers separated by 3 m are driven by the same oscillator as in Figure 18.3. The listener is originally at point O located 8 m from the center line. The listener walks *perpendicular* to the center line a distance of 0.35 m before reaching the *first minimum* in sound intensity. What is the frequency of the oscillator?

Solution The first minimum occurs when the two waves reaching the listener at P are 180° out of phase, or when their path difference equals $\lambda/2$. In order to calculate the path difference, we must first find the path lengths r_1 and r_2. Making use of the two shaded triangles in Figure 18.3, and the given numerical distances, we find the path lengths to be

$$r_1 = \sqrt{(8\text{ m})^2 + (1.15\text{ m})^2} = 8.082\text{ m}$$

$$r_2 = \sqrt{(8\text{ m})^2 + (1.85\text{ m})^2} = 8.211\text{ m}$$

Hence, the path difference is $r_2 - r_1 = 0.129$ m. Since we require that this path difference be equal to $\lambda/2$ for the first minimum, we find that $\lambda = 0.258$ m. To obtain the oscillator frequency, we can use $v = \lambda f$, where v is the speed of sound in air, 344 m/s. This gives

$$f = \frac{v}{\lambda} = \frac{344\text{ m/s}}{0.258\text{ m}} = 1330\text{ Hz}$$

Exercise 1 If the frequency of the oscillator is adjusted such that the listener hears the first minimum at a distance of 0.75 m from 0, what is the new frequency of the oscillator?

Answer 625 Hz.

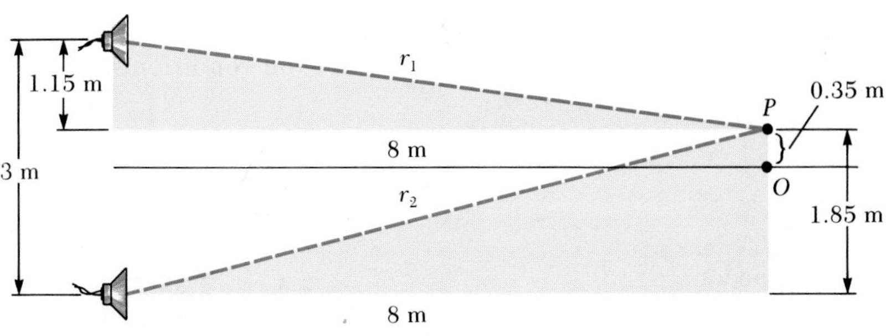

Figure 18.3 (Example 18.1).

18.2 STANDING WAVES

If a stretched string is clamped at both ends, traveling waves are reflected from the fixed ends, creating waves traveling in both directions. The incident and reflected waves combine according to the superposition principle.

Consider two sinusoidal waves in the same medium with the same amplitude, frequency, and wavelength, but traveling in *opposite* directions. Their wave functions can be written

$$y_1 = A_0 \sin(kx - \omega t) \qquad \text{and} \qquad y_2 = A_0 \sin(kx + \omega t)$$

where y_1 represents a wave traveling to the right and y_2 represents a wave traveling to the left. Adding these two functions gives the resultant wave function y:

$$y = y_1 + y_2 = A_0 \sin(kx - \omega t) + A_0 \sin(kx + \omega t)$$

where $k = 2\pi/\lambda$ and $\omega = 2\pi f$, as usual. Using the trigonometric identity $\sin(a \pm b) = \sin a \cos b \pm \cos a \sin b$, this reduces to

Wave function for a standing wave

$$y = (2A_0 \sin kx) \cos \omega t \tag{18.3}$$

This expression represents the wave function of a **standing wave.** From this result, we see that a standing wave has an angular frequency ω and an amplitude given by $2A_0 \sin kx$ (the quantity in the parentheses of Eq. 18.3). That is, every particle of the string vibrates in simple harmonic motion with the same frequency. However, the amplitude of motion of a given particle depends on x. This is in contrast to the situation involving a traveling harmonic wave, in which all particles oscillate with both the same amplitude and the same frequency.

Because the amplitude of the standing wave at any value of x is equal to $2A_0 \sin kx$, we see that the *maximum* amplitude has the value $2A_0$. This occurs when the coordinate x satisfies the condition $\sin kx = 1$, or when

$$kx = \frac{\pi}{2}, \frac{3\pi}{2}, \frac{5\pi}{2}, \ldots$$

Since $k = 2\pi/\lambda$, the positions of maximum amplitude, called **antinodes,** are given by

Position of antinodes

$$x = \frac{\lambda}{4}, \frac{3\lambda}{4}, \frac{5\lambda}{4}, \ldots = \frac{n\lambda}{4} \tag{18.4}$$

where $n = 1, 3, 5, \ldots$. Note that *adjacent antinodes are separated by a distance of $\lambda/2$*. Similarly, the standing wave has a *minimum* amplitude of zero when x satisfies the condition $\sin kx = 0$, or when

$$kx = \pi, 2\pi, 3\pi, \ldots$$

giving

Position of nodes

$$x = \frac{\lambda}{2}, \lambda, \frac{3\lambda}{2}, \ldots = \frac{n\lambda}{2} \tag{18.5}$$

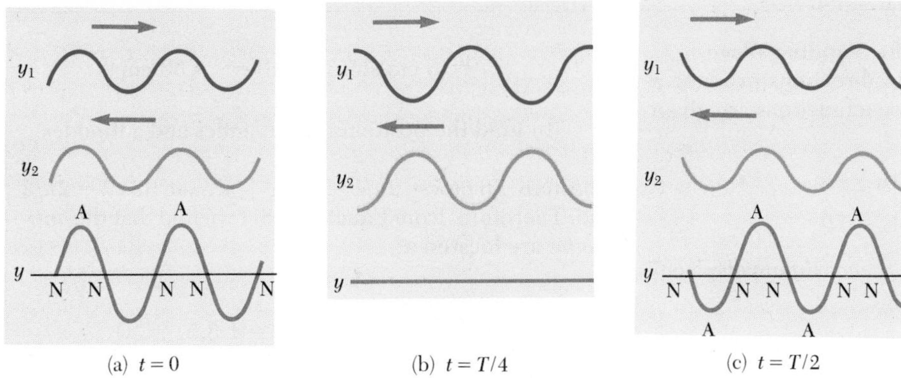

Figure 18.4 Standing wave patterns at various times produced by two waves of equal amplitude traveling in *opposite* directions. For the resultant wave y, the nodes (N) are points of zero displacement, and the antinodes (A) are points of maximum amplitude.

where $n = 1, 2, 3, \ldots$. These points of zero amplitude, called **nodes,** *are also spaced by* $\lambda/2$. The distance between a node and an adjacent antinode is $\lambda/4$.

A graphical description of the standing wave patterns produced at various times by two waves traveling in opposite directions is shown in Figure 18.4. The upper part of each figure represents the individual traveling waves, and the lower part represents the standing wave patterns. The nodes of the standing wave are labeled N, and the antinodes are labeled A. At $t = 0$ (Fig. 18.4a), the two waves are identical spatially, giving a standing wave of maximum amplitude, $2A_0$. One quarter of a period later, at $t = T/4$ (Fig. 18.4b), the individual waves have moved one quarter of a wavelength (one to the right and the other to the left). At this time, the individual displacements are equal and opposite for all values of x, and hence the resultant wave has zero displacement everywhere. At $t = T/2$ (Fig. 18.4c), the individual waves are again identical spatially, producing a standing wave pattern that is inverted relative to the $t = 0$ pattern.

It is instructive to describe the energy associated with the motion of a standing wave. To illustrate this point, consider a standing wave formed on a stretched string fixed at each end, as in Figure 18.5. All points on the string oscillate vertically with the same frequency except for the nodes, which are stationary. Furthermore, the various points have different amplitudes of motion. Figure 18.5 represents snapshots of the standing wave at various times over one half of a cycle. Since the nodal points are stationary, no energy is transmitted along the string across the center nodal point. For this reason, standing waves are often called **stationary waves.** Each point on the string executes simple harmonic motion in the vertical direction. That is, one can view the standing wave as a large number of oscillators vibrating parallel to each other. The energy of the vibrating string continuously alternates between elastic potential energy, at which time the string is momentarily stationary (Fig. 18.5a), and kinetic energy, at which time the string is horizontal and the particles have their maximum speed (Fig. 18.5c). The string particles have both potential energy and kinetic energy at intermediate times (Figs. 18.5b and 18.5d).

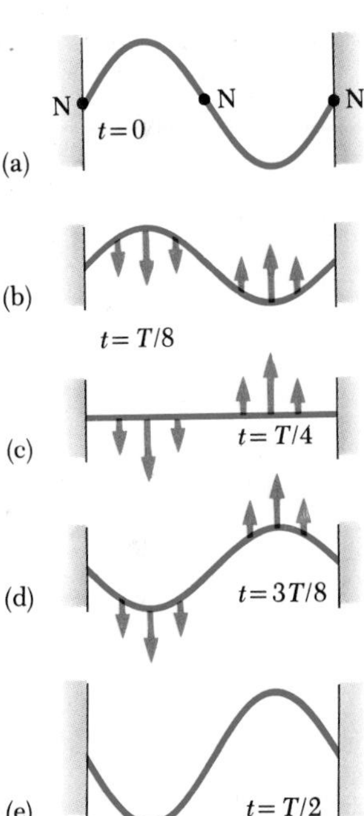

Figure 18.5 A standing wave pattern in a stretched string showing snapshots during one half of a cycle. (a) At $t = 0$, the string is momentarily at rest, and so $K = 0$ and all of the energy is potential energy U associated with the vertical displacements of the string segments (deformation energy). (b) At $t = T/8$, the string is in motion, and the energy is half kinetic and half potential. (c) At $t = T/4$, the string is horizontal (undeformed) and therefore $U = 0$; all of the energy is kinetic. The motion continues as indicated, and ultimately the initial configuration (a) is repeated.

EXAMPLE 18.2 Formation of a Standing Wave □
Two waves traveling in opposite directions produce a standing wave. The individual wave functions are given by

$$y_1 = (4 \text{ cm}) \sin(3x - 2t)$$

$$y_2 = (4 \text{ cm}) \sin(3x + 2t)$$

where x and y are in cm. (a) Find the maximum displacement of the motion at $x = 2.3$ cm.

Solution When the two waves are summed, the result is a standing wave whose function is given by Equation 18.3, with $A_0 = 4$ cm and $k = 3\text{ cm}^{-1}$:

$$y = (2A_0 \sin kx) \cos \omega t = [(8 \text{ cm}) \sin 3x] \cos \omega t$$

Thus, the *maximum* displacement of the motion at the position $x = 2.3$ cm is given by

$$y_{\text{max}} = (8 \text{ cm})\sin 3x]_{x=2.3}$$

$$= (8 \text{ cm})\sin(6.9 \text{ rad}) = \boxed{4.63 \text{ cm}}$$

(b) Find the positions of the nodes and antinodes.

Solution Since $k = 2\pi/\lambda = 3\text{ cm}^{-1}$, we see that $\lambda = 2\pi/3$ cm. Therefore, from Equation 18.4 we find that the *antinodes* are located at

$$x = n\left(\frac{\pi}{6}\right) \text{ cm} \qquad (n = 1,3,5, \ldots)$$

and from Equation 18.5 we find that the *nodes* are located at

$$x = n\frac{\lambda}{2} = n\left(\frac{\pi}{3}\right) \text{ cm} \qquad (n = 1,2,3, \ldots)$$

18.3 STANDING WAVES IN A STRING FIXED AT BOTH ENDS

Consider a string of length L that is fixed at both ends, as in Figure 18.6. Standing waves are set up in the string by a continuous superposition of waves incident on and reflected from the ends. The string has a number of natural patterns of vibration, called **normal modes.** Each of these has a characteristic frequency; the frequencies are easily calculated.

First, note that the ends of the string must be nodes since these points are *fixed.* If the string is displaced at its midpoint and released, the vibration shown in Figure 18.6b is produced, in which the center of the string is an

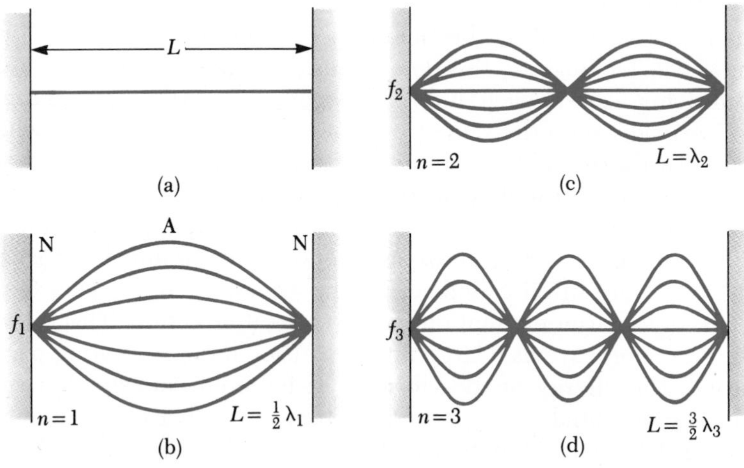

Figure 18.6 (a) Standing waves in a string of length L fixed at both ends. The normal modes of vibration shown as multiple exposures form a harmonic series; (b) the fundamental frequency, or first harmonic; (c) the second harmonic; and (d) the third harmonic.

antinode. For this normal mode, the length of the string equals $\lambda/2$ (the distance between nodes):

$$L = \lambda_1/2 \qquad \text{or} \qquad \lambda_1 = 2L$$

The next normal mode, of wavelength λ_2 (Fig. 18.6c), occurs when the length of the string equals one wavelength, that is, when $\lambda_2 = L$. The third normal mode (Fig. 18.6d) corresponds to the case where the length equals $3\lambda/2$; therefore, $\lambda_3 = 2L/3$. In general, the wavelengths of the various normal modes can be conveniently expressed as

$$\lambda_n = \frac{2L}{n} \qquad (n = 1, 2, 3, \ldots) \tag{18.6}$$

Wavelengths of normal modes

where the index n refers to the nth mode of vibration. The natural frequencies associated with these modes are obtained from the relationship $f = v/\lambda$, where the *wave speed v is the same for all frequencies.* Using Equation 18.6, we find that the frequencies of the normal modes are given by

$$f_n = \frac{v}{\lambda_n} = \frac{n}{2L}v \qquad (n = 1, 2, 3, \ldots) \tag{18.7}$$

Frequencies of normal modes

Because $v = \sqrt{F/\mu}$, where F is the tension in the string and μ is its mass per unit length, we can also express the natural frequencies of a stretched string as[1]

$$f_n = \frac{n}{2L}\sqrt{\frac{F}{\mu}} \qquad (n = 1, 2, 3, \ldots) \tag{18.8}$$

Normal modes of a stretched string fixed at both ends

The lowest frequency, corresponding to $n = 1$, is called the *fundamental* or the **fundamental frequency,** f_1, and is given by

$$f_1 = \frac{1}{2L}\sqrt{\frac{F}{\mu}} \tag{18.9}$$

Fundamental frequency of a stretched string

Clearly, the frequencies of the remaining modes (sometimes called *harmonics*) are integral multiples of the fundamental frequency, that is, $2f_1$, $3f_1$, $4f_1$, and so on. These higher natural frequencies, together with the fundamental frequency, are seen to form a **harmonic series.** The fundamental, f_1, is the first harmonic; the frequency $f_2 = 2f_1$ is the second harmonic; the frequency f_n is the nth harmonic.

We can obtain the above results in an alternative manner. Since we require that the string be fixed at $x = 0$ and $x = L$, the wave function $y(x, t)$ given by Equation 18.3 must be *zero* at these points for *all* times. That is, the boundary conditions require that $y(0, t) = 0$ and $y(L, t) = 0$ for all values of t. Since $y = (2A_0 \sin kx) \cos \omega t$, the first condition, $y(0, t) = 0$, is automatically satisfied because $\sin kx = 0$ at $x = 0$. To meet the second condition, $y(L, t) = 0$, we require that $\sin kL = 0$. This condition is satisfied when the angle kL equals an integral multiple of π (180°). Therefore, the allowed values of k are[2]

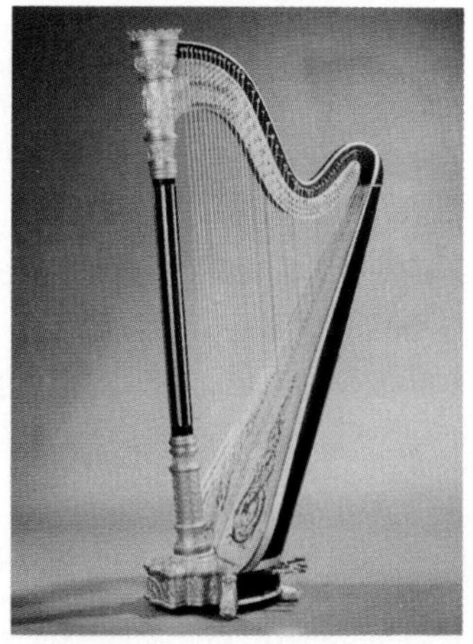

Concert style harp. (Courtesy of Lyon & Healy Harps, Chicago)

[1] The laws governing the sound produced by a vibrating string were first published in 1636 by a Franciscan friar, Pére Mersenne, in a treatise entitled "Harmonie Universelle."

[2] We exclude $n = 0$ since this corresponds to the trivial case where no wave exists ($k = 0$).

$$k_nL = n\pi \qquad (n = 1, 2, 3, \ldots) \tag{18.10}$$

Since $k_n = 2\pi/\lambda_n$, we find that

$$(2\pi/\lambda_n)L = n\pi \qquad \text{or} \qquad \lambda_n = \frac{2L}{n}$$

which is identical to Equation 18.6.

When a stretched string is distorted such that its initial shape corresponds to any one of its harmonics, after being released it will vibrate at the frequency of that harmonic. However, if the string is struck or bowed, the resulting vibration will include frequencies of various harmonics, including the fundamental. In effect, the string "selects" the normal-mode frequencies when disturbed by a nonharmonic disturbance (which happens, for example, when a guitar string is plucked).

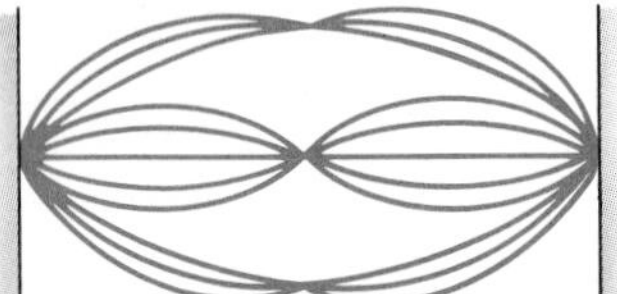

Figure 18.7 Multiple exposures of a stretched string vibrating simultaneously in its first and second harmonics.

Figure 18.7 shows a stretched string vibrating with its first and second harmonics simultaneously. In this figure, the combined vibration is the superposition of the two vibrations shown in Figures 18.6b and 18.6c. The large loop corresponds to the fundamental frequency of vibration, f_1, and the smaller loops correspond to the second harmonic, f_2. In general, the resulting motion, or displacement, can be described by a superposition of the various harmonic wave functions, with different frequencies and amplitudes. Hence, the sound that one hears corresponds to a complex waveform associated with these various modes of vibration. We shall return to this point in Section 18.8.

The frequency of a stringed instrument can be changed either by varying the tension F or by changing the length L. For example, the tension in the strings of guitars and violins is varied by a screw adjustment mechanism or by turning pegs located on the neck of the instrument. As the tension is increased, the frequency of the normal modes increases according to Equation 18.8. Once the instrument is "tuned," the player varies the frequency by moving his or her fingers along the neck, thereby changing the length of the vibrating portion of the string. As the length is shortened, the frequency increases, since the normal-mode frequencies are inversely proportional to string length.

EXAMPLE 18.3 Give Me a C Note

A middle C string of the C-major scale on a piano has a fundamental frequency of 264 Hz, and the A note has a fundamental frequency of 440 Hz. (a) Calculate the frequencies of the next two harmonics of the C string.

Solution Since $f_1 = 264$ Hz, we can use Equations 18.8 and 18.9 to find the frequencies f_2 and f_3:

$$f_2 = 2f_1 = 528 \text{ Hz}$$

$$f_3 = 3f_1 = 792 \text{ Hz}$$

(b) If the two piano strings for the A and C notes are assumed to have the same mass per unit length and the same length, determine the ratio of tensions in the two strings.

Solution Using Equation 18.8 for the two strings vibrating at their fundamental frequencies gives

$$f_{1A} = \frac{1}{2L}\sqrt{F_A/\mu} \qquad \text{and} \qquad f_{1C} = \frac{1}{2L}\sqrt{F_C/\mu}$$

$$f_{1A}/f_{1C} = \sqrt{F_A/F_C}$$

$$F_A/F_C = (f_{1A}/f_{1C})^2 = (440/264)^2 = 2.78$$

(c) While the string densities are, in fact, equal, the A string is 64% as long as the C string. What is the ratio of their tensions?

$$f_{1A}/f_{1C} = (L_C/L_A)\sqrt{F_A/F_C} = (100/64)\sqrt{F_A/F_C}$$

$$F_A/F_C = (0.64)^2(440/264)^2 = 1.14$$

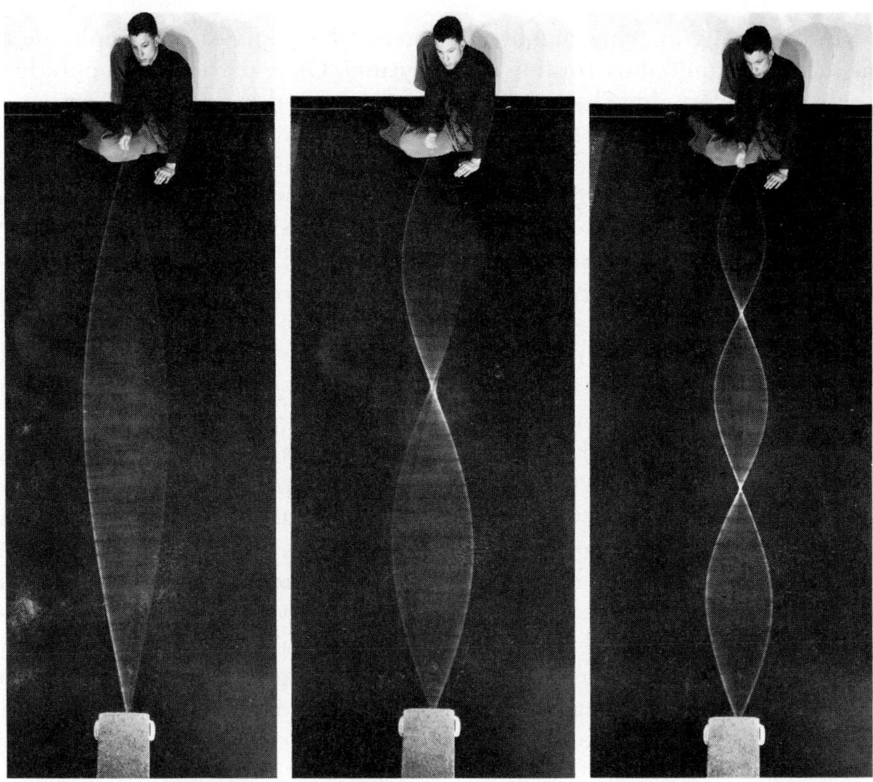

Photographs of standing waves. As one end of the tube is moved from side to side with increasing frequency, patterns with more and more loops are formed; only certain definite frequencies will produce fixed patterns. (Photos, Education Development Center, Newton, Mass.)

18.4 RESONANCE

We have seen that a system such as a stretched string is capable of oscillating in one or more natural modes of vibration. *If a periodic force is applied to such a system, the resulting amplitude of motion of the system will be larger when the frequency of the applied force is equal or nearly equal to one of the natural frequencies of the system* than when the driving force is applied at some other frequency. We have already discussed this phenomenon, known as *resonance*, for mechanical systems.

The corresponding natural frequencies of oscillation of the system are often referred to as **resonant frequencies.** The resonance phenomenon is of great importance in the production of musical sounds. At the atomic level, the electrons and nuclei of atoms and molecules exhibit resonant behavior when exposed to certain frequencies of electromagnetic radiation and applied magnetic fields.

Whenever a system capable of oscillating is driven by a periodic force of constant small amplitude, the resulting amplitude of motion will be large only when the frequency of the driving force is nearly equal to one of the resonant frequencies of the system. Figure 18.8 shows the response of a system to various frequencies, where the peak of the curve represents the resonant frequency, f_0. Note that the amplitude is largest when the frequency of the driving force equals the resonant frequency. When the frequency of the driv-

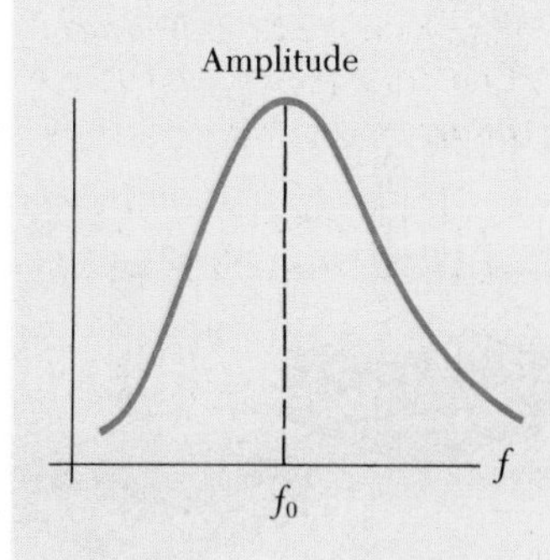

Figure 18.8 The amplitude (response) versus driving frequency for an oscillating system. The amplitude is a maximum at the resonance frequency, f_0.

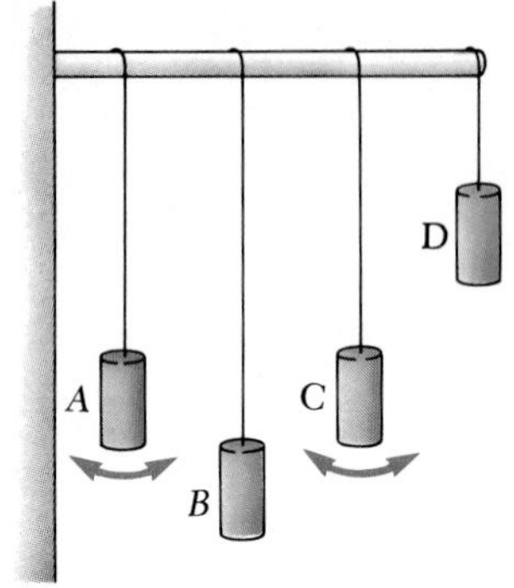

Figure 18.9 If pendulum A is set into oscillation, only pendulum C, whose length is close to the length of A, will eventually oscillate with large amplitude, or resonate.

ing force exactly matches one of the resonant frequencies, the amplitude of the motion is limited by friction in the system. Once maximum amplitude is reached, the work done by the periodic force is used to overcome friction. A system is said to be *weakly damped* when the amount of friction is small. Such a system undergoes a large amplitude of motion when driven at one of its resonant frequencies. The oscillations in such a system will persist for a long time after the driving force is removed. On the other hand, a system with considerable friction, that is, one that is *strongly damped,* will undergo small amplitude oscillations that decrease rapidly with time once the driving force is removed.

Examples of Resonance

A playground swing is a pendulum with a natural frequency that depends on its length. Whenever we push a child in a swing with a series of regular impulses, the swing will go higher if the frequency of the periodic force equals the natural frequency of the swing. One can demonstrate a similar effect by suspending several pendula of different lengths from a horizontal support, as in Figure 18.9. If pendulum A is set into oscillation, the other pendula will soon begin to oscillate as a result of the longitudinal waves transmitted along the beam. However, you will find that a pendulum, such as C, whose length is close to the length of A oscillated with a much larger amplitude than those such as B and D whose lengths are much different from the length of A. This is because the natural frequency of C is nearly the same as the driving frequency associated with A.

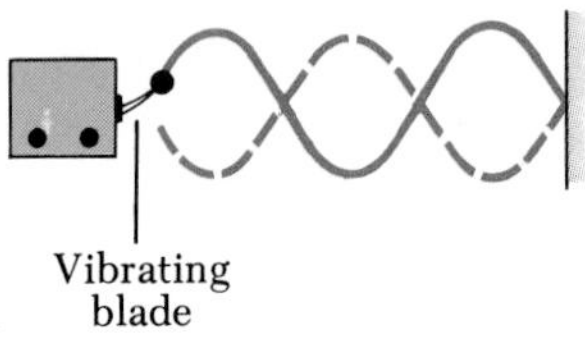

Figure 18.10 Standing waves are set up in a stretched string having one end connected to a vibrating blade when the natural frequencies of the string are nearly the same as those of the vibrating blade.

Next, consider a stretched string fixed at one end and connected at the opposite end to a vibrating blade as in Figure 18.10. As the blade oscillates, transverse waves sent down the string are reflected from the fixed end. As we found in Section 18.3, the string has natural frequencies of vibration that are determined by its length, tension, and mass per unit length (Eq. 18.8). When the frequency of the vibrating blade equals one of the natural frequencies of the string, standing waves are produced and the string vibrates with a large amplitude. In this case, the wave being generated by the vibrating blade is *in phase* with the wave that has been reflected at the fixed end, and so the string absorbs energy from the blade at resonance. Once the amplitude of the standing-wave oscillations reaches a maximum, the energy delivered by the blade and absorbed by the system is lost because of the damping forces. The fixed end is a node, and the point P, which is near the end connected to the vibrating blade, is very nearly a node, since the amplitude of the blade's motion is small compared with that of the string. It is also interesting to note that if the applied frequency differs from one of the natural frequencies, energy is first transferred to the string from the blade. However, later the phase of the wave in the string becomes such that it forces the blade to receive energy from the string, thereby reducing the energy in the string.

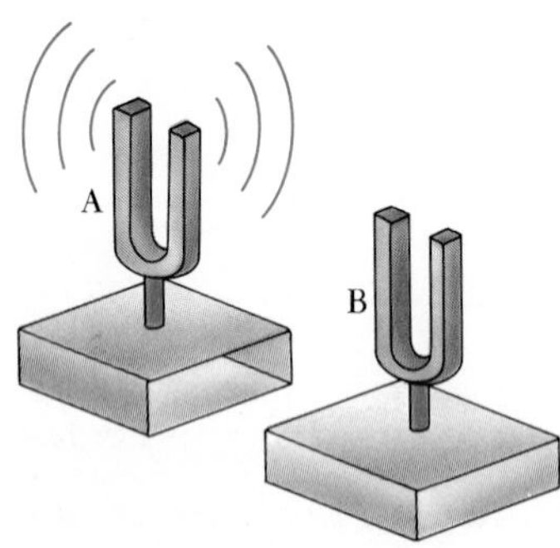

Figure 18.11 If tuning fork A is set into vibration, tuning fork B will eventually vibrate at the same frequency, or resonate, if the two forks are identical.

As a final example of resonance, consider two identical tuning forks mounted on separate hollow boxes (Fig. 18.11). The hollow boxes augment the sound wave power generated by the vibrating tuning forks. If tuning fork A is set into vibration (by striking it, say), tuning fork B will be set into vibration as longitudinal sound waves are received from A. The frequencies of vibration of A and B will be the same, assuming the tuning forks are identical. The energy exchange, or resonance behavior, will not occur if the two have different natural frequencies of vibration.

18.5 STANDING WAVES IN AIR COLUMNS

Standing longitudinal waves can be set up in a tube of air, such as an organ pipe, as the result of interference between longitudinal waves traveling in opposite directions. The phase relationship between the incident wave and the wave reflected from one end depends on whether that end is open or closed. This is analogous to the phase relationships between incident and reflected transverse waves at the ends of a string. *The closed end of an air column is a displacement node,* just as the fixed end of a vibrating string is a displacement node. As a result, at a closed end of a tube of air, the reflected wave is 180° out of phase with the incident wave. Furthermore, since the pressure wave is 90° out of phase with the displacement wave (Section 17.2), *the closed end of an air column corresponds to a pressure antinode* (that is, a point of maximum pressure variation).

If the end of an air column is open to the atmosphere, the air molecules have complete freedom of motion. The wave reflected from an open end is nearly in phase with the incident wave when the tube's diameter is small relative to the wavelength of the sound. Consequently, *the open end of an air column is approximately a displacement antinode and a pressure node.*

Strictly speaking, the open end of an air column is not exactly an antinode. When a condensation reaches an open end, it does not reach full expansion until it passes somewhat beyond the end. For a thin-walled tube of circular cross section, this end correction is about $0.6R$, where R is the tube's radius. Hence, the effective length of the tube is somewhat longer than the true length L.

The first three modes of vibration of a pipe open at both ends are shown in Figure 18.12a. By directing air against an edge at the left, longitudinal stand-

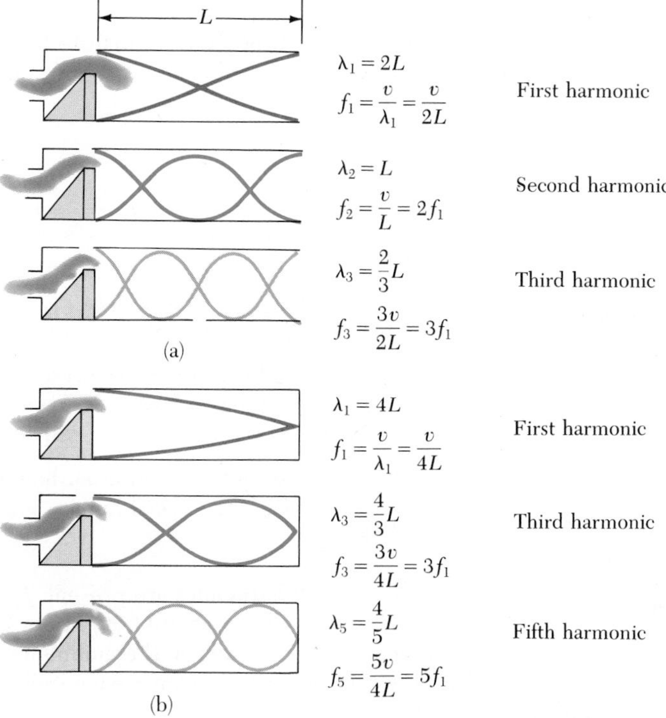

Figure 18.12 (a) Standing longitudinal waves in an organ pipe open at both ends. The natural frequencies which form a harmonic series are f_1, $2f_1$, $3f_1$, ... (b) Standing longitudinal waves in an organ pipe closed at one end. Only the *odd* harmonics are present, and so the natural frequencies are f_1, $3f_1$, $5f_1$, ...

ing waves are formed and the pipe resonates at its natural frequencies. All modes of vibration are excited simultaneously (although not with the same amplitude). Note that the ends are displacement antinodes (approximately). In the fundamental mode, the wavelength is twice the length of the pipe, and hence the frequency of the fundamental, f_1, is given by $v/2L$. Similarly, one finds that the frequencies of the higher harmonics are $2f_1, 3f_1, \ldots$. Thus,

in a pipe open at both ends, the natural frequencies of vibration form a harmonic series, that is, the higher harmonics are integral multiples of the fundamental frequency.

Since all harmonics are present, we can express the natural frequencies of vibration as

Natural frequencies of a pipe open at both ends

$$f_n = n\frac{v}{2L} \qquad (n = 1, 2, 3, \ldots) \tag{18.11}$$

where v is the speed of sound in air.

If a pipe is closed at one end and open at the other, the closed end is a displacement node (Fig. 18.12b). In this case, the wavelength for the fundamental mode is four times the length of the tube. Hence, the fundamental, f_1, is equal to $v/4L$, and the frequencies of the higher harmonics are equal to $3f_1$, $5f_1, \ldots$. That is,

in a pipe closed at one end, only odd harmonics are present, and these are given by

Natural frequencies of a pipe closed at one end

$$f_n = n\frac{v}{4L} \qquad (n = 1, 3, 5, \ldots) \tag{18.12}$$

Standing waves in air columns are primarily responsible for the sounds produced by various wind instruments, as described in the essay that precedes this chapter.

EXAMPLE 18.4 Resonance in a Pipe

A pipe has a length of 1.23 m. (a) Determine the frequencies of the first three harmonics if the pipe is open at each end. Take $v = 344$ m/s as the speed of sound in air.

Solution The first harmonic of an open pipe is

$$f_1 = \frac{v}{2L} = \frac{344 \text{ m/s}}{2(1.23 \text{ m})} = 140 \text{ Hz}$$

Since all harmonics are present, the second and third harmonics are given by $f_2 = 2f_1 = 280$ Hz and $f_3 = 3f_1 = 420$ Hz, respectively.

(b) What are the three frequencies determined in (a) if the pipe is closed at one end?

Solution The fundamental frequency of a pipe closed at one end is

$$f_1 = \frac{v}{4L} = \frac{344 \text{ m/s}}{4(1.23 \text{ m})} = 70 \text{ Hz}$$

In this case, only odd harmonics are present, and so the next two resonances have frequencies given by $f_3 = 3f_1 = 210$ Hz and $f_5 = 5f_1 = 350$ Hz, respectively.

(c) For the case of the open pipe, how many harmonics are present in the normal human hearing range (20 to 20 000 Hz)?

Solution Since all harmonics are present, $f_n = nf_1$. For $f_n = 20\,000$ Hz, we have $n = 20\,000/140 = 142$, so that 142 harmonics are present in the audible range. Actually, only the first few harmonics have sufficient amplitude to be heard.

EXAMPLE 18.5 Measuring the Frequency of a Tuning Fork

A simple apparatus for demonstrating resonance in a tube is described in Figure 18.13a. A long, vertical, open tube is partially submerged in a beaker of water, and a vibrating tuning fork of unknown frequency is placed near the top. The length of the air column, L, is adjusted by moving the tube vertically. The sound waves generated by the fork are reinforced when the length of the column corresponds to one of the resonant frequencies of the tube. The smallest value of L for which a peak occurs in the sound intensity is 9 cm. From this measurement, determine the frequency of the tuning fork and the value of L for the next two resonant modes.

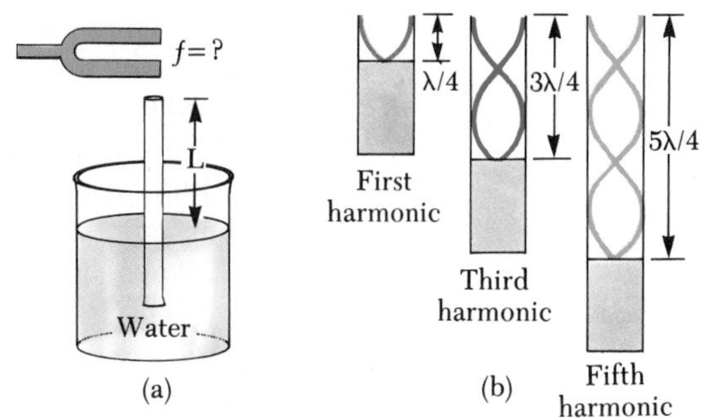

Figure 18.13 (a) Apparatus for demonstrating the resonance of sound waves in a tube closed at one end. The length L of the air column is varied by moving the tube vertically while it is partially submerged in water. (b) The first three normal modes of the system shown in (a).

Solution Since this setup represents a pipe closed at one end, the fundamental has a frequency of $v/4L$ (Fig. 18.13b). Taking $v = 344$ m/s for the speed of sound in air and $L = 0.09$ m, we get

$$f_1 = \frac{v}{4L} = \frac{344 \text{ m/s}}{4(0.09 \text{ m})} = 956 \text{ Hz}$$

From this information about the fundamental mode, we see that the wavelength is given by $\lambda = 4L = 0.36$ m. Since the frequency of the source is constant, we see that the next two resonance modes (Fig. 18.13b) correspond to lengths of $3\lambda/4 = 0.27$ m and $5\lambda/4 = 0.45$ m, respectively.

*18.6 STANDING WAVES IN RODS AND PLATES

Standing wave vibrations can also be set up in rods and plates. If a rod is clamped in the middle and stroked at one end, it will undergo longitudinal vibrations as described in Figure 18.14a. Note that the broken lines in Figure 18.14 represent *longitudinal* displacements of various parts of the rod. The midpoint is a displacement node since it is fixed by the clamp, whereas the ends are displacement antinodes since they are free to vibrate. This is analogous to vibrations set up in a pipe open at each end. The broken lines in Figure 18.14a represent the fundamental mode for which the wavelength is $2L$ and the frequency is $v/2L$, where v is the speed of longitudinal waves in the rod.

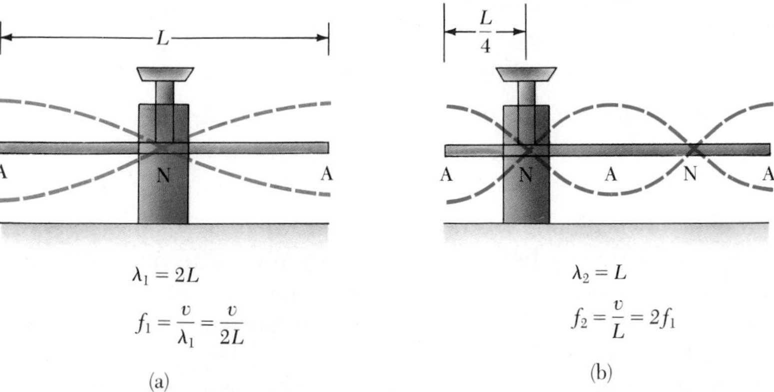

Figure 18.14 Normal longitudinal vibrations of a rod of length L (a) clamped at the middle and (b) clamped at an approximate distance of $L/4$ from one end.

Other modes may be excited by clamping the rod at different points. For example, the second harmonic (Fig. 18.14b) is excited by clamping the rod at a point that is a distance $\lambda/4$ away from one end.

Two-dimensional vibrations can be set up in a flexible membrane stretched over a circular hoop, such as a drumhead. As the membrane is struck at some point, wave pulses traveling toward the fixed boundary are reflected many times. The resulting sound is not melodious, but rather explosive in nature. This is because the vibrating drumhead and the drum's hollow interior produce a disorganized set of waves, which creates a sound of indefinite pitch when they reach a listener's ear. This is in contrast to wind and stringed instruments, which produce sounds of definite pitch.

Some possible normal modes of oscillation of a vibrating, two-dimensional, circular membrane are shown in Figure 18.15. Note that the nodes are curves rather than points, which was the case for a vibrating string. The fixed outer perimeter is one such nodal curve. Some other nodal curves are indicated with arrows. The lowest mode of vibration with frequency f_1 (the fundamental) is a symmetric mode with one nodal curve, the circumference of the membrane. The other possible modes of vibration are *not* integral multiples of f_1; hence the normal frequencies *do not* form a harmonic series. When a drum is struck, many of these modes are excited simultaneously. However, the higher-frequency modes damp out more rapidly. With this information, one can understand why the drum is a nonmelodious instrument.

*18.7 BEATS: INTERFERENCE IN TIME

The interference phenomena we have been dealing with so far involve the superposition of two or more waves with the same frequency traveling in opposite directions. Since the resultant waveform in this case depends on the coordinates of the disturbed medium, we can refer to the phenomenon as *spatial interference.* Standing waves in strings and pipes are common examples of spatial interference.

We now consider another type of interference effect, one that results from the superposition of two waves with slightly *different frequencies* traveling in the *same direction.* In this case, when the two waves are observed at a given point, they are periodically in and out of phase. That is, there is an alternation in time between constructive and destructive interference. Thus, we refer to this phenomenon as *interference in time* or *temporal interference.* For example, if two tuning forks of slightly different frequencies are struck, one hears a sound of pulsating intensity, called a **beat.**

Definition of beats

A **beat** can therefore be defined as the periodic variation in intensity at a given point due to the superposition of two waves having slightly different frequencies.

The number of beats one hears per second, or *beat frequency,* equals the difference in frequency between the two sources. The maximum beat frequency that the human ear can detect is about 20 beats/s. When the beat frequency exceeds this value, it blends indistinguishably with the compound sounds producing the beats.

One can use beats to tune a stringed instrument, such as a piano, by beating a note against a reference tone of known frequency. The string can then be adjusted to equal the frequency of the reference by tightening or loosening it until the beats become too infrequent to notice.

Figure 18.15 (a) Six normal modes of vibration of a circular membrane (drumhead) fixed at its perimeter. Arrows indicate the nodal lines. (From P. M. Morse, *Vibration and Sound*, 2nd ed., New York, McGraw-Hill, 1948, with permission of the publishers.) (b) Representation of some natural modes of vibration on a circular membrane fixed at its perimeter. The frequencies of vibration *do not* form a harmonic series. (From M. L. Warren, *Introductory Physics*, New York, W. H. Freeman, 1979, with permission.)

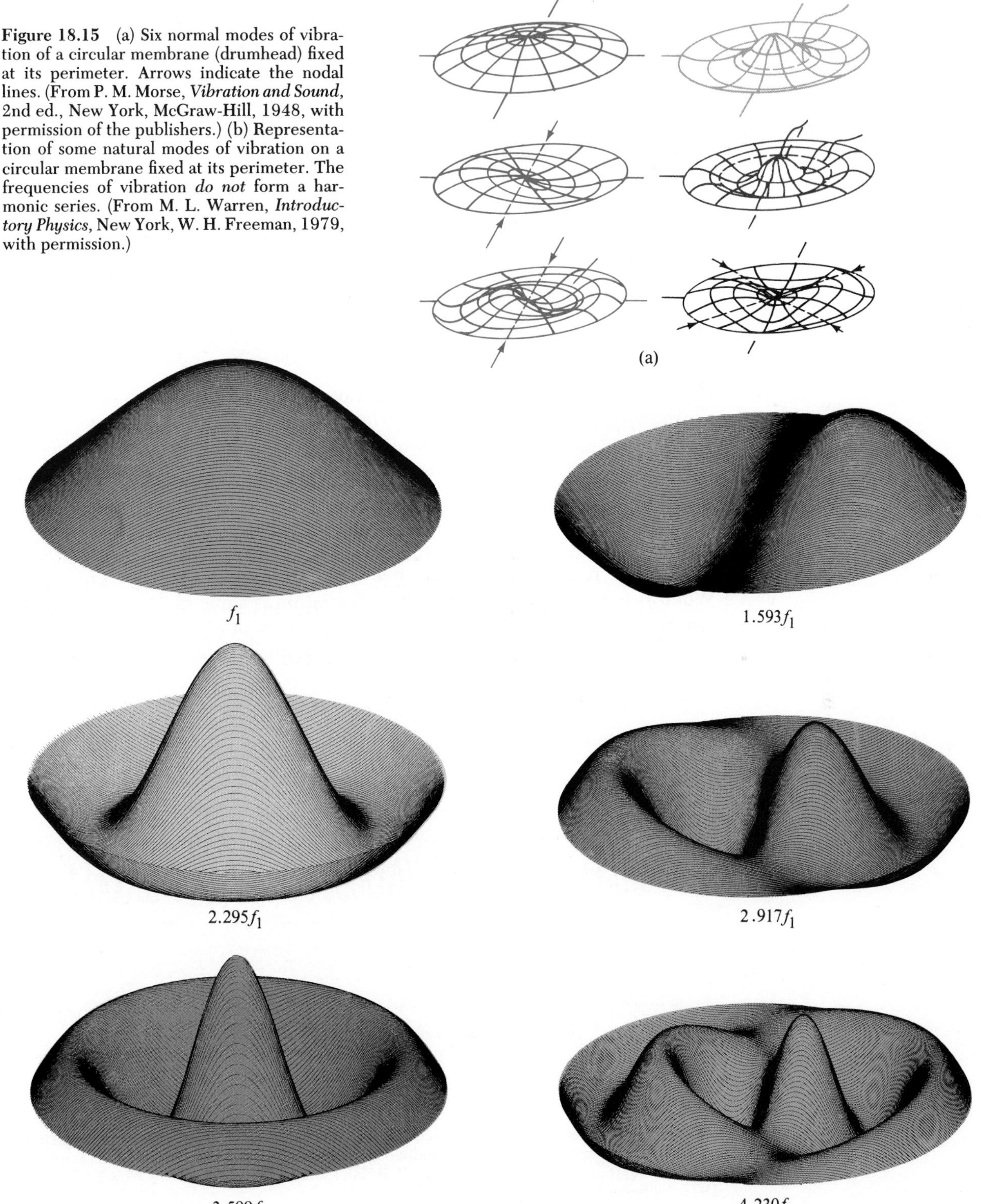

Consider two waves with equal amplitudes traveling through a medium in the *same* direction, but with slightly different frequencies, f_1 and f_2. We can represent the displacement that each wave would produce at a point as

$$y_1 = A_0 \cos 2\pi f_1 t \qquad \text{and} \qquad y_2 = A_0 \cos 2\pi f_2 t$$

Using the superposition principle, we find that the resultant displacement at that point is given by

$$y = y_1 + y_2 = A_0 (\cos 2\pi f_1 t + \cos 2\pi f_2 t)$$

It is convenient to write this in a form that uses the trigonometric identity

$$\cos a + \cos b = 2 \cos \left(\frac{a-b}{2}\right) \cos \left(\frac{a+b}{2}\right)$$

Letting $a = 2\pi f_1 t$ and $b = 2\pi f_2 t$, we find that

Resultant of two waves of different frequencies but equal amplitude

$$y = 2A_0 \cos 2\pi \left(\frac{f_1 - f_2}{2}\right) t \cos 2\pi \left(\frac{f_1 + f_2}{2}\right) t \qquad (18.13)$$

Graphs demonstrating the individual waveforms as well as the resultant wave are shown in Figure 18.16. From the factors in Equation 18.13, we see that the resultant vibration at a point has an effective frequency equal to the average frequency, $(f_1 + f_2)/2$, and an amplitude given by

$$A = 2A_0 \cos 2\pi \left(\frac{f_1 - f_2}{2}\right) t \qquad (18.14)$$

That is, the *amplitude varies in time* with a frequency given by $(f_1 - f_2)/2$. When f_1 is close to f_2, this amplitude variation is slow, as illustrated by the envelope (broken line) of the resultant waveform in Figure 18.16b.

Note that a beat, or a maximum in amplitude, will be detected whenever

$$\cos 2\pi \left(\frac{f_1 - f_2}{2}\right) t = \pm 1$$

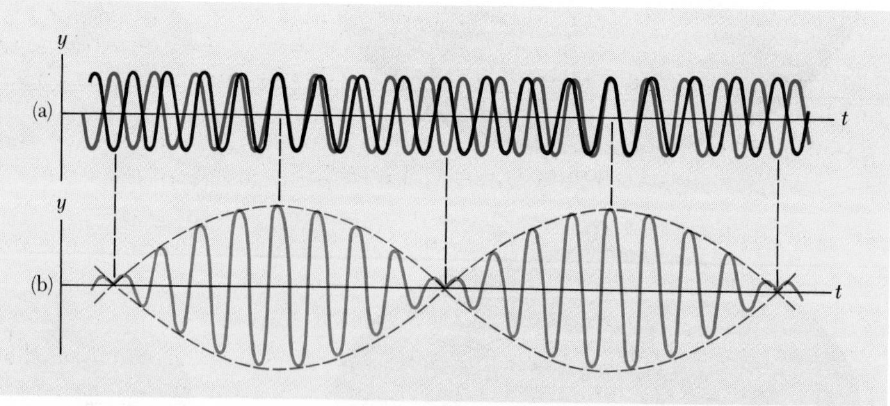

Figure 18.16 Beats are formed by the combination of two waves of slightly different frequencies traveling in the same direction. (a) The individual waves. (b) The combined wave has an amplitude (broken line) that oscillates in time. (From R. Resnick and D. Halliday, *Physics*, New York, Wiley, 1977; by permission of the publisher.)

That is, there are *two* maxima in each cycle. Since the amplitude varies with frequency as $(f_1 - f_2)/2$, the number of beats per second, or the beat frequency f_b, is twice this value. That is,

$$f_b = f_1 - f_2 \quad (18.15)$$

Beat frequency

For instance, if two tuning forks vibrate individually at frequencies of 438 Hz and 442 Hz, respectively, the resultant sound wave of the combination has a frequency of 440 Hz (the musical note A) and a beat frequency of 4 Hz. That is, the listener would hear the 440-Hz sound wave go through an intensity maximum four times every second.

*18.8 COMPLEX WAVES

The sound wave patterns produced by most instruments are very complex. Some characteristic waveforms produced by a tuning fork, a harmonic flute, and a clarinet, each playing the same pitch, are shown in Figure 18.17. Although each instrument has its own characteristic pattern, Figure 18.17 shows that each of the waveforms is periodic in nature. Furthermore, note that a struck tuning fork produces only one harmonic (the fundamental), whereas the flute and clarinet produce many frequencies, which include the fundamental and various harmonics. Thus, the complex waveforms produced by a violin or clarinet, and the corresponding richness of musical tones, are the result of the superposition of various harmonics. This is in contrast to the drum, in which the overtones do not form a harmonic series.

It is interesting to investigate what happens to the frequencies of a flute and a violin during a concert as the temperature rose. A flute goes sharp (increases in frequency) as it warms up since the speed of sound increases. On the other hand, a violin goes flat (decreases in frequency) as the strings expand thermally, which causes their tension to decrease.

(a) Tuning fork

(b) Harmonic flute

(c) Clarinet

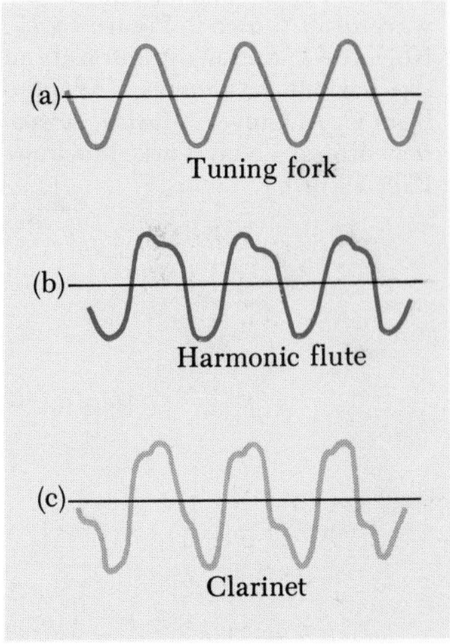

Figure 18.17 Waveform produced by (a) a tuning fork, (b) a harmonic flute, and (c) a clarinet, each at approximately the same frequency. (Adapted from C. A. Culver, *Musical Acoustics*, 4th ed., New York, McGraw-Hill, 1956, p. 128.)

The problem of analyzing complex waveforms appears at first sight to be a rather formidable task. However, if the waveform is periodic, it can be represented with arbitrary precision by the combination of a sufficiently large number of sinusoidal waves that form a harmonic series. In fact, one can represent any periodic function or any finite function as a series of sine and cosine terms by using a mathematical technique based on **Fourier's theorem.**[3] The corresponding sum of terms that represents the periodic waveform is called a **Fourier series.**

Let $y(t)$ be any function that is periodic in time with period T, such that $y(t + T) = y(t)$. Fourier's theorem states that this function can be written

$$y(t) = \sum_n (A_n \sin 2\pi f_n t + B_n \cos 2\pi f_n t) \quad (18.16)$$

Fourier's theorem

where the lowest frequency is $f_1 = 1/T$.

The higher frequencies are integral multiples of the fundamental, so that $f_n = nf_1$. The coefficients A_n and B_n represent the amplitudes of the various waves. The amplitude of the nth harmonic is proportional to $\sqrt{A_n^2 + B_n^2}$, and its intensity is proportional to $A_n^2 + B_n^2$.

[3] Developed by Jean Baptiste Joseph Fourier (1786–1830).

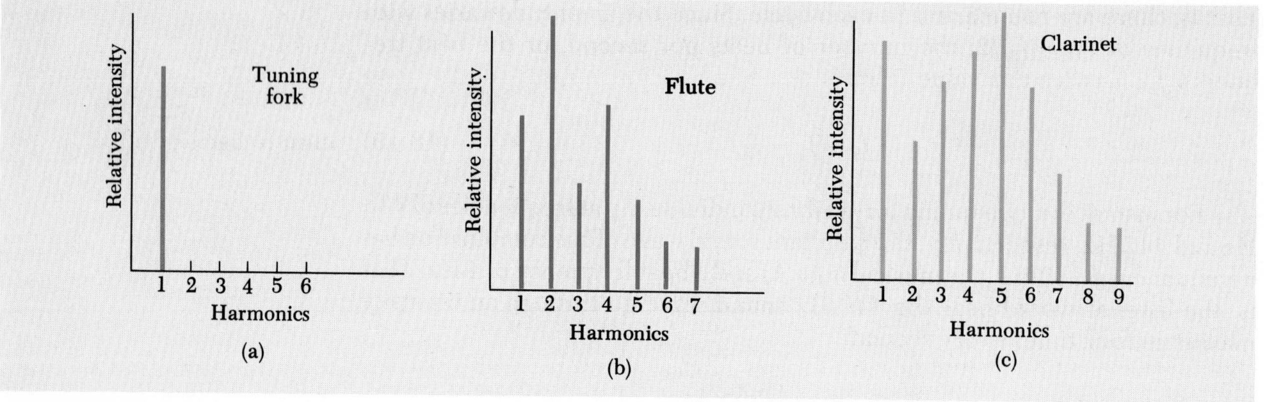

Figure 18.18 Harmonics of the waveforms shown in Figure 18.17. Note the variations in intensity of the various harmonics. (Adapted from C. A. Culver, *Musical Acoustics*, 4th ed., New York, McGraw-Hill, 1956.)

Figure 18.18 represents a harmonic analysis of the waveforms shown in Figure 18.17. Note the variation of relative intensity with harmonic content for the flute and clarinet. In general, any musical sound (of definite pitch) contains components that are members of a harmonic set with varying relative intensities.

As an example of *Fourier synthesis,* consider the periodic square wave shown in Figure 18.19. The square wave is synthesized by a series of *odd* harmonics of the fundamental. The series contains only sine functions (that is, $B_n = 0$ for all n). Only the first four odd harmonics and their respective amplitudes are shown. One obtains a better fit to the true waveform by adding more harmonics.

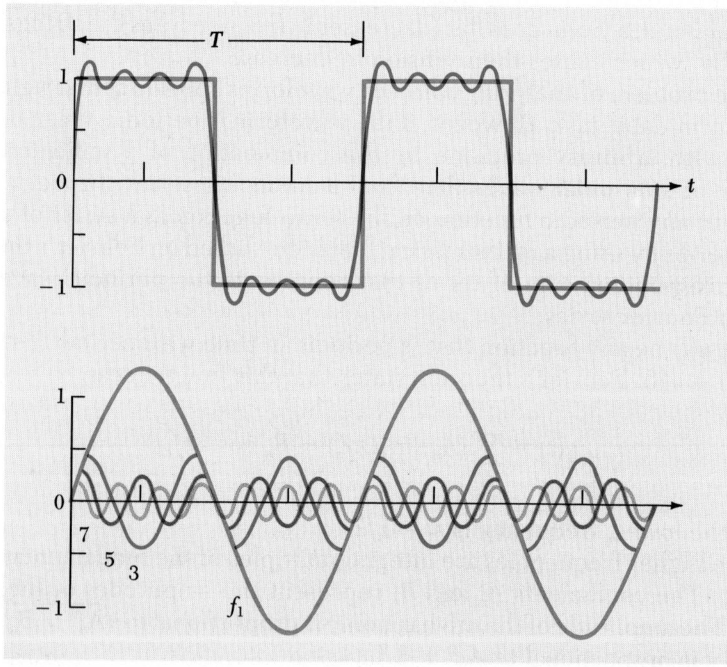

Figure 18.19 Harmonic synthesis of a square wave, which can be represented by the sum of odd harmonics of the fundamental. (From M. L. Warren, *Introductory Physics*, San Francisco, W. H. Freeman, 1979, p. 178; by permission of the publisher.)

Using modern technology, one can generate musical sounds electronically by mixing any number of harmonics with varying amplitudes. These widely used electronic music synthesizers are able to produce an infinite variety of musical tones.

SUMMARY

When two waves with equal amplitudes and frequencies superimpose, the resultant wave has an amplitude that depends on the phase angle ϕ between the two waves. **Constructive interference** occurs when the two waves are *in phase* everywhere, corresponding to $\phi = 0, 2\pi, 4\pi, \ldots$. **Destructive interference** occurs when the two waves are 180° out of phase everywhere, corresponding to $\phi = \pi, 3\pi, 5\pi, \ldots$.

Standing waves are formed from the superposition of two harmonic waves having the same frequency, amplitude, and wavelength, but traveling in *opposite* directions. The resultant standing wave is described by the wave function

Wave function for a standing wave

$$y = (2A_0 \sin kx) \cos \omega t \qquad (18.3)$$

Hence, its amplitude varies as $\sin kx$. The maximum amplitude points (called **antinodes**) occur at $x = n\pi/2k = n\lambda/4$ (for odd n). The points of zero amplitude (called **nodes**) occur at $x = n\pi/k = n\lambda/2$ (for integral values of n).

One can set up standing waves with specific frequencies in such systems as stretched strings, hollow pipes, rods, and drumheads. The natural frequencies of vibration of a stretched string of length L, fixed at both ends, have frequencies given by

Normal modes of a stretched string

$$f_n = \frac{n}{2L}\sqrt{\frac{F}{\mu}} \qquad (n = 1, 2, 3, \ldots) \qquad (18.8)$$

where F is the tension in the string and μ is its mass per unit length. The natural frequencies of vibration form a **harmonic series,** that is, f_1, $2f_1$, $3f_1, \ldots$.

The standing wave patterns for longitudinal waves in a hollow pipe depend on whether the ends of the pipe are open or closed. If the pipe is open at both ends, the natural frequencies of vibration form a harmonic series. If one end is closed, only odd harmonics of the fundamental are present.

A system capable of oscillating is said to be in **resonance** with some driving force whenever the frequency of the driving force matches one of the natural frequencies of the system. When the system is resonating, it responds by oscillating with a relatively large amplitude.

The phenomenon of **beats** occurs from the superposition of two waves of slightly different frequencies, traveling in the same direction. For sound waves at a given point, one hears an alternation in sound intensity with time. Thus, beats correspond to *interference as time passes.*

Any periodic waveform can be represented by the combination of the sinusoidal waves that form a harmonic series. The process is called *Fourier synthesis* and is based upon *Fourier's theorem.*

QUESTIONS

1. For certain positions of the movable section in Figure 18.2, there is no sound detected at the receiver, corresponding to destructive interference. This suggests that perhaps energy is somehow lost! What happens to the energy transmitted by the speaker?
2. Does the phenomenon of wave interference apply only to harmonic waves?
3. When two waves interfere constructively or destructively, is there any gain or loss in energy? Explain.
4. A standing wave is set up on a string as in Figure 18.5. Explain why no energy is transmitted along the string.
5. What is common to *all* points (other than the nodes) on a string supporting a standing wave?
6. Some singers claim to be able to shatter a wine glass by maintaining a certain pitch in their voice over a period of several seconds (see photo). What mechanism causes the glass to break? (The glass must be very clean in order for it to break.)

(Question 6) A wine glass shattered by the amplified sound of a human voice. (Courtesy Memorex Corporation)

7. What limits the amplitude of motion of a real vibrating system that is driven at one of its resonant frequencies?
8. If the temperature of the air in an organ pipe increases, what happens to the resonance frequencies?
9. Explain why your voice seems to sound better than usual when you sing in the shower.
10. What is the purpose of the slide on a trombone or the valves on a trumpet?
11. Explain why all harmonics are present in an organ pipe open at both ends, but only the odd harmonics are present in a pipe closed at one end.
12. Explain how a musical instrument such as a piano may be tuned using the phenomenon of beats.
13. An airplane mechanic notices that the sound from a twin-engine aircraft rapidly varies in loudness when both engines are running. What could be causing this variation from loud to soft?
14. At certain speeds, an automobile driven on a washboard road will vibrate disastrously and lose traction and braking effectiveness. At other speeds, either lesser or greater, the vibration is more manageable. Explain. Why are "rumble strips," which work on this same principle, often used just before stop signs?
15. Why does a vibrating guitar string sound louder when placed on the instrument than it would if allowed to vibrate in the air while off the instrument?
16. When the base of a vibrating tuning fork is placed against a chalk board, the sound becomes louder due to resonance. How does this affect the length of time for which the fork vibrates? Does this agree with conservation of energy?
17. Stereo speakers are supposed to be "phased" when set up. That is, the waves emitted from them should be in phase with each other. What would the sound be like along the center line of the speakers if one speaker were wired up backwards, that is, out of phase?
18. To keep bugs away from their cars, some people mount short thin pipes on the fenders. The pipes give out a high-pitched wail when the cars are moving. How do they create the sound?
19. If you wet your fingers and lightly run them around the rim of a fine wine glass, a high-pitched sound is heard. Why? How could you produce various musical notes with a set of wine glasses?
20. When a bell is rung, standing waves are set up around the bell's circumference. What boundary conditions must be satisfied by the resonant wavelengths? How does a crack in the bell, such as in the Liberty Bell, affect the satisfying of the boundary conditions and the sound emanating from the bell?

PROBLEMS

Section 18.1 Superposition and Interference of Harmonic Waves

1. Two harmonic waves are described by

$$y_1 = (5\text{ m})\sin[\pi(4x - 1200t)]$$
$$y_2 = (5\text{ m})\sin[\pi(4x - 1200t - 0.25)]$$

where x, y_1, and y_2 are in m and t is in s. (a) What is the amplitude of the resultant wave? (b) What is the frequency of the resultant wave?

2. Two harmonic waves are described by

$$y_1 = (6\text{ m})\sin\left(\frac{\pi}{15}x - \frac{\pi}{0.005}t\right)$$

$$y_2 = (6 \text{ m}) \sin\left(\frac{\pi}{15} x - \frac{\pi}{0.005} t - \phi\right)$$

where x, y_1, and y_2 are in m and t is in s. (a) What is the amplitude of the resultant wave when $\phi = (\pi/6)$ rad? (b) For what values of ϕ will the amplitude of the resultant wave have its maximum value?

3. A harmonic wave is described by

$$y_1 = (8 \text{ m}) \sin[2\pi(0.1x - 80t)]$$

where y_1 and x are in m and t is in s. Write an expression for a wave that has the same frequency, amplitude, and wavelength as y_1, but when added to y_1 will give a resultant with an amplitude of $8\sqrt{3}$ m.

4. Two speakers are arranged similarly to those shown in Figure 18.3. The distance between the two speakers is 2 m, and they are driven at a frequency of 1500 Hz. An observer is initially at a point 6 m along the perpendicular bisector of the line joining the two speakers. (a) What distance must the observer move along a line parallel to the line joining the two speakers before reaching the first minimum in intensity? (Use $v = 343$ m/s.) (b) At what distance from the perpendicular bisector will the observer find the first relative maximum in intensity?

5. Two identical sound sources are located along the y axis. Source S_1 is located at (0, 0.1) m and source S_2 is located at (0, −0.1) m. The two sources radiate isotropically at a frequency of 1715 Hz and the amplitude of each wave separately is assumed to be A. A listener is located along the y axis a distance of 5 m from source S_1. (a) What is the phase difference between the sound waves at the position of the listener? (b) What is the amplitude of the resultant wave at the location of the listener? (Use $v = 343$ m/s.)

6. Two identical sound sources are located as described in Problem 5. The frequency of each source is variable. An observer is located at the point (1, 0.5) m. (a) What is the lowest frequency that will produce a relative maximum at the location of the observer? (b) What is the lowest frequency that will produce a relative minimum at the observer's location? (Use $v = 330$ m/s.)

7. Two speakers are driven by a common oscillator at 800 Hz and face each other at a distance of 1.25 m. Locate the points along a line joining the two speakers where relative minima would be expected. (Use $v = 343$ m/s.)

8. For the arrangement shown in Figure 18.2, let the path length $r_1 = 1.20$ m and the path length $r_2 = 0.80$ m. (a) Calculate the three lowest speaker frequencies that will result in intensity maxima at the receiver. (b) What is the highest frequency within the audible range (20–20 000 Hz) that will result in a minimum at the receiver? (Use $v = 330$ m/s.)

Section 18.2 Standing Waves

9. Two harmonic waves are described by

$$y_1 = (3 \text{ cm}) \sin\pi(x + 0.6t)$$

$$y_2 = (3 \text{ cm}) \sin\pi(x - 0.6t)$$

Determine the *maximum* displacement of the motion at (a) $x = 0.25$ m, (b) $x = 0.5$ m, and (c) $x = 1.5$ m. (d) Find the three smallest values of x corresponding to antinodes.

10. Use the trigonometric identity

$$\sin(a \pm b) = \sin a \cos b \pm \cos a \sin b$$

to show that the resultant of two wave functions each of amplitude A_0, angular frequency ω, and propagation number k and traveling in opposite directions can be written

$$y = (2A_0 \sin kx) \cos \omega t$$

11. The wave function for a standing wave in a string is given by

$$y = (0.30 \text{ m}) \sin(0.25x) \cos(120\pi t)$$

where x is in m and t is in s. Determine the wavelength and frequency of the interfering traveling waves.

12. Two harmonic waves traveling in opposite directions interfere to produce a standing wave described by

$$y = (1.50 \text{ m}) \sin(0.4x) \cos(200t)$$

where x is in m and t is in s. Determine the wavelength, frequency, and speed of the interfering waves.

13. A standing wave is formed by the interference of two traveling waves, each of which has an amplitude $A = \pi$ cm, propagation number $k = (\pi/2)\text{ cm}^{-1}$, and angular frequency $\omega = 10\pi$ rad/s. (a) Calculate the distance between the first two antinodes. (b) What is the amplitude of the standing wave at $x = 0.25$ cm?

14. Verify by direct substitution that the wave function for a standing wave given in Equation 18.3,

$$y = 2A_0 \sin kx \cos \omega t,$$

is a solution of the general linear wave equation, Equation 16.26:

$$\frac{\partial^2 y}{\partial x^2} = \frac{1}{v^2} \frac{\partial^2 y}{\partial t^2}$$

15. Two waves which set up a standing wave in a long string are given by

$$y_1 = A \sin(kx - \omega t + \phi)$$

$$y_2 = A \sin(kx + \omega t)$$

Show that the addition of the arbitrary phase angle (a) will change only the positions of the nodes, and (b) that the distance between nodes remains constant.

16. Two waves in a long string are given by

$$y_1 = (0.015 \text{ m}) \cos\left(\frac{x}{2} - 40t\right)$$

$$y_2 = (0.015 \text{ m}) \cos\left(\frac{x}{2} + 40t\right)$$

where the y's and x are in m and t is in s. (a) Determine the positions of the nodes of the resulting standing wave. (b) What is the maximum displacement at the position $x = 0.4$ m?

Section 18.3 Standing Waves in a String Fixed at Both Ends

17. A standing wave is established in a 120-cm-long string fixed at both ends. The string vibrates in four segments when driven at 120 Hz. (a) Determine the wavelength. (b) What is the fundamental frequency?
18. A stretched string is 160 cm long and has a linear density of 0.015 g/cm. What tension in the string will result in a second harmonic of 460 Hz?
19. Consider a tuned guitar string of length L. At what point along the string (fraction of length from one end) should the string be plucked and at what point should the finger be held lightly against the string in order that the second harmonic be the most prominent mode of vibration?
20. A string 50 cm long has a mass per unit length of 20×10^{-5} kg/m. To what tension should this string be stretched if its fundamental frequency is to be (a) 20 Hz and (b) 4500 Hz?
21. Find the fundamental frequency and the next three frequencies that could cause a standing wave pattern on a string that is 30 m long, has a mass per unit length 9×10^{-3} kg/m, and is stretched to a tension of 20 N.
22. A stretched string of length L is observed to vibrate in five equal-length segments when driven by a 630-Hz oscillator. What oscillator frequency will set up a standing wave such that the string vibrates in three segments?
23. A string with $L = 16$ m and $\mu = 0.015$ g/cm is stretched with a tension of 557 N (≈ 125 lb). What is the highest harmonic of this string that is within the typical human's audible range (up to 20 000 Hz)?
24. Two pieces of steel wire having identical cross sections have lengths of L and $2L$. The wires are each fixed at both ends and stretched such that the tension in the longer wire is four times greater than that in the shorter wire. If the fundamental frequency in the shorter wire is 60 Hz, what is the frequency of the second harmonic in the longer wire?
25. A stretched string fixed at each end has a mass of 40 g and a length of 8 m. The tension in the string is 49 N. Determine the position of the nodes and antinodes for the third harmonic.
26. A string of length L, mass per unit length μ, and tension F is vibrating at its fundamental frequency. What effect will the following have on the fundamental frequency? (a) The length of the string is doubled with all other factors held constant. (b) The mass per unit length is doubled with all other factors held constant. (c) The tension is doubled with all other factors held constant.
27. A 60-cm guitar string under a tension of 50 N has a mass per unit length of 0.1 g/cm. What is the highest resonant frequency that can be heard by a person capable of hearing frequencies up to 20 000 Hz?
28. Write an expression for the harmonic number of the nth resonance for a sounding device that has (a) all harmonics present and (b) only odd harmonics present.
29. A steel wire in a piano has a length of 0.70 m and a mass of 4.3×10^{-3} kg. To what tension must this wire be stretched in order that the fundamental vibration correspond to middle C, $f_C = 261.6$ Hz on the chromatic musical scale?
30. A violin string has a length of 0.35 m and is tuned to concert G, $f_G = 392$ Hz. Where must the violinist place her finger to play concert A, $f_A = 440$ Hz? If this position is to remain correct to one half the width of a finger (i.e., to within 0.6 cm), by what fraction may the string tension be allowed to slip?

Section 18.5 Standing Waves in Air Columns

(In this section, unless otherwise indicated, assume that the speed of sound in air is 344 m/s.)

31. A resonance condition is set up in a pipe with a tuning fork whose frequency is f. Write an expression for the length of the pipe that will cause it to resonate in its nth mode if the pipe is (a) open at both ends and (b) closed at one end. (Assume that the speed of sound is v.)
32. If an organ pipe is to resonate at 20 Hz, what is its required length if it is (a) open at both ends and (b) closed at one end?
33. A tuning fork of frequency 512 Hz is placed near the top of the tube shown in Figure 18.13a. The water level is lowered so that the length L slowly increases from an initial value of 20 cm. Determine the next two values of L that correspond to resonant modes.
34. A pipe open at each end has a fundamental frequency of 300 Hz when the speed of sound in air is 333 m/s. (a) What is the length of the pipe? (b) What is the frequency of the second harmonic when the temperature of the air is increased so that the speed of sound in the pipe is 344 m/s?
35. Calculate the minimum length for a pipe that has a fundamental frequency of 240 Hz if the pipe is (a) closed at one end and (b) open at both ends.
36. A tunnel beneath a river is approximately 2 km long. At what frequencies can this tunnel resonate?

37. An organ pipe open at both ends is vibrating in its third harmonic with a frequency of 748 Hz. The length of the pipe is 0.7 m. Determine the speed of sound in air in the pipe.
38. The longest pipe on an organ that has pedal stops is often 16 ft (4.88 m). What is the fundamental frequency (at 0° C) if the nondriven end of the pipe is (a) closed and (b) open? (c) What will be the frequencies at 20°C?
39. The fundamental frequency of an open organ pipe corresponds to middle C (261.6 Hz on the chromatic musical scale). The third resonance of a closed organ pipe has the same frequency. What are the lengths of the two pipes?
40. Determine the frequency corresponding to the first three resonances of a 30-cm pipe when it is (a) open at both ends and (b) closed at one end.
41. An air column 2 m in length is open at both ends. The frequency of a certain harmonic is 410 Hz, and the frequency of the next higher harmonic is 492 Hz. Determine the speed of sound in the air column.
42. At a particular instant, the tube in Figure 18.13a is adjusted so that L, the length above the water surface, is 40 cm. The tuning fork in the figure is replaced by a variable-frequency oscillator that has a frequency range between 20 and 2000 Hz. What are the (a) lowest and (b) highest frequencies within this range that will excite resonant modes in the air column?
43. A glass tube is open at one end and closed at the other (by a movable piston). The tube is filled with 30°C air, and a 384-Hz tuning fork is held at the open end. Resonance is heard when the piston is 22.8 cm from the open end and again when it is 68.3 cm from the open end. (a) What speed of sound is implied by these data? (b) Where would the piston be for the next resonance?
44. The range of a certain pipe organ is from 8 Hz to 30 000 Hz. What range of pipe lengths is necessary if they are (a) open at both ends and (b) closed at one end? (Assume that the speed of sound is 343 m/s.)
45. A shower stall measures 86 cm × 86 cm × 210 cm. When you sing in the shower, which frequencies will sound the richest (resonate), assuming the shower acts as a pipe closed at both ends (nodes at both sides)? Assume also that the human voice ranges from 130 Hz to 2000 Hz (not necessarily one person's voice, however). Let the speed of sound in the hot shower stall be 355 m/s.

*Section 18.6 Standing Waves in Rods and Plates

46. An aluminum rod is clamped at the one-quarter position and set into longitudinal vibration by a variable-frequency driving source. The lowest frequency that produces resonance is 4400 Hz. The speed of sound in aluminum is 5100 m/s. Determine the length of the rod.
47. A 60-cm metal bar that is clamped at one end is struck with a hammer. If the speed of longitudinal (compressional) waves in the bar is 4500 m/s, what is the lowest frequency with which the struck bar will resonate?
48. Longitudinal waves move with a speed v in a bar of length L. Write an expression for the frequencies of the longitudinal vibrations of a metal bar that is (a) clamped at its center, as shown in Figure 18.14a, and (b) clamped at one-fourth the length of the bar from one end, as shown in Figure 18.14b.
49. An aluminum rod 1.6 m long is held at its center. It is stroked with a rosin-coated cloth to set up longitudinal vibrations in the fundamental mode. (a) What is the frequency of the waves established in the rod? (b) What harmonics are set up in the rod held in this manner? (c) What would be the fundamental frequency if the rod were copper?

*Section 18.7 Beats: Interference in Time

50. The tension in the strings of certain instruments normally decreases in time, which results in frequencies lower than intended. It is noted that when a string tuned to 256 Hz is plucked and an "out of tune" string of an identical instrument is plucked simultaneously, beats occur at a rate of 5 beats per second. What is the frequency of the "out of tune" string?
51. Two waves with equal amplitude but with slightly different frequencies are traveling in the same direction through a medium. At a given point the separate displacements are described by

$$y_1 = A_0 \cos \omega_1 t \qquad \text{and} \qquad y_2 = A_0 \cos \omega_2 t$$

Use the trigonometric identity

$$\cos a + \cos b = 2 \cos\left(\frac{a-b}{2}\right)\cos\left(\frac{a+b}{2}\right)$$

to show that the resultant displacement due to the two waves is given by

$$y = 2A_0\left[\cos\left(\frac{\omega_1-\omega_2}{2}\right)t\right]\left[\cos\left(\frac{\omega_1+\omega_2}{2}\right)t\right]$$

52. While attempting to tune a C note at 523 Hz a piano tuner hears 3 beats per second between the oscillator and the string. (a) What are the possible frequencies of the string? (b) By what percentage should the tension in the string be changed to bring the string into tune?
53. A student holds a tuning fork oscillating at 256 Hz. He walks towards a wall at a constant speed of 1.33 m/s. (a) What beat frequency does he observe between the tuning fork and its echo? (b) How fast must he walk away from the wall to observe a beat frequency of 5 Hz?

ADDITIONAL PROBLEMS

54. (a) What is the fundamental frequency of a steel piano wire of 0.005-kg mass and 1-m length, under a tension of 1350 N? (b) What is the fundamental frequency of an organ pipe, 1 m in length, closed at the bottom and open at the top?

55. Two loudspeakers are placed on a wall 2 m apart. A listener stands directly in front of one of the speakers, 3 m from the wall. The speakers are being driven by a single oscillator at a frequency of 300 Hz. (a) What is the phase difference between the two waves when they reach the observer? (b) What is the frequency closest to 300 Hz to which the oscillator may be adjusted such that the observer will hear minimal sound?

56. A variable-length air column as shown in Figure 18.13a is placed just below a vibrating wire fixed at both ends. The length of the air column is gradually increased from zero until the first position of resonance is observed at $L = 34$ cm. The wire is 120 cm in length and is vibrating in its third harmonic. If the speed of sound in air is 344 m/s, what is the speed of transverse waves in the wire?

57. Two speakers are arranged as shown in Figure 18.3. For this problem, assume that point O is 12 m along the center line and the speakers are separated by a distance of 1.5 m. As the listener moves toward point P from point O, a series of alternating minima and maxima is encountered. The distance between the first minimum and the next maximum is 0.4 m. Using 344 m/s as the speed of sound in air, determine the frequency of the speakers. (Use the approximation $\sin\theta \approx \tan\theta$.)

58. Two pipes are each open at one end and are of adjustable length. Each has a fundamental frequency of 480 Hz at 300 K. The air temperature is increased in one pipe to 305 K. (a) If the two pipes are sounded together, what beat frequency will result? (b) By what percent should the length of the 305-K pipe be decreased to again match the frequencies? (Use $v = 331(T/273)^{1/2}$ m/s as the speed of sound in air, where T is the air temperature in K.)

59. The frequency of the third resonance of an organ pipe which is open at both ends is equal to the frequency of the third resonance of another organ pipe which is closed at one end. (a) Find the ratio of the length of the "closed" pipe to the length of the "open" pipe. (b) If the fundamental frequency of the open pipe is 256 Hz, what is the length of each pipe? (Use $v = 340$ m/s.)

60. A speaker at the front of a room and an identical speaker at the rear of the room are being driven by the same oscillator at 456 Hz. A student walks at a uniform rate of 1.5 m/s along the length of the room. How many beats does the student hear per second?

61. To maintain a length of string under tension in a horizontal position, one end of the string is connected to a vibrating blade and the other end is passed over a pulley and attached to a mass. The mass of the string is 10 g, and its *total* length is 1.25 m. (a) When the suspended mass is 10 kg, the string vibrates in three equal length segments. Determine the vibration frequency of the blade. (Assume that the point where the string passes over the pulley and the point where it is attached to the blade are both nodes. Also, ignore the contribution to the tension due to the string's mass.) (b) What mass should be attached to the string if it is to vibrate in four equal segments?

62. While waiting for Stan Speedy to arrive on a late passenger train, Kathy Kool notices beats occurring as a result of two trains blowing their whistles simultaneously. One train is at rest and the other is approaching her at a speed of 20 km/h. Assume that both whistles have the same frequency and that the speed of sound is 344 m/s. If Kathy hears 4 beats per second, what is the frequency of the whistles?

63. A light rope 1.5 m in length lies along the x axis. It is set into vibration with *one* end fixed at $x = 0$. (a) What is the wavelength of the standing wave corresponding to the fundamental mode? (b) If the rope resonates in its fourth harmonic at a frequency of 320 Hz, what is the speed of transverse waves in the rope? (c) Write an expression for the wave function of the standing wave if the displacement at $x = \lambda/4$ is 4 cm.

64. In an arrangement like the one shown in Figure 18.2, paths r_1 and r_2 are each 1.75 m in length. The top portion of the tube (corresponding to r_2) is filled with air at 0°C (273 K). Air in the lower portion is quickly heated to 200°C (473 K). What is the lowest speaker frequency that will produce an intensity maximum at the receiver? (You may determine the speed of sound in air in different temperatures by using the expression $v = 331(T/273)^{1/2}$ m/s, where T is in K.)

65. Two train whistles have identical frequencies of 180 Hz. When one train is at rest in the station sounding its whistle, a beat frequency of 2 Hz is heard from a moving train. What two possible speeds and directions can the moving train have?

66. In a major chord on the physical pitch musical scale, the frequencies are in the ratios 4:5:6:8. A set of pipes, closed at one end, are to be cut so that when sounded in their fundamental mode, they will sound out a major chord. (a) What is the ratio of the lengths of the pipes? (b) What length pipes are needed if the lowest frequency of the chord is 256 Hz? (c) What are the frequencies of this chord?

67. Two wires are welded together. The wires are of the same material, but one is twice the diameter of the other one. They are subjected to a tension of 4.6 N. The thin wire has a length of 40 cm and a linear mass density of 2 g/m. The combination is fixed at both ends and vibrated in such a way that two antinodes are present with the central node being right at the weld. (a) What is the frequency of vibration? (b) How long is the thick wire?

68. Two identical steel wires each fixed at both ends are under equal tension and are vibrating in their third harmonic at 963 Hz. The tension in one wire is increased by 3%. Determine the beat frequency when the two wires now vibrate in their *fundamental* modes.

69. A standing wave is set up in a string of variable length and tension by a vibrator of variable frequency. When the vibrator has a frequency f in a string of length L and tension F there are n antinodes set up in the string. (a) If the length of the string is doubled, by what factor should the frequency be changed to get the same number of antinodes? (b) If the frequency and length are held constant, what tension will produce $n+1$ antinodes? (c) If the frequency is tripled and the length halved, by what factor should the tension be changed to get twice as many antinodes?

70. Radar detects the speed of a car, using the Doppler shift of microwaves that are reflected off the moving car, by beating the received wave with the transmitted wave and measuring the difference. The Doppler shift for light is given by

$$f = f_o \sqrt{\frac{c+v}{c-v}}$$

where f_o is the transmitted frequency, c is the speed of light (3×10^8 m/s) and v the relative speed of the two objects. (a) Show that the wave that reflects back to the source has a frequency given by

$$f = f_o \frac{(c+v)}{(c-v)}$$

(b) Show that the expression for the beat frequency of the microwaves may be written as $f_b = 2\,v/\lambda$. (Since the beat frequency is much smaller than the transmitted frequency, use the approximation $f + f_o = 2f_o$.) (c) What beat frequency is measured for a speed of 30 m/s (67 mph) if the microwaves have a frequency of 10 GHz? (1 GHz = 10^9 Hz.) (d) If the beat frequency measurement is accurate to ± 5 Hz how accurate is the velocity measurement?

CALCULATOR/COMPUTER PROBLEMS

71. Sketch the resultant waveform due to the interference of the two waves y_1 and y_2 in Problem 2 at $t = 0$ s for (a) $\phi = 0$, (b) $\phi = 90°$, and (c) $\phi = 270°$. Let x range over the interval 0 to 30 m.

72. A standing wave is described by the function

$$y = 6 \sin(\pi x/2) \cos(100\pi t)$$

where x and y are in m and t is in s. (a) Plot $y(x)$ versus t for $t = 0$, 0.0005 s, 0.001 s, 0.0015 s, and 0.002 s. (b) What is the frequency of the wave? (c) What is the wavelength λ?

A fiery spray of molten rock, ash, and smoke rises over the small town of Vestmannaeyjar as a volcano erupts on Heimaey, Iceland. (© Pete Turner, The IMAGE Bank)

PART III
Thermodynamics

When dining, I had often observed that some particular dishes retained their Heat much longer than others; and that apple pies, and apples and almonds mixed (a dish in great repute in England) remained hot a surprising length of time. Much struck with this extraordinary quality of retaining Heat, which apples appeared to possess, it frequently occurred to my recollection; and I never burnt my mouth with them, or saw others meet with the same misfortune, without endeavouring, but in vain, to find out some way of accounting, in a satisfactory manner, for this surprising phenomenon.

BENJAMIN THOMPSON
(Count Rumford)

As we saw in the first part of this textbook, Newtonian mechanics explains a wide range of phenomena on a macroscopic scale, such as the motion of baseballs, rockets, and the planets of our solar system. We now turn to the study of thermodynamics, which is concerned with the concepts of heat and temperature. As we shall see, thermodynamics is very successful in explaining the bulk properties of matter and the correlation between these properties and the mechanics of atoms and molecules.

Historically, the development of thermodynamics paralleled the development of the atomic theory of matter. By the middle of the 19th century, chemical experiments provided solid evidence for the existence of atoms. At that time, scientists recognized that there must be a connection between the theory of heat and temperature, and the structure of matter. In 1827, the botanist Robert Brown reported that grains of pollen suspended in a liquid move erratically from one place to another, as if under constant agitation. In 1905, Albert Einstein developed a theory in which he used thermodynamics to explain the cause of this erratic motion, today called Brownian motion. Einstein explained this phenomenon by assuming that the grains of pollen are under constant bombardment by "invisible" molecules in the liquid, which themselves undergo an erratic motion. This important experiment and Einstein's insight gave scientists a means of discovering vital information concerning molecular motion. It also gave reality to the concept of the atomic constituents of matter.

Have you ever wondered how a refrigerator is able to cool its contents or what types of transformations occur in a power plant or in the engine of your automobile or what happens to the kinetic energy of an object when it falls to the ground and comes to rest? The laws of thermodynamics and the concepts of heat and temperature enable us to answer such practical questions.

Many things can happen to an object when it is heated. Its size will change slightly, but it may also melt, boil, ignite, or even explode. The outcome depends upon the composition of the object, the degree to which it is heated, and its environment. In general, thermodynamics must concern itself with the physical and chemical transformations of matter in all of its forms: solid, liquid, gas, and plasma.

19

Temperature, Thermal Expansion, and Ideal Gases

The ceramic material being heated by a flame is a poor conductor of heat but is able to withstand a large temperature gradient; its left side is buried in ice (0°C) while its right side is extremely hot. (Courtesy of Corning Glass Works)

The subject of thermal physics deals with phenomena involving energy transfer between bodies at different temperatures. In the study of mechanics such concepts as mass, force, and kinetic energy were carefully defined in order to make the subject quantitative. Likewise, a quantitative description of thermal phenomena requires a careful definition of the concepts of temperature, heat, and internal energy. The science of thermodynamics is concerned with the study of heat flow from a *macroscopic* viewpoint. The laws of thermodynamics provide us with a relationship between heat flow, work, and internal energy of a system. In practice, suitable observable quantities must be selected to describe the overall behavior of a system. For example, the macroscopic quantities, pressure, volume, and temperature are used to characterize the properties of a gas. Thermal phenomena can also be understood using a *microscopic* approach, which describes what is happening on a molecular scale. For example, the temperature of a gas is a measure of the average kinetic energy of the gas molecules.

The composition of a body is an important factor when dealing with thermal phenomena. For example, liquids and solids expand only slightly when heated. On the other hand, a gas tends to undergo appreciable expansion when heated. If the gas is not free to expand, its pressure rises when heated. Certain substances may melt, boil, burn, or explode, depending on their composition and structure. Thus, the thermal behavior of a substance is closely related to its structure.

It would be far beyond the scope of this book to attempt to present applications of thermodynamics to a wide variety of substances. Instead, we shall examine some rather simple systems, such as a dilute gas and a homogeneous solid. Emphasis will be placed on understanding the key principles of thermodynamics and on providing a basis upon which the thermal behavior of all matter can be understood.

19.1 TEMPERATURE AND THE ZEROTH LAW OF THERMODYNAMICS

When we speak of the temperature of an object, we often associate this concept with the degree of "hotness" or "coldness" of the object when we touch it. Thus, our senses provide us with a qualitative indication of temperature. However, our senses are unreliable and often misleading. For example, if we remove a metal ice tray and a package of frozen vegetables from the freezer, the ice tray feels colder to the hand even though both are at the same temperature. This is because metal is a better conductor of heat than cardboard. What we need is a reliable and reproducible method for establishing the relative "hotness" or "coldness" of bodies. Scientists have developed various types of thermometers for making such quantitative measurements. Some typical thermometers will be described in Section 19.2.

We are familiar with the fact that two objects at different initial temperatures will eventually reach some intermediate temperature when placed in contact with each other. For example, a piece of meat placed on a block of ice in a well-insulated container will eventually reach a temperature near 0°C. Likewise, if an ice cube is dropped into a container of warm water, the ice cube will eventually melt and the water's temperature will decrease. If the process takes place in a thermos bottle, the system (water + ice) is approximately isolated from its surroundings.

In order to understand the concept of temperature, it is useful to first define two often used phrases, *thermal contact* and *thermal equilibrium*. Two objects are in **thermal contact** with each other if energy exchange can occur between them in the absence of macroscopic work done by one on the other. **Thermal equilibrium** is a situation in which two objects in thermal contact with each other cease to have any net energy exchange due to a difference in their temperatures. The time it takes the two objects to reach thermal equilibrium depends on the properties of the objects and on the pathways available for energy exchange.

Now consider two objects, A and B, which are not in thermal contact, and a third object, C, which will be our thermometer. We wish to determine whether or not A and B are in thermal equilibrium with each other. The thermometer (object C) is first placed in thermal contact with A until thermal equilibrium is reached. At that point, the thermometer's reading will remain constant. The thermometer is then placed in thermal contact with B, and its reading is recorded after thermal equilibrium is reached. If the readings after contact with A and B are the same, then A and B are in thermal equilibrium with each other. We can summarize these results in a statement known as the **zeroth law of thermodynamics** *(the law of equilibrium):*

If objects A and B are separately in thermal equilibrium with a third object, C, then A and B are in thermal equilibrium with each other.

This statement, although it may seem obvious, is most fundamental in the field of thermodynamics since it can be used to define temperature. We can think of temperature as the property that determines whether or not an object is in thermal equilibrium with other objects. That is, *two objects in thermal equilibrium with each other are at the same temperature.* Conversely, if two objects have different temperatures, they cannot be in thermal equilibrium with each other at that time.

19.2 THERMOMETERS AND TEMPERATURE SCALES

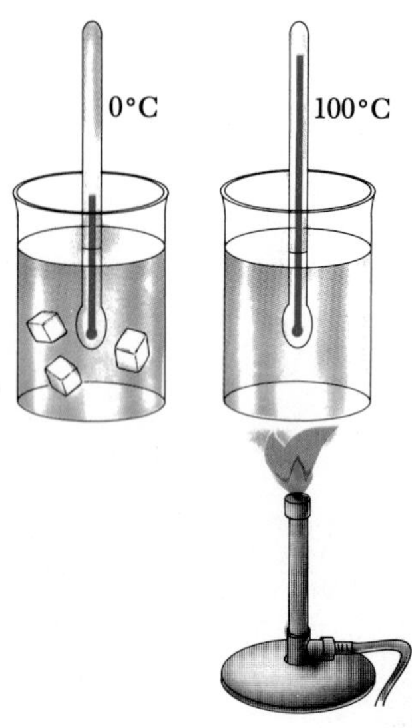

Figure 19.1 Schematic diagram of a mercury thermometer. As a result of thermal expansion, the level of the mercury rises as the mercury is heated from 0°C (the ice point) to 100°C (the steam point).

Thermometers are devices used to define and measure the temperature of a system. A thermometer in thermal equilibrium with a system measures both the temperature of the system and its own temperature. All thermometers make use of the change in some physical property with temperature. Some of these physical properties are (1) the change in volume of a liquid, (2) the change in length of a solid, (3) the change in pressure of a gas at constant volume, (4) the change in volume of a gas at constant pressure, (5) the change in electric resistance of a conductor, and (6) the change in color of a very hot body. A temperature scale can be established for a given substance using any one of these physical quantities.

The most common thermometer in everyday use consists of a glass bulb connected to a glass capillary tube. The glass bulb is filled with a volume of mercury that expands into the capillary tube when heated (Fig. 19.1). Thus, the physical property in this case is the thermal expansion of the mercury. One can now define any temperature change to be proportional to the change in length of the mercury column. The thermometer can be calibrated by placing it in thermal contact with some natural systems that remain at constant temperature (called a *fixed-point temperature*). One of the fixed-point temperatures normally chosen is that of a mixture of water and ice at atmospheric pressure, which is defined to be zero degrees Celsius, written 0°C. (This was formerly called *degrees centigrade.*) Another convenient fixed point is the temperature of a mixture of water and water vapor (steam) in equilibrium at atmospheric pressure. The temperature of this *steam point* is 100°C. Once the mercury levels have been established at these fixed points, the column is divided into 100 equal segments, each denoting a change in temperature of one Celsius degree.

Thermometers calibrated in this way do present problems, however, when extremely accurate readings are needed. For instance, an alcohol thermometer calibrated at the ice and steam points of water might agree with a mercury thermometer only at the calibration points. Because mercury and alcohol have different thermal expansion properties, when one thermometer reads a temperature of 50°C, say, the other will indicate a slightly different value. The discrepancies between thermometers are especially large when the temperatures to be measured are far from the calibration points.[1] An additional practical problem of any thermometer is its limited temperature range. A mercury thermometer, for example, cannot be used below the freezing point of mercury, which is −39°C. What we need is a universal thermometer whose readings are independent of the substance used. The gas thermometer meets this requirement.

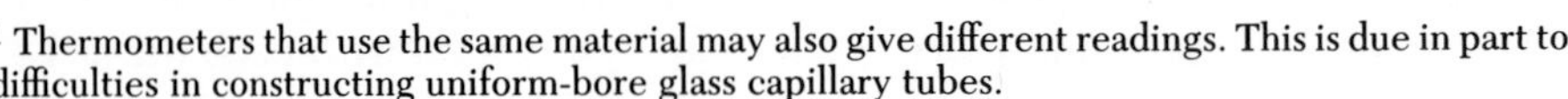
[1] Thermometers that use the same material may also give different readings. This is due in part to difficulties in constructing uniform-bore glass capillary tubes.

19.3 THE CONSTANT-VOLUME GAS THERMOMETER AND THE KELVIN SCALE

In a gas thermometer, the temperature readings are nearly independent of the substance used in the thermometer. One version of this is the constant-volume gas thermometer shown in Figure 19.2. The physical property in this device is the pressure variation with temperature of a fixed volume of gas. As the gas is heated, its pressure increases and the height of the mercury column shown in Figure 19.2 increases. When the gas is cooled, its pressure decreases, so the column height decreases. Thus, we can define temperature in terms of the concept of pressure discussed in Chapter 15. If the variation of temperature, T, with pressure is assumed to be linear, then

$$T = aP + b \tag{19.1}$$

where a and b are constants. These constants can be determined from two fixed points, such as the ice and steam points described in Section 19.2.

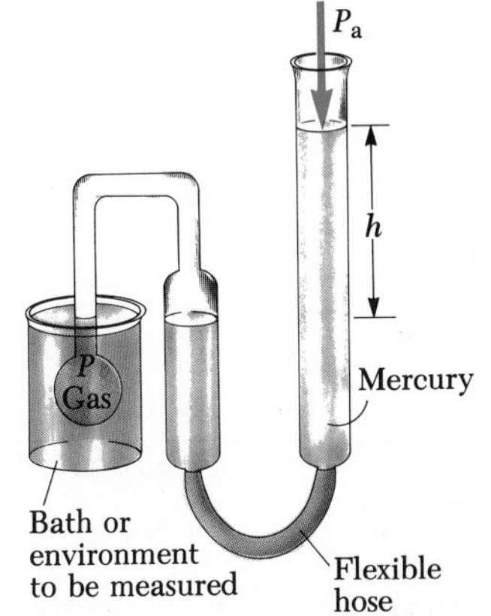

Figure 19.2 A constant-volume gas thermometer measures the pressure of the gas contained in the flask on the left. The volume of gas in the flask is kept constant by raising or lowering the column on the right such that the mercury level on the left remains constant.

Now suppose that temperatures are measured with various gas thermometers containing different gases. Experiments show that the thermometer readings are nearly independent of the type of gas used, so long as the gas pressure is low and the temperature is well above the liquefaction point. The agreement among thermometers using various gases improves as the pressure is reduced. This agreement of all gas thermometers at low pressure and high temperature implies that the intercept b appearing in Equation 19.1 is the same for *all* gases. This fact is illustrated in Figure 19.3. When the pressure versus temperature curve is extrapolated to very low temperatures, one finds that the pressure is zero when the temperature is $-273.15°\text{C}$. This temperature corresponds to the constant b in Equation 19.1. An extrapolation is necessary since all gases liquefy before reaching this temperature.

Early gas thermometers made use of the ice point and steam point as standard temperatures. However, for various technical reasons, these points are experimentally difficult to duplicate. Hence, a new temperature scale based on a single fixed point with b equal to zero was adopted in 1954 by the International Committee on Weights and Measures. The *triple point of water*, which corresponds to the single temperature and pressure at which water, water vapor, and ice can coexist in equilibrium, was chosen as a convenient and reproducible reference temperature for this new scale. The triple point of water occurs at a temperature of about 0.01°C and a pressure of 0.61 kPa. The temperature at the triple point of water on the new scale was set at 273.16

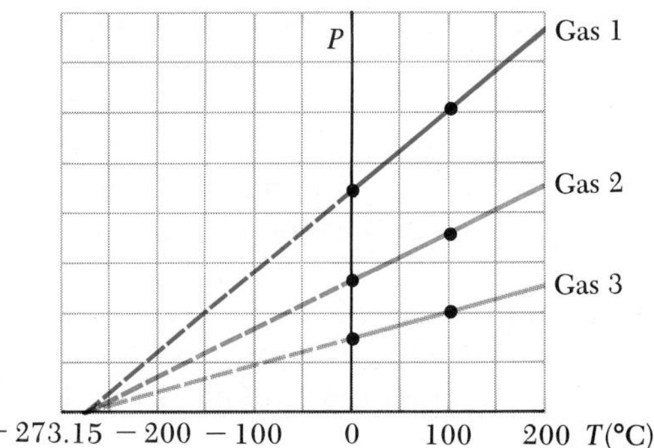

Figure 19.3 Pressure versus temperature for dilute gases. Note that, for all gases, the pressure extrapolates to zero at the unique temperature of $-273.15°\text{C}$.

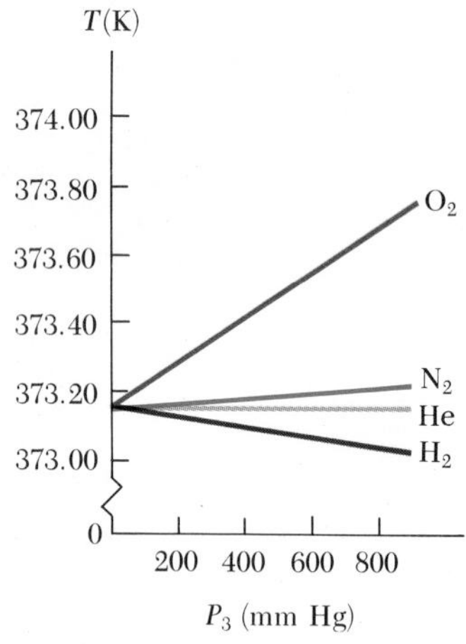

Figure 19.4 The temperature read with a constant-volume gas thermometer versus P_3, the pressure at the steam point of water, for various gases. Note that as the pressure is reduced, the steam-point temperature of water approaches a common value of 373.15 K regardless of which gas is used in the thermometer. Furthermore, the data for helium are nearly independent of pressure, which suggests it behaves like an ideal gas over this range.

kelvin, abbreviated 273.16 K.[2] This choice was made so that the old temperature scale based on the ice- and steam-points would agree closely with the new scale based on the triple point. This new scale is called the **thermodynamic temperature scale** and the SI unit of thermodynamic temperature,

> the **kelvin,** is defined as the fraction 1/273.16 of the temperature of the triple point of water.

If we take $b = 0$ in Equation 19.1 and call P_3 the pressure at the triple-point temperature, $T_3 = 273.16$ K, then we see that $a = (273.16\text{ K})/P_3$. Therefore, the temperature at a measured gas pressure P for a constant-volume gas thermometer is defined to be

$$T = \left(\frac{273.16\text{ K}}{P_3}\right)P \qquad \text{(constant } V) \qquad (19.2)$$

As mentioned earlier, one finds experimentally that as the pressure P_3 decreases, the measured value of the temperature approaches the same value for all gases. An example of such a measurement is illustrated in Figure 19.4, which shows the steam-point temperature measured with a constant-volume gas thermometer using various gases. As P_3 approaches zero, all measurements approach a common value of 373.15 K. Similarly, one finds that the ice-point temperature is 273.15 K.

In the limit of low gas pressures and high temperatures, real gases behave as what is known as an **ideal gas,** which will be discussed in detail in Section 19.6 and Chapter 21. The temperature scale defined in this limit of low gas pressures is called the **ideal gas temperature,** T, given by

Definition of ideal gas temperature

$$T \equiv 273.16\text{ K} \lim_{P_3 \to 0} \frac{P}{P_3} \qquad \text{(constant } V) \qquad (19.3)$$

Thus the constant-volume gas thermometer defines a temperature scale that can be reproduced in laboratories throughout the world. Although the scale depends on the properties of a gas, it is independent of which gas is used. In practice, one can use a gas thermometer down to around 1 K using low-pressure helium gas. Helium liquefies below this temperature; other gases liquefy at even higher temperatures.

It would be convenient to have a temperature scale that is independent of the property of any substance. Such a scale is called an **absolute temperature scale, or kelvin scale.** Later we shall find that the ideal gas scale is identical with the absolute temperature scale for temperatures above 1 K, where gas thermometers can be used. In anticipation of this, we shall also use the symbol T to denote absolute temperature. The absolute temperature scale will be properly defined when we study the second law of thermodynamics in Ch. 22.

Other methods of thermometry calibrated against gas thermometers have been used to provide various other fixed-point temperatures. The "International Practical Temperature Scale of 1968," which was established by international agreement, is based on measurements in various national standard laboratories. The assigned temperatures of particular fixed points associated

[2] A second fixed point at 0 K is implied by Equation 19.1. We shall describe the meaning of this point in Chapter 22 when we discuss the second law of thermodynamics.

TABLE 19.1 Fixed-Point Temperatures

Fixed Point	Temperature (°C)	Temperature (K)
Triple point of hydrogen	−259.34	13.81
Boiling point of hydrogen at 33360.6 N/m^2 pressure	−256.108	17.042
Boiling point of hydrogen	−252.87	20.28
Triple point of neon	−246.048	27.102
Triple point of oxygen	−218.789	54.361
Boiling point of oxygen	−182.962	90.188
Triple point of water	0.01	273.16
Boiling point of water	100.00	373.15
Freezing point of tin	231.9681	505.1181
Freezing point of zinc	419.58	692.73
Freezing point of silver	961.93	1235.08
Freezing point of gold	1064.43	1337.58

All values from National Bureau of Standards Special Publication 420, U. S. Department of Commerce, May 1975. All values at standard atmospheric pressure except as noted.

with various substances are given in Table 19.1. The platinum resistance thermometer (described in the next section) was used to establish all but the last two points in this table. The scale is not defined below 13.81 K.

19.4 THE CELSIUS AND FAHRENHEIT TEMPERATURE SCALES[3]

The Celsius temperature, T_C, is shifted from the absolute (or kelvin) temperature T by 273.15°, since by definition the triple point of water (273.16 K) corresponds to 0.01°C. Therefore,

$$T_C = T - 273.15 \tag{19.4}$$

From this we see that the size of a degree on the kelvin scale is the same as on the Celsius scale. In other words, a temperature difference of 5 Celsius degrees, written 5 C°, is equal to a temperature difference of 5 K. The two scales differ only in the choice of the zero point. Furthermore, the ice point (273.15 K) corresponds to 0.00°C, and the steam point (373.15 K) is equivalent to 100.00°C.

Another temperature scale used in the United States is the *Fahrenheit scale.*[4] The Fahrenheit temperature is related to the Celsius temperature through the relation

$$T_F = \tfrac{9}{5} T_C + 32 \text{ F}° \tag{19.5}$$

From this expression it follows that the ice point (0.00°C) equals 32°F and the steam point (100.00°C) equals 212°F. Figure 19.5 shows on a logarithmic scale the absolute temperatures for various physical processes and structures.

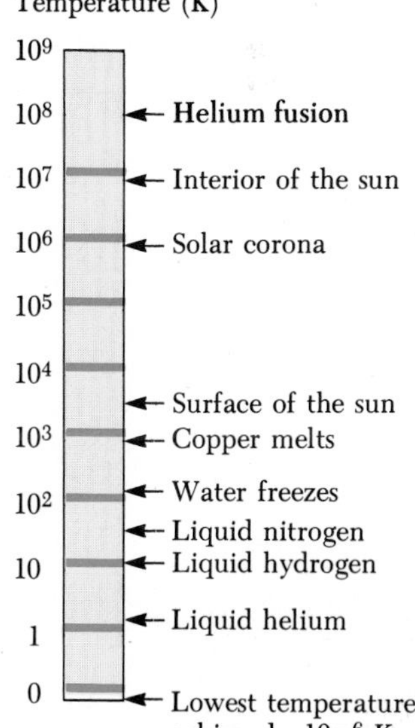

Figure 19.5 Absolute temperatures at which various physical processes take place. Note that the scale is logarithmic.

[3] Named after Anders Celsius (1701–1744), and Gabriel Fahrenheit (1686–1736).

[4] A temperature scale sometimes used in Great Britain is the *Rankine scale,* named after William McQuorn Rankine (1820–1872). The Rankine temperature, T_R, is related to the kelvin temperature through the relation $T_R = \frac{9}{5}T$. Likewise, the Rankine temperature is related to the Fahrenheit temperature through the relation $T_R = T_F + 459.67$.

EXAMPLE 19.1 Converting Temperatures □
An object has a temperature of 50°F. What is its temperature in degrees Celsius and in kelvins?

Solution Substituting $T_F = 50°F$ into Equation 19.5, we get

$$T_C = \frac{5}{9}(T_F - 32) = \frac{5}{9}(50 - 32) = \boxed{10°C}$$

From Equation 19.4, we find that

$$T = T_C + 273.15 = \boxed{283.15\ K}$$

EXAMPLE 19.2 Heating a Pan of Water
A pan of water is heated from 25°C to 80°C. What is the *change* in its temperature on the kelvin scale and on the Fahrenheit scale?

Solution From Equation 19.4, we see that the change in temperature on the Celsius scale equals the change on the kelvin scale. Therefore,

$$\Delta T = \Delta T_C = 80 - 25 = 55\ C° = \boxed{55\ K}$$

From Equation 19.5, we find that the change in temperature on the Fahrenheit scale is greater than the change on the Celsius scale by the factor 9/5. That is,

$$\Delta T_F = \frac{9}{5}\,\Delta T_C = \frac{9}{5}(80 - 25) = \boxed{99\ F°}$$

In other words, 55 C° = 99 F°, where the notations C° and F° refer to temperature *differences,* not to be confused with actual temperatures, which are written °C and °F.

Other Thermometers

A technique that is often used as a temperature standard in thermometry makes use of a pure platinum wire because its electrical resistance changes with temperature. The **platinum resistance thermometer** is essentially a coil of platinum wire mounted in a strain-free glass capsule. The platinum resistance changes by about 0.3% for a temperature change of 1 K. It is commonly used for temperatures ranging from about 14 K to 900 K and can be calibrated to within ±0.0003 K at the triple point of water.

Thermocouple

One of the most useful thermometers for scientific and engineering applications is a device called a **thermocouple.** The thermocouple is essentially a junction formed by two different metals or alloys, labeled A and B in Figure 19.6. The test junction is placed in the material whose temperature is to be

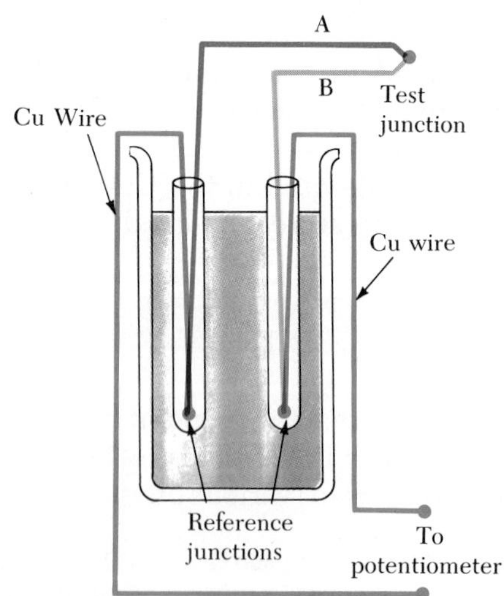

Figure 19.6 Schematic diagram of a thermocouple, which consists of two dissimilar metals, A and B. The reference junction is usually kept at 0°C.

measured, while the opposite ends of the thermocouple wires are maintained at some constant reference temperature (usually in a water-ice mixture) to form two junctions. When the reference temperature is different from the temperature of the test junction, a voltage called the *electromotive force* (emf) appears between the two junctions. The value of this emf is proportional to the temperature difference and therefore can be used to measure an unknown temperature. An instrument called a *potentiometer* is used to measure the emf. In practice, one usually uses junctions for which calibration curves are available.

One advantage of the thermocouple is its small mass, which enables it to quickly reach thermal equilibrium with the material being probed. Some common examples of thermocouple junction materials are copper/constantan (an alloy), which is useful over the temperature range of about −180°C to 400°C, and platinum/platinum-10% rhodium, which is useful over the range from about 0°C to 1500°C. Some typical outputs for various thermocouples are given in Figure 19.7. where the reference junction is at 0°C.

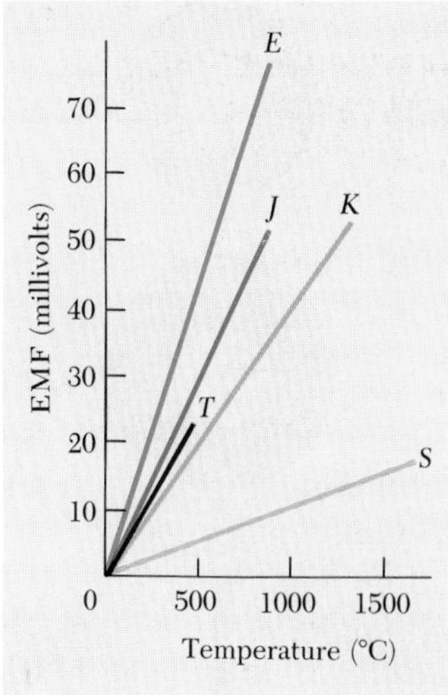

Figure 19.7 Plot of emf (junction voltage) versus temperature for various thermocouples: E, chromel/constantan; J, iron/constantan; T, copper/constantan; K, chromel/alumel; S, platinum/platinum-10% rhodium.

Another thermometer that has extremely high sensitivity is a device called a **thermistor.** This device consists of a small piece of semiconductor material whose electrical resistance changes with temperature. Thermistors are usually fabricated from oxides of various metals, such as nickel, manganese, iron, cobalt, and copper, and can be encapsulated in an epoxy. A careful measurement of the resistance serves as an indicator of temperature, with a typical accuracy of ±0.1 C°. Temperature changes as small as about 10^{-3} C° can be detected with these devices. Most thermistors operate reliably over the temperature range from about −50°C to 100°C. They are often used as clinical thermometers (with digital readout) and in various biological applications.

19.5 THERMAL EXPANSION OF SOLIDS AND LIQUIDS

Most bodies expand as their temperature increases. This phenomenon plays an important role in numerous engineering applications. For example, thermal expansion joints must be included in buildings, concrete highways, railroad tracks, and bridges to compensate for changes in dimensions with temperature variations.

The overall thermal expansion of a body is a consequence of the change in the average separation between its constituent atoms or molecules. To understand this, consider a crystalline solid, which consists of a regular array of atoms held together by electrical forces. We can obtain a mechanical model of these forces by imagining that the atoms are connected by a set of stiff springs as in Figure 19.8. The interatomic forces are taken to be elastic in nature. At ordinary temperatures, the atoms vibrate about their equilibrium positions with an amplitude of about 10^{-11} m and a frequency of about 10^{13} Hz. The average spacing between the atoms is of the order of 10^{-10} m. As the temperature of the solid increases, the atoms vibrate with larger amplitudes and the average separation between them increases.[5] Consequently, the solid as a whole expands with increasing temperature. If the expansion of an object is sufficiently small compared with its initial dimensions, then the change in any

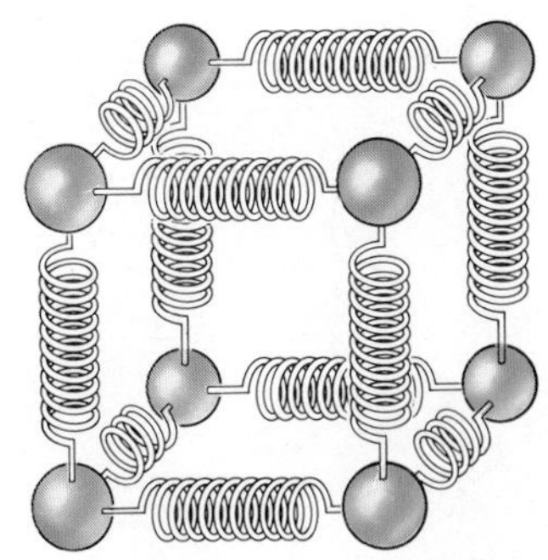

Figure 19.8 A mechanical model of a crystalline solid. The atoms (solid spheres) are imagined to be attached to each other by springs, which reflect the elastic nature of the interatomic forces.

[5] Strictly speaking, thermal expansion arises from the *asymmetric* nature of the potential energy curve for the atoms in a solid. If the oscillators were truly harmonic, the average atomic separations would not change regardless of the amplitude of vibration.

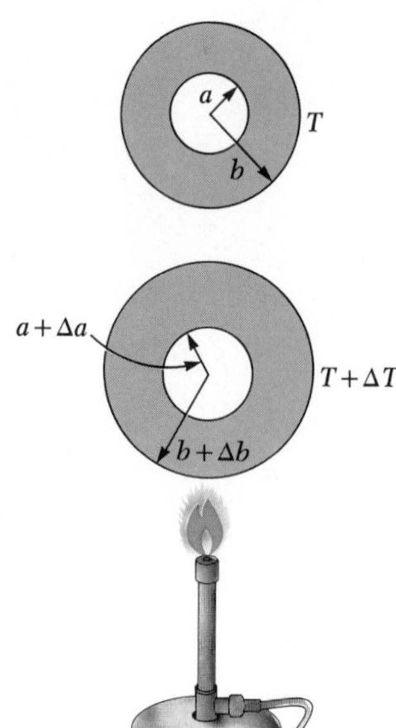

Figure 19.9 Thermal expansion of a homogeneous metal washer. Note that as the washer is heated, all dimensions increase. The expansion is exaggerated.

dimension (length, width, or thickness) is, to a good approximation, a linear function of the temperature.

Suppose the linear dimension of a body along some direction is ℓ at some temperature. The length increases by an amount $\Delta\ell$ for a change in temperature ΔT. Experiments show that the change in length is proportional to the temperature change and to the original length when ΔT is small enough. Thus the basic equation for the expansion of a solid is

$$\Delta\ell = \alpha\ell\Delta T \tag{19.6}$$

where the proportionality constant α is called the **average coefficient of linear expansion** for a given material. From this expression, we see that

$$\alpha = \frac{1}{\ell}\frac{\Delta\ell}{\Delta T} \tag{19.7}$$

In other words, the average coefficient of linear expansion of a solid is the fractional change in length ($\Delta\ell/\ell$) per degree change in temperature. The unit of α is deg^{-1}. For example, an α value of 11×10^{-6} $(\text{C}°)^{-1}$ means that the length of an object changes by 11 parts per million of its original length for every Celsius degree change in temperature. It may be helpful to think of thermal expansion as an effective magnification or as a photographic enlargement of an object when it is heated. For example, as a metal washer is heated (Fig. 19.9) all dimensions increase, including the radius of the hole.

The coefficient of linear expansion generally varies with temperature. Usually this temperature variation is negligible over the temperature range of most everyday measurements. Table 19.2 lists the average coefficient of linear expansion for various materials. Note that α is positive for these materials, indicating an increase in length with increasing temperature. This is not always the case. For example, some single anisotropic crystalline substances, such as calcite ($CaCO_3$), expand along one dimension (positive α) and contract along another (negative α) with increasing temperature.

Because the linear dimensions of a body change with temperature, it follows that the area and volume of a body also change with temperature. The change in volume at constant pressure is proportional to the original volume V and to the change in temperature according to the relation

$$\Delta V = \beta V\,\Delta T \tag{19.8}$$

Thermal expansion: the extreme heat of a July day in Asbury Park, New Jersey, caused these railroad tracks to buckle. (Wide World Photos)

TABLE 19.2 Expansion Coefficients for Some Materials Near Room Temperature

Material	Linear Expansion Coefficient $\alpha(\text{C}°)^{-1}$	Material	Volume Expansion Coefficient $\beta(\text{C}°)^{-1}$
Aluminum	24×10^{-6}	Alcohol, ethyl	1.12×10^{-4}
Brass and bronze	19×10^{-6}	Benzene	1.24×10^{-4}
Copper	17×10^{-6}	Acetone	1.5×10^{-4}
Glass (ordinary)	9×10^{-6}	Glycerine	4.85×10^{-4}
Glass (pyrex)	3.2×10^{-6}	Mercury	1.82×10^{-4}
Lead	29×10^{-6}	Turpentine	9.0×10^{-4}
Steel	11×10^{-6}	Gasoline	9.6×10^{-4}
Invar (Ni-Fe alloy)	0.9×10^{-6}	Air	3.67×10^{-3}
Concrete	12×10^{-6}	Helium	3.665×10^{-3}

where β is the **average coefficient of volume expansion.** *For an isotropic solid, the coefficient of volume expansion is approximately three times the linear expansion coefficient,* or $\beta = 3\alpha$ (An **isotropic solid** is one in which the coefficient of linear expansion is the same in all directions.) Therefore, Equation 19.8 can be written

$$\Delta V = 3\alpha V \, \Delta T \qquad (19.9)$$

Change in volume of an isotropic solid at constant pressure

To show that $\beta = 3\alpha$ for an isotropic solid, consider an object in the shape of a box of dimensions ℓ, w, and h. Its volume at some temperature T is $V = \ell wh$. If the temperature changes to $T + \Delta T$, its volume changes to $V + \Delta V$, where each dimension changes according to Equation 19.6. Therefore,

$$\begin{aligned} V + \Delta V &= (\ell + \Delta\ell)(w + \Delta w)(h + \Delta h) \\ &= (\ell + \alpha\ell \, \Delta T)(w + \alpha w \, \Delta T)(h + \alpha h \, \Delta T) \\ &= \ell wh(1 + \alpha \, \Delta T)^3 \\ &= V[1 + 3\alpha \, \Delta T + 3(\alpha \, \Delta T)^2 + (\alpha \, \Delta T)^3] \end{aligned}$$

Hence the fractional change in volume is

$$\frac{\Delta V}{V} = 3\alpha \, \Delta T + 3(\alpha \, \Delta T)^2 + (\alpha \, \Delta T)^3$$

Since the product $\alpha \, \Delta T$ is small compared with unity for typical values of ΔT (less than $\approx 100\text{C}^\circ$), we can neglect the terms $3(\alpha \, \Delta T)^2$ and $(\alpha \, \Delta T)^3$. In this approximation, we see that

$$\beta = \frac{1}{V} \frac{\Delta V}{\Delta T} = 3\alpha$$

A sheet or flat plate can be described by its area. You should show (Problem 53) that the change in the area of an isotropic plate is given by

$$\Delta A = 2\alpha A \, \Delta T \qquad (19.10)$$

Change in area of an isotropic plate

EXAMPLE 19.3 Expansion of a Railroad Track
A steel railroad track has a length of 30 m when the temperature is 0°C. (a) What is its length on a hot day when the temperature is 40°C?

Solution Making use of Table 19.2 and noting that the change in temperature is 40 C°, we find that the increase in length is

$$\Delta\ell = \alpha\ell \, \Delta T = [11 \times 10^{-6} (\text{C}^\circ)^{-1}](30 \text{ m})(40 \text{ C}^\circ)$$

$$= \; 0.013 \text{ m}$$

Therefore, its length at 40°C is 30.013 m.

(b) Suppose the ends of the rail are rigidly clamped at 0°C so as to prevent expansion. Calculate the thermal stress set up in the rail if its temperature is raised to 40°C.

Solution From the definition of Young's modulus for a solid (Chapter 12), we have

$$\text{Tensile stress} = \frac{F}{A} = Y \frac{\Delta\ell}{\ell}$$

Since Y for steel is 20×10^{10} N/m² we have

$$\frac{F}{A} = \left(20 \times 10^{10} \frac{\text{N}}{\text{m}^2}\right)\left(\frac{0.013 \text{ m}}{30 \text{ m}}\right) = \; 8.67 \times 10^7 \text{ N/m}^2$$

Exercise 1 If the rail has a cross-sectional area of 30 cm², calculate the force of compression in the rail.
Answer 2.60×10^5 N or 58 500 lb!

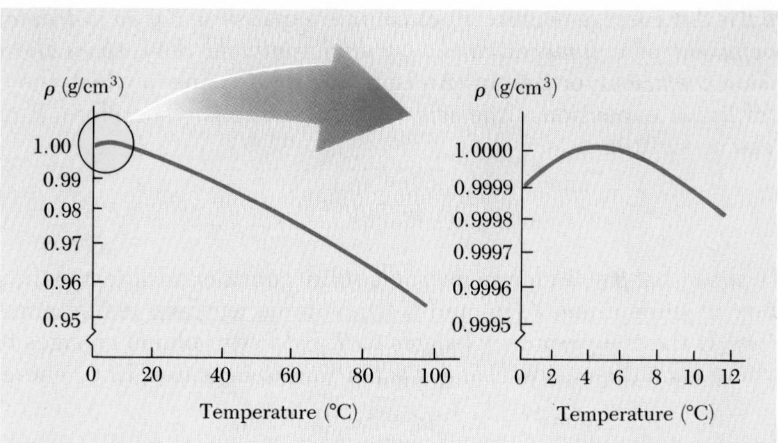

Figure 19.10 The variation of density with temperature for water at atmospheric pressure. The maximum density occurs at 4°C as can be seen in the magnified graph at the right.

Liquids generally increase in volume with increasing temperature, and have volume expansion coefficients about ten times greater than those of solids (Table 19.2). Water is an exception to this rule, as we can see from its density versus temperature curve, shown in Figure 19.10. As the temperature increases from 0°C to 4°C, the water contracts, and thus its density increases. Above 4°C, the water expands with increasing temperature. The density of water reaches a *maximum* value of 1000 kg/m³ at 4°C.

We can explain why a pond or lake freezes at the surface from this unusual thermal expansion behavior of water. As the pond cools, the cooler, denser water at the surface initially flows to the bottom. When the temperature throughout the depth of the pond reaches 4°C, this flow ceases. Consequently, when the surface of the pond is below 4°C, equilibrium is reached when the coldest water is at the surface. As the water freezes at the surface, it remains there since ice is less dense than water. The ice continues to build up at the surface, while water near the bottom remains at 4°C. If this did not happen, fish and other forms of marine life would not survive.

19.6 MACROSCOPIC DESCRIPTION OF AN IDEAL GAS

In this section we shall be concerned with the properties of a gas of mass m confined to a container of volume V at a pressure P and temperature T. It is useful to know how these quantities are related. In general, the equation that interrelates these quantities, called the *equation of state,* is very complicated. However, if the gas is maintained at a very low pressure (or low density), the equation of state is found experimentally to be quite simple. Such a low-density gas is commonly referred to as an **ideal gas.**[6] Most gases at room temperature and atmospheric pressure behave as ideal gases.

It is convenient to express the amount of gas in a given volume in terms of the number of moles, n. By definition, **one mole** of any substance is that mass of the substance that contains a specific number of molecules called Avogadro's number, N_A. The value of N_A is approximately 6.022×10^{23} molecules/mole.

[6] To be more specific, the assumption here is that the temperature of the gas is sufficiently high and its pressure sufficiently low that the gas does not condense into a liquid.

Avogadro's number is defined to be the number of carbon atoms in 12 g of the isotope carbon-12. The number of moles of a substance is related to its mass m through the expression

$$n = \frac{m}{M} \tag{19.11}$$

where M is a quantity called the **molecular weight** of the substance, usually expressed in g/mole. For example, the molecular weight of oxygen, O_2, is 32.0 g/mol. Therefore, the mass of one mole of oxygen is 32.0 g.

Now suppose an ideal gas is confined to a cylindrical container whose volume can be varied by means of a movable piston, as in Figure 19.11. We shall assume that the cylinder does not leak, so the mass (or the number of moles) remains constant. For such a system, experiments provide the following information. First, when the gas is kept at a constant temperature, its pressure is inversely proportional to the volume (Boyle's law). Second, when the pressure of the gas is kept constant, the volume is directly proportional to the temperature (the law of Charles and Gay-Lussac). These observations can be summarized by the following **equation of state for an ideal gas:**

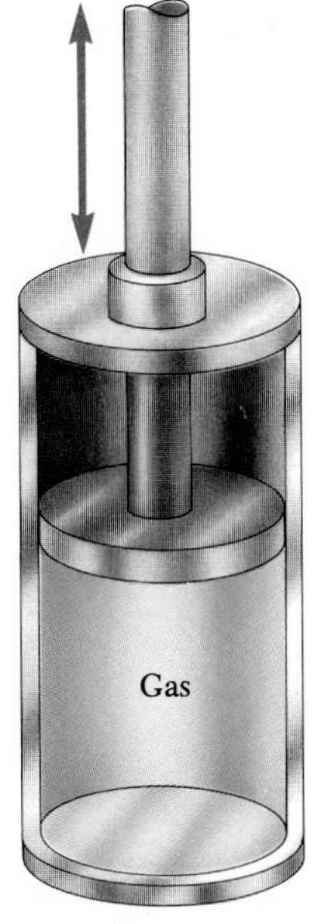

Figure 19.11 An ideal gas contained in a cylinder with a movable piston that allows the volume to be varied. The state of the gas is defined by its pressure, volume, and temperature.

$$PV = nRT \tag{19.12}$$

In this expression, R is a constant for a specific gas, which can be determined from experiments, and T is the absolute temperature in kelvin. Experiments on several gases show that as the pressure approaches zero, the quantity PV/nT approaches the same value of R for all gases. For this reason, R is called the **universal gas constant.** In metric units, where pressure is expressed in Pa and volume in m^3, the product PV has units of N·m, or J, and R has the value

The universal gas constant

$$R = 8.31 \text{ J/mol}\cdot\text{K} \tag{19.13}$$

If the pressure is expressed in atmospheres and the volume in liters ($1 \text{ L} = 10^3 \text{ cm}^3 = 10^{-3} \text{ m}^3$), then R has the value

$$R = 0.0821 \text{ L}\cdot\text{atm/mol}\cdot\text{K}$$

Using this value of R and Equation 19.12, one finds that the volume occupied by 1 mole of any gas at atmospheric pressure and 0°C (273 K) is 22.4 L.

The ideal gas law is often expressed in terms of the total number of molecules, N. Since the total number of molecules equals the product of the number of moles and Avogadro's number, we can write Equation 19.12 as

$$PV = nRT = \frac{N}{N_A}RT$$

$$PV = NkT \tag{19.14}$$

where k is called **Boltzmann's constant,** which has the value

Boltzmann's constant

$$k = \frac{R}{N_A} = 1.38 \times 10^{-23} \text{ J/K} \tag{19.15}$$

We see from Equation 19.14 that the pressure produced by a fixed volume of gas depends only on the temperature and the number of molecules within the volume. As we shall show in Chapter 21, this property of an ideal gas is explained by the kinetic theory of gases.

We have defined an ideal gas as one that obeys the equation of state, $PV = nRT$, under all conditions. In reality, an ideal gas does not exist. However, the concept of an ideal gas is very useful in view of the fact that real gases at low pressures behave as ideal gases. It is common to call quantities such as P, V, and T the **thermodynamic variables** of the system. We note that if the equation of state is known, then one of the variables can always be expressed as some function of the other two thermodynamic variables. That is, given two of the variables, the third can be determined from the equation of state. Other thermodynamic systems are often described with different thermodynamic variables. For example, a wire under tension at constant pressure is described by its length, the tension in the wire, and the temperature.

EXAMPLE 19.4 How Many Gas Molecules Are in a Container?

An ideal gas occupies a volume of 100 cm³ at 20°C and a pressure of 100 Pa. Determine the number of moles of gas in the container.

Solution The quantities given are volume, pressure, and temperature: $V = 100\ \text{cm}^3 = 10^{-4}\ \text{m}^3$, $P = 100$ Pa, and $T = 20°\text{C} = 293$ K. Using Equation 19.12, we get

$$n = \frac{PV}{RT} = \frac{(100\ \text{Pa})(10^{-4}\ \text{m}^3)}{(8.31\ \text{J/mol}\cdot\text{K})(293\ \text{K})}$$

$$= \boxed{4.11 \times 10^{-6}\ \text{moles}}$$

Note that you must express T as an absolute temperature (K) when using the ideal gas law.

Exercise 2 Calculate the number of molecules in the container, using the fact that Avogadro's number is 6.02×10^{23} molecules/mole.
Answer 2.47×10^{18} molecules.

EXAMPLE 19.5 Squeezing a Tank of Gas

Pure helium gas is admitted into a tank containing a movable piston. The initial volume, pressure, and temperature of the gas are $15 \times 10^{-3}\ \text{m}^3$, 200 kPa, and 300 K. If the volume is decreased to $12 \times 10^{-3}\ \text{m}^3$ and the pressure increased to 350 kPa, find the final temperature of the gas. (Assume that helium behaves like an ideal gas.)

Solution If no gas escapes from the tank, the number of moles remains constant; therefore using $PV = nRT$ at the initial and final points gives

$$\frac{P_iV_i}{T_i} = \frac{P_fV_f}{T_f}$$

where i and f refer to the initial and final values. Solving for T_f, we get

$$T_f = \left(\frac{P_fV_f}{P_iV_i}\right)T_i$$

$$= \frac{(350\ \text{kPa})(12 \times 10^{-3}\ \text{m}^3)}{(200\ \text{kPa})(15 \times 10^{-3}\ \text{m}^3)}(300\ \text{K}) = \boxed{420\ \text{K}}$$

EXAMPLE 19.6 Heating a Bottle of Air

A sealed glass bottle containing air at atmospheric pressure (101 kPa) and having a volume of 30 cm³ is at 23°C. It is then tossed into an open fire. When the temperature of the air in the bottle reaches 200°C, what is the pressure inside the bottle? Assume any volume changes of the bottle are negligible.

Solution This example is approached in the same fashion as that used in Example 19.5. We start with the expression

$$\frac{P_iV_i}{T_i} = \frac{P_fV_f}{T_f}$$

Since the initial and final volumes of the gas are assumed equal, this expression reduces to

$$\frac{P_i}{T_i} = \frac{P_f}{T_f}$$

This gives

$$P_f = \left(\frac{T_f}{T_i}\right)(P_i) = \left(\frac{473\text{ K}}{300\text{ K}}\right)(101\text{ kPa}) = \boxed{159\text{ kPa}}$$

Obviously, the higher the temperature, the higher the pressure exerted by the trapped air. Of course, if the pressure rises high enough, the bottle will shatter.

Exercise 3 In this example, we neglected the change in volume of the bottle. If the coefficient of volume expansion for glass is $27 \times 10^{-6}\ (\text{C}°)^{-1}$, find the magnitude of this volume change.
Answer $0.14\ \text{cm}^3$.

SUMMARY

Two bodies are in **thermal equilibrium** with each other if they have the same temperature.

The **zeroth law of thermodynamics** states that if bodies A and B are separately in thermal equilibrium with a third body, C, then A and B are in thermal equilibrium with each other.

The SI unit of thermodynamic temperature is the **kelvin,** which is defined to be the fraction 1/273.16 of the temperature of the triple point of water.

When a substance is heated, it generally expands. The linear expansion of an object is characterized by an **average expansion coefficient,** α, defined by

$$\alpha = \frac{1}{\ell}\frac{\Delta \ell}{\Delta T} \qquad (19.7)$$

Average coefficient of linear expansion

where ℓ is the initial linear dimension of the object and $\Delta\ell$ is the change in linear dimension for a temperature change ΔT. The **average volume expansion coefficient,** β, for a substance is equal to 3α.

An **ideal gas** is one that obeys the *equation of state,*

$$PV = nRT \qquad (19.12)$$

Equation of state for an ideal gas

where n equals the number of moles of gas, V is its volume, R is the universal gas constant (8.31 J/mol·K), and T is the absolute temperature in kelvins. A real gas behaves approximately as an ideal gas if it is far from liquefaction. An ideal gas is used as the working substance in a constant-volume gas thermometer, which defines the absolute temperature scale in kelvins. This absolute temperature T is related to temperatures on the Celsius scale by $T = T_C + 273.15$.

QUESTIONS

1. When a hot body warms a cold one, why don't we say that temperature flows from one to the other?
2. Is it possible for two objects to be in thermal equilibrium if they are not in contact with each other? Explain.
3. A piece of copper is dropped into a beaker of water. If the water's temperature rises, what happens to the temperature of the copper? Under what condition will the water and copper be in thermal equilibrium?
4. In principle, any gas can be used in a gas thermometer. Why is it not possible to use oxygen for temperatures as low as 15 K? What gas would you use? (Look at the data in Table 19.1.)
5. Rubber has a negative coefficient of linear expansion. What happens to the size of a piece of rubber as it is warmed?
6. Why should the amalgam used in dental fillings have the same coefficient of expansion as a tooth? What would occur if they were mismatched?

7. Explain why a column of mercury in a thermometer first descends slightly and then rises when placed in hot water.
8. Explain why the thermal expansion of a spherical shell made of an isotropic solid is equivalent to that of a solid sphere of the same material.
9. A steel wheel bearing has an inside diameter which is 1 mm smaller than an axle. How can it be made to fit onto the axle without removing any material?
10. Markings to indicate length are placed on a steel tape in a room that has a temperature of 22°C. Are measurements made with the tape on a day when the temperature is 27°C too long, too short, or accurate? Defend your answer.
11. What would happen if the glass of a thermometer expanded more upon heating than did the liquid inside?
12. Determine the number of grams in one mole of the following gases: (a) hydrogen, (b) helium, and (c) carbon monoxide.
13. Why is it necessary to use absolute temperature when using the ideal gas law?
14. An inflated rubber balloon filled with air is immersed in a flask of liquid nitrogen that is at 77 K. Describe what happens to the balloon, assuming that it remains flexible while being cooled.
15. Two identical cylinders at the same temperature each contain the same kind of gas. If the volume of cylinder A is three times greater than the volume of cylinder B, what can you say about the relative pressures in the cylinders?
16. The temperature of dry air decreases about 1 C° for every 150 m gain in altitude. Is this what is to be expected according to the gas law?
17. When you let the air out of a car's tires, the air is cool even if the tire is warm. Explain.
18. The suspension of a certain pendulum clock is made of brass. When the temperature increases, does the period of the clock increase, decrease, or remain the same? Explain your answer.
19. An automobile radiator is filled to the brim with water while the engine is cool. What happens to the water when the engine is running and the water is heated? What do modern automobiles have in their cooling systems to prevent the loss of coolants?
20. Metal lids on glass jars can often be loosened by running hot water over them. How is this possible?
21. The blade being heated in the photograph below consists of two different metals bonded together to form a bimetallic strip. Before being heated, the blade was straight. (a) Why does the blade bend when heated? (b) Which way would it bend if it were cooled?

(Question 21). (Courtesy of CENCO)

22. When the metal ring and metal sphere in the photograph below are both at room temperature, the sphere does not fit through the ring. After the ring is heated, the sphere can be passed through the ring. Why does this occur?

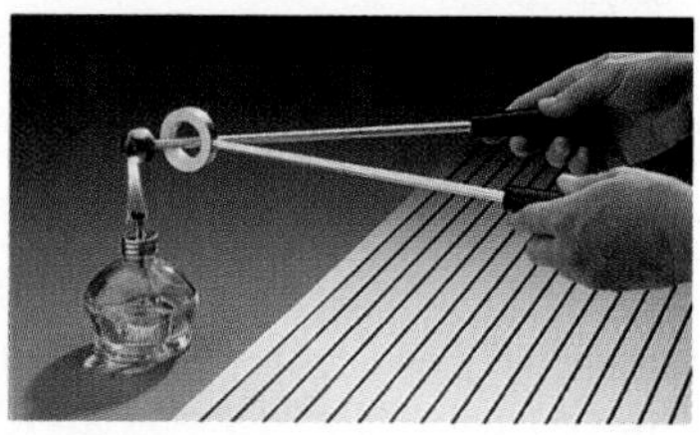

(Question 22). (Courtesy of CENCO)

PROBLEMS

Section 19.3 The Constant-Volume Gas Thermometer and the Kelvin Scale

1. A constant volume gas thermometer is calibrated in dry ice (−80°C) and in boiling ethyl alcohol (78°C). The two pressures are 0.900 atm and 1.635 atm. (a) What value of absolute zero does the calibration yield? (b) What pressures would be found at the freezing and boiling points of water?
2. The gas thermometer shown in Figure 19.2 reads a pressure of 40 mm Hg at the triple-point temperature of water. What pressure will it read at (a) the boiling point of water, and (b) the melting point of gold (1064.43°C)?
3. A constant-volume gas thermometer registers a pressure of 50 mm Hg when it is at a temperature of 450 K. (a) What is the pressure at the triple point of water? (b) What is the temperature when the pressure reads 2 mm Hg?
4. The pressure in a constant volume gas thermometer is 0.700 atm at 100°C and 0.512 atm at 0°C. (a) What is the temperature when the pressure is 0.0400 atm? (b) What is the pressure at 450°C?

5. A constant-volume gas thermometer is filled with helium. When immersed in boiling liquid nitrogen (at a temperature of 77.34 K), the absolute pressure is 25.00 kPa. (a) What is the temperature in degrees Celsius and kelvins when the pressure is 45.00 kPa? (b) What will the pressure be when the thermometer is immersed in boiling liquid hydrogen?

Section 19.4 The Celsius and Fahrenheit Temperature Scales

6. The melting point of gold is 1064°C and the boiling point is 2660°C. (a) Express these temperatures in kelvins. (b) Compute the difference of these temperatures in Celsius degrees and kelvin degrees and compare the two numbers.
7. Liquid nitrogen has a boiling point of −195.81°C at atmospheric pressure. Express this temperature in (a) degrees Fahrenheit, (b) degrees Rankine, and (c) kelvins.
8. The highest recorded temperature on Earth was 136°F, at Azizia, Libya, in 1922. The lowest recorded temperature was −127°F, at Vostok Station, Antarctica, in 1960. Express these temperature extremes in degrees Celsius.
9. The melting point of lead is 327.3°C. Express this in (a) degrees Fahrenheit and (b) kelvins.
10. The temperature of one northeastern state varies from 105°F in the summer to −25°F in winter. Express this range of temperature in degrees Celsius.
11. The boiling point of sulfur is 444.60°C. The melting point is 586.1 F° below the boiling point. (a) Determine the melting point in degrees Celsius. (b) Find the melting and boiling points in degrees Fahrenheit.
12. The normal human body temperature is 98.6°F. A person with a fever may record 102°F. Express these temperatures in degrees Celsius.
13. A substance is heated from −12°F to 150°F. What is its change in temperature on (a) the Celsius scale and (b) the kelvin scale?
14. A certain process cools a body from 350°C to −80°C. Express the change in temperature in (a) kelvins, and (b) Fahrenheit degrees.
15. At what temperature are the readings from a Fahrenheit thermometer and Celsius thermometer the same?

Section 19.5 Thermal Expansion of Solids and Liquids

Use Table 19.2.

16. An aluminum tube is 3.0000 m long at 20°C. What is its length at (a) 100°C and (b) 0°C?
17. A copper telephone wire is strung, with little sag, between two poles that are 35 m apart. How much longer is the wire on a summer day with $T_C = 35$°C than on a winter day with $T_C = -20$°C?
18. What is the change in length of a rod made up of 2.000 m of copper and 1.000 m of Al placed end to end, as it is moved from an ice water bath to a steam bath?
19. A structural steel I-beam is 15 m long when installed at 20°C. How much will its length change over the temperature extremes −30°C to 50°C?
20. The New River Gorge bridge in West Virginia is a steel arch bridge 518 m in length. How much will its length change between temperature extremes of −20°C and 35°C?
21. A steel track is 2 m long. The tracks are laid end to end with a synthetic rubber spacing between them. The coefficient of expansion of this rubber is equal to $-22 \times 10^{-5}(\text{C}°)^{-1}$ (the negative sign means that the rubber contracts when heated.) How thick should the spacer be to contract as much as the tracks expand if they are subjected to a 30 C° temperature spread?
22. A metal rod made of some alloy is to be used as a thermometer. At 0°C its length is 40.000 cm, and at 100°C its length is 40.060 cm. (a) What is the linear expansion coefficient of the alloy? (b) What is the temperature when its length is 39.975 cm?
23. A wire of diameter 0.20 mm is stretched until the tension is 50 N. The wire is then clamped to a stout aluminum bar when both are at a temperature of 20°C. What will be the tension in the wire when the wire and the bar are brought to a temperature of 150°C? [The wire has a Young's modulus of 34×10^{10} Pa and a linear expansion coefficient of $15 \times 10^{-6}\ (\text{C}°)^{-1}$.]
24. A brass ring of diameter 10.00 cm at 20°C is heated and slipped over an aluminum rod of diameter 10.01 cm at 20°C. Assuming the coefficients of linear expansion are constant, to what temperature (a) must this combination be cooled to separate them? Is this attainable? (b) What if the aluminum rod were 10.02 cm in diameter?
25. The concrete sections of a certain superhighway are designed to have a length of 25 m. The sections are poured and cured at 10°C. What minimum spacing should the engineer leave *between the sections* to eliminate "buckling" if the concrete is to reach a temperature of 50°C?
26. A hole with a diameter of 2.00 cm is drilled through a stainless steel plate at room temperature (20°C). By what fraction ($\Delta r/r$) will the hole be larger when the plate is heated to 150°C?
27. At 20°C, an aluminum ring has an inner diameter of 5.000 cm, and a brass rod has a diameter of 5.050 cm. (a) To what temperature must the aluminum ring be heated so that it will just slip over the brass rod? (b) To what temperature must *both* be heated so the aluminum ring will slip off the brass rod? Would this work?
28. A brass washer has an inner diameter of 2.000 cm and an outer diameter of 2.500 cm at 25°C. To what tem-

perature must the washer be heated to just fit over a rod that is 2.005 cm in diameter?

29. A hollow aluminum cylinder 20 cm deep has an internal capacity of 2.000 liters at 20°C. It is completely filled with turpentine, and then warmed to 80°C. (a) How much turpentine overflows? (b) If it is then cooled back to 20°C, how far below the surface of the cylinder's rim is the turpentine surface?

30. Calculate the *fractional* change in the volume ($\Delta V/V$) of an aluminum bar that undergoes a change in temperature of 30 C°. (Note that $\beta = 3\alpha$ for an isotropic substance.)

31. An automobile fuel tank is filled to the brim with 45 liters (11.9 gal) of gasoline at 10°C. Immediately afterward the vehicle is parked in the sun where the temperature is 35°C. How much gasoline overflows from the tank as a result of the expansion? (Neglect the expansion of the tank.)

32. A volumetric Pyrex glass flask is calibrated at 20°C. It is filled to the 100-mL mark with 35°C acetone. (a) What is the volume of the acetone when it is cooled to 20°C? (b) How significant is the change in volume of the flask itself?

33. The active element of a certain laser is made of a glass rod 30 cm long by 1.5 cm in diameter. If the temperature of the rod increases by 65°C, find the increase in (a) its length. (b) its diameter, and (c) its volume. (Take $\alpha = 9 \times 10^{-6}(\mathrm{C}°)^{-1}$.)

Section 19.6 Macroscopic Description of an Ideal Gas

34. An ideal gas is held in a container at constant volume. Initially, its temperature is 10°C and its pressure is 2.5 atm. What is its pressure when its temperature is 80°C?

35. A cylinder with a movable piston contains gas at a temperature of 127°C, a pressure of 30 kPa, and a volume of 4 m^3. What will be its final temperature if the gas is compressed to 2.5 m^3 and the pressure increases to 90 kPa?

36. Gas is contained in an 8-L vessel at a temperature of 20°C and a pressure of 9 atm. (a) Determine the number of moles of gas in the vessel. (b) How many molecules are there in the vessel?

37. Gas is confined in a tank at a pressure of 10.0 atm and a temperature of 15°C. If one half of the gas is withdrawn and the temperature is raised to 65°C, what is the new pressure in the tank?

38. A gas is heated from 27°C to 127°C while maintained at constant pressure in a vessel whose volume increases. By what factor does the volume change?

39. A cylinder of volume 12 liters contains helium gas at a pressure of 136 atm. How many balloons can be filled with this cylinder at atmospheric pressure if each balloon has a volume of 1 liter?

40. A tank with a volume of 0.1 m^3 contains helium gas at a pressure of 150 atm. How many balloons can the tank blow up if each filled balloon is a sphere 30 cm in diameter at an absolute pressure of 1.2 atm?

41. One mole of oxygen gas is at a pressure of 6 atm and a temperature of 27°C. (a) If the gas is heated at constant volume until the pressure triples, what is the final temperature? (b) If the gas is heated such that both the pressure and volume are doubled, what is the final temperature?

42. An automobile tire is inflated using air originally at 10°C and normal atmospheric pressure. During the process, the air is compressed to 28% of its original volume and the temperature is increased to 40°C. What is the tire pressure? After the car is driven at high speed, the tire air temperature rises to 85°C and the interior volume of the tire increases by 2%. What is the new tire pressure? Express each answer in Pa (absolute) and in lb/in.^{2} (gauge). (1 atm = 14.70 lb/in.^{2})

43. A porous balloon has a volume of 2 m^3 at a temperature of 10°C and a pressure of 1.1 atm. When it is warmed to 150°C the volume expands to 2.3 m^3 and it is noted that 5% of the gas leaks out. (a) How much gas was in the balloon at 10°C? (b) What is the pressure in the balloon at 150°C?

44. In state of the art vacuum systems, pressures as low as 10^{-11} mm Hg are being attained. Calculate the number of molecules in a 1 m^3 vessel at this pressure if the temperature is 27°C. (*Note:* One atm of pressure corresponds to 760 mm Hg.)

45. The tire on a bicycle is filled with air to a gauge pressure of 550 kPa (80 lb/in.^{2}) at 20°C. What is the gauge pressure in the tire after a ride on a hot day when the tire air temperature is 40°C? (Assume the volume does not change, and recall that the gauge pressure means absolute pressure in the tire minus atmospheric pressure. Furthermore, assume that the atmospheric pressure remains constant and equal to 101 kPa.)

46. Show that one mole of any gas at atmospheric pressure (1.01×10^5 N/m^2) and standard temperature (273 K) occupies a volume of 22.4 L.

47. A cylindrical diving bell 3 m in diameter and 4 m tall with an open bottom is submerged to a depth of 220 m in the ocean. The surface temperature is 25°C and the temperature 220 m down is 5°C. The density of sea water is 1025 kg/m^3. How high will the sea water rise in the bell when it is submerged?

48. A diving bell in the shape of a cylinder with a height of 2.50 m is closed at the upper end and open at the lower end. The bell is lowered from air into sea water ($\rho = 1.025$ g/cm^3). The air in the bell is initially at a temperature of 20°C. The bell is lowered to a depth (measured to the bottom of the bell) of 45.0 fathoms or 82.3 m. At this depth the water temperature is 4°C, and the bell is in thermal equilibrium with the water. (a) How high will sea water rise in the bell?

(b) To what minimum pressure must the air in the bell be raised to expel the water that entered?

49. A bubble of marsh gas rises from the bottom of a fresh-water lake at a depth of 4.2 m and a temperature of 5°C to the surface where the water temperature is 12°C. What is the ratio of the bubble diameter at the two locations? (Assume that the bubble gas is in thermal equilibrium with the water at each location.)

50. An expandable cylinder has its top connected to a spring of constant 2×10^3 N/m (see Fig. 19.12). The cylinder is filled with 5 L of gas with the spring relaxed at a pressure of 1 atmosphere and a temperature of 20°C. (a) If the lid has a cross-sectional area of 0.01 m^2 and negligible mass, how high will the lid rise when the temperature is raised to 250°C? (b) What is the pressure of the gas at 250°C?

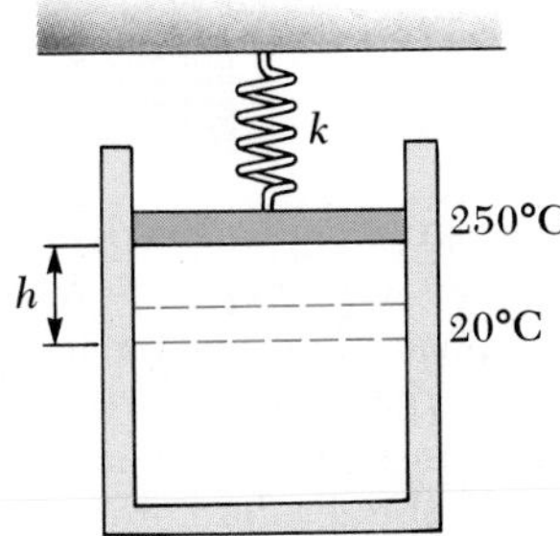

Figure 19.12 (Problem 50).

ADDITIONAL PROBLEMS

51. Precise temperature measurements are often made using the change in electrical resistance of a metal or semiconductor with temperature. The resistance varies approximately according to the expression $R = R_0(1 + AT_C)$, where R_0 and A are constants and T_C is the temperature in degree Celsius. A certain element has a resistance of 50.0 ohms at 0°C and 71.5 ohms at the freezing point of tin (231.97°C). (a) Determine the constants A and R_0. (b) At what temperature is the resistance equal to 89.0 ohms?

52. A pendulum clock with a brass suspension system has a period of 1.000 s at 20°C. If the temperature increases to 30°C, (a) by how much will its period change, and (b) how much time will the clock gain or lose in one week?

53. The rectangular plate shown in Figure 19.13 has an area A equal to ℓw. If the temperature increases by ΔT, show that the increase in area is given by $\Delta A = 2\alpha A\, \Delta T$, where α is the coefficient of linear expansion. What approximation does this expression assume? (*Hint:* Note that each dimension increases according to $\Delta \ell = \alpha \ell\, \Delta T$.)

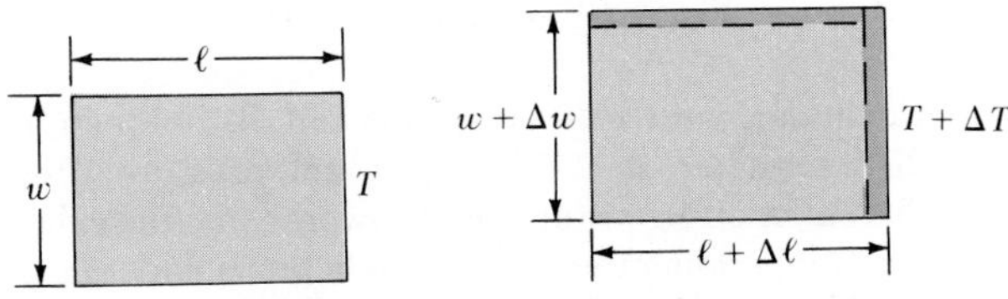

Figure 19.13 (Problem 53).

54. Consider an object with any one of the shapes displayed in Table 10.2. What is the percentage increase in the moment of inertia of the object when it is heated from 0°C to 100°C, if it is composed of (a) copper, (b) aluminum? (See Table 19.2. Assume that the linear expansion coefficients do not vary between 0°C and 100°C.)

55. A mercury thermometer is constructed as in Figure 19.14. The capillary tube has a diameter of 0.004 cm, and the bulb has a diameter of 0.25 cm. Neglecting the expansion of the glass, find the change in height of the mercury column for a temperature change of 30 C°.

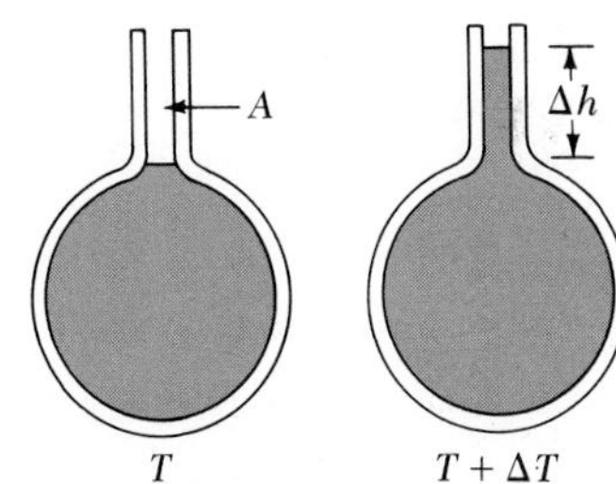

Figure 19.14 (Problems 55 and 57).

56. A fluid has a density ρ. (a) Show that the *fractional* change in density for a change in temperature ΔT is given by $\Delta\rho/\rho = -\beta\, \Delta T$. What does the negative sign signify? (b) Fresh water has a maximum density of 1.000 g/cm^3 at 4°C. At 10°C, its density is 0.9997 g/cm^3. What is β for water over this temperature interval?

57. A liquid with a coefficient of volume expansion β just fills a spherical shell of volume V at a temperature T (Fig. 19.14). The shell is made of a material that has a coefficient of linear expansion of α. The liquid is free to expand into a capillary of cross-sectional area A at the top. (a) If the temperature increases by ΔT, show that the liquid rises in the capillary by an amount Δh given by $\Delta h = \frac{V}{A}(\beta - 3\alpha)\, \Delta T$. (b) For a typical system, such as a mercury thermometer, why is it a good approximation to neglect the expansion of the shell?

58. (a) Derive an expression for the buoyant force on a spherical balloon as it is submerged in water as a function of the depth below the surface, the volume of the balloon at the surface, the pressure at the surface, and the density of the water. (Assume the water temperature to be a constant.) (b) Does the buoyant force increase or decrease as the balloon is submerged? (c) At

what depth will the buoyant force decrease to one-half the surface value?

59. (a) Show that the density of an ideal gas occupying a volume V is given by $\rho = PM/RT$, where M is the molecular weight. (b) Determine the density of oxygen gas at atmospheric pressure and 20°C.

60. (a) Show that the volume coefficient of thermal expansion for an ideal gas at constant pressure is given by $\beta = 1/T$, where T is the kelvin temperature. Start with the definition of β and use the equation of state, $PV = nRT$. (b) What value does this expression predict for β at 0°C? Compare this with the experimental values for helium and air in Table 19.2.

61. Starting with Equation 19.12, show that the total pressure P in a container filled with a mixture of several different ideal gases is given by $P = P_1 + P_2 + P_3 + \cdots$, where P_1, P_2, etc., are the pressures that each gas would exert if it alone filled the container (or the *partial pressures* of the respective gases). This is known as *Dalton's law of partial pressures.*

62. A sample of air with a mass of 100.00 g, collected at sea level, is analyzed and found to consist of the following gases:

$$\text{nitrogen } (N_2) = 75.52 \text{ g}$$
$$\text{oxygen } (O_2) = 23.15 \text{ g}$$
$$\text{argon } (Ar) = 1.28 \text{ g}$$
$$\text{carbon dioxide } (CO_2) = 0.05 \text{ g}$$

plus trace amounts of neon, helium, methane, and other gases. (a) Calculate the partial pressure of each gas when the pressure is 1 atm $= 1.01325 \times 10^5$ Pa. (b) Determine the volume occupied by the 100-g sample at a temperature of 15.00°C and a pressure of 1 atm. What is the density of the air for these conditions? (c) What is the effective molecular weight of the air sample?

63. At $T = 0$°C, each one of three metal bars (two of invar and one of aluminum) is drilled with two holes a distance L apart. Pins are put through the holes to create an equilateral triangle. If the bars are then heated to 100°C, what will be the angle between the two invar bars?

64. A steel ball bearing is 4.000 cm in diameter at 20°C. A bronze plate has a hole in it that is 3.994 cm in diameter at 20°C. What common temperature must they have in order that the ball just squeeze through the hole?

65. At $T = 0$°C, a container is completely full of ethyl alcohol. If the container does not expand when heated (and is pushed out a negligible extent when the ethanol begins to exert an outward pressure), what will be the internal pressure when the temperature is raised to 20°C? Express your answer in Pa and in atmospheres. (Refer to Table 19.2. The bulk modulus of ethyl alcohol is 1×10^9 N/m².)

66. A vertical cylinder of cross-sectional area A is fitted with a tight-fitting, frictionless piston of mass m (Fig. 19.15). (a) If there are n moles of an ideal gas in the cylinder at a temperature T, determine the height h at which the piston will be in equilibrium under its own weight. (b) What is the value for h if $n = 0.2$ mol, $T = 400$ K, $A = 0.008$ m² and $m = 20$ kg?

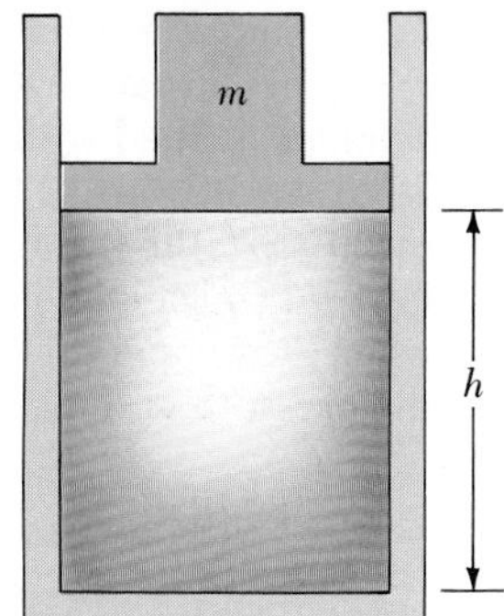

Figure 19.15 (Problem 66).

67. An air bubble originating from a deep sea diver has a radius of 5 mm at some depth h. When the bubble reaches the surface of the water it has a radius of 7 mm. Assuming the temperature of the air in the bubble remains constant, determine (a) the depth h of the diver, and (b) the absolute pressure at this depth.

68. The measurement of the average coefficient of volume expansion for liquids is complicated by the fact that the container also changes size with temperature. Figure 19.16 shows a simple means for overcoming this difficulty. With this apparatus, $\bar{\beta}$ for a liquid is determined by measurements of the column heights, h_0 and h_t, of the liquid columns in the U-tube. One arm of the tube is maintained at 0°C in a water-ice bath, and the other arm is maintained at the temperature T_c in a thermal bath. The connecting tube is horizontal. Derive the expression for $\bar{\beta}$ in terms of h_0 and h_t.

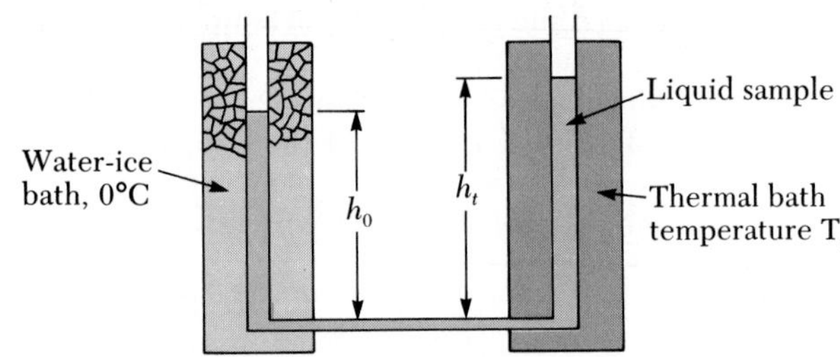

Figure 19.16 (Problem 68).

69. A cylinder, with a 40-cm radius and 50 cm deep, is filled with air at 20°C and 1 atm pressure (Fig. 19.17a). A 20-kg piston is now lowered into the cylinder, compressing the air trapped in the cylinder (Fig. 19.17b). Finally, a 75-kg man stands on the cylinder,

further compressing the air (which remains at 20°C). (a) How far down (Δh) does the piston move when the man steps onto the piston? (b) To what temperature should the gas be heated to raise the piston and man back to h_0?

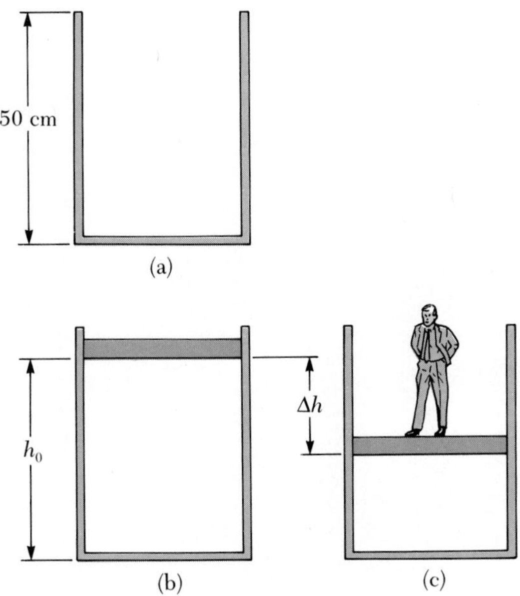

Figure 19.17 (Problem 69).

70. An aluminum pot has the shape of a right circular cylinder. The pot is initially at 4°C, at which temperature it has an inside diameter of 28.00 cm. The pot contains 3.000 gallons of water at 4°C. (a) What is the depth of water in the pot? (1 gallon = 3785 cm^3) (b) The pot and the water in it are heated to a final temperature of 90°C. Allowing for the expansion of the water, but ignoring the expansion of the pot, what is the change in depth of the water? Express the change as percentage of the original depth and also in millimeters. (The density of water is 1.000 g/cm^3 at 4°C and 0.965 g/cm^3 at 90°C.) (c) Modify your solution for part (b) to allow for the expansion of the pot. (Refer to Table 19.2.)

71. A sphere 20 cm in diameter contains an ideal gas at 1 atm pressure and 20°C. As the sphere is heated to 100°C gas is allowed to escape. The valve is closed and the sphere is placed in an ice water bath at 0°C. (a) How many moles of gas escape from the sphere as it warms? (b) What is the pressure in the sphere when it is in the ice water?

72. The relationship $L = L_0(1 + \alpha\,\Delta T)$ is an approximation which works when the coefficient of expansion is small. If α is sizable the relation $dL/dT = \alpha L$ must be integrated to determine the final length. (a) Assuming the coefficient of linear expansion is constant, determine a general expression for the final length. (b) Given a rod of length 1 m and a temperature change of 100 C°, determine the error caused by the approximation when $\alpha = 2 \times 10^{-5}(\text{C}°)^{-1}$ (the normal value for common metals) and when $\alpha = 0.02(\text{C}°)^{-1}$ (an unrealistically large value for comparison).

73. A steel guitar string with a diameter of 1.00 mm is stretched between supports 80 cm apart. The temperature is 0°C. (a) Find the mass per unit length of this string. (Use the value 7.86×10^3 kg/m^3 for the density.) (b) The fundamental frequency of transverse oscillations of the string is 200 Hz. What is the tension in the string? (c) If the temperature is raised to 30°C, find the resulting values of the tension and the fundamental frequency. [Assume that both the Young's modulus (Table 12.1) and the coefficient of thermal expansion (Table 19.2) of steel have constant values between 0°C and 30°C.]

74. A steel wire and a copper wire, each of diameter 2.00 mm, are joined end to end. At 40°C, each has an unstretched length of 2.000 m; they are connected between two fixed supports 4 m apart on a tabletop, so that the steel wire extends from $x = -2.000$ m to $x = 0$, the copper wire extends from $x = 0$ to $x = 2.000$ m, and the tension is negligible. The temperature is then lowered to 20°C, while the supports are held fixed at a separation of 4.000 m. At this lower temperature, find the tension in the wire and the location (x coordinate) of the junction between the steel and copper wires. (Refer to Tables 12.1 and 19.2)

75. A bimetallic bar is made of two thin dissimilar metals bonded together. As they are heated, the one with the larger coefficient of linear expansion will expand more than the other, forcing the bar into an arc, with the outer radius having a larger circumference (see Fig. 19.18). (a) Derive an expression for the angle of bending θ as a function of the initial length of the rods, their coefficients of linear expansion, the change in temperature, and the separation of centers ($\Delta r = r_2 - r_1$). (b) Show that the angle of bending goes to zero when ΔT goes to zero or the two coefficients of expansion become equal. (c) What happens if the bar is cooled?

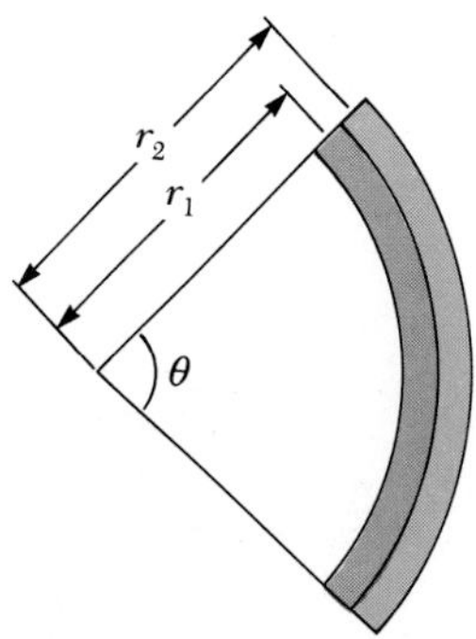

Figure 19.18 (Problem 75).

20

Heat and the First Law of Thermodynamics

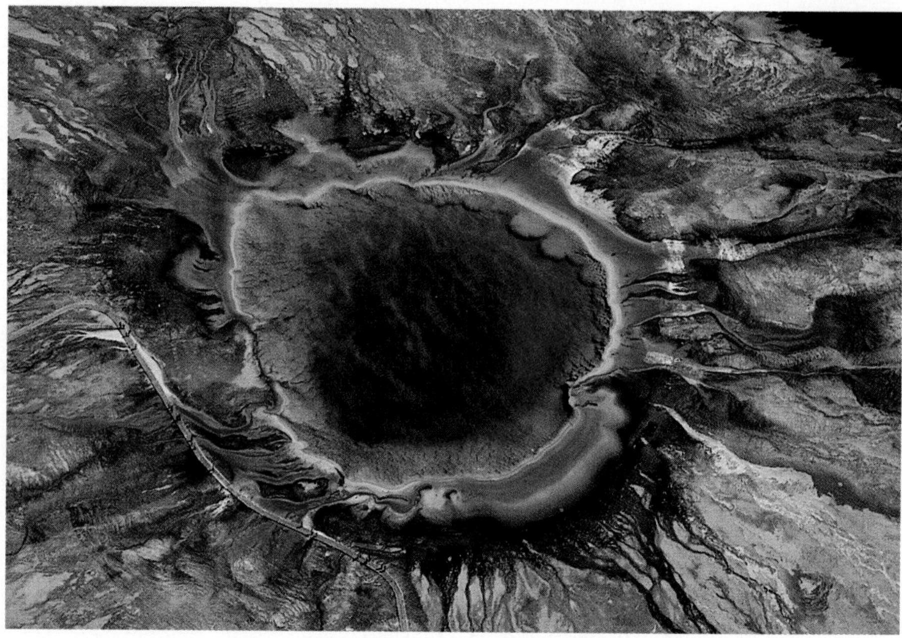

Grand Prismatic Spring in Yellowstone National Park. The water at the surface of this hot spring is nearly 93°C (200°F). Microorganisms thrive in the hot water and are the cause of its vivid colors. A boardwalk at the edge of the spring keeps people at a safe distance. (© Paul Chesley/Photographers Aspen)

It is well known that when two objects at different temperatures are placed in thermal contact with each other, the temperature of the warmer body decreases while the temperature of the cooler body increases. If they are left in contact for some time, they eventually reach a common equilibrium temperature somewhere between the two initial temperatures. When such processes occur, we say that heat is transferred from the warmer to the cooler body. But what is the nature of this heat transfer? Early investigators believed that heat was an invisible material substance called *caloric*, which was transferred from one body to another. According to this theory, caloric could neither be created nor destroyed. Although the caloric theory was successful in describing heat transfer, it eventually was abandoned when various experiments showed that caloric was in fact not conserved.

The first experimental observation suggesting that caloric was not conserved was made by Benjamin Thompson (1753–1814) at the end of the 18th century. Thompson, an American-born scientist, emigrated to Europe during the Revolutionary War because of his Tory sympathies. Following his appointment as director of the Bavarian Arsenal, he was given the title Count Rumford. While supervising the boring of an artillery cannon in Munich, Thompson noticed the great amount of heat generated by the boring tool. The water being used for cooling had to be replaced continuously as it boiled away during the boring process. On the basis of the caloric theory, he reasoned that the ability of the metal filings to retain caloric should decrease as the size of the filings decreased. These heated filings, in turn, presumably transferred caloric to the cooling water, causing it to boil. To his surprise, Thompson discovered that the amount of water boiled away by a blunt boring tool was comparable to the quantity boiled away by a sharper tool for a given turning rate. He then reasoned that if the tool were turned long enough, an almost infinite amount of caloric could be produced from a finite amount of metal filings. For this reason, Thompson rejected the caloric theory and suggested that heat is not a substance, but some form of motion that is transferred from the boring tool to the water. In another experiment, he showed that the heat generated by friction was proportional to the mechanical work done by the boring tool.

Benjamin Thompson (1753–1814). "Being engaged, lately, in superintending the boring of cannon, in the workshops of the military arsenal at Munich, I was struck with the very considerable degree of Heat which a brass gun acquires, in a short time, in being bored; and with the still more intense Heat (much greater than that of boiling water, as I found by experiment) of the metallic chips separated from it by the borer."

There are many other experiments that are at odds with the caloric theory. For example, if you rub two blocks of ice together on a day when the temperature is below 0°C, the blocks will melt. This experiment was first conducted by Sir Humphry Davy (1778–1829). To properly account for this "creation of caloric," we note that mechanical work is done on the system. Thus, we see that the effects of doing mechanical work on a system and of adding heat to it directly, as with a flame, are equivalent. That is, heat and work are both forms of energy transfer.

Although Thompson's observations provided evidence that heat energy is not conserved, it was not until the middle of the 19th century that the modern mechanical model of heat was developed. Before this period, the subjects of heat and mechanics were considered to be two distinct branches of science, and the law of conservation of energy seemed to be a rather specialized result used to describe certain kinds of mechanical systems. After the two disciplines were shown to be intimately related, the law of conservation of energy emerged as a universal law of nature. In this new view, heat is treated as another form of energy that can be transformed into mechanical energy. Experiments performed by the Englishman James Joule (1818–1889) and others in this period showed that whenever heat is gained or lost by a system during some process, the gain or loss can be accounted for by an equivalent quantity of mechanical work done on the system. Thus, by broadening the concept of energy to include heat as a form of energy, the law of energy conservation was extended.

20.1 HEAT AND THERMAL ENERGY

There is a major distinction between the concepts of heat and the internal energy of a substance. The word *heat* should be used only when describing energy transferred from one place to another. That is, *heat flow is an energy transfer that takes place as a consequence of temperature differences only.* On the other hand, *internal energy* is the energy a substance has because of its temperature. In the next chapter, we shall show that the energy of an ideal gas is associated with the internal motion of its atoms and molecules. In other words, the internal energy of a gas is essentially its kinetic energy on a microscopic scale; the higher the temperature of the gas, the greater its internal energy. As an analogy, consider the distinction between work and energy that we discussed in Chapter 7. The work done on (or by) a system is a measure of energy transfer between the system and its surroundings, whereas the mechanical energy (kinetic and/or potential) is a consequence of the motion and

Definition of heat

Biographical Sketch

James Prescott Joule
(1818–1889)

James Prescott Joule, a British physicist, was born in Salford on December 24, 1818 into a wealthy brewing family. He received some formal education in mathematics, philosophy, and chemistry from John Dalton but was in large part self-educated.

Joule's most active research period, from 1837 through 1847, led to the establishment of the principle of conservation of energy and the equivalence of heat and other forms of energy. His study of the quantitative relationship between electrical, mechanical, and chemical effects of heat culminated in his announcement in 1843 of the amount of work required to produce a unit of heat:

> First: that the quantity of heat produced by the friction of bodies, whether solid or liquid, is always proportional to the quantity of energy expended. And second: that the quantity of heat capable of increasing the temperature of water . . . by 1° Fahr requires for its evolution the expenditure of a mechanical energy represented by the fall of 772 lb through the distance of one foot.

This is called the mechanical equivalent of heat (the currently accepted value is equal to 4.186 J/cal).

Much of Joule's later work on the new science of thermodynamics was extended by Lord Kelvin. In 1852 the Joule–Thomson effect showed that when a gas is allowed to expand freely, its temperature drops slightly. This was an important discovery in the field of low-temperature physics.

coordinates of the system. Thus, when you do work on a system, energy is transferred from you to the system. It makes no sense to talk about the work *of* a system — one can refer only to the *work done on or by a system* when some process has occurred in which the system has changed in some way. Likewise, it makes no sense to use the term *heat* unless the thermodynamic variables of the system have undergone a change during some process.

It is also important to recognize that energy can be transferred between two systems even when there is no heat flow. For example, when two objects are rubbed together, their internal energy increases since mechanical work is done on them. When an object slides across a surface and comes to rest as a result of friction, its kinetic energy is transformed into internal energy contained in the block and surface. In such cases, the work done on the system adds energy to the system. The changes in internal energy are measured by corresponding changes in temperature.

This device, called Hero's engine, was invented around 150 B.C. by Hero in Alexandria. When water is boiled in the flask, which is suspended by a cord, steam exits through two tubes at the sides of the flank (in opposite directions), creating a torque that rotates the flask. (Courtesy of CENCO)

Units of Heat

The calorie

Before it was understood that heat was a form of energy, scientists defined heat in terms of the temperature changes it produced in a body. Hence, the **calorie** (cal) was defined as *the amount of heat necessary to raise the temperature of 1 g of water from 14.5°C to 15.5°C.*[1] (Note that the "Calorie," with a capital C, used in describing the chemical energy content of foods, is actually a kilocalorie.) Likewise, the unit of heat in the British system was the **British Thermal Unit** (Btu), defined as *the heat required to raise the temperature of 1 lb of water from 63°F to 64°F.*

Since heat is now recognized as a form of energy, scientists are increasingly using the SI unit of energy for heat, the *joule.* In this textbook, heat will

[1] Originally, the calorie was defined as the heat necessary to raise the temperature of 1 g of water by 1 C°. However, careful measurements showed that energy depends somewhat on temperature; hence, a more precise definition evolved.

usually be given in units of joules. Some useful conversions among the commonly used heat units are as follows:

$$1 \text{ cal} = 4.186 \text{ J} = 3.968 \times 10^{-3} \text{ Btu}$$

$$1 \text{ J} = 0.2389 \text{ cal} = 9.478 \times 10^{-4} \text{ Btu}$$

$$1 \text{ Btu} = 1055 \text{ J} = 252.0 \text{ cal}$$

The Mechanical Equivalent of Heat

When the concept of mechanical energy was introduced in Chapters 7 and 8, we found that whenever friction is present in a mechanical system, some mechanical energy is lost, or is not conserved. Various experiments show that this lost mechanical energy does not simply disappear, but is transformed into thermal energy. Although this connection between mechanical and thermal energy was first suggested by Thompson's crude cannon boring experiment, it was Joule (pronounced "jowl" or "jool") who first established the equivalence of the two forms of energy.

A schematic diagram of Joule's most famous experiment is shown in Figure 20.1. The system of interest is the water in a thermally insulated container. Work is done on the water by a rotating paddle wheel, which is driven by weights falling at a constant speed. The water, which is stirred by the paddles, is warmed due to the friction between it and the paddles. If the energy lost in the bearings and through the walls is neglected, then the loss in potential energy of the weights equals the work done by the paddle wheel on the water. If the two weights fall through a distance h, the loss in potential energy is $2mgh$, and it is this energy that is used to heat the water. By varying the conditions of the experiment, Joule found that the loss in mechanical energy, $2mgh$, is proportional to the increase in temperature of the water, ΔT. The proportionality constant (the specific heat of water) was found to be equal to 4.18 J/g·C°. Hence, 4.18 J of mechanical energy will raise the temperature of 1 g of water from 14.5°C to 15.5°C. One calorie is now defined to be *exactly* 4.186 J without reference to the heating of a substance:

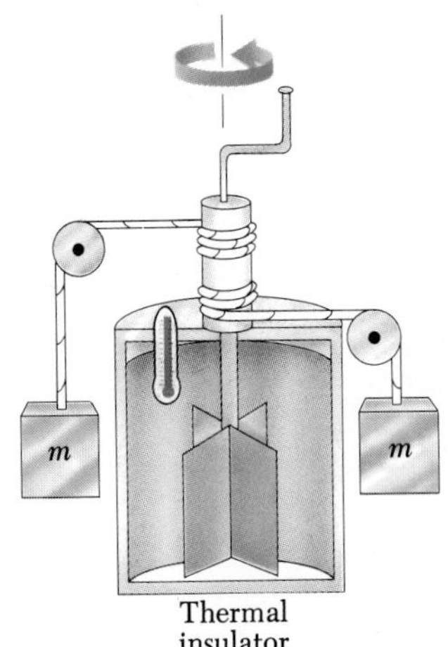

Figure 20.1 An illustration of Joule's experiment for measuring the mechanical equivalent of heat. The falling weights rotate the paddles, causing the temperature of the water to increase.

$$1 \text{ cal} \equiv 4.186 \text{ J} \tag{20.1}$$

Mechanical equivalent of heat

EXAMPLE 20.1 Losing Weight the Hard Way

A student eats a dinner rated at 2000 (food) Calories. He wishes to do an equivalent amount of work in the gymnasium by lifting a 50-kg mass. How many times must he raise the weight to expend this much energy? Assume that he raises the weight a distance of 2 m each time and that no work is done when the weight is dropped to the floor.

Solution Since 1 (food) Calorie = 10^3 cal, the work required is 2×10^6 cal. Converting this to J, we have for the total work required

$$W = (2 \times 10^6 \text{ cal})\,(4.186 \text{ J/cal}) = 8.37 \times 10^6 \text{ J}$$

The work done in lifting the weight once through a distance h is equal to mgh, and the work done in lifting the weight n times is $nmgh$. Equating this to the total work required gives

$$W = nmgh = 8.37 \times 10^6 \text{ J}$$

Since $m = 50$ kg and $h = 2$ m, we get

$$n = \frac{8.37 \times 10^6 \text{ J}}{(50 \text{ kg})(9.80 \text{ m/s}^2)(2 \text{ m})} = 8.54 \times 10^3 \text{ times}$$

If the student is in good shape and lifts the weight, say, once every 5 s, it will take him about 12 h to perform this feat. Clearly, it is much easier to lose weight by dieting.

20.2 HEAT CAPACITY AND SPECIFIC HEAT

The quantity of heat energy required to raise the temperature of a given mass of a substance by some amount varies from one substance to another. For example, the heat required to raise the temperature of 1 kg of water by 1 C° is 4186 J, but the heat required to raise the temperature of 1 kg of copper by 1 C° is only 387 J.

Heat capacity

The **heat capacity**, C, of a particular sample of a substance is defined as the amount of heat energy needed to raise the temperature of that sample by one Celsius degree.

From this definition, we see that if Q units of heat when added to a substance produce a change in temperature of ΔT, then

$$Q = C\,\Delta T \tag{20.2}$$

Specific heat

The heat capacity of any object is proportional to its mass. For this reason, it is convenient to define the heat capacity per unit mass of a substance, c, called the **specific heat:**

$$c = \frac{C}{m} \tag{20.3}$$

Table 20.1 gives specific heat values for various substances measured at room temperature and atmospheric pressure.

TABLE 20.1 Specific Heats of Some Substances at 25°C and Atmospheric Pressure

Substance	Specific heat, c J/kg·C°	Specific heat, c cal/g·C°	Molar heat capacity J/mol·C°
		Elemental Solids	
Aluminum	900	0.215	24.3
Beryllium	1830	0.436	16.5
Cadmium	230	0.055	25.9
Copper	387	0.0924	24.5
Germanium	322	0.077	23.4
Gold	129	0.0308	25.4
Iron	448	0.107	25.0
Lead	128	0.0305	26.4
Silicon	703	0.168	19.8
Silver	234	0.056	25.4
		Other Solids	
Brass	380	0.092	
Wood	1700	0.41	
Glass	837	0.200	
Ice (−5°C)	2090	0.50	
Marble	860	0.21	
		Liquids	
Alcohol (ethyl)	2400	0.58	
Mercury	140	0.033	
Water (15°C)	4186	1.00	

From the definition of heat capacity given by Equation 20.2, we can express the heat energy Q transferred between a substance of mass m and its surroundings for a temperature change $\Delta T = T_f - T_i$ as

$$Q = mc\,\Delta T \tag{20.4}$$

For example, the heat energy required to raise the temperature of 0.5 kg of water by 3 C° is equal to (0.5 kg)(4186 J/kg·C°)(3 C°) = 6280 J. Note that when heat is *added* to the substance, Q and ΔT are both *positive*, and the temperature *increases*. Likewise, when heat is *removed* from the substance, Q and ΔT are both *negative* and the temperature *decreases*.

Molar heat capacity

The **molar heat capacity** of a substance is defined as the heat capacity per mole. Hence, if the substance contains n moles, its molar heat capacity is equal to C/n. Table 20.1 also gives the molar heat capacities of various substances.

It is important to realize that the specific heats of substances vary somewhat with temperature. If the temperature intervals are not too great, the temperature variation can be ignored and c can be treated as a constant.[2] For example, the specific heat of water varies by only about 1% from 0°C to 100°C at atmospheric pressure. Unless stated otherwise, we shall neglect such variations.

When specific heats are measured, the values obtained are also found to depend on the conditions of the experiment. In general, measurements made at constant pressure are different from those made at constant volume. For solids and liquids, the difference between the two values is usually no more than a few percent and is often neglected. The values given in Table 20.1 were measured at atmospheric pressure and room temperature. As we shall see in Chapter 21, the specific heats for gases are quite different under constant pressure conditions compared to constant volume conditions.

It is interesting to note from Table 20.1 that water has the highest specific heat of common earth materials. The high specific heat of water is responsible, in part, for the moderate temperatures found in regions near large bodies of water. As the temperature of a body of water decreases during the winter, heat is transferred from the water to the air, which in turn carries the heat landward when prevailing winds are favorable. For example, the prevailing winds of the western coast of the United States are toward the land (eastward). Hence the heat liberated by the Pacific Ocean as it cools keeps coastal areas much warmer than they would be otherwise. This explains why the western coastal states generally have more favorable winter weather than the eastern coastal states, where the prevailing winds do not tend to carry the heat toward land.

Measuring Specific Heat—Calorimetry

One technique for measuring the specific heat of solids or liquids is simply to heat the substance to some known temperature, place it in a vessel containing

[2] The definition given by Equation 20.4 assumes that the specific heat does not vary with temperature over the interval ΔT. In general, if c varies with temperature over the range T_i to T_f, the correct expression for Q is

$$Q = m\int_{T_i}^{T_f} c\,dT$$

water of known mass and temperature, and measure the temperature of the water after equilibrium is reached. Since a negligible amount of mechanical work is done in the process, the law of conservation of energy requires that the heat that leaves the warmer substance (of unknown specific heat) equals the heat that enters the water.[3] Devices in which this heat transfer occurs are called **calorimeters.** For example, suppose that m_x is the mass of a substance whose specific heat we wish to determine, c_x its specific heat, and T_x its initial temperature. Likewise, let m_w, c_w, and T_w represent corresponding values for the water. If T is the final equilibrium temperature after everything is mixed, then from Equation 20.4, we find that the heat gained by the water is $m_w c_w(T - T_w)$, and the heat lost by the substance of unknown c is $-m_x c_x(T - T_x)$. Assuming that the system (water + unknown) does not lose or gain any heat, it follows that the heat gained by the water must equal the heat lost by the unknown (conservation of energy):

$$m_w c_w(T - T_w) = -m_x c_x(T - T_x)$$

Solving for c_x gives

$$c_x = \frac{m_w c_w(T - T_w)}{m_x(T_x - T)} \tag{20.5}$$

[3] For precise measurements, the container for the water should be included in our calculations, since it also exchanges heat. This would require a knowledge of its mass and composition. However, if the mass of the water is large compared with that of the container, we can neglect the heat gained by the container. Furthermore, precautions must be taken in such measurements to minimize heat transfer between the system and the surroundings.

EXAMPLE 20.2 Cooling a Hot Ingot
A 0.05-kg ingot of metal is heated to 200°C and then dropped into a beaker containing 0.4 kg of water initially at 20°C. If the final equilibrium temperature of the mixed system is 22.4°C, find the specific heat of the metal.

Solution Because the heat lost by the ingot equals the heat gained by the water, we can write

$$m_x c_x(T_i - T_f) = m_w c_w(T_f - T_i)$$

$$(0.05\ \text{kg})(c_x)(200°\text{C} - 22.4°\text{C})$$

$$= (0.4\ \text{kg})(4186\ \text{J/kg}\cdot\text{C°})(22.4°\text{C} - 20°\text{C})$$

from which we find that

$$c_x = 453\ \text{J/kg}\cdot\text{C°}$$

The ingot is most likely iron, as can be seen by comparing this result with the data in Table 20.1.

Exercise 1 What is the total heat transferred to the water in cooling the ingot?
Answer 4020 J

EXAMPLE 20.3 Fun Time for a Cowboy
A cowboy fires a silver bullet of mass 2 g with a muzzle velocity of 200 m/s into the pine wall of a saloon. Assume that all the thermal energy generated by the impact remains with the bullet. What is the temperature change of the bullet?

Solution The kinetic energy of the bullet is

$$\frac{1}{2}mv^2 = \frac{1}{2}(2 \times 10^{-3}\ \text{kg})(200\ \text{m/s})^2 = 40\ \text{J}$$

All of this kinetic energy is transformed into heat, Q, as the bullet stops in the wall. Thus,

$$Q = mc\,\Delta T$$

Since the specific heat of silver is 234 J/kg·C° (Table 20.1), we get

$$\Delta T = \frac{Q}{mc} = \frac{40\ \text{J}}{(2 \times 10^{-3}\ \text{kg})(234\ \text{J/kg}\cdot\text{C°})} = 85.5\ \text{C°}$$

Exercise 2 Suppose the cowboy runs out of silver bullets and fires a lead bullet of the same mass and velocity into the wall. What is the temperature change of the bullet?
Answer 156 C°.

20.3 LATENT HEAT

A substance usually undergoes a change in temperature when heat is transferred between the substance and its surroundings. There are situations, however, where the flow of heat does not result in a change in temperature. This occurs whenever the physical characteristics of the substance change from one form to another, commonly referred to as a **phase change.** Some common phase changes are solid to liquid (melting), liquid to gas (boiling), and a change in crystalline structure of a solid. All such phase changes involve a change in internal energy. The energy required is called the **heat of transformation.**

The heat required to change the phase of a given mass m of a pure substance is given by

$$Q = mL \tag{20.6}$$

where L is called the **latent heat** (hidden heat) of the substance[4] and depends on the nature of the phase change as well as on the properties of the substance. The **heat of fusion, L_f,** is used when the phase change is from a solid to a liquid, and the **heat of vaporization, L_v,** is the latent heat corresponding to the liquid-to-gas phase change.[5] For example, the heat of fusion for water at atmospheric pressure is 3.33×10^5 J/kg, and the latent heat of vaporization of water is 2.26×10^6 J/kg. The latent heats of various substances vary considerably, as is seen in Table 20.2.

Phase changes can be described in terms of a rearrangement of molecules when heat is added or removed from a substance. Consider first the liquid-to-gas phase change. The molecules in a liquid are close together, and the forces between them are stronger than in a gas, where the molecules are far apart. Therefore, work must be done on the liquid against these attractive molecular

[4] The word *latent* is from the Latin *latere,* meaning *hidden or concealed.*

[5] When a gas cools, it eventually returns to the liquid phase, or *condenses.* The heat given up per unit mass is called the *heat of condensation,* which equals the heat of vaporization. Likewise, when a liquid cools it eventually solidifies, and the *heat of solidification* equals the heat of fusion.

TABLE 20.2 Heats of Fusion and Vaporization

Substance	Melting Point (°C)	Heat of Fusion (J/kg)	Boiling Point (°C)	Heat of Vaporization (J/kg)
Helium	−269.65	5.23×10^3	−268.93	2.09×10^4
Nitrogen	−209.97	2.55×10^4	−195.81	2.01×10^5
Oxygen	−218.79	1.38×10^4	−182.97	2.13×10^5
Ethyl alcohol	−114	1.04×10^5	78	8.54×10^5
Water	0.00	3.33×10^5	100.00	2.26×10^6
Sulfur	119	3.81×10^4	444.60	3.26×10^5
Lead	327.3	2.45×10^4	1750	8.70×10^5
Aluminum	660	9.00×10^4	2450	1.14×10^7
Silver	960.80	8.82×10^4	2193	2.33×10^6
Gold	1063.00	6.44×10^4	2660	1.58×10^6
Copper	1083	1.34×10^5	1187	5.06×10^6

forces in order to separate the molecules. The heat of vaporization is the amount of energy that must be added to the liquid to accomplish this.

Similarly, at the melting point of a solid, we imagine that the amplitude of vibration of the atoms about their equilibrium position becomes large enough to overcome the attractive forces binding the atoms into their fixed positions. The heat energy required for total melting of a given mass of solid is equal to the work required to break these bonds and to transform the mass from the ordered solid phase to the disordered liquid phase.

Because the average distance between atoms in the gas phase is much larger than in either the liquid or the solid phase, we could expect that more work is required to vaporize a given mass of a substance then to melt it. Therefore, it is not surprising that the heat of vaporization is much larger than the heat of fusion for a given substance (Table 20.2).

Consider, for example, the heat required to convert a 1-g block of ice at $-30°C$ to steam (water vapor) at 120°C. Figure 20.2 indicates the experimental results obtained when heat is gradually added to the ice. Let us examine each portion of the curve separately.

Part A During this portion of the curve, we are changing the temperature of the ice from $-30°C$ to 0°C. Since the specific heat of ice is 2090 J/kg·C°, we can calculate the amount of heat added as follows:

$$Q = m_i c_i \, \Delta T = (10^{-3} \text{ kg})(2090 \text{ J/kg}\cdot\text{C}°)(30 \text{ C}°) = 62.7 \text{ J}$$

Part B When the ice reaches 0°C, the ice/water mixture remains at this temperature—even though heat is being added—until all the ice melts. The heat required to melt 1 g of ice at 0°C is

$$Q = mL_f = (10^{-3} \text{ kg})(3.33 \times 10^5 \text{ J/kg}) = 333 \text{ J}$$

Part C Between 0°C and 100°C, nothing surprising happens. No phase change occurs in this region. The heat added to the water is being used to increase its temperature. The amount of heat necessary to increase the temperature from 0°C to 100°C is

$$Q = m_w c_w \, \Delta T = (10^{-3} \text{ kg})(4.19 \times 10^3 \text{ J/kg}\cdot\text{C}°)(100 \text{ C}°) = 4.19 \times 10^2 \text{ J}$$

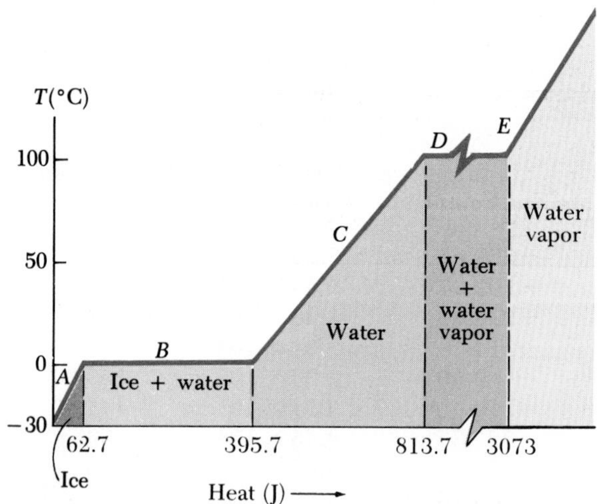

Figure 20.2 A plot of temperature versus heat added when 1 g of ice initially at −30°C is converted to steam.

Part D At 100°C, another phase change occurs as the water changes from water at 100°C to steam at 100°C. We can find the amount of heat required to produce this phase change by using Equation 20.6. In this case, we must set $L = L_v$, the **heat of vaporization.** Since the heat of vaporization for water is 2.26×10^6 J/kg, the amount of heat we must add to convert 1 g of water to steam at 100°C is

$$Q = mL_v = (10^{-3}\text{ kg})(2.26 \times 10^6\text{ J/kg}) = 2.26 \times 10^3\text{ J}$$

Part E On this portion of the curve, heat is being added to the steam with no phase change occurring. Using 2.01×10^3 J/kg·C° for the specific heat of steam, we find that the heat we must add to raise the temperature of the steam to 120°C is

$$Q = m_s c_s\,\Delta T = (10^{-3}\text{ kg})(2.01 \times 10^3\text{ J/kg·C°})(20\text{ C°}) = 40.2\text{ J}$$

The *total amount of heat* that must be added to change one gram of ice at −30°C to steam at 120°C is about 3.11×10^3 J. That is, if we cool one gram of steam at 120°C down to the point at which we have ice at −30°C, we must remove 3.11×10^3 J of heat.

EXAMPLE 20.4 Cooling the Steam

What mass of steam initially at 130°C is needed to warm 200 g of water in a 100-g glass container from 20°C to 50°C?

Solution This is a heat transfer problem in which we must equate the heat lost by the steam to the heat gained by the water and glass container. There are three stages as the steam loses heat. In the first stage, the steam is cooled to 100°C. The heat liberated in the process is

$$Q_1 = m_x c_s\,\Delta T = m_x(2.01 \times 10^3\text{ J/kg·C°})(30\text{ C°})$$
$$= m_x(6.03 \times 10^4\text{ J/kg})$$

In the second stage, the steam is converted to water. In this case, to find the heat removed, we use the heat of vaporization and $Q = mL_v$:

$$Q_2 = m_x(2.26 \times 10^6\text{ J/kg})$$

In the last stage, the temperature of the water is reduced to 50°C. This liberates an amount of heat

$$Q_3 = m_x c_w\,\Delta T = m_x(4.19 \times 10^3\text{ J/kg·C°})(50\text{ C°})$$
$$= m_x(2.09 \times 10^5\text{ J/kg})$$

If we equate the heat lost by the steam to the heat gained by the water and glass and use the given information, we find

$$m_x(6.03 \times 10^4\text{ J/kg}) + m_x(2.26 \times 10^6\text{ J/kg}) + m_x(2.09 \times 10^5\text{ J/kg})$$
$$= (0.2\text{ kg})(4.19 \times 10^3\text{ J/kg·C°})(30\text{ C°}) + (0.1\text{ kg})(837\text{ J/kg·C°})(30\text{ C°})$$

$$m_x = 1.09 \times 10^{-2}\text{ kg} = \boxed{10.9\text{ g}}$$

EXAMPLE 20.5 Boiling Liquid Helium

Liquid helium has a very low boiling point, 4.2 K, and a very low heat of vaporization, 2.09×10^4 J/kg (Table 20.2). A constant power of 10 W (1 W = 1 J/s) is transferred to a container of liquid helium from an immersed electric heater. At this rate, how long does it take to boil away 1 kg of liquid helium? Liquid helium has a density of 0.125 g/cm³, so that 1 kg corresponds to a volume of 8×10^3 cm³ = 8 liters of liquid.

Solution Since $L_v = 2.09 \times 10^4$ J/kg for liquid helium, we must supply 2.09×10^4 J of energy to boil away 1 kg. The power supplied to the helium is 10 W = 10 J/s. That is, in 1 s, 10 J of energy is transferred to the helium. Therefore, the time it takes to transfer 2.09×10^4 J is

$$t = \frac{2.09 \times 10^4\text{ J}}{10\text{ J/s}} = 2.09 \times 10^3\text{ s} \approx \boxed{35\text{ min}}$$

Since 1 kg of helium corresponds to 8 liters, this means a boil-off rate of about 0.23 liter/min. In contrast, 1 kg of liquid nitrogen would boil away in about 5.6 h at the rate of 10 J/s!

Exercise 3 If 10 W of power is supplied to 1 kg of water at 100°C, how long will it take for the water to boil away completely?

Answer 62.8 h.

20.4 WORK AND HEAT IN THERMODYNAMIC PROCESSES

In the macroscopic approach to thermodynamics we describe the *state* of a system with such variables as pressure, volume, temperature, and internal energy. The number of macroscopic variables needed to characterize a system depends on the nature of the system. For a homogeneous system, such as a gas containing only one type of molecule, usually only two variables are needed, such as pressure and volume. However, it is important to note that a *macroscopic state* of an isolated system can be specified only if the system is in thermal equilibrium internally. In the case of a gas in a container, internal thermal equilibrium requires that every part of the container be at the same pressure and temperature.

Consider gas contained in a cylinder fitted with a movable piston (Fig. 20.3). In equilibrium, the gas occupies a volume V and exerts a uniform pressure P on the cylinder walls and piston. If the piston has a cross-sectional area A, the force exerted by the gas on the piston is $F = PA$. Now let us assume that the gas expands **quasi-statically,** that is, slowly enough to allow the system to remain essentially in thermodynamic equilibrium at all times. As the piston moves up a distance dy, the work done by the gas on the piston is

$$dW = F\,dy = PA\,dy$$

Since $A\,dy$ is the increase in volume of the gas dV, we can express the work done as

$$dW = P\,dV \tag{20.7}$$

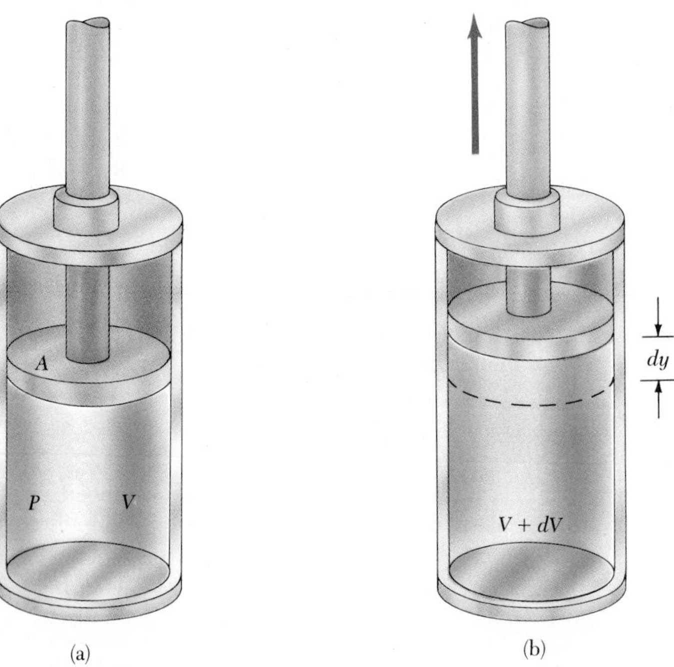

Figure 20.3 Gas contained in a cylinder at a pressure P does work on a moving piston as the system expands from a volume V to a volume $V + dV$.

If the gas expands, as in Figure 20.3b, then dV is positive and the work done by the gas is positive, whereas if the gas is compressed, dV is negative, indicating that the work done by the gas is negative. (In the latter case, negative work can be interpreted as being work done *on* the system.) Clearly, the work done by the system is zero when the volume remains constant. The total work done by the gas as its volume changes from V_i to V_f is given by the integral of Equation 20.7:

$$W = \int_{V_i}^{V_f} P\,dV \tag{20.8}$$

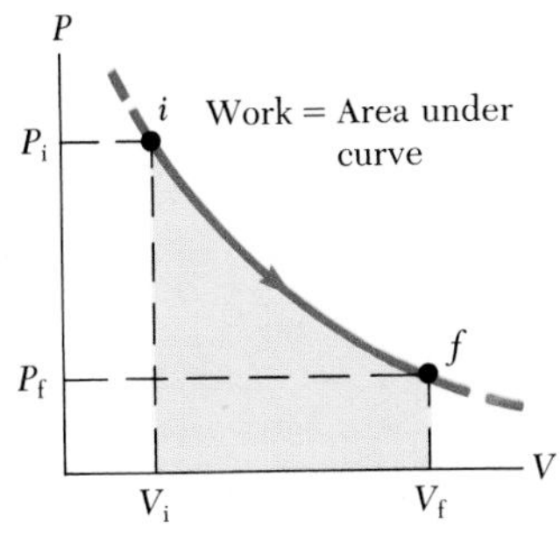

Figure 20.4 A gas expands reversibly (slowly) from state *i* to state *f*. The work done by the gas equals the area under the *PV* curve.

To evaluate this integral, one must know how the pressure varies during the process. (Note that a *process* is *not* specified merely by giving the initial and final states. That is, a process is a *fully specified* change in state of the system.) In general, the pressure is not constant, but depends on the volume and temperature. If the pressure and volume are known at each step of the process, the states of the gas can then be represented as a curve on a *PV* diagram, as in Figure 20.4.

The work done in the expansion from the initial state to the final state is the area under the curve in a *PV* diagram.

Work equals area under the curve in a *PV* diagram

As one can see from Figure 20.4, the work done in the expansion from the initial state, i, to the final state, f, will depend on the specific path taken between these two states. To illustrate this important point, consider several different paths connecting i and f (Fig. 20.5). In the process described in Figure 20.5a, the pressure of the gas is first reduced from P_i to P_f by cooling at constant volume V_i, and the gas then expands from V_i to V_f at constant pressure P_f. The work done along this path is $P_f(V_f - V_i)$. In Figure 20.5b, the gas first expands from V_i to V_f at constant pressure P_i, and then its pressure is reduced to P_f at constant volume V_f. The work done along this path is $P_i(V_f - V_i)$, which is greater than that for the process described in Figure 20.5a. Finally, for the process described in Figure 20.5c, where both P and V change continuously, the work done has some value intermediate between the values obtained in the first two processes. To evaluate the work in this case, the shape of the *PV* curve must be known. Therefore, we see that

the work done by a system depends on the process by which the system goes from the initial to the final state. In other words, the work done depends on the initial, final, and intermediate states of the system.

Work done depends on the path between the initial and final states

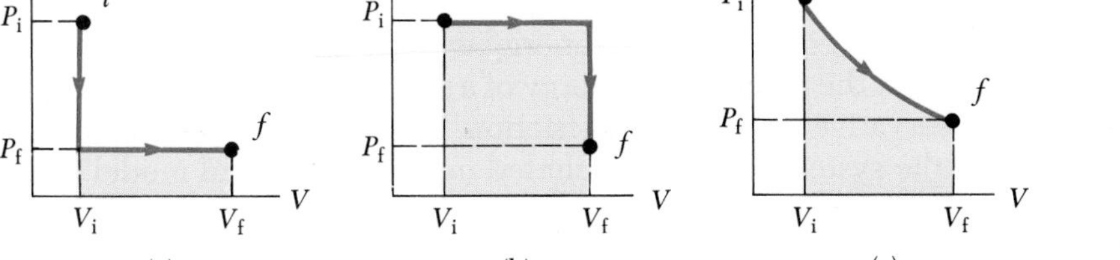
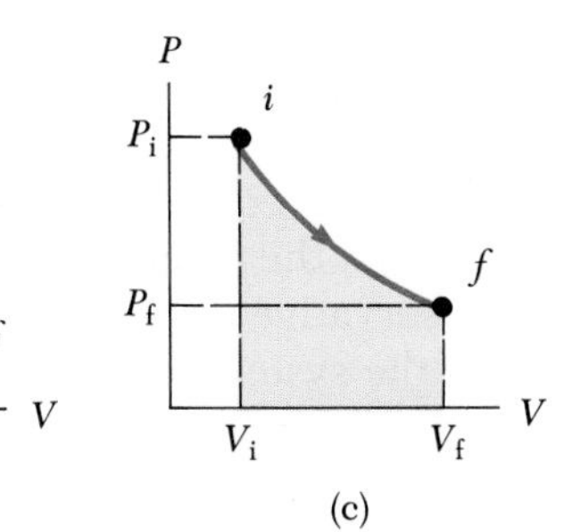

Figure 20.5 The work done by a gas as it is taken from an initial state to a final state depends on the path between these states.

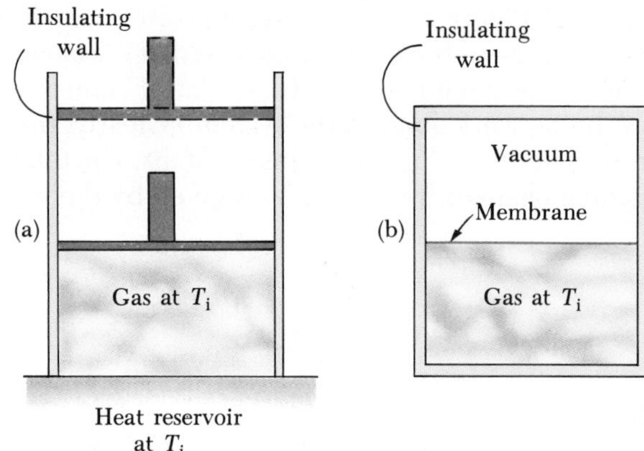

Figure 20.6 (a) A gas at temperature T_i expands slowly by absorbing heat from a reservoir at the same temperature. (b) A gas expands rapidly into an evacuated region after a membrane is broken.

In a similar manner, the heat transferred into or out of the system is also found to depend on the process. This can be demonstrated by considering the situations described in Figure 20.6. In each case, the gas has the same initial volume, temperature, and pressure and is assumed to be ideal. In Figure 20.6a, the gas is in thermal contact with a heat reservoir. If the pressure of the gas is infinitesimally greater than atmospheric pressure, the gas will expand and cause the piston to rise. During this expansion to some final volume V_f, sufficient heat to maintain a constant temperature T_i will be transferred from the reservoir to the gas.

Now consider the thermally insulated system shown in Figure 20.6b. When the membrane is broken, the gas expands rapidly into the vacuum until it occupies a volume V_f. In this case, the gas does no work since there is no movable piston. Furthermore, no heat is transferred through the thermally insulated wall, which we call an *adiabatic wall.* This process is often referred to as **adiabatic free expansion,** or simply *free expansion.* In general, an adiabatic process is one in which no heat is transferred between the system and its surroundings.

Free expansion of a gas

The initial and final states of the ideal gas in Figure 20.6a are identical to the initial and final states in Figure 20.6b, but the paths are different. In the first case, heat is transferred slowly to the gas, and the gas does work on the piston. In the second case, no heat is transferred and the work done is zero. Therefore, we conclude that *heat, like work, depends on the initial, final, and intermediate states of the system.* Furthermore, since heat and work depend on the path, neither quantity is independently conserved during a thermodynamic process.

20.5 THE FIRST LAW OF THERMODYNAMICS

When the law of conservation of energy was first introduced in Chapter 8, it was stated that the mechanical energy of a system is conserved in the absence of nonconservative forces, such as friction. That is, the changes in the internal energy of the system were not included in this mechanical model.

The **first law of thermodynamics** is a generalization of the law of conservation of energy and includes possible changes in internal energy.

It is a universally valid law that can be applied to all kinds of processes. Furthermore, it provides us with a connection between the microscopic and macroscopic worlds.

We have seen that energy can be transferred between a system and its surroundings in two ways. One is work done by (or on) the system. This mode of energy exchange results in measurable changes in the macroscopic variables of the system, such as the pressure, temperature, and volume of a gas. The other is heat transfer, which takes place at the microscopic level.

Change in internal energy

To put these ideas on a more quantitative basis, suppose a thermodynamic system undergoes a change from an initial state to a final state in which Q units of heat are absorbed (or removed) and W is the work done by (or on) the system.[6] For example, the system may be a gas whose pressure and volume change from P_i, V_i to P_f, V_f. If the quantity $Q - W$ is measured for various paths connecting the initial and final equilibrium states (that is, for various *processes*), one finds that $Q - W$ is the same for *all* paths connecting the initial and final states. We conclude that the quantity $Q - W$ is determined completely by the initial and final states of the system, and we call the quantity $Q - W$ the *change in the internal energy of the system.* Although Q and W both depend on the path, the quantity $Q - W$, that is, *the change in internal energy, is independent of the path.* If we represent the internal energy function by the letter U, then the *change* in internal energy, $\Delta U = U_f - U_i$, can be expressed as

$$\Delta U = U_f - U_i = Q - W \tag{20.9}$$

First law of thermodynamics

where all quantities must have the same energy units. Equation 20.9 is known as the **first law of thermodynamics.** When it is used in this form, Q is positive when heat *enters* the system and W is positive when work is done *by* the system.

When a system undergoes an infinitesimal change in state, where a small amount of heat, dQ, is transferred and a small amount of work, dW, is done, the internal energy also changes by a small amount, dU. Thus, for infinitesimal processes we can express the first law as[7]

$$dU = dQ - dW \tag{20.10}$$

First law of thermodynamics for infinitesimal changes

On a microscopic level, the internal energy of a system includes the kinetic and potential energies of the molecules making up the system. In thermodynamics, we do not concern ourselves with the specific form of the internal energy. We simply use Equation 20.9 as a definition of the change in internal energy. One can make an analogy here between the potential energy function associated with a body moving under the influence of gravity without friction. The potential energy function is independent of the path, and it is only its change that is of concern. Likewise, the change in internal energy of a thermodynamic system is what matters, since only differences are defined. Any reference state can be chosen for the internal energy since absolute values are not defined.

[6] We use the convention that Q is positive if the system absorbs heat and negative if it loses heat. Likewise, the work done is positive if the system does work on the surroundings and negative if work is done on the system.

[7] Note that dQ and dW are not true differential quantities, although dU is a true differential. In fact, dQ and dW are *inexact differentials* and are often represented by $đQ$ and $đW$. For further details on this point, see an advanced text in thermodynamics, such as M. W. Zemansky and R. H. Dittman, *Heat and Thermodynamics*, New York, McGraw-Hill, 1981.

Now let us look at some special cases. First consider an *isolated system*, that is, one that does not interact with its surroundings. In this case, there is no heat flow and the work done is zero; hence the internal energy remains constant. That is, since $Q = W = 0$, $\Delta U = 0$, and so $U_i = U_f$. We conclude that

Isolated systems

the internal energy of an isolated system remains constant.

Next consider a process in which the system is taken through a **cyclic process**, that is, one that originates and ends at the same state. In this case, the change in the internal energy is *zero* and the heat added to the system must equal the work done during the cycle. That is, in a cyclic process,

Cyclic process

$$\Delta U = 0 \quad \text{and} \quad Q = W$$

Note that *the net work done per cycle equals the area enclosed by the path representing the process on a PV diagram*. As we shall see in Chapter 22, cyclic processes are very important in describing the thermodynamics of *heat engines*, which are devices in which some part of the heat energy input is extracted as mechanical work.

If a process occurs in which the work done is zero, then the change in internal energy equals the heat entering or leaving the system. If heat enters the system, Q is positive and the internal energy increases. For a gas, we can associate this increase in internal energy with an increase in the kinetic energy of the molecules. On the other hand, if a process occurs in which the heat transferred is zero and work is done by the system, then the magnitude of the change in internal energy equals the negative of the work done by the system. That is, the internal energy of the system decreases. For example, if a gas is compressed with no heat transferred (by a moving piston, say), the work done is negative and the internal energy again increases. This is because kinetic energy is transferred from the moving piston to the gas molecules.

We have seen that there is really no distinction between heat and work on a microscopic scale. Both can produce a change in the internal energy of a system. Although the macroscopic quantities Q and W are *not* properties of a system, they are related to the internal energy of the system through the first law of thermodynamics. Once a process or path is defined, Q and W can be calculated or measured, and the change in internal energy can be found from the first law. One of the important consequences of the first law is that there is a quantity called *internal energy*, the value of which is determined by the state of the system. The internal energy function is therefore called a *state function*.

20.6 SOME APPLICATIONS OF THE FIRST LAW OF THERMODYNAMICS

In order to apply the first law of thermodynamics to specific systems, it is useful to first define some common thermodynamic processes.

Adiabatic process

An **adiabatic process** is defined as a process for which no heat enters or leaves the system, that is, $Q = 0$.

Applying the first law of thermodynamics in this case, we see that

First law for an adiabatic process

$$\Delta U = -W \tag{20.11}$$

An adiabatic process can be achieved either by thermally insulating the system from its surroundings (say, with Styrofoam or an evacuated wall) or by performing the process rapidly. From this result, we see that if a gas expands adiabatically, W is positive, so ΔU is negative and the gas is lowered in temperature. Conversely, a gas is raised in temperature when it is compressed adiabatically.

Adiabatic processes are very important in engineering practice. Some common examples of adiabatic processes include the expansion of hot gases in an internal combustion engine, the liquefaction of gases in a cooling system, and the compression stroke in a diesel engine.

The *free expansion process* described in Figure 20.6b is an adiabatic process in which no work is done on or by the gas. Since $Q = 0$ and $W = 0$, we see from the first law that $\Delta U = 0$ for this process. That is, *the initial and final internal energies of a gas are equal in an adiabatic free expansion.* As we shall see in the next chapter, the internal energy of an ideal gas depends only on its temperature. Thus, we would expect no change in temperature during an adiabatic free expansion. This is in accord with experiments performed at low pressures. Careful experiments at high pressures for real gases show a slight decrease in temperature after the expansion.

Isobaric process

A process that occurs at constant pressure is called an **isobaric process.** When such a process occurs, the heat transferred and the work done are both nonzero. The work done is simply the pressure multiplied by the change in volume, or $P(V_f - V_i)$.

A process that takes place at constant volume is called an **isovolumetric process.** In such a process, the work done is clearly zero.

Hence from the first law we see that

First law for a constant-volume process

$$\Delta U = Q \tag{20.12}$$

This tells us that *if heat is added to a system kept at constant volume, all of the heat goes into increasing the internal energy of the system.* When a mixture of gasoline vapor and air explodes in the cylinder of an engine, the temperature and pressure rise suddenly because the cylinder volume doesn't change appreciably during the short duration of the explosion.

Isothermal process

A process that occurs at constant temperature is called an **isothermal process,** and a plot of P versus V at constant temperature for an ideal gas yields a hyperbolic curve called an *isotherm.* The internal energy of an ideal gas is a function of temperature only. Hence, in an isothermal process of an ideal gas, $\Delta U = 0$.

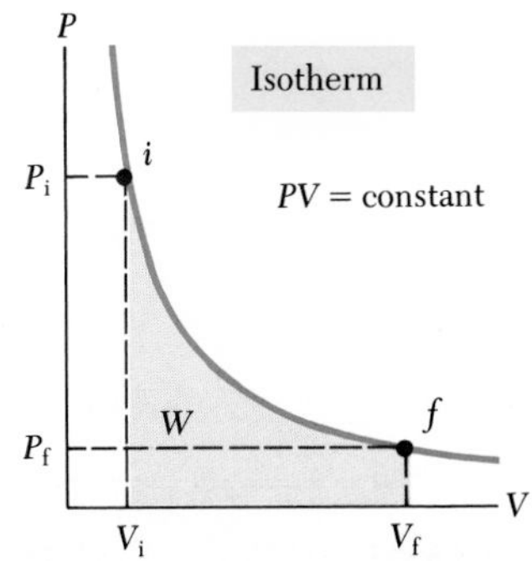

Figure 20.7 The PV diagram for an isothermal expansion of an ideal gas from an initial state to a final state. The curve is a hyperbola.

Isothermal Expansion of an Ideal Gas

Suppose an ideal gas is allowed to expand quasi-statically at constant temperature as described by the PV diagram in Figure 20.7. The curve is a hyperbola with the equation $PV =$ constant. Let us calculate the work done by the gas in the expansion from state i to state f.

The isothermal expansion of the gas can be achieved by placing the gas in good thermal contact with a heat reservoir at the same temperature, as in Figure 20.6a.

The work done by the gas is given by Equation 20.8. Since the gas is ideal and the process is quasi-static, we can apply $PV = nRT$ for each point on the path. Therefore, we have

$$W = \int_{V_i}^{V_f} P\,dV = \int_{V_i}^{V_f} \frac{nRT}{V}\,dV$$

But T is constant in this case; therefore it can be removed from the integral. This gives

$$W = nRT \int_{V_i}^{V_f} \frac{dV}{V} = nRT \ln V \Big]_{V_i}^{V_f}$$

To evaluate the integral, we used $\int \frac{dx}{x} = \ln x$ (Table B.5 in Appendix B). Thus, we find

Work done in an isothermal process

$$W = nRT \ln\left(\frac{V_f}{V_i}\right) \qquad (20.13)$$

Numerically, this work equals the shaded area under the PV curve in Figure 20.7. If the gas expands isothermally, then $V_f > V_i$ and we see that the work done by the gas is positive, as we would expect. If the gas is compressed isothermally, then $V_f < V_i$ and the work done by the gas is negative. (Negative work here implies that positive work must be done *on* the gas by some external agent to compress it.) In the next chapter we shall find that the internal energy of an ideal gas depends only on temperature. Hence, for an isothermal process $\Delta U = 0$, and from the first law we conclude that the heat given up by the reservoir (and transferred to the gas) equals the work done by the gas, or $Q = W$.

The Boiling Process

Suppose that a liquid of mass m vaporizes at constant pressure P. Its volume in the liquid state is V_ℓ, and its volume in the vapor state is V_v. Let us find the work done in the expansion and the change in internal energy of the system.

Since the expansion takes place at constant pressure, the work done by the system is

$$W = \int_{V_\ell}^{V_v} P\,dV = P\int_{V_\ell}^{V_v} dV = P(V_v - V_\ell)$$

The heat that must be transferred to the liquid to vaporize all of it is equal to $Q = mL_v$, where L_v is the heat of vaporization of the liquid. Using the first law and the result above, we get

$$\Delta U = Q - W = mL_v - P(V_v - V_\ell) \qquad (20.14)$$

EXAMPLE 20.6 Work Done During an Isothermal Expansion
Calculate the work done by 1 mole of an ideal gas that is kept at 0°C in an expansion from 3 liters to 10 liters.

Solution Substituting these values into Equation 20.13 gives

$$W = nRT \ln\left(\frac{V_f}{V_i}\right)$$

$$= (1 \text{ mole})(8.31 \text{ J/mole} \cdot \text{K})(273 \text{ K}) \ln\left(\frac{10}{3}\right)$$

$$= \boxed{2.73 \times 10^3 \text{ J}}$$

The heat that must be supplied to the gas from the reservoir to keep T constant is also 2.73×10^3 J.

EXAMPLE 20.7 Boiling Water
One gram of water occupies a volume of 1 cm³ at atmospheric pressure. When this amount of water is boiled, it becomes 1671 cm³ of steam. Calculate the change in internal energy for this process.

Solution Since the heat of vaporization of water is 2.26×10^6 J/kg at atmospheric pressure, the heat required to boil 1 g of water is

$$Q = mL_v = (1 \times 10^{-3} \text{ kg})(2.26 \times 10^6 \text{ J/kg}) = 2260 \text{ J}$$

The work done by the system is positive and equal to

$$W = P(V_v - V_\ell)$$
$$= (1.013 \times 10^5 \text{ N/m}^2)[(1671 - 1) \times 10^{-6} \text{ m}^3]$$
$$= 169 \text{ J}$$

Hence, the change in internal energy is given by

$$\Delta U = Q - W = 2260 \text{ J} - 169 \text{ J} = \boxed{2091 \text{ J}}$$

The internal energy of the system *increases* since ΔU is positive. We see that most of the heat (93%) that is transferred to the liquid goes into increasing the internal energy. Only a small fraction of the heat (7%) goes into external work.

EXAMPLE 20.8 Heat Transferred to a Solid
The internal energy of a solid also increases when heat is transferred to it from its surroundings.

A 1-kg bar of copper is heated at atmospheric pressure. If its temperature increases from 20°C to 50°C, (a) find the work done by the copper.

Solution The change in volume of the copper can be calculated using Equation 19.9 and the volume expansion coefficient for copper taken from Table 19.2 (remembering that $\beta = 3\alpha$):

$$\Delta V = \beta V \, \Delta T = [5.1 \times 10^{-5} \, (\text{C}°)^{-1}](50°\text{C} - 20°\text{C})V$$
$$= 1.5 \times 10^{-3} \, V$$

But the volume is equal to m/ρ, and the density of copper is 8.92×10^3 kg/m³. Hence,

$$\Delta V = (1.5 \times 10^{-3}) \left(\frac{1 \text{ kg}}{8.92 \times 10^3 \text{ kg/m}^3}\right)$$
$$= 1.7 \times 10^{-7} \text{ m}^3$$

Since the expansion takes place at constant pressure, the work done is given by

$$W = P \, \Delta V = (1.013 \times 10^5 \text{ N/m}^2)(1.7 \times 10^{-7} \text{ m}^3)$$

$$= \boxed{1.9 \times 10^{-2} \text{ J}}$$

(b) What quantity of heat is transferred to the copper?

Solution Taking the specific heat of copper from Table 20.1 and using Equation 20.4, we find that the heat transferred is

$$Q = mc \, \Delta T = (1 \text{ kg})(387 \text{ J/kg} \cdot \text{C}°)(30 \text{ C}°)$$

$$= \boxed{1.16 \times 10^4 \text{ J}}$$

(c) What is the increase in internal energy of the copper?

Solution From the first law of thermodynamics, the increase in internal energy is found to be

$$\Delta U = Q - W = \boxed{1.16 \times 10^4 \text{ J}}$$

Note that almost *all* of the heat transferred goes into increasing the internal energy. The fraction of heat energy that is used to do work against the atmosphere is only about 10^{-6}! Hence, in the thermal expansion of a solid or a liquid, the small amount of work done is usually ignored.

20.7 HEAT TRANSFER

In practice, it is important to understand the rate at which heat is transferred between a system and its surroundings and the mechanisms responsible for the heat transfer. You may have used a Thermos bottle or some other thermally insulated vessel to store hot coffee (or ice water) for a length of time. The vessel reduces heat transfer between the outside air and the hot coffee. Ulti-

Melted snow pattern on a parking lot indicates the presence of underground steam pipes used to aid snow removal. Heat from the steam is conducted to the pavement from the pipes, causing the snow to melt. (Courtesy of Dr. Albert A. Bartlett, University of Colorado, Boulder)

mately, of course, the liquid will reach air temperature since the vessel is not a perfect insulator. There will be no heat transfer between a system and its surroundings when they are at the same temperature.

Heat Conduction

The easiest heat transfer process to describe quantitatively is called *heat conduction*. In this process, the heat transfer can be viewed on an atomic scale as an exchange of kinetic energy between molecules, where the less energetic particles gain energy by colliding with the more energetic particles. For example, if you insert a metallic bar into a flame while holding one end, you will find that the temperature of the metal in your hand increases. The heat reaches your hand through conduction. The manner in which heat is transferred from the flame, through the bar, and to your hand can be understood by examining what is happening to the atoms and electrons of the metal. Initially, before the rod is inserted into the flame, the metal atoms and electrons are vibrating about their equilibrium positions. As the flame heats the rod, those metal atoms and electrons near the flame begin to vibrate with larger and larger amplitudes. These, in turn, collide with their neighbors and transfer some of their energy in the collisions. Slowly, metal atoms and electrons farther down the rod increase their amplitude of vibration, until the large-amplitude vibrations arrive at the end being held. The effect of this increased vibration is an increase in temperature of the metal, and possibly a burned hand.

Although the transfer of heat through a metal can be partially explained by atomic vibrations and electron motion, the rate of heat conduction also depends on the properties of the substance being heated. For example, it is possible to hold a piece of asbestos in a flame indefinitely. This implies that very little heat is being conducted through the asbestos. In general, metals are good conductors of heat, and materials such as asbestos, cork, paper, and fiber glass are poor conductors. Gases also are poor heat conductors because of their dilute nature. Metals are good conductors of heat because they contain large numbers of electrons that are relatively free to move through the metal and can transport energy from one region to another. Thus, in a good conductor, such as copper, heat conduction takes place via the vibration of atoms and via the motion of free electrons.

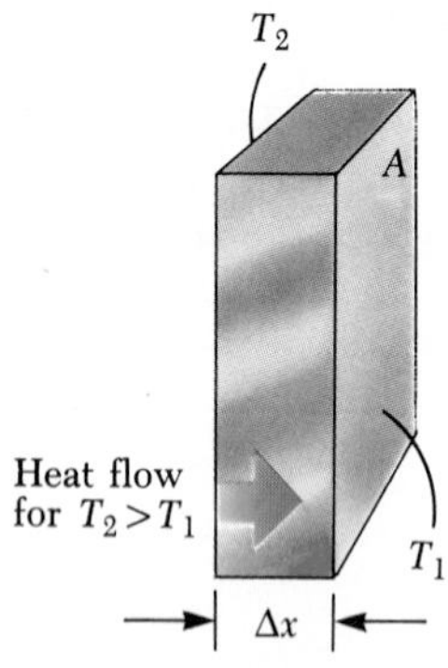

Figure 20.8 Heat transfer through a conducting slab of cross-sectional area A and thickness Δx. The opposite faces are at different temperatures, T_1 and T_2.

The conduction of heat occurs only if there is a difference in temperature between two parts of the conducting medium. Consider a slab of material of thickness Δx and cross-sectional area A with its opposite faces at different temperatures T_1 and T_2, where $T_2 > T_1$ (Fig. 20.8). One finds from experiment that the heat ΔQ transferred in a time Δt flows from the hotter end to the colder end. The rate at which heat flows, $\Delta Q/\Delta t$, is found to be proportional to the cross-sectional area, the temperature difference, and inversely proportional to the thickness. That is,

$$\frac{\Delta Q}{\Delta t} \propto A \frac{\Delta T}{\Delta x}$$

It is convenient to use the symbol H to represent the heat transfer rate. That is, we take $H = \Delta Q/\Delta t$. Note that H has units of watts when ΔQ is in joules and Δt is in seconds (1 W = 1 J/s). For a slab of infinitesimal thickness dx and temperature difference dT, we can write the **law of heat conduction**

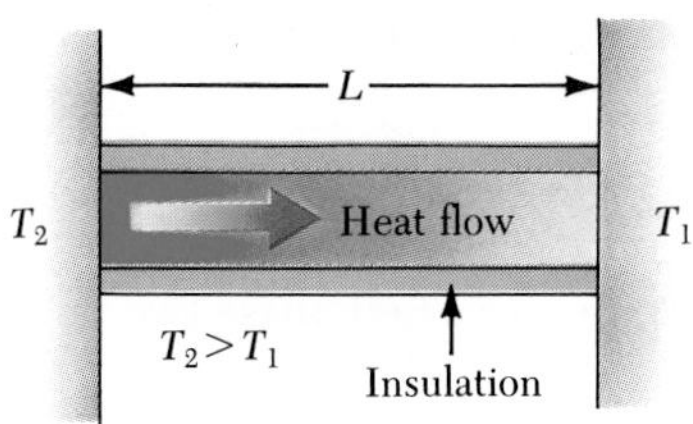

Figure 20.9 Conduction of heat through a uniform, insulated rod of length L. The opposite ends are in thermal contact with heat reservoirs at two different temperatures.

$$H = -kA\frac{dT}{dx} \quad (20.15)$$

Law of heat conduction

where the proportionality constant k is called the **thermal conductivity** of the material, and dT/dx is the **temperature gradient** (the variation of temperature with position). The minus sign in Equation 20.15 denotes the fact that heat flows in the direction of decreasing temperature.

Suppose a substance is in the shape of a long uniform rod of length L, as in Figure 20.9, and is insulated so that no heat can escape from its surface except at the ends, which are in thermal contact with heat reservoirs having temperatures T_1 and T_2. When a steady state has been reached, the temperature at each point along the rod is constant in time. In this case, the temperature gradient is the same everywhere along the rod and is given by $\frac{dT}{dx} = \frac{T_1 - T_2}{L}$. Thus the heat transfer rate is

$$H = kA\frac{(T_2 - T_1)}{L} \quad (20.16)$$

Substances that are good heat conductors have large thermal conductivity values, whereas good thermal insulators have low thermal conductivity values. Table 20.3 lists thermal conductivities for various substances. We see that metals are generally better thermal conductors than nonmetals.

For a compound slab containing several materials of thicknesses L_1, L_2, . . . and thermal conductivities k_1, k_2, . . . , the rate of heat transfer through the slab at steady state is given by

$$H = \frac{A(T_2 - T_1)}{\sum_i (L_i/k_i)} \quad (20.17)$$

where T_1 and T_2 are the temperatures of the outer extremities of the slab (which are held constant) and the summation is over all slabs. The following example is a proof of this result for the case of heat transfer through two slabs.

TABLE 20.3 Thermal Conductivities

Substance	Thermal Conductivity W/m·C°
Metals (at 25°C)	
Aluminum	238
Copper	397
Gold	314
Iron	79.5
Lead	34.7
Silver	427
Gases (at 20°C)	
Air	0.0234
Helium	0.138
Hydrogen	0.172
Nitrogen	0.0234
Oxygen	0.0238
Nonmetals (approximate values)	
Asbestos	0.08
Concrete	0.8
Glass	0.8
Ice	2
Rubber	0.2
Water	0.6
Wood	0.08

EXAMPLE 20.9 Heat Transfer Through Two Slabs □

Two slabs of thickness L_1 and L_2 and thermal conductivities k_1 and k_2 are in thermal contact with each other as in Figure 20.10. The temperatures of their outer surfaces are T_1 and T_2, respectively, and $T_2 > T_1$. Determine the temperature at the interface and the rate of heat transfer through the slabs in the steady-state condition.

Solution If T is the temperature at the interface, then the rate at which heat is transferred through slab 1 is given by

$$(1) \quad H_1 = \frac{k_1 A(T - T_1)}{L_1}$$

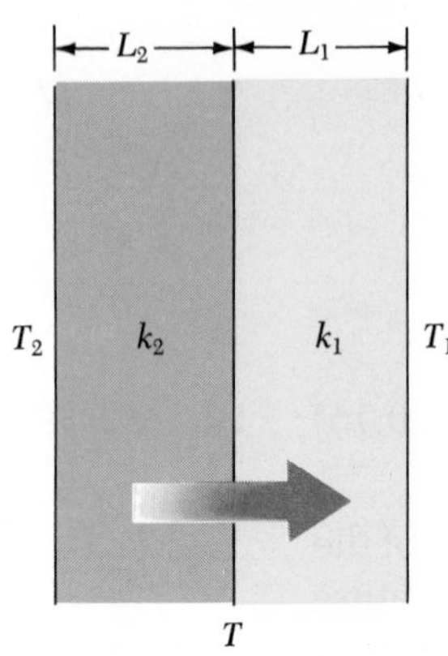

Figure 20.10 Heat transfer by conduction through two slabs in thermal contact with each other. At steady state, the rate of heat transfer through slab 1 equals the rate of heat transfer through slab 2.

Likewise, the rate at which heat is transferred through slab 2 is

$$(2) \qquad H_2 = \frac{k_2A(T_2 - T)}{L_2}$$

When a steady state is reached, these two rates must be equal; hence

$$\frac{k_1A(T - T_1)}{L_1} = \frac{k_2A(T_2 - T)}{L_2}$$

Solving for T gives

$$(3) \qquad T = \frac{k_1L_2T_1 + k_2L_1T_2}{k_1L_2 + k_2L_1}$$

Substituting (3) into either (1) or (2), we get

$$H = \frac{A(T_2 - T_1)}{(L_1/k_1) + (L_2/k_2)}$$

An extension of this model to several slabs of materials leads to Equation 20.17.

*Home Insulation

If you would like to do some calculating to determine whether or not to add insulation to a ceiling or to some other portion of a building, what you have just learned about conduction needs to be modified slightly, for two reasons. (1) The insulating properties of materials used in buildings are usually expressed in engineering rather than SI units. For example, measurements stamped on a package of fiber glass insulating board will be in units such as British thermal units, feet, and degrees Fahrenheit. (2) In dealing with the insulation of a building, we must consider heat conduction through a compound slab, with each portion of the slab having a different thickness and a different thermal conductivity. For example, a typical wall in a house consists of an array of materials, such as wood paneling, dry wall, insulation, sheathing, and wood siding.

In Example 20.9, we showed how to deal with heat conduction through a two-layered slab. A general formula for heat transfer through a compound slab is given by Equation 20.17. For example, if the slab consists of three different materials, the denominator of Equation 20.17 will consist of the sum of three terms. In engineering practice, the term L/k for a particular substance is

The houses in this thermogram look much like jack-o'-lanterns as the heat inside them escapes. This image, made during cold weather, was produced with an infrared scanner. (VANSCAN® Thermogram by Daedalus Enterprises, Inc.)

TABLE 20.4 *R* Values for Some Common Building Materials

Material	*R* value ($ft^2 \cdot F^\circ \cdot h/BTU$)
Hardwood siding (1 in. thick)	0.91
Wood shingles (lapped)	0.87
Brick (4 in. thick)	4.00
Concrete block (filled cores)	1.93
Fiber glass batting (3.5 in. thick)	10.90
Fiber glass batting (6 in. thick)	18.80
Fiber glass board (1 in. thick)	4.35
Cellulose fiber (1 in. thick)	3.70
Flat glass (0.125 in. thick)	0.89
Insulating glass (0.25-in. space)	1.54
Vertical air space (3.5 in. thick)	1.01
Air film	0.17
Dry wall (0.5 in. thick)	0.45
Sheathing (0.5 in. thick)	1.32

referred to as the *R* value of the material. Thus, Equation 20.17 reduces to

$$H = \frac{A(T_2 - T_1)}{\sum_i R_i} \tag{20.18}$$

where $R_i = L_i/R_i$. The *R* values for a few common building materials are given in Table 20.4 (note the units).

Also, near any vertical surface there is a very thin, stagnant layer of air that must be considered when finding the total *R* value for a wall. The thickness of this stagnant layer on an outside wall depends on the velocity of the wind. As a result, heat loss from a house on a day when the wind is blowing hard is greater than heat loss on a day when the wind velocity is zero. A representative *R* value for this stagnant layer of air is given in Table 20.4.

EXAMPLE 20.10 The *R* Value of a Typical Wall
Calculate the total *R* value for a wall constructed as shown in Figure 20.11a. Starting outside the house (to the left in Fig. 20.11a) and moving inward, the wall consists of brick, 0.5 in. of sheathing, a vertical air space 3.5 in. thick, and 0.5 in. of dry wall. Do not forget the dead-air layers inside and outside the house.

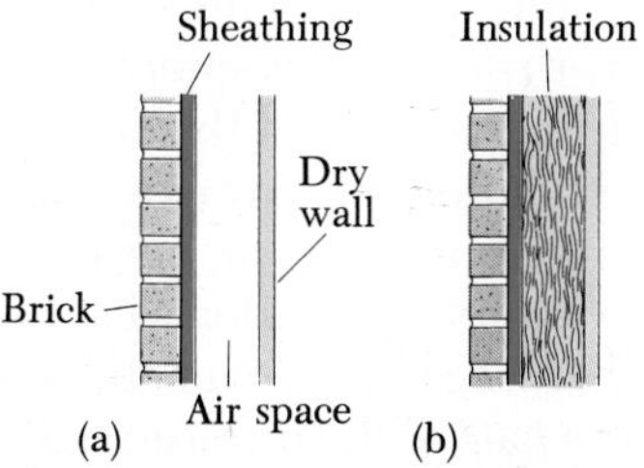

Figure 20.11 (Example 20.10) Cross-sectional view of an exterior wall containing (a) an air space and (b) insulation.

Solution Referring to Table 20.4, we find the total *R* value for the wall as follows:

R_1 (outside air film) $= 0.17\ ft^2 \cdot F^\circ \cdot h/BTU$

R_2 (brick) $= 4.00$

R_3 (sheathing) $= 1.32$

R_4 (air space) $= 1.01$

R_5 (dry wall) $= 0.45$

R_6 (inside air film) $= 0.17$

$R_{total} = 7.12\ ft^2 \cdot F^\circ \cdot h/BTU$

Exercise 4 If a layer of fiberglass insulation 3.5 in. thick is placed inside the wall to replace the air space as in Figure 20.11b, what is the total *R* value of the wall? By what factor is the heat loss reduced?
Answer $R = 17\ ft^2 \cdot F^\circ \cdot h/BTU$; a factor of 2.4.

Convection

At one time or another you probably have warmed your hands by holding them over an open flame. In this situation, the air directly above the flame is heated and expands. As a result, the density of the air decreases and the air rises. This warmed mass of air heats your hands as it flows by. *Heat transferred by the movement of a heated substance is said to have been transferred by* **convection.** When the movement results from differences in density, as in the example of air around a fire, it is referred to as *natural convection.* When the heated substance is forced to move by a fan or pump, as in some hot-air and hot-water heating systems, the process is called *forced convection.*

The circulating pattern of air flow at a beach is an example of convection. Likewise, the mixing that occurs as water is cooled and eventually freezes at its surface (Chapter 19) is an example of convection in nature. Recall that the mixing by convection currents ceases when the water temperature reaches 4°C. Since the water in the lake cannot be cooled by convection below 4°C, and because water is a relatively poor conductor of heat, the water near the bottom remains near 4°C for a long time. As a result, fish have a comfortable temperature in which to live even in periods of prolonged cold weather.

If it were not for convection currents, it would be very difficult to boil water. As water is heated in a teakettle, the lower layers are warmed first. These heated regions expand and rise to the top because their density is lowered. At the same time, the denser cool water replaces the warm water at the bottom of the kettle so that it can be heated.

The same process occurs when a room is heated by a radiator. The hot radiator warms the air in the lower regions of the room. The warm air expands and rises to the ceiling because of its lower density. The denser regions of cooler air from above replace the warm air, setting up the continuous air current pattern shown in Figure 20.12.

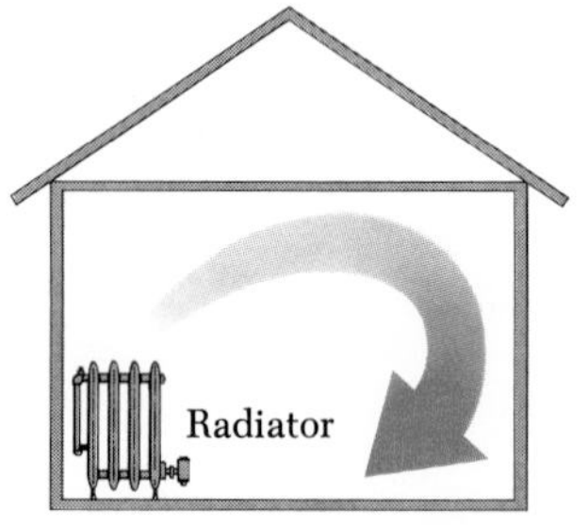

Figure 20.12 Convection currents are set up in a room heated by a radiator.

Radiation

The third way of transferring heat is through **radiation.** All objects radiate energy continuously in the form of electromagnetic waves, which we shall discuss in Chapter 34. The type of radiation associated with the transfer of heat energy from one location to another is referred to as infrared radiation.

Through electromagnetic radiation, approximately 1340 J of heat energy from the sun strikes 1 m^2 of the top of the earth's atmosphere every second. Some of this energy is reflected back into space and some is absorbed by the atmosphere, but enough arrives at the surface of the earth each day to supply all of our energy needs on this planet hundreds of times over—if it could be captured and used efficiently. The growth in the number of solar houses in this country is one example of an attempt to make use of this free energy.

Radiant energy from the sun affects our day-to-day existence in a number of ways. It influences the earth's average temperature, ocean currents, agriculture, rain patterns, and so on. For example, consider what happens to the atmospheric temperature at night. If there is a cloud cover above the earth, the water vapor in the clouds reflects back a part of the infrared radiation emitted by the earth and consequently the temperature remains at moderate levels. In the absence of this cloud cover, however, there is nothing to prevent this radiation from escaping into space, and thus the temperature drops more on a clear night than when it is cloudy.

The rate at which an object emits radiant energy is proportional to the fourth power of its absolute temperature. This is known as **Stefan's law** and is expressed in equation form as

$$P = \sigma A e T^4 \tag{20.19}$$

Stefan's law

where P is the power radiated by the body in watts (or joules per second), σ is a constant equal to $5.6696 \times 10^{-8}\ \mathrm{W/m^2 \cdot K^4}$, A is the surface area of the object in square meters, e is a constant called the **emissivity,** and T is temperature in kelvins. The value of e can vary between zero and unity, depending on the properties of the surface.

An object radiates energy at a rate given by Equation 20.19. At the same time, the object also absorbs electromagnetic radiation. If the latter process did not occur, an object would eventually radiate all of its energy and its temperature would reach absolute zero. The energy that a body absorbs comes from its surroundings, which consists of other objects which radiate energy. If an object is at a temperature T and its surroundings are at a temperature T_0, the net energy gained or lost each second by the object as a result of radiation is given by

$$P_{\mathrm{net}} = \sigma A e (T^4 - T_0{}^4) \tag{20.20}$$

When an object is in *equilibrium* with its surroundings, *it radiates and absorbs energy at the same rate, and so its temperature remains constant.* When an object is hotter than its surroundings, it radiates more energy than it absorbs and so it cools. An *ideal absorber* is defined as an object that absorbs all of the energy incident on it. The emissivity of an ideal absorber is equal to unity. Such an object is often referred to as a **black body.** An ideal absorber is also an ideal radiator of energy. In contrast, an object with an emissivity equal to zero absorbs none of the energy incident on it. Such an object reflects all the incident energy and so is a perfect reflector.

The Dewar Flask

The Thermos bottle, called a *Dewar flask*[8] in the scientific community, is a practical example of a container designed to minimize heat losses by conduction, convection, and radiation. Such a container is used to store either cold or hot liquids for long periods of time. The standard construction (Fig. 20.13) consists of a double-walled pyrex vessel with silvered inner walls. The space between the walls is evacuated to minimize heat transfer by conduction and convection. The silvered surfaces minimize heat transfer by radiation by reflecting most of the radiant heat. Very little heat is lost over the neck of the flask since glass is not a very good conductor, and the cross-section of the glass in the direction of conduction is small. A further reduction in heat loss is obtained by reducing the size of the neck. Dewar flasks are commonly used to store liquid nitrogen (boiling point 77 K) and liquid oxygen (boiling point 90 K).

For another cryogenic liquid, such as liquid helium, which has a very low specific heat (boiling point 4.2 K), it is often necessary to use a double Dewar system in which the Dewar flask containing the liquid is surrounded by a second Dewar flask. The space between the two flasks is filled with liquid nitrogen.

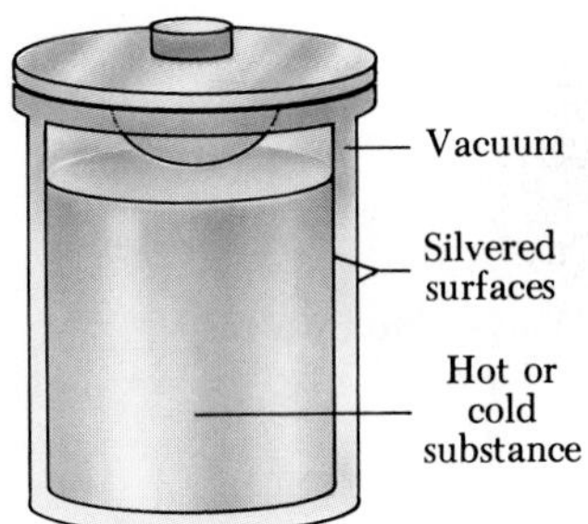

Figure 20.13 A cross-sectional view of a Dewar vessel, used to store hot or cold liquids or other substances.

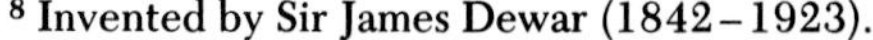

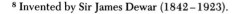

[8] Invented by Sir James Dewar (1842–1923).

EXAMPLE 20.11 Who Turned Down the Thermostat?

An unclothed student is in a room that is 20°C. If the skin temperature of the student is 37°C, how much heat is lost from his body in 10 min, assuming that the emissivity of skin is 0.90? Assume that the surface area of the student is 1.5 m^2.

Solution Using Equation 20.20, the rate of heat loss from the skin is

$$P_{net} = \sigma Ae(T^4 - T_0^4)$$
$$= (5.67 \times 10^{-8}\ W/m^2 \cdot K^4)(1.5\ m^2)$$
$$\times (0.90)[(310\ K)^4 - (293\ K)^4] = 143\ J/s$$

(Why is the temperature given in kelvins?) At this rate of heat loss, the total heat lost by the skin in 10 min is

$$Q = P_{net} \times \text{time} = (143\ J/s)(600\ s) = 8.58 \times 10^4\ J$$

SUMMARY

Heat flow is a form of energy transfer that takes place as a consequence of a temperature difference only. The **internal energy** of a substance is a function of its state and generally increases with increasing temperature.

The **calorie** is the amount of heat necessary to raise the temperature of 1 g of water from 14.5°C to 15.5°C. The **mechanical equivalent of heat** is defined to be 4.186 J/cal.

The **heat capacity,** *C*, of any substance is defined as the amount of heat energy needed to raise the temperature of the substance by one Celsius degree. The heat required to change the temperature of a substance by ΔT is

Heat required to raise the temperature of a substance

$$Q = mc\,\Delta T \tag{20.4}$$

where m is the mass of the substance and c is its **specific heat,** or heat capacity per unit mass.

The heat required to change the phase of a pure substance of mass m is given by

Latent heat

$$Q = mL \tag{20.6}$$

The parameter L is called the **latent heat** of the substance and depends on the nature of the phase change and the properties of the substance.

A **quasi-static process** is one that proceeds slowly enough to allow the system to always be in a state of equilibrium.

The **work done** by a gas as its volume changes from some initial value V_i to some final value V_f is

Work done by a gas

$$W = \int_{V_i}^{V_f} P\,dV \tag{20.8}$$

where P is the pressure, which may vary during the process. In order to evaluate W, the nature of the process must be specified—that is, P and V must be known during each step of the process. Since the work done depends on the initial, final, and intermediate states, it therefore depends on the path taken between the initial and final states.

From the **first law of thermodynamics** we see that when a system undergoes a change from one state to another, the change in its internal energy, ΔU, is given by

First law of thermodynamics

$$\Delta U = Q - W \tag{20.9}$$

where Q is the heat transferred into (or out of) the system and W is the work done by (or on) the system. Although Q and W both depend on the path taken from the initial state to the final state, the quantity ΔU is path-independent.

In a **cyclic process** (one that originates and terminates at the same state), $\Delta U = 0$, and therefore $Q = W$. That is, the heat transferred into the system equals the work done during the cycle.

An **adiabatic process** is one in which no heat is transferred between the system and its surroundings ($Q = 0$). In this case, the first law gives $\Delta U = -W$. That is, the internal energy changes as a consequence of work being done by (or on) the system.

In an **adiabatic free expansion** of a gas, $Q = 0$ and $W = 0$, and so $\Delta U = 0$. That is, the internal energy of the gas does not change in such a process.

An *isovolumetric* (or **isochoric**) **process** is one that occurs at constant volume. No work is done in such a process.

An **isobaric process** is one that occurs at constant pressure. The work done in such a process is simply $P\,\Delta V$.

An **isothermal process** is one that occurs at constant temperature. The work done by an ideal gas during an isothermal process is

Work done in an isothermal process

$$W = nRT \ln\left(\frac{V_f}{V_i}\right) \tag{20.13}$$

Heat may be transferred by three fundamentally distinct mechanisms: conduction, convection, and radiation. The *conduction* process can be viewed as an exchange of kinetic energy between colliding molecules. The rate at which heat flows by conduction through a slab of area A is given by

Law of heat conduction

$$H = -kA\frac{dT}{dx} \tag{20.15}$$

where k is the **thermal conductivity** and $\frac{dT}{dx}$ is the **temperature gradient.**

Convection is a heat transfer process in which the heated substance moves from one place to another.

All bodies radiate and absorb energy in the form of electromagnetic waves. A body that is hotter than its surroundings radiates more energy than it absorbs, whereas a body that is cooler than its surroundings absorbs more than it radiates. An **ideal radiator,** or black body, is one that absorbs all energy incident on it; an ideal radiator emits as much energy as is possible for a body of its size, shape and temperature.

QUESTIONS

1. Ethyl alcohol has about one half the specific heat of water. If equal masses of alcohol and water in separate beakers are supplied with the same amount of heat, compare the temperature increases of the two liquids.
2. Give one reason why coastal regions tend to have a more moderate climate than inland regions.
3. A small crucible is taken from a 200°C oven and immersed in a tub full of water at room temperature (often referred to as *quenching*). What is the approximate final equilibrium temperature?
4. What is the major problem that arises in measuring specific heats if a sample with a temperature above 100°C is placed in water?
5. In a daring lecture demonstration, an instructor dips his wetted fingers into molten lead (327°C) and withdraws them quickly, without getting burned. How is this possible? (This is a dangerous experiment, which you should not attempt.)
6. The pioneers found that a large tub of water placed in a storage cellar would prevent their food from freezing on really cold nights. Explain why this is so.
7. What is wrong with the statement: "Given any two bodies, the one with the higher temperature contains more heat."
8. Why is it possible to hold a lighted match, even when it is burned to within a few millimeters of your fingertips?
9. The photograph below shows the pattern formed by snow on the roof of a barn. What causes the alternating pattern of snow-covered and exposed roof?

(Question 9) Alternating pattern of snow-covered and exposed roof. (Courtesy of Dr. Albert A. Bartlett, University of Colorado, Boulder)

10. Why is a person able to remove a piece of dry aluminum foil from a hot oven with bare fingers, while if there is moisture on the foil, a burn results?
11. A tile floor in a bathroom may feel uncomfortably cold to your bare feet, but a carpeted floor in an adjoining room at the same temperature will feel warm. Why?
12. Why can potatoes be baked more quickly when a skewer has been inserted through them?
13. A Thermos bottle is constructed with double silvered-glass walls, with the space between them evacuated. Give reasons for the silvered walls and the vacuum jacket.
14. A piece of paper is wrapped around a rod made half of wood and half of copper. When held over a flame, the paper in contact with the wood burns but the half in contact with the metal does not. Explain.
15. Why is it necessary to store liquid nitrogen or liquid oxygen in vessels equipped with either Styrofoam insulation or a double-evacuated wall?
16. Why do heavy draperies over the windows help keep a home warm in the winter and cool in the summer?
17. If you wish to cook a piece of meat thoroughly on an open fire, why should you not use a high flame? (Note that carbon is a good thermal insulator.)
18. When insulating a wood-frame house, is it better to place the insulation against the cooler outside wall or against the warmer inside wall? (In either case, there is an air barrier to consider.)
19. In an experimental house, Styrofoam beads were pumped into the air space between the double windows at night in the winter, and pumped out to holding bins during the day. How would this assist in conserving heat energy in the house?
20. Pioneers stored fruits and vegetables in underground cellars. Discuss as fully as possible this choice for a storage site.
21. Why can you get a more severe burn from steam at 100°C than from water at 100°C?
22. Concrete has a higher specific heat than soil. Use this fact to explain (partially) why cities have a higher average night-time temperature than the surrounding countryside. If a city is hotter than the surrounding countryside, would you expect breezes to blow from city to country or from country to city? Explain.
23. When camping in a canyon on a still night, one notices that as soon as the sun strikes the surrounding peaks, a breeze begins to stir. What causes the breeze?
24. Updrafts of air are familiar to all pilots. What causes these currents?
25. If water is a poor conductor of heat, why can it be heated quickly when placed over a flame?
26. The U.S. penny is now made of copper-coated zinc. Can a calorimetric experiment be devised to test for the metal content in a collection of pennies? If so, describe the procedure you would use.

27. If you hold water in a paper cup over a flame, you can bring the water to a boil without burning the cup. How is this possible?
28. When a sealed Thermos bottle full of hot coffee is shaken, what are the changes, if any, in (a) the temperature of the coffee and (b) the internal energy of the coffee?
29. Using the first law of thermodynamics, explain why the *total* energy of an isolated system is always conserved.
30. Is it possible to convert internal energy to mechanical energy? Explain with examples.
31. Suppose you pour hot coffee for your guests, and one of them chooses to drink the coffee after it has been in the cup for several minutes. In order to have the warmest coffee, should the person add the cream just after the coffee is poured or just before drinking? Explain.
32. Two identical cups both at room temperature are filled with the same amount of hot coffee. One cup contains a metal spoon, while the other does not. If you wait for several minutes, which of the two will have the warmer coffee? Which heat transfer process explains your answer?
33. A warning sign often seen on highways just before a bridge is "Caution—Bridge surface freezes before road surface." Which of the three heat transfer processes is most important in causing a bridge surface to freeze before a road surface on very cold days?

PROBLEMS

Section 20.1 Heat and Thermal Energy

1. Consider Joule's apparatus described in Figure 20.1. The two masses are 1.5 kg each, and the tank is filled with 200 g of water. What is the increase in the temperature of the water after the masses fall through a distance of 3 m?
2. An 80-kg weight-watcher wishes to climb a mountain to work off the equivalent of a large piece of chocolate cake rated at 700 (food) Calories. How high must the person climb?
3. Water at the top of Niagara Falls has a temperature of 10°C. If it falls through a distance of 50 m and all of its potential energy goes into heating the water, calculate the temperature of the water at the bottom of the falls.

Section 20.2 Heat Capacity and Specific Heat

4. How many calories of heat are required to raise the temperature of 3 kg of aluminum from 20°C to 50°C?
5. It takes 2 kcal to heat 600 g of an unknown substance from 15°C to 40°C. What is the specific heat of the substance?
6. A 50-g piece of cadmium is at 20°C. If 400 cal of heat is added to the cadmium, what is its final temperature?
7. What is the final equilibrium temperature when 10 g of milk at 10°C is added to 160 g of coffee at 90°C? (Assume the heat capacities of the two liquids are the same as water, and neglect the heat capacity of the container.)
8. Copper pellets, each of mass 1 g, are heated to 100°C. How many pellets must be added to 500 g of water initially at 20°C to make the final equilibrium temperature 25°C? (Neglect the heat capacity of the container.)
9. A 1.5-kg iron horseshoe initially at 600°C is dropped into a bucket containing 20 kg of water at 25°C. What is the final temperature? (Neglect the heat capacity of the container.)
10. A block of ice with a mass $m = 10$ kg is moved back and forth over the flat horizontal top surface of a large block of ice. Both blocks are at 0°C, and the force that produces the back and forth motion acts only horizontally. The coefficient of kinetic friction (wet ice on ice) is $\mu_k = 0.060$. What is the total distance traveled by the upper block relative to the supporting block if 15.2 g of liquid is produced?
11. If 200 g of water is contained in a 300-g aluminum vessel at 10°C and an additional 100 g of water at 100°C is poured into the container, what is the final equilibrium temperature of the system?
12. A 300-g chunk of copper is heated in a furnace and then dropped into a 500-g aluminum calorimeter containing 300 g of water. If the temperature of the water rises from 15°C to 30°C, what was the initial temperature of the copper? (Assume no heat is lost through water flashing to steam.)
13. How much heat must be added to 20 g of aluminum at 20°C to melt it completely?
14. An aluminum calorimeter of mass 100 g contains 250 g of water. They are in thermal equilibrium at 10°C. Two metallic blocks are placed in the water. One is a 50-g piece of copper at 80°C. The other sample has a mass of 70 g and is originally at a temperature of 100°C. The entire system stabilizes at a final temperature of 20°C. (a) Determine the specific heat of the unknown sample. (b) Determine the material of the unknown, using Table 20.1.
15. A Styrofoam cup contains 200 g of mercury at 0°C. To this is added 50 g of ethyl alcohol at 50°C and 100 g of water at 100°C. (a) What is the final temperature of

the mixture? (b) How much heat was gained or lost by the mercury, the alcohol, and the water? (The specific heat of mercury is 0.033 cal/g · C°, that of ethyl alcohol, 0.58 cal/g · C°. The heat capacity of the Styrofoam cup is negligible.)

16. A 1-kg block of copper initially at 20°C is dropped into a large vessel of liquid nitrogen, which is boiling at 77 K. Assuming the vessel is thermally insulated from its surroundings, calculate the number of liters of nitrogen that boils away by the time the copper reaches 77 K. (Note: Nitrogen has a specific heat of 0.21 cal/g · C°, a heat of vaporization of 48 cal/g and a density of 0.8 g/cm^3.)

17. A 20-g ice cube at 0°C is heated until 15 g has become water at 100°C and 5 g has been converted to steam. How much heat was added to do this?

18. One liter of water at 30°C is used to make iced tea. How much ice at 0°C must be added to lower the temperature of the tea to 10°C? (Ice has a specific heat of 0.50 cal/g · C°).

19. A 50 g copper calorimeter contains 250 g of water at 20°C. How much steam must be condensed into the water to make the final temperature of the system 50°C?

20. If 90 g of molten lead at 327.3°C is poured into a 300-g casting made of iron and initially at 20°C, what is the final temperature of the system? (Assume there are no heat losses.)

21. In an insulated vessel, 250 g of ice at 0°C is added to 600 g of water at 18°C. (a) What is the final temperature of the system? (b) How much ice remains?

22. A 40-g block of ice is cooled to −78°C. It is added to 560 g of water in an 80-g copper calorimeter at a temperature of 25°C. Determine the final temperature. (If all the ice does not melt, determine how much ice is left.) Remember that the ice must first warm to 0°C, melt, and then continue warming as water. The specific heat of ice is 0.500 cal/g · C°.

23. Determine the final state when 20 g of 0°C ice and 10 g of 100°C steam are mixed together in an insulated container.

24. A beaker of water sits in the sun until it reaches an equilibrium temperature of 30°C. The beaker is made of 100 g of aluminum and contains 180 g of water. In an attempt to cool this system down, 100 g of ice at 0°C is added to the water. (a) Determine the final temperature. If $T = 0$°C, determine how much ice remains. (b) Repeat this for the case when 50 g of ice is used.

25. A 3-g copper penny at 25°C drops a distance of 50 m to the ground. (a) If 60% of its initial potential energy goes into increasing the internal energy of the penny, determine its final temperature. (b) Does the result depend on the mass of the penny? Explain.

26. A 2-kg aluminum block is given an initial speed of 4 m/s on a rough horizontal surface. Because of friction, it finally comes to rest. (a) If 75% of its initial kinetic energy is absorbed by the block in the form of thermal energy, calculate the increase in temperature of the block. (b) What happens to the remaining energy?

27. A 3-g lead bullet traveling with a speed of 400 m/s is stopped by a large tree. If all of its kinetic energy is converted to thermal energy in the bullet, find the increase in the temperature of the bullet. Is this answer reasonable?

Section 20.4 Work and Heat in Thermodynamic Processes

28. Using the fact that 1 atm = 1.013×10^5 N/m^2, verify the conversion 1 L · atm = 101.3 J = 24.2 cal.

29. A gas expands from I to F along three possible paths as indicated by Figure 20.14. Calculate the work in joules done by the gas along the paths IAF, IF and IBF.

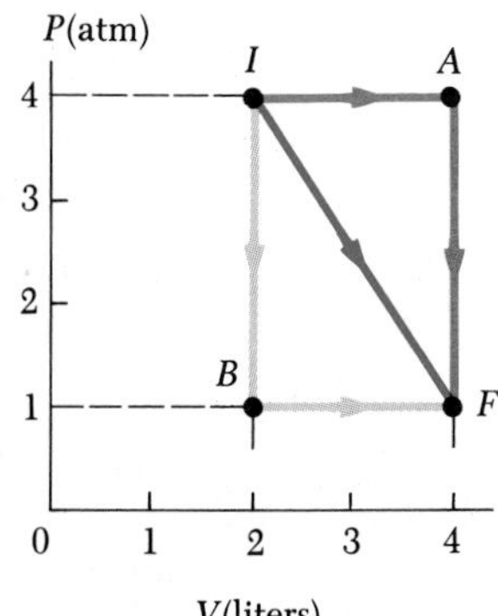

Figure 20.14 (Problems 29 and 36).

30. Gas in a container is at a pressure of 1.5 atm and a volume of 4 m^3. What is the work done by the gas if (a) it expands at constant pressure to twice its initial volume and (b) it is compressed at constant pressure to one quarter its initial volume?

31. An ideal gas is enclosed in a cylinder. There is a movable piston on top of the cylinder. The piston has a mass of 8000 g and an area of 5 cm^2 and is free to slide up and down, keeping the pressure of the gas constant. How much work is done as the temperature of 0.2 moles of the gas is raised from 20°C to 300°C?

32. A 1-mole sample of an ideal gas is carried around the thermodynamic cycle shown in Figure 20.15. The

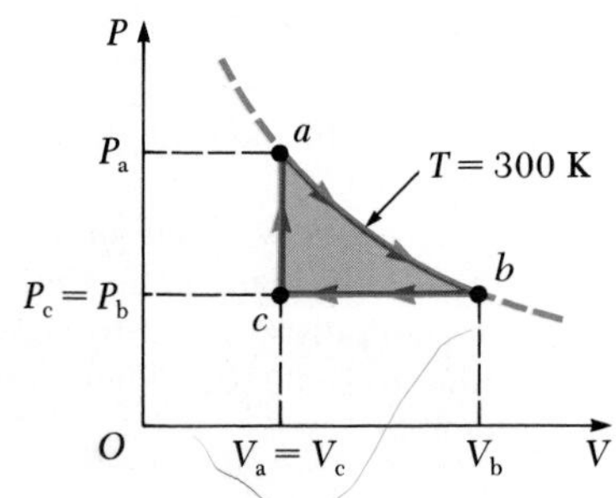

Figure 20.15 (Problem 32).

cycle consists of three parts—an isothermal expansion ($a \to b$), an isobaric compression ($b \to c$), and a constant-volume increase in pressure ($c \to a$). If $T = 300$ K, $P_a = 5$ atm, and $P_b = P_c = 1$ atm, determine the work done by the gas per cycle.

33. A sample of ideal gas is expanded to twice its original volume of 1 m^3 in a quasi-static process for which $P = \alpha V^2$, with $\alpha = 5.0$ atm/m^6, as shown in Figure 20.16. How much work was done by the expanding gas?

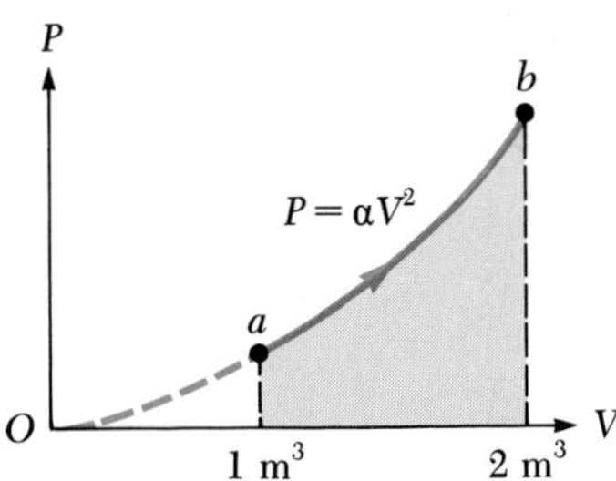

Figure 20.16 (Problem 33).

34. An ideal gas at STP (1 atm and 0°C) is taken through a process where the volume is expanded from 25 L to 80 L. During this process the pressure varies inversely as the volume squared, $P = 0.5\,aV^{-2}$ (a) Determine the constant a in standard SI units. (b) Find the final pressure and temperature. (c) Determine a general expression for the work done by the gas during this process. (d) Compute the actual work in joules done by the gas in this process.

35. One mole of an ideal gas does 3000 J of work on the surroundings as it expands isothermally to a final pressure of 1 atm and volume of 25 L. Determine (a) the initial volume and (b) the temperature of the gas.

Section 20.5 The First Law of Thermodynamics

36. A gas expands from I to F as in Figure 20.14. The heat added to the gas is 400 J when the gas goes from I to F along the diagonal path. (a) What is the change in internal energy of the gas? (b) How much heat must be added to the gas for the indirect path IAF to give the same change in internal energy?

37. A gas is compressed at a constant pressure of 0.8 atm from a volume of 9 L to a volume of 2 L. In the process 400 J of heat energy flows out of the gas. (a) What is the work done by the gas? (b) What is the change in internal energy of the gas?

38. A thermodynamic system undergoes a process in which its internal energy decreases by 500 J. If at the same time, 220 J of work is done on the system, find the heat transferred to or from the system.

39. A gas is taken through the cyclic process described in Figure 20.17. (a) Find the net heat transferred to the system during one complete cycle. (b) If the cycle is reversed, that is, the process goes along $ACBA$, what is the net heat transferred per cycle?

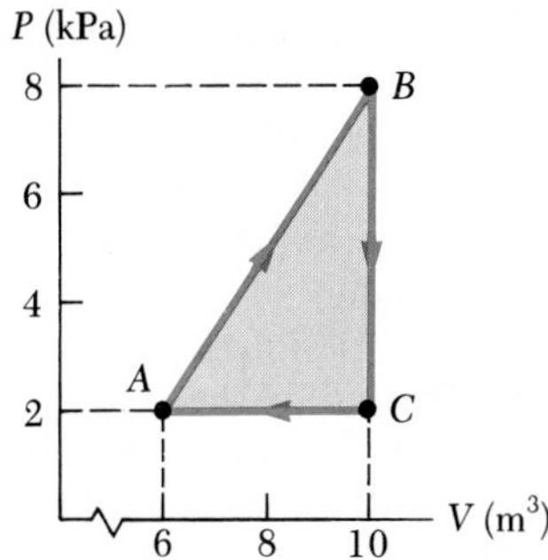

Figure 20.17 (Problems 39 and 40).

40. Consider the cyclic process described by Figure 20.17. If Q is negative for the process BC and ΔU is negative for the process CA, determine the signs of Q, W and ΔU associated with each process.

Section 20.6 Some Applications of the First Law of Thermodynamics

41. Five moles of an ideal gas expands isothermally at 127°C to four times its initial volume. Find (a) the work done by the gas and (b) the heat flow into the system, both in joules.

42. How much work is done by the steam when 1 mole of water at 100°C boils and becomes 1 mole of steam at 100°C at 1 atm pressure? Determine the change in internal energy of the steam as it vaporizes. Consider the steam to be an ideal gas.

43. One mole of gas initially at a pressure of 2 atm and a volume of 0.3 L has an internal energy equal to 91 J. In its final state, the pressure is 1.5 atm, the volume is 0.8 L, and the internal energy equals 182 J. For the three paths IAF, IBF, and IF in Figure 20.18, calculate (a) the work done by the gas and (b) the net heat transferred in the process.

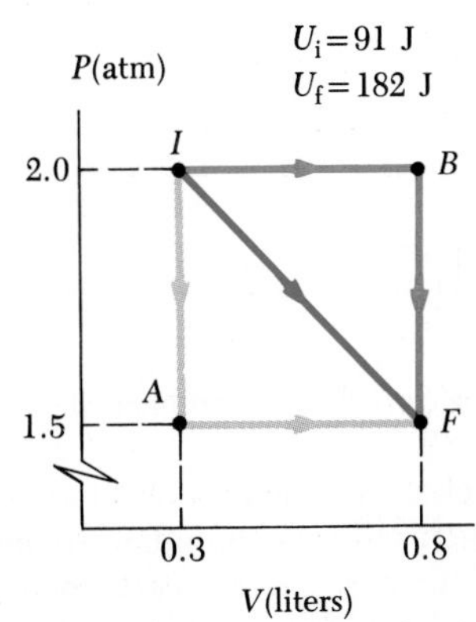

Figure 20.18 (Problem 43).

44. Nitrogen gas ($m = 1.00$ kg) is confined in a cylinder with a movable piston exposed to normal atmospheric pressure. A quantity of heat ($Q = 25\ 000$ cal) is added to the gas in an isobaric process, and the internal energy of the gas increases by 8000 cal. (a) How much work is done by the gas? (b) What is the change in volume?

45. An ideal gas initially at 300 K undergoes an isobaric expansion at a pressure of 2.5 kPa. If the volume increases from 1 m^3 to 3 m^3 and 12 500 J of heat is added to the gas, find (a) the change in internal energy of the gas and (b) its final temperature.

46. Two moles of helium gas initially at a temperature of 300 K and pressure of 0.4 atm is compressed isothermally to a pressure of 1.2 atm. Find (a) the final volume of the gas, (b) the work done by the gas, and (c) the heat transferred. Consider the helium to behave as an ideal gas.

47. One mole of argon is confined in a cylinder with a movable piston at a pressure of 1 atm and at a temperature of 300 K. The gas is heated slowly and isobarically to a temperature of 400 K. The measured value of the molar specific heat at constant pressure for argon in this temperature range is $C_p = 2.5043\ R$, and the measured value of PV/nT is 0.99967 R. Calculate in units of R, carried to two decimals, the following quantities: (a) the work done by the expanding gas; (b) the amount of heat added to the gas; (c) the increase in the internal energy of the gas.

48. One mole of an ideal monatomic gas is carried by a quasi-static isothermal process (at 400 K) to twice its original volume (Fig. 20.19). (a) How much work W_{ab} was done by the gas? (b) How much heat Q_{ab} is supplied to the gas? (c) What is the pressure ratio, P_b/P_a? (d) Suppose that a constant-volume process is used to reduce the original pressure P_a to the same final pressure P_b. Determine the new values for W'_{ab}, Q'_{ab}, and $\Delta U'_{ab}$.

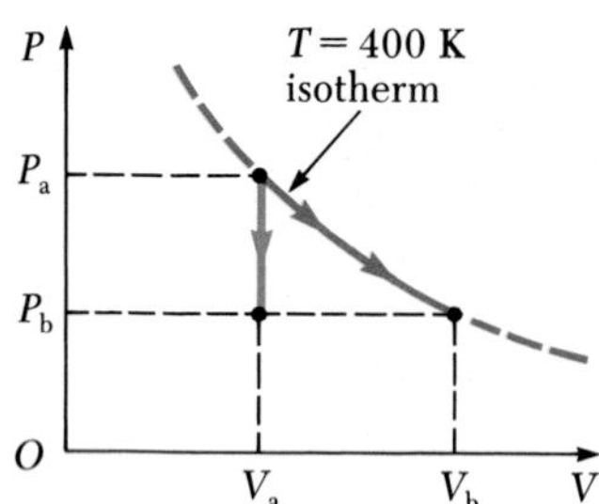

Figure 20.19 (Problem 48).

49. A 1-kg block of aluminum is heated at atmospheric pressure such that its temperature increases from 22°C to 40°C. Find (a) the work done by the aluminum, (b) the heat added to the aluminum, and (c) the change in internal energy of the aluminum.

50. In Figure 20.20, the change in internal energy of a gas that is taken from A to C is +800 J. The work done along path ABC is +500 J. (a) How much heat has to be added to the system as it goes from A through B to C? (b) If the pressure at point A is five times that of point C, what is the work done by the system in going from C to D? (c) What is the heat exchanged with the surroundings as the cycle goes from C to A? (d) If the change in internal energy in going from point D to point A is +500 J, how much heat must be added to the system as it goes from point C to point D?

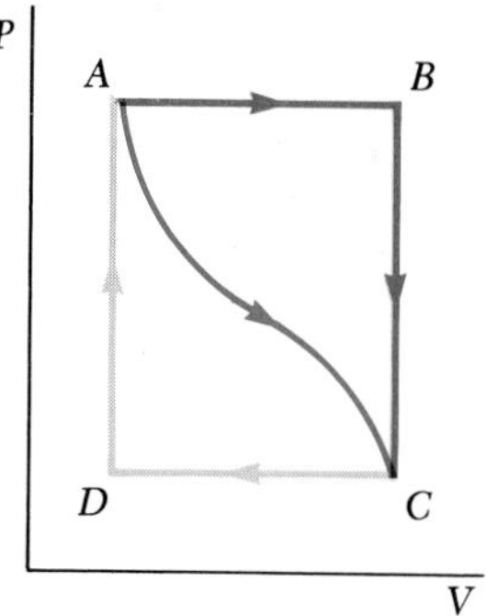

Figure 20.20 (Problems 50 and 51).

51. In the previous problem, (a) determine the difference in internal energy $U_C - U_A$, (b) compute the net work done by the system in going through the complete cycle $ABCDA$, (c) determine how much heat must be added to the system in the same complete cycle, and (d) determine the net change in internal energy for the complete cycle $ABCDA$. Is this consistent with the first law of thermodynamics?

Section 20.7 Heat Transfer

52. A glass window pane has an area of 3 m^2 and a thickness of 0.6 cm. If the temperature difference between its faces is 25°C, how much heat flows through the window per hour?

53. The rod shown in Figure 20.9 is made of iron and has a length of 60 cm and a cross-sectional area of 2 cm^2. One end is maintained at 80°C and the other end is at 20°C. At steady state, find (a) the temperature gradient, (b) the rate of heat transfer, and (c) the temperature in the rod 20 cm from the hot end.

54. The earth's thermal gradient, as measured at the surface, is 30 C°/km, and the earth's radius is 6400 km. Assume that this gradient remains the same all the way to the center of the earth. What is the temperature of the earth at its center if we take the surface temperature to be 0°C? Do you think this is a reasonable answer or is it necessary to refine our assumption of the nature of the thermal gradient with depth?

55. A bar of gold is in thermal contact with a bar of silver of the same length and area (Fig. 20.21). One end of

the compound bar is maintained at 80°C while the opposite end is at 30°C. When the heat flow reaches steady state, find the temperature at the junction.

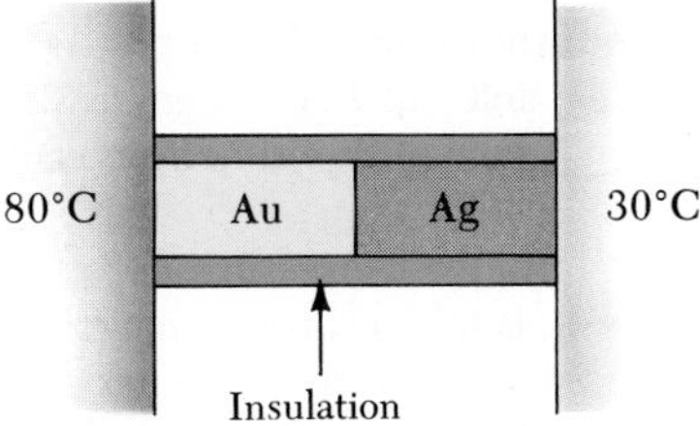

Figure 20.21 (Problem 55).

56. Two rods of the same length but of different materials and cross sectional areas are placed side by side as in Figure 20.22. Determine the rate of heat flow in terms of the thermal conductivity and area of each rod. Generalize this to to several rods.

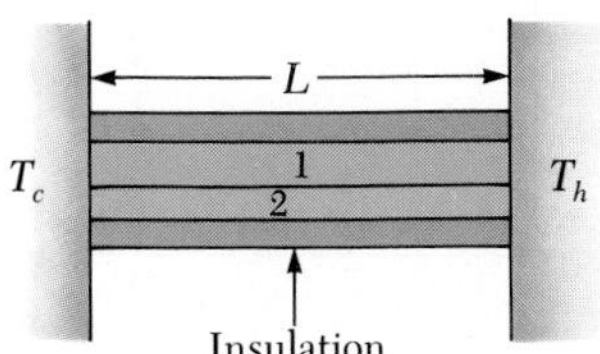

Figure 20.22 (Problem 56).

57. A board of area 2 m^2 and 2 cm thick is used as a thermal barrier between a 20°C room and a 50°C region. How many steel nails 2 cm long and 4 mm in diameter may be driven through the board before the heat flow through the board is doubled?

58. An iron rod 20 cm long with a diameter of 1 cm has one end immersed in an ice water bath while the other end is in a steam tank at 100°C. Assume enough time has elapsed such that a uniform thermal gradient has been established along the rod. (a) Determine the rate of heat flow along the rod. (b) Compute how fast the ice is melting at the cold end. (c) How fast is steam being condensed at the hot end to maintain this uniform gradient? (d) What is the thermal gradient along the rod?

59. A Styrofoam container in the shape of a cube 30 cm on a side is 2 cm thick. The inside is at 5°C and the outside is at 25°C. If it takes 8 h for 5 kg of ice to melt in the container, determine the thermal conductivity of the Styrofoam.

60. A Thermopane window of area 6 m^2 is constructed of two layers of glass, each 4 mm thick separated by an air space of 5 mm. If the inside is at 20°C and the outside is at -30°C, what is the heat loss through the window?

61. A wall is made up of 0.5-inch dry wall, 6-inch fiberglass batting, 0.5-inch sheathing, and lapped wood shingles. If the area of this wall is 20 ft^2, and the temperature difference between the inside and outside temperatures is 30 F°, (a) compute the rate of heat loss. (b) If a 2 ft^2 insulating glass window is installed in the wall, find the new heat loss rate through the wall (i.e., wall plus window).

62. The surface of the sun has a temperature of about 5800 K. Taking the radius of the sun to be equal to 6.96×10^8 m, calculate the total energy radiated by the sun each day. (Assume $e = 1$.)

63. A black thin-walled cylindrical can, 10 cm in diameter and 20 cm tall, is filled with water at 100°C and suspended in a 20°C room. (a) Determine the initial rate of heat transfer due to radiation. (b) Find the initial rate of change of temperature of the water. (Neglect any calorimetric effects of the can.) (c) Repeat this for the case when the can's temperature has decreased to 30°C.

ADDITIONAL PROBLEMS

64. The density of water is 999.17 kg/m^3 at 14.5°C and 999.02 kg/m^3 at 15.5°C. (a) Calculate the work done against the surrounding atmosphere when 2 kg of water expands as its temperature is increased from 14.5°C to 15.5°C. (b) Compare this work with the required heat input of 8370 J.

65. A would-be alchemist places 5 kg of molten lead at 327.3°C and 1 kg of ice at 0°C into an insulated chamber where they reach a common final temperature. (Assume that the specific heats of lead and water are constant throughout the temperature ranges encountered in this problem.) Find (a) the final temperature and (b) the heat transferred during the equilibration process.

66. A *flow calorimeter* is an apparatus used to measure the specific heat of a liquid. The technique is to measure the temperature difference between the input and output points of a flowing stream of the liquid while adding heat at a known rate. In one particular experiment, a liquid of density 0.78 g/cm^3 flows through the calorimeter at the rate of 4 cm^3/s. At steady state, a temperature difference of 4.8 C° is established between the input and output points when heat is supplied at the rate of 30 J/s. What is the specific heat of the liquid?

67. One mole of an ideal gas is contained in a cylinder with a movable piston. The initial pressure, temperature, and volume are P_0, V_0, and T_0, respectively. Find the work done by the gas for the following processes and show each process on a *PV* diagram: (a) an isobaric compresson in which the final volume is one half the initial volume, (b) an isothermal compression in which the final pressure is four times the initial pres-

sure, (c) an isovolumetric process in which the final pressure is triple the initial pressure.

68. A gas expands from a volume of 2 m^3 to a volume of 6 m^3 along two different paths as described in Figure 20.23. The heat added to the gas along the path *IAF* is equal to 4×10^5 cal. Find (a) the work done by the gas along the path *IAF*, (b) the work done along the path *IF*, (c) the change in internal energy of the gas, and (d) the heat transferred in the process along the path *IF*.

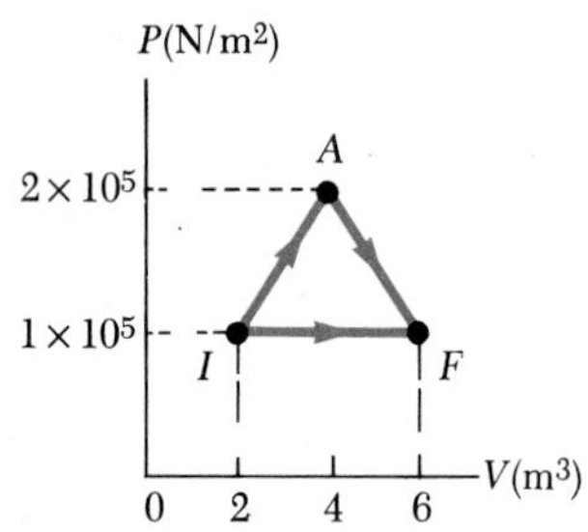

Figure 20.23 (Problem 68).

69. An aluminum rod, 0.5 m in length and of cross-sectional area 2.5 cm^2, is inserted into a thermally insulated vessel containing liquid helium at 4.2 K. The rod is initially at 300 K. (a) If one half of the rod is inserted into the helium, how many liters of helium boil off by the time the inserted half cools to 4.2 K? (Assume the upper half does not cool.) (b) If the upper portion of the rod is maintained at 300 K, what is the approximate boil-off rate of liquid helium after the lower half has reached 4.2 K? (Note that aluminum has a thermal conductivity of 31 J/s · cm · K at 4.2 K, a specific heat of 0.21 cal/g · C°, and a density of 2.7 g/cm^3. See Example 20.5 for data on helium.)

70. Estimate the minimum heat required to transform 300 g of lead at 20°C to a gas at atmospheric pressure. Assume that C_p for lead is constant over the temperature range 20°C to 1750°C. (Use Tables 20.1 and 20.2.)

71. An automobile has a mass of 1500 kg, and its aluminum brakes have an overall mass of 6 kg. (a) Assuming that all of the frictional work done when the car stops is deposited in the brakes, and neglecting heat transfer, how many times could the car be braked to rest from 25 m/s (56 mph) before the brakes melt? (Assume an initial temperature of 20°C.) (b) Identify some effects that are neglected in part (a) but are likely to be important in a more realistic assessment of the heating of brakes.

72. An aluminum kettle has a circular cross section and is 8 cm in radius and 0.3 cm thick. It is placed on a hot plate and filled with 1 kg of water. If the bottom of the kettle is maintained at 101°C and the inside at 100°C, find (a) the rate of heat flow into the water and (b) the time it takes for all of the water to boil away. (Neglect heat transferred from the sides.)

73. Using the data in Example 20.5 and Table 20.2, calculate the change in internal energy when 2 cm^3 of liquid helium gas at 4.2 K is converted to helium gas at 273.15 K and atmospheric pressure. (Assume that the molar heat capacity of helium gas is 24.9 J/mol·K, and note that 1 cm^3 of liquid helium is equivalent to 3.1×10^{-2} moles.)

74. An ideal gas initially at pressure P_0, volume V_0, and temperature T_0 is taken through a cycle as described in Figure 20.24. (a) Find the net work done by the gas per cycle. (b) What is the net heat added to the system per cycle? (c) Obtain a numerical value for the net work done per cycle for one mole of gas initially at 0°C.

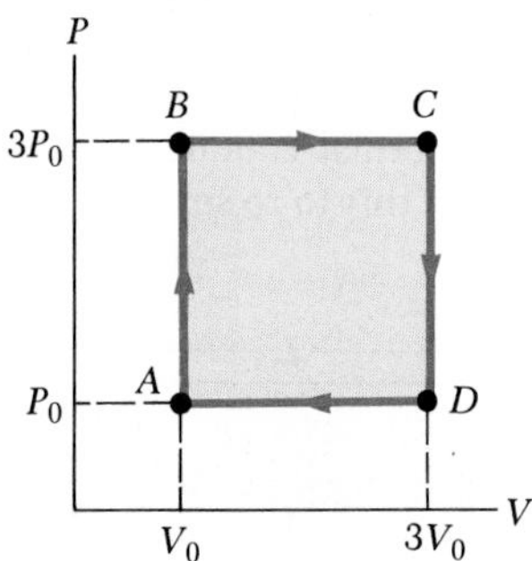

Figure 20.24 (Problem 74).

75. A one-person research submarine has a spherical iron hull 1.50 m in outer radius and 2 cm thick, lined with an equal thickness of rubber. If the submarine is used in arctic waters (temperature 0°C) and the total rate of heat release within the sub (including the occupant's metabolic heat) is 1500 W, find the equilibrium temperature of the interior.

76. An iron plate is held against an iron wheel so that there is a sliding frictional force of 50 N acting between the two pieces of metal. The relative speed at which the two surfaces slide over each other is 40 m/s. (a) Calculate the rate at which mechanical energy is converted to thermal energy. (b) The plate and the wheel have a mass of 5 kg each, and each receives 50% of the thermal energy. If the system is run as described for 10 s and each object is then allowed to reach a uniform internal temperature, what is the resultant temperature increase?

77. A vessel in the shape of a spherical shell has an inner radius a and outer radius b. The wall has a thermal conductivity k. If the inside is maintained at a temperature T_1 and the outside is at a temperature T_2, show that the rate of heat flow between the surfaces is given by

$$\frac{dQ}{dt} = \left(\frac{4\pi kab}{b-a}\right)\left(T_1 - T_2\right)$$

78. The inside of a hollow cylinder is maintained at a temperature T_a while the outside is at a lower temperature, T_b (Fig. 20.25). The wall of the cylinder has a thermal conductivity k. Neglecting end effects, show that the rate of heat flow from the inner to the outer wall in the radial direction is given by

$$\frac{dQ}{dt} = 2\pi Lk\left[\frac{T_a - T_b}{\ln(b/a)}\right]$$

(*Hint:* The temperature gradient is given by dT/dr. Note that a radial heat current passes through a concentric cylinder of area $2\pi rL$.)

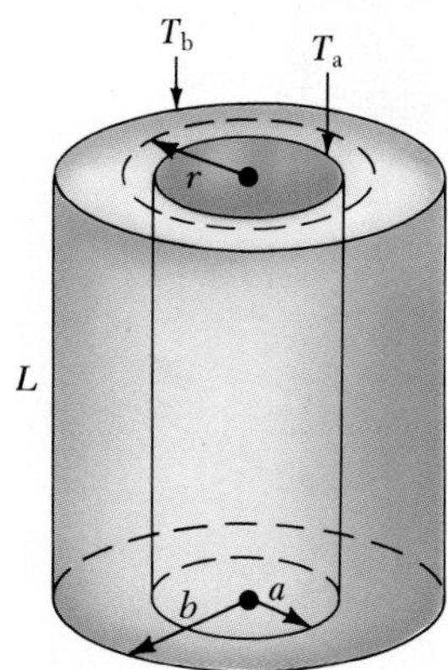

Figure 20.25 (Problem 78).

79. The passenger section of a jet airliner is in the shape of a cylindrical tube of length 35 m and inner radius 2.5 m. Its walls are lined with a 6 cm thickness of insulating material of thermal conductivity 4×10^{-5} cal/s·cm·C°. The inside is to be maintained at 25°C while the outside is at −35°C. What heating rate is required to maintain this temperature difference? (Use the result from Problem 78.)

80. A Thermos bottle in the shape of a cylinder has an inner radius of 4 cm, outer radius of 4.5 cm, and length of 30 cm. The insulating walls have a thermal conductivity equal to 2×10^{-5} cal/s·cm·C°. One liter of hot coffee at 90°C is poured into the bottle. If the outside wall remains at 20°C, how long does it take for the coffee to cool to 50°C? (Neglect end effects and losses by radiation and convection. Use the result from Problem 78 and assume that coffee has the same properties as water.)

81. A "solar cooker" consists of a curved reflecting mirror that focuses sunlight onto the object to be heated (Fig. 20.26). The solar power per unit area reaching the earth at some location is 600 W/m^2, and a small solar cooker has a diameter of 0.6 m. Assuming that 40% of the incident energy is converted into heat energy, how long would it take to completely boil off 0.5 liters of water initially at 20°C? (Neglect the heat capacity of the container.)

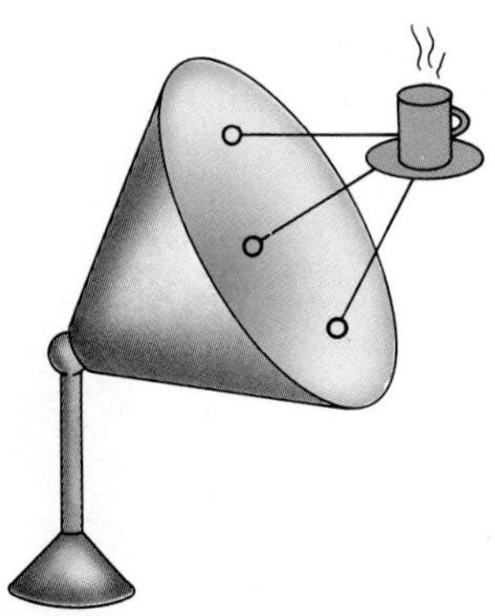

Figure 20.26 (Problem 81).

82. Consider a mass M of liquid that partially fills a cylindrical container of cross-section A. The container has a negligible coefficient of thermal expansion, but the liquid has a volume coefficient of expansion β. (a) Show that the fractional increase $\Delta h/h$ in the depth of the liquid in response to a temperature increase ΔT is given by:

$$\frac{\Delta h}{h} = \beta\,\Delta T$$

(b) Show that the corresponding increase in the potential energy of the liquid (in the gravitational field of the earth) is equal to

$$\frac{Mg\,\Delta h}{2} \quad \text{or} \quad \frac{Mgh\beta\,\Delta T}{2}$$

(c) The nominal heat requirement of 4186 J to raise the temperature of 1 kg of water by 1°C does not include any allowance for energy invested in increased gravitational potential energy. Use the expression given in part (b) to assess this additional energy requirement. Specifically consider water of mass $M = 1$ kg heated from 14.5°C to 15.5°C in a container of cross-section $A = 50$ cm^2, so that $h \approx 20$ cm. At 15°C, $\beta = 1.5 \times 10^{-4}(\text{C°})^{-1}$. Evaluate $(Mg\,\Delta h)/2$ in joules, and also express that energy as a fraction of 4186 J. (d) Does your result suggest that the increase in gravitational potential energy is a significant additional energy requirement when a container of water is heated?

83. A 150-g sample of copper is heated to 500°C. When it is plunged into 700 g of water at 20°C it is noted that 10 g of water immediately flashes to steam and leaves the system. The latent heat of vaporization at 20°C is 585 cal/g. Neglect any effects from the calorimeter. (a) Determine the final temperature of the system less the escaped vapor. (Assume all the water to steam conversion takes place at 20°C.) (b) What would the final temperature have been if no water were converted to steam?

21

The Kinetic Theory of Gases

Hot air balloons over the Rocky Mountains in winter. These colorful balloons rise as a large gas burner heats the air inside them. Because warm air is less dense than cool air, the buoyant force upward can exceed the total force downward, causing the balloons to rise. The vertical motion can also be controlled by venting hot air at the top of the balloon. (© Nicholas De Vore III/Photographers Aspen)

In Chapter 19 we discussed the properties of an ideal gas using such macroscopic variables as pressure, volume, and temperature. We shall now show that such large-scale properties can be described on a microscopic scale, where matter is treated as a collection of molecules. Newton's laws of motion applied in a statistical manner to a collection of particles provide a reasonable description of thermodynamic processes. In order to keep the mathematics relatively simple, we shall consider only the molecular behavior of gases, where the interactions between molecules are much weaker than in liquids or solids. In the current view of gas behavior, called the *kinetic theory*, gas molecules move about in a random fashion, colliding with the walls of their container and with each other. Perhaps the most important consequence of this theory is that it shows the equivalence between the kinetic energy of molecular motion and the internal energy of the system. Furthermore, the kinetic theory provides us with a physical basis upon which the concept of temperature can be understood.

In the simplest model of a gas, each molecule is considered to be a hard sphere that collides elastically with other molecules or with the container wall. The hard-sphere model assumes that the molecules do not interact with each other except during collisions and that they are not deformed by collisions. This description is adequate only for monatomic gases, where the energy is entirely translational kinetic energy. One must modify the theory for more complex molecules, such as O_2 and CO_2, to include the internal energy associated with rotations and vibrations of the molecules.

21.1 MOLECULAR MODEL FOR THE PRESSURE OF AN IDEAL GAS

We begin this chapter by developing a microscopic model of an ideal gas. The model shows that the pressure that a gas exerts on the walls of its container is a consequence of the collisions of the gas molecules with the walls. As we shall see, the model is consistent with the macroscopic description of the preceding chapter. The following assumptions will be made:

Assumptions of the molecular model of an ideal gas

1. *The number of molecules is large, and the average separation between them is large* compared with their dimensions. Therefore, the molecules occupy a negligible volume compared with the volume of the container and are considered to be structureless (that is, point masses).
2. *The molecules obey Newton's laws of motion, but the individual molecules move in a random fashion.* By random fashion, we mean that the molecules move in all directions with equal probability and with various speeds. This distribution of velocities does not change in time, despite the collisions between molecules.
3. *The molecules undergo elastic collisions with each other.* In the collisions both kinetic energy and momentum are conserved.
4. *The forces between molecules are negligible except during a collision.* The forces between molecules are short-range, so that the only time the molecules interact with each other is during a collision.
5. *The gas under consideration is a pure gas.* That is, all molecules are identical.
6. *The gas is in thermal equilibrium with the walls of the container.* Hence, a wall will eject as many molecules as it absorbs, and the ejected molecules will have the same average kinetic energy as the absorbed molecules.

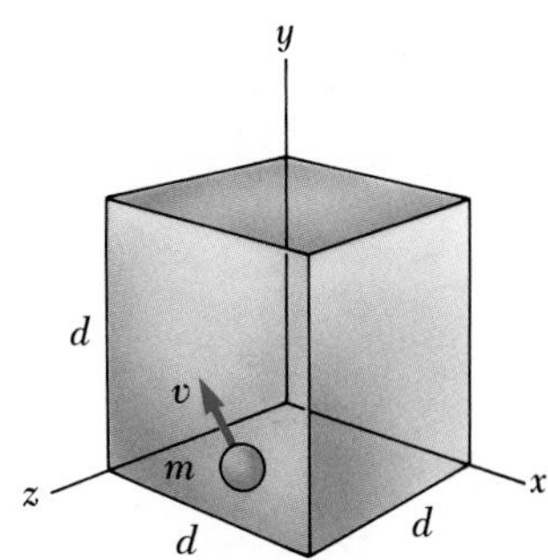

Figure 21.1 A cubical box of sides d containing an ideal gas. The molecule shown moves with velocity $\mathbf{v}$.

Now let us derive an expression for the pressure of an ideal gas consisting of N molecules in a container of volume V. The container is assumed to be in the shape of a cube with edges of length d (Fig. 21.1). Consider the collision of one molecule moving with a velocity $\mathbf{v}$ toward the right-hand face of the box. The molecule has velocity components v_x, v_y, and v_z. As it collides with the wall elastically, its x component of velocity is reversed, while its y and z components of velocity remained unaltered (Fig. 21.2). Since the x component of momentum of the molecule is mv_x before the collision and $-mv_x$ afterward, the *change* in momentum of the molecule is given by

$$\Delta p_x = -mv_x - (mv_x) = -2mv_x$$

The momentum delivered to the wall for each collision is $2mv_x$, since the momentum of the system (molecule + container) is conserved. In order that a molecule make two successive collisions with the same wall, it must travel a distance $2d$ along the x axis in a time Δt. But in a time Δt, the molecule moves a distance $v_x \Delta t$ in the x direction; therefore the time between two successive collisions is $\Delta t = 2d/v_x$. If F is the magnitude of the average force exerted by a molecule on the wall in the time Δt, then from the definition of impulse (which equals change in momentum) we have

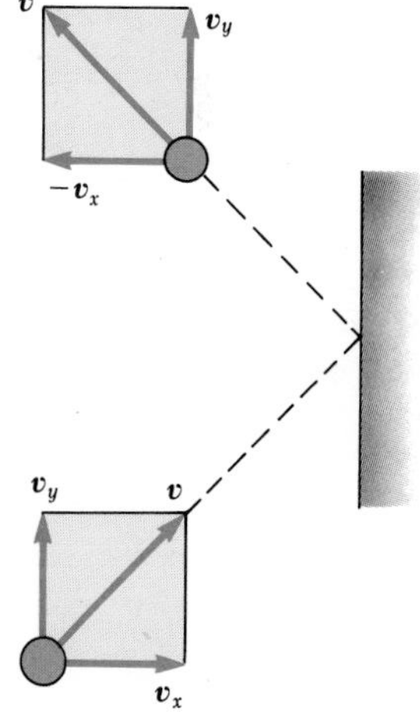

Figure 21.2 A molecule makes an elastic collision with the wall of the container. Its x component of momentum is reversed, thereby imparting momentum to the wall, while its y component remains unchanged. In this construction, we assume that the molecule moves in the xy plane.

$$F\,\Delta t = \Delta p = 2mv_x$$

$$F = \frac{2mv_x}{\Delta t} = \frac{2mv_x}{2d/v_x} = \frac{mv_x^{\,2}}{d} \qquad (21.1)$$

The total force on the wall is the sum of all such terms for all particles. To get the total pressure on the wall, we divide the total force by the area, d^2:

$$P = \frac{\Sigma F}{A} = \frac{m}{d^3}\,(v_{x1}{}^2 + v_{x2}{}^2 + \cdots)$$

where $v_{x1}, v_{x2}, \ldots$ refer to the x components of velocity for particles 1, 2, etc. Since the average value of $v_x{}^2$ is given by

$$\overline{v_x{}^2} = \frac{v_{x1}{}^2 + v_{x2}{}^2 + \cdots}{N}$$

and the volume is given by $V = d^3$, we can express the pressure in the form

$$P = \frac{Nm}{V}\,\overline{v_x{}^2} \tag{21.2}$$

The square of the speed for any one particle is given by

$$v^2 = v_x{}^2 + v_y{}^2 + v_z{}^2$$

Since there is no preferred direction for the molecules, the average values $\overline{v_x{}^2}$, $\overline{v_y{}^2}$, and $\overline{v_z{}^2}$ are equal to each other. Using this fact and the above result, we find that

$$\overline{v_x{}^2} = \overline{v_y{}^2} = \overline{v_z{}^2} = \tfrac{1}{3}\overline{v^2}$$

Hence, the pressure from Equation 21.2 can be expressed as

Pressure and molecular speed

$$P = \tfrac{1}{3}\frac{Nm}{V}\,\overline{v^2} \tag{21.3}$$

The quantity Nm is the total mass of the molecules, which is equal to nM, where n is the number of moles of the gas and M is its molecular weight. Therefore, the pressure can also be expressed in the alternate form

$$P = \tfrac{1}{3}\frac{nM}{V}\,\overline{v^2} \tag{21.4}$$

By rearranging Equation 21.3, we can also express the pressure as

Pressure and molecular kinetic energy

$$P = \tfrac{2}{3}\frac{N}{V}\,(\tfrac{1}{2}m\overline{v^2}) \tag{21.5}$$

This equation tells us that the pressure is proportional to the number of molecules per unit volume and to the average translational kinetic energy per molecule.

With this simplified model of an ideal gas, we have arrived at an important result that relates the macroscopic quantities of pressure and volume to a microscopic quantity, the average molecular speed. Thus we have a key link between the microscopic world of the gas molecules and the macroscopic world as measured, in this case, with a pressure gauge, a meter stick, and a clock.

In the derivation of this result, note that we have not accounted for collisions between gas molecules. When these collisions are considered, the results do not change since collisions will only affect the momenta of the particles, with no net effect on the walls. This is consistent with one of our assumptions, namely: the distribution of velocities does not change in time. In addition, although our result was derived for a cubical container, it is valid for a container of any shape.

Ludwig Boltzmann (1844–1906), Austrian physicist. (Courtesy of AIP Niels Bohr Library, Lande Collection)

21.2 MOLECULAR INTERPRETATION OF TEMPERATURE

We can obtain some insight into the meaning of temperature by first writing Equation 21.5 in the more familiar form

$$PV = \tfrac{2}{3}N\,(\tfrac{1}{2}m\overline{v^2})$$

Let us now compare this with the empirical equation of state for an ideal gas (Eq. 19.14):

$$PV = NkT$$

Recall that the equation of state is based on experimental facts concerning the macroscopic behavior of gases. Equating the right sides of these expressions, we find that

Temperature is proportional to average kinetic energy

$$T = \frac{2}{3k}\,(\tfrac{1}{2}m\overline{v^2}) \qquad (21.6)$$

Since $\frac{1}{2}m\overline{v^2}$ is the average translational kinetic energy per molecule, we see that *temperature is a direct measure of the average molecular kinetic energy.*

By rearranging Equation 21.6, we can relate the translational molecular kinetic energy to the temperature:

Average kinetic energy per molecule

$$\tfrac{1}{2}m\overline{v^2} = \tfrac{3}{2}kT \qquad (21.7)$$

That is, the average translational kinetic energy per molecule is $\frac{3}{2}kT$. Since $\overline{v_x^2} = \frac{1}{3}\overline{v^2}$, it follows that

Equipartition of energy

$$\tfrac{1}{2}m\overline{v_x^2} = \tfrac{1}{2}kT \qquad (21.8)$$

According to this expression, the average translational kinetic energy per molecule associated with motion in the x direction is $\frac{1}{2}kT$. In a similar manner, for the y and z motions it follows that

$$\tfrac{1}{2}m\overline{v_y^2} = \tfrac{1}{2}kT \qquad \text{and} \qquad \tfrac{1}{2}m\overline{v_z^2} = \tfrac{1}{2}kT$$

Thus, each translational degree of freedom contributes an equal amount of energy to the gas, namely, $\frac{1}{2}kT$. (In general, the degrees of freedom refers to the number of independent means by which a molecule can possess energy.)

A generalization of this result, known as **the theorem of equipartition of energy,** says that the energy of a system in thermal equilibrium is equally divided among all degrees of freedom.

We shall return to this important point in Section 21.6.

The total translational kinetic energy of N molecules of gas is simply N times the average energy per molecule, which is given by Equation 21.7:

Total kinetic energy of N molecules

$$E = N\,(\tfrac{1}{2}m\overline{v^2}) = \tfrac{3}{2}NkT = \tfrac{3}{2}nRT \qquad (21.9)$$

where we have used $k = R/N_A$ for Boltzmann's constant and $n = N/N_A$ for the number of moles of gas. This result, together with Equation 21.5, implies that the pressure exerted by an ideal gas depends only on the number of molecules per unit volume and the temperature.

The square root of $\overline{v^2}$ is called the *root mean square* (rms) *speed* of the molecules. From Equation 21.7 we get for the rms speed

Root mean square speed

$$v_{\text{rms}} = \sqrt{\overline{v^2}} = \sqrt{\frac{3kT}{m}} = \sqrt{\frac{3RT}{M}} \tag{21.10}$$

The expression for the rms speed shows that at a given temperature, lighter molecules move faster, on the average, than heavier molecules. For example, hydrogen, with a molecular weight of 2 g/mol, moves four times as fast as oxygen, whose molecular weight is 32 g/mol. The rms speed is not the speed at which a gas molecule will move across a room, since such a molecule undergoes several billion collisions per second with other molecules under standard conditions. We shall describe this in more detail in Section 21.7.

Table 21.1 lists the rms speeds for various molecules at 20°C.

TABLE 21.1 Some rms Speeds

Gas	Molecular Weight (g/mol)	v_{rms} at 20°C (m/s)°
H_2	2.02	1902
He	4.0	1352
H_2O	18	637
Ne	20.1	603
N_2 or CO	28	511
NO	30	494
CO_2	44	408
SO_2	48	390

° All values calculated using Equation 21.10.

EXAMPLE 21.1 A Tank of Helium
A tank of volume 0.3 m³ contains 2 moles of helium gas at 20°C. Assuming the helium behaves like an ideal gas, (a) find the total internal energy of the system.

Solution Using Equation 21.9 with $n = 2$ and $T = 293$ K, we get

$$E = \tfrac{3}{2}nRT = \tfrac{3}{2}(2 \text{ moles})(8.31 \text{ J/mol}\cdot\text{K})(293 \text{ K})$$

$$= 7.30 \times 10^3 \text{ J}$$

(b) What is the average kinetic energy per molecule?

Solution From Equation 21.7, we see that the average kinetic energy per molecule is equal to

$$\tfrac{1}{2}m\overline{v^2} = \tfrac{3}{2}kT = \tfrac{3}{2}(1.38 \times 10^{-23} \text{ J/K})(293 \text{ K})$$

$$= 6.07 \times 10^{-21} \text{ J}$$

Exercise 1 Using the fact that the molecular weight of helium is 4 g/mol, determine the rms speed of the atoms at 20°C.
Answer 1.35×10^3 m/s.

21.3 HEAT CAPACITY OF AN IDEAL GAS

We have found that the temperature of a gas is a measure of the average translational kinetic energy of the gas molecules. This kinetic energy is associated with the motion of the center of mass of each molecule. It does not include the energy associated with the internal motion of the molecule, namely, vibrations and rotations about the center of mass. This should not be surprising, since the simple kinetic theory model assumes a structureless molecule.

In view of this, let us first consider the simplest case of an ideal monatomic gas, that is, a gas containing one atom per molecule, such as helium, neon, or argon. Essentially, all of the kinetic energy of such molecules is associated with the motion of their centers of mass. When energy is added to a monatomic gas in a container of fixed volume (by heating, say) all of the added energy goes into increasing the translational kinetic energy of the atoms.[1] There is no other way to store the energy in a monatomic gas. Therefore, from Equation 21.9 we see that the total internal energy U of N molecules (or n moles) of an ideal monatomic gas is given by

Internal energy of an ideal monatomic gas

$$U = \tfrac{3}{2}NkT = \tfrac{3}{2}nRT \tag{21.11}$$

It is important to note that for an ideal gas (only), U is a function of T only. If heat is transferred to the system at *constant volume*, the work done by the system is zero. That is, $W = \int P\,dV = 0$ for a constant volume process. Hence, from the first law of thermodynamics we see that

$$Q = \Delta U = \tfrac{3}{2}nR\,\Delta T \tag{21.12}$$

In other words, all of the heat transferred goes into increasing the internal energy (and temperature) of the system. The constant-volume process from i to f is described in Figure 21.3, where ΔT is the temperature difference between the two isotherms. Substituting the value for Q given by Equation 20.2 into Equation 21.12, we get

$$nC_v\,\Delta T = \tfrac{3}{2}nR\,\Delta T$$

$$C_v = \tfrac{3}{2}R \tag{21.13}$$

In this notation, C_v is the molar heat capacity of the gas at constant volume. Note that this expression predicts a value of $\tfrac{3}{2}R = 12.5$ J/mol·K for all monatomic gases. This is in excellent agreement with measured values of molar heat capacities for such gases as helium and argon over a wide range of temperatures (Table 21.2).

The change in internal energy for an ideal gas can be expressed as

$$\Delta U = nC_v\,\Delta T \tag{21.14}$$

In the limit of differential changes, we can use Equation 21.12 and the first law of thermodynamics to express the molar heat capacity in the form

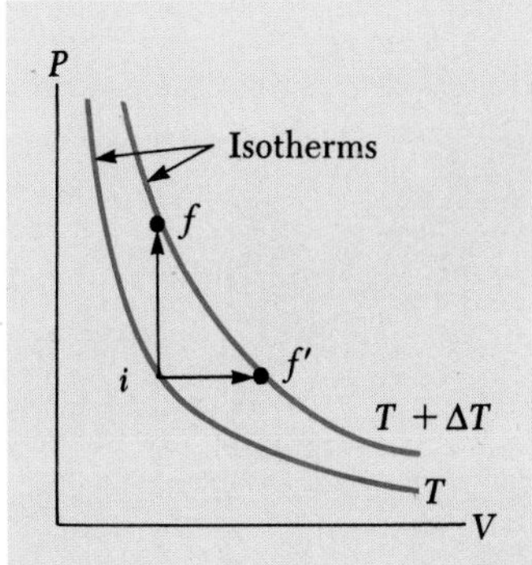

Figure 21.3 Heat is added to an ideal gas in two ways. For the constant-volume path *if*, all the heat added goes into increasing the internal energy of the gas since no work is done. Along the constant-pressure path *if′*, part of the heat added goes into work done by the gas. Note that the internal energy is constant along any isotherm.

[1] If the gas is raised to sufficiently high temperatures, the atom can also be excited or even ionized.

TABLE 21.2 Molar Heat Capacities of Various Gases

	Molar Heat Capacity (J/mol·K)			
	C_p	C_v	$C_p - C_v$	$\gamma = C_p/C_v$
		Monatomic Gases		
He	20.8	12.5	8.33	1.67
Ar	20.8	12.5	8.33	1.67
Ne	20.8	12.7	8.12	1.64
Kr	20.8	12.3	8.49	1.69
		Diatomic Gases		
H_2	28.8	20.4	8.33	1.41
N_2	29.1	20.8	8.33	1.40
O_2	29.4	21.1	8.33	1.40
CO	29.3	21.0	8.33	1.40
Cl_2	34.7	25.7	8.96	1.35
		Polyatomic Gases		
CO_2	37.0	28.5	8.50	1.30
SO_2	40.4	31.4	9.00	1.29
H_2O	35.4	27.0	8.37	1.30
CH_4	35.5	27.1	8.41	1.31

Note: All values obtained at 300 K.

$$C_v = \frac{1}{n}\frac{dU}{dT}$$

For an ideal monatomic gas, where $U = \frac{3}{2}nRT$, Equation 21.14 gives $C_v = \frac{3}{2}R$ in agreement with Equation 21.13.

Now suppose that the gas is taken along the constant-pressure path $i \rightarrow f'$ in Figure 21.3. Along this path, the temperature again increases by ΔT. The heat that must be transferred to the gas in this process is given by $Q = nC_p \Delta T$, where C_p is the molar heat capacity at constant pressure. Since the volume increases in this process, we see that the work done by the gas is $W = P\,\Delta V$. Applying the first law to this process gives

$$\Delta U = Q - W = nC_p\,\Delta T - P\,\Delta V \tag{21.15}$$

In this case, the heat added to the gas is transferred in two forms. Part of it is used to do external work by moving a piston, and the remainder increases the internal energy of the gas. But the change in internal energy for the process $i \rightarrow f'$ is equal to the change for the process $i \rightarrow f$, since U depends only on temperature for an ideal gas and ΔT is the same for each process. In addition, since $PV = nRT$, we note that for a constant-pressure process $P\,\Delta V = nR\,\Delta T$. Substituting this into Equation 21.15 with $\Delta U = nC_v\,\Delta T$ gives

$$nC_v\,\Delta T = nC_p\,\Delta T - nR\,\Delta T$$

or

$$C_p - C_v = R \tag{21.16}$$

This expression applies to *any* ideal gas. It shows that the molar heat capacity of an ideal gas at constant pressure is greater than the molar heat capacity at constant volume by an amount R, the universal gas constant (which has the value 8.31 J/mol·K). This is in good agreement with real gases under standard conditions, that is, 0°C and atmospheric pressure (Table 21.2).

Since $C_v = \frac{3}{2}R$ for a monatomic ideal gas, Equation 21.16 predicts a value $C_p = \frac{5}{2}R = 20.8$ J/mol·K for the molar heat capacity of a monatomic gas at constant pressure. The ratio of these heat capacities is a dimensionless quantity γ given by

$$\gamma = \frac{C_p}{C_v} = \frac{\frac{5}{2}R}{\frac{3}{2}R} = \frac{5}{3} = 1.67 \qquad (21.17)$$

Ratio of heat capacities for an ideal gas.

The values of C_p and γ are in excellent agreement with experimental values for monatomic gases, but in serious disagreement with the values for the more complex gases (Table 21.2). This is not surprising since the value $C_v = \frac{3}{2}R$ was derived for a monatomic ideal gas, and we expect some additional contribution to the specific heat from the internal structure of the more complex molecules. In Section 21.6, we describe the effect of molecular structure on the specific heat of a gas. We shall find that the internal energy and hence the specific heat of a complex gas must include contributions from the rotational and vibrational motions of the molecule.

We have seen that the heat capacities of gases at constant pressure are greater than the heat capacities at constant volume. This difference is a consequence of the fact that in a constant-volume process, no work is done and all of the heat goes into increasing the internal energy (and temperature) of the gas, whereas in a constant-pressure process some of the heat energy is transformed into work done by the gas. In the case of solids and liquids heated at constant pressure, very little work is done since the thermal expansion is small. Consequently C_p and C_v are approximately equal for solids and liquids.

EXAMPLE 21.2 Heating a Cylinder of Helium

A cylinder contains 3 moles of helium gas at a temperature of 300 K. (a) How much heat must be transferred to the gas to increase its temperature to 500 K if the gas is heated at constant volume?

Solution For the constant-volume process, the work done is zero. Therefore from Equation 21.12, we get

$$Q_1 = \tfrac{3}{2}nR\,\Delta T = nC_v\,\Delta T$$

But $C_v = 12.5$ J/mol·K for He and $\Delta T = 200$ K; therefore

$$Q_1 = (3 \text{ moles})(12.5 \text{ J/mol·K})(200 \text{ K})$$

$$= 7.50 \times 10^3 \text{ J}$$

(b) How much heat must be transferred to the gas at constant pressure to raise the temperature to 500 K?

Solution Making use of Table 21.2, we get

$$Q_2 = nC_p\,\Delta T = (3 \text{ moles})(20.8 \text{ J/mol·K})(200 \text{ K})$$

$$= 12.5 \times 10^3 \text{ J}$$

Exercise 2 What is the work done by the gas in this process?
Answer $W = Q_2 - Q_1 = 5.00 \times 10^3$ J.

21.4 ADIABATIC PROCESS FOR AN IDEAL GAS

Definition of an adiabatic process

An **adiabatic process** is one in which there is no heat transfer between the system and its surroundings.

In reality, true adiabatic processes cannot occur since a perfect heat insulator between a system and its surroundings does not exist. However, there are processes that are nearly adiabatic. For example, if a gas is compressed (or expanded) very rapidly, very little heat flows into (or out of) the system, and so the process is nearly adiabatic. Such processes occur in the cycle of a gasoline engine, which we shall discuss in detail in the next chapter.

It is also possible for a process to be both quasi-static and adiabatic. For example, if a gas that is thermally insulated from its surroundings is allowed to expand slowly against a piston, the process is a quasi-static, adiabatic expansion. In general,

a **quasi-static, adiabatic process** is one that is slow enough to allow the system to always be near equilibrium, but fast compared with the time it takes the system to exchange heat with its surroundings.

Suppose that an ideal gas undergoes a *quasi-static, adiabatic* expansion. *At any time during the process, we assume that the gas is in an equilibrium state, so that the equation of state, $PV = nRT$, is valid.* The pressure and volume at any time during the adiabatic process are related by the expression

Relation between P and V for an adiabatic process involving an ideal gas

$$PV^{\gamma} = \text{constant} \tag{21.18}$$

where $\gamma = C_p/C_v$ is assumed to be constant during the process. Thus, we see that all the thermodynamic variables, P, V, and T, change during an adiabatic process. The following represents a proof of Equation 21.18.

Proof that PV^{γ} = constant for an Adiabatic Process

When a gas expands adiabatically in a thermally insulated cylinder, there is no heat transferred between the gas and its surroundings, and so $Q = 0$. Let us take the change in volume to be ΔV and the change in temperature to be ΔT. The work done by the gas is $W = P\,\Delta V$. Since the internal energy of an ideal gas depends only on temperature, the change in internal energy is given by $\Delta U = nC_v\,\Delta T$. Hence, the first law of thermodynamics gives

$$\Delta U = nC_v\,\Delta T = -P\,\Delta V$$

From the equation of state of an ideal gas, $PV = nRT$, we see that

$$P\,\Delta V + V\,\Delta P = nR\,\Delta T$$

Eliminating ΔT from these two equations we find that

$$P\,\Delta V + V\,\Delta P = -\frac{R}{C_v}\,P\,\Delta V$$

Substituting $R = C_p - C_v$ and dividing by PV, we get

$$\frac{\Delta V}{V} + \frac{\Delta P}{P} = -\left(\frac{C_p - C_v}{C_v}\right)\frac{\Delta V}{V} = (1-\gamma)\frac{\Delta V}{V}$$

$$\frac{\Delta P}{P} + \gamma \frac{\Delta V}{V} = 0$$

Taking the limits of differential changes ($\Delta P \rightarrow dP$ and $\Delta V \rightarrow dV$) and integrating, we get

$$\ln P + \gamma \ln V = \text{constant}$$

which is equivalent to Equation 21.18:

$$PV^{\gamma} = \text{constant}$$

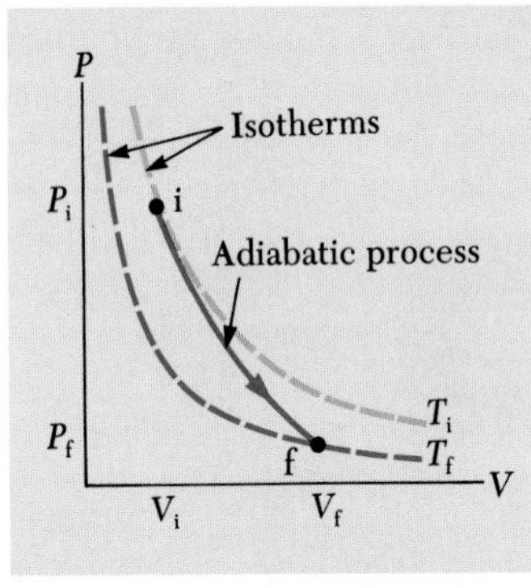

Figure 21.4 The *PV* diagram for an adiabatic expansion. Note that $T_f < T_i$ in this process.

The *PV* diagram for an adiabatic expansion is shown in Figure 21.4. Because $\gamma > 1$, the *PV* curve for the adiabatic expansion is steeper than that for an isothermal expansion. As the gas expands adiabatically, no heat is transferred in or out of the system. Hence, from the first law, we see that ΔU is negative so that ΔT is also negative. Thus, we see that the gas cools ($T_f < T_i$) during an adiabatic expansion. Conversely, the temperature increases if the gas is compressed adiabatically. Applying Equation 21.18 to the initial and final states, we see that

Adiabatic process

$$P_i V_i^{\gamma} = P_f V_f^{\gamma} \tag{21.19}$$

Using $PV = nRT$, it is left as a problem (Problem 26) to show that Equation 21.19 can also be expressed as

$$T_i V_i^{\gamma - 1} = T_f V_f^{\gamma - 1} \tag{21.20}$$

Note that the above analysis is valid only in processes that are slow enough to allow the system to always remain near equilibrium, but fast enough to prevent the system from exchanging heat with its surroundings.

EXAMPLE 21.3 A Diesel Engine Cylinder

Air in the cylinder of a diesel engine at 20°C is compressed from an initial pressure of 1 atm and volume of 800 cm^3 to a volume of 60 cm^3. Assuming that air behaves as an ideal gas ($\gamma = 1.40$) and that the compression is adiabatic, find the final pressure and temperature.

Solution Using Equation 21.19, we find that

$$P_f = P_i (V_i/V_f)^{\gamma} = 1 \text{ atm } (800 \text{ cm}^3/60 \text{ cm}^3)^{1.4}$$
$$= 37.6 \text{ atm}$$

Since $PV = nRT$ is always valid during the process and since no gas escapes from the cylinder,

$$\frac{P_i V_i}{T_i} = \frac{P_f V_f}{T_f}$$

$$T_f = \frac{P_f V_f}{P_i V_i} T_i = \frac{(37.6 \text{ atm})(60 \text{ cm}^3)}{(1 \text{ atm})(800 \text{ cm}^3)} (293 \text{ K})$$
$$= 826 \text{ K} = 553°\text{C}$$

21.5 SOUND WAVES IN A GAS

Most gases are poor heat conductors. Therefore, when a sound wave propagates through a gas, very little heat is transferred between regions of high and low densities. To a good approximation, we can assume that the variations of pressure and volume occur adiabatically, corresponding to no heat transfer between portions of the gas. This is equivalent to assuming that all of the work done in compressing the gas goes into increasing the internal energy of the gas.

We shall use this fact to determine an expression for the speed of sound in a gas. First, recall from Chapter 17 that the speed of a longitudinal wave is given by

$$v = \sqrt{\frac{B}{\rho}}$$

where ρ is the density of the medium and B is its bulk modulus, given by Equation 17.2, $B = -V(\Delta P/\Delta V)$. If ΔP and ΔV are replaced by dP and dV, respectively, we find

$$B = -V\left(\frac{dP}{dV}\right)$$

In Section 21.4, we found that if the gas is ideal, the pressure and volume during an adiabatic process are related by the expression $PV^\gamma = \text{constant}$. Differentiating this with respect to V gives

$$\gamma P V^{\gamma-1} + V^\gamma\left(\frac{dP}{dV}\right) = 0$$

or

$$\frac{dP}{dV} = -\gamma\frac{P}{V}$$

Substituting this into the expression for B gives

$$B_{\text{adiabatic}} = -V\left(-\gamma\frac{P}{V}\right) = \gamma P$$

Therefore the speed of sound in a gas is

Speed of sound in a gas

$$v = \sqrt{\frac{\gamma P}{\rho}} \tag{21.21}$$

Equation 21.21 can be expressed in another useful form, which uses the equation of state of an ideal gas, $PV = nRT$, or

$$P = nRT/V = \rho RT/M$$

where R is the gas constant, M is the molecular weight, and n is the number of moles of gas. Substituting this expression for P into Equation 21.21 gives

Speed of sound in a gas

$$v = \sqrt{\frac{\gamma RT}{M}} \tag{21.22}$$

It is interesting to compare this result with the rms speed of molecules in a gas (Eq. 21.10), where $v_{\text{rms}} = \sqrt{3RT/M}$. The two results differ only by the factors γ and 3. It is known that γ lies between 1 and 1.67; hence the two speeds are nearly the same! Since sound waves propagate through air as a result of collisions between gas molecules, one would expect the wave speed to increase as the temperature (and molecular speed) increase.

EXAMPLE 21.4 The Speed of Sound in Air
Calculate the speed of sound in air at atmospheric pressure and at 0°C, taking $P = 1.01 \times 10^5$ Pa, $\gamma = 1.40$, and $\rho = 1.29$ kg/m^3.

Solution Using Equation 21.21, we find that

$$v_{\text{air}} = \sqrt{(1.4)(1.01) \times 10^5 \text{ Pa})/1.29 \text{ kg/m}^3} = 331 \text{ m/s}$$

This is in excellent agreement with the measured speed of sound in air.

It is interesting to note that the speed of sound in helium is much greater than this because of the lower density of helium. An amusing demonstration of this fact is the variation in the human voice when the vocal cavities are partially filled with helium. The demonstrator talks before and after taking a deep breath of helium, an inert gas. The result is a high-pitched voice sounding a bit like that of Donald Duck. The increase in frequency corresponds to an increase in the speed of sound in helium, since frequency is proportional to velocity.

21.6 THE EQUIPARTITION OF ENERGY

We have found that model predictions based on specific heat agree quite well with the behavior of monatomic gases, but not with the behavior of complex gases (Table 21.2). Furthermore, the value predicted by the model for the quantity $C_p - C_v = R$ is the same for all gases. This is not surprising, since this difference is the result of the work done by the gas, which is independent of its molecular structure.

In order to explain the variations in C_v and C_p in going from monatomic gases to the more complex gases, let us explain the origin of the specific heat. So far, we have assumed that the sole contribution to the internal energy of a gas is the translational kinetic energy of the molecules. However, the internal energy of a gas actually includes contributions from the translational, vibrational, and rotational motion of the molecules. The rotational and vibrational motions of molecules with structure can be activated by collisions and therefore are "coupled" to the translational motion of the molecules. The branch of physics known as *statistical mechanics* has shown that for a large number of particles obeying newtonian mechanics, the available energy is, on the average, shared equally by each independent degree of freedom. Recall that the **equipartition theorem** states that at equilibrium each degree of freedom contributes, on the average, $\frac{1}{2}kT$ of energy per molecule.

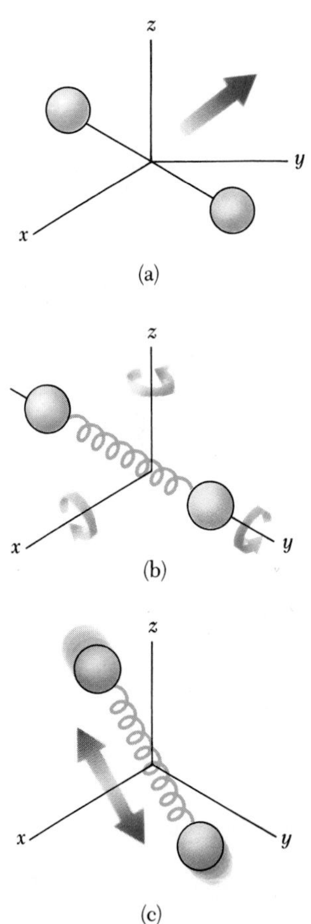

Figure 21.5 Possible motions of a diatomic molecule: (a) translational motion of the center of mass, (b) rotational motion about the various axes, and (c) vibrational motion along the molecular axis.

Let us consider a diatomic gas, which we can visualize as a dumbbell-shaped molecule (Fig. 21.5). In this model, the center of mass of the molecule can translate in the x, y, and z directions (Fig. 21.5a). In addition, the molecule can rotate about three mutually perpendicular axes (Fig. 21.5b). We can neglect the rotation about the y axis since the moment of inertia and the rotational energy, $\frac{1}{2}I\omega^2$, about this axis are negligible compared with those associated with the x and z axes. If the two atoms of the molecule are taken to be point masses, then I_y is identically zero. Thus there are five degrees of freedom: three associated with the translational motion and two associated with the rotational motion. Since *each degree of freedom contributes, on the average, $\frac{1}{2}kT$ of energy per molecule,* the total energy for N molecules is

$$U = 3N(\tfrac{1}{2}kT) + 2N(\tfrac{1}{2}kT) = \tfrac{5}{2}NkT = \tfrac{5}{2}nRT$$

We can use this result and Equation 21.14 to get the molar heat capacity at constant volume:

$$C_v = \frac{1}{n}\frac{dU}{dT} = \frac{1}{n}\frac{d}{dT}(\tfrac{5}{2}nRT) = \tfrac{5}{2}R$$

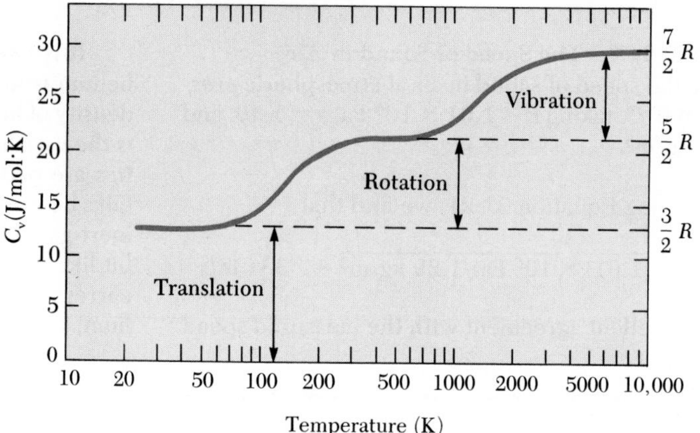

Figure 21.6 The molar heat capacity, C_v, of hydrogen as a function of temperature. The horizontal scale is logarithmic. Note that hydrogen liquefies at 20 K.

From Equations 21.16 and 21.17 we find that

$$C_p = C_v + R = \tfrac{7}{2}R$$

$$\gamma = \frac{C_p}{C_v} = \frac{\frac{7}{2}R}{\frac{5}{2}R} = \frac{7}{5} = 1.40$$

These results agree quite well with most of the data given in Table 21.2 for diatomic molecules. This is rather surprising since we have not yet accounted for the possible vibrations of the molecule. In the vibratory model, the two atoms are joined by an imaginary spring. The vibrational motion adds two more degrees of freedom, corresponding to the kinetic and potential energies associated with vibrations along the length of the molecule. Hence, the equipartition theorem predicts an internal energy of $\frac{7}{2}nRT$ and a higher heat capacity than what is observed. Examination of the experimental data (Table 21.2) suggests that some diatomic molecules, such as H_2 and N_2, do not vibrate at room temperature, and others, such as Cl_2, do. For molecules with more than two atoms, the number of degrees of freedom is even larger and the vibrations are more complex. This results in an even higher predicted heat capacity, which is in qualitative agreement with experiment.

We have seen that the equipartition theorem is successful in explaining some features of the heat capacity of molecules with structure. However, the equipartition theorem does not explain the observed temperature variation in heat capacities. As an example of such a temperature variation, C_v for the hydrogen molecule is $\frac{5}{2}R$ from about 250 K to 750 K and then increases steadily to about $\frac{7}{2}R$ well above 750 K (Fig. 21.6). This suggests that vibrations occur at very high temperatures. At temperatures well below 259 K, C_v has a value of about $\frac{3}{2}R$, suggesting that the molecule has only translational energy at low temperatures.

A Hint of Energy Quantization

The failure of the equipartition theorem to explain such phenomena is due to the inadequacy of classical mechanics when applied to molecular systems. For a more satisfactory description, it is necessary to use a quantum-mechanical model in which the energy of an individual molecule is quantized. The energy separation between adjacent vibrational energy levels for a molecule such as H_2 is about ten times as great as the average kinetic energy of the molecule at room temperature. Consequently, collisions between molecules at low temperatures do not provide enough energy to change the vibrational state of the molecule. It is often stated that such degrees of freedom are "frozen out." This explains why the vibrational energy does not contribute to the heat capacities of molecules at low temperatures.

The rotational energy levels are also quantized, but their spacing at ordinary temperatures is small compared witk kT. Since the spacing between rotational levels is so small compared with kT, the system behaves classically. However, at sufficiently low temperatures (typically less than 50 K), where kT is small compared with the spacing between rotational levels, intermolecular collisions may not be energetic enough to alter the rotational states. This explains why C_v reduces to $\frac{3}{2}R$ for H_2 in the range from 20 K to about 100 K.

Heat Capacities of Solids

Measurements of heat capacities of solids also show a marked temperature dependence. The heat capacities of solids generally decrease in a nonlinear manner with decreasing temperature and approach zero as the absolute temperature approaches zero. At high temperatures (usually above 500 K), the heat capacities of solids approach the value of about $3R \approx 25$ J/mole·K, a result known as the *DuLong-Petit law*. The typical data shown in Figure 21.7 demonstrate the temperature dependence of the heat capacity for two semiconducting solids, silicon and germanium.

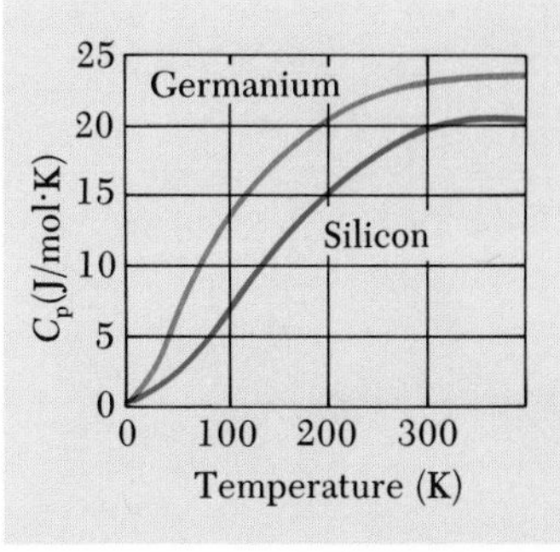

Figure 21.7 Molar heat capacity, C_p, of silicon and germanium. As T approaches zero, the heat capacity also approaches zero. (From C. Kittel, *Introduction to Solid State Physics*, New York, John Wiley, 1971.)

The heat capacity of a solid at high temperatures can be explained using the equipartition theorem. For small displacements of an atom from its equilibrium position, each atom executes simple harmonic motion in the x, y, and z directions. The energy associated with vibrational motion in the x direction is

$$E_x = \tfrac{1}{2}mv_x^2 + \tfrac{1}{2}kx^2$$

There are analogous expressions for E_y and E_z. Therefore, each atom of the solid has six degrees of freedom. According to the equipartition theorem, this corresponds to an average vibrational energy of $6(\frac{1}{2}kT) = 3kT$ per atom. Therefore, the total internal energy of a solid consisting of N atoms is given by

$$U = 3NkT = 3nRT$$

Total internal energy of a solid

From this result, we find that the molar heat capacity

$$C_v = \frac{1}{n}\frac{dU}{dT} = 3R,$$

which agrees with the empirical law of DuLong and Petit. The discrepancies between this model and the experimental data at low temperatures are again due to the inadequacy of classical physics in the microscopic world. One can attribute the decrease in heat capacity with decreasing temperature to a "freezing out" of various vibrational excitations.

*21.7 DISTRIBUTION OF MOLECULAR SPEEDS

Thus far we have neglected the fact that not all molecules in a gas have the same speed and energy. Their motion is extremely chaotic. Any individual molecule is colliding with others at the enormous rate of typically a billion times per second. Each collision results in a change in the speed and direction of motion of each of the participant molecules. From Equation 21.10, we see that average molecular speeds increase with increasing temperature. What we would like to know now is the distribution of molecular speeds. For example, how many molecules of a gas have a speed in the range from, say, 400 to 410 m/s? Intuitively, we expect that the speed distribution depends on temperature. Furthermore, we expect that the distribution peaks in the vicinity of v_{rms}. That is, few molecules are expected to have speeds much less than or much greater than v_{rms}, since these extreme speeds will result only from an unlikely chain of collisions.

The development of a reliable theory for the speed distribution of a large number of particles appears, at first, to be an almost impossible task. However, in 1860 James Clerk Maxwell (1831–1879) derived an expression that describes the distribution of molecular speeds in a very definite manner. His work, and developments by other scientists shortly thereafter, were highly controversial, since experiments at that time were not capable of directly detecting molecules. However, about 60 years later experiments were devised which confirmed Maxwell's predictions.

One experimental arrangement for observing the speed distribution of molecules is illustrated in Figure 21.8. A substance is vaporized in an oven and forms gas molecules, which are permitted to escape through a hole. The molecules enter an evacuated region and pass through a series of slits to form a collimated beam. The beam is incident on two slotted rotating disks separated by a distance s. The slots in the disk are displaced from each other by an angle θ. A molecule passing through the slot in the first disk will pass through the slot in the second disk only if its speed is $v = s\omega/\theta$, where ω is the angular velocity of the disks. Molecules with other speeds will necessarily collide with the second disk and hence will not reach the detector. By varying ω and θ, one can measure the number of molecules in a given range of speeds.

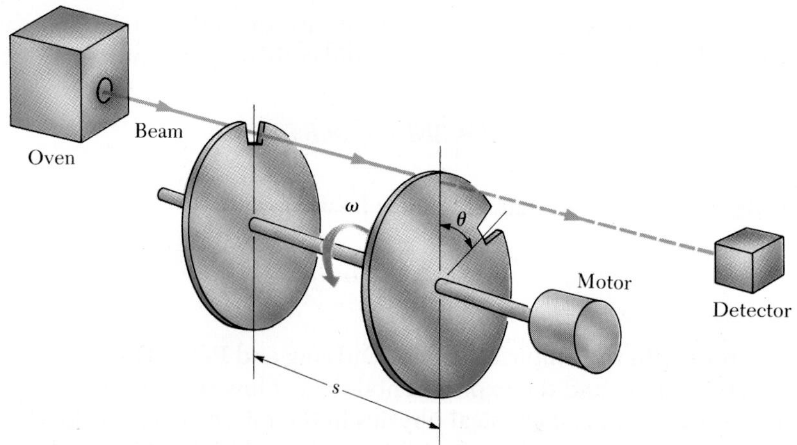

Figure 21.8 A schematic diagram of one apparatus used to measure the speed distribution of gas molecules.

The observed speed distribution of gas molecules in thermal equilibrium is shown in Figure 21.9. The quantity N_v (which is called the *distribution function*) represents the number of molecules per unit interval of speed.

The number of molecules having a speed in the range from v to $v + \Delta v$ is equal to $N_v\,\Delta v$, represented in Figure 21.9 by the area of the shaded rectangle. If N is the total number of molecules, then the fraction of molecules with speeds between v and $v + \Delta v$ is equal to $N_v\,\Delta v/N$. This fraction is also equal to the probability that any given molecule has a speed in the range from v to $v + \Delta v$.

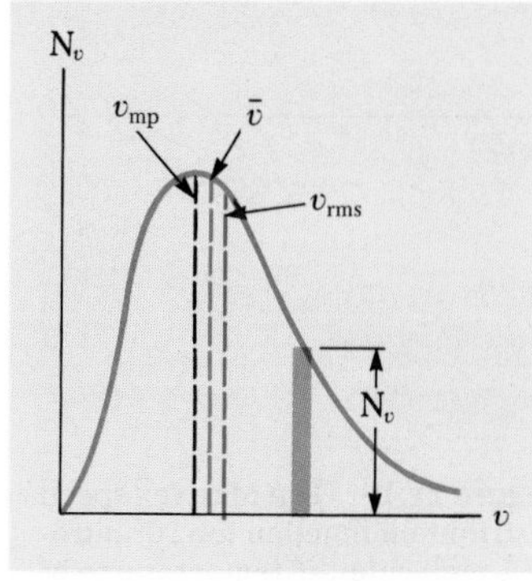

Figure 21.9 The speed distribution of gas molecules at some temperature. The number of molecules in the range Δv is equal to the area of the shaded rectangle, $N_v\,\Delta v$. The function N_v approaches zero as v approaches infinity.

The total number of molecules numerically equals the total area under the speed distribution curve. Since the abscissa ranges from $v = 0$ to $v = \infty$ (classically, all molecular speeds are possible), we can express the total number of particles as the sum of the areas of all shaded rectangles. In the limit $\Delta v \rightarrow 0$, this sum is replaced by an integral:

$$N = \lim_{\Delta v \rightarrow 0} \left(\sum_{v=0}^{\infty} N_v\,\Delta v \right) = \int_0^{\infty} N_v\,dv \qquad (21.23)$$

The fundamental expression (derived by Maxwell) that describes the most probable distribution of speeds of N gas molecules is given by

Maxwell speed distribution function

$$N_v = 4\pi N \left(\frac{m}{2\pi kT} \right)^{3/2} v^2 e^{-mv^2/2kT} \qquad (21.24)$$

where m is mass of a gas molecule, k is Boltzmann's constant, and T is the absolute temperature.[2] The function given by Equation 21.24 satisfies Equation 21.23. Furthermore, N_v approaches zero in the low- and high-speed limits, as expected. Also, the speed distribution for a given gas depends only on temperature.

As indicated in Figure 21.9, the average speed, $\bar{v}$, is somewhat lower than the rms speed. The most probable speed, v_{mp}, is the speed at which the distribution curve reaches a peak. Using Equation 21.24, one finds that

$$v_{rms} = \sqrt{\overline{v^2}} = \sqrt{3kT/m} = 1.73\sqrt{kT/m} \qquad (21.25)$$ rms speed

$$\bar{v} = \sqrt{8kT/\pi m} = 1.60\sqrt{kT/m} \qquad (21.26)$$ Average speed

$$v_{mp} = \sqrt{2kT/m} = 1.41\sqrt{kT/m} \qquad (21.27)$$ Most probable speed

The details of these calculations are left for the student (Problems 50 and 72), but from these equations we see that $v_{rms} > \bar{v} > v_{mp}$.

Figure 21.10 represents specific speed distribution curves for nitrogen molecules. The curves were obtained by using Equation 21.24 to evaluate the distribution function, N_v, at various speeds and at two temperatures (300 K and 900 K). Note that the curve shifts to the right as T increases, indicating that the average speed increases with increasing temperature, as expected. The asymmetric shape of the curves is due to the fact that the lowest speed possible is zero while the upper classical limit of the speed is infinity. Furthermore, as temperature increases the distribution curve broadens and the range of speeds also increases.

Equation 21.24 shows that the distribution of molecular speeds in a gas depends on mass as well as temperature. At a given temperature, the fraction

[2] For the derivation of this expression, see any text on thermodynamics, such as that by M. W. Zemansky and R. H. Dittman, *Heat and Thermodynamics*, New York, McGraw-Hill, 1981.

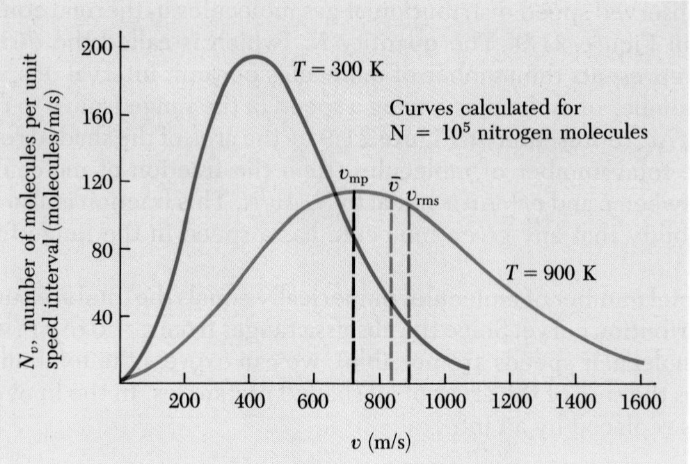

Figure 21.10 The Maxwell speed distribution function for 10^5 nitrogen molecules at temperatures of 300 K and 900 K. The total area under either curve is equal to the total number of molecules, which, in this case, equals 10^5. Note that $v_{rms} > \bar{v} > v_{mp}$.

of particles with speeds exceeding a fixed value increases as the mass decreases. This explains why lighter molecules, such as hydrogen and helium, escape more readily from the earth's atmosphere than heavier molecules, such as nitrogen and oxygen. (See the discussion of escape velocity in Chapter 14. Gas molecules escape even more readily from the moon's surface because the escape velocity on the moon is lower.)

The evaporation process

The speed distribution of molecules in a liquid is similar to that shown in Figure 21.10. The phenomenon of evaporation of a liquid can be understood from this distribution in speeds using the fact that some molecules in the liquid are more energetic than others. Some of the faster-moving molecules in the liquid penetrate the surface and leave the liquid even at temperatures well below the boiling point. The molecules that escape the liquid by evaporation are those that have sufficient energy to overcome the attractive forces of the molecules in the liquid phase. Consequently, the molecules left behind in the liquid phase have a lower average kinetic energy, causing the temperature of the liquid to decrease. Hence evaporation is a cooling process. For example, an alcohol-soaked cloth is often placed on a feverish head to cool and comfort the patient.

EXAMPLE 21.5 A System of Nine Particles □
Nine particles have speeds of 5, 8, 12, 12, 12, 14, 14, 17, and 20 m/s. (a) Find the average speed.

The average speed is the sum of the speeds divided by the total number of particles:

$$\bar{v} = \frac{5+8+12+12+12+14+14+17+20}{9}$$

$$= \boxed{12.7 \text{ m/s}}$$

(b) What is the rms speed?

Solution The average value of the square of the speed is given by

$$\overline{v^2} = \frac{5^2+8^2+12^2+12^2+12^2+14^2+14^2+17^2+20^2}{9}$$

$$= 178 \text{ m}^2/\text{s}^2$$

Hence, the rms speed is

$$v_{rms} = \sqrt{\overline{v^2}} = \sqrt{178 \text{ m}^2/\text{s}^2} = \boxed{13.3 \text{ m/s}}$$

(c) What is the most probable speed of the particles?

Solution Three of the particles have a speed of 12 m/s, two have a speed of 14 m/s, and the remaining have different speeds. Hence, we see that the most probable speed, v_{mp}, is 12 m/s.

*21.8 MEAN FREE PATH

Most of us are familiar with the fact that the strong odor associated with a gas such as ammonia may take several minutes to diffuse through a room. However, since average molecular speeds are typically several hundred meters per second at room temperature, we might expect a time much less than one second. To understand this apparent contradiction, we note that molecules collide with each other, since they are not geometrical points. Therefore, they do not travel from one side of a room to the other in a straight line. Between collisions, the molecules move with constant speed along straight lines.[3] The average distance between collisions is called the **mean free path.** The path of individual molecules is random and resembles that shown in Figure 21.11. As we would expect from this description, the mean free path is related to the diameter of the molecules and the density of the gas.

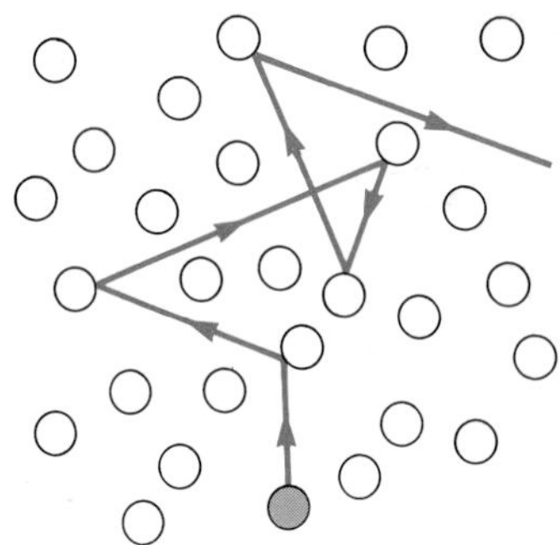

Figure 21.11 A molecule moving through a gas collides with other molecules in a random fashion. This behavior is sometimes referred to as a *random-walk process.* The mean free path increases as the number of molecules per unit volume decreases. Note that the motion is *not* limited to the plane of the paper.

We shall now describe how to estimate the mean free path for a gas molecule. For this calculation we shall assume that the molecules are spheres of diameter d. We see from Figure 21.12a that no two molecules will collide unless their centers are less than a distance d apart as they approach each other. An equivalent description of the collisions is to imagine that one of the molecules has a diameter $2d$ and the rest are geometrical points (Fig. 21.12b). In a time t, the molecule having the speed that we shall take to be the average speed, $\bar{v}$, will travel a distance $\bar{v}t$. In this same time interval, our molecule with equivalent diameter $2d$ will sweep out a cylinder having a cross-sectional area of πd^2 and a length of $\bar{v}t$ (Fig. 21.13). Hence the volume of the cylinder is $\pi d^2 \bar{v}t$. If n_v is the number of particles per unit volume, then the number of particles in the cylinder is $(\pi d^2 \bar{v}t)n_v$. The molecule of equivalent diameter $2d$ will collide with every particle in this cylinder in the time t. Hence, the number of collisions in the time t is equal to the number of particles in the cylinder, which we found was $(\pi d^2 \bar{v}t)n_v$.

The **mean free path,** ℓ, which is the mean distance between collisions, equals the average distance $\bar{v}t$ traveled in a time t divided by the number of collisions that occurs in the time:

$$\ell = \frac{\bar{v}t}{(\pi d^2 \bar{v}t)n_v} = \frac{1}{\pi d^2 n_v}$$

[3] Actually, there is a small curvature in the path because of the force of gravity at the earth's surface. However, this effect is small and can be neglected.

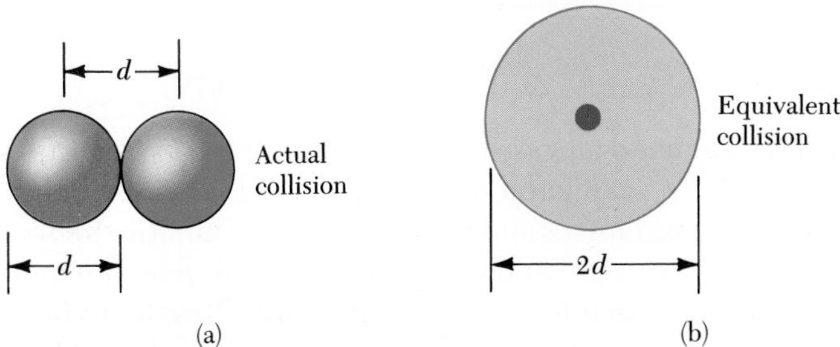

Figure 21.12 (a) Two spherical molecules, each of diameter d, collide if their centers are within a distance d of each other. (b) The collision between the two molecules is equivalent to a point mass colliding with a molecule having an effective diameter of $2d$.

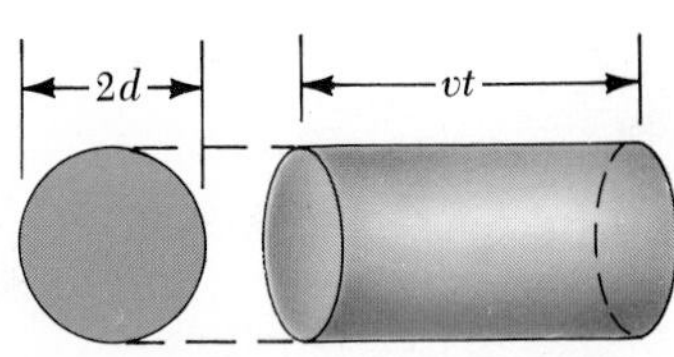

Figure 21.13 In a time t, a molecule of effective diameter $2d$ will sweep out a cylinder of length $\bar{v}t$, where $\bar{v}$ is its average speed. In this time, it will collide with every molecule within this cylinder.

Since the number of collisions in a time t is $(\pi d^2\bar{v}t)n_v$, the number of collisions per unit time, or **collision frequency** f, is given by

$$f = \pi d^2 \bar{v} n_v$$

The inverse of the collision frequency is the average time between collisions, called the **mean free time.**

Our analysis has assumed that particles in the cylinder are stationary. When the motion of the particles is included in the calculation, the correct results are

Mean free path

$$\ell = \frac{1}{\sqrt{2}\,\pi d^2 n_v} \tag{21.28}$$

Collision frequency

$$f = \sqrt{2}\,\pi d^2 \bar{v} n_v = \frac{\bar{v}}{\ell} \tag{21.29}$$

EXAMPLE 21.6 A Collection of Nitrogen Molecules
Calculate the mean free path and collision frequency for nitrogen molecules at a temperature of 20°C and a pressure of 1 atm. Assume a molecular diameter of 2×10^{-10} m.

Solution Assuming the gas is ideal, we can use the equation $PV = NkT$ to obtain the number of molecules per unit volume under these conditions:

$$n_v = \frac{N}{V} = \frac{P}{kT} = \frac{1.01 \times 10^5 \text{ N/m}^2}{(1.38 \times 10^{-23} \text{ J/K})(293 \text{ K})}$$

$$= 2.50 \times 10^{25} \frac{\text{molecules}}{\text{m}^3}$$

Hence, the mean free path is

$$\ell = \frac{1}{\sqrt{2}\,\pi d^2 n_v}$$

$$= \frac{1}{\sqrt{2}\,\pi (2 \times 10^{-10} \text{ m})^2 \left(2.50 \times 10^{25} \dfrac{\text{molecules}}{\text{m}^3}\right)}$$

$$= 2.25 \times 10^{-7} \text{ m}$$

This is about 10^3 times greater than the molecular diameter. Since the average speed of a nitrogen molecule at 20°C is about 511 m/s (Table 21.1), the collision frequency is

$$f = \frac{\bar{v}}{\ell} = \frac{511 \text{ m/s}}{2.25 \times 10^{-7} \text{ m}} = 2.27 \times 10^9/\text{s}$$

The molecule collides with other molecules at the average rate of about two billion times each second!

The mean free path, ℓ, is *not* the same as the average separation between particles. In fact, the average separation, d, between particles is given approximately by $n_v^{-1/3}$. In this example, the average molecular separation is

$$d = \frac{1}{n_v^{1/3}} = \frac{1}{(2.5 \times 10^{25})^{1/3}} = 3.4 \times 10^{-9} \text{ m}$$

*21.9 VAN DER WAALS' EQUATION OF STATE

Thus far we have assumed all gases to be ideal, that is, to obey the equation of state, $PV = nRT$. To a very good approximation, real gases behave as ideal gases at ordinary temperatures and pressures. In the kinetic theory derivation of the ideal-gas law, we neglected the volume occupied by the molecules and assumed that intermolecular forces were negligible. Now let us investigate the qualitative behavior of real gases and the conditions under which deviations from ideal-gas behavior are expected.

Consider a gas contained in a cylinder fitted with a movable piston. As noted in Chapter 20, if the temperature is kept constant while the pressure is measured at various volumes, a plot of P versus V yields a hyperbolic curve (an *isotherm*) as predicted by the ideal-gas law (Fig. 21.14).

Now let us describe what happens to a real gas. Figure 21.15 gives some typical experimental curves taken on a gas at various temperatures. At the higher temperatures, the curves are approximately hyperbolic and the gas behavior is close to ideal. However, as the temperature is lowered, the deviations from the hyperbolic shape are very pronounced.

There are two major reasons for this behavior. First, we must account for the volume occupied by the gas molecules. If V is the volume of the container and b is the volume occupied by the molecules, then $V - b$ is the empty volume available to the gas. The constant b is equal to the number of molecules of gas multiplied by the volume per molecule. As V decreases for a given quantity of gas, the fraction of the volume occupied by the molecules increases.

The second important effect concerns the intermolecular forces when the molecules are close together. At close separations, the molecules attract each other, as we might expect, since gases can condense to form liquids. There is a potential energy associated with this attractive force, hence less of the total energy associated with a particular degree of freedom is kinetic energy. According to the kinetic theory of gases, the decrease in pressure from that of an ideal gas is due the reduction in the average kinetic energy of the molecules colliding with the walls. The net inward force on a molecule near the wall is proportional to the density of molecules, or inversely proportional to the volume. In addition, the pressure at the wall is proportional to the density of molecules. The net pressure is reduced by a factor proportional to the square of the density, which varies as $1/V^2$. Hence, the pressure P is replaced by an effective pressure $P + a/V^2$, where a is a constant.

The two effects just described can now be incorporated into a modified equation of state proposed by J. D. van der Waals (1837–1923) in 1873. For one mole of gas, **van der Waals' equation of state** is given by

$$\left(P + \frac{a}{V^2}\right)(V - b) = RT \tag{21.30}$$

The constants a and b are empirical and are chosen to provide the best fit to the experimental data for a particular gas.

The experimental curves in Figure 21.16 for CO_2 are described quite accurately by van der Waals' equation at the higher temperatures (T_3, T_4, and

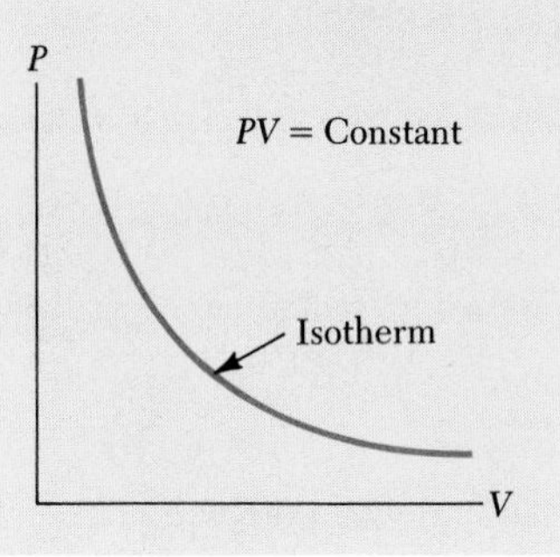

Figure 21.14 The PV diagram of an isothermal process for an ideal gas. In this case, the pressure and volume are related by $PV =$ constant.

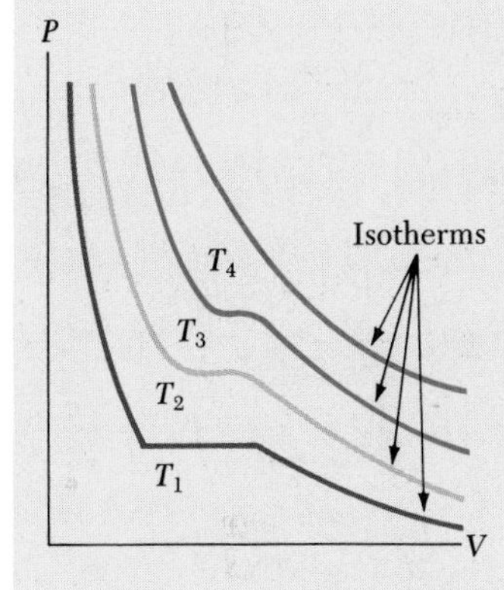

Figure 21.15 Isotherms for a real gas at various temperatures. At higher temperatures, such as T_4, the behavior is nearly ideal. The behavior is not ideal at the lower temperatures.

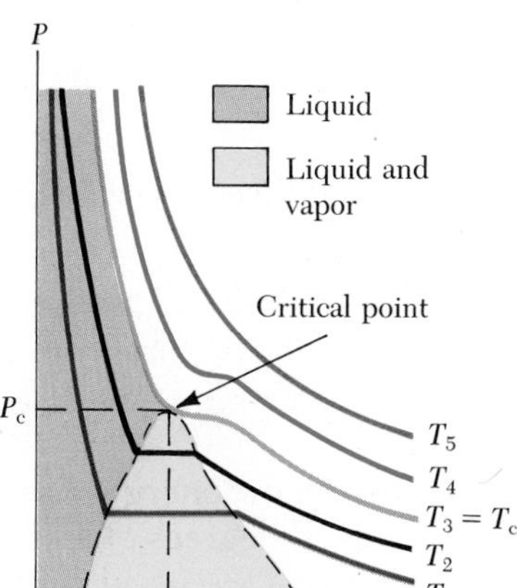

Figure 21.16 Isotherms for CO_2 at various temperatures. Below the critical temperature, T_c, the substance could be in the liquid state, the liquid-vapor equilibrium state, or the gaseous state, depending on the pressure and volume. (Adapted from K. Mendelssohn, *The Quest for Absolute Zero*, New York, McGraw-Hill, World University Library, 1966.)

T_5) and outside the shaded regions. Within the yellow region there are major discrepancies. If the van der Waals equation of state is used to predict the *PV* relationship at a temperature such as T_1, then a nonlinear curve is obtained that is unlike the observed flat portion of the curve in the figure.

The departure from the predictions of van der Waals' equation at the lower temperatures and higher densities is due to the onset of liquefaction. That is, the gas begins to liquefy at the pressure P_c, called the **critical pressure.** In the region within the dotted line below P_c the gas is partially liquefied and the gas vapor and liquid coexist. However, liquefaction cannot occur (even for very large pressures) unless the temperature is below a critical value. In the flat portions of the low-temperature isotherms, as the volume is decreased more gas liquefies and the pressure remains constant. At even lower volumes, the gas is completely liquefied. Any further decrease in volume leads to large increases in pressure because liquids are not easily compressed.

It is now realized, because of the complex nature of the intermolecular forces, that a real gas cannot be rigorously described by any simple equation of state, such as Equation 21.30. Nevertheless, the basic concepts involved in Equation 21.30 are correct. At very low temperatures, the low-energy molecules attract each other and the gas tends to liquefy, or condense. A further pressure increase will accelerate the rate of liquefaction. At the higher temperatures, the average kinetic energy is large enough to overcome the attractive intermolecular forces; hence the molecules do not bind together at the higher temperatures and the gas phase is maintained.

SUMMARY

The **pressure** of N molecules of an ideal gas contained in a volume V is given by

Pressure and molecular kinetic energy

$$P = \frac{2}{3}\frac{N}{V}\left(\tfrac{1}{2}m\overline{v^2}\right) \qquad (21.5)$$

where $\frac{1}{2}m\overline{v^2}$ is the average kinetic energy per molecule.

The **temperature** of an ideal gas is related to the average kinetic energy per molecule through the expression

Temperature is proportional to average kinetic energy

$$T = \frac{2}{3k}\left(\tfrac{1}{2}m\overline{v^2}\right) \qquad (21.6)$$

where k is Boltzmann's constant.

The **average translational kinetic energy per molecule** of a gas is given by

Average kinetic energy per molecule

$$\tfrac{1}{2}m\overline{v^2} = \tfrac{3}{2}kT \qquad (21.7)$$

Each translational degree of freedom (x, y, or z) has $\frac{1}{2}kT$ of energy associated with it.

The **equipartition of energy theorem** states that the energy of a system in thermal equilibrium is equally divided among all degrees of freedom.

The **total energy** of N molecules (or n moles) of an ideal monatomic gas is given by

Internal energy of an ideal monatomic gas

$$U = \tfrac{3}{2}NkT = \tfrac{3}{2}nRT \qquad (21.11)$$

The change in internal energy for n moles of *any* ideal gas that undergoes a change in temperature ΔT is

$$\Delta U = nC_v\,\Delta T \qquad (21.14)$$

where C_v is the **molar** heat capacity at constant volume.

The molar heat capacity of an ideal monatomic gas at constant volume is $C_v = \tfrac{3}{2}R$; the molar heat capacity at constant pressure is $C_p = \tfrac{5}{2}R$. The ratio of heat capacities is $\gamma = C_p/C_v = 5/3$.

An **adiabatic process** is one in which there is no heat transfer between the system and its surroundings.

If an ideal gas undergoes an **adiabatic expansion or compression,** the first law of thermodynamics together with the equation of state, $PV = nRT$, shows that

Adiabatic process

$$PV^\gamma = \text{constant} \qquad (21.18)$$

The speed of sound in a gas of density ρ and at a pressure P is

Speed of sound in a gas

$$v = \sqrt{\frac{\gamma P}{\rho}} \qquad (21.21)$$

The *most probable speed distribution of N* gas molecules at a temperature T is given by **Maxwell's speed distribution function:**

Maxwell speed distribution function

$$N_v = 4\pi N\left(\frac{m}{2\pi kT}\right)^{3/2} v^2 e^{-mv^2/2kT} \qquad (21.24)$$

Using this expression, one can find the rms speed, v_{rms}, the average speed, $\bar{v}$, and the most probable speed, v_{mp}:

$$v_{\text{rms}} = \sqrt{\frac{3kT}{m}} \quad (21.25) \qquad \bar{v} = \sqrt{\frac{8kT}{\pi m}} \quad (21.26) \qquad v_{\text{mp}} = \sqrt{\frac{2kT}{m}} \quad (21.27)$$

The molecules of a gas undergo collisions with each other billions of times each second under standard conditions. If the gas has a volume density n_v and each molecule is assumed to have a diameter d, the average distance between collisions, or **mean free path,** ℓ, is found to be

Mean free path

$$\ell = \frac{1}{\sqrt{2}\,\pi d^2 n_v} \qquad (21.28)$$

Furthermore, the number of collisions per second, or **collision frequency,** f, is given by

Collision frequency

$$f = \sqrt{2}\,\pi d^2 \bar{v} n_v = \frac{\bar{v}}{\ell} \qquad (21.29)$$

QUESTIONS

1. Dalton's law of partial pressures states: *The total pressure of a mixture of gases is equal to the sum of the partial pressures of gases making up the mixture.* Give a convincing argument of this law based on the kinetic theory of gases.
2. One container is filled with helium gas and another with argon gas. If both containers are at the same temperature, which molecules have the higher rms speed?
3. If you wished to manufacture an after-shave lotion with a scent that is less "likely to get there before you do," would you use a high- or low-molecular-weight lotion?
4. A gas consists of a mixture of He and N_2 molecules. Do the lighter He molecules travel faster than the N_2 molecules? Explain.
5. Although the average speed of gas molecules in thermal equilibrium at some temperature is greater than zero, the average velocity is zero. Explain.
6. Why does a fan make you feel cooler on a hot day?
7. Alcohol taken internally makes you feel warmer. Yet when it is rubbed on your body, it lowers body temperature. Explain the latter effect.
8. A liquid partially fills a container. Explain why the temperature of the liquid decreases when the container is partially evacuated. (Using this technique, it is possible to freeze water at temperatures above 0°C.)
9. A vessel containing a fixed volume of gas is cooled. Does the mean free path increase, decrease, or remain constant in the cooling process? What about the collision frequency?
10. A gas is compressed at a constant temperature. What happens to the mean free path of the molecules in this process?
11. If a helium-filled balloon is placed in a freezer, will its volume increase, decrease, or remain the same?
12. What happens to a helium-filled balloon released into the air? Will it expand or contract? Will it stop rising at some height?
13. Which is heavier, standard dry air or air saturated with water vapor? Is this why thunderheads rise?
14. Why does a diatomic gas (Cl_2) have a greater energy content per mole than a mole of monatomic gas (Xe) at the same temperature?
15. What happens to the van der Waals equation (Eq. 21.30) as the volume per mole, V, increases?
16. Ideal gas is contained in a vessel at a temperature of 300 K. If the temperature is increased to 900 K, (a) by what factor would the rms speed of each molecule change? (b) By what factor would the pressure in the vessel change?
17. At room temperature, the average speed of an air molecule is several hundred meters per second. A molecule traveling at this speed should travel across a room in a small fraction of a second. In view of this, why does it take the odor of perfume (or other smells) several seconds to travel across the room?
18. A vessel is filled with gas at some equilibrium pressure and temperature. Can all molecules in the vessel have the same speed?
19. Does it make any sense to use the concept of temperature in describing a vacuum? Explain.
20. In our model of the kinetic theory of gases, molecules were viewed as hard spheres colliding elastically with the walls of the container. Is this model realistic? (A more accurate description is one in which the surface absorbs molecules and emits them later.)
21. In view of the fact that hot air rises, why does it generally become cooler as you climb a mountain? (Note that air is a poor conductor of heat.)

PROBLEMS

Section 21.1 Molecular Model for the Pressure of an Ideal Gas

1. Find the mean square speed of nitrogen molecules under standard conditions, 0°C and 1 atm pressure. Recall that 1 mole of any gas occupies a volume of 22.4 liters under standard conditions.
2. Two moles of oxygen gas are confined to a 5-liter vessel at a pressure of 8 atm. Find the average translational kinetic energy of an oxygen molecule under these conditions. (The mass of an O_2 molecule is 5.31×10^{-26} kg.)
3. In a 1-min interval, a machine gun fires 150 bullets, each of mass 8 g and speed 400 m/s. The bullets strike and become imbedded in a stationary target. If the target has an area of 5 m^2, find the average force and pressure exerted on the target. (*Note:* These are inelastic collisions.)
4. In a period of 1 s, 5×10^{23} nitrogen molecules strike a wall of area 8 cm^2. If the molecules move with a speed of 300 m/s and strike at an angle of 45° to the normal to the wall, find the pressure exerted on the wall. (The mass of an N_2 molecule is 4.65×10^{-26} kg.)
5. A spherical balloon of volume 4000 cm^3 contains helium at an (inside) pressure of 1.2×10^5 N/m^2. How many moles of helium are in the balloon, if each helium atom has an average kinetic energy of 3.6×10^{-22} J?

6. In a 30-s interval, 500 hailstones strike a glass window of area 0.6 m^2 at an angle of 45° to the window surface. Each hailstone has a mass of 5 g and a speed of 8 m/s. If the collisions are assumed to be elastic, find the average force and pressure on the window.

Section 21.2 Molecular Interpretation of Temperature

7. A cylinder contains a mixture of helium and argon gas in equilibrium at a temperature of 150°C. What is the average kinetic energy of each gas molecule?
8. Calculate the root mean square speed of an H_2 molecule at a temperature of 250°C.
9. (a) Determine the temperature at which the rms speed of an He atom equals 500 m/s. (b) What is the rms speed of He on the surface of the sun, where the temperature is 5800 K?
10. Nitrogen molecules have an rms speed of 517 m/s at 300 K. (a) What is the rms speed of nitrogen at 600 K? at 200 K? (b) Construct a graph of v_{rms} versus temperature for helium in intervals of 200 K, over the temperature range 200 K to 2000 K.
11. What is the temperature at which the rms speed of nitrogen molecules equals the rms speed of helium at 20°C?
12. A 5-liter vessel contains nitrogen gas at a temperature of 27°C and a pressure of 3 atm. Find (a) the total translational kinetic energy of the gas molecules and (b) the average kinetic energy per molecule.
13. One mole of xenon gas at 20°C occupies 0.0224 m^3. What is the pressure exerted by the Xe atoms on the walls of a container?
14. (a) How many atoms of helium (He) are required to fill a balloon to diameter 30 cm at room temperature (20°C) and 1 atm? (b) What is the average kinetic energy of each helium atom? (c) What is the average speed of each helium atom?

Section 21.3 Heat Capacity of an Ideal Gas

Use data in Table 21.2.

15. Calculate the change in internal energy of 3 moles of helium gas when its temperature is increased by 2 K.
16. Two moles of oxygen gas are heated from 300 K to 320 K. How much heat is transferred to the gas if the process occurs at (a) constant volume and (b) constant pressure?
17. The heat capacity, C, of a monatomic gas measured at constant pressure is 62.3 J/k. Find (a) the number of moles of gas, (b) the heat capacity at constant volume, and (c) the internal energy of the gas at 350 K.
18. Consider *three* moles of an ideal gas. (a) If the gas is monatomic, find the *total* heat capacity at constant volume and at constant pressure. (b) Repeat (a) for a diatomic gas in which the molecules rotate but do not vibrate.
19. One mole of hydrogen gas is heated at constant pressure from 300 K to 420 K. Calculate (a) the heat transferred to the gas, (b) the increase in internal energy of the gas, and (c) the work done by the gas.
20. In a constant-volume process, 209 J of heat is transferred to 1 mole of an ideal monatomic gas initially at 300 K. Find (a) the increase in internal energy of the gas, (b) the work done by the gas, and (c) the final temperature of the gas.
21. What is the internal energy of 100 g of He gas at 77 K? How much more energy must be supplied to heat this gas to +24°C?
22. How much internal energy is contained in 1 m^3 of air at 0°C?
23. The specific heat at constant volume for neon gas is $c_v = 0.149$ kcal/kg·K. Calculate the mass of a neon atom.
24. How much internal energy is in the air in a 20 m^3 room at (a) 0°C and (b) 20°C? Assume that the pressure remains at 1 atm.

Section 21.4 Adiabatic Process for an Ideal Gas

25. Two moles of an ideal gas ($\gamma = 1.40$) expand quasi-statically and adiabatically from a pressure of 5 atm and a volume of 12 liters to a final volume of 30 liters. (a) What is the final pressure of the gas? (b) What are the initial and final temperatures?
26. Show that Equation 21.20 follows from Equation 21.19 for a quasi-static, adiabatic process. (*Note:* $PV = nRT$ applies during the process.)
27. An ideal gas ($\gamma = 1.40$) expands quasi-statically and adiabatically. If the final temperature is one third the initial temperature, (a) by what factor does its volume change? (b) By what factor does its pressure change?
28. One mole of an ideal monatomic gas ($\gamma = 1.67$) initially at 300 K and 1 atm is compressed quasi-statically and adiabatically to one fourth its initial volume. Find its final pressure and temperature.
29. During the compression stroke of a certain gasoline engine, the pressure increases from 1 atm to 20 atm. Assuming that the process is adiabatic and the gas is ideal with $\gamma = 1.40$, (a) by what factor does the volume change and (b) by what factor does the temperature change?
30. Helium gas at 20°C is compressed without heat loss to 1/5 its initial volume. (a) What is its temperature after compression? (b) What if the gas is dry air (77% N_2, 23% O_2)?
31. Air in a thundercell expands in volume as it rises. If its initial temperature was 300 K, and no heat is lost on expansion, what is its temperature when the initial volume is doubled?
32. How much work is required to compress 5 moles of air at 20°C and 1 atm to 1/10 of the original volume by (a) an isothermal process and (b) an adiabatic process? (c) What are the final pressures for the two cases?

Section 21.5 Sound Waves in a Gas

33. Calculate the speed of sound in methane (CH_4) at 288 K, using the values $\gamma = 1.31$, the molecular weight of CH_4, $M = 16$ kg/kmole, and the universal gas constant $R = 8.314$ J/mol·K.
34. At what temperature will the speed of sound in methane equal the speed of sound in helium at 288 K? For helium, $\gamma = 1.67$ and $M = 4$ kg/kmole.
35. Compare the speed of sound in helium to that in air at 20°C. (*Hint:* Use Table 21.2.)
36. A worker is at one end of a mile-long section (1.61) km) of iron pipeline when an accidental blast occurs at the other end of the section. The worker receives two sound signals from the blast, one transmitted through the pipe and one through the surrounding air. Use values from Table 17.1 to calculate the elapsed time between the two signals. (*Note:* First find the speed of sound in air at 300 K and take the speed of sound in iron at that temperature to be 5200 m/s.)
37. Xenon has a density of 5.9 kg/m^3 at 0°C and 1 atm pressure. Since it is monatomic, $\gamma = 1.67$. (a) Calculate the speed of sound in xenon at 0°C. (b) What is the bulk modulus of xenon?
38. A sound wave propagating in air has a frequency of 4000 Hz. Calculate the percent change in wavelength when the wavefront, initially in a region where $T = 27$°C, enters a region where the air temperature decreases to 10°C.
39. A spelunker attempts to determine the depth of a pit in the floor of a cave by dropping a stone into the pit and measuring the time interval between release and the sound of the stone's hitting bottom. If the measured time interval is 10 s, what is the depth of the pit? (Assume a temperature of 15°C.)

Section 21.6 The Equipartition of Energy

40. If a molecule has f degrees of freedom, show that a gas consisting of such molecules has the following properties: (1) its total internal energy is $fnRT/2$; (2) its molar heat capacity at constant volume is $fR/2$; (3) its molar heat capacity at constant pressure is $(f + 2)R/2$; (4) the ratio $\gamma = C_p/C_v = (f + 2)/f$.
41. Examine the data for polyatomic gases in Table 21.2 and explain why SO_2 has a higher C_v than the other polyatomic gases at 300 K.
42. Inspecting the magnitudes of C_v and C_p for the diatomic and polyatomic gases in Table 21.2, we find that the values increase with increasing molecular mass. Give a qualitative explanation of this observation.
43. In a crude model (Figure 21.17) of a rotating diatomic molecule of chlorine (Cl_2), the two Cl atoms are a distance 2×10^{-10} m apart and rotate about their center-of-mass with angular velocity $\omega = 2 \times 10^{12}$ rad/s. What is the rotational kinetic energy of one molecule of Cl_2, with a molecular weight of 70?

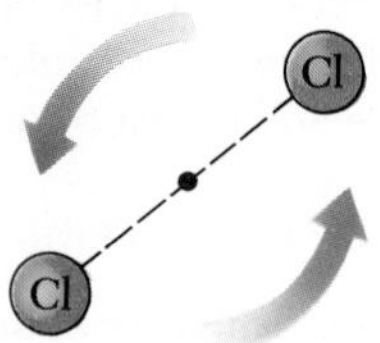

Figure 21.17 (Problem 43).

44. Consider 2 moles of an ideal diatomic gas. Find the *total* heat capacity at constant volume and at constant pressure if (a) the molecules rotate but do not vibrate and (b) the molecules rotate and vibrate.

*Section 21.7 Distribution of Molecular Speeds

45. A vessel containing oxygen gas is at a temperature of 400 K. Find (a) the rms speed, (b) the average speed, and (c) the most probable speed of the gas molecules. (The mass of O_2 is 5.31×10^{-26} kg.)
46. Fifteen identical particles have the following speeds: one has speed 2 m/s; two have speed 3 m/s; three have speed 5 m/s; four have speed 7 m/s; three have speed 9 m/s; two have speed 12 m/s. Find (a) the average speed, (b) the rms speed, and (c) the most probable speed of these particles.
47. Calculate the most probable speed, average speed, and rms speed for nitrogen gas molecules at 900 K. Compare your results with the values obtained from Figure 21.10.
48. Determine the rms velocities of nitrogen (N_2), water vapor (H_2O), and oxygen (O_2) molecules at (a) 20°C and (b) 100°C.
49. Use Figure 21.10 to *estimate* the number of nitrogen molecules with speeds between 400 m/s and 600 m/s at (a) 300 K and (b) 900 K.
50. Show that the most probable speed of a gas molecule is given by Equation 21.27. Note that the most probable speed corresponds to the point where the slope of the speed distribution curve, dN_v/dv, is zero.
51. At what temperature would the average velocity of helium atoms equal (a) the escape velocity from earth, 1.12×10^4 m/s, and (b) the escape velocity from the moon, 2.37×10^3 m/s? (See Chapter 14 for a discussion of escape velocity, and note that the mass of helium is 6.65×10^{-27} kg.)
52. Using the data in Figure 21.10, estimate the *fraction* of N_2 molecules that have speeds in the range 1000 m/s to 1200 m/s at 900 K. The total number of molecules is 10^5.

*Section 21.8 Mean Free Path

53. In an ultrahigh vacuum system, the pressure is measured to be 10^{-10} torr (where 1 torr = 133 N/m^2). If

the gas molecules have a molecular diameter of 3 Å $= 3 \times 10^{-10}$ m and the temperature is 300 K, find (a) the number of molecules in a volume of 1 m^3, (b) the mean free path of the molecules, and (c) the collision frequency, assuming an average speed of 500 m/s.

54. Show that the mean free path for the molecules of an ideal gas is given by

$$\ell = \frac{kT}{\sqrt{2}\pi d^2 P}$$

where d is the molecular diameter.

55. In a tank full of oxygen, how many molecular diameters d (on average) will an oxygen molecule travel (at 1 atm and 20°C) before colliding with another O_2 molecule? (The diameter of the O_2 molecule is about 3.6×10^{-10} m.)

56. Argon gas at atmospheric pressure and at 20°C is confined in a vessel with a volume of 1 m^3. The effective hard-sphere diameter of the argon atom is 3.10×10^{-10} m. (a) Determine the mean free path ℓ. (b) Find the pressure (at 20°C) when $\ell = 1$ m. (c) Find the pressure (at 20°C) when $\ell = 3.1 \times 10^{-10}$ m.

*Section 21.9 Van der Waals' Equation of State

57. The constant b that appears in van der Waals' equation of state for oxygen is measured to be equal to 31.8 cm^3/mole. Assuming a spherical shape, estimate the diameter of the molecule.

58. Use Equation 21.22 to compute the speed of sound in a mixture of 60% oxygen and 40% nitrogen at 40°C.

59. Consider again the situation described in Problem 36. Show that, in general, the elapsed time interval between arrival of the sound signal through the pipe and through the surrounding air is

$$\Delta t = \frac{\ell(v_m - v_a)}{v_m v_a}$$

where v_m is the speed of sound in the metal, v_a is the speed of sound in air, and ℓ is the length of the pipe.

60. Show that the work done in expanding 1 mole of a van der Waals gas from an initial volume V_i to final volume V_f at constant temperature is given by

$$W = RT \ln\left(\frac{V_f - b}{V_i - b}\right) + a(V_f^{-1} - V_i^{-1}).$$

ADDITIONAL PROBLEMS

61. A mixture of two gases will diffuse through a filter at rates proportional to their rms speeds. If the molecules of the two gases have masses m_1 and m_2, show that the ratio of their rms speeds (or the ratio of diffusion rates) is given by

$$\frac{(v_1)_{rms}}{(v_2)_{rms}} = \sqrt{\frac{m_2}{m_1}}$$

This process is used to obtain uranium enriched with the isotope ^{235}U, which is used in nuclear reactors.

62. A cylinder containing n moles of an ideal gas undergoes a quasi-static, adiabatic process. (a) Starting with the expression $W = \int P\,dV$ and using $PV^\gamma =$ constant, show that the work done is given by

$$W = \left(\frac{1}{\gamma - 1}\right)(P_iV_i - P_fV_f)$$

(b) Starting with the first law in differential form, prove that the work done is also equal to $nC_v(T_i - T_f)$. Show that this result is consistent with the equation in (a).

63. Twenty particles, each of mass m and confined to a volume V, have the following speeds: two have speed v; three have speed $2v$; five have speed $3v$; four have speed $4v$; three have speed $5v$; two have speed $6v$; one has speed $7v$. Find (a) the average speed, (b) the rms speed, (c) the most probable speed, (d) the pressure they exert on the walls of the vessel, and (e) the average kinetic energy per particle.

64. A vessel contains 10^4 oxygen molecules at 500 K. (a) Make an accurate graph of the Maxwell speed distribution function, N_v, versus speed with points at speed intervals of 100 m/s. (b) Determine the most probable speed from this graph. (c) Calculate the average and rms speeds for the molecules and label these points on your graph. (d) From the graph, estimate the fraction of molecules with speeds in the range 300 m/s to 600 m/s.

65. A vessel contains 1 mole of helium gas at a temperature of 300 K. Calculate the approximate number of molecules having speeds in the range from 400 m/s to 410 m/s. (*Hint:* This number is approximately equal to $N_v\,\Delta v$, where Δv is the range of speeds.)

66. Argon gas ($n = 3$ mol), initially at a temperature of 20°C, occupies a volume of 10 L. The gas undergoes a slow expansion at constant pressure to a volume of 25 L; then, the gas expands adiabatically until it returns to its initial temperature. (a) Show the process on a P-V diagram. (b) What quantity of heat was supplied to the gas during the entire process? (c) What was the total change in internal energy of the gas? (d) What was the total amount of work done during the process? (e) What is the final volume of the gas?

67. Oxygen at pressures much above 1 atm becomes toxic to lung cells. What ratio, by weight, of helium gas (He) to oxygen (O_2) must be used by a scuba diver who is to descend to an ocean depth of 50 m?

68. The compressibility, κ, of a substance is defined as the fractional change in volume of that substance for a given change in pressure:

$$\kappa = -\frac{1}{V}\frac{dV}{dP}$$

(a) Explain why the negative sign in this expression ensures that κ will always be positive. (b) Show that if an ideal gas is compressed *isothermally*, its compressibility is given by $\kappa_1 = 1/P$. (c) Show that if an ideal gas is compressed *adiabatically*, its compressibility is given by $\kappa_2 = 1/\gamma P$. (d) Determine values for κ_1 and κ_2 for a monatomic ideal gas at a pressure of 2 atm.

69. One mole of a gas obeying van der Waals' equation of state is compressed isothermally. At some critical temperature, T_c, the isotherm has a point of zero slope, as in Figure 21.16. That is, at $T = T_c$,

$$\frac{\partial P}{\partial V} = 0 \quad \text{and} \quad \frac{\partial^2 P}{\partial V^2} = 0$$

Using Equation 21.30 and these conditions, show that at the critical point, the pressure, volume, and temperature are given by $P_c = a/27b^2$; $V_c = 3b$, and $T_c = 8a/27Rb$.

70. Solve for the molar heat capacities C_v and C_p of an ideal gas in terms of the gas constant R and the adiabatic exponent γ.

71. In Equation 21.22 the temperature T must be in degrees kelvin. (a) Starting with this equation, show that the speed of sound in a gas can be expressed in the form $v = v_0 + (v_0/546)t$, where t is the temperature in °C and v_0 is the speed of sound in the gas at 0°C. (*Hint:* Assume that $t \ll 273°\text{C}$ and use the expansion $(1+x)^{1/2} = 1 + \frac{1}{2}x - \frac{1}{8}x^2 + \cdots$ (b) In the case of air, show that this result leads to

$$v = (331 + 0.61t) \text{ m/s}$$

72. Verify Equations 21.25 and 21.26 for the rms and average speed of the molecules of a gas at a temperature T. Note that the average value of v^n is given by

$$\overline{v^n} = \frac{1}{N}\int_0^\infty v^n N_v \, dv$$

and make use of the integrals

$$\int_0^\infty x^3 e^{-ax^2} \, dx = \frac{1}{2a^2}$$

and

$$\int_0^\infty x^4 e^{-ax^2} \, dx = \frac{3}{8a^2}\sqrt{\frac{\pi}{a}}$$

73. The internal energy of a gas consisting of n moles of CO_2 at 300 K is given by $U = anRT + b$, where a and b are constants. (a) From this expression, derive the molar heat capacity at constant volume, C_v. (b) What is C_p for this gas? (c) Use Table 21.2 to obtain a value for the constant a. (d) How many degrees of freedom does the molecule have at this temperature?

74. An ideal monatomic gas undergoes an adiabatic expansion for which the final pressure P_f is related to the initial pressure P_i by $P_f = P_i/10$. Find (a) the ratio of the final volume V_f to the initial volume V_i, (b) the ratio of the final temperature T_f to the initial temperature T_i, (c) the ratio of ℓ_f (the mean free path in the final state) to ℓ_i (the mean free path in the initial state), and (d) the ratio of the collision frequency in the final state to the collision frequency in the initial state.

75. An ideal gas whose constant-volume molar heat capacity is $C_v = \frac{5}{2}R$ undergoes an adiabatic expansion for which $P_f = P_i/10$. Find the ratios requested in parts (a) through (d) of Problem 74.

76. Consider an ideal gas of triatomic molecules. (a) If the molecule is a linear one (such as CO_2) and the gas temperature is low enough that there is negligible vibrational motion, what will be the value of C_v, the molar heat capacity at constant volume? (Give your result as a multiple of R.) (b) At temperatures high enough that vibrations along the length of the molecule are "fully engaged" what will the value of C_v be? (*Note:* A linear triatomic molecule has two distinct patterns or "modes" of vibrational motion along the axis of the molecule. At sufficiently high temperatures *each* of these two modes contributes R ($\frac{1}{2}R$ potential and $\frac{1}{2}R$ kinetic) to the molar heat capacity.) (c) If the molecule is nonlinear (such as H_2O) and the gas temperature is low enough that there is negligible vibrational motion, what will the value of C_v be? (d) Based on your results for parts (a) and (c), how could specific-heat data be used to determine whether a triatomic molecule (of known molecular weight) is linear or nonlinear?

77. An ideal gas mixture consists of n_1 moles of a pure gas whose molar heat capacity at constant volume is $R/(\gamma_1 - 1)$ and n_2 moles of a pure gas whose molar heat capacity at constant volume is $R/(\gamma_2 - 1)$. Find (a) the total heat capacity (at constant volume) of the sample, and (b) the total heat capacity at constant pressure. (*Hint:* The pressure P of the sample obeys the equation of state $PV = (n_1 + n_2)RT$.)

CALCULATOR/COMPUTER PROBLEMS

78. For a Maxwellian gas, find the numerical value of the ratio $\{N_v(v)/N_v(v_{mp})\}$ for the following values of v: $v = (v_{mp}/50)$, $(v_{mp}/10)$, $(v_{mp}/2)$, $2v_{mp}$, $10v_{mp}$, $50v_{mp}$. Give your results to three significant figures.

79. Consider a system of 10^4 oxygen molecules at a temperature T. Write a program that will enable you to calculate the Maxwell distribution function N_v as a function of the speed of the molecules and the temperature. Use your program to evaluate N_v for speeds ranging from $v = 0$ to $v = 2000$ m/s (in intervals of 100 m/s) at temperatures of (a) 300 K and (b) 1000 K. (c) Make graphs of your results (N_v versus v) and use the graph at $T = 1000$ K to calculate the number of molecules having speeds between 800 m/s and 1000 m/s at $T = 1000$ K.

22

Heat Engines, Entropy, and the Second Law of Thermodynamics

A model steam engine equipped with a built-in horizontal boiler. The water is heated electrically, which generates steam that is used to power the electric generator at the right. (Courtesy of CENCO)

The first law of thermodynamics is merely the law of conservation of energy generalized to include heat as a form of energy transfer. This law tells us only that an increase in one form of energy must be accompanied by a decrease in some other form of energy. The first law places no restrictions on the types of energy conversions that can occur. Furthermore, it makes no distinction between heat and work. According to the first law, the internal energy of a body may be increased by either adding heat to it or doing work on it. But there is an important difference between heat and work that is not evident from the first law. For example, it is possible to convert work completely into heat but, in practice, it is impossible to convert heat completely into work without changing the surroundings.

The *second law of thermodynamics* establishes which processes in nature may or may not occur. Of all processes permitted by the first law, only certain types of energy conversions can take place. The following are some examples of processes that are consistent with the first law of thermodynamics but proceed in an order governed by the second law of thermodynamics. (1) When two objects at different temperatures are placed in thermal contact with each other, heat flows from the warmer to the cooler object, but never from the

Lord Kelvin, British physicist and mathematician (1824–1907). Born William Thomson in Belfast, Kelvin was the first to propose the use of an absolute scale of temperature. The kelvin scale, named in honor of the physicist, is discussed in Section 19.3. Kelvin's study of Carnot's theory led to the idea that heat cannot pass spontaneously from a colder body to a hotter body; this is known as the second law of thermodynamics. (The Bettmann Archive)

cooler to the warmer. (2) Salt dissolves spontaneously in water, but extracting salt from salt water requires some external influence. (3) When a rubber ball is dropped to the ground, it bounces several times and eventually comes to rest. The opposite process does not occur. (4) The oscillations of a pendulum will slowly decrease in amplitude because of collisions with air molecules and friction at the point of suspension. Eventually the pendulum will come to rest. Thus, the initial mechanical energy of the pendulum is converted into thermal energy. The reverse transformation of energy does not occur.

These are all examples of *irreversible* processes, that is, processes that occur naturally in only one direction. None of these processes occur in the opposite temporal order; if they did, they would violate the second law of thermodynamics.[1] That is, the one-way nature of thermodynamic processes in fact *establishes* a direction of time.[2] You may have witnessed the humor of an action film running in reverse, which demonstrates the improbable order of events in a time-reversed world.

The second law of thermodynamics, which can be stated in many equivalent ways, has some very practical applications. From an engineering viewpoint, perhaps the most important application is the limited efficiency of heat engines. Simply stated, the second law says that a machine capable of continuously converting thermal energy completely into other forms of energy cannot be constructed.

22.1 HEAT ENGINES AND THE SECOND LAW OF THERMODYNAMICS

The field of thermodynamics developed from a study of heat engines, an application of great importance today. A **heat engine** is a device that converts thermal energy into other useful forms of energy, such as mechanical and electrical energy. More specifically, a heat engine is a device that carries a substance through a cycle during which (1) heat is absorbed from a source at a high temperature, (2) work is done by the engine, and (3) heat is expelled by the engine to a source at a lower temperature. In a typical process for producing electricity in a power plant, coal or some other fuel is burned and the heat produced is used to convert water to steam. This steam is then directed at the blades of a turbine, setting it into rotation. Finally, the mechanical energy associated with this rotation is used to drive an electric generator. The internal combustion engine in your automobile extracts heat from a burning fuel and converts a fraction of this energy to mechanical energy.

As was mentioned above, a heat engine carries some working substance through a cyclic process, defined as one in which the substance eventually returns to its initial state. As an example of a cyclic process, consider the operation of a steam engine in which the working substance is water. The water is carried through a cycle in which it first evaporates into steam in a boiler and then expands against a piston. After the steam is condensed with cooling water, it is returned to the boiler and the process is repeated.

[1] To be more precise, we should say that the set of events in the time-reversed sense is highly improbable. From this viewpoint, events occur with a vastly higher probability in one direction than in the opposite direction.

[2] See, for example, D. Layzer, "The Arrow of Time," *Scientific American*, December 1975.

In the operation of any heat engine, a quantity of heat is extracted from a high-temperature source, some mechanical work is done, and some heat is expelled to a low-temperature reservoir. It is useful to represent a heat engine schematically as in Fig. 22.1. The engine (represented by the circle at the center of the diagram) absorbs a quantity of heat Q_h from the high-temperature reservoir. It does work W and gives up heat Q_c to a lower-temperature heat reservoir. Because the working substance goes through a cycle, its initial and final internal energies are equal, so $\Delta U = 0$. Hence, from the first law of thermodynamics we see that

> the net work W done by the engine equals the net heat flowing into the engine.

As we can see from Figure 19.1, $Q_{net} = Q_h - Q_c$; therefore

$$W = Q_h - Q_c \tag{22.1}$$

where Q_h and Q_c are taken to be positive quantities. If the working substance is a gas, *the net work done for a cyclic process is the area enclosed by the curve representing the process on a PV diagram.* This is shown for an arbitrary cyclic process in Figure 22.2.

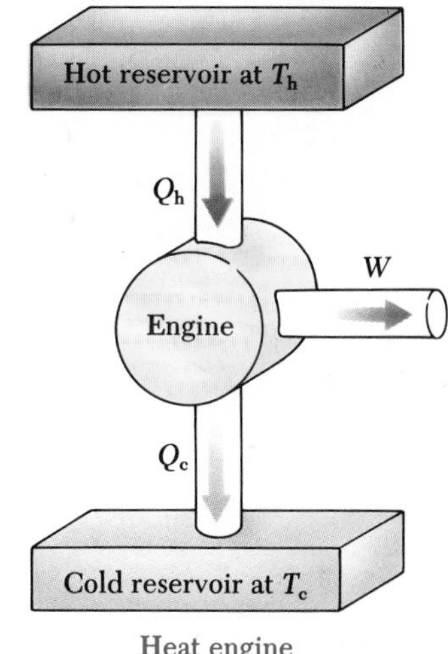

Figure 22.1 Schematic representation of a heat engine. The engine (in the circular area) receives heat Q_h from the hot reservoir, expels heat Q_c to the cold reservoir, and does work W.

The **thermal efficiency,** e, of a heat engine is defined as the ratio of the net work done to the heat absorbed during one cycle:

Thermal efficiency

$$e = \frac{W}{Q_h} = \frac{Q_h - Q_c}{Q_h} = 1 - \frac{Q_c}{Q_h} \tag{22.2}$$

We can think of the efficiency as the ratio of "what you get" (mechanical work) to "what you pay for" (energy). This result shows that a heat engine has 100% efficiency ($e = 1$) only if $Q_c = 0$, that is, if no heat is expelled to the cold reservoir. In other words, a heat engine with perfect efficiency would have to convert all of the absorbed heat energy Q_h into mechanical work. The second law of thermodynamics says that this is impossible.

In practice, it is found that all heat engines convert only a fraction of the absorbed heat into mechanical work. For example, a good automobile engine has an efficiency of about 20%, and diesel engines have efficiencies ranging from 35% to 40%. On the basis of this fact, the **Kelvin-Planck** form of the **second law of thermodynamics** states the following:

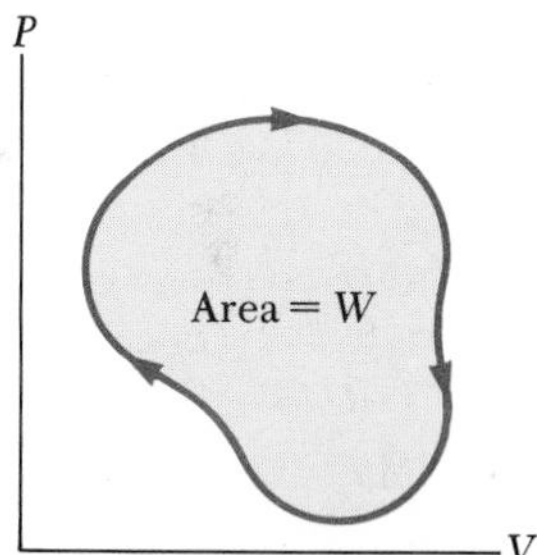

Figure 22.2 The PV diagram for an arbitrary cyclic process. The net work done equals the area enclosed by the curve.

> It is impossible to construct a heat engine that, operating in a cycle, produces no other effect than the absorption of thermal energy from a reservoir and the performance of an equal amount of work.

This is equivalent to stating that *it is impossible to construct a perpetual-motion machine of the second kind,* that is, a machine that would violate the second law.[3] Figure 22.3 is a schematic diagram of the impossible "perfect" heat engine.

A refrigerator (or heat pump) is a heat engine running in reverse. This is shown schematically in Figure 22.4, in which the engine absorbs heat Q_c from

[3] A perpetual-motion machine of the first kind is one that would violate the first law of thermodynamics (energy conservation). It is also impossible to construct this type of machine.

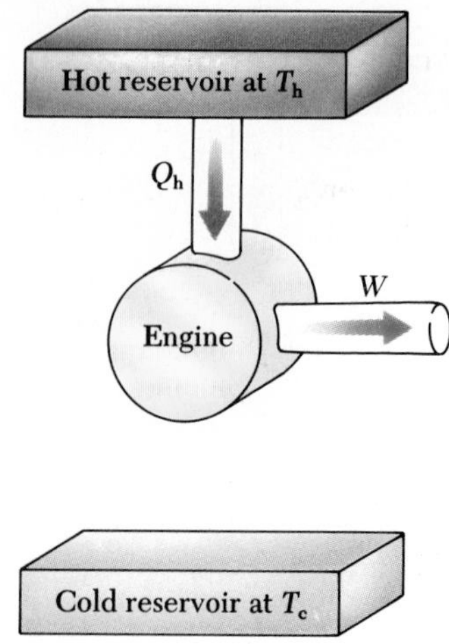

Figure 22.3 Schematic diagram of a heat engine that receives heat Q_h from a hot reservoir and does an equivalent amount of work. It is impossible to construct a perfect engine.

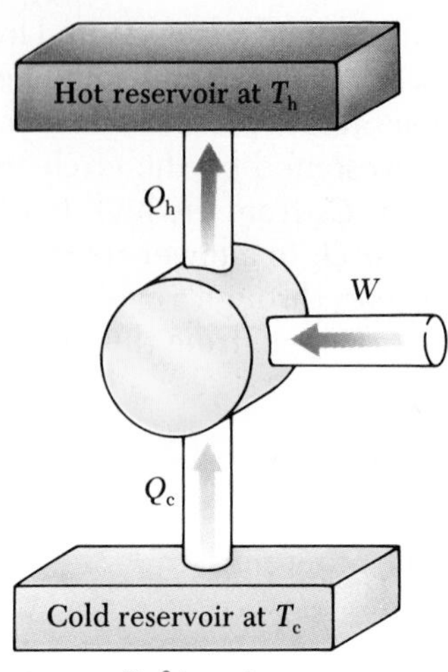

Figure 22.4 Schematic diagram for a refrigerator, which absorbs heat Q_c from the cold reservoir and expels heat Q_h to the hot reservoir. Work W is done *on* the refrigerator.

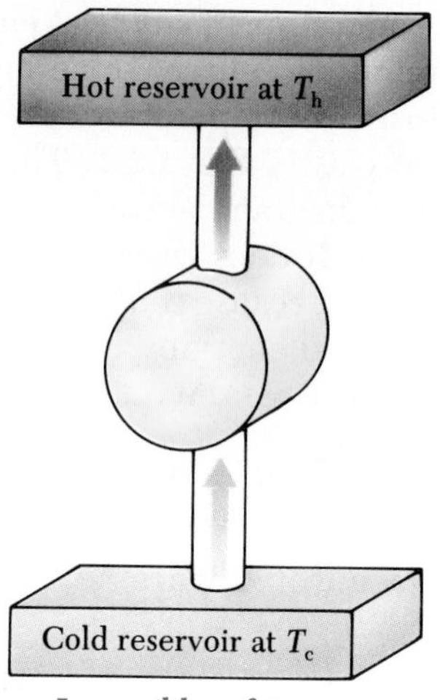

Figure 22.5 Schematic diagram of the impossible refrigerator, that is, one that absorbs heat Q_c from a cold reservoir and expels an equivalent amount of heat to the hot reservoir with $W = 0$.

the cold reservoir and expels heat Q_h to the hot reservoir. This can be accomplished only if work is done *on* the refrigerator. From the first law, we see that the heat given up to the hot reservoir must equal the sum of the work done and the heat absorbed from the cold reservoir. Therefore, we see that the refrigerator transfers heat from a colder body (the contents of the refrigerator) to a hotter body (the room). In practice, it is desirable to carry out this process with a minimum of work. If it could be accomplished without doing any work, we would have a "perfect" refrigerator (Fig. 22.5). Again, this is in violation of the second law of thermodynamics, which in the form of the **Clausius statement**[4] says the following:

> It is impossible to construct a cyclical machine that produces no other effect than to transfer heat continuously from one body to another body at a higher temperature.

In simpler terms, *heat will not flow spontaneously from a cold object to a hot object.* In effect, this statement of the second law governs the direction of heat flow between two bodies at different temperatures. Heat will flow from the colder to the hotter body only if work is done on the system. For example, homes are cooled in summer by pumping heat out; the work done on the air conditioner is supplied by the power company.

The Clausius and Kelvin-Planck statements of the second law appear, at first sight, to be unrelated. They are, in fact, equivalent in all respects. Although we do not prove it here, one can show that if either statement is false, so is the other.[5]

[4] First expressed by Rudolf Clausius (1822–1888).

[5] See, for example, F.W. Sears, *Thermodynamics, The Kinetic Theory of Gases, and Statistical Mechanics,* Reading, Mass., Addison-Wesley, 1953, Chapter 7.

EXAMPLE 22.1 The Efficiency of an Engine
Find the efficiency of an engine that introduces 2000 J of heat during the combustion phase and loses 1500 J at exhaust and through friction.

Solution The efficiency of the engine is given by Equation 22.2 as

$$e = 1 - \frac{Q_c}{Q_h} = 1 - \frac{1500\ \text{J}}{2000\ \text{J}} = 0.25, \text{ or } 25\%$$

If an engine has an efficiency of 20% and loses 3000 J through friction, how much work is done by the engine?
Answer 750 J.

22.2 REVERSIBLE AND IRREVERSIBLE PROCESSES

In our introductory remarks we mentioned that real processes have a preferred direction. Heat flows spontaneously from a hot to a cold body when the two are placed in contact, but the reverse is accomplished only with some external influence. When a block slides on a rough surface, it eventually comes to rest. The mechanical energy of the block is converted into internal energy of the block and table. Such unidirectional processes are called **irreversible** processes. In general, a process is **irreversible** if the system and its surroundings cannot be returned to their initial states.

A process may be **reversible** if the system passes from the initial state to the final state through a succession of equilibrium states. If a real process occurs quasi-statically, that is, slowly enough so that each state departs only infinitesimally from equilibrium, it can be considered reversible. For example, we can imagine compressing a gas quasi-statically by dropping some grains of sand onto a frictionless piston (Fig. 22.6). The pressure, volume, and temperature of the gas are well defined during the isothermal compression. The process is made isothermal by placing the gas in thermal contact with a heat reservoir. Some heat is transferred from the gas to the reservoir during the process. Each time a grain of sand is added to the piston, the volume decreases slightly while the pressure increases slightly. Each added grain of sand represents a change to a new equilibrium state. The process can be reversed by slowly removing grains of sand from the piston.

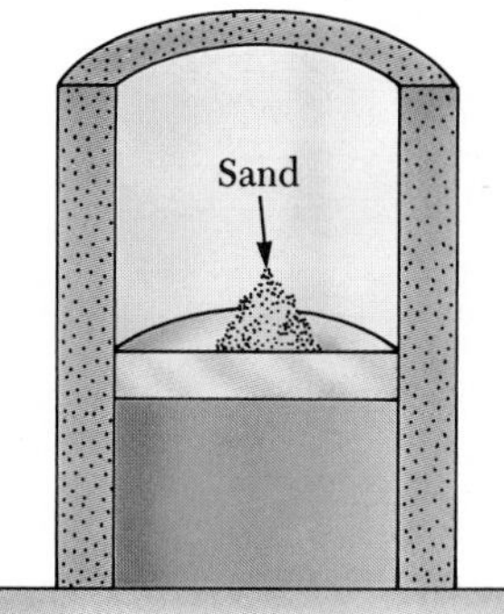

Figure 22.6 A gas in thermal contact with a heat reservoir is compressed slowly by dropping grains of sand onto the piston. The compression is isothermal and reversible.

Since a reversible process is defined by a succession of equilibrium states, it can be represented by a curve on a *PV* diagram, which establishes the path for the process (Fig. 22.7). Each point on this curve represents one of the intermediate equilibrium states. On the other hand, an irreversible process is one that passes from the initial state to the final state through a series of nonequilibrium states. In this case, only the initial and final equilibrium states can be represented on the *PV* diagram. The intermediate, nonequilibrium states may have well-defined volumes, but these states are not characterized by a unique pressure for the entire system. Instead, there are variations in pressure (and temperature) throughout the volume range, and these variations will not persist if left to themselves (i.e., nonequilibrium conditions). For this reason, an irreversible process cannot be represented by a line on a *PV* diagram.

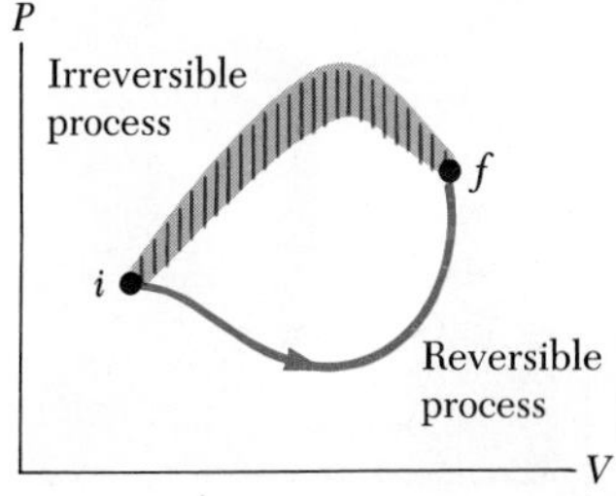

Figure 22.7 A reversible process between the two equilibrium states *i* and *f* can be represented by a line on the *PV* diagram. Each point on this line represents an equilibrium state. An irreversible process passes through a series of nonequilibrium states and cannot be represented by a line on this diagram.

We have stated that a reversible process must take place quasi-statically. In addition, in a reversible process there can be no dissipative effects that produce heat. Other effects that tend to disrupt equilibrium, such as heat conduction resulting from a temperature difference, must not be present. In reality, such effects are impossible to eliminate completely, so it is not surprising that processes in nature are irreversible. Nevertheless, it is possible to approximate reversible processes through carefully controlled procedures. As we shall see in the next section, the concept of reversible processes is especially important in establishing the theoretical limit on the efficiency of a heat engine.

22.3 THE CARNOT ENGINE

In 1824 a French engineer named Sadi Carnot (1796–1832) described a working cycle, now called a **Carnot cycle,** that is of great importance from both a practical and a theoretical viewpoint. Carnot showed that a heat engine operating in this ideal, reversible cycle between two heat reservoirs would be the most efficient engine possible. Such an ideal engine, called a **Carnot engine,** establishes an upper limit on the efficiencies of all engines. That is, the net work done by a working substance taken through the Carnot cycle is the largest possible for a given amount of heat supplied to the working substance. **Carnot's theorem** can be stated as follows:

> No real heat engine operating between two heat reservoirs can be more efficient than a Carnot engine operating between the same two reservoirs.

Let us briefly sketch a proof of this theorem. First, we shall assume that the second law is valid. Next, imagine two heat engines operating between the same heat reservoirs, one which is a Carnot engine with efficiency e_c, and the other whose efficiency e is greater than e_c. If the more efficient engine is used to drive the Carnot engine as a refrigerator, the net result would be a transfer of heat from the cold to the hot reservoir. According to the second law, this is impossible. Hence, the assumption that $e > e_c$ must be false.

To describe the Carnot cycle, we shall assume that the substance working between temperatures T_c and T_h is an ideal gas contained in a cylinder with a

Biographical Sketch

Sadi Carnot
(1796–1832)

Sadi Carnot, a French physicist, was the first to show the quantitative relationship between work and heat. Carnot was born in Paris on June 1, 1796 and was educated at the École Polytechnique in Paris and at the École Genie in Metz. Carnot's many interests include a wide range of study and research in mathematics, tax reform, industrial development, and the fine arts.

In 1824 he published his only work—*Reflections on the Motive Power of Heat*—which reviewed the industrial, political, and economic importance of the steam engine. In it he defined work as "weight lifted through a height." Carnot began to study the physical properties of gases in 1831, particularly the relationship between temperature and pressure.

On August 24, 1832, Sadi Carnot died suddenly of cholera. In accordance with the custom of his time, all of his personal effects were burned. Some of his notes which fortunately escaped destruction indicate that Carnot had arrived at the idea that heat is essentially work, or work that has changed its form. For this reason, he is considered to be the founder of the science of thermodynamics, which states that energy can never disappear; it can only be altered into other forms of energy. Carnot's notes led Lord Kelvin to confirm and extend the science of thermodynamics in 1850.

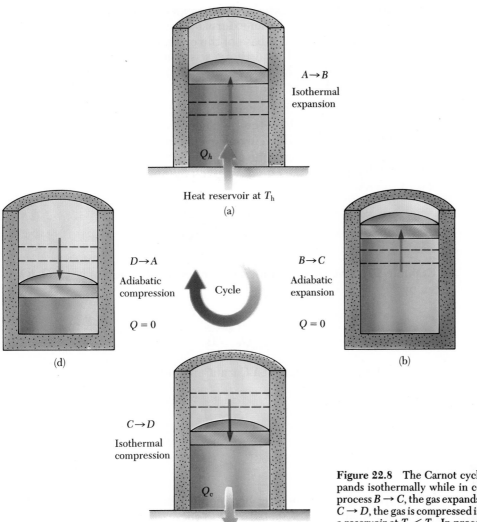

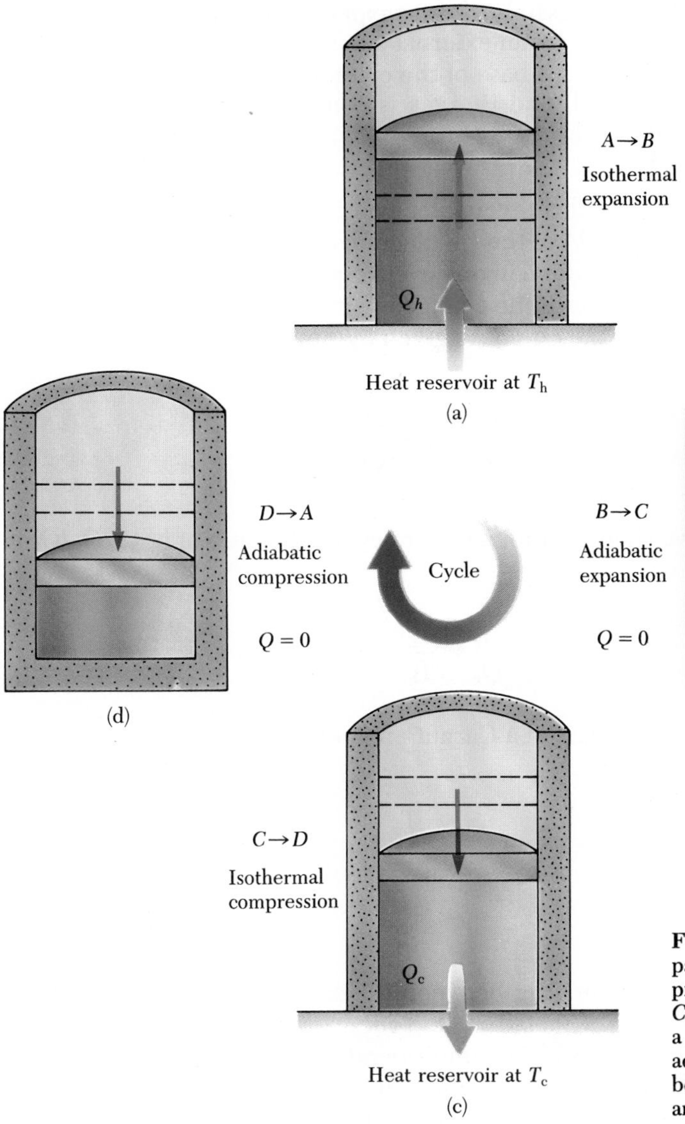

Figure 22.8 The Carnot cycle. In process $A \rightarrow B$, the gas expands isothermally while in contact with a reservoir at T_h. In process $B \rightarrow C$, the gas expands adiabatically ($Q = 0$). In process $C \rightarrow D$, the gas is compressed isothermally while in contact with a reservoir at $T_c < T_h$. In process $D \rightarrow A$, the gas is compressed adiabatically. The upward arrows on the piston indicate sand being removed during the expansions, and the downward arrows indicate the addition of sand during the compressions.

movable piston at one end. The cylinder walls and the piston are thermally nonconducting. Four stages of the Carnot cycle are shown in Figure 22.8, and the *PV* diagram for the cycle is shown in Figure 22.9. The Carnot cycle consists of two adiabatic and two isothermal processes, all reversible.

1. The process $A \rightarrow B$ is an isothermal expansion at temperature T_h, in which the gas is placed in thermal contact with a heat reservoir at temperature T_h (Fig. 22.8a). During the process, the gas absorbs heat Q_h from the reservoir through the base of the cylinder and does work W_{AB} in raising the piston.
2. In the process $B \rightarrow C$, the base of the cylinder is replaced by a thermally nonconducting wall and the gas expands adiabatically, that is, no heat enters or leaves the system (Fig. 22.8b). During the process, the temperature falls from T_h to T_c and the gas does work W_{BC} in raising the piston.
3. In the process $C \rightarrow D$, the gas is placed in thermal contact with a heat reservoir at temperature T_c (Fig. 22.8c) and is compressed isothermally at

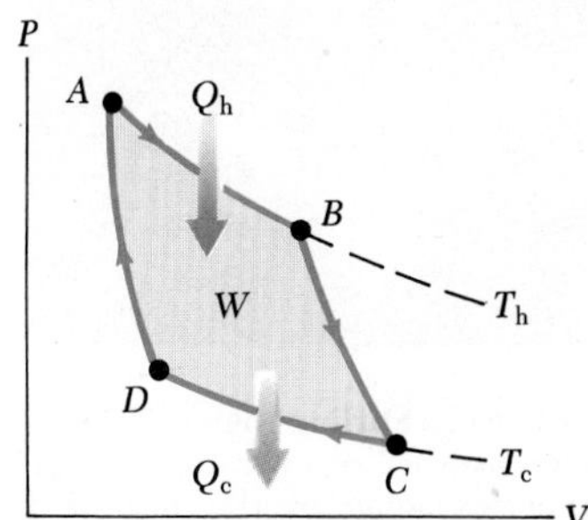

Figure 22.9 The *PV* diagram for the Carnot cycle. The net work done, *W*, equals the net heat received in one cycle, $Q_h - Q_c$. Note that $\Delta U = 0$ for the cycle.

temperature T_c. During this time, the gas expels heat Q_c to the reservoir and the work done on the gas by an external agent is W_{CD}.

4. In the final stage, $D \rightarrow A$, the base of the cylinder is replaced by a nonconducting wall (Fig. 22.8d) and the gas is compressed adiabatically. The temperature of the gas increases to T_h and the work done on the gas by an external agent is W_{DA}.

The net work done in this reversible, cyclic process is equal to the area enclosed by the path *ABCDA* of the *PV* diagram (Fig. 22.9). As we showed in Section 22.1, the net work done in one cycle equals the net heat transferred into the system, $Q_h - Q_c$, since the change in internal energy is zero. Hence, the thermal efficiency of the engine is given by Equation 22.2:

Efficiency of a heat engine

$$e = \frac{W}{Q_h} = 1 - \frac{Q_c}{Q_h}$$

In Example 22.2, we show that for a Carnot cycle, the ratio of heats Q_c/Q_h is given by

Ratio of heats for a Carnot cycle

$$\frac{Q_c}{Q_h} = \frac{T_c}{T_h} \tag{22.3}$$

Hence, the thermal efficiency of a Carnot engine is given by

$$e_c = 1 - \frac{T_c}{T_h} \tag{22.4}$$

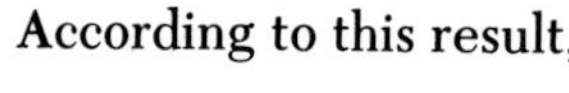

According to this result,

> all Carnot engines operating between the same two temperatures in a reversible manner have the same efficiency. From Carnot's theorem the efficiency of any reversible engine operating in a cycle between two temperatures is greater than the efficiency of any irreversible (real) engine operating between the same two temperatures.[6]

Model of a Stirling engine. This device uses a "regenerator," which recycles some of the heat that would normally be lost back into the system; for this reason, the Stirling cycle has an efficiency nearly equal to that of the Carnot cycle. (Courtesy of CENCO)

Equation 22.4 can be applied to any working substance operating in a Carnot cycle between two heat reservoirs. According to this result, the efficiency is zero if $T_c = T_h$, as one would expect. The efficiency increases as T_c is lowered and as T_h increases. However, the efficiency can only be unity (100%) if $T_c = 0$ K. Such reservoirs are not available, and so the maximum efficiency is always less than unity. In most practical cases, the cold reservoir is near room temperature, about 300 K. Therefore, one usually strives to increase the efficiency by raising the temperature of the hot reservoir. *All real engines are less efficient than the Carnot engine since they are subject to such practical difficulties as friction and heat losses by conduction.*

[6] See, for example, F.W. Sears, *Thermodynamics, The Kinetic Theory of Gases, and Statistical Mechanics*, Reading, Mass., Addison-Wesley, 1953, Chapter 7.

EXAMPLE 22.2 Efficiency of the Carnot Engine
Show that the efficiency of a heat engine operating in a Carnot cycle using an ideal gas is given by Equation 22.4.

Solution During the isothermal expansion, $A \rightarrow B$ (Fig. 22.8a), the temperature does not change and so the internal energy remains constant. The work done by the gas is given by Equation 20.13. According to the first law the heat absorbed, Q_h, equals the work done, so that

$$Q_h = W_{AB} = nRT_h \ln \frac{V_B}{V_A}$$

In a similar manner, the heat rejected to the cold reservoir during the isothermal compression $C \rightarrow D$ is given by

$$Q_c = |W_{CD}| = nRT_c \ln \frac{V_C}{V_D}$$

Dividing these expressions, we find that

$$(1) \qquad \frac{Q_c}{Q_h} = \frac{T_c}{T_h} \frac{\ln(V_C/V_D)}{\ln(V_B/V_A)}$$

We now show that the ratio of the logarithmic quantities is unity by obtaining a relation between the ratio of volumes.

For any quasi-static, adiabatic process, the pressure and volume are related by Equation 21.18:

$$PV^{\gamma} = \text{constant}$$

During any reversible, quasi-static process, the ideal gas must also obey the equation of state, $PV = nRT$. Substituting this into the above expression to eliminate the pressure, we find that

$$TV^{\gamma-1} = \text{constant}$$

Applying this result to the adiabatic processes $B \rightarrow C$ and $D \rightarrow A$, we find that

$$T_h V_B^{\gamma-1} = T_c V_C^{\gamma-1}$$

$$T_h V_A^{\gamma-1} = T_c V_D^{\gamma-1}$$

Dividing these equations, we obtain

$$(V_B/V_A)^{\gamma-1} = (V_C/V_D)^{\gamma-1}$$

$$(2) \qquad \frac{V_B}{V_A} = \frac{V_C}{V_D}$$

Substituting (2) into (1), we see that the logarithmic terms cancel and we obtain the relation

$$\frac{Q_c}{Q_h} = \frac{T_c}{T_h}$$

Using this result and Equation 22.2, the thermal efficiency of the Carnot engine is

$$e_c = 1 - \frac{Q_c}{Q_h} = 1 - \frac{T_c}{T_h} = \frac{T_h - T_c}{T_h}$$

EXAMPLE 22.3 The Steam Engine
A steam engine has a boiler that operates at 500 K. The heat changes water to steam, which drives the piston. The exhaust temperature is that of the outside air, about 300 K. What is the maximum thermal efficiency of this steam engine?

Solution From the expression for the efficiency of a Carnot engine, we find the maximum thermal efficiency for any engine operating between these temperatures:

$$e_c = 1 - \frac{T_c}{T_h} = 1 - \frac{300 \text{ K}}{500 \text{ K}} = \boxed{0.4, \text{ or } 40\%}$$

You should note that this is the highest theoretical efficiency of the engine. In practice, the efficiency will be considerably lower.

Exercise 2 Determine the maximum work the engine can perform in each cycle of operation if it absorbs 200 J of heat from the hot reservoir during each cycle.
Answer 80 J.

EXAMPLE 22.4 The Carnot Efficiency
The highest theoretical efficiency of a gasoline engine, based on the Carnot cycle, is 30%. If this engine expels its gases into the atmosphere, which has a temperature of 300 K, what is the temperature in the cylinder immediately after combustion?

Solution The Carnot efficiency is used to find T_h:

$$e_c = 1 - \frac{T_c}{T_h}$$

$$T_h = \frac{T_c}{1 - e_c} = \frac{300 \text{ K}}{1 - 0.3} = \boxed{429 \text{ K}}$$

Exercise 3 If the heat engine absorbs 837 J of heat from the hot reservoir during each cycle, how much work can it perform in each cycle?
Answer 251 J.

22.4 THE ABSOLUTE TEMPERATURE SCALE

In Chapter 19 we defined temperature scales in terms of observed changes in certain physical properties of materials with temperature. It is desirable to define a temperature scale that is independent of material properties. The Carnot cycle provides us with the basis for such a temperature scale. Equation 22.3 tells us that the ratio Q_c/Q_h depends *only* on the temperatures of the two heat reservoirs. The ratio of the two temperatures, T_c/T_h, can be obtained by operating a reversible heat engine in a Carnot cycle between these two temperatures and carefully measuring the heats Q_c and Q_h. A temperature scale can be determined with reference to some fixed-point temperatures. The *absolute*, or *kelvin*, temperature scale is defined by choosing 273.16 K as the absolute temperature of the triple point of water.

The temperature of any substance can be obtained in the following manner: (1) take the substance through a Carnot cycle; (2) measure the heat Q absorbed or expelled by the system at some temperature T; (3) measure the heat Q_3 absorbed or expelled by the system when it is at the temperature of the triple point of water. From Equation 22.3 and this procedure we find that the unknown temperature is given by

$$T = (273.16\ \text{K})\frac{Q}{Q_3}$$

The absolute temperature scale is identical to the ideal-gas temperature scale and is independent of the property of the working substance. Therefore it can be applied even at very low temperatures.

In the previous section, we found that the thermal efficiency of any Carnot engine is given by $e_c = 1 - (T_c/T_h)$. This result shows that a 100% efficient engine is possible only if a temperature of absolute zero is maintained for T_c. If this were possible, any Carnot engine operating between T_h and $T_c = 0$ K would convert all of the absorbed heat into work.[7] Using this idea, Lord Kelvin defined absolute zero as follows: *Absolute zero is the temperature of a reservoir at which a Carnot engine will expel no heat.*

22.5 THE GASOLINE ENGINE

In this section we shall discuss the efficiency of the common gasoline engine. Five successive processes occur in each cycle, as illustrated in Figure 22.10. During the *intake stroke* of the piston (Fig. 22.10a), air that has been mixed with gasoline vapor in the carburetor is drawn into the cylinder. During the *compression stroke* (Fig. 22.10b), the intake valve is closed and the air-fuel mixture is compressed approximately adiabatically. At this point a spark ignites the air-fuel mixture (Fig. 22.10c), causing a rapid increase in pressure and temperature at nearly constant volume. The burning gases expand and force the piston back, which produces the *power stroke* (Fig. 22.10d). Finally, during the *exhaust stroke* (Fig. 22.10e), the exhaust valve is opened and the rising piston forces most of the remaining gas out of the cylinder. The cycle is repeated after the exhaust valve is closed and the intake valve is opened.

[7] Experimentally, it is not possible to reach absolute zero. Temperatures as low as about 10^{-5} K have been achieved with enormous difficulties using a technique called *nuclear demagnetization*. The fact that absolute zero may be approached but never reached is a consequence of a law of nature known as the *third law of thermodynamics*.

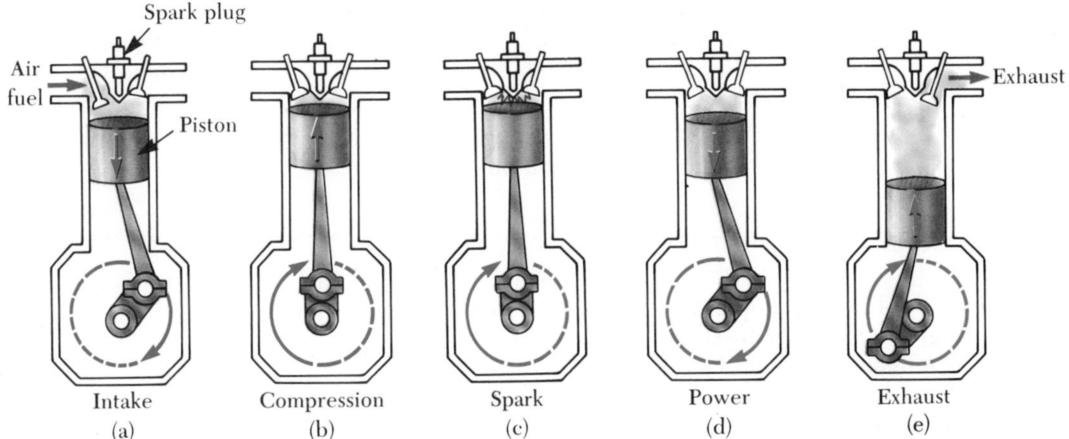

Figure 22.10 The four-stroke cycle of a conventional gasoline engine. (a) In the intake stroke, air is mixed with fuel. (b) The intake valve is then closed, and the air-fuel mixture is compressed by the piston. (c) The mixture is ignited by the spark plug raising it to a higher temperature. (d) In the power stroke, the gas expands against the piston. (e) Finally, the residual gases are expelled and the cycle repeats.

These processes can be approximated by the **Otto cycle,** a *PV* diagram of which is illustrated in Figure 22.11.

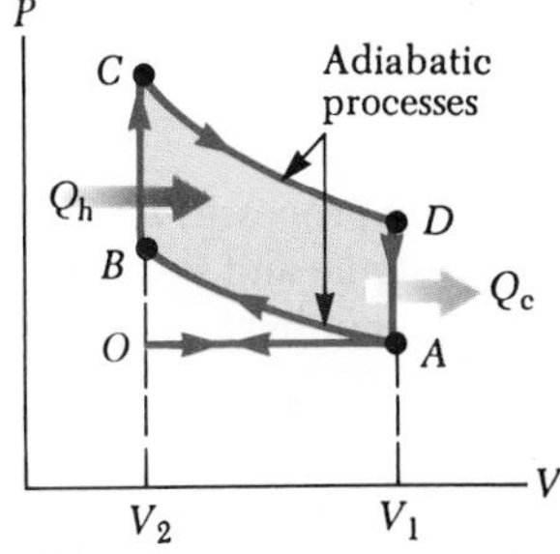

Figure 22.11 The *PV* diagram for the Otto cycle, which approximately represents the processes in the internal combustion engine. No heat is transferred during the adiabatic processes $A \to B$ and $C \to D$.

1. During the intake stroke $0 \to A$ (the horizontal line in Fig. 22.11), air is drawn into the cylinder at atmospheric pressure and the volume increases from V_2 to V_1.
2. In the process $A \to B$ (compression stroke), the air-fuel mixture is compressed adiabatically from volume V_1 to volume V_2, and the temperature increases from T_A to T_B. The work done on the gas is the area under the curve *AB*.
3. In the process $B \to C$, combustion occurs and heat Q_h is added to the gas. This is not an inflow of heat but rather a release of heat from the combustion process. During this time the pressure and temperature rise rapidly, but the volume remains approximately constant. No work is done on the gas.
4. In the process $C \to D$ (power stroke), the gas expands adiabatically from V_2 to V_1, causing the temperature to drop from T_C to T_D. The work done by the gas equals the area under the curve *CD*.
5. In the process $D \to A$, heat Q_c is extracted from the gas as its pressure decreases at constant volume. (Hot gas is replaced by cool gas.) No work is done during this process.
6. In the final process of the exhaust stroke $A \to 0$ (the horizontal line in Fig. 22.11), the residual gases are exhausted at atmospheric pressure, and the volume decreases from V_1 to V_2. The cycle then repeats itself.

If the air-fuel mixture is assumed to be an ideal gas, the efficiency of the Otto cycle is shown in Example 22.5 to be

$$e = 1 - \frac{1}{(V_1/V_2)^{\gamma-1}} \qquad (22.5)$$

Efficiency of the Otto cycle

where γ is the ratio of the molar heat capacities C_p/C_v and V_1/V_2 is called the **compression ratio.** This expression shows that the efficiency increases with increasing compression ratios. For a typical compression ratio of 8 and $\gamma = 1.4$, a theoretical efficiency of 56% is predicted for an engine operating in the idealized Otto cycle. This is much higher than what is achieved in real engines (15% to 20%) because of such effects as friction, heat loss to the cylinder walls, and incomplete combustion of the air-fuel mixture. Diesel engines have higher efficiencies than gasoline engines because of their higher compression ratios (about 16) and higher combustion temperatures.

EXAMPLE 22.5 Efficiency of the Otto Cycle
Show that the thermal efficiency of an engine operating in an idealized Otto cycle (Fig. 22.11) is given by Equation 22.5. Treat the working substance as an ideal gas.

Solution First, let us calculate the work done by the gas during each cycle. No work is done during the processes $B \rightarrow C$ and $D \rightarrow A$. Work is done on the gas during the adiabatic compression $A \rightarrow B$, and work is done by the gas during the adiabatic expansion $C \rightarrow D$. The net work done equals the area bounded by the closed curve in Figure 22.11. Since the change in internal energy is zero for one cycle, we see from the first law that the net work done for each cycle equals the net heat into the system:

$$W = Q_h - Q_c$$

Since the processes $B \rightarrow C$ and $D \rightarrow A$ take place at constant volume and since the gas is ideal, we find from the definition of heat capacity that

$$Q_h = nC_v(T_C - T_B) \qquad \text{and} \qquad Q_c = nC_v(T_D - T_A)$$

Using these expressions together with Equation 22.2, we obtain for the thermal efficiency

$$(1) \qquad e = \frac{W}{Q_h} = 1 - \frac{Q_c}{Q_h} = 1 - \frac{T_D - T_A}{T_C - T_B}$$

We can simplify this expression by noting that the processes $A \rightarrow B$ and $C \rightarrow D$ are adiabatic and hence obey the relation $TV^{\gamma-1} = \text{constant}$. Using this condition, and the facts that $V_A = V_D = V_1$ and $V_B = V_C = V_2$, we find that

$$(2) \qquad \frac{T_D - T_A}{T_C - T_B} = \left[\frac{V_2}{V_1}\right]^{\gamma-1}$$

Substituting (2) into (1) gives for the thermal efficiency

$$(3) \qquad e = 1 - \frac{1}{(V_1/V_2)^{\gamma-1}}$$

This can also be expressed in terms of a ratio of temperatures by noting that since $T_A V_1^{\gamma-1} = T_B V_2^{\gamma-1}$, it follows that

$$\left[\frac{V_2}{V_1}\right]^{\gamma-1} = \frac{T_A}{T_B} = \frac{T_D}{T_C}$$

Therefore (3) becomes

$$(4) \qquad e = 1 - \frac{T_A}{T_B} = 1 - \frac{T_D}{T_C}$$

During this cycle, the lowest temperature is T_A and the highest temperature is T_C. Therefore the efficiency of a Carnot engine operating between reservoirs at these two extreme temperatures $\left(\text{which is given by } e_c = 1 - \frac{T_A}{T_C}\right)$ would be *greater* than the efficiency of the Otto cycle, which is given by (4).

22.6 HEAT PUMPS AND REFRIGERATORS

A **heat pump** is a mechanical device that is finding increased acceptance for use in heating and cooling homes and buildings. In the heating mode, a circulatory fluid absorbs heat from the outside and releases it to the interior of the structure. The circulating fluid is usually in the form of a low pressure vapor when in the coils of a unit exterior to the structure, where it absorbs heat either from the air or the ground. This gas is then compressed and enters the structure as a hot, high-pressure vapor. In an interior unit, the gas condenses to a liquid and releases its stored internal energy. When the heat pump is used as an air conditioner, the cycle above is reversed.

Figure 22.12 is a schematic representation of a heat pump used to heat a structure. The outside temperature is T_c, the inside temperature is T_h, and the heat absorbed by the circulating fluid is Q_c. The compressor does work W on the fluid, and the heat transferred from the pump into the structure is Q_h.

The effectiveness of a heat pump is described in terms of a number called the **coefficient of performance, COP**. This is defined as the ratio of the heat transferred into the hot reservoir and the work required to transfer that heat:

$$\text{COP(heat pump)} \equiv \frac{\text{heat transferred}}{\text{work done by pump}} = \frac{Q_h}{W} \qquad (22.6)$$

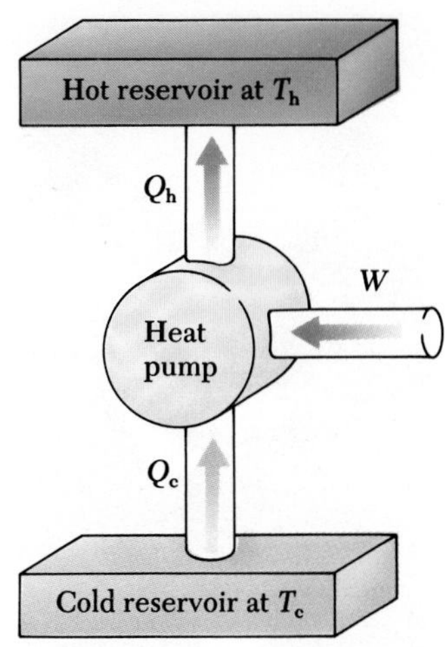

Figure 22.12 Schematic diagram of a heat pump, which absorbs heat Q_c from the cold reservoir and expels heat Q_h to the hot reservoir.

If the outside temperature is 25°F or higher, the COP for a heat pump is about 4. That is, the heat transferred into the house is about four times greater than the work done by the motor in the heat pump. However, as the outside temperature decreases, it becomes more difficult for the heat pump to extract sufficient heat from the air and the COP drops. In fact, the COP can fall below unity for temperatures below the midteens.

Although heat pumps used in buildings are relatively new products in the heating and air conditioning field, the refrigerator has been a standard appliance in homes for years. The refrigerator works much like a heat pump, except that it cools its interior by pumping heat from the food storage compartments into the warmer air outside. During its operation, a refrigerator removes a quantity of heat Q_c from the interior of the refrigerator, and in the process its motor does work W. The coefficient of performance of a refrigerator is given by

$$\text{COP(refrigerator)} = \frac{Q_c}{W} \qquad (22.7)$$

An efficient refrigerator is one that removes the greatest amount of heat from the cold reservoir for the least amount of work. Thus, a good refrigerator should have a high coefficient of performance, typically 5 or 6. The impossible (perfect) refrigerator would have an infinite coefficient of performance.

22.7 ENTROPY

The concept of temperature is involved in the zeroth law of thermodynamics, and the concept of internal energy is involved in the first law. Temperature and internal energy are both state functions. That is, they can be used to describe the thermodynamic state of a system. Another state function related to the second law of thermodynamics is the **entropy function, S**. In this section we define entropy on a macroscopic scale as it was first expressed by Clausius in 1865.

Consider a quasi-static, reversible process between two equilibrium states. If dQ_r is the heat absorbed or expelled by the system during some small interval of the path,

the **change in entropy**, dS, between two equilibrium states is given by the heat transferred, dQ_r, divided by the absolute temperature, T, of the system in this interval. That is,

$$dS = \frac{dQ_r}{T} \qquad (22.8)$$

Clausius definition of change in entropy

Rudolph Clausius (1822–1888). "I propose . . . to call S the entropy of a body, after the Greek word 'transformation.' I have designedly coined the word 'entropy' to be similar to energy, for these two quantities are analogous in their physical significance, that an analogy of denominations seems to be helpful." (AIP Niels Bohr Library, Lande Collection)

The subscript r on the term dQ_r is used to emphasize that the definition applies only to *reversible* processes. When heat is absorbed by the system, dQ_r is positive and hence the entropy increases. When heat is expelled by the system, dQ_r is negative and the entropy decreases. Note that Equation 22.8 does not define entropy, but the *change* in entropy. This is consistent with the fact that a change in state always accompanies heat transfer. Hence, the meaningful quantity in describing a process is the *change* in entropy.

Entropy originally found its place in thermodynamics, but its importance grew tremendously as the field of statistical mechanics developed because this method of analysis provided an alternative way of interpreting the concept of entropy. In statistical mechanics, the behavior of a substance is described in terms of the statistical behavior of the atoms and molecules contained in the substance. One of the main results of this treatment is that

isolated systems tend toward disorder and entropy is a measure of this disorder.

For example, consider the molecules of a gas in the air in your room. If all the gas molecules moved together like soldiers marching in step, this would be a very ordered state. It is also an unlikely state. If you could see the molecules, you would see that they moved haphazardly in all directions, bumping into one another, changing speed upon collision, some going fast, some slowly. This is a highly disordered state, and it is also the most likely state.

All physical processes tend toward the most likely state, and that state is always one in which the disorder increases. Because entropy is a measure of disorder, an alternative way of saying this is

the entropy of the universe increases in all natural processes.

This statement is yet another way of stating the second law of thermodynamics.

To calculate the change in entropy for a finite process, we must recognize that T is generally not constant. If dQ_r is the heat transferred when the system is at a temperature T, then the change in entropy in an arbitrary reversible process between an initial state and a final state is

Changes in entropy for a finite process

$$\Delta S = \int_i^f dS = \int_i^f \frac{dQ_r}{T} \quad \text{(reversible path)} \tag{22.9}$$

Although we do not prove it here, the change in entropy of a system in going from one state to another has the same value for *all* reversible paths connecting the two states.[8] That is,

the change in entropy of a system depends only on the properties of the initial and final equilibrium states.

In the case of a *reversible, adiabatic* process, no heat is transferred between the system and its surroundings, and therefore $\Delta S = 0$ in this case. Since

[8] Note that the quantity dQ_r is called an *inexact differential quantity,* whereas $dQ_r/T = dS$ is a perfect differential. This is because heat is not a property of the system, and hence Q is not a state function. Mathematically, we call $1/T$ the *integrating factor* in this case, since the perfect differential dQ_r/T can be integrated.

there is no change in entropy, such a process is often referred to as an **isentropic process.**

Consider the changes in entropy that occur in a Carnot heat engine operating between the temperatures T_c and T_h. In one cycle, the engine absorbs heat Q_h from a hot reservoir at a temperature T_h and rejects heat Q_c to a cold reservoir at a temperature T_c. Thus, the total change in entropy for one cycle is

$$\Delta S = \frac{Q_h}{T_h} - \frac{Q_c}{T_c}$$

where the negative sign in the second term represents the fact that heat Q_c is expelled by the system. In Example 22.2 we showed that for a Carnot cycle,

$$\frac{Q_c}{Q_h} = \frac{T_c}{T_h}$$

Using this result in the previous expression for ΔS, we find that the total change in entropy for a Carnot engine operating in a cycle is *zero.* That is,

$$\Delta S = 0$$

Change in entropy for a Carnot cycle is zero

Now consider a system taken through an arbitrary reversible cycle. Since the entropy function is a state function and hence depends only on the properties of a given equilibrium state, we conclude that $\Delta S = 0$ for *any* reversible cycle. In general, we can write this condition in the mathematical form

$$\oint \frac{dQ_r}{T} = 0 \qquad (22.10)$$

$\Delta S = 0$ for any reversible cycle

where the symbol $\oint$ indicates that the integration is over a *closed* path.

Another important property of entropy is the fact that

the entropy of the universe remains constant in a reversible process.

This can be understood by noting that two bodies A and B that interact with each other reversibly must always be in thermal equilibrium with each other. That is, their temperatures must always be equal. Therefore, when a small amount of heat dQ is transferred from A to B, the increase in entropy of B is dQ/T, while the corresponding change in entropy of A is $-dQ/T$. Thus the total change in entropy of the system (A + B) is zero, and the entropy of the universe is unaffected by the reversible process.[9]

As a special case, we next show how to calculate the change in entropy for an ideal gas that undergoes a quasi-static, reversible process in which heat is absorbed from a reservoir.

Quasi-static, Reversible Process for an Ideal Gas

An ideal gas undergoes a quasi-static, reversible process from an initial state T_i, V_i to a final state T_f, V_f. Let us calculate the change in entropy for this process.

[9] Alternatively, we can say that since the universe is, by definition, an isolated system, it never gains or loses heat; hence the change in entropy of the universe is zero for a reversible process.

According to the first law, $dQ_r = dU + dW$, where $dW = P\,dV$. For an ideal gas, recall that $dU = nC_v\,dT$ and $P = nRT/V$. Therefore, we can express the heat transferred as

$$dQ_r = dU + P\,dV = nC_v\,dT + nRT\frac{dV}{V} \tag{22.11}$$

We cannot integrate this expression as it stands since the last term contains two variables, T and V. However, if we divide each term by T, we can integrate both terms on the right-hand side:

$$\frac{dQ_r}{T} = nC_v\frac{dT}{T} + nR\frac{dV}{V} \tag{22.12}$$

Assuming that C_v is constant over the interval in question, and integrating Equation 22.12 from T_i, V_i to T_f, V_f, we get

$$\Delta S = \int_i^f \frac{dQ_r}{T} = nC_v \ln\frac{T_f}{T_i} + nR\ln\frac{V_f}{V_i} \tag{22.13}$$

This expression shows that ΔS depends *only on the initial and final states and is independent of the reversible path.* Furthermore, ΔS can be positive or negative depending on whether the gas absorbs or expels heat during the process. Finally, for a cyclic process ($T_i = T_f$ and $V_i = V_f$), we see that $\Delta S = 0$.

EXAMPLE 22.6 Change in Entropy—Melting Process

A solid substance with a heat of fusion L_f melts at a temperature T_m. Calculate the change in entropy that occurs when m grams of this substance is melted.

Solution Let us assume that the melting process occurs so slowly that it can be considered a reversible process. In that case the temperature can be considered to be constant and equal to T_m. Making use of Equations 22.9 and the fact that the heat of fusion $Q = mL_f$, we find that

$$\Delta S = \int \frac{dQ_r}{T} = \frac{1}{T_m}\int dQ = \frac{Q}{T_m} = \frac{mL_f}{T_m} \tag{22.14}$$

Note that we were able to remove T_m from the integral in this case since the process is isothermal. Also, the quantity Q is the total heat required to melt the substance and is equal to mL_f (Section 20.3).

Exercise 4 Calculate the change in entropy when 0.30 kg of lead melts at 327°C. Lead has a heat of fusion equal to 24.5 kJ/kg.
Answer $\Delta S = 12.3$ J/K.

22.8 ENTROPY CHANGES IN IRREVERSIBLE PROCESSES

By definition, the change in entropy for a system can be calculated only for reversible paths connecting the initial and final equilibrium states. In order to calculate changes in entropy for real (irreversible) processes, we must first recognize that the entropy function (like internal energy) depends only on the *state* of the system. That is, entropy is a state function. Hence, the change in entropy of a system between any two equilibrium states depends only on the initial and final states. Experimentally one finds that the entropy change is the same for all processes between the initial and final states. It is possible to show that if this were not the case, the second law of thermodynamics would be violated.

In view of the fact that the entropy of a system depends only on the state of the system, we can now calculate entropy changes for irreversible processes between two equilibrium states. This can be accomplished by devising a reversible process (or series of reversible processes) between the same two equilibrium states and computing $\int dQ_r/T$ for the reversible process. The entropy change for the irreversible process is the same as that of the reversible process between the same two equilibrium states. Let us demonstrate this procedure with a few specific cases.

Heat Conduction

Consider the transfer of heat Q from a hot reservoir at temperature T_h to a cold reservoir at temperature T_c. Since the cold reservoir absorbs heat Q, its entropy increases by Q/T_c. At the same time, the hot reservoir loses heat Q and its entropy decreases by Q/T_h. The increase in entropy of the cold reservoir is greater than the decrease in entropy of the hot reservoir since T_c is less than T_h. Therefore, the total change in entropy of the system (universe) is greater than zero:

$$\Delta S_u = \frac{Q}{T_c} - \frac{Q}{T_h} > 0$$

EXAMPLE 22.7 Which Way Does the Heat Flow?
A cold reservoir is at 273 K, and a hotter reservoir is at 373 K. Show that it is impossible for a small amount of heat energy, say 8 J, to be transferred from the cold reservoir to the hot reservoir without decreasing the entropy of the universe and hence violating the second law.

Solution We assume here that during the heat transfer, the two reservoirs do not undergo a temperature change. The entropy change of the hot reservoir is

$$\Delta S_h = \frac{Q}{T_h} = \frac{8\text{ J}}{373\text{ K}} = 0.0214\text{ J/K}$$

The cold reservoir loses heat, and its entropy change is

$$\Delta S_c = \frac{Q}{T_c} = \frac{-8\text{ J}}{273\text{ K}} = -0.0293\text{ J/K}$$

The net entropy change of the universe is

$$\Delta S_u = \Delta S_c + \Delta S_h = -0.0079\text{ J/K}$$

This is in violation of the concept that the entropy of the universe always increases in natural processes. That is, *the spontaneous transfer of heat from a cold to a hot object cannot occur.*

Free Expansion

An ideal gas in an insulated container initially occupies a volume V_i (Fig. 22.13). A partition separating the gas from another evacuated region is suddenly broken so that the gas expands (irreversibly) to a volume V_f. Let us find the change in entropy of the gas and the universe.

The process is clearly neither reversible nor quasi-static. The work done by the gas against the vacuum is zero, and since the walls are insulating, no heat is transferred during the expansion. That is, $W = 0$ and $Q = 0$. Using the first law, we see that the change in internal energy is zero, therefore $U_i = U_f$, where i and f indicate the initial and final equilibrium states. Since the gas is ideal, U depends on temperature only, so we conclude that $T_i = T_f$.

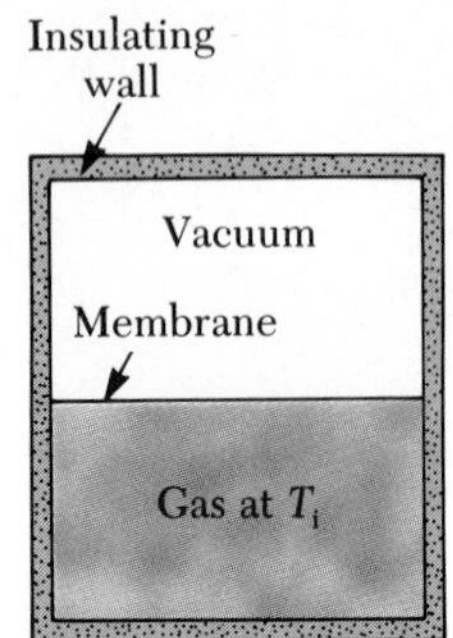

Figure 22.13 Free expansion of a gas. When the partition separating the gas from the evacuated region is ruptured, the gas expands freely and irreversibly so that it occupies a greater final volume. The container is thermally insulated from its surroundings, so $Q = 0$.

We cannot use Equation 22.9 directly to calculate the change in entropy since that equation applies only to reversible processes. In fact, at first sight one might *wrongly* conclude that $\Delta S = 0$ since there is no heat transferred. To calculate the change in entropy, let us imagine a reversible process between the same initial and final equilibrium states. A simple one to choose is an isothermal, reversible expansion in which the gas pushes slowly against a piston. Since T is constant in this process, Equation 22.8 gives

$$\Delta S = \int \frac{dQ_r}{T} = \frac{1}{T}\int_i^f dQ_r$$

But $\int dQ_r$ is simply the work done by the gas during the isothermal expansion from V_i to V_f, which is given by Equation 20.13. Using this result, we find that

$$\Delta S = nR \ln \frac{V_f}{V_i} \qquad (22.15)$$

Since $V_f > V_i$, we conclude that ΔS is positive, and so both the entropy and disorder of the gas (and universe) increase as a result of the irreversible, adiabatic expansion. This result can also be obtained from Equation 22.13, noting that $T_i = T_f$, so $\ln T_f/T_i = \ln(1) = 0$.

EXAMPLE 22.8 Free Expansion of a Gas

Calculate the change in entropy of 2 moles of an ideal gas that undergoes a free expansion to three times its initial volume.

Solution Using Equation 22.15 with $n = 2$ and $V_f = 3V_i$, we find that

$$\Delta S = nR \ln \frac{V_f}{V_i}$$

$$= (2 \text{ moles})(8.31 \text{ J/mole} \cdot \text{K}) \ln 3$$

$$= 18.3 \text{ J/K}$$

Irreversible Heat Transfer

A substance of mass m_1, specific heat c_1, and initial temperature T_1 is placed in thermal contact with a second substance of mass m_2, specific heat c_2, and initial temperature T_2, where $T_2 > T_1$. The two substances are contained in an insulated box so no heat is lost to the surroundings. The system is allowed to reach thermal equilibrium. What is the total entropy change for the system?

First, let us calculate the final equilibrium temperature, T_f. Energy conservation requires that the heat lost by one substance equal the heat gained by the other. Since by definition, $Q = mc\,\Delta T$ for each substance, we get $Q_1 = -Q_2$, or

$$m_1c_1\,\Delta T = -m_2c_2\,\Delta T$$

$$m_1c_1(T_f - T_1) = -m_2c_2(T_f - T_2)$$

Solving for T_f gives

$$T_f = \frac{m_1c_1T_1 + m_2c_2T_2}{m_1c_1 + m_2c_2} \qquad (22.16)$$

Note that $T_1 < T_f < T_2$, as would be expected.

The process is irreversible since the system goes through a series of nonequilibrium states. During such a transformation, the temperature at any time is not well defined. However, we can imagine that the hot body at the initial temperature T_i is slowly cooled to the temperature T_f by placing it in contact with a series of reservoirs differing infinitesimally in temperature, where the first reservoir is at the initial temperature T_i and the last is at T_f. Such a series of very small changes in temperature would approximate a reversible process. Applying Equation 22.9 and noting that $dQ = mc\,dT$ for an infinitesimal change, we get

$$\Delta S = \int_1 \frac{dQ_1}{T} + \int_2 \frac{dQ_2}{T} = m_1c_1 \int_{T_1}^{T_f} \frac{dT}{T} + m_2c_2 \int_{T_2}^{T_f} \frac{dT}{T}$$

where we have assumed that the specific heats remain constant. Integrating, we find

$$\Delta S = m_1c_1 \ln \frac{T_f}{T_1} + m_2c_2 \ln \frac{T_f}{T_2} \qquad (22.17)$$

Change in entropy for a heat transfer process

where T_f is given by Equation 22.16. If Equation 22.16 is substituted into Equation 22.17, you can show that one of the terms in Equation 22.17 will always be positive and the other negative. (You may want to verify this for yourself). However, the positive term will always be larger than the negative term, resulting in a positive value for ΔS. Thus, we conclude that the entropy of the universe (system) increases in this irreversible process.

Finally, you should note that Equation 22.17 is valid when two substances are placed in thermal contact with each other and no mixing occurs. If the substances are liquids, and mixing occurs, the result applies only if the two liquids are identical, as in the following example.

EXAMPLE 22.9 Calculating ΔS for a Mixing Process

One kg of water at 0°C is mixed with an equal mass of water at 100°C. After equilibrium is reached, the mixture has a uniform temperature of 50°C. What is the change in entropy of the system?

Solution The change in entropy can be calculated from Equation 22.17 using the values $m_1 = m_2 = 1$ kg, $c_1 = c_2 = 4186$ J/kg · K, $T_1 = 0°C$ (= 273 K), $T_2 = 100°C$ (= 373 K), and $T_f = 50°C$ (= 323 K).

$$\Delta S = m_1c_1 \ln \frac{T_f}{T_1} + m_2c_2 \ln \frac{T_f}{T_2}$$

$$= (1\text{ kg})(4186\text{ J/kg}\cdot\text{K}) \ln \left(\frac{323\text{ K}}{273\text{ K}}\right)$$

$$+ (1\text{ kg})(4186\text{ J/kg}\cdot\text{K}) \ln \left(\frac{323\text{ K}}{373\text{ K}}\right)$$

$$= 704\text{ J/K} - 602\text{ J/K} = 102\text{ J/K}$$

That is, the increase in entropy of the cold water is greater than the decrease in entropy of the warm water as a result of this irreversible process. Consequently, the increase in entropy of the system is 102 J/K.

The cases just described show that the change in entropy of a system is always positive for an irreversible process. In general, the total entropy (and disorder) always increases in irreversible processes. From these considerations, the second law of thermodynamics can be stated as follows: *The total entropy of an isolated system that undergoes a change cannot decrease.* Furthermore, if the process is *irreversible,* the total entropy of an isolated system always *increases.* On the other hand, in a reversible process, the total entropy of an isolated system remains constant. When dealing with interacting bodies that are not isolated, you must remember that the system refers to the bodies *and* their surroundings. When two bodies interact in an irreversible process, the increase in entropy of one part of the system is greater than the decrease in entropy of the other part. Hence, we conclude that

> the change in entropy of the universe must be greater than zero for an irreversible process and equal to zero for a reversible process.

Ultimately, the entropy of the universe should reach a maximum value. At this point, the universe would be in a state of uniform temperature and density. All physical, chemical, and biological processes would cease, since a state of perfect disorder implies no energy available for doing work. This gloomy state of affairs is sometimes referred to as an ultimate "heat death" of the universe.

*22.9 ENTROPY AND DISORDER

As you look around at the beauties of nature, it is easy to recognize that the events of natural processes have in them a large element of chance. For example, the spacing between trees in a natural forest is quite random. On the other hand, if you were to discover a forest where all the trees were equally spaced, you would probably conclude that the forest was man-made. Likewise, leaves fall to the ground with random arrangements. It would be highly unlikely to find the leaves laid out in perfectly straight rows or in one neat pile. We can express the results of such observations by saying that **a disorderly arrangement is much more probable than an orderly one if the laws of nature are allowed to act without interference.**

One of the main results of statistical mechanics is that **isolated systems tend toward disorder and entropy is a measure of that disorder.** In light of this new view of entropy, Boltzmann found that an alternative method for calculating entropy, S, is through use of the important relation

In real processes, the disorder in the system increases

$$S = k \ln W \tag{22.18}$$

where k is Boltzmann's constant, and W (not to be confused with work) is a number proportional to the probability of the occurrence of a particular event.

Let us explore the meaning of this equation by presenting a specific example. Imagine that you have a bag of 100 marbles, where 50 are red and 50 are green. You are allowed to draw four marbles from the bag according to the following rules. Draw one marble, record its color, return it to the bag and draw again. Continue this process until four marbles have been drawn. Note that because each marble is returned to the bag before the next one is drawn. the probability of drawing a red marble is always the same as that of drawing a green one. The results of all the possible drawing sequences that could occur are shown in Table 22.1. For example, the result RRGR means that you drew a red marble on the first draw, a red one on the second, a green one on the third, and a red one on the fourth. This table indicates that there is only one possible way to draw four red marbles. However, there are four possible sequences that could give one green and three red marbles, six sequences that could give

An illustration from Flammarion's novel *La Fin du Monde,* depicting the "heat-death" of the universe.

TABLE 22.1 Possible Results of Drawing Four Marbles from a Bag

End Result	Possible Draws	Total Number of Same Results
All R	RRRR	1
1G, 3R	RRRG, RRGR, RGRR, GRRR	4
2G, 2R	RRGG, RGRG, GRRG, RGGR, GRGR, GGRR	6
3G, 1R	GGGR, GGRG, GRGG, RGGG	4
All G	GGGG	1

two green and two red, four sequences that produce three green and one red, and one sequence that gives all green. Thus, the most likely occurrence is that you would draw two red and two green marbles, which corresponds to the disordered state. There is a much lower probability that you would draw four red and four green marbles, these being the most ordered states. From Equation 22.18, we see that the state with the greatest disorder has the highest entropy because it is the most probable state. On the other hand, the most ordered states (all red or all green) are least likely to occur, and are states of lowest entropy. In summary, we see that the outcome of the draw can range from a highly ordered state (say all red marbles), which has the lowest entropy, to a highly disordered state (two green and two red marbles), which has the highest entropy. Thus, one can regard entropy as an index of how far the system has progressed from an ordered to a disordered state.

As another example, suppose you were able to measure the velocities of all the air molecules in a room at some instant. It is very unlikely that you would find all molecules moving in the same direction with the same speed. This would, indeed, be a highly ordered state. The most probable situation you would find is a system of molecules moving haphazardly in all directions with a distribution of speeds. This is a highly disordered state and also the most likely. Let us compare this example to that of drawing marbles from a bag. Consider a container of gas consisting of 10^{23} molecules. If all of them were found moving in the same direction with the same speed at some instant, the outcome would be similar to drawing marbles from the bag 10^{23} times and finding a red marble on every draw. This is clearly an unlikely set of events.

Degradation of Energy

The tendency of nature to move toward a state of disorder affects the ability of a system to do work. Consider a ball thrown toward a wall. The ball has kinetic energy, and its state is an ordered one. That is, all of the atoms and molecules of the ball move in unison at the same speed and in the same direction (apart from their random thermal motions). When the ball hits the wall, however, part of this ordered energy is transformed to disordered energy. The temperature of the ball and the wall both increase slightly as part of the ball's kinetic energy is transformed into the random, disordered, thermal motion of the molecules in the ball and the wall. Before the collision, the ball is capable of doing work. It can drive a nail into the wall, for example. When part of the ordered energy is transformed to disordered thermal energy, this capability of doing work is reduced. That is, the ball rebounds with less kinetic energy than it had originally because the collision is inelastic.

Various forms of energy can be converted to thermal energy, as in the collision between the ball and the wall, but the reverse transformation is never complete. In general, if two kinds of energy, A and B, can be completely interconverted, we say that they are the *same grade*. However, if form A can be completely converted to form B and the reverse is never complete, then form A has a *higher grade* of energy than form B. In the case of a ball hitting a wall, the kinetic energy of the ball is of a higher grade than the thermal energy contained in the ball and the wall after the collision. Therefore, high-grade energy which is converted to thermal energy can never be fully recovered as high-grade energy.

This conversion of high-grade energy to thermal energy (the source of thermal pollution) is referred to as the *degradation of energy. The energy is said to be degraded because it takes on a form that is less useful for doing work.* In other words, *in all real processes where heat transfer occurs, the energy available for doing work decreases.*

*22.10 ENERGY CONVERSION AND THERMAL POLLUTION

The main source of thermal pollution is waste heat from electrical power plants. In the United States, about 85% of the electric power is produced by steam engines, which burn either fossil fuels (coal, oil, or natural gas) or nuclear fuels (uranium-235). The remaining 15% of the electric power is generated by water in hydroelectric plants. The overall thermal efficiency of a modern fossil-fuel plant is about 40%. The actual efficiencies of any power plant must be lower than the theoretical efficiencies derived from the second law of thermodynamics. One always seeks the highest efficiency possible for two reasons. First, higher efficiency results in lower fuel costs. Second, thermal pollution of the environment is reduced since there is less waste energy in a highly efficient power plant. Since any power plant involves several steps of energy conversion, the inefficiency will accumulate in steps.

The burning of fossil fuels in an electrical power plant involves three energy-conversion processes: (1) chemical to thermal energy, (2) thermal to mechanical energy, and (3) mechanical to electrical energy. These are indicated schematically in Figure 22.14.

During the first step, heat energy is transferred from the burning fuel to a water supply, which is converted into steam. In this process about 12% of the available energy is lost up the chimney. In the second step, thermal energy in the form of steam at high pressure and temperature passes through a turbine and is converted into mechanical energy. A well-designed turbine has an efficiency of about 47%. Steam, which leaves the turbine at a lower pressure, is then condensed into water and gives up heat in the process. Finally, in the third step, the turbine drives an electrical generator of very high efficiency, typically 99%. Hence, the overall efficiency is the product of the efficiencies of each step, which for the figures given is $(0.88)(0.47)(0.99) = 0.41$, or 41%. The thermal energy transferred to the cooling water amounts to about 47% of the initial fuel energy.

In the case of nuclear power plants, the steam generated by the nuclear reactor is at a lower temperature than that of a fossil-fuel plant. This is due primarily to material limitations in the reactor. Typical water-moderated nuclear power plants have an overall efficiency of about 34%. High temperature

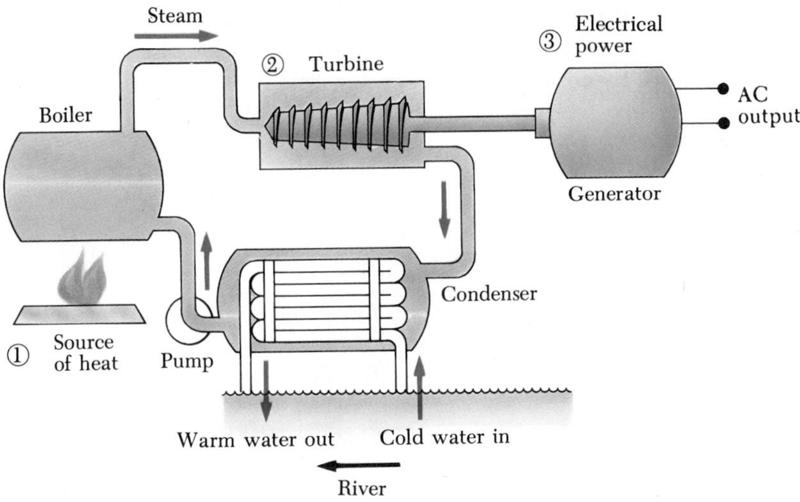

Figure 22.14 Schematic diagram of an electrical power plant.

gas-cooled reactors operate at temperatures and efficiencies comparable to fossil-fuel plants.

The waste heat from electrical power plants can be disposed of in various ways. The method shown in Figure 22.14 involves passing water from a river or lake through a condenser and returning it to that source at a higher temperature. This can raise the water temperature of the river or lake by several degrees, which can produce undesirable ecological effects, such as the increased growth of bacteria, undesirable blue-green algae, and pathogenic organisms. Fish and other marine life are also affected since they require oxygen, and the percentage of dissolved oxygen in the water decreases with increasing temperature. There is further demand for oxygen in the decomposition of organic matter, which also proceeds at a higher rate as the temperature increases.

Cooling towers are also commonly used in disposing of waste heat. These towers usually use the heat to evaporate water, which is then released to the atmosphere. Cooling towers also present environmental problems since evap-

A cooling tower at a nuclear reactor site in Oregon. (Courtesy of U.S. Department of Energy)

orated water can cause increased precipitation, fog, and ice. Another type of cooling tower is the dry cooling tower (nonevaporative), which transfers heat to the atmosphere by conduction. However, this type is more expensive and cannot cool to as low a temperature as an evaporative tower.

SUMMARY

The **first law of thermodynamics** is a generalization of the law of conservation of energy that includes heat transfer in any process.

Real processes proceed in a direction governed by the second law of thermodynamics.

A **heat engine** is a device that converts thermal energy into other useful forms of energy. The net work done by a heat engine in carrying a substance through a cyclic process ($\Delta U = 0$) is given by

Work done by a heat engine

$$W = Q_h - Q_c \tag{22.1}$$

where Q_h is the heat absorbed from a warmer reservoir and Q_c is the heat rejected to a cooler reservoir.

The **thermal efficiency,** e, of a heat engine is defined as the ratio of the net work done to the heat absorbed per cycle:

Thermal efficiency

$$e = \frac{W}{Q_h} = 1 - \frac{Q_c}{Q_h} \tag{22.2}$$

The **second law of thermodynamics** can be stated in many ways:

Statements of the second law

1. No heat engine operating in a cycle can absorb thermal energy from a reservoir and perform an equal amount of work (Kelvin-Planck statement).
2. A perpetual-motion machine of the second kind cannot be built.
3. It is impossible to construct a cyclical machine whose sole effect is to transfer heat continuously from one body to another body at a higher temperature (Clausius statement).

Reversible process

A process is **reversible** if the system passes from the initial to the final state through a succession of equilibrium states. A process can be reversible only if it occurs quasi-statically.

Irreversible process

An **irreversible process** is one in which the system and its surroundings cannot be returned to their initial states. In such a process, the system passes from the initial to the final state through a series of nonequilibrium states.

The *efficiency of a heat engine* operating in the **Carnot cycle** is given by

Efficiency of a Carnot engine

$$e_c = 1 - \frac{T_c}{T_h} \tag{22.4}$$

where T_c is the absolute temperature of the cold reservoir and T_h is the absolute temperature of the hot reservoir.

No real heat engine operating (irreversibly) between the temperatures T_c and T_h can be more efficient than an engine operating reversibly in a Carnot cycle between the same two temperatures.

The second law of thermodynamics states that when real (irreversible) processes occur, the degree of disorder in the system increases. When a process occurs in an isolated system, ordered energy is converted into disordered energy. The measure of disorder in a system is called **entropy,** S.

The **change in entropy,** dS, of a system moving quasi-statically between two equilibrium states is given by

Clausius definition of change in entropy

$$dS = \frac{dQ_r}{T} \tag{22.8}$$

The change in entropy of a system moving reversibly between two equilibrium states is

$$\Delta S = \int_i^f \frac{dQ_r}{T} \tag{22.9}$$

The value of ΔS is the same for all reversible paths connecting the initial and final states.

The change in entropy for any reversible, cyclic process is zero.

In any reversible process, the entropy of the universe remains constant.

The entropy of a system is a state function, that is, it depends on the state of the system. The change in entropy for a system undergoing a real (irreversible) process between two equilibrium states is the same as that of a reversible process between the same states.

In an irreversible process, the total entropy of an isolated system always increases. In general, the total entropy (and disorder) always increases in any irreversible process. Furthermore, the change in entropy of the universe is greater than zero for an irreversible process.

QUESTIONS

1. Distinguish clearly among temperature, heat, and internal energy.
2. When a sealed Thermos bottle full of hot coffee is shaken, what are the changes, if any, in (a) the temperature of the coffee and (b) its internal energy?
3. Use the first law of thermodynamics to explain why the total energy of an isolated system is always conserved.
4. Is it possible to convert internal energy to mechanical energy?
5. What are some factors that affect the efficiency of automobile engines?
6. The statement was made in this chapter that the first law says we cannot get more out of a process than we put in but the second law says that we cannot break even. Explain.
7. Is it possible to cool a room by leaving the door of a refrigerator open? What happens to the temperature of a room in which an air conditioner is left running on a table in the middle of the room?
8. In practical heat engines, which do we have more control of, the temperature of the hot reservoir or the temperature of the cold reservoir? Explain.
9. A steam-driven turbine is one major component of an electric power plant. Why is it advantageous to increase the temperature of the steam as much as possible?
10. Is it possible to construct a heat engine that creates no thermal pollution?
11. Electrical energy can be converted to heat energy with an efficiency of 100%. Why is this number misleading with regard to heating a home? That is, what other factors must be considered in comparing the cost of electric heating with the cost of hot air or hot water heating?
12. Discuss three common examples of natural processes that involve an increase in entropy. Be sure to account for all parts of each system under consideration.
13. Discuss the change in entropy of a gas that expands (a) at constant temperature and (b) adiabatically.

14. A living system, such as a tree, takes unorganized molecules (CO_2, H_2O) and combines them using sunlight to produce leaves and branches. Is this reduction of entropy in the tree a violation of the second law of thermodynamics?
15. A diesel engine operates with a compression ratio of about 15. A gasoline engine has a compression ratio of about 6. Which engine runs hotter? Which engine (at least theoretically) can operate more efficiently?
16. Solar ponds have been constructed in Israel where the sun's energy is concentrated near the bottom of a salty pond. With the proper layering of salt in the water, convection is prevented, and temperatures of 100°C may be reached. Can you make a guess as to the maximum efficiency with which useful energy can be extracted from the pond?
17. The vortex tube (Fig. 22.15) is a T-shaped device that takes in compressed air at 20 atm and 20°C and produces cold air at −20°C out one flared end and +60°C hot air out the other flared end. Does the operation of this device violate the second law of thermodynamics? (For further information, see R. Hilsch, *Rev. Sci. Instr.* **18**, 108, 1947 or Y. Soni and W. Thompson, *Trans. A.S.M.E.*, May 1975, p. 316.)

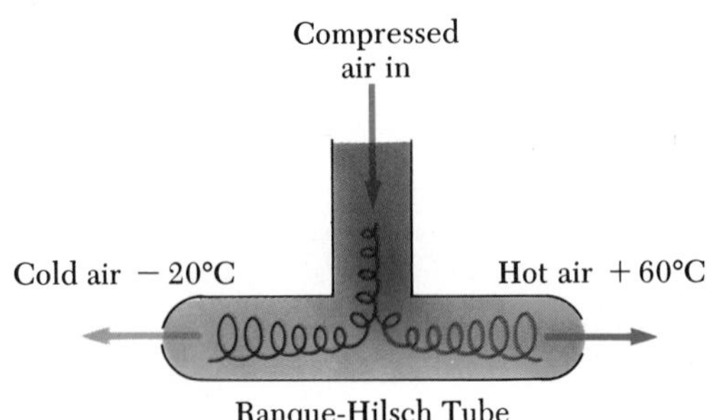

Figure 22.15 (Question 17).

18. Why does your automobile burn more gas in winter than in summer?
19. Can a heat pump have a coefficient of performance less than one? Explain your answer.
20. All natural processes are irreversible. Give some examples of irreversible processes that occur in nature.
21. Give an example of a process in nature that is nearly reversible.
22. A thermodynamic process occurs in which the entropy of a system changes by −8.0 J/K. According to the second law of thermodynamics, what can you conclude about the entropy change of the environment?
23. If a supersaturated sugar solution is allowed to slowly evaporate, sugar crystals form in the container. Hence, sugar molecules go from a disordered form (in solution) to a highly-ordered crystalline form. Does this process violate the second law of thermodynamics? Explain.
24. How could you increase the entropy of one mole of a metal that is at room temperature? How could you decrease its entropy?
25. A heat pump is to be installed in a region where the average outdoor temperature in the winter months is −20°C. In view of this, why would it be advisable to place the outdoor compressor unit deep into the ground? Why are heat pumps not commonly used for heating in cold climates?
26. Suppose your roommate is "Mr. Clean" who tidies up your messy room after a big party. Since more order is being created by your roommate, does this represent a violation of the second law of thermodynamics?
27. Discuss the entropy changes that occur when you (a) bake a loaf of bread, and (b) consume the bread.
28. If you shake a jar full of jelly beans of different sizes, the larger jelly beans tend to appear near the top, while the smaller ones tend to fall to the bottom. Why does this occur? Does the process violate the second law of thermodynamics?
29. The device in the photographs below, called a thermoelectric converter, uses a series of semiconductor cells to convert thermal energy into electrical energy. In the photograph at the left, both "legs" of the device are at the same temperature, and no electrical energy is produced. However, when one leg is at a higher temperature than the other, as in the photograph on the right, electrical energy is produced as the device extracts energy from the hot reservoir and drives a small electric motor. (a) Why does the temperature differential produce electrical energy in this demonstration? (b) In what sense does this intriguing experiment demonstrate the second law of thermodynamics?

Question 29 (Courtesy of PASCO Scientific Co.)

PROBLEMS

Section 22.1 Heat Engines and the Second Law of Thermodynamics

1. A heat engine absorbs 360 J of heat and performs 25 J of work in each cycle. Find (a) the efficiency of the engine and (b) the heat expelled in each cycle.
2. A heat engine performs 200 J of work in each cycle and has an efficiency of 30%. For each cycle of operation, (a) how much heat is absorbed and (b) how much heat is expelled?
3. A refrigerator has a coefficient of performance equal to 5. If the refrigerator absorbs 120 J of heat from a cold reservoir in each cycle, find (a) the work done in each cycle and (b) the heat expelled to the hot reservoir.
4. A particular engine has a power output of 5 kW and an efficiency of 25%. If the engine expels 8000 J of heat in each cycle, find (a) the heat absorbed in each cycle and (b) the time for each cycle.
5. The heat absorbed by an engine is three times greater than the work it performs. (a) What is its thermal efficiency? (b) What fraction of the heat absorbed is expelled to the cold reservoir?
6. In each cycle of its operation, a certain refrigerator absorbs 100 J from the cold reservoir and expels 130 J. (a) What is the power required to operate the refrigerator if it works at 60 cycles/s? (b) What is the coefficient of performance of the refrigerator?
7. An engine absorbs 1600 J from a hot reservoir and expels 1000 J to a cold reservoir in each cycle. (a) What is the efficiency of the engine? (b) How much work is done in each cycle? (c) What is the power output of the engine if each cycle lasts for 0.3 s?

Section 22.3 The Carnot Engine

8. A heat engine operates between two reservoirs at temperatures of 20°C and 300°C. What is the maximum efficiency possible for this engine?
9. The efficiency of a Carnot engine is 30%. The engine absorbs 800 J of heat per cycle from a hot reservoir at 500 K. Determine (a) the heat expelled per cycle and (b) the temperature of the cold reservoir.
10. A Carnot engine has a power output of 150 kW. The engine operates between two reservoirs at 20°C and 500°C. (a) How much heat energy is absorbed per hour? (b) How much heat energy is lost per hour?
11. A power plant has been proposed that would make use of the temperature gradient in the ocean. The system is to operate between 20°C (surface water temperature) and 5°C (water temperature at a depth of about 1 km). (a) What is the maximum efficiency of such a system? (b) If the power output of the plant is 75 MW, how much thermal energy is absorbed per hour? (c) In view of your results to (a), do you think such a system is worthwhile?
12. A heat engine operates in a Carnot cycle between 80°C and 350°C. It absorbs 2×10^4 J of heat per cycle from the hot reservoir. The duration of each cycle is 1 s. (a) What is the maximum power output of this engine? (b) How much heat does it expel in each cycle?
13. One of the most efficient engines ever built operates between 430°C and 1870°C. Its actual efficiency is 42%. (a) What is its maximum theoretical efficiency? (b) How much power does the engine deliver if it absorbs 1.4×10^5 J of heat each second?
14. In a steam turbine, steam at 800°C enters and is exhausted at 120°C. What is the (maximum) efficiency of this turbine?
15. The efficiency of a 1000 MW nuclear power plant is 33%; that is, 2000 MW of heat is rejected to the environment for every 1000 MW of electrical energy produced. If a river of flow rate 10^6 kg/s were used to transport the excess heat away, what would be the average temperature increase of the river?
16. An electrical generating plant has a power output of 500 MW. The plant uses steam at 200°C and exhausts water at 40°C. If the system operates with one half the maximum (Carnot) efficiency, (a) at what rate is heat expelled to the environment? (b) If the waste heat goes into a river whose flow rate is 1.2×10^6 kg/s, what is the rise in temperature of the river?
17. An air conditioner absorbs heat from its cooling coil at 13°C and expels heat to the outside at 30°C. (a) What is the *maximum* coefficient of performance of the air conditioner? (b) If the actual coefficient of performance is one third of the maximum value and if the air conditioner removes 8×10^4 J of heat energy each second, what power must the motor deliver?
18. A heat pump powered by an electric motor absorbs heat from outside at 5°C and exhausts heat inside in the form of hot air at 40°C. (a) What is the maximum coefficient of performance of the heat pump? (b) If the actual coefficient of performance is 3.2, what *fraction* of the theoretical maximum work (electrical energy) is actually done?
19. An ideal gas is taken through a Carnot cycle. The isothermal expansion occurs at 250°C, and the isothermal compression takes place at 50°C. If the gas absorbs 1200 J of heat during the isothermal expansion, find (a) the heat expelled to the cold reservoir in each cycle and (b) the net work done by the gas in each cycle.

Section 22.5 The Gasoline Engine

20. A gasoline engine has a compression ratio of 6 and uses a gas with $\gamma = 1.4$. (a) What is the efficiency of the engine if it operates in an idealized Otto cycle? (b) If the actual efficiency is 15%, what fraction of the fuel is wasted as a result of friction and avoidable heat losses? (Assume complete combustion of the air-fuel mixture.)
21. A gasoline engine using an ideal diatomic gas ($\gamma = 1.4$) operates between temperature extremes of 300 K and 1500 K. Determine its compression ratio if it has an efficiency of 20%. Compare this efficiency to that of a Carnot engine operating between the same temperatures.
22. In a cylinder of an automobile engine, just after combustion, the gas is confined to a volume of 50 cm^3, and has an initial pressure of 3×10^6 N/m^2. The piston moves outward to a final volume of 300 cm^3 and the gas expands without heat loss (adiabatic). If $\gamma = 1.40$ for the gas, what is the final pressure?
23. How much work is done by the gas in Problem 22 in expanding from $V_1 = 50$ cm^3 to $V_2 = 300$ cm^3?

Section 22.6 Heat Pumps and Refrigerators

24. An ideal refrigerator (or heat pump) is equivalent to a Carnot engine running in reverse. That is, heat Q_c is absorbed from a cold reservoir and heat Q_h is rejected to a hot reservoir. (a) Show that the work that must be supplied to run the refrigerator is given by

$$W = \frac{T_h - T_c}{T_c} Q_2$$

(b) Show that the coefficient of performance of the ideal refrigerator is given by

$$\text{COP} = \frac{T_c}{T_h - T_c}$$

25. What is the coefficient of performance of a refrigerator that operates with Carnot efficiency between temperatures -3°C and $+27$°C?
26. What is the coefficient of performance of a heat pump that brings heat from the out-of-doors at -3°C into a $+22$°C house? (*Hint:* the heat pump does work W, which is also available to heat the house.)
27. How much work is required, using an ideal Carnot refrigerator, to remove 1 J of heat energy from helium gas at 4 K and reject this heat to a room-temperature (293 K) environment?

Section 22.7 Entropy

28. What is the change in entropy (ΔS) when one mole of silver (108 g) is melted at 961°C?
29. An avalanche of ice and snow, mass 1000 kg, slides downhill a vertical distance of 200 m. What is the total entropy change if the mountain air temperature is -3°C?
30. What is the entropy decrease in 1 mole of helium gas which is cooled at 1 atm from room temperature 293 K to a final temperature 4 K? (C_p of helium = 21 J/mole · K).
31. Calculate the change in entropy of 250 g of water when it is slowly heated from 20°C to 80°C. (*Hint:* Note that $dQ = mc\, dT$.)
32. An ice tray contains 500 g of water at 0°C. Calculate the change in entropy of the water as it freezes completely and slowly at 0°C.
33. One mole of an ideal monatomic gas is heated quasi-statically at constant volume from 300 K to 400 K. What is the change in entropy of the gas?
34. One kg of mercury is initially at -100°C. What is its change in entropy when heat is slowly added to raise its temperature to 100°C? (Mercury has a heat of fusion of 1.17×10^4 J/kg, a melting temperature of -39°C, and a specific heat of 138 J/kg · C°.)

Section 22.8 Entropy Changes in Irreversible Processes

35. A 1500-kg car traveling at 20 m/s crashes into a concrete wall. If the air temperature is 20°C, what is the total entropy change?
36. The surface of the sun is approximately 5700 K and the temperature of the earth's surface is about 290 K. What entropy change occurs when 1000 J of heat energy is transferred from the sun to the earth?
37. A 100 000-kg iceberg at -5°C breaks away from the polar iceshelf and floats away into the ocean at $+5$°C. What is the final change in the entropy of the system, when the iceberg has completely melted? (The specific heat of ice is 2010 J/kg · C°.)
38. One mole of H_2 gas is contained in the left-hand-side of the container (Figure 22.16, equal volumes left and right). The right-hand-side is evacuated. When the valve is opened, hydrogen gas streams into the right side. What is the final entropy change? Does the temperature of the gas change?

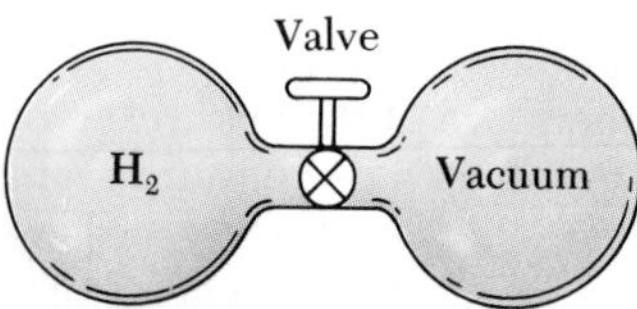

Figure 22.16 (Problem 38).

39. A 2-liter container has a center partition that divides it into two equal parts, as shown in Figure 22.17. The left side contains H_2 gas and the right side contains O_2 gas. Both gases are at room temperature and 1 atmo-

sphere. The partition is removed and the gases are allowed to mix. What is the entropy increase?

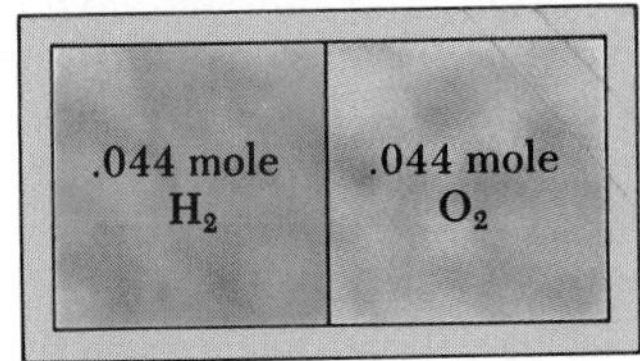

Figure 22.17 (Problem 39).

40. If 3200 J of heat flows from a heat reservoir at 500 K to another reservoir at 300 K through a conducting metal rod, find the change in entropy of (a) the hot reservoir, (b) the cold reservoir, (c) the metal rod, and (d) the universe.

41. A 2-kg block moving with an initial speed of 5 m/s slides on a rough table and is stopped by the force of friction. Assuming the table and air remain at a temperature of 20°C, calculate the entropy change of the universe.

42. A 70-kg log falls from a height of 25 m into a lake. If the log, the lake, and the air are all at 300 K, find the change in entropy of the universe for this process.

43. A 6-kg block of ice at 0°C is dropped into a lake at 27°C. Just after the ice has all melted, and before the ice water has had a chance to warm, what is the change in entropy of (a) the ice, (b) the lake, and (c) the universe?

44. A cyclic heat engine operates between two reservoirs at temperatures of 300 K and 500 K. In each cycle, the engine absorbs 700 J of heat from the hot reservoir and does 160 J of work. Find the entropy change in each cycle for (a) each reservoir, (b) the engine, and (c) the universe.

45. If 200 g of water at 20°C is mixed with 300 g of water at 75°C, find (a) the final equilibrium temperature of the mixture and (b) the change in entropy of the system.

ADDITIONAL PROBLEMS

46. An 18-gram ice cube at 0°C is heated until it vaporizes as steam. (a) How much does the entropy increase? (See Table 20.2.) (b) How much energy was required to vaporize the ice cube?

47. An engine operates in a cycle between the temperatures $T_1 = 180°\text{C}$ and $T_2 = 100°\text{C}$, and exhausts 2×10^4 J of heat per cycle while doing 1.5×10^3 J of work per cycle. Compare the actual efficiency of the engine with that of a reversible engine operating between the same temperatures.

48. What is the minimum amount of work that must be done to extract 400 J of heat from a massive object at 0°C while rejecting heat to a hot reservoir at a temperature of 20°C?

49. A house loses heat through the exterior walls and roof at a rate of 5000 J/s = 5 kW when the interior temperature is 22°C and the outside temperature is −5°C. Calculate the electric power required to maintain the interior temperature at 22°C for the following two cases: (a) The electric power is used in electric resistance heaters (which convert all of the electricity supplied into heat). (b) The electric power is used to drive an electric motor that operates the compressor of a heat pump (which has a coefficient of performance equal to 60 percent of the Carnot cycle value).

50. A 0.30-kg piece of aluminum at a temperature of 85°C is placed in 0.50 L of water at a temperature of 25°C. The system is thermally isolated. Determine the increase in entropy of the system when it has reached equilibrium.

51. One mole of an ideal monatomic gas is taken through the cycle shown in Figure 22.18. The process *AB* is a reversible isothermal expansion. Calculate (a) the net work done by the gas, (b) the heat added to the gas, (c) the heat expelled by the gas, and (d) the efficiency of the cycle.

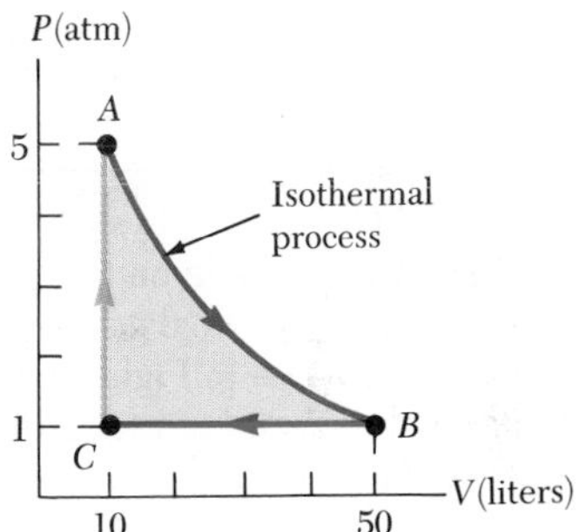

Figure 22.18 (Problem 51).

52. An athlete whose mass is 70 kg drinks 16 ounces (453.6 g) of refrigerated water. The water is at a temperature of 35°F. (a) Neglecting the temperature change of the body resulting from the water intake (so that the body is regarded as a reservoir at 98.6°F), find the entropy increase of the entire system. (b) Assume that the entire body is cooled by the drink and that the average specific heat of a human is equal to the specific heat of liquid water. Neglecting any other heat transfers and any metabolic heat release, find the athlete's temperature after drinking the cold water, given an initial body temperature of 98.6°F. Under *these* assumptions, what is the entropy increase of the entire system? Compare your result with that of (a).

53. Figure 22.19 represents *n* moles of an ideal monatomic gas being taken through a reversible cycle consisting of two isothermal processes at temperatures $3T_0$ and T_0 and two constant-volume processes. For each cycle, determine in terms of n, R, and T_0 (a) the net heat transferred to the gas and (b) the efficiency of an engine operating in this cycle.

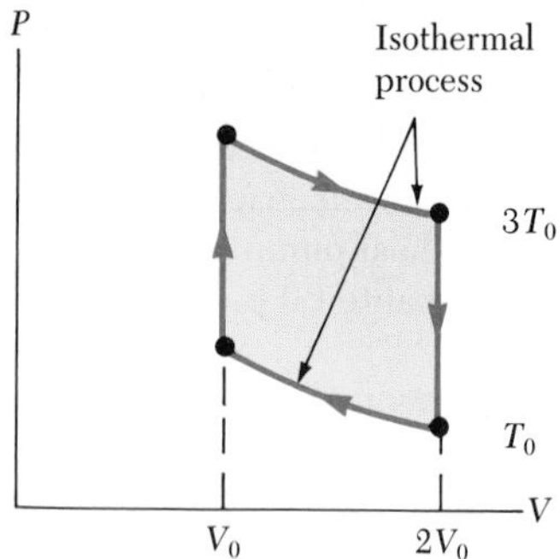

Figure 22.19 (Problem 53).

54. (a) Show that the work done by a system in any reversible cycle is given by $W = \oint T\,dS$, where the integral is over the closed path corresponding to the cyclic process. (*Hint:* Make use of the first law, $dQ = dU + P\,dV$, and the definition of the change in entropy as given by Equation 22.9.) (b) Construct an entropy-temperature plot for the Carnot cycle described in Figure 22.9. Identify the process associated with each line on the graph. What is the physical significance of the area enclosed by the cycle on this plot?

55. One mole of a monatomic ideal gas is taken through the reversible cycle shown in Figure 22.20. At point A, the pressure, volume, and temperature are P_0, V_0, and T_0, respectively. In terms of R and T_0, find (a) the total heat entering the system per cycle, (b) the total heat leaving the system per cycle, (c) the efficiency of an engine operating in this reversible cycle, and (d) the efficiency of an engine operating in a Carnot cycle between the same temperature extremes for this process.

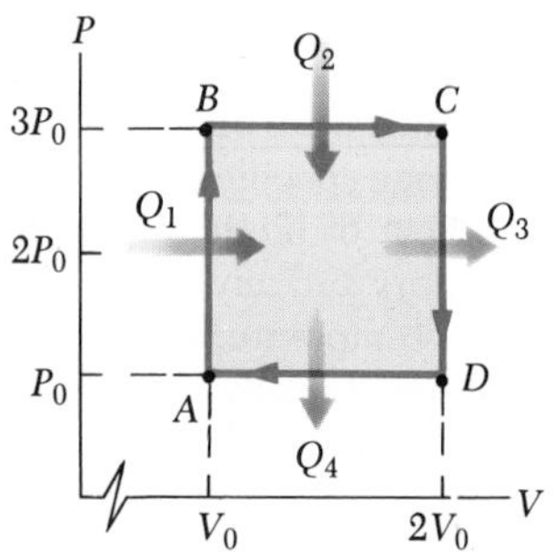

Figure 22.20 (Problem 55).

56. One mole of a diatomic ideal gas at an initial pressure of 4 atm and temperature of 300 K is carried through the following reversible cycle: (1) it expands isothermally until its volume is doubled; (2) it is compressed to its original volume at constant pressure; (3) it is compressed isothermally to a pressure of 4 atm; and (4) it expands at constant pressure to its original volume. (a) Make an accurate plot of the cyclic process on a PV diagram. (b) Calculate the work done by the gas per cycle. (c) Find the gross heat input and efficiency of the engine.

57. A system consisting of n moles of an ideal gas undergoes a reversible, *isobaric* process from a volume V_0 to a volume $3V_0$. Calculate the change in entropy of the gas. (*Hint:* Imagine that the system goes from the initial state to the final state first along an isotherm and then along an adiabatic curve, for which there is no change in entropy.)

58. An electrical power plant has an overall efficiency of 15%. The plant is to deliver 150 MW of power to a city, and its turbines use coal as the fuel. The burning coal produces steam at 190°C, which drives the turbines. This steam is then condensed into water at 25°C by passing it through cooling coils in contact with river water. (a) How many metric tons of coal does the plant consume each day (1 metric ton = 10^3 kg)? (b) What is the total cost of the fuel per year if the delivered price is \$8/metric ton? (c) If the river water is delivered at 20°C, at what minimum rate must it flow over the cooling coils in order that its temperature not exceed 25°C? (*Note:* The heat of combustion of coal is 33 kJ/g.)

59. Two heat engines have efficiencies of e_1 and e_2. The engines operate such that the heat rejected by the engine of efficiency e_1 is used as the input heat for the engine of efficiency e_2. Show that the overall efficiency is given by $e = e_1 + e_2 - e_1 e_2$. (Note that the overall efficiency is equal to the total work done by both engines divided by the heat input to the engine of efficiency e_1.)

60. Two Carnot engines are connected to the same heat reservoir at a temperature of 900 K. The first engine rejects its heat to a reservoir at 400 K, while the second engine rejects its heat to a reservoir at 600 K. Assuming that each engine absorbs the same amount of heat, Q_h, from the hot reservoir, find the overall efficiency of the system. (Note that the overall efficiency is equal to the *total* work done divided by the *total* heat absorbed from the hot reservoir, $2Q_h$.)

61. An idealized Diesel engine operates in a cycle known as the *air-standard Diesel cycle*, shown in Figure 22.21. Fuel is sprayed into the cylinder at the point of maximum compression, B. Combustion occurs during the expansion $B \rightarrow C$, which is approximated as an isobaric process. The rest of the cycle is the same as in

the gasoline engine, described in Figure 22.11. Show that the efficiency of an engine operating in this idealized Diesel cycle is given by

$$e = 1 - \frac{1}{\gamma}\left(\frac{T_D - T_A}{T_C - T_B}\right)$$

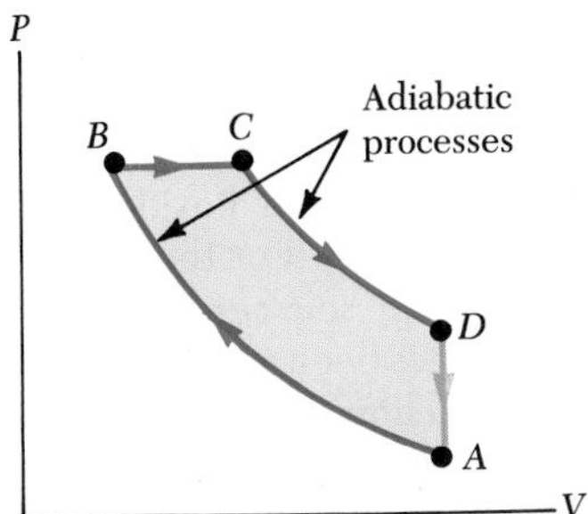

Figure 22.21 (Problem 61).

62. One mole of an ideal gas ($\gamma = 1.4$) is carried through the Carnot cycle described in Figure 22.9. At point A, the pressure is 25 atm and the temperature is 600 K. At point C, the pressure is 1 atm and the temperature is 400 K. (a) Determine the pressures and volumes at points A, B, C, and D. (b) Calculate the net work done per cycle. (c) Determine the efficiency of an engine operating in this cycle.

63. Consider one mole of an ideal gas that undergoes a quasi-static, reversible process in which the heat capacity C remains constant. Show that the pressure and volume of the gas obey the relation

$$PV^{\gamma'} = \text{constant}, \qquad \text{where} \qquad \gamma' = \frac{C - C_p}{C - C_v}$$

(*Hint:* Start with the first law of thermodynamics, $dQ = dU + P\,dV$, and use the fact that $dQ = C\,dT$ and $dU = C_v\,dT$. Furthermore, note that $PV = RT$ applies to the gas, so it follows that $P\,dV + V\,dP = R\,dT$.)

64. A typical human has a mass of 70 kg and produces about 2000 kcal (2×10^6 cal) of metabolic heat per day. (a) Find the rate of heat production in watts, and also express it in cal/h. (b) If none of the metabolic heat were lost, and assuming that the specific heat of the human body is 1 cal/g · C°, find the rate at which body temperature would rise. Give your answer in C° per hour and in F° per hour. (c) The result of (b) indicates that a human who is unable to transfer metabolic heat to the surroundings gets into serious trouble quickly. Assuming now that the heat transfer is rapid enough to maintain a constant body temperature, estimate the rate at which the entropy of the universe increases due to the release and transfer of metabolic heat. Express your result in cal/K/h. In making your estimate, assume that the given heat release is the only significant avenue of entropy production. Given this assumption, notice that there is no entropy increase of the body itself. (Why?) Use 20°C (68°F) for the temperature of the surroundings.

65. A blacksmith plunges a hot iron horseshoe of mass 2 kg into a bucket containing 20 kg of water. The horseshoe is initially at a temperature of 600°C, and the water is initially at a temperature of 20°C. Assuming that no water is vaporized, find: (a) the final temperature of the water and the horseshoe, (b) the change of entropy of the horseshoe, (c) the change of entropy of the water, and (d) the overall change of entropy of the water and the horseshoe. (e) After a time interval, which is long compared with the time for the horseshoe to cool, the horseshoe and the water cool back to the temperature of the surroundings: 20°C. During this process, find the entropy changes of the water, the horseshoe, and the surroundings. (f) Using your results for (d) and (e), find the entropy change of the universe as a result of the entire sequence of events.

66. Consider once more the situation described in Problem 65. A somewhat more realistic assumption is that water is heated to 100°C and vaporized in an amount sufficient to cool the horseshoe to 100°C, and thereafter the horseshoe equilibrates with the remaining water. Under this assumption: (a) How much water is vaporized? (b) What is the final temperature of the horseshoe and the remaining liquid water? (c) Identify where entropy changes occur and whether they are increases or decreases. (You need not give quantitative estimates for those changes.)

"He was working on a theory of entropy, and developed a severe case of it himself."

ESSAY

Alternative Sources of Energy

Laurent Hodges
Iowa State University

Today, most of the world's current energy is obtained from fossil fuels (coal, petroleum, and natural gas) and uranium. Despite their many advantages, these fuels have some significant disadvantages. All have become more costly in recent years; new uranium power plants have become so expensive that none have been built since the late 1970s. Petroleum and natural gas are becoming scarce, with consumption exceeding discoveries, and cannot maintain their dominant position as energy sources for more than a few more decades; the production of these fuels in the United States actually peaked in the early 1970s. Furthermore, all these energy sources have significant environmental disadvantages, particularly air pollution. The most serious environmental problem is the global warming (often referred to as the "greenhouse effect") associated with the increasing amounts of carbon dioxide in the atmosphere; all fossil fuel combustion contributes to this problem. For these and other reasons, society has begun to investigate alternative energy sources.

The term "alternative sources" has sometimes been applied to new fuels derived from old sources, such as hydrocarbon gases and liquids derived from coal or fissionable plutonium derived from uranium-238. However, the term is usually reserved for new primary energy sources which have not been in widespread use in recent years. These include geothermal energy, fusion energy, and some forms of solar energy.

Geothermal Energy

One alternative source is underground heat, or geothermal energy. From a thermodynamic point of view, this energy source exploits the difference between the temperature of underground rocks or water and that of the surface of the earth. The higher underground temperatures are a source of thermal heat, which can be used directly or to generate electricity.

Geothermal energy exists everywhere on earth, but is being used only in places where it is available near the surface (within a few kilometers). These are generally places with natural reservoirs of geothermal steam or hot water. Geothermal heat is being extracted for use as a direct heat source in Iceland (particularly for home heating), in Boise, Idaho and in Klamath Falls, Oregon. Electricity has also been generated from geothermal energy for many years at Larderello, Italy, The Geysers, California, and several other places.

Geothermal energy is not limited to regions with steam or hot water. Research has been carried out at Los Alamos National Laboratory, New Mexico, with the aim of extracting geothermal heat from hot dry rocks deep underground. Two wells are dug, one to bring cold water down from the surface, and the other to bring hot water back to the surface. The underground rocks are fractured with pressurized water, and the water that passes through the fractures is heated. This particular form of geothermal energy would be useful in many parts of the world.

Fusion Energy

Nuclear power plants in use around the world today derive their energy from the fissioning of heavy nuclei such as uranium and plutonium. An alternative is the production of energy from the fusing of light nuclei into heavier nuclei of lower total mass. The decrease in mass in the fusion process appears as energy from Einstein's relationship $E = mc^2$.

An example is the fusion of hydrogen nuclei into a helium nucleus, a process which is the source of the radiant energy in sunlight and most starlight, as well as the energy source in thermonuclear or "hydrogen" bombs.

Because nuclei are positively charged and thus electrically repel each other, fusion reactions require extremely high temperatures—high enough to allow the kinetic energy of the nuclei to overcome their mutual repulsion and collide. It is for this reason that fusion reactions are referred to as thermonuclear reactions.

There is an abundance on earth of naturally occurring isotopes suitable for use in fusion energy reactions, notably deuterium (^2H) and lithium. These could, in theory, produce millions of times as much energy as all the world's fossil fuels combined. The difficulty is in creating conditions suitable for fusion. The temperatures must be very high, the plasma of ionized nuclei must be sufficiently dense, and the length of time during which nuclear reactions occur must be long enough for fusion to take place. Two approaches that have been extensively studied involve confinement of the plasma using magnetic fields or laser radiation.

Controlled fusion has not yet been demonstrated in the laboratory, much less been proved economical, and is unlikely to become a significant source of energy—if ever—until well into the 21st century. Even then, the cost may be too high. Recent claims of successful cold fusion in electrolytic cells are intriguing. However, they seem to contradict some fundamental notions of fusion and require careful verification.

Forms of Solar Energy

The major alternative energy source is the sun. Solar energy comes to the earth in the form of electromagnetic radiation, mainly in the visible and near infrared parts of the spectrum.

Solar energy, broadly defined, includes a great variety of forms which can be classified as shown in Table 1. In addition to direct solar radiation, which can be used to heat or cool buildings, to heat water for domestic use, or to produce electricity, there are indirect forms of solar energy such as water, wind, and photosynthetic power. These are properly classified as solar energy because it is solar radiation that drives the hydrological cycle from which water power is derived and the circulation in the atmosphere from which wind power is derived, and that provides the energy for photosynthesis

Solar energy is often regarded as a possible energy source of the future. In fact, solar energy was the major energy source for humanity throughout almost all of recorded and unrecorded history. Agriculture depended until the 20th century on beasts of burden fed on solar-derived food. The great explorers all used solar energy (in the form of wind energy) on their sea voyages around the world. In the United States, the industrial revolution was powered by solar power—in the form of me-

TABLE 1 Forms and Uses of Solar Energy

Direct Solar Radiation	Indirect Solar Energy
Active solar systems	Water power
Wintertime space heating	Hydroelectricity
Water heating	Mechanical power
Space cooling (air conditioning)	Wind power
Process heat	Ocean transportation
Passive solar space heating	Mechanical power (windmills)
Electric generation systems	Wind-generated electricity
Photovoltaic systems	Photosynthetic power
Solar power towers	Solid fuels (wood)
	Liquid fuels (methanol, ethanol)
	Gaseous fuels (such as methane)
	Ocean thermal energy conversion

chanical water and wind power and wood fuel. Fossil fuels, originally in the form of coal and later in the form of petroleum and natural gas, did not become as important as solar energy in the United States until the 1880s.

Solar energy is still a significant energy source in the United States, ranking behind each of the three fossil fuels but ahead of nuclear energy or other energy sources. Currently solar energy accounts for approximately 10% of the energy used in the United States. The major uses of solar power are hydroelectricity, passive solar energy, and photosynthetic materials (particularly wood).

Hydroelectricity

Hydroelectricity perhaps should not be properly regarded as an alternative energy source because it already accounts for about 10% of the electricity generated in the United States — nearly as much as nuclear fission plants. Hydroelectricity is an interesting example of several steps in energy conversion:

1. Electromagnetic radiation from the sun is absorbed by the oceans and thereby converted into sensible heat.
2. The sensible heat is converted into the latent heat of evaporated water, which is transported throughout the atmosphere and rains down in mountainous regions.
3. The gravitational potential energy of the water is converted into kinetic energy as the water flows down to sea level.
4. The kinetic energy of the water is converted into kinetic energy at the hydroelectricity turbines at the generating plant.
5. The kinetic energy of the turbines is converted into electric energy at the turbine-generator.
6. The electric energy is transported to the end users, who may convert it into heat or light (electromagnetic radiation) or mechanical energy, depending on its intended use.

Where available, hydroelectric power is the cheapest form of electric power. It is a major source of electric systems in several countries (such as Canada and Norway) and in several parts of the United States (such as the Pacific Northwest and in the region served by the Tennessee Valley Authority).

Hydroelectric power has a major environmental impact, since its use requires the damming of a river and often the creation of a large reservoir. Although many good hydroelectric sites remain, most are unlikely to be developed because vast ecosystems would be significantly affected.

Solar Heating of Buildings

The most cost-effective use of solar energy is for space heating, or the winter heating of homes and other buildings. Space heating is the number one use of energy in American homes. Space-heating requirements can be drastically reduced by energy conservation methods such as good wall and roof insulation, multi-pane windows, and air infiltration control by good caulking and weatherstripping. Solar energy can be used to meet a substantial portion of the remaining heating requirements.

A solar heating system requires three major components:

1. *Collection* The function of the collector is to collect the solar radiation and convert it into heat.
2. *Storage* Solar energy comes only during the daylight hours, often at a rate greater than can be immediately used. The function of the storage system is to store the excess heat for later use, such as at night or on a cloudy day.
3. *Distribution* The function of the distribution system is to move the solar heat from where it is to where it is wanted. On a sunny day, the solar heat is distributed from the collector to the living space and, if there is an excess, to storage. During the

night or on a cloudy day, the solar heat must be moved from storage to the living space.

Active Solar Heating

Active solar heating systems use mechanical means to distribute the solar energy. In active air systems, fans and ducts are used to move solar heat in the form of heated air; in active liquid systems, pumps and pipes are used to move solar heat in the form of heated water or antifreeze.

The conversion to active space-heating systems was subsidized by an energy-conscious government in the late 1970s. The systems consisted of a collector, an absorber, and storage. The collector normally consisted of a flatplate collector of one or two panes of glass, behind which was a black absorber. In air systems, the storage was usually provided by a rockbed (a mass of uniformly sized rocks about the size of a small room), and in liquid systems, by a large water tank.

Active space-heating systems, however, were rarely successful. They tended to be large, complex, and expensive, often costing $10,000 or more for a house of typical size. They required expert design and many of their parts required continual maintenance. They also tended to be very conspicuous, which made the homes appear unconventional.

Passive Solar Heating

Passive solar systems have proved much more successful, particularly in new homes where they add little or nothing to the cost. Passive solar systems are those in which the energy transfers are by natural means: natural conduction, natural convection, and natural radiation. Passive solar heating systems typically consist of a large expanse of south-facing windows (which act as the collector) together with interior masonry to serve as thermal storage by providing a large heat capacity. Concrete, stone, brick, and adobe have been used as interior masonry, in the form of walls, floors, or interior partitions. When masonry walls act as thermal storage, they must be insulated on the exterior, not on the interior. Occasionally, water-filled containers have been used instead.

Two common passive solar space heating systems are the direct gain and thermal storage wall systems. In a direct gain system the sun shines directly into the living space through the windows. As long as there is sufficient heat capacity to the thermal storage, the house will not overheat during the day and will cool down very slowly at night. Direct gain homes are well-lit by natural light during the day. As long as there is good air circulation throughout the home, natural distribution of energy is satisfactory, and no active system is required.

A thermal storage wall system has a concrete wall, typically about 30 cm (12 inches) thick, located just behind the south windows. The sunlight shines directly on the wall and is absorbed by its black surface. The wall acts as storage and also distribution, since the heat absorbed is eventually radiated back into the living space behind the wall.

There are several other types of passive heating systems. One popular system is the sunspace system, in which the main collection and storage of solar heat occurs in a separate south-facing room. Sunspaces can be used as sunporches, greenhouses, or swimming pool rooms.

Passive heating systems work very well in many parts of the United States, particularly in areas that have cold but fairly sunny winters. They have been found to work well in Arizona, New Mexico, Colorado, Wyoming, Iowa, Minnesota, and elsewhere. It is not uncommon for well-insulated passive solar homes in cold northern climates to be heated for a whole winter for only $100 to $200 worth of natural gas heat, and my own passive solar home in Iowa requires less natural gas for heating than the family uses for domestic water heating.

Photosynthetic Energy

Photosynthetic energy is a major source of energy in many homes—those that use wood for space heating. Wood and agricultural wastes are sometimes used to generate electricity or industrial process heat, usually by small industries. It is worth remembering that wood was the main energy source in the United States during its first century. Not until the 1880s did coal supplant wood as the leading energy source, and wood remained a significant source through the rest of the 19th century.

Photosynthetic energy in the form of dried animal dung and "biogas" (methane generated from animal wastes) are important energy sources in many rural areas of Asia.

Photovoltaic Cells

Photovoltaic or "solar" cells are made of semiconductor materials that develop a voltage when sunlight strikes them. Originally developed for the space program to provide satellites in orbit with a source of power, they are now widely available on small electronic devices such as solar calculators, but are also sometimes used to power remote installations, traffic lights and electric fences.

A house with photovoltaic cells covering a south-facing roof could collect enough electric energy to meet daily needs but would require a battery system in order to store the energy for use at night. Such systems have been installed on some homes, usually for $20,000 or more, but they are not currently cost-effective except in remote areas where the cost of electric transmission to the home would be excessive.

Research is under way to produce better photovoltaic cells, mainly cells that are higher in efficiency or cheaper to manufacture. It is hoped that they will become economical for many purposes before the end of the century. A more detailed discussion of photovoltaic devices is given in the essay that follows Chapter 43 of the extended version of this text.

Ocean Thermal Energy Conversion

Ocean thermal energy conversion (OTEC) is the production of electricity using the small temperature difference between warm ocean surface water and cold deep ocean water. An OTEC plant would need to use a working substance other than steam; ammonia is one possibility. Although the temperature differences are only a few degrees and the plant efficiency would be very low, very large quantities of water and heat are available, and such a plant might be cost-effective.

A significant problem with OTEC is the transport of the energy back to shore. In some cases electrical cables could be used. Another possibility is the use of the electricity to electrolyze water, with the hydrogen shipped back to shore for use as a fuel.

The OTEC concept has been tested on and off for many decades. The Atlantic Ocean off Florida and the Pacific Ocean near Hawaii look like good locations for such a plant, and research has been carried out in both places. However, OTEC plants are currently far from being technologically and economically feasible and may never actually prove commercially successful.

Power Towers

The solar power tower or central receiver is an electric power plant located at the top of a large tower: its heat source is solar radiation reflected to the tower by a large field of heliostats (mirrors) surrounding it. The heliostats must be computer-controlled so they can reflect the sunlight to the proper place accurately and yet be defocused quickly wherever an emergency arises.

This concept has been thoroughly researched, and in 1982 an experimental version known as "Solar One" went into operation at Barstow, California. Solar One is a small (10 MW) solar electric plant with a steam boiler situated in a field of 1,818 heliostats. The total land area covered is 130 acres.

r towers has not yet been established. The best
e United States would be in the southern states.

nanity in many ways in the past, is currently under
city. Many wind generators were installed on mid-
930s, and a surprising number of them are still in
ıerating systems are commercially available today;
that would provide electricity to several homes or a

higher-than-average wind speeds, a modern wind
a cost higher than the current price of coal-gener-
cost of some nuclear-generated electricity. A wind
ective in the short run but could prove economical
ore.
have been built by utilities in the past, and since the
new designs for large machines have been developed
ully. Several "wind farms" have been built in the
gher-than-average wind speeds, often by knowledge-
ke a good return on their investment over the lifetime
ussion of wind power is provided in Section 15.10 of

, which currently provide most of the world's energy, Society may want to phase them out more quickly to avoid the global warming or a "greenhouse effect" or to otherwise improve environmental quality. Fortunately, there are a number of alternative energy sources available. Many of them (such as most of the solar technologies) are technically feasible today, and some of them (such as passive solar space heating in sunny northern regions) are already economically feasible.

Essay Questions

1. What enables glass to act as a solar collector? Would transparent plastic materials such as Plexiglas work as well?
2. Passive solar homes normally use double- or triple-pane windows. Why not single-pane windows?
3. Does a solar collector have to face exactly south?

Essay Problems

1. The solar radiation reaching the earth, above the atmosphere, averages about 1350 W/m^2. Using the facts that the sun is about 150 million kilometers away and the earth has a diameter of 12.7 thousand kilometers, find: (a) The total power (in watts) of the solar radiation reaching the earth, (b) the total power (in watts) of the solar radiation emitted by the sun, assuming it is emitted equally in all directions, (c) the rate (in kilograms per second) at which the sun is losing mass, making use of Einstein's relation $E = mc^2$, and (d) the fraction of the sun's total mass of 2.0×10^{30} kg being lost each year.
2. At a midlatitude in the United States, the average amount of solar radiation falling on a tilted south-facing roof is about 200 W/m^2, averaging over the whole year, day and night, sunny weather and cloudy. Assuming that a homeowner uses solar cells which cover 80% of the roof and have an efficiency of 10% at converting solar radiation into electricity, how large an area (in square meters) would be needed to produce a total of 6000 kWh of electric energy? Is this a reasonable area to find on a real roof?

This dramatic one-minute exposure captures multiple lightning bolts illuminating Kitt Peak National Observatory in Arizona, illustrating electrical breakdown in the atmosphere. (© Gary Ladd 1972)

PART IV
Electricity and Magnetism

For the sake of persons of . . . different types, scientific truth should be presented in different forms, and should be regarded as equally scientific, whether it appears in the robust form and the vivid coloring of a physical illustration, or in the tenuity and paleness of a symbolic expression.

JAMES CLERK MAXWELL

We now begin the study of that branch of physics which is concerned with electric and magnetic phenomena. The laws of electricity and magnetism play a central role in the operation of various devices such as radios, televisions, electric motors, computers, high-energy accelerators, and a host of electronic devices used in medicine. However, more fundamentally, we now know that the interatomic and intermolecular forces that are responsible for the formation of solids and liquids are electric in origin. Furthermore, such forces as the pushes and pulls between objects and the elastic force in a spring arise from electric forces at the atomic level.

Evidence in Chinese documents suggests that magnetism was known as early as around 2000 B.C. The ancient Greeks observed electric and magnetic phenomena possibly as early as 700 B.C. They found that a piece of amber, when rubbed, becomes electrified and attracts pieces of straw or feathers. The existence of magnetic forces was known from observations that pieces of a naturally occurring stone called *magnetite* (Fe_3O_4) are attracted to iron. (The word *electric* comes from the Greek word for amber, *elecktron*. The work *magnetic* comes from the name of a northern central district of Greece where magnetite was found, *Magnesia*.)

In 1600, William Gilbert discovered that electrification was not limited to amber but is a general phenomenon. Scientists went on to electrify a variety of objects, including chickens and people! Experiments by Charles Coulomb in 1785 confirmed the inverse-square force law for electricity.

It was not until the early part of the 19th century that scientists established that electricity and magnetism are, in fact, related phenomena. In 1820, Hans Oersted discovered that a compass needle is deflected when placed near a circuit carrying an electric current. In 1831, Michael Faraday, and almost simultaneously, Joseph Henry, showed that, when a wire is moved near a magnet (or, equivalently, when a magnet is moved near a wire), an electric current is observed in the wire. In 1873, James Clerk Maxwell used these observations and other experimental facts as a basis for formulating the laws of electromagnetism as we know them today. (*Electromagnetism* is a name given to the combined fields of electricity and magnetism.) Shortly thereafter (around 1888), Heinrich Hertz verified Maxwell's predictions by producing electromagnetic waves in the laboratory. This was followed by such practical developments as radio and television.

Maxwell's contributions to the science of electromagnetism were especially significant because the laws he formulated are basic to *all* forms of electromagnetic phenomena. His work is comparable in importance to Newton's discovery of the laws of motion and the theory of gravitation.

23
Electric Fields

Photograph of a carbon filament incandescent lamp. Thomas Edison's first light bulb also used a piece of carbonized cotton thread as its filament. The carbon filament lamp emits a different light spectrum than that produced by a tungsten lamp because of its composition and because it operates at a lower temperature. (Courtesy of CENCO)

The electromagnetic force between charged particles is one of the fundamental forces of nature. In this chapter, we begin by describing some of the basic properties of electrostatic forces. We then discuss Coulomb's law, which is the fundamental law of force between any two charged particles. The concept of an electric field associated with a charge distribution is then introduced, and its effect on other charged particles is described. The method for calculating electric fields of a given charge distribution from Coulomb's law is discussed, and several examples are given. Then the motion of a charged particle in a uniform electric field is discussed. We conclude the chapter with a brief discussion of the oscilloscope.

23.1 PROPERTIES OF ELECTRIC CHARGES

A number of simple experiments can be performed to demonstrate the existence of electrical forces and charges. For example, after running a comb through your hair, you will find that the comb will attract bits of paper. The attractive force is often strong enough to suspend the pieces of paper. The same effect occurs with other rubbed materials, such as glass or rubber.

Biographical Sketch

Charles Coulomb
(1736–1806)

Charles Coulomb, the great French physicist after whom the unit of electric charge called the *coulomb* was named, was born in Angoulême in 1736. He was educated at the École du Génie in Mézieres, graduating in 1761 as a military engineer with a rank of First Lieutenant. Coulomb served in the West Indies for nine years, where he supervised the building of fortifications in Martinique.

In 1774, Coulomb became a correspondent to the Paris Academy of Science. There he shared the Academy's first prize for his paper on magnetic compasses and also received first prize for his classic work on friction, a study that was unsurpassed for 150 years. During the next 25 years, he presented 25 papers to the Academy on electricity, magnetism, torsion, and applications to the torsion balance, as well as several hundred committee reports on engineering and civil projects.

Coulomb took full advantage of the various positions he held during his lifetime. For example, his experience as an engineer led him to investigate the strengths of materials and determine the forces that affect objects on beams, thereby contributing to the field of structural mechanics. He also contributed to the field of ergonomics. His research provided a fundamental understanding of the ways in which people and animals can best do work and greatly influenced the subsequent research of Gaspard Coriolis (1792–1843).

Coulomb's major contribution to science was in the field of electrostatics and magnetism, in which he made use of the torsion balance he developed (see Fig. 23.2). The paper describing this invention also contained a design for a compass using the principle of torsion suspension. His next paper gave proof of the inverse square law for the electrostatic force between two charges.

Coulomb died in 1806, five years after becoming president of the Institut de France (formerly the Paris Academy of Science). His research on electricity and magnetism brought this area of physics out of traditional natural philosophy and made it an exact science.

(Photograph courtesy of AIP Niels Bohr Library, E. Scott Barr Collection)

Another simple experiment is to rub an inflated balloon with wool. The balloon will then adhere to the wall or the ceiling of a room, often for hours. When materials behave in this way, they are said to be *electrified*, or to have become **electrically charged.** You can easily electrify your body by vigorously rubbing your shoes on a wool rug. The charge on your body can be sensed and removed by lightly touching (and startling) a friend. Under the right conditions, a visible spark is seen when you touch one another, and a slight tingle will be felt by both parties. (Experiments such as these work best on a dry day, since an excessive amount of moisture can lead to a leakage of charge from the electrified body to the earth by various conducting paths.)

In a systematic series of rather simple experiments, one finds that there are two kinds of electric charges, which were given the names **positive** and **negative** by Benjamin Franklin (1706–1790). To demonstrate this fact, consider a hard rubber rod that has been rubbed with fur and then suspended by a nonmetallic thread as in Figure 23.1. When a glass rod that has been rubbed with silk is brought near the rubber rod, the rubber rod is attracted toward the glass rod. On the other hand, if two charged rubber rods (or two charged glass rods) are brought near each other, as in Figure 23.1b, the force between them will be repulsive. This observation shows that the rubber and glass are in two different states of electrification. On the basis of these observations, we conclude that *like charges repel one another and unlike charges attract one another.*

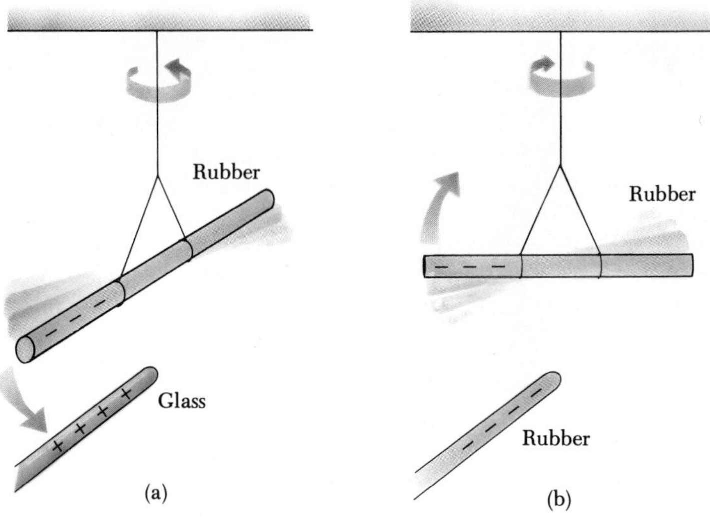

Figure 23.1 (a) A negatively charged rubber rod, suspended by a thread, is attracted to a positively charged glass rod. (b) A negatively charged rubber rod is repelled by another negatively charged rubber rod.

Using the convention suggested by Franklin, the electric charge on the glass rod is called *positive*, and that on the rubber rod is called *negative*. Therefore any charged body that is attracted to a charged rubber rod (or repelled by a charged glass rod) must have a positive charge. Conversely, any charged body that is repelled by a charged rubber rod (or attracted to a charged glass rod) has a negative charge on it.

Charge is conserved

Another important aspect of Franklin's model of electricity is the implication that *electric charge is always conserved*. That is, when one body is rubbed against another, charge is not created in the process. The electrified state is due to a *transfer* of charge from one body to the other. Therefore, one body gains some amount of negative charge while the other gains an equal amount of positive charge. For example, when a glass rod is rubbed with silk, the silk obtains a negative charge that is equal in magnitude to the positive charge on the glass rod. We now know from our understanding of atomic structure *that it is the negatively charged electrons that are transferred* from the glass to the silk in the rubbing process. Likewise, when rubber is rubbed with fur, electrons are transferred from the fur to the rubber, giving the rubber a net negative charge and the fur a net positive charge. This is consistent with the fact that neutral, uncharged matter contains as many positive charges (protons within atomic nuclei) as negative charges (electrons).

Charge is quantized

In 1909, Robert Millikan (1886–1953) discovered that electric charge always occurs as some integral multiple of some fundamental unit of charge, e. In modern terms, the charge q is said to be **quantized.** That is, electric charge exists as discrete "packets." Thus, we can write $q = Ne$, where N is some integer. Other experiments in the same period showed that the electron has a charge $-e$ and the proton has an equal and opposite charge, $+e$. Some elementary particles, such as the neutron, have no charge. A neutral atom must contain as many protons as electrons.

Electric forces between charged objects were measured quantitatively by Coulomb using the torsion balance, which he invented (Fig. 23.2). Using this apparatus, Coulomb confirmed that the electric force between two small charged spheres is proportional to the inverse square of their separation, that is, $F \propto 1/r^2$. The operating principle of the torsion balance is the same as that of the apparatus used by Cavendish to measure the gravitational constant (Section 14.2), with masses replaced by charged spheres. The electric force between the charged spheres produces a twist in the suspended fiber. Since the restoring torque of the twisted fiber is proportional to the angle through which it rotates, a measurement of this angle provides a quantitative measure of the electric force of attraction or repulsion. If the spheres are charged by rubbing, the electrical force between the spheres is very large compared with the gravitational attraction; hence the gravitational force can be neglected.

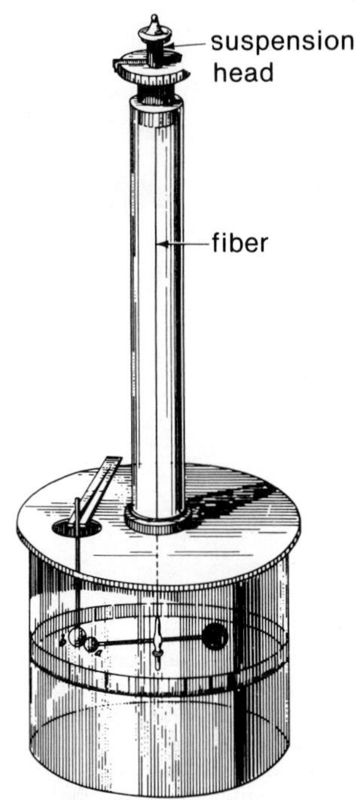

Figure 23.2 Coulomb's torsion balance, which was used to establish the inverse-square law for the electrostatic force between two charges. (Taken from Coulomb's 1785 memoirs to the French Academy of Sciences.)

From our discussion thus far, we conclude that electric charge has the following important properties:

1. There are two kinds of charges in nature, with the property that unlike charges attract one another and like charges repel one another.
2. The force between charges varies as the inverse square of their separation.
3. Charge is conserved.
4. Charge is quantized.

23.2 INSULATORS AND CONDUCTORS

It is convenient to classify substances in terms of their ability to conduct electrical charge.

> **Conductors** are materials in which electric charges move quite freely, whereas **insulators** are materials that do not readily transport charge.

Materials such as glass, rubber, and lucite fall into the category of insulators. When such materials are charged by rubbing, only the area that is rubbed becomes charged and the charge is unable to move to other regions of the material.

Metals are good conductors

In contrast, materials such as copper, aluminum, and silver are good conductors. When such materials are charged in some small region, the charge readily distributes itself over the entire surface of the conductor. If you hold a copper rod in your hand and rub it with wool or fur, it will not attract a small piece of paper. This might suggest that a metal cannot be charged. On the other hand, if you hold the copper rod by a lucite handle and then rub, the rod will remain charged and attract the piece of paper. This is explained by noting that in the first case, the electric charges produced by rubbing will readily move from copper through your body and finally to earth. In the second case, the insulating lucite handle prevents the flow of charge to earth.

Semiconductors are a third class of materials, and their electrical properties are somewhere between those of insulators and conductors. Silicon and germanium are well-known examples of semiconductors commonly used in the fabrication of a variety of electronic devices. The electrical properties of semiconductors can be changed over many orders of magnitude by adding controlled amounts of certain foreign atoms to the materials.

Charging by induction

When a conductor is connected to earth by means of a conducting wire or copper pipe, it is said to be **grounded.** The earth can then be considered an infinite "sink" to which electrons can easily migrate. With this in mind, we can understand how to charge a conductor by a process known as **induction.**

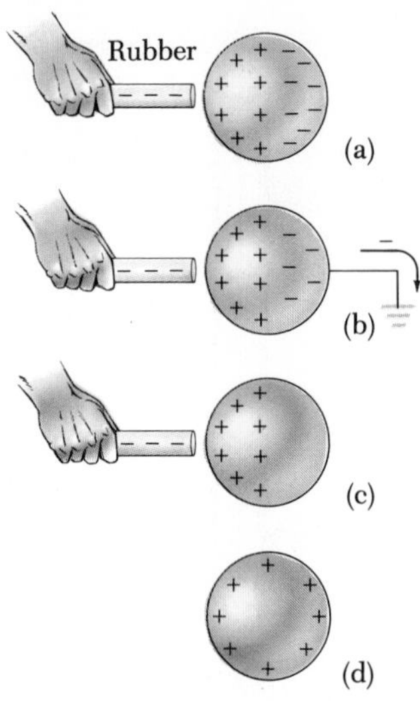

Figure 23.3 Charging a metallic object by induction. (a) The charge on a neutral metallic sphere is redistributed when a charged rubber rod is placed near the sphere. (b) The sphere is grounded, and some of the electrons leave the conductor. (c) The ground connection is removed, and the sphere has a nonuniform positive charge. (d) When the rubber rod is removed, the sphere becomes uniformly charged.

To understand induction, consider a negatively charged rubber rod brought near a neutral (uncharged) conducting sphere insulated from ground. That is, there is no conducting path to ground (Fig. 23.3a). The region of the sphere nearest the negatively charged rod will obtain an excess of positive charge, while the region of the sphere farthest from the rod will obtain an equal excess of negative charge. (That is, electrons in the part of the sphere nearest the rod migrate to the opposite side of the sphere.) If the same experiment is performed with a conducting wire connected from the sphere to ground (Fig. 23.3b), some of the electrons in the conductor will be repelled to earth. If the wire to ground is then removed (Fig. 23.3c), the conducting sphere will contain an excess of *induced* positive charge. Finally, when the rubber rod is removed from the vicinity of the sphere (Fig. 23.3d), the induced positive charge remains on the ungrounded sphere. Note that the charge remaining on the sphere is uniformly distributed over its surface because of the repulsive forces among the like charges. In the process, the electrified rubber rod loses none of its negative charge.

Thus, we see that charging an object by induction requires no contact with the body inducing the charge. This is in contrast to charging an object by rubbing (that is, charging by *conduction*), which does require contact between the two objects.

A process which is very similar to that of charging by induction in conductors also takes place in insulators. In most neutral atoms or molecules the center of positive charge coincides with the center of negative charge. However, in the presence of a charged object, these centers may shift slightly, resulting in more positive charge on one side of the molecule than on the other. This effect, known as **polarization,** will be discussed more completely in Chapter 26. This realignment of charge within individual molecules produces an induced charge on the surface of the insulator as shown in Figure 23.4. With these ideas, you should be able to explain why a comb that has been rubbed through hair will attract bits of neutral paper, or why a balloon that has been rubbed against your clothing is able to stick to a neutral wall.

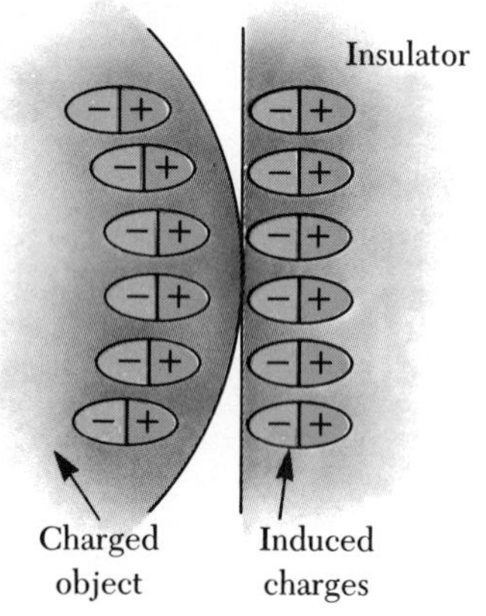

Figure 23.4 The charged object on the left induces charges on the surface of an insulator at the right.

23.3 COULOMB'S LAW

In 1785, Coulomb established the fundamental law of electric force between two stationary, charged particles. Experiments show that an **electric force** has the following properties: (1) The force is inversely proportional to the square of the separation, r, between the two particles and is directed along the line joining the particles. (2) The force is proportional to the product of the charges q_1 and q_2 on the two particles. (3) The force is attractive if the charges are of opposite sign and repulsive if the charges have the same sign. From these observations, we can express the magnitude of the electric force between the two charges as

Coulomb's law

$$F = k\frac{|q_1||q_2|}{r^2} \tag{23.1}$$

where k is a constant called the *Coulomb constant.* In his experiments, Coulomb was able to show that the exponent of r was 2 to within an uncertainty of a few percent. Modern experiments have shown that the exponent is 2 to a precision of a few parts in 10^9.

The constant k in Equation 23.1 has a value that depends on the choice of units. The unit of charge in SI units is the coulomb (C). The coulomb is defined in terms of a unit current called the *ampere* (A), where current equals the rate of flow of charge. (The ampere will be defined in Chapter 27.) When the current in a wire is 1 A, the amount of charge that flows past a given point in the wire in 1 s is 1 C. The Coulomb constant k in SI units has the value

$$k = 8.9875 \times 10^9 \ \mathrm{N \cdot m^2/C^2} \tag{23.2}$$

Coulomb constant

To simplify our calculations, we shall use the approximate value

$$k \cong 9.0 \times 10^9 \ \mathrm{N \cdot m^2/C^2} \tag{23.3}$$

The constant k is also written

$$k = \frac{1}{4\pi\epsilon_0}$$

where the constant ϵ_0 is known as the *permittivity of free space* and has the value

$$\epsilon_0 = 8.8542 \times 10^{-12} \ \mathrm{C^2/N \cdot m^2} \tag{23.4}$$

The smallest unit of charge known in nature is the charge on an electron or proton.[1] The charge of an electron or proton has a magnitude

$$|e| = 1.60219 \times 10^{-19} \ \mathrm{C} \tag{23.5}$$

Charge on an electron or proton

Therefore, 1 C of charge is equal to the charge of 6.3×10^{18} electrons (that is, $1/e$). This can be compared with the number of free electrons in 1 $\mathrm{cm^3}$ of copper,[2] which is of the order of 10^{23}. Note that 1 C is a substantial amount of charge. In typical electrostatic experiments, where a rubber or glass rod is charged by friction, a net charge of the order of 10^{-6} C (= 1 μC) is obtained. In other words, only a very small fraction of the total available charge is transferred between the rod and the rubbing material.

The charges and masses of the electron, proton, and neutron are given in Table 23.1.

When dealing with Coulomb's force law, you must remember that force is a *vector* quantity and must be treated accordingly. Furthermore, note that *Coulomb's law applies exactly only to point charges or particles.* If $\hat{\mathbf{r}}$ is taken to

[1] No unit of charge smaller than e has been detected as a free charge; however, some recent theories have proposed the existence of particles called *quarks* having charges $e/3$ and $2e/3$. Although there is experimental evidence for such particles inside nuclear matter, *free* quarks have never been detected. We shall discuss other properties of quarks in Chapter 47 of the extended version of this text.

[2] A metal atom, such as copper, contains one or more outer electrons, which are weakly bound to the nucleus. When many atoms combine to form a metal, the so-called free electrons are these outer electrons, which are not bound to any one atom. These electrons move about the metal in a manner similar to gas molecules moving in a container.

TABLE 23.1 Charge and Mass of the Electron, Proton, and Neutron

Particle	Charge (C)	Mass (kg)
Electron (e)	$-1.6021917 \times 10^{-19}$	9.1095×10^{-31}
Proton (p)	$+1.6021917 \times 10^{-19}$	1.67261×10^{-27}
Neutron (n)	0	1.67492×10^{-27}

Figure 23.5 Two point charges separated by a distance r exert a force on each other given by Coulomb's law. Note that the force on q_1 is equal to and opposite the force on q_2. (a) When the charges are of the same sign, the force is repulsive. (b) When the charges are of the opposite sign, the force is attractive.

be a unit vector from charge q_2 to charge q_1, as in Figure 23.5, then the electric force on q_1 due to q_2, written $\boldsymbol{F}_{12}$, can be expressed in vector form as

$$\boldsymbol{F}_{12} = k\frac{q_1 q_2}{r^2}\hat{\boldsymbol{r}}_{12} \tag{23.6}$$

where $\hat{\boldsymbol{r}}_{12}$ is a unit vector directed from q_1 to q_2 as in Figure 23.5a. Since Coulomb's law obeys Newton's third law, the electric force on q_2 due to q_1 is equal in magnitude to the force on q_1 due to q_2 and in the opposite direction, that is, $\boldsymbol{F}_{21} = -\boldsymbol{F}_{12}$. Finally, from Equation 23.6 we see that if q_1 and q_2 have the same sign, the product q_1q_2 is positive and the force is repulsive, as in Figure 23.5a. On the other hand, if q_1 and q_2 are of opposite sign, as in Figure 23.5b, the product q_1q_2 is negative and the force is attractive.

When more than two charges are present, the force between any pair of charges is given by Equation 23.6. Therefore, the resultant force on any one of them equals the *vector* sum of the forces due to the various individual charges. This principle of *superposition* as applied to electrostatic forces is an experimentally observed fact. For example, if there are four charges, then the resultant force on particle 1 due to particles 2, 3, and 4 is given by

$$\boldsymbol{F}_1 = \boldsymbol{F}_{12} + \boldsymbol{F}_{13} + \boldsymbol{F}_{14}$$

EXAMPLE 23.1 Find the Resultant Force

Consider three point charges located at the corners of a triangle, as in Figure 23.6, where $q_1 = q_3 = 5\ \mu\text{C}$, $q_2 = -2\ \mu\text{C}$ ($1\ \mu\text{C} = 10^{-6}$ C), and $a = 0.1$ m. Find the resultant force on q_3.

Solution First, note the direction of the individual forces on q_3 due to q_1 and q_2. The force on q_3 due to q_2 is attractive since q_2 and q_3 have opposite signs. The force on q_3 due to q_1 is repulsive since they are both positive.

Now let us calculate the magnitude of the forces on q_3. The magnitude of the force on q_3 due to q_2 is given by

$$F_{32} = k\frac{|q_3||q_2|}{a^2}$$

$$= \left(9.0 \times 10^9\ \frac{\text{N}\cdot\text{m}^2}{\text{C}^2}\right)\frac{(5 \times 10^{-6}\ \text{C})(2 \times 10^{-6}\ \text{C})}{(0.1\ \text{m})^2}$$

$$= 9.0\ \text{N}$$

Note that since q_3 and q_2 are opposite in sign, $\boldsymbol{F}_{32}$ is to the left as shown in Figure 23.6.

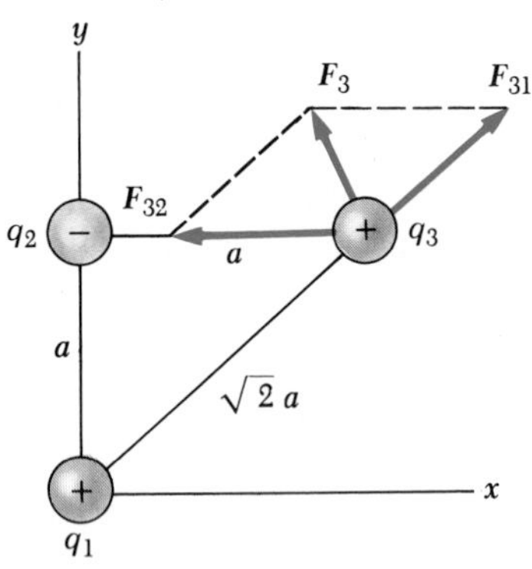

Figure 23.6 (Example 23.1) The force on q_3 due to q_1 is $\boldsymbol{F}_{31}$. The force on q_3 due to q_2 is $\boldsymbol{F}_{32}$. The *total* force, $\boldsymbol{F}_3$, on q_3 is the *vector* sum $\boldsymbol{F}_{31} + \boldsymbol{F}_{32}$.

The magnitude of the force on q_3 due to q_1 is given by

$$F_{31} = k\frac{|q_3||q_1|}{(\sqrt{2}a)^2}$$

$$= \left(9.0 \times 10^9 \frac{\text{N}\cdot\text{m}^2}{\text{C}^2}\right)\frac{(5 \times 10^{-6}\text{ C})(5 \times 10^{-6}\text{ C})}{2(0.1\text{ m})^2}$$

$$= 11\text{ N}$$

The force $\boldsymbol{F}_{31}$ is repulsive and makes an angle of 45° with the x axis. Therefore, the x and y components of $\boldsymbol{F}_{31}$ are equal, with magnitude given by $F_{31}\cos 45° = 7.9$ N. The force $\boldsymbol{F}_{32}$ is in the negative x direction. Hence, the x and y components of the resultant force on q_3 are given by

$$F_x = F_{31x} + F_{32} = 7.9\text{ N} - 9.0\text{ N} = \boxed{-1.1\text{ N}}$$

$$F_y = F_{31y} = \boxed{7.9\text{ N}}$$

We can also express the resultant force on q_3 in unit-vector form as $\boldsymbol{F}_3 = (-1.1\boldsymbol{i} + 7.9\boldsymbol{j})$ N.

Exercise 1 Find the magnitude and direction of the resultant force on q_3.
Answer 8.0 N at an angle of 98° with the x axis.

EXAMPLE 23.2 Where is the Resultant Force Zero?
Three charges lie along the x axis as in Figure 23.7. The positive charge $q_1 = 15\ \mu$C is at $x = 2$ m, and the positive charge $q_2 = 6\ \mu$C is at the origin. Where must a *negative* charge q_3 be placed on the x axis such that the resultant force on it is zero?

Solution Since q_3 is negative and both q_1 and q_2 are positive, the forces $\boldsymbol{F}_{31}$ and $\boldsymbol{F}_{32}$ are both attractive, as indicated in Figure 23.7. If we let x be the coordinate of q_3, then the forces $\boldsymbol{F}_{31}$ and $\boldsymbol{F}_{32}$ have magnitudes given by

$$F_{31} = k\frac{|q_3||q_1|}{(2-x)^2} \quad \text{and} \quad F_{32} = k\frac{|q_3||q_2|}{x^2}$$

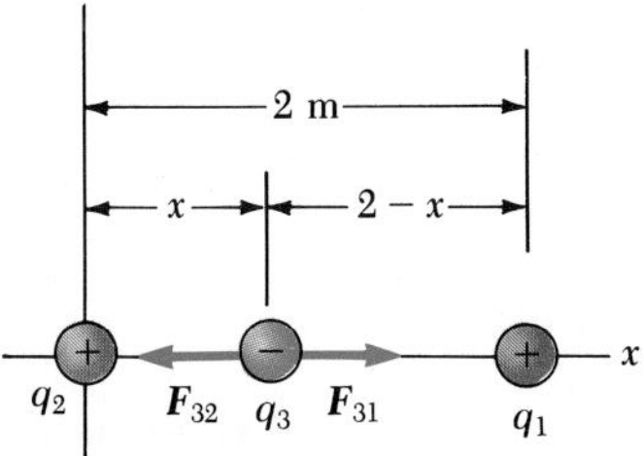

Figure 23.7 (Example 23.2) Three point charges are placed along the x axis. The charge q_3 is negative, whereas q_1 and q_2 are positive. If the net force on q_3 is zero, then the force on q_3 due to q_1 must be equal and opposite to the force on q_3 due to q_2.

If the resultant force on q_3 is zero, then $\boldsymbol{F}_{32}$ must be equal to and opposite $\boldsymbol{F}_{31}$, or

$$k\frac{|q_3||q_2|}{x^2} = k\frac{|q_3||q_1|}{(2-x)^2}$$

Since k and q_3 are common to both sides, we solve for x and find that

$$(2-x)^2|q_2| = x^2|q_1|$$

$$(4 - 4x + x^2)(6 \times 10^{-6}\text{ C}) = x^2(15 \times 10^{-6}\text{ C})$$

Solving this quadratic equation for x, we find that $x = 0.775$ m. Why is the negative root not acceptable?

EXAMPLE 23.3 The Hydrogen Atom
The electron and proton of a hydrogen atom are separated (on the average) by a distance of approximately 5.3×10^{-11} m. Find the magnitude of the electrical force and the gravitational force between the two particles.

Solution From Coulomb's law, we find that the attractive electrical force has the magnitude

$$F_e = k\frac{|e|^2}{r^2} = 9.0 \times 10^9 \frac{\text{N}\cdot\text{m}^2}{\text{C}^2}\frac{(1.6 \times 10^{-19}\text{ C})^2}{(5.3 \times 10^{-11}\text{ m})^2}$$

$$= \boxed{8.2 \times 10^{-8}\text{ N}}$$

Using Newton's universal law of gravity and Table 23.1, we find that the gravitational force has the magnitude

$$F_g = G\frac{m_e m_p}{r^2}$$

$$= \left(6.7 \times 10^{-11}\frac{\text{N}\cdot\text{m}^2}{\text{kg}^2}\right)$$

$$\times \frac{(9.11 \times 10^{-31}\text{ kg})(1.67 \times 10^{-27}\text{ kg})}{(5.3 \times 10^{-11}\text{ m})^2}$$

$$= \boxed{3.6 \times 10^{-47}\text{ N}}$$

The ratio $F_e/F_g \approx 3 \times 10^{39}$. Thus the gravitational force between charged atomic particles is negligible compared with the electrical force.

EXAMPLE 23.4 Find the Charge on the Spheres
Two pieces of small charged spheres, each having a mass of 3×10^{-2} kg, hang in equilibrium as shown in Figure 23.8a. If the length of each string is 0.15 m and the angle $\theta = 5°$, find the magnitude of the charge on each sphere, assuming the spheres have identical charges.

Solution From the right triangle in Figure 23.8a, we see that $\sin\theta = a/L$. From the known length of the string and

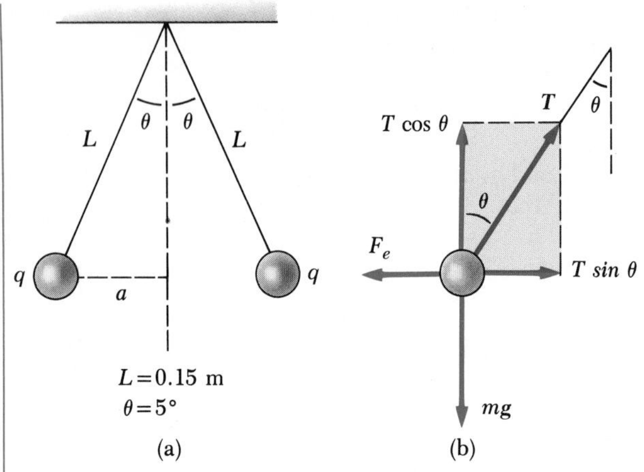

Figure 23.8 (Example 23.4) (a) Two identical spheres, each with the same charge q, suspended in equilibrium by strings. (b) The free-body diagram for the charged spheres on the left side.

the angle the string makes with the vertical, the distance a is calculated to be

$$a = L \sin \theta = (0.15 \text{ m}) \sin 5° = 0.013 \text{ m}$$

Therefore, the separation of the spheres is $2a = 0.026$ m.

The forces acting on one of the spheres are shown in Figure 23.8b. Because the sphere is in equilibrium, the resultants of the forces in the horizontal and vertical directions must separately add up to zero:

$$(1) \qquad \sum F_x = T \sin \theta - F_e = 0$$

$$(2) \qquad \sum F_y = T \cos \theta - mg = 0$$

From (2), we see that $T = mg/\cos \theta$, and so T can be eliminated from (1) if we make this substitution. This gives a value for the electric force, F_e:

$$(3) \qquad F_e = mg \tan \theta$$

$$= (3 \times 10^{-2} \text{ kg})(9.80 \text{ m/s}^2)\tan(5°)$$

$$= 2.57 \times 10^{-2} \text{ N}$$

From Coulomb's law (Eq. 23.1), the electric force between the charges has magnitude given by

$$F_e = k \frac{|q|^2}{r^2}$$

where $r = 2a = 0.026$ m and $|q|$ is the magnitude of the charge on each sphere. Note that the term $|q|^2$ arises here because we have assumed that the charge is the same on both spheres. This equation can be solved for $|q|^2$ to give the charge as follows:

$$|q|^2 = \frac{F_e r^2}{k} = \frac{(2.57 \times 10^{-2} \text{ N})(0.026 \text{ m})^2}{9 \times 10^9 \text{ N} \cdot \text{m}^2/\text{C}^2}$$

$$|q| = 4.4 \times 10^{-8} \text{ C}$$

Exercise 2 If the charge on the foils is negative, how many electrons had to be added to the foils to give a net charge of -4.4×10^{-8} C?
Answer 2.7×10^{11} electrons.

23.4 THE ELECTRIC FIELD

The gravitational field $\mathbf{g}$ at a point in space was defined in Chapter 14 to be equal to the gravitational force $\boldsymbol{F}$ acting on a test mass m_0 divided by the test mass. That is, $\mathbf{g} = \boldsymbol{F}/m_0$. In similar manner, an electric field at a point in space can be defined in terms of the electric force acting on a test charge q_0 placed at that point. To be more precise,

> the electric field vector $\boldsymbol{E}$ at a point in space is defined as the electric force $\boldsymbol{F}$ acting on a positive test charge placed at that point divided by the magnitude of the test charge q_0:

Definition of electric field

$$\boldsymbol{E} \equiv \frac{\boldsymbol{F}}{q_0} \qquad (23.7)$$

Note that $\boldsymbol{E}$ is the field *external* to the test charge—not the field produced by the test charge. The vector $\boldsymbol{E}$ has the SI units of newtons per coulomb (N/C). The direction of $\boldsymbol{E}$ is in the direction of $\boldsymbol{F}$ since we have assumed that $\boldsymbol{F}$ acts on a positive test charge. Thus, we can say that *an electric field exists at a point if a test charge at rest placed at that point experiences an electrical force.* Once the

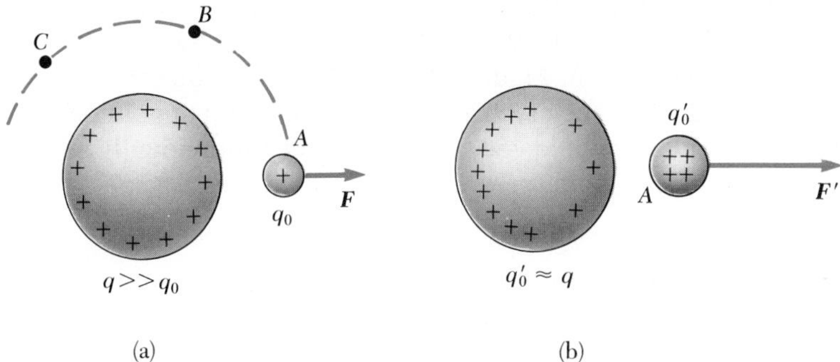

Figure 23.9 (a) When a small test charge q_0 is placed near a conducting sphere of charge q (where $q \gg q_0$), the charge on the conducting sphere remains uniform. (b) If the test charge q_0' is of the order of the charge on the sphere, the charge on the sphere is nonuniform.

electric field is known at some point, the force on *any* charged particle placed at that point can be calculated from Equation 23.7. Furthermore, the electric field is said to exist at some point (even empty space) regardless of whether or not a test charge is located at that point.

When Equation 23.7 is applied, we must assume that the test charge q_0 is small enough such that it does not disturb the charge distribution responsible for the electric field.[3] For instance, if a vanishingly small test charge q_0 is placed near a uniformly charged metallic sphere as in Figure 23.9a, the charge on the metallic sphere, which produces the electric field, will remain uniformly distributed. Furthermore, the force $\mathbf{F}$ on the test charge will have the same magnitude at points A, B, and C, which are equidistant from the sphere. If the test charge is large enough ($q_0' \gg q_0$) as in Figure 23.9b, the charge on the metallic sphere will be redistributed and the ratio of the force to the test charge at point A will be different: ($F'/q_0' \neq F/q_0$). That is, because of this redistribution of charge on the metallic sphere, the electric field at point A set up by the sphere in Figure 23.9b must be different from that of the field at point A in Figure 23.9a. Furthermore, the distribution of charge on the sphere will change as the smaller charge is moved from point A to point B or C.

Consider a point charge q located a distance r from a test charge q_0. According to Coulomb's law, the force on the test charge is given by

$$\mathbf{F} = k\frac{qq_0}{r^2}\,\hat{\mathbf{r}}$$

Since the electric field at the position of the test charge is defined by $\mathbf{E} = \mathbf{F}/q_0$, we find that the electric field *at the position of q_0 due to the charge q* is given by

$$\mathbf{E} = k\frac{q}{r^2}\,\hat{\mathbf{r}} \tag{23.8}$$

[3] To be more precise, the test charge q_0 should be infinitesimally small to ensure that its presence does not affect the original charge distribution. Therefore, strictly speaking, we should replace Equation 23.7 by the expression

$$\mathbf{E} = \lim_{q_0 \to 0} \frac{\mathbf{F}}{q_0}$$

It is impossible to follow this prescription strictly in any experiment since no charges smaller in magnitude than e are known to exist. However, as a practical matter, it is almost always possible to select a sufficiently small test charge to obtain any desired degree of accuracy.

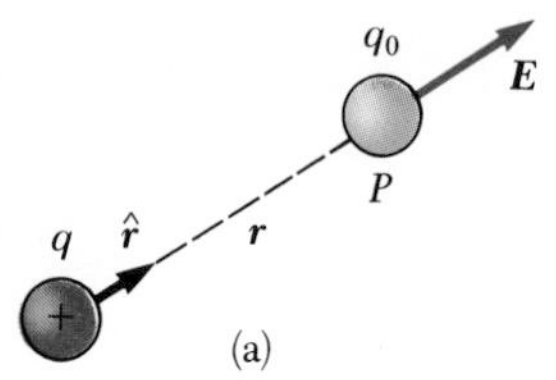

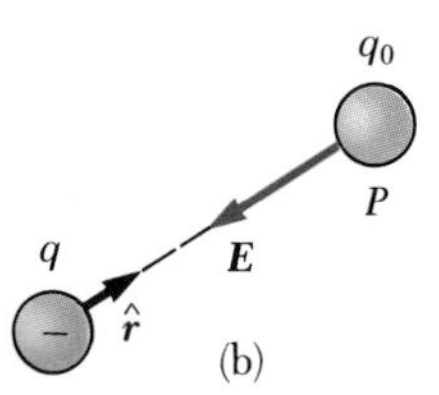

Figure 23.10 A test charge q_0 at the point P is at a distance r from a point charge q. (a) If q is positive, the electric field at P points radially *outward* from q. (b) If q is negative, the electric field at P points radially *inward* toward q.

where $\hat{r}$ is a unit vector that is directed away from q toward q_0 (Fig. 23.10). If q is *positive*, as in Figure 23.10a, the field is directed radially *outward* from this charge. If q is *negative*, as in Figure 23.10b, the field is directed *toward* q.

In order to calculate the electric field due to a group of point charges, we first calculate the electric field vectors at the point P individually using Equation 23.8 and then add them *vectorially*. In other words,

the total electric field due to a group of charges equals the vector sum of the electric fields of all the charges.

This **superposition principle** applied to fields follows directly from the superposition property of electric forces. Thus, the electric field of a group of charges (excluding the test charge q_0) can be expressed as

$$\mathbf{E} = k \sum_i \frac{q_i}{r_i^2} \hat{\mathbf{r}}_i \tag{23.9}$$

where r_i is the distance from the ith charge, q_i, to the point P (the location of the test charge) and $\hat{\mathbf{r}}_i$ is a unit vector directed from q_i toward P.

EXAMPLE 23.5 Electric Force on a Proton

Find the electric force on a proton placed in an electric field of 2×10^4 N/C directed along the positive x axis.

Solution Since the charge on a proton is

$$+e = +1.6 \times 10^{-19}\text{ C},$$

the electric force on it is

$$\begin{aligned} \mathbf{F} = e\mathbf{F} &= (1.6 \times 10^{-19}\text{ C})(2 \times 10^4 \mathbf{i}\text{ N/C}) \\ &= 3.2 \times 10^{-15} \mathbf{i}\text{ N} \end{aligned}$$

where $\mathbf{i}$ is a unit vector in the positive x direction. The weight of the proton is calculated to be equal to $mg = (1.67 \times 10^{-27}\text{ kg})(9.8\text{ m/s}^2) = 1.6 \times 10^{-26}$ N. Hence, we see that the magnitude of the gravitational force in this case is negligible compared with the electric force.

EXAMPLE 23.6 Electric Field Due to Two Charges

A charge $q_1 = 7\ \mu$C is located at the origin, and a second charge $q_2 = -5\ \mu$C is located on the x axis 0.3 m from the origin (Figure 23.11). Find the electric field at the point P with coordinates (0, 0.4) m.

Solution First, let us find the magnitudes of the electric fields due to each charge. The fields $\mathbf{E}_1$ due to the 7-μC charge and $\mathbf{E}_2$ due to the -5-μC charge at P are shown in Figure 23.11. Their magnitudes are given by

$$E_1 = k\frac{|q_1|}{r_1^2} = \left(9.0 \times 10^9\ \frac{\text{N}\cdot\text{m}^2}{\text{C}^2}\right)\frac{(7 \times 10^{-6}\text{ C})}{(0.4\text{ m})^2}$$

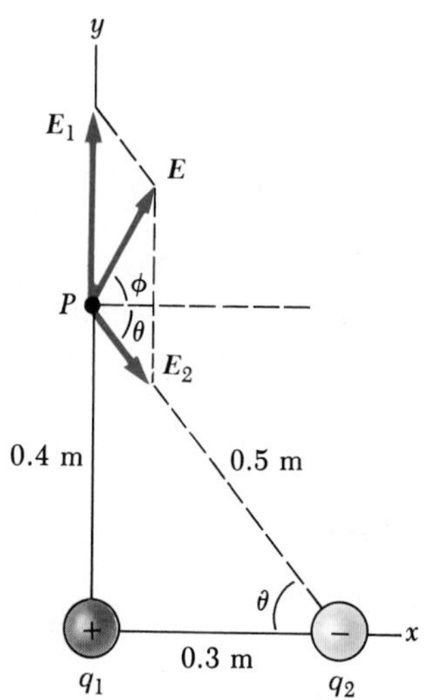

Figure 23.11 (Example 23.6) The total electric field $\mathbf{E}$ at P equals the vector sum $\mathbf{E}_1 + \mathbf{E}_2$, where $\mathbf{E}_1$ is the field due to the positive charge q_1 and $\mathbf{E}_2$ is the field due to the negative charge q_2.

$$= 3.94 \times 10^5\text{ N/C}$$

$$E_2 = k\frac{|q_2|}{r_2^2} = \left(9.0 \times 10^9\ \frac{\text{N}\cdot\text{m}^2}{\text{C}^2}\right)\frac{(5 \times 10^{-6}\text{ C})}{(0.5\text{ m})^2}$$

$$= 1.8 \times 10^5\text{ N/C}$$

The vector $\mathbf{E}_1$ has only a y component. The vector $\mathbf{E}_2$ has an x component given by $E_2 \cos\theta = \frac{3}{5}E_2$ and a negative y

component given by $-E_2 \sin\theta = -\frac{4}{5}E_2$. Hence, we can express the vectors as

$$\mathbf{E}_1 = 3.94 \times 10^5 \mathbf{j} \text{ N/C}$$

$$\mathbf{E}_2 = (1.1 \times 10^5 \mathbf{i} - 1.4 \times 10^5 \mathbf{j}) \text{ N/C}$$

The resultant field $\mathbf{E}$ at P is the superposition of $\mathbf{E}_1$ and $\mathbf{E}_2$:

$$\mathbf{E} = \mathbf{E}_1 + \mathbf{E}_2 = (1.1 \times 10^5 \mathbf{i} + 2.5 \times 10^5 \mathbf{j}) \text{ N/C}$$

From this result, we find that $\mathbf{E}$ has a magnitude of 2.7×10^5 N/C and makes an angle ϕ of 66° with the positive x axis.

Exercise 3 Find the electric force on a test charge of 2×10^{-8} C placed at P.
Answer 5.4×10^{-3} N in the same direction as $\mathbf{E}$.

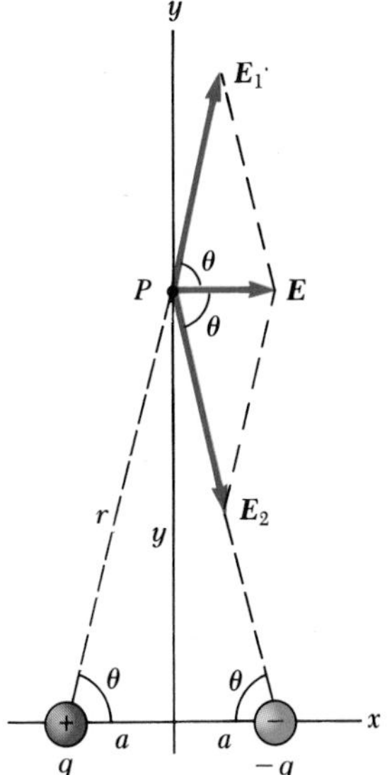

Figure 23.12 (Example 23.7) The total electric field $\mathbf{E}$ at P due to two equal and opposite charges (an electric dipole) equals the vector sum $\mathbf{E}_1 + \mathbf{E}_2$. The field $\mathbf{E}_1$ is due to the positive charge q, and $\mathbf{E}_2$ is the field due to the negative charge $-q$.

EXAMPLE 23.7 Electric Field of a Dipole
An **electric dipole** consists of a positive charge q and a negative charge $-q$ separated by a distance $2a$, as in Figure 23.12. Find the electric field $\mathbf{E}$ due to these charges along the y axis at the point P, which is a distance y from the origin. Assume that $y \gg a$.

Solution At P, the fields $\mathbf{E}_1$ and $\mathbf{E}_2$ due to the two charges are equal in magnitude, since P is equidistant from the two equal and opposite charges. The total field $\mathbf{E} = \mathbf{E}_1 + \mathbf{E}_2$, where the magnitudes of $\mathbf{E}_1$ and $\mathbf{E}_2$ are given by

$$E_1 = E_2 = k\frac{q}{r^2} = k\frac{q}{y^2 + a^2}$$

The y components of $\mathbf{E}_1$ and $\mathbf{E}_2$ cancel each other. The x components are equal since they are both along the x axis. Therefore, $\mathbf{E}$ lies along the x axis and has a magnitude equal to $2E_1 \cos\theta$. From Figure 23.12 we see that $\cos\theta = a/r = a/(y^2 + a^2)^{1/2}$. Therefore,

$$E = 2E_1 \cos\theta = 2k\frac{q}{(y^2 + a^2)}\frac{a}{(y^2 + a^2)^{1/2}}$$

$$= k\frac{2qa}{(y^2 + a^2)^{3/2}}$$

Using the approximation $y \gg a$, we can neglect a^2 in the denominator and write

$$E \approx k\frac{2qa}{y^3} \qquad (23.10)$$

Thus we see that along the y axis the field of a *dipole* at a distant point varies as $1/r^3$, whereas the more slowly varying field of a *point charge* goes as $1/r^2$. This is because at distant points, the fields of the two equal and opposite charges almost cancel each other. The $1/r^3$ variation in E for the dipole is also obtained for a distant point along the x axis (Problem 61) and for a general distant point. The dipole is a good model of many molecules, such as HCl.

As we shall see in later chapters, neutral atoms and molecules behave as dipoles when placed in an external electric field. Furthermore, many molecules, such as HCl, are permanent dipoles. (HCl is essentially an H^+ ion combined with a Cl^- ion.) The effect of such dipoles on the behavior of materials subjected to electric fields will be discussed in Chapter 26.

23.5 ELECTRIC FIELD OF A CONTINUOUS CHARGE DISTRIBUTION

In the previous section, we showed how to calculate the electric field of a point charge using Coulomb's law. The total field of a group of point charges was obtained by taking the vector sum of the individual fields due to all the charges. *This procedure makes use of the superposition principle as applied to the electrostatic field.*

Very often the charges of interest are close together compared with their distances to points of interest. In such situations, the system of charges can be considered to be *continuous.* That is, we imagine that the system of closely spaced charges is equivalent to a total charge that is continuously distributed through a volume or over some surface.

A continuous charge distribution

To evaluate the electric field of a continuous charge distribution, the following procedure is used. First, we divide the charge distribution into small elements each of which contains a small charge Δq, as in Figure 23.13. Next, we use Coulomb's law to calculate the electric field due to one of these elements at a point P. Finally, we evaluate the total field at P due to the charge distribution by summing the contributions of all the charge elements (that is, by applying the superposition principle).

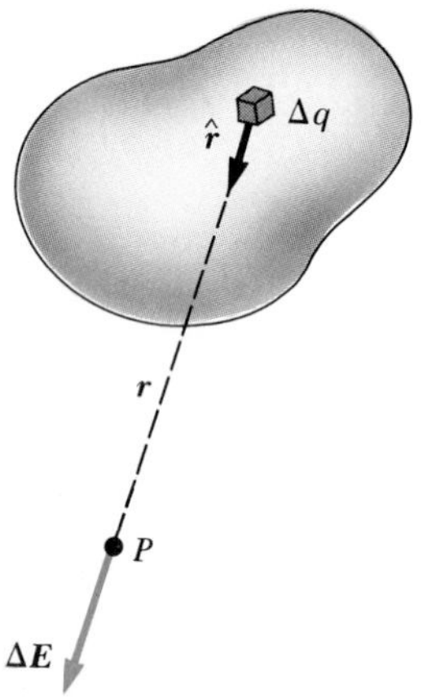

Figure 23.13 The electric field at P due to a continuous charge distribution is the vector sum of the fields due to all the elements Δq of the charge distribution.

The electric field at P due to one element of charge Δq is given by

$$\Delta \mathbf{E} = k \frac{\Delta q}{r^2} \hat{\mathbf{r}}$$

where r is the distance from the element to point P and $\hat{r}$ is a unit vector directed from the charge element toward P. The total electric field at P due to all elements in the charge distribution is approximately given by

$$\mathbf{E} \approx k \sum_i \frac{\Delta q_i}{r_i^2} \hat{\mathbf{r}}_i$$

where the index i refers to the ith element in the distribution. If the separation between elements in the charge distribution is small compared with the distance to P, the charge distribution can be approximated to be continuous. Therefore, the total field at P in the limit $\Delta q_i \rightarrow 0$ becomes

Electric field of a continuous charge distribution

$$\mathbf{E} = k \lim_{\Delta q_i \rightarrow 0} \sum_i \frac{\Delta q_i}{r_i^2} \hat{\mathbf{r}}_i = k \int \frac{dq}{r^2} \hat{\mathbf{r}} \tag{23.11}$$

where the integration is a *vector* operation and must be treated with caution. We shall illustrate this type of calculation with several examples. In these examples, we shall assume that the charge is *uniformly* distributed on a line or a surface or throughout some volume. When performing such calculations, it is convenient to use the concept of a charge density along with the following notations:

If a charge Q is uniformly distributed throughout a volume V, the *charge per unit volume,* ρ, is defined by

Volume charge density

$$\rho \equiv \frac{Q}{V} \tag{23.12}$$

where ρ has units of C/m^3.

If a charge Q is uniformly distributed on a surface of area A, the *surface charge density,* σ, is defined by

Surface charge density

$$\sigma \equiv \frac{Q}{A} \tag{23.13}$$

where σ has units of C/m^2.

Finally, if a charge Q is uniformly distributed along a line of length ℓ, the *linear charge density*, λ, is defined by

$$\lambda \equiv \frac{Q}{\ell} \tag{23.14}$$

Linear charge density

where λ has units of C/m.

If the charge is *nonuniformly* distributed over a volume, surface, or line, we would have to express the charge densities as

$$\rho = \frac{dQ}{dV} \qquad \sigma = \frac{dQ}{dA} \qquad \lambda = \frac{dQ}{d\ell}$$

where dQ is the amount of charge in a small volume, surface, or length element.

EXAMPLE 23.8 The Electric Field Due to a Charged Rod

A rod of length ℓ has a uniform positive charge per unit length λ and a total charge Q. Calculate the electric field at a point P along the axis of the rod, a distance d from one end (Fig. 23.14).

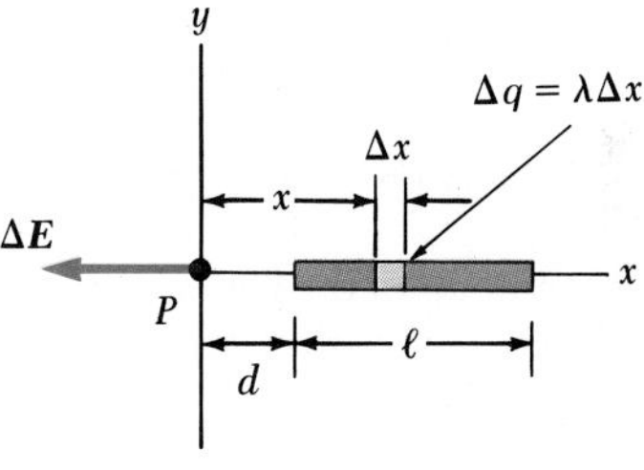

Figure 23.14 (Example 23.8) The electric field at P due to a uniformly charged rod lying along the x axis. The field at P due to the segment of charge Δq is given by $k\,\Delta q/x^2$. The total field at P is the vector sum over all segments of the rod.

Solution For this calculation, the rod is taken to be along the x axis. The ratio of Δq, the charge on the segment to Δx, the length of the segment, is equal to the ratio of the total charge to the total length of the rod. That is, $\Delta q/\Delta x = Q/\ell = \lambda$. Therefore, the charge Δq on the small segment is given by $\Delta q = \lambda\,\Delta x$.

The field ΔE due to this segment at the point P is in the negative x direction, and its magnitude is given by [4]

$$\Delta E = k\frac{\Delta q}{x^2} = k\frac{\lambda\,\Delta x}{x^2}$$

Note that each element produces a field in the negative x direction, and so the problem of summing their contributions is particularly simple in this case. The total field at P due to all segments of the rod, which are at different distances from P, is given by Equation 23.11, which in this case becomes

$$E = \int_d^{\ell+d} k\lambda\frac{dx}{x^2}$$

where the limits on the integral extend from one end of the rod ($x = d$) to the other ($x = \ell + d$). Since k and λ are constants, they can be removed from the integral. Thus, we find that

$$E = k\lambda\int_d^{\ell+d}\frac{dx}{x^2} = k\lambda\left[-\frac{1}{x}\right]_d^{\ell+d}$$

$$= k\lambda\left(\frac{1}{d} - \frac{1}{\ell+d}\right)$$

$$= \frac{kQ}{d(\ell+d)} \tag{23.15}$$

where we have used the fact that the total charge $Q = \lambda\ell$. From this result we see that if the point P is *far* from the rod ($d \gg \ell$), then ℓ in the denominator can be neglected, and $E \approx kQ/d^2$. This is just the form you would expect for a point charge. Therefore, at large distances from the rod, the charge distribution appears to be a point charge of magnitude Q. The use of the limiting technique ($d \to \infty$) is often a good method for checking a theoretical formula.

[4] It is important that you understand the procedure being used to carry out integrations such as this. First, choose an element whose parts are all equidistant from the point at which the field is being calculated. Next, express the charge element Δq in terms of the other variables within the integral (in this example, there is one variable, x.) In examples that have spherical or cylindrical symmetry, the variable will be a radial coordinate.

EXAMPLE 23.9 The Electric Field of a Uniform Ring of Charge

A ring of radius a has a uniform positive charge per unit length, with a total charge Q. Calculate the electric field along the axis of the ring at a point P lying a distance x from the center of the ring (Fig. 23.15a).

Solution The magnitude of the electric field at P due to the segment of charge Δq is

$$\Delta E = k\frac{\Delta q}{r^2}$$

This field has an x component $\Delta E_x = \Delta E \cos\theta$ along the axis of the ring and a component $\Delta E_\perp$ perpendicular to the axis. But as we see in Figure 23.15b, the resultant field at P must lie along the x axis since the perpendicular components sum up to zero. That is, the perpendicular component of any element is canceled by the perpendicular component of an element on the opposite side of the ring. Since $r = (x^2 + a^2)^{1/2}$ and $\cos\theta = x/r$, we find that

$$\Delta E_x = \Delta E \cos\theta = \left(k\frac{\Delta q}{r^2}\right)\frac{x}{r} = \frac{kx}{(x^2+a^2)^{3/2}}\Delta q$$

In this case, all segments of the ring give the *same* contribution to the field at P since they are all equidistant from this point. Thus, we can easily sum over all segments to get the total field at P:

$$E_x = \sum \frac{kx}{(x^2+a^2)^{3/2}}\Delta q = \frac{kx}{(x^2+a^2)^{3/2}}Q \qquad (23.16)$$

This result shows that the field is zero at $x = 0$. Does this surprise you?

Exercise 4 Show that at large distances from the ring ($x \gg a$) the electric field along the axis approaches that of a point charge of magnitude Q.

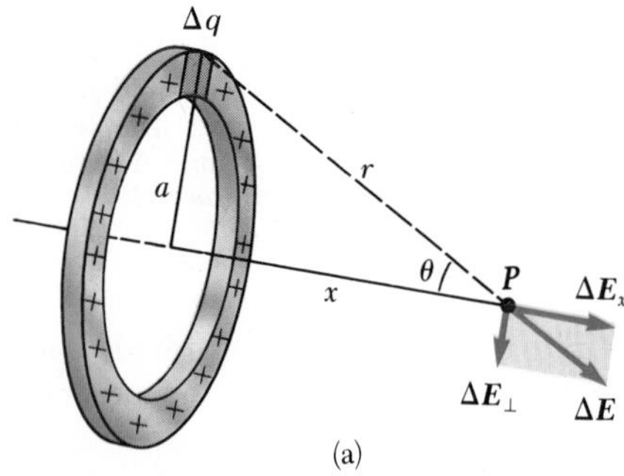

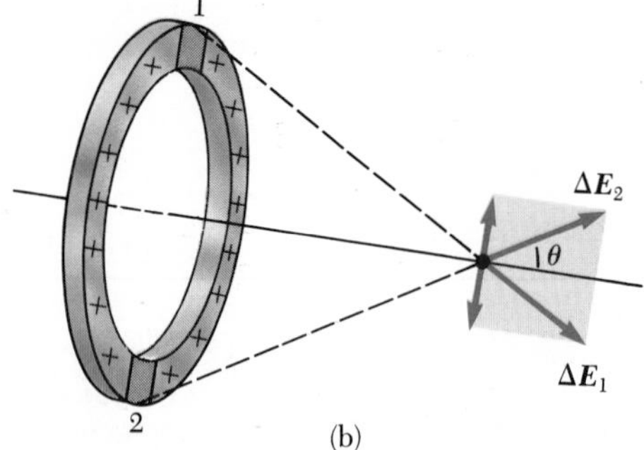

Figure 23.15 (Example 23.9) A uniformly charged ring of radius a. (a) The field at P on the x axis due to an element of charge Δq. (b) The total electric field at P is along the x axis. Note that the perpendicular component of the electric field at P due to segment 1 is canceled by the perpendicular component due to segment 2, which is opposite segment 1.

EXAMPLE 23.10 The Electric Field of a Uniformly Charged Disk □

A disk of radius R has a uniform charge per unit area σ. Calculate the electric field along the axis of the disk, a distance x from its center (Fig. 23.16).

Solution The solution to this problem is straightforward if we consider the disk as a set of concentric rings. We can then make use of Example 23.9, which gives the field of a given ring of radius r, and sum up contributions of all rings making up the disk. By symmetry, the field on an axial point must be parallel to this axis.

The ring of radius r and width dr has an area equal to $2\pi r\,dr$ (Fig. 23.16). The charge dq on this ring is equal to the area of the ring multiplied by the charge per unit area, or $dq = 2\pi\sigma r\,dr$. Using this result in Equation 23.16 (with a replaced by r) gives for the field due to the ring the expression

$$dE = \frac{kx}{(x^2+r^2)^{3/2}}(2\pi\sigma r\,dr)$$

To get the total field at P, we integrate this expression over the limits $r = 0$ to $r = R$, noting that x is a constant, which gives

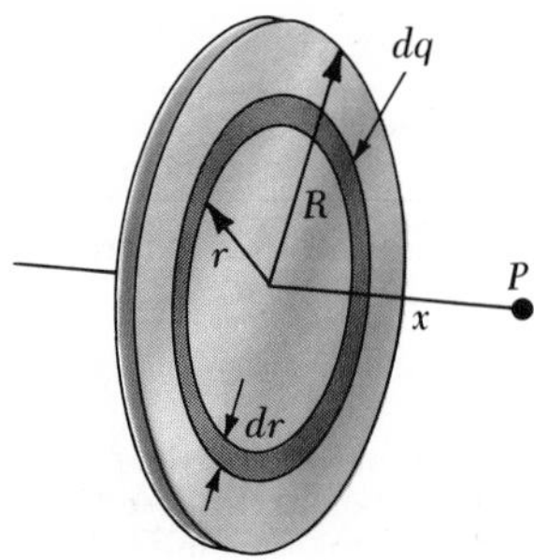

Figure 23.16 (Example 23.10) A uniformly charged disk of radius R. The electric field at an axial point P is directed along this axis, perpendicular to the plane of the disk.

$$E = kx\pi\sigma \int_0^R \frac{2r\,dr}{(x^2 + r^2)^{3/2}}$$

$$= kx\pi\sigma \int_0^R (x^2 + r^2)^{-3/2}\,d(r^2)$$

$$= kx\pi\sigma \left[\frac{(x^2 + r^2)^{-1/2}}{-1/2} \right]_0^R$$

$$= 2\pi k\sigma \left(1 - \frac{x}{(x^2 + R^2)^{1/2}} \right) \qquad (23.17)$$

The field close to the disk along an axial point can also be obtained from Equation 23.17 by letting $x \to 0$ (or $R \to \infty$). This gives

$$E = 2\pi k\sigma = \frac{\sigma}{2\epsilon_0} \qquad (23.18)$$

where ϵ_0 is the permittivity of free space, given by Equation 23.4. As we shall find in the next chapter, the same result is obtained for the field of a uniformly charged infinite sheet.

Exercise 5 Show that at large distances from the disk the electric field along the axis approaches that of a point charge of magnitude $Q = \sigma\pi R^2$.

23.6 ELECTRIC FIELD LINES

A convenient aid for visualizing electric field patterns is to draw lines pointing in the same direction as the electric field vector at any point. These lines, called **electric field lines,** are related to the electric field in any region of space in the following manner:

1. The electric field vector $\boldsymbol{E}$ is *tangent* to the electric field line at each point.
2. The number of lines per unit area through a surface perpendicular to the lines is proportional to the strength of the electric field in that region. Thus $\boldsymbol{E}$ is large when the field lines are close together and small when they are far apart.

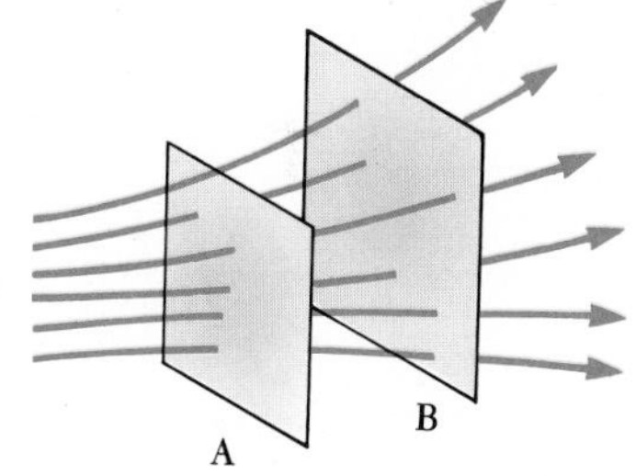

Figure 23.17 Electric field lines penetrating two surfaces. The magnitude of the field is greater on surface A than on surface B.

These properties are illustrated in Figure 23.17. The density of lines through surface A is greater than the density of lines through surface B. Therefore, the electric field is more intense on surface A than on surface B. Furthermore, the field drawn in Figure 23.17 is nonuniform since the lines at different locations point in different directions.

Some representative electric field lines for a single positive point charge are shown in Figure 23.18a. Note that in this two-dimensional drawing we

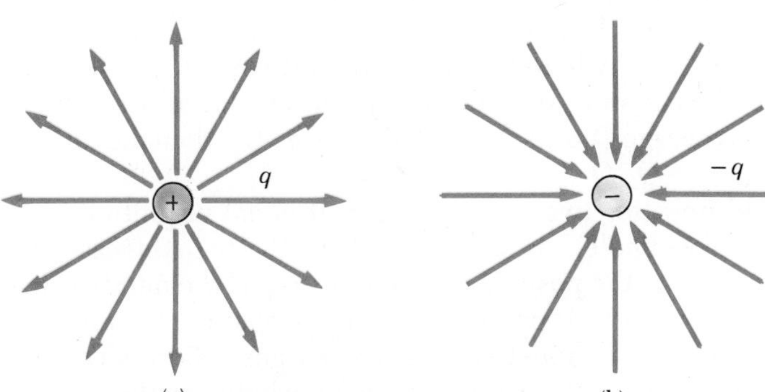

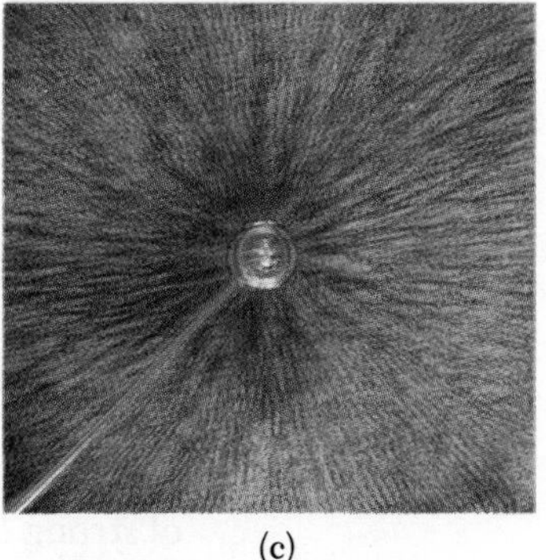

Figure 23.18 The electric field lines for a point charge. (a) For a positive point charge, the lines are radially outward. (b) For a negative point charge, the lines are radially inward. Note that the figures show only those field lines that lie in the plane containing the charge. (c) The dark areas are small pieces of thread suspended in oil, which align with the electric field produced by a small charged conductor at the center. (Photo courtesy of Henry Leap and Jim Lehman)

show only the field lines that lie in the plane containing the point charge. The lines are actually directed radially outward from the charge in *all* directions, somewhat like the needles of a porcupine. Since a positive test charge placed in this field would be repelled by the charge q, the lines are directed radially away from the positive charge. Similarly, the electric field lines for a single negative point charge are directed toward the charge (Fig. 23.18b). In either case, the lines are along the radial direction and extend all the way to infinity. Note that the lines are closer together as they get near the charge, indicating that the strength of the field is increasing.

The rules for drawing electric field lines for any charge distribution are as follows:

Rules for drawing electric field lines

1. The lines must begin on positive charges and terminate on negative charges, or at infinity in the case of an excess of charge.
2. The number of lines drawn leaving a positive charge or approaching a negative charge is proportional to the magnitude of the charge.
3. No two field lines can cross.

Is this visualization of the electric field in terms of field lines consistent with Coulomb's law? To answer this question, consider an imaginary spherical surface of radius r concentric with the charge. From symmetry, we see that the magnitude of the electric field is the same everywhere on the surface of the sphere. The number of lines, N, that emerge from the charge is equal to the number that penetrate the spherical surface. Hence, the number of lines per unit area on the sphere is $N/4\pi r^2$ (where the surface area of the sphere is $4\pi r^2$). Since E is proportional to the number of lines per unit area, we see that E varies as $1/r^2$. This is consistent with the result obtained from Coulomb's law, that is, $E = kq/r^2$.

It is important to note that electric field lines are not material objects. They are used only to provide us with a qualitative description of the electric field. One problem with this model is the fact that one always draws a finite number of lines from each charge, which makes it appear as if the field were quantized and acted only in a certain direction. The field, in fact, is continuous—existing at every point. Another problem with this model is the danger of getting the wrong impression from a two-dimensional drawing of field lines being used to describe a three-dimensional situation.

Since charge is quantized, the number of lines leaving any material object must be $0, \pm C'e, \pm 2C'e, \ldots$, where C' is an arbitrary (but fixed) proportionality constant. Once C' is chosen, the number of lines is not arbitrary. For example, if object 1 has charge Q_1 and object 2 has charge Q_2, then the ratio of number of lines is $N_2/N_1 = Q_2/Q_1$.

The electric field lines for two point charges of equal magnitude, but opposite signs (the electric dipole), are shown in Figure 23.19. In this case, the number of lines that begin at the positive charge must equal the number that terminate at the negative charge. At points very near the charges, the lines are nearly radial. The high density of lines between the charges indicates a region of strong electric field. The attractive nature of the force between the charges can also be seen from Figure 23.19.

Figure 23.20 shows the electric field lines in the vicinity of two equal positive point charges. Again, the lines are nearly radial at points close to either charge. The same number of lines emerge from each charge since the charges are equal in magnitude. At large distances from the charges, the field

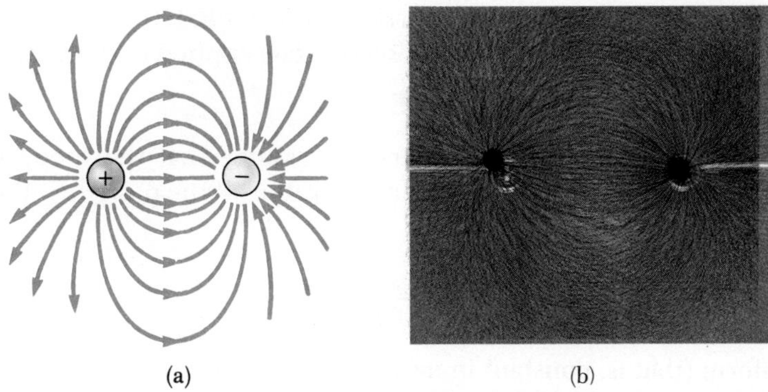

Figure 23.19 (a) The electric field lines for two equal and opposite point charges (an electric dipole). Note that the number of lines leaving the positive charge equals the number terminating at the negative charge. (b) The photograph was taken using small pieces of thread suspended in oil, which align with the electric field. (Photo courtesy of Harold M. Waage, Princeton University)

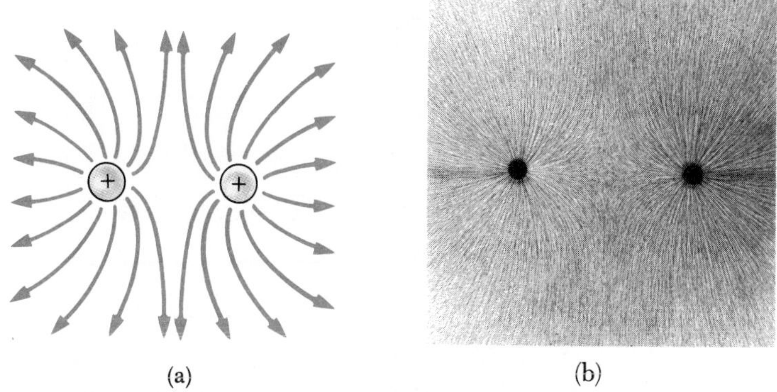

Figure 23.20 (a) The electric field lines for two positive point charges. (b) The photograph was taken using small pieces of thread suspended in oil, which align with the electric field. (Photo courtesy of Harold M. Waage, Princeton University)

is approximately equal to that of a single point charge of magnitude $2q$. The bulging out of the electric field lines between the charges indicates the repulsive nature of the electric force between like charges.

Finally, in Figure 23.21 we sketch the electric field lines associated with a positive charge $+2q$ and a negative charge $-q$. In this case, we see that the number of lines leaving the charge $+2q$ is twice the number entering the charge $-q$. Hence only half of the lines that leave the positive charge enter the negative charge. The remaining half terminate on a negative charge we assume to be located at infinity. At large distances from the charges (large compared with the charge separation), the electric field lines are equivalent to those of a single charge $+q$.

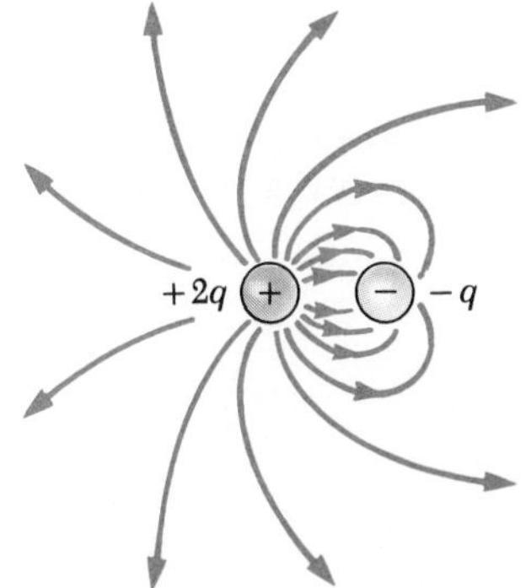

Figure 23.21 The electric field lines for a point charge $+2q$ and a second point charge $-q$. Note that two lines leave the charge $+2q$ for every one that terminates on $-q$.

23.7 MOTION OF CHARGED PARTICLES IN A UNIFORM ELECTRIC FIELD

In this section we describe the motion of a charged particle in a uniform electric field. As we shall see, the motion is equivalent to that of a projectile moving in a uniform gravitational field. When a particle of charge q is placed in an electric field $\boldsymbol{E}$, the electric force on the charge is $q\boldsymbol{E}$. If this is the only force exerted on the charge, then Newton's second law applied to the charge gives

$$\boldsymbol{F} = q\boldsymbol{E} = m\boldsymbol{a}$$

where m is the mass of the charge and we assume that the speed is small compared with the speed of light. The acceleration of the particle is therefore given by

$$\boldsymbol{a} = \frac{q\boldsymbol{E}}{m} \tag{23.19}$$

If $\boldsymbol{E}$ is uniform (that is, constant in magnitude and direction), we see that the acceleration is a constant of the motion. If the charge is positive, the acceleration will be in the direction of the electric field. If the charge is negative, the acceleration will be in the direction *opposite* the electric field.

EXAMPLE 23.11 An Accelerating Positive Charge

A positive point charge q of mass m is released from rest in a uniform electric field $\boldsymbol{E}$ directed along the x axis as in Figure 23.22. Describe its motion.

Solution The acceleration of the charge is constant and given by qE/m. The motion is simple linear motion along the x axis. Therefore, we can apply the equations of kinematics in one dimension (from Chapter 3):

$$x - x_0 = v_0 t + \tfrac{1}{2}at^2 \qquad v = v_0 + at$$

$$v^2 = v_0^2 + 2a(x - x_0)$$

Taking $x_0 = 0$ and $v_0 = 0$ gives

$$x = \tfrac{1}{2}at^2 = \frac{qE}{2m}t^2$$

$$v = at = \frac{qE}{m}t$$

$$v^2 = 2ax = \left(\frac{2qE}{m}\right)x$$

The kinetic energy of the charge after it has moved a distance x is given by

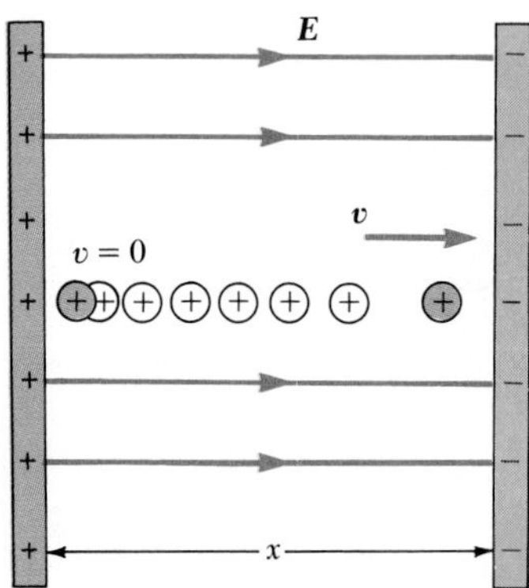

Figure 23.22 (Example 23.11) A positive point charge q in a uniform electric field $\boldsymbol{E}$ undergoes constant acceleration in the direction of the field.

$$K = \tfrac{1}{2}mv^2 = \tfrac{1}{2}m\left(\frac{2qE}{m}\right)x = qEx$$

This result can also be obtained from the work-energy theorem, since the work done by the electric force is $F_e x = qEx$ and $W = \Delta K$.

The electric field in the region between two oppositely charged flat metal plates is approximately uniform (Fig. 23.23). Suppose an electron of charge $-e$ is projected horizontally into this field with an initial velocity $v_0\boldsymbol{i}$. Since the electric field $\boldsymbol{E}$ is in the positive y direction, the acceleration of the electron is in the negative y direction. That is,

$$\boldsymbol{a} = -\frac{eE}{m}\boldsymbol{j} \tag{23.20}$$

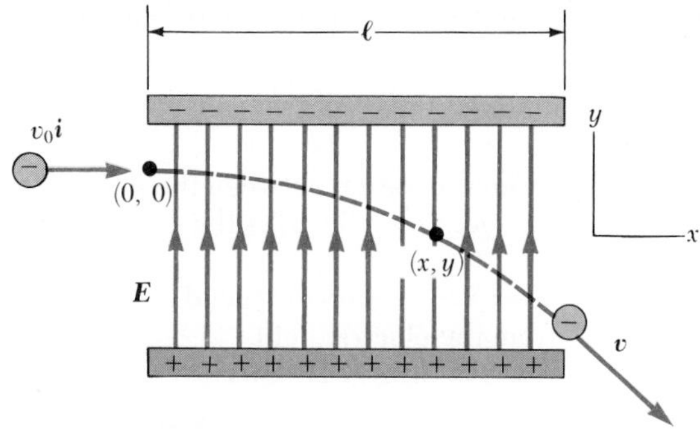

Figure 23.23 An electron is projected horizontally into a uniform electric field produced by two charged plates. The electron undergoes a downward acceleration (opposite $\boldsymbol{E}$), and its motion is parabolic.

Because the acceleration is constant, we can apply the equations of kinematics in two dimensions (from Chapter 4) with $v_{x0} = v_0$ and $v_{y0} = 0$. The components of velocity of the electron after it has been in the electric field a time t are given by

$$v_x = v_0 = \text{constant} \qquad (23.21)$$

$$v_y = at = -\frac{eE}{m} t \qquad (23.22)$$

Likewise, the coordinates of the electron after a time t in the electric field are given by

$$x = v_0 t \qquad (23.23)$$

$$y = \tfrac{1}{2}at^2 = -\tfrac{1}{2}\frac{eE}{m} t^2 \qquad (23.24)$$

Substituting the value $t = x/v_0$ from Equation 23.23 into Equation 23.24, we see that y is proportional to x^2. Hence, the trajectory is a parabola. After the electron leaves the region of uniform electric field, it continues to move in a straight line with a speed $v > v_0$.

Note that we have neglected the gravitational force on the electron. This is a good approximation when dealing with atomic particles. For an electric field of 10^4 N/C, the ratio of the electric force, eE, to the gravitational force, mg, for the electron is of the order of 10^{14}. The corresponding ratio for a proton is of the order of 10^{11}.

EXAMPLE 23.12 An Accelerated Electron

An electron enters the region of a uniform electric field as in Figure 23.23, with $v_0 = 3 \times 10^6$ m/s and $E = 200$ N/C. The width of the plates is $\ell = 0.1$ m. (a) Find the acceleration of the electron while in the electric field.

Since the charge on the electron has a magnitude of 1.60×10^{-9} C and $m = 9.11 \times 10^{-31}$ kg, Equation 23.20 gives

$$\boldsymbol{a} = -\frac{eE}{m}\boldsymbol{j} = -\frac{(1.6 \times 10^{-19}\ \text{C})\ (200\ \text{N/C})}{9.11 \times 10^{-31}\ \text{kg}}\boldsymbol{j}$$

$$= -3.51 \times 10^{13}\boldsymbol{j}\ \text{m/s}^2$$

(b) Find the time it takes the electron to travel through the region of the electric field.

The horizontal distance traveled by the electron while in the electric field is $\ell = 0.1$ m. Using Equation

23.23 with $x = \ell$, we find that the time spent in the electric field is given by

$$t = \frac{\ell}{v_0} = \frac{0.1 \text{ m}}{3 \times 10^6 \text{ m/s}} = 3.33 \times 10^{-8} \text{ s}$$

(c) What is the vertical displacement y of the electron while it is in the electric field?

Using Equation 23.24 and the results from (a) and (b), we find that

$$y = \tfrac{1}{2}at^2 = -\tfrac{1}{2}(3.51 \times 10^{13} \text{ m/s}^2)(3.33 \times 10^{-8} \text{ s})^2$$

$$= -0.0195 \text{ m} = -1.95 \text{ cm}$$

If the separation between the plates is smaller than this, the electron will strike the positive plate.

Exercise 6 Find the speed of the electron as it emerges from the electric field.
Answer 3.22×10^6 m/s.

*23.8 THE OSCILLOSCOPE

The oscilloscope is an electronic instrument widely used in making electrical measurements. The main component of the oscilloscope is the cathode ray tube (CRT), shown in Figure 23.24. This tube is commonly used to obtain a visual display of electronic information for other applications, including radar systems, television receivers, and computers. The CRT is a vacuum tube in which electrons are accelerated and deflected under the influence of electric fields.

The electron beam is produced by an assembly called an *electron gun*, located in the neck of the tube. The assembly shown in Figure 23.24 consists of a heater (H), a cathode (C), and a positively charged anode (A). An electric current maintained in the heater causes its temperature to rise, which in turn heats the cathode. The cathode reaches temperatures high enough to cause electrons to be "boiled off." Although they are not shown in the figure, the electron gun also includes an element that focuses the electron beam and one that controls the number of electrons reaching the anode (that is, a brightness control). The anode has a hole in its center that allows the electrons to pass through without striking the anode. These electrons, if left undisturbed, travel in a straight-line path until they strike the face of the CRT. The screen at the

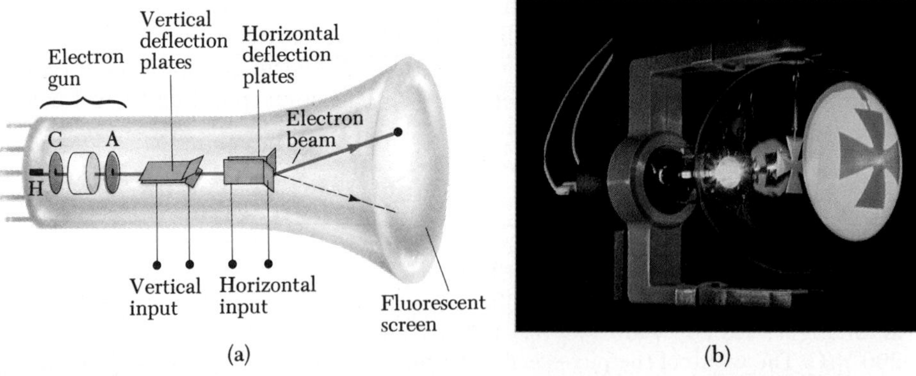

Figure 23.24 (a) Schematic diagram of a cathode ray tube. Electrons leaving the hot cathode C are accelerated to the anode A. The electron gun is also used to focus the beam, and the plates deflect the beam. (b) Photograph of a "Maltese Cross" tube showing the shadow of a beam of cathode rays falling on the tube's luminescent screen. The hot filament also produces a beam of light and a second shadow of the cross. (Courtesy of CENCO)

front of the tube is coated with a material that emits visible light when bombarded with electrons. This results in a visible spot of light on the screen of the CRT.

The electrons are deflected in various directions by two sets of plates placed at right angles to each other in the neck of the tube. In order to understand how the deflection plates operate, first consider the horizontal deflection plates in Figure 23.24a. External electric circuits are used to control and change the amount of charge present on these plates, with positive charge being placed on one plate and negative on the other. (In Chapter 25 we shall see that this can be accomplished by applying a voltage across the plates.) This increasing charge creates an increasing electric field between the plates, which causes the electron beam to be deflected from its straight-line path. The tube face is slightly phosphorescent and therefore glows briefly after the electron beam moves from one point to another on the screen. Slowly increasing the charge on the horizontal plates causes the electron beam to move gradually from the center toward the side of the screen. Because of the phosphorescence, however, one sees a horizontal line extending across the screen instead of the simple movement of the dot. The horizontal line can be maintained on the screen by rapid, repetitive tracing.

The vertical deflection plates act in exactly the same way as the horizontal plates, except that changing the charge in them with external controlling circuits causes a vertical line on the tube face. In practice, the horizontal and vertical deflection plates are used simultaneously. To see how the oscilloscope can display visual information, let us examine how we could observe the sound wave from a tuning fork on the screen. For this purpose, the charge on the horizontal plates changes in such a manner that the beam sweeps across the face of the tube at a constant rate. The tuning fork is then sounded into a microphone, which changes the sound signal to an electric signal that is applied to the vertical plates. The combined effect of the horizontal and vertical plates causes the beam to sweep the tube horizontally and up and down at the same time, with the vertical motion corresponding to the tuning fork signal. A pattern such as that shown in Figure 23.25 is seen on the screen.

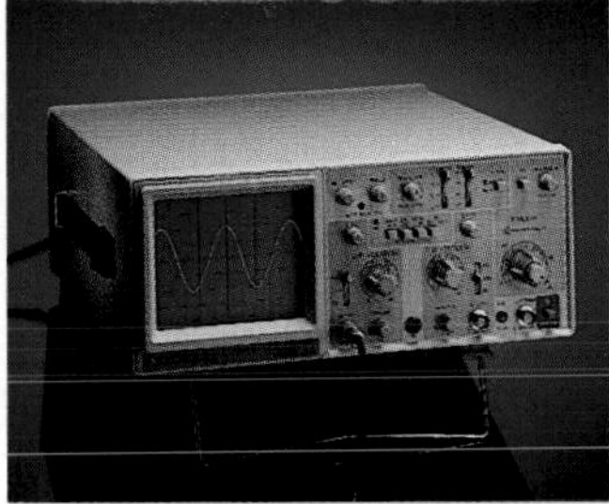

Figure 23.25 A typical laboratory oscilloscope. (Courtesy of PASCO Scientific Co.)

SUMMARY

Properties of electric charges

Electric charges have the following important properties:

1. Unlike charges attract one another and like charges repel one another.
2. Electric charge is always conserved.
3. Charge is quantized, that is, it exists in discrete packets that are some integral multiple of the electronic charge.
4. The force between charged particles varies as the inverse square of their separation.

Conductors are materials in which charges move quite freely. Some examples of good conductors are copper, aluminum, and silver. **Insulators** are materials that do not readily transport charge. Some examples are glass, rubber, and wood.

Coulomb's law

Coulomb's law states that the electrostatic force between two stationary, charged particles separated by a distance r has a magnitude given by

$$F = k\frac{|q_1||q_2|}{r^2} \tag{23.1}$$

Coulomb constant

where the constant k has the value

$$k = 8.9875 \times 10^9\ \text{N}\cdot\text{m}^2/\text{C}^2 \tag{23.2}$$

Charge on an electron or proton

The smallest unit of charge known to exist in nature is the charge on an electron or proton. The magnitude of this charge e is given by

$$|e| = 1.60219 \times 10^{-19}\ \text{C} \tag{23.5}$$

Definition of electric field

The **electric field** $\boldsymbol{E}$ at some point in space is defined as the electric force $\boldsymbol{F}$ that acts on a small positive test charge placed at that point divided by the magnitude of the test charge q_0:

$$\boldsymbol{E} = \frac{\boldsymbol{F}}{q_0} \tag{23.7}$$

Electric field of a point charge q

The electric field due to a point charge q at a distance r from the charge is given by

$$\boldsymbol{E} = k\frac{q}{r^2}\hat{\boldsymbol{r}} \tag{23.8}$$

where $\hat{\boldsymbol{r}}$ is a unit vector directed from the charge to the point in question. The electric field is directed radially outward from a positive charge and is directed *toward* a negative charge.

Electric field of a group of charges

The *electric field* due to a group of charges can be obtained using the **superposition principle.** That is, the total electric field equals the *vector sum* of the electric fields of all the charges at some point:

$$\boldsymbol{E} = k\sum_i \frac{q_i}{r_i^2}\hat{\boldsymbol{r}}_i \tag{23.9}$$

Electric field of a continuous charge distribution

Similarly, the electric field of a continuous charge distribution at some point is given by

$$\boldsymbol{E} = k\int \frac{dq}{r^2}\hat{\boldsymbol{r}} \tag{23.11}$$

where dq is the charge on one element of the charge distribution and r is the distance from the element to the point in question.

Electric field lines are useful for describing the electric field in any region of space. The electric field vector $\boldsymbol{E}$ is always tangent to the electric field lines at every point. Furthermore, the number of lines per unit area through a surface perpendicular to the lines is proportional to the magnitude of $\boldsymbol{E}$ in that region.

Acceleration of a charge in an electric field

A charged particle of mass m and charge q moving in an electric field $\boldsymbol{E}$ has an acceleration $\boldsymbol{a}$ given by

$$\boldsymbol{a} = \frac{q\boldsymbol{E}}{m} \tag{23.19}$$

If the electric field is uniform, the acceleration is constant and the motion of the charge is similar to that of a projectile moving in a uniform gravitational field.

PROBLEM-SOLVING STRATEGY AND HINTS

1. **Units:** When performing calculations that involve the use of the Coulomb constant $k(=1/4\pi\epsilon_0)$ which appears in Coulomb's law, charges must be in coulombs, and distances in meters. If they appear in other units, you must convert them.
2. **Applying Coulomb's law to point charges:** It is important to remember to use the superposition principle properly when dealing with a collection of interacting charges. When several charges are present, the resultant force on any one of the charges is the *vector sum* of the forces due to the individual forces. You must be very careful in the algebraic manipulation of vector quantities. It may be useful to review the material on vector addition in Chapter 2.
3. **Calculating the electric field of point charges:** Remember that the superposition principle can also be applied to electric fields, which are also vector quantities. To find the total electric field at a given point, first calculate the electric field at the point due to each individual charge. The resultant field at the point is the vector sum of the fields due to the individual charges.
4. **Continuous charge distributions:** When you are confronted with problems that involve a continuous distribution of charge, the vector sums for evaluating the total electric field at some point must be replaced by vector integrals. The charge distribution is divided into infinitesimal pieces, and the vector sum is carried out by integrating over the entire charge distribution. You should review Examples 8, 9, and 10, which demonstrate such procedures.
5. **Symmetry:** Whenever dealing with either a distribution of point charges or a continuous charge distribution, you should take advantage of any symmetry in the system to simplify your calculations.
6. **Motion of charged particles in uniform electric fields:** The motion of a charged particle in a uniform electric field (one that is constant in magnitude and direction) can be described by using the equations of projectile motion developed in Chapter 4. In this case, the uniform acceleration is given by $\mathbf{a} = q\mathbf{E}/m$.

QUESTIONS

1. Sparks are often observed (or heard) on a dry day when clothes are removed in the dark. Explain.
2. Explain from an atomic viewpoint why charge is usually transferred by electrons.
3. A balloon is negatively charged by rubbing and then clings to a wall. Does this mean that the wall is positively charged? Why does the balloon eventually fall?
4. A light, uncharged metal sphere suspended from a thread is attracted to a charged rubber rod. After touching the rod, the sphere is repelled by the rod. Explain.
5. Explain what we mean by a neutral atom.
6. If a suspended object A is attracted to object B, which is charged, can we conclude that object A is charged? Explain.
7. A charged comb will often attract small bits of dry paper that fly away when they touch the comb. Explain.
8. Why do some clothes cling together and to your body after being removed from a dryer?
9. A large metal sphere insulated from ground is charged with an electrostatic generator while a person standing on an insulating stool holds the sphere while it is being charged. Why is it safe to do this? Why wouldn't it be safe for another person to touch the sphere after it has been charged?
10. What is the difference between charging an object by induction and charging by conduction?

11. What are the similarities and differences between Newton's Universal Law of Gravitation, $F = \frac{Gm_1m_2}{r^2}$, and Coulomb's Law, $F = \frac{kQ_1Q_2}{r^2}$?
12. Assume that someone proposes a theory that says people are bound to the earth by electric forces rather than by gravity. How could you prove this theory wrong?
13. Would life be different if the electron were positively charged and the proton were negatively charged? Does the choice of signs have any bearing on physical and chemical interactions? Explain.
14. When defining the electric field, why is it necessary to specify that the magnitude of the test charge be very small (i.e., take the limit as $q \to 0$)?
15. Two charged spheres each of radius a are separated by a distance $r > 2a$. Is the force on either sphere given by Coulomb's law? Explain. (*Hint:* Refer back to Chapter 14 on gravitation.)
16. When is it valid to approximate a charge distribution by a "point charge"?
17. Is it possible for an electric field to exist in empty space? Explain.
18. Explain why electric field lines do not form closed loops.
19. Explain why electric field lines never cross. (*Hint:* $\mathbf{E}$ must have a unique direction at all points.)
20. A "free" electron and "free" proton are placed in an identical electric field. Compare the electric forces on each particle. Compare their accelerations.
21. Explain what happens to the magnitude of the electric field of a point charge as r approaches zero.
22. A negative charge is placed in a region of space where the electric field is directed vertically upward. What is the direction of the electric force experienced by this charge?
23. A charge $4q$ is at a distance r from a charge $-q$. Compare the number of electric field lines leaving the charge $4q$ with the number entering the charge $-q$.
24. In Figure 23.21, where do the extra lines leaving the charge $+2q$ end?
25. Consider two equal point charges separated by some distance d. At what point (other than ∞) would a third test charge experience no net force?
26. An uncharged, metallic coated, Ping–Pong ball is placed in the region between two horizontal parallel metal plates. If the two plates are charged, one positive and one negative, describe the motion the Ping–Pong ball will undergo.
27. A negative point charge $-q$ is placed at the point P near the positively charged ring shown in Figure 23.15 of Example 23.9. If $x \ll a$, describe the motion of the point charge if it is released from rest.
28. Explain the differences between linear, surface, and volume charge densities, and give examples of when each would be used.
29. If the electron in Figure 23.23 is projected into the electric field with an arbitrary velocity $\mathbf{v}_0$ (at an angle to $\mathbf{E}$), will its trajectory still be parabolic? Explain.
30. If a metal object receives a positive charge, does its mass increase, decrease, or stay the same? What happens to the mass if the object is given a negative charge?
31. It has been reported that in some instances people near where a lightning bolt strikes the earth have had their clothes thrown off. Explain why this might happen.
32. Why should a ground wire be connected to the metal support rod for a television antenna?
33. Are the occupants of a steel-frame building safer than those in a wood-frame house during an electrical storm or vice versa? Explain.
34. A light piece of aluminum foil is draped over a wooden rod. When a rod with a positive charge is brought close to the foil, two parts of the foil stand apart. Why? What kind of charge is on the foil?
35. Why is it more difficult to charge an object by friction on a humid day than on a dry day?
36. How would you experimentally distinguish an electric field from a gravitational field?

PROBLEMS

Section 23.3 Coulomb's Law

1. Suppose that 1 g of hydrogen is separated into electrons and protons. Suppose also that the protons are placed at the Earth's north pole and the electrons are placed at the south pole. What is the resulting compressional force on the Earth?
2. Calculate the net charge on an arbitrary substance consisting of (a) 8×10^{13} electrons and (b) a combination of 3×10^{14} protons and 7×10^{14} electrons.
3. Two protons in a molecule are separated by a distance of 3.8×10^{-10} m. Find the electrostatic force exerted by one proton on the other.
4. A 6.7-μC charge is located 5.0 m from a -8.4-μC charge. Find the electrostatic force exerted by one charge on the other.
5. A 1.3-μC charge is located on the x axis at $x = -0.5$ m, a 3.2-μC charge is located on the x axis at $x = 1.5$ m, and a 2.5-μC charge is located at the origin. Find the net force on the 2.5-μC charge. All charges are positive.
6. A point charge $q_1 = -4.3\ \mu$C is located on the y axis at $y = 0.18$ m, a charge $q_2 = 1.6\ \mu$C is located at the origin, and a charge $q_3 = 3.7\ \mu$C is located on the x axis at $x = -0.18$ m. Find the resultant force on the charge q_1.

7. Three point charges of 2 μC, 7 μC, and -4 μC are located at the corners of an equilateral triangle as in Figure 23.26. Calculate the net electric force on the 7-μC charge.

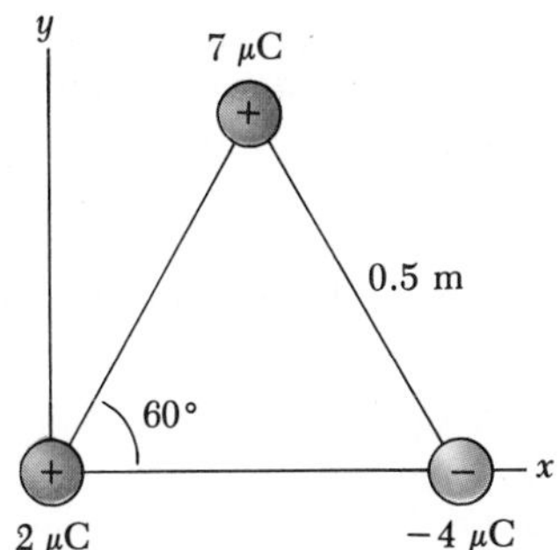

Figure 23.26 (Problems 7 and 24).

8. Four point charges are situated at the corners of a square of sides a as in Figure 23.27. Find the resultant force on the positive charge q.

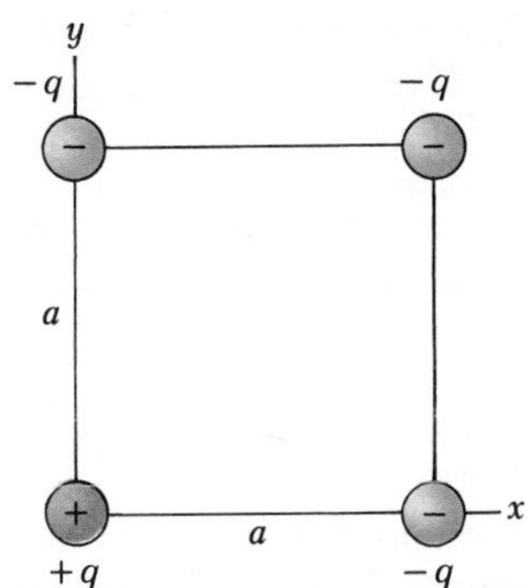

Figure 23.27 (Problems 8 and 25).

9. Four identical point charges ($q = +10$ μC) are located on the corners of a rectangle as shown in Figure 23.28. The dimensions of the rectangle are $L = 60$ cm and $W = 15$ cm. Calculate the magnitude and direction of the net electrostatic force exerted on the charge at the lower left corner of the rectangle by the other three charges.

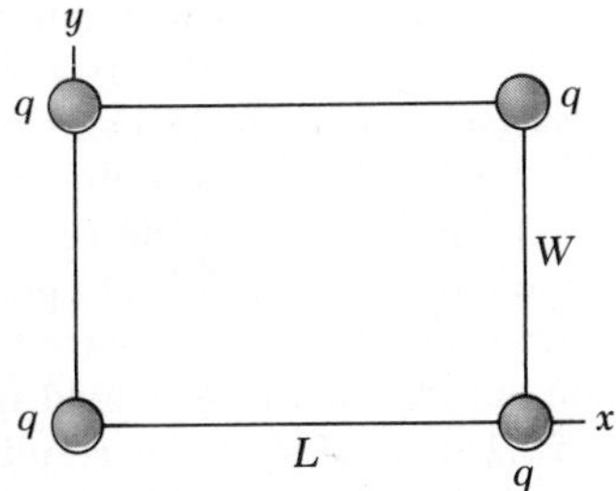

Figure 23.28 (Problems 9 and 20).

10. Four arbitrary pointlike charges are located at the corners of a rigid square plate. Prove that the net torque about the center of the plate is zero.

11. Three point charges lie along the y axis. A charge $q_1 = -9$ μC is at $y = 6.0$ m, and a charge $q_2 = -8$ μC is at $y = -4.0$ m. Where must a third positive charge, q_3, be placed such that the resultant force on it is zero?

12. A charge q_1 of $+3.4$ μC is located at $x = +2$ m, $y = +2$ m and a second charge $q_2 = +2.7$ μC is located at $x = -4$ m, $y = -4$ m. Where must a third charge ($q_3 > 0$) be placed such that the resultant force on q_3 will be zero?

13. Three identical point charges, each of magnitude q, are located on the vertices of an isosceles triangle with its altitude oriented vertically. The altitude of the triangle is 4 cm and the base is 6 cm long. (a) If the resultant electrical force exerted on the charge located at the upper vertex of the triangle has a magnitude of 0.5 N, determine q. (b) If the charge at the lower left vertex of the triangle is replaced by a charge $-q$, determine the magnitude and direction of the resultant force exerted on the charge located at the upper vertex of the triangle.

Section 23.4 The Electric Field

14. The electric force on a point charge of 4.0 μC at some point is 6.9×10^{-4} N in the positive x direction. What is the value of the electric field at that point?

15. What are the magnitude and direction of the electric field that will balance the weight of (a) an electron and (b) a proton? (Use the data in Table 23.1.)

16. An object having a net charge of 24 μC is placed in a uniform electric field of 610 N/C directed vertically. What is the mass of this object if it "floats" in this electric field?

17. A point charge of -5.2 μC is located at the origin. Find the electric field (a) on the x axis at $x = 3$ m, (b) on the y axis at $y = -4$ m, (c) at the point with coordinates $x = 2$ m, $y = 2$ m.

18. Find the total electric field along the line of the two charges shown in Figure 23.29 at the point midway between them.

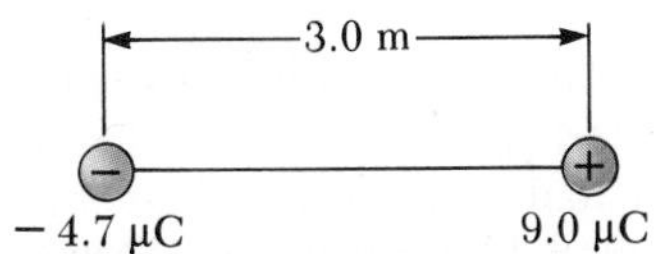

Figure 23.29 (Problems 18 and 27).

19. Two equal point charges each of magnitude 2.0 μC are located on the x axis. One is at $x = 1.0$ m, and the other is at $x = -1.0$ m. (a) Determine the electric field on the y axis at $y = 0.5$ m. (b) Calculate the electric force on a third charge, of -3.0 μC, placed on the y axis at $y = 0.5$ m.

20. Four identical point charges ($q = +6$ μC) are located on a rectangle as shown in Figure 23.28, with $L =$

80 cm and $W = 20$ cm. Calculate the resultant electric field at the center of the rectangle.

21. Three identical point charges ($q = -5\ \mu C$) are located along a circle of 2 m radius at angles of 30°, 150°, and 270° as shown in Figure 23.30. What is the resultant electric field at the center of the circle?

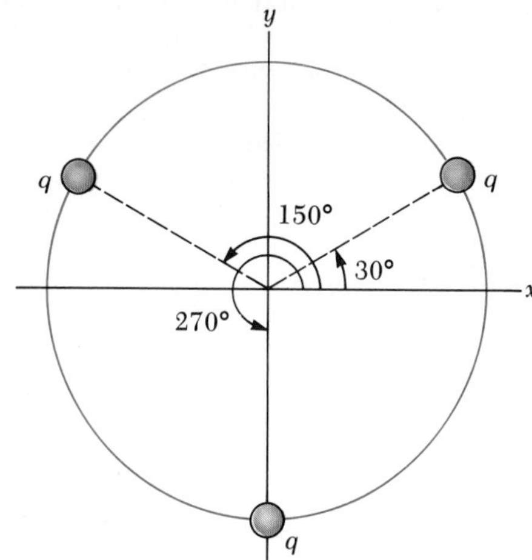

Figure 23.30 (Problem 21).

22. Three identical point charges ($q = +2.7\ \mu C$) are placed on the corners of an equilateral triangle whose sides have a length of 35 cm (see Figure 23.31). What is the magnitude of the resultant electric field at the center of the triangle?

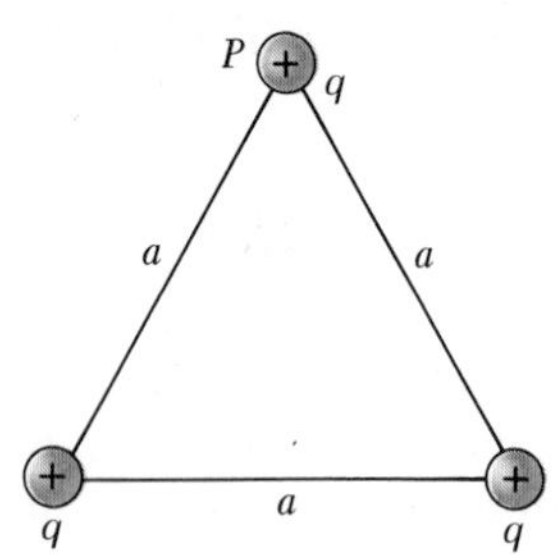

Figure 23.31 (Problems 22 and 23).

23. Three equal positive charges q are at the corners of an equilateral triangle of sides a as in Figure 23.31. (a) At what point in the plane of the charges (other than ∞) is the electric field zero? (b) What are the magnitude and direction of the electric field at the point P due to the two charges at the base of the triangle?

24. Three charges are at the corners of an equilateral triangle as in Figure 23.26. Calculate the electric field intensity at the position of the 2-μC charge due to the 7-μC and -4-μC charges.

25. Four charges are at the corners of a square as in Figure 23.27. (a) Find the magnitude and direction of the electric field at the position of the charge $-q$, the coordinates of which are $x = a$, $y = a$. (b) What is the electric force on this charge?

26. A charge of $-4\ \mu C$ is located at the origin, and a charge of $-5\ \mu C$ is located along the y axis at $y = 2.0$ m. At what point along the y axis is the electric field zero?

27. In Figure 23.29, determine the point (other than ∞) at which the total electric field is zero.

Section 23.5 Electric Field of a Continuous Charge Distribution

28. A rod 14 cm long is uniformly charged and has a total charge of $-22\ \mu C$. Determine the magnitude and direction of the electric field along the axis of the rod, at a point 36 cm from its center.

29. A continuous line of charge lies along the x-axis, extending from $x = +x_0$ to positive infinity. The line carries a uniform linear charge density λ_0. What are the magnitude and direction of the electric field at the origin?

30. A line of charge starts at $x = +x_0$ and extends to positive infinity. If the linear charge density is given by $\lambda = \lambda_0 x_0/x$, determine the electric field at the origin.

31. A uniformly charged ring of radius 10 cm has a total charge of 75 μC. Find the electric field on the *axis* of the ring at (a) 1 cm, (b) 5 cm, (c) 30 cm, and (d) 100 cm from the center of the ring.

32. Show that the maximum field strength E_m along the axis of a uniformly charged ring occurs at $x = a/\sqrt{2}$ (see Fig. 23.15) and has the value $Q/(6\sqrt{3}\pi\epsilon_0 a^2)$.

33. A sphere of radius 4 cm has a net charge of $+39\ \mu C$. (a) If this charge is uniformly distributed throughout the volume of the sphere, what is the volume charge density? (b) If this charge is uniformly distributed on the sphere's surface, what is the surface charge density?

34. A uniformly charged disk of radius 35 cm carries a charge density of $7.9 \times 10^{-3}\ C/m^2$. Calculate the electric field on the *axis* of the disk at (a) 5 cm, (b) 10 cm, (c) 50 cm, and (d) 200 cm from the center of the disk.

35. Example 23.10 derives the exact expression for the electric field at a point on the axis of a uniformly charged disk (see Equation 23.17). Consider a disk of radius $R = 3$ cm, having a uniformly distributed charge of $+5.2\ \mu C$. (a) Using the result of Example 23.10, compute the electric field at a point on the axis and 3 mm from the center. Compare this answer to the field computed from the near-field approximation (Equation 23.18). (b) Using the result of Example 23.10, compute the electric field at a point on the axis and 30 cm from the center of the disk. Compare this to the electric field obtained by treating the disk as a $+5.2\ \mu C$ point charge at a distance of 30 cm.

36. The electric field along the axis of a uniformly charged disk of radius R and total charge Q was calculated in Example 23.10. Show that the electric field at distances x that are large compared with R approaches that of a point charge $Q = \sigma\pi R^2$. (*Hint:* First show that $x/(x^2 + R^2)^{1/2} = (1 + R^2/x^2)^{-1/2}$ and use the binomial expansion $(1 + \delta)^n \approx 1 + n\delta$ when $\delta \ll 1$.)

37. A uniformly charged ring and a uniformly charged disk each have a charge of $+25\ \mu C$ and a radius of 3 cm. For each of these charged objects, determine the electric field at a point along the axis which is 4 cm from the center of the object.

38. A 10 gram piece of Styrofoam carries a net charge of $-0.7\ \mu C$ and "floats" above the center of a very large horizontal sheet of plastic which has a uniform charge density on its surface. What is the charge per unit area on the plastic sheet?

39. A uniformly charged insulating rod of length 14 cm is bent into the shape of a semicircle as in Figure 23.32. If the rod has a total charge of $-7.5\ \mu C$, find the magnitude and direction of the electric field at O, the center of the semicircle.

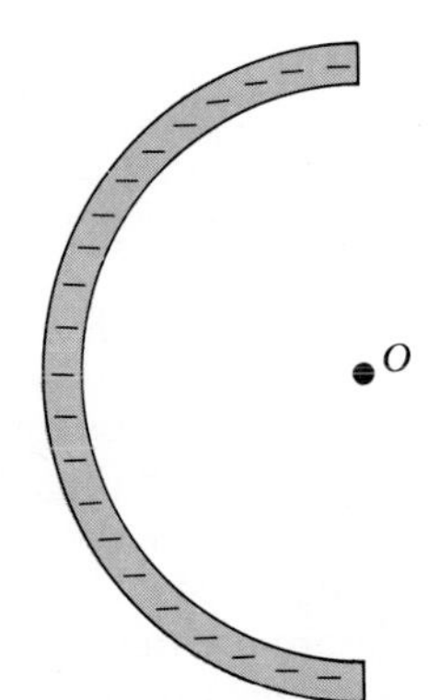

Figure 23.32 (Problem 39).

Section 23.6 Electric Field Lines

40. A positively charged disk has a uniform charge per unit area as described in Example 23.10. Sketch the electric field lines in a plane perpendicular to the plane of the disk passing through its center.

41. A negatively charged rod of finite length has a uniform charge per unit length. Sketch the electric field lines in a plane containing the rod.

42. Four equal positive point charges are at the corners of a square. Sketch the electric field lines in the plane of the square.

43. Figure 23.33 shows the electric field lines for two point charges separated by a small distance. (a) Determine the ratio q_1/q_2. (b) What are the signs of q_1 and q_2?

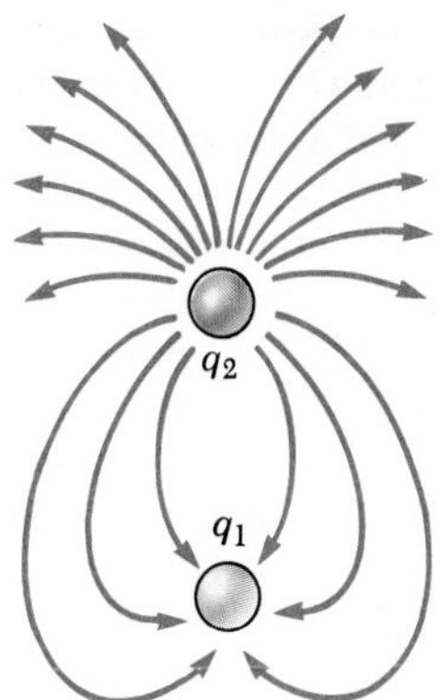

Figure 23.33 (Problem 43).

Section 23.7 Motion of Charged Particles in a Uniform Electric Field

44. An electron and a proton are each placed at rest in an external electric field of 520 N/C. Calculate the speed of each particle after 48 nanoseconds.

45. A proton accelerates from rest in a uniform electric field of 640 N/C. At some later time, its speed is 1.20×10^6 m/s (nonrelativistic since v is much less than the speed of light). (a) Find the acceleration of the proton. (b) How long does it take the proton to reach this velocity? (c) How far has it moved in this time? (d) What is its kinetic energy at this time?

46. Consider an electron which is released from rest in a uniform electric field. (a) If the electron is accelerated to 1% of the speed of light after traveling 2 mm, what is the strength of the electric field? (b) What speed will the electron have after traveling 4 mm from rest?

47. The electrons in a particle beam each have a kinetic energy of 1.6×10^{-17} J. What are the magnitude and direction of the electric field that will stop these electrons in a distance of 10 cm?

48. An electron traveling with an initial velocity equal to $8.6 \times 10^5\mathbf{i}$ m/s enters a region of a uniform electric field given by $\mathbf{E} = 4.1 \times 10^3\mathbf{i}$ N/C. (a) Find the acceleration of the electron. (b) Determine the time it takes for the electron to come to rest after it enters the field. (c) How far does the electron move in the electric field before coming to rest?

49. A proton is projected in the positive x direction into a region of a uniform electric field $\mathbf{E} = -6 \times 10^5\mathbf{i}$ N/C. The proton travels 7 cm before coming to rest. Determine (a) the acceleration of the proton, (b) its initial speed, and (c) the time it takes the proton to come to rest.

50. A proton and an electron both start from rest and from the same point in a uniform electric field of 370 N/C. How far apart are they after 1 μs? (Ignore the attrac-

tion between the electron and the proton. If you like, you might imagine the experiment to be tried with the proton only, and then repeated with the electron only.)

51. A proton has an initial velocity of 4.50×10^5 m/s in the horizontal direction. It enters a uniform electric field of 9.60×10^3 N/C directed vertically. Ignore any gravitational effects and (a) find the time it takes the proton to travel 5.0 cm horizontally, (b) the vertical displacement of the proton after it has traveled 5.0 cm horizontally, and (c) the horizontal and vertical components of the proton's velocity after it has traveled 5.0 cm horizontally.

52. An electron is projected at an angle of 30° above the horizontal at a speed of 8.2×10^5 m/s, in a region of an electric field $\boldsymbol{E} = 390\boldsymbol{j}$ N/C. Neglect gravity and find: (a) the time it takes the electron to return to its initial height, (b) the maximum height reached by the electron, and (c) its horizontal displacement when it reaches its maximum height.

53. Protons are projected with an initial speed, given by $v_0 = 9.55 \times 10^3$ m/s, into a region where a uniform electric field, $\boldsymbol{E} = -720\boldsymbol{j}$ N/C, is present as shown in Figure 23.34. The protons are to hit a target that lies at a horizontal distance of 1.27 mm from the point where the protons are launched. Find (a) the two projection angles θ that will result in a hit, and (b) the total time of flight for each of these two trajectories.

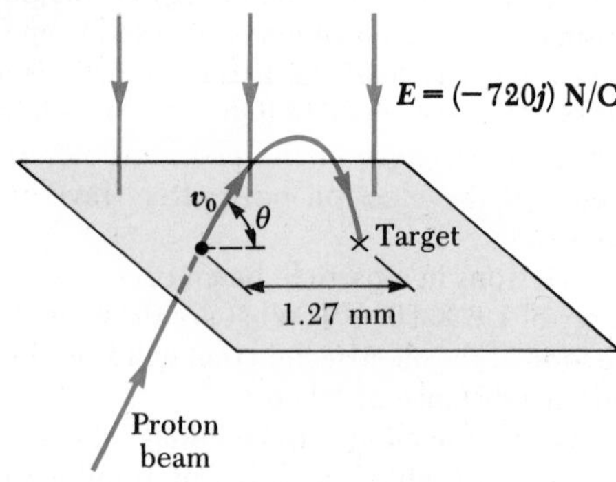

Figure 23.34 (Problem 53).

ADDITIONAL PROBLEMS

54. A small 2-g plastic ball is suspended by a 20-cm long string in a uniform electric field as shown in Figure 23.35. If the ball is in equilibrium when the string makes a 15° angle with the vertical as indicated, what is the net charge on the ball?

55. A charged cork ball of mass 1 g is suspended on a light string in the presence of a uniform electric field as in Figure 23.36. When $\boldsymbol{E} = (3\boldsymbol{i} + 5\boldsymbol{j}) \times 10^5$ N/C, the ball is in equilibrium at $\theta = 37°$. Find (a) the charge on the ball and (b) the tension in the string.

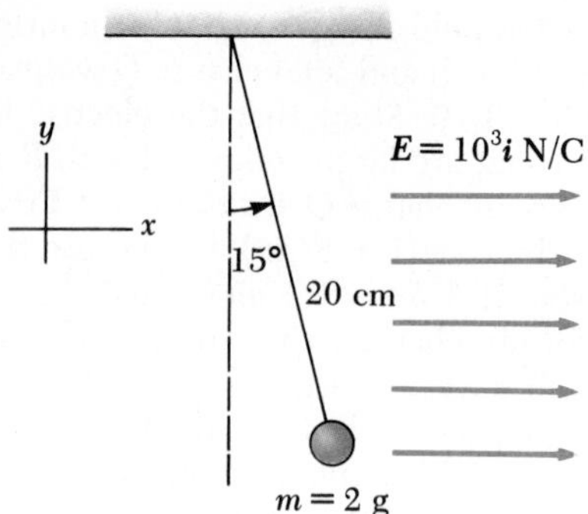

Figure 23.35 (Problem 54).

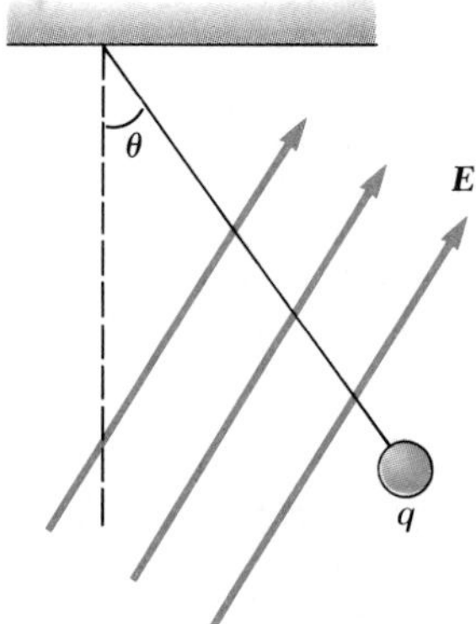

Figure 23.36 (Problem 55).

56. Two small spheres each of mass 2 g are suspended by light strings 10 cm in length (Fig. 23.37). A uniform electric field is applied in the x direction. If the spheres have charges equal to -5×10^{-8} C and $+5 \times 10^{-8}$ C, determine the electric field intensity that enables the spheres to be in equilibrium at an angle of $\theta = 10°$.

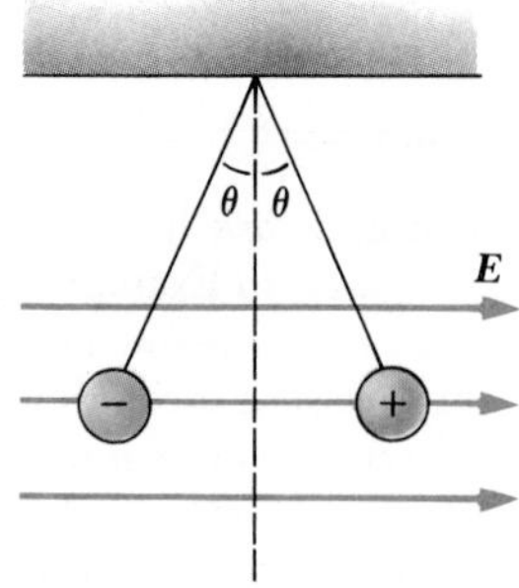

Figure 23.37 (Problem 56).

57. Two small spheres of mass m are suspended from strings of length ℓ that are connected at a common point. One sphere has charge Q; the other has charge $2Q$. Assume the angles, θ_1 and θ_2, that the strings make with the vertical are small. (a) How are θ_1 and θ_2

related? (b) Show that the distance r between the spheres is

$$r \simeq \left(\frac{4kQ^2\ell}{mg}\right)^{1/3}$$

58. Three charges are located at the corners of an equilateral triangle as in Figure 23.38. (a) Determine the magnitude and direction of the net electric field at the *center* of the triangle. (b) Find the magnitude and direction of the resultant force on the charge $-q$.

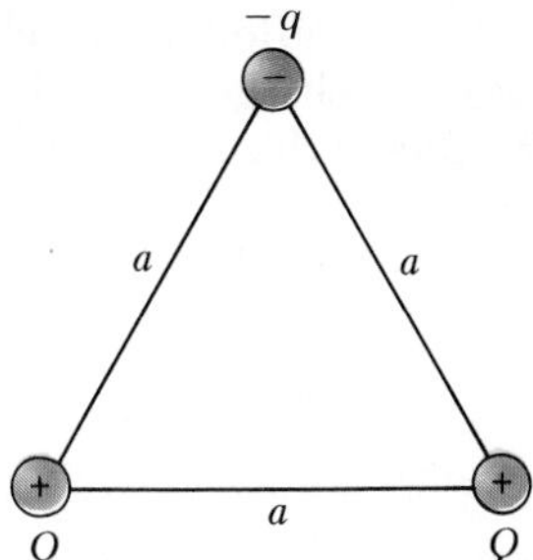

Figure 23.38 (Problem 58).

59. Three identical small Styrofoam balls ($m = 2$ g) are suspended from a fixed point by three nonconducting threads, each with a length of 50 cm and with negligible mass. At equilibrium the three balls form an equilateral triangle with sides of 30 cm. What is the common charge q that is carried by each ball?

60. Two small spheres of charge Q are suspended from strings of length ℓ that are connected at a common point. One sphere has mass m; the other has mass $2m$. Assume the angles, θ_1 and θ_2, that the strings make with the vertical are small. (a) How are θ_1 and θ_2 related? (b) Show that the distance r between the spheres is

$$r \simeq \left(\frac{3kQ^2\ell}{2mg}\right)^{1/3}$$

61. Consider the electric dipole shown in Figure 23.39. Show that the electric field at a *distant* point along the x axis is given by $E_x \cong 2kp/x^3$, where $p = 2qa$ is the dipole moment.

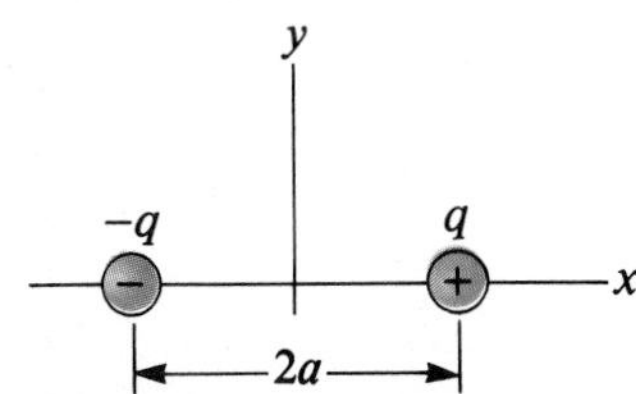

Figure 23.39 (Problem 61).

62. Two equal positive charges q are located on the x axis at $x = a$ and $x = -a$. (a) Show that the field at a point on the y axis is in the y direction and is given by $E_y = 2kqy(y^2 + a^2)^{-3/2}$. (b) Determine the field at a point on the y axis for $y \gg a$, and explain your result. (c) Show that the field is a maximum at $y = \pm a/\sqrt{2}$ (*Hint:* When E_y is a maximum, $dE_y/dy = 0$.)

63. Three charges of equal magnitude q reside at the corners of an equilateral triangle of side length a. Two of the charges are negative, and the other is positive, as shown in Figure 23.40. (a) Find the magnitude and direction of the electric field at point P, midway between the negative charges, in terms of k, q, and a. (b) Where must a $-4q$ charge be placed so that any charge located at point P will experience no net electrostatic force ($\mathbf{F}_e = 0$)? In (b) let the distance between the $+q$ charge and point P be one meter.

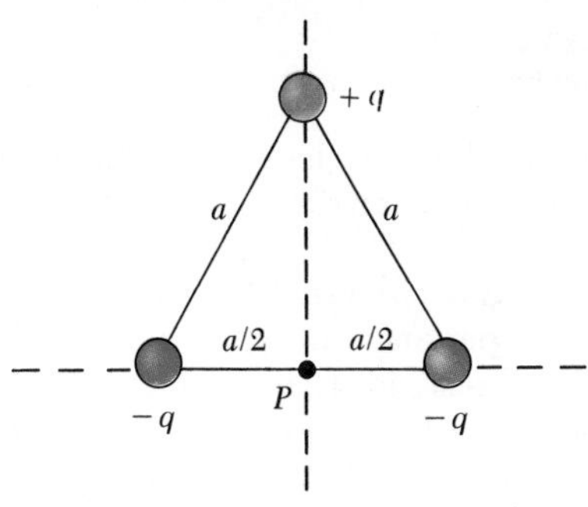

Figure 23.40 (Problem 63).

64. Two rings each of radius R have their axes oriented along the same line and are separated by a distance $2R$. One ring has uniform linear charge density $+\lambda$ and the other ring has uniform linear charge density $-\lambda$. Find (a) the electric field at point P_1 midway between the two rings on the common axis and (b) the electric field at point P_2 on the axis a distance R outside the negatively charged ring.

65. A rod of length ℓ with uniform charge per unit length λ is placed a distance d from the origin along the x axis. A similar rod with the same charge is placed along the y axis (Fig. 23.41). Determine the net electric field intensity at the origin. (*Hint:* See Example 23.8.)

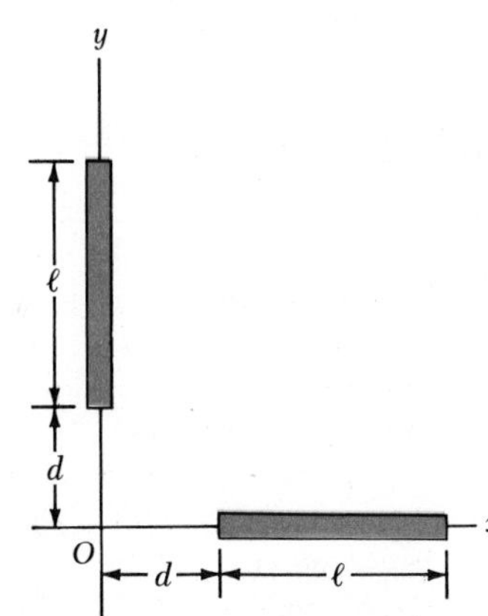

Figure 23.41 (Problem 65).

66. A uniformly charged rod with charge per unit length λ is bent into the shape of a circular arc of radius R as in Figure 23.42. The arc subtends an angle 2θ at the center of the circle. Show that the electric field at the center of the circle is in the y direction and is given by $E = (2k\lambda \sin \theta)/R$.

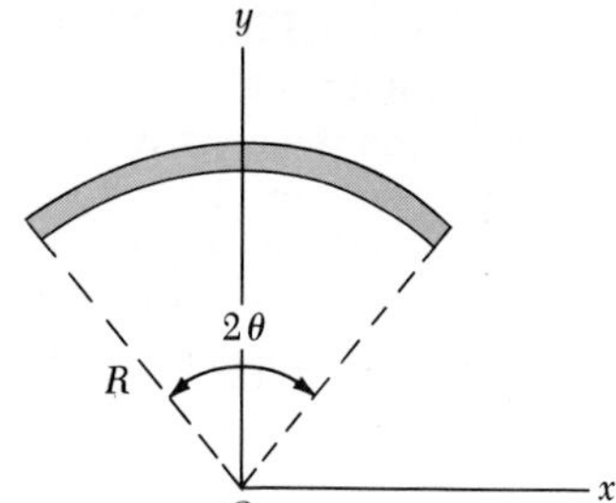

Figure 23.42 (Problem 66).

67. Air will break down (lose its insulating quality) and sparking will result if the field strength is increased to about 3×10^6 N/C. (This field strength is also expressed as 3×10^6 V/m.) What acceleration will an electron experience in such a field? If the electron starts from rest, in what distance will it acquire a speed equal to 10 percent of the speed of light?

68. A line charge of length ℓ and oriented along the x axis as in Figure 23.14 has a charge per unit length λ, which varies with x as $\lambda = \lambda_0(x - d)/d$, where d is the distance of the rod from the origin (point P in the figure) and λ_0 is a constant. Find the electric field at the origin. (*Hint:* An infinitesimal element has a charge $dq = \lambda\, dx$, but note that λ is *not* a constant.)

69. A thin rod of length ℓ and uniform charge per unit length λ lies along the x axis as shown in Figure 23.43. (a) Show that the electric field at the point P, a distance y from the rod, along the perpendicular bisector has no x component and is given by $E = 2k\lambda \sin \theta_0/y$. (b) Using your result to (a), show that the field of a rod of *infinite* length is given by $E = 2k\lambda/y$. (*Hint:* First calculate the field at P due to an element of length dx, which has a charge $\lambda\, dx$. Then change variables from x to θ using the facts that $x = y \tan \theta$ and $dx = y \sec^2 \theta\, d\theta$ and integrate over θ.)

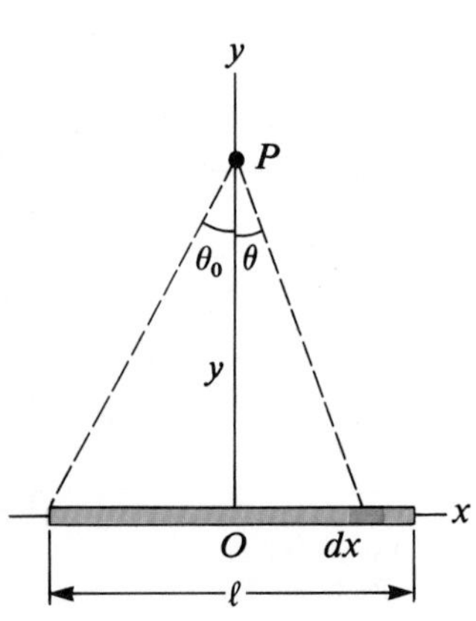

Figure 23.43 (Problem 69).

70. A thin insulating rod is bent into a semicircle of radius r. The rod has nonuniform charge density $\lambda = \lambda_0 \sin \theta$, where λ_0 is a constant and θ is an angle measured from either end of the rod using the center of the semicircle as the vertex. Find the magnitude and direction of the electric field at the center of the semicircle.

71. A set of eight point charges, each of magnitude $+q$, is located on the corners of a cube of side s as shown in Figure 23.44. (a) Determine x, y, and z components of the resultant force exerted on the charge located at point A by the other charges. (b) What are the magnitude and direction of this resultant force?

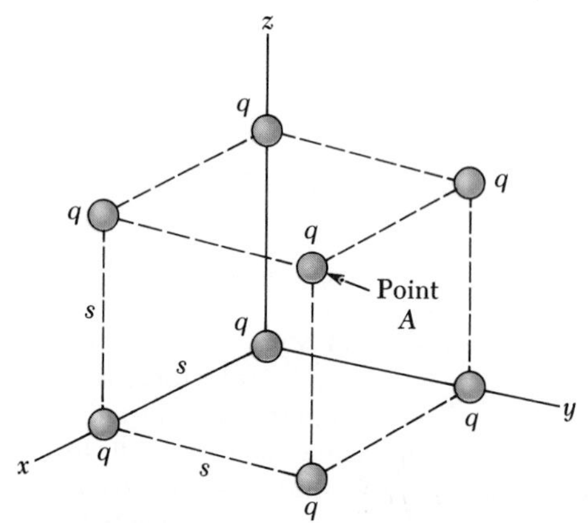

Figure 23.44 (Problems 71 and 72).

72. Consider the charge distribution shown in Figure 23.44. (a) Show that the magnitude of the electric field at the center of any face of the cube has a value of $2.18\ kq/s^2$. (b) What is the direction of the electric field at the center of the top face of the cube?

73. Three point charges q, $-2q$, and q are located along the x axis as in Figure 23.45. Show that the electric field at the distant point P ($y \gg a$) along the y axis is given by

$$\mathbf{E} = -k\frac{3qa^2}{y^4}\,\mathbf{j}$$

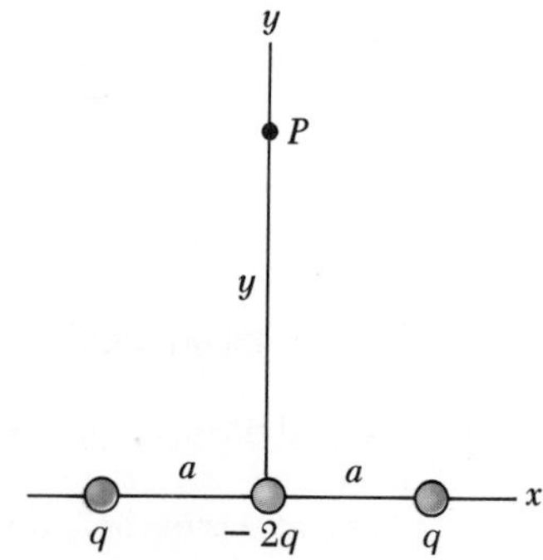

Figure 23.45 (Problem 73).

This charge distribution, which is essentially that of two electric dipoles, is called an *electric quadrupole.* Note that E varies as r^{-4} for the quadrupole, compared with variations of r^{-3} for the dipole and r^{-2} for the monopole (a single charge).

74. An electric dipole in a uniform electric field is displaced slightly from its equilibrium position, as in Figure 23.46, where θ is small. The dipole moment is $p = 2qa$ and the moment of inertia of the dipole is I. If the dipole is released from this position, show that it exhibits simple harmonic motion with a frequency given by

$$f = \frac{1}{2\pi}\sqrt{\frac{pE}{I}}$$

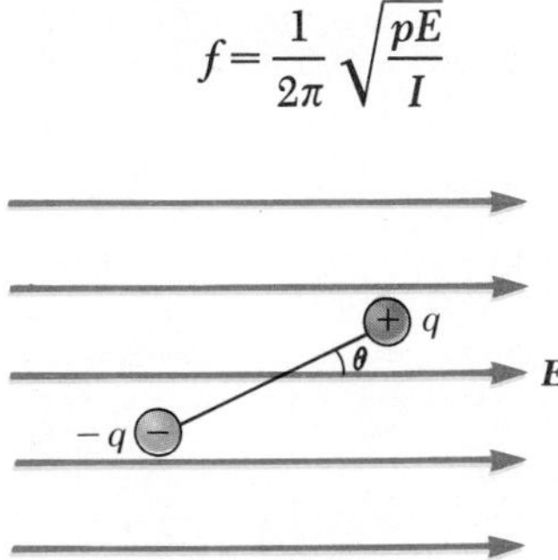

Figure 23.46 (Problem 74).

75. A *negatively* charged particle $-q$ is placed at the center of a uniformly charged ring, where the ring has a total positive charge Q as in Example 23.9. The particle, confined to move along the x axis, is displaced a *small* distance x along the axis (where $x \ll a$) and released. Show that the particle oscillates with simple harmonic motion along the x axis with a frequency given by

$$f = \frac{1}{2\pi}\left(\frac{kqQ}{ma^3}\right)^{1/2}$$

76. The *cathode-ray oscilloscope* operates on the following principle. An electron with charge $-e$ and mass m is projected with a speed v_0 at right angles to a uniform electric field (Fig. 23.47) and is deflected as shown. A screen is placed a distance L from the charged plates. Ignoring the effects of gravity, (a) show that the equation of the path followed by the charge *in the field* is given by $y = (eE/2mv_0^2)x^2$. (b) If $L \gg d$, show that the charge-to-mass ratio is given by $e/m = hv_0^2/ELd$. (This problem also suggests a means of measuring e/m for other charged particles.)

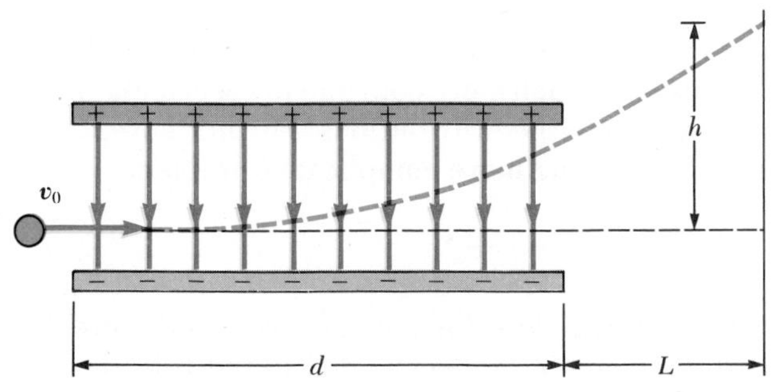

Figure 23.47 (Problem 76).

77. Consider an infinite sheet of charge with uniform charge density σ situated in the yz plane. Find the electric field at a point P along the x axis a distance x from the origin. (*Hint:* Use a ring of charge as the differential element. Change all variables to terms of θ, where θ is the angle indicated in Figure 23.15. Sum the contributions of all rings by evaluating the integral for E from $\theta = 0$ to $\theta = \pi/2$.)

78. A charge distribution with linear charge density λ and radius r, is a semicircle oriented in the xz plane with its bisector coincident with the negative x axis and with its center at the origin. Show that the electric field at a distance y from the origin along the y axis is

$$\mathbf{E} = \frac{k\lambda}{(r^2 + y^2)^{1/2}}\left(2\mathbf{i} + \pi\frac{y}{r}\mathbf{j}\right)$$

CALCULATOR/COMPUTER PROBLEMS

79. A continuous charge is distributed along a rod lying along the x axis as in Figure 23.14. The total charge on the rod is $Q = +16 \times 10^{-10}$ C, $d = 1.0$ m, and $\ell = 2.0$ m. Estimate the electric field at $x = 0$ by approximating the rod to be (a) a point charge at $x = 2.0$ m, (b) two point charges (each of charge 8×10^{-10} C) at $x = 1.5$ m and $x = 2.5$ m, and (c) four point charges (each of charge 4×10^{-10} C) at $x = 1.25$ m, $x = 1.75$ m, $x = 2.25$ m, and $x = 2.75$ m. (d) Write a program that will enable you to extend your calculations to 256 equally spaced point charges, and compare your result with that given by the exact expression, Equation 23.15.

80. Consider a uniform ring of charge located in the yz plane as in Figure 23.15a, where $Q = +16 \times 10^{-10}$ C, and the radius $a = 1$ m. Estimate the electric field along the x axis at $x = 3$ m by approximating the ring to be (a) a point charge at $x = 0$, (b) two point charges (each of charge 8×10^{-10} C) diametrically opposite each other on the ring, and (c) four point charges (each of charge 4×10^{-10} C) symmetrically spaced on the ring. (d) Write a program that will enable you to extend your calculations to 64 point charges equally spaced on the ring, and compare your result with that given by the exact expression, Equation 23.16.

24
Gauss' Law

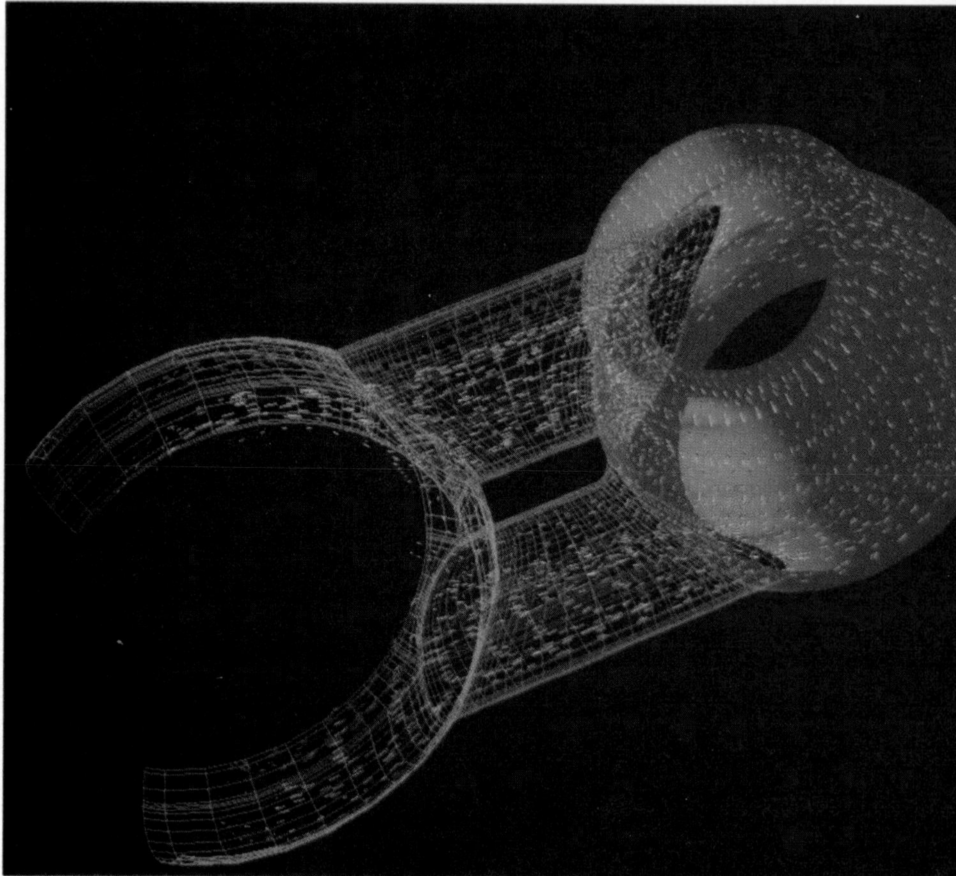

This computer-generated picture shows the surfaces of a manifold in a space shuttle's main engine—just one of the many kinds of surfaces that crop up in scientific and mathematical problems. Gaussian surfaces, described in this chapter, are hypothetical closed surfaces, useful for solving many problems in electrostatics. (© Dale E. Boyer, Science Source/Photo Researchers)

In the preceding chapter we showed how to calculate the electric field of a given charge distribution from Coulomb's law. This chapter describes an alternative procedure for calculating electric fields known as *Gauss' law*. This formulation is based on the fact that the fundamental electrostatic force between point charges is an inverse-square law. Although Gauss' law is a consequence of Coulomb's law, it is much more convenient for calculating the electric field of highly symmetric charge distributions. Furthermore, Gauss' law serves as a guide for understanding more complicated problems.

24.1 ELECTRIC FLUX

The concept of electric field lines was described qualitatively in the previous chapter. We shall now use the concept of electric flux to put this idea on a quantitative basis. *Electric flux is a measure of the number of electric field lines penetrating some surface.* When the surface being penetrated encloses some net charge, the net number of lines that go through the surface is proportional to the net charge within the surface. The number of lines counted is independent of the shape of the surface enclosing the charge. This is essentially a statement of Gauss' law, which we describe in the next section.

First consider an electric field that is uniform in both magnitude and direction, as in Figure 24.1. The electric field lines penetrate a rectangular surface of area A, which is perpendicular to the field. Recall that the number of lines per unit area is proportional to the magnitude of the electric field. Therefore, the number of lines penetrating the surface of area A is proportional to the product EA. The product of the electric field strength, E, and a surface area A perpendicular to the field is called the **electric flux, Φ**:

$$\Phi = EA \tag{24.1}$$

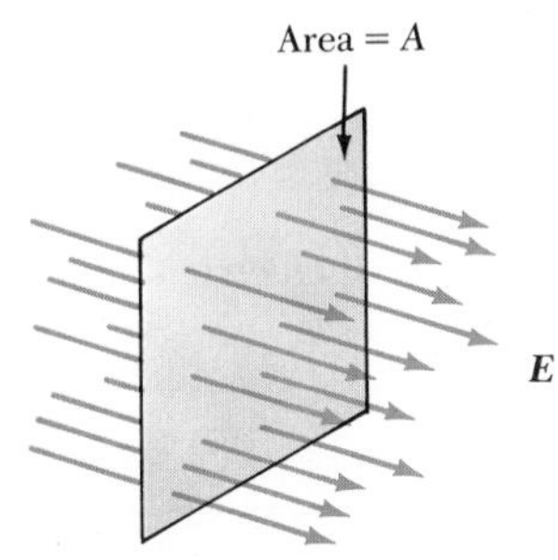

Figure 24.1 Field lines of a uniform electric field penetrating a plane of area A perpendicular to the field. The electric flux, Φ, through this area is equal to EA.

From the SI units of E and A, we see that electric flux has the units of $N \cdot m^2/C$.

If the surface under consideration is not perpendicular to the field, the number of lines (or the flux) through it must be less than that given by Equation 24.1. This can be easily understood by considering Figure 24.2, where the normal to the surface of area A is at an angle θ to the uniform electric field. Note that the number of lines that cross this area is equal to the number that cross the projected area A', which is perpendicular to the field. From Figure 24.2 we see that the two areas are related by $A' = A \cos\theta$. Since the flux through the area A equals the flux through A', we conclude that the desired flux is given by

$$\Phi = EA\cos\theta \tag{24.2}$$

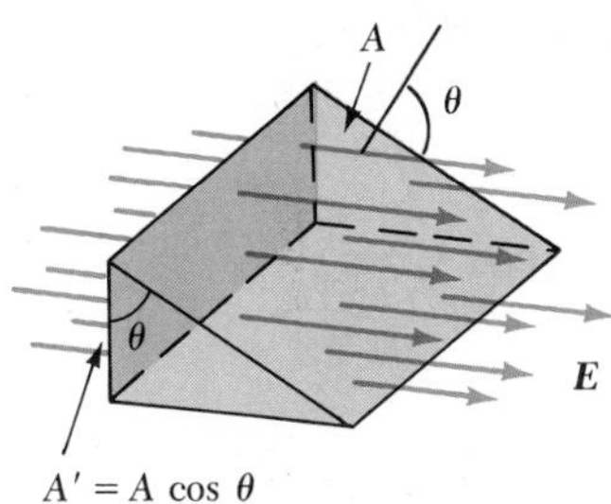

Figure 24.2 Field lines for a uniform electric field through an area A that is at an angle θ to the field. Since the number of lines that go through the shaded area A' is the same as the number that go through A, we conclude that the flux through A' is equal to the flux through A and is given by $\Phi = EA \cos\theta$.

From this result, we see that the flux through a surface of fixed area has the maximum value, EA, when the surface is perpendicular to the field (or when the *normal* to the surface is parallel to the field, that is, $\theta = 0°$); the flux is zero when the surface is parallel to the field (or when the normal to the surface is perpendicular to the field, that is, $\theta = 90°$).

In more general situations, the electric field may vary over the surface in question. Therefore, our definition of flux given by Equation 24.2 has meaning only over a small element of area. Consider a general surface divided up into a large number of small elements, each of area ΔA. The variation in the electric field over the element can be neglected if the element is small enough. It is convenient to define a vector $\Delta\boldsymbol{A}_i$ whose magnitude represents the area of the ith element and whose direction is *defined to be perpendicular* to the surface, as in Figure 24.3. The electric flux $\Delta\Phi_i$ through this small element is given by

$$\Delta\Phi_i = E_i\,\Delta A_i \cos\theta = \boldsymbol{E}_i \cdot \Delta\boldsymbol{A}_i$$

where we have used the definition of the scalar product of two vectors ($\boldsymbol{A} \cdot \boldsymbol{B} = AB\cos\theta$). By summing the contributions of all elements, we obtain the total flux through the surface.[1] If we let the area of each element approach

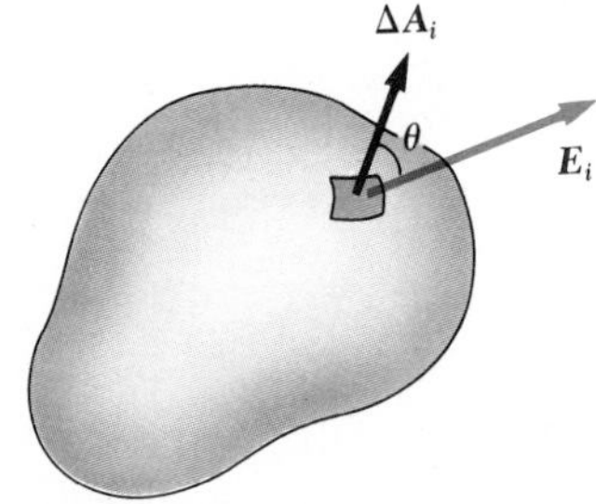

Figure 24.3 A small element of a surface of area ΔA_i. The electric field makes an angle θ with the normal to the surface (the direction of $\Delta\boldsymbol{A}_i$), and the flux through the element is equal to $E_i\,\Delta A_i \cos\theta$.

[1] It is important to note that drawings with field lines have their inaccuracies, since a small area (depending on its location) may happen to have too many or too few penetrating lines. At any rate, it is stressed that the basic definition of electric flux is $\int \boldsymbol{E} \cdot d\boldsymbol{A}$. The use of lines is only an aid for visualizing the concept.

zero, then the number of elements approaches infinity and the sum is replaced by an integral. Therefore *the general definition of electric flux is*

Definition of electric flux

$$\Phi \equiv \lim_{\Delta A_i \to 0} \sum \boldsymbol{E}_i \cdot \Delta \boldsymbol{A}_i = \int_{\text{surface}} \boldsymbol{E} \cdot d\boldsymbol{A} \tag{24.3}$$

Equation 24.3 is a surface integral, which must be evaluated over the hypothetical surface in question. In general, the value of Φ depends both on the field pattern and on the specified surface.

We shall usually be interested in evaluating the flux through a *closed surface.* (A closed surface is defined as a surface which divides space into an inside and an outside region, so that one cannot move from one region to the other without crossing the surface. The surface of a sphere, for example, is a closed surface.) Consider the closed surface in Figure 24.4. Note that the vectors $\Delta \boldsymbol{A}_i$ point in different directions for the various surface elements. At each point, these vectors are *normal* to the surface and, by convention, always point *outward.* At the elements labeled ① and ②, $\boldsymbol{E}$ is outward and $\theta < 90°$; hence the flux $\Delta\Phi = \boldsymbol{E} \cdot \Delta \boldsymbol{A}$ through these elements is positive. On the other hand, for elements such as ③, where the field lines are directed into the surface, $\theta > 90°$ and the flux becomes negative with $\cos\theta$. The total, or net, flux through the surface is proportional to the net number of lines penetrating the surface (where the net number means *the number leaving the volume surrounding the surface minus the number entering the surface*). If there are more lines leaving the surface than entering, the net flux is positive. If more lines enter than leave the surface, the net flux is negative. Using the symbol $\oint$ to represent an *integral over a closed surface,* we can write the net flux, Φ_c, through a closed surface

$$\Phi_c = \oint \boldsymbol{E} \cdot d\boldsymbol{A} = \oint E_n \, dA \tag{24.4}$$

where E_n represents the component of the electric field perpendicular, or normal, to the surface and the subscript c denotes a closed surface. Evaluating the net flux through a closed surface could be very cumbersome. However, if the field is normal to the surface at each point and constant in magnitude, the calculation is straightforward. The following example illustrates this point.

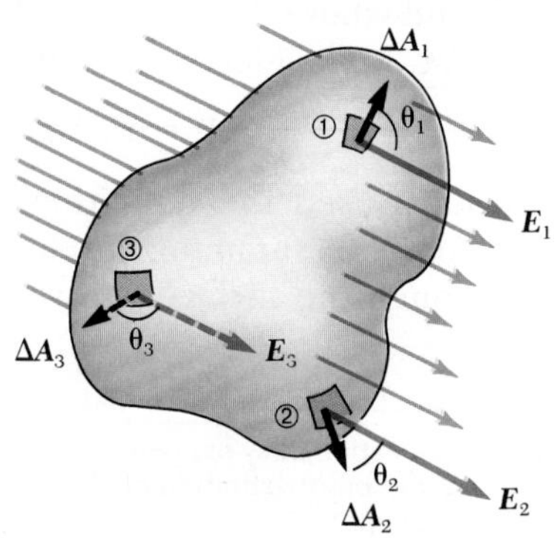

Figure 24.4 A closed surface in an electric field. The area vectors $\Delta \boldsymbol{A}_i$ are, by convention, normal to the surface and point outward. The flux through an area element can be positive (elements ① and ②) or negative (element ③).

EXAMPLE 24.1 Flux Through a Cube

Consider a uniform electric field $\mathbf{E}$ oriented in the x direction. Find the net electric flux through the surface of a cube of edges ℓ oriented as shown in Figure 24.5.

Solution The net flux can be evaluated by summing up the fluxes through each face of the cube. First, note that the flux through *four* of the faces is zero, since $\mathbf{E}$ is perpendicular to $d\mathbf{A}$ on these faces. In particular, the orientation of $d\mathbf{A}$ is perpendicular to $\mathbf{E}$ for the two faces labeled ③ and ④ in Figure 24.5. Therefore, $\theta = 90°$, so that $\mathbf{E}\cdot d\mathbf{A} = E\,dA\cos 90° = 0$. The fluxes through the planes parallel to the yx plane are also zero for the same reason.

Now consider the faces labeled ① and ②. The net flux through these faces is given by

$$\Phi_c = \int_1 \mathbf{E}\cdot d\mathbf{A} + \int_2 \mathbf{E}\cdot d\mathbf{A}$$

For the face labeled ①, $\mathbf{E}$ is constant and inward while $d\mathbf{A}$ is outward ($\theta = 180°$), so that we find that the flux through this face is

$$\int_1 \mathbf{E}\cdot d\mathbf{A} = \int_1 E\,dA\cos 180° = -E\int_1 dA$$

$$= -EA = -E\ell^2$$

since the area of each face is $A = \ell^2$.

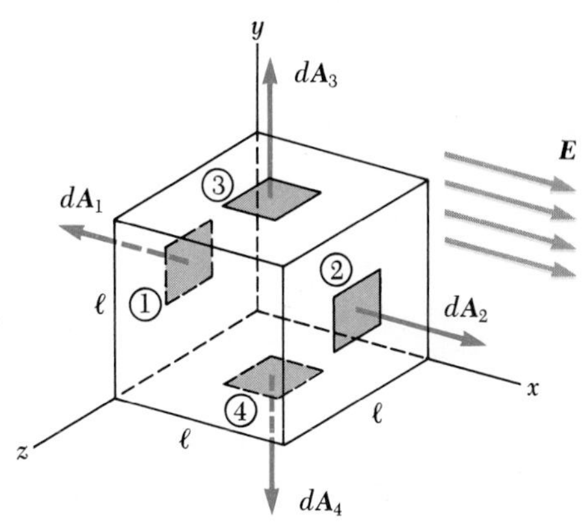

Figure 24.5 (Example 24.1) A hypothetical surface in the shape of a cube in a uniform electric field parallel to the x axis. The net flux through the surface is zero.

Likewise, for the face labeled ②, $\mathbf{E}$ is constant and outward and in the same direction as $d\mathbf{A}$ ($\theta = 0°$), so that the flux through this face is

$$\int_2 \mathbf{E}\cdot d\mathbf{A} = \int_2 E\,dA\cos 0° = E\int_2 dA = +EA = E\ell^2$$

Hence, the net flux over all faces is zero, since

$$\Phi_c = -E\ell^2 + E\ell^2 = 0$$

24.2 GAUSS' LAW

In this section we describe a general relation between the net electric flux through a closed surface (often called a *gaussian surface*) and the charge *enclosed* by the surface. This relation, known as *Gauss' law*, is of fundamental importance in the study of electrostatic fields.

First, let us consider a positive point charge q located at the center of a sphere of radius r as in Figure 24.6. From Coulomb's law we know that the magnitude of the electric field everywhere on the surface of the sphere is $E = kq/r^2$. Furthermore, the field lines are radial outward, and hence are perpendicular (or normal) to the surface at each point. That is, at each point E is parallel to the vector $\Delta\mathbf{A}_i$ representing the local element of area ΔA_i. Therefore

$$\mathbf{E}\cdot\Delta\mathbf{A}_i = E_n\,\Delta A_i = E\,\Delta A_i$$

and from Equation 24.4 we find that the net flux through the gaussian surface is given by

$$\Phi_c = \oint E_n\,dA = \oint E\,dA = E\oint dA$$

since E is constant over the surface and given by $E = kq/r^2$. Furthermore, for a spherical gaussian surface, $\oint dA = A = 4\pi r^2$ (the surface area of a sphere).

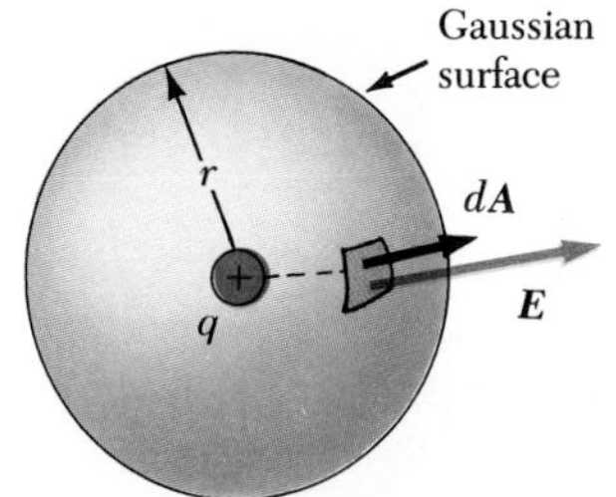

Figure 24.6 A spherical surface of radius r surrounding a point charge q. When the charge is at the center of the sphere, the electric field is normal to the surface and constant in magnitude everywhere on the surface.

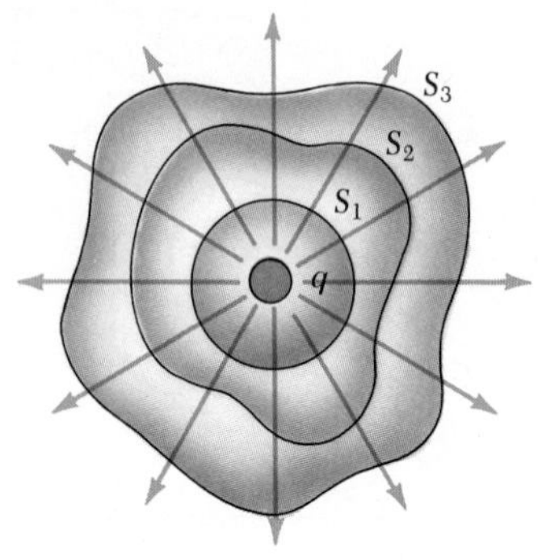

Figure 24.7 Closed surfaces of various shapes surrounding a charge q. Note that the net electric flux through each surface is the same.

Hence the net flux through the gaussian surface is

$$\Phi_c = \frac{kq}{r^2}(4\pi r^2) = 4\pi kq$$

Recalling that $k = 1/4\pi\epsilon_0$, we can write this in the form

$$\Phi_c = \frac{q}{\epsilon_0} \tag{24.5}$$

Note that this result, which is independent of r, says that the net flux through a spherical gaussian surface is proportional to the charge q *inside* the surface. The fact that the flux is independent of the radius is a consequence of the inverse-square dependence of the electric field given by Coulomb's law. That is, E varies as $1/r^2$, but the area of the sphere varies as r^2. Their combined effect produces a flux that is independent of r.

Now consider several closed surfaces surrounding a charge q as in Figure 24.7. Surface S_1 is spherical, whereas surfaces S_2 and S_3 are nonspherical. The flux that passes through surface S_1 has the value q/ϵ_0. As we discussed in the previous section, the flux is proportional to the number of electric field lines passing through that surface. The construction in Figure 24.7 shows that the number of electric field lines through the spherical surface S_1 is equal to the number of electric field lines through the nonspherical surfaces S_2 and S_3. Therefore, it is reasonable to conclude that the net flux through any closed surface is independent of the shape of that surface. (One can prove that this is the case if $E \propto 1/r^2$.) In fact, *the net flux through any closed surface surrounding a point charge q is given by q/ϵ_0.*

Now consider a point charge located *outside* a closed surface of arbitrary shape, as in Figure 24.8. As you can see from this construction, some electric field lines enter the surface, and others leave the surface. However, *the number of electric field lines entering the surface equals the number leaving the surface.* Therefore, we conclude that *the net electric flux through a closed surface that surrounds no charge is zero.* If we apply this result to Example 24.1, we can easily see that the net flux through the cube is zero, since it was assumed there was no charge inside the cube.

The net flux through a closed surface is zero if there is no charge inside

Let us extend these arguments to the generalized case of many point charges, or a continuous distribution of charge. We shall make use of the **superposition principle,** which says that *the electric field due to many charges is the vector sum of the electric fields produced by the individual charges.* That is, we can express the flux through any closed surface as

$$\oint \mathbf{E}\cdot d\mathbf{A} = \oint (\mathbf{E}_1 + \mathbf{E}_2 + \mathbf{E}_3)\cdot d\mathbf{A}$$

where $\mathbf{E}$ is the total electric field at any point on the surface and $\mathbf{E}_1$, $\mathbf{E}_2$, and $\mathbf{E}_3$ are the fields produced by the individual charges at that point. Consider the system of charges shown in Figure 24.9. The surface S surrounds only one charge, q_1; hence the net flux through S is q_1/ϵ_0. The flux through S due to the charges outside it is zero since each electric field line that enters S at one point leaves it at another. The surface S′ surrounds charges q_2 and q_3; hence the net flux through S′ is $(q_2 + q_3)/\epsilon_0$. Finally, the net flux through surface S″ is zero

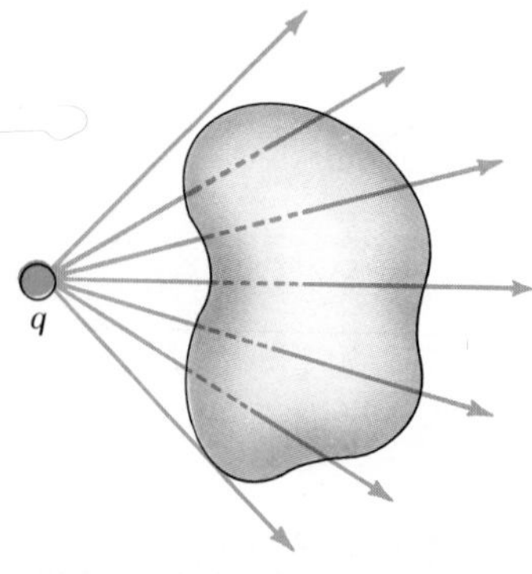

Figure 24.8 A point charge located *outside* a closed surface. In this case, note that the number of lines entering the surface equals the number leaving the surface.

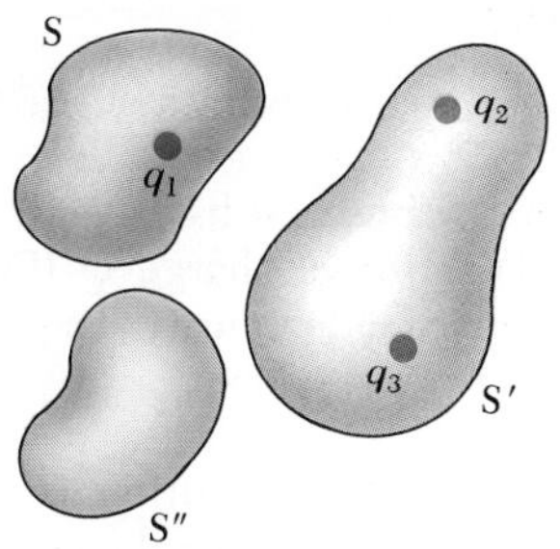

Figure 24.9 The net electric flux through any closed surface depends only on the charge *inside* that surface. The net flux through surface S is q_1/ϵ_o, the net flux through surface S′ is $(q_2 + q_3)/\epsilon_o$, and the net flux through surface S″ is zero.

since there is no charge inside this surface. That is, *all* lines that enter S″ at one point leave S″ at another.

Gauss' law, which is a generalization of the above discussion, states that the net flux through *any* closed surface is given by

$$\Phi_c = \oint \mathbf{E} \cdot d\mathbf{A} = \frac{q_{in}}{\epsilon_o} \tag{24.6}$$

where q_{in} represents the *net charge inside* the gaussian surface and $\mathbf{E}$ represents the electric field at any point on the gaussian surface. In words,

Gauss' law

Gauss' law states that the net electric flux through any closed gaussian surface is equal to the net charge inside the surface divided by ϵ_o.

A formal proof of Gauss' law is presented in Section 24.6. When using Equation 24.6, you should note that although the charge q_{in} is the net charge inside the gaussian surface, the $\mathbf{E}$ that appears in Gauss' law represents the *total electric field*, which includes contributions from charges both inside and outside the gaussian surface. This point is often neglected or misunderstood.

In principle, Gauss' law can always be used to calculate the electric field of a system of charges or a continuous distribution of charge. However, in practice, *the technique is useful only in a limited number of situations where there is a high degree of symmetry*. As we shall see in the next section, *Gauss' law can be used to evaluate the electric field for charge distributions that have spherical, cylindrical, or plane symmetry*. If one carefully chooses the gaussian surface surrounding the charge distribution, the integral in Equation 24.6 will be easy to evaluate. You should also note that a gaussian surface is a mathematical surface and need not coincide with any real physical surface.

24.3 APPLICATION OF GAUSS' LAW TO CHARGED INSULATORS

In this section we give some examples of how to use Gauss' law to calculate $\mathbf{E}$ for a given charge distribution. It is important to recognize that *Gauss' law is useful when there is a high degree of symmetry in the charge distribution, as in the case of uniformly charged spheres, long cylinders, and plane sheets*. In such cases, it is possible to find a simple gaussian surface over which the surface integral given by Equation 24.6 is easily evaluated.

Gauss' law is useful for evaluating E when the charge distribution has symmetry

The surface should always be chosen such that it has the same symmetry as that of the charge distribution.

The following examples should clarify this procedure.

EXAMPLE 24.2 The Electric Field Due to a Point Charge

Starting with Gauss' law, calculate the electric field due to an isolated point charge q and show that Coulomb's law follows from this result.

Solution For this situation we choose a spherical gaussian surface of radius r and centered on the point charge, as in Figure 24.10. The electric field of a positive point charge is radial outward by symmetry, and is therefore normal to the surface at every point. That is, $\mathbf{E}$ is parallel to $d\mathbf{A}$ at each point, and so $\mathbf{E}\cdot d\mathbf{A} = E\,dA$ and Gauss' law gives

$$\Phi_c = \oint \mathbf{E}\cdot d\mathbf{A} = \oint E\,dA = \frac{q}{\epsilon_0}$$

By symmetry, E is constant everywhere on the surface, and so it can be removed from the integral. Therefore,

$$\oint E\,dA = E\oint dA = E(4\pi r^2) = \frac{q}{\epsilon_0}$$

where we have used the fact that the surface area of a sphere is $4\pi r^2$. Hence, the magnitude of the field at a distance r from the charge q is

$$E = \frac{q}{4\pi\epsilon_0 r^2} = k\frac{q}{r^2}$$

If a second point charge q_0 is placed at a point where the field is E, the electrostatic force on this charge has a magnitude given by

$$F = q_0E = k\frac{qq_0}{r^2}$$

This, of course, is Coulomb's law. Note that this example is logically circular. It does, however, demonstrate the equivalence of Coulomb's law and Gauss' law.

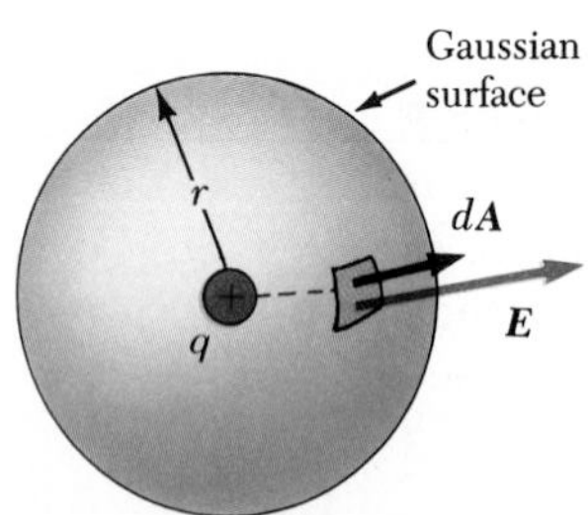

Figure 24.10 (Example 24.2) The point charge q is at the center of the spherical gaussian surface, and $\mathbf{E}$ is parallel to $d\mathbf{A}$ at every point on the surface.

EXAMPLE 24.3 A Spherically Symmetric Charge Distribution □

An insulating sphere of radius a has a uniform charge density ρ and a total positive charge Q (Fig. 24.11). (a) Calculate the electric field intensity at a point *outside* the sphere, that is, for $r > a$.

Solution Since the charge distribution is spherically symmetric, we again select a spherical gaussian surface of radius r, concentric with the sphere, as in Figure 24.11a. Following the line of reasoning given in Example 24.2, we find that

$$E = k\frac{Q}{r^2} \quad \text{(for } r > a\text{)} \tag{24.7}$$

Note that this result is identical to that obtained for a point charge. Therefore, we conclude that, for a uniformly charged sphere, the field in the region external to the sphere is *equivalent* to that of a point charge located at the center of the sphere.

(b) Find the electric field intensity at a point *inside* the sphere, that is, for $r < a$.

Solution In this case we select a spherical gaussian surface with radius $r < a$, concentric with the charge distribution (Fig. 24.11b). To apply Gauss' law in this situation, it is important to recognize that the charge q_{in} *within* the gaussian surface of volume V' is a quantity *less* than the total charge Q. To calculate the charge q_{in}, we use the fact that $q_{in} = \rho V'$, where ρ is the charge per unit volume and V' is the volume enclosed by the gaussian surface, given by $V' = \frac{4}{3}\pi r^3$ for a sphere. Therefore,

$$q_{in} = \rho V' = \rho\,(\tfrac{4}{3}\pi r^3)$$

As in Example 24.2, the electric field is constant in magnitude everywhere on the spherical gaussian surface and

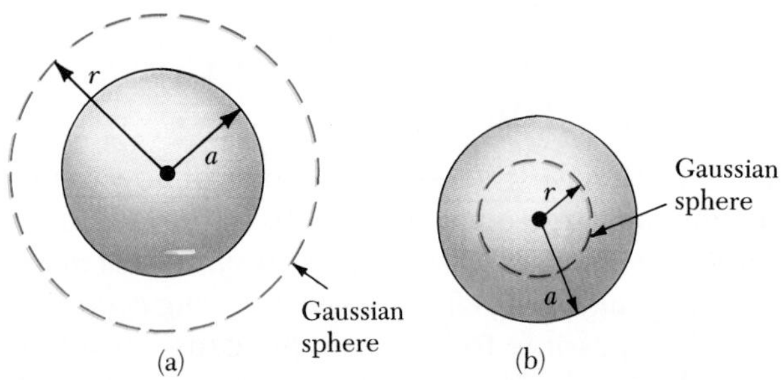

Figure 24.11 (Example 24.3) A uniformly charged insulating sphere of radius a and total charge Q. (a) The field at a point exterior to the sphere is kQ/r^2. (b) The field inside the sphere is due only to the charge *within* the gaussian surface and is given by $(kQ/a^3)r$.

is normal to the surface at each point. Therefore, Gauss' law in the region $r < a$ gives

$$\oint E\,dA = E\oint dA = E(4\pi r^2) = \frac{q_{\text{in}}}{\epsilon_0}$$

Solving for E gives

$$E = \frac{q_{\text{in}}}{4\pi\epsilon_0 r^2} = \frac{\rho\,\frac{4}{3}\pi r^3}{4\pi\epsilon_0 r^2} = \frac{\rho}{3\epsilon_0}\,r$$

Since by definition $\rho = Q/\frac{4}{3}\pi a^3$, this can be written

$$E = \frac{Qr}{4\pi\epsilon_0 a^3} = \frac{kQ}{a^3}\,r \qquad \text{(for } r < a\text{)} \qquad (24.8)$$

Note that this result for E differs from that obtained in (a). It shows that $E \rightarrow 0$ as $r \rightarrow 0$, as you might have guessed based on the spherical symmetry of the charge distribution. Therefore, the result fortunately eliminates the singularity that would exist at $r = 0$ if E varied as $1/r^2$ inside the sphere. That is, if $E \propto 1/r^2$, the field would be infinite at $r = 0$, which is clearly a physically impossible situation. A plot of E versus r is shown in Figure 24.12.

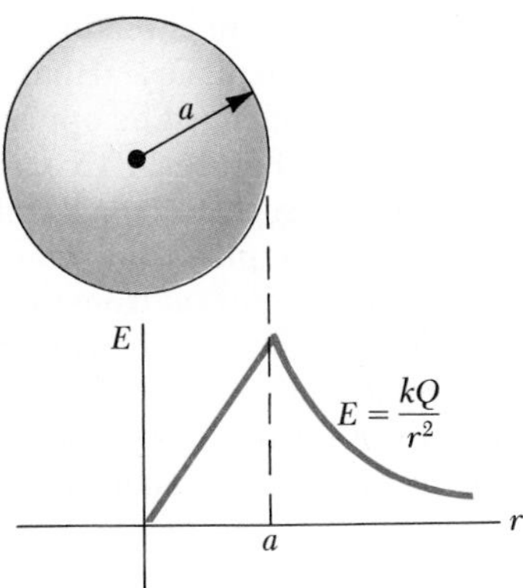

Figure 24.12 (Example 24.3) A plot of E versus r for a uniformly charged insulating sphere. The field inside the sphere ($r < a$) varies linearly with r. The field outside the sphere ($r > a$) is the same as that of a point charge Q located at the origin.

EXAMPLE 24.4 The E Field of a Thin Spherical Shell

A thin *spherical shell* of radius a has a total charge Q distributed uniformly over its surface (Fig. 24.13). Find the electric field at points inside and outside the shell.

Solution The calculation of the field outside the shell is identical to that already carried out for the solid sphere in Example 24.3a. If we construct a spherical gaussian surface of radius $r > a$, concentric with the shell, then the charge inside this surface is Q. Therefore, the field at

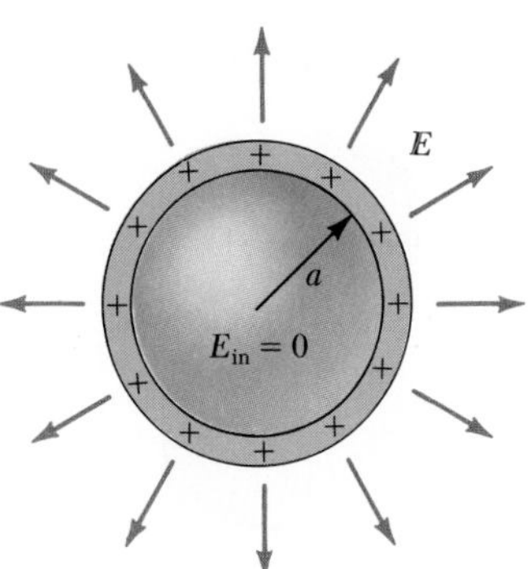

Figure 24.13 (Example 24.4) The electric field inside a uniformly charged spherical shell is *zero*. The field outside is the same as that of a point charge having a total charge Q located at the center of the shell.

a point outside the shell is equivalent to that of a point charge Q at the center:

$$E = k\frac{Q}{r^2} \qquad \text{(for } r > a\text{)}$$

The electric field inside the spherical shell is zero. This also follows from Gauss' law applied to a spherical surface of radius $r < a$. Since the net charge inside the surface is zero, and because of the spherical symmetry of the charge distribution, application of Gauss' law shows that $E = 0$ in the region $r < a$.

The same results can be obtained using Coulomb's law and integrating over the charge distribution. This calculation is rather complicated and will be omitted.

EXAMPLE 24.5 A Cylindrically Symmetric Charge Distribution

Find the electric field at a distance r from a uniform positive line charge of infinite length whose charge per unit length is λ = constant (Fig. 24.14).

Solution The symmetry of the charge distribution shows that **E** must be perpendicular to the line charge and directed outward as in Figure 24.14a. The end view of the line charge shown in Figure 24.14b should help visualize the directions of the electric field lines. In this situation, we select a cylindrical gaussian surface of radius r and length ℓ that is coaxial with the line charge. For the curved part of this surface, **E** is constant in magnitude and perpendicular to the surface at each point. Furthermore, the flux through the *ends* of the gaussian cylinder is *zero* since **E** is *parallel* to these surfaces.

The total charge inside our gaussian surface is $\lambda\ell$, where λ is the charge per unit length and ℓ is the length of the cylinder. Applying Gauss' law and noting that **E** is

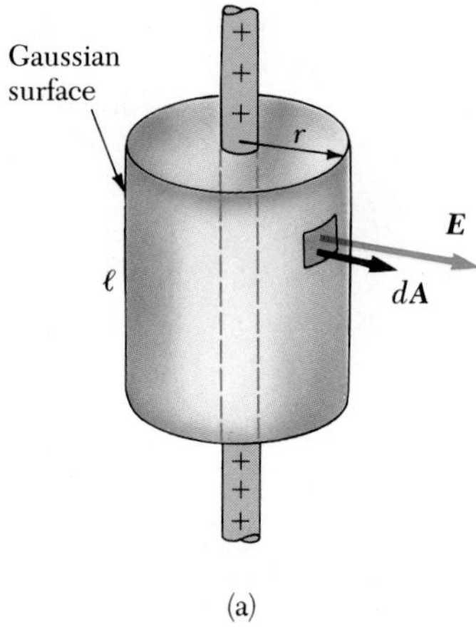

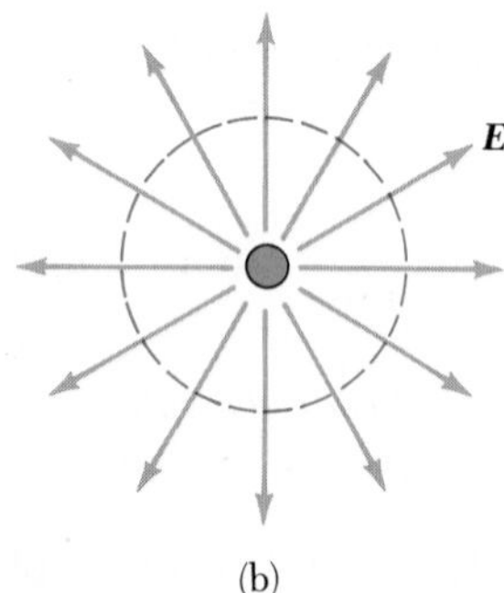

Figure 24.14 (Example 24.5) (a) An infinite line of charge surrounded by a cylindrical gaussian surface concentric with the line charge. (b) The field on the cylindrical surface is constant in magnitude and perpendicular to the surface.

parallel to $d\mathbf{A}$ everywhere on the cylindrical surface, we find that

$$\Phi_c = \oint \mathbf{E} \cdot d\mathbf{A} = E \oint dA = \frac{q_{\text{in}}}{\epsilon_0} = \frac{\lambda \ell}{\epsilon_0}$$

But the area of the curved surface is $A = 2\pi r \ell$; therefore

$$E(2\pi r \ell) = \frac{\lambda \ell}{\epsilon_0}$$

$$E = \frac{\lambda}{2\pi \epsilon_0 r} = 2k \frac{\lambda}{r} \qquad (24.9)$$

Thus, we see that the field of a cylindrically symmetric charge distribution varies as $1/r$, whereas the field external to a spherically symmetric charge distribution varies as $1/r^2$. Equation 24.9 can also be obtained using Coulomb's law and integration; however, the mathematical techniques necessary for this calculation are more cumbersome.

If the line charge has a finite length, the result for E is *not* the same as that given by Equation 24.9. For points close to the line charge and far from the ends, Equation 24.9 gives a good approximation of the actual value of the field. It turns out that Gauss' law is *not useful* for calculating E for a finite line charge. This is because the electric field is no longer constant in magnitude over the surface of the gaussian cylinder. Furthermore, $\mathbf{E}$ is not perpendicular to the cylindrical surface at all points. When there is little symmetry in the charge distribution, as in this situation, it is necessary to calculate $\mathbf{E}$ using Coulomb's law.

It is left as a problem (Problem 35) to show that the $\mathbf{E}$ field *inside* a uniformly charged rod of finite thickness is proportional to r.

EXAMPLE 24.6 A Nonconducting Plane Sheet of Charge

Find the electric field due to a nonconducting, infinite plane with uniform charge per unit area σ.

Solution The symmetry of the situation shows that $\mathbf{E}$ must be perpendicular to the plane and that the direction of $\mathbf{E}$ on one side of the plane must be opposite its direction on the other side, as in Figure 24.15. It is convenient to choose for our gaussian surface a small cylinder whose axis is perpendicular to the plane and whose ends each have an area A and are equidistant from the plane. Here we see that since $\mathbf{E}$ is parallel to the cylindrical surface, there is no flux through this surface. The flux out of *each* end of the cylinder is EA (since $\mathbf{E}$ is perpendicular to the ends); hence the *total* flux through our gaussian surface is

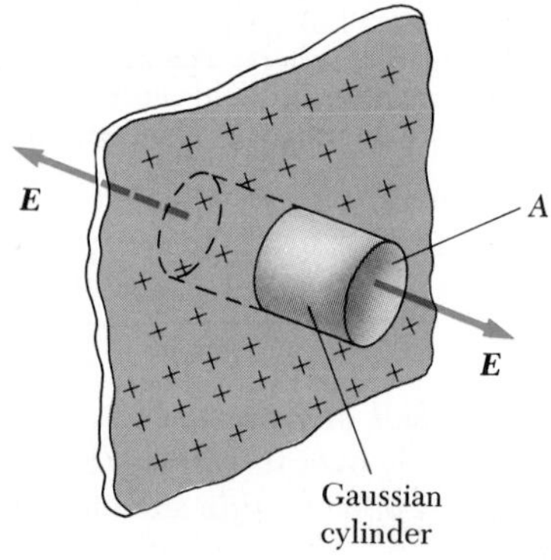

Figure 24.15 (Example 24.6) A cylindrical gaussian surface penetrating an infinite sheet of charge. The flux through each end of the gaussian surface is EA. There is no flux through the cylinder's surface.

$2EA$. Noting that the total charge *inside* the surface is σA, we use Gauss' law to get

$$\Phi_c = 2EA = \frac{q_{in}}{\epsilon_0} = \frac{\sigma A}{\epsilon_0}$$

$$E = \frac{\sigma}{2\epsilon_0} \qquad (24.10)$$

Since the distance of the surfaces from the plane does not appear in Equation 24.10, we conclude that $E = \sigma/2\epsilon_0$ at *any* distance from the plane. That is, the field is *uniform* everywhere.

An important configuration related to this example is the case of two parallel sheets of charge, with charge densities σ and $-\sigma$, respectively (Problem 58). In this situation, the electric field is σ/ϵ_0 *between* the sheets and approximately zero elsewhere.

24.4 CONDUCTORS IN ELECTROSTATIC EQUILIBRIUM

A good electrical conductor, such as copper, contains charges (electrons) that are not bound to any atom and are free to move about within the material. When there is no *net* motion of charge within the conductor, the conductor is in **electrostatic equilibrium.** As we shall see, *a conductor in electrostatic equilibrium* has the following properties:

Properties of a conductor in electrostatic equilibrium

1. The electric field is zero everywhere inside the conductor.
2. Any excess charge on an isolated conductor must reside entirely on its surface.
3. The electric field just outside a charged conductor is perpendicular to the conductor's surface and has a magnitude σ/ϵ_0, where σ is the charge per unit area at that point.
4. On an irregularly shaped conductor, charge tends to accumulate at locations where the radius of curvature of the surface is the smallest, that is, at sharp points.

The first property can be understood by considering a conducting slab placed in an external field $\mathbf{E}$ (Fig. 24.16). In electrostatic equilibrium, the electric field *inside* the conductor must be zero. If this were not the case, the free charges would accelerate under the action of an electric field. Before the external field is applied, the electrons are uniformly distributed throughout the conductor. When the external field is applied, the free electrons accelerate to the left, causing a buildup of negative charge on the left surface (excess electrons) and of positive charge on the right (where electrons have been removed). These charges create their own electric field, which *opposes* the external field. The surface charge density increases until the magnitude of the electric field set up by these charges equals that of the external field, giving a net field of zero *inside* the conductor. In a good conductor, the time it takes the conductor to reach equilibrium is of the order of 10^{-16} s, which for most purposes can be considered instantaneous.

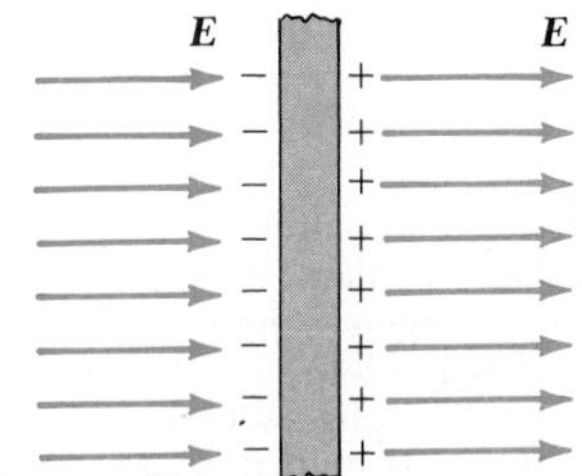

Figure 24.16 A conducting slab in an external electric field $\mathbf{E}$. The charges induced on the surfaces of the slab produce an electric field which opposes the external field, giving a resultant field of zero in the conductor.

We can use Gauss' law to verify the second and third properties of a conductor in electrostatic equilibrium. Figure 24.17 shows an arbitrarily shaped insulated conductor. A gaussian surface is drawn inside the conductor as close to the surface as we wish. As we have just shown, the electric field everywhere inside the conductor is zero when it is in electrostatic equilibrium. Since the electric field is also zero at *every* point on the gaussian surface, we see that the net flux through this surface is zero. From this result and Gauss' law, we conclude that the net charge inside the gaussian surface is zero. Since

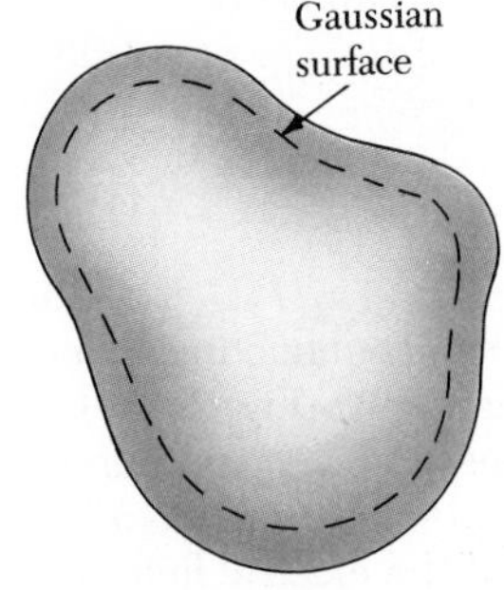

Figure 24.17 An insulated conductor of arbitrary shape. The broken line represents a gaussian surface just inside the conductor.

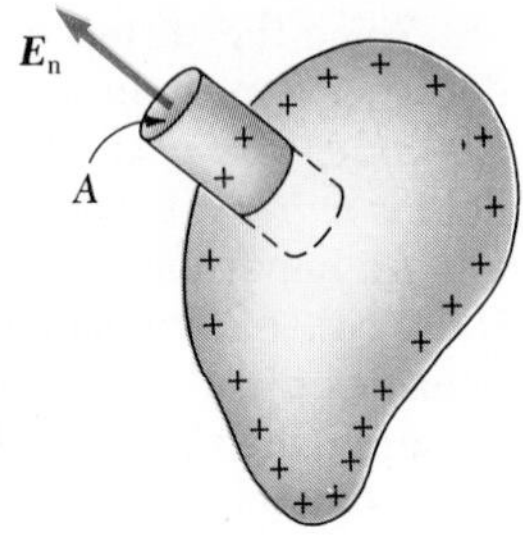

Figure 24.18 A gaussian surface in the shape of a small cylinder is used to calculate the electric field just outside a charged conductor. The flux through the gaussian surface is E_nA. Note that $\boldsymbol{E}$ is zero inside the conductor.

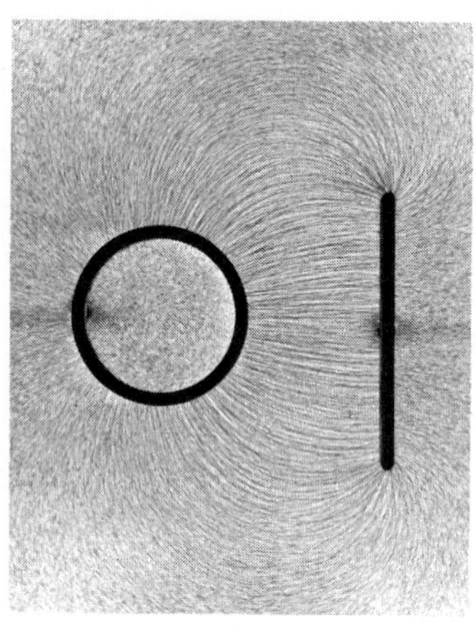

Electric field pattern of a charged conducting plate near an oppositely charged conducting cylinder. Small pieces of thread suspended in oil align with the electric field lines. Note that (1) the electric field lines are perpendicular to the conductors and (2) there are no lines inside the cylinder ($E = 0$). (Courtesy of Harold M. Waage, Princeton University)

there can be no net charge inside the gaussian surface (which is arbitrarily close to the conductor's surface), *any net charge on the conductor must reside on its surface.* Gauss' law does *not* tell us how this excess charge is distributed on the surface. In Section 25.6 we shall prove the fourth property of a conductor in electrostatic equilibrium.

We can use Gauss' law to relate the electric field just outside the surface of a charged conductor in equilibrium to the charge distribution on the conductor. To do this, it is convenient to draw a gaussian surface in the shape of a small cylinder with end faces parallel to the surface (Fig. 24.18). Part of the cylinder is just outside the conductor, and part is inside. There is no flux through the face on the inside of the cylinder since $\boldsymbol{E} = 0$ inside the conductor. Furthermore, the field is normal to the surface. If $\boldsymbol{E}$ had a tangential component, the free charges would move along the surface creating surface currents, and the conductor would not be in equilibrium. There is no flux through the cylindrical face of the gaussian surface since $\boldsymbol{E}$ is tangent to this surface. Hence, the net flux through the gaussian surface is E_nA, where E_n is the electric field just outside the conductor. Applying Gauss' law to this surface gives

$$\Phi_c = \oint E_n\, dA = E_nA = \frac{q_{in}}{\epsilon_0} = \frac{\sigma A}{\epsilon_0}$$

We have used the fact that the charge inside the gaussian surface is $q_{in} = \sigma A$, where A is the area of the cylinder's face and σ is the (local) charge per unit area. Solving for E_n gives

Electric field just outside a charged conductor

$$E_n = \frac{\sigma}{\epsilon_0} \tag{24.11}$$

EXAMPLE 24.7 A Sphere Inside a Spherical Shell

A solid conducting sphere of radius a has a net positive charge $2Q$ (Fig. 24.19). A conducting spherical shell of inner radius b and outer radius c is concentric with the solid sphere and has a *net* charge $-Q$. Using Gauss' law, find the electric field in the regions labeled ①, ②, ③, and ④ and the charge distribution on the spherical shell.

Solution First note that the charge distribution on both spheres has spherical symmetry, since they are concentric. To determine the electric field at various distances r from the center, we construct spherical gaussian surfaces of radius r.

To find E inside the solid sphere of radius a (region ①), we construct a gaussian surface of radius $r < a$. Since there can be no charge inside a conductor in electrostatic equilibrium, we see that $q_{in} = 0$, and so from Gauss' law $E_1 = 0$ for $r < a$. Thus we conclude that the net charge $2Q$ on the solid sphere is distributed on its outer surface.

In region ② between the spheres, where $a < r < b$, we again construct a spherical gaussian surface of radius r and note that the charge inside this surface is $+2Q$ (the

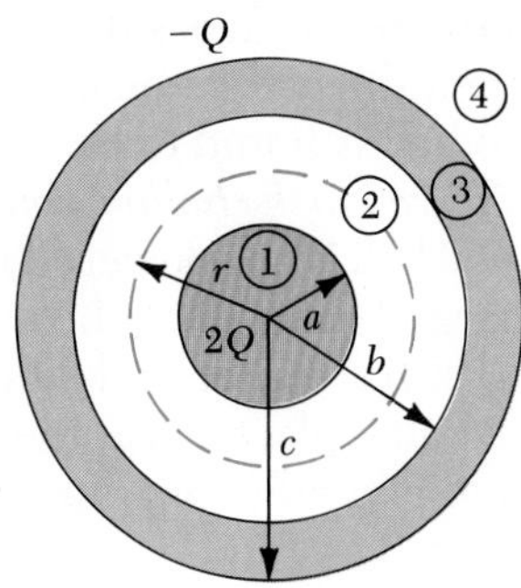

Figure 24.19 (Example 24.7) A solid conducting sphere of radius a and charge $2Q$ surrounded by a conducting spherical shell of charge $-Q$.

charge on the inner sphere). Because of the spherical symmetry, the electric field lines must be radial outward and constant in magnitude on the gaussian surface. Following Example 24.2 and using Gauss' law, we find that

$$E_2A = E_2(4\pi r^2) = \frac{q_{in}}{\epsilon_0} = \frac{2Q}{\epsilon_0}$$

$$E_2 = \frac{2Q}{4\pi\epsilon_0 r^2} = \frac{2kQ}{r^2} \qquad (\text{for } a < r < b)$$

In region ④ outside both spheres, where $r > c$, the spherical gaussian surface surrounds a *total* charge of $q_{in} = 2Q + (-Q) = Q$. Therefore, Gauss' law applied to this surface gives

$$E_4 = \frac{kQ}{r^2} \qquad (\text{for } r > c)$$

Finally, consider region ③, where $b < r < c$. The electric field must be *zero* in this region since the spherical shell is also a conductor in equilibrium. If we construct a Gaussian surface of this radius, we see that q_{in} must be zero since $E_3 = 0$. From this argument, we conclude that the charge on the *inner surface* of the *spherical shell* must be $-2Q$ to cancel the charge $+2Q$ on the solid sphere. (The charge $-2Q$ is induced by the charge $+2Q$ on the solid sphere.) Furthermore, since the net charge on the spherical shell is $-Q$, we conclude that the outer surface of the shell must have a charge equal to $+Q$.

*24.5 EXPERIMENTAL PROOF OF GAUSS' LAW AND COULOMB'S LAW

When a net charge is placed on a conductor, the charge distributes itself on the surface in such a way that the electric field inside is zero. Since $\mathbf{E} = 0$ inside a conductor in electrostatic equilibrium, Gauss' law shows that there can be no net charge inside the conductor. We have seen that Gauss' law is a consequence of Coulomb's law (Example 24.2). Hence, it should be possible to test the validity of the inverse-square law of force by attempting to detect a net charge inside a conductor. If a net charge is detected anywhere but on the conductor's surface, Coulomb's law, and hence Gauss' law, is invalid. Many experiments, including early work by Faraday, Cavendish, and Maxwell, have been performed to show that the net charge on a conductor resides on its surface. In all reported cases, no electric field could be detected in a closed conductor. The most recent and precise experiments by Williams, Faller, and Hill in 1971 showed that the exponent of r in Coulomb's law is $(2 + \delta)$, where $\delta = (2.7 \pm 3.1) \times 10^{-16}$!

The following experiment can be performed to verify that the net charge on a conductor resides on its surface. A positively charged metal ball at the end of a silk thread is lowered into an uncharged, hollow conductor through a small opening[2] (Fig. 24.20a). The hollow conductor is insulated from ground. The charged ball induces a negative charge on the inner wall of the hollow conductor, leaving an equal positive charge on the outer wall (Fig. 24.20b). The presence of positive charge on the outer wall is indicated by the deflection of an electrometer (a device used to measure charge). The deflection of the

[2] The experiment is often referred to as *Faraday's ice-pail experiment,* since it was first performed by Faraday using an ice pail for the hollow conductor.

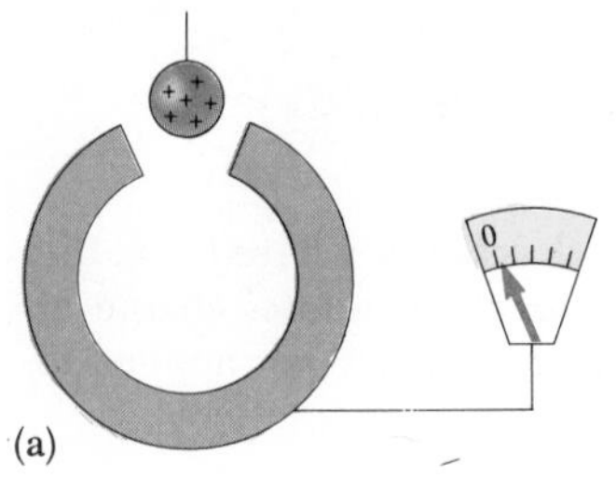

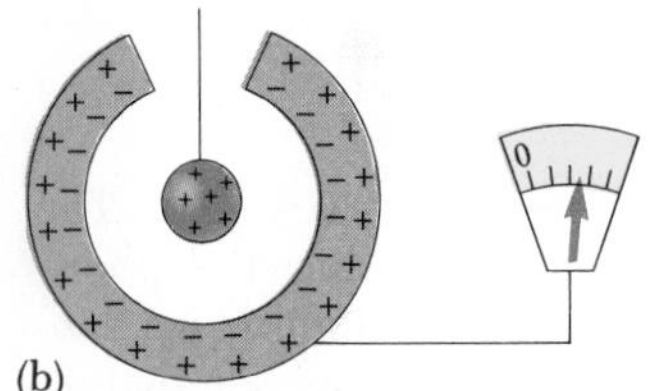

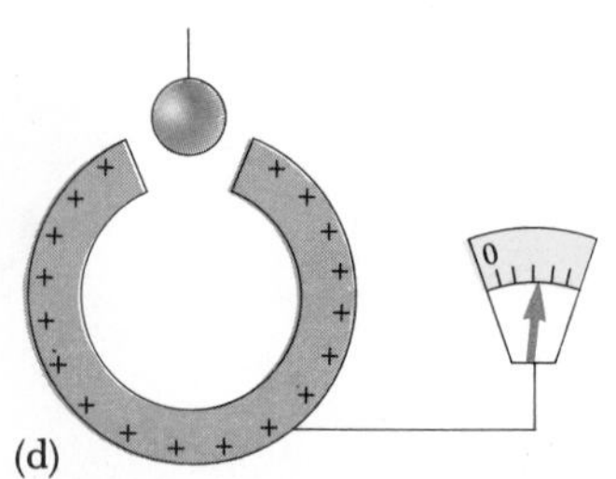

Figure 24.20 An experiment showing that any charge transferred to a conductor resides on its surface in electrostatic equilibrium. The hollow conductor is insulated from ground, and the small metal ball is supported by an insulating thread.

electrometer remains unchanged when the ball touches the inner surface of the hollow conductor (Fig. 24.20c). When the ball is removed, the electrometer reading remains the same and the ball is found to be uncharged (Fig. 24.20d). This shows that *the charge transferred to the hollow conductor resides on its outer surface.* If a small charged metal ball is now lowered into the *center* of the charged hollow conductor, the charged ball will not be attracted to the hollow conductor. This shows that $\mathbf{E} = 0$ at the center of the hollow conductor. On the other hand, if a small charged ball is placed near the outside of the conductor, the ball will be repelled by the conductor, showing that $\mathbf{E} \neq 0$ outside the conductor.

*24.6 DERIVATION OF GAUSS' LAW

One method that can be used to derive Gauss' law involves the concept of the *solid angle.* Consider a spherical surface of radius r containing an area element ΔA. The solid angle $\Delta\Omega$ subtended by this element at the center of the sphere is defined to be

$$\Delta\Omega \equiv \frac{\Delta A}{r^2}$$

From this expression, we see that $\Delta\Omega$ has no dimensions, since ΔA and r^2 both have the dimension of L^2. The unit of a solid angle is called the **steradian.** Since the total surface area of a sphere is $4\pi r^2$, the total solid angle subtended by the sphere at the center is given by

$$\Omega = \frac{4\pi r^2}{r^2} = 4\pi \text{ steradians}$$

Now consider a point charge q surrounded by a closed surface of arbitrary shape (Fig. 24.21). The total flux through this surface can be obtained by evaluating $\mathbf{E}\cdot\Delta\mathbf{A}$ for each element of area and summing over all elements of the surface. The flux through the element of area ΔA is

$$\Delta\Phi = \mathbf{E}\cdot\Delta\mathbf{A} = E\cos\theta\,\Delta A = kq\frac{\Delta A\cos\theta}{r^2}$$

where we have used the fact that $E = kQ/r^2$ for a point charge. But the quantity $\Delta A\cos\theta/r^2$ is equal to the solid angle $\Delta\Omega$ subtended at the charge q by the surface element ΔA. From Figure 24.22 we see that $\Delta\Omega$ is equal to the solid angle subtended by the element of a spherical surface of radius r. Since the

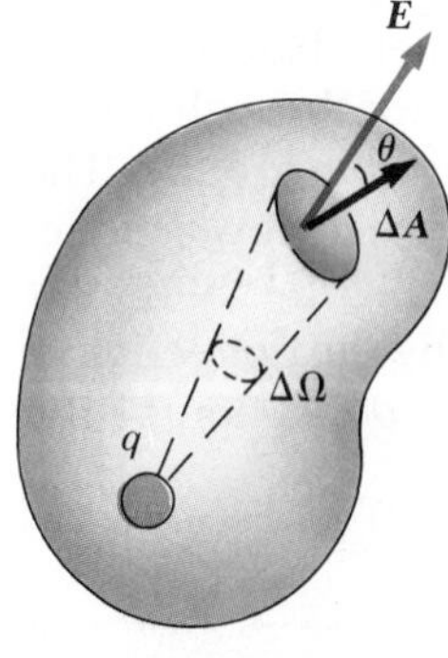

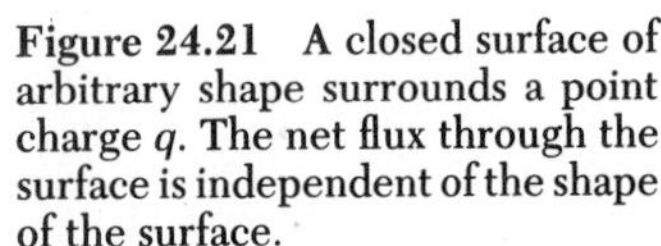
Figure 24.21 A closed surface of arbitrary shape surrounds a point charge q. The net flux through the surface is independent of the shape of the surface.

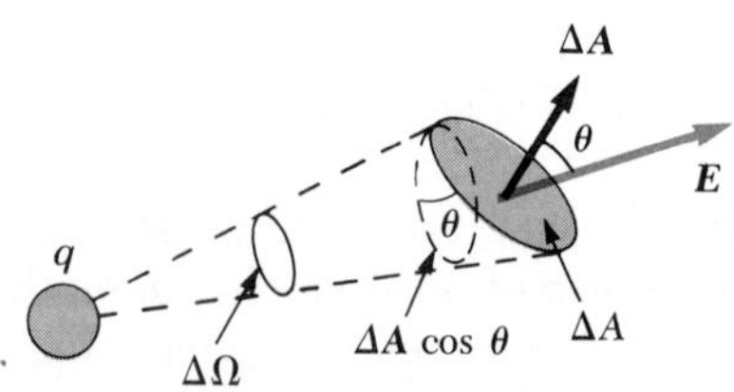

Figure 24.22 The area element ΔA subtends a solid angle $\Delta\Omega = (\Delta A\cos\theta)/r^2$ at the charge q.

total solid angle at a point is 4π steradians, we see that the total flux through the closed surface is

$$\Phi_c = kq \oint \frac{dA \cos\theta}{r^2} = kq \oint d\Omega = 4\pi kq = \frac{q}{\epsilon_o}$$

Thus we have derived Gauss' law, Equation 24.6. Note that this result is independent of the shape of the closed surface and independent of the position of the charge within the surface.

SUMMARY

Electric flux is a measure of the number of electric field lines that penetrate a surface. If the electric field is uniform and makes an angle θ with the normal to the surface, the electric flux through the surface is

Flux through a surface in a uniform electric field

$$\Phi = EA \cos\theta \qquad (24.2)$$

In general, the electric flux through a surface is defined by the expression

Definition of electric flux

$$\Phi = \int_{\text{surface}} \mathbf{E} \cdot d\mathbf{A} \qquad (24.3)$$

Gauss' law says that the net electric flux, Φ_c, through any closed gaussian surface is equal to the *net* charge *inside* the surface divided by ϵ_o:

Gauss' law

$$\Phi_c = \oint \mathbf{E} \cdot d\mathbf{A} = \frac{q_{\text{in}}}{\epsilon_o} \qquad (24.6)$$

Using Gauss' law, one can calculate the electric field due to various symmetric charge distributions. Table 24.1 lists some typical results.

TABLE 24.1 Typical Electric Field Calculations Using Gauss' Law

Charge Distribution	Electric Field	Location
Insulating sphere of radius R, uniform charge density, and total charge Q	$k\frac{Q}{r^2}$	$r > R$
	$k\frac{Q}{R^3}r$	$r < R$
Thin spherical shell of radius R and total charge Q	$k\frac{Q}{r^2}$	$r > R$
	0	$r < R$
Line charge of infinite length and charge per unit length λ	$2k\frac{\lambda}{r}$	Outside the line charge
Nonconducting, infinite charged plane with charge per unit area σ	$\frac{\sigma}{2\epsilon_o}$	Everywhere outside the plane
Conductor of surface charge per unit area σ	$\frac{\sigma}{\epsilon_o}$	Just outside the conductor
	0	Inside the conductor

A conductor in electrostatic equilibrium has the following properties:

Properties of a conductor in electrostatic equilibrium

1. The electric field is zero everywhere inside the conductor.
2. Any excess charge on an isolated conductor must reside entirely on its surface.
3. The electric field just outside the conductor is perpendicular to its surface and has a magnitude σ/ϵ_0, where σ is the charge per unit area at that point.
4. On an irregularly shaped conductor, charge tends to accumulate where the radius of curvature of the surface is the smallest, that is, at sharp points.

PROBLEM-SOLVING STRATEGY AND HINTS

Gauss' law may seem mysterious to you, and it is usually one of the most difficult concepts to understand in introductory physics. However, as we have seen, Gauss' law is very powerful in solving problems having a high degree of symmetry. In this chapter, you will only encounter problems with three kinds of symmetry: plane symmetry, cylindrical symmetry, and spherical symmetry. It is important to review Examples 2 through 7 and to use the following procedure:

1. First, select a gaussian surface which *has the same symmetry as the charge distribution.* For point charges or spherically symmetric charge distributions, the gaussian surface should be a sphere centered on the charge as in Examples 2, 3, 4 and 7. For uniform line charges or uniformly charged cylinders, your choice of a gaussian surface should be a cylindrical surface that is coaxial with the line charge or cylinder as in Example 5. For sheets of charge having plane symmetry, the gaussian surface should be a "pillbox" that straddles the sheet as in Example 6. Note that in all cases, the gaussian surface is selected such that the electric field has the same magnitude everywhere on the surface, and is directed perpendicular to the surface. This enables you to easily evaluate the surface integral that appears on the left side of Gauss' law, which represents the total electric flux through that surface.
2. Now evaluate the right side of Gauss' law, which amounts to calculating the total electric charge, q_{in}, *inside* the gaussian surface. If the charge density is uniform as is usually the case (that is, if λ, σ, or ρ is constant), simply multiply that charge density by the length, area, or volume enclosed by the gaussian surface. However, if the charge distribution is *nonuniform,* you must integrate the charge density over the region enclosed by the gaussian surface. For example, if the charge is distributed along a line, you would integrate the expression $dq = \lambda\, dx$, where dq is the charge on an infinitesimal element dx and λ is the charge per unit length. For a plane of charge, you would integrate $dq = \sigma\, dA$, where σ is the charge per unit area and dA is an infinitesimal element of area. Finally, for a volume of charge you would integrate $dq = \rho\, dV$, where ρ is the charge per unit volume and dV is an infinitesimal element of volume.
3. Once the left and right sides of Gauss's law have been evaluated, you can proceed to calculate the electric field on the gaussian surface assuming the charge distribution is given in the problem. Conversely, if the electric field is known, you can calculate the charge distribution that produces the field.

QUESTIONS

1. If the net flux through a gaussian surface is zero, which of the following statements are true? (a) There are no charges inside the surface. (b) The net charge inside the surface is zero. (c) The electric field is zero everywhere on the surface. (d) The number of electric field lines entering the surface equals the number leaving the surface.
2. If the electric field in a region of space is zero, can you conclude there are no electric charges in that region? Explain.
3. A spherical gaussian surface surrounds a point charge q. Describe what happens to the flux through the surface if (a) the charge is tripled, (b) the volume of the sphere is doubled, (c) the shape of the surface is changed to that of a cube, and (d) the charge is moved to another position inside the surface.
4. If there are more electric field lines leaving a gaussian surface than there are entering the surface, what can you conclude about the *net* charge enclosed by that surface?
5. A uniform electric field exists in a region of space in which there are no charges. What can you conclude about the *net* electric flux through a gaussian surface placed in this region of space?
6. Explain why Gauss' law cannot be used to calculate the electric field of (a) an electric dipole, (b) a charged disk, (c) a charged ring, and (d) three point charges at the corners of a triangle.
7. If the total charge inside a closed surface is known but the distribution of the charge is unspecified, can you use Gauss' law to find the electric field? Explain.
8. Explain why the electric flux through a closed surface with a given enclosed charge is independent of the size or shape of the surface.
9. Consider the electric field due to a nonconducting infinite plane with a uniform charge density. Explain why the electric field does not depend on the distance from the plane in terms of the spacing of the electric field lines.
10. Use Gauss' law to explain why electric field lines must begin and end on electric charges. (*Hint:* Change the size of the gaussian surface.)
11. A point charge is placed at the center of an uncharged metallic spherical shell insulated from ground. As the point charge is moved off center, describe what happens to (a) the total induced charge on the shell and (b) the distribution of charge on the interior and exterior surfaces of the shell.
12. Explain why excess charge on a isolated conductor must reside on its surface, using the repulsive nature of the force between like charges and the freedom of motion of charge within the conductor.
13. A person is placed in a large hollow metallic sphere that is insulated from ground. If a large charge is placed on the sphere, will the person be harmed upon touching the inside of the sphere? Explain what will happen if the person also has an initial charge whose sign is opposite to that of the charge on the sphere.
14. How would the observations described in Figure 24.20 differ if the hollow conductor were grounded? How would they differ if the small charged ball were an insulator rather than a conductor?
15. What other experiment might be performed on the ball in Figure 24.20 to show that its charge was transferred to the hollow conductor?
16. What would happen to the electrometer reading if the charged ball in Figure 24.20 touched the inner wall of the conductor? the outer wall?
17. Two solid spheres, both of radius R, carry identical total charges, Q. One sphere is a good conductor while the other is an insulator. If the charge on the insulating sphere is uniformly distributed throughout its interior volume, how do the electric fields outside these two spheres compare? Are the fields identical inside the two spheres?

PROBLEMS

Section 24.1 Electric Flux

1. A flat surface having an area of 3.2 m^2 is rotated in a uniform electric field of intensity $E = 6.2 \times 10^5$ N/C. Calculate the electric flux through this area when the electric field is (a) perpendicular to the surface, (b) parallel to the surface, and (c) makes an angle of 75° with the plane of the surface.
2. An electric field of intensity 3.5×10^3 N/C is applied along the x axis. Calculate the electric flux through a rectangular plane 0.35 m wide and 0.70 m long if (a) the plane is parallel to the yz plane, (b) the plane is parallel to the xy plane, and (c) the plane contains the y axis and its normal makes an angle of 40° with the x axis.
3. A uniform electric field $a\mathbf{i} + b\mathbf{j}$ intersects a surface of area A. What is the flux through this area if the surface lies (a) in the yz plane? (b) in the xz plane? (c) in the xy plane?
4. Consider a closed triangular box resting within a horizontal electric field $E = 7.8 \times 10^4$ N/C as shown in Figure 24.23. Calculate the electric flux through

(a) the left-hand vertical surface (A'), (b) the slanted surface (A), and (c) the entire surface of the box.

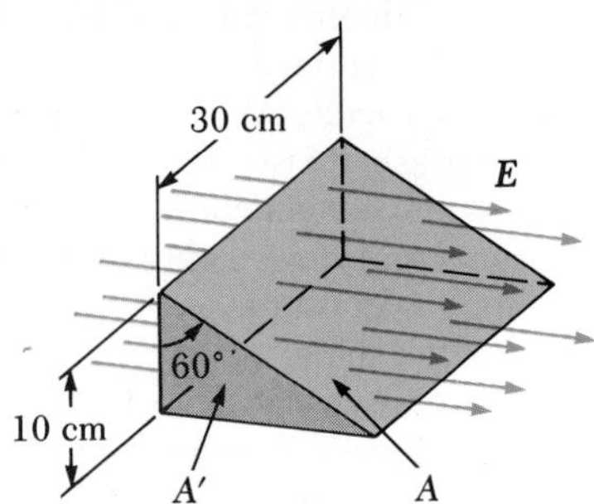

Figure 24.23 (Problem 4).

5. A 40-cm diameter loop is rotated in a uniform electric field until the position of maximum electric flux is found. The flux in this position is measured to be 5.2×10^5 Nm^2/C. What is the electric field strength?
6. A nonuniform electric field is given by the expression $\mathbf{E} = ay\mathbf{i} + bz\mathbf{j} + cx\mathbf{k}$, where a, b, and c are constants. Determine the electric flux through a rectangular surface in the xy plane, extending from $x = 0$ to $x = w$ and from $y = 0$ to $y = h$.
7. An electric field is given by $\mathbf{E} = az\mathbf{i} + bx\mathbf{k}$, where a and b are constants. Determine the electric flux through the triangular surface shown in Figure 24.24.

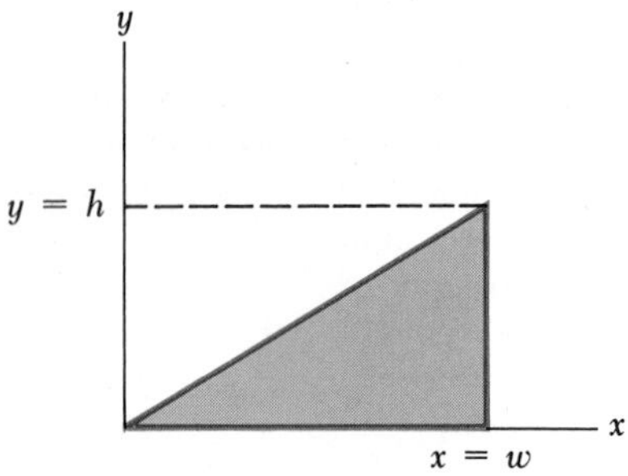

Figure 24.24 (Problem 7).

8. A cone with a circular base of radius R stands upright so that its axis is vertical. A uniform electric field $\mathbf{E}$ is applied in the vertical direction. Show that the flux through the cone's surface (not counting its base) is given by $\pi R^2 E$.
9. A pyramid with a 6-m square base and height of 4 m is placed in a vertical electric field of 52 N/C. Calculate the total electric flux through the pyramid's four slanted surfaces.

Section 24.2 Gauss' Law

10. Four closed surfaces, S_1 through S_4, together with the charges $-2Q$, $+Q$, and $-Q$ are sketched in Figure 24.25. Find the electric flux through each surface.

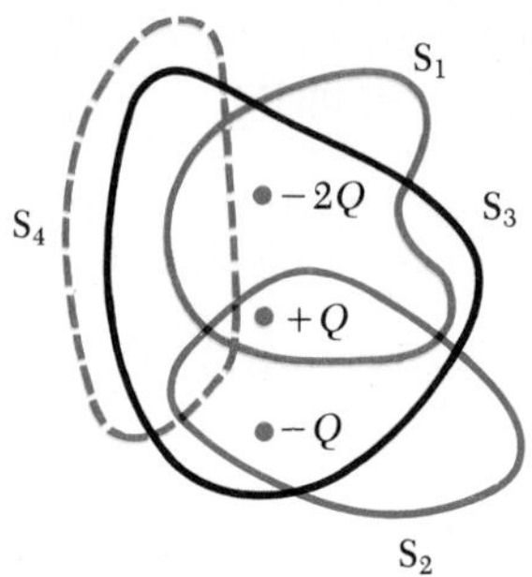

Figure 24.25 (Problem 10).

11. A point charge of $+5\ \mu$C is located at the center of a sphere with a radius of 12 cm. What is the electric flux through the surface of this sphere?
12. On a clear day, the electric field near the earth's surface is 100 N/C pointing radially inward. If the same electric field existed everywhere on the earth's surface, determine the total charge that would be stored in the earth.
13. A point charge of 12 μC is placed at the center of a *spherical* shell of radius 22 cm. What is the total electric flux through (a) the entire surface of the shell and (b) any hemispherical surface of the shell? (c) Do the results depend on the radius? Explain.
14. (a) Two charges of 8 μC and $-5\ \mu$C are inside a cube of sides 0.45 m. What is the total electric flux through the cube? (b) Repeat (a) if the same two charges are inside a spherical shell of radius 0.45 m.
15. The following charges are located inside a submarine: $+5\ \mu$C, $-9\ \mu$C, $+27\ \mu$C, and $-84\ \mu$C. Calculate the net electric flux through the submarine. Compare the number of electric field lines leaving the submarine with the number entering it.
16. The electric field on the surface of an uncharged, conducting hollow sphere is directed outward but varies in magnitude over the sphere's surface. A point charge within the sphere has a magnitude Q. (a) What is the sign of the charge? (b) What can you say about the position of the charge? (c) What can you say about the surroundings? (d) What is the net flux through the sphere?
17. Five charges are placed in a closed box. Each charge (except the first) has a magnitude which is twice that of the previous one placed in the box. If all charges have the same sign and if (after all charges have been placed in the box) the net electric flux through the box

is 4.8×10^7 Nm^2/C, what is the magnitude of the smallest charge in the box? Does the answer depend on the size of the box?

18. The electric field everywhere on the surface of a hollow sphere of radius 0.75 m is measured to be equal to 8.90×10^2 N/C and points radially toward the center of the sphere. (a) What is the net charge within the sphere's surface? (b) What can you conclude about the nature and distribution of the charge inside the sphere?

19. The electric field everywhere on the surface of a hollow sphere of radius 11 cm is measured to be equal to 3.8×10^4 N/C and points radially outward from the center of the sphere. (a) What is the electric flux through this surface? (b) How much charge is enclosed by this surface?

20. A charge of 170 μC is at the center of a cube of sides 80 cm. (a) Find the total flux through each face of the cube. (b) Find the flux through the whole surface of the cube. (c) Would your answers to (a) or (b) change if the charge were not at the center? Explain.

21. The total electric flux through a closed surface in the shape of a cylinder is 8.60×10^4 N·m^2/C. (a) What is the net charge within the cylinder? (b) From the information given, what can you say about the charge within the cylinder? (c) How would your answers to (a) and (b) change if the net flux were -8.60×10^4 N·m^2/C?

22. A cube of sides 10 cm is centered at the origin. A point charge of 2 μC is located on the y axis at $y = 20$ cm. (a) Sketch the electric lines for the point charge. (b) What is the net flux through the surface of the cube? (c) Repeat (a) and (b) if a second point charge of 4 μC is located at the center of the cube. (Neglect the lines that go through the edges and corners.)

23. A point charge Q is located just above the center of the flat face of a hemisphere of radius R as shown in Figure 24.26. (a) What is the electric flux through the curved surface of this hemisphere? (b) What is the electric flux through the flat face of this hemisphere?

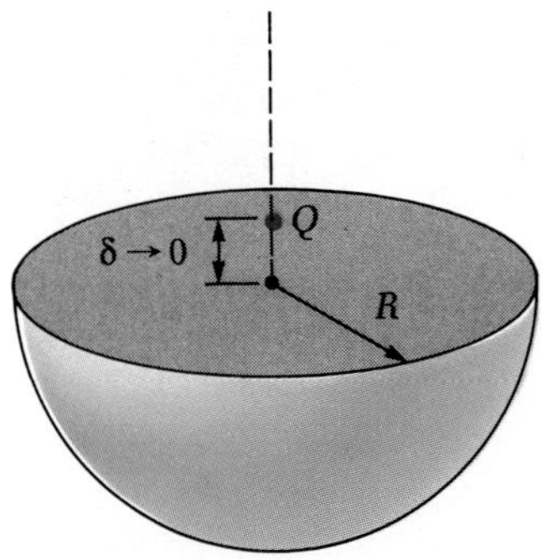

Figure 24.26 (Problem 23).

Section 24.3 Application of Gauss' Law to Charged Insulators

24. Show that the electric field intensity at a point just outside a uniformly charged spherical shell is σ/ϵ_0, where σ is the charge per unit area on the shell.

25. Consider a thin spherical shell of radius 14 cm with a total charge of 32 μC distributed uniformly on its surface. Find the electric field for the following distances from the center of the charge distribution: (a) $r =$ 10 cm and (b) $r = 20$ cm.

26. An inflated balloon in the shape of a sphere of radius 12 cm has a total charge of 7 μC uniformly distributed on its surface. Calculate the electric field intensity at the following distances from the center of the balloon: (a) 10 cm, (b) 12.5 cm, (c) 30 cm.

27. An insulating sphere is 8 cm in diameter, and carries a $+5.7$ μC charge uniformly distributed throughout its interior volume. Calculate the charge enclosed by a concentric spherical surface with the following radii: (a) $r = 2$ cm and (b) $r = 6$ cm.

28. An insulating sphere of radius 10 mm has a uniform charge density 6×10^{-3} C/m^3. Calculate the electric flux through a concentric spherical surface with the following radii: (a) $r = 5$ mm, (b) $r = 10$ mm, and (c) $r = 25$ mm.

29. A solid sphere of radius 40 cm has a total positive charge of 26 μC uniformly distributed throughout its volume. Calculate the electric field intensity at the following distances from the center of the sphere: (a) 0 cm, (b) 10 cm, (c) 40 cm, (d) 60 cm.

30. An insulating sphere of 10 cm radius has a uniform charge density throughout its volume. If the magnitude of the electric field at a distance of 5 cm from the center is 8.6×10^4 N/C, what is the magnitude of the electric field at 15 cm from the center?

31. A spherically symmetric charge distribution has a charge density given by $\rho = a/r$ where a is constant. Find the electric field as a function of r. [See the note in Problem 54.]

32. The charge per unit length on a *long*, straight filament is -90 μC/m. Find the electric field at the following distances from the filament: (a) 10 cm, (b) 20 cm, (c) 100 cm.

33. A uniformly charged, straight filament 7 m in length has a total positive charge of 2 μC. An uncharged cardboard cylinder 2 cm in length and 10 cm in radius surrounds the filament at its center, with the filament as the axis of the cylinder. Using any reasonable approximations, find (a) the electric field at the surface of the cylinder and (b) the total electric flux through the cylinder.

34. A cylindrical shell of radius 7 cm and length 240 cm has its charge uniformly distributed on its surface. The electric field intensity at a point 19 cm radially

outward from its axis (measured from the midpoint of the shell) is 3.6×10^4 N/C. Use approximate relations to find (a) the net charge on the shell and (b) the electric field at a point 4 cm from the axis, measured from the midpoint.

35. Consider a long cylindrical charge distribution of radius R with a uniform charge density ρ. Find the electric field at distance r from the axis where $r < R$.

36. A nonconducting wall carries a uniform charge density of 8.6 μC/cm^2. What is the electric field at a distance of 7 cm from the wall? Does your result change as the distance from the wall is varied?

37. A large plane sheet of charge has a charge per unit area of 9.0 μC/m^2. Find the electric field intensity *just above the surface* of the sheet, measured from the sheet's midpoint.

Section 24.4 Conductors in Electrostatic Equilibrium

38. A conducting spherical shell of radius 15 cm carries a net charge of -6.4 μC uniformly distributed on its surface. Find the electric field at points (a) just outside the shell and (b) inside the shell.

39. A long, straight metal rod has a radius of 5 cm and a charge per unit length of 30 nC/m. Find the electric field at the following distances from the axis of the rod: (a) 3 cm, (b) 10 cm, (c) 100 cm.

40. A square plate of copper of sides 50 cm is placed in an extended electric field of 8×10^4 N/C directed *perpendicular* to the plate. Find (a) the charge density of each face of the plate and (b) the total charge on each face.

41. A thin conducting plate 50 cm on a side lies in the xy plane. If a total charge of 4×10^{-8} C is placed on the plate, find (a) the charge density on the plate, (b) the electric field just above the plate, and (c) the electric field just below the plate.

42. A very large, thin, flat plate of aluminum of area A has a total charge Q uniformly distributed over its surfaces. If the same charge is spread uniformly over the *upper* surface of an otherwise identical glass plate, compare the electric fields just above the center of the upper surface of each plate.

43. A solid copper sphere 15 cm in radius has a total charge of 40 nC. Find the electric field at the following distances measured from the center of the sphere: (a) 12 cm, (b) 17 cm, (c) 75 cm. (d) How would your answers change if the sphere were hollow?

44. A hollow (but not necessarily empty) conducting sphere has a uniform charge per unit area of $+\sigma$ on its outer surface and $-\sigma$ on its inner surface. From this information, (a) what can you conclude about the charge in the region *inside* the hollow sphere? (b) What can you say about the electric field just outside the sphere?

45. A solid conducting sphere of radius 2 cm has a positive charge of $+8$ μC. A conducting spherical shell of inner radius 4 cm and outer radius 5 cm is concentric with the solid sphere and has a net charge of -4 μC. Find the electric field at the following distances from the center of this charge configuration: (a) $r = 1$ cm, (b) $r = 3$ cm, (c) $r = 4.5$ cm, and (d) $r = 7$ cm.

46. Consider the data given in Problem 45. Calculate the net charge enclosed by a concentric spherical gaussian surface with the following radii: (a) $r = 1$ cm, (b) 3 cm, (c) $r = 4.5$ cm, and (d) $r = 7$ cm.

47. A *long*, straight wire is surrounded by a hollow metallic cylinder whose axis coincides with that of the wire. The solid wire has a charge per unit length of $+\lambda$, and the hollow cylinder has a *net* charge per unit length of $+2\lambda$. From this information, use Gauss' law to find (a) the charge per unit length on the inner and outer surfaces of the hollow cylinder and (b) the electric field outside the hollow cylinder, a distance r from the axis.

48. The electric field on the surface of an irregularly shaped conductor varies from 5.6×10^4 N/C to 2.8×10^4 N/C. Calculate the local surface charge density at the point on the surface where the radius of curvature of the surface is (a) greatest and (b) smallest.

Section 24.6 Derivation of Gauss' Law

49. A sphere of radius R surrounds a point charge Q, located at its center. (a) Show that the electric flux through a circular cap of half-angle θ (see Figure 24.27) is given by

$$\Phi = \frac{Q}{2\epsilon_0}(1 - \cos\theta)$$

(b) What is the flux for $\theta = 90°$? (c) What is the flux for $\theta = 180°$?

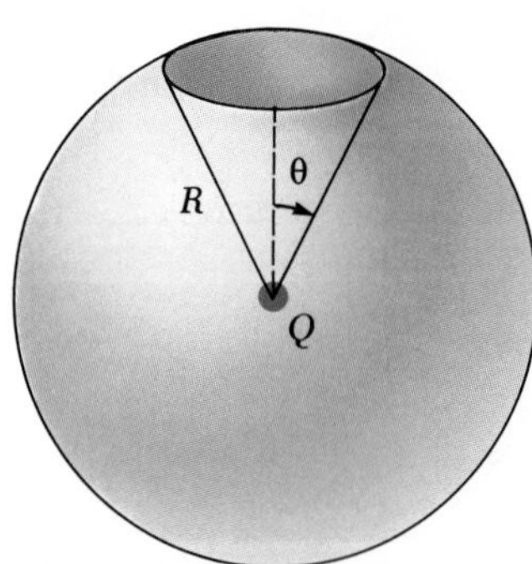

Figure 24.27 (Problem 49).

ADDITIONAL PROBLEMS

50. For the configuration shown in Figure 24.28, suppose that $a = 5$ cm, $b = 20$ cm, and $c = 25$ cm. Furthermore, suppose that the electric field at a point 10 cm from the center is measured to be 3.6×10^3 N/C radially *inward* while the electric field at a point 50 cm from the center is 2.0×10^2 N/C radially *outward.* From this information, find (a) the charge on the insulating sphere, (b) the net charge on the hollow conducting sphere, and (c) the total charge on the inner and outer *surfaces*, respectively, of the hollow conducting sphere.

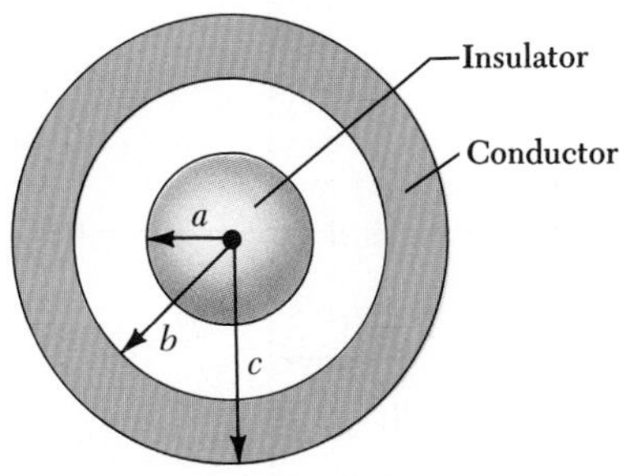

Figure 24.28 (Problems 50 and 51).

51. A solid *insulating* sphere of radius a has a uniform charge density ρ and a total charge Q. Concentric with this sphere is an *uncharged, conducting* hollow sphere whose inner and outer radii are b and c, as in Figure 24.28. (a) Find the electric field intensity in the regions $r < a$, $a < r < b$, $b < r < c$, and $r > c$. (b) Determine the induced charge per unit area on the inner and outer surfaces of the hollow sphere.
52. Consider an insulating sphere of radius R and having a *uniform* volume charge density ρ. Plot the magnitude of the electric field, E, as a function of the distance from the center of the sphere, r. Let r range over the interval $0 < r < 3R$ and plot E in units of $\rho R/\epsilon_0$.
53. A *hollow* insulating sphere has a uniform charge density ρ. Its inner and outer radii are a and b, respectively. Use Gauss' law to find expressions for the electric field in the regions (a) $r < a$, (b) $a < r < b$, (c) $r > b$.
54. Consider a solid insulating sphere of radius b with nonuniform charge density $\rho = Cr$. Find the charge contained within the radius when (a) $r < b$ and (b) $r > b$. (*Note:* The volume element dV for a spherical shell of radius r and thickness dr is equal to $4\pi r^2\, dr$.)
55. A solid insulating sphere of radius R has a *nonuniform* charge density that varies with r according to the expression $\rho = Ar^2$, where A is a constant and $r < R$ is measured from the center of the sphere. (a) Show that the electric field *outside* ($r > R$) the sphere is given by the expression $E = AR^5/5\epsilon_0 r^2$. (b) Show that the electric field *inside* ($r < R$) the sphere is given by $E = Ar^3/5\epsilon_0$. (*Hint:* Note that the total charge Q on the sphere is equal to the integral of $\rho\, dV$, where r extends from 0 to R; also note that the charge q within a radius $r < R$ is *less* than Q. To evaluate the integrals, note that the volume element dV for a spherical shell of radius r and thickness dr is equal to $4\pi r^2\, dr$.)
56. An infinitely long insulating cylinder of radius R has a volume charge density that varies with the radius as

$$\rho = \rho_0\left(a - \frac{r}{b}\right),$$

where ρ_0, a, and b are positive constants and r is the distance from the axis of the cylinder. Use Gauss' law to determine the magnitude of the electric field at radial distances (a) $r < R$ and (b) $r > R$.
57. The flux of *any* vector field **V** through a surface can be defined as

$$\Phi_{\mathbf{V}} = \int_S \mathbf{V}\cdot d\mathbf{A}$$

Radial fields whose magnitudes vary inversely as the square of the distance, such as the gravitational field

$$\mathbf{g} = \frac{\mathbf{F}_g}{m}$$

also obey Gauss' Law. Using Gauss' (gravitational) law, find the gravitational field **g** at a point distance r from the center of the earth where $r < R_e$. Pretend that the earth's density is uniform.
58. Two infinite, nonconducting sheets of charge are parallel to each other as in Figure 24.29. The sheet on the left has a uniform surface charge density σ, and the one on the right has a uniform charge density $-\sigma$. Calculate the value of the electric field at points (a) to the left of, (b) in between, and (c) to the right of the two sheets. (*Hint:* See Example 24.6.)

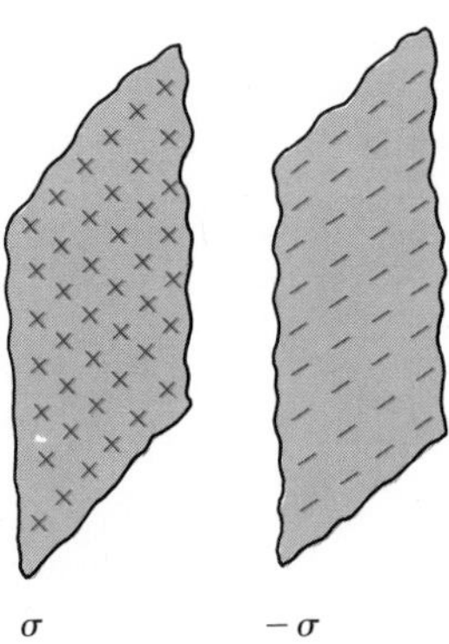

Figure 24.29 (Problems 58 and 59).

59. Repeat the calculations for Problem 58 when both sheets have *positive* uniform charge densities of value σ.

60. A closed surface with dimensions $a = b = 0.4$ m and $c = 0.6$ m is located as shown in Figure 24.30. The electric field throughout the region is *nonuniform* and given by

$$\mathbf{E} = (3 + 2x^2)\mathbf{i} \text{ N/C}$$

where x is in meters. Calculate the net electric flux leaving the closed surface. What net charge is enclosed by the surface?

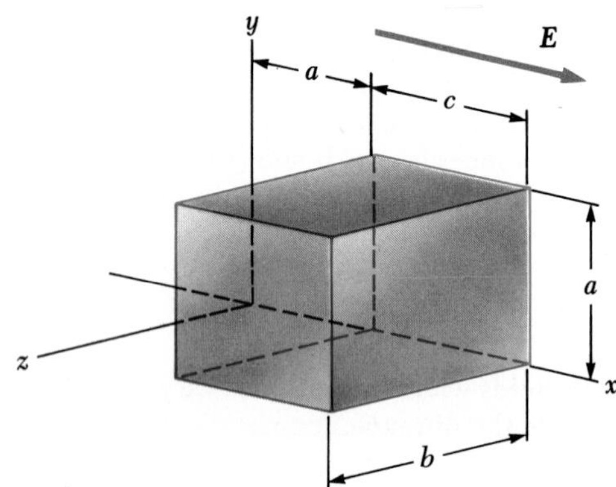

Figure 24.30 (Problem 60).

61. A slab of insulating material (infinite in two of its three dimensions) has a uniform positive charge density ρ as in the edge view of Figure 24.31. (a) Show that the electric field a distance x from its center and inside the slab is $E = \rho x/\epsilon_0$. (b) Suppose an electron of charge $-e$ and mass m is placed inside the slab. If it is released from rest at a distance x from the center, show that the electron exhibits simple harmonic motion with a frequency given by

$$f = \frac{1}{2\pi}\sqrt{\frac{\rho e}{m\epsilon_0}}$$

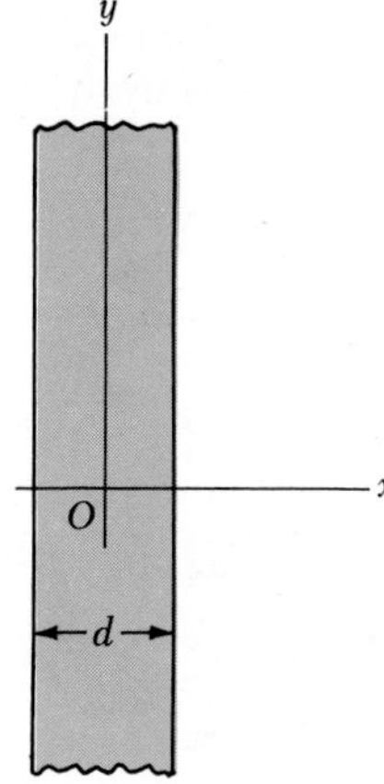

Figure 24.31 (Problems 61 and 62).

62. A slab of insulating material has a *nonuniform* positive charge density given by $\rho = Cx^2$, where x is measured from the center of the slab as in Figure 24.31, and C is a constant. The slab is infinite in the y and z directions. Derive expressions for the electric field in (a) the exterior regions and (b) the interior region of the slab $(-d/2 < x < d/2)$.

25

Electric Potential

Jennifer is holding onto a charged sphere that reaches a potential of about 100,000 volts. The device that generates this high potential is called a Van de Graaff generator. Why do you suppose Jennifer's hair stands on end like the needles of a porcupine? Why is it important that she stand on a pedestal insulated from ground? (Courtesy of Henry Leap and Jim Lehman)

The concept of potential energy was first introduced in Chapter 8 in connection with such conservative forces as the force of gravity and the elastic force of a spring. By using the law of energy conservation, we were often able to avoid working directly with forces when solving various mechanical problems. In this chapter we shall see that the energy concept is also of great value in the study of electricity. Since the electrostatic force given by Coulomb's law is conservative, one can conveniently describe electrostatic phenomena in terms of an electrical potential energy. This idea enables us to define a scalar quantity called *electric potential.* Because the potential is a scalar function of position, it offers a simpler way of describing electrostatic phenomena than does the electric field. In later chapters we shall see that the concept of the electric potential is of great practical value. In fact, the measured voltage between any two points in an electrical circuit is simply the difference in electric potential between the points.

25.1 POTENTIAL DIFFERENCE AND ELECTRIC POTENTIAL

In Chapter 14, we showed that the gravitational force is conservative. Since the electrostatic force, given by Coulomb's law, is of the same form as the universal law of gravity, it follows that the electrostatic force is also conservative. Therefore, it is possible to define a potential energy function associated with this force.

When a test charge q_o is placed in an electrostatic field $\mathbf{E}$, the electric force on the test charge is $q_o\mathbf{E}$. The force $q_o\mathbf{E}$ is the vector sum of the individual forces exerted on q_o by the various charges producing the field $\mathbf{E}$. It follows that the force $q_o\mathbf{E}$ is conservative, since the individual forces governed by Coulomb's law are conservative. The work done by the force $q_o\mathbf{E}$ is equal to the negative of the work done by an external agent. Furthermore, the work done by the electric force $q_o\mathbf{E}$ on the test charge for an infinitesimal displacement $d\mathbf{s}$ is given by

$$dW = \mathbf{F}\cdot d\mathbf{s} = q_o\mathbf{E}\cdot d\mathbf{s} \tag{25.1}$$

By definition, the work done by a conservative force equals the negative of the change in potential energy, dU; therefore, we see that

$$dU = -q_o\mathbf{E}\cdot d\mathbf{s} \tag{25.2}$$

For a finite displacement of the test charge between points A and B, the **change in the potential energy** is given by

Change in potential energy

$$\Delta U = U_B - U_A = -q_o \int_A^B \mathbf{E}\cdot d\mathbf{s} \tag{25.3}$$

The integral in Equation 25.3 is performed along the path by which q_o moves from A to B and is called a *path integral,* or *line integral.* Since the force $q_o\mathbf{E}$ is conservative, *this integral does not depend on the path taken between A and B.*

The **potential difference,** $V_B - V_A$, between the points A and B is defined as the change in potential energy divided by the test charge q_o:

Potential difference

$$V_B - V_A = \frac{U_B - U_A}{q_o} = -\int_A^B \mathbf{E}\cdot d\mathbf{s}$$

Potential difference should not be confused with potential energy. The potential difference is *proportional* to the potential energy, and we see from Equation 25.4 that the two are related by $\Delta U = q_o\,\Delta V$. Because potential energy is a scalar, electric potential is also scalar quantity. Note that the change in the potential energy of the charge is the negative of the work done by the electric force. Hence, we see that

the potential difference $V_B - V_A$ equals the work per unit charge that an external agent must perform to move a test charge from A to B without a change in kinetic energy.

Equation 25.4 defines potential differences only. That is, only *differences* in V are meaningful. The electric potential function is often taken to be zero at some convenient point. We shall usually choose the potential to be zero for a point at infinity (that is, a point infinitely remote from the charges producing the electric field). With this choice, we can say that the *electric potential at an arbitrary point equals the work required per unit charge to bring a positive test charge from infinity to that point.* Thus, if we take $V_A = 0$ at infinity in Equation 25.4, then the potential at any point P is given by

$$V_P = -\int_{\infty}^{P} \mathbf{E} \cdot d\mathbf{s} \qquad (25.5)$$

In reality, V_P represents the potential difference between the point P and a point at infinity. (Equation 25.5 is a special case of Eq. 25.4.)

Since potential difference is a measure of energy per unit charge, the SI unit of potential is joules per coulomb, defined to be equal to a unit called the **volt** (V):

Definition of a volt

$$1 \text{ V} \equiv 1 \text{ J/C}$$

That is, 1 J of work must be done to take a 1-C charge through a potential difference of 1 V. Equation 25.4 shows that the potential difference also has units of electric field times distance. From this, it follows that the SI unit of electric field (N/C) can also be expressed as volts per meter:

$$1 \text{ N/C} = 1 \text{ V/m}$$

A unit of energy commonly used in atomic and nuclear physics is the **electron volt,** which is defined as *the energy that an electron (or proton) gains when moving through a potential difference of magnitude 1 V.* Since 1 V = 1 J/C and since the fundamental charge is equal to 1.6×10^{-19} C, we see that the electron volt (eV) is related to the joule through the relation

The electron volt

$$1 \text{ eV} = 1.6 \times 10^{-19} \text{ C} \cdot \text{V} = 1.6 \times 10^{-19} \text{ J} \qquad (25.6)$$

For instance, an electron in the beam of a typical TV picture tube (or cathode ray tube) has a speed of 5×10^7 m/s. This corresponds to a kinetic energy of 1.1×10^{-15} J, which is equivalent to 7.1×10^3 eV. Such an electron has to be accelerated from rest through a potential difference of 7.1 kV to reach this speed.

25.2 POTENTIAL DIFFERENCES IN A UNIFORM ELECTRIC FIELD

In this section, we shall describe the potential difference between any two points in a *uniform* electric field. The potential difference is independent of the path between these two points; that is, the work done in taking a test charge from point A to point B is the same along all paths. This confirms that a static, uniform electric field is conservative. By definition, a force is conservative if it has this property (see Section 8.1).

First, consider a uniform electric field directed along the x axis, as in Figure 25.1. Let us calculate the potential difference between two points,

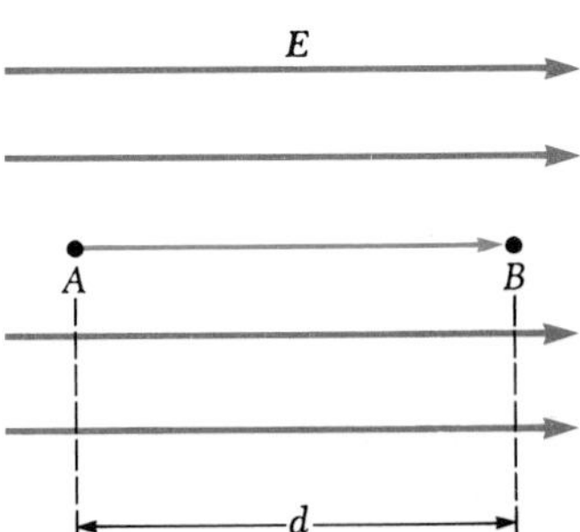

Figure 25.1 The displacement of a charged particle from A to B in the presence of a uniform electric field $\mathbf{E}$.

A and B, separated by a distance d, where d is measured parallel to the field lines. If we apply Equation 25.4 to this situation, we get

$$V_B - V_A = \Delta V = -\int_A^B \mathbf{E}\cdot d\mathbf{s} = -\int_A^B E\cos 0^\circ\, ds = -\int_A^B E\, ds$$

Since E is constant, it can be removed from the integral sign, giving

Potential difference in a uniform E field

$$\Delta V = -E\int_A^B ds = -Ed \qquad (25.7)$$

The minus sign results from the fact that point B is at a lower potential than point A; that is, $V_B < V_A$.

Now suppose that a test charge q_o moves from A to B. The change in its potential energy can be found from Equations 25.4 and 25.7:

$$\Delta U = q_o\,\Delta V = -q_o Ed \qquad (25.8)$$

From this result, we see that if q_o is positive, ΔU is negative. This means that *a positive charge will lose electric potential energy when it moves in the direction of the electric field.* This is analogous to a mass losing gravitational potential energy when it moves to lower elevations in the presence of gravity. If a positive test charge is released from rest in this electric field, it experiences an electric force $q_o\mathbf{E}$ in the direction of $\mathbf{E}$ (to the right in Fig. 25.1). Therefore, it accelerates to the right, gaining kinetic energy. *As it gains kinetic energy, it loses an equal amount of potential energy.*

On the other hand, if the test charge q_o is negative, then ΔU is positive and the situation is reversed. *A negative charge gains electric potential energy when it moves in the direction of the electric field.* If a negative charge is released from rest in the field $\mathbf{E}$, it accelerates in a direction opposite the electric field.[1]

Now consider the more general case of a charged particle moving between any two points in a uniform electric field directed along the x axis, as in Figure 25.2. If $\mathbf{d}$ represents the displacement vector between points A and B, Equation 25.4 gives

$$\Delta V = -\int_A^B \mathbf{E}\cdot d\mathbf{s} = -\mathbf{E}\cdot\int_A^B d\mathbf{s} = -\mathbf{E}\cdot\mathbf{d} \qquad (25.9)$$

where again, we are able to remove $\mathbf{E}$ from the integral since it is constant. Further, the change in potential energy of the charge is

$$\Delta U = q_o\,\Delta V = -q_o\,\mathbf{E}\cdot\mathbf{d} \qquad (25.10)$$

Finally, our results show that all points in a plane *perpendicular* to a uniform electric field are at the same potential. This can be seen in Figure 25.2, where the potential difference $V_B - V_A$ is *equal* to $V_C - V_A$. Therefore, $V_B = V_C$.

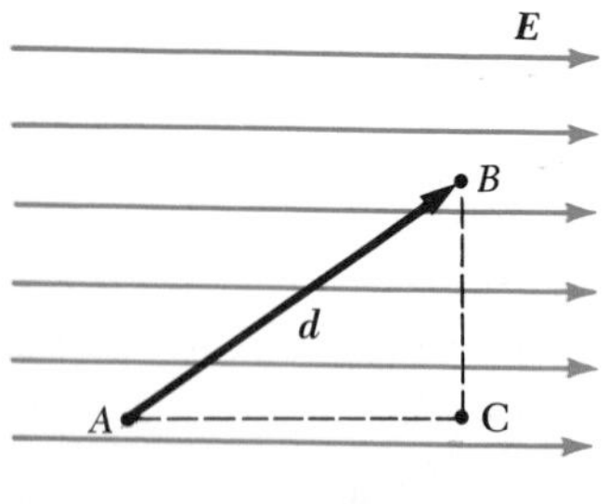

Figure 25.2 A uniform electric field directed along the positive x axis. Point B is at a lower potential than point A. Points B and C are at the *same* potential.

[1] Note that when a charged particle accelerates, it actually loses energy by radiating electromagnetic waves. However, in most cases encountered in this course, there is no *net* loss of energy. Furthermore, the radiation emitted by an accelerating charge is not predicted by the nonrelativistic physics we have studied so far.

The name **equipotential surface** is given to any surface consisting of a continuous distribution of points having the same potential.

An equipotential surface

Note that since $\Delta U = q_0\,\Delta V$, *no* work is done in moving a test charge between any two points on an equipotential surface. The equipotential surfaces of a uniform electric field consist of a family of planes all *perpendicular* to the field (Fig. 25.2). Equipotential surfaces for fields with other symmetries will be described in later sections.

EXAMPLE 25.1 The Field Between Two Parallel Plates of Opposite Charge

A 12-V battery is connected between two parallel plates as in Figure 25.3. The separation between the plates is 0.3 cm, and the electric field is assumed to be uniform. (This assumption is reasonable if the plate separation is small compared to the plate size and if we do not consider points near the edges of the plates.) Find the electric field between the plates.

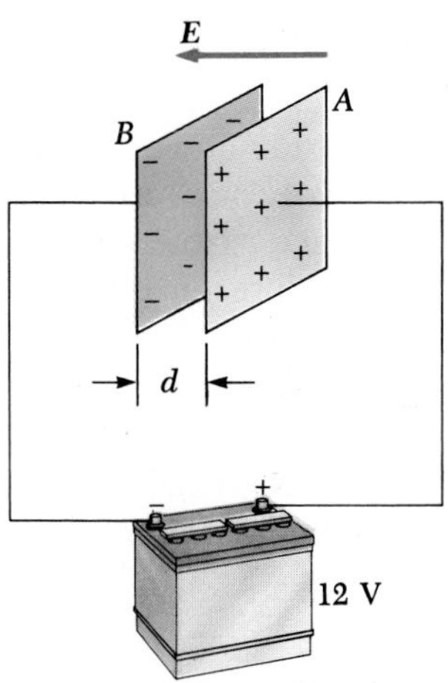

Figure 25.3 (Example 25.1) A 12-V battery connected to two parallel plates. The electric field between the plates has a magnitude given by the potential difference divided by the plate separation *d*.

Solution The electric field is directed from the positive plate toward the negative plate. We see that the positive plate (at the right) is at a *higher* potential than the negative plate. Note that the potential difference between plates *B* and *A* must equal the potential difference between the battery terminals. This can be understood by noting that all points on a conductor in equilibrium are at the same potential,[2] and hence there is no potential difference between a terminal of the battery and any portion of the plate to which it is connected. Therefore, the magnitude of the electric field between the plates is

$$E = \frac{|V_B - V_A|}{d} = \frac{12\ \text{V}}{0.3 \times 10^{-2}\ \text{m}} = 4.0 \times 10^3\ \text{V/m}$$

This configuration, which is called a *parallel-plate capacitor,* will be examined in more detail in the next chapter.

EXAMPLE 25.2 Motion of a Proton in a Uniform Electron Field □

A proton is released from *rest* in a uniform electric field of 8×10^4 V/m directed along the positive x axis (Fig. 25.4). The proton undergoes a displacement of 0.5 m in the direction of $\mathbf{E}$. (a) Find the *change* in the electric potential between the points *A* and *B*.

Using Equation 25.4 and noting that the displacement is *in the direction* of the field, we have

$$\Delta V = V_B - V_A = -\int \mathbf{E}\cdot d\mathbf{s} = -\int_0^d E\,dx = -E\int_0^d dx$$

$$= -Ed = -\left(8 \times 10^4 \frac{\text{V}}{\text{m}}\right)(0.5\ \text{m})$$

$$= -4 \times 10^4\ \text{V}$$

Thus, the electric potential of the proton *decreases* as it moves from *A* to *B*.

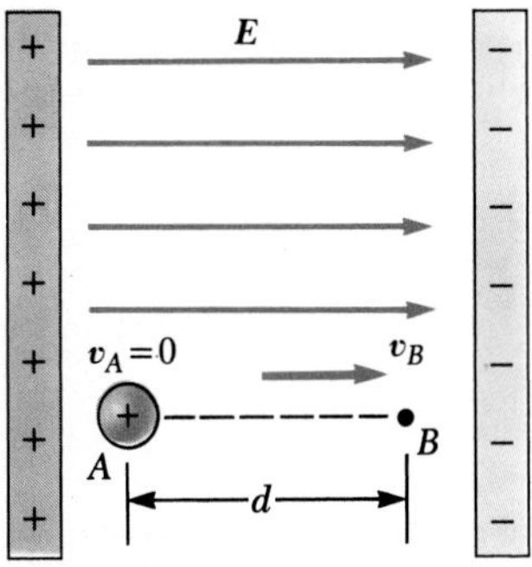

Figure 25.4 (Example 25.2) A proton accelerates from *A* to *B* in the direction of the electric field.

[2] The electric field vanishes within a conductor in electrostatic equilibrium, and so the path integral $\int \mathbf{E}\cdot d\mathbf{s}$ between any two points within the conductor must be zero. A fuller discussion of this point is given in Section 25.6.

(b) Find the change in potential energy of the proton for this displacement.

$$\Delta U = q_0\,\Delta V = e\,\Delta V$$
$$= (1.6 \times 10^{-19}\text{ C})(-4 \times 10^4\text{ V})$$
$$= -6.4 \times 10^{-15}\text{ J}$$

The negative sign here means that the potential energy of the proton decreases as it moves in the direction of **E**. This makes sense since as the proton *accelerates* in the direction of **E**, it gains kinetic energy and at the same time loses electrical potential energy (the total energy is conserved).

(c) Find the speed of the proton after it has been displaced from rest by 0.5 m.

If there are no forces acting on the proton other than the conservative electric force, we can apply the principle of conservation of mechanical energy in the form $\Delta K + \Delta U = 0$; that is, *the decrease in potential energy must be accompanied by an equal increase in kinetic energy*. Because the mass of the proton is given by the value $m_p = 1.67 \times 10^{-27}$ kg, we find

$$\Delta K + \Delta U = (\tfrac{1}{2}m_p v^2 - 0) - 6.4 \times 10^{-15}\text{ J} = 0$$

$$v^2 = \frac{2(6.4 \times 10^{-15})\text{ J}}{1.67 \times 10^{-27}\text{ kg}} = 7.66 \times 10^{12}\text{ m}^2/\text{s}^2$$

$$v = 2.77 \times 10^6\text{ m/s}$$

If an electron were accelerated under the same circumstances, its speed would approach the speed of light and the problem would have to be treated by relativistic mechanics (Chapter 39).

25.3 ELECTRIC POTENTIAL AND POTENTIAL ENERGY DUE TO POINT CHARGES

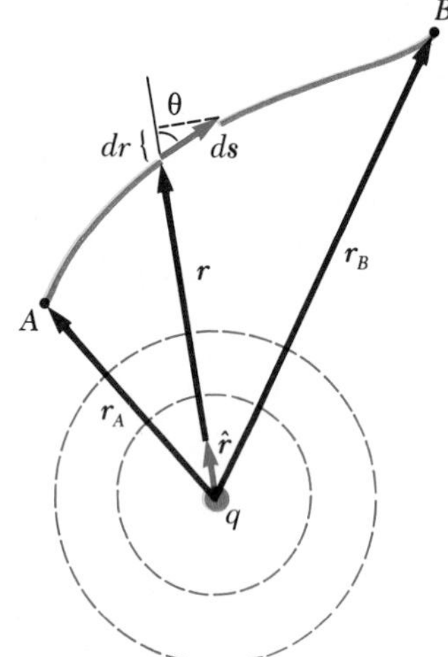

Figure 25.5 The potential difference between points A and B due to a point charge q depends *only* on the initial and final radial coordinates, r_A and r_B, respectively.

Consider an isolated positive point charge q as in Figure 25.5. Recall that such a charge produces an electric field that is radially outward from the charge. In order to find the electric potential at a distance r from the charge, we begin with the general expression for the potential difference given by

$$V_B - V_A = -\int_A^B \mathbf{E}\cdot d\mathbf{s}$$

Since the electric field due to the point charge is given by $\mathbf{E} = kq\hat{\mathbf{r}}/r^2$, where $\hat{\mathbf{r}}$ is a unit vector directed from the charge to the field point, the quantity $\mathbf{E}\cdot d\mathbf{s}$ can be expressed as

$$\mathbf{E}\cdot d\mathbf{s} = k\frac{q}{r^2}\,\hat{\mathbf{r}}\cdot d\mathbf{s}$$

The dot product $\hat{\mathbf{r}}\cdot d\mathbf{r} = ds\cos\theta$, where θ is the angle between $\hat{\mathbf{r}}$ and $d\mathbf{s}$ as in Figure 25.5. Furthermore, note that $ds\cos\theta$ is the projection of $d\mathbf{s}$ onto $\mathbf{r}$, so that $ds\cos\theta = dr$. That is, any displacement $d\mathbf{s}$ produces a change dr in the magnitude of $\mathbf{r}$. With these substitutions, we find that $\mathbf{E}\cdot d\mathbf{s} = (kq/r^2)\,dr$, so the expression for the potential difference becomes

$$V_B - V_A = -\int E_r\,dr = -kq\int_{r_A}^{r_B}\frac{dr}{r^2} = \frac{kq}{r}\Bigg]_{r_A}^{r_B}$$

$$V_B - V_A = kq\left[\frac{1}{r_B} - \frac{1}{r_A}\right] \qquad (25.11)$$

The integral $-\int_A^B \mathbf{E}\cdot d\mathbf{s}$ is *independent* of the path between A and B, as it must be. (We had already concluded that the electric field of a point charge is a conservative field, by analogy with the gravitational field of a point mass.)

Furthermore, Equation 25.11 expresses the important result that the potential difference between any two points A and B depends *only* on the *radial* coordinates r_A and r_B. It is customary to choose the reference of potential to be zero at $r_A = \infty$. (This is quite natural since $V \propto 1/r_A$ and as $r_A \rightarrow \infty$, $V \rightarrow 0$.) With this choice, the electric potential due to a point charge at any distance r from the charge is given by

$$V = k\frac{q}{r} \qquad (25.12)$$

Potential of a point charge

From this we see that V is constant on a spherical surface of radius r. Hence, we conclude that *the equipotential surfaces (surfaces on which V remains constant) for an isolated point charge consist of a family of spheres concentric with the charge,* as shown in Figure 25.5. Note that the equipotential surfaces are perpendicular to the lines of electric force, as was the case for a uniform electric field.

The electric potential of two or more point charges is obtained by applying the superposition principle. That is, the total potential at some point P due to several point charges is the sum of the potentials due to the individual charges. For a group of charges, we can write the total potential at P in the form

$$V = k\sum_i \frac{q_i}{r_i} \qquad (25.13)$$

The potential of several point charges

where the potential is again taken to be zero at infinity and r_i is the distance from the point P to the charge q_i. Note that the sum in Equation 25.13 is an *algebraic sum* of scalars rather than a vector sum (which is used to calculate the electric field of a group of charges). Thus, it is much easier to evaluate V than to evaluate $\mathbf{E}$.

We now consider the potential energy of interaction of a system of charged particles. If V_1 is the electric potential due to charge q_1 at a point P, then the work required to bring a second charge, q_2, from infinity to the point P without acceleration is given by q_2V_1. By definition, this work equals the potential energy, U, of the two-particle system when the particles are separated by a distance r_{12} (Fig. 25.6).

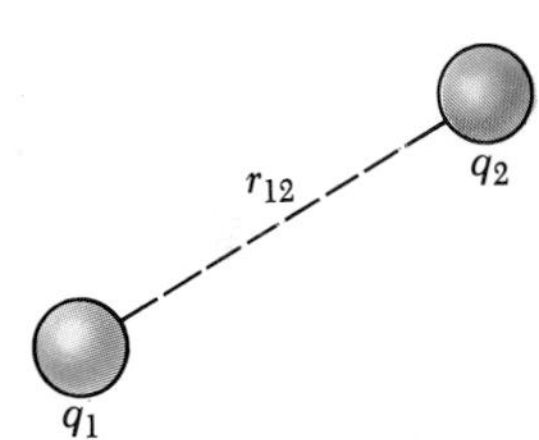

Figure 25.6 If two point charges are separated by a distance r_{12}, the potential energy of the pair of charges is given by kq_1q_2/r_{12}.

Therefore, we can express the potential energy as

$$U = q_2V_1 = k\frac{q_1q_2}{r_{12}} \qquad (25.14)$$

Electric potential energy of two charges

Note that if the charges are of the same sign, U is positive.[3] This is consistent with the fact that like charges repel, and so positive work must be done *on* the system to bring the two charges near one another. Conversely, if the charges are of opposite sign, the force is attractive and U is negative. This means that negative work must be done to bring the unlike charges near one another.

[3] The expression for the electric potential energy for two point charges, Equation 25.14, is of the *same* form as the gravitational potential energy of two point masses given by Gm_1m_2/r (Chapter 14). The similarity is not surprising in view of the fact that both are derived from an inverse-square force law.

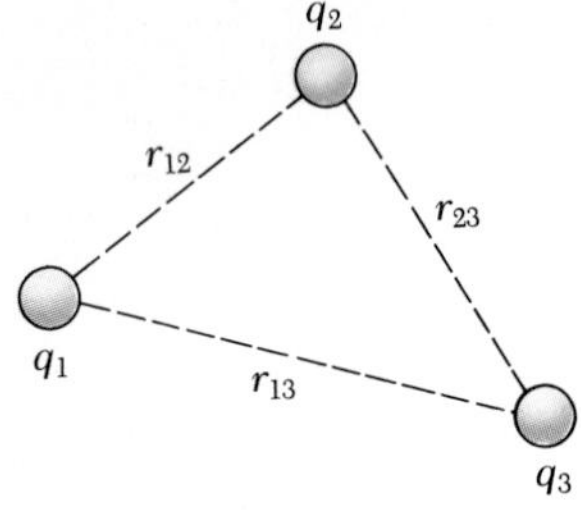

Figure 25.7 Three point charges are fixed at the positions shown. The potential energy of this system of charges is given by Equation 25.15.

If there are more than two charged particles in the system, the total potential energy can be obtained by calculating U for every *pair* of charges and summing the terms algebraically. As an example, the total potential energy of the three charges shown in Figure 25.7 is given by

$$U = k\left(\frac{q_1 q_2}{r_{12}} + \frac{q_1 q_3}{r_{13}} + \frac{q_2 q_3}{r_{23}}\right) \tag{25.15}$$

Physically, we can interpret this as follows: Imagine that q_1 is fixed at the position shown in Figure 25.7, but q_2 and q_3 are at infinity. The work required to bring q_2 from infinity to its position near q_1 is kq_1q_2/r_{12}, which is the first term in Equation 25.15. The last two terms in Equation 25.15 represent the work required to bring q_3 from infinity to its position near q_1 and q_2. (You should show that the result is independent of the order in which the charges are transported.)

EXAMPLE 25.3 The Potential Due to Two Point Charges

A 5-μC point charge is located at the origin, and a second point charge of $-2\ \mu$C is located on the x axis at the position (3, 0) m, as in Figure 25.8a. (a) If the potential is taken to be zero at infinity, find the total electric potential due to these charges at the point P, whose coordinates are (0, 4) m.

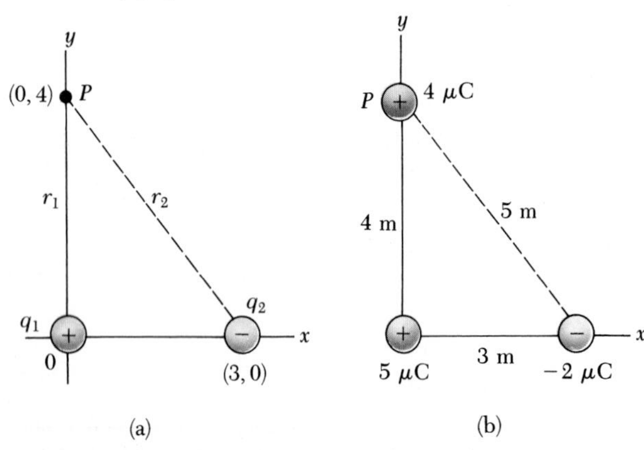

Figure 25.8 (Example 25.3) The electric potential at the point P due to the two point charges q_1 and q_2 is the algebraic sum of the potentials due to the individual charges.

The total potential at P due to the two charges is given by

$$V_P = k\left(\frac{q_1}{r_1} + \frac{q_2}{r_2}\right)$$

Since $r_1 = 4$ m and $r_2 = 5$ m, we get

$$V_P = 9 \times 10^9 \frac{\text{N}\cdot\text{m}^2}{\text{C}^2}\left(\frac{5 \times 10^{-6}\ \text{C}}{4\ \text{m}} - \frac{2 \times 10^{-6}\ \text{C}}{5\ \text{m}}\right)$$

$$= \boxed{7.65 \times 10^3\ \text{V}}$$

(b) How much work is required to bring a third point charge of 4 μC from infinity to the point P?

$$W = q_3 V_P = (4 \times 10^{-6}\ \text{C})(7.65 \times 10^3\ \text{V})$$

Since 1 V = 1 J/C, W reduces to

$$W = \boxed{3.06 \times 10^{-2}\ \text{J}}$$

Exercise 1 Find the *total* potential energy of the system of three charges in the configuration shown in Figure 25.8b.

Answer 6.0×10^{-4} J.

25.4 ELECTRIC POTENTIAL DUE TO CONTINUOUS CHARGE DISTRIBUTIONS

The electric potential due to a continuous charge distribution can be calculated in two ways. If the charge distribution is known, we can start with Equation 25.12 for the potential of a point charge. We then consider the potential due to a small charge element dq, treating this element as a point charge (Figure 25.9). The potential dV at some point P due to the charge element dq is given by

$$dV = k\frac{dq}{r} \tag{25.16}$$

where r is the distance from the charge element to the point P. To get the total potential at P, we integrate Equation 25.16 to include contributions from all elements of the charge distribution. Since each element is, in general, at a different distance from P and since k is a constant, we can express V as

$$V = k\int \frac{dq}{r} \tag{25.17}$$

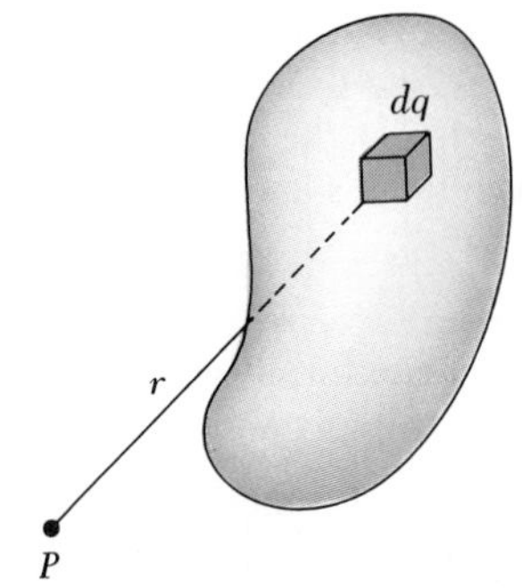

Figure 25.9 The electric potential at the point P due to a continuous charge distribution can be calculated by dividing the charged body into segments of charge dq and summing the potential contributions over all segments.

In effect, we have replaced the sum in Equation 25.13 by an integral. Note that this expression for V uses a particular choice of reference: the potential is taken to be zero for point P located infinitely far from the charge distribution.

The second method for calculating the potential of a continuous charge distribution makes use of Equation 25.4. This procedure is useful when the electric field is already known from other considerations, such as Gauss' law. If the charge distribution is highly symmetric, we first evaluate $\boldsymbol{E}$ at any point using Gauss' law and then substitute the value obtained into Equation 25.4 to determine the potential difference between any two points. We then choose V to be zero at *any* convenient point. Let us illustrate both methods with several examples.

EXAMPLE 25.4 Potential Due to a Uniformly Charged Ring

Find the electric potential at a point P located on the axis of a uniformly charged ring of radius a and total charge Q. The plane of the ring is chosen perpendicular to the x axis (Fig. 25.10).

Solution Let us take the point P to be at a distance x from the center of the ring, as in Figure 25.10. The charge element dq is at a distance equal to $\sqrt{x^2 + a^2}$ from the point P. Hence, we can express V as

$$V = k\int \frac{dq}{r} = k\int \frac{dq}{\sqrt{x^2 + a^2}}$$

In this case, *each* element dq is at the *same distance* from the point P. Therefore, the term $\sqrt{x^2 + a^2}$ can be removed from the integral and V reduces to

$$V = \frac{k}{\sqrt{x^2 + a^2}}\int dq = \frac{kQ}{\sqrt{x^2 + a^2}} \tag{25.18}$$

The only variable that appears in this expression for V is x. This is not surprising, since our calculation is valid only for points along the x axis, where y and z are both zero. From the symmetry of the situation, we see that

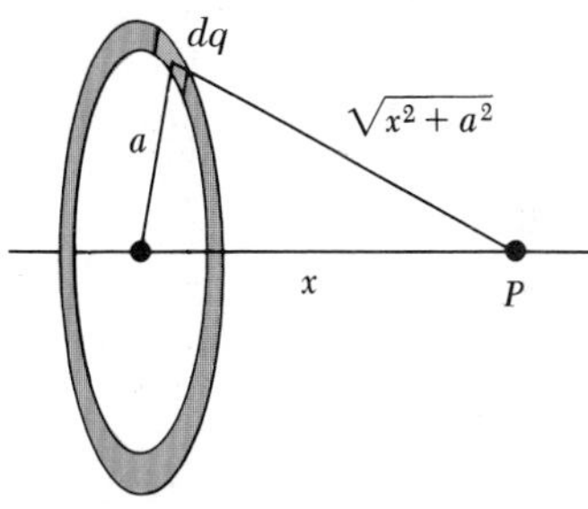

Figure 25.10 (Example 25.4) A uniformly charged ring of radius a, whose plane is perpendicular to the x axis. All segments of the ring are at the same distance from any axial point P.

along the x axis $\boldsymbol{E}$ can have only an x component. Therefore, we can use the expression $E_x = -dV/dx$, which we shall derive in Section 25.5, to find the electric field at P:

$$E_x = -\frac{dV}{dx} = -kQ\frac{d}{dx}(x^2 + a^2)^{-1/2}$$

$$= -kQ(-\tfrac{1}{2})(x^2 + a^2)^{-3/2}(2x)$$

$$= \frac{kQx}{(x^2 + a^2)^{3/2}} \tag{25.19}$$

This result agrees with that obtained by direct integration (see Example 23.9). Note that $E_x = 0$ at $x = 0$ (the center of the ring). Could you have guessed this from Coulomb's law?

Exercise 2 What is the electric potential at the center of the uniformly charged ring? What does the field at the center imply about this result?

Answer $V = kQ/a$ at $x = 0$. Because $E = 0$, V must have a maximum or minimum value; it is in fact a maximum.

EXAMPLE 25.5 Potential of a Uniformly Charged Disk

Find the electric potential along the axis of a uniformly charged disk of radius a and charge per unit area σ (Fig. 25.11).

Solution Again we choose the point P to be at a distance x from the center of the disk and take the plane of the disk perpendicular to the x axis. The problem is simplified by dividing the disk into a series of charged rings. The potential of each ring is given by Equation 25.18 in Example 25.4. Consider one such ring of radius r and width dr, as indicated in Figure 25.11. The area of the ring is $dA = 2\pi r\,dr$ (the circumference multiplied by the width), and the charge on the ring is $dq = \sigma\,dA = \sigma 2\pi r\,dr$. Hence, the potential at the point P due to this ring is given by

$$dV = \frac{k\,dq}{\sqrt{r^2+x^2}} = \frac{k\sigma 2\pi r\,dr}{\sqrt{r^2+x^2}}$$

To find the *total* potential at P, we sum over all rings making up the disk. That is, we integrate dV from $r = 0$ to $r = a$:

$$V = \pi k\sigma \int_0^a \frac{2r\,dr}{\sqrt{r^2+x^2}} = \pi k\sigma \int_0^a (r^2+x^2)^{-1/2} 2r\,dr$$

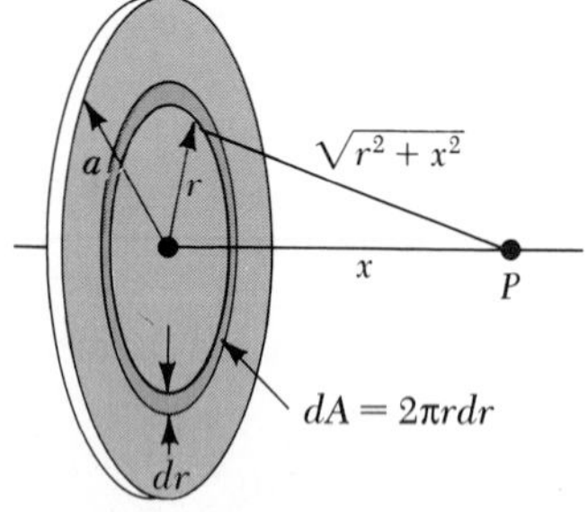

Figure 25.11 (Example 25.5) A uniformly charged disk of radius a, whose plane is perpendicular to the x axis. The calculation of the potential at an axial point P is simplified by dividing the disk into rings of area $2\pi r\,dr$.

This integral is of the form $u^n\,du$ and has the value $u^{n+1}/(n+1)$, where $n = -\frac{1}{2}$ and $u = r^2 + x^2$. This gives the result

$$V = 2\pi k\sigma[(x^2+a^2)^{1/2} - x] \qquad (25.20)$$

As in Example 25.4, we can find the electric field at any axial point by taking the negative of the derivative of V with respect to x. This gives

$$E_x = -\frac{dV}{dx} = 2\pi k\sigma\left(1 - \frac{x}{\sqrt{x^2+a^2}}\right) \qquad (25.21)$$

The calculation of V and $\mathbf{E}$ for an arbitrary point off the axis is more difficult to perform.

EXAMPLE 25.6 Potential of a Finite Line Charge

A rod of length ℓ located along the x axis has a uniform charge per unit length and a total charge Q. Find the electric potential at a point P along the y axis at a distance d from the origin (Figure 25.12).

Solution The element of length dx has a charge dq given by $\lambda\,dx$, where λ is the charge per unit length, Q/ℓ. Since this element is at a distance $r = \sqrt{x^2+d^2}$ from the point P, we can express the potential at P due to this element as

$$dV = k\frac{dq}{r} = k\frac{\lambda\,dx}{\sqrt{x^2+d^2}}$$

To get the total potential at P, we integrate this expression over the limits $x = 0$ to $x = \ell$. Noting that k, λ and d are constants, we find that

$$V = k\lambda \int_0^\ell \frac{dx}{\sqrt{x^2+d^2}} = k\frac{Q}{\ell}\int_0^\ell \frac{dx}{\sqrt{x^2+d^2}}$$

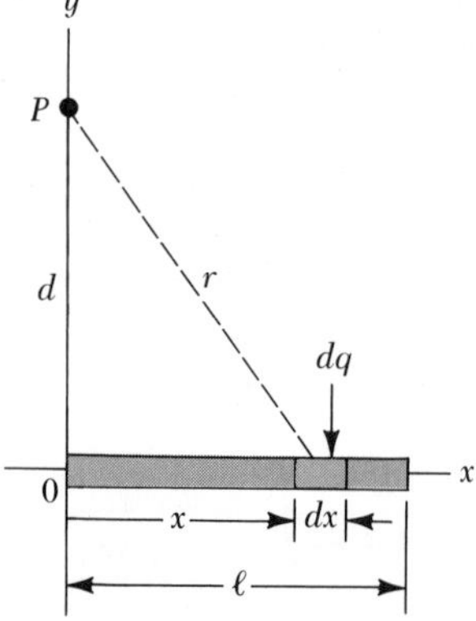

Figure 25.12 (Example 25.6) A uniform line charge of length ℓ located along the x axis. To calculate the potential at P, the line charge is divided into segments each of length dx, having a charge $dq = \lambda\,dx$.

This integral, found in most integral tables, has the value

$$\int \frac{dx}{\sqrt{x^2 + d^2}} = \ln(x + \sqrt{x^2 + d^2})$$

Evaluating V, we find that

$$V = \frac{kQ}{\ell} \ln\left(\frac{\ell + \sqrt{\ell^2 + d^2}}{d}\right) \qquad (25.22)$$

EXAMPLE 25.7 Potential of a Uniformly Charged Sphere

An insulating solid sphere of radius R has a uniform positive charge density with total charge Q (Fig. 25.13). (a) Find the electric potential at a point *outside* the sphere, that is, for $r > R$. Take the potential to be zero at $r = \infty$.

Solution In Example 24.3, we found from Gauss' law that the magnitude of the electric field *outside* a uniformly charged sphere is given by

$$E_r = k\frac{Q}{r^2} \qquad (\text{for } r > R)$$

where the field is directed radially outward when Q is positive. To obtain the potential at an exterior point, such as B in Figure 25.13, we substitute this expression for E into Equation 25.5. Since $\mathbf{E} \cdot d\mathbf{s} = E_r\,dr$ in this case, we get

$$V_B = -\int_{\infty}^{r} E_r\,dr = -kQ\int_{\infty}^{r} \frac{dr}{r^2}$$

$$V_B = k\frac{Q}{r} \qquad (\text{for } r > R)$$

Note that the result is identical to that for the electric potential due to a point charge. Since the potential must be continuous at $r = R$, we can use this expression to obtain the potential at the surface of the sphere. That is, the potential at a point such as C in Figure 25.13 is given by

$$V_C = k\frac{Q}{R} \qquad (\text{for } r = R)$$

(b) Find the potential at a point *inside* the charged sphere, that is, for $r < R$.

Solution In Example 24.3 we found that the electric field inside a uniformly charged sphere is given by

$$E_r = \frac{kQ}{R^3}r \qquad (\text{for } r < R)$$

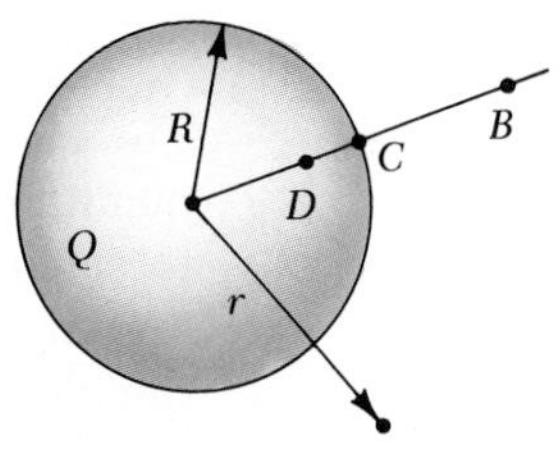

Figure 25.13 (Example 25.7) A uniformly charged insulating sphere of radius R and total charge Q. The electric potential at points B and C is equivalent to that of a point charge Q located at the center of the sphere.

We can use this result and Equation 25.4 to evaluate the potential difference $V_D - V_C$, where D is an interior point:

$$V_D - V_C = -\int_R^r E_r\,dr = -\frac{kQ}{R^3}\int_R^r r\,dr = \frac{kQ}{2R^3}(R^2 - r^2)$$

Substituting $V_C = kQ/R$ into this expression and solving for V_D, we get

$$V_D = \frac{kQ}{2R}\left(3 - \frac{r^2}{R^2}\right) \qquad (\text{for } r < R) \quad (25.23)$$

At $r = R$, this expression gives a result that agrees with that for the potential at the surface, that is, V_C. A plot of V versus r for this charge distribution is given in Figure 25.14.

Exercise 3 What is the electric field at the center of a uniformly charged sphere? What is the electric potential at this point?
Answer At $r = 0$, $E = 0$ and $V_0 = 3kQ/2R$.

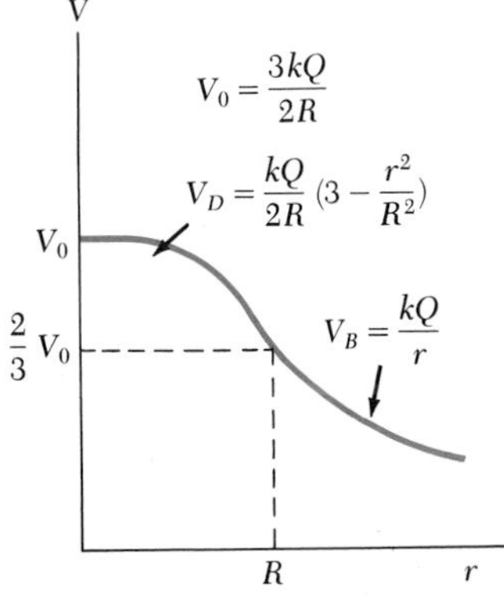

Figure 25.14 (Example 25.7) A plot of the electric potential V versus the distance r from the center of a uniformly charged, insulating sphere of radius R. The curve for V_D inside the sphere is parabolic and joins smoothly with the curve for V_B outside the sphere, which is a hyperbola. The potential has a maximum value V_0 at the center of the sphere.

*25.5 OBTAINING *E* FROM THE ELECTRIC POTENTIAL

The electric field $\mathbf{E}$ and the potential V are related by Equation 25.4. Both quantities are determined by a specific charge distribution. We now show how to calculate the electric field if the electric potential is known in a certain region. As we shall see, the electric field is simply the negative derivative of the electric potential.

From Equation 25.4 we can express the potential difference dV between two points a distance ds apart as

$$dV = -\mathbf{E} \cdot d\mathbf{s} \tag{25.24}$$

If the electric field has only *one* component, E_x, then $\mathbf{E} \cdot d\mathbf{s} = E_x\, dx$. Therefore, Equation 25.24 becomes $dV = -E_x\, dx$, or

$$E_x = -\frac{dV}{dx} \tag{25.25}$$

That is,

> the electric field is equal to the negative of the derivative of the potential with respect to some coordinate.

Note that the potential change is zero for any displacement perpendicular to the electric field. This is consistent with the notion of equipotential surfaces being perpendicular to the field, as in Figure 25.15a.

If the charge distribution has *spherical symmetry, where the charge density depends only on the radial distance r,* then the electric field is radial. In this case, $\mathbf{E} \cdot d\mathbf{s} = E_r\, dr$, and so we can express dV in the form $dV = -E_r\, dr$. Therefore,

Equipotential surfaces are always perpendicular to the electric field lines

$$E_r = -\frac{dV}{dr} \tag{25.26}$$

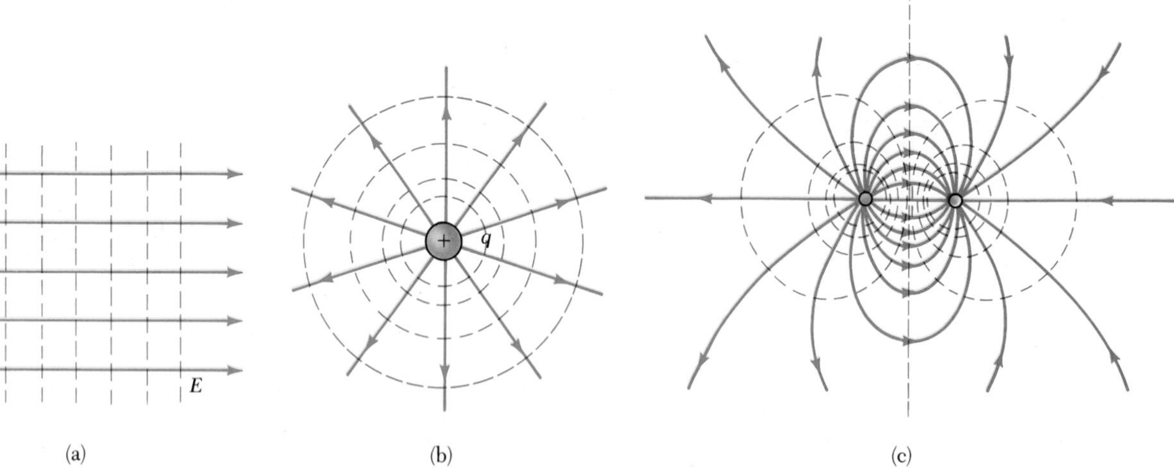

Figure 25.15 Equipotential surfaces (dashed blue lines) and electric field lines (red lines) for (a) a uniform electric field produced by an infinite sheet of charge, (b) a point charge, and (c) an electric dipole. In all cases, the equipotential surfaces are *perpendicular* to the electric field lines at every point.

Note that the potential changes only in the radial direction, not in a direction perpendicular to r. Thus V (like E_r) is a function only of r. Again, this is consistent with the idea that *equipotential surfaces are perpendicular to field lines.* In this case the equipotential surfaces are a family of spheres concentric with the spherically symmetric charge distribution (Figure 25.15b). The equipotential surfaces for the electric dipole are sketched in Figure 25.15c.

When a test charge is displaced by a vector $d\boldsymbol{s}$ that lies *within* any equipotential surface, then by definition $dV = -\boldsymbol{E}\cdot d\boldsymbol{s} = 0$. This shows that the equipotential surfaces must *always* be *perpendicular* to the electric field lines.

In general, the electric potential is a function of all three spatial coordinates. If $V(r)$ is given in terms of the rectangular coordinates, the electric field components E_x, E_y, and E_z can readily be found from $V(x, y, z)$. The field components are given by

$$E_x = -\frac{\partial V}{\partial x} \qquad E_y = -\frac{\partial V}{\partial y} \qquad E_z = -\frac{\partial V}{\partial z}$$

In these expressions, the derivatives are called *partial derivatives.* This means that in the operation $\partial V/\partial x$, *one takes a derivative with respect to x while y and z are held constant.* For example, if $V = 3x^2y + y^2 + yz$, then

$$\frac{\partial V}{\partial x} = \frac{\partial}{\partial x}(3x^2y + y^2 + yz) = \frac{\partial}{\partial x}(3x^2y) = 3y\frac{d}{dx}(x^2) = 6xy$$

In vector notation, $\boldsymbol{E}$ is often written $\boldsymbol{E} = -\nabla V = -\left(\boldsymbol{i}\frac{\partial}{\partial x} + \boldsymbol{j}\frac{\partial}{\partial y} + \boldsymbol{k}\frac{\partial}{\partial z}\right)V$, where ∇ is called the *gradient operator.*

EXAMPLE 25.8 The Point Charge Revisited

Let us use the potential function for a point charge q to derive the electric field at a distance r from the charge.

Solution The potential of a point charge is given by Equation 25.12:

$$V = k\frac{q}{r}$$

Since the potential is a function of r only, it has spherical symmetry and we can apply Equation 25.26 directly to obtain the electric field:

$$E_r = -\frac{dV}{dr} = -\frac{d}{dr}\left(k\frac{q}{r}\right) = -kq\frac{d}{dr}\left(\frac{1}{r}\right)$$

$$E_r = \frac{kq}{r^2}$$

Thus, the electric field is radial and the result agrees with that obtained using Gauss' law.

EXAMPLE 25.9 The Electric Potential of a Dipole

An electric dipole consists of two equal and opposite charges separated by a distance $2a$, as in Figure 25.16. Calculate the electric potential and the electric field at the point P on the x axis and located a distance x from the center of the dipole.

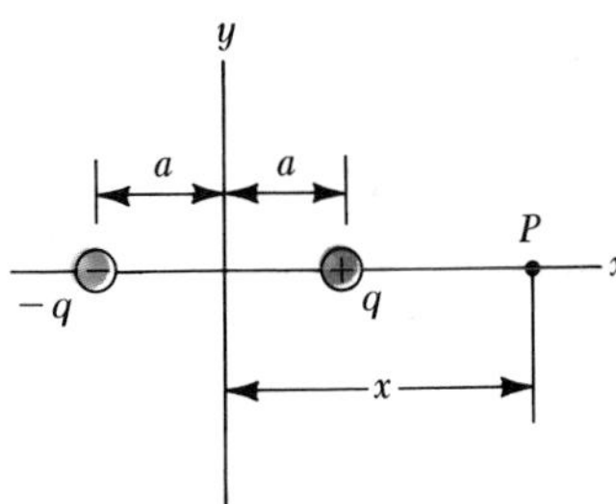

Figure 25.16 (Example 25.9) An electric dipole located on the x axis.

Solution

$$V = k\sum\frac{q_i}{r_i} = k\left(\frac{q}{x-a} - \frac{q}{x+a}\right) = \frac{2kqa}{x^2 - a^2}$$

If the point P is far from the dipole, so that $x \gg a$, then a^2 can be neglected in the term $x^2 - a^2$ and V becomes

$$V \approx \frac{2kqa}{x^2} \qquad (x \gg a)$$

Using Equation 25.25 and this result, the electric field at P is given by

$$E = -\frac{dV}{dx} = \frac{4kqa}{x^3} \qquad \text{for } x \gg a$$

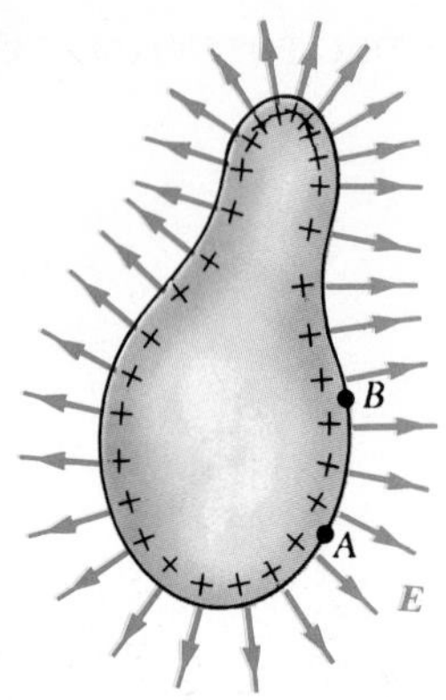

Figure 25.17 An arbitrarily shaped conductor with an excess positive charge. When the conductor is in electrostatic equilibrium, all of the charge resides at the surface, $\mathbf{E} = 0$ inside the conductor, and the electric field just outside the conductor is perpendicular to the surface. The potential is constant inside the conductor and is equal to the potential at the surface. The surface charge density is nonuniform.

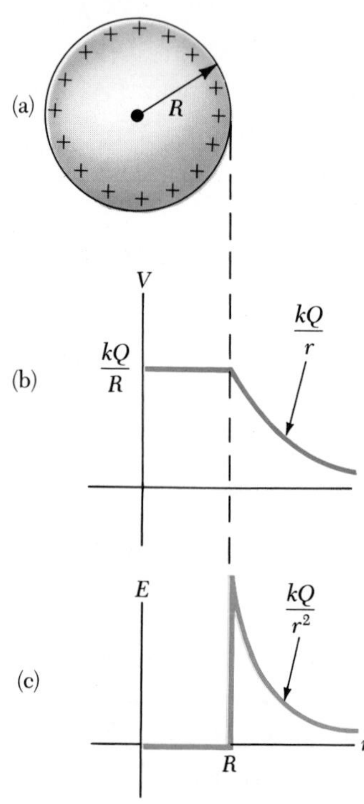

Figure 25.18 (a) The excess charge on a conducting sphere of radius R is uniformly distributed on its surface. (b) The electric potential versus the distance r from the center of the charged conducting sphere. (c) The electric field intensity versus the distance r from the center of the charged conducting sphere.

25.6 POTENTIAL OF A CHARGED CONDUCTOR

In the previous chapter we found that when a solid conductor in equilibrium carries a net charge, the charge resides on the outer surface of the conductor. Furthermore, we showed that the electric field just outside the surface of a conductor in equilibrium is perpendicular to the surface while the field *inside* the conductor is zero. If the electric field had a component parallel to the surface, this would cause surface charges to move, creating a current and a nonequilibrium situation.

We shall now show that *every point on the surface of a charged conductor in equilibrium is at the same potential.* Consider two points A and B on the surface of a charged conductor, as in Figure 25.17. Along a surface path connecting these points, $\mathbf{E}$ is always perpendicular to the displacement $d\mathbf{s}$; therefore $\mathbf{E} \cdot d\mathbf{s} = 0$. Using this result and Equation 25.4, we conclude that the potential difference between A and B is necessarily zero. That is,

$$V_B - V_A = -\int_A^B \mathbf{E} \cdot d\mathbf{s} = 0$$

This result applies to *any* two points on the surface. Therefore, V is constant everywhere on the surface of a charged conductor in equilibrium. That is,

> the surface of any charged conductor in equilibrium is an equipotential surface. Furthermore, since the electric field is zero inside the conductor, we conclude that the potential is constant everywhere inside the conductor and equal to its value at the surface.

Therefore, no work is required to move a test charge from the interior of a charged conductor to its surface. (Note that the potential is *not zero* inside the conductor even though the electric field is zero.)

For example, consider a solid metal sphere of radius R and total positive charge Q, as shown in Figure 25.18a. The electric field outside the charged sphere is given by kQ/r^2 and points radially outward. Following Example 25.7, we see that the potential at the interior and surface of the sphere must be kQ/R relative to infinity. The potential outside the sphere is given by kQ/r. Figure 25.18b is a plot of the potential as a function of r, and Figure 25.18c shows the variations of the electric field with r.

When a net charge is placed on a spherical conductor, the surface charge density is uniform, as indicated in Figure 25.18a. However, if the conductor is nonspherical, as in Figure 25.17, the surface charge density is high where the radius of curvature is small and convex and low where the radius of curvature is small and concave. Since the electric field just outside a charged conductor is proportional to the surface charge density, σ, we see that *the electric field is large near points having a small convex radius of curvature and reaches very high values at sharp points.*

Figure 25.19 shows the electric field lines around two spherical conductors, one with a net charge Q and one with zero net charge. In this case, the surface charge density is *not* uniform on either conductor. The larger sphere (on the right), with zero net charge, has negative charges induced on its side that faces the charged sphere and positive charge on its side opposite the charged sphere. The blue lines in Figure 25.19 represent the boundaries of the equipotential surfaces for this charge configuration. Again, you should notice that the field lines are perpendicular to the conducting surfaces. Fur-

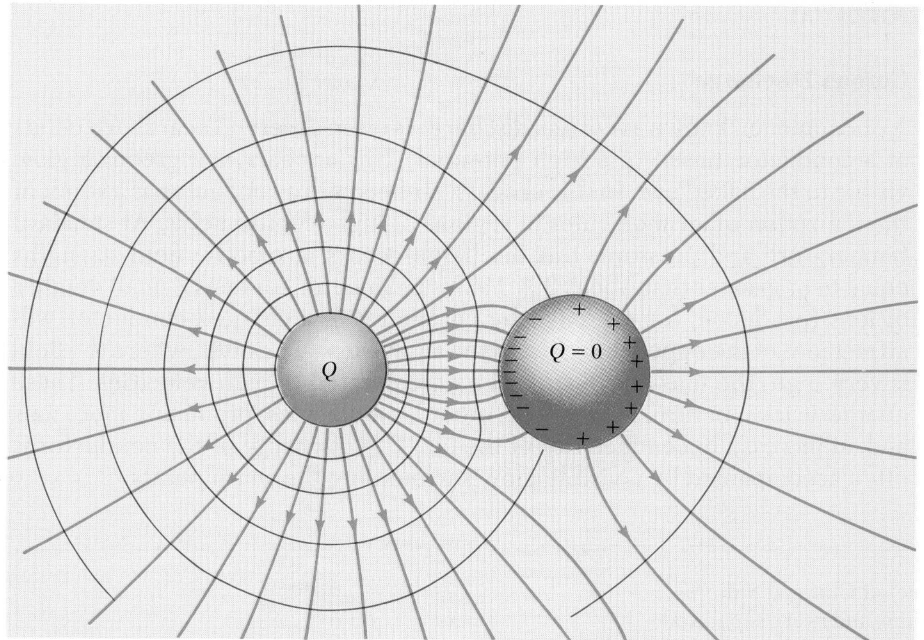

Figure 25.19 The electric field lines (in red) around two spherical conductors. The smaller sphere on the left has a net charge Q, and the sphere on the right has zero net charge. The blue lines represent the edges of the equipotential surfaces. (From E. Purcell, *Electricity and Magnetism*, New York, McGraw-Hill, 1965, with permission of the Education Development Center, Inc.)

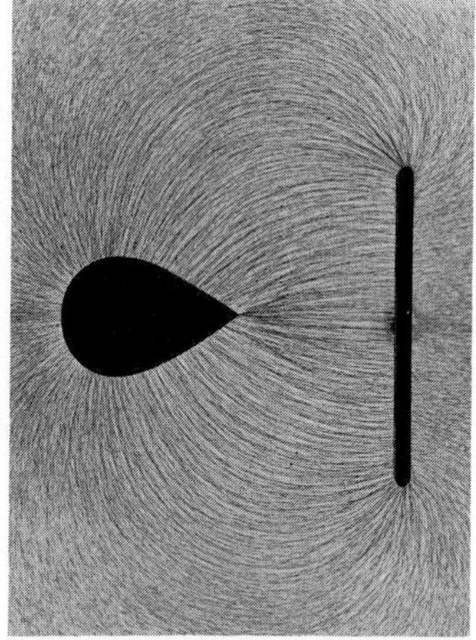

Electric field pattern of a charged conducting plate near an oppositely charged pointed conductor. Small pieces of thread suspended in oil align with the electric field lines. Note that the electric field is most intense near the pointed part of the conductor and at other points where the radius of curvature is small. (Courtesy of Harold M. Waage, Princeton University)

thermore, the equipotential surfaces are perpendicular to the field lines at the boundaries of the conductor and everywhere else in space.

A Cavity Within a Conductor

Now consider a conductor of arbitrary shape containing a cavity as in Figure 25.20. Let us assume there are no charges *inside* the cavity. We shall show that *the electric field inside the cavity must be zero,* regardless of the charge distribution on the *outside* surface of the conductor. Furthermore, the field in the cavity is zero even if an electric field exists outside the conductor.

In order to prove this point, we shall use the fact that every point on the conductor is at the same potential, and therefore any two points A and B on the surface of the cavity must be at the same potential. Now *imagine* that a field $\mathbf{E}$ exists in the cavity, and evaluate the potential difference $V_B - V_A$ defined by the expression

$$V_B - V_A = -\int_A^B \mathbf{E} \cdot d\mathbf{s}$$

If $\mathbf{E}$ is non-zero, we can always find a path between A and B for which $\mathbf{E} \cdot d\mathbf{s}$ is always a positive number (a path along the direction of $\mathbf{E}$), and so the integral must be positive. However, since $V_B - V_A = 0$, the integral must also be zero. This contradiction can be reconciled only if $\mathbf{E} = 0$ inside the cavity. Thus, we conclude that a cavity surrounded by conducting walls is a field-free region as long as there are no charges inside the cavity.

This result has some interesting applications. For example, it is possible to shield an electronic circuit or even an entire laboratory from external fields by surrounding it with conducting walls. Shielding is often necessary when performing highly sensitive electrical measurements.

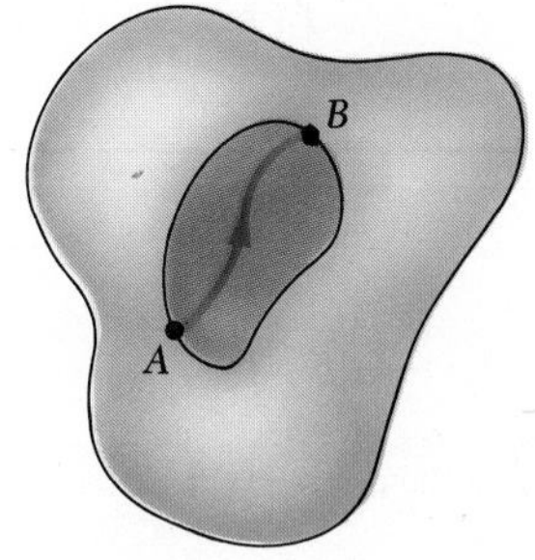

Figure 25.20 A conductor in electrostatic equilibrium containing an empty cavity. The electric field in the cavity is *zero*, regardless of the charge on the conductor.

Corona Discharge

A phenomenon known as **corona discharge** is often observed near sharp points of a conductor raised to a high potential. This appears as a greenish glow visible to the naked eye. In this process, air becomes a conductor as a result of the ionization of air molecules in regions of high electric fields. At standard temperature and pressure, this discharge occurs at electric field strengths equal to or greater than about 3×10^6 V/m. Since air contains a small number of ions (produced, for example, by cosmic rays), a charged conductor will attract ions of the opposite sign from the air. Near sharp points, where the field is very high, the ions in the air will be accelerated to high velocities. These energetic ions, in turn, collide with other air molecules, producing more ions and an increase in conductivity of the air. The discharge of the conductor is often accompanied by a visible glow surrounding the sharp points.

EXAMPLE 25.10 Two Connected Charged Spheres

Two spherical conductors of radii r_1 and r_2 are separated by a distance much larger than the radius of either sphere. The spheres are connected by a conducting wire as in Figure 25.21. If the charges on the spheres in equilibrium are q_1 and q_2, respectively, find the ratio of the field strengths at the surfaces of the spheres.

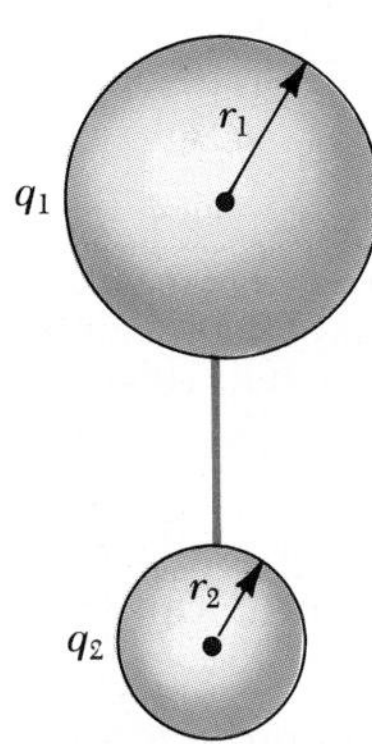

Figure 25.21 (Example 25.10) Two charged spherical conductors connected by a conducting wire. The spheres are at the *same* potential, *V*.

Solution Since the spheres are connected by a conducting wire, they must both be at the *same* potential V, given by

$$V = k\frac{q_1}{r_1} = k\frac{q_2}{r_2}$$

Therefore, the ratio of charges is

$$(1) \qquad \frac{q_1}{q_2} = \frac{r_1}{r_2}$$

Since the spheres are very far apart, their surfaces are uniformly charged and we can express the electric fields at their surfaces as

$$E_1 = k\frac{q_1}{{r_1}^2} \quad \text{and} \quad E_2 = k\frac{q_2}{{r_2}^2}$$

Taking the ratio of these two fields and making use of (1), we find that

$$(2) \qquad \frac{E_1}{E_2} = \frac{r_2}{r_1}$$

Hence, the field is more intense in the vicinity of the smaller sphere.

*25.7 THE MILLIKAN OIL-DROP EXPERIMENT

During the period 1909 to 1913, Robert Andrews Millikan (1868–1953) performed a brilliant set of experiments at the University of Chicago in which he measured the elementary charge on an electron, e, and demonstrated the quantized nature of the electronic charge. The apparatus used by Millikan, diagrammed in Figure 25.22, contains two parallel metal plates. Oil droplets charged by friction in an atomizer are allowed to pass through a small hole in the upper plate. A light beam directed horizontally is used to illuminate the oil droplets, which are veiwed by a telescope whose axis is at right angles to the

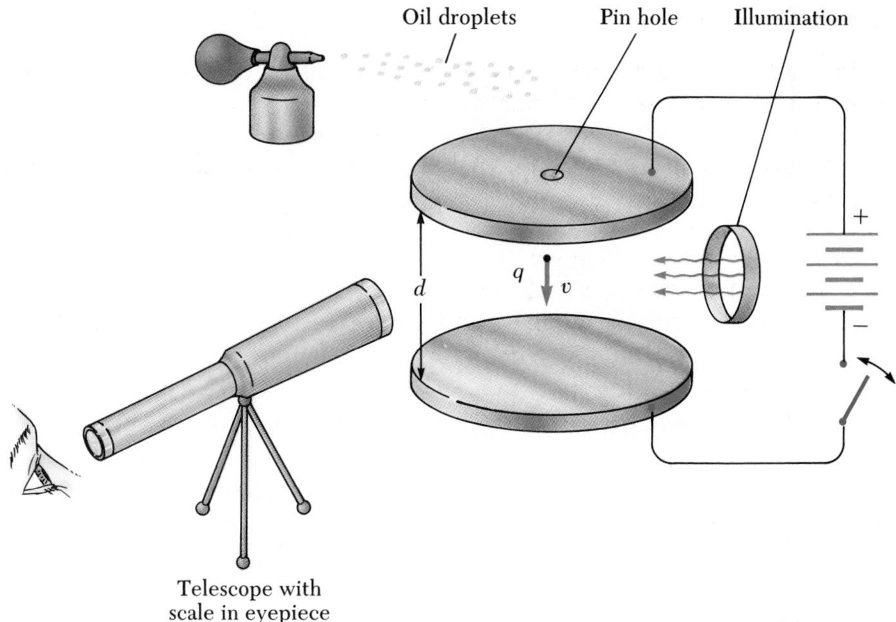

Figure 25.22 A schematic view of the Millikan oil-drop apparatus.

beam. When the droplets are viewed in this manner, they appear as shining stars against a dark background, and the rate of fall of individual drops may be determined.[4]

Let us assume a single drop having a mass m and carrying a charge q is being viewed, and that its charge is negative. If there is no electric field present between the plates, the two forces acting on the charge are its weight, mg, acting downward, and an upward viscous drag force D, as indicated in Figure 25.23a. The drag force is proportional to the speed of the drop. When the drop reaches its terminal speed v, the two forces balance each other ($mg = D$).

Now suppose that an electric field is set up between the plates by connecting a battery such that the upper plate is at the higher potential. In this case, a third force $q\mathbf{E}$ acts on the charged drop. Since q is negative and $\mathbf{E}$ is downward, the electric force is *upward* as in Figure 25.23b. If this force is large enough, the drop will move upward and the drag force D' will act downward. When the upward electric force, qE, balances the sum of the weight and the drag force both acting downward, the drop reaches a new terminal speed v'.

With the field turned on, a drop moves slowly upward, typically at rates of *hundredths* of a centimeter per second. The rate of fall in the absence of a field is comparable. Hence, a single droplet with constant mass and radius may be followed for hours, alternately rising and falling, by simply turning the electric field on and off.

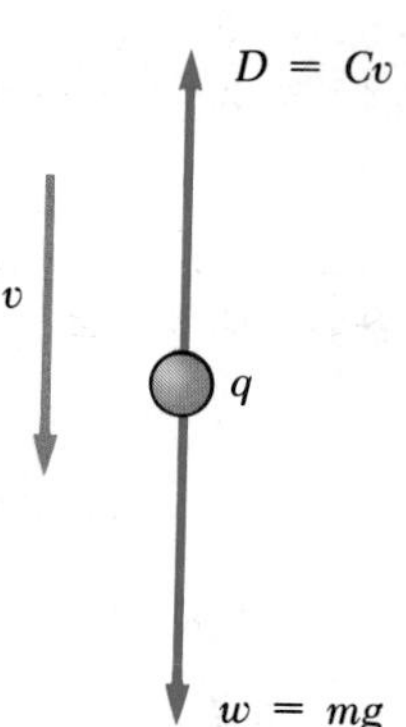

(a) Field off

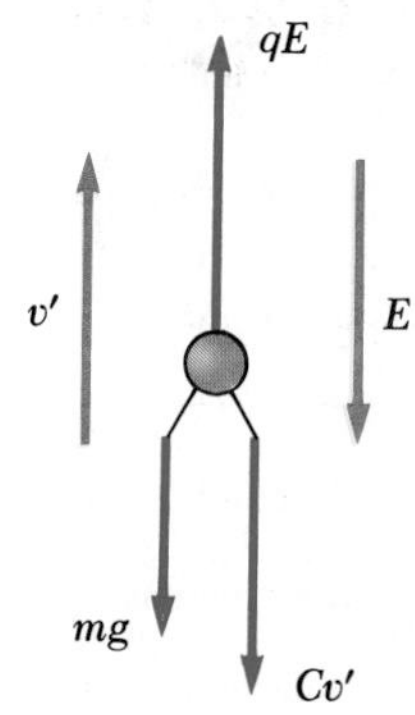

(b) Field on

Figure 25.23 The forces on a charged oil droplet in the Millikan experiment.

[4] At one time, the oil droplets were termed "Millikan's Shining Stars." Perhaps this description has lost its popularity because of the generations of physics students who have experienced hallucinations, near blindness, migraine headaches, etc., while repeating his experiment!

After making measurements on thousands of droplets, Millikan and his coworkers found that all drops, to within about 1% precision, had a charge equal to some integer multiple of the elementary charge e. That is,

$$q = ne \qquad n = 0, \pm 1, \pm 2, \pm 3, \ldots \tag{25.27}$$

where $e = 1.60 \times 10^{-19}$ C. Millikan's experiment is conclusive evidence that charge is quantized. He was awarded the Nobel Prize in physics in 1923 for this work.

*25.8 APPLICATIONS OF ELECTROSTATICS

The principles of electrostatics have been used in various applications, a few of which we shall briefly discuss in this section. Some of the more practical applications include electrostatic precipitators, used to reduce the level of atmospheric pollution from coal-burning power plants, and the xerography process, which has revolutionized imaging process technology. Scientific applications of electrostatic principles include electrostatic generators for accelerating elementary charged particles and the field-ion microscope, which is used to image atoms on the surface of metallic samples.

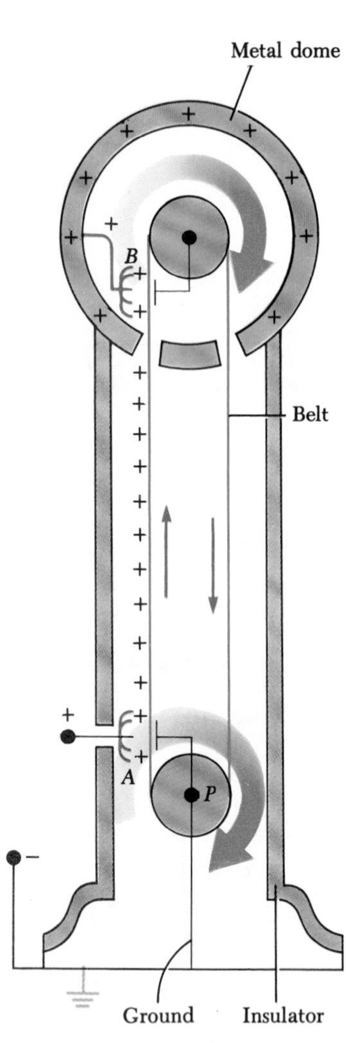

Figure 25.24 Schematic diagram of a Van de Graaff generator. Charge is transferred to the hollow conductor at the top by means of a moving belt. The charge is deposited on the belt at point A and is transferred to the hollow conductor at point B.

The Van de Graaff Generator

In the previous chapter we described an experiment that demonstrates a method for transferring charge to a hollow conductor (the Faraday ice-pail experiment). When a charged conductor is placed in contact with the inside of a hollow conductor, all of the charge of the first conductor is transferred to the hollow conductor. In principle, the charge on the hollow conductor and its potential can be increased without limit by repeating the process.

In 1929 Robert J. Van de Graaff used this principle to design and build an electrostatic generator. This type of generator is used extensively in nuclear physics research. The basic idea of the Van de Graaff generator is described in Figure 25.24. Charge is delivered continuously to a high-voltage electrode on a moving belt of insulating material. The high-voltage electrode is a hollow conductor mounted on an insulating column. The belt is charged at A by means of a corona discharge between comb-like metallic needles and a grounded grid. The needles are maintained at a positive potential of typically 10^4 V. The positive charge on the moving belt is transferred to the high-voltage electrode by a second comb of needles at B. Since the electric field inside the hollow conductor is negligible, the positive charge on the belt easily transfers to the high-voltage electrode, regardless of its potential. In practice, it is possible to increase the potential of the high-voltage electrode until electrical discharge occurs through the air. Since the "breakdown" voltage of air is equal to about 3×10^6 V/m, a sphere 1 m in radius can be raised to a maximum potential of 3×10^6 V. The potential can be increased further by increasing the radius of the hollow conductor and by placing the entire system in a container filled with high-pressure gas.

Van de Graaff generators can produce potential differences as high as 20 million volts. Protons accelerated through such potential differences receive enough energy to initiate nuclear reactions between the protons and various target nuclei.

The Electrostatic Precipitator

One important application of electrical discharge in gases is a device called an *electrostatic precipitator.* This device is used to remove particulate matter from combustion gases, thereby reducing air pollution. They are especially useful in coal-burning power plants and in industrial operations that generate large quantities of smoke. Current systems are able to eliminate more than 99% of the ash and dust (by weight) from the smoke. Figure 25.25 shows the basic idea of the electrostatic precipitator. A high voltage (typically 40 kV to 100 kV) is maintained between a wire running down the center of a duct and the outer wall, which is grounded. The wire is maintained at a negative potential with respect to the walls, and so the electric field is directed toward the wire. The electric field near the wire reaches high enough values to cause a corona discharge around the wire and the formation of positive ions, electrons, and negative ions, such as O_2^-. As the electrons and negative ions are accelerated toward the outer wall by the nonuniform electric field, the dirt particles in the streaming gas become charged by collisions and ion capture. Since most of the charged dirt particles are negative, they are also drawn to the outer wall by the electric field. By periodically shaking the duct, the particles fall loose and are collected at the bottom.

In addition to reducing the level of particulate matter in the atmosphere, the electrostatic precipitator also recovers valuable materials from the stack in the form of metal oxides.

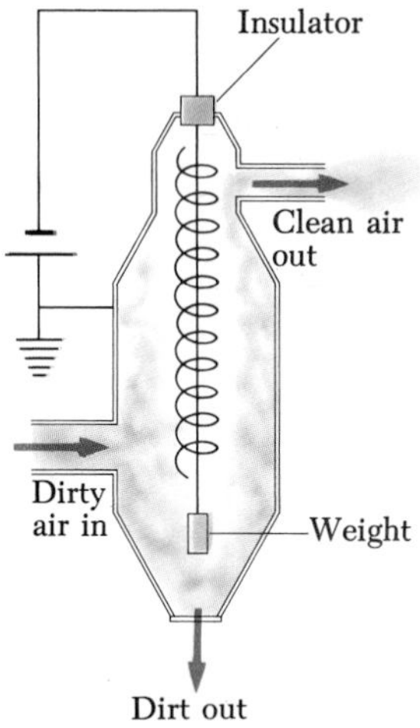

Figure 25.25 Schematic diagram of an electrostatic precipitator. The high negative voltage maintained on the central wire creates an electrical discharge in the vicinity of the wire.

Xerography

The process of xerography is widely used for making photocopies of letters, documents, and other printed materials. The basic idea for the process was developed by Chester Carlson, for which he was granted a patent in 1940. In 1947, the Xerox Corporation launched a full-scale program to develop automated duplicating machines using this process. The huge success of this development is quite evident; today, practically all modern offices and libraries have one or more duplicating machines, and the capabilities of modern machines are on the increase.

Some features of the xerographic process involve simple concepts from electrostatics and optics. However, the one idea that makes the process unique is the use of a photoconductive material to form an image. (A photoconductor is a material that is a poor conductor in the dark but becomes a good electrical conductor when exposed to light.)

The sequence of steps used in the xerographic process is illustrated in Figure 25.26. First, the surface of a plate or drum is coated with a thin film of the photoconductive material (usually selenium or some compound of selenium), and the photoconductive surface is given a positive electrostatic charge in the dark. The page to be copied is then projected onto the charged surface. The photoconducting surface becomes conducting only in areas where light strikes. In these areas, the light produces charge carriers in the photoconductor, which neutralize the positively charged surface. However, the charges remain on those areas of the photoconductor not exposed to light, leaving a latent (hidden) image of the object in the form of a positive surface charge distribution.

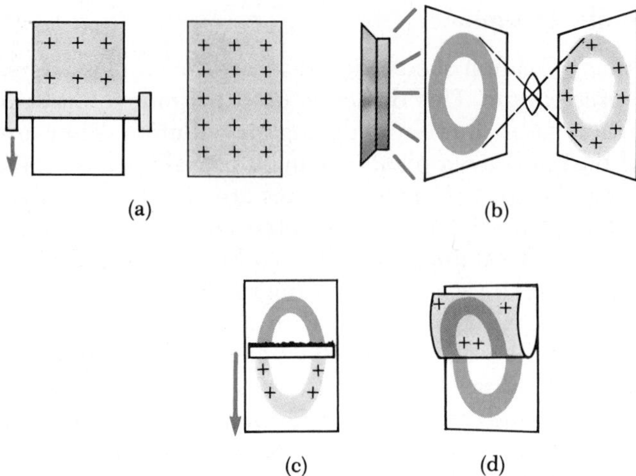

Figure 25.26 The xerographic process: (a) The photoconductive surface is positively charged. (b) Through the use of a light source and lens, an image is formed on the surface in the form of hidden positive charges. (c) The surface containing the image is covered with a charged powder, which adheres only to the image area. (d) A piece of paper is placed over the surface and given a charge. This transfers the visible image to the paper, which is finally heat-treated to "fix" the powder to the paper.

Next, a negatively charged powder called a *toner* is dusted onto the photoconducting surface. The charged powder adheres only to those areas of the surface that contain the positively charged image. At this point, the image becomes visible. The image is then transferred to the surface of a sheet of positively charged paper.

Finally, the toner material is "fixed" to the surface of the paper through the application of heat. This results in a permanent copy of the original.

The Field-Ion Microscope

In Section 25.6 we pointed out that the electric field intensity can be very high in the vicinity of a sharp point on a charged conductor. A device that makes use of this intense field is the *field-ion microscope,* which was invented in 1956 by E. W. Mueller of the Pennsylvania State University.

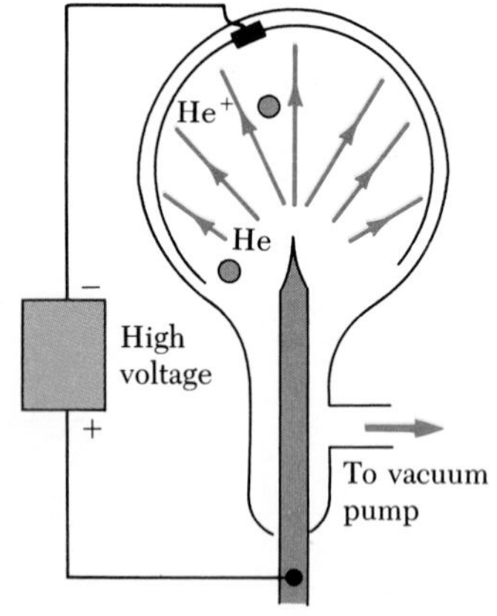

Figure 25.27 Schematic diagram of a field-ion microscope. The electric field is very intense at the tip of the needle-shaped specimen.

The basic construction of the field-ion microscope is shown in Figure 25.27. A specimen to be studied is fabricated from a fine wire, and a sharp tip is formed, usually by etching the wire in an acid. Typically, the diameter of the tip is about 0.1 μm (= 100 nm). The specimen is placed at the center of an evacuated glass tube containing a fluorescent screen. Next, a small amount of helium is introduced into the vessel. A very high potential difference is applied between the needle and the screen, producing a very intense electric field near the tip of the needle. It is important to cool the tip to at least the temperature of liquid nitrogen to obtain stable pictures. The helium atoms in the vicinity of this high-field region are ionized by the loss of an electron, which leaves the helium positively charged. The positively charged He^+ ions then accelerate to the negatively charged fluorescent screen. This results in a pattern on the screen that represents an image of the tip of the specimen.

Under the proper conditions (low specimen temperature and high vacuum), the images of the individual atoms on the surface of the sample are

Figure 25.28 Field ion microscope image of the surface of a platinum crystal with a magnification of 1 000 000×. Individual atoms can be seen on surface layers using this technique. (Courtesy of Prof. T.T. Tsong, The Pennsylvania State University)

visible, and the atomic arrangement on the surface can be studied. Unfortunately, the high electric fields also set up large mechanical stresses near the tip of the specimen, which limits the application of the technique to strong metallic elements, such as tungsten and rhenium. Figure 25.28 represents a typical field-ion microscope pattern of a platinum crystal.

SUMMARY

When a positive test charge q_o is moved between points A and B in an electrostatic field $\mathbf{E}$, the **change in the potential energy** is given by

Change in potential energy

$$\Delta U = -q_o \int_A^B \mathbf{E} \cdot d\mathbf{s} \tag{25.3}$$

The **potential difference** ΔV between points A and B in an electrostatic field $\mathbf{E}$ is defined as the change in potential energy divided by the test charge q_o:

Potential difference

$$\Delta V = \frac{\Delta U}{q_o} = -\int_A^B \mathbf{E} \cdot d\mathbf{s} \tag{25.4}$$

where the electric potential V is a scalar and has the units of J/C, defined to be 1 volt (V).

The potential difference between two points A and B in a *uniform* electric field $\mathbf{E}$ is given by

$$\Delta V = -Ed \tag{25.7}$$

where d is the displacement in the direction *parallel* to $\mathbf{E}$.

Equipotential surfaces are surfaces on which the electric potential remains constant. Equipotential surfaces are *perpendicular* to the electric field lines.

The potential due to a point charge q at any distance r from the charge is given by

Potential of a point charge

$$V = k\frac{q}{r} \tag{25.12}$$

The potential due to a group of point charges is obtained by summing the potentials due to the individual charges. Since V is a scalar, the sum is a simple algebraic operation.

The **potential energy of a pair of point charges** separated by a distance r_{12} is given by

Electric potential energy of two charges

$$U = k\frac{q_1 q_2}{r_{12}} \tag{25.14}$$

This represents the work required to bring the charges from an infinite separation to the separation r_{12}. The potential energy of a distribution of point charges is obtained by summing terms like Equation 25.14 over all *pairs* of particles.

TABLE 25.1 Potentials Due to Various Charge Distributions

Charge Distribution	Electric Potential	Location
Uniformly charged ring of radius a	$V = k\dfrac{Q}{\sqrt{x^2 + a^2}}$	Along the axis of the ring, a distance x from its center
Uniformly charged disk of radius a	$V = 2\pi k\sigma[(x^2 + a^2)^{1/2} - x]$	Along the axis of the disk, a distance x from its center
Uniformly charged, *insulating* solid sphere of radius R and total charge Q	$V = k\dfrac{Q}{r}$	$r \geq R$
	$V = \dfrac{kQ}{2R}\left(3 - \dfrac{r^2}{R^2}\right)$	$r < R$
Isolated *conducting* sphere of total charge Q and radius R	$V = k\dfrac{Q}{R}$	$r \leq R$
	$V = k\dfrac{Q}{r}$	$r > R$

The **electric potential due to a continuous charge distribution** is given by

$$V = k\int \frac{dq}{r} \qquad (25.17)$$

Electric potential due to a continuous charge distribution

If the electric potential is known as a function of coordinates x, y, z the components of the electric field can be obtained by taking the negative derivative of the potential with respect to the coordinates. For example, the x component of the electric field is given by

$$E_x = -\frac{dV}{dx} \qquad (25.25)$$

Every point on the surface of a charged conductor in electrostatic equilibrium is at the same potential. Furthermore, the potential is constant everywhere inside the conductor and equal to its value at the surface. Table 25.1 lists potentials due to several charge distributions.

PROBLEM-SOLVING STRATEGY AND HINTS

1. When working problems involving electric potential, remember that potential is a *scalar quantity* (rather than a vector quantity like the electric field), so there are no components to worry about. Therefore, when using the superposition principle to evaluate the electric potential at a point due to a system of point charges, you simply take the algebraic sum of the potentials due to each charge. However, you must keep track of signs. The potential for each positive charge ($V = kq/r$) is positive, while the potential for each negative charge is negative.
2. Just as in mechanics, only *changes* in potential are significant, hence the point where you choose the potential to be zero is arbitrary. When dealing with point charges or a finite-sized charge distribution, we usually define $V = 0$ to be at a point infinitely far from the charges. However, if the charge distribution itself extends to infinity, some other nearby point must be selected as the reference point.
3. The electric potential at some point P due to a continuous distribution of charge can be evaluated by dividing the charge distribution into infinitesimal elements of charge dq located at a distance r from the point P. You then treat this element as a point charge, so that the potential at P due to the element is $dV = k\,dq/r$. The total potential at P is obtained by integrating dV over the entire charge distribution. In performing the integration for most problems, it is necessary to express dq and r in terms of a single variable. In order to simplify the integration, it is important to give careful consideration of the geometry involved in the problem. You should review Examples 25.4 through 25.6 as guides for using this method.
4. Another method that can be used to obtain the potential due to a finite continuous charge distribution is to start with the definition of the potential difference given by Equation 25.4. If $\mathbf{E}$ is known or can be obtained easily (say from Gauss' law), then the line integral of $\mathbf{E} \cdot d\mathbf{s}$ can be evaluated. An example of this method is given in Example 25.7.
5. Once you know the electric potential at a point, it is possible to obtain the electric field at that point by remembering that *the electric field is equal to the negative of the derivative of the potential with respect to some coordinate.* Examples 25.4 and 25.5 illustrate how to use this procedure.

QUESTIONS

1. In your own words, distinguish between electric potential and electrical potential energy.
2. A negative charge moves in the direction of a uniform electric field. Does its potential energy increase or decrease? Does the electric potential increase or decrease?
3. If a proton is released from rest in a uniform electric field, does its electric potential increase or decrease? What about its potential energy?
4. Give a physical explanation of the fact that the potential energy of a pair of like charges is positive whereas the potential energy of a pair of unlike charges is negative.
5. A uniform electric field is parallel to the x axis. In what direction can a charge be displaced in this field without any external work being done on the charge?
6. Explain why equipotential surfaces are always perpendicular to electric field lines.
7. Describe the equipotential surfaces for (a) an infinite line of charge and (b) a uniformly charged sphere.
8. Explain why, under static conditions, all points in a conductor must be at the same electric potential.
9. If the electric potential at some point is zero, can you conclude that there are no charges in the vicinity of that point? Explain.
10. If the potential is constant in a certain region, what is the electric field in that region?
11. The electric field inside a hollow, uniformly charged sphere is zero. Does this imply that the potential is zero inside the sphere? Explain.
12. The potential of a point charge is defined to be zero at an infinite distance. Why can we not define the potential of an infinite line of charge to be zero at $r = \infty$?
13. Two charged conducting spheres of different radii are connected by a conducting wire as in Figure 25.21. Which sphere has the greater charge density?
14. What determines the maximum potential to which the dome of a Van de Graaff generator can be raised?
15. In what type of weather would a car battery be more likely to discharge and why?
16. Explain the origin of the glow that is sometimes observed around the cables of a high-voltage power line.
17. Why is it important to avoid sharp edges, or points, on conductors used in high-voltage equipment?
18. How would you shield an electronic circuit or laboratory from stray electric fields? Why does this work?
19. Why is it relatively safe to stay in an automobile with a metal body during a severe thunderstorm?
20. Walking across a carpet and then touching someone can result in a shock. Explain why this occurs.

PROBLEMS

Section 25.1 Potential Difference and Electric Potential

1. Concentric spherical surfaces surrounding a point charge at their center are *equipotential surfaces.* The intersections of these surfaces with a plane through their common center are *equipotential lines.* How much work is done in moving a charge q a distance s along an arc of an equipotential of circular shape and of radius R?
2. What change in potential energy does a 12-μC charge experience when it is moved between two points for which the potential difference is 65 V? Express the answer in eV.
3. (a) Calculate the speed of a proton that is accelerated from rest through a potential difference of 120 V. (b) Calculate the speed of an electron that is accelerated through the same potential difference.
4. Through what potential difference would one need to accelerate an electron in order for it to achieve a velocity of 40% of the velocity of light, starting from rest? ($c = 3.0 \times 10^8$ m/s.)
5. A deuteron (a nucleus which consists of one proton and one neutron) is accelerated through a 2.7-kV potential difference. (a) How much energy does it gain? (b) How fast would it be going if it started from rest?
6. What potential difference is needed to stop an electron with an initial speed of 4.2×10^5 m/s?
7. An ion accelerated through a potential difference of 115 V experiences an increase in potential energy of 7.37×10^{-17} J. Calculate the charge on the ion.
8. How much energy is gained by a charge of 75 μC moving through a potential difference of 90 V? Express your answer in joules and electron volts.
9. A positron, when accelerated from rest between two points at a fixed potential difference, acquires a speed of 30% of the speed of light. What speed will be achieved by a *proton* if accelerated from rest between the same two points?

Section 25.2 Potential Differences in a Uniform Electric Field

10. Consider two points in an electric field. The potential at point P_1 is $V_1 = -30$ V, and the potential at point P_2 is $V_2 = +150$ V. How much work is done by an external force in moving a charge $q = -4.7$ μC from P_2 to P_1?
11. How much work is done (by a battery, generator, or other source of electrical energy) in moving Avogadro's number of electrons from an initial point where the electric potential is 9 V to a point where the po-

tential is −5 V? (The potential in each case is measured relative to a common reference point.)

12. A capacitor consists of two parallel plates separated by a distance of 0.3 mm. If a 20-V potential difference is maintained between those plates, calculate the electric field strength in the region between the plates.

13. The electric field between two charged parallel plates separated by a distance of 1.8 cm has a uniform value of 2.4×10^4 N/C. Find the potential difference between the two plates. How much kinetic energy would be gained by a deuteron in accelerating from the positive to the negative plate?

14. Suppose an electron is released from rest in a uniform electric field whose strength is 5.9×10^3 V/m. (a) Through what potential difference will it have passed after moving 1 cm? (b) How fast will the electron be moving after it has traveled 1 cm?

15. An electron moving parallel to the x axis has an initial velocity of 3.7×10^6 m/s at the origin. The velocity of the electron is reduced to 1.4×10^5 m/s at the point $x = 2$ cm. Calculate the potential difference between the origin and the point $x = 2$ cm. Which point is at the higher potential?

16. A positron has the same charge as a proton, but the same mass as an electron. Suppose a positron moves 5.2 cm in the direction of a uniform 480 V/m electric field. (a) How much potential energy does it gain or lose? (b) How much kinetic energy does it gain or lose?

17. A proton moves in a region of a uniform electric field. The proton experiences an increase in kinetic energy of 5×10^{-18} J after being displaced 2 cm in a direction parallel to the field. What is the magnitude of the electric field?

18. A uniform electric field of magnitude 325 V/m is directed in the *negative y* direction in Figure 25.29. The coordinates of point A are $(-0.2, -0.3)$ m, and those of point B are $(0.4, 0.5)$ m. Calculate the potential $V_B - V_A$ using the blue path.

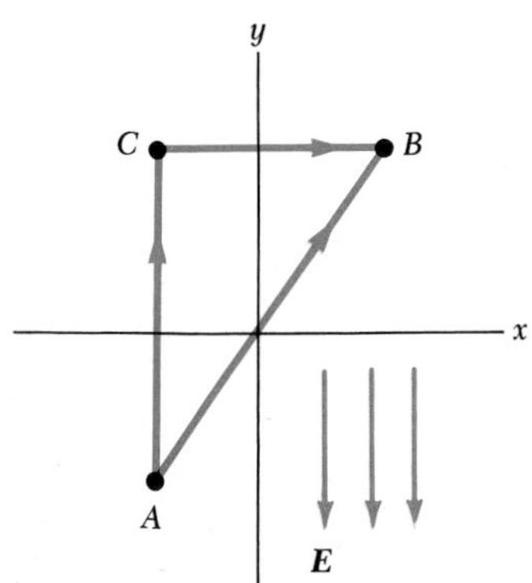

Figure 25.29 (Problems 18 and 19).

19. For the situation described in Problem 18, calculate the change in electric potential while going from point A to point B along the direct red path AB. Which point is at the higher potential?

20. A uniform electric field of magnitude 250 V/m is directed in the positive x direction. (a) Suppose a $+12\ \mu$C charge moves from the origin to the point $(x, y) = (20$ cm, 50 cm$)$. Through what potential difference did it move? (b) What was the change in its potential energy?

Section 25.3 Electric Potential and Potential Energy Due to Point Charges

Note: Assume a reference level for potential as $V = 0$ at $r = \infty$ unless the statement of the problem requires otherwise.

21. At what distance from a point charge of 8 μC would the potential equal 3.6×10^4 V?

22. A small spherical object carries a charge of 8 nC. At what distance from the center of the object is the potential equal to 100 V? 50 V? 25 V? Is the spacing of the equipotentials proportional to the change in V?

23. At a distance r away from a point charge q, the electrical potential is $V = 400$ V and the magnitude of the electric field is $E = 150$ N/C. Determine the value of q and r.

24. Two point charges are located as shown in Figure 25.30, where $q_1 = +4\ \mu$C, $q_2 = -2\ \mu$C, $a = 0.30$ m, and $b = 0.90$ m. Calculate the value of the electrical potential at points P_1 and P_2. Which point is at the higher potential?

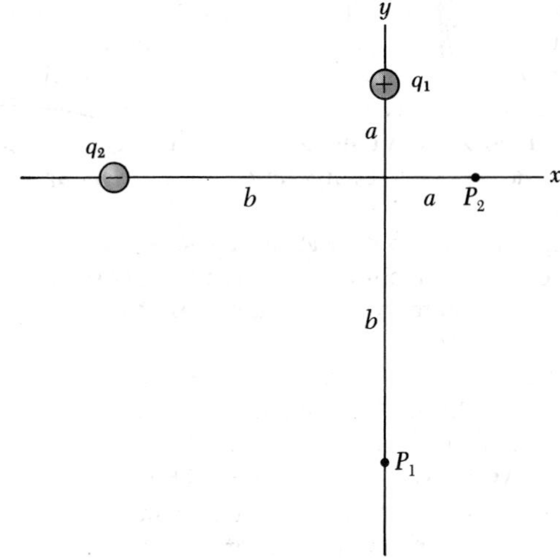

Figure 25.30 (Problem 24).

25. A $+2.8\ \mu$C charge is located on the y axis at $y = +1.6$ m and a $-4.6\ \mu$C charge is located at the origin. Calculate the net electric potential at the point (0.4 m, 0).

26. Two point charges, $Q_1 = +5\ \mu C$ and $Q_2 = -2\ \mu C$, are separated by 50 cm. (a) Where, along a line that passes through both charges, is the potential zero (in addition to $r = \infty$)? (b) Determine the electric field strength at the point found in (a). If $V = 0$ at this point, why is E not also equal to zero?

27. The three charges shown in Figure 25.31 are at the vertices of an isosceles triangle. Calculate the electric potential at the *midpoint of the base*, taking $q = 7\ \mu C$.

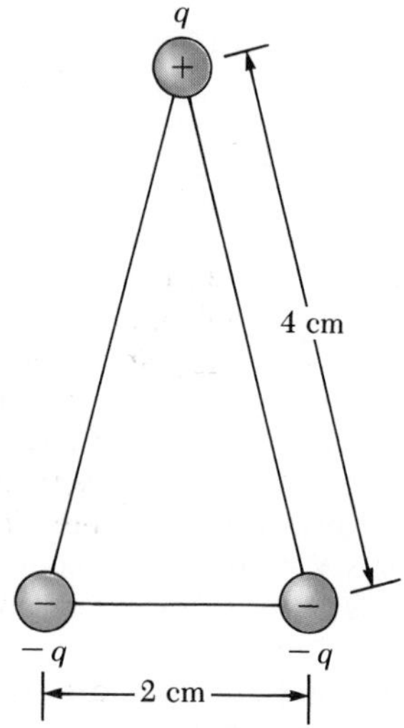

Figure 25.31 (Problem 27).

28. Calculate the value of the electric potential at point P due to the charge configuration shown in Figure 25.32. Use the values $q_1 = 5\ \mu C$, $q_2 = -10\ \mu C$, $a = 0.4$ m, and $b = 0.50$ m.

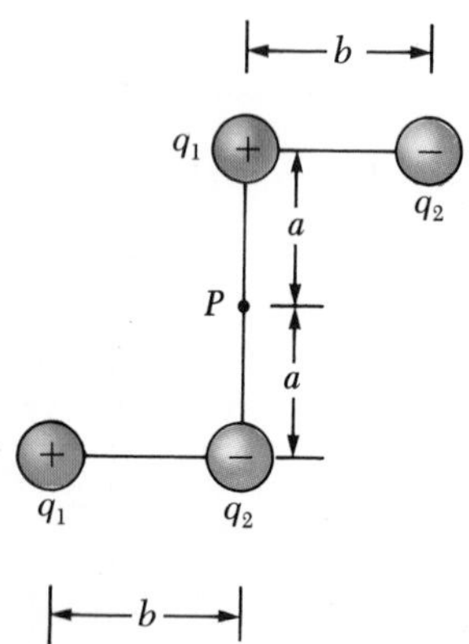

Figure 25.32 (Problem 28).

29. Two point charges, $Q_1 = +5$ nC and $Q_2 = -3$ nC, are separated by 35 cm. (a) What is the potential energy of the pair? What is the significance of the algebraic sign of your answer? (b) What is the electric potential at a point midway between the charges?

30. A charge $q_1 = -9\ \mu C$ is located at the origin, and a second charge $q_2 = -1\ \mu C$ is located on the x axis at $x = 0.7$ m. Calculate the electric potential energy of this pair of charges.

31. The Bohr model of the hydrogen atom states that the electron can only exist in certain allowed orbits. The radius of each Bohr orbit is given by the expression $r = n^2(0.0529 \text{ nm})$ where $n = 1, 2, 3, \ldots$. Calculate the electric potential energy of a hydrogen atom when the electron is in the (a) first allowed orbit, $n = 1$, (b) second allowed orbit, $n = 2$, and (c) when the electron has escaped from the atom, $r = \infty$. Express your answers in electron volts.

32. Calculate the energy required to assemble the array of charges shown in Figure 25.33, where $a = 0.20$ m, $b = 0.40$ m, and $q = 6\ \mu C$.

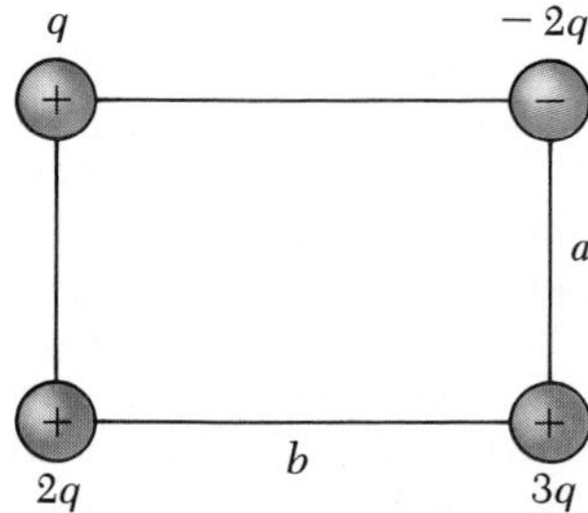

Figure 25.33 (Problem 32).

33. Show that the amount of work required to assemble four identical point charges of magnitude Q at the corners of a square of side s is given by $5.41kQ^2/s$.

34. Four equal point charges of charge $q = +5\ \mu C$ are located at the corners of a 30 cm by 40 cm rectangle. Calculate the electric potential energy stored in this charge configuration.

35. Four charges are located at the corners of a rectangle as in Figure 25.34. How much energy would be expended in removing the two 4-μC charges to infinity?

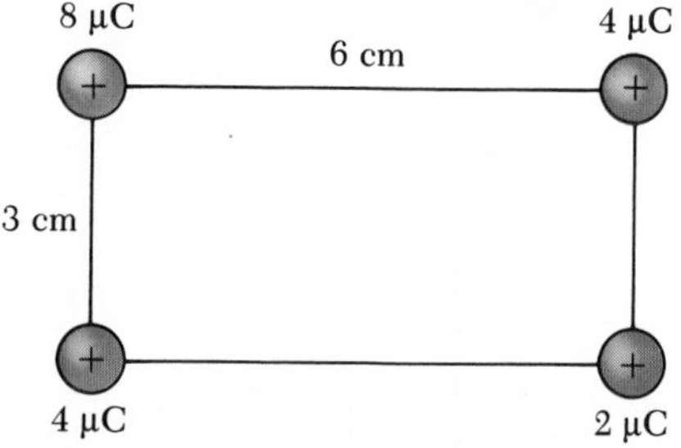

Figure 25.34 (Problem 35).

36. How much work is required to assemble eight identical point charges, each of magnitude q, at the corners of a cube of side s?

Section 25.4 Electric Potential Due to Continuous Charge Distributions

37. Consider a ring of radius R with total charge Q spread uniformly over its perimeter. What is the potential difference between the point at the center of the ring and a point on the axis of the ring at a distance $2R$ from the center of the ring?

38. Consider a Helmholtz pair consisting of two coaxial rings of 30 cm radius, separated by a distance of 30 cm. (a) Calculate the electric potential at a point on their common axis midway between the two rings, assuming that each ring carries a uniformly distributed charge of $+5\ \mu C$. (b) What is the potential at this point if the two rings carry equal and opposite charges?

39. A rod of length L (Fig. 25.35) lies along the x axis with its left end at the origin and has a *nonuniform* charge density $\lambda = \alpha x$ (where α is a positive constant). (a) What are the units of the constant α? (b) Calculate the electric potential at point A, a distance d from the left end of the rod.

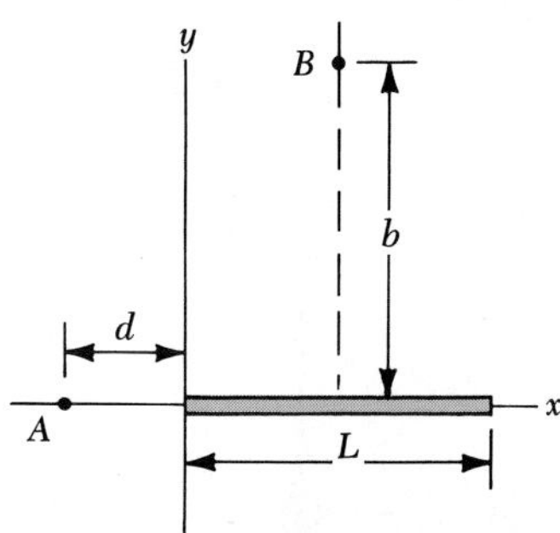

Figure 25.35 (Problems 39 and 40).

40. For the arrangement described in the previous problem, calculate the electric potential at point B on the perpendicular bisector of the rod a distance b above the x axis. Note that the rod has a *nonuniform* charge density $\lambda = \alpha x$.

41. Calculate the electric potential at point P on the axis of the annulus shown in Figure 25.36, which has a uniform charge density σ and inner and outer radii a and b, respectively.

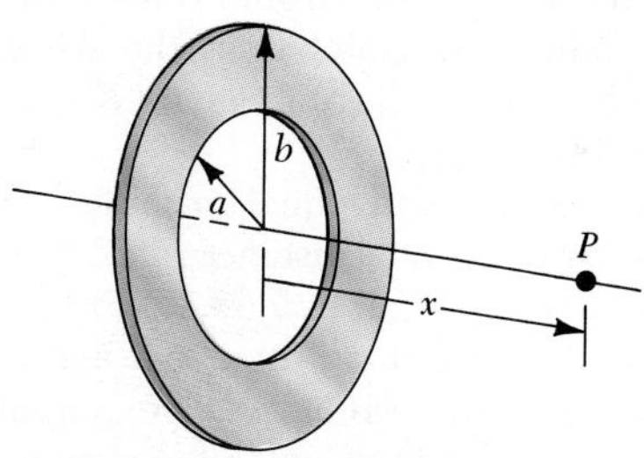

Figure 25.36 (Problem 41).

*Section 25.5 Obtaining *E* from the Electric Potential

42. The electric potential over a certain region of space is given by $V = 3x^2y - 4xz - 5y^2$ volts. Find (a) the electric potential and (b) the components of the electric field at the point $(+1, 0, +2)$ where all distances are in meters.

43. Over a certain region of space, the electric potential is given by $V = 5x - 3x^2y + 2yz^2$. Find the expressions for the x, y, and z components of the electric field over this region. What is the magnitude of the field at the point P, which has coordinates (in meters) $(1, 0, -2)$?

44. The electric potential in a certain region is given by $V = 4xz - 5y + 3z^2$ volts. Find the magnitude of the electric field at the point $(+2, -1, +3)$ where all distances are in meters.

45. The electric potential over a particular region is given by $V = 3xy^2 - 4zx$. Determine the angle between the direction of the electric field, $\mathbf{E}$, and the direction of the positive x axis at the point P, which has coordinates (in meters) $(1, 0, 1)$.

46. Two parallel plates are perpendicular to the x axis. The negative plate is in the yz plane and the positive plate is at the point $x = x_0$. The potential at some point between the plates at $x < x_0$ is given by $V(x) = bx$, where b is a constant. Find an expression for the electric field $\mathbf{E}$ between the two plates.

47. The electric potential inside a charged spherical conductor of radius R is given by $V = kQ/R$ and outside the potential is given by $V = kQ/r$. Using $E_r = -\dfrac{dV}{dr}$, derive the electric field both (a) inside ($r < R$) and (b) outside ($r > R$) this charge distribution.

48. The electric potential inside a uniformly charged spherical insulator of radius R is given by

$$V = \frac{kQ}{2R}\left(3 - \frac{r^2}{R^2}\right)$$

and outside by

$$V = \frac{kQ}{r}$$

Use $E_r = -\dfrac{dV}{dr}$ to derive the electric field both (a) inside ($r < R$) and (b) outside ($r > R$) this charge distribution.

49. Use the exact result from Example 25.9 to evaluate V at $x = 3a$ for the described dipole when $a = 2$ mm and $q = 3\ \mu C$. Compare this answer to that which you would obtain if you used the approximate formula valid when $x \gg a$.

50. Use the exact result for the dipole potential from Example 25.9 to find the electric field at any point where $x > a$ on the axis of the dipole. Evaluate E at $x = 3a$ if $a = 2$ mm and $q = 3\ \mu C$.

Section 25.6 Potential of a Charged Conductor

51. How many electrons should be removed from an initially uncharged spherical conductor of radius 0.3 m to produce a potential of 7.5 kV at the surface?
52. Calculate the surface charge density, σ (in C/m^2), for a solid spherical conductor of radius $R = 0.25$ m if the potential at a distance 0.5 m from the center of the sphere is 1300 V.
53. A spherical conductor has a radius of 14 cm and has a charge of $+26\ \mu$C. Calculate the electric field and the electric potential at the following distances from the center of this conductor: (a) $r = 10$ cm, (b) $r = 20$ cm, and (c) $r = 14$ cm.
54. Two spherical conductors of radii r_1 and r_2 are connected by a conducting wire as shown in Figure 25.21. If $r_1 = 0.94$ m, $r_2 = 0.47$ m, and the field at the surface of the smaller sphere is 890 N/C, calculate the total charge on the larger sphere assuming it is initially uncharged.
55. Two charged spherical conductors are connected by a long conducting wire. A total charge of $+20\ \mu$C is placed on this combination of two spheres. (a) If one has a radius of 4 cm and the other has a radius of 6 cm, what is the electric field near the surface of each sphere? (b) What is the electrical potential of each sphere?
56. An egg-shaped conductor has a charge of $+43$ nC placed on its surface. It has a total surface area of 38 cm^2. (a) What is the average surface charge density? (b) What is the electric field inside the conductor? (c) What is the (average) electric field just outside the conductor?

°Section 25.8 Applications of Electrostatics

57. Consider the Van de Graaff generator with a 30-cm diameter dome operating in dry air. (a) What is the maximum potential of the dome? (b) What is the maximum charge on the dome?
58. (a) Calculate the *largest* amount of charge possible on the surface of a Van de Graaff generator with a 40-cm diameter dome surrounded by air. (b) What is the potential of this spherical dome when it has the charge calculated above?
59. What charge would have to be placed on the surface of a Van de Graaff generator, whose dome has a radius of 15 cm, to produce a spark across a 10-cm air gap?
60. What power in watts must a Van de Graaff generator deliver if it produces a 100-μA beam of protons at an energy of 12 MeV?

ADDITIONAL PROBLEMS

61. At a certain distance from a point charge the field intensity is 500 V/m and the potential is -3000 V. (a) What is the distance to the charge? (b) What is the magnitude of the charge?
62. All of the corners, except one, of a 1-m cube are occupied by charges of $+1\ \mu$C. What is the electric potential at the empty corner?
63. Three point charges of magnitude $+8\ \mu$C, $-3\ \mu$C, and $+5\ \mu$C are located on the corners of a triangle whose sides are each 9 cm long. Calculate the electric potential at the center of this triangle.
64. Consider an array of eight equal negative charges located so as to define the corners of a *cube* of edge length $\ell = 0.15$ m. If each of the eight charges has a charge $q = -6\ \mu$C, determine the potential at the *center* of the cube.
65. Equal charges ($q = +2\ \mu$C) are placed at 30° intervals around the equator of a sphere with a radius of 1.2 m. What is the electrical potential (a) at the center of the sphere and (b) at the north pole of the sphere?
66. The charge distribution shown in Figure 25.37 is referred to as a linear quadrupole. (a) Find the exact expression for the potential at a point on the x axis where $x > d$. (b) Show that the expression obtained in (a) reduces to

$$V = \frac{2kQd^2}{x^3}$$

when $x \gg d$.

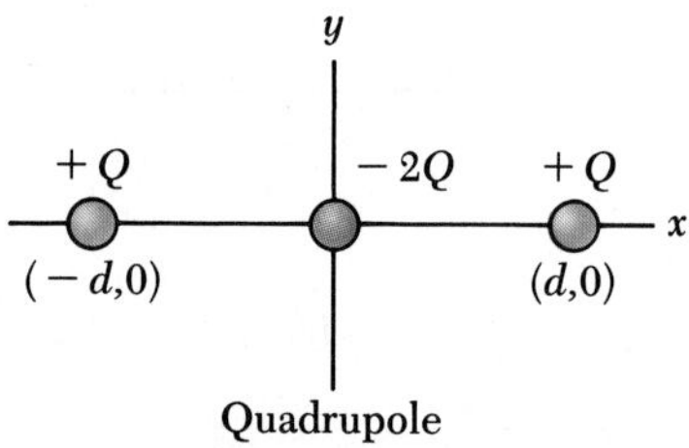

Figure 25.37 (Problems 66, 67, and 75).

67. Use the exact result from Problem 66 to evaluate the potential for the linear quadrupole at $x = 3d$ if $d =$ 2 mm and $q = 3\ \mu$C. Compare this answer to what you obtain when you use the approximate result valid when $x \gg d$.
68. (a) Use the exact result from Problem 66 to find the electric field at any point along the axis of the linear quadrupole for $x > d$. (b) Evaluate E at $x = 3d$ if $d =$ 2 mm and $Q = 3\ \mu$C.
69. Two point charges of equal magnitude are located along the y axis at equal distances above and below the x axis, as shown in Figure 25.38. (a) Plot a graph of the potential at points along the x axis over the interval $-3a < x < 3a$. You should plot the potential in units of kQ/a, where k is the Coulomb constant. (b) Let the charge located at $-a$ be *negative* and plot the potential along the y axis over the interval $-4a < y < 4a$.

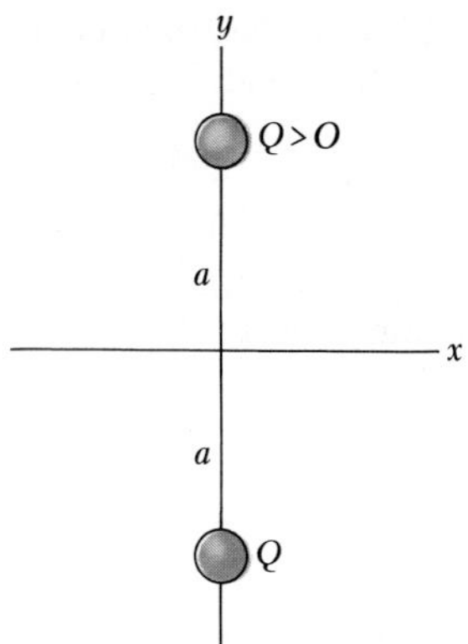

Figure 25.38 (Problem 69).

70. A net positive charge Q is placed on a solid conducting sphere of radius R. Plot a graph of the electric potential V as a function of r, the distance from the center of the sphere. Let r range over the interval $0 < r < 3R$, and plot V in units of kQ/R.
71. A spherical drop of water 2 mm in radius has an electric potential of 300 V at its surface. (a) What is the charge on the drop? (b) If two such drops of equal charge and radius unite to form a single spherical drop, what is the potential at the surface of the resulting drop? (Assume no charge is lost when the two drops unite.)
72. A Van de Graaff generator is operating so that the potential difference between the high-voltage electrode and the charging needles (points B and A in Figure 25.24) is 1.5×10^4 V. Calculate the power required to drive the belt (against electrical forces) at an instant when the effective current delivered to the high-voltage electrode is 500 μA.
73. Calculate the work that must be done to charge a spherical shell of radius R to a total charge Q.
74. A conducting sphere of radius r is concentric with a larger conducting spherical shell of radius R. For such an arrangement, show that if the charges on the spheres have values of q and Q, respectively, the potential difference between the two spheres will be

$$V_r - V_R = \frac{q}{4\pi\epsilon_0}\left(\frac{1}{r} - \frac{1}{R}\right)$$

75. How much work is required to assemble the linear quadrupole shown in Figure 25.37 if $d = 2$ mm and $Q = 3\ \mu$C?
76. A large rectangle of length $4a$ and width $3a$ contains equal positive charges of magnitude $q_1 = 4\ \mu$C at opposite vertices, as shown in Figure 25.39. A small rectangle of length $2a$ and width a has equal positive charges $q_2 = 6\ \mu$C located at two vertices as shown in the figure. How much work must be done against electrostatic forces in order to rotate the small rectangle about its long side to the position shown by the broken line? Let $a = 0.1$ m.

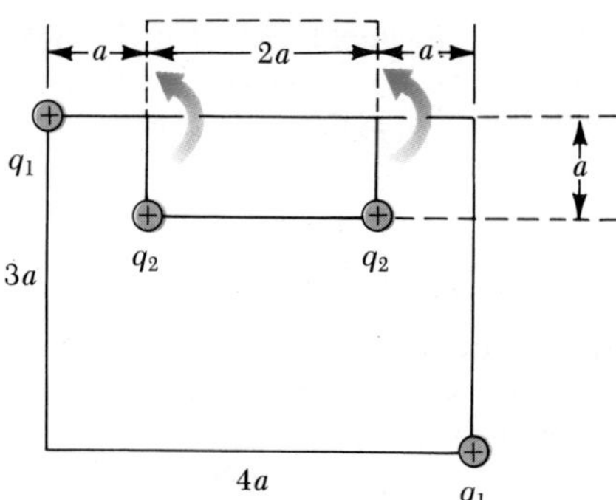

Figure 25.39 (Problem 76).

77. A square sheet of sides L has a uniform charge density σ and is located in the xy plane as in Figure 25.40. Set up the integral expression necessary to calculate the electric potential at a point P on a perpendicular axis through the center of the sheet. Assume that the point P is a distance d from the sheet.

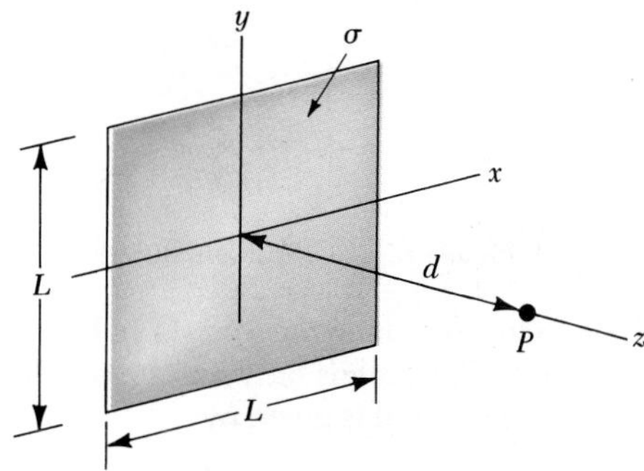

Figure 25.40 (Problem 77).

78. The thin, uniformly charged rod shown in Figure 25.41 has a length L and a linear charge density λ. Find an expression for the electric potential at point P, a distance b along the positive y axis.

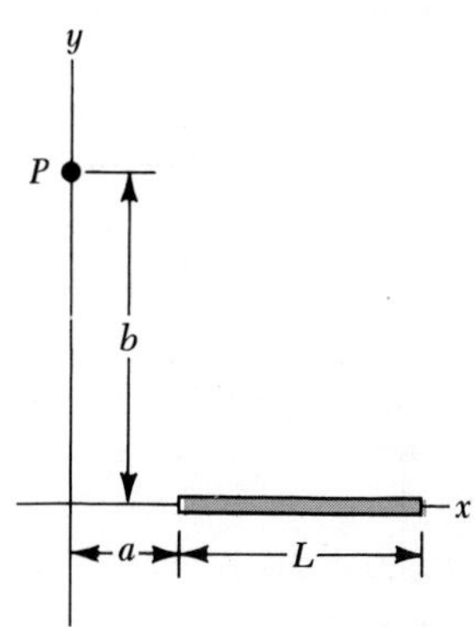

Figure 25.41 (Problem 78).

79. It is shown in Example 25.6 that the potential at a point P a distance d above one end of a uniformly charged rod of length ℓ lying along the x axis is given by

$$V = \frac{kQ}{\ell} \ln\left(\frac{\ell + \sqrt{\ell^2 + d^2}}{d}\right)$$

Use this result to derive an expression for the y component of the electric field at the point P. (*Hint:* Replace d with y.)

80. Figure 25.42 shows several equipotential lines each labeled by its potential in volts. The distance between the lines of the square grid represents 1 cm. (a) Is the magnitude of the **E** field bigger at A or B? Why? (b) What is **E** at B? (c) Represent what the **E** field looks like by drawing at least 8 field lines.

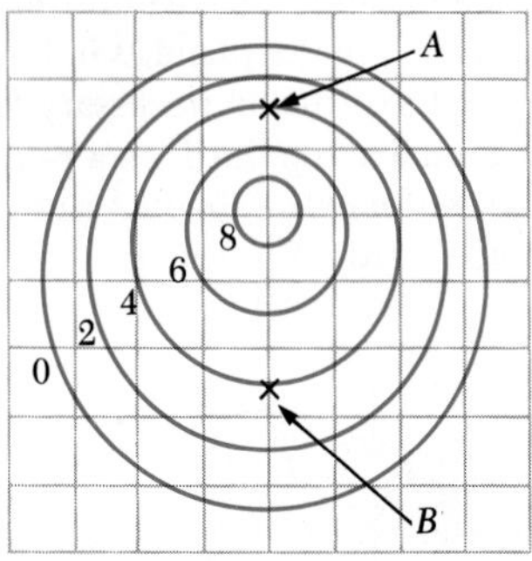

Figure 25.42 (Problem 80).

81. A dipole is located along the y axis as in Figure 25.43. (a) At a point P, which is far from the dipole ($r \gg a$), the electric potential is given by

$$V = k\frac{p \cos \theta}{r^2}$$

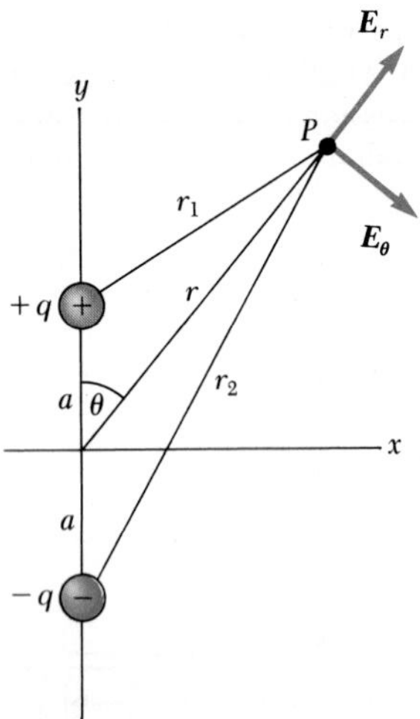

Figure 25.43 (Problem 81).

where $p = 2qa$. Calculate the radial component of the associated electric field, E_r, and the perpendicular component, E_θ. Note that $E_\theta = -\frac{1}{r}\left(\frac{\partial V}{\partial \theta}\right)$. Do these results seem reasonable for $\theta = 90°$ and $0°$? for $r = 0$? (b) For the dipole arrangement shown, express V in terms of rectangular coordinates using $r = (x^2 + y^2)^{1/2}$ and

$$\cos \theta = \frac{y}{(x^2 + y^2)^{1/2}}$$

Using these results and taking $r \gg a$, calculate the field components E_x and E_y.

82. A disk of radius R has a nonuniform surface charge density $\sigma = Cr$, where C is a constant and r is measured from the center of the disk (Fig. 25.44). Find (by direct integration) the potential at an axial point P a distance x from the disk.

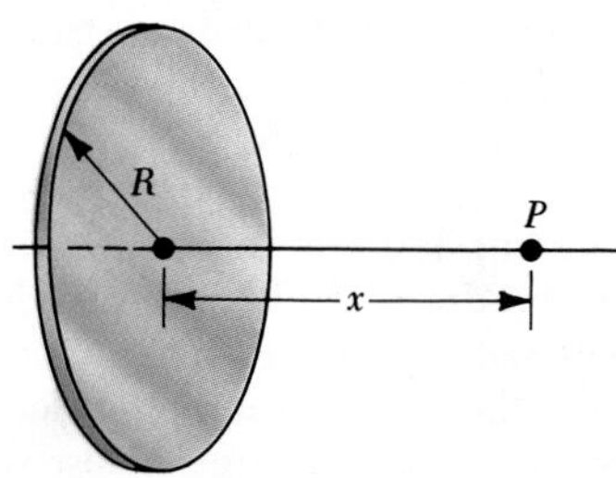

Figure 25.44 (Problem 82).

83. A solid sphere of radius R has a *uniform* charge density ρ and *total* charge Q. Derive an expression for the total electric potential energy of the charged sphere. (*Hint:* Imagine that the sphere is constructed by adding successive layers of concentric shells of charge $dq = (4\pi r^2\, dr)\rho$ and use $dU = V\, dq$.)

84. A Geiger-Müller counter is a type of radiation detector that essentially consists of a hollow cylinder (the cathode) of inner radius r_a and a coaxial cylindrical wire (the anode) of radius r_b (Fig. 25.45). The charge per unit length on the anode is λ, while the charge per unit length on the cathode is $-\lambda$. (a) Show that the

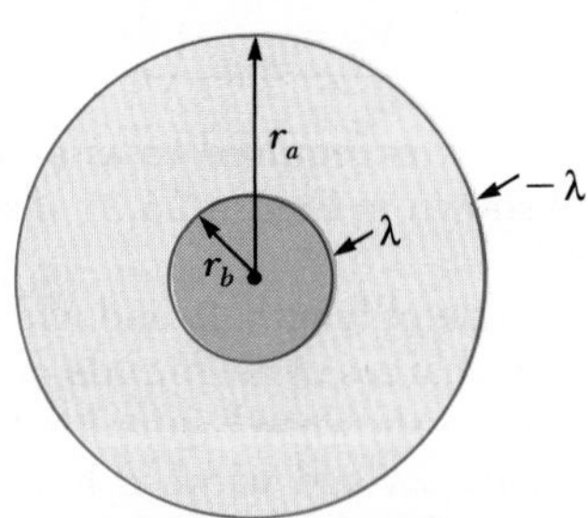

Figure 25.45 (Problem 84).

potential difference between the wire and the cylinder in the sensitive region of the detector is given by

$$V = 2k\lambda \ln\left(\frac{r_a}{r_b}\right)$$

(b) Show that the magnitude of the electric field over that region is given by

$$E = \frac{V}{\ln(r_a/r_b)}\left(\frac{1}{r}\right)$$

where r is the distance from the center of the anode to the point where the field is to be calculated.

CALCULATOR/COMPUTER PROBLEMS

85. A uniformly charged rod is located along the x axis as in Figure 23.14. The total charge on the rod is $+16 \times 10^{-10}$ C, $d = 1.0$ m, and $\ell = 2.0$ m. Estimate the electrical potential at $x = 0$ by approximating the rod to be (a) a point charge at $x = 2.0$ m, (b) two point charges (each of charge $+8 \times 10^{-10}$ C) at $x = 1.5$ m and $x = 2.5$ m, and (c) four point charges (each of charge $+4 \times 10^{-10}$ C) at $x = 1.25$ m, $x = 1.75$ m, $x = 2.25$ m, and $x = 2.75$ m. (d) Write a program that will enable you to extend your calculations to 256 equally spaced point charges, and compare your result with that given by the exact expression

$$V = k\frac{Q}{\ell}\ln\left(\frac{\ell + d}{d}\right)$$

86. A ring of radius 1 m has a uniform charge per unit length and a total charge of $+16 \times 10^{-10}$ C. The ring lies in the yz plane, and its center is at $x = 0$, as in Figure 25.10. Estimate the electric potential along the x axis at $x = 2$ m by approximating the ring to be (a) a point charge located at the origin, (b) two point charges (each of charge $+8 \times 10^{-10}$ C) diametrically opposite each other on the ring, and (c) four point charges (each of charge $+4 \times 10^{-10}$ C) symmetrically spaced on the ring. (d) Write a program that will enable you to extend your calculations to 64 point charges equally spaced on the ring, and compare your result with that given by the exact expression, Equation 25.18.

26
Capacitance and Dielectrics

Very large capacitors are used as components in high-voltage power generators. In this photograph, a technician installs a safety switch in the "oil section" that contains the Marx generators seen at the right of the picture. This is the main high-voltage mechanical switch linking the Marx generators to accelerator modules. (Courtesy of Sandia National Laboratories)

This chapter is concerned with the properties of capacitors, devices that store charge. Capacitors are commonly used in a variety of electrical circuits. For instance, they are used (1) to tune the frequency of radio receivers, (2) as filters in power supplies, (3) to eliminate sparking in automobile ignition systems, and (4) as energy-storing devices in electronic flash units.

A capacitor basically consists of two conductors separated by an insulator. We shall see that the capacitance of a given device depends on its geometry and on the material separating the charged conductors, called a *dielectric*. A dielectric is an insulating material having distinctive electrical properties that can best be understood as a consequence of the properties of atoms.

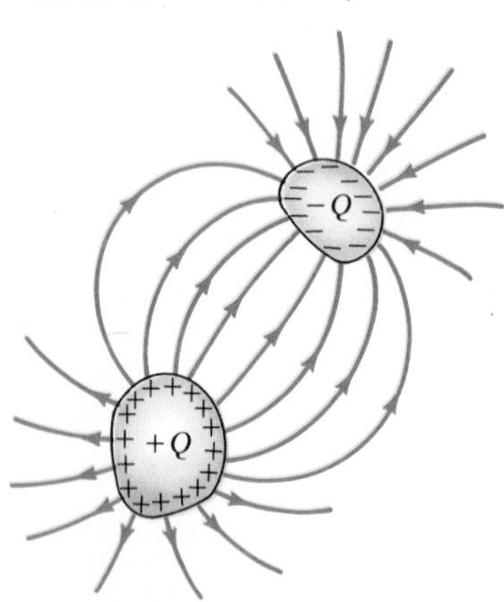

Figure 26.1 A capacitor consists of two conductors isolated from each other and their surroundings. Once the capacitor is charged, the two conductors carry equal but opposite charges.

26.1 DEFINITION OF CAPACITANCE

Consider two conductors having a potential difference V between them. Let us assume that the conductors have equal and opposite charges as in Figure 26.1. This can be accomplished by connecting the two uncharged conductors to the terminals of a battery. Such a combination of two conductors is called a *capacitor*. The potential difference V is found to be proportional to the magnitude of the charge Q on the capacitor.[1]

The **capacitance,** C, of a capacitor is defined as the ratio of the magnitude of the charge on either conductor to the magnitude of the potential difference between them:

$$C \equiv \frac{Q}{V} \tag{26.1}$$

Definition of capacitance

Note that by definition *capacitance is always a positive quantity*. Furthermore, since the potential difference increases as the stored charge increases, the ratio Q/V is constant for a given capacitor. Therefore, the capacitance of a device is a measure of its ability to store charge and electrical potential energy.

From Equation 26.1, we see that capacitance has **SI** units of coulombs per volt. The SI unit of capacitance is the **farad** (F), in honor of Michael Faraday. That is,

$$[\text{Capacitance}] = 1\text{ F} = 1\text{ C/V}$$

The farad is a very large unit of capacitance. In practice, typical devices have capacitances ranging from microfarads ($1\ \mu\text{F} = 10^{-6}$ F) to picofarads ($1\text{ pF} = 10^{-12}$ F). As a practical note, capacitors are often labeled mF for microfarads and mmF for micromicrofarads (picofarads).

As we shall show in the next section, the capacitance of a device depends on the geometrical arrangement of the conductors. To illustrate this point, let us calculate the capacitance of an isolated spherical conductor of radius R and charge Q. (The second conductor can be taken as a concentric hollow conducting sphere of infinite radius.) Since the potential of the sphere is simply kQ/R (where $V = 0$ at infinity), its capacitance is given by

$$C = \frac{Q}{V} = \frac{Q}{kQ/R} = \frac{R}{k} = 4\pi\epsilon_0 R \tag{26.2}$$

This shows that the capacitance of an isolated charged sphere is proportional to its radius and is independent of both the charge and the potential difference. For example, an isolated metallic sphere of radius 0.15 m has a capacitance of

$$C = 4\pi\epsilon_0 R = 4\pi(8.85 \times 10^{-12}\ \text{C}^2/\text{N}\cdot\text{m}^2)(0.15\text{ m}) = 17\text{ pF}$$

26.2 CALCULATION OF CAPACITANCE

The capacitance of a pair of oppositely charged conductors can be calculated in the following manner. A convenient charge of magnitude Q is assumed, and the potential difference is calculated using the techniques described in the previous chapter. One then simply uses $C = Q/V$ to evaluate the capacitance.

[1] The proportionality between the potential difference and charge on the conductors can be proved from Coulomb's law or by experiment.

As you might expect, the calculation is relatively easy to perform if the geometry of the capacitor is simple.

Let us illustrate this with three geometries that we are all familiar with, namely, two parallel plates, two concentric cylinders, and two concentric spheres. In these examples, we shall assume that the charged conductors are separated by a vacuum. The effect of a dielectric material between the conductors will be treated in Section 26.5.

The Parallel-Plate Capacitor

Two parallel plates of equal area A are separated by a distance d as in Figure 26.2. One plate has a charge $+Q$, the other, charge $-Q$. The charge per unit area on either plate is $\sigma = Q/A$. If the plates are very close together (compared with their length and width), we can neglect end effects and assume that the electric field is uniform between the plates and zero elsewhere. According to Example 24.6, the electric field between the plates is given by

$$E = \frac{\sigma}{\epsilon_0} = \frac{Q}{\epsilon_0 A}$$

The potential difference between the plates equals Ed; therefore

$$V = Ed = \frac{Qd}{\epsilon_0 A}$$

Substituting this result into Equation 26.1, we find that the capacitance is given by

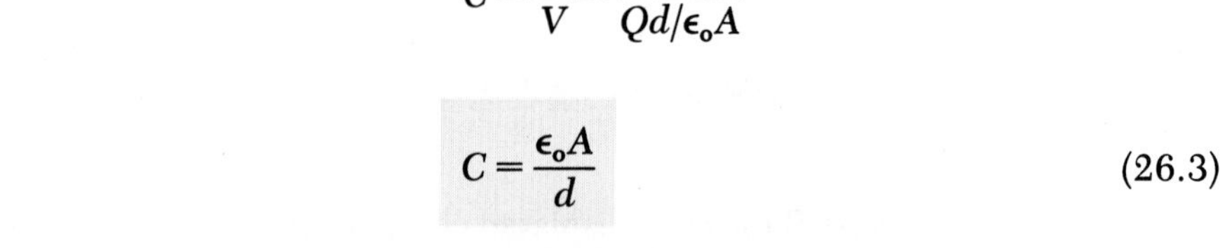

$$C = \frac{Q}{V} = \frac{Q}{Qd/\epsilon_0 A}$$

$$C = \frac{\epsilon_0 A}{d} \qquad (26.3)$$

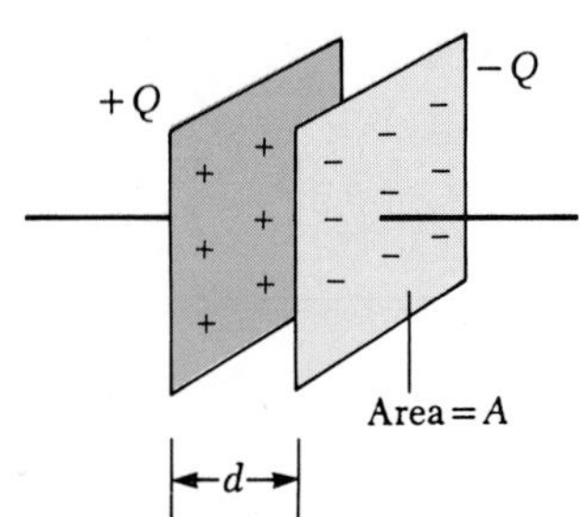

Figure 26.2 A parallel-plate capacitor consists of two parallel plates each of area A, separated by a distance d. The plates carry equal and opposite charges.

That is, *the capacitance of a parallel-plate capacitor is proportional to the area of its plates and inversely proportional to the plate separation.*

As you can see from the definition of capacitance, $C = Q/V$, the amount of charge a given capacitor is able to store for a given potential difference across its plates increases as the capacitance increases. Therefore, it seems reasonable that a capacitor constructed from plates having a large area should be able to store a large charge. The amount of charge needed to produce a given potential difference increases with decreasing plate separation.

A careful inspection of the electric field lines for a parallel-plate capacitor reveals that the field is uniform in the central region between the plates as in Figure 26.3a. However, the field is nonuniform at the edges of the plates. Figure 26.3b is a photograph of the electric field pattern of a parallel-plate capacitor showing the nonuniform field lines at its edges.

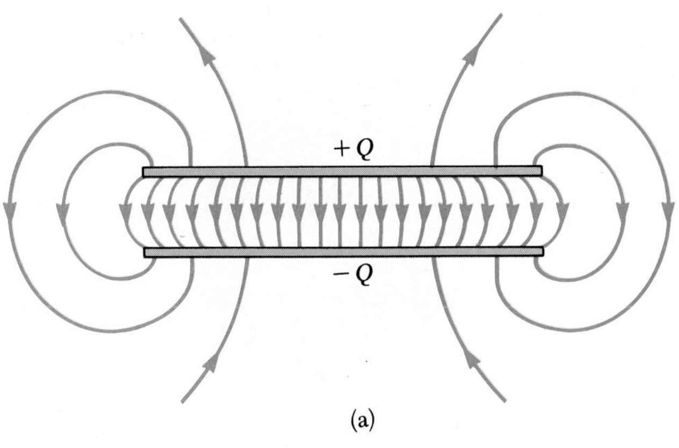

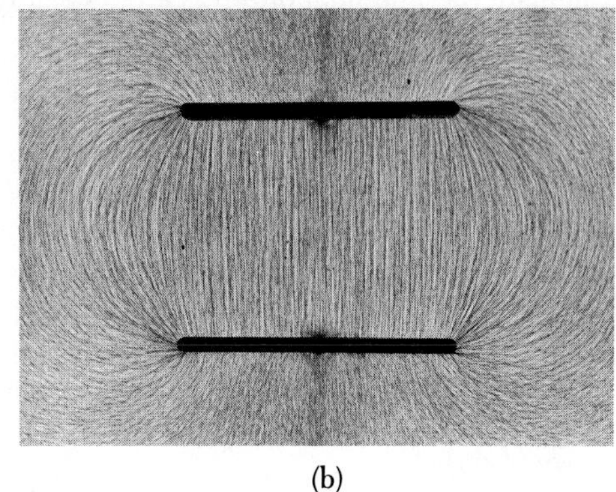

Figure 26.3 (a) The electric fields between the plates of a parallel-plate capacitor is uniform near its center, but is nonuniform near its edges. (b) Electric field pattern of two oppositely charged conducting parallel plates. Small pieces of thread on an oil surface align with the electric field. Note the nonuniform nature of the electric field at the ends of the plates. Such end effects can be neglected if the plate separation is small compared to the length of the plates. (Courtesy of Harold M. Waage, Princeton University)

EXAMPLE 26.1 Parallel-Plate Capacitor

A parallel-plate capacitor has an area of $A = 2\ \text{cm}^2 = 2 \times 10^{-4}\ \text{m}^2$ and a plate separation of $d = 1\ \text{mm} = 10^{-3}\ \text{m}$. Find its capacitance.

Solution From Equation 26.3, we find

$$C = \epsilon_0 \frac{A}{d} = \left(8.85 \times 10^{-12}\ \frac{\text{C}^2}{\text{N}\cdot\text{m}^2}\right)\left(\frac{2 \times 10^{-4}\ \text{m}^2}{1 \times 10^{-3}\ \text{m}}\right)$$

$$= 1.77 \times 10^{-12}\ \text{F} = \boxed{1.77\ \text{pF}}$$

Exercise 1 If the plate separation of this capacitor is increased to 3 mm, find its capacitance. Answer: 0.59 pF.

EXAMPLE 26.2 The Cylindrical Capacitor

A cylindrical conductor of radius a and charge $+Q$ is concentric with a larger cylindrical shell of radius b and charge $-Q$ (Fig. 26.4a). Find the capacitance of this cylindrical capacitor if its length is ℓ.

Solution If we assume that ℓ is long compared with a and b, we can neglect end effects. In this case, the field is perpendicular to the axis of the cylinders and is confined to the region between them (Fig. 26.4b). We must

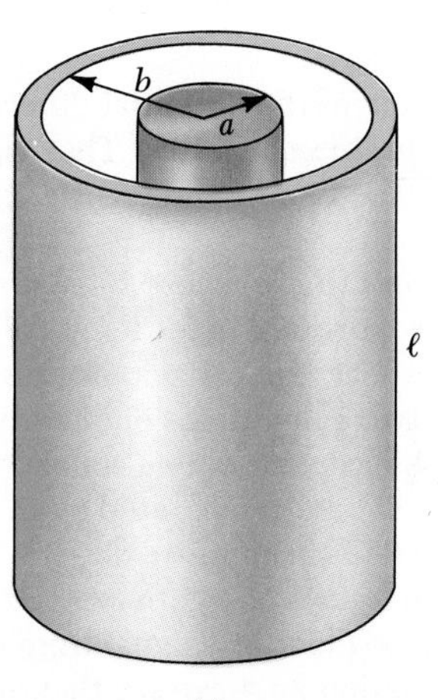

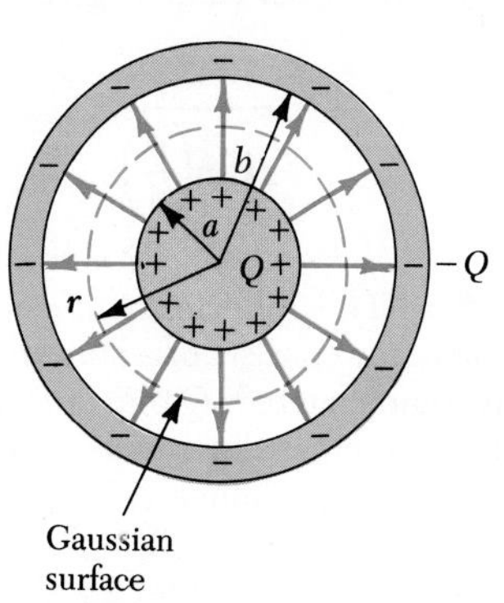

Figure 26.4 (Example 26.2) (a) A cylindrical capacitor consists of a cylindrical conductor of radius a and length ℓ surrounded by a coaxial cylindrical shell of radius b. (b) The end view of a cylindrical capacitor. The blue broken line represents the end of the cylindrical gaussian surface of radius r and length ℓ.

first calculate the potential difference between the two cylinders, which is given in general by

$$V_b - V_a = -\int_a^b \mathbf{E} \cdot d\mathbf{s}$$

where $\mathbf{E}$ is the electric field in the region $a < r < b$. In Chapter 24, we showed using Gauss' law that the electric field of a cylinder of charge per unit length λ is given by $2k\lambda/r$. The same result applies here, since the outer cylinder does not contribute to the electric field inside it. Using this result and noting that $\mathbf{E}$ is along r in Figure 26.4b, we find that

$$V_b - V_a = -\int_a^b E_r\,dr = -2k\lambda \int_a^b \frac{dr}{r} = -2k\lambda \ln\left(\frac{b}{a}\right)$$

Substituting this into Equation 26.1 and using the fact that $\lambda = Q/\ell$, we get

$$C = \frac{Q}{V} = \frac{Q}{\dfrac{2kQ}{\ell}\ln\left(\dfrac{b}{a}\right)} = \frac{\ell}{2k\ln\left(\dfrac{b}{a}\right)} \tag{26.4}$$

V is the magnitude of the potential difference given by $2k\lambda \ln(b/a)$, a *positive* quantity. That is, $V = V_a - V_b$ is *positive* since the inner cylinder is at the higher potential.

Our result for C makes sense since it shows that the capacitance is proportional to the length of the cylinders. As you might expect, the capacitance also depends on the radii of the two cylindrical conductors. As an example, a coaxial cable consists of two concentric cylindrical conductors of radii a and b separated by an insulator. The cable carries currents in opposite directions in the inner and outer conductors. Such a geometry is especially useful for shielding an electrical signal from external influences. From Equation 26.4, we see that the capacitance per unit length of a coaxial cable is given by

$$\frac{C}{\ell} = \frac{1}{2k\ln\left(\dfrac{b}{a}\right)}$$

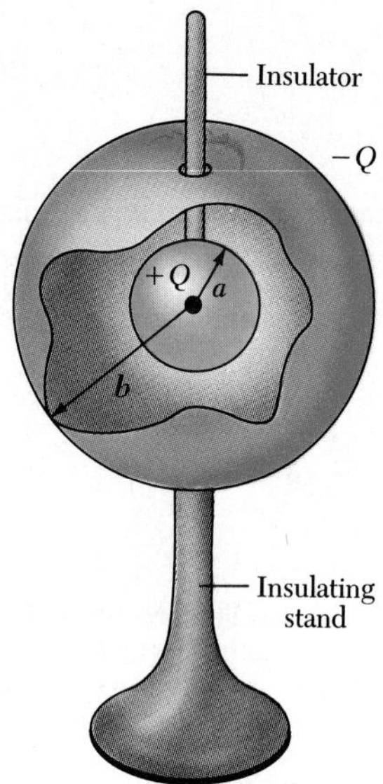

Figure 26.5 (Example 26.3) A spherical capacitor consists of an inner sphere of radius a surrounded by a concentric spherical shell of radius b. The electric field between the spheres is radial outward if the inner sphere is positively charged.

EXAMPLE 26.3 The Spherical Capacitor

A spherical capacitor consists of a spherical conducting shell of radius b and charge $-Q$ that is concentric with a smaller conducting sphere of radius a and charge $+Q$ (Fig. 26.5). Find its capacitance.

Solution As we showed in Chapter 24, the field outside a spherically symmetric charge distribution is radial and given by kQ/r^2. In this case, this corresponds to the field between the spheres ($a < r < b$). (The field is zero elsewhere.) From Gauss' law we see that only the inner sphere contributes to this field. Thus, the potential difference between the spheres is given by

$$V_b - V_a = -\int_a^b E_r\,dr = -kQ\int_a^b \frac{dr}{r^2} = kQ\left[\frac{1}{r}\right]_a^b$$

$$= kQ\left(\frac{1}{b} - \frac{1}{a}\right)$$

The magnitude of the potential difference is given by

$$V = V_a - V_b = kQ\frac{(b-a)}{ab}$$

Substituting this into Equation 26.1, we get

$$C = \frac{Q}{V} = \frac{ab}{k(b-a)} \tag{26.5}$$

Exercise 2 Show that as the radius b of the outer sphere approaches infinity, the capacitance approaches the value $a/k = 4\pi\epsilon_0 a$. This is consistent with the result obtained earlier (Eq. 26.2).

26.3 COMBINATIONS OF CAPACITORS

Two or more capacitors are often combined in circuits in several ways. The equivalent capacitance of certain combinations can be calculated using methods described in this section. The circuit symbols for capacitors and batteries, together with their color codes, are given in Figure 26.6. The positive terminal of the battery is at the higher potential and is represented by the longer vertical line in the battery symbol.

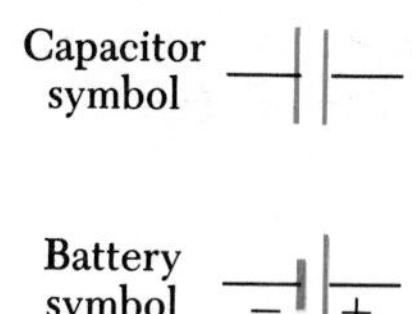

Figure 26.6 Circuit symbols for capacitors and batteries. Note that capacitors are in blue, while batteries are in red.

Parallel Combination

Two capacitors connected as shown in Figure 26.7a are known as a *parallel combination* of capacitors. The left plates of the capacitors are connected by a conducting wire to the positive terminal of the battery and are therefore at the same potential. Likewise, the right plates are connected to the negative terminal of the battery. When the capacitors are first connected in the circuit, electrons are transferred through the battery from the left plates to the right plates, leaving the left plates positively charged and the right plates negatively charged. The energy source for this charge transfer is the internal chemical energy stored in the battery, which is converted to electrical energy. The flow of charge ceases when the voltage across the capacitors is equal to that of the battery. The capacitors reach their maximum charge when the flow of charge ceases. Let us call the maximum charges on the two capacitors Q_1 and Q_2. Then the *total charge*, Q, stored by the two capacitors is

$$Q = Q_1 + Q_2 \tag{26.6}$$

Suppose we wish to replace these two capacitors by one equivalent capacitor having a capacitance C_{eq}. This equivalent capacitor must have exactly the

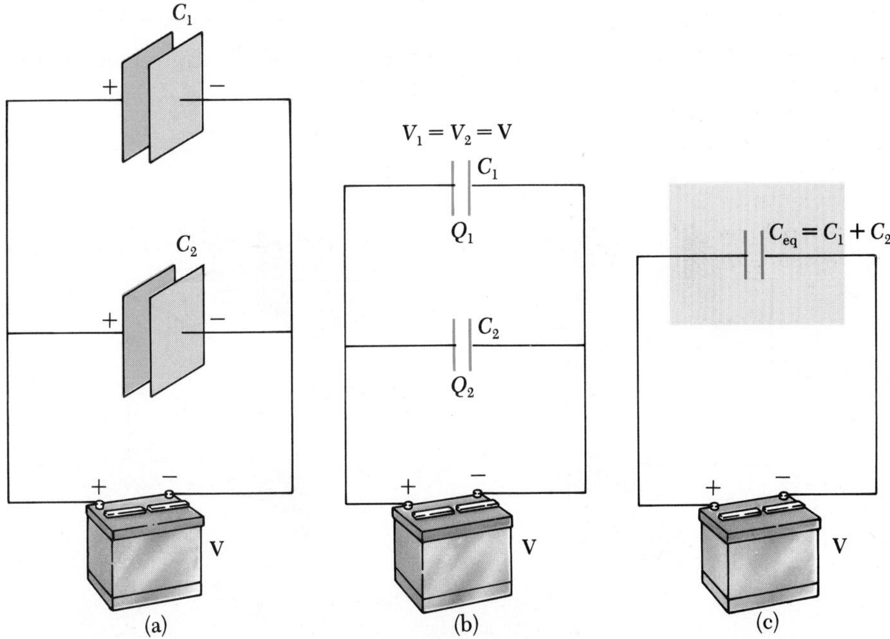

Figure 26.7 (a) A parallel connection of two capacitors. (b) The circuit diagram for the parallel combination. (c) The potential difference is the same across each capacitor, and the equivalent capacitance is $C_{eq} = C_1 + C_2$.

same external effect on the circuit as the original two. That is, it must store Q units of charge. We also see from Figure 26.7b that

> the potential difference across each capacitor in the parallel circuit is the same and is equal to the voltage of the battery, V.

From Figure 26.7c, we see that the voltage across the equivalent capacitor is also V. Thus, we have

$$Q_1 = C_1 V \qquad Q_2 = C_2 V$$

and, for the equivalent capacitor,

$$Q = C_{eq} V$$

Substituting these relations into Equation 26.6 gives

$$C_{eq} V = C_1 V + C_2 V$$

or

$$C_{eq} = C_1 + C_2 \qquad \begin{pmatrix}\text{parallel} \\ \text{combination}\end{pmatrix} \qquad (26.7)$$

If we extend this treatment to three or more capacitors connected in parallel, the equivalent capacitance is found to be

$$C_{eq} = C_1 + C_2 + C_3 + \cdots \qquad \begin{pmatrix}\text{parallel} \\ \text{combination}\end{pmatrix} \qquad (26.8)$$

Thus we see that *the equivalent capacitance of a parallel combination of capacitors is larger than any of the individual capacitances.*

Series Combination

Now consider two capacitors connected in *series,* as illustrated in Figure 26.8a.

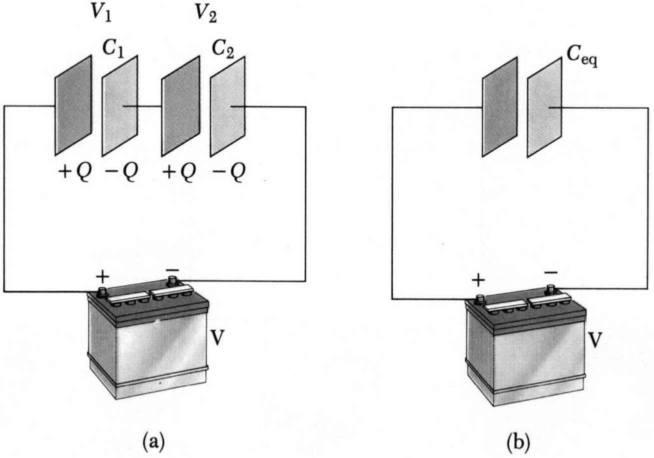

Figure 26.8 A series connection of two capacitors. The charge on each capacitor is the same, and the equivalent capacitance can be calculated from the relation $\dfrac{1}{C_{eq}} = \dfrac{1}{C_1} + \dfrac{1}{C_2}$.

For this series combination of capacitors, the magnitude of the charge must be the same on all the plates.

To see why this must be true, let us consider the charge transfer process in some detail. We start with uncharged capacitors and follow what happens just after a battery is connected to the circuit. When the battery is connected, electrons are transferred from the left plate of C_1 to the right plate of C_2 through the battery. As this negative charge accumulates on the right plate of C_2, an equivalent amount of negative charge is forced off the left plate of C_2, leaving it with an excess positive charge. The negative charge leaving the left plate of C_2 accumulates on the right plate of C_1, where again an equivalent amount of negative charge leaves the left plate. The result of this is that *all of the right plates gain a charge of* $-Q$ *while all of the left plates have a charge of* $+Q$.

Suppose an equivalent capacitor performs the same function as the series combination. After it is fully charged, *the equivalent capacitor must end up with a charge of* $-Q$ *on its right plate and* $+Q$ *on its left plate.* By applying the definition of capacitance to the circuit shown in Figure 26.8b, we have

$$V = \frac{Q}{C_{eq}}$$

where V is the potential difference between the terminals of the battery and C_{eq} is the equivalent capacitance. From Figure 26.8a, we see that

$$V = V_1 + V_2 \qquad (26.9)$$

where V_1 and V_2 are the potential differences across capacitors C_1 and C_2. In general, the potential difference across any number of capacitors in series is equal to the sum of the potential differences across the individual capacitors. Since $Q = CV$ can be applied to each capacitor, the potential difference across each is given by

$$V_1 = \frac{Q}{C_1} \qquad V_2 = \frac{Q}{C_2}$$

Substituting these expressions into Equation 26.9, and noting that $V = Q/C_{eq}$, we have

$$\frac{Q}{C_{eq}} = \frac{Q}{C_1} + \frac{Q}{C_2}$$

Cancelling Q, we arrive at the relationship

$$\frac{1}{C_{eq}} = \frac{1}{C_1} + \frac{1}{C_2} \qquad \begin{pmatrix}\text{series}\\ \text{combination}\end{pmatrix} \qquad (26.10)$$

If this analysis is applied to three or more capacitors connected in series, the equivalent capacitance is found to be

$$\frac{1}{C_{eq}} = \frac{1}{C_1} + \frac{1}{C_2} + \frac{1}{C_3} + \cdots \qquad \begin{pmatrix}\text{series}\\ \text{combination}\end{pmatrix} \qquad (26.11)$$

This shows that *the equivalent capacitance of a series combination is always less than any individual capacitance in the combination.*

EXAMPLE 26.4 Equivalent Capacitance
Find the equivalent capacitance between *a* and *b* for the combination of capacitors shown in Figure 26.9a. All capacitances are in μF.

Solution Using Equations 26.8 and 26.11, we reduce the combination step by step as indicated in the figure. The 1-μF and 3-μF capacitors are in *parallel* and combine according to $C_{eq} = C_1 + C_2$. Their equivalent capacitance is 4 μF. Likewise, the 2-μF and 6-μF capacitors are also in *parallel* and have an equivalent capacitance of 8 μF. The upper branch in Figure 26.9b now consists of two 4-μF capacitors in *series*, which combine according to

$$\frac{1}{C_{eq}} = \frac{1}{C_1} + \frac{1}{C_2} = \frac{1}{4\ \mu\text{F}} + \frac{1}{4\ \mu\text{F}} = \frac{1}{2\ \mu\text{F}}$$

$$C_{eq} = 2\ \mu\text{F}$$

Likewise, the lower branch in Figure 26.9b consists of two 8-μF capacitors in *series*, which give an equivalent of 4 μF. Finally, the 2-μF and 4-μF capacitors in Figure 26.9c are in *parallel* and have an equivalent capacitance of 6 μF. Hence, the equivalent capacitance of the circuit is 6 μF.

Exercise 3 Consider three capacitors having capacitances of 3 μF, 6 μF, and 12 μF. Find their equivalent capacitance if they are connected (a) in parallel, (b) in series.
Answer (a) 21 μF, (b) 1.71 μF.

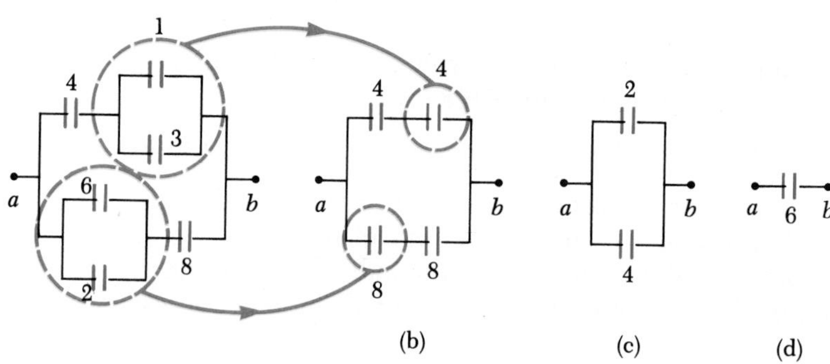

Figure 26.9 (Example 26.4) To find the equivalent combination of the capacitors in (a), the various combinations are reduced in steps as indicated in (b), (c), and (d), using the series and parallel rules described in the text.

26.4 ENERGY STORED IN A CHARGED CAPACITOR

Almost everyone that works with electronic equipment has at some time verified that a capacitor is able to store energy. If the plates of a charged capacitor are connected together by a conductor, such as a wire, charge will transfer from one plate to the other until the two are uncharged. The discharge can often be observed as a visible spark. If you should accidentally touch the opposite plates of a charged capacitor, your fingers would act as a pathway by which the capacitor could discharge, which would result in an electric shock. The degree of shock you would receive depends on the capacitance and voltage applied to the capacitor. Such a shock could be fatal where high voltages are present, such as in the power supply of a television set.

Consider a parallel-plate capacitor that is initially uncharged, so that the initial potential difference across the plates is zero. Now imagine that the capacitor is connected to a battery and develops a maximum charge Q. We shall assume that the capacitor is charged *slowly* so that the problem can be considered as an electrostatic system. The final potential difference across the capacitor is $V = Q/C$. Since the initial potential difference is zero, the *average* potential difference during the charging process is $V/2 = Q/2C$. From this we might conclude that the work needed to charge the capacitor is given by $W = QV/2 = Q^2/2C$. Although this result is correct, a more detailed proof is desirable and is now given.

Suppose that q is the charge on the capacitor at some instant during the charging process. At the same instant, the potential difference across the capacitor is $V = q/C$. The work necessary[2] to transfer an increment of charge dq from the plate of charge $-q$ to the plate of charge q (which is at the higher potential) is given by

$$dW = V\,dq = \frac{q}{C}\,dq$$

Thus, the total work required to charge the capacitor from $q = 0$ to some final charge $q = Q$ is given by

$$W = \int_0^Q \frac{q}{C}\,dq = \frac{Q^2}{2C}$$

But the work done in charging the capacitor can be considered as potential energy U stored in the capacitor. Using $Q = CV$, we can express the electrostatic energy stored in a charged capacitor in the following alternative forms:

$$U = \frac{Q^2}{2C} = \tfrac{1}{2}QV = \tfrac{1}{2}CV^2 \qquad (26.12)$$

Energy stored in a charged capacitor

This result applies to *any* capacitor, regardless of its geometry. We see that the stored energy increases as C increases and as the potential difference increases. In practice, there is a limit to the maximum energy (or charge) that can be stored. This is because electrical discharge will ultimately occur between the plates of the capacitor at a sufficiently large value of V. For this reason, capacitors are usually labeled with a maximum operating voltage.

The energy stored in a capacitor can be considered as being stored in the electric field created between the plates as the capacitor is charged. This description is reasonable in view of the fact that the electric field is proportional to the charge on the capacitor. For a parallel-plate capacitor, the potential difference is related to the electric field through the relationship $V = Ed$. Furthermore, its capacitance is given by $C = \epsilon_0 A/d$. Substituting these expressions into Equation 26.12 gives

$$U = \tfrac{1}{2}\frac{\epsilon_0 A}{d}(E^2 d^2) = \tfrac{1}{2}(\epsilon_0 A d)E^2 \qquad (26.13)$$

Energy stored in a parallel-plate capacitor

Since the volume of a parallel-plate capacitor that is occupied by the electric field is Ad, the *energy per unit volume* $u = U/Ad$, called the *energy density*, is

$$u = \tfrac{1}{2}\epsilon_0 E^2 \qquad (26.14)$$

Energy density in an electric field

[2] One mechanical analog of this process is the work required to raise a mass through some vertical distance in the presence of gravity.

Although Equation 26.14 was derived for a parallel-plate capacitor, the expression is generally valid. That is, the *energy density in any electrostatic field is proportional to the square of the electric field intensity at a given point.* (A formal proof of this statement is given in intermediate and advanced courses in electricity and magnetism.)

EXAMPLE 26.5 Rewiring Two Charged Capacitors □

Two capacitors C_1 and C_2 (where $C_1 > C_2$) are charged to the same potential difference V_0, but with opposite polarity. The charged capacitors are removed from the battery, and their plates are connected as shown in Figure 26.10a. The switches S_1 and S_2 are then closed as in Figure 26.10b. (a) Find the final potential difference between a and b after the switches are closed.

Solution The charges on the left-hand plates of the capacitors *before* the switches are closed are given by

$$Q_1 = C_1V_0 \qquad \text{and} \qquad Q_2 = -C_2V_0$$

The negative sign for Q_2 is necessary since this capacitor's polarity is *opposite* that of capacitor C_1. After the switches are closed, the charges on the plates redistribute until the total charge Q shared by both capacitors is

$$Q = Q_1 + Q_2 = (C_1 - C_2)V_0$$

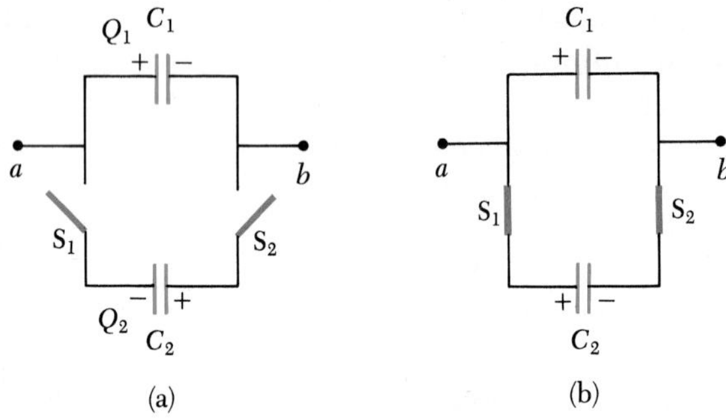

Figure 26.10 (Example 26.5).

The two capacitors are now in *parallel*, so the final potential difference across each is the *same* and given by

$$V = \frac{Q}{C_1 + C_2} = \left(\frac{C_1 - C_2}{C_1 + C_2}\right)V_0$$

(b) Find the total energy stored in the capacitors before and after the switches are closed.

Solution Before the switches are closed, the total energy stored in the capacitors is given by

$$U_i = \tfrac{1}{2}C_1V_0^2 + \tfrac{1}{2}C_2V_0^2 = \tfrac{1}{2}(C_1 + C_2)V_0^2$$

After the switches are closed and the capacitors have reached an equilibrium charge, the total energy stored in the capacitors is given by

$$U_f = \tfrac{1}{2}C_1V^2 + \tfrac{1}{2}C_2V^2 = \tfrac{1}{2}(C_1 + C_2)V^2$$

$$= \tfrac{1}{2}(C_1 + C_2)\left(\frac{C_1 - C_2}{C_1 + C_2}\right)^2 V_0^2 = \left(\frac{C_1 - C_2}{C_1 + C_2}\right)^2 U_i$$

Therefore, the ratio of the final to the initial energy stored is

$$\frac{U_f}{U_i} = \left(\frac{C_1 - C_2}{C_1 + C_2}\right)^2$$

This shows that the final energy is *less* than the initial energy. At first, you might think that energy conservation has been violated, but this is not the case since we have assumed that the circuit is ideal. Part of the missing energy appears as heat energy in the connecting wires, which have resistance, and part of the energy is radiated away in the form of electromagnetic waves (Chapter 34).

26.5 CAPACITORS WITH DIELECTRICS

A *dielectric* is a nonconducting material, such as rubber, glass, or waxed paper. When a dielectric material is inserted between the plates of a capacitor, the capacitance increases. If the dielectric completely fills the space between the plates, the capacitance increases by a dimensionless factor κ, called the **dielectric constant.**

The following experiment can be performed to illustrate the effect of a dielectric in a capacitor. Consider a parallel-plate capacitor of charge Q_0 and capacitance C_0 in the absence of a dielectric. The potential difference across the capacitor as measured by a voltmeter is $V_0 = Q_0/C_0$ (Fig. 26.11a). Notice

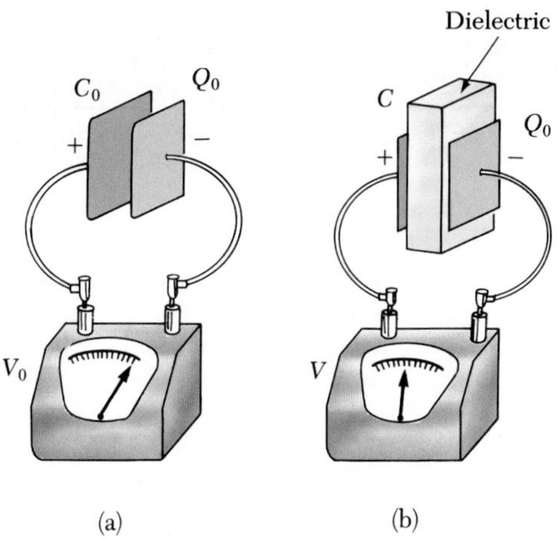

Figure 26.11 When a dielectric is inserted between the plates of a charged capacitor, the charge on the plates remains unchanged, but the potential difference as recorded by an electrostatic voltmeter is reduced from V_0 to $V = V_0/\kappa$. Thus, the capacitance *increases* in the process by the factor κ.

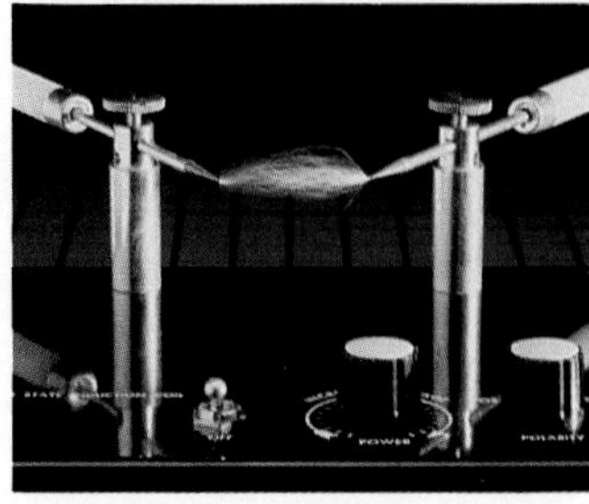

This photograph illustrates dielectric breakdown in air. Sparks are produced when a large alternating voltage is applied across the electrodes using a high-voltage induction coil power supply. (Courtesy of CENCO)

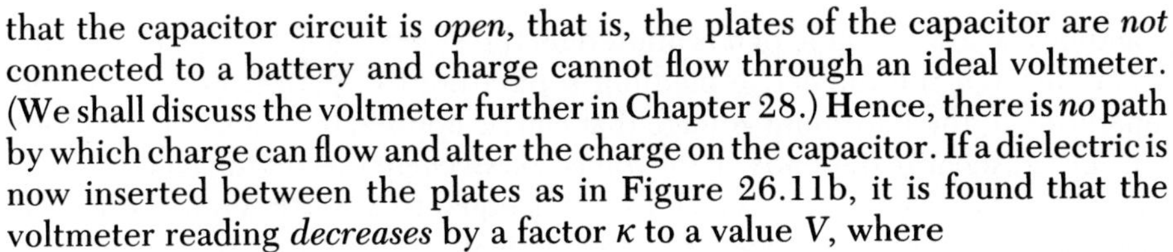

that the capacitor circuit is *open*, that is, the plates of the capacitor are *not* connected to a battery and charge cannot flow through an ideal voltmeter. (We shall discuss the voltmeter further in Chapter 28.) Hence, there is *no* path by which charge can flow and alter the charge on the capacitor. If a dielectric is now inserted between the plates as in Figure 26.11b, it is found that the voltmeter reading *decreases* by a factor κ to a value V, where

$$V = \frac{V_0}{\kappa}$$

Since $V < V_0$, we see that $\kappa > 1$.

Since the charge Q_0 on the capacitor *does not change*, we conclude that the capacitance must change to the value

$$C = \frac{Q_0}{V} = \frac{Q_0}{V_0/\kappa} = \kappa \frac{Q_0}{V_0}$$

$$C = \kappa C_0 \qquad (26.15)$$

where C_0 is the capacitance in the absence of the dielectric. That is, the capacitance *increases* by the factor κ when the dielectric completely fills the region between the plates.[3] For a parallel-plate capacitor, where $C_0 = \epsilon_0 A/d$, we can express the capacitance when the capacitor is filled with a dielectric as

$$C = \kappa \frac{\epsilon_0 A}{d} \qquad (26.16)$$

The capacitance of a filled capacitor is greater than that of an empty one by a factor κ.

From Equations 26.3 and 26.16, it would appear that the capacitance could be made very large by decreasing d, the distance between the plates. In

[3] If another experiment is performed in which the dielectric is introduced while the potential difference remains constant by means of a battery, the charge increases to a value $Q = \kappa Q_0$. The additional charge is supplied by the battery and the capacitance still increases by the factor κ.

TABLE 26.1 Dielectric Constants and Dielectric Strengths of Various Materials at Room Temperature

Material	Dielectric Constant κ	Dielectric Strength[a] (V/m)
Vacuum	1.00000	—
Air	1.00059	3×10^6
Bakelite	4.9	24×10^6
Fused quartz	3.78	8×10^6
Pyrex glass	5.6	14×10^6
Polystyrene	2.56	24×10^6
Teflon	2.1	60×10^6
Neoprene rubber	6.7	12×10^6
Nylon	3.4	14×10^6
Paper	3.7	16×10^6
Strontium titanate	233	8×10^6
Water	80	—
Silicone oil	2.5	15×10^6

[a] The dielectric strength equals the maximum electric field that can exist in a dielectric without electrical breakdown.

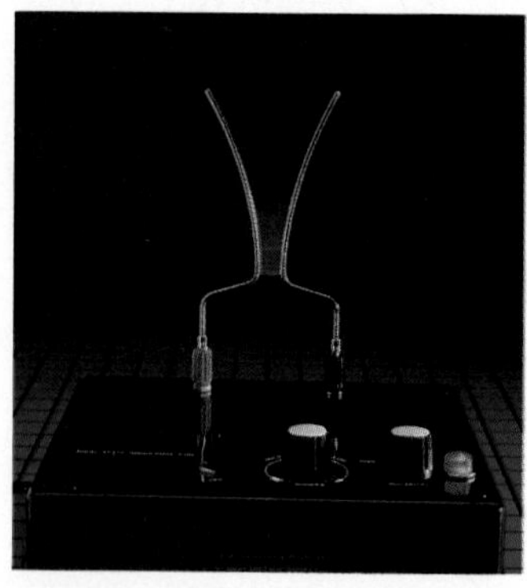

A continuous electrical discharge is produced between two electrodes when the applied voltage produces an electric field that exceeds the dielectric strength of air. What happens to the air in the vicinity of this discharge? As the discharge continues, the "sparks" rise to the top. Can you explain this behavior? (Courtesy CENCO)

practice, the lowest value of d is limited by the electrical discharge that could occur through the dielectric medium separating the plates. For any given separation d, the maximum voltage that can be applied to a capacitor without causing a discharge depends on the *dielectric strength* (maximum electric field intensity) of the dielectric, which for air is equal to 3×10^6 V/m. If the field strength in the medium exceeds the dielectric strength, the insulating properties will break down and the medium will begin to conduct. Most insulating materials have dielectric strengths and dielectric constants greater than that of air, as Table 26.1 indicates. Thus, we see that a dielectric provides the following advantages:

1. A dielectric increases the capacitance of a capacitor.
2. A dielectric increases the maximum operating voltage of a capacitor.
3. A dielectric may provide mechanical support between the conducting plates.

Types of Capacitors

Commercial capacitors are often made using metal foil interlaced with thin sheets of paraffin-impregnated paper or mylar, which serves as the dielectric material. These alternate layers of metal foil and dielectric are then rolled into the shape of a cylinder to form a small package (Fig. 26.12a). High-voltage capacitors commonly consist of a number of interwoven metal plates immersed in silicone oil (Fig. 26.12b). Small capacitors are often constructed from ceramic materials. Variable capacitors (typically 10 to 500 pF) usually consist of two interwoven sets of metal plates, one fixed and the other movable, with air as the dielectric.

An electrolytic capacitor is often used to store large amounts of charge at relatively low voltages. This device, shown in Figure 26.12c, consists of a metal foil in contact with an electrolyte — a solution that conducts electricity by virtue of the motion of ions contained in the solution. When a voltage is applied between the foil and the electrolyte, a thin layer of metal oxide (an

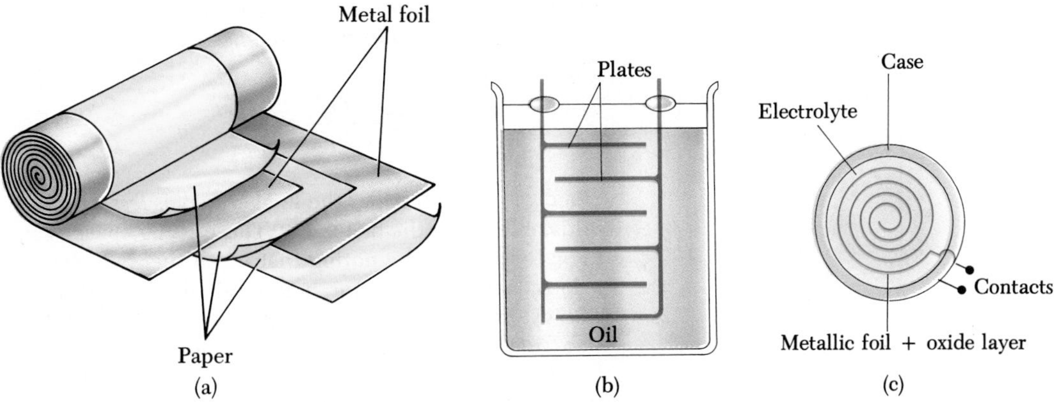

Figure 26.12 Three commercial capacitor designs. (a) A tubular capacitor whose plates are separated by paper and then rolled into a cylinder. (b) A high-voltage capacitor consists of many parallel plates separated by insulating oil. (c) An electrolytic capacitor.

insulator) is formed on the foil, and this layer serves as the dielectric. Very large values of capacitance can be obtained because the dielectric layer is very thin.

When electrolytic capacitors are used in circuits, the polarity (the plus and minus signs on the device) must be installed properly. If the polarity of the applied voltage is opposite to what is intended, the oxide layer will be removed and the capacitor will conduct electricity rather than store charge.

EXAMPLE 26.6 A Paper-Filled Capacitor
A parallel-plate capacitor has plates of dimensions 2 cm × 3 cm. The plates are separated by a 1-mm thickness of paper. (a) Find the capacitance of this device.

Solution Since $\kappa = 3.7$ for paper (Table 26.1), we get

$$C = \kappa\frac{\epsilon_0 A}{d} = 3.7\left(8.85 \times 10^{-12}\,\frac{\text{C}^2}{\text{N}\cdot\text{m}^2}\right)\left(\frac{6 \times 10^{-4}\ \text{m}^2}{1 \times 10^{-3}\ \text{m}}\right)$$

$$= 19.6 \times 10^{-12}\ \text{F} = 19.6\ \text{pF}$$

(b) What is the maximum charge that can be placed on the capacitor?

Solution From Table 26.1 we see that the dielectric strength of paper is 16×10^6 V/m. Since the thickness of the paper is 1 mm, the maximum voltage that can be applied before breakdown occurs is

$$V_{\text{max}} = E_{\text{max}}d = \left(16 \times 10^6\ \frac{\text{V}}{\text{m}}\right)(1 \times 10^{-3}\ \text{m})$$

$$= 16 \times 10^3\ \text{V}$$

Hence, the maximum charge is given by

$$Q_{\text{max}} = CV_{\text{max}} = (19.6 \times 10^{-12}\ \text{F})(16 \times 10^3\ \text{V})$$

$$= 0.31\ \mu\text{C}$$

Exercise 4 What is the maximum energy that can be stored in the capacitor?
Answer 2.5×10^{-3} J.

EXAMPLE 26.7 Energy Stored Before and After
A parallel-plate capacitor is charged with a battery to a charge Q_0, as in Figure 26.13a. The battery is then removed, and a slab of dielectric constant κ is inserted between the plates, as in Figure 26.13b. Find the energy stored in the capacitor before and after the dielectric is inserted.

Solution The energy stored in the capacitor in the absence of the dielectric is

$$U_0 = \tfrac{1}{2}C_0V_0^2$$

Since $V_0 = Q_0/C_0$, this can be expressed as

$$U_0 = \frac{Q_0^2}{2C_0}$$

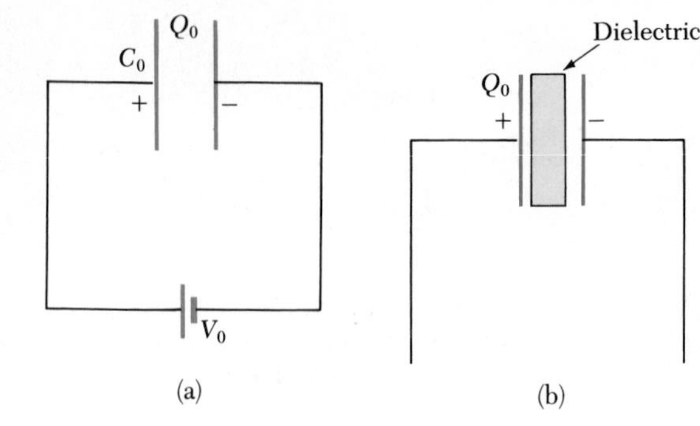

Figure 26.13 (Example 26.7).

After the battery is removed and the dielectric is inserted between the plates, the *charge on the capacitor remains the same.* Hence, the energy stored in the presence of the dielectric is given by

$$U = \frac{Q_0^2}{2C}$$

But the capacitance in the presence of the dielectric is given by $C = \kappa C_0$, and so U becomes

$$U = \frac{Q_0^2}{2\kappa C_0} = \frac{U_0}{\kappa}$$

Since $\kappa > 1$, we see that the final energy is *less* than the initial energy by the factor $1/\kappa$. This missing energy can be accounted for by noting that when the dielectric is inserted into the capacitor, it gets pulled into the device. The external agent must do negative work to keep the slab from accelerating. This work is simply the difference $U - U_0$. (Alternatively, the positive work done by the system on the external agent is given by $U_0 - U$.)

Exercise 5 Suppose that the capacitance in the absence of a dielectric is 8.50 pF, and the capacitor is charged to a potential difference of 12.0 V. If the battery is disconnected, and a slab of polystyrene ($\kappa = 2.56$) is inserted between the plates, calculate the energy difference $U - U_0$.
Answer 373 pJ

As we have seen, the energy of a capacitor is lowered when a dielectric is inserted between the plates, which means that work is done on the dielectric. This, in turn, implies that a force must act on the dielectric which draws it into the capacitor. This force originates from the nonuniform nature of the electric field of the capacitor near its edges as indicated in Figure 26.14. The horizontal component of this fringe field acts on the induced charges on the surface of the dielectric, producing a net horizontal force directed into the capacitor.

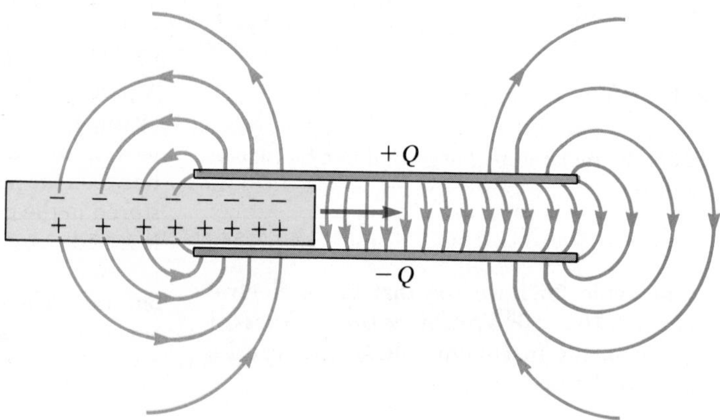

Figure 26.14 The nonuniform electric field near the edges of a parallel-plate capacitor causes a dielectric to be pulled into the capacitor. Note that the field acts on the induced surface charges on the dielectric which are nonuniformly distributed.

*26.6 ELECTRIC DIPOLE IN AN EXTERNAL ELECTRIC FIELD

The electric dipole, discussed briefly in Example 23.7, consists of two equal and opposite charges separated by a distance $2a$, as in Figure 26.15. Let us define the **electric dipole moment** of this configuration as the vector $\boldsymbol{p}$ whose magnitude is $2aq$ (that is, the separation $2a$ multiplied by the charge q).

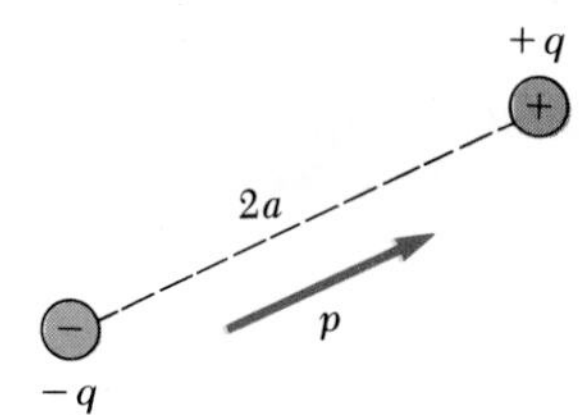

Figure 26.15 An electric dipole consists of two equal and opposite charges separated by a distance $2a$.

$$p \equiv 2aq \qquad (26.17)$$

Now suppose an electric dipole is placed in a uniform *external* electric field $\boldsymbol{E}$ as in Figure 26.16, where the dipole moment makes an angle θ with the field. The forces on the two charges are equal and opposite as shown, each having a magnitude of

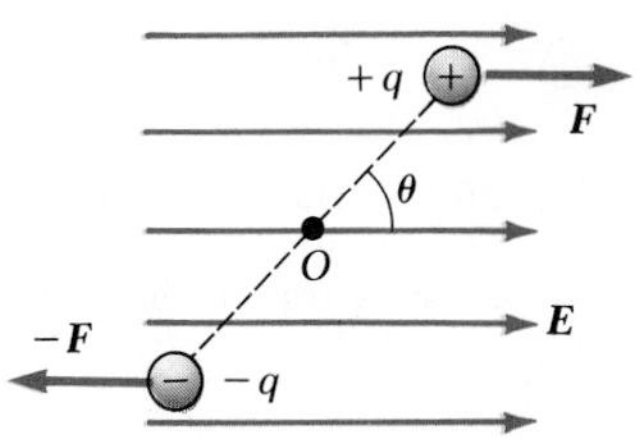

Figure 26.16 An electric dipole in a uniform electric field. The dipole moment $\boldsymbol{p}$ is at an angle θ with the field, and the dipole experiences a torque.

$$F = qE$$

Thus, we see that the net force on the dipole is *zero*. However, the two forces produce a net torque on the dipole, and the dipole tends to rotate such that its axis is aligned with the field. The torque due to the force on the positive charge about an axis through O is given by $Fa \sin\theta$, where $a \sin\theta$ is the moment arm of F about O. In Figure 26.16, this force tends to produce a clockwise rotation. The torque on the negative charge about O is also $Fa \sin\theta$, so the net torque about O is given by

$$\tau = 2Fa \sin\theta$$

Because $F = qE$ and $p = 2aq$, we can express τ as

$$\tau = 2aqE \sin\theta = pE \sin\theta \qquad (26.18)$$

It is convenient to express the torque in vector form as the cross product of the vectors $\boldsymbol{p}$ and $\boldsymbol{E}$:

Torque on an electric dipole in an extended electric field

$$\boldsymbol{\tau} = \boldsymbol{p} \times \boldsymbol{E} \qquad (26.19)$$

We can also determine the potential energy of an electric dipole as a function of its orientation with respect to the external electric field. In order to do this, you should recognize that work must be done by an external agent to rotate the dipole through a given angle in the field. The work done is then stored as potential energy in the system, that is, the dipole and the external field. The work dW required to rotate the dipole through an angle $d\theta$ is given by $dW = \tau\, d\theta$ (Chapter 10). Because $\tau = pE \sin\theta$, and because the work is transformed into potential energy U, we find that for a rotation from θ_0 to θ, the change in potential energy is

$$U - U_0 = \int_{\theta_0}^{\theta} \tau\, d\theta = \int_{\theta_0}^{\theta} pE \sin\theta\, d\theta = pE \int_{\theta_0}^{\theta} \sin\theta\, d\theta$$

$$U - U_0 = pE[-\cos\theta]_{\theta_0}^{\theta} = pE(\cos\theta_0 - \cos\theta)$$

The term involving $\cos\theta_0$ is a constant that depends on the initial orientation of the dipole. It is convenient to choose $\theta_0 = 90°$, so that $\cos\theta_0 = \cos 90° = 0$. Furthermore, let us choose $U_0 = 0$ at $\theta_0 = 90°$ as our reference of potential energy. Hence, we can express U as

$$U = -pE \cos\theta \qquad (26.20)$$

This can be written as the dot product of the vectors $\boldsymbol{p}$ and $\boldsymbol{E}$:

Potential energy of an electric dipole in an external electric field

$$U = -\boldsymbol{p} \cdot \boldsymbol{E} \tag{26.21}$$

Molecules are said to be polarized when there is a separation between the "center of gravity" of the negative charges and that of the positive charges on the molecule. In some molecules, such as water, this condition is always present. This can be understood by inspecting the geometry of the water molecule. The molecule is arranged so that the oxygen atom is bonded to the hydrogen atoms with an angle of 105° between the two bonds (Fig. 26.17). The center of negative charge is near the oxygen atom, and the center of positive charge lies at a point midway along the line joining the hydrogen atoms (point x in the diagram). Materials composed of molecules that are permanently polarized in this fashion have large dielectric constants. For example, the dielectric constant of water is quite large ($\kappa = 80$).

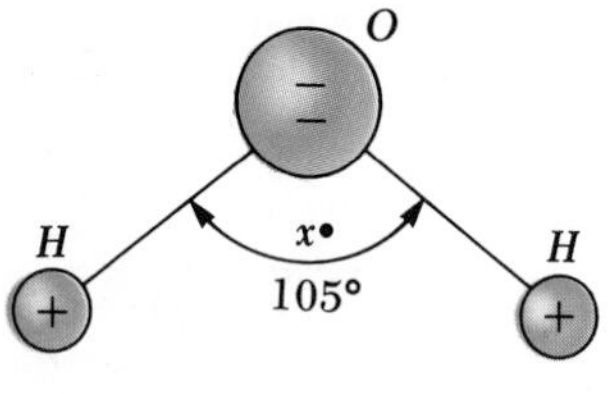

A symmetrical molecule might have no permanent polarization, but a polarization can be induced by an external electric field. For example, if a linear molecule lies along the x axis, an external electric field in the positive x direction would cause the center of positive charge to shift to the right from its initial position and the center of negative charge to shift to the left. This *induced polarization* is the effect that predominates in most materials used as dielectrics in capacitors.

Figure 26.17 The water molecule, H_2O, has a permanent polarization resulting from its bent geometry.

EXAMPLE 26.8 The H_2O Molecule

The H_2O molecule has a dipole moment of 6.3×10^{-30} C · m. A sample contains 10^{21} such molecules, whose dipole moments are all oriented in the direction of an electric field of 2.5×10^5 N/C. How much work is required to rotate the dipoles from this orientation ($\theta = 0°$) to one in which all of the moments are perpendicular to the field ($\theta = 90°$)?

Solution The work required to rotate *one* molecule by 90° is equal to the difference in potential energy between the 90° orientation and the 0° orientation. Using Equation 26.20 gives

$$W = U_{90} - U_0 = (-pE\cos 90°) - (-pE\cos 0°)$$
$$= pE = (6.3 \times 10^{-30}\ \text{C}\cdot\text{m})(2.5 \times 10^5\ \text{N/C})$$
$$= 1.6 \times 10^{-24}\ \text{J}$$

Since there are 10^{21} molecules in the sample, the *total* work required is given by

$$W_{\text{total}} = (10^{21})(1.6 \times 10^{-24}\ \text{J}) = 1.6 \times 10^{-3}\ \text{J}$$

*26.7 AN ATOMIC DESCRIPTION OF DIELECTRICS

In Section 26.5 we found that the potential difference between the plates of a capacitor is reduced by the factor κ when a dielectric is introduced. Since the potential difference between the plates equals the product of the electric field and the separation d, the electric field is also reduced by the factor κ. Thus, if $\boldsymbol{E}_0$ is the electric field without the dielectric, the field in the presence of a dielectric is

$$\boldsymbol{E} = \frac{\boldsymbol{E}_0}{\kappa} \tag{26.22}$$

This can be understood by noting that a dielectric can be polarized. At the atomic level, a polarized material is one in which the positive and negative charges are slightly separated. If the molecules of the dielectric possess permanent electric dipole moments in the absence of an electric field, they are called *polar molecules* (water is an example). The dipoles are randomly oriented in the absence of an electric field, as shown in Figure 26.18a. When an external field is applied, a torque is exerted on the dipoles, causing them to be partially aligned with the field, as in Figure 26.18b. The degree of alignment depends on temperature and on the magnitude of the applied field. In general, the alignment increases with decreasing temperature and with increasing electric field strength. The partially aligned dipoles produce an internal electric field that *opposes* the external field, thereby causing a reduction of the original field.

If the molecules of the dielectric do not possess a permanent dipole moment, they are called **nonpolar molecules.** In this case, an external electric field produces some charge separation, and the resulting dipole moments are said to be *induced.* These induced dipole moments tend to align with the external field, causing a reduction in the internal electric field.

With these ideas in mind, consider a slab of dielectric material in a uniform electric field $\boldsymbol{E}_0$ as in Figure 26.19a. Positive portions of the molecules are shifted in the direction of the electric field, and negative portions are shifted in the opposite direction. Hence, the applied electric field polarizes the dielectric. The net effect on the dielectric is the formation of an "induced" positive surface charge density σ_i on the right face and an equal negative surface charge density on the left face, as shown in Figure 26.19b. These induced surface charges on the dielectric give rise to an induced electric field $\boldsymbol{E}_i$, which *opposes* the external field $\boldsymbol{E}_0$. Therefore, the net electric field $\boldsymbol{E}$ in the dielectric has a magnitude given by

$$E = E_0 - E_i \tag{26.23}$$

In the parallel-plate capacitor shown in Figure 26.20, the external field E_0 is related to the free charge density σ on the plates through the relation $E_0 = \sigma/\epsilon_0$. The induced electric field in the dielectric is related to the induced

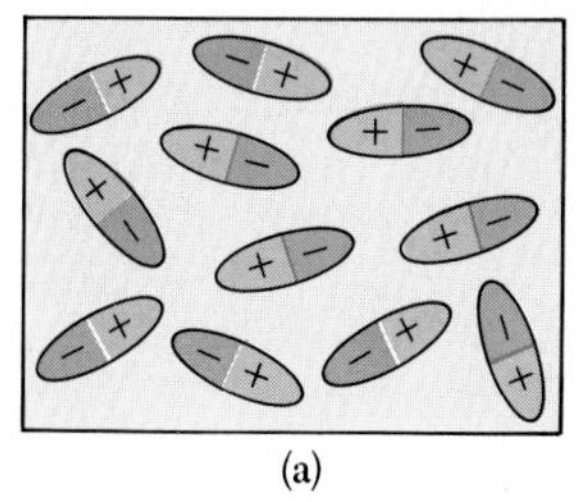
(a)

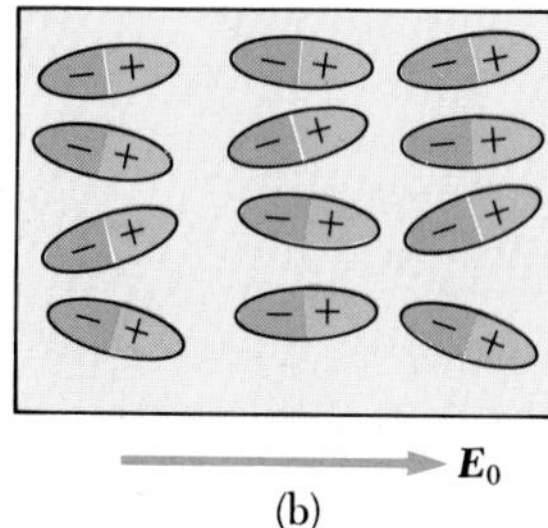

(b)

Figure 26.18 (a) Molecules with a permanent dipole moment are randomly oriented in the absence of an external electric field. (b) When an external field is applied, the dipoles are partially aligned with the field.

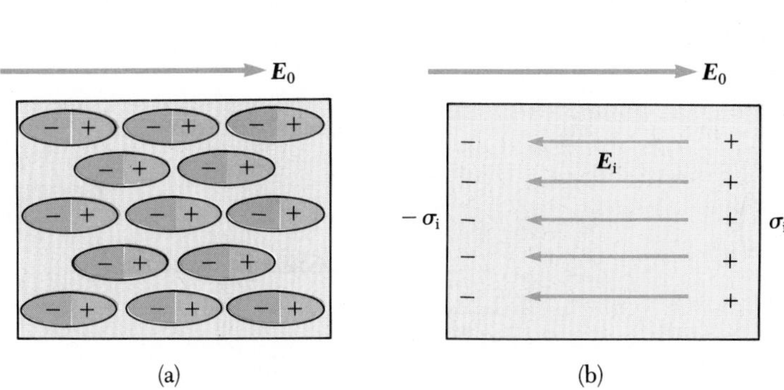

(a) (b)

Figure 26.19 (a) When a dielectric is polarized, the molecular dipole moments in the dielectric are aligned with the external field $\boldsymbol{E}_0$. (b) This polarization causes an induced negative surface charge on one side of the dielectric and an equal positive surface charge on the opposite side. This results in a reduction in the electric field within the dielectric.

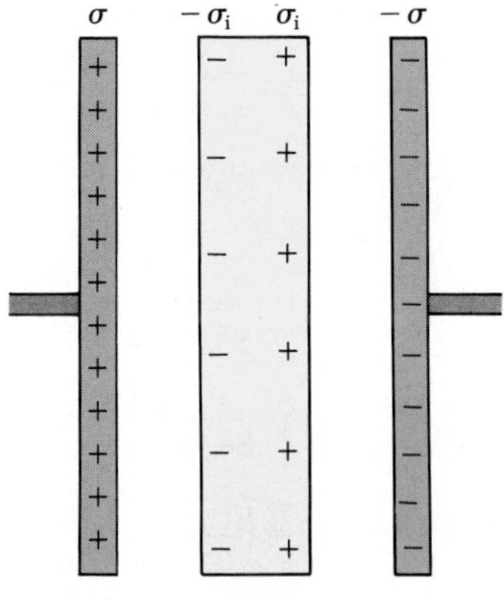

Figure 26.20 Induced charge on a dielectric placed between the plates of a charged capacitor. Note that the induced charge density on the dielectric is *less* than the free charge density on the plates.

charge density σ_i through the relation $E_i = \sigma_i/\epsilon_o$. Since $E = E_0/\kappa = \sigma/\kappa\epsilon_o$, substitution into Equation 26.23 gives

$$\frac{\sigma}{\kappa\epsilon_o} = \frac{\sigma}{\epsilon_o} - \frac{\sigma_i}{\epsilon_0}$$

$$\sigma_i = \left(\frac{\kappa - 1}{\kappa}\right)\sigma \qquad (26.24)$$

Because $\kappa > 1$, this shows that the charge density σ_i induced on the dielectric is *less* than the free charge density σ on the plates. For instance, if $\kappa = 3$, we see that the induced charge density on the dielectric is two thirds the free charge density on the plates. If there is no dielectric present, $\kappa = 1$ and $\sigma_i = 0$ as expected. However, if the dielectric is replaced by a *conductor*, for which $E = 0$, then Equation 26.23 shows that $E_0 = E_i$, corresponding to $\sigma_i = \sigma$. That is, the surface charge induced on the conductor will be equal to and opposite that on the plates, resulting in a net field of *zero* in the conductor.

EXAMPLE 26.9 A Partially Filled Capacitor

A parallel-plate capacitor has a capacitance C_0 in the absence of a dielectric. A slab of dielectric material of dielectric constant κ and thickness $\frac{1}{3}d$ is inserted between the plates (Fig. 26.21a). What is the new capacitance when the dielectric is present?

Solution This capacitor is equivalent to two parallel-plate capacitors of the same area A connected in series, one with a plate separation $d/3$ (dielectric filled) and the other with a plate separation $2d/3$ (Fig. 26.21b). (This step is permissible since there is no potential difference between the lower plate of C_1 and the upper plate of C_2.)[4]

From Equations 26.3 and 26.15, the two capacitances are given by

$$C_1 = \frac{\kappa\epsilon_o A}{d/3} \quad \text{and} \quad C_2 = \frac{\epsilon_o A}{2d/3}$$

Using Equation 26.10 for two capacitors combined in series, we get

$$\frac{1}{C} = \frac{1}{C_1} + \frac{1}{C_2} = \frac{d/3}{\kappa\epsilon_o A} + \frac{2d/3}{\epsilon_o A}$$

$$\frac{1}{C} = \frac{d}{3\epsilon_o A}\left(\frac{1}{\kappa} + 2\right) = \frac{d}{3\epsilon_o A}\left(\frac{1 + 2\kappa}{\kappa}\right)$$

$$C = \left(\frac{3\kappa}{2\kappa + 1}\right)\frac{\epsilon_o A}{d}$$

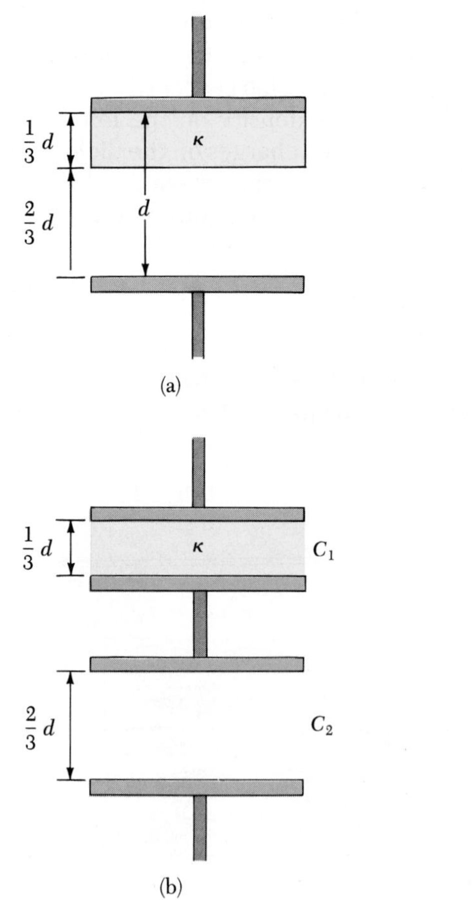

Figure 26.21 (Example 26.9) (a) A parallel-plate capacitor of plate separation d partially filled with a dielectric of thickness $d/3$. (b) The equivalent circuit of the capacitor consists of two capacitors connected in series.

[4] You could also imagine placing two thin metallic plates (with a coiled-up conducting wire between them) at the lower surface of the dielectric in Figure 26.21a and then pulling the assembly out until it becomes like Figure 26.21b.

Since the capacitance *without* the dielectric is given by $C_0 = \epsilon_0 A/d$, we see that

$$C = \left(\frac{3\kappa}{2\kappa + 1}\right)C_0$$

EXAMPLE 26.10 Effect of a Metal Slab

A parallel-plate capacitor has a plate separation d and plate area A. An uncharged *metal* slab of thickness a is inserted midway between the plates, as shown in Figure 26.22a. Find the capacitance of the device.

Solution This problem can be solved by noting that whatever charge appears on one plate of the capacitor must induce an *equal* and *opposite* charge on the metal slab, as shown in Figure 26.22a. Consequently, the net charge on the metal slab remains zero, and the field inside the slab is zero. Hence, the capacitor is equivalent to two capacitors in *series*, each having a plate separation $(d - a)/2$ as shown in Figure 26.22b. Using the rule for adding two capacitors in series we get

$$\frac{1}{C} = \frac{1}{C_1} + \frac{1}{C_2} = \frac{1}{\dfrac{\epsilon_0 A}{(d-a)/2}} + \frac{1}{\dfrac{\epsilon_0 A}{(d-a)/2}}$$

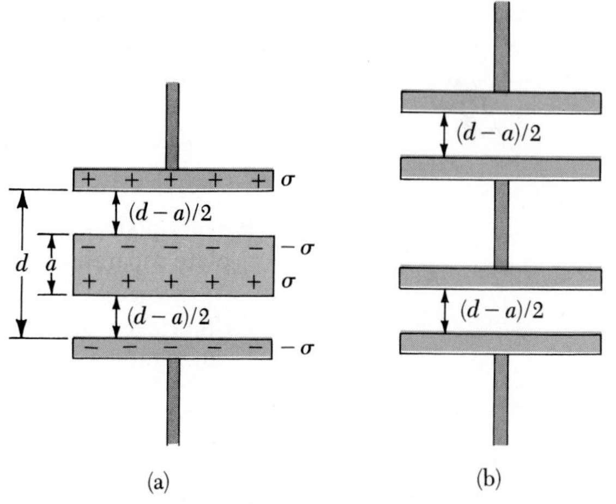

Figure 26.22 (Example 26.10) (a) A parallel-plate capacitor of plate separation d partially filled with a metal slab of thickness a. (b) The equivalent circuit of the device in (a) consists of two capacitors in series, each with a plate separation $(d - a)/2$.

Solving for C gives

$$C = \frac{\epsilon_0 A}{d - a}$$

Note that C approaches infinity as a approaches d. Why?

SUMMARY

A *capacitor* consists of two equal and oppositely charged conductors spaced very close together compared to their size with a potential difference V between them. **The capacitance C of any capacitor is defined to be the ratio of the magnitude of the charge Q on either conductor to the magnitude of the potential difference V:**

$$C \equiv \frac{Q}{V} \qquad (26.1)$$

Definition of capacitance

The SI unit of capacitance is coulomb per volt, or farad (F), and 1 F = 1 C/V.

The capacitance of several capacitors is summarized in Table 26.2. The formulas apply when the charged conductors are separated by a vacuum.

If two or more capacitors are connected in *parallel*, the potential difference across them must be the same. The equivalent capacitance of a parallel combination of capacitors is given by

$$C_{eq} = C_1 + C_2 + C_3 + \cdots \qquad (26.8)$$

Parallel combinations

TABLE 26.2 Capacitance and Geometry

Geometry	Capacitance	Equation
Isolated charged sphere of radius R	$C = 4\pi\epsilon_0 R$	(26.2)
Parallel-plate capacitor of plate area A and plate separation d	$C = \epsilon_0 \dfrac{A}{d}$	(26.3)
Cylindrical capacitor of length ℓ and inner and outer radii a and b, respectively	$C = \dfrac{\ell}{2k \ln\left(\dfrac{b}{a}\right)}$	(26.4)
Spherical capacitor with inner and outer radii a and b, respectively	$C = \dfrac{ab}{k(b-a)}$	(26.5)

If two or more capacitors are connected in *series*, the charge on them is the same, and the equivalent capacitance of the series combination is given by

Series combination

$$\frac{1}{C_{eq}} = \frac{1}{C_1} + \frac{1}{C_2} + \frac{1}{C_3} + \cdots \tag{26.11}$$

Work is required to charge a capacitor, since the charging process consists of transferring charges from one conductor at a lower potential to another conductor at a higher potential. The work done in charging the capacitor to a charge Q equals the electrostatic potential energy U stored in the capacitor, where

Energy stored in a charged capacitor

$$U = \frac{Q^2}{2C} = \tfrac{1}{2}QV = \tfrac{1}{2}CV^2 \tag{26.12}$$

When a dielectric material is inserted between the plates of a capacitor, the capacitance generally increases by a dimensionless factor κ called the **dielectric constant.** That is,

$$C = \kappa C_0 \tag{26.15}$$

where C_0 is the capacitance in the absence of the dielectric. The increase in capacitance is due to a decrease in the electric field in the presence of the dielectric and to a corresponding decrease in the potential difference between the plates—assuming the charging battery is removed from the circuit before the dielectric is inserted. The decrease in $\mathbf{E}$ arises from an internal electric field produced by aligned dipoles in the dielectric. This internal field produced by the dipoles opposes the original applied field, and this results in a reduction in the net electric field.

An *electric dipole* consists of two equal and opposite charges separated by a distance $2a$. The **electric dipole moment** $\mathbf{p}$ of this configuration has a magnitude given by

Electric dipole moment

$$p \equiv 2aq \tag{26.17}$$

The **torque** acting on an electric dipole in a uniform electric field $\mathbf{E}$ is given by

$$\tau = \mathbf{p} \times \mathbf{E} \qquad (26.19)$$

Torque on an electric dipole in an extended electric field

The **potential energy** of an electric dipole in a uniform external electric field $\mathbf{E}$ is given by

$$U = -\mathbf{p} \cdot \mathbf{E} \qquad (26.21)$$

Potential energy of an electric dipole in an external electric field

PROBLEM-SOLVING STRATEGY AND HINTS

1. Be careful with your choice of units. To calculate capacitance in farads, make sure that distances are in meters and use the SI value of ϵ_0. When checking consistency of units, remember that the units for electric fields can be either N/C or V/m.
2. When two or more unequal capacitors are connected in *series,* they carry the same charge, but the potential differences are not the same. Their capacitances add as reciprocals, and the equivalent capacitance of the combination is always *less* than the smallest individual capacitor.
3. When two or more capacitors are connected in *parallel,* the potential difference across each is the same. The charge on each capacitor is proportional to its capacitance, hence the capacitances add directly to give the equivalent capacitance of the parallel combination.
4. The effect of a dielectric on a capacitor is to *increase* its capacitance by a factor κ (the dielectric constant) over its empty capacitance. The reason for this is that induced surface charges on the dielectric reduce the electric field inside the material from E to E/κ.
5. Be careful about problems in which you may be connecting or disconnecting a battery to a capacitor. It is important to note whether modifications to the capacitor are being made while the capacitor is connected to the battery or after it is disconnected. If the capacitor remains connected to the battery, the *voltage* across the capacitor necessarily remains the *same* (equal to the battery voltage), and the charge will be proportional to the capacitance *however it may be modified* (say by inserting a dielectric). On the other hand, if you disconnect the capacitor from the battery *before* making any modifications to the capacitor, then its *charge* remains the same. In this case, as you vary the capacitance, the voltage across the plates will change in an inverse proportion to capacitance according to $V = Q/C$.

QUESTIONS

1. What happens to the charge on a capacitor if the potential difference between the conductors is doubled?
2. The plates of a capacitor are connected to a battery. What happens to the charge on the plates if the connecting wires are removed from the battery? What happens to the charge if the wires are removed from the battery and connected to each other?
3. A farad is a very large unit of capacitance. Calculate the length of one side of a square, air-filled capacitor with a plate separation of 1 meter. Assume it has a capacitance of 1 farad.
4. A pair of capacitors are connected in parallel while an identical pair are connected in series. Which pair would be more dangerous to handle after being connected to the same voltage source? Explain.
5. If you are given 3 different capacitors C_1, C_2, C_3, how many different combinations of capacitance can you produce?
6. What advantage might there be in using 2 identical capacitors in parallel connected in series with another identical parallel pair, rather than using a single capacitor by itself?

7. Is it always possible to reduce a combination of capacitors to one equivalent capacitor with the rules we have just developed? Explain your answer.
8. Since the net charge in a capacitor is always zero, what does a capacitor store?
9. Since the charges on the plates of a parallel-plate capacitor are equal and opposite, they attract each other. Hence, it would take positive work to increase the plate separation. What happens to the external work done in this process?
10. Explain why the work needed to move a charge, Q, through a potential, V, is given by $W = QV$ whereas the energy stored in a charged capacitor is $U = \frac{1}{2}QV$. Where does the $\frac{1}{2}$ factor come from?
11. If the potential difference across a capacitor is doubled, by what factor does the energy stored change?
12. Why is it dangerous to touch the terminals of a high-voltage capacitor even after the applied voltage has been turned off? What can be done to make the capacitor safe to handle after the voltage source has been removed?
13. If you want to increase the maximum operating voltage of a parallel-plate capacitor, describe how you can do this for a fixed plate separation.
14. An air-filled capacitor is charged, then disconnected from the power supply, and finally connected to a voltmeter. Explain how and why the voltage reading changes when a dielectric is inserted between the plates of the capacitor.
15. Using the polar molecule description of a dielectric, explain how a dielectric affects the electric field inside a capacitor.
16. Explain why a dielectric increases the maximum operating voltage of a capacitor although the physical size of the capacitor does not change.
17. What is the difference between dielectric strength and the dielectric constant?
18. Where in a coaxial cable will electrical breakdown first occur if the cable is connected to an excessive potential difference?
19. Explain why a water molecule is permanently polarized. What type of molecule has no permanent polarization?
20. If a dielectric-filled capacitor is heated, how will its capacitance change? (Neglect thermal expansion and assume that the dipole orientations are temperature-dependent.)
21. In terms of induced charges, explain why a charged comb attracts small bits of paper.
22. If you were asked to design a capacitor where small size and large capacitance were required, what factors would be important in your design?

PROBLEMS

Section 26.1 Definition of Capacitance

1. The excess charge on each conductor of a parallel-plate capacitor is 53 μC. What is the potential difference between the conductors if the capacitance of the system is 4×10^{-3} μF?
2. Show that the units $C^2/N \cdot m$ equal 1 F.
3. Two parallel wires are suspended in a vacuum. When the potential difference between the two wires is 52 V, each wire has a charge of 73 pC (the two charges are of opposite sign). Calculate the capacitance of the parallel-wire system.
4. Two conductors insulated from each other are charged by transferring electrons from one conductor to the other. After 1.6×10^{12} electrons have been transferred, the potential difference between the conductors is found to be 14 V. What is the capacitance of the system?
5. A parallel-plate capacitor has a capacitance of 19 μF. What charge on each plate will produce a potential difference of 36 V between the plates of the capacitor?
6. An isolated conducting sphere can be considered as one element of a capacitor (the other element being a concentric sphere of infinite radius). (a) If the capacitance of this system is 9.1×10^{-11} F, what is the radius of the sphere? (b) If the potential at the surface of the sphere is 2.8×10^4 V, what is the corresponding surface charge density?
7. An isolated charged conducting sphere of radius 12 cm creates an electric field of 4.9×10^4 N/C at a distance of 21 cm from its center. (a) What is its surface charge density? (b) What is its capacitance?
8. Two conducting spheres with diameters of 0.40 m and 1.0 m are separated by a distance that is large compared with the diameters. The spheres are connected by a thin wire and are charged to 7 μC. (a) How is this total charge shared between the spheres? (Neglect any charge on the wire.) (b) What is the potential of the system of spheres relative to $V = 0$ at $r = \infty$?
9. Two spherical conductors with radii R_1 and R_2 are separated by a sufficiently large distance that induction effects are negligible. The spheres are connected by a thin conducting wire and are brought to the same potential V relative to $V = 0$ at $r = \infty$. (a) Determine the capacitance C of the system, where $C = (Q_1 + Q_2)/V$. (b) What is the charge ratio Q_1/Q_2?

Section 26.2 Calculation of Capacitance

10. An air-filled parallel plate capacitor is to have a capacitance of 1 F. If the distance between the plates is 1 mm, calculate the required surface area of each plate. Convert your answer to square miles.
11. A parallel-plate capacitor has a plate area of 12 cm^2 and a capacitance of 7 pF. What is the plate separation?
12. The plates of a parallel-plate capacitor are separated by 0.2 mm. If the space between the plates is air, what plate area is required to provide a capacitance of 9 pF?
13. When a potential difference of 150 V is applied to the plates of a parallel-plate capacitor, the plates carry a surface charge density of 30 nC/cm^2. What is the spacing between the plates?
14. A small object with a mass of 350 mg carries a charge of 30 nC and is suspended by a thread between the vertical plates of a parallel-plate capacitor. The plates are separated by 4 cm. If the thread makes an angle of 15° with the vertical, what is the potential difference between the plates?
15. An air-filled capacitor consists of two parallel plates, each with an area of 7.6 cm^2, separated by a distance of 1.8 mm. If a 20-V potential difference is applied to these plates, calculate (a) the electric field between the plates, (b) the surface charge density, (c) the capacitance, and (d) the charge on each plate.
16. Calculate the electric field inside a parallel-plate 75-pF capacitor with an applied voltage of 0.34 V. Assume the surface area of the plates to be 1 cm^2.
17. A circular parallel-plate capacitor with a spacing $d =$ 3 mm is charged to produce an electric field strength of 3×10^6 V/m. What plate radius R is required if the stored charge Q is 1 μC?
18. A capacitor is constructed of interlocking plates as shown in Figure 26.23 (a cross-sectional view). The separation between adjacent plates is 0.8 mm, and the effective area of overlap of adjacent plates is 7 cm^2. Ignoring side effects, calculate the capacitance of the unit.

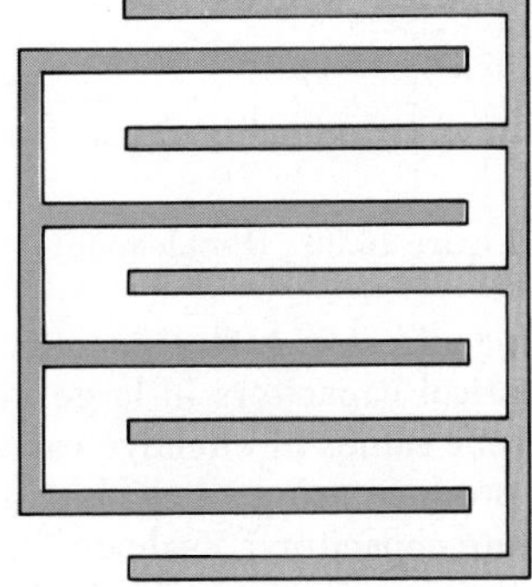

Figure 26.23 (Problems 18 and 87).

19. One plate of a parallel-plate capacitor (with area $A =$ 100 cm^2) is grounded. The ungrounded plate carries a fixed charge Q. When the ungrounded plate is moved 0.5 cm farther away from the grounded plate, the potential difference between the plates increases by 200 V. Determine the magnitude of the charge Q.
20. An air-filled *cylindrical* capacitor has a capacitance of 10 pF and is 6 cm in length. If the radius of the outside conductor is 1.5 cm, what is the required radius of the inner conductor?
21. A 50-m length of coaxial cable has an inner conductor with a diameter of 2.58 mm and a charge of +8.1 μC. The surrounding conductor has an inner diameter of 7.27 mm and a charge of −8.1 μC. (a) What is the capacitance of this cable? (b) What is the potential difference between the two conductors?
22. A cylindrical capacitor has outer and inner conductors whose radii are in the ratio of $b/a = 4/1$. The inner conductor is to be replaced by a wire whose radius is one half of the original inner conductor. By what factor should the length be increased in order to obtain a capacitance equal to that of the original capacitor?
23. An air-filled spherical capacitor is constructed with inner and outer shell radii of 7 and 14 cm, respectively. (a) Calculate the capacitance of the device. (b) What potential difference between the spheres will result in a charge of 4 μC on each conductor?
24. An air-filled spherical capacitor has an outer spherical conductor of radius 0.25 m. If the capacitance of the device is to be 1 μF, calculate the required value for the radius of the inner spherical conductor.
25. A spherical capacitor consists of a conducting ball with a diameter of 10 cm that is centered inside a grounded conducting spherical shell with an inner diameter of 12 cm. What capacitor charge is required to achieve a potential of 1000 V on the ball?
26. Compare the capacitance of a spherical capacitor with radii $a = 7$ cm and $b = 8$ cm with that of a cylindrical capacitor having the same radii and an axial length of 15 cm. (In each case the outer conductor is grounded.) Why are the capacitance values nearly equal?

Section 26.3 Combinations of Capacitors

27. Two capacitors, $C_1 = 2\ \mu$F and $C_2 = 16\ \mu$F, are connected in parallel. What is the value of the equivalent capacitance of the combination?
28. Calculate the equivalent capacitance of the two capacitors in the previous exercise if they are connected in series.
29. (a) Determine the equivalent capacitance for the capacitor network shown in Figure 26.24. (b) If the network is connected to a 12-V battery, calculate the potential difference across each capacitor and the charge on each capacitor.

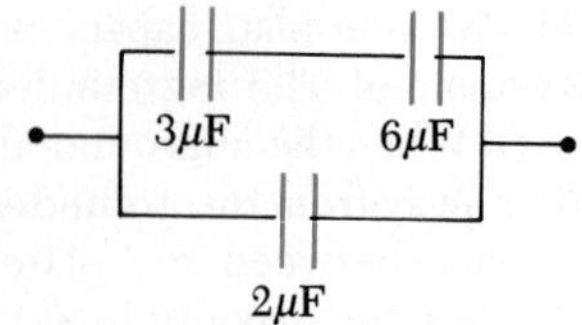

Figure 26.24 (Problem 29).

30. Evaluate the effective capacitance of the configuration shown in Figure 26.25. Each of the capacitors is identical and has capacitance C.

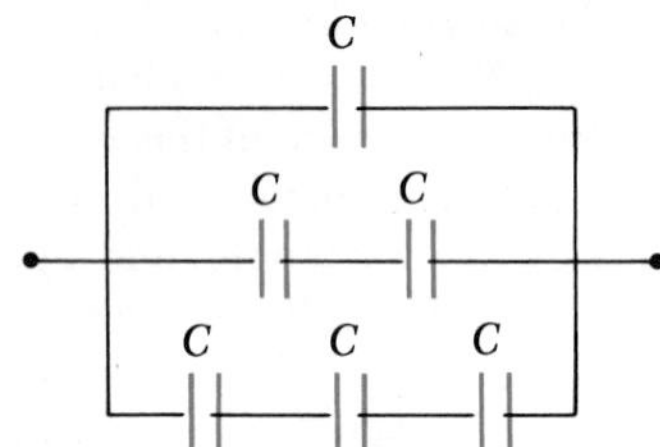

Figure 26.25 (Problem 30).

31. Four capacitors are connected as shown in Figure 26.26. (a) Find the equivalent capacitance between points a and b. (b) Calculate the charge on each capacitor if $V_{ab} = 15$ V.

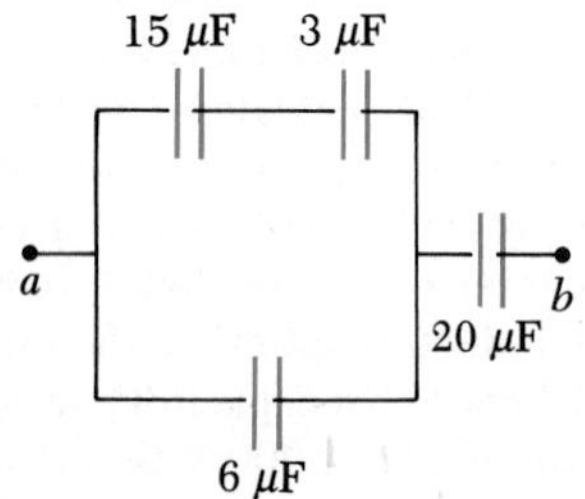

Figure 26.26 (Problems 31 and 44).

32. (a) Figure 26.27 shows a network of capacitors between the terminals a and b. Reduce this network to a single equivalent capacitor. (b) Determine the charge on the 4-μF and 8-μF capacitors when the capacitors are fully charged by a 12-V battery connected to the terminals. (c) Determine the potential difference across each capacitor.

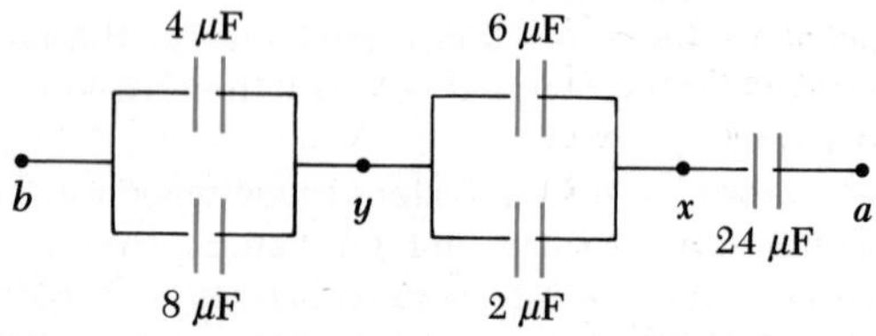

Figure 26.27 (Problem 32).

33. Consider the circuit shown in Figure 26.28, where $C_1 = 6\ \mu$F, $C_2 = 3\ \mu$F, and $V = 20$ V. C_1 is first charged by the closing of switch S_1. Switch S_1 is then opened, and the charged capacitor is connected to the uncharged capacitor by the closing of S_2. Calculate the initial charge acquired by C_1 and the final charge on each of the two capacitors.

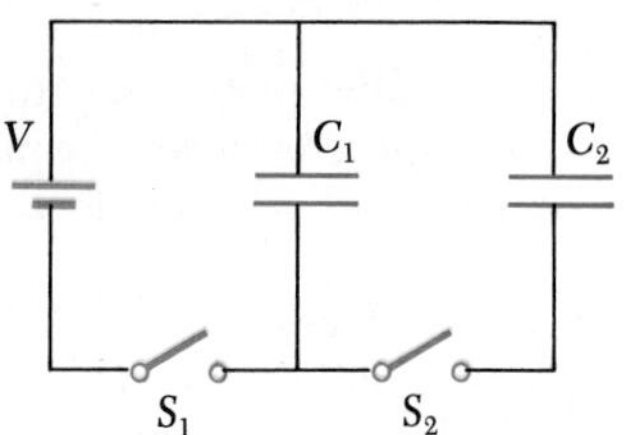

Figure 26.28 (Problem 33).

34. Consider the group of capacitors shown in Figure 26.29. (a) Find the equivalent capacitance between points a and b. (b) Determine the charge on each capacitor when the potential difference between a and b is 12 V.

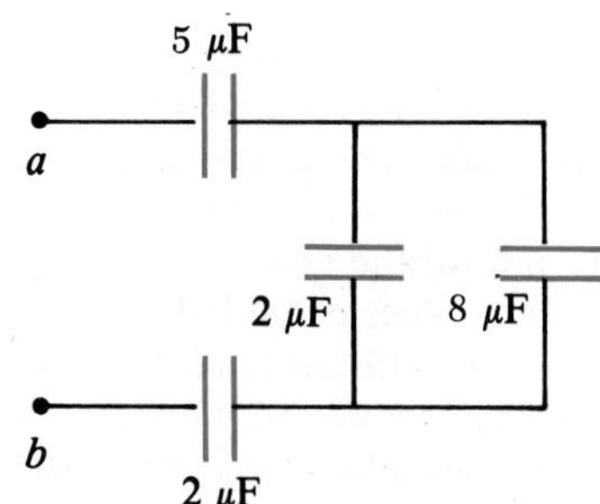

Figure 26.29 (Problem 34).

35. Consider the combination of capacitors shown in Figure 26.30. (a) What is the equivalent capacitance between points a and b? (b) Determine the charge on each capacitor if $V_{ab} = 4.8$ V.

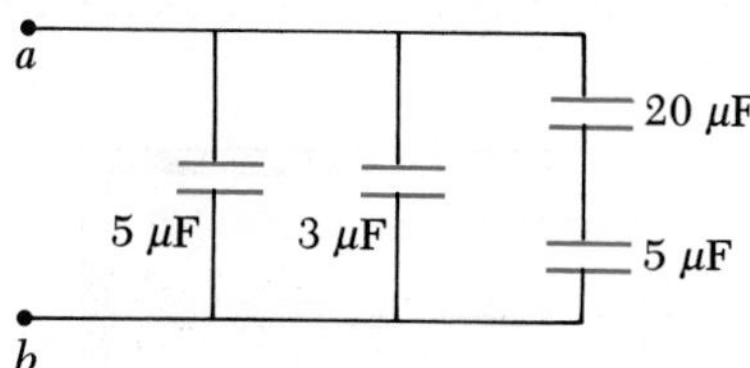

Figure 26.30 (Problem 35).

36. A purchasing director gets a quantity discount for buying identical capacitors in large numbers. How many different values of effective capacitance can a technician produce using all combinations of one to three separate capacitors? Evaluate the capacitance, in terms of the capacitance C of a single capacitor, for each combination.

37. A 4-μF capacitor charged to 400 V and a 6-μF capacitor charged to 600 V are connected to each other, with the positive plate of each connected to the negative plate of the other. (a) What is the final value of the charge that resides on each capacitor? (b) What is the potential difference across each capacitor after they have been connected?
38. Find the equivalent capacitance between points *a* and *b* for the group of capacitors connected as shown in Figure 26.31 if $C_1 = 5\ \mu$F, $C_2 = 10\ \mu$F, and $C_3 = 2\ \mu$F.

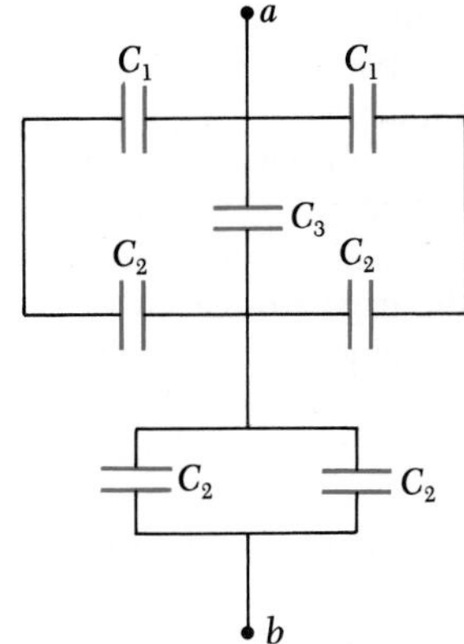

Figure 26.31 (Problems 38 and 39).

39. For the circuit described in the previous exercise, if the potential between points *a* and *b* is 60 V, what charge is stored on the capacitor C_3?
40. Find the equivalent capacitance between points *a* and *b* in the capacitor network shown in Figure 26.32.

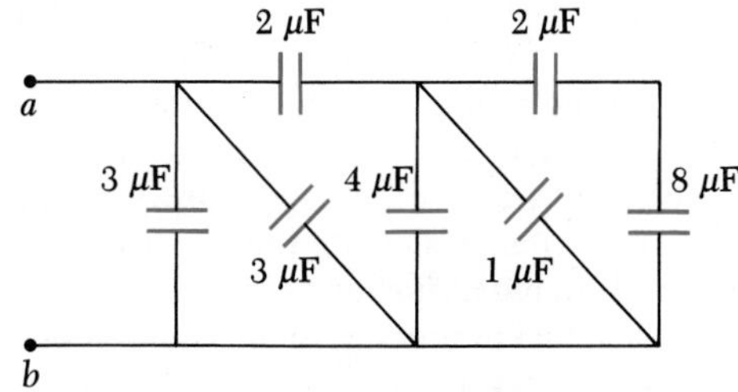

Figure 26.32 (Problem 40).

41. Find the equivalent capacitance between points *a* and *b* in the combination of capacitors shown in Figure 26.33.

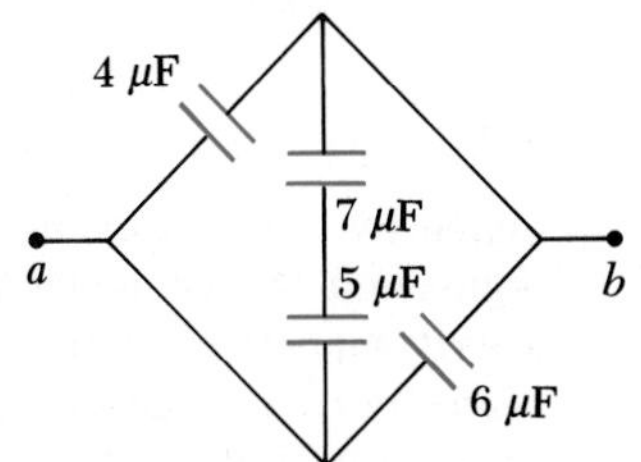

Figure 26.33 (Problem 41).

42. How should four 2-μF capacitors be connected to have a total capacitance of (a) 8 μF, (b) 2 μF, (c) 1.5 μF, and (d) 0.5 μF?
43. A conducting slab with a thickness *d* and area *A* is inserted into the space between the plates of a parallel-plate capacitor with spacing *s* and area *A*, as shown in Figure 26.34. What is the value of the capacitance of the system?

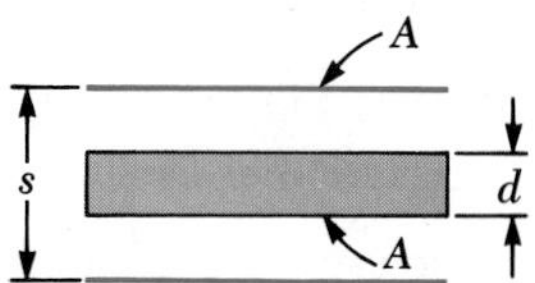

Figure 26.34 (Problem 43).

Section 26.4 Energy Stored in a Charged Capacitor

44. What total energy is stored in the group of capacitors shown in Figure 26.26 if $V_{ab} = 15$ V?
45. Calculate the energy stored in a 18-μF capacitor when it is charged to a potential of 100 V.
46. A parallel-plate capacitor is charged and then disconnected from a battery. By what fraction does the stored energy change (increase or decrease) when the plate separation is doubled?
47. Suppose that a particular capacitor has a capacitance *C* and can withstand a maximum potential difference *V*. Compare the maximum energy that can be stored in a series combination of two of these capacitors to the maximum energy that can be stored in a parallel combination of two of these capacitors.
48. Two capacitors, $C_1 = 25\ \mu$F and $C_2 = 5\ \mu$F, are connected in parallel and charged with a 100-V power supply. (a) Calculate the total energy stored in the two capacitors. (b) What potential difference would be required across the same two capacitors connected in series in order that the combination store the same energy as in (a)?
49. A 16-pF parallel-plate capacitor is charged by a 10-V battery. If each plate of the capacitor has an area of 5 cm^2, what is the energy stored in the capacitor? What is the energy density (energy per unit volume) in the electric field of the capacitor if the plates are separated by air?
50. The energy density in a parallel-plate capacitor is given as 2.1×10^{-9} J/m^3. What is the value of the electric field in the region between the plates?
51. A parallel-plate capacitor has a charge *Q* and plates of area *A*. Show that the force exerted on each plate by the other is given by $F = Q^2/2\epsilon_0 A$. (*Hint:* Let $C = \epsilon_0 A/x$ for an arbitrary plate separation *x*; then require that the work done in separating the two charged plates be $W = \int F\,dx$.)

52. A uniform electric field $E = 3000$ V/m exists within a certain region. What volume of space would contain an energy equal to 10^{-7} J? Express your answer in cubic meters and in liters.
53. Show that the energy associated with a conducting sphere of radius R and charge Q surrounded by a vacuum is given by $U = kQ^2/2R$.
54. Use Equation 26.14 to make an explicit calculation of the energy stored in the field of a simple spherical capacitor. Show that $U = Q^2/2C$.

*Section 26.5 Capacitors with Dielectrics and *Section 26.7 An Atomic Description of Dielectrics

55. Determine (a) the capacitance and (b) the maximum voltage which can be applied to a Teflon-filled parallel-plate capacitor having a plate area of 175 cm² and insulation thickness of 0.04 mm.
56. A parallel-plate capacitor is to be constructed using paper as a dielectric. If a maximum voltage before breakdown of 2500 V is desired, what thickness of dielectric is needed?
57. A parallel-plate capacitor has a plate area of 0.64 cm². When the plates are in a vacuum, the capacitance of the device is 4.9 pF. (a) Calculate the value of the capacitance if the space between the plates is filled with nylon. (b) What is the maximum potential difference that can be applied to the plates without causing dielectric breakdown, or discharge?
58. A capacitor is constructed from two square metal plates of side length L and separated by a distance d (Fig. 26.35). One half of the space between the plates (top to bottom) is filled with polystyrene ($\kappa = 2.56$), and the other half is filled with neoprene rubber ($\kappa = 6.7$). Calculate the capacitance of the device, taking $L = 2$ cm and $d = 0.75$ mm. (*Hint:* The capacitor can be considered as two capacitors connected in parallel.)

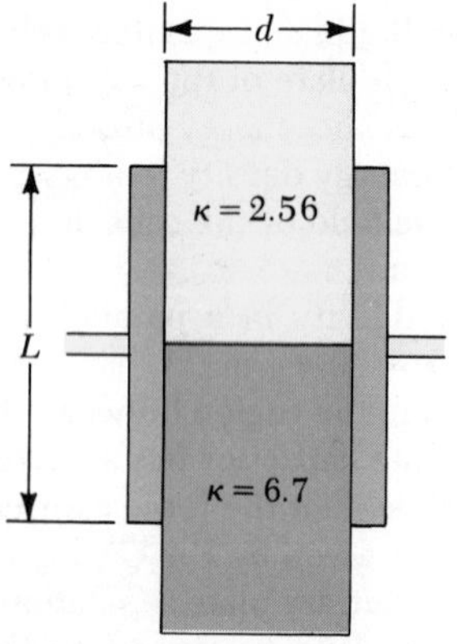

Figure 26.35 (Problem 58).

59. A commercial capacitor is constructed as shown in Figure 26.12a. This particular capacitor is "rolled" from two strips of aluminum separated by two strips of paraffin-coated paper. Each strip of foil and paper is 7 cm wide. The foil is 0.004 mm thick, and the paper is 0.025 mm thick and has a dielectric constant of 3.7. What length should the strips be if a capacitor of 9.5×10^{-8} F is desired? (Use the parallel-plate formula.)
60. A coaxial cable has an outer conductor of inside diameter 0.9 cm, and is filled with polyethylene, which has a dielectric constant of 2.3. If the capacitance per unit length of this cable is 40 pF/m, what is the diameter of the inner conductor?
61. A capacitor with air between the plates is charged to 120 V and then disconnected from the battery. When a piece of glass is placed between the plates, the voltage across the capacitor drops to 20 V. What is the dielectric constant of the glass? (Assume the glass completely fills the space between the plates.)
62. A parallel-plate capacitor can store a maximum energy of U_a when the space between the plates is filled with air. What maximum energy can this capacitor store if the space between the plates is filled with Teflon?
63. A parallel-plate capacitor can be charged to 3 C before breakdown occurs when air is between the plates. What maximum charge can the plates carry if the space between the plates is filled with (a) glass ($\kappa = 5$) or (b) polyethylene ($\kappa = 2.3$)?
64. Let Q_0 be the greatest charge that can be placed on the plates of an air-filled parallel-plate capacitor without causing electrical breakdown. Let Q be the greatest charge that can be placed on the plates of this same capacitor when the gap between the plates is filled with neoprene rubber. What is the ratio Q/Q_0?
65. A sheet of 0.1-mm thick paper is inserted between the plates of a 340-pF air-filled capacitor with a plate separation of 0.4 mm. Calculate the new capacitance.
66. A wafer of titanium dioxide ($\kappa = 173$) has an area of 1 cm² and a thickness of 0.10 mm. Aluminum is evaporated on the parallel faces to form a parallel-plate capacitor. (a) Calculate the capacitance. (b) When the capacitor is charged with a 12-V battery, what is the magnitude of charge delivered to each plate? (c) For the situation in (b), what are the free and induced surface charge densities? (d) What is the electric field strength E?

ADDITIONAL PROBLEMS

67. When two capacitors are connected in parallel, the equivalent capacitance is 4 μF. If the same capacitors are reconnected in series, the equivalent capacitance is one fourth the capacitance of one of the two capacitors. Determine the two capacitances.

68. For the system of capacitors shown in Figure 26.36, find (a) the equivalent capacitance of the system, (b) the potential across each capacitor, (c) the charge on each capacitor, and (d) the total energy stored by the group.

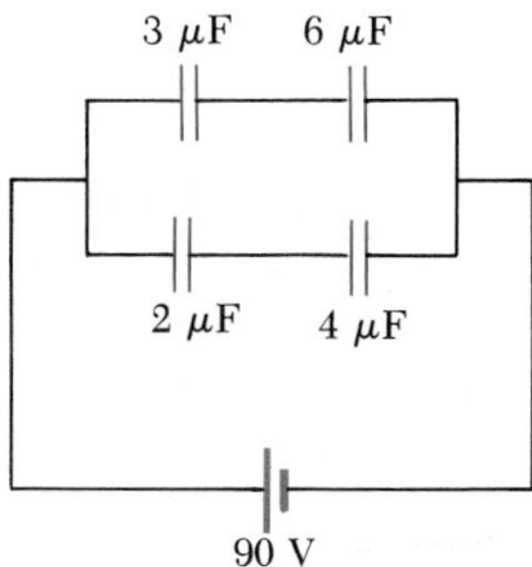

Figure 26.36 (Problem 68).

69. When the voltage applied to a capacitor increases from 80 V to 110 V, the charge on the capacitor increases by 9.0×10^{-5} C. Determine the capacitance.

70. (a) What is the capacitance of a square parallel-plate capacitor measuring 5 cm on a side with a 0.2 mm gap between the plates if this gap is filled with Teflon? (b) What is the maximum voltage this capacitor can withstand? (c) What is the maximum energy this capacitor can store?

71. When a certain air-filled parallel-plate capacitor is connected across a battery, it acquires a charge (on each plate) of 150 μC. While the battery connection is maintained, a dielectric slab is inserted into and fills the region between the plates. This results in the accumulation of an *additional* charge of 200 μC on each plate. What is the dielectric constant of the dielectric slab?

72. The energy stored in a 52-μF capacitor is used to melt a 6-mg sample of lead. To what voltage must the capacitor be initially charged, assuming the initial temperature of the lead is 20°C?

73. Three capacitors of 8 μF, 10 μF, and 14 μF are connected to the terminals of a 12-volt battery. How much energy does the battery supply if the capacitors are connected (a) in series and (b) in parallel?

74. An isolated metal sphere of 50 pF capacitance is charged to a potential of 10^3 V. (a) How much energy is stored on the sphere? (b) An identical uncharged sphere is brought in contact with the charged sphere; next, the second sphere is removed, so in effect, both spheres are isolated. Calculate the total energy stored on the two spheres. (c) Explain any difference between the results of (a) and (b).

75. Capacitors $C_1 = 4\ \mu$F and $C_2 = 2\ \mu$F are charged as a series combination across a 100-V battery. The two capacitors are disconnected from the battery and from each other. They are then connected positive plate to positive plate and negative plate to negative plate. Calculate the resulting charge on each capacitor.

76. Find 9 combinations of unique effective capacitance which can be obtained by using 4 identical capacitors, each having a capacitance C.

77. Two identical capacitors are connected in parallel. Initially they are charged to a potential V_0 and each acquired a charge Q_0. The battery is then disconnected, and the gap between the plates of *one* capacitor is filled with a dielectric ($\kappa = 3$). (a) Calculate the charge, in terms of Q_0, on each capacitor after the dielectric is inserted. (b) Calculate the new potential difference across the plates of each capacitor in terms of V_0.

78. A capacitor $C_1 = 4\ \mu$F is charged to a potential difference of 800 V. The capacitor is then removed from the charging source, and each plate of the charged capacitor is connected to a corresponding plate of an *uncharged* capacitor $C_2 = 6\ \mu$F. (a) What is the resulting charge on each capacitor? (b) What is the total electrostatic energy associated with the two capacitors *before* and *after* they are connected?

79. A parallel-plate capacitor is to be constructed using nylon as a dielectric. If the capacitance of the device is to be 0.6 μF and it is to be operated at 4000 V, (a) calculate the minimum required plate area. (b) What is the energy stored in the capacitor at the operating voltage?

80. A parallel-plate capacitor is constructed using three different dielectric materials, as shown in Figure 26.37. (a) Find an expression for the capacitance of the device in terms of the plate area A and d, κ_1, κ_2, and κ_3. (b) Calculate the capacitance using the values $A = 1\ \text{cm}^2$, $d = 2$ mm, $\kappa_1 = 4.9$, $\kappa_2 = 5.6$, and $\kappa_3 = 2.1$.

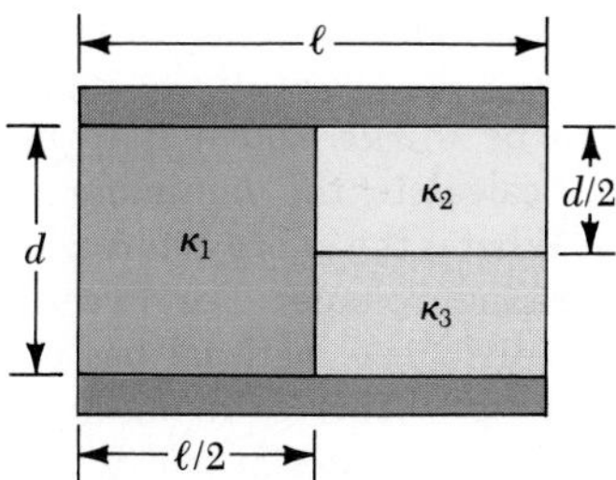

Figure 26.37 (Problem 80).

81. In the arrangement shown in Figure 26.38, a potential V is applied, and C_1 is adjusted so that the electrostatic voltmeter between points b and d reads zero. This "balance" occurs when $C_1 = 4\ \mu$F. If $C_3 = 9\ \mu$F and $C_4 = 12\ \mu$F, calculate the value of C_2.

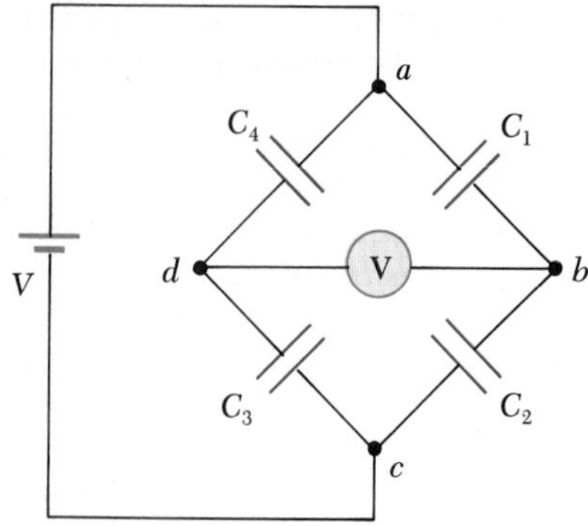

Figure 26.38 (Problem 81).

82. It is possible to obtain large potential differences by first charging a group of capacitors connected in parallel and then activating a switch arrangement that in effect disconnects the capacitors from the charging source and from each other and reconnects them in a *series* arrangement. The group of charged capacitors is then discharged in *series*. What is the maximum potential difference that can be obtained in this manner by using ten capacitors each of 500 μF and a charging source of 800 V?

83. A parallel-plate capacitor of plate separation d is charged to a potential difference V_0. A dielectric slab of thickness d and dielectric constant κ is introduced between the plates *while the battery remains connected to the plates.* (a) Show that the ratio of energy stored after the dielectric is introduced to the energy stored in the empty capacitor is given by $U/U_0 = \kappa$. Give a physical explanation for this increase in stored energy. (b) What happens to the charge on the capacitor? (Note that this situation is not the same as Example 26.7, in which the battery was removed from the circuit before introducing the dielectric.)

84. A parallel-plate capacitor is to be constructed using Pyrex glass as a dielectric. If the capacitance of the device is to be 0.2 μF and it is to be operated at 6000 V, (a) calculate the minimum plate area required. (b) What is the energy stored in the capacitor at the operating voltage? For Pyrex, use $\kappa = 5.6$. (*Note:* Each dielectric material has a characteristic dielectric strength. This is the maximum voltage per unit thickness the material can withstand without electrical breakdown or rupture. For Pyrex, the dielectric strength is 14×10^6 V/m.)

85. A capacitor is constructed from two square plates of sides ℓ and separation d. A material of dielectric constant κ is inserted a distance x into the capacitor, as in Figure 26.39. (a) Find the equivalent capacitance of the device. (b) Calculate the energy stored in the capacitor if the potential difference is V. (c) Find the direction and magnitude of the force exerted on the dielectric, assuming a constant potential difference V. Neglect friction and edge effects. (d) Obtain a numerical value for the force assuming that $\ell = 5$ cm, $V = 2000$ V, $d = 2$ mm, and the dielectric is glass ($\kappa = 4.5$). (*Hint:* The system can be considered as two capacitors connected in *parallel.*)

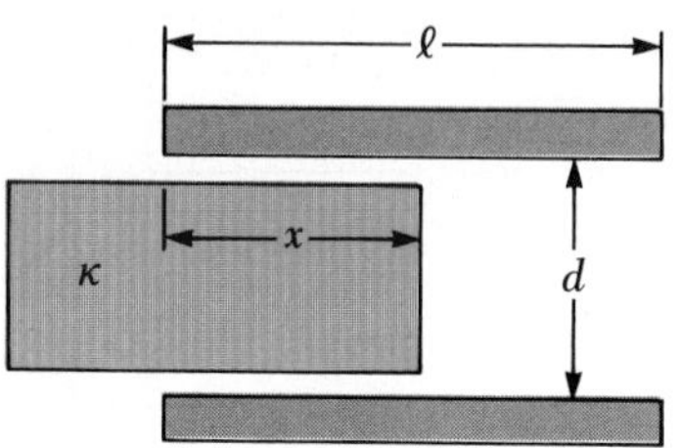

Figure 26.39 (Problem 85).

86. Capacitors $C_1 = 6\ \mu$F and $C_2 = 2\ \mu$F are charged as a parallel combination across a 250-V battery. The capacitors are disconnected from the battery and from each other. They are then connected positive plate to negative plate and negative plate to positive plate. Calculate the resulting charge on each capacitor.

87. A stack of N plates has alternate plates connected to form a capacitor similar to Figure 26.23. Adjacent plates are separated by a dielectric of thickness d. The dielectric constant is κ and the area of overlap of adjacent plates is A. Show that the capacitance of this stack of plates is $C = \dfrac{\kappa\epsilon_0 A}{d}(N-1)$.

88. A coulomb balance is constructed of two parallel plates, each being 10 cm square. The upper plate is movable. A 25-mg mass is placed on the upper plate and the plate is observed to lower and the mass is then removed. When a potential difference is applied to the plates, it is found that the applied voltage must be 375 V to cause the upper plate to lower the same amount as it lowered when the mass was on it. If the force exerted on each plate by the other is given by $F = Q^2/2\epsilon_0 A$, calculate the following, assuming an applied voltage of 375 V: (a) The charge on the plates. (b) The electric field between the plates. (c) The separation distance of the plates. (d) The capacitance of this capacitor.

89. The inner conductor of a coaxial cable has a diameter of 0.8 mm and the outer conductor's inside diameter is 3.0 mm. The space between the conductors is filled with polyethylene, which has a dielectric constant of 2.3 and a dielectric strength of 18×10^6 V/m. What is the maximum potential difference that this cable can withstand?

90. You are optimizing coaxial cable design for a major manufacturer. Show that for a given outer conductor radius b, maximum potential difference capability is

attained when the radius of the inner conductor is given by $a = b/e$ where e is the base of natural logarithms.

91. Calculate the equivalent capacitance between the points a and b in Figure 26.40. Note that this is not a simple series or parallel combination. (*Hint:* Assume a potential difference V between points a and b. Write expressions for V_{ab} in terms of the charges and capacitances for the various possible pathways from a to b, and require conservation of charge for those capacitor plates that are connected to each other.)

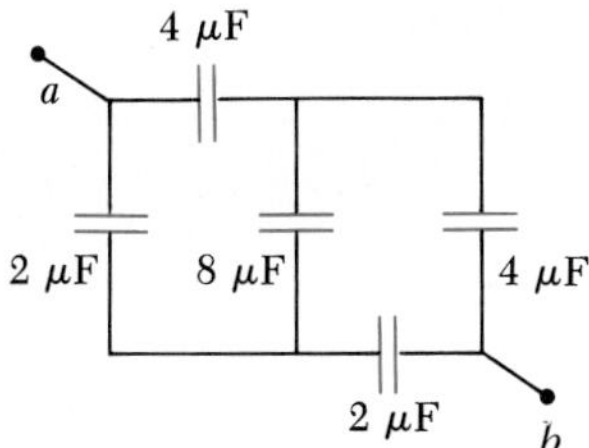

Figure 26.40 (Problem 91).

92. Determine the effective capacitance of the combination shown in Figure 26.41. (*Hint:* Consider the symmetry involved!)

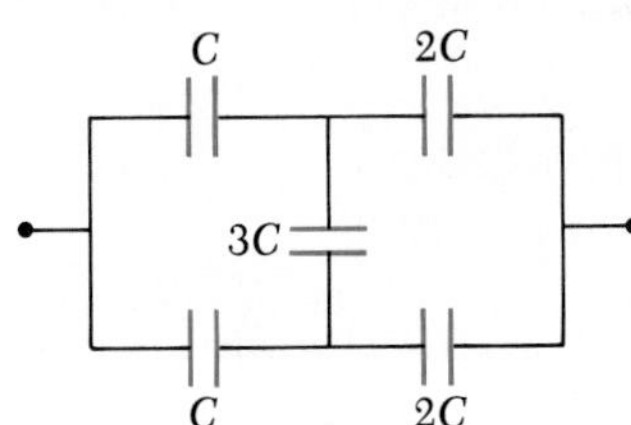

Figure 26.41 (Problem 92).

93. Consider two *long,* parallel, and oppositely charged wires of radius d with their centers separated by a distance D. Assuming the charge is distributed uniformly on the surface of each wire, show that the capacitance per unit length of this pair of wires is given by the following expression:

$$\frac{C}{\ell} = \frac{\pi\epsilon_0}{\ln\left(\frac{D-d}{d}\right)}.$$

94. An air-dielectric capacitor is formed by two *nonparallel* plates, each of area A. An edge view of the arrangement is shown in Figure 26.42. Note that the top plate is tilted relative to the bottom plate so that on one edge the plate separation is $d + \Delta d$, while on the other edge it is $d - \Delta d$. Assuming that $\Delta d \ll d$ and that d is small compared with the length of the plate, show that $C = \frac{\epsilon_0 A}{d}\left[1 + \frac{1}{3}\left(\frac{\Delta d}{d}\right)^2\right]$.

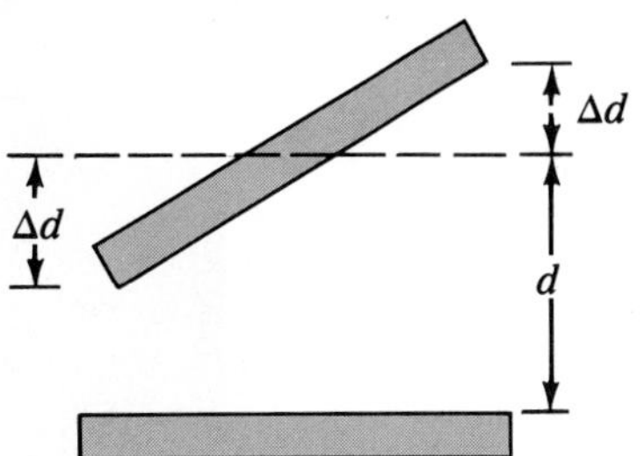

Figure 26.42 (Problem 94).

"But we just don't have the technology to carry it out."

27
Current and Resistance

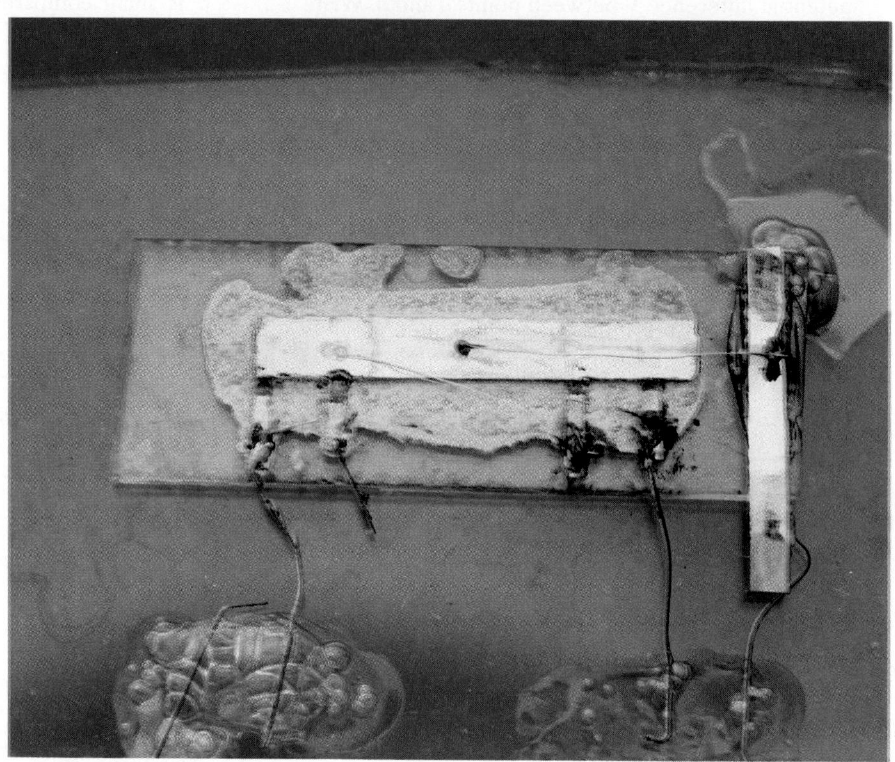

The first integrated circuit, tested on September 12, 1958. (Courtesy of Texas Instruments)

Thus far our discussion of electrical phenomena has been confined to charges at rest, or electrostatics. We shall now consider situations involving electric charges in motion. The term *electric current,* or simply *current,* is used to describe the rate of flow of charge through some region of space. Most practical applications of electricity deal with electric currents. For example, the battery of a flashlight supplies current to the filament of the bulb when the switch is turned on. A variety of home appliances operate on alternating current. In these common situations, the flow of charge takes place in a conductor, such as a copper wire. However, it is possible for currents to exist outside of a conductor. For instance, a beam of electrons in a TV picture tube constitutes a current.

In this chapter we shall first discuss the battery, one source of continuous current, followed by a definition of current and current density. A microscopic description of current will be given, and some of the factors that contribute to the resistance to the flow of charge in conductors will be discussed. Mechanisms responsible for the electrical resistance of various materials depend on the composition of the material and on temperature. A classical model is used to describe electrical conduction in metals, and some of the limitations of this model are pointed out.

27.1 THE BATTERY

Although electrical phenomena were known before 1800, electrical machines of that era were limited to devices that could produce static charge and large potential differences by means of friction. Such machines were capable of producing large sparks, but were of little practical value.

The electric battery, invented in 1800 by Alessandro Volta (1745–1827), was one of the most important practical discoveries in science. This invention represented the basis for a wide range of subsequent developments in electrical technology.

It is interesting to describe briefly some important events that led to Volta's invention. In 1786, Luigi Galvani (1737–1798) found that when a copper hook was inserted into the spinal cord of a frog, which in turn was hung from an iron railing, the leg muscles contracted. Galvani observed the same effect when other dissimilar metals were used. In reporting this unusual phenomenon, he proposed that the source of the charge was the muscle or nerve of the frog. Hence, he termed the source "animal electricity."

After hearing of Galvani's results, Volta proceeded to confirm and expand these experiments. He then offered the idea that the source of the charge was not the animal, but the contact between the two dissimilar metals, iron and copper. During his investigations, Volta recognized that the contact between the two metals required a moist conductor (such as the frog's muscle) to obtain a sizable effect. He eventually proved his point conclusively by showing that the effect occurred (although weakly) when the frog muscle was replaced by an inorganic substance. Further, he showed that certain pairs of metals produced a larger effect than others.

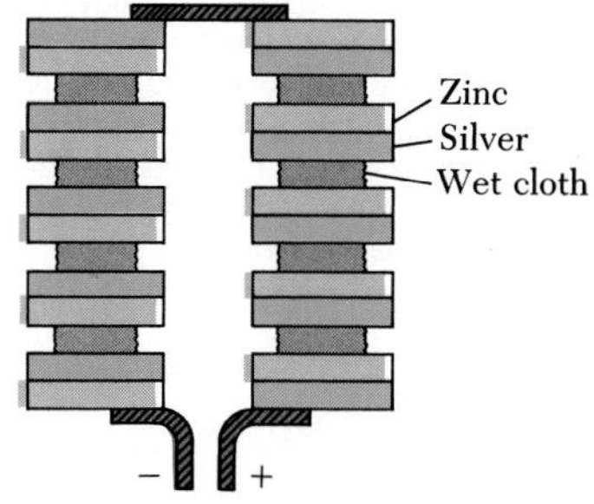

Figure 27.1 Diagram of Volta's original pile. The cloth separating the plates is soaked in a salt solution. A potential difference is produced between the two end plates.

Volta then proceeded to invent a continuous source of electricity, the first battery. His original device, called the Voltaic pile, consisted of alternate disks of silver and zinc, as in Figure 27.1. Adjacent layers were separated by a cloth that had been soaked in a salt solution or dilute acid. The layered structure provided a continuous potential difference between the two ends, with an excess of positive charge at the silver end and an equal amount of negative charge at the zinc end. In effect, the pile was an energy converter, where internal chemical energy was converted into electric potential energy. Although this battery produced small potential differences compared to those produced by friction machines, it was able to provide a large electric charge, and hence proved to be of great practical importance. These early sources were very important for experiments because they provided a nearly constant potential difference.

There are many different kinds of batteries in use today. One of the most common types is the ordinary flashlight battery. These batteries are produced in a variety of shapes and sizes, but they all work in basically the same way. Figure 27.2 is a diagram of the interior of such a battery. In this particular battery, often referred to as a dry cell, the zinc case serves as the negative terminal, while the carbon rod down its center serves as the positive terminal. The space between the two terminals contains a paste-like mixture of manganese dioxide, ammonium chloride, and carbon.

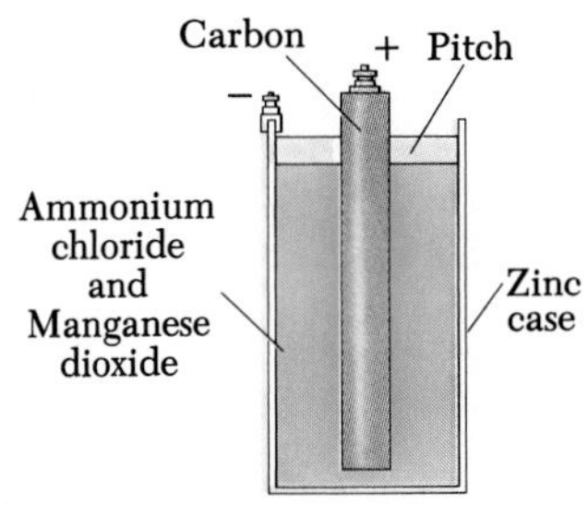

Figure 27.2 Cross-sectional view of a dry cell.

When these materials are assembled in this fashion, two chemical reactions take place; one occurs at the zinc case, the other at the manganese dioxide layer surrounding the carbon rod. Positive charged zinc ions (Zn^{2+}) leave the case and enter the ammonium chloride paste, where they combine with chloride ions (Cl^-). (The chloride ions are present because a small percentage of the ammonium chloride dissociates, leaving some free chloride ions in the solution.) As each zinc ion is removed from the case, it leaves behind two electrons. As additional zinc ions leave the case, more electrons accumulate, leaving the zinc case with a net negative charge.

When a chloride ion breaks free from the ammonium chloride molecule, the remnant portion of the molecule becomes singly ionized. This positively charged ion is neutralized by the manganese dioxide, which supplies the needed electrons. As a result, the carbon rod surrounded by its manganese dioxide layer ends up with a net positive charge.

These chemical reactions and thus the charge separation do not continue without limit. The zinc case ultimately achieves such a strong negative charge that the zinc ions can no longer escape. A similar charge saturation occurs at the carbon rod.

27.2 ELECTRIC CURRENT

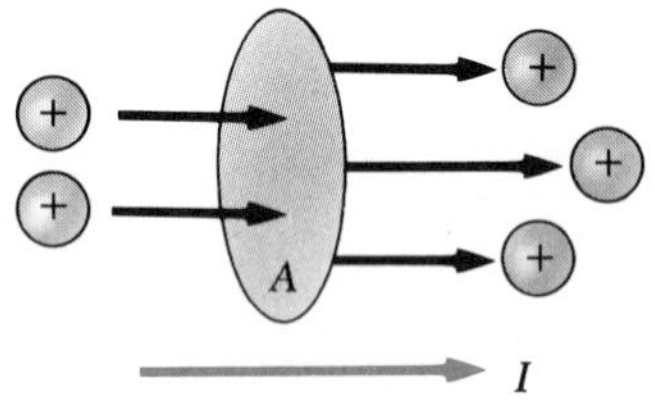

Whenever electric charges of like sign move, a *current* is said to exist. To define current more precisely, suppose the charges are moving perpendicular to a surface of area A as in Figure 27.3. This area could be the cross-sectional area of a wire, for example. The **current** is *the rate at which charge flows through this surface.* If ΔQ is the amount of charge that passes through this area in a time interval Δt, the **average current, I_{av},** is equal to the ratio of the charge to the time interval:

Figure 27.3 Charges in motion through an area A. The time rate of flow of charge through the area is defined as the current I. The direction of the current is in the direction in which positive charge would flow if free to do so.

$$I_{av} = \frac{\Delta Q}{\Delta t} \tag{27.1}$$

If the rate at which charge flows varies in time, the current also varies in time and we define the **instantaneous current,** I, as the differential limit of the expression above:

Electric current

$$I \equiv \frac{dQ}{dt} \tag{27.2}$$

The SI unit of current is the **ampere** (A), where

The direction of the current

$$1\ \text{A} = 1\ \text{C/s} \tag{27.3}$$

That is, 1 A of current is equivalent to 1 C of charge passing through the surface in 1 s. In practice, smaller units of current are often used, such as the milliampere (1 mA $= 10^{-3}$ A) and the microampere (1 μA $= 10^{-6}$ A).

When charges flow through the surface in Figure 27.3, they can be positive, negative, or both. *It is conventional to choose the direction of the current to be in the direction of flow of positive charge.* In a conductor such as copper, the current is due to the motion of the negatively charged electrons. Therefore, when we speak of current in an ordinary conductor, such as a copper wire, *the direction of the current will be opposite the direction of flow of electrons.* On the other hand, if one considers a beam of positively charged protons in an accelerator, the current is in the direction of motion of the protons. In some cases, the current is the result of the flow of both positive and negative charges. This occurs, for example, in semiconductors and electrolytes. It is common to refer to a moving charge (whether it is positive or negative) as a mobile *charge carrier.* For example, the charge carriers in a metal are electrons.

It is instructive to relate current to the motion of the charged particles. To illustrate this point, consider the current in a conductor of cross-sectional area A (Fig. 27.4). The volume of an element of the conductor of length Δx (the shaded region in Fig. 27.4) is $A\,\Delta x$. If n represents the number of mobile charge carriers per unit volume, then the number of mobile charge carriers in the volume element is given by $nA\,\Delta x$. Therefore, the charge ΔQ in this element is given by

$$\Delta Q = \text{number of charges} \times \text{charge per particle} = (nA\,\Delta x)q$$

where q is the charge on each particle. If the charge carriers move with a speed v_d, the distance they move in a time Δt is given by $\Delta x = v_d\,\Delta t$. Therefore, we can write ΔQ in the form

$$\Delta Q = (nAv_d\,\Delta t)q$$

If we divide both sides of this equation by Δt, we see that the current in the conductor is given by

$$I = \frac{\Delta Q}{\Delta t} = nqv_dA \qquad (27.4)$$

Current in a conductor

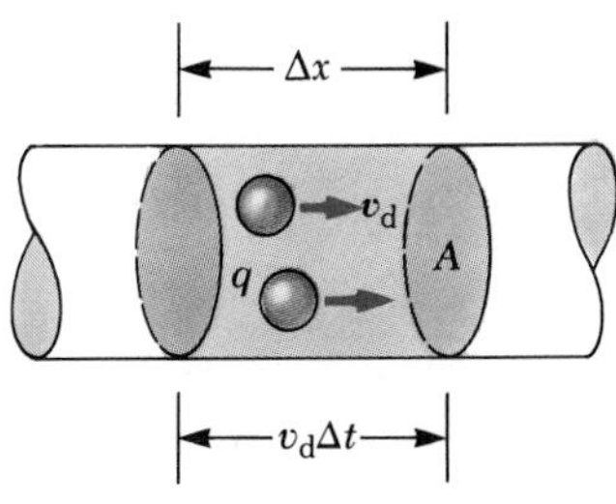

Figure 27.4 A section of a uniform conductor of cross-sectional area A. The charge carriers move with a speed v_d, and the distance they travel in a time Δt is given by $\Delta x = v_d\,\Delta t$. The number of mobile charge carriers in the section of length Δx is given by $nAv_d\,\Delta t$, where n is the number of mobile carriers per unit volume.

The velocity of the charge carriers, v_d, is actually an average velocity and is called the **drift velocity.** To understand the meaning of drift velocity, consider a conductor in which the charge carriers are free electrons. In an isolated conductor, these electrons undergo random motion similar to that of gas molecules. When a potential difference is applied across the conductor (say, by means of a battery), an electric field is set up in the conductor, which creates an electric force on the electrons and hence a current. In reality, the electrons do not simply move in straight lines along the conductor. Instead, they undergo repeated collisions with the metal atoms, which results in a complicated zigzag motion (Fig. 27.5). The energy transferred from the electrons to the metal atoms causes an increase in the vibrational energy of the atoms and a corresponding increase in the temperature of the conductor. However, despite the collisions, the electrons move slowly along the conductor (in a direction opposite $\mathbf{E}$) with an average velocity called the drift velocity, $\mathbf{v}_d$. The field does work on the electrons that exceeds the average loss due to collisions, which results in a net current. As we shall see in an example that follows, drift velocities are *much* smaller than the average speed between collisions. We shall discuss this model in more detail in Section 27.6. One can think of the collisions of the electrons within a conductor as being an effective internal friction (or drag force), similar to that experienced by the molecules of a liquid flowing through a pipe stuffed with steel wool.

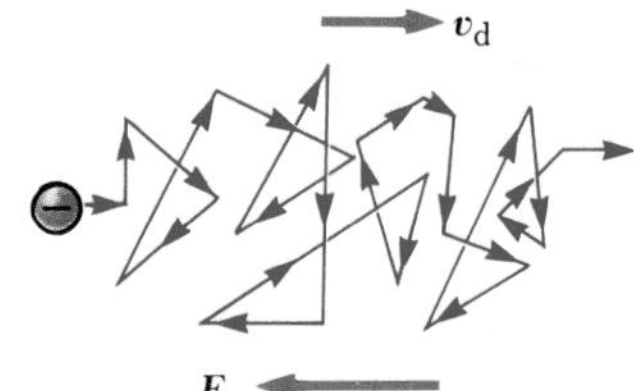

Figure 27.5 A schematic representation of the zigzag motion of a charge carrier in a conductor. The changes in direction are due to collisions with atoms in the conductor. Note that the net motion of electrons is opposite the direction of the electric field.

The following quotation is an interesting and amusing description by W.F.G. Swann of electronic conduction in telephone cables.[1]

> Think of the cables which carry the telephone current in the form of electrons. In the absence of the current the electrons are moving in all directions. As many are

[1] W.F.G. Swann, *Physics Today,* June 1951, p. 9.

moving from left to right as are moving from right to left; and the nothingness which is there is composed of two equal and opposite halves, about a million million amperes per square centimeter in one direction, and a million million amperes per square centimeter in the other direction. The telephone current constitutes an upsetting of the balance to the extent of one hundredth of a millionth of an ampere per square centimeter, or about one part in a hundred million million million. Then if this one part in a hundred million million million is at fault by one part in a thousand, we ring up the telephone company and complain that the quality of the speech is faulty.

EXAMPLE 27.1 The Drift Velocity in a Copper Wire
A copper wire of cross-sectional area $3 \times 10^{-6}\ \text{m}^2$ carries a current of 10 A. Find the drift velocity of the electrons in this wire. The density of copper is $8.95\ \text{g/cm}^3$.

Solution From the periodic table of the elements, we find that the atomic weight of copper is 63.5 g/mole. Recall that one atomic mass of any substance contains Avogadro's number of atoms, 6.02×10^{23} atoms. Knowing the density of copper enables us to calculate the volume occupied by 63.5 g of copper:

$$V = \frac{m}{\rho} = \frac{63.5\ \text{g}}{8.95\ \text{g/cm}^3} = 7.09\ \text{cm}^3$$

If we now assume that each copper atom contributes one free electron to the body of the material, we have

$$n = \frac{6.02 \times 10^{23}\ \text{electrons}}{7.09\ \text{cm}^3}$$

$$= 8.48 \times 10^{22}\ \text{electrons/cm}^3$$

$$= \left(8.48 \times 10^{22}\ \frac{\text{electrons}}{\text{cm}^3}\right)\left(10^6\ \frac{\text{cm}^3}{\text{m}^3}\right)$$

$$= 8.48 \times 10^{28}\ \text{electrons/m}^3$$

From Equation 27.4, we find that the drift velocity is

$$v_d = \frac{I}{nqA}$$

$$= \frac{10\ \text{C/s}}{(8.48 \times 10^{28}\ \text{m}^{-3}/\text{m}^3)(1.6 \times 10^{-19}\ \text{C})(3 \times 10^{-6}\ \text{m}^2)}$$

$$= 2.46 \times 10^{-4}\ \text{m/s}$$

Example 27.1 shows that typical drift velocities are very small. In fact, the drift velocity is much smaller than the average velocity between collisions. For instance, electrons traveling with this velocity would take about 68 min to travel 1 m! In view of this low speed, you might wonder why a light turns on almost instantaneously when a switch is thrown. This can be explained by considering the flow of water through a pipe. If a drop of water is forced in one end of a pipe that is already filled with water, a drop must be pushed out the other end of the pipe. While it may take individual drops of water a long time to make it through the pipe, a flow initiated at one end produces a similar flow at the other end very quickly. In a conductor, the electric field that drives the free electrons travels through the conductor with a speed close to that of light. Thus, when you flip a light switch, the message for the electrons to start moving through the wire (the electric field) reaches them at a speed of the order of 10^8 m/s.

27.3 RESISTANCE AND OHM'S LAW

Earlier, we found that there can be no electric field inside a conductor. However, this statement is true *only* if the conductor is in static equilibrium. The purpose of this section is to describe what happens when the charges are allowed to move in the conductor.

Charges move in a conductor to produce a current under the action of an electric field inside the conductor. An electric field can exist in the conductor in this case since we are dealing with charges in motion, a *nonelectrostatic* situation. This is in contrast with the situation in which a conductor in *electrostatic equilibrium* (where the charges are at rest) can have no electric field inside.

Consider a conductor of cross-sectional area A carrying a current I. The **current density** J in the conductor is defined to be the current per unit area. Since $I = nqv_dA$, the current density is given by

$$J \equiv \frac{I}{A} = nqv_d \tag{27.5}$$

Current density

where J has SI units of A/m^2. In general, the current density is a *vector quantity*. That is,

$$\mathbf{J} = nq\mathbf{v}_d \tag{27.6}$$

From this definition, we see once again that the current density is in the direction of motion of the charges for positive charge carriers and opposite the direction of motion for negative charge carriers.

A current density $\mathbf{J}$ *and an electric field* $\mathbf{E}$ *are established in a conductor when a potential difference is maintained across the conductor.* If the potential difference is constant, the current in the conductor will also be constant. Very often, the current density in a conductor is proportional to the electric field in the conductor. That is,

$$\mathbf{J} = \sigma\mathbf{E} \tag{27.7}$$

Ohm's law

where the constant of proportionality σ is called the **conductivity** of the conductor.[2] Materials that obey Equation 27.7 are said to follow Ohm's law, named after Georg Simon Ohm (1787–1854). More specifically,

> Ohm's law states that for many materials (including most metals), the ratio of the current density and electric field is a constant, σ, which is independent of the electric field producing the current.

Materials that obey Ohm's law, and hence demonstrate this linear behavior between $\mathbf{E}$ and $\mathbf{J}$, are said to be *ohmic.* The electrical behavior of most materials is quite linear for *very small changes* in the current. Experimentally, one finds that not all materials have this property. Materials that do not obey Ohm's law are said to be *nonohmic.* Ohm's law is *not* a fundamental law of nature, but an empirical relationship valid only for certain materials.

A form of Ohm's law that is more directly useful in practical applications can be obtained by considering a segment of a straight wire of cross-sectional area A and length ℓ, as in Figure 27.6. A potential difference $V_b - V_a$ is maintained across the wire, creating an electric field in the wire and a current. If the electric field in the wire is assumed to be uniform, the potential difference $V = V_b - V_a$ is related to the electric field through the relationship[3]

$$V = E\ell$$

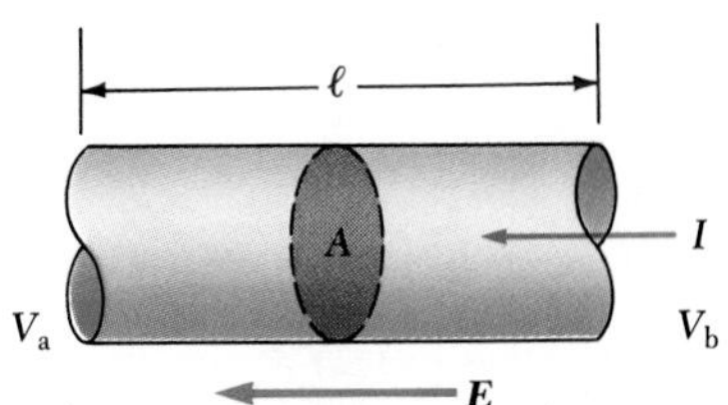

Figure 27.6 A uniform conductor of length ℓ and cross-sectional area A. A potential difference $V_b - V_a$ maintained across the conductor sets up an electric field $\mathbf{E}$ in the conductor, and this field produces a current I.

[2] Do not confuse the conductivity σ with the surface charge density, for which the same symbol is used.

[3] This result follows from the definition of potential difference:

$$V_b - V_a = -\int_a^b \mathbf{E} \cdot d\mathbf{s} = E\int_0^\ell dx = E\ell$$

Therefore, we can express the magnitude of the current density in the wire as

$$J = \sigma E = \sigma \frac{V}{\ell}$$

Since $J = I/A$, the potential difference can be written

$$V = \frac{\ell}{\sigma} J = \left(\frac{\ell}{\sigma A}\right) I$$

The quantity $\ell/\sigma A$ is called the **resistance** R of the conductor:

Resistance of a conductor

$$R = \frac{\ell}{\sigma A} = \frac{V}{I} \tag{27.8}$$

From this result we see that resistance has SI units of volts per ampere. One volt per ampere is defined to be one ohm (Ω):

$$1\ \Omega \equiv 1\ \text{V/A}$$

That is, if a potential difference of 1 V across a conductor causes a current of 1 A, the resistance of the conductor is 1 Ω. For example, if an electrical appliance connected to a 120-V source carries a current of 6A, its resistance is 20 Ω.

The inverse of the conductivity of a material is called the **resistivity** ρ:

Resistivity

$$\rho \equiv \frac{1}{\sigma} \tag{27.9}$$

Using this definition and Equation 27.8, the resistance can be expressed as

Resistance of a uniform conductor

$$R = \rho \frac{\ell}{A} \tag{27.10}$$

where ρ has the units ohm-meters ($\Omega \cdot$m). (The symbol ρ for resistivity should not be confused with the same symbol used earlier in the text for mass density or charge density.) Every ohmic material has a characteristic resistivity, a parameter that depends on the properties of the material and on temperature. On the other hand, as you can see from Equation 27.10, the resistance of a substance depends on simple geometry as well as on the resistivity of the substance. Good electrical conductors have very low resistivity (or high conductivity), and good insulators have very high resistivity (low conductivity). Table 27.1 gives the resistivities of a variety of materials at 20°C.

Equation 27.10 shows that the resistance of a given cylindrical conductor is proportional to its length and inversely proportional to its cross-sectional area. Therefore, if the length of a wire is doubled, its resistance doubles. Furthermore, if its cross-sectional area is doubled, its resistance drops by one-half. The situation is analogous to the flow of a liquid through a pipe. As the length of the pipe is increased, the resistance to liquid flow increases. As its cross-sectional area is increased, the pipe can more readily transport liquid.

All electric appliances such as toasters, heaters, and light bulbs have a fixed resistance. Most electric circuits make use of devices called **resistors** to control the current level in the various parts of the circuit. Two common types of resistors are the "composition" resistor containing carbon, which is a semi-

TABLE 27.1 Resistivities and Temperature Coefficients of Resistivity for Various Materials

Material	Resistivity[a] ($\Omega \cdot m$)	Temperature Coefficient α [(C°)$^{-1}$]
Silver	1.59×10^{-8}	3.8×10^{-3}
Copper	1.7×10^{-8}	3.9×10^{-3}
Gold	2.44×10^{-8}	3.4×10^{-3}
Aluminum	2.82×10^{-8}	3.9×10^{-3}
Tungsten	5.6×10^{-8}	4.5×10^{-3}
Iron	10×10^{-8}	5.0×10^{-3}
Platinum	11×10^{-8}	3.92×10^{-3}
Lead	22×10^{-8}	3.9×10^{-3}
Nichrome[b]	150×10^{-8}	0.4×10^{-3}
Carbon	3.5×10^{-5}	-0.5×10^{-3}
Germanium	0.46	-48×10^{-3}
Silicon	640	-75×10^{-3}
Glass	$10^{10} - 10^{14}$	
Hard rubber	$\approx 10^{13}$	
Sulfur	10^{15}	
Quartz (fused)	75×10^{16}	

[a] All values at 20°C.
[b] A nickel-chromium alloy commonly used in heating elements.

conductor, and the "wire-wound" resistor, which consists of a coil of wire. Resistors are normally color-coded to give their values in ohms, as shown in Figure 27.7. Table 27.2 will enable you to translate from the color code to a specific value of resistance.

Ohmic materials, such as copper, have a linear current-voltage relationship over a large range of applied voltage (Fig. 27.8a). The slope of the I versus V curve in the linear region yields a value for R. Nonohmic materials have a nonlinear current-voltage relationship. One common semiconducting device that has nonlinear I versus V characteristics is the diode (Fig. 27.8b). The

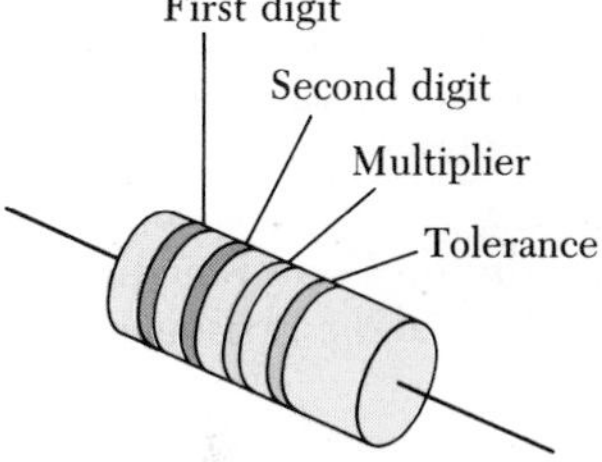

Figure 27.7 The colored bands on a resistor represent a code for determining the value of its resistance. The first two colors give the first two digits in the resistance value. The third color represents the power of ten for the multiplier of the resistance value. The last color is the tolerance of the resistance value. As an example, if the four colors are orange, blue, yellow, and gold, the resistance value is $36 \times 10^4\ \Omega$ or 360 kΩ, with a tolerance value of 18 kΩ (5%).

TABLE 27.2 Color Code for Resistors

Color	Number	Multiplier	Tolerance (%)
Black	0	1	
Brown	1	10^1	
Red	2	10^2	
Orange	3	10^3	
Yellow	4	10^4	
Green	5	10^5	
Blue	6	10^6	
Violet	7	10^7	
Gray	8	10^8	
White	9	10^9	
Gold		10^{-1}	5%
Silver		10^{-2}	10%
Colorless			20%

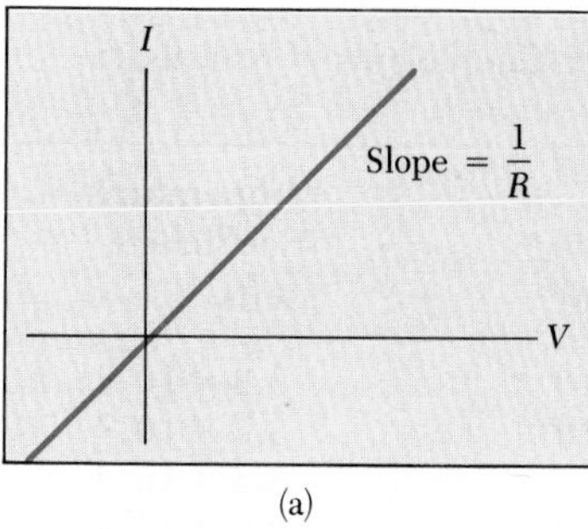

(a)

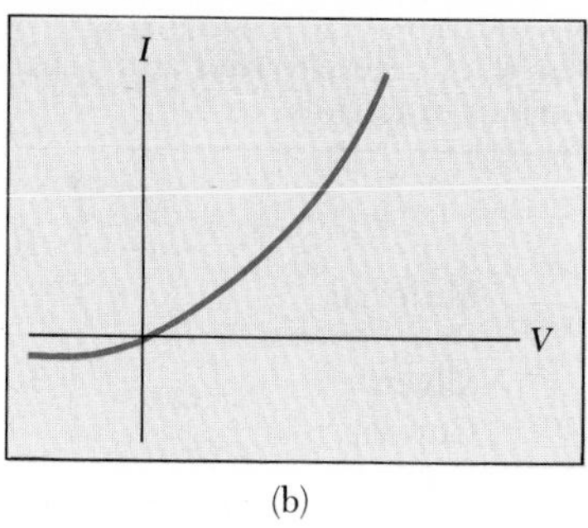

(b)

Figure 27.8 (a) The current-voltage curve for an ohmic material. The curve is linear, and the slope gives the resistance of the conductor. (b) A nonlinear current-voltage curve for a semiconducting diode. This device does not obey Ohm's law.

effective resistance of this device (inversely proportional to the slope of its I versus V curve) is small for currents in one direction (positive V) and large for currents in the reverse direction (negative V). In fact, most modern electronic devices, such as transistors, have nonlinear current-voltage relationships; their proper operation depends on the particular way in which they violate Ohm's law.

EXAMPLE 27.2 The Resistance of a Conductor
Calculate the resistance of a piece of aluminum that is 10 cm long and has a cross-sectional area of 10^{-4} m². Repeat the calculation for a piece of glass of resistivity $10^{10}\ \Omega \cdot$ m.

Solution From Equation 27.10 and Table 27.1, we can calculate the resistance of the aluminum bar:

$$R = \rho \frac{L}{A} = (2.82 \times 10^{-8}\ \Omega \cdot \text{m})\left(\frac{0.1\ \text{m}}{10^{-4}\ \text{m}^2}\right)$$

$$= \boxed{2.82 \times 10^{-5}\ \Omega}$$

Similarly, for glass we find

$$R = \rho \frac{L}{A} = (10^{10}\ \Omega \cdot \text{m})\left(\frac{0.1\ \text{m}}{10^{-4}\ \text{m}^2}\right) = \boxed{10^{13}\ \Omega}$$

As you might expect, aluminum has a much lower resistance than glass. This is the reason that aluminum is a good conductor and glass is a poor conductor.

EXAMPLE 27.3 The Resistance of Nichrome Wire
(a) Calculate the resistance per unit length of a 22-gauge nichrome wire of radius 0.321 mm.

Solution The cross-sectional area of this wire is

$$A = \pi r^2 = \pi(0.321 \times 10^{-3}\ \text{m})^2 = 3.24 \times 10^{-7}\ \text{m}^2$$

The resistivity of nichrome is $1.5 \times 10^{-6}\ \Omega \cdot$ m (Table 27.1). Thus, we can use Equation 27.10 to find the resistance per unit length:

$$\frac{R}{\ell} = \frac{\rho}{A} = \frac{1.5 \times 10^{-6}\ \Omega \cdot \text{m}}{3.24 \times 10^{-7}\ \text{m}^2} = \boxed{4.6\ \Omega/\text{m}}$$

(b) If a potential difference of 10 V is maintained across a 1-m length of the nichrome wire, what is the current in the wire?

Solution Since a 1-m length of this wire has a resistance of 4.6 Ω, Ohm's law gives

$$I = \frac{V}{R} = \frac{10\ \text{V}}{4.6\ \Omega} = \boxed{2.2\ \text{A}}$$

Note that the resistance of the nichrome wire is about 100 times larger than that of the copper wire. Therefore, a copper wire of the same radius would have a resistance per unit length of only 0.052 Ω/m. A 1-m length of copper wire of the same radius would carry the same current (2.2 A) with an applied voltage of only 0.11 V.

Because of its high resistivity and its resistance to oxidation, nichrome is often used for heating elements in toasters, irons, and electric heaters.

Exercise 1 What is the resistance of a 6-m length of 22-gauge nichrome wire? How much current does it carry when connected to a 120-V source?
Answer 28 Ω, 4.3 A.

Exercise 2 Calculate the current density and electric field in the wire assuming that it carries a current of 2.2 A.
Answer 6.7×10^6 A/m^2; 10 N/C.

EXAMPLE 27.4 The Resistance of a Coaxial Tube □
The gap between a pair of coaxial tubes is completely filled with silicon as in Figure 27.9a. The inner radius of the tube is $a = 0.500$ cm, the outer radius, $b = 1.75$ cm, and its length, $L = 15.0$ cm. Calculate the total resistance of the silicon when measured between the inner and outer tubes.

Solution The resistivity of copper is small compared with that of silicon, so its resistance can be neglected. In this type of problem, we must divide the conductor into elements of infinitesimal thickness over which the area may be considered constant. We can start by using the differential form of Equation 27.10, which is $dR = \rho\, d\ell/A$, where dR is the resistance of a section of the conductor of thickness $d\ell$ and area A. In this example, we take as our element a hollow cylinder of thickness dr and length L as in Figure 27.9b. Any current that passes between the inner and outer tubes must pass radially through such elements, and the area through which it passes is $A = 2\pi rL$. (This would be the surface area of our hollow cylinder neglecting the area of its ends.) Hence, we can write the resistance of our hollow cylinder as

$$dR = \frac{\rho}{2\pi rL}\, dr$$

Since we wish to know the total resistance of the silicon, we must integrate this expression over dr from $r = a$ to $r = b$. This gives

$$R = \int_a^b dR = \frac{\rho}{2\pi L}\int_a^b \frac{dr}{r} = \frac{\rho}{2\pi L}\ln\left(\frac{b}{a}\right)$$

Substituting in the values given, and using $\rho = 640\ \Omega \cdot \text{m}$ for silicon gives

$$R = \frac{640\ \Omega \cdot \text{m}}{2\pi(0.150\ \text{m})}\ln\left(\frac{1.75\ \text{cm}}{0.500\ \text{cm}}\right) = 851\ \Omega$$

Exercise 3 If a potential difference of 12.0 V is applied between the inner and outer copper tube, calculate the total current that passes between them.
Answer 14.1 mA

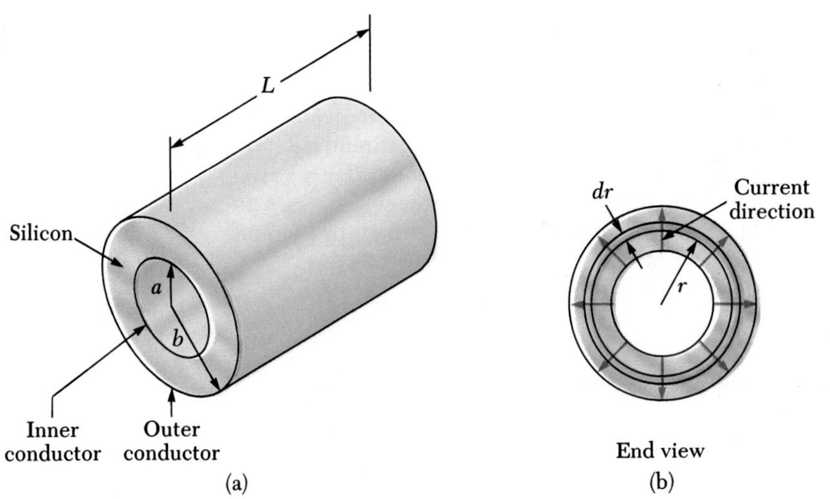

Figure 27.9 (Example 27.4).

27.4 THE RESISTIVITY OF DIFFERENT CONDUCTORS

The resistivity of a conductor depends on a number of factors, one of which is temperature. For most metals, resistivity increases with increasing temperature. The resistivity of a conductor varies in an approximately linear fashion with temperature over a limited temperature range according to the expression

$$\rho = \rho_0[1 + \alpha(T - T_0)] \qquad (27.11)$$

Variation of ρ with temperature

where ρ is the resistivity at some temperature T (in °C), ρ_o is the resistivity at some reference temperature T_o (usually taken to be 20°C), and α is called the **temperature coefficient of resistivity.** From Equation 27.11, we see that the temperature coefficient of resistivity can also be expressed as

Temperature coefficient of resistivity

$$\alpha = \frac{1}{\rho_o}\frac{\Delta\rho}{\Delta T} \tag{27.12}$$

where $\Delta\rho = \rho - \rho_o$ is the change in resistivity in the temperature interval $\Delta T = T - T_o$.

The resistivities and temperature coefficients for various materials are given in Table 27.1. Note the enormous range in resistivities, from very low values for good conductors, such as copper and silver, to very high values for good insulators, such as glass and rubber. An ideal, or "perfect," conductor would have zero resistivity, and an ideal insulator would have infinite resistivity.

Since the resistance of a conductor is proportional to the resistivity according to Equation 27.10, the temperature variation of the resistance can be written

$$R = R_o[1 + \alpha(T - T_o)] \tag{27.13}$$

Precise temperature measurements are often made using this property, as shown in the following example.

EXAMPLE 27.5 A Platinum Resistance Thermometer
A resistance thermometer made from platinum has a resistance of 50.0 Ω at 20°C. When immersed in a vessel containing melting indium, its resistance increases to 76.8 Ω. From this information, find the melting point of indium. For platinum, $\alpha = 3.92 \times 10^{-3}\ (\text{C}°)^{-1}$.

Solution Using Equation 27.13 and solving for ΔT, we get

$$\Delta T = \frac{R - R_o}{\alpha R_o} = \frac{76.8\ \Omega - 50.0\ \Omega}{[3.92 \times 10^{-3}\ (\text{C}°)^{-1}](50.0\ \Omega)}$$

$$= 137\ \text{C}°$$

Since $\Delta T = T - T_o$ and $T_o = 20°\text{C}$, we find that $T = 157°\text{C}$.

As mentioned above, many ohmic materials have resistivities that increase linearly with increasing temperature, as shown in Figure 27.10. In reality, however, there is always a nonlinear region at very low temperatures, and the resistivity usually approaches some finite value near absolute zero (see magnified insert in Fig. 27.10). This residual resistivity near absolute zero is due primarily to collisions of electrons with impurities and imperfections in the metal. In contrast, the high-temperature resistivity (the linear region) is dominated by collisions of electrons with the metal atoms. We shall describe this process in more detail in Section 27.6.

Semiconductors, such as silicon and germanium, have intermediate values of resistivity. The resistivity of semiconductors generally decreases with increasing temperature, corresponding to a negative temperature coefficient of resistivity (Fig. 27.11). This is due to the increase in the density of charge carriers at the higher temperatures. Since the charge carriers in a

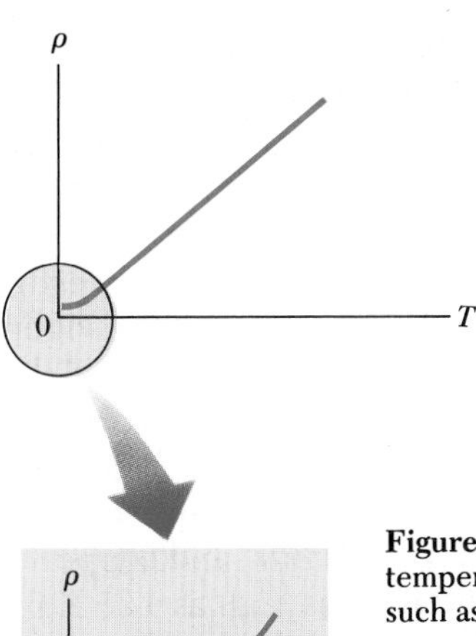

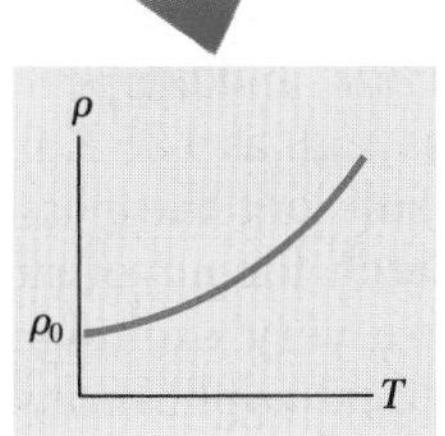

Figure 27.10 Resistivity versus temperature for a normal metal, such as copper. The curve is linear over a wide range of temperatures, and ρ increases with increasing temperature. As T approaches absolute zero (insert), the resistivity approaches a finite value ρ_0.

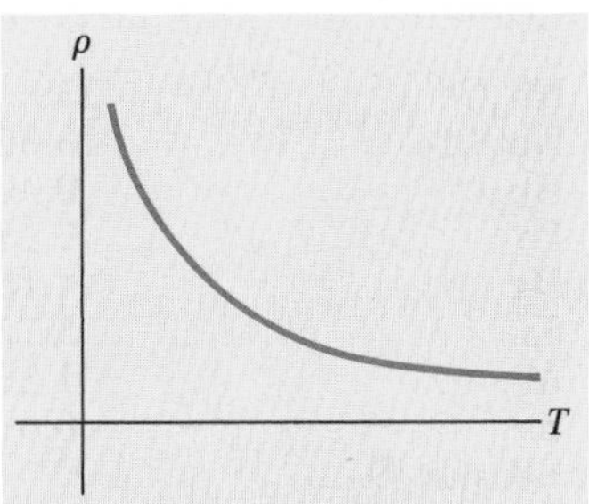

Figure 27.11 Resistivity versus temperature for a pure semiconductor, such as silicon or germanium.

semiconductor are often associated with impurity atoms, the resistivity is very sensitive to the type and concentration of such impurities. The **thermistor** is a semiconducting thermometer that makes use of the large changes in its resistivity with temperature. We shall return to the study of semiconductors in the extended version of this text, Chapter 43.

27.5 SUPERCONDUCTORS

There is a class of metals and compounds whose resistance goes virtually to *zero* below a certain temperature, T_c, called the *critical temperature*. These materials are known as **superconductors.** The resistance-temperature graph for a superconductor follows that of a normal metal at temperatures above T_c (Fig. 27.12). When the temperature is at or below T_c, the resistivity drops suddenly to zero. This phenomenon was discovered in 1911 by the Dutch physicist H. Kamerlingh-Onnes when he was working with mercury, which is a superconductor below 4.2 K. Recent measurements have shown that the resistivities of superconductors below T_c are less than $4 \times 10^{-25}\ \Omega \cdot \text{m}$, which is around 10^{17} times smaller than the resistivity of copper and considered to be *zero* in practice.

Today there are thousands of known superconductors. Such common metals as aluminum, tin, lead, zinc, and indium are superconductors. Table 27.3 lists the critical temperatures of several superconductors. The value of T_c is sensitive to chemical composition, pressure, and crystalline structure. It is interesting to note that copper, silver, and gold, which are excellent conductors, do not exhibit superconductivity.

One of the truly remarkable features of superconductors is the fact that once a current is set up in them, it will persist *without any applied voltage* (since $R = 0$). In fact, steady currents have been observed to persist in superconducting loops for several years with no apparent decay!

One of the most important recent developments in physics that has created much excitement in the scientific community has been the discovery of high temperature copper-oxide-based superconductors. The excitement

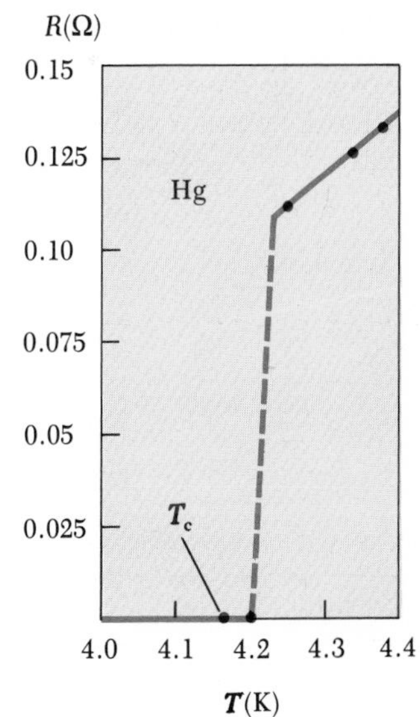

Figure 27.12 Resistance versus temperature for mercury. The graph follows that of a normal metal above the critical temperature, T_c. The resistance drops to zero at the critical temperature, which is 4.15 K for mercury.

TABLE 27.3 Critical Temperatures for Various Superconductors

Material	T_c (K)
Nb_3Ge	23.2
Nb_3Sn	18.05
Nb	9.46
Pb	7.18
Hg	4.15
Sn	3.72
Al	1.19
Zn	0.88
$YBa_2Cu_3O_{7-\delta}$	92
Bi-Sr-Ca-Cu-O	105
Tl-Ba-Ca-Cu-O	125

began with a 1986 publication by Georg Bednorz and K. Alex Müller, two scientists working at the IBM Zurich Research Laboratory in Switzerland, who reported evidence for superconductivity at a temperature near 30 K in an oxide of barium, lanthanum, and copper. Bednorz and Müller were awarded the Nobel Prize in 1987 for their remarkable and important discovery. Shortly thereafter, a new family of compounds was open for investigation, and research activity in the field of superconductivity proceeded vigorously. In early 1987, groups at the University of Alabama at Huntsville and the University of Houston announced the discovery of superconductivity at about 92 K in an oxide of yttrium, barium, and copper ($YBa_2Cu_3O_7$). Late in 1987, teams of scientists from Japan and the United States reported superconductivity at 105 K in an oxide of bismuth, strontium, calcium, and copper. Most recently, scientists have reported superconductivity as high as 125 K in an oxide containing thallium. At this point, one cannot rule out the possibility of room temperature superconductivity, and the search for novel superconducting materials continues. These developments are very exciting and important both for scientific reasons and because practical applications become more probable and widespread as the critical temperature is raised.

An important and useful application of superconductivity has been the construction of superconducting magnets in which the magnetic field strengths are about ten times greater than those of the best normal electromagnets. Such superconducting magnets are being considered as a means of storing energy. The idea of using superconducting power lines for transmitting power efficiently is also receiving some consideration. Modern superconducting electronic devices consisting of two thin-film superconductors separated by a thin insulator have been constructed. These devices include magnetometers (a magnetic-field measuring device) and various microwave devices.

We shall return to the subject of superconductivity in more depth in Chapter 44 of the extended version of this text. This same material is also available as a supplement to the standard version of the text.

27.6 A MODEL FOR ELECTRICAL CONDUCTION

In this section we describe a classical model of electrical conduction in metals. This model leads to Ohm's law and shows that resistivity can be related to the motion of electrons in metals.

Consider a conductor as a regular array of atoms containing free electrons (sometimes called *conduction* electrons). These electrons are free to move through the conductor and are approximately equal in number to the number of atoms. In the absence of an electric field, the free electrons move in random directions through the conductor with average speeds of the order of 10^6 m/s. (These speeds can be properly calculated only if a quantum mechanical description is used.) The situation is similar to the motion of gas molecules confined in a vessel. In fact, some scientists refer to conduction electrons in a metal as an *electron gas*. The conduction electrons are not totally "free" since they are confined to the interior of the conductor and undergo frequent collisions with the array of atoms. These collisions are the predominant mechanism for the resistivity of a metal at normal temperatures. Note that there is no current through a conductor in the absence of an electric field since the

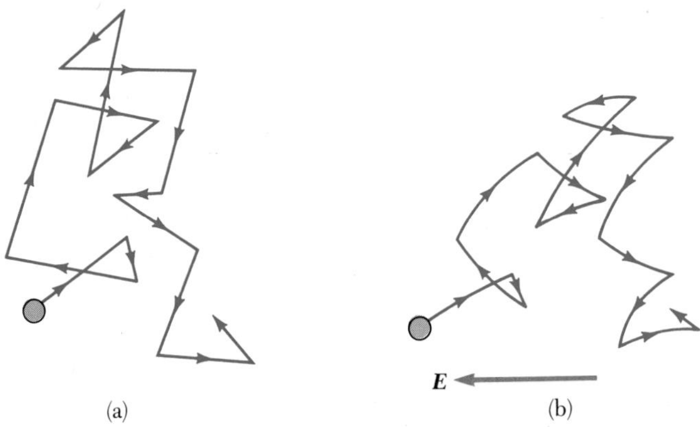

Figure 27.13 (a) A schematic diagram of the random motion of a charge carrier in a conductor in the absence of an electric field. Note that the drift velocity is zero. (b) The motion of a charge carrier in a conductor in the presence of an electric field. Note that the random motion is modified by the field, and the charge carrier has a drift velocity.

average velocity of the free electrons is zero. That is, on the average, just as many electrons move in one direction as in the opposite direction, and so there is no net flow of charge.

The situation is modified when an electric field is applied to the metal. In addition to the random thermal motion just described, the free electrons drift slowly in a direction opposite that of the electric field, with an average drift speed v_d, which is much smaller (typically 10^{-4} m/s) than the average speed between collisions (typically 10^6 m/s). Figure 27.13 provides a crude description of the motion of free electrons in a conductor. In the absence of an electric field, there is no net displacement after many collisions (Fig. 27.13a). An electric field $\mathbf{E}$ modifies the random motion and causes the electrons to drift in a direction opposite that of $\mathbf{E}$ (Fig. 27.13b). The slight curvature in the paths in Figure 27.13b results from the acceleration of the electrons between collisions, caused by the applied field. One mechanical system somewhat analogous to this situation is a ball rolling down a slightly inclined plane through an array of closely spaced, fixed pegs (Fig. 27.14). The ball represents a conduction electron, the pegs represent defects in the crystal lattice, and the component of the gravitational force along the incline represents the electric force $e\mathbf{E}$.

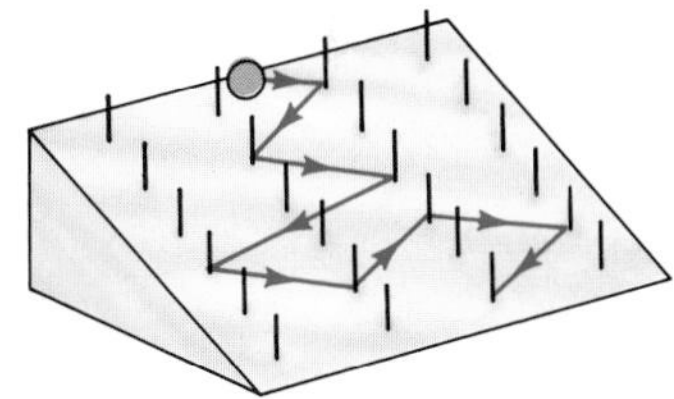

Figure 27.14 A mechanical system somewhat analogous to the motion of charge carriers in the presence of an electric field. The collisions of the ball with the pegs represent the resistance to the ball's motion down the incline.

In our model, we shall assume that the excess energy acquired by the electrons in the electric field is lost to the conductor in the collision process. The energy given up to the atoms in the collisions increases the vibrational energy of the atoms, causing the conductor to heat up. The model also assumes that the motion of an electron after a collision is independent of its motion before the collision.

We are now in a position to obtain an expression for the drift velocity. When a mobile charged particle of mass m and charge q is subjected to an electric field $\mathbf{E}$, it experiences a force $q\mathbf{E}$. Since $\mathbf{F} = m\mathbf{a}$, we conclude that the acceleration of the particle is given by

$$\mathbf{a} = \frac{q\mathbf{E}}{m} \tag{27.14}$$

This acceleration, which occurs for only a short time between collisions, enables the electron to acquire a small drift velocity. If t is the time since the last collision and $\boldsymbol{v}_0$ is the initial velocity, then the velocity of the electron after a time t is given by

$$\boldsymbol{v} = \boldsymbol{v}_0 + \boldsymbol{a}t = \boldsymbol{v}_0 + \frac{q\mathbf{E}}{m}t \qquad (27.15)$$

We now take the average value of $\boldsymbol{v}$ over all possible times t and all possible values of $\boldsymbol{v}_0$. If the initial velocities are assumed to be randomly distributed in space, we see that the average value of $\boldsymbol{v}_0$ is zero. The term $(q\mathbf{E}/m)t$ is the velocity added by the field at the end of one trip between atoms. If the electron starts with zero velocity, the average value of the second term of Equation 27.15 is $(q\mathbf{E}/m)\tau$, where τ is the *average time between collisions.* Because the average of $\boldsymbol{v}$ is equal to the drift velocity,[4] we have

Drift velocity

$$\boldsymbol{v}_d = \frac{q\mathbf{E}}{m}\tau \qquad (27.16)$$

Substituting this result into Equation 27.6, we find that the magnitude of the current density is given by

Current density

$$J = nqv_d = \frac{nq^2E}{m}\tau \qquad (27.17)$$

Comparing this expression with Ohm's law, $J = \sigma E$, we obtain the following relationships for the conductivity and resistivity:

Conductivity

$$\sigma = \frac{nq^2\tau}{m} \qquad (27.18)$$

Resistivity

$$\rho = \frac{1}{\sigma} = \frac{m}{nq^2\tau} \qquad (27.19)$$

The average time between collisions is related to the average distance between collisions ℓ (the mean free path) and the average thermal speed $\bar{v}$ through the expression[5]

$$\tau = \frac{\ell}{\bar{v}} \qquad (27.20)$$

According to this classical model, the conductivity and resistivity do not depend on the electric field. This feature is characteristic of a conductor obeying Ohm's law. The model shows that the conductivity can be calculated from a knowledge of the density of the charge carriers, their charge and mass, and the average time between collisions.

[4] Since the collision process is random, each collision event is *independent* of what happened earlier. This is analogous to the random process of throwing a die. The probability of rolling a particular number on one throw is independent of the result of the previous throw. On the average, it would take six throws to come up with that number, starting at any arbitrary time.

[5] Recall that the thermal speed is the speed a particle has as a consequence of the temperature of its surroundings (Chapter 20).

EXAMPLE 27.6 Electron Collisions in Copper
(a) Using the data and results from Example 27.1 and the classical model of electron conduction, estimate the average time between collisions for electrons in copper at 20°C.

From Equation 27.19 we see that

$$\tau = \frac{m}{nq^2\rho}$$

where $\rho = 1.7 \times 10^{-8}\ \Omega \cdot \text{m}$ for copper and the carrier density $n = 8.48 \times 10^{28}$ electrons/m^3 for the wire described in Example 27.1. Substitution of these values into the expression above gives

$$\tau = \frac{(9.11 \times 10^{-31}\ \text{kg})}{(8.48 \times 10^{28}\ \text{m}^{-3})(1.6 \times 10^{-19}\ \text{C})^2(1.7 \times 10^{-8}\ \Omega \cdot \text{m})}$$

$$= \boxed{2.5 \times 10^{-14}\ \text{s}}$$

(b) Assuming the mean thermal speed for free electrons in copper to be 1.6×10^6 m/s and using the result from (a), calculate the mean free path for electrons in copper.

$$\ell = \bar{v}\tau = (1.6 \times 10^6\ \text{m/s})(2.5 \times 10^{-14}\ \text{s})$$

$$= \boxed{4.0 \times 10^{-8}\ \text{m}}$$

which is equivalent to 40 nm (compared with atomic spacings of about 0.2 nm). Thus, although the time between collisions is very short, the electrons travel about 200 atomic distances before colliding with an atom.

Although this classical model of conduction is consistent with Ohm's law, it is not satisfactory for explaining some important phenomena. For example, classical calculations for $\bar{v}$ using the ideal-gas model are about a factor of 10 smaller than the true values. Furthermore, according to Equations 27.17 and 27.18, the temperature variation of the resistivity is predicted to vary as $\bar{v}$, which, according to an ideal-gas model (Chapter 21), is proportional to $\sqrt{T}$. This is in disagreement with the linear dependence of resistivity with temperature for pure metals (Fig. 27.8). It is possible to account for such observations only by using a quantum mechanical model, which we shall describe briefly.

According to quantum mechanics, electrons have wavelike properties. If the array of atoms is regularly spaced (that is, periodic), the wavelike character of the electrons makes it possible for them to move freely through the conductor, and a collision with an atom is unlikely. For an idealized conductor, there would be no collisions, the mean free path would be infinite, and the resistivity would be zero. Electron waves are scattered only if the atomic arrangement is irregular (not periodic) as a result of, for example, structural defects or impurities. At low temperatures, the resistivity of metals is dominated by scattering caused by collisions between the electrons and impurities. At high temperatures, the resistivity is dominated by scattering caused by collisions between the electrons and the atoms of the conductor, which are continuously displaced as a result of thermal agitation. The thermal motion of the atoms causes the structure to be irregular (compared with an atomic array at rest), thereby reducing the electron's mean free path.

27.7 ELECTRICAL ENERGY AND POWER

If a battery is used to establish an electric current in a conductor, there is a continuous transformation of chemical energy stored in the battery to kinetic energy of the charge carriers. This kinetic energy is quickly lost as a result of collisions between the charge carriers and the lattice ions, resulting in an increase in the temperature of the conductor. Therefore, we see that the chemical energy stored in the battery is continuously transformed into thermal energy.

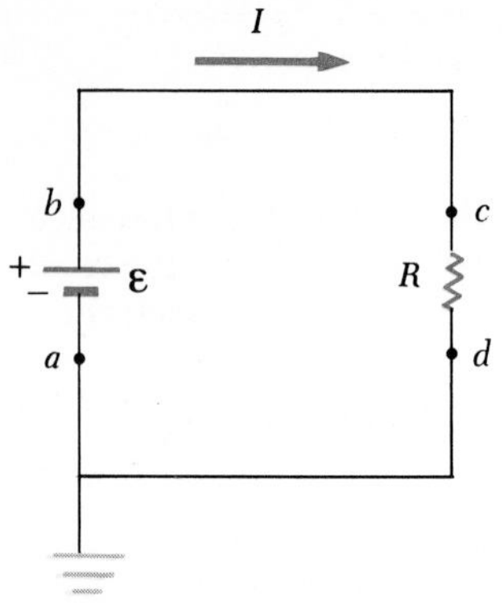

Figure 27.15 A circuit consisting of a battery of emf $\mathcal{E}$ and resistance R. Positive charge flows in the clockwise direction, from the negative to the positive terminal of the battery. Points a and d are grounded.

Consider a simple circuit consisting of a battery whose terminals are connected to a resistor R, as shown in Figure 27.15. The symbol ⊣⊢ is used to designate a battery (or any other direct current source), and resistors are designated by the symbol —\/\/\/—. The positive terminal of the battery is at the higher potential. Now imagine following a positive quantity of charge ΔQ moving around the circuit from point a through the battery and resistor and back to a. Point a is a reference point that is grounded (ground symbol ⏚), and its potential is taken to be zero. As the charge moves from a to b through the battery, its electrical potential energy *increases* by an amount $V\,\Delta Q$ (where V is the potential at b) while the chemical potential energy in the battery *decreases* by the same amount. (Recall from Chapter 25 that $\Delta U = q\,\Delta V$.) However, as the charge moves from c to d through the resistor, it *loses* this electrical potential energy as it undergoes collisions with atoms in the resistor, thereby producing thermal energy. Note that if we neglect the resistance of the interconnecting wires there is no loss in energy for paths bc and da. When the charge returns to point a, it must have the same potential energy (zero) as it had at the start.[6]

The rate at which the charge ΔQ *loses* potential energy in going through the resistor is given by

$$\frac{\Delta U}{\Delta t} = \frac{\Delta Q}{\Delta t} V = IV$$

where I is the current in the circuit. Of course, the charge regains this energy when it passes through the battery. Since the rate at which the charge loses energy equals the power P lost in the resistor, we have

Power

$$P = IV \tag{27.21}$$

In this case, the power is supplied to a resistor by a battery. However, Equation 27.21 can be used to determine the power transferred to *any* device carrying a current I and having a potential difference V between its terminals.

Using Equation 27.21 and the fact that $V = IR$ for a resistor, we can express the power dissipated in the alternative forms

Power loss in a conductor

$$P = I^2R = \frac{V^2}{R} \tag{27.22}$$

When I is in amperes, V in volts, and R in ohms, the SI unit of power is the watt (W). The power lost as heat in a conductor of resistance R is called *joule heating*[7]; however, it is often referred to as an *I^2R loss.*

A battery or any other device that provides electrical energy is called an *electromotive force,* usually referred to as an *emf.* The concept of emf will be discussed in more detail in Chapter 28. (The phrase *electromotive force* is an unfortunate one, since it does not describe a force but actually refers to a potential difference in volts.) *Neglecting the internal resistance of the battery, the potential difference between points a and b is equal to the emf $\mathcal{E}$ of the*

[6] Note that when the current reaches its steady-state value, there is *no* change with time in the kinetic energy associated with the current.

[7] It is called *joule heating* even though its dimensions are *energy per unit time,* which are dimensions of power.

battery. That is, $V = V_b - V_a = \mathcal{E}$, and the current in the circuit is given by $I = V/R = \mathcal{E}/R$. Since $V = \mathcal{E}$, the power supplied by the emf can be expressed as $P = I\mathcal{E}$, which, of course, equals the power lost in the resistor, I^2R.

EXAMPLE 27.7 Power in an Electric Heater

An electric heater is constructed by applying a potential difference of 110 V to a nichrome wire of total resistance 8 Ω. Find the current carried by the wire and the power rating of the heater.

Solution Since $V = IR$, we have

$$I = \frac{V}{R} = \frac{110\text{ V}}{8\ \Omega} = 13.8\text{ A}$$

We can find the power rating using $P = I^2R$:

$$P = I^2R = (13.8\text{ A})^2(8\ \Omega) = 1.52\text{ kW}$$

If we were to double the applied voltage, the current would double but the power would quadruple.

EXAMPLE 27.8 Electrical Rating of a Lightbulb

A light bulb is rated at 120 V/75 W. That is, its operating voltage is 120 V and it has a power rating of 75 W. The bulb is powered by a 120-V direct-current power supply. Find the current in the bulb and its resistance.

Solution Since the power rating of the bulb is 75 W and the operating voltage is 120 V, we can use $P = IV$ to find the current:

$$I = \frac{P}{V} = \frac{75\text{ W}}{120\text{ V}} = 0.625\text{ A}$$

Using Ohm's law, $V = IR$, the resistance is calculated to be

$$R = \frac{V}{I} = \frac{120\text{ V}}{0.625\text{ A}} = 192\ \Omega$$

Exercise 4 What would the resistance be in a lamp rated at 120 V and 100 W?
Answer 144 Ω.

*27.8 ENERGY CONVERSION IN HOUSEHOLD CIRCUITS

The heat generated when current passes through a resistive material is used in many common devices. A cross-sectional view of the spiral heating element of an electric range is shown in Figure 27.16a. The material through which the current passes is surrounded by an insulating substance in order to prevent the current from flowing through the cook to the earth when he or she touches the pan. A material that is a good conductor of heat surrounds the insulator.

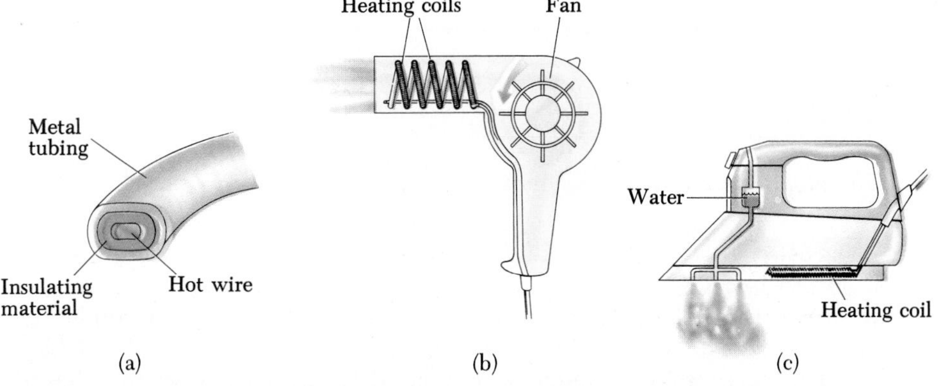

Figure 27.16 (a) The cross-section of a heating element used in an electric range. (b) In a hair dryer, warm air is produced by blowing air from a fan past the heating coils. (c) In a steam iron, water is turned into steam by heat from a heating coil.

Figure 27.16b shows a common hair dryer, in which a fan blows air past heating coils. In this case the warm air can be used to dry hair, but on a broader scale this same principle is used to dry clothes and to heat buildings.

A final example of a household appliance that uses the heating effect of electric currents is the steam iron shown in Figure 27.16c. A heating coil warms the bottom of the iron and simultaneously turns water to steam, which is sprayed from jets located in the bottom of the iron.

The unit of energy the electric company uses to calculate energy consumption, the **kilowatt-hour,** is defined in terms of the unit of power. One kilowatt-hour (kWh) is the energy converted or consumed in 1 h at the constant rate of 1 kW. The numerical value of 1 kWh is

$$1 \text{ kWh} = (10^3 \text{ W})(3600 \text{ s}) = 3.6 \times 10^6 \text{ J} \tag{27.23}$$

On your electric bill, the amount of electricity used is usually stated in multiples of kWh.

EXAMPLE 27.9 The Cost of Operating a Lightbulb
How much does it cost to burn a 100-W lightbulb for 24 h if electricity costs eight cents per kilowatt-hour?

Solution A 100-W lightbulb is equivalent to a 0.1-kW bulb. Since the energy consumed equals power × time, the amount of energy you must pay for, expressed in kWh, is

$$\text{Energy} = (0.10 \text{ kW})(24 \text{ h}) = 2.4 \text{ kWh}$$

If energy is purchased at eight cents per kWh, the cost is

$$\text{Cost} = (2.4 \text{ kWh})(\$0.08/\text{kWh}) = \$0.19$$

That is, it will cost 19 cents to operate the lightbulb for one day. This is a small amount, but when larger and more complex devices are being used, the costs go up rapidly.

Demands on our energy supplies have made it necessary to be aware of the energy requirements of our electric devices. This is true not only because they are becoming more expensive to operate but also because, with the dwindling of the coal and oil resources that ultimately supply us with electrical energy, increased awareness of conservation becomes necessary. On every electric appliance is a label that contains the information you need to calculate the power requirements of the appliance. The power consumption in watts is often stated directly, as on a lightbulb. In other cases, the amount of current used by the device and the voltage at which it operates are given. This information and Equation 27.21 are sufficient to calculate the operating cost of any electric device.

Exercise 5 If electricity costs eight cents per kilowatt-hour, what does it cost to operate an electric oven, which operates at 20 A and 220 V, for 5 h?
Answer $1.76.

SUMMARY

The **electric current** I in a conductor is defined as

Electric current

$$I \equiv \frac{dQ}{dt} \tag{27.2}$$

where dQ is the charge that passes through a cross section of the conductor in a time dt. The SI unit of current is the ampere (A), where 1 A = 1 C/s.

The current in a conductor is related to the motion of the charge carriers through the relationship

Current in a conductor

$$I = nqv_dA \tag{27.4}$$

where n is the density of charge carriers, q is their charge, v_d is the drift velocity, and A is the cross-sectional area of the conductor.

The **current density** J in a conductor is defined as the current per unit area:

$$J = nqv_d \qquad (27.6)$$

The current density in a conductor is proportional to the electric field according to the expression

Ohm's law

$$J = \sigma E \qquad (27.7)$$

The constant σ is called the **conductivity** of the material. The inverse of σ is called the **resistivity,** ρ. That is, $\rho = 1/\sigma$.

A material is said to obey Ohm's law if its conductivity is independent of the applied field.

The **resistance** R of a conductor is defined as the ratio of the potential difference across the conductor to the current:

Resistance of a conductor

$$R \equiv \frac{V}{I} \qquad (27.8)$$

If the resistance is independent of the applied voltage, the conductor obeys Ohm's law.

If the conductor has a uniform cross-sectional area A and a length ℓ, its resistance is given by

Resistance of a uniform conductor

$$R = \frac{\ell}{\sigma A} = \rho \frac{\ell}{A} \qquad (27.10)$$

The SI unit of resistance is volt per ampere, which is defined to be 1 ohm (Ω). That is, $1\ \Omega = 1$ V/A.

The resistivity of a conductor varies with temperature in an approximately linear fashion, that is

Variation of ρ with temperature

$$\rho = \rho_o[1 + \alpha(T - T_o)] \qquad (27.11)$$

where α is the temperature coefficient of resistivity and ρ_o is the resistivity at some reference temperature T_o.

In a classical model of electronic conduction in a metal, the electrons are treated as molecules of a gas. In the absence of an electric field, the average velocity of the electrons is zero. When an electric field is applied, the electrons move (on the average) with a **drift velocity** $\mathbf{v}_d$, which is opposite the electric field. The drift velocity is given by

Drift velocity

$$\mathbf{v}_d = \frac{q\mathbf{E}}{m} \tau \qquad (27.16)$$

where τ is the average time between collisions with the atoms of the metal. The resistivity of the material according to this model is given by

Resistivity

$$\rho = \frac{m}{nq^2\tau} \qquad (27.19)$$

where n is the number of free electrons per unit volume.

If a potential difference V is maintained across a resistor, the **power**, or rate at which energy is supplied to the resistor, is given by

Power

$$P = IV \tag{27.21}$$

Since the potential difference across a resistor is given by $V = IR$, we can express the power dissipated in a resistor in the form

Power loss in a resistor

$$P = I^2R = \frac{V^2}{R} \tag{27.22}$$

The electrical energy supplied to a resistor appears in the form of internal energy (thermal energy) in the resistor.

QUESTIONS

1. Explain the chemistry involved in the operation of a "dry cell" battery.
2. In an analogy between traffic flow and electrical current, what would correspond to the charge Q? What would correspond to the current I?
3. What factors affect the resistance of a conductor?
4. What is the difference between resistance and resistivity?
5. We have seen that an electric field must exist inside a conductor that carries a current. How is this possible in view of the fact that in *electrostatics*, we concluded that E must be zero inside a conductor?
6. Two wires A and B of circular cross-section are made of the same metal and have equal lengths, but the resistance of wire A is three times greater than that of wire B. What is the ratio of their cross-sectional areas? How do their radii compare?
7. What is required in order to maintain a steady current in a conductor?
8. Do all conductors obey Ohm's law? Give examples to justify your answer.
9. When the voltage across a certain conductor is doubled, the current is observed to increase by a factor of 3. What can you conclude about the conductor?
10. In the water analogy of an electric circuit, what corresponds to the power supply, resistor, charge, and potential difference?
11. Why might a "good" electrical conductor also be a "good" thermal conductor?
12. Use the atomic theory of matter to explain why the resistance of a material should increase as its temperature increases.
13. How does the resistance change with temperature for copper and silicon? Why are they different?
14. Explain how a current can persist in a superconductor without any applied voltage.
15. What single experimental requirement makes superconducting devices expensive to operate? In principle, can this limitation be overcome?
16. What would happen to the drift velocity of the electrons in a wire and to the current in the wire if the electrons could move freely without resistance through the wire?
17. If charges flow very slowly through a metal, why does it not require several hours for a light to come on when you throw a switch?
18. In a conductor, the electric field that drives the electrons through the conductor propagates with a speed close to the speed of light, although the drift velocity of the electrons is very small. Explain how these can both be true. Does the same electron move from one end of the conductor to the other?
19. Two conductors of the same length and radius are connected across the same potential difference. One conductor has twice the resistance of the other. Which conductor will dissipate more power?
20. When incandescent lamps burn out, they usually do so just after they are switched on. Why?
21. If you were to design an electric heater using nichrome wire as the heating element, what parameters of the wire could you vary to meet a specific power output, such as 1000 W?
22. Two light bulbs both operate from 110 V, but one has a power rating of 25 W and the other of 100 W. Which bulb has the higher resistance? Which bulb carries the greater current?
23. A typical monthly utility rate structure might go something like this: $1.60 for the first 16 kWh, 7.05 cents/kWh for the next 34 kWh used, 5.02 cents/kWh for the next 50 kWh, 3.25 cents/kWh for the next 100 kWh, 2.95 cents/kWh for the next 200 kWh, 2.35 cents/kWh for all in excess of 400 kWh. Based on these rates, what would be the charge for 327 kWh? From the standpoint of encouraging conservation of energy, what is wrong with this pricing method?

PROBLEMS

Section 27.2 Electric Current

1. Calculate the current in the case for which 3×10^{12} electrons pass a given cross section of a conductor each second.
2. In a particular cathode ray tube, the measured beam current is 30 μA. How many electrons strike the tube screen every 40 s?
3. A small sphere that carries a charge of 8 nC is whirled in a circle at the end of an insulating string. The rotation frequency is 100π rad/s. What average current does this rotating charge represent?
4. The quantity of charge q (in C) passing through a surface of area 2 cm^2 varies with time as $q = 4t^3 + 5t + 6$, where t is in s. (a) What is the instantaneous current through the surface at $t = 1.0$ s? (b) What is the value of the current density?
5. The current I (in A) in a conductor depends on time as $I = 2t^2 - 3t + 7$, where t is in s. What quantity of charge moves across a section through the conductor during the interval $t = 2$ s to $t = 4$ s?
6. Suppose that the current through a conductor decreases exponentially with time according to

$$I(t) = I_0 e^{-t/\tau}$$

where I_0 is the intial current (at $t = 0$), and τ is a constant having dimensions of time. Consider a fixed observation point within the conductor. (a) How much charge passes this point between $t = 0$ and $t = \tau$? (b) How much charge passes this point between $t = 0$ and $t = 10\tau$? (c) How much charge passes this point between $t = 0$ and $t = \infty$?

7. Calculate the number of free electrons per cubic meter for gold, assuming one free electron per atom.
8. Calculate the drift velocity of the electrons in a conductor that has a cross-sectional area of 8×10^{-6} m^2 and carries a current of 8 A. Take the concentration of free electrons to be 5×10^{28} electrons/m^3.
9. A copper bus bar has a cross section of 5 cm $\times$ 15 cm and carries a current with a density of 2000 A/cm^2. (a) What is the total current in the bus bar? (b) What amount of charge passes a given point in the bar per hour?
10. Figure 27.17 represents a section of a circular conductor of nonuniform diameter carrying a current of 5 A. The radius of cross section A_1 is 0.4 cm. (a) What is the magnitude of the current density across A_1? (b) If the current density across A_2 is one fourth the value across A_1, what is the radius of the conductor at A_2?

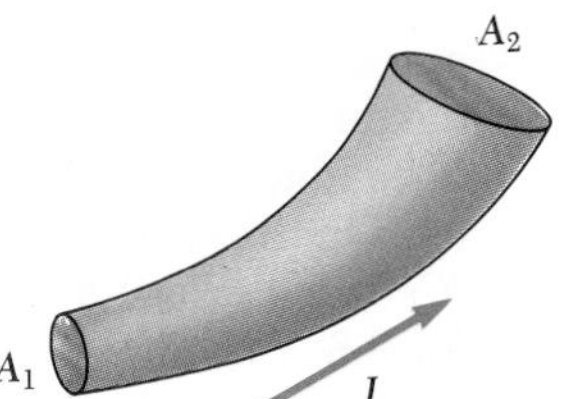

Figure 27.17 (Problem 10).

11. A coaxial conductor with a length of 20 m consists of an inner cylinder with a radius of 3.0 mm and a concentric outer cylindrical tube with an inside radius of 9.0 mm. A uniformly distributed leakage current of 10 μA flows between the two conductors. Determine the leakage current density (in A/m^2) through a cylindrical surface (concentric with the conductors) that has a radius of 6.0 mm.

Section 27.3 Resistance and Ohm's Law

12. A conductor of uniform radius 1.2 cm carries a current of 3 A produced by an electric field of 120 V/m. What is the resistivity of the material?
13. An electric field of 2100 V/m is applied to a section of silver of uniform cross section. Calculate the resulting current density if the specimen is at a temperature of 20°C.
14. If the current density in a copper wire is equal to 5.8×10^6 A/m^2, calculate the drift velocity of the free electrons in this wire.
15. Calculate the resistance at 20°C of a 40-m length of silver wire having a cross-sectional area of 0.4 mm^2.
16. Eighteen-gauge wire has a diameter of 1.024 mm. Calculate the resistance of 15 m of 18-gauge copper wire at 20°C.
17. A 2.4-m length of wire that is 0.031 cm^2 in cross section has a measured resistance of 0.24 Ω. Calculate the conductivity of the material.
18. What is the resistance of a tungsten filament 15 cm in length and 0.002 cm in diameter at 20°C?
19. What diameter copper wire has a resistance per unit length of 3.28×10^{-3} Ω/m at 20°C?
20. Determine the resistance at 20°C of a 1.5-m length of platinum wire that has a diameter of 0.10 mm.
21. Suppose that a uniform wire of resistance R is stretched uniformly to three times its original length. What will be its new resistance, assuming its density and resistivity remain constant?
22. Aluminum and copper wires of equal length are found to have the same resistance. What is the ratio of their radii?
23. Suppose that you wish to fabricate a uniform wire out of 1 g of copper. If the wire is to have a resistance of $R = 0.5$ Ω, and all of the copper is to be used, what will be (a) the length and (b) the diameter of this wire?
24. What is the resistance of a device that operates with a current of 7 A when the applied voltage is 110 V?

25. A 0.9-V potential difference is maintained across a 1.5-m length of tungsten wire that has a cross-sectional area of 0.6 mm^2. What is the current in the wire?
26. Twenty-two gauge wire has a diameter of 0.644 mm. Calculate the voltage drop across a 5-m length of 22-gauge aluminum wire when it carries a 0.5 A current.
27. A 22-V power supply is connected to a 4.4-Ω resistor in the shape of a cylinder of length 2 cm and diameter 0.8 cm. Calculate (a) the current in the resistor, and (b) the electric field in the resistor. (c) What happens to these two values if the 4.4-Ω resistor is replaced with a 2.2-Ω resistor with the same dimensions?
28. A potential difference of 6 V is found to produce a current of 0.14 A in a 2.6-m length of conductor having a uniform radius of 0.3 cm. (a) Calculate the resistivity of the material. (b) What is the resistance of the conductor?

Section 27.4 The Resistivity of Different Conductors

29. An aluminum wire with a diameter of 0.1 mm has a uniform electric field of 0.2 V/m imposed along its entire length. The temperature of the wire is 50°C. Assume one free electron per atom. (a) Use the information in Table 27.1 and determine the resistivity. (b) What is the current density in the wire? (c) What is the total current in the wire? (d) What is the drift speed of the conduction electrons? (e) What potential difference must exist between the ends of a 2-m length of the wire to produce the stated electric field strength?
30. Calculate the percentage change in the resistance of a carbon filament when it is heated from room temperature to 160°C.
31. What is the fractional change in the resistance of an iron filament when its temperature changes from 25°C to 50°C?
32. Assuming that Equation 27.13 holds over a large temperature range (which it doesn't), calculate the temperature at which copper would have no resistance. Why doesn't the resistance of copper go to zero at very low temperatures?
33. If a copper wire has a resistance of 18 Ω at 20°C, what resistance will it have at 60°C? (Neglect any change in length or cross-sectional area due to the change in temperature.)
34. A wire 3 m in length and 0.45 mm^2 in cross-sectional area has a resistance of 41 Ω at 20°C. If the resistance of the wire increases to 41.4 Ω at 29°C, what is the temperature coefficient of resistivity?
35. At what temperature will tungsten have a resistivity four times that of copper? (Assume that the copper is at 20°C.)
36. A segment of nichrome wire is initially at 20°C. Using the data from Table 27.1, calculate the temperature to which the wire must be heated to double its resistance.
37. At 45°C, the resistance of a segment of gold wire is 85 Ω. When the wire is placed in a liquid bath, the resistance decreases to 80 Ω. What is the temperature of the bath?
38. Calculate the resistivity of copper from the following data: A potential difference of 1.0 V produces a current of 3.4 A in a 125-m length of copper wire that is 0.30 cm in diameter and at a temperature of 20°C.

Section 27.6 A Model for Electrical Conduction

39. Calculate the current density in a gold wire in which an electric field of 0.74 V/m exists.
40. If the drift velocity of free electrons in a copper wire is 7.84×10^{-4} m/s, calculate the electric field in the conductor.
41. Use data from Example 27.6 to calculate the collision mean free path of electrons in copper if the average thermal speed of conduction electrons is 8.6×10^5 m/s.
42. If the current through a given conductor is doubled, what happens to the (a) charge carrier density? (b) current density? (c) electron drift velocity? (d) average time between collisions?

Section 27.7 Electrical Energy and Power

43. A 10-V battery is connected to a 120-Ω resistor. Neglecting the internal resistance of the battery, calculate the power dissipated in the resistor.
44. How much current is being supplied by a 200-V generator delivering 100 kW of power?
45. Suppose that a voltage surge produces 140 V for a moment. By what percentage will the output of a 120-V, 100-W light bulb increase, assuming its resistance does not change?
46. A particular type of automobile storage battery is characterized as "360-ampere-hour, 12 V." What total energy can the battery deliver?
47. If a 55-Ω resistor is rated at 125 W (the maximum allowed power), what is the maximum allowed operating voltage?
48. In a hydroelectric installation, a turbine delivers 1500 hp to a generator, which in turn converts 80% of the mechanical energy into electrical energy. Under these conditions, what current will the generator deliver at a terminal potential difference of 2000 V?
49. Two conductors made of the same material are connected across a common potential difference. Conductor A has twice the diameter and twice the length of conductor B. What is the ratio of the power delivered to the two conductors?
50. A current of 0.9 A is maintained in a 160-Ω resistor for 3 h. Calculate the heat energy developed in the resistor. Give the answer in both J and cal.

Section 27.8 Energy Conservation in Household Circuits

51. What is the required resistance of an immersion heater that will increase the temperature of 1.5 kg of water from 10°C to 50°C in 10 min while operating at 110 V?
52. If an electric stove is capable of heating 500 cm³ of water from room temperature to boiling in 10 minutes, what is the power rating of the stove? Assume 50% efficiency.
53. Compute the cost per day of operating a lamp that draws 1.7 A from a 110-V line if the cost of electrical energy is 6 cents/kWh.
54. An electric heater operating at full power draws a current of 8 A from a 110-V circuit. (a) What is the resistance of the heater? (b) Assuming constant R, how much current should the heater draw in order to dissipate 750 W?
55. A 110-V motor produces 2.5 hp of mechanical power. If this motor is 90% efficient in converting electrical power into mechanical power, find (a) the current drawn by the motor, and (b) the total electrical energy used by this motor after running for one hour. (c) If electrical energy costs 8 cents/kWh, how much did it cost to run the motor for this hour?

ADDITIONAL PROBLEMS

56. Many problems in Chapter 27 deal with resistors of odd shapes and sizes. To verify the results experimentally, a voltage may be applied to the indicated surfaces and the resulting current may be measured. The resistance could then be calculated by Ohm's law. Describe a method to assure that the potential applied will be uniform over the surface.
57. The potential difference across the filament of a lamp is maintained at a constant level while equilibrium temperature is being reached. It is observed that the steady-state current in the lamp is only one tenth of the current drawn by the lamp when it is first turned on. If the temperature coefficient of resistivity for the lamp at 20°C is 0.0045 $(\mathrm{C}^\circ)^{-1}$, and if the resistance increases linearly with increasing temperature, what is the final operating temperature of the filament?
58. The current in a resistor decreases by 3 A when the voltage applied across the resistor decreases from 12 V to 6 V. Find the resistance of the resistor.
59. (a) A 115-g mass of aluminum is formed into a right circular cylinder shaped so that the diameter of the cylinder equals its height. Calculate the resistance between the top and bottom faces of the cylinder at 20°C. (b) Calculate the resistance between opposite faces if the same mass of aluminum is formed into a cube.
60. (a) A sheet of copper ($\rho = 1.7 \times 10^{-8}\ \Omega \cdot \mathrm{m}$) is 2 mm thick and has surface dimensions of 8 cm × 24 cm. If the long edges are joined to form a hollow tube 24 cm in length, what is the resistance between the ends? (b) What mass of copper would be required to manufacture a spool of copper cable 1500 m in length and having a total resistance of 4.5 Ω?
61. A Wheatstone bridge can be used to measure the strain ($\Delta L/L_0$) of a wire (see Section 12.4), where L_0 is the length before stretching, L is the length after stretching, and $\Delta L = L - L_0$. Let $\alpha = \Delta L/L_0$. Show that the resistance is $R = R_0\ (1 + 2\alpha + \alpha^2)$ for any length where $R_0 = \dfrac{\rho L_0}{A_0}$. Assume that the resistivity and volume of the wire stay constant.
62. A cylindrical tungsten conductor has an initial length L_1 and cross-sectional area A_1. The metal is drawn uniformly to a final length $L_2 = 10L_1$ and is then annealed. If the resistance of the conductor at the new length is 75 Ω, what is the initial value of R?
63. A resistor is constructed by forming a material of resistivity ρ into the shape of a hollow cylinder of length L and inner and outer radii r_a and r_b, respectively (Fig. 27.18). In use, a potential difference is applied between the ends of the cylinder, producing a current parallel to the axis. (a) Find a general expression for the resistance of such a device in terms of L, ρ, r_a, and r_b. (b) Obtain a numerical value for R when $L = 4$ cm, $r_a = 0.5$ cm, $r_b = 1.2$ cm, and the resistivity $\rho = 3.5 \times 10^5\ \Omega \cdot \mathrm{m}$.

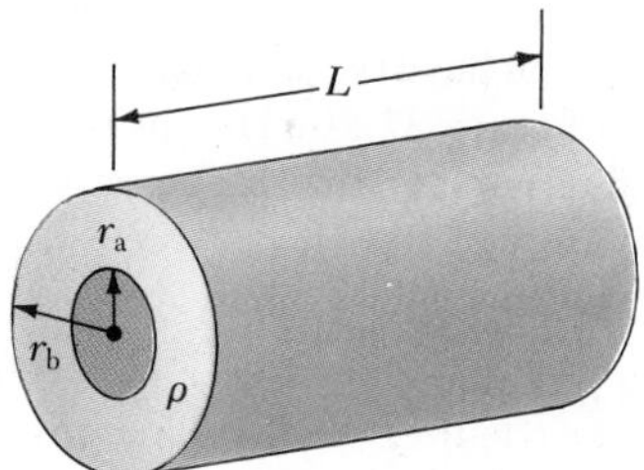

Figure 27.18 (Problems 63 and 64).

64. Consider the device described in Problem 63. Suppose now that the potential difference is applied between the inner and outer surfaces so that the resulting current flows radially outward. (a) Find a general expression for the resistance of the device in terms of L, ρ, r_a, and r_b. (b) Calculate the value of R using the parameter values given in (b) of Problem 63.
65. Type 58u coaxial cable consists of an inner cylindrical conductor of radius 0.81 mm surrounded by an insulating sheath of polyethylene of outside diameter 2.95 mm with an outer conductor formed by a braided sheath around the insulator. Find the leakage current between conductors for a 10-m length of this coaxial cable if the potential difference is 1000 V. Take the resistivity of polyethylene to be equal to $1.6 \times 10^{11}\ \Omega \cdot \mathrm{m}$.

66. Two concentric spherical shells with inner and outer radii r_a and r_b, respectively, form a resistive element when the region between the two surfaces contains a material of resistivity ρ. Show that the resistance of the device is given by

$$R = \frac{\rho}{4\pi}\left(\frac{1}{r_a} - \frac{1}{r_b}\right)$$

67. Material with uniform resistivity ρ is formed into a wedge as shown in Figure 27.19. Show that the resistance between face *A* and face *B* of this wedge is given by

$$R = \rho \frac{L}{w(y_2 - y_1)} \ln\left(\frac{y_2}{y_1}\right)$$

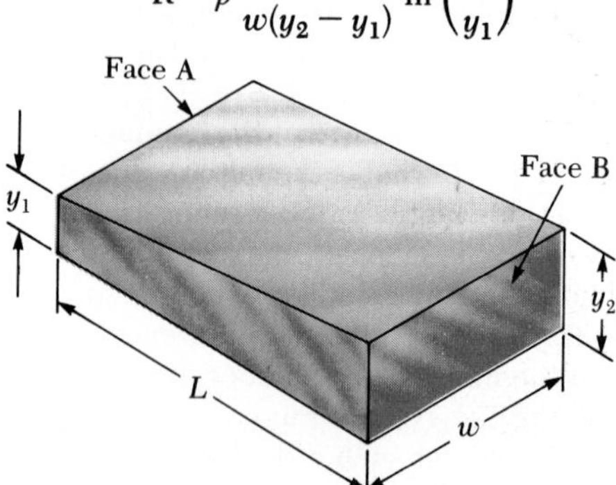

Figure 27.19 (Problem 67).

68. Find the resistance between face *A* and face *B* of a quarter ring that has a rectangular cross-sectional area as shown in Figure 27.20. The inside radius is *b*, the outside radius is *a*, and the thickness is *t*. The material has resistivity ρ.

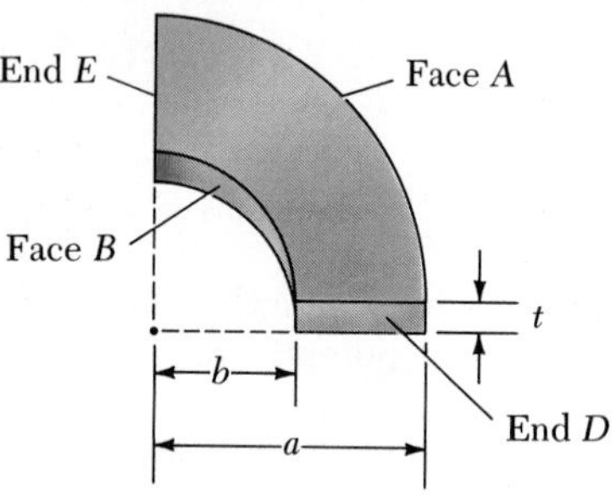

Figure 27.20 (Problems 68 and 69).

69. Find the resistance between the ends *D* and *E* of the quarter ring shown in Figure 27.20 and described in Problem 68.

70. A material of resistivity ρ is formed into the shape of a truncated cone of altitude *h* as in Figure 27.21. The bottom end has a radius *b* and the top end has a radius *a*. Assuming a uniform current density through any circular cross section of the cone, show that the resistance between the two ends is

$$R = \frac{\rho}{\pi}\left(\frac{h}{ab}\right)$$

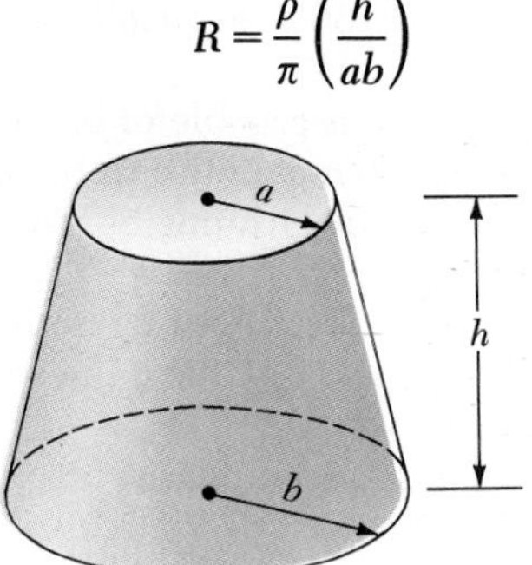

Figure 27.21 (Problem 70).

71. A sphere of uniform resistivity ρ and radius *R* has opposite sides ground off to produce identical parallel flat faces. The distance between these flat faces is *D*. (a) Find the expression for the resistance between the two flat faces. (b) What happens to the resistance as *D* goes to zero? (c) What happens to the resistance as *D* goes to a value of 2*R*?

72. A semiconducting rod of radius *a* and length *L* has diffused impurities so that its conductivity varies along its length as $\sigma = \sigma_0 e^{-bx}$ where σ_0 and *b* are constants. Find the resistance of the rod.

73. An engineer is in need of a resistor that is to have zero overall temperature coefficient of resistance at 20°C. The design is a composite of right circular cylinders of two materials, as in Figure 27.22. The ratio of the resistivities of the two materials is $\rho_1/\rho_2 = 3.2$, and the ratio of the lengths of the sections is $\ell_1/\ell_2 = 2.6$. The radius *r* is uniform throughout. Assuming that the temperature of the two sections remains equal, calculate α_1/α_2, the required ratio of temperature coefficients of resistivity of the two materials. You should also assume that the total resistance is given by $R = R_1 + R_2$ (to be proved in Ch. 28).

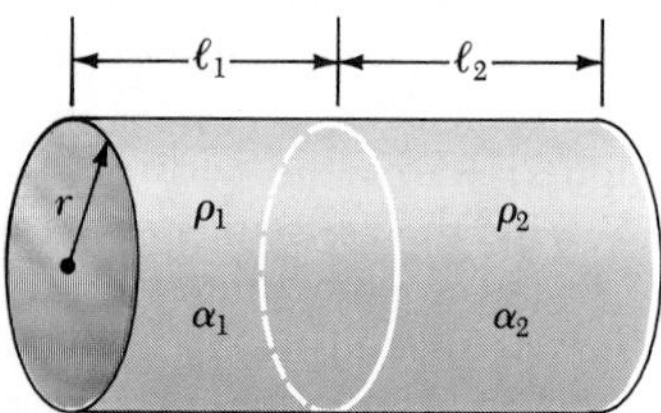

Figure 27.22 (Problem 73).

28

Direct Current Circuits

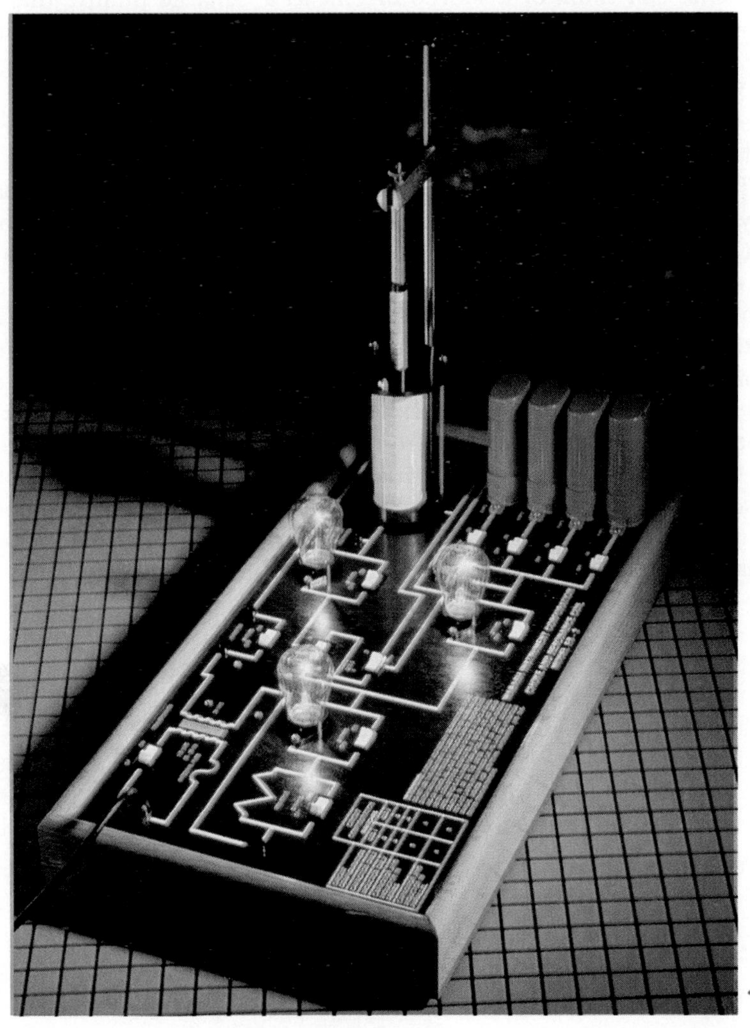

This versatile circuit enables the experimenter to examine the properties of circuit elements such as capacitors and resistors and their effect on circuit behavior. (Courtesy of CENCO)

This chapter is concerned with the analysis of some simple circuits whose elements include batteries, resistors, and capacitors in various combinations. The analysis of these circuits is simplified by the use of two rules known as *Kirchhoff's rules.* These rules follow from the laws of conservation of energy and conservation of charge. Most of the circuits analyzed are assumed to be in *steady state,* where the currents are constant in magnitude and direction. In one section we discuss circuits containing resistors and capacitors, for which the current varies with time. Finally, a number of common electrical devices and techniques are described for measuring current, potential differences, resistance, and emfs.

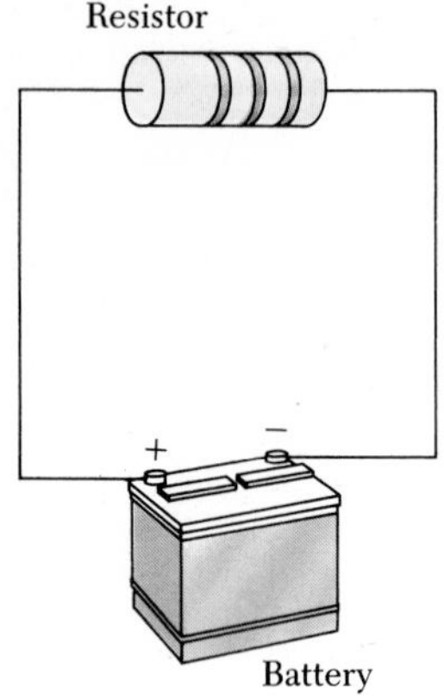

Figure 28.1 A circuit consisting of a resistor connected to the terminals of a battery.

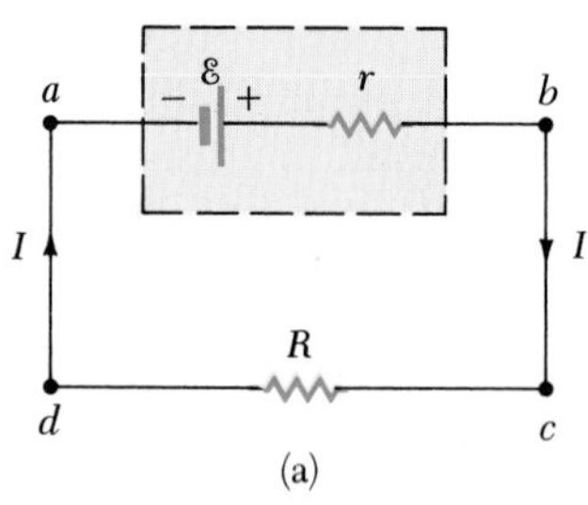

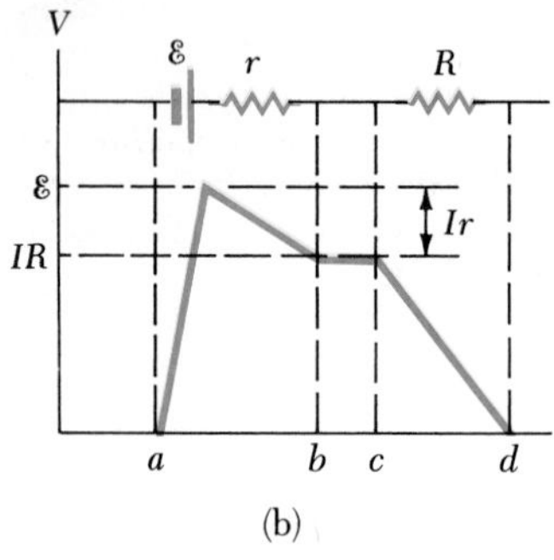

Figure 28.2 (a) Circuit diagram of a source of emf $\mathcal{E}$ of internal resistance r connected to an external resistor R. (b) Graphical representation showing how the potential changes as the series circuit in (a) is traversed clockwise.

28.1 ELECTROMOTIVE FORCE

In the previous chapter we found that a constant current can be maintained in a closed circuit through the use of a source of energy, called an **electromotive force** (abbreviated *emf*). A source of emf is any device (such as a battery or generator) that will *increase* the potential energy of charges circulating in a circuit. One can think of a source of emf as a "charge pump" that forces electrons to move in a direction opposite the electrostatic field inside the source. The emf, $\mathcal{E}$, of a source describes the work done per unit charge, and hence the SI unit of emf is the volt.

Consider the circuit shown in Figure 28.1, consisting of a battery connected to a resistor. We shall assume that the connecting wires have no resistance. The positive terminal of the battery is at a higher potential than the negative terminal. If we were to neglect the internal resistance of the battery itself, then the potential difference across the battery (the terminal voltage) would equal the emf of the battery. However, because a real battery always has some internal resistance r, the terminal voltage is not equal to the emf of the battery. The circuit shown in Figure 28.1 can be described by the circuit diagram in Figure 28.2a. The battery within the dotted rectangle is represented by a source of emf, $\mathcal{E}$, in series with the internal resistance, r. Now imagine a positive charge moving from a to b in Figure 28.2a. As the charge passes from the negative to the positive terminal of the battery, its potential *increases* by $\mathcal{E}$. However, as it moves through the resistance r, its potential *decreases* by an amount Ir, where I is the current in the circuit. Thus, the terminal voltage of the battery, $V = V_b - V_a$, is given by[1]

$$V = \mathcal{E} - Ir \tag{28.1}$$

From this expression, note that $\mathcal{E}$ is equivalent to the **open-circuit voltage,** that is, the *terminal voltage when the current is zero.* Figure 28.2b is a graphical representation of the changes in potential as the circuit is traversed in the clockwise direction. By inspecting Figure 28.2a we see that the terminal voltage V must also equal the potential difference across the external resistance R, often called the **load resistance.** That is, $V = IR$. Combining this with Equation 28.1, we see that

$$\mathcal{E} = IR + Ir \tag{28.2}$$

Solving for the current gives

$$I = \frac{\mathcal{E}}{R + r} \tag{28.3}$$

This shows that the current in this simple circuit depends on both the resistance external to the battery and the internal resistance. If the load resistance R is much greater than the internal resistance r, we can neglect r in this analysis. In many circuits we shall ignore this internal resistance.

If we multiply Equation 28.2 by the current I, the following expression is obtained:

$$I\mathcal{E} = I^2R + I^2r \tag{28.4}$$

[1] The terminal voltage in this case is less than the emf by an amount Ir. In some situations, the terminal voltage may *exceed* the emf by an amount Ir. This happens when the direction of the current is *opposite* that of the emf, as in the case of charging a battery with another source of emf.

This equation tells us that the total power output of the source of emf, $I\mathcal{E}$, is converted into power dissipated as joule heat in the load resistance, I^2R, *plus* power dissipated in the internal resistance, I^2r. Again, if $r \ll R$, then most of the power delivered by the battery is transferred to the load resistance.

EXAMPLE 28.1 Terminal Voltage of a Battery

A battery has an emf of 12 V and an internal resistance of 0.05 Ω. Its terminals are connected to a load resistance of 3 Ω. (a) Find the current in the circuit and the terminal voltage of the battery.

Using Equations 28.1 and 28.3, we get

$$I = \frac{\mathcal{E}}{R + r} = \frac{12\ \text{V}}{3.05\ \Omega} = 3.93\ \text{A}$$

$$V = \mathcal{E} - Ir = 12\ \text{V} - (3.93\ \text{A})(0.05\ \Omega) = 11.8\ \text{V}$$

As a check of this result, we can calculate the voltage drop across the load resistance R. This gives

$$V = IR = (3.93\ \text{A})(3\ \Omega) = 11.8\ \text{V}$$

(b) Calculate the power dissipated in the load resistor, the power dissipated by the internal resistance of the battery, and the power delivered by the battery.

The power dissipated by the load resistor is

$$P_R = I^2R = (3.93\ \text{A})^2(3\ \Omega) = 46.3\ \text{W}$$

The power dissipated by the internal resistance is

$$P_r = I^2r = (3.93\ \text{A})^2(0.05\ \Omega) = 0.772\ \text{W}$$

Hence, the power delivered by the battery is the sum of these quantities, or 47.1 W. This can be checked using the expression $P = I\mathcal{E}$.

EXAMPLE 28.2 Matching the Load

Show that the *maximum* power lost in the load resistance R in Figure 28.2a occurs when $R = r$, that is, when the load resistance *matches* the internal resistance.

Solution The power dissipated in the load resistance is equal to I^2R, where I is given by Equation 28.3:

$$P = I^2R = \frac{\mathcal{E}^2R}{(R + r)^2}$$

When P is plotted versus R as in Figure 28.3, we find that P reaches a *maximum* value of $\mathcal{E}^2/4r$ at $R = r$. This can also be proved by differentiating P with respect to R, setting the result equal to zero, and solving for R. The details are left as a problem (Problem 79).

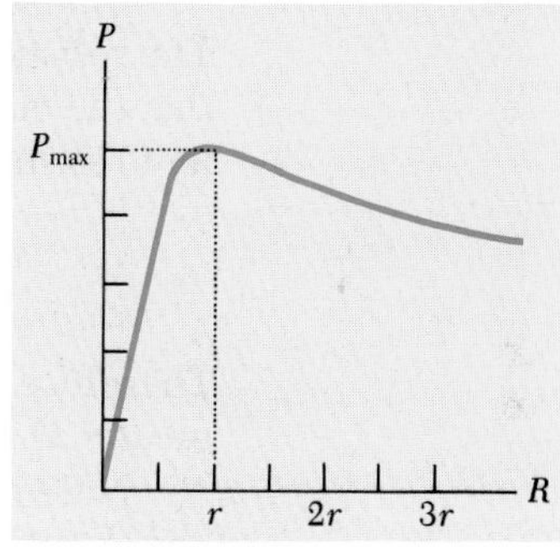

Figure 28.3 Graph of the power P delivered to a load resistor as a function of R. Note that the power into R is a maximum when R equals r, the internal resistance of the battery.

28.2 RESISTORS IN SERIES AND IN PARALLEL

When two or more resistors are connected together such that they have only one common point per pair, they are said to be in *series*. Figure 28.4 shows two resistors connected in series. Note that

For a series connection of resistors, the current is the same in each resistor

> the current is the same through each resistor since any charge that flows through R_1 must equal the charge that flows through R_2.

Since the potential drop from a to b in Figure 28.4b equals IR_1 and the potential drop from b to c equals IR_2, the potential drop from a to c is given by

$$V = IR_1 + IR_2 = I(R_1 + R_2)$$

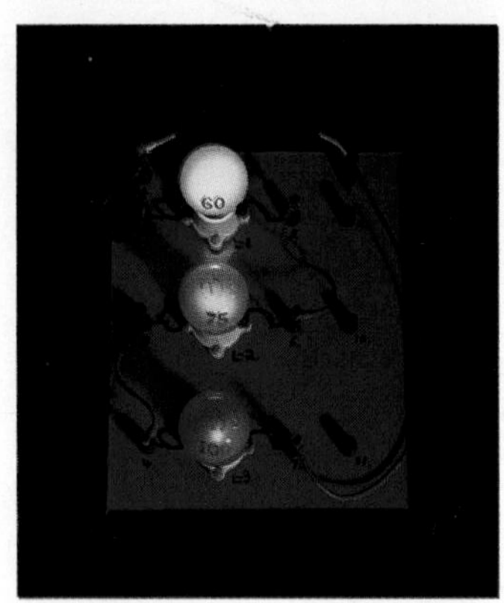

A series connection of three lamps, all rated at 120 V, with power ratings of 60 W, 75 W, and 150 W. Why are the intensities of the lamps different? Which lamp has the greatest resistance? How would their relative intensities differ if they were connected in parallel? (Courtesy of Henry Leap and Jim Lehman)

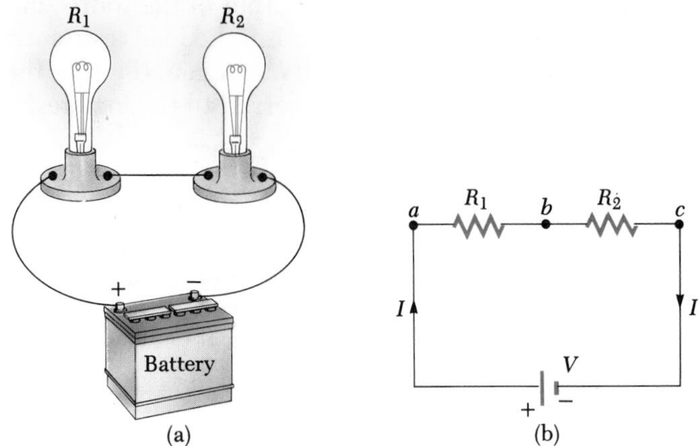

Figure 28.4 Series connection of two resistors, R_1 and R_2. The current in each resistor is the same.

Therefore, we can replace the two resistors in series by a single *equivalent resistance* R_{eq} whose value is the *sum* of the individual resistances:

$$R_{eq} = R_1 + R_2 \tag{28.5}$$

The resistance R_{eq} is equivalent to the series combination $R_1 + R_2$ in the sense that the circuit current is unchanged when R_{eq} replaces $R_1 + R_2$. The equivalent resistance of three or more resistors connected in series is simply

$$R_{eq} = R_1 + R_2 + R_3 + \cdots \tag{28.6}$$

Therefore, *the equivalent resistance of a series connection of resistors is always greater than any individual resistance.*

Note that if the filament of one light bulb in Figure 28.4 were to break, or "burn out," the circuit would no longer be complete (an open-circuit condition) and the second bulb would also go out. Some Christmas tree light sets (especially older ones) are connected in this way, and the agonizing experience of determining which bulb is burned out is a familiar one. Frustrating experiences such as this illustrate how inconvenient it would be to have all appliances in a house connected in series. In many circuits, fuses are used in series with other circuit elements for safety purposes. The conductor in the fuse is designed to melt and open the circuit at some maximum current, the value of which depends on the nature of the circuit. If a fuse is not used, excessive currents could damage circuit elements, overheat wires, and perhaps cause a fire. In modern home construction, circuit breakers are used in place of fuses. When the current in a circuit exceeds some value (typically 15 A), the circuit breaker acts as a switch and opens the circuit.

Now consider two resistors connected in *parallel* as shown in Figure 28.5.

In this case, there is an equal potential difference across each resistor.

However, the current in each resistor is in general not the same. When the current I reaches point a (called a *junction*), it splits into two parts, I_1 going

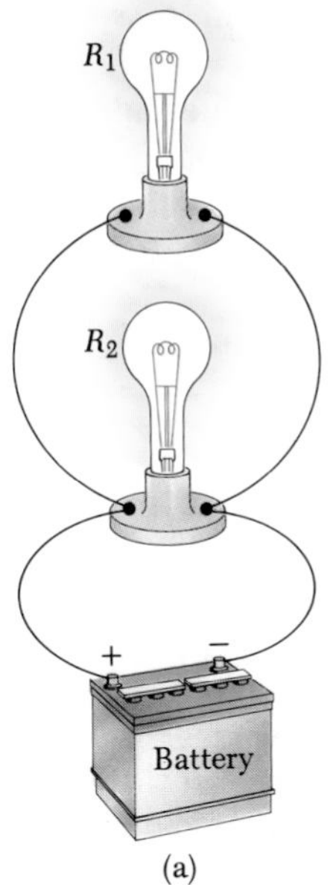

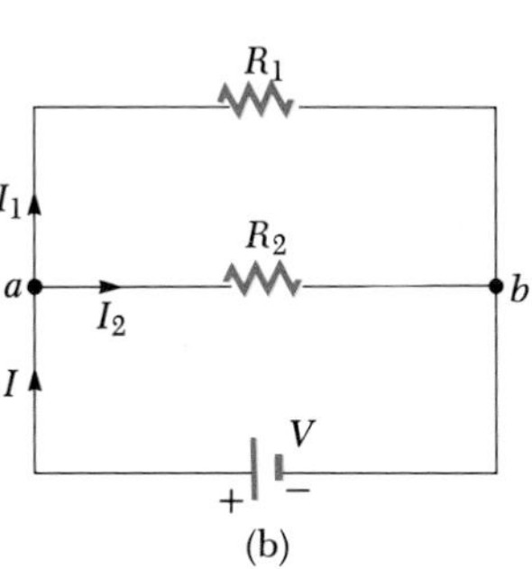

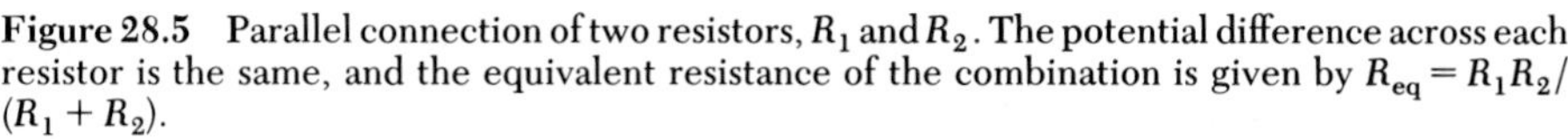
Figure 28.5 Parallel connection of two resistors, R_1 and R_2. The potential difference across each resistor is the same, and the equivalent resistance of the combination is given by $R_{eq} = R_1R_2/(R_1 + R_2)$.

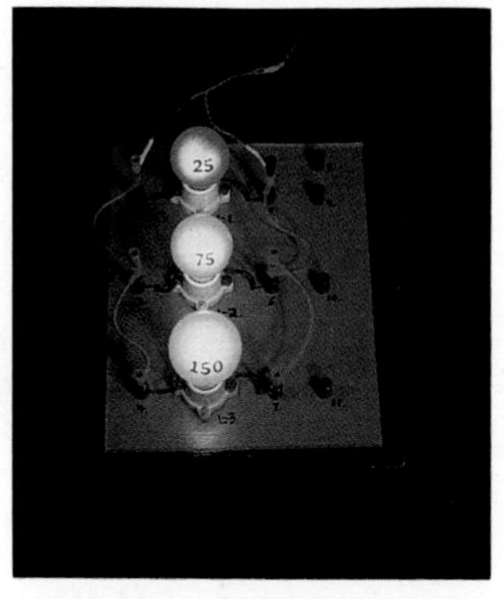

Three incandescent lamps with power ratings of 25 W, 75 W, and 150 W, connected in parallel to a voltage source of about 100 V. All lamps are rated at the same voltage. Why do the intensities of the lamps differ? Which lamp draws the most current? Which has the least resistance? (Courtesy of Henry Leap and Jim Lehman)

through R_1 and I_2 going through R_2. If R_1 is greater than R_2, then I_1 will be less than I_2. That is, the charge will tend to take the path of least resistance. Clearly, since charge must be conserved, the current I that enters point a must equal the total current leaving this point, $I_1 + I_2$:

$$I = I_1 + I_2$$

Since the potential drop across each resistor must be the *same*, Ohm's law gives

$$I = I_1 + I_2 = \frac{V}{R_1} + \frac{V}{R_2} = V\left(\frac{1}{R_1} + \frac{1}{R_2}\right) = \frac{V}{R_{eq}}$$

From this result, we see that the equivalent resistance of two resistors in parallel is given by

$$\frac{1}{R_{eq}} = \frac{1}{R_1} + \frac{1}{R_2} \qquad (28.7)$$

This can be rearranged to give

$$R_{eq} = \frac{R_1R_2}{R_1 + R_2}$$

Georg Simon Ohm (1787–1854), German physicist. (Courtesy of AIP Niels Bohr Library, E. Scott Barr Collection)

An extension of this analysis to three or more resistors in parallel gives the following general expression:

Several resistors in parallel

$$\frac{1}{R_{eq}} = \frac{1}{R_1} + \frac{1}{R_2} + \frac{1}{R_3} + \cdots \qquad (28.8)$$

It can be seen from this expression that the equivalent resistance of two or more resistors connected in parallel is always *less* than the smallest resistance in the group.

Household circuits are always wired such that the light bulbs (or appliances, etc.) are connected in parallel, as in Figure 28.5a. In this manner, each device operates independently of the others, so that if one is switched off, the others remain on. Equally important, each device operates on the same voltage.

Finally, it is interesting to note that parallel resistors combine in the same way that series capacitors combine, and vice versa.

EXAMPLE 28.3 Find the Equivalent Resistance
Four resistors are connected as shown in Figure 28.6a. (a) Find the equivalent resistance between a and c.

The circuit can be reduced in steps as shown in Figure 28.6. The 8-Ω and 4-Ω resistors are in series, and so the equivalent resistance between a and b is 12 Ω (Eq. 28.5). The 6-Ω and 3-Ω resistors are in parallel, and so from Equation 28.7 we find that the equivalent resistance from b to c is 2 Ω. Hence, the equivalent resistance from a to c is 14 Ω.

(b) What is the current in each resistor if a potential difference of 42 V is maintained between a and c?

The current I in the 8-Ω and 4-Ω resistors is the same since they are in series. Using Ohm's law and the results from (a), we get

$$I = \frac{V_{ac}}{R_{eq}} = \frac{42\ \text{V}}{14\ \Omega} = 3\ \text{A}$$

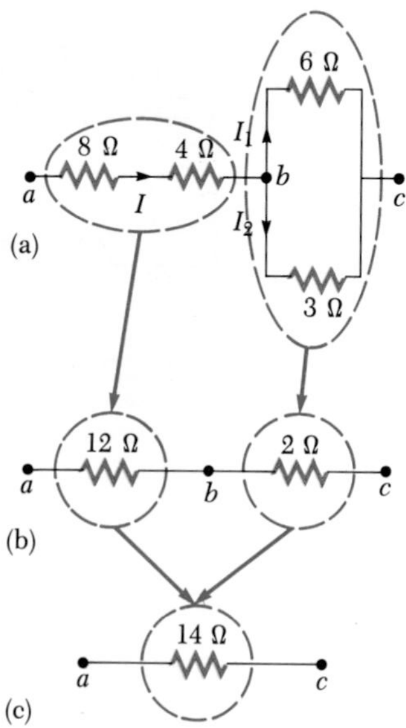

Figure 28.6 (Example 28.3) The equivalent resistance of the four resistors shown in (a) can be reduced in steps to an equivalent 14-Ω resistor.

When this current enters the junction at b, it splits and part of the current passes through the 6-Ω resistor (I_1) and part goes through the 3-Ω resistor (I_2). Since the potential difference across these resistors, V_{bc}, is the *same* (they are in parallel), we see that $6I_1 = 3I_2$, or $I_2 = 2I_1$. Using this result and the fact that $I_1 + I_2 = 3$ A, we find that $I_1 = 1$ A and $I_2 = 2$ A. We could have guessed this from the start by noting that the current through the 3-Ω resistor has to be twice the current through the 6-Ω resistor in view of their relative resistances and the fact that the same voltage is applied to each of them.

As a final check, note that $V_{bc} = 6I_1 = 3I_2 = 6$ V and $V_{ab} = 12I = 36$ V; therefore, $V_{ac} = V_{ab} + V_{bc} = 42$ V, as it must.

EXAMPLE 28.4 Three Resistors in Parallel
Three resistors are connected in parallel as in Figure 28.7. A potential difference of 18 V is maintained between points a and b. (a) Find the current in each resistor.

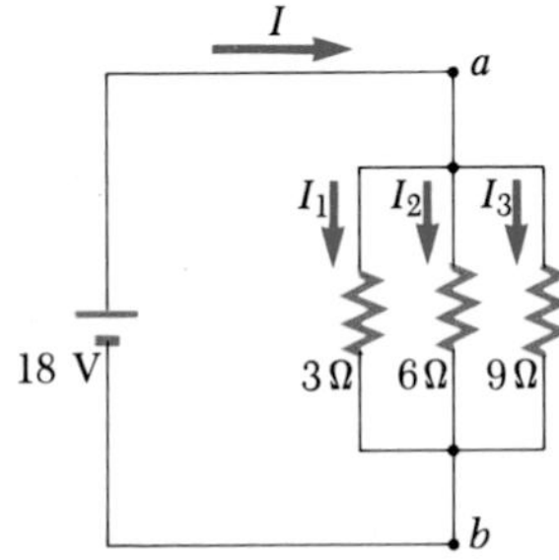

Figure 28.7 (Example 28.4) Three resistors connected in parallel. The voltage across each resistor is 18 V.

The resistors are in parallel, and the potential difference across each is 18 V. Applying $V = IR$ to each resistor gives

$$I_1 = \frac{V}{R_1} = \frac{18\ \text{V}}{3\ \Omega} = \boxed{6\ \text{A}}$$

$$I_2 = \frac{V}{R_2} = \frac{18\ \text{V}}{6\ \Omega} = \boxed{3\ \text{A}}$$

$$I_3 = \frac{V}{R_3} = \frac{18\ \text{V}}{9\ \Omega} = \boxed{2\ \text{A}}$$

(b) Calculate the power dissipated by each resistor and the total power dissipated by the three resistors.

Applying $P = I^2R$ to each resistor gives

3-Ω: $P_1 = I_1{}^2R_1 = (6\ \text{A})^2(3\ \Omega) = \boxed{108\ \text{W}}$

6-Ω: $P_2 = I_2{}^2R_2 = (3\ \text{A})^2(6\ \Omega) = \boxed{54\ \text{W}}$

9-Ω: $P_3 = I_3{}^2R_3 = (2\ \text{A})^2(9\ \Omega) = \boxed{36\ \text{W}}$

This shows that the smallest resistor dissipates the most power since it carries the most current. (Note that you can also use $P = V^2/R$ to find the power dissipated by each resistor.) Summing the three quantities gives a total power of 198 W.

(c) Calculate the equivalent resistance of the three resistors, and from this result find the total power dissipated.

We can use Equation 28.7 to find R_{eq}:

$$\frac{1}{R_{eq}} = \frac{1}{3} + \frac{1}{6} + \frac{1}{9}$$

$$R_{eq} = \boxed{\frac{18}{11}\ \Omega}$$

Exercise 1 Use the result for R_{eq} to calculate the total power dissipated in the circuit.

Answer 198 W.

EXAMPLE 28.5 Finding R_{eq} by Symmetry Arguments

Consider the five resistors connected as shown in Figure 28.8a. Find the equivalent resistance of the combination of resistors between points *a* and *b*.

Solution In this type of problem, it is convenient to assume a current entering junction *a* and then apply symmetry arguments. Because of the symmetry in the circuit (all 1-Ω resistors in the outside loop), the currents in branches *ac* and *ad* must be equal; hence, the potentials at points *c* and *d* must be equal. Since $V_c = V_d$, points *c* and *d* may be connected together without affecting the circuit, as in Figure 28.8b. Thus, the 5-Ω resistor may be removed from the circuit, and the circuit may be reduced as shown in Figures 28.8c and 28.8d. From this reduction, we see that the equivalent resistance of the combination is 1 Ω. Note that the result is 1 Ω regardless of what resistor is connected between *c* and *d*.

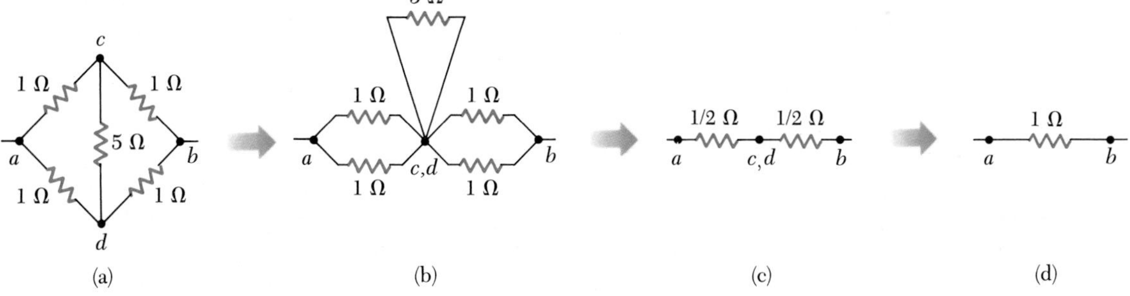

Figure 28.8 (Example 28.5). Because of the symmetry in this circuit, the 5-Ω resistor does not contribute to the resistance between points *a* and *b* and can be disregarded.

28.3 KIRCHHOFF'S RULES

As we saw in the previous section, simple circuits can be analyzed using Ohm's law and the rules for series and parallel combinations of resistors. Very often it is not possible to reduce a circuit to a single loop. The procedure for analyzing more complex circuits is greatly simplified by the use of two simple rules called **Kirchhoff's rules:**

1. The sum of the currents entering any junction must equal the sum of the currents leaving that junction. (A **junction** is any point in the circuit where a current can split.)

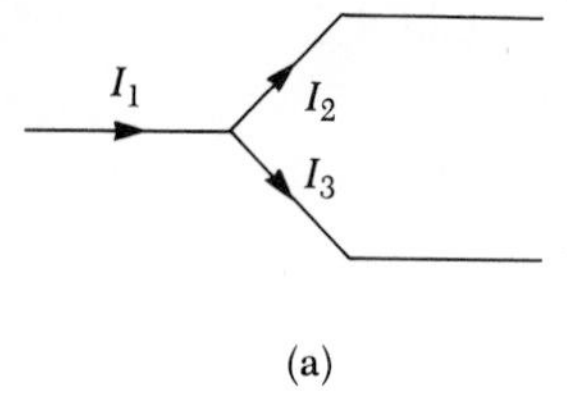

(a)

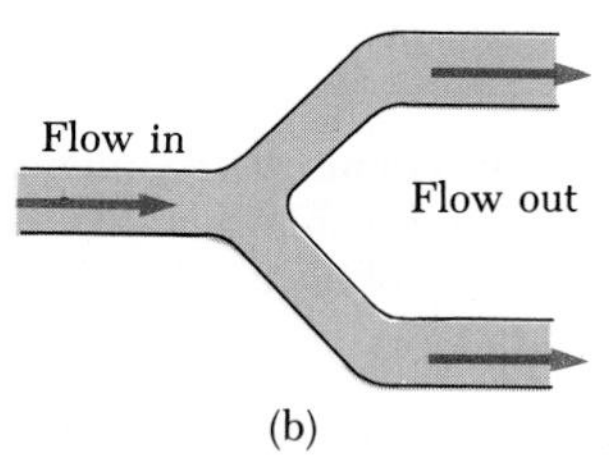

(b)

Figure 28.9 (a) A schematic diagram illustrating Kirchhoff's junction rule. Conservation of charge requires that whatever current enters a junction must leave that junction. Therefore, in this case, $I_1 = I_2 + I_3$. (b) A mechanical analog of the junction rule: the flow out must equal the flow in.

2. The algebraic sum of the changes in potential across all of the elements around any closed circuit loop must be *zero*.

The first rule is a statement of **conservation of charge.** That is, whatever current enters a given point in a circuit must leave that point, since charge cannot build up at a point. If we apply this rule to the junction shown in Figure 28.9a, we get

$$I_1 = I_2 + I_3$$

Figure 28.9b represents a mechanical analog to this situation, in which water flows through a branched pipe with no leaks. The flow rate into the pipe equals the total flow rate out of the two branches.

The second rule follows from **conservation of energy.** That is, any charge that moves around *any* closed loop in a circuit (it starts and ends at the same point) must gain as much energy as it loses. Its energy may decrease in the form of a potential drop, $-IR$, across a resistor or as the result of having the charge go the reverse direction through a source of emf. In a practical application of the latter case, electrical energy is converted into chemical energy when a battery is charged; similarly, electrical energy may be converted into mechanical energy for operating a motor.

As an aid in applying the second rule, the following calculational tools should be noted. These points are summarized in Fig. 28.10.

1. If a resistor is traversed in the direction of the current, the change in potential across the resistor is $-IR$ (Fig. 28.10a).
2. If a resistor is traversed in the direction *opposite* the current, the change in potential across the resistor is $+IR$ (Figure 28.10b).
3. If a source of emf is traversed in the direction of the emf (from $-$ to $+$ on the terminals), the change in potential is $+\mathcal{E}$ (Fig. 28.10c).
4. If a source of emf is traversed in the direction opposite the emf (from $+$ to $-$ on the terminals), the change in potential is $-\mathcal{E}$ (Fig. 28.10d).

(a)

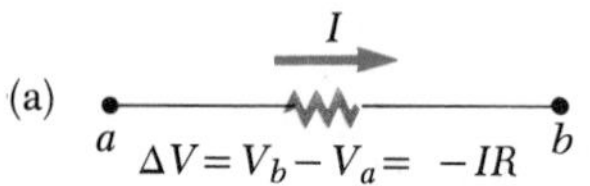

(b) I a b $\Delta V = V_b - V_a = +IR$

(c) $\mathcal{E}$ a b $\Delta V = V_b - V_a = +\mathcal{E}$

(d) $\mathcal{E}$ a b $\Delta V = V_b - V_a = -\mathcal{E}$

Figure 28.10 Rules for determining the potential changes across a resistor and a battery, assuming the battery has no internal resistance.

There are limitations on the number of times you can use the junction rule and the loop rule. The junction rule can be used as often as needed so long as each time you write an equation, you include in it a current that has not been used in a previous junction rule equation. In general, the number of times the junction rule must be used is one fewer than the number of junction points in the circuit. The loop rule can be used as often as needed so long as a new circuit element (resistor or battery) or a new current appears in each new equation. In general, *the number of independent equations you need must at least equal the number of unknowns in order to solve a particular circuit problem.*

Complex networks with many loops and junctions generate large numbers of independent, linear equations and a corresponding large number of unknowns. Such situations can be handled formally using matrix algebra. Computer programs can also be written to solve for the unknowns.

The following examples illustrate the use of Kirchhoff's rules in analyzing circuits. In all cases, it is assumed that the circuits have reached steady-state conditions, that is, the currents in the various branches are constant. If a capacitor is included as an element in one of the branches, *it acts as an open circuit,* that is, the current in the branch containing the capacitor will be zero under steady-state conditions.

PROBLEM-SOLVING STRATEGY AND HINTS: KIRCHHOFF'S RULES

1. First, draw the circuit diagram and assign labels and symbols to all the known and unknown quantities. You must assign a *direction* to the currents in each part of the circuit. Do not be alarmed if you guess the direction of a current incorrectly; the result will have a negative value, but *its magnitude will be correct.* Although the assignment of current directions is arbitrary, you must adhere *rigorously* to the assigned directions when applying Kirchhoff's rules.
2. Apply the junction rule (Kirchhoff's first rule) to any junction in the circuit which provides a relation between the various currents. (This step is easy!)
3. Now apply Kirchhoff's second rule to as many loops in the circuit as are needed to solve for the unknowns. In order to apply this rule, you must correctly identify the change in potential as you cross each element in traversing the closed loop (either clockwise or counterclockwise). Watch out for signs! We suggest that you follow the four "rules-of-thumb" listed above and summarized in Figure 28.10.
4. Finally, you must solve the equations simultaneously for the unknown quantities. Be careful in your algebraic steps, and check your numerical answers for consistency.

EXAMPLE 28.6 A Single-Loop Circuit

A single-loop circuit contains two external resistors and two sources of emf as shown in Figure 28.11. The internal resistances of the batteries have been neglected. (a) Find the current in the circuit.

There are no junctions in this single-loop circuit, and so the current is the same in all elements. Let us assume that the current is in the clockwise direction as shown in Figure 28.11. Traversing the circuit in the clockwise direction, starting at point a, we see that $a \rightarrow b$ represents a potential increase of $+\mathcal{E}_1$, $b \rightarrow c$ represents a potential decrease of $-IR_1$, $c \rightarrow d$ represents a potential decrease of $-\mathcal{E}_2$, and $d \rightarrow a$ represents a potential decrease of $-IR_2$. Applying Kirchhoff's second rule gives

$$\sum_i \Delta V_i = 0$$

$$\mathcal{E}_1 - IR_1 - \mathcal{E}_2 - IR_2 = 0$$

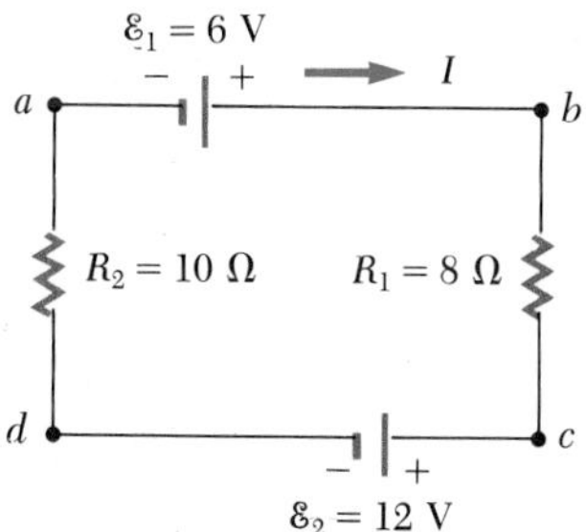

Figure 28.11 (Example 28.6) A series circuit containing two batteries and two resistors, where the polarities of the batteries are in opposition to each other.

Solving for I and using the values given in Figure 28.11, we get

$$I = \frac{\mathcal{E}_1 - \mathcal{E}_2}{R_1 + R_2} = \frac{6\ \text{V} - 12\ \text{V}}{8\ \Omega + 10\ \Omega} = -\frac{1}{3}\ \text{A}$$

The negative sign for I indicates that the direction of the current is *opposite* the assumed direction, or *counterclockwise.*

(b) What is the power lost in each resistor?

$$P_1 = I^2R_1 = (\tfrac{1}{3}\text{A})^2(8\ \Omega) = \frac{8}{9}\ \text{W}$$

$$P_2 = I^2R_2 = (\tfrac{1}{3}\text{A})^2(10\ \Omega) = \frac{10}{9}\ \text{W}$$

Hence, the total power lost is $P_1 + P_2 = 2$ W. Note that the 12-V battery delivers power $I\mathcal{E}_2 = 4$ W. Half of this power is delivered to the external resistors. The other half is delivered to the 6-V battery, which is being charged by the 12-V battery. If we had included the internal resistances of the batteries, some of the power would be dissipated as heat in the batteries, so that *less* power would be delivered to the 6-V battery.

EXAMPLE 28.7 Applying Kirchhoff's Rules

Find the currents I_1, I_2, and I_3 in the circuit shown in Figure 28.12.

We shall choose the directions of the currents as shown in Figure 28.12. Applying Kirchhoff's first rule to junction c gives

$$(1) \qquad I_1 + I_2 = I_3$$

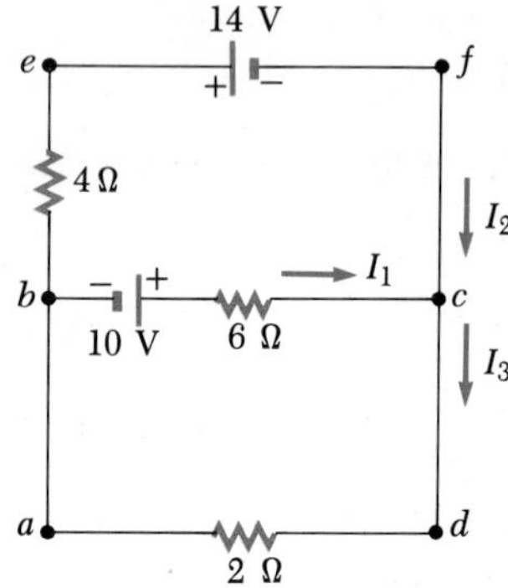

Figure 28.12 (Example 28.7) A circuit containing three loops.

There are *three* loops in the circuit, *abcda, befcb,* and *aefda* (the outer loop). We need only *two* loop equations to determine the unknown currents. The third loop equation would give no new information. Applying Kirchhoff's second rule to loops *abcda* and *befcb* and traversing these loops in the clockwise direction, we obtain the following expressions:

(2) Loop *abcda:* $10\text{ V} - (6\ \Omega)I_1 - (2\ \Omega)I_3 = 0$

(3) Loop *befcb:* $-14\text{ V} - 10\text{ V} + (6\ \Omega)I_1 - (4\ \Omega)I_2 = 0$

Note that in loop *befcb*, a positive sign is obtained when traversing the 6-Ω resistor since the direction of the path is opposite the direction of the current I_1. A third loop equation for *aefda* gives $-14 = 2I_3 + 4\ I_2$, which is just the sum of (2) and (3). Expressions (1), (2), and (3) represent three linear, independent equations with three unknowns. We can solve the problem as follows: Substituting (1) into (2) gives

$$10 - 6I_1 - 2(I_1 + I_2) = 0$$

(4) $$10 = 8I_1 + 2I_2$$

Dividing each term in (3) by 2 and rearranging the equation gives

(5) $$-12 = -3I_1 + 2I_2$$

Subtracting (5) from (4) eliminates I_2, giving

$$22 = 11I_1$$

$$I_1 = 2\text{ A}$$

Using this value of I_1 in (5) gives a value for I_2:

$$2I_2 = 3I_1 - 12 = 3(2) - 12 = -6$$

$$I_2 = -3\text{ A}$$

Finally, $I_3 = I_1 + I_2 = -1$ A. Hence, the currents have the values

$I_1 = 2\text{ A}$ $\quad I_2 = -3\text{ A}$ $\quad I_3 = -1\text{ A}$

The fact that I_2 and I_3 are both negative indicates only that we chose the *wrong* direction for these currents. However, the numerical values are correct.

Exercise 2 Find the potential difference between points *b* and *c*.
Answer $V_b - V_c = 2$ V.

EXAMPLE 28.8 A Multi-Loop Circuit

The multiloop circuit in Figure 28.13 contains three resistors, three batteries, and one capacitor. (a) Under steady-state conditions, find the unknown currents.

First note that *the capacitor represents an open circuit, and hence there is no current along path ghab under steady-state conditions.* Therefore, $I_{gf} = I_1$. Labeling the currents as shown in Figure 28.13 and applying Kirchhoff's first rule to junction *c*, we get

(1) $$I_1 + I_2 = I_3$$

Kirchhoff's second rule applied to loops *defcd* and *cfgbc* gives

(2) Loop *defcd:* $4\text{ V} - (3\ \Omega)I_2 - (5\ \Omega)I_3 = 0$

(3) Loop *cfgbc:* $8\text{ V} - (5\ \Omega)I_1 + (3\ \Omega)I_2 = 0$

From (1) we see that $I_1 = I_3 - I_2$, which when substituted into (3) gives

(4) $$8\text{ V} - (5\ \Omega)I_3 + (8\ \Omega)I_2 = 0$$

Subtracting (4) from (2), we eliminate I_3 and find

$$I_2 = -\tfrac{4}{11}\text{ A} = -0.364\text{ A}$$

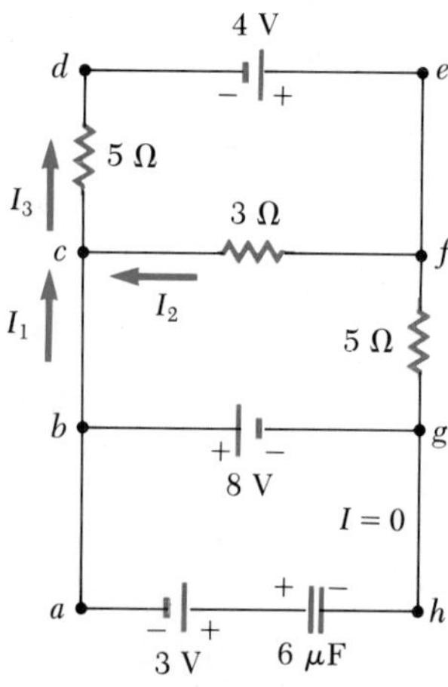

Figure 28.13 (Example 28.8) A multiloop circuit. Note that Kirchhoff's loop equation can be applied to *any* closed loop, including one containing the capacitor.

Since I_2 is negative, we conclude that I_2 is from c to f through the 3-Ω resistor. Using this value of I_2 in (3) and (1) gives the following values for I_1 and I_3:

$$I_1 = 1.38 \text{ A} \qquad I_3 = 1.02 \text{ A}$$

Under state-steady conditions, the capacitor represents an *open* circuit, and so there is no current in the branch *ghab*.

(b) What is the charge on the capacitor?

We can apply Kirchhoff's second rule to loop *abgha* (or any loop that contains the capacitor) to find the potential difference V_c across the capacitor:

$$-8 \text{ V} + V_c - 3 \text{ V} = 0$$

$$V_c = 11.0 \text{ V}$$

Since $Q = CV_c$, we find that the charge on the capacitor is equal to

$$Q = (6\ \mu\text{F})(11.0 \text{ V}) = 66.0\ \mu\text{C}$$

Why is the left side of the capacitor positively charged?

Exercise 3 Find the voltage across the capacitor by traversing any other loop, such as the outside loop.
Answer 11.0 V.

28.4 *RC* CIRCUITS

So far we have been concerned with circuits with constant currents, or so-called *steady-state circuits.* We shall now consider circuits containing capacitors, in which the currents may vary in time. When a potential difference is first applied across a capacitor, the rate at which it charges depends on its capacitance and on the resistance in the circuit.

Charging a Capacitor

Consider the series circuit shown in Figure 28.14. Let us assume that the capacitor is initially uncharged. There is no current when the switch S is open (Fig. 28.14b). If the switch is closed at $t = 0$, charges will begin to flow, setting up a current in the circuit, and the capacitor will begin to charge (Fig. 28.14c). Note that during the charging process, charges do not jump across the plates of the capacitor since the gap between the plates represents an open circuit. Instead, charge is transferred from one plate to the other through the resistor, switch, and battery until the capacitor is fully charged. The value of the maximum charge depends on the emf of the battery. Once the maximum charge is reached, the current in the circuit is zero.

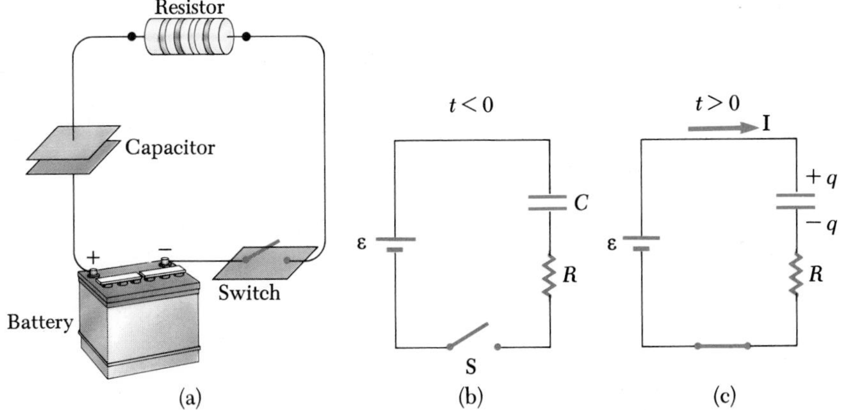

Figure 28.14 (a) A capacitor in series with a resistor, battery, and switch. (b) Circuit diagram representing this system before the switch is closed, $t < 0$. (c) Circuit diagram after the switch is closed, $t > 0$.

To put this discussion on a quantitative basis, let us apply Kirchhoff's second rule to the circuit *after* the switch is closed. This gives

$$\mathcal{E} - IR - \frac{q}{C} = 0 \tag{28.9}$$

where IR is the potential drop across the resistor and q/C is the potential drop across the capacitor. Note that q and I are *instantaneous* values of the charge and current, respectively, as the capacitor is being charged.

We can use Equation 28.9 to find the initial current in the circuit and the maximum charge on the capacitor. At $t = 0$, when the switch is closed, the charge on the capacitor is zero, and from Equation 28.9 we find that the initial current in the circuit, I_0, is a maximum and equal to

Maximum current

$$I_0 = \frac{\mathcal{E}}{R} \qquad \text{(current at } t = 0\text{)} \tag{28.10}$$

At this time, *the potential drop is entirely across the resistor.* Later, when the capacitor is charged to its maximum value Q, charges cease to flow, the current in the circuit is zero, and *the potential drop is entirely across the capacitor.* Substituting $I = 0$ into Equation 28.9 gives the following expression for Q:

Maximum charge on the capacitor

$$Q = C\mathcal{E} \qquad \text{(maximum charge)} \tag{28.11}$$

To determine analytical expressions for the time dependence of the charge and current, we must solve Equation 28.9, a single equation containing two variables, q and I. In order to do this, let us differentiate Equation 28.9 with respect to time. Since $\mathcal{E}$ is a constant, $d\mathcal{E}/dt = 0$ and we get

$$\frac{d}{dt}\left(\mathcal{E} - \frac{q}{C} - IR\right) = 0 - \frac{1}{C}\frac{dq}{dt} - R\frac{dI}{dt} = 0$$

Recalling that $I = dq/dt$, we can express this equation in the form

$$R\frac{dI}{dt} + \frac{I}{C} = 0$$

$$\frac{dI}{I} = -\frac{1}{RC}\,dt \tag{28.12}$$

Since R and C are constants, this can be integrated using the initial condition that at $t = 0$, $I = I_0$:

$$\int_{I_0}^{I} \frac{dI}{I} = -\frac{1}{RC}\int_0^t dt$$

$$\ln\left(\frac{I}{I_0}\right) = -\frac{t}{RC}$$

Current versus time

$$I(t) = I_0\, e^{-t/RC} = \frac{\mathcal{E}}{R}\, e^{-t/RC} \tag{28.13}$$

where e is the base of the natural logarithm and $I_0 = \mathcal{E}/R$ is the initial current.

In order to find the charge on the capacitor as a function of time, we can substitute $I = dq/dt$ into Equation 28.13 and integrate once more:

$$\frac{dq}{dt} = \frac{\mathcal{E}}{R} e^{-t/RC}$$

$$dq = \frac{\mathcal{E}}{R} e^{-t/RC}\, dt$$

We can integrate this expression using the condition that $q = 0$ at $t = 0$:

$$\int_0^q dq = \frac{\mathcal{E}}{R} \int_0^t e^{-t/RC}\, dt$$

In order to integrate the right side of this expression, we use the fact that $\int e^{-ax}\, dx = -\frac{1}{a} e^{-ax}$. The result of the integration gives

Charge versus time for a capacitor being charged

$$q(t) = C\mathcal{E}[1 - e^{-t/RC}] = Q[1 - e^{-t/RC}] \quad (28.14)$$

where $Q = C\mathcal{E}$ is the *maximum* charge on the capacitor.

Plots of Equations 28.13 and 28.14 are shown in Figure 28.15. Note that the charge is zero at $t = 0$ and approaches the maximum value of $C\mathcal{E}$ as $t \to \infty$ (Fig. 28.15a). Furthermore, the current has its maximum value of $I_0 = \mathcal{E}/R$ at $t = 0$ and decays exponentially to zero as $t \to \infty$ (Fig. 28.15b). The quantity *RC*, which appears in the exponential of Equations 28.13 and 28.14, is called the **time constant,** τ, of the circuit. It represents the time it takes the current to decrease to $1/e$ of its initial value; that is, in a time τ, $I = e^{-1}I_0 = 0.37I_0$. In a time 2τ, $I = e^{-2}I_0 = 0.135I_0$, and so forth. Likewise, in a time τ the charge will increase from zero to $C\mathcal{E}[1 - e^{-1}] = 0.63C\mathcal{E}$.

The following dimensional analysis shows that τ has the unit of time:

$$[\tau] = [RC] = \left[\frac{V}{I} \times \frac{Q}{V}\right] = \left[\frac{Q}{Q/T}\right] = [T]$$

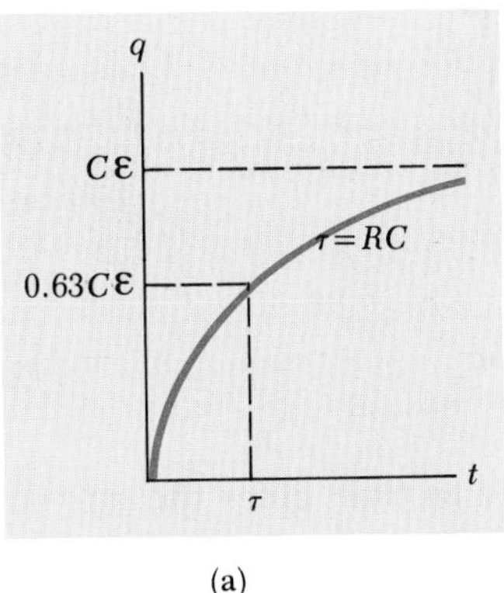

(a)

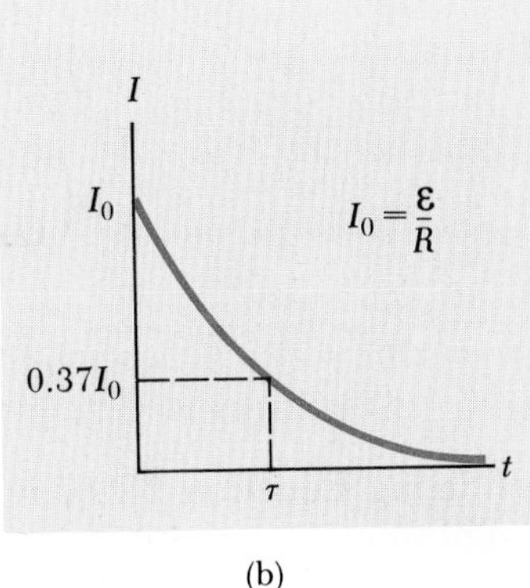

(b)

Figure 28.15 (a) Plot of capacitor charge versus time for the circuit shown in Figure 28.14. After one time constant, τ, the charge is 63% of the maximum value, $C\mathcal{E}$. The charge approaches its maximum value as t approaches infinity. (b) Plot of current versus time for the *RC* circuit shown in Figure 28.14. The current has its maximum value, $I_0 = \mathcal{E}/R$, at $t = 0$ and decays to zero exponentially as t approaches infinity. After one time constant, τ, the current decreases to 37% of its initial value.

The work done by the battery during the charging process is $Q\mathcal{E} = C\mathcal{E}^2$. After the capacitor is fully charged, the energy stored in the capacitor is $\frac{1}{2}Q\mathcal{E} = \frac{1}{2}C\mathcal{E}^2$, which is just half the work done by the battery. It is left as a problem to show that the remaining half of the energy supplied by the battery goes into joule heat in the resistor (Problem 82).

Discharging a Capacitor

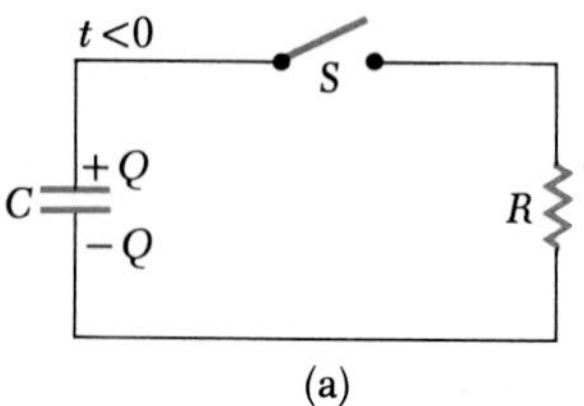

(a)

t>0 S I +q C R −q

(b)

Figure 28.16 (a) A charged capacitor connected to a resistor and a switch, which is open at $t < 0$. (b) After the switch is closed, a nonsteady current is set up in the direction shown and the charge on the capacitor decreases exponentially with time.

Now consider the circuit in Figure 28.16, consisting of a capacitor with an initial charge Q, a resistor, and a switch. When the switch is open (Fig. 28.16a), there is a potential difference of Q/C across the capacitor and zero potential difference across the resistor since $I = 0$. If the switch is closed at $t = 0$, the capacitor begins to discharge through the resistor. At some time during the discharge, the current in the circuit is I and the charge on the capacitor is q (Fig. 28.16b). From Kirchhoff's second rule, we see that the potential drop across the resistor, IR, must equal the potential difference across the capacitor, q/C:

$$IR = \frac{q}{C} \tag{28.15}$$

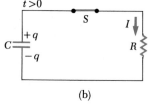

However, the current in the circuit must equal the rate of *decrease* of charge on the capacitor. That is, $I = -dq/dt$, and so Equation 28.15 becomes

$$-R\frac{dq}{dt} = \frac{q}{C}$$

$$\frac{dq}{q} = -\frac{1}{RC}\,dt \tag{28.16}$$

Integrating this expression using the fact that $q = Q$ at $t = 0$ gives

$$\int_Q^q \frac{dq}{q} = -\frac{1}{RC}\int_0^t dt$$

$$\ln\left(\frac{q}{Q}\right) = -\frac{t}{RC}$$

Charge versus time for a discharging capacitor

$$q(t) = Q\,e^{-t/RC} \tag{28.17}$$

Differentiating Equation 28.17 with respect to time gives the current as a function of time:

$$I(t) = -\frac{dq}{dt} = \frac{Q}{RC}\,e^{-t/RC} = I_0\,e^{-t/RC} \tag{28.18}$$

where the initial current $I_0 = Q/RC$. Therefore, we see that both the charge on the capacitor and the current decay exponentially at a rate characterized by the time constant $\tau = RC$.

EXAMPLE 28.9 Charging a Capacitor in an *RC* Circuit

An uncharged capacitor and a resistor are connected in series to a battery as in Figure 28.17. If $\mathcal{E} = 12$ V, $C = 5\ \mu$F, and $R = 8 \times 10^5\ \Omega$, find the time constant of the circuit, the maximum charge on the capacitor, the maximum current in the circuit, and the charge and current as a function of time.

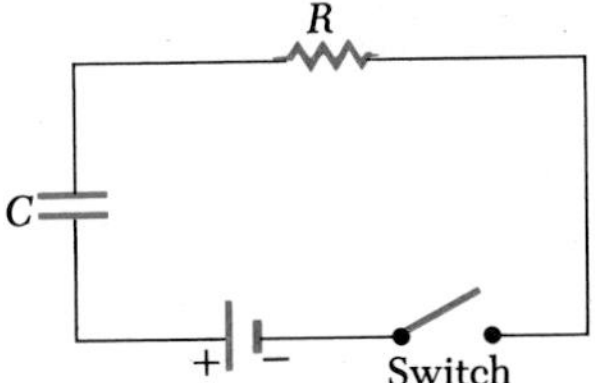

Figure 28.17 (Example 28.9) The switch of this series *RC* circuit is closed at $t = 0$.

Solution The time constant of the circuit is $\tau = RC = (8 \times 10^5\ \Omega)(5 \times 10^{-6}\ \text{F}) = 4$ s. The maximum charge on the capacitor is $Q = C\mathcal{E} = (5 \times 10^{-6}\ \text{F})(12\ \text{V}) = 60\ \mu$C. The maximum current in the circuit is $I_0 = \mathcal{E}/R = (12\ \text{V})/(8 \times 10^5\ \Omega) = 15\ \mu$A. Using these values and Equations 28.13 and 28.14, we find that

$$q(t) = 60[1 - e^{-t/4}]\ \mu\text{C}$$

$$I(t) = 15\, e^{-t/4}\ \mu\text{A}$$

Graphs of these functions are given in Figure 28.18.

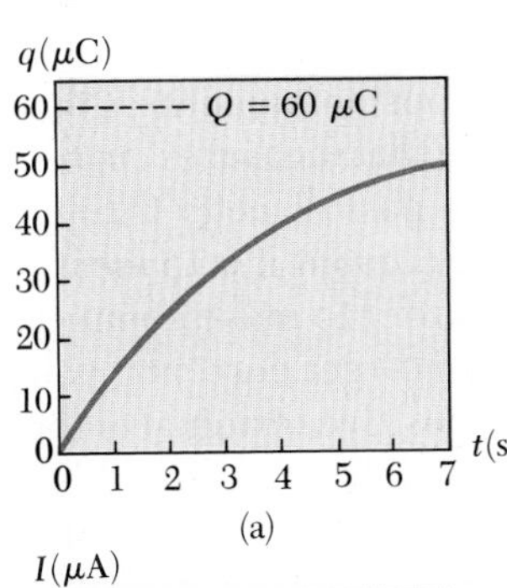

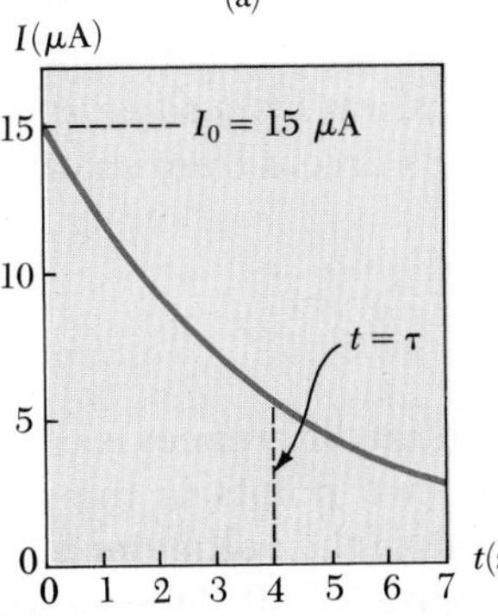

Figure 28.18 (Example 28.9) Plots of (a) charge versus time and (b) current versus time for the *RC* circuit shown in Figure 28.17, with $\mathcal{E} = 12$ V, $R = 8 \times 10^5\ \Omega$, and $C = 5\ \mu$F.

Exercise 4 Calculate the charge on the capacitor and the current in the circuit after one time constant has elapsed.

Answer 37.9 μC, 5.52 μA.

EXAMPLE 28.10 Discharging a Capacitor in an *RC* Circuit □

Consider a capacitor C being discharged through a resistor R as in Figure 28.16. (a) After how many time constants will the charge on the capacitor drop to one-fourth of its initial value?

Solution The charge on the capacitor varies with time according to Equation 28.17,

$$q(t) = Qe^{-t/RC}$$

where Q is the initial charge on the capacitor. To find the time it takes the charge q to drop to one-fourth of its initial value, we substitute $q(t) = Q/4$ into this expression and solve for t:

$$\tfrac{1}{4}Q = Qe^{-t/RC}$$

or

$$\tfrac{1}{4} = e^{-t/RC}$$

Taking logarithms of both sides, we find

$$-\ln 4 = -\frac{t}{RC}$$

or

$$t = RC \ln 4 = 1.39RC$$

(b) The energy stored in the capacitor decreases with time as it discharges. After how many time constants will this stored energy reduce to one-fourth of its initial value?

Solution Using Equations 26.12 and 28.17, we can express the energy stored in the capacitor at any time t as

$$U = \frac{q^2}{2C} = \frac{Q^2}{2C} e^{-2t/RC} = U_0 e^{-2t/RC}$$

where U_0 is the initial energy stored in the capacitor. As in part (a), we now set $U = U_0/4$ and solve for t:

$$\tfrac{1}{4}U_0 = U_0 e^{-2t/RC}$$

$$\tfrac{1}{4} = e^{-2t/RC}$$

Again, taking logarithms of both sides and solving for t gives

$$t = \tfrac{1}{2}RC \ln 4 = 0.693RC$$

Exercise 5 After how many time constants will the current in the RC circuit drop to one-half of its initial value?
Answer $0.693RC$

EXAMPLE 28.11 Heat Loss in a Resistor
A 5-μF capacitor is charged to a potential difference of 800 V and is then discharged through a 25-kΩ resistor, as in Figure 28.16. What is the total energy which has been lost as joule heating in the resistor when the capacitor is fully discharged?

Solution We shall solve this problem in *two* ways. The first method, which is actually quite simple, is to note that the initial energy in the system equals the energy stored in the capacitor, given by $C\mathcal{E}^2/2$. Once the capacitor is fully discharged, the energy stored in it is zero. Since energy is conserved, the initial energy stored in the capacitor is transformed into thermal energy dissipated in the resistor. Using the given values of C and $\mathcal{E}$, we find

$$\text{Energy} = \tfrac{1}{2}C\mathcal{E}^2 = \tfrac{1}{2}(5 \times 10^{-6}\ \text{F})(800\ \text{V})^2 = 1.60\ \text{J}$$

The second method, which is more difficult but perhaps more instructive, is to note that as the capacitor discharges through the resistor, the rate at which heat is generated in the resistor (or the power loss) is given by RI^2, where I is the instantaneous current given by Equation 28.18. Since the power is defined as the rate of change of energy, we conclude that the energy lost in the resistor in the form of heat must equal the time integral of $RI^2\,dt$. That is

$$\text{Energy} = \int_0^\infty RI^2\,dt = \int_0^\infty R(I_0 e^{-t/RC})^2\,dt$$

To evaluate this integral, we note that the initial current $I_0 = \mathcal{E}/R$, and all parameters are constants except for t. Thus, we find

$$\text{Energy} = \frac{\mathcal{E}^2}{R}\int_0^\infty e^{-2t/RC}\,dt$$

The integral in the expression above has a value of $RC/2$, so we find

$$\text{Energy} = \tfrac{1}{2}C\mathcal{E}^2$$

which agrees with the simpler approach, as it must. Note that this second approach can be used to find the energy lost as heat at *any* time after the switch is closed by simply replacing the upper limit in the integral by that specific value of t.

Exercise 6 Show that the integral given in this example has the value of $RC/2$.

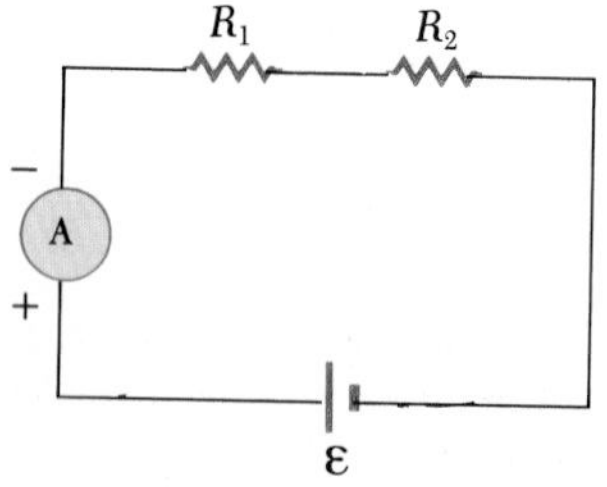

Figure 28.19 The current in a circuit can be measured with an ammeter connected in series with the resistor and battery. An ideal ammeter has zero resistance.

*28.5 ELECTRICAL INSTRUMENTS

The Ammeter

Current is one of the most important quantities that one would like to measure in an electric circuit. A device that measures current is called an **ammeter.** The current to be measured must pass directly through the ammeter, so that the ammeter is in series with the current it is to measure, as in Figure 28.19. The wires usually must be cut in order to make connections to the ammeter. When using an ammeter to measure direct currents, you must be sure to connect it such that current enters the positive terminal of the instrument and exits at the negative terminal. **Ideally, an ammeter should have zero resistance so as not to alter the current being measured.** In the circuit shown in Figure 28.19, this condition requires that the ammeter's resistance be small compared to $R_1 + R_2$. Since any ammeter always has some resistance, its presence in the circuit will slightly reduce the current from its value when the ammeter is not present.

The Voltmeter

A device that measures potential differences is called a **voltmeter.** The potential difference between any two points in the circuit can be measured by simply attaching the terminals of the voltmeter between these points without breaking the circuit, as in Figure 28.20. The potential difference across resistor R_2 is measured by connecting the voltmeter in parallel with R_2. Again, it is necessary to observe the polarity of the instrument. The positive terminal of the voltmeter must be connected to the end of the resistor at the higher potential, and the negative terminal to the low-potential end of the resistor. **An ideal voltmeter has infinite resistance so that no current will pass through it.** In Figure 28.20, this condition requires that the voltmeter must have a resistance that is very large compared to R_2. In practice, if this condition is not met, one should make corrections for the known resistance of the voltmeter.

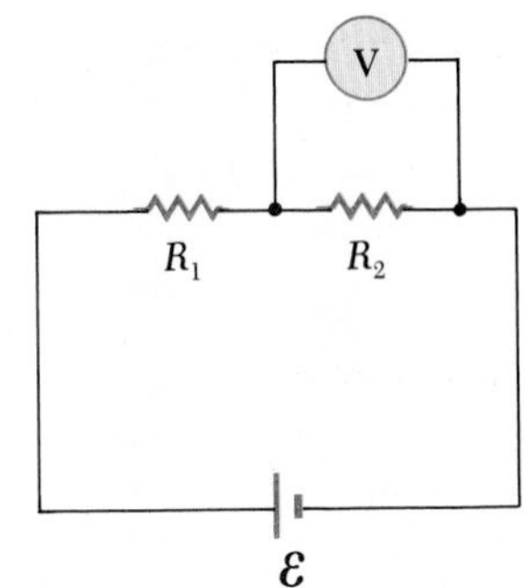

Figure 28.20 The potential difference across a resistor can be measured with a voltmeter connected in parallel with the resistor. An ideal voltmeter has infinite resistance and does not affect the circuit.

The Galvanometer

The **galvanometer** is the main component used in the construction of ammeters and voltmeters. The essential features of a common type, called the *D'Arsonval galvanometer*, are shown in Figure 28.21. It consists of a coil of wire mounted such that it is free to rotate on a pivot in a magnetic field provided by a permanent magnet. The basic operation of the galvanometer makes use of the fact that a torque acts on a current loop in the presence of a magnetic field. (The reason for this is discussed in detail in Chapter 29.) The torque experienced by the coil is proportional to the current through it. This means that the larger the current, the larger the torque and the more the coil will rotate before the spring tightens enough to stop the rotation. Hence, the amount of deflection is proportional to the current. Once the instrument is properly calibrated, it can be used in conjunction with other circuit elements to measure either currents or potential differences.

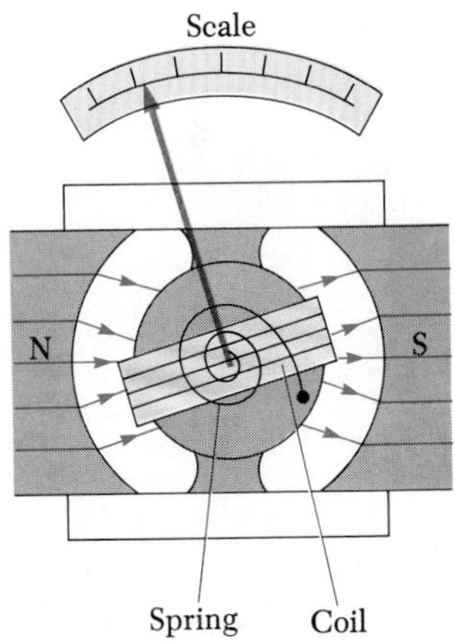

Figure 28.21 The principal components of a D'Arsonval galvanometer. When current passes through the coil, situated in a magnetic field, the magnetic torque causes the coil to twist. The angle through which the coil rotates is proportional to the current through it because of the spring's torque.

A typical off-the-shelf galvanometer is often not suitable for use as an ammeter. One of the main reasons for this is that a typical galvanometer has a resistance of about 60 Ω. An ammeter resistance this large would considerably alter the current in the circuit in which it is placed. This can easily be understood by considering the following example. Suppose you were to construct a simple series circuit containing a 3-V battery and a 3-Ω resistor. The current in such a circuit is 1 A. However, if you insert a 60-Ω galvanometer in the circuit to measure the current, the total resistance of the circuit would now be 63 Ω, and the current would be reduced to 0.048 A.

A second factor that limits the use of a galvanometer as an ammeter is the fact that a typical galvanometer will give a full-scale deflection for very low currents, of the order of 1 mA or less. Consequently, such a galvanometer cannot be used directly to measure currents greater than this. However, one can convert a galvanometer into an ammeter by simply placing a resistor, R_p, in *parallel* with the galvanometer as in Figure 28.22a. The value of R_p, some-

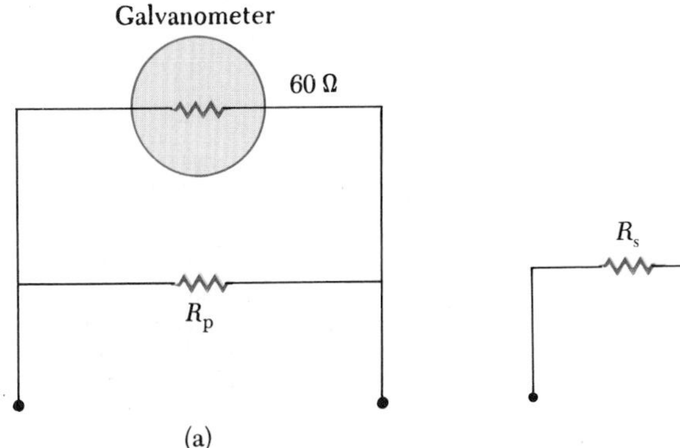

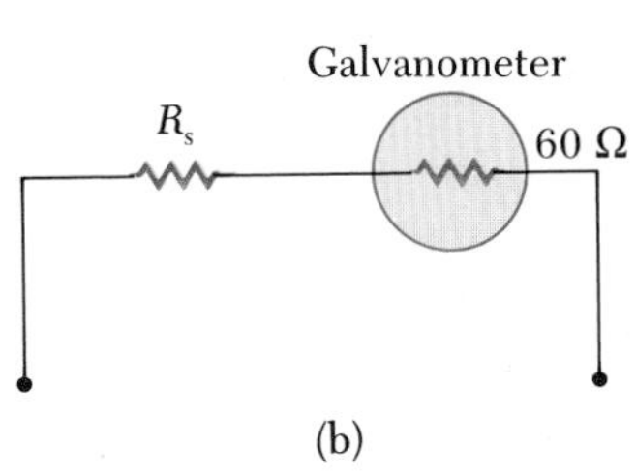

Figure 28.22 (a) When a galvanometer is to be used as an ammeter, a resistor, R_p, is connected in parallel with the galvanometer. (b) When the galvanometer is used as a voltmeter, a resistor, R_s, is connected in series with the galvanometer.

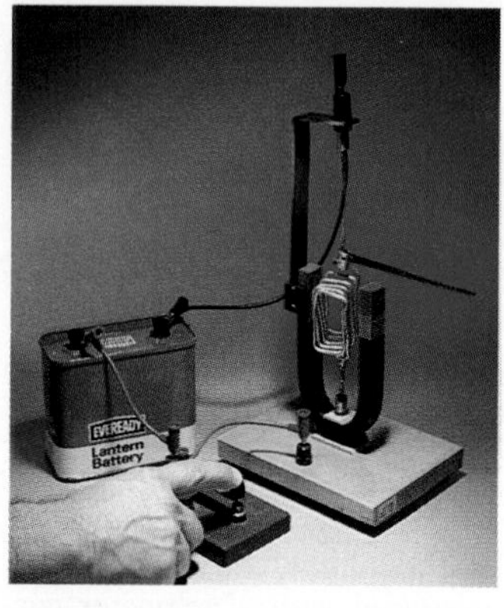

Large-scale model of a galvanometer movement. Why does the coil rotate about the vertical axis after the switch is closed? (Courtesy of Henry Leap and Jim Lehman)

times called the *shunt resistor*, must be very small compared to the resistance of the galvanometer so that most of the current to be measured passes through the shunt resistor. For example, if you wish to measure a current of 2 A with a galvanometer whose resistance is 60 Ω, the shunt resistance should have a value of about 0.03 Ω.

A galvanometer can also be used as a voltmeter by adding an external resistor, R_s, in series with it, as in Figure 28.22b. In this case, the external resistor must have a value which is very large compared to the resistance of the galvanometer. This will insure that the galvanometer will not significantly alter the voltage to be measured. For example, if you wish to measure a maximum voltage of 100 V with a galvanometer whose resistance is 60 Ω, the external series resistor should have a value of about 10^5 Ω.

When a voltmeter is constructed with several available ranges, one selects various values of R_s by using a switch that can be connected to a preselected set of resistors. The required value of R_s increases as the maximum voltage to be measured increases.

*28.6 THE WHEATSTONE BRIDGE

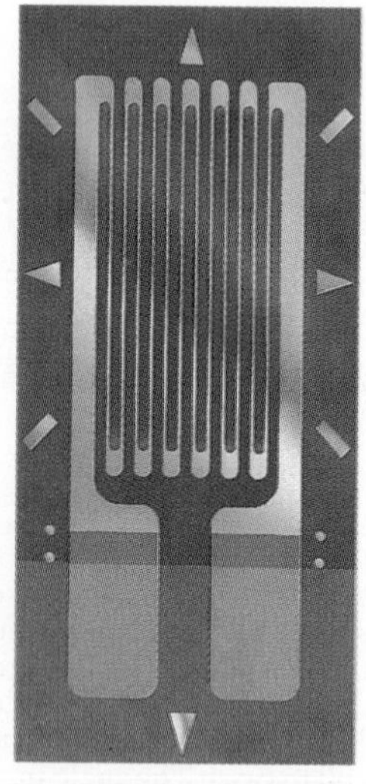

The strain gage, a device used for experimental stress analysis, consists of a thin coiled wire bonded to a flexible plastic backing. Stresses are measured by detecting changes in resistance of the coil as the strip bends. Resistance measurements are made with the gage as one element of a Wheatstone bridge. These devices are commonly used in modern electronic balances to measure the mass of an object.

Unknown resistances can be accurately measured using a circuit known as a **Wheatstone bridge** (Fig. 28.23). This circuit consists of the unknown resistance, R_x, three known resistors, R_1, R_2, and R_3 (where R_1 is a calibrated variable resistor), a galvanometer, and a source of emf. The principle of its operation is quite simple. The known resistor R_1 is varied until the galvanometer reading is zero, that is, until there is no current from a to b. Under this condition the bridge is said to be balanced. Since the potential at point a must equal the potential at point b when the bridge is balanced, the potential difference across R_1 must equal the potential difference across R_2. Likewise, the potential difference across R_3 must equal the potential difference across R_x. From these considerations, we see that

$$(1) \qquad I_1R_1 = I_2R_2$$

$$(2) \qquad I_1R_3 = I_2R_x$$

Dividing (1) by (2) eliminates the currents, and solving for R_x we find

$$R_x = \frac{R_2R_3}{R_1} \tag{28.19}$$

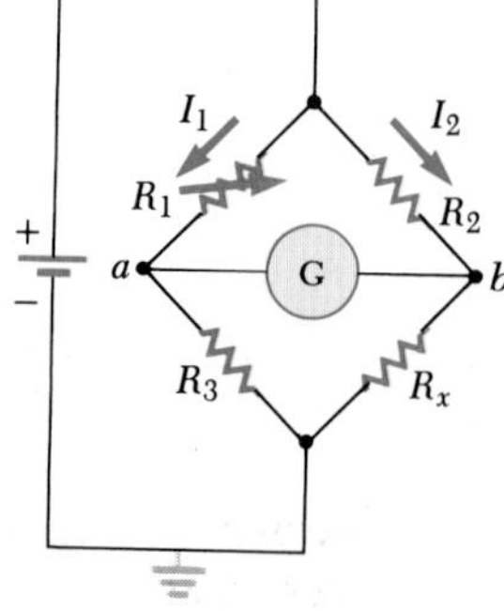

Figure 28.23 Circuit diagram for a Wheatstone bridge. This circuit is often used to measure an unknown resistance R_x in terms of known resistances R_1, R_2, and R_3. When the bridge is balanced, there is no current in the galvanometer.

Since R_1, R_2, and R_3 are known quantities, R_x can be calculated. There are a number of similar devices that use the null measurement, such as a capacitance bridge (used to measure unknown capacitances). These devices do not require the use of calibrated meters and can be used with any source of emf.

When very high resistances are to be measured (above $10^5\ \Omega$), the Wheatstone bridge method becomes difficult for technical reasons. As a result of recent advances in the technology of such solid state devices as the field-effect transistor, modern electronic instruments are capable of measuring resistances as high as $10^{12}\ \Omega$. Such instruments are designed to have an extremely high effective resistance between their input terminals. For example, input resistances of $10^{10}\ \Omega$ are common in most digital multimeters.

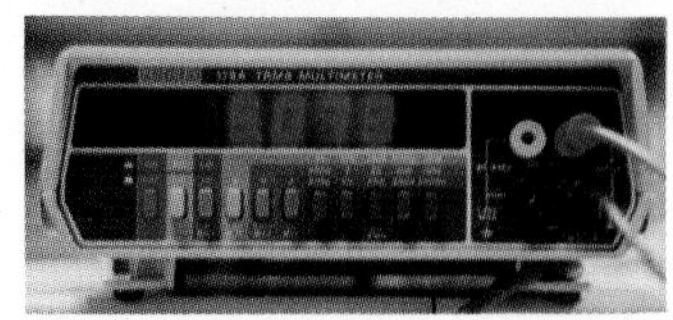

Voltages, currents, and resistances are frequently measured by digital multimeters like the one shown in this photograph. (Courtesy of Henry Leap and Jim Lehman)

*28.7 THE POTENTIOMETER

A **potentiometer** is a circuit that is used to measure an unknown emf, $\mathcal{E}_x$, by comparison with a known emf. Figure 28.24 shows the essential components of the potentiometer. Point d represents a sliding contact used to vary the resistance (and hence the potential difference) between points a and d. In a common version of the potentiometer, called a **slide-wire potentiometer,** the variable resistor is a wire with the contact point d at some position on the wire. The other required components in this circuit are a galvanometer, a power source with emf $\mathcal{E}_0$, a standard reference battery, and the unknown emf, $\mathcal{E}_x$.

With the currents in the directions shown in Fig. 28.24, we see from Kirchhoff's first rule that the current through the resistor R_x is $I - I_x$, where I is the current in the lower branch (through the battery of emf $\mathcal{E}_0$) and I_x is the current in the upper branch. Kirchhoff's second rule applied to loop *abcd* gives

$$-\mathcal{E}_x + (I - I_x)R_x = 0$$

where R_x is the resistance between points a and d. The sliding contact at d is now adjusted until the galvanometer reads zero (a balanced circuit). Under this condition, the current in the galvanometer and in the unknown cell is *zero* ($I_x = 0$), and the potential difference between a and d equals the unknown emf, $\mathcal{E}_x$. That is,

$$\mathcal{E}_x = IR_x$$

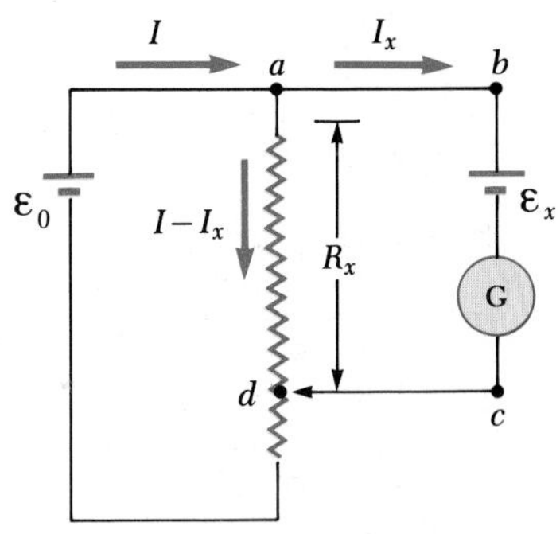

Figure 28.24 Circuit diagram for a potentiometer. The circuit is used to measure an unknown emf $\mathcal{E}_x$ in terms of a known emf $\mathcal{E}_s$, provided by a standard cell.

Next, the cell of unknown emf is replaced by a standard cell of known emf, $\mathcal{E}_s$, and the above procedure is repeated. That is, the moving contact at d is varied until a balance is obtained. If R_s is the resistance between a and d when balance is achieved, then

$$\mathcal{E}_s = IR_s$$

where it is assumed that I remains the same.

Combining this expression with the previous equation, $\mathcal{E}_x = IR_x$, we see that

$$\mathcal{E}_x = \frac{R_x}{R_s}\mathcal{E}_s \tag{28.20}$$

This result shows that the unknown emf can be determined from a knowledge of the standard-cell emf and the ratio of the two resistances.

If the resistor is a wire of resistivity ρ, its resistance can be varied using sliding contacts to vary the length of the circuit. With the substitutions $R_s = \rho L_s/A$ and $R_x = \rho L_x/A$, Equation 28.20 reduces to

$$\mathcal{E}_x = \frac{L_x}{L_s}\mathcal{E}_s \tag{28.21}$$

According to this result, the unknown emf can be obtained from a measurement of the two wire lengths and the magnitude of the standard emf.

*28.8 HOUSEHOLD WIRING AND ELECTRICAL SAFETY

Household circuits represent a very practical application of some of the ideas we have presented in this chapter concerning circuit analysis. In our world of electrical appliances, it is useful to understand the power requirements and limitations of conventional electrical systems and the safety measures that should be practiced to prevent accidents.

In a conventional installation, the utilities company distributes electrical power to individual homes with a pair of power lines. Each user is connected in parallel to these lines, as shown in Figure 28.25. The potential difference between these wires is about 120 V. The voltage alternates in time, but for the present discussion we shall assume a steady direct current (dc) voltage. (Alternating voltages and currents will be discussed in Chapter 33.) One of the wires is connected to ground, and the potential on the "live" wire oscillates relative to ground.[2]

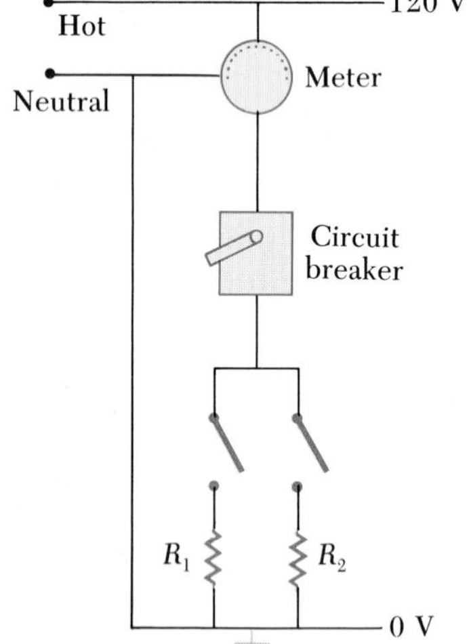

Figure 28.25 Wiring diagram for a household circuit. The resistances R_1 and R_2 represent appliances or other electric devices which operate with an applied voltage of 120 V.

A meter and circuit breaker (or in older installations, a fuse) are connected in series with the wire entering the house as indicated in Figure 28.25. The circuit breaker is a device that protects against too large a current, which can cause overheating and fires. When the current exceeds some safe value (typically 15 A or 30 A), the circuit breaker disconnects the voltage source from the load. Some circuit breakers make use of the principle of the bimetallic strip discussed in Chapter 19.

The wire and circuit breaker are carefully selected to meet the current demands for that circuit. If a circuit is to carry currents as large as 30 A, a heavy wire and appropriate circuit breaker must be selected to handle this current. Other individual household circuits, which are normally used to power lamps and small appliances, often require only 15 A. Therefore, each circuit has its own circuit breaker to accommodate various load conditions.

As an example, consider a circuit in which a toaster, a microwave oven, and a heater are in the same circuit (corresponding to $R_1, R_2, \ldots$ in Fig. 28.25). We can calculate the current through each appliance using the expression $P = IV$. The toaster, rated at 1000 W, would draw a current of 1000/120 = 8.33 A. The microwave oven, rated at 800 W, would draw a current of 6.67 A, and the electric heater, rated at 1300 W, would draw a current of 10.8 A. If the three appliances are operated simultaneously, they will draw a total current of 25.8 A. Therefore, the circuit should be wired to handle at least this much current. In order to accommodate a small additional load, such as a 100-W lamp, a 30-A circuit should be installed. Alternatively, one could

[2] The phrase *live wire* is common jargon for a conductor whose potential is above or below ground.

operate the toaster and microwave oven on one 20-A circuit and the heater on a separate 20-A circuit.

Many heavy-duty appliances, such as electric ranges and clothes dryers, require 240 V for their operation. The power company supplies this voltage by providing a third live wire, which is 120 V *below* ground potential (Fig. 28.26). Therefore, the potential difference between this wire and the other live wire (which is 120 V above ground potential) is 240 V. An appliance that operates from a 240-V line requires half the current of one operating from a 120-V line; therefore smaller wires can be used in the higher-voltage circuit without overheating becoming a problem.

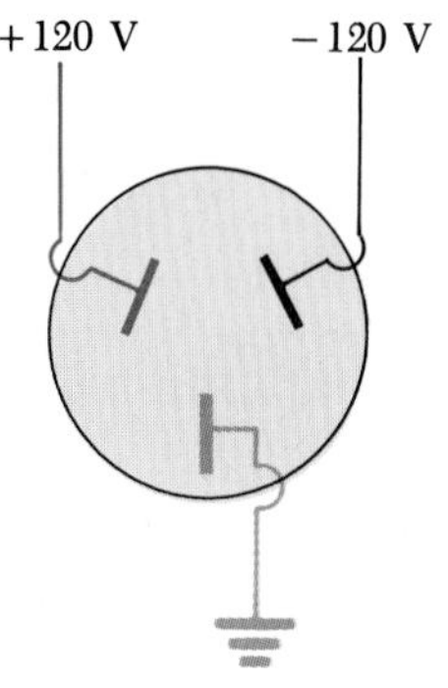

Figure 28.26 Power connections for a 240-V appliance.

Electrical Safety

When the live wire of an electrical outlet is connected directly to ground, the circuit is completed and a short-circuit condition exists. When this happens accidentally, a properly operating circuit breaker will "break" the circuit. On the other hand, a person can be electrocuted by touching the live wire of a frayed cord or other exposed conductors while in contact with ground. An exceptionally good ground contact might be made either by the person's touching a water pipe (which is normally at ground potential) or by standing on ground with wet feet, since water is not a good insulator. Such situations should be avoided at all costs.

Electrical shock can result in fatal burns, or it can cause the muscles of vital organs, such as the heart, to malfunction. The degree of damage to the body depends on the magnitude of the current, the length of time it acts, the location of the contact, and the part of the body through which the current passes. Currents of 5 mA or less can cause a sensation of shock, but ordinarily do little or no damage. If the current is larger than about 10 mA, the hand muscles contract and the person may be unable to release the live wire. If a current of about 100 mA passes through the body for only a few seconds, the result could be fatal. Such large currents will paralyze the respiratory muscles and prevent breathing. In some cases, currents of about 1 A through the body can produce serious (and sometimes fatal) burns. In practice, no contact with live wires should be regarded as safe.

Many 120-V outlets are designed to take a three-pronged power cord. (This feature is required in all new electrical installations.) One of these prongs is the live wire and two are common with ground. The additional ground connection is provided as a safety feature. Many appliances contain a three-pronged 120-V power cord with one of the ground wires connected directly to the casing of the appliance. If the live wire is accidentally shorted to the casing (which often occurs when the wire insulation wears off), the current will take the low-resistance path through the appliance to ground. In contrast, if the casing of the appliance is not properly grounded and a short occurs, anyone in contact with the appliance will experience an electric shock since his or her body will provide a low-resistance path to ground.

Special power outlets called ground-fault interrupters (GFI's) are now being used in kitchens, bathrooms, basements, and other hazardous areas of new homes. These devices are designed to protect persons from electrical shock by sensing small currents (≈ 5 mA) leaking to ground. When an excessive leakage current is detected, the current is shut off (interrupted) in less than one millisecond.

SUMMARY

The **emf** of a battery is equal to the voltage across its terminals when the current is zero. That is, the emf is equivalent to the open-circuit voltage of the battery.

The **equivalent resistance** of a set of resistors connected in **series** is given by

Resistors in series

$$R_{eq} = R_1 + R_2 + R_3 + \cdots \tag{28.6}$$

The **equivalent resistance** of a set of resistors connected in *parallel* is given by

Resistors in parallel

$$\frac{1}{R_{eq}} = \frac{1}{R_1} + \frac{1}{R_2} + \frac{1}{R_3} + \cdots \tag{28.8}$$

Kirchhoff's rules

Complex circuits involving more than one loop are conveniently analyzed using two simple rules called **Kirchhoff's rules:**

1. The sum of the currents entering any junction must equal the sum of the currents leaving that junction.
2. The sum of the potential differences across each element around any closed-circuit loop must be *zero.*

The first rule is a statement of **conservation of charge.** The second rule is equivalent to a statement of **conservation of energy.**

When a resistor is traversed in the direction of the current, the change in potential, ΔV, across the resistor is $-IR$. If a resistor is traversed in the direction opposite the current, $\Delta V = +IR$.

If a source of emf is traversed in the direction of the emf (negative to positive) the change in potential is $+\mathcal{E}$. If it is traversed opposite the emf (positive to negative), the change in potential is $-\mathcal{E}$.

If a capacitor is charged with a battery of emf $\mathcal{E}$ through a resistance R, the current in the circuit and charge on the capacitor vary in time according to the expressions

Current versus time

$$I(t) = \frac{\mathcal{E}}{R} e^{-t/RC} \tag{28.13}$$

Charge versus time

$$q(t) = Q[1 - e^{-t/RC}] \tag{28.14}$$

where $Q = C\mathcal{E}$ is the *maximum* charge on the capacitor. The product RC is called the **time constant** of the circuit.

If a charged capacitor is discharged through a resistance R, the charge and current decrease exponentially in time according to the expressions

$$q(t) = Q\, e^{-t/RC} \tag{28.17}$$

$$I(t) = I_0\, e^{-t/RC} \tag{28.18}$$

where $I_0 = Q/RC$ is the initial current in the circuit and Q is the initial charge on the capacitor.

A **Wheatstone bridge** is a particular circuit that can be used to measure an unknown resistance.

A **potentiometer** is a circuit that can be used to measure an unknown emf.

QUESTIONS

1. Explain the difference between load resistance and internal resistance for a battery.
2. Under what condition does the potential difference across the terminals of a battery equal its emf? Can the terminal voltage ever exceed the emf? Explain.
3. Is the direction of current through a battery always from negative to positive on the terminals? Explain.
4. Two different sets of Christmas-tree lights are available. For set A, when one bulb is removed (or burns out), the remaining bulbs remain illuminated. For set B, when one bulb is removed, the remaining bulbs will not operate. Explain the difference in wiring for the two sets of lights.
5. How would you connect resistors in order for the equivalent resistance to be larger than the individual resistances? Give an example.
6. How would you connect resistors in order for the equivalent resistance to be smaller than the individual resistances? Give an example.
7. Given three lightbulbs and a battery, sketch as many different electric circuits as you can.
8. When resistors are connected in series, which of the following will be the same, for each resistor: potential difference, current, power?
9. When resistors are connected in parallel, which of the following will be the same for each resistor: potential difference, current, power?
10. What advantage might there be in using two identical resistors in parallel connected in series with another identical parallel pair, rather than just using a single resistor?
11. Are the two headlights on a car wired in series or in parallel? How can you tell?
12. An incandescent lamp connected to a 120-V source with a short extension cord will provide more illumination than if it were connected to the same source with a very long extension cord. Explain.
13. Embodied in Kirchhoff's rules are two conservation laws. What are they?
14. When can the potential difference across a resistor be positive?
15. With reference to Figure 28.13, suppose the wire between points g and h is replaced by a 10-Ω resistor. Explain why this change will *not* affect the currents calculated in Example 28.8.
16. With reference to Figure 28.27, describe what happens to the light bulb after the switch is closed. Assume the capacitor has a large capacitance and is initially uncharged, and assume that the light will illuminate when connected directly across the battery terminals.
17. What would be the internal resistance of an ideal ammeter and voltmeter? Why would these become ideal meters?

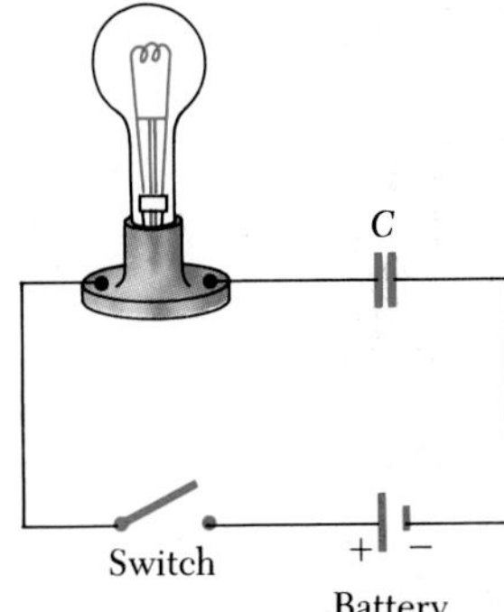

Figure 28.27 (Question 16).

18. What are a Wheatstone bridge and potentiometer used to measure?
19. Although the internal resistance of the unknown and known emfs was neglected in the treatment of the potentiometer (Section 28.7), it is really not necessary to make this assumption. Explain why the internal resistances play no role in this measurement.
20. Why is it dangerous to turn on a light when you are in the bathtub?
21. Why is it possible for a bird to sit on a high-voltage wire without being electrocuted?
22. Suppose you fall from a building and on the way down grab a high-voltage wire. Assuming that the wire holds you, will you be electrocuted? If the wire then breaks, should you continue to hold onto an end of the wire as you fall?
23. Would a fuse work successfully if it were placed in parallel with the device it is supposed to protect?
24. What advantage does 120-V operation offer over 240 V? What disadvantages?
25. When electricians work with potentially live wires, they often use the backs of their hands or fingers to move wires. Why do you suppose they use this technique?
26. What procedure would you use to try to save a person who is "frozen" to a live high-voltage wire without endangering your own life?
27. At what levels of current do you experience (a) the sensation of shock, (b) involuntary muscle contractions, (c) paralysis of respiratory muscle, and (d) serious and possibly fatal burns? In practice what current level is regarded as safe?
28. If it is the current flowing through the body that determines how serious a shock will be, why do we see warnings of high voltage rather than high current near electric equipment?
29. Suppose you are flying a kite when it strikes a high-voltage wire. What factors determine how great a shock you receive?

30. A series circuit consists of three identical lamps connected to a battery as in the circuit shown at the right. When the switch *S* is closed, (a) what happens to the intensities of lamps *A* and *B*? (b) What happens to the intensity of lamp *C*? (c) What happens to the current in the circuit? (d) What happens to the voltage drop across the three lamps? (e) Does the power dissipated in the circuit increase, decrease, or remain the same?

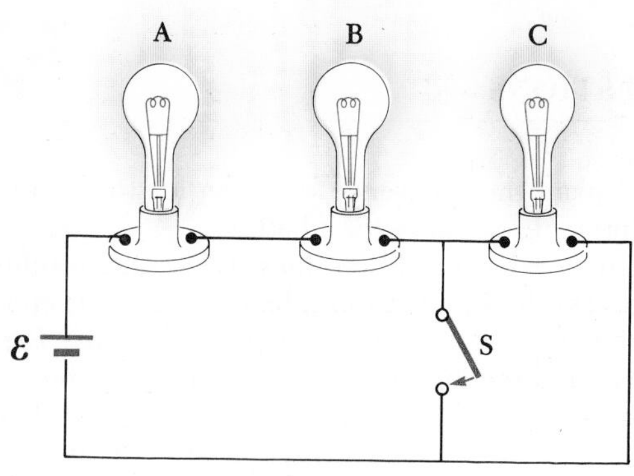

(Question 30).

PROBLEMS

Section 28.1 Electromotive Force

1. A battery with an emf of 12 V and internal resistance of 0.9 Ω is connected across a load resistor *R*. If the current in the circuit is 1.4 A, what is the value of *R*?
2. What power is dissipated in the internal resistance of the battery in the circuit described in Problem 1?
3. (a) What is the current in a 5.6-Ω resistor connected to a battery with an 0.2-Ω internal resistance if the terminal voltage of the battery is 10 V? (b) What is the emf of the battery?
4. If the emf of a battery is 15 V and a current of 60 A is measured when the battery is shorted, what is the internal resistance of the battery?
5. The current in a loop circuit that has a resistance of R_1 is 2 A. The current is reduced to 1.6 A when an additional resistor $R_2 = 3\ \Omega$ is added in series with R_1. What is the value of R_1?
6. A battery has an emf of 6 V and an internal resistance of 0.2 Ω. When its terminals are connected to a load resistor the current in the circuit is 2 A. What is the value of the load resistor?
7. A battery has an emf of 15 V. The terminal voltage of the battery is 11.6 V when it is delivering 20 W of power to an external load resistor *R*. (a) What is the value of *R*? (b) What is the internal resistance of the battery?
8. What potential difference will be measured across a 18-Ω load resistor when it is connected across a battery of emf 5 V and internal resistance 0.45 Ω?
9. A certain battery has an open-circuit voltage of 42 V. A load resistance of 12 Ω reduces the terminal voltage to 35 V. What is the value of the internal resistance of the battery?

Section 28.2 Resistors in Series and in Parallel

10. Two circuit elements with fixed resistances R_1 and R_2 are connected in *series* with a 6-V battery and a switch. The battery has an internal resistance of 5 Ω, $R_1 = 132\ \Omega$, and $R_2 = 56\ \Omega$. (a) What is the current through R_1 when the switch is closed? (b) What is the voltage across R_2 when the switch is closed?
11. The components of Problem 10 are reconnected with R_1 and R_2 in *parallel* across the battery. (a) What is the voltage across R_1 when the switch is closed? (b) What is the current in each resistor?
12. A technician has a box full of resistors, each having the same resistance *R*. How many different values of effective resistance could the technician make using all combinations of one to three separate resistors? Express the effective resistance of each combination in terms of *R*.
13. Three resistors (10 Ω, 20 Ω, and 30 Ω) are connected in parallel. The total current through this network is 5 A. (a) What is the voltage drop across the network? (b) What is the current in each resistor?
14. Find the equivalent resistance between points *a* and *b* in Figure 28.28.

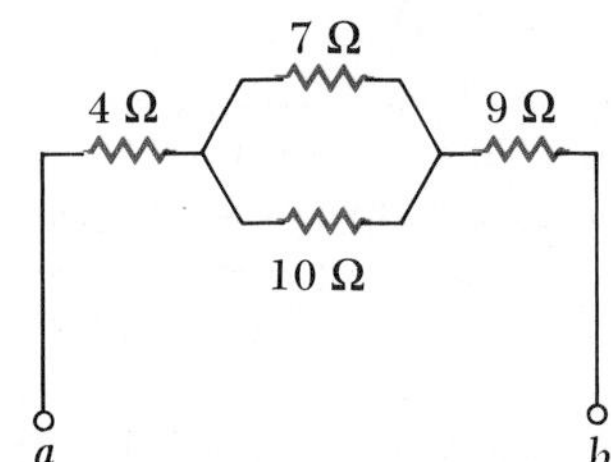

Figure 28.28 (Problems 14 and 15).

15. A potential difference of 34 V is applied between points *a* and *b* in Figure 28.28. Calculate the current in each resistor.
16. Find the equivalent resistance between points *a* and *b* in Figure 28.29.

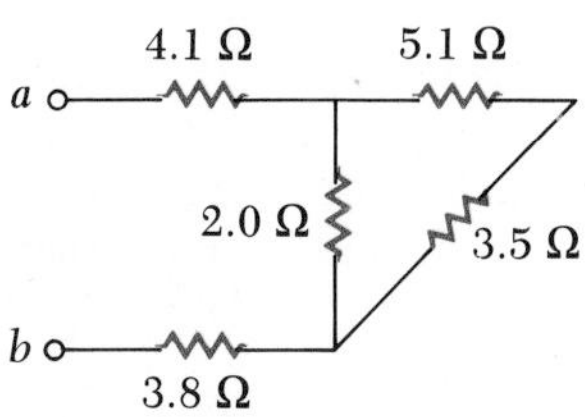

Figure 28.29 (Problem 16).

17. Evaluate the effective resistance of the network of identical resistors, each having resistance R, shown in Figure 28.30.

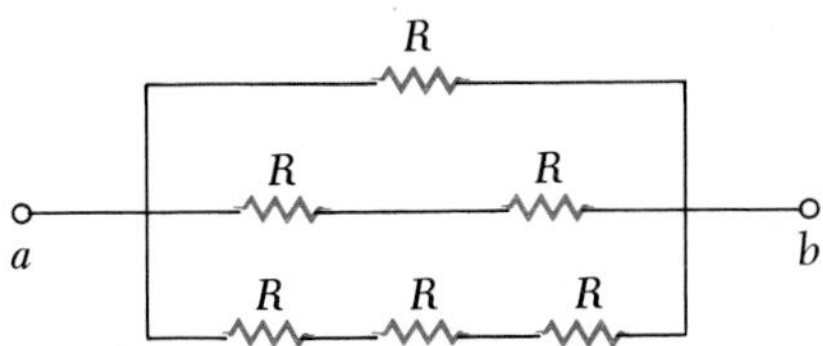

Figure 28.30 (Problem 17).

18. In Figures 28.4 and 28.5, let $R_1 = 11\ \Omega$, $R_2 = 22\ \Omega$, and the battery have a terminal voltage of 33 V. (a) In the parallel circuit shown in Figure 28.5, which resistor uses more power? (b) Verify that the sum of the power (I^2R) used by each resistor equals the power supplied by the battery (IV). (c) In the series circuit, Figure 28.4, which resistor uses more power? (d) Verify that the sum of the power (I^2R) used by each resistor equals the power supplied by the battery ($P = IV$). (e) Which circuit configuration uses more power?

19. Calculate the power dissipated in each resistor in the circuit of Figure 28.31.

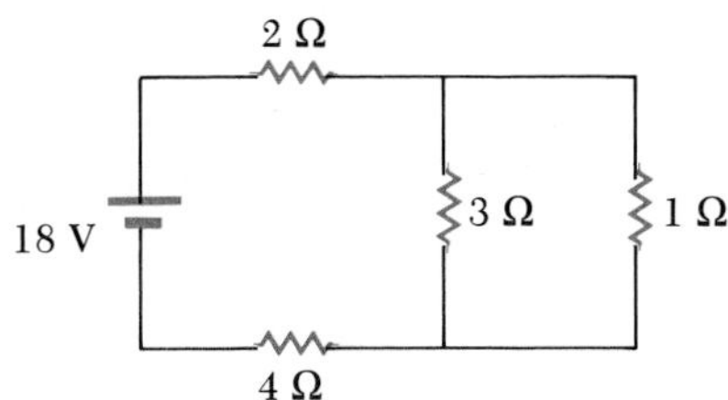

Figure 28.31 (Problem 19).

20. Determine the equivalent resistance between the terminals a and b for the network illustrated in Figure 28.32.

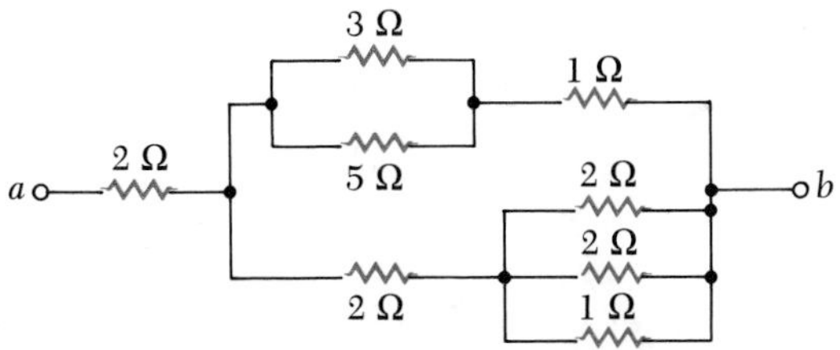

Figure 28.32 (Problem 20).

21. Consider the circuit shown in Figure 28.33. Find (a) the current in the 20-Ω resistor and (b) the potential difference between points a and b.

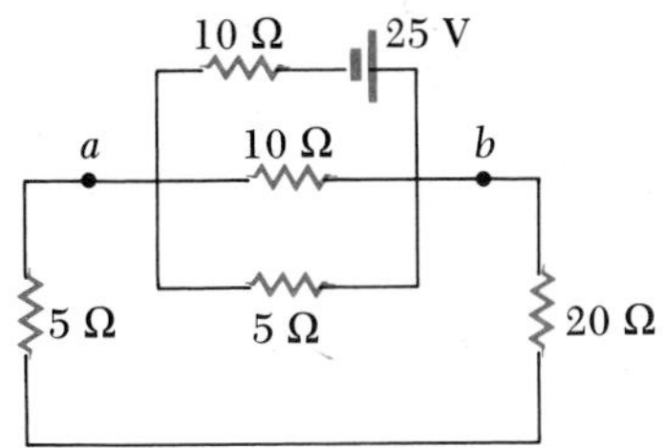

Figure 28.33 (Problem 21).

22. Consider the combination of resistors in Figure 28.34. (a) Find the resistance between points a and b. (b) If the current in the 5-Ω resistor is 1 A, what is the potential difference between points a and b?

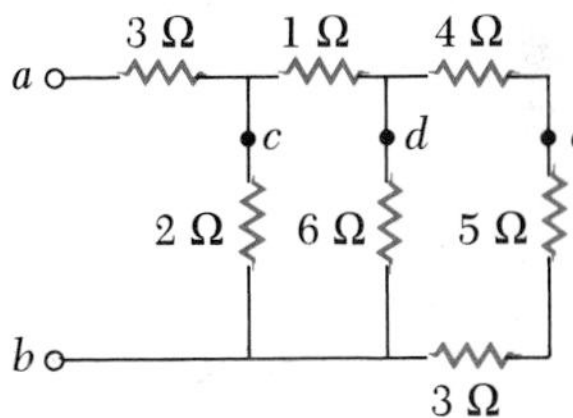

Figure 28.34 (Problems 22, 23, and 71).

23. In Figure 28.34, connect points c, d, and e with a conductor and then find the equivalent resistance between points a and b. (*Hint:* Redraw the circuit.)

24. Three 100-Ω resistors are connected as shown in Figure 28.35. The maximum power that can be dissipated in any one of the resistors is 25 W. (a) What is the maximum voltage that can be applied to the terminals a and b? (b) For the voltage determined in (a), what is the power dissipation in each resistor? What is the total power dissipation?

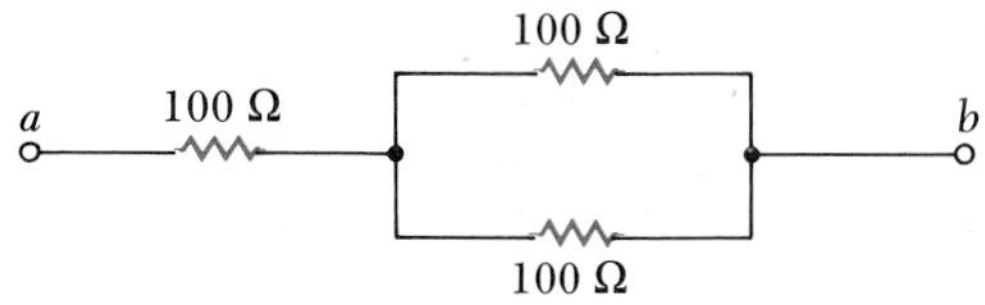

Figure 28.35 (Problem 24).

Section 28.3 Kirchhoff's Rules

The currents are not necessarily in the direction shown for some circuits.

25. Find the potential difference between points a and b in the circuit in Figure 28.36.

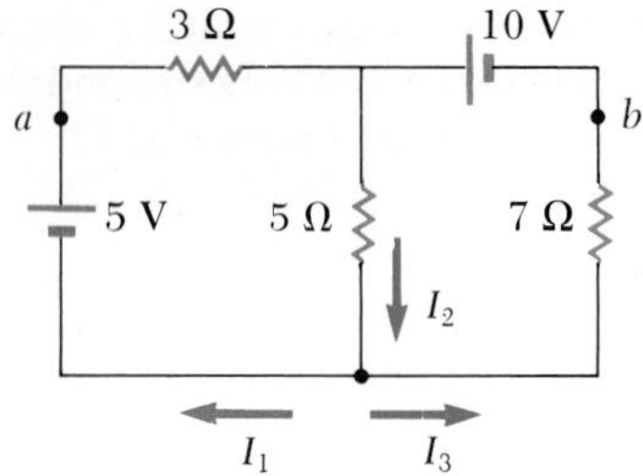

Figure 28.36 (Problems 25 and 26).

26. Find the currents I_1, I_2, and I_3 in the circuit shown in Figure 28.36.

27. Determine the current in each of the branches of the circuit shown in Figure 28.37.

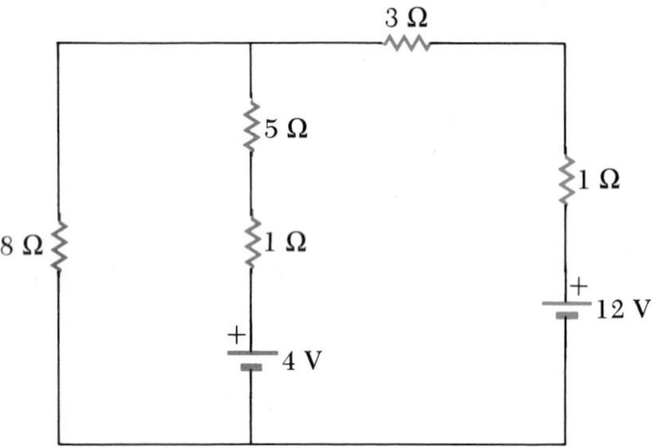

Figure 28.37 (Problem 27).

28. Determine the potential difference V_{ab} for the circuit shown in Figure 28.38.

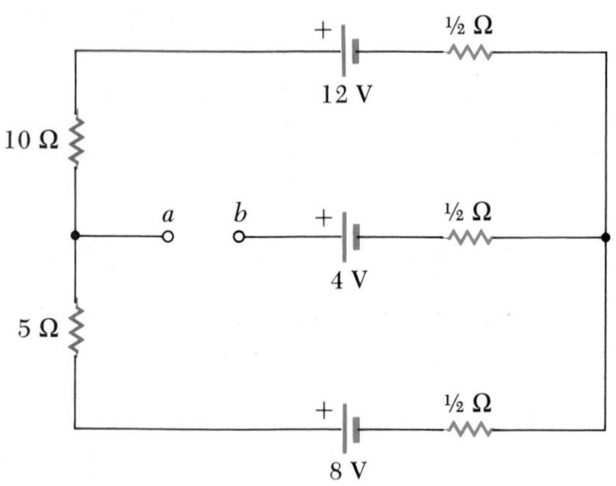

Figure 28.38 (Problem 28).

29. The two branch currents in the circuit shown in Figure 28.39 are $I_1 = \frac{1}{3}$A and $I_2 = \frac{1}{2}$A. Determine the emfs, $\mathcal{E}_1$ and $\mathcal{E}_2$.

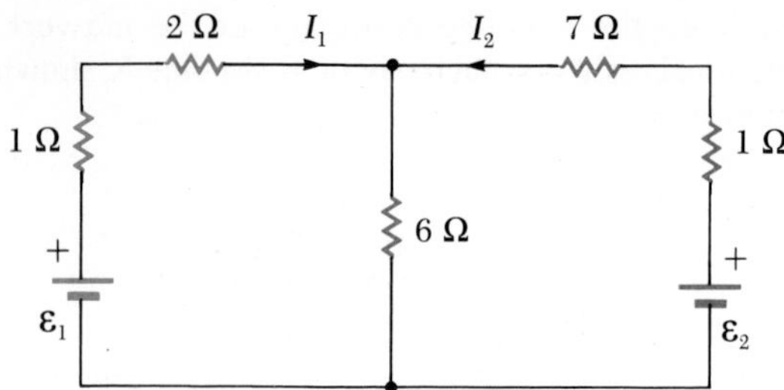

Figure 28.39 (Problem 29).

30. (a) Calculate the value of R for the circuit shown in Figure 28.40. (b) Determine the currents in the 6-Ω and 4-Ω resistors.

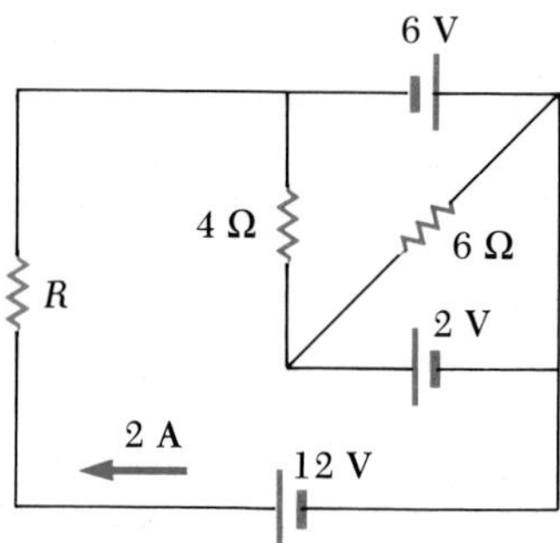

Figure 28.40 (Problem 30).

31. Calculate each of the unknown currents I_1, I_2, and I_3 for the circuit of Figure 28.41.

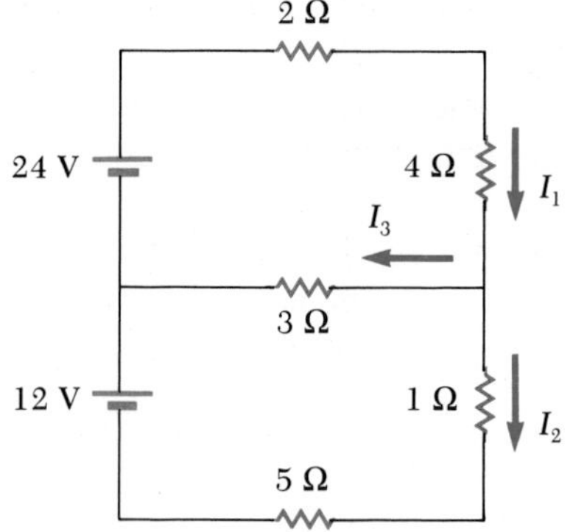

Figure 28.41 (Problem 31).

32. The ammeter in the circuit shown in Figure 28.42 reads 2 A. Find the currents I_1 and I_2 and the value of $\mathcal{E}$.

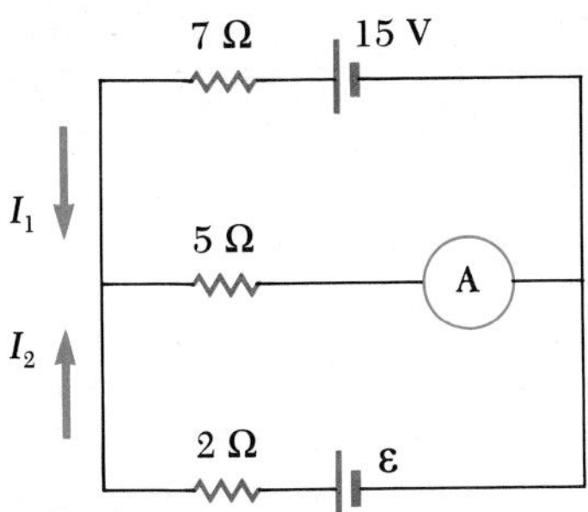

Figure 28.42 (Problem 32).

33. What is the emf $\mathcal{E}_1$ of the battery in the circuit shown in Figure 28.43?

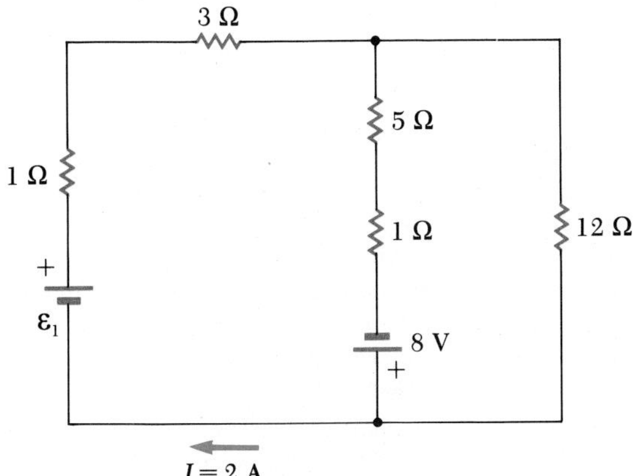

Figure 28.43 (Problem 33).

34. Consider the circuit shown in Figure 28.44. Find the value of I_1, I_2, and I_3.

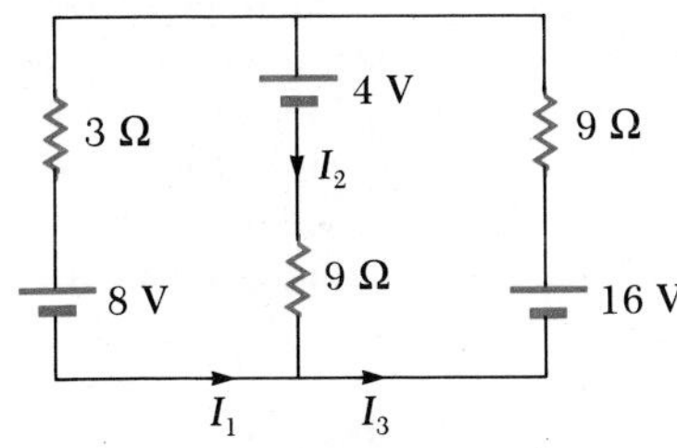

Figure 28.44 (Problems 34 and 35).

35. (a) Find the value of I_1 and I_3 in the circuit of Figure 28.44 if the 4-V battery is replaced by a 5-μF capacitor. (b) Determine the charge on the 5-μF capacitor.
36. Determine the current that flows through each of the batteries in the circuit shown in Figure 28.45.

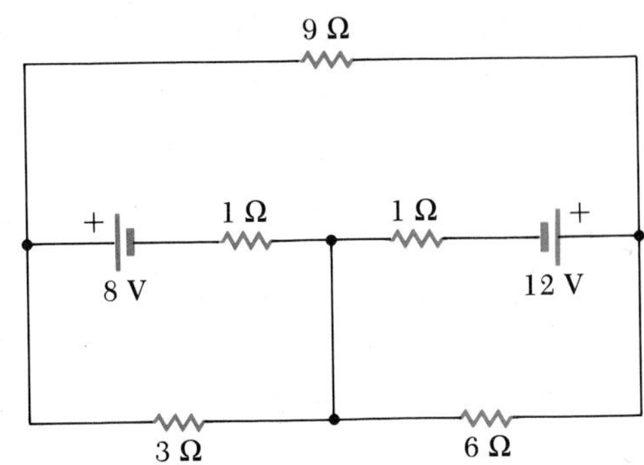

Figure 28.45 (Problem 36).

37. For the circuit shown in Figure 28.46, calculate (a) the current in the 2-Ω resistor and (b) the potential difference between points a and b.

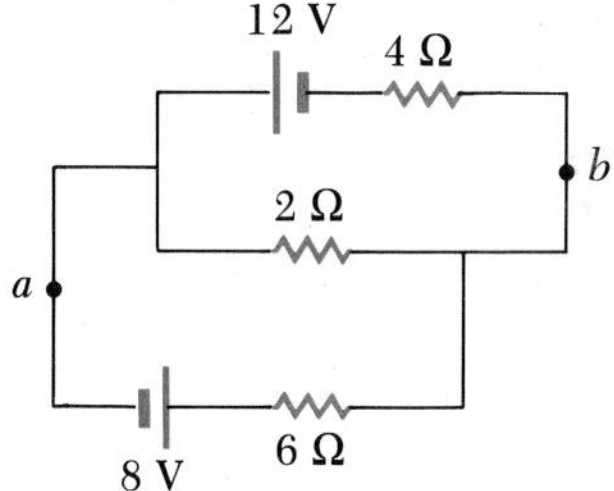

Figure 28.46 (Problem 37).

38. Determine the value of the current in each of the four resistors shown in the circuit of Figure 28.47.

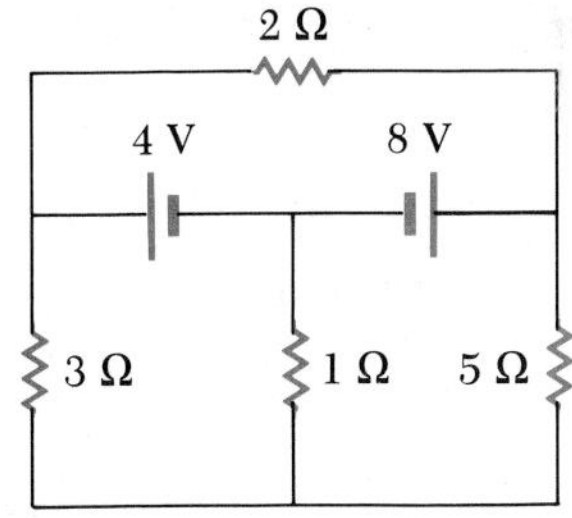

Figure 28.47 (Problem 38).

39. Calculate the power dissipated in each resistor in the circuit shown in Figure 28.48.

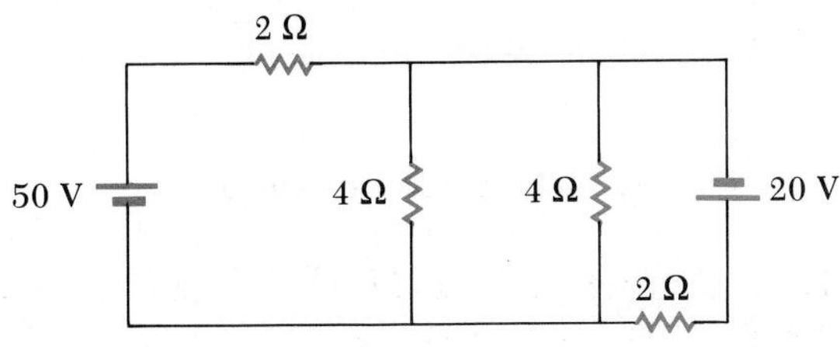

Figure 28.48 (Problem 39).

40. Calculate the power dissipated in each resistor in the circuit of Figure 28.49.

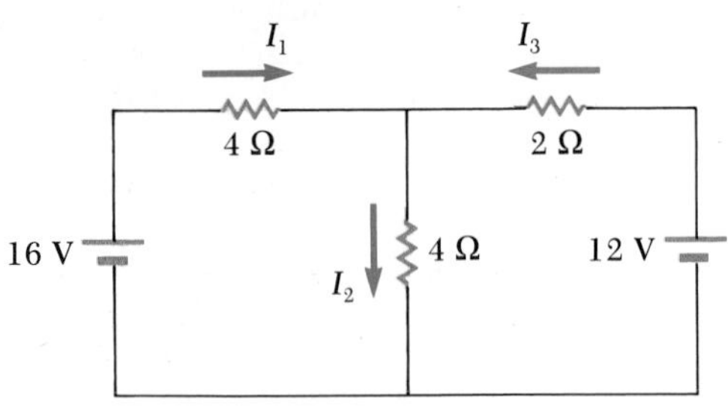

Figure 28.49 (Problem 40).

Section 28.4 *RC* Circuits

41. Consider an *RC* circuit in which the capacitor is being charged by a battery connected in the circuit. In five time constants, what percentage of the *final* charge is on the capacitor?

42. Consider a series *RC* circuit (Fig. 28.14) for which $R = 1$ MΩ, $C = 5$ μF, and $\mathcal{E} = 30$ V. Find (a) the time constant of the circuit and (b) the *maximum* charge on the capacitor after the switch is closed.

43. The switch in the *RC* circuit described in Problem 42 is closed at $t = 0$. Find the current in the resistor *R* at a time 10 s after the switch is closed.

44. At $t = 0$, an uncharged capacitor of capacitance *C* is connected through a resistance *R* to a battery of constant emf $\mathcal{E}$ (Figure 28.50). (a) How long does it take for the capacitor to reach one half of its final charge? (b) How long does it take for the capacitor to become fully charged?

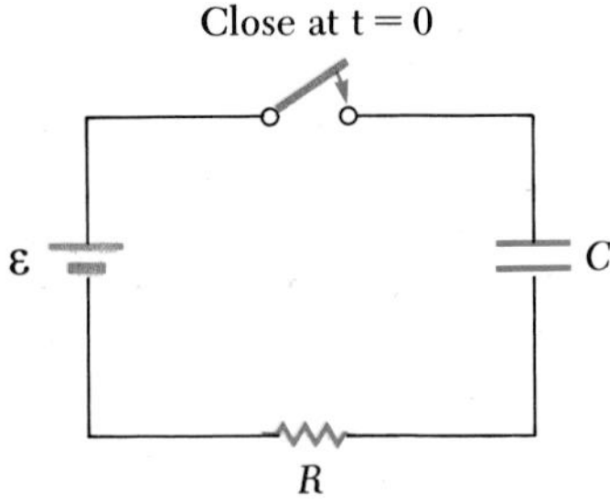

Figure 28.50 (Problem 44).

45. A 4-MΩ resistor and a 3-μF capacitor are connected in series with a 12-V power supply. (a) What is the time constant for the circuit? (b) Express the current in the circuit and the charge on the capacitor as functions of time.

46. A 750-pF capacitor has an initial charge of 6 μC. It is then connected to a 150-MΩ resistor and allowed to discharge through the resistor. (a) What is the time constant for the circuit? (b) Express the current in the circuit and the charge on the capacitor as functions of time.

47. The circuit has been connected as shown in Figure 28.51 a "long" time. (a) What is the voltage across the capacitor? (b) If the battery is disconnected, how long does it take for the capacitor to discharge to 1/10 of its initial voltage?

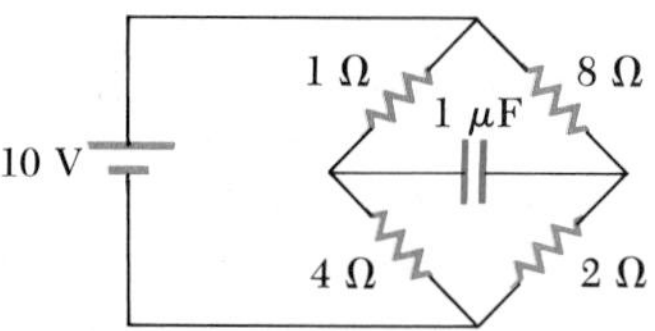

Figure 28.51 (Problem 47).

48. A 2×10^{-3}-μF capacitor with an initial charge of 5.1 μC is discharged through a 1300-Ω resistor. (a) Calculate the current through the resistor 9 μs after the resistor is connected across the terminals of the capacitor. (b) What charge remains on the capacitor after 8 μs? (c) What is the maximum current through the resistor?

49. Consider the capacitor-resistor combination described in Problem 48. (a) How much energy is stored initially in the charged capacitor? (b) If the capacitor is completely discharged through the resistor, how much energy will be dissipated as heat in the resistor?

50. A capacitor in an *RC* circuit is charged to 60% of its maximum value in 0.9 s. What is the time constant of the circuit?

51. Dielectric materials used in the manufacture of capacitors are characterized by conductivities that are small but not zero. Therefore, a charged capacitor will slowly lose its charge by "leaking" across the dielectric. If a certain 3.6-μF capacitor leaks charge such that the potential difference decreases to half its initial value in 4 s, what is the equivalent resistance of the dielectric?

°Section 28.5 Electrical Instruments

52. A typical galvanometer, which requires a current of 1.5 mA for full-scale deflection and has a resistance of 75 Ω, may be used to measure currents of much larger values. To enable the measuring of large currents without damage to the sensitive meter, a relatively small shunt resistor is wired in parallel with the meter movement similar to Figure 28.22a. Most of the cur-

rent will then flow through the shunt resistor. Calculate the value of the shunt resistor that enables the meter to be used to measure a current of 1 A at full-scale deflection. (*Hint:* Use Kirchhoff's laws.)

53. The same galvanometer movement as used in the previous problem may be used to measure voltages. In this case a large resistor is wired in series with the meter movement similar to Figure 28.22b, which in effect limits the current that flows through the movement when large voltages are applied. Most of the potential drop occurs across the resistor placed in series. Calculate the value of the resistor that enables the movement to measure an applied voltage of 25 V at full-scale deflection.

54. Consider a galvanometer with an internal resistance of 60 Ω. If the galvanometer deflects full scale when it carries a current of 0.5 mA, what value series resistance must be connected to the galvanometer if this combination is to be used as a voltmeter having a full-scale deflection for a potential difference of 1.0 V?

55. Assume that a galvanometer has an internal resistance of 60 Ω and requires a current of 0.5 mA to produce full-scale deflection. What value of resistance must be connected in parallel with the galvanometer if the combination is to serve as an ammeter with a full-scale deflection for a current of 0.1 A?

56. A meter movement has a resistance of 100 Ω and gives a full scale deflection when it carries a current of 25 μA. (a) What resistor would you connect in parallel with this movement in order to construct an ammeter which gives a full-scale deflection for a current of 1 A? (b) What is the equivalent resistance of this ammeter?

57. A meter movement has a resistance of 50 Ω and gives a full-scale deflection when it carries a current of 50 μA. (a) What resistor would you connect in series with this movement in order to construct a voltmeter that gives a full-scale deflection for 20 V? (b) What is the equivalent resistance of this voltmeter?

58. A current of 2.5 mA causes a given galvanometer movement to deflect full scale. The resistance of the movement is 200 Ω. (a) Show by means of a circuit diagram, using two resistors and three external jacks, how the meter movement may be made into a dual-range voltmeter. (b) Determine the values of the resistors needed to make the high range 0–200 V and the low range 0–20 V. Indicate these values on the diagram.

59. The same meter movement is given as in the previous problem. (a) Show by means of a circuit diagram, using two resistors and three external jacks, how the meter movement may be made into a dual-range ammeter. (b) Determine the values of the resistors needed to make the high range 0–10 A and the low range 0–1 A. Indicate these values on the diagram.

*Section 28.6 The Wheatstone Bridge

60. A Wheatstone bridge of the type shown in Figure 28.23 is used to make a precise measurement of the resistance of a wire connector. The resistor shown in the circuit as R_3 is 1 kΩ. If the bridge is balanced by adjusting R_1 such that $R_1 = 2.5R_2$, what is the resistance of the wire connector, R_x?

61. Consider the case when the Wheatstone bridge shown in Figure 28.23 is *unbalanced.* Calculate the current through the galvanometer when $R_x = R_3 = 7\ \Omega$, $R_2 = 21\ \Omega$, and $R_1 = 14\ \Omega$. Assume the voltage across the bridge is 70 V, and neglect the galvanometer's resistance.

62. Consider the Wheatstone bridge shown in Figure 28.23. When the Wheatstone bridge is balanced the voltage drop across R_x is 3.2 V and $I_1 = 200\ \mu$A. If the total current drawn from the power supply is 500 μA, what is the resistance, R_x?

63. The Wheatstone bridge in Figure 28.23 is balanced when $R_1 = 10\ \Omega$, $R_2 = 20\ \Omega$, and $R_3 = 30\ \Omega$. Calculate the value of R_x.

*Section 28.7 The Potentiometer

64. Consider the potentiometer circuit shown in Figure 28.24. When a standard cell of emf 1.0186 V is used in the circuit, and the resistance between *a* and *d* is 36 Ω, the galvanometer reads zero. When the standard cell is replaced by an unknown emf, the galvanometer reads zero when the resistance is adjusted to 48 Ω. What is the value of the unknown emf?

*Section 28.8 Household Wiring and Electrical Safety

65. An electric heater is rated at 1500 W, a toaster is rated at 750 W, and an electric grill is rated at 1000 W. The three appliances are connected to a common 120-V circuit. (a) How much current does each appliance draw? (b) Is a 25-A circuit sufficient in this situation? Explain.

66. A 1000-W toaster, an 800-W microwave oven, and a 500-W coffee pot are all plugged into the same 120-V outlet. If the circuit is protected by a 20-A fuse, will the fuse blow if all these appliances are used at once?

67. An 8-foot extension cord has two 18-gauge copper wires, each having a diameter of 1.024 mm. (a) How much power does this cord dissipate when carrying a current of 1 A? (b) How much power does this cord dissipate when carrying a current of 10 A?

68. Sometimes aluminum wiring is used instead of copper for economic reasons. According to the National Electrical Code, the maximum allowable current for 12-gauge copper wire with rubber insulation is 20 A. What should be the maximum allowable current in a 12-gauge aluminum wire if it is to dissipate the same power per unit length as the copper wire?

69. A 4-kW heater is wired for 240-V operation with nichrome wire having a total mass M. (a) How much current does the heater require? (b) How much current would a 120-V, 4-kW heater require? (c) If a 240-V, 4-kW heater and a 120-V, 4-kW heater have the same length resistance wires in them, how does the mass of the resistance wire in the 120-V heater compare to the mass of the resistance wire in the 240-V heater?

ADDITIONAL PROBLEMS

70. What must be the value of R_3 to make the equivalent resistance between the terminals a and b equal to R_1 in the circuit of Figure 28.52?

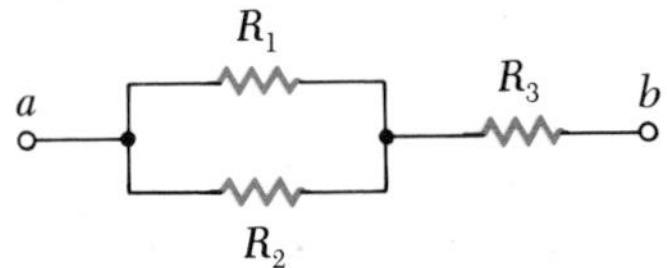

Figure 28.52 (Problem 70).

71. A voltage of 50.4 V is applied across points a and b to the circuit as shown in Figure 28.34. What voltage difference will be measured by a voltmeter connected (a) between points a and c, (b) between points a and e, and (c) between points c and e?

72. A cylindrical rod of material with a uniform resistivity ρ_a has a radius of a and length L. A hollow cylindrical sleeve, of inner radius a and outer radius b, of the same length L, and of material having a uniform resistivity ρ_b, fits over the rod. (See Figure 28.53.) (a) Find an expression for the effective resistance between the ends of the combination. (b) Show that if $\rho_a = \rho_b = \rho$, your expression reduces to

$$R = \frac{\rho L}{\pi b^2}$$

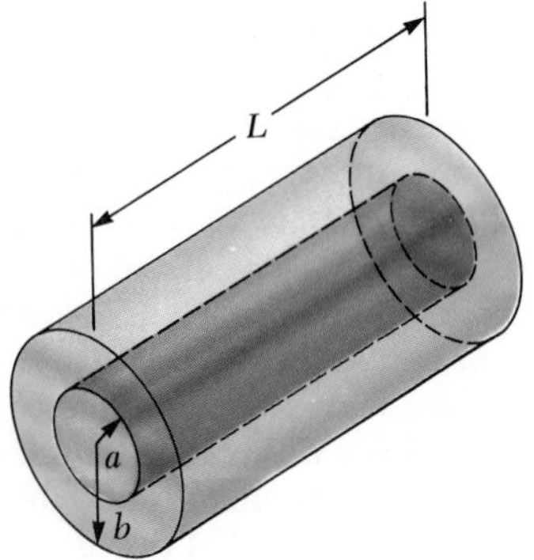

Figure 28.53 (Problem 72).

73. Three resistors, each of value 3 Ω, are arranged in two different arrangements as shown in Figure 28.54. If the maximum allowable power for each individual resistor is 48 W, calculate the maximum power that can be dissipated by (a) the circuit shown in Figure 28.54a and (b) the circuit shown in Figure 28.54b.

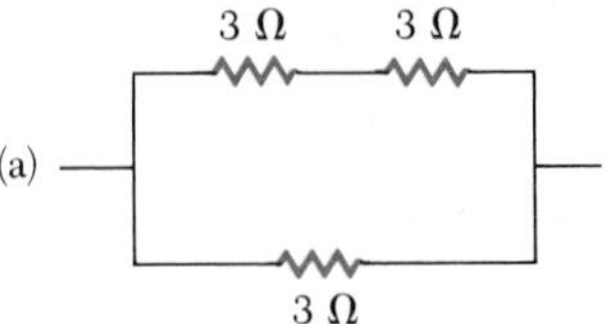

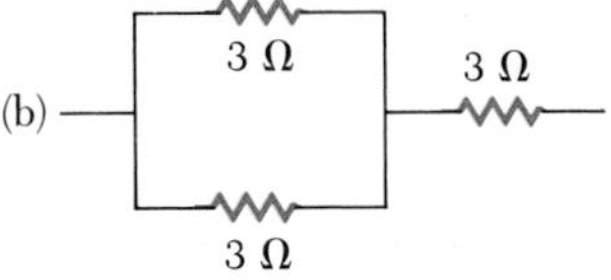

Figure 28.54 (Problem 73).

74. (a) Calculate the current through the 6-V battery in Figure 28.55. (b) Determine the potential difference between points a and b.

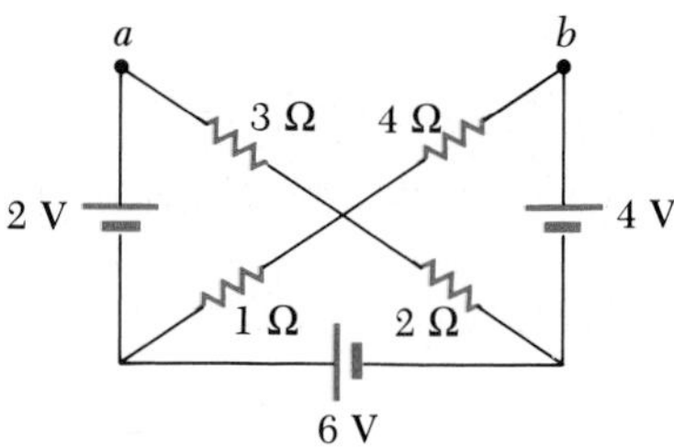

Figure 28.55 (Problem 74).

75. Consider the circuit shown in Figure 28.56. What are the expected readings of the ammeter and voltmeter?

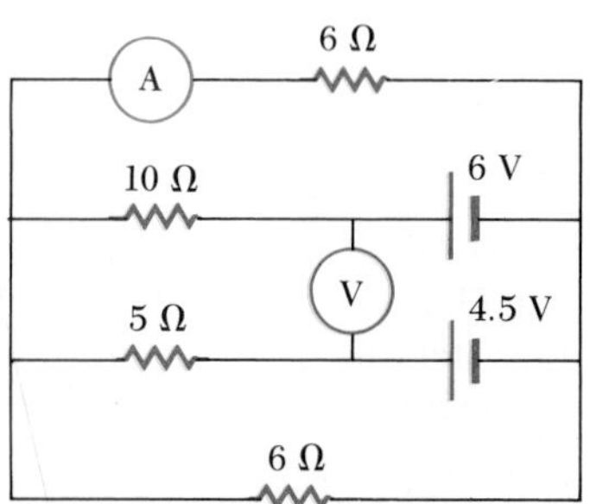

Figure 28.56 (Problem 75).

76. Consider the circuit shown in Figure 28.57. Calculate (a) the current in the 4-Ω resistor, (b) the potential difference between points a and b, (c) the terminal potential difference of the 4-V battery, and (d) the thermal energy expended in the 3-Ω resistor during 10 min of operation of the circuit.

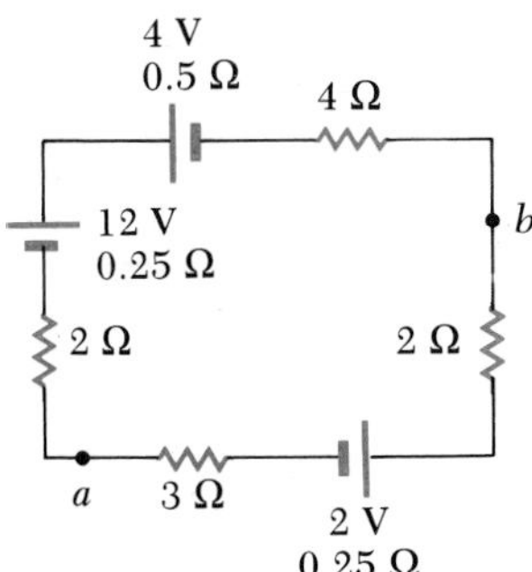

Figure 28.57 (Problem 76).

77. The value of a resistor R is to be determined using the ammeter-voltmeter setup shown in Figure 28.58. The ammeter has a resistance of 0.5 Ω, and the voltmeter has a resistance of 20 000 Ω. Within what range of actual values of R will the measured values be correct to within 5% if the measurement is made using the circuit shown in (a) Figure 28.58a and (b) Figure 28.58b?

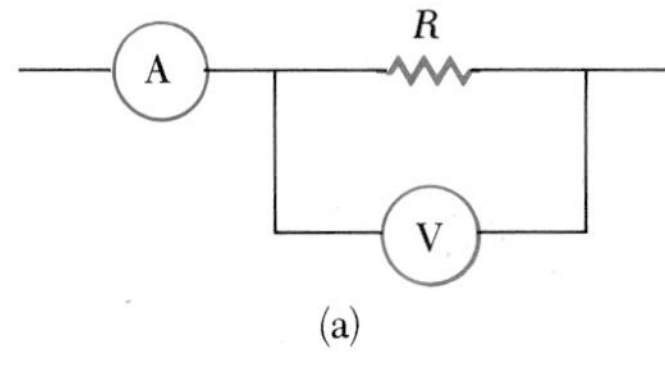

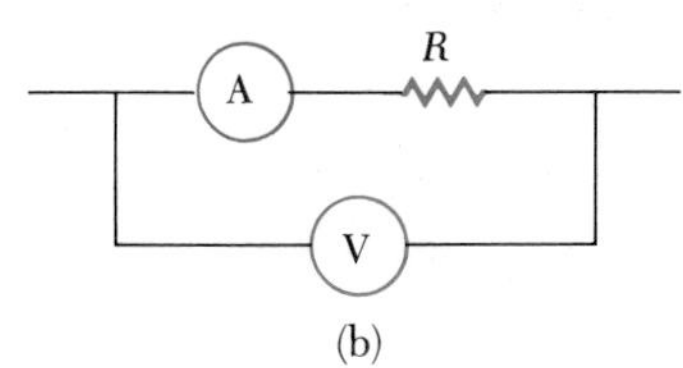

Figure 28.58 (Problem 77).

78. A dc power supply has an open-circuit emf of 40 V and an internal resistance of 2 Ω. It is used to charge two storage batteries connected in series, each having an emf of 6 V and internal resistance of 0.3 Ω. If the charging current is to be 4 A, (a) what additional resistance should be added in series? (b) Find the power lost in the supply, the batteries, and the added series resistance. (c) How much power is converted to chemical energy in the batteries?

79. A battery has an emf $\mathcal{E}$ and internal resistance r. A variable resistor R is connected across the terminals of the battery. Find the value of R such that (a) the potential difference across the terminals is a maximum, (b) the current in the circuit is a maximum, (c) the power delivered to the resistor is a maximum.

80. Consider the circuit shown in Figure 28.59. (a) Calculate the current in the 5-Ω resistor. (b) What power is dissipated by the entire circuit? (c) Determine the potential difference between points a and b. Which point is at the higher potential?

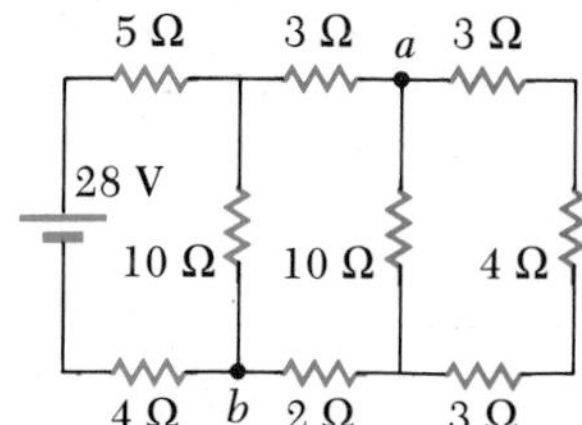

Figure 28.59 (Problem 80).

81. The values of the components in a simple RC circuit (Fig. 28.14) are as follows: $C = 1\ \mu\text{F}$, $R = 2 \times 10^6\ \Omega$, and $\mathcal{E} = 10$ V. At the instant 10 s after the switch in the circuit is closed, calculate (a) the charge on the capacitor, (b) the current in the resistor, (c) the rate at which energy is being stored in the capacitor, and (d) the rate at which energy is being delivered by the battery.

82. A battery is used to charge a capacitor through a resistor, as in Figure 28.14. Show that in the process of charging the capacitor, half of the energy supplied by the battery is dissipated as heat in the resistor and half is stored in the capacitor.

83. The switch in the circuit shown in Figure 28.60a closes when $V_c \geq 2V/3$ and opens when $V_c \leq V/3$.

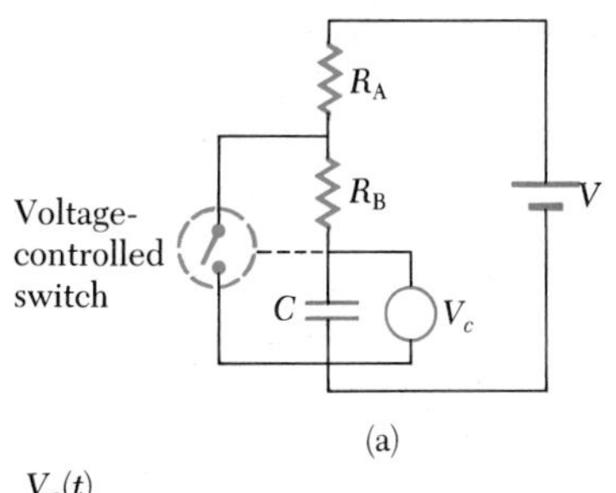

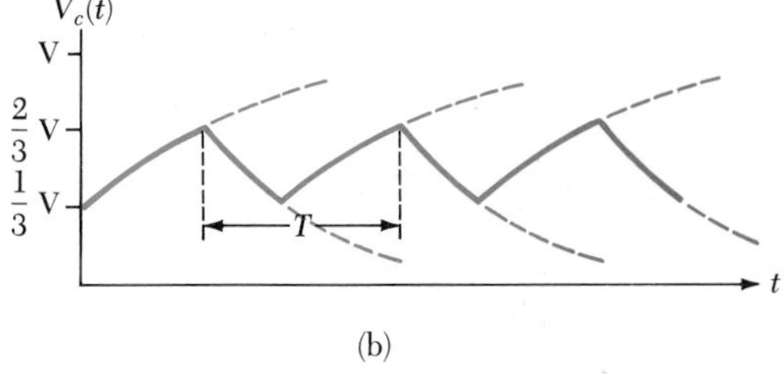

Figure 28.60 (Problem 83).

The voltmeter will show a voltage as plotted in Figure 28.60b. What is the period, T, of the waveform in terms of R_A, R_B, and C?

84. Design a multirange dc voltmeter that is capable of a full-scale deflection for the following divisions of voltage: (a) 20 V, (b) 50 V, and (c) 100 V. Assume a meter movement which has a coil resistance of 60 Ω and gives a full-scale deflection for a current of 1 mA.

85. Design a multirange dc ammeter that is capable of a full-scale deflection for the following divisions of current: (a) 25 mA, (b) 50 mA, and (c) 100 mA. Assume a meter movement which has a coil resistance of 25 Ω and gives a full-scale deflection for a current of 1 mA.

86. A particular galvanometer serves as a 2-V full-scale voltmeter when a 2500-Ω resistor is connected in series with it. It serves as a 0.5-A full-scale ammeter when a 0.22-Ω resistor is connected in parallel with it. Determine the internal resistance of the galvanometer and the current required to produce full-scale deflection.

Hint for Problems 87 through 90: Assume a current, use symmetry to identify points of equal potential, connect these points and redraw.

87. Find the equivalent resistance between any two vertices of a pyramid of six equal resistors of value R. (See Fig. 28.61.)

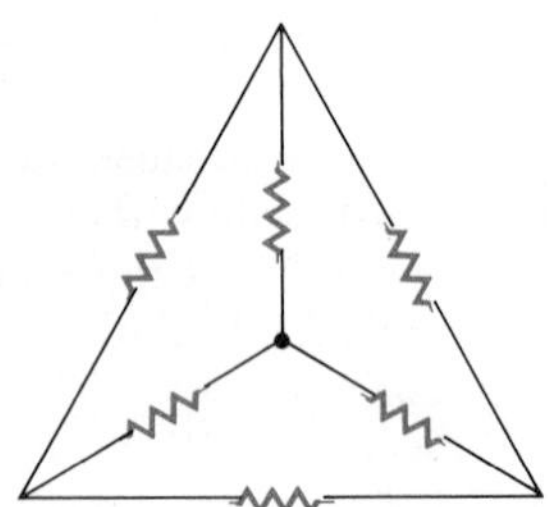

Figure 28.61 (Problems 87 and 88).

88. Three vertices of the pyramid configuration of resistors R, shown in Figure 28.61, are connected to a common ground. Find the equivalent resistance between the fourth vertex and the common ground.

89. Find the equivalent resistance between the center connection A and any one of the six vertex connections around the hexagon of resistors, as shown in Figure 28.62.

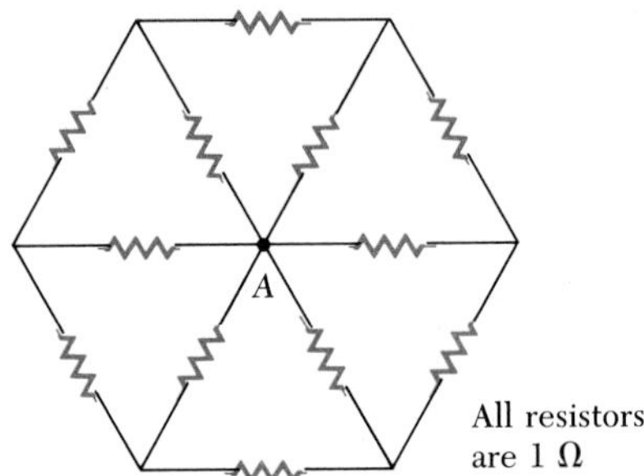

Figure 28.62 (Problem 89).

90. Twelve resistors, each of value 1 Ω, are connected so that each is along one edge of a cube, as shown in Figure 28.63. Find the resistance between the points (a) a and b, (b) a and c, and (c) b and c.

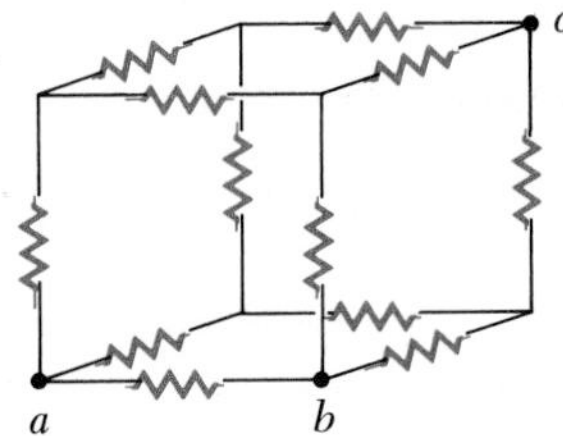

Figure 28.63 (Problem 90).

ESSAY

Exponential Growth

Albert A. Bartlett
University of Colorado

If something is growing at a constant rate, such as $P = 5\%$ per year, we refer to the growth as steady growth. It is also called **exponential growth** because the size, N, of the growing quantity at some time t in the future is related to its present size, N_0 (at time $t = 0$), by the exponential function

$$N = N_0 e^{kt} \tag{E.1}$$

where $e = 2.718 \ldots$ is the base of the natural logarithims and k is the annual percent growth rate P divided by 100:

$$k = \frac{P}{100} \tag{E.2}$$

Note that if k is positive, N increases exponentially with time (exponential growth). If k is negative, N decreases exponentially with time (exponential decay). Some exam-

Figure 1 If you try to draw an ordinary graph of the size versus time of anything that is growing steadily, the graph will go right through the ceiling.

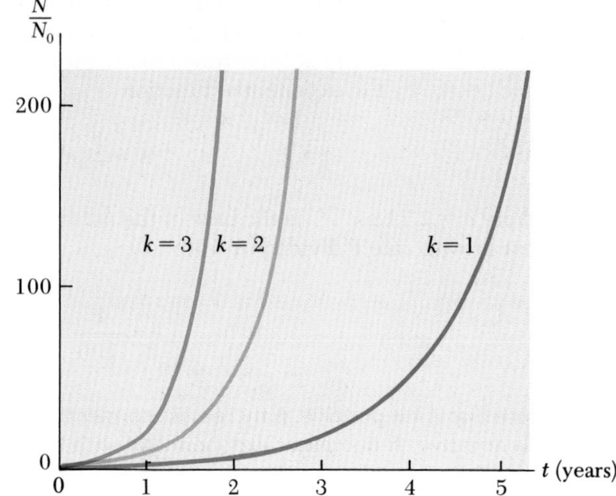

Figure 2 Exponential growth curve $N/N_0 = e^{kt}$. Note, N/N_0 has the value 1 at $t = 0$.

ples of exponential decay are radioactive decay (Section 45.4) and the decay of the quantity of charge on the plates of a capacitor as the capacitor is discharging through a resistor (Section 28.4). Figures 1 and 2 are representative graphs of exponential growth. This essay describes several examples of exponential growth.

The condition of steady growth is represented by the simple differential equation

$$\frac{dN}{dt} = kN \tag{E.3}$$

In words, this equation says that the rate of change of the quantity N is proportional to N. We can rearrange Equation E.3 to give

$$\frac{1}{N}\frac{dN}{dt} = k$$

In this form we see that the fractional change of N per unit time is constant. An example would be the case where the quantity N is growing 6% per year. In this case $P = 6\%$ per year and $k = 0.06$ per year.

The solution of the Equation E.3 is Equation E.1. If we can show that a quantity N changes with time according to Equation E.3 it follows automatically that N will obey Equation E.1.

For example, the fundamental concept of compound interest on a savings account in the bank is that the interest ΔN added to the number N of dollars in the account in the time interval Δt is proportional to the number of dollars in the account. The constant of proportionality is the interest rate. If $N = \$250$, $P = 8\%$ per year, and $\Delta t = 1$ year, then we have simple compounding once a year. The interest in that year is $\Delta N = 0.08 \times \$250 = \$20$ so that the value of N at the end of the year would be $\$250 + \$20 = \$270$. In the next year $\Delta N = 0.08 \times \$270 = \$21.60$, and at the end of the second year, $N = \$270 + \$21.60 = \$291.60$. Suppose the interest were 8% per year, compounded semiannually. In the first half year $\Delta N = (0.08/2) \times \$250 = \$10$, and at the end of the first half-year, $N = \$260$. In the second half year, $\Delta N = 0.04 \times \$260 = \$10.40$, and at the end of the first full year, $N = \$270.40$. At the end of the first year, compounding once gave $N = \$270$. Compounding twice gave $N = \$270.40$. This suggests a very fundamental fact. The more frequently we compound the interest, the more rapid is the increase in the size of N. Equation E.3 represents the limiting case where Δt approaches zero and the interest is *compounded continuously*. In this case at the end of one year we use Equation E.1:

$$N = \$250e^{0.08\times 1} = \$250 \times 1.08329 = \$270.82$$

Many banks calculate interest by compounding continuously. This leads to newpaper ads such as one that says "9.54% annual yield" which corresponds to a "9.11% annual rate." What this means is that the rate of 9.11% *compounded continuously* for one year gives the same result (yield) as a rate of 9.54% compounded once. In other words

$$e^{0.0911\times 1} = 1.0954 \qquad \text{(E.4)}$$

It has been shown that the number of miles of highway in the United States obeys Equation E.3 so that the number of miles of highway will grow exponentially according to Equation E.1.

In steady growth, it takes a fixed length of time for a quantity to grow by a fixed fraction such as 5%. From this it follows that it takes a fixed longer length of time for that quantity to grow by 100%. Let us calculate the time required for the quantity N to double in value, which is called the **doubling time, T_2**. We can obtain an expression for T_2 by writing Equation E.1 as $N/N_0 = e^{kt}$ and taking the natural logarithm of each side:

$$\ln\left(\frac{N}{N_0}\right) = kt$$

If we set $N = 2N_0$ (that is, we double N_0), then T_2 (which is the time t when $N = 2N_0$) is

$$T_2 = \frac{\ln(2N_0/N_0)}{k} = \frac{\ln 2}{k} = \frac{0.693}{k}$$

Since $k = P/100$, this becomes

$$T_2 \approx \frac{70}{P} \qquad \text{(E.5)}$$

Likewise, if we wanted the time for N to *triple* in size, we would use the natural logarithm of 3, to find

$$T_3 = \frac{100 \ln 3}{P} \approx \frac{110}{P}$$

EXAMPLE E.1 Compound Interest—The Eighth Wonder

Suppose you put \$15 in a savings account at 9% annual interest to be compounded continuously. How large a sum of money would be in the account at the end of 200 years?

Solution $N_0 = \$15.$; $k = 9/100 = 0.09$ per year and $t = 200$ years. Therefore, from Equation E.1 we have

$$N = \$15 \times e^{(0.09\times 200)} = \$15 \times e^{18} = \$15 \times 6.57 \times 10^7 = \$985 \text{ million!}$$

Now you can see why a famous financier once said that he could not name the seven wonders of the ancient world but surely the eighth wonder would have to be compound interest!

(Continued)

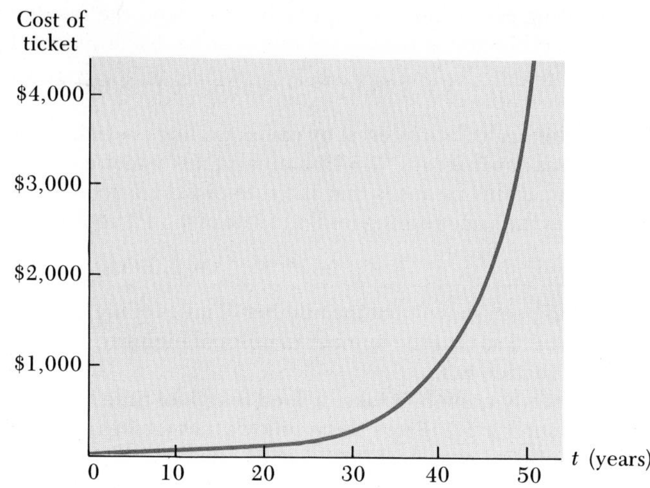

Figure 3 Price of a $4 ticket with 14% annual inflation. The cost at $t = 0$ is $N_0 = \$4$.

EXAMPLE E.2 The Consequences of Inflation

Let us use the doubling time instead of the quantity e to estimate the consequences of an annual inflation rate of 14% that continued for 50 years.

Solution First, we calculate T_2 from Equation E.5

$$T_2 = \frac{70}{14} = 5 \text{ years}$$

This inflation rate will cause prices to double every five years!

In the next step, we calculate the number of doubling times in 50 years:

$$\text{Number of doublings} = \frac{50 \text{ years}}{5 \text{ years/doubling}} = 10$$

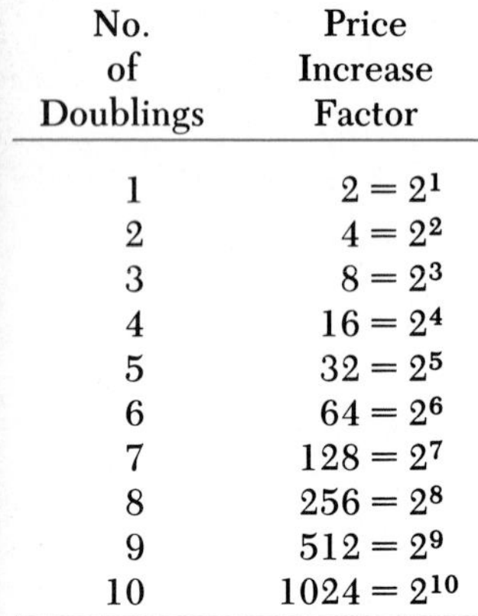

No. of Doublings	Price Increase Factor
1	$2 = 2^1$
2	$4 = 2^2$
3	$8 = 2^3$
4	$16 = 2^4$
5	$32 = 2^5$
6	$64 = 2^6$
7	$128 = 2^7$
8	$256 = 2^8$
9	$512 = 2^9$
10	$1024 = 2^{10}$

Finally, we count up the consequences of each doubling by making use of a table, which shows us that, in 10 doubling times, prices will increase by a factor of 1024, which is approximately 1000. Thus, in 50 years of 14% annual inflation, the cost of a $4 ticket to the movies would increase to roughly $4000! (See Fig. 3.)

It is very convenient to remember that 10 doublings give an increase by a factor of approximately 10^3, that 20 doublings give an increase of a factor of approximately 10^6, that 30 doublings give an increase by a factor of approximately 10^9, and so on.

EXAMPLE E.3 The Increasing Rate of Energy Consumption

For many years before 1975, consumption of electrical energy in the United States grew steadily at a rate of about 7% per year. By what factor would consumption increase if this growth rate continued for 40 years?

Solution In this case, $P = 7$, and so the doubling time from Equation E.5 is

$$T_2 = \frac{70}{7} = 10 \text{ years}$$

and the number of doublings in 40 years is

$$\text{Number of doubings} = \frac{40 \text{ years}}{10 \text{ years/doubling}} = 4$$

Therefore, in 40 years the amount of power consumed would be $2^4 = 16$ times the amount used today. That is, 40 years from now we would need 16 times as many electric generating plants as we have at the present. Furthermore, if those additional plants are similar to today's, then each day they would consume 16 times as much fuel as our present plants use, and there would be 16 times as much pollution and waste heat to contend with!

EXAMPLE E.4 Annual Increase in World Population

Populations tend to grow steadily. In July of 1987 we saw reports that the population of the earth had reached 5×10^9 people. The world birth rate was estimated to be 28 per 1000 each year while the annual death rate was estimated to be 11 per 1000. Thus, for every 1000 people, the population increase each year is $28 - 11 = 17$. For this growth rate we find

$$k = \frac{17}{1000} = 0.017 \text{ per year}$$

$$P = 100\,k = 1.7\% \text{ per year}$$

This growth rate seems so small that many people regard it as trivial and inconsequential. A proper perspective of this rate appears only when we calculate the doubling time:

$$T_2 = \frac{70}{1.7} = 41 \text{ years}$$

This simple calculation indicates that it is most likely that the world population will double within the life expectancy of today's students! At the most elemental level, this means that we have approximately 41 years to double world food production.

What is the annual increase in the earth's population? Since for one year $\Delta N \ll N$ we can get a good answer from Equation E.3:

$$\frac{\Delta N}{\Delta t} = 0.017 \times 5 \times 10^9 = 85 \text{ million per year}$$

This annual increase in the world population is roughly one third of the population of the United States.

Some illuminating calculations can be made based on the assumption that this rate of growth has been constant and will remain constant. These calculations will demonstrate that the growth rate has not been constant at this value in the past and cannot remain this high for very long.

EXAMPLE E.5 When did Adam and Eve Live?

When we use Equation E.1, setting $N = 5 \times 10^9$, $N_0 = 2$ (Adam and Eve) and $k = 0.017$ (from Example E.4), we have

$$5 \times 10^9 = 2e^{0.017t}$$

This gives $t = 1273$ years ago, or about 714 A.D.! This result proves that through essentially all of human history the population growth rate was very much smaller than it is today. It must have been near zero through most of human history.

EXAMPLE E.6 Growth of Population Density

The land area of the continents (excluding Antarctica) is $1.24 \times 10^{14}\ \text{m}^2$. If this modest annual growth rate of 1.7% were to continue steadily in the future, how long

(Continued)

would it take for the population to reach a density of one person per square meter on the continents?

$$1.24 \times 10^{14} = 5 \times 10^{9} e^{0.017t}$$

Solving, we find t is slightly less than 600 years.

EXAMPLE E.7 Growth of the Mass of People

If this very low rate of growth continued, how long would it take for the mass of people to equal the mass of the earth (5.98×10^{24} kg)? (Assume that the mass of an average person is 65 kg.)

$$5.98 \times 10^{24} = (5 \times 10^{9} \times 65) e^{0.017t}$$

This gives a value of t of about 1800 years! We have assumed that the mass of a person is 65 kg.

The last two examples prove that the growth rate of world population cannot stay as high as it presently is for any extended period of time. Although world agricultural production has been just barely keeping pace with world population growth, millions are malnourished and many people are starving. However, we will not have to double food production in 41 years if we can lower the worldwide birth rate. If we fail to double world food production in 41 years, then the death rate will rise. Dramatic increases in world food production in recent decades are due almost exclusively to the rapid growth of the use of petroleum for powering machinery and for manufacturing fertilizers and insecticides. Indeed, it has been noted that "modern agriculture is the use of land to convert petroleum into food." The student must wonder how much longer we can continue the long history of approximately steady population growth when our food supplies are tied so closely to dwindling supplies of petroleum.

This brief introduction to the arithmetic of steady growth enables us to understand that, in all biological systems, the normal condition is the steady-state condition, where the birth rate is equal to the death rate. Growth is a short-term transient phenomenon that can never continue for more than a short period of time. Yet in the United States, business and government leaders at all levels, from local communities to Washington, D.C., would have us believe that steady growth forever is a goal we can achieve. They would have us believe that we should continue our population growth (the U.S. population increases by about 2 million people per year) and the growth in our rates of consumption of natural resources. We now hear about "sustainable growth" as though the addition of the adjective "sustainable" would render inoperable the laws of nature.

In contrast to all this optimism, please remember that someone once noted that "The greatest shortcoming of the human race is our inability to understand the exponential function."

Suggested Readings

Bartlett, A. A., *Civil Engineering,* December 1969, pp. 71–72.

Bartlett, A. A., "The Exponential Function," *The Physics Teacher,* October 1976 to January 1979.

Bartlett, A. A., "The Forgotten Fundamentals of the Energy Crisis," *Am. J. Physics* 46:876, 1978.

Kerr, R. A., "Another Oil Resource Warning," *Science,* January 27, 1984, p. 382.

Essay Problems

1. In the year 1626 Manhattan Island was purchased for $24. Assuming a continually compounded interest rate of 4.4%, calculate the current land valuation of the island.
2. The following "mystery" was taken from Deborah Hughes-Hallett: *Elementary Functions,* W. W. Norton, 1980, p. 264.

   The police were baffled by what seemed to be the perfect murder of a girl who had been found, apparently suffocated, in her kitchen. Finally, Sherlock Holmes was called in. With the aid of Dr. Watson's knowledge of botany, the mystery was solved and the following story told. The girl had been making bread in her kitchen, whose dimensions were 10 ft by 10 ft. She had formed the dough into a ball of volume 1/6 cubic feet and turned away to wash some dishes. At that moment Holmes' enemy, Professor Moriarty, had added a particularly virulent strain of yeast to the bread. As a result, the bread immediately started to rise, tripling in volume every 4 minutes. Before long, the dough filled the room, stopping the clock at 3:48 and squashing the girl to death against the wall. By the time Inspector Lestrade of Scotland Yard reached the scene the next day, the yeast had worked itself out and the dough returned to its original size.

At what time did Professor Moriarty add the yeast?

29

Magnetic Fields

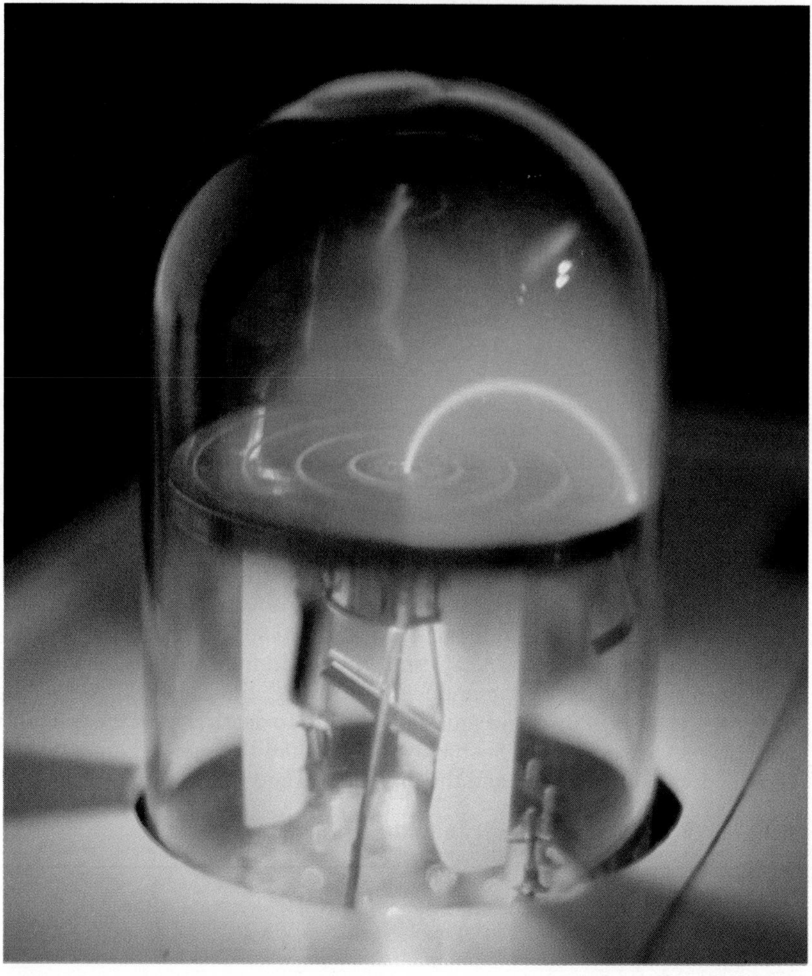

The blue arc in this photograph indicates the circular path followed by an electron beam moving in a magnetic field. The vessel contains gas at very low pressure, and the beam is made visible as the electrons collide with the gas atoms, which in turn emit visible light. The magnetic field is produced by two coils (not shown). The apparatus can be used to measure the ratio of e/m for the electron. (Courtesy of CENCO)

The behavior of bar magnets is well known to anyone who has studied science. Permanent magnets, which are usually made of alloys containing iron, will attract or repel other magnets. Furthermore, they will attract other bits of iron, which in turn can become magnetized. The list of important technological applications of magnetism is extensive. For instance, large electromagnets are used to pick up heavy loads. Magnets are also used in such devices as meters, transformers, motors, particle accelerators, and loudspeakers. Magnetic tapes are routinely used in sound recording, TV recording, and computer memories. Intense magnetic fields generated by superconducting magnets are currently being used as a means of containing the plasmas (heated to temperatures of the order of 10^8 K) used in controlled nuclear fusion research.

29.1 INTRODUCTION

The phenomenon of magnetism was known to the Greeks as early as around 800 B.C. They discovered that certain stones, now called *magnetite* (Fe_3O_4), attract pieces of iron. Legend ascribes the name *magnetite* to the shepherd Magnes, "the nails of whose shoes and the tip of whose staff stuck fast in a magnetic field while he pastured his flocks." In 1269 Pierre de Maricourt, using a spherical natural magnet, mapped out the directions taken by a needle when placed at various points on the surface of the sphere. He found that the directions formed lines that encircle the sphere passing through two points diametrically opposite each other, which he called the *poles* of the magnet. Subsequent experiments showed that every magnet, regardless of its shape, has two poles, called *north* and *south poles,* which exhibit forces on each other in a manner analogous to electrical charges. That is, like poles repel each other and unlike poles attract each other.

In 1600 William Gilbert extended these experiments to a variety of materials. Using the fact that a compass needle orients in preferred directions, he suggested that the earth itself is a large permanent magnet. In 1750 John Michell (1724–1793) used a torsion balance to show that magnetic poles exert attractive or repulsive forces on each other and that these forces vary as the inverse square of their separation. Although the force between two magnetic poles is similar to the force between two electric charges, there is an important difference. Electric charges can be isolated (witness the electron or proton), whereas *magnetic poles cannot be isolated.* That is, *magnetic poles are always found in pairs.* All attempts thus far to detect an isolated magnetic monopole have been unsuccessful. No matter how many times a permanent magnet is cut, each piece will always have a north and a south pole.

The relationship between magnetism and electricity was discovered in 1819 when, during a lecture demonstration, the Danish scientist Hans Oersted found that an electric current in a wire deflected a nearby compass

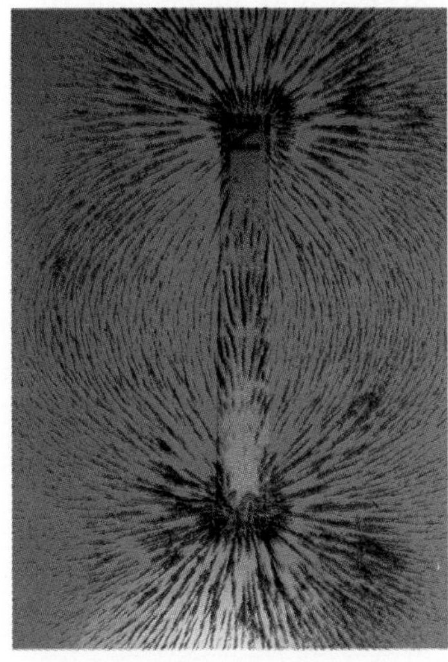

Magnetic field pattern of a bar magnet as displayed by iron filings on a sheet of paper. (Courtesy of Henry Leap and Jim Lehman)

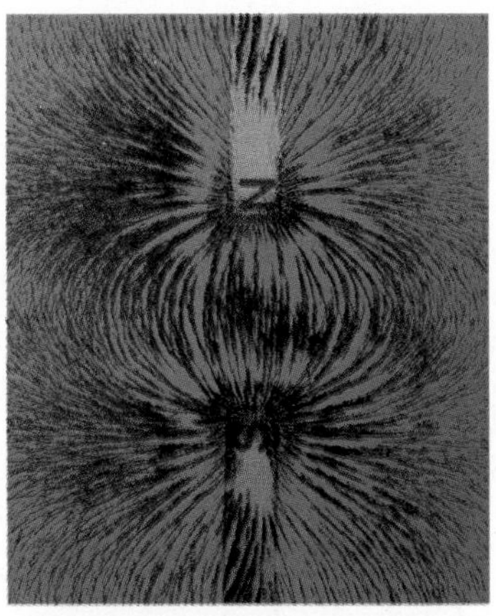

Magnetic field patterns surrounding two bar magnets as displayed with iron filings. This demonstrates the magnetic field pattern between unlike poles. (Courtesy of Henry Leap and Jim Lehman)

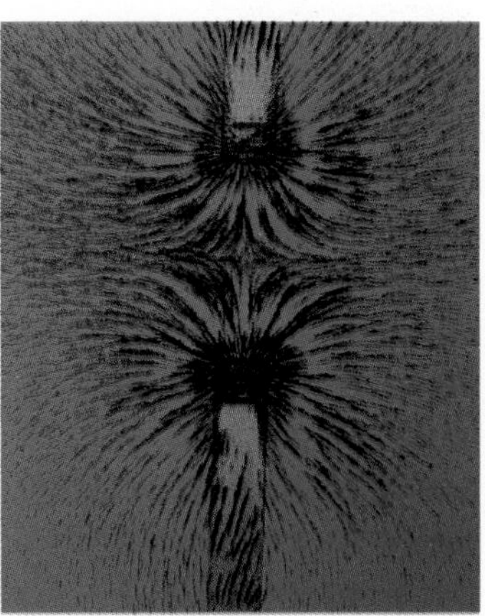

This demonstrates the magnetic field pattern between two like poles. (Courtesy of Henry Leap and Jim Lehman)

Hans Christian Oersted (1777–1851), Danish physicist. (The Bettmann Archive)

needle.[1] Shortly thereafter, André Ampère (1775–1836) obtained quantitative laws of magnetic force between current-carrying conductors. He also suggested that electric current loops of molecular size are responsible for *all* magnetic phenomena. This idea is the basis for the modern theory of magnetism.

In the 1820's, further connections between electricity and magnetism were demonstrated by Faraday and independently by Joseph Henry (1797–1878). They showed that an electric current could be produced in a circuit either by moving a magnet near the circuit or by changing the current in another, nearby circuit. These observations demonstrate that a changing magnetic field produces an electric field. Years later, theoretical work by Maxwell showed that a changing electric field gives rise to a magnetic field.

This chapter examines forces on moving charges and on current-carrying wires in the presence of a magnetic field. The source of the magnetic field itself will be described in Chapter 30.

29.2 DEFINITION AND PROPERTIES OF THE MAGNETIC FIELD

The electric field $\boldsymbol{E}$ at a point in space has been defined as the electric force per unit charge acting on a test charge placed at that point. Similarly, the gravitational field $\boldsymbol{g}$ at a point in space is the gravitational force per unit mass acting on a test mass.

We now define a magnetic field vector $\boldsymbol{B}$ (sometimes called the *magnetic induction* or *magnetic flux density*) at some point in space in terms of a magnetic force that would be exerted on an appropriate test object. Our test object is taken to be a charged particle moving with a velocity $\boldsymbol{v}$. For the time being, let us assume that there are no electric or gravitational fields present in the region of the charge. Experiments on the motion of various charged particles moving in a magnetic field give the following results:

Properties of the magnetic force on a charge moving in a $\boldsymbol{B}$ field

1. The magnetic force is proportional to the charge q and speed v of the particle.
2. The magnitude and direction of the magnetic force depend on the velocity of the particle and on the magnitude and direction of the magnetic field.
3. When a charged particle moves in a direction *parallel* to the magnetic field vector, the magnetic force $\boldsymbol{F}$ on the charge is *zero*.
4. When the velocity vector makes an angle θ with the magnetic field, the magnetic force acts in a direction perpendicular to both $\boldsymbol{v}$ and $\boldsymbol{B}$; that is, $\boldsymbol{F}$ is perpendicular to the plane formed by $\boldsymbol{v}$ and $\boldsymbol{B}$ (Fig. 29.1a).
5. The magnetic force on a positive charge is in the direction opposite the direction of the force on a negative charge moving in the same direction (Fig. 29.1b).
6. If the velocity vector makes an angle θ with the magnetic field, the magnitude of the magnetic force is proportional to $\sin\theta$.

These observations can be summarized by writing the magnetic force in the form

Magnetic force on a charged particle moving in a magnetic field

$$\boldsymbol{F} = q\boldsymbol{v} \times \boldsymbol{B} \tag{29.1}$$

[1] It is interesting to note that the same discovery was reported in 1802 by an Italian jurist, Gian Dominico Romognosi, but was overlooked, probably because it was published in a newspaper, *Gazetta de Trentino*, rather than in a scholarly journal.

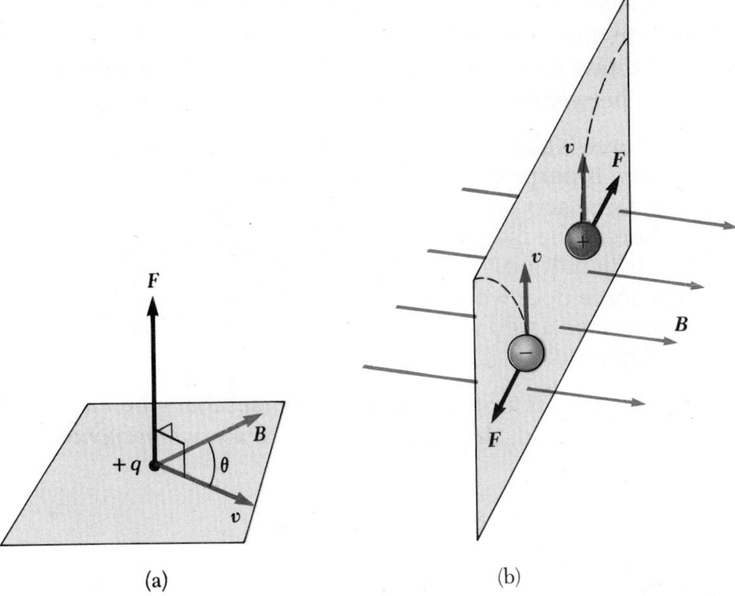

Figure 29.1 The direction of the magnetic force on a charged particle moving with a velocity $\boldsymbol{v}$ in the presence of a magnetic field. (a) When $\boldsymbol{v}$ is at an angle θ to $\boldsymbol{B}$, the magnetic force is perpendicular to both $\boldsymbol{v}$ and $\boldsymbol{B}$. (b) In the presence of a magnetic field, the moving charged particles are deflected as indicated by the dotted lines.

where the direction of the magnetic force is in the direction of $\boldsymbol{v} \times \boldsymbol{B}$, which, by definition of the cross product, is perpendicular to both $\boldsymbol{v}$ and $\boldsymbol{B}$.

Figure 29.2 gives a brief review of the right-hand rule for determining the direction of the cross product $\boldsymbol{v} \times \boldsymbol{B}$. You point the four fingers of your right hand along the direction of $\boldsymbol{v}$, and then turn them until they point along the direction of $\boldsymbol{B}$. The thumb then points in the direction of $\boldsymbol{v} \times \boldsymbol{B}$. Since $\boldsymbol{F} = q\boldsymbol{v} \times \boldsymbol{B}$, $\boldsymbol{F}$ is in the direction of $\boldsymbol{v} \times \boldsymbol{B}$ if q is positive (Fig. 29.2a) and *opposite* the direction of $\boldsymbol{v} \times \boldsymbol{B}$ if q is negative (Fig. 29.2b). The magnitude of the magnetic force has the value

$$F = qvB \sin \theta \tag{29.2}$$

where θ is the angle between $\boldsymbol{v}$ and $\boldsymbol{B}$. From this expression, we see that F is *zero* when $\boldsymbol{v}$ is parallel to $\boldsymbol{B}$ ($\theta = 0$ or 180°). Furthermore, the force has its *maximum* value, $F = qvB$, when $\boldsymbol{v}$ is perpendicular to $\boldsymbol{B}$ ($\theta = 90°$).

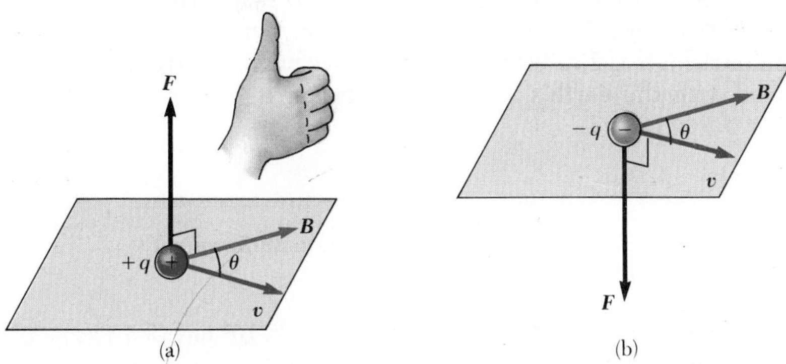

Figure 29.2 The right-hand rule for determining the direction of the magnetic force $\boldsymbol{F}$ acting on a charge q moving with a velocity $\boldsymbol{v}$ in a magnetic field $\boldsymbol{B}$. If q is positive, $\boldsymbol{F}$ is upward in the direction of the thumb. If q is negative, $\boldsymbol{F}$ is downward.

We can regard Equation 29.1 as an operational definition of the magnetic field at a point in space. That is, the magnetic field is defined in terms of a *sideways* force acting on a moving charged particle. There are several important differences between electric and magnetic forces:

Differences between electric and magnetic fields

1. The electric force is always in the direction of the electric field, whereas the magnetic force is perpendicular to the magnetic field.
2. The electric force acts on a charged particle independent of the particle's velocity, whereas the magnetic force acts on a charged particle only when the particle is in motion.
3. The electric force does work in displacing a charged particle, whereas the magnetic force associated with a steady magnetic field does *no* work when a particle is displaced.

This last statement is a consequence of the fact that when a charge moves in a steady magnetic field, the magnetic force is always *perpendicular* to the displacement. That is,

$$\boldsymbol{F}\cdot d\boldsymbol{s} = (\boldsymbol{F}\cdot\boldsymbol{v})\, dt = 0$$

since the magnetic force is a vector perpendicular to $\boldsymbol{v}$. From this property and the work-energy theorem, we conclude that the kinetic energy of a charged particle *cannot* be altered by a magnetic field alone. In other words,

A magnetic field cannot change the speed of a particle

when a charge moves with a velocity $\boldsymbol{v}$, an applied magnetic field can alter the direction of the velocity vector, but it cannot change the speed of the particle.

The SI unit of the magnetic field is the **weber per square meter** (Wb/m^2), also called the **tesla** (T). This unit can be related to the fundamental units by using Equation 29.1: a 1-coulomb charge moving through a field of 1 tesla with a velocity of 1 m/s perpendicular to the field experiences a force of 1 newton:

$$[\text{B}] = \text{T} = \frac{\text{Wb}}{\text{m}^2} = \frac{\text{N}}{\text{C}\cdot\text{m/s}} = \frac{\text{N}}{\text{A}\cdot\text{m}} \tag{29.3}$$

In practice, the cgs unit for magnetic field, called the **gauss** (G), is often used. The gauss is related to the tesla through the conversion

$$1\ \text{T} = 10^4\ \text{G} \tag{29.4}$$

Conventional laboratory magnets can produce magnetic fields as large as about 25 000 G, or 2.5 T. Superconducting magnets that can generate magnetic fields as high as 250 000 G, or 25 T, have been constructed. This can be compared with the earth's magnetic field near its surface, which is about 0.5 G, or 0.5×10^{-4} T.

EXAMPLE 29.1 A Proton Moving in a Magnetic Field

A proton moves with a speed of 8×10^6 m/s along the x axis. It enters a region where there is a field of magnitude 2.5 T, directed at an angle of 60° to the x axis and lying in the xy plane (Fig. 29.3). Calculate the initial magnetic force and acceleration of the proton.

Solution From Equation 29.2, we get

$$\begin{aligned} F &= qvB \sin\theta \\ &= (1.6 \times 10^{-19}\ \text{C})(8 \times 10^6\ \text{m/s})(2.5\ \text{T})(\sin 60°) \\ &= 2.77 \times 10^{-12}\ \text{N} \end{aligned}$$

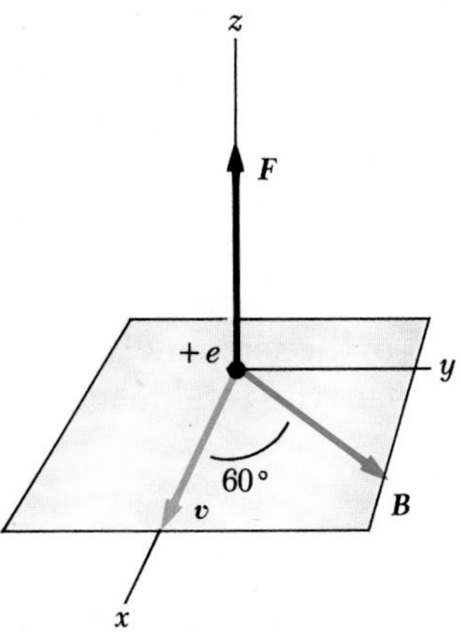

Figure 29.3 (Example 29.1) The magnetic force $\boldsymbol{F}$ on a proton is in the positive z direction when $\boldsymbol{v}$ and $\boldsymbol{B}$ lie in the xy plane.

Since $\boldsymbol{v} \times \boldsymbol{B}$ is in the positive z direction and since the charge is positive, the force $\boldsymbol{F}$ is in the positive z direction.

Since the mass of the proton is 1.67×10^{-27} kg, its initial acceleration is

$$a = \frac{F}{m} = \frac{2.77 \times 10^{-12}\ \text{N}}{1.67 \times 10^{-27}\ \text{kg}} = 1.66 \times 10^{15}\ \text{m/s}^2$$

in the positive z direction.

Exercise 1 Verify that the units of $\boldsymbol{F}$ in the above calculation for the magnetic force reduce to newtons.

29.3 MAGNETIC FORCE ON A CURRENT-CARRYING CONDUCTOR

If a force is exerted on a single charged particle when it moves through a magnetic field, it should not surprise you that a current-carrying wire also experiences a force when placed in a magnetic field. This follows from the fact that the current represents a collection of many charged particles in motion; hence, the resultant force on the wire is due to the sum of the individual forces on the charged particles.

The force on a current-carrying conductor can be demonstrated by hanging a wire between the faces of a magnet as in Figure 29.4. In this figure, the magnetic field is directed into the page and covers the region within the shaded circle. When the current in the wire is zero, the wire remains vertical as in Figure 29.4a. However, when a current is set up in the wire directed upwards as in Figure 29.4b, the wire deflects to the left. If we reverse the current, as in Figure 29.4c, the wire deflects to the right.

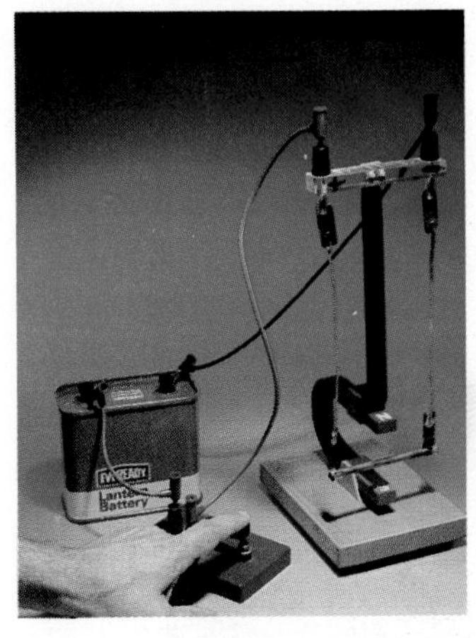

This apparatus demonstrates the force on a current-carrying conductor in an external magnetic field. Why does the bar swing *away* from the magnet after the switch is closed? (Courtesy of Henry Leap and Jim Lehman)

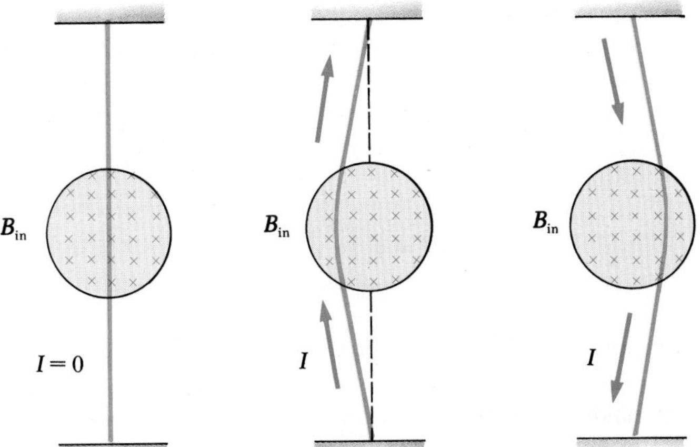

Figure 29.4 A flexible vertical wire which is partially stretched between the faces of a magnet with the field (blue crosses) directed into the paper. (a) When there is no current in the wire, it remains vertical. (b) When the current is upwards, the wire deflects to the left. (c) When the current is downwards, the wire deflects to the right.

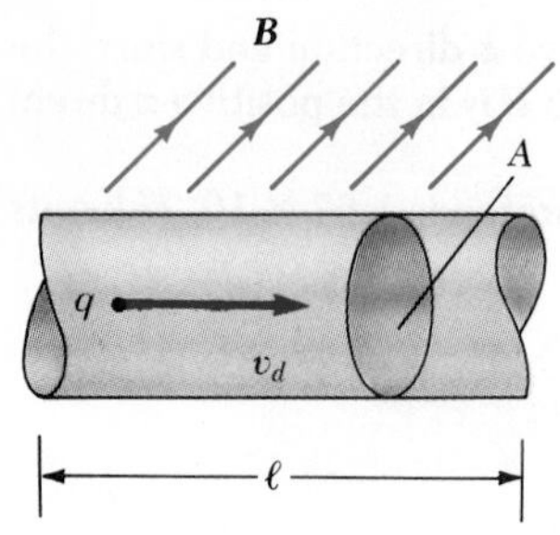

Figure 29.5 A section of a wire containing moving charges in an external magnetic field $\boldsymbol{B}$. The magnetic force on each charge is $q\boldsymbol{v}_d \times \boldsymbol{B}$, and the net force on a straight element is $I\boldsymbol{\ell} \times \boldsymbol{B}$.

Let us quantify this discussion by considering a straight segment of wire of length ℓ and cross-sectional area A, carrying a current I in a uniform *external* magnetic field $\boldsymbol{B}$ as in Figure 29.5. The magnetic force on a charge q moving with a drift velocity v_d is given by $q\boldsymbol{v}_d \times \boldsymbol{B}$. The force on the charge carriers is transmitted to the "bulk" of the wire through collisions with the atoms making up the wire. To find the total force on the wire, we multiply the force on one charge, $q\boldsymbol{v}_d \times \boldsymbol{B}$, by the number of charges in the segment. Since the volume of the segment is $A\ell$, the number of charges in the segment is $nA\ell$, where n is the number of charges per unit volume. Hence, the total magnetic force on the wire of length ℓ is

$$\boldsymbol{F} = (q\boldsymbol{v}_d \times \boldsymbol{B})nA\ell$$

This can be written in a more convenient form by noting that, from Equation 27.4, the current in the wire is given by $I = nqv_dA$. Therefore, $\boldsymbol{F}$ can be expressed as

$$\boldsymbol{F} = I\boldsymbol{\ell} \times \boldsymbol{B} \tag{29.5}$$

where $\boldsymbol{\ell}$ is a vector in the direction of the current I; the magnitude of $\boldsymbol{\ell}$ equals the length ℓ of the segment. Note that this expression applies only to a straight segment of wire in a uniform external magnetic field. Furthermore, we have neglected the field produced by the current itself. (In fact, the wire cannot produce a force on itself.)

Now consider an arbitrarily shaped wire of uniform cross section in an external magnetic field, as in Figure 29.6. It follows from Equation 29.5 that the magnetic force on a very small segment ds in the presence of a field $\boldsymbol{B}$ is given by

$$d\boldsymbol{F} = I\, d\boldsymbol{s} \times \boldsymbol{B} \tag{29.6}$$

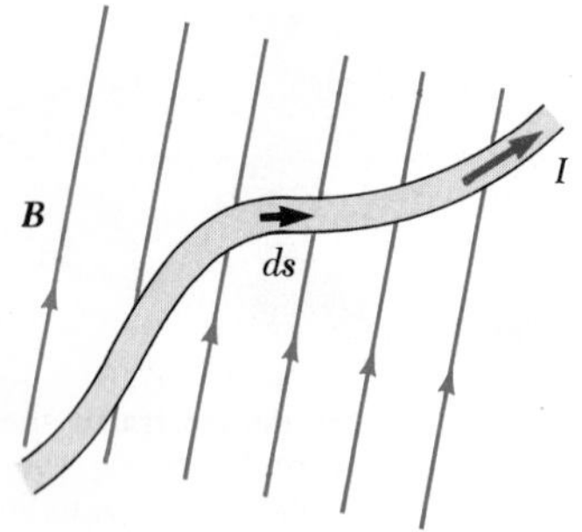

Figure 29.6 A wire of arbitrary shape carrying a current I in an external magnetic field $\boldsymbol{B}$ experiences a magnetic force. The force on any segment $d\boldsymbol{s}$ is given by $I\, d\boldsymbol{s} \times \boldsymbol{B}$ and is directed *out* of the page.

where $d\boldsymbol{F}$ is directed out of the page for the directions assumed in Figure 29.6. We can consider Equation 29.6 as an alternative definition of $\boldsymbol{B}$. That is, the field $\boldsymbol{B}$ can be defined in terms of a measurable force on a current element, where the force is a maximum when $\boldsymbol{B}$ is perpendicular to the element and zero when $\boldsymbol{B}$ is parallel to the element.

To get the total force $\boldsymbol{F}$ on the wire, we integrate Equation 29.6 over the length of the wire:

$$\boldsymbol{F} = I \int_a^b d\boldsymbol{s} \times \boldsymbol{B} \tag{29.7}$$

In this expression, a and b represent the end points of the wire. When this integration is carried out, the magnitude of the magnetic field and the direction the field makes with the vector $d\boldsymbol{s}$ (that is, the element orientation) may vary at each point.

Now let us consider two special cases involving the application of Equation 29.7. In both cases, the external magnetic field is taken to be constant in magnitude and direction.

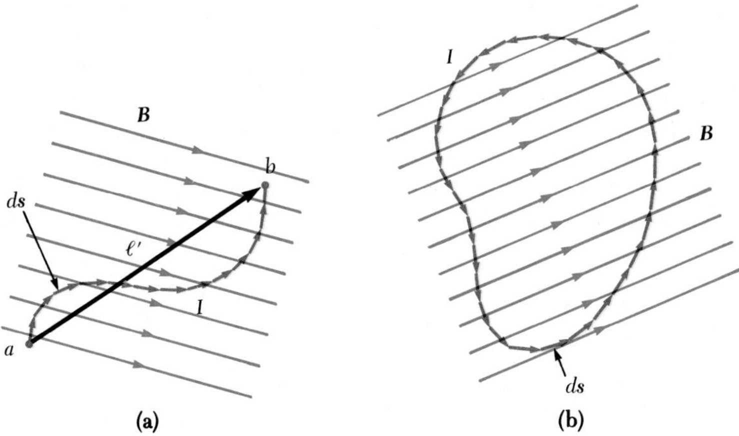

Figure 29.7 (a) A curved conductor carrying a current I in a uniform magnetic field. The magnetic force on the conductor is equivalent to the force on a straight segment of length ℓ' running between the ends of the wire, a and b. (b) A current-carrying loop of arbitrary shape in a uniform magnetic field. The net magnetic force on the loop is 0.

Case I

Consider a curved wire carrying a current I; the wire is located in a uniform external magnetic field $\mathbf{B}$ as in Figure 29.7a. Since the field is assumed to be uniform (that is, $\mathbf{B}$ has the same value over the region of the conductor), $\mathbf{B}$ can be taken outside the integral in Equation 29.7, and we get

$$\mathbf{F} = I\left(\int_a^b d\mathbf{s}\right) \times \mathbf{B} \tag{29.8}$$

But the quantity $\int_a^b d\mathbf{s}$ represents the *vector sum* of all the displacement elements from a to b as described in Figure 29.6. From the law of addition of many vectors (Chapter 2), the sum equals the vector $\boldsymbol{\ell}'$, which is directed from a to b. Therefore, Equation 29.8 reduces to

$$\mathbf{F} = I\boldsymbol{\ell}' \times \mathbf{B} \tag{29.9}$$

Force on a wire in a uniform field

Case II

An arbitrarily shaped, closed loop carrying a current I is placed in a uniform external magnetic field $\mathbf{B}$ as in Figure 29.7b. Again, we can express the force in the form of Equation 29.8. In this case, the vector sum of the displacement vectors must be taken over the closed loop. That is,

$$\mathbf{F} = I\left(\oint d\mathbf{s}\right) \times \mathbf{B}$$

Since the set of displacement vectors forms a *closed polygon* (Fig. 29.7b), the vector sum must be *zero*. This follows from the graphical procedure of adding vectors by the *polygon method* (Chapter 2). Since $\oint d\mathbf{s} = 0$, we conclude that

$$\mathbf{F} = 0 \tag{29.10}$$

That is,

> the total magnetic force on any closed current loop in a uniform magnetic field is zero.

EXAMPLE 29.2 Force on a Semicircular Conductor □

A wire bent into the shape of a semicircle of radius R forms a closed circuit and carries a current I. The circuit lies in the xy plane, and a uniform magnetic field is present along the positive y axis as in Figure 29.8. Find the magnetic forces on the straight portion of the wire and on the curved portion.

Solution The force on the straight portion of the wire has a magnitude given by $F_1 = I\ell B = 2IRB$, since $\ell = 2R$ and the wire is perpendicular to $\mathbf{B}$. The direction of $\mathbf{F}_1$ is *out* of the paper since $\boldsymbol{\ell} \times \mathbf{B}$ is outward. (That is, $\boldsymbol{\ell}$ is to the right in the direction of the current, and so by the rule of cross products, $\boldsymbol{\ell} \times \mathbf{B}$ is outward.)

To find the force on the curved part, we must first write an expression for the force $d\mathbf{F}_2$ on the element ds. If θ is the angle between $\mathbf{B}$ and $d\mathbf{s}$ in Figure 29.8, then the magnitude of $d\mathbf{F}_2$ is given by

$$dF_2 = I|d\mathbf{s} \times \mathbf{B}| = IB \sin \theta \, ds$$

where ds is the length of the small element measured along the circular arc. In order to integrate this expression, we must express ds in terms of the variable θ. Since $s = R\theta$, $ds = R\, d\theta$, and the expression for dF_2 can be written

$$dF_2 = IRB \sin \theta \, d\theta$$

To get the *total* force F_2 on the curved portion, we can integrate this expression to account for contributions from *all* elements. Note that the direction of the force on every element is the same: *into* the paper (since $d\mathbf{s} \times \mathbf{B}$ is inward). Therefore, the resultant force $\mathbf{F}_2$ on the curved wire must also be *into* the paper. Integrating dF_2 over the limits $\theta = 0$ to $\theta = \pi$ (that is, the entire semicircle) gives

$$F_2 = IRB \int_0^{\pi} \sin \theta \, d\theta = IRB[-\cos \theta]_0^{\pi}$$

$$= -IRB(\cos \pi - \cos 0) = -IRB(-1-1) = 2IRB$$

Since $F_2 = 2IRB$ and is directed *into* the paper while the force on the straight wire $F_1 = 2IRB$ is *out* of the paper, we see that the *net* force on the closed loop is *zero*. This result is consistent with Case II as described above and given by Equation 29.10.

Exercise 2 Find the force $\mathbf{F}_2$ on the semicircular part of the wire by making use of the more direct method, Case I, discussed in the text (Eq. 29.9).

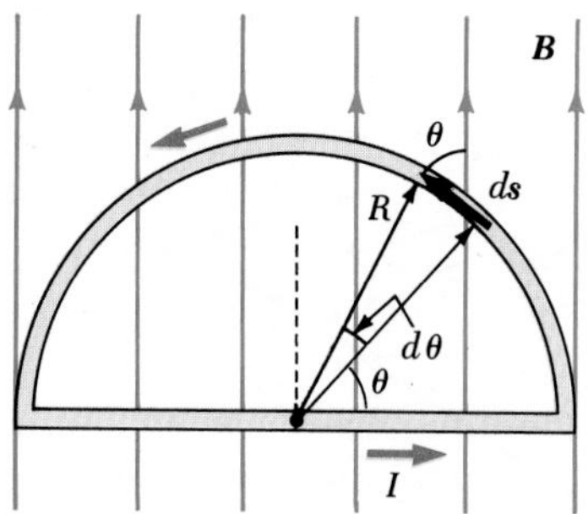

Figure 29.8 (Example 29.2) The net force on a closed current loop in a uniform magnetic field is *zero*. In this case, the force on the straight portion is $2IRB$ and outward, while the force on the curved portion is also $2IRB$ and inward.

29.4 TORQUE ON A CURRENT LOOP IN A UNIFORM MAGNETIC FIELD

In the previous section we showed how a force is exerted on a current-carrying conductor when the conductor is placed in an external magnetic field. With this as a starting point, we shall show that a torque is exerted on a current loop placed in a magnetic field. The results of this analysis will be of great practical value when we discuss motors in a future chapter.

Consider a rectangular loop carrying a current I in the presence of a uniform magnetic field *in the plane of the loop*, as in Figure 29.9a. The forces on the sides of length a are zero since these wires are parallel to the field; hence $d\mathbf{s} \times \mathbf{B} = 0$ for these sides. The magnitude of the forces on the sides of length b, however, is given by

$$F_1 = F_2 = IbB$$

The direction of $\mathbf{F}_1$, the force on the left side of the loop, is out of the paper and that of $\mathbf{F}_2$, the force on the right side of the loop, is into the paper. If we were to view the loop from an end view, as in Figure 29.9b, we would see the forces directed as shown. If we assume that the loop is pivoted so that it can rotate about point O, we see that these two forces produce a torque about O that rotates the loop clockwise. The magnitude of this torque, τ_{max}, is

$$\tau_{max} = F_1 \frac{a}{2} + F_2 \frac{a}{2} = (IbB) \frac{a}{2} + (IbB) \frac{a}{2} = IabB$$

where the moment arm about O is $a/2$ for each force. Since the area of the loop is $A = ab$, the torque can be expressed as

$$\tau = IAB \tag{29.11}$$

Remember that this result is valid only when the field $\boldsymbol{B}$ is parallel to the plane of the loop. The sense of the rotation is clockwise when viewed from the bottom end, as indicated in Figure 29.9b. If the current were reversed, the forces would reverse their directions and the rotational tendency would be counterclockwise.

Now suppose the uniform magnetic field makes an angle θ with respect to a line perpendicular to the plane of the loop, as in Figure 29.10a. For convenience, we shall assume that the field $\boldsymbol{B}$ is perpendicular to the sides of length b. In this case, the magnetic forces $\boldsymbol{F}_3$ and $\boldsymbol{F}_4$ on the sides of length a cancel each other and produce no torque since they pass through a common origin. However, the forces $\boldsymbol{F}_1$ and $\boldsymbol{F}_2$ acting on the sides of length b form a couple and hence produce a torque about *any point*. Referring to the end view shown in Figure 29.10b, we note that the moment arm of the force $\boldsymbol{F}_1$ about the point O is equal to $(a/2) \sin \theta$. Likewise, the moment arm of $\boldsymbol{F}_2$ about O is also $(a/2) \sin \theta$. Since $F_1 = F_2 = IbB$, the net torque about O has a magnitude given by

$$\tau = F_1 \frac{a}{2} \sin \theta + F_2 \frac{a}{2} \sin \theta$$

$$= IbB \left(\frac{a}{2} \sin \theta\right) + IbB \left(\frac{a}{2} \sin \theta\right) = IabB \sin \theta$$

$$= IAB \sin \theta$$

where $A = ab$ is the area of the loop. This result shows that the torque has the *maximum* value IAB when the field is parallel to the plane of the loop ($\theta = 90°$)

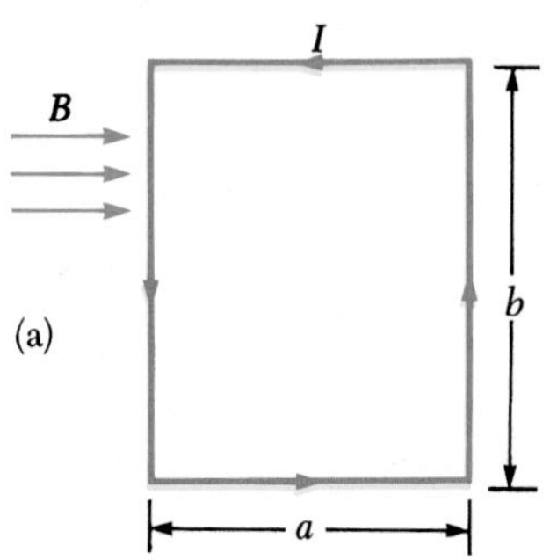

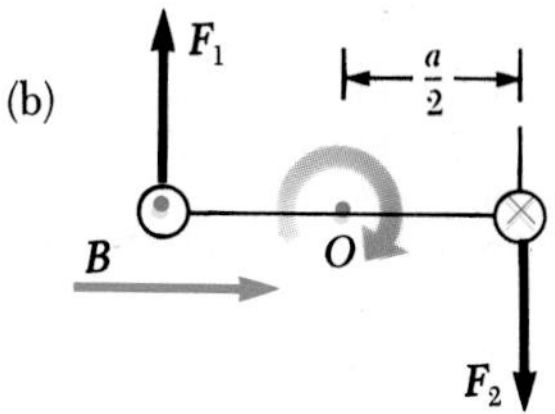

Figure 29.9 (a) Front view of a rectangular loop in a uniform magnetic field. There are no forces on the sides of width a parallel to $\boldsymbol{B}$, but there are forces acting on the sides of length b. (b) Bottom view of the rectangular loop shows that the forces $\boldsymbol{F}_1$ and $\boldsymbol{F}_2$ on the sides of length b create a torque that tends to twist the loop clockwise as shown.

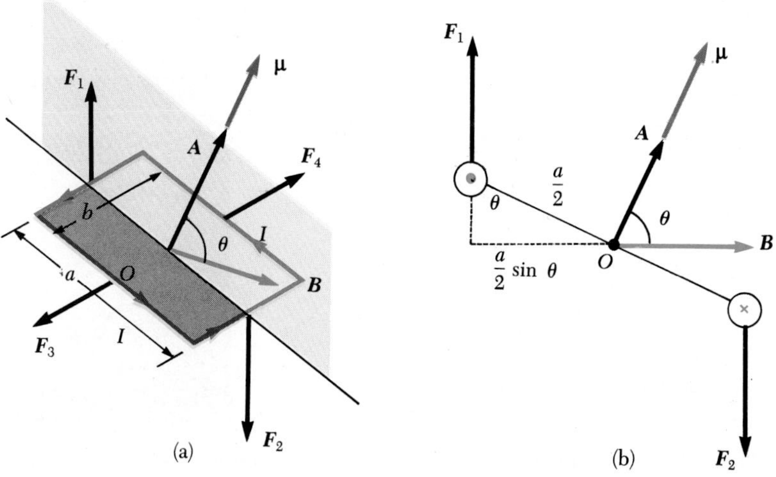

Figure 29.10 (a) A rectangular current loop whose normal makes an angle θ with a uniform magnetic field. The forces on the sides of length a cancel while the forces on the sides of width b create a torque on the loop. (b) An end view of the loop. The magnetic moment $\boldsymbol{\mu}$ is in the direction normal to the plane of the loop.

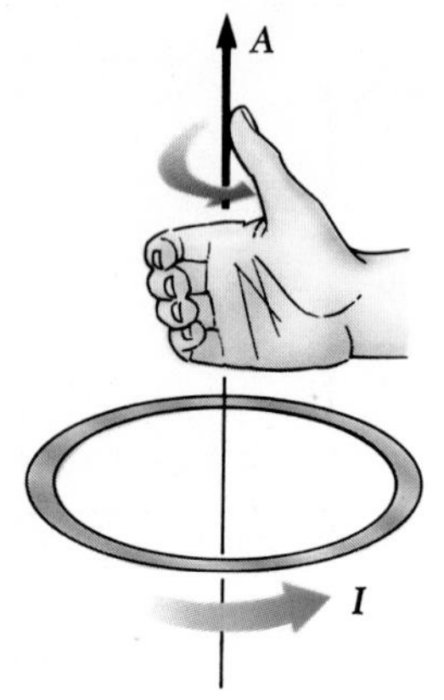

Figure 29.11 A right-hand rule for determining the direction of the vector **A**. The magnetic moment μ is also in the direction of **A**.

and is *zero* when the field is perpendicular to the plane of the loop ($\theta = 0$). As we see in Figure 29.10, the loop tends to rotate to smaller values of θ (that is, such that the normal to the plane of the loop rotates toward the direction of the magnetic field).

A convenient vector expression for the torque is the following cross-product relationship:

$$\boldsymbol{\tau} = I\,\mathbf{A} \times \mathbf{B} \tag{29.12}$$

where **A**, a vector perpendicular to the plane of the loop, has a magnitude equal to the area of the loop. The sense of **A** is determined by the right-hand rule as described in Figure 29.11. By rotating the four fingers of the right hand in the direction of the current in the loop, the thumb points in the direction of **A**. The product IA is defined to be the **magnetic moment** $\boldsymbol{\mu}$ of the loop. That is,

$$\boldsymbol{\mu} = I\,\mathbf{A} \tag{29.13}$$

The SI unit of magnetic moment is ampere-meter^2 ($A \cdot m^2$). Using this definition, the torque can be expressed as

Torque on a current loop

$$\boldsymbol{\tau} = \boldsymbol{\mu} \times \mathbf{B} \tag{29.14}$$

Note that this result is analogous to the torque acting on an electric dipole moment $\mathbf{p}$ in the presence of an external electric field $\mathbf{E}$, where $\boldsymbol{\tau} = \mathbf{p} \times \mathbf{E}$ (Section 26.6). If a coil has N turns all of the same dimensions, the magnetic moment and the torque on the coil will clearly be N times greater than in a single loop.

Although the torque was obtained for a particular orientation of $\mathbf{B}$ with respect to the loop, the equation $\boldsymbol{\tau} = \boldsymbol{\mu} \times \mathbf{B}$ is valid for any orientation. Furthermore, although the torque expression was derived for a rectangular loop, the result is valid for a loop of *any* shape.

It is interesting to note the similarity between the tendency for rotation of a current loop in an external magnetic field and the motion of a compass needle (or pivoted bar magnet) in such a field. Like the current loop, the compass needle and bar magnet can be regarded as magnetic dipoles. The similarity in their magnetic field lines is described in Figure 29.12. Note that one face of

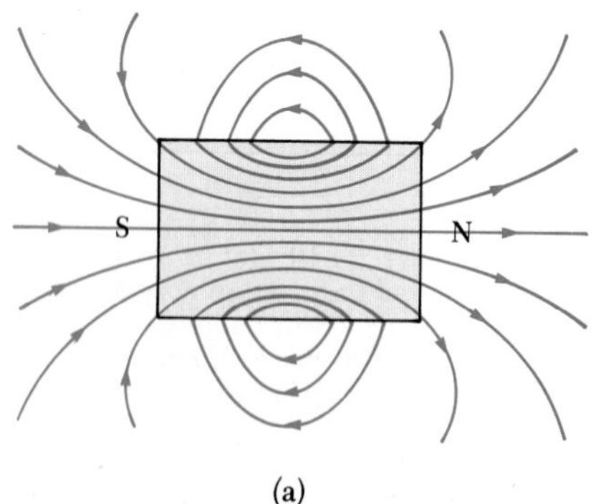

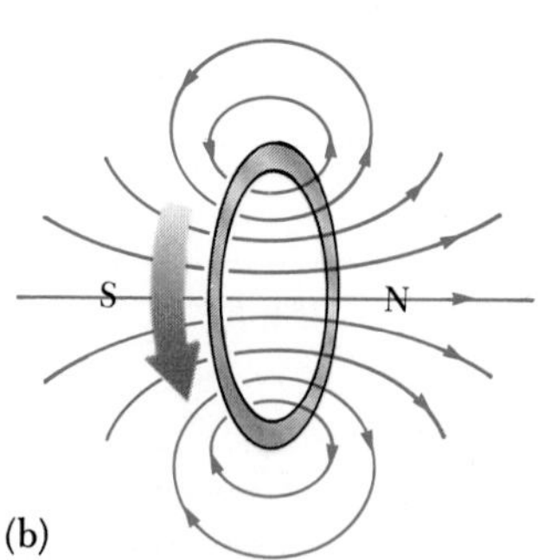

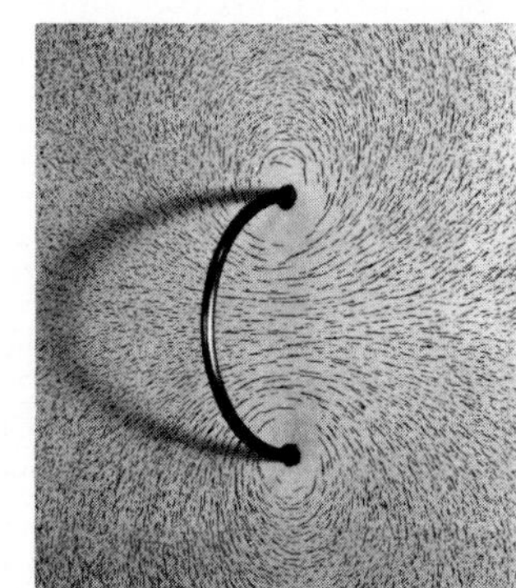

Figure 29.12 The similarity between the magnetic field patterns of (a) a bar magnet and (b) a current loop. (c) Photograph courtesy of Education Development Center, Newton, MA.

the current loop behaves as the north pole of a bar magnet while the opposite face behaves as the south pole. The field lines shown in Figure 29.12 are the patterns due to the bar magnet (Fig. 29.12a) and the current loop (Fig. 29.12b). There is *no* external field present in these diagrams. Furthermore, the diagrams are a simplified, two-dimensional description of the field lines.

EXAMPLE 29.3 The Magnetic Moment of a Coil
A rectangular coil of dimensions 5.40 cm × 8.50 cm consists of 25 turns of wire. The coil carries a current of 15 mA. (a) Calculate the magnitude of the magnetic moment of the coil.

Solution The magnitude of the magnetic moment of a current loop is given by $\mu = IA$ (see Eq. 29.13), where A is the area of the loop. In this case, $A = (0.0540\text{ m})(0.0850\text{ m}) = 4.59 \times 10^{-3}\text{ m}^2$. Since the coil has 25 turns, and assuming that each turn has the same area A, we have

$$\mu_{\text{coil}} = NIA = (25)(15 \times 10^{-3}\text{ A})(4.59 \times 10^{-3}\text{ m}^2)$$

$$= 1.72 \times 10^{-3}\text{ A}\cdot\text{m}^2 = 1.72 \times 10^{-3}\text{ J/T}$$

(b) Suppose a magnetic field of magnitude 0.350 T is applied parallel to the plane of the loop. What is the magnitude of the torque acting on the loop?

Solution In general, the torque is given by $\boldsymbol{\tau} = \boldsymbol{\mu} \times \mathbf{B}$, where the vector $\boldsymbol{\mu}$ is directed perpendicular to the plane of the loop. In this case, $\mathbf{B}$ is *perpendicular* to μ_{coil}, so that

$$\tau = \mu_{\text{coil}}B = (1.72 \times 10^{-3}\text{ J/T})(0.350\text{ T})$$

$$= 6.02 \times 10^{-4}\text{ J}$$

Note that this is the basic principle behind the operation of a galvanometer coil discussed in Chapter 28.

Exercise 3 Calculate the magnitude of the torque on the coil when the 0.350 T magnetic field makes angles of (a) 60° and (b) 0° with $\boldsymbol{\mu}$.
Answer (a) 5.21×10^{-4} J (b) zero

29.5 MOTION OF A CHARGED PARTICLE IN A MAGNETIC FIELD

In Section 29.2 we found that the magnetic force acting on a charged particle moving in a magnetic field is always perpendicular to the velocity of the particle. From this property, it follows that

> the work done by the magnetic force is zero since the displacement of the charge is always perpendicular to the magnetic force. Therefore, a static magnetic field changes the direction of the velocity but does not affect the speed or kinetic energy of the charged particle.

Consider the special case of a positively charged particle moving in a uniform external magnetic field with its initial velocity vector *perpendicular* to the field. Let us assume that the magnetic field is *into* the page (this is indicated by the crosses in Fig. 29.13). The crosses are used to represent the *tail* of $\mathbf{B}$, since $\mathbf{B}$ is directed *into* the page. Later, we shall use dots to represent the *tip* of a vector directed *out* of the page. Figure 29.13 shows that the

> charged particle moves in a circle whose plane is perpendicular to the magnetic field.

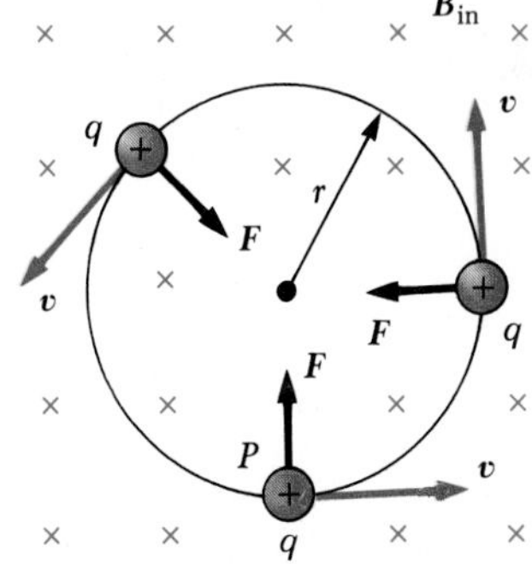

Figure 29.13 When the velocity of a charged particle is perpendicular to a uniform magnetic field, the particle moves in a circular path whose plane is perpendicular to **B**, which is directed into the page (the blue crosses represent the tail of the vector). The magnetic force, **F**, on the charge is always directed toward the center of the circle.

This is because the magnetic force $\mathbf{F}$ is at right angles to $\mathbf{v}$ and $\mathbf{B}$ and has a constant magnitude equal to qvB. As the force $\mathbf{F}$ deflects the particle, the directions of $\mathbf{v}$ and $\mathbf{F}$ change continuously, as shown in Figure 29.13. Therefore the force $\mathbf{F}$ is a *centripetal force*, which changes only the direction of $\mathbf{v}$ while the speed remains constant. The sense of the rotation, as shown in

Figure 29.13, is counterclockwise for a positive charge. If q were negative, the sense of the rotation would be reversed, or clockwise. Since the resultant force $\boldsymbol{F}$ in the radial direction has a magnitude of qvB, we can equate this to the required centrifugal force, which is the mass m multiplied by the centripetal acceleration v^2/r. From Newton's second law, we find that

$$F = qvB = \frac{mv^2}{r}$$

Radius of the circular orbit

$$r = \frac{mv}{qB} \tag{29.15}$$

That is, the radius of the path is proportional to the momentum mv of the particle and is inversely proportional to the magnetic field. The angular frequency of the rotating charged particle is given by

Cyclotron frequency

$$\omega = \frac{v}{r} = \frac{qB}{m} \tag{29.16}$$

The period of its motion (the time for one revolution) is equal to the circumference of the circle divided by the speed of the particle:

$$T = \frac{2\pi r}{v} = \frac{2\pi}{\omega} = \frac{2\pi m}{qB} \tag{29.17}$$

These results show that the angular frequency and period of the circular motion do not depend on the speed of the particle or the radius of the orbit. The angular frequency ω is often referred to as the **cyclotron frequency** since charged particles circulate at this frequency in one type of accelerator called a *cyclotron*, which will be discussed in Section 29.6.

If a charged particle moves in a uniform magnetic field with its velocity at some arbitrary angle to $\boldsymbol{B}$, its path is a helix. For example, if the field is in the x direction as in Figure 29.14, there is no component of force in the x direction, and hence $a_x = 0$ and the x component of velocity, v_x, remains constant. On the other hand, the magnetic force $q\boldsymbol{v} \times \boldsymbol{B}$ causes the components v_y and v_z to

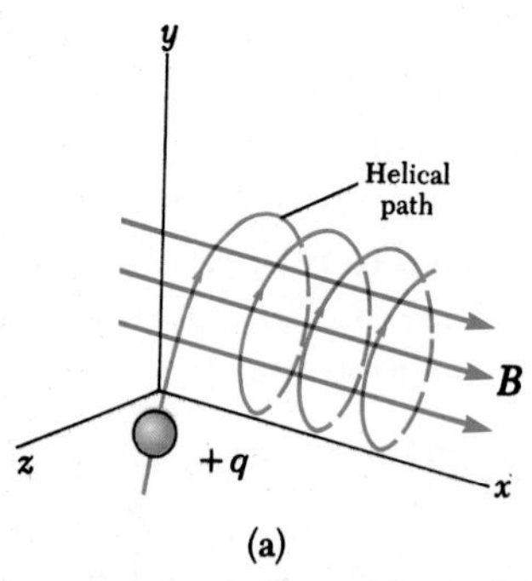

(a)

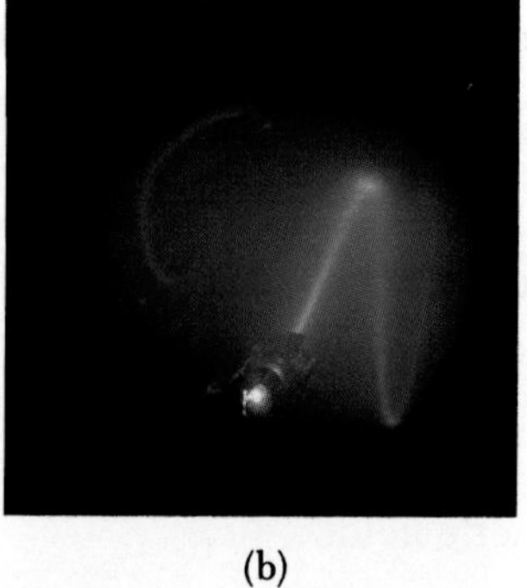

(b)

Figure 29.14 (a) A charged particle having a velocity vector that has a component parallel to a uniform magnetic field moves in a helical path. (b) Photograph of the helical path followed by an electron beam when the beam is directed at an arbitrary angle to the magnetic field. (Photo courtesy of Henry Leap and Jim Lehman) change in time, and the resulting motion is a helix having its axis parallel to the **B** field. The projection of the path onto the yz plane (viewed along the x axis) is a circle. (The projections of the path onto the xy and xz planes are sinusoids!) Equations 29.15 to 29.17 still apply, provided that v is replaced by $v_\perp = \sqrt{v_y^2 + v_z^2}$.

EXAMPLE 29.4 A Proton Moving Perpendicular to a Uniform Magnetic Field

A proton is moving in a circular orbit of radius 14 cm in a uniform magnetic field of magnitude 0.35 T directed perpendicular to the velocity of the proton. Find the oribital speed of the proton.

Solution From Equation 29.15, we get

$$v = \frac{qBr}{m} = \frac{(1.60 \times 10^{-19}\ \text{C})(0.35\ \text{T})(14 \times 10^{-2}\ \text{m})}{1.67 \times 10^{-27}\ \text{kg}}$$

$$= 4.69 \times 10^{6}\ \text{m/s}$$

Exercise 4 If an electron moves perpendicular to the same magnetic field with this speed, what is the radius of its circular orbit?

Answer 7.63×10^{-5} m.

EXAMPLE 29.5 The Bending of an Electron Beam

In an experiment designed to measure the strength of a uniform magnetic field produced by a set of coils, the electrons are accelerated from rest through a potential difference of 350 V, and the beam associated with the electrons is measured to have a radius of 7.5 cm as in Figure 29.15. Assuming the magnetic field is perpendicular to the beam, (a) what is the magnitude of the magnetic field?

Solution First, we must calculate the speed of electrons using the fact that the increase in kinetic energy of the electrons must equal the change in their potential energy, $|e|V$ (because of conservation of energy). Since $K_i = 0$ and $K_f = mv^2/2$, we have

$$\tfrac{1}{2}mv^2 = |e|V$$

$$v = \sqrt{\frac{2|e|V}{m}} = \sqrt{\frac{2(1.60 \times 10^{-19}\ \text{C})(350\ \text{V})}{9.11 \times 10^{-31}\ \text{kg}}}$$

$$= 1.11 \times 10^{7}\ \text{m/s}$$

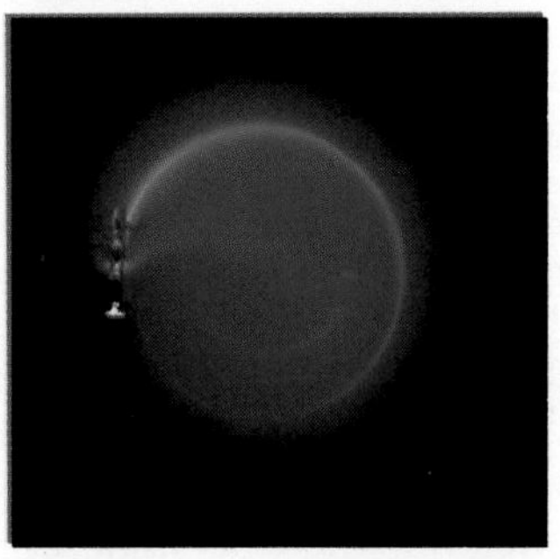

Figure 29.15 The bending of an electron beam in an external magnetic field. The tube contains gas at very low pressure, and the beam is made visible as the electrons collide with the gas atoms, which in turn emit visible light. The apparatus used to take this photograph is part of a system used to measure the ratio e/m. (Courtesy of Henry Leap and Jim Lehman)

We can now use Equation 29.15 to find the strength of the magnetic field:

$$B = \frac{mv}{|e|r} = \frac{(9.11 \times 10^{-31}\ \text{kg})(1.11 \times 10^{7}\ \text{m/s})}{(1.60 \times 10^{-19}\ \text{C})(0.075\ \text{m})}$$

$$= 8.43 \times 10^{-4}\ \text{T}$$

(b) What is the angular frequency of revolution of the electrons?

Solution Using Equation 29.16, we find

$$\omega = \frac{v}{r} = \frac{1.11 \times 10^{7}\ \text{m/s}}{0.075\ \text{m}} = 1.48 \times 10^{8}\ \text{Hz}$$

$$= 148\ \text{MHz}$$

Exercise 5 What is the period of revolution of the electrons?

Answer $T = 6.76$ ns

When charged particles move in a nonuniform magnetic field, the motion is rather complex. For example, in a magnetic field that is strong at the ends and weak in the middle, as in Figure 29.16, the particles can oscillate back and forth between the end points. Such a field can be produced by two current

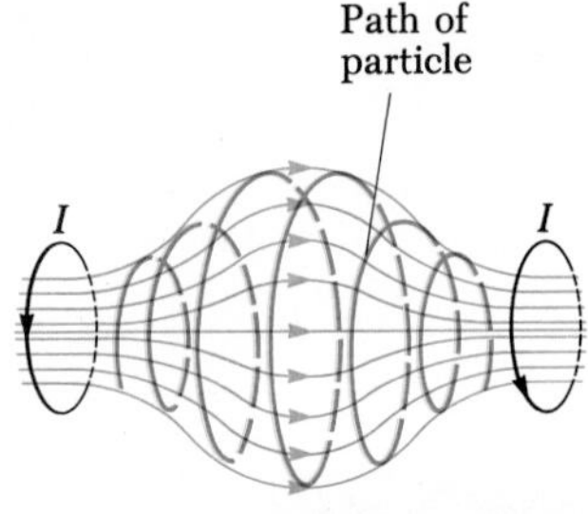

Figure 29.16 A charged particle moving in a nonuniform magnetic field represented by the blue lines (a magnetic bottle) spirals about the field (red path) and oscillates between the end points.

loops as in Figure 29.16. In this case, a charged particle starting at one end will spiral along the field lines until it reaches the other end, where it reverses its path and spirals back. This configuration is known as a *magnetic bottle* because charged particles can be trapped in it. This concept has been used to confine very hot gases (T greater than 10^6 K) consisting of electrons and positive ions, known as **plasmas.** Such a plasma-confinement scheme could play a crucial role in achieving a controlled nuclear fusion process, which could supply us with an almost endless source of energy. Unfortunately, the magnetic bottle has its problems. If a large number of particles is trapped, collisions between the particles cause them to eventually "leak" from the system.

The Van Allen radiation belts consist of charged particles (mostly electrons and protons) surrounding the earth in doughnut-shaped regions (Fig. 29.17a). These radiation belts were discovered in 1958 by a team of researchers under the direction of James Van Allen, using data gathered by instrumentation aboard the Explorer I satellite. The charged particles, trapped by the earth's nonuniform magnetic field, spiral around the earth's field lines from pole to pole. These particles originate mainly from the sun, but some come from stars and other heavenly objects. For this reason, these particles are given the name *cosmic rays.* Most cosmic rays are deflected by the earth's magnetic field and never reach the earth. However, some become trapped, and these make up the Van Allen belts. When these charged particles are in the earth's atmosphere over the poles, they often collide with other atoms, causing them to emit visible light. This is the origin of the beautiful Aurora Borealis, or Northern Lights (Fig. 29.17b). A similar phenomenon seen in the southern hemisphere is called the Aurora Australis.

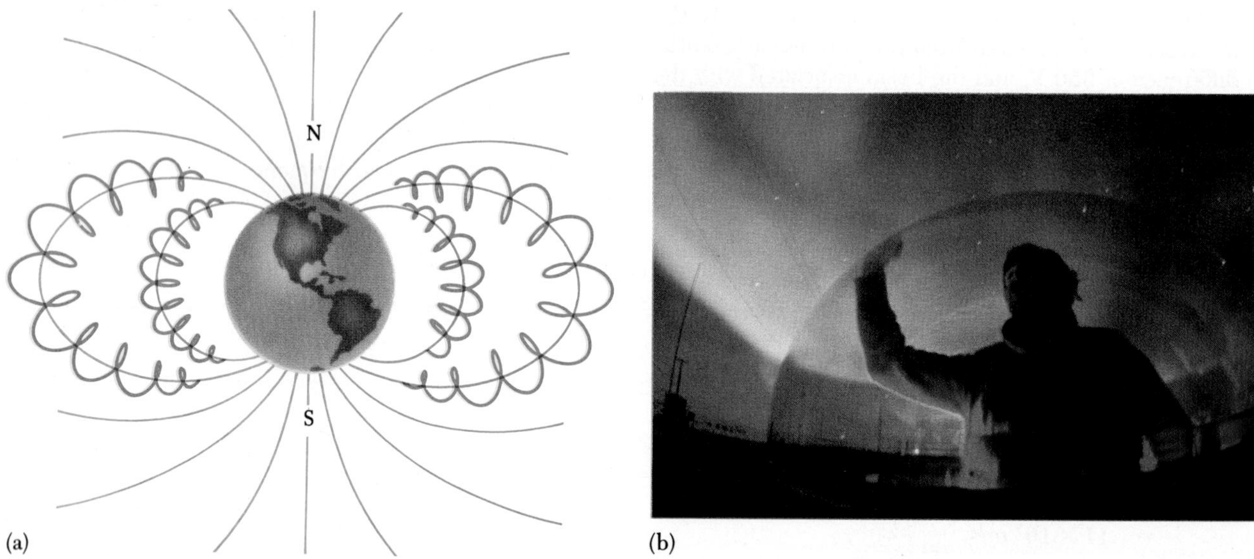

Figure 29.17 (a) The Van Allen belts are made up of charged particles (electrons and protons) trapped by the earth's nonuniform magnetic field. The field lines are in blue and the particle paths in red. (b) A scientist inside a heated dome on the western shore of Hudson Bay near Churchill, Manitoba, observes an Aurora Borealis, or Northern Lights. Auroras occur when cosmic rays—electrically charged particles originating mainly from the sun—become trapped in the earth's atmosphere over earth's magnetic poles and collide with other atoms, resulting in the emission of visible light. (© David Hiser/Photographers, Aspen)

*29.6 APPLICATIONS OF THE MOTION OF CHARGED PARTICLES IN A MAGNETIC FIELD

In this section we describe some important devices that involve the motion of charged particles in uniform magnetic fields. For many situations, the charge under consideration will be moving with a velocity $\boldsymbol{v}$ in the presence of both an electric field $\boldsymbol{E}$ and a magnetic field $\boldsymbol{B}$. Therefore, the charge will experience both an electric force $q\boldsymbol{E}$ and a magnetic force $q\boldsymbol{v} \times \boldsymbol{B}$, and so the total force on the charge will be given by

$$\boldsymbol{F} = q\boldsymbol{E} + q\boldsymbol{v} \times \boldsymbol{B} \tag{29.18}$$

Lorentz force

The force described by Equation 29.18 is known as the **Lorentz force.**

Velocity Selector

In many experiments involving the motion of charged particles, it is important to have a source of particles that move with essentially the same velocity. This can be achieved by applying a combination of an electric field and a magnetic field oriented as shown in Figure 29.18. A uniform electric field vertically downward is provided by a pair of charged parallel plates, while a uniform magnetic field is applied perpendicular to the page (indicated by the crosses). Assuming that q is positive, we see that the magnetic force $q\boldsymbol{v} \times \boldsymbol{B}$ is upward and the electric force $q\boldsymbol{E}$ is downward. If the fields are chosen such that the electric force balances the magnetic force, the particle will move in a straight horizontal line and emerge from the slit at the right. If we equate the upward magnetic force qvB to the downward electric force qE, we find $qvB = qE$, from which we get

$$v = \frac{E}{B} \tag{29.19}$$

Note that only those particles having this velocity will pass undeflected through the perpendicular electric and magnetic fields. In practice, $\boldsymbol{E}$ and $\boldsymbol{B}$ are adjusted to provide this specific velocity. The magnetic force acting on particles with velocities greater than this will be stronger than the electric force, and these particles will be deflected upward. Those with velocities less than this will be deflected downward.

Bubble chamber photograph. The spiral tracks at the bottom of the photograph are an electron-positron pair (left and right, respectively) formed by a gamma ray interacting with a hydrogen nucleus. An applied magnetic field causes the electron and the positron to be deflected in opposite directions. The track leaving from the cusp between the two spirals is an additional electron knocked out of a hydrogen atom during this interaction. (G. Holton, F.J. Rutherford, F.G. Watson, *Project Physics,* New York: HRW, 1981.)

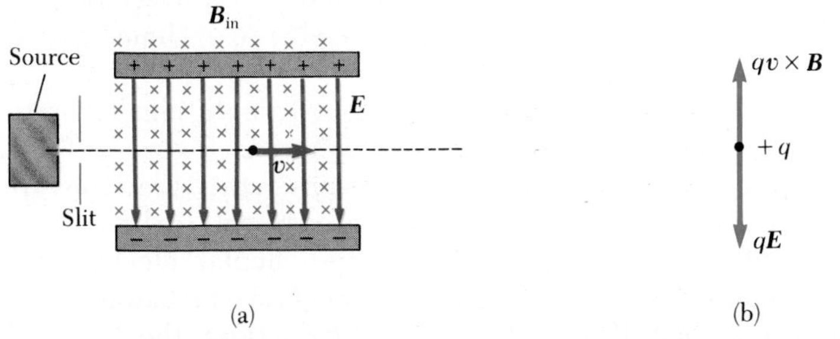

Figure 29.18 (a) A velocity selector. When a positively charged particle is in the presence of both an inward magnetic field as indicated by the blue crosses, and a downward electric field indicated by the red arrows, it experiences both an electric force $q\boldsymbol{E}$ downward and a magnetic force $q\boldsymbol{v} \times \boldsymbol{B}$ upward. (b) When these forces balance each other as shown here, the particle moves in a horizontal line through the fields.

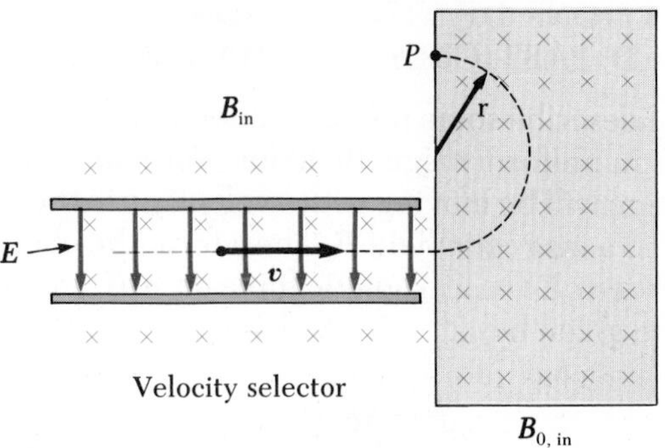

Figure 29.19 A mass spectrometer. Charged particles are first sent through a velocity selector. They then enter a region where the magnetic field $\boldsymbol{B}_0$ (inward) causes positive ions to move in a semicircular path and strike a photographic film at *P*.

The Mass Spectrometer

The **mass spectrometer** is an instrument that separates atomic and molecular ions according to their mass-to-charge ratio. In one version, known as the *Bainbridge mass spectrometer*, a beam of ions first passes through a velocity selector and then enters a uniform magnetic field $\boldsymbol{B}_0$ directed into the paper (Fig. 29.19). Upon entering the magnetic field $\boldsymbol{B}_0$, the ions move in a semicircle of radius r before striking a photographic plate at *P*. From Equation 29.15, we can express the ratio m/q as

$$\frac{m}{q} = \frac{rB_0}{v} \tag{29.20}$$

Assuming that the magnitude of the magnetic field in the region of the velocity selector is B and using Equation 29.19, which gives the speed of the particle, we find that

$$\frac{m}{q} = \frac{rB_0B}{E} \tag{29.21}$$

Therefore, one can determine m/q by measuring the radius of curvature and knowing the fields B, B_0, and E. In practice, one usually measures the masses of various isotopes of a given ion with the same charge q. Hence, the mass ratios can be determined even if q is unknown.

A variation of this technique was used by Joseph John Thomson (1856–1940) in 1897 to measure the ratio e/m for electrons. Figure 29.20a shows the basic apparatus used by Thomson in his measurements. Electrons are accelerated from the cathode to the anodes, collimated by slits in the anodes, and then allowed to drift into a region of crossed (perpendicular) electric and magnetic fields. The simultaneously applied $\boldsymbol{E}$ and $\boldsymbol{B}$ fields are first adjusted to produce an undeflected beam. If the $\boldsymbol{B}$ field is then turned off, the $\boldsymbol{E}$ field alone produces a measurable beam deflection on the phosphorescent screen. From the size of the deflection and the measured values of $\boldsymbol{E}$ and $\boldsymbol{B}$, the charge to mass ratio, e/m, may be determined. The results of this crucial experiment represent the discovery of the electron as a fundamental particle of nature.

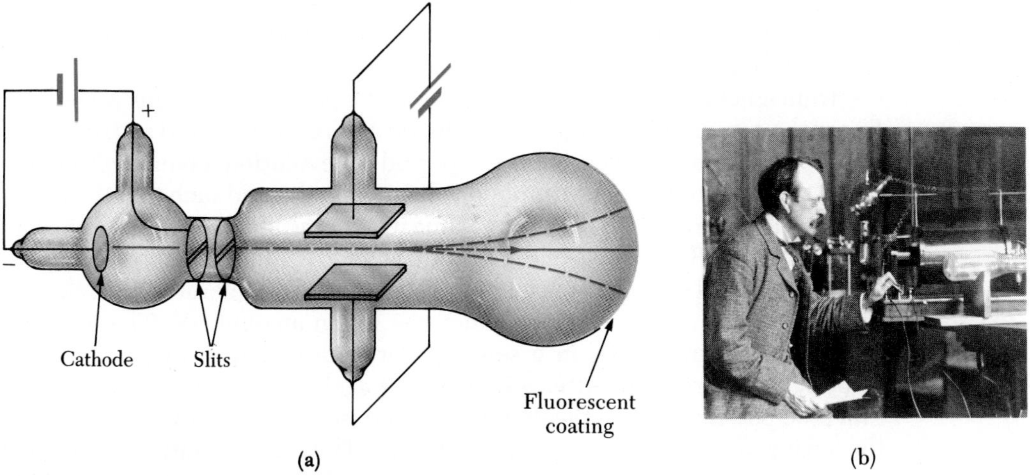

Figure 29.20 (a) Thomson's apparatus for measuring q/m. Electrons are accelerated from the cathode, pass through two slits, and are deflected by both an electric field and a magnetic field (not shown, but directed into the paper). The deflected beam then strikes a phosphorescent screen. (b) J.J. Thomson in the Cavendish Laboratory, University of Cambridge.

The Cyclotron

The **cyclotron,** invented in 1934 by E. O. Lawrence and M. S. Livingston, is a machine that can accelerate charged particles to very high velocities. Both electric and magnetic forces play a key role in the operation of the cyclotron. The energetic particles that emerge from the cyclotron are used to bombard other nuclei; this bombardment in turn produces nuclear reactions of interest to researchers. A number of hospitals use cyclotron facilities to produce radioactive substances that can be used in diagnosis and treatment.

A schematic drawing of a cyclotron is shown in Figure 29.21. Motion of the charges occurs in two semicircular containers, D_1 and D_2, referred to as

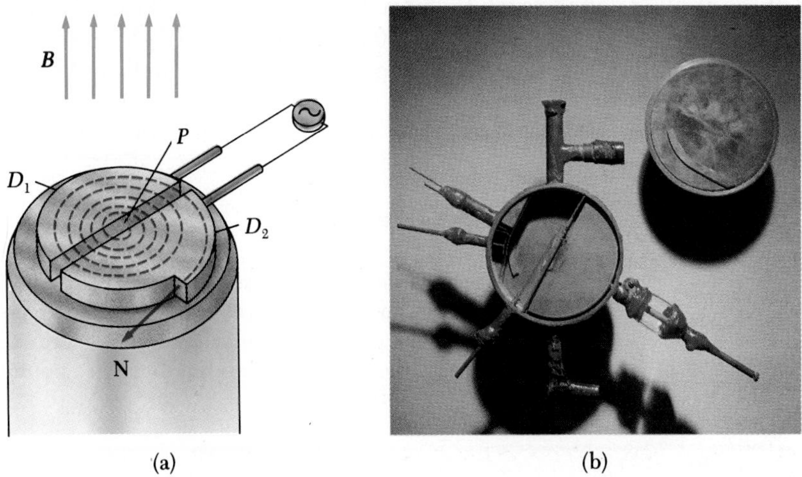

Figure 29.21 (a) The cyclotron consists of an ion source, two dees across which an alternating voltage is applied, and a uniform magnetic field provided by an electromagnet. (The south pole of the magnet is not shown.) (b) The first cyclotron, invented by E.O. Lawrence and M.S. Livingston in 1934. (Courtesy of Lawrence Berkeley Laboratory, University of California)

dees. The dees are evacuated in order to minimize energy losses resulting from collisions between the ions and air molecules. A high-frequency alternating voltage is applied to the dees, and a uniform magnetic field provided by an electromagnet is directed perpendicular to the dees. Positive ions released at P near the center of the magnet move in a semicircular path and arrive back at the gap in a time $T/2$, where T is the period of revolution, given by Equation 29.17. The frequency of the applied voltage V is adjusted such that the polarity of the dees is reversed in the same time it takes the ions to complete one half of a revolution. If the phase of the applied voltage is adjusted such that D_2 is at a *lower* potential than D_1 by an amount V, the ion will accelerate across the gap to D_2 and its kinetic energy will increase by an amount qV. The ion then continues to move in D_2 in a semicircular path of larger radius (since its velocity has increased). After a time $T/2$, it again arrives at the gap. By this time, the potential across the dees is reversed (so that D_1 is now negative) and the ion is given another "kick" across the gap. The motion continues such that for each half revolution, the ion gains additional kinetic energy equal to qV. When the radius of its orbit is nearly that of the dees, the energetic ions leave the system through an exit slit as shown in Figure 29.21.

It is important to note that the operation of the cyclotron is based on the fact that the time for one revolution is *independent* of the speed (or radius) of the ion.

We can obtain the maximum kinetic energy of the ion when it exits from the cyclotron in terms of the radius R of the dees. From Equation 29.15 we find that $v = qBR/m$. Hence, the kinetic energy is given by

$$K = \tfrac{1}{2}mv^2 = \frac{q^2B^2R^2}{2m} \tag{29.22}$$

When the energy of the ions exceeds about 20 MeV, relativistic effects come into play and the masses of the ions no longer remain constant. (Such effects will be discussed in Chapter 39.) For this reason, the period of the orbit increases and the rotating ions do not remain in phase with the applied voltage. Accelerators have been built which solve this problem by modifying the period of the applied voltage such that it remains in phase with the rotating ion. In 1977, protons were accelerated to 400 GeV (1 GeV $= 10^9$ eV) in an accelerator in Batavia, Illinois. The system incorporates 954 magnets and has a circumference of 6.3 km (4.1 miles)!

EXAMPLE 29.6 A Proton Accelerator
Calculate the maximum kinetic energy of protons in a cyclotron of radius 0.50 m in a magnetic field of 0.35 T.

Solution Using Equation 29.22, we find that

$$K = \frac{q^2B^2R^2}{2m} = \frac{(1.6 \times 10^{-19}\ \text{C})^2(0.35\ \text{T})^2(0.50\ \text{m})^2}{2(1.67 \times 10^{-27}\ \text{kg})}$$

$$K = 2.34 \times 10^{-13}\ \text{J} = 1.46\ \text{MeV}$$

In this calculation, we have used the conversions 1 eV $= 1.6 \times 10^{-19}$ J and 1 MeV $= 10^6$ eV. The kinetic energy acquired by the protons is equivalent to the energy they would gain if they were accelerated through a potential difference of 1.46 MV!

*29.7 THE HALL EFFECT

In 1879 Edwin Hall discovered that when a current-carrying conductor is placed in a magnetic field, a voltage is generated in a direction perpendicular to both the current and the magnetic field. This observation, known as the *Hall*

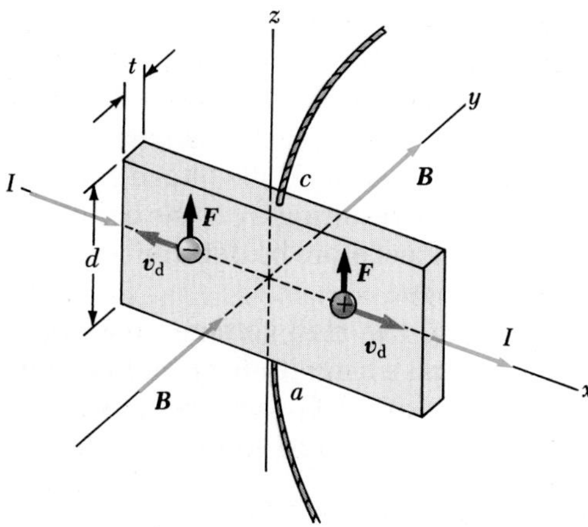

Figure 29.22 To observe the Hall effect, a magnetic field is applied to a current-carrying conductor. When I is in the x direction and B in the y direction as shown, both positive and negative charge carriers are deflected upward in the magnetic field. The Hall voltage is measured between points a and c.

effect, arises from the deflection of charge carriers to one side of the conductor as a result of the magnetic force experienced by the charge carriers. A proper analysis of experimental data gives information regarding the sign of the charge carriers and their density. The effect also provides a convenient technique for measuring magnetic fields.

The arrangement for observing the Hall effect consists of a conductor in the form of a flat strip carrying a current I in the x direction as in Figure 29.22. A uniform magnetic field $\boldsymbol{B}$ is applied in the y direction. If the charge carriers are electrons moving in the negative x direction with a velocity $\boldsymbol{v}_d$, they will experience an *upward* magnetic force $\boldsymbol{F}$. Hence, the electrons will be deflected upward, accumulating at the upper edge and leaving an excess positive charge at the lower edge (Fig. 29.23a). This accumulation of charge at the edges will continue until the electrostatic field set up by this charge separation balances the magnetic force on the charge carriers. When this equilibrium condition is reached, the electrons will no longer be deflected upward. A sensitive voltmeter or potentiometer connected across the sample as shown in Figure 29.23 can be used to measure the potential difference generated across

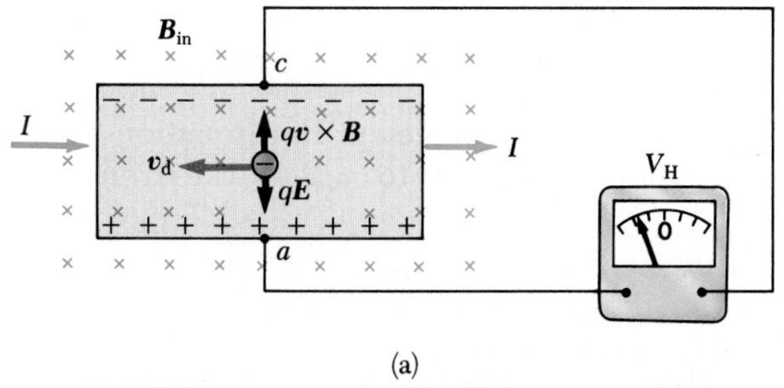

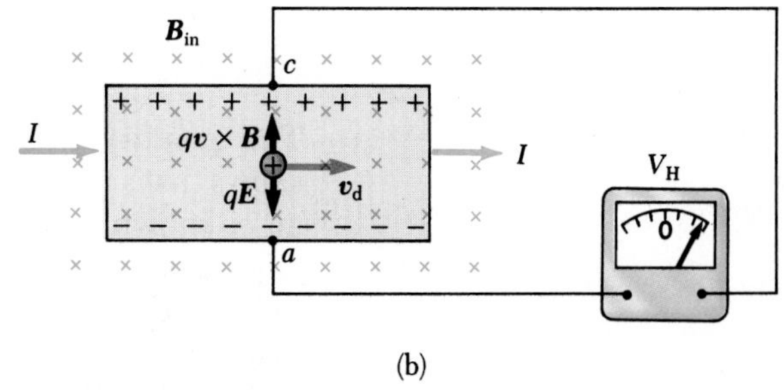

Figure 29.23 (a) When the charge carriers are negative, the upper edge becomes negatively charged, and c is at a lower potential than a. (b) When the charge carriers are positive, the upper edge becomes positively charged, and c is at a higher potential than a. In either case, the charge carriers are no longer deflected when the edges become fully charged, that is, when there is a balance between the electrostatic force trying to combine the charges and the magnetic deflection force.

the conductor, known as the **Hall voltage** V_H. If the charge carriers are positive, and hence move in the positive x direction as in Figure 29.23b, they will also experience an *upward* magnetic force $q\mathbf{v}_d \times \mathbf{B}$. This produces a buildup of positive charge on the upper edge and leaves an excess of negative charge on the lower edge. Hence, the sign of the Hall voltage generated in the sample is opposite the sign of the voltage resulting from the deflection of electrons. The sign of the charge carriers can therefore be determined from a measurement of the polarity of the Hall voltage.

To find an expression for the Hall voltage, first note that the magnetic force on the charge carriers has a magnitude qv_dB. In equilibrium, this force is balanced by the electrostatic force qE_H, where E_H is the electric field due to the charge separation (sometimes referred to as the *Hall field*). Therefore,

$$qv_dB = qE_H$$

$$E_H = v_dB$$

If d is taken to be the width of the conductor, then the Hall voltage V_H measured by the potentiometer is equal to E_Hd, or

$$V_H = E_Hd = v_dBd \tag{29.23}$$

Thus, we see that the measured Hall voltage gives a value for the drift velocity of the charge carriers if d and B are known.

The number of charge carriers per unit volume (or charge density), n, can be obtained by measuring the current in the sample. From Equation 27.4, the drift velocity can be expressed as

$$v_d = \frac{I}{nqA} \tag{29.24}$$

where A is the cross-sectional area of the conductor. Substituting Equation 29.24 into Equation 29.23 we obtain

The Hall voltage

$$V_H = \frac{IBd}{nqA} \tag{29.25}$$

Since $A = td$, where t is the thickness of the sample, we can also express Equation 29.25 as

$$V_H = \frac{IB}{nqt} \tag{29.26}$$

The quantity $1/nq$ is referred to as the **Hall coefficient** R_H. Equation 29.26 shows that a properly calibrated sample can be used to measure the strength of an unknown magnetic field.

Since all quantities appearing in Equation 29.26 other than nq can be measured, a value for the Hall coefficient is readily obtained. The sign and magnitude of R_H give the sign of the charge carriers and their density. In most metals, the charge carriers are electrons and the charge density determined from Hall effect measurements is in good agreement with calculated values for monovalent metals, such as Li, Na, Cu, and Ag, where n is approximately equal to the number of valence electrons per unit volume. However, this classical model is not valid for metals such as Fe, Bi, and Cd or for semiconductors such as silicon and germanium. These discrepancies can be explained only by using a model based on the quantum nature of solids.

EXAMPLE 29.7 The Hall Effect for Copper
A rectangular copper strip 1.5 cm wide and 0.1 cm thick carries a current of 5 A. A 1.2-T magnetic field is applied perpendicular to the strip as in Figure 29.23. Find the resulting Hall voltage.

Solution If we assume there is one electron per atom available for conduction, then we can take the charge density to be $n = 8.48 \times 10^{28}$ electrons/m^3 (Example 27.1). Substituting this value and the given data into Equation 29.26 gives

$$V_H = \frac{IB}{nqt}$$

$$= \frac{(5\text{ A})(1.2\text{ T})}{(8.48 \times 10^{28}\text{ m}^{-3})(1.6 \times 10^{-19}\text{ C})(0.1 \times 10^{-2}\text{ m})}$$

$$V_H = 0.442\ \mu\text{V}$$

Hence, the Hall voltage is quite small in good conductors. Note that the width of this sample is not needed in this calculation.

In semiconductors, where n is much smaller than in monovalent metals, one finds a larger Hall voltage since V_H varies as the inverse of n. Current levels of the order of 1 mA are generally used for such materials. Consider a piece of silicon with the same dimensions as the copper strip, with $n = 10^{20}$ electrons/m^3. Taking $B = 1.2$ T and $I = 0.1$ mA, we find that $V_H = 7.5$ mV. Such a voltage is readily measured with a potentiometer.

*29.8 THE QUANTUM HALL EFFECT

Although the Hall effect was discovered over one hundred years ago, it continues to be one of the most valuable techniques for helping scientists understand the electronic properties of metals and semiconductors. For example, in 1980, scientists reported that at low temperatures and very strong magnetic fields, a two-dimensional system of electrons in a semiconductor exhibits a conductivity given by $\sigma = i(e^2/h)$, where i is a small integer, e is the electronic charge, and h is an atomic constant called Planck's constant. This behavior manifests itself as a series of plateaus in the Hall voltage as the applied magnetic field is varied. The quantized nature of this two-dimensional conductivity (or resistivity) was totally unanticipated. As its discoverer, Klaus von Klitzing stated, "It is quite astonishing that it is the total macroscopic conductance of the Hall device which is quantized rather than some idealized microscopic conductivity."

One of the important consequences of the quantum Hall effect is the ability to measure the ratio of fundamental constants, e^2/h, to an accuracy of at least one part in 10^5. This provides a very accurate measure of the dimensionless fine structure constant, given by $\alpha = e^2/hc \approx 1/137$, since c is an *exactly* defined quantity (the speed of light). In addition, the quantum Hall effect provides scientists with a new and convenient standard of resistance. The 1985 Nobel Prize in physics was awarded to von Klitzing for this fundamental discovery.

Another great surprise occurred in 1982 when scientists announced that in some nearly ideal samples at very low temperatures, the Hall conductivity could take on both integer values of e^2/h and *fractional* values of e^2/h. Undoubtedly, future discoveries in this and related areas of science will continue to improve our understanding of the nature of matter.

SUMMARY

Magnetic force on a charged particle moving in a magnetic field

The **magnetic force** that acts on a charge q moving with a velocity $\boldsymbol{v}$ in an external magnetic field $\boldsymbol{B}$ is given by

$$\boldsymbol{F} = q\boldsymbol{v} \times \boldsymbol{B} \tag{29.1}$$

That is, the magnetic force is in a direction perpendicular both to the velocity of the particle and to the field. The *magnitude* of the magnetic force is given by

$$F = qvB \sin \theta \tag{29.2}$$

where θ is the angle between $\boldsymbol{v}$ and $\boldsymbol{B}$. From this expression, we see that $F = 0$ when $\boldsymbol{v}$ is parallel to (or opposite) $\boldsymbol{B}$. Furthermore, $F = qvB$ when $\boldsymbol{v}$ is perpendicular to $\boldsymbol{B}$.

The SI unit of $\boldsymbol{B}$ is the **weber per square meter** (Wb/m^2), also called the **tesla** (T), where

$$[\mathrm{B}] = \mathrm{T} = \frac{\mathrm{Wb}}{\mathrm{m}^2} = \frac{\mathrm{N}}{\mathrm{A} \cdot \mathrm{m}} \tag{29.3}$$

If a straight conductor of length ℓ carries a current I, the force on that conductor when placed in a uniform *external* magnetic field $\boldsymbol{B}$ is given by

Force on a straight wire carrying a current

$$\boldsymbol{F} = I\boldsymbol{\ell} \times \boldsymbol{B} \tag{29.5}$$

where the direction of $\boldsymbol{\ell}$ is in the direction of the current and $|\boldsymbol{\ell}| = \ell$.

If an arbitrarily shaped wire carrying a current I is placed in an *external* magnetic field, the force on a very small segment $d\boldsymbol{s}$ is given by

Force on a current element

$$d\boldsymbol{F} = I\, d\boldsymbol{s} \times \boldsymbol{B} \tag{29.6}$$

To determine the total force on the wire, one has to integrate Equation 29.6, keeping in mind that both $\boldsymbol{B}$ and $d\boldsymbol{s}$ may vary at each point.

The net magnetic force on any *closed* loop carrying a current in a uniform *external* magnetic field is *zero.*

The force on a current-carrying conductor of arbitrary shape in a uniform magnetic field is given by

Force on a wire in a uniform field

$$\boldsymbol{F} = I\boldsymbol{\ell}' \times \boldsymbol{B} \tag{29.9}$$

where $\boldsymbol{\ell}'$ is a vector directed from one end of the conductor to the opposite end.

The **magnetic moment** $\boldsymbol{\mu}$ of a current loop carrying a current I is

Magnetic moment of a current loop

$$\boldsymbol{\mu} = I\boldsymbol{A} \tag{29.13}$$

where $\boldsymbol{A}$ is perpendicular to the plane of the loop and $|\boldsymbol{A}|$ is equal to the area of the loop. The SI unit of $\boldsymbol{\mu}$ is A · m^2.

The torque $\boldsymbol{\tau}$ on a current loop when the loop is placed in a uniform *external* magnetic field $\boldsymbol{B}$ is given by

Torque on a current loop

$$\boldsymbol{\tau} = \boldsymbol{\mu} \times \boldsymbol{B} \tag{29.14}$$

When a charged particle moves in an external magnetic field, the work done by the magnetic force on the particle is *zero* since the displacement is always *perpendicular* to the direction of the magnetic force. The external magnetic field can alter the direction of the velocity vector, but it cannot change the speed of the particle.

If a charged particle moves in a uniform external magnetic field such that its initial velocity is *perpendicular* to the field, the particle will move in a circle whose plane is *perpendicular* to the magnetic field. The radius r of the circular path is given by

$$r = \frac{mv}{qB} \tag{29.15}$$

where m is the mass of the particle and q is its charge. The angular frequency (cyclotron frequency) of the rotating charged particle is given by

$$\omega = \frac{qB}{m} \tag{29.16}$$

Cyclotron frequency

If a charged particle is moving in the presence of both a magnetic field and an electric field, the total force on the charge is given by the **Lorentz force,**

$$\mathbf{F} = q\mathbf{E} + q\mathbf{v} \times \mathbf{B} \tag{29.18}$$

Lorentz force

That is, the charge experiences both an electric force $q\mathbf{E}$ and a magnetic force $q\mathbf{v} \times \mathbf{B}$.

QUESTIONS

1. At a given instant, a proton moves in the positive x direction in a region where there is a magnetic field in the negative z direction. What is the direction of the magnetic force? Will the proton continue to move in the positive x direction? Explain.
2. Two charged particles are projected into a region where there is a magnetic field perpendicular to their velocities. If the charges are deflected in opposite directions, what can you say about them?
3. If a charged particle moves in a straight line through some region of space, can you say that the magnetic field in that region is zero?
4. Suppose an electron is chasing a proton up this page when suddenly a magnetic field is formed perpendicular to the page. What will happen to the particles?
5. Why does the picture on a TV screen become distorted when a magnet is brought near the screen?
6. How can the motion of a moving charged particle be used to distinguish between a magnetic field and an electric field? Give a specific example to justify your argument.
7. List several similarities and differences in electric and magnetic forces.
8. Justify the following statement: "It is impossible for a constant (i.e., time independent) magnetic field to alter the speed of a charged particle."
9. In view of the above statement, what is the role of a magnetic field in a cyclotron?
10. A current-carrying conductor experiences no magnetic force when placed in a certain manner in a uniform magnetic field. Explain.
11. Is it possible to orient a current loop in a uniform magnetic field such that the loop will not tend to rotate? Explain.
12. How can a current loop be used to determine the presence of a magnetic field in a given region of space?
13. What is the *net* force on a compass needle in a uniform magnetic field?
14. What type of magnetic field is required to exert a resultant force on a magnetic dipole? What will be the direction of the resultant force?
15. A proton moving horizontally enters a region where there is a uniform magnetic field perpendicular to the proton's velocity, as shown in Figure 29.24. Describe

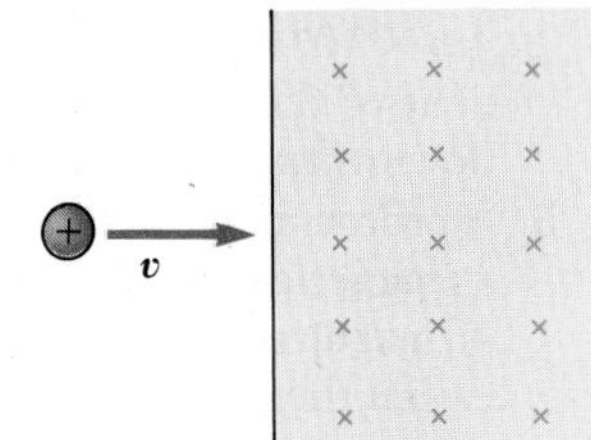

Figure 29.24 (Question 15).

its subsequent motion. How would an electron behave under the same circumstances?

16. In a magnetic bottle, what reverses the direction of the velocity of the confined charged particles at the ends of the bottle? (*Hint:* Find the direction of the magnetic force on these particles in a region where the field becomes stronger and the field lines converge.)
17. In the cyclotron, why do particles of differing velocities take the same amount of time to complete one half of a revolution?
18. The *bubble chamber* is a device used for observing tracks of particles that pass through the chamber, which is immersed in a magnetic field. If some of the tracks are spirals and others are straight lines, what can you say about the particles?
19. Can a magnetic field set a resting electron into motion? If so, how?
20. You are designing a magnetic probe that uses the Hall effect to measure magnetic fields. Assume that you are restricted to using a given material and that you have already made the probe as thin as possible. What, if anything, can be done to increase the Hall voltage produced for a given magnetic field strength?
21. The electron beam in the photograph below is projected to the right. The beam deflects downward in the presence of a magnetic field produced by a pair of current-carrying coils. (a) What is the direction of the magnetic field? (b) What would happen to the beam if the current in the coils were reversed?

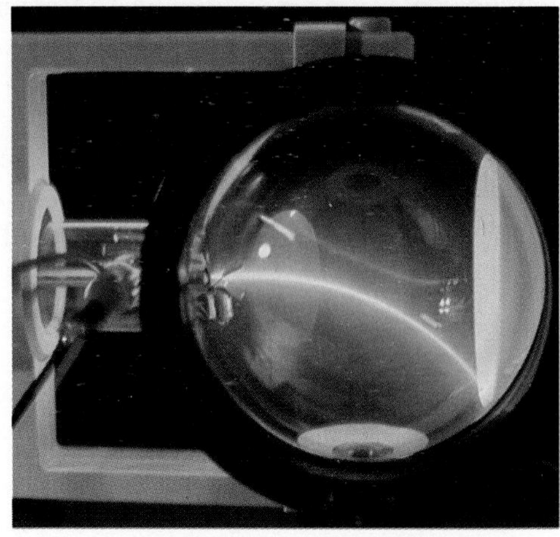

(Courtesy of CENCO)

PROBLEMS

Section 29.2 Definition and Properties of the Magnetic Field

1. Consider an electron near the magnetic equator. In which direction will it tend to be deflected if its velocity is directed (a) downward, (b) northward, (c) westward, or (d) southeastward?
2. An electron moving along the positive x axis perpendicular to a magnetic field experiences a magnetic deflection in the negative y direction. What is the direction of the magnetic field over this region?
3. An alpha particle (which is the nucleus of a helium atom) is moving with a northward velocity of 3.8×10^5 m/s in a region where the magnetic field is 1.9 T and points horizontally to the east. What are the magnitude and direction of the magnetic force on this alpha particle?
4. What force of magnetic origin is experienced by a proton moving north to south with a speed equal to 4.8×10^6 m/s at a location where the vertical component of the earth's magnetic field is 75 μT directed downward? In what direction is the proton deflected?
5. A proton moving with a speed of 4×10^6 m/s through a magnetic field of 1.7 T experiences a magnetic force of magnitude 8.2×10^{-13} N. What is the angle between the proton's velocity and the field?
6. An electron is accelerated through 2400 V and then enters a region where there is a uniform 1.7-T magnetic field. What are (a) the maximum and (b) the minimum values of the magnetic force this charge can experience?
7. The magnetic field over a certain region is given by $\boldsymbol{B} = (4\boldsymbol{i} - 11\boldsymbol{j})$ T. An electron moves in the field with a velocity $\boldsymbol{v} = (-2\boldsymbol{i} + 3\boldsymbol{j} - 7\boldsymbol{k})$ m/s. Write out in unit-vector notation the force exerted on the electron by the magnetic field.
8. An electron is projected into a uniform magnetic field given by $\boldsymbol{B} = (1.4\boldsymbol{i} + 2.1\boldsymbol{j})$ T. Find the vector expression for the force on the electron when its velocity is $\boldsymbol{v} = 3.7 \times 10^5\boldsymbol{j}$ m/s.
9. A proton moves with a velocity of $\boldsymbol{v} = (2\boldsymbol{i} - 4\boldsymbol{j} + \boldsymbol{k})$ m/s in a region in which the magnetic field is given by $\boldsymbol{B} = (\boldsymbol{i} + 2\boldsymbol{j} - 3\boldsymbol{k})$ T. What is the magnitude of the magnetic force this charge experiences?
10. A proton moves perpendicular to a uniform magnetic field $\boldsymbol{B}$ with a speed of 10^7 m/s and experiences an acceleration of 2×10^{13} m/s^2 in the $+x$ direction when its velocity is in the $+z$ direction. Determine the magnitude and direction of the field.
11. Show that the work done by the magnetic force on a charged particle moving in a magnetic field is zero for any displacement of the particle.

Section 29.3 Magnetic Force on a Current-Carrying Conductor

12. Calculate the magnitude of the force per unit length exerted on a conductor carrying a current of 22 A in a region where a uniform magnetic field has a magnitude of 0.77 T and is directed perpendicular to the conductor.
13. A wire carries a steady current of 2.4 A. A straight section of the wire, with a length of 0.75 m along the x axis, lies within a uniform magnetic field, $\mathbf{B} = (1.6\ \mathbf{k})$T. If the current flows in the $+x$ direction, what is the magnetic force on the section of wire?
14. A conductor suspended by two flexible wires as in Figure 29.25 has a mass per unit length of 0.04 kg/m. What current must exist in the conductor in order for the tension in the supporting wires to be zero if the magnetic field over the region is 3.6 T into the page? What is the required direction for the current?

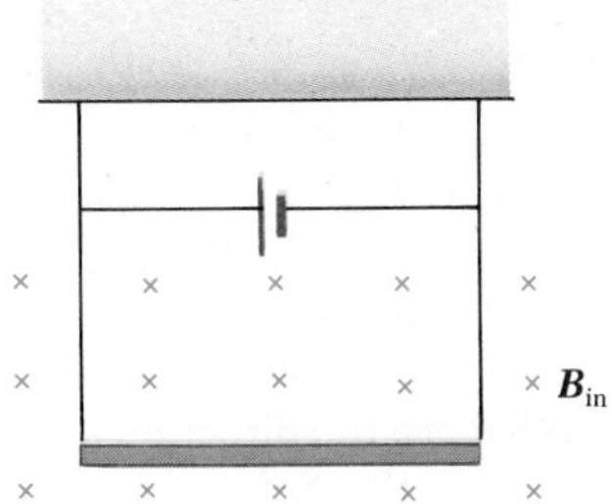

Figure 29.25 (Problem 14).

15. A wire with a mass of 0.5 g/cm carries a 2-A current horizontally to the south. What are the direction and magnitude of the magnetic field needed to lift this wire vertically upward?
16. A rectangular loop with dimensions 10 cm × 20 cm is suspended by a string, and the lower horizontal section of the loop is immersed in a magnetic field confined to a circular region (Fig. 29.26). If a current of 3 A is maintained in the loop in the direction shown, what are the direction and magnitude of the magnetic field required to produce a tension of 4×10^{-2} N in the supporting string? (Neglect the mass of the loop.)

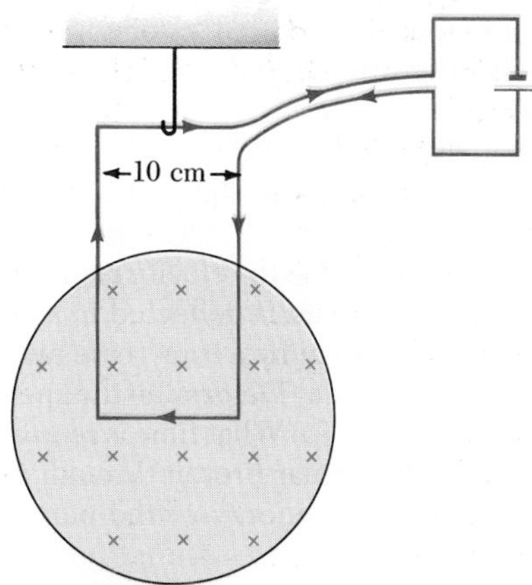

Figure 29.26 (Problem 16).

17. A wire 2.8 m in length carries a current of 5 A in a region where a uniform magnetic field has a magnitude of 0.39 T. Calculate the magnitude of the magnetic force on the wire if the angle between the magnetic field and the direction of the current in the wire is (a) 60°, (b) 90°, (c) 120°.
18. The segment of conductor shown in Figure 29.27 carries a current $I = 0.2$ A. The shorter section is 0.80 m long, and the longer section is 1.6 m long. Determine the magnitude and direction of the magnetic force on the conductor if there is a uniform magnetic field given by $\mathbf{B} = 1.9\mathbf{i}$ T over the region.

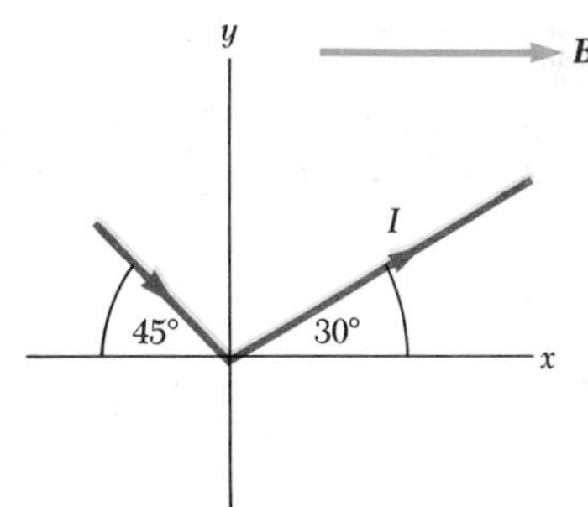

Figure 29.27 (Problem 18).

19. A current $I = 15$ A is directed along the positive x axis in a wire perpendicular to a magnetic field. The current experiences a magnetic force per unit length of 0.63 N/m in the negative y direction. Calculate the magnitude and direction of the magnetic field in the region through which the current passes.
20. The Earth has a magnetic field of 0.6×10^{-4} T, pointing 75° below the horizontal in a north-south plane. A 10-m-long straight wire carries a 15-A current. (a) If the current is directed horizontally toward the east, what are the magnitude and direction of the magnetic force on the wire? (b) What are the magnitude and direction of the force if the current is directed vertically upward?
21. A strong magnet is placed under a horizontal conducting ring of radius r which carries a current I as shown in Figure 29.28. If the magnetic lines of force make an angle θ with the vertical at the ring's location, what are the magnitude and direction of the resultant force on the ring?

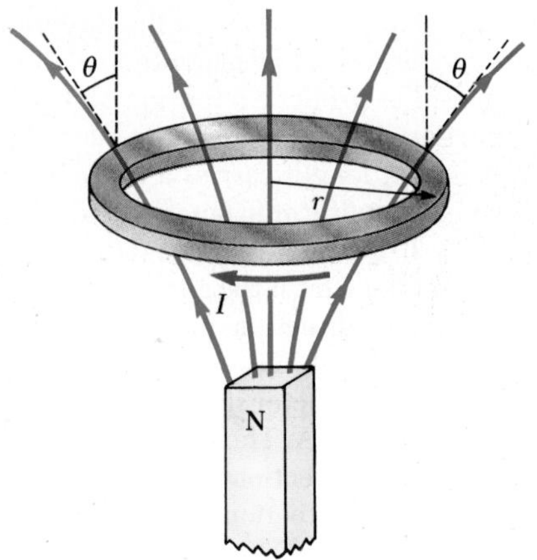

Figure 29.28 (Problem 21).

Section 29.4 Torque on a Current Loop in a Uniform Magnetic Field

22. A current of 17 mA is maintained in a single circular loop of 2 m circumference. An external magnetic field of 0.8 T is directed parallel to the plane of the loop. (a) Calculate the magnetic moment of the current loop. (b) What is the magnitude of the torque exerted on the loop by the magnetic field?

23. A rectangular loop consists of 100 closely wrapped turns and has dimensions 0.4 m by 0.3 m. The loop is hinged along the y axis, and the plane of the coil makes an angle of 30° with the x axis (Fig. 29.29). What is the magnitude of the torque exerted on the loop by a uniform magnetic field of 0.8 T directed along the x axis when the current in the windings has a value of 1.2 A in the direction shown? What is the expected direction of rotation of the loop?

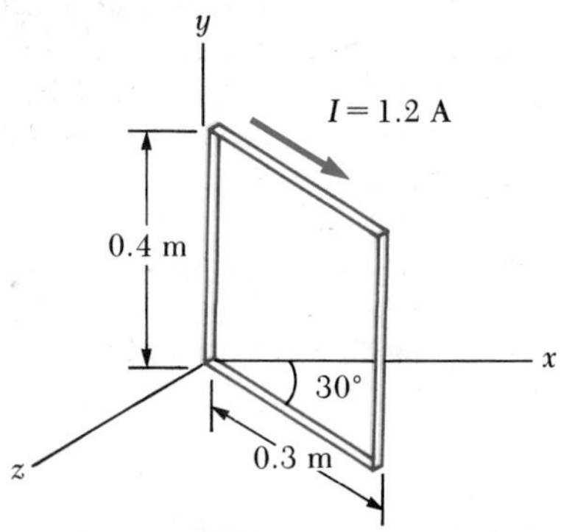

Figure 29.29 (Problem 23).

24. A small bar magnet is suspended in a uniform 0.25-T magnetic field. The maximum torque experienced by the bar magnet is 4.6×10^{-3} N · m. Calculate the magnetic moment of the bar magnet.

25. A rectangular coil of 225 turns and area 0.45 m^2 is in a uniform magnetic field of 0.21 T. Measurements indicate that the maximum torque exerted on the loop by the field is 8×10^{-3} N · m. (a) Calculate the current in the coil. (b) Would the value found for the required current be different if the 225 turns of wire were used to form a single-turn coil with the same shape of larger area? Explain.

26. A circular loop is made from a 1-m-long wire carrying a 70-mA current. (a) Calculate the magnetic moment of this wire. (b) Wrap this same wire into 50 circular turns and calculate its magnetic moment.

27. A circular coil of 100 turns has a radius of 0.025 m and carries a current of 0.1 A while in a uniform external magnetic field of 1.5 T. How much work must be done to rotate the coil from a position where the magnetic moment is parallel to the field to a position where the magnetic moment is opposite the field?

Section 29.5 Motion of a Charged Particle in a Magnetic Field

28. The magnetic field of the earth at a certain location is directed vertically downward and has a magnitude of 0.5×10^{-4} T. A proton is moving horizontally towards the west in this field with a velocity of 6.2×10^6 m/s. (a) What are the direction and magnitude of the magnetic force the field exerts on this charge? (b) What is the radius of the circular arc followed by this proton?

29. A singly charged positive ion has a mass of 3.2×10^{-26} kg. After being accelerated through a potential difference of 833 V, the ion enters a magnetic field of 0.92 T along a direction perpendicular to the direction of the field. Calculate the radius of the path of the ion in the field.

30. Consider a particle of mass m and charge q moving with a velocity $\boldsymbol{v}$. The particle enters a region perpendicular to a magnetic field $\boldsymbol{B}$. Show that while in the region of the magnetic field the kinetic energy of the particle is proportional to the square of the radius of its orbit.

31. What magnetic field would be required to constrain an electron whose energy is 725 eV to a circular path of radius 0.5 m?

32. A beam of protons (all with velocity $\boldsymbol{v}$) emerges from a particle accelerator and is deflected in a circular arc with a radius of 0.45 m by a transverse uniform magnetic field of 0.80 T. (a) Determine the speed v of the protons in the beam. (b) What time is required for the deflection of a particular proton through an angle of 90°? (c) What is the energy of the particles in the beam?

33. A proton, a deuteron, and an alpha particle ($^4_2$He nucleus) are accelerated through a common potential difference V. The particles enter a uniform magnetic

field $\boldsymbol{B}$ along a direction perpendicular to $\boldsymbol{B}$. The proton moves in a circular path of radius r_p. Find the value of the radii of the orbits of the deuteron, r_d, and the alpha particle, r_α, in terms of r_p.

34. Calculate the cyclotron frequency of a proton in a magnetic field of 5.2 T.

35. A cosmic-ray proton in interstellar space has an energy of 10 MeV and executes a circular orbit with a radius equal to that of Mercury's orbit around the Sun (5.8×10^{10} m). What is the galactic magnetic field in that region of space?

36. A singly charged ion of mass m is accelerated from rest by a potential difference V. It is then deflected by a uniform magnetic field (perpendicular to the ion's velocity) into a semicircle of radius R. Now a doubly-charged ion of mass m' is accelerated through the same potential difference and deflected by the same magnetic field into a semicircle of radius $R' = 2R$. What is the ratio of the ions' masses?

37. A singly charged positive ion moving with a speed of 4.6×10^5 m/s leaves a spiral track of radius 7.94 mm in a photograph along a direction perpendicular to the magnetic field of a bubble chamber. The magnetic field applied for the photograph has a magnitude of 1.8 T. Compute the mass (in atomic mass units) of this particle, and, from that, identify the particle.

38. The accelerating voltage that is applied to an electron gun is 15 kV, and the horizontal distance from the gun to a viewing screen is 35 cm. What is the deflection caused by the vertical component of the Earth's magnetic field (4×10^{-5} T), assuming that any change in the horizontal component of the beam velocity is negligible?

°Section 29.6 Applications of the Motion of Charged Particles in a Magnetic Field

39. A crossed-field velocity selector has a magnetic field of 10^{-2} T. What electric field strength is required if 10-keV electrons are to pass through undeflected?

40. An alpha particle with velocity $\boldsymbol{v} = 6.2 \times 10^5 \boldsymbol{i}$ m/s enters a region where the magnetic field has a value $\boldsymbol{B} = 1.1\boldsymbol{k}$ T. Determine the required magnitude and direction of an electric field $\boldsymbol{E}$ that will allow the alpha particle to continue to move along the x axis.

41. Assume that a proton and an electron, each having a kinetic energy of 200 eV, enter a uniform magnetic field of strength 0.01 T in a direction perpendicular to the field. What are the radii of the circular paths these two particles will follow?

42. Singly charged uranium ions are accelerated through a potential difference of 2 kV and enter a uniform magnetic field of 1.2 T directed perpendicular to their velocities. Determine the radius of the circular path followed by these ions assuming that they are (a) U^{238} ions and (b) U^{235} ions. How does the ratio of these path radii depend on the accelerating voltage and the magnetic field strength?

43. Consider the mass spectrometer shown schematically in Figure 29.19. The electric field between the plates of the velocity selector is 950 V/m, and the magnetic field in both the velocity selector and the deflection chamber has a magnitude of 0.93 T. Calculate the radius of the path in the system for a singly charged ion with a mass $m = 2.18 \times 10^{-26}$ kg.

44. What is the required radius of a cyclotron designed to accelerate protons to energies of 34 MeV using a magnetic field of 5.2 T?

45. What is the minimum size of a cyclotron designed to accelerate protons to an energy of 18 MeV with a cyclotron frequency of 3×10^7 Hz?

46. A certain cyclotron designed to accelerate alpha particles has a diameter of 1.52 m and operates at a frequency of 9.37 MHz. What is the maximum kinetic energy of the alpha particles?

47. A cyclotron designed to accelerate protons is provided with a magnetic field of 0.45 T and has a radius of 1.2 m. (a) What is the cyclotron frequency? (b) What is the maximum speed acquired by the protons?

48. (a) What must be the magnetic field strength within a 60-inch diameter cyclotron if that cyclotron is to accelerate protons to a maximum kinetic energy of 10.5 MeV? (b) At what frequency must the oscillator in the cyclotron operate? (c) If the frequency of the oscillator is maintained at the value found in (b), to what value must the magnetic field strength be altered if the cyclotron is to accelerate deuterons? (Note: A deuteron is a deuterium nucleus, consisting of a proton and a neutron bound together.) (d) After the cyclotron is adjusted to accelerate deuterons, what is the maximum kinetic energy of the deuterons produced?

49. The picture tube in a television uses magnetic deflection coils rather than electric deflection plates. Suppose an electron beam is accelerated through a 50-kV potential difference and then passes through a uniform magnetic field produced by these coils for 1 cm. The screen is located 10 cm from the center of the coils and is 50 cm wide. When the field is turned off, the electron beam hits the center of the screen. What field strength is necessary to deflect the beam to the side of the screen?

°Section 29.7 The Hall Effect

50. A flat ribbon of silver with a thickness $t = 0.20$ mm is used for a Hall-effect measurement of a uniform magnetic field that is perpendicular to the ribbon, as shown in Figure 29.30. The Hall coefficient for silver is $R_H = -0.84 \times 10^{-10}$ m^3/C. (a) What is the effective density of charge carriers, n, in silver? (b) If a

current $I = 20$ A produces a Hall voltage $V_H = 15\ \mu$V, what is the magnitude of the applied magnetic field?

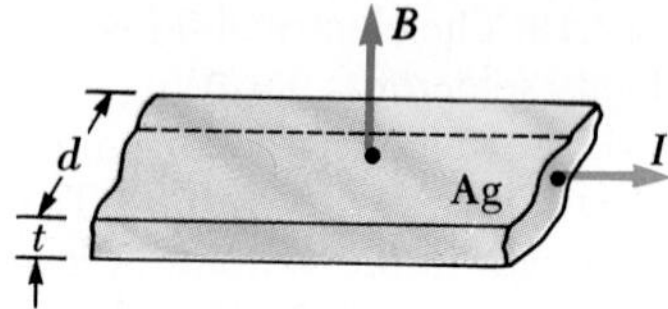

Figure 29.30 (Problem 50).

51. A section of conductor 0.4 cm in thickness is used as the experimental specimen in a Hall effect measurement. If a Hall voltage of 35 μV is measured for a current of 21 A in a magnetic field of 1.8 T, calculate the Hall coefficient for the conductor.
52. Suppose a silver foil carries a current of 5 A. If the current is directed perpendicular to a uniform 1.2-T magnetic field, what must be the thickness of the foil to produce a Hall voltage of (a) 1.0 μV, and (b) 1.0 mV?
53. In an experiment designed to measure the earth's magnetic field using the Hall effect, a copper bar 0.5 m in thickness is positioned along an east-west direction. If a current of 8 A in the conductor results in a measured Hall voltage of 5.1×10^{-12} V, what is the calculated value of the earth's magnetic field? (Assume that $n = 8.48 \times 10^{28}$ electrons/m³ and that the plane of the bar is rotated to be perpendicular to the direction of $\boldsymbol{B}$.)
54. A flat copper ribbon with a thickness of $\frac{1}{3}$ mm carries a steady current of 50 A and is located within a uniform magnetic field of 1.3 T that is directed perpendicular to the plane of the ribbon. If a Hall voltage of 9.6 μV is measured across the ribbon, what is the charge density of free electrons in the copper ribbon? What effective number of free electrons per atom does this result indicate?
55. The Hall effect can be used to measure the number of conduction electrons per unit volume n for an unknown sample. The sample is 15 mm thick, and when placed in a 1.8-T magnetic field produces a Hall voltage of 0.122 μV while carrying a 12-A current. What is the value of n?
56. A 2-mm thick sample of silver is used to measure the magnetic field in a certain region of space. Silver has approximately $n = 5.86 \times 10^{28}$ electrons/m³. If a current of 15 A in the sample results in a Hall voltage of 0.24 μV, what is the magnetic field strength?

ADDITIONAL PROBLEMS

57. A wire with a mass of 1 g/cm is placed on a horizontal surface with a coefficient of friction of 0.2. The wire carries a current of 1.5 A toward the east, and moves horizontally to the north. What are the magnitude and the direction of the *smallest* magnetic field that enables the wire to move in this fashion?
58. Indicate the initial direction of the deflection of the charged particles as they enter the magnetic fields as shown in Figure 29.31.

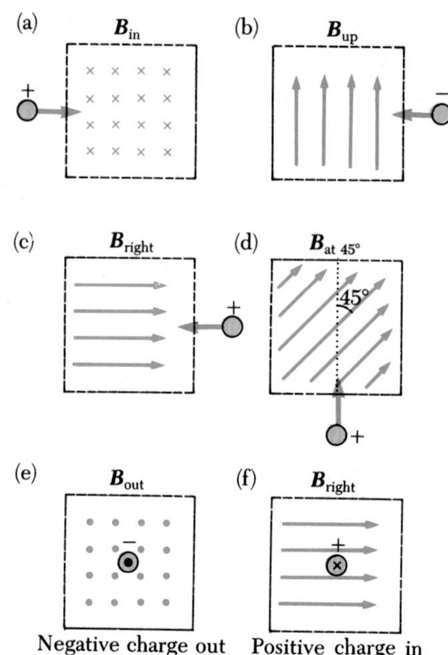

Figure 29.31 (Problem 58).

59. A positive charge $q = 3.2 \times 10^{-19}$ C moves with a velocity $\boldsymbol{v} = (2\boldsymbol{i} + 3\boldsymbol{j} - \boldsymbol{k})$ m/s through a region where both a uniform magnetic field and a uniform electric field exist. (a) Calculate the total force on the moving charge (in unit-vector notation) if $\boldsymbol{B} = (2\boldsymbol{i} + 4\boldsymbol{j} + \boldsymbol{k})$ T and $\boldsymbol{E} = (4\boldsymbol{i} - \boldsymbol{j} - 2\boldsymbol{k})$ V/m. (b) What angle does the force vector make relative to the positive x axis?
60. A conducting wire of circular cross section formed of a material that has a mass density of 2.7 g/cm³ is placed in a uniform magnetic field with the axis of the wire perpendicular to the direction of the field. A current density of 2.4×10^6 A/m² is established in the wire and the magnetic field increased until the magnetic force on the wire just balances the gravitational force. Calculate the value of B when this condition is met.
61. A straight wire of mass 10 g and length 5 cm is suspended from two identical springs which, in turn, form a closed circuit (Fig. 29.32). The springs stretch a distance of 0.5 cm under the weight of the wire. The circuit has a *total* resistance of 12 Ω. When a magnetic field is turned on, directed *out* of the page (indicated by the dots in Fig. 29.32), the springs are observed to stretch an *additional* 0.3 cm. What is the strength of the magnetic field? (The upper portion of the circuit is fixed.)

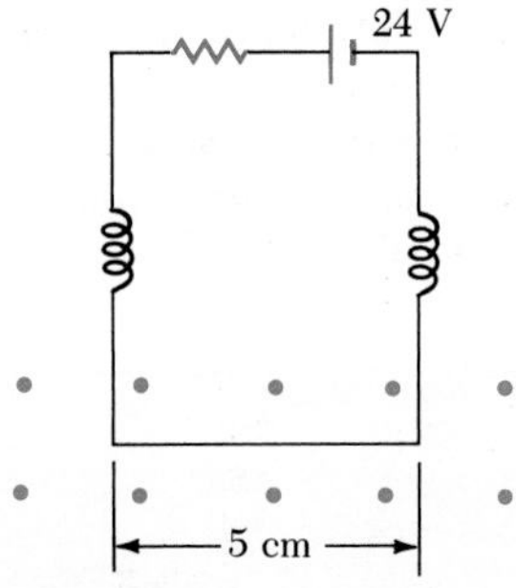

Figure 29.32 (Problem 61).

62. A rectangular loop of wire carrying a current of 2 A is suspended vertically and attached to the right arm of a balance. After the system is balanced, an external magnetic field is introduced. The field threads the lower end of the loop in a direction perpendicular to the wire. If the width of the loop is 20 cm and it takes 13.5 g of added mass on the left arm to rebalance the system, determine $\boldsymbol{B}$.

63. A conducting wire formed into the shape of an M with dimensions as shown in Figure 29.33 carries a current of 15 A. An external magnetic field $B = 2.5$ T is directed as shown throughout the region occupied by the conductor. Calculate the magnitude and direction of the net force exerted on the conductor by the magnetic field.

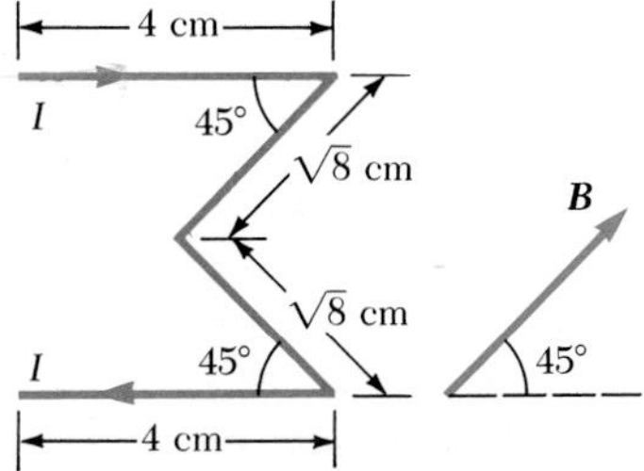

Figure 29.33 (Problem 63).

64. A metal rod of mass M and length L is pivoted at its upper end and is free to swing about the pivot. The rod hangs within a uniform horizontal magnetic field $\boldsymbol{B}$. When the rod carries a steady current I, the rod swings out to an angle θ with respect to the vertical. What is the magnitude of $\boldsymbol{B}$, expressed in terms of M, g, I, L, and θ? (*Hint:* The magnetic force, which is uniformly distributed along the rod, may be treated as a single force acting at the center of the rod.)

65. A mass spectrometer of the Bainbridge type is used to examine the isotopes of uranium. Ions in the beam emerge from the velocity selector with a speed equal to 3×10^5 m/s and enter a uniform magnetic field of 0.6 T directed perpendicular to the velocity of the ions. What is the distance between the impact points formed on the photographic plate by singly charged ions of ^{235}U and ^{238}U?

66. A cyclotron designed to accelerate deuterons has a magnetic field with a uniform intensity of 1.5 T over a region of radius 0.45 m. If the alternating potential applied between the dees of the cyclotron has a maximum value of 15 kV, what time is required for the deuterons to acquire maximum attainable energy?

67. A singly charged heavy ion is observed to complete five revolutions in a uniform magnetic field of magnitude 5×10^{-2} T in 1.50 ms. Calculate the (approximate) mass of the ion in kg.

68. A uniform magnetic field of 0.15 T is directed along the positive x axis. A positron moving with a speed of 5×10^6 m/s enters the field along a direction that makes an angle of 85° with the x axis (Fig. 29.34). The motion of the particle is expected to be a helix, as described in Section 29.5. Calculate (a) the pitch p and (b) the radius r of the trajectory.

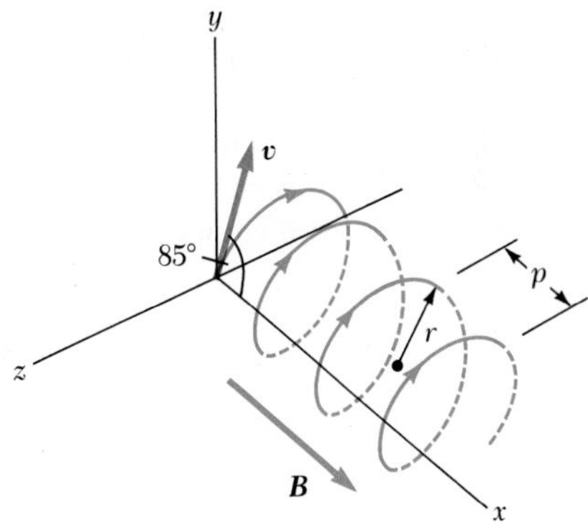

Figure 29.34 (Problem 68).

69. Consider an electron orbiting a proton and maintained in a fixed circular path of radius equal to $R = 5.29 \times 10^{-11}$ m by the Coulomb force of mutual attraction. Treating the orbiting charge as a current loop, calculate the resulting torque when the system is in an external magnetic field of 0.4 T directed perpendicular to the magnetic moment of the orbiting electron.

70. A proton moving in the plane of the page has kinetic energy of 6 MeV. It enters a magnetic field $B = 1$ T (into the page) at an angle $\theta = 45°$ to the linear boundary of the field as shown in Figure 29.35.

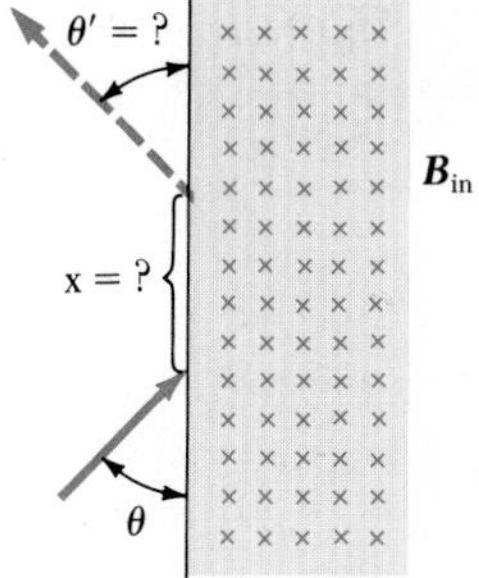

Figure 29.35 (Problem 70).

(a) Find x, the distance from the point of entry to where the proton will leave the field. (b) Determine θ', the angle between the boundary and the proton's velocity vector as it leaves the field.

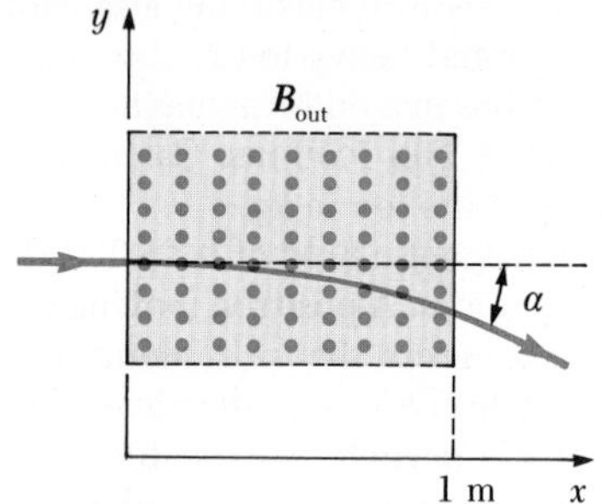

Figure 29.36 (Problem 71).

71. Protons with kinetic energy of 50 MeV are moving in the positive x direction and enter a magnetic field $\boldsymbol{B} = (0.5\text{ T})\boldsymbol{k}$ directed out of the plane of the page and extending from $x = 0$ to $x = 1$ m as shown in Figure 29.36. (a) Calculate the y component of the protons' momentum as they leave the magnetic field at $x = 1$ m. (b) Find the angle α between the initial velocity vector of the proton beam and the velocity vector after the beam emerges from the field. (Note that $1\text{ eV} = 1.60 \times 10^{-19}$ J.)

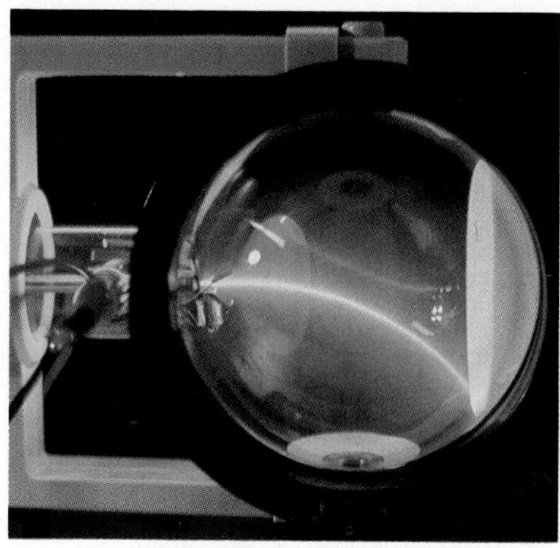

(Courtesy of CENCO)

30

Sources of the Magnetic Field

An assortment of commercially available magnets. The four red magnets and the large black magnet on the left are made of an alloy of iron, aluminum, and cobalt. The six horseshoe magnets on the right are made of different nickel alloy steels. The rectangular magnets on the lower right are ceramics made of iron, nickel, and beryllium oxides. (Courtesy of CENCO)

The preceding chapter treated a class of problems involving the magnetic force on a charged particle moving in a magnetic field. To complete the description of the magnetic interaction, this chapter deals with the origin of the magnetic field, namely, moving charges or electric currents. We begin by showing how to use the law of Biot and Savart to calculate the magnetic field produced at a point by a current element. Using this formalism and the superposition principle, we then calculate the total magnetic field due to a distribution of currents for several geometries. Next, we show how to determine the force between two current-carrying conductors, which leads to the definition of the ampere. We shall also introduce Ampère's law, which is very useful for calculating the magnetic field of highly symmetric configurations carrying steady currents. We apply Ampère's law to determine the magnetic field for several current configurations, including that of a solenoid.

This chapter is also concerned with some aspects of the complex processes that occur in magnetic materials. All magnetic effects in matter can be explained on the basis of effective current loops associated with atomic magnetic dipole moments. These atomic magnetic moments can arise both from the orbital motion of the electrons and from an intrinsic, or "built-in," property of the electrons known as *spin*. Our description of magnetism in matter will be based in part on the experimental fact that the presence of bulk matter generally modifies the magnetic field produced by currents. For example, when a material is placed inside a current-carrying solenoid, the material sets up its own magnetic field, which adds (vectorially) to the field previously present.

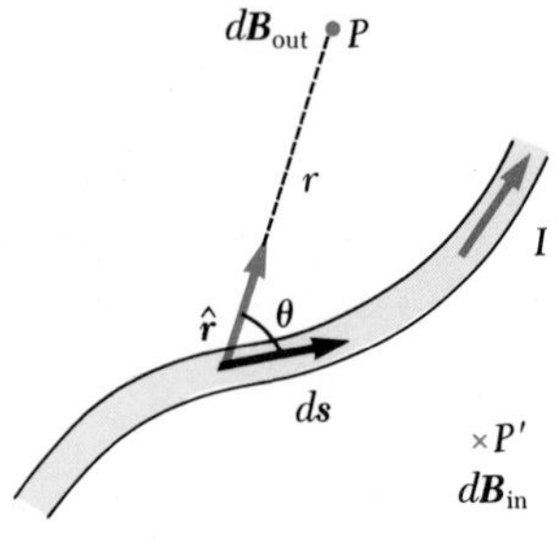

Figure 30.1 The magnetic field $d\mathbf{B}$ at a point P due to a current element $d\mathbf{s}$ is given by the Biot-Savart law, Equation 30.1. The field is out of the paper at P.

30.1 THE BIOT-SAVART LAW

Shortly after Oersted's discovery in 1819 that a compass needle is deflected by a current-carrying conductor, Jean Baptiste Biot and Felix Savart reported that a conductor carrying a steady current produces a force on a magnet. From their experimental results, Biot and Savart were able to arrive at an expression that gives the magnetic field at some point in space in terms of the current that produces the field. The *Biot-Savart law* says that if a wire carries a steady current I, the magnetic field $d\mathbf{B}$ at a point P associated with an element $d\mathbf{s}$ (Fig. 30.1) has the following properties:

Properties of the magnetic field due to a current element

1. The vector $d\mathbf{B}$ is perpendicular both to $d\mathbf{s}$ (which is in the direction of the current) and to the unit vector $\hat{\mathbf{r}}$ directed from the element to the point P.
2. The magnitude of $d\mathbf{B}$ is inversely proportional to r^2, where r is the distance from the element to the point P.
3. The magnitude of $d\mathbf{B}$ is proportional to the current and to the length ds of the element.
4. The magnitude of $d\mathbf{B}$ is proportional to $\sin\theta$, where θ is the angle between the vectors $d\mathbf{s}$ and $\hat{\mathbf{r}}$.

The **Biot-Savart law** can be summarized in the following convenient form:

Biot-Savart law

$$d\mathbf{B} = k_m \frac{I\, d\mathbf{s} \times \hat{\mathbf{r}}}{r^2} \tag{30.1}$$

where k_m is a constant that in SI units is exactly 10^{-7} Wb/A·m. The constant k_m is usually written $\mu_0/4\pi$, where μ_0 is another constant, called the **permeability of free space.** That is,

$$\frac{\mu_0}{4\pi} = k_m = 10^{-7}\ \text{Wb/A}\cdot\text{m} \tag{30.2}$$

Permeability of free space

$$\mu_0 = 4\pi k_m = 4\pi \times 10^{-7}\ \text{Wb/A}\cdot\text{m} \tag{30.3}$$

Hence, the Biot-Savart Law, Equation 30.1, can also be written

Biot-Savart law

$$d\mathbf{B} = \frac{\mu_0}{4\pi} \frac{I\, d\mathbf{s} \times \hat{\mathbf{r}}}{r^2} \tag{30.4}$$

It is important to note that the Biot-Savart law gives the magnetic field at a point only for a small element of the conductor. To find the *total* magnetic field $\mathbf{B}$ at some point due to a conductor of finite size, we must sum up contributions from all current elements making up the conductor. That is, we must evaluate $\mathbf{B}$ by integrating Equation 30.4:

$$\mathbf{B} = \frac{\mu_0 I}{4\pi} \int \frac{d\mathbf{s} \times \hat{\mathbf{r}}}{r^2} \tag{30.5}$$

where the integral is taken over the entire conductor. This expression must be handled with special care since the integrand is a vector quantity.

There are interesting similarities between the Biot-Savart law of magnetism and Coulomb's law of electrostatics. That is, the current element $I\, d\mathbf{s}$

produces a magnetic field, whereas a point charge q produces an electric field. Furthermore, *the magnitude of the magnetic field varies as the inverse square of the distance from the current element,* as does the electric field due to a point charge.

However, the directions of the two fields are quite different. The electric field due to a point charge is radial. In the case of a positive point charge, $\mathbf{E}$ is directed from the charge to the field point. On the other hand, the magnetic field due to a current element is perpendicular to both the current element and the radius vector. Hence, if the conductor lies in the plane of the paper, as in Figure 30.1, $d\mathbf{B}$ points *out* of the paper at the point P and into the paper at P'.

The examples that follow illustrate how to use the Biot-Savart law for calculating the magnetic induction of several important geometric arrangements. It is important that you recognize that the magnetic field described in these calculations is *the field due to a given current-carrying conductor.* This is not to be confused with any *external* field that may be applied to the conductor.

EXAMPLE 30.1 Magnetic Field of a Thin Straight Conductor

Consider a thin, straight wire carrying a constant current I and placed along the x axis as in Figure 30.2. Let us calculate the total magnetic field at the point P located at a distance a from the wire.

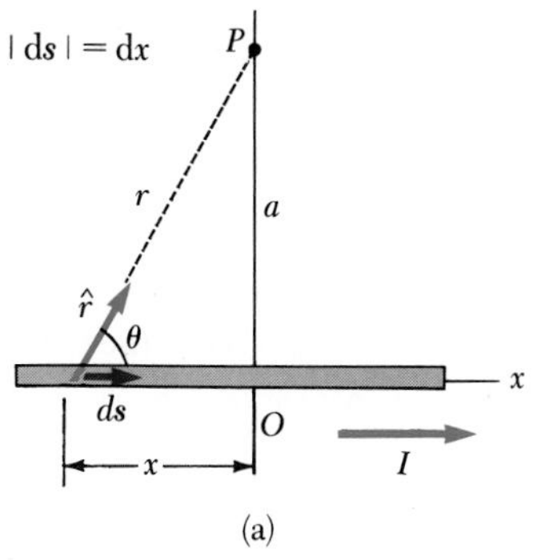

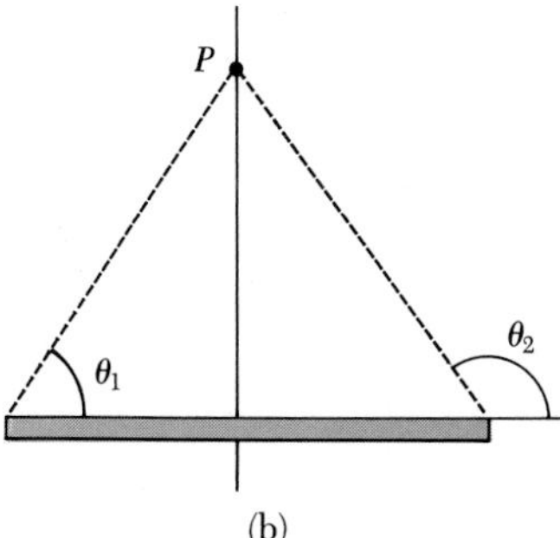

Figure 30.2 (Example 30.1) (a) A straight wire segment carrying a current I. The magnetic field at P due to each element ds is out of the paper, and so the net field is also out of the paper. (b) The limiting angles θ_1 and θ_2 for this geometry.

Solution An element $d\mathbf{s}$ is at a distance r from P. The direction of the field at P due to this element is out of the paper, since $d\mathbf{s} \times \hat{\mathbf{r}}$ is out of the paper. In fact, *all* elements give a contribution directly out of the paper at P. Therefore, we have only to determine the magnitude of the field at P. In fact, taking the origin at O and letting P be along the *positive* y axis, with $\mathbf{k}$ being a unit vector pointing *out* of the paper, we see that

$$d\mathbf{s} \times \hat{\mathbf{r}} = \mathbf{k}|d\mathbf{s} \times \hat{\mathbf{r}}| = \mathbf{k}(dx \sin \theta)$$

Substitution into Equation 30.4 gives $d\mathbf{B} = \mathbf{k}\, dB$, with

$$(1) \qquad dB = \frac{\mu_0 I}{4\pi} \frac{dx \sin \theta}{r^2}$$

In order to integrate this expression, we must relate the variables θ, x, and r. One approach is to express x and r in terms of θ. From the geometry in Figure 30.2a and some simple differentiation, we obtain the following relationship:

$$(2) \qquad r = \frac{a}{\sin \theta} = a \csc \theta$$

Since $\tan \theta = -a/x$ from the right triangle in Figure 30.2a,

$$x = -a \cot \theta$$

$$(3) \qquad dx = a \csc^2 \theta \, d\theta$$

Substitution of (2) and (3) into (1) gives

$$(4) \qquad dB = \frac{\mu_0 I}{4\pi} \frac{a \csc^2 \theta \sin \theta \, d\theta}{a^2 \csc^2 \theta} = \frac{\mu_0 I}{4\pi a} \sin \theta \, d\theta$$

Thus, we have reduced the expression to one involving only the variable θ. We can now obtain the total field at P by integrating (4) over all elements subtending angles ranging from θ_1 to θ_2 as defined in Figure 30.2b. This gives

$$B = \frac{\mu_0 I}{4\pi a}\int_{\theta_1}^{\theta_2} \sin\theta \, d\theta = \frac{\mu_0 I}{4\pi a}(\cos\theta_1 - \cos\theta_2) \qquad (30.6)$$

We can apply this result to find the magnetic field of any straight wire if we know the geometry and hence the angles θ_1 and θ_2.

Consider the special case of an infinitely long, straight wire. In this case, $\theta_1 = 0$ and $\theta_2 = \pi$, as can be seen from Figure 30.2b, for segments ranging from $x = -\infty$ to $x = +\infty$. Since $(\cos\theta_1 - \cos\theta_2) = (\cos 0 - \cos\pi) = 2$, Equation 30.6 becomes

$$B = \frac{\mu_0 I}{2\pi a} \qquad (30.7)$$

A three-dimensional view of the direction of $\boldsymbol{B}$ for a long, straight wire is shown in Figure 30.3. *The field lines are circles concentric with the wire and are in a plane perpendicular to the wire.* The magnitude of $\boldsymbol{B}$ is constant on any circle of radius a and is given by Equation 30.7. A convenient rule for determining the direction of $\boldsymbol{B}$ is to grasp the wire with the right hand, with the thumb along the direction of the current. The four fingers wrap in the direction of the magnetic field.

Our result shows that the magnitude of the magnetic field is proportional to the current and decreases as the distance from the wire increases, as one might intuitively expect. Notice that Equation 30.7 has the same mathematical form as the expression for the magnitude of the electric field due to a long charged wire (Eq. 24.9).

Exercise 1 Calculate the magnetic field of a long, straight wire carrying a current of 5 A, at a distance of 4 cm from the wire.
Answer 2.5×10^{-5} T.

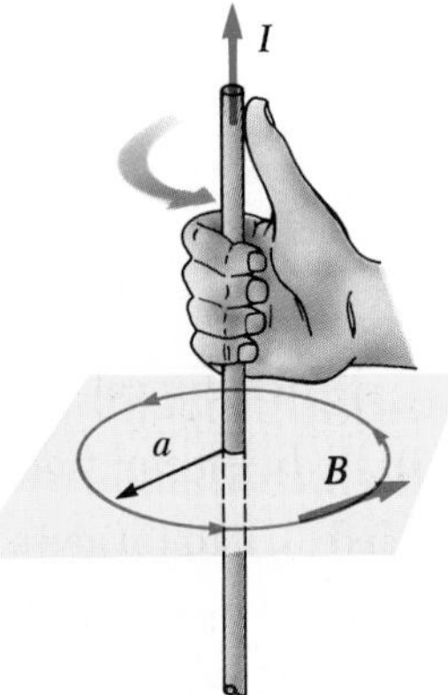

Figure 30.3 The right-hand rule for determining the direction of the magnetic field due to a long, straight wire. Note that the magnetic field lines form circles around the wire.

EXAMPLE 30.2 Field of a Current Loop

Calculate the magnetic field at the point O for the closed current loop shown in Figure 30.4. The loop consists of two straight portions and a circular arc of radius R, which subtends an angle θ at the center of the arc.

Solution First, note that the magnetic field at O due to the straight segments OA and OC is identically *zero*, since $d\boldsymbol{s}$ is parallel to $\hat{\boldsymbol{r}}$ along these paths and therefore $d\boldsymbol{s} \times \hat{\boldsymbol{r}} = 0$. This simplifies the problem because now we need to be concerned only with the magnetic field at O due to the curved portion AC. Note that each element along the path AC is at the same distance R from O, and each gives a contribution $d\boldsymbol{B}$, which is directed into the paper at O. Furthermore, at every point on the path AC, we see that $d\boldsymbol{s}$ is perpendicular to $\hat{\boldsymbol{r}}$, so that $|d\boldsymbol{s} \times \hat{\boldsymbol{r}}| = ds$. Using this information and Equation 30.4, we get the following expression for the field at O due to the segment ds:

$$dB = \frac{\mu_0 I}{4\pi}\frac{ds}{R^2}$$

Since I and R are constants, we can easily integrate this expression, which gives

$$B = \frac{\mu_0 I}{4\pi R^2}\int ds = \frac{\mu_0 I}{4\pi R^2}s = \frac{\mu_0 I}{4\pi R}\theta \qquad (30.8)$$

where we have used the fact that $s = R\theta$, where θ is measured in *radians*. The direction of $\boldsymbol{B}$ is *into* the paper at O since $d\boldsymbol{s} \times \hat{\boldsymbol{r}}$ is into the paper for every segment.

For example, if an arc subtends an angle $\theta = \pi/2$ rad, we find from Equation 30.8 that $B = \mu_0 I/8R$.

Exercise 2 A current loop in the form of a full circle of radius R carries a current I. What is the magnitude of the magnetic field at its center?
Answer $\mu_0 I/2R$.

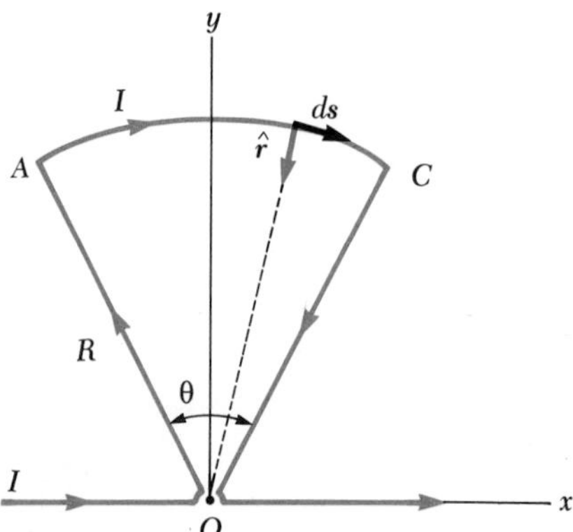

Figure 30.4 (Example 30.2) The magnetic field at O due to the closed current loop is into the paper. Note that the contribution to the field at O due to the straight segments OA and OC is zero.

EXAMPLE 30.3 Magnetic Field on the Axis of a Circular Current Loop

Consider a circular loop of wire of radius R located in the yz plane and carrying a steady current I, as in Figure 30.5. Let us calculate the magnetic field at an axial point P a distance x from the center of the loop.

Solution In this situation, note that any element $d\mathbf{s}$ is perpendicular to $\hat{\mathbf{r}}$. Furthermore, all elements around the loop are at the same distance r from P, where $r^2 = x^2 + R^2$. Hence, the *magnitude* of $d\mathbf{B}$ due to the element $d\mathbf{s}$ is given by

$$dB = \frac{\mu_0 I}{4\pi}\frac{|d\mathbf{s} \times \hat{\mathbf{r}}|}{r^2} = \frac{\mu_0 I}{4\pi}\frac{ds}{(x^2 + R^2)}$$

The direction of the magnetic field $d\mathbf{B}$ due to the element $d\mathbf{s}$ is perpendicular to the plane formed by $\hat{\mathbf{r}}$ and $d\mathbf{s}$, as shown in Figure 30.5. The vector $d\mathbf{B}$ can be resolved into a component dB_x, along the x axis, and a component dB_y, which is perpendicular to the x axis. When the components perpendicular to the x axis are summed over the whole loop, the result is *zero*. That is, by symmetry any element on one side of the loop will set up a perpendicular component that cancels the component set up by an element diametrically opposite it. Therefore, we see that *the resultant field at P must be along the x axis* and can be found by integrating the components $dB_x = dB \cos\theta$, where this expression is obtained from resolving the vector $d\mathbf{B}$ into its components as shown in Figure 30.5. That is, $\mathbf{B} = \mathbf{i}B_x$, where B_x is given by

$$B_x = \oint dB \cos\theta = \frac{\mu_0 I}{4\pi}\oint \frac{ds \cos\theta}{x^2 + R^2}$$

where the integral must be taken over the entire loop.

Since θ, x, and R are constants for all elements of the loop and since $\cos\theta = R/(x^2 + R^2)^{1/2}$, we get

$$B_x = \frac{\mu_0 IR}{4\pi(x^2 + R^2)^{3/2}}\oint ds = \frac{\mu_0 IR^2}{2(x^2 + R^2)^{3/2}} \qquad (30.9)$$

where we have used the fact that $\oint ds = 2\pi R$ (the circumference of the loop).

To find the magnetic field at the *center* of the loop, we set $x = 0$ in Equation 30.9. At this special point, this gives

$$B = \frac{\mu_0 I}{2R} \qquad (\text{at } x = 0) \qquad (30.10)$$

It is also interesting to determine the behavior of the magnetic field at large distances from the loop, that is, when x is large compared with R. In this case, we can neglect the term R^2 in the denominator of Equation 30.9 and get

$$B \approx \frac{\mu_0 IR^2}{2x^3} \qquad (\text{for } x \gg R) \qquad (30.11)$$

Since the magnitude of the magnetic dipole moment μ of the loop is defined as the product of the current and the area (Eq. 29.13), $\mu = I(\pi R^2)$ and we can express Equation 30.11 in the form

$$B = \frac{\mu_0}{2\pi}\frac{\mu}{x^3} \qquad (30.12)$$

This result is similar in form to the expression for the electric field due to an electric dipole, $E = kp/y^3$ (Eq. 23.10), where p is the electric dipole moment. The pattern of the magnetic field lines for a circular loop is shown in Figure 30.6. For clarity, the lines are drawn only for one plane which contains the axis of the loop. The field pattern is axially symmetric.

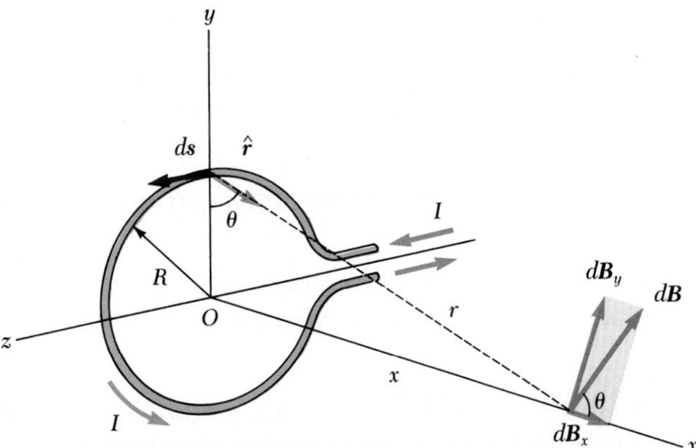

Figure 30.5 (Example 30.3) The geometry for calculating the magnetic field at an axial point P for a current loop. Note that by symmetry the total field $\mathbf{B}$ is along the x axis.

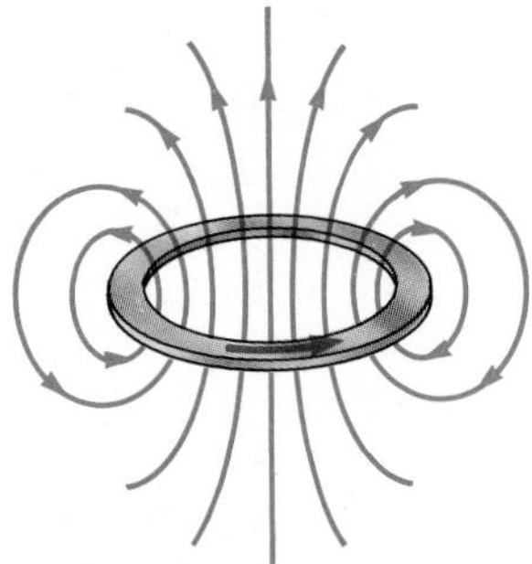

Figure 30.6 Magnetic field lines for a current loop. Far from the loop, the field lines are identical in form to those of an electric dipole.

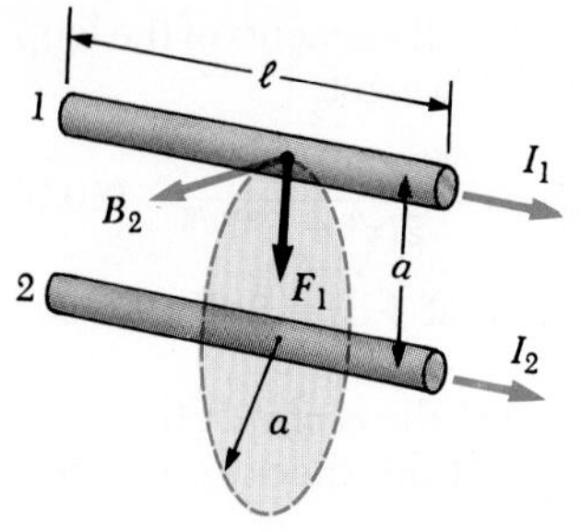

Figure 30.7 Two parallel wires each carrying a steady current exert a force on each other. The field $\boldsymbol{B}_2$ at wire 1 due to wire 2 produces a force on wire 1 given by $F_1 = I_1 \ell B_2$. The force is attractive if the currents are parallel as shown and repulsive if the currents are antiparallel.

30.2 THE MAGNETIC FORCE BETWEEN TWO PARALLEL CONDUCTORS

In the previous chapter we described the magnetic force that acts on a current-carrying conductor when the conductor is placed in an external magnetic field. Since a current in a conductor sets up its own magnetic field, it is easy to understand that two current-carrying conductors will exert magnetic forces upon each other. As we shall see, such forces can be used as the basis for defining the ampere and the coulomb. Consider two long, straight, parallel wires separated by a distance a and carrying currents I_1 and I_2 in the same direction, as in Figure 30.7. We can easily determine the force on one wire due to a magnetic field set up by the other wire. Wire 2, which carries a current I_2, sets up a magnetic field $\boldsymbol{B}_2$ at the position of wire 1. The direction of $\boldsymbol{B}_2$ is *perpendicular* to the wire, as shown in Figure 30.7. According to Equation 29.5, the magnetic force on a length ℓ of wire 1 is $\boldsymbol{F}_1 = I_1 \boldsymbol{\ell} \times \boldsymbol{B}_2$. Since $\boldsymbol{\ell}$ is perpendicular to $\boldsymbol{B}_2$, the magnitude of $\boldsymbol{F}_1$ is given by $F_1 = I_1 \ell B_2$. Since the field due to wire 2 is given by Equation 30.7,

$$B_2 = \frac{\mu_0 I_2}{2\pi a}$$

we see that

$$F_1 = I_1 \ell B_2 = I_1 \ell \left(\frac{\mu_0 I_2}{2\pi a} \right) = \frac{\ell \mu_0 I_1 I_2}{2\pi a}$$

We can rewrite this in terms of the force per unit length as

$$\frac{F_1}{\ell} = \frac{\mu_0 I_1 I_2}{2\pi a} \tag{30.13}$$

The direction of $\boldsymbol{F}_1$ is downward, toward wire 2, since $\boldsymbol{\ell} \times \boldsymbol{B}_2$ is downward. If one considers the field set up at wire 2 due to wire 1, the force $\boldsymbol{F}_2$ on wire 2 is found to be equal to and opposite $\boldsymbol{F}_1$. This is what one would expect, because Newton's third law of action-reaction must be obeyed.[1] When the currents are in opposite directions, the forces are reversed and the wires repel each other. Hence, we find that

> parallel conductors carrying currents in the same direction attract each other, whereas parallel conductors carrying currents in opposite directions repel each other.

The force between two parallel wires each carrying a current is used to define the **ampere** as follows:

> If two long, parallel wires 1 m apart carry the same current and the force per unit length on each wire is 2×10^{-7} N/m, then the current is defined to be 1 A.

The numerical value of 2×10^{-7} N/m is obtained from Equation 30.13, with $I_1 = I_2 = 1$ A and $a = 1$ m. Therefore, a mechanical measurement can be used

[1] Although the total force on wire 1 is equal to and opposite the total force on wire 2, Newton's third law does not apply when one considers two small elements of the wires that are not opposite each other. This apparent violation of Newton's third law and of conservation of momentum is described in more advanced treatments on electricity and magnetism.

to standardize the ampere. For instance, the National Bureau of Standards uses an instrument called a *current balance* for primary current measurements. These results are then used to standardize other, more conventional instruments, such as ammeters.

The SI unit of charge, the **coulomb**, can now be defined in terms of the ampere as follows:

If a conductor carries a steady current of 1 A, then the quantity of charge that flows through a cross section of the conductor in 1 s is 1 C.

30.3 AMPÈRE'S LAW

A simple experiment first carried out by Oersted in 1820 clearly demonstrates the fact that a current-carrying conductor produces a magnetic field. In this experiment, several compass needles are placed in a horizontal plane near a long vertical wire, as in Figure 30.8a. When there is no current in the wire, all compasses in the loop point in the same direction (that of the earth's field), as one would expect. However, when the wire carries a strong, steady current, the compass needles will all deflect in a direction tangent to the circle, as in Figure 30.8b. These observations show that the direction of $\boldsymbol{B}$ is consistent with the right-hand rule described in Section 30.1.

If the wire is grasped in the right hand with the thumb in the direction of the current, the fingers will wrap (or curl) in the direction of $\boldsymbol{B}$.

When the current is reversed, the compass needles in Figure 30.8b will also reverse.

Since the compass needles point in the direction of $\boldsymbol{B}$, we conclude that the lines of $\boldsymbol{B}$ form circles about the wire, as we discussed in the previous section. By symmetry, the magnitude of $\boldsymbol{B}$ is the same everywhere on a circular path that is centered on the wire and lying in a plane that is perpendicular to the wire. By varying the current and distance r from the wire, one finds that $\boldsymbol{B}$ is proportional to the current and inversely proportional to the distance from the wire.

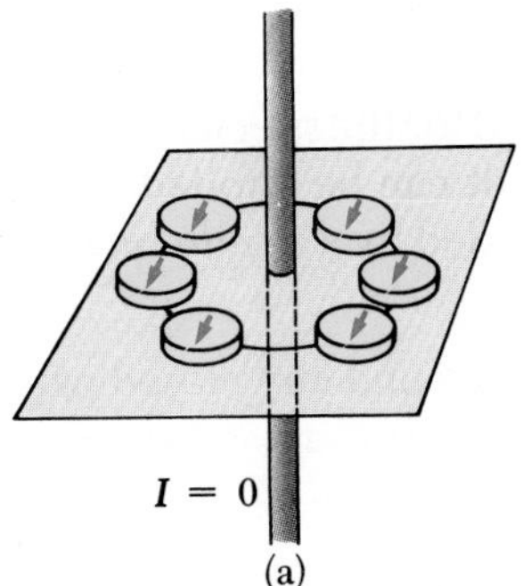

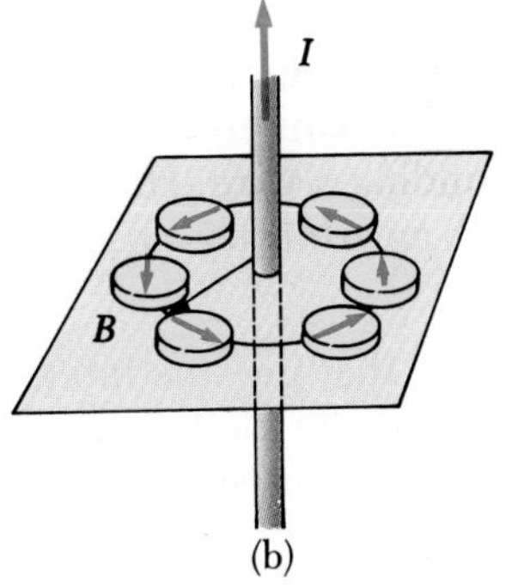

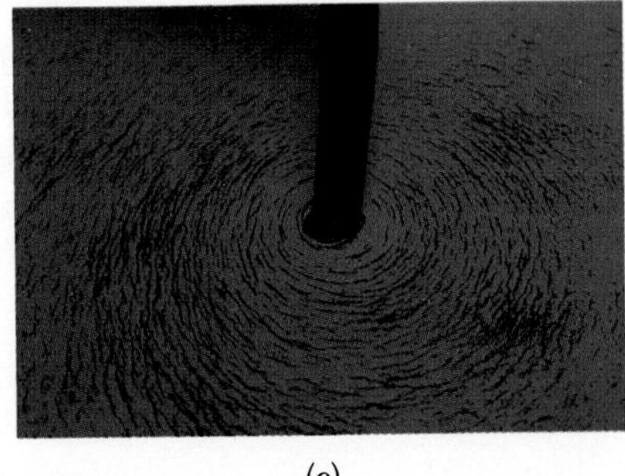

Figure 30.8 (a) When there is no current in the vertical wire, all compass needles point in the same direction. (b) When the wire carries a strong current, the compass needles deflect in a direction tangent to the circle, which is the direction of $\boldsymbol{B}$ due to the current. (c) Circular magnetic field line surrounding a current-carrying conductor as displayed with iron filings. The photograph was taken using 30 parallel wires each carrying a current of $\frac{1}{2}$ A. (Photo courtesy of Henry Leap and Jim Lehman)

Biographical Sketch

André-Marie Ampère
(1775–1836)

André-Marie Ampère was a French mathematician, chemist, and philosopher who founded the science of electrodynamics. The unit of measure for electric current was named in his honor.

Ampère's genius, particularly in mathematics, became evident early in his life: he had mastered advanced mathematics by the age of 12. In his first publication, *Considerations on the Mathematical Theory of Games,* an early contribution to the theory of probability, he proposed the inevitability of a player's losing a game of chance to a player with greater financial resources.

Ampère is credited with the discovery of electromagnetism—the relationship between electric current and magnetic fields. His work in this field was influenced by the findings of Danish physicist Hans Christian Oersted. Ampère presented a series of papers expounding the theory and basic laws of electromagnetism, which he called electrodynamics, to differentiate it from the study of stationary electric forces, which he called electrostatics.

The culmination of Ampère's studies came in 1827 when he published his *Mathematical Theory of Electrodynamic Phenomena Deduced Solely from Experiment,* in which he derived precise mathematical formulations of electromagnetism, notably Ampère's law.

Many stories are told of Ampère's absent-mindedness, a trait he shared with Newton. In one instance, he forgot to honor an invitation to dine with the Emperor Napoleon.

Ampère's personal life was filled with tragedy. His father, a wealthy city official, was guillotined during the French Revolution, and his wife's death in 1803 was a major blow. Ampère died at the age of 63 of pneumonia. His judgment of his life is clear from the epitaph he chose for his gravestone: *Tandem felix* (Happy at last).

(Photo courtesy of AIP Niels Bohr Library)

Now let us evaluate the product $\mathbf{B}\cdot d\mathbf{s}$ and sum these products over the closed circular path centered on the wire. Along this path, the vectors $d\mathbf{s}$ and $\mathbf{B}$ are parallel at each point (Fig. 30.8b), so that $\mathbf{B}\cdot d\mathbf{s} = B\,ds$. Furthermore, $\mathbf{B}$ is constant in magnitude on this circle and given by Equation 30.7. Therefore, the sum of the products $B\,ds$ over the closed path, which is equivalent to the line integral of $\mathbf{B}\cdot d\mathbf{s}$, is given by

$$\oint \mathbf{B}\cdot d\mathbf{s} = B\oint ds = \frac{\mu_0 I}{2\pi r}(2\pi r) = \mu_0 I \tag{30.14}$$

where $\oint ds = 2\pi r$ is the circumference of the circle.

This result, known as **Ampère's law,** was calculated for the special case of a circular path surrounding a wire. However, the result can be applied in the general case in which an arbitrary closed path is threaded by a *steady current.* That is,

Ampère's law says that the line integral of $\mathbf{B}\cdot d\mathbf{s}$ around any closed path equals $\mu_0 I$, where I is the total steady current passing through any surface bounded by the closed path.

Ampère's law

$$\oint \mathbf{B}\cdot d\mathbf{s} = \mu_0 I \tag{30.15}$$

Ampère's law is valid only for steady currents. Furthermore, *Ampère's law is useful only for calculating the magnetic field of current configurations with a high degree of symmetry*, just as Gauss' law is useful only for calculating the electric field of highly symmetric charge distributions. The following examples illustrate some symmetric current configurations for which Ampère's law is useful.

EXAMPLE 30.4 The B Field of a Long Wire

A long, straight wire of radius R carries a steady current I_0 that is uniformly distributed through the cross section of the wire (Figure 30.9). Calculate the magnetic field at a distance r from the center of the wire in the regions $r \geq R$ and $r < R$.

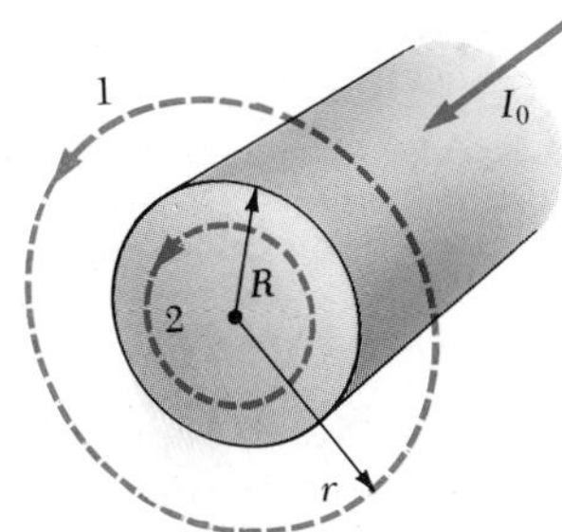

Figure 30.9 (Example 30.4) A long, straight wire of radius R carrying a steady current I_0 uniformly distributed across the wire. The magnetic field at any point can be calculated from Ampère's law using a circular path of radius r, concentric with the wire.

Solution In region 1, where $r \geq R$, let us choose a circular path of radius r centered at the wire. From symmetry, we see that $\mathbf{B}$ must be constant in magnitude and parallel to $d\mathbf{s}$ at every point on the path. Since the total current passing through by path 1 is I_0, Ampère's law applied to the path gives

$$\oint \mathbf{B} \cdot d\mathbf{s} = B \oint ds = B(2\pi r) = \mu_0 I_0$$

$$B = \frac{\mu_0 I_0}{2\pi r} \qquad \text{(for } r \geq R) \qquad (30.16)$$

which is identical in meaning to Equation 30.7.

Now consider the interior of the wire, that is, region 2, where $r < R$. In this case, note that the current I enclosed by the path is *less* than the total current, I_0. Since the current is assumed to be uniform over the cross section of the wire, we see that the fraction of the current enclosed by the path of radius $r < R$ must equal the ratio of the area πr^2 enclosed by path 2 and the cross-sectional area πR^2 of the wire.[2] That is,

$$\frac{I}{I_0} = \frac{\pi r^2}{\pi R^2}$$

$$I = \frac{r^2}{R^2} I_0$$

Following the same procedure as for path 1, we can now apply Ampère's law to path 2. This gives

$$\oint \mathbf{B} \cdot d\mathbf{s} = B(2\pi r) = \mu_0 I = \mu_0 \left(\frac{r^2}{R^2} I_0 \right)$$

$$B = \left(\frac{\mu_0 I_0}{2\pi R^2} \right) r \qquad \text{(for } r < R) \qquad (30.17)$$

The magnetic field versus r for this configuration is sketched in Figure 30.10. Note that inside the wire, $B \rightarrow 0$ as $r \rightarrow 0$. This result is similar in form to that of the electric field inside a uniformly charged rod.

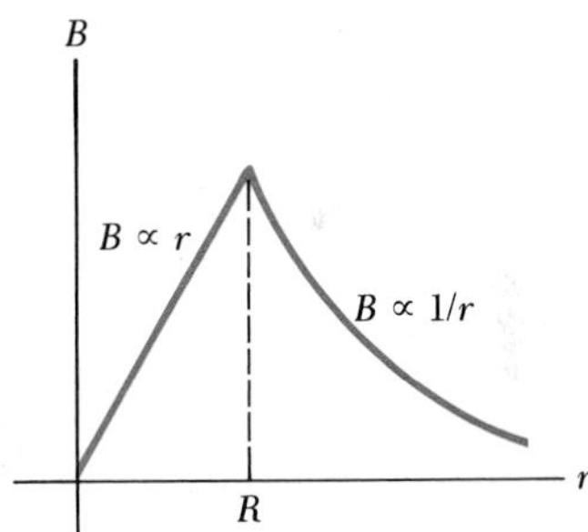

Figure 30.10 A sketch of the magnetic field versus r for the wire described in Example 30.4. The field is proportional to r inside the wire and varies as $1/r$ outside the wire.

EXAMPLE 30.5 The Magnetic Field of a Toroidal Coil

The *toroidal coil* consists of N turns of wire wrapped around a doughnut-shaped structure as in Figure 30.11. Assuming that the turns are closely spaced, calculate the magnetic field *inside* the coil, a distance r from the center.

Solution To calculate the field inside the coil, we evaluate the line integral of $\mathbf{B} \cdot d\mathbf{s}$ over a circle of radius r. By symmetry, we see that the magnetic field is constant in magnitude on this path and tangent to it, so that $\mathbf{B} \cdot d\mathbf{s} = B\,ds$. Furthermore, note that the closed path threads N

[2] Alternatively, the current linked by path 2 must equal the product of the current density, $J = I_0/\pi R^2$, and the area πr^2 enclosed by path 2.

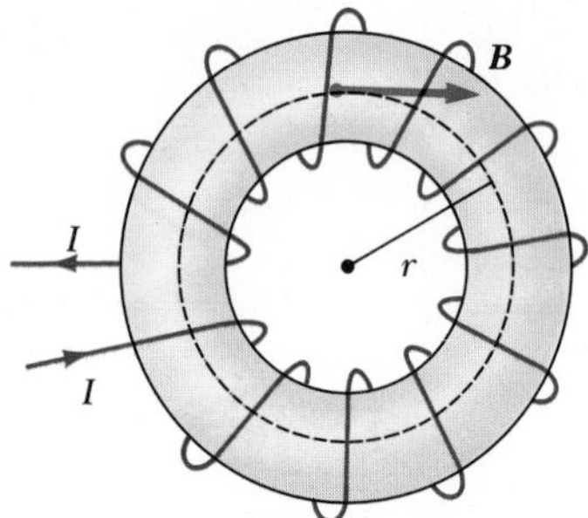

Figure 30.11 (Example 30.6) A toroidal coil consists of many turns of wire wrapped around a doughnut-shaped structure (torus). If the coils are closely spaced, the field inside the toroidal coil is tangent to the dashed circular path and varies as $1/r$, and the exterior field is zero.

loops of wire, each of which carries a current I. Therefore, the right side of Ampère's law, Equation 30.15, is $\mu_0 NI$ in this case. Ampère's law applied to this path then gives

$$\oint \mathbf{B} \cdot d\mathbf{s} = B \oint ds = B(2\pi r) = \mu_0 NI$$

$$B = \frac{\mu_0 NI}{2\pi r} \qquad (30.18)$$

This result shows that B varies as $1/r$ and hence is nonuniform within the coil. However, if r is large compared with a, where a is the cross-sectional radius of the toroid, then the field will be approximately uniform inside the coil. Furthermore, for the ideal toroidal coil, where the turns are closely spaced, the external field is *zero*. This can be seen by noting that the *net* current threaded by any circular path lying outside the toroidal coil is zero (including the region of the "hole in the doughnut"). Therefore, from Ampère's law one finds that $\mathbf{B} = 0$ in the regions exterior to the toroidal coil. In reality, the turns of a toroidal coil forms a helix rather than circular loops (the ideal case). As a result, there is always a small field external to the coil.

EXAMPLE 30.6 Magnetic Field of an Infinite Current Sheet

An infinite sheet lying in the yz plane carries a surface current of density $\mathbf{J}_s$. The current is in the y direction, and J_s represents the current per unit length measured along the z axis. Find the magnetic field near the sheet.

Solution To evaluate the line integral in Ampère's law, let us take a rectangular path around the sheet as in Figure 30.12. The rectangle has dimensions ℓ and w, where the sides of length ℓ are parallel to the surface. The *net* current through the loop is $J_s\ell$ (that is, the net current equals the current per unit length multiplied by the length of the rectangle). Hence, applying Ampère's law over the loop and noting that the paths of length w do not contribute to the line integral (because the component of $\mathbf{B}$ along the direction of these paths is zero), we get

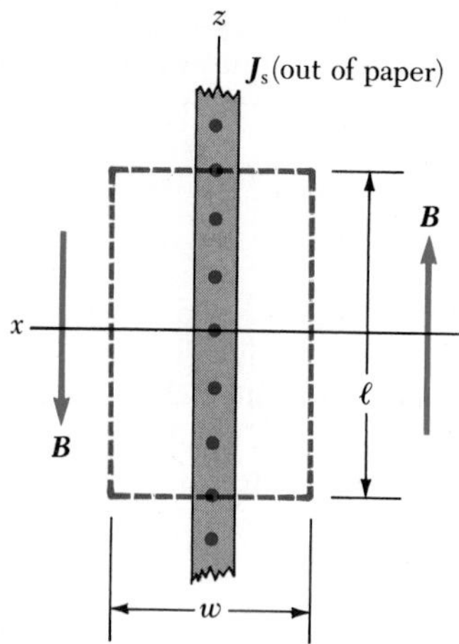

Figure 30.12 A top view of an infinite current sheet lying in the yz plane, where the current is in the y direction (out of the paper). This view shows the direction of $\mathbf{B}$ on both sides of the sheet.

$$\oint \mathbf{B} \cdot d\mathbf{s} = \mu_0 I = \mu_0 J_s \ell$$

$$2B\ell = \mu_0 J_s \ell$$

$$B = \mu_0 \frac{J_s}{2} \qquad (30.19)$$

The result shows that *the magnetic field is independent of the distance from the current sheet.* In fact, the magnetic field is uniform and is everywhere parallel to the plane of the sheet. This is reasonable since we are dealing with an *infinite* sheet of current. The result is analogous to the uniform electric field associated with an infinite sheet of charge. (Example 24.6.)

EXAMPLE 30.7 The Magnetic Force on a Current Segment

A long straight wire oriented along the y axis carries a steady current I_1 as in Figure 30.13. A rectangular circuit located to the right of the wire carries a current I_2. Find the magnetic force on the *upper horizontal segment* of the circuit that runs from $x = a$ to $x = a + b$.

Solution In this problem, you may be tempted to use Equation 30.13 to obtain the force. However, this result applies *only* to two *parallel* wires, and cannot be used here. The correct approach is to start with the force on a small segment of the conductor given by $d\mathbf{F} = I\, d\mathbf{s} \times \mathbf{B}$

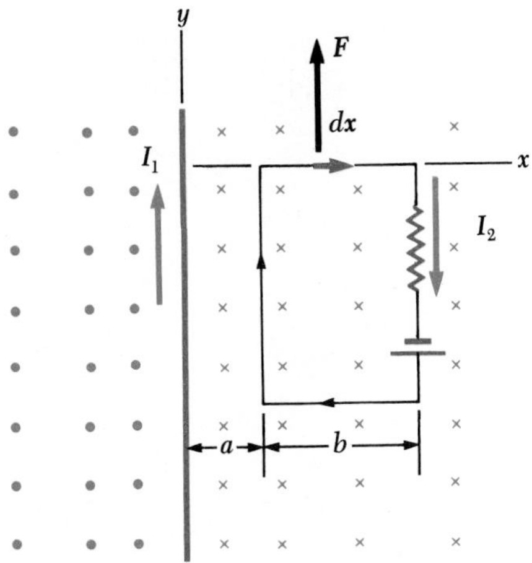

Figure 30.13 Example 30.7.

(Eq. 29.6), where in this case, $I = I_2$ and $\boldsymbol{B}$ is the magnetic field due to the long straight wire at the position of the segment of length ds. From Ampere's law, the field at a distance x from the straight wire is given by

$$\boldsymbol{B} = \frac{\mu_0 I_1}{2\pi x}(-\boldsymbol{k})$$

where the field points *into* the page as indicated by the unit vector notation $(-\boldsymbol{k})$. Taking the length of our segment as $d\boldsymbol{s} = dx\,\boldsymbol{i}$, we find

$$d\boldsymbol{F} = \boldsymbol{B} = \frac{\mu_0 I_1 I_2}{2\pi x}[\boldsymbol{i} \times (-\boldsymbol{k})]\,dx = \frac{\mu_0 I_1 I_2}{2\pi}\frac{dx}{x}\boldsymbol{j}$$

Integrating this equation over the limits $x = a$ to $x = a + b$ gives

$$\boldsymbol{F} = \frac{\mu_0 I_1 I_2}{2\pi}\ln x\bigg]_a^{a+b}\boldsymbol{j} = \frac{\mu_0 I_1 I_2}{2\pi}\ln\left(1 + \frac{b}{a}\right)\boldsymbol{j}$$

The force points *upward* as indicated by the notation $\boldsymbol{j}$, and as shown in Figure 30.13.

Exercise 3 What is the force on the lower horizontal segment of the circuit?
Answer The force has the same magnitude as the force on the upper horizontal segment, but is directed *downward*.

30.4 THE MAGNETIC FIELD OF A SOLENOID

A solenoid is a long wire wound in the form of a helix. With this configuration, one can produce a reasonably uniform magnetic field within a small volume of the solenoid's interior region if the consecutive turns are closely spaced. When the turns are closely spaced, each can be regarded as a circular loop, and the net magnetic field is the vector sum of the fields due to all the turns.

Figure 30.14 shows the magnetic field lines of a loosely wound solenoid. Note that the field lines inside the coil are nearly parallel, uniformly distributed, and close together. This indicates that the field inside the solenoid is uniform. The field lines between the turns tend to cancel each other. The field outside the solenoid is both nonuniform and weak. The field at exterior points, such as P, is weak since the field due to current elements on the upper portions tends to cancel the field due to current elements on the lower portions.

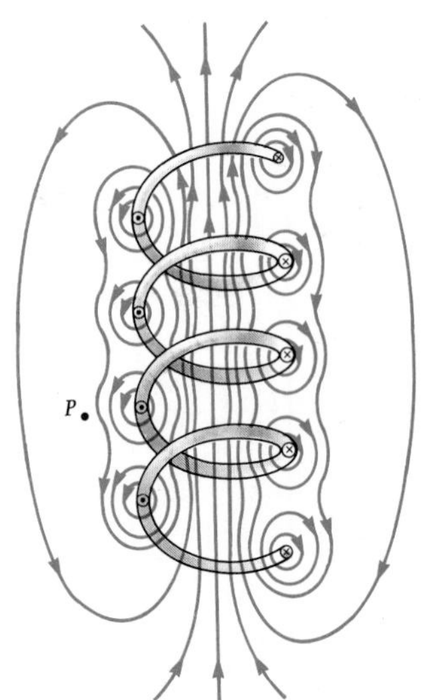

Figure 30.14 The magnetic field lines for a loosely wound solenoid. Adapted from D. Halliday and R. Resnick, *Physics*, New York, Wiley, 1978.

If the turns are closely spaced and the solenoid is of finite length, the field lines are as shown in Figure 30.15. In this case, the field lines diverge from one end and converge at the opposite end. An inspection of this field distribution exterior to the solenoid shows a similarity with the field of a bar magnet. Hence, one end of the solenoid behaves like the north pole of a magnet while the opposite end behaves like the south pole. As the length of the solenoid increases, the field within it becomes more and more uniform. One approaches the case of an *ideal solenoid* when the turns are closely spaced and the length is long compared with the radius. In this case, the field outside the solenoid is weak compared with the field inside the solenoid, and the field inside is uniform over a large volume.

We can use Ampère's law to obtain an expression for the magnetic field inside an ideal solenoid. A longitudinal cross section of part of our ideal sole-

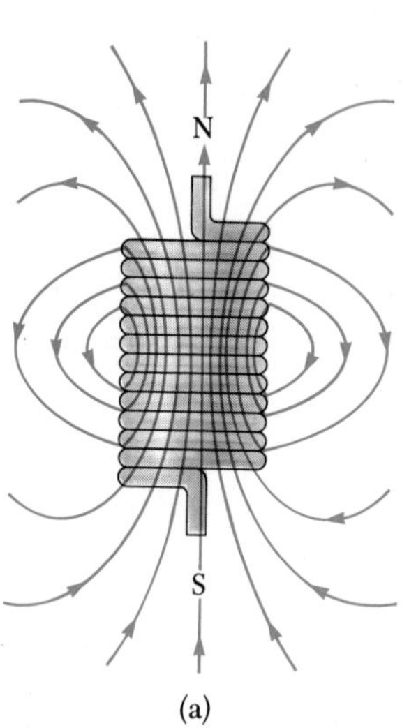

(a)

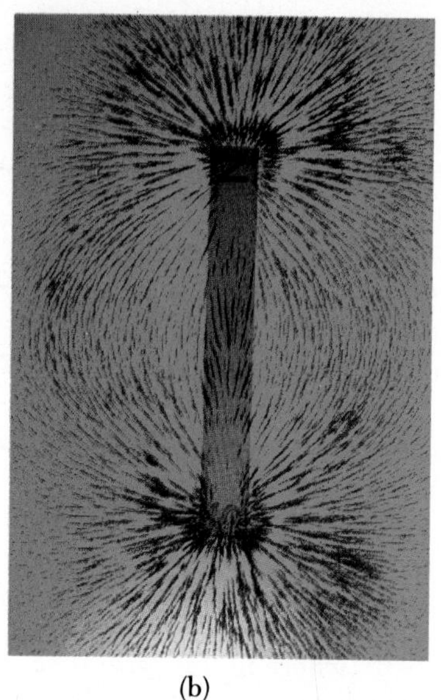

(b)

Figure 30.15 (a) Magnetic field lines for a tightly wound solenoid of finite length carrying a steady current. The field inside the solenoid is nearly uniform and strong. Note that the field lines resemble those of a bar magnet, so that the solenoid effectively has north and south poles. (b) Magnetic field pattern of a bar magnet, as displayed by small iron filings on a sheet of paper. (Courtesy of Henry Leap and Jim Lehman)

noid (Fig. 30.16) carries a current I. For the ideal solenoid, $\mathbf{B}$ inside the solenoid is uniform and parallel to the axis and $\mathbf{B}$ outside is zero. Consider a rectangular path of length ℓ and width w as shown in Figure 30.16. We can apply Ampère's law to this path by evaluating the integral of $\mathbf{B} \cdot d\mathbf{s}$ over each of the four sides of the rectangle. The contribution along side 3 is clearly zero, since $\mathbf{B} = 0$ in this region. The contributions from sides 2 and 4 are both zero since $\mathbf{B}$ is perpendicular to $d\mathbf{s}$ along these paths. Side 1, whose length is ℓ, gives a contribution $B\ell$ to the integral since $\mathbf{B}$ is uniform and parallel to $d\mathbf{s}$ along this path. Therefore, the integral over the closed rectangular path has the value

$$\oint \mathbf{B} \cdot d\mathbf{s} = \int_{\text{path 1}} \mathbf{B} \cdot d\mathbf{s} = B \int_{\text{path 1}} ds = B\ell$$

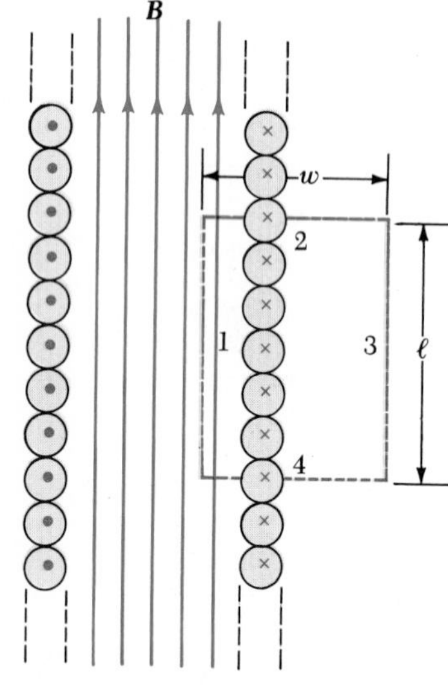

Figure 30.16 A cross-sectional view of a tightly wound solenoid. If the solenoid is long relative to its radius, we can assume that the field inside is uniform and the field outside is zero. Ampère's law applied to the red dashed rectangular path can then be used to calculate the field inside the solenoid.

The right side of Ampère's law involves the *total* current that passes through the area bound by the path of integration. In our case, the total current through the rectangular path equals the current through each turn multiplied by the number of turns. If N is the number of turns in the length ℓ, then the total current through the rectangle equals NI. Therefore, Ampère's law applied to this path gives

$$\oint \mathbf{B} \cdot d\mathbf{s} = B\ell = \mu_0 NI$$

$$B = \mu_0 \frac{N}{\ell} I = \mu_0 nI \tag{30.20}$$

where $n = N/\ell$ is the number of turns *per unit length* (not to be confused with N).

We also could obtain this result in a simpler manner by reconsidering the magnetic field of a toroidal coil (Example 30.5). If the radius r of the toroidal coil containing N turns is large compared with its cross-sectional radius a, then a short section of the toroidal coil approximates a solenoid with $n = N/2\pi r$. In this limit, we see that Equation 30.18 derived for the toroidal coil agrees with Equation 30.20.

Equation 30.20 is valid only for points near the center of a very long solenoid. As you might expect, the field near each end is smaller than the value given by Equation 30.20. At the very end of a long solenoid, the magnitude of the field is about one half that of the field at the center. The field at arbitrary axial points of the solenoid is derived in Section 30.5.

*30.5 THE MAGNETIC FIELD ALONG THE AXIS OF A SOLENOID

Consider a solenoid of length ℓ and radius R containing N closely spaced turns and carrying a steady current I. Let us determine an expression for the magnetic field at an axial point P inside the solenoid, as indicated in Figure 30.17.

Perhaps the simplest way to obtain the desired result is to consider the solenoid as a distribution of current loops. The field of any one loop along the axis is given by Equation 30.9. Hence, the net field in the solenoid is the superposition of fields from all loops. The number of turns in a length dx of the solenoid is $(N/\ell)\ dx$; therefore the total current in a width dx is given by $I(N/\ell)\ dx$. Then, using Equation 30.9, we find that the field at P due to the section dx is given by

$$dB = \frac{\mu_0 R^2}{2(x^2 + R^2)^{3/2}}\, I\left(\frac{N}{\ell}\right) dx \qquad (30.21)$$

This expression contains the variable x, which can be expressed in terms of the variable ϕ, defined in Figure 30.17. That is, $x = R \tan \phi$, so that we have $dx = R \sec^2 \phi\ d\phi$. Substituting these expressions into Equation 30.21 and integrating from ϕ_1 to ϕ_2, we get

$$B = \frac{\mu_0 NI}{2\ell} \int_{\phi_1}^{\phi_2} \cos \phi\ d\phi = \frac{\mu_0 NI}{2\ell} (\sin \phi_2 - \sin \phi_1) \qquad (30.22)$$

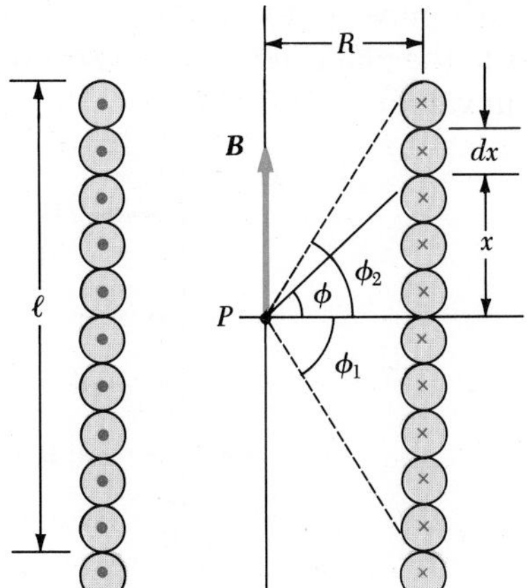

Figure 30.17 The geometry for calculating the magnetic field at an axial point P inside a tightly wound solenoid.

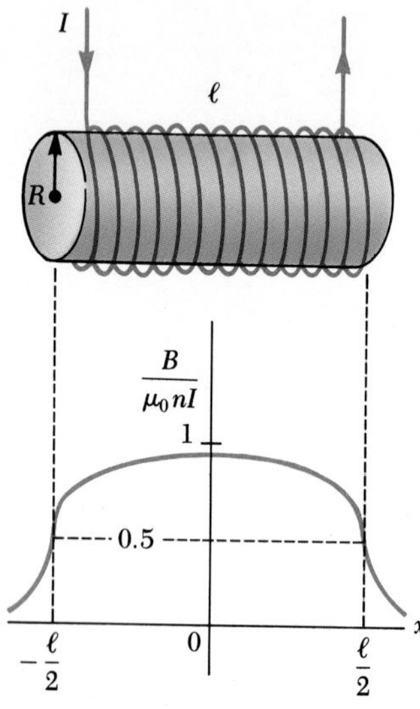

Figure 30.18 A sketch of the magnetic field along the axis versus x for a long, tightly wound solenoid. Note that the magnitude of the field at the ends is about one half the value of the center.

If P is at the *midpoint* of the solenoid and if we assume that the solenoid is long compared with R, then $\phi_2 \approx 90°$ and $\phi_1 \approx -90°$; therefore

$$B \approx \frac{\mu_0 NI}{2\ell}(1+1) = \frac{\mu_0 NI}{\ell} = \mu_0 nI \qquad \text{(at the center)}$$

which is in agreement with our previous result, Equation 30.20.

If P is a point at the end of a long solenoid (say, the bottom), then $\phi_1 \approx 0°$, $\phi_2 \approx 90°$, and

$$B \approx \frac{\mu_0 NI}{2\ell}(1+0) = \tfrac{1}{2}\mu_0 nI \qquad \text{(at the ends)}$$

This shows that the field at each end of a solenoid approaches *one half* the value at the solenoid's center as the length ℓ approaches infinity.

A sketch of the field at axial points versus x for a solenoid is shown in Figure 30.18. If the length ℓ is large compared with R, the axial field will be quite uniform over most of the solenoid and the curve will be quite flat except at points near the ends. On the other hand, if ℓ is comparable to R, then the field will have a value somewhat less than $\mu_0 nI$ at the middle and will be uniform only over a small region of the solenoid.

30.6 MAGNETIC FLUX

The flux associated with a magnetic field is defined in a manner similar to that used to define the electric flux. Consider an element of area dA on an arbitrarily shaped surface, as in Figure 30.19. If the magnetic field at this element is $\boldsymbol{B}$, then the magnetic flux through the element is $\boldsymbol{B}\cdot d\boldsymbol{A}$, where $d\boldsymbol{A}$ is a vector perpendicular to the surface whose magnitude equals the area dA. Hence, the total magnetic flux Φ_m through the surface is given by

Magnetic flux

$$\Phi_m = \int \boldsymbol{B}\cdot d\boldsymbol{A} \tag{30.23}$$

Consider the special case of a plane of area A and a uniform field $\boldsymbol{B}$, which makes an angle θ with the vector $d\boldsymbol{A}$. The magnetic flux through the plane in this case is given by

$$\Phi_m = BA\cos\theta \tag{30.24}$$

If the magnetic field lies in the plane as in Figure 30.20a, then $\theta = 90°$ and the flux is zero. If the field is perpendicular to the plane as in Figure 30.20b, then $\theta = 0°$ and the flux is BA (the maximum value).

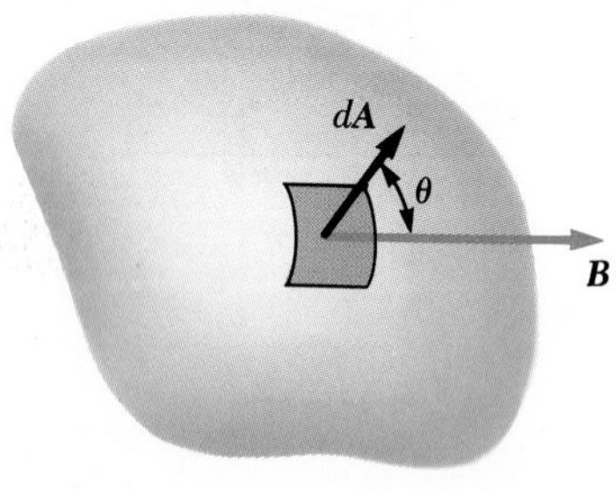

Figure 30.19 The magnetic flux through an area element dA is given by $\boldsymbol{B}\cdot d\boldsymbol{A} = BdA\cos\theta$. Note that $d\boldsymbol{A}$ is perpendicular to the surface.

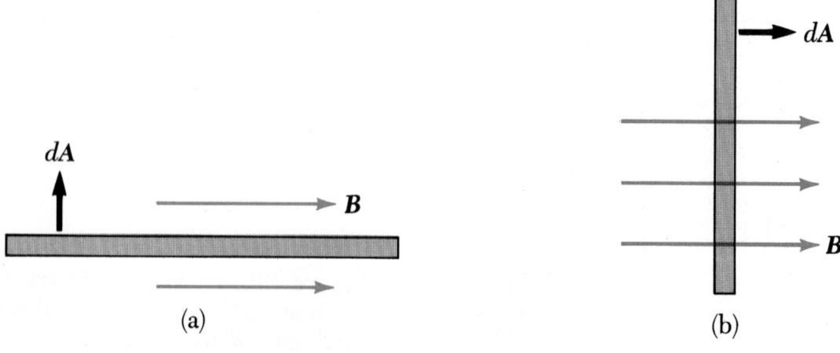

Figure 30.20 (a) The flux is zero when the magnetic field is parallel to the surface of the plane (an edge view). (b) The flux is a maximum when the magnetic field is perpendicular to the plane.

Since B has units of Wb/m^2, or T, the unit of flux is the weber (Wb), where $1\ Wb = 1\ T \cdot m^2$.

EXAMPLE 30.8 Flux Through a Rectangular Loop
A rectangular loop of width a and length b is located a distance c from a long wire carrying a current I (Fig. 30.21). The wire is parallel to the long side of the loop. Find the total magnetic flux through the loop.

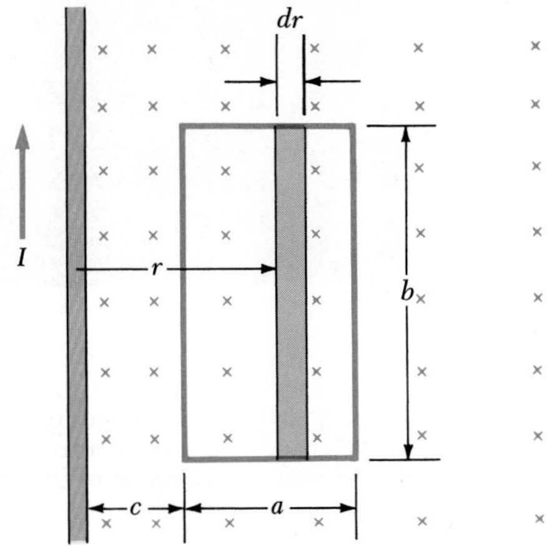

Figure 30.21 (Example 30.8) The magnetic field due to the wire carrying a current I is *not* uniform over the rectangular loop.

Solution From Ampère's law, we found that the magnetic field due to the wire at a distance r from the wire is given by

$$B = \frac{\mu_0 I}{2\pi r}$$

That is, the field *varies* over the loop and is directed *into* the page as shown in Figure 30.21. Since $\mathbf{B}$ is parallel to $d\mathbf{A}$, we can express the magnetic flux through an area element dA as

$$\Phi_m = \int B\, dA = \int \frac{\mu_0 I}{2\pi r}\, dA$$

Note that since $\mathbf{B}$ *is not uniform, but depends on r, it cannot be removed from the integral.* In order to carry out the integration, we first express the area element (the blue region in Fig. 30.21) as $dA = b\, dr$. Since r is the only variable that now appears in the integral, the expression for Φ_m becomes

$$\Phi_m = \frac{\mu_0 I}{2\pi} b \int_c^{a+c} \frac{dr}{r} = \frac{\mu_0 I b}{2\pi} \ln r \Big]_c^{a+c}$$

$$= \frac{\mu_0 I b}{2\pi} \ln\left(\frac{a+c}{c}\right)$$

30.7 GAUSS' LAW IN MAGNETISM

In Chapter 24 we found that the flux of the electric field through a closed surface surrounding a net charge is proportional to that charge (Gauss' law). In other words, the number of electric field lines leaving the surface depends only on the net charge within it. This property is based in part on the fact that electric field lines originate on electric charges.

The situation is quite different for magnetic fields, which are continuous and form closed loops. Magnetic field lines due to currents do not begin or end at any point. The magnetic field lines of the bar magnet in Figure 30.22 illustrate this point. Note that for any closed surface, the number of lines entering that surface equals the number leaving that surface, and so the net magnetic flux is *zero*. This is in contrast to the case of a surface surrounding one charge of an electric dipole (Fig. 30.23), where the net electric flux is not zero.

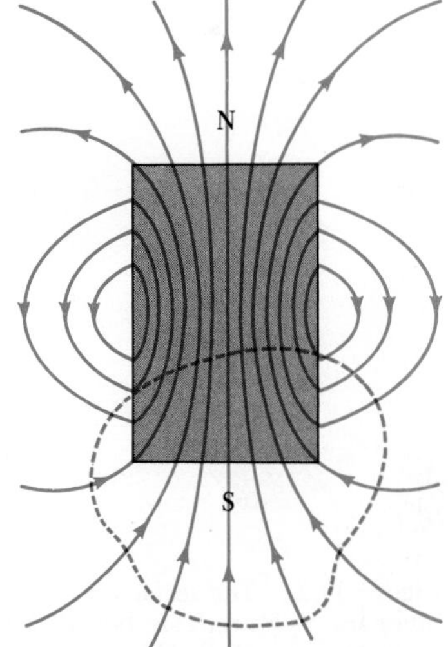

Figure 30.22 The magnetic field lines of a bar magnet form closed loops. Note that the net flux through the closed surface surrounding one of the poles (or any other closed surface) is zero.

Gauss' law in magnetism states that the net magnetic flux through any closed surface is always zero:

$$\oint \mathbf{B} \cdot d\mathbf{A} = 0 \qquad (30.25)$$

This statement is based on the experimental fact that *isolated magnetic poles (or monopoles) have not been detected, and perhaps do not even exist.* The only

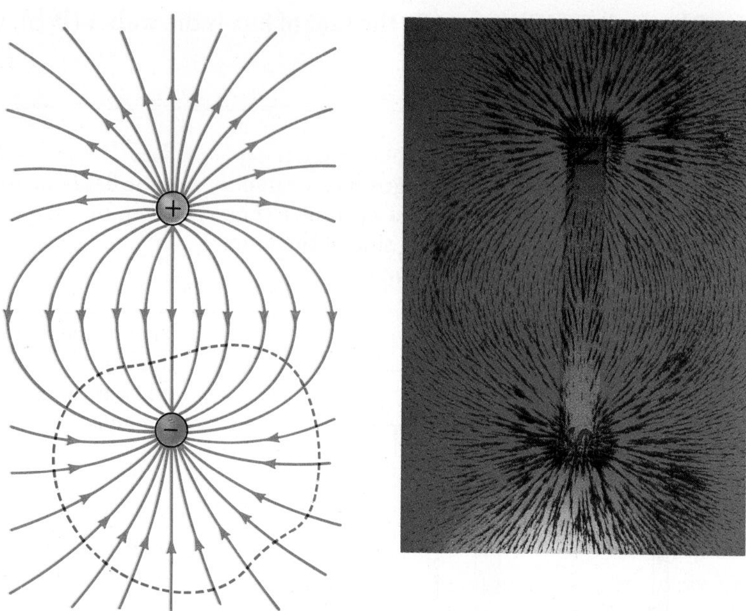

Figure 30.23 (a) The electric field lines of an electric dipole begin on the positive charge and terminate on the negative charge. The electric flux through a closed surface surrounding one of the charges is *not* zero. (b) Magnetic field pattern of a bar magnet. (Courtesy of Henry Leap and Jim Lehman)

known sources of magnetic fields are magnetic dipoles (current loops), even in magnetic materials. In fact, all magnetic effects in matter can be explained in terms of magnetic dipole moments (effective current loops) associated with electrons and nuclei. This will be discussed further in Section 30.9.

30.8 DISPLACEMENT CURRENT AND THE GENERALIZED AMPÈRE'S LAW

We have seen that charges in motion, or currents, produce magnetic fields. When a current-carrying conductor has high symmetry, we can calculate the magnetic field using Ampère's law, given by Equation 30.15:

$$\oint \boldsymbol{B} \cdot d\boldsymbol{s} = \mu_0 I$$

where the line integral is over *any closed path through which the conduction current passes.* If Q is the charge on the capacitor at any instant, the conduction current is defined by

$$I \equiv \frac{dQ}{dt}$$

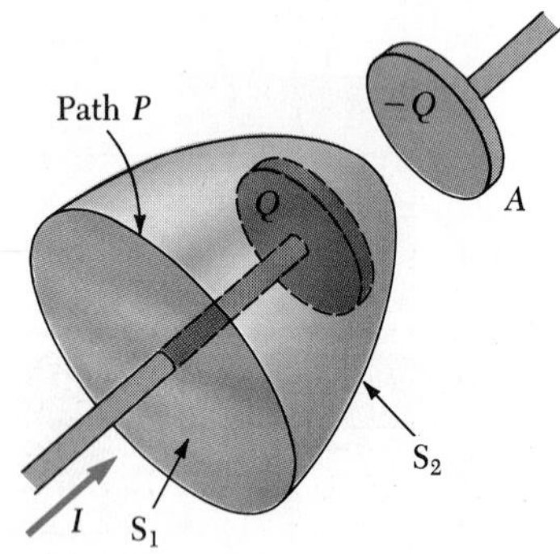

Figure 30.24 The surfaces S_1 (in blue) and S_2 (in red) are bounded by the same path P. The conduction current passes only through S_1. This leads to a contradictory situation in Ampère's law which is resolved only if one postulates a displacement current through S_2.

We shall now show that *Ampère's law in this form is valid only if the conduction is constant in time.* Maxwell recognized this limitation and modified Ampère's law to include all possible situations.

We can understand this problem by considering a capacitor being charged as in Figure 30.24. The argument given here is equivalent to Maxwell's original reasoning. When the current I changes with time (for example, when an ac voltage source is used), the charge on the plate changes, but *no conduction current passes between the plates*. Now consider the two surfaces S_1 and S_2 bounded by the same path P. Ampère's law says that the line integral of $\boldsymbol{B}\cdot d\boldsymbol{s}$ around this path must equal $\mu_0 I$, where I is the total current through any surface bounded by the path P.

When the path P is considered to bound S_1, the result of the integral is $\mu_0 I$ since the current passes through S_1. However when the path bounds S_2, the result is *zero* since no conduction current passes through S_2. Thus, we have a contradictory situation which arises from the discontinuity of the current! Maxwell solved this problem by postulating an additional term on the right side of Equation 30.15, called the **displacement current,** I_d, defined as

$$I_d \equiv \epsilon_0 \frac{d\Phi_e}{dt} \qquad (30.26)$$

Displacement current

Recall that Φ_e is the flux of the electric field, defined as $\Phi_e = \int \boldsymbol{E}\cdot d\boldsymbol{A}$.

As the capacitor is being charged (or discharged), the *changing* electric field between the plates may be thought of as a sort of current that bridges the discontinuity in the conduction current. When this expression for the current (Eq. 30.26) is added to the right side of Ampère's law, the difficulty represented by Figure 30.24 is resolved. No matter what surface bounded by the path P is chosen, some combination of conduction and displacement current will pass through it. With this new term I_d, we can express the generalized form of Ampère's law (sometimes called the **Ampère-Maxwell law**) as[3]

$$\oint \boldsymbol{B}\cdot d\boldsymbol{s} = \mu_0(I + I_d) = \mu_0 I + \mu_0\epsilon_0 \frac{d\Phi_e}{dt} \qquad (30.27)$$

Ampère-Maxwell law

The meaning of this expression can be understood by referring to Figure 30.25. The electric flux through S_2 is $\Phi_e = \int \boldsymbol{E}\cdot d\boldsymbol{A} = EA$, where A is the area of the plates and E is the uniform electric field strength between the plates. If Q is the charge on the plates at any instant, then one finds that $E = Q/\epsilon_0 A$ (Section 26.2). Therefore, the electric flux through S_2 is simply

$$\Phi_e = EA = \frac{Q}{\epsilon_0}$$

Hence, the displacement current I_d through S_2 is

$$I_d = \epsilon_0 \frac{d\Phi_e}{dt} = \frac{dQ}{dt} \qquad (30.28)$$

That is, the displacement current is precisely equal to the conduction current I passing through S_1!

The central point of this formalism is the fact that

magnetic fields are produced both by conduction currents and by changing electric fields.

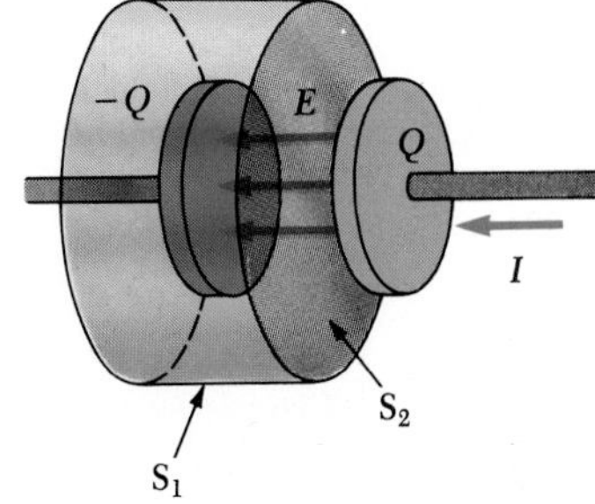

Figure 30.25 The conduction current $I = dQ/dt$ passes through S_1. The displacement current $I_d = \epsilon_0\, d\Phi_e/dt$ passes through S_2. The two currents must be equal for continuity. In general, the total current through any surface bounded by some path is $I + I_d$.

[3] Strictly speaking, this expression is valid only in a vacuum. If a magnetic material is present, one must also include a magnetizing current I_m on the right side of Equation 30.27 to make Ampère's law fully general. On a microscopic scale, I_m is a current that is as real as the conduction current I.

EXAMPLE 30.9 Displacement Current in a Capacitor
An ac voltage is applied directly across an 8-μF capacitor. The frequency of the source is 3 kHz, and the voltage amplitude is 30 V. Find the displacement current between the plates of the capacitor.

Solution The angular frequency of the source is given by $\omega = 2\pi f = 2\pi(3 \times 10^3 \text{ Hz}) = 6\pi \times 10^3 \text{ s}^{-1}$. Hence, the voltage across the capacitor in terms of t is

$$V = V_m \sin \omega t = (30 \text{ V}) \sin(6\pi \times 10^3 t)$$

We can make use of Equation 30.28 and of the fact that the charge on the capacitor is given by $Q = CV$ to find the displacement current:

$$I_d = \frac{dQ}{dt} = \frac{d}{dt}(CV) = C\frac{dV}{dt}$$

$$= (8 \times 10^{-6}) \frac{d}{dt} [30 \sin(6\pi \times 10^3 t)]$$

$$= (4.52 \text{ A}) \cos(6\pi \times 10^3 t)$$

Hence, the displacement current varies sinusoidally with time and has a *maximum* value of 4.52 A.

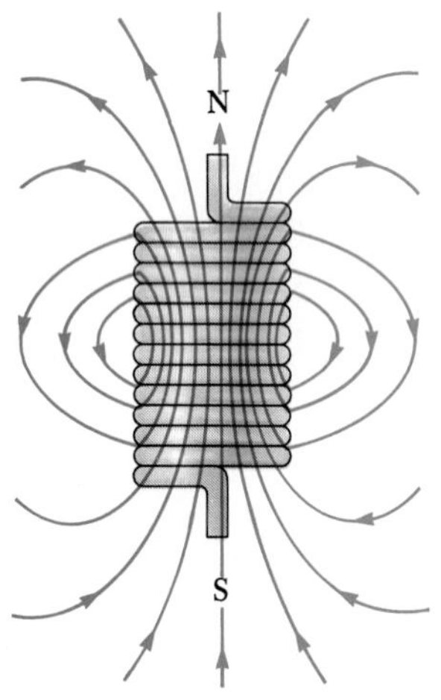

Figure 30.26 Magnetic field lines for a tightly wound solenoid of finite length carrying a steady current.

30.9 MAGNETISM IN MATTER

The magnetic field produced by a current in a coil of wire gives us a hint as to what might cause certain materials to exhibit strong magnetic properties. Earlier we found that a coil like that shown in Figure 30.26 has a north and a south pole. In general, any current loop has a magnetic field and a corresponding magnetic moment. Similarly, the magnetic moments in a magnetized substance are associated with internal atomic currents. One can view these currents as arising from electrons orbiting around the nucleus and protons orbiting about each other inside the nucleus.

We shall begin this section with a brief discussion of the magnetic moments due to electrons. As we shall see, the net magnetic moment of an electron is due to a combination of its orbital motion and an intrinsic property called *spin*. The mutual forces between these magnetic dipole moments and their interaction with an external magnetic field are of fundamental importance in understanding the behavior of magnetic materials. We shall describe three categories of materials, paramagnetic, ferromagnetic, and diamagnetic. **Paramagnetic** and **ferromagnetic** materials are those that have atoms with permanent magnetic dipole moments. **Diamagnetic** materials are those whose atoms have no permanent magnetic dipole moments. For materials whose atoms have permanent magnetic moments, the diamagnetic contribution to the magnetism is usually overshadowed by paramagnetic or ferromagnetic effects.

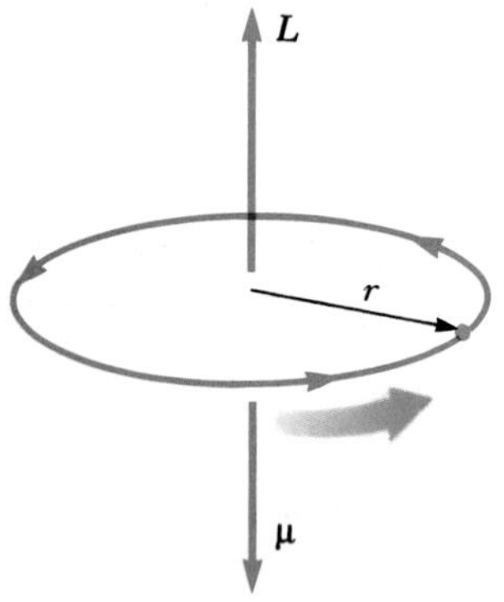

Figure 30.27 An electron moving in a circular orbit of radius r has an angular momentum $\mathbf{L}$ and a magnetic moment $\boldsymbol{\mu}$ which are in *opposite* directions.

The Magnetic Moments of Atoms

It is instructive to begin our discussion with a classical model of the atom in which electrons are assumed to move in circular orbits about the much more massive nucleus. In this model, an orbiting electron is viewed as a tiny current loop, and the atomic magnetic moment is associated with this orbital motion. Although this model has many deficiencies, its predictions are in good agreement with the correct theory from quantum physics.

Consider an electron moving with constant speed v in a circular orbit of radius r about the nucleus, as in Figure 30.27. Since the electron travels a distance of $2\pi r$ (the circumference of the circle) in a time T, where T is the time for one revolution, the orbital speed of the electron is $v = 2\pi r/T$. The effective

Oxygen, a paramagnetic substance, is attracted to a magnetic field. The liquid oxygen in this photograph is suspended between the poles of the magnet. (Courtesy of Leon Lewandowski)

current associated with this orbiting electron equals its charge divided by the time for one revolution:

$$I = \frac{e}{T} = \frac{e}{2\pi r/T} = \frac{ev}{2\pi r}$$

The magnetic moment associated with this effective current loop is given by $\mu = IA$, where $A = \pi r^2$ is the area of the orbit. Therefore,

Orbital magnetic moment

$$\mu = IA = \left(\frac{ev}{2\pi r}\right)\pi r^2 = \tfrac{1}{2}\, evr \tag{30.29}$$

Since the magnitude of the orbital angular momentum of the electron is given by $L = mvr$, the magnetic moment can be written as

$$\mu = \left(\frac{e}{2m}\right) L \tag{30.30}$$

This result says that *the magnetic moment of the electron is proportional to its orbital angular momentum.* Note that since the electron is negatively charged, the vectors $\boldsymbol{\mu}$ and $\mathbf{L}$ point in *opposite* directions. Both vectors are perpendicular to the plane of the orbit as indicated in Figure 30.27.

Angular momentum is quantized

A fundamental outcome of quantum physics is that the orbital angular momentum must be *quantized,* and is always some integer multiple of $\hbar = h/2\pi = 1.06 \times 10^{-34}$ J·s, where h is Planck's constant. That is

$$L = 0, \hbar, 2\hbar, 3\hbar, \ldots$$

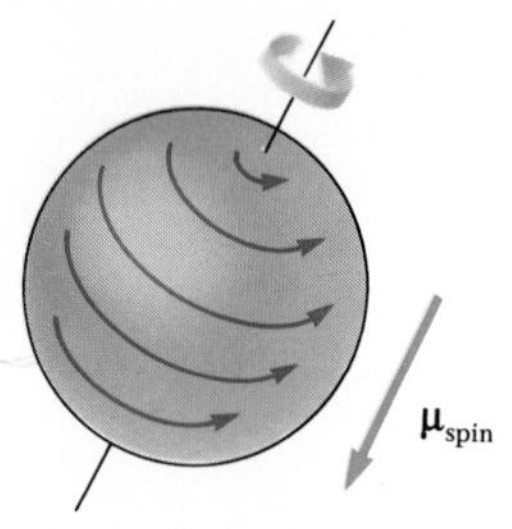

Figure 30.28 Model of a spinning electron. The magnetic moment μ_{spin} can be viewed as arising from the effective current loops associated with a spinning charged sphere.

Hence, the smallest nonzero value of the magnetic moment is

$$\mu = \frac{e}{2m}\hbar \qquad (30.31)$$

Since all substances contain electrons, you may wonder why all substances are not magnetic. The main reason is that in most substances, the magnetic moment of one electron in an atom is canceled by the moment of another electron in the atom orbiting in the opposite direction. The net result is that *the magnetic effect produced by the orbital motion of the electrons is either zero or very small for most materials.*

So far we have only considered the contribution to the magnetic moment of an atom from the orbital motion of the electron. However, an electron has another intrinsic property called *spin* which also contributes to the magnetic moment. In this regard, one can view the electron as a sphere of charge spinning about its axis as it orbits the nucleus, as in Figure 30.28. (This classical description of a spinning electron should not be taken literally. The property of spin can be understood only through a quantum mechanical model.) This spinning motion produces an effective current loop and hence a magnetic moment which is of the same order of magnitude as that due to the orbital motion. The magnitude of the spin angular momentum predicted by quantum theory is

Spin angular momentum

$$S = \frac{\hbar}{2} = 5.2729 \times 10^{-35}\ \text{J}\cdot\text{s}$$

The intrinsic magnetic moment associated with the spin of an electron has the value

Bohr magneton

$$\mu_B = \frac{e}{2m}\hbar = 9.27 \times 10^{-24}\ \text{J/T} \qquad (30.32)$$

which is called the **Bohr magneton.**

In atoms or ions containing many electrons, the electrons usually pair up with their spins *opposite* each other, which results in a cancellation of the spin magnetic moments. However, atoms with an *odd* number of electrons must have at least one "unpaired" electron and a corresponding spin magnetic moment. The total magnetic moment of an atom is the vector sum of the orbital and spin magnetic moments. The magnetic moments of a few atoms and ions are given in Table 30.1. Note that some atoms such as helium and neon have zero moments because their individual moments cancel.

The nucleus of an atom also has a magnetic moment associated with its constituent protons and neutrons. However, the magnetic moment of a proton or neutron is small compared to the magnetic moment of an electron and can usually be neglected. This can be understood by inspecting Equation 30.32. Since the masses of the proton and neutron are much greater than that of the electron, their magnetic moments are much smaller by a factor of about 10^3.

TABLE 30.1 Magnetic Moments of Some Atoms and Ions

Atom (or ion)	Magnetic moment (10^{-24} J/T)
H	9.27
He	0
Li	9.27
O	13.9
Ne	0
Ce^{3+}	19.8
Yb^{3+}	37.1

Magnetization and Magnetic Field Strength

The magnetic state of a substance is described by a quantity called the **magnetization vector, $\boldsymbol{M}$.** *The magnitude of the magnetization vector is equal to the magnetic moment per unit volume of the substance.* As you might expect, the total magnetic field in a substance depends on both the applied (external) field and the magnetization of the substance.

Magnetization

Consider a region where there exists a magnetic field $\boldsymbol{B}_0$ produced by a current-carrying conductor, such as the interior of a toroidal winding. If we now fill that region with a magnetic substance, the *total* field $\boldsymbol{B}$ in that region will be given by $\boldsymbol{B} = \boldsymbol{B}_0 + \boldsymbol{B}_{\text{m}}$ where $\boldsymbol{B}_{\text{m}}$ is the field produced by the magnetic substance. This contribution can be expressed in terms of the magnetization vector as $\boldsymbol{B}_{\text{m}} = \mu_0 \boldsymbol{M}$; hence the total field in the substance becomes

$$\boldsymbol{B} = \boldsymbol{B}_0 + \mu_0 \boldsymbol{M} \tag{30.33}$$

Magnetic field strength

It is convenient to introduce another field quantity $\boldsymbol{H}$, called the **magnetic field strength.** This vector quantity is defined by the relation $\boldsymbol{H} = (\boldsymbol{B}/\mu_0) - \boldsymbol{M}$, or

$$\boldsymbol{B} = \mu_0(\boldsymbol{H} + \boldsymbol{M}) \tag{30.34}$$

In SI units, the dimensions of both $\boldsymbol{H}$ and $\boldsymbol{M}$ are A/m.

To better understand these expressions, consider the region inside a toroidal coil which carries a current I. If the interior region is a vacuum, then $\boldsymbol{M} = 0$, and $\boldsymbol{B} = \boldsymbol{B}_0 = \mu_0 \boldsymbol{H}$. Since $B_0 = \mu_0 nI$ inside a toroid, where n is the number of turns per unit length in its windings, then $H = B_0/\mu_0 = \mu_0 nI/\mu_0$, or

$$H = nI \tag{30.35}$$

That is, the magnetic field strength inside the toroid is due to the current in the its windings.

If the toroid core is now filled with some substance, and the current I is kept constant, then $\boldsymbol{H}$ inside the substance will remain *unchanged,* with a magnitude nI. This is because the magnetic field strength $\boldsymbol{H}$ is due *solely* to the current in the coil. The total field $\boldsymbol{B}$, however, changes when the substance is introduced. From Equation 30.34, we see that part of $\boldsymbol{B}$ arises from the term $\mu_0 \boldsymbol{H}$ associated with the toroidal current; the second contribution to $\boldsymbol{B}$ is the term $\mu_0 \boldsymbol{M}$ due to the magnetization of the substance.

For a large class of substances, specifically paramagnetic and diamagnetic substances, the magnetization $\boldsymbol{M}$ is proportional to the magnetic field strength $\boldsymbol{H}$. In these linear substances, we can write

$$\boldsymbol{M} = \chi \boldsymbol{H} \tag{30.36}$$

Magnetic susceptibility

where χ is a dimensionless factor called the **magnetic susceptibility.** If the sample is paramagnetic, χ is positive, in which case $\boldsymbol{M}$ is in the same direction as $\boldsymbol{H}$. If the substance is diamagnetic, χ is negative, and $\boldsymbol{M}$ is opposite $\boldsymbol{H}$. It is important to note that *this linear relationship does not apply to ferromagnetic substances.* The susceptibilities of some substances are given in Table 30.2.

TABLE 30.2 Magnetic Susceptibilities of Some Paramagnetic and Diamagnetic Substances at 300 K

Paramagnetic Substance	χ	Diamagnetic Substance	χ
Aluminum	2.3×10^{-5}	Bismuth	-1.66×10^{-5}
Calcium	1.9×10^{-5}	Copper	-9.8×10^{-6}
Chromium	2.7×10^{-4}	Diamond	-2.2×10^{-5}
Lithium	2.1×10^{-5}	Gold	-3.6×10^{-5}
Magnesium	1.2×10^{-5}	Lead	-1.7×10^{-5}
Niobium	2.6×10^{-4}	Mercury	-2.9×10^{-5}
Oxygen (STP)	2.1×10^{-6}	Nitrogen (STP)	-5.0×10^{-9}
Platinum	2.9×10^{-4}	Silver	-2.6×10^{-5}
Tungsten	6.8×10^{-5}	Silicon	-4.2×10^{-6}

Substituting Equation 30.36 for $\boldsymbol{M}$ into Equation 30.34 gives

$$\boldsymbol{B} = \mu_0(\boldsymbol{H} + \boldsymbol{M}) = \mu_0(\boldsymbol{H} + \chi\boldsymbol{H}) = \mu_0(1 + \chi)\boldsymbol{H}$$

or

$$\boldsymbol{B} = \kappa_m \boldsymbol{H} \tag{30.37}$$

where the constant κ_m is called the **permeability** of the substance and has the value[4]

Permeability

$$\kappa_m = \mu_0(1 + \chi) \tag{30.38}$$

Substances may also be classified in terms of how their permeability κ_m compares to μ_0 (the permeability of free space) as follows

Paramagnetic	$\kappa_m > \mu_0$
Diamagnetic	$\kappa_m < \mu_0$
Ferromagnetic	$\kappa_m \gg \mu_0$

Since χ is very small for paramagnetic and diamagnetic substances (see Table 30.2), κ_m is nearly equal to μ_0 in these cases. For ferromagnetic substances, however, κ_m is typically several thousand times larger than μ_0. Although Equation 30.37 provides a simple relation between $\boldsymbol{B}$ and $\boldsymbol{H}$, it must be interpreted with care when dealing with ferromagnetic substances. As mentioned earlier, $\boldsymbol{M}$ is not a linear function of $\boldsymbol{H}$ for ferromagnetic substances such as iron, nickel, and cobalt. This is because the value of κ_m *is not a characteristic of the substance,* but depends on the previous state and treatment of the sample.

[4] The symbol μ is often used for permeability, but we have already used this same symbol for magnetic moment. For this reason, we use the symbol κ_m for permeability.

EXAMPLE 30.10 An Iron-Filled Toroid

A toroidal winding carrying a current of 5 A is wound with 60 turns/m of wire. The core is iron, which has a magnetic permeability of $5000\mu_0$ under the given conditions. Find H and B inside the iron core.

Solution Using Equations 30.35 and 30.37, we get

$$H = nI = \left(60\ \frac{\text{turns}}{\text{m}}\right)(5\text{ A}) = 300\ \frac{\text{A}\cdot\text{turns}}{\text{m}}$$

$$B = \kappa_m H = 5000\mu_0 H$$

$$= 5000\left(4\pi \times 10^{-7}\ \frac{\text{Wb}}{\text{A}\cdot\text{m}}\right)\left(300\ \frac{\text{A}\cdot\text{turns}}{\text{m}}\right)$$

$$= 1.88\ \text{T}$$

This value of B is 5000 times larger than the field in the absence of iron!

Exercise 4 Determine the magnitude and direction of the magnetization inside the iron core.
Answer $M = 1.5 \times 10^6$ A/m; $\boldsymbol{M}$ is in the direction of $\boldsymbol{H}$.

Ferromagnetism

Iron, cobalt, nickel, gadolinium, and dysprosium are strongly magnetic materials and are said to be ferromagnetic. Ferromagnetic substances are used to fabricate permanent magnets. Such substances contain atomic magnetic moments that tend to align parallel to each other even in a weak external magnetic field. Once the moments are aligned, the substance will remain magnetized after the external field is removed. This permanent alignment is due to a strong coupling between neighboring moments, which can only be understood in quantum mechanical terms.

All ferromagnetic materials contain microscopic regions called **domains,** within which all magnetic moments are aligned. These domains have volumes of about 10^{-12} to 10^{-8} m^3 and contain 10^{17} to 10^{21} atoms. The boundaries between the various domains having different orientations are called **domain walls.** In an unmagnetized sample, the domains are randomly oriented such that the net magnetic moment is zero as shown in Figure 30.29a. When the sample is placed in an external magnetic field, the domains tend to align with the field by rotating slightly, which results in a magnetized sample, as in Figure 30.29b. Observations show that domains initially oriented along the external field will grow in size at the expense of the less favorably oriented domains. When the external field is removed, the sample may retain a net magnetization in the direction of the original field.[5] At ordinary temperatures, thermal agitation is not sufficiently high to disrupt this preferred orientation of magnetic moments.

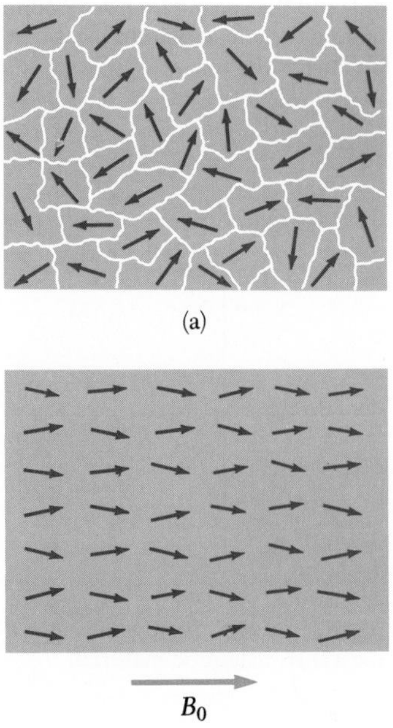

A typical experimental arrangement used to measure the magnetic properties of a ferromagnetic material consists of a toroid-shaped sample wound with N turns of wire, as in Figure 30.30. This configuration is sometimes referred to as the **Rowland ring.** A secondary coil connected to a galvanometer is used to measure the magnetic flux. The magnetic field $\boldsymbol{B}$ within the core of the toroid is measured by increasing the current in the toroid coil from zero to I. As the current changes, the magnetic flux through the secondary coil changes by BA, where A is the cross-sectional area of the toroid. Because of this changing flux, an emf is induced in the secondary coil that is proportional to the rate of change in magnetic flux. If the galvanometer in the secondary circuit is properly calibrated, one can obtain a value for $\boldsymbol{B}$ corresponding to any value of the current in the toroidal coil. The magnetic field $\boldsymbol{B}$ is measured first in the empty coil and then with the same coil filled with the magnetic substance. The magnetic properties of the substance are then obtained from a comparison of the two measurements.

(a)

B_0
(b)

Figure 30.29 (a) Random orientation of atomic magnetic dipoles in an unmagnetized substance. (b) When an external field $\boldsymbol{B}_0$ is applied, the atomic magnetic dipoles tend to align with the field, giving the sample a net magnetization $\boldsymbol{M}$.

[5] It is possible to observe the domain walls directly and follow their motion under a microscope. In this technique, a liquid suspension of finely powdered ferromagnetic substance is applied to the sample. The fine particles tend to accumulate at the domain walls and shift with them.

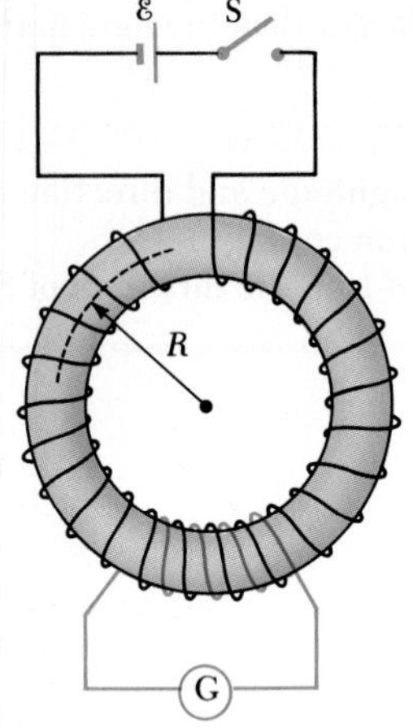

Figure 30.30 Cross section of a toroidal winding arrangement used to measure the magnetic properties of a substance. The material under study fills the core of the toroid, and the secondary circuit containing the galvanometer measures the magnetic flux.

Now consider a toroidal coil whose core consists of unmagnetized iron. If the current in the windings is increased from zero to some value I, the field intensity H increases linearly with I according to the expression $H = nI$. Furthermore, the total field B also increases with increasing current as shown in Figure 30.31. At point O, the domains are randomly oriented, corresponding to $B_m = 0$. As the external field increases, the domains become more aligned until all are nearly aligned at point a. At this point, the iron core is approaching saturation. (The condition of saturation corresponds to the case where all domains are aligned in the same direction.) Next, suppose the current is reduced to zero, thereby eliminating the external field. The B versus H curve, called a **magnetization curve,** now follows the path ab shown in Figure 30.31. Note that at point b, the field $\mathbf{B}$ is not zero, although the external field is $\mathbf{B}_0 = 0$. This is explained by the fact that the iron core is now magnetized due to the alignment of a large number of domains (that is, $\mathbf{B} = \mathbf{B}_m$). At this point, the iron is said to have a *remanent magnetization.* If the external field is reversed in direction and increased in strength by reversing the current, the domains reorient until the sample is again unmagnetized at point c, where $\mathbf{B} = 0$. A further increase in the reverse current causes the iron to be magnetized in the opposite direction, approaching saturation at point d. A similar sequence of events occurs as the current is reduced to zero and then increased in the original (positive) direction. In this case, the magnetization curve follows the path def. If the current is increased sufficiently, the magnetization curve returns to point a, where the sample again has its maximum magnetization.

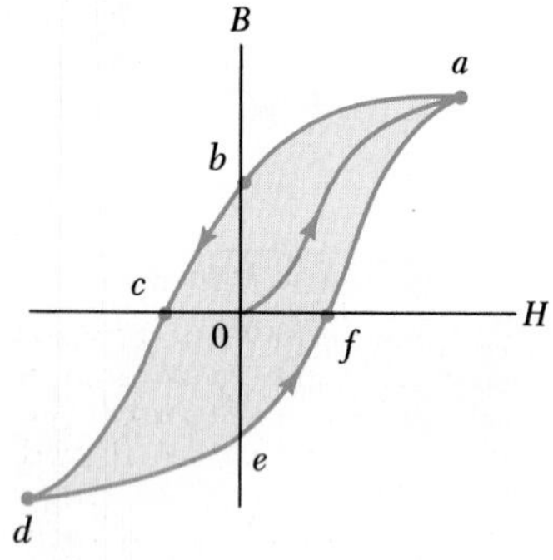

Figure 30.31 Hysteresis curve for a ferromagnetic material.

The effect just described, called **magnetic hysteresis,** shows that the magnetization of a ferromagnetic substance depends on the history of the substance as well as the strength of the applied field. (The word *hysteresis* literally means to "lag behind.") One often says that a ferromagnetic substance has a "memory" since it remains magnetized after the external field is removed. The closed loop in Figure 30.31 is referred to as a *hysteresis loop.* Its shape and size depend on the properties of the ferromagnetic substance and on the strength of the maximum applied field. The hysteresis loop for "hard" ferromagnetic materials (used in permanent magnets) is characteristically wide as in Figure 30.32a, corresponding to a large remanent magnetization. Such materials cannot be easily demagnetized by an external field. This is in contrast to "soft" ferromagnetic materials, such as iron, that have a very narrow

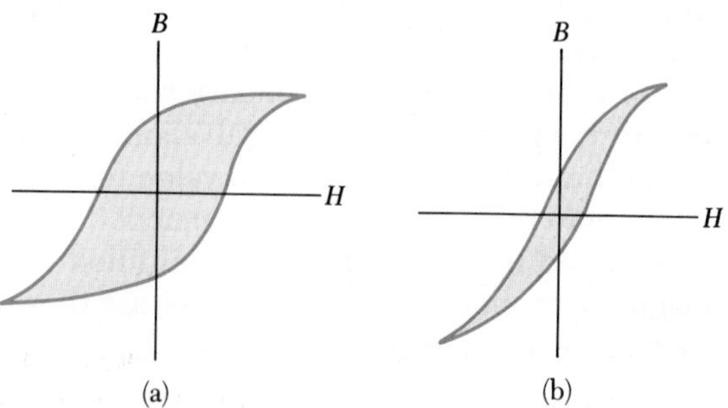

Figure 30.32 Hysteresis curves for (a) a hard ferromagnetic material and (b) a soft ferromagnetic material.

hysteresis loop and a small remanent magnetization (Fig. 30.32b.) Such materials are easily magnetized and demagnetized. An ideal soft ferromagnet would exhibit no hysteresis and hence would have no remanent magnetization. A ferromagnetic substance can be demagnetized by carrying the substance through successive hysteresis loops, gradually decreasing the applied field as in Figure 30.33.

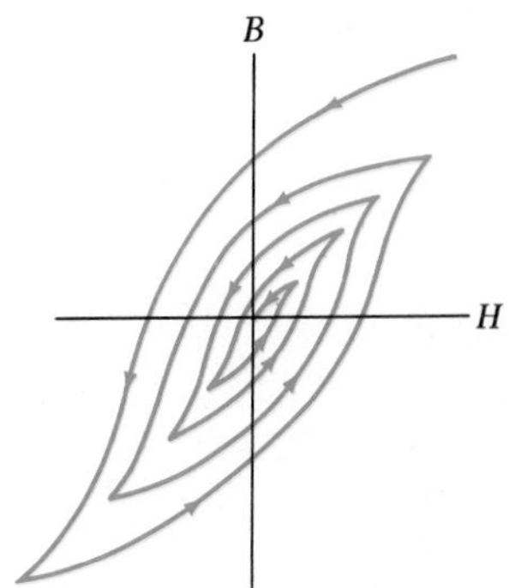

Figure 30.33 Demagnetizing a ferromagnetic material by carrying it through successive hysteresis loops.

The magnetization curve is useful for another reason. *The area enclosed by the magnetization curve represents the work required to take the material through the hysteresis cycle.* The energy acquired by the sample in the magnetization process originates from the source of the external field, that is, the emf in the circuit of the toroidal coil. When the magnetization cycle is repeated, dissipative processes within the material due to realignment of the domains result in a transformation of magnetic energy into internal thermal energy, which raises the temperature of the substance. For this reason, devices subjected to alternating fields (such as transformers) use cores made of soft ferromagnetic substances, which have narrow hysteresis loops and a correspondingly small energy loss per cycle.

Paramagnetism

Paramagnetic substances have a positive but small susceptibility ($0 < \chi \ll 1$), which is due to the presence of atoms (or ions) with *permanent* magnetic dipole moments. These dipoles interact only weakly with each other and are randomly oriented in the absence of an external magnetic field. When the substance is placed in an external magnetic field, its atomic dipoles tend to line up with the field. However, this alignment process must compete with the effects of thermal motion, which tends to randomize the dipole orientations.

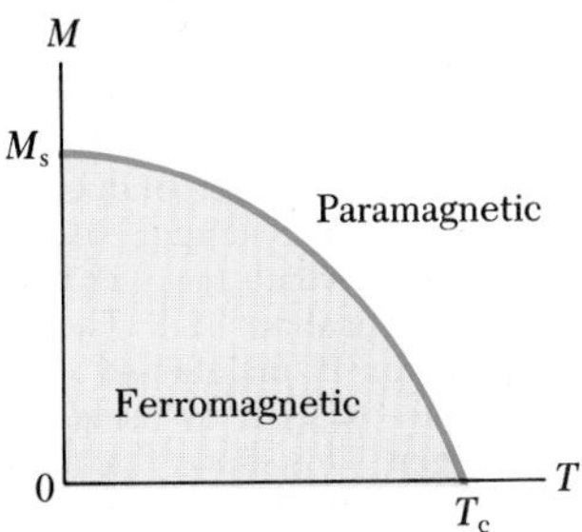

Figure 30.34 Plot of the magnetization versus absolute temperature for a ferromagnetic substance. The magnetic moments are aligned (ordered) below the Curie temperature T_c, where the substance is ferromagnetic. The substance becomes paramagnetic (disordered) above T_c.

Experimentally, one finds that the magnetization of a paramagnetic substance is proportional to the applied field and inversely proportional to the absolute temperature under a wide range of conditions. That is,

$$M = C\frac{B}{T} \tag{30.39}$$

This is known as **Curie's law** after its discoverer Pierre Curie (1859–1906), and the constant C is called **Curie's constant.** This shows that the magnetization increases with increasing applied field and with decreasing temperature. When $B = 0$, the magnetization is zero, corresponding to a random orientation of dipoles. At very high fields or very low temperatures, the magnetization approaches its maximum, or saturation, value corresponding to a complete alignment of its dipoles and Equation 30.39 is no longer valid.

It is interesting to note when the temperature of a ferromagnetic substance reaches or exceeds a critical temperature, called the **Curie temperature,** the substance loses its spontaneous magnetization and becomes paramagnetic (see Fig. 30.34). Below the Curie temperature, the magnetic moments are aligned and the substance is ferromagnetic. Above the Curie temperature, the thermal energy is large enough to cause a random orientation of dipoles, hence the substance becomes paramagnetic. For example, the Curie temperature for iron is 1043 K. A list of Curie temperatures for several ferromagnetic substances is given in Table 30.3.

TABLE 30.3 Curie Temperature for Several Ferromagnetic Substances

Substance	T_c (K)
Iron	1043
Cobalt	1394
Nickel	631
Gadolinium	317
Fe_2O_3	893

Diamagnetism

A diamagnetic substance is one whose atoms have no permanent magnetic dipole moment. When an external magnetic field is applied to a diamagnetic substance such as bismuth or silver, a weak magnetic dipole moment is *induced* in the direction opposite the applied field. Although the effect of diamagnetism is present in all matter, it is weak compared to paramagnetism or ferromagnetism.

Magnetic levitation of a permanent magnet above a superconductor. A small permanent magnet "floats" above a disk of superconducting yttrium barium copper oxide ($YBa_2Cu_3O_7$), which has a critical temperature of about 92 K. The superconductor is maintained at a temperature of 77 K by using liquid nitrogen. This shows that a superconductor is diamagnetic; that is, it expels magnetic fields. (Courtesy of Argonne National Laboratory)

We can obtain some understanding of diamagnetism by considering two electrons of an atom orbiting the nucleus in opposite directions but with the same speed. The electrons remain in these circular orbits because of the attractive electrostatic force (the centripetal force) of the positively charged nucleus. Because the magnetic moments of the two electrons are equal in magnitude and opposite in direction, they cancel each other and the dipole moment of the atom is zero. When an external magnetic field is applied, the electrons experience an additional force $q\boldsymbol{v} \times \boldsymbol{B}$. This added force modifies the centripetal force so as to increase the orbital speed of the electron whose magnetic moment is antiparallel to the field and decreases the speed of the electron whose magnetic moment is parallel to the field. As a result, the magnetic moments of the electrons no longer cancel, and the substance acquires a net dipole moment that opposes the applied field.

As you recall from Chapter 27, superconductors are substances whose dc resistance is *zero* below some critical temperature characteristic of the substance. Certain types of superconductors also exhibit *perfect diamagnetism* in the superconducting state. As a result, an applied magnetic field is *expelled* by the superconductor so that the field is *zero* in its interior. This phenomenon of flux expulsion is known as the **Meissner effect.** If a permanent magnet is brought near a superconductor, the two substances will *repel* each other. This is illustrated in the photograph, which shows a small permanent magnet levitated above a superconductor maintained at 77 K. A more detailed description of the unusual properties of superconductors is presented in Chapter 44 of the extended version of this text.

EXAMPLE 30.11 Saturation Magnetization

Estimate the *maximum magnetization* in a long cylinder of iron, assuming there is one unpaired electron spin per atom.

Solution The maximum magnetization, called the *saturation magnetization,* is obtained when all the magnetic moments in the sample are aligned. If the sample contains n atoms per unit volume, then the saturation magnetization M_s has the value

$$M_s = n\mu$$

where μ is the magnetic moment per atom. Since the molecular weight of iron is 55 g/mole and its density is 7.9 g/cm^3, the value of n is 8.5×10^{28} atoms/m^3. Assuming each atom contributes one Bohr magneton (due to one unpaired spin) to the magnetic moment, we get

$$M_s = \left(8.5 \times 10^{28} \frac{\text{atoms}}{\text{m}^3}\right)\left(9.27 \times 10^{-24} \frac{\text{A}\cdot\text{m}^2}{\text{atom}}\right)$$

$$= 7.9 \times 10^5 \text{ A/m}$$

This is about one half the experimentally determined saturation magnetization for annealed iron, which indicates that there are actually *two* unpaired electron spins per atom.

*30.10 MAGNETIC FIELD OF THE EARTH

When we speak of a small bar magnet's having a north and a south pole, we should more properly say that it has a "north-seeking" and a "south-seeking" pole. By this we mean that if such a magnet is used as a compass, one end will seek, or point to, the north geographic pole of the earth. Thus, we conclude that *a north magnetic pole is located near the south geographic pole, and a south magnetic pole is located near the north geographic pole.* In fact, the configuration of the earth's magnetic field, pictured in Figure 30.35, is very much like that which would be achieved by burying a bar magnet deep in the interior of the earth.

If a compass needle is suspended in bearings that allow it to rotate in the vertical plane as well as in the horizontal plane, the needle is horizontal with respect to the earth's surface only near the equator. As the device is moved northward, the needle rotates such that it points more and more toward the surface of the earth. Finally, at a point just north of Hudson Bay in Canada, the north pole of the needle would point directly downward. This location, first found in 1832, is considered to be the location of the south-seeking magnetic pole of the earth. This site is approximately 1300 mi from the earth's geographic north pole and varies with time. Similarly, the north-seeking magnetic pole of the earth is about 1200 miles away from the earth's geographic south pole. Thus, it is only approximately correct to say that a compass needle points north. The difference between true north, defined as the geographic north pole, and north indicated by a compass varies from point to point on the earth, and the difference is referred to as *magnetic declination.* For example, along a line through Florida and the Great Lakes, a compass indicates true north, whereas in Washington state, it aligns 25° east of true north.

Although the magnetic field pattern of the earth is similar to that which would be set up by a bar magnet deep within the earth, it is easy to understand why the source of the earth's field cannot be large masses of permanently magnetized material. The earth does have large deposits of iron ore deep beneath its surface, but the high temperatures in the earth's core prevent the iron from retaining any permanent magnetization. It is considered more likely that the true source is charge-carrying convection currents in the earth's core.

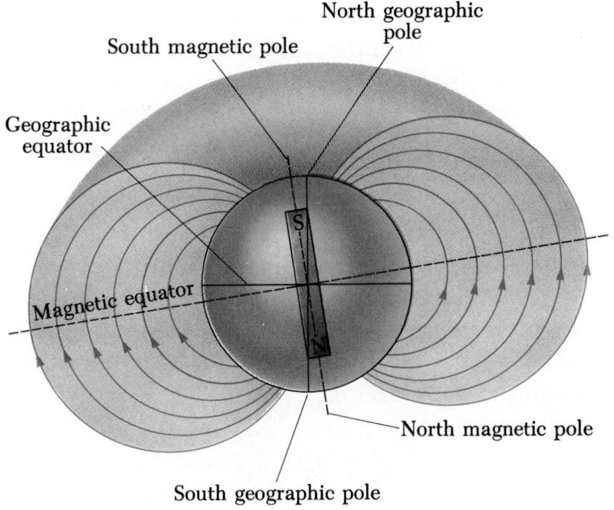

Figure 30.35 The earth's magnetic field lines. Note that a magnetic south pole is at the north geographic pole and a magnetic north pole is at the south geographic pole.

Charged ions or electrons circling in the liquid interior could produce a magnetic field, just as a current in a loop of wire produces a magnetic field. There is also strong evidence to indicate that the strength of a planet's field is related to the planet's rate of rotation. For example, Jupiter rotates faster than the earth, and recent space probes indicate that Jupiter's magnetic field is stronger than ours. Venus, on the other hand, rotates more slowly than the earth, and its magnetic field is found to be weaker. Investigation into the cause of the earth's magnetism remains open.

There is an interesting sidelight concerning the earth's magnetic field. It has been found that the direction of the field has been reversed several times during the last million years. Evidence for this is provided by basalt (a type of rock that contains iron) that is spewed forth by volcanic activity on the ocean floor. As the lava cools, it solidifies and retains a picture of the earth's magnetic field direction. The rocks can be dated by other means to provide the evidence for these periodic reversals of the magnetic field.

SUMMARY

The **Biot-Savart law** says that the magnetic field $d\mathbf{B}$ at a point P due to a current element $d\mathbf{s}$ carrying a steady current I is

Biot-Savart law

$$d\mathbf{B} = k_m \frac{I\, d\mathbf{s} \times \hat{\mathbf{r}}}{r^2} \qquad (30.1)$$

where $k_m = 10^{-7}$ Wb/A·m and r is the distance from the element to the point P. To find the total field at P due to a current-carrying conductor, we must integrate this vector expression over the entire conductor.

The **magnetic field** at a distance a from a long, straight wire carrying a current I is given by

Magnetic field of an infinitely long wire

$$B = \frac{\mu_0 I}{2\pi a} \qquad (30.7)$$

where $\mu_0 = 4\pi \times 10^{-7}$ Wb/A·m is the **permeability of free space.** The field lines are circles concentric with the wire.

The force per unit length between two parallel wires separated by a distance a and carrying currents I_1 and I_2 has a magnitude given by

Force per unit length between two wires

$$\frac{F}{\ell} = \frac{\mu_0 I_1 I_2}{2\pi a} \qquad (30.13)$$

The force is attractive if the currents are in the same direction and repulsive if they are in opposite directions.

Ampère's law says that the line integral of $\mathbf{B} \cdot d\mathbf{s}$ around any closed path equals $\mu_0 I$, where I is the total steady current passing through any surface bounded by the closed path. That is,

Ampère's law

$$\oint \mathbf{B} \cdot d\mathbf{s} = \mu_0 I \qquad (30.15)$$

Using Ampère's law, one finds that the fields inside a toroid and solenoid are given by

Magnetic field inside a toroid

$$B = \frac{\mu_0 NI}{2\pi r} \tag{30.18}$$

Magnetic field inside a solenoid

$$B = \mu_0 \frac{N}{\ell} I = \mu_0 nI \tag{30.20}$$

where N is the total number of turns.

The **magnetic flux** Φ_m through a surface is defined by the surface integral

Magnetic flux

$$\Phi_m \equiv \int \mathbf{B} \cdot d\mathbf{A} \tag{30.23}$$

Gauss' law of magnetism states that the net magnetic flux through any closed surface is zero. That is, isolated magnetic poles (or magnetic monopoles) do not exist.

A displacement current I_d arises from a time-varying electric flux and is defined by

Displacement current

$$I_d \equiv \epsilon_0 \frac{d\Phi_e}{dt} \tag{30.26}$$

The **generalized form of Ampère's law,** which includes the displacement current, is given by

Ampère-Maxwell law

$$\oint \mathbf{B} \cdot d\mathbf{s} = \mu_0 I + \mu_0 \epsilon_0 \frac{d\Phi_e}{dt} \tag{30.27}$$

This law describes the fact that magnetic fields are produced both by conduction currents and by changing electric fields.

The fundamental sources of all magnetic fields are the magnetic dipole moments associated with atoms. The atomic dipole moments can arise both from the orbital motions of the electrons and from an intrinsic property of electrons known as *spin*.

The magnetic properties of substances can be described in terms of their response to an external field. In a broad sense, materials can be described as being *ferromagnetic, paramagnetic,* or *diamagnetic*. The atoms of **ferromagnetic** and **paramagnetic** materials have permanent magnetic moments. **Diamagnetic** materials consist of atoms with no permanent magnetic moments.

When a paramagnetic or ferromagnetic material is placed in an external magnetic field, its dipoles tend to align parallel to the field, and this aligning in turn increases the net field. The increase in the field is quite small in the case of paramagnetic substances. This is because the magnetic dipoles in paramagnetic materials are randomly oriented in the absence of a magnetic field. The dipoles are partially aligned in the presence of an applied field.

QUESTIONS

1. Is the magnetic field due to a current loop uniform? Explain.
2. A current in a conductor produces a magnetic field which can be calculated using the Biot-Savart law. Since current is defined as the rate of flow of charge, what can you conclude about the magnetic field due to stationary charges? What about moving charges?
3. Two parallel wires carry currents in opposite directions. Describe the nature of the resultant magnetic field due to the two wires at points (a) between the wires and (b) outside the wires in a plane containing the wires.
4. Explain why two parallel wires carrying currents in opposite directions repel each other.
5. Two wires carrying equal and opposite currents are twisted together in the construction of a circuit. Why does this technique reduce stray magnetic fields?
6. Is Ampère's law valid for all closed paths surrounding a conductor? Why is it not useful for calculating $\boldsymbol{B}$ for all such paths?
7. Compare Ampère's law with the Biot-Savart law. Which is the more general method for calculating $\boldsymbol{B}$ for a current-carrying conductor?
8. Is the magnetic field inside a toroidal coil uniform? Explain.
9. Describe the similarities between Ampère's law in magnetism and Gauss' law in electrostatics.
10. A hollow copper tube carries a current. Why is $\boldsymbol{B} = 0$ inside the tube? Is $\boldsymbol{B}$ nonzero outside the tube?
11. Why is $\boldsymbol{B}$ nonzero outside a solenoid? Why is $\boldsymbol{B} = 0$ outside a toroid? (The lines of $\boldsymbol{B}$ must form closed paths.)
12. Describe the change in the magnetic field inside a solenoid carrying a steady current I if (a) the length of the solenoid is doubled, but the number of turns remains the same and (b) the number of turns is doubled, but the length remains the same.
13. A plane conducting loop is located in a uniform magnetic field that is directed along the x axis. For what orientation of the loop is the flux through it a maximum? For what orientation is the flux a minimum?
14. What new concept did Maxwell's generalized form of Ampère's circuital law include?
15. A magnet attracts a piece of iron. The iron can then attract another piece of iron. On the basis of alignment of the domains, explain what happens in each piece of iron.
16. You are an astronaut stranded on a planet with no test equipment or minerals around. The planet does not even have a magnetic field. You have two bars of iron in your possession; one is magnetized, one is not. How could you determine which is magnetized?
17. Why will hitting a magnet with a hammer cause its magnetism to be reduced?
18. Will a nail be attracted to either pole of a magnet? Explain what is happening inside the nail.
19. The north-seeking pole of a magnet is attracted toward the geographic north pole of the earth. Yet, like poles repel. What is the way out of this dilemma?
20. A Hindu ruler once suggested that he be entombed in a magnetic coffin with the polarity arranged such that he would be forever suspended between heaven and earth. Is such magnetic levitation possible? Discuss.
21. Why is $\boldsymbol{M} = 0$ in a vacuum? What is the relationship between $\boldsymbol{B}$ and $\boldsymbol{H}$ in a vacuum?
22. Explain why some atoms have permanent magnetic dipole moments and others do not.
23. What factors can contribute to the total magnetic dipole moment of an atom?
24. Why is the susceptibility of a diamagnetic substance negative?
25. Why can the effect of diamagnetism be neglected in a paramagnetic substance?
26. Explain the significance of the Curie temperature for a ferromagnetic substance.
27. Discuss the difference between ferromagnetic, paramagnetic, and diamagnetic substances.
28. What is the difference between hard and soft ferromagnetic materials?
29. Should the surface of a computer disk be make from a "hard" or a "soft" ferromagnetic substance?
30. Explain why it is desirable to use hard ferromagnetic materials to make permanent magnets.
31. Why is an ordinary, unmagnetized steel nail attracted to a permanent magnet?
32. Would you expect the tape from a tape recorder to be attracted to a magnet? (Try it, but not with a recording you wish to save.)
33. Given only a strong magnet and a screwdriver, (a) how would you magnetize the screwdriver, and (b) how would you then demagnetize the screwdriver?
34. The photograph below shows two permanent magnets with holes through their centers. Note that the

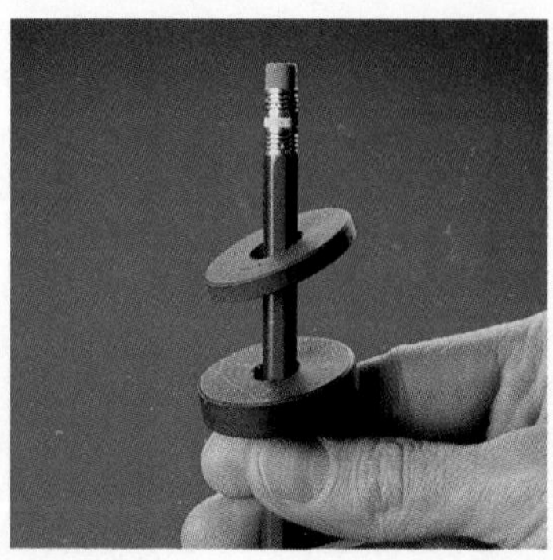

Question 34. Magnetic levitation using two ceramic magnets. (Courtesy of CENCO)

upper magnet is levitated above the lower magnet. (a) How does this occur? (b) What purpose does the pencil serve? (c) What can you say about the poles of the magnets from this observation? (d) If the upper magnet were inverted, what do you suppose would happen?

PROBLEMS

Section 30.1 The Biot-Savart Law

1. Calculate the magnitude of the magnetic field at a point 100 cm from a long, thin conductor carrying a current of 1 A.
2. A long, thin conductor carries a current of 10 A. At what distance from the conductor is the magnitude of the resulting magnetic field equal to 10^{-4} T?
3. A wire in which there is a current of 5 A is to be formed into a circular loop of one turn. If the required value of the magnetic field at the *center* of the loop is 10 μT, what is the required radius of the loop?
4. In Neils Bohr's 1913 model of the hydrogen atom, an electron circles the proton at a distance of 5.3×10^{-11} m with a speed of 2.2×10^{6} m/s. Compute the magnetic field strength produced by the electron's motion at the location of the proton.
5. A conductor in the shape of a square of edge length $\ell = 0.4$ m carries a current $I = 10$ A (Fig. 30.36). Calculate the magnitude and direction of the magnetic field produced at the *center* of the square.

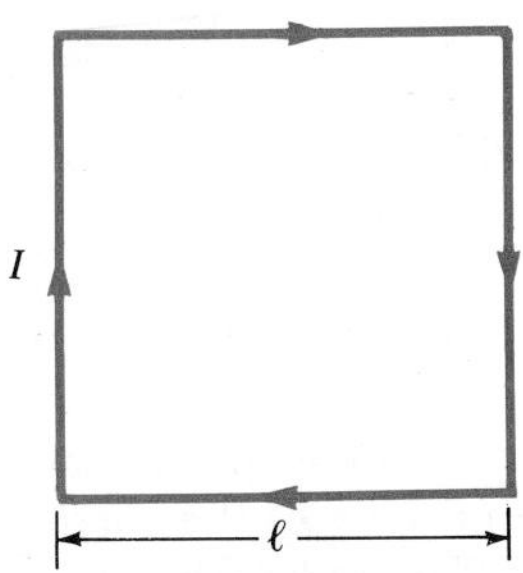

Figure 30.36 (Problems 5 and 7).

6. A 12 cm $\times$ 16 cm rectangular loop of superconducting wire carries a current of 30 A. What is the magnetic field at the center of the loop?
7. If the total length of the conductor in Problem 5 is formed into a single *circular* turn with the *same* current, what is the value of the magnetic field at the center of the turn?
8. How many turns should be in a flat circular coil of radius 0.1 m in order for a current of 10 A to produce a magnetic field of 3×10^{-3} T at its center?
9. Determine the magnetic field at a point P that is a distance x from the corner of an infinitely long wire that is bent at a right angle, as shown in Figure 30.37. The wire carries a steady current I.

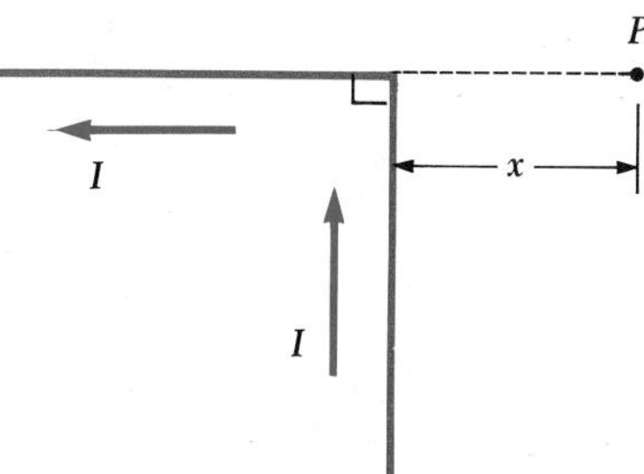

Figure 30.37 (Problem 9).

10. A segment of wire of total length $4r$ is formed into a shape as shown in Figure 30.38 and carries a current $I = 6$ A. Find the magnitude and direction of the magnetic field at point P when $r = 2\pi$ cm.

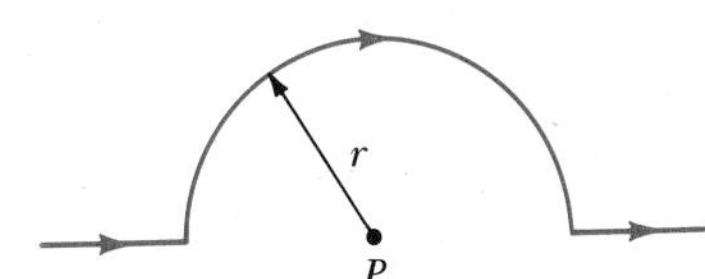

Figure 30.38 (Problem 10).

11. A closed current path shaped as shown in Figure 30.39 produces a magnetic field at P, the center of the arc. If the arc subtends an angle of 30° and the total length of wire in the closed path is 1.2 m, what are the magnitude and direction of the field produced at P if the current in the loop is 3 A?

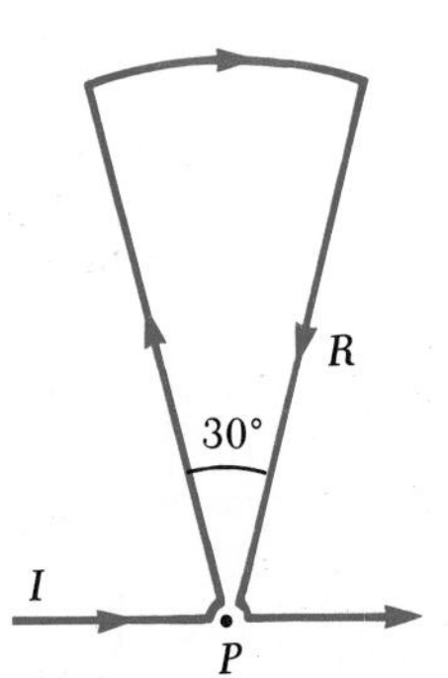

Figure 30.39 (Problem 11).

12. Use the Biot-Savart law to calculate the magnitude and direction of the magnetic field at a point P located at the center of concentric semicircles of radii $a = 5$ cm and $b = 8$ cm (Fig. 30.40) when a current $I = 2$ A is maintained in the loop circuit.

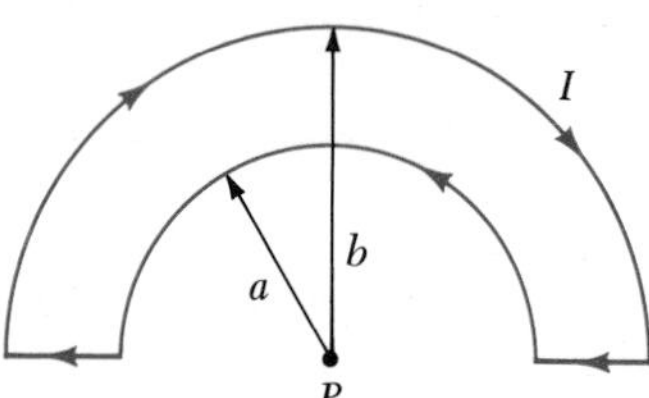

Figure 30.40 (Problem 12).

13. Consider a flat circular coil of radius R and carrying a current I. If the coil lies in the yz plane with its center at the origin, at what x coordinates (expressed as multiples of R) will the magnitude of the magnetic field be (a) one-half of that at the origin? (b) one-tenth of that at the origin?

14. Recalling that the current density $J = nqv_d$ (Eq. 27.6), show that the Biot-Savart law can be written

$$d\mathbf{B} = \frac{\mu_0}{4\pi}\frac{q\mathbf{v}_d \times \hat{\mathbf{r}}}{r^2}\, n\, dV$$

where dV is the volume element of the conductor and the drift velocity $\mathbf{v}_d$ is as defined in Chapter 27.

Section 30.2 The Magnetic Force Between Two Parallel Conductors

15. Two long parallel conductors, separated by a distance $a = 10$ cm, carry currents in the same direction (Fig. 30.41). If $I_1 = 5$ A and $I_2 = 8$ A, what is the force per unit length exerted on each conductor by the other?

16. For the arrangement of parallel conductors described in Problem 15 and shown in Figure 30.41, calculate the magnitude and direction of the magnetic field at point P, located 2 cm to the left of the conductor carrying current I_2.

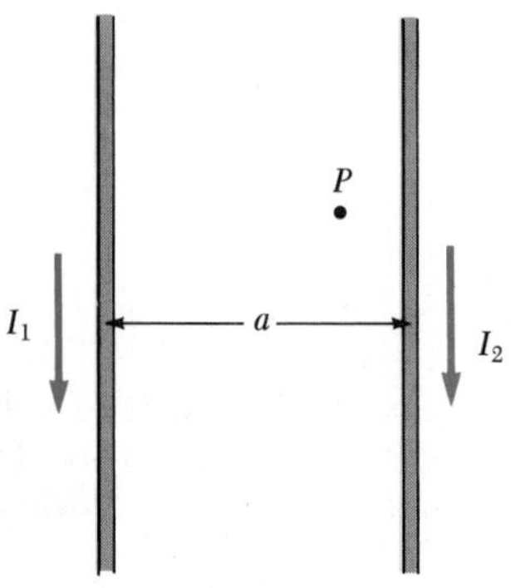

Figure 30.41 (Problems 15 and 16).

17. Two parallel conductors are each 0.5 m long and carry 10 A currents in opposite directions. (a) What center-to-center separation must the conductors have if they are to repel each other with a force of 1.0 N? (b) Is this physically possible?

18. Compute the magnetic force per unit length between two adjacent windings of a solenoid if each carries a current $I = 100$ A, and the center-to-center distance between the wires is 4 mm.

19. For the arrangement shown in Figure 30.42, the current in the long, straight conductor has the value $I_1 = 5$ A and lies in the plane of the rectangular loop, which carries a current $I_2 = 10$ A. The dimensions are $c = 0.1$ m, $a = 0.15$ m, and $\ell = 0.45$ m. Find the magnitude and direction of the *net force* exerted on the rectangle by the magnetic field of the straight current-carrying conductor.

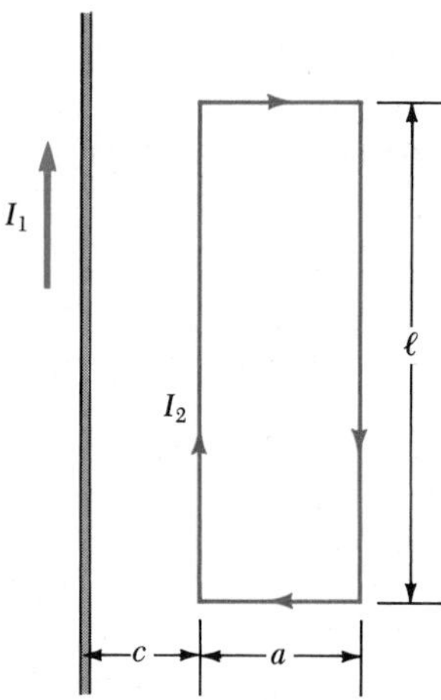

Figure 30.42 (Problem 19).

20. Four long, parallel conductors carry equal currents $I = 5$ A. An end view of the conductors is shown in Figure 30.43. The current direction is out of the page at points A and B (indicated by the dots) and into the page at points C and D (indicated by the crosses). Calculate the magnitude and direction of the magnetic field at point P, located at the center of the square of edge length 0.2 m.

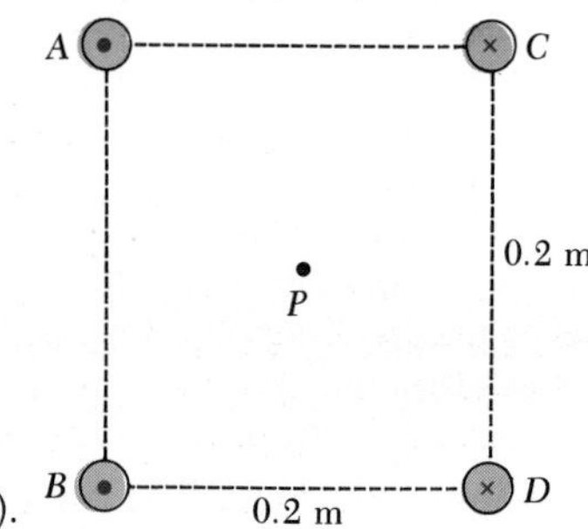

Figure 30.43 (Problem 20).

21. Two long, parallel conductors carry currents $I_1 = 3$ A and $I_2 = 3$ A, both directed into the page in Figure 30.44. The conductors are separated by a distance of 13 cm. Determine the magnitude and direction of the resultant magnetic field at point P, located 5 cm from I_1 and 12 cm from I_2.

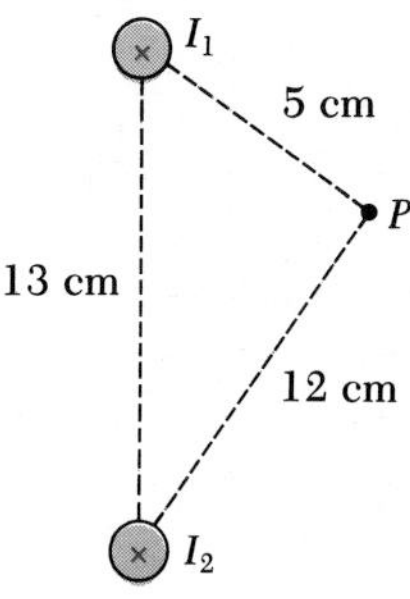

Figure 30.44 (Problem 21).

Section 30.3 Ampère's Law and Section 30.4 The Magnetic Field of a Solenoid

22. A closely wound long solenoid of overall length 30 cm has a magnetic field $B = 5 \times 10^{-4}$ T at its center due to a current $I = 1$ A. How many turns of wire are on the solenoid?
23. A superconducting solenoid is to be designed to generate a magnetic field of 10 T. (a) If the solenoid winding has 2000 turns/meter, what is the required current? (b) What force per unit length is exerted on the solenoid windings by this magnetic field?
24. What current is required in the windings of a long solenoid that has 1000 turns uniformly distributed over a length of 0.4 m in order to produce a magnetic field of magnitude 1.0×10^{-4} T at the center of the solenoid?
25. Some superconducting alloys at very low temperatures can carry very high currents. For example, Nb_3Sn wire at 10 K can carry 10^3 A and maintain its superconductivity. Determine the maximum B field which can be achieved in a solenoid of length 25 cm if 1000 turns of Nb_3Sn wire are wrapped on the outside surface.
26. A toroidal winding (Fig. 30.11) has a total of 400 turns on a core with inner radius $a = 4$ cm and outer radius $b = 6$ cm. Calculate the magnitude of the magnetic field at a point midway between the inner and outer walls of the core when there is a current of 0.5 A maintained in the windings.
27. The magnetic coils of a Tokamak fusion reactor are in the shape of a toroid having an inner radius of 0.7 m and outer radius of 1.3 m. Inside the toroid is the plasma. If the toroid has 900 turns of large diameter wire, each of which carries a current of 14 000 A, find the magnetic field strength along (a) the inner radius of the toroid and (b) the outer radius of the toroid.
28. A cylindrical conductor of radius $R = 2.5$ cm carries a current $I = 2.5$ A along its length; this current is uniformly distributed throughout the cross section of the conductor. Calculate the magnetic field midway along the radius of the wire (that is, at $r = R/2$).
29. For the conductor described in Problem 28, find the distance beyond the surface of the conductor at which the magnitude of the magnetic field has the same value as the magnitude of the field at $r = R/2$.
30. Niobium metal becomes a superconductor (with electrical resistance equal to zero) when cooled below 9 K. If superconductivity is destroyed when the surface magnetic field exceeds 0.1 T, determine the maximum current a 2-mm diameter niobium wire can carry and remain superconducting.
31. A *packed bundle* of 100 long straight insulated wires forms a cylinder of radius $R = 0.5$ cm. (a) If each wire carries a 2 A current, what are the magnitude and direction of the magnetic force per unit length acting on a wire located 0.2 cm from the center of the bundle? (b) Would a wire on the outer edge of the bundle experience a greater or a smaller force compared to the wire 0.2 cm from the center?
32. Consider a coaxial arrangement with a wire of radius a along the axis of a thin cylindrical shell of radius b, as in Figure 30.45. Current is directed *into* the page along the center wire and returns out of the page along the cylinder. If $I = 5$ A, $a = 0.6$ cm, and $b = 1.2$ cm, calculate the magnetic field (a) at point P_1, a distance $r_1 = 1$ cm, and (b) at point P_2, a distance $r_2 = 2.4$ cm from the center of the wire.

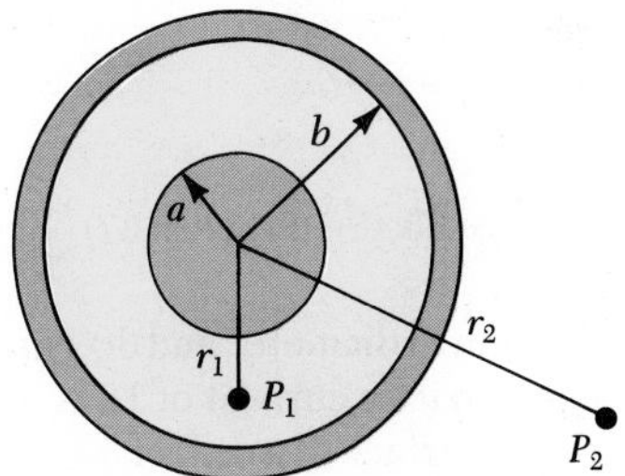

Figure 30.45 (Problem 32).

*Section 30.5 The Magnetic Field Along the Axis of a Solenoid

33. A short solenoid, with a length of 10 cm and a radius of 5 cm, consists of 200 turns of fine wire that carries a current of 15 A. What is the magnetic induction at the center of the solenoid? For the same number of turns per unit length, what value of B would result for $\ell \rightarrow \infty$?

34. A solenoid has 900 turns, carries a current of 3 A, has a length of 80 cm, and a radius of 2.5 cm. Calculate the magnetic field along its axis at (a) its center and (b) a point near the end.
35. A solenoid has 500 turns, a length of 50 cm, a radius of 5 cm, and carries a current of 4 A. Calculate the magnetic field at an axial point, a distance of 15 cm from the center (that is, 10 cm from one end).

Section 30.6 Magnetic Flux

36. A toroid is constructed from N *rectangular* turns of wire. Each turn has height h. The toroid has an inner radius a and outer radius b. (a) If the toroid carries a current I, show that the total magnetic flux through the turns of the toroid is proportional to $\ln(b/a)$. (b) Evaluate this flux if $N = 200$ turns, $h = 1.5$ cm, $a = 2$ cm, $b = 5$ cm, and $I = 2$ A.
37. A cube of edge length $\ell = 2.5$ cm is positioned as shown in Figure 30.46. There is a uniform magnetic field throughout the region given by the expression $\mathbf{B} = (5\mathbf{i} + 4\mathbf{j} + 3\mathbf{k})$ T. (a) Calculate the flux through the shaded face of the cube. (b) What is the total flux through the six faces of the cube?

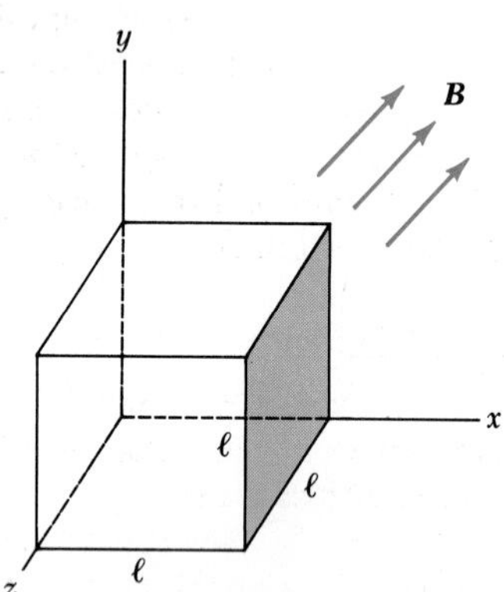

Figure 30.46 (Problem 37).

38. A solenoid 2.5 cm in diameter and 30 cm in length has 300 turns and carries a current of 12 A. Calculate the flux through the surface of a disk of 5-cm radius that is positioned perpendicular to and centered on the axis of the solenoid, as in Figure 30.47.

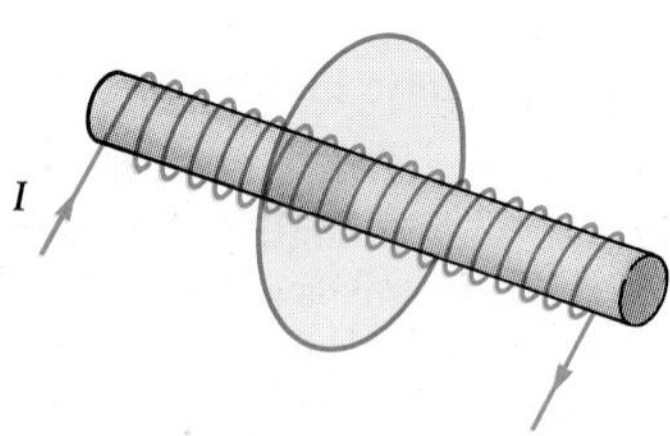

Figure 30.47 (Problem 38).

39. Figure 30.48 shows an enlarged *end* view of the solenoid described in Problem 38. Calculate the flux through the blue area defined by an annulus with an inner radius of 0.4 cm and outer radius of 0.8 cm.

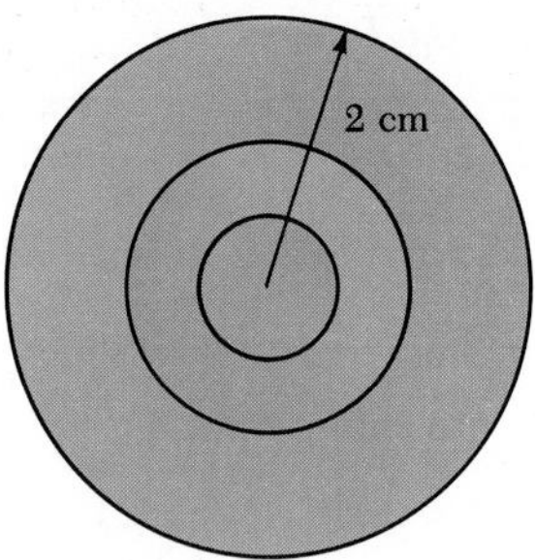

Figure 30.48 (Problem 39).

40. A circular loop of wire of radius R is placed in a uniform magnetic field $\mathbf{B}$ and is then spun at constant angular velocity ω about an axis through its diameter. Determine the magnetic flux through the loop as a function of time if the axis of rotation is (a) perpendicular to $\mathbf{B}$, and (b) parallel to $\mathbf{B}$.

Section 30.8 Displacement Current and the Generalized Ampère's Law

41. The applied voltage across the plates of a 4-μF capacitor varies in time according to the expression

$$V_{\text{app}} = (8 \text{ V})(1 - e^{-t/4})$$

where t is in s. Calculate (a) the displacement current as a function of time and (b) the value of the current at $t = 4$ s.
42. A capacitor of capacitance C has a charge Q at $t = 0$. At that time, a resistor of resistance R is connected to the plates of the charged capacitor. (a) Find the displacement current in the dielectric between the plates of the capacitor as a function of time. (b) Evaluate this displacement current at time $t = 0.1$ s if $C = 2.0$ μF, $Q = 20$ μC, and $R = 500$ kΩ. (c) At what rate is the electric flux between the capacitor plates changing at $t = 0.1$ s?
43. A 0.1 A current is charging a capacitor with square plates, 5 cm on a side. If the plate separation is 4 mm, find (a) the rate of change of electric flux $\frac{d\Phi_E}{dt}$ between the plates and (b) the displacement current I_d between the plates.
44. A 0.2-A current is charging a capacitor with circular plates, 10 cm in radius. If the plate separation is 4 mm, (a) what is the rate of increase of electric field dE/dt between the plates? (b) What is the magnetic field between the plates at a radius of 5 cm from the center?

Section 30.9 Magnetism In Matter

45. What is the relative permeability of a material that has a magnetic susceptibility of 10^{-4}?
46. An iron-core toroid is wrapped with 250 turns of wire per meter of its length. The current in the winding is 8 A. Taking the magnetic permeability of iron to be $\kappa_m = 5000\mu_0$, calculate (a) the magnetic field strength, $\boldsymbol{H}$ and (b) the magnetic flux density, $\boldsymbol{B}$.
47. A toroidal winding with a mean radius of 20 cm and 630 turns (as in Figure 30.30) is filled with powdered steel whose magnetic susceptibility χ is 100. If the current in the windings is 3 A, find B (assumed uniform) inside the toroid.
48. A toroid has an average radius of 9 cm. The current in the coil is 0.5 A. How many turns are required to produce a magnetic field strength of 700 A·turns/m within the toroid?
49. A magnetic field of flux density 1.3 T is to be set up in an iron-core toroid. The toroid has a mean radius of 10 cm, and magnetic permeability of $5000\mu_0$. What current is required if there are 470 turns of wire in the winding?
50. A toroidal solenoid has an average radius of 10 cm and a cross-sectional area of 1 cm^2. There are 400 turns of wire on the soft iron core, which has a permeability of $800\mu_0$. Calculate the current necessary to produce a magnetic flux of 5×10^{-4} Wb through a cross section of the core.
51. A coil of 500 turns is wound on an iron ring ($\kappa_m = 750\mu_0$) of 20 cm mean radius and 8 cm^2 cross-sectional area. Calculate the magnetic flux Φ in this Rowland ring when the current in the coil is 0.5 A.
52. Show that the product of magnetic field strength H and magnetic flux density B has SI units of J/m^3.
53. In the text, we found that an alternative description for magnetic field $\boldsymbol{B}$ in terms of magnetic field strength $\boldsymbol{H}$ and magnetization $\boldsymbol{M}$ is $\boldsymbol{B} = \mu_0\boldsymbol{H} + \mu_0\boldsymbol{M}$. Relate the magnetic susceptibility χ to $|\boldsymbol{H}|$ and $|\boldsymbol{M}|$ for paramagnetic or diamagnetic materials.
54. Calculate the magnetic field strength H of a magnetized substance characterized by a magnetization of 0.88×10^6 A·turns/m and a magnetic field of flux density 4.4 T. (*Hint:* See Problem 53.)
55. A magnetized cylinder of iron has a magnetic field $B = 0.04$ T in its interior. The magnet is 3 cm in diameter and 20 cm long. If the same magnetic field is to be produced by a 5-A current carried by an air-core solenoid having the same dimensions as the cylindrical magnet, how many turns of wire must be on the solenoid?
56. In Bohr's 1913 model of the hydrogen atom, the electron is in a circular orbit of radius 5.3×10^{-11} m, and its speed is 2.2×10^6 m/s. (a) What is the magnitude of the magnetic moment due to the electron's motion? (b) If the electron orbits counterclockwise in a horizontal circle, what is the direction of this magnetic moment vector?
57. At saturation, the alignment of spins in iron can contribute as much as 2 Tesla to the total magnetic field $\boldsymbol{B}$. If each electron contributes a magnetic moment of 9.27×10^{-24} A·m^2 (one Bohr magneton), how many electrons per atom contribute to the saturated field of iron? (*Hint:* There are 8.5×10^{28} iron atoms/m^3.)

°Section 30.10 Magnetic Field of the Earth

58. A circular coil of 5 turns and a diameter of 30 cm is oriented in a vertical plane with its axis perpendicular to the horizontal component of the earth's magnetic field. A horizontal compass placed at the center of the coil is made to deflect 45° from magnetic North by a current of 0.60 A in the coil. (a) What is the horizontal component of the earth's magnetic field? (b) If a compass "dip" needle oriented in a vertical north-south plane makes an angle of 13° from the vertical, what is the total strength of the earth's magnetic field at this location?
59. The magnetic moment of the earth is approximately 8.7×10^{22} A·m^2. (a) If this were caused by the complete magnetization of a huge iron deposit, how many unpaired electrons would this correspond to? (b) At 2 unpaired electrons per iron atom, how many kilograms of iron would this correspond to? (The density of iron is 7900 kg/m^3, and there are approximately 8.5×10^{28} iron atoms/m^3.)

ADDITIONAL PROBLEMS

60. A lightning bolt may carry a current of 10^4 A for a short period of time. What is the resulting magnetic field at a point 100 m from the bolt?
61. Measurements of the magnetic field of a large tornado were made at the Geophysical Observatory in Tulsa, Oklahoma in 1962. If the tornado's field was $B = 1.5 \times 10^{-8}$ T pointing north when the tornado was 9 km east of the observatory, what current was carried up/down the funnel of the tornado?
62. At the Fermilab Accelerator, in Weston, Illinois, protons with momentum 4.8×10^{-16} kg·m/s are confined to a circular orbit of radius 1 km by an upward magnetic field. What vertical B field must be used to maintain the protons in this orbit?
63. Two long, parallel conductors are carrying currents in the same direction as in Figure 30.49. Conductor A carries a current of 150 A and is held firmly in position. Conductor B carries a current I_B and is allowed to slide freely up and down (parallel to A) between a set of nonconducting guides. If the linear density of conductor B is 0.10 g/cm, what value of current I_B will result in equilibrium when the distance between the two conductors is 2.5 cm?

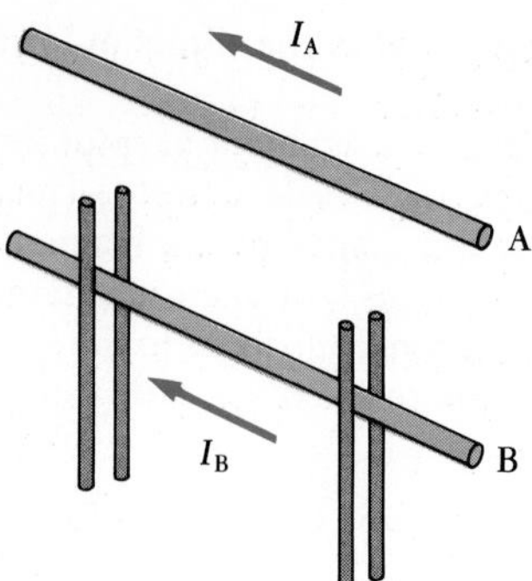

Figure 30.49 (Problem 63).

64. Two parallel conductors carry current in opposite directions as shown in Figure 30.50. One conductor carries a current of 10 A. Point A is at the *midpoint* between the wires and point C is a distance $d/2$ to the right of the 10-A current. If $d = 18$ cm and I is adjusted so that the magnetic field at C is zero, find (a) the value of the current I and (b) the value of the magnetic field at A.

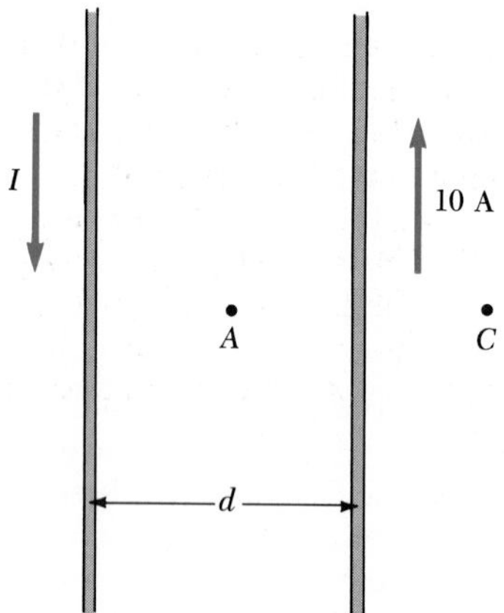

Figure 30.50 (Problem 64).

65. A very long, thin strip of metal of width w carries a current I along its length as in Figure 30.51. Find the magnetic field in the *plane* of the strip (at an external point P) a distance b from one edge.

66. A large nonconducting belt with a uniform surface charge density σ moves with a speed v on a set of rollers as shown in Figure 30.52. Consider a point *just above* the surface of the moving belt. (a) Find an expression for the magnitude of the magnetic field $\mathbf{B}$ at this point. (b) If the belt is positively charged, what is the direction of $\mathbf{B}$? (Note that the belt may be considered as an infinite sheet.)

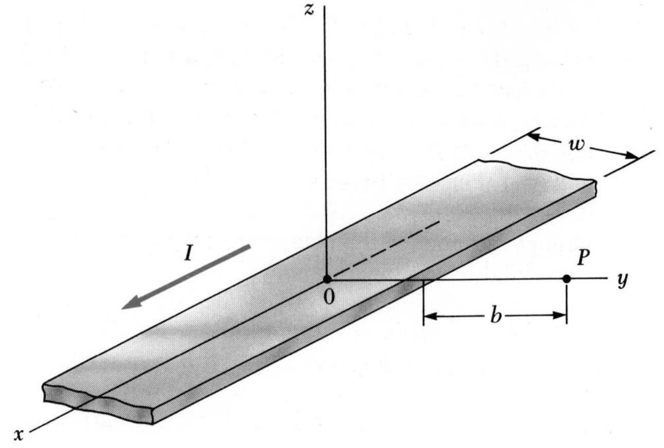

Figure 30.51 (Problem 65).

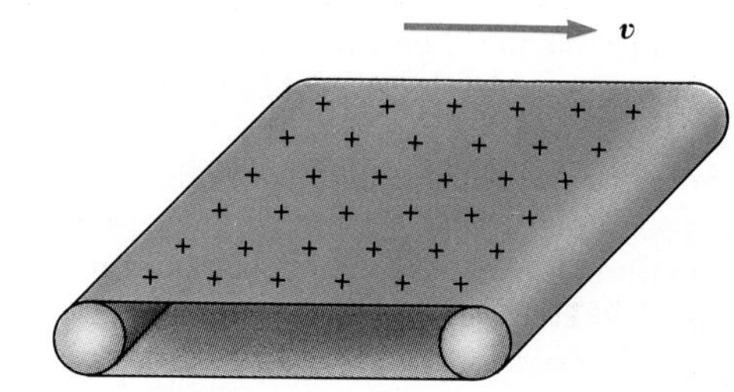

Figure 30.52 (Problem 66).

67. A straight wire located at the equator is oriented parallel to the earth along the east-west direction. The earth's magnetic field at this point is horizontal and has a magnitude of 3.3×10^{-5} T. If the mass per unit length of the wire is 2×10^{-3} kg/m, what current must the wire carry in order that the magnetic force balance the weight of the wire?

68. The earth's magnetic field at either pole is about 0.7 G $= 7 \times 10^{-5}$ T. Using a model in which you assume that this field is produced by a current loop around the equator, determine the current that would generate such a field. ($R_e = 6.37 \times 10^6$ m.)

69. A nonconducting ring of radius R is uniformly charged with a total positive charge q. The ring rotates at a constant angular velocity ω about an axis through its center, perpendicular to the plane of the ring. If $R =$ 0.1 m, $q = 10\ \mu$C, and $\omega = 20$ rad/s, what is the resulting magnetic field on the axis of the ring a distance of 0.05 m from the center?

70. Consider a thin disk of radius R mounted to rotate about the x axis in the yz plane. The disk has a positive uniform surface charge density σ and angular velocity ω. Show that the magnetic field at the center of the disk is given by $B = \frac{1}{2}\mu_0\sigma\omega R$.

71. Two circular coils of radius R are each perpendicular to a common axis. The coil centers are a distance R apart and a steady current I flows in the same direction around each coil as shown in Figure 30.53. (a) Show that the magnetic field on the axis at a distance x from the center of one coil is

$$B = \frac{\mu_0 I R^2}{2}\left[\frac{1}{(R^2 + x^2)^{3/2}} + \frac{1}{(2R^2 + x^2 - 2Rx)^{3/2}}\right]$$

(b) Show that dB/dx and $\frac{d^2B}{dx^2}$ are both zero at a point *midway* between the coils. This means the magnetic field in the region midway between the coils is *uniform*. Coils in this configuration are called **Helmholtz coils.**

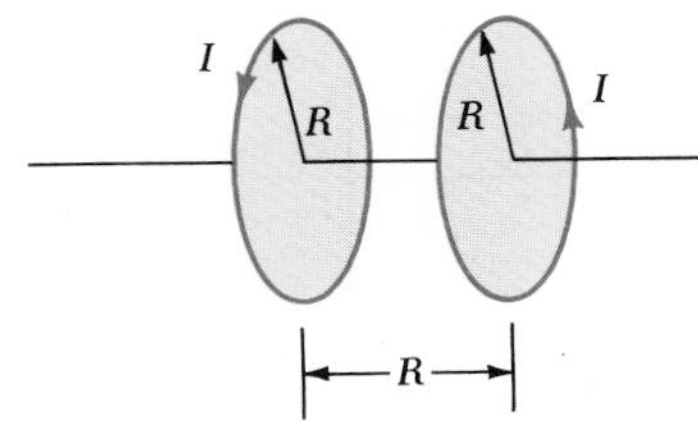

Figure 30.53 (Problem 71).

72. Two identical, flat, circular coils of wire each have 100 turns and a radius of 0.50 m. If these coils are arranged as a set of Helmholtz coils and each coil carries a current of 10 A, determine the magnitude of the magnetic field at a point half way between the coils and on the axis of the coils. (See Figure 30.53.)

73. A long cylindrical conductor of radius R carries a current I as in Figure 30.54. The current density J, however, is *not* uniform over the cross section of the conductor but is a function of the radius according to $J = br$, where b is a constant. Find an expression for the magnetic field B (a) at a distance $r_1 < R$ and (b) at a distance $r_2 > R$, measured from the axis.

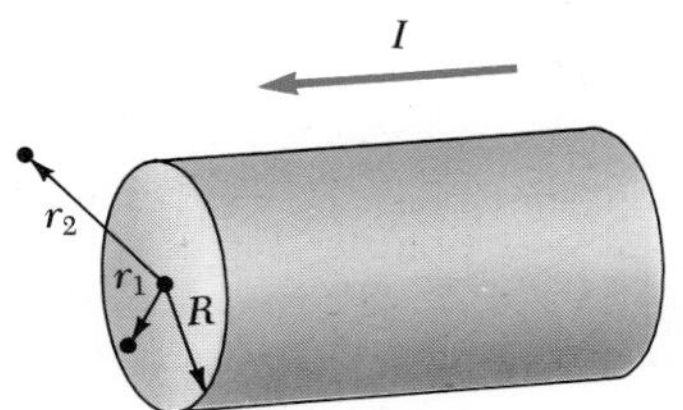

Figure 30.54 (Problem 73).

74. Shown in Figure 30.55 is the cross section of a long nonconducting cylinder that has N wires parallel to the axis of the cylinder and uniformly spaced around the curved surface. The cylinder has a radius R and the current in *each* conductor is I and directed *out* of the plane of the figure. Assuming N to be a very large number and the radius of each wire to be small compared with the radius of the cylinder, find an expression for the magnetic field B (a) at $r_1 < R$ and (b) at $r_2 > R$. (c) Obtain a numerical value for B when $N = 100$, $R = 5$ cm, $I = 10$ A, and $r_2 = 15$ cm.

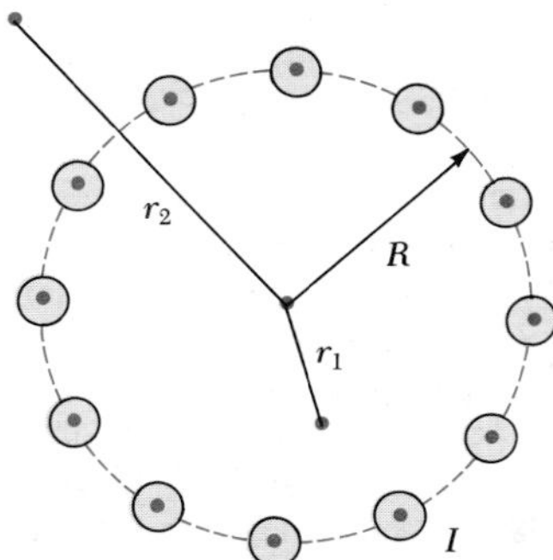

Figure 30.55 (Problem 74).

75. A current I flows around a closed path in the horizontal plane of the circuit shown in Figure 30.56. The path consists of six arcs with alternating radii r_1 and r_2 connected by radial segments. Each segment of arc subtends an angle of 60° at the common center P, with $r_2/r_1 = 2/3$. This current path produces a magnetic field B at P. If the path is *modified* so that the ratio $r_2/r_1 = 1/3$, by what factor must I be multiplied in order that the field at P remain the same?

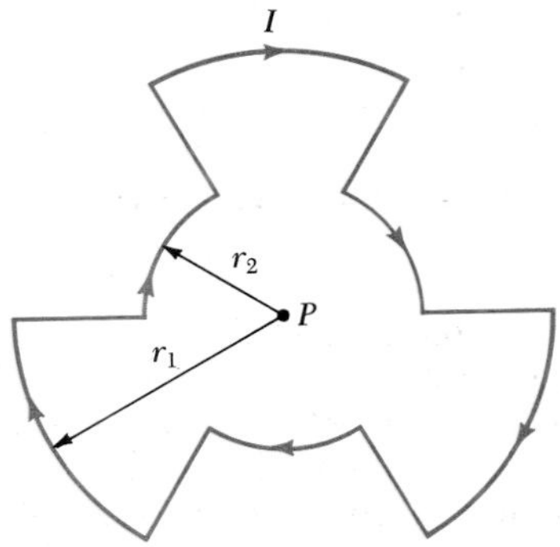

Figure 30.56 (Problem 75).

76. A 4-m length of wire of diameter 2 mm is wrapped in a single layer around a hollow cardboard cylinder of radius 2 cm. Each turn is placed in contact with the adjacent winding. (Assume that adjacent wire turns

are electrically insulated from each other, but neglect the thickness of the insulating cover.) A steady current of 11 A is maintained in the wire. (a) Calculate the value of the magnetic field at the center of the solenoidal array. (b) How would the answer to (a) be changed if the total length of wire were used to wrap a *single*-layer solenoid of radius 1 cm?

77. A toroidal winding filled with a magnetic substance carries a steady current of 2 A. The coil contains a total of 1505 turns, has an average radius of 4 cm, and the core has a cross-sectional area of 1.21 cm². The total magnetic flux through a cross-section for the toroid is measured as 3×10^{-5} Wb. Assume the flux density is constant. (a) What is the magnetic field strength H within the core? (b) Determine the permeability of the core material.

78. A paramagnetic substance achieves 10% of its saturation magnetization when placed in a magnetic field of 5.0 T at a temperature of 4.0 K. The density of magnetic atoms in the sample is 8×10^{27} atoms/m³, and the magnetic moment per atom is 5 Bohr magnetons. Calculate the Curie constant for this substance.

79. The density of a specimen of a suspected new element is determined to be 4.15 g/cm³. The saturation magnetism of the material is found to be 7.6×10^4 A/m, and the measured atomic magnetic moment is 1.2 Bohr magnetons. Calculate the expected value of the atomic weight of the element based on these values.

80. The force on a magnetic dipole M aligned with a nonuniform magnetic field in the x direction is given by $F_x = M\dfrac{dB}{dx}$. Suppose that 2 flat loops of wire each have radius R and carry current I. (a) If the loops are arranged coaxially and separated by a large variable distance x, show that the magnetic force between them varies as $1/x^4$. (b) Evaluate the magnitude of this force if $I = 10$ A, $R = 0.5$ cm, and $x = 5.0$ cm.

81. A wire is formed into the shape of a square of edge length L (Fig. 30.57). Show that when the current in the loop is I, the magnetic field at point P a distance x from the center of the square along its axis is given by

$$B = \frac{\mu_0 I L^2}{2\pi\left(x^2 + \dfrac{L^2}{4}\right)\sqrt{x^2 + \dfrac{L^2}{2}}}$$

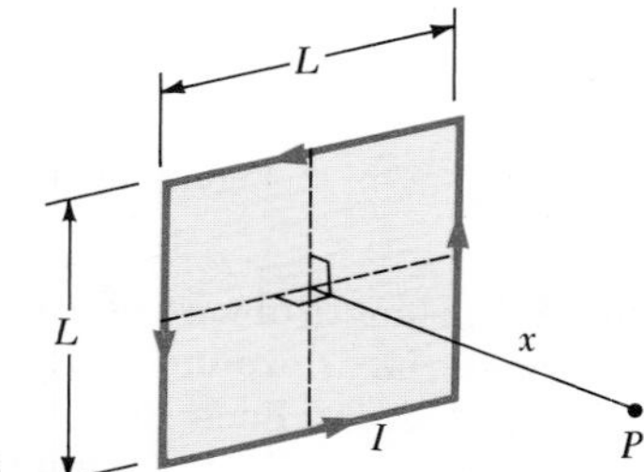

Figure 30.57 (Problem 81).

82. A wire is bent into the shape shown in Figure 30.58a, and the magnetic field is measured at P_1 when the current in the wire is I. The same wire is then formed into the shape shown in Figure 30.58b, and the magnetic field measured at point P_2 when the current is again I. If the *total* length of wire is the same in each case, what is the ratio of B_1/B_2?

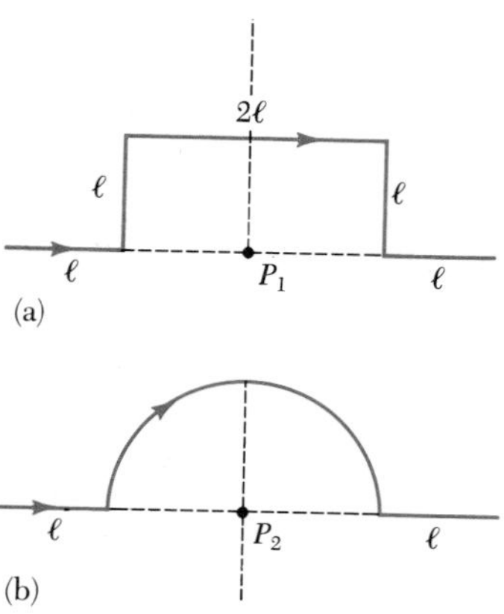

Figure 30.58 (Problem 82).

83. A wire carrying a current I is bent into the shape of an exponential spiral, $r = e^\theta$, from $\theta = 0$ to $\theta = 2\pi$ as in Figure 30.59. To complete a loop, the ends of the spiral are connected by a straight wire along the x axis. Find the magnitude and direction of $\boldsymbol{B}$ at the origin. (*Hints:* Use the Biot-Savart Law. The angle β between a radial line and its tangent line at any point on the curve $r = f(\theta)$ is related to the function in the following way:

$$\tan \beta = \frac{r}{dr/d\theta}$$

In this case $r = e^\theta$, thus $\tan \beta = 1$ and $\beta = \pi/4$. Therefore, the angle between $d\boldsymbol{s}$ and $\hat{\boldsymbol{r}}$ is $\pi - \beta = 3\pi/4$. Also

$$ds = \frac{dr}{\sin \pi/4} = \sqrt{2}\, dr$$

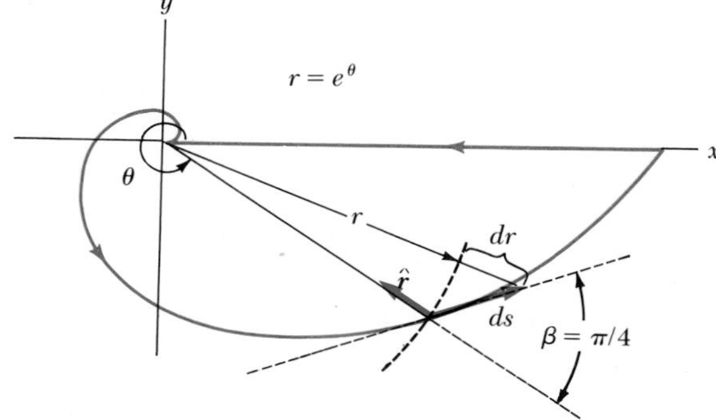

Figure 30.59 (Problem 83).

84. A long cylindrical conductor of radius a has two cylindrical cavities of diameter a through its entire length, as shown in Figure 30.60. A current, I, is directed out of the page and is uniform through a cross-section of the conductor. Find the magnitude and direction of the magnetic field at point P_1 in terms of μ_0, I, r, and a.

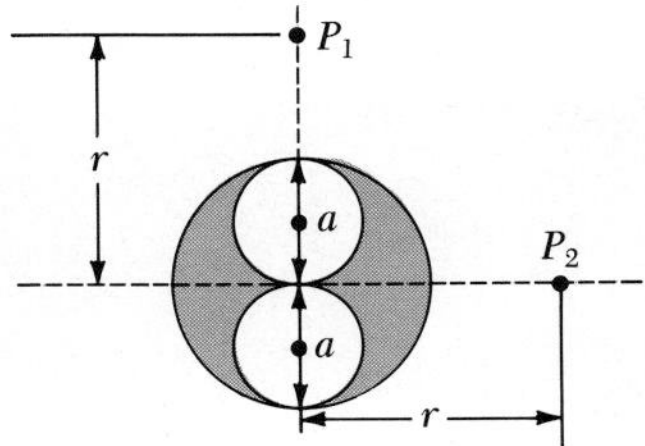

Figure 30.60 (Problems 84 and 85).

85. Given the same conductor as described in Problem 84, find the magnitude and direction of the magnetic field at point P_2 as shown in Figure 30.60 in terms of μ_0, I, r, and a.

CALCULATOR/COMPUTER PROBLEMS

86. Consider a flat circular current loop of radius R carrying current I. Choose the x axis to be along the axis of the loop with the origin at the center of the loop. Plot a graph of the ratio of the magnitude of the magnetic field at coordinate x to that at the origin for $x = 0$ to $x = 5R$.

31
Faraday's Law

Michael Faraday in his laboratory at the Royal Institute, London, in 1870. (Courtesy of The Bettmann Archive)

Our studies so far have been concerned with the electric fields due to stationary charges and the magnetic fields produced by moving charges. This chapter deals with electric fields that originate from changing magnetic fields.

Experiments conducted by Michael Faraday in England in 1831 and independently by Joseph Henry in the United States that same year showed that an electric current could be induced in a circuit by a changing magnetic field. The results of these experiments led to a very basic and important law of electromagnetism known as *Faraday's law of induction.* This law says that the magnitude of the emf induced in a circuit equals the time rate of change of the magnetic flux through the circuit.

As we shall see, an induced emf can be produced in many ways. For instance, an induced emf and an induced current can be produced in a closed loop of wire when the wire moves into a magnetic field. We shall describe such experiments along with a number of important applications that make use of the phenomenon of electromagnetic induction.

With the treatment of Faraday's law, we complete our introduction to the fundamental laws of electromagnetism. These laws can be summarized in a set of four equations called *Maxwell's equations.* Together with the Lorentz force law, which we shall discuss briefly, they represent a complete theory for describing the interaction of charged objects. Maxwell's equations relate electric and magnetic fields to each other and to their ultimate source, namely, electric charges.

Biographical Sketch

Joseph Henry
(1797–1878)

Joseph Henry, an American physicist who carried out early experiments in electrical induction, was born in Albany, New York, in 1797. The son of a laborer, Henry had little schooling and was forced to go to work at a very young age. After working his way through Albany Academy to study medicine, then engineering, Henry became professor of mathematics and physics in 1826. He later became professor of natural philosophy at New Jersey College (now Princeton University).

In 1848, Henry became the first director of the Smithsonian Institute, where he introduced a weather-forecasting system based on meteorological information received by the electric telegraph. He was also the first president of The Academy of Natural Science, a position he held until his death in 1878.

Many of Henry's early experiments were with electromagnetism. He improved the electromagnet of William Sturgeon and made one of the first electromagnetic motors. By 1830, Henry had made powerful electromagnets by using many turns of fine insulated wire wound around iron cores. He discovered the phenomenon of self-induction but failed to publish his findings; as a result, credit was given to Michael Faraday.

Henry's contribution to science was ultimately recognized: in 1893 the unit of inductance was named the henry.

(Courtesy of AIP Niels Bohr Library, E. Scott Barr Collection)

31.1 FARADAY'S LAW OF INDUCTION

We begin by describing two simple experiments that demonstrate that a current can be produced by a changing magnetic field. First, consider a loop of wire connected to a galvanometer as in Figure 31.1. If a magnet is moved toward the loop, the galvanometer needle will deflect in one direction, as in Figure 31.1a. If the magnet is moved away from the loop, the galvanometer needle will deflect in the opposite direction, as in Figure 31.1b. If the magnet is held stationary relative to the loop, no deflection is observed. Finally, if the magnet is held stationary and the coil is moved either toward or away from the magnet, the needle will also deflect. From these observations, one concludes

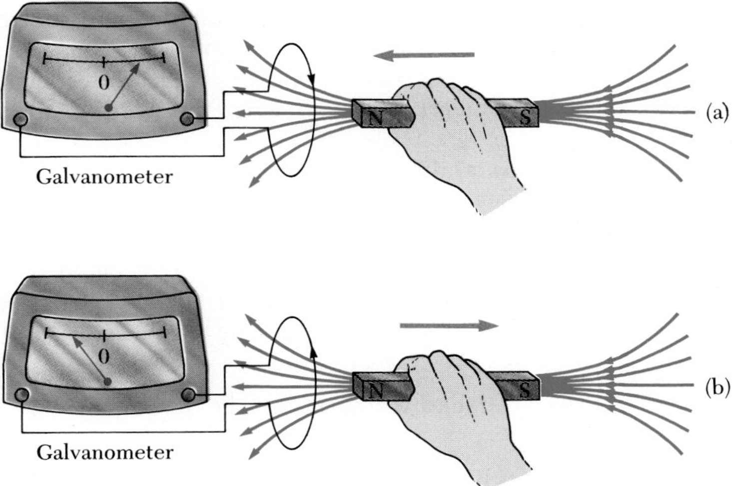

Figure 31.1 (a) When a magnet is moved toward a loop of wire connected to a galvanometer, the galvanometer deflects as shown. This shows that a current is induced in the loop. (b) When the magnet is moved away from the loop, the galvanometer deflects in the opposite direction, indicating that the induced current is opposite that shown in (a).

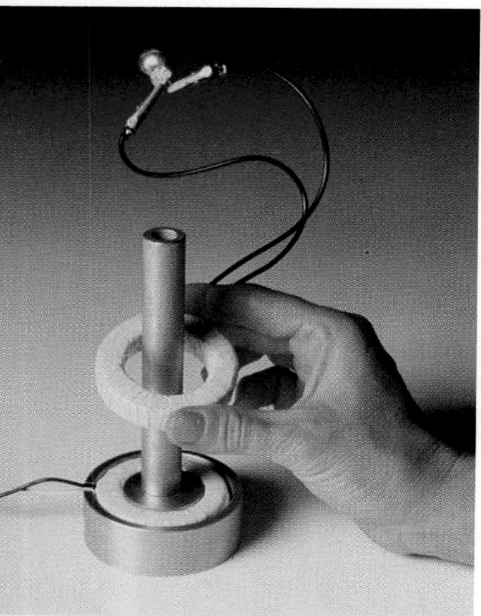

A demonstration of electromagnetic induction. An ac voltage is applied to the lower coil. A voltage is induced in the upper coil as indicated by the illuminated lamp connected to the upper coil. What do you think happens to the lamp's intensity as the upper coil is moved over the vertical tube? (Courtesy of CENCO)

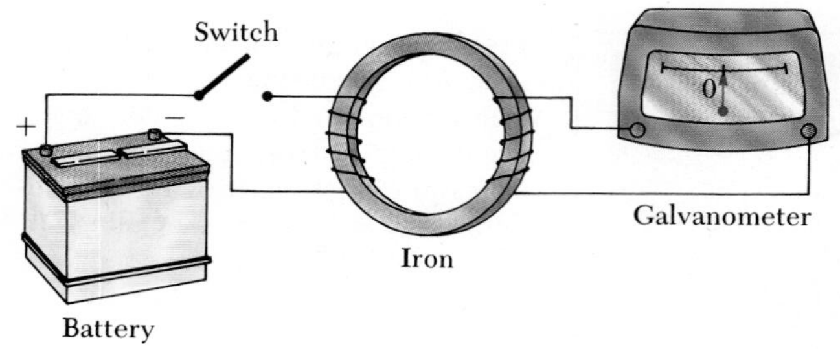

Figure 31.2 Faraday's experiment. When the switch in the primary circuit at the left is closed, the galvanometer in the secondary circuit at the right deflects momentarily. The emf induced in the secondary circuit is caused by the changing magnetic field through the coil in this circuit.

that *a current is set up in the circuit as long as there is relative motion between the magnet and the coil.*[1]

These results are quite remarkable in view of the fact that *a current is set up in the circuit even though there are no batteries in the circuit!* We call such a current an *induced current,* which is produced by an *induced emf.*

Now let us describe an experiment, first conducted by Faraday, that is illustrated in Figure 31.2. Part of the apparatus consists of a coil connected to a switch and a battery. We shall refer to this coil as the *primary coil* and to the corresponding circuit as the primary circuit. The coil is wrapped around an iron ring to intensify the magnetic field produced by the current through the coil. A second coil, at the right, is also wrapped around the iron ring and is connected to a galvanometer. We shall refer to this as the *secondary coil* and to the corresponding circuit as the secondary circuit. There is no battery in the secondary circuit and the secondary coil is not connected to the primary coil. The only purpose of this circuit is to detect any current that might be produced by a change in the magnetic field.

At first sight, you might guess that no current would ever be detected in the secondary circuit. However, something quite amazing happens when the switch in the primary circuit is suddenly closed or opened. At the instant the switch in the primary circuit is closed, the galvanometer in the secondary circuit deflects in one direction and then returns to zero. When the switch is opened, the galvanometer deflects in the opposite direction and again returns to zero. Finally, the galvanometer reads zero when there is a steady current in the primary circuit.

As a result of these observations, Faraday concluded that *an electric current can be produced by a changing magnetic field.* A current cannot be produced by a steady magnetic field. The current that is produced in the secondary circuit occurs for only an instant while the magnetic field through the secondary coil is changing. In effect, the secondary circuit behaves as though there were a source of emf connected to it for a short instant. It is customary to say that

> an induced emf is produced in the secondary circuit by the changing magnetic field.

These two experiments have one thing in common. In both cases, an emf is induced in a circuit when the *magnetic flux* through the circuit *changes with*

[1] The exact magnitude of the current depends on the particular resistance of the circuit, but the existence of the current (or the algebraic sign) does *not.*

time. In fact, a general statement that summarizes such experiments involving induced currents and emfs is as follows:

> The emf induced in a circuit is directly proportional to the time rate of change of magnetic flux through the circuit.

This statement, known as **Faraday's law of induction**, can be written

$$\mathcal{E} = -\frac{d\Phi_m}{dt} \quad (31.1)$$

Faraday's law

where Φ_m is the magnetic flux threading the circuit (Section 30.6), which can be expressed as

$$\Phi_m = \int \mathbf{B} \cdot d\mathbf{A} \quad (31.2)$$

The integral given by Equation 31.2 is taken over the area bounded by the circuit. The meaning of the negative sign in Equation 31.1 is a consequence of Lenz's law and will be discussed in Section 31.3. If the circuit is a coil consisting of N loops all of the same area and if the flux threads all loops, the induced emf is given by

$$\mathcal{E} = -N\frac{d\Phi_m}{dt} \quad (31.3)$$

Suppose the magnetic field is uniform over a loop of area A lying in a plane as in Figure 31.3. In this case, the flux through the loop is equal to $BA \cos \theta$; hence the induced emf can be expressed as

$$\mathcal{E} = -\frac{d}{dt}(BA \cos \theta) \quad (31.4)$$

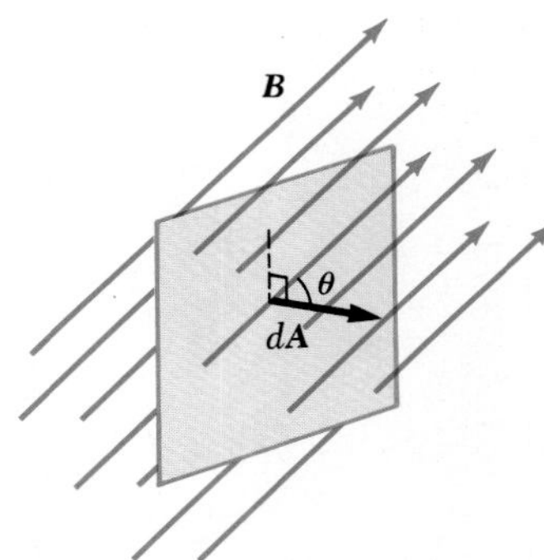

Figure 31.3 A conducting loop of area A in the presence of a uniform magnetic field $\mathbf{B}$, which is at an angle θ with the normal to the loop.

From this expression, we see that an emf can be induced in the circuit in several ways: (1) The magnitude of $\mathbf{B}$ can vary with time; (2) the area of the circuit can change with time; (3) the angle θ between $\mathbf{B}$ and the normal to the plane can change with time; and (4) any combination of these can occur.

The following examples illustrate cases where an emf is induced in a circuit as a result of a time variation of the magnetic field.

EXAMPLE 31.1 Application of Faraday's Law

A coil is wrapped with 200 turns of wire on the perimeter of a square frame of sides 18 cm. Each turn has the same area, equal to that of the frame, and the total resistance of the coil is 2 Ω. A uniform magnetic field is turned on perpendicular to the plane of the coil. If the field changes *linearly* from 0 to 0.5 Wb/m^2 in a time of 0.8 s, find the magnitude of the induced emf in the coil while the field is changing.

Solution The area of the loop is $(0.18 \text{ m})^2 = 0.0324 \text{ m}^2$. The magnetic flux through the loop at $t = 0$ is zero since $B = 0$. At $t = 0.8$ s, the magnetic flux through the loop is $\Phi_m = BA = (0.5 \text{ Wb/m}^2)(0.0324 \text{ m}^2) = 0.0162$ Wb. Therefore, the magnitude of the induced emf is

$$|\mathcal{E}| = \frac{N\,\Delta\Phi_m}{\Delta t} = \frac{200(0.0162 \text{ Wb} - 0 \text{ Wb})}{0.8 \text{ s}} = 4.05 \text{ V}$$

(Note that 1 Wb = 1 V·s.)

Exercise 1 What is the magnitude of the induced current in the coil while the field is changing?
Answer 2.03 A.

EXAMPLE 31.2 An Exponentially Decaying *B* Field
A plane loop of wire of area A is placed in a region where the magnetic field is *perpendicular* to the plane. The magnitude of $\mathbf{B}$ varies in time according to the expression $B = B_0e^{-at}$. That is, at $t = 0$ the field is B_0, and for $t > 0$, the field decreases exponentially in time (Fig. 31.4). Find the induced emf in the loop as a function of time.

Solution Since $\mathbf{B}$ is perpendicular to the plane of the loop, the magnetic flux through the loop at time $t > 0$ is given by

$$\Phi_m = BA = AB_0e^{-at}$$

Also, since the coefficient AB_0 and the parameter a are constants, the induced emf can be calculated from Equation 31.1:

$$\mathcal{E} = -\frac{d\Phi_m}{dt} = -AB_0 \frac{d}{dt} e^{-at} = aAB_0e^{-at}$$

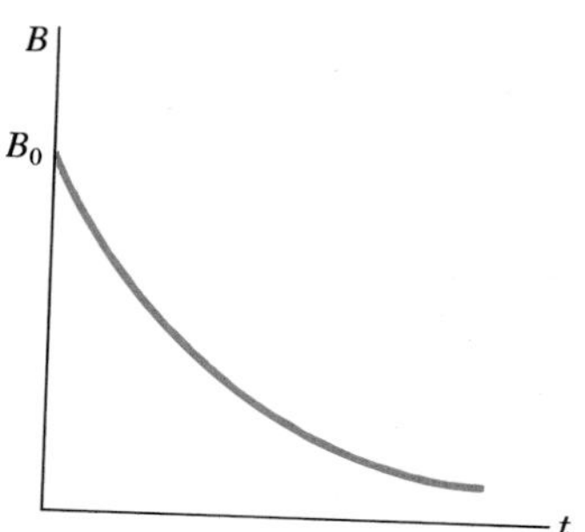

Figure 31.4 (Example 31.2) Exponential decrease of the magnetic field with time. The induced emf and induced current have similar time variations.

That is, the induced emf decays exponentially in time. Note that the maximum emf occurs at $t = 0$, where $\mathcal{E}_{max} = aAB_0$. Why is this true? The plot of $\mathcal{E}$ versus t is similar to the B versus t curve shown in Figure 31.4.

Biographical Sketch

Michael Faraday
(1791–1867)

Michael Faraday was a British physicist and chemist who is often regarded as the greatest experimental scientist of the 1800s. His many contributions to the study of electricity include the invention of the electric motor, electric generator, and transformer, as well as the discovery of electromagnetic induction, the laws of electrolysis, the discovery of benzene, and the theory that the plane of polarization of light is rotated in an electric field.

Faraday was born in 1791 in rural England, but his family moved to London shortly thereafter. One of ten children and the son of a blacksmith, Faraday received a minimal education and became apprenticed to a bookbinder at age 14. He was fascinated by articles on electricity and chemistry and was fortunate to have an employer who allowed him to read books and attend scientific lectures. He received some education in science from the City Philosophical Society.

When Faraday finished his apprenticeship in 1812, he expected to devote himself to bookbinding rather than to science. That same year, Faraday attended a lecture by Humphry Davy, who made many contributions in the field of heat and thermodynamics. Faraday sent 386 pages of notes, bound in leather, to Davy; Davy was impressed and appointed Faraday his permanent assistant at the Royal Institution. Faraday toured France and Italy from 1813 to 1815 with Davy, visiting leading scientists of the time such as Volta and Vauquelin.

Despite his limited mathematical ability, Faraday succeeded in making the basic discoveries on which virtually all our uses of electricity depend. He conceived the fundamental nature of magnetism and, to a degree, that of electricity and light.

A modest man who was content to serve science as best he could, Faraday declined a knighthood and an offer to become president of the Royal Society. He was also a moral man; he refused to take part in the preparation of poison gas for use in the Crimean War.

Faraday died in 1867. His many achievements are recognized by the use of his name. The Faraday constant is the quantity of electricity required to deliver a standard amount of substance in electrolysis, and the SI unit of capacitance is the farad.

(Courtesy of AIP Niels Bohr Library)

31.2 MOTIONAL EMF

In Examples 31.1 and 31.2, we considered cases in which an emf is produced in a circuit when the magnetic field changes with time. In this section we describe the so-called **motional emf,** which is the emf induced in a conductor moving through a magnetic field.

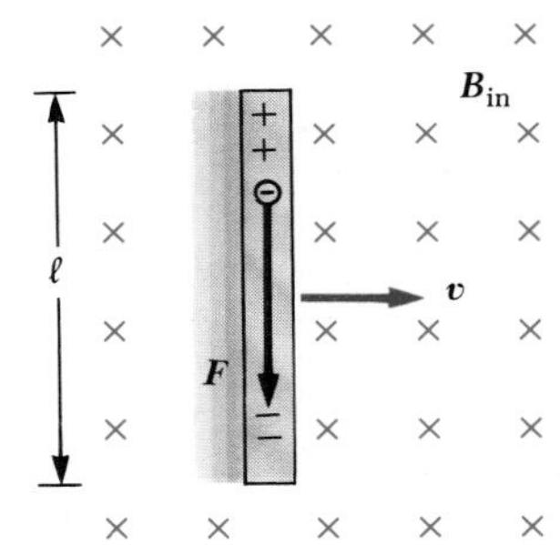

Figure 31.5 A straight conducting bar of length ℓ moving with a velocity $\boldsymbol{v}$ through a uniform magnetic field $\boldsymbol{B}$ directed perpendicular to $\boldsymbol{v}$. An emf equal to $B\ell v$ is induced between the ends of the bar.

First, consider a straight conductor of length ℓ moving with constant velocity through a uniform magnetic field directed into the paper as in Figure 31.5. For simplicity, we shall assume that the conductor is moving perpendicular to the field. The electrons in the conductor will experience a force along the conductor given by $\boldsymbol{F} = q\boldsymbol{v} \times \boldsymbol{B}$. Under the influence of this force, the electrons will move to the *lower* end and accumulate there, leaving a net positive charge at the upper end. An electric field is therefore produced within the conductor as a result of this charge separation. The charge at the ends builds up until the magnetic force qvB is balanced by the electric force qE. At this point, charge stops flowing and the condition for equilibrium requires that

$$qE = qvB \qquad \text{or} \qquad E = vB$$

Since the electric field is constant, the electric field produced in the conductor is related to the potential difference across the ends according to the relation $V = E\ell$. Thus,

$$V = E\ell = B\ell v$$

where the upper end is at a higher potential than the lower end. Thus, *a potential difference is maintained as long as there is motion through the field. If the motion is reversed, the polarity of V is also reversed.*

A more interesting situation occurs if we now consider what happens when the moving conductor is part of a closed conducting path. This situation is particularly useful for illustrating how a changing magnetic flux can cause an induced current in a closed circuit. Consider a circuit consisting of a conducting bar of length ℓ sliding along two fixed parallel conducting rails as in Figure 31.6a. For simplicity, we assume that the moving bar has zero resistance and that the stationary part of the circuit has a resistance R. A uniform and constant magnetic field $\boldsymbol{B}$ is applied perpendicular to the plane of the circuit. As the bar is pulled to the right with a velocity $\boldsymbol{v}$, under the influence of an applied force $\boldsymbol{F}_{app}$, free charges in the bar experience a magnetic force along the length of the bar. This force, in turn, sets up an induced current since the charges are free to move in a closed conducting path. In this case, the rate of change of magnetic flux through the loop and the corresponding induced emf across the moving bar are proportional to the change in area of the loop as the bar moves through the magnetic field. As we shall see, if the bar is pulled to the right with a constant velocity, the work done by the applied force is dissipated in the form of joule heating in the circuit's resistive element.

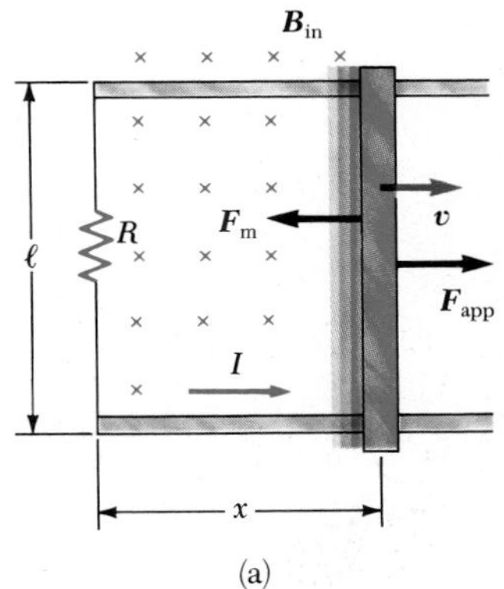

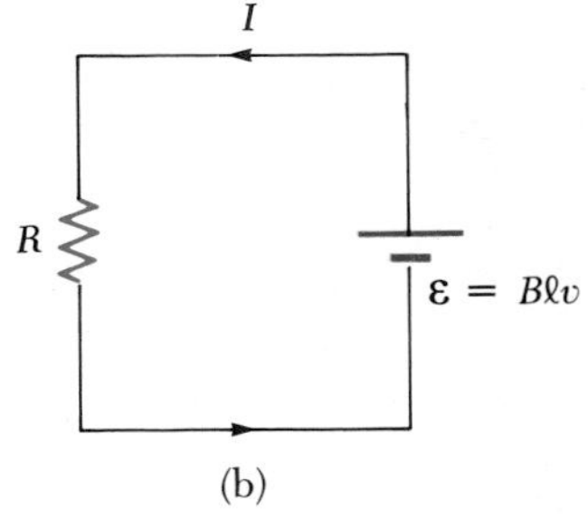

Figure 31.6 (a) A conducting bar sliding with a velocity $\boldsymbol{v}$ along two conducting rails under the action of an applied force $\boldsymbol{F}_{app}$. The magnetic force $\boldsymbol{F}_m$ opposes the motion, and a counterclockwise current is induced in the loop. (b) The equivalent circuit of (a).

Since the area of the circuit at any instant is ℓx, the external magnetic flux through the circuit is given by

$$\Phi_m = B\ell x$$

where x is the width of the circuit, which changes with time. Using Faraday's law, we find that the induced emf is

$$\mathcal{E} = -\frac{d\Phi_m}{dt} = -\frac{d}{dt}(B\ell x) = -B\ell\frac{dx}{dt}$$

$$\mathcal{E} = -B\ell v \tag{31.5}$$

If the resistance of the circuit is R, the magnitude of the induced current is given by

$$I = \frac{|\mathcal{E}|}{R} = \frac{B\ell v}{R} \tag{31.6}$$

The equivalent circuit diagram for this example is shown in Figure 31.6b.

Let us examine the system using energy considerations. Since there is no real battery in the circuit, one might wonder about the origin of the induced current and the electrical energy in the system. We can understand this by noting that the external force does work on the conductor, thereby moving charges through a magnetic field. This causes the charges to move along the conductor with some average drift velocity, and hence a current is established. From the viewpoint of energy conservation, the total work done by the applied force during some time interval should equal the electrical energy that the induced emf supplies in that same period. Furthermore, if the bar moves with constant speed, the work done must equal the energy dissipated as heat in the resistor in this time interval.

As the conductor of length ℓ moves through the uniform magnetic field $\mathbf{B}$, it experiences a magnetic force $\mathbf{F}_m$ of magnitude $I\ell B$ (Section 29.3). The direction of this force is opposite the motion of the bar, or to the left in Figure 31.6a.

If the bar is to move with a *constant* velocity, the applied force must be equal to and opposite the magnetic force, or to the right in Figure 31.6a. If the magnetic force acted in the direction of motion, it would cause the bar to accelerate once it was in motion, thereby increasing its velocity. This state of affairs would represent a violation of the principle of energy conservation. Using Equation 31.6 and the fact that $F_{app} = I\ell B$, we find that the power delivered by the applied force is

$$P = F_{app}v = (I\ell B)v = \frac{B^2\ell^2v^2}{R} \tag{31.7}$$

This power is equal to the rate at which energy is dissipated in the resistor, I^2R, as we would expect. It is also equal to the power $I\mathcal{E}$ supplied by the induced emf. This example is a clear demonstration of the conversion of mechanical energy into electrical energy and finally into thermal energy (joule heating).

EXAMPLE 31.3 Emf Induced in a Rotating Bar

A conducting bar of length ℓ rotates with a constant angular velocity ω about a pivot at one end. A uniform magnetic field $\mathbf{B}$ is directed perpendicular to the plane of rotation, as in Figure 31.7. Find the emf induced between the ends of the bar.

Solution Consider a segment of the bar of length dr, whose velocity is $\mathbf{v}$. According to Equation 31.5, the emf induced in a conductor of this length moving perpendicular to a field $\mathbf{B}$ is given by

$$(1) \qquad d\mathcal{E} = Bv\, dr$$

Each segment of the bar is moving perpendicular to $\mathbf{B}$, so there is an emf generated across each segment; the value of this emf is given by (1). Summing up the emfs induced across all elements, which are in series, gives the total emf between the ends of the bar. That is,

$$\mathcal{E} = \int Bv\, dr$$

In order to integrate this expression, note that the linear

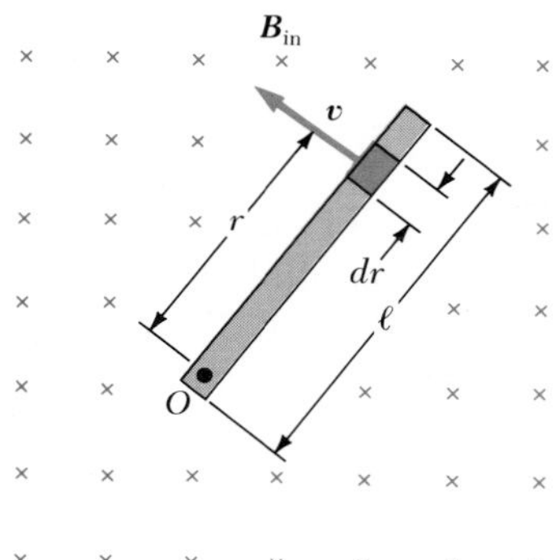

Figure 31.7 (Example 31.3) A conducting bar rotating about a pivot at one end in a uniform magnetic field that is perpendicular to the plane of rotation. An emf is induced across the ends of the bar.

speed of an element is related to the angular speed ω through the relationship $v = r\omega$. Therefore, since B and ω are constants, we find that

$$\mathcal{E} = B\int v\,dr = B\omega\int_0^{\ell} r\,dr = \tfrac{1}{2}B\omega\ell^2$$

EXAMPLE 31.4 Magnetic Force on a Sliding Bar □

A bar of mass m and length ℓ moves on two frictionless parallel rails in the presence of a uniform magnetic field directed into the paper (Fig. 31.8). The bar is given an initial velocity $\mathbf{v}_0$ to the right and is released. Find the velocity of the bar as a function of time.

Solution First note that the induced current is counterclockwise and the magnetic force is $\mathbf{F}_m = -I\ell\mathbf{B}$, where the negative sign denotes that the force is to the left and *retards* the motion. This is the *only* horizontal force acting on the bar, and hence Newton's second law applied to motion in the horizontal direction gives

$$F_x = ma = m\frac{dv}{dt} = -I\ell B$$

Since the induced current is given by Equation 31.6, $I = B\ell v/R$, we can write this expression as

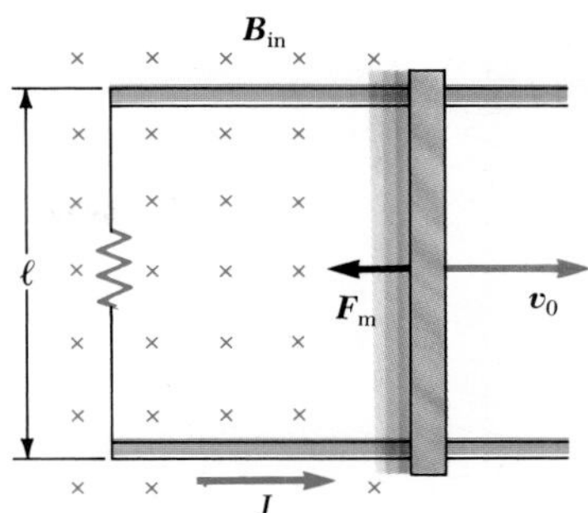

Figure 31.8 (Example 31.4) A conducting bar of length ℓ sliding on two fixed conducting rails is given an initial velocity $\mathbf{v}_0$ to the right.

$$m\frac{dv}{dt} = -\frac{B^2\ell^2}{R}v$$

$$\frac{dv}{v} = -\left(\frac{B^2\ell^2}{mR}\right)dt$$

Integrating this last equation using the initial condition that $v = v_0$ at $t = 0$, we find that

$$\int_{v_0}^{v}\frac{dv}{v} = \frac{-B^2\ell^2}{mR}\int_0^t dt$$

$$\ln\left(\frac{v}{v_0}\right) = -\left(\frac{B^2\ell^2}{mR}\right)t = -\frac{t}{\tau}$$

where the constant $\tau = mR/B^2\ell^2$. From this, we see that the velocity can be expressed in the exponential form

$$v = v_0 e^{-t/\tau}$$

Therefore, the velocity of the bar decreases exponentially with time under the action of the magnetic retarding force. Furthermore, if we substitute this result into Equations 31.5 and 31.6, we find that the induced emf and induced current also decrease exponentially with time. That is,

$$I = \frac{B\ell v}{R} = \frac{B\ell v_0}{R}e^{-t/\tau}$$

$$\mathcal{E} = IR = B\ell v_0 e^{-t/\tau}$$

31.3 LENZ'S LAW

The direction of the induced emf and induced current can be found from **Lenz's law,**[2] which can be stated as follows:

> The polarity of the induced emf is such that it tends to produce a current that will create a magnetic flux to oppose the change in magnetic flux through the loop.

[2] Developed by the German physicist Heinrich Lenz (1804–1865).

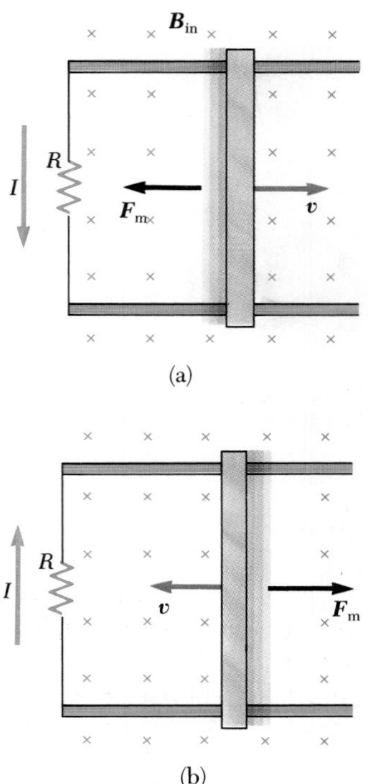

Figure 31.9 (a) As the conducting bar slides on the two fixed conducting rails, the magnetic flux through the loop increases in time. By Lenz's law, the induced current must be *counterclockwise* so as to produce a counteracting flux *out* of the paper. (b) When the bar moves to the left, the induced current must be *clockwise*. Why?

That is, the induced current tends to keep the original flux through the circuit from changing. The interpretation of this statement depends on the circumstances. As we shall see, this law is a consequence of the law of conservation of energy.

In order to obtain a better understanding of Lenz's law, let us return to the example of a bar moving to the right on two parallel rails in the presence of a uniform magnetic field directed into the paper (Fig. 31.9a). As the bar moves to the right, the magnetic flux through the circuit increases with time since the area of the loop increases. Lenz's law says that the induced current must be in a direction such that the flux *it* produces opposes the change in the external magnetic flux. Since the flux due to the external field is increasing *into* the paper, the induced current, if it is to oppose the change, must produce a flux *out* of the paper. Hence, the induced current must be counterclockwise when the bar moves to the right to give a counteracting flux out of the paper in the region *inside* the loop. (Use the right-hand rule to verify this direction.) On the other hand, if the bar is moving to the left, as in Figure 31.9b, the magnet flux through the loop decreases with time. Since the flux is into the paper, the induced current has to be clockwise to produce a flux into the paper inside the loop. In either case, the induced current tends to maintain the original flux through the circuit.

Let us look at this situation from the viewpoint of energy considerations. Suppose that the bar is given a slight push to the right. In the above analysis, we found that this motion leads to a counterclockwise current in the loop. Let us see what happens if we assume that the current is clockwise. For a clockwise current I, the direction of the magnetic force on the sliding bar would be to the right. This force would accelerate the rod and increase its velocity. This, in turn, would cause the area of the loop to increase more rapidly, thus increasing the induced current, which would increase the force, which would increase the current, which would. . . . In effect, the system would acquire energy with no additional input energy. This is clearly inconsistent with all experience and with the law of conservation of energy. Thus, we are forced to conclude that the current must be counterclockwise.

Consider another situation, one in which a bar magnet is moved to the right toward a stationary loop of wire, as in Figure 31.10a. As the magnet moves to the right toward the loop, the magnetic flux through the loop in-

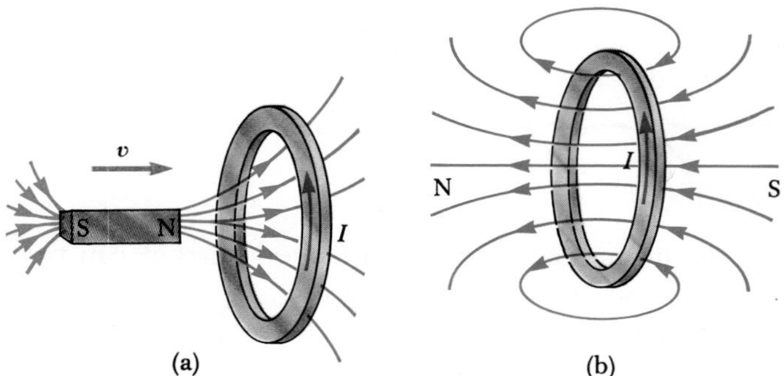

Figure 31.10 (a) When the magnet is moved toward the stationary conducting loop, a current is induced in the direction shown. (b) This induced current produces its own flux to the left to counteract the increasing external flux to the right.

creases with time. To counteract this increase in flux to the right, the induced current produces a flux to the left, as in Figure 31.10b; hence the induced current is in the direction shown. Note that the magnetic field lines associated with the induced current oppose the motion of the magnet. Therefore, the left face of the current loop is a north pole and the right face is a south pole.

On the other hand, if the magnet were moving to the left, its flux through the loop, which is toward the right, would decrease in time. Under these circumstances, the induced current in the loop would be in a direction such as to set up a field through the loop directed from left to right in an effort to maintain a constant number of flux lines. Hence, the induced current in the loop would be opposite that shown in Figure 31.10b. In this case, the left face of the loop would be a south pole and the right face would be a north pole.

EXAMPLE 31.5 Application of Lenz's Law

A coil of wire is placed near an electromagnet as shown in Figure 31.11a. Find the direction of the induced current in the coil (a) at the instant the switch is closed, (b) after the switch has been closed for several seconds, and (c) when the switch is opened.

Solution (a) When the switch is closed, the situation changes from a condition in which no lines of flux pass through the coil to one in which lines of flux pass through in the direction shown in Figure 31.11b. To counteract this change in the number of lines, the coil must set up a field from left to right in the figure. This requires a current directed as shown in Figure 31.11b.

(b) After the switch has been closed for several seconds, there is no change in the number of lines through the loop; hence the induced current is zero.

(c) Opening the switch causes the magnetic field to change from a condition in which flux lines thread through the coil from right to left to a condition of zero flux. The induced current must then be as shown in Figure 31.11c, so as to set up its own field from right to left.

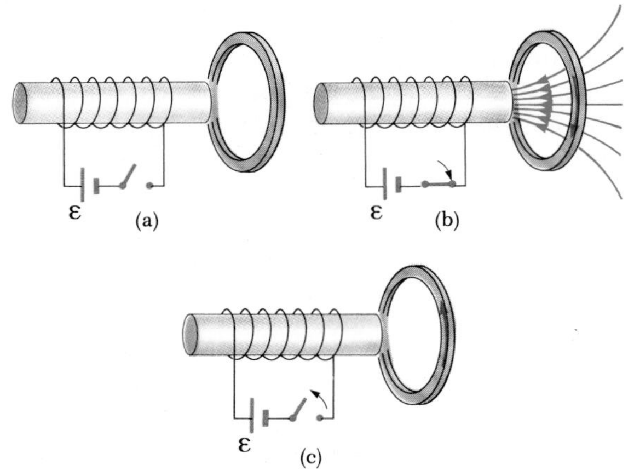

Figure 31.11 (Example 31.5).

EXAMPLE 31.6 A Loop Moving Through a *B* Field

A rectangular loop of dimensions ℓ and w and resistance R moves with constant speed v to the right, as in Figure 31.12a. It continues to move with this speed through a region containing a uniform magnetic field $\mathbf{B}$ directed into the paper and extending a distance $3w$. Plot the flux, the induced emf, and the external force acting on the loop as a function of the position of the loop in the field.

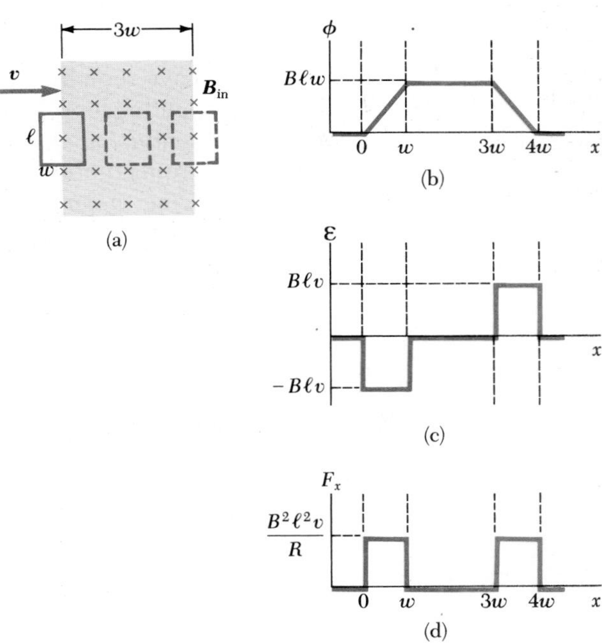

Figure 31.12 (Example 31.6) (a) A conducting rectangular loop of width w and length ℓ moving with a velocity $\mathbf{v}$ through a uniform magnetic field extending a distance $3w$. (b) A plot of the flux as a function of the position of the loop. (c) A plot of the induced emf versus the position of the leading edge. (d) A plot of the force versus position such that the velocity of the loop remains constant.

Solution Figure 31.12b shows the flux through the loop as a function of loop position. Before the loop enters the field, the flux is zero. As it enters the field, the flux increases linearly with position. Finally, the flux decreases linearly to zero as the loop leaves the field.

Before the loop enters the field, there is no induced emf since there is no field present (Fig. 31.12c). As the right side of the loop enters the field, the flux inward begins to increase. Hence, according to Lenz's law, the induced current is counterclockwise and the induced emf is given by $-B\ell v$. This motional emf arises from the magnetic force experienced by charges in the right side of the loop. When the loop is entirely in the field, the *change* in flux is zero, and hence the induced emf vanishes.

From another point of view, the right and left sides of the loop experience magnetic forces that tend to set up currents that cancel one another. As the right side of the loop leaves the field, the flux inward begins to decrease, a clockwise current is induced, and the induced emf is $B\ell v$. As soon as the left side leaves the field, the emf drops to zero.

The external force that must act on the loop to maintain this motion is plotted in Figure 31.12d. When the loop is not in the field, there is no magnetic force on it; hence the external force on it must be zero if v is constant. When the right side of the loop enters the field, the external force necessary to maintain constant speed must be equal to and opposite the magnetic force on that side, given by $F_m = -I\ell B = -B^2\ell^2 v/R$. When the loop is entirely in the field, the flux through the loop is not changing with time. Hence, the net emf induced in the loop is zero, and the current is also zero. Therefore, no external force is needed to maintain the motion of the loop. (From another point of view, the right and left sides of the loop experience equal and opposite forces; hence, the net force is zero.) Finally, as the right side leaves the field, the external force must be equal to and opposite the magnetic force on the left side of the loop. From this analysis, we conclude that power is supplied only when the loop is either entering or leaving the field. Furthermore, this example shows that the induced emf in the loop can be zero even when there is motion through the field! Again, it is emphasized that an emf is induced in the loop *only* when the magnetic flux through the loop *changes* in time.

31.4 INDUCED EMFS AND ELECTRIC FIELDS

We have seen that a changing magnetic flux induces an emf and a current in a conducting loop. We therefore must conclude that *an electric field is created in the conductor as a result of the changing magnetic flux.* In fact, the law of electromagnetic induction shows that *an electric field is always generated by a changing magnetic flux,* even in free space where no charges are present. However, this induced electric field has properties that are quite different from those of an electrostatic field *produced by stationary charges.*

We can illustrate this point by considering a conducting loop of radius r situated in a uniform magnetic field that is perpendicular to the plane of the loop, as in Figure 31.13. If the magnetic field changes with time, then Faraday's law tells us that an emf given by $\mathcal{E} = -d\Phi_m/dt$ is induced in the loop. The induced current that is produced implies the presence of an induced electric field $\boldsymbol{E}$, which must be tangent to the loop since all points on the loop are equivalent. The work done in moving a test charge q once around the loop is equal to $q\mathcal{E}$. Since the electric force on the charge is $q\boldsymbol{E}$, the work done by this force in moving the charge once around the loop is given by $qE(2\pi r)$, where $2\pi r$ is the circumference of the loop. These two expressions for the work must be equal; therefore we see that

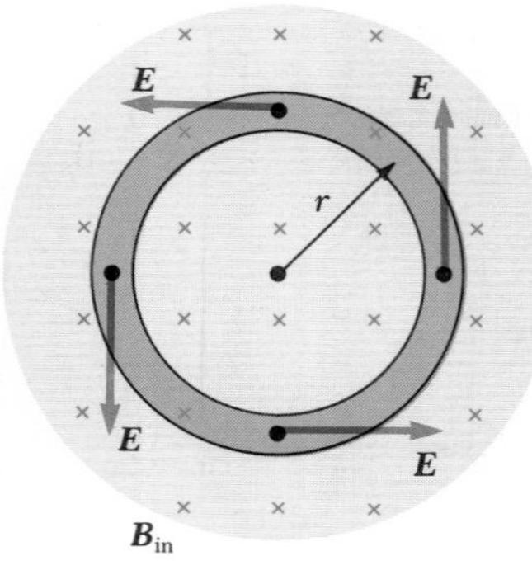

Figure 31.13 A loop of radius r in a uniform magnetic field perpendicular to the plane of the loop. If $\boldsymbol{B}$ changes in time, an electric field is induced in a direction tangent to the loop.

$$q\mathcal{E} = qE(2\pi r)$$

$$E = \frac{\mathcal{E}}{2\pi r}$$

Using this result, Faraday's law, and the fact that $\Phi_m = BA = \pi r^2 B$ for a circular loop, we find that the induced electric field can be expressed as

$$E = -\frac{1}{2\pi r}\frac{d\Phi_m}{dt} = -\frac{r}{2}\frac{dB}{dt} \qquad (31.8)$$

If the time variation of the magnetic field is specified, the induced electric field can easily be calculated from Equation 31.8. The negative sign indicates that the induced electric field **E** *opposes* the change in the magnetic field. It is important to understand that *this result is also valid in the absence of a conductor.* That is, a free charge placed in a changing magnetic field will also experience the same electric field.

In general, the emf for any closed path can be expressed as the line integral of $\mathbf{E}\cdot d\mathbf{s}$ over that path. Hence, the general form of Faraday's law of induction is given by

$$\mathcal{E} = \oint \mathbf{E}\cdot d\mathbf{s} = -\frac{d\Phi_m}{dt} \qquad (31.9)$$

Faraday's law in general form

It is important to recognize that *the induced electric field* **E** *that appears in Equation 31.9 is a nonconservative, time-varying field that is generated by a changing magnetic field.* The field **E** that satisfies Equation 31.9 could not possibly be an electrostatic field for the following reason. If the field were electrostatic, and hence conservative, the line integral of $\mathbf{E}\cdot d\mathbf{s}$ over a closed loop would be zero, contrary to Equation 31.9.

EXAMPLE 31.7 Electric Field Due to a Solenoid

A long solenoid of radius R has n turns per unit length and carries a time-varying current that varies sinusoidally as $I = I_0 \cos \omega t$, where I_0 is the maximum current and ω is the angular frequency of the current source (Fig. 31.14). (a) Determine the electric field outside the solenoid, a distance r from its axis.

First, let us consider an external point and take the path for our line integral to be a circle centered on the solenoid, as in Figure 31.14. By symmetry we see that the magnitude of **E** is constant on this path and tangent to it. The magnetic flux through this path is given by $BA = B(\pi R^2)$, and hence Equation 31.9 gives

$$\oint \mathbf{E}\cdot d\mathbf{s} = -\frac{d}{dt}[B(\pi R^2)] = -\pi R^2 \frac{dB}{dt}$$

$$E(2\pi r) = -\pi R^2 \frac{dB}{dt}$$

Since the magnetic field inside a long solenoid is given by Equation 30.20, $B = \mu_0 nI$, and $I = I_0 \cos \omega t$, we find that

$$E(2\pi r) = -\pi R^2 \mu_0 n I_0 \frac{d}{dt}(\cos \omega t) = \pi R^2 \mu_0 n I_0 \omega \sin \omega t$$

$$E = \frac{\mu_0 n I_0 \omega R^2}{2r} \sin \omega t \qquad (\text{for } r > R)$$

Hence, the electric field varies sinusoidally with time, and its amplitude falls off as $1/r$ outside the solenoid.

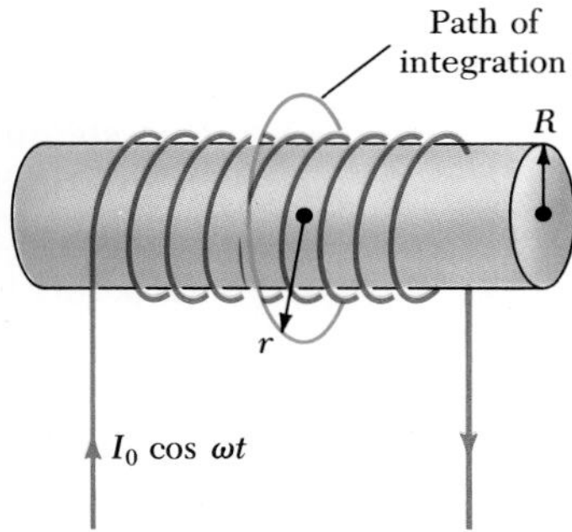

Figure 31.14 (Example 31.7) A long solenoid carrying a time-varying current given by $I = I_0 \cos \omega t$. An electric field is induced both inside and outside the solenoid.

(b) What is the electric field inside the solenoid, a distance r from its axis?

For an interior point ($r < R$), the flux threading an integration loop is given by $B(\pi r^2)$. Using the same procedure as in (a), we find that

$$E(2\pi r) = -\pi r^2 \frac{dB}{dt} = \pi r^2 \mu_0 n I_0 \omega \sin \omega t$$

$$E = \frac{\mu_0 n I_0 \omega}{2} r \sin \omega t \qquad (\text{for } r < R)$$

This shows that the amplitude of the electric field *inside* the solenoid increases linearly with r and varies sinusoidally with time.

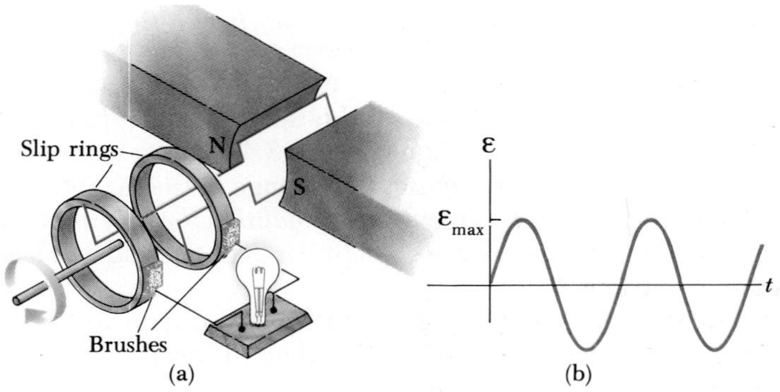

Figure 31.15 (a) Schematic diagram of an ac generator. An emf is induced in a coil that rotates by some external means in a magnetic field. (b) The alternating emf induced in the loop plotted versus time.

*31.5 GENERATORS AND MOTORS

Generators and motors are important devices that operate on the principle of electromagnetic induction. First, let us consider the **alternating current generator** (or ac generator), a device that converts mechanical energy to electrical energy. In its simplest form, the ac generator consists of a loop of wire rotated by some external means in a magnetic field (Fig. 31.15a). In commercial power plants, the energy required to rotate the loop can be derived from a variety of sources. For example, in a hydroelectric plant, falling water directed against the blades of a turbine produces the rotary motion; in a coal-fired plant, the heat produced by burning coal is used to convert water to steam and this steam is directed against the turbine blades. As the loop rotates, the magnetic flux through it changes with time, inducing an emf and a current in an external circuit. The ends of the loop are connected to slip rings that rotate with the loop. Connections to the external circuit are made by stationary brushes in contact with the slip rings.

To put our discussion of the generator on a quantitative basis, suppose that the loop has N turns (a more practical situation), all of the same area A, and suppose that the loop rotates with a constant angular velocity ω. If θ is the angle between the magnetic field and the normal to the plane of the loop as in Figure 31.16, then the magnetic flux through the loop at any time t is given by

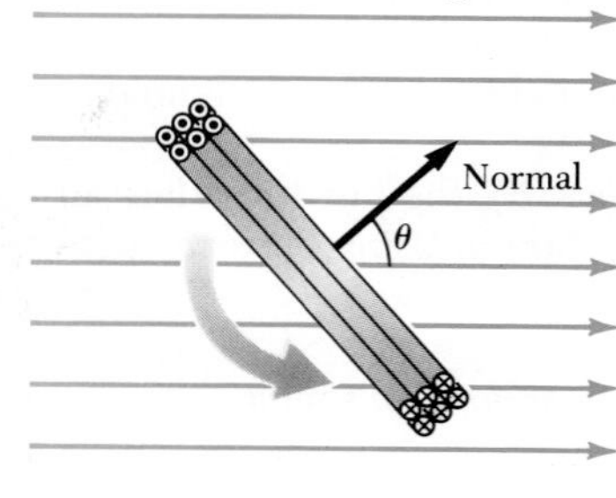

Figure 31.16 A loop of area A containing N turns, rotating with constant angular velocity ω in the presence of a magnetic field. The emf induced in the loop varies sinusoidally in time.

$$\Phi_m = BA \cos \theta = BA \cos \omega t$$

where we have used the relationship between angular displacement and angular velocity, $\theta = \omega t$. (We have set the clock so that $t = 0$ when $\theta = 0$.) Hence, the induced emf in the coil is given by

$$\mathcal{E} = -N \frac{d\Phi_m}{dt} = -NAB \frac{d}{dt} (\cos \omega t) = NAB\omega \sin \omega t \qquad (31.10)$$

This result shows that the emf varies sinusoidally with time, as plotted in Figure 31.15b. From Equation 31.10 we see that the maximum emf has the value

$$\mathcal{E}_{max} = NAB\omega \qquad (31.11)$$

which occurs when $\omega t = 90°$ or $270°$. In other words, $\mathcal{E} = \mathcal{E}_{max}$ when the magnetic field is in the plane of the coil, and the time rate of change of flux is a maximum. Furthermore, the emf is *zero* when $\omega t = 0$ or $180°$, that is, when $\boldsymbol{B}$ is perpendicular to the plane of the coil, and the time rate of change of flux is zero. The frequency for commercial generators in the United States and Canada is 60 Hz, whereas in some European countries, 50 Hz is used. (Recall that $\omega = 2\pi f$, where f is the frequency in hertz.)

EXAMPLE 31.8 Emf Induced in a Generator

An ac generator consists of 8 turns of wire each of area $A = 0.09\ \text{m}^2$ and total resistance $12\ \Omega$. The loop rotates in a magnetic field $B = 0.5$ T at a constant frequency of 60 Hz. (a) Find the maximum induced emf.

First note that $\omega = 2\pi f = 2\pi(60\ \text{Hz}) = 377\ \text{s}^{-1}$. Using Equation 31.11 with the appropriate numerical values gives

$$\mathcal{E}_{max} = NAB\omega = 8(0.09\ \text{m}^2)(0.5\ \text{T})(377\ \text{s}^{-1}) = 136\ \text{V}$$

(b) What is the maximum induced current?

From Ohm's law and the results to (a), we find that the maximum induced current is

$$I_{max} = \frac{\mathcal{E}_{max}}{R} = \frac{136\ \text{V}}{12\ \Omega} = 11.3\ \text{A}$$

Exercise 2 Determine the time variation of the induced emf and induced current when the output terminals are connected by a low-resistance conductor.

Answers:

$$\mathcal{E} = \mathcal{E}_{max} \sin \omega t = (136\ \text{V}) \sin 377t$$

$$I = I_{max} \sin \omega t = (11.3\ \text{A}) \sin 377t$$

The **direct current** (dc) **generator** is illustrated in Figure 31.17a. Such generators are used, for instance, to charge storage batteries used in older style cars. The components are essentially the same as those of the ac generator, except that the contacts to the rotating loop are made using a split ring, or commutator.

In this configuration, the output voltage always has the same polarity and the current is a pulsating direct current as in Figure 31.17b. The reason for this can be understood by noting that the contacts to the split ring reverse their roles every half cycle. At the same time, the polarity of the induced emf reverses; hence the polarity of the split ring (which is the same as the polarity of the output voltage) remains the same.

A pulsating dc current is not suitable for most applications. To obtain a more steady dc current, commercial dc generators use many armature coils

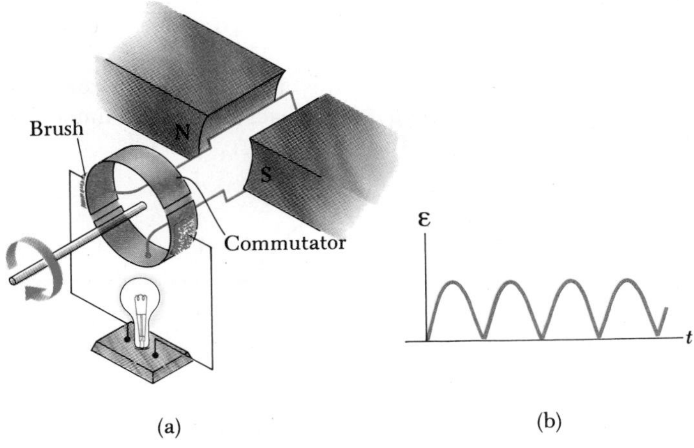

Figure 31.17 (a) Schematic diagram of a dc generator. (b) The emf versus time fluctuates in magnitude but always has the same polarity.

and commutators distributed so that the sinusoidal pulses from the various coils are out of phase. When these pulses are superimposed, the dc output is almost free of fluctuations.

Motors are devices that convert electrical energy into mechanical energy. Essentially, *a motor is a generator operating in reverse.* Instead of generating a current by rotating a loop, a current is supplied to the loop by a battery and the torque acting on the current-carrying loop causes it to rotate.

Useful mechanical work can be done by attaching the rotating armature to some external device. However, as the loop rotates, the changing magnetic flux induces an emf in the loop; this induced emf *always* acts to reduce the current in the loop. If this were not the case, Lenz's law would be violated. The back emf increases in magnitude as the rotational speed of the armature increases. (The phrase *back emf* is used to indicate an emf that tends to reduce the supplied current.) Since the voltage available to supply current equals the difference between the supply voltage and the back emf, the current through the armature coil is limited by the back emf.

When a motor is first turned on, there is initially no back emf and the current is very large because it is limited only by the resistance of the coil. As the coils begin to rotate, the induced back emf opposes the applied voltage and the current in the coils is reduced. If the mechanical load increases, the motor will slow down, which causes the back emf to decrease. This reduction in the back emf increases the current in the coils and therefore also increases the power needed from the external voltage source. For this reason, the power requirements are greater for starting a motor and for running it under heavy loads. If the motor is allowed to run under no mechanical load, the back emf reduces the current to a value just large enough to overcome energy losses due to heat and friction.

EXAMPLE 31.9 The Induced Current in a Motor

Assume that a motor having coils with a resistance of 10 Ω is supplied by a voltage of 120 V. When the motor is running at its maxiumum speed, the back emf is 70 V. Find the current in the coils (a) when the motor is first turned on and (b) when the motor has reached maximum speed.

Solution (a) When the motor is first turned on, the back emf is zero. (The coils are motionless.) Thus the current in the coils is a maximum and equal to

$$I = \frac{\mathcal{E}}{R} = \frac{120\ \text{V}}{10\ \Omega} = 12\ \text{A}$$

(b) At the maximum speed, the back emf has its maximum value. Thus, the effective supply voltage is now that of the external source minus the back emf. Hence, the current is reduced to

$$I = \frac{\mathcal{E} - \mathcal{E}_{\text{back}}}{R} = \frac{120\ \text{V} - 70\ \text{V}}{10\ \Omega} = \frac{50\ \text{V}}{10\ \Omega} = 5\ \text{A}$$

Exercise 3 If the current in the motor is 8 A at some instant, what is the back emf at this time?

Answer 40 V.

*31.6 EDDY CURRENTS

As we have seen, an emf and a current are induced in a circuit by a changing magnetic flux. In the same manner, circulating currents called **eddy currents** are set up in bulk pieces of metal moving through a magnetic field. This can easily be demonstrated by allowing a flat metal plate at the end of a rigid bar to swing as a pendulum through a magnetic field (Fig. 31.18). The metal should be a material such as aluminum or copper. As the plate enters the field, the changing flux creates an induced emf in the plate, which in turn causes the free electrons in the metal to move, producing the swirling eddy currents. According to Lenz's law, the direction of the eddy currents must oppose the change that causes them. For this reason, the eddy currents must produce effective magnetic poles on the plate, which are repelled by the poles of the magnet, thus giving rise to a repulsive force that opposes the motion of the pendulum. (If the opposite were true, the pendulum would accelerate and its energy would increase after each swing, in violation of the law of energy conservation.) Alternatively, the retarding force can be "felt" by pulling a metal sheet through the field of a strong magnet.

As indicated in Figure 31.19, with $\boldsymbol{B}$ into the paper the eddy current is counterclockwise as the swinging plate *enters* the field in position 1. This is because the external flux into the paper is increasing, and hence by Lenz's law the induced current must provide a flux out of the paper. The opposite is true as the plate leaves the field in position 2, where the current is clockwise. Since the induced eddy current always produces a retarding force $\boldsymbol{F}$ when the plate enters or leaves the field, the swinging plate eventually comes to rest.

If slots are cut in the metal plate as in Figure 31.20, the eddy currents and the corresponding retarding force are greatly reduced. This can be understood since the cuts in the plate result in open circuits for any large current loops that might otherwise be formed.

The braking systems on many subway and rapid transit cars make use of electromagnetic induction and eddy currents. An electromagnet, which can be energized with a current, is positioned near the steel rails. The braking action occurs when a large current is passed through the electromagnet. The relative motion of the magnet and rails induces eddy currents in the rails, and the direction of these currents produces a drag force on the moving vehicle. The loss in mechanical energy of the vehicle is transformed into joule heat. Since the eddy currents decrease steadily in magnitude as the vehicle slows down, the braking effect is quite smooth. Eddy current brakes are also used in some mechanical balances and in various machines.

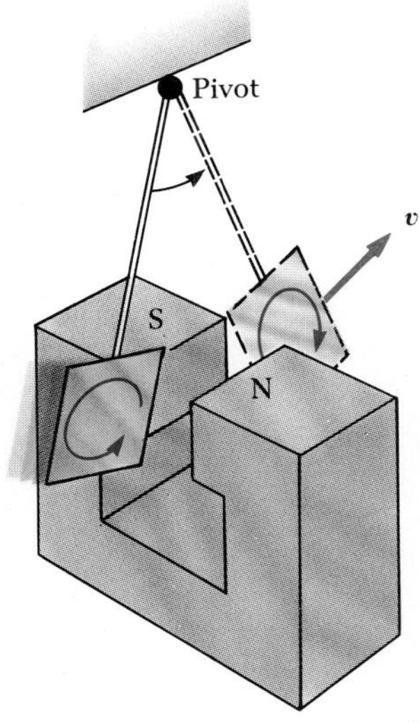

Figure 31.18 An apparatus that demonstrates the formation of eddy currents in a conductor moving through a magnetic field. As the plate enters or leaves the field, the changing flux sets up an induced emf, which causes the eddy currents.

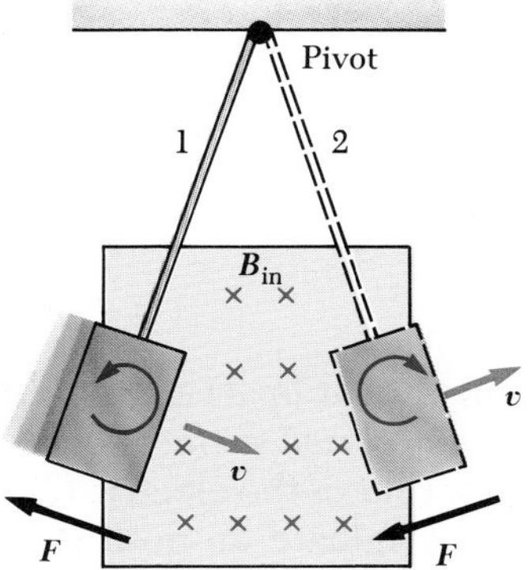

Figure 31.19 As the conducting plate enters the field in position 1, the eddy currents are counterclockwise. However, in position 2, the currents are clockwise. In either case, the plate is repelled by the magnet and eventually comes to rest.

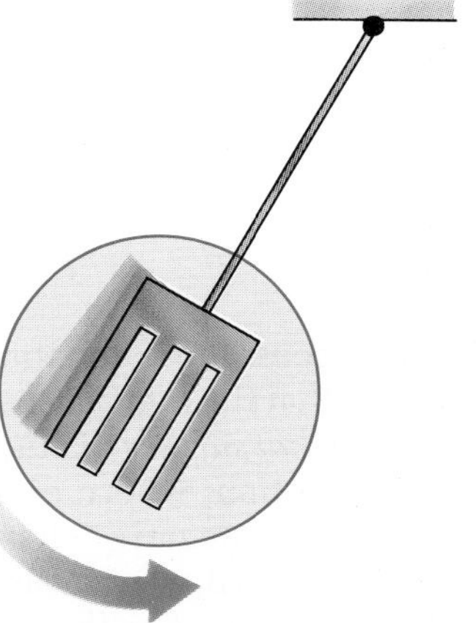

Figure 31.20 When slots are cut in the conducting plate, the eddy currents are reduced and the plate swings more freely through the magnetic field.

Eddy currents are often undesirable since they dissipate energy in the form of heat. To reduce this energy loss, the moving conducting parts are often laminated, that is, built up in thin layers separated by a nonconducting material such as lacquer or a metal oxide. This layered structure increases the resistance of the possible paths of the eddy currents and effectively confines the currents to individual layers. Such a laminated structure is used in the cores of transformers and motors to minimize eddy currents and thereby increase the efficiency of these devices.

31.7 MAXWELL'S WONDERFUL EQUATIONS

We conclude this chapter by presenting four equations that can be regarded as the basis of all electrical and magnetic phenomena. These equations, known as Maxwell's equations, after James Clerk Maxwell, are as fundamental to electromagnetic phenomena as Newton's laws are to the study of mechanical phenomena. In fact, the theory developed by Maxwell was more far-reaching than even he imagined at the time, because it turned out to be in agreement with the special theory of relativity, as Einstein showed in 1905. As we shall see, Maxwell's equations represent laws of electricity and magnetism that have already been discussed. However, the equations have additional important consequences. In Chapter 34 we shall show that these equations predict the existence of electromagnetic waves (traveling patterns of electric and magnetic fields), which travel with a speed $c = 1/\sqrt{\mu_0\epsilon_0} \approx 3 \times 10^8$ m/s, the speed of light. Furthermore, the theory shows that such waves are radiated by accelerating charges.

For simplicity, we present **Maxwell's equations** as applied to *free space*, that is, in the absence of any dielectric or magnetic material. The four equations are:

Gauss' law

$$\oint \boldsymbol{E} \cdot d\boldsymbol{A} = \frac{Q}{\epsilon_0} \tag{31.12}$$

Gauss' law in magnetism

$$\oint \boldsymbol{B} \cdot d\boldsymbol{A} = 0 \tag{31.13}$$

Faraday's law

$$\oint \boldsymbol{E} \cdot d\boldsymbol{s} = -\frac{d\Phi_m}{dt} \tag{31.14}$$

Ampère-Maxwell law

$$\oint \boldsymbol{B} \cdot d\boldsymbol{s} = \mu_0 I + \epsilon_0\mu_0 \frac{d\Phi_e}{dt} \tag{31.15}$$

Let us discuss these equations one at a time. Equation 31.12 is *Gauss' law*, which states that the *total electric flux through any closed surface equals the net charge inside that surface divided by* ϵ_0. This law relates the electric field to the charge distribution, where electric field lines originate on positive charges and terminate on negative charges.

Equation 31.13, which can be considered *Gauss' law in magnetism*, says that *the net magnetic flux through a closed surface is zero.* That is, the number of magnetic field lines that enter a closed volume must equal the number that leave that volume. This implies that magnetic field lines cannot begin or end at any point. If they did, this would mean that isolated magnetic monopoles existed at those points. The fact that isolated magnetic monopoles have not been observed in nature can be taken as a confirmation of Equation 31.13.

Equation 31.14 is *Faraday's law of induction,* which describes the relationship between an electric field and a changing magnetic flux. This law states that *the line integral of the electric field around any closed path (which equals the emf) equals the rate of change of magnetic flux through any surface area bounded by that path.* One consequence of Faraday's law is the current induced in a conducting loop placed in a time-varying magnetic field.

Equation 31.15 is the generalized form of Ampère's law, which describes a relationship between magnetic and electric fields and electric currents. That is, *the line integral of the magnetic field around any closed path is determined by the sum of the net conduction current through that path and the rate of change of electric flux through any surface bounded by that path.*

Once the electric and magnetic fields are known at some point in space, the force on a particle of charge q can be calculated from the expression

$$\boldsymbol{F} = q\boldsymbol{E} + q\boldsymbol{v} \times \boldsymbol{B} \tag{31.16}$$

The Lorentz force

This is called the **Lorentz force.** Maxwell's equations, together with this force law, give a complete description of all classical electromagnetic interactions.

It is interesting to note the symmetry of Maxwell's equations. Equations 31.12 and 31.13 are symmetric, apart from the absence of a magnetic monopole term in Equation 31.13. Furthermore, Equations 31.14 and 31.15 are symmetric in that the line integrals of $\boldsymbol{E}$ and $\boldsymbol{B}$ around a closed path are related to the rate of change of magnetic flux and electric flux, respectively. "Maxwell's wonderful equations," as they were called by John R. Pierce,[3] are of fundamental importance not only to electronics but to all of science. Heinrich Hertz once wrote, "One cannot escape the feeling that these mathematical formulas have an independent existence and an intelligence of their own, that they are wiser than we are, wiser even than their discoverers, that we get more out of them than we put into them."

[3] John R. Pierce, *Electrons and Waves,* New York, Doubleday Science Study Series, 1964. Chapter 6 of this interesting book is recommended as supplemental reading.

SUMMARY

Faraday's law of induction states that the emf induced in a circuit is directly proportional to the time rate of change of magnetic flux through the circuit. That is,

$$\mathcal{E} = -\frac{d\Phi_m}{dt} \tag{31.1}$$

Faraday's law

where Φ_m is the magnetic flux, given by

$$\Phi_m = \int \boldsymbol{B} \cdot d\boldsymbol{A}$$

When a conducting bar of length ℓ moves through a magnetic field $\boldsymbol{B}$ with a speed v such that $\boldsymbol{B}$ is perpendicular to the bar, the emf induced in the bar (the so-called **motional emf**) is given by

Motional emf

$$\mathcal{E} = -B\ell v \tag{31.5}$$

Lenz's law states that the induced current and induced emf in a conductor are in such a direction as to oppose the change that produced them.

A general form of **Faraday's law of induction** is

Faraday's law in general form

$$\mathcal{E} = \oint \boldsymbol{E} \cdot d\boldsymbol{s} = -\frac{d\Phi_m}{dt} \tag{31.9}$$

where $\boldsymbol{E}$ is a nonconservative, time-varying electric field that is produced by the changing magnetic flux.

When used with the Lorentz force law, $\boldsymbol{F} = q\boldsymbol{E} + q\boldsymbol{v} \times \boldsymbol{B}$, **Maxwell's equations,** given below in integral form, describe *all* electromagnetic phenomena:

Gauss' law (electricity)

$$\oint \boldsymbol{E} \cdot d\boldsymbol{A} = \frac{Q}{\epsilon_0} \tag{31.12}$$

Gauss' law (magnetism)

$$\oint \boldsymbol{B} \cdot d\boldsymbol{A} = 0 \tag{31.13}$$

Faraday's law

$$\oint \boldsymbol{E} \cdot d\boldsymbol{s} = -\frac{d\Phi_m}{dt} \tag{31.14}$$

Ampère-Maxwell law

$$\oint \boldsymbol{B} \cdot d\boldsymbol{s} = \mu_0 I + \mu_0 \epsilon_0 \frac{d\Phi_e}{dt} \tag{31.15}$$

The last two equations are of particular importance for the material discussed in this chapter. Faraday's law describes how an electric field can be induced by a changing magnetic flux. Similarly, the Ampère-Maxwell law describes how a magnetic field can be produced by both a conduction current and a changing electric flux.

QUESTIONS

1. What is the difference between magnetic flux and magnetic field?
2. A circular loop is located in a uniform and constant magnetic field. Describe how an emf can be induced in the loop in this situation.
3. A loop of wire is placed in a uniform magnetic field. For what orientation of the loop is the magnetic flux a maximum? For what orientation is the flux zero?
4. As the conducting bar in Figure 31.21 moves to the right, an electric field is set up directed downward. If the bar were moving to the left, explain why the electric field would be upward.
5. As the bar in Figure 31.21 moves perpendicular to the field, is an external force required to keep it moving with constant velocity?

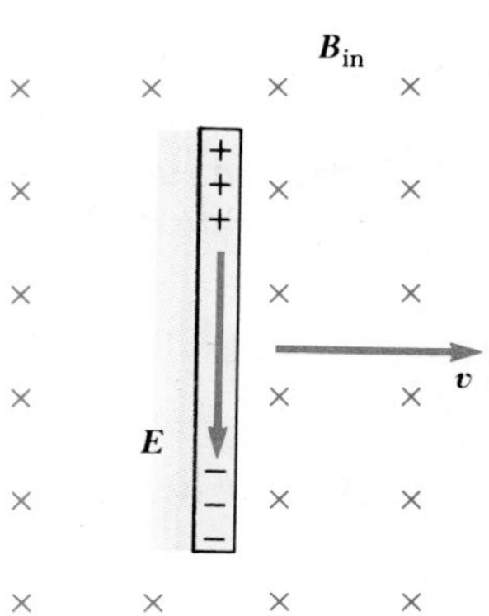

Figure 31.21 (Questions 4 and 5).

6. The bar in Figure 31.22 moves on rails to the right with a velocity $\boldsymbol{v}$, and the uniform, constant magnetic field is *outward*. Why is the induced current clockwise? If the bar were moving to the left, what would be the direction of the induced current?

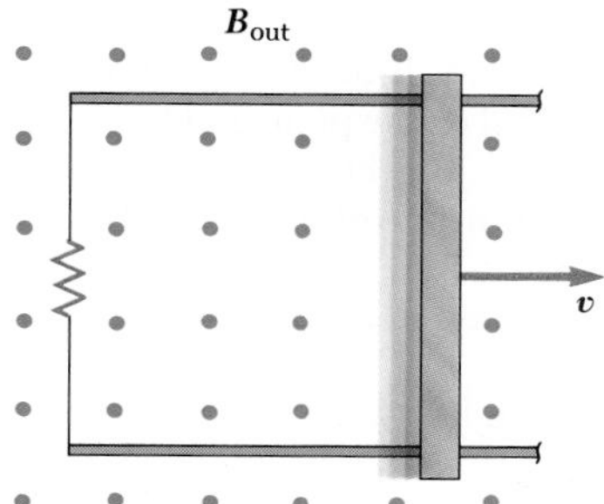

Figure 31.22 (Questions 6 and 7).

7. Explain why an external force is necessary to keep the bar in Figure 31.22 moving with a constant velocity.
8. A large circular loop of wire lies in the horizontal plane. A bar magnet is dropped through the loop. If the axis of the magnet remains horizontal as it falls, describe the emf induced in the loop. How is the situation altered if the axis of the magnet remains vertical as it falls?
9. When a small magnet is moved toward a solenoid, an emf is induced in the coil. However, if the magnet is moved around inside a toroidal coil, there is *no* induced emf. Explain.
10. Will dropping a magnet down a long copper tube produce a current in the tube? Explain.
11. How is electrical energy produced in dams (that is, how is the energy of motion of the water converted into ac electricity)?
12. In a beam balance scale, an aluminum plate is sometimes used to slow the oscillations of the beam near equilibrium. The plate is mounted at the end of the beam, and moves between the poles of a small horseshoe magnet, attached to the frame. Why are the oscillations of the beam strongly damped near equilibrium?
13. What happens when the coil of a generator is rotated at a faster rate?
14. Could a current be induced in a coil by rotating a magnet inside the coil? If so, how?
15. When the switch in the circuit shown in Figure 31.23a is closed, a current is set up in the coil and the metal ring springs upward (see Fig. 31.23b). Explain this behavior.

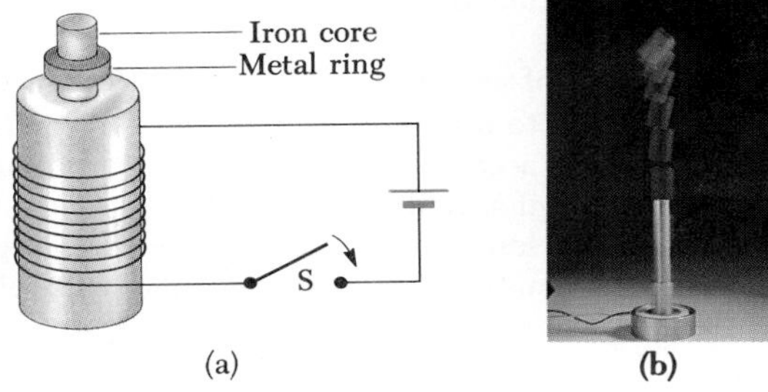

Figure 31.23 (Questions 15 and 16). (Courtesy of CENCO)

16. Assume that the battery in Figure 31.23a is replaced by an alternating current source and the switch S is held closed. If the metal ring on top of the solenoid is held down, it will become *hot*. Why?
17. Identify the individual generally associated with each of Maxwell's four equations.
18. Do Maxwell's equations (Section 31.7) allow for the existence of magnetic "charges," that is, isolated N or S poles?

PROBLEMS

Section 31.1 Faraday's Law of Induction

1. A 50-turn rectangular coil of dimensions 5 cm × 10 cm is "dropped" from a position where $B = 0$ to a new position where $B = 0.5$ T and is directed perpendicular to the plane of the coil. Calculate the resulting average emf induced in the coil if the displacement occurs in 0.25 s.
2. A plane loop of wire consisting of a single turn of cross-sectional area 8.0 cm^2 is perpendicular to a magnetic field that increases uniformly in magnitude from 0.5 T to 2.5 T in a time of 1.0 s. What is the resulting induced current if the coil has a total resistance of 2 Ω?
3. A powerful electromagnet has a field of 1.6 T and a cross sectional area of 0.2 m^2. If we place a coil of 200 turns with a total resistance of 20 Ω around the electromagnet, and then turn off the power to the electromagnet in 0.02 s, what will be the induced current in the coil?
4. A square, single-turn coil 0.20 m on a side is placed with its plane perpendicular to a constant magnetic field. An emf of 18 mV is induced in the winding when the area of the coil decreases at a rate of 0.1 m^2/s. What is the magnitude of the magnetic field?
5. The plane of a rectangular coil of dimensions 5 cm by 8 cm is perpendicular to the direction of a magnetic field $\boldsymbol{B}$. If the coil has 75 turns and a total resistance of 8 Ω, at what rate must the magnitude of $\boldsymbol{B}$ change in order to induce a current of 0.1 A in the windings of the coil?

6. A tightly wound circular coil has 50 turns, each of radius 0.1 m. A uniform magnetic field is turned on along a direction perpendicular to the plane of the coil. If the field increases linearly from 0 to 0.6 T in a time of 0.2 s, what emf is induced in the windings of the coil?
7. A 30-turn circular coil of radius 4 cm and resistance 1 Ω is placed in a magnetic field directed perpendicular to the plane of the coil. The magnitude of the magnetic field varies in time according to the expression $B = 0.01t + 0.04t^2$, where t is in s and B is in T. Calculate the induced emf in the coil at $t = 5$ s.
8. A plane loop of wire of 10 turns, each of area 14 cm^2, is perpendicular to a magnetic field whose magnitude changes in time according to $B = (0.5\text{ T})\sin(60\pi t)$. What is the induced emf in the loop as a function of time?
9. A plane loop of wire of area 14 cm^2 with 2 turns is perpendicular to a magnetic field whose magnitude decays in time according to $B = (0.5\text{ T})e^{-t/7}$. What is the induced emf as a function of time?
10. A rectangular loop of area A is placed in a region where the magnetic field is perpendicular to the plane of the loop. The magnitude of the field is allowed to vary in time according to $B = B_0 e^{-t/\tau}$, where B_0 and τ are constants. The field has a value of B_0 at $t \leq 0$. (a) Use Faraday's law to show that the emf induced in the loop is given by

$$\mathcal{E} = \frac{AB_0}{\tau} e^{-t/\tau}$$

(b) Obtain a numerical value for $\mathcal{E}$ at $t = 4$ s when $A = 0.16$ m^2, $B_0 = 0.35$ T, and $\tau = 2$ s. (c) For the values of A, B_0, and τ given in (b), what is the *maximum* value of $\mathcal{E}$?
11. A long solenoid has n turns per meter and carries a current $I = I_0(1 - e^{-\alpha t})$, with $I_0 = 30$ A and $\alpha = 1.6$ s^{-1}. Inside the solenoid and coaxial with it is a loop that has a radius $R = 6$ cm and consists of a total of N turns of fine wire. What emf is induced in the loop by the changing current? Take $n = 400$ turns/m and $N = 250$ turns. (See Figure 31.24.)

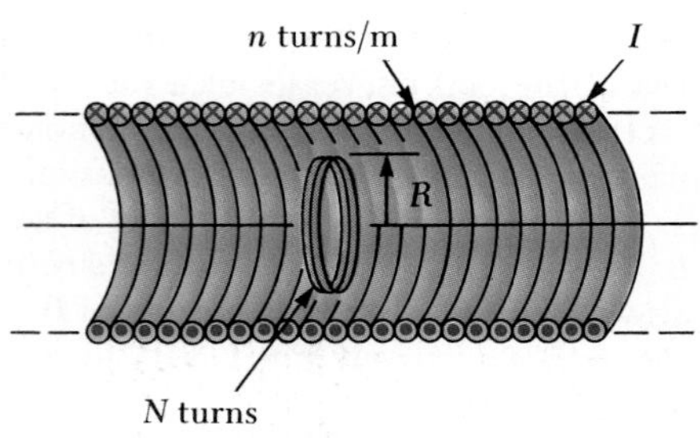

Figure 31.24 (Problem 11).

12. A magnetic field of 0.2 T exists within a solenoid of 500 turns and a diameter of 10 cm. How rapidly (that is, within what period of time) must the field be reduced to zero magnitude if the average magnitude of the induced emf within the coil during this time interval is to be 10 kV?
13. A coil, formed by wrapping 50 turns of wire in the shape of a square, is positioned in a magnetic field so that the normal to the plane of the coil makes an angle of 30° with the direction of the field. It is observed that if the magnitude of the magnetic field is increased uniformly from 200 μT to 600 μT in 0.4 s, an emf of 80 mV is induced in the coil. What is the total length of the wire?
14. A long straight wire carries a current $I = I_0 \sin(\omega t + \delta)$ and lies in the plane of a rectangular loop of N turns of wire, as shown in Figure 31.25. The quantities I_0, ω, and δ are all constants. Determine the emf induced in the loop by the magnetic field due to the current in the straight wire. Assume $I_0 = 50$ A, $\omega = 200\pi$ s^{-1}, $N = 100$, $a = b = 5$ cm, and $\ell = 20$ cm.

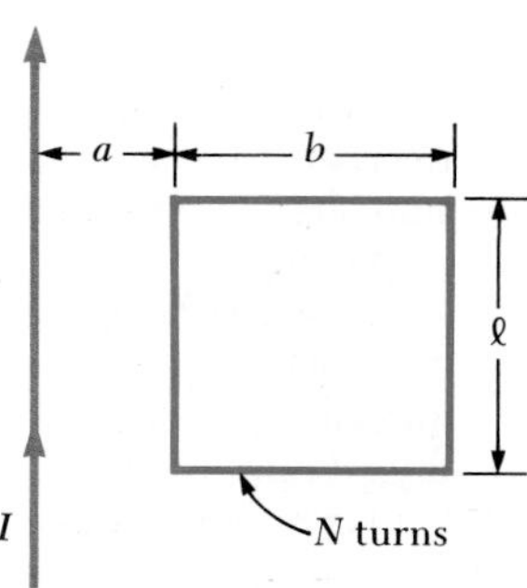

Figure 31.25 (Problem 14).

15. A circular loop with a radius R consists of N tight turns of wire and is penetrated by an external magnetic field directed perpendicular to the plane of the loop. The magnitude of the field in the plane of the loop is $B = B_0(1 - r/2R)\cos\omega t$, where R is the radius of the loop, and where r is measured from the center of the loop, as shown in Figure 31.26. Determine the induced emf in the loop.

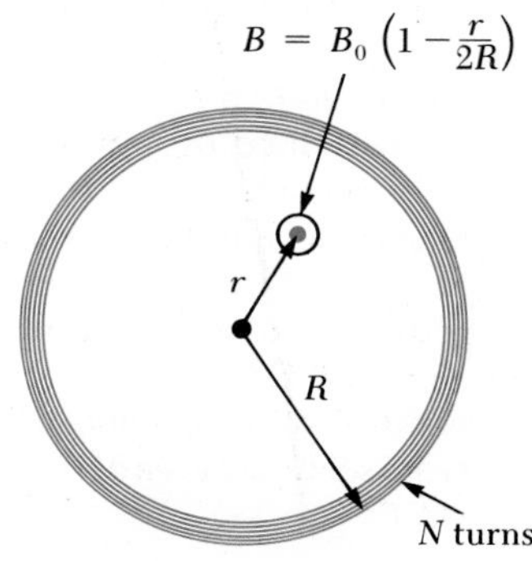

Figure 31.26 (Problem 15).

16. A toroid with a rectangular cross section ($a = 2$ cm by $b = 3$ cm) and inner radius $R = 4$ cm consists of 500 turns of wire that carries a current $I = I_0 \sin \omega t$, with $I_0 = 50$ A and a frequency $f = \omega/2\pi = 60$ Hz. A loop that consists of 20 turns of wire links the toroid, as shown in Figure 31.27. Determine the emf induced in the loop by the changing current I.

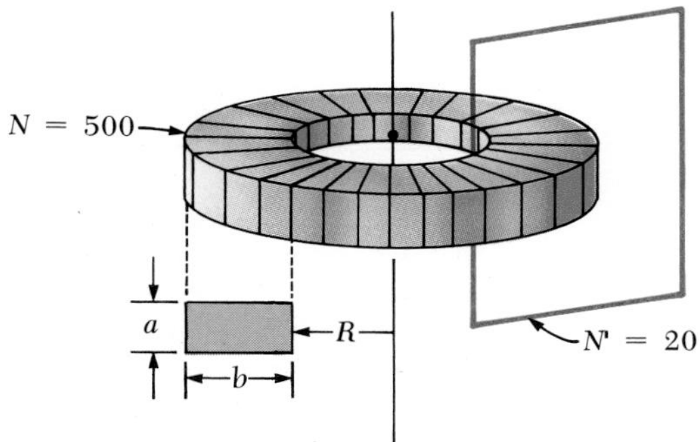

Figure 31.27 (Problem 16).

Section 31.2 Motional emf and Section 31.3 Lenz's Law

17. A measured average emf of 30 mV is induced in a small circular coil of 600 turns and 4 cm diameter under the following condition: It is rotated in a uniform magnetic field in 0.05 s from a position where the plane of the coil is parallel to the field to a position where the plane of the coil is at an angle of 45° to the field. What is the value of B within the region where the measurement is made?

18. Consider the arrangement shown in Figure 31.28. Assume that $R = 6\ \Omega$, $\ell = 1.2$ m, and that a uniform 2.5-T magnetic field is directed *into* the page. At what speed should the bar be moved to produce a current of 0.5 A in the resistor?

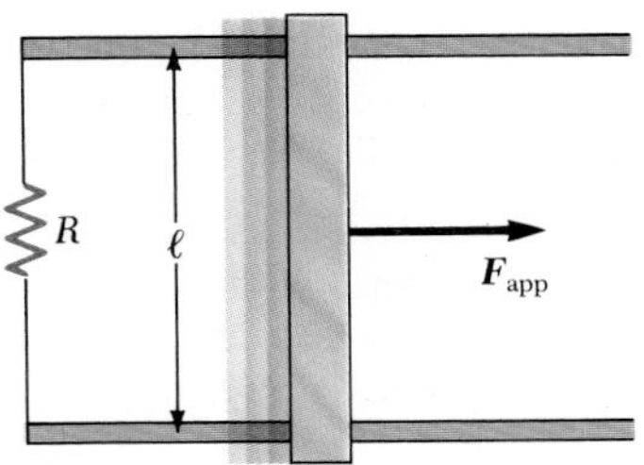

Figure 31.28 (Problems 18, 19, and 20).

19. In the arrangement shown in Figure 31.28, a conducting bar moves to the right along parallel, frictionless conducting rails connected on one end by a 6-Ω resistor. A 2.5-T magnetic field is directed *into* the paper. Let $\ell = 1.2$ m and neglect the mass of the bar. (a) Calculate the applied force required to move the bar to the right at a *constant* speed of 2 m/s. (b) At what rate is energy dissipated in the resistor?

20. A conducting rod of length ℓ moves on two horizontal frictionless rails as shown in Figure 31.28. If a constant force of 1 N moves the bar at 2 m/s through a magnetic field $\boldsymbol{B}$ which is into the paper, (a) what is the current through an 8-Ω resistor R? (b) What is the rate of energy dissipation in the resistor? (c) What is the mechanical power delivered by the force $\boldsymbol{F}$?

21. Over a region where the *vertical* component of the earth's magnetic field is 40 μT directed downward, a 5-m length of wire is held along an east-west direction and moved horizontally to the north with a speed of 10 m/s. Calculate the potential difference between the ends of the wire and determine which end is positive.

22. A small airplane with a wing span of 14 m is flying due north at a speed of 70 m/s over a region where the vertical component of the earth's magnetic field is 1.2 μT. (a) What potential difference is developed between the wing tips? (b) How would the answer to (a) change if the plane were flying due east?

23. A helicopter has blades of length 3 m, rotating at 2 rev/s about a central hub. If the vertical component of the earth's magnetic field is 0.5×10^{-4} T, what is the emf induced between the blade tip and the center hub?

24. A metal blade spins at a constant rate in the magnetic field of the earth as in Figure 31.7. The rotation occurs in a region where the component of the earth's magnetic field perpendicular to the plane of rotation is 3.3×10^{-5} T. If the blade is 1.0 m in length and its angular velocity is 5π rad/s, what potential difference is developed between its ends?

25. A rigid thin conducting rod of length L is mechanically rotated at constant angular velocity ω about an axis perpendicular to the rod and *through its center*. If a uniform magnetic field exists *parallel* to the rotation axis of the rod, (a) show that the induced emf between the center of the rod and one of its ends is proportional to L^2. (b) Evaluate the magnitude of this emf for $L = 0.2$ m, $\omega = 60$ rad/s, and $|\boldsymbol{B}| = 1.2$ T.

26. A 200-turn circular coil of radius 10 cm is located in a uniform magnetic field of 0.8 T such that the plane of the coil is perpendicular to the direction of the field. The coil is rotated at a constant rate (uniform angular velocity) through 90° in a time of 1.5 s, so that the plane of the coil is finally parallel to the direction of the field. (a) Calculate the *average* emf induced in the coil as a result of the rotation. (b) What is the instantaneous value of the emf in the coil at the moment the plane of the coil makes an angle of 45° with the magnetic field?

27. Use Lenz's law to answer the following questions concerning the direction of induced currents. (a) What is the direction of the induced current in resistor R in

Figure 31.29a when the bar magnet is moved to the left? (b) What is the direction of the current induced in the resistor R right after the switch S in the circuit of Figure 31.29b is closed? (c) What is the direction of the induced current in R when the current I in Figure 31.29c decreases rapidly to zero? (d) A copper bar is moved to the right while its axis is maintained perpendicular to a magnetic field, as in Figure 31.29d. If the top of the bar becomes positive relative to the bottom, what is the direction of the magnetic field?

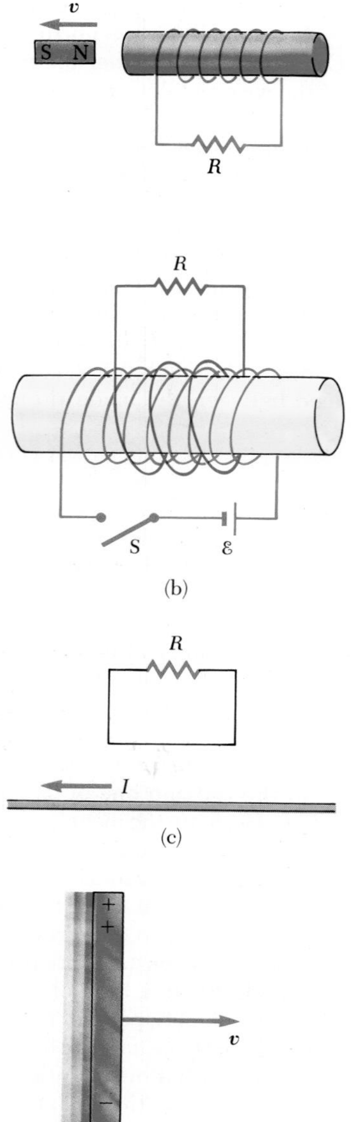

Figure 31.29 (Problem 27).

28. A conducting rectangular loop of mass M, resistance R, and dimensions of w wide by ℓ long falls from rest into a magnetic field $\mathbf{B}$ as shown in Figure 31.30. The loop accelerates until it reaches terminal speed, v_t. (a) Show that

$$v_t = \frac{MgR}{B^2w^2}$$

(b) Why is v_t proportional to R? (c) Why is it inversely proportional to B^2?

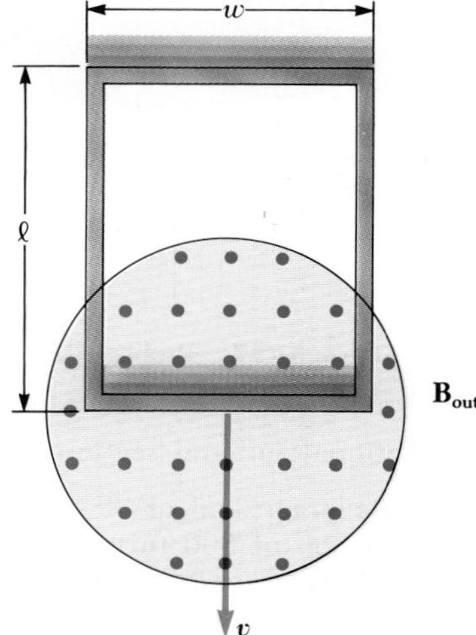

Figure 31.30 (Problems 28 and 29).

29. A 0.15-kg wire in the shape of a closed rectangle 1 m wide and 1.5 m long has a total resistance of 0.75 Ω. The rectangle is allowed to fall through a magnetic field directed perpendicular to the direction of motion of the wire (Fig. 31.30). The rectangle accelerates downward until it acquires a *constant* speed of 2 m/s with the top of the rectangle not yet in that region of the field. Calculate the magnitude of B.

Section 31.4 Induced emfs and Electric Fields

30. The current in a solenoid is increasing at a rate of 10 A/s. The cross-sectional area of the solenoid is π cm² and there are 300 turns on its 15-cm length. What is the induced emf which acts to oppose the increasing current?

31. A single-turn, circular loop of radius R is coaxial with a long solenoid of radius r and length ℓ and having N turns (Fig. 31.31). The variable resistor is changed so that the solenoid current decreases linearly from 7.2 A to 2.4 A in 0.3 s. If $r = 0.03$ m, $\ell = 0.75$ m, and $N = 1500$ turns, calculate the induced emf in the circular loop.

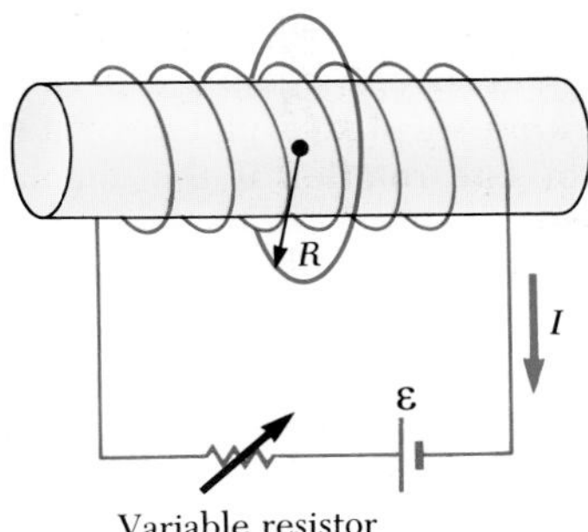

Figure 31.31 (Problem 31).

32. A coil of 15 turns and radius 10 cm surrounds a long solenoid of radius 2 cm and 10^3 turns/meter. If the current in the inner solenoid changes as $I = (5\text{ A}) \sin(120t)$, what is the induced emf in the 15-turn coil?

33. A magnetic field directed into the page changes with time according to $B = (0.03t^2 + 1.4)$ T, where t is in s. The field has a circular cross-section of radius $R = 2.5$ cm (Fig. 31.32). What are the magnitude and direction of the electric field at point P_1 when $t = 3$ s and $r_1 = 0.02$ m?

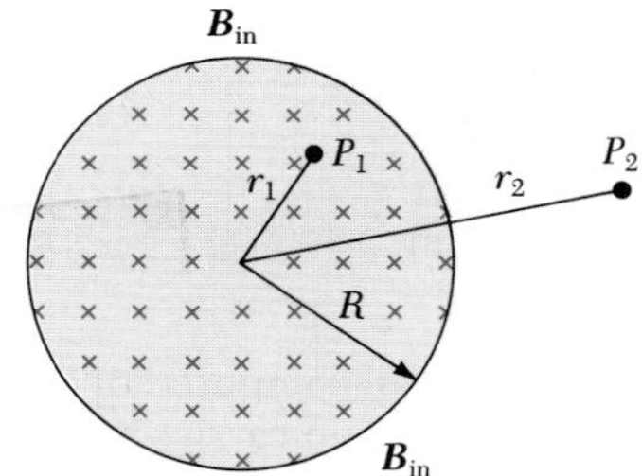

Figure 31.32 (Problems 33 and 34).

34. For the situation described in Figure 31.32, the magnetic field varies as $B = (2t^3 - 4t^2 + 0.8)$ T, and $r_2 = 2R = 5$ cm. (a) Calculate the magnitude and direction of the force exerted on an electron located at point P_2 when $t = 2$ s. (b) At what time is the force equal to zero?

35. A long solenoid with 1000 turns/meter and radius 2 cm carries an oscillating current $I = (5\text{ A}) \sin(100\pi t)$. What is the electric field induced at a radius $r = 1$ cm from the axis of the solenoid? What is the direction of this electric field when the current is *increasing* counterclockwise in the coil?

36. A solenoid has a radius of 2 cm and has 1000 turns/m. The current varies with time according to the expression $I = 3e^{0.2t}$, where I is in A and t is in s. Calculate the electric field at a distance of 5 cm from the axis of the solenoid at $t = 10$ s.

37. An aluminum ring of radius 5 cm and resistance $3 \times 10^{-4}\ \Omega$ is placed on top of a long air-core solenoid with 1000 turns per meter and radius 3 cm as shown in Figure 31.33. At the location of the ring, the magnetic field due to the current in the solenoid is one-half that at the center of the solenoid. If the current in the solenoid is *increasing* at a rate of 270 A/s, (a) what is the induced current in the ring? (b) At the center of the ring, what is the magnetic field produced by the induced current in the ring? (c) What is the direction of the field in (b)?

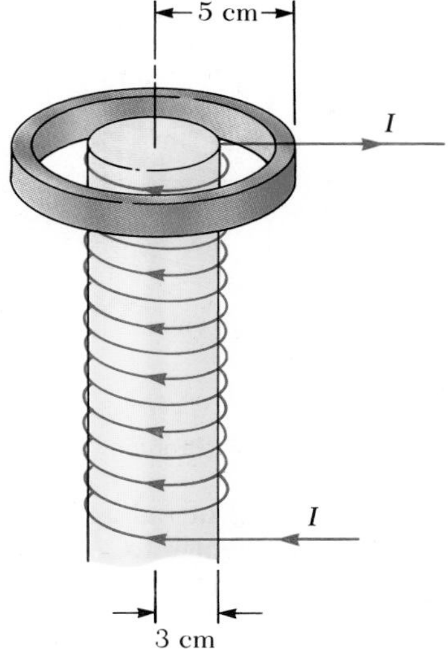

Figure 31.33 (Problem 37).

38. A circular coil, enclosing an area of 100 cm², is made of 200 turns of copper wire as shown in Figure 31.34. Initially, a 1.1-T uniform magnetic field points perpendicularly *upward* through the plane of the coil. The direction of the field then reverses so that the final magnetic field has a magnitude of 1.1 T pointing *downward* through the coil. During the time the field is changing, how much charge flows through the coil if the coil is connected to a 5-Ω resistor as shown?

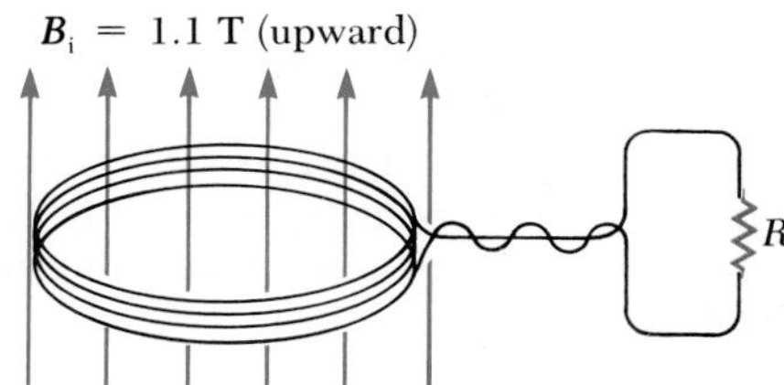

Figure 31.34 (Problem 38).

*Section 31.5 Generators and Motors

39. A square coil (20 cm × 20 cm) that consists of 100 turns of wire rotates about a vertical axis at 1500 rpm, as indicated in Figure 31.35. The horizontal component of the earth's magnetic field at the location of the loop is 2×10^{-4} T. Calculate the maximum emf induced in the coil by the earth's field.

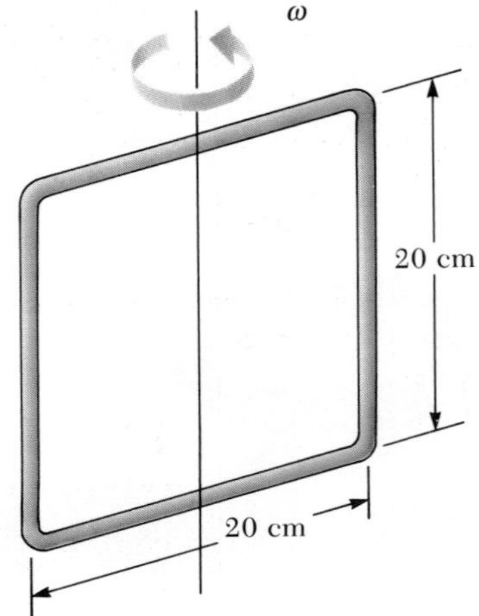

Figure 31.35 (Problem 39).

40. A 400-turn circular coil of radius 15 cm is rotating about an axis perpendicular to a magnetic field of 0.02 T. What angular velocity will produce a maximum induced emf of 4 V?
41. A loop of area 0.1 m² is rotating at 60 rev/s with the axis of rotation perpendicular to a 0.2-T magnetic field. (a) If there are 1000 turns on the loop, what is the maximum voltage induced in the loop? (b) When the maximum induced voltage occurs, what is the orientation of the loop with respect to the magnetic field?
42. The coil of a simple ac generator develops a sinusoidal emf of maximum value 90.4 V and frequency 60 Hz. If the coil has dimensions of 10 cm by 20 cm and rotates in a magnetic field of 1.0 T, how many turns are in the winding?
43. A long solenoid, whose axis coincides with the x axis, consists of 200 turns per meter of wire that carries a steady current of 15 A. A coil is formed by wrapping 30 turns of thin wire around a frame that has a radius of 8 cm. The coil is placed inside the solenoid and mounted on an axis that is a diameter of the coil and coincides with the y axis. The coil is then rotated with an angular speed of 4π radians per second. (The plane of the coil is in the yz plane at $t = 0$.) Determine the emf developed in the coil.
44. Consider a 100-turn rectangular coil of cross-sectional area 0.06 m² rotating with an angular velocity of 20 rad/s about an axis perpendicular to a magnetic field of 2.5 T. (a) Plot the induced voltage as a function of time over one complete period of rotation. (b) On the same set of axes, plot the voltage induced in the coil for one rotation if the angular velocity is 40 rad/s.
45. (a) What is the maximum torque delivered by an electric motor if it has 80 turns of wire wrapped on a rectangular coil, of dimensions 2.5 cm by 4 cm? Assume that the motor uses 10 A of current and that a uniform 0.8-T magnetic field exists within the motor. (b) If the motor rotates at 3600 rev/min, what is the (peak) power produced by the motor?
46. A semicircular conductor of radius $R = 0.25$ m is rotated about the axis AC at a constant rate of 120 revolutions per minute (Fig. 31.36). A uniform magnetic field in all of the lower half of the figure is directed *out* of the plane of rotation and has a magnitude of 1.3 T. (a) Calculate the *maximum* value of the emf induced in the conductor. (b) What is the value of the *average* induced emf for each complete rotation? (c) How would the answers to (a) and (b) change if the uniform field **B** were allowed to extend a distance R above the axis of rotation? (d) Sketch the emf versus time in each case.

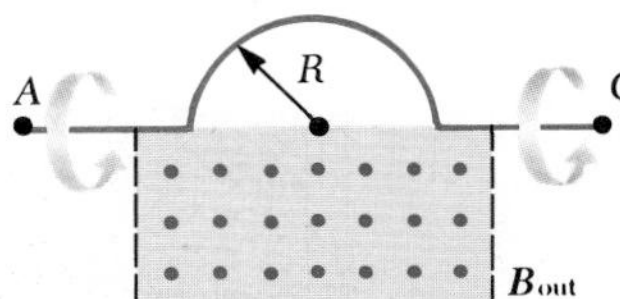

Figure 31.36 (Problem 46).

47. A small rectangular coil composed of 50 turns of wire has an area of 30 cm², and carries a current of 1.5 A. When the plane of the coil makes an angle of 30° with a uniform magnetic field, the torque on the coil is 0.1 N · m. What is the strength of the magnetic field?
48. A bar magnet is spun at constant angular velocity ω about an axis as shown in Figure 31.37. A flat rectan-

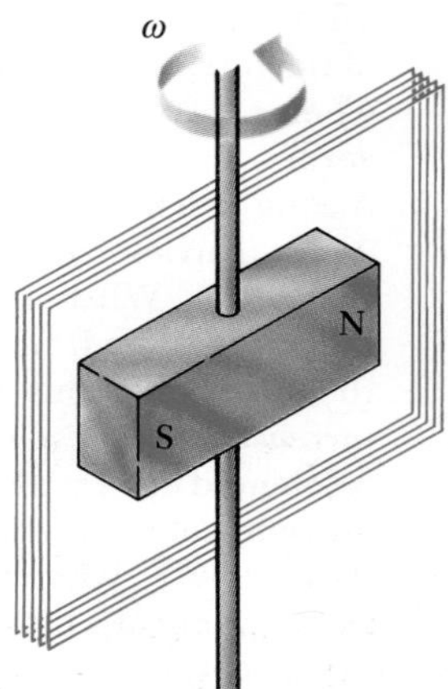

Figure 31.37 (Problem 48).

gular conducting loop surrounds the magnet, and at $t = 0$, the magnet is oriented as shown. Sketch the induced current in the loop as a function of time, plotting counterclockwise currents as positive and clockwise currents as negative.

*Section 31.6 Eddy Currents

49. A rectangular loop with resistance R has N turns, each of length ℓ and width w as shown in Figure 31.38. The loop moves into a uniform magnetic field $\boldsymbol{B}$ with velocity $\boldsymbol{v}$. What is the magnitude and direction of the resultant force on the loop (a) as it enters the magnetic field? (b) as it moves within the magnetic field? (c) as it leaves the magnetic field?

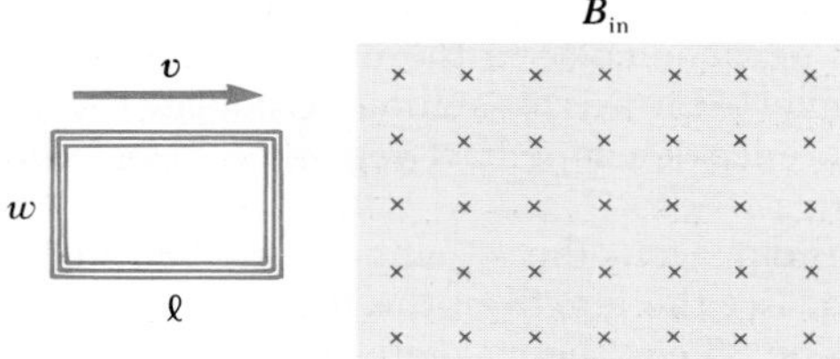

Figure 31.38 (Problem 49).

50. A dime is suspended from a thread and hung between the poles of a strong horseshoe magnet as shown in Figure 31.39. The dime rotates at constant angular speed ω about a vertical axis. Let θ be the angle between the direction of $\boldsymbol{B}$ and the normal to the face of the dime, and sketch a graph of the torque due to induced currents as a function of θ for $0 \le \theta \le 2\pi$.

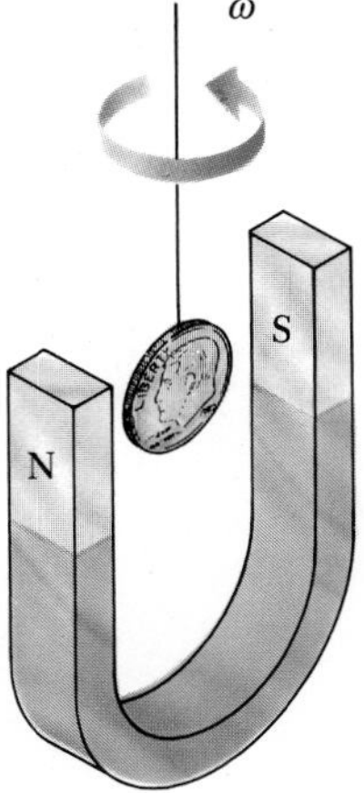

Figure 31.39 (Problem 50).

Section 31.7 Maxwell's Wonderful Equations

51. A proton moves through a uniform electric field $\boldsymbol{E} = 50\boldsymbol{j}$ V/m and a uniform magnetic field $\boldsymbol{B} = (0.2\boldsymbol{i} + 0.3\boldsymbol{j} + 0.4\boldsymbol{k})$ T. Determine the acceleration of the proton when it has a velocity of $\boldsymbol{v} = 200\boldsymbol{i}$ m/s.

52. An electron moves through a uniform electric field $\boldsymbol{E} = (2.5\boldsymbol{i} + 5.0\boldsymbol{j})$ V/m and a uniform magnetic field $\boldsymbol{B} = 0.4\boldsymbol{k}$ T. Determine the acceleration of the electron when it has a velocity of $\boldsymbol{v} = 10\boldsymbol{i}$ m/s.

ADDITIONAL PROBLEMS

53. A conducting rod moves with a constant velocity $\boldsymbol{v}$ perpendicular to a long, straight wire carrying a current I as in Figure 31.40. Show that the emf generated between the ends of the rod is given by

$$|\mathcal{E}| = \frac{\mu_0 v I}{2\pi r}\,\ell$$

In this case, note that the emf decreases with increasing r, as you might expect.

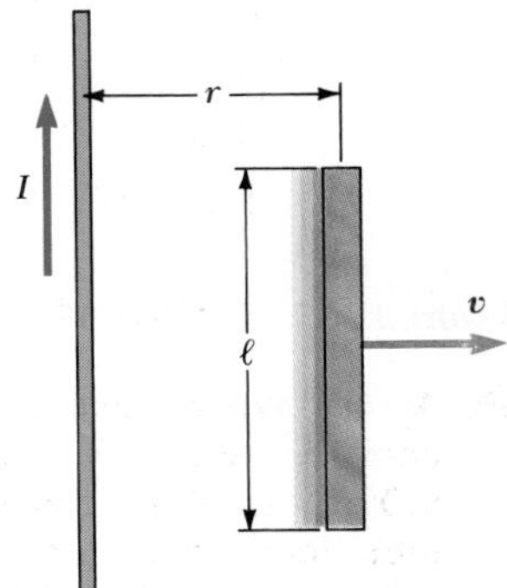

Figure 31.40 (Problem 53).

54. Shown in Figure 31.41 is a graph of the induced emf versus time for a coil of N turns rotating with angular velocity ω in a uniform magnetic field directed perpendicular to the axis of rotation of the coil. Copy this sketch (to a larger scale) and on the same set of axes show the graph of $\mathcal{E}(t)$ versus t when (a) the number of turns in the coil is doubled, (b) the angular velocity is doubled, and (c) the angular velocity is doubled while the number of turns in the coil is halved.

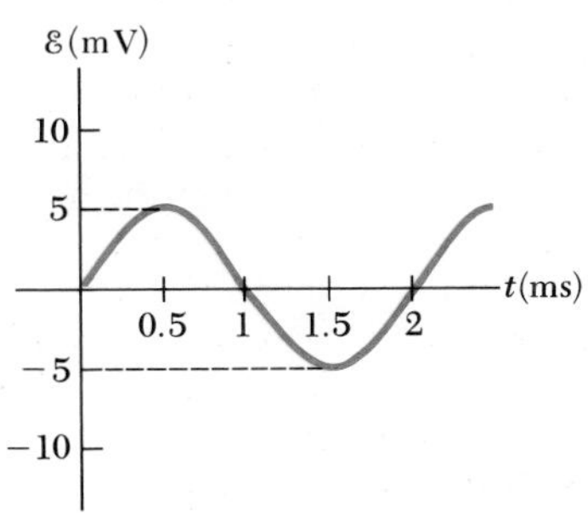

Figure 31.41 (Problem 54).

55. Consider a long solenoid of length ℓ containing a core of permeability μ. The core material is magnetized by increasing the current in the coil, so as to produce a changing field dB/dt. (a) Show that the rate at which work done against the induced emf in the coil is given by

$$\frac{dW}{dt} = I\mathcal{E} = HA\ell \frac{dB}{dt}$$

where A is the area of the solenoid. (*Hint:* Use Faraday's law to find $\mathcal{E}$ and make use of Equation 30.34.) (b) Use the result of part (a) to show that the total work done in a complete hysteresis cycle equals the area enclosed by the B versus H curve (Fig. 30.31).

56. Indicate the direction of the electric field between the plates of the capacitor shown in Figure 31.42 if the magnetic field is increasing linearly in time *into* the paper. Give a brief explanation.

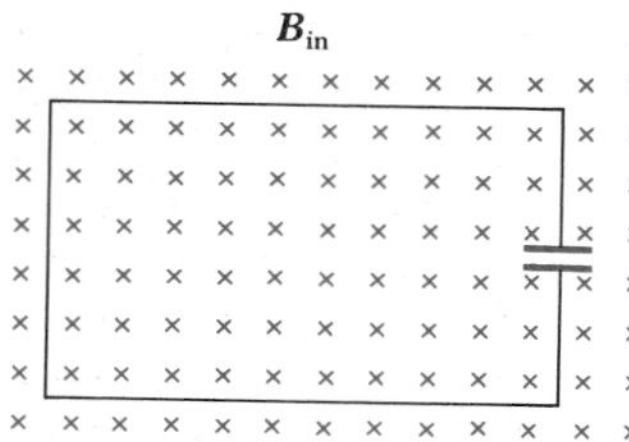

Figure 31.42 (Problem 56).

57. A rectangular coil of N turns, dimensions ℓ and w, and total resistance R rotates with angular velocity ω about the y axis in a region where a uniform magnetic field $\mathbf{B}$ is directed along the x axis (Fig. 31.15a). The rotation is initiated so that the plane of the coil is perpendicular to the direction of $\mathbf{B}$ at $t = 0$. Let $\omega =$ 30 rad/s, $N = 60$ turns, $\ell = 0.10$ m, $w = 0.20$ m, $R =$ 10 Ω, and $B = 1.0$ T. Calculate (a) the maximum induced emf in the coil, (b) the maximum rate of change of magnetic flux through the coil, (c) the value of the induced emf at $t = 0.05$ s, and (d) the torque exerted on the loop by the magnetic field at the instant when the emf is a maximum.

58. A horizontal wire is free to slide on the vertical rails of a conducting frame as in Figure 31.43. The wire has a mass m and length ℓ, and the resistance of the circuit is R. If a uniform magnetic field is directed perpendicular to the frame, what is the *terminal* velocity of the wire as it falls under the force of gravity? (Neglect mechanical friction.)

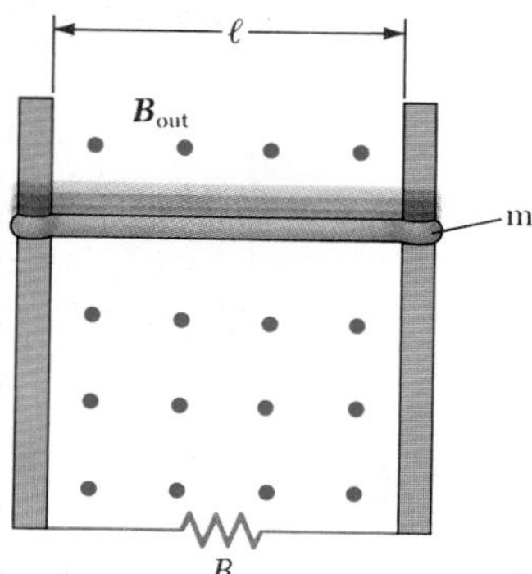

Figure 31.43 (Problem 58).

59. A solenoid wound with 2000 turns/m is supplied with current that varies in time according to $I = 4\sin(120\pi t)$, where I is in A and t is in s. A small coaxial circular coil of 40 turns and radius $r = 5$ cm is located inside the solenoid near its center. (a) Derive an expression that describes the manner in which the emf in the small coil varies in time. (b) At what average rate is energy dissipated in the small coil if the windings have a total resistance of 8 Ω?

60. A conducting disk of radius 0.3 m rotates about an axle through its center with a constant angular speed of $\omega = 15$ rad/s. A uniform magnetic field of 0.8 T acts perpendicular to the plane of the disk (that is, parallel to the axis of rotation). Calculate the potential difference developed between the rim and axle of the disk.

61. Consider again the situation described in Example 31.4. Let the length of the bar between the sliding rails be 25 cm, the magnetic field $B = 0.8$ T, the stationary resistor $R = 5\ \Omega$, and the initial velocity $\mathbf{v}_0 =$ 1.0 m/s. If the velocity of the moving bar decreases to $0.4\mathbf{v}_0$ in 3 s, what is the value of m_1, the mass of the bar?

62. A long straight wire is parallel to one edge and is in the plane of a single turn rectangular loop as in Figure 31.44. (a) If the current in the long wire varies in time as $I = I_0e^{-t/\tau}$, show that the induced emf in the loop is given by

$$\mathcal{E} = \frac{\mu_0 b}{2\pi}\frac{I}{\tau}\ln\left(1 + \frac{a}{d}\right)$$

(b) Calculate the value for the induced emf at $t = 5$ s taking $I_0 = 10$ A, $d = 3$ cm, $a = 6$ cm, $b = 15$ cm, and $\tau = 5$ s.

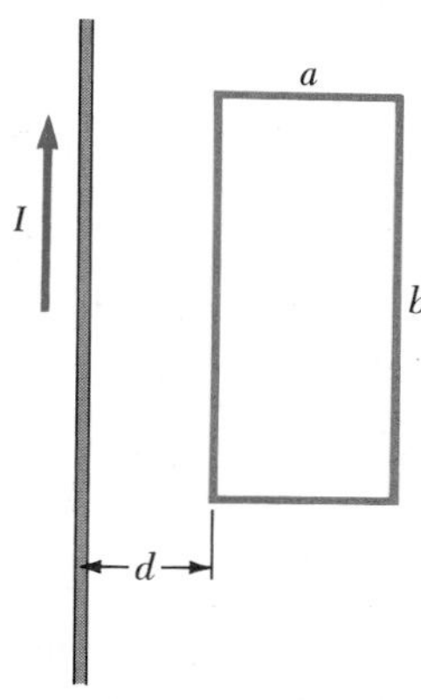

Figure 31.44 (Problem 62).

63. A conducting rod of length ℓ moves with velocity $\boldsymbol{v}$ along a direction parallel to a long wire carrying a steady current I. The axis of the rod is maintained perpendicular to the wire with the near end a distance r away, as shown in Figure 31.45. Show that the emf induced in the rod is given by

$$|\mathcal{E}| = \frac{\mu_0 I}{2\pi} v \ln\left(1 + \frac{\ell}{r}\right)$$

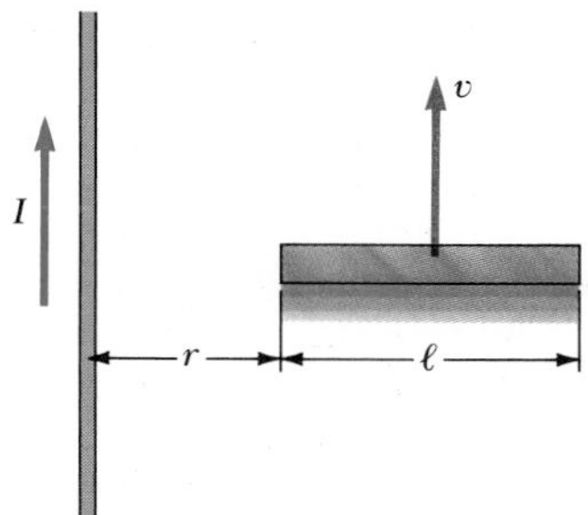

Figure 31.45 (Problem 63).

64. A rectangular loop of dimensions ℓ and w moves with a constant velocity $\boldsymbol{v}$ away from a long wire that carries a current I in the plane of the loop (Fig. 31.46). The total resistance of the loop is R. Derive an expression that gives the current in the loop at the instant the near side is a distance r from the wire.

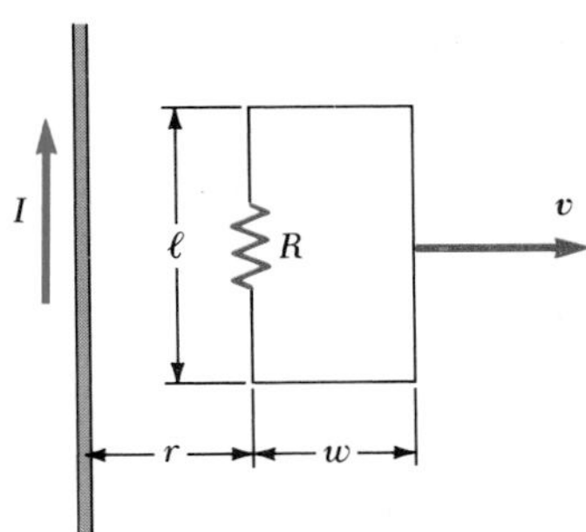

Figure 31.46 (Problem 64).

65. A square loop of wire with edge length $a = 0.2$ m is perpendicular to the earth's magnetic field at a point where $B = 15\ \mu$T, as in Figure 31.47. The total resistance of the loop and the wires connecting the loop to the galvanometer is 0.5 Ω. If the loop is suddenly collapsed by horizontal forces as shown, what total charge will pass through the galvanometer?

66. Magnetic field values are often determined by using a device known as a *search coil*. This technique depends on the measurement of the total charge passing through a coil in a time interval during which the magnetic flux linking the windings changes either because of the motion of the coil or because of a change in the value of B. (a) Show that if the flux through the coil changes from Φ_1 to Φ_2, the charge transferred through the coil between t_1 and t_2 will be given by $Q = N(\Phi_2 - \Phi_1)/R$ where R is the resistance of the coil and associated circuitry (galvanometer). (b) As a specific example, calculate B when a 100-turn coil of resistance 200 Ω and cross-sectional area 40 cm² produces the following results: A total charge of 5×10^{-4} C passes through the coil when it is rotated in a uniform field from a position where the plane of the coil is perpendicular to the field into a position where the coil's plane is parallel to the field.

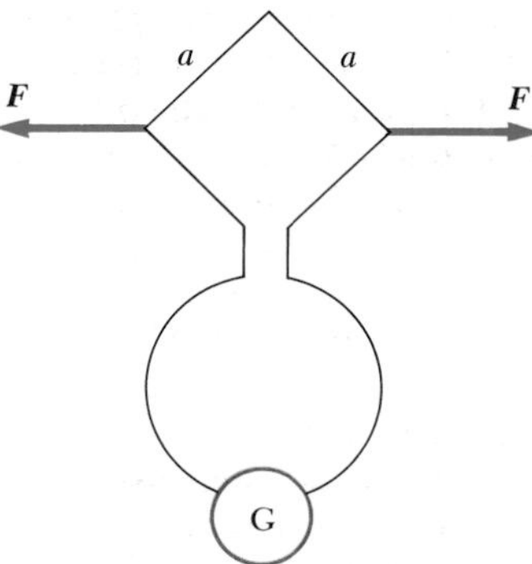

Figure 31.47 (Problem 65).

67. A flat circular coil of N turns has a diameter D and total resistance R. The coil is oriented with its axis parallel to a uniform magnetic field $\boldsymbol{B}$ and the ends of the wire making up the coil are connected to a device capable of measuring the total charge which passes through it. If, when the coil is quickly flipped over (i.e., rotated 180° about an axis perpendicular to the magnetic field), the metering device measures a total charge Q, express the magnitude of the magnetic field in terms of N, D, R, and Q.

68. Figure 31.48 illustrates an arrangement similar to that discussed in Example 31.4, except in this case the bar is pulled horizontally across the set of parallel rails by a string (assumed massless) that passes over an

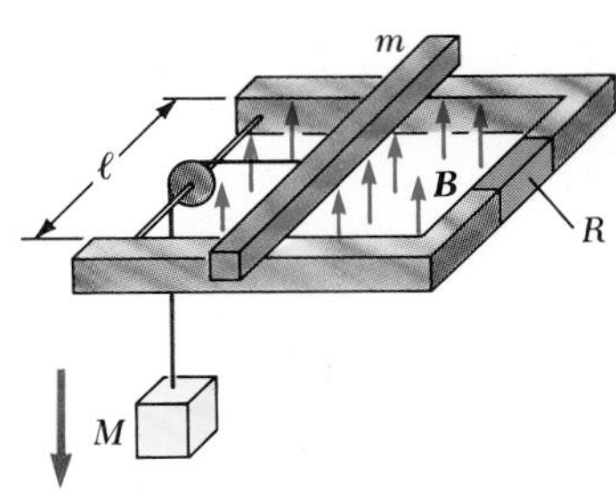

Figure 31.48 (Problem 68).

ideal pulley and is attached to a freely suspended mass M. The uniform magnetic field has a magnitude B, the sliding bar has mass m, and the distance between the rails is ℓ. The rails are connected at one end by a load resistor R. Derive an expression that gives the value of the horizontal speed of the bar as a function of time, assuming that the suspended mass was released with the bar at rest at $t = 0$. Assume no friction between the rails and the bar.

69. A thin metal strip is allowed to slide down parallel frictionless rails of negligible resistance connected at the bottom end and elevated at an angle θ above the horizontal. A uniform magnetic field **B** is directed vertically upward throughout the region as shown in Figure 31.49. The strip has a mass $m = 35$ g, resistance $R = 20\ \Omega$, and length between the rails $\ell = 0.3$ m. (a) Derive a general expression for the terminal speed of the strip. (b) Calculate the terminal speed achieved by the strip sliding along the incline if $\theta = 30°$ and $B = 1.5$ T.

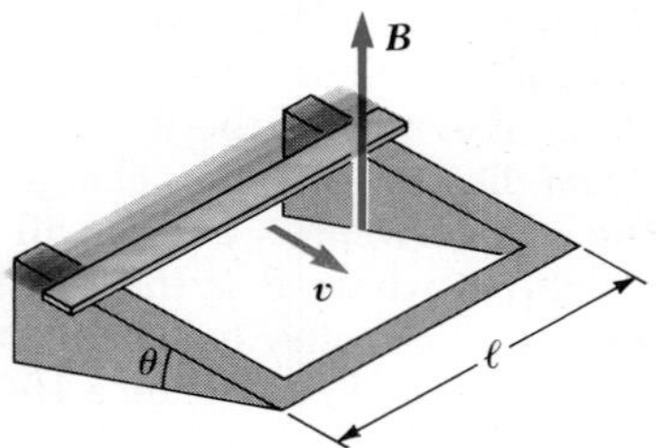

Figure 31.49 (Problem 69).

70. Two infinitely long solenoids (seen in cross section) thread the circuit as shown in Figure 31.50. The solenoids have radii of 0.1 m and 0.15 m, respectively. The magnitude of **B** inside each is the same and is increasing at the rate of 100 T/s. What is the current in each resistor?

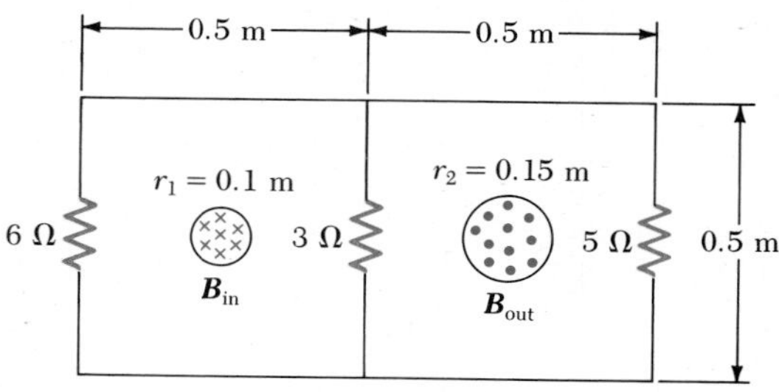

Figure 31.50 (Problem 70).

32

Inductance

Hydroelectric power is generated using the mechanical energy of water and electromagnetic induction. (© Tomas D. Friedmann 1971/Photo Researchers, Inc.)

In the previous chapter, we saw that currents and emfs are induced in a circuit when the magnetic flux through the circuit changes with time. This phenomenon of electromagnetic induction has some practical consequences, which we shall describe in this chapter. First, we shall describe an effect known as *self-induction,* in which a time-varying current in a conductor induces an emf in the conductor that opposes the external emf that set up the current. This phenomenon is the basis of the element known as the *inductor*, which plays an important role in circuits with time-varying currents. We shall discuss the concepts of the energy stored in the magnetic field of an inductor and the energy density associated with a magnetic field.

Next, we shall study how an emf can be induced in a circuit as a result of a changing flux produced by an external circuit, which is the basic principle of *mutual induction*. Finally, we shall examine the characteristics of circuits containing inductors, resistors, and capacitors in various combinations. For example, we shall find that in a circuit containing only an inductor and a capacitor, the charge and current both oscillate in simple harmonic fashion. These oscillations correspond to a continuous transfer of energy between the electric field of the charged capacitor and the magnetic field of the current-carrying inductor.

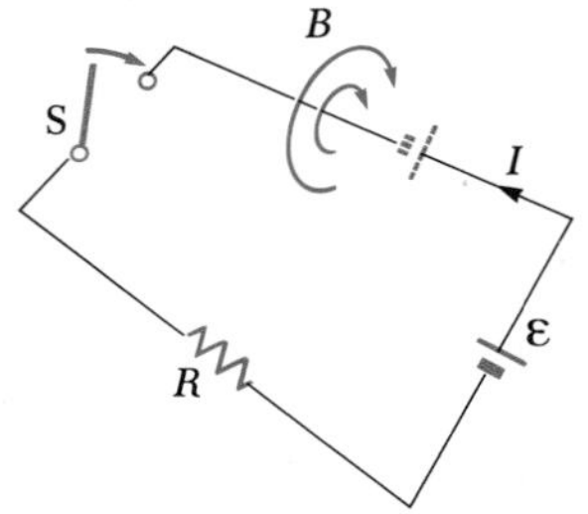

Figure 32.1 After the switch in the circuit is closed, the current produces a magnetic flux through the loop. As the current increases toward its equilibrium value, the flux changes in time and induces an emf in the loop.

32.1 SELF-INDUCTANCE

Consider an isolated circuit consisting of a switch, resistor, and source of emf, as in Figure 32.1. When the switch is closed, the current doesn't immediately jump from zero to its maximum value, $\mathcal{E}/R$. The law of electromagnetic induction (Faraday's law) prevents this from occurring. What happens is the following: As the current increases with time, the magnetic flux through the loop due to this current also increases with time. This increasing flux induces an emf in the circuit that opposes the change in the net magnetic flux through the loop. By Lenz's law, the induced electric field in the wires must therefore be opposite the direction of the conventional current, and this opposing emf results in a *gradual* increase in the current. This effect is called *self-induction* since the changing flux through the circuit arises from the circuit itself. The emf that is set up in this case is called a **self-induced emf.** Later, in Section 32.4, we shall describe a related effect called *mutual induction* in which an emf is induced in one circuit as a result of a changing magnetic flux set up by another circuit.

To obtain a quantitative description of self-induction, we recall from Faraday's law that the induced emf is given by the negative time rate of change of the magnetic flux. The magnetic flux is proportional to the magnetic field, which in turn is proportional to the current in the circuit. Therefore, *the self-induced emf is always proportional to the time rate of change of the current.* For a closely spaced coil of N turns of fixed geometry (a toroidal coil or the ideal solenoid), we find that

Induced emf

$$\mathcal{E} = -N\frac{d\Phi_m}{dt} = -L\frac{dI}{dt} \tag{32.1}$$

where L is a proportionality constant, called the **inductance** of the device, that depends on the geometric features of the circuit and other physical characteristics. From this expression, we see that the inductance of a coil containing N turns is given by

Inductance of an N-turn coil

$$L = \frac{N\Phi_m}{I} \tag{32.2}$$

where it is assumed that the same flux passes through each turn. Later we shall use this equation to calculate the inductance of some special current geometries.

From Equation 32.1, we can also write the inductance as the ratio

Inductance

$$L = -\frac{\mathcal{E}}{dI/dt} \tag{32.3}$$

This is usually taken to be the defining equation for the inductance of any coil, regardless of its shape, size, or material characteristics. Just as resistance is a measure of the opposition to current, inductance is a measure of the opposition to the *change* in current.

The SI unit of inductance is the **henry** (H), which, from Equation 32.3, is seen to be equal to 1 volt-second per ampere:

$$1\ \text{H} = 1\ \frac{\text{V}\cdot\text{s}}{\text{A}}$$

As we shall see, *the inductance of a device depends on its geometry.* However, the calculation of a device's inductance can be quite difficult for complicated geometries. The examples below involve rather simple situations for which inductances are easily evaluated.

EXAMPLE 32.1 Inductance of a Solenoid

Find the inductance of a uniformly wound solenoid with N turns and length ℓ. Assume that ℓ is long compared with the radius and that the core of the solenoid is air.

Solution In this case, we can take the interior field to be uniform and given by Equation 30.20:

$$B = \mu_0 n I = \mu_0 \frac{N}{\ell} I$$

where n is the number of turns per unit length, N/ℓ. The flux through each turn is given by

$$\Phi_m = BA = \mu_0 \frac{NA}{\ell} I$$

where A is the cross-sectional area of the solenoid. Using this expression and Equation 32.2 we find that

$$L = \frac{N\Phi_m}{I} = \frac{\mu_0 N^2 A}{\ell} \qquad (32.4)$$

This shows that L depends on geometric factors and is proportional to the square of the number of turns. Since $N = n\ell$, we can also express the result in the form

$$L = \mu_0 \frac{(n\ell)^2}{\ell} A = \mu_0 n^2 A\ell = \mu_0 n^2 (\text{volume}) \qquad (32.5)$$

where $A\ell$ is the volume of the solenoid.

EXAMPLE 32.2 Calculating Inductance and Emf

(a) Calculate the inductance of a solenoid containing 300 turns if the length of the solenoid is 25 cm and its cross-sectional area is 4 cm^2 = 4×10^{-4} m^2.

Solution Using Equation 32.4 we get

$$L = \frac{\mu_0 N^2 A}{\ell} = (4\pi \times 10^{-7}\ \text{Wb/A}\cdot\text{m}) \frac{(300)^2 (4 \times 10^{-4}\ \text{m}^2)}{25 \times 10^{-2}\ \text{m}}$$

$$= 1.81 \times 10^{-4}\ \text{Wb/A} = \boxed{0.181\ \text{mH}}$$

(b) Calculate the self-induced emf in the solenoid described in (a) if the current through it is *decreasing* at the rate of 50 A/s.

Solution Using Equation 32.1 and given that $dI/dt =$ 50 A/s, we get

$$\mathcal{E} = -L\frac{dI}{dt} = -(1.81 \times 10^{-4}\ \text{H})(-50\ \text{A/s})$$

$$= \boxed{9.05\ \text{mV}}$$

32.2 *RL* CIRCUITS

A circuit that contains a coil, such as a solenoid, has a self-inductance that prevents the current from increasing or decreasing instantaneously. A circuit element that has a large inductance is called an **inductor.** The circuit symbol for an inductor is ꝏꝏ. We shall always assume that the self-inductance of the remainder of the circuit is negligible compared with that of the inductor.

Consider the circuit consisting of a resistor, inductor, and battery shown in Figure 32.2. The internal resistance of the battery will be neglected. Suppose the switch S is closed at $t = 0$. The current will begin to increase, and due to the increasing current, the inductor will produce an emf (sometimes referred to as a *back emf*) that opposes the increasing current. In other words, the inductor acts like a battery whose polarity is opposite that of the real battery in the circuit. The back emf produced by the inductor is given by

$$\mathcal{E}_L = -L\frac{dI}{dt}$$

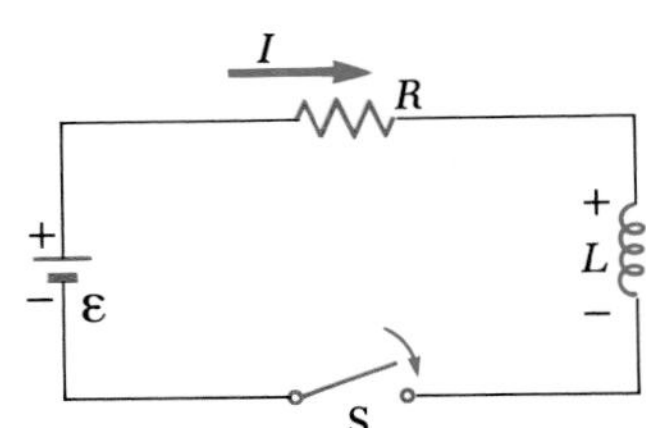

Figure 32.2 A series *RL* circuit. As the current increases toward its maximum value, the inductor produces an emf that opposes the increasing current.

Since the current is increasing, dI/dt is positive; therefore $\mathcal{E}_L$ is negative. This corresponds to the fact that there is a potential drop in going from a to b across the inductor. For this reason, point a is at a higher potential than point b, as illustrated in Figure 32.2.

With this in mind, we can apply Kirchhoff's loop equation to this circuit:

$$\mathcal{E} - IR - L\frac{dI}{dt} = 0 \tag{32.6}$$

where IR is the voltage drop across the resistor. We must now look for a solution to this differential equation, which is seen to be similar in form to that of the RC circuit (Section 28.4).

To obtain a mathematical solution of Equation 32.6, it is convenient to change variables by letting $x = \frac{\mathcal{E}}{R} - I$, so that $dx = -dI$. With these substitutions, Equation 32.6 can be written

$$x + \frac{L}{R}\frac{dx}{dt} = 0$$

$$\frac{dx}{x} = -\frac{R}{L}\,dt$$

Integrating this last expression gives

$$\ln\frac{x}{x_0} = -\frac{R}{L}t$$

where the integrating constant is taken to be $-\ln x_0$. Taking the antilog of this result gives

$$x = x_0 e^{-Rt/L}$$

Since at $t = 0$, $I = 0$, we note that $x_0 = \mathcal{E}/R$. Hence, the last expression is equivalent to

$$\frac{\mathcal{E}}{R} - I = \frac{\mathcal{E}}{R}e^{-Rt/L}$$

$$I = \frac{\mathcal{E}}{R}(1 - e^{-Rt/L})$$

which represents the solution of Equation 32.6.

This mathematical solution of Equation 32.6, which represents the current as a function of time, can also be written:

$$I(t) = \frac{\mathcal{E}}{R}(1 - e^{-Rt/L}) = \frac{\mathcal{E}}{R}(1 - e^{-t/\tau}) \tag{32.7}$$

where the constant τ is the **time constant** of the RL circuit:

Time constant of the *RL* circuit

$$\tau = L/R \tag{32.8}$$

It is left as an exercise to show that the dimension of τ is time. Physically, τ is the time it takes the current to reach $(1 - e^{-1}) = 0.63$ of its final value, $\mathcal{E}/R$.

Figure 32.3 represents a graph of the current versus time, where $I = 0$ at $t = 0$. Note that the final equilibrium value of the current, which occurs at $t = \infty$, is given by $\mathcal{E}/R$. This can be seen by setting dI/dt equal to zero in Equation 32.6 (at equilibrium, the change in the current is zero) and solving for the current. Thus, we see that the current rises very fast initially and then gradually approaches the equilibrium value $\mathcal{E}/R$ as $t \rightarrow \infty$.

One can show that Equation 32.7 is a solution of Equation 32.6 by computing the derivative dI/dt and requiring that $I = 0$ at $t = 0$. Taking the first time derivative of Equation 32.7, we get

$$\frac{dI}{dt} = \frac{\mathcal{E}}{L} e^{-t/\tau} \tag{32.9}$$

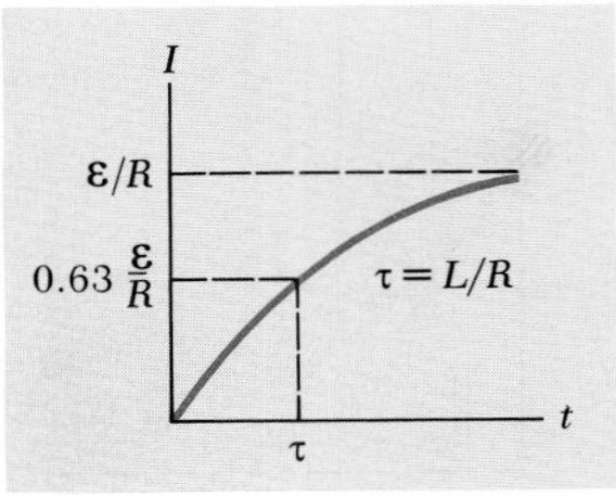

Figure 32.3 Plot of the current versus time for the *RL* circuit shown in Figure 32.2. The switch is closed at $t = 0$, and the current increases toward its maximum value, $\mathcal{E}/R$. The time constant τ is the time it takes I to reach 63% of its maximum value.

Substitution of this result together with Equation 32.7 will indeed verify that our solution satisfies Equation 32.6. That is,

$$\mathcal{E} - IR - L\frac{dI}{dt} = 0$$

$$\mathcal{E} - \frac{\mathcal{E}}{R}(1 - e^{-t/\tau})R - L\left(\frac{\mathcal{E}}{L} e^{-t/\tau}\right) = 0$$

$$\cancel{\mathcal{E}} - \cancel{\mathcal{E}} + \cancel{\mathcal{E}e^{-t/\tau}} - \cancel{\mathcal{E}e^{-t/\tau}} = 0$$

and the solution is verified.

From Equation 32.9 we see that the rate of increase of current, dI/dt, is a *maximum* (equal to $\mathcal{E}/L$) at $t = 0$ and falls off exponentially to zero as $t \rightarrow \infty$ (Fig. 32.4).

Now consider the *RL* circuit arranged as shown in Figure 32.5. The circuit contains two switches that operate such that when one is closed, the other is opened. Now suppose that S_1 is closed for a long enough time to allow the current to reach its equilibrium value, $\mathcal{E}/R$. If S_1 is now opened and S_2 is closed at $t = 0$, we have a circuit with no battery ($\mathcal{E} = 0$). If we apply Kirchhoff's circuit law to the upper loop containing the resistor and inductor, we obtain the expression

$$IR + L\frac{dI}{dt} = 0 \tag{32.10}$$

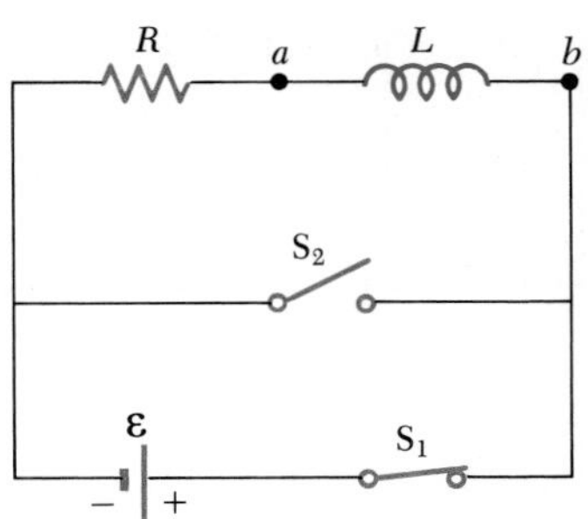

Figure 32.5 An *RL* circuit containing two switches. When S_1 is closed and S_2 is open as shown, the battery is in the circuit. At the instant S_2 is closed, S_1 is opened and the battery is removed from the circuit.

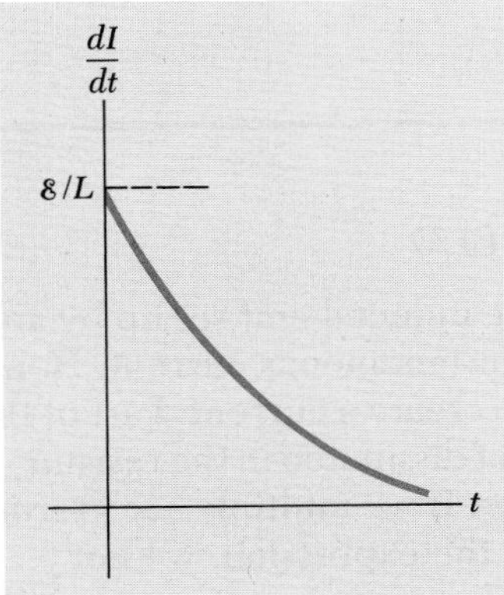

Figure 32.4 Plot of dI/dt versus time for the *RL* circuit shown in Figure 32.2. The rate of change of current is a maximum at $t = 0$ when the switch is closed. The rate dI/dt decreases exponentially with time as I increases toward its maximum value.

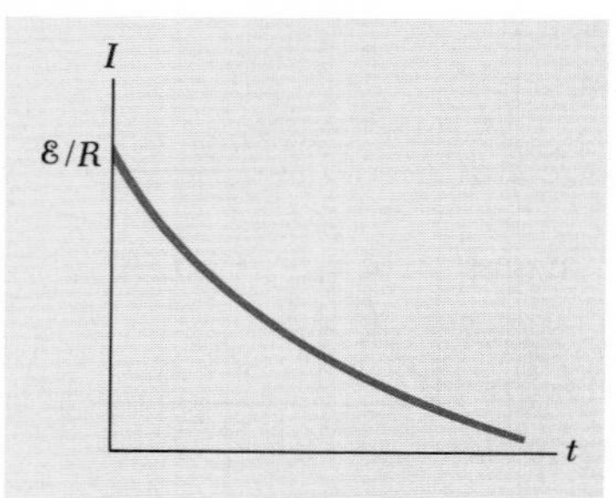

Figure 32.6 Current versus time for the circuit shown in Figure 32.5. At $t < 0$, S_1 is closed and S_2 is open. At $t = 0$, S_2 is closed, S_1 is open, and the current has its maximum value $\mathcal{E}/R$.

It is left as a problem (Problem 18) to show that the solution of this differential equation is

$$I(t) = \frac{\mathcal{E}}{R}e^{-t/\tau} = I_0 e^{-t/\tau} \tag{32.11}$$

where the current at $t = 0$ is given by $I_0 = \mathcal{E}/R$ and $\tau = L/R$.

The graph of the current versus time (Fig. 32.6) shows that the current is continuously decreasing with time, as one would expect. Furthermore, note that the slope, dI/dt, is always negative and has its maximum value at $t = 0$. The negative slope signifies that $\mathcal{E}_L = -L\,(dI/dt)$ is now *positive;* that is, point *a* is at a lower potential than point *b* in Figure 32.5.

EXAMPLE 32.3 Time Constant of an *RL* Circuit □ The circuit shown in Figure 32.7a consists of a 30-mH inductor, a 6-Ω resistor, and a 12-V battery. The switch is closed at $t = 0$. (a) Find the time constant of the circuit.

Solution The time constant is given by Equation 32.8

$$\tau = \frac{L}{R} = \frac{30 \times 10^{-3}\ \text{H}}{6\ \Omega} = 5.00\ \text{ms}$$

(b) Calculate the current in the circuit at $t = 2$ ms.

Solution Using Equation 32.7 for the current as a function of time (with t and τ in ms), we find that at $t = 2$ ms

$$I = \frac{\mathcal{E}}{R}(1 - e^{-t/\tau}) = \frac{12\ \text{V}}{6\ \Omega}(1 - e^{-0.4}) = 0.659\ \text{A}$$

A plot of Equation 32.7 for this circuit is given in Figure 32.7b.

Exercise 1 Calculate the current in the circuit and the voltage across the resistor after one time constant has elapsed.

Answer 1.26 A, 7.56 V.

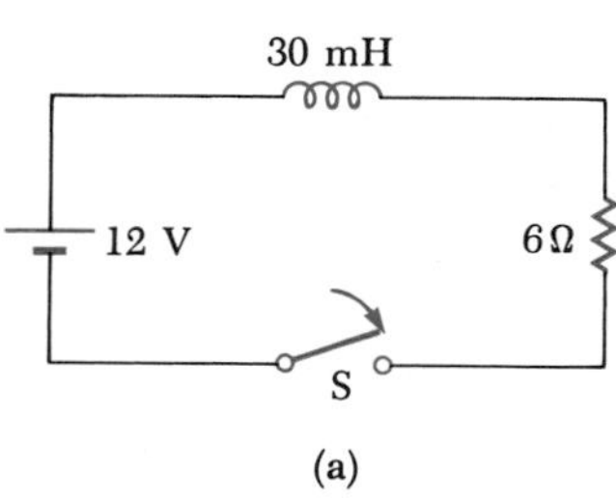

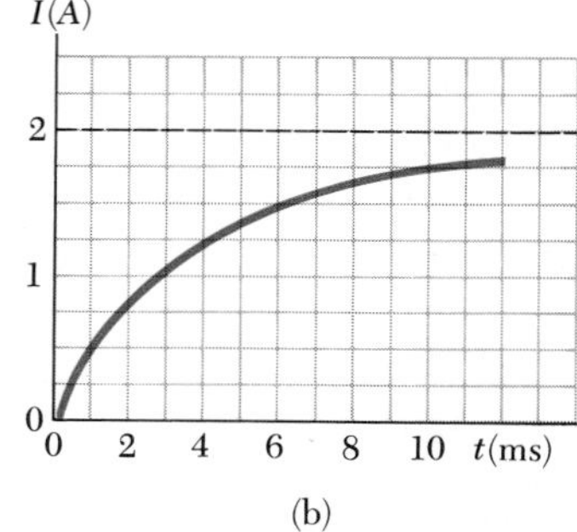

Figure 32.7 (Example 32.3) (a) The switch in this *RL* circuit is closed at $t = 0$. (b) A graph of the current versus time for the circuit in (a).

32.3 ENERGY IN A MAGNETIC FIELD

In the previous section we found that the induced emf set up by an inductor prevents a battery from establishing an instantaneous current. Hence, a battery has to do work against an inductor to create a current. Part of the energy supplied by the battery goes into joule heat dissipated in the resistor, while the remaining energy is stored in the inductor. If we multiply each term in Equation 32.6 by the current I and rearrange the expression, we get

$$I\mathcal{E} = I^2R + LI\frac{dI}{dt} \quad (32.12)$$

This expression tells us that the rate at which energy is supplied by the battery, $I\mathcal{E}$, equals the sum of the rate at which joule heat is dissipated in the resistor, I^2R, and the rate at which energy is stored in the inductor, $LI\,(dI/dt)$. Thus, Equation 32.12 is simply an expression of energy conservation. If we let U_m denote the energy stored in the inductor at any time, then the rate dU_m/dt at which energy is stored in the inductor can be written

$$\frac{dU_m}{dt} = LI\frac{dI}{dt}$$

To find the total energy stored in the inductor, we can rewrite this expression as $dU_m = LI\,dI$ and integrate:

$$U_m = \int_0^{U_m} dU_m = \int_0^I LI\,dI$$

$$U_m = \tfrac{1}{2}LI^2 \quad (32.13)$$

Energy stored in an inductor

where L is constant and has been removed from the integral. Equation 32.13 represents the energy stored as magnetic energy in the field of the inductor when the current is I. Note that it is similar in form to the equation for the energy stored in the electric field of a capacitor, $Q^2/2C$. In either case, we see that it takes work to establish a field.

We can also determine the energy per unit volume, or energy density, stored in a magnetic field. For simplicity, consider a solenoid whose inductance is given by Equation 32.5:

$$L = \mu_0 n^2 A\ell$$

The magnetic field of a solenoid is given by

$$B = \mu_0 nI$$

Substituting the expression for L and $I = B/\mu_0 n$ into Equation 32.13 gives

$$U_m = \tfrac{1}{2}LI^2 = \tfrac{1}{2}\mu_0 n^2 A\ell\left(\frac{B}{\mu_0 n}\right)^2 = \frac{B^2}{2\mu_0}(A\ell) \quad (32.14)$$

Because $A\ell$ is the volume of the solenoid, the energy stored per unit volume in a magnetic field is given by

$$u_m = \frac{U_m}{A\ell} = \frac{B^2}{2\mu_0} \quad (32.15)$$

Magnetic energy density

Although Equation 32.15 was derived for the special case of a solenoid, *it is valid for any region of space in which a magnetic field exists.* Note that Equation 32.15 is similar in form to the equation for the energy per unit volume stored in an electric field, given by $\tfrac{1}{2}\epsilon_0 E^2$. In both cases, the energy density is proportional to the *square* of the field strength.

EXAMPLE 32.4 What Happens to the Energy in the Inductor?
Consider once again the *RL* circuit shown in Figure 32.5, in which switch S_2 is closed at the instant S_1 is opened (at $t = 0$). Recall that the current in the upper loop decays exponentially with time according to the expression $I = I_0e^{-t/\tau}$, where $I_0 = \mathcal{E}/R$ is the initial current in the circuit and $\tau = L/R$ is the time constant. The energy stored in the magnetic field of the inductor gradually dissipates as thermal energy in the resistor. Let us show explicitly that all the energy stored in the inductor gets dissipated as heat in the resistor.

Solution The rate at which energy is dissipated in the resistor, dU/dt, (or the power) is equal to I^2R, where I is the instantaneous current. That is,

$$\frac{dU}{dt} = I^2R = (I_0e^{-Rt/L})^2R = I_0{}^2Re^{-2Rt/L} \qquad (1)$$

To find the total energy dissipated in the resistor, we integrate this expression over the limits $t = 0$ to $t = \infty$. (The upper limit of ∞ is used because it takes an infinite time for the current to reach zero.) Hence,

$$U = \int_0^\infty I_0{}^2Re^{-2Rt/L}\,dt = I_0{}^2R\int_0^\infty e^{-2Rt/L}\,dt \qquad (2)$$

The value of the definite integral is $L/2R$, so U becomes

$$U = I_0{}^2R\left(\frac{L}{2R}\right) = \frac{1}{2}LI_0{}^2$$

Note that this is equal to the initial energy stored in the magnetic field of the inductor, given by Equation 32.13, as we set out to prove.

Exercise 2 Evaluate the integral on the right hand side of Equation (2) and show that it has the value $L/2R$.

EXAMPLE 32.5 The Coaxial Cable
A long coaxial cable consists of two concentric cylindrical conductors of radii a and b and length ℓ, as in Figure 32.8. The inner conductor is assumed to be a thin cylindrical shell. Each conductor carries a current I (the outer one being a return path). (a) Calculate the self-inductance L of this cable.

Solution To obtain L, we must know the magnetic flux through any cross section between the two conductors. From Ampère's law (Section 30.3), it is easy to see that the magnetic field *between* the conductors is given by $B = \mu_0I/2\pi r$. Furthermore, the field is zero outside the conductors and zero inside the inner hollow conductor. The field is zero outside since the *net* current through a circular path surrounding both wires is zero, and hence

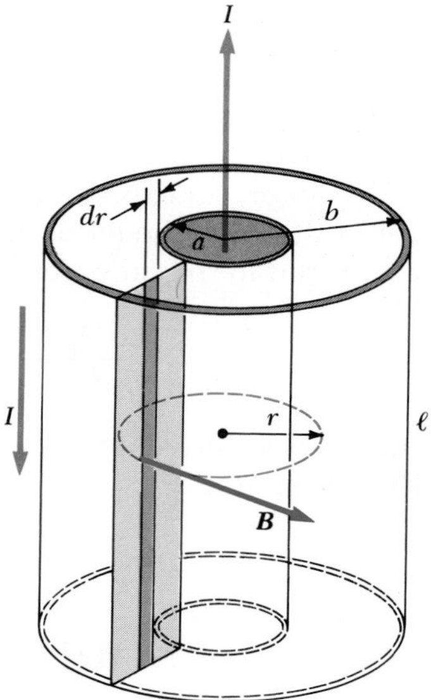

Figure 32.8 (Example 32.5) Section of a long coaxial cable. The inner and outer conductors carry equal and opposite currents.

from Ampère's law, $\oint \mathbf{B}\cdot d\mathbf{s} = 0$. The field is zero inside the inner conductor since it is hollow and there is no current within a radius $r < a$.

The magnetic field is *perpendicular* to the shaded rectangular strip of length ℓ and width $(b - a)$. This is the cross section of interest. Dividing this rectangle into strips of width dr, we see that the area of each strip is ℓdr and the flux through each strip is $B\,dA = B\ell\,dr$. Hence, the *total* flux through any cross section is

$$\Phi_m = \int B\,dA = \int_a^b \frac{\mu_0I}{2\pi r}\ell\,dr = \frac{\mu_0I\ell}{2\pi}\int_a^b \frac{dr}{r} = \frac{\mu_0I\ell}{2\pi}\ln\left(\frac{b}{a}\right)$$

Using this result, we find that the self-inductance of the cable is

$$L = \frac{\Phi_m}{I} = \frac{\mu_0\ell}{2\pi}\ln\left(\frac{b}{a}\right)$$

Furthermore, the self-inductance per unit length is given by $(\mu_0/2\pi)\ln(b/a)$.

(b) Calculate the total energy stored in the magnetic field of the cable.

Solution Using Equation 32.14 and the results to (a), we get

$$U_m = \tfrac{1}{2}LI^2 = \frac{\mu_0\ell I^2}{4\pi}\ln\left(\frac{b}{a}\right)$$

*32.4 MUTUAL INDUCTANCE

Very often the magnetic flux through a circuit varies with time because of varying currents in nearby circuits. This gives rise to an induced emf through a process known as **mutual induction,** so called because it depends on the interaction of two circuits.

Consider two closely wound coils as shown in the cross-sectional view of Figure 32.9. The current I_1 in coil 1, which has N_1 turns, creates magnetic field lines, some of which pass through coil 2, which has N_2 turns. The corresponding flux through coil 2 produced by coil 1 is represented by Φ_{21}. We define the **mutual inductance** M_{21} of coil 2 with respect to coil 1 as the ratio of $N_2\Phi_{21}$ and the current I_1:

$$M_{21} = \frac{N_2\Phi_{21}}{I_1} \tag{32.16}$$

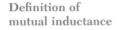

Definition of mutual inductance

$$\Phi_{21} = \frac{M_{21}}{N_2} I_1$$

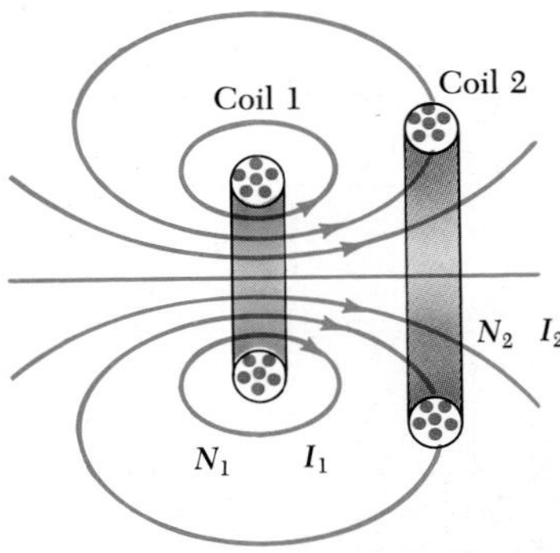

Figure 32.9 A cross-sectional view of two adjacent coils. A current in coil 1 sets up a flux, part of which passes through coil 2.

The mutual inductance depends on the geometry of both circuits and on their orientation with respect to one another. Clearly, as the circuit separation increases, the mutual inductance decreases since the flux linking the circuits decreases.

If the current I_1 varies with time, we see from Faraday's law and Equation 32.16 that the emf induced in coil 2 by coil 1 is given by

$$\mathcal{E}_2 = -N_2\frac{d\Phi_{21}}{dt} = -M_{21}\frac{dI_1}{dt} \tag{32.17}$$

Similarly, if the current I_2 varies with time, the induced emf in coil 1 due to coil 2 is given by

$$\mathcal{E}_1 = -M_{12}\frac{dI_2}{dt} \tag{32.18}$$

These results are similar in form to the expression for the self-induced emf $\mathcal{E} = -L\,(dI/dt)$. *The emf induced by mutual induction in one coil is always proportional to the rate of current change in the other coil.* If the rates at which the currents change with time are equal (that is, if $dI_1/dt = dI_2/dt$), then one finds that $\mathcal{E}_1 = \mathcal{E}_2$. Although the proportionality constants M_{12} and M_{21} appear to be different, one can show that they are equal. Thus, taking $M_{12} = M_{21} = M$, Equations 32.17 and 32.18 become

$$\mathcal{E}_2 = -M\frac{dI_1}{dt} \quad \text{and} \quad \mathcal{E}_1 = -M\frac{dI_2}{dt} \tag{32.19}$$

The unit of mutual inductance is also the henry.

EXAMPLE 32.6 Mutual Inductance of Two Solenoids

A long solenoid of length ℓ has N_1 turns, carries a current I, and has a cross-sectional area A. A second coil containing N_2 turns is wound around the center of the first coil, as in Figure 32.10. Find the mutual inductance of the system.

Solution If the solenoid carries a current I_1, the magnetic field at its center is given by

$$B = \frac{\mu_0 N_1 I_1}{\ell}$$

Since the flux Φ_{21} through coil 2 due to coil 1 is BA, the mutual inductance is

$$M = \frac{N_2 \Phi_{21}}{I_1} = \frac{N_2 BA}{I_1} = \mu_0 \frac{N_1 N_2 A}{\ell}$$

For example, if $N_1 = 500$ turns, $A = 3 \times 10^{-3}\ \text{m}^2$, $\ell = 0.5$ m, and $N_2 = 8$ turns, we get

$$M = \frac{(4\pi \times 10^{-7}\ \text{Wb/A}\cdot\text{m})(500)(8)(3 \times 10^{-3}\ \text{m}^2)}{0.5\ \text{m}}$$

$$\approx 30.2 \times 10^{-6}\ \text{H} = 30.2\ \mu\text{H}$$

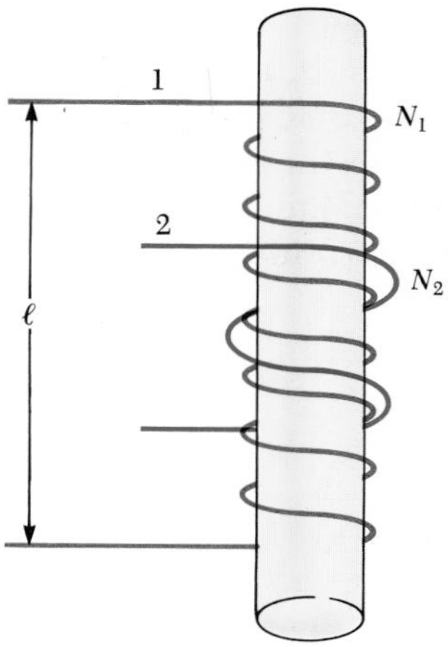

Figure 32.10 (Example 32.6) A small coil of N_2 turns wrapped around the center of a long solenoid of N_1 turns.

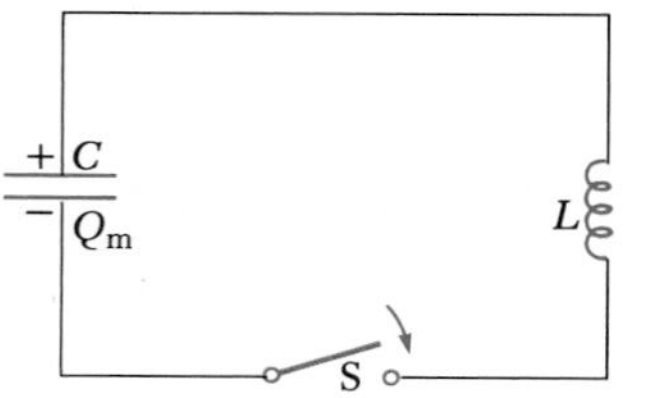

Figure 32.11 A simple LC circuit. The capacitor has an initial charge Q_m, and the switch is closed at $t = 0$.

32.5 OSCILLATIONS IN AN LC CIRCUIT

When a *charged* capacitor is connected to an inductor as in Figure 32.11 and the switch is then closed, oscillations will occur in the current and charge on the capacitor. If the resistance of the circuit is zero, no energy is dissipated as joule heat and the oscillations will persist. In this section we shall neglect the resistance in the circuit.

In the following analysis, let us assume that the capacitor has an initial charge Q_m and that the switch is closed at $t = 0$. It is convenient to describe what happens from an energy viewpoint.

When the capacitor is fully charged, the total energy U in the circuit is stored in the electric field of the capacitor and is equal to $Q_m{}^2/2C$. At this time, the current is zero and so there is no energy stored in the inductor. As the capacitor begins to discharge, the energy stored in its electric field decreases. At the same time, the current increases and some energy is now stored in the magnetic field of the inductor. Thus, we see that energy is transferred from the electric field of the capacitor to the magnetic field of the inductor. When the capacitor is fully discharged, it stores no energy. At this time, the current reaches its maximum value and all of the energy is now stored in the inductor. The process then repeats in the reverse direction. The energy continues to transfer between the inductor and capacitor indefinitely, corresponding to oscillations in the current and charge.

A graphical description of this energy transfer is shown in Figure 32.12. The circuit behavior is analogous to the oscillating mass-spring system studied in Chapter 13. The potential energy stored in a stretched spring, $\frac{1}{2}kx^2$, is analogous to the potential energy stored in the capacitor, $Q_m{}^2/2C$. The kinetic energy of the moving mass, $\frac{1}{2}mv^2$, is analogous to the energy stored in the inductor, $\frac{1}{2}LI^2$, which requires the presence of moving charges. In Figure 32.12a, all of the energy is stored as potential energy in the capacitor at $t = 0$ (since $I = 0$). In Figure 32.12b, all of the energy is stored as "kinetic" energy

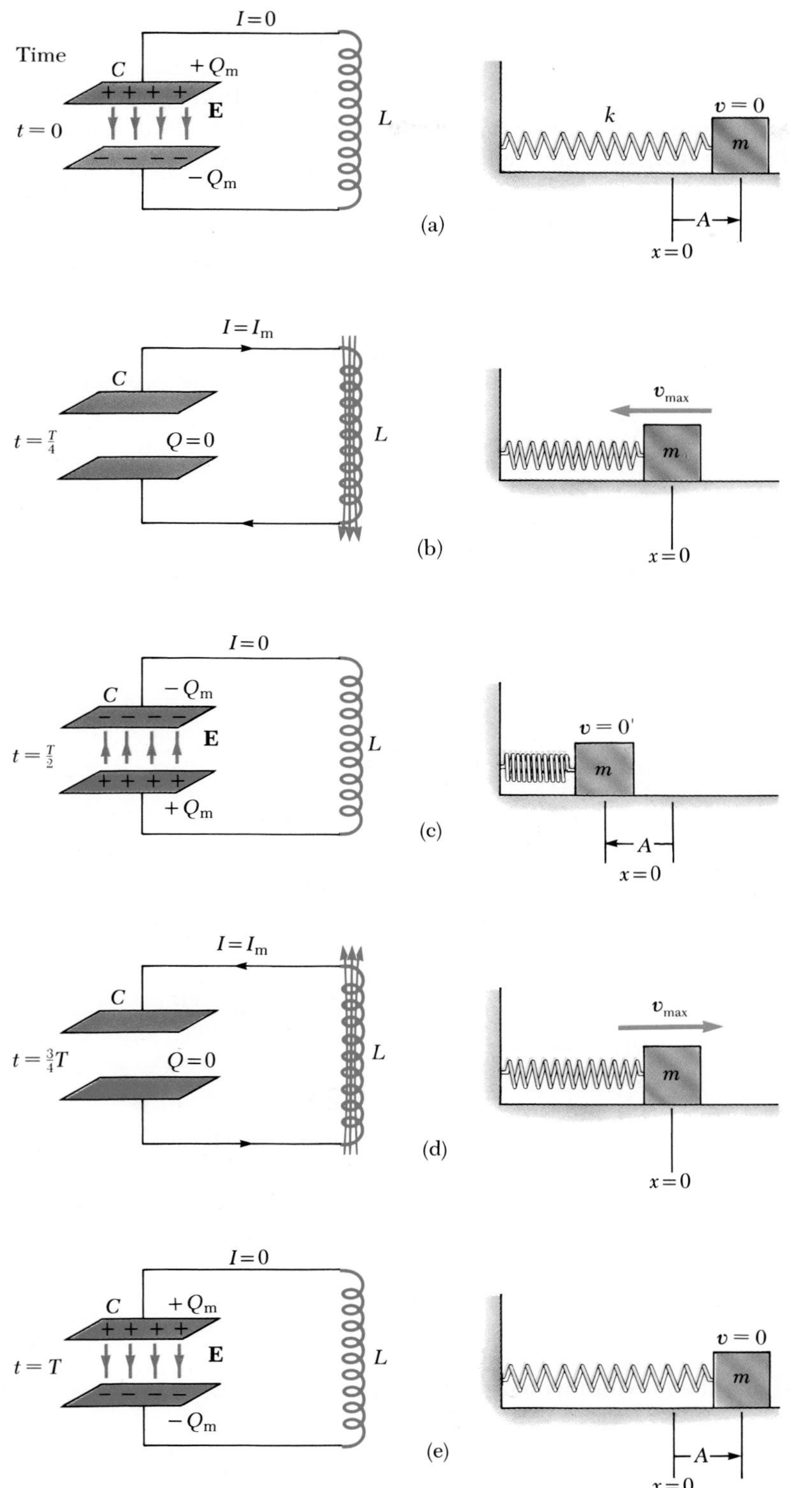

Figure 32.12 Energy transfer in a resistanceless LC circuit. The capacitor has a charge Q_m at $t = 0$ when the switch is closed. The mechanical analog of this circuit, the mass-spring system, is shown at the right.

in the inductor, $\frac{1}{2}LI_m^2$, where I_m is the maximum current. At intermediate points, part of the energy is potential energy and part is kinetic energy.

Consider some arbitrary time t after the switch is closed, such that the capacitor has a charge Q and the current is I. At this time, both elements store energy, but the sum of the two energies must equal the total initial energy U stored in the fully charged capacitor at $t = 0$. That is,

Total energy stored in the LC circuit

$$U = U_C + U_L = \frac{Q^2}{2C} + \tfrac{1}{2}LI^2 \qquad (32.20)$$

Since we have assumed the circuit resistance to be zero, no energy is dissipated as joule heat and hence *the total energy must remain constant in time.* This means that $dU/dt = 0$. Therefore, by differentiating Equation 32.20 with respect to time while noting that Q and I vary with time, we get

The total energy in an LC circuit remains constant; therefore, $dU/dt = 0$

$$\frac{dU}{dt} = \frac{d}{dt}\left(\frac{Q^2}{2C} + \tfrac{1}{2}LI^2\right) = \frac{Q}{C}\frac{dQ}{dt} + LI\frac{dI}{dt} = 0 \qquad (32.21)$$

We can reduce this to a differential equation in one variable by using the relationship between Q and I, namely, $I = dQ/dt$. From this, it follows that $dI/dt = d^2Q/dt^2$. Substitution of these relationships into Equation 32.21 gives

$$L\frac{d^2Q}{dt^2} + \frac{Q}{C} = 0$$

$$\frac{d^2Q}{dt^2} = -\frac{1}{LC}Q \qquad (32.22)$$

We can solve for the function Q by noting that Equation 32.22 is of the *same* form as that of the mass-spring system (simple harmonic oscillator) studied in Chapter 13. For this system, the equation of motion is given by

$$\frac{d^2x}{dt^2} = -\frac{k}{m}x = -\omega^2 x$$

where k is the spring constant, m is the mass, and $\omega = \sqrt{k/m}$. The solution of this equation has the general form

$$x = A\cos(\omega t + \delta)$$

where ω is the angular frequency of the simple harmonic motion, A is the amplitude of motion (the maximum value of x), and δ is the phase constant; the values of A and δ depend on the initial conditions. Since Equation 32.22 is of the same form as the differential equation of the simple harmonic oscillator, its solution is

Charge versus time for the LC circuit

$$Q = Q_m\cos(\omega t + \delta) \qquad (32.23)$$

where Q_m is the maximum charge of the capacitor and the angular frequency ω is given by

Angular frequency of oscillation

$$\omega = \frac{1}{\sqrt{LC}} \qquad (32.24)$$

Note that *the angular frequency of the oscillations depends solely on the inductance and capacitance of the circuit.*

Since Q varies periodically, the current also varies periodically. This is easily shown by differentiating Equation 32.23 with respect to time, which gives

$$I = \frac{dQ}{dt} = -\omega Q_m \sin(\omega t + \delta) \tag{32.25}$$

Current versus time for the LC current

To determine the value of the phase angle δ, we examine the initial conditions, which in our situation require that at $t = 0$, $I = 0$ and $Q = Q_m$. Setting $I = 0$ at $t = 0$ in Equation 32.25 gives

$$0 = -\omega Q_m \sin \delta$$

which shows that $\delta = 0$. This value for δ is also consistent with Equation 32.23 and the second condition that $Q = Q_m$ at $t = 0$. Therefore, in our case, the time variation of Q and that of I are given by

$$Q = Q_m \cos \omega t \tag{32.26}$$

$$I = -\omega Q_m \sin \omega t = -I_m \sin \omega t \tag{32.27}$$

where $I_m = \omega Q_m$ is the *maximum* current in the circuit.

Graphs of Q versus t and I versus t are shown in Figure 32.13. Note that the charge on the capacitor oscillates between the extreme values Q_m and $-Q_m$, and the current oscillates between I_m and $-I_m$. Furthermore, the current is 90° out of phase with the charge. That is, when the charge reaches an extreme value, the current is zero, and when the charge is zero, the current has an extreme value.

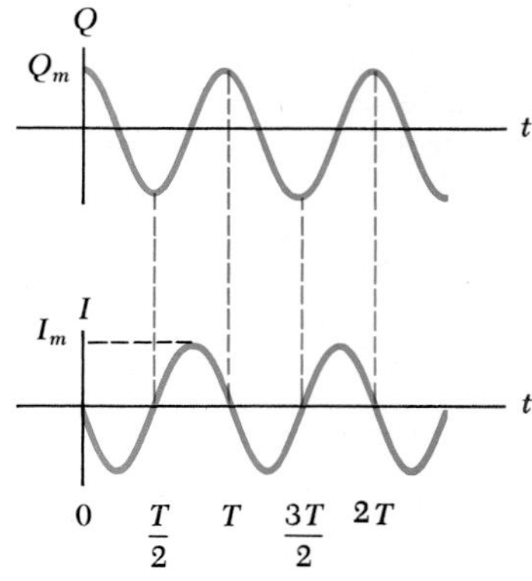

Figure 32.13 Graphs of charge versus time and current versus time for a resistanceless LC circuit. Note that Q and I are 90° out of phase with each other.

Let us return to the energy of the LC circuit. Substituting Equations 32.26 and 32.27 in Equation 32.20, we find that the total energy is given by

$$U = U_C + U_L = \frac{Q_m^2}{2C} \cos^2 \omega t + \frac{LI_m^2}{2} \sin^2 \omega t \tag{32.28}$$

This expression contains all of the features that were described qualitatively at the beginning of this section. It shows that the energy of the system continuously oscillates between energy stored in the electric field of the capacitor and energy stored in the magnetic field of the inductor. When the energy stored in the capacitor has its maximum value, $Q_m^2/2C$, the energy stored in the inductor is zero. When the energy stored in the inductor has its maximum value, $\frac{1}{2}LI_m^2$, the energy stored in the capacitor is zero.

Plots of the time variations of U_C and U_L are shown in Figure 32.14. Note that the sum $U_C + U_L$ is a constant and equal to the total energy, $Q_m^2/2C$. An analytical proof of this is straightforward. Since the maximum energy stored in the capacitor (when $I = 0$) must equal the maximum energy stored in the inductor (when $Q = 0$),

$$\frac{Q_m^2}{2C} = \frac{1}{2}LI_m^2 \tag{32.29}$$

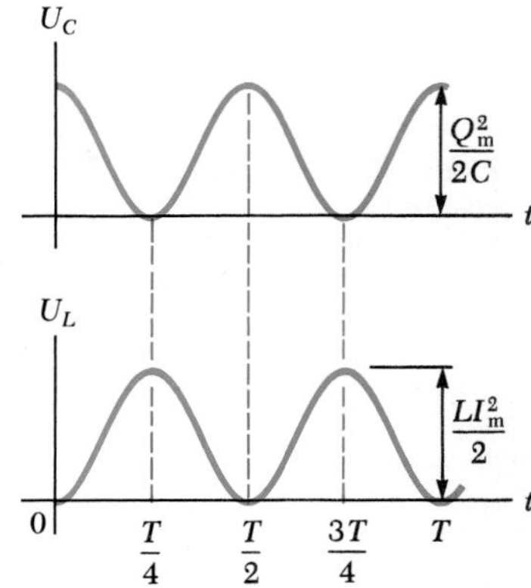

Figure 32.14 Plots of U_C versus t and U_L versus t for a resistanceless LC circuit. The sum of the two curves is a constant and equal to the total energy stored in the circuit.

Substitution of this into Equation 32.28 for the total energy gives

$$U = \frac{Q_m^{\ 2}}{2C}(\cos^2 \omega t + \sin^2 \omega t) = \frac{Q_m^{\ 2}}{2C} \tag{32.30}$$

because $\cos^2 \omega t + \sin^2 \omega t = 1$.

You should note that the total energy U remains constant *only* if energy losses are neglected. In actual circuits, there will always be some resistance and so energy will be lost in the form of heat. (In fact, even when the energy losses due to wire resistance are neglected, energy will also be lost in the form of electromagnetic waves radiated by the circuit.) In our idealized situation, the oscillations in the circuit persist indefinitely.

EXAMPLE 32.7 An Oscillatory *LC* Circuit

An *LC* circuit has an inductance of 2.81 mH and a capacitance of 9 pF (Fig. 32.15). The capacitor is initially charged with a 12-V battery when the switch S_1 is open and switch S_2 is closed. S_1 is then closed at the same instant that S_2 is opened so that the capacitor is shorted across the inductor. (a) Find the frequency of oscillation.

Solution Using Equation 32.24 gives for the frequency

$$f = \frac{\omega}{2\pi} = \frac{1}{2\pi\sqrt{LC}}$$

$$= \frac{1}{2\pi[(2.81 \times 10^{-3}\ \text{H})(9 \times 10^{-12}\ \text{F})]^{1/2}} = 10^6\ \text{Hz}$$

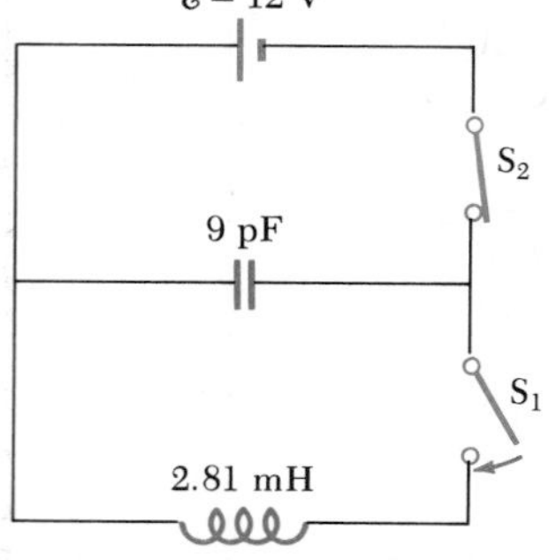

Figure 32.15 (Example 32.7) First the capacitor is fully charged with the switch S_1 open and S_2 closed. Then, S_1 is closed at the same time that S_2 is being opened.

(b) What are the maximum values of charge on the capacitor and current in the circuit?

Solution The initial charge on the capacitor equals the maximum charge, and since $C = Q/V$, we get

$$Q_m = CV = (9 \times 10^{-12}\ \text{F})(12\ \text{V}) = 1.08 \times 10^{-10}\ \text{C}$$

From Equation 32.27, we see that the maximum current is related to the maximum charge:

$$I_m = \omega Q_m = 2\pi f Q_m$$

$$= (2\pi \times 10^6\ \text{s}^{-1})(1.08 \times 10^{-10}\ \text{C})$$

$$= 6.79 \times 10^{-4}\ \text{A}$$

(c) Determine the charge and current as functions of time.

Solution Equations 32.26 and 32.27 give the following expressions for the time variation of Q and I:

$$Q = Q_m \cos \omega t = (1.08 \times 10^{-10}\ \text{C}) \cos \omega t$$

$$I = -I_m \sin \omega t = (-6.79 \times 10^{-4}\ \text{A}) \sin \omega t$$

where

$$\omega = 2\pi f = 2\pi \times 10^6\ \text{rad/s}$$

Exercise 3 What is the total energy stored in the circuit?

Answer 6.48×10^{-10} J.

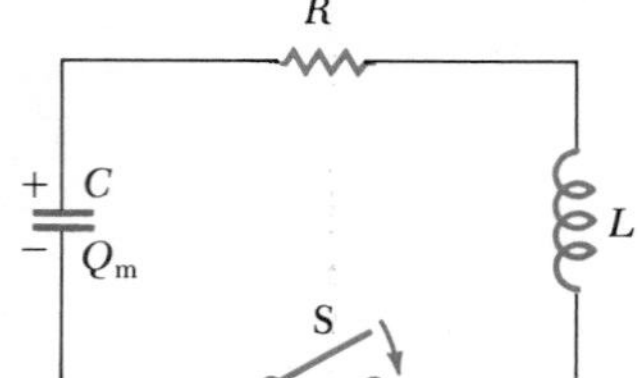

Figure 32.16 A series *RLC* circuit. The capacitor has a charge Q_m at $t = 0$ when the switch is being closed.

*32.6 THE *RLC* CIRCUIT

We now turn our attention to a more realistic circuit consisting of an inductor, a capacitor, and a resistor connected in series, as in Figure 32.16. We shall assume that the capacitor has an initial charge Q_m before the switch is closed. Once the switch is closed and a current is established, the total energy stored in the circuit at any time is given, as before, by Equation 32.20. That is, the energy stored in the capacitor is $Q^2/2C$, and the energy stored in the inductor is $\frac{1}{2}LI^2$. However, the total energy is no longer constant, as it was in the *LC* circuit, because of the presence of a resistor, which dissipates energy as heat. Since the rate of energy dissipation through a resistor is I^2R, we have

$$\frac{dU}{dt} = -I^2R \tag{32.31}$$

where the negative sign signifies that U is *decreasing* in time. Substituting this result into the time derivative of Equation 32.20 gives

$$LI\frac{dI}{dt} + \frac{Q}{C}\frac{dQ}{dt} = -I^2R \tag{32.32}$$

Using the fact that $I = dQ/dt$ and $dI/dt = d^2Q/dt^2$, and dividing Equation 32.32 by I, we get

$$L\frac{d^2Q}{dt^2} + R\frac{dQ}{dt} + \frac{Q}{C} = 0 \tag{32.33}$$

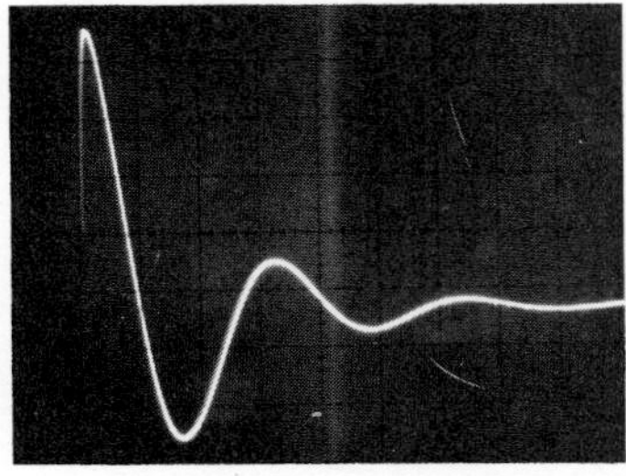

Oscilloscope pattern showing the decay in the oscillations of an *RLC* circuit. The parameters used were $R = 75\ \Omega$, $L = 10$ mH, $C = 0.19\ \mu$F, and $f = 300$ Hz. (Courtesy of J. Rudmin)

Note that the *RLC* circuit is analogous to the damped harmonic oscillator discussed in Section 13.6 and illustrated in Figure 32.17. The equation of motion for this mechanical system is

$$m\frac{d^2x}{dt^2} + b\frac{dx}{dt} + kx = 0 \tag{32.34}$$

Comparing Equations 32.33 and 32.34, we see that Q corresponds to x, L corresponds to m, R corresponds to the damping constant b, and C corresponds to $1/k$, where k is the force constant of the spring.

The analytical solution of Equation 32.33 is rather cumbersome and is usually covered in courses dealing with differential equations. Therefore, we shall give only a qualitative description of the circuit behavior.

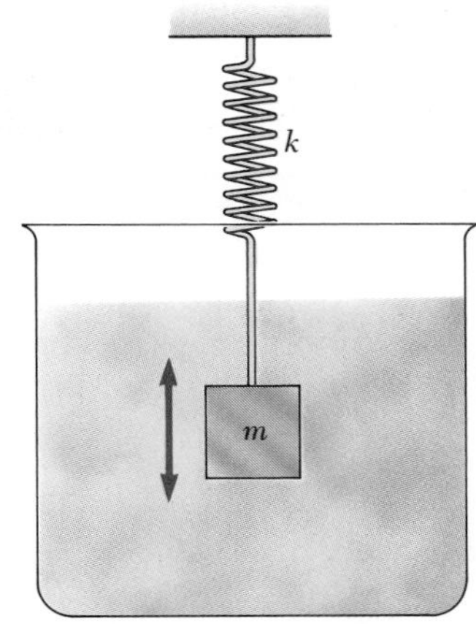

Figure 32.17 A mass-spring system moving in a viscous medium with damped harmonic motion is analogous to an *RLC* circuit.

In the simplest case, when $R = 0$, Equation 32.33 reduces to that of a simple *LC* circuit, as expected, and the charge and the current oscillate sinusoidally in time.

Next consider the situation where R is reasonably small. In this case, the solution of Equation 32.33 is given by

$$Q = Q_m e^{-Rt/2L} \cos \omega_d t \tag{32.35}$$

where

$$\omega_d = \left[\frac{1}{LC} - \left(\frac{R}{2L}\right)^2\right]^{1/2} \tag{32.36}$$

That is, the charge will oscillate with *damped harmonic motion* in analogy with a mass-spring system moving in a viscous medium. From Equation 32.35, we see that when $R \ll \sqrt{4L/C}$ the frequency ω_d of the damped oscillator will be close to that of the undamped oscillator, $1/\sqrt{LC}$. Since $I = dQ/dt$, it follows that the current will also undergo damped harmonic motion. A plot of the charge versus time for the damped oscillator is shown in Figure 32.18. Note that the maximum value of Q decreases after each oscillation, just as the amplitude of a damped harmonic oscillator decreases in time.

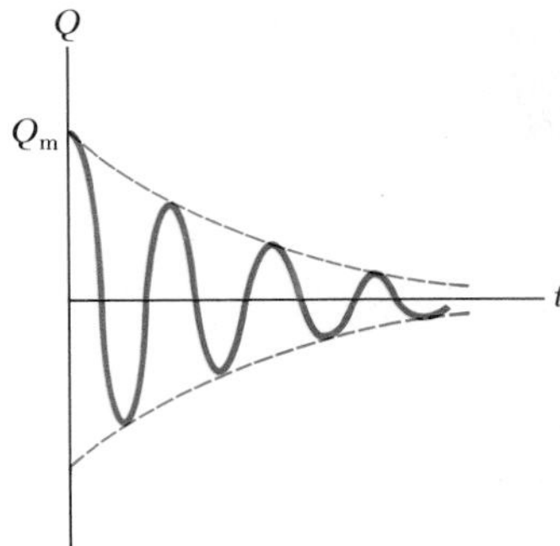

Figure 32.18 Charge versus time for a damped *RLC* circuit. This occurs for $R \ll \sqrt{4L/C}$. The Q versus t curve represents a plot of Equation 32.35.

When we consider larger values of R, we find that the oscillations damp out more rapidly; in fact, there exists a critical resistance value R_c above which

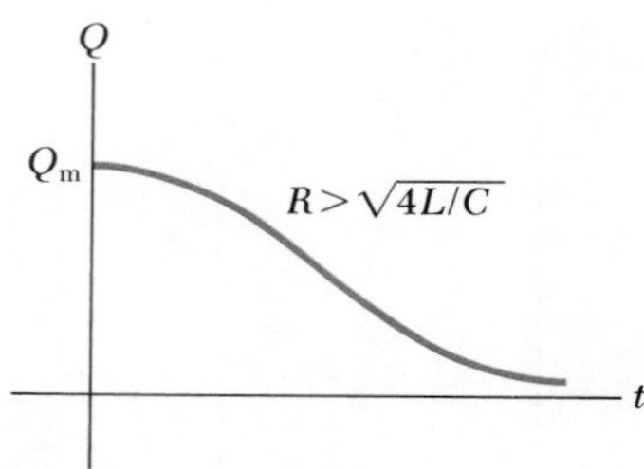

Figure 32.19 Plot of Q versus t for an overdamped RLC circuit, which occurs for values of $R > \sqrt{4L/C}$.

no oscillations occur. The critical value is given by $R_c = \sqrt{4L/C}$. A system with $R = R_c$ is said to be *critically damped*. When R exceeds R_c, the system is said to be *overdamped* (Fig. 32.19).

SUMMARY

When the current in a coil changes with time, an emf is induced in the coil according to Faraday's law. The **self-induced emf** is defined by the expression

Induced emf

$$\mathcal{E} = -L\frac{dI}{dt} \tag{32.1}$$

where L is the *inductance* of the coil. Inductance is a measure of the opposition of a device to a *change* in current. Inductance has the SI unit of **henry** (H), where 1 H = 1 V·s/A.

The **inductance** *of any coil*, such as a solenoid or toroid, is given by the expression

Inductance of an N-turn coil

$$L = \frac{N\Phi_m}{I} \tag{32.2}$$

where Φ_m is the magnetic flux through the coil and N is the total number of turns.

The inductance of a device depends on its geometry. For example, the *inductance of a solenoid* (whose core is a vacuum), as calculated from Equation 32.2, is given by

Inductance of a solenoid

$$L = \frac{\mu_0 N^2 A}{\ell} \tag{32.4}$$

where N is the number of turns, A is the cross-sectional area, and ℓ is the length of the solenoid.

If a resistor and inductor are connected in series to a battery of emf $\mathcal{E}$ as shown in Figure 32.2 and a switch in the circuit is closed at $t = 0$, the current in the circuit varies in time according to the expression

Current in an RL circuit

$$I(t) = \frac{\mathcal{E}}{R}(1 - e^{-t/\tau}) \tag{32.7}$$

where $\tau = L/R$ is the *time constant* of the RL circuit. That is, the current rises to an equilibrium value of $\mathcal{E}/R$ after a time that is long compared to τ.

If the battery is removed from an RL circuit, as in Figure 32.5 with S_1 open and S_2 closed, the current decays exponentially with time according to the expression

$$I(t) = \frac{\mathcal{E}}{R}e^{-t/\tau} \tag{32.11}$$

where $\mathcal{E}/R$ is the initial current in the circuit.

The **energy** *stored in the magnetic field of an inductor* carrying a current I is given by

Energy stored in an inductor

$$U_m = \tfrac{1}{2}LI^2 \tag{32.13}$$

This result is obtained by applying the principle of energy conservation to the *RL* circuit.

The **energy per unit volume** (or energy density) at a point where the magnetic field is B is given by

$$u_m = \frac{B^2}{2\mu_0} \tag{32.15}$$

Magnetic energy density

That is, the energy density is proportional to the square of the field at that point.

If two coils are close to each other, a changing current in one coil can induce an emf in the other coil. If dI_1/dt is rate of change of current in the first coil, the emf induced in the second is given by

$$\mathcal{E}_2 = -M\frac{dI_1}{dt} \tag{32.19}$$

where M is a constant called the **mutual inductance** of one coil with respect to the other.

If Φ_{21} is the magnetic flux through coil 2 due to the current I_1 in coil 1 and N_2 is the number of turns in coil 2, then the **mutual inductance** of coil 2 is given by

$$M_{21} = \frac{N_2\Phi_{21}}{I_1} \tag{32.16}$$

Mutual inductance

In an *LC* circuit with zero resistance, the charge on the capacitor and the current in the circuit vary in time according to the expressions

$$Q = Q_m \cos(\omega t + \delta) \tag{32.23}$$

$$I = \frac{dQ}{dt} = -\,\omega Q_m \sin(\omega t + \delta) \tag{32.25}$$

Charge and current versus time in an *LC* circuit

where Q_m is the maximum charge on the capacitor, δ is a phase constant, and ω is the angular frequency of oscillation, given by

$$\omega = \frac{1}{\sqrt{LC}} \tag{32.24}$$

Frequency of oscillation in an *LC* circuit

The energy in an *LC* circuit continuously transfers between energy stored in the capacitor and energy stored in the inductor. The **total energy** of the *LC* circuit at any time t is given by

$$U = U_C + U_L = \frac{{Q_m}^2}{2C}\cos^2 \omega t + \frac{L{I_m}^2}{2}\sin^2 \omega t \tag{32.28}$$

Energy of an *LC* circuit

where I_m is the maximum current in the circuit. At $t = 0$, all of the energy is stored in the electric field of the capacitor ($U = {Q_m}^2/2C$). Eventually, all of this energy is transferred to the inductor ($U = L{I_m}^2/2$). However, the *total* energy remains constant since the energy losses are neglected in the ideal *LC* circuit.

The charge and current in an *RLC* circuit exhibit a damped harmonic behavior for small values of R. This is analogous to the damped harmonic motion of a mass-spring system in which friction is present.

QUESTIONS

1. Why is the induced emf that appears in an inductor called a "counter" or "back" emf?
2. A circuit containing a coil, resistor, and battery is in steady state, that is, the current has reached a constant value. Does the coil have an inductance? Does the coil affect the value of the current in the circuit?
3. Does the inductance of a coil depend on the current in the coil? What parameters affect the inductance of a coil?
4. How can a long piece of wire be wound on a spool so that it has a negligible self-inductance?
5. A long fine wire is wound as a solenoid with a self-inductance L. If this is connected directly across the terminals of a battery, how does the maximum current depend on L?
6. For the series RL circuit shown in Figure 32.20, can the back emf ever be greater than the battery emf? Explain.

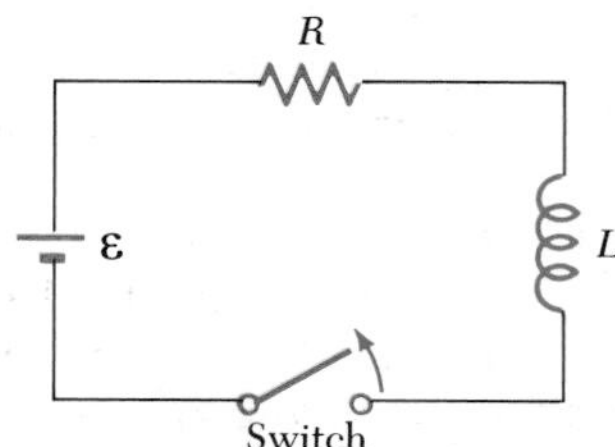

Figure 32.20 (Questions 6 and 7).

7. Suppose the switch in the RL circuit in Figure 32.20 has been closed for a long time and is suddenly opened. Does the current instantaneously drop to zero? Why does a spark tend to appear at the switch contacts when the switch is opened?
8. If the current in an inductor is doubled, by what factor does the stored energy change?
9. Discuss the similarities between the energy stored in the electric field of a charged capacitor and the energy stored in the magnetic field of a current-carrying coil.
10. What is the effective inductance of two isolated inductors, connected in series?
11. Discuss how the mutual inductance arises between the primary and secondary coils in a transformer.
12. The centers of two circular loops are separated by a fixed distance. For what relative orientation of the loops will their mutual inductance be a maximum? For what orientation will it be a minimum?
13. Two solenoids are connected in series such that each carries the same current at any instant. Is mutual induction present? Explain.
14. In the LC circuit shown in Figure 32.12, the charge on the capacitor is sometimes zero, even though there is current in the circuit. How is this possible?
15. If the resistance of the wires in an LC circuit were not zero, would the oscillations persist? Explain.
16. How can you tell whether an RLC circuit will be over- or underdamped?
17. What is the significance of "critical damping" in an RLC circuit?

PROBLEMS

Section 32.1 Self-Inductance

1. A 2-H inductor carries a steady current of 0.5 A. When the switch in the circuit is opened, the current disappears in 0.01 s. What is the induced emf that appears in the inductor during this time?
2. A "Slinky toy" spring has a radius of 4 cm and an inductance of 125 μH when extended to a length of 2 m. What is the total number of turns in the spring?
3. What is the inductance of a 510-turn solenoid that has a radius of 8 cm and an overall length of 1.4 m?
4. A small air-core solenoid has a length of 4 cm and a radius of 0.25 cm. If the inductance is to be 0.06 mH, how many turns per cm are required?
5. Show that the two expressions for inductance given by

$$L = \frac{N\Phi_m}{I} \quad \text{and} \quad L = -\frac{\mathcal{E}}{dI/dt}$$

have the same units.

6. Calculate the magnetic flux through a 300-turn, 7.2-mH coil when the current in the coil is 10 mA.
7. A 40-mA current is carried by a uniformly wound air-core solenoid with 450 turns, a 15-mm diameter, and 12-cm length. Compute (a) the magnetic field inside the solenoid, (b) the magnetic flux through each turn, and (c) the inductance of the solenoid. (d) Which of these quantities depends on the current?
8. A 0.388-mH inductor has a length that is four times its diameter. If it is wound with 22 turns per centimeter, what is its length?
9. An emf of 36 mV is induced in a 400-turn coil at an instant when the current has a value of 2.8 A and is changing at a rate of 12 A/s. What is the total magnetic flux through the coil?
10. The current in a 0.02-H inductor varies in time as $I = 3t^2 - 4t$, where I is in A and t is in s. (a) Calculate the magnitude of the induced emf at $t = 1$ and $t = 5$ s. (b) For what value of t will the induced emf be zero?

11. A current $I = I_0 \sin \omega t$, with $I_0 = 5$ A and $\omega/2\pi = 60$ Hz, flows through an inductor whose inductance is 10 mH. What is the back emf as a function of time?

12. Three solenoidal windings of 300, 200, and 100 turns are wrapped at well-spaced positions along a cardboard tube of radius 1 cm. Each winding extends for 5 cm along the cylindrical surface. What is the equivalent inductance of the 600 turns when the three sets of windings are connected in series?

13. Two coils, A and B, are wound using *equal lengths* of wire. Each coil has the same number of turns per unit length, but coil A has twice as many turns as coil B. What is the ratio of the self-inductance of A to the self-inductance of B? (*Note:* The radii of the two coils are not equal.)

14. A toroid has a major radius R and a minor radius r, and is tightly wound with N turns of wire, as shown in Figure 32.21. If $R \gg r$, the magnetic field inside the toroid is essentially that of a long solenoid that has been bent into a large circle of radius R. Using the uniform field of a long solenoid, show that the self-inductance of such a toroid is given (approximately) by

$$L \cong \frac{\mu_0 N^2 A}{2\pi R}$$

(An exact expression for the inductance of a toroid with a rectangular cross-section is derived in Problem 78.)

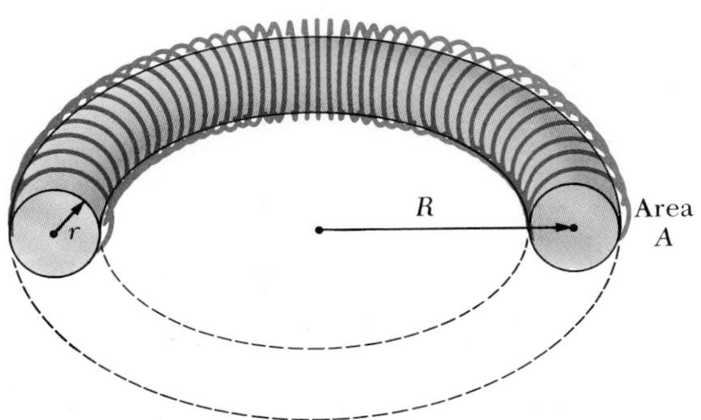

Figure 32.21 (Problem 14).

15. A solenoid has 120 turns uniformly wrapped around a wooden core, which has a diameter of 10 mm and a length of 9 cm. (a) Calculate the inductance of this solenoid. (b) The wooden core is replaced with a soft iron rod with the same dimensions, but a magnetic permeability $\kappa_m = 800\mu_0$. What is the new inductance?

16. A solenoidal inductor contains 420 turns, is 16 cm in length, and has a cross-sectional area of 3 cm². What uniform rate of decrease of current through the inductor will produce an induced emf of 175 μV?

Section 32.2 *RL* Circuits

17. Verify by direct substitution that the expression for current given in Equation 32.7 is a solution of Kirchhoff's loop equation for the *RL* circuit as given by Equation 32.6.

18. Show that $I = I_0 e^{-t/\tau}$ is a solution of the differential equation

$$IR + L\frac{dI}{dt} = 0,$$

where $\tau = L/R$ and $I_0 = \mathcal{E}/R$ is the value of the current at $t = 0$.

19. Calculate the inductance in an *RL* circuit in which $R = 0.5\ \Omega$ and the current increases to one fourth its final value in 1.5 s.

20. A 12-V battery is connected in series with a resistor and an inductor. The circuit has a time constant of 500 μs, and the maximum current is 200 mA. What is the value of the inductance?

21. Show that the inductive time constant τ has SI units of seconds.

22. An inductor with an inductance of 15 H and resistance of 30 Ω is connected across a 100-V battery. (a) What is the *initial* rate of increase of current in the circuit? (b) At what rate is the current changing at $t = 1.5$ s?

23. A 12-V battery is about to be connected to a series circuit containing a 10-Ω resistor and a 2-H inductor. (a) How long will it take the current to reach 50% of its final value? (b) How long will it take to reach 90% of its final value?

24. Consider the circuit shown in Figure 32.22, taking $\mathcal{E} = 6$ V, $L = 8$ mH, and $R = 4\ \Omega$. (a) What is the inductive time constant of the circuit? (b) Calculate the current in the circuit at a time 250 μs after the switch S_1 is closed. (c) What is the value of the final steady-state current? (d) How long does it take the current to reach 80% of its maximum value?

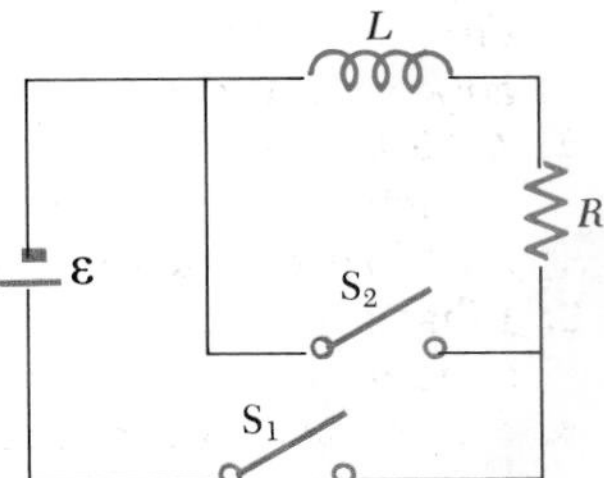

Figure 32.22 (Problems 24, 27, and 28).

25. A 140-mH inductor and a 4.9-Ω resistor are connected with a switch to a 6-V battery as shown in Figure 32.23. (a) If the switch is thrown to the left (connecting the battery), how much time elapses before the current reaches 220 mA? (b) What is the current through the inductor 10 s after the switch is

closed? (c) Now the switch is quickly thrown from A to B. How much time elapses before the current falls to 160 mA?

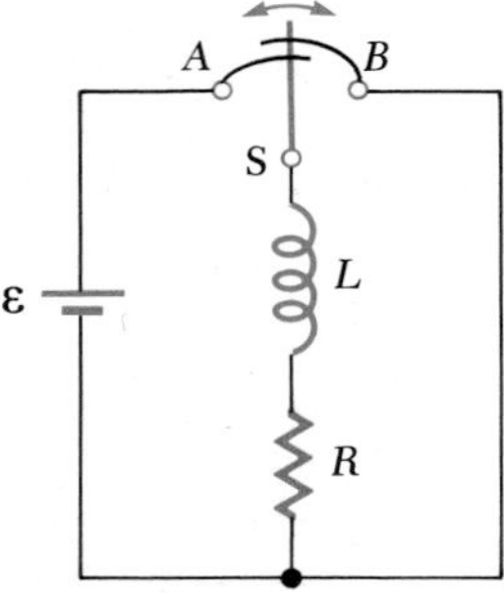

Figure 32.23 (Problem 25).

26. When the switch in Figure 32.24 is closed, the current takes 3.0 ms to reach 98% of its final value. If $R = 10\ \Omega$, what is the inductance L?

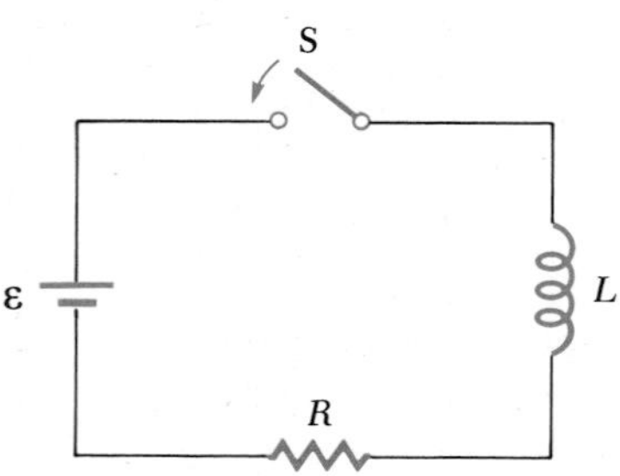

Figure 32.24 (Problems 26, 29, 30, 35).

27. Let the following values be assigned to the components in the circuit shown in Figure 32.22. $\mathcal{E} = 6$ V, $L = 24$ mH, and $R = 10\ \Omega$. (a) Calculate the current in the circuit at a time 0.5 ms after switch S_1 is closed. (b) What is the maximum value of the current in the circuit?

28. Assume that switch S_1 in the circuit of Figure 32.22 has been closed long enough for the current to reach its *maximum* value. If switch S_1 is now opened and switch S_2 closed at $t = 0$, after what time interval will the current in R be 25% of the maximum value? (Use the numerical values given in Problem 27.)

29. For the RL circuit shown in Figure 32.24, let $L = 3$ H, $R = 8\ \Omega$, and $\mathcal{E} = 36$ V. (a) Calculate the ratio of the potential difference across the resistor to that across the inductor when $I = 2$ A. (b) Calculate the voltage across the inductor when $I = 4.5$ A.

30. In the circuit shown in Figure 32.24 let $L = 7$ H, $R = 9\ \Omega$, and $\mathcal{E} = 120$ V. What is the self-induced emf 0.2 s after the switch is closed?

31. One application of an RL circuit is the generation of high-voltage transients from a low-voltage dc source, as shown in Figure 32.25. (a) What is the current in the circuit a long time after the switch has been in position A? (b) Now the switch is thrown quickly from A to B. Compute the initial voltage across each resistor and the inductor. (c) How much time elapses before the voltage across the inductor drops to 12 V?

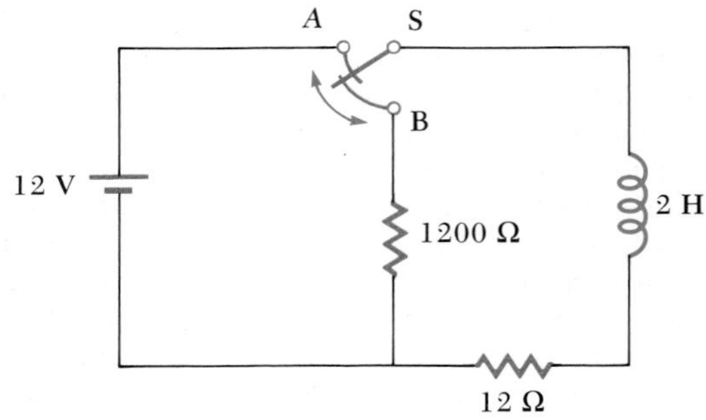

Figure 32.25 (Problem 31).

Section 32.3 Energy in a Magnetic Field

32. Calculate the energy associated with the magnetic field of a 200-turn solenoid in which a current of 1.75 A produces a flux of 3.7×10^{-4} Wb in each turn.

33. An air-core solenoid with 68 turns is 8 cm long and has a diameter of 1.2 cm. How much energy is stored in its magnetic field when it carries a current of 0.77 A?

34. Consider the circuit shown in Figure 32.26. What energy is stored in the inductor when the current reaches its final equilibrium value after the switch is closed?

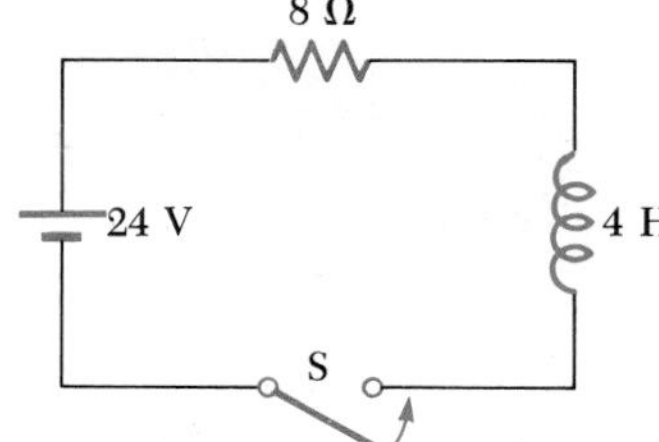

Figure 32.26 (Problems 34 and 42).

35. In the circuit shown in Figure 32.24, let $\mathcal{E} = 6.0$ V, $L = 350$ mH, and $R = 5\ \Omega$. Determine the following quantities 0.1 s after the switch is closed: (a) the rate at which energy is being stored in the magnetic field of the inductor; (b) the instantaneous power delivered to the resistor; and (c) the instantaneous rate at which energy is being drawn from the battery. (d) What is the algebraic sum of the individual rates of energy transfer in this circuit?

36. (a) Calculate the energy stored in the magnetic field of the inductor of Problem 10 at $t = 0.5$ s. (b) At what rate is energy being added to the magnetic field at the time $t = 0.5$ s?

37. Calculate the magnetic energy density near the center of a closely wound solenoid of 990 turns/m when the current in the solenoid is 2.8 A.

rent [as found in (a)] and a maximum potential difference [as found in (b)]?

73. When the current I in the portion of circuit shown in Figure 32.28 is 2 A and *increasing* at a rate of 1 A/s, the measured potential difference $V_{ab} = 8$ V. However, when the current $I = 2$ A and is *decreasing* at a rate of 1 A/s, the measured potential difference $V_{ab} = 4$ V. Calculate the value of L and R.

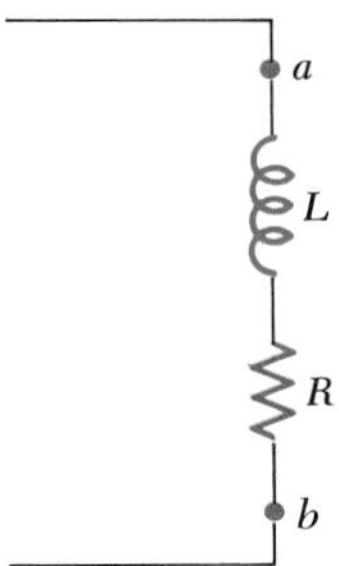

Figure 32.28 (Problem 73).

74. A platinum wire 2.5 mm in diameter is connected in series to a 100-μF capacitor and a 1.2×10^{-3}-μH inductor to form an *RLC* circuit. The resistivity of platinum is $11 \times 10^{-8}\ \Omega \cdot$ m. Calculate the *maximum* length of wire for which the current in the circuit will oscillate.

75. Assume that the switch in the circuit shown in Figure 32.29 is initially in position 1. Show that if the switch is thrown from position 1 to position 2, all the energy stored in the magnetic field of the inductor will be dissipated as thermal energy in the resistor.

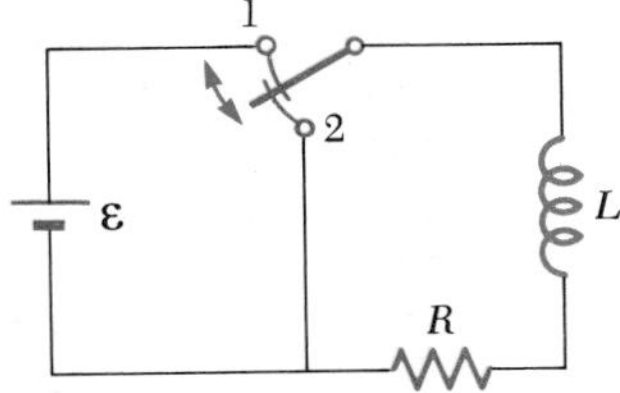

Figure 32.29 (Problems 75, 85, and 86).

76. (a) Determine the time constant of the circuit shown in Figure 32.30. (b) How much energy is stored in the 30-mH inductor when the total energy stored in the circuit is 50% of the maximum possible value? (Neglect mutual inductance between the coils.)

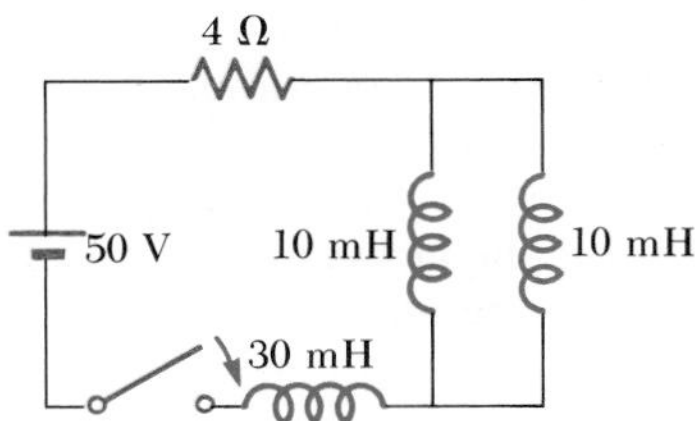

Figure 32.30 (Problem 76).

77. At $t = 0$, the switch in Figure 32.31 is closed. By using Kirchhoff's laws for the instantaneous currents and voltages in this two-loop circuit, show that the current through the inductor is given by

$$I(t) = \frac{\mathcal{E}}{R_1}\left[1 - e^{-(R'/L)t}\right]$$

where $R' = R_1R_2/(R_1 + R_2)$.

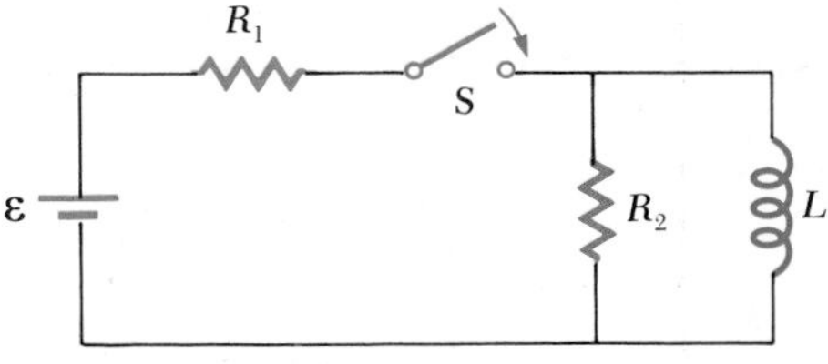

Figure 32.31 (Problem 77).

78. The toroidal coil shown in Figure 32.32 consists of N turns and has a rectangular cross section. Its inner and outer radii are a and b, respectively. (a) Show that the self inductance of the coil is given by

$$L = \frac{\mu_0 N^2 h}{2\pi} \ln(b/a)$$

(b) Using this result, compute the self-inductance of a 500-turn toroid with $a = 10$ cm, $b = 12$ cm, and $h = 1$ cm. (c) In Problem 14, an approximate formula for the inductance of a toroid with $R \gg r$ was derived. To get a feel for the accuracy of this result, use the expression in Problem 14 to compute the (approximate) inductance of the toroid described in part (b).

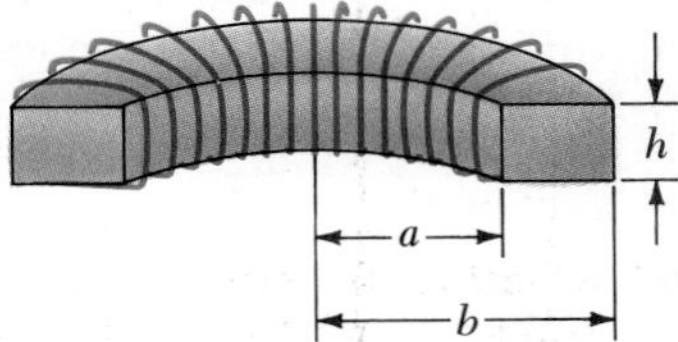

Figure 32.32 (Problem 78).

79. The battery in the circuit shown in Figure 32.33 has an emf $\mathcal{E} = 24$ V. (a) What current is being delivered

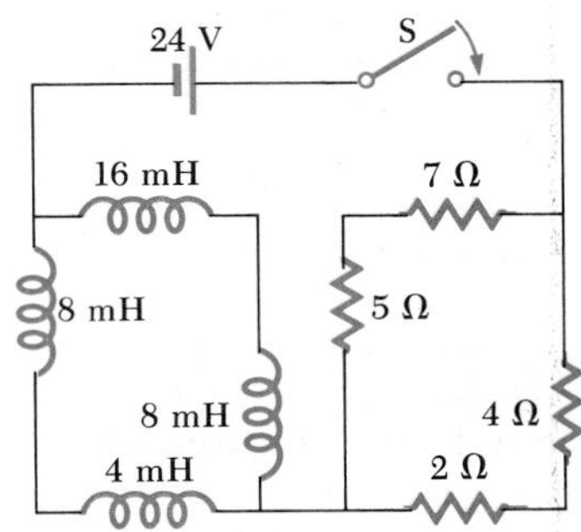

Figure 32.33 (Problem 79).

by the battery 1 ms after the switch is closed? (b) Determine the potential difference across the 5-Ω resistor 3 ms after the switch is closed. (Neglect the mutual inductance of the coils.)

80. In previous problems, when two or more coils were present in a circuit, it was assumed that they were located so that the flux from one coil did not link the turns of the other coils. Now consider the situation shown in Figure 32.34, in which two coaxial solenoids are positioned so that some of the flux from one links the windings of the other. (a) Use Kirchhoff's loop theorem to show that the effective inductance of the pair is given by

$$L_{\text{eff}} = L_1 + L_2 \pm 2M$$

where M is the mutual inductance of the two solenoids. (b) Why is it necessary to allow the choice of + and − signs for the $2M$ term?

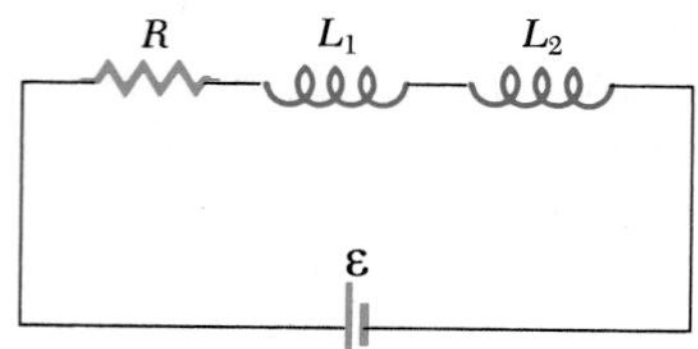

Figure 32.34 (Problem 80).

81. Two long parallel wires, each of radius a, have their centers a distance d apart and carry *equal* currents in *opposite* directions. Neglecting the flux within the wires themselves, calculate the inductance per unit length of such a pair of wires.

82. An air-core solenoid 0.5 m in length contains 1000 turns and has a cross-sectional area of 1 cm². (a) Neglecting end effects, what is the self-inductance? (b) A secondary winding wrapped around the center of the solenoid has 100 turns. What is the mutual inductance? (c) A constant current of 1 A flows in the secondary winding, and the solenoid is connected to a load of 10^3 Ω. The constant current is suddenly stopped. How much charge flows through the load resistor?

83. To prevent damage from arcing in an electric motor, a discharge resistor is sometimes placed in parallel with the armature. If the motor is suddenly unplugged while running, this resistor limits the voltage that appears across the armature coils. Consider a 12-V dc motor that has an armature with a resistance of 7.5 Ω and an inductance of 450 mH. Assume the counter emf in the armature coils is 10 V when the motor is running at normal speed. (The equivalent circuit for the armature is shown in Figure 32.35.) Calculate the maximum resistance R that will limit the voltage across the armature to 80 V when the motor is unplugged.

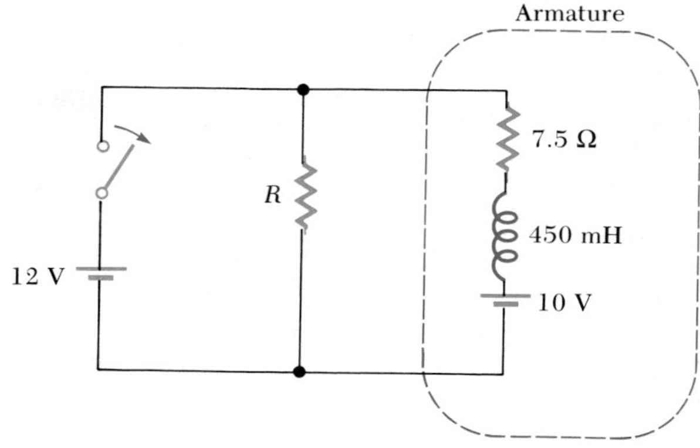

Figure 32.35 (Problem 83).

84. A long wire of radius R carries a steady current I, which is distributed evenly over the cross section of the conductor. (a) What is the magnetic energy density at $r_1 = R/2$? (b) Calculate the total magnetic energy per unit length stored in the wire. (c) What is the magnetic energy density at $r_2 = 3R$? (d) Calculate the total magnetic energy stored per unit length within a cylindrical shell of inner and outer radii $2R$ and $2.5R$, respectively, when the current-carrying wire is along the axis of the cylindrical shell.

85. The switch in the circuit of Figure 32.29 is placed in position 1 at time $t = 0$ and then switched to position 2 at $t = 0.3$ s. The values of the circuit parameters are $\mathcal{E} = 24$ V, $R = 3$ Ω, and $L = 1.2$ H. (a) Calculate the current in R at $t = 0.6$ s. (b) For what value of $t > 0.3$ s does the current in R have the same value that it has at $t = 0.15$ s?

86. Let the values of L, R, and $\mathcal{E}$ in the circuit shown in Figure 32.29 be 3.6 mH, 6 Ω, and 36 V, respectively. Consider the situation when the switch S is placed in position 1 at $t = 0$ and left there indefinitely. Let t_1 equal the time required for the current in R to increase from zero to 9% of its *maximum* value and let t_2 equal the time required for the current in R to increase from 90% to 99% of its maximum value. (a) What is the ratio of t_2 to t_1? (b) At what time will the rate of change of current equal one half of the *initial* rate of change of current? (c) At what time will the potential difference across the resistor have the same magnitude as the potential difference across the inductor? (d) At what time will the power delivered to the inductor equal the power dissipated in the resistor? (e) How is the answer to (d) related to the time constant of the circuit?

33
Alternating Current Circuits

High-voltage transmission lines at a power station in Oxfordshire, England. (© E. Nagele, FPG International)

In this chapter, we shall describe the basic principles of simple alternating current (ac) circuits. We shall investigate the characteristics of circuits containing familiar elements and driven by a sinusoidal voltage. Our discussion will be limited to analyzing simple series circuits containing a resistor, inductor, and capacitor, both individually and in combination with each other. We shall make use of the fact that these elements respond linearly; that is, the ac current through each element is proportional to the instantaneous ac voltage across the element. We shall find that when the applied voltage of the generator is sinusoidal, the current in each element is also sinusoidal, but not necessarily in phase with the applied voltage. We conclude the chapter with two sections concerning the characteristics of *RC* filters, transformers, and power transmission.

33.1 AC SOURCES AND PHASORS

An ac circuit consists of combinations of circuit elements and a generator which provides the alternating current. The basic principles of the ac generator were described in Section 31.5. By rotating a coil in a magnetic field with constant angular velocity ω, a sinusoidal voltage (emf) is induced in the coil. This instantaneous voltage, v, is given by

$$v = V_m \sin \omega t \tag{33.1}$$

where V_m is the *peak voltage of the ac generator*, or the **voltage amplitude.** The angular frequency, ω, is given by

$$\omega = 2\pi f = \frac{2\pi}{T}$$

where f is the frequency of the source and T is the period. Commercial electric-power plants in the United States use a frequency of $f = 60$ Hz (cycles per second), which corresponds to an angular frequency of $\omega = 377$ rad/s.

The primary aim of this chapter can be summarized as follows: Consider an ac generator connected to a series circuit containing R, L, and C elements. If the voltage amplitude and frequency of the generator are given, together with the values of R, L, and C, find the resulting current as specified by its amplitude and its phase constant. In order to simplify our analysis of more complex circuits containing two or more elements, we shall use graphical constructions called *phasor diagrams.* In these constructions, alternating quantities, such as current and voltage, are represented by rotating vectors called **phasors.** The length of the phasor represents the amplitude (maximum value) of the quantity, while the projection of the phasor onto the vertical axis represents the instantaneous value of that quantity. Phasors rotate counterclockwise. As we shall see, the method of combining several sinusoidally varying currents or voltages with different phases is greatly simplified using this procedure. We will also use phasor diagrams in Chapters 37 and 38, which concern interference and diffraction of light waves.

33.2 RESISTORS IN AN AC CIRCUIT

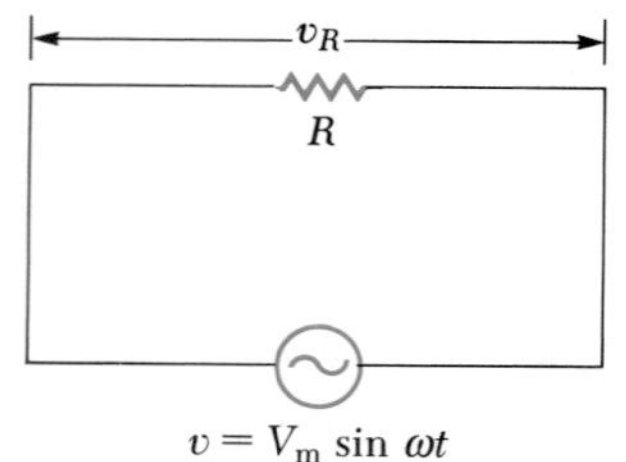

Figure 33.1 A circuit consisting of a resistor R connected to an ac generator, designated by the symbol —⊙—.

Consider a simple ac circuit consisting of a resistor and an ac generator (designated by the symbol —⊙—), as in Figure 33.1. At any instant, the algebraic sum of the potential increases and decreases around a closed loop in a circuit must be zero (Kirchhoff's loop equation). Therefore, $v - v_R = 0$, or

$$v = v_R = V_m \sin \omega t \tag{33.2}$$

where v_R is the *instantaneous voltage drop across the resistor.* Therefore, the instantaneous current is equal to

$$i_R = \frac{v}{R} = \frac{V_m}{R} \sin \omega t = I_m \sin \omega t \tag{33.3}$$

where I_m is the *peak current,* given by

Maximum current in a resistor

$$I_m = \frac{V_m}{R} \tag{33.4}$$

From Equations 33.2 and 33.3, we see that the instantaneous voltage drop across the resistor is

$$v_R = I_m R \sin \omega t \tag{33.5}$$

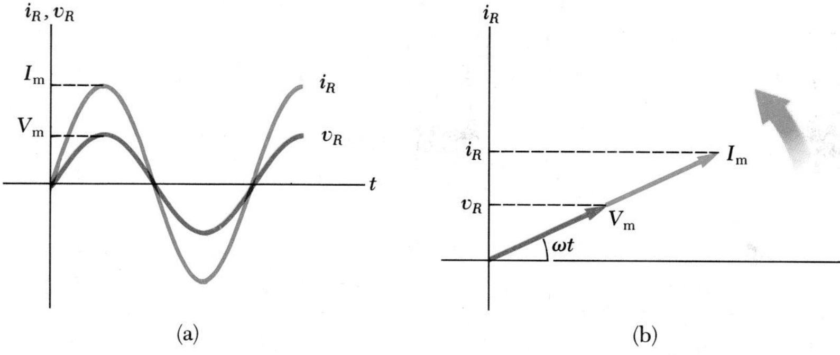

Figure 33.2 (a) Plots of the current and voltage across a resistor as functions of time. The current is in phase with the voltage. (b) A phasor diagram for the resistive circuit, showing that the current is *in phase* with the voltage. The projections of the rotating arrows (phasors) onto the vertical axis represent the instantaneous values v_R and i_R.

Since i_R and v_R both vary as sin ωt and reach their peak values at the *same time*, they are said to be *in phase*. Graphs of the voltage and current as functions of time (Fig. 33.2a) show that they each reach their peak and zero values at the same instant.

A phasor diagram may be used to represent the phase relationship between current and voltage. The lengths of the arrows correspond to V_m and L_m. The projections of the arrows onto the vertical axis give v_R and i_R. In the case of the single-loop resistive circuit, the current and voltage phasors lie along the same line, as in Figure 33.2b, since i_R and v_R are *in phase with each other*.

The current is in phase with the voltage for a resistor

Note that *the average value of the current over one cycle is zero*. That is, the current is maintained in one direction (the positive direction) for the same amount of time and at the same magnitude as in the opposite direction (the negative direction). However, the direction of the current has no effect on the behavior of the resistor in the circuit. This can be understood by realizing that collisions between electrons and the fixed atoms of the resistor result in an increase in the temperature of the resistor. Although this temperature increase depends on the magnitude of the current, it is independent of the direction of the current.

This discussion can be made quantitative by recalling that the rate at which electrical energy is converted to heat in a resistor, which is the power P, is given by

Power

$$P = i^2R$$

where i is the instantaneous current in the resistor. Since the heating effect of a current is proportional to the *square* of the current, it makes no difference whether the current is direct or alternating, that is, whether the sign associated with the current is positive or negative. However, the heating effect produced by an alternating current having a maximum value of I_m *is not the same* as that produced by a direct current of the same value. This is because the alternating current is at this maximum value for only a very brief instant of time during a cycle. What is of importance in an ac circuit is an average value of current referred to as the rms current. The term **rms** refers to *root mean square*, which simply means that one takes the square root of the average value of the square of the current. I^2 varies as $\sin^2 \omega t$, and one can show[1] that the average value of i^2 is $\frac{1}{2}I_m^2$ (Fig. 33.3). Therefore, the rms current, I_{rms}, is related to the peak value of the alternating current, I_m, as

rms current

$$I_{rms} = \frac{I_m}{\sqrt{2}} = 0.707 I_m \tag{33.6}$$

This equation says that an alternating current whose maximum value is 2A will produce the same heating effect in a resistor as a direct current of $(0.707)(2) = 1.414$ A. Thus, we can say that the average power dissipated in a resistor that carries an alternating current is $P_{av} = I_{rms}^2 R$.

Alternating voltages are also best discussed in terms of rms voltages, and the relationship here is identical to the above, that is, the rms voltage, V_{rms}, is related to the peak value of the alternating voltage, V_m, as

rms voltage

$$V_{rms} = \frac{V_m}{\sqrt{2}} = 0.707 V_m \tag{33.7}$$

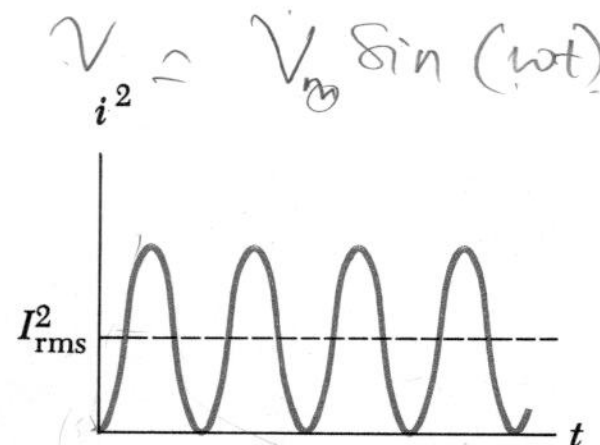

Figure 33.3 Plot of the square of the current in a resistor versus time. The rms current is the square root of the average of the square of the current.

When one speaks of measuring an ac voltage of 120 V from an electric outlet, one is really referring to an rms voltage of 120 V. A quick calculation using Equation 33.7 shows that such an ac voltage actually has a peak value of about 170 V. In this chapter we shall use rms values when discussing alternating currents and voltages. One reason for this is that ac ammeters and voltmeters are designed to read rms values. Furthermore, we shall find that if we use rms values, many of the equations we use will have the same form as those used in the study of direct current (dc) circuits. Table 33.1 summarizes the notation that will be used in this chapter.

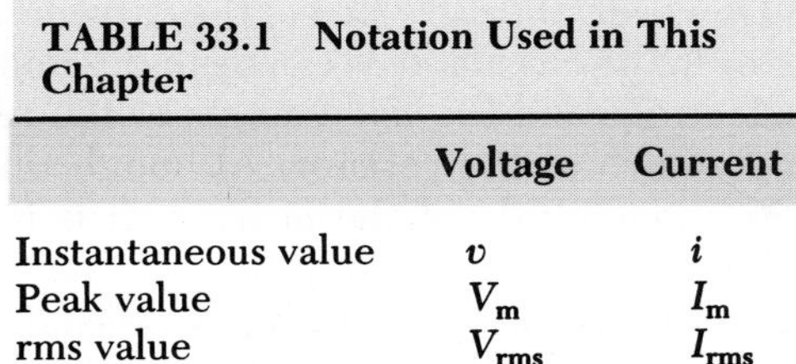

TABLE 33.1 Notation Used in This Chapter

	Voltage	Current
Instantaneous value	v	i
Peak value	V_m	I_m
rms value	V_{rms}	I_{rms}

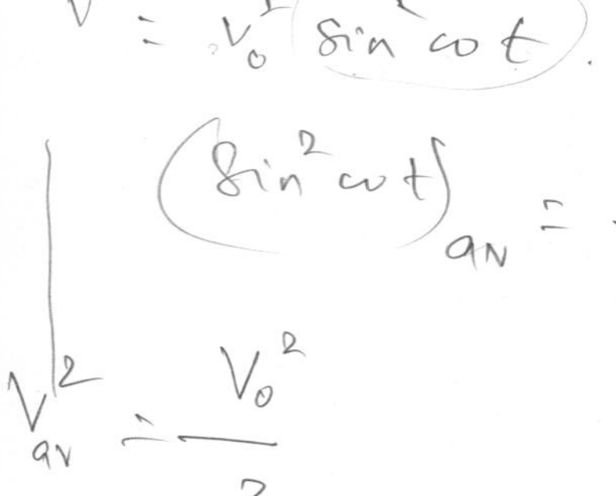

[1] The fact that the square root of the average value of the square of the current is equal to $I_m/\sqrt{2}$ can be shown as follows. The current in the circuit varies with time according to the expression $i = I_m \sin \omega t$, so that $i^2 = I_m^2 \sin^2 \omega t$. Therefore we can find the average value of i^2 by calculating the average value of $\sin^2 \omega t$. Note that a graph of $\cos^2 \omega t$ versus time is identical to a graph of $\sin^2 \omega t$ versus time, except that the points are shifted on the time axis. Thus, the time average of $\sin^2 \omega t$ is equal to the time average of $\cos^2 \omega t$ when taken over one or more complete cycles. That is,

$$(\sin^2 \omega t)_{av} = (\cos^2 \omega t)_{av}$$

With this fact and the trigonometric identity $\sin^2 \theta + \cos^2 \theta = 1$, we get

$$(\sin^2 \omega t)_{av} + (\cos^2 \omega t)_{av} = 2(\sin^2 \omega t)_{av} = 1$$

or

$$(\sin^2 \omega t)_{av} = \tfrac{1}{2}$$

When this result is substituted in the expression $i^2 = I_m^2 \sin^2 \omega t$, we get $(i^2)_{av} = I_{rms}^2 = I_m^2/2$, or $I_{rms} = I_m/\sqrt{2}$, where I_{rms} is the rms current. The factor of $1/\sqrt{2}$ is only valid for sinusoidally varying currents. Other waveforms such as sawtooth variations have different factors.

EXAMPLE 33.1 What is the rms Current?
An ac voltage source has an output given by the expression $v = 200 \sin \omega t$. This source is connected to a 100-Ω resistor as in Figure 33.1. Find the rms current in the circuit.

Solution Compare the expression for the voltage output given above with the general form, $v = V_m \sin \omega t$. We see that the peak output voltage of the device is 200 V. Thus, the rms voltage output of the source is

$$V_{rms} = \frac{V_m}{\sqrt{2}} = \frac{200 \text{ V}}{\sqrt{2}} = 141 \text{ V}$$

Ohm's law can be used in resistive ac circuits as well as in dc circuits. The calculated rms voltage can be used with Ohm's law to find the rms current in the circuit:

$$I_{rms} = \frac{V_{rms}}{R} = \frac{141 \text{ V}}{100 \ \Omega} = 1.41 \text{ A}$$

Exercise 1 Find the peak current in the circuit.
Answer 2 A.

33.3 INDUCTORS IN AN AC CIRCUIT

Now consider an ac circuit consisting only of an inductor connected to the terminals of an ac generator as in Figure 33.4. If v_L is the *instantaneous voltage drop across the inductor,* then Kirchhoff's loop rule applied to this circuit gives $v + v_L = 0$, or

$$v - L\frac{di}{dt} = 0$$

When we rearrange this equation and substitute $v = V_m \sin \omega t$, we get

$$L\frac{di}{dt} = V_m \sin \omega t \tag{33.8}$$

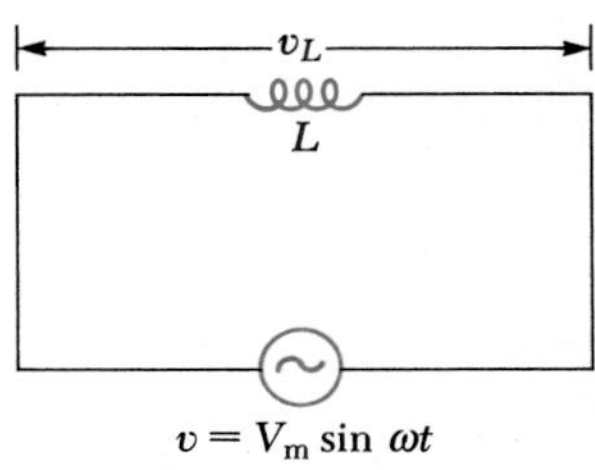

Figure 33.4 A circuit consisting of an inductor L connected to an ac generator.

Integrating this expression[2] gives the current as a function of time:

$$i_L = \frac{V_m}{L}\int \sin \omega t \, dt = -\frac{V_m}{\omega L}\cos \omega t \tag{33.9}$$

Using the trigonometric identity $\cos \omega t = -\sin(\omega t - \pi/2)$, Equation 33.9 can also be expressed as

$$i_L = \frac{V_m}{\omega L}\sin\left(\omega t - \frac{\pi}{2}\right) \tag{33.10}$$

Comparing this result with Equation 33.8 clearly shows that the current is out of phase with the voltage by $\pi/2$ rad, or 90°. A plot of the voltage and current versus time is given in Figure 33.5a. The voltage reaches its peak value at a time that is one quarter of the oscillation period *before* the current reaches its peak value. The corresponding phasor diagram for this circuit is shown in Figure 33.5b. Thus we see that

> for a sinusoidal applied voltage, the current always lags behind the voltage across an inductor by 90°.

The current in an inductor lags the voltage by 90°

This can be understood by noting that since the voltage across the inductor is proportional to di/dt, the value of v_L is largest when the current is changing most rapidly. Since i versus t is a sinusoidal curve, di/dt (the slope) is maximum when the curve goes through zero. This shows that v_L reaches its maximum value when the current is zero.

[2] The constant of integration is neglected here since it depends on the initial conditions, which are not important for this situation.

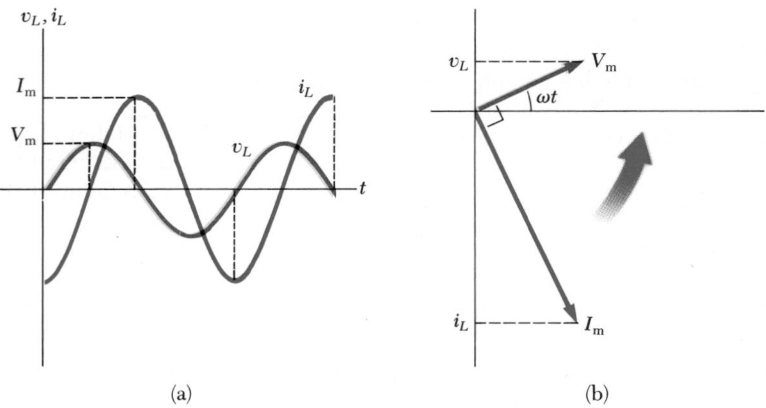

Figure 33.5 (a) Plots of the current and voltage across the inductor as functions of time. The voltage *leads* the current by 90°. (b) The phasor diagram for the inductive circuit. Projections of the phasors onto the vertical axis give the instantaneous values v_L and i_L.

From Equation 33.9 we see that the peak current I_m is

Maximum current in an inductor

$$I_m = \frac{V_m}{\omega L} = \frac{V_m}{X_L} \quad (33.11)$$

where the quantity X_L, called the **inductive reactance,** is given by

Inductive reactance

$$X_L = \omega L \quad (33.12)$$

The rms current is given by an expression similar to Equation 33.11, with V_m replaced by V_{rms}.

The term *reactance* is used so that it is not confused with *resistance.* Reactance is distinguished from resistance by the phase difference between v and i. Recall that i and v are always in phase in a purely resistive circuit, whereas i lags behind v by 90° in a purely inductive circuit.

Using Equations 33.8 and 33.11, we find that the instantaneous voltage drop across the inductor can be expressed as

$$v_L = V_m \sin \omega t = I_m X_L \sin \omega t \quad (33.13)$$

We can think of Equation 33.13 as Ohm's law for an inductive circuit. It is left as a problem (Problem 8) to show that X_L has the SI unit of ohm.

Note that the reactance of an inductor increases with increasing frequency. This is because at higher frequencies, the current must change more rapidly, which in turn causes an increase in the induced emf associated with a given peak current.

EXAMPLE 33.2 A Purely Inductive AC Circuit

In a purely inductive ac circuit (Fig. 33.4), $L = 25$ mH and the rms voltage is 150 V. Find the inductive reactance and rms current in the circuit if the frequency is 60 Hz.

Solution First, note that $\omega = 2\pi f = 2\pi(60) = 377\ s^{-1}$. Equation 33.12 then gives

$$X_L = \omega L = (377\ s^{-1})(25 \times 10^{-3}\ H) = 9.43\ \Omega$$

The rms current is given by

$$I_{\rm rms} = \frac{V_L}{X_L} = \frac{150\ \rm V}{9.43\ \Omega} = 15.9\ \rm A$$

Exercise 2 Calculate the inductive reactance and rms current in the circuit if the frequency is 6 kHz.
Answers $X_L = 943\ \Omega$, $I_{\rm rms} = 0.159$ A.

33.4 CAPACITORS IN AN AC CIRCUIT

Figure 33.6 shows an ac circuit consisting of a capacitor connected across the terminals of an ac generator. Kirchhoff's loop rule applied to this circuit gives $v - v_C = 0$, or

$$v = v_C = V_{\rm m} \sin \omega t \tag{33.14}$$

where v_C is the *instantaneous voltage drop across the capacitor.* But from the definition of capacitance, $v_C = Q/C$, which when substituted into Equation 33.14 gives

$$Q = CV_{\rm m} \sin \omega t \tag{33.15}$$

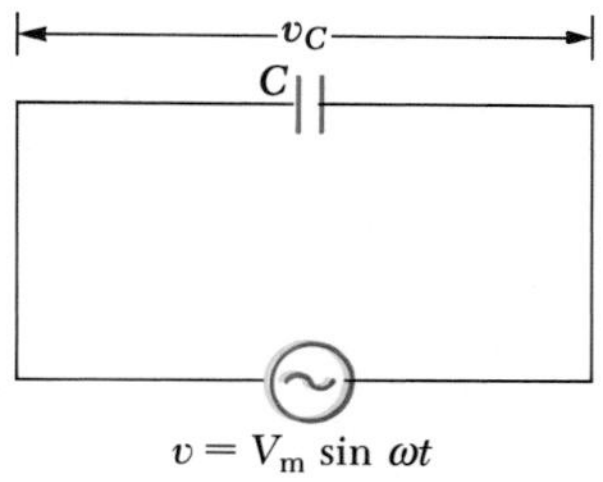

Figure 33.6 A circuit consisting of a capacitor C connected to an ac generator.

Since $i = dQ/dt$, differentiating Equation 33.15 gives the instantaneous current:

$$i_C = \frac{dQ}{dt} = \omega CV_{\rm m} \cos \omega t \tag{33.16}$$

Again, we see that the current is not in phase with the voltage drop across the capacitor, given by Equation 33.14. Using the trigonometric identity $\cos \omega t = \sin\left(\omega t + \frac{\pi}{2}\right)$, we can express Equation 33.16 in the alternative form

$$i_C = \omega CV_{\rm m} \sin\left(\omega t + \frac{\pi}{2}\right) \tag{33.17}$$

Comparing this expression with Equation 33.14, we see that the current is 90° out of phase with the voltage across the inductor. A plot of the current and voltage versus time (Fig. 33.7a) shows that the current reaches its peak value

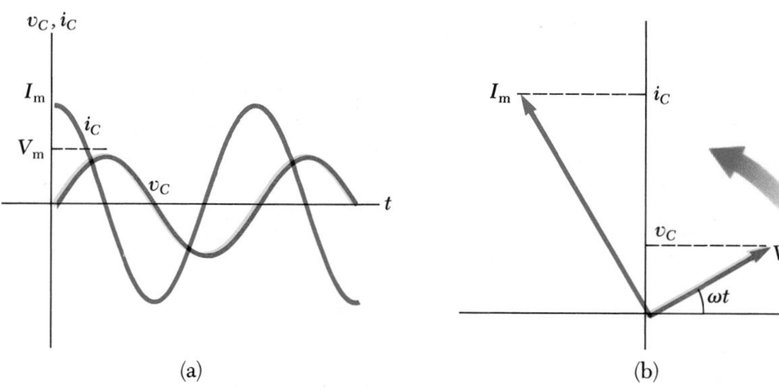

Figure 33.7 (a) Plots of the current and voltage across the capacitor as functions of time. The voltage *lags behind* the current by 90°. (b) Phasor diagram for the purely capacitive circuit. Projections of the phasors onto the vertical axis gives the instantaneous values v_C and i_C.

one quarter of a cycle sooner than the voltage reaches it peak value. The corresponding phasor diagram in Figure 33.7b also shows that

The current leads the voltage across the capacitor by 90°

for a sinusoidally applied emf, the current always leads the voltage across a capacitor by 90°.

From Equation 33.17, we see that the peak current in the circuit is

$$I_m = \omega C V_m = \frac{V_m}{X_C} \tag{33.18}$$

where X_C is called the **capacitive reactance.**

Capacitive reactance

$$X_C = \frac{1}{\omega C} \tag{33.19}$$

The rms current is given by an expression similar to Equation 33.18, with V_m replaced by V_{rms}.

Combining Equations 33.14 and 33.18, we can express the instantaneous voltage drop across the capacitor as

$$v_C = I_m X_C \sin \omega t \tag{33.20}$$

The SI unit of X_C is also the ohm. As the frequency of the circuit increases, the current increases but the reactance decreases. For a given maximum applied voltage V_m, the current increases as the frequency increases. As the frequency approaches zero, the capacitive reactance approaches infinity. Therefore, the current approaches zero. This makes sense since the circuit approaches dc conditions as $\omega \rightarrow 0$. Of course, a capacitor passes no current under steady-state dc conditions.

EXAMPLE 33.3 A Purely Capacitive AC Circuit
An 8-μF capacitor is connected to the terminals of an ac generator whose rms voltage is 150 V and whose frequency is 60 Hz. Find the capacitive reactance and the rms current in the circuit.

Solution Using Equation 33.19 and the fact that $\omega = 2\pi f = 377\ s^{-1}$ gives

$$X_C = \frac{1}{\omega C} = \frac{1}{(377\ s^{-1})(8 \times 10^{-6}\ F)} = 332\ \Omega$$

Hence, the rms current is

$$I_{rms} = \frac{V_{rms}}{X_C} = \frac{150\ V}{332\ \Omega} = 0.452\ A$$

Exercise 3 If the frequency is doubled, what happens to the capacitive reactance and the current?
Answer X_C is halved, and I is doubled.

33.5 THE *RLC* SERIES CIRCUIT

In the previous sections, we examined the effects that an inductor, a capacitor, and a resistor have when placed separately across an ac-voltage source. We shall now consider what happens when combinations of these devices are used.

Figure 33.8a shows a circuit containing a resistor, an inductor, and a capacitor connected in series across an ac-voltage source. As before, we assume that the applied voltage varies sinusoidally with time. It is convenient to assume that the applied voltage is given by

$$v = V_m \sin \omega t$$

while the current varies as

$$i = I_m \sin(\omega t - \phi)$$

The quantity ϕ is called the **phase angle** between the current and the applied voltage. Our aim is to determine ϕ and I_m. Figure 33.8b shows the voltage versus time across each element in the circuit and their phase relations.

In order to solve this problem, we must construct and analyze the phasor diagram for this circuit. First, note that since the elements are in series, the current everywhere in the circuit must be the same at any instant. That is, *the ac current at all points in a series ac circuit has the same amplitude and phase.* Therefore, as we found in the previous sections, the voltage across each element will have *different* amplitudes and phases, as summarized in Figure 33.9. In particular, the voltage across the resistor is in phase with the current (Figure 33.9a), the voltage across the inductor leads the current by 90° (Fig. 33.9b), and finally, the voltage across the capacitor lags behind the current by 90° (Fig. 33.9c). Using these phase relationships, we can express the *instantaneous* voltage drops across the three elements as

$$v_R = I_m R \sin \omega t = V_R \sin \omega t \qquad (33.21)$$

$$v_L = I_m X_L \sin\left(\omega t + \frac{\pi}{2}\right) = V_L \cos \omega t \qquad (33.22)$$

$$v_C = I_m X_C \sin\left(\omega t - \frac{\pi}{2}\right) = -V_C \cos \omega t \qquad (33.23)$$

where V_R, V_L, and V_C are the *peak* voltages across each element, given by

$$V_R = I_m R \qquad (33.24)$$

$$V_L = I_m X_L \qquad (33.25)$$

$$V_C = I_m X_C \qquad (33.26)$$

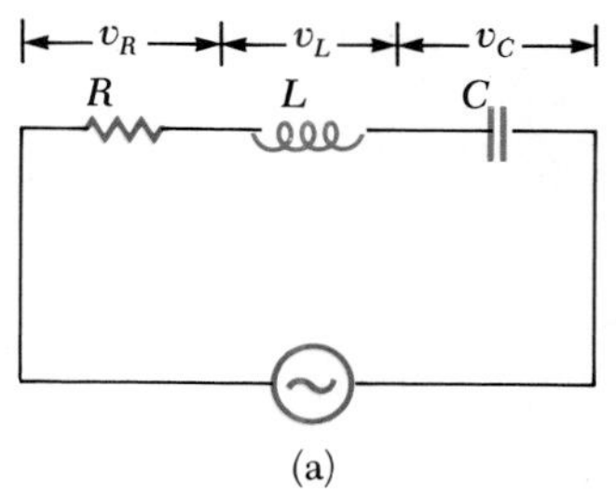

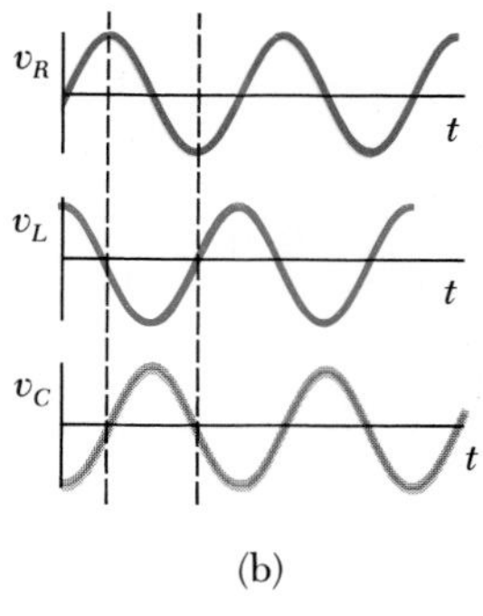

Figure 33.8 (a) A series circuit consisting of a resistor, an inductor, and a capacitor connected to an ac generator. (b) Phase relations in the series *RLC* circuit shown in part (a).

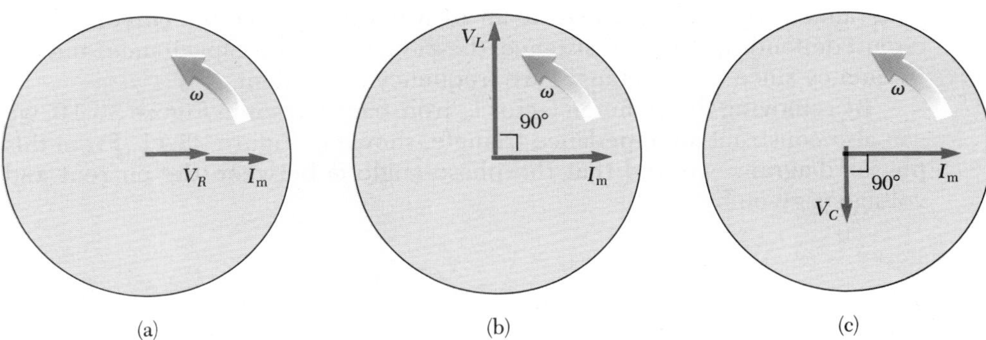

Figure 33.9 Phase relationships between the peak voltage and current phasors for (a) a resistor, (b) an inductor, and (c) a capacitor.

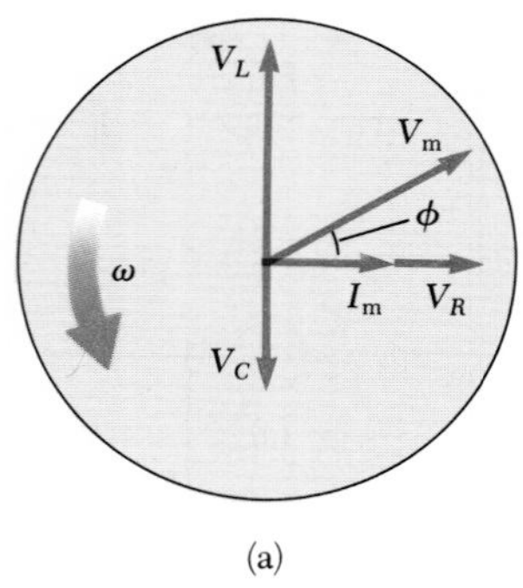

(a)

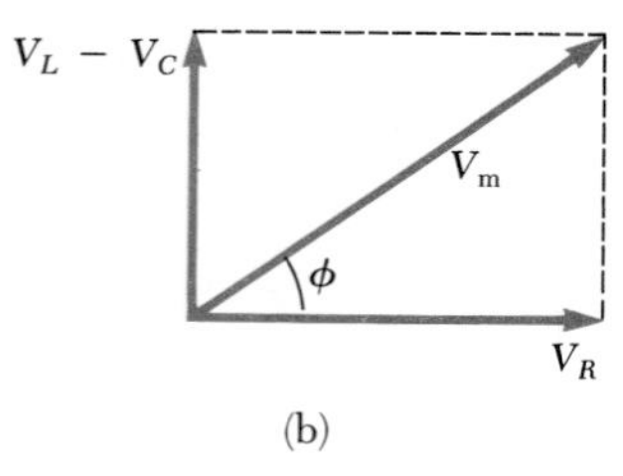

(b)

Figure 33.10 (a) The phasor diagram for the series *RLC* circuit shown in Figure 33.8. Note that the phasor V_R is *in phase* with the current phasor I_m, the phasor V_L *leads* the phasor I_m by 90°, and the phasor V_C *lags behind* the phasor I_m by 90°. The total voltage V_m makes an angle ϕ with the current phasor I_m. (b) Simplified version of the phasor diagram shown in (a).

At this point, we could proceed by noting that the instantaneous voltage v across the three elements equals the sum

$$v = v_R + v_L + v_C \tag{33.27}$$

Although this analytical approach is correct, it is simpler to obtain the sum by examining the phasor diagram.

Because the current in each element is the same at any instant, we can obtain the resulting phasor diagram by combining the three phasors shown in Figure 33.9. This gives the diagram shown in Figure 33.10a, where a single phasor I_m is used to represent the current in each element. To obtain the vector sum of these voltages, it is convenient to redraw the phasor diagram as in Figure 33.10b. From this diagram, we see that the *vector* sum of the voltage amplitudes V_R, V_L, and V_C equals a phasor whose length is the peak applied voltage, V_m, where the phasor V_m makes an angle ϕ with the current phasor, I_m. Note that the voltage phasors V_L and V_C are in opposite directions along the same line, and hence we are able to construct the difference phasor $V_L - V_C$, which is perpendicular to the phasor V_R. From the right triangle in Figure 33.10b, we see that

$$V_m = \sqrt{V_R^2 + (V_L - V_C)^2} = \sqrt{(I_m R)^2 + (I_m X_L - I_m X_C)^2}$$

$$V_m = I_m \sqrt{R^2 + (X_L - X_C)^2} \tag{33.28}$$

where $X_L = \omega L$ and $X_C = 1/\omega C$. Therefore, we can express the maximum current as

$$I_m = \frac{V_m}{\sqrt{R^2 + (X_L - X_C)^2}}$$

The **impedance** Z of the circuit is defined to be

$$Z \equiv \sqrt{R^2 + (X_L - X_C)^2} \tag{33.29}$$

Therefore, we can write Equation 33.28 in the form

$$V_m = I_m Z \tag{33.30}$$

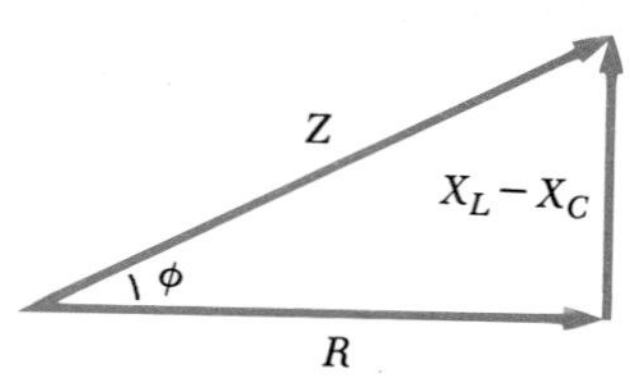

Figure 33.11 The impedance triangle for a series *RLC* circuit which gives the relationship $Z = \sqrt{R^2 + (X_L - X_C)^2}$.

Impedance also has the SI unit of ohm. We can regard Equation 33.30 as a generalized Ohm's law applied to an ac circuit. Note that the current in the circuit depends upon the resistance, the inductance, the capacitance, and the frequency since the reactances are frequency dependent.

By removing the common factor I_m from each phasor in Figure 33.10, we can also construct an impedance triangle, shown in Figure 33.11. From this phasor diagram, we find that the phase angle ϕ between the current and voltage is given by

Phase angle ϕ

$$\tan \phi = \frac{X_L - X_C}{R} \tag{33.31}$$

For example, when $X_L > X_C$ (which occurs at high frequencies), the phase angle is positive, signifying that the current lags behind the applied voltage, as

Circuit Elements	Impedance, Z	Phase angle, ϕ
R	R	$0°$
C	X_C	$-90°$
L	X_L	$+90°$
R C	$\sqrt{R^2+X_C^2}$	Negative, between $-90°$ and $0°$
R L	$\sqrt{R^2+X_L^2}$	Positive, between $0°$ and $90°$
R L C	$\sqrt{R^2+(X_L-X_C)^2}$	Negative if $X_C>X_L$ Positive if $X_C<X_L$

Figure 33.12 The impedance values and phase angles for various circuit element combinations. In each case, an ac voltage (not shown) is applied across the combination of elements (that is, across the dots).

in Figure 33.10. On the other hand, if $X_L < X_C$, the phase angle is negative, signifying that the current leads the applied voltage. Finally, when $X_L = X_C$, the phase angle is zero. In this case, the ac impedance equals the resistance and the current has its *peak* value, given by V_m/R. The frequency at which this occurs is called the *resonance frequency,* which will be described further in Section 33.7.

Figure 33.12 gives impedance values and phase angles for various series circuits containing different combinations of circuit elements.

EXAMPLE 33.4 Analyzing a Series *RLC* Circuit

Analyze a series *RLC* ac circuit for which $R = 250\ \Omega$, $L = 0.6$ H, $C = 3.5\ \mu$F, $\omega = 377\ \text{s}^{-1}$, and $V_m = 150$ V.

Solution The reactances are given by $X_L = \omega L = 226\ \Omega$ and $X_C = 1/\omega C = 758\ \Omega$. Therefore, the impedance is equal to

$$Z = \sqrt{R^2 + (X_L - X_C)^2}$$

$$= \sqrt{(250\ \Omega)^2 + (226\ \Omega - 758\ \Omega)^2} = 588\ \Omega$$

The maximum current is given by

$$I_m = \frac{V_m}{Z} = \frac{150\text{ V}}{588\ \Omega} = 0.255\text{ A}$$

The phase angle between the current and voltage is

$$\phi = \tan^{-1}\left(\frac{X_L - X_C}{R}\right) = \tan^{-1}\left(\frac{226 - 758}{250}\right)$$

$$= -64.8°$$

Since the circuit is more capacitive than inductive, ϕ is negative and the current leads the applied voltage.

The *peak* voltages across each element are given by

$$V_R = I_mR = (0.255\text{ A})(250\ \Omega) = 63.8\text{ V}$$

$$V_L = I_mX_L = (0.255\text{ A})(226\ \Omega) = 57.6\text{ V}$$

$$V_C = I_mX_C = (0.255\text{ A})(758\ \Omega) = 193\text{ V}$$

Using Equations 33.21, 33.22, and 33.23, we find that the *instantaneous* voltages across the three elements can be written

$$v_R = (63.8\text{ V})\sin 377t$$

$$v_L = (57.6\text{ V})\cos 377t$$

$$v_C = (-193\text{ V})\cos 377t$$

and the applied voltage is $v = 150\sin(\omega t - 64.8°)$. The sum of the *peak* voltages is $V_R + V_L + V_C = 314$ V, which is much larger than the maximum voltage of the generator, 150 V. The former is a meaningless quantity. This is because when harmonically varying quantities are added, *both their amplitudes and their phases* must be taken into account and we know that the peak voltages across the different circuit elements occur at *different* times. That is, the voltages must be added in a way that takes account of the different phases. When this is done, Equation 33.28 is satisfied. You should verify this result.

33.6 POWER IN AN AC CIRCUIT

As we shall see in this section, *there are no power losses associated with pure capacitors and pure inductors in an ac circuit.* (A pure inductor is defined as one with no resistance or capacitance.) First, let us analyze the power dissipated in an ac circuit containing only a generator and a capacitor.

When the current begins to increase in one direction in an ac circuit, charge begins to accumulate on the capacitor and a voltage drop appears across it. When the voltage drop across the capacitor reaches its peak value, the energy stored in the capacitor is $\frac{1}{2}CV_m^2$. However, this energy storage is only momentary. The capacitor is charged and discharged twice during each cycle. In this process, charge is delivered to the capacitor during two quarters of the cycle, and is returned to the voltage source during the remaining two quarters. Therefore, *the average power supplied by the source is zero.* In other words, *a capacitor in an ac circuit does not dissipate energy.*

Similarly, the source must do work against the back emf of the inductor, which carries a current. When the current reaches its peak value, the energy stored in the inductor is a maximum and is given by $\frac{1}{2}LI_m^2$. When the current begins to decrease in the circuit, this stored energy is returned to the source as the inductor attempts to maintain the current in the circuit.

When we studied dc circuits in Chapter 27, we found that the power delivered by a battery to an external circuit is equal to the product of the current and the emf of the battery. Likewise, the instantaneous power delivered by an ac generator to any circuit is the product of the generator current and the applied voltage. For the *RLC* circuit shown in Figure 33.8, we can express the instantaneous power P as

$$P = iv = I_m \sin(\omega t - \phi)[V_m \sin \omega t]$$

$$= I_m V_m \sin \omega t \sin(\omega t - \phi) \quad (33.32)$$

Clearly this result is a complicated function of time and, in itself, is not very useful from a practical viewpoint. What is generally of interest is the average power over one or more cycles. Such an average can be computed by first using the trigonometric identity $\sin(\omega t - \phi) = \sin \omega t \cos \phi - \cos \omega t \sin \phi$. Substituting this into Equation 33.32 gives

$$P = I_m V_m \sin^2 \omega t \cos \phi - I_m V_m \sin \omega t \cos \omega t \sin \phi \quad (33.33)$$

We now take the time average of P over one or more cycles, noting that I_m, V_m, ϕ, and ω are all constants. The time average of the first term on the right of Equation 33.33 involves the average value of $\sin^2 \omega t$, which is $\frac{1}{2}$, as shown in footnote 1. The time average of the second term on the right of Equation 33.33 is identically zero because $\sin \omega t \cos \omega t = \frac{1}{2} \sin 2\omega t$, whose average value is zero.

Therefore, we can express the **average power** P_{av} as

$$P_{av} = \tfrac{1}{2} I_m V_m \cos \phi \quad (33.34)$$

It is convenient to express the average power in terms of the rms current and rms voltage defined by Equations 33.6 and 33.7. Using these defined quantities, the average power becomes

Average power

$$P_{av} = I_{rms} V_{rms} \cos \phi \quad (33.35)$$

where the quantity $\cos \phi$ is called the **power factor.** By inspecting Figure 33.10, we see that the maximum voltage drop across the resistor is given by $V_R = V_m \cos \phi = I_m R$. Using Equations 33.6 and 33.7 and the fact that $\cos \phi = I_m R/V_m$, we find that P_{av} can be expressed as

$$P_{av} = I_{rms} V_{rms} \cos \phi = I_{rms} \left(\frac{V_m}{\sqrt{2}}\right) \frac{I_m R}{V_m} = I_{rms} \frac{I_m R}{\sqrt{2}}$$

$$P_{av} = I_{rms}^2 R \qquad (33.36)$$

In other words, the *average power delivered by the generator is dissipated as heat in the resistor,* just as in the case of a dc circuit. *There is no power loss in an ideal inductor or capacitor.* When the load is purely resistive, then $\phi = 0$, $\cos \phi = 1$, and from Equation 33.35 we see that $P_{av} = I_{rms} V_{rms}$.

EXAMPLE 33.5 Average Power in a *RLC* Series Circuit

Calculate the average power delivered to the series *RLC* circuit described in Example 33.4.

Solution First, let us calculate the rms voltage and rms current:

$$V_{rms} = \frac{V_m}{\sqrt{2}} = \frac{150 \text{ V}}{\sqrt{2}} = 106 \text{ V}$$

$$I_{rms} = \frac{I_m}{\sqrt{2}} = \frac{V_m/Z}{\sqrt{2}} = \frac{0.255 \text{ A}}{\sqrt{2}} = 0.180 \text{ A}$$

Since $\phi = -64.8°$, the power factor, $\cos \phi$, is 0.426, and hence the average power is

$$P_{av} = I_{rms} V_{rms} \cos \phi = (0.180 \text{ A})(106 \text{ V})(0.426)$$

$$= 8.13 \text{ W}$$

The same result can be obtained using Equation 33.36.

33.7 RESONANCE IN A SERIES *RLC* CIRCUIT

A series *RLC* circuit is said to be in *resonance* when the current has its peak value. In general, the rms current can be written

$$I_{rms} = \frac{V_{rms}}{Z} \qquad (33.37)$$

where Z is the impedance. Substituting Equation 33.29 into 33.37 gives the relationship

$$I_{rms} = \frac{V_{rms}}{\sqrt{R^2 + (X_L - X_C)^2}} \qquad (33.38)$$

Because the impedance depends on the frequency of the source, we see that the current in the *RLC* circuit will also depend on the frequency. Note that the current reaches its peak when $X_L = X_C$, corresponding to $Z = R$. The frequency ω_0 at which this occurs is called the **resonance frequency** of the circuit. To find ω_0, we use the condition $X_L = X_C$, from which we get

$$\omega_0 L = \frac{1}{\omega_0 C}$$

$$\omega_0 = \frac{1}{\sqrt{LC}} \qquad (33.39)$$

Resonance frequency

Note that this frequency also corresponds to the natural frequency of oscillation of an *LC* circuit (Section 32.5). Therefore, the current in a series *RLC* circuit reaches its *peak* value when the frequency of the applied voltage matches the natural oscillator frequency, which depends only on L and C. Furthermore, at this frequency the current is in phase with the applied voltage.

A plot of the rms current versus frequency for a series *RLC* circuit is shown in Figure 33.13a. The data that are plotted assume a constant rms voltage of 5 mV, $L = 5\ \mu\text{H}$, and $C = 2$ nF. The three curves correspond to three different values of R. Note that in each case, the current reaches its peak value at the resonance frequency, ω_0. Furthermore, the curves become narrower and taller as the resistance decreases.

By inspecting Equation 33.38, one must conclude that the current would become infinite at resonance when $R = 0$. Although the equation predicts this, real circuits always have some resistance, which limits the value of the current. Mechanical systems can also exhibit resonances. For example, when an undamped mass-spring system is driven at its natural frequency of oscillation, its amplitude increases with time, as we discussed in Chapter 13. Large-amplitude mechanical vibrations can be disastrous, as in the case of the Tacoma Narrows Bridge collapse. (See the essay in Chapter 13.)

It is also interesting to calculate the average power as a function of frequency for a series *RLC* circuit. Using Equation 33.36 together with Equation 33.37, we find that

$$P_{\text{av}} = I_{\text{rms}}^2 R = \frac{V_{\text{rms}}^2}{Z^2} R = \frac{V_{\text{rms}}^2 R}{R^2 + (X_L - X_C)^2} \tag{33.40}$$

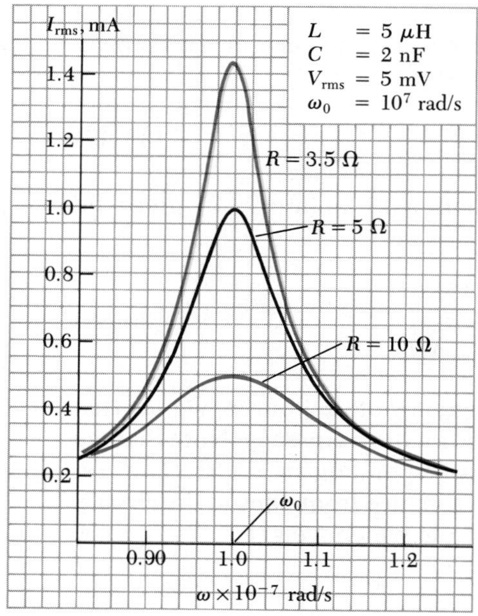

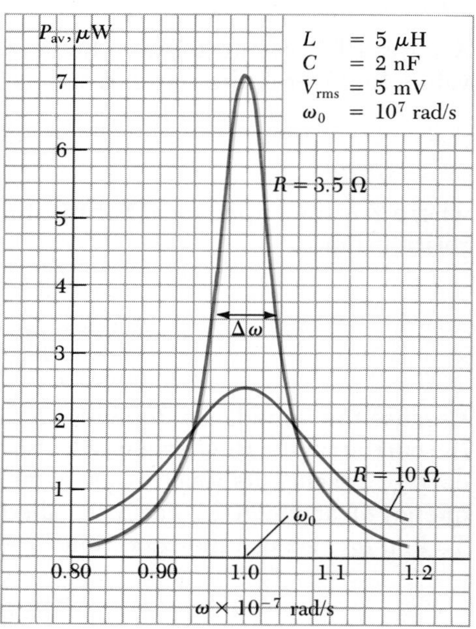

Figure 33.13 (a) Plots of the rms current versus frequency for a series *RLC* circuit for three different values of R. Note that the current reaches its peak value at the resonance frequency ω_0. (b) Plots of the average power versus frequency for the series *RLC* circuit for two different values of R.

Since $X_L = \omega L$, $X_C = \dfrac{1}{\omega C}$, and $\omega_0{}^2 = 1/LC$, the factor $(X_L - X_C)^2$ can be expressed as

$$(X_L - X_C)^2 = \left(\omega L - \frac{1}{\omega C}\right)^2 = \frac{L^2}{\omega^2}(\omega^2 - \omega_0{}^2)^2$$

Using this result in Equation 33.40 gives

$$P_{av} = \frac{V_{rms}^2 R\omega^2}{R^2\omega^2 + L^2(\omega^2 - \omega_0{}^2)^2} \quad (33.41)$$

Power in an *RLC* circuit

This expression shows that at resonance, when $\omega = \omega_0$, the *average power is a maximum* and has the value $V_{rms}{}^2/R$. A plot of the average power versus the frequency ω of the applied voltage is shown in Figure 33.13b for the series *RLC* circuit described in Figure 33.13a, taking $R = 3.5\ \Omega$ and $R = 10\ \Omega$. As the resistance is made smaller, the curve becomes sharper in the vicinity of the resonance. The sharpness of the curve is usually described by a dimensionless parameter known as the **quality factor,** denoted by Q_0 (not to be confused with the symbol for charge), which is given by the ratio[3]

$$Q_0 = \frac{\omega_0}{\Delta\omega} \quad (33.42)$$

where $\Delta\omega$ is the width of the curve measured between the two values of ω for which P_{av} has *half* its maximum value (half-power points, see Figure 33.13b). It is left as a problem (Problem 75) to show that the width at the half-power points has the value $\Delta\omega = R/L$, so that

$$Q_0 = \frac{\omega_0 L}{R} \quad (33.43)$$

Quality factor

That is, Q_0 is equal to the ratio of the inductive reactance to the resistance evaluated at the resonance frequency, ω_0. Note that Q_0 is a dimensionless quantity.

The curves plotted in Figure 33.14 show that a high-Q_0 circuit responds to a very narrow range of frequencies, whereas a low-Q_0 circuit responds to a much broader range of frequencies. Typical values of Q_0 in electronic circuits range from 10 to 100. For example, $Q_0 = 14.3$ for the circuit described in Figure 33.13 when $R = 3.5\ \Omega$.

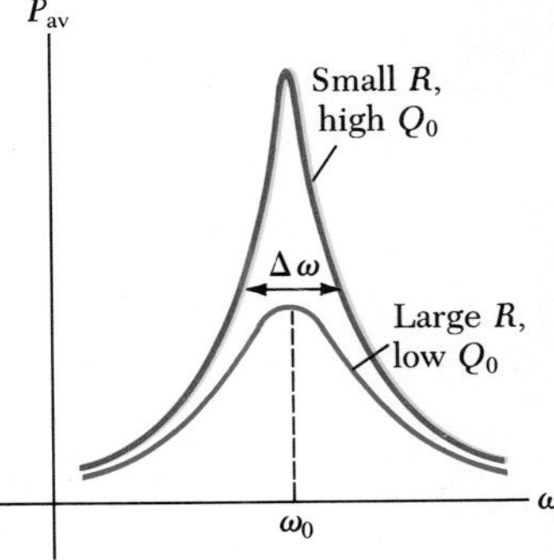

Figure 33.14 Plots of the average power versus frequency for a series *RLC* circuit (see Eq. 33.41). The upper, narrow curve is for a small value of *B*, and the lower, broad curve is for a large value of *R*. The width $\Delta\omega$ of each curve is measured between points where the power is half its maximum value. The power is a maximum at the resonance frequency, ω_0.

The receiving circuit of a radio is an important application of a resonant circuit. The radio is tuned to a particular station (which transmits a specific radio frequency signal) by varying a capacitor, which changes the resonant frequency of the receiving circuit. When the resonance frequency of the circuit matches that of the incoming radio wave, the current in the receiving circuit increases. This signal is then amplified and fed to a speaker. Since many signals are often present over a range of frequencies, it is important to design a high-Q_0 circuit in order to eliminate unwanted signals. In this manner, stations whose frequencies are near but not at the resonance frequency will give

[3] The quality factor is also defined as the ratio $2\pi E/\Delta E$, where E is the energy stored in the oscillating system and ΔE is the energy lost per cycle of oscillation. One can also define the quality factor for a mechanical system such as a damped oscillator.

negligibly small signals at the receiver relative to the one that matches the resonance frequency.

EXAMPLE 33.6 A Resonating Series *RLC* Circuit
Consider a series *RLC* circuit for which $R = 150\ \Omega$, $L = 20$ mH, $V_{rms} = 20$ V, and $\omega = 5000\ s^{-1}$. Determine the value of the capacitance for which the current has its peak value.

Solution The current has its peak value at the resonance frequency ω_0, which should be made to match the "driving" frequency of $5000\ s^{-1}$ in this problem:

$$\omega_0 = 5 \times 10^3\ s^{-1} = \frac{1}{\sqrt{LC}}$$

$$C = \frac{1}{(25 \times 10^6\ s^{-2})L}$$

$$= \frac{1}{(25 \times 10^6\ s^{-2})(20 \times 10^{-3}\ H)} = 2.00\ \mu F$$

Exercise 4 Calculate the maximum value of the rms current in the circuit.
Answer 0.133 A.

*33.8 FILTER CIRCUITS

In this section, we give a brief description of *RC* filters, which are commonly used in ac circuits to modify the characteristics of a time-varying signal. A filter circuit can be used to smooth out or eliminate a time-varying voltage. For example, radios are usually powered by a 60-Hz ac voltage. The ac voltage is converted to dc using a *rectifier circuit.* After rectification, however, the voltage will still contain a small ac component at 60 Hz (sometimes called *ripple*), which must be filtered. This 60-Hz ripple must be reduced to a value much smaller than the audio signal to be amplified. Without filtering, the resulting audio signal includes an annoying hum at 60 Hz.

First, consider the simple series *RC* circuit shown in Figure 33.15a. The input voltage is across the two elements and is represented by $V_m \sin \omega t$. Since we shall be interested only in peak values, we can use Equation 33.30, which shows that the peak input voltage is related to the peak current by

$$V_{in} = I_m Z = I_m \sqrt{R^2 + \left(\frac{1}{\omega C}\right)^2}$$

If the voltage across the resistor is considered to be the output voltage, V_{out}, then from Ohm's law the peak output voltage is given by

$$V_{out} = I_m R$$

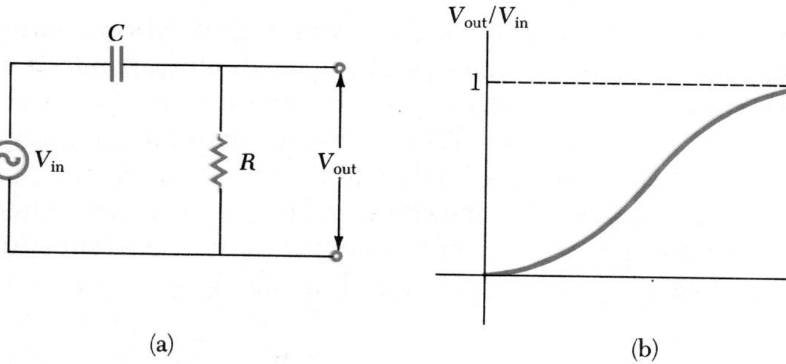

Figure 33.15 (a) A simple *RC* high-pass filter. (b) Ratio of the output voltage to the input voltage for an *RC* high-pass filter.

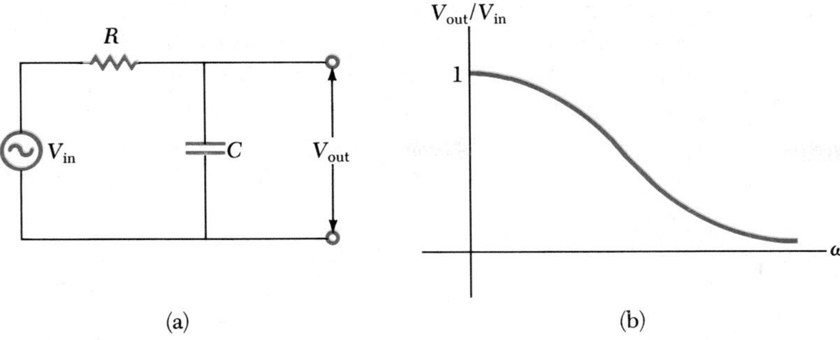

Figure 33.16 (a) A simple *RC* low-pass filter. (b) Ratio of the output voltage to the input voltage for an *RC* low-pass filter.

Therefore, the ratio of the output voltage to the input voltage is given by

$$\frac{V_{\text{out}}}{V_{\text{in}}} = \frac{R}{\sqrt{R^2 + \left(\frac{1}{\omega C}\right)^2}} \tag{33.44}$$

High-pass filter

A plot of Equation 33.34, given in Figure 33.15b, shows that at low frequencies, V_{out} is small compared with V_{in}, whereas at high frequencies the two voltages are equal. Since the circuit preferentially passes signals of higher frequency while low frequencies are filtered (or attenuated), the circuit is called an *RC high-pass filter*. Physically, the high-pass filter is a result of the "blocking action" of the capacitor to direct current or low frequencies.

Now consider the *RC* series circuit shown in Figure 33.16a, where the output voltage is taken across the capacitor. In this case, the peak voltage equals the voltage across the capacitor. Since the impedance across the capacitor is $X_C = 1/\omega C$,

$$V_{\text{out}} = I_{\text{m}} X_C = \frac{I_{\text{m}}}{\omega C}$$

Therefore, the ratio of the output voltage to the input voltage is given by

$$\frac{V_{\text{out}}}{V_{\text{in}}} = \frac{1/\omega C}{\sqrt{R^2 + \left(\frac{1}{\omega C}\right)^2}} \tag{33.45}$$

Low-pass filter

This ratio, plotted in Figure 33.16b, shows that in this case the circuit preferentially passes signals of low frequency. Hence, the circuit is called an *RC low-pass filter.*

We have considered only two simple filters. One can also use a series *RL* circuit as a high-pass or low-pass filter. It is also possible to design filters, called *band-pass filters,* that pass only a narrow range of frequencies.

*33.9 THE TRANSFORMER AND POWER TRANSMISSION

When electrical power is transmitted over large distances, it is economical to use a high voltage and low current to minimize the I^2R heating loss in the transmission lines. For this reason, 350-kV lines are common, and in many areas even higher-voltage (765 kV) lines are under construction. Such high-voltage transmission systems have met with considerable public resistance because of the potential safety and environmental problems they pose. At the receiving end of such lines, the consumer requires power at a low voltage and high current (for safety and efficiency in design) to operate such things as appliances and motor-driven machines. Therefore, a device is required that will increase (or decrease) the ac voltage V and current I without causing appreciable changes in the product IV. The *ac transformer* is the device used for this purpose.

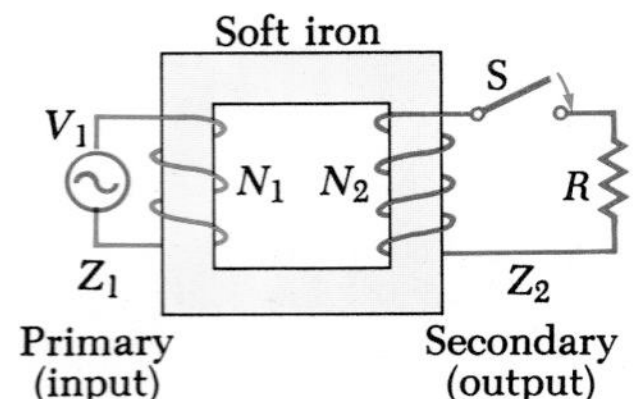

Figure 33.17 An ideal transformer consists of two coils wound on the same soft iron core. An ac voltage V_1 is applied to the primary coil, and the output voltage V_2 is across the load resistance R.

In its simplest form, the ac transformer consists of two coils of wire wound around a core of soft iron as in Figure 33.17. The coil on the left, which is connected to the input ac voltage source and has N_1 turns, is called the *primary* winding (or primary). The coil on the right, consisting of N_2 turns and connected to a load resistor R, is called the *secondary*. The purpose of the common iron core is to increase the magnetic flux and to provide a medium in which nearly all the flux through one coil passes through the other coil. Eddy current losses are reduced by using a laminated iron core.[4] Soft iron is used as the core material to reduce hysteresis losses. Joule heat losses due to the finite resistance of the coil wires are usually quite small. Typical transformers have power efficiencies ranging from 90% to 99%. In what follows, we shall assume an ideal transformer, for which there are no power losses.

First, let us consider what happens in the primary circuit when the switch in the secondary circuit of Figure 33.17 is open. If we assume that the resistance of the primary coil is negligible relative to its inductive reactance, then the primary circuit is equivalent to a simple circuit consisting of an inductor connected to an ac generator (described in Section 33.3). Since the current is 90° out of phase with the voltage, the power factor, cos ϕ, is zero, and hence the average power delivered from the generator to the primary circuit is zero. Faraday's law tells us that the voltage V_1 across the primary coil is given by

$$V_1 = -N_1 \frac{d\Phi_m}{dt} \tag{33.46}$$

where Φ_m is the magnetic flux through each turn. If we assume that no flux leaks out of the iron core, then the flux through each turn of the primary equals the flux through each turn of the secondary. Hence, the voltage across the secondary coil is given by

$$V_2 = -N_2 \frac{d\Phi_m}{dt} \tag{33.47}$$

Since $d\Phi_m/dt$ is common to Equations 33.46 and 33.47, we find that

$$V_2 = \frac{N_2}{N_1} V_1 \tag{33.48}$$

When N_2 is greater than N_1, the output voltage V_2 exceeds the input voltage V_1. This is referred to as a *step-up transformer*. When N_2 is less than N_1,

[4] Losses in the core are present even under the condition of no load, that is, when the secondary circuit is open. Most of the power loss in this case is in the form of hysteresis losses as the core is magnetized cyclically.

the output voltage is less than the input voltage, and we speak of a *step-down transformer*.

When the switch in the secondary circuit is closed, a current I_2 is induced in the secondary. If the load in the secondary circuit is a pure resistance, R_L, the induced current will be in phase with the induced voltage. The power supplied to the secondary circuit must be provided by the ac generator that is connected to the primary circuit, as in Figure 33.18. An *ideal transformer* with a resistive load is one in which the energy losses in the transformer windings and core can be neglected. In an ideal transformer, the power supplied by the generator, I_1V_1, is equal to the power in the secondary circuit, I_2V_2. That is,

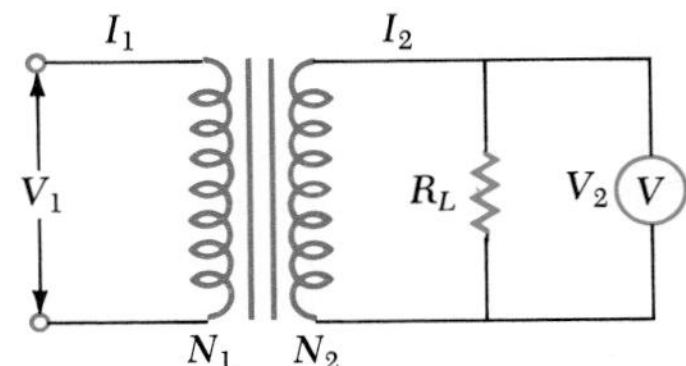

Figure 33.18 Conventional circuit diagram for a transformer.

$$I_1V_1 = I_2V_2 \tag{33.49}$$

Clearly, the value of the load resistance R determines the value of the secondary current, since $I_2 = V_2/R$. Furthermore, the current in the primary is $I_1 = V_1/R_{eq}$, where R_{eq} is the equivalent resistance of the load resistance R when viewed from the primary side, given by

$$R_{eq} = \left(\frac{N_1}{N_2}\right)^2 R \tag{33.50}$$

From this analysis, we see that a transformer may be used to match resistances between the primary circuit and the load. In this manner, one can achieve *maximum* power transfer between a given power source and the load resistance.

In real transformers, the power in the secondary is typically between 90% and 99% of the primary power. The energy losses are due mainly to hysteresis losses in the transformer core, and thermal energy losses from currents induced in the core and the coil windings themselves.

We can now understand why transformers are useful for transmitting power over long distances. By stepping up the generator voltage, the current in the transmission line is reduced, thereby reducing I^2R losses. In practice, the voltage is stepped up to around 230 000 V at the generating station, then stepped down to around 20 000 V at a distributing station, and finally stepped down to 110–220 V at the customer's utility poles. The power is supplied by a three-wire cable. In the United States, two of these wires are "hot," with voltages of 110 V with respect to a common ground wire. Most home appliances operating on 110 V are connected in parallel between one of the hot wires and ground. Larger appliances, such as electric stoves and clothes dryers, require 220 V. This is obtained across the two hot wires, which are 180° out of phase so that the voltage difference between them is 220 V.

There is a practical upper limit to the voltages one can use in transmission lines. Excessive voltages could ionize the air surrounding the transmission lines, which could result in a conducting path to ground or to other objects in the vicinity. This, of course, would present a serious hazard to any living creatures. For this reason, a long string of insulators is used to keep high-voltage wires away from their supporting metal towers. Other insulators are used to maintain separation between wires.

EXAMPLE 33.7 A Step-up Transformer
A generator produces 10 A (rms) of current at 400 V. The voltage is stepped up to 4500 V by an ideal transformer and transmitted a long distance through a power line of total resistance 30 Ω. (a) Determine the percentage of power lost when the voltage is stepped up.

Solution From Equation 33.49, we find that the current in the transmission line is

$$I_2 = \frac{I_1 V_1}{V_2} = \frac{(10\ \text{A})(400\ \text{V})}{4500\ \text{V}} = 0.89\ \text{A}$$

Hence, the power lost in the transmission line is

$$P_{\text{lost}} = I_2{}^2 R = (0.89\ \text{A})^2(30\ \Omega) = 24\ \text{W}$$

Since the output power of the generator is $P = IV = (10\ \text{A})(400\ \text{V}) = 4000$ W, we find that the percentage of power lost is

$$\%\ \text{power lost} = \left(\frac{24}{4000}\right) \times 100 = 0.6\%$$

(b) What percentage of the original power would be lost in the transmission line if the voltage were not stepped up?

Solution If the voltage were not stepped up, the current in the transmission line would be 10 A and the power lost in the line would be $I^2R = (10\ \text{A})^2(30\ \Omega) = 3000$ W. Hence, the percentage of power lost would be

$$\%\ \text{power lost} = \frac{3000}{4000} \times 100 = 75\%$$

This example illustrates the advantage of high-voltage transmission lines.

Exercise 5 If the transmission line is cooled so that the resistance is reduced to 5 Ω, how much power will be lost in the line if it carries a current of 0.89 A?
Answer 4 W.

SUMMARY

If an ac circuit consists of a generator and a resistor, the current in the circuit is in phase with the voltage. That is, the current and voltage reach their peak values at the same time.

The **rms current** and **rms voltage** in an ac circuit in which the voltages and current vary sinusoidally are given by the relations

$$I_{\text{rms}} = \frac{I_{\text{m}}}{\sqrt{2}} = 0.707 I_{\text{m}} \qquad (33.6)$$

$$V_{\text{rms}} = \frac{V_{\text{m}}}{\sqrt{2}} = 0.707 V_{\text{m}} \qquad (33.7)$$

where I_{m} and V_{m} are the peak values of the current and voltage, respectively.

If an ac circuit consists of a generator and an inductor, the current *lags behind* the voltage by 90°. That is, the voltage reaches its peak value one quarter of a period before the current reaches its peak value.

If an ac circuit consists of a generator and a capacitor, the current *leads* the voltage by 90°. That is, the current reaches its peak value one quarter of a period before the voltage reaches its peak value.

In ac circuits that contain inductors and capacitors, it is useful to define the **inductive reactance** X_L and **capacitive reactance** X_C as

Inductive reactance

$$X_L = \omega L \qquad (33.12)$$

Capacitive reactance

$$X_C = \frac{1}{\omega C} \qquad (33.19)$$

where ω is the angular frequency of the ac generator. The SI unit of reactance is the ohm.

The quantity in the denominator of Equation 33.38 is defined as the **impedance** Z of the circuit, which also has the unit of ohm:

Impedance

$$Z \equiv \sqrt{R^2 + (X_L - X_C)^2} \tag{33.29}$$

In an RLC series ac circuit, the applied voltage and current are out of phase. The **phase angle** ϕ between the current and voltage is given by

Phase angle

$$\tan \phi = \frac{X_L - X_C}{R} \tag{33.31}$$

The sign of ϕ can be positive or negative, depending on whether X_L is greater or less than X_C. The phase angle is zero when $X_L = X_C$.

The **average power** delivered by the generator in an RLC ac circuit is given by

Average power

$$P_{av} = I_{rms} V_{rms} \cos \phi \tag{33.35}$$

An equivalent expression for the average power is

Average power

$$P_{av} = I_{rms}^2 R \tag{33.36}$$

The average power delivered by the generator is dissipated as heat in the resistor. There is no power loss in an ideal inductor or capacitor.

The rms current in a series RLC circuit is

$$I_{rms} = \frac{V_{rms}}{\sqrt{R^2 + (X_L - X_C)^2}} \tag{33.38}$$

where V_{rms} is the rms value of the applied voltage.

A series RLC circuit is in resonance when the inductive reactance equals the capacitive reactance. When this condition is met, the current given by Equation 33.38 reaches its peak value. Setting $X_L = X_C$, one finds that the **resonance frequency** ω_0 of the circuit has the value

Resonance frequency

$$\omega_0 = 1/\sqrt{LC} \tag{33.39}$$

The current in a series RLC circuit reaches its peak value when the frequency of the generator equals ω_0, that is, when the "driving" frequency matches the resonance frequency.

A transformer is a device designed to raise or lower an ac voltage and current without causing an appreciable change in the product IV. In its simplest form, it consists of a primary coil of N_1 turns and a secondary coil of N_2 turns, both wound on a common soft iron core. When a voltage V_1 is applied across the primary, the voltage V_2 across the secondary is given by

$$V_2 = \frac{N_2}{N_1} V_1 \tag{33.48}$$

In an ideal transformer, the power delivered by the generator must equal the power dissipated in the load. If a load resistor R is connected across the secondary coil, this means that

$$I_1 V_1 = I_2 V_2 = \frac{{V_2}^2}{R} \tag{33.49}$$

QUESTIONS

1. What is meant by the statement "the voltage across an inductor leads the current by 90°"?
2. Explain why the reactance of a capacitor decreases with increasing frequency, whereas the reactance of an inductor increases with increasing frequency.
3. Why does a capacitor act as a short circuit at high frequencies? Why does it act as an open circuit at low frequencies?
4. Explain how the acronym "ELI the ICE man" can be used to recall whether current leads voltage or voltage leads current in *RLC* circuits.
5. Why is the sum of the peak voltages across each of the elements in a series *RLC* circuit usually greater than the peak applied voltage? Doesn't this violate Kirchhoff's voltage law?
6. Does the phase angle depend on frequency? What is the phase angle when the inductive reactance equals the capacitive reactance?
7. In a series *RLC* circuit, what is the possible range of values for the phase angle?
8. If the frequency is doubled in a series *RLC* circuit, what happens to the resistance, the inductive reactance, and the capacitive reactance?
9. Energy is delivered to a series *RLC* circuit by a generator. This energy is dissipated as heat in the resistor. What is the source of this energy?
10. Explain why the average power delivered to an *RLC* circuit by the generator depends on the phase between the current and applied voltage.
11. A particular experiment requires a beam of light of very stable intensity. Why would an ac voltage be unsuitable for powering the light source?
12. What is the impedance of an *RLC* circuit at the resonance frequency?
13. Consider a series *RLC* circuit in which *R* is an incandescent lamp, *C* is some fixed capacitor, and *L* is a *variable* inductance. The source is 110 V ac. Explain why the lamp glows brightly for some values of *L* and does not glow at all for other values.
14. What is the advantage of transmitting power at high voltages?
15. What determines the peak voltage that can be used on a transmission line?
16. Why do power lines carry electrical energy at several thousand volts potential, but it is always stepped down to 240 V or 120 V as it enters your home?
17. Will a transformer operate if a battery is used for the input voltage across the primary? Explain.
18. How can the average value of a current be zero and yet the square root of the average squared current not be zero?
19. What is the time average of a sinusoidal potential with amplitude V_m? What is its rms voltage?
20. What is the time average of the "square-wave" potential shown in Figure 33.19? What is its rms voltage?

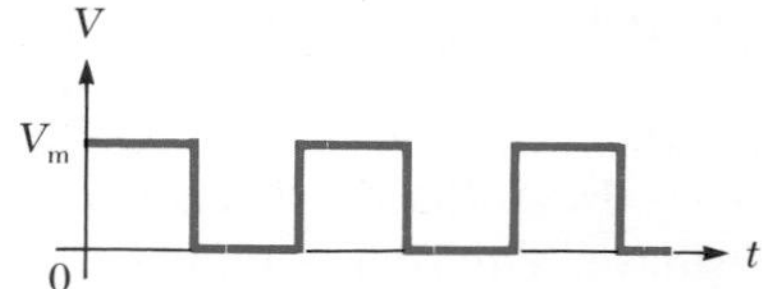

Figure 33.19
(Question 20).

21. Do ac ammeters and voltmeters read peak, rms, or average values?
22. Is the voltage applied to a circuit always in phase with the current through a resistor in the circuit?
23. Would an inductor and a capacitor used together in an ac circuit dissipate any power?
24. Show that the peak current in an *RLC* circuit occurs when the circuit is in resonance.
25. Explain how the quality factor is related to the response characteristics of a receiver. Which variable most strongly determines the quality factor?
26. List some applications for a filter circuit.
27. The approximate efficiency of an incandescent lamp for converting electrical energy into heat is (a) 30%, (b) 60%, (c) 100%, or (d) 10%.
28. A night-watchman is fired by his boss for being wasteful and keeping all the lights on in the building. The night-watchman defends himself by claiming that the building is electrically heated, so his boss's claim is unfounded. Who should win the argument if this were to end up in a court of law?
29. Why are the primary and secondary coils of a transformer wrapped on an iron core that passes through both coils?
30. With reference to Figure 33.20, explain why the capacitor prevents a dc voltage from passing between A and B, yet allows an ac signal to pass from A to B. (The circuits are said to be capacitively coupled.)

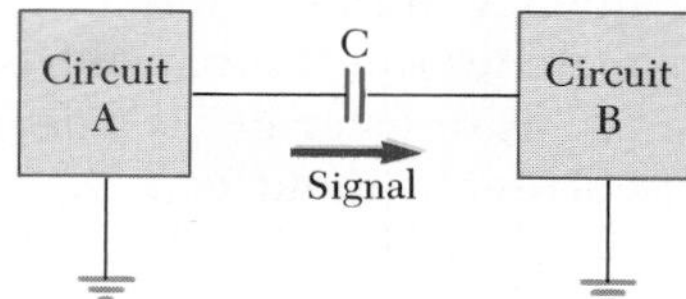

Figure 33.20
(Question 30).

31. With reference to Figure 33.21, one finds that if *C* is made sufficiently large, an ac signal passes from A to ground rather than into B. Hence, the capacitor acts as a filter. Explain.

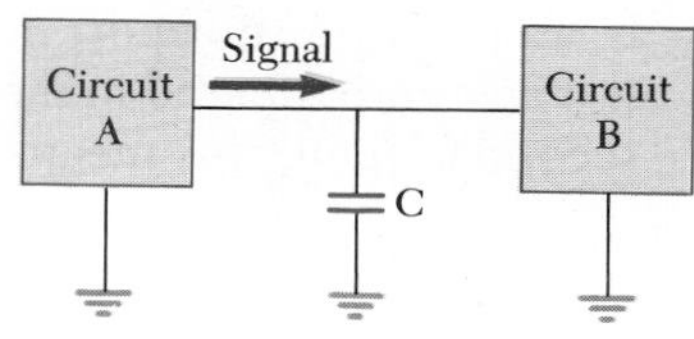

Figure 33.21
(Question 31).

PROBLEMS

Assume all AC voltages and currents are sinusoidal, unless stated otherwise.

Section 33.2 Resistors in an AC Circuit

1. Show that the rms value for the sawtooth voltage shown in Figure 33.22 is given by $V_m/\sqrt{3}$.
2. (a) What is the resistance of a lightbulb that uses an average power of 75 W when connected to a 60-Hz power source with a peak voltage of 170 V? (b) What is the resistance of a 100-W bulb?

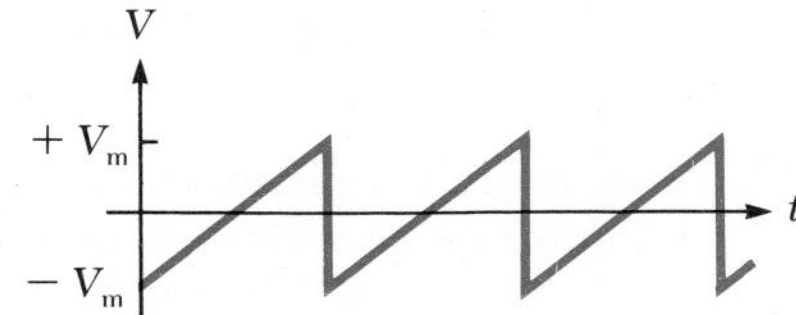

Figure 33.22 (Problem 1).

3. An ac power supply produces a peak voltage $V_m = 100$ V. This power supply is connected to a 24-Ω resistor, and the current and resistor voltage are measured with an ideal ac ammeter and voltmeter, as shown in Figure 33.23. What does each meter read?

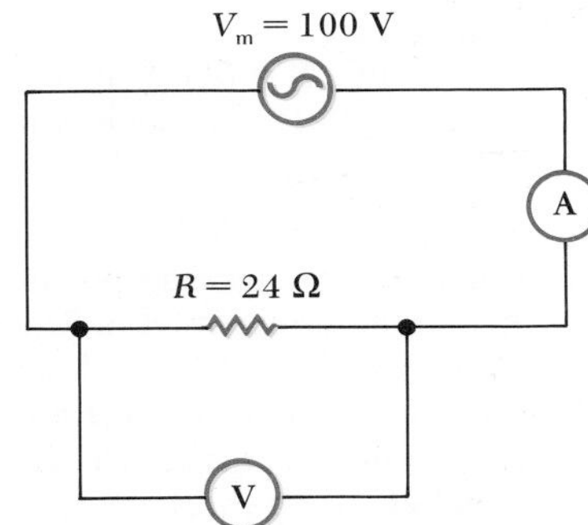

Figure 33.23 (Problem 3).

4. In the simple ac circuit shown in Figure 33.1, let $R = 60\ \Omega$, $V_m = 100$ V, and the frequency of the generator $f = 50$ Hz. Assume that the voltage across the resistor $V_R = 0$ when $t = 0$. Calculate (a) the peak current in the resistor and (b) the angular frequency of the generator.
5. Use the values given in Problem 4 for the circuit of Figure 33.1 to calculate the current through the resistor at (a) $t = \frac{1}{75}$ s and (b) $t = \frac{1}{150}$ s.
6. In the simple ac circuit shown in Figure 33.1, $R = 70\ \Omega$. (a) If $V_R = 0.25V_m$ at $t = 0.01$ s, what is the angular frequency of the generator? (b) What is the next value of t for which V_R will be $0.25V_m$?
7. The current in the circuit shown in Figure 33.1 equals 60% of the peak current at $t = 0.007$ s. What is the smallest frequency of the generator that gives this current?

Section 33.3 Inductors in an AC Circuit

8. Show that the inductive reactance X_L has the SI unit of ohm.
9. In a purely inductive ac circuit, as in Figure 33.4, $V_m = 100$ V. (a) If the peak current is 7.5 A at a frequency of 50 Hz, calculate the inductance L. (b) At what angular frequency ω will the maximum current be reduced to 2.5 A?
10. When a particular inductor is connected to a sinusoidal voltage with a 110-V amplitude, a peak current of 3 A appears in the inductor. (a) What will be the peak current if the frequency of the applied voltage is doubled? (b) What is the inductive reactance at these two frequencies?
11. An inductor is connected to a 20-Hz power supply that produces a 50-V rms voltage. What inductance is needed to keep the instantaneous current in the circuit below 80 mA?
12. An inductor has a 54-Ω reactance at 60 Hz. What will be the peak current if this inductor is connected to a 50-Hz source that produces a 100-V rms voltage?
13. For the circuit shown in Figure 33.4, $V_m = 80$ V, $\omega = 65\pi$ rad/s, and $L = 70$ mH. Calculate the current in the inductor at $t = 0.0155$ s.
14. (a) If $L = 310$ mH and $V_m = 130$ V in the circuit of Figure 33.4, at what frequency will the inductive reactance equal 40 Ω? (b) Calculate the peak value of the current in the circuit at this frequency.
15. What is the inductance of a coil that has an inductive reactance of 63 Ω at an angular frequency of 820 rad/s?

Section 33.4 Capacitors in an AC Circuit

16. Show that the SI unit of capacitive reactance is the ohm.
17. (a) For what linear frequencies does a 22-μF capacitor have a reactance below 175 Ω? (b) Over this same frequency range, what would be the reactance of a 44-μF capacitor?
18. Calculate the capacitive reactance of a 10-μF capacitor when connected to an ac generator having an angular frequency of 95π rad/s.
19. A 98-pF capacitor is connected to a 60-Hz power supply that produces a 20-V rms voltage. What is the maximum charge that appears on either of the capacitor plates?
20. A sinusoidal voltage $v(t) = V_m \cos \omega t$ is applied to a capacitor as shown in Figure 33.24. (a) Write an expression for the instantaneous charge on the capacitor in terms of V_m, C, t, and ω. (b) What is the instantaneous current in the circuit?

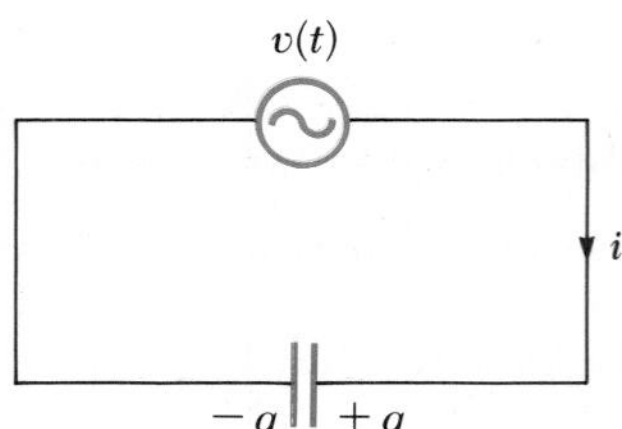

Figure 33.24 (Problem 20).

21. What peak current will be delivered by an ac generator with $V_m = 48$ V and $f = 90$ Hz when connected across a 3.7-μF capacitor?

22. A variable-frequency ac generator with $V_m = 18$ V is connected across a 9.4×10^{-8}-F capacitor. At what frequency should the generator be operated to provide a peak current of 5 A?

23. The generator in a purely capacitive ac circuit (Fig. 33.6) has an angular frequency of 100π rad/s and $V_m = 220$ V. If $C = 20$ μF, what is the current in the circuit at $t = 0.004$ s?

Section 33.5 The *RLC* Series Circuit

24. At what frequency will the inductive reactance of a 57-μH inductance equal the capacitive reactance of a 57-μF capacitor?

25. A series ac circuit contains the following components: $R = 150\ \Omega$, $L = 250$ mH, $C = 2\ \mu$F and a generator with $V_m = 210$ V operating at 50 Hz. Calculate the (a) inductive reactance, (b) capacitive reactance, (c) impedance, (d) peak current, and (e) phase angle.

26. A sinusoidal voltage $v(t) = (40\text{ V}) \sin(100t)$ is applied to a series *RLC* circuit with $L = 160$ mH, $C = 99\ \mu$F, and $R = 68\ \Omega$. (a) What is the impedance of the circuit? (b) What is the current amplitude? (c) Determine the numerical values for I_m, ω, and ϕ in the equation $i(t) = I_m \sin(\omega t - \phi)$.

27. An *RLC* circuit consists of a 150-Ω resistor, a 21-μF capacitor, and a 460-mH inductor, connected in series with a 120-V, 60-Hz power supply. (a) What is the phase angle between the current and the applied voltage? (b) Does the current or voltage reach its peak earlier?

28. A resistor ($R = 900\ \Omega$), a capacitor ($C = 0.25\ \mu$F), and an inductor ($L = 2.5$ H) are connected in series across a 240-Hz ac source for which $V_m = 140$ V. Calculate the (a) impedance of the circuit, (b) peak current delivered by the source, and (c) phase angle between the current and voltage. (d) Is the current leading or lagging behind the voltage?

29. A coil with an inductance of 18.1 mH and a resistance of 7 Ω is connected to a *variable*-frequency ac generator. At what frequency will the voltage across the coil lead the current by 45°?

30. A 400-Ω resistor, an inductor, and a capacitor are in series with a generator. When the frequency is adjusted to $600/\pi$ Hz, the inductive reactance is 700 Ω. What is the *minimum* value of capacitance that will result in a circuit impedance of 910 Ω?

31. An ac source with $V_m = 150$ V and $f = 50$ Hz is connected between points *a* and *d* in Figure 33.25. Calculate the *peak* voltages between points (a) *a* and *b*, (b) *b* and *c*, (c) *c* and *d*, (d) *b* and *d*.

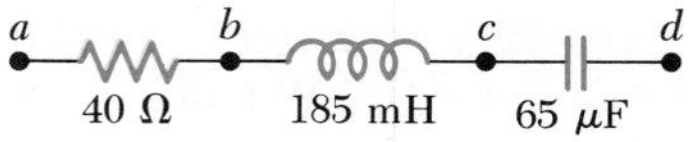

Figure 33.25 (Problem 31).

32. Draw to scale a phasor diagram showing Z, X_L, X_C, and ϕ for an ac series circuit for which $R = 300\ \Omega$, $C = 11\ \mu$F, $L = 0.2$ H, and $f = 500/\pi$ Hz.

33. An inductor ($L = 400$ mH), a capacitor ($C = 4.43\ \mu$F), and a resistor ($R = 500\ \Omega$) are connected in series. A 50-Hz ac generator produces a peak current of 250 mA in the circuit. (a) Calculate the required peak voltage V_m. (b) Determine the angle by which the current in the circuit leads or lags behind the applied voltage.

34. Repeat Problem 33 for the case where $L = 130$ mH, $C = 75\ \mu$F, $R = 50\ \Omega$, if an applied voltage of frequency 100 Hz produces a peak current of 1.7 A.

Section 33.6 Power in an AC Circuit

35. Calculate the average power delivered to the series *RLC* circuit described in Problem 25.

36. Consider the circuit described in Problem 28. (a) What is the power factor of the circuit? (b) What is the rms current in the circuit? (c) What average power is delivered by the source?

37. For the circuit described in Problem 26, find (a) the average power delivered to the circuit by the power supply, and (b) the average power dissipated by the resistor.

38. A series *RLC* circuit is driven by a sinusoidal voltage. (a) What is the phase angle between the current and the applied voltage at resonance? (b) What is the power factor at resonance? (c) What phase angle(s) will produce $P_{av} = \frac{1}{2}V_{rms}I_{rms}$, where V_{rms} and I_{rms} are the rms values of the applied voltage and the current, respectively? (d) In terms of R, what impedance Z will produce this "half-power" condition?

39. The rms terminal voltage of an ac generator is 200 V. The operating frequency is 100 Hz. Write the equation giving the output voltage as a function of time.

40. The average power in a circuit for which the rms current is 5 A is 450 W. Calculate the resistance of the circuit.

41. In a certain series *RLC* circuit, $I_{rms} = 9$ A, $V_{rms} =$ 180 V, and the current leads the voltage by 37°. (a) What is the total resistance of the circuit? (b) Calculate the reactance of the circuit $(X_L - X_C)$.

42. A series *RLC* circuit has a resistance of 45 Ω and an impedance of 75 Ω. What average power will be delivered to this circuit when $V_{rms} = 210$ V?

Section 33.7 Resonance in a Series *RLC* Circuit

43. Calculate the resonance frequency of a series *RLC* circuit for which $C = 8.4\ \mu$F and $L = 120$ mH.

44. (a) Compute the quality factor for each of the circuits described in Problems 26 and 27. (b) Which of these two circuits has the sharper resonance?

45. An *RLC* circuit is used in a radio to tune into an FM station broadcasting at 99.7 MHz. The resistance in the circuit is 12 Ω and the inductance is 1.40 μH. What capacitance should be used?

46. The resonant frequency of a series *RLC* circuit is found to be 4.2×10^4 Hz. If the capacitance of the circuit is 300 μF, what is the value of the inductance?

47. A coil of resistance 35 Ω and inductance 20.5 H is in series with a capacitor and a 200-V (rms), 100-Hz source. The current in the circuit is 4 A (rms). (a) Calculate the capacitance in the circuit. (b) What is V_{rms} across the coil?

48. A series *RLC* circuit has $L = 156$ mH, $C = 0.2\ \mu$F, and $R = 88$ Ω. The circuit is driven by a generator with $V_m = 110$ V and a *variable* angular frequency. Calculate (a) the resonance frequency of the circuit, (b) the quality factor of the circuit, and (c) the two values of ω at which the average power has one half its maximum value.

49. What average power is delivered to the circuit described in Problem 48 when the frequency is adjusted to 600 Hz?

•Section 33.8 Filter Circuits

50. Consider the circuit shown in Figure 33.15, with $R =$ 800 Ω and $C = 0.09\ \mu$F. Calculate the ratio V_{out}/V_{in} for (a) $\omega = 300\ s^{-1}$ and (b) $\omega = 7 \times 10^5\ s^{-1}$.

51. The *RC* high-pass filter shown in Figure 33.15 has a resistance $R = 0.50$ Ω. (a) What capacitance will give an output signal with one-half the amplitude of a 300-Hz input signal? (b) What is the gain (V_{out}/V_{in}) for a 600-Hz signal?

52. The *RC* low-pass filter shown in Figure 33.16 has a resistance $R = 90$ Ω and a capacitance $C = 8000$ pF. Calculate the gain (V_{out}/V_{in}) for an input frequency (a) $f = 600$ Hz, and (b) $f = 600$ kHz.

53. Assign the values of *R* and *C* given in Problem 50 to the circuit shown in Figure 33.16 and calculate V_{out}/V_{in} for (a) $\omega = 300\ s^{-1}$ and (b) $\omega = 7 \times 10^5\ s^{-1}$.

54. (a) For the circuit shown in Figure 33.26, show that the maximum possible value of the ratio V_{out}/V_{in} is unity. (b) At what frequency (expressed in terms of *R*, *L*, and *C*) does this occur?

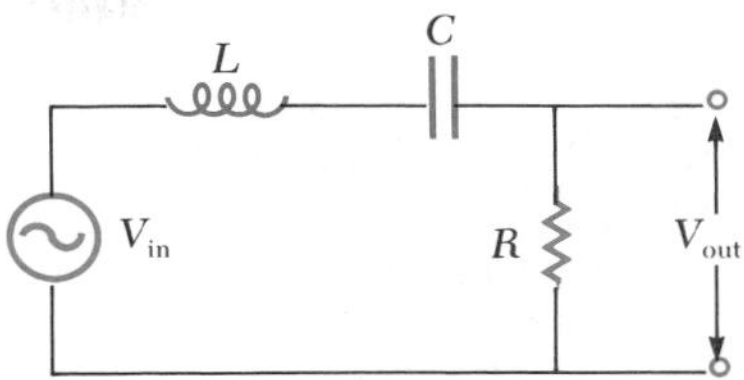

Figure 33.26 (Problems 54 and 55).

55. The circuit shown in Figure 33.26 can be used as a filter to pass signals that lie in a certain frequency band. (a) Show that the gain (V_{out}/V_{in}) for an input voltage of frequency ω is given by

$$\frac{V_{out}}{V_{in}} = \frac{1}{\sqrt{1 + \left[\frac{(\omega^2/\omega_0{}^2) - 1}{\omega RC}\right]^2}}$$

(b) Let $R = 100$ Ω, $C = 0.050\ \mu$F, and $L = 0.127$ H. Compute the gain of this circuit for input frequencies $f_1 = 1.5$ kHz, $f_2 = 2.0$ kHz, and $f_3 = 2.5$ kHz.

56. Show that two successive high-pass filters with the same values of *R* and *C* give a combined gain

$$\frac{V_{out}}{V_{in}} = \frac{1}{1 + (1/\omega RC)^2}$$

57. Consider a low-pass filter followed by a high-pass filter, as shown in Figure 33.27. If $R = 1000$ Ω and $C = 0.050\ \mu$F, determine V_{out}/V_{in} for a 2.0-kHz input frequency.

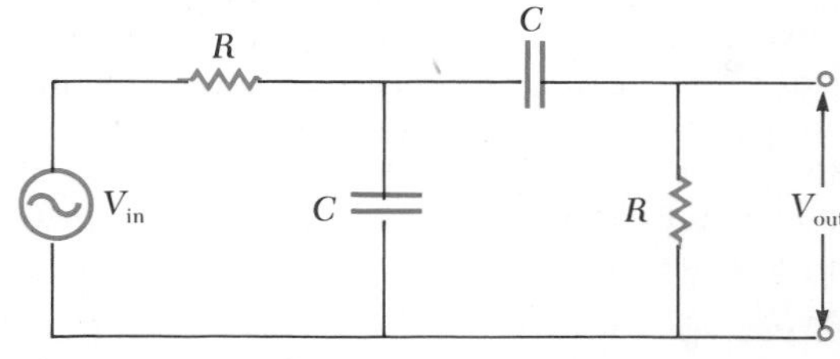

Figure 33.27 (Problem 57).

•Section 33.9 The Transformer and Power Transmission

58. An ideal transformer has 240 turns on the primary winding and 20 turns on the secondary. If the primary is connected across a 110-V (rms) generator, what is the rms output voltage?

59. A transformer has $N_1 = 350$ turns and $N_2 = 2000$ turns. If the input voltage is $v(t) = (170\text{ V}) \cos \omega t$, what rms voltage is developed across the secondary coil?

60. Consider an ideal transformer with N_1 primary and N_2 secondary windings. Show that a step-up transformer (one with $N_2 > N_1$) actually reduces the current in the output by a factor of N_1/N_2.

61. A particular transformer is 95% efficient and has twice as many secondary windings as primary windings. If the primary windings carry a 5 A current at an rms voltage of 120 V, what are the secondary current and rms voltage?

62. A step-up transformer is designed to have an output voltage of 2200 V (rms) when the primary is connected across a 110-V (rms) source. (a) If there are 80 turns on the primary winding, how many turns are required on the secondary? (b) If a load resistor across the secondary draws a current of 1.5 A, what is the current in the primary, assuming ideal conditions?

63. If the transformer in Problem 62 has an efficiency of 95%, what is the current in the primary when the secondary current is 1.2 A?

64. The primary current of an ideal transformer is 8.5 A when the primary voltage is 77 V. Calculate the voltage across the secondary when a current of 1.4 A is delivered to a load resistor.

ADDITIONAL PROBLEMS

65. A series *RLC* circuit consists of an 8-Ω resistor, a 5-μF capacitor, and a 50-mH inductor. A variable frequency source of 400 V (rms) is applied across the combination. Determine the power delivered to the circuit when the frequency is equal to one half of the resonance frequency.

66. The *RLC* circuit shown in Figure 33.8 has $L = 10$ mH, $C = 1\ \mu$F, $R = 3.3\ \Omega$, and $V_m = 1$ V. (a) Determine the resonant frequency of this circuit. (b) Determine the Q of this circuit, and find the frequency interval between the two half-power points. (c) At the resonant frequency ω_0, calculate the current amplitude and determine the average power dissipated. (d) At the frequency $0.95\omega_0$, calculate the current amplitude and determine the average power dissipated.

67. In a series ac circuit, $R = 21\ \Omega$, $L = 25$ mH, $C = 17\ \mu$F, $V_m = 150$ V, and $\omega = \dfrac{2000}{\pi}\ \text{s}^{-1}$. (a) Calculate the peak current in the circuit. (b) Determine the peak voltage across each of the three elements. (c) What is the power factor for the circuit? (d) Show X_L, X_C, R, and ϕ in a phasor diagram for the circuit.

68. A 1000-Ω resistor is connected in series to a 0.6-H inductor and a 2.5-μF capacitor. This *RLC* combination is then connected across a voltage source that varies as $v = (80\text{ V})\sin\left(\dfrac{1000}{\pi}t\right)$. Calculate (a) the peak current, (b) the phase angle, (c) the power factor, (d) V_{rms} across the inductor, and (e) the average power delivered to the circuit.

69. For a certain series *RLC* circuit, $R = 100\ \Omega$, $I = (2\sqrt{3}\text{ A})\sin(200t - \phi)$, $v = (400\sqrt{3}\text{ V})\sin(200t)$, and $X_C = \sqrt{3} \times 10^2\ \Omega$. Find (a) the impedance, (b) the inductive reactance of the circuit, and (c) the phase angle.

70. A series *RLC* circuit has a resonance frequency of $2000/\pi$ Hz. When operating at a frequency $\omega > \omega_0$, $X_L = 12\ \Omega$ and $X_C = 8\ \Omega$. Calculate the values of L and C for the circuit.

71. A transmission line with a resistance per unit length of $4.5 \times 10^{-4}\ \Omega$/m is to be used to transmit 5000 kW of power over a distance of 400 miles (6.44×10^5 m). The terminal voltage of the generator is 4500 V. (a) What is the line loss if a transformer is used to step up the voltage to 500 kV? (b) What fraction of the input power is lost to the line under these circumstances? (c) What difficulties would be encountered on attempting to transmit the 5000 kW of power at the generator voltage of 4500 V?

72. A small transformer is used to supply 6 V ac to a model railroad lighting circuit. The primary has 220 turns and is connected to a standard 110-V, 60-Hz line. Although the resistance of the primary may be neglected, it has an inductance of 150 mH. (a) How many turns are required on the secondary winding? (b) If the transformer is left plugged in, what current will be drawn by the primary when the secondary is open? (c) What power will be drawn by the primary when the secondary is open?

73. *LC* filters are used as both high- and low-pass filters as were the *RC* filters in Section 33.8. However, all real inductors have resistance, as indicated in Figure 33.28, which must be taken into account. (a) Deter-

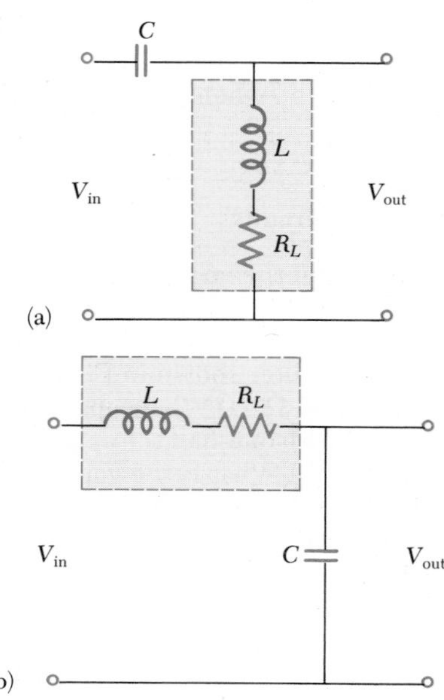

Figure 33.28 (Problem 73).

mine which circuit in Figure 33.28 is the high-pass filter and which is the low-pass filter. (b) Derive the output/input formulas for each circuit following the procedure used for the RC filters in Section 33.8.

74. A resistor of 80 Ω and a 200-mH inductor are connected in *parallel* across a 100-V (rms), 60-Hz source. (a) What is the rms current in the resistor? (b) By what angle does the total current lead or lag behind the voltage?

75. The quality factor of a series RLC circuit is discussed in Section 33.7. With reference to Equation 33.42, show that $\Delta\omega = R/L$ and therefore $Q_0 = \omega L/R$.

76. The average power delivered to a series RLC circuit at frequency ω (Section 33.7) is given by Equation 33.41. (a) Show that the peak current can be written

$$I_m = \omega V_m[L^2(\omega_0^2 - \omega^2)^2 + (\omega R)^2]^{-1/2}$$

where ω is the operating frequency of the circuit and ω_0 is the resonance frequency. (b) Show that the phase angle can be expressed as

$$\phi = \tan^{-1}\left[\frac{L}{R}\left(\frac{\omega_0^2 - \omega^2}{\omega}\right)\right]$$

77. A series RLC circuit consists of a 30-Ω resistor, a 133-μF capacitor, and a 159-mH inductor, which also has a resistance of 25 Ω. The voltage source for the circuit is 180 V (rms) at an angular frequency of 120π rad/s. Calculate the rms voltages across (a) the inductor, (b) the capacitor, and (c) the resistor.

78. A series RLC circuit as shown in Figure 33.8 is connected to a 60-Hz voltage source. The measured voltage across each of the circuit elements is 100 V (rms). The current through the resistor is 1 A (rms). (a) Explain the conditions under which these measurements would occur. (b) What is the value of V (rms) across the capacitor and inductor combination? Find (c) the resistance, (d) the capacitance, (e) the inductance, and (f) the average power delivered to this circuit.

79. *Impedance matching:* A transformer may be used to provide maximum power transfer between two ac circuits that have different impedances. (a) Show that the ratio of turns N_1/N_2 needed to meet this condition is given by

$$\frac{N_1}{N_2} = \sqrt{\frac{Z_1}{Z_2}}$$

(b) Suppose you want to use a transformer as an impedance-matching device between an audio amplifier that has an output impedance of 8000 Ω and a speaker that has an input impedance of 8 Ω. What should be the ratio of primary to secondary turns on the transformer?

80. A power transmission line consists of two parallel wires each having a cross-sectional area A. (One serves as a return line, so the effective length is 2ℓ when connected to a load.) Show that the power lost as heat in the conductors is given by

$$P_{\text{loss}} = \frac{2\rho\ell P_L^2}{AV^2\cos^2\phi}$$

where $\cos\phi$ is the power factor, P_L is the power supplied to the load, and ρ is the resistivity of the conductor. This result shows that power losses may be minimized by minimizing ϕ and increasing the operating voltage V. (Note that V represents an rms value in this problem.)

81. Figure 33.29a shows a parallel RLC circuit, and the corresponding phasor diagram is given in Figure 33.29b. The instantaneous voltage (and rms voltage) across each of the three circuit elements is the same, and each is in phase with the current through the resistor. The currents in C and L lead (or lag behind) the current in the resistor, as shown in Figure 33.29b. (a) Show that the rms current delivered by the source is given by

$$I_{\text{rms}} = V_{\text{rms}}\left[\frac{1}{R^2} + \left(\omega C - \frac{1}{\omega L}\right)^2\right]^{1/2}$$

(b) Show that the phase angle ϕ between V_{rms} and I_{rms} is given by

$$\tan\phi = R\left(\frac{1}{X_C} - \frac{1}{X_L}\right)$$

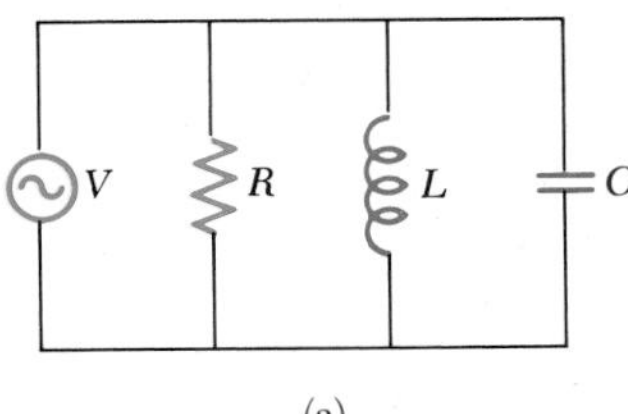

(a)

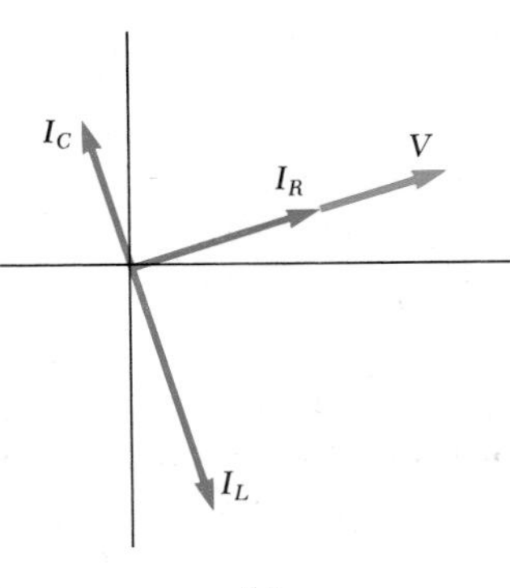

(b)

Figure 33.29 (Problem 81).

82. An 80-Ω resistor, a 200-mH inductor, and a 0.15-μF capacitor are connected in parallel across a 120-V (rms) source operating at an angular frequency of 374 rad/s. (a) What is the resonant frequency of the circuit? (b) Calculate the rms current in the resistor, inductor, and capacitor. (c) What is the rms current delivered by the source? (d) Is the current leading or lagging behind the voltage? By what angle?

CALCULATOR/COMPUTER PROBLEMS

83. A series *RLC* circuit is operating at 2000 Hz. At this frequency, $X_L = X_C = 1884\ \Omega$. The resistance of the circuit is 40 Ω. (a) Prepare a table showing the values of X_L, X_C, and Z for $f = 300$, 600, 800, 1000, 1500, 2000, 3000, 4000, 6000, and 10 000 Hz. (b) Plot on the same set of axes X_L, X_C, and Z as a function of ln *f*.

84. Suppose the high-pass filter shown in Figure 33.15 has $R = 1000\ \Omega$ and $C = 0.050\ \mu F$. (a) At what frequency does $V_{out}/V_{in} = \frac{1}{2}$? (b) Plot $\log_{10}(V_{out}/V_{in})$ versus $\log_{10}(f)$ over the frequency range from 1 Hz to 1 MHz. (This log–log plot of gain versus frequency is known as a **Bode plot.**)

85. Suppose the low-pass filter shown in Figure 33.16 has $R = 1000\ \Omega$ and $C = 0.050\ \mu F$. (a) At what frequency does $V_{out}/V_{in} = \frac{1}{2}$? (b) Plot $\log_{10}(V_{out}/V_{in})$ versus $\log_{10}(f)$ over the frequency range from 1 Hz to 1 MHz.

34

Electromagnetic Waves

An old-style rancher in St. Augustine Plains, New Mexico, rides past one of the 27 radio telescopes that comprise the Very Large Array (VLA). Arranged in a Y-shaped configuration on a system of railroad tracks, the radio telescopes of the VLA capture and focus electromagnetic waves from space. (© Danny Lehman)

The waves we have described in Chapters 16, 17, and 18 are mechanical waves. Such waves correspond to the disturbance of a medium. By definition, mechanical disturbances such as sound waves, water waves, and waves on a string require the presence of a medium. This chapter is concerned with the properties of electromagnetic waves that (unlike mechanical waves) can propagate through empty space.

In Section 31.7 we gave a brief description of Maxwell's equations, which form the theoretical basis of all electromagnetic phenomena. The consequences of Maxwell's equations are far reaching and very dramatic for the history of physics. One of Maxwell's equations, the Ampère-Maxwell law, predicts that a time-varying electric field produces a magnetic field just as a time-varying magnetic field produces an electric field (Faraday's law). From this generalization, Maxwell introduced the concept of displacement current, a new source of a magnetic field. Thus, Maxwell's theory provided the final important link between electric and magnetic fields.

Astonishingly, Maxwell's formalism also predicts the existence of electromagnetic waves that propagate through space with the speed of light. This prediction was confirmed experimentally by Hertz, who first generated and

Biographical Sketch

James Clerk Maxwell
(1831–1879)

James Clerk Maxwell is generally regarded as the greatest theoretical physicist of the 19th century. Born in Edinburgh to a well-known Scottish family, he entered The University of Edinburgh at age 15, around the time that he discovered an original method for drawing a perfect oval. Maxwell was appointed to his first professorship in 1856 at Aberdeen. This was the beginning of a career during which Maxwell would develop the electromagnetic theory of light, the kinetic theory of gases, and explanations of the nature of Saturn's rings and of color vision.

Maxwell's development of the electromagnetic theory of light took many years and began with the paper "On Faraday's Lines of Force," in which Maxwell expanded upon Faraday's theory that electric and magnetic effects result from fields of lines of force surrounding conductors and magnets. His next publication, "On Physical Lines of Force," included a series of papers on the nature of electromagnetism. By considering how the motion of the vortices and cells could produce magnetic and electric effects, Maxwell was successful in explaining all the known effects of electromagnetism. He effectively showed that the lines of force behaved in a similar way.

Maxwell's other important contributions to theoretical physics were made in the area of the kinetic theory of gases. Here, he furthered the work of Rudolf Clausius, who in 1858 had shown that a gas must consist of molecules in constant motion colliding with one another and the walls of the container. This resulted in Maxwell's distribution of molecular velocities in addition to important applications of the theory to viscosity, conduction of heat, and diffusion of gases.

Maxwell's successful interpretation of Faraday's concept of the electromagnetic field resulted in the field equation bearing Maxwell's name. Formidable mathematical ability combined with great insight enabled Maxwell to lead the way in the study of the two most important areas of physics at that time. Maxwell died of cancer before he was 50.

(Photo courtesy of AIP Niels Bohr Library)

detected electromagnetic waves. This discovery has led to many practical communication systems, including radio, television, and radar. On a conceptual level, Maxwell unified the subjects of light and electromagnetism by developing the idea that light is a form of electromagnetic radiation.

Electromagnetic waves are generated by accelerating electric charges. The radiated waves consist of oscillating electric and magnetic fields, which are *at right angles to each other* and also *at right angles to the direction of wave propagation.* Thus, electromagnetic waves are transverse in nature. Maxwell's theory shows that the electric and magnetic field amplitudes, E and B, in an electromagnetic wave are related by $E = cB$. At large distances from the source of the waves, the amplitudes of the oscillating fields diminish with distance, in proportion to $1/r$. The radiated waves can be detected at great distances from the oscillating charges. Furthermore, electromagnetic waves carry energy and momentum and hence exert pressure on a surface.

Electromagnetic waves cover a wide range of frequencies. For example, radio waves (frequencies of about 10^7 Hz) are electromagnetic waves produced by oscillating currents in a radio tower's transmitting antenna. Light waves are a high-frequency form of electromagnetic radiation (about 10^{14} Hz) produced by oscillating electrons within atomic systems.

34.1 MAXWELL'S EQUATIONS AND HERTZ'S DISCOVERIES

The fundamental laws governing the behavior of electric and magnetic fields are Maxwell's equations, which were discussed in Section 31.7.[1] In this unified theory of electromagnetism, Maxwell showed that electromagnetic waves are a natural consequence of these fundamental laws. Recall that *Maxwell's equations* in free space are given by

$$\oint \mathbf{E} \cdot d\mathbf{A} = \frac{Q}{\epsilon_0} \tag{34.1}$$

$$\oint \mathbf{B} \cdot d\mathbf{A} = 0 \tag{34.2}$$

$$\oint \mathbf{E} \cdot d\mathbf{s} = -\frac{d\Phi_m}{dt} \tag{34.3}$$

$$\oint \mathbf{B} \cdot d\mathbf{s} = \mu_0 I + \mu_0 \epsilon_0 \frac{d\Phi_e}{dt} \tag{34.4}$$

As we shall see in the next section, one can combine Equations 34.3 and 34.4 and obtain a wave equation for both the electric and the magnetic field. In empty space ($Q = 0, I = 0$), these equations permit a wavelike solution, where the *wave velocity* $(\mu_0\epsilon_0)^{-1/2}$ *equals the measured speed of light.* This result led Maxwell to the prediction that light waves are, in fact, a form of electromagnetic radiation.

Electromagnetic waves were first generated and detected in 1887 by Hertz, using electrical sources. His experimental apparatus is shown schematically in Figure 34.1. An induction coil is connected to two spherical electrodes with a narrow gap between them (the transmitter). The coil provides short voltage surges to the spheres, making one positive, the other negative. A spark is generated between the spheres when the voltage between them reaches the breakdown voltage for air. As the air in the gap is ionized, it conducts more readily and the discharge between the spheres becomes oscillatory. From an electrical circuit viewpoint, this is equivalent to an LC circuit, where the inductance is that of the loop and the capacitance is due to the spherical electrodes.

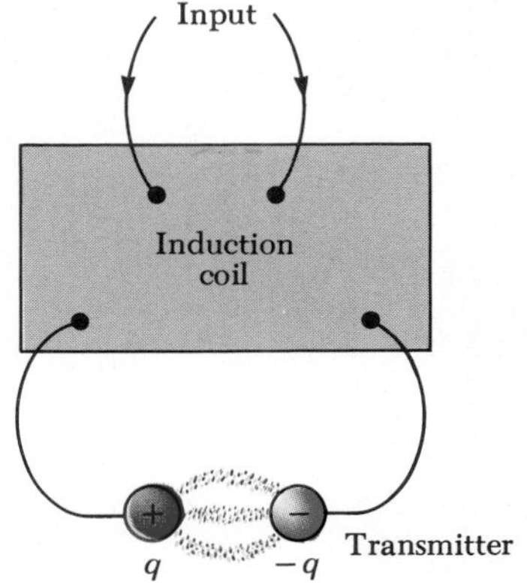

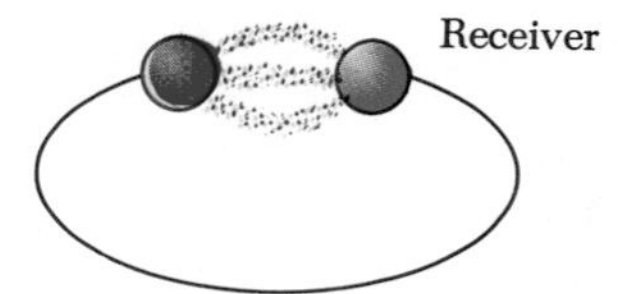

Figure 34.1 Schematic diagram of Hertz's apparatus for generating and detecting electromagnetic waves. The transmitter consists of two spherical electrodes connected to an induction coil, which provides short voltage surges to the spheres, setting up oscillations in the discharge. The receiver is a nearby loop containing a second spark gap.

Since L and C are quite small, the frequency of oscillation is very high, ≈ 100 MHz. (Recall that $\omega = 1/\sqrt{LC}$ for an LC circuit.) Electromagnetic waves are radiated at this frequency as a result of the oscillation (and hence acceleration) of free charges in the loop. Hertz was able to detect these waves using a single loop of wire with its own spark gap (the receiver). This loop, placed several meters from the transmitter, has its own effective inductance, capacitance, and natural frequency of oscillation. Sparks were induced across the gap of the receiving electrodes when the frequency of the receiver was adjusted to match that of the transmitter. Thus, Hertz demonstrated that the oscillating current induced in the receiver was produced by electromagnetic waves radiated by the transmitter. Hertz's experiment is analogous to the mechanical phenomenon in which a tuning fork picks up the vibrations from another, identical oscillating tuning fork.

[1] The reader should review Section 31.7 as a background for the material in this chapter.

Biographical Sketch

Heinrich Rudolf Hertz
(1857–1894)

Heinrich Hertz was born in 1857 in Hamburg, Germany. He studied physics under Helmholtz and Kirchhoff at the University of Berlin. In 1885, Hertz accepted the position of Professor of Physics at Karlsruhe; it was here that he discovered radio waves in 1888, his most important accomplishment.

In 1889 Hertz succeeded Rudolf Clausius as Professor of Physics at the University of Bonn, where his experiments involving the cathode ray's penetration through certain metal films led him to the conclusion that cathode rays were waves rather than particles.

Discovering radio waves, demonstrating their generation, and determining their velocity are among Hertz's many achievements. After finding that the velocity of a radio wave was the same as that of light, Hertz showed that radio waves, like light waves, could be reflected, refracted, and diffracted.

Hertz died of blood poisoning at the age of 36. During his short life, he made many contributions to science. The hertz, equal to one complete vibration or cycle per second, is named after him.

In a series of experiments, Hertz also showed that the radiation generated by his spark-gap device exhibited the wave properties of interference, diffraction, reflection, refraction, and polarization, all of which are properties exhibited by light. Thus, it became evident that the radio-frequency waves had properties similar to light waves and differed only in frequency and wavelength.

Perhaps the most convincing experiment performed by Hertz was the measurement of the velocity of the radio-frequency waves. Radio-frequency waves of known frequency were reflected from a metal sheet and created an interference pattern whose nodal points (where $\boldsymbol{E}$ was zero) could be detected. The measured distance between the nodal points allowed determination of the wavelength λ. Using the relation $v = \lambda f$, Hertz found that v was close to 3×10^8 m/s, the known speed of visible light.

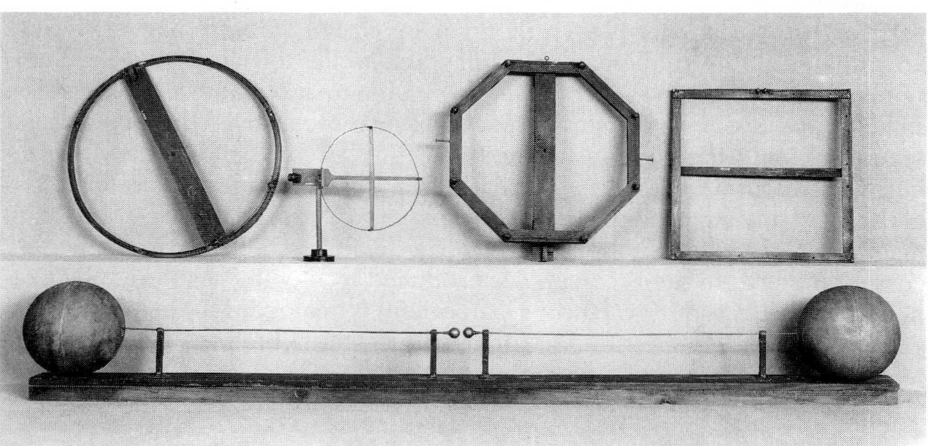

Large oscillator as well as circular and square resonators used by Heinrich Hertz, 1886–88. (Photo Deutsches Museum Munich)

34.2 PLANE ELECTROMAGNETIC WAVES

The properties of electromagnetic waves can be deduced from Maxwell's equations. One approach that can be used to derive such properties would be to solve the second-order differential equation that can be obtained from Maxwell's third and fourth equations. A rigorous mathematical treatment of this sort is beyond the scope of this text. To circumvent this problem, we shall assume that the electric and magnetic vectors have a specific space-time behavior that is consistent with Maxwell's equations.

First, we shall assume that the electromagnetic wave is a *plane wave*, that is, one that travels in one direction. The plane wave we are describing has the following properties. The wave travels in the x direction (the direction of propagation), the electric field $\mathbf{E}$ is in the y direction, and the magnetic field $\mathbf{B}$ is in the z direction, as in Figure 34.2. Waves in which the electric and magnetic fields are restricted to being parallel to certain lines in the yz plane are said to be **linearly polarized waves.**[2] Furthermore, we assume that E and B at any point P depend upon x and t and not upon the y or z coordinates of the point P.

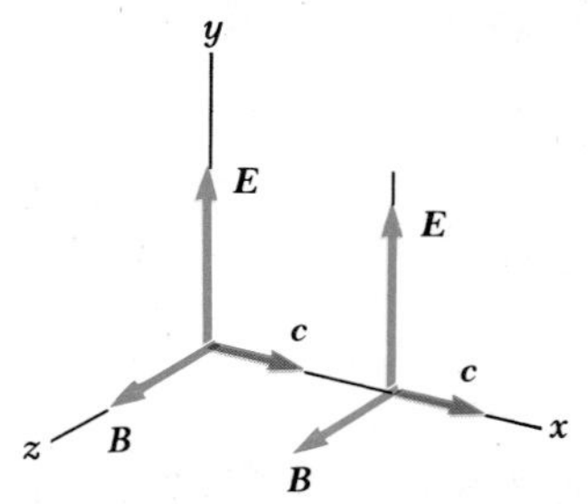

Figure 34.2 A plane polarized electromagnetic wave traveling in the positive x direction. The electric field is along the y direction, and the magnetic field is along the z direction. These fields depend only on x and t.

We can relate E and B to each other by using Maxwell's third and fourth equations (Eqs. 34.3 and 34.4). In empty space, where $Q = 0$ and $I = 0$, these equations become

$$\oint \mathbf{E} \cdot d\mathbf{s} = -\frac{d\Phi_m}{dt} \tag{34.5}$$

$$\oint \mathbf{B} \cdot d\mathbf{s} = \epsilon_0 \mu_0 \frac{d\Phi_e}{dt} \tag{34.6}$$

Using these expressions and the plane wave assumption, one obtains the following differential equations relating E and B. For simplicity of notation, we have dropped the subscripts on the components E_y and B_z:

$$\frac{\partial E}{\partial x} = -\frac{\partial B}{\partial t} \tag{34.7}$$

$$\frac{\partial B}{\partial x} = -\mu_0 \epsilon_0 \frac{\partial E}{\partial t} \tag{34.8}$$

Note that the derivatives here are partial derivatives. For example, when $\partial E/\partial x$ is evaluated, we assume that t is constant. Likewise, when evaluating $\partial B/\partial t$, x is held constant. We shall derive these expressions from Maxwell's equations later in this section. Taking the derivative of Equation 34.7 and combining this with Equation 34.8 we get

$$\frac{\partial^2 E}{\partial x^2} = -\frac{\partial}{\partial x}\left(\frac{\partial B}{\partial t}\right) = -\frac{\partial}{\partial t}\left(\frac{\partial B}{\partial x}\right) = -\frac{\partial}{\partial t}\left(\frac{-\mu_0 \epsilon_0 \partial E}{\partial t}\right) \tag{34.9}$$

$$\frac{\partial^2 E}{\partial x^2} = \mu_0 \epsilon_0 \frac{\partial^2 E}{\partial t^2} \tag{34.10}$$

Wave equations for electromagnetic waves in free space

[2] Waves with other particular patterns of vibrations of $\mathbf{E}$ and $\mathbf{B}$ include *circularly polarized waves*. The most general polarization pattern is *elliptical*.

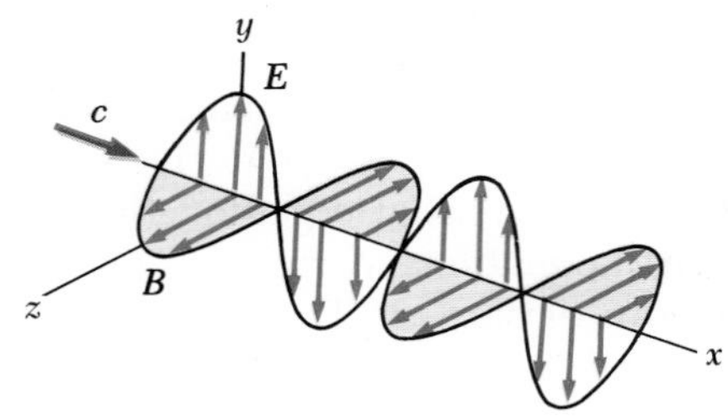

Figure 34.3 Representation of a sinusoidal, plane polarized electromagnetic wave moving in the positive x direction with a speed c. The drawing represents a snapshot, that is, the wave at some instant. Note the sinusoidal variations of E and B with x.

In the same manner, taking a derivative of Equation 34.8 and combining this with Equation 34.10, we get

$$\frac{\partial^2 B}{\partial x^2} = \mu_0 \epsilon_0 \frac{\partial^2 B}{\partial t^2} \tag{34.11}$$

Equations 34.10 and 34.11 both have the form of the general wave equation,[3] with a speed c given by

$$c = \frac{1}{\sqrt{\mu_0 \epsilon_0}} \tag{34.12}$$

Taking $\mu_0 = 4\pi \times 10^{-7}$ Wb/A·m and $\epsilon_0 = 8.85418 \times 10^{-12}$ C^2/N·m^2 in Equation 34.12, we find that

The speed of electromagnetic waves

$$c = 2.99792 \times 10^8 \text{ m/s} \tag{34.13}$$

Since this speed is precisely the same as the speed of light in empty space,[4] one is led to believe (correctly) that light is an electromagnetic wave.

The simplest plane wave solution is a sinusoidal wave, for which the field amplitudes E and B vary with x and t according to the expressions

Sinusoidal electric and magnetic fields

$$E = E_m \cos(kx - \omega t) \tag{34.14}$$

$$B = B_m \cos(kx - \omega t) \tag{34.15}$$

where E_m and B_m are the *maximum* values of the fields. The constant $k = 2\pi/\lambda$, where λ is the wavelength, and the angular frequency $\omega = 2\pi f$, where f is the number of cycles per second. The ratio ω/k equals the speed c, since

$$\frac{\omega}{k} = \frac{2\pi f}{2\pi/\lambda} = \lambda f = c$$

Figure 34.3 is a pictorial representation at one instant of a sinusoidal, linearly polarized plane wave moving in the positive x direction.

[3] The general wave equation is of the form $(\partial^2 f/\partial x^2) = (1/v^2)(\partial^2 f/\partial t^2)$, where v is the speed of the wave and f is the wave amplitude. The wave equation was first introduced in Chapter 16, and it would be useful for the reader to review this material.

[4] Because of the redefinition of the meter in 1983, the speed of light is now a *defined* quantity with an *exact* value of $c = 2.99792458 \times 10^8$ m/s.

Taking partial derivatives of Equations 34.14 and 34.15, we find that

$$\frac{\partial E}{\partial x} = -kE_m \sin(kx - \omega t)$$

$$-\frac{\partial B}{\partial t} = -\omega B_m \sin(kx - \omega t)$$

Since these must be equal, according to Equation 34.7, we find that at any instant

$$kE_m = \omega B_m$$

$$\frac{E_m}{B_m} = \frac{\omega}{k} = c$$

The minus sign is ignored here since we are interested only in comparing the amplitudes. Using these results together with Equations 34.14 and 34.15, we see that

$$\frac{E_m}{B_m} = \frac{E}{B} = c \qquad (34.16)$$

That is, *at every instant the ratio of the electric field to the magnetic field of an electromagnetic wave equals the speed of light.*

Finally, one should note that electromagnetic waves obey the *superposition principle,* since the differential equations involving E and B are *linear* equations. For example, two waves traveling in opposite directions with the same frequency could be added by simply adding the wave fields algebraically. Furthermore, we now have a theoretical value for c, given by the relation $c = 1/\sqrt{\mu_0\epsilon_0}$.

Let us summarize the properties of electromagnetic waves as we have described them:

Properties of electromagnetic waves

1. The solutions of Maxwell's third and fourth equations are wavelike, where both E and B satisfy the same wave equation.
2. Electromagnetic waves travel through empty space with the speed of light, $c = 1/\sqrt{\epsilon_0\mu_0}$.
3. The electric and magnetic field components of plane electromagnetic waves are perpendicular to each other and also perpendicular to the direction of wave propagation. The latter property can be summarized by saying that electromagnetic waves are transverse waves.
4. The relative magnitudes of $\boldsymbol{E}$ and $\boldsymbol{B}$ in empty space are related by $E/B = c$.
5. Electromagnetic waves obey the principle of superposition.

EXAMPLE 34.1 An Electromagnetic Wave

A plane electromagnetic sinusoidal wave of frequency 40 MHz travels in free space in the x direction, as in Figure 34.4. At some point and at some instant, the electric field $\boldsymbol{E}$ has its *maximum* value of 750 N/C and is along the y axis. (a) Determine the wavelength and period of the wave.

Since $c = \lambda f$ and $f = 40 \text{ MHz} = 4 \times 10^7 \text{ s}^{-1}$, we get

$$\lambda = \frac{c}{f} = \frac{3 \times 10^8 \text{ m/s}}{4 \times 10^7 \text{ s}^{-1}} = 7.50 \text{ m}$$

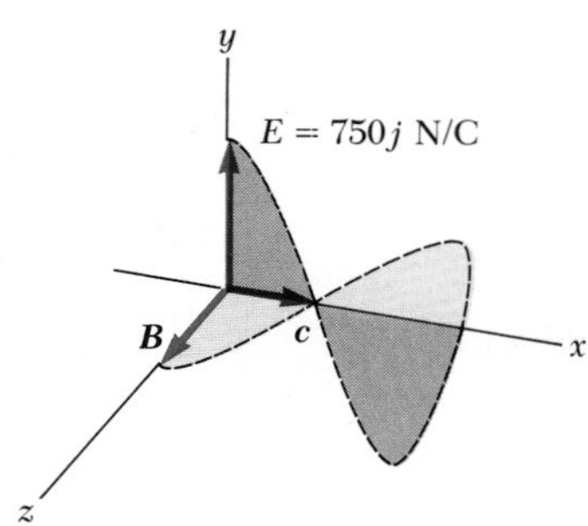

Figure 34.4 (Example 34.1) At some instant, a plane electromagnetic wave moving in the x direction has a maximum electric field of 750 N/C in the positive y direction. The corresponding magnetic field at that point has a magnitude E/c and is in the z direction.

The period of the wave T equals the inverse of the frequency, and so

$$T = \frac{1}{f} = \frac{1}{4 \times 10^7 \text{ s}^{-1}} = \boxed{2.5 \times 10^{-8} \text{ s}}$$

(b) Calculate the magnitude and direction of the magnetic field $\mathbf{B}$ when $\mathbf{E} = 750\mathbf{j}$ N/C.

From Equation 34.16 we see that

$$B_\text{m} = \frac{E_\text{m}}{c} = \frac{750 \text{ N/C}}{3 \times 10^8 \text{ m/s}} = \boxed{2.50 \times 10^{-6} \text{ T}}$$

Since $\mathbf{E}$ and $\mathbf{B}$ must be perpendicular to each other and both must be perpendicular to the direction of wave propagation (x in this case), we conclude that $\mathbf{B}$ is in the z direction.

(c) Write expressions for the space-time variation of the electric and magnetic field components for this plane wave.

We can apply Equations 34.14 and 34.15 directly:

$$E = E_\text{m} \cos(kx - \omega t) = (750 \text{ N/C}) \cos(kx - \omega t)$$

$$B = B_\text{m} \cos(kx - \omega t)$$
$$= (2.50 \times 10^{-6}) \cos(kx - \omega t)$$

where

$$\omega = 2\pi f = 2\pi(4 \times 10^7 \text{ s}^{-1}) = 8\pi \times 10^7 \text{ rad/s}$$

$$k = \frac{2\pi}{\lambda} = \frac{2\pi}{7.5 \text{ m}} = 0.838 \text{ m}^{-1}$$

We shall now give derivations for Equations 34.7 and 34.8. To derive Equation 34.7, we start with Faraday's law, that is, Equation 34.5:

$$\oint \mathbf{E} \cdot d\mathbf{s} = -\frac{d\Phi_\text{m}}{dt}$$

Again, let us assume that the electromagnetic plane wave travels in the x direction with the electric field $\mathbf{E}$ in the positive y direction and the magnetic field $\mathbf{B}$ in the positive z direction.

Consider a thin rectangle lying in the xy plane. The dimensions of the rectangle are width dx and height ℓ, as in Figure 34.5. To apply Equation 34.5, we must first evaluate the line integral of $\mathbf{E} \cdot d\mathbf{s}$ around this rectangle. The contributions from the top and bottom of this rectangle are zero since $\mathbf{E}$ is perpendicular to $d\mathbf{s}$ for these paths. We can express the electric field on the right side of the rectangle as

$$E(x + dx, t) \approx E(x, t) + \left.\frac{dE}{dx}\right]_{t \text{ constant}} dx = E(x, t) + \frac{\partial E}{\partial x}\, dx$$

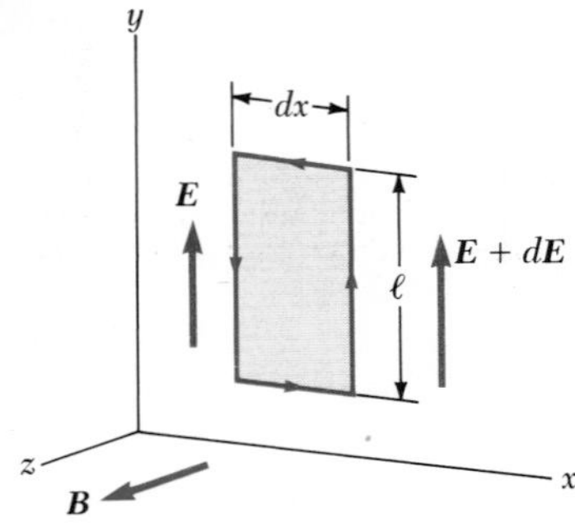

Figure 34.5 As a plane wave passes through a rectangular path of width dx lying in the xy plane, the electric field in the y direction varies from $\mathbf{E}$ to $\mathbf{E} + d\mathbf{E}$. This spatial variation in $\mathbf{E}$ gives rise to a time-varying magnetic field along the z direction, according to Equation 34.19.

while the field on the left side is simply $E(x, t)$. Therefore, the line integral over this rectangle becomes approximately[5]

$$\oint \mathbf{E} \cdot d\mathbf{s} = E(x + dx, t) \cdot \ell - E(x, t) \cdot \ell \approx (\partial E/\partial x)\, dx \cdot \ell \qquad (34.17)$$

[5] Since dE/dx means the change in E with x at a given instant t, dE/dx is equivalent to the partial derivative $\partial E/\partial x$. Likewise, dB/dt means the change in B with time at a particular position x, and so we can replace dB/dt by $\partial B/\partial t$.

Since the magnetic field is in the z direction, the magnetic flux through the rectangle of area $\ell\, dx$ is approximately

$$\Phi_{\mathrm{m}} = B\ell dx$$

(This assumes that dx is small compared with the wavelength of the wave.) Taking the time derivative of the flux gives

$$\frac{d\Phi_{\mathrm{m}}}{dt} = \ell\, dx \left.\frac{dB}{dt}\right]_{x\ \text{constant}} = \ell\, dx \frac{\partial B}{\partial t} \tag{34.18}$$

Substituting Equations 34.17 and 34.18 into Equation 34.5 gives

$$\left(\frac{\partial E}{\partial x}\right) dx \cdot \ell = -\ell\, dx \frac{\partial B}{\partial t}$$

$$\frac{\partial E}{\partial x} = -\frac{\partial B}{\partial t} \tag{34.19}$$

Thus, we see that Equation 34.19 is equivalent to Equation 34.7.

In a similar manner, we can verify Equation 34.8 by starting with Maxwell's fourth equation in empty space (Eq. 34.6):

$$\oint \boldsymbol{B} \cdot d\boldsymbol{s} = \mu_0 \epsilon_0 \frac{d\Phi_{\mathrm{e}}}{dt}$$

In this case, we evaluate the line integral of $\boldsymbol{B} \cdot d\boldsymbol{s}$ around a rectangle lying in the yz plane and having width dx and length ℓ, as in Figure 34.6, where the magnetic field is in the z direction. Using the sense of the integration shown, and noting that the magnetic field changes from $B(x, t)$ to $B(x + dx, t)$ over the width dx, we get

$$\oint \boldsymbol{B} \cdot d\boldsymbol{s} = B(x, t) \cdot \ell - B(x + dx, t) \cdot \ell = -(\partial B/\partial x)\, dx \cdot \ell \tag{34.20}$$

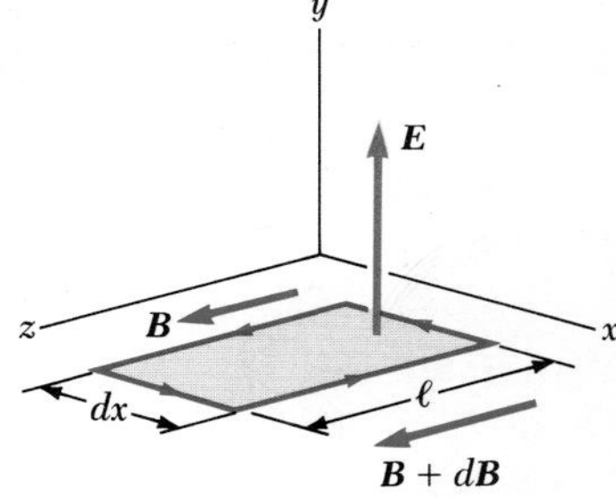

Figure 34.6 As a plane wave passes through a rectangular curve of width dx lying in the xz plane, the magnetic field along z varies from $\boldsymbol{B}$ to $\boldsymbol{B} + d\boldsymbol{B}$. This spatial variation in $\boldsymbol{B}$ gives rise to a time-varying electric field along the y direction, according to Equation 34.22.

The electric flux through the rectangle is

$$\Phi_{\mathrm{e}} = E\ell\, dx$$

which when differentiated with respect to time gives

$$\frac{\partial \Phi_{\mathrm{e}}}{\partial t} = \ell\, dx \frac{\partial E}{\partial t} \tag{34.21}$$

Substituting Equations 34.20 and 34.21 into Equation 34.6 gives

$$-(\partial B/\partial x) dx \cdot \ell = \mu_0 \epsilon_0 \ell\, dx (\partial E/\partial t)$$

$$\frac{\partial B}{\partial x} = -\mu_0 \epsilon_0 \frac{\partial E}{\partial t} \tag{34.22}$$

which is equivalent to Equation 34.8.

34.3 ENERGY CARRIED BY ELECTROMAGNETIC WAVES

Electromagnetic waves carry energy, and as they propagate through space they can transfer energy to objects placed in their path. The rate of flow of energy in an electromagnetic wave is described by a vector **S**, called the **Poynting vector,** defined by the expression

Poynting vector

$$\mathbf{S} \equiv \frac{1}{\mu_0} \mathbf{E} \times \mathbf{B} \tag{34.23}$$

The magnitude of the Poynting vector represents the rate at which energy flows through a unit surface area perpendicular to the flow.

The direction of **S** is along the direction of wave propagation (Fig. 34.7). The SI units of the Poynting vector are $\mathrm{J/s \cdot m^2} = \mathrm{W/m^2}$. (These are the units **S** must have since it represents the power per unit area, where the unit area is oriented at right angles to the direction of wave propagation.)

As an example, let us evaluate the magnitude of **S** for a plane electromagnetic wave where $|\mathbf{E} \times \mathbf{B}| = EB$. In this case

Poynting vector for a plane wave

$$S = \frac{EB}{\mu_0} \tag{34.24}$$

Since $B = E/c$, we can also express this as

$$S = \frac{E^2}{\mu_0 c} = \frac{c}{\mu_0} B^2 \tag{34.25}$$

These equations for S apply at any instant of time.

What is of more interest for a sinusoidal plane electromagnetic wave is the time average of S taken over one or more cycles, called the *wave intensity*, *I*. When this average is taken, one obtains an expression involving the time average of $\cos^2(kx - \omega t)$, which equals $\frac{1}{2}$. Hence, the average value of S (or the intensity of the wave) is

Wave intensity

$$I = S_{\text{av}} = \frac{E_{\text{m}} B_{\text{m}}}{2\mu_0} = \frac{E_{\text{m}}^{\ 2}}{2\mu_0 c} = \frac{c}{2\mu_0} B_{\text{m}}^{\ 2} \tag{34.26}$$

where it is important to note that E_{m} and B_{m} represent *maximum* values of the fields. The constant $\mu_0 c$, called the **impedance of free space**, has the SI units of ohms and has the value

Impedance of free space

$$\mu_0 c = \sqrt{\frac{\mu_0}{\epsilon_0}} = 377\ \Omega$$

Recall that the energy per unit volume u_{e}, the instantaneous energy density associated with an electric field (Section 26.4), is given by

$$u_{\text{e}} = \tfrac{1}{2}\epsilon_0 E^2 \tag{26.14}$$

and that the instantaneous energy density u_{m} associated with a magnetic field (Section 32.3) is given by

$$u_{\text{m}} = \frac{B^2}{2\mu_0} \tag{32.15}$$

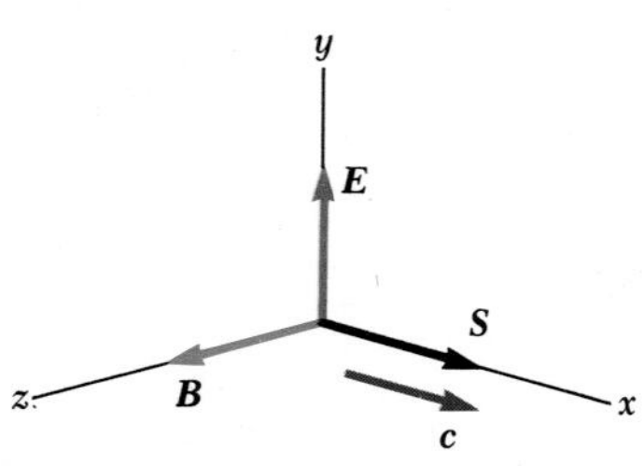

Figure 34.7 The Poynting vector **S** for a plane electromagnetic wave moving in the x direction is along the direction of propagation.

Because E and B vary with time for an electromagnetic wave, we see that the energy densities also vary with time. Using the relationships $B = E/c$ and $c = 1/\sqrt{\epsilon_0 \mu_0}$, Equation 32.15 becomes

$$u_m = \frac{(E/c)^2}{2\mu_0} = \frac{\epsilon_0\mu_0}{2\mu_0}E^2 = \tfrac{1}{2}\epsilon_0 E^2$$

Comparing this result with Equation 32.15, we see that

$$u_m = u_e = \tfrac{1}{2}\epsilon_0 E^2 = \frac{B^2}{2\mu_0} \quad (34.27)$$

That is, *for an electromagnetic wave the instantaneous energy density associated with the magnetic field equals the instantaneous energy density associated with the electric field.* Hence, in a given volume the energy is equally shared by the two fields.

The **total instantaneous energy density** u is equal to the sum of the energy densities associated with the electric and magnetic fields:

$$u = u_e + u_m = \epsilon_0 E^2 = \frac{B^2}{\mu_0} \quad (34.28)$$

Total energy density

When this is averaged over one or more cycles of an electromagnetic wave, we again get a factor of $\frac{1}{2}$. Hence, the total *average* energy per unit volume of an electromagnetic wave is given by

$$u_{av} = \epsilon_0 (E^2)_{av} = \tfrac{1}{2}\epsilon_0 E_m{}^2 = \frac{B_m{}^2}{2\mu_0} \quad (34.29)$$

Average energy density of an electromagnetic wave

Comparing this result with Equation 34.26 for the average value of S, we see that

$$I = S_{av} = cu_{av} \quad (34.30)$$

In other words, *the intensity of an electromagnetic wave equals the average energy density multiplied by the speed of light.*

EXAMPLE 34.2 Fields Due to a Point Source

A point source of electromagnetic radiation has an average power output of 800 W. Calculate the *maximum* values of the electric and magnetic fields at a point 3.50 m from the source.

Solution Recall from Chapter 17 that the wave intensity, I, at a distance r from a point source is given by

$$I = \frac{P_{av}}{4\pi r^2}$$

where P_{av} is the average power output of the source and $4\pi r^2$ is the area of a sphere of radius r centered on the source. Since the intensity of an electromagnetic wave is also given by Equation 34.26, we have

$$I = \frac{P_{av}}{4\pi r^2} = \frac{E_m{}^2}{2\mu_0 c}$$

Solving for the maximum electric field, E_m, gives

$$E_m = \sqrt{\frac{\mu_0 c P_{av}}{2\pi r^2}}$$

$$= \sqrt{\frac{(4\pi \times 10^{-7}\ \text{N/A}^2)(3.00 \times 10^8\ \text{m/s})(800\ \text{W})}{2\pi(3.50\ \text{m})^2}}$$

$$= 62.6\ \text{V/m}$$

We can easily calculate the maximum value of the magnetic field using the result above and the relation $B_m = E_m/c$ (Eq. 34.16):

$$B_m = \frac{E_m}{c} = \frac{62.6\ \text{V/m}}{3.00 \times 10^8\ \text{m/s}} = 2.09 \times 10^{-7}\ \text{T}$$

Exercise 1 Calculate the value of the energy density at the point 3.50 m from the point source.
Answer $1.73 \times 10^{-8}\ \text{J/m}^3$

34.4 MOMENTUM AND RADIATION PRESSURE

Electromagnetic waves transport linear momentum as well as energy. Hence, it follows that pressure (radiation pressure) is exerted on a surface when an electromagnetic wave impinges on it. In what follows, we shall assume that the electromagnetic wave transports a total energy U to a surface in a time t. If the surface *absorbs all* the incident energy U in this time, Maxwell showed that the total momentum $\mathbf{p}$ delivered to this surface has a magnitude given by

Momentum delivered to an absorbing surface

$$p = \frac{U}{c} \qquad \text{(complete absorption)} \tag{34.31}$$

Furthermore, if the Poynting vector of the wave is $\mathbf{S}$, the *radiation pressure* P (force per unit area) exerted on the perfect absorbing surface is given by

Radiation pressure exerted on a perfect absorbing surface

$$P = \frac{S}{c} \tag{34.32}$$

We can apply these results to a perfect black body, where *all* of the incident energy is absorbed (none is reflected).

On the other hand, if the surface is a perfect reflector (for example, a mirror with a 100% reflecting surface), then the momentum delivered in a time t for normal incidence is *twice* that given by Equation 34.31, or $2U/c$. That is, a momentum equal to U/c is delivered by the incident wave and U/c is delivered by the reflected wave, in analogy with a ball colliding elastically with a wall. Therefore,

$$p = \frac{2U}{c} \qquad \text{(complete reflection)} \tag{34.33}$$

The momentum delivered to an arbitrary surface has a value between U/c and $2U/c$, depending on the properties of the surface. Finally, the radiation pressure exerted on a perfect reflecting surface for normal incidence of the wave is given by[6]

$$P = \frac{2S}{c} \tag{34.34}$$

Although radiation pressures are very small (about 5×10^{-6} N/m^2 for direct sunlight), they have been measured using torsion balances such as the one shown in Figure 34.8. Light is allowed to strike either a mirror or a black disk, both of which are suspended from a fine fiber. Light striking the black disk is completely absorbed, and so all of its momentum is transferred to the disk. Light striking the mirror (normal incidence) is totally reflected, hence the momentum transfer is twice as great as that transferred to the disk. The radiation pressure is determined by measuring the angle through which the horizontal portion rotates. The apparatus must be placed in a high vacuum to eliminate the effects of air currents.

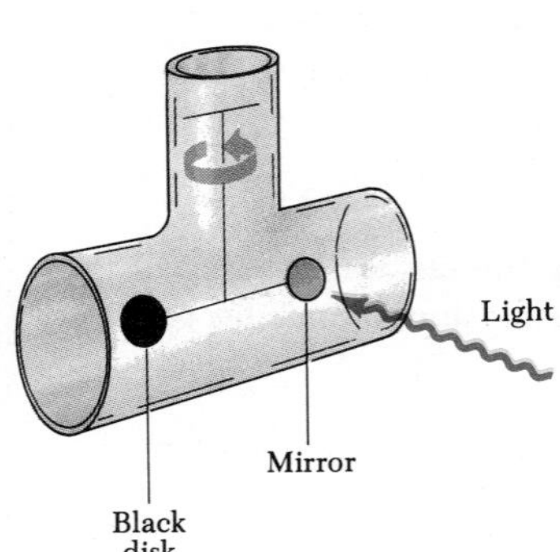

Figure 34.8 An apparatus for measuring the pressure of light. In practice, the system is contained in a high vacuum.

[6] For *oblique* incidence, the momentum transferred is $2U \cos\theta/c$ and the pressure is given by $P = 2S\cos\theta/c$, where θ is the angle between the normal to the surface and the direction of propagation.

EXAMPLE 34.3 Solar Energy
The sun delivers about 1000 W/m^2 of electromagnetic flux to the earth's surface. (a) Calculate the total power that is incident on a roof of dimensions 8 m × 20 m.

Solution The Poynting vector has a magnitude of $S = 1000\ W/m^2$, which represents the power per unit area, or the light intensity. Assuming the radiation is incident *normal* to the roof (sun directly overhead), we get

$$\text{Power} = SA = (1000\ W/m^2)(8 \times 20\ m^2)$$
$$= 1.60 \times 10^5\ W$$

If this power could *all* be converted into electrical energy, it would provide more than enough power for the average home. However, solar energy is not easily harnessed, and the prospects for large-scale conversion are not as "bright" as they may appear from this simple calculation. For example, the conversion efficiency from solar to electrical energy is far less than 100% (typically, 10% for photovoltaic cells). Roof systems for converting solar energy to *thermal* energy have been built with efficiencies of around 50%; however, there are other practical problems with solar energy that must be considered, such as overcast days, geographic location, and energy storage.

(b) Determine the radiation pressure and radiation force on the roof assuming the roof covering is a perfect absorber.

Solution Using Equation 34.32 with $S = 1000\ W/m^2$, we find that the radiation pressure is

$$P = \frac{S}{c} = \frac{1000\ W/m^2}{3 \times 10^8\ m/s} = 3.33 \times 10^{-6}\ N/m^2$$

Because pressure equals force per unit area, this corresponds to a radiation force of

$$F = PA = (3.33 \times 10^{-6}\ N/m^2)(160\ m^2)$$
$$= 5.33 \times 10^{-4}\ N$$

Of course, this "load" is *far* less than the other loads one must contend with on roofs, such as the roof's own weight or a layer of snow.

Exercise 2 How much solar energy (in joules) is incident on the roof in 1 h?
Answer 5.76×10^8 J.

EXAMPLE 34.4 Poynting Vector for a Wire
A long, straight wire of resistance R, radius a, and length ℓ carries a constant current I as in Figure 34.9. Calculate the Poynting vector for this wire.

Solution First, let us find the electric field $\mathbf{E}$ along the wire. If V is the potential difference across the ends of the wire, then $V = IR$ and

$$E = V/\ell = IR/\ell$$

Recall that the magnetic field at the surface of the wire (Example 30.4) is given by

$$B = \mu_0 I/2\pi a$$

The vectors $\mathbf{E}$ and $\mathbf{B}$ are mutually *perpendicular*, as shown in Figure 34.9, and therefore $|\mathbf{E} \times \mathbf{B}| = EB$. Hence, the Poynting vector $\mathbf{S}$ is directed radially *inward* and has a magnitude

$$S = \frac{EB}{\mu} = \frac{1}{\mu}\frac{IR}{\ell}\frac{\mu_0 I}{2\pi a} = \frac{I^2R}{2\pi a\ell} = \frac{I^2R}{A}$$

where $A = 2\pi a\ell$ is the *surface* area of the wire, and the total area through which S passes. From this result, we see that

$$SA = I^2R$$

where SA has units of power (J/s = W). That is, *the rate at which electromagnetic energy flows into the wire, SA, equals the rate of energy (or power) dissipated as joule heat,* I^2R.

Exercise 3 A heater wire of radius 0.3 mm and resistance 5 Ω carries a current of 2 A. Determine the magnitude and direction of the Poynting vector for this wire.
Answer $1.06 \times 10^4\ W/m^2$ directed radially inward.

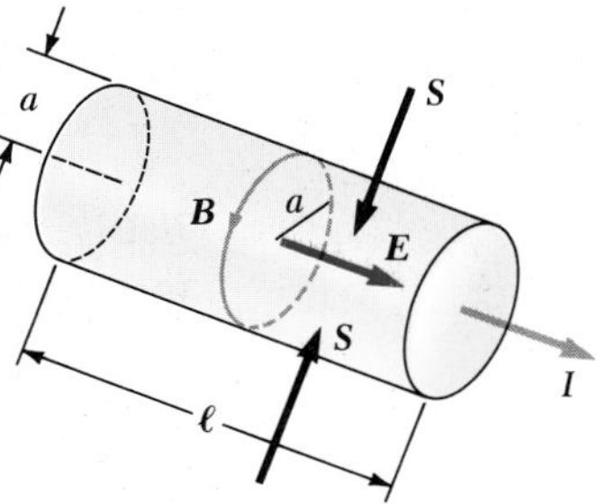

Figure 34.9 (Example 34.4) A wire of length ℓ, resistance R, and radius a carrying a current I. The Poynting vector $\mathbf{S}$ is directed radially *inward*.

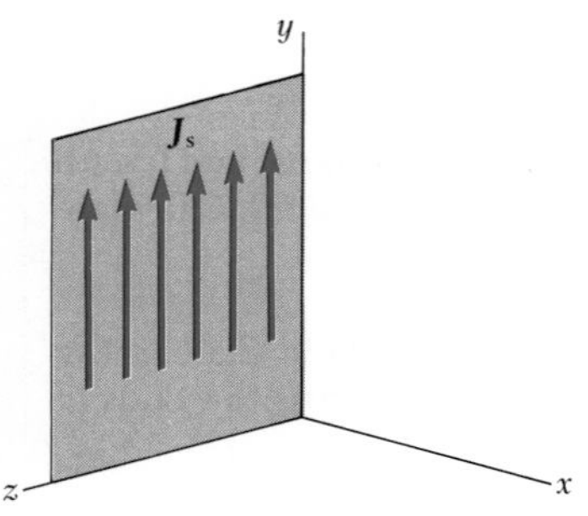

Figure 34.10 An infinite current sheet lying in the yz plane. The current density is sinusoidal and given by $J_s = J_0 \cos \omega t$. The magnetic field is everywhere parallel to the sheet and lies along z.

*34.5 RADIATION FROM AN INFINITE CURRENT SHEET

In this section, we shall describe the fields radiated by a conductor carrying a time-varying current. The plane geometry we shall treat reduces the mathematical complexities one would encounter in a lower-symmetry situation, such as an oscillating electric dipole.

Consider an *infinite* conducting sheet lying in the yz plane and carrying a *surface current per unit length* $\mathbf{J}_\mathrm{S}$ in the y direction, as in Figure 34.10. Let us assume that J_S varies sinusoidally with time as

$$J_\mathrm{S} = J_0 \cos \omega t$$

A similar problem for the case of a steady current was treated in Example 30.6, where we found that the magnetic field outside the sheet is everywhere parallel to the sheet and lies along the z axis. The magnetic field was found to have a magnitude

Radiated magnetic field

$$B_z = -\mu_0 \frac{J_\mathrm{S}}{2}$$

In the present situation, where J_S varies with time, this equation for B_z is valid only for distances *close* to the sheet. That is,

$$B_z = -\frac{\mu_0}{2} J_0 \cos \omega t \qquad \text{(for small values of } x\text{)}$$

To obtain the expression for B_z for *arbitrary values* of x, we can investigate the following solution:[7]

$$B_z = -\frac{\mu_0 J_0}{2} \cos(kx - \omega t) \tag{34.35}$$

There are two things to note about this solution, which is unique to the geometry under consideration. First, it agrees with our original solution for small values of x. Second, it satisfies the wave equation as it is expressed in Equation 34.11. Hence, we conclude that the magnetic field lies along the z axis and is characterized by a transverse traveling wave having an angular frequency ω, wave number $k = 2\pi/\lambda$, and wave speed c.

We can obtain the radiated electric field that accompanies this varying magnetic field by using Equation 34.16:

Radiated electric field

$$E_y = cB_z = -\frac{\mu_0 J_0 c}{2} \cos(kx - \omega t) \tag{34.36}$$

That is, the electric field is in the y direction, perpendicular to $\mathbf{B}$, and has the same space and time dependences.

These expressions for B_z and E_y show that the radiation field of an infinite current sheet carrying a sinusoidal current is a plane electromagnetic wave propagating with a speed c along the x axis, as shown in Figure 34.11.

We can calculate the Poynting vector for this wave by using Equation 34.24 together with Equations 34.35 and 34.36:

[7] Note that the solution could also be written in the form $\cos(\omega t - kx)$, which is equivalent to $\cos(kx - \omega t)$. That is, $\cos \theta$ is an even function, which means that $\cos(-\theta) = \cos \theta$.

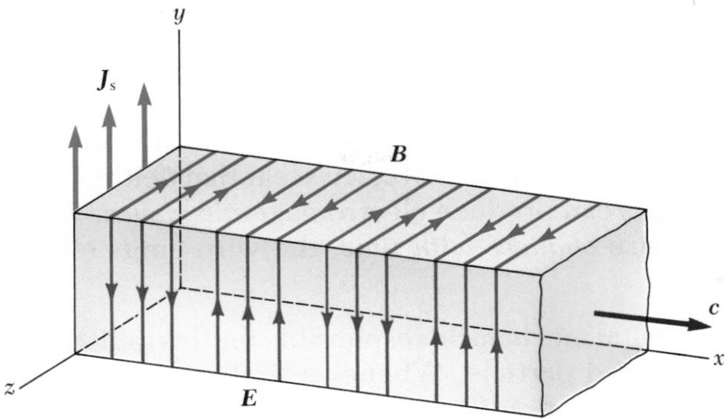

Figure 34.11 Representation of the plane electromagnetic wave radiated by the infinite current sheet lying in the yz plane. Note that $\mathbf{B}$ is in the z direction, $\mathbf{E}$ is in the y direction, and the direction of wave motion is along x. Both vectors have a $\cos(kx - \omega t)$ behavior.

$$S = \frac{EB}{\mu_0} = \frac{\mu_0 J_0^2 c}{4} \cos^2(kx - \omega t) \quad (34.37)$$

The intensity of the wave, which equals the average value of S, is

$$S_{av} = \frac{\mu_0 J_0^2 c}{8} \quad (34.38)$$

The intensity given by Equation 34.38 represents the average intensity of the outgoing wave on each side of the sheet. The total rate of energy emitted per unit area of the conductor is $2S_{av} = \mu_0 J_0^2 c/4$.

EXAMPLE 34.5 An Infinite Sheet Carrying a Sinusoidal Current

An infinite current sheet lying in the yz plane carries a sinusoidal current density that has a maximum value of 5 A/m. (a) Find the *maximum* values of the radiated magnetic field and electric field.

Solution From Equations 34.35 and 34.36, we see that the *maximum* values of B_z and E_y are given by

$$B_m = \frac{\mu_0 J_0}{2} \quad \text{and} \quad E_m = \frac{\mu_0 J_0 c}{2}$$

Using the values $\mu_0 = 4\pi \times 10^{-7}$ Wb/A · m, $J_0 = 5$ A/m, and $c = 3 \times 10^8$ m/s, we get

$$B_m = \frac{(4\pi \times 10^{-7}\ \text{Wb/A}\cdot\text{m})(5\ \text{A/m})}{2} = 3.14 \times 10^{-6}\ \text{T}$$

$$E_m = \frac{(4\pi \times 10^{-7}\ \text{Wb/A}\cdot\text{m})(5\ \text{A/m})(3 \times 10^8\ \text{m/s})}{2}$$

$$= 942\ \text{V/m}$$

(b) What is the average power incident on a second plane surface that is parallel to the sheet and has an area of 3 m²? (The length and width of the plate are both much larger than the wavelength of the light.)

Solution The power per unit area (the average value of the Poynting vector) radiated in each direction by the current sheet is given by Equation 34.38. Multiplying this by the area of the plane in question gives the incident power:

$$P = \left(\frac{\mu_0 J_0^2 c}{8}\right) A$$

$$= \frac{(4\pi \times 10^{-7}\ \text{Wb/A}\cdot\text{m})(5\ \text{A/m})^2(3 \times 10^8\ \text{m/s})}{8} (3\ \text{m}^2)$$

$$= 3.54 \times 10^3\ \text{W}$$

The result is *independent of the distance from the current sheet* since we are dealing with a plane wave.

*34.6 THE PRODUCTION OF ELECTROMAGNETIC WAVES BY AN ANTENNA

Electromagnetic waves arise as a consequence of two effects: (1) a changing magnetic field produces an electric field and (2) a changing electric field produces a magnetic field. Therefore, it is clear that neither stationary charges nor steady currents can produce electromagnetic waves. Whenever the current through a wire *changes with time,* the wire emits electromagnetic radiation.

Accelerating charges produce *EM* radiation

The fundamental mechanism responsible for this radiation is the acceleration of a charged particle. Whenever a charged particle undergoes an acceleration, it must radiate energy.

An alternating voltage applied to the wires of an antenna forces an electric charge in the antenna to oscillate. This is a common technique for accelerating charged particles and is the source of the radio waves emitted by the antenna of a radio station.

Figure 34.12 illustrates the production of an electromagnetic wave by oscillating electric charges in an antenna. Two metal rods are connected to an ac generator, which causes charges to oscillate between the two rods. The output voltage of the generator is sinusoidal. At $t = 0$, the upper rod is given a maximum positive charge and the bottom rod an equal negative charge, as in Figure 34.12a. The electric field near the antenna at this instant is also shown in Figure 34.12a. As the charges oscillate, the rods become less charged, the field near the rods decreases in strength, and the downward-directed maximum electric field produced at $t = 0$ moves away from the rod. When the charges are neutralized, as in Figure 34.12b, the electric field has dropped to zero. This occurs at a time equal to one quarter of the period of oscillation.

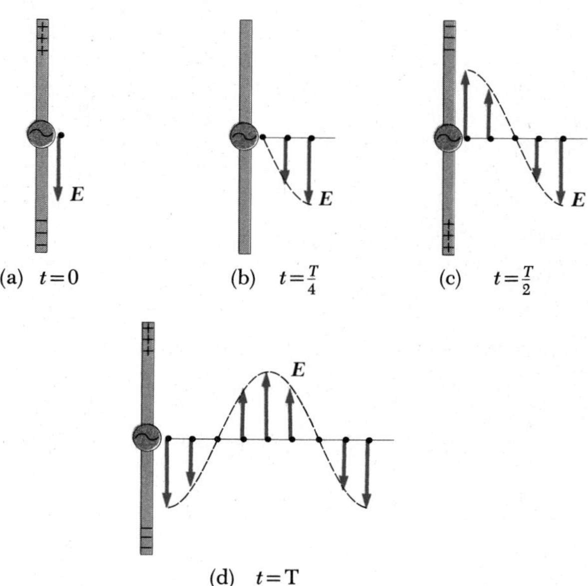

Figure 34.12 The electric field set up by oscillating charges in an antenna. The field moves away from the antenna with the speed of light.

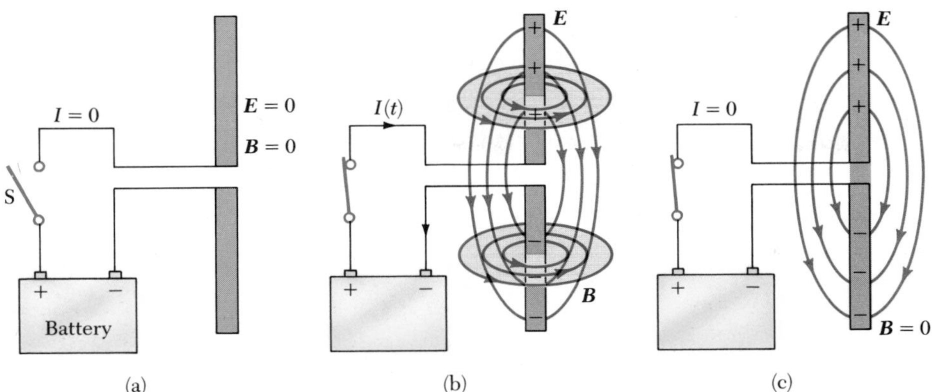

Figure 34.13 A pair of metal rods connected to a battery. (a) When the switch is open and there is no current, the electric and magnetic fields are both zero. (b) After the switch is closed and the rods are being charged (so that a current exists), the rods generate changing electric and magnetic fields. (c) When the rods are fully charged, the current is zero, the electric field is a maximum, and the magnetic field is zero.

Continuing in this fashion, the upper rod soon obtains a maximum negative charge and the lower rod becomes positive, as in Figure 34.12c, resulting in an electric field directed upward. This occurs after a time equal to one half the period of oscillation. The oscillations continue as indicated in Figure 34.12d. A magnetic field oscillating perpendicular to the diagram in Figure 34.12 accompanies the oscillating electric field, but is not shown for clarity. The electric field near the antenna oscillates in phase with the charge distribution. That is, the field points down when the upper rod is positive and up when the upper rod is negative. Furthermore, the magnitude of the field at any instant depends on the amount of charge on the rods at that instant.

As the charges continue to oscillate (and accelerate) between the rods, the electric field set up by the charges moves away from the antenna at the speed of light. Figure 34.12 shows the electric field pattern at various times during the oscillation cycle. As you can see, one cycle of charge oscillation produces one full wavelength in the electric field pattern.

Next, consider what happens when two conducting rods are connected to the opposite ends of a battery (Fig. 34.13). Before the switch is closed, the current is zero, so there are no fields present (Fig. 34.13a). Just after the switch is closed, charge of opposite signs begins to build up on the rods (Fig. 34.13b), which corresponds to a time-varying current, $I(t)$. The changing charge causes the electric field to change, which in turn produces a magnetic field around the rods.[8] Finally, when the rods are fully charged, the current is zero and there is no magnetic field (Fig. 34.13c).

Now let us consider the production of electromagnetic waves by a *half-wave antenna*. In this arrangement, two conducting rods, each one quarter of a wavelength long, are connected to a source of alternating emf (such as an *LC* oscillator), as in Figure 34.14. The oscillator forces charges to accelerate back and forth between the two rods. Figure 34.14 shows the field configuration at some instant when the current is upward. The electric field lines resemble those of an electric dipole, that is, two equal and opposite charges. Since these

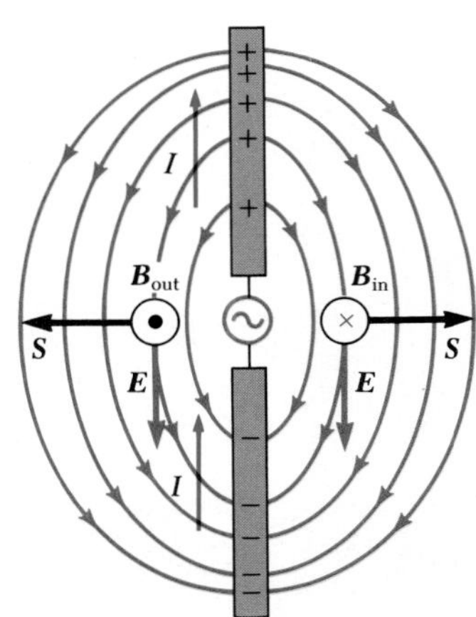

Figure 34.14 A half-wave (dipole) antenna consists of two metal rods connected to an alternating voltage source. The diagram shows ***E*** and ***B*** at an instant when the current is upward. Note that the electric field lines resemble those of a dipole.

[8] We have neglected the field due to the wires leading to the rods. This is a good approximation if the circuit dimensions are small relative to the length of the rods.

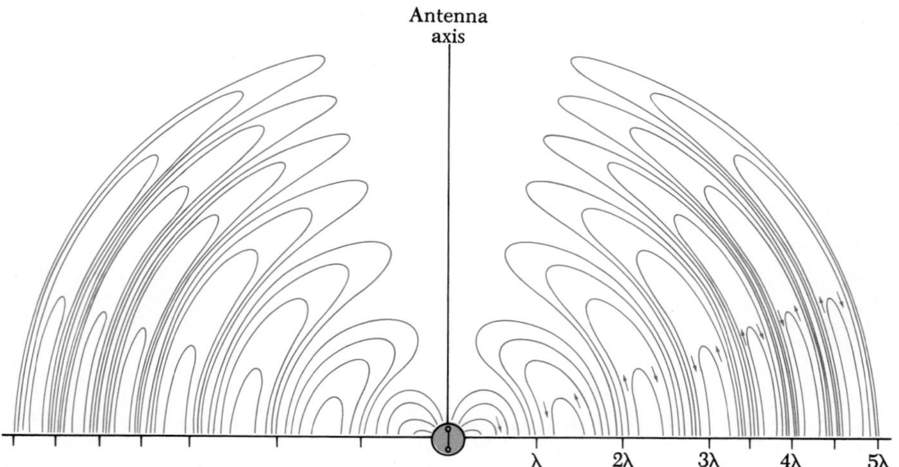

Figure 34.15 Electric field lines surrounding an oscillating dipole at a given instant. The radiation fields propagate outward from the dipole with a speed c.

charges are continuously oscillating between the two rods, the antenna can be approximated by an oscillating electric dipole. The magnetic field lines form concentric circles about the antenna and are perpendicular to the electric field lines at all points. The magnetic field is zero at all points along the axis of the antenna. Furthermore, $\boldsymbol{E}$ and $\boldsymbol{B}$ are 90° out of phase in time, that is, $\boldsymbol{E}$ at some point reaches its maximum value when $\boldsymbol{B}$ is zero and vice versa. This is because when the charges at the ends of the rods are at a maximum, the current is zero.

At the two points shown in Figure 34.14, Poynting's vector $\boldsymbol{S}$ is radially outward. This indicates that energy is flowing away from the antenna at this instant. At later times, the fields and Poynting's vector change direction as the current alternates. Since $\boldsymbol{E}$ and $\boldsymbol{B}$ are 90° out of phase at points near the dipole, the net energy flow is zero. From this, we might conclude (incorrectly) that no energy is radiated by the dipole.

Since the dipole fields fall off as $1/r^3$ (as in the case of a static dipole), they are not important at large distances from the antenna. However, at these large distances, another effect produces the radiation field. The source of this radiation is the continuous induction of an electric field by a time-varying magnetic field and the induction of a magnetic field by a time-varying electric field. These are predicted by two of Maxwell's equations (Eqs. 34.3 and 34.4). The electric and magnetic fields produced in this manner are in phase with each other and vary as $1/r$. The result is an outward flow of energy at all times.

The electric field lines produced by an oscillating dipole at some instant are shown in Figure 34.15. Note that the intensity (and the power radiated) are a maximum in a plane that is perpendicular to the antenna and passing through its midpoint. Furthermore, the power radiated is zero along the antenna's axis. A mathematical solution to Maxwell's equations for the oscillating dipole shows that the intensity of the radiation field varies as $\sin^2 \theta/r^2$, where θ is measured from the axis of the antenna. The angular dependence of the radiation intensity (power per unit area) is sketched in Figure 34.16.

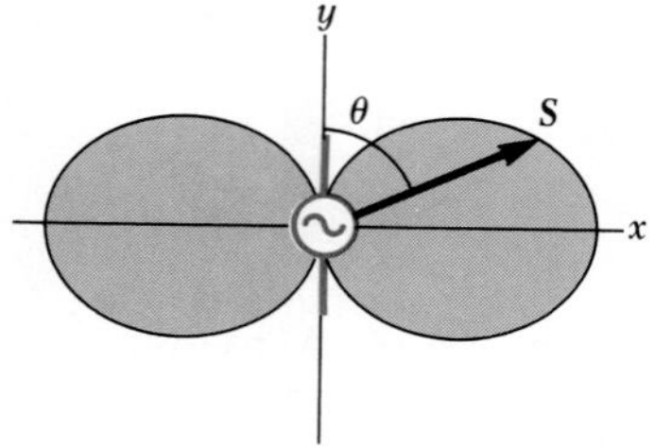

Figure 34.16 Angular dependence of the intensity of radiation produced by an oscillating electric dipole.

Electromagnetic waves can also induce currents in a *receiving antenna.* The response of a dipole receiving antenna at a given position will be a maximum when its axis is parallel to the electric field at that point and zero when its axis is perpendicular to the electric field.

34.7 THE SPECTRUM OF ELECTROMAGNETIC WAVES

We have seen that all electromagnetic waves travel in a vacuum with the speed of light, c. These waves transport energy and momentum from some source to a receiver. In 1887, Hertz successfully generated and detected the radio-frequency electromagnetic waves predicted by Maxwell.[9] At this time, the only electromagnetic waves recognized were radio waves and visible light. It is now known that other forms of electromagnetic waves exist which are distinguished by their frequency and wavelength.

Since all electromagnetic waves travel through vacuum with a speed c, their frequency f and wavelength λ are related by the important expression

$$c = f\lambda \tag{34.39}$$

The various types of electromagnetic waves are listed in Figure 34.17. Note the wide range of frequencies and wavelengths. For instance, a radio wave of frequency 5 MHz (a typical value) has a wavelength given by

[9] Following Hertz's discoveries, Marconi succeeded in developing a practical, long-range radio communication system. However, Hertz must be recognized as the true inventor of radio communication.

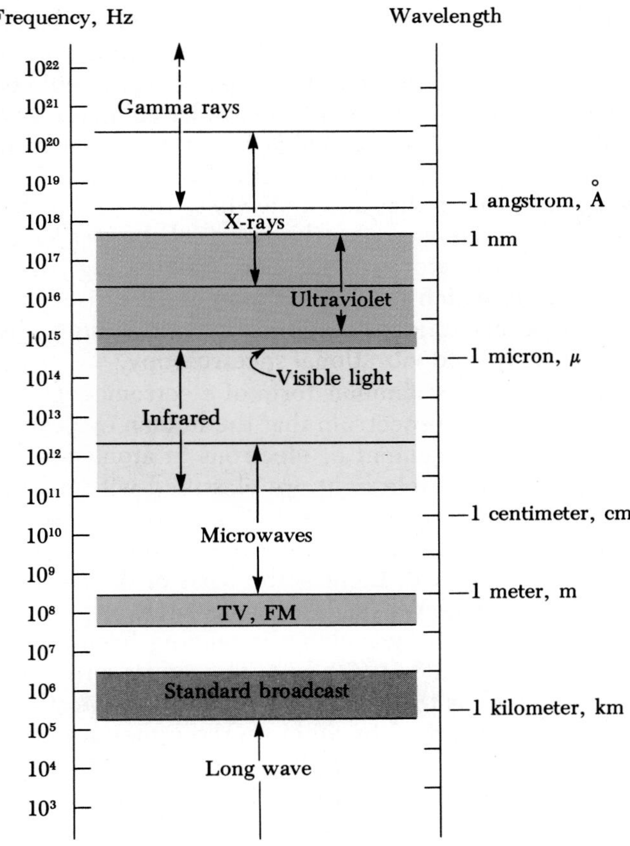

Figure 34.17 The electromagnetic spectrum. Note the overlap between one type of wave and the next.

$$\lambda = \frac{c}{f} = \frac{3 \times 10^8 \text{ m/s}}{5 \times 10^6 \text{ s}^{-1}} = 60 \text{ m}$$

The following abbreviations are often used to designate short wavelengths and distances:

$$1 \text{ micrometer } (\mu\text{m}) = 10^{-6} \text{ m}$$
$$1 \text{ nanometer (nm)} = 10^{-9} \text{ m}$$
$$1 \text{ angstrom (Å)} = 10^{-10} \text{ m}$$

For example, the wavelengths of visible light range from 0.4 to 0.7 μm, or 400 to 700 nm, or 4000 to 7000 Å.

Let us give a brief description of these various waves in order of decreasing wavelength. There is no sharp dividing point between one kind of wave and the next. It should be noted that all forms of radiation are produced by accelerating charges.

Radio waves

Radio waves, which were discussed in the previous section, are the result of charges accelerating through conducting wires. They are generated by such electronic devices as *LC* oscillators and are used in radio and television communication systems.

Microwaves

Microwaves (short-wavelength radio waves) have wavelengths ranging between about 1 mm and 30 cm and are also generated by electronic devices. Because of their short wavelength, they are well suited for the radar systems used in aircraft navigation and for studying the atomic and molecular properties of matter. Microwave ovens represent an interesting domestic application of these waves. It has been suggested that solar energy could be harnessed by beaming microwaves down to earth from a solar collector in space.[10]

Infrared waves

Infrared waves (sometimes called *heat waves*) have wavelengths ranging from about 1 mm to the longest wavelength of visible light, 7×10^{-7} m. These waves, produced by hot bodies and molecules, are readily absorbed by most materials. The infrared energy absorbed by a substance appears as heat since the energy agitates the atoms of the body, increasing their vibrational or translational motion, which results in a temperature rise. Infrared radiation has many practical and scientific applications, including physical therapy, infrared photography, and vibrational spectroscopy.

Visible waves

Visible light, the most familiar form of electromagnetic waves, may be defined as that part of the spectrum that the human eye can detect. Light is produced by the rearrangement of electrons in atoms and molecules. The various wavelengths of visible light are classified with colors ranging from violet ($\lambda \approx 4 \times 10^{-7}$ m) to red ($\lambda \approx 7 \times 10^{-7}$ m). The eye's sensitivity is a function of wavelength, the sensitivity being a maximum at a wavelength of about 5.6×10^{-7} m (yellow-green). Light is the basis of the science of optics and optical instruments, which we shall deal with later.

Ultraviolet waves

Ultraviolet light covers wavelengths ranging from about 3.8×10^{-7} m (380 nm) down to 6×10^{-8} m (60 nm). The sun is an important source of ultraviolet light, which is the main cause of suntans. Most of the ultraviolet light from the sun is absorbed by atoms in the upper atmosphere, or stratosphere. This is fortunate since uv light in large quantities produces harmful effects on humans. One important constituent of the stratosphere is ozone (O_3), which results from reactions of oxygen with ultraviolet radiation. This

[10] P. Glaser, "Solar Power from Satellites," *Physics Today*, February, 1977, p. 30.

ozone shield converts lethal high-energy ultraviolet radiation into heat, which in turn warms the stratosphere. Recently, there has been a great deal of controversy concerning the possible depletion of the protective ozone layer as a result of the use of the freons used in aerosal spray cans and as refrigerants.

X-rays

X-rays are electromagnetic waves with wavelengths in the range of about 10^{-8} m (10 nm) down to 10^{-13} m (10^{-4} nm). The most common source of x-rays is the deceleration of high-energy electrons bombarding a metal target. X-rays are used as a diagnostic tool in medicine and as a treatment for certain forms of cancer. Since x-rays damage or destroy living tissues and organisms, care must be taken to avoid unnecessary exposure or overexposure. X-rays are also used in the study of crystal structure, since x-ray wavelengths are comparable to the atomic separation distances (≈ 0.1 nm) in solids.

Gamma rays

Gamma rays are electromagnetic waves emitted by radioactive nuclei (such as ^{60}Co and ^{137}Cs) and during certain nuclear reactions. They have wavelengths ranging from about 10^{-10} m to less than 10^{-14} m. They are highly penetrating and produce serious damage when absorbed by living tissues. Consequently, those working near such dangerous radiation must be protected with heavily absorbing materials, such as thick layers of lead.

SUMMARY

Electromagnetic waves, which are predicted by Maxwell's equations, have the following properties:

1. The electric and magnetic fields satisfy the following wave equations, which can be obtained from Maxwell's third and fourth equations:

Wave equations

$$\frac{\partial^2 E}{\partial x^2} = \mu_0 \epsilon_0 \frac{\partial^2 E}{\partial t^2} \qquad (34.10)$$

$$\frac{\partial^2 B}{\partial x^2} = \mu_0 \epsilon_0 \frac{\partial^2 B}{\partial t^2} \qquad (34.11)$$

2. Electromagnetic waves travel through a vacuum with the speed of light c, where

The speed of electromagnetic waves

$$c = \frac{1}{\sqrt{\mu_0 \epsilon_0}} = 3.00 \times 10^8 \text{ m/s} \qquad (34.12)$$

3. The electric and magnetic fields of an electromagnetic wave are perpendicular to each other and perpendicular to the direction of wave propagation. (Hence, they are transverse waves.)
4. The instantaneous magnitudes of $|\boldsymbol{E}|$ and $|\boldsymbol{B}|$ in an electromagnetic wave are related by the expression

$$\frac{E}{B} = c \qquad (34.16)$$

5. Electromagnetic waves carry energy. The rate of flow of energy crossing a unit area is described by the Poynting vector **S**, where

Poynting vector

$$\boldsymbol{S} \equiv \frac{1}{\mu_0} \boldsymbol{E} \times \boldsymbol{B} \qquad (34.23)$$

6. Electromagnetic waves carry momentum and hence can exert pressure on surfaces. If an electromagnetic wave whose Poynting vector is **S** is completely absorbed by a surface upon which it is normally incident, the radiation pressure on that surface is

$$P = \frac{S}{c} \quad \text{(complete absorption)} \tag{34.32}$$

If the surface totally reflects a normally incident wave, the pressure is doubled.

The electric and magnetic fields of a sinusoidal plane electromagnetic wave propagating in the positive x direction can be written

Sinusoidal electric and magnetic fields

$$E = E_m \cos(kx - \omega t) \tag{34.14}$$

$$B = B_m \cos(kx - \omega t) \tag{34.15}$$

where ω is the angular frequency of the wave and k is the wave number. These equations represent special solutions to the wave equations for $\mathbf{E}$ and $\mathbf{B}$. Since $\omega = 2\omega f$ and $k = 2\pi/\lambda$, where f and λ are the frequency and wavelength, respectively, one finds that

$$\frac{\omega}{k} = \lambda f = c$$

The average value of the Poynting vector for a plane electromagnetic wave has a magnitude given by

Wave intensity

$$S_{av} = \frac{E_m B_m}{2\mu_0} = \frac{E_m^{\ 2}}{2\mu_0 c} = \frac{c}{2\mu_0} B_m^{\ 2} \tag{34.26}$$

The intensity of a sinusoidal plane electromagnetic wave equals the average value of the Poynting vector taken over one or more cycles.

The fundamental sources of electromagnetic waves are *accelerating electric charges.* For instance, radio waves emitted by an antenna arise from the continuous oscillation (and hence acceleration) of charges within the antenna structure.

The electromagnetic spectrum includes waves covering a broad range of frequencies and wavelengths. The frequency f and wavelength λ of a given wave are related by

$$c = f\lambda \tag{34.39}$$

QUESTIONS

1. For a given incident energy of electromagnetic wave, why is the radiation pressure on a perfect reflecting surface twice as large as the pressure on a perfect absorbing surface?
2. In your own words, describe the physical significance of the Poynting vector.
3. Do all current-carrying conductors emit electromagnetic waves? Explain.
4. What is the fundamental source of electromagnetic radiation?
5. Electrical engineers often speak of the *radiation resistance* of an antenna. What do you suppose they mean by this phrase?
6. If a high-frequency current is passed through a solenoid containing a metallic core, the core will heat up by induction. This process also cooks foods in micro-

wave ovens. Explain why the materials heat up in these situations.

7. Certain orientations of the receiving antenna on a TV give better reception than others. Furthermore, the best orientation varies from station to station. Explain these observations.
8. Does a wire connected to a battery emit an electromagnetic wave? Explain.
9. If you charge a comb by running it through your hair and then hold the comb next to a bar magnet, do the electric and magnetic fields produced constitute an electromagnetic wave?
10. An empty plastic or glass dish removed from a microwave oven is cool to the touch. How can this be possible?
11. Often when you touch the indoor antenna on a television receiver, the reception instantly improves. Why?
12. Explain how the (dipole) VHF antenna of a television set works.
13. Explain how the (loop) UHF antenna of a television set works.
14. Explain why the voltage induced in a UHF antenna depends on the frequency of the signal, while the voltage in a VHF antenna does not.
15. List as many similarities and differences as you can between sound waves and light waves.
16. What does a radio wave do to the charges in the receiving antenna to provide a signal for your car radio?
17. What is meant by the terms "amplitude modulation" and "frequency modulation?"
18. What determines the height of an AM radio station's broadcast antenna?
19. Some radio transmitters use a "phased array" of antennas. What is their purpose?
20. What happens to the radio reception in an airplane as it flies over the (vertical) dipole antenna of the control tower?
21. When light (or other electromagnetic radiation) travels across a given region, what is it that moves?
22. Why should an infrared photograph of a person look different from a photograph taken with visible light?

PROBLEMS

Section 34.2 Plane Electromagnetic Waves

1. Verify that Equation 34.12 gives c with dimensions of length per unit time.
2. An electromagnetic wave in vacuum has an electric field amplitude of 220 V/m. Calculate the amplitude of the corresponding magnetic field.
3. (a) Use the relationship $B = \mu_0 H$ described in Equation 30.37 (for free space where $\kappa_m = \mu_0$) together with the properties of E and B described in Section 34.2 to show that $E/H = \sqrt{\mu_0/\epsilon_0}$. (b) Calculate the numerical value of this ratio and show that it has SI units of ohms. (The ratio E/H is referred to as the *impedance of free space.*)
4. Calculate the maximum value of the magnetic field in a region where the measured maximum value of the electric field is 5.3 mV/m.
5. The magnetic field amplitude of an electromagnetic wave is 5.4×10^{-7} T. Calculate the electric field amplitude if the wave is traveling (a) in free space and (b) in a medium in which the speed of the wave is $0.8c$.
6. The speed of an electromagnetic wave traveling in a transparent substance is given by $v = 1/\sqrt{\kappa\mu_0\epsilon_0}$ where κ is the dielectric constant of the substance. Determine the speed of light in water, which has a dielectric constant of 1.78 at optical frequencies.
7. Verify that the following pair of equations for E and B are solutions of Equations 34.7 and 34.8:

$$E = \frac{A}{\sqrt{\epsilon_0\mu_0}} e^{a(x-ct)} \quad \text{and} \quad B = Ae^{a(x-ct)}$$

8. A plane electromagnetic wave has a frequency of 6.4×10^{14} Hz and an electric field amplitude of 320 N/C. Write equations in the form $E(x, t) = E_m \cos(kx - \omega t)$ and $B(x, t) = L_m \cos(kx - \omega t)$ that describe the electric and magnetic fields in this wave. Give numerical values for E_m, B_m, k, and ω.
9. Figure 34.3 shows a plane electromagnetic sinusoidal wave propagating in the x direction. The wavelength is 50 m, and the electric field vibrates in the xy plane with an amplitude of 22 V/m. Calculate (a) the sinusoidal frequency and (b) the magnitude and direction of $\mathbf{B}$ when the electric field has its maximum value in the negative y direction. (c) Write an expression for $\mathbf{B}$ in the form

$$B = B_m \cos(kx - \omega t)$$

with numerical values for B_m, k, and ω.

10. Verify that the following equations are solutions to Equations 34.10 and 34.11, respectively:

$$E = E_m \cos(kx - \omega t)$$

$$B = B_m \cos(kx - \omega t).$$

11. Verify that the following pair of equations satisfies Equations 34.7 and 34.8:

$$E = E_m \sin(kx - \omega t)$$

$$B = B_m \sin(kx - \omega t).$$

Section 34.3 Energy Carried by Electromagnetic Waves

12. Compute the average energy content of a liter of sunlight as it reaches the top of the earth's atmosphere, where its intensity is 1340 W/m^2.
13. At what distance from 100-W isotropic electromagnetic wave power source will $E_m = 15$ V/m?
14. What power must be radiated (isotropically) by a source if the amplitude of the electric field is 55 V/m at a distance of 20 m?
15. What is the average magnitude of the Poynting vector at a distance of 5 miles from an isotropic radio transmitter, broadcasting with an average power of 250 kW?
16. The sun radiates electromagnetic energy at the rate of $P_{sun} = 3.85 \times 10^{26}$ W. (a) At what distance from the sun does the intensity of its radiation fall to 1000 W/m^2? (Compare this distance to the radius of the earth's orbit.) (b) At the distance you just found, what is the average energy density of the sun's radiation?
17. A radio transmitter broadcasts isotropically with an average power of 15 kW. Compute the maximum value of the electric field in its radio wave at the following distances from the transmitter: (a) 1 km; (b) 10 km; (c) 100 km.
18. The amplitude of the magnetic field in an electromagnetic wave is $B_m = 4.1 \times 10^{-8}$ T. What is the average intensity of this wave?
19. The filament of an incandescent lamp has a 150-Ω resistance, and carries a dc current of 1 A. The filament is 8 cm long and 0.9 mm in radius. (a) Calculate the Poynting vector at the surface of the filament. (b) Find the magnitude of the electric and magnetic fields at the surface of the filament.
20. An incandescent lamp is radiating isotropically (i.e., identically in all directions) at 100 W. Calculate the root-mean-square amplitudes of the electric and magnetic fields at distances of (a) 2 m and (b) 8 m from the source.
21. A helium-neon laser intended for instructional use operates at a typical power of 5.0 mW. (a) Determine the maximum value of the electric field at a point where the cross section of the beam is 4 mm^2. (b) Calculate the electromagnetic energy in a 1-m length of the beam.
22. At a particular point, the electric field in an electromagnetic wave has the instantaneous value $E =$ 240 V/m. Compute the instantaneous magnitudes of the following quantities at the same point and time: (a) the magnetic field; (b) the Poynting vector; (c) the energy density of the electric field; (d) the energy density of the magnetic field.
23. At one location on the earth, the rms value of the magnetic field due to solar radiation is 1.8 μT. From this value calculate (a) the average electric field due to solar radiation, (b) the average energy density of the solar component of electromagnetic radiation at this location, and (c) the magnitude of the Poynting vector for the sun's radiation. (d) Compare the value found in (c) to the value of the solar flux given in Example 34.3.
24. A long wire has a radius of 1 mm and a resistance per unit length of 2 Ω/m. Determine the current required if the Poynting vector at the surface of the wire equals 2.68×10^3 W/m^2.

Section 34.4 Momentum and Radiation Pressure

25. A radio wave transmits 25 W/m^2 of power per unit area. A plane surface of area A is perpendicular to the direction of propagation of the wave. Calculate the radiation pressure on the surface if the surface is a perfect absorber.
26. Let the planar surface in Problem 25 have dimensions of 2.4 m × 0.7 m. Calculate the momentum delivered to the surface per second.
27. Determine the momentum per unit volume in a low-power helium-neon laser (2 mW) if the beam diameter is 2.5 mm.
28. Sunlight reaches the earth's surface with an average intensity of 650 W/m^2. Assuming normal incidence, what force does sunlight exert on a perfectly reflecting 40 cm × 80 cm mirror?
29. A plane electromagnetic wave has an energy flux of 750 W/m^2. A flat, rectangular surface of dimensions 50 cm × 100 cm is placed perpendicular to the direction of the plane wave. If the surface absorbs half of the energy and reflects half (that is, it is a 50% reflecting surface), calculate (a) the total energy absorbed by the surface in a time of 1 min and (b) the momentum absorbed in this time.
30. A disk 1.0 cm in diameter is located 2.5 m from a 100-W light bulb. The surface of the disk is a perfect reflector, and the normal to the plane of the disk makes an angle of 37° with the outward radial direction from the bulb. Calculate the radiation force on the disk.

Section 34.5 Radiation from an Infinite Current Sheet

31. A rectangular surface of dimensions 120 cm × 40 cm is parallel to and 4.4 m from a very large conducting sheet in which there is a sinusoidally varying surface current which has a maximum value of 10 A/m. (a) Calculate the average power incident on the smaller sheet. (b) What power per unit area is radiated by the current-carrying sheet?
32. A large current-carrying sheet is expected to radiate in each direction (normal to the plane of the sheet) at a rate equal to 570 W/m^2 (approximately one half of the solar constant). What maximum value of sinusoidal current density is required?

Section 34.6 The Production of Electromagnetic Waves by an Antenna

33. What is the length of a half-wave antenna designed to broadcast 20 MHz radio waves?
34. An AM radio station broadcasts isotropically with an average power of 4 kW. A dipole receiving antenna, 65 cm long, is located 4 miles from the transmitter. Compute the emf induced by this signal between the ends of the receiving antenna.
35. A television set uses a dipole receiving antenna for VHF channels and a loop antenna for UHF channels. The UHF antenna produces a voltage from the changing *magnetic* flux through the loop. (a) Using Faraday's Law, derive an expression for the amplitude of the voltage that appears in a single-turn circular loop antenna with a radius r. The TV station broadcasts a signal with a frequency f, and the signal has an electric field amplitude E_m and magnetic field amplitude B_m at the receiving antenna's location. (b) If the electric field in the signal points vertically, what should be the orientation of the loop for best reception?
36. Figure 34.14 shows a Hertz antenna (also known as a half-wave antenna, since its length is $\lambda/2$). The Hertz antenna is located far enough from the ground that reflections do not significantly affect its radiation pattern. Most AM radio stations, however, use a Marconi antenna, which consists of the top half of a Hertz antenna. The lower end of this (quarter-wave) antenna is connected to earth ground, and the ground itself serves as the missing lower half. (a) What is the antenna height for a radio station broadcasting at the lower end of the AM band, at 560 kHz? (b) What is the height for a station broadcasting at 1600 kHz?
37. Two hand-held radio transceivers with dipole antennas are separated by a large fixed distance. Assuming a vertical transmitting antenna, what fraction of the maximum received power will occur in the receiving antenna when it is inclined from the vertical by (a) 15°; (b) 45°; (c) 90°?
38. Two radio-transmitting antennas are separated by half the broadcast wavelength and are driven in phase with each other. (a) In which direction is the strongest signal radiated? (b) In which direction is the weakest signal radiated?

Section 34.7 The Spectrum of Electromagnetic Waves

39. What is the wavelength of an electromagnetic wave in free space that has a frequency of (a) 5×10^{19} Hz and (b) 4×10^{9} Hz?
40. The eye is most sensitive to light having a wavelength of 5.5×10^{-7} m, which is in the green-yellow region of the electromagnetic spectrum. What is the frequency of this light?
41. Compute the frequency of the following electromagnetic waves: (a) microwaves ($\lambda = 1$ cm); (b) infrared radiation ($\lambda = 1\ \mu$m); (c) yellow light ($\lambda = 580$ nm); (d) ultraviolet light ($\lambda = 100$ nm); and (e) x-rays ($\lambda = 1$ pm).
42. What are the wavelength ranges in (a) the AM radio band (540–1600 kHz), and (b) the FM radio band (88–108 MHz)?
43. There are 12 VHF channels (Channels 2–13) that lie in a frequency range from 54 MHz to 216 MHz. Each channel has a width of 6 MHz, with the two ranges 72–76 MHz and 88–174 MHz reserved for non-TV purposes. (Channel 2, for example, lies between 54 and 60 MHz.) Calculate the *wavelength* range for (a) Channel 4; (b) Channel 6; (c) Channel 8.

ADDITIONAL PROBLEMS

44. Assume that the solar radiation incident on the earth is 1340 W/m^2. (This is the value of the solar flux above the earth's atmosphere.) (a) Calculate the total power radiated by the sun, taking the average earth-sun separation to be 1.49×10^{11} m. (b) Determine the maximum values of the electric and magnetic fields at the earth's surface due to solar radiation.
45. A community plans to build a facility to convert solar radiation into electrical power. They require 1 MW of power (10^6 W), and the system to be installed has an efficiency of 30% (that is, 30% of the solar energy incident on the surface is converted to electrical energy). What must be the effective area of a perfectly absorbing surface used in such an installation, assuming a constant energy flux of 1000 W/m^2?
46. A microwave source produces pulses of 20-GHz radiation, with each pulse lasting 1.0 ns. A parabolic reflector ($R = 6$ cm) is used to focus these into a parallel beam of radiation, as shown in Figure 34.18. The average power during each pulse is 25 kW. (a) What is the wavelength of these microwaves? (b) What is the total energy contained in each pulse? (c) Compute the average energy density inside each pulse. (d) Determine the amplitude of the electric and magnetic fields in these microwaves. (e) If this pulsed beam strikes an absorbing surface, compute the force exerted on the surface during the 1-ns duration of each pulse.

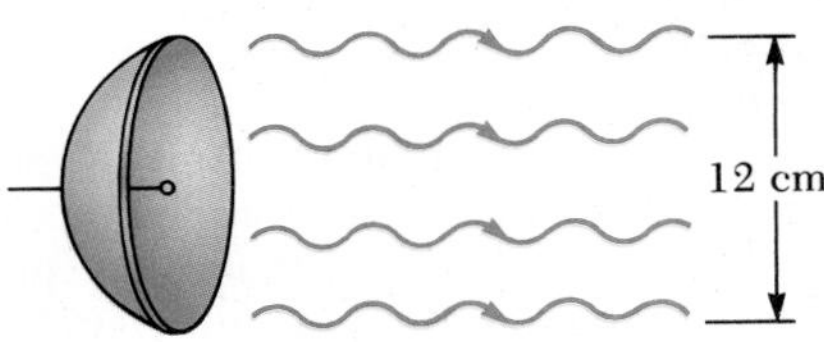

Figure 34.18 (Problem 46).

47. A dish antenna with a diameter of 20 m receives (at normal incidence) a radio signal from a distant source, as shown in Figure 34.19. The radio signal is a continuous sinusoidal wave with amplitude $E_0 = 0.2\ \mu\text{V/m}$. Assume the antenna absorbs all the radiation that falls on the dish. (a) What is the amplitude of the magnetic field in this wave? (b) What is the intensity of the radiation received by this antenna? (c) What is the power received by the antenna? (d) What force is exerted on the antenna by the radio waves?

Figure 34.19 (Problem 47).

48. Show that the instantaneous value of the Poynting vector has a magnitude given by

$$S = \frac{c}{2}(\epsilon_0 E^2 + \mu_0 H^2)$$

49. A thin tungsten filament of length 1 m radiates 60 W of energy in the form of electromagnetic waves. A perfectly absorbing surface in the form of a hollow cylinder of radius 5 cm and length 1 m is placed concentric with the filament. Calculate the radiation pressure acting on the cylinder. (Assume that the radiation is emitted in the radial direction, and neglect end effects.)

50. A group of astronauts plan to propel a spaceship by using a "sail" to reflect solar radiation. The sail is totally reflecting, oriented with its plane perpendicular to the direction to the sun, and 1 km × 1.5 km in size. What is the maximum acceleration that can be expected for a spaceship of 4 metric tons (4000 kg)? (Use the solar radiation data from Problem 44 and neglect gravitational forces.)

51. A microwave transmitter emits monochromatic electromagnetic waves. The maximum electric field at a distance 1 km from the transmitter is 6.0 V/m. Assuming the transmitter is a point source and neglecting waves reflected from the earth, calculate (a) the maximum magnetic field at this distance and (b) the total power emitted by the transmitter.

52. The magnetic field of a linearly-polarized electromagnetic wave is described by the equation

$$B = (1.5 \times 10^{-6}\ T)\sin\left[2\pi\left(\frac{x}{20} - \frac{t \times 10^8}{6.7}\right)\right]$$

where x is in m and t in s. (a) Calculate the velocity of the wave and (b) write the equation for the associated electric field.

53. An astronaut in a spacecraft moving with constant velocity wishes to increase the speed of the craft by using a laser beam attached to the spaceship. The laser beam emits 100 J of electromagnetic energy per pulse, and the laser is pulsed at the rate of 0.2 pulse/s. If the mass of the spaceship plus its contents is 5000 kg, for how long a time must the beam be on in order to increase the speed of the vehicle by 1 m/s in the direction of its initial motion? In what direction should the beam be pointed to achieve this?

54. Consider a small, spherical particle of radius r located in space a distance R from the sun. (a) Show that the ratio $F_{\text{rad}}/F_{\text{grav}} \propto 1/r$, where F_{rad} = the force due to solar radiation and F_{grav} = the force of gravitational attraction. (b) The result of (a) means that for a sufficiently small value of r the force exerted on the particle due to solar radiation will exceed the force of gravitational attraction. Calculate the value of r for which the particle will be in equilibrium under the two forces. (Assume that the particle has a perfectly absorbing surface and a mass density of 1.5 g/cm^3. Let the particle be located 3.75×10^{11} m from the sun and use 214 W/m^2 as the value of the solar flux at that point.)

55. The torsion balance shown in Figure 34.8 is used in an experiment to measure radiation pressure. The torque constant (elastic restoring torque) of the suspension fiber is 1×10^{-11} N·m/deg, and the length of the horizontal rod is 6 cm. The beam from a 3-mW helium-neon laser is incident on the black disk, and the mirror disk is completely shielded. Calculate the angle between the *equilibrium* positions of the horizontal bar when the beam is switched from "off" to "on."

56. Monoenergetic x-rays move through a material with a speed of $0.95c$. The photon flux on a surface perpendicular to the x-ray beam is 10^{13} photons/m^2·s.

(a) Calculate the density of photons (number per unit volume) in the material. (b) If each photon has an energy of 8.88 keV, what is the energy density within the material?

57. A linearly polarized microwave of wavelength 1.5 cm is directed along the positive x axis. The electric field vector has a maximum value of 175 V/m and vibrates in the xy plane. (a) Assume that the magnetic field component of the wave can be written in the form $B = B_0 \sin(kx - \omega t)$ and give values for B_0, k, and ω. Also, determine in which plane the magnetic field vector vibrates. (b) Calculate the Poynting vector for this wave. (c) What radiation pressure would this wave exert if directed at normal incidence onto a perfectly reflecting sheet? (d) What acceleration would be imparted to a 500-g sheet (perfectly reflecting and at normal incidence) with dimensions 1 m $\times$ 0.75 m?

58. A police radar unit transmits at a frequency of 10.525 GHz, better known as the "X-band." (a) Show that when this radar wave is reflected from the front of a moving car, the reflected wave is shifted in frequency by the amount $\Delta f \cong 2fv/c$, where v is the speed of the car. (*Hint:* Treat the reflection of a wave as two separate processes: First, the motion of the car produces a "moving-observer" Doppler shift in the frequency of the waves striking the car, and then these Doppler-shifted waves are re-emitted by a moving source. Also, automobile speeds are low enough that you can use the binomial expansion $(1 + x)^n \cong 1 + nx$ in the acoustic Doppler formulas derived in Chapter 17.) (b) The unit is usually calibrated before and after an arrest for speeds of 35 mph and 80 mph. Calculate the frequency shift produced by these two speeds.

White light directed through a triangular glass prism creats a "Phaenoma of Colours" on an old manuscript. (© Eric Lessing, Magnum Photos)

PART V
Light and Optics

Scientists have long been intrigued by the nature of light, and philosophers have had endless arguments concerning the proper definition and perception of light. It is important to understand the nature of light because it is one of the basic ingredients of life on earth. Plants convert light energy from the sun to chemical energy through photosynthesis. Light is the principle means by which we are able to transmit and receive information from objects around us and throughout the universe.

The nature and properties of light have been a subject of great interest and speculation since ancient times. The Greeks believed that light consisted of tiny particles (corpuscles) that were emitted by a light source and then stimulated the perception of vision upon striking the observer's eye. Newton used this corpuscular theory to explain the reflection and refraction of light. In 1670, one of Newton's contemporaries, the Dutch scientist Christian Huygens, was able to explain many properties of light by proposing that light was wave-like in character. In 1803, Thomas Young showed that light beams can interfere with one another, giving strong support to the wave theory. In 1865, Maxwell developed a brilliant theory that electromagnetic waves travel with the speed of light (Chapter 34). By this time, the wave theory of light seemed to be on firm ground.

However, at the beginning of the 20th century, Max Planck returned to the corpuscular theory of light in order to explain the radiation emitted by hot objects. Albert Einstein used the same concept to explain the electrons emitted by a metal exposed to light (the photoelectric effect). We shall discuss these and other topics in the last part of this text, which is concerned with modern physics.

Today, scientists view light as having a dual nature. Sometimes light behaves as if it were particle-like, and sometimes it behaves as if it were wave-like.

In this part of the book, we shall concentrate on those aspects of light that are best understood through the wave model. First, we shall discuss the reflection of light at the boundary between two media and the refraction (bending) of light as it travels from one medium into another. We shall use these ideas to study the refraction of light as it passes through lenses and the reflection of light from surfaces. Then we shall describe how lenses and mirrors can be used to view objects with such instruments as cameras, telescopes, and microscopes. Finally, we shall study the phenomena of diffraction, polarization, and interference as they apply to light.

"I procured me a Triangular glass-Prisme to try therewith the celebrated Pheaenomena of Colours. . . . *I placed my Prisme at his entrance (the sunlight), that it might thereby be refracted to the opposite wall. It was a very pleasing divertisement to view the vivid and intense colours produced thereby; . . . I have often with admiration beheld that all the colours of the Prisme being made to converge, and thereby to be again mixed, as they were in the light before it was incident upon the Prisme, reproduced light, entirely and perfectly white, and not at all sensibly differing from the direct light of the sun. . . ."*

ISAAC NEWTON

35
The Nature of Light and the Laws of Geometric Optics

Rainbow over the Potala Palace, Lhasa, Tibet. (© Galen Rowell/Mountain Light)

35.1 THE NATURE OF LIGHT

Before the beginning of the 19th century, light was considered to be a stream of particles that were emitted by a light source and stimulated the sense of sight upon entering the eye. The chief architect of this particle theory of light was, once again, Isaac Newton.[1] With this theory, Newton was able to provide a simple explanation of some known experimental facts concerning the nature of light, namely, the laws of reflection and refraction.

Most scientists accepted Newton's particle theory of light. However, during Newton's lifetime another theory was proposed — one that argued that light might be some sort of wave motion. In 1678, a Dutch physicist and astronomer, Christian Huygens (1629–1695), showed that a wave theory of light could explain the laws of reflection and refraction. The wave theory did not receive immediate acceptance for several reasons. All the waves known at the time (sound, water, etc.) traveled through some sort of medium. On the other hand, light could travel to us from the sun through the vacuum of space. Furthermore, it was argued that if light were some form of wave motion, the

[1] Isaac Newton, *Opticks*, 1704. [The fourth edition (1730) was reprinted by Dover Publications, New York, 1952.]

waves would be able to bend around obstacles; hence, we should be able to see around corners. It is now known that light does indeed bend around the edges of objects. This phenomenon, known as *diffraction,* is not easy to observe because light waves have such short wavelengths. Thus, although experimental evidence for the diffraction of light was discovered by Francesco Grimaldi (1618–1663) around 1660, most scientists rejected the wave theory and adhered to Newton's particle theory for more than a century. This was, for the most part, due to Newton's great reputation as a scientist.

The first clear demonstration of the wave nature of light was provided in 1801 by Thomas Young (1773–1829), who showed that, under appropriate conditions, light exhibits interference behavior. That is, at certain points in the vicinity of two sources, light waves can combine and cancel each other by destructive interference. Such behavior could not be explained at that time by a particle theory because there was no conceivable way by which two or more particles could come together and cancel one another. Several years later, a French physicist, Augustin Fresnel (1788–1829), performed a number of detailed experiments dealing with interference and diffraction phenomena. In 1850, Jean Foucault (1791–1868) provided further evidence of the inadequacy of the particle theory by showing that the speed of light in liquids is less than in air. According to the particle model of light, the speed of light would be higher in glasses and liquids than in air. Additional developments during the 19th century led to the general acceptance of the wave theory of light.

The most important development concerning the theory of light was the work of Maxwell, who in 1873 asserted that light was a form of high-frequency electromagnetic wave (Chapter 34). His theory predicted that these waves should have a speed of about 3×10^8 m/s. Within experimental error, this value is equal to the speed of light. As discussed in Chapter 34, Hertz provided experimental confirmation of Maxwell's theory in 1887 by producing and detecting electromagnetic waves. Furthermore, Hertz and other investigators showed that these *waves exhibited reflection, refraction, and all the other characteristic properties of waves.*

Although the classical theory of electricity and magnetism was able to explain most known properties of light, some subsequent experiments could not be explained by assuming that light was a wave. The most striking of these is the *photoelectric effect,* also discovered by Hertz. The photoelectric effect is the ejection of electrons from a metal whose surface is exposed to light. As one example of the difficulties that arose, experiments showed that the kinetic energy of an ejected electron is *independent* of the light intensity. This was in contradiction of the wave theory, which held that a more intense beam of light should add more energy to the electron. An explanation of this phenomenon was proposed by Einstein in 1905. Einstein's theory used the concept of quantization developed by Max Planck (1858–1947) in 1900. The quantization model assumes that the energy of a light wave is present in bundles of energy called *photons;* hence the energy is said to be *quantized.* (Any quantity that appears in discrete bundles is said to be quantized. For example, electric charge is quantized because it always appears in bundles equal to the elementary charge of 1.6×10^{-19} C.) According to Einstein's theory, the energy of a photon is proportional to the frequency of the electromagnetic wave:

$$E = hf \tag{35.1}$$

Energy of a photon

where $h = 6.63 \times 10^{-34}$ J · s is **Planck's constant.** It is important to note that this theory retains some features of both the wave theory and the particle theory of light. As we shall discuss later, the photoelectric effect is the result of energy transfer from a single photon to an electron in the metal. That is, the electron interacts with one photon of light as if the electron had been struck by a particle. Yet this photon has wave-like characteristics because its energy is determined by the frequency (a wave-like quantity).

In view of these developments, one must regard light as having a *dual nature.* That is, *in some cases light acts like a wave and in others it acts like a particle.* Classical electromagnetic wave theory provides an adequate explanation of light propagation and of the effects of interference, whereas the photoelectric effect and other experiments involving the interaction of light with matter are best explained by assuming that light is a particle. Light is light, to be sure. However, the question, "Is light a wave or a particle?" is an inappropriate one. Sometimes it acts like a wave, sometimes like a particle. In the next few chapters, we shall investigate the wave nature of light.

35.2 MEASUREMENTS OF THE SPEED OF LIGHT

Light travels at such a high speed ($c \approx 3 \times 10^8$ m/s) that early attempts to measure its speed were unsuccessful. Galileo attempted to measure the speed of light by positioning two observers in towers separated by about 5 mi. Each observer carried a shuttered lantern. One observer would open his lantern first, and then the other would open his lantern at the moment he saw the light from the first lantern. The velocity could then be obtained, in principle, knowing the transit time of the light beams between lanterns. The results were inconclusive. Today, we realize (and as Galileo himself concluded) that it is impossible to measure the speed of light in this manner because the transit time of the light is very small compared with the reaction time of the observers.

We now describe two methods for determining the speed of light.

Roemer's Method

The first successful estimate of the speed of light was made in 1675 by the Danish astronomer Ole Roemer (1644–1710). His technique involved astronomical observations of one of the moons of Jupiter, called Io. At that time, 4 of Jupiter's 14 moons had been discovered, and the periods of their orbits were known. Io, the innermost moon, has a period of about 42.5 h. This was measured by observing the eclipse of Io as it passed behind Jupiter (Fig. 35.1). The period of Jupiter is about 12 years, so as the earth moves through 180° about the sun, Jupiter revolves through only 15°.

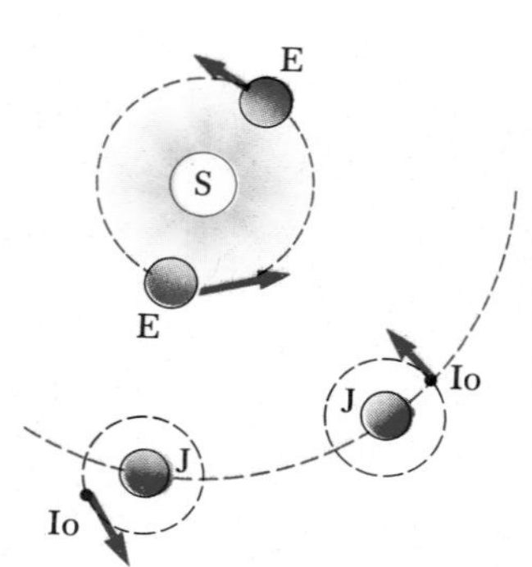

Figure 35.1 Roemer's method for measuring the speed of light.

Using the orbital motion of Io as a clock, one would expect a constant period in its orbit over long time intervals. However, Roemer observed a systematic variation in Io's period during a year's time. He found that the periods were larger than average when the earth receded from Jupiter and smaller than average when the earth was approaching Jupiter. For example, if Io had a constant period, Roemer should have been able to see an eclipse occurring at a particular instant and be able to predict when an eclipse should begin at a later time in the year. However, when Roemer checked to see if the second eclipse did occur at the predicted time, he found that if the earth was receding from Jupiter, the eclipse was late. In fact, if the interval between observations was three months, the delay was approximately 600 s. Roemer attributed this variation in period to the fact that the distance between the earth and Jupiter was changing between the observations. In three months (one quarter of the period of the earth), the light from Jupiter has to travel an additional distance equal to the radius of the earth's orbit.

Using Roemer's data, Huygens estimated the lower limit for the speed of light to be about 2.3×10^8 m/s. This experiment is important historically because it demonstrated that light does have a finite speed and gave an estimate of this speed.

Fizeau's Technique

The first successful method for measuring the speed of light using purely terrestrial techniques was developed in 1849 by Armand H. L. Fizeau (1819–1896). Figure 35.2 represents a simplified diagram of his apparatus.[2] The basic idea is to measure the total time it takes light to travel from some point to a distant mirror and back. If d is the distance between the light source and the mirror and if the transit time for one round trip is t, then the speed of light is $c = 2d/t$. To measure the transit time, Fizeau used a rotating toothed wheel, which converts an otherwise continuous beam of light into a series of light pulses. Additionally, the rotation of the wheel controls what an observer at the light source sees. For example, if the light passing the opening at point A in Figure 35.2 should return at the instant that tooth B had rotated into position to cover the return path, the light would not reach the observer. At a faster rate of rotation, the opening at point C could move into position to allow the reflected beam to pass and reach the observer. Knowing the distance d, the number of teeth in the wheel, and the angular velocity of the wheel, Fizeau arrived at a value of $c = 3.1 \times 10^8$ m/s. Similar measurements made by subsequent investigators yielded more precise values for c, approximately 2.9977×10^8 m/s.

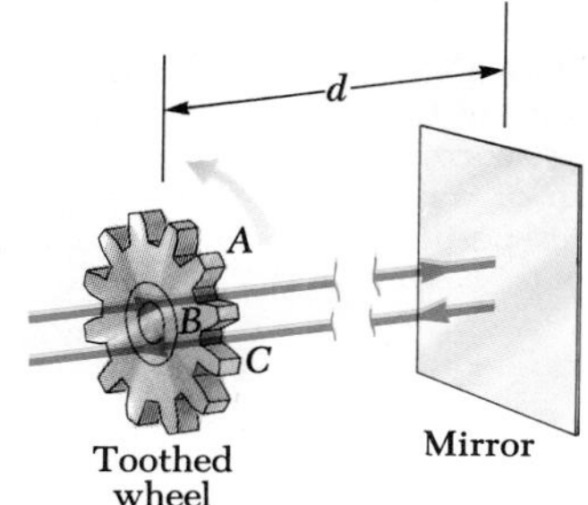

Figure 35.2 Fizeau's method for measuring the speed of light using a rotating toothed wheel.

[2] The actual apparatus involved several lenses and mirrors which we have omitted for the sake of clarity. For more details, see F. W. Sears, *Optics*, 3rd ed., Reading, Mass., Addison-Wesley, 1958, Chapter 1.

EXAMPLE 35.1 Measuring the Speed of Light with Fizeau's Toothed Wheel □

Assume the toothed wheel of the Fizeau experiment has 360 teeth and is rotating with a speed of 27.5 rev/s when the light from the source is extinguished, that is, when a burst of light passing through opening A in Figure 35.2 is blocked by tooth B on return. If the distance to the mirror is 7500 m, find the speed of light.

Solution If the wheel has 360 teeth, it will turn through an angle of 1/720 rev in the time that passes while the light makes its round trip. From the definition of angular velocity, we see that the time is

$$t = \frac{\theta}{\omega} = \frac{(1/720)\text{ rev}}{27.5\text{ rev/s}} = 5.05 \times 10^{-5}\text{ s}$$

Hence, the speed of light is

$$c = \frac{2d}{t} = \frac{2\,(7500\text{ m})}{5.05 \times 10^{-5}\text{ s}} = 2.97 \times 10^8\text{ m/s}$$

35.3 THE RAY APPROXIMATION IN GEOMETRIC OPTICS

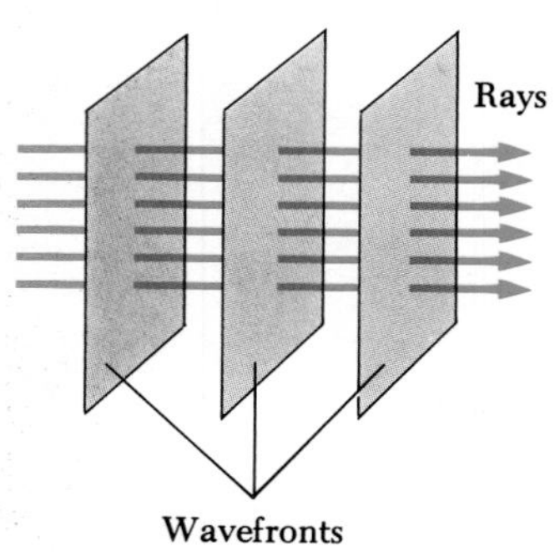

Figure 35.3 A plane wave propagating to the right. Note that the rays, corresponding to the direction of wave motion, are straight lines perpendicular to the wavefronts.

In studying geometric optics here and in Chapter 36, we shall use what is called the *ray approximation*. To understand this approximation, first recall that the direction of energy flow of a wave, corresponding to the direction of wave propagation, is called a ray. The rays of a given wave are straight lines that are perpendicular to the wavefronts, as illustrated in Figure 35.3 for a plane wave. In the ray approximation, we assume that a wave moving through a medium travels in a straight line in the direction of its rays. That is, a ray is a line drawn in the direction in which the light is traveling. For example, a beam of sunlight passing through a darkened room traces out the path of a ray.

If the wave meets a barrier with a circular opening whose diameter is large relative to the wavelength, as in Figure 35.4a, the wave emerging from the opening continues to move in a straight line (apart from some small edge effects); hence, the ray approximation continues to be valid. On the other hand, if the diameter of the opening of the barrier is of the order of the wavelength, as in Figure 35.4b, the waves spread out from the opening in all directions. We say that the outgoing wave is noticeably *diffracted*. Finally, if the opening is small relative to the wavelength, the opening can be approximated as a point source of waves (Fig. 35.4c). Thus, the effect of diffraction is more pronounced as the ratio d/λ approaches zero. Similar effects are seen when waves encounter an opaque circular object. In this case, when $\lambda \ll d$, the object casts a sharp shadow.

The ray approximation and the assumption that $\lambda \ll d$ will be used here and in Chapter 36, both dealing with geometric optics. This approximation is very good for the study of mirrors, lenses, prisms, and associated optical instruments, such as telescopes, cameras, and eyeglasses. We shall return to the subject of diffraction (where $\lambda \geq d$) in Chapter 38.

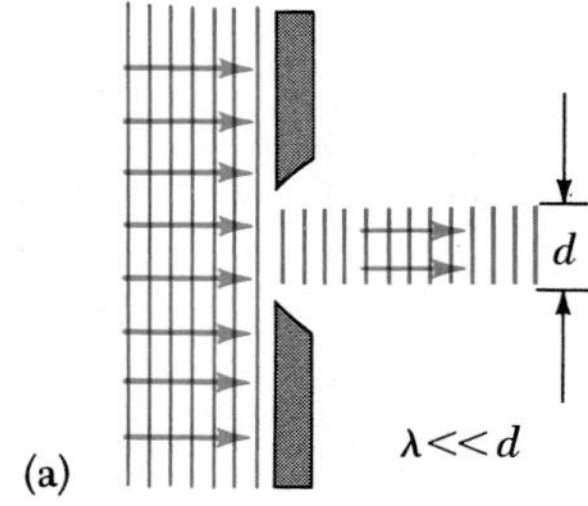

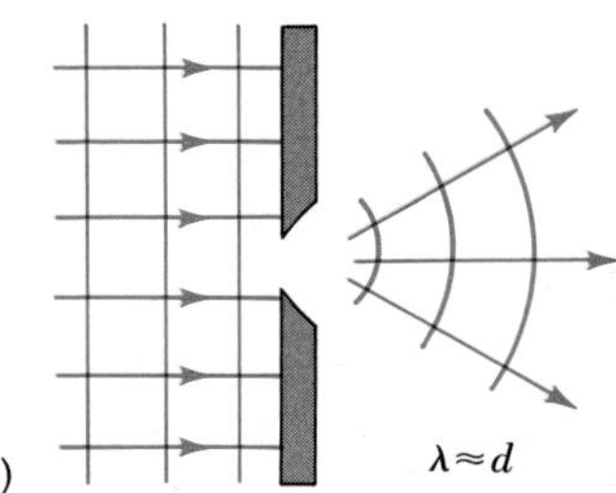

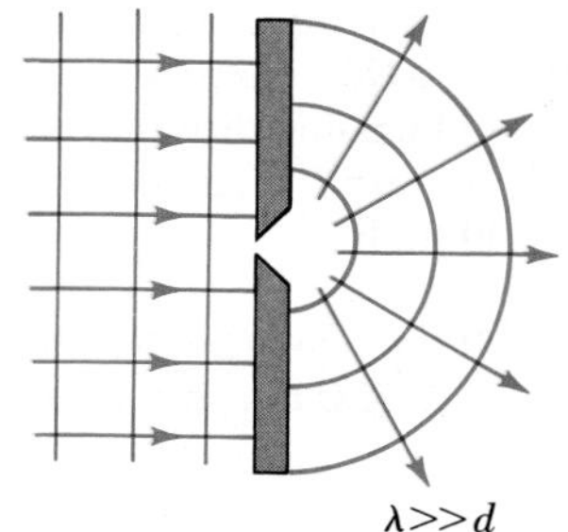

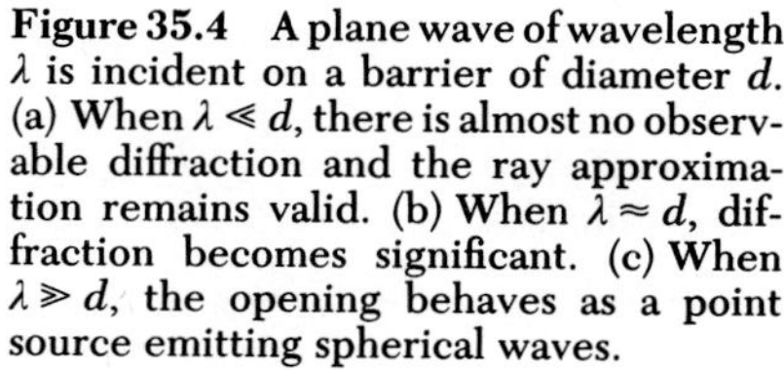

Figure 35.4 A plane wave of wavelength λ is incident on a barrier of diameter d. (a) When $\lambda \ll d$, there is almost no observable diffraction and the ray approximation remains valid. (b) When $\lambda \approx d$, diffraction becomes significant. (c) When $\lambda \gg d$, the opening behaves as a point source emitting spherical waves.

35.4 REFLECTION AND REFRACTION

Reflection of Light

When a light ray traveling in a medium encounters a boundary leading into a second medium, part of the incident ray is reflected back into the first medium. Figure 35.5a shows several rays of a beam of light incident on a smooth, mirror-like, reflecting surface. The reflected rays are parallel to each other, as indicated in the figure. Reflection of light from such a smooth surface is called **specular reflection.** On the other hand, if the reflecting surface is rough, as in Figure 35.5b, the surface will reflect the rays in various directions. Reflection from any rough surface is known as **diffuse reflection.** A surface will behave as a smooth surface as long as the surface variations are small compared with the wavelength of the incident light. Photographs of specular reflection and diffuse reflection using laser light are shown in Figures 35.5c and 35.5d.

For instance, consider the two types of reflection one can observe from a road's surface while driving a car at night. When the road is dry, light from oncoming vehicles is scattered off the road in different directions (diffuse reflection) and the road is quite visible. On a rainy night, when the road is wet, the road irregularities are filled with water. Because the water surface is quite smooth, the light undergoes specular reflection. In this book, we shall concern ourselves only with specular reflection, and we shall use the term *reflection* to mean specular reflection.

Consider a light ray traveling in air and incident at an angle on a flat, smooth surface, as in Figure 35.6. The incident and reflected rays make angles

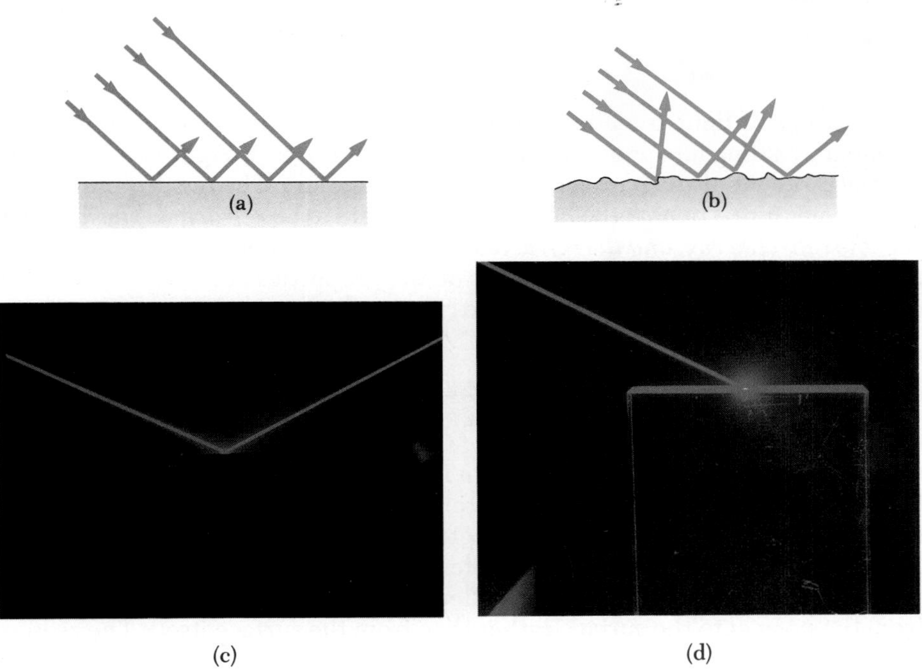

Figure 35.5 Schematic representation of (a) specular reflection, where the reflected rays are all parallel to each other, and (b) diffuse reflection, where the reflected rays travel in random directions. (c) and (d) Photographs of specular and diffuse reflection using laser light. (Photographs courtesy of Henry Leap and Jim Lehman)

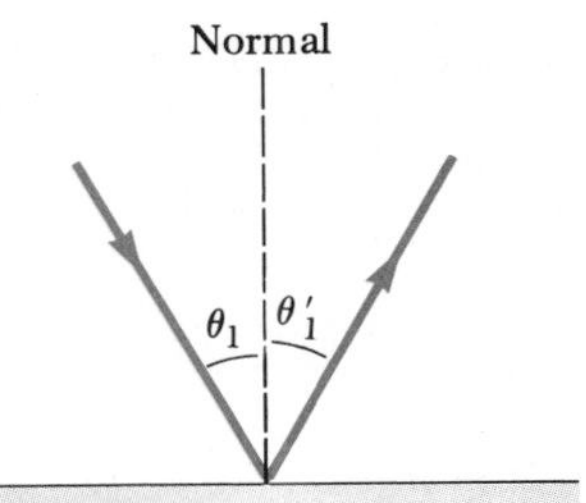

Figure 35.6 According to the law of reflection, $\theta_1 = \theta'_1$.

θ_1 and θ_1', respectively, with a line drawn perpendicular to the surface at the point where the incident ray strikes the surface. We shall call this line the *normal* to the surface. Experiments show that *the angle of reflection equals the angle of incidence,* that is,

Law of reflection

$$\theta_1' = \theta_1 \tag{35.2}$$

EXAMPLE 35.2 The Double-Reflected Light Ray

Two mirrors make an angle of 120° with each other, as in Figure 35.7. A ray is incident on mirror M_1 at an angle of 65° to the normal. Find the direction of the ray after it is reflected from mirror M_2.

Solution From the law of reflection, we see that the first ray also makes an angle of 65° with the normal. Thus, it follows that this same ray makes an angle of 90° − 65°, or 25°, with the horizontal. From the triangle made by the first reflected ray and the two mirrors, we see that the first reflected ray makes an angle of 35° with M_2 (since the sum of the interior angles of any triangle is 180°). This means that this ray makes an angle of 55° with the normal to M_2. Hence, from the law of reflection, it follows that the second reflected ray makes an angle of 55° with the normal to M_2. Finally, by comparing the direction of the ray incident on M_1 with its direction after reflecting from M_2, we see that the ray is rotated through 120°, which corresponds to the angle between the mirrors.

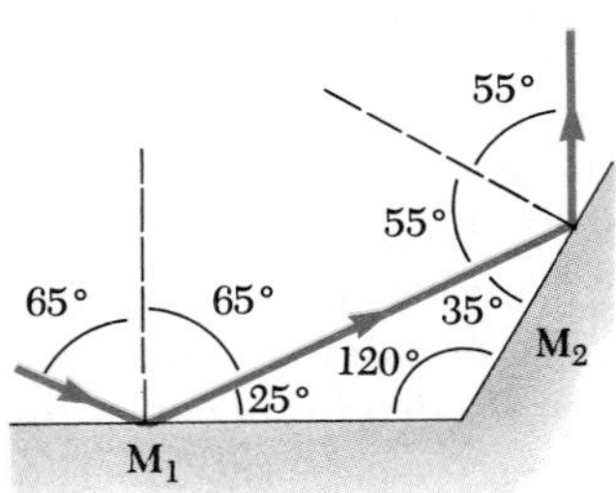

Figure 35.7 (Example 35.2) Mirrors M_1 and M_2 make an angle of 120° with each other.

Refraction of Light

When a ray of light traveling through a transparent medium encounters a boundary leading into another transparent medium, as in Figure 35.8, part of the ray is reflected and part enters the second medium. The ray that enters the second medium is bent at the boundary and is said to be *refracted. The incident ray, the reflected ray, and the refracted ray all lie in the same plane.* The **angle of refraction,** θ_2 in Figure 35.8, depends on the properties of the two media and on the angle of incidence through the relationship

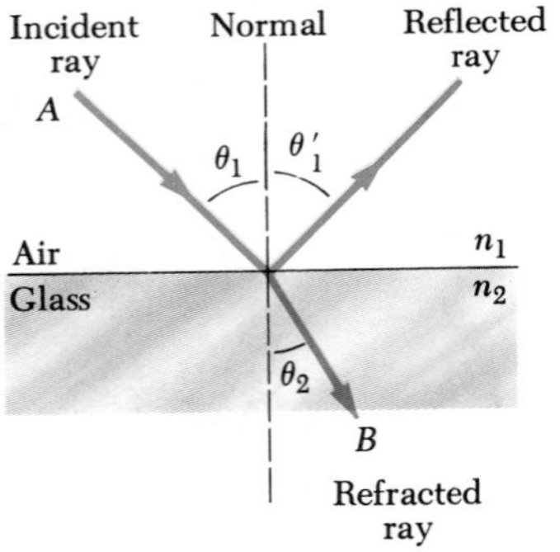

$$\frac{\sin\theta_2}{\sin\theta_1} = \frac{v_2}{v_1} = \text{constant} \tag{35.3}$$

Figure 35.8 A ray obliquely incident on an air-glass interface. The refracted ray is bent toward the normal since $n_2 > n_1$ and $v_2 > v_1$.

where v_1 is the speed of light in medium 1 and v_2 is the speed of light in medium 2. The experimental discovery of this relationship is usually credited to Willebrord Snell (1591–1627) and is therefore known as **Snell's law.**[3] In Section 35.7, we shall verify the laws of reflection and refraction using Huygens' principle. The angles of incidence, reflection, and refraction are *all* measured from the *normal* to the surface rather than from the surface itself. The reason for this convention is that there is no unique angle that a light ray makes with the surface of a three-dimensional object.

[3] This law was also deduced from the particle theory of light by René Descartes (1596–1650) and hence is known as *Descartes' law* in France.

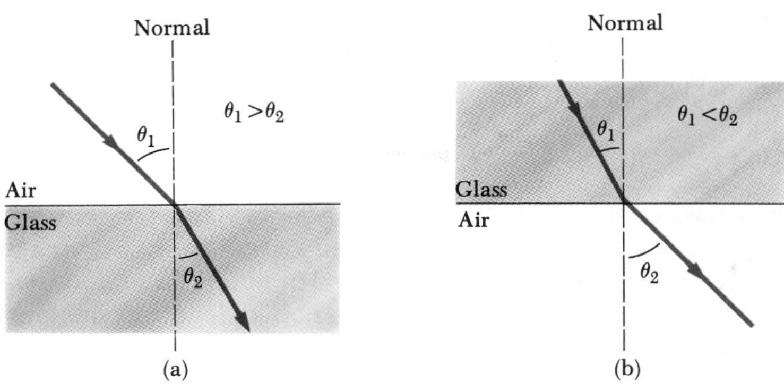

Figure 35.9 (a) When the light beam moves from air into glass, its path is bent toward the normal. (b) When the beam moves from glass into air, its path is bent away from the normal.

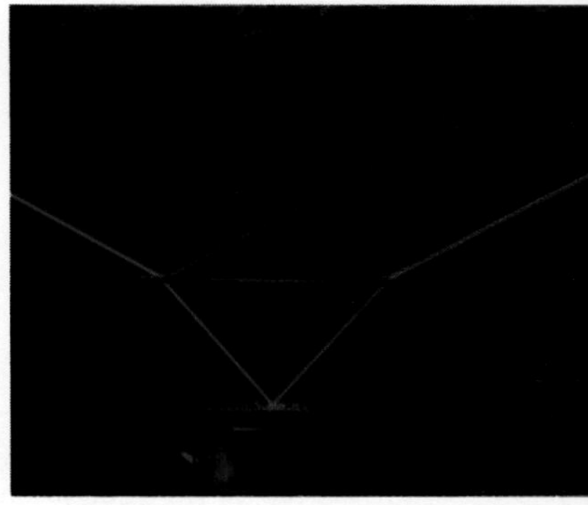

The laser beam in this photograph is incident from the left on a container of water. Part of the beam is reflected, and part is transmitted. The refracted ray then reflects off a mirror at the bottom and emerges from the water at the right. Why is the transmitted ray more intense than the reflected ray? (Courtesy of Henry Leap and Jim Lehman)

It is found that *the path of a light ray through a refracting surface is reversible.* For example, the ray in Figure 35.8 travels from point A to point B. If the ray originated at B, it would follow the same path to reach point A. In the latter case, however, the reflected ray would be in the glass.

When light moves from a material in which its speed is high to a material in which its speed is lower, the angle of refraction, θ_2, is less than the angle of incidence, as shown in Figure 35.9a. If the ray moves from a material in which it moves slowly to a material in which it moves more rapidly, it is bent away from the normal, as shown in Figure 35.9b.

The behavior of light as it passes from air into another substance and then re-emerges into air is often a source of confusion to students. Let us take a look at what happens and see why this behavior is so different from other occurrences in our daily lives. When light travels in air, its speed is equal to 3×10^8 m/s, and its speed is reduced to about 2×10^8 m/s upon entering a block of glass. When the light re-emerges into air, its speed instantaneously increases to its original value of 3×10^8 m/s. This is far different from what happens, for example, when a bullet is fired through a block of wood. In this case, the speed of the bullet is reduced as it moves through the wood because some of its original energy is used to tear apart the fibers of the wood. When the bullet enters the air once again, it emerges at the speed it had just before leaving the block of wood.

In order to see why light behaves as it does, consider Figure 35.10, which represents a beam of light entering a piece of glass from the left. Once inside the glass, the light may encounter an electron bound to an atom, indicated as point A in the figure. Let us assume that light is absorbed by the atom, which causes the electron to oscillate. The oscillating electron then acts as an antenna and radiates the beam of light toward an atom at point B, where the light is again absorbed by an atom at that point. The details of these absorptions and emissions are best explained in terms of quantum mechanics, a subject we shall study in the extended version of this text. For now, it is sufficient to think of the process as one in which the light passes from one atom to another through the glass. (The situation is somewhat analogous to a relay race in which a baton is passed between runners on the same team.) Although light travels from one atom to another with a speed of 3×10^8 m/s, the processes of absorption and emission of light by the atoms take time. Enough time is required, in fact, to

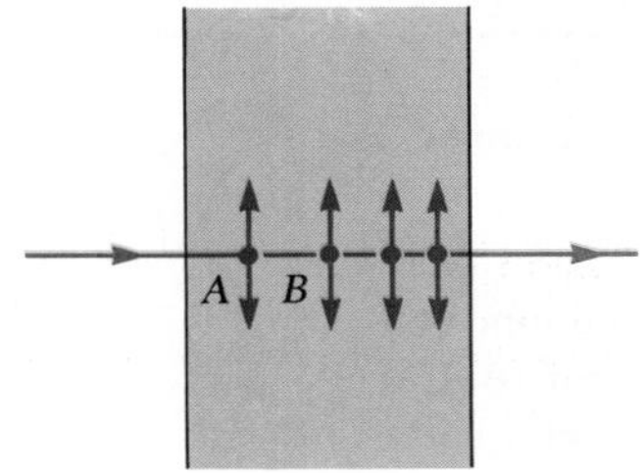

Figure 35.10 Light passing from one atom to another in a medium.

lower the speed of light in one type of glass to 2×10^8 m/s. Once the light emerges into the air, the absorptions and emissions cease and its speed returns to the original value.

The Law of Refraction

When light passes from one medium to another, it is refracted because the speed of light is different in the two media. In general, one finds that the speed of light in any material is less than the speed of light in vacuum. In fact, *light travels at its maximum speed in vacuum.* It is convenient to define the **index of refraction,** n, of a medium to be the ratio

Index of refraction

$$n = \frac{\text{speed of light in vacuum}}{\text{speed of light in a medium}} = \frac{c}{v} \tag{35.4}$$

From this definition, we see that the index of refraction is a dimensionless number greater than unity since v is always less than c. Furthermore, n is equal to unity for vacuum. The indices of refraction for various substances measured with respect to vacuum are listed in Table 35.1.

As light travels from one medium to another, *the frequency of the light does not change.* To see why this is so, consider Figure 35.11. Wavefronts pass an observer at point A in medium 1 with a certain frequency and are incident on the boundary between medium 1 and medium 2. The frequency with which the wavefronts pass an observer at point B in medium 2 must equal the frequency at which they arrive at point A in medium 1. If this were not the case, either wavefronts would be piling up at the boundary or they would be destroyed or created at the boundary. Since there is no mechanism for this to occur, the frequency must be a constant as a light ray passes from one medium into another.

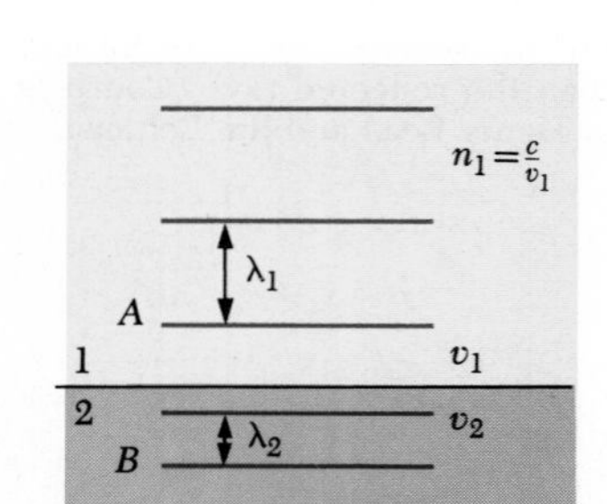

Figure 35.11 As a wavefront moves from medium 1 to medium 2, its wavelength changes but its frequency remains constant.

Therefore, because the relation $v = f\lambda$ must be valid in both media and because $f_1 = f_2 = f$, we see that

$$v_1 = f\lambda_1 \qquad \text{and} \qquad v_2 = f\lambda_2$$

TABLE 35.1 Index of Refraction for Various Substances Measured with Light of Vacuum Wavelength $\lambda_0 = 589$ nm

Substance	Index of Refraction	Substance	Index of Refraction
Solids at 20°C		Liquids at 20°C	
Diamond (C)	2.419	Benzene	1.501
Fluorite (CaF_2)	1.434	Carbon disulfide	1.628
Fused quartz (SiO_2)	1.458	Carbon tetrachloride	1.461
Glass, crown	1.52	Ethyl alcohol	1.361
Glass, flint	1.66	Glycerine	1.473
Ice (H_2O)	1.309	Water	1.333
Polystyrene	1.49	Gases at 0°C, 1 atm	
Sodium chloride (NaCl)	1.544	Air	1.000293
Zircon	1.923	Carbon dioxide	1.00045

where the subscripts refer to the two media. A relationship between index of refraction and wavelength can be obtained by dividing these two equations and making use of the definition of the index of refraction given by Equation 35.4:

$$\frac{\lambda_1}{\lambda_2} = \frac{v_1}{v_2} = \frac{c/n_1}{c/n_2} = \frac{n_2}{n_1} \qquad (35.5)$$

which gives

$$\lambda_1 n_1 = \lambda_2 n_2 \qquad (35.6)$$

If medium 1 is vacuum, or for all practical purposes air, then $n_1 = 1$. Hence, it follows from Equation 35.5 that the index of refraction of any medium can be expressed as the ratio

$$n = \frac{\lambda_0}{\lambda_n} \qquad (35.7)$$

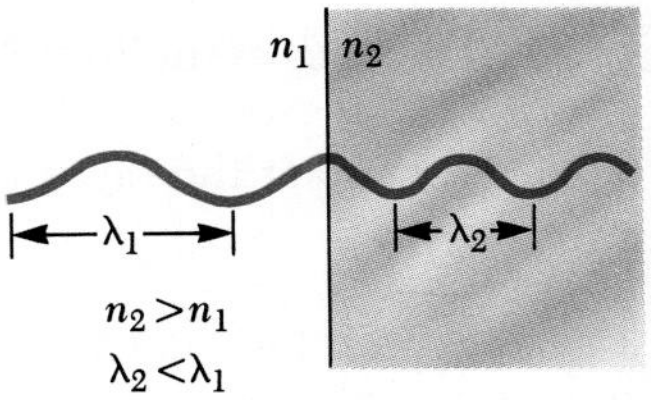

where λ_0 is the wavelength of light in vacuum and λ_n is the wavelength in the medium whose index of refraction is n. A schematic representation of this reduction in wavelength is shown in Figure 35.12.

Figure 35.12 Schematic diagram of the *reduction* in wavelength when light travels from a medium of low index of refraction to one of higher index of refraction.

We are now in a position to express Snell's law (Eq. 35.3) in an alternative form. If we substitute Equation 35.5 into Equation 35.3, we get

Snell's law

$$n_1 \sin\theta_1 = n_2 \sin\theta_2 \qquad (35.8)$$

This is the most widely used and practical form of **Snell's law.**

EXAMPLE 35.3 An Index of Refraction Measurement
A beam of light of wavelength 550 nm traveling in air is incident on a slab of transparent material. The incident beam makes an angle of 40° with the normal, and the refracted beam makes an angle of 26° with the normal. Find the index of refraction of the material.

Solution Snell's law of refraction (Eq. 35.8) with the given data, $\theta_1 = 40°$, $n_1 = 1.00$ for air, and $\theta_2 = 26°$, gives

$$n_1 \sin\theta_1 = n_2 \sin\theta_2$$

$$n_2 = \frac{n_1 \sin\theta_1}{\sin\theta_2} = (1.00)\frac{(\sin 40°)}{\sin 26°} = \frac{0.643}{0.438} = 1.47$$

If we compare this value with the data in Table 35.1, we see that the material is probably fused quartz.

Exercise 1 What is the wavelength of light in the material?
Answer 374 nm

EXAMPLE 35.4 Angle of Refraction for Glass
A light ray of wavelength 589 nm (produced by a sodium lamp) traveling through air is incident on a smooth, flat slab of crown glass at an angle of 30° to the normal, as sketched in Figure 35.13. Find the angle of refraction.

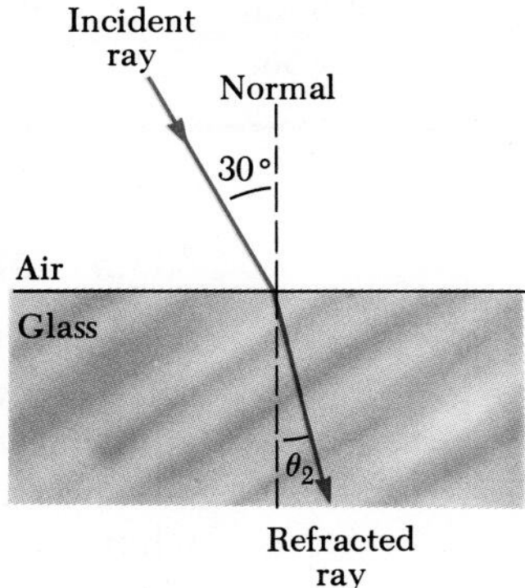

Figure 35.13 (Example 35.4) Refraction of light by glass.

Solution Snell's law given by Equation 35.8 can be rearranged as

$$\sin\theta_2 = \frac{n_1}{n_2}\sin\theta_1$$

From Table 35.1, we find that $n_1 = 1.00$ for air and $n_2 = 1.52$ for glass. Therefore, the unknown refraction angle is

$$\sin\theta_2 = \left(\frac{1.00}{1.52}\right)(\sin 30°) = 0.329$$

$$\theta_2 = \sin^{-1}(0.329) = 19.2°$$

Thus we see that the ray is bent *toward* the normal, as expected.

Exercise 2 If the light ray moves from inside the glass toward the glass-air interface at an angle of 30° to the normal, determine the angle of refraction.
Answer 49.5° *away* from the normal.

EXAMPLE 35.5 The Speed of Light in Quartz
Light of wavelength 589 nm in vacuum passes through a piece of fused quartz ($n = 1.458$). (a) Find the speed of light in quartz.

Solution The speed of light in quartz can be easily obtained from Equation 35.4:

$$v = \frac{c}{n} = \frac{3\times10^8 \text{ m/s}}{1.458} = 2.058\times10^8 \text{ m/s}$$

(b) What is the wavelength of this light in quartz?

Solution We can use $\lambda_n = \lambda_0/n$ (Eq. 35.7) to calculate the wavelength in quartz, noting that we are given the wavelength in vacuum to be $\lambda_0 = 589$ nm:

$$\lambda_n = \frac{\lambda_0}{n} = \frac{589 \text{ nm}}{1.458} = 404 \text{ nm}$$

Exercise 3 Find the frequency of the light passing through the quartz.
Answer 5.09×10^{14} Hz.

EXAMPLE 35.6 Light Passing Through a Slab
A light beam passes from medium 1 to medium 2 through a thick slab of material whose index of refraction is n_2 (Fig. 35.14). Show that the emerging beam is parallel to the incident beam.

Solution First, let us apply Snell's law to the upper surface:

$$(1) \qquad \sin\theta_2 = \frac{n_1}{n_2}\sin\theta_1$$

Applying Snell's law to the lower surface gives

$$(2) \qquad \sin\theta_3 = \frac{n_2}{n_1}\sin\theta_2$$

Substituting (1) into (2) gives

$$\sin\theta_3 = \frac{n_2}{n_1}\left(\frac{n_1}{n_2}\sin\theta_1\right) = \sin\theta_1$$

That is, $\theta_3 = \theta_1$, and so the layer does not alter the direction of the beam. It does, however, produce a displacement of the beam. The same result is obtained when light passes through multiple layers of materials.

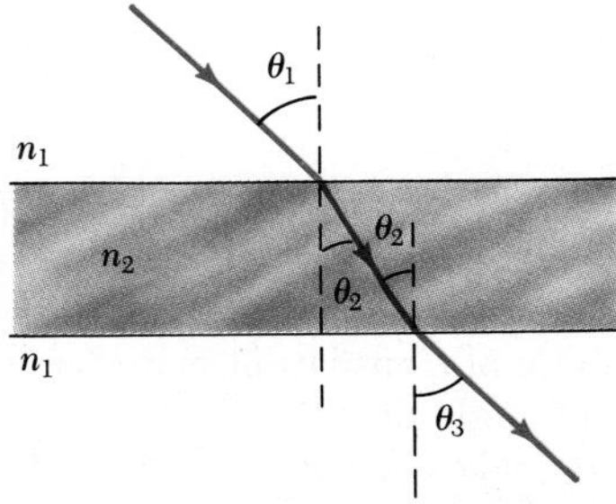

Figure 35.14 (Example 35.6) When light passes through a flat slab of material, the emerging beam is parallel to the incident beam, and therefore $\theta_1 = \theta_3$.

*35.5 DISPERSION AND PRISMS

An important property of the index of refraction is that it is different for different wavelengths of light. A graph of the index of refraction for three materials is shown in Figure 35.15. Since n is a function of wavelength, Snell's law indicates that light of *different wavelengths* will be bent at *different angles* when incident on a refracting material. As we see from Figure 35.15, the index of refraction decreases with increasing wavelength. This means that blue light will bend more than red light when passing into a refracting material.

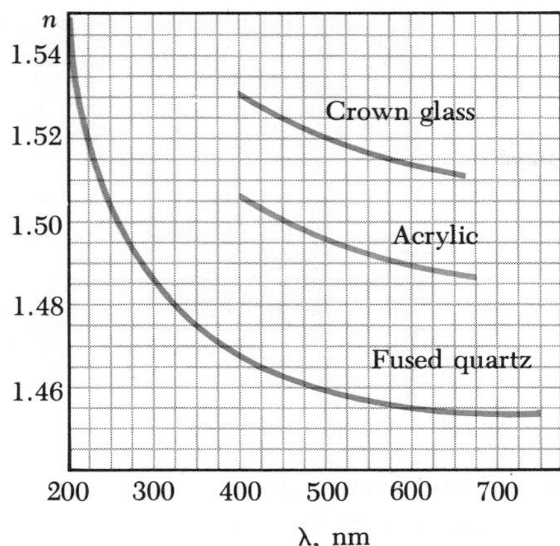

Figure 35.15 Variation of index of refraction with wavelength for three materials.

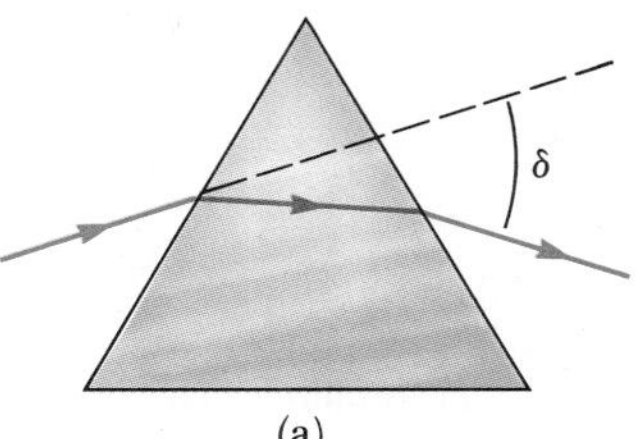

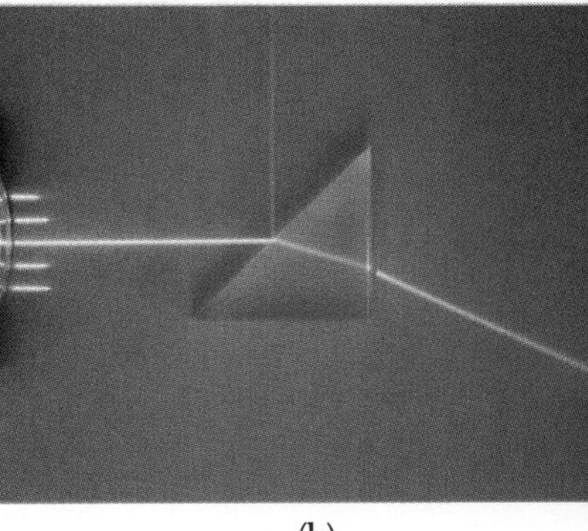

Figure 35.16 (a) A prism refracts a light ray and deviates the light through an angle δ. (b) The light is refracted as it passes through the prism. (Courtesy of Jim Lehman, James Madison University)

Any substance in which n varies with wavelength is said to exhibit **dispersion.** To understand the effects that dispersion can have on light, let us consider what happens when light strikes a prism, as in Figure 35.16. A single ray of light incident on the prism from the left emerges bent away from its original direction of travel by an angle δ, called the **angle of deviation.** Now suppose a beam of white light (a combination of all visible wavelengths) is incident on a prism, as in Figure 35.17. The rays that emerge from the second face spread out in a series of colors known as a **spectrum.** These colors, in order of decreasing wavelength, are red, orange, yellow, green, blue, indigo, and violet. Newton showed that each color had a particular angle of deviation, that the spectrum could not be further broken down, and that the colors could be recombined to form the original white light. Clearly, the angle of deviation, δ, depends on the wavelength of a given color. Violet light deviates the most, red light deviates the least, and the remaining colors in the visible spectrum fall between these extremes. When light is spread out by a substance such as the prism, the light is said to be dispersed into a spectrum.

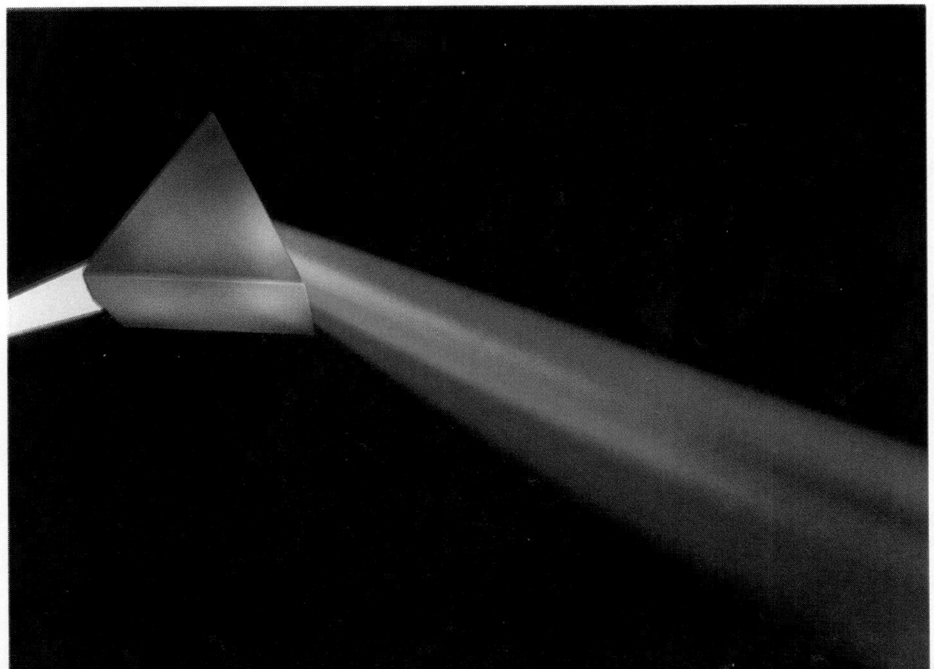

Figure 35.17 Dispersion of white light. White light is dispersed into a spectrum of visible colors by a prism. The various colors are refracted at different angles because the index of refraction of the glass depends on wavelength. Refraction at the first surface produces a different path for each color passing into the glass. The path separation between the various colors is enhanced following refraction from the second surface as the light exits the prism. The blue light deviates the most, while red light deviates the least. (Courtesy of Bausch and Lomb)

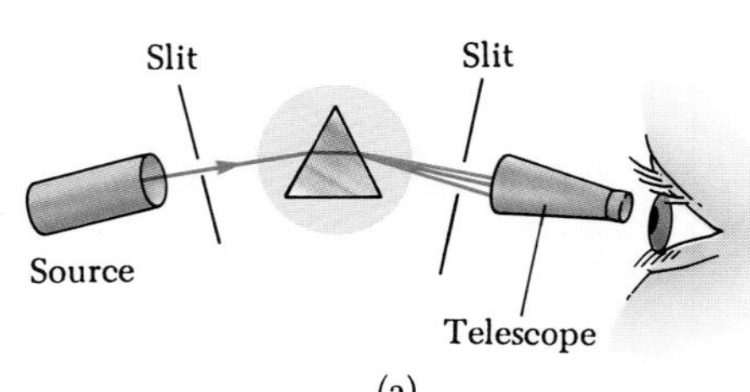

(a)

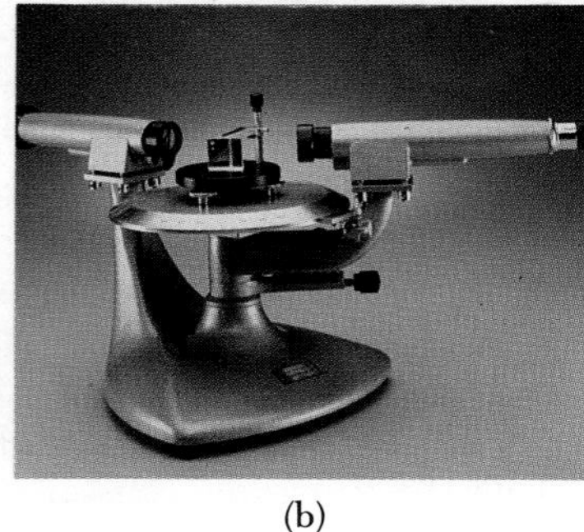
(b)

Figure 35.18 (a) Diagram of a prism spectrometer. The various colors in the spectrum are viewed through a telescope. (b) Photograph of a prism spectrometer. (Courtesy of CENCO)

A prism is often used in an instrument known as a **prism spectrometer,** the essential elements of which are shown in Figure 35.18a. The instrument is commonly used to study the wavelengths emitted by a light source, such as a sodium vapor lamp. Light from the source is sent through a narrow, adjustable slit to produce a parallel, or collimated, beam. The light then passes through the prism and is dispersed into a spectrum. The refracted light is observed through a telescope. The experimenter sees an image of the slit through the eyepiece of the telescope. The telescope can be moved or the prism can be rotated in order to view the various images formed by different wavelengths at different angles of deviation.

All hot, low-pressure gases emit their own characteristic spectra. Thus, one use of a prism spectrometer is to identify gases. For example, sodium emits two wavelengths in the visible spectrum, which appear as two closely spaced yellow lines. Thus, a gas emitting these colors can be identified as having sodium as one of its constituents. Likewise, mercury vapor has its own characteristic spectrum, consisting of four prominent wavelengths — orange, green, blue, and violet lines — along with some wavelengths of lower intensity. The particular wavelengths emitted by a gas serve as "fingerprints" of that gas.

EXAMPLE 35.7 Measuring *n* Using a Prism

A prism is often used to measure the index of refraction of a transparent solid. Although we do not prove it here, one finds that the *minimum angle of deviation,* δ_m, occurs at the angle of incidence θ_1 where the refracted ray inside the prism makes the same angle α with the normal to the two prism faces,[4] as in Figure 35.19. Let us obtain an expression for the index of refraction of the prism material.

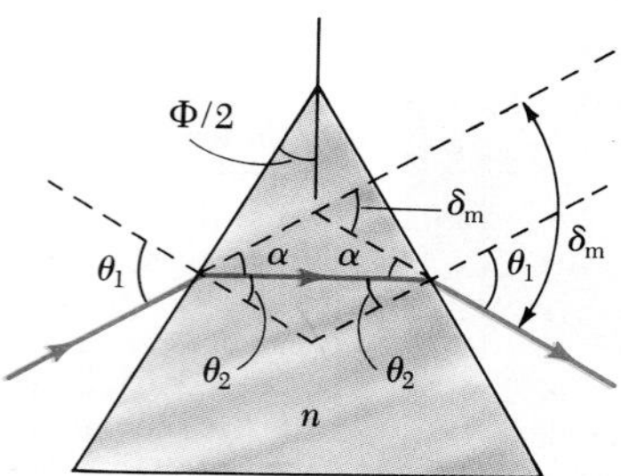

Figure 35.19 (Example 35.7) A light ray passing through a prism at the minimum angle of deviation, δ_m.

Using the geometry shown, one finds that $\theta_2 = \Phi/2$ and

$$\theta_1 = \theta_2 + \alpha = \frac{\Phi}{2} + \frac{\delta_m}{2} = \frac{\Phi + \delta_m}{2}$$

From Snell's law,

$$\sin\,\theta_1 = n\,\sin\,\theta_2$$

$$\sin\left(\frac{\Phi + \delta_m}{2}\right) = n\,\sin(\Phi/2)$$

$$n = \frac{\sin\left(\dfrac{\Phi + \delta_m}{2}\right)}{\sin(\Phi/2)} \qquad (35.9)$$

Hence, knowing the apex angle Φ of the prism and measuring δ_m, one can calculate the index of refraction of the prism material. Furthermore, one can use a hollow prism to determine the values of n for various liquids.

[4] For details, see F. A. Jenkins and H. E. White, *Fundamentals of Optics,* New York, McGraw-Hill, 1976, Chapter 2.

35.6 HUYGENS' PRINCIPLE

In this section, we shall develop the laws of reflection and refraction by using a geometric method proposed by Huygens in 1678. Huygens assumed that light is some form of wave motion rather than a stream of particles. He had no knowledge of the nature of light or of its electromagnetic character. Nevertheless, his simplified wave model is adequate for understanding many practical aspects of the propagation of light.

Huygens' principle is a geometric construction for determining the position of a new wavefront at some instant from the knowledge of an earlier wavefront. In Huygens' construction,

> all points on a given wavefront are taken as point sources for the production of spherical secondary waves, called wavelets, which propagate outward with speeds characteristic of waves in that medium. After some time has elapsed, the new position of the wavefront is the surface tangent to the wavelets.

Figure 35.20 illustrates two simple examples of Huygens' construction. First, consider a plane wave moving through free space, as in Figure 35.20a. At $t = 0$, the wavefront is indicated by the plane labeled AA'. (A wavefront is a surface passing through those points of a wave that are behaving identically. For instance, a wavefront could be a surface over which the wave is at a crest.) In Huygens' construction, each point on this wavefront is considered a point source. For clarity, only a few points on AA' are shown. With these points as sources for the wavelets, we draw circles each of radius $c\,\Delta t$, where c is the speed of light in free space and Δt is the time of propagation from one wavefront to the next. The surface drawn tangent to these wavelets is the plane BB', which is parallel to AA'. In a similar manner, Figure 35.20b shows Huygens' construction for an outgoing spherical wave.

A convincing demonstration of Huygens' principle is obtained with water waves in a shallow tank (called a ripple tank), as in Figure 35.21. Plane waves produced below the slit emerge above the slit as two-dimensional circular waves propagating outward.

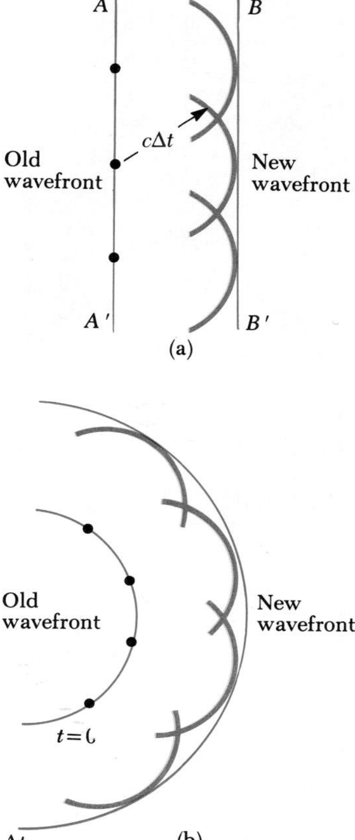

Figure 35.20 Huygens' construction for (a) a plane wave propagating to the right and (b) a spherical wave.

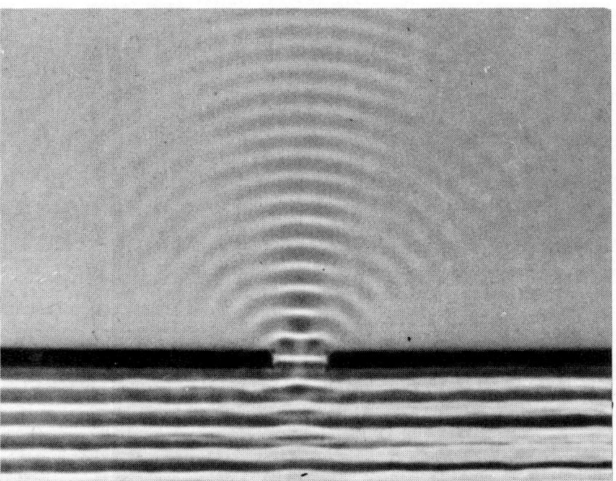

Figure 35.21 Water waves in a ripple tank, which demonstrates Huygens' wavelets. A plane wave is incident on a barrier with a small opening. The opening acts as a source of circular wavelets. (Photography courtesy of Education Development Center, Newton, Mass.)

Biographical Sketch

Christian Huygens
(1629–1695)

Christian Huygens was a Dutch physicist and astronomer who is best known for his contributions to the fields of dynamics and optics. As a physicist, his accomplishments include invention of the pendulum clock and the first exposition of a wave theory of light. As an astronomer, Huygens was the first to recognize the rings around Saturn and to discover Titan, a satellite of Saturn.

Huygens was born in 1629 into a prominent family in The Hague. He was the son of Constantin Huygens, one of the most important figures of the Renaissance in Holland. Educated at the University of Leyden, Christian became a close friend of René Descartes, who was a frequent guest at the Huygens home. Huygens published his first paper in 1651 on the subject of the quadrature of various curves.

Huygens' reputation in optics and dynamics spread throughout Europe, and in 1663 he was elected a charter member of the Royal Society. Louis XIV lured Huygens to France in 1666, in accordance with the king's policy of collecting scholars for the glory of his regime. While in France, Huygens became one of the founders of the French Academy of Science.

In 1673, in Paris, Huygens published *Horologium Oscillatorium.* In this work he described a solution to the problem of the compound pendulum, for which he calculated the equivalent simple pendulum length. In the same publication he also derived a formula for computing the period of oscillation of a simple pendulum, and he explained his laws of centrifugal force for uniform motion in a circle.

Huygens returned to Holland in 1681, constructed some lenses of large focal lengths and invented the achromatic eyepiece for telescopes. Shortly after returning from a visit to England, where he met Isaac Newton, Huygens published his treatise on the wave theory of light. To Huygens, light was a vibratory motion in the ether, spreading out and producing the sensation of light when impinging on the eye. On the basis of this theory, he was able to deduce the laws of reflection and refraction and to explain the phenomenon of double refraction.

Huygens, second only to Newton among the greatest scientists in the second half of the 17th century, was the first to proceed in the field of dynamics beyond the point reached by Galileo and Descartes. It was Huygens who essentially solved the problem of centrifugal force. A solitary man, Huygens did not attract students or disciples and was very slow in publishing his findings. Huygens died in 1695 after a long illness.

(Rijksmuseum voor de Geschiedenis der Natuurwetenschappen. Courtesy AIP Niels Bohr Library)

Huygens' Principle Applied to Reflection and Refraction

The laws of reflection and refraction were stated earlier in this chapter without proof. We shall now derive these laws using Huygens' principle. Figure 35.22a will be used in our consideration of the law of reflection. The line AA' represents a wavefront of the incident light. As ray 3 travels from A' to C, ray 1 reflects from A and produces a spherical wavelet of radius AD. (Recall that the radius of a Huygens wavelet is equal to vt.) Since the two wavelets having radii $A'C$ and AD are in the same medium, they have the same velocity, v, and thus $AD = A'C$. Meanwhile, the spherical wavelet centered at B has spread only half as far as the one centered at A since ray 2 strikes the surface later than ray 1.

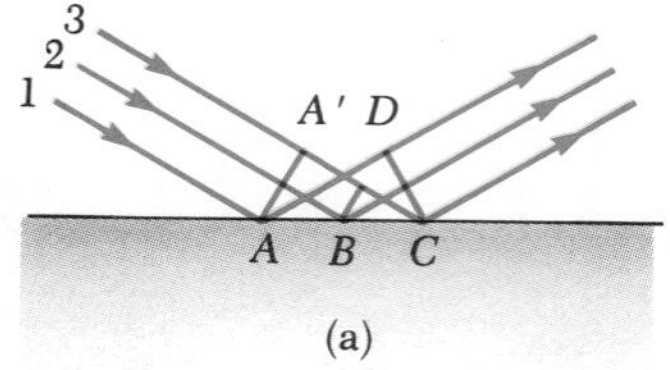

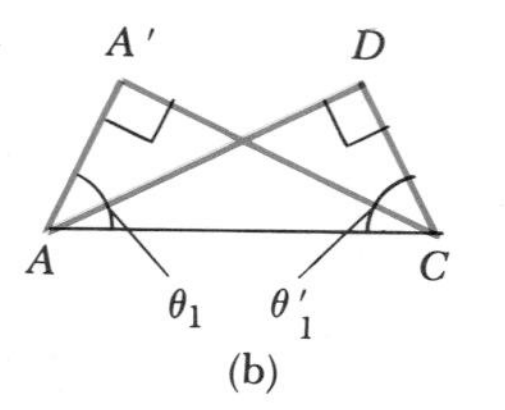

Figure 35.22 (a) Huygens' construction for proving the law of reflection. (b) Triangle ADC is identical to triangle $AA'C$.

From Huygens' principle, we find that the reflected wavefront is CD, a line tangent to all the outgoing spherical wavelets. The remainder of the analysis depends upon geometry, as summarized in Figure 35.22b. Note that the right triangles ADC and $AA'C$ are congruent because they have the same hypotenuse, AC, and because $AD = A'C$. From Figure 35.22b we have

$$\sin\theta_1 = \frac{A'C}{AC} \qquad \text{and} \qquad \sin\theta_1' = \frac{AD}{AC}$$

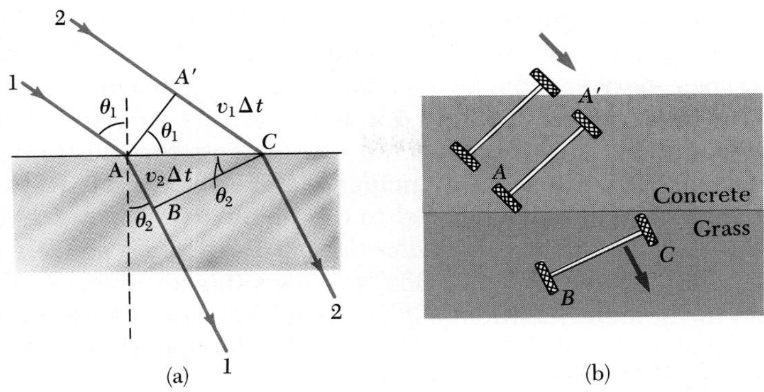

Figure 35.23 (a) Huygens' construction for proving the law of refraction. (b) A mechanical analog of refraction.

Thus,

$$\sin \theta_1 = \sin \theta_1'$$

$$\theta_1 = \theta_1'$$

which is the law of reflection.

Now let us use Huygens' principle and Figure 35.23a to derive Snell's law of refraction. Note that in the time interval Δt, ray 1 moves from A to B and ray 2 moves from A' to C. The radius of the outgoing spherical wavelet centered at A is equal to $v_2 \Delta t$. The distance $A'C$ is equal to $v_1 \Delta t$. Geometric considerations show that angle $A'AC$ equals θ_1 and angle ACB equals θ_2. From triangles $AA'C$ and ACB, we find that

$$\sin \theta_1 = \frac{v_1 \Delta t}{AC} \quad \text{and} \quad \sin \theta_2 = \frac{v_2 \Delta t}{AC}$$

If we divide these two equations, we get

$$\frac{\sin \theta_1}{\sin \theta_2} = \frac{v_1}{v_2}$$

But from Equation 35.4 we know that $v_1 = c/n_1$ and $v_2 = c/n_2$. Therefore,

$$\frac{\sin \theta_1}{\sin \theta_2} = \frac{c/n_1}{c/n_2} = \frac{n_2}{n_1}$$

$$n_1 \sin \theta_1 = n_2 \sin \theta_2$$

which is the law of refraction.

A mechanical analog of refraction is shown in Figure 35.23b. The wheels on a device such as a wagon change their direction as they move from a concrete surface to a grass surface.

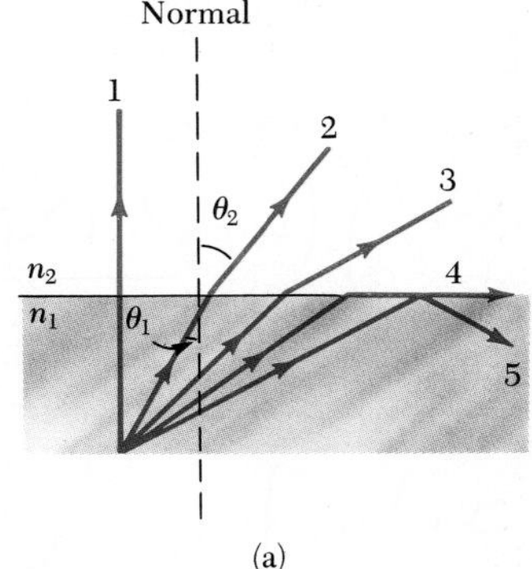

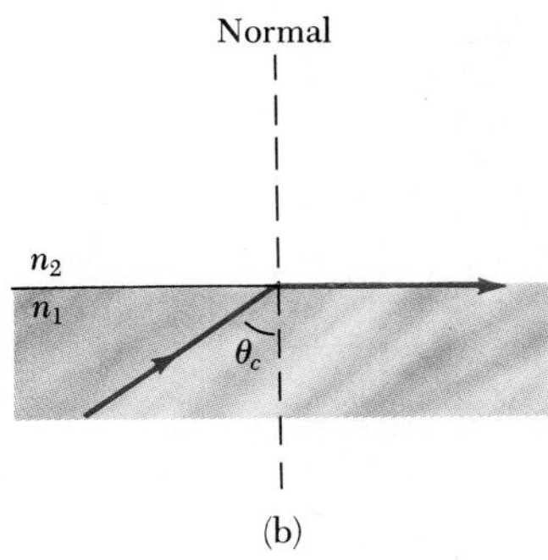

Figure 35.24 (a) A ray from a medium of index of refraction n_1 to a medium of index of refraction n_2, where $n_1 > n_2$. As the angle of incidence increases, the angle of refraction increases until θ_2 is 90° (ray 4). For even larger angles of incidence, total internal reflection occurs (ray 5). (b) The angle of incidence producing an angle of refraction equal to 90° is often called the *critical angle*, θ_c.

35.7 TOTAL INTERNAL REFLECTION

An interesting effect called *total internal reflection* can occur when light attempts to move from a medium having a given index of refraction to one having a *lower* index of refraction. Consider a light beam traveling in medium 1 and meeting the boundary between medium 1 and medium 2, where n_1 is greater than n_2 (Fig. 35.24). Various possible directions of the beam are indicated by rays 1 through 5. The refracted rays are bent away from the normal because n_1 is greater than n_2. Furthermore, you should remember that when light refracts at the interface between the two media, it is also partially reflected. For example, the rays labeled 2, 3, and 4 in Figure 35.24 are *partially* reflected back into medium 1, but the reflected components for these rays are not shown. At some particular angle of incidence, θ_c, called the **critical angle,** the refracted light ray will move parallel to the boundary so that $\theta_2 = 90°$ (Fig. 35.24b). *For angles of incidence greater than* θ_c, the beam is entirely reflected at the boundary. Ray 5 in Figure 35.24a shows this occurrence. This ray is reflected at the boundary as though it had struck a perfectly reflecting surface. This ray, and all those like it, obey the law of reflection; that is, the angle of incidence equals the angle of reflection.

We can use Snell's law to find the critical angle. When $\theta_1 = \theta_c$, $\theta_2 = 90°$ and Snell's law (Eq. 35.8) gives

$$n_1 \sin \theta_c = n_2 \sin 90° = n_2$$

Critical angle

$$\sin \theta_c = \frac{n_2}{n_1} \qquad (\text{for } n_1 > n_2) \qquad (35.10)$$

This equation can be used only when n_1 is greater than n_2. That is,

> total internal reflection occurs only when light attempts to move from a medium of given index of refraction to a medium of lower index of refraction.

If n_1 were less than n_2, Equation 35.10 would give $\sin \theta_c > 1$, which is an absurd result because the sine of an angle can never be greater than unity.

The critical angle for a substance in air is small for substances with a large index of refraction, such as diamond, where $n = 2.42$ and $\theta_c = 24°$. For crown glass, $n = 1.52$ and $\theta_c = 41°$. In fact, this property combined with proper faceting causes diamonds and crystal glass to sparkle.

One can use a prism and the phenomenon of total internal reflection to alter the direction of travel of a light beam. Two such possibilities are illustrated in Figure 35.25. In one case the light beam is deflected by 90° (Fig. 35.25a), and in the second case the path of the beam is reversed (Fig. 35.25b). A common application of total internal reflection is in a submarine periscope. In this device, two prisms are arranged as in Figure 35.25c so that an incident beam of light follows the path shown and one is able to "see around corners."

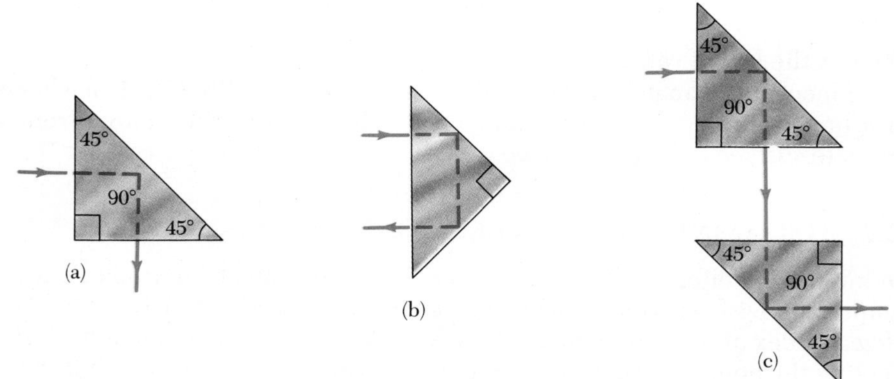

Figure 35.25 Internal reflection in a prism. (a) The ray is deviated by 90°. (b) The direction of the ray is reversed. (c) Two prisms used as a periscope.

EXAMPLE 35.8 A View from the Fish's Eye
(a) Find the critical angle for a water-air boundary if the index of refraction of water is 1.33.

Solution Applying Equation 35.10, we find the critical angle to be

$$\sin \theta_c = \frac{n_2}{n_1} = \frac{1}{1.33} = 0.752$$

$$\theta_c = 48.8°$$

(b) Use the results of (a) to predict what a fish will see if it looks upward toward the water surface at an angle of 40°, 49°, and 60°.

Solution Because the path of a light ray is reversible, the fish can see out of the water if it looks toward the surface at an angle less than the critical angle. Thus, at 40°, the fish can see into the air above the water. At an angle of 49°, the critical angle for water, the light that reaches the fish has to skim along the water surface before being refracted to the fish's eye. At angles greater than the critical angle, the light reaching the fish comes via internal reflection at the surface. Thus, at 60°, the fish sees a reflection of some object on the bottom of the pool.

Fiber Optics

Another interesting application of total internal reflection is the use of glass or transparent plastic rods to "pipe" light from one place to another. As indicated in Figure 35.26, light is confined to traveling within the rods, even around gentle curves, as the result of successive internal reflections. Such a "light pipe" will be flexible if thin fibers are used rather than thick rods. If a bundle of parallel fibers is used to construct an optical transmission line, images can be transferred from one point to another.

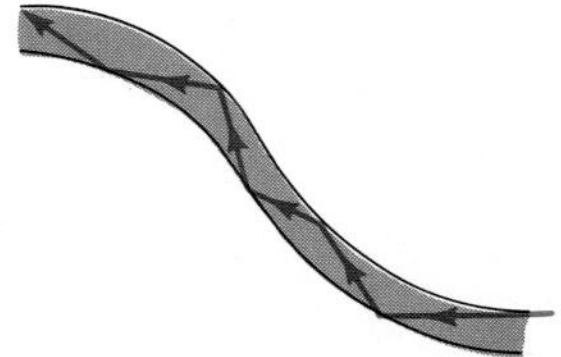

Figure 35.26 Light travels in a curved transparent rod by multiple internal reflections.

This technique is used in a sizable industry known as *fiber optics* (see the essay at the end of this chapter). There is very little light intensity lost in these fibers as a result of reflections on the sides. Any loss in intensity is due essentially to reflections from the two ends and absorption by the fiber material. These devices are particularly useful when one wishes to view an image produced at inaccessible locations. For example, physicians often use this technique to examine internal organs of the body. The field of fiber optics is finding increasing use in telecommunications, since the fibers can carry a much higher volume of telephone calls or other forms of communication than electrical wires. The essay in this chapter discusses the use of fiber optics in the expanding field of telecommunications.

*35.8 FERMAT'S PRINCIPLE

A general principle that can be used for determining the actual paths of light rays was developed by Pierre de Fermat (1601–1665). **Fermat's principle** states that

> when a light ray travels between any two points *P* and *Q*, its actual path will be the one that requires the least time.

Spray of glass fiber optics consisting of 2000 individual strands, each measuring 60 mμ thick. (Adam Hart, Davis/Science Photo Library/Photo Researchers, Inc.)

Fermat's principle is sometimes called the *principle of least time.* An obvious consequence of this principle is that when the rays travel in a single, homogeneous medium, the paths are straight lines because a straight line is the shortest distance between two points. Let us illustrate how to use Fermat's principle to derive the law of refraction.

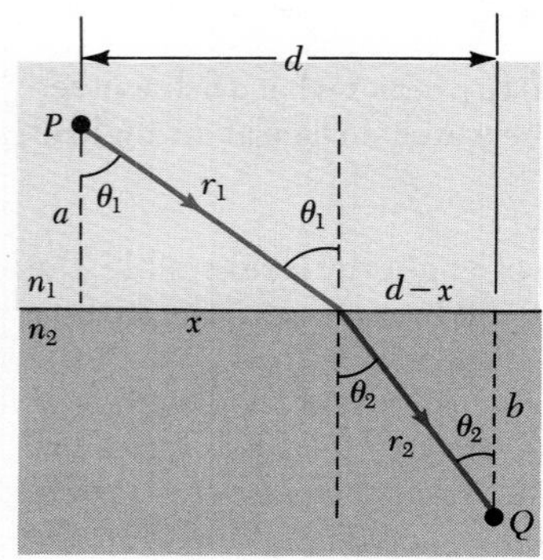

Figure 35.27 Geometry for deriving the law of refraction using Fermat's principle.

Suppose a light ray is to travel from P to Q, where P is in medium 1 and Q is in medium 2 (Fig. 35.27). The points P and Q are at perpendicular distances a and b, respectively, from the interface. The speed of light is c/n_1 in medium 1 and c/n_2 in medium 2. Using the geometry of Figure 35.27, we see that the time it takes the ray to travel from P to Q is

$$t = \frac{r_1}{v_1} + \frac{r_2}{v_2} = \frac{\sqrt{a^2 + x^2}}{c/n_1} + \frac{\sqrt{b^2 + (d - x)^2}}{c/n_2}$$

We obtain the least time, or the minimum value of t, by taking the derivative of t with respect to x (the variable) and setting the derivative equal to zero. Using this procedure, we get

$$\frac{dt}{dx} = \frac{n_1}{c}\frac{d}{dx}(a^2 + x^2)^{1/2} + \frac{n^2}{c}\frac{d}{dx}[b^2 + (d - x)^2]^{1/2}$$

$$= \frac{n_1}{c}\left(\frac{1}{2}\right)\frac{2x}{(a^2 + x^2)^{1/2}} + \frac{n_2}{c}\left(\frac{1}{2}\right)\frac{2(d - x)(-1)}{[b^2 + (d - x)^2]^{1/2}}$$

$$\frac{dt}{dx} = \frac{n_1 x}{c(a^2 + x^2)^{1/2}} - \frac{n_2(d - x)}{c[b^2 + (d - x)^2]^{1/2}} = 0$$

From Figure 35.27 and recognizing $\sin\theta_1$ and $\sin\theta_2$ in this equation, we find that

$$n_1 \sin\theta_1 = n_2 \sin\theta_2$$

which is Snell's law of refraction.

It is a simple matter to use a similar procedure to derive the law of reflection. The calculation is left for you to carry out (Problem 46).

SUMMARY

In geometric optics, we use the so-called **ray approximation,** in which we assume that a wave travels through a medium in straight lines in the direction of the rays. Furthermore, we neglect diffraction effects, which is a good approximation as long as the wavelength is short compared with any aperture dimensions.

The basic laws of geometric optics are the *laws of reflection and refraction* for light rays. The **law of reflection** states that the angle of reflection, θ_1', equals the angle of incidence, θ_1. The **law of refraction, or Snell's Law,** states that

Snell's law of refraction

$$n_1 \sin\theta_1 = n_2 \sin\theta_2 \qquad (35.8)$$

where θ_2 is the angle of refraction and n_1 and n_2 are the indices of refraction in the two media. The incident ray, the reflected ray, the refracted ray, and the normal to the surface all lie in the same plane.

The **index of refraction** of a medium, n, is defined by the ratio

Index of refraction

$$n \equiv \frac{c}{v} \qquad (35.4)$$

where c is the speed of light in a vacuum and v is the speed of light in the medium. In general, n varies with wavelength and is given by

Index of refraction and wavelength

$$n = \frac{\lambda_0}{\lambda_n} \tag{35.7}$$

where λ_0 is the vacuum wavelength and λ_n is the wavelength in the medium.

Huygens' principle

Huygens' principle states that all points on a wavefront can be taken as point sources for the production of secondary wavelets. At some later time, the new position of the wavefront is the surface tangent to these secondary wavelets.

Total internal reflection can occur when light travels from a medium of high index of refraction to one of lower index of refraction. The minimum angle of incidence, θ_c, for which total reflection occurs at an interface is given by

Critical angle for total internal reflection

$$\sin \theta_c = \frac{n_2}{n_1} \qquad \text{(where } n_1 > n_2\text{)} \tag{35.10}$$

Fermat's principle

Fermat's principle states that when a light ray travels between two points, its path will be the one that requires the least time.

QUESTIONS

1. Light of wavelength λ is incident on a slit of width d. Under what conditions is the ray approximation valid? Under what circumstances will the slit produce significant diffraction?
2. Sound waves have much in common with light waves, including the properties of reflection and refraction. Give examples of such phenomena for sound waves.
3. Does a light ray traveling from one medium into another always bend toward the normal as in Figure 35.8? Explain.
4. As light travels from one medium to another, does its wavelength change? Does its frequency change? Does its velocity change? Explain.
5. A laser beam passing through a nonhomogeneous sugar solution is observed to follow a curved path. Explain.
6. A laser beam ($\lambda = 632.8$ nm) is incident on a piece of Lucite as in Figure 35.28. Part of the beam is reflected and part is refracted. What information can you get from this photograph?
7. Suppose blue light were used instead of red light in the experiment shown in Figure 35.28. Would the refracted beam be bent at a larger or smaller angle?
8. The level of water in a clear, colorless glass is easily observed with the naked eye. The level of liquid helium in a clear glass vessel is extremely difficult to see with the naked eye. Explain.
9. Describe an experiment in which internal reflection is used to determine the index of refraction of a medium.
10. Why does a diamond show flashes of color when observed under ordinary white light?
11. Explain why a diamond shows more "sparkle" than a glass crystal of the same shape and size.
12. Explain why an oar in the water appears bent.

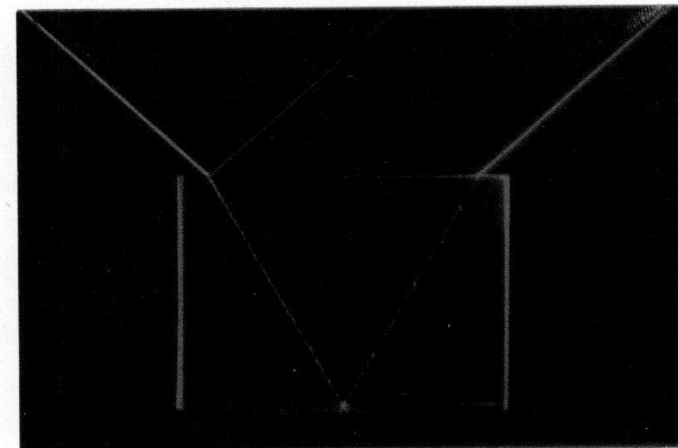

Figure 35.28 (Questions 6 and 7) Light from a helium-neon laser beam ($\lambda = 632.8$ nm) is incident on a block of Lucite. The photograph shows both reflected and refracted rays. Can you identify the incident, reflected, and refracted rays? From this photograph, estimate the index of refraction of Lucite at this wavelength. (Courtesy of Henry Leap and Jim Lehman)

13. Redesign the periscope of Figure 35.25c so that it can show you where you have been rather than where you are going.
14. Under certain circumstances, sound can be heard over extremely long distances. This frequently happens over a body of water, where the air near the water surface is cooler than the air higher up. Explain how the refraction of sound waves in such a situation could increase the distance over which the sound can be heard.
15. Why do astronomers looking at distant galaxies talk about looking backward in time?
16. A solar eclipse occurs when the moon gets between the earth and the sun. Use a diagram to show why some areas of the earth see a total eclipse, other areas see a partial eclipse, and most areas see no eclipse.
17. Some department stores have their windows slanted slightly inward at the bottom. This is to decrease the glare from streetlights or the sun, which would make it difficult for shoppers to see the display inside. Draw a sketch of a light ray reflecting off such a window to show how this technique works.
18. Suppose you are told only that two colors of light (X and Y) are sent through a glass prism and that X is bent more than Y. Which color travels more slowly in the prism?
19. Figure 35.29 represents sunlight striking a drop of water in the atmosphere. Use the laws of refraction and reflection and the fact that sunlight consists of a wide range of wavelengths to discuss the formation of rainbows.
20. Why does the arc of a rainbow appear with red colors on top and violet hues on the bottom?
21. How is it possible that a complete *circular* rainbow can sometimes be seen from an airplane?
22. You can make a corner reflector by placing three plane mirrors in the corner of a room where the ceiling meets the walls. Show that no matter where you are in the room, you can see yourself reflected in the mirrors—upside down.
23. Several corner reflectors were left on the moon's Sea of Tranquility by the astronauts of Apollo 11. How can scientists utilize a laser beam sent from Earth even today to determine the precise distance from the Earth to the moon?
24. What are the conditions for the production of a mirage? On a hot day, what is it that we are seeing when we observe "water on the road"?
25. The "professor in the box" shown in Figure 35.30 appears to be balancing himself on a few fingers with his feet elevated from the floor. How do you suppose this illusion was created? (The same fellow has also been known to "fly" at times.)

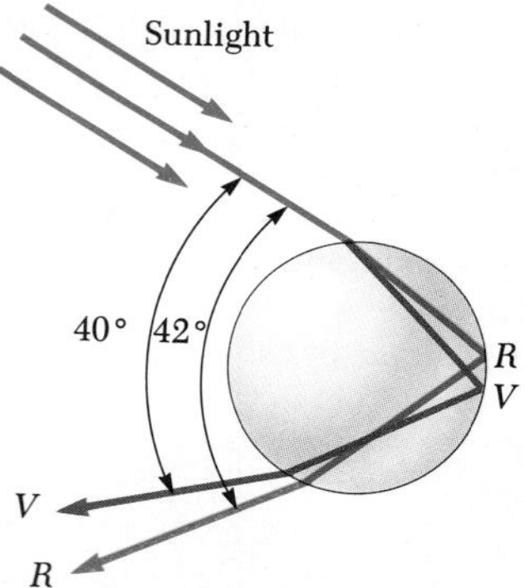

Figure 35.29 (Question 19) Refraction of sunlight by a spherical raindrop.

Figure 35.30 (Question 25). (Photograph courtesy of Henry Leap and Jim Lehman)

PROBLEMS

Section 35.2 Measurements of the Speed of Light

1. Experimenters at the National Bureau of Standards have made precise measurements of the speed of light using the property of electromagnetic waves that in vacuum the phase velocity of the waves is $c = \sqrt{1/\mu_0\epsilon_0}$, where μ_0 (permeability constant) $= 4\pi \times 10^{-7}$ N $\cdot$ s^2/C^2 and ϵ_0 (permittivity constant) $= 8.854 \times 10^{-12}$ C^2/N $\cdot$ m^2. What value (to four significant figures) does this give for the speed of light in vacuum?
2. As a result of his observations, Roemer concluded that the time interval between successive eclipses of the moon Io by the planet Jupiter increased by 22 min during a 6-month period as the earth moved from a point in its orbit, where its motion is *toward* Jupiter, to a diametrically opposite point where it moves *away*

ceiling and forms a spot on the floor 10.0 m below. (a) At what speed (in cm/min) does the spot move across the (flat) floor? (b) If a mirror is placed on the floor to intercept the light, at what speed will the reflected spot move across the ceiling?

45. A drinking glass is 4 cm wide at the bottom, as shown in Figure 35.32. When an observer's eye is placed as shown, the observer sees the edge of the bottom of the glass. When this glass is filled with water, the observer sees the center of the bottom of the glass. Find the height of the glass.

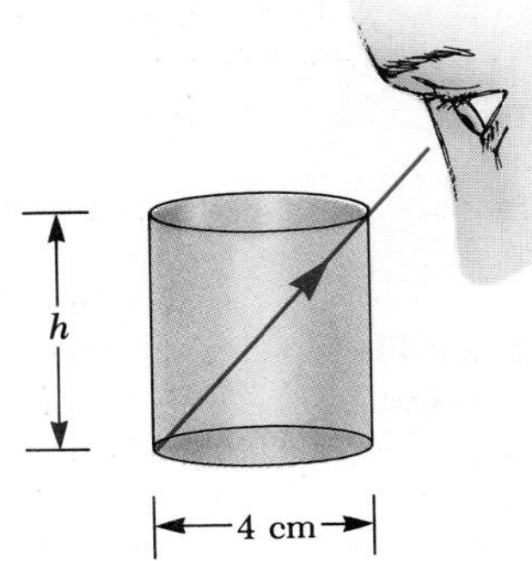

Figure 35.32 (Problem 45).

46. Show that the time required for light to travel from a point source A in air a distance h_1 above a water surface to a point B at h_2 below the water surface is given by

$$t = \frac{h_1 \sec \theta_1}{c} + \frac{h_2 n \sec \theta_2}{c}$$

where n is the index of refraction of water, θ_1 is the angle of incidence, and θ_2 is the angle of refraction.

47. Derive the law of reflection (Eq. 35.2) from Fermat's principle of least time. (See the procedure outlined in Section 35.9 for the derivation of the law of refraction from Fermat's principle.)

48. A thick plate of flint glass ($n = 1.66$) rests on top of a thick plate of Lucite ($n = 1.50$). A beam of light is incident on the top surface of the flint glass at an angle θ_i. The beam passes through the glass and the Lucite and emerges from the Lucite at an angle of 40° with respect to the normal. Calculate the value of θ_i. A sketch of the light path through the two plates of refracting materials might be helpful.

49. A light ray of wavelength 589 nm is incident at an angle θ on the top surface of a block of polystyrene, as shown in Figure 35.33. (a) Find the maximum value of θ for which the refracted ray will undergo *total* internal reflection at the left vertical face of the block. (b) Repeat the calculation for the case in which the polystyrene block is immersed in water. (c) What happens if the block is immersed in carbon disulfide?

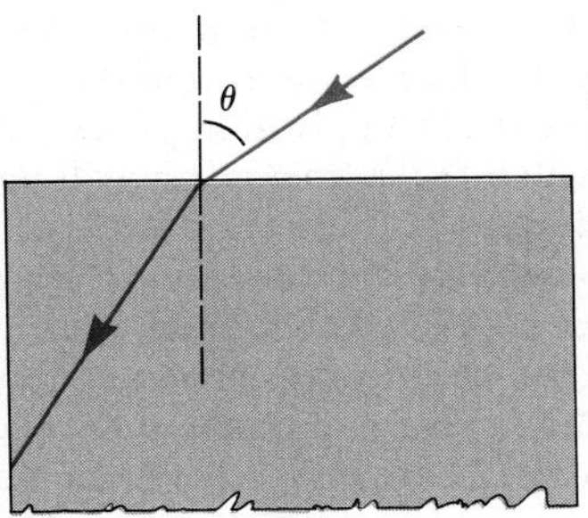

Figure 35.33 (Problem 49).

50. A narrow beam of light passes through a plate of glass with a thickness $s = 1$ cm and a refractive index $n = 1.52$. The beam enters from air at an angle $\theta_1 = 30°$. Calculate the deviation d of the ray, as indicated in Figure 35.34.

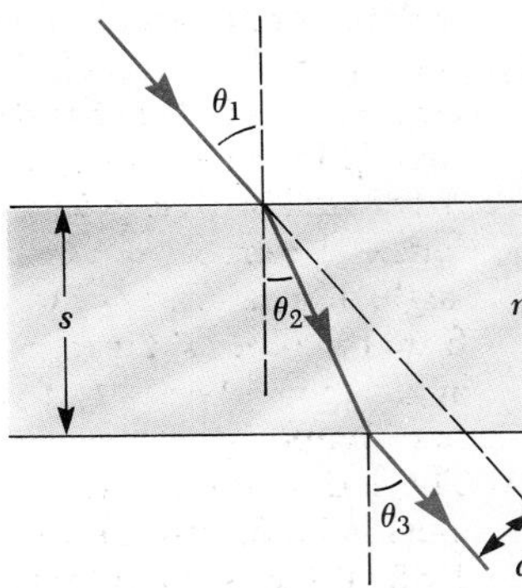

Figure 35.34 (Problem 50).

51. A hiker stands on a mountain peak near sunset and observes a (primary) rainbow caused by water droplets in the air about 8 km away. The valley is 2 km below the mountain peak and is entirely flat. What fraction of the complete circular arc of the rainbow is visible to the hiker?

52. A fish is at a depth d under water. Show that when viewed from an angle of incidence θ_1, the *apparent depth* z of the fish is

$$z = \frac{3d \cos \theta_1}{\sqrt{7 + 9 \cos^2 \theta_1}}$$

53. A light ray is incident on a prism and refracted at the first surface as shown in Figure 35.35. Let Φ repre-

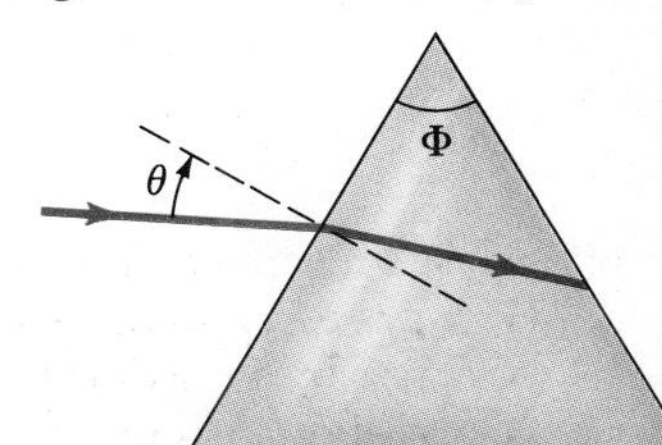

Figure 35.35 (Problem 53).

sent the apex angle of the prism and n its index of refraction. Find in terms of n and Φ the smallest allowed value of the angle of incidence at the first surface for which the refracted ray will *not* undergo internal reflection at the second surface.

54. When a polychromatic (multiwavelength) light ray is incident on a prism, the various wavelength components are deviated by different amounts, as shown in Figure 35.17. This dispersion of the incident light is due to the variation of index of refraction with wavelength (shown for several types of glass in Fig. 35.15). The *dispersive power* of a refracting material is defined by

$$\omega \equiv \frac{n_V - n_R}{n_Y - 1}$$

where n_V, n_R, and n_Y are the respective indices of refraction of three reference wavelengths λ_V, λ_R, and λ_Y. For flint glass, $n_R = 1.644$, $n_Y = 1.650$, and $n_V = 1.665$. For crown glass, $n_R = 1.517$, $n_Y = 1.520$, and $n_V = 1.527$. Calculate the dispersion power of (a) flint glass and (b) crown glass. (c) For small apex angles, the dispersive power can also be written

$$\omega = \frac{\delta_V - \delta_R}{\delta_Y}$$

where the angles of deviation δ are as shown in Figure 35.36. Show that these two forms for ω are equivalent.

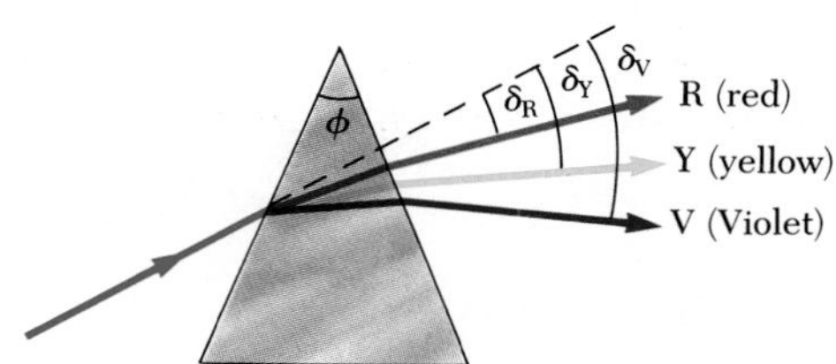

Figure 35.36 (Problem 54).

ESSAY

Fiber Optics: The Ultimate in Telecommunications

Edward A. Lacy

In the 1980s light wave communications—photonics—came of age. The United States and other industrial nations went on a "high-fiber" diet, installing so much fiber-optic cable that the decade was called the "decade of glass." More than 3 million km of these cables were installed in the United States alone. In the process, thousands of miles of copper cables—both coaxial and twisted-pair—were made obsolete as far as long distance telecommunications were concerned. Copper cables have been displaced simply because they do not have the tremendous information-carrying capacity, called *bandwidth,* of fiber-optic cables.

With such bandwidth, fiber-optic systems can transmit thousands of telephone conversations, dozens of television programs, and numerous computer data signals over one or two flexible, hair-thin threads of ultrapure glass called *optical fibers.* Theoretically, the bandwidth of a fiber-optic cable is 50 THz (50 million MHz). For comparison, a television signal, the biggest user of bandwidth, requires 6 MHz of bandwidth, a telephone conversation just a few kilohertz.

Currently, such phenomenal bandwidths are not available because the input and output equipment to the cables are not capable of such speed. But even present systems, with their bandwidths of a few gigahertz, are able to transmit video conference signals and high definition television at economical prices.

These and other broadband services are not available to homes and businesses at present because of cost and government regulations. The long distance fiber-optic trunks across the nation are like superhighways. While most homes are within 50 miles of these super trunks, the connection between the trunks and homes and businesses is copper cable. In effect, the superhighways of telecommunications are connected to your home with dirt footpaths. To get the advantage of broadband services, the subscriber loop must also be replaced with fiber optics.

Running fiber all the way to a customer's home is presently too expensive. In addition, cable television companies and telephone companies are battling to see who will control the fiber to the home. Obviously, only one fiber would be needed per home. This fiber would carry telephone signals as well as cable TV and other signals. Government bodies will have to decide whether the phone company or the cable TV company will provide the service.

Meanwhile, local area networks (LANs) that interconnect computers at a common geographic site are being connected with optical fibers. In some cases these LANs do not use fiber optics because they have low bandwidth requirements. However, LANs at university campuses throughout the world are using fiber optics to

Figure 1 Ultra-pure glass optical fibers carry voice, video and data signals in high-capacity telecommunications networks. (Photo courtesy of Corning Incorporated)

(Continued on next page)

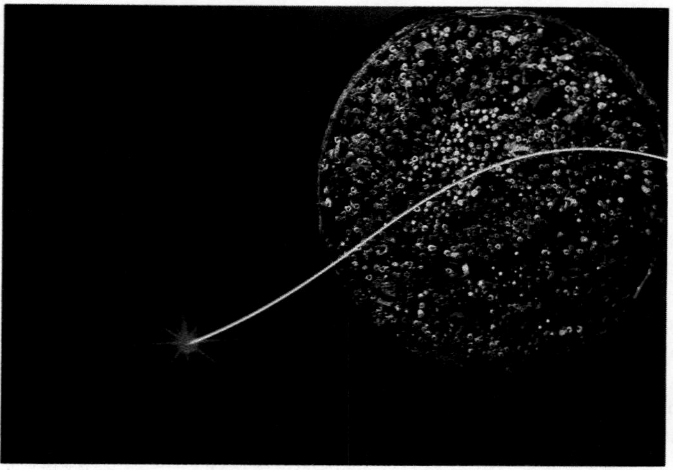

Figure 2 Optical fiber, the transmission medium of choice for telephone company interoffice trunks, proves in against copper for new and rebuild feeder systems. (Photo courtesy of Corning Incorporated)

connect host computers with terminals at dozens of buildings and hundreds of dormitory rooms.

In addition to giving extremely high bandwidth, fiber-optic systems have significantly smaller, lighter-weight cables than conventional telephone cables. An optical fiber with its protective jacket may be, typically, 0.635 cm in diameter, yet it can replace a 7.62-cm diameter bundle of copper wires now used to carry the same number of telephone conversations and other signals. The importance of this dramatic decrease in size is not obvious in uncongested rural areas; in major cities, however, where telephone cables must be placed underground, conduits are so crammed that they can scarcely accommodate a single additional copper cable.

Along with the size reduction, there is a corresponding weight reduction. For example, 94.5 kg of copper wire can be replaced with 3.6 kg of optical fiber. Weight reduction is important for the military services as it allows faster deployment of communication cables on battlefields. On the civilian side, it is important in huge jet aircraft, which use a surprising amount of copper cables between the various equipment and instrumentation on board. By replacing these cables with optical cables, up to 1 000 lb of weight may be saved, thereby giving better fuel consumption.

Perhaps of equal importance, fiber-optic systems are immune to electromagnetic interference such as lightning and arcs created in some factories. Because of this immunity, fiber-optic systems give accurate transmission of data—about 100 times better than transmissions over copper cables. For information transmitted in bits, the accuracy is typically one error in 100 million bits.

Still another advantage of fiber-optic systems is that it is difficult to tap them, as compared to telephone cables, for eavesdropping. In fact, fiber-optic systems are generally considered to be "secure," an important feature for the military. There are numerous other advantages that are important but discussion of them must be reserved for other texts.°

In a fiber-optic communication system, information is carried by lightwaves (photons) rather than by the movement of electrons as in metallic systems.

A fiber-optic communication system has three major components: a transmitter that converts electrical signals to light signals, an optical fiber for transmitting the signals, and a receiver that captures the signals at the other end of the fiber and converts them to electrical signals.

The key part of the transmitter is a light source—either a semiconductor laser diode or a light-emitting diode (LED). In general, laser diodes have greater capabili-

° E.A. Lacy, *Fiber Optics,* Englewood Cliffs, N.J., Prentice-Hall, 1982.

ties than LEDs but cost more, require more complex circuitry, and are less reliable. Laser diodes are generally used for long-haul routes, LEDs for short-haul loops. With either device, the light emitted is an invisible infrared signal with a wavelength of 1.31 μm (1310 nm). Wavelengths that are either higher or lower are significantly attenuated as they cannot pass through the "window" at 1.31 μm. Older systems had a wavelength of 0.85 μm; future systems may have a wavelength of 1.55 μm. Visible and ultraviolet light are impractical for fiber-optic systems because of high losses in the optical fiber. The laser diodes and LEDs used in this application are miniture units less than half the size of a thumbnail in order to couple the light into the tiny fibers effectively. Even so, a glass lens is frequently used to transfer the emitted light effectively to the fiber.

To transmit an audio, television, or computer data signal by light waves, it is necessary to change or *modulate* the light waves in accordance with the information in these signals. By varying the intensity of the light beam from the laser diode or the LED, *analog modulation* is achieved. By flashing the laser diode or LED on and off at an extremely fast rate, *digital modulation* is achieved.

In digital modulation, a pulse of light represents the number 1, and the absence of light represents 0. In a sense, instead of flashes of light traveling down the fiber, 1s and 0s are moving down the path. With computer-type equipment, any communication can be represented by a particular pattern or code of 1s and 0s. If the receiver is programmed to recognize such digital patterns, it can reconstruct the original signal from the 1s and 0s it receives.

Digital modulation is expressed in bits (short for "binary digit") per second, megabits (1 000 000 bits) per second, or gigabits (1 000 000 000 bits) per second. Engineers have demonstrated a fiber-optic system that can transmit 27 gigabits per second. At this rate, 400 000 phone conversations could be held simultaneously over this system.

As remarkable as these bandwidths are, engineers are pursuing other techniques to give even greater bandwidth. In *wavelength division multiplexing* (also called *color* multiplexing) the outputs of two or more lasers with different wavelengths are combined and sent through one optical fiber. In effect, this doubles the bandwidth. In *coherent modulation and detection* a laser sends a continuous beam whose frequency is varied by the messages that are being transmitted. At the receiver, this light beam is combined with a lightbeam that has been generated at the receiver. The frequencies are designed to be slightly different so that when they are combined there will be a new signal which is the difference of the two. The new signal can be processed, as in superheterodyne radio and television sets, much more easily than the original incoming lightbeam.

As you might suspect, the equipment used in digital modulation, such as encoders, is much more complicated than that used in analog modulation. Digital modulation also requires more bandwidth than analog modulation to send the same message. The former is, however, far more popular because it allows greater transmission distance with the same power and less expensive switching equipment. Thus, even though digital telecommunication is only a minority of present telecommunication now, it is rapidly replacing analog transmission which is used mainly for television signals.

Even though an optical fiber is usually made of glass, it is surprisingly tough; in fact, it can be bent and twisted just like wire. Splicing optical fiber, however, can be difficult: the ends of the fibers can be joined by fusion or with mechanical splices, but not by twisting and soldering as with copper cables. In splicing, great care must be taken to ensure precision mating of the tiny fibers. Otherwise the system will have enough losses to make it inoperable.

Optical fibers have very low transmission losses because of their ultra purity. If a window pane 1 km thick were to be made of such glass, it would be as transparent as an ordinary pane of glass. Despite this purity, the light waves eventually become dim

(Continued on next page)

Figure 3 Interference contrast light micrograph of a fiber optics wave guide of the type used for signal transmission in all aspects of communications. This particular type of fiber is known as a monomode. The faint impression of the innermost core of the fiber is visible as a darker shade of red (its refractive index is different from the surrounding material): it is along this inner core that the light signals travel. (Courtesy of Science Photo Library/Photo Researchers)

or attenuated because of absorption and scattering. *Absorption* occurs within the fiber when the light waves encounter impurities and are turned into heat. *Scattering* occurs primarily at splices or junctions in the fiber where light leaves the fiber because of imprecise connections.

Attenuation is measured in decibels per kilometer (dB/km). In most long haul circuits, the attenuation is less than 1 dB/km for signals being transmitted at a wavelength of 1310 nm. When equipment and fibers now being developed to transmit at a wavelength of 1550 nm are perfected, the losses are expected to be less than 0.25 dB/km.

Because of attenuation, the light signals must eventually be regenerated at intervals by devices called *repeaters.* A repeater is a combination of a fiber-optic receiver and a fiber-optic transmitter. The receiver decodes the signal and triggers the transmitter to produce an identical version, only now the signal has greater strength and purity. In the case of digital signals, it is also in better time synchronism. Repeaters are typically placed about 30 km apart, but in the newer systems they may be separated as much as 200 km or more. Whereas present repeaters convert the light signals to electrical signals and then back to light signals, *all-optical* repeaters being developed will convert weak signals directly to strong light signals, skipping the conversion from light signals to electric signals to light signals. By doing this, these repeaters will theoretically be much more sensitive and thus can be placed farther apart. In addition, they will be smaller, cost less, and have greater reliability.

Each fiber has three parts. At the center of the fiber is the *core,* which carries the light signal. A concentric layer of glass about 125 μm in diameter, called the *cladding,* surrounds the core. Because the cladding has a different index of refraction than the core, total internal reflection occurs in the core, keeping the light in the core. Surrounding the cladding is a polyurethane *jacket* that protects the fiber from abrasion, crushing, and chemicals. From one to several hundred fibers are grouped to form a cable.

In large-core fibers, typically with core diameter of 62.5 μm, light pulses can take numerous paths (called *modes*) as they bounce back and forth down the fiber. Because the different paths are not equal in length, some pulses will take longer to travel down the fiber, causing some of the pulses to overlap and thereby cause distortion. These *multimode* fibers, which are less expensive than other fibers, were widely used in the early days of fiber optics but now are used mainly for certain short-distance links such as local area networks.

In the newer fibers, the core is much smaller: only 8 μm in diameter, about one sixth the thickness of a sheet of paper. Because of the small diameter, only one light path is possible—straight down the core with no zigzagging. As there is only one path, there is much less distortion, giving these *single-mode* fibers a much higher bandwidth than multimode fibers.

At the end of the fiber, a photodiode converts the light signals to electric signals, which are then amplified and decoded, if necessary, to reform the signals originally transmitted.

While long-distance fiber-optic systems have received most of the attention, fiber-optic links are also useful for very short distances, such as *between* large computer mainframes and their peripheral terminals and printers. *Within* computers, fiber optics is being used to carry signals between circuit boards.

In an entirely different use of fiber optics, in a nontransmission application, optical fibers are being used as sensors to detect strain, pressure, temperature, and other stimulus. In this function, the fibers offer the advantages of compactness, sensitivity, and immunity to hostile environments, as compared to other sensors. For example, a fiber-optic sensor is being used to measure the temperature of a volcano. In another use, by embedding fiber-optic strain sensors in polymer composites, used to replace metal on some aircraft, it may be possible to give an aircraft a "smart skin" that would warn the pilot of dangerous strains in the wings or fuselage.

In one type of sensor, short lengths of optical fibers are made with intentional small lateral deformations called *microbends.* At these microbends some of the light radiates from the fiber. The behavior of this light can be influenced by temperature, acceleration, and other parameters to be sensed. In the optical interferometer type of sensor, light from a laser is transmitted down two paths: a reference path and a measuring path. The perturbation being measured will cause the light through the measuring path to be out of phase with the light through the reference path when they are recombined. This phase difference can then be converted to an amplitude change.

While most optical fibers are made of glass, *plastic* fibers serve a useful role in short-range data links, electronic billboards, automobile displays, and automobile electrical systems. Plastic fibers have excessive losses when used for distances greater than 1 km, but for short distances they have the advantages of being more flexible than glass fibers and less expensive.

Suggested Readings

Books

Lacy, Edward A., *Fiber Optics,* Prentice-Hall, Inc. Englewood Cliffs, NJ, 1982 (an introductory text).

Basch, E.E. (Editor-in-chief), *Optical-Fiber Transmission;* Indianapolis, Howard W. Sams & Co., 1987 (for the more advanced student).

Chaffee, C. David, *The Rewiring of America: The Fiber Optics Revolution,* San Diego, Academic Press, CA, 1987.

Trade Magazines

Lightwave, The Journal of Fiber Optics
Laser focus/Electro-optics
Lasers and Applications

Journals

I.E.E.E. Spectrum
I.E.E.E. Journal of Lightwave Technology

Essay Questions

1. How will business and society be affected if the cost of transferring information becomes insignificant?
2. Why is fiber optics useful as invasive medical probes?
3. Why would unauthorized "tapping" of optical fibers be difficult, if not impossible?
4. If economical superconductors are developed, will they displace optical fibers?
5. In addition to optical fibers, what devices had to be developed in order for lightwave communication to be practical?

Essay Problems

1. What is the percentage of weight reduction of fiber-optic cables in comparison with copper cables?
2. If a telephone conversation requires 3 kHz, what is the theoretical maximum number of conversations that could be carried on a fiber-optic system without using special modulation techniques?
3. When fiber-optic systems start using the 1550 nm wavelength, how much further apart, relatively, can the repeaters be placed, as compared with spacing for the 1310-nm wavelength systems?

36
Geometric Optics

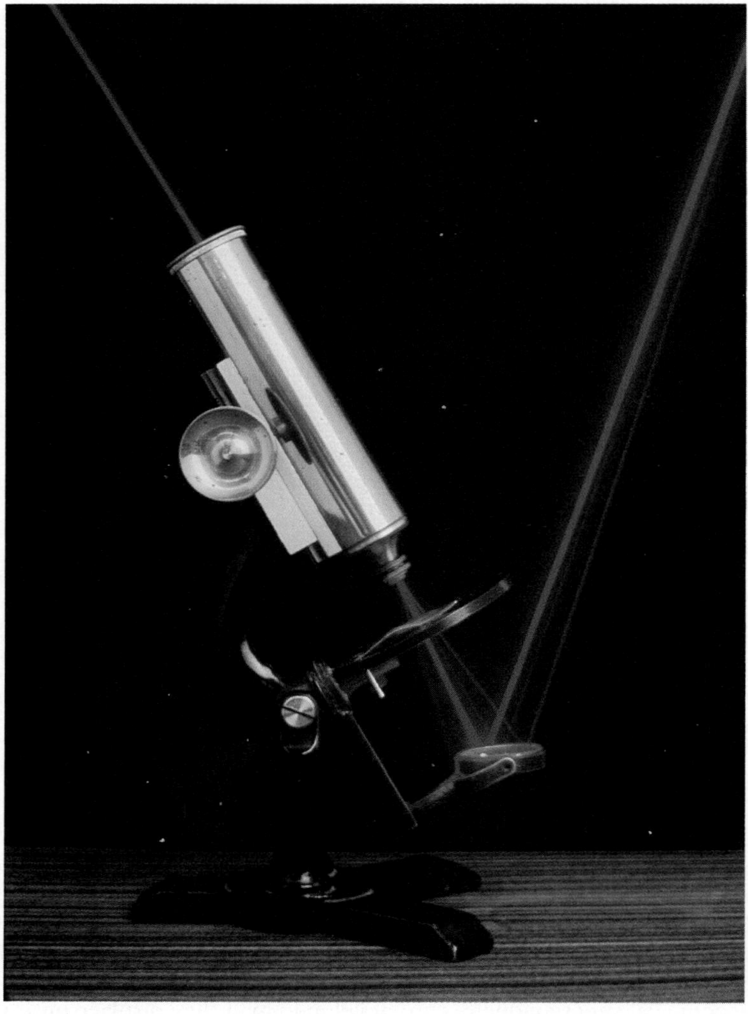

A microscope is a common device used to view small objects with the assistance of a light source. This particular microscope is accompanied by a laser beam. (Garry Gay © THE IMAGE BANK)

This chapter is concerned with the study of the formation of images when spherical waves fall on plane and spherical surfaces. We shall find that images can be formed by reflection or by refraction. From a practical viewpoint, mirrors and lenses are devices that work on the basis of image formation by reflection and refraction. Such devices, commonly used in optical instruments and systems, will be described in some detail. We shall continue to use the ray approximation and to assume that light travels in straight lines. This corresponds to the field of geometric optics. In subsequent chapters, we shall concern ourselves with interference and diffraction effects, or the field of wave optics.

36.1 IMAGES FORMED BY PLANE MIRRORS

One of the objectives of this chapter will be to discuss the manner in which optical elements such as lenses and mirrors form images. We shall begin this investigation by considering the simplest possible mirror, the plane mirror. Throughout our discussion of both mirrors and lenses, we shall use the ray model of light.

Consider a point source of light placed at O in Figure 36.1, a distance s in front of a plane mirror. The distance s is called the **object distance.** Light rays leave the source and are reflected from the mirror. After reflection, the rays diverge (spread apart), but they appear to the viewer to come from a point I located behind the mirror. Point I is called the **image** of the object at O. Regardless of the system under study, images are always formed in the same way. *Images are formed at the point where rays of light actually intersect or at the point from which they appear to originate.* Since the rays in Figure 36.1 appear to originate at I, which is a distance s' behind the mirror, this is the location of the image. The distance s' is called the **image distance.**

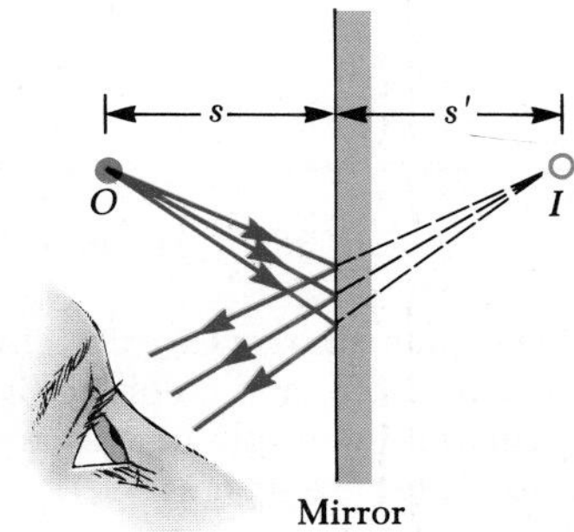

Figure 36.1 An image formed by reflection from a plane mirror. The image point, I, is located behind the mirror at a distance s', which is equal to the object distance, s.

Images are classified as real or virtual. A **real image** *is one in which light actually intersects, or passes through, the image point;* a **virtual image** *is one in which the light does not really pass through the image point but appears to diverge from that point.* The image formed by the plane mirror in Figure 36.1 is a virtual image. The images seen in plane mirrors *are always virtual* for real objects. Real images can usually be displayed on a screen (as at a movie), but virtual images cannot be displayed on a screen.

We shall examine some of the properties of the images formed by plane mirrors by using the simple geometric techniques shown in Figure 36.2. In order to find out where an image is formed, it is always necessary to follow at least two rays of light as they reflect from the mirror. One of those rays starts at P, follows a horizontal path to the mirror, and reflects back on itself. The second ray follows the oblique path PR and reflects as shown. An observer to the left of the mirror would trace the two reflected rays back to the point from which they appear to have originated, that is, point P'. A continuation of this process for points on the object other than P would result in a virtual image (drawn as a yellow arrow) to the right of the mirror. Since triangles PQR and $P'QR$ are congruent, $PQ = P'Q$. Hence, we conclude that the *image formed by an object placed in front of a plane mirror is as far behind the mirror as the object is in front of the mirror.* Geometry also shows that the object height, h, equals the image height, h'. Let us define **lateral magnification,** M, as follows:

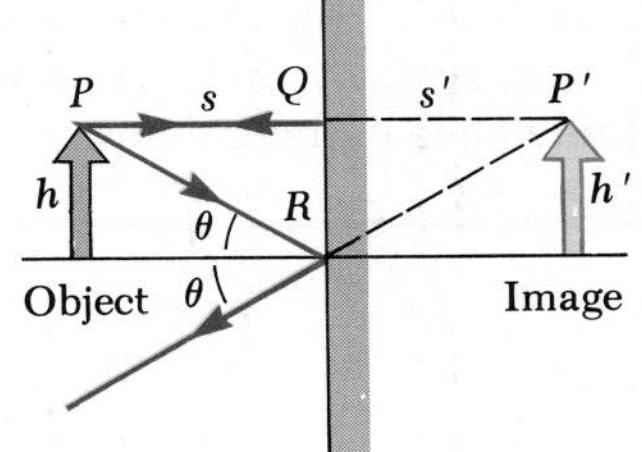

Figure 36.2 Geometric construction used to locate the image of an object placed in front of a plane mirror. Because the triangles PQR and $P'QR$ are congruent, $s = s'$ and $h = h'$.

$$M \equiv \frac{\text{image height}}{\text{object height}} = \frac{h'}{h} \tag{36.1}$$

Magnification

This is a general definition of the lateral magnification of any type of mirror. $M = 1$ for a plane mirror because $h' = h$ in this case.

The image formed by a plane mirror has one more important property, that of right-left reversal between image and object. This reversal can be seen by standing in front of a mirror and raising your right hand. The image you see raises its left hand. Likewise, your hair appears to be parted on the opposite side and a mole on your right cheek appears to be on your left cheek.

Thus, we conclude that the image formed by a plane mirror has the following properties:

1. The image is as far behind the mirror as the object is in front.
2. The image is unmagnified, virtual, and erect. (By erect we mean that, if the object arrow points upward as in Figure 36.2, so does the image arrow.)
3. The image has right-left reversal.

EXAMPLE 36.1 Multiple Images Formed by Two Mirrors

Two plane mirrors are at right angles to each other, as in Figure 36.3, and an object is placed at point O. In this situation, multiple images are formed. Locate the positions of these images.

Solution The image of the object is at I_1, in mirror 1 and at I_2 in mirror 2. In addition, a third image is formed at I_3, which will be considered to be the image of I_1 in mirror 2 or, equivalently, the image of I_2 in mirror 1. That is, the image at I_1 (or I_2) serves as the object for I_3. When viewing I_3, note that the rays reflect twice after leaving the object at O.

Exercise 1 Sketch the rays corresponding to viewing the images at I_1 and I_2 and show that the light is reflected only once in these cases.

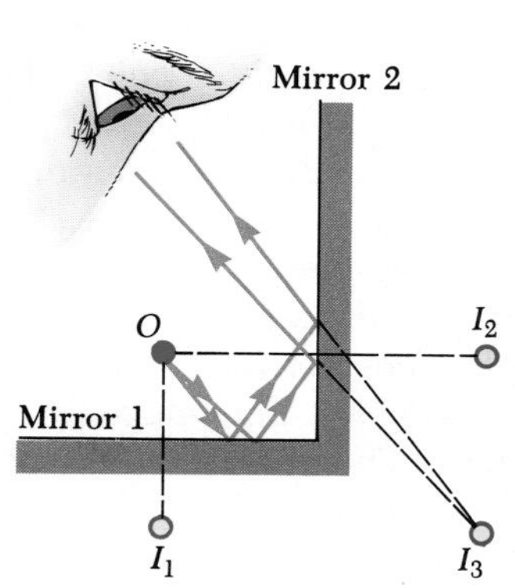

Figure 36.3 (Example 36.1) When an object is placed in front of two mutually perpendicular mirrors as shown, three images are formed.

36.2 IMAGES FORMED BY SPHERICAL MIRRORS

Concave Mirrors

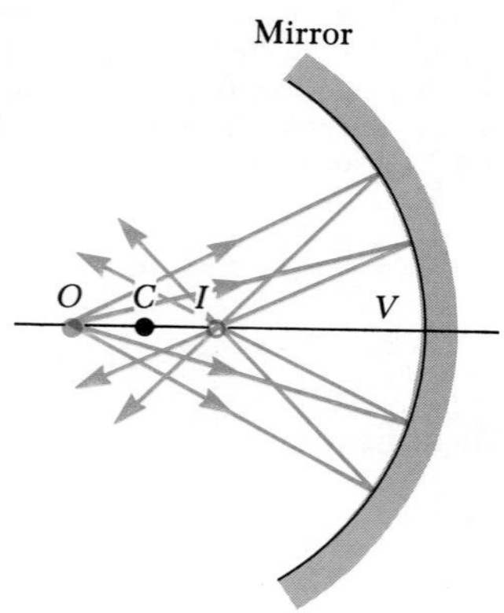

Figure 36.4 A point object placed at O, outside the center of curvature of a concave spherical mirror, forms a real image at I. If the rays diverge from O at small angles, they reflect through the same image point.

A spherical mirror, as its name implies, has the shape of a segment of a sphere. Figure 36.4 shows the cross section of a spherical mirror with light reflecting from its surface, represented by the solid curved line. Such a mirror, in which light is reflected from the inner, concave surface, is called a **concave mirror.** The mirror has a radius of curvature R, and its center of curvature is located at point C. Point V is the center of the spherical segment, and a line drawn from C to V is called the **principal axis** of the mirror.

Now consider a point source of light placed at point O in Figure 36.4, located on the principal axis and outside point C. Several diverging rays originating at O are shown. After reflecting from the mirror, these rays converge and meet at I, called the **image point.** The rays then continue to diverge from I as if there were an object there. As a result, we have a real image formed. *Real images are always formed at a point when reflected light actually passes through the point.*

In what follows, we shall assume that all rays that diverge from the object make a small angle with the principal axis. Such rays are called **paraxial rays.** All such rays reflect through the image point, as in Figure 36.4. Rays that are far from the principal axis, as in Figure 36.5, converge to other points on the principal axis, producing a blurred image. This effect, called **spherical aberra-**

tion, is present to some extent for any spherical mirror and will be discussed in Section 36.5.

We can use the geometry shown in Figure 36.6 to calculate the image distance, s', from a knowledge of the object distance, s, and radius of curvature, R. By convention, these distances are measured from point V. Figure 36.6 shows two rays of light leaving the tip of the object. One of these rays passes through the center of curvature, C, of the mirror, hitting the mirror head on (perpendicular to the mirror surface) and reflecting back on itself. The second ray strikes the mirror at the center, point V, and reflects as shown, obeying the law of reflection. The image of the tip of the arrow is located at the point where these two rays intersect. From the largest triangle in Figure 36.6 we see that $\tan \theta = h/s$, while the smallest triangle gives $\tan \theta = -h'/s'$. The negative sign signifies that the image is inverted, so h' is negative. Thus, from Equation 36.1 and these results, we find that the magnification of the mirror is

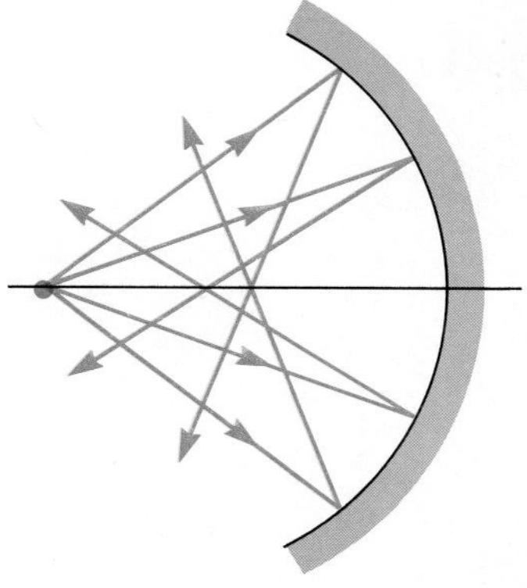

Figure 36.5 Rays at large angles from the horizontal axis reflect from a spherical concave mirror to intersect the principal axis at different points, resulting in a blurred image. This is called *spherical aberration.*

$$M = \frac{h'}{h} = -\frac{s'}{s} \tag{36.2}$$

We also note from two other triangles in the figure that

$$\tan \alpha = \frac{h}{s-R} \qquad \text{and} \qquad \tan \alpha = -\frac{h'}{R-s'}$$

from which we find that

$$\frac{h'}{h} = -\frac{R-s'}{s-R} \tag{36.3}$$

If we compare Equations 36.2 and 36.3 we see that

$$\frac{R-s'}{s-R} = \frac{s'}{s}$$

Simple algebra reduces this to

$$\frac{1}{s} + \frac{1}{s'} = \frac{2}{R} \tag{36.4}$$

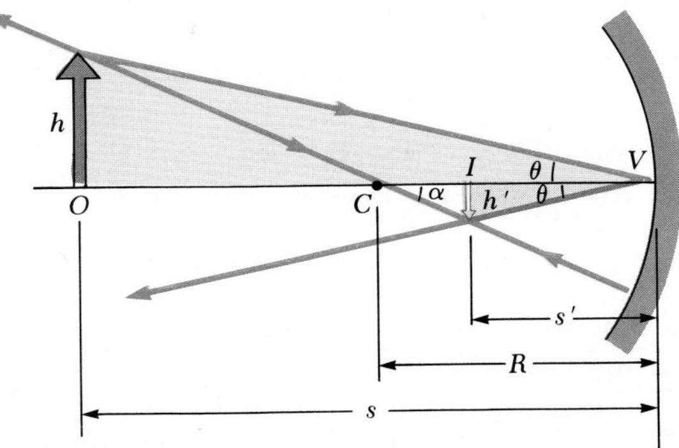

Figure 36.6 The image formed by a spherical concave mirror where the object O lies outside the center of curvature, C.

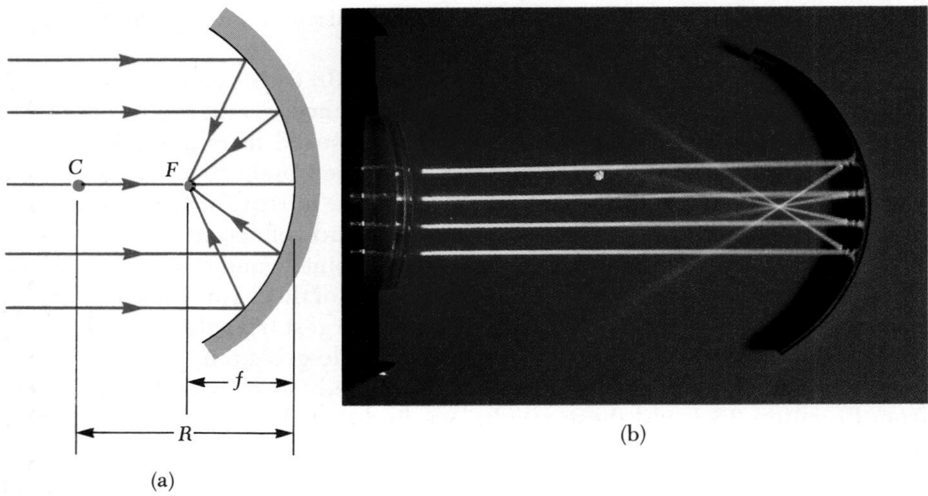

Figure 36.7 (a) Light rays from a distant object ($s = \infty$) reflect from a concave mirror through the focal point, F. In this case, the image distance $s' = R/2 = f$, where f is the focal length of the mirror. (b) Photograph of the reflection of parallel rays from a concave mirror. (Courtesy of Henry Leap and Jim Lehman)

This expression is called the **mirror equation.** This equation is applicable only to paraxial rays.

If the object is very far from the mirror, that is, if the object distance, s, is large enough compared with R that s can be said to approach infinity, then $1/s \approx 0$, and we see from Equation 36.4 that $s' \approx R/2$. That is, when the object is very far from the mirror, *the image point is halfway between the center of curvature and the center of the mirror,* as in Figure 36.7a. The rays are essentially parallel in this figure because the source is assumed to be very far from the mirror. We call the image point in this special case the **focal point,** F, and the image distance the **focal length,** f, where

Focal length

$$f = \frac{R}{2} \tag{36.5}$$

The mirror equation can therefore be expressed in terms of the focal length:

Mirror equation

$$\frac{1}{s} + \frac{1}{s'} = \frac{1}{f} \tag{36.6}$$

Convex Mirrors

Figure 36.8 shows the formation of an image by a **convex mirror,** that is, one silvered such that light is reflected from the outer, convex surface. This is sometimes called a **diverging mirror** because the rays from any point on a real object diverge after reflection as though they were coming from some point behind the mirror. The image in Figure 36.8 is virtual rather than real because it lies behind the mirror at the location from which the reflected rays appear to

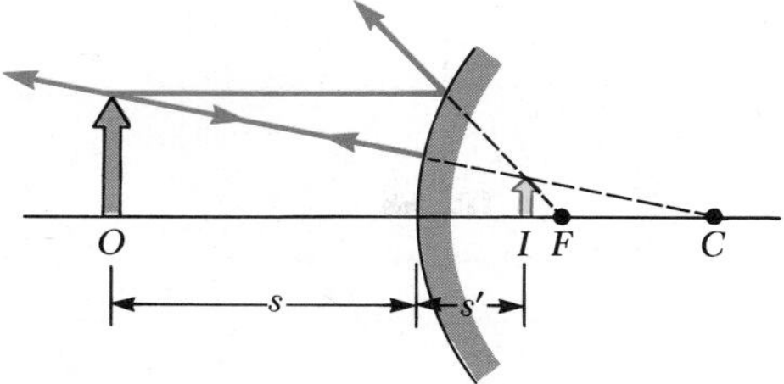

Figure 36.8 Formation of an image by a spherical convex mirror. The image formed by the real object is virtual and erect.

originate. Furthermore, the image will always be erect, virtual, and smaller than the object, as shown in the figure.

We shall not derive any equations for convex spherical mirrors. Such derivations show that the equations developed for concave mirrors can be used if we adhere to a particular sign convention.

We can use Equations 36.2, 36.4, and 36.6 for either concave or convex mirrors if we adhere to the following procedure. Let us refer to the region in which light rays move as the *front side* of the mirror and the other side, where virtual images are formed, as the *back side.* For example, in Figures 36.6 and 36.8, the side to the left of the mirrors is the front side and the side to the right of the mirrors is the back side. Table 36.1 summarizes the sign conventions for all the necessary quantities.

Ray Diagrams for Mirrors

The position and size of images formed by mirrors can be conveniently determined by using *ray diagrams.* These graphical constructions tell us the total nature of the image and can be used to check parameters calculated from the mirror and magnification equations. In these diagrams, one needs to know the position of the object and the location of the center of curvature. In order to locate the image, three rays are then constructed, as shown by the various examples in Figure 36.9. These rays all start from any object point (chosen to be the top) and are drawn as follows:

1. The first, labeled ray 1, is drawn from the top of the object parallel to the optical axis and is reflected back through the focal point, F.

TABLE 36.1 Sign Convention for Mirrors

s is + if the object is in front of the mirror (real object).
s is − if the object is in back of the mirror (virtual object).

s' is + if the image is in front of the mirror (real image).
s' is − if the image is in back of the mirror (virtual image).

Both f and R are + if the center of curvature is in front of the mirror (concave mirror).
Both f and R are − if the center of curvature is in back of the mirror (convex mirror).

If M is positive, the image is erect.
If M is negative, the image is inverted.

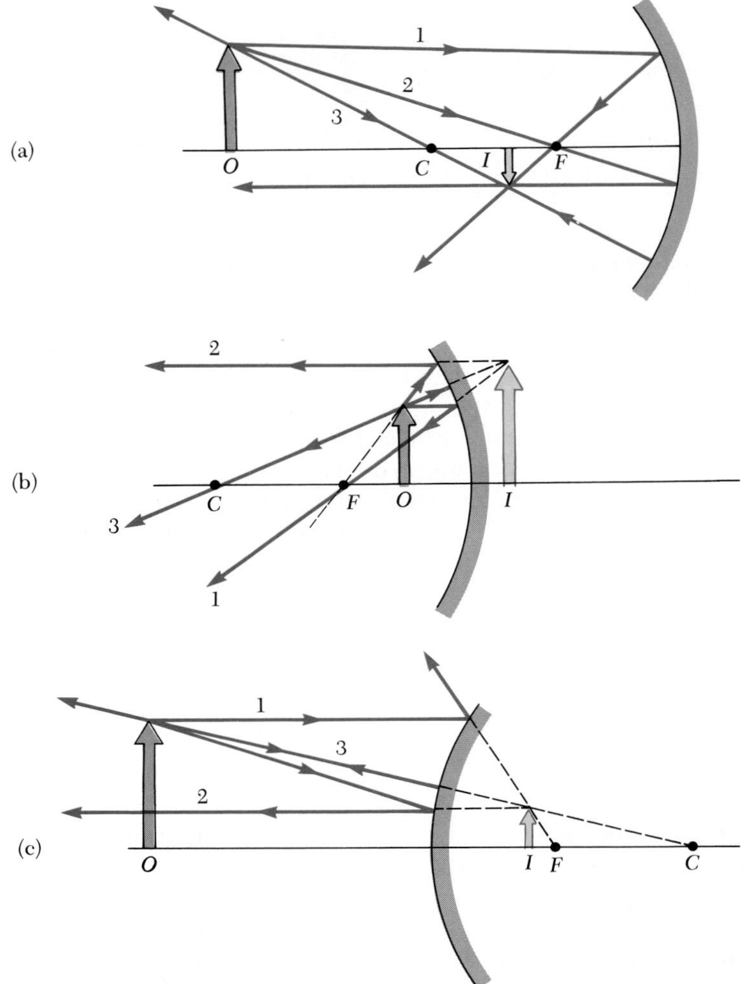

Figure 36.9 Ray diagrams for spherical mirrors. (a) The object is located outside the center of curvature of a spherical concave mirror. (b) The object is located between the spherical concave mirror and the focal point, F. (c) The object is located in front of a spherical convex mirror.

2. The second, labeled ray 2, is drawn from the top of the object to the vertex of the mirror and is reflected with the angle of incidence equal to the angle of reflection.
3. The third, labeled ray 3, is drawn from the top of the object through the center of curvature, C, which is reflected back on itself.

The intersection of any *two* of these rays at a point locates the image. The third ray serves as a check on your construction. The image point obtained in this fashion must always agree with the value of s' calculated from the mirror formula.

In the case of a concave mirror, note what happens as the object is moved closer to the mirror. The real, inverted image in Figure 36.9a moves to the left as the object approaches the focal point. When the object is at the focal point, the image is infinitely far to the left. However, when the object lies between the focal point and the vertex, as in Figure 36.9b, the image is virtual and erect. Finally, for the convex mirror shown in Figure 36.9c, the image of a real object is always virtual and erect. In this case, as the object distance increases, the virtual image decreases in size and approaches the focal point as s approaches infinity. You should construct other diagrams to verify the variation of image position with object position.

EXAMPLE 36.2 The Image for a Concave Mirror
Assume that a certain concave spherical mirror has a focal length of 10 cm. Find the location of the image for object distances of (a) 25 cm, (b) 10 cm, and (c) 5 cm. Describe the image in each case.

Solution (a) For an object distance of 25 cm, we find the image distance using the mirror equation:

$$\frac{1}{s}+\frac{1}{s'}=\frac{1}{f}$$

$$\frac{1}{25\text{ cm}}+\frac{1}{s'}=\frac{1}{10\text{ cm}}$$

$$s' = 16.7\text{ cm}$$

The magnification is given by Equation 36.2:

$$M=-\frac{s'}{s}=-\frac{16.7\text{ cm}}{25\text{ cm}}=-0.668$$

Thus, the image is smaller than the object. Furthermore, the image is inverted because M is negative. Finally, because s' is positive, the image is located on the front side of the mirror and is real. This situation is pictured in Figure 36.9a.

(b) When the object distance is 10 cm, the object is located at the focal point. Substituting the values $s = 10$ cm and $f = 10$ cm into the mirror equation, we find

$$\frac{1}{10\text{ cm}}+\frac{1}{s'}=\frac{1}{10\text{ cm}}$$

$$s' = \infty$$

Thus, we see that rays of light originating from an object located at the focal point of a mirror are reflected such that the image is formed at an infinite distance from the mirror; that is, the rays travel parallel to one another after reflection.

(c) When the object is at the position $s = 5$ cm, it is inside the focal point of the mirror. In this case, the mirror equation gives

$$\frac{1}{5\text{ cm}}+\frac{1}{s'}=\frac{1}{10\text{ cm}}$$

$$s' = -10\text{ cm}$$

That is, the image is virtual since it is located behind the mirror. The magnification is

$$M=-\frac{s'}{s}=-\left(\frac{-10\text{ cm}}{5\text{ cm}}\right)=2$$

From this, we see that the image is magnified by a factor of 2, and the positive sign indicates that the image is erect (Fig. 36.9b).

Note the characteristics of the images formed by a concave spherical mirror. When the object is outside the focal point, the image is inverted and real; at the focal point, the image is formed at infinity; inside the focal point, the image is erect and virtual.

Exercise 2 If the object distance is 20 cm, find the image distance and the magnification of the mirror.
Answer $s' = 20$ cm, $M = -1$.

EXAMPLE 36.3 The Image for a Convex Mirror
An object 3 cm high is placed 20 cm from a convex mirror having a focal length of 8 cm. Find (a) the position of the final image and (b) the magnification of the mirror.

Solution (a) Since the mirror is convex, its focal length is negative. To find the image position, we use the mirror equation:

$$\frac{1}{s}+\frac{1}{s'}=\frac{1}{f}=-\frac{1}{8\text{ cm}}$$

$$\frac{1}{s'}=-\frac{1}{8\text{ cm}}-\frac{1}{20\text{ cm}}$$

$$s' = -5.71\text{ cm}$$

The negative value of s' indicates that the image is virtual, or behind the mirror, as in Figure 36.9c.

(b) The magnification of the mirror is

$$M=-\frac{s'}{s}=-\left(\frac{-5.71\text{ cm}}{20\text{ cm}}\right)=0.286$$

The image is erect because M is positive.

Exercise 3 Find the height of the image.
Answer 0.857 cm.

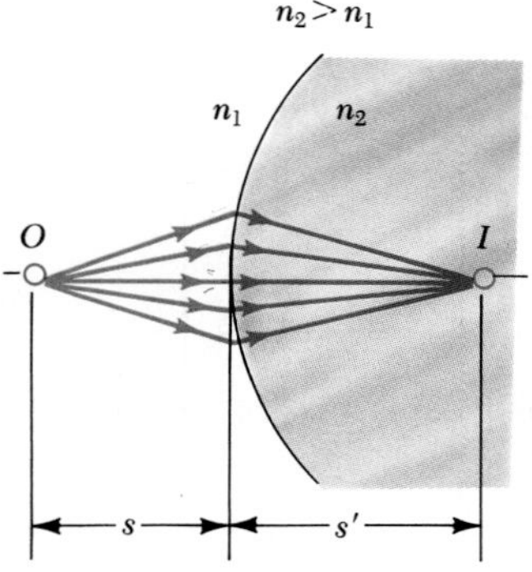

Figure 36.10 Image formed by refraction at a spherical surface. Rays making small angles with the optic axis diverge from a point object at O and pass through the image point, I.

36.3 IMAGES FORMED BY REFRACTION

In this section, we shall describe how images are formed by the refraction of rays at a spherical surface of a transparent material. Consider two transparent media with indices of refraction n_1 and n_2, where the boundary between the two media is a spherical surface of radius R (Fig. 36.10). We shall assume that the object at point O is in the medium whose index of refraction is n_1. Furthermore, let us consider paraxial rays leaving the point O that make a small angle with the axis and with each other. As we shall see, all such rays originating at the object point will be refracted at the spherical surface and focus at a single point I, the image point.

Let us proceed by considering the geometric construction in Figure 36.11, which shows a single ray leaving point O and focusing at point I. Snell's law applied to this refracted ray gives

$$n_1 \sin \theta_1 = n_2 \sin \theta_2$$

Because the angles θ_1 and θ_2 are assumed to be small, we can use the approximations $\sin \theta_1 \approx \theta_1$ and $\sin \theta_2 \approx \theta_2$ (angles in radians). Therefore, Snell's law becomes

$$n_1\theta_1 = n_2\theta_2$$

Now we make use of the fact that an exterior angle of any triangle equals the sum of the two opposite interior angles. Applying this to the triangles OPC and PIC in Figure 36.11 gives

$$\theta_1 = \alpha + \beta$$

$$\beta = \theta_2 + \gamma$$

If we combine the last three relations, and eliminate θ_1 and θ_2, we find

$$n_1\alpha + n_2\gamma = (n_2 - n_1)\beta \tag{36.7}$$

Again, in the small angle approximation, $\tan \theta \approx \theta$, so we can write the approximate relations

$$\alpha = \frac{d}{s}, \qquad \beta = \frac{d}{R}, \qquad \gamma = \frac{d}{s'}$$

where d is the distance shown in Figure 36.11. We substitute these into Equation 36.7 and divide through by d to give

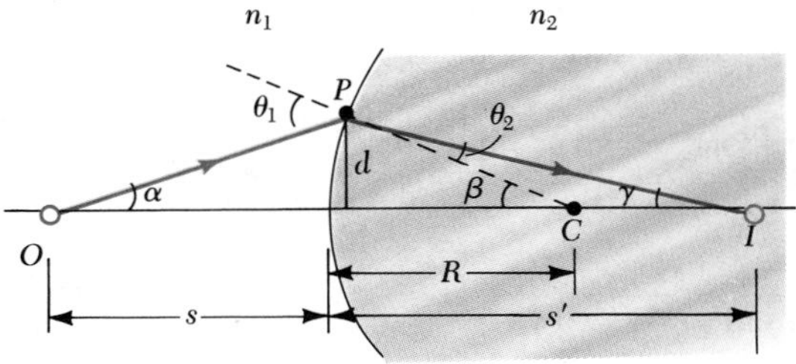

Figure 36.11 Geometry used to derive Equation 36.8.

TABLE 36.2 Sign Convention for Refracting Surfaces

s is + if the object is in front of the surface (real object). s is − if the object is in back of the surface (virtual object).
s' is + if the image is in back of the surface (real image). s' is − if the image is in front of the surface (virtual image).
R is + if the center of curvature is in back of the surface. R is − if the center of curvature is in front of the surface.

$$\frac{n_1}{s} + \frac{n_2}{s'} = \frac{n_2 - n_1}{R} \tag{36.8}$$

For a fixed object distance s, the image distance s' is independent of the angle that the ray makes with the axis. This tells us that all paraxial rays focus at the same point I.

As was the case for mirrors, we must use a sign convention if we are to apply this equation to a variety of circumstances. First note that real images are formed on the side of the surface that is *opposite* the side from which the light comes, in contrast to mirrors, whereas real images are formed on the *same* side of the reflecting surface. Therefore, *the sign convention for spherical refracting surfaces is similar to the convention for mirrors, recognizing the change in sides of the surface for real and virtual images.* For example, in Figure 36.11, s, s', and R are all positive.

The sign convention for spherical refracting surfaces is summarized in Table 36.2. The same sign convention will be used for thin lenses, which will be discussed in the next section. As with mirrors, we assume that the front of the refracting surface is the side from which the light approaches the surface.

Plane Refracting Surfaces

If the refracting surface is a plane, then R approaches infinity and Equation 36.8 reduces to

$$\frac{n_1}{s} = -\frac{n_2}{s'}$$

or

$$s' = -\frac{n_2}{n_1}s \tag{36.9}$$

The ratio n_2/n_1 represents the index of refraction of medium 2 *relative* to that of medium 1. From Equation 36.9 we see that the sign of s' is opposite that of s. Thus, *the image formed by a plane refracting surface is on the same side of the surface as the object.* This is illustrated in Figure 36.12 for the situation in which n_1 is greater than n_2, where a virtual image is formed between the object and the surface. If n_1 is less than n_2, the image will still be virtual but will be formed to the left of the object.

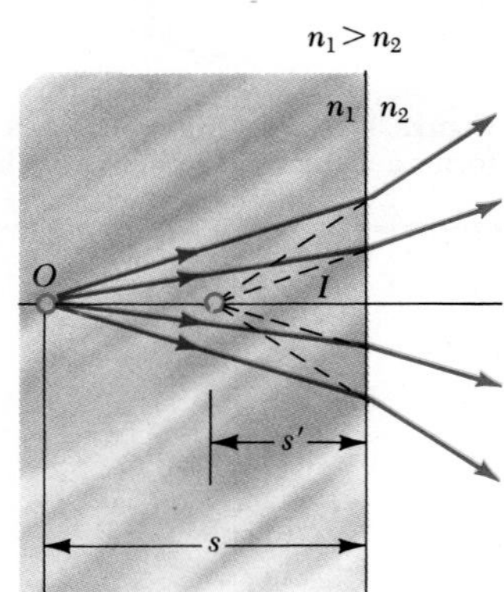

Figure 36.12 The image formed by a plane refracting surface is virtual, that is, it forms to the left of the refracting surface.

EXAMPLE 36.4 Gaze into the Crystal Ball
A coin 2 cm in diameter is embedded in a solid glass ball of radius 30 cm (Fig. 36.13). The index of refraction of the ball is 1.5, and the coin is 20 cm from the surface. Find the position and height of the image.

Solution The rays originating from the object are refracted away from the normal at the surface and diverge outward. Hence, the image is formed in the glass and is virtual. Applying Equation 36.8 and taking $n_1 = 1.5$, $n_2 = 1$, $s = 20$ cm, and $R = -30$ cm, we get

$$\frac{n_1}{s} + \frac{n_2}{s'} = \frac{n_2 - n_1}{R}$$

$$\frac{1.5}{20 \text{ cm}} + \frac{1}{s'} = \frac{1 - 1.5}{-30 \text{ cm}}$$

$$s' = -17.1 \text{ cm}$$

The negative sign indicates that the image is in the same medium as the object (the side of incident light), in agreement with our ray diagram. Since the image is in the same medium as the object, it must be virtual.

Exercise 4 Find the diameter of the coin's image.
Answer 1.7 cm.

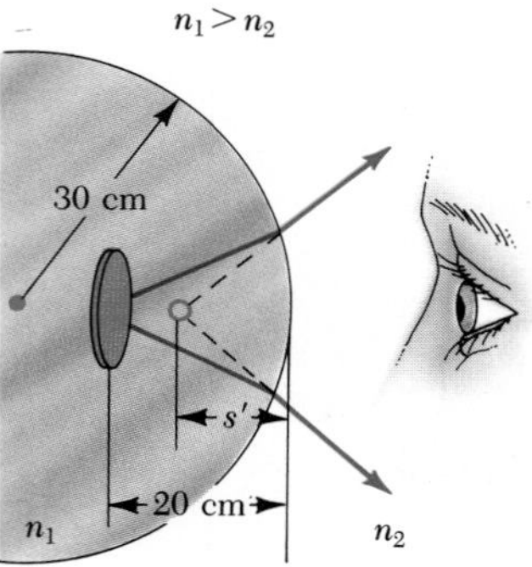

Figure 36.13 (Example 36.4) A coin embedded in a glass ball forms a virtual image between the coin and the glass surface.

EXAMPLE 36.5 The One That Got Away
A small fish is swimming at a depth d below the surface of a pond (Fig. 36.14). What is the *apparent depth* of the fish as viewed from directly overhead?

Solution In this example, the refracting surface is a plane, and so R is infinite. Hence, we can use Equation 36.9 to determine the location of the image. Using the facts that $n_1 = 1.33$ for water and $s = d$ gives

$$s' = -\frac{n_2}{n_1} s = -\frac{1}{1.33} d = -0.750d$$

Again, since s' is negative, the image is virtual, as indicated in Figure 36.14. The apparent depth is three fourths the actual depth. For instance, if $d = 4$ m, then $s' = -3$ m.

Exercise 5 If the fish is 10 cm long, how long is its image?
Answer 10 cm.

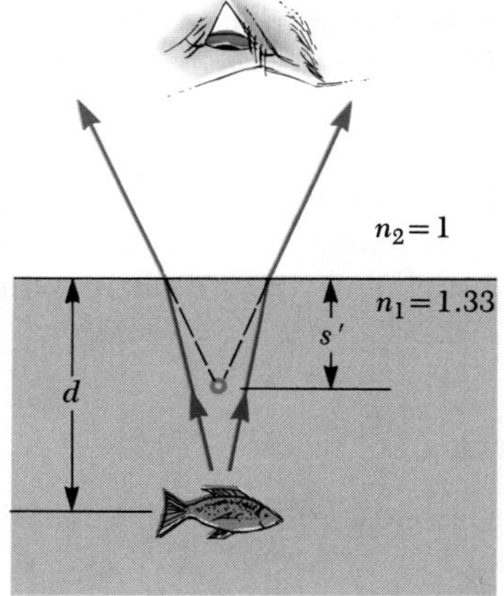

Figure 36.14 (Example 36.5) The apparent depth, s', of the fish is less than the true depth, d.

36.4 THIN LENSES

Lenses are commonly used to form images by refraction in optical instruments, such as cameras, telescopes, and microscopes. The methods discussed in the previous section will be used here to locate the image position. The essential idea in locating the final image of a lens is to *use the image formed by one refracting surface as the object for the second surface.*

Consider a lens having an index of refraction n and two spherical surfaces of radii of curvature R_1 and R_2, as in Figure 36.15. An object is placed at point O at a distance s_1 in front of the first refracting surface. For this example, s_1 has been chosen so as to produce a virtual image I_1, located to the left of the

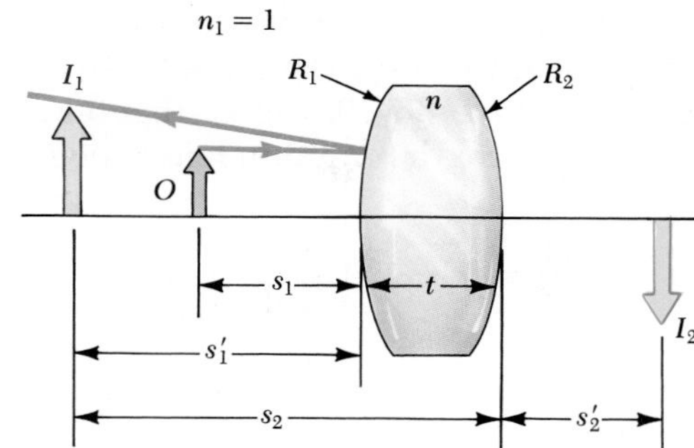

Figure 36.15 To locate the image of a lens, the image at I_1 formed by the first surface is used as the object for the second surface. The final image is at I_2.

lens. This image is used as the object for the second surface, of radius R_2, which results in a real image I_2.

Using Equation 36.8 and assuming $n_1 = 1$, we find that the image formed by the first surface satisfies the equation

$$(1) \qquad \frac{1}{s_1} + \frac{n}{s_1'} = \frac{n-1}{R_1}$$

Now we apply Equation 36.8 to the second surface, taking $n_1 = n$ and $n_2 = 1$. That is, light approaches the second refracting surface *as if* it had come from the image, I_1, formed by the first refracting surface. Taking s_2 as the object distance and s_2' as the image distance for the second surface gives

$$(2) \qquad \frac{n}{s_2} + \frac{1}{s_2'} = \frac{1-n}{R_2}$$

But $s_2 = -s_1' + t$, where t is the thickness of the lens. (Remember s_1' is a negative number and s_2 must be positive by our sign convention.) For a thin lens, we can neglect t. In this approximation and from Figure 36.15, we see that $s_2 = -s_1'$. Hence, (2) becomes

$$(3) \qquad -\frac{n}{s_1'} + \frac{1}{s_2'} = \frac{1-n}{R_2}$$

Adding (1) and (3), we find that

$$(4) \qquad \frac{1}{s_1} + \frac{1}{s_2'} = (n-1)\left(\frac{1}{R_1} - \frac{1}{R_2}\right)$$

For the thin lens, we can omit the subscripts on s_1 and s_2' in (4) and call the object distance s and the image distance s', as in Figure 36.16. Hence, we can write (4) in the form

$$\frac{1}{s} + \frac{1}{s'} = (n-1)\left(\frac{1}{R_1} - \frac{1}{R_2}\right) \qquad (36.10)$$

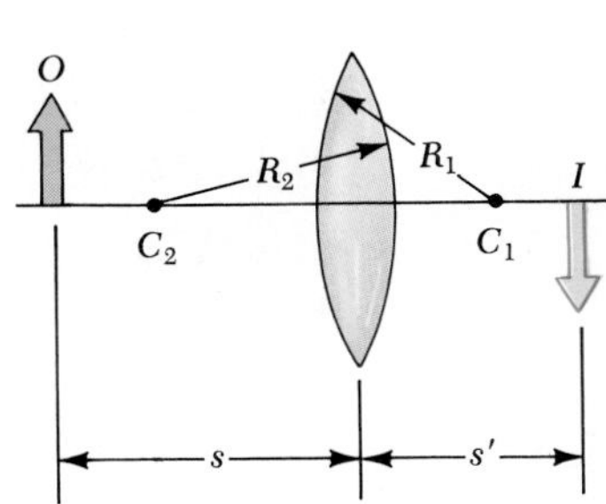

Figure 36.16 The biconvex lens.

This expression relates the image distance s' of a thin lens to the object distance s and to the thin lens properties (index of refraction and radii of curvature). It is valid only for paraxial rays and only when the lens thickness is small relative to the radii R_1 and R_2.

We now define the focal length f of a thin lens as the image distance that corresponds to an infinite object distance, as we did with mirrors. According to this definition and from Equation 36.10, we see that for $s \rightarrow \infty$, $f = s'$; therefore, the inverse of the focal length for a thin lens is given by

Lens makers' equation

$$\frac{1}{f} = (n - 1)\left(\frac{1}{R_1} - \frac{1}{R_2}\right) \tag{36.11}$$

Using this result, we can write Equation 36.10 in an alternate form identical to Equation 36.6 for mirrors:

$$\frac{1}{s} + \frac{1}{s'} = \frac{1}{f} \tag{36.12}$$

Equation 36.11 is called the **lens makers' equation,** since it enables one to calculate f from the known properties of the lens. It can also be used to determine the values of R_1 and R_2 needed for a given index of refraction and desired focal length. A thin lens has *two* focal points, corresponding to incident parallel light rays traveling from the left or right. This is illustrated in Figure 36.17 for a biconvex lens (converging, positive f) and a biconcave lens (diverging, negative f). Focal point F_1 is sometimes called the *primary focal point,* and F_2 is called the *secondary focal point.*

Table 36.3 lists the signs of the quantities appearing in the thin lens equation. Note that the sign convention for thin lenses is the same as for refracting surfaces discussed in the previous section.

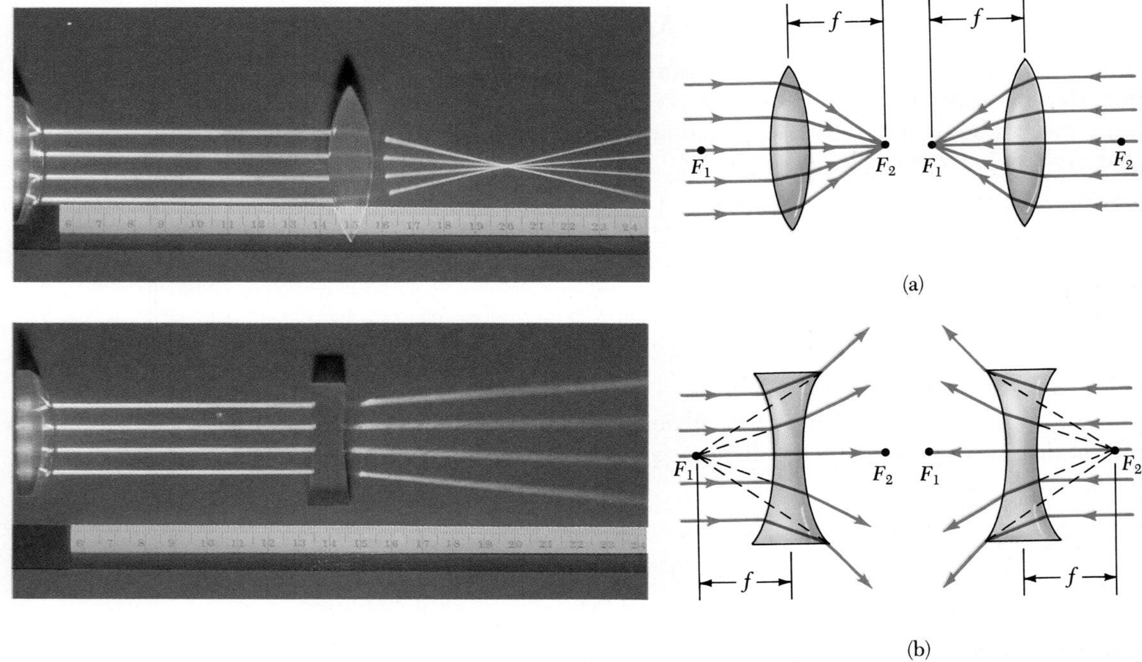

Figure 36.17 (*Left*) Photographs of the effect of converging and diverging lenses on parallel rays. (Courtesy of Henry Leap and Jim Lehman) (*Right*) The principal focal points of (a) the biconvex lens and (b) the biconcave lens.

TABLE 36.3 Sign Convention for Thin Lenses

s is + if the object is in front of the lens.
s is − if the object is in back of the lens.
s' is + if the image is in back of the lens.
s' is − if the image is in front of the lens.
R_1 and R_2 are + if the center of curvature is in back of the lens.
R_1 and R_2 are − if the center of curvature is in front of the lens.

Applying these rules to the *converging* lens, we see that when $s > f$, the quantities s, s', and R_1 are positive and R_2 is negative. Therefore, in the case of a converging lens, where a real object forms a real image, s, s', and f are all positive. Likewise, for a *diverging* lens, s and R_2 are positive and s' and R_1 are negative. Thus, f is negative for a diverging lens.

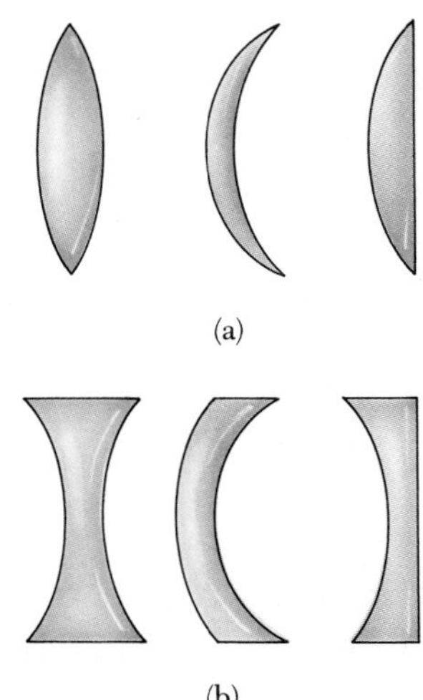

Figure 36.18 Various lens shapes: (a) Converging lenses have a positive focal length and are thickest at the middle. (b) Diverging lenses have a negative focal length and are thickest at the edges.

Sketches of various lens shapes are shown in Figure 36.18. In general, note that a converging lens (positive f) is thicker at the center than at the edge, whereas a diverging lens (negative f) is thinner at the center than at the edge.

Consider a single thin lens illuminated by a *real* object, so that $s > 0$. As with mirrors, the *lateral magnification* of a thin lens is defined as the ratio of the image height h' to the object height h. That is,

$$M = \frac{h'}{h} = -\frac{s'}{s}$$

From this expression, it follows that when M is positive, the image is erect and on the same side of the lens as the object. When M is negative, the image is inverted and on the side of the lens opposite the object.

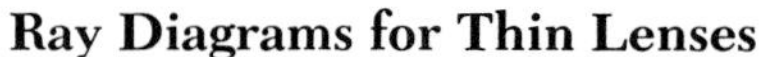

Ray Diagrams for Thin Lenses

Graphical methods, or ray diagrams, are very convenient for determining the image of a thin lens or a system of lenses. Such constructions should also help clarify the sign conventions that have been discussed. Figure 36.19 illustrates

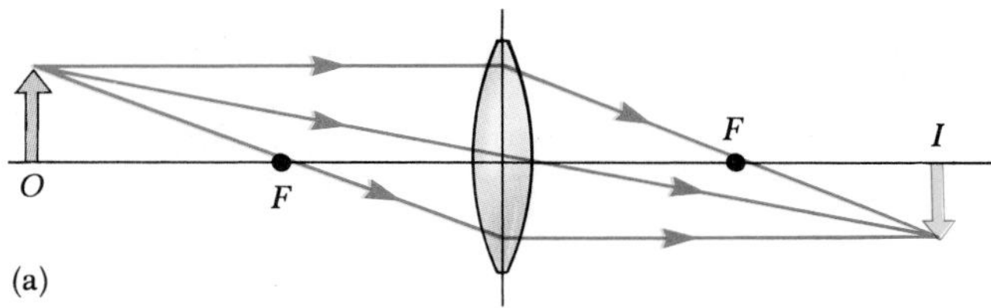

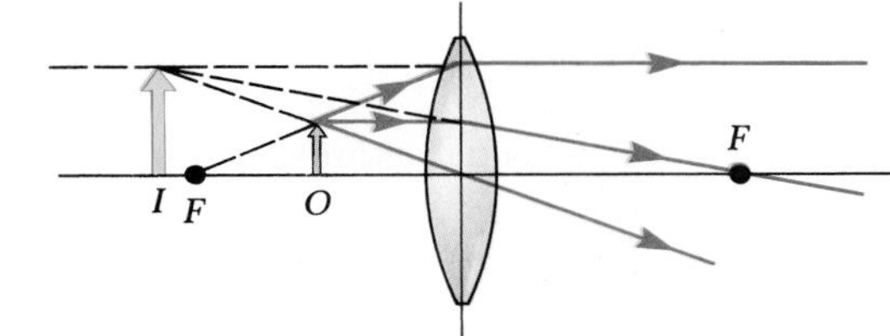

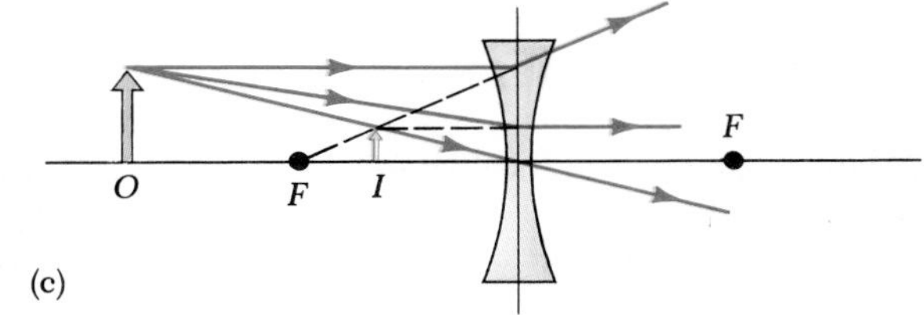

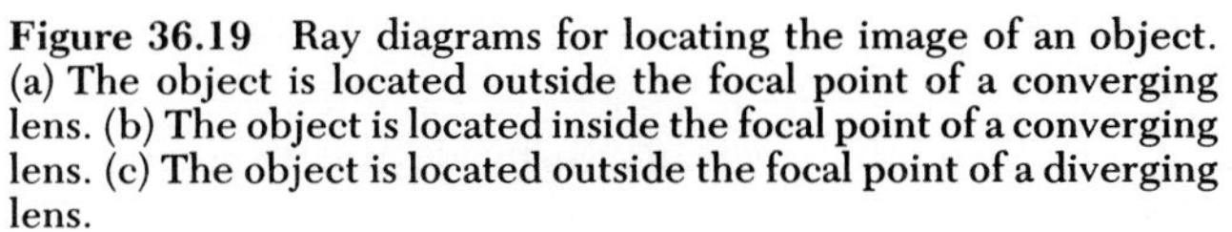

Figure 36.19 Ray diagrams for locating the image of an object. (a) The object is located outside the focal point of a converging lens. (b) The object is located inside the focal point of a converging lens. (c) The object is located outside the focal point of a diverging lens.

this method for three different single-lens situations. To locate the image of a converging lens (Figs. 36.19a and 36.19b), the following three rays are drawn from the top of the object:

1. The first ray is drawn parallel to the optic axis. After being refracted by the lens, this ray passes through (or appears to come from) one of the focal points.
2. The second ray is drawn through the center of the lens. This ray continues in a straight line.
3. The third ray is drawn through the focal point F, and emerges from the lens parallel to the optic axis.

A similar construction is used to locate the image of a diverging lens, as shown in Figure 36.19c. The point of intersection of *any two* of the rays in these diagrams can be used to locate the image. The third ray serves as a check on your construction.

For the converging lens in Figure 36.19a, where the object is *outside* the front focal point ($s > f$), the image is real and inverted. On the other hand, when the real object is *inside* the front focal point ($s < f$), as in Figure 36.19b, the image is virtual, erect, and enlarged. Finally, for the diverging lens shown in Figure 36.19c, the image is always virtual and erect. These geometric constructions are reasonably accurate only if the distance between the rays and the principal axis is small compared to the radii of the lens surfaces.

EXAMPLE 36.6 An Image Formed by a Diverging Lens

A diverging lens has a focal length of -20 cm. An object 2 cm in height is placed 30 cm in front of the lens. Locate the position of the image.

Solution Using the thin lens equation with $s = 30$ cm and $f = -20$ cm, we get

$$\frac{1}{30\text{ cm}} + \frac{1}{s'} = -\frac{1}{20\text{ cm}}$$

$$s' = -12.0\text{ cm}$$

Thus, the image is virtual, as indicated in Figure 36.19c.

Exercise 6 Find the magnification of the lens and the height of the image.
Answer $M = 0.400$, $h' = 0.800$ cm.

EXAMPLE 36.7 An Image Formed by a Converging Lens

A converging lens of focal length 10 cm forms an image of an object placed (a) 30 cm, (b) 10 cm, and (c) 5 cm from the lens. Find the image distance and describe the image in each case.

Solution (a) The thin lens equation, Equation 36.12, can be used to find the image distance:

$$\frac{1}{s} + \frac{1}{s'} = \frac{1}{f}$$

$$\frac{1}{30\text{ cm}} + \frac{1}{s'} = \frac{1}{10\text{ cm}}$$

$$s' = 15.0\text{ cm}$$

The positive sign for the image distance tells us that the image is on the real side of the lens. The magnification of the lens is

$$M = -\frac{s'}{s} = -\frac{15\text{ cm}}{30\text{ cm}} = -0.500$$

Thus, the image is reduced in size by one half, and the negative sign for M tells us that the image is inverted. The situation is like that pictured in Figure 36.19a.

(b) No calculation should be necessary for this case because we know that, when the object is placed at the focal point, the image will be formed at infinity. This is readily verified by substituting $s = 10$ cm into the lens equation.

(c) We now move inside the focal point, to an object distance of 5 cm. In this case, the lens equation gives

$$\frac{1}{5\text{ cm}} + \frac{1}{s'} = \frac{1}{10\text{ cm}}$$

$$s' = -10\text{ cm}$$

and

$$M = -\frac{s'}{s} = -\left(\frac{-10\text{ cm}}{5\text{ cm}}\right) = 2$$

The negative image distance tells us that the image is formed on the side of the lens from which the light is incident. The image is enlarged, and the positive sign for M tells us that the image is erect, as shown in Figure 36.19b.

There are two general cases for a converging lens. When the real object is outside the focal point ($s > f$), the image is real, inverted, and smaller than the object. When the real object is inside the focal point ($s < f$), the image is virtual, erect, and enlarged.

EXAMPLE 36.8 A Lens Under Water

A converging glass lens ($n = 1.52$) has a focal length of 40 cm in air. Find its focal length when it is immersed in water, which has an index of refraction of 1.33.

Solution We can use the lens makers' formula (Eq. 36.11) in both cases, noting that R_1 and R_2 remain the same in air and water. In air, we have

$$\frac{1}{f_a} = (n-1)\left(\frac{1}{R_1} - \frac{1}{R_2}\right)$$

where $n = 1.52$. In water we get

$$\frac{1}{f_w} = (n'-1)\left(\frac{1}{R_1} - \frac{1}{R_2}\right)$$

where n' is the index of refraction of glass *relative* to water. That is, $n' = 1.52/1.33 = 1.14$. Dividing the two equations gives

$$\frac{f_w}{f_a} = \frac{n-1}{n'-1} = \frac{1.52-1}{1.14-1} = 3.71$$

Since $f_a = 40$ cm, we find that

$$f_w = 3.71\,f_a = 3.71(40\text{ cm}) = 148\text{ cm}$$

In fact, the focal length of *any* glass lens is *increased* by the factor $(n-1)/(n'-1)$ when immersed in water.

Combination of Thin Lenses

If two thin lenses are used to form an image, the system can be treated in the following manner. First, the image of the first lens is calculated as if the second lens were not present. The light then approaches the second lens *as if* it had come from the image formed by the first lens. Hence, the image of the first lens is treated as the object of the second lens. The image of the second lens is the final image of the system. If the image of the first lens lies to the right of the second lens, then the image is treated as a virtual object for the second lens (that is, s negative). The same procedure can be extended to a system of three or more lenses. The overall magnification of a system of thin lenses equals the *product* of the magnification of the separate lenses.

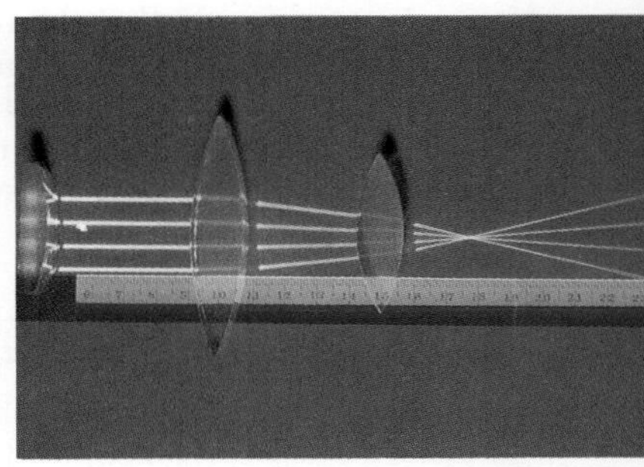

Light from a distant object brought into focus by two converging lenses. Can you estimate the overall focal length of this combination from the photo? (Courtesy of Henry Leap and Jim Lehman)

Now suppose two thin lenses of focal lengths f_1 and f_2 are placed in contact with each other. If s is the object distance for the combination, then application of the thin lens equation to the first lens gives

$$\frac{1}{s} + \frac{1}{s_1'} = \frac{1}{f_1}$$

where s_1' is the image distance for the first lens. Treating this image as the object for the second lens, we see that the object distance for the second lens must be $-s_1'$. Therefore, for the second lens

$$-\frac{1}{s_1'} + \frac{1}{s'} = \frac{1}{f_2}$$

where s' is the final image distance from the second lens. Adding these equations eliminates s_1' and gives

$$\frac{1}{s} + \frac{1}{s'} = \frac{1}{f_1} + \frac{1}{f_2}$$

Focal length of two thin lenses in contact

$$\frac{1}{f} = \frac{1}{f_1} + \frac{1}{f_2} \tag{36.13}$$

If the two *thin* lenses are in contact with one another, then s' is also the distance of the final image from the first lens. Therefore, *two thin lenses in contact are equivalent to a single thin lens whose focal length is given by Equation 36.13.*

EXAMPLE 36.9 Where is the Final Image?

Two thin converging lenses of focal lengths 10 cm and 20 cm are separated by 20 cm, as in Figure 36.20. An object is placed 15 cm in front of the first lens. Find the position of the final image and the magnification of the system.

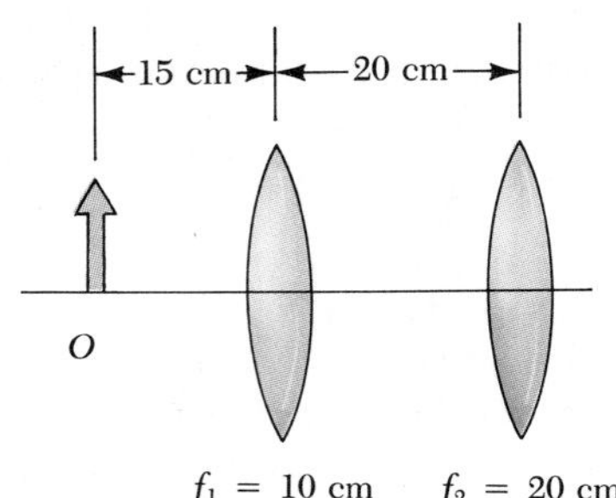

Figure 36.20 (Example 36.9) A combination of two converging lenses.

Solution First we find the image position for the first lens while neglecting the second lens:

$$\frac{1}{s_1} + \frac{1}{s_1'} = \frac{1}{15 \text{ cm}} + \frac{1}{s_1'} = \frac{1}{10 \text{ cm}}$$

$$s_1' = 30.0 \text{ cm}$$

where s'_1 is measured from the first lens.

Since s_1' is greater than the separation between the two lenses, we see that the image of the first lens lies 10 cm to the *right* of the second lens. We take this as the object distance for the second lens. That is, we apply the thin lens equation to the second lens with $s_2 = -10$ cm, where distances are now measured from the second lens, whose focal length is 20 cm:

$$\frac{1}{s_2} + \frac{1}{s_2'} = \frac{1}{f_2}$$

$$\frac{1}{-10 \text{ cm}} + \frac{1}{s_2'} = \frac{1}{20 \text{ cm}}$$

Solving for s_2' gives $s_2' = (20/3)$ cm. That is, the final image lies (20/3) cm to the *right* of the second lens.

The magnification of each lens separately is given by

$$M_1 = \frac{-s_1'}{s_1} = -\frac{30 \text{ cm}}{15 \text{ cm}} = -2$$

$$M_2 = \frac{-s_2'}{s_2} = -\frac{(20/3) \text{ cm}}{-10 \text{ cm}} = \frac{2}{3}$$

The total magnification M of the two lenses is the product $M_1M_2 = (-2)(2/3) = -4/3$. Hence, the final image is real, inverted, and enlarged.

*36.5 LENS ABERRATIONS

One of the basic problems of lenses and lens systems is the imperfect quality of the images. Such imperfect images are largely the result of defects in the shape and form of the lenses. The simple theory of mirrors and lenses assumes that rays make small angles with the optic axis. In this simple model, all rays leaving a point source focus at a single point, producing a sharp image. Clearly, this is not always true. For those cases where the approximations used in this theory do not hold, imperfect images are formed.

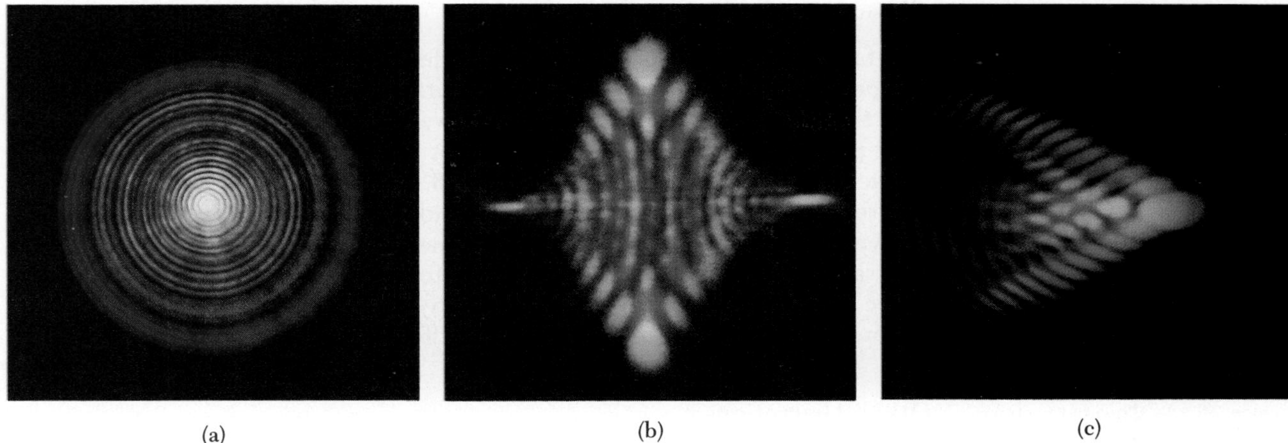

(a) (b) (c)

Lenses can produce various forms of aberrations, as shown by the blurred images of a point source in these photos. (a) Spherical aberration occurs when light passing through the lens at different distances from the optical axis is focused at different points. (b) Astigmatism is an aberration that occurs for objects that are not located on the optical axis of the lens. (c) Coma. This aberration occurs as light passing through the lens far from the optical axis focuses at a different part of the focal plane from light passing near the center of the lens. (Photos by Norman Goldberg)

If one wishes to perform a precise analysis of image formation, it is necessary to trace each ray using Snell's law at each refracting surface. This procedure shows that the rays from a point object do *not* focus at a single point. That is, there is no single point image; instead, the image is *blurred*. The departures of real (imperfect) images from the ideal image predicted by the simple theory are called **aberrations.** Two types of aberrations will now be described.

Spherical Aberrations

Spherical aberrations result from the fact that the focal points of light rays far from the optic axis of a spherical lens (or mirror) are different from the focal points of rays of the same wavelength passing near the center. Figure 36.21 illustrates spherical aberration for parallel rays passing through a converging lens. Rays near the middle of the lens are imaged at greater distances from the lens than rays at the edges. Hence, there is no single focal length for a lens. Many cameras are equipped with an adjustable aperture to control the light intensity and reduce spherical aberration when possible. (An aperture is an opening that controls the amount of light transmitted through the lens.) Sharper images are produced as the aperture size is reduced, since only the central portion of the lens is exposed to the incident light. At the same time, however, less light is imaged. To compensate for this, a longer exposure time is used on the photographic film. A good example is the sharp image produced by a "pin-hole" camera, whose aperture size is approximately 1 mm.

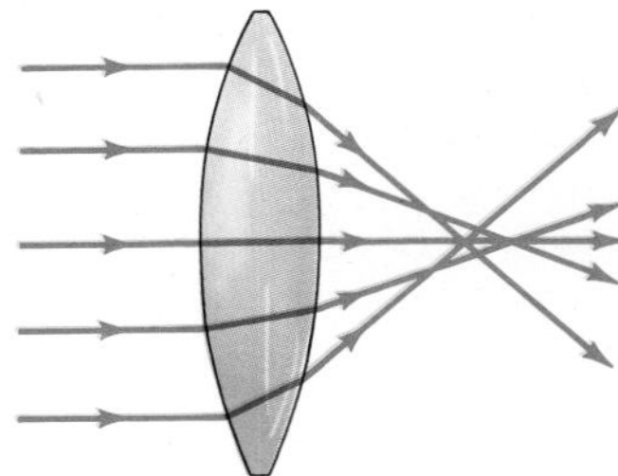

Figure 36.21 Spherical aberration caused by a converging lens. Does a diverging lens cause spherical aberration?

In the case of mirrors used for very distant objects, one can eliminate, or at least minimize, spherical aberration by using a parabolic surface rather than a spherical surface. Parabolic surfaces are not often used, however, because those with high-quality optics are very expensive to make. Parallel light rays incident on such a surface focus at a common point. Parabolic reflecting surfaces are used in many astronomical telescopes in order to enhance the image quality. They are also used in flashlights, where a nearly parallel light beam is produced from a small lamp placed at the focus of the surface.

Chromatic Aberrations

The fact that different wavelengths of light refracted by a lens focus at different points gives rise to *chromatic aberrations.* In Chapter 35, we described how the index of refraction of a material varies with wavelength. When white light passes through a lens, one finds, for example, that violet light rays are refracted more than red light rays (Fig. 36.22). From this we see that the focal length is larger for red light than for violet light. Other wavelengths (not shown in Fig. 36.22) would have intermediate focal points. The chromatic aberration for a diverging lens is opposite that for a converging lens. Chromatic aberration can be greatly reduced by using a combination of a converging and diverging lens made from two different types of glass.

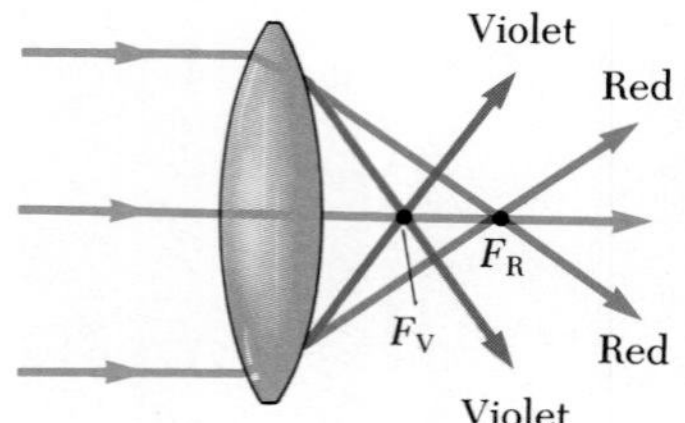

Figure 36.22 Chromatic aberration caused by a converging lens. Rays of different wavelengths focus at different points.

Other Aberrations

Several other defects occur as the result of object points being off the optical axis. *Astigmatism* results when a point object off the axis produces two line images at different points. A defect called *coma* is usually found in lenses with large spherical aberration. For this defect, an off-axis object produces a comet-shaped image. *Distortion* in an image exists for an extended object since magnification of off-axis points differs from magnification of those points near the axis. In high-quality optical systems, these defects are minimized by using properly designed, nonspherical surfaces or specific lens combinations.

*36.6 THE CAMERA

The photographic **camera** is a simple optical instrument whose essential features are shown in Figure 36.23. It consists of a light-tight box, a converging lens that produces a real image, and a film behind the lens to receive the image. Focusing is accomplished by varying the distance between lens and film with an adjustable bellows in older style cameras or some other mechanical arrangement in modern cameras. For proper focusing, or sharp images, the lens-to-film distance will depend on the object distance as well as on the focal length of the lens. The shutter, located behind the lens, is a mechanical device that is opened for selected time intervals. With this arrangement, one can photograph moving objects by using short exposure times or dark scenes (low light levels) by using long exposure times. If this arrangement were not avail-

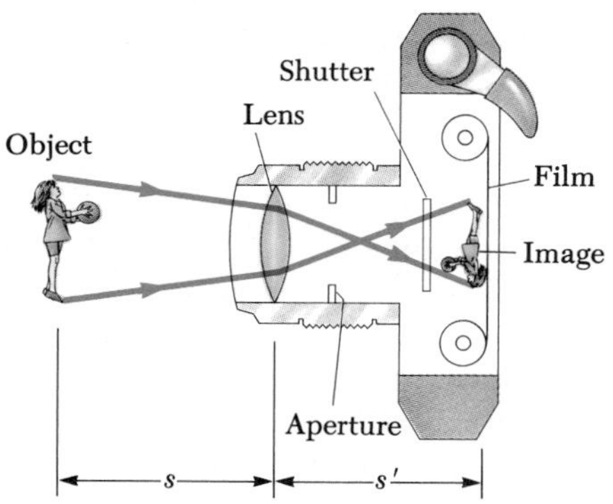

Figure 36.23 Cross-sectional view of a simple camera.

able, it would be impossible to take stop-action photographs. For example, a rapidly moving vehicle could move enough in the time that the shutter was open to produce a blurred image. Another major cause of blurred image is the movement of the camera while the shutter is open. For this reason, you should use short exposure times, or use a tripod, even for stationary objects. Typical shutter speeds are 1/30, 1/60, 1/125, and 1/250 s. A stationary object is normally shot with a shutter speed of 1/60 s.

More expensive cameras also have an aperture of adjustable diameter either behind or in between the lenses to provide further control of the intensity of the light reaching the film. When an aperture of small diameter is used, only light from the central portion of the lens reaches the film and so the aberration is reduced somewhat.

The brightness (or energy flux) of the image focused on the film depends on the focal length of the lens and on the diameter D of the lens. Clearly, the light intensity I will be proportional to the area of the lens. Since the area is proportional to D^2, we conclude that $I \propto D^2$. Furthermore, the intensity is a measure of the energy received by the film per unit area of the image. Since the area of the image is proportional to $(s')^2$, and $s' \approx f$ (for objects with $s \gg f$), we conclude that the intensity is also proportional to $1/f^2$, so that $I \propto D^2/f^2$. The ratio f/D is defined to be the *f-number* of a lens:

$$f\text{-number} \equiv \frac{f}{D} \qquad (36.14)$$

Hence, the intensity of light incident on the film can be expressed as

$$I \propto \frac{1}{(f/D)^2} \propto \frac{1}{(f\text{-number})^2} \qquad (36.15)$$

The f-number is a measure of the "light-concentrating" power and determines the "speed" of the lens. A "fast" lens has a small f-number. Fast lenses, with an f-number as low as about 1.4, are more expensive, because it is more difficult to keep aberrations acceptably small. Camera lenses are often marked with various f-numbers such as $f/2.8, f/4, f/5.6, f/8, f/11, f/16$. The various f-numbers are obtained by adjusting the aperture, which effectively changes D. The smallest f-number corresponds to the case where the aperture is wide open and the full lens area is in use. Simple cameras for routine snapshots usually have a fixed focal length and fixed aperture size, with an f-number of about $f/11$.

EXAMPLE 36.10 Finding the Correct Exposure Time

The lens of a certain 35-mm camera (where 35 mm is the width of the film strip) has a focal length of 55 mm and a speed of $f/1.8$. The correct exposure time for this speed under certain conditions is known to be (1/500) s. (a) Determine the diameter of the lens.

Solution From Equation 36.14, we find that

$$D = \frac{f}{f\text{-number}} = \frac{55\text{ mm}}{1.8} = 30.6\text{ mm}$$

(b) Calculate the correct exposure time if the f-number is changed to $f/4$ under the same lighting conditions.

Solution The total light energy received by each part of the image is proportional to the product of the flux and the exposure time. If I is the light intensity reaching the film, then in a time t, the energy received by the film is It. Comparing the two situations, we require that $I_1t_1 = I_2t_2$, where t_1 is the correct exposure time for $f/1.8$ and t_2 is the correct exposure time for some other f-number. Using this result, together with Equation 36.15, we find that

$$\frac{t_1}{(f_1\text{-number})^2} = \frac{t_2}{(f_2\text{-number})^2}$$

$$t_2 = \left(\frac{f_2\text{-number}}{f_1\text{-number}}\right)^2 t_1 = \left(\frac{4}{1.8}\right)^2\left(\frac{1}{500}\text{ s}\right) \approx \frac{1}{100}\text{ s}$$

That is, as the aperture is reduced in size, the exposure time must increase.

*36.7 THE EYE

The eye is an extremely complex part of the body, and because of its complexity, certain defects often arise that can cause the impairment of vision. In these cases, external aids, such as eyeglasses, are often used. In this section we shall describe the parts of the eye, their purpose, and some of the corrections that can be made when the eye does not function properly. You will find that the eye has much in common with the camera. Like the camera, a normal eye focuses light and produces a sharp image. However, the mechanisms by which the eye controls the amount of light admitted and adjusts itself to produce correctly focused images are far more complex, intricate, and effective than those in the most sophisticated camera. In all respects, the eye is an architectural wonder.

Figure 36.24 shows the essential parts of the eye. The front is covered by a transparent membrane called the *cornea*. This is followed by a clear liquid region (the *aqueous humor*), a variable aperture (the *iris* and *pupil*), and the *crystalline lens*. Most of the refraction occurs in the cornea because the liquid medium surrounding the lens has an average index of refraction close to that of the lens. The iris, which is the colored portion of the eye, is a muscular diaphragm that controls the size of the pupil. The iris regulates the amount of light entering the eye by dilating the pupil in light of low intensity and contracting the pupil in high-intensity light. The f-number range of the eye is from about $f/2.8$ to $f/16$.

Light entering the eye is focused by the cornea-lens system onto the back surface of the eye, called the *retina*. The surface of the retina consists of millions of sensitive structures called *rods* and *cones*. When stimulated by

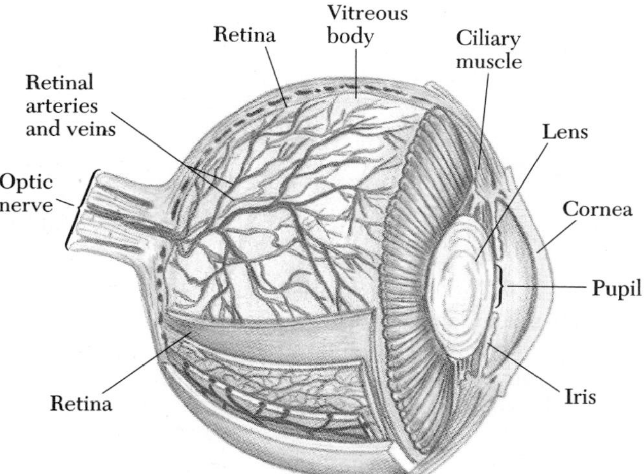

Figure 36.24 Essential parts of the eye. Note the similarity between the eye and the simple camera. Can you correlate the parts of the eye with those of the camera?

light, these receptors send impulses via the optic nerve to the brain, where an image is perceived. By this process, a distinct image of an object is observed when the image falls on the retina.

The eye focuses on a given object by varying the shape of the pliable crystalline lens through an amazing process called **accommodation.** An important component in accommodation is the *ciliary muscle,* which is attached to the lens. When the eye is focused on distant objects, the ciliary muscle is relaxed. For an object distance of infinity, the focal length of the eye (the distance between the lens and the retina) is about 1.7 cm. The eye focuses on nearby objects by tensing the ciliary muscle. This action effectively decreases the focal length by slightly decreasing the radius of curvature of the lens, which allows the image to be focused on the retina. This lens adjustment takes place so swiftly that we are not even aware of the change. Again in this respect, even the finest electronic camera is a toy compared with the eye. It is evident that there is a limit to accommodation because objects that are very close to the eye produce blurred images.

The **near point** represents the closest distance for which the lens will produce a sharp image on the retina. This distance usually increases with age and has an average value of around 25 cm.

Typically, at age ten the near point of the eye is about 18 cm. This increases to about 25 cm at age 20, to 50 cm at age 40, and to 500 cm or greater at age 60.

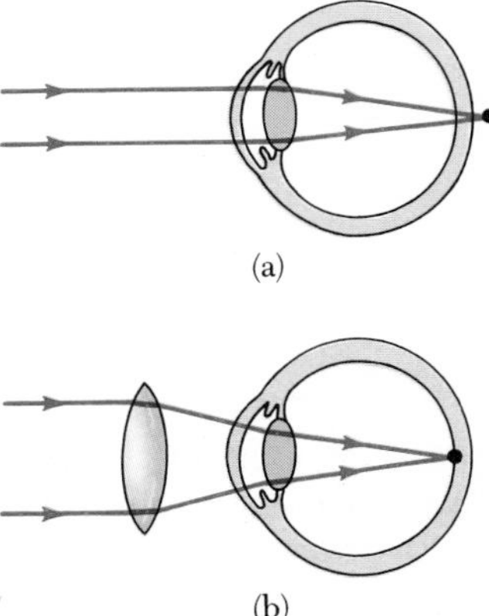

Figure 36.25 (a) A hyperopic eye (farsightedness) is slightly shorter than normal; hence the image of a distant object focuses behind the retina. (b) The condition can be corrected with a converging lens.

Conditions of the Eye

Although the eye is one of the most remarkable organs in the body, it may have several abnormalities, which can often be corrected with eyeglasses, contact lenses, or surgery.

When the relaxed eye produces an image of a distant object *behind* the retina, as in Figure 36.25a, the condition is known as **hyperopia,** and the person is said to be farsighted. With this defect, distant objects are seen clearly but nearby objects are blurred. Either the hyperopic eye is too short or the ciliary muscle is unable to change the shape of the lens enough to properly focus the image. The condition can be corrected with a converging lens, as shown in Figure 36.25b.

Another condition, known as **myopia,** or nearsightedness, occurs either when the eye is longer than normal or when the maximum focal length of the lens is insufficient to produce a clearly formed image on the retina. In this case, light from a distant object is focused in front of the retina (Fig. 36.26a). The distinguishing feature of this condition is that distant objects are not seen clearly. Nearsightedness can be corrected with a diverging lens, as in Figure 36.26b.

Beginning with middle age, most people lose some of their accommodation power as a result of a weakening of the ciliary muscle and a hardening of the lens. This causes an individual to become farsighted. Fortunately, the condition can be corrected with converging lenses.

A person may also have an eye defect known as **astigmatism,** in which light from a point source produces a line image on the retina. This condition arises either when the cornea or the crystalline lens or both are not perfectly spherical. Astigmatism can be corrected with lenses having different curvatures in two mutually perpendicular directions.

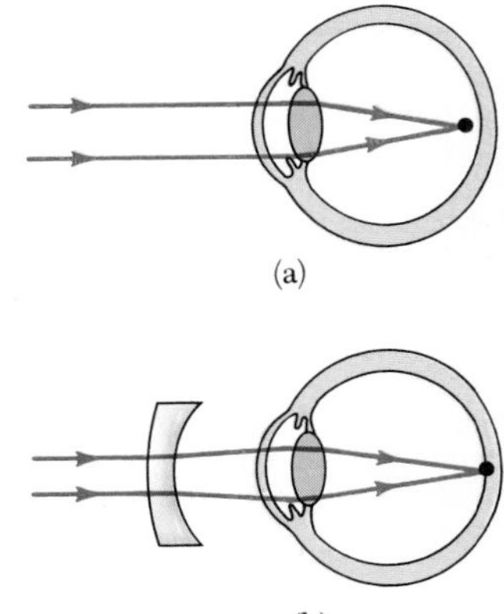

Figure 36.26 (a) A myopic eye (nearsightedness) is slightly longer than normal; hence the image of a distant object focuses in front of the retina. (b) The condition can be corrected with a diverging lens.

The eye is also subject to several diseases. One disease, which usually occurs in old age, is the formation of **cataracts,** where the lens becomes partially or totally opaque. One remedy for cataracts is surgical removal of the lens. Another disease, called **glaucoma,** arises from an abnormal increase in fluid pressure inside the eyeball. The pressure increase can lead to a swelling of the lens and to strong myopia. There is a chronic form of glaucoma in which the pressure increase causes a reduction in blood supply to the retina. This can eventually lead to blindness because the nerve fibers of the retina eventually die. If the disease is discovered early enough, it can be treated with medicine or surgery.

Optometrists and ophthalmologists usually prescribe lenses measured in **diopters.**

The **power,** P, of a lens in diopters equals the inverse of the focal length in meters, that is, $P = 1/f$.

For example, a converging lens whose focal length is +20 cm has a power of +5 diopters, and a diverging lens whose focal length is −40 cm has a power of −2.5 diopters.

EXAMPLE 36.11 A Case of Nearsightedness

A particular nearsighted person is unable to see objects clearly when they are beyond 50 cm (the far point of the eye). What should the focal length of the lens prescribed to correct this problem be?

Solution The purpose of the lens in this instance is to "move" an object from infinity to a distance where it can be seen clearly. This is accomplished by having the lens produce an image at the far point of the eye. From the thin lens equation, we have

$$\frac{1}{s} + \frac{1}{s'} = \frac{1}{\infty} - \frac{1}{50 \text{ cm}} = \frac{1}{f}$$

$$f = -50 \text{ cm}$$

Why did we use a negative sign for the image distance? As you should have suspected, the lens must be a diverging lens (negative focal length) to correct nearsightedness.

Exercise 6 What is the power of this lens?
Answer −2 diopters.

*36.8 THE SIMPLE MAGNIFIER

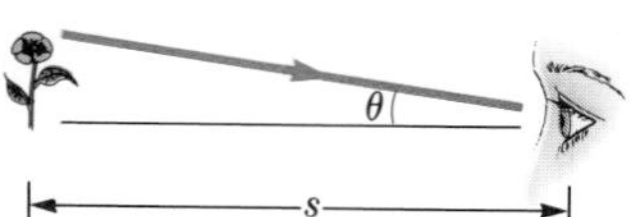

Figure 36.27 The size of the image formed on the retina depends on the angle θ subtended at the eye.

The simple magnifier is one of the simplest and most basic of all optical instruments because it consists of only a single converging lens. As the name implies, this device is used to increase the apparent size of an object. Suppose an object is viewed at some distance s from the eye, as in Figure 36.27. Clearly, the size of the image formed at the retina depends on the angle θ subtended by the object at the eye. As the object moves closer to the eye, θ increases and a larger image is observed.[1] However, an average normal eye is unable to focus on an object closer than about 25 cm, the near point (Fig. 36.28a). Try it! Therefore, θ is maximum at the near point.

To further increase the apparent angular size of an object, a converging lens can be placed in front of the eye with the object located at point O, just inside the focal point of the lens, as in Figure 36.28b. At this location, the lens

[1] Regular eyeglasses give some magnification because the lenses are not located at the lens of the eye. On the other hand, contact lenses minimize this effect because of their close proximity to the lens of the eye.

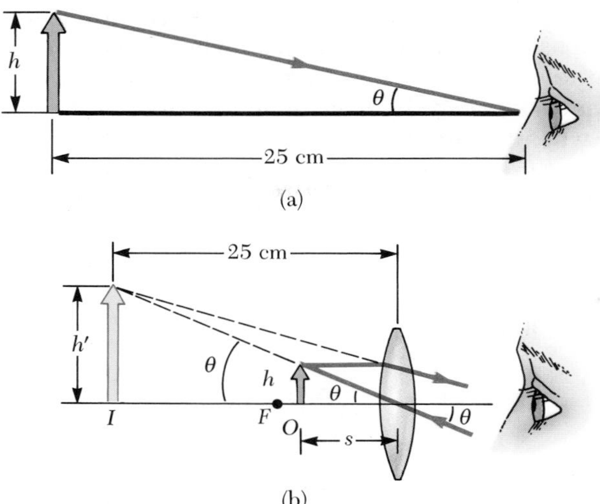

Figure 36.28 (a) An object placed at the near point of the eye ($s = 25$ cm) subtends an angle θ_0 at the eye, where $\theta_0 \approx h/25$. (b) An object placed near the focal point of a converging lens produces a magnified image, which subtends an angle $\theta \approx h'/25$ at the eye.

forms a virtual, erect, and enlarged image, as shown. Clearly, the lens increases the angular size of the object. We define the **angular magnification,** m, as the ratio of the angle subtended by an object with a lens in use (angle θ in Figure 36.28b) to that subtended by the object when it is placed at the near point with no lens (angle θ_0 in Fig. 36.28a):

$$m \equiv \frac{\theta}{\theta_0} \tag{36.16}$$

Angular magnification

The angular magnification is a maximum when the image is at the near point of the eye, that is, when $s' = -25$ cm. The object distance corresponding to this image distance can be calculated from the thin lens formula:

$$\frac{1}{s} + \frac{1}{-25 \text{ cm}} = \frac{1}{f}$$

$$s = \frac{25f}{25 + f}$$

where f is the focal length in centimeters. Let us now make the small angle approximation as follows:

$$\theta_0 \approx \frac{h}{25} \quad \text{and} \quad \theta \approx \frac{h}{s} \tag{36.17}$$

Thus, Equation 36.16 becomes

$$m = \frac{\theta}{\theta_0} = \frac{h/s}{h/25} = \frac{25}{s} = \frac{25}{25f/(25 + f)}$$

$$m = 1 + \frac{25 \text{ cm}}{f} \tag{36.18}$$

The magnification given by Equation 36.18 is the ratio of the angular size seen with the lens to the angular size seen when the object is viewed at the near point of the eye with no lens. Actually, the eye can focus on an image formed anywhere between the near point and infinity. However, the eye is more relaxed when the image is at infinity (Section 36.7). In order for the image formed by the magnifying lens to appear at infinity, the object has to be placed at the focal point of the lens. In this case, the equations in 36.17 become

$$\theta_0 \approx \frac{h}{25} \quad \text{and} \quad \theta \approx \frac{h}{f}$$

and the magnification is

$$m = \frac{\theta}{\theta_0} = \frac{25 \text{ cm}}{f} \tag{36.19}$$

With a single lens, it is possible to obtain angular magnifications up to about 4 without serious aberrations. Magnifications up to about 20 can be achieved by using one or two additional lenses to correct for aberrations.

EXAMPLE 36.12 Maximum Magnification of a Lens
What is the maximum magnification of a lens having a focal length of 10 cm, and what is the magnification of this lens when the eye is relaxed?

Solution The maximum magnification occurs when the image formed by the lens is located at the near point of the eye. Under these circumstances, Equation 36.18 gives us the magnification as

$$m = 1 + \frac{25 \text{ cm}}{f} = 1 + \frac{25 \text{ cm}}{10 \text{ cm}} = 3.50$$

When the eye is relaxed, the image is at infinity. In this case, we use Equation 36.19:

$$m = \frac{25}{f} = \frac{25 \text{ cm}}{10 \text{ cm}} = 2.50$$

*36.9 THE COMPOUND MICROSCOPE

A simple magnifier provides only limited assistance in inspecting the minute details of an object. Greater magnification can be achieved by combining two lenses in a device called a compound microscope, a schematic diagram of which is shown in Figure 36.29a. It consists of an objective lens with a very short focal length, f_o (where $f_o < 1$ cm), and an ocular, or eyepiece lens, having a focal length, f_e, of a few centimeters. The two lenses are separated by a distance L, where L is much greater than either f_o or f_e. The object, which is placed just outside the focal length of the objective, forms a real, inverted image at I_1, which is at or close to the focal point of the eyepiece. The eyepiece, which serves as a simple magnifier, produces at I_2 an image of the image at I_1, and this image at I_2 is virtual and inverted. The lateral magnification, M_1, of the first image is $-s_1'/s_1$. Note from Figure 36.29a that s_1' is approximately equal to L, and recall that the object is very close to the focal point of the objective; thus, $s_1 \approx f_o$. This gives a magnification for the objective of

$$M_1 \approx -\frac{L}{f_o}$$

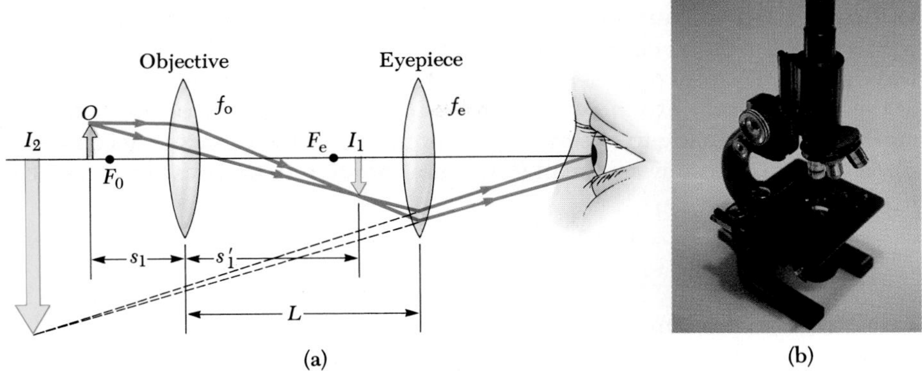

Figure 36.29 (a) Diagram of a compound microscope, which consists of an objective lens and an eyepiece, or ocular, lens. (b) An old-fashioned compound microscope. The three-objective turret allows the user to switch to several different powers of magnification. Combinations of oculars with different focal lengths and different objectives can produce a wide range of magnifications. (Courtesy of Henry Leap and Jim Lehman)

The angular magnification of the eyepiece for an object (corresponding to the image at I_1) placed at the focal point is found from Equation 36.19 to be

$$m_e = \frac{25 \text{ cm}}{f_e}$$

The overall magnification of the compound microscope is defined as the product of the lateral and angular magnifications:

$$M = M_1 m_e = -\frac{L}{f_o}\left(\frac{25 \text{ cm}}{f_e}\right) \tag{36.20}$$

The negative sign indicates that the image is inverted.

The microscope has extended our vision to include the previously unknown details of incredibly small objects. The capabilities of this instrument have steadily increased with improved techniques in the precision grinding of lenses. A question that is often asked about microscopes is, "If you were extremely patient and careful, would it be possible to construct a microscope that would enable you to see an atom?" The answer to this question is no, as long as light is used to illuminate the object. The reason is that, in order to be seen, the object under a microscope must be at least as large as a wavelength of light. An atom is many times smaller than the wavelengths of visible light, and so its mysteries have to be probed using other types of "microscopes."

The wavelength dependence of the "seeing" ability of a wave can be illustrated by water waves set up in a bathtub in the following manner. Suppose you vibrate your hand in the water until waves having a wavelength of about 15 cm are moving along the surface. If you fix a small object, such as a toothpick, in the path of the waves, you will find that the waves are not disturbed appreciably by the toothpick but instead continue along their path, oblivious of the small object. Now suppose you fix a larger object, such as a toy sailboat, in the path of the waves. In this case, the waves are considerably "disturbed" by the object. In the first case, the toothpick is smaller than the wavelength of the waves, and as a result the waves do not "see" the toothpick. (The intensity of the scattered waves is low.) In the second case, the toy sailboat is about the same size as the wavelength of the waves and hence the sailboat creates a disturbance. That is, the object acts as the source of scattered waves that appear to come from it. Light waves behave in this same general way. The ability of an optical microscope to view an object depends on the size of the object relative to the wavelength of the light used to observe it. Hence, one will never be able to observe atoms or molecules with such a microscope, since their dimensions are small (≈0.1 nm) relative to the wavelength of the light (≈500 nm).

*36.10 THE TELESCOPE

There are two fundamentally different types of **telescopes,** both designed to aid in viewing distant objects, such as the planets in our solar system. The two classifications are (1) the **refracting telescope,** which uses a combination of lenses to form an image, and (2) the **reflecting telescope,** which uses a curved mirror and a lens to form an image.

The telescope sketched in Figure 36.30a is a refracting telescope. The two lenses are arranged such that the objective forms a real, inverted image of the distant object very near the focal point of the eyepiece. Furthermore, the image at I_1 is formed at the focal point of the objective because the object is essentially at infinity. Hence, the two lenses are separated by a distance $f_o + f_e$, which corresponds to the length of the telescope's tube. The eyepiece finally forms, at I_2, an enlarged, inverted image of the image at I_1.

The angular magnification of the telescope is given by θ/θ_o, where θ_o is the angle subtended by the object at the objective and θ is the angle subtended by the final image. From the triangles in Figure 36.30a, and for small angles, we have

$$\theta \approx -\frac{h'}{f_e} \quad \text{and} \quad \theta_o \approx \frac{h'}{f_o}$$

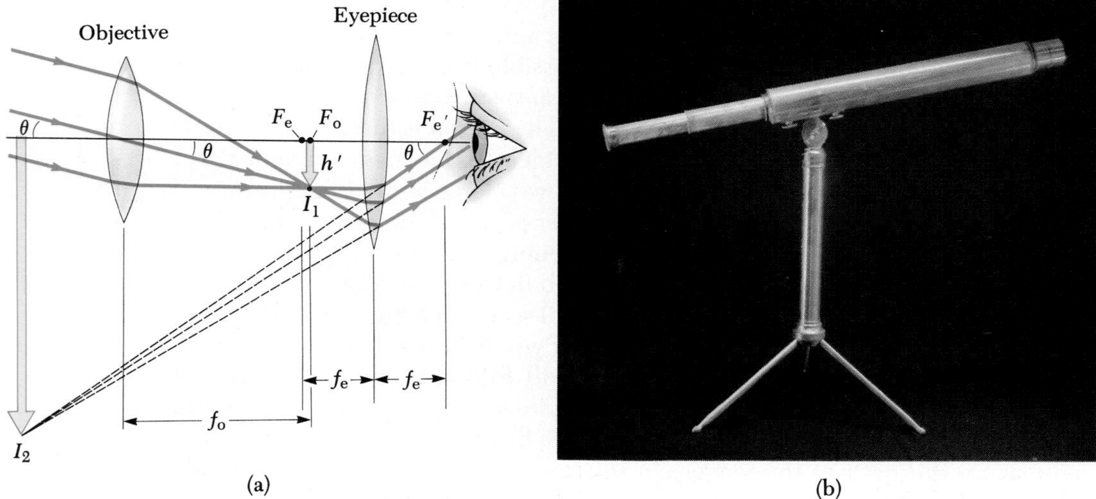

Figure 36.30 (a) Diagram of a refracting telescope, with the object at infinity. (b) Photograph of a refracting telescope. (Photo courtesy of Henry Leap and Jim Lehman)

Hence, the angular magnification of the telescope can be expressed as

$$m = \frac{\theta}{\theta_o} = \frac{-h'/f_e}{h'/f_o} = -\frac{f_o}{f_e} \qquad (36.21)$$

The minus sign indicates that the image is inverted. This says that the angular magnification of a telescope equals the ratio of the objective focal length to the eyepiece focal length. Here again, the magnification is the ratio of the angular size seen with the telescope to the angular size seen with the unaided eye.

In some applications, such as observing nearby objects like the sun, moon, or planets, magnification is important. However, stars are so far away that they always appear as small points of light regardless of how much magnification is used. Large research telescopes used to study very distant objects must have a large diameter in order to gather as much light as possible. It is difficult and expensive to manufacture large lenses for refracting telescopes. Another difficulty with large lenses is that their large weight leads to sagging, which is an additional source of aberration. These problems can be partially overcome by replacing the objective lens with a reflecting, concave mirror. Figure 36.31 shows the design for a typical reflecting telescope. Incoming light rays pass down the barrel of the telescope and are reflected by a parabolic mirror at the base. These rays converge toward point A in the figure, where an image would be formed. However, before this image is formed, a small flat mirror at point M reflects the light toward an opening in the side of the tube that passes into an eyepiece. This particular design is said to have a Newtonian focus because it was Newton who developed it. Note that the light never passes through glass in the reflecting telescope (except through the small eyepiece). As a result, problems associated with chromatic aberration are virtually eliminated.

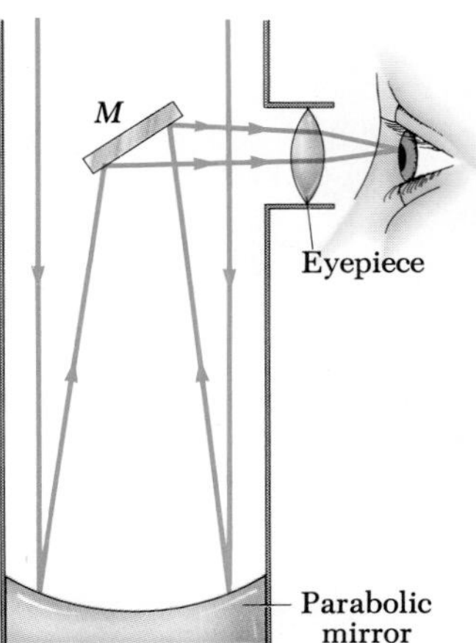

Figure 36.31 A reflecting telescope with a Newtonian focus.

The largest telescope in the world is the 6-m-diameter reflecting telescope on Mount Pastukhov in the Caucasus, Soviet Union. The largest reflecting telescope in the United States is the 5-m-diameter instrument on Mount Palomar in California. In contrast, the largest refracting telescope in the world, which is located at the Yerkes Observatory in Williams Bay, Wisconsin, has a diameter of only 1 m.

SUMMARY

The **magnification** M of a mirror or lens is defined as the ratio of the image height h' to the object height h:

Magnification of a mirror

$$M = \frac{h'}{h} = -\frac{s'}{s} \qquad (36.2)$$

In the paraxial ray approximation, the object distance s and image distance s' for a spherical mirror of radius R are related by the **mirror equation**

Mirror equation

$$\frac{1}{s} + \frac{1}{s'} = \frac{2}{R} = \frac{1}{f} \qquad (36.4, 36.6)$$

where $f = R/2$ is the **focal length** of the mirror.

An image can be formed by refraction from a spherical surface of radius R. The object and image distances for refraction from such a surface are related by

Formation of an image by refraction

$$\frac{n_1}{s} + \frac{n_2}{s'} = \frac{n_2 - n_1}{R} \tag{36.8}$$

where the light is incident in the medium of index of refraction n_1 and is refracted in the medium whose index of refraction is n_2.

The inverse of the **focal length** f of a thin lens in air is given by

Lens makers' equation

$$\frac{1}{f} = (n-1)\left(\frac{1}{R_1} - \frac{1}{R_2}\right) \tag{36.11}$$

Converging lenses have positive focal lengths, and **diverging lenses** have negative focal lengths.

For a thin lens, and in the paraxial ray approximation, the object and image distances are related by the **thin lens equation:**

Thin lens formula

$$\frac{1}{s} + \frac{1}{s'} = \frac{1}{f} \tag{36.12}$$

Aberrations are responsible for the formation of imperfect images by lenses and mirrors. **Spherical aberration** is due to the variation in focal points for parallel incident rays that strike the lens at various distances from the optical axis. **Chromatic aberration** arises from the fact that light of different wavelengths focuses at different points when refracted by a lens.

QUESTIONS

1. When you look in a mirror, the image of your left and right sides is reversed, yet the image of your head and legs is not reversed. Explain.
2. Using a simple ray diagram, as in Figure 36.2, show that a plane mirror whose top is at eye level need not be as long as your height in order for you to see your entire body.
3. Consider a concave spherical mirror with a real object. Is the image always inverted? Is the image always real? Give conditions for your answers.
4. Repeat the previous question for a convex spherical mirror.
5. It is well known that distant objects viewed under water with the naked eye appear blurred and out of focus. On the other hand, the use of goggles provides the swimmer with a clear view of objects. Explain this, using the fact that the indices of refraction of the cornea, water, and air are 1.376, 1.333, and 1.00029, respectively.
6. Why does a clear stream always appear to be shallower than it actually is?
7. A person spearfishing in a boat sees a fish apparently located 3 m from the boat at a depth of 1 m. In order to hit the fish with his spear, should the person aim at the fish, above the fish, or below the fish? Explain.
8. Consider the image formed by a thin converging lens. Under what conditions will the image be (a) inverted, (b) erect; (c) real, (d) virtual, (e) larger than the object, and (f) smaller than the object?
9. Repeat Question 8 for a thin diverging lens.
10. If a cylinder of solid glass or clear plastic is placed above the words LEAD OXIDE and viewed from the side as shown in Figure 36.32, the word "LEAD" appears inverted but the word "OXIDE" does not. Explain.

Figure 36.32 (Question 10) (Courtesy of Henry Leap and Jim Lehman)

11. Describe two types of aberration common in a spherical lens.

12. Explain why a mirror cannot give rise to chromatic aberration.
13. What is the magnification of a plane mirror? What is its focal length?
14. Why do some emergency vehicles have the symbol ƎƆИA⅃UᗺMA written on the front?
15. Explain why a fish in a spherical goldfish bowl appears larger than it really is.
16. Lenses used in eyeglasses, whether converging or diverging, are always designed such that the middle of the lens curves away from the eye, like the center lenses of Figures 36.18a and 36.18b. Why?
17. A mirage is formed when the air gets gradually cooler as the height above the ground increases. What might happen if the air grows gradually warmer as the height is increased? This often happens over bodies of water or snow-covered ground: the effect is called looming.
18. Consider a spherical concave mirror, with the object located to the left of the mirror beyond the focal point. Using ray diagrams, show that the image of the object moves to the left as the object approaches the focal point.
19. In a Jules Verne novel, a piece of ice is shaped to form a magnifying lens to focus sunlight to start a fire. Is this possible?
20. The "*f*-number" of a camera is the focal length of the camera lens divided by its aperture (or diameter). How can the "*f*-number" of the lens be changed? How does this change the required exposure time?
21. A solar furnace can be constructed by using a concave mirror to reflect and focus sunlight into a furnace enclosure. What factors in the design of the reflecting mirror would guarantee that very high temperatures may be achieved?

PROBLEMS

Section 36.1 Images Formed by Plane Mirrors

1. In a physics laboratory experiment, a torque is applied to a small-diameter wire that is suspended vertically under tensile stress. It is necessary to measure accurately the small angle through which the wire turns as a consequence of the net torque. This is accomplished by attaching a small mirror to the wire and reflecting a beam of light off the mirror and onto a circular scale. Such an arrangement is known as an *optical lever* and is shown from a top view in Figure 36.33. Show that when the mirror turns through an angle θ, the reflected beam is rotated by an angle 2θ.

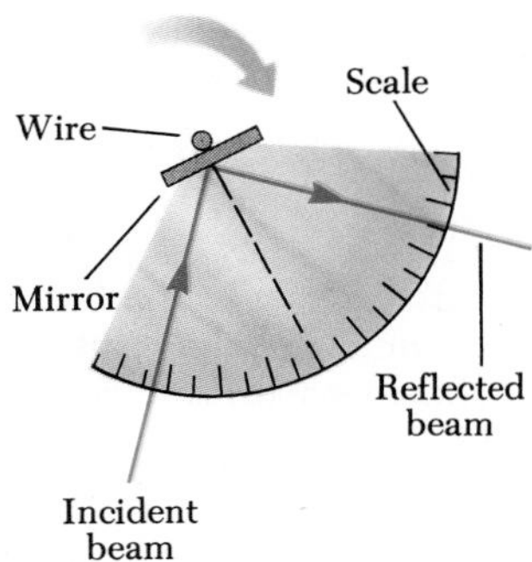

Figure 36.33 (Problem 1).

2. Two plane mirrors, A and B, are in contact along one edge, and the planes of the two mirrors are at an angle of 45° with respect to each other (Fig. 36.34). A point object is placed at P along the bisector of the angle between the two mirrors. Make a sketch similar to Figure 36.34 to a suitable scale and locate graphically (a) the image of P in mirror A and the image of P in mirror B. (b) Label the images found in (a) P'_A and P'_B, respectively, and locate the image of P'_A in mirror B and the image of P'_B in mirror A. (c) Determine the total number of images for the arrangement described.

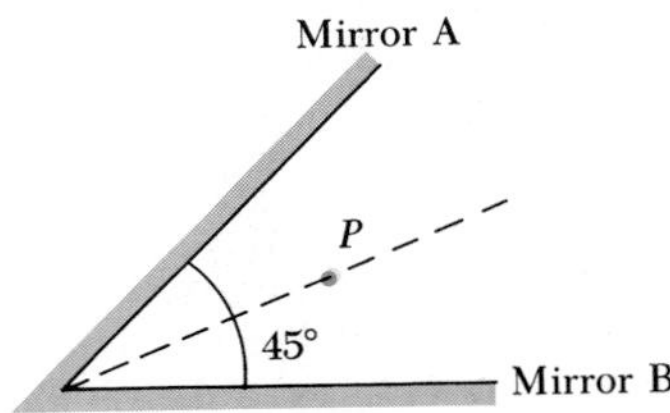

Figure 36.34 (Problem 2).

3. Determine the minimum height of a vertical plane mirror in which a person 5′10″ in height can see his or her full image. (A ray diagram would be helpful.)
4. Consider the case in which a light ray A is incident on mirror 1 in Figure 36.3. The reflected ray is incident on mirror 2 and subsequently reflected as ray B. Let the angle of incidence (with respect to the normal) on mirror 1 equal 53° and the point of incidence be located 20 cm from the edge of contact between the two mirrors. Determine the angle between ray A and ray B.

Section 36.2 Images Formed by Spherical Mirrors

5. A concave mirror has a focal length of 40 cm. Determine the object position for which the resulting image will be erect and four times the size of the object.

6. A convex mirror has a focal length of -20 cm. Determine an object location for which the image will be one half the size of the object.
7. A concave mirror has a radius of curvature of 60 cm. Calculate the image position and magnification of an object placed in front of the mirror at distances of (a) 90 cm and (b) 20 cm. (c) Draw ray diagrams to obtain the image in each case.
8. The real-image height of a concave mirror is observed to be four times larger than the object height when the object is 30 cm in front of the mirror. (a) What is the radius of curvature of the mirror? (b) Use a ray diagram to locate the image position corresponding to the given object position and the radius of curvature calculated in (a).
9. Calculate the image position and magnification for an object placed (a) 20 cm and (b) 60 cm in front of a convex mirror of focal length -40 cm. (c) Use ray diagrams to locate image positions corresponding to the object positions in (a) and (b).
10. A candle is 49 cm in front of a convex spherical mirror, of radius of curvature 70 cm. (a) Where is the image? (b) What is the magnification?
11. An object 2 cm high is placed 10 cm in front of a mirror. What type of mirror and what radius of curvature is needed for an upright image that is 4 cm high?
12. Use a ray diagram to demonstrate that the image of a real object placed in front of a spherical mirror is always virtual and erect when $s < |f|$.
13. A spherical mirror is to be used to form an image five times the size of an object on a screen located 5 m from the object. (a) Describe the type of mirror required. (b) Where should the mirror be positioned relative to the object?
14. A real object is located at the zero end of a meter stick. A concave mirror located at the 100-cm end of the meter stick forms an image of the object at the 70-cm position. A convex mirror placed at the 60-cm position forms a final image at the 10-cm point. What is the radius of curvature of the convex mirror?
15. A spherical convex mirror has a radius of 40 cm. Determine the position of the virtual image and magnification of the mirror for object distances of (a) 30 cm and (b) 60 cm. (c) Are the images erect or inverted?
16. An object is 15 cm from the surface of a reflective spherical Christmas-tree ornament 6 cm in diameter. What are the magnification and position of the image?

Section 36.3 Images Formed by Refraction

17. A smooth block of ice ($n = 1.309$) rests on the floor with one face parallel to the floor. The block has a vertical thickness of 50 cm. Find the location of the image of a pattern in the floor covering as formed by rays that are nearly perpendicular to the block.
18. One end of a long glass rod ($n = 1.5$) is formed into the shape of a *convex* surface of radius 6 cm. An object is located in air along the axis of the rod. Find the image positions corresponding to the object at distances of (a) 20 cm, (b) 10 cm, and (c) 3 cm from the end of the rod.
19. Calculate the image positions corresponding to the object positions stated in Problem 18 if the end of the rod has the shape of a *concave* surface of radius 8 cm.
20. Repeat Problem 18 if the object is in water surrounding the glass rod instead of in air.
21. A glass sphere ($n = 1.50$) of radius 15 cm has a tiny air bubble located 5 cm from the center. The sphere is viewed along a direction parallel to the radius containing the bubble. What is the apparent depth of the bubble below the surface of the sphere?
22. A flint glass plate ($n = 1.66$) rests on the bottom of an aquarium tank. The plate is 8 cm thick (vertical dimension) and is covered with water ($n = 1.33$) to a depth of 12 cm. Calculate the apparent thickness of the plate as viewed from above the water. (Assume nearly normal incidence.)
23. A glass hemisphere is used as a paperweight with its flat face resting on a stack of papers. The radius of the circular cross section is 4 cm, and the index of refraction of the glass is 1.55. The center of the hemisphere is directly over a letter "O" that is 2.5 mm in diameter. What is the diameter of the image of the letter as seen looking along a vertical radius?
24. A goldfish is swimming in water inside a spherical plastic bowl of index of refraction 1.33. If the goldfish is 10 cm from the wall of the 15-cm-radius bowl, where does the goldfish appear to an observer outside the bowl?
25. A transparent sphere of unknown composition is observed to form an image of the sun on the surface of the sphere opposite the sun. What is the refractive index of the sphere material?

Section 36.4 Thin Lenses

26. An object located 32 cm in front of a lens forms an image on a screen 8 cm behind the lens. (a) Find the focal length of the lens. (b) Determine the magnification. (c) Is the lens converging or diverging?
27. The left face of a biconvex lens has a radius of curvature of 12 cm, and the right face has a radius of curvature of 18 cm. The index of refraction of the glass is 1.44. (a) Calculate the focal length of the lens. (b) Calculate the focal length if the radii of curvature of the two faces are interchanged.
28. A thin converging lens has a focal length f. Find the object distance if the image is (a) real and twice as large as the object and (b) virtual and twice the size of the object.

29. What is the image distance of an object 1 m in front of a converging lens of focal length 20 cm? What is the magnification of the object? (See Figure 36.19a for a ray diagram.)
30. A convex lens forms a real image of an object at a point located 12 cm to the right of the lens. The object is positioned 50 cm to the left of the lens. (a) Calculate the focal length of the lens. (b) What is the ratio of the height of the image to the height of the object? (c) Is the image erect or inverted? Real or virtual?
31. Construct a ray diagram for the arrangement described in Problem 30.
32. A converging lens has a focal length of 40 cm. Calculate the size of the real image of an object 4 cm in height for the following object distances: (a) 50 cm, (b) 60 cm, (c) 80 cm, (d) 100 cm, (e) 200 cm, (f) ∞.
33. A real object is located 20 cm to the left of a diverging lens of focal length $f = -32$ cm. Determine (a) the location and (b) the magnification of the image.
34. Construct a ray diagram for the arrangement described in Problem 33.
35. A thin-walled, hollow convex lens is immersed in a tank of water. The hollow lens has $R_1 = 20$ cm and $R_2 = 30$ cm. Calculate the focal length of this "air lens" surrounded by water ($n = 1.33$). Use the derivation of Equation 36.11 as a guide.
36. A diverging lens is used to form a virtual image of a real object. The object is positioned 80 cm to the left of the lens, and the image is located 40 cm to the left of the lens. (a) Determine the focal length of the lens. (b) If the surfaces of the lens have radii of curvature of $R_1 = -40$ cm and $R_2 = -50$ cm, what is the value of the index of refraction of the lens?
37. The thin lens shown in Fig. 36.16 is made from glass ($n = 1.520$) and has radii of curvature $R_1 = 10$ cm and $R_2 = -4$ cm. (a) Determine the focal length f of the lens. (b) An object with a height $h = 2$ cm is located 30 cm in front of the lens. Determine the position, the character, and the size of the image.

*Section 36.6 The Camera and *Section 36.7 The Eye

38. A camera is found to give proper film exposure when it is set at $f/16$ and the shutter is open for (1/32) s. Determine the correct exposure time if a setting of $f/8$ is used. (Assume the lighting conditions are unchanged.)
39. A camera is being used with correct exposure at $f/4$ and a shutter speed of (1/16) s. In order to "stop" a fast-moving subject, the shutter speed is changed to (1/128) s. Find the new f-number setting that should be used to maintain satisfactory exposure.
40. A 1.7-m tall woman stands 5 m in front of a camera equipped with a 50-mm focal length lens. What is the size of the image formed on the film?
41. A nearsighted person cannot see objects clearly beyond 25 cm (the far point). If the patient has no astigmatism and contact lenses are prescibed, what are the power and type of lens required to correct the patient's vision?
42. What is the unaided near point for a person required to wear lenses with a power of $+1.5$ diopters to read at 25 cm?
43. If the aqueous humor of the eye has an index of refraction of 1.34 and the distance from the vertex of the cornea to the retina is 2.2 cm, what is the radius of curvature of the cornea for which distant objects will be focused on the retina? (Assume all refraction occurs in the aqueous humor.)
44. Assume that the camera shown in Figure 36.23 has a fixed focal length of 6.5 cm and is adjusted to properly focus the image of a distant object. By how much and in what direction must the lens be moved in order to focus the image of an object at a distance of 2 m?
45. Figure 36.25a illustrates the case of a farsighted person who can focus clearly on objects that are more distant than 90 cm from the eye. Determine the power of lenses that will enable this person to read comfortably at a normal near point of 25 cm.
46. The eye of a nearsighted person is illustrated in Figure 36.26a. In this case, the person cannot focus clearly on objects that are more distant than 200 cm from the eye. Determine the power of lenses that will enable this person to see distant objects clearly.

*Section 36.8 The Simple Magnifier, *Section 36.9 The Compound Microscope, and *Section 36.10 The Telescope

47. A philatelist examines the printing detail on a stamp using a convex lens of focal length 10 cm as a simple magnifier. The lens is held close to the eye, and the lens-to-object distance is adjusted so that the virtual image is formed at the normal near point (25 cm). Calculate the expected magnification.
48. A lens with focal length 5 cm is used as a magnifying glass. (a) To obtain maximum magnification, where should the object be placed? (b) What is the magnification?
49. The Yerkes refracting telescope has a 1-m diameter objective lens of focal length 20 m and an eyepiece of focal length 2.5 cm. (a) Determine the magnification of the planet Mars as seen through this telescope. (b) Are the polar caps right side up, or upside down?
50. The Palomar reflecting telescope has a parabolic mirror, with an 80-m focal length. Determine the magnification achieved when an eyepiece of 2.5 cm focal length is used.
51. An astronomical telescope has an objective with focal length 75 cm and an eyepiece with focal length 4 cm. What is the magnifying power of the instrument?

52. The distance between the eyepiece and the objective lens in a certain compound microscope is 23 cm. The focal length of the eyepiece is 2.5 cm, and the focal length of the objective is 0.4 cm. What is the overall magnification of the microscope?

53. The desired overall magnification of a compound microscope is 140×. The objective alone produces a lateral magnification of 12×. Determine the required focal length of the eyepiece lens.

ADDITIONAL PROBLEMS

54. An object placed 10 cm from a concave spherical mirror produces a real image 8 cm from the mirror. If the object is moved to a new position 20 cm from the mirror, what is the position of the image? Is the final image real or virtual?

55. An object is located in a fixed position in front of a screen. A thin lens, placed between the object and the screen, produces a sharp image on the screen when it is in either of two positions that are 10 cm apart. The sizes of images in the two situations are in the ratio 3 : 2. (a) What is the focal length of the lens? (b) What is the distance from the screen to the object?

56. Show that, for a converging lens, unit magnification ($m = -1$) results when $s = 2f$. Under what condition is $m = +1$ approached?

57. Construct a graph of magnification m as a function of s for (a) a converging lens and (b) a diverging lens.

58. An object is located 36 cm to the left of a biconvex lens of index of refraction 1.5. The left surface of the lens has a radius of curvature of 20 cm. The right surface of the lens is to be shaped so that a real image will be formed 72 cm to the right of the lens. What is the required radius of curvature of the second surface?

59. A parallel beam of light enters a glass hemisphere perpendicular to the flat face, as shown in Figure 36.35. The radius is $R = 6$ cm and the index of refraction is $n = 1.560$. Determine the point at which the beam is focused. (Assume paraxial rays.)

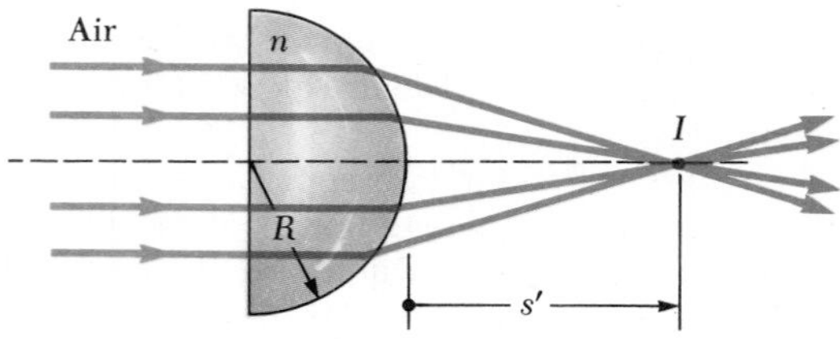

Figure 36.35 (Problem 59).

60. A thin lens with refractive index n is immersed in a liquid with index n'. Show that the focal length f of the lens is given by

$$\frac{1}{f} = \left(\frac{n}{n'} - 1\right)\left(\frac{1}{R_1} - \frac{1}{R_2}\right)$$

61. A thin lens of focal length 20 cm lies on a horizontal front-surfaced mirror. How far above the lens should an object be held if its image is to coincide with the object?

62. Find the object distances (in terms of f) of a thin converging lens of focal length f if (a) the image is real and the image distance is four times the focal length and (b) the image is virtual and the image distance is three times the focal length. (c) Calculate the magnification of the lens for cases (a) and (b).

63. A cataract-impaired lens may be surgically removed and replaced by a manufactured lens. The focal length of this "implant" is determined by the lens-to-retina distance, which is measured by a "sonar-like" device and by the requirement that the implant provide for correct distant vision (objects which are far from the eye). (a) If the distance from the lens to the retina is 22.4 mm, calculate the power of the implanted lens in diopters. (b) Since there is no accommodation and the implant allows for correct distant vision, a corrective lens for close work or reading must be used. Assume a reading distance of 33 cm and calculate the power of the lens in the reading glasses.

64. A concave mirror has a radius of 40 cm. (a) Calculate the image distance s' for an arbitrary real object distance s. (b) Obtain values of s' for object distances of 5 cm, 10 cm, 40 cm, and 60 cm. (c) Make a plot of s' versus s using the results of (b).

65. An object is placed 12 cm to the left of a diverging lens of focal length −6 cm. A converging lens of focal length 12 cm is placed a distance d to the right of the diverging lens. Find the distance d such that the final image is at infinity. Draw a ray diagram for this case.

66. A converging lens has a focal length of 20 cm. Find the position of the image for a real object at distances of (a) 50 cm, (b) 30 cm, (c) 10 cm. (d) Determine the magnification of the lens for these object distances and whether the image is erect or inverted. (e) Draw ray diagrams to locate the images for these object distances.

67. Repeat Problem 66 if the lens is diverging and has a focal length of 15 cm.

68. The image formed by the objective lens of an astronomical telescope is called the "prime-focus image." Show that (a) this image is located at the rear focal point of the objective lens and (b) the prime-focus image size is related to the astronomical object's angular size (measured in degrees of arc) and the focal length of the objective lens as

$$h' = f_o \frac{\theta_o}{57.3°}$$

69. A colored marble is dropped in a large tank filled with benzene ($n = 1.50$). (a) What is the depth of the tank if the apparent depth of the marble when viewed from

directly above the tank is 35 cm? (b) If the marble has a diameter of 1.5 cm, what is its apparent diameter when viewed from directly above, outside the tank?

70. An object 1 cm in height is placed 4 cm to the left of a converging lens of focal length 8 cm. A diverging lens of focal length −16 cm is located 6 cm to the right of the converging lens. Find the position and size of the final image. Is the image inverted or erect? Real or virtual?

71. The disk of the sun subtends an angle of 0.5 degrees at the earth. What are the position and diameter of the solar image formed by a concave spherical mirror of radius 3 m?

72. A reflecting telescope with 2-m focal length and an eyepiece of 10-cm focal length is used to view the moon. Calculate the size of the image formed at the viewer's near point, 25 cm from the eye. (The diameter of the moon = 3.5×10^3 km; the distance from the earth to the moon = 3.84×10^5 km.)

73. The cornea of the eye has a radius of curvature of 0.80 cm. (a) What is the focal length of the reflecting surface of the eye? (b) If a $20 gold piece 3.4 cm in diameter is held 25 cm from the cornea, what are the size and location of the reflected image?

74. Because the index of refraction for all substances depends on the wavelength of the incident light, the focal length of a lens is different for different colors of light. This effect, called chromatic aberration, causes blurring of the colors in an image. What is the fractional difference in focal length $\Delta f/f$ between yellow and red light for a lens made from fused quartz? Between yellow and violet light? Use λ (red) = 750 nm, λ (yellow) = 600 nm, and λ (violet) = 450 nm; refer to Figure 35.17.

75. In a darkened room a burning candle is placed 1.5 m from a white wall. A lens is placed between the candle and wall at a location that causes a larger, inverted image of the candle to form on the wall. When the lens is moved 90 cm toward the wall, another image of the candle is formed. Find (a) the two object distances that produce the images stated above and (b) the focal length of the lens. (c) Characterize the second image.

76. A converging lens of focal length 20 cm is separated by 50 cm from a converging lens of focal length 5 cm. (a) Find the final position of the image of an object placed 40 cm in front of the first converging lens. (b) If the height of the object is 2 cm, what is the height of the final image? Is it real or virtual? (c) If the two lenses are placed in contact, what is the focal length of the combination? (d) Determine the image position of an object placed 5 cm in front of the two lenses in contact.

77. Each face of a double convex lens has radius of curvature R. The lens has index of refraction n_L. The two faces are in contact with different substances with indices of refraction n_1 and n_2. (a) Show that an object must be placed at distance

$$s_1 = \frac{R(n_1 + n_2)}{2n_L - (n_1 + n_2)}$$

from the lens in the first medium with index of refraction n_1 to form an image the same distance from the lens in the second medium with index of refraction n_2. (That is, $s_1 = s_2'$.) (b) In the above case, show that the image height h_2' is related to the object height h_1 in the following manner:

$$h_2' = -\frac{n_1}{n_2} h_1$$

78. Figure 36.36 shows the longitudinal section of a shape formed from glass of index of refraction $n = 1.5$. The ends are hemispheres of radius r and $2r$, and the centers of the hemispherical ends are separated by a distance $4r$. A point object is located in air a distance $0.5r$ from the left end of the glass form. Find the location of the image of the object due to refraction at the two spherical surfaces when $r = 2$ cm.

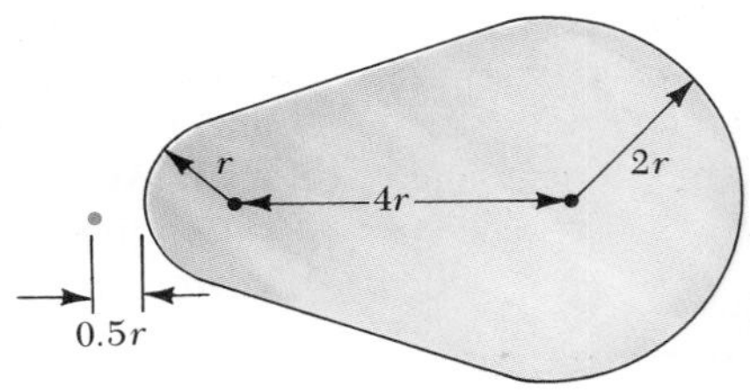

Figure 36.36 (Problem 78).

37

Interference of Light Waves

Interference in soap bubbles. White light incident on soap bubbles (thin films of soap) forms a beautiful pattern of colors as a result of interference in the films, as described in Section 37.6. (Peter Aprahamian, Science Photo Library)

In the previous chapter on geometric optics, we used the concept of light rays to examine what happens when light passes through a lens or reflects from a mirror. The next two chapters are concerned with the subject of *wave optics,* which deals with the phenomena of interference, diffraction, and polarization of light. These phenomena cannot be adequately explained with ray optics, but we shall describe how the wave theory of light leads to a satisfying description of such phenomena. This chapter is aimed at explaining various types of interference effects associated with light.

37.1 CONDITIONS FOR INTERFERENCE

In our discussion of wave interference in Chapter 18, we found that two waves could add together constructively or destructively. In constructive interference, the amplitude of the resultant wave is greater than that of either of the individual waves, while in destructive interference, the resultant amplitude is less than that of either of the individual waves. Light waves also undergo interference. Fundamentally, all interference associated with light waves arises as a result of combining the fields that constitute the individual waves.

Interference effects in light waves are not easy to observe because of the short wavelengths involved (about 4×10^{-7} m to about 7×10^{-7} m). In order to observe sustained interference in light waves, the following conditions must be met:

Conditions for interference

1. The sources must be **coherent,** that is, they must maintain a constant phase with respect to each other.
2. The sources must be **monochromatic,** that is, of a single wavelength.
3. The superposition principle must apply.

We shall now describe the characteristics of **coherent sources.** As we have said, two sources (producing two traveling waves) are needed to create interference. However, in order to produce a stable interference pattern, *the individual waves must maintain a constant phase relationship with one another.* When this situation prevails, the sources are said to be **coherent.** As an example, the sound waves emitted by two side-by-side loudspeakers driven by a single amplifier can produce interference because the two speakers respond to the amplifier in the same way at the same time.

Now, if two separate light sources are placed side by side, no interference effects are observed because in this case the light waves emitted by each of the sources are emitted independently of the other source; hence, their emissions do not maintain a constant phase relationship with each other over the time of observation. Light from an ordinary light source undergoes such random changes about once every 10^{-8} s. Therefore the conditions for constructive interference, destructive interference, or some intermediate state last for times of the order of 10^{-8} s. The result is that no interference effects are observed since the eye cannot follow such short-term changes. Such light sources are said to be **incoherent.**

A common method for producing two coherent light sources is to use one monochromatic source to illuminate a screen containing two small openings (usually in the shape of slits). The light emerging from the two slits is coherent because a single source produces the original light beam and the two slits serve only to separate the original beam into two parts (which, after all, is what was done to the sound signal discussed above). A random change in the light emitted by the source will occur in the two separate beams at the same time, and interference effects can still be observed.

37.2 YOUNG'S DOUBLE-SLIT EXPERIMENT

The phenomenon of interference in light waves from two sources was first demonstrated by Thomas Young in 1801. A schematic diagram of the apparatus used in this experiment is shown in Figure 37.1a. Light is incident on a screen, which is provided with a narrow slit S_0. The waves emerging from this slit arrive at a second screen, which contains two narrow, parallel slits, S_1 and S_2. These two slits serve as a pair of coherent light sources because waves emerging from them originate from the same wavefront and therefore maintain a constant phase relationship. The light from the two slits produces a visible pattern on screen C; the pattern consists of a series of bright and dark parallel bands called **fringes** (Fig. 37.1b). When the light from slits S_1 and S_2 arrives at a point on screen C such that constructive interference occurs at that location, a bright line appears. When the light from two slits combines de-

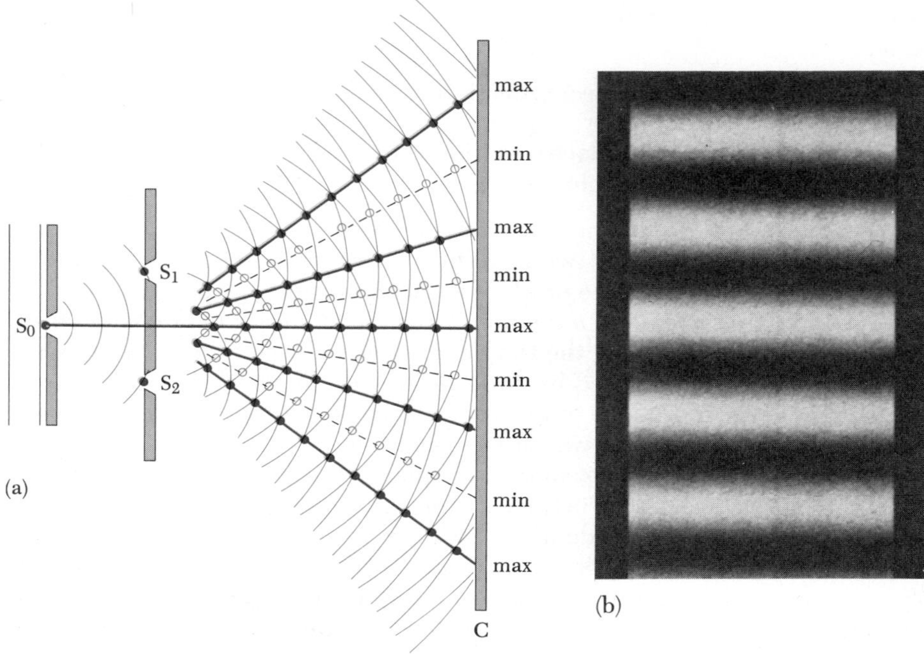

Figure 37.1 (a) Schematic diagram of Young's double-slit experiment. The narrow slits act as sources of waves. Slits S_1 and S_2 behave as coherent sources which produce an interference pattern on screen C. (Note that this drawing is not to scale.) (b) The fringe pattern formed on screen C could look like this.

structively at any location on the screen, a dark line results. Figure 37.2 is a photograph of an interference pattern produced by two coherent vibrating sources in a water tank.

Figure 37.2 Water waves set up by two vibrating sources produce an effect similar to Young's double-slit interference. The waves interfere constructively at X and destructively at Y.

Figure 37.3 is a schematic diagram of some of the ways the two waves can combine at the screen. In Figure 37.3a, the two waves, which leave the two slits in phase, strike the screen at the central point *P*. Since these waves travel an equal distance, they arrive in phase at *P*, and as a result constructive interference occurs at this location and a bright area is observed. In Figure 37.3b, the two light waves again start in phase, but the upper wave has to travel one wavelength farther to reach point *Q* on the screen. Since the upper wave falls behind the lower one by exactly one wavelength, they still arrive in phase at *Q*, and so a second bright light appears at this location. Now consider point *R*, midway between *P* and *Q* in Figure 37.3c. At this location, the upper wave has fallen half a wavelength behind the lower wave. This means that the trough from the bottom wave overlaps the crest from the upper wave, giving rise to destructive interference at *R*. For this reason, one observes a dark region at this location.

We can obtain a quantitative description of Young's experiment with the help of Figure 37.4. Consider a point *P* on the viewing screen; the screen is located a perpendicular distance *L* from the screen containing slits S_1 and S_2, which are separated by a distance *d*. Let us assume that the source is monochromatic. Under these conditions, the waves emerging from S_1 and S_2 have the same frequency and amplitude and are in phase. The light intensity on the screen at *P* is the resultant of the light coming from both slits. Note that a wave

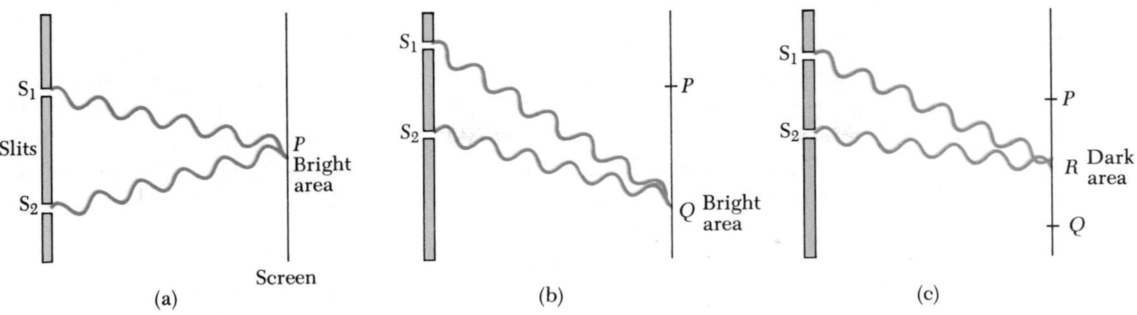

Figure 37.3 (a) Constructive interference occurs at P when the waves combine. (b) Constructive interference also occurs at Q. (c) Destructive interference occurs at R because the wave from the upper slit falls half a wavelength behind the wave from the lower slit. (Note that these figures are not drawn to scale.)

from the lower slit travels farther than a wave from the upper slit by an amount equal to $d \sin \theta$. This distance is called the **path difference,** δ, where

$$\delta = r_2 - r_1 = d \sin \theta \qquad (37.1)$$

Path difference

This equation assumes that r_1 and r_2 are parallel, which is approximately true because L is much greater than d. As noted earlier, the value of this path difference will determine whether or not the two waves are in phase when they arrive at P. If the path difference is either zero or some integral multiple of the wavelength, the two waves are in phase at P and constructive interference results. Therefore, the condition for bright fringes, or **constructive interference,** at P is given by

$$\delta = d \sin \theta = m\lambda \qquad (m = 0, \pm 1, \pm 2, \ldots) \qquad (37.2)$$

Constructive interference

The number m is called the **order number.** The central bright fringe at $\theta = 0$ ($m = 0$) is called the *zeroth-order maximum.* The first maximum on either side, when $m = \pm 1$, is called the *first-order maximum,* and so forth.

Similarly, when the path difference is an odd multiple of $\lambda/2$, the two waves arriving at P will be 180° out of phase and will give rise to destructive interference. Therefore, the condition for dark fringes, or **destructive interference,** at P is given by

$$\delta = d \sin \theta = (m + \tfrac{1}{2})\lambda \qquad (m = 0, \pm 1, \pm 2, \ldots) \qquad (37.3)$$

Destructive interference

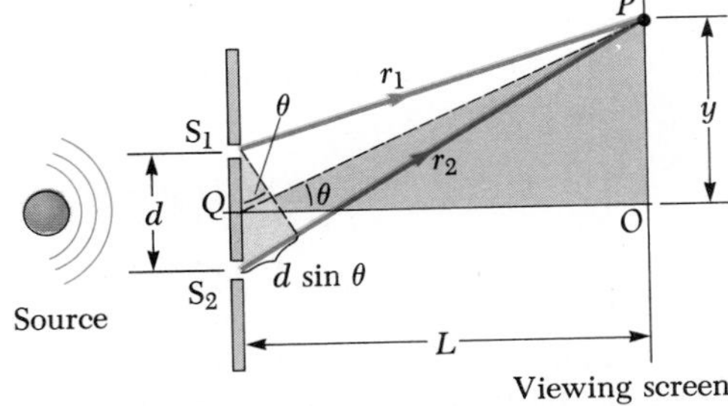

Figure 37.4 Geometric construction for describing Young's double-slit experiment. The path difference between the two rays is $r_2 - r_1 = d \sin \theta$. (Note that this figure is not drawn to scale.)

It is useful to obtain expressions for the positions of the bright and dark fringes measured vertically from O to P. In addition to our assumption that $L \gg d$, we shall assume that $d \gg \lambda$; that is, the distance between the two slits is much larger than the wavelength. This situation prevails in practice because L is often of the order of 1 m while d is a fraction of a millimeter and λ is a fraction of a micrometer for visible light. Under these conditions, θ is small, and so we can use the approximation $\sin\theta \approx \tan\theta$. From the triangle OPQ in Figure 37.4, we see that

$$\sin\theta \approx \tan\theta = \frac{y}{L} \tag{37.4}$$

Using this result together with Equation 37.2, we see that the positions of the bright fringes measured from O are given by

$$y_{\text{bright}} = \frac{\lambda L}{d} m \tag{37.5}$$

Similarly, using Equations 37.3 and 37.4, we find that the *dark fringes* are located at

$$y_{\text{dark}} = \frac{\lambda L}{d} (m + \tfrac{1}{2}) \tag{37.6}$$

As we shall demonstrate in Example 37.1, Young's double-slit experiment provides a method for measuring the wavelength of light. In fact, Young used this technique to make the first measurement of the wavelength of light. Additionally, the experiment gave the wave model of light a great deal of credibility. It was inconceivable that particles of light coming through the slits could cancel each other in a way that would explain the regions of darkness. Today we still use the phenomenon of interference to describe wave-like behavior in many observations.

EXAMPLE 37.1 Measuring the Wavelength of a Light Source

A screen is separated from a double-slit source by 1.2 m. The distance between the two slits is 0.03 mm. The second-order bright fringe ($m = 2$) is measured to be 4.5 cm from the center line. (a) Determine the wavelength of the light.

Solution We can use Equation 37.5, with $m = 2$, $y_2 = 4.5 \times 10^{-2}$ m, $L = 1.2$ m, and $d = 3 \times 10^{-5}$ m:

$$\lambda = \frac{dy_2}{mL} = \frac{(3 \times 10^{-5}\ \text{m})(4.5 \times 10^{-2}\ \text{m})}{2 \times 1.2\ \text{m}}$$

$$= 5.62 \times 10^{-7}\ \text{m} = 560\ \text{nm}$$

(b) Calculate the distance between adjacent bright fringes.

Solution From Equation 37.5 and the results to (a), we get

$$y_{m+1} - y_m = \frac{\lambda L(m+1)}{d} - \frac{\lambda Lm}{d}$$

$$= \frac{\lambda L}{d} = \frac{(5.62 \times 10^{-7}\ \text{m})(1.2\ \text{m})}{3 \times 10^{-5}\ \text{m}}$$

$$= 2.25 \times 10^{-2}\ \text{m} = 2.25\ \text{cm}$$

EXAMPLE 37.2 The Distance Between Bright Fringes

A light source emits light of two wavelengths in the visible region, given by $\lambda = 430$ nm and $\lambda' = 510$ nm. The source is used in a double-slit interference experiment in which $L = 1.5$ m and $d = 0.025$ mm. Find the separation between the third-order bright fringes corresponding to these wavelengths.

Solution Using Equation 37.5, with $m = 3$ for the third-order bright fringes, we find that the values of the fringe positions corresponding to these two wavelengths are given by

$$y_3 = \frac{\lambda L}{d} m = 3 \frac{\lambda L}{d} = 7.74 \times 10^{-2} \text{ m}$$

$$y'_3 = \frac{\lambda' L}{d} m = 3 \frac{\lambda' L}{d} = 9.18 \times 10^{-2} \text{ m}$$

Hence, the separation between the two fringes is given by

$$\Delta y = y'_3 - y_3 = \frac{3(\lambda' - \lambda)}{d} L$$

$$= 1.44 \times 10^{-2} \text{ m} = \boxed{1.44 \text{ cm}}$$

37.3 INTENSITY DISTRIBUTION OF THE DOUBLE-SLIT INTERFERENCE PATTERN

We shall now calculate the distribution of light intensity associated with the double-slit interference pattern. Again, suppose that the two slits represent coherent sources of sinusoidal waves. Hence, they have the same angular frequency ω and a constant phase difference ϕ. The total electric field intensity at the point P on the screen in Figure 37.5 is the *vector superposition* of the two waves from slits S_1 and S_2. Assuming the two waves have the same amplitude E_0, we can write the electric field intensities at P due to each wave separately as

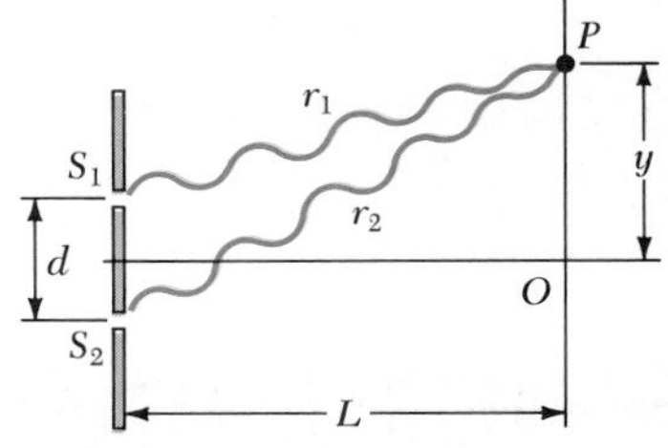

Figure 37.5 Construction for analyzing the double-slit interference pattern. A bright region, or intensity maximum, is observed at O.

$$E_1 = E_0 \sin \omega t \qquad \text{and} \qquad E_2 = E_0 \sin(\omega t + \phi) \tag{37.7}$$

Although the waves have equal phase at the slits, *their phase difference ϕ at P depends on the path difference $\delta = r_2 - r_1 = d \sin \theta$.* Since a path difference of λ corresponds to a phase difference of 2π radians (constructive interference), while a path difference of $\lambda/2$ corresponds to a phase difference of π radians (destructive interference), we obtain the ratio

$$\frac{\delta}{\phi} = \frac{\lambda}{2\pi}$$

$$\phi = \frac{2\pi}{\lambda} \delta = \frac{2\pi}{\lambda} d \sin \theta \tag{37.8}$$

Phase difference

This equation gives the precise dependence of the phase difference ϕ on the angle θ.

Using the superposition principle and Equation 37.7, we can obtain the resultant electric field at the point P:

$$E_P = E_1 + E_2 = E_0[\sin \omega t + \sin(\omega t + \phi)] \tag{37.9}$$

To simplify this expression, we use the following trigonometric identity:

$$\sin A + \sin B = 2 \sin\left(\frac{A + B}{2}\right)\cos\left(\frac{A - B}{2}\right)$$

Taking $A = \omega t + \phi$ and $B = \omega t$, we can write Equation 37.9 in the form

$$E_P = 2E_0 \cos\left(\frac{\phi}{2}\right)\sin\left(\omega t + \frac{\phi}{2}\right) \tag{37.10}$$

Hence, the electric field at P has the same frequency ω, but its amplitude is multiplied by the factor $2 \cos(\phi/2)$. To check the consistency of this result, note that if $\phi = 0, 2\pi, 4\pi, \ldots$, the amplitude at P is $2E_0$, corresponding to the condition for constructive interference. Referring to Equation 37.8, we find that our result is consistent with Equation 37.2. Likewise, if $\phi = \pi, 3\pi, 5\pi, \ldots$, the amplitude at P is zero, which is consistent with Equation 37.3 for destructive interference.

Finally, to obtain an expression for the light intensity at P, recall that *the intensity of a wave is proportional to the square of the resultant electric field at that point* (Section 34.3). Using Equation 37.10, we can therefore express the intensity at P as

$$I \propto E_P^2 = 4E_0^2 \cos^2(\phi/2) \sin^2\left(\omega t + \frac{\phi}{2}\right)$$

Since most light-detecting instruments measure the time average light intensity and the time average value of $\sin^2(\omega t + \phi/2)$ over one cycle is 1/2, we can write the average intensity at P as

$$I_{av} = I_0 \cos^2(\phi/2) \tag{37.11}$$

where I_0 is the *maximum* possible time average light intensity. [You should note that $I \propto (E_0 + E_0)^2 = (2E_0)^2 = 4E_0^2$.] Substituting Equation 37.8 into Equation 37.11, we find that

$$I_{av} = I_0 \cos^2\left(\frac{\pi d \sin\theta}{\lambda}\right) \tag{37.12}$$

Alternatively, since $\sin\theta \approx y/L$ for small values of θ, we can write Equation 37.12 in the form

$$I_{av} = I_0 \cos^2\left(\frac{\pi d}{\lambda L} y\right) \tag{37.13}$$

Constructive interference, which produces intensity maxima, occurs when the quantity $(\pi y d/\lambda L)$ is an integral multiple of 2π, corresponding to $y = (\lambda L/d)m$. This is consistent with Equation 37.5. A plot of the intensity distribution versus θ is given in Figure 37.6a. Note that the interference pattern consists of equally spaced fringes of equal intensity. However, the result is valid only if the slit-to-screen distance L is large relative to the slit separation, and only for small values of θ.

We have seen that the interference phenomena arising from two sources depend on the relative phase of the waves at a given point. Furthermore, the phase difference at a given point depends on the *path difference* between the two waves. The *resultant intensity at a point is proportional to the square of the resultant amplitude.* That is, the intensity is proportional to $(E_1 + E_2)^2$. It would be *incorrect* to calculate the resultant intensity by adding the intensities of the individual waves. This procedure would give a different quantity, namely, $E_1^2 + E_2^2$. Finally, $(E_1 + E_2)^2$ has the same *average* value as $E_1^2 + E_2^2$, when the average is taken over all values of the phase difference between E_1 and E_2. Hence, there is no violation of energy conservation.

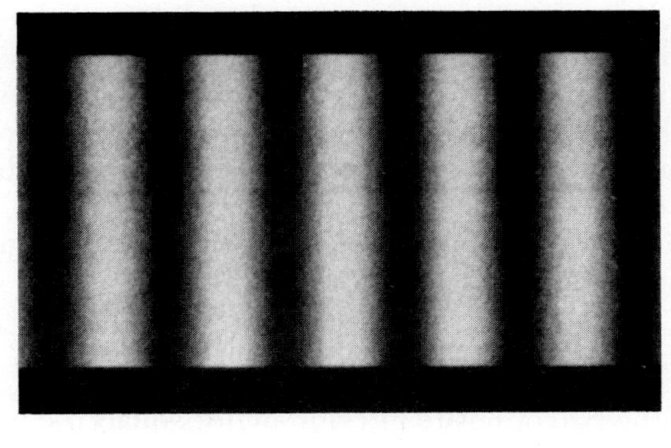

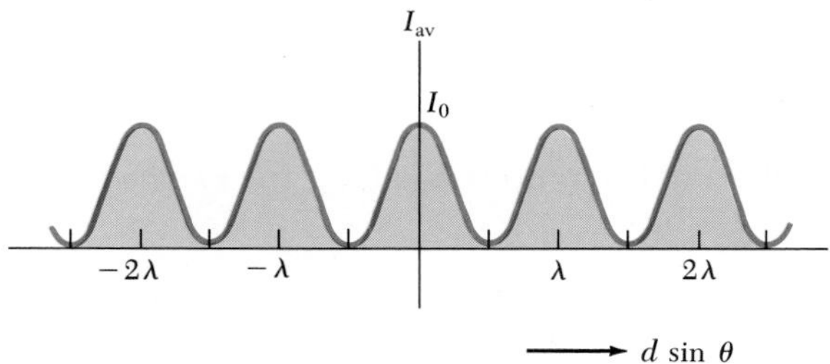

Figure 37.6 Intensity distribution versus $d \sin \theta$ for the double-slit pattern when the screen is far from two slits ($L \gg d$). (Photo from M. Cagnet, M. Francon, and J.C. Thierr, *Atlas of Optical Phenomena,* Berlin, Springer-Verlag, 1962)

37.4 PHASOR ADDITION OF WAVES

In the previous section we combined two waves algebraically to obtain the resultant wave amplitude at some point on a screen. Unfortunately, this analytical procedure becomes rather cumbersome when several wave amplitudes have to be added. Since we shall eventually be interested in combining a large number of waves, we now describe a graphical procedure for this purpose.

Again, consider a sinusoidal wave whose electric field component is given by the expression

$$E_1 = E_0 \sin \omega t$$

where E_0 is the wave amplitude and ω is the angular frequency. This wave disturbance can be represented graphically with a vector of magnitude E_0, *rotating* about the origin in a counterclockwise direction with an angular frequency ω, as in Figure 37.7a. Such a rotating vector is called a *phasor* and is commonly used in the field of electrical engineering (see Chapter 33). Note that the phasor makes an angle of ωt with the horizontal axis. The projection of the phasor on the vertical axis represents E_1, the magnitude of the wave disturbance at some time t. Hence, as the phasor rotates in a circle, the projection E oscillates along the vertical axis about the origin.

Now consider a second sinusoidal wave whose electric field is given by

$$E_2 = E_0 \sin(\omega t + \phi)$$

That is, this wave has the same amplitude and frequency as E_1, but its phase is ϕ with respect to the wave E_1. The phasor representing the wave E_2 is shown in Figure 37.7b. The resultant wave, which is the sum of E_1 and E_2, can be obtained graphically by redrawing the phasors end to end, as in Figure 37.7c,

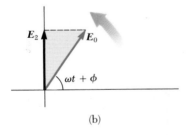

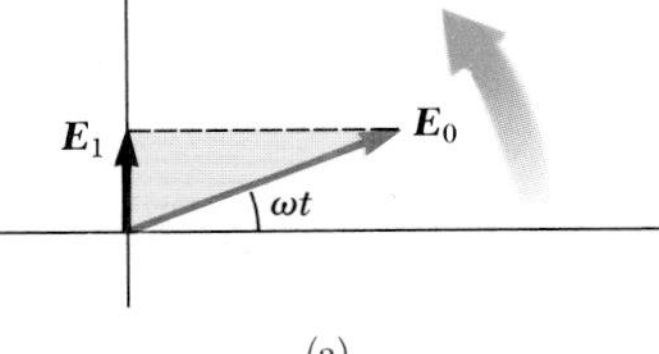

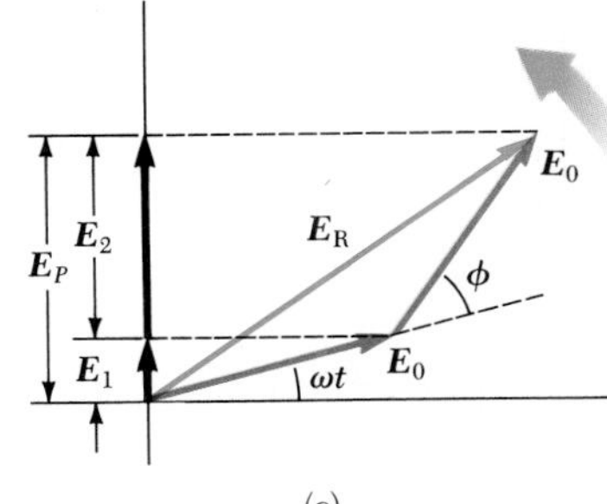

Figure 37.7 (a) Phasor diagram for the wave disturbance $E_1 = E_0 \sin \omega t$. The phasor is a vector of length E_0 rotating counterclockwise. (b) Phasor diagram for the wave $E_2 = E_0 \sin(\omega t + \phi)$. (c) E_R is the resultant phasor formed from the individual phasors shown in (a) and (b).

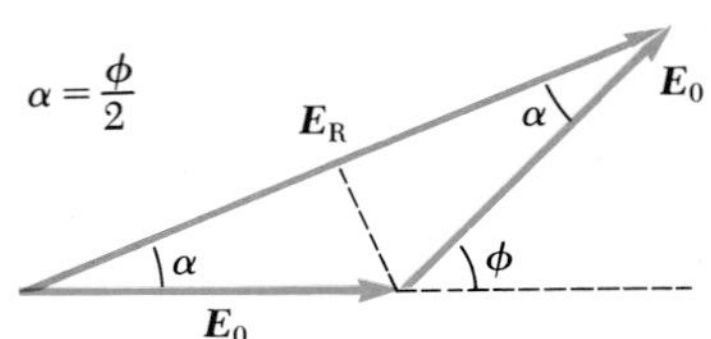

Figure 37.8 Reconstruction of the resultant phasor $\boldsymbol{E}_\mathbf{R}$. From the geometry, note that $\alpha = \phi/2$.

where the tail of the second phasor is placed at the tip of the first phasor. As with vector addition, the resultant phasor $\boldsymbol{E}_\mathbf{R}$ runs from the tail of the first phasor to the tip of the second phasor. Furthermore, $\boldsymbol{E}_\mathbf{R}$ rotates along with the two individual phasors at the same angular frequency ω. The projection of $\boldsymbol{E}_\mathbf{R}$ along the vertical axis equals the sum of the projections of the two phasors. That is, $E_P = E_1 + E_2$.

It is convenient to construct the phasors at $t = 0$ as in Figure 37.8. From the geometry of the triangle, we see that

$$E_\mathrm{R} = E_0 \cos \alpha + E_0 \cos \alpha = 2E_0 \cos \alpha$$

Since the sum of the two opposite interior angles equals the exterior angle ϕ, we see that $\alpha = \phi/2$, so that

$$E_\mathrm{R} = 2E_0 \cos(\phi/2)$$

Hence, the projection of the phasor $\boldsymbol{E}_\mathbf{R}$ along the *vertical axis* at any time t is given by

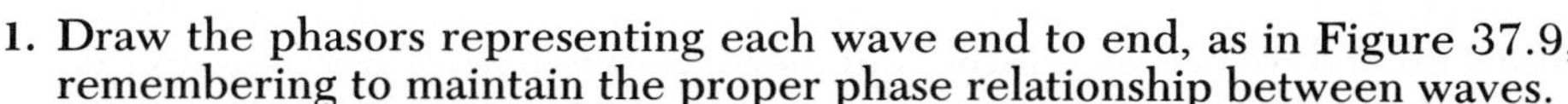

$$E_P = E_\mathrm{R} \sin\left(\omega t + \frac{\phi}{2}\right) = 2E_0 \cos(\phi/2)\sin\left(\omega t + \frac{\phi}{2}\right)$$

This is consistent with the result obtained algebraically, Equation 37.10. The resultant phasor has an amplitude $2E_0 \cos(\phi/2)$ and makes an angle of $\phi/2$ with the first phasor. Furthermore, the average intensity at P, which varies as $E_P{}^2$, is proportional to $\cos^2(\phi/2)$, as described previously in Equation 37.11.

We can now describe how to obtain the resultant of several waves which have the same frequency:

1. Draw the phasors representing each wave end to end, as in Figure 37.9, remembering to maintain the proper phase relationship between waves.
2. The resultant represented by the phasor $\boldsymbol{E}_\mathbf{R}$ is the vector sum of the individual phasors. At each instant, the projection of $\boldsymbol{E}_\mathbf{R}$ along the vertical axis represents the time variation of the resultant wave. The phase angle α of the resultant wave is the angle between $\boldsymbol{E}_\mathbf{R}$ and the first phasor. From the construction in Figure 37.9, drawn for four phasors, we see that the phasor of the resultant wave is given by $E_P = E_\mathrm{R} \sin(\omega t + \alpha)$.

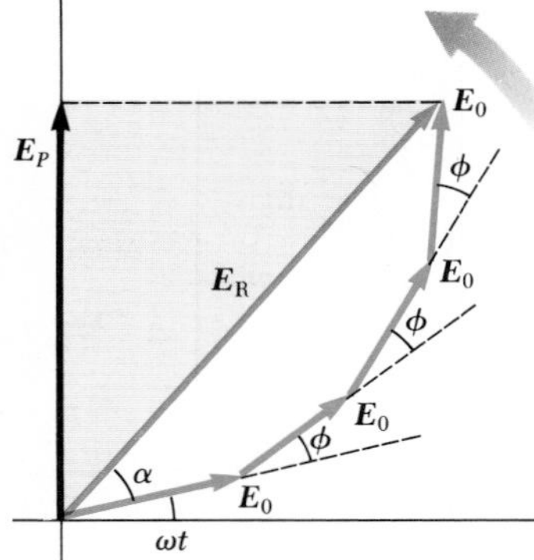

Figure 37.9 The phasor $\boldsymbol{E}_\mathbf{R}$ is the resultant of four phasors of equal amplitude E_0. The phase of $\boldsymbol{E}_\mathbf{R}$ is α with respect to the first phasor.

Phasor Diagrams for Two Coherent Sources

As an example of the phasor method, consider the interference pattern produced by two coherent sources, which was discussed in the previous section. Figure 37.10 represents the phasor diagrams for various values of the phase difference ϕ, and the corresponding values of the path difference δ, which are obtained using Equation 37.8.

From Figure 37.10, we see that the intensity at a point will be a maximum when $\boldsymbol{E}_\mathbf{R}$ is a maximum. This occurs at values of ϕ equal to 0, 2π, 4π, etc. Likewise, we see that the intensity at some observation point will be zero when $\boldsymbol{E}_\mathbf{R}$ is zero. The first zero-intensity point occurs at $\phi = 180°$, corresponding to $\delta = \lambda/2$, while the other zero points (not shown) occur at values of δ equal to $3\lambda/2$, $5\lambda/2$, etc. These results are in complete agreement with the analytical procedure described in the previous section.

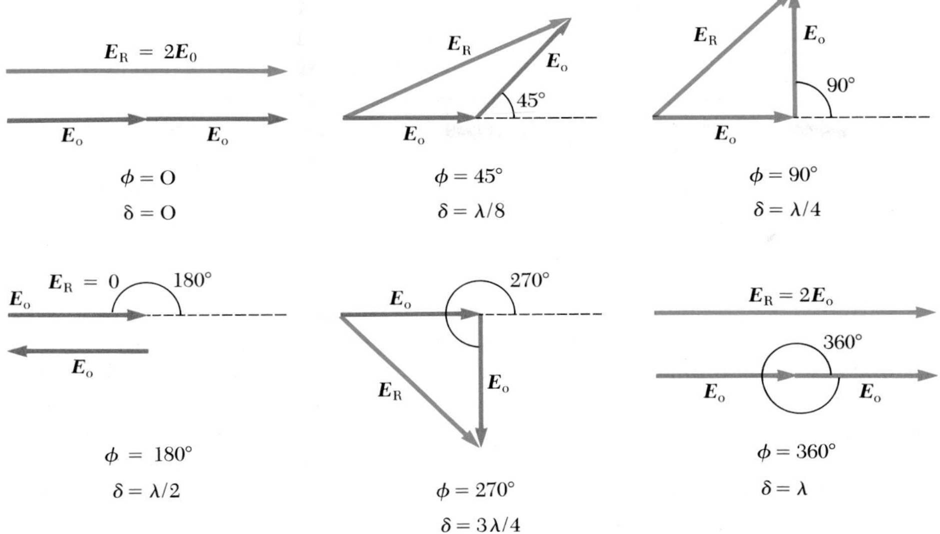

Figure 37.10 Phasor diagrams for the double-slit interference pattern. The resultant phasor E_R is a maximum when $\phi = 0, 2\pi, 4\pi, \ldots$ and is zero when $\phi = \pi, 3\pi, 5\pi, \ldots$.

Three-Slit Interference Pattern

Using phasor diagrams, let us analyze the interference pattern caused by three equally spaced slits. The electric fields at a point P on the screen due to waves from the individual slits can be expressed as

$$E_1 = E_0 \sin \omega t$$

$$E_2 = E_0 \sin(\omega t + \phi)$$

$$E_3 = E_0 \sin(\omega t + 2\phi)$$

where ϕ is the phase difference between waves from adjacent slits. Hence, the resultant field at P can be obtained by using a phasor diagram as shown in Figure 37.11.

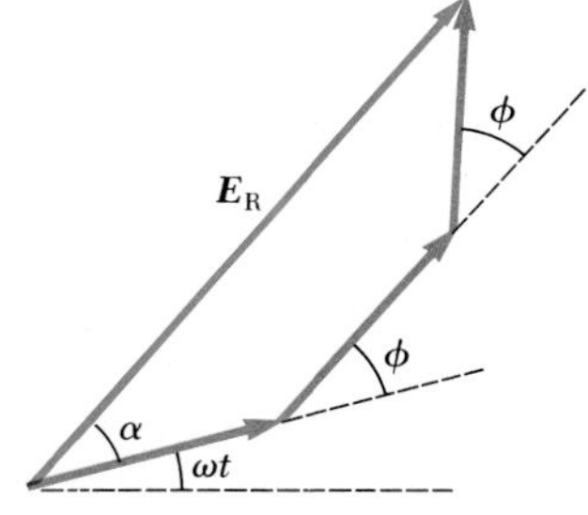

Figure 37.11 Phasor diagram for three equally spaced slits.

The phasor diagrams for various specific values of ϕ for the three slits are shown in Figure 37.12. Note that the resultant amplitude at P has a maximum

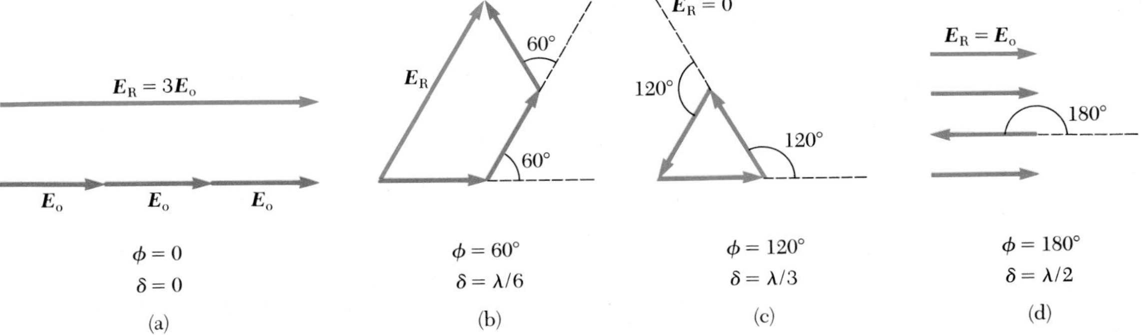

Figure 37.12 Phasor diagrams for three equally spaced slits at various values of ϕ. Note that there are primary maxima of amplitude $3E_0$ and secondary maxima of amplitude E_0.

value of $3E_0$ (called the primary maximum) when $\phi = 0, \pm 2\pi, \pm 4\pi, \ldots$. This corresponds to the case where the three individual phasors are aligned as in Figure 37.12a. However, we also find that secondary maxima of amplitude E_0 occur between the primary maxima when $\phi = \pm\pi, \pm 3\pi, \ldots$. For these points, the wave from one slit exactly cancels that from another slit (Fig. 37.12d), which results in a total amplitude of E_0. Total destructive interference occurs whenever the three phasors form a closed triangle as in Figure 37.12c. These points where $E_0 = 0$ correspond to $\phi = \pm 2\pi/3, \pm 4\pi/3, \ldots$. You should be able to construct other phasor diagrams for values of ϕ greater than π.

Figure 37.13 shows multiple-slit interference patterns for a number of configurations. These patterns represent plots of the intensity for the various primary and secondary maxima. For the case of three slits, note that the primary maxima are nine times more intense than the secondary maxima. This is because the intensity varies as $E_R{}^2$. Figure 37.13 also shows that as the number of slits increases, the number of secondary maxima also increases. In fact, the number of secondary maxima is always equal to N − 2, where N is the number of slits. Finally, as the number of slits increases, the primary maxima

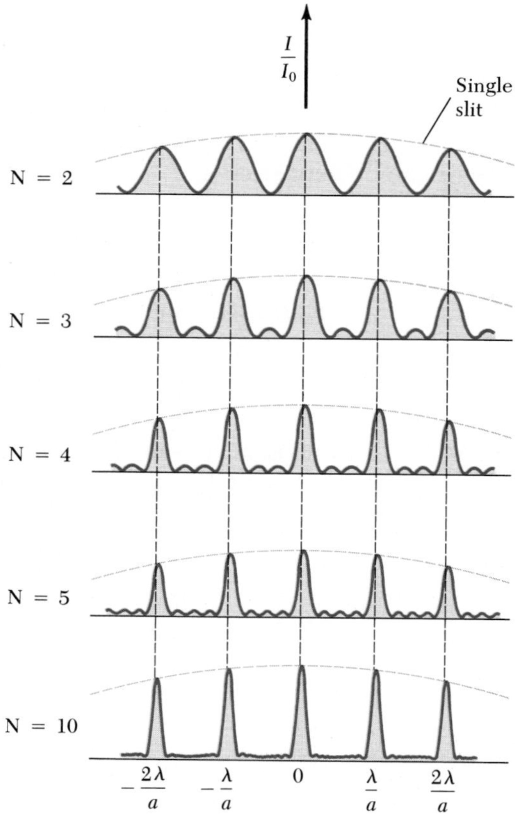

Figure 37.13 Multiple-slit interference patterns. Note that as the number of slits is increased, the primary maxima become narrower but remain fixed in position. Furthermore, the number of secondary maxima increases as the number of slits is increased. The decrease in intensity of the maxima, as indicated by the dashed lines, is due to the phenomenon of diffraction, which is discussed in Chapter 38.

increase in intensity and become narrower, while the secondary maxima decrease in intensity.

37.5 CHANGE OF PHASE DUE TO REFLECTION

We have described interference effects produced by two or more coherent light sources. Young's method for producing two coherent light sources involves illuminating a pair of slits with a single source. Another simple, yet ingenious, arrangement for producing an interference pattern with a single light source is known as *Lloyd's mirror*. A light source is placed at S close to a mirror, as illustrated in Figure 37.14. Waves can reach the viewing point P either by the direct path SP or by the path involving reflection from the mirror. The reflected ray can be treated as a ray originating from a source at S', located behind the mirror. S', which is the image of S, can be considered a virtual source. Hence, at observation points far from the source, one would expect an interference pattern due to waves from S and S' just as is observed for two real coherent sources. An interference pattern is indeed observed. However, the positions of the dark and bright fringes are *reversed* relative to the pattern of two real coherent sources (Young's experiment). This is because the coherent sources at S and S' differ in phase by 180°. This 180° phase change is produced upon reflection. To illustrate this further, consider the point P', where the mirror meets the screen. This point is equidistant from S and S'. If path difference alone were responsible for the phase difference, one would expect to see a bright fringe at P' (since the path difference is zero for this point), corresponding to the central fringe of the two-slit interference pattern. Instead, one observes a *dark* fringe at P' because of the 180° phase change produced by reflection. In general,

> an electromagnetic wave undergoes a phase change of 180° upon reflection from a medium that is optically more dense than the one in which it was traveling. There is also a 180° phase change upon reflection from a conducting surface.

It is useful to draw an analogy between reflected light waves and the reflections of a transverse wave on a stretched string when the wave meets the boundary (Section 16.6). The reflected pulse on a string undergoes a phase change of 180° when it is reflected from a denser medium, such as a heavier string. On the other hand, there is no phase change if the pulse reflects from a less dense medium. Similarly, electromagnetic waves undergo a 180° phase change when reflected from a boundary leading to an optically denser medium. There is no phase change when the wave is reflected from a boundary leading to a less dense medium. In either case, the transmitted wave under-

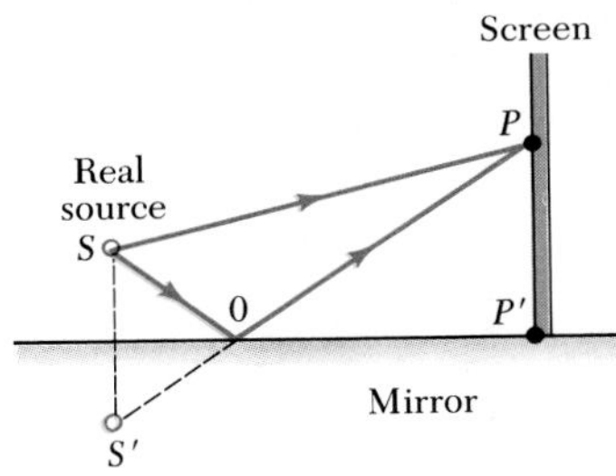

Figure 37.14 Lloyd's mirror. An interference pattern is produced on a screen at P as a result of the combination of the direct ray and the reflected ray. The reflected ray undergoes a phase change of 180°.

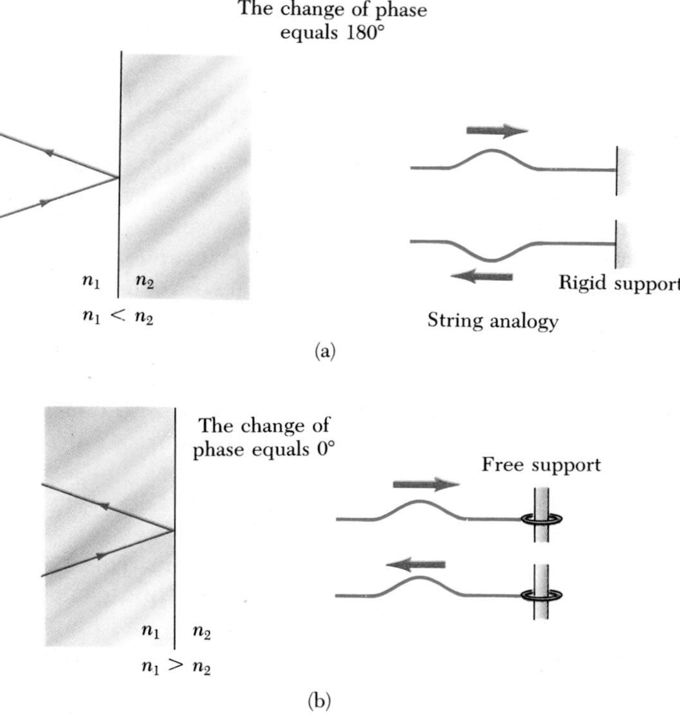

Figure 37.15 (a) A ray reflecting from a medium of higher refractive index undergoes a 180° phase change. The right side shows the analogy with a reflected pulse on a string. (b) A ray reflecting from a medium of lower refractive index undergoes *no* phase change.

goes no phase change. These rules, summarized in Figure 37.15, can be deduced from Maxwell's equations, but the treatment is beyond the scope of this text.

37.6 INTERFERENCE IN THIN FILMS

Interference effects are commonly observed in thin films, such as thin layers of oil on water and soap bubbles. The various colors that are observed with ordinary white light result from the interference of waves reflected from the opposite surfaces of the film.

Consider a film of uniform thickness t and index of refraction n, as in Figure 37.16. Let us assume that the light rays traveling in air are nearly normal to the surface. To determine whether the reflected rays interfere constructively or destructively, we must first note the following facts:

1. *A wave traveling in a medium of low refractive index (air) undergoes a 180° phase change upon reflection from a medium of higher refractive index.* There is no phase change in the reflected wave if it reflects from a medium of lower refractive index.
2. The wavelength of light λ_n in a medium whose refraction index is n (Section 35.4) is given by

$$\lambda_n = \frac{\lambda}{n} \qquad (37.14)$$

where λ is the wavelength of light in free space.

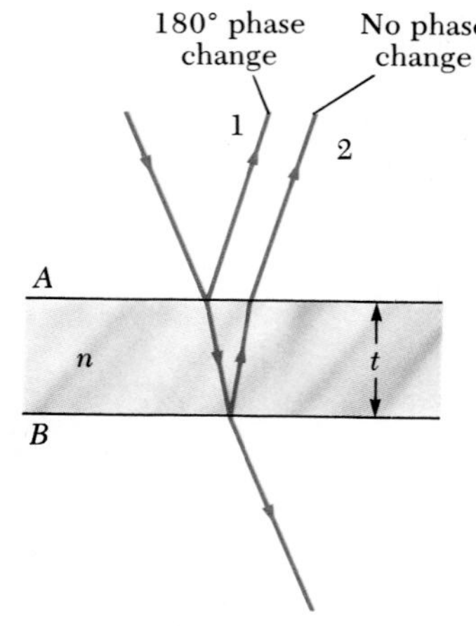

Figure 37.16 Interference in light reflected from a thin film is due to a combination of rays reflected from the upper and lower surfaces.

Thin film interference. A thin film of oil on water displays interference, as shown by the pattern of colors when white light is incident on the film. The film thickness varies, thereby producing the interesting color pattern. (© Tom Branch 1984, Photo Researchers)

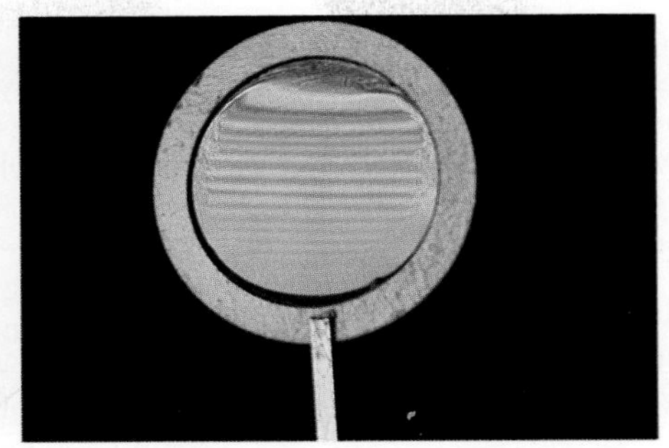

Interference in a vertical soap film of variable thickness. The top of the film appears darkest where the film is thinnest. (© 1983 Larry Mulvehill, Photo Researchers)

Let us apply these rules to the film described in Figure 37.16. According to the first rule, ray 1, which is reflected from the upper surface (A), undergoes a phase change of 180° with respect to the incident wave. On the other hand, ray 2, which is reflected from the lower surface (B), undergoes no phase change with respect to the incident wave. Therefore, ray 1 is 180° out of phase with respect to ray 2, which is equivalent to a path difference of $\lambda_n/2$. However, we must also consider that ray 2 travels an extra distance equal to $2t$ before the waves recombine. For example, if $2t = \lambda_n/2$, rays 1 and 2 will recombine in phase and constructive interference will result. In general, the condition for constructive interference can be expressed as

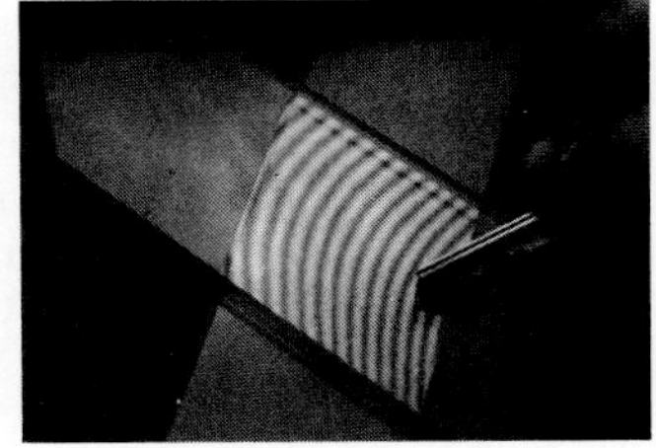

The thin film of air between two glass plates is responsible for the interference pattern. The lines are curved because pressure from the key bends the glass slightly, thus changing the thickness of the air film.

$$2t = (m + \tfrac{1}{2})\lambda_n \qquad (m = 0, 1, 2, \ldots) \tag{37.15}$$

This condition takes into account two factors: (a) the difference in optical path length for the two rays (the term $m\lambda_n$) and (b) the 180° phase change upon reflection (the term $\lambda_n/2$). Since $\lambda_n = \lambda/n$, we can write Equation 37.15 in the form

$$2nt = (m + \tfrac{1}{2})\lambda \qquad (m = 0, 1, 2, \ldots) \tag{37.16}$$

If the extra distance $2t$ traveled by ray 2 corresponds to a multiple of λ_n, the two waves will combine out of phase and destructive interference will result. The general equation for destructive interference is

$$2nt = m\lambda \qquad (m = 0, 1, 2, \ldots) \tag{37.17}$$

These conditions for constructive and destructive interference are valid only when the film is surrounded by a common medium. The surrounding medium may have a refractive index less than or greater than that of the film. In either case, the rays reflected from the two surfaces will be out of phase by 180°. On the other hand, if the film is located between two *different* media, one of lower refractive index and one of higher refractive index, the conditions for constructive and destructive interference are *reversed*. In this case, either there is a phase change of 180° for both ray 1 reflecting from surface A and ray 2 reflecting from surface B or there is no phase change for either ray; hence, the net change in relative phase due to the reflections is *zero*.

EXAMPLE 37.3 Interference in a Soap Film
Calculate the minimum thickness of a soap bubble film ($n = 1.33$) that will result in constructive interference in the reflected light if the film is illuminated with light whose wavelength in free space is 600 nm.

Solution The minimum film thickness for constructive interference in the reflected light corresponds to $m = 0$ in Equation 37.16. This gives $2nt = \lambda/2$, or

$$t = \frac{\lambda}{4n} = \frac{600 \text{ nm}}{4(1.33)} = 113 \text{ nm}$$

Exercise 1 What other film thicknesses will produce constructive interference?
Answer 338 nm, 564 nm, 789 nm, and so on.

EXAMPLE 37.4 Nonreflecting Coatings for Solar Cells
Solar cells are often coated with transparent thin film, such as silicon monoxide (SiO, $n = 1.45$), in order to minimize reflective losses from the surface (Fig. 37.17). A silicon solar cell ($n = 3.5$) is coated with a thin film of silicon monoxide for this purpose. Determine the minimum thickness of the film that will produce the least reflection at a wavelength of 550 nm, which is the center of the visible spectrum.

Solution The reflected light is a minimum when rays 1 and 2 meet the condition of destructive interference. Note that *both* rays undergo a 180° phase change upon reflection in this case, one from the upper and one from the lower surface. Hence, the net change in phase is zero due to reflection, and the condition for a reflection *minimum* requires a path difference of $\lambda_n/2$; hence $2t = \lambda/2n$ or

$$t = \frac{\lambda}{4n} = \frac{550 \text{ nm}}{4(1.45)} = 94.8 \text{ nm}$$

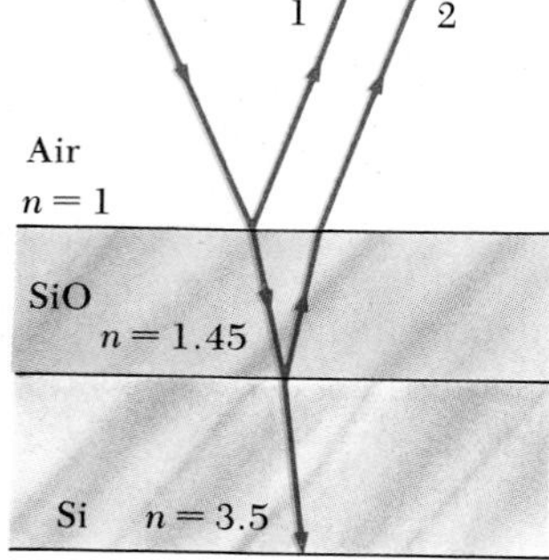

Figure 37.17 (Example 37.4) Reflective losses from a silicon solar cell are minimized by coating it with a thin film of silicon monoxide.

Typically, such antireflecting coatings reduce the reflective loss from 30% (with no coating) to 10% (with coating). Such coatings therefore increase the cell's efficiency since more light is available to create charge carriers in the cell. In reality, the coating is never perfectly nonreflecting because the required thickness is wavelength-dependent and the incident light covers a wide range of wavelengths.

Glass lenses used in cameras and other optical instruments are usually coated with a transparent thin film, such as magnesium fluoride (MgF_2), to reduce or eliminate unwanted reflection. More important, such coatings enhance the transmission of light through the lenses.

EXAMPLE 37.5 Interference in a Wedge-Shaped Film
A thin, wedge-shaped film of refractive index n is illuminated with monochromatic light of wavelength λ, as illustrated in Figure 37.18. Describe the interference pattern observed for this case.

Solution The interference pattern is that of a thin film of variable thickness surrounded by air. Hence, the pattern will be a series of alternating bright and dark parallel bands. A dark band corresponding to destructive interference appears at point O, the apex, since the upper reflected ray undergoes a 180° phase change while the lower one does not. According to Equation 37.17, other dark bands appear when $2nt = m\lambda$, so that $t_1 = \lambda/2n$, $t_2 = \lambda/n$, $t_3 = 3\lambda/2n$, and so on. Similarly, bright bands will be observed when the thickness satisfies the condition $2nt = (m + \frac{1}{2})\lambda$, corresponding to thicknesses of $\lambda/4n$, $3\lambda/4n$, $5\lambda/4n$, and so on. If white light is used, bands of different colors will be observed at different points, corresponding to the different wavelengths of light.

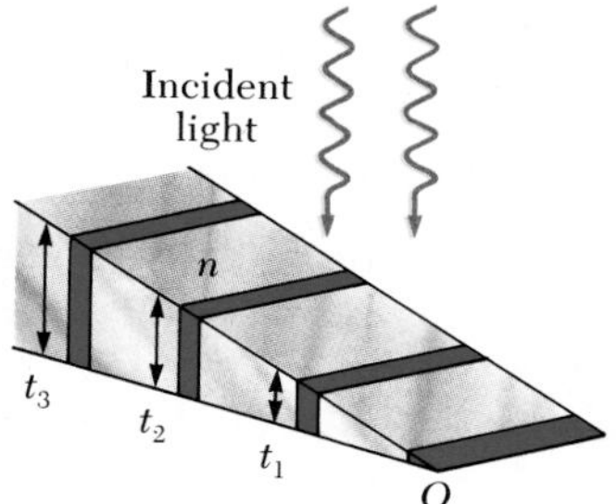

Figure 37.18 (Example 37.5) Interference bands in reflected light can be observed by illuminating a wedge-shaped film with monochromatic light. The blue areas correspond to positions of destructive interference.

Newton's Rings

Another method for observing interference of light waves is to place a plano-convex lens (one having one plane side and one convex side) on top of a plane glass surface as in Figure 37.19a. With this arrangement, the air film between the glass surfaces varies in thickness from zero at the point of contact to some value t at P. If the radius of curvature of the lens R is very large compared with the distance r, and if the system is viewed from above using light of wavelength λ, a pattern of light and dark rings is observed. A photograph of such a pattern is shown in Figure 37.19b. These circular fringes, discovered by Newton, are called **Newton's rings.** Newton's particle model of light could not explain the origin of the rings.

The interference effect is due to the combination of ray 1, reflected from the plane glass plate, with ray 2, reflected from the lower part of the lens. Ray 1 undergoes a phase change of 180° upon reflection, since it is reflected from a medium of higher refractive index, whereas ray 2 undergoes no phase change. Hence, the conditions for constructive and destructive interference are given by Equations 37.16 and 37.17, respectively, with $n = 1$ since the "film" is air. Here again, one might guess that the contact point O would be bright, corresponding to constructive interference. Instead, the contact point is dark, as seen in Figure 37.19b, because ray 1, reflected from the plane surface, undergoes a 180° phase change with respect to ray 2. Using the geometry shown in Figure 37.19a, one can obtain expressions for the radii of the bright and dark bands in terms of the radius of curvature R and wavelength λ. For example, the dark rings have radii given by $r \approx \sqrt{m\lambda R/n}$. The details are left as a problem for the reader (Problem 62). By measuring the radii of the rings, one can obtain the wavelength, provided R is known. Conversely, if the wavelength is accurately known, one can use this effect to obtain R.

One of the important uses of Newton's rings is in the testing of optical lenses. A circular pattern like that pictured in Figure 37.19b is obtained only

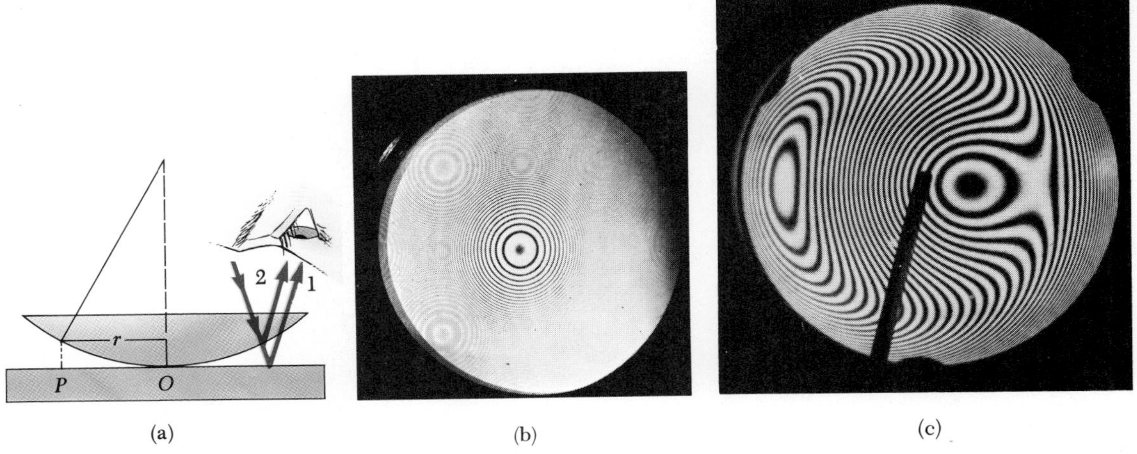

Figure 37.19 (a) The combination of rays reflected from the glass plate and the curved surface of the lens give rise to an interference pattern known as Newton's rings. (b) Photograph of Newton's rings. (Courtesy of Bausch and Lomb Optical Co.) (c) This asymmetrical interference pattern indicates imperfections in the lens. (From Physical Science Study Committee, *College Physics*, Lexington, Mass., Heath, 1968)

when the lens is ground to a perfectly symmetric curvature. Variations from such symmetry might produce a pattern like that in Figure 37.19c. These variations give an indication of how the lens must be ground and polished in order to remove the imperfections.

*37.7 THE MICHELSON INTERFEROMETER

The **interferometer,** invented by the American physicist A. A. Michelson (1852–1931), is an ingenious device that splits a light beam into two parts and then recombines them to form an interference pattern after they have traveled over different paths. The device can be used for obtaining accurate measurements of wavelengths or for precise length measurements.

A schematic diagram of the interferometer is shown in Figure 37.20. A beam of light provided by a monochromatic source is split into two rays by a partially silvered mirror M inclined at 45° relative to the incident light beam. One ray is reflected vertically upward toward mirror M_1 while the second ray is transmitted horizontally through M toward mirror M_2. Hence, the two rays travel separate paths L_1 and L_2. After reflecting from mirrors M_1 and M_2, the two rays eventually recombine to produce an interference pattern, which can be viewed through a telescope. The glass plate P, equal in thickness to mirror M, is placed in the path of the horizontal ray in order to insure that the two rays travel the same distance through glass.

The interference condition for the two rays is determined by the difference in their optical path lengths. When the two rays are viewed as shown, the

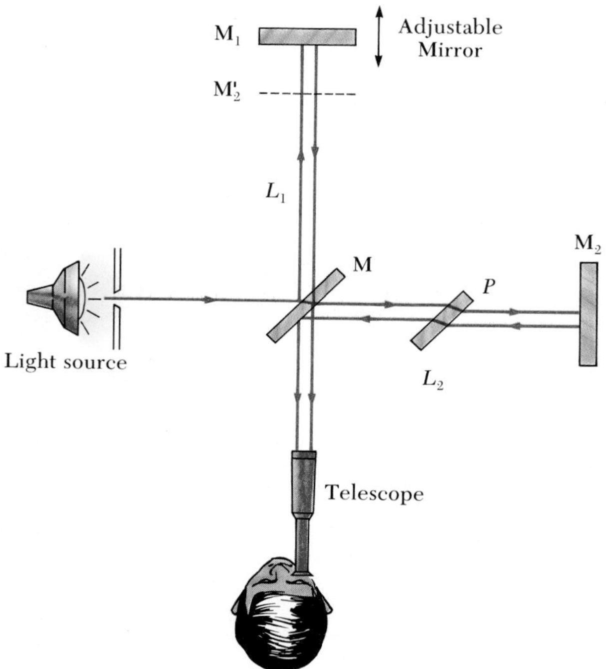

Figure 37.20 Diagram of the Michelson interferometer. A single beam is split into two rays by the partially silvered mirror M. The path difference between the two rays is varied with the adjustable mirror M_1.

image of M_2 is at M_2' parallel to M_1. Hence, M_2' and M_1 form the equivalent of a parallel air film. The effective thickness of the air film is varied by moving mirror M_1 parallel to itself with a finely threaded screw. Under these conditions, the interference pattern is a series of bright and dark circular rings which resemble Newton's rings. If a dark circle appears at the center of the pattern, the two rays interfere destructively. If the mirror M_1 is then moved a distance of $\lambda/4$, the path difference changes by $\lambda/2$ (twice the separation between M_1 and M_2'). The two rays will now interfere constructively, giving a bright circle in the middle. As M_1 is moved an additional distance of $\lambda/4$, a dark circle will appear once again. Thus, we see that successive dark or bright circles are formed each time M_1 is moved a distance of $\lambda/4$. The wavelength of light is then measured by counting the number of fringe shifts for a given displacement of M_1. Conversely, if the wavelength is accurately known (as with a laser beam), mirror displacements can be measured to within a fraction of the wavelength.

Since the interferometer can accurately measure displacement, it is often used to make highly precise measurements of the length of mechanical components. The interferometer makes such precise wavelength measurements possible.

SUMMARY

Interference of light waves is the result of the linear superposition of two or more waves at a given point. A sustained interference pattern is observed if (1) the sources are coherent (that is, they maintain a constant relative phase), (2) the sources are monochromatic (of a single wavelength), and (3) the linear superposition principle is applicable.

In Young's double-slit experiment, two slits separated by a distance d are illuminated by a monochromatic light source. An interference pattern consisting of bright and dark fringes is observed on a screen a distance L from the slits. The condition for bright fringes in the double-slit experiment (**constructive interference**) is given by

Conditions for constructive interference

$$d \sin \theta = m\lambda \qquad (m = 0, \pm 1, \pm 2, \ldots) \tag{37.2}$$

The condition for dark fringes in the double-slit experiment (**destructive interference**) is

Conditions for destructive interference

$$d \sin \theta = (m + \tfrac{1}{2})\lambda \qquad (m = 0, \pm 1, \pm 2, \ldots) \tag{37.3}$$

The index number m is called the **order number** of the fringe.

The **average intensity** of the double-slit interference pattern is given by

$$I_{av} = I_0 \cos^2 \frac{\pi d \sin \theta}{\lambda} \tag{37.12}$$

where I_0 is the maximum intensity on the screen.

When a series of N slits is illuminated, a diffraction pattern is produced that can be viewed as interference arising from the superposition of a large number of waves. It is convenient to use phasor diagrams to simplify the analysis of interference from three or more equally spaced slits.

An electromagnetic wave undergoes a phase change of 180° upon reflection from an optically more dense medium or from any conducting surface.

The wavelength of light λ_n in a medium whose refractive index is n is given by

$$\lambda_n = \frac{\lambda}{n} \tag{37.14}$$

where λ is the wavelength of light in free space.

The condition for constructive interference in a film of thickness t and refractive index n with a common medium on both sides of the film is given by

Constructive interference in thin films

$$2nt = (m + \tfrac{1}{2})\lambda \qquad (m = 0, 1, 2, \ldots) \tag{37.16}$$

Similarly, the condition for destructive interference in thin films is

Destructive interference in thin films

$$2nt = m\lambda \qquad (m = 0, 1, 2, \ldots) \tag{37.17}$$

QUESTIONS

1. What is the necessary condition on the path length difference between two waves that interfere (a) constructively and (b) destructively?
2. Explain why two flashlights held close together will not produce an interference pattern on a distant screen.
3. If Young's double-slit experiment were performed under water, how would the observed interference pattern be affected?
4. What is the difference between interference and diffraction?
5. In Young's double-slit experiment, why do we use monochromatic light? If white light is used, how would the pattern change?
6. In the process of evaporation, a soap bubble appears black just before it breaks. Explain this phenomenon in terms of the phase changes that occur upon reflection from the two surfaces.
7. If an oil film is observed on water, the film appears brightest at the outer regions, where it is thinnest. From this information, what can you say about the index of refraction of the oil relative to that of water?
8. If a soap film on a wire loop is held in air, it appears black in the thinnest regions when observed by reflected light and shows a variety of colors in thicker regions, as in the photograph at the right. Explain.
9. A simple way of observing an interference pattern is to look at a distant light source through a stretched handkerchief or an opened umbrella. Explain how this works.
10. In order to observe interference in a thin film, why must the film not be very thick (on the order of a few wavelengths)? (*Hint:* How far apart are the two reflected waves when they attempt to interfere if the film is thick?)
11. A lens with outer radius of curvature R and index of refraction n rests on a flat glass plate. It is illuminated with white light from above. Is there a dark spot or a light spot at the center of the lens? What does it mean if the observed rings are noncircular?
12. Why is the lens on a good quality camera coated with a thin film?
13. Would it be possible to place a nonreflective coating on an airplane to cancel radar waves of wavelength 3 cm?
14. How is it that good quality pictures can be taken with a pinhole camera (if the pinhole is the right diameter)? Why isn't a lens required to get good focus?
15. Why is it so much easier to perform interference experiments with a laser than an ordinary light source?

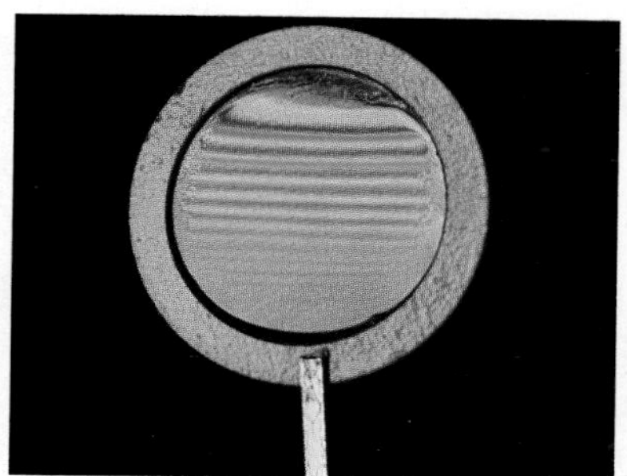

Question 8.

PROBLEMS

Section 37.2 Young's Double-Slit Experiment

1. A pair of narrow, parallel slits separated by 0.25 mm are illuminated by the green component from a mercury vapor lamp ($\lambda = 546.1$ nm). The interference pattern is observed on a screen located 1.2 m from the plane of the parallel slits. Calculate the distance (a) from the central maximum to the first bright region on either side of the central maximum and (b) between the first and second dark bands in the interference pattern.
2. A laser beam ($\lambda = 632.8$ nm) is incident upon two slits 0.2 mm apart. Approximately how far apart will the bright interference lines be on a screen 5 m from the double slits?
3. A Young's interference experiment is performed with blue-green argon laser light. The separation between the slits is 0.50 mm and the interference pattern on a screen 3.3 m away shows the first maximum at a distance of 3.4 mm from the center of the pattern. What is the wavelength of argon laser light?
4. The slits in a Young's interference apparatus are illuminated with monochromatic light. The third dark band is 9.5 mm from the central maximum. The two slits are 0.15 mm apart, and the screen is 90 cm away from the slits. Calculate the wavelength of the light used.
5. The yellow component of light from a helium discharge tube ($\lambda = 587.5$ nm) is allowed to fall on a plane containing parallel slits that are 0.2 mm apart. A screen is located so that the second bright band in the interference pattern is at a distance equal to 10 slit spacings from the central maximum. What is the distance between the source plane and the screen?
6. One of the bright bands in a Young's interference pattern is located 12 mm from the central maximum. The screen is located 119 cm from the pair of slits that serve as secondary sources. The slits are 0.241 mm apart and are illuminated by the blue light from a hydrogen discharge tube ($\lambda = 486$ nm). How many bright lines are there *between* the central maximum and the 12-mm position?
7. A narrow slit is cut into each of two overlapping opaque squares. The slits are parallel, and the distance between them is adjustable. Monochromatic light of wavelength 600 nm illuminates the slits, and an interference pattern is formed on a screen 80 cm away. The third dark band is located 1.2 cm from the central bright band. What is the distance between the center of the central bright band and the center of the first bright band on either side of the central band?
8. In a double-slit interference experiment, the slits are illuminated with light of wavelength 680 nm. If the second bright fringe is 3.5 cm from the central line and the slits are 2 m from the observing screen, calculate (a) the slit separation and (b) the position of the second dark fringe.
9. In deriving Equation 37.5, it is assumed as stated in Equation 37.4 that $\sin\theta \approx \tan\theta$. The validity of this assumption depends on the requirement that $d \gg \lambda$. In the arrangement of Figure 37.4, let $L = 40$ cm, $d = 0.5$ mm, and $\lambda = 656.3$ nm (the red line in hydrogen). Under the assumption stated in Equation 37.4, the ninth-order dark band would be located 4.4631 mm on either side of the central maximum. What percent error is introduced by the assumption that $d \gg \lambda$?
10. A third-order maximum is located 4.2 mm above the central maximum in a Young's interference pattern. The distance between the slits equals 200 wavelengths of the incident light. What is the distance between the source plane and the screen?
11. Light of wavelength 546 nm (the intense green line from a mercury discharge tube) produces a Young's interference pattern in which the second-order minimum is along a direction that makes an angle of 18 minutes of arc relative to the direction to the central maximum. What is the distance between the parallel slits?
12. In a double-slit arrangement as illustrated in Figure 37.4, $d = 0.15$ mm, $L = 140$ cm, $\lambda = 643$ nm, and $y = 1.8$ cm. (a) What is the path difference δ for the two slits at the point P? (b) Express this path difference in terms of the wavelength. (c) Will point P correspond to a maximum, a minimum, or an intermediate condition?

Section 37.3 Intensity Distribution of the Double-Slit Interference Pattern

13. In the arrangement of Figure 37.4, let $L = 120$ cm and $d = 0.25$ cm. The slits are illuminated with light of wavelength 600 nm. Calculate the distance y above the central maximum for which the average intensity on the screen will be 75% of the maximum.
14. Two slits are separated by a distance of 0.18 mm. An interference pattern is formed on a screen 80 cm away by the H_α line in hydrogen ($\lambda = 656.3$ nm). Calculate the fraction of the maximum intensity that would be measured at a point 0.6 cm above the central maximum.
15. In a double-slit interference experiment (Fig. 37.4), $d = 0.2$ mm, $L = 160$ cm, and $y = 1$ mm. What wavelength will result in an average intensity at P that is 36% of the maximum?
16. In an arrangement similar to that illustrated in Figure 37.4, let $L = 140$ cm and $y = 8$ mm. Find the value of

the ratio d/λ for which the average intensity at point P will be 60% of the maximum intensity.

17. Make a plot of I/I_0 as a function of θ (see Fig. 37.4) for the interference pattern produced by the arrangement described in Problem 1. Let θ range over the interval from 0 to 0.3°.

18. In Figure 37.4 let $L = 1.2$ m and $d = 0.12$ mm and assume that the slit system is illuminated with monochromatic light of wavelength 500 nm. Calculate the phase difference between the two wavefronts arriving at point P from S_1 and S_2 when (a) $\theta = 0.5°$ and (b) $y = 5$ mm.

19. For the situation described in Problem 18, what is the value of θ for which (a) the phase difference will be equal to 0.333 rad and (b) the path difference will be $\lambda/4$?

20. The intensity on the screen at a certain point in a double-slit interference pattern is 64% of the maximum value. (a) What is the minimum phase difference (in radians) between sources that will produce this result? (b) Express the phase difference calculated in (a) as a path difference if the wavelength of the incident light is 486.1 nm (H_β line).

21. At a particular location in a Young's interference pattern, the intensity on the screen is 6.4% of maximum. (a) Calculate the minimum phase difference in this case. (b) If the wavelength of light is 587.5 nm (from a helium discharge tube), determine the path difference.

22. Light from a helium-neon laser ($\lambda = 632.8$ nm) is incident on two parallel slits 0.2 mm apart. What is the distance to the first maximum and its intensity (relative to the central maximum) on a screen 2 m beyond the slits?

23. Two narrow parallel slits are separated by 0.85 mm and are illuminated by light with $\lambda = 6000$ Å. (a) What is the phase difference between the two interfering waves on a screen (2.8 m away) at a point 2.50 mm from the central bright fringe? (b) What is the ratio of the intensity at this point to the intensity at the center of a bright fringe?

Section 37.4 Phasor Addition of Waves

24. The electric fields from three coherent sources are described by $E_1 = E_0 \sin \omega t$, $E_2 = E_0 \sin(\omega t + \phi)$, and $E_3 = E_0 \sin(\omega t + 2\phi)$. Let the resultant field be represented by $E_P = E_R \sin(\omega t + \alpha)$. Use the phasor method to find E_R and α when (a) $\phi = 20°$, (b) $\phi = 60°$, (c) $\phi = 120°$.

25. Repeat Problem 24 when $\phi = (3\pi/2)$ radians.

26. Use the method of phasors to find the resultant (magnitude and phase angle) of two fields represented by $E_1 = 12 \sin \omega t$ and $E_2 = 18 \sin(\omega t + 60°)$. (Note that in this case the amplitudes of the two fields are unequal.)

27. Determine the resultant of the two waves $E_1 = 6 \sin(100\,\pi t)$ and $E_2 = 8 \sin(100\,\pi t + \pi/2)$.

28. You are given that $E_1 = 5.77 \sin \omega t$, $E_2 = E_0 \sin(\omega t + \phi)$, and $E_P = E_1 + E_2$. Find ϕ and E_0 such that $E_P = 10 \sin(\omega t + \pi/6)$. (Use the method of phasor addition.)

29. When illuminated, four equally spaced parallel slits act as multiple coherent sources, each differing in phase from the adjacent one by an angle ϕ. Use a phasor diagram to determine the smallest value of ϕ for which the resultant of the four waves (assumed to be of equal amplitude) will be zero.

30. Sketch a phasor diagram to illustrate the resultant of $E_1 = E_{01} \sin \omega t$ and $E_2 = E_{02} \sin(\omega t + \phi)$, where $E_{02} = 1.5E_{01}$ and $\pi/6 \le \phi \le \pi/3$. Use the sketch and the law of cosines to show that, for two coherent waves, the resultant *intensity* can be written in the form $I_R = I_1 + I_2 + 2\sqrt{I_1 I_2} \cos \phi$.

31. Consider N coherent sources described by $E_1 = E_0 \sin(\omega t + \phi)$, $E_2 = E_0 \sin(\omega t + 2\phi)$, $E_3 = E_0 \sin(\omega t + 3\phi)$, . . . , $E_N = E_0 \sin(\omega t + N\phi)$. Find the minimum value of ϕ for which $E_R = E_1 + E_2 + E_3 + \cdots E_N$ will be zero.

Section 37.6 Interference in Thin Films

32. A material having an index of refraction of 1.30 is used to coat a piece of glass ($n = 1.50$). What should be the minimum thickness of this film in order to minimize reflected light at a wavelength of 500 nm?

33. A thin film of MgF_2 10^{-5} cm thick ($n = 1.38$) is used to coat a camera lens. Will any wavelengths in the visible spectrum be intensified in the reflected light?

34. A soap bubble of index of refraction 1.33 strongly reflects both red and green colors in white light. What thickness of soap bubble allows this to happen? (In air, λ red = 700 nm, λ green = 500 nm.)

35. Determine the minimum thickness of a soap film ($n = 1.33$) that will result in constructive interference of (a) the red H_α line ($\lambda = 656.3$ nm) and (b) the blue H_γ line ($\lambda = 434.0$ nm).

36. A thin layer of oil ($n = 1.25$) is floating on water. How thick is the oil in the region that reflects green light ($\lambda = 525$ nm)?

37. A thin layer of liquid methylene iodide ($n = 1.756$) is sandwiched between two flat parallel plates of glass. What must be the thickness of the liquid layer if normally incident light with $\lambda = 600$ nm is to be strongly reflected?

38. A thin film of glass ($n = 1.52$) with a thickness of 0.42 μm is viewed in white light at near-normal incidence. Visible light with what wavelength is most strongly reflected by the film?

39. A 500-nm-thick oil film in air is illuminated with white light in the direction perpendicular to the film. What

wavelengths will be strongly reflected in the range 300–700 nm? (Take $n = 1.46$ for oil.)

40. Repeat Problem 39 if the oil film is placed on a thick piece of glass ($n = 1.50$).

41. Let the film shown in Figure 37.16 have an index of refraction of 1.36 and a thickness of 7×10^{-5} cm. A beam of sunlight is incident in air on the top surface of the film. Determine the wavelengths (within the range 400–700 nm) that will be strongly reflected by the film.

42. Two rectangular, optically flat glass plates ($n = 1.52$) are in contact along one end and are separated along the other end by a sheet of paper that is 4×10^{-3} cm thick (Fig. 37.21). The top plate is illuminated by monochromatic light ($\lambda = 546.1$ nm). Calculate the number of dark parallel bands crossing the top plate (include the dark band at zero thickness along the edge of contact between the two plates).

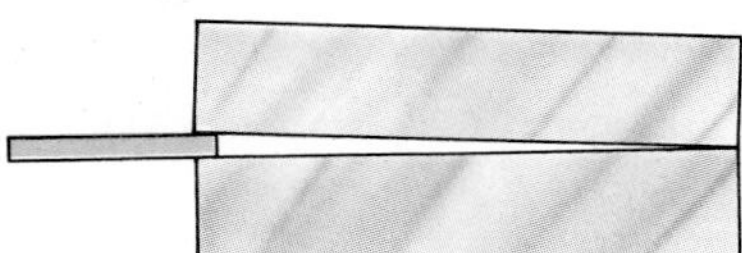

Figure 37.21 (Problems 42 and 43).

43. An air wedge is formed between two thick glass plates in a manner similar to that described in Problem 42. Light of wavelength 434 nm is incident vertically on the top plate. In this case, there are 20 bright parallel interference fringes across the top plate. Calculate the thickness of the paper separating the plates.

44. Suppose that the wedge shape shown in Figure 37.18 has a length (measured along the incline) of 12 cm and that the angle between the two faces is 3.5×10^{-4} rad (2×10^{-2} deg). The wedge has an index of refraction of 1.41 and is illuminated (along the vertical to the top face) by light of wavelength 600 nm. How many bright bands appear in the interference pattern of the reflected light?

°Section 37.7 The Michelson Interferometer

45. The mirror on one arm of a Michelson interferometer is displaced a distance ΔL. During this displacement, 250 fringe shifts (formation of *successive dark or bright line fringes*) are counted. The light being used has a wavelength of 632.8 nm. Calculate the displacement ΔL.

46. A Michelson interferometer is used to measure an unknown wavelength. The mirror in one arm of the instrument is moved 0.12 mm as 481 *dark* fringes are counted. Determine the wavelength of the light used.

47. Light of wavelength 550.5 nm is used to calibrate a Michelson interferometer. By use of a micrometer screw, the platform on which one mirror is mounted is moved 0.18 mm. How many *dark* fringe shifts are counted?

48. A Michelson interferometer (Fig. 37.20) is illuminated with green light ($\lambda = 530$ nm). (a) How many fringes will be seen to pass the cross-hair in the viewing telescope as the mirror M_1 is moved from one end of a 1.0-cm gauge block to the other end? (b) A new light source with a different color is substituted and the observer counts 36,200 fringes as the mirror is returned to its original position. What is the wavelength of the light from the new source?

ADDITIONAL PROBLEMS

49. The calculation of Example 37.1 shows that the double-slit arrangement produced fringe separations of 2.2 cm for $\lambda = 560$ nm. Calculate the fringe separation for this same arrangement if the apparatus is submerged in a tank containing a 30% sugar solution ($n = 1.38$).

50. Interference effects are produced at point P on a screen as a result of direct rays from a source of wavelength 500 nm and reflected rays off the mirror, as shown in Figure 37.22. If the source is 100 m to the left of the screen, and 1 cm above the mirror, find the distance y (in mm) to the first dark band above the mirror.

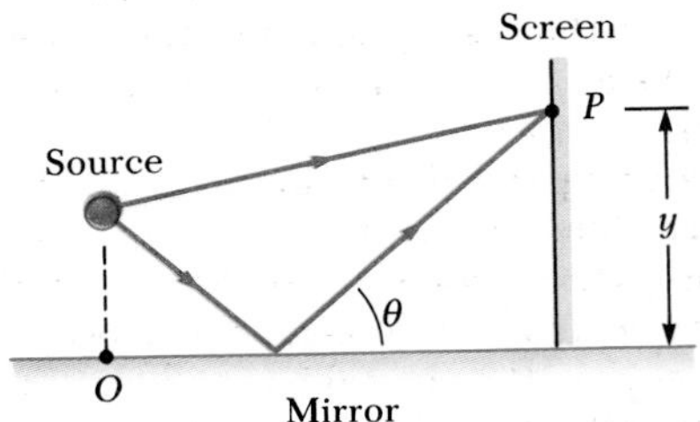

Figure 37.22 (Problem 50).

51. In an application of interference effects to radioastronomy, Australian astronomers observed a 60-MHz radiosource both directly from the source and from its reflection from the sea. If the receiving dish is 20 m above sea level, what is the angle of the radiosource above the horizon at first maximum?

52. Waves from a radio station have a wavelength of 300 m. They arrive at a home receiver 20 km away from the transmitter by two paths. One is a direct path, and the second is by reflection from a mountain directly behind the home receiver. What is the minimum distance from the mountain to the receiver such that destructive interference occurs at the receiver?

53. Measurements are made of the intensity distribution in a Young's interference pattern (as illustrated in Figure 37.6). At a particular value of y (distance from the center of the screen), it is found that $I/I_0 = 0.81$ when light of wavelength 600 nm is used. What wavelength of light should be used to reduce the relative intensity at the same location to 64%?

54. The light source used to illuminate the parallel-slit system illustrated in Figure 37.4 emits two wavelengths, the longer of which is 700 nm. The fifth dark fringe of the long-wavelength pattern occupies the same position as the fifth bright fringe (not counting the central maximum) of the short-wavelength pattern. Determine the wavelength of the second component.

55. In a Young's interference experiment, the two slits are separated by 0.15 mm and the incident light includes light of wavelengths $\lambda_1 = 540$ nm and $\lambda_2 = 450$ nm. The overlapping interference patterns are formed on a screen 1.4 m from the slits. Calculate the minimum distance from the center of the screen to the point where a bright line of the λ_1 light coincides with a bright line of the λ_2 light.

56. Two sinusoidal vectors of the same amplitude A and frequency ω have a phase difference ϕ. Calculate the resultant amplitude of the two vectors both graphically and analytically if ϕ equals (a) 0, (b) 60°, (c) 90°.

57. Young's double-slit experiment is performed with sodium yellow light ($\lambda = 5890$ Å) and with a slits-to-screen distance of 2.0 m. The tenth interference minimum (dark fringe) is observed to be 7.26 mm from the central maximum. Determine the spacing of the slits.

58. Young's experiment is performed by using white light. The two narrow slits are separated by 0.40 mm. The screen is at a distance of 3.0 m. (a) Locate the first-order maximum for yellow-green light ($\lambda = 550$ nm). (b) What is the separation between the first-order maxima for red light ($\lambda = 700$ nm) and violet light ($\lambda = 400$ nm)? (c) At what point nearest the central maximum will a maximum for yellow light ($\lambda = 600$ nm) coincide with a maximum for violet light ($\lambda = 400$ nm)? Identify the order of each maximum.

59. A hair is placed at one edge between two flat glass plates 8 cm long. When this arrangement is illuminated with yellow light of wavelength 600 nm, a total of 121 dark bands are counted, starting at the point of contact of the two plates. How thick is the hair?

60. A glass plate ($n = 1.61$) is covered with a thin uniform layer of oil ($n = 1.2$). A nonmonochromatic light beam in air is incident normally on the oil surface. Observation of the reflected beam shows destructive interference at 500 nm and constructive interference at 750 nm with no intervening maxima or minima. Calculate the thickness of the oil film.

61. A piece of transparent material having an index of refraction n is cut into the shape of a wedge as shown in Figure 37.23. The angle of the wedge is small, and monochromatic light of wavelength λ is normally incident from above. If the height of the wedge is h and the width is ℓ, show that bright fringes occur at the positions $x = \lambda\ell(m + \frac{1}{2})/2hn$ and dark fringes occur at the positions $x = \lambda\ell m/2hn$, where $m = 0, 1, 2, \ldots$ and x is measured as shown.

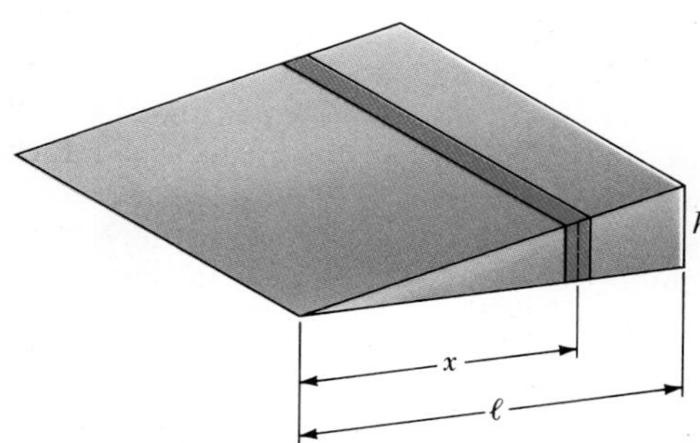

Figure 37.23 (Problem 61).

62. Refer to Figure 37.19a, which illustrates an arrangement for observing Newton's rings. Assume that n is the index of refraction of the material in the gap between the lens and the glass plate. Use the geometry of the figure to show that when $r \ll R$, (a) the radii of the dark fringes are given by $r \approx \sqrt{m\lambda R/n}$ and (b) the radii of the bright fringes are given by $r \approx \sqrt{(m + \frac{1}{2})\lambda R/n}$. (c) When Newton's rings are formed using sodium light ($\lambda = 590$ nm), the diameters of two successive dark rings are 2 mm and 2.23 mm. Calculate the radius of curvature of the convex lens surface.

63. The condition for constructive interference by reflection from a thin film in air as developed in Section 37.6 assumes nearly normal incidence. (a) Show that if the light is incident on the film at an angle $\phi_1 \gg 0$ (relative to the normal), then the condition for constructive interference is given by $2nt \cos \theta_2 = (m + \frac{1}{2})\lambda$, where θ_2 is the angle of refraction. (b) Calculate the minimum thickness for constructive interference if sodium light ($\lambda = 5.9 \times 10^{-5}$ cm) is incident at an angle of 30° on a film with index of refraction 1.38.

64. Use the method of phasor addition to find the resultant amplitude and phase constant when the following three harmonic functions are combined: $E_1 = \sin(\omega t + \pi/6)$, $E_2 = 3\sin(\omega t + 7\pi/2)$, $E_3 = 6\sin(\omega t + 4\pi/3)$.

65. A fringe pattern is established in the field of view of a Michelson interferometer using light of wavelength 580 nm. A parallel-faced sheet of transparent material 2.5 μm thick is placed in front of one of the mirrors

perpendicular to the incident and reflected light beams. An observer counts a fringe shift of six *dark* fringes. What is the index of refraction of the sheet?

66. A soap film ($n = 1.33$) is contained within a rectangular wire frame. The frame is held vertically so that the film drains downward due to gravity and becomes thicker at the bottom than at the top, where the thickness is essentially zero. The film is viewed in white light with near-normal incidence, and the first violet ($\lambda = 420$ nm) interference band is observed 3 cm from the top edge of the film. (a) Locate the first red ($\lambda = 680$ nm) interference band. (b) Determine the film thickness at the positions of the violet and red bands. (c) What is the wedge angle of the film?

67. A soap film such as that in Problem 66 is viewed in yellow light ($\lambda = 589$ nm) with near-normal incidence. Interference fringes with a uniform spacing of 4.5 mm are observed. What is the variation in thickness of the film per centimeter of vertical distance?

68. Figure 37.24 illustrates the formation of an interference pattern by the Lloyd's mirror method (see also Section 37.5 and Fig. 37.14). In the case shown here, the actual source S and the virtual source S′ are in a plane 25 cm to the left of the mirror and the screen is a distance $L = 120$ cm to the right of this plane. The source S is a distance $d = 2.5$ mm above the top surface of the mirror (arranged for reflection at glancing incidence), and the light is monochromatic with $\lambda = 620$ nm. (a) Show that, in general, the separation between bright (or dark) fringes on the screen is given by $\Delta y = L\lambda/2d$. (b) Determine the distance of the first bright fringe above the top surface of the mirror.

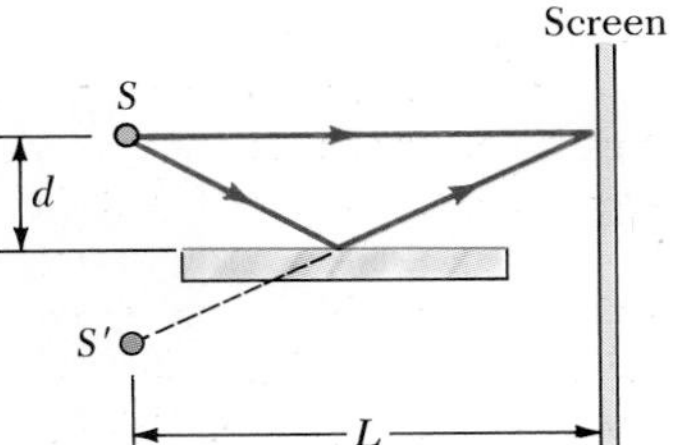

Figure 37.24 (Problem 68).

69. (a) Both sides of a uniform film of index of refraction n and thickness d are in contact with air. For normal incidence of light, an intensity minimum is observed in the reflected light at λ_2 and an intensity maximum is observed at λ_1, where $\lambda_1 > \lambda_2$. If there are no intensity minima observed between λ_1 and λ_2, show that the integer m that appears in Equations 37.16 and 37.17 is given by $m = \lambda_1/2(\lambda_1 - \lambda_2)$. (b) Determine the thickness of the film if $n = 1.40$, $\lambda_1 = 500$ nm, and $\lambda_2 = 370$ nm.

38

Diffraction and Polarization

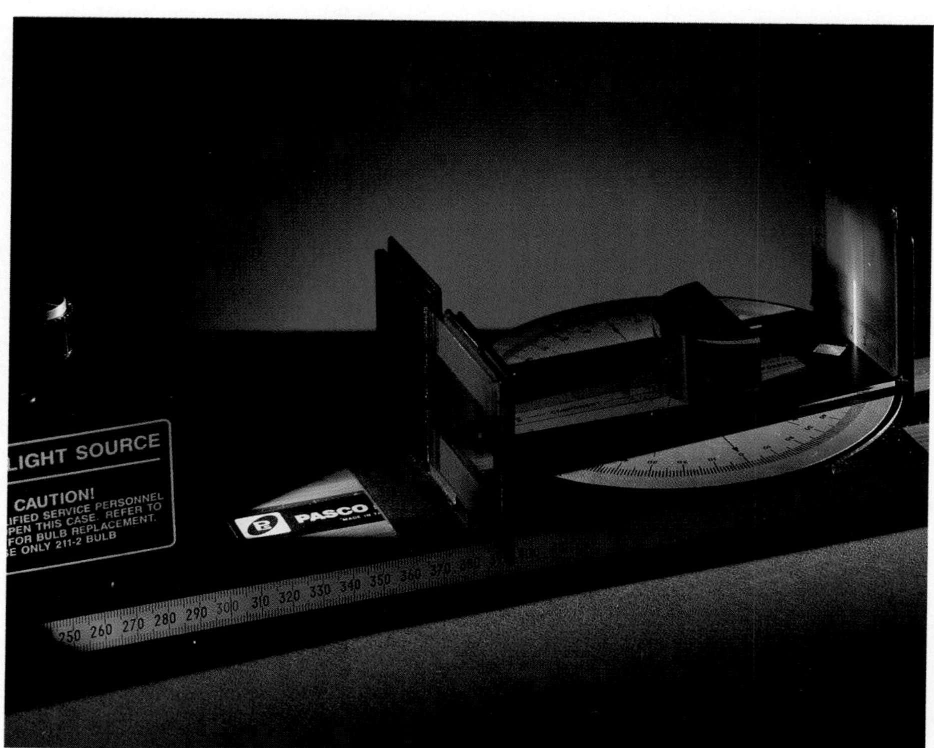

White light is dispersed into its various spectral components using a diffraction grating, as shown in this photograph. (Courtesy of PASCO Scientific)

In this chapter, we continue our treatment of wave optics with the discussion of diffraction and polarization phenomena. When light waves pass through a small aperture, an interference pattern is observed rather than a sharp spot of light cast by the aperture. This shows that light spreads in various directions beyond the aperture into regions where a shadow would be expected if light traveled in straight lines. Other waves, such as sound waves and water waves, also have this property of being able to bend around corners. This phenomenon, known as *diffraction*, can be regarded as a consequence of interference from many coherent wave sources. In other words, the phenomena of diffraction and interference are basically equivalent.

In Chapter 34, we discussed the properties of electromagnetic waves and the fact that they are transverse in nature. That is, the electric and magnetic field vectors associated with the wave are perpendicular to the direction of propagation. Under certain conditions, light waves can be plane-polarized. Although ordinary light is usually not polarized, we shall discuss various methods for producing polarized light, such as by using polarizing sheets.

38.1 INTRODUCTION TO DIFFRACTION

Suppose a light beam is incident on two slits, as in Young's double-slit experiment. If the light truly traveled in straight-line paths after passing through the slits, as in Figure 38.1a, the waves would not overlap and no interference pattern would be seen. Instead, Huygens' principle requires that the waves spread out from the slits as shown in Figure 38.1b. In other words, the light deviates from a straight-line path and enters the region that would otherwise be shadowed. This divergence of light from its initial line of travel is called **diffraction.**

In general, diffraction occurs when waves pass through small openings, around obstacles, or by relatively sharp edges. As an example of diffraction, consider the following. When an opaque object is placed between a point source of light and a screen, as in Figure 38.2a, the boundary between the shadowed and illuminated regions on the screen is not sharp. A careful inspection of the boundary shows that a small amount of light bends into the shadowed region. The region outside the shadow contains alternating light and dark bands, as in Figure 38.2b. The intensity in the first bright band is greater than the intensity in the region of uniform illumination. Effects of this type were first reported in the 17th century by Francesco Grimaldi.

Figure 38.3 shows the shadow of the diffraction pattern of a penny. There is a bright spot at the center, circular fringes near the shadow's edge, and another set of fringes outside the shadow. This particular type of diffraction pattern was first observed in 1818 by Dominique Arago. The bright spot at the center of the shadow can be explained only through the use of the wave theory of light, which predicts constructive interference at this point. From the viewpoint of geometrical optics, the center of the pattern would be screened by the object, hence one would never expect to observe a central bright spot. It is interesting to point out a historical incident that occurred shortly before the central bright spot was observed. One of the supporters of geometrical optics, Simeon Poisson, argued that if Augustin Fresnel's wave theory of light were valid, then a central bright spot should be observed. Because the spot was

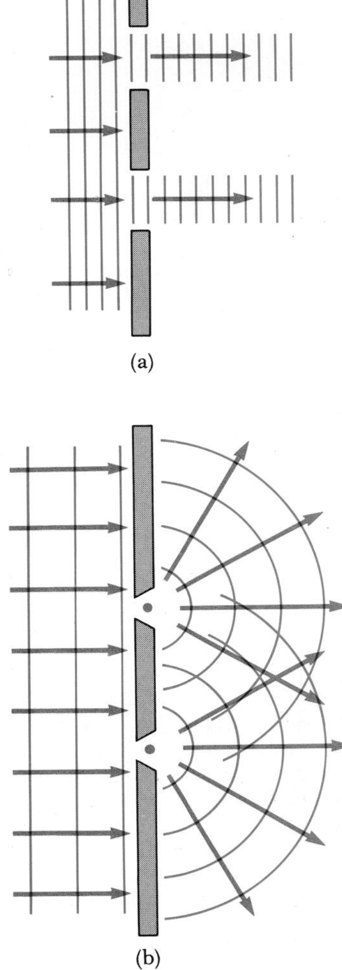

Figure 38.1 (a) If light waves did not spread out after passing through the slits, no interference would occur. (b) The light waves from the two slits overlap as they spread out, producing interference fringes.

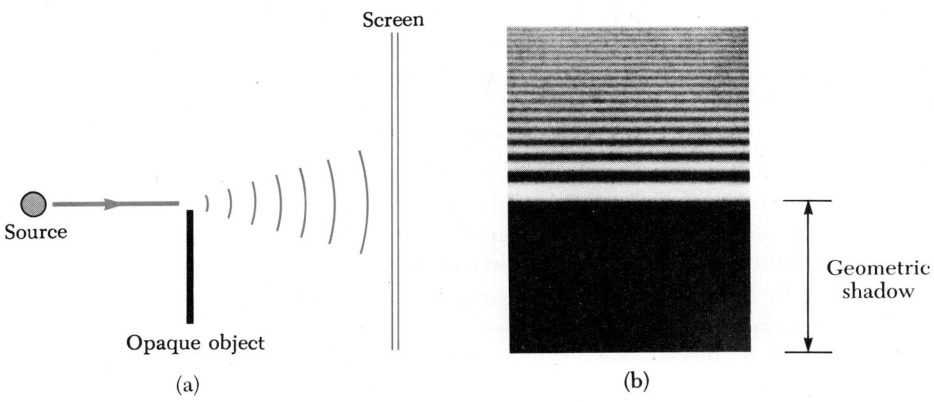

Figure 38.2 (a) Light bends around the opaque object. (b) The result is a series of dark and light bands in the area that would be completely in the shadow if light did not bend around the object.

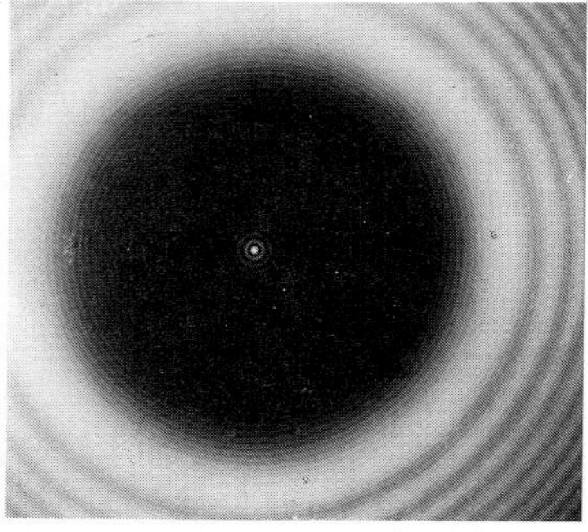

Figure 38.3 Diffraction pattern of a penny, taken with the penny midway between screen and source. (Courtesy of P.M. Rinard, from *Am. J. Phys.* 44:70, 1976)

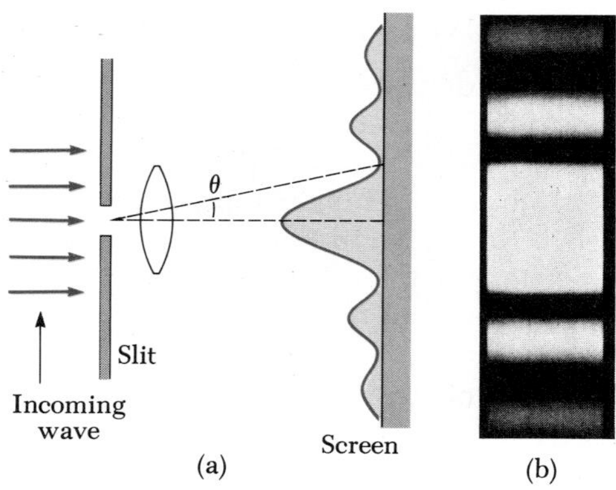

Figure 38.4 (a) Fraunhofer diffraction pattern of a single slit. The parallel rays are brought into focus on the screen with a converging lens. The pattern consists of a central bright region flanked by much weaker maxima. (Note that this is not to scale.) (b) Photograph of a single-slit Fraunhofer diffraction pattern. (From M. Cagnet, M. Francon, and J.C. Thierr, *Atlas of Optical Phenomena*, Berlin, Springer-Verlag, 1962, plate 18)

observed shortly thereafter, Poisson's prediction *reinforced* the wave theory rather than disproved it. This was certainly a most dramatic experimental proof of the wave nature of light.

Diffraction phenomena are usually classified as being of two types, which are named after the men who first explained them. The first type, called **Fraunhofer diffraction,** occurs when the rays reaching a point are approximately parallel. This can be achieved experimentally either by placing the observing screen far from the opening or by using a converging lens to focus the parallel rays on the screen, as in Figure 38.4a. A bright fringe is observed along the axis at $\theta = 0$, with alternating dark and bright fringes on either side of the central bright fringe. Figure 38.4b is a photograph of a single-slit Fraunhofer diffraction pattern.

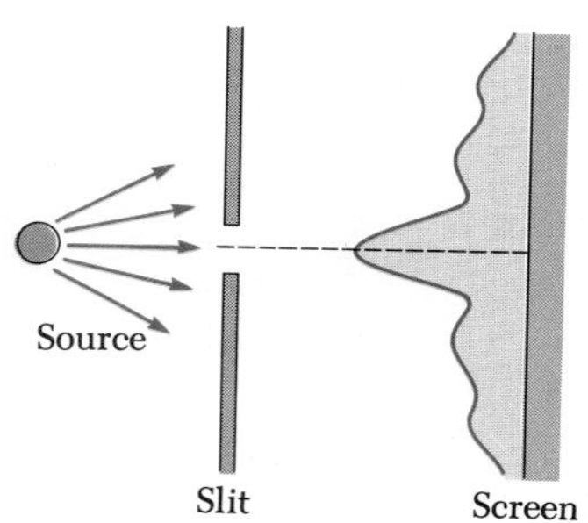

Figure 38.5 A Fresnel diffraction pattern of a single slit is observed when the incident rays are not parallel and the observing screen is a finite distance from the slit. (Note that this is not to scale.)

When the observing screen is placed a finite distance from the slit and no lens is used to focus parallel rays, the observed pattern is called a **Fresnel diffraction** pattern (Fig. 38.5). The diffraction patterns shown in Figures 38.2b and 38.3 are examples of Fresnel diffraction. Fresnel diffraction is rather complex to treat quantitatively. Therefore, the following discussion will be restricted to Fraunhofer diffraction.

38.2 SINGLE-SLIT DIFFRACTION

Up until now, we have assumed that the slits are point sources of light. In this section, we shall determine how their finite width is the basis for understanding the nature of the Fraunhofer diffraction pattern produced by a single slit.

We can deduce some important features of this problem by examining waves coming from various portions of the slit, as shown in Figure 38.6. According to Huygens' principle, *each portion of the slit acts as a source of waves.* Hence, *light from one portion of the slit can interfere with light from another portion,* and the resultant intensity on the screen will depend on the direction θ.

To analyze the diffraction pattern, it is convenient to divide the slit in two halves, as in Figure 38.6. All the waves that originate from the slit are in phase. Consider waves 1 and 3, which originate from the bottom and center of the slit, respectively. Wave 1 travels farther than wave 3 by an amount equal to the path difference $(a/2)\sin\theta$, where a is the width of the slit. Similarly, the path difference between waves 2 and 4 is also $(a/2)\sin\theta$. If this path difference is exactly one half of a wavelength (corresponding to a phase difference of 180°), the two waves cancel each other and destructive interference results. This is true, in fact, for any two waves that originate at points separated by half the slit width because the phase difference between two such points is 180°. Therefore, waves from the upper half of the slit interfere *destructively* with waves from the lower half of the slit when

$$\frac{a}{2}\sin\theta = \frac{\lambda}{2}$$

or when

$$\sin\theta = \frac{\lambda}{a}$$

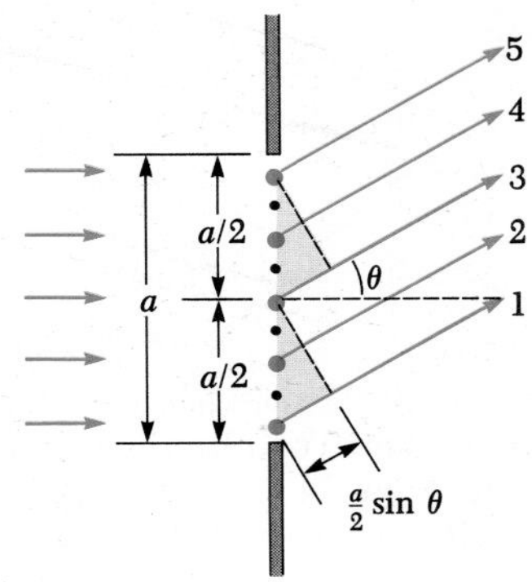

Figure 38.6 Diffraction of light by a narrow slit of width *a*. Each portion of the slit acts as a point source of waves. The path difference between rays 1 and 3 or between rays 2 and 4 is equal to $(a/2)\sin\theta$. (Note that this is not to scale.)

If we divide the slit into four parts rather than two and use similar reasoning, we find that the screen is also dark when

$$\sin\theta = \frac{2\lambda}{a}$$

Likewise, we can divide the slit into six parts and show that darkness occurs on the screen when

$$\sin\theta = \frac{3\lambda}{a}$$

Therefore, the general condition for **destructive interference** is

Condition for destructive interference

$$\sin\theta = m\frac{\lambda}{a} \qquad (m = \pm 1, \pm 2, \pm 3, \ldots) \tag{38.1}$$

Equation 38.1 gives the values of θ for which the diffraction pattern has zero intensity, that is, where a dark fringe is formed. However, it tells us nothing about the variation in intensity along the screen. The general features of the intensity distribution along the screen are shown in Figure 38.7. A broad central bright fringe is observed, flanked by much weaker, alternating

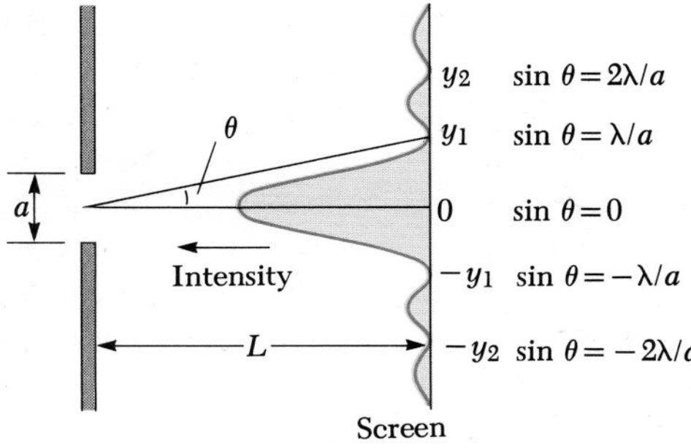

Figure 38.7 Positions of the minima for the Fraunhofer diffraction pattern of a single slit of width *a*. (Note that this is not to scale.)

bright fringes. The various dark fringes (points of zero intensity) occur at the values of θ that satisfy Equation 38.1. The position of the points of constructive interference lie approximately halfway between the dark fringes. Note that the central bright fringe is twice as wide as the weaker maxima.

EXAMPLE 38.1 Where Are The Dark Fringes? Light of wavelength 580 nm is incident on a slit of width 0.30 mm. The observing screen is placed 2 m from the slit. Find the positions of the first dark fringes and the width of the central bright fringe.

Solution The first dark fringes that flank the central bright fringe correspond to $m = \pm 1$ in Equation 38.1. Hence, we find that

$$\sin\theta = \pm\frac{\lambda}{a} = \pm\frac{5.8\times 10^{-7}\text{ m}}{0.3\times 10^{-3}\text{ m}} = \pm 1.93\times 10^{-3}$$

From the triangle in Figure 38.7, note that $\tan\theta = y_1/L$. Since θ is very small, we can use the approximation $\sin\theta \approx \tan\theta$, so that $\sin\theta \approx y_1/L$. Therefore, the positions of the first minima measured from the central axis are given by

$$y_1 \approx L\sin\theta = \pm L\frac{\lambda}{a} = \pm 3.87\times 10^{-3}\text{ m}$$

The positive and negative signs correspond to the dark fringes on either side of the central bright fringe. Hence, the width of the central bright fringe is equal to $2|y_1| = 7.73\times 10^{-3}\text{ m} = 7.73$ mm. Note that this value is *much larger* than the width of the slit. However, as the width of the slit is *increased*, the diffraction pattern will *narrow*, corresponding to smaller values of θ. In fact, for large values of a, the various maxima and minima are so closely spaced that only a large central bright area is observed, which resembles the geometric image of the slit. This matter is of great importance in the design of lenses used in telescopes, microscopes, and other optical instruments.

Exercise 1 Determine the width of the first order bright fringe.
Answer 3.87 mm

Intensity of the Single-Slit Diffraction Pattern

We can make use of phasors to determine the intensity distribution for the single-slit diffraction pattern. Imagine that a slit is divided into a large number of small zones, each of width Δy as in Figure 38.8. Each zone acts as a source of coherent radiation, and each contributes an incremental electric field amplitude ΔE at some point P on the screen. The total electric field amplitude E at the point P is obtained by summing the contributions from each zone. Note that the incremental electric field amplitudes between adjacent zones are out

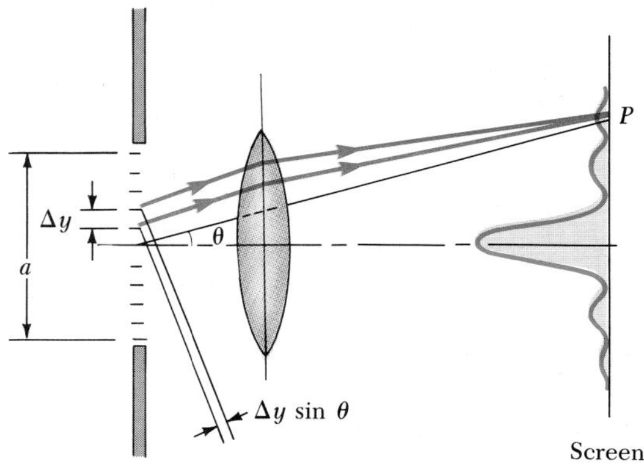

Figure 38.8 Fraunhofer diffraction by a single slit. The intensity at the point P on the screen is the resultant of all the incremental fields from zones of width Δy leaving the slit.

of phase with one another by an amount $\Delta\beta$, corresponding to a path difference of $\Delta y \sin\theta$ (see Fig. 38.8). The ratio

$$\frac{\Delta\beta}{2\pi} = \frac{\Delta y \sin\theta}{\lambda}$$

can be written as

$$\Delta\beta = \frac{2\pi}{\lambda}\,\Delta y \sin\theta \tag{38.2}$$

To find the total electric field amplitude on the screen at any angle θ, we sum the incremental amplitudes due to each zone. For small values of θ, we can assume that the amplitudes ΔE due to each zone are the same. It is convenient to use the phasor diagrams for various angles as in Figure 38.9. When $\theta = 0$, all phasors are aligned as in Figure 38.9a, since the waves from each zone are in phase. In this case, the total amplitude at the center of the screen is $E_0 = N\,\Delta E$, where N is the number of zones. The amplitude E_θ at some small angle θ is shown in Figure 38.9b, where each phasor differs in phase from an adjacent one by an amount $\Delta\beta$. In this case, note that E_θ is the *vector sum* of the incremental amplitudes, and hence is given by the length of the chord. Therefore, $E_\theta < E_0$. The total phase difference between waves from the top and bottom portions of the slit is

$$\beta = N\,\Delta\beta = \frac{2\pi}{\lambda}\,N\,\Delta y \sin\theta = \frac{2\pi}{\lambda}\,a \sin\theta \tag{38.3}$$

where $a = N\,\Delta y$ is the width of the slit.

As θ increases, the chain of phasors eventually forms a closed path as in Figure 38.9c. At this point, the vector sum is zero, so $E_\theta = 0$, corresponding to the first minimum on the screen. Noting that $\beta = N\,\Delta\beta = 2\pi$ in this situation, we see from Equation 38.3

$$2\pi = \frac{2\pi}{\lambda}\,a \sin\theta$$

or

$$\sin\theta = \frac{\lambda}{a}$$

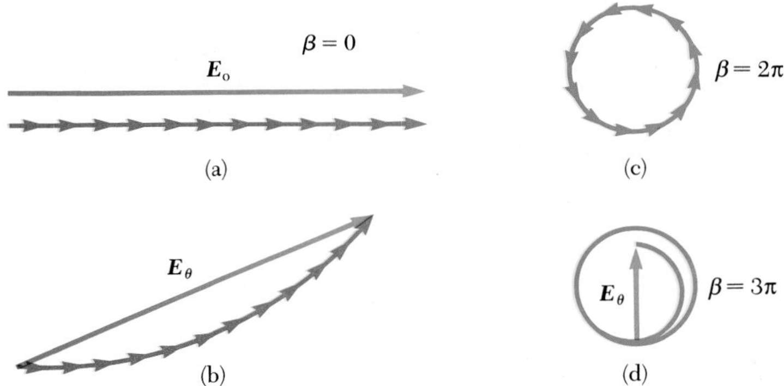

Figure 38.9 Phasor diagrams for obtaining the various maxima and minima of the single slit diffraction pattern.

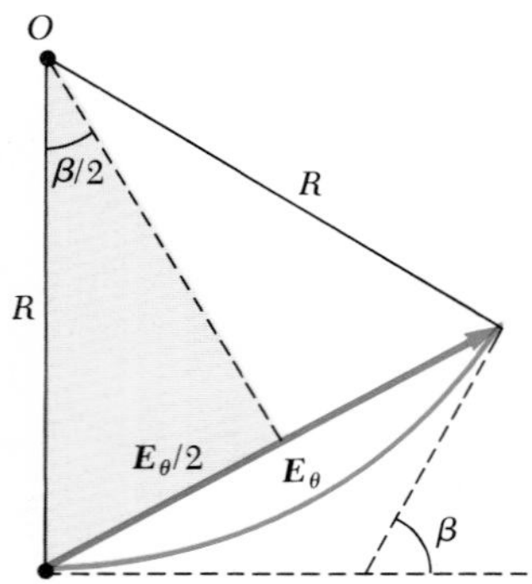

Figure 38.10 Phasor diagram for a large number of coherent sources. Note that the ends of the phasors lie on a circular arc of radius R. The resultant amplitude E_θ equals the length of the chord.

That is, the first minimum in the diffraction pattern occurs when $\sin\theta = \lambda/a$, which agrees with Equation 38.1.

At even larger values of θ, the spiral chain of phasors continues. For example, Figure 38.9d represents the situation corresponding to the second maximum, which occurs when $\beta \simeq 360° + 180° = 540°$ (3π rad). The second minimum (two complete spirals not shown) corresponds to $\beta = 720°$ (4π rad), which satisfies the condition $\sin\theta = 2\lambda/a$.

The total amplitude and intensity at any point on the screen can now be obtained by considering the limiting case where Δy becomes infinitesimal (dy) and $N \rightarrow \infty$. In this limit, the phasor diagrams in Figure 38.9 become smooth curves, as in Figure 38.10. From this figure, we see that at some angle θ, the wave amplitude on the screen, E_θ, is equal to the chord length, while E_0 is the arc length. From the triangle whose angle is $\beta/2$, we see that

$$\sin\frac{\beta}{2} = \frac{E_\theta/2}{R}$$

where R is the radius of curvature. But the arc length E_0 is equal to the product $R\beta$, where β is in radians. Combining this with the expression above gives

$$E_\theta = 2R\sin\frac{\beta}{2} = 2\left(\frac{E_0}{\beta}\right)\sin\frac{\beta}{2}$$

or

$$E_\theta = E_0\left[\frac{\sin(\beta/2)}{\beta/2}\right]$$

Since the resultant intensity I_θ at P is proportional to the square of the amplitude E_θ, we find

$$I_\theta = I_0\left[\frac{\sin(\beta/2)}{\beta/2}\right]^2 \tag{38.4}$$

where I_0 is the intensity at $\theta = 0$ (the central maximum), and $\beta = 2\pi a\sin\theta/\lambda$. Substitution of this expression for β into Equation 38.4 gives

Intensity of a single-slit Fraunhofer diffraction pattern

$$I_\theta = I_0\left[\frac{\sin(\pi a\sin\theta/\lambda)}{\pi a\sin\theta/\lambda}\right]^2 \tag{38.5}$$

From this result, we see that minima occur when

$$\frac{\pi a\sin\theta}{\lambda} = m\pi$$

or

Condition for intensity minima

$$\sin\theta = m\frac{\lambda}{a}$$

where $m = \pm 1, \pm 2, \pm 3, \ldots$. This is in agreement with our earlier result, given by Equation 38.1.

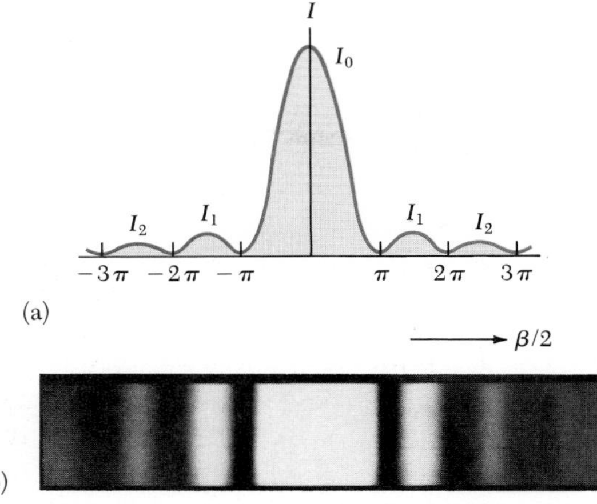

Figure 38.11 (a) A plot of the intensity I versus β for the single-slit Fraunhofer diffraction pattern. (b) Photograph of a single-slit Fraunhofer diffraction pattern. (From M. Cagnet, M. Francon, and J.C. Thierr, *Atlas of Optical Phenomena,* Berlin, Springer-Verlag, 1962, plate 18)

Figure 38.11a represents a plot of Equation 38.5, and a photograph of a single-slit Fraunhofer diffraction pattern is shown in Figure 38.11b. Most of the light intensity is concentrated in the central bright fringe.

EXAMPLE 38.2 Relative Intensities of the Maxima

Find the ratio of intensities of the secondary maxima to the intensity of the central maximum for the single-slit Fraunhofer diffraction pattern.

Solution To a good approximation, we can assume that the secondary maxima lie midway between the zero points. From Figure 38.11a, we see that this corresponds to $\beta/2$ values of $3\pi/2$, $5\pi/2$, $7\pi/2$, ... Substituting these into Equation 38.4 gives for the first two ratios

$$\frac{I_1}{I_0} = \left[\frac{\sin(3\pi/2)}{(3\pi/2)}\right]^2 = \frac{1}{9\pi^2/4} = 0.045$$

$$\frac{I_2}{I_0} = \left[\frac{\sin(5\pi/2)}{5\pi/2}\right]^2 = \frac{1}{25\pi^2/4} = 0.016$$

That is, the secondary maximum that is adjacent to the central maximum has an intensity of 4.5% that of the central bright fringe, and the next secondary maximum has an intensity of 1.6% that of the central bright fringe.

Exercise 2 Determine the intensity of the secondary maximum corresponding to $m = 3$ relative to the central maximum.

Answer 0.0083.

38.3 RESOLUTION OF SINGLE-SLIT AND CIRCULAR APERTURES

The ability of optical systems such as microscopes and telescopes to distinguish between closely spaced objects is limited because of the wave nature of light. To understand this difficulty, consider Figure 38.12, which shows two light sources far from a narrow slit of width a. The sources can be considered as two point sources, S_1 and S_2, that are *not* coherent. For example, they could be two distant stars. If no diffraction occurred, one would observe two distinct bright spots (or images) on the screen at the right in the figure. However, because of diffraction, each source is imaged as a bright central region flanked by weaker bright and dark rings. What is observed on the screen is the sum of two diffraction patterns, one from S_1, and the other from S_2.

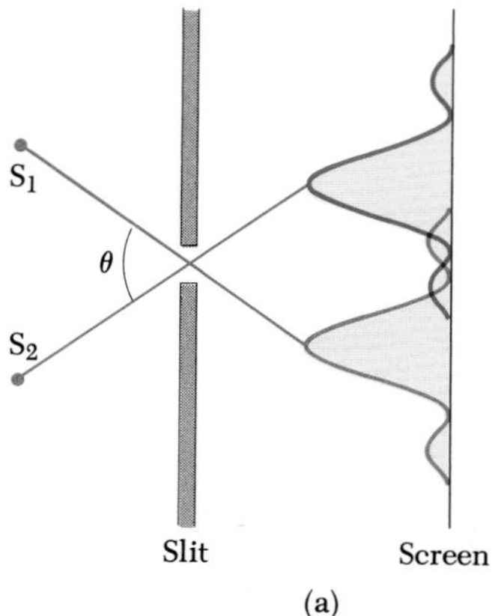

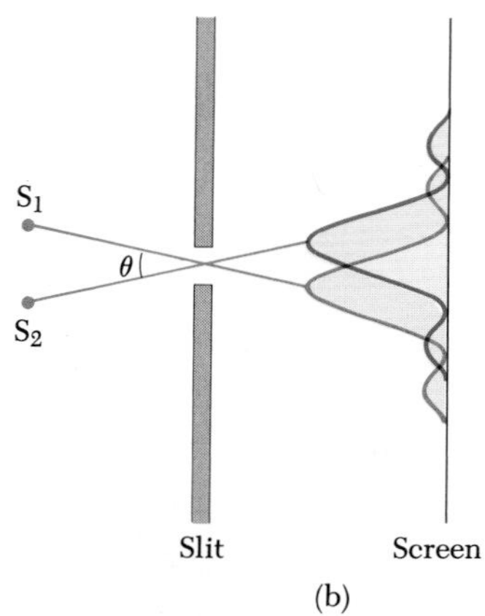

Figure 38.12 Two point sources at some distance from a small aperture each produce a diffraction pattern. (a) The angle subtended by the sources at the aperture is large enough so that the diffraction patterns are distinguishable. (b) The angle subtended by the sources is so small that their diffraction patterns overlap and the images are not well resolved. (Note that the angles are greatly exaggerated.)

If the two sources are separated such that their central maxima do not overlap, as in Figure 38.12a, their images can be distinguished and are said to be *resolved.* If the sources are close together, however, as in Figure 38.12b, the two central maxima may overlap and the images are *not resolved.* To decide when two images are resolved, the following condition is often used:

When the central maximum of one image falls on the first minimum of another image, the images are said to be just resolved. This limiting condition of resolution is known as **Rayleigh's criterion.**

Figure 38.13 shows the diffraction patterns for three situations. When the objects are far apart, their images are well resolved (Fig. 38.13a). The images are just resolved when their angular separation satisfies Rayleigh's criterion (Fig. 38.13b). Finally, the images are not resolved in Figure 38.13c.

From Rayleigh's criterion, we can determine the minimum angular separation, θ_m, subtended by the source at the slit such that their images will be just resolved. In Section 38.2, we found that the first minimum in a single-slit diffraction pattern occurs at the angle that satisfies the relationship

$$\sin \theta = \frac{\lambda}{a}$$

where a is the width of the slit. According to Rayleigh's criterion, this expression gives the smallest angular separation for which the two images will be resolved. Because $\lambda \ll a$ in most situations, $\sin \theta$ is small and we can use the approximation $\sin \theta \approx \theta$. Therefore, the limiting angle of resolution for a slit of width a is

Limiting angle of resolution for a slit

$$\theta_m = \frac{\lambda}{a} \tag{38.6}$$

where θ_m is expressed in radians. Hence, the angle subtended by the two sources at the slit must be *greater* than λ/a if the images are to be resolved.

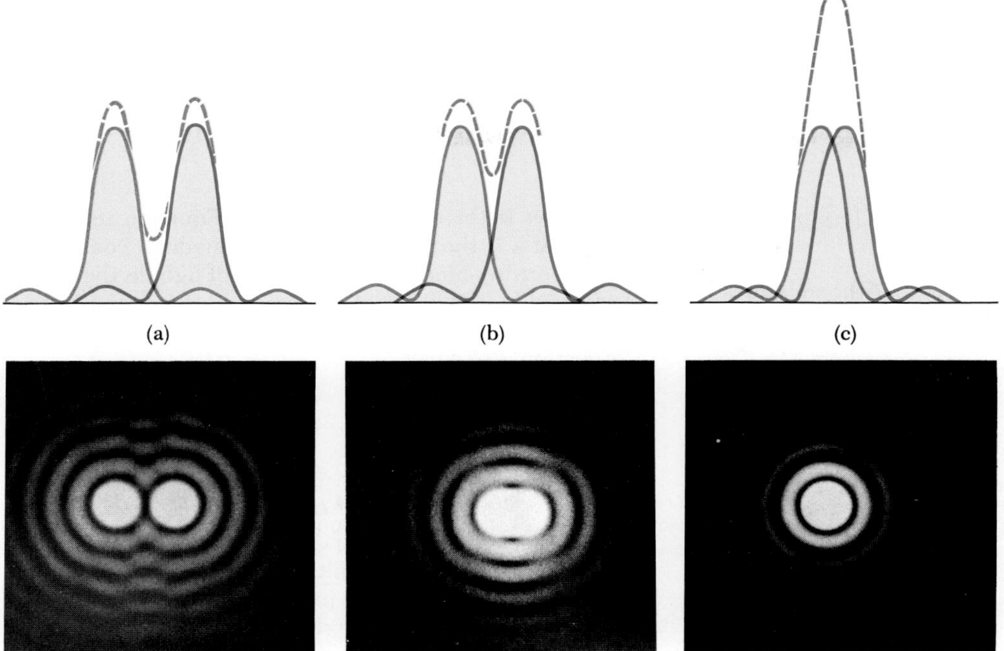

Figure 38.13 The diffraction patterns of two point sources (solid curves) and the resultant pattern (dashed curves), for various angular separations of the sources. In each case, the dashed curve is the sum of the two solid curves. (a) The sources are far apart, and the patterns are well resolved. (b) The sources are closer together, and the patterns are just resolved. (c) The sources are so close together that the patterns are not resolved. (From M. Cagnet, M. Francon, and J.C. Thierr, *Atlas of Optical Phenomena*, Berlin, Springer-Verlag, 1962, plate 16)

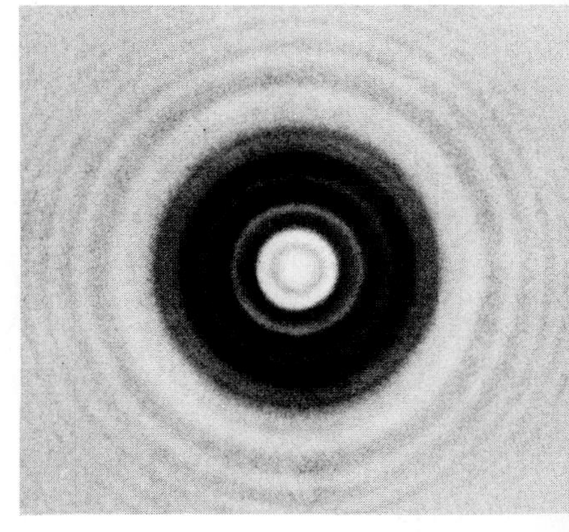

Figure 38.14 The Fresnel diffraction pattern of a circular aperture consists of a central bright disk surrounded by concentric bright and dark rings. (From M. Cagnet, M. Francon, and J.C. Thierr, *Atlas of Optical Phenomena*, Berlin, Springer-Verlag, 1962, plate 34)

Many optical systems use circular apertures rather than slits. The diffraction pattern of a circular aperture, illustrated in Figure 38.14, consists of a central circular bright disk surrounded by progressively fainter rings. Analysis shows that the limiting angle of resolution of the circular aperture is

$$\theta_m = 1.22\frac{\lambda}{D} \tag{38.7}$$

where D is the diameter of the aperture. Note that Equation 38.7 is similar to Equation 38.6 except for the factor of 1.22, which arises from a complex mathematical analysis of diffraction from the circular aperture.

EXAMPLE 38.3 Limiting Resolution of a Microscope
Sodium light of wavelength 589 nm is used to view an object under a microscope. If the aperture of the objective has a diameter of 0.9 cm, (a) find the limiting angle of resolution. (b) Using visible light of any wavelength you desire, what is the maximum limit of resolution for this microscope? (c) Suppose water of index of refraction 1.33 fills the space between the object and the objective. What effect would this have on the resolving power of the microscope?

Solution (a) From Equation 38.7, we find the limiting angle of resolution to be

$$\theta_m = 1.22\left(\frac{589 \times 10^{-9}\text{ m}}{0.9 \times 10^{-2}\text{ m}}\right) = 7.98 \times 10^{-5}\text{ rad}$$

This means that any two points on the object subtending an angle less than 8×10^{-5} rad at the objective cannot be distinguished in the image.

(b) To obtain the smallest angle corresponding to the maximum limit of resolution, we have to use the shortest wavelength available in the visible spectrum. Violet light of wavelength 400 nm gives us a limiting angle of resolution of

$$\theta_m = 1.22\left(\frac{400 \times 10^{-9}\ \text{m}}{0.9 \times 10^{-2}\ \text{m}}\right) = \boxed{5.42 \times 10^{-5}\ \text{rad}}$$

(c) In this case, the wavelength of the sodium light in the water is found by $\lambda_w = \lambda_a/n$ (Chap. 35). Thus, we have

$$\lambda_w = \frac{\lambda_a}{n} = \frac{589\ \text{nm}}{1.33} = 443\ \text{nm}$$

The limiting angle of resolution at this wavelength is

$$\theta_m = 1.22\left(\frac{443 \times 10^{-9}\ \text{m}}{0.9 \times 10^{-2}\ \text{m}}\right) = \boxed{6.00 \times 10^{-5}\ \text{rad}}$$

EXAMPLE 38.4 Resolution of a Telescope

The Hale telescope at Mount Palomar has a diameter of 200 in. What is its limiting angle of resolution at a wavelength of 600 nm?

Solution Because $D = 200$ in. $= 5.08$ m and the wavelength $\lambda = 6 \times 10^{-7}$ m, Equation 38.7 gives

$$\theta_m = 1.22\frac{\lambda}{D} = 1.22\left(\frac{6 \times 10^{-7}\ \text{m}}{5.08\ \text{m}}\right)$$

$$= 1.44 \times 10^{-7}\ \text{rad} = 0.03\ \text{s of arc}$$

Therefore, any two stars that subtend an angle greater than or equal to this value will be resolved (assuming ideal atmospheric conditions).

The Hale telescope can never reach its diffraction limit. Instead, its limiting angle of resolution is always set by atmospheric blurring. This seeing limit is usually about 1 s of arc and is never smaller than about 0.1 s of arc. (This is one of the reasons for the current interest in a large space telescope.)

Exercise 3 The large radio telescope at Arecibo, Puerto Rico, has a diameter of 305 m, and is designed to detect radio waves at a wavelength of 0.75 m. Calculate the minimum angle of resolution for this telescope, and compare your answer with that of the Hale telescope.
Answer 3×10^{-3} rad (10 min 19 s of arc), which is more than 10 000 times larger than the calculated minimum angle for the Hale telescope.

EXAMPLE 38.5 Resolution of the Eye

Calculate the limiting angle of resolution for the eye, assuming a pupil diameter of 2 mm, a wavelength of 500 nm in air, and an index of refraction for the eye equal to 1.33.

Solution In this example, we can use Equation 38.7, noting that λ is the wavelength in the medium containing the aperture. Since the wavelength of light in the eye is reduced by the index of refraction of the eye medium, we find that $\lambda = (500\ \text{nm})/1.33 = 376$ nm. Therefore, Equation 38.7 gives

$$\theta_m = 1.22\frac{\lambda}{D} = 1.22\left(\frac{3.76 \times 10^{-7}\ \text{m}}{2 \times 10^{-3}\ \text{m}}\right)$$

$$= 2.29 \times 10^{-4}\ \text{rad} = \boxed{0.0131^\circ}$$

We can use this result to calculate the minimum separation d between two point sources that the eye can distinguish if they are at a distance L from the observer (Fig. 38.15). Since θ_m is small, we see that

$$\sin\theta_m \approx \theta_m \approx d/L$$

$$d = L\theta_m$$

For example, if the objects are located at a distance of 25 cm from the eye (the near point), then

$$d = (25\ \text{cm})(2.29 \times 10^{-4}\ \text{rad}) = 5.73 \times 10^{-3}\ \text{cm}$$

This is approximately equal to the thickness of a human hair.

Exercise 4 If the eye is dilated to a diameter of 5 mm, what is the minimum distance between two point sources that the eye can distinguish at a distance of 40 cm from the eye?
Answer 3.67×10^{-3} cm

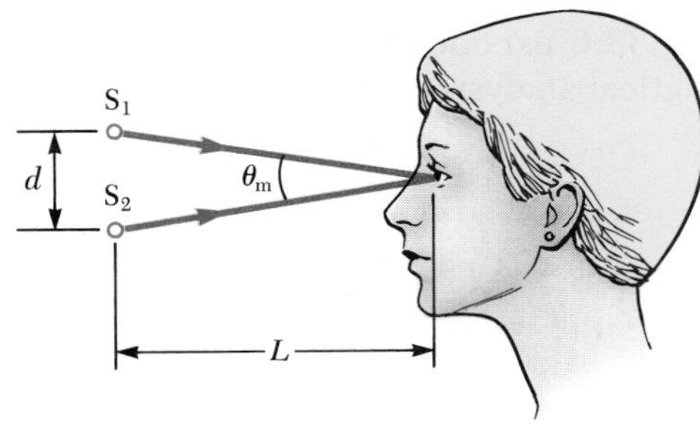

Figure 38.15 (Example 38.5) Two point sources separated by a distance d as observed by the eye.

38.4 THE DIFFRACTION GRATING

The diffraction grating, a very useful device for analyzing light sources, consists of a large number of equally spaced parallel slits. A grating can be made by engraving parallel lines on a glass plate with a precision machining technique. The spaces between each line are transparent to the light and hence act as separate slits. A typical grating contains several thousand lines per centimeter. For example, a grating ruled with 5000 lines/cm has a slit spacing d equal to the inverse of this number, or $d = (1/5000)\text{ cm} = 2 \times 10^{-4}\text{ cm}$.

A schematic diagram of a section of plane diffraction grating is illustrated in Figure 38.16. A plane wave is incident from the left, normal to the plane of the grating. A converging lens can be used to bring the rays together at the point P. The intensity of the observed pattern on the screen is the result of the combined effects of interference and diffraction. Each slit produces diffraction, as was described in the previous section. The diffracted beams in turn interfere with each other to produce the final pattern. Moreover, each slit acts as a source of waves, where all waves start at the slits in phase. However, for some arbitrary direction θ measured from the horizontal, the waves must travel *different* path lengths before reaching a particular point P on the screen. From Figure 38.16, note that the path difference between waves from any two adjacent slits is equal to $d \sin \theta$. If this path difference equals one wavelength or some integral multiple of a wavelength, waves from all slits will be in phase at P and a bright line will be observed. Therefore, the condition for **maxima** in the interference pattern at the angle θ is

$$d \sin \theta = m\lambda \qquad (m = 0, 1, 2, 3, \ldots) \qquad (38.8)$$

Condition for interference maxima for a grating

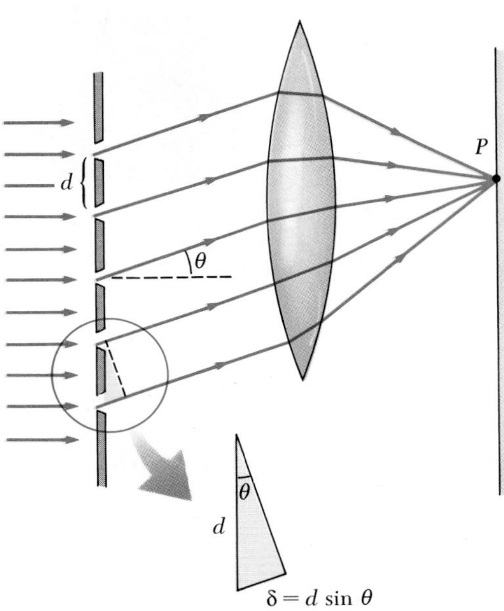

Figure 38.16 Side view of a diffraction grating. The slit separation is d, and the path difference between adjacent slits is $d \sin \theta$.

This expression can be used to calculate the wavelength from a knowledge of the grating spacing and the angle of deviation θ. The integer m represents the *order number* of the diffraction pattern. If the incident radiation contains several wavelengths, the mth order maximum for each wavelength occurs at a specific angle. All wavelengths are seen at $\theta = 0$, corresponding to $m = 0$. This is called the *zeroth-order maximum*. The *first-order maximum*, corresponding to $m = 1$, is observed at an angle that satisfies the relationship $\sin\theta = \lambda/d$; the *second-order maximum*, corresponding to $m = 2$, is observed at a larger angle θ, and so on.

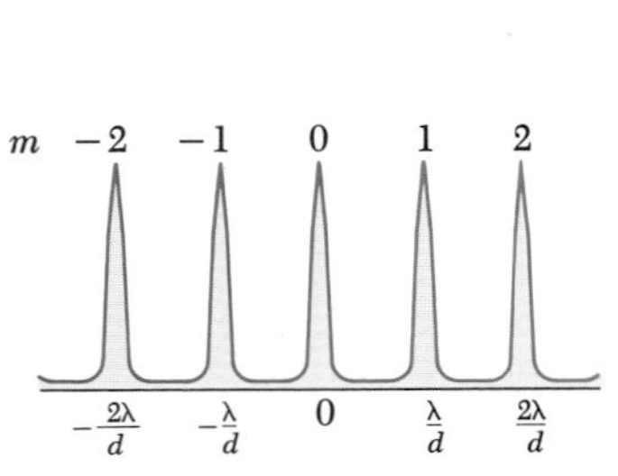

Figure 38.17 Intensity versus $\sin\theta$ for a diffraction grating. The zeroth-, first-, and second-order maxima are shown.

A sketch of the intensity distribution for the diffraction grating is shown in Figure 38.17. If the source contains various wavelengths, a spectrum of lines will be observed at different positions for each order number. Note the sharpness of the principal maxima and the broad range of dark areas. This is in contrast to the broad bright fringes characteristic of the two-slit interference pattern (Fig. 37.1).

A simple arrangement that can be used to measure various orders of the diffraction pattern is shown in Figure 38.18. This is a form of a diffraction grating spectrometer. The light to be analyzed passes through a slit, and a parallel beam of light exits from the collimator, which is perpendicular to the grating. The diffracted light leaves the grating at angles that satisfy Equation 38.8. A telescope is used to view the image of the slit. The wavelength can be determined by measuring the precise angles at which the images of the slit appear for the various orders.

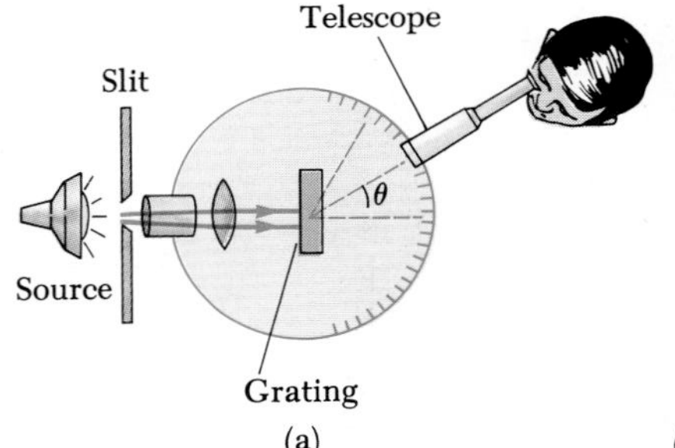

Figure 38.18 (a) Diagram of a diffraction grating spectrometer. The collimated beam incident on the grating is diffracted into the various orders at the angles θ that satisfy the equation $d\sin\theta = m\lambda$, where $m = 0, 1, 2, \ldots$ (b) Photograph of a diffraction grating. (Courtesy of PASCO Scientific)

EXAMPLE 38.6 The Orders of a Diffraction Grating
Monochromatic light from a helium-neon laser ($\lambda = 632.8$ nm) is incident normally on a diffraction grating containing 6000 lines/cm. Find the angles at which one would observe the first-order maximum, the second-order maximum, and so forth.

Solution First, we must calculate the slit separation, which is equal to the inverse of the number of lines per cm:

$$d = (1/6000)\text{ cm} = 1.667 \times 10^{-4}\text{ cm} = 1667\text{ nm}$$

For the first-order maximum ($m = 1$), we get

$$\sin\theta_1 = \frac{\lambda}{d} = \frac{632.8\text{ nm}}{1667\text{ nm}} = 0.3797$$

$$\theta_1 = 22.31°$$

Likewise, for $m = 2$, we find that

$$\sin\theta_2 = \frac{2\lambda}{d} = \frac{2(632.8\text{ nm})}{1667\text{ nm}} = 0.7592$$

$$\theta_2 = 49.41°$$

However, for $m = 3$ we find that $\sin\theta_3 = 1.139$. Since $\sin\theta$ cannot exceed unity, this does not represent a realistic solution. Hence, only zeroth-, first-, and second-order maxima will be observed for this situation.

Resolving Power of the Diffraction Grating

The diffraction grating is most useful for taking accurate wavelength measurements. Like the prism, the diffraction grating can be used to disperse a spectrum into its components. Of the two devices, the grating is more precise if one wants to distinguish between two closely spaced wavelengths. We say that the grating spectrometer has a "higher resolution" than a prism spectrometer. If λ_1 and λ_2 are the two nearly equal wavelengths between which the spectrometer can just barely distinguish, the **resolving power** R of the grating is defined as

$$R \equiv \frac{\lambda}{\lambda_2 - \lambda_1} = \frac{\lambda}{\Delta\lambda} \tag{38.9}$$

Resolving power

where $\lambda \approx \lambda_1 \approx \lambda_2$ and $\Delta\lambda = \lambda_2 - \lambda_1$. Thus, we see that a grating with a high resolving power can distinguish small differences in wavelength. Furthermore, if N lines of the grating are illuminated, it can be shown that the resolving power in the mth order diffraction equals the product Nm:

$$R = Nm \tag{38.10}$$

Resolving power of a grating

The derivation of Equation 38.10 is left as a problem (Problem 71). Thus, the resolving power increases with increasing order number. Furthermore, R is large for a grating with a large number of illuminated slits. Note that for $m = 0$, $R = 0$, which signifies that *all wavelengths are indistinguishable* for the zeroth-order maximum. However, consider the second-order diffraction pattern ($m = 2$) of a grating that has 5000 rulings illuminated by the light source. The resolving power of such a grating in second-order is $R = 5000 \times 2 = 10\ 000$. Therefore, the *minimum* wavelength separation between two spectral lines that can be just resolved, assuming a mean wavelength of 600 nm, is given by $\Delta\lambda = \lambda/R = 6 \times 10^{-2}$ nm. For the third-order principal maximum, we find $R = 15\ 000$ and $\Delta\lambda = 4 \times 10^{-2}$ nm, and so on.

EXAMPLE 38.7 Resolving the Sodium Spectral Lines

Two strong lines in the spectrum of sodium have wavelengths of 589.00 nm and 589.59 nm. (a) What must the resolving power of the grating be in order to distinguish these wavelengths?

$$R = \frac{\lambda}{\Delta\lambda} = \frac{589.30\ \text{nm}}{589.59\ \text{nm} - 589.00\ \text{nm}} = \frac{589.30}{0.59} = 999$$

(b) In order to resolve these lines in the second-order spectrum ($m = 2$), how many lines of the grating must be illuminated?

From Equation 38.10 and the results to (a), we find that

$$N = \frac{R}{m} = \frac{999}{2} = 500\ \text{lines}$$

*38.5 DIFFRACTION OF X-RAYS BY CRYSTALS

We have seen that the wavelength of light can be measured with a diffraction grating having a known number of rulings per unit length. In principle, the wavelength of any electromagnetic wave can be determined if a grating of the proper spacing (of the order of λ) is available. X-rays, discovered by W. Roentgen (1845–1923) in 1895, are electromagnetic waves with very short wavelengths (of the order of 1 Å $= 10^{-10}$ m $= 0.1$ nm). Obviously, it would be impossible to construct a grating with such a small spacing. However, the atomic spacing in a solid is known to be about 10^{-10} m. In 1913, Max von Laue (1879–1960) suggested that the regular array of atoms in a crystal could act as a three-dimensional diffraction grating for x-rays. Subsequent experiments confirmed this prediction. The diffraction patterns that one observes are rather complicated because of the three-dimensional nature of the crystal. Nevertheless, x-ray diffraction has proved to be an invaluable technique for elucidating crystalline structures and for understanding the structure of matter.[1]

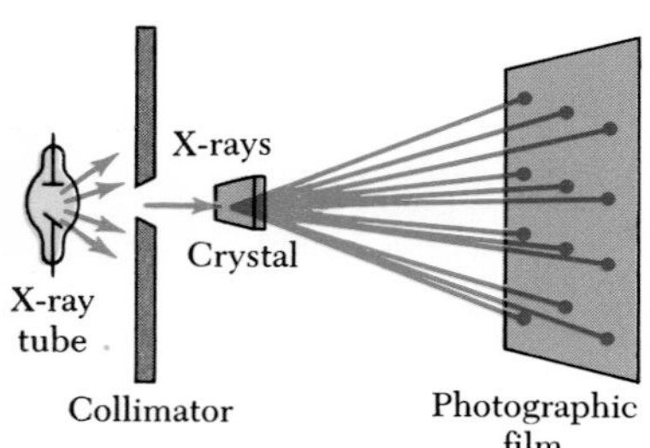

Figure 38.19 Schematic diagram of the technique used to observe the diffraction of x-rays by a single crystal. The array of spots formed on the film by the strongly diffracted beams is called a Laue pattern.

Figure 38.19 is one experimental arrangement for observing x-ray diffraction from a crystal. A collimated beam of x-rays with a continuous range of wavelengths is incident on a crystal, such as one of sodium chloride, for example. The diffracted beams are very intense in certain directions, corresponding to constructive interference from waves reflected from layers of atoms in the crystal. The diffracted beams can be detected by a photographic film, and they form an array of spots known as a "Laue pattern." The crystalline structure is deduced by analyzing the positions and intensities of the various spots in the pattern.

The arrangement of atoms in a crystal of NaCl is shown in Figure 38.20. The smaller, dark spheres represent Na^+ ions and the larger, hollow spheres represent Cl^- ions. Note that the ions are located at the corners of a cube; for this reason, the structure is said to have *cubic symmetry*.

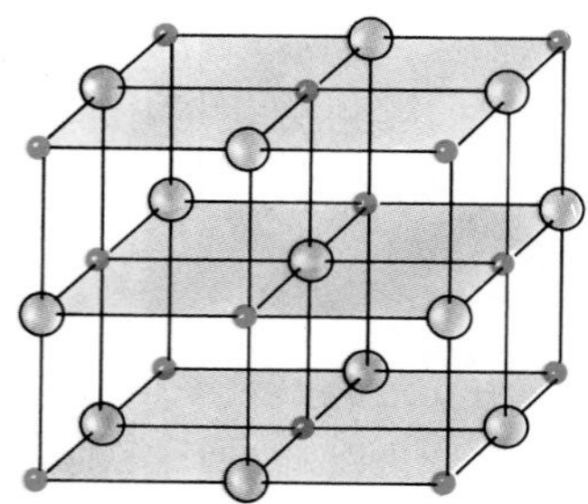

Figure 38.20 A model of the cubic crystalline structure of sodium chloride. The larger blue spheres represent the Cl^- ions, and the smaller red spheres represent the Na^+ ions. The length of the cube edge is $a = 0.562737$ nm.

A careful examination of the NaCl structure shows that the ions appear to lie in various planes. The shaded areas in Figure 38.20 represent one example in which the atoms lie in equally spaced planes. Now suppose an incident x-ray beam makes an angle θ with one of the planes, as in Figure 38.21. The beam can be reflected from both the upper and the lower plane of atoms. However, the geometric construction in Figure 38.21 shows that the beam reflected from the lower surface travels farther than the beam reflected from the upper surface. The effective path difference between the two beams is $2d \sin \theta$. The two beams will reinforce each other (constructive interference) when this path difference equals some integral multiple of the wavelength λ. The same is

[1] For more details on this subject, see Sir Lawrence Bragg, "X-Ray Crystallography," *Scientific American*, July 1968.

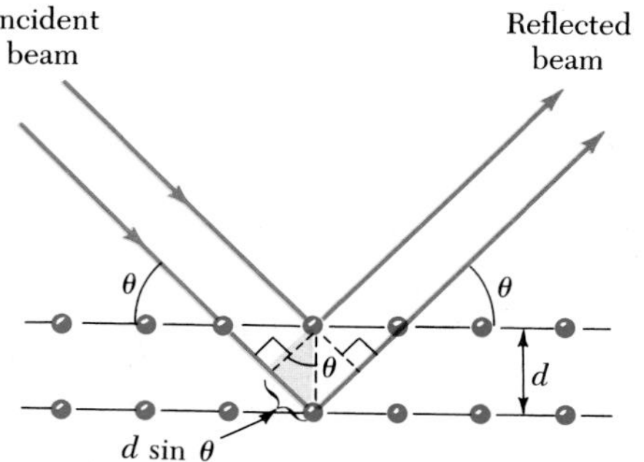

Figure 38.21 A two-dimensional description of the reflection of an x-ray beam from two parallel crystalline planes separated by a distance d. The beam reflected from the lower plane travels farther than the one reflected from the upper plane by an amount equal to $2d \sin \theta$.

true for reflection from the entire family of parallel planes. Hence, the condition for constructive interference (maxima in the reflected wave) is given by

$$2d \sin \theta = m\lambda \qquad (m = 1, 2, 3, \ldots) \tag{38.11}$$

Bragg's law

This condition is known as **Bragg's law** after W. L. Bragg (1890–1971), who first derived the relationship. If the wavelength and diffraction angle are measured, Equation 38.11 can be used to calculate the spacing between atomic planes.

38.6 POLARIZATION OF LIGHT WAVES

The wave nature of light has been used to explain the phenomena of interference and diffraction. In Chapter 34 we described the transverse nature of light waves and, in fact, of all electromagnetic waves. Figure 38.22 shows that the electric and magnetic vectors associated with an electromagnetic wave are at right angles to each other and also to the direction of wave propagation. The phenomenon of polarization, which will be described in this section, is firm evidence of the transverse nature of electromagnetic waves.

An ordinary beam of light consists of a large number of waves emitted by the atoms or molecules of the light source. Each atom produces a wave with its own orientation of $\boldsymbol{E}$, as in Figure 38.22, corresponding to the direction of atomic vibration. The direction of polarization of the electromagnetic wave is defined to be the direction in which $\boldsymbol{E}$ is vibrating. However, since all directions of vibration are possible, the resultant electromagnetic wave is a superposition of waves produced by the individual atomic sources. The result is an **unpolarized** light wave, described in Figure 38.23a. The direction of wave propagation in this figure is perpendicular to the page. Note that *all* directions of the electric field vector are equally probable all lying in a plane perpendicular to the direction of propagation. At any given point and at some instant of time, there is only one resultant electric field, hence you should not be misled by the meaning of Figure 38.23a. A wave is said to be **linearly polarized** if $\boldsymbol{E}$ vibrates in the same direction *at all times* at a particular point, as in Figure 38.23b. (Sometimes such a wave is described as *plane-polarized,* or simply *polarized.*)

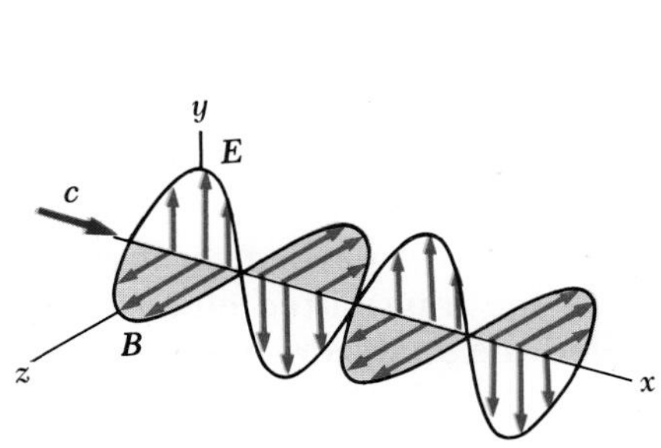

Figure 38.22 Schematic diagram of an electromagnetic wave propagating in the x direction. The electric field vector $\boldsymbol{E}$ vibrates in the xy plane, and the magnetic field vector $\boldsymbol{B}$ vibrates in the xz plane.

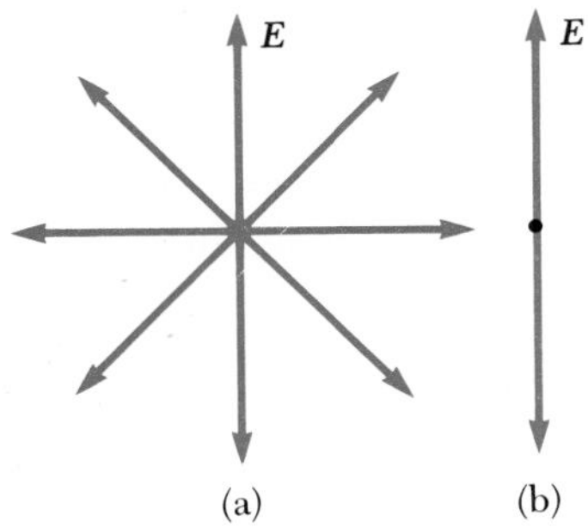

Figure 38.23 (a) An unpolarized light beam viewed along the direction of propagation (perpendicular to the page). The transverse electric field vector can vibrate in any direction with equal probability. (b) A linearly polarized light beam with the electric field vector vibrating in the vertical direction.

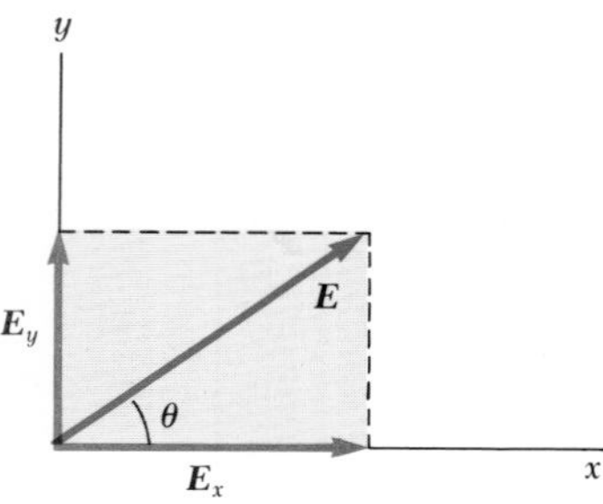

Figure 38.24 A linearly polarized wave with $\boldsymbol{E}$ at an angle θ to x has components $E_x = E \cos\theta$ and $E_y = E \sin\theta$.

Suppose a light beam traveling in the z direction has an electric field vector that is at an angle θ with the x axis at some instant, as in Figure 38.24. The vector has components E_x and E_y as shown. Obviously, the light is linearly polarized if one of these components is always zero or if the angle θ remains constant in time. However, if the tip of the vector $\boldsymbol{E}$ rotates in a circle with time, the wave is said to be **circularly polarized.** This occurs when the magnitudes of E_x and E_y are *equal*, but differ in phase by 90°. On the other hand, if the magnitudes of E_x and E_y are *not* equal, but differ in phase by 90°, the tip of $\boldsymbol{E}$ moves in an ellipse. Such a wave is said to be **elliptically polarized.** Finally, if E_x and E_y are, on the average, equal in magnitude, but have a randomly varying phase difference, the light beam is unpolarized.

It is possible to obtain a linearly polarized beam from an unpolarized beam by removing all waves from the beam except those whose electric field vectors oscillate in a single plane. We shall now discuss four different physical processes for producing polarized light from unpolarized light. These are (1) selective absorption, (2) reflection, (3) double refraction, and (4) scattering.

Polarization by Selective Absorption

The most common technique for obtaining polarized light is to use a material that transmits waves whose electric field vectors vibrate in a plane parallel to a certain direction and absorbs those waves whose electric field vectors vibrate in other directions. Any substance that has the property of transmitting light with the electric field vector vibrating in only one direction is called a **dichroic substance.** In 1938, E. H. Land discovered a material, which he called **polaroid,** that polarizes light through selective absorption by oriented molecules. This material is fabricated in thin sheets of long-chain hydrocarbons, such as polyvinyl alcohol. The sheets are stretched during manufacture so that the long-chain molecules align.[2] After a sheet is dipped into a solution containing iodine, the molecules become conducting. However, the conduction takes place primarily along the hydrocarbon chains since the valence electrons of the molecules can move easily only along the chains (recall that valence electrons are "free" electrons that can readily move through the conductor). As a result, the molecules readily *absorb* light whose electric field vector is parallel to their length and *transmit* light whose electric field vector is perpendicular to their length. It is common to refer to the direction perpendicular to the molecular chains as the **transmission axis.** In an ideal polarizer, all light with $\boldsymbol{E}$ parallel to the transmission axis is transmitted, and all light with $\boldsymbol{E}$ perpendicular to the transmission axis is absorbed.

Figure 38.25 represents an unpolarized light beam incident on the first polarizing sheet, called the **polarizer,** where the transmission axis is indicated by the straight lines on the polarizer. The light that is passing through this sheet is polarized vertically as shown, where the transmitted electric field vector is $\boldsymbol{E}_0$. A second polarizing sheet, called the **analyzer,** intercepts this beam with its transmission axis at an angle θ to the axis of the polarizer. The component of $\boldsymbol{E}_0$ perpendicular to the axis of the analyzer is completely absorbed, and the component of $\boldsymbol{E}_0$ parallel to the axis of the analyzer is $E_0 \cos\theta$

[2] An earlier version of a Polaroid material developed by Land consisted of oriented dichroic crystals of quinine sulfate periodide imbedded in a plastic film.

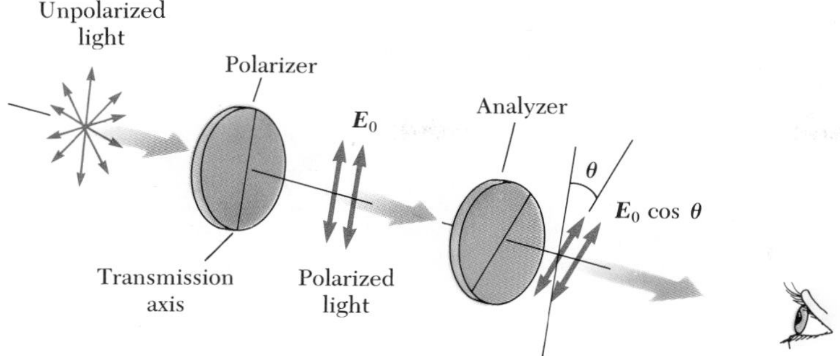

Figure 38.25 Two polarizing sheets whose transmission axes make an angle θ with each other. Only a fraction of the polarized light incident on the analyzer is transmitted.

and since the transmitted intensity varies as the *square* of the transmitted amplitude, we conclude that the transmitted intensity varies as

$$I = I_0 \cos^2 \theta \tag{38.12}$$

where I_0 is the intensity of the polarized wave incident on the analyzer. This expression, known as **Malus's law,**[3] applies to any two polarizing materials whose transmission axes are at an angle θ to each other. From this expression, note that the transmitted intensity is a maximum when the transmission axes are parallel ($\theta = 0$ or $180°$). In addition, the transmitted intensity is zero (complete absorption by the analyzer) when the transmission axes are perpendicular to each other. This variation in transmitted intensity through a pair of polarizing sheets is illustrated in Figure 38.26.

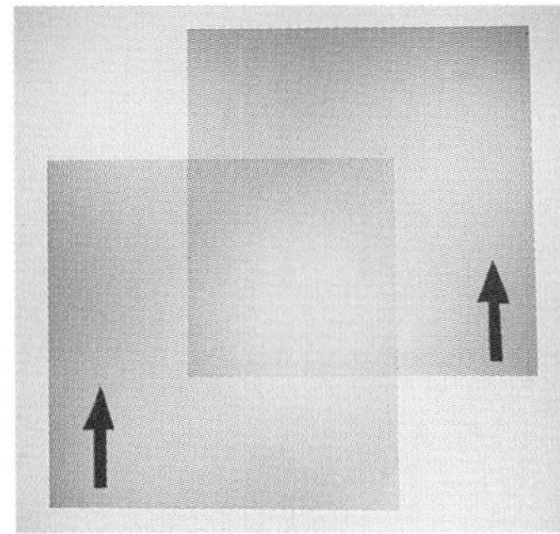

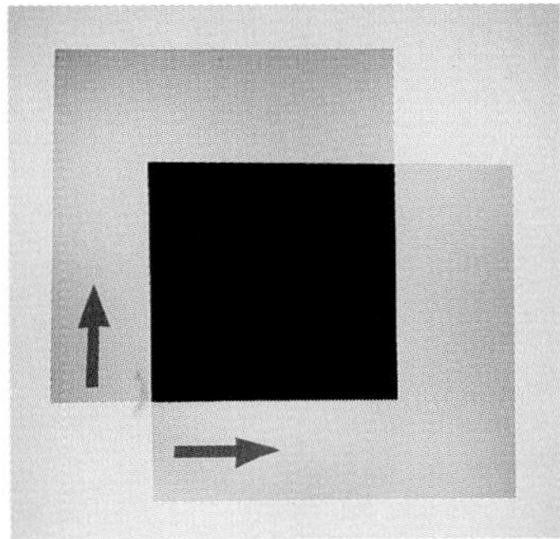

Figure 38.26 *(Top)* Two polarizers whose transmission axes are aligned with each other. *(Bottom)* Two crossed polarizers, whose transmission axes are *perpendicular* to each other. (Courtesy of Henry Leap and Jim Lehman)

Polarization by Reflection

Another method for obtaining polarized light is by reflection. When an unpolarized light beam is reflected from a surface, the reflected light is completely polarized, partially polarized, or unpolarized, depending on the angle of incidence. If the angle of incidence is either 0 or 90° (normal or grazing angles), the reflected beam is unpolarized. However, for intermediate angles of incidence, the reflected light is polarized to some extent. In fact, for one particular angle of incidence, the reflected light is completely polarized.

Suppose an unpolarized light beam is incident on a surface as in Figure 38.27a. The beam can be described by two electric field components, one parallel to the surface (the dots) and the other perpendicular to the first and to the direction of propagation (the arrows). It is found that the parallel component reflects more strongly than the other component, and this results in a partially polarized beam (Fig. 38.27a). Furthermore, the refracted ray is also partially polarized.

Now suppose the angle of incidence, θ_1, is varied until the angle between the reflected and refracted beams is 90° (Fig. 38.27b). At this particular angle of incidence, experiments show that the reflected beam is completely polarized with its electric field vector parallel to the surface, while the refracted

[3] Named after its discoverer, E. L. Malus (1775–1812). Actually, Malus first discovered that reflected light was polarized by viewing it through a calcite ($CaCO_3$) crystal.

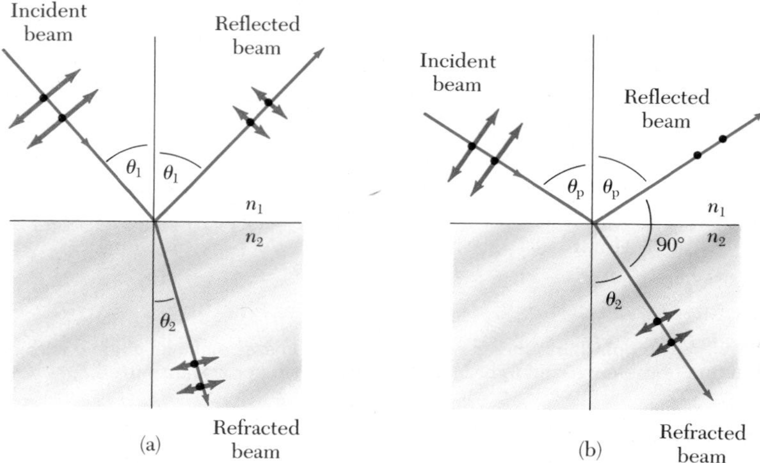

Figure 38.27 (a) When unpolarized light is incident on a reflecting surface, the reflected and refracted beams are partially polarized. (b) The reflected beam is completely polarized when the angle of incidence equals the polarizing angle, θ_p, which satisfies the equation $n = \tan\theta_p$.

The polarizing angle

beam is partially polarized. The angle of incidence at which this occurs is called the **polarizing angle,** θ_p.

An expression relating the polarizing angle to the index of refraction, n, of the reflecting substance can be obtained by using Figure 38.27b. From this figure, we see that at the polarizing angle, $\theta_p + 90° + \theta_2 = 180°$, so that $\theta_2 = 90° - \theta_p$. Using Snell's law, we have

$$n = \frac{\sin\theta_1}{\sin\theta_2} = \frac{\sin\theta_p}{\sin\theta_2}$$

Because $\sin\theta_2 = \sin(90° - \theta_p) = \cos\theta_p$, the expression for n can be written

$$n = \frac{\sin\theta_p}{\cos\theta_p} = \tan\theta_p \tag{38.13}$$

Brewster's law

This expression is called **Brewster's law,** and the polarizing angle θ_p is sometimes called **Brewster's angle,** after its discoverer, Sir David Brewster (1781–1868). For example, the Brewster's angle for crown glass ($n = 1.52$) has the value $\theta_p = \tan^{-1}(1.52) = 56.7°$. Because n varies with wavelength for a given substance, the Brewster's angle is also a function of the wavelength.

Polarization by reflection is a common phenomenon. Sunlight reflected from water, glass, and snow is partially polarized. If the surface is horizontal, the electric field vector of the reflected light will have a strong horizontal component. Sunglasses made of polarizing material reduce the glare of reflected light. The transmission axes of the lenses are oriented vertically so as to absorb the strong horizontal component of the reflected light.

Polarization by Double Refraction

When light travels through an amorphous material, such as glass, it travels with a speed that is the same in all directions. That is, glass has a single index of refraction. However, in certain crystalline materials, such as calcite and quartz, the speed of light is *not* the same in all directions. Such materials are characterized by two indices of refraction. Hence, they are often referred to as **double-refracting** or **birefringent** materials.[4]

When an unpolarized beam of light enters a calcite crystal, the beam splits into two plane-polarized rays that travel with different velocities, corresponding to two different angles of refraction, as in Figure 38.28. The two rays are polarized in two mutually perpendicular directions, as indicated by the dots and arrows. One ray, called the **ordinary (O) ray,** is characterized by an index of refraction, n_O, that is the *same* in all directions. This means that if one could place a point source of light inside the crystal, as in Figure 38.29, the ordinary waves would spread out from the source as spheres.

The second plane-polarized ray, called the **extraordinary (E) ray,** travels with *different* speeds in different directions and hence is characterized by an index of refraction, n_E, that *varies* with the direction of propagation. A point source of light inside such a crystal would send out an extraordinary wave having wavefronts that are elliptical in cross section (Fig. 38.29). Note from Figure 38.29 that there is one direction, called the **optic axis,** along which the ordinary and extraordinary rays have the *same* velocity, corresponding to the direction for which $n_O = n_E$. The difference in velocity for the two rays is a maximum in the direction perpendicular to the optic axis. For example, in calcite $n_O = 1.658$ at a wavelength of 589.3 nm and n_E varies from 1.658 along the optic axis to 1.486 perpendicular to the optic axis. Values for n_O and n_E for various double-refracting crystals are given in Table 38.1.

If one places a piece of calcite on a sheet of paper and then looks through the crystal at any writing on the paper, two images of the writing are seen, as shown in Figure 38.30. As can be seen from Figure 38.28, these two images correspond to one formed by the ordinary ray and the second formed by the extraordinary ray. If the two images are viewed through a sheet of rotating polarizing glass, they will alternately appear and disappear because the ordinary and extraordinary rays are plane-polarized along mutually perpendicular directions.

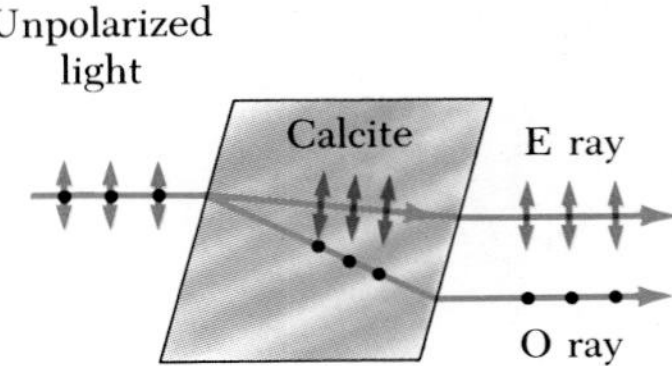

Figure 38.28 An unpolarized light beam incident on a calcite crystal splits into an ordinary (O) ray and an extraordinary (E) ray. These two rays are polarized in mutually perpendicular directions. (Note that this is not to scale.)

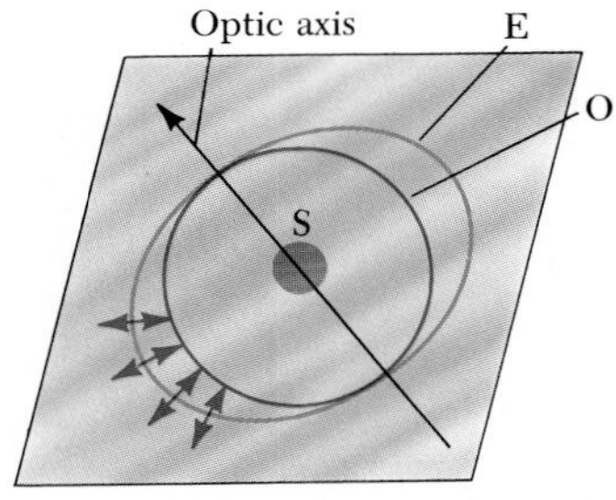

Figure 38.29 A point source, S, inside a double refracting crystal produces a spherical wavefront corresponding to the ordinary ray and an elliptical wavefront corresponding to the extraordinary ray. The two waves propagate with the same velocity along the optic axis.

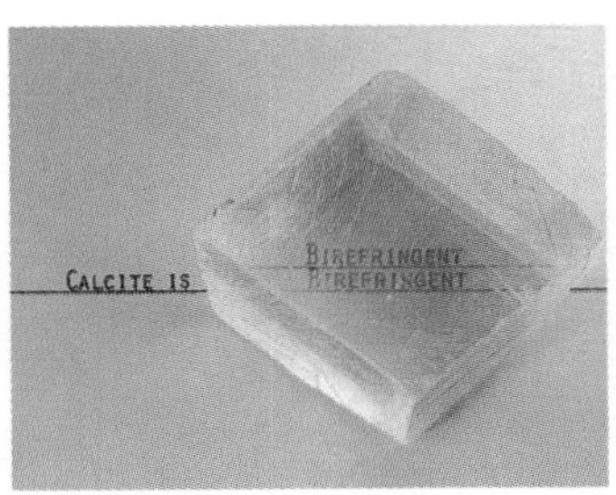

Figure 38.30 A calcite crystal produces a double image because it is a birefringent (double-refracting) material. (Courtesy of Henry Leap and Jim Lehman)

Polarization by Scattering

When light is incident on a system of particles, such as a gas, the electrons in the medium can absorb and reradiate part of the light. The absorption and reradiation of light by the medium, called **scattering,** is what causes sunlight

[4] For a lucid treatment of this topic, see Elizabeth A. Wood, *Crystals and Light,* New York, Van Nostrand (Momentum), 1964, Chapter 12.

TABLE 38.1 Indices of Refraction for Some Double-Refracting Crystals at a Wavelength of 589.3 nm

Crystal	n_O	n_E	n_O/n_E
Calcite ($CaCO_3$)	1.658	1.486	1.116
Quartz (SiO_2)	1.544	1.553	0.994
Sodium nitrate ($NaNO_3$)	1.587	1.336	1.188
Sodium sulfite ($NaSO_3$)	1.565	1.515	1.033
Zinc chloride ($ZnCl_2$)	1.687	1.713	0.985
Zinc sulfide (ZnS)	2.356	2.378	0.991

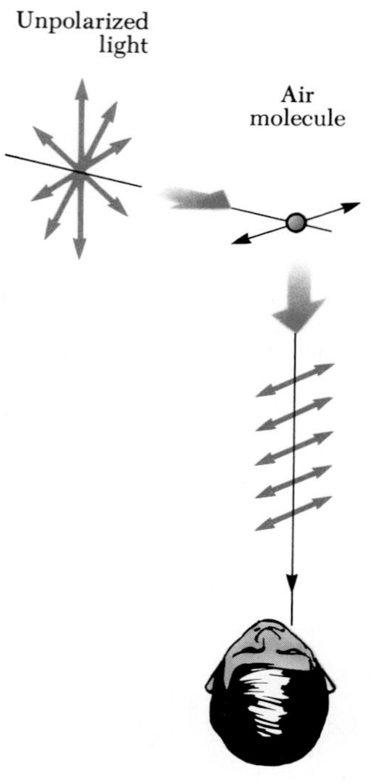

Figure 38.31 The scattering of unpolarized sunlight by air molecules. The light observed at right angles is plane-polarized because the vibrating molecule has a horizontal component of vibration.

reaching an observer on the earth from straight overhead to be partially polarized. You can observe this effect by looking directly up through a pair of sunglasses whose lenses are made of polarizing material. At certain orientations of the lenses, less light passes through than at others.

Figure 38.31 illustrates how the sunlight becomes partially polarized. The left side of the figure shows an incident unpolarized beam of sunlight traveling in the horizontal direction on the verge of striking an air molecule. When this beam strikes the air molecule, it sets the electrons of the molecule into vibration. These vibrating charges act like the vibrating charges in an antenna, except that these charges are vibrating in a complicated pattern. The horizontal part of the electric field vector in the incident wave causes the charges to vibrate horizontally, and the vertical part of the vector simultaneously causes them to vibrate vertically. A horizontally polarized wave is emitted by the electrons as a result of their horizontal motion, and a vertically polarized wave is emitted parallel to the earth as a result of their motion.

Some phenomena involving the scattering of light in the atmosphere can be understood as follows: When light of various wavelengths λ is incident on air molecules of size d, where $d \ll \lambda$, the relative intensity of the scattered light varies as $1/\lambda^4$. The condition $d \ll \lambda$ is satisfied for scattering from O_2 and N_2 molecules in the atmosphere, whose diameters are about 0.2 nm. Hence shorter wavelengths (blue light) are scattered more efficiently than longer wavelengths (red light). Because sunlight contains wavelengths from the entire visible spectrum, the sky appears to be blue. (This also explains why the sky is black in outer space, where there is no atmosphere to scatter the sunlight!)

Optical Activity

Many important practical applications of polarized light involve the use of certain materials which display the property of **optical activity.** A substance is said to be optically active if it rotates the plane of polarization of transmitted light. The angle through which the light is rotated by the material depends on the length of the sample and on the concentration if the substance is in solution. One optically active material is a solution of common sugar dextrose. A standard method for determining the concentration of sugar solutions is to measure the rotation produced by a fixed length of the solution.

Optical activity occurs in a material because of an asymmetry in the shape of its constituent molecules. For example, some proteins are optically active because of their spiral shape. Other materials, such as glass and plastic, become optically active when placed under stress. If polarized light is passed through an unstressed piece of plastic and then through an analyzer whose axis is perpendicular to that of the polarizer, none of the polarized light is transmitted. However, if the plastic is placed under stress, the regions of greatest stress produce the largest angles of rotation of polarized light. Hence, one observes a series of bright and dark bands in the transmitted light, with the bright bands corresponding to regions of greatest stress. Engineers often use this technique, called *optical stress analysis,* to assist in the design of various structures ranging from bridges to small tools and machine parts. A plastic model is built and analyzed under different load conditions to determine regions of potential weakness and failure under stress. Some examples of a plastic model under stress are shown in the photographs on the next page.

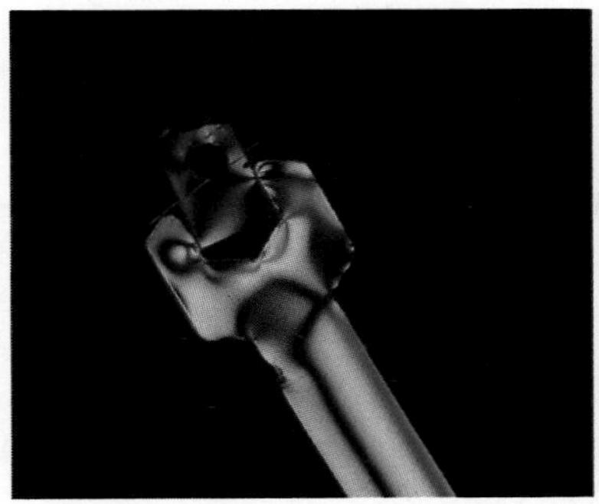

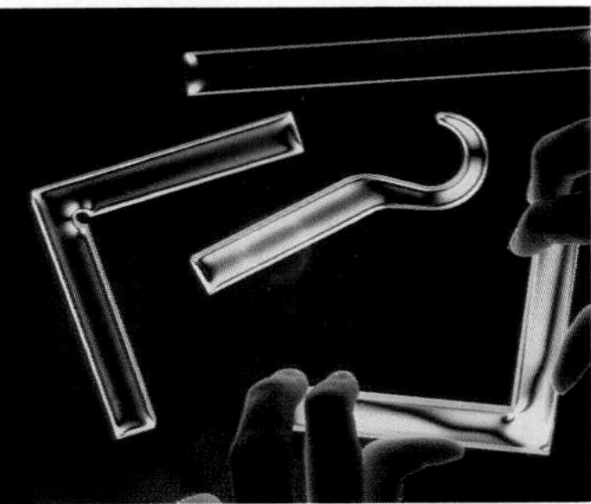

(Left) The strain distribution in a plastic wrench under stress. The test piece is viewed between two crossed polarizers. (Courtesy of Henry Leap and Jim Lehman) *(Right)* The photograph shows that the strain pattern disappears when the stress is released. (Courtesy of CENCO)

SUMMARY

The phenomenon of **diffraction** arises from the interference of a large number or continuous distribution of coherent sources. Diffraction accounts for the deviation of light from a straight-line path when the light passes through an aperture or around obstacles.

The **Fraunhofer diffraction pattern** produced by a *single slit* of width a on a distant screen consists of a central, bright maximum and alternating bright and dark regions of much lower intensities. The angles θ at which the diffraction pattern has *zero* intensity are given by

Condition for intensity minima in the single-slit diffraction pattern

$$\sin\theta = m\frac{\lambda}{a} \qquad (m = \pm 1, \pm 2, \pm 3, \ldots) \tag{38.1}$$

where $|m| \leq a/\lambda$.

The variation of intensity I with angle θ is given by

Intensity of a single-slit Fraunhofer diffraction pattern

$$I_\theta = I_0\left[\frac{\sin(\beta/2)}{\beta/2}\right]^2 \tag{38.4}$$

where $\beta = 2\pi a \sin\theta/\lambda$ and I_0 is the intensity at $\theta = 0$.

Rayleigh's criterion

Rayleigh's criterion, which is a limiting condition of resolution, says that two images formed by an aperture are just distinguishable if the central maximum of the diffraction pattern for one image falls on the first minimum of the other image. The limiting angle of resolution for a slit of width a is given by $\theta_m = \lambda/a$, and the limiting angle of resolution for a circular aperture of diameter D is given by $\theta_m = 1.22\lambda/D$.

A diffraction grating consists of a large number of equally spaced, identical slits. The condition for intensity maxima in the interference pattern of a diffraction grating is given by

Condition for intensity maxima for a grating

$$d\sin\theta = m\lambda \qquad (m = 0, 1, 2, 3, \ldots) \tag{38.8}$$

where d is the spacing between adjacent slits and m is the *order number* of the diffraction pattern. The resolving power of a diffraction grating in the mth order of the diffraction pattern is given by $R = Nm$, where N is the number of rulings in the grating.

Unpolarized light can be polarized by four processes: (1) selective absorption, (2) reflection, (3) double refraction, and (4) scattering.

When polarized light of intensity I_0 is incident on a polarizing film, the light transmitted through the film has an intensity equal to $I_0 \cos^2 \theta$, where θ is the angle between the transmission axis of the polarizer and the electric field vector of the incident light.

In general, light reflected from a dielectric material, such as glass, is partially polarized. However, the reflected light is completely polarized when the angle of incidence is such that the angle between the reflected and refracted beams is 90°. This angle of incidence, called the **polarizing angle** θ_p, satisfies **Brewster's law,** given by

Brewster's law

$$n = \tan \theta_p \tag{38.13}$$

where n is the index of refraction of the reflecting medium.

QUESTIONS

1. If you place your thumb and index finger very close together and view light passing between them when they are a few cm in front of your eye, dark lines parallel to your thumb and finger will appear. Explain.
2. Observe the shadow of your book or some other straight edge when it is held a few inches above a table with a lamp several feet above the book. Why is the shadow of the book somewhat fuzzy at the edges?
3. What is the difference between Fraunhofer and Fresnel diffraction?
4. Although we can hear around corners, we cannot see around corners. How can you explain this in view of the fact that sound and light are both waves?
5. Describe the change in width of the central maximum of the single-slit diffraction pattern as the width of the slit is made smaller.
6. Assuming that the headlights of a car are point sources, estimate the maximum distance from an observer to the car at which the headlights are distinguishable from each other.
7. A laser beam is incident at a shallow angle on a machinist's ruler that has a finely calibrated scale. The rulings on the scale give rise to a diffraction pattern on a screen. Discuss how you can use this technique to obtain a measure of the wavelength of the laser light.
8. Certain sunglasses use a polarizing material to reduce the intensity of light reflected from shiny surfaces, such as water or the hood of a car. What orientation of polarization should the material have in order to be most effective?
9. Why is the sky black when viewed from the moon?
10. The diffraction grating effect is easily observed with everyday equipment. For example, a compact disk can be held so that light is reflected from it at a glancing angle. When held this way, various colors in the reflected light can be seen. Explain how this works.

(Question 10). Compact disks act as diffraction gratings when observed under white light. (© Bobbie Kingsley, Photo Researchers, Inc.)

11. The path of a light beam from a helium-neon laser can be made visible by placing chalk dust in the air (perhaps by shaking a blackboard eraser in the path of the light beam). Explain why the beam can be seen under these circumstances.
12. Is light from the sky polarized? Why is it that clouds seen through Polaroid glasses stand out in bold contrast to the sky?
13. If a coin is glued to a glass sheet and this arrangement is held in front of a helium-neon laser, the projected shadow shows diffraction rings around the coin and a bright spot in the center of the shadow. How is this possible?
14. If a fine wire is stretched across the path of a laser beam, is it possible to produce a diffraction pattern?
15. How would one determine the index of refraction of a flat piece of dark obsidian glass?

PROBLEMS

Section 38.2 Single-Slit Diffraction

1. In Figure 38.7, let the slit width $a = 0.5$ mm and assume incident monochromatic light of wavelength 460 nm. (a) Find the value of θ corresponding to the second dark fringe beyond the central maximum. (b) If the observing screen is located 120 cm in front of the slit, what is the distance y from the center of the central maximum to the second dark fringe?
2. A Fraunhofer diffraction pattern is produced on a screen 140 cm from a single slit. The distance from the center of the central maximum to the first secondary maximum is $10^4\lambda$. Calculate the slit width.
3. The second bright fringe in a single-slit diffraction pattern is located 1.4 mm beyond the center of the central maximum. The screen is 80 cm from a slit opening of 0.8 mm. Assuming monochromatic incident light, calculate the wavelength.
4. Helium-neon laser light ($\lambda = 632.8$ nm) is sent through a 0.3-mm-wide single slit. What is the width of the central maximum on a screen 1 m in back of the slit?
5. The pupil of a cat's eye narrows to a slit of width 0.5 mm in daylight. What is the angular resolution? (Let the wavelength of light in the cat's eye be 500 nm.)
6. Light of wavelength 587.5 nm illuminates a single slit 0.75 mm in width. (a) At what distance from the slit should a screen be located if the first minimum in the diffraction pattern is to be 0.85 mm from the center of the screen? (b) What is the width of the central maximum?
7. A screen is placed 50 cm from a single slit, which is illuminated with light of wavelength 690 nm. If the distance between the first and third minima in the diffraction pattern is 3.0 mm, what is the width of the slit?
8. In equation 38.4, let $\beta/2 \equiv \phi$ and show that $I = 0.5I_0$ when $\sin\phi = \phi/\sqrt{2}$.
9. The equation $\sin\phi = \phi/\sqrt{2}$ found in Problem 8 is known as a *transcendental equation*. One method of solving such an equation is the graphical method. To illustrate this, let $\phi = \beta/2$, $y_1 = \sin\phi$, and $y_2 = \phi/\sqrt{2}$. Plot y_1 and y_2 on the same set of axes over a range from $\phi = 1$ rad to $\phi = \pi/2$ rad. Determine ϕ from the point of intersection of the two curves.
10. A slit of width 1 mm is illuminated with monochromatic light of wavelength 580 nm. A diffraction pattern is formed on a screen 50 cm away. Calculate the intensity on the screen as a fraction of the maximum intensity (I/I_0) at an angular distance of 1.7 min of arc from the center of the diffraction pattern.
11. A diffraction pattern is formed on a screen 120 cm away from a 0.4-mm-wide slit. Monochromatic light of 546.1 nm is used. Calculate the fractional intensity I/I_0 at a point on the screen 4.1 mm from the center of the principal maximum.
12. In the calculation of Example 38.2, it is assumed that the secondary maxima in the diffraction pattern lie midway between points of zero intensity. Verify that this is a reasonable approximation by (a) calculating I/I_0 for $\beta/2 = 1.4303\pi$ and $\beta/2 = 2.4590\pi$. (These are the first two values of $\beta/2$ that satisfy the equation $\tan(\beta/2) = \beta/2$.) (b) Compare the results of (a) to the intensity ratios calculated in Example 38.2.

Section 38.3 Resolution of Single-Slit and Circular Apertures

13. In Example 38.5, the limiting angle of resolution of the eye at a wavelength of 500 nm in air is calculated to be $\theta_{min} = 2.3 \times 10^{-4}$ rad. What is the maximum distance from the eye at which two points separated by 1 cm could be resolved?
14. The *resolving power* of a telescope is expressed as $R = 1/\theta_m$, where θ_m is given by Equation 38.7. Determine the resolving power of a 25-in.-diameter telescope at a wavelength of 550 nm.
15. What is the minimum distance between two points that will permit them to be resolved at 1 km (a) using a terrestrial telescope with a 6.5-cm-diameter objective (assume $\lambda = 550$ nm) and (b) using the unaided eye (assume a pupil diameter of 2.5 mm)?
16. If we were to send a ruby laser beam ($\lambda = 694.3$ nm) outward from the barrel of a 2.7-m-diameter telescope, what would be the diameter of the big red spot when the beam hit the moon 384 000 km away? (Neglect atmospheric dispersion.)
17. The 2.4-m diameter Hubble space telescope will be placed into earth orbit by the space shuttle in 1989 or 1990. What angular resolution will this telescope achieve for visible light ($\lambda = 500$ nm)?
18. Find the radius of a star-image formed on the retina of the eye if the aperture diameter (the pupil) at night is 0.7 cm, and the length of the eye is 3 cm. Assume the wavelength of starlight in the eye is 500 nm.
19. At what distance could one theoretically distinguish two automobile headlights separated by 1.4 m? Assume a pupil diameter of 6 mm and yellow headlights ($\lambda = 580$ nm). The index of refraction in the eye is approximately 1.33.
20. A binary star system in the constellation Orion has an angular separation between the two stars of 10^{-5} radians. If $\lambda = 500$ nm, what is the smallest aperture (diameter) telescope that can just resolve the two stars?

21. The Impressionist painter Georges Seurat created paintings with an enormous number of dots of pure pigment about 2 mm in diameter. The idea was to have colors such as red and green next to each other to form a scintillating canvas. Outside what distance would one be unable to discern individual dots on the canvas? (Assume $\lambda = 500$ nm within the eye and a pupil diameter of 4 mm.)
22. Calculate the angular separation between two points which are just resolvable by a 100-in.-diameter telescope at an average wavelength of 550 nm.
23. A radar installation operates at a frequency equal to 9×10^9 Hz and uses an antenna with a diameter of 15 m. Two objects are 150 m apart. At what distance from the antenna would they be at the limit of resolution?
24. Two motorcycles separated laterally by 2 m are approaching an observer who is holding an infrared "snooper scope" (effective $\lambda = 885$ nm). What aperture diameter is required if the two headlights are to be resolved at a distance of 10 km?

Section 38.4 The Diffraction Grating

25. Collimated light from a hydrogen discharge tube is incident normally on a diffraction grating. The incident light includes four wavelength components: $\lambda_1 = 410.1$ nm, $\lambda_2 = 434.0$ nm, $\lambda_3 = 486.1$ nm, and $\lambda_4 = 656.3$ nm. There are 410 lines/mm in the grating. Calculate the angle between (a) λ_1 and λ_4 in the first-order spectrum and (b) λ_1 and λ_3 in the third-order spectrum.
26. Monochromatic light is incident on a grating that is 75 mm wide and ruled with 50 000 lines. The line is imaged at 32.5° in the second-order spectrum. Determine the wavelength of the incident light.
27. A grating with 250 lines/mm is used with an incandescent light source. Assume the visible spectrum to range in wavelength from 400 to 700 nm. In how many orders can one see (a) the entire visible spectrum and (b) the short-wavelength region?
28. A grating of width 4 cm is illuminated with monochromatic light of wavelength 577 nm. The second-order maximum is formed at an angle of 41.25°. What is the total number of lines in the grating?
29. The full width of a 3-cm-wide grating is illuminated by a sodium discharge tube. The lines in the grating are uniformly spaced at 775 nm. Calculate the angular separation in the first-order spectrum between the two wavelengths forming the sodium doublet ($\lambda_1 = 589.0$ nm and $\lambda_2 = 589.6$ nm).
30. Determine the minimum wavelength difference that can be resolved at 600 nm by the grating described in Problem 29. (Assume that the full grating width is illuminated.)
31. (a) Determine the minimum number of lines in a grating that will allow resolution of the sodium doublet in the second order (see Problem 29 for wavelength values). (b) Calculate the width of the grating if the second-order spectrum is to be formed at an angle of 15°.
32. The 546.1-nm line in mercury is measured at an angle of 81° in the third-order spectrum of a diffraction grating. Calculate the number of lines per mm for the grating.
33. The 501.5-nm line in helium is observed at an angle of 30° in the second-order spectrum of a diffraction grating. Calculate the angular deviation associated with the 667.8-nm line in helium in the first-order spectrum for the same grating.
34. A helium-neon laser ($\lambda = 632.8$ nm) is used to calibrate a diffraction grating. If the first-order maximum occurs at 20.5°, what is the line spacing, *d*?
35. White light is spread out into spectral hues by a diffraction grating. If the grating has 2000 lines per cm, at what angle will red light ($\lambda = 640$ nm) appear in first order?
36. Two spectral lines in a mixture of hydrogen (H_2) and deuterium (D_2) gas have wavelengths of 656.30 nm and 656.48 nm, respectively. What is the minimum number of lines a diffraction grating must have to resolve these two wavelengths in first order?
37. Monochromatic light from a He-Ne laser ($\lambda = 632.8$ nm) is incident on a diffraction grating containing 4000 lines/cm. Determine the angle of the first-order maximum.

*Section 38.5 Diffraction of X-Rays by Crystals

38. Potassium iodide (KI) has the same crystalline structure as that of NaCl, with $d = 0.353$ nm. A monochromatic x-ray beam shows a diffraction maximum when the grazing angle is 7.6°. Calculate the x-ray wavelength. (Assume first order.)
39. Monochromatic x-rays of the K_α line of potassium from a nickel target ($\lambda = 0.166$ nm) are incident on a KCl crystal surface. The interplanar distance in KCl is 0.314 nm. At what angle (relative to the surface) should the beam be directed in order that a second-order maximum be observed?
40. A monochromatic x-ray beam is incident on a NaCl crystal surface which has an interplanar spacing of 0.281 mm. The second-order maximum in the reflected beam is found when the angle between the incident beam and the surface is 20.5°. Determine the wavelength of the x-rays.
41. Show why the Bragg condition expressed by Equation 38.11 cannot be satisfied in cases where the wavelength is greater than $2d$ (the length of the unit cell).
42. The dimension labeled *a* in Figure 38.20 is the edge length of the unit cell of NaCl. This is twice the dis-

tance between adjacent ions, which is the parameter d in Equation 38.11. Calculate d for the NaCl crystal from the following data: density $\rho = 2.164$ g/cm^3, molecular mass $M = 58.45$ g/mol, and Avogadro's number $N_A = 6.022 \times 10^{23}$ atoms/mol.

43. X-rays of wavelength 0.14 nm are reflected from a NaCl crystal, and the first-order maximum occurs at an angle of 14.4°. (a) What value does this give for the interplane spacing of NaCl? (b) Compare the value found in (a) with the value calculated in Problem 42.
44. A wavelength of 0.129 nm characterizes K_β x-rays from zinc. When a beam of these x-rays is incident on the surface of a crystal whose structure is similar to that of NaCl, a first-order maximum is observed at an angle of 8.15°. Calculate the interplanar spacing based on this information.
45. If the interplanar spacing of NaCl is 0.281 nm, what is the predicted angle at which x-rays of wavelength 0.14 nm will be diffracted in a first-order maximum?
46. In an x-ray diffraction experiment using x-rays of $\lambda = 0.5 \times 10^{-10}$ m, a first-order maximum occurred at 5°. Find the crystal plane spacing.

Section 38.6 Polarization of Light Waves

47. The angle of incidence of a light beam onto a reflecting surface is continuously variable. The reflected ray is found to be completely polarized when the angle of incidence is 48°. What is the index of refraction of the reflecting material?
48. Light is reflected from a smooth ice surface, and the reflected ray is completely polarized. Determine the angle of incidence. ($n = 1.309$ for ice.)
49. A light beam is incident on heavy flint glass ($n = 1.65$) at the polarizing angle. Calculate the angle of refraction for the transmitted ray.
50. How far above the horizon is the moon when its image reflected in calm water is completely polarized? ($n_{water} = 1.33$.)
51. Unpolarized light passes through two polaroid sheets. The axis of the first is vertical, the second is at 30° to the vertical. What fraction of the initial light is transmitted?
52. Vertically polarized light is passed through three successive polaroid filters. The transmission axes are at 30°, 60°, and 90° to the vertical. What percentage of the light gets through?
53. Plane-polarized light is incident on a single polarizing disk with the direction of E_0 parallel to the direction of the transmission axis. Through what angle should the disk be rotated so that the intensity in the transmitted beam will be reduced by a factor of (a) 3, (b) 5, (c) 10?
54. Three polarizing disks whose planes are parallel are centered on a common axis. The direction of the transmission axis in each case is shown, in Figure 38.32, relative to the common vertical direction. A plane-polarized beam of light with E_0 parallel to the vertical reference direction is incident from the left on the first disk with intensity $I_i = 10$ units (arbitrary). Calculate the transmitted intensity I_f when (a) $\theta_1 = 20°$, $\theta_2 = 40°$, and $\theta_3 = 60°$; (b) $\theta_1 = 0°$, $\theta_2 = 30°$, and $\theta_3 = 60°$.

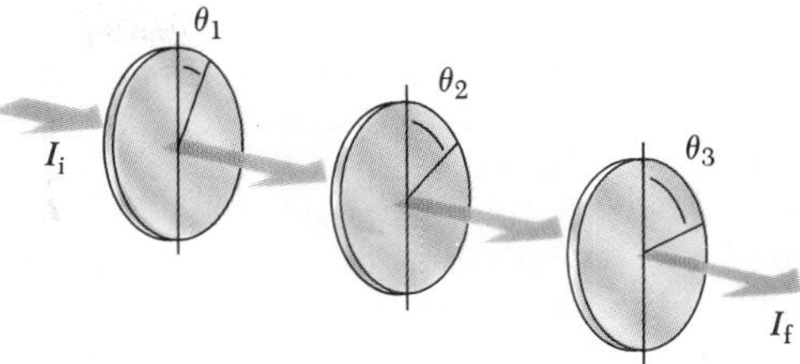

Figure 38.32 (Problems 54 and 72).

55. The critical angle for sapphire surrounded by air is 34.4°. Calculate the polarizing angle for sapphire.
56. For a particular transparent medium surrounded by air, show that the critical angle for internal reflection and the polarizing angle are related by $\cot \theta_p = \sin \theta_c$.
57. If the polarizing angle for cubic zirconia (ZrO_2) is 65.6°, what is the index of refraction for this material?

ADDITIONAL PROBLEMS

58. A solar eclipse is projected through a pinhole of diameter 0.5 mm and strikes a screen 2 meters away. (a) What is the diameter of the projected image? (b) What is the radius of the first diffraction minimum? Both the sun and moon have angular diameters of very nearly 0.5 degrees. (Assume $\lambda = 550$ nm.)
59. The hydrogen spectrum has a red line at 656 nm, and a blue line at 434 nm. What is the angular separation between two spectral lines obtained with a diffraction grating with 4500 lines/cm?
60. What are the approximate dimensions of the smallest object on earth that the astronauts can resolve by eye at 250 km height from the space shuttle? Assume $\lambda = 500$ nm light in the eye, and a pupil diameter of 0.005 m.
61. A spy satellite circles the earth at an altitude of 200 km and carries out surveillance with a special high-resolution telescopic camera having a lens diameter of 35 cm. If the angular resolution of this camera is limited by diffraction, estimate the separation of two small objects on the earth's surface that are just resolved in yellow-green light ($\lambda = 5500$ Å).
62. An unpolarized beam of light is reflected from a glass surface. It is found that the reflected beam is linearly polarized when the light is incident from air at an angle of 58.6°. What is the refractive index of the glass?

63. Light consisting of two wavelength components is incident on a grating. The shorter-wavelength component has a wavelength of 440 nm. The third-order image of this component is coincident with the second-order image of the longer-wavelength component. Determine the value of the longer-wavelength component.

64. Sunlight is incident on a diffraction grating which has 2750 lines/cm. The second-order spectrum over the visible range (400–700 nm) is to be limited to 1.75 cm along a screen a distance L from the grating. What is the required value of L?

65. A diffraction grating of length 4 cm contains 6000 rulings over a width of 2 cm. (a) What is the resolving power of this grating in the first three orders? (b) If two monochromatic waves incident on this grating have a mean wavelength of 400 nm, what is their wavelength separation if they are just resolved in the third order?

66. A single slit of width 0.20 mm is illuminated with light of wavelength 500 nm. The observing screen is placed 80 cm from the slit. (a) Calculate the width of the central bright fringe and of the secondary maxima of the diffraction pattern. (b) What is the distance between the first and fourth minima?

67. Light of wavelength 500 nm is incident normally on a diffraction grating. If the third-order maximum of the diffraction pattern is observed at an angle of 32°, (a) what is the number of rulings per cm for the grating? (b) Determine the total number of primary maxima that can be observed in this situation.

68. Consider the case of a light beam containing two discrete wavelength components whose difference in wavelength $\Delta\lambda$ is small relative to the mean wavelength λ incident on a diffraction grating. A useful measure of the angular separation of the maxima corresponding to the two wavelengths is the *dispersion* D, given by $D = d\theta/d\lambda$. (The dispersion of a grating should not be confused with its resolving power R, given by Equations 38.9 and 38.10.) (a) Starting with Equation 38.8 show that the dispersion can be written

$$D = \frac{\tan\theta}{\lambda}$$

(b) Calculate the dispersion in the third order for the grating described in Problem 45. Give the answer in units of deg/nm.

69. An *unpolarized* beam of light is incident on a stack of polarizing disks. Find the fraction by which the intensity of the transmitted beam is reduced under each of the following conditions. (Note that in each case the angle between the directions of the transmission axes of the first and last disk is 90°.) (a) There are three disks in the stack, and each disk has its transmission axis at an angle of 45° relative to the preceding disk. (b) There are four disks in the stack, and each disk has its transmission axis at an angle of 30° relative to the preceding one. (c) There are seven disks in the stack, and each disk has its axis at an angle of 15° relative to the preceding one. (d) Comment on the different values of I found in (a), (b), and (c).

70. Figure 38.33a is a three-dimensional sketch of a birefringent crystal. The dotted lines illustrate how one could cut a thin parallel-faced slab of material from the larger specimen with the optic axis of the crystal parallel to the faces of the plate. A section cut from the crystal in this manner is known as a *retardation plate.* When a beam of light is incident on the plate perpendicular to the direction of the optic axis, as shown in Figure 38.33b, the O ray and the E ray travel along a single straight line, but with different speeds. (a) Let the thickness of the plate be d and show that the phase difference between the O ray and the E ray in the transmitted beam is given by

$$\theta = \frac{2\pi d}{\lambda_0}|n_O - n_E|$$

where λ_0 is the wavelength in air. (Recall that the *optical path length* in a material is the product of the geometric path and the index of refraction.) (b) If in a particular case the incident light has a wavelength of 550 nm, find the minimum value of d for a quartz plate for which $\theta = \pi/2$. Such a plate is called a *quarter-wave plate.* (Use values of n_O and n_E from Table 38.1.)

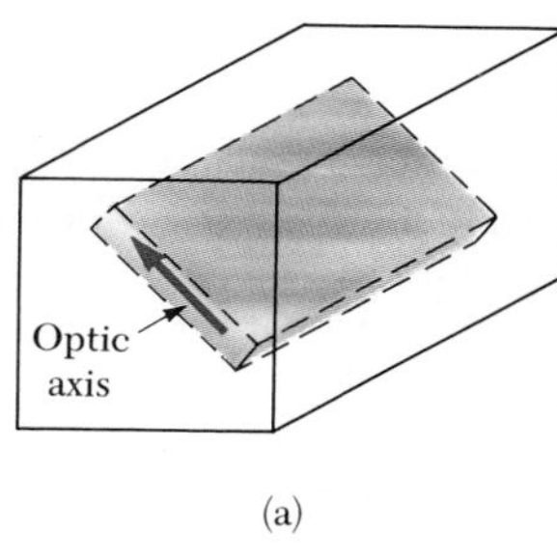

(a)

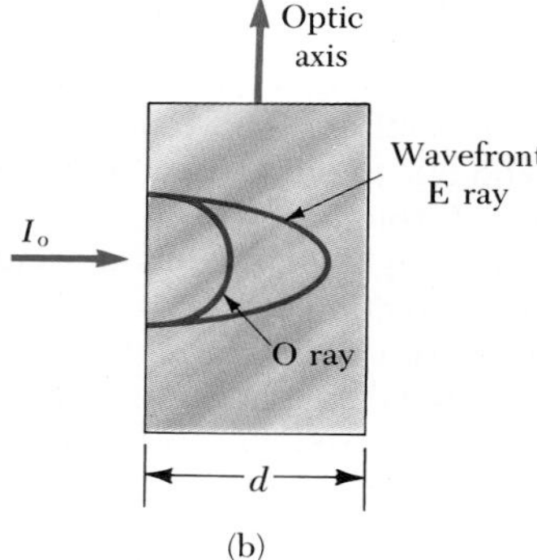

(b)

Figure 38.33 (Problem 70).

71. Derive Equation 38.10 for the resolving power of a grating, $R = Nm$, where N is the number of lines illuminated and m is the order in the diffraction pattern. Remember that Rayleigh's criterion (Section 38.3) states that two wavelengths will be resolved when the principal maximum for one falls on the first minimum for the other.

72. In Figure 38.32, suppose that the left and right polarizing disks have their transmission axes perpendicular to each other. Also, let the center disk be rotated on the common axis with an angular velocity ω. Show that if *unpolarized* light is incident on the left disk with an intensity I_0, the intensity of the beam emerging from the right disk will be

$$I = \frac{1}{16} I_0(1 - \cos 4\omega t)$$

This means that the intensity of the emerging beam will be modulated at a rate that is four times the rate of rotation of the center disk. (*Hint:* Use the trigonometric identities $\cos^2\theta = (1 + \cos 2\theta)/2$ and $\sin^2\theta = (1 - \cos 2\theta)/2$, and recall that $\theta = \omega t$.)

73. Suppose that the single slit opening in Figure 38.7 is 6 cm wide and is placed in front of a microwave source operating at a frequency of 7.5 GHz. (a) Calculate the angle subtended by the first minimum in the diffraction pattern. (b) What is the relative intensity I/I_0 at $\theta = 15°$? (c) Consider the case when there are *two* such sources, separated laterally by 20 cm, behind the single slit. What is the maximum distance between the plane of the sources and the slit if the diffraction patterns are to be resolved? (In this case, the approximation that $\sin\theta \approx \tan\theta$ may not be valid because of the relatively small value of the ratio a/λ.)

CALCULATOR/COMPUTER PROBLEMS

74. Figure 38.11 shows the relative intensity of a single-slit Fraunhofer diffraction pattern as a function of the parameter $\beta/2 = (\pi a \sin\theta)/\lambda$. Make a plot of the relative intensity I/I_0 as a function of θ, the angle subtended by a point on the screen at the slit, when (a) $\lambda = a$, (b) $\lambda = 0.5a$, (c) $\lambda = 0.1a$, (d) $\lambda = 0.05a$. Let θ range over the interval from 0 to 20° and choose a number of steps appropriate for each case.

75. From Equation 38.4 show that, in the Fraunhofer diffraction pattern of a single slit, the angular width of the central maximum at the point where $I = 0.5I_0$ is $\Delta\theta = 0.886\lambda/a$. (*Hint:* In Equation 38.4, let $\beta/2 = \phi$ and solve the resulting transcendental equation graphically; see Problem 9.)

76. Another method to solve the equation $\phi = \sqrt{2}\sin\phi$ in Problem 9 is to use a scientific calculator, guess a first value of ϕ, see if it fits, and continue to update your estimate until the equation balances. How many steps (iterations) did this take? [Another approach is to apply the Newton-Raphson method to find the roots of $f(\phi) = \phi - \sqrt{2}\sin\phi$. In this approach, $\phi_2 = \phi_1 - f(\phi_1)/f'(\phi_1)$, where f' is the first derivative of f.]

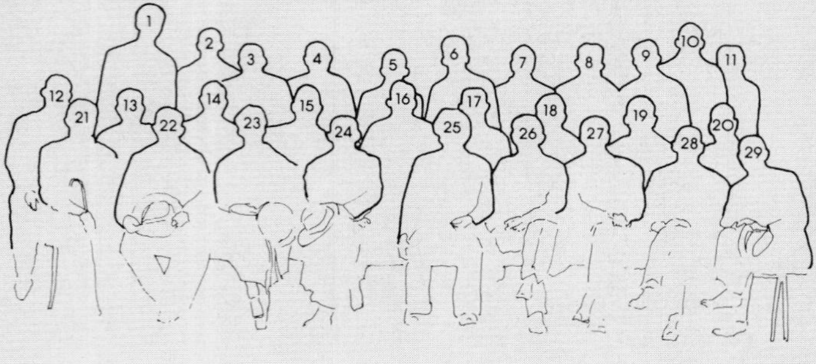

1. A. Piccard
2. E. Henriot
3. P. Ehrenfest
4. E. Herzen
5. Th. de Donder
6. E. Schroedinger
7. E. Verschaffelt
8. W. Pauli
9. W. Heisenberg
10. R.H. Fowler
11. L. Brillouin
12. P. Debye
13. M. Knudsen
14. W.L. Bragg
15. H.A. Kramers
16. P.A.M. Dirac
17. A.H. Compton
18. L.V. de Broglie
19. M. Born
20. N. Bohr
21. I. Langmuir
22. M. Planck
23. M. Curie
24. H.A. Lorentz
25. A. Einstein
26. P. Langevin
27. C.E. Guye
28. C.T.R. Wilson
29. O.W. Richardson

PART VI
Modern Physics

"The new order of science makes possible, for the first time, a cooperative effort in which everyone is the gainer and no one the loser. The advent of modern science is the most important social event in history."

KARL TAYLOR-COMPTON

At the end of the 19th century, scientists believed that they had learned most of what there was to know about physics. Newton's laws of motion and his universal theory of gravitation, Maxwell's theoretical work in unifying electricity and magnetism, and the laws of thermodynamics and kinetic theory were highly successful in explaining a wide variety of phenomena.

However, at the turn of the 20th century, a major revolution shook the world of physics. In 1900 Planck provided the basic ideas that led to the formulation of the quantum theory, and in 1905 Einstein formulated his brilliant special theory of relativity. The excitement of the times is captured in Einstein's own words: "It was a marvelous time to be alive." Both ideas were to have a profound effect on our understanding of nature. Within a few decades, these theories inspired new developments and theories in the fields of atomic physics, nuclear physics, and condensed matter physics.

The discussion of modern physics in this last part of the text will begin with a treatment of the special theory of relativity in Chapter 39. Although the concepts underlying this theory often violate our common sense, the theory provides us with a new and deeper view of physical laws. Next we shall discuss various developments in quantum theory (Chapter 40), which provides us with a successful model for understanding electrons, atoms, and molecules. The extended version of this text includes seven additional chapters on selected topics in modern physics. In these chapters, we cover the basic concepts of quantum mechanics, its application to atomic and molecular physics, and introductions to solid state physics, superconductivity, nuclear physics, particle physics, and cosmology.

You should keep in mind that, although modern physics has been developed during this century and has led to a multitude of important technological achievements, the story is still incomplete. Discoveries will continue to evolve during our lifetime, many of which will deepen or refine our understanding of nature and the world around us. It is still a "marvelous time to be alive."

The "architects" of modern physics. This unique photograph shows many eminent scientists who participated in the fifth international congress of physics held in 1927 by the Solvay Institute in Brussels. At this and similar conferences, held regularly from 1911 on, scientists were able to discuss and share the many dramatic developments in atomic and nuclear physics. This elite company of scientists includes fifteen Nobel Prize winners in physics and three in chemistry. (Photograph courtesy of AIP Niels Bohr Library)

39
Relativity

Albert Einstein, German physicist. Einstein is best known for his special theory of relativity. (Color added by John Austin.) (Photograph courtesy of Roger Richman Agency)

Light waves and other forms of electromagnetic radiation travel through free space with a speed of $c = 3.00 \times 10^8$ m/s. As we shall see in this chapter, the speed of light is an upper limit for the speeds of particles and mechanical waves.

39.1 INTRODUCTION

Most of our everyday experiences and observations deal with objects that move at speeds much less than the speed of light. Newtonian mechanics, and the early ideas on space and time, were formulated to describe the motion of such objects. This formalism is very successful for describing a wide range of phenomena. Although Newtonian mechanics works very well at low speeds, it fails when applied to particles whose speeds approach that of light. Experimentally, one can test the predictions of the theory at large speeds by accelerating an electron through a large electric potential difference. For example, it is possible to accelerate an electron to a speed of $0.99c$ by using a potential difference of several million volts. According to Newtonian mechanics, if the potential difference (as well as the corresponding energy) is increased by a factor of 4, then the speed of the electron should be doubled to $1.98c$. However, experiments show that the speed of the electron always remains *less* than the speed of light, regardless of the size of the accelerating voltage. Since Newtonian mechanics places no upper limit on the speed that a particle can attain, it is contrary to modern experimental results and is clearly a limited theory.

In 1905, at the age of only 26, Einstein published his *special theory of relativity:*

> The relativity theory arose from necessity, from serious and deep contradictions in the old theory from which there seemed no escape. The strength of the new theory lies in the consistency and simplicity with which it solves all these difficulties, using only a few very convincing assumptions. . . .[1]

Although Einstein made many important contributions to science, the theory of relativity alone represents one of the greatest intellectual achievements of the 20th century. With this theory, one can correctly predict experimental observations over the range of speeds from $v = 0$ to velocities approaching the speed of light. Newtonian mechanics, which was accepted for over 200 years, is in fact a specialized case of Einstein's generalized theory. This chapter gives an introduction to the special theory of relativity, with emphasis on some of the consequences of the theory. A discussion of general relativity and some of its consequences and experimental tests is presented in the essay written by Clifford Will.

As we shall see, the special theory of relativity is based on two basic postulates:

The postulates of the special theory of relativity

1. The laws of physics are the same in all inertial reference systems. That is, basic laws of physics have the same mathematical form for all observers moving at constant velocity with respect to each other.
2. The speed of light in vacuum is always measured to be 3×10^8 m/s, and the measured value is independent of the motion of the observer or of the motion of the source of light. That is, the speed of light is the same for *all* inertial observers.

Special relativity covers phenomena such as the slowing down of clocks and the contraction of lengths in moving reference frames as measured by a stationary observer. We shall also discuss the relativistic forms of momentum and energy, and some consequences of the famous mass-energy equivalence formula, $E = mc^2$. You may want to consult a number of excellent books on relativity for more details on the subject.[2]

Although it is well known that relativity plays an essential role in contemporary theoretical physics, it also has many practical applications, including the design of accelerators and other devices that utilize high-speed particles. We shall have occasion to use relativity in some subsequent chapters of this text, but often only the outcome of relativistic effects will be presented.

[1] A. Einstein and L. Infeld, *The Evolution of Physics,* New York, Simon and Schuster, 1961.

[2] The following books are recommended for more details on the theory of relativity at the introductory level: E. F. Taylor and J. A. Wheeler, *Spacetime Physics,* San Francisco, W. H. Freeman, 1963; R. Resnick, *Introduction to Special Relativity,* New York, Wiley, 1968; A. P. French, *Special Relativity,* New York, Norton, 1968; other suggested readings are selected reprints on "Special Relativity Theory" published by the American Institute of Physics.

39.2 THE PRINCIPLE OF RELATIVITY

In order to describe a physical event, it is necessary to establish a frame of reference, such as one that is fixed in the laboratory. You should recall from your studies in mechanics that Newton's laws are valid in *all* inertial frames of reference. Since an inertial frame of reference is defined as one in which Newton's first law is valid, one can say that *an inertial system is a system in which a free body exhibits no acceleration.* Furthermore, any system moving with constant velocity with respect to an inertial system is also an inertial system. There is no preferred frame. This means that the results of an experiment performed in a vehicle moving with uniform velocity will be identical to the results of the same experiment performed in the stationary laboratory.

Inertial frame of reference

According to the **principle of Newtonian relativity,** the laws of mechanics are the same in all inertial frames of reference.

For example, if you perform an experiment while at rest in a laboratory, and an observer in a passing car moving with constant velocity also observes your experiment, the laboratory coordinate system and the coordinate system of the moving car are both inertial reference frames. Therefore, if you find the laws of mechanics to be true in the lab, the person in the moving car must agree with your observation. This also implies that no mechanical experiment can detect any difference between the two inertial frames. The only thing that can be detected is the relative motion of one frame with respect to the other. That is, the notion of *absolute* motion through space is meaningless.

Suppose that some physical phenomenon, which we call an *event,* occurs in an inertial system. The event's location and time of occurrence can be specified by the coordinates (x, y, z, t). We would like to be able to transform the space and time coordinates of the event from one inertial system to another moving with uniform relative velocity. This is accomplished by using a so-called *Galilean transformation.*

Consider two inertial systems S and S′, as in Figure 39.1. The system S′ moves with a constant velocity v along the xx' axes, where v is measured relative to the system S. We assume that an event occurs at the point P. The event might be the "explosion" of a flashbulb or a heartbeat. An observer in system S would describe the event with space-time coordinates (x, y, z, t), while an observer in system S′ would use (x', y', z', t') to describe the same event. As we can see from Figure 39.1, these coordinates are related by the equations

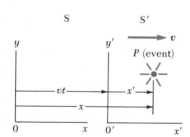

Figure 39.1 An event occurs at a point P. The event is observed by two observers in inertial frames S and S′, where S′ moves with a velocity v relative to S.

Galilean transformation of coordinates

$$\begin{aligned} x' &= x - vt \\ y' &= y \\ z' &= z \\ t' &= t \end{aligned} \qquad (39.1)$$

These equations constitute what is known as a **Galilean transformation of coordinates.** Note that the fourth coordinate, time, is *assumed* to be the same in both inertial systems. That is, within the framework of classical mechanics, clocks are universal, so that the time of an event for an observer in S is the same as the time for the same event in S′. Consequently, the time interval between two successive events should be the same for both observers. Although this assumption may seem obvious, it turns out to be *incorrect* when treating situations in which v is comparable to the speed of light. In fact, this point represents one of the most profound differences between Newtonian concepts and the ideas contained in Einstein's theory of relativity.

Now suppose two events are separated by a distance Δx and a time interval Δt as measured by an observer in S. It follows from Equation 39.1 that the corresponding displacement $\Delta x'$ measured by an observer in S′ is given by $\Delta x' = \Delta x - v\,\Delta t$, where Δx is the displacement measured by an observer in S. Since $\Delta t = \Delta t'$, we find that

$$\frac{\Delta x'}{\Delta t} = \frac{\Delta x}{\Delta t} - v$$

If the two events correspond to the passage of a moving object (or wave pulse) past two "milestones" in frame S, then $\Delta \boldsymbol{x}/\Delta t$ is the time-averaged velocity in S and $\Delta \boldsymbol{x}'/\Delta t$ is the average velocity in S′. In the time limit $\Delta t \rightarrow 0$, this becomes

$$\boldsymbol{u}_x' = \boldsymbol{u}_x - \boldsymbol{v} \qquad (39.2)$$

Galilean addition law for velocities

where $\boldsymbol{u}_x$ and $\boldsymbol{u}_x'$ are the instantaneous velocities of the object relative to S and S′, respectively.

This result, which is called the **Galilean addition law for velocities** (or Galilean velocity transformation), is used in everyday observations and is consistent with our intuitive notion of time and space. However, as we shall soon see, these results lead to serious contradictions when applied to electromagnetic waves.

The Speed of Light

It is quite natural to ask whether the concept of Newtonian relativity in mechanics also applies to experiments in electricity, magnetism, optics, and other areas. For example, if one assumes that the laws of electricity and magnetism are the same in all inertial frames, a paradox concerning the velocity of light immediately arises. This can be understood by recalling that according to Maxwell's equations in electromagnetic theory, the velocity of light always has the fixed value of $(\mu_0\epsilon_0)^{-1/2} \approx 3.00 \times 10^8$ m/s. But this is in direct contradiction to what one would expect based on the Galilean addition law for velocities. According to this law, the velocity of light should *not* be the same in all inertial frames. For example, suppose a light pulse is sent out by an observer in a boxcar moving with a velocity $\boldsymbol{v}$ (Fig. 39.2). The light pulse has a velocity $\boldsymbol{c}$

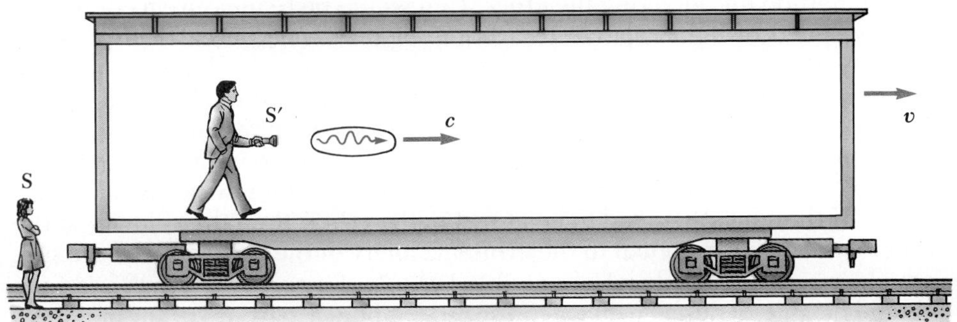

Figure 39.2 A pulse of light is sent out by a person in a moving boxcar. According to Newtonian relativity, the speed of the pulse should be $c + v$ relative to a stationary observer.

relative to observer S′ in the boxcar. According to the ideas of Newtonian relativity, the velocity of the pulse relative to the stationary observer S outside the boxcar should be $\mathbf{c} + \mathbf{v}$. This is in obvious contradiction to Einstein's theory, which postulates that the velocity of the light pulse is the same for all observers.

In order to resolve this paradox, one must conclude that either (1) the Galilean addition law for velocities is incorrect, or else (2) the laws of electricity and magnetism are not the same in all inertial frames. If the Galilean addition law for velocities were incorrect, we would be forced to abandon the seemingly "obvious" notions of absolute time and absolute length that form the basis for the Galilean transformations.

If we assume that the second alternative is true, then a preferred reference frame must exist in which the speed of light has the value c, and that it is greater or less than this value in any other reference frame in accordance with the Galilean addition law for velocities. It is useful to draw an analogy with sound waves, which propagate through a medium such as air. The speed of sound in air is about 330 m/s when measured in a reference frame in which the air is stationary. However, the measured speed of sound is greater or less than this value when measured from a reference frame that is moving with respect to the source of sound.

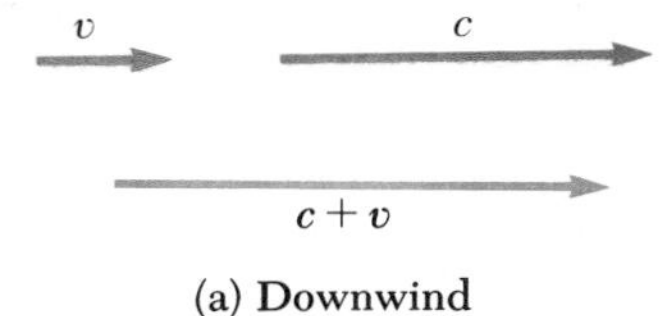

(a) Downwind

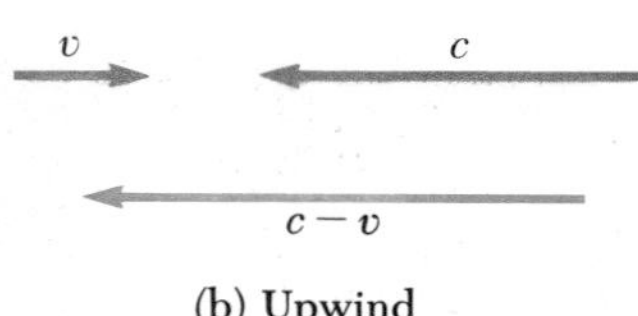

(b) Upwind

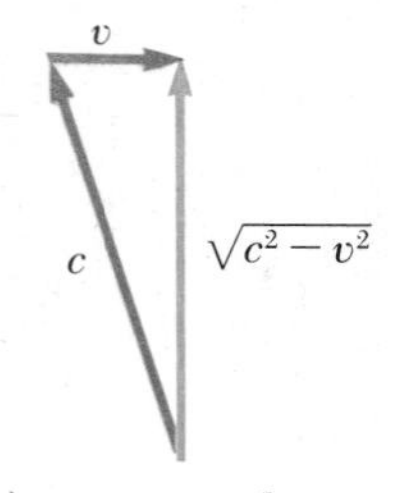

(c) Across wind

Figure 39.3 If the velocity of the ether wind relative to the earth is $\mathbf{v}$, and $\mathbf{c}$ is the velocity of light relative to the ether, the speed of light relative to the earth is (a) $c + v$ in the downwind direction, (b) $c - v$ in the upwind direction, and (c) $(c^2 - v^2)^{1/2}$ in the direction perpendicular to the wind.

In the case of light signals (electromagnetic waves), recall that Maxwell's theory predicted that such waves must propagate through free space with a speed equal to the speed of light. However, Maxwell's theory does not require the presence of a medium for the wave propagation. This is in contrast to mechanical waves, such as water or sound waves, which do require a medium to support the disturbances. In the 19th century, physicists thought that electromagnetic waves also required a medium in order to propagate. They proposed that such a medium existed, and they gave it the name **luminiferous ether.** The ether was assumed to be present everywhere, even in free space, and light waves were viewed as ether oscillations. Furthermore, the ether had to have the unusual properties of being a massless but rigid medium, and would have no effect on the motion of planets or other objects. Indeed, this is a strange concept. Additionally, it was found that the troublesome laws of electricity and magnetism would take on their simplest form in a frame of reference at *rest* with respect to this ether. This frame was called the *absolute frame.* The laws of electricity and magnetism would be valid in this absolute frame, but they would have to be modified in any reference frame moving with respect to the ether frame.

As a result of the importance attached to this absolute frame, it became of considerable interest in physics to prove by experiment that it existed. A direct method for detecting the ether wind would be to measure its influence on the speed of light relative to a frame of reference on earth. If $\mathbf{v}$ is the velocity of the ether relative to the earth, then the speed of light should have its maximum value, $c + v$, when propagating downwind as shown in Figure 39.3a. Likewise, the speed of light should have its minimum value, $c - v$, when propagating upwind as in Figure 39.3b, and some intermediate value, $(c^2 - v^2)^{1/2}$, in the direction perpendicular to the ether wind as in Figure 39.3c. If the sun is assumed to be at rest in the ether, then the velocity of the ether wind would be equal to the orbital velocity of the earth around the sun, which has a magnitude of about 3×10^4 m/s. Since $c = 3 \times 10^8$ m/s, one should be able to detect a change in speed of about 1 part in 10^4 for measurements in the upwind or downwind directions. However, as we shall see in the next section, all attempts to detect such changes and establish the existence of the ether (and hence the absolute frame) proved futile!

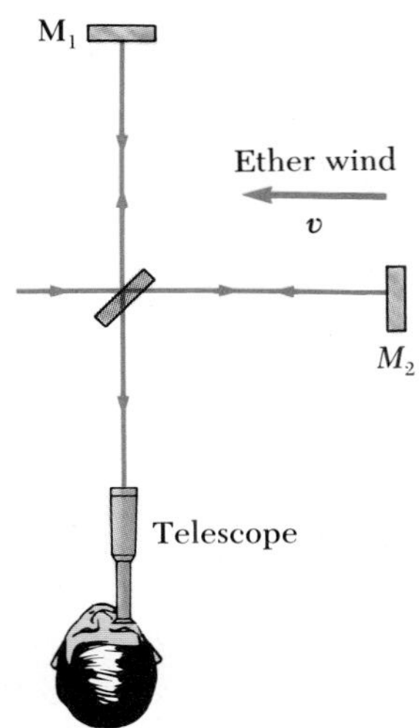

Figure 39.4 According to the ether wind theory, the speed of light should be $c - v$ as the beam approaches mirror M_2 and $c + v$ after reflection.

39.3 THE MICHELSON-MORLEY EXPERIMENT

The most famous experiment that was designed to detect small changes in the speed of light was performed in 1887 by A.A. Michelson (1852–1931) and E.W. Morley (1838–1923).[3] We should state at the outset that the outcome of the experiment was *negative*, thus contradicting the ether hypothesis. The experiment was designed to determine the velocity of the earth with respect to the hypothetical ether. The experimental tool used was the Michelson interferometer shown in Figure 39.4. Suppose one of the arms of the interferometer is aligned along the direction of the motion of the earth through space. The motion of the earth through the ether would be equivalent to the ether flowing past the earth in the opposite direction. This ether wind blowing in the opposite direction should cause the speed of light as measured in the earth's frame of reference to be $c - v$ as it approaches the mirror M_2 in Figure 39.4 and $c + v$ after reflection. The speed v is the speed of the earth through space, and hence the speed of the ether wind, while c is the speed of light in the ether frame. The two beams of light reflected from M_1 and M_2 recombine, and an interference pattern consisting of alternating dark and bright bands or fringes is formed. During the experiment, the interference pattern was observed while the interferometer was rotated through an angle of 90 degrees. This rotation would change the speed of the ether wind along the direction of the arms of the interferometer. The effect of this rotation should have been to cause the fringe pattern to shift slightly but measurably. Measurements failed to show any change in the interference pattern! The Michelson-Morley experiment was repeated by other researchers under various conditions and at different locations, but the results were always the same: *no fringe shift of the magnitude required was ever observed.*

The negative results of the Michelson-Morley experiment meant that it was impossible to measure the absolute velocity of the earth (the orbital velocity) with respect to the ether frame. However, as we shall see in the next section, Einstein developed a postulate for his theory of relativity that places quite a different interpretation on these null results. In later years, when more was known about the nature of light, the idea of an ether that permeates all of space was relegated to the ash heap of worn out concepts. Light is now understood to be *an electromagnetic wave, which requires no medium for its propagation.* As a result, the idea of having an ether in which these waves could travel became unnecessary.

Albert A. Michelson (1852–1931). (AIP Niels Bohr Library)

Details of the Michelson-Morley Experiment

In order to understand the outcome of the Michelson-Morley experiment, let us assume that the interferometer shown in Figure 39.4 has two arms of equal length L. First consider the beam traveling parallel to the direction of the ether wind, which is taken to be horizontal in Figure 39.4. As the beam moves to the right, its speed is reduced by the wind and its speed with respect to the

[3] A. A. Michelson and E. W. Morley, *Am. J. Sci* **134**:333 (1887).

earth is $c - v$. On its return journey, as the light beam moves to the left downwind, its speed with respect to the earth is $c + v$. Thus, the time of travel to the right is $L/(c - v)$ and the time of travel to the left is $L/(c + v)$. The total time of travel for the round trip along the horizontal path is

$$t_1 = \frac{L}{c+v} + \frac{L}{c-v} = \frac{2Lc}{c^2 - v^2} = \frac{2L}{c}\left(1 - \frac{v^2}{c^2}\right)^{-1}$$

Now consider the light beam traveling perpendicular to the wind, which is the vertical direction in Figure 39.4. Since the speed of the beam relative to the earth is $(c^2 - v^2)^{1/2}$ in this case (see Fig. 39.3), then the time of travel for each half of this trip is $L/(c^2 - v^2)^{1/2}$, and the total time of travel for the round trip is

$$t_2 = \frac{2L}{(c^2 - v^2)^{1/2}} = \frac{2L}{c}\left(1 - \frac{v^2}{c^2}\right)^{-1/2}$$

Thus, the time difference between the light beam traveling horizontally and the beam traveling vertically is

$$\Delta t = t_1 - t_2 = \frac{2L}{c}\left[\left(1 - \frac{v^2}{c^2}\right)^{-1} - \left(1 - \frac{v^2}{c^2}\right)^{-1/2}\right]$$

Since $v^2/c^2 \ll 1$, this expression can be simplified by using the following binomial expansions after dropping all terms higher than second order:

$$(1 - x)^n \approx 1 - nx \qquad (\text{for } x \ll 1)$$

In our case, $x = v^2/c^2$, and we find

$$t_1 - t_2 = \Delta t = \frac{Lv^2}{c^3} \tag{39.3}$$

Because the speed of the earth in its orbital path is approximately 3×10^4 m/s, the speed of the wind should be at least this great. The two light beams start out in phase and return to form an interference pattern. Let us assume that the interferometer is adjusted for parallel fringes and that a telescope is focused on one of these fringes. The time difference between the two light beams gives rise to a phase difference between the beams, producing an interference pattern when they combine at the position of the telescope. A difference in the pattern should be detected by rotating the interferometer through 90° in a horizontal plane, such that the two beams exchange roles. This results in a net time difference of twice that given by Equation 39.3. Thus, the path difference which corresponds to this time difference is

$$\Delta d = c(2\ \Delta t) = \frac{2Lv^2}{c^2}$$

The corresponding phase difference $\Delta\phi$ is equal to $2\pi/\lambda$ times this path difference, where λ is the wavelength of light. That is,

$$\Delta\phi = \frac{4\pi L v^2}{\lambda c^2} \tag{39.4}$$

In the first experiments by Michelson and Morley, each light beam was reflected by mirrors many times to give an increased effective path length L of about 11 meters. Using this value, and taking v to be equal to 3×10^4 m/s, gives a path difference of

$$\Delta d = \frac{2(11 \text{ m})(3 \times 10^4 \text{ m/s})^2}{(3 \times 10^8 \text{ m/s})^2} = 2.2 \times 10^{-7} \text{ m}$$

This extra distance of travel should produce a noticeable shift in the fringe pattern. Specifically, calculations show that if one views the pattern while the interferometer is rotated through 90°, a shift of about 0.4 fringe should be observed. The instrument used by Michelson and Morley had the capability of detecting a shift in the fringe pattern as small as 0.01 fringe. However, *they detected no shift in the fringe pattern.* Since then, the experiment has been repeated many times by various scientists under various conditions, and no fringe shift has ever been detected. Thus, it was concluded that one cannot detect the motion of the earth with respect to the ether.

Many efforts were made to explain the null results of the Michelson-Morley experiment. For example, perhaps the earth drags the ether with it in its motion through space. To test this assumption, interferometer measurements were made at various altitudes, but again no fringe shift was detected. In the 1890s, G. F. Fitzgerald and H. A. Lorentz independently tried to explain the null results by making the following ad hoc assumption. They proposed that the length of an object moving along the direction of the ether wind would contract by a factor of $\sqrt{1 - v^2/c^2}$. The net result of this contraction would be a change in length of one of the arms of the interferometer such that no path difference would occur as it was rotated.

No experiment in the history of physics has received such valiant efforts to try to explain the absence of an expected result as did the Michelson-Morley experiment. The stage was set for the brilliant Albert Einstein, who solved the problem in 1905 with his special theory of relativity.

39.4 EINSTEIN'S PRINCIPLE OF RELATIVITY

In the previous section we noted the serious contradiction between the invariance of the speed of light and the Galilean addition law for velocities. In 1905, Albert Einstein proposed a theory that would resolve this contradiction but at the same time would completely alter our notion of space and time.[4] Einstein based his special theory of relativity on the following general hypothesis, which is called the **Principle of Relativity:**

All the laws of physics are the same in all inertial reference frames.

The postulates of special relativity

In Einstein's own words, ". . . the same laws of electrodynamics and optics will be valid for all frames of reference for which the equations of mechanics hold good." This simple statement is a generalization of Newton's principle of relativity, and is the foundation of the special theory of relativity.

An immediate consequence of the Principle of Relativity is as follows:

The speed of light in a vacuum has the same value, $c = 3.00 \times 10^8$ m/s, in all inertial reference frames.

[4] A. Einstein, "On the Electrodynamics of Moving Bodies," *Ann. Physik* **17:** 891 (1905). For an English translation of this article and other publications by Einstein, see the book by H. Lorentz, A. Einstein, H. Minkowski, and H. Weyl, *The Principle of Relativity,* Dover, 1958.

Biographical Sketch

Albert Einstein
(1879–1955)

(Photograph Courtesy of AIP Niels Bohr Library)

Albert Einstein, one of the greatest physicists of all times, was born in Ulm, Germany. He showed little intellectual promise as a youngster, and left the highly disciplined German school system after one teacher stated "You will never amount to anything, Einstein." Following a vacation in Italy, he completed his education at the Swiss Federal Polytechnic School in 1901. Although Einstein attended very few lectures, he was able to pass the courses with the help of excellent lecture notes taken by a friend. Unable to find an academic position, Einstein accepted a position as a junior official in the Swiss Patent Office in Berne. In this setting, and during his "spare time," he continued his independent studies in theoretical physics. In 1905, at the age of 26, he published four scientific papers that revolutionized physics. (In that same year, he earned his Ph.D.) One of these papers, for which he was awarded the 1921 Nobel prize in physics, dealt with the photoelectric effect. Another was concerned with Brownian motion, the irregular motion of small particles suspended in a liquid. The remaining two papers were concerned with what is now considered his most important contribution of all, the special theory of relativity. In 1915, Einstein published his work on the general theory of relativity, which relates gravity to the structure of space and time. The most dramatic prediction of this theory is the degree to which light would be deflected by a gravitational field. Measurements made by astronomers on bright stars in the vicinity of the eclipsed sun in 1919 confirmed Einstein's prediction, and Einstein suddenly became a world celebrity. This and other predictions of the general theory of relativity are discussed in the delightful essay by Clifford Will that follows this chapter.

In 1913, following academic appointments in Switzerland and Czechoslovakia, Einstein accepted a special position created for him at the Kaiser Wilhelm Institute in Berlin. This made it possible for him to be able to devote all of his time to research, free of financial troubles and routine duties. Einstein left Germany in 1933, which was then under Hitler's power, thereby escaping the fate of millions of other European Jews. In the same year he accepted a special position at the Institute for Advanced Study in Princeton where he remained for the rest of his life. He became an American citizen in 1940. Although he was a pacifist, Einstein was persuaded by Leo Szilard to write a letter to President Franklin D. Roosevelt urging him to initiate a program to develop a nuclear bomb. The result was the successful six-year Manhattan project and two nuclear explosions in Japan which ended World War II in 1945.

Einstein made many important contributions to the development of modern physics, including the concept of the light quantum and the idea of stimulated emission of radiation, which led to the invention of the laser 40 years later. Einstein was deeply disturbed by the development of quantum mechanics in the 1920s despite his own role as a scientific revolutionary. In particular, he could never accept the probabilistic view of events in nature which is a central feature of the highly successful quantum theory. He once said, "God does not play dice with nature." The last few decades of his life were devoted to an unsuccessful search for a unified theory that would combine gravitation and electromagnetism into one picture.

In other words, anyone who measures the speed of light will get the same value, c. This implies that the ether simply does not exist. The Principle of Relativity, together with the above statement, are often referred to as the **two postulates of special relativity.**

Although the Michelson-Morley experiment was performed before Einstein published his work on relativity, it is not clear that Einstein was aware of the details of the experiment. Nonetheless, the null result of the experiment can be readily understood within the framework of Einstein's theory. According to his Principle of Relativity, the premises of the Michelson-Morley experiment were incorrect. In the process of trying to explain the expected results, we stated that when light traveled against the ether wind its speed was $c - v$, in accordance with the Galilean addition law for velocities. However, if the state of motion of the observer or of the source has no influence on the value found for the speed of light, one will always measure the value to be c. Likewise, the light makes the return trip after reflection from the mirror at a speed of c, and not with the speed $c + v$. Thus, the motion of the earth should not influence the fringe pattern observed in the Michelson-Morley experiment and a null result should be expected.

The Michelson-Morley experiment has been repeated many times, always giving the same null results. Modern versions of the experiment have compared the frequencies of resonant laser cavities of identical length oriented at right angles to each other. More recently, Doppler shift experiments using gamma rays emitted by a radioactive sample of ^{57}Fe placed an upper limit of about 5 cm/s on the ether wind velocity. These results have shown quite conclusively that the motion of the earth has no effect on the speed of light!

If we accept Einstein's theory of relativity, we must conclude that relative motion is unimportant when measuring the speed of light. At the same time, we must alter our common-sense notion of space and time and be prepared for some rather bizarre consequences.

39.5 DESCRIBING EVENTS IN RELATIVITY

Before we discuss the consequences of special relativity, we must first understand how an observer located in an inertial reference frame describes an event. As we mentioned earlier, an event is an occurrence which is described by three space coordinates and one time coordinate. Different observers in different inertial frames will, in general, describe the same event with different spacetime coordinates. The reference frame which is used to describe an event consists of a coordinate grid and a set of clocks located at the grid intersections as shown in Figure 39.5 in a two-dimensional representation. It is necessary that the array of clocks in the grid be synchronized. This can be accomplished with the help of light signals in many ways. For example, suppose the observer located at the origin of coordinates with his master clock sends out a pulse of light at $t = 0$. The light pulse takes a time r/c to reach a second clock located a distance r from the origin. Hence, the second clock will

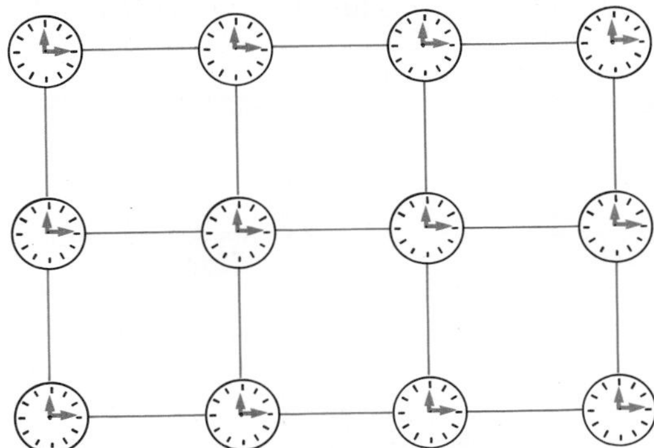

Figure 39.5 In relativity, we use a reference frame consisting of a coordinate grid and a set of synchronized clocks.

be synchronized with the clock at the origin if the second clock reads a time r/c at the instant the pulse reaches it. This procedure of synchronization assumes that the speed of light has the same value in all directions and in all inertial frames. Furthermore, the procedure concerns an event recorded by an observer in a specific inertial reference frame. An observer in some other inertial frame would assign different spacetime coordinates to events they observe using their own coordinate grid and array of clocks.

Almost everyone who has dabbled even superficially with science is aware of some of the startling predictions that arise because of Einstein's approach to relative motion. As we examine some of the consequences of relativity in the following three sections, we shall find that they conflict with our basic notions of space and time. We shall restrict our discussion to the concepts of length, time, and simultaneity, which are quite different in relativistic mechanics than in Newtonian mechanics. For example, we shall see that *the distance between two points and the time interval between two events depend on the frame of reference in which they are measured.* That is, *there is no such thing as absolute length or absolute time in relativity.* Furthermore, *events at different locations that occur simultaneously in one frame are not simultaneous in another frame.*

39.6 SIMULTANEITY

A basic premise of Newtonian mechanics is that a universal time scale exists which is the same for all observers. In fact, Newton wrote that "Absolute, true, and mathematical time, of itself, and from its own nature, flows equably without relation to anything external." Thus, Newton and his followers simply took simultaneity for granted. In his special theory of relativity, Einstein abandoned this assumption. According to Einstein, *time interval measurements depend on the reference frame in which the measurement is made.*

Einstein devised the following thought experiment to illustrate this point. A boxcar moves with uniform velocity, and two lightning bolts strike the ends of the boxcar, as in Figure 39.6a, leaving marks on the boxcar and ground. The marks left on the boxcar are labeled A' and B', while those on the ground are labeled A and B. An observer at O' moving with the boxcar is midway between A' and B', while a ground observer at O is midway between A and B. The events recorded by the observers are the light signals from the lightning bolts.

Let us assume that the two light signals reach the observer at O at the same time, as indicated in Figure 39.6b. This observer realizes that the light signals

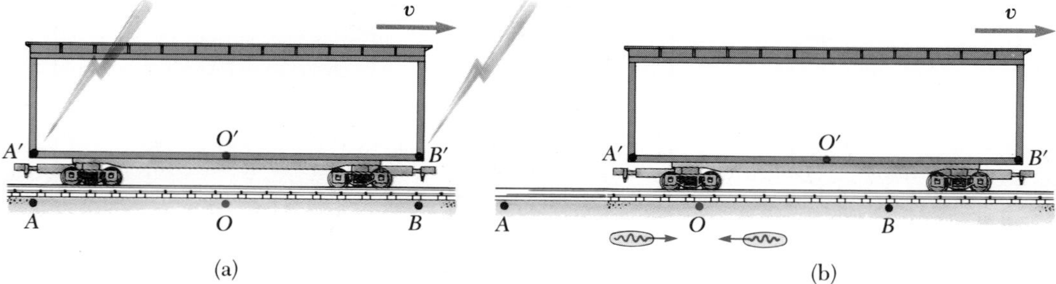

Figure 39.6 Two lightning bolts strike the end of a moving boxcar. (a) The events appear to be simultaneous to the stationary observer at O, who is midway between A and B. (b) The events do not appear to be simultaneous to the observer at O', who claims that the front of the train is struck *before* the rear.

have traveled at the same speed over distances of equal length. Thus, the observer at O rightly concludes that the events at A and B occurred simultaneously. Now consider the same events as viewed by the observer on the boxcar at O'. By the time the light has reached the observer at O, the observer at O' has moved as indicated in Figure 39.6b. Thus, the light signal from B' has already swept past O', while the light from A' has not yet reached O'. According to Einstein, the observer at O' must find that light travels at the same speed as that measured by the observer at O. Therefore, the observer at O' concludes that the light reaches the front of the boxcar before it reaches the back. This thought experiment clearly demonstrates that the two events, which appear to be simultaneous to the observer at O, do not appear to be simultaneous to the observer at O'. In other words,

> two events that are simultaneous in one reference frame are in general not simultaneous in a second frame moving with respect to the first. That is, simultaneity is not an absolute concept but one that depends upon the state of motion of the observer.

At this point, you might wonder which observer is right concerning the two events. The answer is that *both are correct,* because the Principle of Relativity states that *there is no preferred inertial frame of reference.* Although the two observers reach different conclusions, both are correct in their own reference frame because the concept of simultaneity is not absolute. This, in fact, is the central point of relativity—any uniformly moving frame of reference can be used to describe events and do physics. There is nothing wrong with the clocks and meter sticks used to perform measurements. It is simply that time intervals and length measurements depend on the state of motion of the observer. Observers in different inertial frames of reference will always measure different time intervals with their clocks and different distances with their meter sticks. However, they will both agree on the laws of physics in their respective frames, since they must be the same for all observers in uniform motion. It is the alteration of time and space that allows the laws of physics (including Maxwell's equations) to be the same for all observers in uniform motion.

39.7 THE RELATIVITY OF TIME

As we have seen, observers in different inertial frames will always measure different time intervals between a pair of events. This can be illustrated by considering a vehicle moving to the right with a speed v as in Figure 39.7a. A mirror is fixed to the ceiling of the vehicle and an observer Liz at O' at rest in this system holds a laser a distance d below the mirror. At some instant, the laser emits a pulse of light directed towards the mirror (event 1), and at some later time after reflecting from the mirror, the pulse arrives back at the laser (event 2). Liz carries a clock C' which she uses to measure the time interval $\Delta t'$ between these two events. Because the light pulse has a speed c, the time it takes the pulse to travel from Liz to the mirror and back to Liz can be found from the definition of speed:

$$\Delta t' = \frac{\text{distance traveled}}{\text{speed}} = \frac{2d}{c} \qquad (39.5)$$

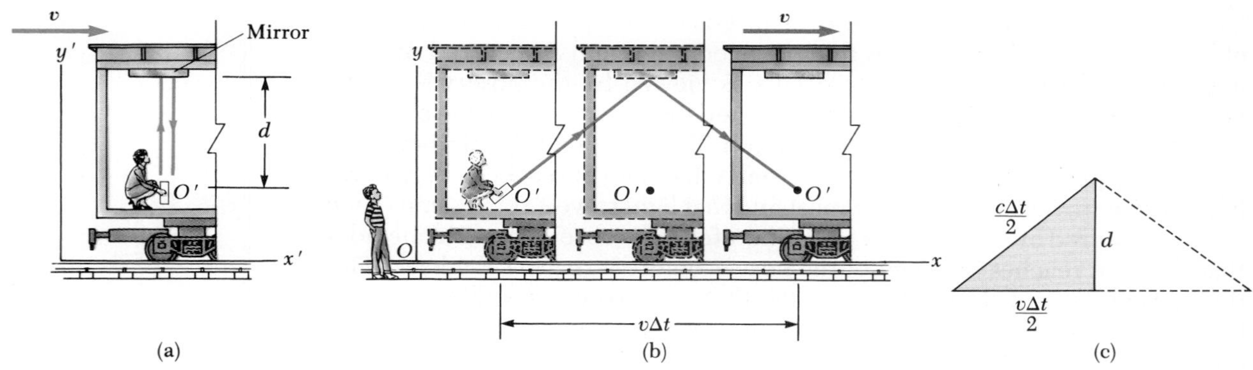

Figure 39.7 (a) A mirror is fixed to a moving vehicle, and a light pulse leaves O' at rest in the vehicle. (b) Relative to a stationary observer on earth, the mirror and O' move with a speed v. Note that the distance the pulse travels is greater than $2d$ as measured by the stationary observer. (c) The right triangle for calculating the relationship between Δt and $\Delta t'_{O'}$.

This time interval $\Delta t'$ measured by Liz who is at rest in the moving vehicle requires only a *single* clock C' located at the *same place* in this frame.

Now consider the same pair of events as viewed by an observer named Mark at O in a stationary frame as in Figure 39.7b. According to Mark, the mirror and laser are moving to the right with a speed v. By the time the light pulse reaches the mirror, the mirror moves a distance $v\,\Delta t/2$, where Δt is the time it takes the light to travel from O' to the mirror and back to O' as measured by Mark. In other words, Mark concludes that, because of the motion of the vehicle, if the light is to hit the mirror, it must leave the laser at an angle with respect to the vertical direction. Comparing Figures 39.7a and 39.7b, we see that the light must travel farther for Mark than it does for Liz.

According to the second postulate of special relativity, the speed of light must be c as measured by both observers. Since the light travels a farther distance for Mark, it follows that the time interval Δt measured by Mark in the stationary frame is *longer* than the time interval $\Delta t'$ measured by Liz in the moving frame. To obtain a relationship between these two time intervals, it is convenient to use the right triangle shown in Figure 39.7c. The Pythagorean theorem applied to this triangle gives

$$\left(\frac{c\,\Delta t}{2}\right)^2 = \left(\frac{v\,\Delta t}{2}\right)^2 + d^2$$

Solving for Δt gives

$$\Delta t = \frac{2d}{\sqrt{c^2 - v^2}} = \frac{2d}{c\sqrt{1 - \dfrac{v^2}{c^2}}} \tag{39.6}$$

Because $\Delta t' = 2d/c$, we can express Equation 39.6 as

Time dilation

$$\Delta t = \frac{\Delta t'}{\sqrt{1 - \dfrac{v^2}{c^2}}} = \gamma\,\Delta t' \tag{39.7}$$

where $\gamma = (1 - v^2/c^2)^{-1/2}$. The two events observed by Mark occur at *different* positions. In order to measure the time interval Δt, Mark must use *two synchronized clocks* located at *different places* in his reference frame. Thus, their situations are not symmetrical.

From Equation 39.7, we see that the time interval Δt measured by Mark in the stationary frame is *longer* than the time interval $\Delta t'$ measured by Liz in the moving frame (because γ is always greater than unity). That is, $\Delta t > \Delta t'$.

According to a stationary observer, a moving clock runs **slower** than an identical stationary clock. This effect is known as **time dilation.**

The time interval $\Delta t'$ in Equation 39.7 is called the **proper time.** In general, proper time is defined as *the time interval between two events as measured by an observer who sees the events occur at the same place.* In our case, the observer at O' measures the proper time. That is, *proper time is always the time measured with a single clock at rest in that frame.*

We have seen that moving clocks run slow by a factor of γ^{-1}. This is true for ordinary mechanical clocks as well as for the light clock just described. In fact, we can generalize these results by stating that *all physical processes, including chemical reactions and biological processes, slow down relative to a stationary clock when they occur in a moving frame.* For example, the heartbeat of an astronaut moving through space would have to keep time with a clock inside the spaceship. Both the astronaut's clock and his heartbeat are slowed down relative to a stationary clock. The astronaut would not have any sensation of life slowing down in the spaceship.

Time dilation is a very real phenomenon that has been verified by various experiments. For example, muons are unstable elementary particles, which have a charge equal to that of an electron and a mass 207 times that of the electron. Muons can be produced by the absorption of cosmic radiation high in the atmosphere. These unstable particles have a lifetime of only 2.2 μs when measured in a reference frame at rest with respect to them. If we take 2.2 μs as the average lifetime of a muon and assume that their speed is close to the speed of light, we would find that these particles could travel a distance of only about 600 m before they decayed into something else (Fig. 39.8a). Hence, they could not reach the earth from the upper atmosphere where they are produced. However, experiments show that a large number of muons *do* reach the earth. The phenomenon of time dilation explains this effect. Relative to an observer on earth, the muons have a lifetime equal to $\gamma\tau$, where $\tau = 2.2$ μs is the lifetime in a frame of reference traveling with the muons. For example, for $v = 0.99c$, $\gamma \approx 7.1$ and $\gamma\tau \approx 16$ μs. Hence, the average distance traveled as measured by an observer on earth is $\gamma v\tau \approx 4800$ m, as indicated in Figure 39.8b.

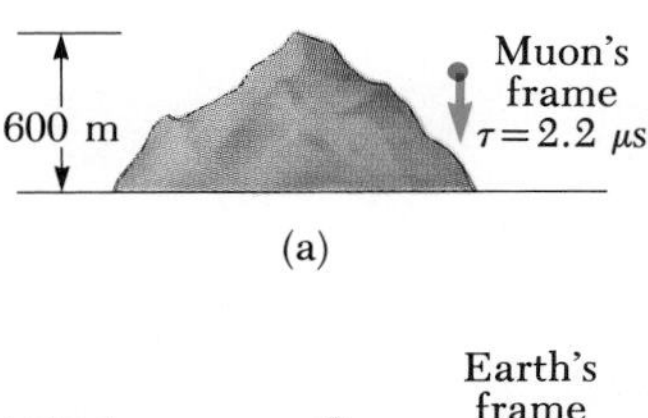

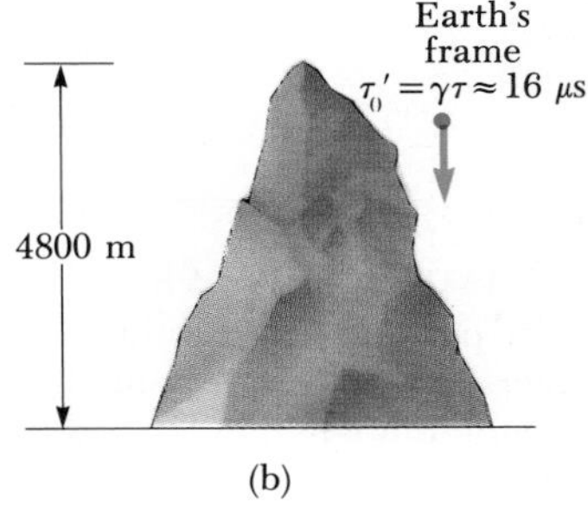

Figure 39.8 (a) Muons traveling with a speed of 0.99*c* travel only about 600 m as measured in the muons' reference frame, where their lifetime is about 2.2 μs. (b) The muons travel about 4800 m as measured by an observer on earth. Because of time dilation, the muons' lifetime is longer as measured by the earth observer.

In 1976, experiments with muons were conducted at the laboratory of the European Council for Nuclear Research (CERN) in Geneva. Muons were injected into a large storage ring, reaching speeds of about 0.9994*c*. Electrons produced by the decaying muons were detected by counters around the ring, enabling scientists to measure the decay rate, and hence the lifetime of the muons. The lifetime of the moving muons was measured to be about 30 times as long as that of the stationary muon (see Fig. 39.9), in agreement with the prediction of relativity to within two parts in a thousand.

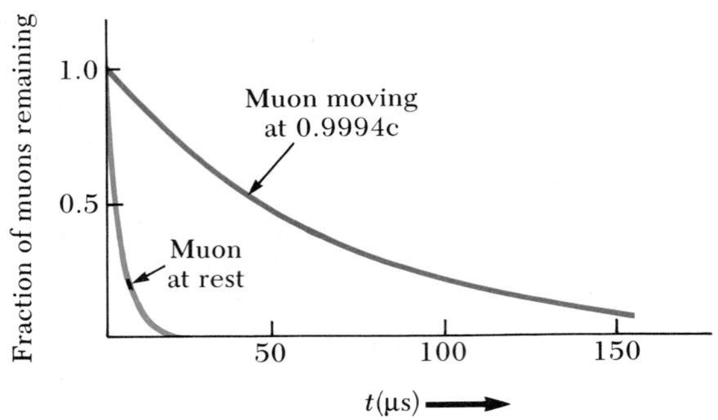

Figure 39.9 Decay curves for muons traveling at a speed of $0.9994c$ and for muons at rest.

The results of an experiment reported by Hafele and Keating provided direct evidence for the phenomenon of time dilation.[5] The experiment involved the use of very stable cesium beam atomic clocks. Time intervals measured with four such clocks in jet flight were compared with time intervals measured by reference atomic clocks located at the U.S. Naval Observatory. (Because of the earth's rotation about its axis, a ground-based clock is not in a true inertial frame.) In order to compare these results with the theory, many factors had to be considered, including periods of acceleration and deceleration relative to the earth, variations in direction of travel, and the weaker gravitational field experienced by the flying clocks compared to the earth-based clock. Their results were in good agreement with the predictions of the special theory of relativity. In their paper, Hafele and Keating report the following: "Relative to the atomic time scale of the U.S. Naval Observatory, the flying clocks lost 59 ± 10 ns during the eastward trip and gained 273 ± 7 ns during the westward trip. . . . These results provide an unambiguous empirical resolution of the famous clock paradox with macroscopic clocks."

EXAMPLE 39.1 What is the Period of the Pendulum?
The period of a pendulum is measured to be 3.0 s in the inertial frame of the pendulum. What is the period of the pendulum when measured by an observer moving at a speed of $0.95c$ with respect to the pendulum?

Solution In this case, the proper time is equal to 3 s. We can use Equation 39.7 to calculate the period measured by the moving observer. This gives

$$T = \gamma T' = \frac{1}{\sqrt{1 - \frac{(0.95c)^2}{c^2}}} T' = (3.2)(3.0 \text{ s}) = \boxed{9.6 \text{ s}}$$

That is, the observer moving at a speed of $0.95c$ observes the pendulum as oscillating more slowly.

The Twin Paradox

An interesting consequence of time dilation is the so-called twin paradox. Consider a controlled experiment involving 20-year-old twin brothers Speedo and Goslo. Speedo, the more venturesome twin, sets out on a journey toward a star located 30 lightyears from earth. His spaceship is able to acceler-

[5] J. C. Hafele and R. E. Keating, "Around the World Atomic Clocks: Relativistic Time Gains Observed," *Science*, July 14, 1972, p. 168.

ate to a speed close to the speed of light. After reaching the star, Speedo becomes very homesick and immediately returns to earth at the same high speed. Upon his return, he is shocked to find that many things have changed. Old cities have expanded and new cities have appeared. Lifestyles, people's appearances, and transportation systems have changed dramatically. Speedo's twin brother, Goslo, has aged to about 80 years old and is now wiser, feeble, and somewhat hard of hearing. Speedo, on the other hand, has aged only about ten years. This is because his bodily processes slowed down during his travels in space.

It is quite natural to raise the question, "Which twin actually traveled at a speed close to the speed of light, and therefore would be the one who did not age?" Herein lies the paradox: From Goslo's frame of reference, he was at rest while his brother Speedo traveled at a high velocity. On the other hand, according to the space traveler Speedo, it was he who was at rest while his brother zoomed away from him on earth and then returned. This leads to contradictions as to which twin actually aged.

In order to resolve this paradox, it should be pointed out that the trip is not as symmetrical as we may have led you to believe. Speedo, the space traveler, had to experience a series of accelerations and decelerations during his journey to the star and back home, and therefore is not always in uniform motion. This means that Speedo has been in a noninertial frame during a large part of his trip, so that predictions based on special relativity are not valid in his frame. On the other hand, the brother on earth has been in an inertial frame, and he can make reliable predictions based on the special theory. The situation is not symmetrical since Speedo experiences forces when his spaceship turns around, whereas Goslo is not subject to such forces. Therefore, the space traveler will indeed be younger upon returning to earth.

39.8 THE RELATIVITY OF LENGTH

We have seen that measured time intervals are not absolute, that is, the time interval between two events depends on the frame of reference in which it is measured. Likewise, the measured distance between two points depends on the frame of reference. The **proper length** of an object is defined as *the length of the object measured in the reference frame in which the object is at rest.* You should note that proper length is defined similarly to proper time, in that proper time is the time measured by a clock at rest relative to both events. However, proper length is *not* the distance between two points measured at the same time. The length of an object measured in a reference frame in which the object is moving is always less than the proper length. This effect is known as **length contraction.**

To understand length contraction quantitatively, let us consider a spaceship traveling with a speed v from one star to another, and two observers. An observer at rest on earth (and also assumed to be at rest with respect to the two stars) measures the distance between the stars to be L', where L' is the proper length. According to this observer, the time it takes the spaceship to complete the voyage is $\Delta t = L'v$. What does an observer in the moving spaceship measure for the distance between the stars? Because of time dilation, the space traveler measures a smaller time of travel: $\Delta t' = \Delta t/\gamma$. The space traveler claims to be at rest and sees the destination star as moving toward the spaceship with speed v. Since the space traveler reaches the star in the time $\Delta t'$, he or she concludes that the distance, L, between the stars is shorter than L'. This distance measured by the space traveler is given by

$$L = v\,\Delta t' = v\,\frac{\Delta t}{\gamma}$$

Since $L' = v\,\Delta t$, we see that $L = L'/\gamma$ or

Length contraction

$$L = L'\left(1 - \frac{v^2}{c^2}\right)^{1/2} \qquad (39.8)$$

According to this result,

> if an observer at rest with respect to an object measures its length to be L', an observer moving with a relative speed v with respect to the object will find it to be shorter than its rest length by the factor $(1 - v^2/c^2)^{1/2}$.

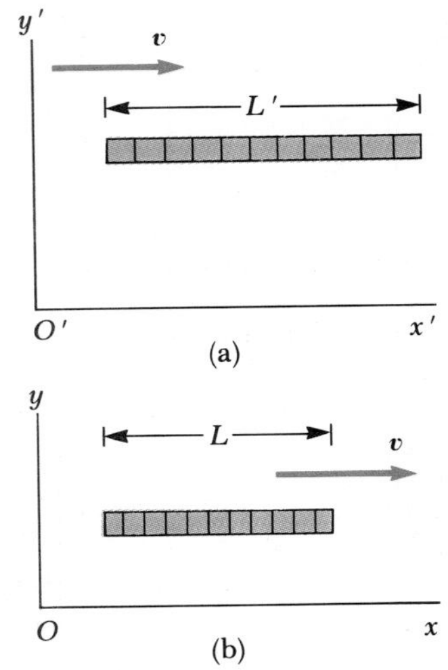

You should note that *the length contraction takes place only along the direction of motion.* For example, suppose a stick moves past a stationary earth observer with a speed v as in Figure 39.10. The length of the stick as measured by an observer in the frame attached to it is the proper length L', as illustrated in Figure 39.10a. The length of the stick, L, as measured by the earth observer in the stationary frame is shorter than L' by the factor $(1 - v^2/c^2)^{1/2}$. Note that length contraction is a symmetrical effect: if the stick were at rest on earth, an observer in the moving frame would also measure its length to be shorter by the same factor $(1 - v^2/c^2)^{1/2}$.

Figure 39.10 (a) A stick as viewed by an observer in a frame attached to the stick (i.e., both have the same velocity). (b) The stick as seen by an observer in a frame in which the stick has a velocity $\mathbf{v}$ relative to the frame. The length is shorter than the proper length, L', by a factor $(1 - v^2/c^2)^{1/2}$.

It is important to emphasize that proper length and proper time are measured in *different* reference frames. As an example of this point, let us return to the example of the decaying muons moving at speeds close to the speed of light. An observer in the muon's reference frame would measure the proper lifetime, while an earth-based observer measures the proper height of the mountain in Figure 39.8. In the muon's reference frame, there is no time dilation, but the distance of travel is observed to be shorter when measured in this frame. Likewise, in the earth observer's reference frame, there is time dilation, but the distance of travel is measured to be the actual height of the mountain. Thus, when calculations on the muon are performed in both frames, one sees the effect of "offsetting penalties" and the outcome of the experiment is the same!

If an object in the shape of a box passing by could be photographed, its image would show length contraction, but its shape would also be distorted. This is illustrated in the computer-simulated drawings shown in Figure 39.11

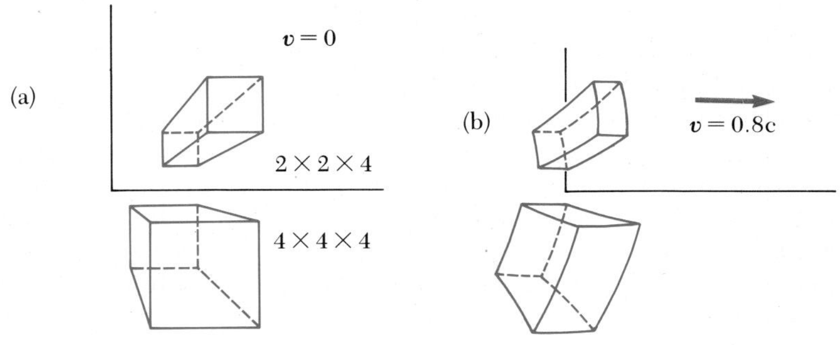

Figure 39.11 Computer-simulated photographs of a set of two rectangular boxes (a) at rest relative to the camera and (b) moving at a velocity $\mathbf{v} = 0.8\mathbf{c}$ relative to the camera. [From G. D. Scott and M. R. Viner, *Am. J. Phys.* 33, 534 (1965).]

for a box moving past an observer with a speed $v = 0.8c$. When the shutter of the camera is opened, it records the shape of the object at the instant light from it began its journey to the camera. Since light from different parts of the object must arrive at the camera at the same time (when the photograph is taken), light from more distant parts of the object must start its journey earlier than light from closer parts. Hence, the photograph records different parts of the object at different times. This results in a highly distorted image.

EXAMPLE 39.2 The Contraction of a Spaceship
A spaceship is measured to be 100 m long while it is at rest with respect to an observer. If this spaceship now flies by the observer with a speed of $0.99c$, what length will the observer find for the spaceship?

Solution From Equation 39.8, the length measured by an observer in the spaceship is

$$L = L'\sqrt{1 - \frac{v^2}{c^2}} = (100\text{ m})\sqrt{1 - \frac{(0.99c)^2}{c^2}} = 14\text{ m}$$

Exercise 1 If the ship moves past the observer with a speed of $0.01c$, what length will the observer measure?
Answer 99.99 m.

EXAMPLE 39.3 How High is the Spaceship?
An observer on earth sees a spaceship at an altitude of 435 m moving downward toward the earth with a speed of $0.970c$. What is the altitude of the spaceship as measured by an observer in the spaceship?

Solution The moving observer in the ship finds the altitude to be

$$L = L'\sqrt{1 - \frac{v^2}{c^2}} = (435\text{ m})\sqrt{1 - \frac{(0.970c)^2}{c^2}}$$

$$= 106\text{ m}$$

EXAMPLE 39.4 The Triangular Spaceship
A spaceship in the form of a triangle flies by an observer with a speed of $0.950c$. When the ship is at rest (Fig. 39.12a), the distances x and y are found to be 50.0 m and 25 m, respectively. What is the shape of the ship as seen by an observer at rest when the ship is in motion along the direction shown in Figure 39.12b?

Solution The observer sees the horizontal length of the ship to be contracted to a length of

$$L = L'\sqrt{1 - \frac{v^2}{c^2}} = (50.0\text{ m})\sqrt{1 - \frac{(0.950c)^2}{c^2}}$$

$$= 15.6\text{ m}$$

The 25-m vertical height is unchanged because it is perpendicular to the direction of relative motion between the observer and the spaceship. Figure 39.12b represents the shape of the spaceship as seen by the observer at rest.

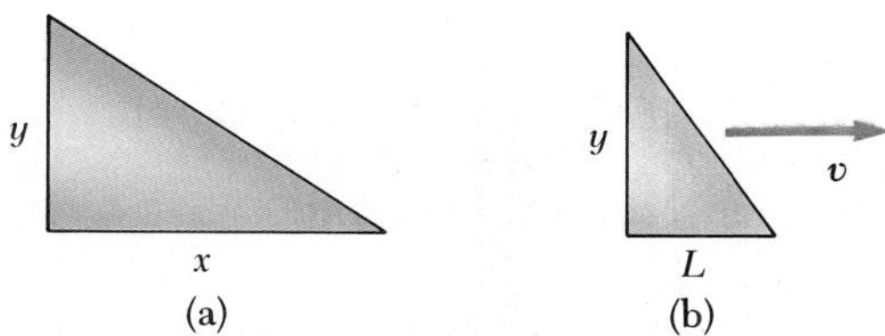

Figure 39.12 (Example 39.4) (a) When the spaceship is at rest, its shape is as shown. (b) The spaceship appears to look like this when it moves to the right with a speed v. Note that only its x dimension is contracted in this case.

39.9 THE LORENTZ TRANSFORMATION EQUATIONS

An event such as a flash of light is specified by three space coordinates and one time coordinate. Suppose that an event that occurs at some point P is reported by two observers, one at rest in the S frame and another in the S′ frame moving to the right with a speed v as in Figure 39.13. The observer in S reports the event with spacetime coordinates (x, y, z, t), while the observer in S' reports the same event using the coordinates (x', y', z', t'). We would like to find a relation between these coordinates that would be valid for all speeds. In Section 39.2, we found that the Galilean transformation of coordinates, given by Equation 39.1, does not agree with experiment at speeds comparable to the speed of light.

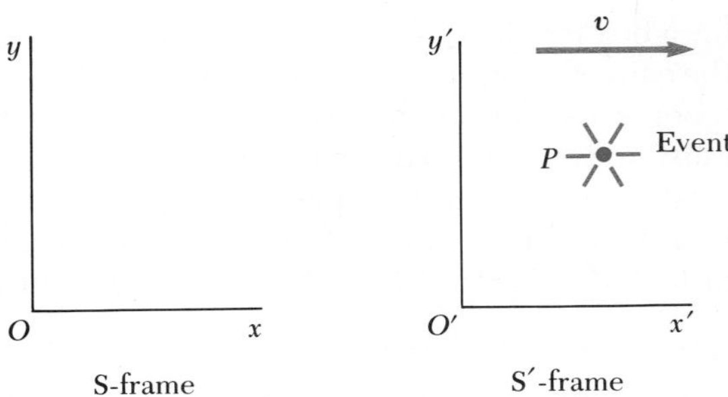

Figure 39.13 Representation of an event that occurs at some point P as observed by an observer at rest in the S frame and another in the S′ frame, which is moving to the right with a speed v.

The correct equations which are valid for speeds ranging from $v = 0$ to $v = c$, and enable us to transform some S to S′, are given by the **Lorentz transformation equations:**

Lorentz transformation for S → S′

$$\begin{aligned} x' &= \gamma(x - vt) \\ y' &= y \\ z' &= z \\ t' &= \gamma\left(t - \frac{v}{c^2}x\right) \end{aligned} \tag{39.9}$$

where the symbol γ (gamma) is defined as

$$\gamma \equiv \frac{1}{\sqrt{1 - \frac{v^2}{c^2}}} \tag{39.10}$$

The Lorentz transformation equations were originally derived by H. A. Lorentz (1853–1928) in 1890. However, it was Einstein who recognized their physical significance and took the bold step of interpreting them within the framework of the theory of relativity.

We see that the value for t' assigned to an event by observer O' depends both on the time t and on the coordinate x as measured by observer O. This is consistent with the notion that an event is characterized by four space-time coordinates (x, y, z, t). In other words, space and time are not actually separate concepts, but are closely interwoven in relativity. This is unlike the case of the Galilean transformation in which $t = t'$.

If we wish to transform coordinates in the S′ frame to coordinates in the S frame, we simply replace v by $-v$ and interchange the primed and unprimed coordinates in Equation 39.9. The resulting transformation is given by

Inverse Lorentz transformation for S′ → S

$$\begin{aligned} x &= \gamma(x' + vt') \\ y &= y' \\ z &= z' \\ t &= \gamma\left(t' + \frac{v}{c^2}x'\right) \end{aligned} \tag{39.11}$$

When $v \ll c$, the Lorentz transformation should reduce to the Galilean transformation. To check this, note that as $v \rightarrow 0$, $v/c^2 \ll 1$ and $v^2/c^2 \ll 1$, so that Equation 39.9 reduces in this limit to the Galilean coordinate transformation equations, given by

$$x' = x - vt \qquad y' = y \qquad z' = z \qquad t' = t$$

In many situations, we would like to know the *difference* in coordinates between two events or the time *interval* between two events as seen by observer O and observer O'. This can be accomplished by writing the Lorentz equations in a form suitable for describing pairs of events. From Equations 39.9 and 39.11, we can express the differences between the four variables x, x', t, and t' in the form

$$\begin{aligned} \Delta x' &= \gamma(\Delta x - v\,\Delta t) \\ \Delta t' &= \gamma\left(\Delta t - \frac{v}{c^2}\,\Delta x\right) \end{aligned} \qquad S \rightarrow S' \tag{39.12}$$

$$\begin{aligned} \Delta x &= \gamma(\Delta x' + v\Delta t') \\ \Delta t &= \gamma\left(\Delta t' + \frac{v}{c^2}\,\Delta x'\right) \end{aligned} \qquad S \rightarrow S' \tag{39.13}$$

where $\Delta x' = x_2' - x_1'$ and $\Delta t' = t_2' - t_1$ are the differences measured by observer O', while $\Delta x = x_2 - x_1$ and $\Delta t = t_2 - t_1$ are the differences measured by observer O. We have not included the expressions for relating the y and z coordinates since they are unaffected by motion along the x direction.

EXAMPLE 39.5 Simultaneity and Time Dilation Revisited

Use the Lorentz transformation equations in difference form to show that (a) simultaneity is not an absolute concept and (b) moving clocks run slower than stationary clocks.

(a) **Simultaneity:** Suppose that two events are simultaneous according to the moving observer O', so that $\Delta t' = 0$. From the expression for Δt given in Equation 39.13, we see that in this case, $\Delta t = \gamma v\,\Delta x'/c^2$. That is, the time interval for the same two events as measured by observer O is nonzero, so they will not appear to be simultaneous in O.

(b) **Time Dilation:** Suppose that observer O' finds that two events occur at the same place ($\Delta x' = 0$), but at different times ($\Delta t' \neq 0$). In this situation, the expression for Δt given in Equation 39.13 becomes $\Delta t = \gamma\,\Delta t'$. This is the equation for time dilation found earlier, Equation 39.7, where $\Delta t' = \Delta t$ is the proper time measured by the single clock in O'.

Exercise 2 Use the Lorentz transformation equations in difference form to confirm length contraction. That is, $L = L'/\gamma$.

39.10 LORENTZ VELOCITY TRANSFORMATION

Let us now derive the Lorentz velocity transformation, which is the relativistic counterpart of the Galilean velocity transformation. Suppose that an unaccelerated object is observed in the S′ frame at x_1' at time t_1' and at x_2' at time t_2'. Its speed u_x' measured in S′ is given by

$$u_x' = \frac{x_2' - x_1'}{t_2' - t_1'} = \frac{dx'}{dt'} \tag{39.14}$$

Using Equation 39.9, we have

$$dx' = \gamma(dx - v\,dt)$$

$$dt' = \gamma\left(dt - \frac{v}{c^2}\,dx\right)$$

Substituting these into Equation 39.14 gives

$$u'_x = \frac{dx'}{dt'} = \frac{dx - v\,dt}{dt - \frac{v}{c^2}\,dx} = \frac{\frac{dx}{dt} - v}{1 - \frac{v}{c^2}\frac{dx}{dt}}$$

But dx/dt is just the velocity component u_x of the object measured in S, and so this expression becomes

Lorentz velocity transformation for $S \to S'$

$$u'_x = \frac{u_x - v}{1 - \frac{u_x v}{c^2}} \tag{39.15}$$

Similarly, if the object has velocity components along y and z, the components in S′ are given by

$$u'_y = \frac{u_y}{\gamma\left(1 - \frac{u_x v}{c^2}\right)} \quad \text{and} \quad u'_z = \frac{u_z}{\gamma\left(1 - \frac{u_x v}{c^2}\right)} \tag{39.16}$$

When u_x and v are both much smaller than c (the nonrelativistic case), the denominator of Equation 39.15 approaches unity, and $u'_x \approx u_x - v$. This corresponds to the Galilean velocity transformations. In the other extreme, when $u_x = c$, Equation 39.14 becomes

$$u'_x = \frac{c - v}{1 - \frac{cv}{c^2}} = \frac{c\left(1 - \frac{v}{c}\right)}{1 - \frac{v}{c}} = c$$

From this result, we see that an object moving with a speed c relative to an observer in S also has a speed c relative to an observer in S′—*independent* of the relative motion of S and S′. Note that this conclusion is consistent with Einstein's second postulate, namely, that the speed of light must be c with respect to all inertial frames of reference. Furthermore, the speed of an object can never exceed c. That is, the speed of light is the "ultimate" speed. We shall return to this point later when we consider the energy of a particle.

To obtain u_x in terms of u'_x, we replace v by $-v$ in Equation 39.15 and interchange the roles of u_x and u'_x. This gives

Inverse Lorentz velocity transformation for $S' \to S$

$$u_x = \frac{u'_x + v}{1 + \frac{u'_x v}{c^2}} \tag{39.17}$$

EXAMPLE 39.6 Relative Velocity of Spaceships
Two spaceships A and B are moving in *opposite* directions, as in Figure 39.14. An observer on the earth measures the speed of A to be $0.75c$ and the speed of B to be $0.85c$. Find the velocity of B with respect to A.

Solution This problem can be solved by taking the S′ frame as being attached to spacecraft A, so that $\mathbf{v} = 0.75\mathbf{c}$ relative to an observer on the earth (the S frame). Spacecraft B can be considered as an object moving with a velocity $\mathbf{u}_x = -0.85\mathbf{c}$ relative to the earth observer. Hence, the velocity of B with respect to A can be obtained using Equation 39.15:

$$u'_x = \frac{u_x - v}{1 - \dfrac{u_x v}{c^2}} = \frac{-0.85c - 0.75c}{1 - \dfrac{(-0.85c)(0.75c)}{c^2}}$$

$$= -0.98c$$

The negative sign for u'_x indicates that spaceship B is moving in the negative x direction as observed by A. Note that the result is less than c. That is, a body whose speed is less than c in one frame of reference must have a speed less than c in *any other* frame. If the Galilean velocity transformation were used in this example, we would find that $u'_x = u_x - v = -0.85c - 0.75c = -1.6c$, which is greater than c.

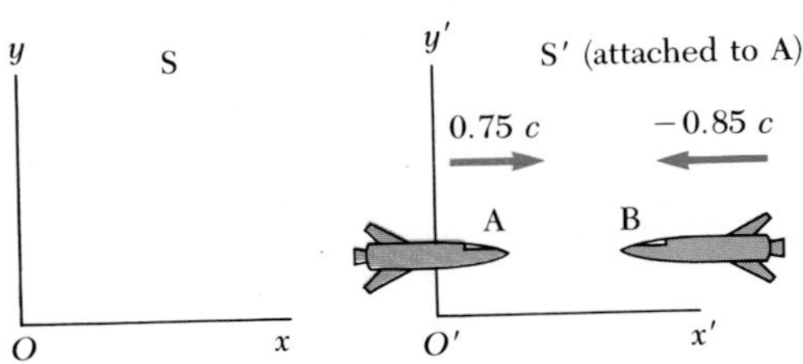

Figure 39.14 (Example 39.6) Two spaceships *A* and *B* move in *opposite* directions. The speed of *B* relative to *A* is *less* than *c* and is obtained by using the relativistic velocity transformation.

EXAMPLE 39.7 The Speeding Motorcycle
Imagine a motorcycle rider moving with a speed of $0.8c$ past a stationary observer, as shown in Figure 39.15. If the rider tosses a ball in the forward direction with a speed of $0.7c$ relative to himself, what is the speed of the ball as seen by the stationary observer?

Solution In this situation, the velocity of the motorcycle with respect to the stationary observer is $v = 0.8c$. The velocity of the ball in the frame of reference of the motorcyclist is $0.7c$. Therefore, the velocity, u_x, of the ball relative to the stationary observer is

$$u_x = \frac{u'_x + v}{1 + \dfrac{u'_x v}{c^2}} = \frac{0.7c + 0.8c}{1 + \dfrac{(0.7c)(0.8c)}{c^2}} = 0.962c$$

Exercise 3 Suppose that the motorcyclist moving with a speed $0.8c$ turns on a beam of light that moves away from him with a speed of c in the same direction as the moving motorcycle. What would the stationary observer measure for the speed of the beam of light?
Answer c.

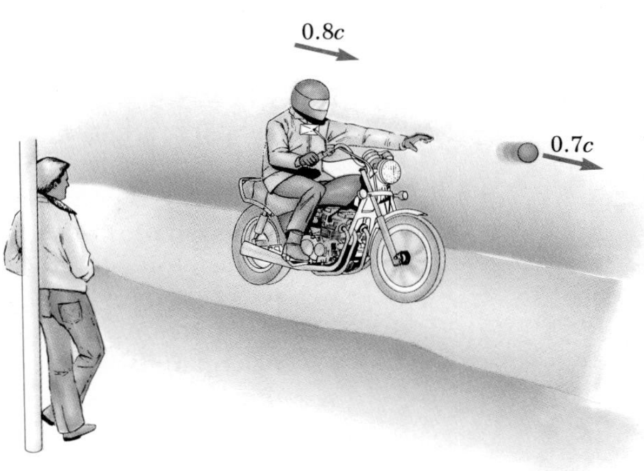

Figure 39.15 (Example 39.7) A motorcyclist moves past a stationary observer with a speed of $0.8c$ and throws a ball in the direction of motion with a speed of $0.7c$ relative to himself.

39.11 RELATIVISTIC MOMENTUM

We have seen that the principle of relativity is satisfied if the Galilean transformation is replaced by the more general Lorentz transformation. Therefore, in order to properly describe the motion of material particles within the framework of special relativity, we must generalize Newton's laws and the definition of momentum and energy. These generalized definitions of momentum and energy will reduce to the classical (nonrelativistic) definitions for $v \ll c$.

First, recall that the conservation of momentum states that when two bodies collide, the total momentum remains constant, assuming the bodies are isolated (that is, they interact only with each other). Suppose the collision is described in a reference frame S in which the momentum is conserved. If the velocities in a second reference frame S′ are calculated using the Lorentz transformation and the classical definition of momentum, $\mathbf{p} = m\mathbf{u}$, one finds that momentum *is not* conserved in the second reference frame. However, since the laws of physics are the same in all inertial frames, the momentum must be conserved in all systems. In view of this condition and assuming the Lorentz transformation is correct, we must modifiy the definition of momentum.

Our definition of relativistic momentum $\mathbf{p}$ must satisfy the following conditions:

1. The relativistic momentum must be conserved in all collisions.
2. The relativistic momentum must approach the classical value $m\mathbf{u}$ as $\mathbf{u} \rightarrow 0$.

The correct relativistic equation for the momentum that satisfies these conditions is given by the expression

Definition of relativistic momentum

$$\mathbf{p} \equiv \frac{m\mathbf{u}}{\sqrt{1 - \frac{u^2}{c^2}}} = \gamma m\mathbf{u} \tag{39.18}$$

where $\mathbf{u}$ is the velocity of the particle. We use the symbol $\mathbf{u}$ for particle velocity rather than $\mathbf{v}$, which is used for the relative velocity of two reference frames. The proof of this generalized expression for $\mathbf{p}$ is beyond the scope of this text. When u is much less than c, the denominator of Equation 39.18 approaches unity, so that $\mathbf{p}$ approaches $m\mathbf{u}$. Therefore, the relativistic equation for $\mathbf{p}$ reduces to the classical expression when u is small compared with c.

In many situations, it is useful to interpret Equation 39.18 as the product of a relativistic mass, γm, and the velocity of the object. Using this description, one can say that *the observed mass of an object increases with speed according to the relation*

$$\text{Relativistic mass} = \gamma m = \frac{m}{\sqrt{1 - \frac{v^2}{c^2}}} \tag{39.19}$$

where the relativistic mass is the mass of the object moving at a speed v with respect to the observer and m is the mass of the object as measured by an observer at rest with respect to the object. The quantity m is often referred to as the **rest mass** of the object. (In future chapters, we shall sometimes use the symbol m_0 to represent rest mass.) Note that Equation 39.19 indicates that *the mass of an object increases as its speed increases.* This prediction has been borne out through many observations on elementary particles, such as energetic electrons.

The relativistic force $\mathbf{F}$ on a particle whose momentum is $\mathbf{p}$ is defined by the expression

$$\boldsymbol{F} \equiv \frac{d\boldsymbol{p}}{dt} \qquad (39.20)$$

where $\boldsymbol{p}$ is given by Equation 39.18. This expression is identical to the classical statement of Newton's second law, which says that force equals the time rate of change of momentum.

EXAMPLE 39.8 Momentum of an Electron
An electron, which has a rest mass of 9.11×10^{-31} kg, moves with a speed of $0.750c$. Find its relativistic momentum and compare this with the momentum calculated from the classical expression.

Solution Using Equation 39.18 with $u = 0.75c$, we have

$$p = \frac{mu}{\sqrt{1 - \frac{u^2}{c^2}}}$$

$$p = \frac{(9.11 \times 10^{-31}\text{ kg})(0.750 \times 3 \times 10^8\text{ m/s})}{\sqrt{1 - \frac{(0.750c)^2}{c^2}}}$$

$$= 3.10 \times 10^{-22}\text{ kg}\cdot\text{m/s}$$

The incorrect classical expression would give

$$\text{Momentum} = mu = 2.05 \times 10^{-22}\text{ kg}\cdot\text{m/s}$$

Hence, the correct relativistic result is 50% greater than the classical result!

It is left as a problem (Problem 55) to show that the acceleration $\boldsymbol{a}$ of a particle decreases under the action of a constant force, in which case $a \propto (1 - u^2/c^2)^{3/2}$. Furthermore, as the speed approaches c, the acceleration caused by any finite force approaches zero. Hence, it is impossible to accelerate a particle from rest to a speed equal to or greater than c.

39.12 RELATIVISTIC ENERGY

We have seen that the definition of momentum and the laws of motion required generalization to make them compatible with the principle of relativity. This implies that the relation between work and energy must also be modified.

In order to derive the relativistic form of the work-energy theorem, let us start with the definition of work done by a force F and make use of the definition of relativistic force, Equation 39.20. That is,

$$W = \int_{x_1}^{x_2} F\,dx = \int_{x_1}^{x_2} \frac{dp}{dt}\,dx$$

where we have assumed that the force and motion are along the x axis. In order to perform this integration, we make repeated use of the chain rule for derivatives in the following manner:

$$\left(\frac{dp}{dt}\right) dx = \left(\frac{dp}{du}\frac{du}{dt}\right) dx = \frac{dp}{du}\left(\frac{du}{dx}\frac{dx}{dt}\right) dx$$

$$= \frac{dp}{du}\,u\,\frac{du}{dx}\,dx = \frac{dp}{du}\,u\,du$$

Since p depends on u according to Equation 39.18, we have

$$\frac{dp}{du} = \frac{d}{du}\frac{mu}{\sqrt{1-\frac{u^2}{c^2}}} = \frac{m}{\left(1-\frac{u^2}{c^2}\right)^{3/2}}$$

Using these results, we can express the work as

$$W = \int_0^u \frac{dp}{du}\, u\, du = \int_0^u \frac{mu \quad du}{\left(1-\frac{u^2}{c^2}\right)^{3/2}}$$

where we have assumed that the particle is accelerated from rest to some final velocity u. Evaluating the integral, we find that

$$W = \frac{mc^2}{\sqrt{1-\frac{u^2}{c^2}}} - mc^2 \qquad (39.21)$$

Recall that the work-energy theorem states the work done by a force acting on a particle equals the change in kinetic energy of the particle. Since the initial kinetic energy is zero, we conclude that the work W is equivalent to the relativistic kinetic energy K, that is,

Relativistic kinetic energy

$$K = \frac{mc^2}{\sqrt{1-\frac{u^2}{c^2}}} - mc^2 \qquad (39.22)$$

This equation has been confirmed by experiments using high-energy particle accelerators. At low speeds, where $u/c \ll 1$, Equation 39.22 should reduce to the classical expression $K = \frac{1}{2}mu^2$. We can check this by using the binomial expansion $(1 - x^2)^{-1/2} \approx 1 + \frac{1}{2}x^2 + \cdots$ for $x \ll 1$, where the higher-order powers of x are neglected in the expansion. In our case, $x = u/c$, so that

$$\frac{1}{\sqrt{1-\frac{u^2}{c^2}}} = \left(1-\frac{u^2}{c^2}\right)^{-1/2} \approx 1 + \frac{1}{2}\frac{u^2}{c^2} + \cdots$$

Substituting this into Equation 39.22 gives

$$K \approx mc^2\left(1 + \frac{1}{2}\frac{u^2}{c^2} + \cdots\right) - mc^2 = \frac{1}{2}mu^2$$

which agrees with the classical result. A graph comparing the relativistic and nonrelativistic expressions is given in Figure 39.16. In the relativistic case, the particle speed never exceeds c, regardless of the kinetic energy. The two curves are in good agreement when $v \ll c$.

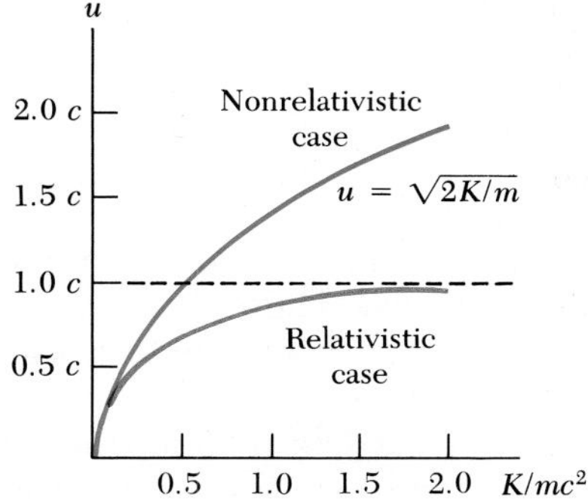

Figure 39.16 A graph comparing relativistic and nonrelativistic kinetic energy. The energies are plotted versus speed. In the relativistic case, u is always *less* than c.

It is useful to write the relativistic kinetic energy in the form

$$K = \gamma mc^2 - mc^2 \tag{39.23}$$

where

$$\gamma = \frac{1}{\sqrt{1 - \dfrac{u^2}{c^2}}}$$

The constant term mc^2, which is independent of the speed, is called the **rest energy** of the particle E_0. The term γmc^2, which depends on the particle speed, is therefore the sum of the kinetic and rest energies. We define γmc^2 to be the **total energy** E, that is,

$$E = \gamma mc^2 = K + mc^2 \tag{39.24}$$

Definition of total energy

or

$$E = \frac{mc^2}{\sqrt{1 - \dfrac{u^2}{c^2}}} \tag{39.25}$$

Energy-mass equivalence

This, of course, is Einstein's famous mass-energy equivalence equation. The relation $E = \gamma mc^2 = \gamma E_0$ shows that *mass is a form of energy*. Furthermore, this result shows that a small mass corresponds to an enormous amount of energy. This concept has revolutionized the field of nuclear physics.

In many situations, the momentum or energy of a particle is known rather than its speed. It is therefore useful to have an expression relating the total energy E to the relativistic momentum p. This is accomplished by using the expressions $E = \gamma mc^2$ and $p = \gamma mu$. By squaring these equations and subtracting, we can eliminate u (Problem 23). The result, after some algebra, is

$$E^2 = p^2c^2 + (mc^2)^2 \tag{39.26}$$

Energy-momentum relation

When the particle is at rest, $p = 0$, and so we see that $E = E_0 = mc^2$. That is, the total energy equals the rest energy. As we shall discuss in the next chapter, it is well established that there are particles that have zero mass, such as photons (quanta of electromagnetic radiation). If we set $m = 0$ in Equation 39.26, we see that

$$E = pc \tag{39.27}$$

This equation is an *exact* expression relating energy and momentum for photons and neutrinos, which always travel at the speed of light. In these cases, the particles have a characteristic wavelength λ, and their momentum is given by $p = h/\lambda$.

Finally, note that since the mass m of a particle is independent of its motion, m must have the same value in all reference frames. On the other hand, the total energy and momentum of a particle depend on the reference frame in which they are measured, since they both depend on velocity. Since m is a constant, then according to Equation 39.26 the quantity $E^2 - p^2c^2$ must have the same value in all reference frames. That is, $E^2 - p^2c^2$ is *invariant* under a Lorentz transformation.

When dealing with electrons or other subatomic particles, it is convenient to express their energy in electron volts (eV), since the particles are usually given this energy by acceleration through a potential difference. The conversion factor is

$$1 \text{ eV} = 1.60 \times 10^{-19} \text{ J}$$

For example, the mass of an electron is 9.11×10^{-31} kg. Hence, the rest energy of the electron is

$$mc^2 = (9.11 \times 10^{-31} \text{ kg})(3.00 \times 10^8 \text{ m/s})^2 = 8.20 \times 10^{-14} \text{ J}$$

Converting this to eV, we have

$$mc^2 = (8.20 \times 10^{-14} \text{ J})(1 \text{ eV}/1.60 \times 10^{-19} \text{ J}) = 0.511 \text{ MeV}$$

where 1 MeV $= 10^6$ eV.

EXAMPLE 39.9 The Energy of a Speedy Electron

An electron moves with a speed $u = 0.850c$. Find its total energy and kinetic energy in eV.

Solution Using the fact that the rest energy of the electron is 0.511 MeV together with Equation 39.25 gives

$$E = \frac{mc^2}{\sqrt{1 - \frac{u^2}{c^2}}} = \frac{0.511 \text{ MeV}}{\sqrt{1 - \frac{(0.85c)^2}{c^2}}}$$

$$= 1.90(0.511 \text{ MeV}) = \boxed{0.970 \text{ MeV}}$$

The kinetic energy is obtained by subtracting the rest energy from the total energy:

$$K = E - mc^2 = 0.970 \text{ MeV} - 0.511 \text{ MeV}$$

$$= \boxed{0.459 \text{ MeV}}$$

EXAMPLE 39.10 The Energy of a Speedy Proton

The total energy of a proton is three times its rest energy. (a) Find the proton's rest energy in eV.

Solution

$$\text{Rest energy} = mc^2 = (1.67 \times 10^{-27} \text{ kg})(3 \times 10^8 \text{ m/s})^2$$

$$= (1.50 \times 10^{-10} \text{ J})(1 \text{ eV}/1.60 \times 10^{-19} \text{ J})$$

$$= \boxed{938 \text{ MeV}}$$

(b) With what speed is the proton moving?

Solution Since the total energy E is three times the rest energy, Equation 39.25 gives

$$E = 3mc^2 = \frac{mc^2}{\sqrt{1 - \frac{u^2}{c^2}}}$$

$$3 = \frac{1}{\sqrt{1 - \frac{u^2}{c^2}}}$$

Solving for u gives

$$\left(1 - \frac{u^2}{c^2}\right) = \frac{1}{9} \quad \text{or} \quad \frac{u^2}{c^2} = \frac{8}{9}$$

$$u = \frac{\sqrt{8}}{3} c = \boxed{2.83 \times 10^8 \text{ m/s}}$$

(c) Determine the kinetic energy of the proton in eV.

Solution

$$K = E - mc^2 = 3mc^2 - mc^2 = 2mc^2$$

Since $mc^2 = 938$ MeV, $K = \boxed{1876 \text{ MeV}}$

(d) What is the proton's momentum?

Solution We can use Equation 39.26 to calculate the momentum with $E = 3mc^2$:

$$E^2 = p^2c^2 + (mc^2)^2 = (3mc^2)^2$$

$$p^2c^2 = 9(mc^2)^2 - (mc^2)^2 = 8(mc^2)^2$$

$$p = \sqrt{8}\,\frac{mc^2}{c} = \sqrt{8}\,\frac{(938 \text{ MeV})}{c} = \boxed{2653 \frac{\text{MeV}}{c}}$$

The unit of momentum is written MeV/c for convenience.

39.13 CONFIRMATIONS AND CONSEQUENCES OF RELATIVITY THEORY

The special theory of relativity has been confirmed by a number of experiments. One important experiment concerned with the decay of muons, and time dilation in the muon's reference frame, was discussed in Section 39.7. This section describes further evidence of Einstein's special theory of relativity.

One of the first predictions of relativity that was experimentally confirmed is the variation of momentum with velocity. Experiments were performed as early as 1909 on electrons, which can easily be accelerated to speeds close to c through the use of electric fields. When an energetic electron enters a magnetic field $\mathbf{B}$ with its velocity vector perpendicular to $\mathbf{B}$, a magnetic force is exerted on the electron, causing it to move in a circle of radius r. In this situation, the relativistic momentum of the electron is given by $p = eB/r$. From the relativistic equivalent of Newton's second law, $\mathbf{F} = d\mathbf{p}/dt$, the variation in momentum with kinetic energy can be checked experimentally. The results of such experiments on electrons and other charged particles support the relativistic expressions.

The release of enormous quantities of energy in nuclear fission and fusion processes is a manifestation of the equivalence of mass and energy. The conversion of mass into energy is, of course, the basis of atomic and hydrogen bombs, the most powerful and destructive weapons ever constructed. In fact, all reactions that release energy do so at the expense of mass (including chemical reactions). In a conventional nuclear reactor, the uranium nucleus undergoes fission, a reaction that results in several lighter fragments having considerable kinetic energy. In the case of ^{235}U (the parent nucleus), which undergoes spontaneous fission, the fragments are two lighter nuclei and two neutrons. The total mass of the fragments is *less* than that of the parent nucleus by some amount Δm. The corresponding energy Δmc^2 associated with this mass difference is *exactly* equal to the total kinetic energy of the fragments. This kinetic energy is then used to produce heat and steam for the generation of electrical power.

Next, consider the basic fusion reaction in which two deuterium atoms combine to form one helium atom. This reaction is of major importance in current research and development of controlled-fusion reactors. The decrease in mass which results from the creation of one helium atom from two deuterium atoms is calculated to be $\Delta m = 4.25 \times 10^{-29}$ kg. Hence, the corresponding excess energy which results from one fusion reaction is given by the expression $\Delta mc^2 = 3.83 \times 10^{-12}$ J $= 23.9$ MeV. To appreciate the magnitude of this result, if 1 g of deuterium is converted into helium, the energy released is about 10^{12} J! At the 1988 cost of electrical energy, this would be worth about \$50 000.

EXAMPLE 39.11 Binding Energy of the Deuteron
The mass of the deuteron, which is the nucleus of "heavy hydrogen," is not equal to the sum of the masses of its constituents, which are the proton and neutron. Calculate this mass difference and determine its energy equivalence.

Solution Using atomic mass units (u), we have

$$m_p = \text{mass of proton} = 1.007276 \text{ u}$$

$$m_n = \text{mass of neutron} = 1.008665 \text{ u}$$

$$m_p + m_n = 2.015941 \text{ u}$$

Since the mass of the deuteron is 2.013553 u, we see that the mass difference Δm is 0.002388 u. By definition, 1 u $= 1.66 \times 10^{-27}$ kg, and therefore

$$\Delta m = 0.002388 \text{ u} = 3.96 \times 10^{-30} \text{ kg}$$

Using $E = \Delta mc^2$, we find that

$$E = \Delta mc^2 = (3.96 \times 10^{-30} \text{ kg})(3.00 \times 10^8 \text{ m/s})^2$$
$$= 3.56 \times 10^{-13} \text{ J}$$
$$= 2.23 \text{ MeV}$$

Therefore, the minimum energy required to separate the proton from the neutron of the deuterium nucleus (the binding energy) is 2.23 MeV.

SUMMARY

The two basic postulates of the ***special theory of relativity*** are as follows:

1. All the laws of physics must be the same for all observers moving at constant velocity with respect to each other.
2. In particular, the speed of light must be the same for all inertial observers, independent of their relative motion.

In order to satisfy these postulates, the Galilean transformations must be replaced by the **Lorentz transformations** given by

$$\begin{aligned} x' &= \gamma(x - vt) \\ y' &= y \\ z' &= z \\ t' &= \gamma\left(t - \frac{v}{c^2}x\right) \end{aligned} \tag{39.9}$$

where

$$\gamma = \frac{1}{\sqrt{1 - \dfrac{v^2}{c^2}}} \tag{39.10}$$

In these equations, it is assumed that the primed system moves with a speed v along the xx' axes.

The relativistic form of the **velocity transformation** is

$$u'_x = \frac{u_x - v}{1 - \dfrac{u_x v}{c^2}} \tag{39.15}$$

where u_x is the speed of an object as measured in the S frame and u'_x is its speed measured in the S′ frame.

Some of the consequences of the special theory of relativity are as follows:

1. Clocks in motion relative to an observer appear to be slowed down by a factor γ. This is known as **time dilation.**
2. Lengths of objects in motion appear to be contracted in the direction of motion.
3. Events that are simultaneous for one observer are not simultaneous for another observer in motion relative to the first.

These three statements can be summarized by saying that duration, length, and simultaneity are not absolute concepts in relativity.

The relativistic expression for the **momentum** of a particle moving with a velocity $\boldsymbol{u}$ is

$$\boldsymbol{p} \equiv \frac{m\boldsymbol{u}}{\sqrt{1 - \dfrac{u^2}{c^2}}} = \gamma m\boldsymbol{u} \tag{39.18}$$

where

$$\gamma = \frac{1}{\sqrt{1 - \dfrac{u^2}{c^2}}}$$

The relativistic expression for the **kinetic energy** of a particle is

$$K = \gamma mc^2 - mc^2 \tag{39.23}$$

where mc^2 is called the **rest energy** of the particle.

The total energy E of a particle is related to the mass through the famous **energy-mass equivalence** expression:

$$E = \gamma mc^2 = \frac{mc^2}{\sqrt{1 - \dfrac{u^2}{c^2}}} \tag{39.25}$$

Finally, the relativistic momentum is related to the total energy through the equation

$$E^2 = p^2c^2 + (mc^2)^2 \tag{39.26}$$

QUESTIONS

1. What two speed measurements will two observers in relative motion *always* agree upon?
2. A spaceship in the shape of a sphere moves past an observer on earth with a speed $0.5c$. What shape will the observer see as the spaceship moves past?
3. An astronaut moves away from the earth at a speed close to the speed of light. If an observer on earth could make measurements of the astronaut's size and pulse rate, what changes (if any) would he or she measure? Would the astronaut measure any changes?
4. Two identically constructed clocks are synchronized. One is put in orbit around the earth while the other remains on earth. Which clock runs slower? When the moving clock returns to earth, will the two clocks still be synchronized?
5. Two lasers situated on a moving spacecraft are triggered simultaneously. An observer on the spacecraft claims to see the pulses of light simultaneously. What condition is necessary in order that a stationary observer agree that the two pulses are emitted simultaneously?
6. When we say that a moving clock runs slower than a stationary one, does this imply that there is something physically unusual about the moving clock?
7. When we speak of time dilation, do we mean that time passes more slowly in moving systems or that it simply appears to do so?
8. List some ways our day-to-day lives would change if the speed of light were only 50 m/s.
9. Give a physical argument which shows that it is impossible to accelerate an object of mass m to the speed of light, even with a continuous force acting on it.
10. It is said that Einstein, in his teenage years, asked the question, "What would I see in a mirror if I carried it in my hands and ran at the speed of light?" How would you answer this question?
11. Since mass is a form of energy, can we conclude that a compressed spring has more mass than the same spring when it is not compressed? On the basis of your answer, which has more mass, a spinning planet or an otherwise identical but nonspinning planet?
12. Suppose astronauts were paid according to the time spent traveling in space. After a long voyage at a speed near that of light, a crew of astronauts return and open their pay envelopes. What will their reaction be?
13. What happens to the density of an object as its speed increases?
14. Consider the incorrect statement, "Matter can neither be created nor destroyed." How would you correct this statement in view of the special theory of relativity?
15. Some of the distant stars, called quasars, are receding from us at half the speed of light (or greater). What is the speed of light we receive from these quasars?
16. How is it possible that photons of light, with zero mass, have momentum?
17. The expression $E = mc^2$ is often given in popular descriptions of Einstein's work. Is this expression strictly correct? For example, does it accurately take into account the kinetic energy of a moving mass?
18. With regard to reference frames, how does general relativity differ from special relativity?
19. Imagine an astronaut on a trip to Sirius, which lies 8 lightyears from the earth. Upon arrival at Sirius, the astronaut finds that the trip lasted 6 years. If the trip was made at a constant speed of $0.8c$, how can the 8-lightyear distance be reconciled with the six-year duration?

PROBLEMS

Section 39.2 The Principle of Relativity

1. In a laboratory frame of reference, an observer notes that Newton's second law is valid. Show that it is also valid for an observer moving at a constant speed relative to the laboratory frame.
2. Show that Newton's second law is not valid in a reference frame moving past the laboratory frame of Problem 1 with a constant acceleration.
3. A 2000-kg car moving with a speed of 20 m/s collides with and sticks to a 1500-kg car at rest at a stop sign. Show that momentum is conserved in a reference frame moving with a speed of 10 m/s in the direction of the moving car.
4. A billiard ball of mass 0.3 kg moves with a speed of 5 m/s and collides elastically with a ball of mass 0.2 kg moving in the opposite direction with a speed of 3 m/s. Show that momentum is conserved in a frame of reference moving with a speed of 2 m/s in the direction of the second ball.
5. A ball is thrown at 20 m/s inside a boxcar moving along the tracks at 40 m/s. What is the speed of the ball with respect to the ground if it is thrown (a) forward, (b) backward, (c) out the side door?

Section 39.7 The Relativity of Time

Section 39.8 The Relativity of Length

6. With what speed will a clock have to be moving in order to run at a rate that is one half the rate of a clock at rest?
7. How fast must a meter stick be moving if its length is observed to shrink to 0.5 m?

8. A clock on a moving spacecraft runs 1 minute slower per day relative to an identical clock on earth. What is the speed of the spacecraft? (*Hint:* For $v/c \ll 1$, note that $\gamma \approx 1 + v^2/2c^2$.)
9. An atomic clock is placed in a jet airplane. The clock measures a time interval of 3600 s when the jet moves with a speed of 300 m/s. What corresponding time interval does an identical clock held by an observer on the ground measure? (See hint in Problem 8.)
10. A meter stick moving in a direction parallel to its length appears to be only 75 cm long to an observer. What is the speed of the meter stick relative to the observer?
11. A spacecraft moves at a speed of $0.9c$. If its length is L_0 when measured from inside the spacecraft, what is its length measured by a ground observer?
12. The cosmic rays of highest energy are protons, with kinetic energy of 10^{13} MeV. (a) How long would it take a proton of this energy to travel across the Milky Way galaxy, of diameter 10^5 lightyears, as measured in the proton's frame? (b) From the point of view of the proton, how many kilometers across is our galaxy?
13. A muon formed high in the earth's atmosphere travels at speed $v = 0.99c$ for a distance 4.6 k before it decays into an electron, a neutrino, and an antineutrino ($\mu^- \rightarrow e^- + \nu + \bar{\nu}$). (a) How long does the muon live as measured in its reference frame? (b) How far does the muon travel, as measured in its frame?
14. If astronauts could travel at $v = 0.95c$, we on earth would say it takes $(4.2/0.95) = 4.4$ years to reach Alpha Centauri, 4.2 lightyears away. The astronauts disagree. (a) How much time passes on the astronauts' clocks? (b) What is the distance to Alpha Centauri, as measured by the astronauts?

Section 39.9 The Lorentz Transformation Equations

Section 39.10 Lorentz Velocity Transformation

15. For what value of v does $\gamma = 1.01$?
16. An observer sees two particles moving in opposite directions, each with speed $0.9c$. What is the speed of one particle with respect to the other?
17. Two spaceships approach each other, each moving with the *same* speed as measured by an observer on the earth. If their *relative* speed is $0.7c$, what is the speed of each spaceship?
18. Show that the Lorentz velocity transformation equations for u_y' and u_z' are given by Equations 39.16. (*Hint:* Make use of Equations 39.9.)
19. A cube of steel has a volume of 1 cm^3 and a mass of 8 g when at rest on the earth. If this cube is now given a speed $v = 0.9c$, what is its density as measured by a stationary observer?

Section 39.11 Relativistic Momentum

20. Calculate the momentum of a proton moving with a speed of (a) $0.01c$, (b) $0.5c$, (c) $0.9c$.
21. Find the momentum of a proton in MeV/c units if its total energy is twice its rest energy.
22. An electron has a momentum that is 90% larger than its classical momentum. (a) Find the speed of the electron. (b) How would your result change if the particle were a proton?

Section 39.12 Relativistic Energy

23. Show that the energy-momentum relationship given by $E^2 = p^2c^2 + (mc^2)^2$ follows from the expressions $E = \gamma mc^2$ and $p = \gamma mu$.
24. Show that the kinetic energy K, momentum p, and rest mass m of a relativistic particle are related by $p^2c^2 = 2Kmc^2 + K^2$.
25. A proton moves with the speed of $0.95c$. Calculate its (a) rest energy, (b) total energy, and (c) kinetic energy.
26. Find the speed at which the relativistic kinetic energy equals two times the nonrelativistic value.
27. Find the speed of a particle whose total energy is twice that of its rest energy.
28. A proton in a high-energy accelerator is given a kinetic energy of 50 GeV. Determine the (a) momentum and (b) speed of the proton.
29. Determine the energy required to accelerate an electron from (a) $0.50c$ to $0.9c$ and (b) $0.90c$ to $0.99c$.
30. In a typical color television tube, the electrons are accelerated through a potential difference of 25 000 volts. (a) What speed do the electrons have when they strike the screen? (b) What is their kinetic energy (in Joules)?
31. Electrons are accelerated to an energy of 2×10^{10} eV in the 3-km long Stanford Linear Accelerator. (a) What is the γ factor for the electrons? (b) What is the speed of the 20-GeV electrons? (c) How long does the accelerator appear to a 20-GeV electron?
32. A spaceship of mass 10^6 kg is to be accelerated to $0.6c$. (a) How much energy does this require? (b) How many kilograms of matter and antimatter (in equal proportions) will it take to provide this much energy?
33. A pion at rest ($m_\pi = 270m_e$) decays into a muon ($m_\mu = 206m_e$) and an antineutrino ($m_\nu = 0$) according to the following: $\pi^- \rightarrow \mu^- + \bar{\nu}$. Find the kinetic energy of the muon and the antineutrino in MeV. (*Hint:* Relativistic momentum is conserved.)

Section 39.13 Confirmations and Consequences of Relativity Theory

34. A radium isotope decays by emitting an α particle to a radon isotope according to the scheme ${}^{222}_{88}Ra \rightarrow {}^{222}_{86}Rn + {}^{4}_{2}He$. The masses of the atoms are 226.0254 (Ra), 222.0175 (Rn), and 4.0026 (He). How much energy is released as the result of this decay?

35. Consider the decay $^{55}_{24}\text{Cr} \rightarrow ^{55}_{25}\text{Mn} + \text{e}$, where e is an electron. The ^{55}Cr nucleus has a mass of 54.9279 u, and the ^{55}Mn nucleus has a mass of 54.9244 u. (a) Calculate the mass difference between the two nuclei in MeV. (b) What is the maximum kinetic energy of the emitted electron?

36. Calculate the binding energy in MeV per nucleon (proton or neutron) in the isotope $^{12}_{6}$C. Note that the mass of this isotope is exactly 12 u, and the masses of the proton and neutron are 1.007276 u and 1.008665 u, respectively.

37. The power output of the sun is 3.8×10^{26} W. How much matter is converted into energy in the sun each second?

38. A gamma ray (a very high energy photon of light) can produce an electron and an antielectron when it enters the electric field of a heavy nucleus: ($\gamma \rightarrow e^+ + e^-$). What is the minimum energy of the γ-ray to accomplish this task? (*Hint:* The masses of the electron and the antielectron are equal.)

ADDITIONAL PROBLEMS

39. An astronaut is traveling in a space vehicle that has a speed of $0.50c$ relative to the earth. The astronaut measures his pulse rate to be 75 per minute. Signals generated by the astronaut's pulse are radioed to earth when the vehicle is moving perpendicular to a line that connects the vehicle with an earth observer. What pulse rate does the earth observer measure? What would be the pulse rate if the speed of the space vehicle were increased to $0.99c$?

40. An astronaut is traveling in a space vehicle that has a speed of $0.50c$ relative to the earth. The astronaut and an earth observer compare the length of the standard meter stick that each has in his rest frame with the length of the other's meter stick. The measurements are made when each meter stick is held parallel to the relative velocity $\mathbf{u}$. By what amount does each claim that the other's meter stick is shorter?

41. The dominant nuclear reaction inside the sun is $4\text{p} \rightarrow \text{He}^4 + \Delta E$. If the rest mass of each proton is 938.2 MeV and the rest mass of the He^4 nucleus is 3727 MeV, calculate the percentage of the starting mass that is converted into energy.

42. The annual energy requirement for the United States is on the order of 10^{20} J. How many kilograms of matter would have to be completely converted to energy to meet this requirement?

43. An electron has a speed of $0.75c$. Find the speed of a proton which has (a) the same kinetic energy as the electron and (b) the same momentum as the electron.

44. How fast would a motorist have to be going to make a red light appear green? ($\lambda_{\text{red}} = 650$ nm, $\lambda_{\text{green}} = 550$ nm.) In computing this, use the correct relativistic formula for Doppler shift:

$$\frac{\Delta\lambda}{\lambda} + 1 = \sqrt{\frac{c - v}{c + v}}$$

where v is the approach velocity and λ is the source wavelength.

45. Determine the velocity of recession of the quasar 3C9 if its redshift $\Delta\lambda/\lambda = 2$. Use the relativistic formula for Doppler shift:

$$\frac{\Delta\lambda}{\lambda} + 1 = \sqrt{\frac{c + u}{c - u}}$$

where u = recessional velocity and $\Delta\lambda$ is the wavelength shift.

46. The average lifetime of a pi meson in its own frame of reference is 2.6×10^{-8} s. If the meson moves with a speed of $0.95c$, what is (a) its mean lifetime as measured by an observer on earth and (b) the average distance it travels before decaying, as measured by an observer on earth?

47. An observer on earth observes two spacecraft moving in the *same* direction toward the earth. Spacecraft A appears to have a speed of $0.5c$, and spacecraft B appears to have a speed of $0.8c$. What is the speed of spacecraft A measured by an observer in spacecraft B?

48. A rod of length L_0 moves with a speed v along the horizontal direction. The rod makes an angle of θ_0 with respect to the x' axis. (a) Show that the length of the rod as measured by a stationary observer is given by $L = L_0[1 - (v^2/c^2)\cos^2\theta_0]^{1/2}$. (b) Show that the angle that the rod makes with the x axis is given by the expression $\tan\theta = \gamma\tan\theta_0$. These results show that the rod is both contracted and rotated. (Take the lower end of the rod to be at the origin of the primed coordinate system.)

49. An antiproton $\bar{\text{p}}$ can be created in a large synchrotron by an energetic moving proton striking a stationary proton: $\text{p} + \text{p} \rightarrow \text{p} + \text{p} + \text{p} + \bar{\text{p}}$. What minimum kinetic energy must the incoming proton have to produce an antiproton? The rest mass of the p and $\bar{\text{p}}$ is 938 MeV/c^2. (*Hint:* The incoming proton has speed v, the outgoing ensemble of 4 particles leaves with speed u.)

50. The muon is an unstable particle that spontaneously decays into an electron and two neutrinos. If the number of muons at $t = 0$ is N_0, the number at time t is given by $N = N_0 e^{-t/\tau}$ where τ is the mean lifetime, equal to 2.2 μs. Suppose the muons move at a speed of $0.95c$ and there are 5×10^4 muons at $t = 0$. (a) What is the observed lifetime of the muons? (b) How many muons remain after traveling a distance of 3 km?

51. *Speed of light in a moving medium.* The motion of a medium such as water influences the speed of light. This effect was first observed by Fizeau in 1851. Consider a light beam passing through a horizontal column of water moving with a speed v. (a) Show that if the beam travels in the same direction as the flow of

water, the speed of light measured in the laboratory frame is given by

$$u = \frac{c}{n}\left(\frac{1 + nv/c}{1 + v/nc}\right)$$

where n is the index of refraction of the water. (*Hint:* Use the velocity transformation relation, Equation 39.17, and note that the velocity of the water with respect to the *moving* frame is given by c/n.) (b) Show that for $v \ll c$, the expression above is, to a good approximation,

$$u \approx \frac{c}{n} + v - \frac{v}{n^2}$$

52. *Time dilation in an atom.* The atoms in a gas move in random directions with thermal velocities. Because of their motion with respect to a laboratory observer, the frequency at which they radiate is shifted by time dilation. (a) If an atom radiates at a frequency f_0 in its rest frame, show that the observed frequency in the laboratory frame is given by $f = (1 - v^2/c^2)^{1/2} f_0$. (b) If $v \ll c$, show that the fractional change in frequency is given by $\Delta f/f = -v^2/2c^2$. (c) Evaluate the fractional change in frequency for a hydrogen atom at a temperature of 300 K. (*Hint:* From the kinetic theory of gases, $Mv^2/2 = 3kT/2$.)

53. *Doppler effect for light.* If a light source moves with a speed v relative to an observer, there is a shift in the observed frequency analogous to the Doppler effect for sound waves. Show that the observed frequency f_0 is related to the true frequency f through the expression

$$f_0 = \sqrt{\frac{c \pm v_s}{c \mp v_s}}\, f$$

where the upper signs correspond to the source approaching the observer and the lower signs correspond to the source receding from the observer. [*Hint:* In the moving frame S′, the period is the proper time interval and is given by $T = 1/f$. Furthermore, the wavelength measured by the observer is $\lambda_0 = (c - v_s)T_0$, where T_0 is the period measured in s.]

54. *The red shift.* A light source recedes from an observer with a speed v_s, which is small compared with c. (a) Show that the fractional shift in the measured wavelength is given by the approximate expression

$$\frac{\Delta\lambda}{\lambda} \approx \frac{v_s}{c}$$

This result is known as the *red shift*, since the visible light is shifted toward the red. (Note that the proper period and measured period are approximately equal in this case.) (b) Spectroscopic measurements of light at $\lambda = 397$ nm coming from a galaxy in Ursa Major reveal a red shift of 20 nm. What is the recessional speed of the galaxy?

55. A charged particle moves along a straight line in a uniform electric field $\boldsymbol{E}$ with a speed v. If the motion and the electric field are both in the x direction, (a) show that the acceleration of the charge q in the x direction is given by

$$a = \frac{dv}{dt} = \frac{qE}{m}\left(1 - \frac{v^2}{c^2}\right)^{3/2}$$

(b) Discuss the significance of the dependence of the acceleration on the speed. (c) If the particle starts from the rest at $x = 0$ at $t = 0$, how would you proceed to find the speed of the particle and its position after a time t has elapsed?

ESSAY

The Renaissance of General Relativity

Clifford M. Will
McDonnell Center for the Space Sciences, Washington University

During the two decades from 1960 to 1980, the subject of general relativity experienced a rebirth. Despite its enormous influence on scientific thought in its early years, by the late 1950s general relativity had become a sterile, formalistic subject, cut off from the mainstream of physics. It was thought to have very little observational contact, and was believed to be an extremely difficult subject to learn and comprehend.

Yet by 1970, general relativity had become one of the most active and exciting branches of physics. It took on new roles both as a theoretical tool of the astrophysicist and as a playground for the elementary-particle physicist. New experimental tests verified its predictions in unheard-of-ways, and to remarkable levels of precision. Fields of study were created, such as "black-hole physics" and "gravitational-wave astronomy," that brought together the efforts of theorists and experimentalists. One of the most remarkable and important aspects of this renaissance of relativity was the degree to which experiment and observation motivated and complemented theoretical advances.

This was not always the case. In deriving general relativity during the final months of 1915, Einstein himself was not particularly motivated by a desire to account for observational results. Instead, he was driven by purely theoretical ideas of elegance and simplicity. His goal was to produce a theory of gravitation that incorporated in a natural way both the special theory of relativity that dealt with physics in inertial frames, and the principle of equivalence, the proposal that physics in a frame falling freely in a gravitational field was in some sense equivalent to physics in an inertial frame.

Once the theory was formulated, however, he did try to confront it with experiment by proposing three tests. One of these tests was an immediate success — the explanation of the anomalous advance in the perihelion of Mercury of 43 arcseconds per century, a problem that had bedeviled celestial mechanicians of the latter part of the 19th century. The next test, the deflection of light by the Sun, was such a success that it produced what today would be called a "media event." The measurements of the deflection, amounting to 1.75 arcseconds for a ray that grazes the Sun, by two teams of British astronomers in 1919, made Einstein an instant international celebrity. However, these measurements were not all that accurate, and subsequent measurements weren't much better. The third test, actually proposed by Einstein in 1907, was the gravitational redshift of light, but it was a test that remained unfulfilled until 1960, by which time it was no longer viewed as a true test of general relativity.

Cosmology was the other area where general relativity was believed to have observational relevance. Although the general relativistic picture of the expansion of the universe from a "big bang" was compatible with observations, there were problems. As late as the middle 1950s the measured values of the universal expansion rate implied that the universe was younger than the Earth! Even though this "age" problem had been resolved by 1960, cosmological observations were still in their infancy, and could not distinguish between various alternative models.

The turning point for general relativity came in the early 1960s, when discoveries of unusual astronomical objects such as quasars demonstrated that the theory would have important applications in astrophysical situations. Theorists found new ways to understand the theory and its observable consequences. Finally, the technological revolution of the last quarter century, together with the development of the interplanetary space program, provided new high-precision tools to carry out experimental tests of general relativity.

After 1960, the pace of research in general relativity and in an emerging field called "relativistic astrophysics" began to accelerate. New advances, both theoretical and observational, came at an ever increasing rate. They included the discovery of the cosmic background radiation; the analysis of the synthesis of helium from hydrogen in the big bang; observations of pulsars and of black-hole candidates; the development of the theory of relativistic stars and black holes; the theoretical study of gravitational radiation and the beginning of an experimental program to detect it; improved versions of old tests of general relativity, and brand new tests, discovered after 1960; the discovery of the binary pulsar, which provided evidence for gravity waves; the analysis of quantum effects outside black holes and of black hole evaporation; the discovery of a gravitational lens; and the beginnings of a unification of gravitation theory with the other interactions and with quantum mechanics.

During the two decades following 1960, general relativity rejoined the world of physics and astronomy. Research in relativity took on an increasingly interdisciplinary flavor, spanning such subjects as celestial mechanics, pure mathematics, experimental physics, quantum mechanics, observational astronomy, particle physics, and theoretical astrophysics.

Einstein's Equivalence Principle

The foundations of general relativity are actually quite old, dating back to the equivalence principle of Galileo and Newton. In Newton's view, the principle of equivalence stated that all objects accelerate at the same rate in a gravitational field regardless of their mass or composition. This equality has been verified abundantly over the years, including classic experiments by the Hungarian physicist Baron Lorand von Eötvös around 100 years ago and recent experiments at Princeton and Moscow State Universities. The accuracy of these tests is better than one part in 10^{11}. Einstein's insight was the recognition that, to an observer inside a freely falling laboratory, not only should objects float as if gravity were absent as a consequence of this equality, but also *all* laws of nongravitational physics, such as electromagnetism and quantum mechanics, should behave as if gravity were truly absent.

This idea is now known as the Einstein equivalence principle, and it was a key step, because it implied the converse: that in a reference frame where gravity is felt, such as in a laboratory in the Earth's surface, the effects of gravitation on physical laws can be obtained simply by mathematically transforming the laws from a freely falling frame to the laboratory frame. According to the branch of mathematics known as differential geometry, this is the same as saying that space-time is curved; in other words, that the effects of gravity are *indistinguishable* from the effects of being in curved space-time.

Gravitational Redshift

An immediate consequence of Einstein's principle of equivalence is the gravitational redshift effect. It is not a consequence of general relativity itself, although Einstein believed that it was. One version of a gravitational redshift experiment measures the frequency shift between two identical clocks (meaning any device that produces a signal at a well-defined, steady frequency), placed at rest at different heights in a gravitational field. To derive the frequency shift from the equivalence principle, one imagines a freely falling frame that is momentarily at rest with respect to one clock at the moment the clock emits its signal. Because special relativity is valid in that frame, the frequency of the signal is unaffected by gravity as it travels from one clock to the other. However, by the time the signal reaches the second clock, the frame has picked up a downward velocity because it is in free fall in the gravitational field, and therefore, as seen by the falling frame, the second clock is moving upward. Thus the frequency seen by the second clock will appear to be shifted from the standard value by the Doppler effect. For small differences in height h between clocks, the shift in the frequency Δf is given by

$$\frac{\Delta f}{f} = \frac{v_{\text{fall}}}{c} = \frac{gh}{c^2}$$

(continued on next page)

where g is the local gravitational acceleration and c is the speed of light. If the receiver is at a lower height than the emitter, the received signal is shifted to higher frequencies ("blueshift"), while if the receiver is higher, the signal is shifted to lower frequencies ("redshift"). The generic name for the effect is "gravitational redshift."

The first and most famous high-precision redshift measurement was the Pound-Rebka experiment of 1960, which measured the frequency shift of gamma-ray photons from the decay of iron-57 (^{57}Fe) as they ascended or descended the Jefferson Physical Laboratory tower at Harvard University.

The most precise gravitational redshift experiment performed to date was a rocket experiment carried out in June 1976. A "hydrogen maser" atomic clock was flown on a Scout D rocket to an altitude of 10 000 km, and its frequency was compared to that of a similar clock on the ground by radio signals. After the effects of the rocket's motion were taken into account, the observations confirmed the gravitational redshift to 0.02 percent.

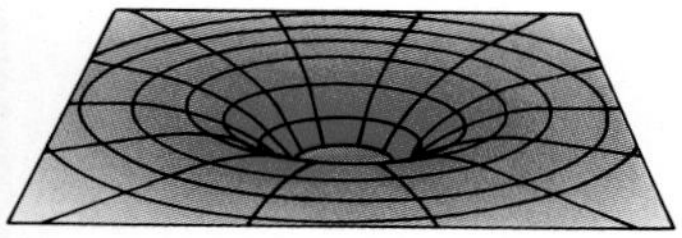

Because of these kinds of experiments, physicists are now convinced that spacetime *is* curved, and that the correct theory of gravity must be based on curved spacetime (see Fig. 1). That does not automatically imply general relativity, however, since it is not the only such theory. To test the specific predictions of general relativity for the amount and nature of space-time curvature, other experiments are needed.

Figure 1 The gravitation curvature of space. According to general relativity, gravity curves spacetime. This drawing shows how space is curved around a massive object such as the sun or a star. The red shaded region in the center indicates the location of the star. The greatest curvature is found immediately above the star's surface. Far from the star, where gravity is weak, spacetime is almost perfectly flat.

Deflection of Light

The first of these is a test that made Einstein's name a household word: the deflection of light. According to general relativity, a light ray that passes the Sun at a distance d is deflected by an angle $\Delta\theta = 1''.75/d$, where d is measured in units of the solar radius, and the notation $''$ denotes seconds of arc (see Fig. 2). Half of the deflection can be calculated by appealing only to the principle of equivalence. The idea is to determine what observers in a sequence of free-falling frames all along the trajectory of the photon would find for the deflection from one frame to the next. It turns out that the net deflection over all the frames is $0''\ 875/d$. The same result can be obtained by calculating the deflection of a particle by a massive body using ordinary Newtonian theory, and then taking the limit in which the particle's velocity becomes that of light. But the result of both versions is the deflection of light only relative to local straight lines, as defined for example by rigid rods laid end-to-end. However, because of the curvature of space, a local straight line defined by rods passing close to the Sun is itself bent relative to straight lines far from the Sun by an amount that yields the remaining half of the deflection.

The prediction of the bending of light by the Sun was one of the great successes of general relativity. Confirmation by the British astronomers Eddington and Crommelin of the bending of optical starlight observed during a total solar eclipse in the first

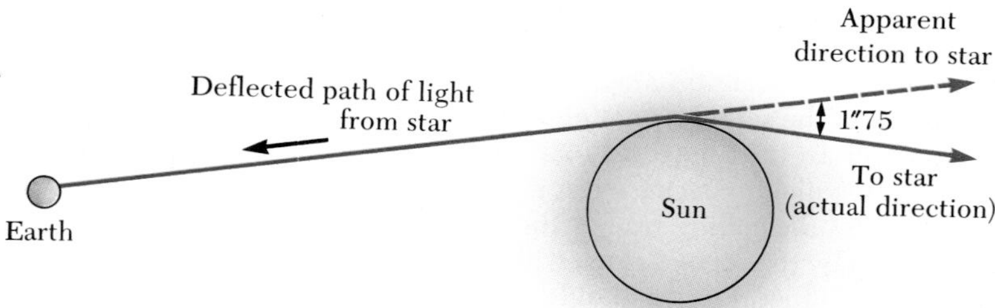

Figure 2 Deflection of starlight passing near the sun. Because of this effect, the sun and other remote objects can act as a *gravitational lens*. In his general relativity theory, Einstein calculated that starlight just grazing the sun's surface should be deflected by an angle of $1''.75$.

months following World War I helped make Einstein a celebrity. However, those measurements had only 30 percent accuracy, and succeeding eclipse experiments weren't much better; the results were scattered between one half and twice the Einstein value, and the accuracies were low.

However, the development of long-baseline radio interferometry produced a method for greatly improved determinations of the deflection of light. Radio interferometry is a technique of combining widely separated radio telescopes in such a way that the direction of a source of radio waves can be measured by determining the difference in phase of the signal received at the different telescopes. Modern interferometry has the ability to measure angular separations and changes in angles as small as 10^{-4} arcseconds. Coupled with this technological advance is a series of heavenly coincidences: each year groups of strong quasars pass near the Sun (as seen from the Earth). The idea is to measure the differential deflection of radio waves from one quasar relative to those from another as they pass near the Sun. A number of measurements of this kind were made almost annually over the period 1969–1975, yielding a confirmation of the predicted deflection to 1.5 percent.

The 1979 discovery of the "double" quasar Q0957 + 561 converted the deflection of light from a test of relativity to an important effect in astronomy and cosmology. The "double" quasar was found to be a multiple image of a single quasar caused by the gravitational lensing effect of a galaxy or a cluster of galaxies along the line of sight between us and the quasar. Several other such lenses have been found.

Closely related to light deflection is the "Shapiro time delay," a retardation of light signals that pass near the Sun. For instance, for a signal that grazes the Sun on a round trip from Earth to Mars at superior conjunction (when Mars is on the far side of the Sun), the round trip travel time is increased over what Newtonian theory would give by about 250 μs. The effect decreases with increasing distance of the signal from the Sun.

In the two decades following radio-astronomer Irwin Shapiro's 1964 discovery of this effect, several high-precision measurements have been made using the technique of radar-ranging to planets and spacecraft. Three types of targets were employed: planets such as Mercury and Venus, free-flying spacecraft such as Mariners 6 and 7, and combinations of planets and spacecraft, known as "anchored spacecraft," such as the Mariner 9 Mars orbiter and the 1976 Viking Mars landers and orbiters. The Viking experiments produced dramatic results, agreeing with the general relativistic prediction to one part in a thousand. This corresponded to a measurement accuracy in the Earth-Mars distance of 30 meters.

Perihelion Shift

The explanation of the anomalous perihelion shift of Mercury's orbit was another of the triumphs of general relativity (see Figs. 3 and 4). This had been an unsolved problem in celestial mechanics for over half a century, since the announcement by Le Verrier in 1859 that, after the perturbing effects of the planets on Mercury's orbit had been accounted for, there remained in the data an unexplained advance in the perihelion of Mercury. The modern value for this discrepancy is 42.98 arcseconds per century. Many ad hoc proposals were made in an attempt to account for this excess, including the existence of a new planet Vulcan near the Sun, or a deviation from the inverse square law of gravitation. But each was doomed to failure. General relativity accounted for the anomalous shift in a natural way. Radar measurements of the orbit of Mercury since 1966 have led to improved accuracy, so that the relativistic advance is known to about 0.5 percent.

Other measurements carried out since 1960 have given further support to general relativity. Observations of the motion of the Moon using laser ranging to a collection of specially designed reflectors, deposited on the lunar surface during the

(continued on next page)

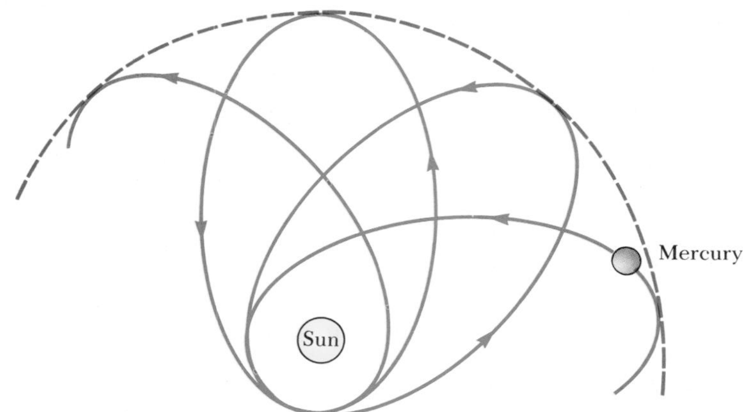

Figure 3 Advance of perihelion of Mercury. The elliptical orbit of Mercury about the sun rotates very slowly relative to the system connected with the sun. General relativity successfully explains this small effect, which predicts that the direction of the perihelion should change by only 43 arcseconds per century.

Apollo and Luna missions, have shown that the Moon and the Earth accelerate toward the Sun equally to a part in 10^{11}, an important confirmation of the equivalence principle applied to planetary bodies. Other measurements of planetary and lunar orbits have shown that the gravitational constant is a true constant of nature; it does not vary with time as the universe ages. Ambitious experimenters are currently designing an experiment using supercooled quartz gyroscopes to be placed in Earth orbit, to try to detect an important general relativistic effect called the "dragging of inertial frames," caused by the rotation of the Earth.

General relativity has passed every experimental test to which it has been put, and many alternative theories have fallen by the wayside. Most physicists now take the theory for granted, and look to see how it can be used as a practical tool in physics and astronomy.

Gravitational Radiation

One of these new tools is gravitational radiation, a subject that is almost as old as general relativity itself. By 1916, Einstein had succeeded in showing that the field equations of general relativity admitted wavelike solutions analogous to those of electromagnetic theory. For example, a dumbbell rotating about an axis passing at right angles through its handle will emit gravitational waves that travel at the speed of light. They also carry energy away from the rotating dumbbell, just as electromagnetic waves carry energy away from a light source. That was about all that was heard on the subject for over 40 years, primarily because the effects associated with gravitational waves were extremely tiny, unlikely (it was thought) ever to be of experimental or observational interest.

But in 1968 the idea of gravitational radiation was resurrected by the stunning announcement by Joseph Weber that he had detected gravitational radiation of extraterrestrial origin by using massive aluminum bars as detectors. A passing gravitational wave acts as an oscillating gravitational force field that alternately compresses and extends the bar in the lengthwise direction. However, subsequent observations by other researchers using bars with sensitivity that was claimed to be better than Weber's failed to confirm Weber's results. His results are now generally regarded as a false alarm, although there is still no good explanation for the "events" that he recorded in his bars.

Nevertheless, Weber's experiments did initiate the program of gravitational-wave detection, and inspired other groups to build better detectors. Currently a dozen laboratories around the world are engaged in building and improving upon the

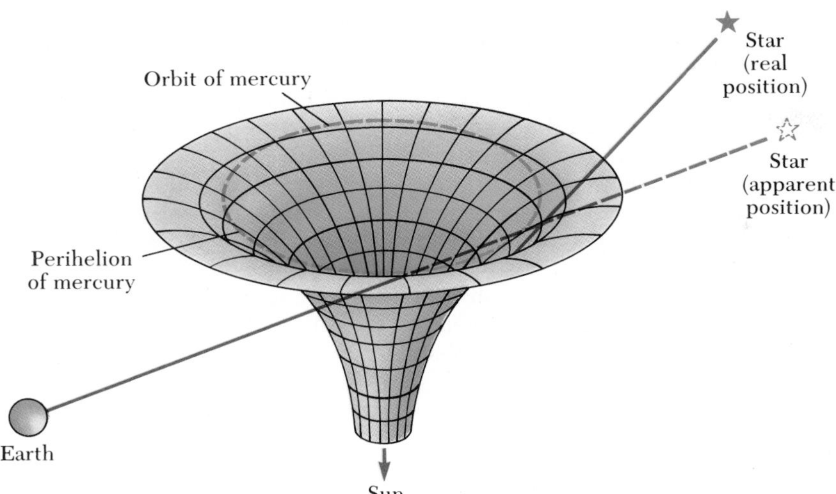

Figure 4 Under the general theory of relativity, the presence of a massive body essentially warps the space nearby. This can account for both the bending of light near the sun and the advance of the perihelion point of Mercury by 43 arcseconds per century more than would otherwise be expected. The diagram shows how a two-dimensional surface warped into three dimensions can change the direction of a "straight" line that is constrained to its surface; the warping of space is analogous, although with a greater number of dimensions to consider. The effect is similar to the golfer putting on a warped green. Though the ball is hit in a straight line, we see it appear to curve. (Adapted from Pasachoff, Jay M., *Astronomy,* 3rd ed., Saunders College Publishing, 1987).

basic "Weber bar" detector, some using bigger bars, some using smaller bars, some working at room temperature, and some working near absolute zero. The goal in all cases is to reduce noise from thermal, electrical, and environmental sources as much as possible to have a hope of detecting the very weak oscillations produced by a gravitational wave. For a bar of one meter length, the challenge is to detect a variation in length smaller than 10^{-20} meters, or 10^{-5} of the radius of a proton.

Another important type of detector is the laser interferometer, currently under development at several laboratories. This device operates on the same principle as the interferometer used by Michelson and Morley, but uses a laser as the light source. A passing gravitational wave will change the length of one arm of the apparatus relative to the other, and cause the interference pattern to vary. One ambitious proposal is to build two independent interferometers, each with arm lengths of 4 km, and separated by 1000 km.

It is hoped that these detectors will eventually be sensitive enough to detect gravitational waves from many sources, both in our galaxy and in distant galaxies, such as collapsing stars, double star systems, colliding black holes, and possibly even gravitational waves left over from the big bang. When this happens, a new field of "gravitational-wave astronomy" will be born.

Although gravitational radiation has not been detected directly, we know that it exists, through a remarkable system known as the binary pulsar. Discovered in 1974 by radio astronomers Russell Hulse and Joseph Taylor, it consists of a pulsar (which is a rapidly spinning neutron star) and a companion star in orbit around each other. Although the companion has not been seen directly, it is also believed to be a neutron star. The two bodies are circling so closely—their orbit is about the size of the Sun, and the period of the orbit is eight hours—that the effects of general relativity are very large. For example, the periastron shift (the analogue of the perihelion shift for Mercury) is more than 4 degrees per *year*.

(continued on next page)

Furthermore, the pulsar acts as an extremely stable clock, its pulse period of approximately 59 milliseconds drifting by only a quarter of a nanosecond per year. By measuring the arrival times of radio pulses at Earth, observers were able to determine the motion of the pulsar about its companion with amazing accuracy. For example, the accurate value for the orbital period is 27906.98163 seconds, and the orbital eccentricity is 0.617127.

Like a rotating dumbbell, an orbiting binary system should emit gravitational radiation, and in the process lose some of its orbital energy. This energy loss will cause the pulsar and its companion to spiral in toward each other, and the orbital period to shorten. According to general relativity, the predicted decrease in the orbital period is 75 microseconds per year. The observed decrease rate is in agreement with the prediction to about 4 percent. This confirms the existence of gravitational radiation and the general relativistic equations that describe it.

Black Holes

One of the most important and exciting aspects of the relativity renaissance is the study of and search for black holes. The subject began in the early 1960s as astrophysicists looked for explanations of the quasars, warmed up in the early 1970s following the possible detection of a black hole in an X-ray source, became white hot in the middle 1970s when black holes were found theoretically to evaporate, and continues today as one of the most active branches of general relativity.

However, the first glimmerings of the black hole idea date back to the 18th century, in the writings of a British amateur astronomer, the Reverend John Michell. Reasoning on the basis of the corpuscular theory that light would be attracted by gravity in the same way that ordinary matter is attracted, he noted that light emitted from the surface of a body such as the Earth or the Sun would be reduced in velocity by the time it reached great distances (Michell of course did not know special relativity). How large would a body of the same density as the Sun have to be in order that light emitted from it would be stopped and pulled back before reaching infinity? The answer he obtained was 500 times the diameter of the Sun. Light could therefore never escape from such a body. In today's language, such an object would be a supermassive black hole of about 100 million solar masses.

Although the general relativistic solution for a nonrotating black hole was discovered by Karl Schwarzschild in 1916, and a calculation of gravitational collapse to a black hole state was performed by J. Robert Oppenheimer and Hartland Snyder in 1939, black hole physics didn't really begin until the middle 1960s, when astronomers confronted the problem of the energy output of the quasars, and the mathematician Roy Kerr discovered the rotating black-hole solution.

A black hole is formed when a star has exhausted the thermonuclear fuel necessary to produce the heat and pressure that support it against gravity. The star begins to collapse, and if it is massive enough, it continues to collapse until the radius of the star approaches a value called the gravitational radius or Schwarzschild radius. In the nonrotating spherical case, this radius has a value given by $2GM/c^2$, where M is the mass of the star. For a body of one solar mass, the gravitational radius is about 3 km; for a body of the mass of the Earth, it is about 9 mm. An observer sitting on the surface of the star sees the collapse continue to smaller and smaller radii, until both star and observer reach the origin $r = 0$, with consequences too horrible to describe in detail. On the other hand, an observer at great distances observes the collapse to slow down as the radius approaches the gravitional radius, a result of the gravitational redshift of the light signals sent outward. However, in this case, the redshifting or slowing down becomes so extreme that the star appears almost to stop just outside the gravitational radius. The distant observer never sees any signals emitted by the falling observer once the latter is inside the gravitational radius. Any signal emitted inside can never escape the sphere bounded by the gravitational radius, called the "event horizon."

Since the middle 1960s the laws of "black hole physics" have been established and codified. For example, it was discovered that the Kerr (rotating) and Schwarzschild (nonrotating) solutions are unique; there are no other black-hole solutions in general relativity. Inside every black hole resides a "singularity," a pathological region of spacetime where the gravitational forces become infinite, where time comes to an end for any observer unfortunate enough to hit the singularity, indeed where all the laws of physics break down. Fortunately the event horizon of the black hole prevents any bizarre phenomena that such singularities might cause from reaching the outside world (this notion has been dubbed "cosmic censorship"). In 1974, Stephen Hawking discovered that the laws of quantum mechanics applied to the physics outside a black hole required it to evaporate by the creation of particles with a thermal energy spectrum, and to have an associated temperature and entropy. The temperature of a Schwarzschild black hole is $T = hc^3/8\pi kGM$, where h is Planck's constant and k is Boltzmann's constant. This discovery demonstrated a remarkable connection between gravity, thermodynamics, and quantum mechanics, which helped renew the theoretical quest for a grand synthesis of all the fundamental interactions. For black holes of astronomical masses, however, the evaporation is completely negligible, since for a solar-mass black hole, $T \approx 10^{-6}$ K.

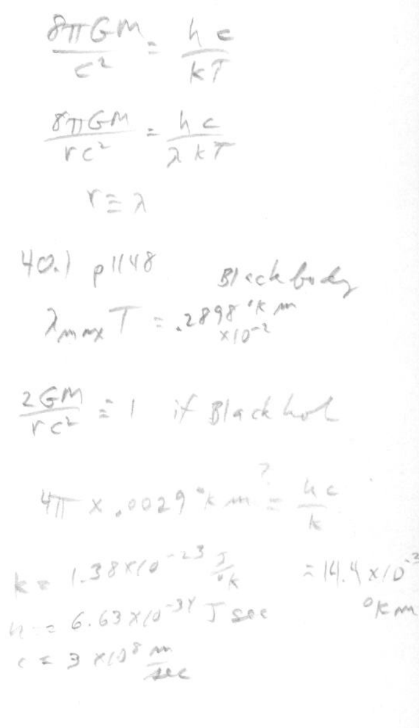

Although a great deal is known about black holes in theory, rather less is known about them observationally. There are several instances in which the evidence for the existence of black holes is impressive, but in all cases it is indirect. For instance, in the X-ray source Cygnus X1, the source of the X-rays is believed to be a collapsed object with a mass larger than about 6 solar masses in orbit around a giant star (see Figs. 5 and 6). The object cannot be a neutron star, because general relativity predicts that neutron stars must be lighter than about 3 solar masses. Thus the object must be a black hole. The X-rays are emitted by matter pulled from the surface of the companion star and sent into a spiralling orbit around the black hole. Similarly, there is evidence in the centers of certain galaxies, such as M87 and possibly even our own, of

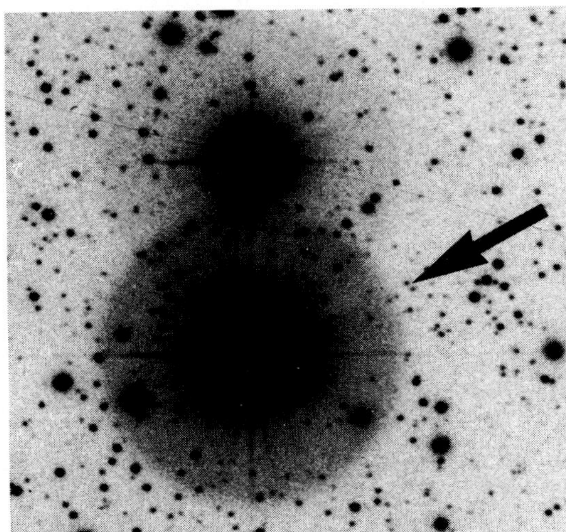

Figure 5 The overexposed dark object in the center of this negative print is the blue supergiant star HDE 226868, which thought to be the companion of the first black hole to be discovered, Cygnus X-1. The black hole in Cygnus X-1 that is thought to be orbiting the supergiant star is not visible. The image of the supergiant appears so large because it is overexposed on the film. (From Pasachoff, *Astronomy*.)

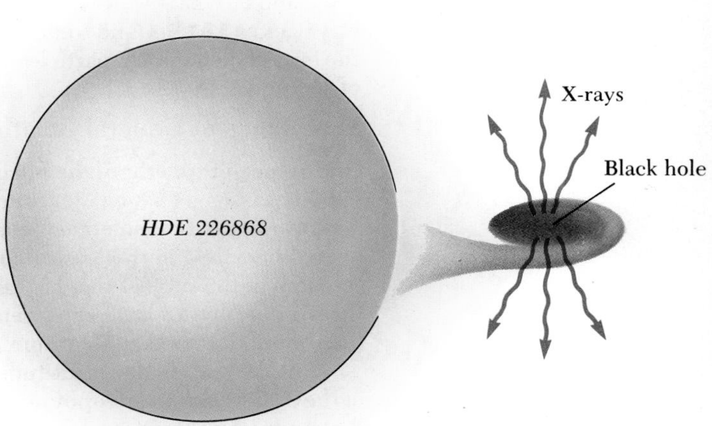

Figure 6 The Cygnus X-1 black hole. The stellar wind from HDE 226868 pours matter onto a huge disk around its black hole companion. The infalling gases are heated to enormous temperatures as they spiral toward the black hole, the gases are so hot that they emit vast quantities of x-rays.

collapsed objects of between 10^2 and 10^8 solar masses. A black hole model is consistent with the observations, although it is not necessarily required. Accretion of matter onto supermassive rotating black holes may produce the jets of outflowing matter that are observed in many quasars and active galactic nuclei. These and other astrophysical processes that might aid in the detection of black holes are being studied by relativists and astrophysicists.

Cosmology

The other area in which there has been a renaissance for general relativity is cosmology. Although Einstein in 1917 first used general relativity to calculate a model for the universe as a whole, the subject was not considered a serious branch of physics until the 1960s, when astronomical observations lent firm credence to the idea that the universe was expanding from a "big bang." For instance, observations of the rates at which galaxies are receding from us, coupled with improved determinations of their distances, implied that the universe was at least 10 billion years old, a value that was consistent with other observations such as the age of the Earth and of old star clusters.

In 1965 came the discovery of the cosmic background radiation by Arno Penzias and Robert Wilson. This radiation is the remains of the hot electromagnetic blackbody radiation that dominated the universe in its earlier phase, now cooled to 3 kelvins by the subsequent expansion of the universe. Next came calculations of the amount of helium that would be synthesized from hydrogen by thermonuclear fusion in the very early universe, around 1000 seconds after the big bang. The amount, approximately 25 percent by weight, was in agreement with the abundances of helium observed in stars and in interstellar space. This was an important confirmation of the hot big bang picture, because the amount of helium believed to be produced by fusion in the interiors of stars is woefully inadequate to explain the observed abundances.

Today, the general relativistic hot big-bang model of the universe has broad acceptance, and cosmologists now focus their attention on more detailed issues, such as how galaxies and other large-scale structures formed out of the hot primordial soup, and on what the universe might have been like earlier than 1000 seconds, all the way back to 10^{-36} s (and some brave cosmologists are going back even further) when the laws of elementary-particle physics may have played a major role in the evolution of the universe.

Acceptance of General Relativity

One of the outgrowths of the renaissance of general relativity that has occurred since 1960 has been a change in attitude about the importance and use of the theory. Its importance as a fundamental theory of the nature of spacetime and gravitation has not been diminished in the least; if anything, it has been enhanced by the flowering of research in the subject that has taken place. Its importance as a foundation for other theories of physics has been strengthened by current searches for unified quantum theories of nature that incorporate gravity along with the other interactions.

But the real change in attitude about general relativity has been in its use as a tool in the real world. In astrophysical situations, general relativity plays a central role in the study of neutron stars, black holes, gravitational lenses, relativistic binary star systems, and the universe as a whole. Gravitational radiation may one day provide a completely new observational tool for exploring and examining the cosmos.

Relativity even plays a role in everyday life. For example, the gravitational redshift effect on clocks *must* be taken into account in satellite-based navigation systems, such as the US Air Force's Global Positioning System, in order to achieve the required positional accuracy of a few meters or time transfer accuracy of a few nanoseconds.

To general relativists, always eager to find practical consequences of their subject, these have been very welcome developments!

Suggested Readings

Davies, P. C. W. *The Search for Gravity Waves,* Cambridge, England, Cambridge University Press, 1980.

Greenstein, G. *Frozen Star: Of Pulsars, Black Holes and the Fate of Stars,* New York, Freundlich Books, 1984.

Weinberg, S. *The First Three Minutes: A Modern View of the Origin of the Universe,* New York, Basic Books, 1977.

Will, C. M. *Was Einstein Right? Putting General Relativity to the Test,* New York, Basic Books, 1986.

Essay Questions

1. Two stars are a certain distance from each other as seen in the night sky. Half a year later, the stars are now overhead during the day, and the Sun is now located midway between the two stars. Because of the deflection of light by the Sun, do the stars now appear closer together or farther apart?
2. When gravitational waves are emitted by the binary pulsar, the double-star system loses energy. As a consequence, its orbital velocity increases, and thus its kinetic energy increases. How is this possible?
3. The temperature of a black hole is inversely proportional to its mass, so that when it gains energy (which is equivalent to mass), its temperature goes down, and when it loses energy, its temperature goes up. What do you conclude about the sign of its heat capacity? Can a black hole ever be in thermal equilibrium with an infinite reservoir at a fixed temperature?

Essay Problems

1. What is the gravitational frequency shift between two atomic clocks separated by 1000 km, both at sea level?
2. A pair of identical twin scientists set out to test the gravitational redshift effect in the following way. One climbs to the top of Mount Baldy and lives there for a year, while the other remains behind in the laboratory at the base of the mountain. When they are reunited after one year, one of the twins is slightly older than the other. Which one is older and why? If Mount Baldy is 5000 m high, what is the difference in age between the twins after one year?
3. Calculate the deflection of light for light rays that pass the Sun at distances of 2 solar radii, 10 solar radii, and 100 solar radii.
4. Gravitational waves travel with the same speed as light, 3×10^{10} cm/s. Calculate the wavelength of gravitational waves emitted by the following sources: (a) a star collapsing to form a black hole, emitting waves with a frequency around 1000 Hz, and (b) the binary pulsar, emitting waves with a period of four hours (one half the orbital period).
5. What is the fractional decrease in the orbital period of the binary pulsar per year? Estimate how long the binary system will continue before gravitational radiation will reduce the orbital period to zero (in other words will bring the stars so close together that they will merge into one object).
6. Calculate the mass of a black hole (in solar mass units) whose Schwarzschild radius is equal to the radius of the Earth's orbit around the Sun (1.5×10^8 km). One solar mass is about 2×10^{33} g.
7. Show that a billion-ton black hole has a Schwarzschild radius comparable to the radius of a proton (1 ton $\approx 10^6$ g, proton radius $\approx 10^{-13}$ cm). What kind of object on Earth weighs a billion tons?
8. Determine the mass and radius of a black hole whose apparent density is equal to that of water, 1 g/cm^3. Assume that the volume of a black hole is given by $4\pi R^3/3$, and express the mass in solar mass units.
9. Calculate the mass of a black hole whose temperature is room temperature.

40
Introduction to Quantum Physics

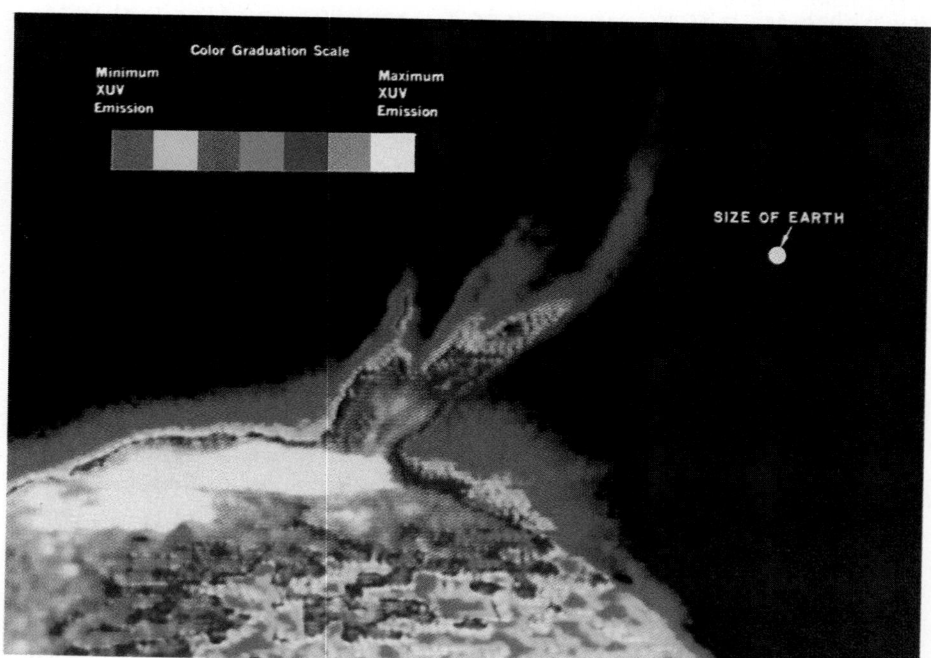

A solar prominence recorded in wavelengths of extreme ultraviolet light. The temperatures of the churning gases at the sun's surface are about 4 000°C, but they increase rapidly to about 500 000°C in the chromosphere and up to some 3 000 000°C in the corona (black area of picture; the colors on the sun's surface represent various intensities of extreme ultraviolet radiation). When the energy of a solar explosion heads toward the earth, taking several days to reach us, it can produce so-called geomagnetic storms, which may disrupt broadcast communications and power transmission lines and cause navigational systems to go haywire. Such storms are also the cause of the beautiful aurora borealis. (Courtesy of NASA)

In the previous chapter, we discussed the fact that Newtonian mechanics must be replaced by Einstein's special theory of relativity when we are dealing with particle velocities comparable to the speed of light. Although many problems were indeed resolved by the theory of relativity in the early part of the 20th century, many experimental and theoretical problems remained unanswered. Attempts to apply the laws of classical physics to explain the behavior of matter on the atomic scale were totally unsuccessful. Various phenomena, such as blackbody radiation, the photoelectric effect, and the emission of sharp spectral lines by atoms in a gas discharge, could not be understood within the framework of classical physics. We shall describe these phenomena because of their importance in subsequent developments.

Another revolution took place in physics between 1900 and 1930. This was the era of a new and more general scheme called *quantum mechanics*. This new approach was highly successful in explaining the behavior of atoms, molecules, and nuclei. Moreover, the quantum theory reduces to classical physics when applied to macroscopic systems. As with relativity, the quantum theory requires a modification of our ideas concerning the physical world.

The basic ideas of quantum theory were first introduced by Max Planck, but most of the subsequent mathematical developments and interpretations were made by a number of distinguished physicists, including Einstein, Bohr, Schrödinger, de Broglie, Heisenberg, Born, and Dirac. Despite the great success of the quantum theory, Einstein frequently played the role of critic, especially with regard to the manner in which the theory was interpreted. In

Biographical Sketch

Max Planck
(1858–1947)

(Courtesy of AIP Niels Bohr Library, W.F. Meggers Collection)

Max Planck was born in Kiel and received his college education in Munich and Berlin. He joined the faculty at Munich in 1880 and five years later received a professorship at Kiel University. He replaced Kirchhoff at the University of Berlin in 1889 and remained there until 1926. He introduced the concept of a "quantum of action" (Planck's constant h) in an attempt to explain the spectral distribution of blackbody radiation, which laid the foundations for quantum theory. He was awarded the Nobel prize in 1918 for this discovery of the quantized nature of energy. The work leading to the "lucky" blackbody radiation formula was described by Planck in his Nobel prize acceptance speech (1920): "But even if the radiation formula proved to be perfectly correct, it would after all have been only an interpolation formula found by lucky guesswork and thus, would have left us rather unsatisfied. I therefore strived from the day of its discovery, to give it a real physical interpretation and this led me to consider the relations between entropy and probability according to Boltzmann's ideas."

Planck's life was filled with personal tragedies. One of his sons was killed in action in World War I, and two daughters were lost in childbirth in the same period. His house was destroyed by bombs in World War II, and his son Erwin was executed by the Nazis in 1944, after being accused of plotting to assassinate Hitler.

Planck became president of the Kaiser Wilhelm Institute of Berlin in 1930. The institute was renamed the Max Planck Institute in his honor after World War II. Although Planck remained in Germany during the Hitler regime, he openly protested the Nazi treatment of his Jewish colleagues, and consequently was forced to resign his presidency in 1937. Following World War II, he was renamed the president of the Max Planck Institute. He spent the last two years of his life in Göttingen as an honored and respected scientist and humanitarian.

particular, Einstein did not accept Heisenberg's uncertainty principle, which says that it is impossible to obtain a precise simultaneous measurement of the position and the velocity of a particle. According to this principle, it is possible to predict only the *probability* of the future of a system, contrary to the deterministic view held by Einstein.[1]

An extensive study of quantum theory is certainly beyond the scope of this book. This chapter is simply an introduction to the underlying ideas of quantum theory. We shall also discuss some simple applications of quantum theory, including the photoelectric effect, the interpretation of atomic spectra, and the Compton effect.

40.1 BLACKBODY RADIATION AND PLANCK'S HYPOTHESIS

An object at any temperature is known to emit radiation sometimes referred to as **thermal radiation.** The characteristics of this radiation depend on the temperature and properties of the object. At low temperatures, the wavelengths of the thermal radiation are mainly in the infrared region and hence are not observed by the eye. As the temperature of the object is increased, it eventually begins to glow red. At sufficiently high temperatures, it appears to be white, as in the glow of the hot tungsten filament of a lightbulb. A careful study of thermal radiation shows that it consists of a continuous distribution of wavelengths from the infrared, visible, and ultraviolet portions of the spectrum.

[1] Einstein's views on the probabilistic nature of quantum theory are brought out in his statement, "God does not play dice with the universe."

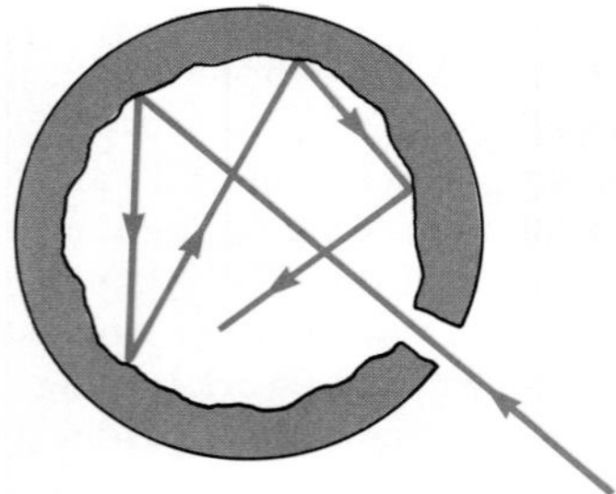

Figure 40.1 The opening in the cavity of a body is a good approximation to a blackbody. As light enters the cavity through the small opening, part is reflected and part is absorbed on each reflection from the interior walls. After many reflections, essentially all of the incident energy is absorbed.

From a classical viewpoint, thermal radiation originates from accelerated charged particles near the surface of the object; those charges emit radiation much like small antennas. The thermally agitated charges can have a distribution of accelerations, which accounts for the continuous spectrum of radiation emitted by the object. By the end of the 19th century, it had become apparent that the classical theory of thermal radiation was inadequate. The basic problem was in understanding the observed distribution of wavelengths in the radiation emitted by a black body. By definition, a black body is an ideal system that absorbs all radiation incident on it. A good approximation to a black body is the inside of a hollow object, as shown in Figure 40.1. The nature of the radiation emitted through a small hole leading to the cavity depends only on the temperature of the cavity walls.

Experimental data for the distribution of energy for blackbody radiation at three different temperatures are shown in Figure 40.2. The radiated energy varies with wavelength and temperature. As the temperature of the black body increases, the total amount of energy it emits increases. Also, with increasing temperatures, the peak of the distribution shifts to shorter wavelengths. This shift was found to obey the following relationship, called **Wien's displacement law:**

$$\lambda_{max} T = 0.2898 \times 10^{-2} \text{ m} \cdot \text{K} \tag{40.1}$$

where λ_{max} is the wavelength at which the curve peaks and T is the absolute temperature of the object emitting the radiation.

Early attempts to explain these results based on classical theories failed. To describe the radiation spectrum, it is useful to define $I(\lambda, T)\, d\lambda$ to be the power per unit area emitted in the wavelength interval $d\lambda$. The result of a calculation based on a classical model of blackbody radiation known as the **Rayleigh-Jeans law** is

$$I(\lambda, T) = \frac{2\pi ckT}{\lambda^4} \tag{40.2}$$

where k is Boltzmann's constant. In this classical model of blackbody radiation, the atoms in the cavity walls are treated as a set of oscillators that emit electromagnetic waves at all wavelengths. This model leads to an average energy per oscillator that is proportional to T.

An experimental plot of the blackbody radiation spectrum is shown in Figure 40.3, together with the theoretical prediction of the Rayleigh-Jeans law. At long wavelengths, the Rayleigh-Jeans law is in reasonable agreement with experimental data. However, at short wavelengths there is major disagreement. This can be seen by noting that as λ approaches zero, the function $I(\lambda, T)$ given by Equation 40.2 approaches infinity; that is, short-wavelength radiation should predominate. This is contrary to the experimental data plotted in Figure 40.3, which shows that as λ approaches zero, $I(\lambda, T)$ also approaches zero. This contradiction is often called the **ultraviolet catastrophe.** Another major problem with classical theory is that it predicts an *infinite total energy density* since all wavelengths are possible.[2] Physically, an infinite energy in the electromagnetic field is an impossible situation.

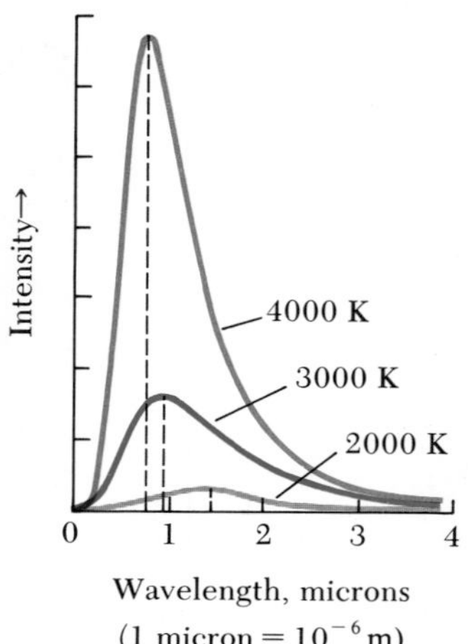

Figure 40.2 Intensity of blackbody radiation versus wavelength at three temperatures. Note that the amount of radiation emitted (the area under a curve) increases with increasing temperature.

[2] The total power per unit area $I = \int_0^{\infty} I(\lambda, T)\, d\lambda$ diverges to ∞ when all wavelengths are allowed.

In 1900 Max Planck discovered a formula for blackbody radiation that was in complete agreement with experiment at all wavelengths. Figure 40.3 shows that Planck's law fits the experimental results quite well. The empirical function proposed by Planck is given by

$$I(\lambda, T) = \frac{2\pi hc^2}{\lambda^5(e^{hc/\lambda kT} - 1)} \tag{40.3}$$

where h is a constant that can be adjusted to fit the data. The current value of h, known as **Planck's constant**, is given by

$$h = 6.626 \times 10^{-34}\ \mathrm{J \cdot s} \tag{40.4}$$

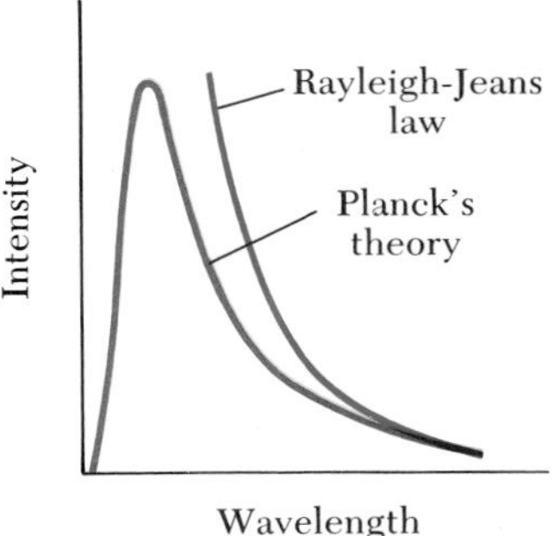

Figure 40.3 Comparison between Planck's theory and classical theory for the distribution of blackbody radiation.

You should show that at long wavelengths, Planck's expression, Equation 40.3, reduces to the Rayleigh-Jeans expression given by Equation 40.2 . Furthermore, at short wavelengths, Planck's law predicts an exponential decrease in $I(\lambda, T)$ with decreasing wavelength, in agreement with experimental results.

In his theory, Planck made two bold and controversial assumptions concerning the nature of the oscillating molecules of the cavity walls:

Quantization of energy

1. The oscillating molecules that emit the radiation could have only *discrete* units of energy E_n given by

$$E_n = nhf \tag{40.5}$$

   where n is a positive integer called a **quantum number** and f is the frequency of vibration of the molecules. The energies of the molecules are said to be *quantized*, and the allowed energy states are called *quantum states*.
2. The molecules emit or absorb energy in discrete units of light energy called **quanta** (or **photons**, as they are now called). They do so by "jumping" from one quantum state to another. If the quantum number n changes by one unit, Equation 40.5 shows that the amount of energy radiated or absorbed by the molecule equals hf. Hence, the energy of a photon corresponding to the energy difference between two adjacent quantum states is given by

$$E = hf \tag{40.6}$$

The molecule will radiate or absorb energy only when it changes quantum states. If it remains in one quantum state, no energy is absorbed or emitted. Figure 40.4 shows the quantized energy levels and allowed transitions proposed by Planck.

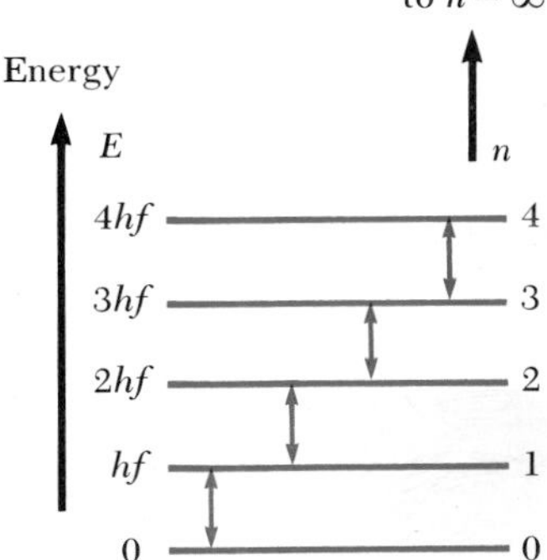

Figure 40.4 Allowed energy levels for an oscillator of natural frequency f. Allowed transitions are indicated by vertical arrows.

The key point in Planck's theory is the radical assumption of quantized energy states. This development marked the birth of the quantum theory. We should emphasize that Planck's work involved more than clever mathematical manipulation. In fact, Planck spent more than six years attempting to derive the blackbody distribution curve. In his own words, the emission problem "represents something absolute, and since I had always regarded the search for the absolute as the loftiest goal of all scientific activity, I eagerly set to work." This work was to occupy most of his life as he continued in his search for a physical interpretation of his formula and to reconcile the quantum concept with classical theory.

EXAMPLE 40.1 Thermal Radiation from the Human Body
The temperature of the skin is approximately 35°C. What is the wavelength at which the peak occurs in the radiation emitted from the skin?

Solution From Wien's displacement law, we have

$$\lambda_{max}T = 0.2898 \times 10^{-2}\ \text{m}\cdot\text{K}$$

Solving for λ_{max}, noting that 35°C corresponds to an absolute temperature of 308 K, we have

$$\lambda_{max} = \frac{0.2898 \times 10^{-2}\ \text{m}\cdot\text{K}}{308\ \text{K}} = 940\ \mu\text{m}$$

This radiation is in the infrared region of the spectrum.

EXAMPLE 40.2 The Quantized Oscillator
A 2-kg mass is attached to a massless spring of force constant $k = 25$ N/m. The spring is stretched 0.4 m from its equilibrium position and released. (a) Find the total energy and frequency of oscillation according to classical calculations. (b) Assume that the energy is quantized and find the quantum number, n, for the system. (c) How much energy would be carried away in a one-quantum change?

Solution (a) The total energy of a simple harmonic oscillator having an amplitude A is $\frac{1}{2}kA^2$. Therefore,

$$E = \tfrac{1}{2}kA^2 = \tfrac{1}{2}(25\ \text{N/m})(0.40\ \text{m})^2 = 2.0\ \text{J}$$

The frequency of oscillation is

$$f = \frac{1}{2\pi}\sqrt{\frac{k}{m}} = \frac{1}{2\pi}\sqrt{\frac{25\ \text{N/m}}{2.0\ \text{kg}}} = 0.56\ \text{Hz}$$

(b) If the energy is quantized, we have $E_n = nhf$, and from the result of (a) we have

$$E_n = nhf = n(6.626 \times 10^{-34}\ \text{J}\cdot\text{s})(0.56\ \text{Hz}) = 2.0\ \text{J}$$

Therefore,

$$n = 5.4 \times 10^{33}$$

(c) The energy carried away in a one-quantum change of energy is

$$E = hf = (6.63 \times 10^{-34}\ \text{J}\cdot\text{s})(0.56\ \text{Hz})$$

$$= 3.7 \times 10^{-34}\ \text{J}$$

The energy carried away by a one-quantum change in energy is such a small fraction of the total energy of the oscillator that we could not expect to see such a small change in the system. Thus, even though the decrease in energy of a spring-mass system is quantized and does decrease by small quantum jumps, our senses perceive the decrease as continuous. Quantum effects become important and measurable only on the submicroscopic level of atoms and molecules.

EXAMPLE 40.3 The Energy of a "Yellow" Photon
What is the energy carried by a quantum of light whose frequency equals 6.00×10^{-14} Hz (yellow light)?

Solution The energy carried by one quantum of light is given by Equation 40.6:

$$E = hf = (6.626 \times 10^{-34}\ \text{J}\cdot\text{s})(6.00 \times 10^{14}\ \text{Hz})$$

$$= 3.98 \times 10^{-19}\ \text{J} = 2.45\ \text{eV}$$

Exercise 1 What is the wavelength of this light?
Answer 500 nm.

40.2 THE PHOTOELECTRIC EFFECT

In the latter part of the 19th century, experiments showed that, when light is incident on certain metallic surfaces, electrons are emitted from the surfaces. This phenomenon is known as the **photoelectric effect,** and the emitted electrons are called **photoelectrons.** The first discovery of this phenomenon was made by Hertz, who was also the first to produce the electromagnetic waves predicted by Maxwell.

Figure 40.5 is a schematic diagram of an apparatus in which the photoelectric effect can occur. An evacuated glass or quartz tube contains a metal plate, C, connected to the negative terminal of a battery. Another metal plate, A, is maintained at a positive potential by the battery. When the tube is kept in the dark, the ammeter reads zero, indicating that there is no current in the circuit. However, when monochromatic light of the appropriate wavelength shines on plate C, a current is detected by the ammeter, indicating a flow of charges across the gap between C and A. The current associated with this process arises from electrons emitted from the negative plate and collected at the positive plate.

A plot of the photoelectric current versus the potential difference, V, between A and C is shown in Figure 40.6 for two light intensities. Note that for large values of V, the current reaches a maximum value, corresponding to the case where all photoelectrons are collected at A. In addition, the current increases as the incident light intensity increases, as you might expect. Finally, when V is negative, that is, when the battery in the circuit is reversed to make C positive and A negative, the photoelectrons are repelled by the negative plate A. Only those electrons having a kinetic energy greater than eV will reach A, where e is the charge on the electron. Furthermore, if V is less than or equal to V_s, called the **stopping potential,** no electrons will reach A and the current will be zero. The stopping potential is *independent* of the radiation intensity. The maximum kinetic energy of the photoelectrons is related to the stopping potential through the relation

$$K_{\text{max}} = eV_s \quad (40.7)$$

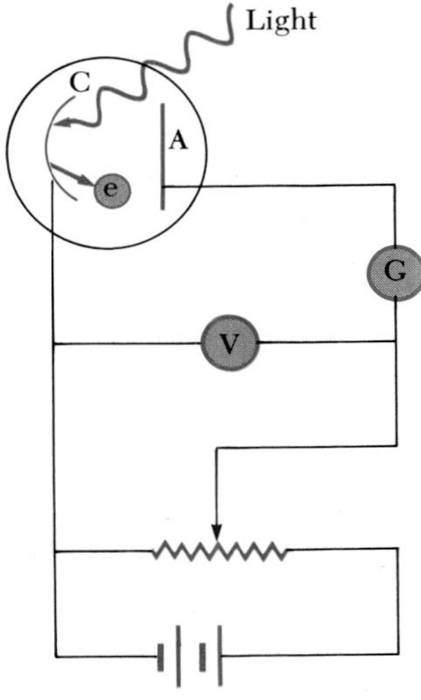

Figure 40.5 Circuit diagram for observing the photoelectric effect. When light strikes plate C (the emitter), photoelectrons are ejected from the plate. The flow of electrons to plate A (the collector) constitutes a current in the circuit.

Several features of the photoelectric effect could not be explained with classical physics or with the wave theory of light. The major observations that were not understood are as follows:

1. No electrons are emitted if the incident light frequency falls below some **cutoff frequency,** f_c, which is characteristic of the material being illuminated. For example, in the case of sodium, $f_c = 5.50 \times 10^{14}$ Hz. This is inconsistent with the wave theory, which predicts that the photoelectric effect should occur at any frequency, provided the light intensity is high enough.
2. If the light frequency exceeds the cutoff frequency, a photoelectric effect is observed and the number of photoelectrons emitted is proportional to the light intensity. However, the maximum kinetic energy of the photoelectrons is independent of light intensity, a fact that cannot be explained by the concepts of classical physics.
3. The maximum kinetic energy of the photoelectrons increases with increasing light frequency.
4. Electrons are emitted from the surface almost instantaneously (less than 10^{-9} s after the surface is illuminated), even at low light intensities. Classically, one would expect that the electrons would require some time to absorb the incident radiation before they acquired enough kinetic energy to escape from the metal.

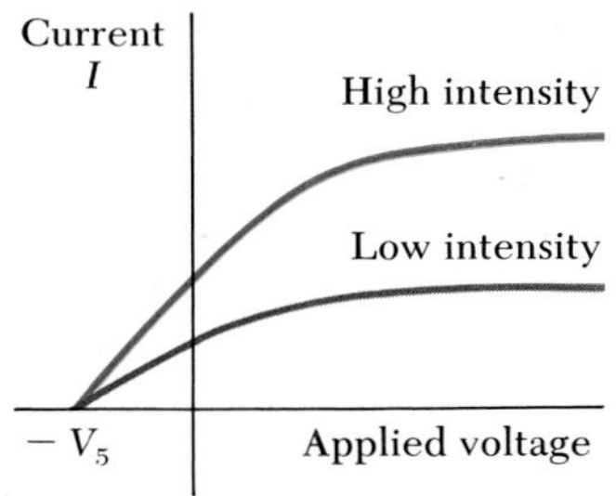

Figure 40.6 Photoelectric current versus applied voltage for two light intensities. The current increases with intensity but reaches a saturation level for large values of V. At voltages equal to or less than $-V_s$, the current is zero.

A successful explanation of the photoelectric effect was given by Einstein in 1905, the same year he published his special theory of relativity. As part of a general paper on electromagnetic radiation, for which he received the Nobel prize in 1921, Einstein extended Planck's concept of quantization to electromagnetic waves. He assumed that light (or any electromagnetic wave) of frequency f can be considered a stream of photons. Each photon has an energy E, given by

$$E = hf$$

Energy of a photon

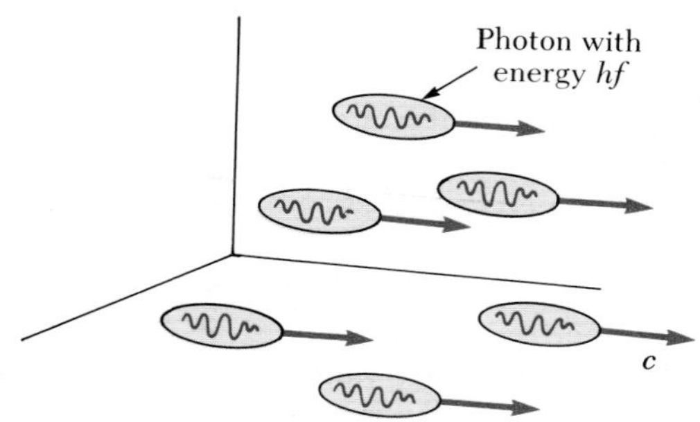

Figure 40.7 Einstein's photon picture of "a traveling light wave."

where h is Planck's constant. Einstein maintained that the energy of light is not distributed evenly over the classical wave front, but is concentrated in discrete regions (or in "bundles") called quanta or photons. A suggestive image of photons, not to be taken too literally, is shown in Figure 40.7. Einstein's simple view of the photoelectric effect was that a photon gives *all* its energy, hf, to a single electron in the metal. Electrons emitted from the surface of the metal possess the maximum kinetic energy, K_{max}. According to Einstein, the maximum kinetic energy for these liberated electrons is

Photoelectric effect equation

$$K_{max} = hf - \phi \tag{40.9}$$

where ϕ is called the **work function** of the metal. The work function represents the minimum energy with which an electron is bound in the metal, and is of the order of a few electron volts. Table 40.1 lists values of work functions measured for different metals.

With the photon theory of light, one can explain the features of the photoelectric effect that cannot be understood using classical concepts. These are briefly described in the order they were introduced earlier:

1. The fact that the photoelectric effect is not observed below a certain cutoff frequency follows from the fact that the energy of the photon must be greater than or equal to ϕ. If the energy of the incoming photon is not equal to or greater than ϕ, the electrons will never be ejected from the surface, regardless of the intensity of the light.
2. The fact that K_{max} is independent of the light intensity can be understood with the following argument. If the light intensity is doubled, the number of photons is doubled, which doubles the number of photoelectrons emitted. However, their kinetic energy, which equals $hf - \phi$, depends only on the light frequency and the work function, not on the light intensity.
3. The fact that K_{max} increases with increasing frequency is easily understood with Equation 40.9.
4. Finally, the fact that the electrons are emitted almost instantaneously is consistent with the particle theory of light, in which the incident energy appears in small packets and there is a one-to-one interaction between photons and electrons. This is in contrast to having the energy of the photons distributed uniformly over a large area.

TABLE 40.1 Work Functions of Selected Metals

Metal	ϕ (eV)
Na	2.28
Al	4.08
Cu	4.70
Zn	4.31
Ag	4.73
Pt	6.35
Pb	4.14
Fe	4.50

A final confirmation of Einstein's theory is a test of the prediction of a linear relationship between f and K_{max}. Indeed, such a linear relationship is observed, as sketched in Figure 40.8. The slope of such a curve gives a value for h. The intercept on the horizontal axis gives the cutoff frequency, which is related to the work function through the relation $f_c = \phi/h$. This corresponds to a **cutoff wavelength** of

$$\lambda_c = \frac{c}{f_c} = \frac{c}{\phi/h} = \frac{hc}{\phi} \tag{40.10}$$

where c is the speed of light (3.00×10^8 m/s). Wavelengths *greater* than λ_c incident on a material with a work function ϕ do not result in the emission of photoelectrons.

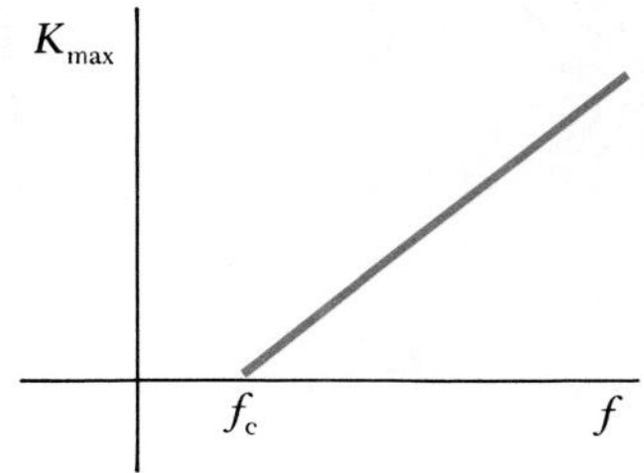

Figure 40.8 A sketch of K_{max} versus frequency of incident light for photoelectrons in a typical photoelectric effect experiment. Photons with frequency less than f_c do not have sufficient energy to eject an electron from the metal.

EXAMPLE 40.4 The Photoelectric Effect for Sodium
A sodium surface is illuminated with light of wavelength 300 nm. The work function for sodium metal is 2.46 eV. Find (a) the kinetic energy of the ejected photoelectrons and (b) the cutoff wavelength for sodium.

Solution (a) The energy of the illuminating light beam is

$$E = hf = \frac{hc}{\lambda} = \frac{(6.626 \times 10^{-34}\ \mathrm{J \cdot s})(3.00 \times 10^8\ \mathrm{m/s})}{300 \times 10^{-9}\ \mathrm{m}}$$

$$= 6.63 \times 10^{-19}\ \mathrm{J} = \frac{6.63 \times 10^{-19}\ \mathrm{J}}{1.60 \times 10^{-19}\ \mathrm{J/eV}} = 4.14\ \mathrm{eV}$$

where we have used the conversion of units $1\ \mathrm{eV} = 1.60 \times 10^{-19}$ J. Using Equation 40.9 gives

$$K_{max} = hf - \phi = 4.14\ \mathrm{eV} - 2.46\ \mathrm{eV} = 1.68\ \mathrm{eV}$$

(b) The cutoff wavelength can be calculated from Equation 40.10 after we convert ϕ from electron volts to joules:

$$\phi = 2.46\ \mathrm{eV} = (2.46\ \mathrm{eV})(1.60 \times 10^{-19}\ \mathrm{J/eV})$$

$$= 3.94 \times 10^{-19}\ \mathrm{J}$$

Hence

$$\lambda_c = \frac{hc}{\phi} = \frac{(6.626 \times 10^{-34}\ \mathrm{J \cdot s})(3.00 \times 10^8\ \mathrm{m/s})}{3.94 \times 10^{-19}\ \mathrm{J}}$$

$$= 5.05 \times 10^{-7}\ \mathrm{m} = 505\ \mathrm{nm}$$

This wavelength is in the green region of the visible spectrum.

Exercise 2 Calculate the maximum speed of the photoelectrons under the conditions described in this example.
Answer 7.68×10^5 m/s.

40.3 THE COMPTON EFFECT

In 1919, Einstein concluded that a photon of energy E travels in a single direction (unlike a spherical wave) and carries a momentum equal to E/c or hf/c. In his own words, "if a bundle of radiation causes a molecule to emit or absorb an energy packet hf, then momentum of quantity hf/c is transferred to the molecule, directed along the line of the bundle for absorption and opposite the bundle for emission." In 1923, Arthur Holly Compton (1892–1962) and Peter Debye independently carried Einstein's idea of photon momentum farther. They realized that the scattering of x-ray photons from electrons could be explained by treating photons as point-like particles with energy hf and momentum hf/c, and by conserving energy and momentum of the photon-electron pair in a collision.

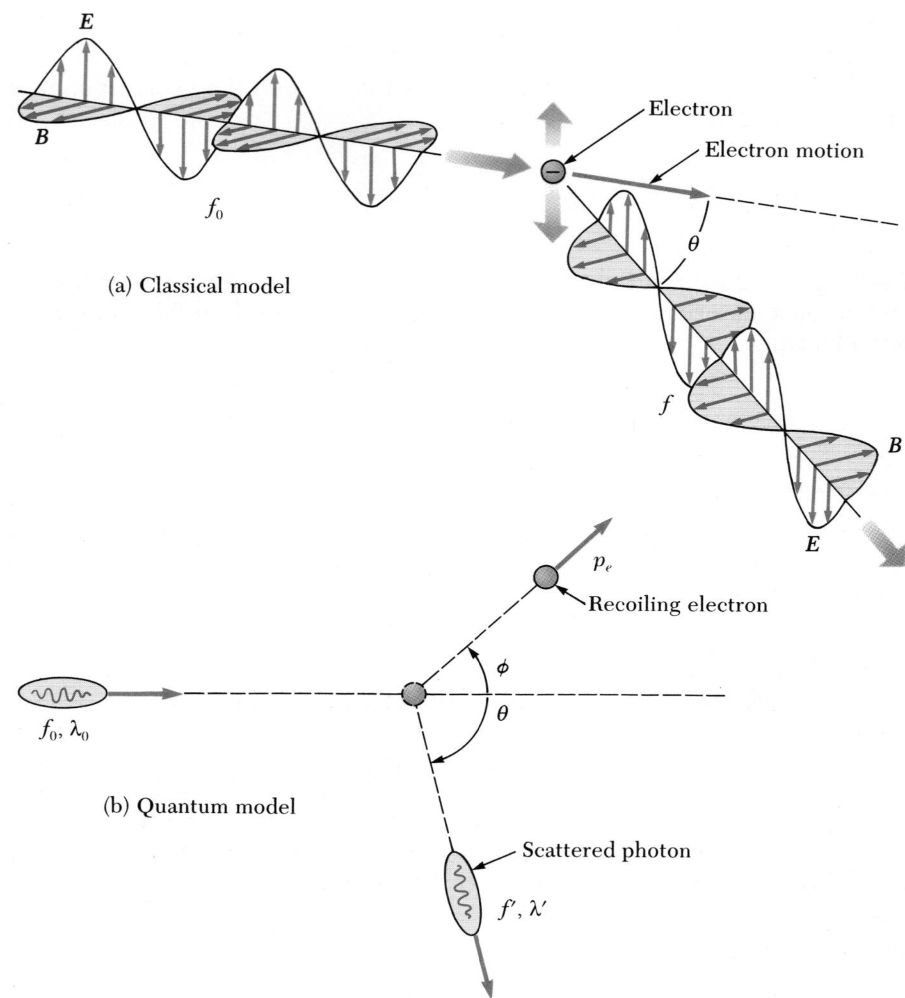

Figure 40.9 X-ray scattering from an electron: (a) the classical model, (b) the quantum model.

Arthur Holly Compton (1892–1962), American physicist. (Courtesy of AIP Niels Bohr Library)

Prior to 1922, Compton and his coworkers had accumulated evidence that showed that classical wave theory failed to explain the scattering of x-rays from electrons. According to classical wave theory, incident electromagnetic waves of frequency f_0 should accelerate electrons, forcing them to oscillate and reradiate at a frequency $f < f_0$ as in Figure 40.9a. Furthermore, according to classical theory, the frequency or wavelength of the scattered radiation should depend on the time of exposure of the sample to the incident radiation as well as the intensity of the incident radiation. Contrary to these predictions, Compton's experimental results showed that the wavelength shift of x-rays scattered at a given angle depends *only on the scattering angle.* Figure 40.9b shows the quantum picture of the transfer of momentum and energy between an individual x-ray photon and an electron.

A schematic diagram of the apparatus used by Compton is shown in Figure 40.10a. In his original experiment Compton measured the dependence of scattered x-ray intensity on wavelength at three different scattering angles. The wavelength was measured with a rotating crystal spectrometer using carbon as the target, and the intensity was determined by an ioni-

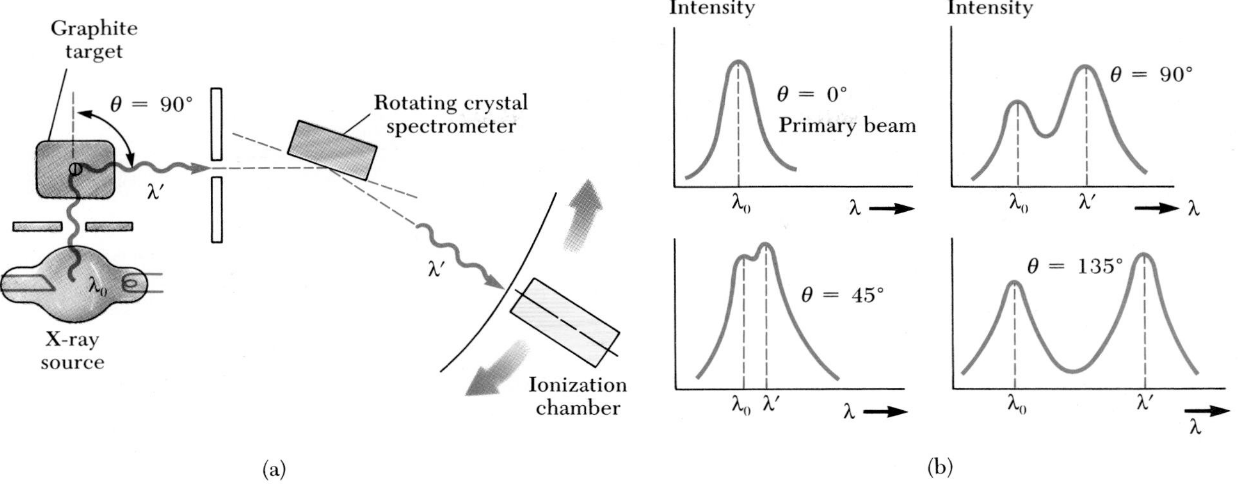

Figure 40.10 (a) Schematic diagram of Compton's apparatus. (b) Scattered intensity versus wavelength of Compton scattering at $\theta = 0°$, 45°, 90°, and 135°.

zation chamber that generated a current proportional to the x-ray intensity. The incident beam consisted of monochromatic x-rays of wavelength $\lambda_0 = 0.071$ nm. The experimental intensity versus wavelength plots observed by Compton for scattering angles of 0°, 45°, 90°, and 135° are shown in Figure 40.10b. They show two peaks, one at λ_0 and a shifted peak at λ'. The peak at λ_0 is caused by x-rays scattered from electrons that are tightly bound to the target atoms, while the shifted peak is caused by scattering of x-rays from free electrons in the target. In his analysis, Compton predicted that the shifted peak should depend on scattering angle θ as

$$\lambda' - \lambda_0 = \frac{h}{mc}(1 - \cos\theta) \qquad (40.11)$$

Compton shift equation

In this expression, known as the **Compton shift equation,** m is the mass of the electron; h/mc is called the **Compton wavelength** λ_c of the electron; and has a currently accepted value of

$$\lambda_c = \frac{h}{mc} = 0.00243 \text{ nm}$$

Compton wavelength

Compton's measurements were in excellent agreement with the predictions of Equation 40.11. It is fair to say that these were the first experimental results to convince most physicists of the fundamental validity of the quantum theory!

Derivation of the Compton Shift Equation

We can derive the Compton shift expression, Equation 40.11, by assuming that the photon exhibits particle-like behavior and collides elastically with a free electron initially at rest, as in Figure 40.11. In this model, the photon is treated as a particle of energy $E = hf = hc/\lambda$, with a rest mass of zero. In the scattering process, the total energy and total momentum of the system must be

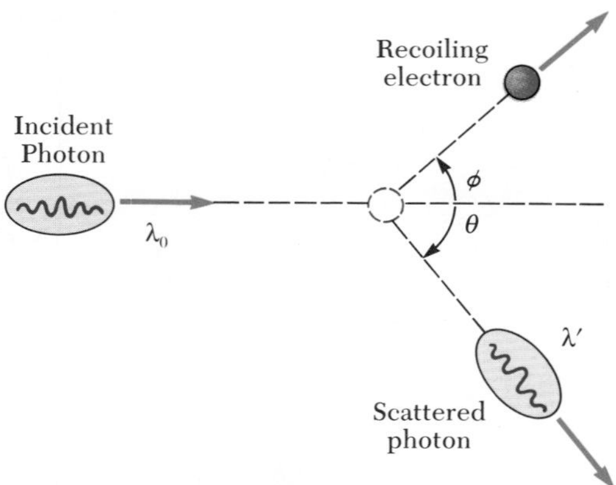

Figure 40.11 Diagram representing Compton scattering of a photon by an electron. The scattered photon has less energy (or longer wavelength) than the incident photon.

conserved. Applying conservation of energy to this process gives

$$\frac{hc}{\lambda_0} = \frac{hc}{\lambda'} + K_e$$

where hc/λ_0 is the energy of the incident photon, hc/λ' is the energy of the scattered photon, and K_e is the kinetic energy of the recoiling electron. Since the electron may recoil at speeds comparable to the speed of light, we must use the relativistic expression for K_e, given by $K_e = \gamma mc^2 - mc^2$ (Equation 39.23). Therefore,

$$\frac{hc}{\lambda_0} = \frac{hc}{\lambda'} + \gamma mc^2 - mc^2 \qquad (40.12)$$

where $\gamma = 1/\sqrt{1 - v^2/c^2}$.

Next, we can apply the law of conservation of momentum to this collision, noting that *both the x and y components of momentum are conserved.* Since the momentum of a photon has a magnitude given by $p = E/c = h/\lambda$, and since the relativistic expression for the momentum of the recoiling electron is $p_e = \gamma mv$ (Eq. 39.18), we obtain the following expressions for the x and y components of linear momentum:

$$x \text{ component:} \qquad \frac{h}{\lambda_0} = \frac{h}{\lambda'} \cos\theta + \gamma mv \cos\phi \qquad (40.13)$$

$$y \text{ component:} \qquad 0 = \frac{h}{\lambda'} \sin\theta - \gamma mv \sin\phi \qquad (40.14)$$

By eliminating v and ϕ from Equations 40.12 to 40.14, we obtain a single expression that relates the remaining three variables (λ', λ_0, and θ). After some algebra (Problem 68), one obtains the Compton shift equation:

$$\Delta\lambda = \lambda' - \lambda_0 = \frac{h}{mc}(1 - \cos\theta)$$

EXAMPLE 40.5 Compton Scattering at 45°
X-rays of wavelength $\lambda_0 = 0.20$ nm are scattered from a block of material. The scattered x-rays are observed at an angle of 45° to the incident beam. Calculate the wavelength of the scattered x-rays at this angle.

Solution The shift in wavelength of the scattered x-rays is given by Equation 40.11. Taking $\theta = 45°$, we find that

$$\Delta\lambda = \frac{h}{mc}(1 - \cos\theta)$$

$$\Delta\lambda = \frac{6.626 \times 10^{-34}\ \text{J}\cdot\text{s}}{(9.11 \times 10^{-31}\ \text{kg})(3.00 \times 10^{8}\ \text{m/s})}(1 - \cos 45°)$$

$$= 7.10 \times 10^{-13}\ \text{m} = 0.000710\ \text{nm}$$

Hence, the wavelength of the scattered x-ray at this angle is

$$\lambda' = \Delta\lambda + \lambda_0 = 0.200710\ \text{nm}$$

Exercise 3 Find the fraction of energy lost by the photon in this collision.
Answer Fraction $= \Delta\lambda/\lambda_0 = 0.00354$.

40.4 ATOMIC SPECTRA

As already pointed out in Section 40.1, all substances at some temperature emit thermal radiation which is characterized by a continuous distribution of wavelengths. The shape of the distribution depends on the temperature and properties of the substance. In sharp contrast to this continuous distribution spectrum is the discrete **line spectrum** emitted by a low-pressure gas subject to electric discharge. When the light from such a low-pressure gas discharge is examined with a spectroscope, it is found to consist of a few bright lines of pure color on a generally dark background. This contrasts sharply with the continuous rainbow of colors seen when a glowing solid is viewed through a spectroscope. Furthermore, as you can see from Figure 40.12a, the wavelengths

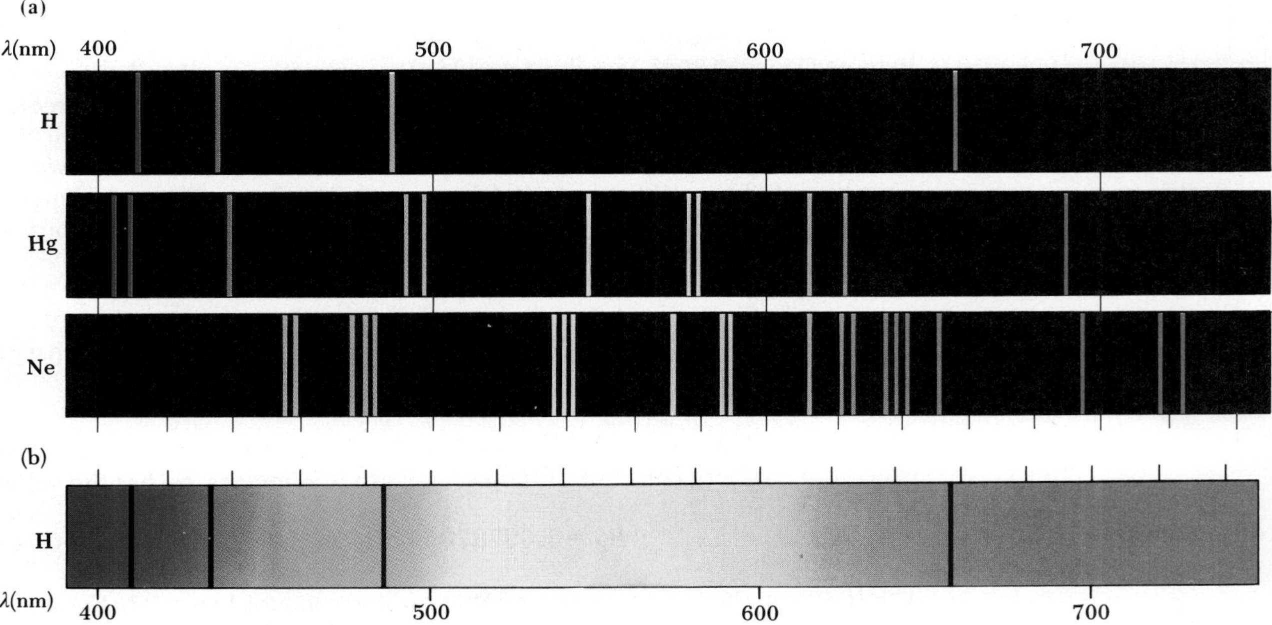

Figure 40.12 Visible spectra (a) Line spectra produced by emission in the visible range for the elements hydrogen, helium, and neon. (b) The absorption spectrum for hydrogen. The dark absorption lines occur at the same wavelengths as the emission lines for hydrogen shown in (a). (Whitten, K.W., K.D. Gailey, and R.E. Davis, *General Chemistry*, 3rd ed., Saunders College Publishing, 1987).

contained in a given line spectrum are characteristic of the particular element emitting the light. The simplest line spectrum is observed for atomic hydrogen, and we shall describe this spectrum in detail. Other atoms such as mercury and neon emit completely different line spectra. Since no two elements emit the same line spectrum, this phenomenon represents a practical and sensitive technique for identifying the elements present in unknown samples.

Another form of spectroscopy which is very useful in analyzing substances is **absorption spectroscopy.** An absorption spectrum is obtained by passing light from a continuous source through a gas or dilute solution of the element being analyzed. The absorption spectrum consists of a series of dark lines superimposed on the otherwise continuous spectrum of that same element, as shown in Figure 40.12b for atomic hydrogen. In general, not all of the emission lines are present in the absorption spectrum.

The absorption spectrum of an element has many practical applications. For example, the continuous spectrum of radiation emitted by the sun must pass through the cooler gases of the solar atmosphere and through the earth's atmosphere. The various absorption lines observed in the solar spectrum have been used to identify elements in the solar atmosphere. In early studies of the solar spectrum, some lines were found that did not correspond to any known element. A new element had been discovered! The new element was named helium, after the Greek word for sun, *helios*. Scientists are able to examine the light from stars other than our sun in this fashion, but elements other than those present on earth have never been detected. Atomic absorption spectroscopy has also been a useful technique in analyzing heavy metal contamination of the food chain. For example, the first determination of high levels of mercury in tuna fish was made with atomic absorption.

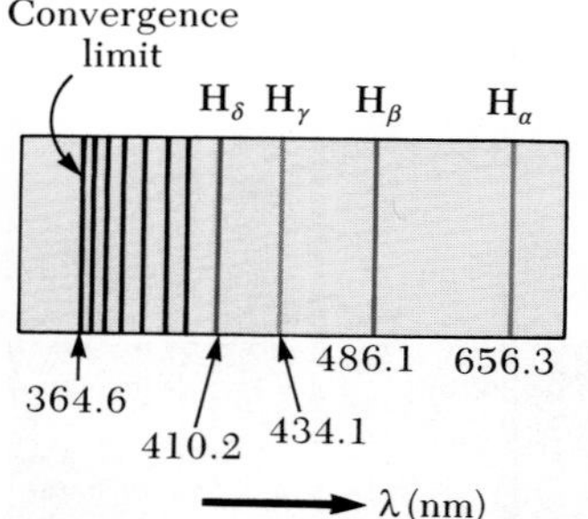

Figure 40.13 The Balmer series of spectral lines for hydrogen (emission spectrum).

From 1860 to 1885, scientists had accumulated a great deal of data using spectroscopic measurements. In 1885, a Swiss school teacher, Johann Jacob Balmer (1825–1898) found a formula that correctly predicted the wavelengths of four visible emission lines of hydrogen: H_α(red), H_β(green), H_γ(blue), and H_δ(violet). Figure 40.13 shows these and other lines in the emission spectrum of hydrogen. The four visible lines occur at the wavelengths 656.3 nm, 486.1 nm, 434.1 nm, and 410.2 nm. The wavelengths of these lines (called the Balmer series) can be described by the empirical equation

Balmer series

$$\frac{1}{\lambda} = R_H\left(\frac{1}{2^2} - \frac{1}{n^2}\right) \qquad (40.15)$$

where n may have integral values of 3, 4, 5, . . . and R_H is a constant, now called the **Rydberg constant.** If the wavelength is in meters, R_H has the value

Rydberg constant

$$R_H = 1.0973732 \times 10^7 \text{ m}^{-1} \qquad (40.16)$$

The H_α line in the Balmer series at 656.3 nm corresponds to $n = 3$ in Equation 40.15; the H_β line at 486.1 nm corresponds to $n = 4$, and so on. Much to Balmer's delight, the measured spectral lines agreed with his empirical formula to within 0.1%!

Other line spectra for hydrogen were found following Balmer's discovery. These spectra are called the Lyman, Paschen, and Brackett series after their discoverers. The wavelengths of the lines in these series can be calculated by the following empirical formulas:

$$\frac{1}{\lambda} = R_H\left(1 - \frac{1}{n^2}\right) \qquad n = 2, 3, 4, \ldots \qquad (40.17)$$ Lyman series

$$\frac{1}{\lambda} = R_H\left(\frac{1}{3^2} - \frac{1}{n^2}\right) \qquad n = 4, 5, 6, \ldots \qquad (40.18)$$ Paschen series

$$\frac{1}{\lambda} = R_H\left(\frac{1}{4^2} - \frac{1}{n^2}\right) \qquad n = 5, 6, 7, \ldots \qquad (40.19)$$ Brackett series

40.5 BOHR'S QUANTUM MODEL OF THE ATOM

At the beginning of the 20th century, scientists were perplexed by the failure of classical physics in explaining the characteristics of atomic spectra. Why did hydrogen emit only certain lines in the visible part of the spectrum? Furthermore, why did hydrogen absorb only those wavelengths which it emitted? In 1913, the Danish scientist Niels Bohr (1885–1963) provided an explanation of atomic spectra that included some features present in the currently accepted theory. Bohr's theory contained a combination of ideas from Planck's original quantum theory, Einstein's photon theory of light, and Rutherford's model of the atom. Bohr's model of the hydrogen atom contains some classical features as well as some revolutionary postulates that could not be justified within the framework of classical physics. The Bohr model can be applied quite successfully to such hydrogen-like ions as singly ionized helium and doubly ionized lithium. However, the theory does not properly describe the spectra of more complex atoms and ions.

The basic ideas of the Bohr theory as it applies to an atom of hydrogen are as follows:

Assumptions of the Bohr theory

1. The electron moves in circular orbits about the proton under the influence of the Coulomb force of attraction, as in Figure 40.14. So far nothing new!
2. Only certain orbits are stable. These stable orbits are ones in which the electron does not radiate. Hence the energy is fixed or stationary, and classical mechanics may be used to describe the electron's motion.
3. Radiation is emitted by the atom when the electron "jumps" from a more energic initial stationary state to a lower state. This "jump" cannot be visualized or treated classically. In particular, the frequency f of the photon emitted in the jump *is independent of the frequency of the electron's orbital motion.* The frequency of the light emitted is related to the change in the atom's energy and is given by the Planck-Einstein formula

$$E_i - E_f = hf \qquad (40.20)$$

where E_i is the energy of the initial state, E_f is the energy of the final state, and $E_i > E_f$.
4. The size of the allowed electron orbits is determined by an additional quantum condition imposed on the electron's orbital angular momentum. Namely, the allowed orbits are those for which the electron's orbital angular momentum about the nucleus is an integral multiple of $\hbar = h/2\pi$,

$$mvr = n\hbar \qquad n = 1, 2, 3, \ldots \qquad (40.21)$$

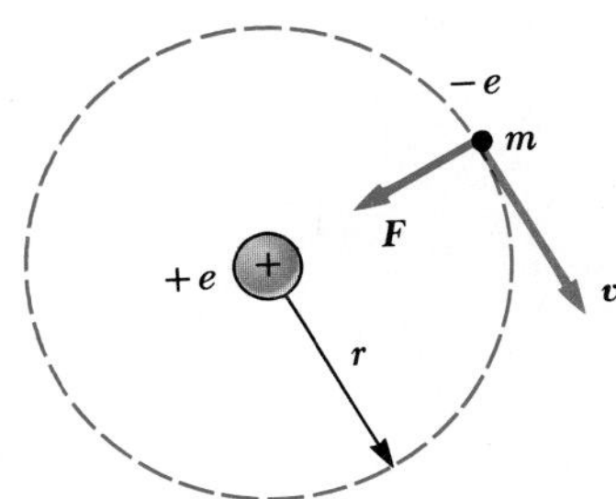

Figure 40.14 Diagram representing Bohr's model of the hydrogen atom, in which the orbiting electron is allowed to be only in specific orbits of discrete radii.

Biographical Sketch

Niels Bohr
(1885–1962)

Niels Bohr, a Danish physicist, proposed the first quantum model of the atom. He was an active participant in the early development of quantum mechanics and provided much of its philosophical framework. He made many other important contributions to theoretical nuclear physics, including the development of the liquid-drop model of the nucleus and work in nuclear fission. He was awarded the 1922 Nobel prize in physics for his investigation of the structure of atoms and of the radiation emanating from them.*

Bohr spent most of his life in Copenhagen, Denmark, and received his doctorate at the University of Copenhagen in 1911. The following year he traveled to England where he worked under J. J. Thomson at Cambridge, and then went to Manchester where we worked under Ernest Rutherford. He was married in 1912 and returned to the University of Copenhagen in 1916 as professor of physics. During the 1920s and 1930s, Bohr headed the Institute for Advanced Studies in Copenhagen, under the support of the Carlsberg brewery. (This was certainly the greatest benefit provided by beer to the field of theoretical physics.) The institute was a magnet for many of the world's best physicists and provided a forum for the exchange of ideas. Bohr, a firm believer in doing physics on a "man-to-man" basis, was always the initiator of probing questions, reflections, and discussions with his guests.

When Bohr visited the United States in 1939 to attend a scientific conference, he brought news that the fission of uranium had been discovered by Hahn and Strassman in Berlin. The results, confirmed by other scientists shortly thereafter, were the foundations of the atomic bomb developed in the United States during World War II. He returned to Denmark and was there during the German occupation in 1940. He escaped to Sweden in 1943 to avoid imprisonment, and helped to arrange the escape of many imperiled Danish citizens.

Although Bohr himself worked on the Manhattan project at Los Alamos until 1945, he strongly felt that openness between nations concerning nuclear weapons should be the first step in establishing their control. After the war, Bohr committed himself to many human issues including the development of peaceful uses of atomic energy. He was awarded the Atoms for Peace award in 1957. As John Archibald Wheeler summarizes, "Bohr was a great scientist. He was a great citizen of Denmark, of the World. He was a great human being."

(Photograph courtesy of AIP Niels Bohr Library, Margarethe Bohr Collection)

* For several interesting articles concerning Bohr, read the special Niels Bohr centennial issue of *Physics Today*, October 1985.

Using these four assumptions, we can now calculate the allowed energy levels and emission wavelengths of the hydrogen atom. Recall that the electrical potential energy of the system shown in Figure 40.14 is given by $U = qV = -ke^2/r$, where k (the Coulomb constant) has the value $1/4\pi\epsilon_0$. Thus, the total energy of the atom, which contains both kinetic and potential energy terms, is given by

$$E = K + U = \tfrac{1}{2}mv^2 - k\frac{e^2}{r} \tag{40.22}$$

Applying Newton's second law to this system, we see that the Coulomb attractive force on the electron, ke^2/r^2, must equal the mass times the centripetal acceleration of the electron, or

$$\frac{ke^2}{r^2} = \frac{mv^2}{r}$$

From this expression, we immediately find the kinetic energy to be

$$K = \frac{mv^2}{2} = \frac{ke^2}{2r} \qquad (40.23)$$

Substituting this value of K into Equation 40.22, we find that the total energy of the atom is

$$E = -\frac{ke^2}{2r} \qquad (40.24)$$

Note that the total energy is negative, indicating a bound electron-proton system. This means that energy in the amount of $ke^2/2r$ must be added to the atom to just remove the electron and make the total energy zero. An expression for r, the radius of the stationary orbits, may be obtained by solving Equations 40.21 and 40.23 for v and equating the results:

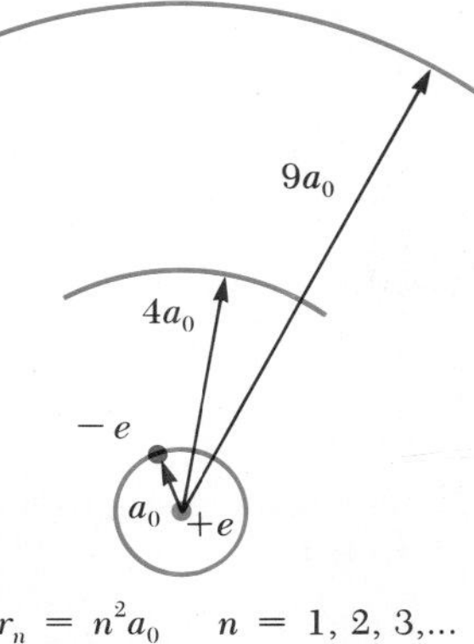

Figure 40.15 The first three Bohr orbits for hydrogen.

Radii of Bohr orbits in hydrogen

$$r_n = \frac{n^2\hbar^2}{mke^2} \qquad n = 1, 2, 3, \ldots \qquad (40.25)$$

The orbit for which $n = 1$ has the smallest radius; it is called the **Bohr radius,** a_0, and has the value

$$a_0 = \frac{\hbar^2}{mke^2} = 0.529 \text{ Å} = 0.0529 \text{ nm} \qquad (40.26)$$

The fact that Bohr's theory gave an accurate value for the radius of hydrogen from first principles without any empirical calibration of orbit size was considered a striking triumph for this theory. The first three Bohr orbits are shown to scale in Figure 40.15.

The quantization of the orbit radii immediately leads to energy quantization. This can be seen by substituting $r_n = n^2a_0$ into Equation 40.24, giving for the allowed energy levels

$$E_n = -\frac{ke^2}{2a_0}\left(\frac{1}{n^2}\right) \qquad n = 1, 2, 3, \ldots \qquad (40.27)$$

Inserting numerical values into Equation 40.27 gives

$$E_n = -\frac{13.6}{n^2} \text{ eV} \qquad n = 1, 2, 3, \ldots \qquad (40.28)$$

The lowest stationary or non-radiating state is called the **ground state,** has $n = 1$, and has an energy $E_1 = -13.6$ eV. The next state or **first excited state** has $n = 2$, and an energy $E_2 = E_1/2^2 = -3.4$ eV. An energy level diagram showing the energies of these discrete energy states and the corresponding quantum numbers is shown in Figure 40.16. The uppermost level, corresponding to $n = \infty$ (or $r = \infty$) and $E = 0$, represents the state for which the electron is removed from the atom. The minimum energy required to ionize the atom (that is, to completely remove an electron in the ground state from the proton's influence) is called the **ionization energy.** As can be seen from Figure 40.16, the ionization energy for hydrogen based on Bohr's calculation is 13.6 eV. This constituted another major achievement for the Bohr theory, since the ionization energy for hydrogen had already been measured to be precisely 13.6 eV.

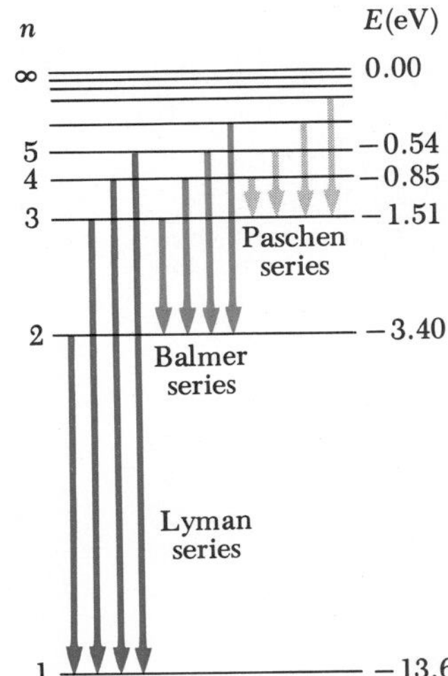

Figure 40.16 An energy level diagram for hydrogen. In such diagrams the discrete allowed energies are plotted on the vertical axis. Nothing is plotted on the horizontal axis, but the horizontal extent of the diagram is made large enough to show allowed transitions. Note that the quantum numbers are given on the left.

Equation 40.27 together with Bohr's third postulate can be used to calculate the frequency of the photon emitted when the electron jumps from an outer orbit to an inner orbit:

$$f = \frac{E_i - E_f}{h} = \frac{ke^2}{2a_0h}\left(\frac{1}{n_f^2} - \frac{1}{n_i^2}\right) \tag{40.29}$$

Since the quantity actually measured is wavelength, it is convenient to convert frequency to wavelength using $c = f\lambda$ to get

Emission wavelengths of hydrogen

$$\frac{1}{\lambda} = \frac{f}{c} = \frac{ke^2}{2a_0hc}\left(\frac{1}{n_f^2} - \frac{1}{n_i^2}\right) \tag{40.30}$$

The remarkable fact is that the *theoretical* expression, Equation 40.30, is identical to a generalized form of the empirical relations discovered by Balmer and others, given by Equations 40.16 to 40.19,

$$\frac{1}{\lambda} = R_H\left(\frac{1}{n_f^2} - \frac{1}{n_i^2}\right) \tag{40.31}$$

provided that the combination of constants $ke^2/2a_0hc$ is equal to the experimentally determined Rydberg constant, $R_H = 1.0973732 \times 10^7 \text{ m}^{-1}$. After Bohr demonstrated the agreement of these two quantities to a precision of about 1%,it was soon recognized as the crowning achievement of his new theory of quantum mechanics. Furthermore, Bohr showed that all of the spectral series for hydrogen have a natural interpretation in his theory. These spectral series are shown as transitions between energy levels in Figure 40.16.

Bohr immediately extended his model for hydrogen to other elements in which all but one electron had been removed. Ionized elements such as He^+, Li^{++}, and Be^{+++} were suspected to exist in hot stellar atmospheres, where frequent atomic collisions occurred with enough energy to completely remove one or more atomic electrons. Bohr showed that many mysterious lines observed in the sun and several stars could not be due to hydrogen, but were correctly predicted by his theory if attributed to singly ionized helium. In general, to describe a single electron orbiting a fixed nucleus of charge $+Ze$, Bohr's theory gives

$$r_n = (n^2)\frac{a_0}{Z} \tag{40.32}$$

and

$$E_n = -\frac{ke^2}{2a_0}\left(\frac{Z^2}{n^2}\right) \qquad n = 1, 2, 3, \ldots \tag{40.33}$$

EXAMPLE 40.6 Spectral Lines from the Star ζ-Puppis The mysterious lines observed by Pickering in 1896 in the spectrum of the star ζ-Puppis fit the empirical formula

$$\frac{1}{\lambda} = R_H\left(\frac{1}{(n_f/2)^2} - \frac{1}{(n_i/2)^2}\right)$$

Show that these lines can be explained by the Bohr theory as originating from He^+.

Solution He^+ has $Z = 2$. Thus the allowed energy levels are given by Equation 40.33 as

$$E_n = -\frac{ke^2}{2a_0}\left(\frac{4}{n^2}\right)$$

Using $hf = E_i - E_f$ we find

$$f = \frac{E_i - E_f}{h} = \frac{ke^2}{2a_0 h}\left(\frac{4}{n_f^2} - \frac{4}{n_i^2}\right)$$
$$= \frac{ke^2}{2a_0 h}\left(\frac{1}{(n_f/2)^2} - \frac{1}{(n_i/2)^2}\right)$$

or

$$\frac{1}{\lambda} = \frac{f}{c} = \frac{ke^2}{2a_0 hc}\left(\frac{1}{(n_f/2)^2} - \frac{1}{(n_i/2)^2}\right)$$

This is the desired solution, since $R_H \equiv ke^2/2a_0hc$.

EXAMPLE 40.7 An Electronic Transition in Hydrogen

The electron in the hydrogen atom makes a transition from the $n = 2$ energy state to the ground state (corresponding to $n = 1$). Find the wavelength and frequency of the emitted photon.

Solution We can use Equation 40.31 directly to obtain λ, with $n_i = 2$ and $n_f = 1$:

$$\frac{1}{\lambda} = R_H\left(\frac{1}{n_f^2} - \frac{1}{n_i^2}\right)$$
$$\frac{1}{\lambda} = R_H\left(\frac{1}{1^2} - \frac{1}{2^2}\right) = \frac{3R_H}{4}$$
$$\lambda = \frac{4}{3R_H} = \frac{4}{3(1.097 \times 10^7\ \text{m}^{-1})}$$
$$= 1.215 \times 10^{-7}\ \text{m} = 121.5\ \text{nm}$$

This wavelength lies in the ultraviolet region.

Since $c = f\lambda$, the frequency of the photon is

$$f = \frac{c}{\lambda} = \frac{3.00 \times 10^8\ \text{m/s}}{1.215 \times 10^{-7}\ \text{m}} = 2.47 \times 10^{15}\ \text{Hz}$$

Exercise 4 What is the wavelength of the photon emitted by hydrogen when the electron makes a transition from the $n = 3$ state to the $n = 1$ state?

Answer $\frac{9}{8R_H} = 102.6$ nm.

EXAMPLE 40.8 The Balmer Series for Hydrogen

The Balmer series for the hydrogen atom corresponds to electronic transitions that terminate in the state of quantum number $n = 2$, as shown in Figure 40.16. (a) Find the longest-wavelength photon emitted and determine its energy.

Solution The longest-wavelength photon in the Balmer series results from the transition from $n = 3$ to $n = 2$. Using Equation 40.31 gives

$$\frac{1}{\lambda} = R_H\left(\frac{1}{n_f^2} - \frac{1}{n_i^2}\right)$$
$$\frac{1}{\lambda_{max}} = R_H\left(\frac{1}{2^2} - \frac{1}{3^2}\right) = \frac{5}{36}R_H$$
$$\lambda_{max} = \frac{36}{5R_H} = \frac{36}{5(1.097 \times 10^7\ \text{m}^{-1})} = 656.3\ \text{nm}$$

This wavelength is in the red region of the visible spectrum.

The energy of this photon is

$$E_{photon} = hf = \frac{hc}{\lambda_{max}}$$
$$= \frac{(6.626 \times 10^{-34}\ \text{J}\cdot\text{s})(3.00 \times 10^8\ \text{m/s})}{656.3 \times 10^{-9}\ \text{m}}$$
$$= 3.03 \times 10^{-19}\ \text{J} = 1.89\ \text{eV}$$

We could also obtain the energy of the photon by using the expression $hf = E_3 - E_2$, where E_2 and E_3 are the energy levels of the hydrogen atom, which can be calculated from Equation 40.28. Note that this is the lowest-energy photon in this series since it involves the smallest energy change.

(b) Find the shortest-wavelength photon emitted in the Balmer series.

Solution The shortest-wavelength photon in the Balmer series is emitted when the electron makes a transition from $n = \infty$ to $n = 2$. Therefore

$$\frac{1}{\lambda_{min}} = R_H\left(\frac{1}{2^2} - \frac{1}{\infty}\right) = \frac{R_H}{4}$$
$$\lambda_{min} = \frac{4}{R_H} = \frac{4}{1.097 \times 10^7\ \text{m}^{-1}} = 364.6\ \text{nm}$$

This wavelength is in the ultraviolet region and corresponds to the series limit.

Exercise 5 Find the energy of the shortest-wavelength photon emitted in the Balmer series for hydrogen.

Answer 3.40 eV.

Although the theoretical derivation of the line spectrum was a remarkable feat in itself, the scope, breadth, and impact of Bohr's monumental achievement is truly seen only when it is realized what else he treated in his three-part paper of 1913:

(a) He explained the limited number of lines seen in the absorption spectrum of hydrogen compared to the emission spectrum.
(b) He explained the emission of x-rays from atoms.
(c) He explained the chemical properties of atoms in terms of the electron shell model.
(d) He explained how atoms associate to form molecules.

Bohr's Correspondence Principle

In our study of relativity, we found that newtonian mechanics cannot be used to describe phenomena that occur at speeds approaching the speed of light. Newtonian mechanics is a special case of relativistic mechanics and is usable only when v is much less than c. Similarly, *quantum mechanics is in agreement with classical physics when the quantum numbers are very large.* This principle, first set forth by Bohr, is called the **correspondence principle.**

For example, consider an electron orbiting the hydrogen atom with $n >$ 10 000. For such large values of n, the energy differences between adjacent levels approach zero and the levels are nearly continuous. Consequently, the classical model is reasonably accurate in describing the system for large values of n. According to the classical picture, the frequency of the light emitted by the atom is equal to the frequency of revolution of the electron in its orbit about the nucleus. Calculations show that for $n > 10\ 000$, this frequency is different from that predicted by quantum mechanics by less than 0.015%.

SUMMARY

The characteristics of *blackbody radiation* cannot be explained using classical concepts. Planck first introduced the *quantum concept* when he assumed that the atomic oscillators responsible for this radiation existed only in discrete states.

The **photoelectric effect** is a process whereby electrons are ejected from a metallic surface when light is incident on that surface. Einstein provided a successful explanation of this effect by extending Planck's quantum hypothesis to the electromagnetic field. In this model, light is viewed as a stream of particles called *photons*, each with energy $E = hf$, where f is the frequency and h is Planck's constant. The kinetic energy of the ejected photoelectron is given by

Photoelectric effect equation

$$K_{\text{max}} = hf - \phi \tag{40.9}$$

where ϕ is the work function of the metal.

X-rays from an incident beam are scattered at various angles by electrons in a target such as carbon. In such a scattering event, a shift in wavelength is observed for the scattered x-rays, and the phenomenon is known as the **Compton effect.** Classical physics does not explain this effect. If the x-ray is treated as a photon, conservation of energy and momentum applied to the photon-electron collisions yields the following expression for the Compton shift:

Compton shift equation

$$\Delta\lambda = \frac{h}{mc}(1 - \cos\theta) \tag{40.11}$$

where m is the mass of the electron, c is the speed of light, and θ is the scattering angle.

The Bohr model of the atom is successful in describing the spectra of atomic hydrogen and hydrogen-like ions. One of the basic assumptions of the model is that the electron can exist only in discrete orbits such that the angular momentum mvr is an integral multiple of $h/2\pi = \hbar$. Assuming circular orbits and a simple coulombic attraction between the electron and proton, the energies of the quantum states for hydrogen are calculated to be

Allowed energies of the hydrogen atom

$$E_n = -\frac{ke^2}{2a_0}\left(\frac{1}{n^2}\right) \qquad (40.27)$$

where k is the Coulomb constant, e is electronic charge, n is an integer called a *quantum number*, and $a_0 = 0.0529$ nm is the **Bohr radius.**

If the electron in the hydrogen atom makes a transition from an orbit whose quantum number is n_i to one whose quantum number is n_f, where $n_f < n_i$, a photon is emitted by the atom whose frequency is given by

Frequency of a photon emitted from hydrogen

$$f = \frac{ke^2}{2a_0h}\left(\frac{1}{n_f^2} - \frac{1}{n_i^2}\right) \qquad (40.29)$$

Using $E = hf = hc/\lambda$, one can calculate the wavelengths of the photons for various transitions in which there is a change in quantum number, $n_i \rightarrow n_f$. The calculated wavelengths are in excellent agreement with observed line spectra.

QUESTIONS

1. What assumptions were made by Planck in dealing with the problem of blackbody radiation? Discuss the consequences of these assumptions.
2. The classical model of blackbody radiation given by the Rayleigh-Jeans law has two major flaws. Identify them and explain how Planck's law deals with these two issues.
3. What is the difference between a "regular" electron and a photoelectron?
4. If the photoelectric effect is observed for one metal, can you conclude that the effect will also be observed for another metal under the same conditions? Explain.
5. In the photoelectric effect, explain why the photocurrent depends on the intensity of the light source but not on the frequency.
6. In the photoelectric effect, explain why the stopping potential depends on the frequency of light but not on the intensity.
7. Explain why the results of the photoelectric effect experiment may vary if an incandescent light with filters is used instead of using gas tubes.
8. Suppose the photoelectric effect occurs in a gaseous target rather than a solid. Will photoelectrons be produced at *all* frequencies of the incident photon? Explain.
9. How does the Compton effect differ from the photoelectric effect?
10. What assumptions were made by Compton in dealing with the scattering of a photon from an electron?
11. The Bohr theory of the hydrogen atom is based upon several assumptions. Discuss these assumptions and their significance. Do any of these assumptions contradict classical physics?
12. Suppose that the electron in the hydrogen atom obeyed classical mechanics rather than quantum mechanics. Why should such a "hypothetical" atom emit a continuous spectrum rather than the observed line spectrum?
13. Can the electron in the ground state of hydrogen absorb a photon of energy (a) *less* than 13.6 eV and (b) *greater* than 13.6 eV?
14. By studying the relative intensities of various spectral lines emitted by a gas, the temperature, pressure, and density of the gas can be determined. Explain how these three variables might affect the various shells the electrons could occupy about the nucleus of the atom.
15. Why aren't the relative intensities of the various spectral lines in a gas all the same?

16. Why would the spectral lines of the diatomic hydrogen be different from those of monatomic hydrogen?
17. Explain the significance behind the fact that the total energy of the atom in the Bohr model is negative.
18. What are the limitations on the Bohr model for the hydrogen atom?
19. An x-ray photon is scattered by an electron. What happens to the frequency of the scattered photon relative to that of the incident photon?
20. Why does the existence of a cutoff frequency in the photoelectric effect favor a particle theory for light rather than a wave theory?
21. Using Wien's law, calculate the wavelength of highest intensity given off by a human body. Using this information, explain why an infrared detector would be a useful alarm for security work.
22. All objects radiate energy. Why, then, are we not able to see all objects in a dark room?
23. Which has more energy, a photon of ultraviolet radiation or a photon of yellow light?
24. What effect, if any, would you expect the temperature of a material to have on the ease with which electrons can be ejected from it in the photoelectric effect?
25. Some stars are observed to be reddish, and some are blue. Which stars have the higher surface temperature? Explain.

PROBLEMS

Section 40.1 Blackbody Radiation and Planck's Hypothesis

1. Calculate the energy of a photon whose frequency is (a) 6.2×10^{14} Hz, (b) 3.1 GHz, (c) 46 MHz. Express your answers in eV.
2. Determine the corresponding wavelengths for the photons described in Problem 1.
3. An FM radio transmitter has a power output of 150 kW and operates at a frequency of 99.7 MHz. How many photons per second does the transmitter emit?
4. The average power generated by the sun is equal to 3.74×10^{26} W. Assuming the average wavelength of the sun's radiation to be 500 nm, find the number of photons emitted by the sun in 1 s.
5. Consider the mass-spring system described in Example 40.2. If the quantum number n changes by unity, calculate the *fractional* change in energy of the oscillator.
6. A sodium-vapor lamp has a power output of 10 W. Using 589.3 nm as the average wavelength of the source, calculate the number of photons emitted per second.
7. Using Wien's displacement law, calculate the surface temperature of a red giant star that radiates with a peak wavelength of $\lambda_{max} = 650$ nm.
8. Using the Rayleigh-Jeans law, calculate the total power radiated per unit area by a 300°C blackbody in the wavelength range between 7.00 mm and 7.50 mm.
9. What is the peak wavelength emitted by the human body? Assume a body temperature of 98.6°F and use the Wien displacement law. In what part of the electromagnetic spectrum does this wavelength lie?
10. A tungsten filament is heated to a temperature of 800°C. What is the wavelength of the most intense radiation?
11. The human eye is most sensitive to light with a wavelength $\lambda = 560$ nm. What temperature blackbody would radiate most intensely at this wavelength?
12. Show that at *long* wavelengths, Planck's radiation law (Eq. 40.3) reduces to the Rayleigh-Jeans law (Eq. 40.2).
13. Show that at *short* wavelengths or *low* temperatures, Planck's radiation law (Eq. 40.3) predicts an exponential decrease in $I(\lambda, T)$ given by *Wien's radiation law:*

$$I(\lambda, T) = \frac{2\pi hc^2}{\lambda^5} e^{-hc/\lambda kT}$$

Section 40.2 The Photoelectric Effect

14. The photocurrent of a photocell is stopped by a retarding potential of 0.54 V for radiation of wavelength 750 nm. Find the work function for the material.
15. The work function for potassium is 2.24 eV. If potassium metal is illuminated with light of wavelength 480 nm, find (a) the maximum kinetic energy of the photoelectrons and (b) the cutoff wavelength.
16. Molybdenum has a work function of 4.2 eV. (a) Find the cutoff wavelength and threshold frequency for the photoelectric effect. (b) Calculate the stopping potential if the incident light has a wavelength of 180 nm.
17. When cesium metal is illuminated with light of wavelength 500 nm, the photoelectrons emitted have a maximum kinetic energy of 0.57 eV. Find (a) the work function of cesium and (b) the stopping potential if the incident light has a wavelength of 600 nm.
18. Lithium has a work function $\phi = 2.3$ eV. (a) Calculate the cutoff frequency for lithium. (b) Calculate the stopping potential for 500-nm light falling on a clean lithium surface.
19. Two light sources are used in a photoelectric experiment to determine the work function for a particular metal surface. When the green light from a mercury

lamp ($\lambda = 546.1$ nm) is used, a retarding potential of 1.70 V reduces the photocurrent to zero. Based on this measurement, what is the work function for this metal? (b) What stopping potential would be observed when using the yellow light from a helium discharge tube ($\lambda = 587.5$ nm)?

20. Electrons are ejected from a metallic surface with speeds ranging up to 4.6×10^5 m/s when light with a wavelength $\lambda = 625$ nm is used. (a) What is the work function of the surface? (b) What is the cutoff frequency for this surface?

21. Consider the metals lithium, beryllium, and mercury, which have work functions of 2.3 eV, 3.9 eV, and 4.5 eV, respectively. If light of wavelength 400 nm is incident on each of these metals, determine (a) which metals exhibit the photoelectric effect and (b) the maximum kinetic energy for the photoelectron in each case.

22. Light of wavelength 300 nm is incident on a metallic surface. If the stopping potential for the photoelectric effect is 1.2 V, find (a) the maximum energy of the emitted electrons, (b) the work function, and (c) the cutoff wavelength.

23. The active material in a photocell has a work function of 3.1 eV. Under reverse-bias conditions, the cutoff wavelength is found to be 270 nm. What is the value of the bias voltage?

24. A light source of wavelength λ illuminates a metal and ejects photoelectrons with a maximum kinetic energy of 1.6 eV. A second light source with half the wavelength of the first ejects photoelectrons with a maximum kinetic energy of 5.2 eV. What is the work function of the metal?

Section 40.3 The Compton Effect

25. Calculate the energy and momentum of a photon of wavelength 700 nm.

26. X-rays of wavelength 0.200 nm are scattered from a block of carbon. If the scattered radiation is detected at 60° to the incident beam, find (a) the Compton shift $\Delta\lambda$ and (b) the kinetic energy imparted to the recoiling electron.

27. What scattering angle will shift the wavelength of a 1-nm x-ray by 0.02%?

28. A 0.45-nm x-ray photon is deflected through a 23° angle after scattering from a free electron. (a) What is the kinetic energy of the recoiling electron? (b) What is its speed?

29. A 0.03-nm x-ray photon is scattered from a free electron. (a) If the *shift* in the x-ray wavelength equals the Compton wavelength of the electron, what is the electron's kinetic energy after the collision? (b) What is its speed?

30. X-rays with an energy of 300 keV undergo Compton scattering from a target. If the scattered rays are detected at 37° relative to the incident rays, find (a) the Compton shift at this angle, (b) the energy of the scattered x-ray, and (c) the energy of the recoiling electron.

31. After a 0.80-nm x-ray photon scatters from a free electron, the electron recoils with a speed equal to 1.4×10^6 m/s. (a) What was the Compton shift in the photon's wavelength? (b) Through what angle was the photon scattered?

32. A beam of 0.68-nm photons undergoes Compton scattering from free electrons. What are the energy and momentum of the photons that emerge at a 45° angle with respect to the incident beam?

33. X-rays with a wavelength of 0.12 nm undergo Compton scattering. (a) Find the wavelength of photons scattered at angles of 30, 60, 90, 120, 150, and 180°. (b) Find the energy of the scattered electrons corresponding to these scattered x-rays. (c) Which one of the given scattering angles provides the electron with the greatest energy?

34. A 0.5-nm x-ray photon is deflected through a 134° angle in a Compton scattering event from a free electron. At what angle (with respect to the incident beam) is the recoiling electron found?

35. A 0.0016-nm photon scatters from a free electron. For what (photon) scattering angle will the recoiling electron and scattered photon have the same kinetic energy?

Section 40.4 Atomic Spectra

36. Calculate the wavelengths of the first three lines in the Paschen series for hydrogen using Equation 40.18.

37. Calculate the wavelengths of the first three lines in the Lyman series for hydrogen using Equation 40.17.

38. (a) Compute the shortest wavelength in each of these hydrogen spectral series: Lyman, Balmer, Paschen, and Brackett. (b) Compute the energy (in eV) of the highest-energy photon produced in each of these series.

39. (a) What value of n is associated with the Lyman series line in hydrogen whose wavelength is 94.96 nm? (b) Could this wavelength be associated with the Paschen or Brackett series?

40. Consider a large number of hydrogen atoms, with electrons all initially in the $n = 4$ state. (a) How many different wavelengths would be observed in the emission spectrum of these atoms? (b) What is the longest wavelength that could be observed? To which series does it belong?

Section 40.5 Bohr's Quantum Model of the Atom

41. Use Equation 40.25 to calculate the radius of the first, second, and third Bohr orbits of hydrogen.

42. For a hydrogen atom in its ground state, use the Bohr model to compute (a) the orbital velocity of the electron, (b) the kinetic energy (in eV) of the electron, and (c) the electrical potential energy (in eV) of the atom.

43. (a) Construct an energy level diagram for the He^+ ion, for which $Z = 2$. (b) What is the ionization energy for He^+?

44. Construct an energy level diagram for the Li^{2+} ion, for which $Z = 3$.

45. What is the radius of the first Bohr orbit in (a) He^+, (b) Li^{2+}, and (c) Be^{3+}?

46. A hydrogen atom initially in its ground state ($n = 1$) absorbs a photon and ends up in the state for which $n = 4$. (a) What is the energy of the absorbed photon? (b) If the atom returns to the ground state, what photon energies could the atom emit?

47. A photon is emitted from a hydrogen atom which undergoes a transition from the state $n = 6$ to the state $n = 2$. Calculate (a) the energy, (b) the wavelength, and (c) the frequency of the emitted photon.

48. What is the energy of the photon that could cause (a) an electronic transition from the $n = 3$ state to the $n = 5$ state and (b) an electronic transition from the $n = 5$ state to the $n = 7$ state?

49. Four possible transitions for a hydrogen atom are listed below.

(A) $n_i = 2;\ n_f = 5$

(B) $n_i = 5;\ n_f = 3$

(C) $n_i = 7;\ n_f = 4$

(D) $n_i = 4;\ n_f = 7$

(a) Which transition will emit the shortest wavelength photon? (b) For which transition will the atom gain the most energy? (c) For which transition(s) does the atom lose energy?

50. (a) Using Equation 40.19 calculate the longest and shortest wavelengths for the Brackett series. (b) Determine the photon energies corresponding to these wavelengths.

51. Find the potential energy and kinetic energy of an electron in the first excited state of the hydrogen atom.

ADDITIONAL PROBLEMS

52. A hydrogen atom is in its first excited state ($n = 2$). Using the Bohr theory of the atom, calculate (a) the radius of the orbit, (b) the linear momentum of the electron, (c) the angular momentum of the electron, (d) the kinetic energy, (e) the potential energy, and (f) the total energy.

53. *Positronium* is a hydrogen-like atom consisting of a positron (a positively charged electron) and an electron revolving around each other. Using the Bohr model, find the allowed radii (relative to the center of mass of the two particles) and the allowed energies of the system.

54. Figure 40.17 shows the stopping potential versus incident photon frequency for the photoelectric effect for sodium. Use these data points to find (a) the work function, (b) the ratio h/e, and (c) the cutoff wavelength. (Data taken from R. A. Millikan, *Phys. Rev.* 7:362 [1916].)

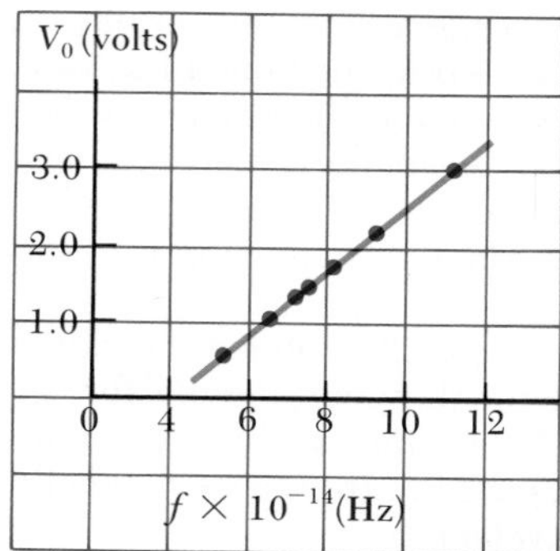

Figure 40.17 (Problem 54).

55. *The Auger process.* An electron in chromium makes a transition from the $n = 2$ state to the $n = 1$ state without emitting a photon. Instead, the excess energy is transferred to an outer electron (in the $n = 4$ state), which is ejected by the atom. (This is called an *Auger process,* and the ejected electron is referred to as an *Auger electron.*) Use the Bohr theory to find the kinetic energy of the Auger electron.

56. Photons of wavelength 450 nm are incident on a metal. The most energetic electrons ejected from the metal are bent into a circular arc of radius 20 cm by a magnetic field whose strength is equal to 2×10^{-5} T. What is the work function of the metal?

57. Gamma rays (high-energy photons) of energy 1.02 MeV are scattered from electrons that are initially at rest. If the scattering is *symmetric,* that is, if $\theta = \phi$, find (a) the scattering angle θ and (b) the energy of the scattered photons.

58. If the maximum energy given to an electron during Compton scattering is 30 keV, what is the wavelength of the incident photon?

59. A *muon* is a particle with a charge of $-e$ and a mass equal to 207 times the mass of an electron. Muonic lead is formed when a ^{208}Pb nucleus captures a muon. According to the Bohr theory, what are the radius and energy of the ground state of muonic lead?

60. Use Bohr's model of the hydrogen atom to show that when the atom makes a transition from the state n to the state $n - 1$, the frequency of the emitted light is given by

$$f = \frac{2\pi^2 mk^2e^4}{h^3}\left[\frac{2n-1}{(n-1)^2n^2}\right]$$

Show that as $n \rightarrow \infty$, the expression above varies as $1/n^3$ and reduces to the classical frequency one would expect the atom to emit. (*Hint:* To calculate the classical frequency, note that the frequency of revolution is $v/2\pi r$, where r is given by Eq. 40.25.) This is an example of the correspondence principle, which requires that the classical and quantum models agree for large values of n.

61. A muon (Problem 59) is captured by a deuteron to form a muonic atom. (a) Find the energy of the ground state and the first excited state. (b) What is the wavelength of the photon emitted when the atom makes a transition from the first excited state to the ground state?
62. Show that a photon cannot transfer all of its energy to a free electron. (*Hint:* Note that energy and momentum must be conserved.)
63. A photon of initial energy 0.1 MeV undergoes Compton scattering at an angle of 60°. Find (a) the energy of the scattered photon, (b) the recoil energy of the electron, and (c) the recoil angle of the electron.
64. The total power per unit area radiated by a blackbody at a temperature T is given by the area under the $I(\lambda, T)$ versus λ curve, as in Figure 40.2. (a) Show that this power per unit area is given by

$$\int_0^\infty I(\lambda, T)\, d\lambda = \sigma T^4$$

where $I(\lambda, T)$ is given by Planck's radiation law and σ is a constant independent of T. This result is known as the *Stefan-Boltzmann law* (see Eq. 20.11). To carry out the integration, you should make the change of variable $x = hc/\lambda kT$ and use the fact that

$$\int_0^\infty \frac{x^3\, dx}{e^x - 1} = \frac{\pi^4}{15}$$

(b) Show that the Stefan-Boltzmann constant σ has the value

$$\sigma = \frac{2\pi^5 k^4}{15c^2h^3} = 5.7 \times 10^{-8}\ \text{W/m}^2\cdot\text{K}^4$$

65. (a) Use the results of Problem 64 to calculate the total power radiated per unit area by a tungsten filament at a temperature of 3000 K. (Assume that the filament is an ideal radiator.) (b) If the tungsten filament of a light bulb is rated at 75 W, what is the surface area of the filament? (Assume that the main energy loss is due to radiation.) (c) Use the results of Problem 64 to find the surface temperature of the sun, assuming the sun is an ideal radiator. Take the radius of the sun to have the value 6.96×10^8 m and the power radiated by the sun to be 3.74×10^{26} W.

66. Wavelengths of spectral lines depend, to some extent, on the nuclear mass. Determine the shift in the first line of the Balmer series (relative to hydrogen) for (a) deuterium (^{2}H) and (b) tritium (^{3}H).
67. Show that the ratio of the Compton wavelength λ_c to the de Broglie wavelength λ for a *relativistic* electron is given by

$$\frac{\lambda_c}{\lambda} = \left[\left(\frac{E}{mc^2}\right)^2 - 1\right]^{1/2}$$

where E is the total energy of the electron and m is its mass.

68. Derive the formula for the Compton shift (Eq. 40.11) from Equations 40.12, 40.13, and 40.14.
69. In the Compton scattering event illustrated in Figure 40.7, the scattered photon has an energy of 120 keV and the recoiling electron has an energy of 40 keV. Find (a) the wavelength of the incident photon, (b) the angle θ at which the photon is scattered, and (c) the recoil angle ϕ of the electron.
70. Consider the problem of the distribution of blackbody radiation described in Figure 40.2. Note that as T *increases*, the wavelength λ_{max} at which $I(\lambda, T)$ reaches a maximum shifts toward shorter wavelengths. (a) Show that there is a general relationship between temperature and λ_{max} (known as *Wien's displacement law*), which states that $T\lambda_{max}$ = constant. (b) Obtain a numerical value for this constant. [*Hint:* Start with Planck's radiation law and note that the slope of $I(\lambda, T)$ versus λ is zero when $\lambda = \lambda_{max}$.]
71. An electron initially in the $n = 3$ state of a one-electron atom of mass M undergoes a transition to the $n = 1$ ground state. Show that the recoil velocity is given approximately by

$$v = \frac{8hR}{9M}$$

"Actually I started out in quantum mechanics, but somewhere along the way I took a wrong turn."

Appendix A

TABLE A.1 Conversion Factors

Length

	m	cm	km	in.	ft	mi
1 meter	1	10^{2}	10^{-3}	39.37	3.281	6.214×10^{-4}
1 centimeter	10^{-2}	1	10^{-5}	0.3937	3.281×10^{-2}	6.214×10^{-6}
1 kilometer	10^{3}	10^{5}	1	3.937×10^{4}	3.281×10^{3}	0.6214
1 inch	2.540×10^{-2}	2.540	2.540×10^{-5}	1	8.333×10^{-2}	1.578×10^{-5}
1 foot	0.3048	30.48	3.048×10^{-4}	12	1	1.894×10^{-4}
1 mile	1609	1.609×10^{5}	1.609	6.336×10^{4}	5280	1

Mass

	kg	g	slug	u
1 kilogram	1	10^{3}	6.852×10^{-2}	6.024×10^{26}
1 gram	10^{-3}	1	6.852×10^{-5}	6.024×10^{23}
1 slug (lb/g)	14.59	1.459×10^{4}	1	8.789×10^{27}
1 atomic mass unit	1.660×10^{-27}	1.660×10^{-24}	1.137×10^{-28}	1

Time

	s	min	h	day	year
1 second	1	1.667×10^{-2}	2.778×10^{-4}	1.157×10^{-5}	3.169×10^{-8}
1 minute	60	1	1.667×10^{-2}	6.994×10^{-4}	1.901×10^{-6}
1 hour	3600	60	1	4.167×10^{-2}	1.141×10^{-4}
1 day	8.640×10^{4}	1440	24	1	2.738×10^{-3}
1 year	3.156×10^{7}	5.259×10^{5}	8.766×10^{3}	365.2	1

Speed

	m/s	cm/s	ft/s	mi/h
1 meter/second	1	10^{2}	3.281	2.237
1 centimeter/second	10^{-2}	1	3.281×10^{-2}	2.237×10^{-2}
1 foot/second	0.3048	30.48	1	0.6818
1 mile/hour	0.4470	44.70	1.467	1

Note: 1 mi/min = 60 mi/h = 88 ft/s.

Force

	N	dyn	lb
1 newton	1	10^{5}	0.2248
1 dyne	10^{-5}	1	2.248×10^{-6}
1 pound	4.448	4.448×10^{5}	1

TABLE A.1 (Continued)

Work, Energy, Heat

	J	erg	ft · lb
1 joule	1	10^{7}	0.7376
1 erg	10^{-7}	1	7.376×10^{-8}
1 ft · lb	1.356	1.356×10^{7}	1
1 eV	1.602×10^{-19}	1.602×10^{-12}	1.182×10^{-19}
1 cal	4.186	4.186×10^{7}	3.087
1 Btu	1.055×10^{3}	1.055×10^{10}	7.779×10^{2}
1 kWh	3.600×10^{6}	3.600×10^{13}	2.655×10^{6}

	eV	cal	Btu	kWh
1 joule	6.242×10^{18}	0.2389	9.481×10^{-4}	2.778×10^{-7}
1 erg	6.242×10^{11}	2.389×10^{-8}	9.481×10^{-11}	2.778×10^{-14}
1 ft · lb	8.464×10^{18}	0.3239	1.285×10^{-3}	3.766×10^{-7}
1 eV	1	3.827×10^{-20}	1.519×10^{-22}	4.450×10^{-26}
1 cal	2.613×10^{19}	1	3.968×10^{-3}	1.163×10^{-6}
1 Btu	6.585×10^{21}	2.520×10^{2}	1	2.930×10^{-4}
1 kWh	2.247×10^{25}	8.601×10^{5}	3.413×10^{2}	1

Pressure

	Pa	dyn/cm²	atm
1 pascal	1	10	9.869×10^{-6}
1 dyne/centimeter²	10^{-1}	1	9.869×10^{-7}
1 atmosphere	1.013×10^{5}	1.013×10^{6}	1
1 centimeter mercury*	1.333×10^{3}	1.333×10^{4}	1.316×10^{-2}
1 pound/inch²	6.895×10^{3}	6.895×10^{4}	6.805×10^{-2}
1 pound/foot²	47.88	4.788×10^{2}	4.725×10^{-4}

	cm Hg	lb/in.²	lb/ft²
1 newton/meter²	7.501×10^{-4}	1.450×10^{-4}	2.089×10^{-2}
1 dyne/centimeter²	7.501×10^{-5}	1.450×10^{-5}	2.089×10^{-3}
1 atmosphere	76	14.70	2.116×10^{3}
1 centimeter mercury*	1	0.1943	27.85
1 pound/inch²	5.171	1	144
1 pound/foot²	3.591×10^{-2}	6.944×10^{-3}	1

* At 0°C and at a location where the acceleration due to gravity has its "standard" value, 9.80665 m/s².

TABLE A.2 Symbols, Dimensions, and Units of Physical Quantities

Quantity	Common Symbol	Unit*	Dimensions†	Unit in Terms of Base SI Units
Acceleration	a	m/s²	L/T²	m/s²
Amount of substance	n	mole		mol
Angle	θ, ϕ	radian (rad)	1	
Angular acceleration	α	rad/s²	T⁻²	s⁻²
Angular frequency	ω	rad/s	T⁻¹	s⁻¹
Angular momentum	L	kg · m²/s	ML²/T	kg · m²/s
Angular velocity	ω	rad/s	T⁻¹	s⁻¹
Area	A	m²	L²	m²

(Table continues)

TABLE A.2 (Continued)

Quantity	Common Symbol	Unit*	Dimensions†	Unit in Terms of Base SI Units
Atomic number	Z			
Capacitance	C	farad (F)(=C/V)	Q^2T^2/ML^2	$A^2 \cdot s^4/kg \cdot m^2$
Charge	q, Q, e	coulomb (C)	Q	$A \cdot s$
Charge density				
Line	λ	C/m	Q/L	$A \cdot s/m$
Surface	σ	C/m^2	Q/L^2	$A \cdot s/m^2$
Volume	ρ	C/m^3	Q/L^3	$A \cdot s/m^3$
Conductivity	σ	$1/\Omega \cdot m$	Q^2T/ML^3	$A^2 \cdot s^3/kg \cdot m^3$
Current	I	AMPERE	Q/T	A
Current density	J	A/m^2	Q/T^2	A/m^2
Density	ρ	kg/m^3	M/L^3	kg/m^3
Dielectric constant	κ			
Displacement	s	METER	L	m
Distance	d, h			
Length	ℓ, L			
Electric dipole moment	p	$C \cdot m$	QL	$A \cdot s \cdot m$
Electric field	E	V/m	ML/QT^2	$kg \cdot m/A \cdot s^3$
Electric flux	Φ	$V \cdot m$	ML^3/QT^2	$kg \cdot m^3/A \cdot s^3$
Electromotive force	$\mathcal{E}$	volt (V)	ML^2/QT^2	$kg \cdot m^2/A \cdot s^3$
Energy	E, U, K	joule (J)	ML^2/T^2	$kg \cdot m^2/s^2$
Entropy	S	J/K	$ML^2/T^2 {}^\circ K$	$kg \cdot m^2/s^2 \cdot K$
Force	F	newton (N)	ML/T^2	$kg \cdot m/s^2$
Frequency	f, ν	hertz (Hz)	T^{-1}	s^{-1}
Heat	Q	joule (J)	ML^2/T^2	$kg \cdot m^2/s^2$
Inductance	L	henry (H)	ML^2/Q^2	$kg \cdot m^2/A^2 \cdot s^2$
Magnetic dipole moment	μ	$N \cdot m/T$	QL^2/T	$A \cdot m^2$
Magnetic field	B	tesla (T)(=Wb/m^2)	M/QT	$kg/A \cdot s^2$
Magnetic flux	Φ_m	weber (Wb)	ML^2/QT	$kg \cdot m^2/A \cdot s^2$
Mass	m, M	KILOGRAM	M	kg
Molar specific heat	C	$J/mol \cdot K$		$kg \cdot m^2/s^2 \cdot kmol \cdot K$
Moment of inertia	I	$kg \cdot m^2$	ML^2	$kg \cdot m^2$
Momentum	p	$kg \cdot m/s$	ML/T	$kg \cdot m/s$
Period	T	s	T	s
Permeability of space	μ_o	N/A^2 (=H/m)	ML/Q^2	$kg \cdot m/A^2 \cdot s^2$
Permittivity of space	ϵ_o	$C^2/N \cdot m^2$ (= F/m)	Q^2T^2/ML^3	$A^2 \cdot s^4/kg \cdot m^3$
Potential (voltage)	V	volt (V)(=J/C)	ML^2/QT^2	$kg \cdot m^2/A \cdot s^3$
Power	P	watt (W)(=J/s)	ML^2/T^3	$kg \cdot m^2/s^3$
Pressure	P, p	pascal (Pa) = (N/m^2)	M/LT^2	$kg/m \cdot s^2$
Resistance	R	ohm (Ω)(=V/A)	ML^2/Q^2T	$kg \cdot m^2/A^2 \cdot s^3$
Specific heat	c	$J/kg \cdot K$	$L^2/T^2 {}^\circ K$	$m^2/s^2 \cdot K$
Temperature	T	KELVIN	°K	K
Time	t	SECOND	T	s
Torque	τ	$N \cdot m$	ML^2/T^2	$kg \cdot m^2/s^2$
Speed	v	m/s	L/T	m/s
Volume	V	m^3	L^3	m^3
Wavelength	λ	m	L	m
Work	W	joule (J)(=$N \cdot m$)	ML^2/T^2	$kg \cdot m^2/s^2$

* The base SI units are given in upper case letters.

† The symbols M, L, T, and Q denote mass, length, time, and charge, respectively.

TABLE A.3 Table of Selected Atomic Masses*

Atomic Number Z	Element	Symbol	Mass Number, A	Atomic Mass†	Percent Abundance, or Decay Mode (if radioactive)‡	Half-Life (if radioactive)
0	(Neutron)	n	1	1.008665	β^-	10.6 min
1	Hydrogen	H	1	1.007825	99.985	
	Deuterium	D	2	2.014102	0.015	
	Tritium	T	3	3.016049	β^-	12.33 y
2	Helium	He	3	3.016029	0.00014	
			4	4.002603	≈ 100	
3	Lithium	Li	6	6.015123	7.5	
			7	7.016005	92.5	
4	Beryllium	Be	7	7.016930	EC, γ	53.3 days
			8	8.005305	2α	6.7×10^{-17} s
			9	9.012183	100	
5	Boron	B	10	10.012938	19.8	
			11	11.009305	80.2	
6	Carbon	C	11	11.011433	β^+, EC	20.4 min
			12	12.000000	98.89	
			13	13.003355	1.11	
			14	14.003242	β^-	5730 y
7	Nitrogen	N	13	13.005739	β^+	9.96 min
			14	14.003074	99.63	
			15	15.000109	0.37	
8	Oxygen	O	15	15.003065	β^+, EC	122 s
			16	15.994915	99.759	
			18	17.999159	0.204	
9	Fluorine	F	19	18.998403	100	
10	Neon	Ne	20	19.992439	90.51	
			22	21.991384	9.22	
11	Sodium	Na	22	21.994435	β^+, EC, γ	2.602 y
			23	22.989770	100	
			24	23.990964	β^-, γ	15.0 h
12	Magnesium	Mg	24	23.985045	78.99	
13	Aluminum	Al	27	26.981541	100	
14	Silicon	Si	28	27.976928	92.23	
			31	30.975364	β^-, γ	2.62 h
15	Phosphorus	P	31	30.973763	100	
			32	31.973908	β^-	14.28 days
16	Sulfur	S	32	31.972072	95.0	
			35	34.969033	β^-	87.4 days
17	Chlorine	Cl	35	34.968853	75.77	
			37	36.965903	24.23	
18	Argon	Ar	40	39.962383	99.60	
19	Potassium	K	39	38.963708	93.26	
			40	39.964000	β^-, EC, γ, β^+	1.28×10^9 y
20	Calcium	Ca	40	39.962591	96.94	
21	Scandium	Sc	45	44.955914	100	
22	Titanium	Ti	48	47.947947	73.7	
23	Vanadium	V	51	50.943963	99.75	

(Table continues)

TABLE A.3 (Continued)

Atomic Number Z	Element	Symbol	Mass Number, A	Atomic Mass†	Percent Abundance, or Decay Mode (if radioactive)‡	Half-Life (if radioactive)
24	Chromium	Cr	52	51.940510	83.79	
25	Manganese	Mn	55	54.938046	100	
26	Iron	Fe	56	55.934939	91.8	
27	Cobalt	Co	59	58.933198	100	
			60	59.933820	β^-, γ	5.271 y
28	Nickel	Ni	58	57.935347	68.3	
			60	59.930789	26.1	
			64	63.927968	0.91	
29	Copper	Cu	63	62.929599	69.2	
			64	63.929766	β^-, β^+	12.7 h
			65	64.927792	30.8	
30	Zinc	Zn	64	63.929145	48.6	
			66	65.926035	27.9	
31	Gallium	Ga	69	68.925581	60.1	
32	Germanium	Ge	72	71.922080	27.4	
			74	73.921179	36.5	
33	Arsenic	As	75	74.921596	100	
34	Selenium	Se	80	79.916521	49.8	
35	Bromine	Br	79	78.918336	50.69	
36	Krypton	Kr	84	83.911506	57.0	
			89	88.917563	β^-	3.2 min
37	Rubidium	Rb	85	84.911800	72.17	
38	Strontium	Sr	86	85.909273	9.8	
			88	87.905625	82.6	
			90	89.907746	β^-	28.8 y
39	Yttrium	Y	89	88.905856	100	
40	Zirconium	Zr	90	89.904708	51.5	
41	Niobium	Nb	93	92.906378	100	
42	Molybdenum	Mo	98	97.905405	24.1	
43	Technetium	Tc	98	97.907210	β^-, γ	4.2×10^6 y
44	Ruthenium	Ru	102	101.904348	31.6	
45	Rhodium	Rh	103	102.90550	100	
46	Palladium	Pd	106	105.90348	27.3	
47	Silver	Ag	107	106.905095	51.83	
			109	108.904754	48.17	
48	Cadmium	Cd	114	113.903361	28.7	
49	Indium	In	115	114.90388	95.7; β^-	5.1×10^{14} y
50	Tin	Sn	120	119.902199	32.4	
51	Antimony	Sb	121	120.903824	57.3	
52	Tellurium	Te	130	129.90623	34.5; β^-	2×10^{21} y
53	Iodine	I	127	126.904477	100	
			131	130.906118	β^-, γ	8.04 days

(Table continues)

* Data are taken from *Chart of the Nuclides*, 12th ed., General Electric, 1977, and from C. M. Lederer and V. S. Shirley, eds., *Table of Isotopes*, 7th ed., New York, John Wiley & Sons, Inc., 1978.

† The masses given in column (5) are those for the neutral atom, including the Z electrons.

‡ The process EC stands for "electron capture."

TABLE A.3 (Continued)

Atomic Number Z	Element	Symbol	Mass Number, A	Atomic Mass†	Percent Abundance, or Decay Mode (if radioactive)‡	Half-Life (if radioactive)
54	Xenon	Xe	132	131.90415	26.9	
			136	135.90722	8.9	
55	Cesium	Cs	133	132.90543	100	
56	Barium	Ba	137	136.90582	11.2	
			138	137.90524	71.7	
			144	143.922673	β^-	11.9 s
57	Lanthanum	La	139	138.90636	99.911	
58	Cerium	Ce	140	139.90544	88.5	
59	Praseodymium	Pr	141	140.90766	100	
60	Neodymium	Nd	142	141.90773	27.2	
61	Promethium	Pm	145	144.91275	EC, α, γ	17.7 y
62	Samarium	Sm	152	151.91974	26.6	
63	Europium	Eu	153	152.92124	52.1	
64	Gadolinium	Gd	158	157.92411	24.8	
65	Terbium	Tb	159	158.92535	100	
66	Dysprosium	Dy	164	163.92918	28.1	
67	Holmium	Ho	165	164.93033	100	
68	Erbium	Er	166	165.93031	33.4	
69	Thulium	Tm	169	168.93423	100	
70	Ytterbium	Yb	174	173.93887	31.6	
71	Lutecium	Lu	175	174.94079	97.39	
72	Hafnium	Hf	180	179.94656	35.2	
73	Tantalum	Ta	181	180.94801	99.988	
74	Tungsten (wolfram)	W	184	183.95095	30.7	
75	Rhenium	Re	187	186.95577	62.60, β^-	4×10^{10} y
76	Osmium	Os	191	190.96094	β^-, γ	15.4 days
			192	191.96149	41.0	
77	Iridium	Ir	191	190.96060	37.3	
			193	192.96294	62.7	
78	Platinum	Pt	195	194.96479	33.8	
79	Gold	Au	197	196.96656	100	
80	Mercury	Hg	202	201.97063	29.8	
81	Thallium	Tl	205	204.97441	70.5	
82	Lead	Pb	210	209.990069	β^-	1.3 min
			204	203.973044	β^-, 1.48	1.4×10^{17} y
			206	205.97446	24.1	
			207	206.97589	22.1	
			208	207.97664	52.3	
			210	209.98418	α, β^-, γ	22.3 y
			211	210.98874	β^-, γ	36.1 min
			212	211.99188	β^-, γ	10.64 h
			214	213.99980	β^-, γ	26.8 min
83	Bismuth	Bi	209	208.98039	100	
			211	210.98726	α, β^-, γ	2.15 min

(Table continues)

TABLE A.3 (Continued)

Atomic Number Z	Element	Symbol	Mass Number, A	Atomic Mass†	Percent Abundance, or Decay Mode (if radioactive)‡	Half-Life (if radioactive)
84	Polonium	Po	210	209.98286	α, γ	138.38 days
			214	213.99519	α, γ	164 μs
85	Astatine	At	218	218.00870	α, β^-	$\approx$2 s
86	Radon	Rn	222	222.017574	α, γ	3.8235 days
87	Francium	Fr	223	223.019734	α, β^-, γ	21.8 min
88	Radium	Ra	226	226.025406	α, γ	1.60×10^3 y
			228	228.031069	β^-	5.76 y
89	Actinium	Ac	227	227.027751	α, β^-, γ	21.773 y
90	Thorium	Th	228	228.02873	α, γ	1.9131 y
			232	232.038054	100, α, γ	1.41×10^{10} y
91	Protactinium	Pa	231	231.035881	α, γ	3.28×10^4 y
92	Uranium	U	232	232.03714	α, γ	72 y
			233	233.039629	α, γ	1.592×10^5 y
			235	235.043925	0.72; α, γ	7.038×10^8 y
			236	236.045563	α, γ	2.342×10^7 y
			238	238.050786	99.275; α, γ	4.468×10^9 y
			239	239.054291	β^-, γ	23.5 min
93	Neptunium	Np	239	239.052932	β^-, γ	2.35 days
94	Plutonium	Pu	239	239.052158	α, γ	2.41×10^4 y
95	Americium	Am	243	243.061374	α, γ	7.37×10^3 y
96	Curium	Cm	245	245.065487	α, γ	8.5×10^3 y
97	Berkelium	Bk	247	247.07003	α, γ	1.4×10^3 y
98	Californium	Cf	249	249.074849	α, γ	351 y
99	Einsteinium	Es	254	254.08802	α, γ, β^-	276 days
100	Fermium	Fm	253	253.08518	EC, α, γ	3.0 days
101	Mendelevium	Md	255	255.0911	EC, α	27 min
102	Nobelium	No	255	255.0933	EC, α	3.1 min
103	Lawrencium	Lr	257	257.0998	α	$\approx$35 s
104	Unnilquadium	Rf	261	261.1087	α	1.1 min
105	Unnilpentium	Ha	262	262.1138	α	0.7 min
106	Unnilhexium		263	263.1184	α	0.9 s
107	Unnilseptium		261	261	α	1–2 ms

° Data are taken from *Chart of the Nuclides*, 12th ed., General Electric, 1977, and from C. M. Lederer and V. S. Shirley, eds., *Table of Isotopes*, 7th ed., New York, John Wiley & Sons, Inc., 1978.

† The masses given in column (5) are those for the neutral atom, including the Z electrons.

‡ The process EC stands for "electron capture."

Appendix B

Mathematics Review

These appendices in mathematics are intended as a brief review of operations and methods. Early in this course, you should be totally familiar with basic algebraic techniques, analytic geometry, and trigonometry. The appendices on differential and integral calculus are more detailed and are intended for those students who have difficulties in applying calculus concepts to physical situations.

B.1 SCIENTIFIC NOTATION

Many quantities that scientists deal with often have very large or very small values. For example, the speed of light is about 300 000 000 m/s and the ink required to make the dot over an *i* in this textbook has a mass of about 0.000 000 001 kg. Obviously, it is very cumbersome to read, write, and keep track of numbers such as these. We avoid this problem by using a method dealing with powers of the number 10:

$$10^0 = 1$$

$$10^1 = 10$$

$$10^2 = 10 \times 10 = 100$$

$$10^3 = 10 \times 10 \times 10 = 1000$$

$$10^4 = 10 \times 10 \times 10 \times 10 = 10\,000$$

$$10^5 = 10 \times 10 \times 10 \times 10 \times 10 = 100\,000$$

and so on. The number of zeros corresponds to the power to which 10 is raised, called the **exponent** of 10. For example, the speed of light, 300 000 000 m/s, can be expressed as 3×10^8 m/s.

For numbers less than one, we note the following:

$$10^{-1} = \frac{1}{10} = 0.1$$

$$10^{-2} = \frac{1}{10 \times 10} = 0.01$$

$$10^{-3} = \frac{1}{10 \times 10 \times 10} = 0.001$$

$$10^{-4} = \frac{1}{10 \times 10 \times 10 \times 10} = 0.0001$$

$$10^{-5} = \frac{1}{10 \times 10 \times 10 \times 10 \times 10} = 0.00001$$

In these cases, the number of places the decimal point is to the left of the digit 1 equals the value of the (negative) exponent. Numbers that are expressed as some power of 10 multiplied by another number between 1 and 10 are said to be in **scientific notation.** For example, the scientific notation for 5 943 000 000 is 5.943×10^9 and that for 0.0000832 is 8.32×10^{-5}.

When numbers expressed in scientific notation are being multiplied, the following general rule is very useful:

$$10^n \times 10^m = 10^{n+m} \tag{B.1}$$

where n and m can be *any* numbers (not necessarily integers). For example, $10^2 \times 10^5 = 10^7$. The rule also applies if one of the exponents is negative. For example, $10^3 \times 10^{-8} = 10^{-5}$.

When dividing numbers expressed in scientific notation, note that

$$\frac{10^n}{10^m} = 10^n \times 10^{-m} = 10^{n-m} \tag{B.2}$$

EXERCISES

With help from the above rules, verify the answers to the following:

1. $86{,}400 = 8.64 \times 10^4$
2. $9{,}816{,}762.5 = 9.8167625 \times 10^6$
3. $0.0000000398 = 3.98 \times 10^{-8}$
4. $(4 \times 10^8)(9 \times 10^9) = 3.6 \times 10^{18}$
5. $(3 \times 10^7)(6 \times 10^{-12}) = 1.8 \times 10^{-4}$
6. $\dfrac{75 \times 10^{-11}}{5 \times 10^{-3}} = 1.5 \times 10^{-7}$
7. $\dfrac{(3 \times 10^6)(8 \times 10^{-2})}{(2 \times 10^{17})(6 \times 10^5)} = 2 \times 10^{-18}$

B.2 ALGEBRA

Some Basic Rules

When algebraic operations are performed, the laws of arithmetic apply. Symbols such as x, y, and z are usually used to represent quantities that are not specified, what are called the **unknowns.**

First, consider the equation

$$8x = 32$$

If we wish to solve for x, we can divide (or multiply) each side of the equation by the same factor without destroying the equality. In this case, if we divide both sides by 8, we have

$$\frac{8x}{8} = \frac{32}{8}$$

$$x = 4$$

Next consider the equation

$$x + 2 = 8$$

In this type of expression, we can add or subtract the same quantity from each side. If we subtract 2 from each side, we get

$$x + 2 - 2 = 8 - 2$$
$$x = 6$$

In general, if $x + a = b$, then $x = b - a$.

Now consider the equation

$$\frac{x}{5} = 9$$

If we multiply each side by 5, we are left with x on the left by itself and 45 on the right:

$$\left(\frac{x}{5}\right)(5) = 9 \times 5$$
$$x = 45$$

In all cases, *whatever operation is performed on the left side of the equality must also be performed on the right side.*

The following rules for multiplying, dividing, adding, and subtracting fractions should be recalled, where a, b, and c are three numbers:

	Rule	Example
Multiplying	$\left(\frac{a}{b}\right)\left(\frac{c}{d}\right) = \frac{ac}{bd}$	$\left(\frac{2}{3}\right)\left(\frac{4}{5}\right) = \frac{8}{15}$
Dividing	$\frac{(a/b)}{(c/d)} = \frac{ad}{bc}$	$\frac{2/3}{4/5} = \frac{(2)(5)}{(4)(3)} = \frac{10}{12}$
Adding	$\frac{a}{b} \pm \frac{c}{d} = \frac{ad \pm bc}{bd}$	$\frac{2}{3} - \frac{4}{5} = \frac{(2)(5) - (4)(3)}{(3)(5)} = -\frac{2}{15}$

EXERCISES

In the following exercises, solve for x:

	Answers
1. $a = \frac{1}{1 + x}$	$x = \frac{1 - a}{a}$
2. $3x - 5 = 13$	$x = 6$
3. $ax - 5 = bx + 2$	$x = \frac{7}{a - b}$
4. $\frac{5}{2x + 6} = \frac{3}{4x + 8}$	$x = -\frac{11}{7}$

Powers

When powers of a given quantity x are multiplied, the following rule applies:

$$x^n x^m = x^{n+m} \tag{B.3}$$

For example, $x^2x^4 = x^{2+4} = x^6$.

When dividing the powers of a given quantity, the rule is

$$\frac{x^n}{x^m} = x^{n-m} \tag{B.4}$$

For example, $x^8/x^2 = x^{8-2} = x^6$.

A power that is a fraction, such as $\frac{1}{3}$, corresponds to a root as follows:

$$x^{1/n} = \sqrt[n]{x} \tag{B.5}$$

For example, $4^{1/3} = \sqrt[3]{4} = 1.5874$. (A scientific calculator is useful for such calculations.)

Finally, any quantity x^n that is raised to the mth power is

$$(x^n)^m = x^{nm} \tag{B.6}$$

Table B.1 summarizes the rules of exponents.

TABLE B.1 Rules of Exponents

$x^0 = 1$
$x^1 = x$
$x^n x^m = x^{n+m}$
$x^n/x^m = x^{n-m}$
$x^{1/n} = \sqrt[n]{x}$
$(x^n)^m = x^{nm}$

EXERCISES

Verify the following:

1. $3^2 \times 3^3 = 243$
2. $x^5x^{-8} = x^{-3}$
3. $x^{10}/x^{-5} = x^{15}$
4. $5^{1/3} = 1.709975$ (Use your calculator.)
5. $60^{1/4} = 2.783158$ (Use your calculator.)
6. $(x^4)^3 = x^{12}$

Factoring

Some useful formulas for factoring an equation are

$ax + ay + az = a(x + y + x)$ common factor

$a^2 + 2ab + b^2 = (a + b)^2$ perfect square

$a^2 - b^2 = (a + b)(a - b)$ differences of squares

Quadratic Equations

The general form of a quadratic equation is

$$ax^2 + bx + c = 0 \tag{B.7}$$

where x is the unknown quantity and a, b, and c are numerical factors referred to as **coefficients** of the equation. This equation has two roots, given by

$$x = \frac{-b \pm \sqrt{b^2 - 4ac}}{2a} \tag{B.8}$$

If $b^2 \geq 4ac$, the roots will be real.

EXAMPLE 1

The equation $x^2 + 5x + 4 = 0$ has the following roots corresponding to the two signs of the square-root term:

$$x = \frac{-5 \pm \sqrt{5^2 - (4)(1)(4)}}{2(1)} = \frac{-5 \pm \sqrt{9}}{2} = \frac{-5 \pm 3}{2}$$

that is,

$$x_+ = \frac{-5+3}{2} = -1 \qquad x_- = \frac{-5-3}{2} = -4$$

where x_+ refers to the root corresponding to the positive sign and x_- refers to the root corresponding to the negative sign.

EXERCISES

Solve the following quadratic equations:

	Answers	
1. $x^2 + 2x - 3 = 0$	$x_+ = 1$	$x_- = -3$
2. $2x^2 - 5x + 2 = 0$	$x_+ = 2$	$x_- = \frac{1}{2}$
3. $2x^2 - 4x - 9 = 0$	$x_+ = 1 + \sqrt{22}/2$	$x_- = 1 - \sqrt{22}/2$

Linear Equations

A linear equation has the general form

$$y = ax + b \tag{B.9}$$

where a and b are constants. This equation is referred to as being linear because the graph of y versus x is a straight line, as shown in Figure B.1. The constant b, called the **intercept**, represents the value of y at which the straight line intersects the y axis. The constant a is equal to the **slope** of the straight line and is also equal to the tangent of the angle that the line makes with the x axis. If any two points on the straight line are specified by the coordinates (x_1, y_1) and (x_2, y_2), as in Figure B.1, then the **slope** of the straight line can be expressed as

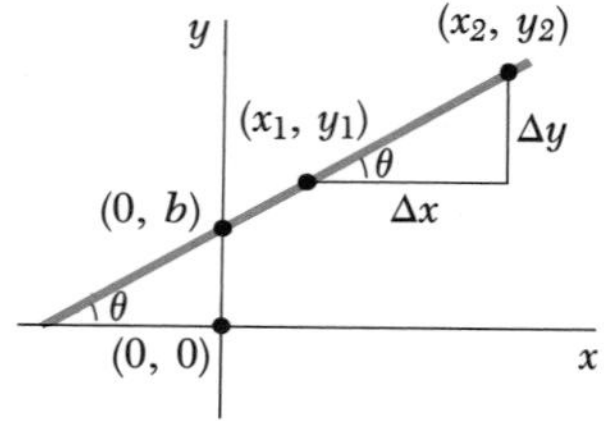

Figure B.1

$$\text{Slope} = \frac{y_2 - y_1}{x_2 - x_1} = \frac{\Delta y}{\Delta x} = \tan\theta \tag{B.10}$$

Note that a and b can have either positive or negative values. If $a > 0$, the straight line has a *positive* slope, as in Figure B.1. If $a < 0$, the straight line has a *negative* slope. In Figure B.1, both a and b are positive. Three other possible situations are shown in Figure B.2: $a > 0, b < 0$; $a < 0, b > 0$; and $a < 0, b < 0$.

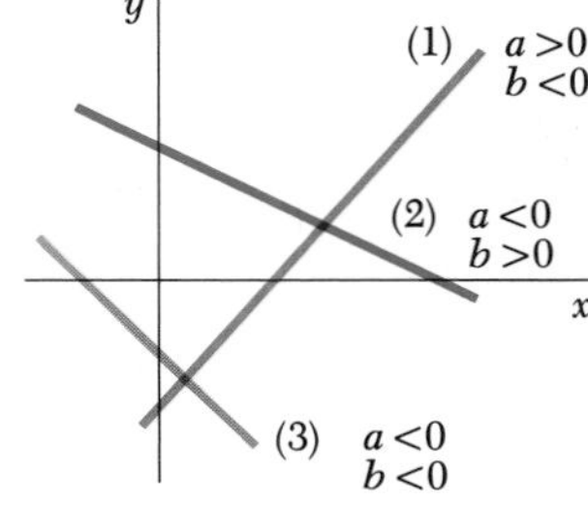

Figure B.2

EXERCISES

1. Draw graphs of the following straight lines:
(a) $y = 5x + 3$ (b) $y = -2x + 4$ (c) $y = -3x - 6$
2. Find the slopes of the straight lines described in Exercise 1.
Answers (a) 5 (b) -2 (c) -3
3. Find the slopes of the straight lines that pass through the following sets of points:
(a) $(0, -4)$ and $(4, 2)$, (b) $(0, 0)$ and $(2, -5)$, and (c) $(-5, 2)$ and $(4, -2)$
Answers (a) 3/2 (b) $-5/2$ (c) $-4/9$

Solving Simultaneous Linear Equations

Consider an equation such as $3x + 5y = 15$, which has two unknowns, x and y. Such an equation does not have a unique solution. That is, $(x = 0,\ y = 3)$, $(x = 5,\ y = 0)$, and $(x = 2,\ y = 9/5)$ are all solutions to this equation.

If a problem has two unknowns, a unique solution is possible only if we have *two* equations. In general, if a problem has n unknowns, its solution requires n equations. In order to solve two simultaneous equations involving two unknowns, x and y, we solve one of the equations for x in terms of y and substitute this expression into the other equation.

EXAMPLE 2
Solve the following two simultaneous equations:

$$(1)\ \ 5x + y = -8$$

$$(2)\ \ 2x - 2y = 4$$

Solution From (2), $x = y + 2$. Substitution of this into (1) gives

$$5(y + 2) + y = -8$$
$$6y = -18$$
$$y = -3$$
$$x = y + 2 = -1$$

Alternate Solution Multiply each term in (1) by the factor 2 and add the result to (2):

$$10x + 2y = -16$$
$$\underline{2x - 2y = 4}$$
$$12x = -12$$
$$x = -1$$
$$y = x - 2 = -3$$

Two linear equations with two unknowns can also be solved by a graphical method. If the straight lines corresponding to the two equations are plotted in

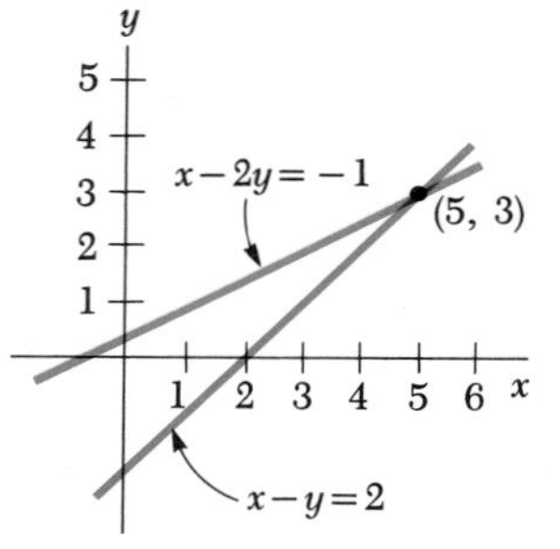

Figure B.3

a conventional coordinate system, the intersection of the two lines represents the solution. For example, consider the two equations

$$x - y = 2$$

$$x - 2y = -1$$

These are plotted in Figure B.3. The intersection of the two lines has the coordinates $x = 5$, $y = 3$. This represents the solution to the equations. You should check this solution by the analytical technique discussed above.

EXERCISES

Solve the following pairs of simultaneous equations involving two unknowns:

		Answers
1.	$x + y = 8$ $x - y = 2$	$x = 5$, $y = 3$
2.	$98 - T = 10a$ $T - 49 = 5a$	$T = 65$, $a = 3.27$
3.	$6x + 2y = 6$ $8x - 4y = 28$	$x = 2$, $y = -3$

Logarithms

Suppose that a quantity x is expressed as a power of some quantity a:

$$x = a^y \tag{B.11}$$

The number a is called the **base** number. The **logarithm** of x with respect to the base a is equal to the exponent to which the base must be raised in order to satisfy the expression $x = a^y$:

$$y = \log_a x \tag{B.12}$$

Conversely, the **antilogarithm** of y is the number x:

$$x = \text{antilog}_a\, y \tag{B.13}$$

In practice, the two bases most often used are base 10, called the *common* logarithm base, and base $e = 2.718\ldots$, called the *natural* logarithm base. When common logarithms are used,

$$y = \log_{10} x \qquad (\text{or } x = 10^y) \tag{B.14}$$

When natural logarithms are used,

$$y = \ln_e x \qquad (\text{or } x = e^y) \tag{B.15}$$

For example, $\log_{10} 52 = 1.716$, so that $\text{antilog}_{10}\, 1.716 = 10^{1.716} = 52$. Likewise, $\ln_e 52 = 3.951$, so $\text{antiln}_e\, 3.951 = e^{3.951} = 52$.

In general, note that you can convert between base 10 and base e with the equality

$$\ln_e x = (2.302585) \log_{10} x \tag{B.16}$$

Finally, some useful properties of logarithms are as follows:

$$\log(ab) = \log a + \log b$$
$$\log(a/b) = \log a - \log b$$
$$\log(a^n) = n \log a$$
$$\ln e = 1$$
$$\ln e^a = a$$
$$\ln\left(\frac{1}{a}\right) = -\ln a$$

B.3 GEOMETRY

The **distance** d between two points whose coordinates are (x_1, y_1) and (x_2, y_2)

$$d = \sqrt{(x_2 - x_1)^2 + (y_2 - y_1)^2} \tag{B.17}$$

The **radian measure:** the arc length s of a circular arc (Fig. B.4) is proportional to the radius r for a fixed value of θ (in radians)

$$s = r\theta$$
$$\theta = \frac{s}{r} \tag{B.18}$$

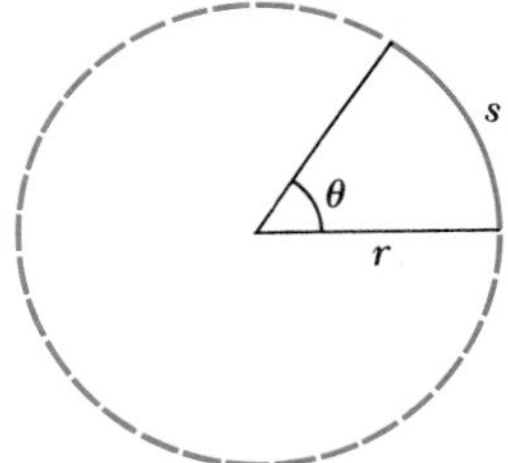

Figure B.4

Table B.2 gives the areas and volumes for several geometric shapes used throughout this text:

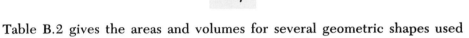

TABLE B.2 Useful Information for Geometry

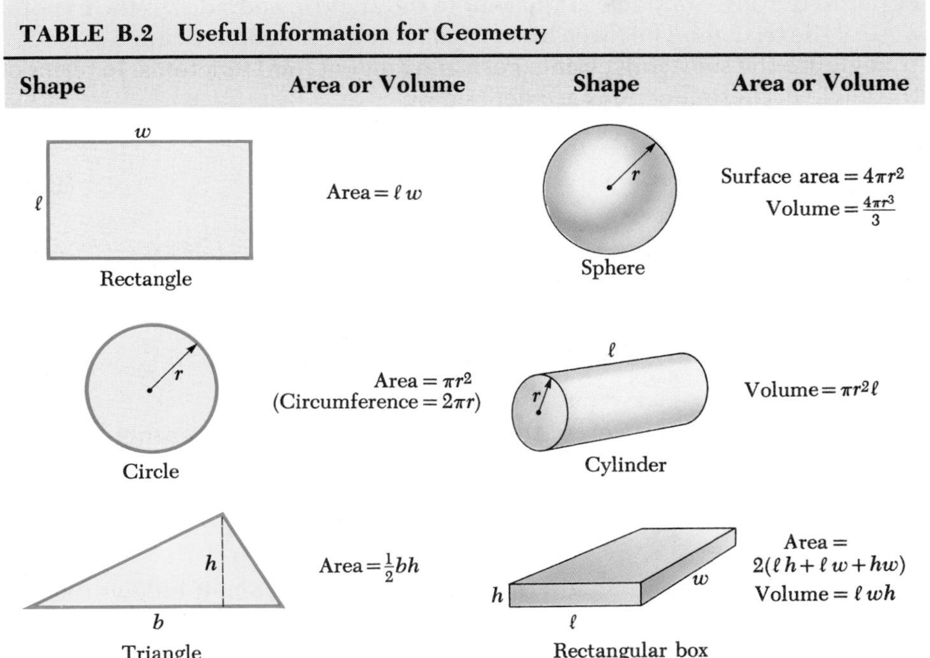

Shape	Area or Volume	Shape	Area or Volume
Rectangle	Area $= \ell w$	Sphere	Surface area $= 4\pi r^2$ Volume $= \frac{4\pi r^3}{3}$
Circle	Area $= \pi r^2$ (Circumference $= 2\pi r$)	Cylinder	Volume $= \pi r^2 \ell$
Triangle	Area $= \frac{1}{2}bh$	Rectangular box	Area $=$ $2(\ell h + \ell w + hw)$ Volume $= \ell wh$

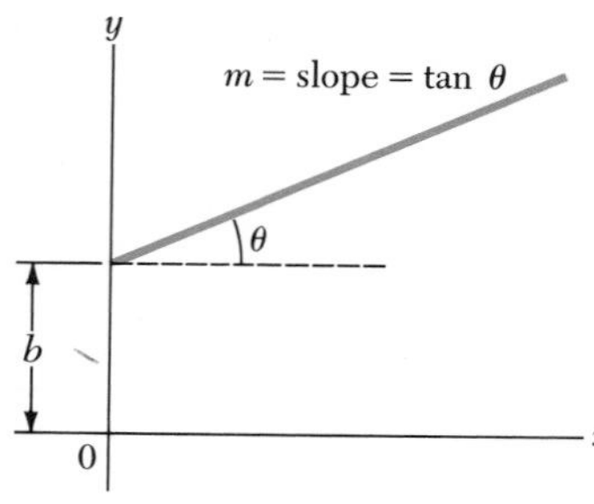
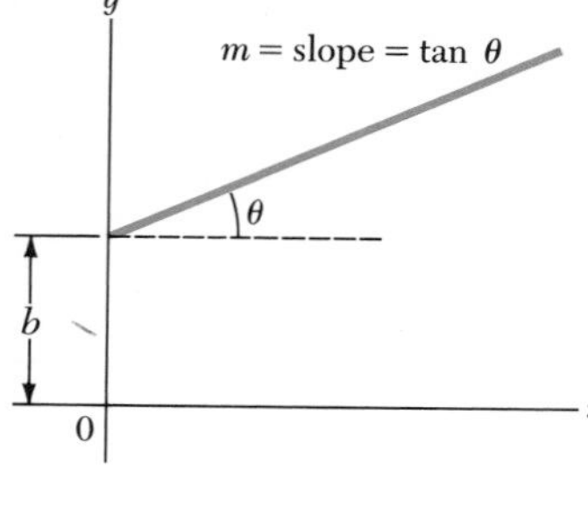

Figure B.5

The equation of a **straight line** (Fig. B.5) is given by

$$y = mx + b \tag{B.19}$$

where b is the y intercept and m is the slope of the line.

The equation of a **circle** of radius R centered at the origin is

$$x^2 + y^2 = R^2 \tag{B.20}$$

The equation of an **ellipse** with the origin at its center (Fig. B.6) is

$$\frac{x^2}{a^2} + \frac{y^2}{b^2} = 1 \tag{B.21}$$

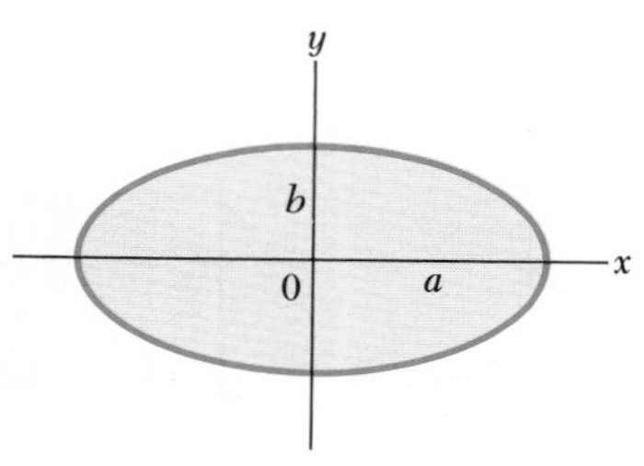

Figure B.6

where a is the length of the semi-major axis and b is the length of the semi-minor axis.

The equation of a **parabola** whose vertex is at $y = b$ (Fig. B.7) is

$$y = ax^2 + b \tag{B.22}$$

The equation of a **rectangular hyperbola** (Fig. B.8) is

$$xy = \text{constant} \tag{B.23}$$

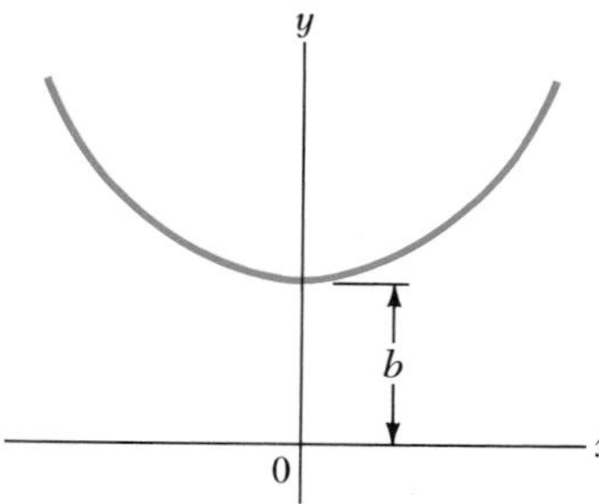

Figure B.7

B.4 TRIGONOMETRY

That portion of mathematics based on the special properties of the right triangle is called trigonometry. By definition, a right triangle is one containing a 90° angle. Consider the right triangle shown in Figure B.9, where side a is opposite the angle θ, side b is adjacent to the angle θ, and side c is the hypotenuse of the triangle. The three basic trigonometric functions defined by such a triangle are the sine (sin), cosine (cos), and tangent (tan) functions. In terms of the angle θ, these functions are defined by

$$\sin\theta \equiv \frac{\text{side opposite } \theta}{\text{hypotenuse}} = \frac{a}{c} \tag{B.24}$$

$$\cos\theta \equiv \frac{\text{side adjacent to } \theta}{\text{hypotenuse}} = \frac{b}{c} \tag{B.25}$$

$$\tan\theta \equiv \frac{\text{side opposite } \theta}{\text{side adjacent to } \theta} = \frac{a}{b} \tag{B.26}$$

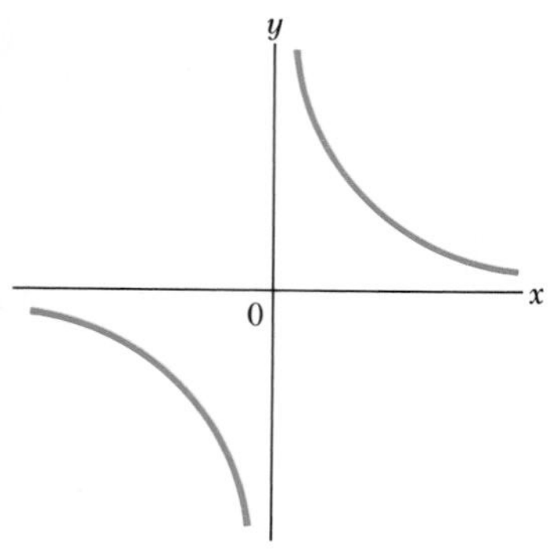

Figure B.8

The Pythagorean theorem provides the following relationship between the sides of a triangle:

$$c^2 = a^2 + b^2 \tag{B.27}$$

From the above definitions and the Pythagorean theorem, it follows that

$$\sin^2\theta + \cos^2\theta = 1$$

$$\tan\theta = \frac{\sin\theta}{\cos\theta}$$

The cosecant, secant, and cotangent functions are defined by

$$\csc\theta \equiv \frac{1}{\sin\theta} \qquad \sec\theta \equiv \frac{1}{\cos\theta} \qquad \cot\theta \equiv \frac{1}{\tan\theta}$$

The relations below follow directly from the right triangle shown in Figure B.9:

$$\begin{cases} \sin\theta = \cos(90^\circ - \theta) \\ \cos\theta = \sin(90^\circ - \theta) \\ \cot\theta = \tan(90^\circ - \theta) \end{cases}$$

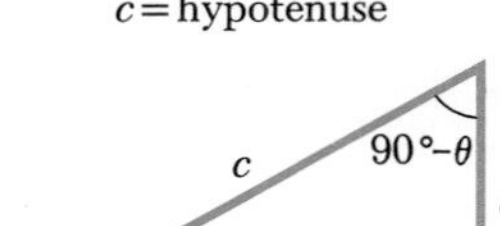

Figure B.9

Some properties of trigonometric functions are as follows:

$$\begin{cases} \sin(-\theta) = -\sin\theta \\ \cos(-\theta) = \cos\theta \\ \tan(-\theta) = -\tan\theta \end{cases}$$

The following relations apply to *any* triangle as shown in Figure B.10:

$$\alpha + \beta + \gamma = 180^\circ$$

$$\text{Law of cosines} \quad \begin{cases} a^2 = b^2 + c^2 - 2bc\cos\alpha \\ b^2 = a^2 + c^2 - 2ac\cos\beta \\ c^2 = a^2 + b^2 - 2ab\cos\gamma \end{cases}$$

$$\text{Law of sines} \quad \left\{ \frac{a}{\sin\alpha} = \frac{b}{\sin\beta} = \frac{c}{\sin\gamma} \right.$$

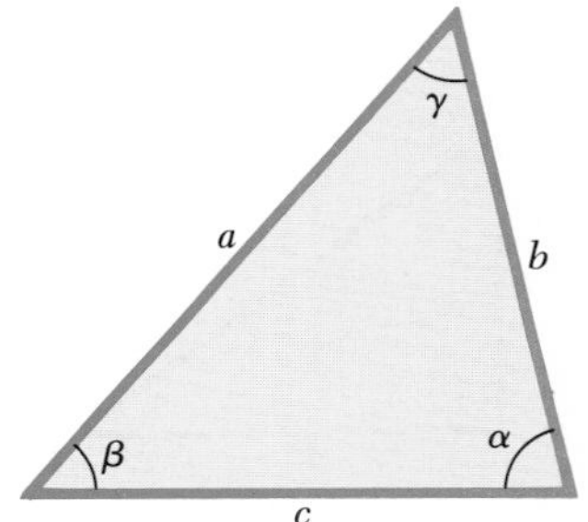

Figure B.10

Table B.3 lists a number of useful trigonometric identities.

TABLE B.3 Some Trigonometric Identities

$\sin^2\theta + \cos^2\theta = 1$	$\csc^2\theta = 1 + \cot^2\theta$
$\sec^2\theta = 1 + \tan^2\theta$	$\sin^2\frac{\theta}{2} = \frac{1}{2}(1 - \cos\theta)$
$\sin 2\theta = 2\sin\theta\cos\theta$	$\cos^2\frac{\theta}{2} = \frac{1}{2}(1 + \cos\theta)$
$\cos 2\theta = \cos^2\theta - \sin^2\theta$	$1 - \cos\theta = 2\sin^2\frac{\theta}{2}$
$\tan 2\theta = \frac{2\tan\theta}{1 - \tan^2\theta}$	$\tan\frac{\theta}{2} = \sqrt{\frac{1 - \cos\theta}{1 + \cos\theta}}$
$\sin(A \pm B) = \sin A\cos B \pm \cos A\sin B$	
$\cos(A \pm B) = \cos A\cos B \mp \sin A\sin B$	
$\sin A \pm \sin B = 2\sin[\frac{1}{2}(A \pm B)]\cos[\frac{1}{2}(A \mp B)]$	
$\cos A + \cos B = 2\cos[\frac{1}{2}(A + B)]\cos[\frac{1}{2}(A - B)]$	
$\cos A - \cos B = 2\sin[\frac{1}{2}(A + B)]\sin[\frac{1}{2}(B - A)]$	

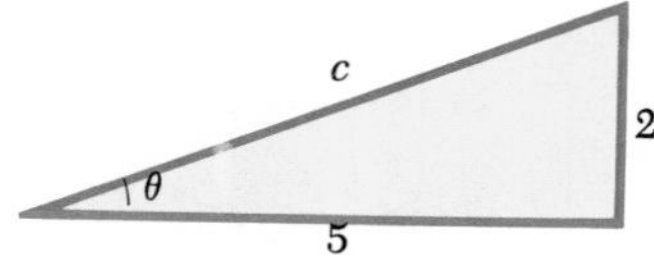

Figure B.11

EXAMPLE 3
Consider the right triangle in Figure B.11, in which $a = 2$, $b = 5$, and c is unknown. From the Pythagorean theorem, we have

$$c^2 = a^2 + b^2 = 2^2 + 5^2 = 4 + 25 = 29$$

$$c = \sqrt{29} = 5.39$$

To find the angle θ, note that

$$\tan \theta = \frac{a}{b} = \frac{2}{5} = 0.400$$

From a table of functions or from a calculator, we have

$$\theta = \tan^{-1}(0.400) = 21.8°$$

where $\tan^{-1}(0.400)$ is the notation for "angle whose tangent is 0.400," sometimes written as arctan(0.400).

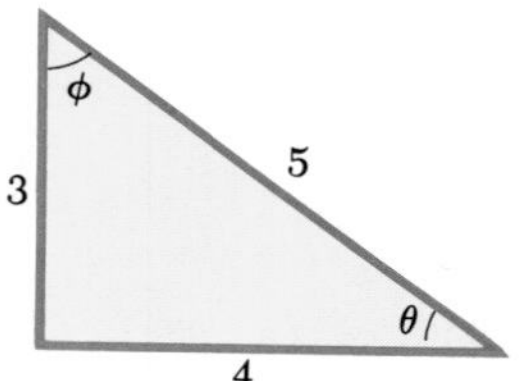

Figure B.12

EXERCISES

1. In Figure B.12, find (a) the side opposite θ, (b) the side adjacent to ϕ, (c) $\cos \theta$, (d) $\sin \phi$, and (e) $\tan \phi$.
Answers (a) 3, (b) 3, (c) $\frac{4}{5}$, (d) $\frac{4}{5}$, and (e) $\frac{4}{3}$
2. In a certain right triangle, the two sides that are perpendicular to each other are 5 m and 7 m long. What is the length of the third side of the triangle?
Answer 8.60 m
3. A right triangle has a hypotenuse of length 3 m, and one of its angles is 30°. What is the length of (a) the side opposite the 30° angle and (b) the side adjacent to the 30° angle?
Answers (a) 1.5 m and (b) 2.60 m

B.5 SERIES EXPANSIONS

$$(a + b)^n = a^n + \frac{n}{1!} a^{n-1}b + \frac{n(n-1)}{2!} a^{n-2}b^2 + \cdots$$

$$(1 + x)^n = 1 + nx + \frac{n(n-1)}{2!} x^2 + \cdots$$

$$e^x = 1 + x + \frac{x^2}{2!} + \frac{x^3}{3!} + \cdots$$

$$\ln(1 \pm x) = \pm x - \tfrac{1}{2}x^2 \pm \tfrac{1}{3}x^3 - \cdots$$

$$\left.\begin{aligned} \sin x &= x - \frac{x^3}{3!} + \frac{x^5}{5!} - \cdots \\ \cos x &= 1 - \frac{x^2}{2!} + \frac{x^4}{4!} - \cdots \\ \tan x &= x + \frac{x^3}{3} + \frac{2x^5}{15} + \cdots \quad |x| < \pi/2 \end{aligned}\right\} x \text{ in radians}$$

For $x \ll 1$, the following approximations can be used:

$$(1 + x)^n \approx 1 + nx \qquad \sin x \approx x$$

$$e^x \approx 1 + x \qquad \cos x \approx 1$$

$$\ln(1 \pm x) \approx \pm x \qquad \tan x \approx x$$

B.6 DIFFERENTIAL CALCULUS

In various branches of science, it is sometimes necessary to use the basic tools of calculus, first invented by Newton, to describe physical phenomena. The use of calculus is fundamental in the treatment of various problems in newtonian mechanics, electricity, and magnetism. In this section, we simply state some basic properties and "rules of thumb" that should be a useful review to the student.

First, a **function** must be specified that relates one variable to another (such as a coordinate as a function of time). Suppose one of the variables is called y (the dependent variable), the other x (the independent variable). We might have a function relation such as

$$y(x) = ax^3 + bx^2 + cx + d$$

If a, b, c, and d are specified constants, then y can be calculated for any value of x. We usually deal with continuous functions, that is, those for which y varies "smoothly" with x.

The **derivative** of y with respect to x is defined as the limit of the slopes of chords drawn between two points on the y versus x curve as Δx approaches zero. Mathematically, we write this definition as

$$\frac{dy}{dx} = \lim_{\Delta x \to 0} \frac{\Delta y}{\Delta x} = \lim_{\Delta x \to 0} \frac{y(x + \Delta x) - y(x)}{\Delta x} \tag{B.28}$$

where Δy and Δx are defined as $\Delta x = x_2 - x_1$ and $\Delta y = y_2 - y_1$ (see Fig. B.13).

A useful expression to remember when $y(x) = ax^n$, where a is a *constant* and n is *any* positive or negative number (integer or fraction), is

$$\frac{dy}{dx} = nax^{n-1} \tag{B.29}$$

If $y(x)$ is a polynomial or algebraic function of x, we apply Equation B.29 to *each* term in the polynomial and take $da/dx = 0$. It is important to note that

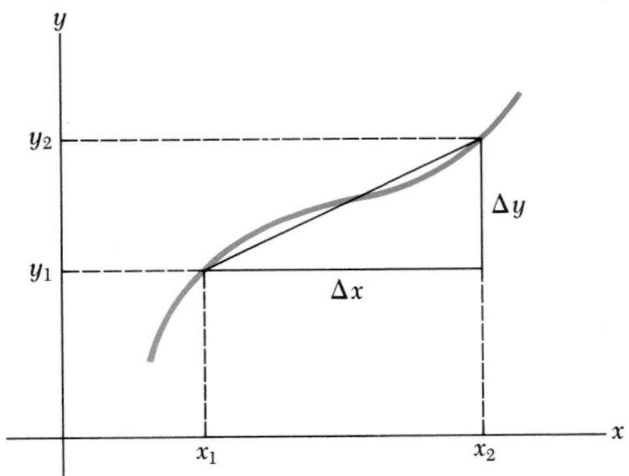

Figure B.13

dy/dx *does not* mean dy divided by dx, but is simply a notation of the limiting process of the derivative as defined by Equation B.28. In Examples 4 through 7, we evaluate the derivatives of several well-behaved functions.

EXAMPLE 4

Suppose $y(x)$ (that is, y as a function of x) is given by

$$y(x) = ax^3 + bx + c$$

where a and b are constants. Then it follows that

$$y(x + \Delta x) = a(x + \Delta x)^3 + b(x + \Delta x) + c$$

$$y(x + \Delta x) = a(x^3 + 3x^2\,\Delta x + 3x\,\Delta x^2 + \Delta x^3) + b(x + \Delta x) + c$$

so

$$\Delta y = y(x + \Delta x) - y(x) = a(3x^2\,\Delta x + 3x\,\Delta x^2 + \Delta x^3) + b\,\Delta x$$

Substituting this into Equation B.28 gives

$$\frac{dy}{dx} = \lim_{\Delta x \to 0} \frac{\Delta y}{\Delta x} = \lim_{\Delta x \to 0} [3ax^2 + 3x\,\Delta x + \Delta x^2] + b$$

$$\frac{dy}{dx} = 3ax^2 + b$$

EXAMPLE 5

$$y(x) = 8x^5 + 4x^3 + 2x + 7$$

Solution Applying Equation B.29 to each term independently, and remembering that d/dx (constant) $= 0$, we have

$$\frac{dy}{dx} = 8(5)x^4 + 4(3)x^2 + 2(1)x^0 + 0$$

$$\frac{dy}{dx} = 40x^4 + 12x^2 + 2$$

Special Properties of the Derivative

A. **Derivative of the Product of Two Functions** If a function y is given by the product of two functions, say, $g(x)$ and $h(x)$, then the derivative of y is defined as

$$\frac{d}{dx} f(x) = \frac{d}{dx} [g(x)h(x)] = g\frac{dh}{dx} + h\frac{dg}{dx} \tag{B.30}$$

B. **Derivative of the Sum of Two Functions** If a function y is equal to the sum of two functions, then the derivative of the sum is equal to the sum of the derivatives:

$$\frac{d}{dx} f(x) = \frac{d}{dx} [g(x) + h(x)] = \frac{dg}{dx} + \frac{dh}{dx} \tag{B.31}$$

C. **Chain Rule of Differential Calculus** If $y = f(x)$ and x is a function of some other variable z, then dy/dx can be written as the product of two derivatives:

$$\frac{dy}{dx} = \frac{dy}{dz}\frac{dz}{dx} \qquad (B.32)$$

D. The Second Derivative The second derivative of y with respect to x is defined as the derivative of the function dy/dx (or, the derivative of the derivative). It is usually written

$$\frac{d^2y}{dx^2} = \frac{d}{dx}\left(\frac{dy}{dx}\right) \qquad (B.33)$$

EXAMPLE 6

Find the first derivative $y(x) = x^3/(x+1)^2$ with respect to x.

Solution We can rewrite this function as $y(x) = x^3(x+1)^{-2}$ and apply Equation B.30 directly:

$$\frac{dy}{dx} = (x+1)^{-2}\frac{d}{dx}(x^3) + x^3\frac{d}{dx}(x+1)^{-2}$$

$$= (x+1)^{-2}\,3x^2 + x^3(-2)(x+1)^{-3}$$

$$\frac{dy}{dx} = \frac{3x^2}{(x+1)^2} - \frac{2x^3}{(x+1)^3}$$

EXAMPLE 7

A useful formula that follows from Equation B.30 is the derivative of the quotient of two functions. Show that the expression is given by

$$\frac{d}{dx}\left[\frac{g(x)}{h(x)}\right] = \frac{h\dfrac{dg}{dx} - g\dfrac{dh}{dx}}{h^2}$$

Solution We can write the quotient as gh^{-1} and then apply Equations B.29 and B.30:

$$\frac{d}{dx}\left(\frac{g}{h}\right) = \frac{d}{dx}(gh^{-1}) = g\frac{d}{dx}(h^{-1}) + h^{-1}\frac{d}{dx}(g)$$

$$= -gh^{-2}\frac{dh}{dx} + h^{-1}\frac{dg}{dx}$$

$$= \frac{h\dfrac{dg}{dx} - g\dfrac{dh}{dx}}{h^2}$$

Some of the more commonly used derivatives of functions are listed in Table B.4.

TABLE B.4 Derivatives for Several Functions

$\frac{d}{dx}(a) = 0$
$\frac{d}{dx}(ax^n) = nax^{n-1}$
$\frac{d}{dx}(e^{ax}) = ae^{ax}$
$\frac{d}{dx}(\sin ax) = a\cos ax$
$\frac{d}{dx}(\cos ax) = -a\sin ax$
$\frac{d}{dx}(\tan ax) = a\sec^2 ax$
$\frac{d}{dx}(\cot ax) = -a\csc^2 ax$
$\frac{d}{dx}(\sec x) = \tan x\sec x$
$\frac{d}{dx}(\csc x) = -\cot x\csc x$
$\frac{d}{dx}(\ln ax) = \frac{a}{x}$

Note: The letters a and n are constants.

B.7 INTEGRAL CALCULUS

We think of integration as the inverse of differentiation. As an example, consider the expression

$$f(x) = \frac{dy}{dx} = 3ax^2 + b$$

which was the result of differentiating the function

$$y(x) = ax^3 + bx + c$$

in Example 4. We can write the first expression $dy = f(x)dx = (3ax^2 + b)dx$ and obtain $y(x)$ by "summing" over all values of x. Mathematically, we write this inverse operation

$$y(x) = \int f(x)dx$$

For the function $f(x)$ given above,

$$y(x) = \int (3ax^2 + b)dx = ax^3 + bx + c$$

where c is a constant of the integration. This type of integral is called an *indefinite integral* since its value depends on the choice of the constant c.

A general **indefinite integral** $I(x)$ is defined as

$$I(x) = \int f(x)dx \tag{B.34}$$

where $f(x)$ is called the *integrand* and $f(x) = \frac{dI(x)}{dx}$.

For a *general continuous* function $f(x)$, the integral can be described as the area under the curve bounded by $f(x)$ and the x axis, between two specified values of x, say, x_1 and x_2, as in Figure B.14.

The area of the shaded element is approximately $f_i \Delta x_i$. If we sum all these area elements from x_1 to x_2 and take the limit of this sum as $\Delta x_i \rightarrow 0$, we obtain the *true* area under the curve bounded by $f(x)$ and x, between the limits x_1 and x_2:

$$\text{Area} = \lim_{\Delta x \rightarrow 0} \sum_i f_i(x)\Delta x_i = \int_{x_1}^{x_2} f(x)dx \tag{B.35}$$

Integrals of the type defined by Equation B.35 are called **definite integrals.**

One of the common types of integrals that arise in practical situations has the form

$$\int x^n \, dx = \frac{x^{n+1}}{n+1} + c \qquad (n \neq -1) \tag{B.36}$$

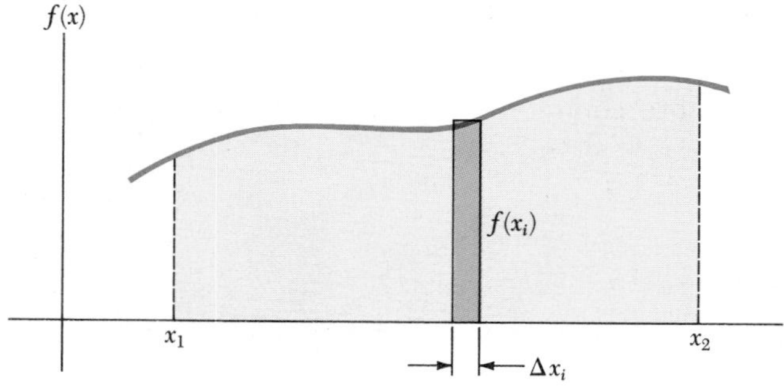

Figure B.14

This result is obvious since differentiation of the right-hand side with respect to x gives $f(x) = x^n$ directly. If the limits of the integration are known, this integral becomes a *definite integral* and is written

$$\int_{x_1}^{x_2} x^n\, dx = \frac{x_2^{\,n+1} - x_1^{\,n+1}}{n+1} \qquad (n \neq -1) \tag{B.37}$$

Examples

1. $\displaystyle\int_0^a x^2\, dx = \frac{x^3}{3}\bigg]_0^a = \frac{a^3}{3}$
2. $\displaystyle\int_0^b x^{3/2}\, dx = \frac{x^{5/2}}{5/2}\bigg]_0^b = \frac{2}{5}\, b^{5/2}$
3. $\displaystyle\int_3^5 x\, dx = \frac{x^2}{2}\bigg]_3^5 = \frac{5^2 - 3^2}{2} = 8$

Partial Integration

Sometimes it is useful to apply the method of *partial integration* to evaluate certain integrals. The method uses the property that

$$\int u\, dv = uv - \int v\, du \tag{B.38}$$

where u and v are *carefully* chosen so as to reduce a complex integral to a simpler one. In many cases, several reductions have to be made. Consider the example

$$I(x) = \int x^2 e^x\, dx$$

This can be evaluated by integrating by parts twice. First, if we choose $u = x^2$, $v = e^x$, we get

$$\int x^2 e^x\, dx = \int x^2\, d(e^x) = x^2 e^x - 2\int e^x x\, dx + c_1$$

Now, in the second term, choose $u = x$, $v = e^x$, which gives

$$\int x^2 e^x\, dx = x^2 e^x - 2xe^x + 2\int e^x\, dx + c_1$$

or

$$\int x^2 e^x\, dx = x^2 e^x - 2xe^x + 2e^x + c_2$$

The Perfect Differential

Another useful method to remember is the use of the *perfect differential*. That is, we should sometimes look for a change of variable such that the differential of the function is the differential of the independent variable appearing in the integrand. For example, consider the integral

$$I(x) = \int \cos^2 x \sin x\, dx$$

This becomes easy to evaluate if we rewrite the differential as $d(\cos x) = -\sin x\, dx$. The integral then becomes

$$\int \cos^2 x \sin x\, dx = -\int \cos^2 x\, d(\cos x)$$

If we now change variables, letting $y = \cos x$, we get

$$\int \cos^2 x \sin x\, dx = -\int y^2 dy = -\frac{y^3}{3} + c = -\frac{\cos^3 x}{3} + c$$

Table B.5 lists some useful indefinite integrals. Table B.6 gives Gauss' probability integral and other definite integrals. A more complete list can be found in various handbooks, such as *The Handbook of Chemistry and Physics*, CRC Press.

TABLE B.5 Some Indefinite Integrals (an arbitrary constant should be added to each of these integrals)

$$\int x^n\, dx = \frac{x^{n+1}}{n+1} \quad \text{(provided } n \neq -1\text{)}$$

$$\int \frac{dx}{x} = \int x^{-1}\, dx = \ln x$$

$$\int \frac{dx}{a + bx} = \frac{1}{b} \ln(a + bx)$$

$$\int \frac{dx}{(a + bx)^2} = -\frac{1}{b(a + bx)}$$

$$\int \frac{dx}{a^2 + x^2} = \frac{1}{a} \tan^{-1} \frac{x}{a}$$

$$\int \frac{dx}{a^2 - x^2} = \frac{1}{2a} \ln \frac{a + x}{a - x} \quad (a^2 - x^2 > 0)$$

$$\int \frac{dx}{x^2 - a^2} = \frac{1}{2a} \ln \frac{x - a}{x + a} \quad (x^2 - a^2 > 0)$$

$$\int \frac{x\, dx}{a^2 \pm x^2} = \pm \tfrac{1}{2} \ln(a^2 \pm x^2)$$

$$\int \frac{dx}{\sqrt{a^2 - x^2}} = \sin^{-1} \frac{x}{a} = -\cos^{-1} \frac{x}{a} \quad (a^2 - x^2 > 0)$$

$$\int \frac{dx}{\sqrt{x^2 \pm a^2}} = \ln(x + \sqrt{x^2 \pm a^2})$$

$$\int \frac{x\, dx}{\sqrt{a^2 - x^2}} = -\sqrt{a^2 - x^2}$$

$$\int \frac{x\, dx}{\sqrt{x^2 \pm a^2}} = \sqrt{x^2 \pm a^2}$$

$$\int \sqrt{a^2 - x^2}\, dx = \tfrac{1}{2}\left(x\sqrt{a^2 - x^2} + a^2 \sin^{-1} \frac{x}{a}\right)$$

$$\int x\sqrt{a^2 - x^2}\, dx = -\tfrac{1}{3}(a^2 - x^2)^{3/2}$$

$$\int x e^{ax}\, dx = \frac{e^{ax}}{a^2}(ax - 1)$$

$$\int \frac{dx}{a + be^{cx}} = \frac{x}{a} - \frac{1}{ac} \ln(a + be^{cx})$$

$$\int \sin ax\, dx = -\frac{1}{a} \cos ax$$

$$\int \cos ax\, dx = \frac{1}{a} \sin ax$$

$$\int \tan ax\, dx = -\frac{1}{a} \ln(\cos ax) = \frac{1}{a} \ln(\sec ax)$$

$$\int \cot ax\, dx = \frac{1}{a} \ln(\sin ax)$$

$$\int \sec ax\, dx = \frac{1}{a} \ln(\sec ax + \tan ax) = \frac{1}{a} \ln\left[\tan\left(\frac{ax}{2} + \frac{\pi}{4}\right)\right]$$

$$\int \csc ax\, dx = \frac{1}{a} \ln(\csc ax - \cot ax) = \frac{1}{a} \ln\left(\tan \frac{ax}{2}\right)$$

$$\int \sin^2 ax\, dx = \frac{x}{2} - \frac{\sin 2ax}{4a}$$

$$\int \cos^2 ax\, dx = \frac{x}{2} + \frac{\sin 2ax}{4a}$$

$$\int \frac{dx}{\sin^2 ax} = -\frac{1}{a} \cot ax$$

$$\int \frac{dx}{\cos^2 ax} = \frac{1}{a} \tan ax$$

$$\int \tan^2 ax\, dx = \frac{1}{a} (\tan ax) - x$$

$$\int \cot^2 ax\, dx = -\frac{1}{a} (\cot ax) - x$$

(Table continues)

TABLE B.5 (Continued)

$\int \sqrt{x^2 \pm a^2}\,dx = \frac{1}{2}[x\sqrt{x^2 \pm a^2} \pm a^2 \ln(x + \sqrt{x^2 \pm a^2})]$	$\int \sin^{-1} ax\,dx = x(\sin^{-1} ax) + \frac{\sqrt{1 - a^2x^2}}{a}$
$\int x\,(\sqrt{x^2 \pm a^2})\,dx = \frac{1}{3}(x^2 \pm a^2)^{3/2}$	$\int \cos^{-1} ax\,dx = x(\cos^{-1} ax) - \frac{\sqrt{1 - a^2x^2}}{a}$
$\int e^{ax}\,dx = \frac{1}{a}\,e^{ax}$	$\int \frac{dx}{(x^2 + a^2)^{3/2}} = \frac{x}{a^2\sqrt{x^2 + a^2}}$
$\int \ln ax\,dx = (x \ln ax) - x$	$\int \frac{x\,dx}{(x^2 + a^2)^{3/2}} = \frac{1}{a^2\sqrt{x^2 + a^2}}$

TABLE B.6 Gauss' Probability Integral and Related Integrals

$$I_0 = \int_0^\infty e^{-\alpha x^2}\,dx = \frac{1}{2}\sqrt{\frac{\pi}{\alpha}} \qquad \text{(Gauss' probability integral)}$$

$$I_1 = \int_0^\infty x e^{-\alpha x^2}\,dx = \frac{1}{2\alpha}$$

$$I_2 = \int_0^\infty x^2 e^{-\alpha x^2}\,dx = -\frac{dI_0}{d\alpha} = \frac{1}{4}\sqrt{\frac{\pi}{\alpha^3}}$$

$$I_3 = \int_0^\infty x^3 e^{-\alpha x^2}\,dx = -\frac{dI_1}{d\alpha} = \frac{1}{2\alpha^2}$$

$$I_4 = \int_0^\infty x^4 e^{-\alpha x^2}\,dx = \frac{d^2 I_0}{d\alpha^2} = \frac{3}{8}\sqrt{\frac{\pi}{\alpha^5}}$$

$$I_5 = \int_0^\infty x^5 e^{-\alpha x^2}\,dx = \frac{d^2 I_1}{d\alpha^2} = \frac{1}{\alpha^3}$$

$$\vdots$$

$$I_{2n} = (-1)^n \frac{d^n}{d\alpha^n} I_0$$

$$I_{2n+1} = (-1)^n \frac{d^n}{d\alpha^n} I_1$$

Appendix C

Periodic Table of the Elements*

Key: Symbol — Ca 20 — Atomic number; Atomic mass† — 40.08; $4s^2$ — Electron configuration

Group I	Group II	Transition elements						
H 1 1.0080 $1s^1$								
Li 3 6.94 $2s^1$	Be 4 9.012 $2s^2$							
Na 11 22.99 $3s^1$	Mg 12 24.31 $3s^2$							
K 19 39.102 $4s^1$	Ca 20 40.08 $4s^2$	Sc 21 44.96 $3d^1 4s^2$	Ti 22 47.90 $3d^2 4s^2$	V 23 50.94 $3d^3 4s^2$	Cr 24 51.996 $3d^5 4s^1$	Mn 25 54.94 $3d^5 4s^2$	Fe 26 55.85 $3d^6 4s^2$	Co 27 58.93 $3d^7 4s^2$
Rb 37 85.47 $5s^1$	Sr 38 87.62 $5s^2$	Y 39 88.906 $4d^1 5s^2$	Zr 40 91.22 $4d^2 5s^2$	Nb 41 92.91 $4d^4 5s^1$	Mo 42 95.94 $4d^5 5s^1$	Tc 43 (99) $4d^5 5s^2$	Ru 44 101.1 $4d^7 5s^1$	Rh 45 102.91 $4d^8 5s^1$
Cs 55 132.91 $6s^1$	Ba 56 137.34 $6s^2$	57 - 71‡	Hf 72 178.49 $5d^2 6s^2$	Ta 73 180.95 $5d^3 6s^2$	W 74 183.85 $5d^4 6s^2$	Re 75 186.2 $5d^5 6s^2$	Os 76 190.2 $5d^6 6s^2$	Ir 77 192.2 $5d^7 6s^2$
Fr 87 (223) $7s^1$	Ra 88 (226) $7s^2$	89 - 103§	Rf 104 (261) $6d^2 7s^2$	Ha 105 (262) $6d^3 7s^2$	Unh 106 (263)	Uns 107 (261)		

‡Lanthanide series	La 57 138.91 $5d^1 6s^2$	Ce 58 140.12 $5d^1 4f^1 6s^2$	Pr 59 140.91 $4f^3 6s^2$	Nd 60 144.24 $4f^4 6s^2$	Pm 61 (147) $4f^6 6s^2$	Sm 62 150.4 $4f^6 6s^2$
§Actinide series	Ac 89 (227) $6d^1 7s^2$	Th 90 (232) $6d^2 7s^2$	Pa 91 (231) $5f^2 d^1 7s^2$	U 92 (238) $5f^3 6d^1 7s^2$	Np 93 (239) $5f^3 6d^0 7s^2$	Pu 94 (239) $5f^6 6d^0 7s^2$

* Atomic mass values given are averaged over isotopes in the percentages in which they exist in nature.

† For an unstable element, mass number of the most stable known isotope is given in parentheses.

			Group III	Group IV	Group V	Group VI	Group VII	Group 0
							H 1 1.0080 $2s^4$	He 2 4.0026 $1s^2$
			B 5 10.81 $2p^2$	C 6 12.011 $2p^2$	N 7 14.007 $2p^3$	O 8 15.999 $2p^4$	F 9 18.998 $2p^5$	Ne 10 20.18 $2p^6$
			Al 13 26.98 $3p^1$	Si 14 28.09 $3p^2$	P 15 30.97 $3p^3$	S 16 32.06 $3p^4$	Cl 17 35.453 $3p^5$	Ar 18 39.948 $3p^6$
Ni 28 58.71 $3d^84s^2$	Cu 29 63.54 $3d^{10}4s^1$	Zn 30 65.37 $3d^{10}4s^2$	Ga 31 69.72 $4p^1$	Ge 32 72.59 $4p^2$	As 33 74.92 $4p^3$	Se 34 78.96 $4p^4$	Br 35 79.91 $4p^5$	Kr 36 83.80 $4p^6$
Pd 46 106.4 $4d^{10}5s^6$	Ag 47 107.87 $4d^{10}5s^1$	Cd 48 112.40 $4d^{10}5s^2$	In 49 114.82 $5p^1$	Sn 50 118.69 $5p^2$	Sb 51 121.75 $5p^3$	Te 52 127.60 $5p^4$	I 53 126.90 $5p^5$	Xe 54 131.30 $5p^6$
Pt 78 195.09 $5d^96s^1$	Au 79 196.97 $5d^{10}6s^1$	Hg 80 200.59 $5d^{10}6s^2$	Tl 81 204.37 $6p^2$	Pb 82 207.2 $6p^2$	Bi 83 208.98 $6p^3$	Po 84 (210) $6p^4$	At 85 (218) $6p^5$	Rn 86 (222) $6p^6$

Eu 63 152.0 $4f^76s^2$	Gd 64 157.25 $5d^14f^76s^2$	Tb 65 158.92 $5d^14f^86s^2$	Dy 66 162.50 $4f^{10}6s^2$	Ho 67 164.93 $4f^{11}6s^2$	Er 68 167.26 $4f^{12}6s^2$	Tm 69 168.93 $4f^{13}6s^2$	Yb 70 173.04 $4f^{14}6s^2$	Lu 71 174.97 $5d^14f^{14}6s^2$
Am 95 (243) $5f^76d^07s^2$	Cm 96 (245) $5f^76d^17s^2$	Bk 97 (247) $5f^86d^17s^2$	Cf 98 (249) $5f^{10}6d^07s^2$	Es 99 (254) $5s^{11}6d^07s^2$	Fm 100 (253) $5f^{12}6d^07s^2$	Md 101 (255) $5f^{13}6d^07s^2$	No 102 (255) $6d^07s^2$	Lr 103 (257) $6d^17s^2$

Appendix D

SI Units

TABLE D.1 SI Base Units

Base Quantity	SI Base Unit	
	Name	Symbol
Length	Meter	m
Mass	Kilogram	kg
Time	Second	s
Electric current	Ampere	A
Temperature	Kelvin	K
Amount of substance	Mole	mol
Luminous intensity	Candela	cd

TABLE D.2 Some Derived SI Units

Quantity	Name	Symbol	Expression in Terms of Base Units	Expression in Terms of Other SI Units
Plane angle	Radian	rad	m/m	
Frequency	Hertz	Hz	s^{-1}	
Force	Newton	N	$kg \cdot m/s^2$	J/m
Pressure	Pascal	Pa	$kg/m \cdot s^2$	N/m^2
Energy: work	Joule	J	$kg \cdot m^2/s^2$	$N \cdot m$
Power	Watt	W	$kg \cdot m^2/s^3$	J/s
Electric charge	Coulomb	C	$A \cdot s$	
Electric potential (emf)	Volt	V	$kg \cdot m^2/A \cdot s^3$	W/A
Capacitance	Farad	F	$A^2 \cdot s^4/kg \cdot m^2$	C/V
Electric resistance	Ohm	Ω	$kg \cdot m^2/A^2 \cdot s^3$	V/A
Magnetic flux	Weber	Wb	$kg \cdot m^2/A \cdot s^2$	$V \cdot s$
Magnetic field intensity	Tesla	T	$kg/A \cdot s^2$	Wb/m^2
Inductance	Henry	H	$kg \cdot m^2/A^2 \cdot s^3$	Wb/A

Appendix E

Nobel Prizes

All Nobel Prizes in physics are listed (and marked with a P), as well as relevant Nobel Prizes in Chemistry (C). The key dates for some of the scientific work are supplied; they often antedate the prize considerably.

1901 (P) *Wilhelm Roentgen* for discovering x-rays (1895).

1902 (P) *Hendrik A. Lorentz* for predicting the Zeeman effect and *Pieter Zeeman* for discovering the Zeeman effect, the splitting of spectral lines in magnetic fields.

1903 (P) *Antoine-Henri Becquerel* for discovering radioactivity (1896) and *Pierre* and *Marie Curie* for studying radioactivity.

1904 (P) *Lord Rayleigh* for studying the density of gases and discovering argon.
(C) *William Ramsay* for discovering the inert gas elements helium, neon, xenon, and krypton, and placing them in the periodic table.

1905 (P) *Philipp Lenard* for studying cathode rays, electrons (1898–1899).

1906 (P) *J. J. Thomson* for studying electrical discharge through gases and discovering the electron (1897).

1907 (P) *Albert A. Michelson* for inventing optical instruments and measuring the speed of light (1880s).

1908 (P) *Gabriel Lippmann* for making the first color photographic plate, using interference methods (1891).
(C) *Ernest Rutherford* for discovering that atoms can be broken apart by alpha rays and for studying radioactivity.

1909 (P) *Guglielmo Marconi* and *Carl Ferdinand Braun* for developing wireless telegraphy.

1910 (P) *Johannes D. van der Waals* for studying the equation of state for gases and liquids (1881).

1911 (P) *Wilhelm Wien* for discovering Wien's law giving the peak of a blackbody spectrum (1893).
(C) *Marie Curie* for discovering radium and polonium (1898) and isolating radium.

1912 (P) *Nils Dalén* for inventing automatic gas regulators for lighthouses.

1913 (P) *Heike Kamerlingh Onnes* for the discovery of superconductivity and liquefying helium (1908).

1914 (P) *Max T. F. von Laue* for studying x-rays from their diffraction by crystals, showing that x-rays are electromagnetic waves (1912).
(C) *Theodore W. Richards* for determining the atomic weights of sixty elements, indicating the existence of isotopes.

1915 (P) *William Henry Bragg* and *William Lawrence Bragg*, his son, for studying the diffraction of x-rays in crystals.

1917 (P) *Charles Barkla* for studying atoms by x-ray scattering (1906).

1918 (P) *Max Planck* for discovering energy quanta (1900).

1919 (P) *Johannes Stark*, for discovering the Stark effect, the splitting of spectral lines in electric fields (1913).

1920 (P) *Charles-Édouard Guillaume* for discovering invar, a nickel-steel alloy with low coefficient of expansion.
(C) *Walther Nernst* for studying heat changes in chemical reactions and formulating the third law of thermodynamics (1918).
1921 (P) *Albert Einstein* for explaining the photoelectric effect and for his services to theoretical physics (1905).
(C) *Frederick Soddy* for studying the chemistry of radioactive substances and discovering isotopes (1912).
1922 (P) *Niels Bohr* for his model of the atom and its radiation (1913).
(C) *Francis W. Aston* for using the mass spectrograph to study atomic weights, thus discovering 212 of the 287 naturally occurring isotopes.
1923 (P) *Robert A. Millikan* for measuring the charge on an electron (1911) and for studying the photoelectric effect experimentally (1914).
1924 (P) *Karl M. G. Siegbahn* for his work in x-ray spectroscopy.
1925 (P) *James Franck* and *Gustav Hertz* for discovering the Franck-Hertz effect in electron-atom collisions.
1926 (P) *Jean-Baptiste Perrin* for studying Brownian motion to validate the discontinuous structure of matter and measure the size of atoms.
1927 (P) *Arthur Holly Compton* for discovering the Compton effect on x-rays, their change in wavelength when they collide with matter (1922), and *Charles T. R. Wilson* for inventing the cloud chamber, used to study charged particles (1906).
1928 (P) *Owen W. Richardson* for studying the thermionic effect and electrons emitted by hot metals (1911).
1929 (P) *Louis Victor de Broglie* for discovering the wave nature of electrons (1923).
1930 (P) *Chandrasekhara Venkata Raman* for studying Raman scattering, the scattering of light by atoms and molecules with a change in wavelength (1928).
1932 (P) *Werner Heisenberg* for creating quantum mechanics (1925).
1933 (P) *Erwin Schrödinger* and *Paul A. M. Dirac* for developing wave mechanics (1925) and relativistic quantum mechanics (1927).
(C) *Harold Urey* for discovering heavy hydrogen, deuterium (1931).
1935 (P) *James Chadwick* for discovering the neutron (1932).
(C) *Irène* and *Frédéric Joliot-Curie* for synthesizing new radioactive elements.
1936 (P) *Carl D. Anderson* for discovering the positron in particular and antimatter in general (1932) and *Victor F. Hess* for discovering cosmic rays.
(C) *Peter J. W. Debye* for studying dipole moments and diffraction of x-rays and electrons in gases.
1937 (P) *Clinton Davisson* and *George Thomson* for discovering the diffraction of electrons by crystals, confirming de Broglie's hypothesis (1927).
1938 (P) *Enrico Fermi* for producing the transuranic radioactive elements by neutron irradiation (1934–1937).
1939 (P) *Ernest O. Lawrence* for inventing the cyclotron.
1943 (P) *Otto Stern* for developing molecular-beam studies (1923), and using them to discover the magnetic moment of the proton (1933).
1944 (P) *Isidor I. Rabi* for discovering nuclear magnetic resonance in atomic and molecular beams.
(C) *Otto Hahn* for discovering nuclear fission (1938).

1945 (P) *Wolfgang Pauli* for discovering the exclusion principle (1924).

1946 (P) *Percy W. Bridgman* for studying physics at high pressures.

1947 (P) *Edward V. Appleton* for studying the ionosphere.

1948 (P) *Patrick M. S. Blackett* for studying nuclear physics with cloud-chamber photographs of cosmic-ray interactions.

1949 (P) *Hideki Yukawa* for predicting the existence of mesons (1935).

1950 (P) *Cecil F. Powell* for developing the method of studying cosmic rays with photographic emulsions and discovering new mesons.

1951 (P) *John D. Cockcroft* and *Ernest T. S. Walton* for transmuting nuclei in an accelerator (1932).
(C) *Edwin M. McMillan* for producing neptunium (1940) and *Glenn T. Seaborg* for producing plutonium (1941) and further transuranic elements.

1952 (P) *Felix Bloch* and *Edward Mills Purcell* for discovering nuclear magnetic resonance in liquids and gases (1946).

1953 (P) *Frits Zernike* for inventing the phase-contrast microscope, which uses interference to provide high contrast.

1954 (P) *Max Born* for interpreting the wave function as a probability (1926) and other quantum-mechanical discoveries and *Walther Bothe* for developing the coincidence method to study subatomic particles (1930–1931), producing, in particular, the particle interpreted by Chadwick as the neutron.

1955 (P) *Willis E. Lamb, Jr.* for discovering the Lamb shift in the hydrogen spectrum (1947) and *Polykarp Kusch* for determining the magnetic moment of the electron (1947).

1956 (P) *John Bardeen, Walter H. Brattain,* and *William Shockley* for inventing the transistor (1956).

1957 (P) *T.-D. Lee* and *C.-N. Yang* for predicting that parity is not conserved in beta decay (1956).

1958 (P) *Pavel A. Čerenkov* for discovering Čerenkov radiation (1935) and *Ilya M. Frank* and *Igor Tamm* for interpreting it (1937).

1959 (P) *Emilio G. Segrè* and *Owen Chamberlain* for discovering the antiproton (1955).

1960 (P) *Donald A. Glaser* for inventing the bubble chamber to study elementary particles (1952).
(C) *Willard Libby* for developing radiocarbon dating (1947).

1961 (P) *Robert Hofstadter* for discovering internal structure in protons and neutrons and *Rudolf L. Mössbauer* for discovering the Mössbauer effect of recoilless gamma-ray emission (1957).

1962 (P) *Lev Davidovich Landau* for studying liquid helium and other condensed matter theoretically.

1963 (P) *Eugene P. Wigner* for applying symmetry principles to elementary-particle theory and *Maria Goeppert Mayer* and *J. Hans D. Jensen* for studying the shell model of nuclei (1947).

1964 (P) *Charles H. Townes, Nikolai G. Basov,* and *Alexandr M. Prokhorov* for developing masers (1951–1952) and lasers.

1965 (P) *Sin-itiro Tomonaga, Julian S. Schwinger,* and *Richard P. Feynman* for developing quantum electrodynamics (1948).

1966 (P) *Alfred Kastler* for his optical methods of studying atomic energy levels.

1967 (P) *Hans Albrecht Bethe* for discovering the routes of energy production in stars (1939).

1968 (P) *Luis W. Alvarez* for discovering resonance states of elementary particles.

1969 (P) *Murray Gell-Mann* for classifying elementary particles (1963).

1970 (P) *Hannes Alfvén* for developing magnetohydrodynamic theory and *Louis Eugène Félix Néel* for discovering antiferromagnetism and ferrimagnetism (1930s).

1971 (P) *Dennis Gabor* for developing holography (1947).
(C) *Gerhard Herzberg* for studying the structure of molecules spectroscopically.

1972 (P) *John Bardeen, Leon N. Cooper,* and *John Robert Schrieffer* for explaining superconductivity (1957).

1973 (P) *Leo Esaki* for discovering tunneling in semiconductors, *Ivar Giaever* for discovering tunneling in superconductors, and *Brian D. Josephson* for predicting the Josephson effect, which involves tunneling of paired electrons (1958–1962).

1974 (P) *Anthony Hewish* for discovering pulsars and *Martin Ryle* for developing radio interferometry.

1975 (P) *Aage N. Bohr, Ben R. Mottelson,* and *James Rainwater* for discovering why some nuclei take asymmetric shapes.

1976 (P) *Burton Richter* and *Samuel C. C. Ting* for discovering the J/psi particle, the first charmed particle (1974).

1977 (P) *John H. Van Vleck, Nevill F. Mott,* and *Philip W. Anderson* for studying solids quantum-mechanically.
(C) *Ilya Prigogine* for extending thermodynamics to show how life could arise in the face of the second law.

1978 (P) *Arno A. Penzias* and *Robert W. Wilson* for discovering the cosmic background radiation (1965) and *Pyotr Kapitsa* for his studies of liquid helium.

1979 (P) *Sheldon L. Glashow, Abdus Salam,* and *Steven Weinberg* for developing the theory that unified the weak and electromagnetic forces (1958–1971).

1980 (P) *Val Fitch* and *James W. Cronin* for discovering CP (charge-parity) violation (1964), which possibly explains the cosmological dominance of matter over antimatter.

1981 (P) *Nicolaas Bloembergen* and *Arthur L. Schawlow* for developing laser spectroscopy and *Kai M. Siegbahn* for developing high-resolution electron spectroscopy (1958).

1982 (P) *Kenneth G. Wilson* for developing a method of constructing theories of phase transitions to analyze critical phenomena.

1983 (P) *William A. Fowler* for theoretical studies of astrophysical nucleosynthesis and *Subramanyan Chandrasekhar* for studying physical processes of importance to stellar structure and evolution, including the prediction of white dwarf stars (1930).

1984 (P) *Carlo Rubbia* for discovering the W and Z particles, verifying the electroweak unification, and *Simon van der Meer,* for developing the method of stochastic cooling of the CERN beam that allowed the discovery (1982–1983).

1985 (P) *Klaus von Klitzing* for the quantized Hall effect, relating to conductivity in the presence of a magnetic field (1980).

1986 (P) *Ernst Ruska* for inventing the electron microscope (1931), and *Gerd Binnig* and *Heinrich Rohrer* for inventing the scanning-tunneling electron microscope (1981).

1987 (P) *J. Georg Bednorz* and *Karl Alex Müller* for the discovery of high temperature superconductivity (1986).

1988 (P) *Leon M. Lederman, Melvin Schwartz,* and *Jack Steinberger* for a collaborative experiment that led to the development of a new tool for studying the weak nuclear force, which affects the radioactive decay of atoms.

Appendix F

Physics Problem Solving with a Spreadsheet

D.A. Stetser, Miami University-Middletown

INTRODUCTION

Spreadsheet computer solutions of problems for a first course in physics provide the student with an effective means to increase understanding of complex phenomena. The year or semester of preparation to learn a high level language, such as BASIC or FORTRAN, reduces to a few hours of spreadsheet introduction. The spreadsheet with internal programming capability quickly models physical phenomena. Most spreadsheets allow immediate graphical representations of the variables. Spreadsheet modeling details the physical phenomena without the burden of writing cumbersome language code.

Spreadsheet modeling for this text is introduced by providing a floppy disk containing worksheet files to be retrieved into the Lotus 1-2-3 software program. While the worksheets provided require the Lotus 1-2-3 software, other spreadsheets, QUATTRO, TWIN, EXCEL, are equally applicable to problem solving. The spreadsheet programs are designed for use with IBM PC microcomputers (or IBM-compatible machines). The Lotus 1-2-3 software provides the blank row and column format. The rows and columns define cells on this electronic simulation of a sheet of paper with a grid system. Labels, numerical values and equations may be inserted into these cells. The equations acquire values in cells, algebraically manipulate these values, and compute results. The spreadsheet recalculates immediately after the keyboard insertion of a new parameter to facilitate a "what if" approach to the modeling of a phenomenon. Let us examine the details of one of the worksheet files on the disk.

WORKSHEET EXAMPLE

The worksheet RAPGRO.wk1 models the solution to Question 3.15.

Question 3.15 A rapidly growing plant doubles in height each week. At the end of the 25th day, the plant reaches the height of the building. At what time was the plant one-fourth the height of the building?

The *.wk1 files on the floppy disk represent models to examples and problems from the text. Additional problems extend the scope of the text material. The worksheets are used to model the problem with interactive input into the cells representing constant parameters. Suggestions are made for modifications to the worksheet in order to model other problems in the text. After examination of the provided spreadsheets, the student should modify the worksheets and create original worksheets. If you do not save a modified worksheet under the original name, it can be retrieved in its original form.

If you wish to save a modified worksheet, rename it RAPGRO1.wk1, for example.

With the Lotus 1-2-3 spreadsheet activated, the worksheet RAPGRO is retrieved by the keystrokes /FR and the selection of METHOD and then RAPGRO from the menu. The screen displays the worksheet.

```
RAPGRO  Q3.15  A Rapidly Growing Plant

DT=1 days               DH=k*H*Dt        Hanal=Ho*EXP(k*t)
 k=0.1041 1/time          H=H+DH              k=LN(2)/7
Ho=1 ft                                       k=0.099021
TAB ──→ for Instructions
```

t days	H ft	Hanal ft
0	1.000	1
1	1.104	1.104089
2	1.219	1.219013
3	1.346	1.345900
4	1.486	1.485994
5	1.641	1.640670
6	1.812	1.811447
7	2.000	2
8	2.208	2.208179
9	2.438	2.438027
10	2.692	2.691800

Instructions

1. Enter the time interval, 1, into cell B3. By trial and error, adjust the constant k until the height = 2 m at $t = 7$ days.
2. Enter $Dt = 2.5$ days, and determine the height of the plant at $t = 25$ days. This is the height of the building. Divide this value by 4, and find the time that corresponds to this height. Adjust the time interval to achieve an H value that is equal to 1/4 the height of the building. This is a rough estimate.
3. The analytical solution is an exponential growth. The growth constant k is determined analytically from:

```
H=Ho*EXP(k*t)
2=EXP(k*7)
LN(2)=k*7
k=LN(2)/7=0.099021
```

4. Notice that the approximate numerical solution is in best agreement with analytical solution when $k = 0.099021$ and small time intervals, $Dt = 0.1$ are entered.
5. A satisfactory numerical solution uses a time interval, $Dt = 1$. Larger values create errors. Copy the last row of the worksheet to include values for times extending to $t = 25$ days.

This example demonstrates the method of numerical analysis. The key difference between the analytical technique using calculus and numerical analysis is that the change in the independent variable time is infinitesimally small, using calculus, and finite, using numerical analysis. The dependent variable, *H*, assumes instantaneous values with calculus, and the *H* value is approximated by a constant value over the finite interval, *Dt*, using numerical analysis. The time rate of change in the height of the plant, *DH/Dt*, is proportional to the height. The change, *DH*, during this time interval equals k*H*Dt. The proportionality constant, *k*, is a growth constant. The worksheet calculates this change for each time interval and adds this change to the value of height for the previous interval. These *H* values are tabulated in a column which corresponds to time values in the first column. The analytical solution is tabulated in the last column with respect to times in the first column. The *k* value for the analytical equation

$$H=H_0*e^{k*t}$$

is calculated automatically in a cell above the column containing the analytical solution values.

The graph of height versus time is viewed by pressing F10. The graph is constructed by selecting the graph type as XY and the horizontal and vertical variable ranges. In this example the height determined by numerical analysis and the height determined by analytical solution are chosen as two dependent

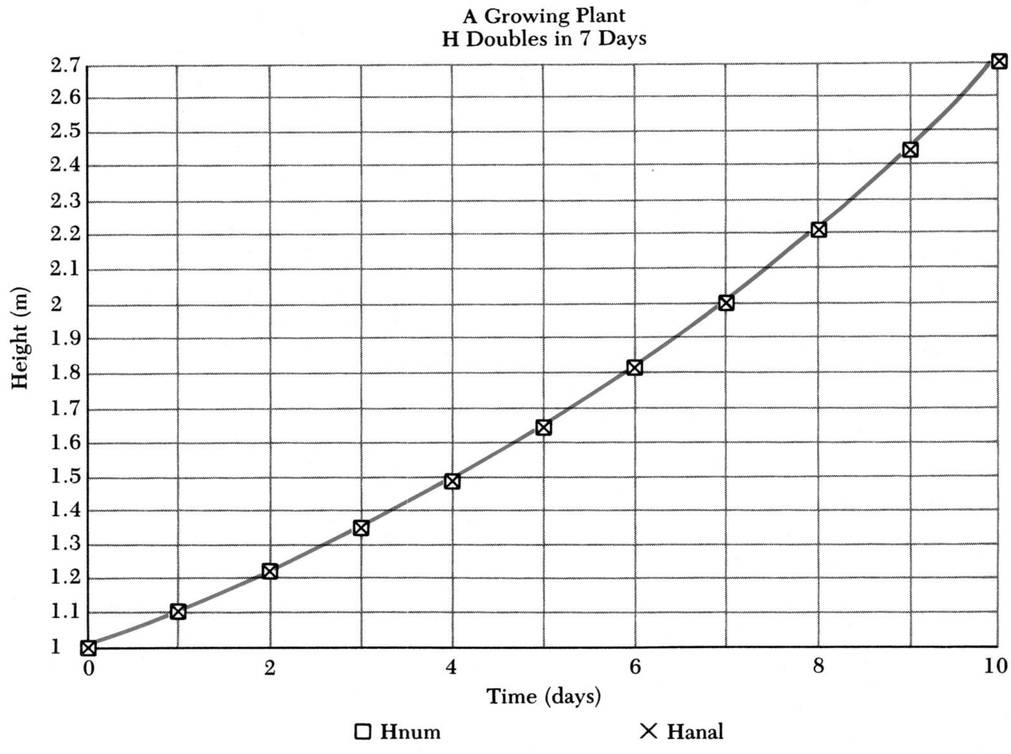

Figure 1 HEIGHT versus TIME.

variables on the vertical ranges, and the independent variable time is placed on the horizontal range.

Modeling provides understanding of physical concepts without the constraints of simplifying assumptions, necessary for an analytical solution by the techniques of calculus.

BASICS

A spreadsheet worksheet is similar in purpose to a program written in programming languages, such as BASIC, FORTRAN or Pascal. The spreadsheet software requires only a few hours of familiarization to begin creating models of physical phenomena. The graphic abilities of a spreadsheet are easily accessed compared to the code required for programming languages. The spreadsheet is ideally suited to satisfy the computational requirements for an introductory text in physics.

Your introduction to spreadsheet modeling is incorporated into text examples. Worksheets are provided on disk to expand upon the presentation in the text. You are requested to enter data or constants into the worksheet, observe and interpret the resulting computations and modify the spreadsheet to perform other calculations. Examine these text examples to determine the technique of their creation. As soon as possible, try to create original worksheets. Spreadsheet modeling is a tool to replace the electronic calculator, which replaced the slide rule that hung from my belt many years ago.

Problems from each chapter are suggested as appropriate for spreadsheet modeling. Many others are appropriate as well, and you should develop worksheets for any and all purposes. Worksheet solutions to these and other problems of a more difficult nature are provided on another disk offered to instructors of the course.

The Worksheet

An electronic grid of rows and columns appears on your computer screen. The rows are labeled with numbers and the columns are labeled with letters. The specification of a column and a row indicates an individual cell. A1 is the cell address of the upper left corner. The cells are accessed by moving the cursor to the desired cell.

Entries to the cell are made from the keyboard.

1. A number is referred to as a value. This value is formatted as desired (Fixed, scientific, currency, etc.).

   0.238, 2.38E-1, $0.24

2. Text is referred to as a label. The labels may be right- or left-justified or centered. Values cannot be adjusted within the cell; if ′, ″, or ^ precedes a value, the computer recognizes the combination as a label.

   ′text—right justified label
   ″text—left justified label
   ^text—centered label

3. Formulas with values and cell addresses.

   The entry 0.238*A10 represents the value 0.238 multiplying the value contained in cell A10.

   If a letter is the first character, a symbol (+,−,() must precede the letter: +A10*0.238.

4. A spreadsheet function: mathematical, statistical, logic. The functions are always preceded by the @ symbol. A list of all functions is obtained from electronic HELP(F1) or from the software documentation.

   The function @SQRT(A10) produces the value 10, if A10 contains the value 100.

   The function @SIN(3.14/6) produces the value 0.5. The parentheses are required for function notation.

   The function @AVG(A10..A20) finds the average value of the values in the range of cells A10 to and including A20.

   The function @IF(A10 > 1, 1, 0) returns the value 1 to the cell if the value in cell A10 is greater than 1 and returns the value 0 to the cell if the value in the cell A10 is less than 1.

The spreadsheet command menu is accessed by the / key. Command choices appear across the top of the screen. The commands are in a tree structure that leads the user to the desired result. Experimentation is the best method for learning the commands. A useful command is the COPY command. Either moving the cursor to COPY and pressing **ENTER** or pressing the key **C** starts the copy sequence. The computer responds **FROM**, and the cell to copy from is highlighted with the cursor: you should then press **ENTER**. The computer responds **TO**, and the cell to which the first cell is to be copied is highlighted: press **ENTER**. A more useful copy sequence is for copying a formula in a cell to a range of cells. Cells may be copied in a relative or absolute fashion. A cell with an address (A10) is copied relatively. A cell with an address ($A$10) is copied in an absolute fashion. An example follows.

In the first row, the number in the first column is multiplied by the value 0.238. The result is shown in the second and third columns. The second column has the cell entry 0.238*E1. Copying this cell down over the four rows computes the values shown. This is described as relative copying. The third column has the cell entry 0.238*$E$1. Copying this cell down over the four rows computes the values in the third column. This is an absolute cell copy. The cell text entries are listed to the right of the numbers.

```
1    0.24    0.238        1 0.238*E1  0.238*$E$1
2    0.48    0.238        2 0.238*E2  0.238*$E$1
3    0.71    0.238        3 0.238*E3  0.238*$E$1
4    0.95    0.238        4 0.238*E4  0.238*$E$1
5    1.19    0.238        5 0.238*E5  0.238*$E$1
```

The simplest method to further introduce spreadsheet technique is to detail the construction of a worksheet. If possible, sit in front of a computer with the Lotus 1-2-3 spreadsheet installed. Insert a copy of the disk provided into the B floppy drive of the computer and enter the keystrokes:

/FD (sets the current directory)
B:\method (specifies the drive and the desired directory)
/FR (selects a group file for retrieval)

Move the cursor to EX1FF and press **ENTER.** The screen displays the worksheet.

```
EX1FF  Free  Fall                         Numerical  Analysis
  A=g=        9.80  m/s/s                  Yex=Vo*t+0.5*A*t^2
   Vo=        0.00  m/s                    V=V+DV
   Dt=        1.00  s                      Y=Y+V*Dt

    t          V(t+0.5t)                    Y(t)          Yex(t)
    s            m/s                          m              m

  0.00           4.90                        0.00           0.00
  1.00          14.70                        4.90           4.90
  2.00          24.50                       19.60          19.60
  3.00          34.30                       44.10          44.10
  4.00          44.10                       78.40          78.40
  5.00          53.90                      122.50         122.50
  6.00          63.70                      176.40         176.40
  7.00          73.50                      240.10         240.10
  8.00          83.30                      313.60         313.60
  9.00          93.10                      396.90         396.90
 10.00         102.90                      490.00         490.00
 11.00         112.70                      592.90         592.90
 12.00                                     705.60         705.60
```

This worksheet computes the velocity at the middle of the time intervals defined in the first column. The acceleration, entered as a constant in cell B3, multiplied by the time interval *Dt*, entered as a constant in cell B5, calculates the change in velocity from the beginning of the interval to the midpoint of the time interval. Adding this velocity change to the initial value of velocity computes the velocity at the middle of the first time interval. Repeating the process for a whole time interval gives the velocity at the middle of the time intervals for the remainder of the column.

The distance fallen by the object is computed by adding the change in distance, DY = V*Dt, to the initial value, Y = Y + DY. The last column computes the exact result of an analytical solution, $Y = 0.5*A*t^2$. These equations are copied down over the range of the worksheet. The computational process approximates the phenomena of a freely falling body by assuming the velocity is a constant over the brief time interval *Dt*. The change in position over this interval is calculated by

$$Y_{final} - Y_{initial} = Velocity*time\ interval$$

The worksheet equations appear as follows:

```
EX1FF Free Fall                              Numerical Analysis
   A=g=          9.8 m/s/s                       Yex=Vo*t+0.5*A*t^2
    Vo=            0 m/s                         Y=Y+V*Dt
    Dt=            1 s
```

t sec	V(t+0.5Dt) m/s	Y(t) m	Yex(t) m
	0+B$4+B$3*B$5/2		00.5*(A8^2)*B$3
+A8+$B$5	+B8+B$3*B$5	+C8+B8*B$5	0.5*(A9^2)*B$3
+A9+$B$5	+B9+B$3*B$5	+C9+B9*B$5	0.5*(A10^2)*B$3
+A10+$B$5	+B10+B$3*B$5	+C10+B10*B$5	0.5*(A11^2)*B$3
+A11+$B$5	+B11+B$3*B$5	+C11+B11*B$5	0.5*(A12^2)*B$3
+A12+$B$5	+B12+B$3*B$5	+C12+B12*B$5	0.5*(A13^2)*B$3
+A13+$B$5	+B13+B$3*B$5	+C13+B13*B$5	0.5*(A14^2)*B$3
+A14+$B$5	+B14+B$3*B$5	+C14+B14*B$5	0.5*(A15^2)*B$3
+A15+$B$5	+B15+B$3*B$5	+C15+B15*B$5	0.5*(A16^2)*B$3
+A16+$B$5	+B16+B$3*B$5	+C16+B16*B$5	0.5*(A17^2)*B$3
+A17+$B$5	+B17+B$3*B$5	+C17+B17*B$5	0.5*(A18^2)*B$3
+A18+$B$5	+B18+B$3*B$5	+C18+B18*B$5	0.5*(A19^2)*B$3
+A19+$B$5		+C19+B19*B$5	0.5*(A20^2)*B$3

The Distance versus Time Graph

1. Key **/GT** and select type with cursor at **XY** and press **ENTER.**
2. Key **X** to define the time range by moving the cursor to the top value in the time column, press period (.) to "pin" the top of the range and move the cursor to the bottom of the range and press **ENTER** to select this range.
3. Key **A** to define the distance range in the third column. Create the range just as you did for the time.
4. Titles and grids are created from the **OPTIONS** command.
5. View the graph with the **VIEW** command if you are in the graph command menu or **F10** from the **READY** mode.

The graph displays the quadratic dependence of distance on time for the falling object. The graph as well as the spreadsheet automatically adjust to new constants. Change the acceleration to that of the moon by placing the cursor on the cell with the acceleration value and press **F2** (the edit function). Edit this value by entering **/6** after the current value 9.8 and press **ENTER.** The worksheet recalculates for this new value of acceleration.

This worksheet is easily modified to include the effects of an initial velocity, Vo. The upward direction is now considered to be represented by a positive number. Downward directions are negative. Let the initial velocity equal +49 m/s and let the acceleration due to gravity equal −9.8 m/s/s. Enter these values into the cells B3 and B4. Your modified worksheet appears as follows:

```
EX2FFVO Free Fall with Initial Velocity Upward
A=g=       -9.80 m/s/s                Yex=Vo*t+0.5*A*t^2
 Vo=       49.00 m/s                     V=V+DV
 Dt=        1.00 s                       Y=Y+V*Dt

  t        V(t+0.5Dt)                    Y(t)       Yex(t)
  s           m/s                         m           m

 0.00       44.10                        0.00        0.00
 1.00       34.30                       44.10       44.10
 2.00       24.50                       78.40       78.40
 3.00       14.70                      102.90      102.90
 4.00        4.90                      117.60      117.60
 5.00       -4.90                      122.50      122.50
 6.00      -14.70                      117.60      117.60
 7.00      -24.50                      102.90      102.90
 8.00      -34.30                       78.40       78.40
 9.00      -44.10                       44.10       44.10
10.00                                    0.00        0.00
```

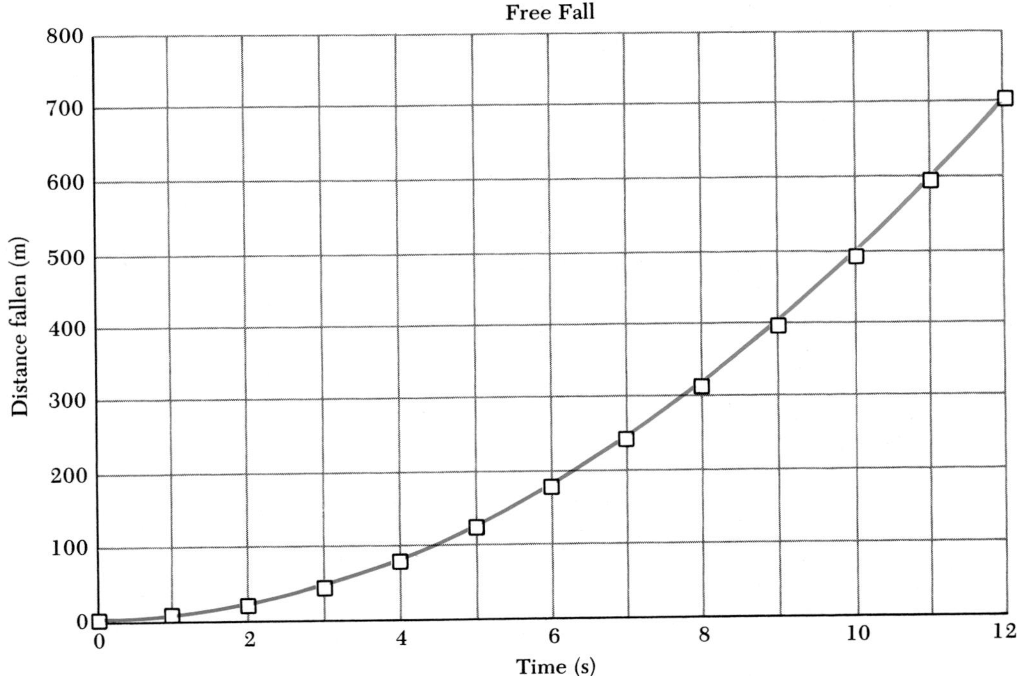

Figure 2 DISTANCE versus TIME.

Your screen does not exactly look like this. You must clean up your new worksheet. Move the cursor to A1, type in the new file name **EX2FFVO** and press **ENTER**. Type in the new title "Free Fall with Initial Velocity Upward" in cell A2. Find the cell with "Numerical Analysis," and erase the contents by the keystrokes **/RE** and **ENTER** with the cursor specifying the cell. Erase the extraneous cells in the worksheet. The graph must be altered by adding a velocity range column, changing the titles and the legend. The ranges of the X and A variables must also be changed. Try to modify your worksheet to match the worksheet above.

The new graph appears as follows. The values for velocity represent the velocities at the times t + 0.5*Dt even though the velocity points appear graphically at t.

Notice that the velocity decreases, goes through zero, and becomes negative in a symmetric fashion. The object rises to a maximum height and returns to its starting height also in a symmetric fashion. The time up is equal to the time down. Experiment with different values of velocity and observe the results.

A further modification of the free fall worksheet describes the effect of air resistance on the object. Air resistance provides a force that is opposite in direction to the velocity, and the magnitude of this force is proportional to the velocity squared.

$$F = k*v^2$$

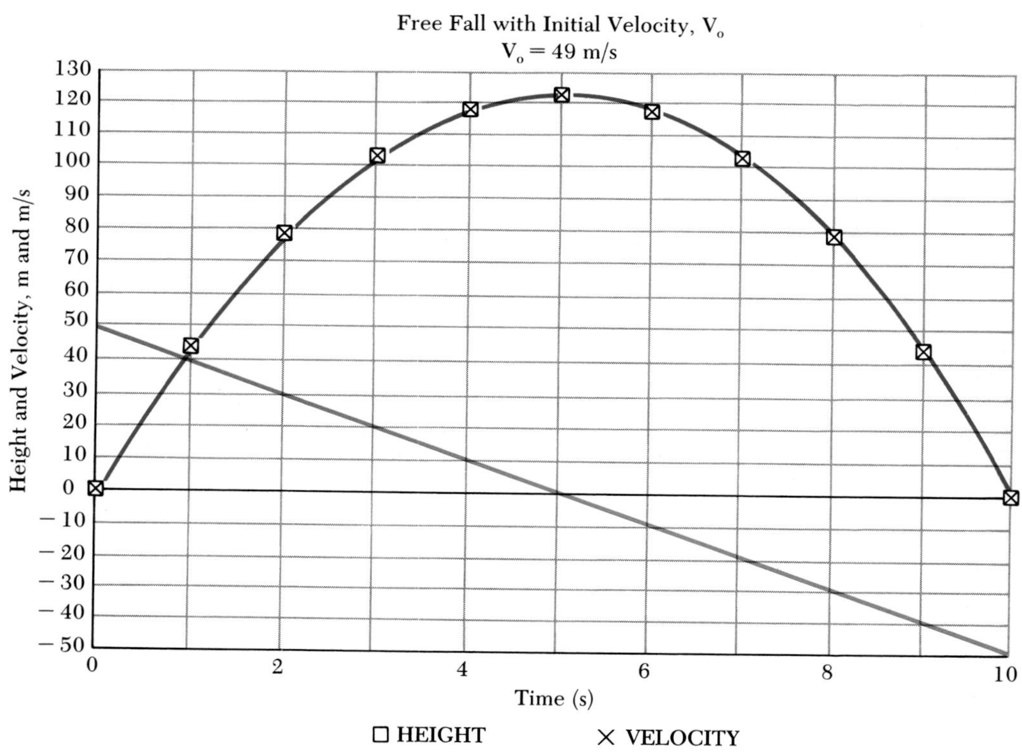

Figure 3 Ht and V versus TIME.

The net acceleration for the rising object is:

$$a_r=-g-(k/m)*v^2$$

And the net acceleration for the falling object is:

$$a_f=-g+(k/m)*v^2$$

The acceleration is no longer a constant. It is a function of the velocity. Let us construct a worksheet that models an object thrown upward under the influence of gravity and air resistance. The key element in this worksheet is the calculation of a variable acceleration, dependent upon the velocity squared.

$$a=-9.8-v*@ABS(v)$$

The term −v*@ABS(v) gives a magnitude equal to v^2 and a direction opposite to the velocity. (@ABS(v) gives the absolute value.)

Retrieve this worksheet by the keystrokes **/FR**, select **METHOD** and press **ENTER**, select EX3VOAR and press **ENTER.** The following worksheet should appear on your screen.

```
EX3VOAR Free Fall with an Initial Velocity and
Air Resistance
 k=     0.15  m      DV=A*Dt          DY=Vav*Dt
 Vo=   20.00 m/s     V=V+DV           Y=Y+DY
 Dt=    0.10  s      Vav=(VO+V)/2     A=-g-k*V*@ABS(V)
```

t s	A m/s/s	DV m/s	V m/s	Vav m	Y m
0.00	-69.80		20.00		0.00
0.10	-35.23	-6.98	13.02	16.51	1.65
0.20	-23.33	-3.52	9.50	11.26	2.78
0.30	-17.50	-2.33	7.16	8.33	3.61
0.40	-14.20	-1.75	5.41	6.29	4.24
0.50	-12.19	-1.42	3.99	4.70	4.71
0.60	-10.96	-1.22	2.78	3.38	5.05
0.70	-10.22	-1.10	1.68	2.23	5.27
0.80	-9.86	-1.02	0.66	1.17	5.39
0.90	-9.78	-0.99	-0.33	0.16	5.40
1.00	-9.54	-0.98	-1.31	-0.82	5.32
1.10	-9.03	-0.95	-2.26	-1.78	5.14
1.20	-8.30	-0.90	-3.17	-2.71	4.87
1.30	-7.41	-0.83	-3.99	-3.58	4.51
1.40	-6.44	-0.74	-4.74	-4.37	4.08
1.50	-5.46	-0.64	-5.38	-5.06	3.57
1.60	-4.53	-0.55	-5.93	-5.65	3.01
1.70	-3.70	-0.45	-6.38	-6.15	2.39
1.80	-2.97	-0.37	-6.75	-6.56	1.74
1.90	-2.35	-0.30	-7.05	-6.90	1.05
2.00	-1.85	-0.24	-7.28	-7.16	0.33

Let us get a fresh start and see if we can reconstruct this worksheet. I will lead you through the construction. First clear the screen. The keystrokes **/WEY** denote worksheet—erase—yes. You are not permanently erasing the file. Since it is saved on disk, you may retrieve it at any time by **/FR** and **ENTER** with the cursor on the filename.

Type the filename in cell A1, **EX3VOAR.** In cell B1 type the worksheet title, **Free Fall with an Initial Velocity and Air Resistance.** Cells, A4..A6, contain the labels for the constants, **"k=, "Vo=, "Dt=**. The values of the constants, **0.15, 20, 0.10** are entered in cells **B4..B6**. Type the units for the constants, **m, m/s, s** in cells **C4..C6**. The equations to the right of the constants represent the equations typed in the main body of the worksheet. Columns representing values of time, acceleration, change in velocity, velocity, average velocity and vertical position represent the main body of the worksheet. Rows 8 and 9 contain labels for the column headings and their units. Enter these as centered (^) labels.

The column for time starts with **0** entered in cell **A10** for the initial time value. In cell **A11** type the equation **+A10+b$6.** This will add the time interval DT to the value in cell A10. Copy this downward from **A11** to **A11..A30**. This creates time intervals from 0 to 2.0 in increments, Dt = 0.10 s.

The acceleration is computed in the **B** column. Enter the equation **−9.8 − B$4*D10*@ABS(D10)** in cell **B10**. As yet there is nothing in cell D10; it will represent the initial velocity. Column C is for the change in velocity, Dv = a*Dt. Type the equation **+B10*B$6** in cell **C11** to compute the change in velocity for the interval from t = 0 to t = 0.10 s. The **$** in **B$6** is to have that term copy absolutely (always as cell B6, which represents the con-

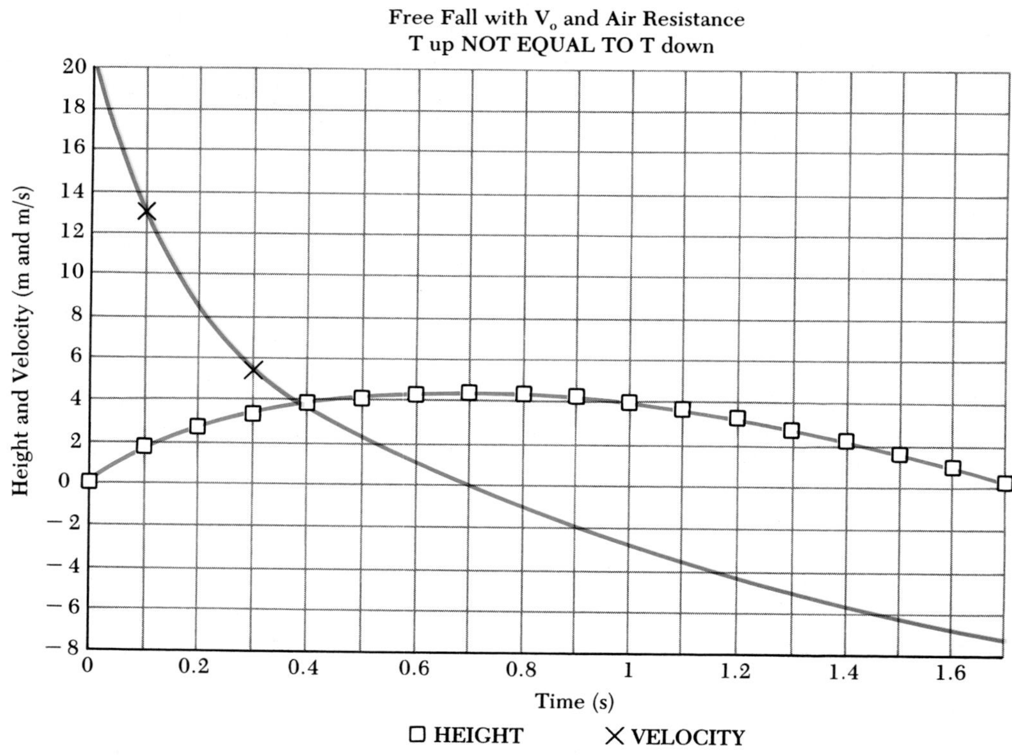

Figure 4 V & Y versus TIME.

stant Dt). Column D starts out with the initial velocity value in cell B5; enter **+B$5** in cell D10. The equation **+D10+C11,** representing V + DV, is entered in cell D11. An average velocity is computed in column E by entering the equation **(D10 + D11)/2** in cell E11. These values are used to compute the vertical position in Column F. The vertical position is calculated from Y = Yo + v_{ave}*Dt. Type **0** in cell F10 and type **+F10 + E11*B$6** in cell F11.

The spreadsheet now takes form. Copy row 11 downward to row 30. Press **/C,** move the cursor to cell **A11**, press **.** to "pin" the from range and move the cursor to stretch the highlighted range **A11..F11**. Press **ENTER,** stretch the cursor over the total copy range **A11..F30**, at the command TO and again press **ENTER.** If you are very lucky, the worksheet takes form. The copy commands are like learning to ride a bicycle, only difficult until you do it once. If all else fails, retrieve the file from disk by **/FR,** selecting **EX3VOAR** and pressing **ENTER.**

Figure 4 uses an **XY** plot of velocity (**A** variable) and position (**B** variable) versus the independent variable time. Use the **/G** commands to create the graph. The graph is very informative in accessing the effects of air resistance. Notice the time up is not equal to the time down. Why not? Change the time interval to 0.20 s, and note what happens. Why does the velocity become constant? What happens with the acceleration as time progresses? Try to construct a qualitative picture of the quantitative results of the spreadsheet modeling.

These examples indicate the applicability of spreadsheet modeling to the solution of physics problems. With practice the spreadsheet software and the computer will enlarge your understanding of the physical laws describing natural phenomena. The simplifying assumptions, sometimes necessary for analytical solutions, are usually not needed. The computer leads us into more realistic descriptions of our world. The value of spreadsheet modeling is the ease of application, as compared to language programming, afforded to you to *create your own* worksheets.

Index of Worksheets

Chapter	Example	Filename	Problems	Filenames
1 Measurement			P1.58,59	MTHNOT.wk1
2 Vectors	2.3	VECTORS.wk1	P2.17	ROLCOS.wk1
3 Motion 1D	3.2	MOTION1D.wk1		GEOSER.wk1
			P3.83	VARACC.wk1
4 Motion 2D	4.1	MOTION2D.wk1		GOLFDR2.wk1
			P4.66	HOMER.wk1
			P4.82	SAFZON.wk1
			P4.84	COYOTE.wk1
5 Laws of Motion	5.10	LAWSMOT.wk1		
6 Circular Motion	6.5	CIRCMOT.wk1	P6.54	ADRAG.wk1
7 Work & Energy			P7.84	WVARF.wk1
8 PE & Cons. of E	8.1	BAFRFL.wk1		CONSME.wk1
			P8.56	PEG&PEND.wk1
			P8.66	FALROP.wk1
9 P & Collisions	9.14	ROCKET.wk1		
10 Rotation	10.6	RBROT.wk1	P10.6	ROTWHL.wk1
11 Angular Momentum			Q11.15	ROMORB.wk1
13 Oscillatory Mo.	13.1	OSCBOD.wk1		TORPEND.wk1
				DAMPOSC.wk1
14 Universal Grav.	14.2	GVERSH.wk1		CENFOR1.wk1
	14.8	GFORCE.wk1		
15 Solids & Fluids	15.6	FORDAM.wk1	P15.76	PVERSH.wk1
16 Wave Motion	16.1	PULSE.wk1	P16.52	TWVEL.wk1
17 Sound Waves	17.4	SNDWAV.wk1	P17.9	SNDSPD.wk1
18 Standing Waves	18.2	STAWAV.wk1	P18.51	BEATS.wk1
19 Temperature	19.1	CONVTEMP.wk1		COOLER.wk1
20 Heat & 1st Law	20.9	HTTRAN.wk1	P20.79	HTRATE.wk1
21 Kinetic Theory	21.5	MAXDIS.wk1		
22 Entropy				ENTROP.wk1
			P23.28	EFLDRD.wk1
23 Electric Fields	23.10	CHGDSK.wk1	P23.79	RODCHG.wk1
			P23.80	RNGCHG.wk1
24 Gauss' Law	24.3	SPSYCD.wk1		

Chapter	Example	Filename	Problems	Filenames
25 Elec. Potential	25.2	MOPEFLD.wk1	P25.40	CHGLIN.wk1
26 Capacitance	26.5	REWCC.wk1		
27 I & R	27.4	RESCOAX.wk1	P27.70	RESTOR.wk1
28 D. C. Circuits	28.10	DISCAP.wk1		
29 Mag. Fields	29.2	FSEMIC.wk1		
30 Sources Mag Fd			P30.86	CURLOOP.wk1
31 Faraday's Law	31.4	BFORBAR.wk1		
32 Inductance	32.3	RLCIR.wk1		
33 AC Circuits				LCRCIR.wk1
				LCRCIRD.wk1
34 E.M. Waves			P34.54	FG&FR.wk1
35 Nature of Light	35.1	FIZEAU.wk1	P35.5,6	FIZEAU.wk1
36 Geo. Optics				
37 Interference				
38 Diff. & Polar.				
39 Relativity				
40 Intro. Q. Phys.				

Answers to Odd-numbered Problems

CHAPTER 1

1. 2.80 g/cm^3
3. 184 g
5. (a) 4 u $= 6.64 \times 10^{-27}$ kg
(b) 56 u $= 9.30 \times 10^{-26}$ kg
(c) 207 u $= 3.44 \times 10^{-25}$ kg
7. 1.033; 24.3% larger
9. It is.
11. It is.
13. (b) only
15. L/T^3
17. (a) ML/T^2 (b) kg·m/s^2
19. 1.39×10^3 m^2
21. 8.32×10^{-4} m/s
23. 11.4×10^3 kg/m^3
25. 4.1×10^9 years
27. 6.71×10^8 mi
29. (a) 1 mi/h = 1.609 km/h (b) 88.5 km/h
(c) 16.1 km/h
31. 294.5 m^3
33. 1.00×10^{10} lb
35. 8.45 years
37. 5.95×10^{24} kg
39. 3.11×10^3 lb
41. $\approx 10^6$
43. $\approx 10^{10}$ containers; 10^5 tons
45. $\approx 10^4$
47. $\approx 10^3$
49. $\approx 10^2$
51. 67.9 cm^2
53. 195.8 cm^2 $\pm$ 0.7%
55. 3, 4, 3, 2
57. 5.2 m^3 $\pm$ 3%
59. It is not.

CHAPTER 2

1. (a) 8.60 m (b) 4.47 m at 297°; 4.24 m at 135°
3. (−2.75 m, −4.76 m)
5. 2.24 m
7. (2.05 m, 1.43 m)
9. 196 cm at 345°
11. (a) 5.00 at 307° (b) 5.00 at 53.1°
13. (a) 10.0 m (b) 15.7 m (c) 0
15. (a) 5.20 m at 60.0° (b) 3.00 m at 330°
(c) 3.00 m at 150° (d) 5.20 m at 300°
17. 421 ft at −2.63°
19. 86.6 m, −50.0 m
21. 82.5 m at 147°
23. 47.2 units at 122°
25. See Figure 2.14
27. 7.21 m at 56.3°
29. (a) $2\boldsymbol{i} - 6\boldsymbol{j}$ (b) $4\boldsymbol{i} + 2\boldsymbol{j}$ (c) 6.32 (d) 4.47
(e) 288°; 26.6°
31. (a) $(-11.1 \text{ m})\boldsymbol{i} + (6.40 \text{ m})\boldsymbol{j}$
(b) $(1.65 \text{ cm})\boldsymbol{i} + (2.86 \text{ cm})\boldsymbol{j}$
(c) $(-18.0 \text{ in.})\boldsymbol{i} - (12.6 \text{ in.})\boldsymbol{j}$
33. 9.48 m at 166°
35. 1320 mi at 17.0°
37. $(2.60 \text{ m})\boldsymbol{i} + (4.50 \text{ m})\boldsymbol{j}$
39. (a) $8\boldsymbol{i} + 12\boldsymbol{j} - 4\boldsymbol{k}$ (b) $2\boldsymbol{i} + 3\boldsymbol{j} - \boldsymbol{k}$
(c) $-24\boldsymbol{i} - 36\boldsymbol{j} + 12\boldsymbol{k}$
41. (a) 7.00 at 217° (b) 95.3 at 253°
43. (a) $-3\boldsymbol{i} + 2\boldsymbol{j}$ (b) 3.61 at 146° (c) $3\boldsymbol{i} - 6\boldsymbol{j}$
45. 5.83 m at 149°
47. (a) 49.5, 27.1 (b) 56.4 at 28.7°
49. 240 m at 237°
51. (a) $(-3 \text{ m})\boldsymbol{i} - (5 \text{ m})\boldsymbol{j}$; $(-1 \text{ m})\boldsymbol{i} + (8 \text{ m})\boldsymbol{j}$
(b) $(2 \text{ m})\boldsymbol{i} + (13 \text{ m})\boldsymbol{j}$

CHAPTER 3

1. (a) 2.3 m/s (b) 16.1 m/s (c) 11.5 m/s
3. (a) 5.0 m/s (b) 1.2 m/s (c) −2.5 m/s
(d) −3.3 m/s (e) 0
5. (a) 2.50 m/s (b) −2.27 m/s (c) 0
7. 50.0 km/h
9. (a) −2.5 m/s (b) −3.4 m/s (c) 4.2 s
11. (b) 1.60 m/s
13. (a) 5.0 m/s (b) −2.5 m/s (c) 0 (d) 5.0 m/s
15. −4.00 m/s^2
17. 20.0 m/s; 6.00 m/s^2
19. (b) 1.6 m/s^2; 0.80 m/s^2
21. (a) 2.00 m (b) −3.00 m/s (c) −2.00 m/s^2
23. (b) yes. (c) 1.0 m/s^2
25. −16.0 cm/s^2
27. (a) 4.76×10^{-5} m (b) 15.9 ns
29. (a) 12.7 m/s (b) −2.30 m/s
31. (a) 20.0 s (b) No.
33. (a) −10.0 m/s^2 (b) 125 m
35. (a) 0.444 m/s^2 (b) 1.33 m/s (c) 2.12 s
(d) 0.943 m/s
37. (a) −3.20 m/s^2 (b) 5.00 s
39. (a) $x = 30t - t^2$, $v = 30 - 2t$ (b) 225 m
41. −32.3 ft/s^2
43. (a) −662 ft/s^2 (b) 649 ft
45. (a) −96 ft/s (b) 3.08×10^3 ft/s^2
(c) 3.12×10^{-2} s
47. (a) 10.0 m/s (b) −4.68 m/s

49. (a) 1.53 s (b) 11.5 m (c) -4.60 m/s, -9.80 m/s^2
51. (a) 0.510 s (b) 0.497 m
53. (a) 0.800 m/s (b) 18.0 m
55. (a) 0.206 s (b) -1.76 m/s, 23.0 m/s (c) 2.04 s, 5.10 s
57. (a) m/s^3, m/s^4 (b) $v = Bt^2/2 - Ct^3/3$ (c) $T = B/C$ (d) $B^3/6C^2$
59. (a) 10.0 m/s^2 (b) $20 + 5t + 5t^2$ (c) 5.00 m/s
61. 3.25 m/s, 3.12 m/s, 3.02 m/s, 3.002 m/s, 3.0002 m/s
63. (a) 3.00 s (b) -15.3 m/s (c) -31.4 m/s, -34.8 m/s
65. (a) -6.26 m/s (b) 6.02 m/s (c) 1.25 s
67. 0.815 s
69. (a) 6.46 s (b) 334 ft (c) 103 ft/s, 89.6 ft/s
71. -1.80 m/s^2, 5 m + $20t$ m/s $- 0.9t^2$ m/s^2, 22.2 s
73. (a) 79.3 ft/s (b) -113 ft/s (c) -16.7 ft/s
75. (a) 5500 ft (b) 367 ft/s
77. (a) 5.43 m/s^2, 3.83 m/s^2 (b) 10.9 m/s, 11.5 m/s (c) Maggie, 2.62 m
79. 155 s; 129 s
81. $0.577v$
83. (a) 2, 4, 3, 5, 5, 2, 3, 4, 2, 5 m/s^2 (b) 0, 0.25, 1.25, 3.12, 6, 10.1, 15.1, 20.8, 27.2, 34.5, 42.6 m
85. 44.3 m, 19.8 m/s

CHAPTER 4

1. $(8a + 2b)\mathbf{i} + 2c\mathbf{j}$
3. 7.86×10^{-3} cm/s at 225°
5. (a) $(2\mathbf{i} + 3\mathbf{j})$ m/s^2 (b) $(3t + t^2)\mathbf{i} + (-2t + 1.5t^2)\mathbf{j}$
7. (a) $(0.8\mathbf{i} - 0.3\mathbf{j})$ m/s^2 (b) at 339° (c) $(360\mathbf{i} - 72.8\mathbf{j})$ m; at 345°
9. (a) $\mathbf{r} = (5t\mathbf{i} + 1.5t^2\mathbf{j})$ m, $\mathbf{v} = (5\mathbf{i} + 3t\mathbf{j})$ m/s (b) (10 m, 6 m), 7.81 m/s
11. (a) 3.34 m/s at 0° (b) 309°
13. (a) $(12\mathbf{i} - 15.2\mathbf{j})$ m/s (b) (29.4 m, 29.4 m)
15. (a) 2.67 s (b) $29.9\mathbf{i}$ m/s (c) $(29.9\mathbf{i} - 26.2\mathbf{j})$ m/s
17. (a) clears by 0.888 m (b) while falling 13.3 m/s
19. (a) 27.1 m/s (b) 3.91 s
21. (a) 277 km (b) 284 s
23. 53.1°
25. 2.70 m/s^2
27. 32.0 m/s^2
29. 284 m/s^2
31. (a) 1.02 km/s (b) 2.72 mm/s^2
33. (a) 12.6 m/s (b) 395 m/s^2
35. (a) 13.0 m/s^2 (b) 5.70 m/s (c) 7.50 m/s^2
37. 1.48 m/s^2
39. (a) $-30.8\mathbf{j}$ m/s^2 (b) $70.4\mathbf{j}$ m/s^2
41. (a) 110 km/h north (b) 110 km/h south
43. 2.02×10^3 s; 21.0% longer
45. 12.0 s
47. 153 km/h at 11.3° N of W
49. (a) 57.7 km/h at 210° (b) 28.9 km/h down
51. (a) 10.1 m/s^2 S at 75.7° below horizontal (b) 9.80 m/s^2 down
53. (a) $4.00\mathbf{i}$ m/s (b) (4.00 m, 6.00 m)
55. (a) 20.0 m/s, 5.00 s (b) 31.4 m/s at 301° (c) 6.53 s (d) 24.5 m away
57. 3.14 m
61. (a) 0.60 m (b) 0.40 m (c) 1.87 m/s^2 toward center (d) 9.80 m/s^2 down
63. 4.12 m
65. 87.3 m upstream or 567 m downstream
67. 10.8 m
69. (a) 6.80 km (b) directly above explosion (c) 66.2°
71. (a) $(-7.05$ cm$)\mathbf{j}$ (b) $(7.61$ cm$)\mathbf{i} - (6.48$ cm$)\mathbf{j}$ (c) $(10.0$ cm$)\mathbf{i} - (7.05$ cm$)\mathbf{j}$
73. (a) 22.2° or 67.8° (b) 235 m (c) 235 m
75. (a) east (b) 15.0 min (c) 6.40 km/h at 51.3° N of E (d) 1.25 km
77. (a) 5.15 s (b) 4.85 m/s at 74.5° N of W (c) 19.4 m
79. (a) 43.2 m (b) $(9.66$ m/s$)\mathbf{i} - (25.6$ m/s$)\mathbf{j}$
81. (18.8 m, -17.4 m)
83. (a) 46.5 m/s (b) $-77.6°$ (c) 6.34 s
85. He must long jump to reach safety, with 3.18 ft to spare.

CHAPTER 5

1. (a) 1/3 (b) 0.750 m/s^2
3. (a) 5.00 m/s^2 (b) 19.6 N (c) 10.0 m/s^2
5. 312 N
7. $(6\mathbf{i} + 15\mathbf{j})$ N, 16.2 N
9. (a) 534 N (b) 54.5 kg
11. 70.6 kg
13. 1.72 kg; 5.81 m/s^2
15. (c) Forces of brick on spring and spring on brick; of spring on table and table on spring; of table on earth and earth on table; of brick on earth and of earth on brick; of spring on earth and of earth on spring.
17. (a) 4.27 m/s^2 at 321° (b) $(26.7$ m$)\mathbf{i} - (21.3$ m$)\mathbf{j}$
19. 3.00 N at 180°
21. (a) $(2.50$ N$)\mathbf{i} + (5.00$ N$)\mathbf{j}$ (b) 5.59 N
23. 1.24 ft/s^2
25. (a) 15.0 lb up (b) 5.00 lb up (c) 0
27. $T_1 = 294$ N, $T_2 = 196$ N, $T_3 = 98$ N
29. 100 N
31. (b) 316 N, 343 N, 200 N
33. 1.36 kN
35. 188 m
37. (a) $F_x > 19.6$ N (b) $F_x \leqq -78.4$ N
39. (a) 36.8 N (b) 2.45 m/s^2 (c) 1.22 m
41. 3.73 m
43. (a) 706 N (b) 814 N (c) 706 N (d) 648 N
45. 74.0 m/s
47. (a) mg (b) $0.6mg$ (c) m has $a = g$; $4m$ has $a > g$
49. 19.6 N
51. 0.159
53. (a) 0.161 (b) 1.01 m/s^2
55. (b) 16.7 N, 0.687 m/s^2

57. (a) $1.78\ m/s^2$ (b) 0.368 (c) 9.37 N (d) 2.67 m/s
59. 0.727, 0.577
61. (a) 35.4 N (b) 0.601
63. (a) 0.548 (b) $0.250\ m/s^2$
65. (a) $Mg/2$, $Mg/2$, $Mg/2$, $3Mg/2$, Mg (b) $Mg/2$
67. 1.66×10^6 N
69. (a) $2.00\ m/s^2$ (b) 4.00 N, 6.00 N, 8.00 N (c) 14.0 N between m_1 and m_2, 8.00 N between m_2 and m_3.
71. (a) F forward (b) $3F/2Mg$
(c) $F/(M + m_1 + m_2 + m_3)$
(d) $m_1F/(M + m_1 + m_2 + m_3)$,
$(m_1 + m_2)F/(M + m_1 + m_2 + m_3)$,
$(m_1 + m_2 + m_3)F/(M + m_1 + m_2 + m_3)$
(e) $m_2F/(M + m_1 + m_2 + m_3)$
73. (a) Friction between the two blocks (b) 34.7 N (c) 0.306
75. (a) $0.408\ m/s^2$ (b) 83.3 N
77. 304 N, 290 N, 152 N, 138 N
81. (a) $(-45\mathbf{i} + 15\mathbf{j})$ m/s (b) at 162° (c) $(-225\mathbf{i} + 75\mathbf{j})$ m (d) (−227 m, 79 m)
83. $(M + m_1 + m_2)m_2\,g/m_1$
85. (a) 78.0 N, 35.9 N (b) 0.655
87. (a) $3.12\ m/s^2$ (b) 17.5 N
89. (a) 33.2 N (b) 37.4 N (c) 35.5% larger; 11.8% smaller
91. 6.00 cm

CHAPTER 6

1. (a) 8.0 m/s (b) 3.02 N
3. 60.0 N
5. $0 < v < 8.08$ m/s
7. (a) 6.05×10^6 m (b) 2.21 h
9. $v \le 14.3$ m/s
11. (a) 9.80 N (b) 9.80 N (c) 6.26 m/s
13. (a) static friction (b) 0.085
15. 3.13 m/s
17. (a) 4.81 m/s (b) 700 N up
19. 2.87 N down
21. No.
23. (c) The tension increases as the friction increases.
25. (a) 17.0° (b) 5.12 N
27. (a) $m(g - a)$ (b) $m(g - a)$
29. (a) $6.27\ m/s^2$ down (b) 784 N up (c) 283 N up
31. (a) $3.47 \times 10^{-2}\ s^{-1}$ (b) 2.50 m/s (c) $a = -cv$
33. (a) 1.47 N·s/m (b) 2.04×10^{-3} s (c) 2.94×10^{-2} N
35. (a) 8.32×10^{-8} N (b) $9.13 \times 10^{22}\ m/s^2$ (c) 6.61×10^{15} rev/s
37. (a) 1.22 N (b) 2.42 m/s
39. (a) The true weight is greater than the apparent weight.
(b) $w = w' = 735$ N at the poles; $w' = 732.4$ N at the equator
41. 20.3 m/s
43. 12.8 N
45. (a) 967 lb (b) 647 lb up
47. $T_1 = 108$ N; $T_2 = 55.7$ N
49. (b) 2.54 s, 23.6 rev/min
51. (a) $4\pi^2 mR/T^2$ (b) $\cos^{-1}(T^2g/4\pi^2R)$ (c) $m[(4\pi^2R/T^2)^2 - g^2]^{1/2}$
53. $v = v_0e^{-(b/m)t}$

CHAPTER 7

1. 30.6 m
3. 15.0 MJ
5. 5.88 kJ
7. (a) 900 J (b) −900 J (c) 0.383 J
9. 8.75 m
11. (a) 79.4 N (b) 1.49 kJ (c) −1.49 kJ
13. (a) 3.61 (b) 33.7°
17. (a) 16.0 J (b) 36.9°
19. 68.0°
21. (a) 11.3° (b) 156° (c) 82.3°
23. (a) 7.50 J (b) 15.0 J (c) 7.50 J (d) 30.0 J
25. (a) 575 N/m (b) 46.0 J
27. (a) 0.938 cm (b) 1.25 J
29. 1.59 kJ
31. 1.50 kJ
33. (a) 33.8 J (b) 135 J
35. (a) 2.00 m/s (b) 200 N
37. (a) 650 J (b) −588 J (c) 62.0 J (d) 1.76 m/s
39. (a) 2.83 m/s; 24.0 J (b) 3.46 m/s; 36.0 J (c) 3.32 m/s; 33.0 J
41. (a) 400 J (b) −294 J (c) 5.94 m/s
43. 1.25 m/s
45. 2.04 m
47. (a) 168 J (b) 184 J (c) 500 J (d) 148 J (e) 5.65 m/s
49. (a) 4.51 m (b) No, since $f > mg \sin\theta$.
51. (a) 63.9 J (b) −35.4 J (c) −9.51 J (d) 19.0 J
53. 875 W
55. 3.27 kW
57. (a) 7.5×10^4 J (b) 2.50×10^4 W (33.5 hp) (c) 3.33×10^4 W (44.7 hp)
59. 6.47×10^3 N (≈1450 lb)
61. (a) 1.31×10^5 J (b) 1.63 kW (2.19 hp)
63. (a) 29.8 kW (b) 37.3 kW
65. 0.111 gal
67. 5.90 km/liter
69. (a) 5.37×10^{-11} J (b) 1.33×10^{-9} J
71. 3.70 m/s
75. (a) $(2 + 24t^2 + 72t^4)$ J (b) $a = 12t\ m/s^2$; $F = 48t$ N (c) $(48t + 288t^3)$ W (d) 1.25×10^3 J
77. (a) 20.0 J (b) 6.71 m/s
79. (a) 4.12 m (b) 3.35 m
81. (a) −5.6.0 J (b) 0.152 (c) 2.29 rev
83. (a) $W = mgh$ (b) $\Delta K = mgh$ (c) $K_f = mgh + mv_0^2/2$
85. 1.94 kJ
87. (a) 11.0 m/s (b) 5.49 m/s (c) 4.62 m
89. (c) 72.9 MJ, 19.7 kW (d) 0.132

CHAPTER 8

1. (a) -196 J (b) -196 J (c) -196 J
The force is conservative
3. (a) -30.0 J (b) -51.2 J (c) -42.4 J
(d) Friction is a nonconservative force.
5. (a) 125 J (b) 50.0 J (c) 66.7 J
(d) nonconservative, since W is a path-dependent
7. (a) 45.0 J (b) -45.0 J (c) 67.5 J
9. (a) -9.00 J; No. A constant force is conservative.
(b) 3.39 m/s (c) 9.00 J
11. $v_A = \sqrt{3gR}$; 0.098 N downward
13. (a) $v = (gh + v_0^2)^{1/2}$ (b) $v_x = 0.6v_0$;
$v_y = -(0.64v_0^2 + gh)^{1/2}$
15. (a) 64 J, 0, 64 J (b) 39.5 J, 24.5 J, 64.0 J
(c) 0 J, 64 J, 64 J (d) 13.1 m
17. (a) 4.43 m/s (b) 5.00 m
19. (a) -160 J (b) 73.5 J (c) 28.8 N (d) 0.679
21. (a) 8.85 m/s (b) 54.1%
23. (a) 2.29 m/s (b) -15.6 J
25. 3.74 m/s
27. 0.721 m/s
29. (a) 0.400 J (b) 0.225 J (c) 0 J
31. (a) -28.0 J (b) 0.446 m
33. 10.2 m
35. 0.327
37. (a) $F_r = A/r^2$
39. $\boldsymbol{F} = (7 - 9x^2y)\boldsymbol{i} - 3x^3\boldsymbol{j}$
41.

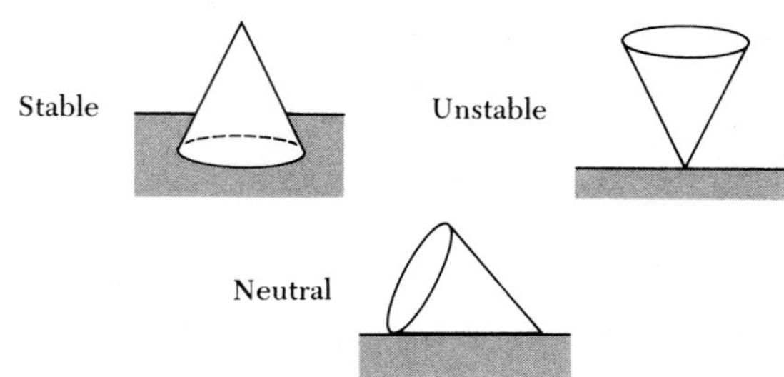

43. (a) $v_B = 5.94$ m/s; $v_C = 7.67$ m/s (b) 147 J
45. (a) 1.50×10^{-10} J (b) 1.07×10^{-9} J
(c) 9.15×10^{-10} J
47. (a) 0.225 J (b) 0.363 J (c) No. The normal force varies with position, and so the frictional force also varies.
49. (a) $v_0 = (Rg - 2gh/3)^{1/2}$ (b) $h' = R/2 + 2h/3$
51. (a) 349 J, 676 J, 741 J (b) 174 N, 338 N, 370 N (c) yes
53. (a) $\Delta U = -\dfrac{ax^2}{2} - \dfrac{bx^3}{3}$ (b) $\Delta U = \dfrac{A}{\alpha}(1 - e^{\alpha x})$
55. 0.115
59. 1.24 m/s
61. 140 kW, yes
63. (a) 7.46 kJ (b) -1.57 kJ (c) -5.90 kJ
65. (a) 0.400 mm (b) 4.10 m/s (c) It reaches the top
67. 3.92 kJ
69. (a) 0.378 m (b) 2.30 m/s (c) 1.08 m

CHAPTER 9

1. $(9.00\boldsymbol{i} - 12.0\boldsymbol{j})$kg·m/s, 15.0 kg·m/s
3. (a) 1.70×10^4 kg·m/s in the northwesterly direction (b) 5.66×10^3 N @ 135° from the East
5. (a) 12.0 kg·m/s (b) 6.00 m/s (c) 4.00 m/s
7. (a) 1.35×10^4 kg·m/s (b) 9.00×10^3 N
(c) 18.0×10^3 N
9. (a) 8.35×10^{-21} kg·m/s (b) 7.50 kg·m/s
(c) 900 kg·m/s (d) 1.78×10^{29} kg·m/s
11. (a) 7.50 kg·m/s (b) 375 N
13. (a) 13.5 kg·m/s toward the pitcher
(b) 6.75×10^3 N toward the pitcher
15. 260 N toward the left in the diagram
17. 0.400 m/s to the west
19. The boy moves westward with a speed of 2.67 m/s.
21. (a) 1.15 m/s (b) -0.346 m/s
23. 4.01×10^{-20} m/s
25. 314 m/s
27. (a) 20.9 m/s east (b) 8.74 kJ into thermal energy
29. (a) 0.284, or 28.4% (b) $K_n = 1.15 \times 10^{-13}$ J,
$K_c = 4.54 \times 10^{-14}$ J
31. (a) -10.0 cm/s, 20.0 cm/s (b) 0.889
33. (a) 1.40 m/s (b) 39.2 J
35. 91.2 m/s
37. 0.556 m
39. 1000 m/s
41. $\boldsymbol{v} = (3.00\boldsymbol{i} - 1.20\boldsymbol{j})$ m/s
43. (a) $v_x = -9.33 \times 10^6$ m/s, $v_y = -8.33 \times 10^6$ m/s
(b) 4.39×10^{-13} J
45. 3.01 m/s, 3.99 m/s
47. From Newton's third law, $\boldsymbol{F}_{12} = -\boldsymbol{F}_{21}$.
$\boldsymbol{F}_{net} = d\boldsymbol{p}/dt = \boldsymbol{F}_{12} + \boldsymbol{F}_{21} = 0$. Thus $\boldsymbol{p}$ is conserved.
51. -0.429 m
53. C.M. = 454 km, well within the sun.
55. $3R/2$
57. 11.7 cm, 13.3 cm
59. (a) $(1.40\boldsymbol{i} + 2.40\boldsymbol{j})$m/s (b) $(7.00\boldsymbol{i} + 12.0\boldsymbol{j})$kg·m/s
61. (a) 2.10 m/s, 0.900 m/s (b) 6.30×10^{-3} kg·m/s,
-6.30×10^{-3} kg·m/s
63. 200 kN
65. -523 kg/s (b) -1.57×10^3 kg/s (c) 1.57 MN, 4.70 MN
67. 0.595 m^3/s
69. 291 N
71. (a) 1.80 m/s to the left (b) 257 N to the left
(c) larger than part b
73. (a) 4160 N (b) 4.17 m/s
75. 32.0 kN; 7.13 MW
77. (a) 6.81 m/s (b) 1.00 m
79. 240 s
81. $(3Mgx/L)\boldsymbol{j}$
83. (a) As the child walks to the right, the boat moves to the left, but the center of mass remains fixed.
(b) 5.55 m from the pier (c) Since the turtle is 7 m from the pier, the boy will not be able to reach the turtle, even with a 1 m reach.
85. (a) 100 m/s (b) 374 J

CHAPTER 10

1. (a) 4.00 rad/s^2 (b) 18.0 rad
3. (a) 1.99×10^{-7} rad/s (b) 2.66×10^{-6} rad/s
5. (a) 5.24 s (b) 27.4 rad
7. -12.7 rad/s^2, 3.14 s
9. (a) 0.18 rad/s (b) 8.10 m/s^2 toward the center of the track
11. (a) 8.00 rad/s (b) 8.00 m/s, $a_r = -64.0$ m/s^2, $a_t = 4.00$ m/s^2 (c) 9.00 rad
13. (a) 126 rad/s (b) 3.77 m/s (c) 1.26 km/s^2 (d) 20.1 m
15. 29.4 m/s^2, 9.80 m/s^2
17. (a) 143 $kg \cdot m^2$ (b) 2.57×10^3 J
19. (a) 92.0 $kg \cdot m^2$, 184 J (b) 6.00 m/s, 4.00 m/s, 8.00 m/s, 184 J
23. (a) $(3/2)MR^2$ (b) $(7/5)MR^2$
25. -3.55 N·m
27. 2.79%
29. 12.6 kW (16.8 hp)
31. (a) 2.00 rad/s^2 (b) 6.00 rad/s (c) 90.0 J (d) 4.50 m
33. (a) $2(Rg/3)^{1/2}$ (b) $4(Rg/3)^{1/2}$ (c) $(Rg)^{1/2}$
35. (a) 11.4 N, 7.57 m/s^2, 9.53 m/s down, (b) 9.53 m/s
37. (a) 3.00 N·m (b) 12.0 rad/s^2 (c) 3.00 m/s^2 (d) 24.0 rad/s (e) 6.00 m/s (f) 72.0 J (g) 72.0 J (h) 24.0 rad (i) 6.00 m
39. (a) $\dfrac{Mmg}{M+4m}$ (b) $\dfrac{4mg}{M+4m}$ (c) $\dfrac{1}{R}\sqrt{\dfrac{8mgh}{M+4m}}$
41. (a) 4.00 J (b) 1.60 s (c) Yes.
43. (a) $\omega = \sqrt{3g/L}$ (b) $\alpha = 3g/2L$ (c) $-\frac{3}{8}g\boldsymbol{i} - \frac{3}{4}g\boldsymbol{j}$ (d) $-\frac{3}{2}Mg\boldsymbol{i} + \frac{1}{4}Mg\boldsymbol{j}$
45. (a) $0.707R$ (b) $0.289L$ (c) $0.632R$
49. $247MR^2/512$
51. (a) 118 N, 156 N (b) 1.17 $kg \cdot m^2$
53. (a) 0.800 N·m, 40.0 $rads/s^2$ (b) 1.25 rad (c) $a_t = 4.00$ m/s^2, $a_r = 10.0$ m/s^2 (d) 1.00 J

CHAPTER 11

1. (a) 500 J (b) 250 J (c) 750 J
3. (a) $a_c = \frac{2}{3}g \sin\theta$ (disk), $a_c = \frac{1}{2}g \sin\theta$ (hoop) (b) $\frac{1}{3}\tan\theta$
5. $0.750Mv^2$
7. (a) $-17\boldsymbol{k}$ (b) 161°
9. (a) negative z direction (b) positive z direction
11. 45.0°
13. $|\boldsymbol{F}_3| = |\boldsymbol{F}_1| + |\boldsymbol{F}_2|$, no
15. $(17.5\ kg \cdot m^2/s)\boldsymbol{k}$
17. $(60\ kg \cdot m^2/s)\boldsymbol{k}$
19. (a) mvd (out of plane) (b) $-2mvd$ (into plane) (c) zero
21. (a) 7.06×10^{33} $kg \cdot m^2/s$ (b) 2.64×10^{40} $kg \cdot m^2/s$
23. (a) zero (b) $[-mv_0^3 \sin^2\theta \cos\theta/2g]\boldsymbol{k}$ (c) $[-2mv_0^3 \sin^2\theta \cos\theta/g]\boldsymbol{k}$ (d) The downward force of gravity exerts a torque in the $-z$ direction.
25. (a) 0.433 $kg \cdot m^2/s$ (b) 1.73 $kg \cdot m^2/s$
27. (a) $\omega = \omega_0 I_1/(I_1 + I_2)$ (b) $I_1/(I_1 + I_2)$
29. (a) 0.360 rad/s in the counterclockwise direction (b) 99.9 J
31. (a) 6.05 rad/s (b) 114 J
33. (a) $mv\ell$ down (b) $M/(M + m)$
35. (a) 2.19×10^6 m/s (b) 2.18×10^{-18} J (c) 4.13×10^{16} rad/s
37. (a) 2.63×10^{33} J (b) 2.57×10^{29} J (c) 1.03×10^4
41. (a) The net torque about this axis is zero. (b) Since $\boldsymbol{\tau} = 0$, $\boldsymbol{L} =$ const. But initially, $\boldsymbol{L} = 0$, hence it remains zero throughout the motion. Consequently, the monkey and bananas move upward with the same speed at any instant. The distance between the monkey and bananas stays constant. Hence, the monkey will not reach the bananas.
43. (a) 4.44 rad/s^2 (b) 8.89 rad/s
45. $(-60t\boldsymbol{k})$ $kg \cdot m^2/s$, $(-60.0\ \boldsymbol{k})$ N·m
47. (a) $\tau_x = yF_z - zF_y$, $\tau_y = zF_x - xF_z$, $\tau_z = xF_y - yF_x$
49. (a) $v_0 r_0/r$ (b) $T = (mv_0^2 r_0^2)r^{-3}$ (c) $\dfrac{1}{2}mv_0^2\left(\dfrac{r_0^2}{r^2} - 1\right)$ (d) 4.50 m/s, 10.1 N, 0.450 J
51. (a) $F_y = \dfrac{W}{L}\left(d - \dfrac{ah}{g}\right)$ (b) 0.306 m (c) $(-306\boldsymbol{i} + 553\boldsymbol{j})$ N
55. 5.29×10^3 rad/s, 2.34×10^{44} J
57. (c) $(8Fd/3M)^{1/2}$
61. $v_0 = [ag(16/3)(\sqrt{2} - 1)]^{1/2}$

CHAPTER 12

1. $F_1 + F_2 - W_1 - W_2 = 0$, $F_2\ell - W_1 d_1 - W_2 d_2 = 0$
3. $[(W_1 + W)d + W_1\ell/2]/W_2$
5. $F_W = 480$ N, $F_v = 1200$ N
7. The y coordinate of the center of mass is 15.3 cm from the bottom. The x coordinate is 8.00 cm from the left side of the "tee."
9. -1.50 m, -1.50 m
11. 0.25 m to the right of the point of support
13. 0.789
15. $F_f = 4410$ N, $F_r = 2940$ N
17. $2R/5$
19. $\frac{1}{3}$ by the left string, $\frac{2}{3}$ by the right string
21. 1.76 m
23. 400 N
25. 2.23×10^{-5}
27. 2.80×10^7 N/m^2
29. (a) 3.14×10^4 N (b) 62.8 kN
31. 7.71 kg/m^3
33. $N_A = 5.98 \times 10^5$ N, $N_B = 4.80 \times 10^5$ N
35. (b) 69.8 N (c) 0.877ℓ
37. (b) $T = 213$ N, $R_x = 184$ N, $R_y = 188$ N
39. (a) $T = W(\ell + d)/\sin\theta(2\ell + d)$, and (b) $R_x = W(\ell + d)\cot\theta/(2\ell + d)$; $R_y = W\ell/(2\ell + d)$
41. (a) $F_x = 268$ N, $F_y = 1300$ N (b) 0.324
43. 5.08 kN, $R_x = 4.77$ kN, $R_y = 8.26$ kN

45. $T = 2710$ N, $R_x = 2640$ N, $R_y = -12$ N
47. (a) 20.1 cm to the left of the front edge, $\mu = 0.571$ (b) 0.501 m
49. (a) $W = \frac{w}{2}\left(\frac{2\mu_s \sin\theta - \cos\theta}{\cos\theta - \mu_s \sin\theta}\right)$
(b) $R = (w + W)\sqrt{1 + \mu_s^2}$, $F = \sqrt{W^2 + \mu_s^2(w + W)^2}$
51. (a) 133 N (b) $N_A = 429$ N, $N_B = 257$ N
(c) $R_x = 133$ N, $R_y = -257$ N
53. $\boldsymbol{F}_{\text{floor}} = \left(100\text{ N} + \frac{T}{2}\right)\boldsymbol{j}$

$\boldsymbol{F}_{\text{hinge}} = -T\boldsymbol{i} + \left(100\text{ N} - \frac{T}{2}\right)\boldsymbol{j}$
55. 2.81 m
57. 7.51×10^6 N/m²
59. (a) 4500 N (b) 4.50×10^6 N/m²
(c) This is more than sufficient to break the board.

CHAPTER 13

1. (a) 1.50 Hz, 0.667 s (b) 4.00 m (c) π rad
(d) -4.00 m
3. (a) 4.33 cm (b) -5.00 cm/s (c) -17.3 cm/s²
(d) π s, 5.00 cm
5. (a) 13.9 cm/s, 16.0 cm/s² (b) 16.0 cm/s, 0.263 s
(c) 32.0 cm/s², 1.05 s
7. (b) 6π cm/s, 0.333 s (c) $18\pi^2$ cm/s², 0.500 s
(d) 12.0 cm
9. 3.33 cm
11. 3.95 N/m
13. (a) 2.40 s (b) 0.417 Hz (c) 2.62 rad/s
15. (a) 0.400 m/s, 1.60 m/s² (b) ± 0.320 m/s, -0.960 m/s² (c) 0.232 s
17. (a) 0.750 m (b) $x = -(0.75\text{ m})\sin(2.0t)$
19. (a) 0.153 J (b) 0.783 m/s (c) 17.5 m/s²
21. (a) quadrupled (b) doubled (c) doubled
(d) no change
23. ± 2.60 cm
25. (a) 1.55 m (b) 6.06 s
27. (a) halved (b) doubled
29. (a) 0.820 m/s (b) 2.57 rad/s² (c) 0.641 N
31. increases by 1.78×10^{-3} s
33. 8.50×10^{-2} kg·m²
35. (a) 5.00×10^{-7} kg·m² (b) 3.16×10^{-4} N·m/rad
39. 1.00×10^{-3} s⁻¹

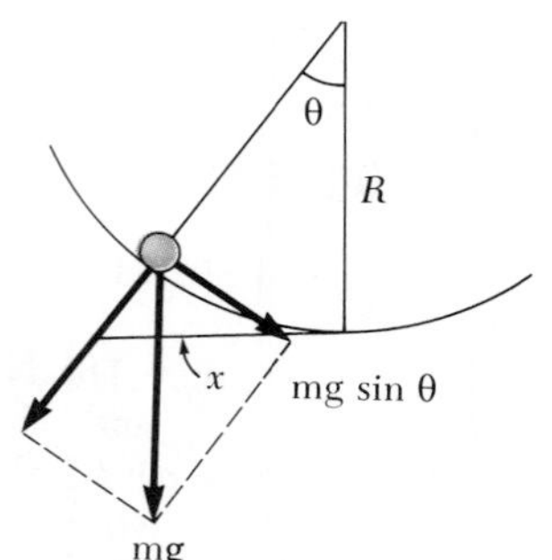

Problem 49

41. (a) 1.42 Hz (b) 0.407 Hz
43. 318 N
45. 1.57 s
47. (a) $E = \frac{1}{2}mv^2 + mgL(1 - \cos\theta)$ (b) $U = \frac{1}{2}m\omega^2 s^2$
49. Referring to the sketch, we have $F = -mg \sin\theta$ and $\tan\theta = x/R$. For small displacements, $\tan\theta \approx \sin\theta$ and $F = -(mg/R)x = -kx$ and $\omega = (k/m)^{1/2} = (g/R)^{1/2}$.
51. (a) $2Mg$, $T_p = Mg\left(1 + \frac{y}{L}\right)$ (b) $\frac{4\pi}{3}\sqrt{\frac{2L}{g}} = 2.68$ s
55. $f = \frac{1}{2\pi}\sqrt{\frac{MgL + kh^2}{ML^2}}$
57. 0.0662 m
59. (a) 250 N/m (b) 0.281 s, 3.56 Hz, 22.4 rad/s
(c) 5.00 cm, $\delta = 3\pi/2$ (d) 1.12 m/s, 25.0 m/s²
(e) 0.312 J (f) 4.62 cm, 0.428 m/s, +23.1 m/s²
61. 9.19×10^{13} Hz
63. (a) 15.8 rad/s (b) 5.23 cm (c) 1.31 cm, π
67. 1.1 s

CHAPTER 14

1. 2.96×10^{-9} N
3. 4.62×10^{-8} N toward the center of the triangle
5. $F_x = Gm^2\left[\frac{2}{b^2} + \frac{3b}{(a^2 + b^2)^{3/2}}\right]$,

$F_y = Gm^2\left[\frac{2}{a^2} + \frac{3a}{(a^2 + b^2)^{3/2}}\right]$
7. $\boldsymbol{F}_6 = (12.6\boldsymbol{i} + 1.92\boldsymbol{j}) \times 10^{-11}$ N,
$F_6 = 12.7 \times 10^{-11}$ N
9. 26.9 m/s²
11. 12.6×10^{31} kg
13. 6.01×10^{24} kg
15. 1.90×10^{27} kg
17. 5.90×10^4 m/s
19. 2.98×10^4 m/s
21. $2GMr/(r^2 + a^2)^{3/2}$, to the left
23. 3.84×10^4 km from the moon's center
25. (a) -1.67×10^{-14} J (b) at the center of the triangle
29. 1.66×10^4 m/s
31. (a) 1.88×10^{11} J (b) 100 kW
35. 1.58×10^{10} J
37. (a) 42.1 km/s relative to the sun (b) 2.20×10^{11} m
39. $Gm\lambda_0 L[d(L + d)]^{-1} + GmAL$ to the right
41. (a) 7.41×10^{-10} N (b) 1.04×10^{-8} N
(c) 5.21×10^{-9} N
43. 2.26×10^{-7}
45. 0.0572 rad/s = 32.7 rev/h
47. (a) -2.33×10^{-10} N
(b) 1.00×10^{-10} N along the positive x axis
49. (a) $k = \frac{GmM_e}{R_e^3}$, $A = \frac{L}{2}$

(b) $\frac{L}{2}\left(\frac{GM_e}{R_e^3}\right)^{1/2}$, at the middle of the tunnel

(c) 1.55×10^3 m/s

51. 2.67×10^{-7} m/s^2
53. 2.99×10^3 rev/min
55. (a) $v_1 = m_2 \left[\frac{2G}{d(m_1 + m_2)}\right]^{1/2}$

$v_2 = m_1 \left[\frac{2G}{d(m_1 + m_2)}\right]^{1/2}$

$v_{rel} = \left[\frac{2G(m_1 + m_2)}{d}\right]^{1/2}$

(b) $K_1 = 1.07 \times 10^{32}$ J, $K_2 = 2.67 \times 10^{31}$ J
57. (a) 7.34×10^{22} kg (b) 1.63×10^3 m/s (c) 1.32×10^{10} J
59. -5.37×10^{33} J
61. (a) 5300 s (b) 7.79 km/s (c) 6.45×10^9 J
63. (a) $M/\pi R^4$ (b) $-GmM/r^2$ (c) $-(GmM/R^4)r^2$
65. 1.48×10^{22} kg
67. (b) 981 kg/m^3
69. (b) $GMm/2R$

CHAPTER 15

1. 0.111 kg
3. 3.99×10^{17} kg/m^3. Matter is mostly free space.
5. 6.24×10^6 Pa
7. 4.77×10^{17} kg/m^3
9. 20.6 m
11. 1.62 m
13. 77.4 cm^2
15. 2.47%
17. 9.12 MPa
19. 10.3 m
21. 10.5 m; no, a little alcohol and water evaporate
23. 1.08 cm
25. 10.67% above
27. (a) 7.00 cm (b) 2.80 kg
29. 0.424 kg
31. 0.611 kg
33. 1.07 m^2
35. 1430 m^3
37. (a) 17.7 m/s (b) 1.73 mm
39. (a) 8.00 m/s (b) 1.50×10^4 Pa
41. 31.6 m/s
43. (a) 28.0 m/s (b) 392 kPa
45. 4.31×10^4 N
47. 103 m/s
49. $2[h(h_0 - h)]^{1/2}$
51. 4.83 kW at 20°C
53. (b) $\frac{1}{2}\rho Av^3$ if the mill could make the air stop; the same
55. 6.54×10^{-4} Pa·s; water
57. 1.40×10^8 Pa
63. 0.443 cm
65. 2.01×10^6 N
69. 3.17 m
71. 5.02 GW
73. (a) $(\rho_1 h_1 + \rho_2 h_2)/(h_1 + h_2)$
(b) $(\rho_1 h_1 + \rho_2 h_2)/\rho_w$
(c) $d' = d$ (d) $\Delta U = (\rho_2 - \rho_1)s^2 h_1 h_2 g$
75. (a) $T_\ell = \left[\frac{\rho_b - \rho_w}{\rho_b}\right] M_b g$
(b) $T_u = T_\ell + \frac{1}{4}[\rho_c - \rho_w]\pi d^2 hg$
(c) $T'_\ell = M_b g$; $T'_u = T'_\ell + \frac{1}{4}\rho_c \pi d^2 hg$
(d) $T_\ell = 1.14 \times 10^4$ N; $T_u = 1.35 \times 10^4$ N
$T'_\ell = 1.96 \times 10^4$ N; $T'_u = 2.20 \times 10^4$ N
77. (a) 18.3 mm (b) 14.3 mm (c) 8.56 mm

CHAPTER 16

1. $y = \frac{6}{(x - 4.5t)^2 + 3}$
7. (a) 5.00 rad (b) 0.858 cm
9. (a) Wave 1 travels in the $+x$ direction; wave 2 travels in the $-x$ direction. (b) 0.75 s
(c) $x = 1.00$ m
11. 25.8 m/s
13. 13.5 N
15. 586 m/s
17. 0.329 s
19. (b) 0.125 s
21. 0.319 m
23. (a) $y = A \sin k(x - vt)$

(b) $y = A \sin \frac{2\pi}{\lambda}(x - vt)$

(c) $y = A \sin 2\pi \left(\frac{x}{\lambda} - ft\right)$

(d) $y = A \sin\left[2\pi f\left(\frac{x}{v} - t\right)\right]$
25. (a) 0.200 m (b) 4π rad/s (c) 5.03 m^{-1}
(d) 1.25 m (e) 2.50 m/s (f) to the left
27. (a) $y = (0.0800 \text{ m}) \sin(7.85x + 6\pi t)$
(b) $y = (0.0800 \text{ m}) \sin(7.85x + 6\pi t - 0.785)$
29. (a) 2.15 cm (b) 0.379 rad (c) 541 cm/s
(d) $y(x, t) = (2.15 \text{ cm})\cos(80\pi t + 8\pi x/3 + 0.379)$
31. (a) 125 m/s (b) 2.40 m/s
33. (a) $y = (0.0200 \text{ m})\sin(7.85x - 1005t)$
(b) $y = (0.0200 \text{ m})\sin[2\pi(x/0.8 - 160t)]$ where x and y are in m and t is in s.
35. 1.07 kW
37. (a) remains constant (b) remains constant
(c) remains constant (d) p is quadrupled
39. (a) 62.5 m/s (b) 7.85 m (c) 7.96 Hz
(d) 21.1 W
43. Both functions in (a) and (b) are solutions to the wave equation.
45. (a) 3.33 m/s in the positive x direction
(b) -5.48 cm (c) 0.667 m, 5.00 Hz
(d) 11.0 m/s
47. (a) 179 m/s (b) 17.7 kW
49. 6.67 cm
51. (a) 25 m^{-1} (b) 12.0 rad/s (c) 0.340 rad
(d) 0.262 s (e) 0.0136 m
53. (a) $v = \sqrt{\frac{F}{\rho\,(10^{-3}x + 10^{-2})\text{cm}^2}}$ (b) 94.3 m/s, 66.7 m/s

55. 3.86×10^{-4} (at 5.93°C)
57. $(2L/g)^{1/2}$; $L/4$
59. (a) $\frac{\mu\omega^3}{2k} A_0^2 e^{-2bx}$ (b) $\frac{\mu\omega^3}{2k} A_0^2$ (c) e^{-2bx}

CHAPTER 17

1. 5.56 km
3. 1430 m/s
5. Pa
7. 1.988 km
9. (a) 27.2 s (b) 25.7 s It is shorter by 5.30%.
11. 1.55×10^{-10} m
13. 5.81 m
15. (a) 2.00 μm, 0.400 m, 54.6 m/s (b) -0.433 μm (c) 1.72 mm/s
17. (0.2 Pa) $\sin(62.8x - 2.16 \times 10^4 t)$
19. 66.0 dB
21. (a) 1.94×10^{-2} W/m² (b) 103 dB
23. 28.7 Pa
25. (a) 84.0 dB (b) 356 m
27. 241 W
29. (a) 30.0 m (b) 9.49×10^5 m
31. (a) 112 dB, 148 μm (b) 112 dB, 11.1 μm (c) 112 dB, 0.221 μm
33. 46.4°
35. 41.8°
37. 26.4 m/s
39. (a) 339 Hz (b) 483 Hz
41. 2.82×10^8 m/s
43. (a) 56.3 s (b) (56.6 km)$\boldsymbol{i}$ + (20.0 km)$\boldsymbol{j}$ from the observer
45. 130 m/s, 1.73 km
47. (a) 10.0 (b) 3.01 dB
49. 1204 Hz
51. (a) 0.948° (b) 4.40°
53. 1.34×10^4 N
55. 34.3 W
57. (a) 55.8 m/s (b) 2500 Hz
59. (a) 6.45 (b) 0
61. 1.60

CHAPTER 18

1. (a) 9.24 m (b) 600 Hz
3. $y_2 = (8\text{ m})\sin[2\pi(0.1x - 80t - 0.167)]$
5. (a) 6.04 rad (b) 1.98A
7. at 0.0891 m, 0.303 m, 0.518 m, 0.732 m, 0.947 m, and 1.16 m
9. (a) 4.24 cm (b) 6.00 cm (c) 6.00 cm (d) $x = 0.5$ cm, 1.5 cm, 2.5 cm
11. 25.1 m, 60.0 Hz
13. (a) 2.00 cm (b) 2.40 cm
15. It does.
17. (a) 60.0 cm (b) 30.0 Hz
19. $L/4$, $L/2$
21. 0.786 Hz, 1.57 Hz, 2.36 Hz, 3.14 Hz
23. 19.995 kHz
25. Nodes at 0, 2.67, 5.33, 8.00 m; Antinodes at 1.33, 4.00, 6.67 m.
27. 19,976 kHz
29. 824 N
31. (a) $L = nv/2f$ (b) $L = (2n - 1)v/4f$, $n = 1, 2, 3, \ldots$
33. 50.4 cm, 84.0 cm
35. 35.8 cm, 71.7 cm
37. 349 m/s
39. 65.7 cm, 164 cm
41. 328 m/s
43. (a) 350 m/s (b) 114 cm
45. n(206 Hz) and n(84.5 Hz), where $n = 1, 2, 3, .$
47. 1.88 kHz
49. (a) 1.59 kHz (b) odd (c) 1.11 kHz
51. It is.
53. (a) 1.99 Hz (b) 3.38 m/s
55. (a) 3.33 rad (b) 283 Hz
57. 3.40 kHz
59. (a) 0.833 (b) 0.664 m, 0.553 m
61. (a) 133 Hz (b) 5.62 kg
63. (a) 6.00 m (b) 274 m/s (c) (4 cm)$\sin(7.33x)\cos(2010t)$
65. 3.87 m/s *away* from the station *or* 3.78 m/s *toward* the station
67. (a) 59.9 Hz (b) 20 cm
69. (a) 0.5 (b) $\frac{n^2F}{(n+1)^2}$ (c) $\frac{F'}{F} = \frac{9}{16}$

CHAPTER 19

1. (a) -273.5°C (b) 1.27 atm, 1.74 atm
3. (a) 30.4 mm Hg (b) 18.0 K
5. (a) 139 K, -134°C (b) 6.56 kPa
7. (a) -321°F (b) 139°R (c) 77.3 K
9. (a) 621°F (b) 600 K
11. (a) 119.0°C (b) 246.2°F and 832.3°F
13. (a) 90.0°C (b) 90.0 K
15. -40.0°C
17. 3.27 cm
19. 1.32 cm
21. 10.0 cm
23. 62.5 N
25. 1.20 cm
27. (a) 437°C (b) 2100°C. No; they melt first.
29. (a) 99.4 cm³ (b) 1.00 cm
31. 1.08 L
33. (a) 0.176 mm (b) 8.78 μm (c) 93.0 mm³
35. 750 K
37. 5.87 atm
39. 1.63×10^3
41. (a) 900 K (b) 1200 K
43. (a) 94.7 mol (b) 1.36 atm
45. 594 kPa
47. 3.84 m
49. 1.13
51. (a) $A = 1.85 \times 10^{-3}$ (1/C°), $R_0 = 50.0\ \Omega$ (b) 421°C

53. $\alpha \Delta T \ll 1$
55. 3.55 cm
57. (b) For most solids α is small compared to β for most liquids.
59. (b) 1.33 kg/m^3
63. 60.2°
65. 2.24 MPa = 22.1 atm
67. (a) 18.0 m (b) 277 kPa
69. (a) 7.06 mm (b) 297 K
71. (a) 0.0374 mol (b) 0.732 atm
73. (a) 6.17×10^{-3} kg/m (b) 632 N (c) 580 N; 192 Hz
75. (a) $(\alpha_2 - \alpha_1) L \Delta T/(r_2 - r_1)$ (c) It bends the other way.

CHAPTER 20

1. 0.105°C
3. 10.117°C
5. 0.133 cal/g·C°
7. 85.3°C
9. 29.6°C
11. 34.7°C
13. 13.3 kJ
15. (a) 84.4°C (b) 557 cal, 998 cal, 1560 cal
17. 26.3 kJ
19. 12.9 g
21. (a) 0°C (b) 115 g
23. 26.7 g liquid and 3.35 g vapor at 100°C
25. (a) 25.760°C (b) No.
27. 434°C. No; heat will very soon be lost to the wood.
29. 810 J, 506 J, 203 J
31. 466 J
33. 1.18 MJ
35. (a) 7.65 L (b) 305 K
37. (a) −567 J (b) 167 J
39. (a) 12.0 kJ (b) −12.0 kJ
41. (a) 23.1 kJ (b) 23.1 kJ
43. (a) 76.0 J, 101 J, 88.6 J (b) 167 J, 192 J, 180 J
45. (a) 7.50 kJ (b) 900 K
47. (a) 99.97R (b) 250.43R (c) 150.46R
49. (a) 48.6 mJ (b) 16.2 kJ (c) 16.2 kJ
51. (a) −800 J (b) 400 J (c) 400 J (d) 0, yes
53. (a) 100°C/m (b) 1.59 W (c) 60.0°C
55. 51.0°C
57. 160
59. 0.143 W/mC°
61. (a) 27.5 BTU/h (b) 56.7 BTU/h
63. (a) 53.4 W (b) -8.12×10^{-3} C°/s (c) 4.72 W, -7.17×10^{-4} C°/s
65. (a) 0°C (b) 332 kJ
67. (a) $-0.500 P_0 V_0$ (b) $-1.39 P_0 V_0$ (c) 0
69. (a) 16.8 L (b) 0.351 L/s
71. (a) 7 (b) heat transfer from the brakes into the environment
73. 285 J
75. 5.46°C
79. 9.32 kW
81. 5.31 h
83. (a) 21.4°C (b) 29.3°C

CHAPTER 21

1. 2.43×10^5 m^2/s^2
3. 8.00 N, 1.60 Pa
5. 3.32 mol
7. 8.76×10^{-21} J
9. (a) 40.1 K (b) 6.01 km/s
11. 2.05×10^3 K
13. 109 kPa
15. 75.0 J
17. (a) 3.00 mol (b) 37.4 J/K (c) 13.1 kJ
19. (a) 3.46 kJ (b) 2.45 kJ (c) 1.01 kJ
21. 24.0 kJ; 68.7 kJ
23. 3.33×10^{-26} kg
25. (a) 1.39 atm (b) 366 K, 254 K
27. (a) 15.6 (b) 0.0214
29. (a) 8.50 (b) 2.35
31. 227 K
33. 443 m/s
35. 2.94 times higher
37. (a) 169 m/s (b) 169 kPa
39. 385 m
41. SO_2 has the highest molecular weight of the polyatomic gases listed. The frequencies of some vibrational modes are low enough so that some vibration is excited at 300 K.
43. 2.32×10^{-21} J
45. (a) 558 m/s (b) 515 m/s (c) 456 m/s
47. 731 m/s, 825 m/s, 895 m/s
49. (a) 36 000 (b) 18 000
51. (a) 2.37×10^4 K (b) 1.06×10^3 K
53. (a) 3.21×10^{12} molecules (b) 778 km (c) 6.42×10^{-4} s^{-1}
55. 193
57. 4.65×10^{-10} m
63. (a) $3.65v$ (b) $3.99v$ (c) $3.00v$ (d) $106mv^2/V$ (e) $7.98mv^2$
65. 1.40×10^{21}
67. 0.605
73. (a) aR (b) $(a+1)R$ (c) 3.43 (d) 7
75. (a) 5.18 (b) 0.518 (c) 5.18 (d) 0.139
77. (a) $n_1R/(\gamma_1 - 1) + n_2R/(\gamma_2 - 1)$ (b) $\gamma_1 n_1 R/(\gamma_1 - 1) + \gamma_2 n_2 R/(\gamma_2 - 1)$

CHAPTER 22

1. (a) 6.94% (b) 335 J
3. (a) 24.0 J (b) 144 J
5. (a) 0.333 (b) 0.667
7. (a) 0.375 (b) 600 J (c) 2.00 kW
9. (a) 560 J (b) 350 K
11. (a) 5.12% (b) 5.27 TJ
13. (a) 0.672 (b) 58.8 kW
15. 0.478°C
17. (a) 16.8 (b) 14.3 kW
19. (a) 741 J (b) 459 J

21. 1.75; $\frac{1}{4}$ as large
23. 192 J
25. 9.00
27. 72.2 J
29. 7.26×10^3 J/K
31. 195 J/K
33. 3.59 J/K
35. 1.02×10^3 J/K
37. 1.34×10^8 J/K
39. 507 J/K
41. 0.0853 J/K
43. (a) 7.34 kJ/K (b) −6.68 kJ/K (c) 661 J/K
45. (a) 53.0°C (b) 7.34 J/K
47. 0.395 times as high
49. (a) 5.00 kW (b) 763 W
51. (a) 4.10 kJ (b) 14.2 kJ (c) 10.1 kJ (d) 28.8%
53. (a) $2nRT_0 \ln 2$ (b) 0.273
55. (a) $10.5nRT_0$ (b) $8.5nRT_0$ (c) 0.190 (d) 0.833
57. $nC_p \ln 3$
65. (a) 26.1°C (b) −960 J/K (c) 1.74 kJ/K (d) 777 J/K (e) −1.74 kJ/K, −18.6 J/K, +1.77 kJ/K (f) 777 J/K

CHAPTER 23

1. 5.14×10^3 N
3. 1.60×10^{-9} N repelling one another
5. $(8.50 \times 10^{-2}\text{ N})\boldsymbol{i}$
7. 0.873 N at 330°
9. 40.9 N at 263°
11. (0, 0.853 m)
13. (a) 295 nC (b) $(-0.375\text{ N})\boldsymbol{i}$
15. (a) $(-5.58 \times 10^{-11}\text{ N/C})\boldsymbol{j}$ (b) $(1.02 \times 10^{-7}\text{ N/C})\boldsymbol{j}$
17. (a) $(-5.20 \times 10^3\text{ N/C})\boldsymbol{i}$ (b) $(2.93 \times 10^3\text{ N/C})\boldsymbol{j}$ (c) 5.85×10^3 N/C at 225°
19. (a) $(1.29 \times 10^4\text{ N/C})\boldsymbol{j}$ (b) $(-3.87 \times 10^{-2}\text{ N})\boldsymbol{j}$
21. 0
23. (a) at the center (b) $\left(\frac{\sqrt{3}\,kq}{a^2}\right)\boldsymbol{j}$
25. (a) $0.914 \frac{kq}{a^2}$ at 225° (b) $0.914 \frac{kq^2}{a^2}$ at 45°
27. 7.82 m to the left of the negative charge
29. $-\left(\frac{k\lambda_0}{x_0}\right)\boldsymbol{i}$
31. (a) $(6.65 \times 10^6\text{ N/C})\boldsymbol{i}$ (b) $(2.42 \times 10^7\text{ N/C})\boldsymbol{i}$ (c) $(6.40 \times 10^6\text{ N/C})\boldsymbol{i}$ (d) $(6.65 \times 10^5\text{ N/C})\boldsymbol{i}$
33. (a) 0.145 C/m³ (b) 1.94×10^{-3} C/m²
35. (a) 9.37×10^7 N/C away from the center; 1.04×10^8 N/C is 11.0% larger (b) 516 kN/C away from the center; 520 kN/C is 0.749% larger
37. 7.20×10^7 N/C away from the center; 1.00×10^8 N/C axially away
39. $(-21.6\text{ MN/C})\boldsymbol{i}$
41.

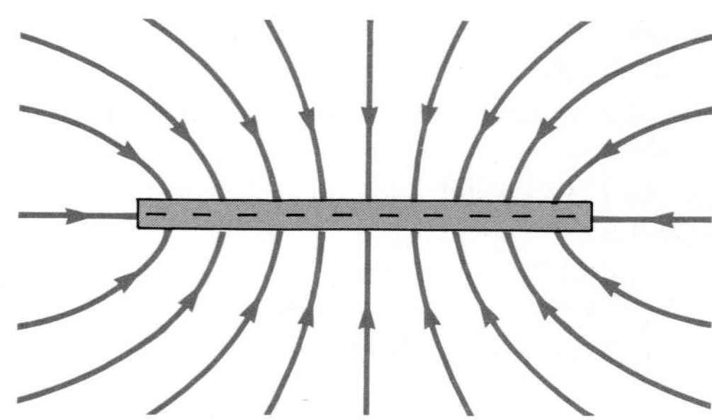

43. (a) $\frac{q_1}{q_2} = -1/3$ (b) q_1 is negative and q_2 is positive
45. (a) 6.14×10^{10} m/s² (b) 19.5 μs (c) 11.7 m (d) 1.20 fJ
47. 1.00×10^3 N/C in the direction of the beam
49. (a) $(-5.75 \times 10^{13}\text{ m/s}^2)\boldsymbol{i}$ (b) 2.84×10^6 m/s (c) 49.4 ns
51. (a) 111 ns (b) 5.67 mm (c) $(450\text{ km/s})\boldsymbol{i} + (102\text{ km/s})\boldsymbol{j}$
53. (a) 36.9°, 53.1° (b) 167 ns, 221 ns
55. (a) 10.9 nC (b) 5.43×10^{-3} N
57. (a) $\theta_1 = \theta_2$
59. 204 nC
63. (a) $-\left(\frac{4kq}{3a^2}\right)\boldsymbol{j}$ (b) (0, 2.00 m)
65. $\boldsymbol{E} = -\frac{k\lambda\ell}{d(d+\ell)}(\boldsymbol{i}+\boldsymbol{j})$
67. 5.27×10^{17} m/s²; 0.854 mm
71. (a) $\boldsymbol{F} = \frac{kq^2}{s^2}(1.90)(\boldsymbol{i}+\boldsymbol{j}+\boldsymbol{k})$

(b) $F = 3.29 \frac{kq^2}{s^2}$ in a direction away from the vertex diagonally opposite to it
77. $\left(\frac{\sigma}{2\epsilon_0}\right)\boldsymbol{i}$ for $x > 0$ and $-\left(\frac{\sigma}{2\epsilon_0}\right)\boldsymbol{i}$ for $x < 0$

CHAPTER 24

1. (a) 1.98×10^6 Nm²/C (b) 0 (c) 1.92×10^6 Nm²/C
3. (a) aA (b) bA (c) 0
5. 4.14×10^6 N/C
7. $bhw^2/3$
9. 1.87×10^3 Nm²/C
11. 5.65×10^5 Nm²/C
13. (a) 1.36×10^6 Nm²/C (b) 6.78×10^5 Nm²/C (c) No, the same field lines go through spheres of all sizes.
15. -6.89×10^6 Nm²/C. The number of lines entering exceeds the number leaving by 2.91 times or more.
17. 13.7 μC, No
19. (a) 5.78×10^3 Nm²/C (b) 51.1 nC
21. (a) 761 nC (b) It may have any distribution. Any and all point and smeared-out charges, positive and

negative, must add algebraically to +761 nC.
(c) Total charge is −761 nC

23. (a) $\frac{Q}{2\epsilon_0}$ (out of the volume enclosed)
(b) $-\frac{Q}{2\epsilon_0}$ (into it).

25. (a) 0 (b) 7.20×10^6 N/C away from the center
27. (a) 0.713 μC (b) 5.7 μC
29. (a) 0 (b) $(3.66 \times 10^5 \text{ N/C})\hat{\boldsymbol{r}}$
(c) $(1.46 \times 10^6 \text{ N/C})\hat{\boldsymbol{r}}$ (d) $(6.50 \times 10^5 \text{ N/C})\hat{\boldsymbol{r}}$
31. $\boldsymbol{E} = (a/2\epsilon_0)\hat{\boldsymbol{r}}$
33. (a) 5.14×10^4 N/C outward (b) 646 Nm^2/C
35. $\boldsymbol{E} = (\rho r/2\epsilon_0)\hat{\boldsymbol{r}}$
37. 5.08×10^5 N/C up.
39. (a) 0 (b) 5.40×10^3 N/C
(c) 540 N/C, both radially outward
41. (a) 80.0 nC/m^2 on each face (b) $(9.04 \times 10^3 \text{ N/C})\boldsymbol{k}$
(c) $(-9.04 \times 10^3 \text{ N/C})\boldsymbol{k}$
43. (a) 0 (b) $(125 \times 10^4 \text{ N/C})\hat{\boldsymbol{r}}$ (c) $(639 \text{ N/C})\hat{\boldsymbol{r}}$
(d) No change
45. (a) 0 (b) $(8.00 \times 10^7 \text{ N/C})\hat{\boldsymbol{r}}$ (c) 0
(d) $(7.35 \times 10^6 \text{ N/C})\hat{\boldsymbol{r}}$
47. (a) $-\lambda$, $+3\lambda$ (b) $\left(\frac{3\lambda}{2\pi\epsilon_0 r}\right)\hat{\boldsymbol{r}}$
49. (b) $\frac{Q}{2\epsilon_0}$ (c) $\frac{Q}{\epsilon_0}$
51. (a) $\boldsymbol{E} = \left(\frac{\rho r}{3\epsilon_0}\right)\hat{\boldsymbol{r}}$ for $r < a$; $\boldsymbol{E} = \left(\frac{kQ}{r^2}\right)\hat{\boldsymbol{r}}$ for $a < r < b$;
$\boldsymbol{E} = 0$ for $b < r < c$; $\boldsymbol{E} = \left(\frac{kQ}{r^2}\right)\hat{\boldsymbol{r}}$ for $r > c$
(b) $\sigma_1 = -\frac{Q}{4\pi b^2}$ inner; $\sigma_2 = +\frac{Q}{4\pi c^2}$ outer
53. (a) $\boldsymbol{E} = 0$ (b) $\boldsymbol{E} = \frac{\rho(r^3 - a^3)}{3\epsilon_0 r^2}\hat{\boldsymbol{r}}$
(c) $\boldsymbol{E} = \frac{\rho(b^3 - a^3)}{3\epsilon_0 r^2}\hat{\boldsymbol{r}}$
57. $\boldsymbol{g} = -\left(\frac{GMr}{R_e^3}\right)\hat{\boldsymbol{r}}$
59. (a) σ/ϵ_0 to the left (b) zero (c) σ/ϵ_0 to the right

CHAPTER 25

1. 0
3. (a) 152 km/s (b) 6.50×10^6 m/s
5. (a) 2.7 keV (b) 509 km/s
7. 6.41×10^{-19} C
9. 2.10×10^6 m/s
11. 1.35 MJ
13. 432 V; 432 eV
15. −38.9 V; the origin
17. 1.56×10^3 N/C
19. 260 V, B
21. 2.00 m
23. 119 nC, 2.67 m
25. −88.2 kV
27. −11.0 MV
29. (a) −386 nJ. Positive binding energy would have to be put in to separate them. (b) 103 V
31. (a) −27.3 eV (b) −6.81 eV (c) 0
35. 20.1 J
37. $-(0.553)\frac{kQ}{R}$
39. (a) C/m^2 (b) $k\alpha\left[L - d\ln\frac{d+L}{d}\right]$
41. $V = \frac{\sigma}{2\epsilon_0}(\sqrt{x^2+b^2} - \sqrt{x^2+a^2})$
43. $\boldsymbol{E} = (6xy - 5)\boldsymbol{i} + (3x^2 - 2z^2)\boldsymbol{j} - 4yz\boldsymbol{k}$; 7.07 N/C
45. 45.0°
47. (a) 0 (b) $\left(\frac{kQ}{r^2}\right)\hat{\boldsymbol{r}}$
49. 3.38 MV; 3.00 MV is 11.1% smaller
51. 1.56×10^{12} electrons removed
53. (a) 0, 1.67 MV (b) 5.85×10^6 N/C away, 1.17 MV (c) 1.19×10^7 N/C away, 1.67 MV
55. (a) 4.50×10^7 N/C outward, 30.0 MN/C outward
(b) 1.80 MV
57. (a) 450 kV (b) 7.50 μC
59. 5.00 μC
61. (a) 6.00 m (b) −2.00 μC
63. 1.73 MV
65. (a) 180 kV (b) 127 kV
67. 1.12 MV; 1.00 MV is smaller by 11.1%
69.

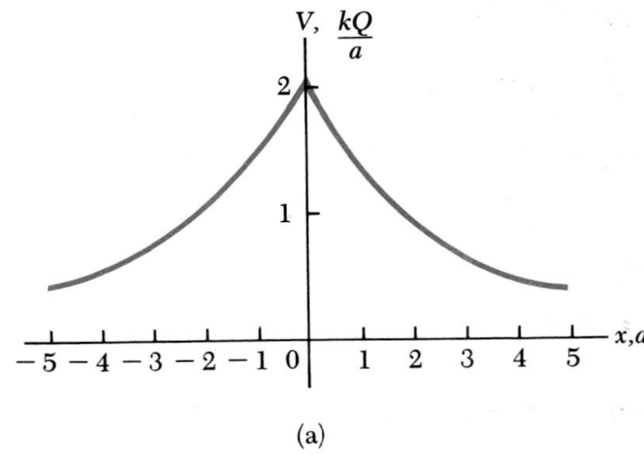

(a)

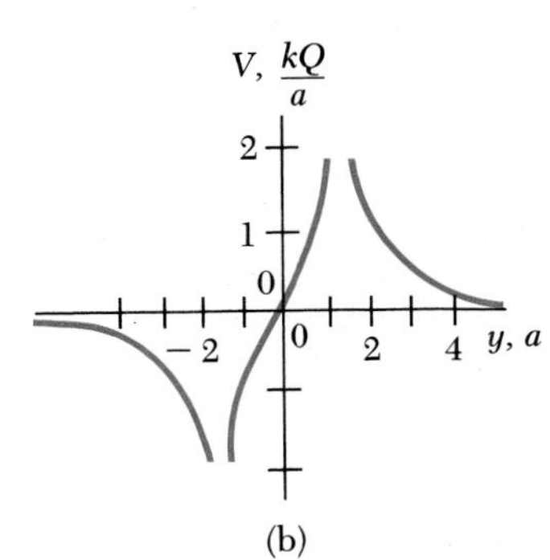

(b)

71. (a) 66.7 pC (b) 476 V

73. $k\dfrac{Q^2}{2R}$

75. −142 J

77. $V = k\sigma \int_{+L/2}^{+L/2} \int_{-L/2}^{-L/2} \dfrac{dx\,dy}{\sqrt{x^2+y^2+z^2}}$

79. $E_y = k\dfrac{Q}{\ell y}\left[1 - \dfrac{y^2}{\ell^2 + y^2 + \ell\sqrt{\ell^2+y^2}}\right]$

81. (a) $E_r = \dfrac{2kp\cos\theta}{r^3}$; $E_\theta = \dfrac{kp\sin\theta}{r^3}$; yes; no

(b) $\mathbf{E} = \dfrac{3kpxy\mathbf{i}}{(x^2+y^2)^{5/2}} + \dfrac{kp(2y^2-x^2)\mathbf{j}}{(x^2+y^2)^{5/2}}$

83. $\dfrac{3}{5}\left(\dfrac{kQ^2}{R}\right)$

CHAPTER 26

1. 13.3 kV
3. 1.40 pF
5. 684 μC
7. (a) 1.33 μC/m^2 (b) 13.3 pF
9. (a) $4\pi\epsilon_0(R_1 + R_2)$ (b) $\dfrac{Q_1}{Q_2} = \dfrac{R_1}{R_2}$
11. 1.52 mm
13. 4.42 μm
15. (a) 1.11×10^4 N/C toward the negative plate (b) 98.3 nC/m^2 (c) 3.74 pF (d) 74.8 pC
17. 109 mm
19. 3.54 nC
21. (a) 2.68 nF (b) 3.02 kV
23. (a) 15.6 pF (b) 256 kV
25. 66.7 nC
27. 18.0 μF
29. (a) 4.00 μF (b) 8.00 V, 4.00 V, 12.0 V, 24.0 μC, 24.0 μC, 24.0 μC
31. (a) 5.96 μF (b) 89.2 μC, 63.1 μC, 26.4 μC, 26.4 μC.
33. 120 μC; 80.0 μC and 40.0 μC
35. (a) 12.0 μF (b) 24.0 μC, 14.4 μC, 19.2 μC, 19.2 μC
37. (a) 800 μC, 1200 μC (b) 200 V
39. 83.6 μC
41. 12.9 μF
43. $\dfrac{\epsilon_0 A}{(s-d)}$
45. 90.0 mJ
47. The same
49. 800 pJ, 5.79 mJ/m^3
55. (a) 8.13 nF (b) 2.40 kV
57. 16.7 pF, 1.62 kV
59. 1.04 m
61. 6.00
63. (a) 70 C (b) 55.2 C
65. 416 pF
67. 1.00 μF and 3.00 μF
69. 3.00 μF
71. 2.33
73. (a) 243 μJ (b) 2.30 mJ
75. 177 μC and 88.6 μC
77. (a) $\frac{3}{2}Q_0$ and $\frac{1}{2}Q_0$ (b) $\frac{1}{2}V_0$
79. 1.73 m^2, 3.60 J
81. 3.00 μF
83. (b) $Q/Q_0 = \kappa$
85. (a) $\dfrac{\epsilon_0}{d}[\ell^2 + \ell x(\kappa - 1)]$ (b) $\dfrac{\epsilon_0 V^2}{2d}[\ell x(\kappa-1) + \ell^2]$ (c) $\dfrac{\epsilon_0 V^2}{2d}\ell(\kappa - 1)$ to the right (d) 1.55×10^{-3} N
89. 19.0 kV
91. 3.00 μF

CHAPTER 27

1. 481 nA
3. 400 nA
5. 33.3 C
7. 5.90×10^{28}/m^3
9. (a) 1.50×10^5 A (b) 5.40×10^8 C
11. 13.3 μA/m^2
13. 1.32×10^{11} A/m^2
15. 1.59 Ω
17. $3.23 \times 10^6/\Omega\cdot$m
19. 2.57 mm
21. 9.00R
23. (a) 1.82 m (b) 280 μm
25. 6.43 A
27. (a) 5.00 A (b) 1.10 kV/m (c) Current doubles for constant field.
29. (a) $3.18 \times 10^{-8}\ \Omega\cdot$m (b) 6.29×10^6 A/m^2 (c) 49.9 mA (d) 6.59×10^{-4} m/s (assume 1 conduction electron per atom) (e) 0.400 V
31. 0.125
33. 20.8 Ω
35. 67.6°C
37. 26.2°C
39. 3.03×10^7 A/m^2
41. 21.2 nm
43. 0.833 W
45. 36.1%
47. 82.9 V
49. $\dfrac{P_A}{P_B} = 2$
51. 28.9 Ω
53. 26.9 cents/day
55. (a) 18.8 A (b) 7.46 MJ (c) 16.6 cents
57. 2020°C
59. (a) $9.49 \times 10^{-7}\ \Omega$ (b) $8.07 \times 10^{-7}\ \Omega$
63. (a) $R = \dfrac{\rho L}{\pi({r_b}^2 - {r_a}^2)}$ (b) 37.4 MΩ

65. 304 nA

69. $R = \dfrac{\pi\rho}{2t\ln(a/b)}$

71. (a) $R = \dfrac{\rho}{\pi R}\ln\left(\dfrac{2R+D}{2R-D}\right)$ (b) $\lim_{D\to 0} R = 0$

(c) $\lim_{D\to 2R} R = \infty$

73. $\dfrac{\alpha_1}{\alpha_2} = -0.120$

CHAPTER 28

1. $R = 7.67\ \Omega$
3. (a) 1.79 A (b) 10.4 V
5. 12.0 Ω
7. (a) 6.73 Ω (b) 1.98 Ω
9. $r = 2.40\ \Omega$
11. (a) 5.32 V (b) 40.3 mA, 95.1 mA
13. (a) 27.3 V (b) 2.73 A, 1.37 A, 0.91 A
15. 1.99 A, 1.17 A, 0.819 A
17. $\frac{6}{11}R$
19. 14.3 W, 28.5 W, 1.33 W, 4.00 W
21. (a) 0.227 A (b) 5.68 V
23. 4.26 Ω
25. −10.4 V
27. $\frac{11}{13}$ A, $\frac{6}{13}$ A, $\frac{17}{13}$ A
29. 6.00 V, 9.00 V
31. 3.50 A, 2.50 A, 1.00 A
33. 10.7 V
35. (a) 0.667 A (b) 50.0 μC
37. (a) 909 mA (b) −1.82 V
39. 800 W, 450 W, 25.0 W, 25.0 W
41. 99.3%
43. 4.06 μA
45. (a) 12.0 s (b) $i(t) = (3.00\ \mu\text{A})e^{-t/12.1}$ $q(t) = (36.0\ \mu\text{C})[1 - e^{-t/12.1}]$
47. (a) 6.00 V (b) 8.29 μs
49. (a) 6.50 mJ (b) 6.50 mJ
51. 1.60 MΩ
53. 16.6 kΩ
55. 0.302 Ω
57. (a) 400 kΩ (b) 400 kΩ
59. (b) 0.0501 Ω, 0.451 Ω
61. 0.588 A
63. 15.0 Ω
65. (a) 12.5 A, 6.25 A, 8.33 A (b) 27.1 A; No, it would not be sufficient since the current drawn is greater than 25 A.
67. (a) 0.101 W (b) 10.1 W
69. (a) 16.7 A (b) 33.3 A (c) The 120-V heater requires four times as much mass.
71. (a) 34.1 V (b) 41.7 V (c) 7.57 V
73. (a) 72.0 W (b) 72.0 W
75. 395 mA; 1.50 V
77. (a) $R \le 1050\ \Omega$ (b) $R \ge 10.0\ \Omega$
79. (a) $R \to \infty$ (b) $R \to 0$ (c) $R \to r$
81. (a) 9.93 μC (b) 3.37×10^{-8} A (c) 3.34×10^{-7} W (d) 3.37×10^{-7} W
83. $T = (R_A + 2R_B)C \ln 2$
85. $R = 0.521\ \Omega$, 0.260 Ω, 0.260 Ω, assuming resistors are in series with the galvanometer
87. $R/2$
89. 0.450 Ω

CHAPTER 29

1. (a) West (b) zero deflection (c) up (d) down
3. 2.31×10^{-13} N
5. 48.8° or 131°
7. $(12.3\boldsymbol{i} + 4.48\boldsymbol{j} - 1.60\boldsymbol{k}) \times 10^{-18}$ N
9. 2.34×10^{-18} N
11. zero
13. $(-2.88\ \text{N})\boldsymbol{j}$
15. 0.245 T west
17. (a) 4.73 N (b) 5.46 N (c) 4.73 N
19. $(0.042\ \text{T})\boldsymbol{k}$
21. $F = 2\pi rIB \sin\theta$, up
23. 9.98 N·m, clockwise as seen looking in the negative y-direction
25. (a) 376 μA (b) 1.67 μA
27. 58.9 mJ
29. 1.98 cm
31. 182 μT
33. $r_\alpha = r_d = \sqrt{2}r_p$
35. 7.88 pT
37. 2.99 u; ${}^3_1\text{H}^+$ or ${}^3_2\text{He}^+$
39. 5.93×10^5 N/C
41. 20.4 cm and 4.77 mm
43. 150 μm
45. 31.2 cm
47. (a) 4.31×10^7 rad/s (b) 5.17×10^7 m/s
49. 70.1 mT
51. 3.70×10^{-9} m³/C
53. 4.32×10^{-5} T
55. 7.37×10^{28} electrons/m³
57. 126 mT at 78.7° below the horizon
59. (a) $(3.52\boldsymbol{i} - 1.60\boldsymbol{j}) \times 10^{-18}$ N (b) 24.4°
61. 0.588 T
63. $(1.06\ \text{N})\boldsymbol{k}$
65. 3.13 cm
67. 3.82×10^{-25} kg
69. 3.70×10^{-24} N·m
71. (a) $(-8.34 \times 10^{-20}\ \text{kg}\cdot\text{m/s})\boldsymbol{j}$ (b) 29.4°

CHAPTER 30

1. 200 nT
3. 31.4 cm
5. 28.3 μT
7. 24.6 μT
9. $\dfrac{\mu_0 I}{4\pi x}$ into the plane of the paper
11. 330 nT

13. (a) $0.766R$ (b) $1.91R$
15. 80.0 μN/m
17. (a) 10.0 μm (b) silver wire, yes; copper wire, no
19. $(-27.0\ \mu\text{N})\boldsymbol{i}$
21. 13.0 μT at 270°
23. (a) 3.98 kA (b) ≈ 0
25. 5.03 T
27. (a) 3.60 T (b) 1.94 T
29. 5.00 cm
31. (a) 6.34×10^{-3} N/m, inward (b) $\boldsymbol{F}$ is greatest at the outer surface
33. 26.7 mT, 37.7 mT
35. 4.74 mT
37. (a) 3.13 mWb (b) zero
39. 2.27 μWb
41. (a) $(8.00\ \mu\text{A})e^{-t/4}$ (b) 2.94 μA
43. (a) 11.3×10^9 V·m/s (b) 100 mA
45. 1.0001
47. 191 mT
49. 277 mA
51. 150 μWb
53. M/H
55. 1.27×10^3 turns
57. 2.02
59. (a) 1.88×10^{46} (b) 8.74×10^{20} kg
61. 675 A down
63. 81.7 A
65. $\boldsymbol{B} = \dfrac{\mu_0 I}{2\pi w} \ln\left(\dfrac{b+w}{b}\right) \boldsymbol{k}$
67. 594 A east
69. 1.43×10^{-10} T directed away from the center
73. (a) $B = \dfrac{1}{3}\mu_0 b r_1{}^2$ (b) $B = \dfrac{\mu_0 b R^3}{3r_2}$
75. 5/4
77. (a) 12.0 kA-turns/m (b) 2.07×10^{-5} T·m/A
79. 366 g/mol
83. $\dfrac{\mu_0 I}{4\pi}(1 - e^{-2\pi})$ perpendicularly out of the paper
85. $\dfrac{\mu_0 I}{\pi r}\left(\dfrac{2r^2 + a^2}{4r^2 + a^2}\right)$ up

CHAPTER 31

1. 500 mV
3. 160 A
5. 2.67 T/s
7. 61.8 mV
9. $(200\ \mu\text{V})e^{-t/7}$
11. $Nn\pi R^2 \mu_0 I_0 \alpha e^{-\alpha t} = (68.2\ \text{mV})e^{-1.6t}$ counterclockwise
13. 272 m
15. $\frac{2}{3}\pi R^2 N B_0 \omega \sin \omega t$ counterclockwise looking to the left
17. 2.81 mT
19. (a) 3.00 N to the right (b) 6.00 W
21. 2.00 mV; the west end is positive in the northern hemisphere
23. 2.83 mV
25. (a) $\frac{1}{8} B\omega L^2$ (b) 360 mV
27. (a) to the right (b) to the right (c) to the right (d) into the plane of the paper
29. 0.742 T
31. 114 μV clockwise
33. 1.80×10^{-3} N/C counterclockwise
35. (a) $(9.87 \times 10^{-3}\ \text{V/m})\cos(100\pi t)$ (b) clockwise
37. (a) 1.60 A counterclockwise (b) 20.1 μT (c) up
39. 12.6 mV
41. (a) 7.54 kV (b) $\boldsymbol{B}$ is parallel to the plane of the loop.
43. $(28.6\ \text{mV})\sin(4\pi t)$
45. (a) 0.640 N·m (b) 241 W
47. 0.513 T
49. (a) $F = \dfrac{N^2 B^2 w^2 v}{R}$ to the left (b) 0
(c) $F = \dfrac{N^2 B^2 w^2 v}{R}$ to the left
51. $(-2.87\boldsymbol{j} + 5.75\boldsymbol{k}) \times 10^9\ \text{m/s}^2$
53. It is, with the top end in the picture positive.
57. (a) 36.0 V (b) 600 mWb/s (c) 35.9 V (d) 4.32 N·m
59. (a) $(1.19\ \text{V})\cos(120\pi t)$ (b) 88.5 mW
61. 26.2 g
63. It is, with the left end in the picture positive
65. 1.20 μC
67. $B = \dfrac{2QR}{\pi N D^2}$
69. (a) $\dfrac{Rmg \sin\theta}{B^2 \ell^2 \cos^2\theta}$ (b) 22.6 m/s

CHAPTER 32

1. 100 V
3. 4.69 mH
7. (a) 188 μT (b) 33.3 nWb (c) 375 μH (d) field and flux
9. 21.0 μWb
11. $(18.8\ \text{V})\cos(377t)$
13. $\frac{1}{2}$
15. (a) 15.8 μH (b) 12.6 mH
19. 2.61 H
23. (a) 139 ms (b) 461 ms
25. (a) 5.66 ms (b) 1.22 A (c) 58.1 ms
27. (a) 113 mA (b) 600 mA
29. (a) 0.800 (b) 0
31. (a) 1.00 A (b) 12.0 V, 1.20 kV, 1.21 kV (c) 7.62 ms
33. 2.44 μJ
35. (a) 1.31 W (b) 4.16 W (c) 5.47 W (d) zero
37. 4.83 J/m³
39. 44.3 nJ/m³ in the $\boldsymbol{E}$ field; 995 μJ/m³ in the $\boldsymbol{B}$ field
41. 30.7 μJ and 72.2 μJ
43. (a) 500 mJ (b) 17.0 W (c) 11.0 W
45. $(-0.168t + 0.112)$V
47. 80.0 mH
49. 553 μH

51. (a) 18.0 mH (b) 34.3 mH (c) −9.00 mV
53. 400 mA
55. (a) 5.19 mA (b) 1.72 MHz (c) 7.68 nJ
57. 2.60 pF
59. (a) 36.0 μF (b) 8.00 ms
61. (a) 6.03 J (b) 0.529 J (c) 6.56 J
63. (a) 2.51 kHz (b) 69.9 Ω
65. (a) 4.47 krad/s (b) 4.36 krad/s (c) 2.53%
67. (a) True. (b) There is no ratio, for the *RLC* circuit is overdamped and has no frequency of oscillation.
69. 979 mH
71.

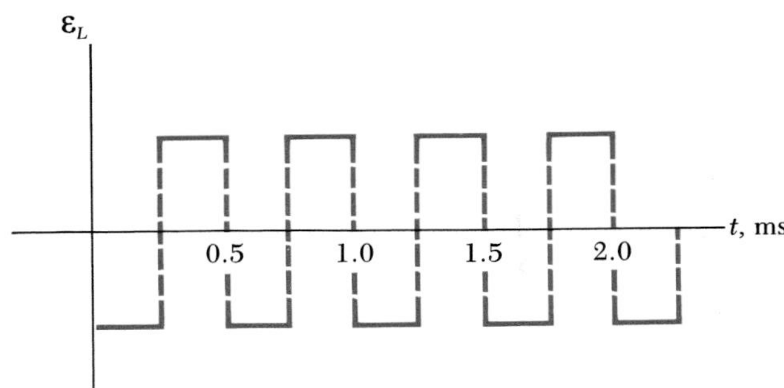

73. 2.00 H, 3.00 Ω
79. (a) 2.36 A (b) 7.77 V
81. $\dfrac{\mu_0}{\pi} \ln\left(\dfrac{d-a}{a}\right)$
83. 300 Ω
85. (a) 1.99 A (b) 0.509 s

CHAPTER 33

3. 2.95 A, 70.7 V
5. (a) −1.44 A (b) 1.44 A
7. 14.6 Hz
9. (a) 42.4 mH (b) 943 Hz
11. 7.03 H
13. 5.60 A
15. 76.8 mH
17. (a) $f > 41.3$ Hz (b) $X_C < 87.5\ \Omega$
19. 2.77 nC
21. 100 mA
23. 0.427 A
25. (a) 78.5 Ω (b) 1.59 kΩ (c) 1.52 kΩ (d) 138 mA (e) −84.3°
27. (a) 17.4° (b) voltage leads the current
29. 61.6 Hz
31. (a) 146 V (b) 213 V (c) 179 V (d) 33.4 V
33. (a) 194 V (b) current leads by 33.3°
35. 1.43 W
37. (a) 4.59 W (b) 4.59 W
39. $v(t) = (283\text{ V})\sin(628t)$
41. (a) 16.0 Ω (b) −12.0 Ω
43. 159 Hz
45. 1.82 pF
47. (a) 124 nF (b) 51.5 kV
49. 0.964 W
51. (a) 613 μF (b) 0.756
53. (a) 0.9998 (b) 0.0199
55. (a) True, if $\omega_0 = (LC)^{-1/2}$ (b) 0.107, 0.999, 0.137
57. 0.317
59. 687 V
61. 2.38 A, 240 V
63. 25.3 A
65. 56.7 W
67. (a) 1.89 A (b) $V_R = 39.7$ V, $V_L = 30.1$ V, $V_C = 175$ V (c) $\cos\phi = 0.265$
(d)

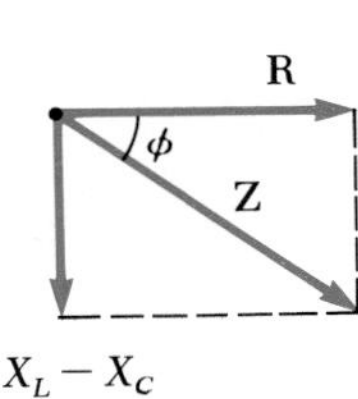

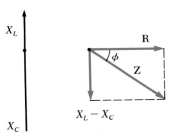

69. (a) 200 Ω (b) 346 Ω (c) 60.0°
71. (a) 2.90 kW (b) 5.80×10^{-3}
(c) If the generator is limited to 4500 V, no more than 16 kW could be delivered to the load, never 5000 kW.
73. (a) Circuit (a) is a high-pass filter, (b) is a low-pass filter.
(b) $\dfrac{V_{\text{out}}}{V_{\text{in}}} = \dfrac{\sqrt{R_L{}^2 + X_L{}^2}}{\sqrt{R_L{}^2 + (X_L - X_C)^2}}$ for circuit (a);
$\dfrac{V_{\text{out}}}{V_{\text{in}}} = \dfrac{X_C}{\sqrt{R_L{}^2 + (X_L - X_C)^2}}$ for circuit (b)
77. (a) 172 V (b) 52.7 V (c) 79.4 V
79. (b) 31.6

CHAPTER 34

3. (b) 377 Ω
5. (a) 162 N/C (b) 130 N/C
9. (a) 6.00 MHz (b) $(73.3\text{ nT})(-\mathbf{k})$
(c) $\mathbf{B} = (73.3\text{ nT})\cos(0.126x - 3.77 \times 10^7 t)(-\mathbf{k})$
13. 5.16 m
15. 307 μW/m^2
17. (a) 0.949 V/m
(b) 0.0949 V/m
(c) 0.00949 V/m
19. (a) 332 kW/m^2 (b) 1.88 kV/m, 222 μT
21. (a) 971 V/m (b) 16.7 pJ
23. (a) 540 N/C (b) 1.29 μJ/m^3 (c) 387 W/m^2
(d) This is 38.7% of the flux mentioned in Example 34.3 and 28.9% of the 1340 W/m^2 intensity above

the atmosphere. It may be cloudy at this location, or the sun may be setting.

25. 83.3 nPa
27. 4.53×10^{-15} kg/m²s
29. (a) 11.3 kJ (b) 5.65×10^{-5} kg·m/s
31. (a) 2.26 kW (b) 4.71 kW/m²
33. 7.50 m
35. (a) $\mathcal{E}_m = 2\pi^2 r^2 f B_m \cos\theta$, where θ is the angle between the magnetic field and the normal to the loop. (b) Vertical, with its plane pointing toward the broadcast antenna
37. (a) 0.933 (b) 0.500 (c) 0
39. (a) 6.00 pm (b) 7.50 cm
41. (a) 30.0 GHz (b) 3.00×10^{14} Hz (c) 5.17×10^{14} Hz (d) 3.00×10^{15} Hz (e) 3.00×10^{20} Hz
43. (a) 4.17 m to 4.55 m (b) 3.41 to 3.66 m (c) 1.61 m to 1.67 m
45. 3.33×10^3 m²
47. (a) 6.67×10^{-16} T (b) 5.31×10^{-17} W/m² (c) 1.67×10^{-14} W (d) 5.57×10^{-23} N
49. 6.37×10^{-7} Pa
51. (a) 20.0 nT (b) 600 kW
53. 7.50×10^{10} s = 2370 y; out the back
55. 3.00×10^{-2} degrees
57. (a) $B_0 = 583$ nT, $k = 419$ rad/m, $\omega = 1.26 \times 10^{11}$ rad/s; xz (b) 40.6 W/m² in average value (c) 271 nPa (d) 406 nm/s²

CHAPTER 35

1. 2.998×10^8 m/s
3. 1.98×10^8 km
5. 114 rad/s
7. 20 μs 60 μs
9. (a) 4.74×10^{14} Hz (b) 422 nm (c) 2.00×10^8 m/s
11. 70.5° from the vertical
13. 61.3°
15. 19.5°, 19.5°, 30.0°.
17. (a) 1.97×10^8 m/s (b) 3.88×10^{-5} m
19. 0.890
21. 30.0°, 19.5° at entry; 40.5°, 77.1° at exit
23. 7.89°
27. 18.4°
29. 86.8°
31. 4.61°
33. 1.89
35. (a) 24.4° (b) 37.0° (c) 49.8°
37. 1.00008
39. $\theta < 48.2°$
41. 53.6°
43. 2.27 m
45. 2.37 cm
49. 90°, 30°, No
51. 62.2%
53. $\sin^{-1}[(n^2 - 1)^{1/2} \sin\phi - \cos\phi]$ If $n \sin\phi \leq 1$, $\theta = 0$

CHAPTER 36

3. 2′11″
5. 30 cm
7. (a) $s' = 45.0$ cm, $M = -\frac{1}{2}$ (b) $s' = -60.0$ cm, $M = 3.00$ (c) Similar to Figures 36.6 and 36.9b
9. (a) $s' = -13.3$ cm, $M = 0.667$ (b) $s' = -24.0$ cm, $M = 0.400$ (c) See Figure 36.8
11. concave with radius 40.0 cm
13. (a) 2.08 m (concave) (b) 1.25 m in front of the object
15. (a) $s' = -12.0$ cm, $M = 0.400$ (b) $s' = -15.0$ cm, $M = 0.250$ (c) The images are erect
17. 11.8 cm above the floor
19. (a) −13.33 cm (b) −9.23 cm (c) −3.79 cm
21. 8.57 cm
23. 3.88 mm
25. 2
27. (a) 16.4 cm (b) 16.4 cm
29. 25.0 cm; −0.250
31.

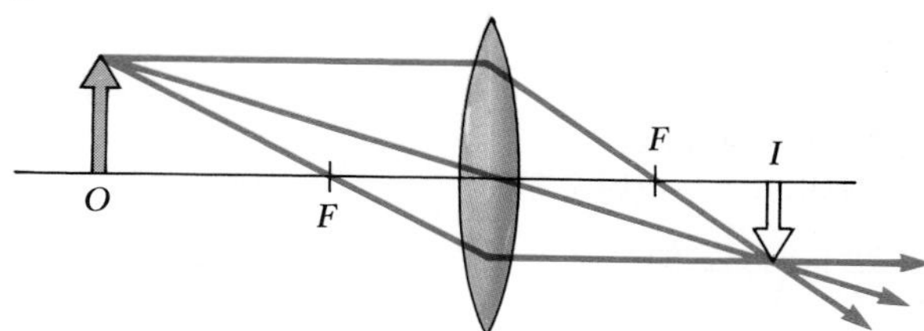

33. (a) −12.3 cm, to left of lens (b) −0.615
35. −48.4 cm
37. (a) +5.49 cm (b) $s' = 6.73$ cm, real and inverted, diminished, $h' - 0.448$ cm
39. $\frac{f}{1.41}$
41. −4 diopters, a diverging lens
43. 0.558 cm
45. −2.89 diopters
47. 3.5
49. −800, image is inverted
51. −18.8
53. 2.14 cm
55. (a) 24.5 cm (b) 99.0 cm
57.

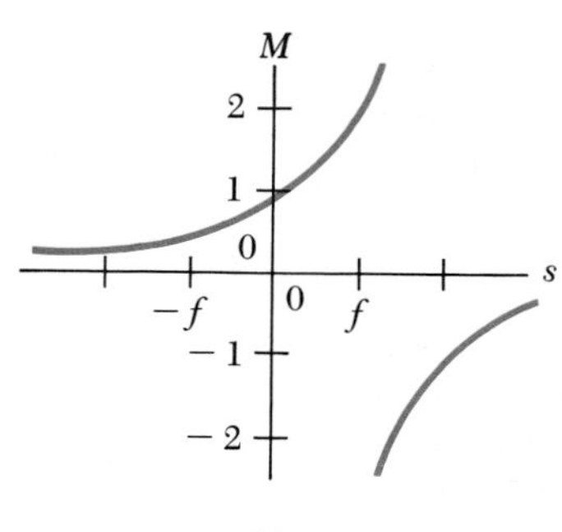

(a)

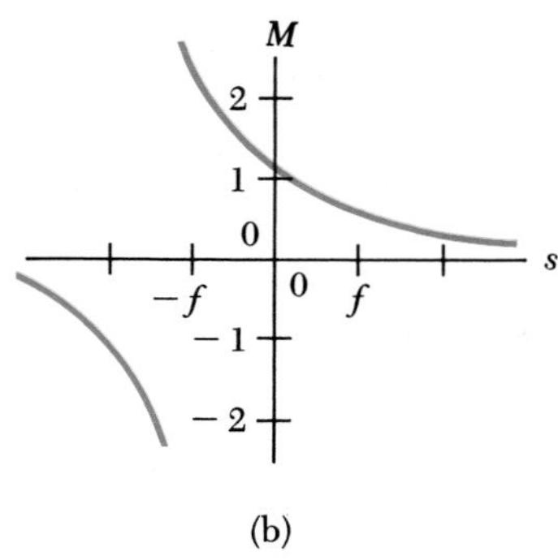

(b)

59. 10.7 cm beyond the curved surface
61. 20.0 cm
63. (a) 44.6 diopters (b) 3.03 diopters
65. $d = 8$ cm
67. (a) −11.5 cm (b) −10.0 cm (c) −6.00 cm
(d) 0.231, upright; 0.333, upright; 0.600, upright
(e) All similar to Figure 36.19c
69. (a) 52.5 cm (b) 1.50 cm
71. $s' = 1.5$ m, $h' = -13.1$ mm
73. (a) 0.400 cm (b) $s' = -3.94$ mm, $h' = 535\ \mu$m
75. (a) 30.0 cm and 120 cm (b) 24.0 cm
(c) real, inverted, diminished

CHAPTER 37

1. (a) 2.62×10^{-3} m (b) 2.62×10^{-3} m
3. 515 nm
5. 0.340 m
7. 4.80×10^{-3} m
9. 4.48×10^{-3}%
11. 1.56×10^{-4} m
13. 4.80×10^{-5} m
15. 423.5 nm
17.

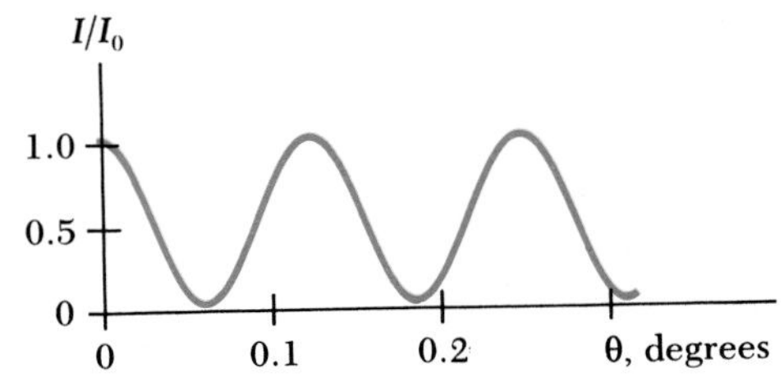

19. (a) 1.27×10^{-2} degrees (b) 5.97×10^{-2} degrees
21. (a) 2.63 rad (b) 246 nm
23. (a) 7.95 rad (b) 0.453
25. $E_R = 1.00E_0$, $\alpha = 3\pi/2$
27. $10 \sin(100\pi t + 0.93)$
29. $\pi/2$
31. 360°/N
33. No reflection maxima in the visible spectrum
35. (a) 123 nm (b) 81.6 nm
37. 85.4 nm, or 256 nm, or 427 nm . . .
39. 584 nm, 417 nm, 324 nm
41. 544 nm, 423.1 nm
43. 4.23×10^{-6} m
45. 3.96×10^{-5} m
47. 654 dark fringes
49. 0.0162 m
51. 3.58°
53. 421 nm
55. 2.52 cm
57. 1.54 mm
59. 3.6×10^{-5} m
61. $x_{\text{bright}} = \dfrac{\lambda \ell (m + \frac{1}{2})}{2hn}$, $x_{\text{dark}} = \dfrac{\lambda \ell m}{2hn}$
63. (b) 9.96×10^{-8} m
65. 1.7
67. For each centimeter of vertical displacement, the thickness varies by 492 nm.
69. (b) 254 nm

CHAPTER 38

1. (a) 0.105° (b) 2.20 mm
3. 560 nm
5. $\cong 10^{-3}$ rad
7. 0.230 mm
9. $\phi = 1.392$ rad

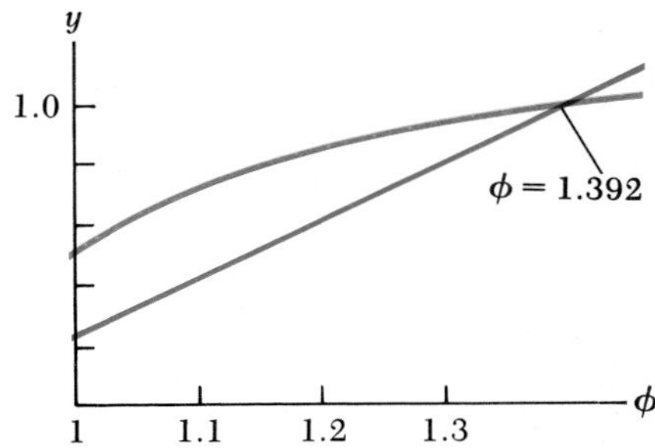

11. 1.62×10^{-2}
13. 43.5 m
15. (a) 1.03 cm (b) 26.8 cm
17. 2.54×10^{-7} rad
19. 15.7 km
21. 13 m
23. 55.4 km
25. (a) 5.93° (b) 6.43°
27. (a) 5.7; 5 orders is the maximum
(b) 10; 10 orders to be in the short-wavelength region.
29. 0.0683°
31. (a) 491 lines (b) 2.23 mm
33. 19.5°
35. 7.35°
37. 14.7°
39. 31.9°
41. It is not possible.
43. (a) 0.281 nm (b) 0.18%
45. 14.4°
47. 1.11
49. 31.2°

51. $\frac{3}{8} I_0$
53. (a) 54.7° (b) 63.4° (c) 71.6°
55. 60.5°
57. 2.2
59. 5.9°
61. 38.3 cm
63. 660 nm
65. (a) 12 000, 24 000, 36 000 (b) 0.0111 nm
67. (a) 3.53×10^3 lines/cm
(b) Eleven maxima can be observed.
69. (a) $I_0(0.125)$ (b) $I_0(0.211)$ (c) $I_0(0.33)$
(d) The overall decrease in intensity is less when the number of steps in the 90° rotation of the plane of polarization is increased.
73. (a) 41.8° (b) 0.593 (c) 0.262 m

CHAPTER 39

5. (a) 60 m/s (b) 20 m/s (c) 44.7 m/s
7. $0.866c$
9. $3600 \text{ s} + 1.8 \times 10^{-9} \text{ s}$
11. $0.436L_0$
13. (a) 2.18 μs (b) 649 m
15. $0.1404c$
17. $0.4086c$
19. 42 g/cm^3
21. 1625 MeV/c
25. (a) 939.4 MeV (b) 3.008×10^3 MeV
(c) 2.069×10^3 MeV
27. $0.864c$
29. (a) 0.582 MeV (b) 2.45 MeV
31. (a) 3.91×10^4 (b) $0.9999999997c$ (c) 7.66 cm
33. 4 MeV and 29 MeV
35. (a) 3.29 MeV (b) 2.77 MeV
37. 4.2×10^9 kg/s
39. 65/min, 10.6/min.
41. 0.7%
43. (a) $0.0236c$ (b) $6.18 \times 10^{-4}\, c$
45. $0.8c$
47. $-0.5c$
49. 5600 MeV
55. (b) For $v \ll c$, $a = qE/m$ as in the classical description. As $v \to c$, $a \to 0$, describing how the particle can never reach the speed of light.
(c) Perform $\int_0^v \left(1 - \frac{v^2}{c^2}\right)^{-3/2} dv = \int_0^t \frac{qE}{m}\, dt$ to

obtain $v = \dfrac{qEct}{\sqrt{m^2c^2 + q^2E^2t^2}}$ and then

$$\int_0^x dx = \int_0^t \frac{qEtc}{\sqrt{m^2c^2 + q^2E^2t^2}}\, dt$$

CHAPTER 40

1. (a) 2.57 eV (b) 1.28×10^{-5} eV
(c) 1.91×10^{-7} eV
3. 2.27×10^{30} photons/s
5. 1.86×10^{-34}
7. 4.46×10^3 K
9. 9.35 μm; infrared
11. 5.18×10^3 K
15. (a) 0.350 eV (b) 555 nm
17. (a) 1.92 eV (b) 0.159 V
19. (a) 0.571 eV (b) 1.54 V
21. (a) only lithium (b) 0.808 eV
23. 1.50 V
25. 1.78 eV, 9.47×10^{-28} kg·m/s
27. 23.4°
29. (a) 3.10 keV (b) $0.110c = 32.9 \times 10^6$ m/s
31. (a) 2.88×10^{-12} m (b) 101°
33. (a) 0.1203 nm, 0.1212 nm, 0.1224 nm, 0.1236 nm, 0.1245 nm, 0.1248 nm (b) 28.0 eV, 104 eV, 205 eV, 305 eV, 377 eV, 403 eV (c) 180°
35. 70.1°
37. 121 nm, 103 nm, 97.2 nm
39. (a) 5 (b) No; No
41. 0.529 Å, 2.12 Å, 4.77 Å
43. (a) $E_n = -54.4 \dfrac{\text{eV}}{n^2}$, $n = 1, 2, 3, \ldots$
(b) -54.4 eV
45. (a) 0.265 Å (b) 0.177 Å (c) 0.132 Å
47. (a) 3.03 eV (b) 411 nm
(c) 7.32×10^{14} Hz
49. (a) B (b) D (c) B and C
51. -6.80 eV, $+3.40$ eV
53. $r_n = (1.06 \text{ Å})n^2$, $E_n = -6.80 \text{ eV}/n^2$, $n = 1, 2, 3, \ldots$
55. 5.39 keV
57. (a) 43.5° (b) 622 keV
59. (a) 3.12 fm (b) -18.9 MeV
61. (a) -6.67 keV, -668 eV (b) 0.622 nm
63. (a) 91.1 keV (b) 8.90 keV
(c) 57.5°
65. (a) 4.62 MW/m^2 (b) 1.62×10^{-5} m^2
(c) 5.73×10^3 K
69. (a) 7.77 pm (b) 93.9° (c) 35.5°

Index

Page numbers in *italics* indicate illustrations, page numbers followed by "n" indicate footnotes, page numbers followed by "t" indicate tables.

Standard Abbreviations and Symbols for Units

Abbreviation	Unit	Abbreviation	Unit
A	ampere	in.	inch
Å	angstrom	J	joule
u	atomic mass unit	K	kelvin
atm	atmosphere	kcal	kilocalorie
Btu	British thermal unit	kg	kilogram
C	coulomb	kmol	kilomole
°C	degree Celsius	lb	pound
cal	calorie	m	meter
deg	degree (angle)	min	minute
eV	electron volt	N	newton
°F	degree Fahrenheit	Pa	pascal
F	farad	rev	revolution
ft	foot	s	second
G	gauss	T	tesla
g	gram	V	volt
H	henry	W	watt
h	hour	Wb	weber
hp	horsepower	μm	micrometer
Hz	hertz	Ω	ohm

Mathematical Symbols Used in the Text and Their Meaning

Symbol	Meaning
$=$	is equal to
$\equiv$	is defined as
$\neq$	is not equal to
$\propto$	is proportional to
$>$	is greater than
$<$	is less than
$\gg$ ($\ll$)	is much greater (less) than
$\approx$	is approximately equal to
Δx	the change in x
$\sum_{i=1}^{N} x_i$	the sum of all quantities x_i from $i = 1$ to $i = N$
$\|x\|$	the magnitude of x (always a positive quantity)
$\Delta x \rightarrow 0$	Δx approaches zero
$\frac{dx}{dt}$	the derivative of x with respect to t
$\frac{\partial x}{\partial t}$	the partial derivative of x with respect to t
$\int$	integral